CHEMISTRY

STRUCTURE AND PROPERTIES

Nivaldo J. Tro
WESTMONT COLLEGE

PEARSON

Boston Columbus Indianapolis New York San Francisco Upper Saddle River
Amsterdam Cape Town Dubai London Madrid Milan Munich Paris Montréal Toronto
Delhi Mexico City São Paulo Sydney Hong Kong Seoul Singapore Taipei Tokyo

Editor in Chief: Adam Jaworski
Senior Acquisitions Editor: Terry Haugen
Director of Development: Jennifer Hart
Executive Marketing Manager: Jonathan Cottrell
Senior Market Development Manager: Michelle Cadden
Associate Team Lead, Program Management, Chemistry and Geosciences: Jessica Moro
Development Editor: Erin Mulligan
Editorial Assistant: Fran Falk/Caitlin Falco
Marketing Assistant: Nicola Houston
Team Lead, Project Management, Chemistry and Geosciences: Gina M. Cheselka
Project Manager: Beth Sweeten
Production Management: codeMantra, LLC
Compositor: codeMantra, LLC
Illustrator: Precision Graphics
Image Lead: Maya Melenchuk
Photo Researcher: Peter Jardim, PreMedial Global
Text Permissions Manager: Alison Bruckner
Text Permission Researcher: Haydee Hidalgo, Electronic Publishing Services Inc.
Design Manager: Derek Bacchus
Interior Designer: Elise Lansdon
Cover Designer: Elise Lansdon
Operations Specialist: Christy Hall
Cover Art: Quade Paul

Credits and acknowledgments borrowed from other sources and reproduced, with permission, in this textbook appear on the appropriate page within text or on page C1.

Library of Congress Cataloging-in-Publication Data

Tro, Nivaldo J.
Chemistry : structure and properties / Nivaldo J. Tro, Westmont College.
pages cm
ISBN 978-0-321-83468-3
1. Chemistry—Textbooks. I. Title.

QD33.2.T7595 2015
540—dc23 2013035237

3 16

www.pearsonhighered.com

Student Edition: ISBN 10: 0-321-83468-2; ISBN 13: 978-0-321-83468-3
Instructor's Review Copy: ISBN 10: 0-321-93904-2; ISBN 13: 978-0-321-93904-3

About the Author

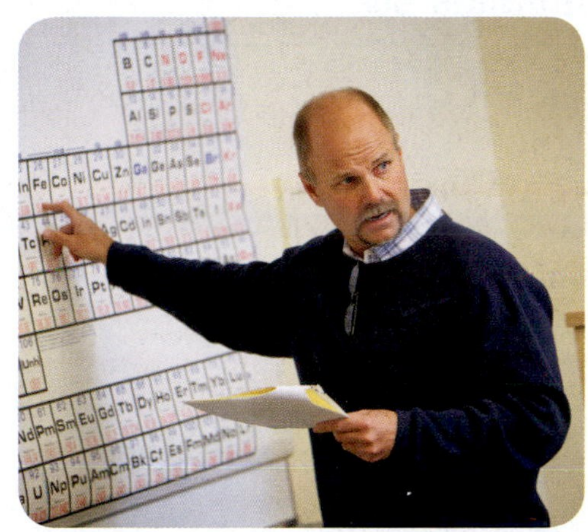

Nivaldo Tro is a professor of chemistry at Westmont College in Santa Barbara, California, where he has been a faculty member since 1990. He received his Ph.D. in chemistry from Stanford University for work on developing and using optical techniques to study the adsorption and desorption of molecules to and from surfaces in ultrahigh vacuum. He then went on to the University of California at Berkeley, where he did postdoctoral research on ultrafast reaction dynamics in solution. Since coming to Westmont, Professor Tro has been awarded grants from the American Chemical Society Petroleum Research Fund, from the Research Corporation, and from the National Science Foundation to study the dynamics of various processes occurring in thin adlayer films adsorbed on dielectric surfaces. He has been honored as Westmont's outstanding teacher of the year three times and has also received the college's outstanding researcher of the year award. Professor Tro lives in Santa Barbara with his wife, Ann, and their four children, Michael, Ali, Kyle, and Kaden. In his leisure time, Professor Tro enjoys mountain biking, surfing, reading to his children, and being outdoors with his family.

To Ann, Michael, Ali, Kyle, and Kaden

Brief Contents

1 Atoms 2

2 Measurement, Problem Solving, and the Mole Concept 34

3 The Quantum-Mechanical Model of the Atom 62

4 Periodic Properties of the Elements 100

5 Molecules and Compounds 144

6 Chemical Bonding I: Drawing Lewis Structures and Determining Molecular Shapes 188

7 Chemical Bonding II: Valence Bond Theory and Molecular Orbital Theory 232

8 Chemical Reactions and Chemical Quantities 270

9 Introduction to Solutions and Aqueous Reactions 300

10 Thermochemistry 342

11 Gases 390

12 Liquids, Solids, and Intermolecular Forces 440

13 Phase Diagrams and Crystalline Solids 480

14 Solutions 508

15 Chemical Kinetics 554

16 Chemical Equilibrium 608

17 Acids and Bases 654

18 Aqueous Ionic Equilibrium 708

19 Free Energy and Thermodynamics 766

20 Electrochemistry 812

21 Radioactivity and Nuclear Chemistry 860

22 Organic Chemistry 902

23 Transition Metals and Coordination Compounds 954

Appendix I The Units of Measurement A-1

Appendix II Significant Figure Guidelines A-6

Appendix III Common Mathematical Operations in Chemistry A-11

Appendix IV Useful Data A-17

Appendix V Answers to Selected End-of-Chapter Problems A-29

Appendix VI Answers to In-Chapter Practice Problems A-61

Glossary G-1

Credits C-1

Index I-1

Contents

Preface xvii

Atoms 2

1.1 A Particulate View of the World: Structure Determines Properties 3

1.2 Classifying Matter: A Particulate View 4
The States of Matter: Solid, Liquid, and Gas 5 Elements, Compounds, and Mixtures 6

1.3 The Scientific Approach to Knowledge 7
The Importance of Measurement in Science 8 Creativity and Subjectivity in Science 8

1.4 Early Ideas about the Building Blocks of Matter 9

1.5 Modern Atomic Theory and the Laws That Led to It 10
The Law of Conservation of Mass 10 The Law of Definite Proportions 11 The Law of Multiple Proportions 12 John Dalton and the Atomic Theory 13

1.6 The Discovery of the Electron 13
Cathode Rays 13 Millikan's Oil Drop Experiment: The Charge of the Electron 14

1.7 The Structure of the Atom 16

1.8 Subatomic Particles: Protons, Neutrons, and Electrons 18
Elements: Defined by Their Numbers of Protons 18 Isotopes: When the Number of Neutrons Varies 20 Ions: Losing and Gaining Electrons 22

1.9 Atomic Mass: The Average Mass of an Element's Atoms 22
Mass Spectrometry: Measuring the Mass of Atoms and Molecules 24

1.10 The Origins of Atoms and Elements 25

REVIEW Self-Assessment Quiz 26 Key Learning Outcomes 27 Key Terms 27 Key Concepts 27 Key Equations and Relationships 28

EXERCISES Review Questions 28 Problems by Topic 29 Cumulative Problems 32 Challenge Problems 32 Conceptual Problems 33 Answers to Conceptual Connections 33

2 Measurement, Problem Solving, and the Mole Concept 34

2.1 The Metric Mix-up: A $125 Million Unit Error 35

2.2 The Reliability of a Measurement 36
Reporting Measurements to Reflect Certainty 36 Precision and Accuracy 37

2.3 Density 38

2.4 Energy and Its Units 40
The Nature of Energy 40 Energy Units 41 Quantifying Changes in Energy 42

2.5 Converting between Units 43

2.6 Problem-Solving Strategies 45
Units Raised to a Power 47 Order-of-Magnitude Estimations 49

2.7 Solving Problems Involving Equations 49

2.8 Atoms and the Mole: How Many Particles? 51
The Mole: A Chemist's "Dozen" 51 Converting between Number of Moles and Number of Atoms 52 Converting between Mass and Amount (Number of Moles) 52

REVIEW Self-Assessment Quiz 56 Key Learning Outcomes 56 Key Terms 57 Key Concepts 57 Key Equations and Relationships 57

EXERCISES Review Questions 58 Problems by Topic 58 Cumulative Problems 59 Challenge Problems 60 Conceptual Problems 61 Answers to Conceptual Connections 61

3

The Quantum-Mechanical Model of the Atom 62

3.1 Schrödinger's Cat 63

3.2 The Nature of Light 64
The Wave Nature of Light 64 The Electromagnetic Spectrum 66 Interference and Diffraction 68 The Particle Nature of Light 68

3.3 Atomic Spectroscopy and the Bohr Model 73
Atomic Spectra 73 The Bohr Model 74 Atomic Spectroscopy and the Identification of Elements 75

3.4 The Wave Nature of Matter: The de Broglie Wavelength, the Uncertainty Principle, and Indeterminacy 77
The de Broglie Wavelength 78 The Uncertainty Principle 79 Indeterminacy and Probability Distribution Maps 80

3.5 Quantum Mechanics and the Atom 81
Solutions to the Schrödinger Equation for the Hydrogen Atom 82 Atomic Spectroscopy Explained 84

3.6 The Shapes of Atomic Orbitals 87
s Orbitals ($l = 0$) 87 *p* Orbitals ($l = 1$) 90 *d* Orbitals ($l = 2$) 90 *f* Orbitals ($l = 3$) 90 The Phase of Orbitals 92 The Shape of Atoms 92

REVIEW Self-Assessment Quiz 93 Key Learning Outcomes 93 Key Terms 94 Key Concepts 94 Key Equations and Relationships 95

EXERCISES Review Questions 95 Problems by Topic 96 Cumulative Problems 97 Challenge Problems 98 Conceptual Problems 99 Answers to Conceptual Connections 99

4

Periodic Properties of the Elements 100

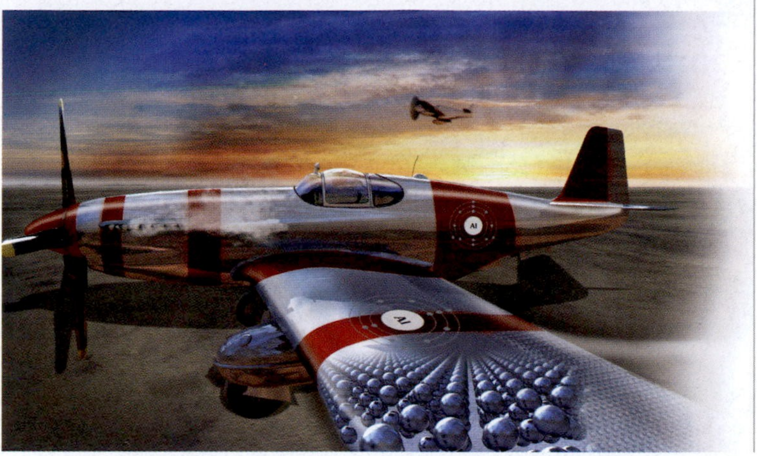

4.1 Aluminum: Low-Density Atoms Result in Low-Density Metal 101

4.2 Finding Patterns: The Periodic Law and the Periodic Table 102

4.3 Electron Configurations: How Electrons Occupy Orbitals 105
Electron Spin and the Pauli Exclusion Principle 105 Sublevel Energy Splitting in Multi-electron Atoms 106 Electron Configurations for Multi-electron Atoms 109

4.4 Electron Configurations, Valence Electrons, and the Periodic Table 112
Orbital Blocks in the Periodic Table 113 Writing an Electron Configuration for an Element from Its Position in the Periodic Table 114 The Transition and Inner Transition Elements 115

4.5 How the Electron Configuration of an Element Relates to Its Properties 116
Metals and Nonmetals 116 Families of Elements 117 The Formation of Ions 118

4.6 Periodic Trends in the Size of Atoms and Effective Nuclear Charge 119
Effective Nuclear Charge 121 Atomic Radii and the Transition Elements 122

4.7 Ions: Electron Configurations, Magnetic Properties, Ionic Radii, and Ionization Energy 124
Electron Configurations and Magnetic Properties of Ions 124 Ionic Radii 126 Ionization Energy 128 Trends in First Ionization Energy 128 Exceptions to Trends in First Ionization Energy 131 Trends in Second and Successive Ionization Energies 131

4.8 Electron Affinities and Metallic Character 132
Electron Affinity 132 Metallic Character 133

REVIEW Self-Assessment Quiz 136 Key Learning Outcomes 137 Key Terms 137 Key Concepts 138 Key Equations and Relationships 138

EXERCISES Review Questions 139 Problems by Topic 140 Cumulative Problems 141 Challenge Problems 142 Conceptual Problems 143 Answers to Conceptual Connections 143

5 Molecules and Compounds 144

5.1 Hydrogen, Oxygen, and Water 145

5.2 Types of Chemical Bonds 146

5.3 Representing Compounds: Chemical Formulas and Molecular Models 148
Types of Chemical Formulas 148 Molecular Models 150

5.4 The Lewis Model: Representing Valence Electrons with Dots 150

5.5 Ionic Bonding: The Lewis Model and Lattice Energies 152
Ionic Bonding and Electron Transfer 152 Lattice Energy: The Rest of the Story 153 Ionic Bonding: Models and Reality 154

5.6 Ionic Compounds: Formulas and Names 155
Writing Formulas for Ionic Compounds 155 Naming Ionic Compounds 156 Naming Binary Ionic Compounds Containing a Metal That Forms Only One Type of Cation 156 Naming Binary Ionic Compounds Containing a Metal That Forms More than One Kind of Cation 157 Naming Ionic Compounds Containing Polyatomic Ions 158 Hydrated Ionic Compounds 160

5.7 Covalent Bonding: Simple Lewis Structures 161
Single Covalent Bonds 161 Double and Triple Covalent Bonds 162 Covalent Bonding: Models and Reality 162

5.8 Molecular Compounds: Formulas and Names 163

5.9 Formula Mass and the Mole Concept for Compounds 165
Molar Mass of a Compound 165 Using Molar Mass to Count Molecules by Weighing 166

5.10 Composition of Compounds 167
Mass Percent Composition as a Conversion Factor 168 Conversion Factors from Chemical Formulas 170

5.11 Determining a Chemical Formula from Experimental Data 172
Calculating Molecular Formulas for Compounds 174 Combustion Analysis 175

5.12 Organic Compounds 177

REVIEW Self-Assessment Quiz 179 Key Learning Outcomes 180 Key Terms 180 Key Concepts 181 Key Equations and Relationships 181

EXERCISES Review Questions 182 Problems by Topic 182 Cumulative Problems 186 Challenge Problems 186 Conceptual Problems 187 Answers to Conceptual Connections 187

6 Chemical Bonding I: Drawing Lewis Structures and Determining Molecular Shapes 188

6.1 Morphine: A Molecular Imposter 189

6.2 Electronegativity and Bond Polarity 190
Electronegativity 191 Bond Polarity, Dipole Moment, and Percent Ionic Character 192

6.3 Writing Lewis Structures for Molecular Compounds and Polyatomic Ions 194
Writing Lewis Structures for Molecular Compounds 194 Writing Lewis Structures for Polyatomic Ions 196

6.4 Resonance and Formal Charge 196
Resonance 196 Formal Charge 199

6.5 Exceptions to the Octet Rule: Odd-Electron Species, Incomplete Octets, and Expanded Octets 201
Odd-Electron Species 202 Incomplete Octets 202 Expanded Octets 203

6.6 Bond Energies and Bond Lengths 204
Bond Energy 205 Bond Length 206

6.7 VSEPR Theory: The Five Basic Shapes 207
Two Electron Groups: Linear Geometry 207 Three Electron Groups: Trigonal Planar Geometry 208 Four Electron Groups: Tetrahedral Geometry 208 Five Electron Groups: Trigonal Bipyramidal Geometry 209 Six Electron Groups: Octahedral Geometry 210

6.8 VSEPR Theory: The Effect of Lone Pairs 211
Four Electron Groups with Lone Pairs 211 Five Electron Groups with Lone Pairs 213 Six Electron Groups with Lone Pairs 214

6.9 VSEPR Theory: Predicting Molecular Geometries 215
Representing Molecular Geometries on Paper 218 Predicting the Shapes of Larger Molecules 218

6.10 Molecular Shape and Polarity 219
Vector Addition 221

REVIEW Self-Assessment Quiz 224 Key Learning Outcomes 225 Key Terms 225 Key Concepts 226 Key Equations and Relationships 226

EXERCISES Review Questions 226 Problems by Topic 227 Cumulative Problems 229 Challenge Problems 231 Conceptual Problems 231 Answers to Conceptual Connections 231

7 Chemical Bonding II: Valence Bond Theory and Molecular Orbital Theory 232

7.1 Oxygen: A Magnetic Liquid 233

7.2 Valence Bond Theory: Orbital Overlap as a Chemical Bond 234

7.3 Valence Bond Theory: Hybridization of Atomic Orbitals 236
sp^3 Hybridization 237 sp^2 Hybridization and Double Bonds 239 sp Hydridization and Triple Bonds 243 sp^3d and sp^3d^2 Hybridization 244 Writing Hybridization and Bonding Schemes 245

7.4 Molecular Orbital Theory: Electron Delocalization 248
Linear Combination of Atomic Orbitals (LCAO) 249 Second-Period Homonuclear Diatomic Molecules 252 Second-Period Heteronuclear Diatomic Molecules 258

7.5 Molecular Orbital Theory: Polyatomic Molecules 259

7.6 Bonding in Metals and Semiconductors 261
Bonding in Metals: The Electron Sea Model 261 Semiconductors and Band Theory 261 Doping: Controlling the Conductivity of Semiconductors 262

REVIEW Self-Assessment Quiz 263 Key Learning Outcomes 264 Key Terms 264 Key Concepts 264 Key Equations and Relationships 265

EXERCISES Review Questions 265 Problems by Topic 265 Cumulative Problems 267 Challenge Problems 268 Conceptual Problems 269 Answers to Conceptual Connections 269

8 Chemical Reactions and Chemical Quantities 270

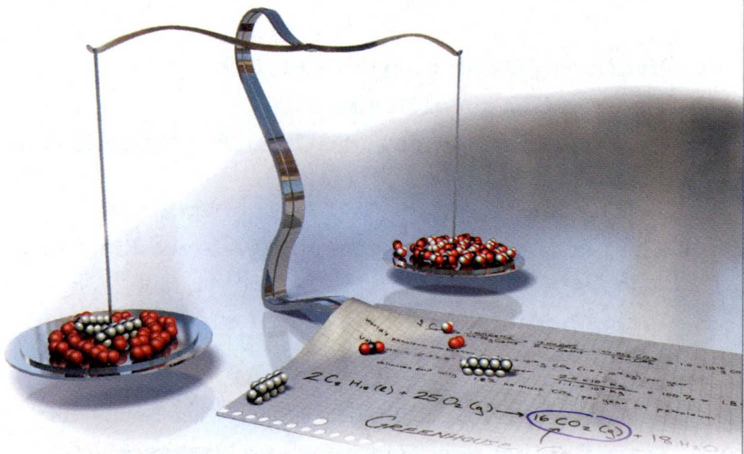

8.1 Climate Change and the Combustion of Fossil Fuels 271

8.2 Chemical Change 273

8.3 Writing and Balancing Chemical Equations 274

8.4 Reaction Stoichiometry: How Much Carbon Dioxide? 279
Making Pizza: The Relationships among Ingredients 279 Making Molecules: Mole-to-Mole Conversions 279 Making Molecules: Mass-to-Mass Conversions 280

8.5 Limiting Reactant, Theoretical Yield, and Percent Yield 283

8.6 Three Examples of Chemical Reactions: Combustion, Alkali Metals, and Halogens 289
Combustion Reactions 289 Alkali Metal Reactions 290 Halogen Reactions 290

REVIEW Self-Assessment Quiz 292 Key Learning Outcomes 292 Key Terms 293 Key Concepts 293 Key Equations and Relationships 293

EXERCISES Review Questions 293 Problems by Topic 294 Cumulative Problems 297 Challenge Problems 298 Conceptual Problems 299 Answers to Conceptual Connections 299

9 Introduction to Solutions and Aqueous Reactions 300

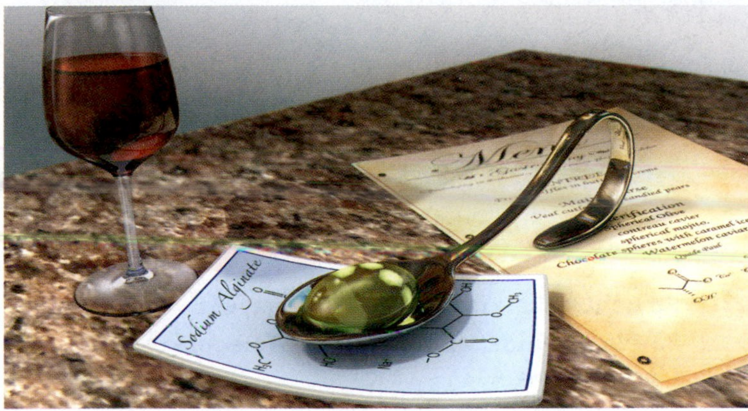

9.1 Molecular Gastronomy 301

9.2 Solution Concentration 302
Quantifying Solution Concentration 302 Using Molarity in Calculations 303 Solution Dilution 304

9.3 Solution Stoichiometry 307

9.4 Types of Aqueous Solutions and Solubility 308
Electrolyte and Nonelectrolyte Solutions 309 The Solubility of Ionic Compounds 311

9.5 Precipitation Reactions 313

9.6 Representing Aqueous Reactions: Molecular, Ionic, and Complete Ionic Equations 318

9.7 Acid–Base Reactions 319
Properties of Acids and Bases 320 Naming Oxyacids 322 Acid–Base Reactions 322 Acid–Base Titrations 324

9.8 Gas-Evolution Reactions 327

9.9 Oxidation–Reduction Reactions 328
Oxidation States 330 Identifying Redox Reactions 332

REVIEW Self-Assessment Quiz 335 Key Learning Outcomes 335 Key Terms 336 Key Concepts 336 Key Equations and Relationships 337

EXERCISES Review Questions 337 Problems by Topic 338 Cumulative Problems 340 Challenge Problems 340 Conceptual Problems 341 Answers to Conceptual Connections 341

10 Thermochemistry 342

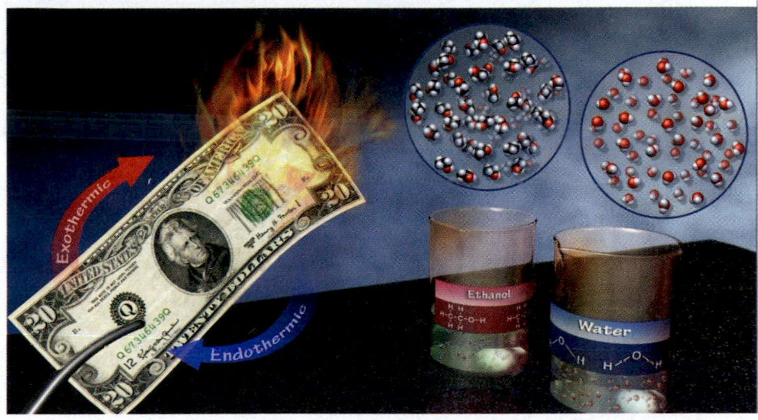

10.1 On Fire, But Not Consumed 343

10.2 The Nature of Energy: Key Definitions 344

10.3 The First Law of Thermodynamics: There Is No Free Lunch 346

10.4 Quantifying Heat and Work 349
Heat 349 Work: Pressure–Volume Work 353

10.5 Measuring ΔE for Chemical Reactions: Constant-Volume Calorimetry 355

10.6 Enthalpy: The Heat Evolved in a Chemical Reaction at Constant Pressure 358
Exothermic and Endothermic Processes: A Molecular View 360 Stoichiometry Involving ΔH: Thermochemical Equations 360

10.7 Measuring ΔH for Chemical Reactions: Constant-Pressure Calorimetry 362

10.8 Relationships Involving ΔH_{rxn} 364

10.9 Determining Enthalpies of Reaction from Bond Energies 367

10.10 Determining Enthalpies of Reaction from Standard Enthalpies of Formation 370
Standard States and Standard Enthalpy Changes 370 Calculating the Standard Enthalpy Change for a Reaction 372

10.11 Lattice Energies for Ionic Compounds 375
Calculating Lattice Energy: The Born–Haber Cycle 375 Trends in Lattice Energies: Ion Size 377 Trends in Lattice Energies: Ion Charge 377

REVIEW Self-Assessment Quiz 379 Key Learning Outcomes 380 Key Terms 381 Key Concepts 381 Key Equations and Relationships 382

EXERCISES Review Questions 382 Problems by Topic 383 Cumulative Problems 386 Challenge Problems 388 Conceptual Problems 388 Answers to Conceptual Connections 389

11 Gases 390

11.1 Supersonic Skydiving and the Risk of Decompression 391

11.2 Pressure: The Result of Particle Collisions 392
Pressure Units 393 The Manometer: A Way to Measure Pressure in the Laboratory 394

11.3 The Simple Gas Laws: Boyle's Law, Charles's Law, and Avogadro's Law 395
Boyle's Law: Volume and Pressure 395 Charles's Law: Volume and Temperature 397 Avogadro's Law: Volume and Amount (in Moles) 400

11.4 The Ideal Gas Law 401

11.5 Applications of the Ideal Gas Law: Molar Volume, Density, and Molar Mass of a Gas 404
Molar Volume at Standard Temperature and Pressure 404 Density of a Gas 404 Molar Mass of a Gas 406

11.6 Mixtures of Gases and Partial Pressures 407
Deep-Sea Diving and Partial Pressures 409 Collecting Gases over Water 412

11.7 A Particulate Model for Gases: Kinetic Molecular Theory 414
Kinetic Molecular Theory, Pressure, and the Simple Gas Laws 415 Kinetic Molecular Theory and the Ideal Gas Law 416

11.8 Temperature and Molecular Velocities 417

11.9 Mean Free Path, Diffusion, and Effusion of Gases 420

11.10 Gases in Chemical Reactions: Stoichiometry Revisited 422
Molar Volume and Stoichiometry 423

11.11 Real Gases: The Effects of Size and Intermolecular Forces 425
The Effect of the Finite Volume of Gas Particles 425 The Effect of Intermolecular Forces 426 Van der Waals Equation 427 Real Gases 427

REVIEW Self-Assessment Quiz 429 Key Learning Outcomes 430 Key Terms 430 Key Concepts 431 Key Equations and Relationships 431

EXERCISES Review Questions 432 Problems by Topic 433 Cumulative Problems 436 Challenge Problems 438 Conceptual Problems 438 Answers to Conceptual Connections 439

12 Liquids, Solids, and Intermolecular Forces 440

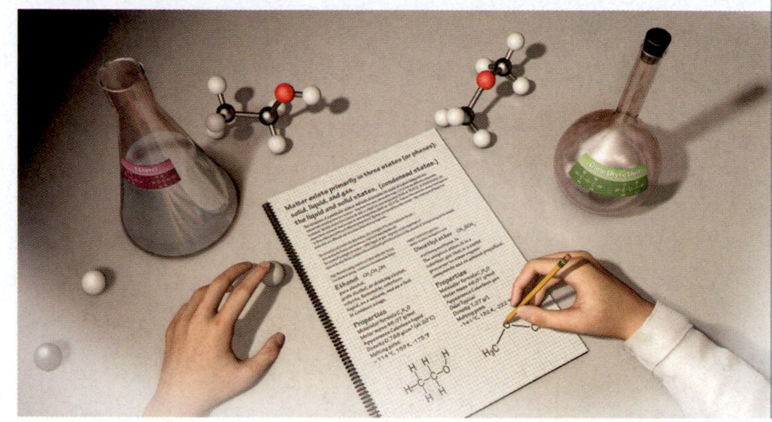

12.1 Structure Determines Properties 441

12.2 Solids, Liquids, and Gases: A Molecular Comparison 442
Changes between States 444

12.3 Intermolecular Forces: The Forces That Hold Condensed States Together 445
Dispersion Force 446 Dipole–Dipole Force 448 Hydrogen Bonding 450 Ion–Dipole Force 453

12.4 Intermolecular Forces in Action: Surface Tension, Viscosity, and Capillary Action 454
Surface Tension 454 Viscosity 455 Capillary Action 455

12.5 Vaporization and Vapor Pressure 456
The Process of Vaporization 456 The Energetics of Vaporization 457 Vapor Pressure and Dynamic Equilibrium 459 Temperature Dependence of Vapor Pressure and Boiling Point 461 The Critical Point: The Transition to an Unusual State of Matter 465

12.6 Sublimation and Fusion 466
Sublimation 466 Fusion 466 Energetics of Melting and Freezing 467

12.7 Heating Curve for Water 468

12.8 Water: An Extraordinary Substance 470

REVIEW Self-Assessment Quiz 472 Key Learning Outcomes 473 Key Terms 473 Key Concepts 473 Key Equations and Relationships 474

EXERCISES Review Questions 474 Problems by Topic 475 Cumulative Problems 477 Challenge Problems 478 Conceptual Problems 478 Answers to Conceptual Connections 479

13 Phase Diagrams and Crystalline Solids 480

13.1 Sliding Glaciers 481

13.2 Phase Diagrams 482
The Major Features of a Phase Diagram 482 Navigation within a Phase Diagram 483 The Phase Diagrams of Other Substances 484

13.3 Crystalline Solids: Determining Their Structure by X-Ray Crystallography 485

13.4 Crystalline Solids: Unit Cells and Basic Structures 487
The Unit Cell 488 Closest-Packed Structures 493

13.5 Crystalline Solids: The Fundamental Types 495
Molecular Solids 495 Ionic Solids 495 Atomic Solids 495

13.6 The Structures of Ionic Solids 497

13.7 Network Covalent Atomic Solids: Carbon and Silicates 498
Carbon 499 Silicates 501

REVIEW Self-Assessment Quiz 502 Key Learning Outcomes 503 Key Terms 503 Key Concepts 503 Key Equations and Relationships 504

EXERCISES Review Questions 504 Problems by Topic 504 Cumulative Problems 506 Challenge Problems 507 Conceptual Problems 507 Answers to Conceptual Connections 507

14 Solutions 508

14.1 Antifreeze in Frogs 509

14.2 Types of Solutions and Solubility 510
Nature's Tendency toward Mixing: Entropy 511 The Effect of Intermolecular Forces 511

14.3 Energetics of Solution Formation 514
Energy Changes during Solution Formation 515 Aqueous Solutions and Heats of Hydration 516

14.4 Solution Equilibrium and Factors Affecting Solubility 518
The Effect of Temperature on the Solubility of Solids 519 Factors Affecting the Solubility of Gases in Water 520

14.5 Expressing Solution Concentration 522
Molarity 523 Molality 524 Parts by Mass and Parts by Volume 524 Mole Fraction and Mole Percent 525

14.6 Colligative Properties: Vapor Pressure Lowering, Freezing Point Depression, Boiling Point Elevation, and Osmotic Pressure 528
Vapor Pressure Lowering 528 Vapor Pressures of Solutions Containing a Volatile (Nonelectrolyte) Solute 530 Freezing Point Depression and Boiling Point Elevation 533 Osmotic Pressure 537

14.7 Colligative Properties of Strong Electrolyte Solutions 539
Strong Electrolytes and Vapor Pressure 540 Colligative Properties and Medical Solutions 541

REVIEW Self-Assessment Quiz 543 Key Learning Outcomes 544 Key Terms 545 Key Concepts 545 Key Equations and Relationships 546

EXERCISES Review Questions 546 Problems by Topic 547 Cumulative Problems 550 Challenge Problems 551 Conceptual Problems 552 Answers to Conceptual Connections 553

15
Chemical Kinetics 554

15.1 Catching Lizards 555

15.2 Rates of Reaction and the Particulate Nature of Matter 556
The Concentration of the Reactant Particles 556 The Temperature of the Reactant Mixture 557 The Structure and Orientation of the Colliding Particles 557

15.3 Defining and Measuring the Rate of a Chemical Reaction 557
Defining Reaction Rate 558 Measuring Reaction Rates 561

15.4 The Rate Law: The Effect of Concentration on Reaction Rate 563
Determining the Order of a Reaction 564 Reaction Order for Multiple Reactants 565

15.5 The Integrated Rate Law: The Dependence of Concentration on Time 568
Integrated Rate Laws 569 The Half-Life of a Reaction 573

15.6 The Effect of Temperature on Reaction Rate 576
The Arrhenius Equation 576 Arrhenius Plots: Experimental Measurements of the Frequency Factor and the Activation Energy 578 The Collision Model: A Closer Look at the Frequency Factor 581

15.7 Reaction Mechanisms 583
Rate Laws for Elementary Steps 583 Rate-Determining Steps and Overall Reaction Rate Laws 584 Mechanisms with a Fast Initial Step 585

15.8 Catalysis 588
Homogeneous and Heterogeneous Catalysis 590 Enzymes: Biological Catalysts 591

REVIEW Self-Assessment Quiz 593 Key Learning Outcomes 595 Key Terms 596 Key Concepts 596 Key Equations and Relationships 597

EXERCISES Review Questions 597 Problems by Topic 598 Cumulative Problems 603 Challenge Problems 606 Conceptual Problems 607 Answers to Conceptual Connections 607

16
Chemical Equilibrium 608

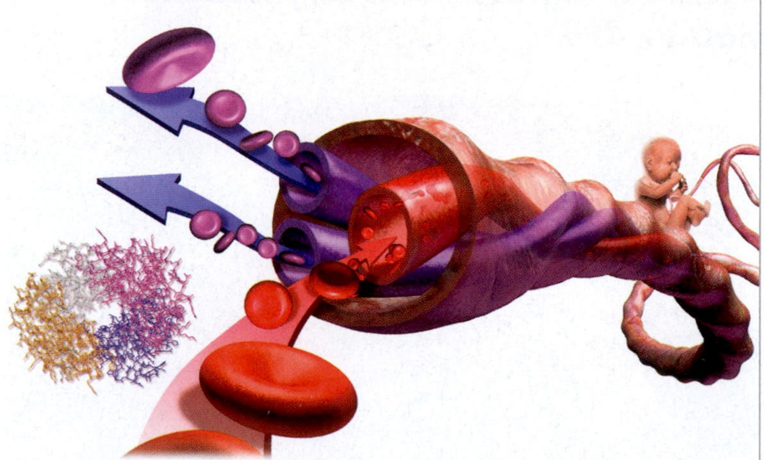

16.1 Fetal Hemoglobin and Equilibrium 609

16.2 The Concept of Dynamic Equilibrium 611

16.3 The Equilibrium Constant (*K*) 612
Expressing Equilibrium Constants for Chemical Reactions 614 The Significance of the Equilibrium Constant 614 Relationships between the Equilibrium Constant and the Chemical Equation 615

16.4 Expressing the Equilibrium Constant in Terms of Pressure 617
Units of *K* 619

16.5 Heterogeneous Equilibria: Reactions Involving Solids and Liquids 620

16.6 Calculating the Equilibrium Constant from Measured Equilibrium Concentrations 621

16.7 The Reaction Quotient: Predicting the Direction of Change 623

16.8 Finding Equilibrium Concentrations 626
Finding Equilibrium Concentrations from the Equilibrium Constant and All but One of the Equilibrium Concentrations of the Reactants and Products 626 Finding Equilibrium Concentrations from the Equilibrium Constant and Initial Concentrations or Pressures 627 Simplifying Approximations in Working Equilibrium Problems 632

16.9 Le Châtelier's Principle: How a System at Equilibrium Responds to Disturbances 636
The Effect of a Concentration Change on Equilibrium 636 The Effect of a Volume (or Pressure) Change on Equilibrium 638 The Effect of a Temperature Change on Equilibrium 641

REVIEW Self-Assessment Quiz 644 Key Learning Outcomes 645 Key Terms 645 Key Concepts 646 Key Equations and Relationships 646

EXERCISES Review Questions 647 Problems by Topic 647 Cumulative Problems 651 Challenge Problems 652 Conceptual Problems 653 Answers to Conceptual Connections 653

17

Acids and Bases 654

17.1 Batman's Basic Blunder 655

17.2 The Nature of Acids and Bases 656

17.3 Definitions of Acids and Bases 658
The Arrhenius Definition 658 The Brønsted–Lowry Definition 659

17.4 Acid Strength and Molecular Structure 661
Binary Acids 661 Oxyacids 662

17.5 Acid Strength and the Acid Ionization Constant (K_a) 663
Strong Acids 663 Weak Acids 664 The Acid Ionization Constant (K_a) 664

17.6 Autoionization of Water and pH 666
Specifying the Acidity or Basicity of a Solution: The pH Scale 668 pOH and Other p Scales 669

17.7 Finding the [H_3O^+] and pH of Strong and Weak Acid Solutions 670
Strong Acids 670 Weak Acids 671 Percent Ionization of a Weak Acid 676 Mixtures of Acids 678

17.8 Finding the [OH^-] and pH of Strong and Weak Base Solutions 680
Strong Bases 680 Weak Bases 681 Finding the [OH^-] and pH of Basic Solutions 682

17.9 The Acid–Base Properties of Ions and Salts 684
Anions as Weak Bases 684 Cations as Weak Acids 688 Classifying Salt Solutions as Acidic, Basic, or Neutral 689

17.10 Polyprotic Acids 691
Finding the pH of Polyprotic Acid Solutions 693 Finding the Concentration of the Anions for a Weak Diprotic Acid Solution 695

17.11 Lewis Acids and Bases 696
Molecules That Act as Lewis Acids 697 Cations That Act as Lewis Acids 697

REVIEW Self-Assessment Quiz 698 Key Learning Outcomes 699 Key Terms 699 Key Concepts 700 Key Equations and Relationships 700

EXERCISES Review Questions 701 Problems by Topic 701 Cumulative Problems 705 Challenge Problems 707 Conceptual Problems 707 Answers to Conceptual Connections 707

18

Aqueous Ionic Equilibrium 708

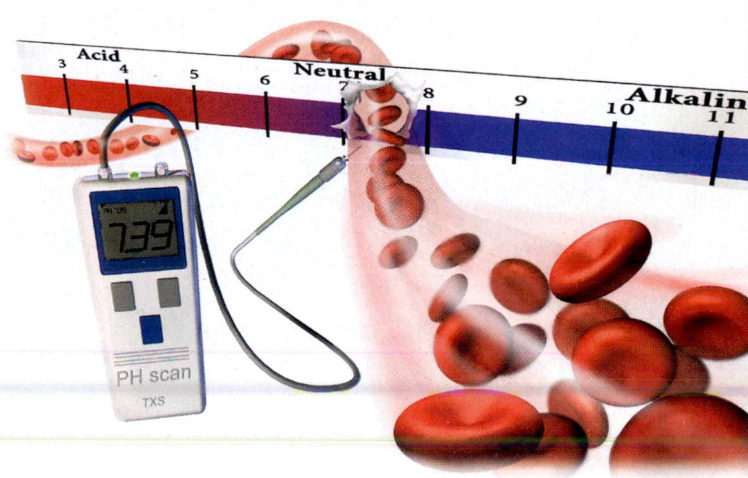

18.1 The Danger of Antifreeze 709

18.2 Buffers: Solutions That Resist pH Change 710
Calculating the pH of a Buffer Solution 712 The Henderson–Hasselbalch Equation 713 Calculating pH Changes in a Buffer Solution 716 Buffers Containing a Base and Its Conjugate Acid 720

18.3 Buffer Effectiveness: Buffer Range and Buffer Capacity 722
Relative Amounts of Acid and Base 722 Absolute Concentrations of the Acid and Conjugate Base 722 Buffer Range 723 Buffer Capacity 724

18.4 Titrations and pH Curves 725
The Titration of a Strong Acid with a Strong Base 726 The Titration of a Weak Acid with a Strong Base 730 The Titration of a Weak Base with a Strong Acid 735 The Titration of a Polyprotic Acid 736 Indicators: pH-dependent Colors 737

18.5 Solubility Equilibria and the Solubility-Product Constant 739
K_{sp} and Molar Solubility 740 K_{sp} and Relative Solubility 742 The Effect of a Common Ion on Solubility 743 The Effect of pH on Solubility 744

18.6 Precipitation 745
Selective Precipitation 747

18.7 Complex Ion Equilibria 748
The Effect of Complex Ion Equilibria on Solubility 750 The Solubility of Amphoteric Metal Hydroxides 752

REVIEW Self-Assessment Quiz 754 Key Learning Outcomes 755 Key Terms 756 Key Concepts 756 Key Equations and Relationships 757

EXERCISES Review Questions 757 Problems by Topic 758 Cumulative Problems 763 Challenge Problems 764 Conceptual Problems 764 Answers to Conceptual Connections 765

19

Free Energy and Thermodynamics 766

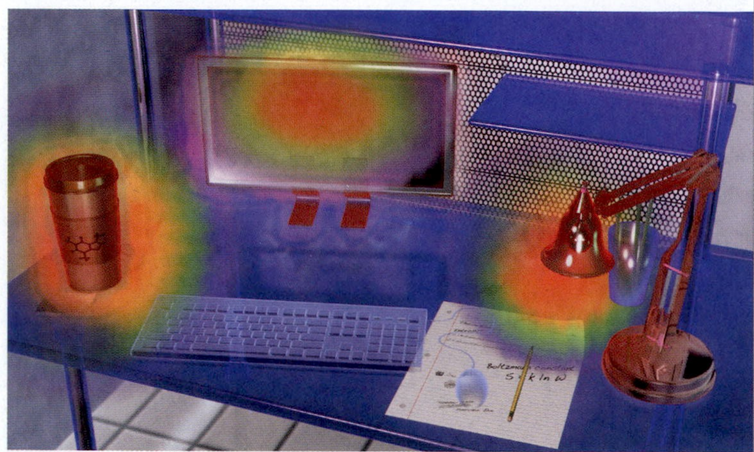

19.1 Energy Spreads Out 767

19.2 Spontaneous and Nonspontaneous Processes 768

19.3 Entropy and the Second Law of Thermodynamics 769
Entropy 770 The Second Law of Thermodynamics 771 Macrostates and Microstates 771 The Units of Entropy 773

19.4 Predicting Entropy and Entropy Changes for Chemical Reactions 774
The Entropy Change Associated with a Change in State 774 The Entropy Change Associated with a Chemical Reaction (ΔS°_{rxn}) 776 Standard Molar Entropies (S°) and the Third Law of Thermodynamics 776 Calculating the Standard Entropy Change (ΔS°_{rxn}) for a Reaction 779

19.5 Heat Transfer and Entropy Changes of the Surroundings 780
The Temperature Dependence of ΔS_{surr} 781 Quantifying Entropy Changes in the Surroundings 782

19.6 Gibbs Free Energy 784
The Effect of ΔH, ΔS, and T on Spontaneity 785

19.7 Free Energy Changes in Chemical Reactions: Calculating ΔG°_{rxn} 788
Calculating Standard Free Energy Changes with $\Delta G^\circ_{rxn} = \Delta H^\circ_{rxn} - T\Delta S^\circ_{rxn}$ 788 Calculating ΔG°_{rxn} with Tabulated Values of Free Energies of Formation 790 Calculating ΔG°_{rxn} for a Stepwise Reaction from the Changes in Free Energy for Each of the Steps 791 Making a Nonspontaneous Process Spontaneous 793 Why Free Energy Is "Free" 793

19.8 Free Energy Changes for Nonstandard States: The Relationship between ΔG°_{rxn} and ΔG_{rxn} 794

19.9 Free Energy and Equilibrium: Relating ΔG°_{rxn} to the Equilibrium Constant (K) 797
The Temperature Dependence of the Equilibrium Constant 799

REVIEW Self-Assessment Quiz 801 Key Learning Outcomes 802 Key Terms 803 Key Concepts 803 Key Equations and Relationships 803

EXERCISES Review Questions 804 Problems by Topic 805 Cumulative Problems 808 Challenge Problems 809 Conceptual Problems 810 Answers to Conceptual Connections 811

20

Electrochemistry 812

20.1 Lightning and Batteries 813

20.2 Balancing Oxidation–Reduction Equations 814

20.3 Voltaic (or Galvanic) Cells: Generating Electricity from Spontaneous Chemical Reactions 817
Electrochemical Cell Notation 820

20.4 Standard Electrode Potentials 822
Predicting the Spontaneous Direction of an Oxidation–Reduction Reaction 827 Predicting Whether a Metal Will Dissolve in Acid 829

20.5 Cell Potential, Free Energy, and the Equilibrium Constant 829
The Relationship between ΔG° and E°_{cell} 830 The Relationship between E°_{cell} and K 832

20.6 Cell Potential and Concentration 833
Concentration Cells 836

20.7 Batteries: Using Chemistry to Generate Electricity 838
Dry-Cell Batteries 838 Lead–Acid Storage Batteries 838 Other Rechargeable Batteries 839 Fuel Cells 840

20.8 Electrolysis: Driving Nonspontaneous Chemical Reactions with Electricity 841
Predicting the Products of Electrolysis 843 Stoichiometry of Electrolysis 847

20.9 Corrosion: Undesirable Redox Reactions 848

REVIEW Self-Assessment Quiz 851 Key Learning Outcomes 852 Key Terms 853 Key Concepts 853 Key Equations and Relationships 854

EXERCISES Review Questions 854 Problems by Topic 855 Cumulative Problems 857 Challenge Problems 859 Conceptual Problems 859 Answers to Conceptual Connections 859

21 Radioactivity and Nuclear Chemistry 860

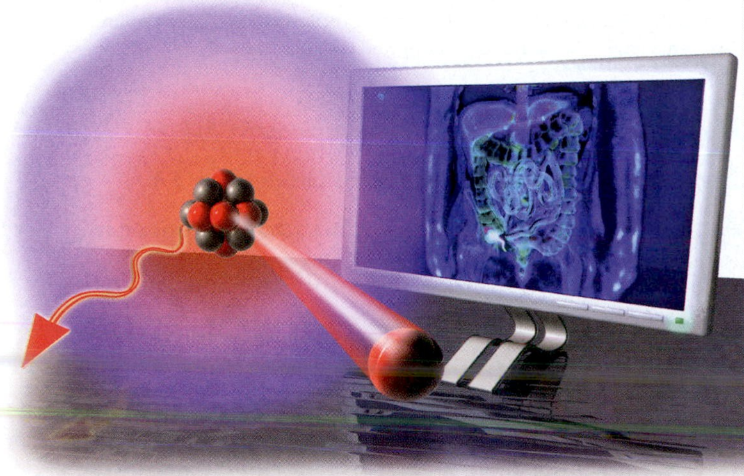

21.1 Diagnosing Appendicitis 861

21.2 The Discovery of Radioactivity 862

21.3 Types of Radioactivity 863
Alpha (α) Decay 864 Beta (β) Decay 865 Gamma (γ) Ray Emission 866 Positron Emission 866 Electron Capture 867

21.4 The Valley of Stability: Predicting the Type of Radioactivity 869
Magic Numbers 870 Radioactive Decay Series 871

21.5 Detecting Radioactivity 871

21.6 The Kinetics of Radioactive Decay and Radiometric Dating 872
The Integrated Rate Law 873 Radiocarbon Dating: Using Radioactivity to Measure the Age of Fossils and Artifacts 875 Uranium/Lead Dating 877

21.7 The Discovery of Fission: The Atomic Bomb and Nuclear Power 879
The Atomic Bomb 880 Nuclear Power: Using Fission to Generate Electricity 880

21.8 Converting Mass to Energy: Mass Defect and Nuclear Binding Energy 883
The Conversion of Mass to Energy 883 Mass Defect and Nuclear Binding Energy 884

21.9 Nuclear Fusion: The Power of the Sun 886

21.10 Nuclear Transmutation and Transuranium Elements 887

21.11 The Effects of Radiation on Life 888
Acute Radiation Damage 889 Increased Cancer Risk 889 Genetic Defects 889 Measuring Radiation Exposure 889

21.12 Radioactivity in Medicine and Other Applications 891
Diagnosis in Medicine 891 Radiotherapy in Medicine 892 Other Applications 893

REVIEW Self-Assessment Quiz 894 Key Learning Outcomes 895 Key Terms 895 Key Concepts 895 Key Equations and Relationships 896

EXERCISES Review Questions 897 Problems by Topic 897 Cumulative Problems 899 Challenge Problems 900 Conceptual Problems 900 Answers to Conceptual Connections 901

22 Organic Chemistry 902

22.1 Fragrances and Odors 903

22.2 Carbon: Why It Is Unique 904
Carbon's Tendency to Form Four Covalent Bonds 904 Carbon's Ability to Form Double and Triple Bonds 905 Carbon's Tendency to Catenate 905

22.3 Hydrocarbons: Compounds Containing Only Carbon and Hydrogen 905
Drawing Hydrocarbon Structures 906 Stereoisomerism and Optical Isomerism 909

22.4 Alkanes: Saturated Hydrocarbons 912
Naming Alkanes 913

22.5 Alkenes and Alkynes 916
Naming Alkenes and Alkynes 918 Geometric (Cis–Trans) Isomerism in Alkenes 920

22.6 Hydrocarbon Reactions 921
Reactions of Alkanes 922 Reactions of Alkenes and Alkynes 923

22.7 Aromatic Hydrocarbons 924
Naming Aromatic Hydrocarbons 925 Reactions of Aromatic Compounds 926

22.8 Functional Groups 928

22.9 Alcohols 929
Naming Alcohols 929 About Alcohols 929 Alcohol Reactions 930

22.10 Aldehydes and Ketones 931
Naming Aldehydes and Ketones 932 About Aldehydes and Ketones 932 Aldehyde and Ketone Reactions 933

22.11 Carboxylic Acids and Esters 934
Naming Carboxylic Acids and Esters 934 About Carboxylic Acids and Esters 934 Carboxylic Acid and Ester Reactions 935

22.12 Ethers 936
Naming Ethers 936 About Ethers 937

22.13 Amines 937
Amine Reactions 937

22.14 Polymers 937

REVIEW Self-Assessment Quiz 940 Key Learning Outcomes 941 Key Terms 941 Key Concepts 941 Key Equations and Relationships 942

EXERCISES Review Questions 943 Problems by Topic 944 Cumulative Problems 950 Challenge Problems 952 Conceptual Problems 953 Answers to Conceptual Connections 953

23 Transition Metals and Coordination Compounds 954

23.1 The Colors of Rubies and Emeralds 955

23.2 Properties of Transition Metals 956
Electron Configurations 956 Atomic Size 958 Ionization Energy 958 Electronegativity 959 Oxidation States 959

23.3 Coordination Compounds 960
Ligands 960 Coordination Numbers and Geometries 962 Naming Coordination Compounds 963

23.4 Structure and Isomerization 965
Structural Isomerism 965 Stereoisomerism 966

23.5 Bonding in Coordination Compounds 970
Valence Bond Theory 970 Crystal Field Theory 970

23.6 Applications of Coordination Compounds 975
Chelating Agents 975 Chemical Analysis 975 Coloring Agents 975 Biomolecules 975

REVIEW Self-Assessment Quiz 979 Key Learning Outcomes 980 Key Terms 980 Key Concepts 980 Key Equations and Relationships 981

EXERCISES Review Questions 981 Problems by Topic 981 Cumulative Problems 983 Challenge Problems 983 Conceptual Problems 984 Answers to Conceptual Connections 984

Appendix I The Units of Measurement A-1

Appendix II Significant Figure Guidelines A-6

Appendix III Common Mathematical Operations in Chemistry A-11
- A Scientific Notation A-11
- B Logarithms A-13
- C Quadratic Equations A-15
- D Graphs A-15

Appendix IV Useful Data A-17
- A Atomic Colors A-17
- B Standard Thermodynamic Quantities for Selected Substances at 25 °C A-17
- C Aqueous Equilibrium Constants at 25 °C A-23
- D Standard Reduction Half-Cell Potentials at 25 °C A-27
- E Vapor Pressure of Water at Various Temperatures A-28

Appendix V Answers to Selected End-of-Chapter Problems A-29

Appendix VI Answers to In-Chapter Practice Problems A-61

Glossary G-1

Credits C-1

Index I-1

Preface

To the Student

In this book, I tell the story of chemistry, a field of science that has not only revolutionized how we live (think of drugs designed to cure diseases or fertilizers that help feed the world), but also helps us to understand virtually everything that happens all around us all the time. The core of the story is simple: Matter is composed of particles, and the structure of those particles determines the properties of matter. Although these ideas may seem familiar to you as a 21st-century student, they were not so obvious as recently as 200 years ago. Yet, they are among the most powerful ideas in all of science. You need not look any further than the advances in biology over the last half-century to see how the particulate view of matter drives understanding. In that time, we have learned how even living things derive much of what they are from the particles (especially proteins and DNA) that compose them. I invite you to join the story as you read this book. Your part in its unfolding is yet to be determined, but I wish you the best as you start your journey.

Nivaldo J. Tro
tro@westmont.edu

To the Professor

In recent years, some chemistry professors have begun teaching their General Chemistry courses with what is now called an *atoms-first* approach. In a practical sense, the main thrust of this approach is a reordering of topics so that atomic theory and bonding models come much earlier than in the traditional approach. A primary rationale for this approach is that students should understand the theory and framework behind the chemical "facts" they are learning. For example, in the traditional approach students learn early that magnesium atoms tend to form ions with a charge of 2+. However, they don't understand *why* until much later (when they get to quantum theory). In an *atoms-first* approach, students learn quantum theory first and understand immediately why magnesium atoms form ions with a charge of 2+. In this way, students see chemistry as a more coherent picture and not just a jumble of disjointed facts.

From my perspective, the *atoms-first* movement is better understood—not in terms of topic order—but in terms of emphasis. Professors who teach with an *atoms-first* approach generally emphasize: (1) the particulate nature of matter; and (2) the connection between the *structure* of atoms and molecules and their *properties* (or their function). The result of this emphasis is that the topic order is rearranged to make these connections earlier, stronger, and more often than is possible with the traditional approach. Consequently, I have chosen to name this book *Chemistry: Structure and Properties*, and I have not included the phrase *atoms-first* in the title. From my perspective, the topic order grows out of the particulate emphasis, not the other way around.

In addition, by making the relationship between structure and properties the emphasis of the book, I extend that emphasis beyond just the topic order in the first half of the book. For example, in the chapter on acids and bases, a more traditional approach puts the relationship between the structure of an acid and its acidity toward the end of the chapter, and many professors even skip this material. In contrast, in this book, I cover this relationship early in the chapter, and I emphasize its importance in the continuing story of structure and properties. Similarly, in the chapter on free energy and thermodynamics, a traditional approach does not put much emphasis on the relationship between molecular structure and entropy. In this book, however, I emphasize this relationship and use it to tell the overall story of entropy and its ultimate importance in determining the direction of chemical reactions.

Throughout the course of writing this book and in conversations with many of my colleagues, I have also come to realize that the *atoms-first* approach has some unique challenges. For example, how do you teach quantum theory and bonding (with topics like bond energies) when you have not covered thermochemistry? Or how do you find laboratory activities for the first few weeks if you have not covered chemical quantities and stoichiometry? I have sought to develop solutions to these challenges in this book. For example, I have included a section on energy and its units in Chapter 2. This section introduces changes in energy and the concepts of exothermicity and endothermicity. These topics are therefore in place when you need them to discuss the energies of orbitals and spectroscopy in Chapter 3 and bond energies in Chapter 6. Similarly, I have introduced the mole concept in Chapter 2; this placement allows not only for a more even distribution of quantitative homework problems, but also for laboratory exercises that require the use of the mole concept. In addition, because I strongly support the efforts of my colleagues at the Examinations Institute of the American Chemical Society, and because I have sat on several committees that write the ACS General Chemistry exam, I have ordered the chapters in this book so that they can be used with those exams in their present form. The end result is a table of contents that emphasizes structure and properties, while still maintaining the overall traditional division of first- and second-semester topics.

For those of you who have used my other General Chemistry book (*Chemistry: A Molecular Approach*), you will find that this book is a bit shorter and more focused and streamlined. I have shortened some chapters, divided others in half, and completely eliminated three chapters (Biochemistry, Chemistry of the Nonmetals, and Metals and Metallurgy). These topics are simply not being taught much in most General Chemistry courses. *Chemistry: Structure and Properties* is a leaner and more efficient book that fits well with current trends that emphasize depth over breadth. Nonetheless, the main features that have made *Chemistry: A Molecular Approach* a success continue in this book. For example, strong problem-solving pedagogy, clear and concise writing, mathematical and chemical rigor, and dynamic art are all vital components of this book.

I hope that this book supports you in your vocation of teaching students chemistry. I am increasingly convinced of the importance of our task. Please feel free to e-mail me with any questions or comments about the book.

Nivaldo J. Tro
tro@westmont.edu

The Development Story

A great textbook starts with an author's vision, but that vision and its implementation must be continuously tested and refined to ensure that the book meets its primary goal—to teach the material in new ways that result in improved student learning. The development of a first edition textbook is an arduous process, typically spanning several years. This process is necessary to ensure that the content and pedagogical framework meet the educational needs of those who are in the classroom: *both* instructors and students.

The development of Dr. Tro's *Structure and Properties* was accomplished through a series of interlocking feedback loops. Each chapter was drafted by the author and subjected to an initial round of internal developmental editing, with a focus on making sure that the author's goal of "emphasizing the particulate nature of matter" was executed in a clear and concise way.

The chapters were then revised by the author and exposed to intensive reviewer scrutiny. We asked over 150 reviewers across the country to define what teaching with an *atoms-first* approach meant to them and to focus on how that philosophy was executed in *Chemistry: Structure and Properties.* They were also asked to analyze the table of contents and to read each chapter carefully. We asked them to evaluate the breadth and depth of coverage, the execution of the art program, the worked examples, and the overall pedagogical effectiveness of each chapter. The author and the development editor then worked closely together to analyze the feedback and determine which changes were necessary to improve each chapter.

In addition to reviews, we hosted six focus groups where professors scrutinized the details of several chapters and participated in candid group discussions with the author and editorial team. These group meetings not only focused on the content within the book, but also provided the author and participants with an opportunity to discuss the challenges they face each day in the classroom and what the author and the publisher could do to address these concerns in the book and within our media products. These sessions generated valuable insights that would have been difficult to obtain in any other way and were the inspiration for some significant ideas and improvements.

Class-Tested and Approved

General Chemistry students across the country also contributed to the development of *Chemistry: Structure and Properties.* Over 2000 students provided feedback through extensive class testing prior to publication. We asked students to use the chapters in place of, or alongside, their current textbook during their course. We then asked them to evaluate numerous aspects of the text, including how it explains difficult topics; how clear and understandable the writing style is; if the text helped them to see the "big picture" of chemistry through its macroscopic-to-microscopic organization of the material; and how well the Interactive Worked Examples helped them further understand the examples in the book. Through these student reviews, the strengths of *Chemistry: Structure and Properties* were put to the test, and it passed. Overwhelmingly, the majority of students who class tested would prefer to use *Chemistry: Structure and Properties* over their current textbook in their General Chemistry course!

In addition, our market development team interviewed over 75 General Chemistry instructors, gathering feedback on how well the *atoms-first* approach is carried out throughout the text; how well the text builds conceptual understanding; and how effective the end-of-chapter and practice material is. The team also reported on the accuracy and depth of the content overall. All comments, suggestions, and corrections were provided to the author and editorial team to analyze and address prior to publication.

Acknowledgments

The book you hold in your hands bears my name on the cover, but I am really only one member of a large team that carefully crafted this book. Most importantly, I thank my editor, Terry Haugen. Terry is a great editor and friend who really gets the *atoms-first* approach. He gives me the right balance of freedom and direction and always supports my efforts. Thanks, Terry, for all you have done for me and for the progression of the *atoms-first* movement throughout the world. I am also grateful for my project editor, Jessica Moro, who gave birth to her baby girl at about the same time that we gave birth to this book. Thanks Jessica for your hard labor on this project and congratulations on your beautiful baby! Thanks also to Coleen Morrison who capably filled in while Jessica was on maternity leave.

Thanks to Jennifer Hart, who has now worked with me on multiple editions of several books. Jennifer, your guidance, organizational skills, and wisdom are central to the success of my projects, and I am eternally grateful.

I also thank Erin Mulligan, who has now worked with me on several editions of multiple projects. Erin is an outstanding developmental editor, a great thinker, and a good friend. We work together almost seamlessly now, and I am lucky and grateful to have Erin on my team. I am also grateful to Adam Jaworski. His skills and competence have led the chemistry team at Pearson since he took over as editor-in-chief. And, of course, I am continually grateful to Paul Corey, with whom I have now worked for over 13 years and on 10 projects. Paul is a man of incredible energy and vision, and it is my great privilege to work with him. Paul told me many years ago (when he first signed me on to the Pearson team) to dream big, and then he provided the resources I needed to make those dreams come true. *Thanks, Paul.*

I would also like to thank my marketing manager, Jonathan Cottrell. Jonathan is wise, thoughtful, and outstanding at what he does. He knows how to convey ideas clearly and has done an amazing job at marketing and promoting this book. I am continually grateful for Quade and Emiko Paul, who make my ideas come alive with their art. We have also worked together on many projects over many editions, and I am continually impressed by their creativity and craftsmanship. I owe a special debt of gratitude to them. I am also grateful to Derek Bacchus and Elise Lansdon for their efforts in the design of this book.

Special thanks to Beth Sweeten and Gina Cheselka, whose skill and diligence gave this book its physical existence. I also appreciate the expertise and professionalism of my copy editor, Betty Pessagno, as well as the skill and diligence of Francesca Monaco and her colleagues at codeMantra. I am a picky author, and they always accommodate my seemingly endless requests. Thank you, Francesca.

I acknowledge the great work of my colleague Kathy Thrush Shaginaw, who put countless hours into developing the solutions manual. She is exacting, careful, and consistent, and I am so grateful for her hard work. I acknowledge the help of my colleagues Allan Nishimura,

Kristi Lazar, David Marten, Stephen Contakes, Michael Everest, and Carrie Hill who have supported me in my department while I worked on this book. I am also grateful to Gayle Beebe (President of Westmont College) and Mark Sargent (Provost of Westmont College) for giving me the time and space to work on my books. Thank you, Gayle and Mark, for allowing me to pursue my gifts and my vision.

I am also grateful to those who have supported me personally. First on that list is my wife, Ann. Her patience and love for me are beyond description, and without her, this book would never have been written. I am also indebted to my children, Michael, Ali, Kyle, and Kaden, whose smiling faces and love of life always inspire me. I come from a large Cuban family whose closeness and support most people would envy. Thanks to my parents, Nivaldo and Sara; my siblings, Sarita, Mary, and Jorge; my siblings-in-law, Jeff, Nachy, Karen, and John; my nephews and nieces, Germain, Danny, Lisette, Sara, and Kenny. These are the people with whom I celebrate life.

I would like to thank all of the General Chemistry students who have been in my classes throughout my 23 years as a professor at Westmont College. You have taught me much about teaching that is now in this book. I am especially grateful to Michael Tro who put in many hours proofreading my manuscript, working problems and quiz questions, and organizing art codes and appendices. Michael, you are an amazing kid—it is my privilege to have you work with me on this project. I would also like to express my appreciation to Katherine Han, who was a tremendous help with proofreading and self-assessment quizzes.

I would like to thank Brian Woodfield and Ed McCulloph for helping me create the interactive worked examples and Key Concept Videos.

Lastly, I am indebted to the many reviewers, listed on the following pages, whose ideas are imbedded throughout this book. They have corrected me, inspired me, and sharpened my thinking on how best to emphasize structure and properties while teaching chemistry. I deeply appreciate their commitment to this project. Last but by no means least, I would like to thank Alyse Dilts, Brian Gute, Jim Jeitler, Milt Johnston, Jessica Parr, Binyomin Abrams, and Allison Soult for their help in reviewing page proofs.

Faculty Advisory Board

Stacey Brydges, *University of California—San Diego*
Amina El-Ashmawy, *Collin College*
Lee Friedman, *University of Maryland*
Margie Haak, *Oregon State University*
Willem Leenstra, *University of Vermont*
Douglas Mulford, *Emory University*
Dawn Richardson, *Collin College*
Ali Sezer, *California University of Pennsylvania*

Focus Group Participants

We would like to thank the following professors for contributing their valuable time to meet with the author and the publishing team in order to provide a meaningful perspective on the most important challenges they face in teaching General Chemistry and give us insight into creating a new General Chemistry text that successfully responds to those challenges.

Focus Group 1

Stacey Brydges, *University of California—San Diego*
Amina El-Ashmawy, *Collin College*
Tracy Hamilton, *University of Alabama, Birmingham*
David Jenkins, *University of Tennessee*
Daniel Knauss, *Colorado School of Mines*
Willem Leenstra, *University of Vermont*
Daniel Moriarty, *Siena College*
Clifford Murphy, *Roger Williams University*
Jodi O'Donnell, *Siena College*
Ali Sezer, *California University of Pennsylvania*
Mark Watry, *Franciscan University of Steubenville*
Paul Wine, *Georgia Institute of Technology*
Lin Zhu, *Indiana University Purdue University Indianapolis*

Focus Group 2

David Boatright, *University of West Georgia*
Jon Camden, *University of Tennessee, Knoxville*
Kathleen Carrigan, *Portland Community College*
Sandra Chimon-Peszek, *DePaul University*
Amina El-Ashmawy, *Collin College*
Nicole Grove, *Western Wyoming Community College*
Margie Haak, *Oregon State University*
Antony Hascall, *Northern Arizona University*
Richard Jew, *University of North Carolina, Charlotte*
Willem Leenstra, *University of Vermont*
Douglas Mulford, *Emory University*
Daphne Norton, *University of Georgia*
Allison Wind, *Middle Tennessee State University*
Lioudmila Woldman, *Florida State College, Jacksonville*

Focus Group 3

Cynthia Judd, *Palm Beach State College*
Farooq Khan, *University of West Georgia*
Zhengrong Li, *Southeastern Louisiana University*
Tracy McGill, *Emory University*
David Perdian, *Broward College*
Thomas Sommerfeld, *Southern Louisiana University*
Shane Street, *University of Alabama*
Carrie Shepler, *Georgia Institute of Technology*

Focus Group 4

William Cleaver, *University of Texas at Arlington*
Deanna Dunlavy, *New Mexico State University*
Susan Hendrickson, *University of Colorado, Boulder*
Christian Madu, *Collin College*
Dawn Richardson, *Collin College*
Alan Van Orden, *Colorado State University*
Kristin Ziebart, *Oregon State University*

Focus Group 5

Mary Jo Bojan, *Pennsylvania State University*
Leslie Farris, *University of Massachusetts, Lowell*
Amy Irwin, *Monroe Community College*
Janet Schrenk, *University of Massachusetts, Lowell*
Lori Van Der Sluys, *Pennsylvania State University*
Michael Vannatta, *West Virginia University*
Joshua Wallach, *Old Dominion University*
Suzanne Young, *University of Massachusetts, Lowell*

Focus Group 6

Bryan Breyfogle, *Missouri State University*
Gregory Ferrence, *Illinois State University*
Brian Gute, *University of Minnesota, Duluth*
Daniel Kelly, *Indiana University Northwest*
Vanessa McCaffrey, *Albion College*
Yasmin Patell, *Kansas State University*
Lynmarie Posey, *Michigan State University*
Jen Snyder, *Ozark Technical College*
Cathrine Southern, *DePaul University*
Hongqiu Zhao, *Indiana University-Purdue University*

Accuracy Reviewers

Alyse Dilts, *Harrisburg Area Community College*
Brian Gute, *University of Minnesota—Duluth*
Jim Jeitler, *Marietta College*
Milton Johnston, *University of South Florida*
Jessica Parr, *University of Southern California*
Allison Soult, *University of Kentucky*

Chapter Reviewers

Binyomin Abrams, *Boston University*
David Ballantine, *Northern Illinois University*
Mufeed Basti, *North Carolina A&T State University*
Sharmistha Basu-Dutt, *University of West Georgia*
Shannon Biros, *Grand Valley State University*
John Breen, *Providence College*
Nicole Brinkman, *University of Notre Dame*
Mark Campbell, *United States Naval Academy*
Sandra Chimon-Peszek, *DePaul University*
Margaret Czerw, *Raritan Valley Community College*
Richard Farrer, *Colorado State University—Pueblo*
Debbie Finocchio, *University of San Diego*
Andrew Frazer, *University of Central Florida*
Kenneth Friedrich, *Portland Community College*
Anthony Gambino, *State College of Florida*
Harold Harris, *University of Missouri—St. Louis*
David Henderson, *Trinity College*
Jim Jeitler, *Marietta College*
Milton Johnston, *University of South Florida*
Scott Kennedy, *Anderson University*
Farooq Khan, *University of West Georgia*
Angela King, *Wake Forest University*
John Kiser, *Western Piedmont Community College*
Robert LaDuca, *Michigan State University*
Joseph Lanzafame, *Rochester Institute of Technology*
Rita Maher, *Richland College*
Marcin Majda, *University of California—Berkeley*
Tracy McGill, *Emory University*
Vanessa McCaffrey, *Albion College*
Gail Meyer, *University of Tennessee—Chattanooga*
Daniel Moriarty, *Siena College*
Gary Mort, *Lane Community College*
Richard Mullins, *Xavier University*
Clifford Murphy, *Roger Williams University*
Anne-Marie Nickel, *Milwaukee School of Engineering*
Chifuru Noda, *Bridgewater State University*
Stacy O'Reilly, *Butler University*
Edith Osborne, *Angelo State University*
Jessica Parr, *University of Southern California*
Yasmin Patell, *Kansas State University*
Thomas Pentecost, *Grand Valley State University*
Robert Pike, *College of William and Mary*
Karen Pressprich, *Clemson University*
Robert Rittenhouse, *Central Washington University*
Al Rives, *Wake Forest University*
Steven Rowley, *Middlesex Community College—Edison*
Raymond Sadeghi, *University of Texas—San Antonio*
Jason Schmeltzer, *University of North Carolina—Asheville*
Sarah Siegel, *Gonzaga University*
Jacqueline Smits, *Bellevue College*
David Son, *Southern Methodist University*
Kimberly Stieglitz, *Roxbury Community College*
John Stubbs, *University of New England*
Steven Tait, *Indiana University*
Dennis Taylor, *Clemson University*
Stephen Testa, *University of Kentucky*
Thomas Ticich, *Centenary College of Louisiana*
Paula Weiss, *Oregon State University*
Wayne Wesolowski, *University of Arizona*
Kimberly Woznack, *California University of Pennsylvania*
Dan Wright, *Elon University*
Darrin York, *Rutgers University*
Lin Zhu, *Indiana University, Purdue University Indianapolis*

Class Test Participants

Keith Baessler, *Suffolk County Community College*
James Bann, *Wichita State University*
Ericka Barnes, *Southern Connecticut State University*
Sharmistha Basu-Dutt, *University of West Georgia*
Richard Bell, *Pennsylvania State University—Altoona*
David Boatright, *University of West Georgia*
Shannon Biros, *Grand Valley State University*
Charles Burns Jr., *Wake Technical Community College*
Sarah Dimick Gray, *Metropolitan State University*
Tara Carpenter, *University of Maryland—Baltimore Country*
David Dearden, *Brigham Young University*
Barrett Eichler, *Augustana College*
Amina El-Ashmawy, *Collin College*
Mark Ellison, *Ursinus College*
Robin Ertl, *Marywood University*
Sylvia Esjornson, *Southwestern Oklahoma State University*
Renee Falconer, *Colorado School of Mines*
Richard Farrer, *Colorado State University—Pueblo*
Christine Gaudinski, *AIMS Community College*
Nicole Grove, *Western Wyoming Community College*
Alexander Grushow, *Rider University*
Brian Gute, *University of Minnesota—Duluth*
Janet Haff, *Suffolk County Community College*
Eric Hawrelak, *Bloomsburg State University of Pennsylvania*
Renee Henry, *University of Colorado—Colorado Springs*
Deborah Hokien, *Marywood University*
Donna Iannotti, *Brevard College*
Milton Johnston, *University of South Florida*
Jason Kahn, *University of Maryland*
Rick Karpeles, *University of Massachusetts—Lowell*
Daniel Kelly, *Indiana University— Northwest*
Vivek Kumar, *Suffolk County Community College*
Fiona Lihs, *Estrella Mountain Community College*
Douglas Linder, *Southwestern Oklahoma State University*
Daniel Moriarty, *Siena College*
Douglas Mulford, *Emory University*
Maureen Murphy, *Huntingdon College*
Chifuru Noda, *Bridgewater State University*
Jodi O' Donnell, *Siena College*
Stacy O'Reilly, *Butler University*
John Ondov, *University of Maryland*
Robert Pike, *College of William and Mary*
Curtis Pulliam, *Utica College*
Jayashree Ranga, *Salem State University*
Patricia Redden, *Saint Peter's University*
Michael Roper, *Frontrange Community College*
Sharadha Sambasivan, *Suffolk County Community College*
Stephen Schvaneveldt, *Clemson University*
Carrie Shepler, *Georgia Institute of Technology*
Kim Shih, *University of Massachusetts—Lowell*
Janet Schrenk, *University of Massachusetts—Lowell*
Sarah Siegel, *Gonzaga University*
Gabriela Smeureanu, *Hunter College*
Thomas Sorensen, *University of Wisconsin—Milwaukee*
Allison Soult, *University of Kentucky*
Katherine Swanson, *University of Minnesota—Duluth*
Dennis Taylor, *Clemson University*
Nicolay Tsarevsky, *Southern Methodist University*
Col. Michael Van Valkenburg, *United States Air Force Academy*
Jeffrey Webb, *Southern Connecticut State University*
David Zax, *Cornell University*
Hongquin Zhao, *Indiana University, Purdue University Indianapolis*
Lin Zhu, *Indiana University, Purdue University Indianapolis*
Brian Zoltowski, *Southern Methodist University*
James Zubricky, *University of Toledo*

Dear Colleague:

In recent years, many chemistry professors have begun teaching their General Chemistry courses with what is now called an *atoms-first* approach. On the surface, this approach may seem like a mere reordering of topics, so that atomic theory and bonding theories come much earlier than in the traditional approach. A rationale for this reordering is that students should understand the theory and framework behind the chemical "facts" they are learning. For example, in the traditional approach students learn early that magnesium atoms tend to form ions with a charge of 2+. However, they don't understand why until much later (when they get to quantum theory). In an atoms-first approach, students learn quantum theory first and understand immediately why magnesium atoms form ions with a charge of 2+. In this way, students see chemistry as a more coherent picture and not just a jumble of disjointed facts.

From my perspective, however, the *atoms-first* movement is much more than just a reordering or topics. To me, the *atoms-first* movement is a result of the growing emphasis in chemistry courses on the two main ideas of chemistry: a) *that matter is particulate*, and b) *that the structure of those particles determines the properties of matter.* In other words, the atoms-first movement is—at its core—an attempt to tell the story of chemistry in a more unified and thematic way. As a result, an atoms-first textbook must be more than a rearrangement of topics: it must tell the story of chemistry through the lens of the particulate model of matter. That is the book that I present to you here. The table of contents reflects the ordering of an atoms-first approach, but more importantly, the entire book is written and organized so that the theme—*structure determines properties*—unifies and animates the content. My hope is that students will see the power and beauty of the simple ideas that lie at the core of chemistry, and that they may learn to apply them to see and understand the world around them in new ways.

"My hope is that students will see the power and beauty of the simple ideas that lie at the core of chemistry."

Niva

150 **Peer reviewers**

who scrutinized each chapter and provided feedback on everything from content and organization to art and pedagogy.

75 **Instructors**

who tested chapters in their own classrooms and advised how students interacted with and learned from the content.

50 **Focus Group Participants**

who joined Dr. Tro and the editorial team for in-person candid discussions on the challenges they face in their classrooms and how we could address those challenges in the book and within our media products.

Structure and Properties was developed with the goal of presenting the story of chemistry in a unified way. To ensure that the book consistently emphasizes the theme—*structure determines properties*—Dr. Tro consulted a community of general chemistry instructors teaching with an atoms-first approach.

What Instructors are Saying:

This book is exactly what I have been looking for in a book. It has what I would consider the perfect order of topics. It has a true atoms-first approach.

Ken Friedrich — Portland Community College

Chemistry: Structures and Properties is a student-friendly text, offering a pedagogically sound treatment of an atoms first approach to chemistry. With its well-written text, supporting figures and worked examples, students have access to a text possessing the potential to maximize their learning.

Christine Mina Kelly — University of Colorado

It is an outstanding, very well written text that nails the "atoms-first" approach. The book is clear, concise and entertaining to read.

Richard Mullins — Xavier University

Dr. Tro takes excellent artwork, excellent worked examples, and excellent explanations and combines them in an Atoms First General Chemistry book that raises the bar for others to follow.

John Kiser — Western Piedmont Community College

Niva Tro presents the science of chemistry using a very warm writing style and approach that connects well with both the student and scientist reader.

Amina El-Ashwamy/Collin County CC

2,000

Student Class Testers

In addition to peer reviews, general chemistry students across the country also contributed to the development of *Chemistry: Structure and Properties*. Students were asked to use chapters in place of, or alongside, their current textbook during their course and provide feedback to the author and editorial team.

What Students are Saying:

"This sample is really unlike any chemistry book I've ever seen. The examples and breakdowns of problems were awesome. The concepts are clear and down to earth. This book just makes it seem like the author really wants you to get it."

Kenneth Bell — Colorado School of Mines

"It is the best text I've read that clearly and concisely presents chemistry concepts in a fun and organized way!"

Peter Inirio — Marywood University

"I think that sometimes in chemistry, it's very hard to see the "big picture." I thought that this textbook did a great job with that by organizing the material and making me think about how it relates to real life."

Megan Little — University of Massachusetts Lowell

"I really enjoyed how this chapter/author doesn't assume your knowledge of prerequisite material. Going from macro to micro allows the reader/student to truly conceptualize all aspects of the material. The organization and step-by-step approach delivers the chapter in a simple yet thorough manner. This booklet helped me tremendously, thank you."

Meghan Berthold — Collin County Community College

"Students need to learn chemistry in a way that is not intimidating. My current textbook had language that was too advanced for a beginner. This book was a fresh breath of air that made me relax and understand the topics better than ever before."

Megan Van Doren — Bloomsburg University

"It was very similar to a classroom format, giving me the confidence to solve problems on my own."

Zachary Ghalayini — University of South Florida

Unifying Theme of Structure and Properties

Section 1.1 – Introduction to the theme

1.1 A Particulate View of the World: Structure Determines Properties

A good novel usually has a strong *premise*—a short statement that describes the central idea of the story. The story of chemistry as described in this book also has a strong premise, which consists of two simple statements:

1. Matter is particulate—it is composed of particles.
2. The structure of those particles determines the properties of matter.

Matter is anything that occupies space and has mass. Most things you can think of—such as this book, your desk, and even your body—are composed of matter. The particulate nature of matter—first

Section 4.1 – How the structure of Al atoms determines the density of aluminum metal

The densities of elements and the radii of their atoms are examples of *periodic properties*. A **periodic property** is one that is generally predictable based on an element's position within the periodic table. In this chapter, we examine several periodic properties of elements, including atomic radius, ionization energy, and electron affinity. As we do, we will see that these properties—as well as the overall arrangement of the periodic table—are explained by quantum-mechanical theory, which we first examined in Chapter 3. *Quantum-mechanical theory explains the electronic structure of atoms—this in turn determines the properties of those atoms.*

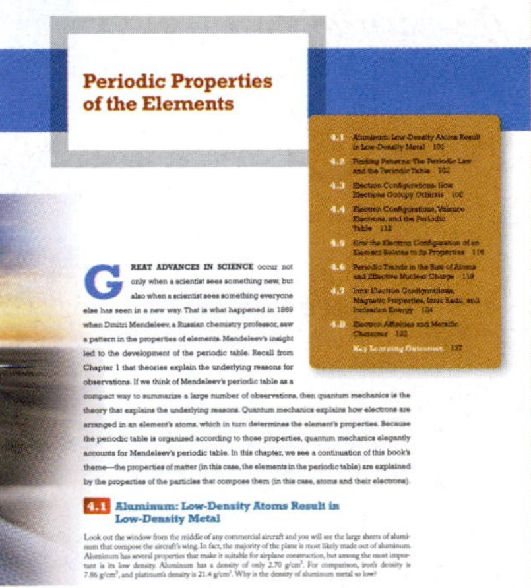

Periodic Properties of the Elements

4.1 Aluminum: Low-Density Atoms Result in Low-Density Metal 101
4.2 Finding Patterns: The Periodic Law and the Periodic Table 102
4.3 Electron Configurations: How Electrons Occupy Orbitals 106
4.4 Electron Configurations, Valence Electrons, and the Periodic Table 112
4.5 How the Electron Configuration of an Element Relates to Its Properties 116
4.6 Periodic Trends in the Size of Atoms and Effective Nuclear Charge 119
4.7 Ions: Electron Configurations, Magnetic Properties, Ionic Radii, and Ionization Energy 124
4.8 Electron Affinities and Metallic Character 132
Key Learning Outcomes 137

GREAT ADVANCES IN SCIENCE occur not only when a scientist sees something new, but also when a scientist sees something everyone else has seen in a new way. That is what happened in 1869 when Dmitri Mendeleev, a Russian chemistry professor, saw a pattern in the properties of elements. Mendeleev's insight led to the development of the periodic table. Recall from Chapter 1 that theories explain the underlying reasons for observations. If we think of Mendeleev's periodic table as a compact way to summarize a large number of observations, then quantum mechanics is the theory that explains the underlying reasons. Quantum mechanics explains how electrons are arranged in an element's atoms, which in turn determines the element's properties. Because the periodic table is organized according to those properties, quantum mechanics elegantly accounts for Mendeleev's periodic table. In this chapter, we see a continuation of this book's theme—the properties of matter (in this case, the elements in the periodic table) are explained by the properties of the particles that compose them (in this case, atoms and their electrons).

4.1 Aluminum: Low-Density Atoms Result in Low-Density Metal

Look out the window from the middle of any commercial aircraft and you will see the large sheets of aluminum that compose the aircraft's wing. In fact, the majority of the plane is most likely made out of aluminum. Aluminum has several properties that make it suitable for airplane construction, but among the most important is its low density. Aluminum has a density of only 2.70 g/cm^3. For comparison, iron's density is 7.86 g/cm^3, and platinum's density is 21.4 g/cm^3. Why is the density of aluminum metal so low?

Section 4.5 – How atomic structure determines the properties of the elements

4.5 How the Electron Configuration of an Element Relates to Its Properties

As we discussed in Section 4.4, *the chemical properties of elements are largely determined by the number of valence electrons they contain.* The properties of elements are periodic because the number of valence electrons is periodic. Mendeleev grouped elements into families (or columns) based on observations about their properties. We now know that elements in a family have the same number of valence electrons. In other words, elements in a family have similar properties because they have the same number of valence electrons.

Section 6.1 – How the structure of morphine allows it to be a molecular imposter for the body's natural endorphins

Morphine binds to opioid receptors because it fits into a special pocket (called the active site) on the opioid receptor protein (just as a key fits into a lock) that normally binds endorphins. Certain parts of the morphine molecule have a similar enough shape to endorphins that they fit the lock (even though they are not the original key). In other words, morphine is a *molecular imposter*, mimicking the action of endorphins because of similarities in shape.

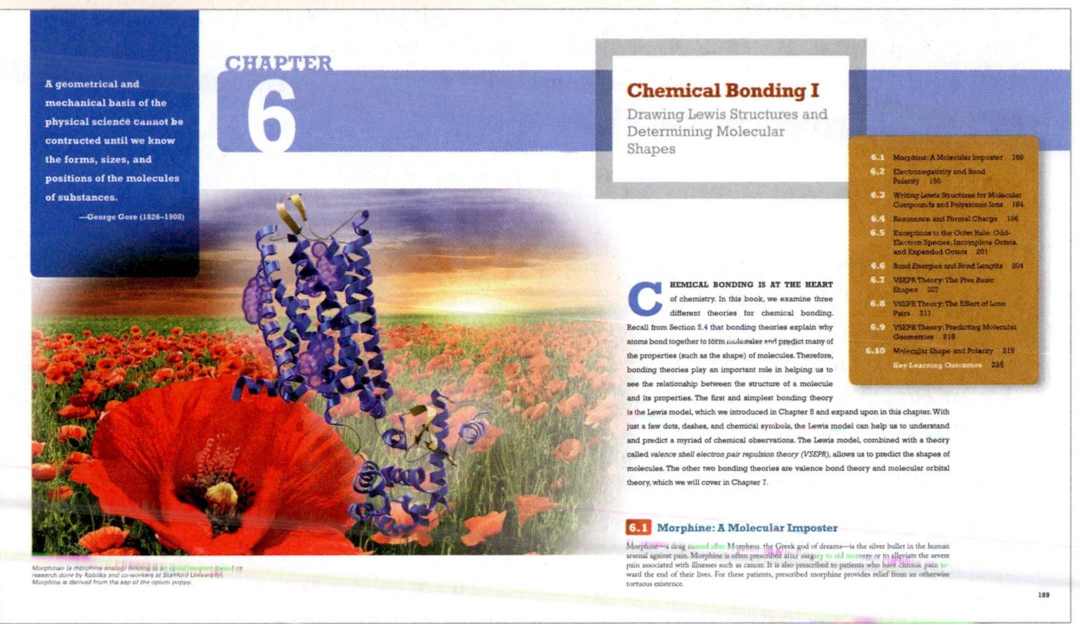

A geometrical and mechanical basis of the physical science cannot be contructed until we know the forms, sizes, and positions of the molecules of substances.

—George Gore (1826–1908)

CHAPTER 6

Chemical Bonding I

Drawing Lewis Structures and Determining Molecular Shapes

6.1 Morphine: A Molecular Imposter 189
6.2 Electronegativity and Bond Polarity 190
6.3 Writing Lewis Structures for Molecular Compounds and Polyatomic Ions 194
6.4 Resonance and Formal Charge 196
6.5 Exceptions to the Octet Rule: Odd-Electron Species, Incomplete Octets, and Expanded Octets 201
6.6 Bond Energies and Bond Lengths 204
6.7 VSEPR Theory: The Five Basic Shapes 207
6.8 VSEPR Theory: The Effect of Lone Pairs 211
6.9 VSEPR Theory: Predicting Molecular Geometries 216
6.10 Molecular Shape and Polarity 219
Key Learning Outcomes 226

CHEMICAL BONDING IS AT THE HEART of chemistry. In this book, we examine three different theories for chemical bonding. Recall from Section 5.4 that bonding theories explain why atoms bond together to form molecules and predict many of the properties (such as the shape) of molecules. Therefore, bonding theories play an important role in helping us to see the relationship between the structure of a molecule and its properties. The first and simplest bonding theory is the Lewis model, which we introduced in Chapter 5 and expand upon in this chapter. With just a few dots, dashes, and chemical symbols, the Lewis model can help us to understand and predict a myriad of chemical observations. The Lewis model, combined with a theory called *valence shell electron pair repulsion theory (VSEPR)*, allows us to predict the shapes of molecules. The other two bonding theories are valence bond theory and molecular orbital theory, which we will cover in Chapter 7.

6.1 Morphine: A Molecular Imposter

Morphine—a drug named after Morpheus, the Greek god of dreams—is the silver bullet in the human arsenal against pain. Morphine is often prescribed after surgery to aid recovery or to alleviate the severe pain associated with illnesses such as cancer. It is also prescribed to patients who have chronic pain toward the end of their lives. For these patients, prescribed morphine provides relief from an otherwise tortuous existence.

Morphinan (a morphine analog) binding to an opioid receptor (based on research done by Kobilka and co-workers at Stanford University). Morphine is derived from the sap of the opium poppy.

189

Section 6.10 – How molecular structure determines whether a substance is polar or nonpolar

6.10 Molecular Shape and Polarity

In Section 6.2, we discussed polar bonds. Entire molecules can also be polar, depending on their shape and the nature of their bonds. For example, if a diatomic molecule has a polar bond, the molecule as a whole will be polar.

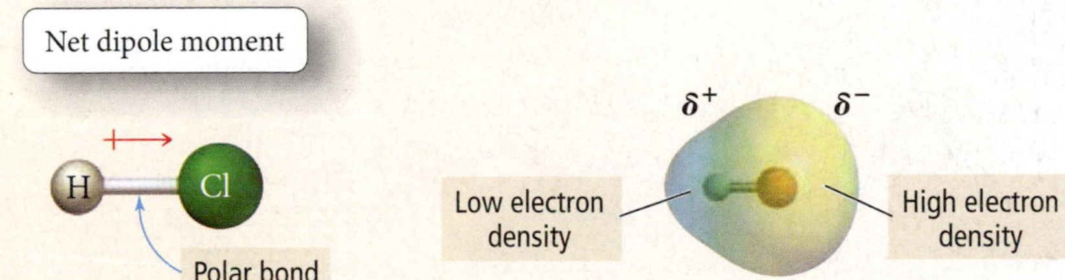

In the figure shown here the image to the right is an electrostatic potential map of HCl. In these maps, red areas indicate electron-rich regions in the molecule and the blue areas indicate electron-poor regions. Yellow indicates moderate electron density. Notice that the region around the more

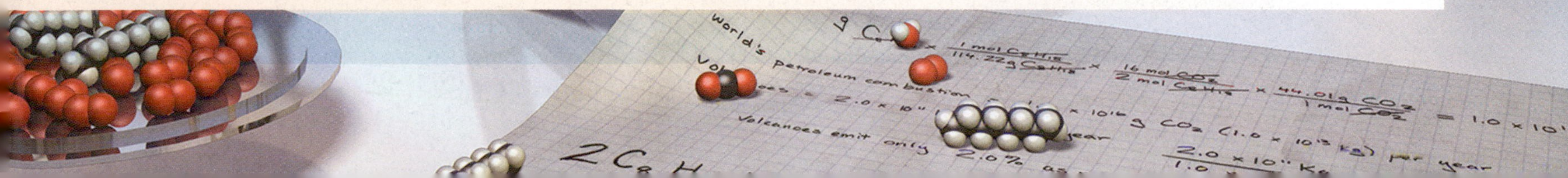

Structure and Properties: Unified Theme Carries through the Second Semester

Section 12.1 – How ethanol and dimethyl ether are composed of exactly the same atoms, but their different structures result in different properties

12.1 Structure Determines Properties

Ethanol and dimethyl ether are isomers—they have the same chemical formula, C_2H_6O but are different compounds. In ethanol, the nine atoms form a molecule that is a liquid at room temperature (boils at 78.3 °C). In dimethyl ether, the atoms form a molecule that is a gas at room temperature (boils at −22.0 °C). How can the same nine atoms bond together to form molecules with such different properties? By now, you should know the answer—the structures of these two molecules are different, and *structure determines properties*.

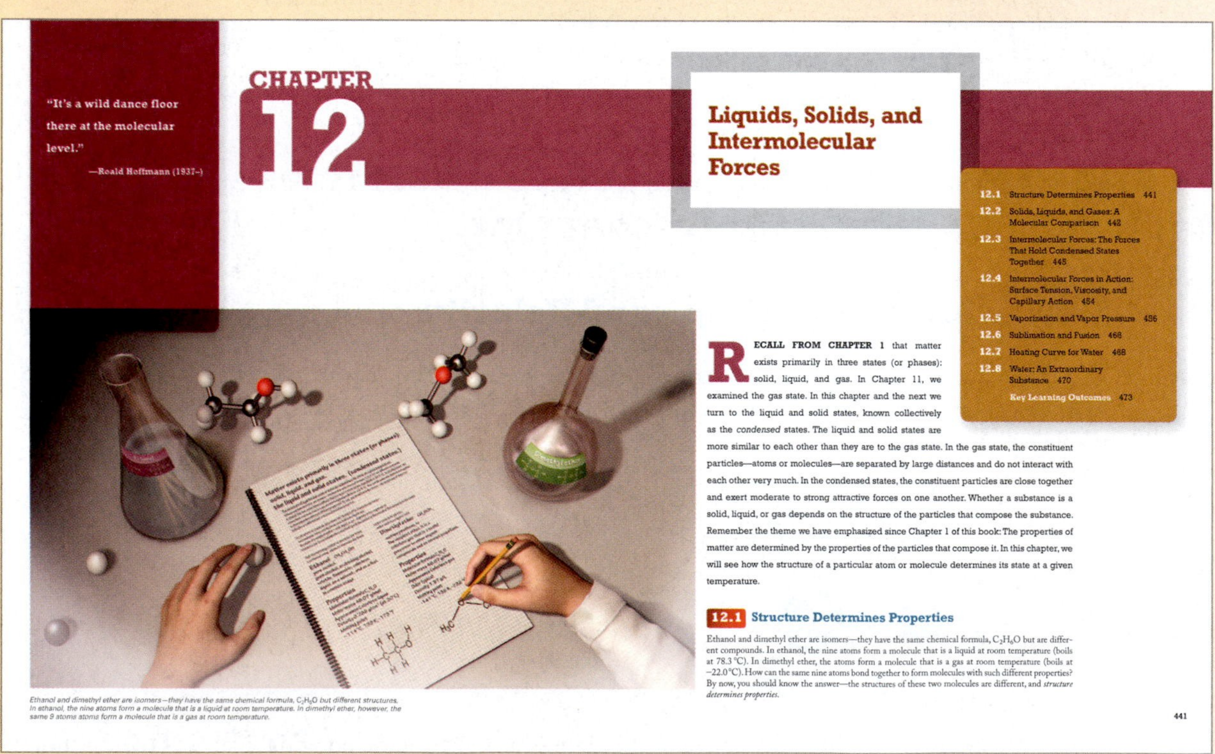

"It's a wild dance floor there at the molecular level."

—Roald Hoffmann (1937–)

CHAPTER 12

Liquids, Solids, and Intermolecular Forces

12.1 Structure Determines Properties 441

12.2 Solids, Liquids, and Gases: A Molecular Comparison 442

12.3 Intermolecular Forces: The Forces That Hold Condensed States Together 445

12.4 Intermolecular Forces in Action: Surface Tension, Viscosity, and Capillary Action 454

12.5 Vaporization and Vapor Pressure 456

12.6 Sublimation and Fusion 466

12.7 Heating Curve for Water 468

12.8 Water: An Extraordinary Substance 470

Key Learning Outcomes 473

RECALL FROM CHAPTER 1 that matter exists primarily in three states (or phases): solid, liquid, and gas. In Chapter 11, we examined the gas state. In this chapter and the next we turn to the liquid and solid states, known collectively as the *condensed* states. The liquid and solid states are more similar to each other than they are to the gas state. In the gas state, the constituent particles—atoms or molecules—are separated by large distances and do not interact with each other very much. In the condensed states, the constituent particles are close together and exert moderate to strong attractive forces on one another. Whether a substance is a solid, liquid, or gas depends on the structure of the particles that compose the substance. Remember the theme we have emphasized since Chapter 1 of this book: The properties of matter are determined by the properties of the particles that compose it. In this chapter, we will see how the structure of a particular atom or molecule determines its state at a given temperature.

12.1 Structure Determines Properties

Ethanol and dimethyl ether are isomers—they have the same chemical formula, C_2H_6O but are different compounds. In ethanol, the nine atoms form a molecule that is a liquid at room temperature (boils at 78.3 °C). In dimethyl ether, the atoms form a molecule that is a gas at room temperature (boils at −22.0 °C). How can the same nine atoms bond together to form molecules with such different properties? By now, you should know the answer—the structures of these two molecules are different, and *structure determines properties*.

Ethanol and dimethyl ether are isomers—they have the same chemical formula, C_2H_6O but different structures. In ethanol, the nine atoms form a molecule that is a liquid at room temperature. In dimethyl ether, however, the same 9 atoms atoms form a molecule that is a gas at room temperature.

441

Section 15.2 – How reaction rates depend of the structure of the reacting particles

15.2 Rates of Reaction and the Particulate Nature of Matter

We have seen throughout this book that matter is composed of particles (atoms, ions, and molecules). The simplest way to begin to understand the factors that influence a reaction rate is to think of a chemical reaction as the result of a collision between these particles, which is the basis of *the collision model* (which we cover in more detail in Section 15.6). For example, consider the following simple generic reaction occurring in the gaseous state:

$$A{-}A + B \longrightarrow A{-}B + A$$

According to the collision model, the reaction occurs as a result of a collision between A-A particles and B particles.

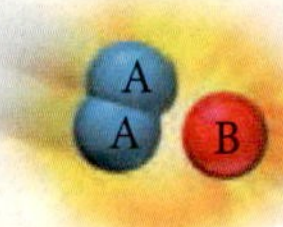

Section 17.4 – How the structure of an acid determines its strength

17.4 Acid Strength and Molecular Structure

We have learned that a Brønsted–Lowry acid is a proton (H^+) donor. Now we explore why some hydrogen-containing molecules act as proton donors while others do not. In other words, we want to explore *how the structure of a molecule affects its acidity*. Why is H_2S acidic while CH_4 is not? Or why is HF a weak acid while HCl is a strong acid? We divide our discussion about these issues into two categories: binary acids (those containing hydrogen and only one other element) and oxyacids (those containing hydrogen bonded to an oxygen atom that is bonded to another element).

Section 19.4 – How the structure of a molecule determines its entropy

19.4 Predicting Entropy and Entropy Changes for Chemical Reactions

We now turn our attention to predicting and quantifying entropy and entropy changes in a sample of matter. As we examine this topic, we again encounter the theme of this book: *structure determines properties.* In this case, the property we are interested in is entropy. In this section we see how the structure of the particles that compose a particular sample of matter determines the entropy that the sample possesses at a given temperature and pressure.

Key Concept Videos

Key Concept Videos and Interactive Worked Examples digitally bring Dr. Tro's award winning teaching directly to students.

In these highly conceptual videos, the author visually explains key concepts within each chapter and engages students in the learning process by asking them to answer embedded questions.

Scan this QR code (located on the back cover of the textbook) with your smartphone to access the Key Concept videos.

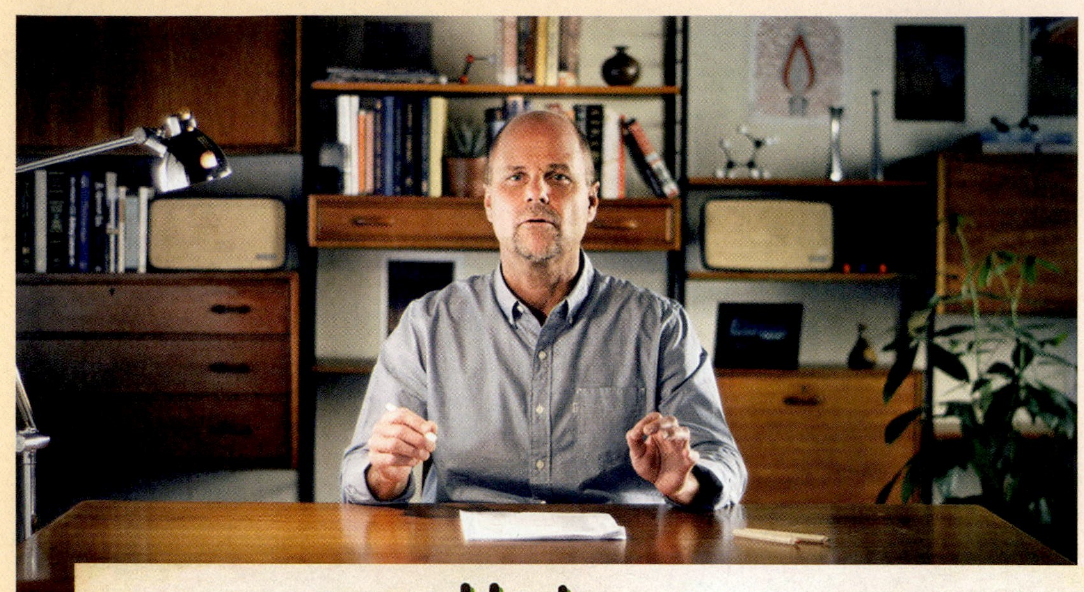

The Mole Concept

26.98 g aluminum = 1 mol aluminum = 6.022×10^{23} Al atoms

Al

12.01 g carbon = 1 mol carbon = 6.022×10^{23} C atoms

C

Interactive Worked Examples

Interactive Worked Examples are digital versions of the text's worked examples that make Tro's unique problem-solving strategies interactive, bringing his award-winning teaching directly to all students using his text. In these digital versions, students are instructed how to break down problems using Tro's proven technique.

These examples and videos are often paired and can be accessed by scanning the QR code on the back cover allowing students to quickly access an office-hour type experience. These problems are incorporated into MasteringChemistry® as assignable activities, and are also available for download via the Instructor Resource Center for instructional and classroom use.

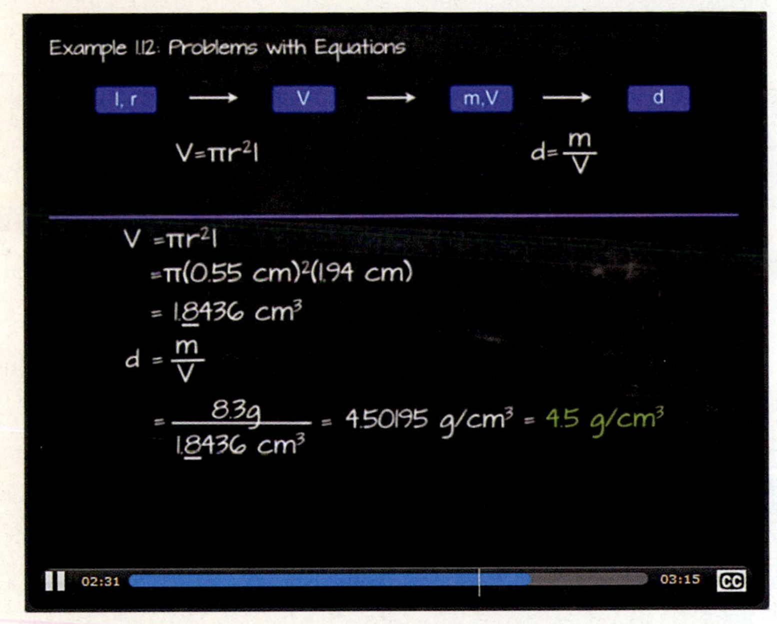

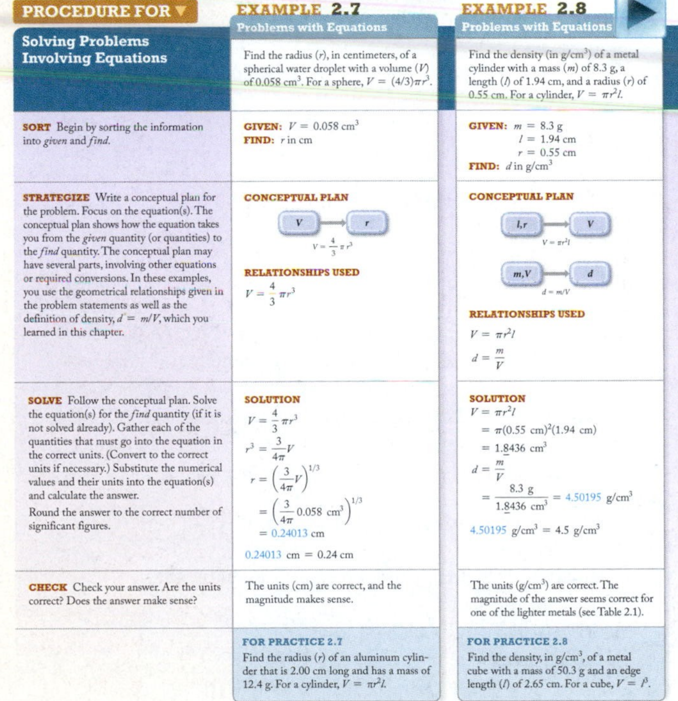

PROCEDURE FOR Solving Problems Involving Equations	EXAMPLE 2.7 Problems with Equations	EXAMPLE 2.8 Problems with Equations
	Find the radius (r), in centimeters, of a spherical water droplet with a volume (V) of 0.058 cm^3. For a sphere, $V = (4/3)\pi r^3$.	Find the density (in g/cm^3) of a metal cylinder with a mass (m) of 8.3 g, a length (l) of 1.94 cm, and a radius (r) of 0.55 cm. For a cylinder, $V = \pi r^2 l$.
SORT Begin by sorting the information into *given* and *find*.	**GIVEN:** $V = 0.058\ cm^3$ **FIND:** r in cm	**GIVEN:** $m = 8.3$ g, $l = 1.94$ cm, $r = 0.55$ cm **FIND:** d in g/cm^3
STRATEGIZE Write a conceptual plan for the problem. Focus on the equation(s). The conceptual plan shows how the equation takes you from the *given* quantity (or quantities) to the *find* quantity. The conceptual plan may have several parts, involving other equations or required conversions. In these examples, you use the geometrical relationships given in the problem statements as well as the definition of density, $d = m/V$, which you learned in this chapter.	**CONCEPTUAL PLAN** V → r ($V = \frac{4}{3}\pi r^3$) **RELATIONSHIPS USED** $V = \frac{4}{3}\pi r^3$	**CONCEPTUAL PLAN** l,r → V ($V = \pi r^2 l$) m,V → d ($d = m/V$) **RELATIONSHIPS USED** $V = \pi r^2 l$ $d = \frac{m}{V}$
SOLVE Follow the conceptual plan. Solve the equation(s) for the *find* quantity (if it is not solved already). Gather each of the quantities that must go into the equation in the correct units. (Convert to the correct units if necessary.) Substitute the numerical values and their units into the equation(s) and calculate the answer. Round the answer to the correct number of significant figures.	**SOLUTION** $V = \frac{4}{3}\pi r^3$ $r^3 = \frac{3}{4\pi}V$ $r = \left(\frac{3}{4\pi}V\right)^{1/3}$ $= \left(\frac{3}{4\pi}0.058\ cm^3\right)^{1/3}$ $= 0.24013$ cm $0.24013\ cm = 0.24\ cm$	**SOLUTION** $V = \pi r^2 l$ $= \pi(0.55\ cm)^2(1.94\ cm)$ $= 1.8436\ cm^3$ $d = \frac{m}{V}$ $= \frac{8.3\ g}{1.8436\ cm^3} = 4.50195\ g/cm^3$ $4.50195\ g/cm^3 = 4.5\ g/cm^3$
CHECK Check your answer. Are the units correct? Does the answer make sense?	The units (cm) are correct, and the magnitude makes sense.	The units (g/cm^3) are correct. The magnitude of the answer seems correct for one of the lighter metals (see Table 2.1).
	FOR PRACTICE 2.7 Find the radius (r) of an aluminum cylinder that is 2.00 cm long and has a mass of 12.4 g. For a cylinder, $V = \pi r^2 l$.	**FOR PRACTICE 2.8** Find the density, in g/cm^3, of a metal cube with a mass of 50.3 g and an edge length (l) of 2.65 cm. For a cube, $V = l^3$.

Linking the Conceptual with the Quantitative

Self-Assessment Quizzes

Niva Tro actively participates on the ACS Exams Committee for Gen Chem I, Gen Chem II and full year exams. Tro's Self-Assessment Quizzes at the end of each chapter contain 10-15 multiple-choice questions that are similar to those found on the ACS exam and on other standardized exams. The Self-Assessment Quizzes are also assignable in MasteringChemistry®.

SELF-ASSESSMENT Quiz

1. Which wavelength of light has the highest frequency?
   a) 10 nm **b)** 10 mm **c)** 1 nm **d)** 1 mm

2. Which kind of electromagnetic radiation contains the greatest energy per photon?
   a) Microwaves **b)** Gamma rays
   c) X-rays **d)** Visible light

3. How much energy (in J) is contained in 1.00 mole of 552-nm photons?
   a) 3.60×10^{-19} J **b)** 2.17×10^{5} J
   c) 3.60×10^{-28} J **d)** 5.98×10^{-43} J

4. Light from three different lasers (A, B, and C), each with a different wavelength, is shined onto the same metal surface. Laser A produces no photoelectrons. Lasers B and C both produce photoelectrons, but the photoelectrons produced by laser B have a greater velocity than those produced by laser C. Arrange the lasers in order of increasing wavelength.
   a) $A < B < C$ **b)** $B < C < A$
   c) $C < B < A$ **d)** $A < C < B$

5. Calculate the frequency of an electron traveling at 1.85×10^{7} m/s.
   a) $1.31 \times 10^{-19}\ s^{-1}$ **b)** $1.18 \times 10^{-2}\ s^{-1}$
   c) $3.93 \times 10^{-11}\ s^{-1}$ **d)** $7.63 \times 10^{18}\ s^{-1}$

6. Which set of three quantum numbers *does not* specify an orbital in the hydrogen atom?
   a) $n = 2; l = 1; m_l = -1$ **b)** $n = 3; l = 3; m_l = -2$
   c) $n = 2; l = 0; m_l = 0$ **d)** $n = 3; l = 2; m_l = 2$

7. Calculate the wavelength of light emitted when an electron in the hydrogen makes a transition from an orbital with $n = 5$ to an orbital with $n = 3$.
   a) 1.28×10^{-6} m **b)** 6.04×10^{-7} m
   c) 2.28×10^{-6} m **d)** 1.55×10^{-19} m

8. Which electron transition produces light of the highest frequency in the hydrogen atom?
   a) $5p \longrightarrow 1s$ **b)** $4p \longrightarrow 1s$
   c) $3p \longrightarrow 1s$ **d)** $2p \longrightarrow 1s$

9. How much time (in seconds) does it take light to travel 1.00 billion km?
   a) 3.00×10^{17} s **b)** 3.33 s
   c) 3.33×10^{3} s **d)** 3.00×10^{20} s

10. Which figure represents a *d* orbital?
    a) 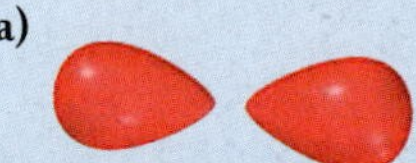 **b)** 
    c)  **d)** None of the above

Answers: 1:c; 2:b; 3:b; 4:b; 5:d; 6:b; 7:a; 8:a; 9:c; 10:b

Two-Column Example

The **general procedure** is shown in the left column.

The **right column** shows the implementation of the steps explained in the left column

A four-part structure ("**Sort, Strategize, Solve, Check**") provides you with a framework for analyzing and solving problems.

Every Worked Example is followed by "For Practice" Problems that you can try to solve on your own. Answers to "For Practice" Problems are in Appendix VI

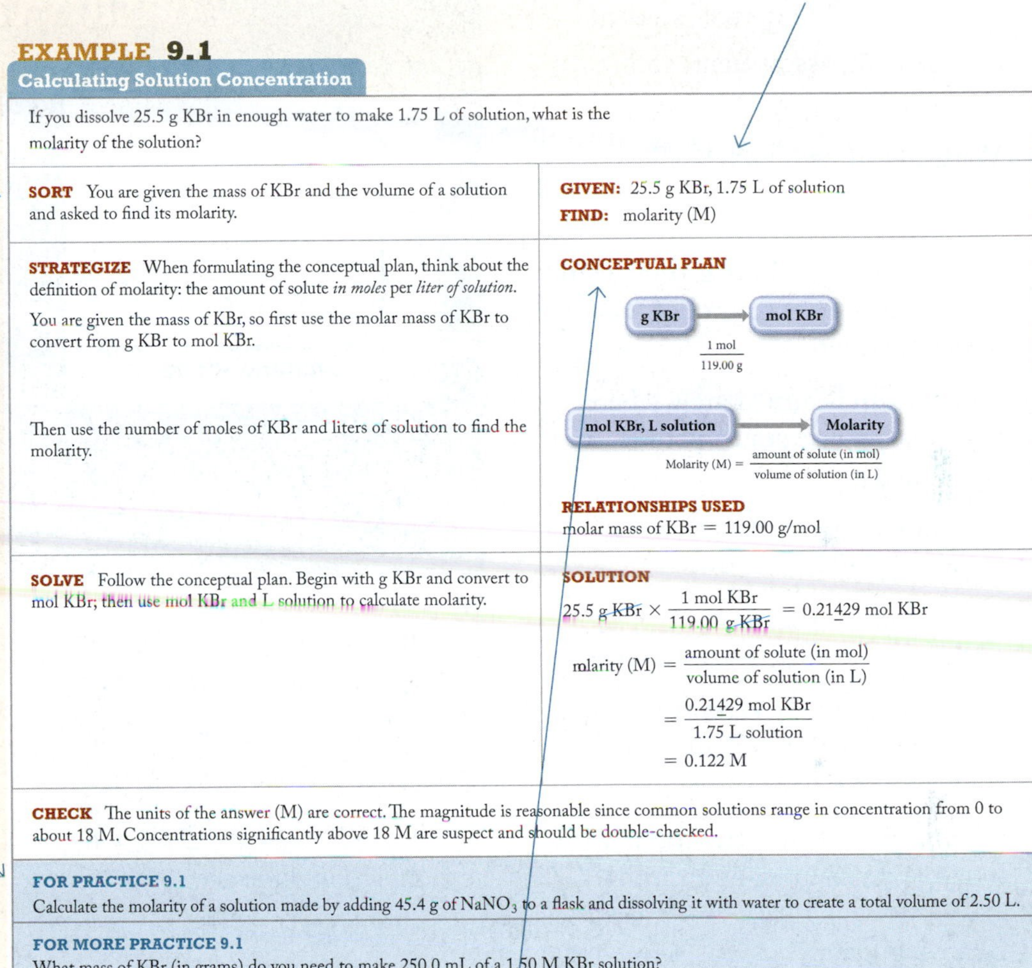

EXAMPLE 9.1

Calculating Solution Concentration

If you dissolve 25.5 g KBr in enough water to make 1.75 L of solution, what is the molarity of the solution?

SORT You are given the mass of KBr and the volume of a solution and asked to find its molarity.

GIVEN: 25.5 g KBr, 1.75 L of solution

FIND: molarity (M)

STRATEGIZE When formulating the conceptual plan, think about the definition of molarity: the amount of solute *in moles* per *liter of solution*.

You are given the mass of KBr, so first use the molar mass of KBr to convert from g KBr to mol KBr.

Then use the number of moles of KBr and liters of solution to find the molarity.

CONCEPTUAL PLAN

g KBr → mol KBr

$\frac{1 \text{ mol}}{119.00 \text{ g}}$

mol KBr, L solution → Molarity

$\text{Molarity (M)} = \frac{\text{amount of solute (in mol)}}{\text{volume of solution (in L)}}$

RELATIONSHIPS USED

molar mass of KBr = 119.00 g/mol

SOLVE Follow the conceptual plan. Begin with g KBr and convert to mol KBr; then use mol KBr and L solution to calculate molarity.

SOLUTION

$$25.5 \text{ g KBr} \times \frac{1 \text{ mol KBr}}{119.00 \text{ g KBr}} = 0.21429 \text{ mol KBr}$$

$$\text{molarity (M)} = \frac{\text{amount of solute (in mol)}}{\text{volume of solution (in L)}} = \frac{0.21429 \text{ mol KBr}}{1.75 \text{ L solution}} = 0.122 \text{ M}$$

CHECK The units of the answer (M) are correct. The magnitude is reasonable since common solutions range in concentration from 0 to about 18 M. Concentrations significantly above 18 M are suspect and should be double-checked.

FOR PRACTICE 9.1

Calculate the molarity of a solution made by adding 45.4 g of $NaNO_3$ to a flask and dissolving it with water to create a total volume of 2.50 L.

FOR MORE PRACTICE 9.1

What mass of KBr (in grams) do you need to make 250.0 mL of a 1.50 M KBr solution?

Many problems are solved with a **conceptual plan** that provides a visual outline of the steps leading from the given information to the solution.

Active and Adaptive

Learning Catalytics™

Learning Catalytics™ is a "bring your own device" student engagement, assessment, and classroom intelligence system. With Learning Catalytics™ you can:

- Assess students in real time, using open-ended tasks to probe student understanding.
- Understand immediately where students are and adjust your lecture accordingly.
- Improve your students' critical-thinking skills.
- Access rich analytics to understand student performance.
- Add your own questions to make Learning Catalytics™ fit your course exactly.
- Manage student interactions with intelligent grouping and timing.

Learning Catalytics™ is a technology that has grown out of twenty years of cutting edge research, innovation, and implementation of interactive teaching and peer instruction.

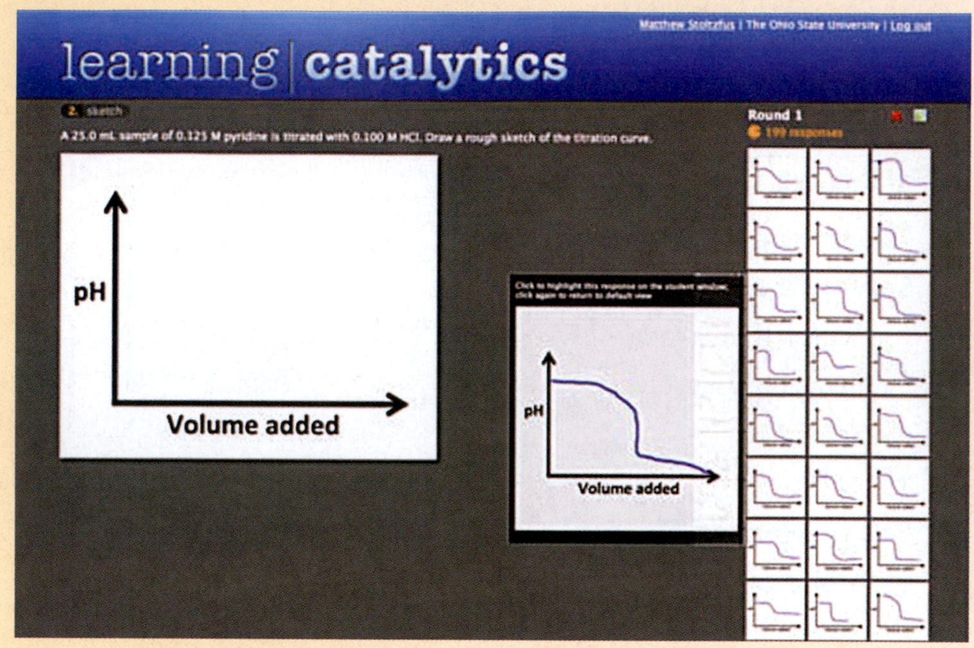

Learning Catalytics™ is included with the purchase of Mastering with eText. Students purchasing Mastering without eText will be able to upgrade their Mastering accounts to include access to Learning Catalytics™.

Instructors using Learning Catalytics™ in conjunction with MasteringChemistry® will be able to select publisher provided questions specific to each course.

Adaptive Follow-up Assignments in MasteringChemistry®

Instructors are given the ability to assign adaptive follow-up assignments to students for *Chemistry: Structure and Properties*. Content delivered to students as part of adaptive learning will automatically be personalized for each individual based on strengths and weaknesses as identified by his or her performance on Mastering parent assignments.

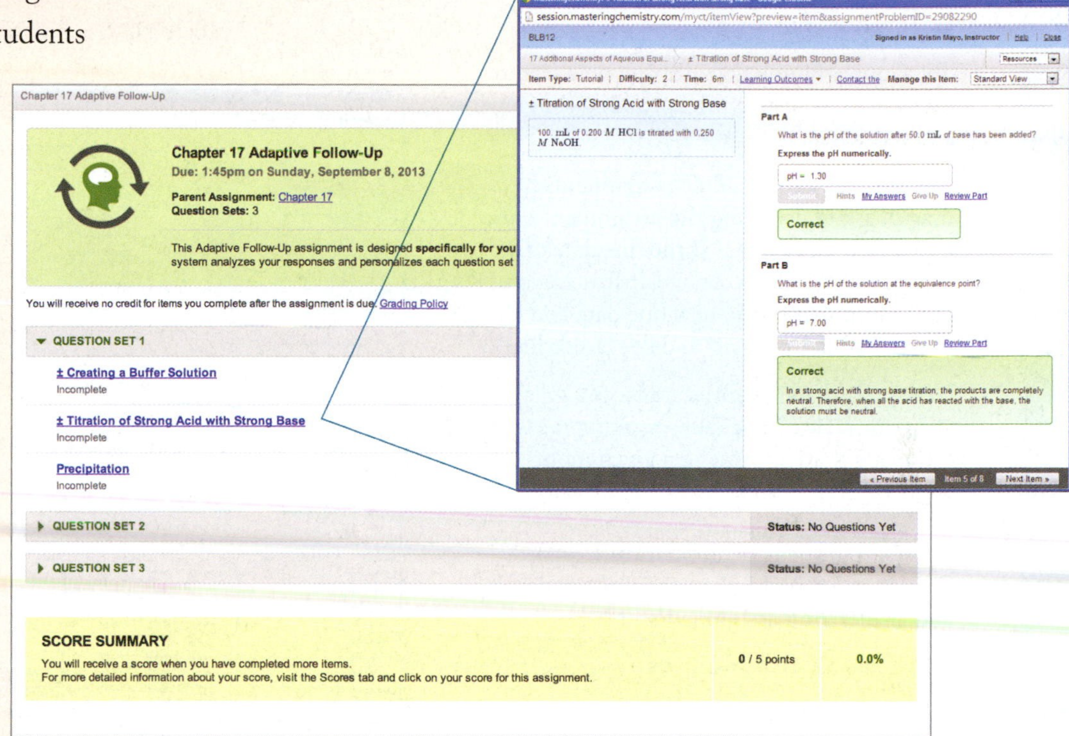

Dynamic Study Modules

NEW! Dynamic Study Modules, designed to enable students to study effectively on their own as well as help students quickly access and learn the nomenclature they need to be more successful in chemistry. These modules can be accessed on smartphones, tablets, and computers and results can be tracked in the MasteringChemistry® Gradebook.

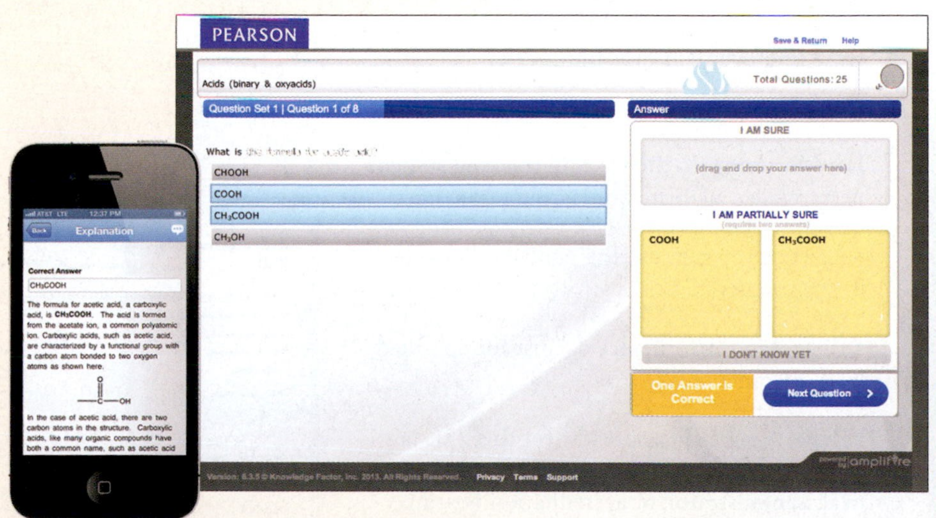

MasteringChemistry® for Instructors

www.masteringchemistry.com

The Mastering platform was developed by scientists for science students and instructors. Mastering has been refined from data-driven insights derived from over a decade of real-world use by faculty and students.

Calendar Features

The Course Home default page now features a calendar view displaying upcoming assignments and due dates.

- Instructors can schedule assignments by dragging and dropping the assignment onto a date in the calendar. If the due date of an assignment needs to change, instructors can drag the assignment to the new due date and change the "available from and to dates" accordingly.
- The calendar view gives students a syllabus-style overview of due dates, making it easy to see all assignments due in a given month.

Gradebook

Every assignment is automatically graded. Shades of red highlight struggling students and challenging assignments.

Gradebook Diagnostics

This screen provides you with your favorite diagnostics. With a single click, charts summarize the most difficult problems, vulnerable students, grade distribution, and even score improvement over the course.

Learning Outcomes

Let Mastering do the work in tracking student performance against your learning outcomes:

- Add your own or use the publisher provided learning outcomes.
- View class performance against the specified learning outcomes.
- Export results to a spreadsheet that you can further customize and share with your chair, deal, administrator, or accreditation board.

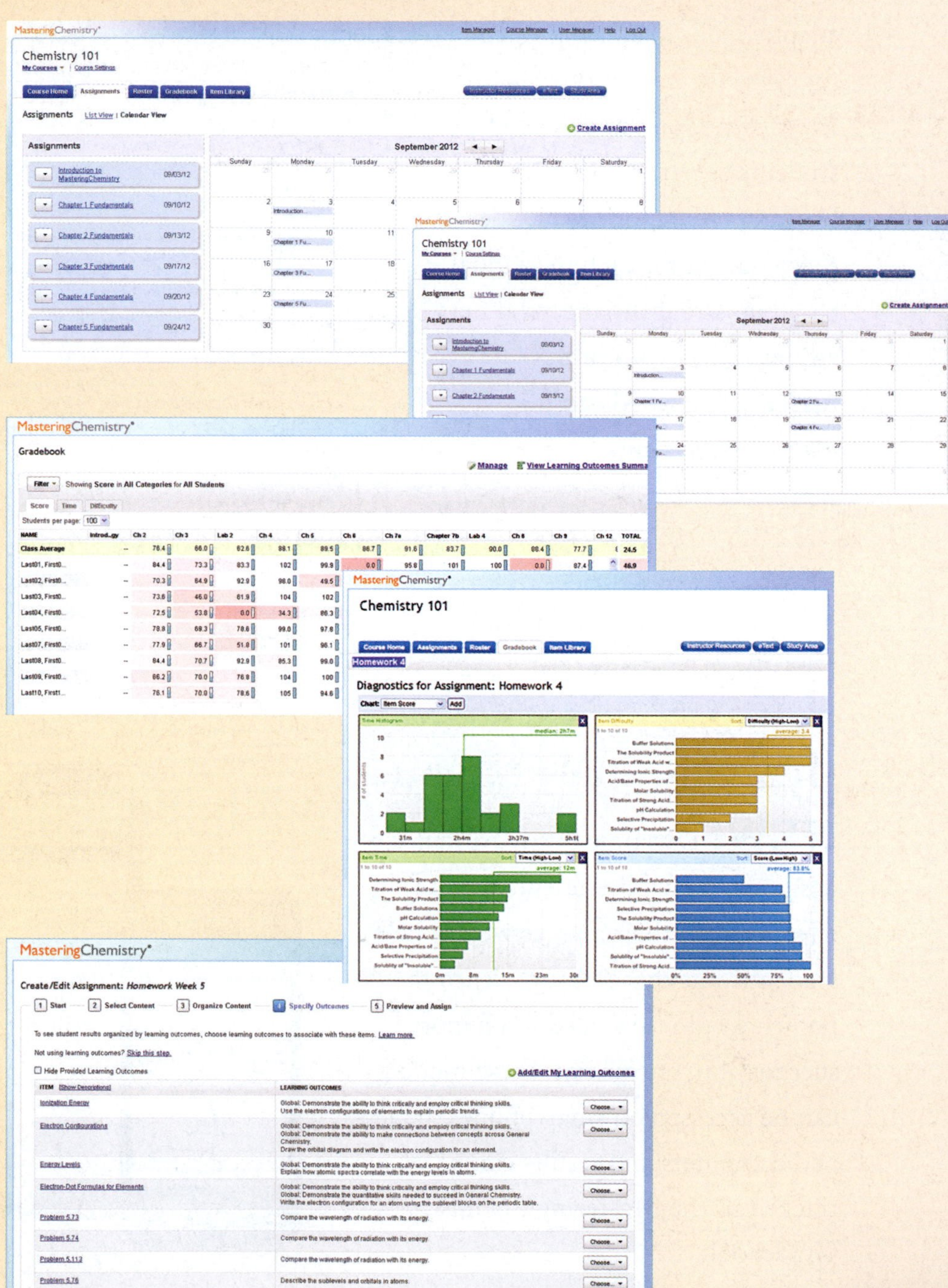

Labs Designed for S&P

Laboratory Manual for Chemistry: Structure and Properties
0321869079 / 9780321869074

The Tro/Norton Lab Manual is authored by Daphne Norton from the University of Georgia. Written to correspond with teaching using an atoms-first approach, this author emphasizes critical thinking and problem-solving skills while fostering student engagement in real world applications.

Students will be exposed to recent advances in science by presenting labs in an investigative context. Emphasis is placed on data collection and analysis versus mere step-by-step instruction.

Lab Manual Table of Contents

1 Liquid Crystals
2 Atomic Emission Spectra: Comparing Experimental Results to Bohr's Theoretical Model
3 Energy & Electromagnetism: Irradiance Measurements
4 Structure of Molecules
5 A Gravimetric Analysis of Phosphorus in Fertilizer
6 Recycling Aluminum
7 Qualitative Analysis- The Detection of Anions
8 Qualitative Analysis- Detection of Metal Cations
9 Qualitative Analysis- Identification of the Single Salt
10 Qualitative Analysis of Household Chemicals
11 Iron Deficiency Analysis
12 Gasimetric Analysis of a Carbonate
13 Calorimetry: Heat of Fusion and Specific Heat
14 Chromatography: Isolation and Characterization of Yellow Dye No. 5
15 Freezing Point Depression or A Lesson in Making Ice Cream
16 Alternative Fuel Project
17 The Green Fades Away
18 Chemical Kinetics
19 Analysis of Phosphoric Acid in Coca-Cola Classic: Spectrophotometric Analysis
20 Analysis of Phosphoric Acid in Coca-Cola Classic: pH Titration
21 Borax Solubility: Investigating the Relationship between Thermodynamics and Equilibrium
22 Electrochemical Preparation of Nickel Nanowires
23 Synthesis of $K_3Fe(C_2O_4)_3 \cdot 3\ H_2O$
24 Analysis of Oxalate in $K_3Fe(C_2O_4)_3 \cdot 3\ H_2O$

Supplements

For Students

Study Guide for Chemistry: Structure and Properties
0321965612 / 9780321965615
This Study Guide was written specifically to assist students using Structure and Properties. It presents the major concepts, theories, and applications discussed in the text in a comprehensive and accessible manner for students. It contains learning objectives, chapter summaries and outlines, as well as examples, self-tests and concept questions.

Student's Selected Solutions Manual for Chemistry: Structure and Properties
0321965388 / 9780321965387
The selected solution manual for students contains complete, step-by-step solutions to selected odd-numbered end-of-chapter problems.

For Instructors

Instructor Supplements

MasteringChemistry® with Pearson eText—Instant Access—for Chemistry: Structure and Properties
0321834666 / 9780321834669
http://www.masteringchemistry.com

This includes all of the resources of MasteringChemistry® in addition to Pearson eText content.

MasteringChemistry®—Instant Access—for Chemistry: Structure and Properties
0321933648 / 9780321933645
http://www.masteringchemistry.com
MasteringChemistry® from Pearson is the leading online homework, tutorial, and assessment product designed to improve results by helping students quickly master concepts. Students benefit from self-paced tutorials, featuring specific wrong-answer feedback, hints, and a vast variety of educationally effective content to keep them engaged and on track. Robust diagnostics and unrivalled gradebook reporting allow instructors to pinpoint the weaknesses and misconceptions of a student or class to provide timely intervention.

Solutions Manual for Chemistry: Structure and Properties
0321965299 / 9780321965295
The solution manual contains complete, step-by-step solutions to end-of-chapter problems and can be made available for purchase with instructor approval.

Instructor's Resource Manual (Download only) for Chemistry: Structure and Properties
0321965396 / 9780321965394
Organized by chapter, this useful guide includes objectives, lecture outlines, and references to figures and worked examples, as well as teaching tips.

Online Instructor Resource Center for Chemistry: Structure and Properties
0321965108 / 9780321965103

This resource contains the following:

- All illustrations, tables, and photos from the text in JPEG format
- Four pre-built PowerPoint™ Presentations (lecture, worked examples, images, CRS/clicker questions)
- Interactive animations, movies, and 3-D molecules
- TestGen computerized software with the TestGen version of the Testbank
- Word files of the Test Item File

Test Bank (Download Only) for Chemistry: Structure and Properties
032196523X / 9780321965233
The Testbank is downloadable directly from the Instructor Resource Center in either Microsoft Word or TestGen formats.

Tro | Chemistry: Structure and Properties

CHAPTER 1

"It will be found that everything depends on the composition of the forces with which the particles of matter act upon one another; and from these forces...all phenomena of nature take their origin."

—Roger Joseph Boscovich (1711–1787)

Water, like all matter, is composed of atoms. The atoms are bound together to form a molecule. The structure of the molecule determines the properties of water.

Atoms

1.1 A Particulate View of the World: Structure Determines Properties 3

1.2 Classifying Matter: A Particulate View 4

1.3 The Scientific Approach to Knowledge 7

1.4 Early Ideas about the Building Blocks of Matter 9

1.5 Modern Atomic Theory and the Laws That Led to It 10

1.6 The Discovery of the Electron 13

1.7 The Structure of the Atom 16

1.8 Subatomic Particles: Protons, Neutrons, and Electrons 18

1.9 Atomic Mass: The Average Mass of an Element's Atoms 22

1.10 The Origins of Atoms and Elements 25

Key Learning Outcomes 27

WHAT DO YOU THINK is the most powerful idea in all of human knowledge? There are, of course, many possible answers to this question—some practical, some philosophical, and some scientific. If we limit ourselves only to scientific answers, mine would be this: *The properties of matter are determined by the structure of the atoms and molecules that compose it.* Atoms and molecules determine how matter behaves—if they were different, matter would be different. The structure of helium atoms determines how helium behaves; the structure of water molecules determines how water behaves; and the structures of the molecules that compose our bodies determine how our bodies behave. The understanding of matter at the particulate level gives us unprecedented control over that matter. For example, our understanding of the details of the molecules that compose living organisms has revolutionized biology over the last 50 years.

1.1 A Particulate View of the World: Structure Determines Properties

A good novel usually has a strong *premise*—a short statement that describes the central idea of the story. The story of chemistry as described in this book also has a strong premise, which consists of two simple statements:

1. Matter is particulate—it is composed of particles.
2. The structure of those particles determines the properties of matter.

Matter is anything that occupies space and has mass. Most things you can think of—such as this book, your desk, and even your body—are composed of matter. The particulate nature of matter—first

In chemistry, atoms are often portrayed as colored spheres, with each color representing a different kind of atom. For example, a black sphere represents a carbon atom, a red sphere represents an oxygen atom, and a white sphere represents a hydrogen atom. For a complete color code of atoms, see Appendix IV A.

conceived in ancient Greece, but widely accepted only about 200 years ago—is the foundation of chemistry and the premise of this book.

As an example of this premise, consider water, the familiar substance we all know and depend on for life. The particles that compose water are *water molecules*, which we can represent like this:

Water molecule

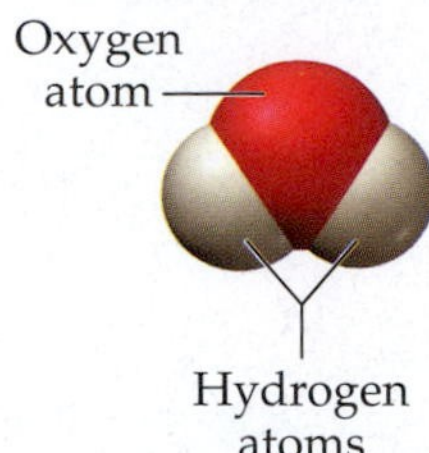

Atoms themselves, as we discuss later in this chapter, are composed of even smaller particles.

A water molecule is composed of three *atoms*: one oxygen atom and two hydrogen atoms. **Atoms** are the basic particles that compose ordinary matter, and about 91 different types of atoms naturally exist. Atoms often bind together in specific geometrical arrangements to form **molecules**, as we see in water.

The first thing you should know about water molecules—and all molecules—is that they are extremely small, much too small to see with even the strongest optical microscope. The period at the end of this sentence has a diameter of about one-fifth of a millimeter (less than one-hundredth of an inch); yet a spherical drop of water with the same diameter as this period contains over 100 million billion water molecules.

The second thing you should know about water molecules is that their structure determines the properties of water. The water molecule is *bent*: The two hydrogen atoms and the oxygen atom are not in a straight line. If the atoms were in a straight line, water itself would be different. For example, suppose that the water molecule were linear instead of bent:

Hypothetical linear water molecule

If water had this hypothetical structure, it would be a different substance. First of all, linear water would have a lower boiling point than normal water (and may even be a gas at room temperture). Just this change in shape would cause the attractive forces between water molecules to weaken so that the molecules would have less of a tendency to clump together as a liquid and more of a tendency to evaporate into a gas. In its liquid form, linear water would be quite different than the water we know. It would feel more like gasoline or paint thinner than water. Substances that normally dissolve easily in water—such as sugar or salt—would probably not dissolve in linear water.

The key point here is that the properties of the substances around us radically depend on the structure of the particles that compose them—a small change in structure, such as a different shape, results in a significant change in properties. If we want to understand the substances around us, we must understand the particles that compose them—and that is the central goal of chemistry. A good simple definition of **chemistry** is:

> **Chemistry—the science that seeks to understand the properties of matter by studying the structure of the particles that compose it.**

1.2 Classifying Matter: A Particulate View

Recall from Section 1.1 that matter is anything that occupies space and has mass. A specific instance of matter—such as air, water, or sand—is a **substance**. We can begin to understand the particulate view of matter by classifying matter based on the particles that compose it. The first classification—the **state** of matter—depends on the *relative positions* of the particles and *how strongly they interact* with one another (relative to temperature). The second classification—the **composition** of matter—depends on the *types* of particles.

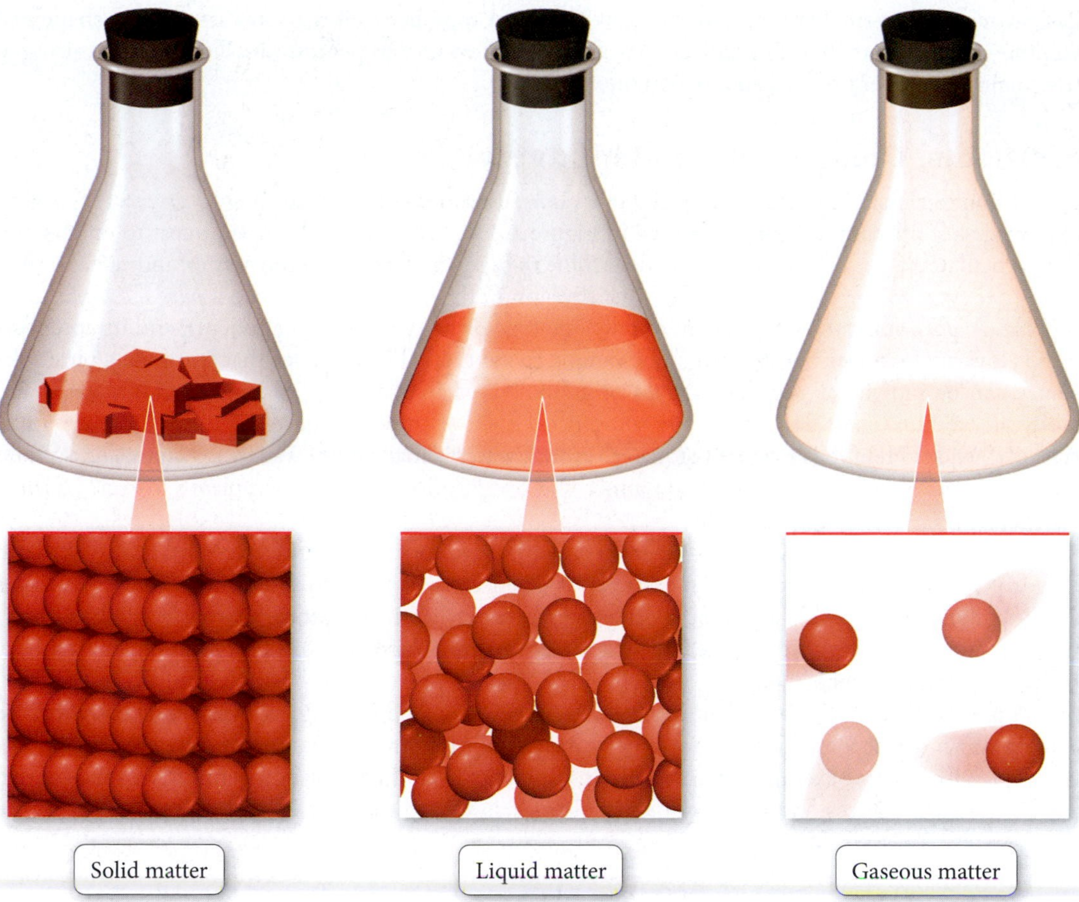

◀ **FIGURE 1.1 The States of Matter** In a solid, the composite particles are fixed in place and can only vibrate. In a liquid, although the particles are closely packed, they can move past one another, allowing the liquid to flow and assume the shape of its container. In a gas, the particles are widely spaced, making gases compressible as well as fluid (able to flow).

The States of Matter: Solid, Liquid, and Gas

Matter can exist in three different states: **solid**, **liquid**, and **gas** (Figure 1.1 ▲). The particles that compose *solid matter* attract one another strongly and therefore pack close to each other in fixed locations. Although the particles vibrate, they do not move around or past each other. Consequently, a solid has a fixed volume and rigid shape. Ice, aluminum, and diamond are good examples of solids.

The particles that compose *liquid matter* pack about as closely as particles do in solid matter, but slightly weaker attractions between the particles allow them to move relative to each other, giving liquids a fixed volume but not a fixed shape. Liquids assume the shape of their container. Water, alcohol, and gasoline are examples of substances that are liquids at room temperature.

The particles that compose *gaseous matter* attract each other only very weakly—so weakly that they do not clump together as particles do in a liquid or solid. Instead the particles are free to move large distances before colliding with one another. The large spaces between the particles make gases *compressible* (Figure 1.2 ▼). When you squeeze a balloon or sit down on an air mattress, you force the

The state of matter changes from solid to liquid to gas with increasing temperature.

The discussion here assumes that the three samples of matter are all at the same fixed temperature. At this temperature, strong attractions between particles favor the solid state and weak attractions between particles favor the gas state.

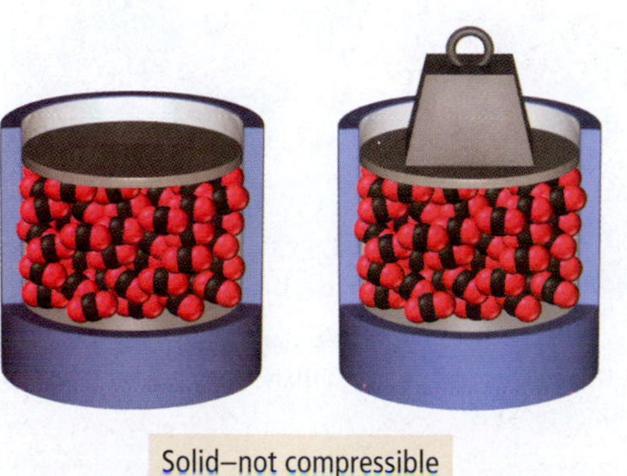

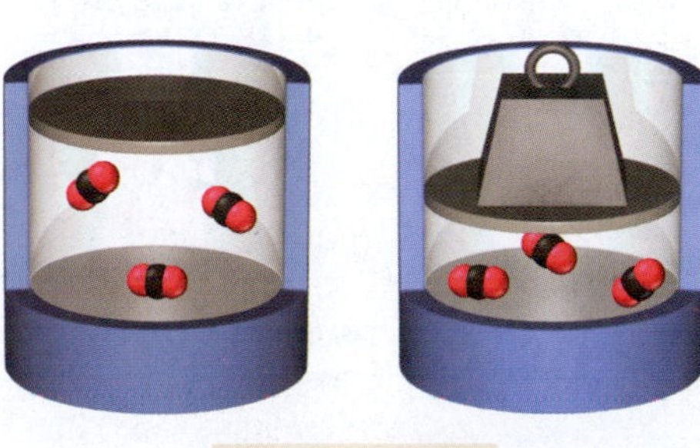

◀ **FIGURE 1.2 The Compressibility of Gases** Gases can be compressed—squeezed into a smaller volume—because there is so much empty space between atoms or molecules in the gaseous state.

gas particles into a smaller space, so that they are closer together. Gases always assume the shape *and* volume of their container. Substances that are gases at room temperature include helium, nitrogen (the main component of air), and carbon dioxide.

Elements, Compounds, and Mixtures

In addition to classifying matter according to its *state*, we can classify it according to *the types of particles that compose it* (its composition), as shown in Figure 1.3 ▼. In other words, in our quest to understand the particulate nature of matter we must determine the *types* of particles in the matter, and *whether there is only one type or more than one type*. The first division in this scheme is between a *pure substance* and a *mixture*. A **pure substance** is made up of only one type of particle (one component), and its composition is invariant (it does not vary from one sample to another). The particles that compose a pure substance can be individual atoms, or groups of atoms joined together. For example, helium, water, and table salt (sodium chloride) are all pure substances. Each of these substances is made up of only one type of particle: Helium is made up of helium atoms; water is made up of water molecules; and sodium chloride is made up of sodium chloride units. The composition of a pure sample of any one of these substances is always exactly the same.

A **mixture**, by contrast, is a substance composed of two or more particles in proportions that can vary from one sample to another. For example, sweetened tea, composed primarily of water molecules and sugar molecules (with a few other substances mixed in), is a mixture. It can be slightly sweet (a small proportion of sugar to water) or very sweet (a large proportion of sugar to water) or any level of sweetness in between.

A pure substance can be either an *element* or a *compound*, depending on whether or not it can be broken down (or decomposed) into simpler substances. Helium, which we just noted is a pure substance, is also a good example of an **element**, a substance that cannot be chemically broken down into simpler substances. Water, also a pure substance, is a good example of a **compound**, a substance

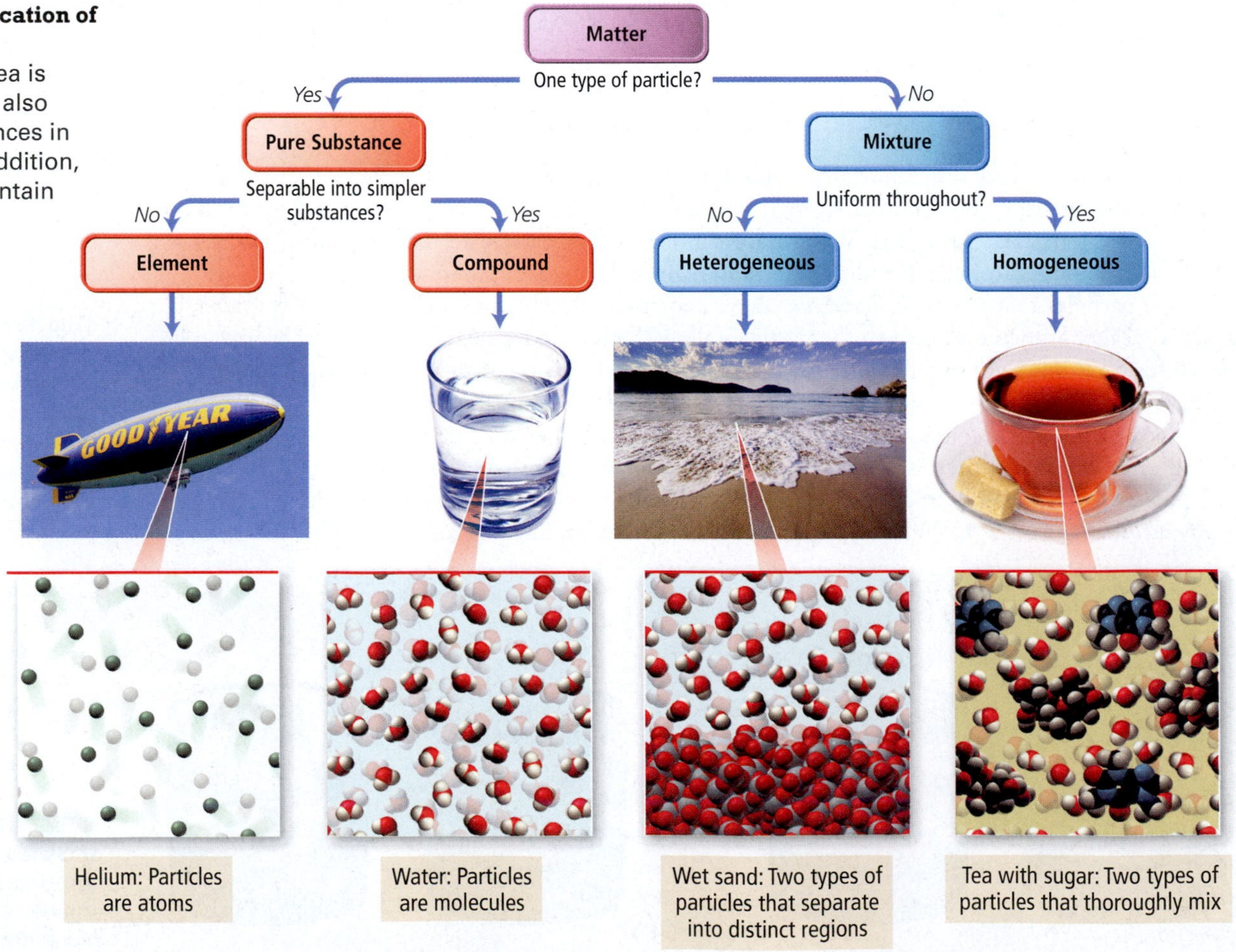

▶ **FIGURE 1.3 The Classification of Matter According to Its Composition** Sweetened tea is mostly sugar and water, but also contains a few other substances in much smaller amounts. In addition, the tea is assumed to not contain any solid impurities.

Matter
One type of particle?
Yes
No
Pure Substance
Mixture
Separable into simpler substances?
No
Yes
Uniform throughout?
No
Yes
Element
Compound
Heterogeneous
Homogeneous
Helium: Particles are atoms
Water: Particles are molecules
Wet sand: Two types of particles that separate into distinct regions
Tea with sugar: Two types of particles that thoroughly mix

composed of two or more elements (in this case hydrogen and oxygen) in fixed, definite proportions. On Earth, compounds are more common than pure elements because most elements combine with other elements to form compounds.

A mixture can be either heterogeneous or homogeneous, depending on how *uniformly* the particles that compose the mixture combine. Water and sand is a **heterogeneous mixture**, one in which the composition varies from one region of the mixture to another—the different particles that compose water and sand *do not* mix uniformly. Sweetened tea is a **homogeneous mixture**, one with the same composition throughout—the particles that compose sweetened tea mix uniformly.

Classifying a substance according to its composition is not always obvious and requires that you either know the true composition of the substance or are able to test it in a laboratory. For now, we will focus on relatively common substances which you are likely to have encountered. During this course, you will gain the knowledge to understand the composition of a larger variety of substances.

Pure Substances and Mixtures

1.1 Cc Conceptual Connection

Using a small circle to represent each atom of one type of element and a small square to represent each atom of a second type of element, make a drawing of: (a) a pure substance (a compound) composed of the two elements (in a one-to-one ratio); (b) a homogeneous mixture composed of the two elements; and (c) a heterogeneous mixture composed of the two elements.

Note: Answers to Conceptual Connections can be found at the end of each chapter.

1.3 The Scientific Approach to Knowledge

The particulate model of matter introduced in Section 1.1 is not obvious to a casual observer of matter. In fact, early influential thinkers rejected it. Nonetheless, it came to be accepted because *scientists were driven to it by the data*. When thinkers applied the *scientific approach to knowledge* to understanding matter, the only explanation consistent with their observations was that matter is particulate.

Although some Greek philosophers, such as Aristotle (384–322 B.C.E.), did use observation to attain knowledge, they did not emphasize experiment and measurement to the extent that modern science does.

The scientific approach to knowledge is a fairly recent phenomenon, only finding broad acceptance in the last 400 years or so. Greek thinkers were heavily influenced by the Greek philosopher Plato (427–347 B.C.E.), who thought that the best way to learn about reality was not through the senses, but through reason. Plato believed that the physical world was an imperfect representation of a perfect and transcendent world (a world beyond space and time). For him, true knowledge came, not through observing the real physical world, but through reasoning and thinking about the ideal one.

The *scientific* approach to knowledge is exactly the opposite of Plato's approach. Scientific knowledge is empirical—it is based on *observation* and *experiment*. Scientists observe and perform experiments on the physical world to learn about it. Some observations and experiments are qualitative (noting or describing how a process happens), but many are quantitative (measuring or quantifying something about the process). For example, Antoine Lavoisier (1743–1794), a French chemist who studied combustion (burning), made careful measurements of the mass of objects before and after burning them in closed containers. He noticed that there was no change in the total mass of material within the container after combustion. In doing so, Lavoisier made an important *observation* about the physical world.

▲ A painting of the French chemist Antoine Lavoisier with his wife, Marie, who helped him in his work by illustrating his experiments and translating scientific articles from English. Lavoisier, who also made significant contributions to agriculture, industry, education, and government administration, was executed during the French Revolution. (The Metropolitan Museum of Art)

Observations often lead a scientist to formulate a **hypothesis**, a tentative interpretation or explanation of the observations. For example, Lavoisier explained his observations on combustion by hypothesizing that when a substance burns, it combines with a component of air. A good hypothesis is *falsifiable*, which means that it makes predictions that can be confirmed or refuted by further observations. Scientists test hypotheses by **experiments**, highly controlled procedures designed to generate observations that can confirm or refute a hypothesis. The results of an experiment may support a hypothesis or prove it wrong, in which case the scientist must modify or discard the hypothesis. In some cases, a series of similar observations can lead to the development of a **scientific law**, a brief statement that summarizes past observations and predicts future ones. Lavoisier summarized his observations on combustion with the **law of conservation of mass**, which states, "In a chemical reaction, matter is neither created nor destroyed." This statement summarized his observations on chemical reactions and predicted the outcome of future observations on reactions. Laws, like hypotheses, are also subject to experiments, which can support them or prove them wrong.

Scientific laws are not *laws* in the same sense as civil or governmental laws. Nature does not follow laws in the way that we obey the laws against speeding or running a stop sign. Rather, scientific laws *describe* how nature behaves—they are generalizations about what nature does. For that reason, some people find it more appropriate to refer to them as *principles* rather than *laws*.

One or more well-established hypotheses may form the basis for a scientific **theory**. A scientific theory is a model for the way nature is and tries to explain not merely *what* nature does but *why*. As such, well-established theories are the pinnacle of scientific knowledge, often predicting behavior far beyond the observations or laws from which they were developed. The particulate view of matter grows out of the **atomic theory** proposed by English chemist John Dalton (1766–1844). Dalton explained the law of conservation of mass, as well as other laws and observations of the time, by proposing that matter is composed of small, indestructible particles called atoms. Since these particles are merely rearranged in chemical reactions (and not created or destroyed), the total amount of mass remains the same. Dalton's theory is a model for the physical world—it gives us insight into how nature works, and therefore *explains* our laws and observations.

In Dalton's time, people thought atoms were indestructible. Today, because of nuclear reactions, we know that atoms can be broken apart into their smaller components.

The scientific approach always returns to observation to test theories. For example, scientists have tested the atomic theory by isolating single atoms and even imaging them, providing strong validation for the theory. Nonetheless, theories are never proven because some new observation or experiment always has the potential to reveal a flaw.

Established theories with strong experimental support are the most powerful kind of scientific knowledge. You may have heard the phrase, "That is just a theory," as if theories are easily dismissible. Such a statement reveals a deep misunderstanding of the nature of a scientific theory. Well-established theories are as close to truth as we get in science. The idea that all matter is made of atoms is "just a theory," but it has over 200 years of experimental evidence to support it. It is a powerful piece of scientific knowledge on which many other scientific ideas have been built.

1.2 Cc Conceptual Connection

Laws and Theories

Which statement best explains the difference between a law and a theory?

(a) A law is truth whereas a theory is mere speculation.

(b) A law summarizes a series of related observations, while a theory gives the underlying reasons for them.

(c) A theory describes *what* nature does; a law describes *why* nature does it.

The Importance of Measurement in Science

Scientific observations are often quantifiable; that is, they can be expressed with numbers. In fact, much of the power of science stems from the ability to assign an accurate number to a particular observation. For example, it is one thing to say that one sample of matter is hot and another cold; it is quite another to say that one sample of matter has a temperature of 115.2 °C and another is 10.5 °C. By assigning a number to an observation, we more clearly and accurately specify the details of an observation and how it differs from other related observations.

To assign a number to an observation we use *units*. A **unit** is a standard, agreed-upon quantity by which to specify a measurement. The two most common unit systems are the **metric system**, used in most of the world, and the **English system**, used in the United States. Scientists use the **International System of Units (SI)**, which is based on the metric system. Table 1.1 shows the standard SI base units. The most important for our purposes are the *meter*, the standard unit of length; the *kilogram*, the standard unit of mass; the *second*, the standard unit of time; the *kelvin*, the standard unit of temperature, and the *mole* the standard unit of amount. A complete description of the first four of these units, as well as commonly used prefix multipliers, is included in Appendix I of this book. The mole is first introduced in Section 2.8.

Creativity and Subjectivity in Science

As we have discussed, empiricism is the hallmark of science. However, that does not mean that creativity, subjectivity, and even a bit of luck do not also play important roles. Novices imagine science to be a strict set of rules and procedures that automatically lead to inarguable, objective facts. But this is not the case. Even our discussion of the scientific approach to knowledge is only an idealization of real science, useful to help us see the key distinctions of science. Real science requires creativity and hard work. Scientific theories do not just arise out of data—men and women of great genius and creativity craft theories. A great theory is not unlike a master painting, and many see a similar kind of beauty in both.

The Structure of Scientific Revolutions, a book by Thomas Kuhn (1922–1996), published in 1962, details the history of science and highlights the creative and subjective aspects of the scientific approach to knowledge. In the book, Kuhn argues that scientific history does not support the idea that science progresses in the smooth cumulative way one might expect of a wholly objective linear enterprise. Instead, Kuhn shows how science goes through fairly quiet periods that he calls *normal science.* In these periods, scientists make their data fit the reigning theory. Small inconsistencies are swept aside during periods of normal science. However, when too many inconsistencies and anomalies develop, a crisis emerges. The crisis brings about a *revolution* and a new reigning theory. According to Kuhn, the new theory is usually quite different from the old one; it not only helps us to make sense of new or anomalous information, but also enables us to see accumulated data from the past in a dramatically new way.

The abbreviation *SI* comes from the French, *Système International d'Unités.*

Kuhn further contends that theories are held for reasons that are not always logical or unbiased, and that theories are not *true* models—in the sense of a one-to-one mapping—of the physical world. Because new theories are often so different from the ones they replace, he argues, and because old theories always make good sense to those holding them, they must not be "True" with a capital *T*; otherwise "truth" would be constantly changing.

Kuhn's ideas created a controversy among scientists and science historians that continues to this day. Some, especially postmodern philosophers of science, have taken Kuhn's ideas one step further. They argue that scientific knowledge is *completely* biased and lacks any objectivity. Most scientists, including Kuhn, would disagree. Although Kuhn pointed out that scientific knowledge has *arbitrary elements*, he also said that, "Observation . . . can and must drastically restrict the range of admissible scientific belief else there would be no science." In other words, saying that science has arbitrary elements is quite different from saying that science is arbitrary.

TABLE 1.1 SI Base Units

Quantity	Unit	Symbol
Length	Meter	m
Mass	Kilogram	kg
Time	Second	s
Temperature	Kelvin	K
Amount of substance	Mole	mol
Electric current	Ampere	A
Luminous intensity	Candela	cd

1.4 Early Ideas about the Building Blocks of Matter

The first people to propose that matter was composed of small, indestructible particles were Leucippus (fifth century B.C.E., exact dates unknown) and his student Democritus (460–370 B.C.E.). These Greek philosophers theorized that matter was ultimately composed of small, indivisible particles they named *atomos*. Democritus wrote, "Nothing exists except atoms and empty space; everything else is opinion." Leucippus and Democritus proposed that many different kinds of atoms existed, each different in shape and size, and that they moved randomly through empty space. As we discussed in Section 1.3, other influential Greek thinkers of the time, such as Plato and Aristotle, did not embrace the atomic ideas of Leucippus and Democritus. Instead, they held that matter had no smallest parts (that it was continuous) and that different substances were composed of various proportions of fire, air, earth, and water. Since there was no experimental way to test the relative merits of the competing ideas, Aristotle's view prevailed, largely because he was so influential. The idea that matter was composed of atoms took a back seat for nearly 2000 years.

In the sixteenth century modern science began to emerge. A greater emphasis on observation led Nicolaus Copernicus (1473–1543) to publish *On the Revolution of the Heavenly Orbs* in 1543. The publication of that book—which proposed that the sun, not Earth, was at the center of the universe—marks the beginning of what we now call the *scientific revolution*. The next 200 years—and the work of scientists such as Francis Bacon (1561–1626), Johannes Kepler (1571–1630), Galileo Galilei (1564–1642), Robert Boyle (1627–1691), and Isaac Newton (1642–1727)—brought rapid advancement as the scientific approach became the established way to learn about the physical world. By the early 1800s, certain observations led the English chemist John Dalton (introduced in Section 1.3) to offer convincing evidence that supported the early atomic ideas of Leucippus and Democritus and to propose his atomic theory.

1.5 Modern Atomic Theory and the Laws That Led to It

Like most theories, Dalton's theory that all matter is composed of atoms grew out of observations and laws. The three most important laws that led to the development and acceptance of the atomic theory are the law of conservation of mass, the law of definite proportions, and the law of multiple proportions.

The Law of Conservation of Mass

Recall from Section 1.3 that in 1789 Antoine Lavoisier studied combustion and formulated the law of conservation of mass, which states:

In a chemical reaction, matter is neither created nor destroyed.

We will see in Chapter 21 that this law is a slight oversimplification. However, the changes in mass in ordinary chemical processes are so minute that we can ignore them for all practical purposes.

A **chemical reaction** (discussed more fully in Chapter 8) is a process in which one or more substances are converted into one or more different substances. The law of conservation of mass states that when a chemical reaction occurs, the total mass of the substances involved in the reaction does not change. For example, consider the reaction between sodium and chlorine to form sodium chloride shown here:

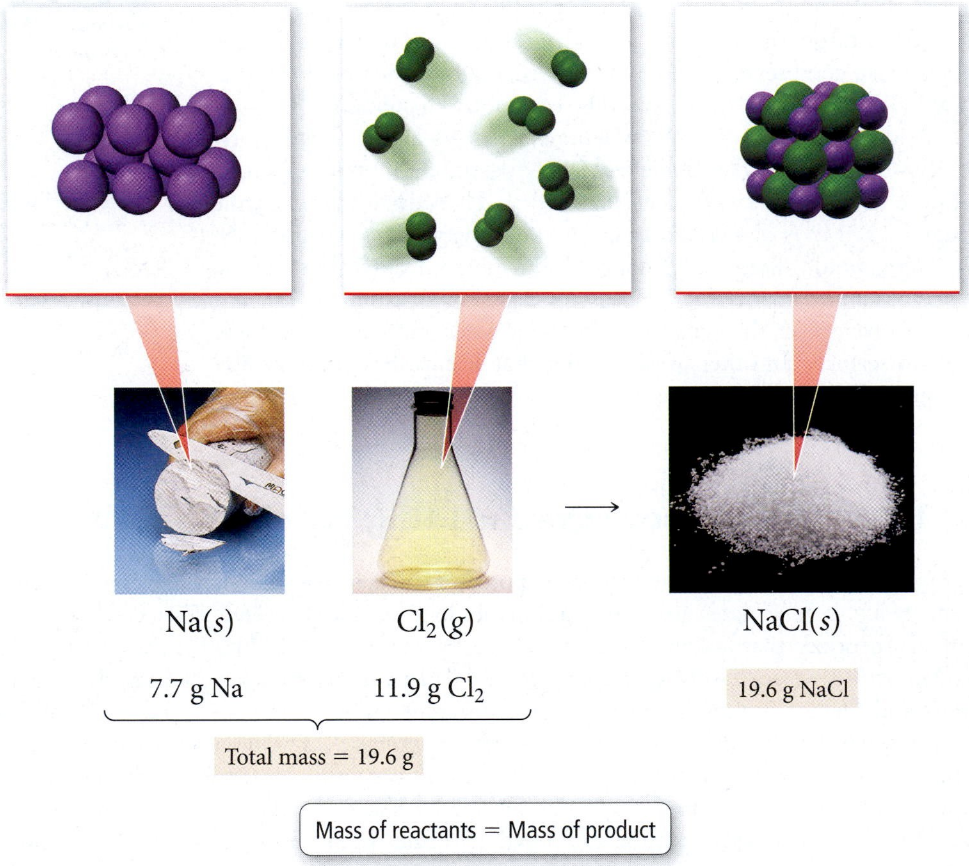

$Na(s)$ + $Cl_2(g)$ → $NaCl(s)$

7.7 g Na, 11.9 g Cl_2 — Total mass = 19.6 g

19.6 g NaCl

Mass of reactants = Mass of product

The combined mass of the sodium and chlorine that react (the reactants) exactly equals the mass of the sodium chloride that forms (the product). This law is consistent with the idea that matter is composed of small, indestructible particles. The particles rearrange during a chemical reaction, but the number of particles is conserved because the particles themselves are indestructible (at least by chemical means).

1.3 Cc Conceptual Connection

The Law of Conservation of Mass

When a log completely burns in a campfire, the mass of the ash is much less than the mass of the log. What happens to the matter that composed the log?

The Law of Definite Proportions

In 1797, a French chemist named Joseph Proust (1754–1826) made observations on the composition of compounds. He found that the elements composing a given compound always occur in fixed (or definite) proportions in all samples of the compound. In contrast, the components of a mixture can be present in any proportions whatsoever. He summarized his observations in the **law of definite proportions**:

> **All samples of a given compound, regardless of their source or how they were prepared, have the same proportions of their constituent elements.**

The law of definite proportions is sometimes called the law of constant composition.

For example, the decomposition of 18.0 g of water results in 16.0 g of oxygen and 2.0 g of hydrogen, or an oxygen-to-hydrogen mass ratio of:

$$\text{Mass ratio} = \frac{16.0 \text{ g O}}{2.0 \text{ g H}} = 8.0 \text{ or } 8{:}1$$

This ratio holds for any sample of pure water, regardless of its origin. The law of definite proportions applies to all compounds. Consider ammonia, a compound composed of nitrogen and hydrogen. Ammonia contains 14.0 g of nitrogen for every 3.0 g of hydrogen, resulting in a nitrogen-to-hydrogen mass ratio of:

$$\text{Mass ratio} = \frac{14.0 \text{ g N}}{3.0 \text{ g H}} = 4.7 \text{ or } 4.7{:}1$$

Again, this ratio is the same for every sample of ammonia. The law of definite proportions hints at the idea that matter is composed of atoms. Compounds have definite proportions of their constituent elements because the atoms that compose them, each with its own specific mass, occur in a definite ratio. Because the ratio of atoms is the same for all samples of a particular compound, the ratio of masses is also the same.

EXAMPLE 1.1
Law of Definite Proportions

Two samples of carbon dioxide decompose into their constituent elements. One sample produces 25.6 g of oxygen and 9.60 g of carbon, and the other produces 21.6 g of oxygen and 8.10 g of carbon. Show that these results are consistent with the law of definite proportions.

SOLUTION

To show this, calculate the mass ratio of one element to the other for both samples by dividing the mass of one element by the mass of the other. For convenience, divide the larger mass by the smaller one.	For the first sample: $\frac{\text{Mass oxygen}}{\text{Mass carbon}} = \frac{25.6}{9.60} = 2.67 \text{ or } 2.67{:}1$ For the second sample: $\frac{\text{Mass oxygen}}{\text{Mass carbon}} = \frac{21.6}{8.10} = 2.67 \text{ or } 2.67{:}1$

The ratios are the same for the two samples, so these results are consistent with the law of definite proportions.

FOR PRACTICE 1.1

Two samples of carbon monoxide decompose into their constituent elements. One sample produces 17.2 g of oxygen and 12.9 g of carbon, and the other sample produces 10.5 g of oxygen and 7.88 g of carbon. Show that these results are consistent with the law of definite proportions.

Answers to For Practice and For More Practice Problems are in Appendix VI.

The Law of Multiple Proportions

In 1804, John Dalton published his **law of multiple proportions:**

> **When two elements (call them A and B) form two different compounds, the masses of element B that combine with 1 g of element A can be expressed as a ratio of small whole numbers.**

Dalton already suspected that matter was composed of atoms, so that when two elements A and B combine to form more than one compound, an atom of A combines with either one, two, three, or more atoms of B (AB_1, AB_2, AB_3, etc.). Therefore the masses of B that react with a fixed mass of A are always related to one another as small whole-number ratios. Consider the compounds carbon monoxide and carbon dioxide, two compounds composed of the same two elements: carbon and oxygen. We saw in Example 1.1 that the mass ratio of oxygen to carbon in carbon dioxide is 2.67:1; therefore, 2.67 g of oxygen reacts with 1 g of carbon. In carbon monoxide, however, the mass ratio of oxygen to carbon is 1.33:1, or 1.33 g of oxygen to every 1 g of carbon.

The ratio of these two masses is a small whole number.

$$\frac{\text{Mass oxygen to 1 g carbon in carbon dioxide}}{\text{Mass oxygen to 1 g carbon in carbon monoxide}} = \frac{2.67}{1.33} = 2$$

With the help of the molecular models, we can see why the ratio is 2:1—carbon dioxide contains two oxygen atoms to every carbon atom while carbon monoxide contains only one. Of course, neither John Dalton nor Joseph Proust had access to any kind of modern instrumentation that could detect individual atoms—Dalton supported his atomic ideas primarily by using the *weights* of samples. *But the weights implied that matter was ultimately particulate; what else would explain why these ratios were always whole numbers?*

EXAMPLE 1.2
Law of Multiple Proportions

Nitrogen forms several compounds with oxygen, including nitrogen dioxide and dinitrogen monoxide. Measurements of the masses of nitrogen and oxygen that form upon decomposing these compounds show that nitrogen dioxide contains 2.28 g oxygen to every 1.00 g nitrogen while dinitrogen monoxide contains 0.570 g oxygen to every 1.00 g nitrogen. Show that these results are consistent with the law of multiple proportions.

SOLUTION

To show this, calculate the ratio of the mass of oxygen from one compound to the mass of oxygen in the other. Always divide the larger of the two masses by the smaller one.

$$\frac{\begin{array}{c}\text{Mass oxygen to 1 g}\\ \text{nitrogen in nitrogen dioxide}\end{array}}{\begin{array}{c}\text{Mass oxygen to 1 g}\\ \text{nitrogen in dinitrogen monoxide}\end{array}} = \frac{2.28}{0.570} = 4.00$$

The ratio is a small whole number (4); these results are consistent with the law of multiple proportions.

FOR PRACTICE 1.2

Hydrogen and oxygen form both water and hydrogen peroxide. The decomposition of a sample of water forms 0.125 g hydrogen to every 1.00 g oxygen. The decomposition of a sample of hydrogen peroxide forms 0.0625 g hydrogen to every 1.00 g oxygen. Show that these results are consistent with the law of multiple proportions.

The Laws of Definite and Multiple Proportions

Explain the difference between the law of definite proportions and the law of multiple proportions.

John Dalton and the Atomic Theory

In 1808, John Dalton explained the laws discussed in this section with his **atomic theory**, which states that:

1. Each element is composed of tiny, indestructible particles called atoms.
2. All atoms of a given element have the same mass and other properties that distinguish them from the atoms of other elements.
3. Atoms combine in simple, whole-number ratios to form compounds.
4. Atoms of one element cannot change into atoms of another element. In a chemical reaction, atoms only change the way that they are *bound together* with other atoms.

The most important idea presented in this section is that measurements of the relative weights of matter samples in three categories (of samples before and after a reaction; of different samples of the same compound; and of different compounds composed of the same elements) indicate that matter is particulate. These scientists—with the help of a balance—gathered and interpreted data that settled an age-old question: Is matter continuous or particulate? Thanks to the careful observations they made, the evidence was in and the particulate view was confirmed.

1.6 The Discovery of the Electron

By the end of the nineteenth century, scientists were convinced that matter is made up of atoms—permanent, supposedly indestructible building blocks that compose everything. However, further experiments revealed that the atom itself is composed of even smaller, more fundamental particles.

Cathode Rays

In the late 1800s an English physicist named J. J. Thomson (1856–1940), working at Cambridge University, performed experiments to probe the properties of **cathode rays.** Thomson constructed a partially evacuated glass tube called a **cathode ray tube**, shown in Figure 1.4 ▼. Thomson then applied a high electrical voltage between two electrodes at either end of the tube. He found that a beam of particles, called cathode rays, traveled from the negatively charged electrode (called the cathode) to the positively charged one (called the anode).

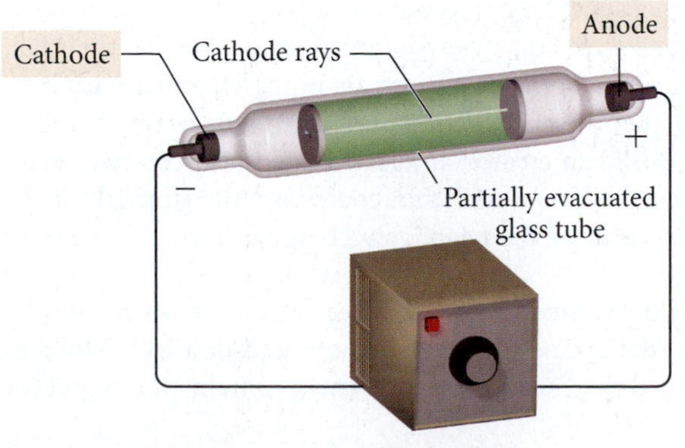

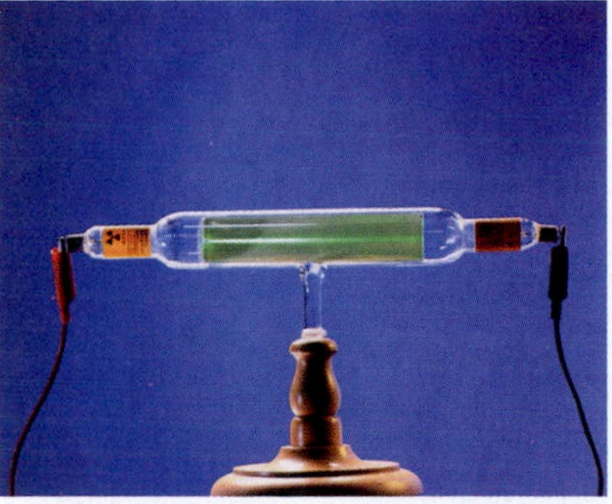

▲ **FIGURE 1.4 Cathode Ray Tube**

▶ **FIGURE 1.5 Thomson's Measurement of the Charge-to-Mass Ratio of the Electron** J. J. Thomson used electric and magnetic fields to deflect the electron beam in a cathode ray tube. By measuring the strengths at which the effects of the two fields (electric and magnetic) canceled exactly, leaving the beam undeflected, he was able to calculate the charge-to-mass ratio of the electron.

Charge-to-Mass Ratio of the Electron

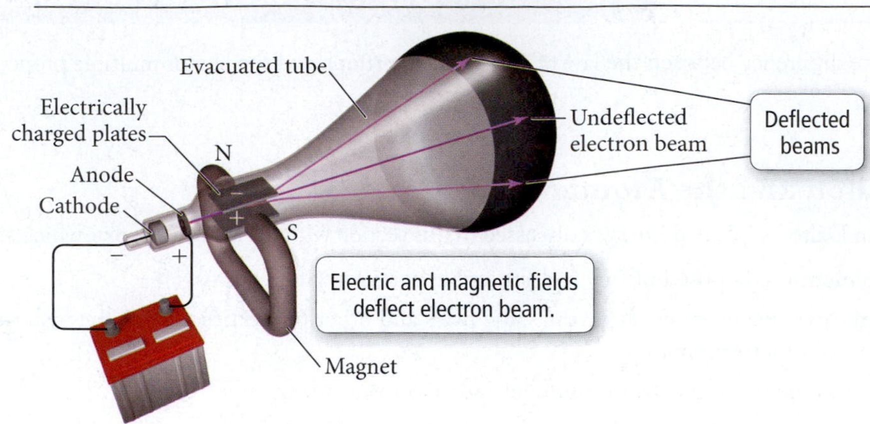

Properties of Electrical Charge

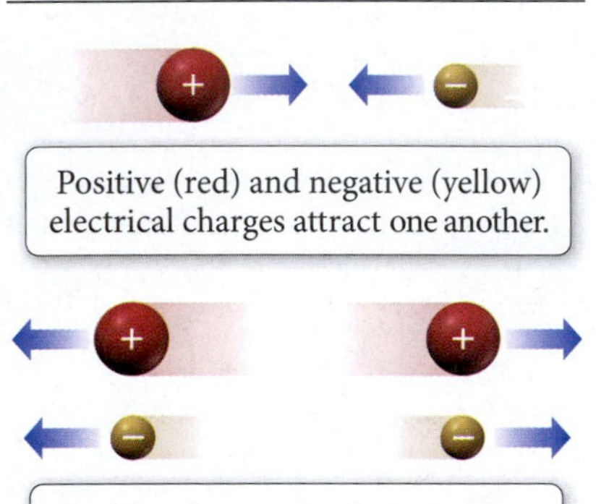

Positive (red) and negative (yellow) electrical charges attract one another.
Positive charges repel one another.
Negative charges repel one another.

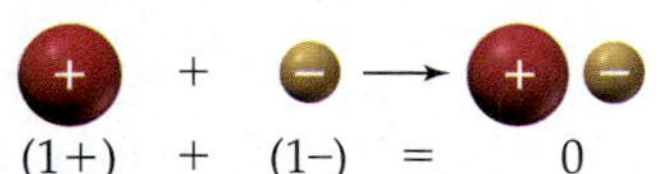

Positive and negative charges of exactly the same magnitude sum to zero when combined.

For a full explanation of electrical voltage, see Chapter 20. The coulomb (C) is the SI unit for charge.

Thomson observed that the particles that compose the cathode ray have the following properties: They travel in straight lines; they are independent of the composition of the material from which they originate (the cathode); and they carry a negative **electrical charge**. Electrical charge is a fundamental property of some of the particles that compose atoms that results in attractive and repulsive forces—called *electrostatic forces*—between those particles. The area around a charged particle where these forces exist is called an *electric field*. The characteristics of electrical charge are summarized in the figure in the margin. You have probably experienced excess electrical charge when brushing your hair on a dry day. The brushing action causes the accumulation of charged particles in your hair, which repel each other, making your hair stand on end.

J. J. Thomson measured the charge-to-mass ratio of the cathode ray particles by deflecting them using electric and magnetic fields, as shown in Figure 1.5 ▲. The value he measured, -1.76×10^8 coulombs (C) per gram, implied that the cathode ray particle was about 2000 times lighter (less massive) than hydrogen, the lightest known atom. These results were incredible—the indestructible atom could apparently be chipped!

J. J. Thomson had discovered the **electron**, a negatively charged, low-mass particle present within all atoms. He wrote, "We have in the cathode rays matter in a new state, a state in which the subdivision of matter is carried very much further . . . a state in which all matter . . . is of one and the same kind; this matter being the substance from which all the chemical elements are built up."

Millikan's Oil Drop Experiment: The Charge of the Electron

In 1909, American physicist Robert Millikan (1868–1953), working at the University of Chicago, performed his now famous oil drop experiment in which he deduced the charge of a single electron. The apparatus for the oil drop experiment is shown in Figure 1.6 ▶.

In his experiment, Millikan sprayed oil into fine droplets using an atomizer. The droplets were allowed to fall under the influence of gravity through a small hole into the lower portion of the apparatus where Millikan viewed them with the aid of a light source and a viewing microscope. During their fall, the drops acquired electrons that had been produced by bombarding the air in the chamber with ionizing radiation (a kind of energy described in Chapter 3). The electrons imparted a negative charge to the drops. In the lower portion of the apparatus, Millikan created an electric field between two metal plates. Since the lower plate was negatively charged, and since Millikan could vary the strength of the electric field, he could slow or even reverse the free fall of the negatively charged drops. (Remember that like charges repel each other.)

By measuring the strength of the electric field required to halt the free fall of the drops, and by figuring out the masses of the drops themselves (determined from their radii and density), Millikan calculated the charge of each drop. He then reasoned that, since each drop must contain an integral (or

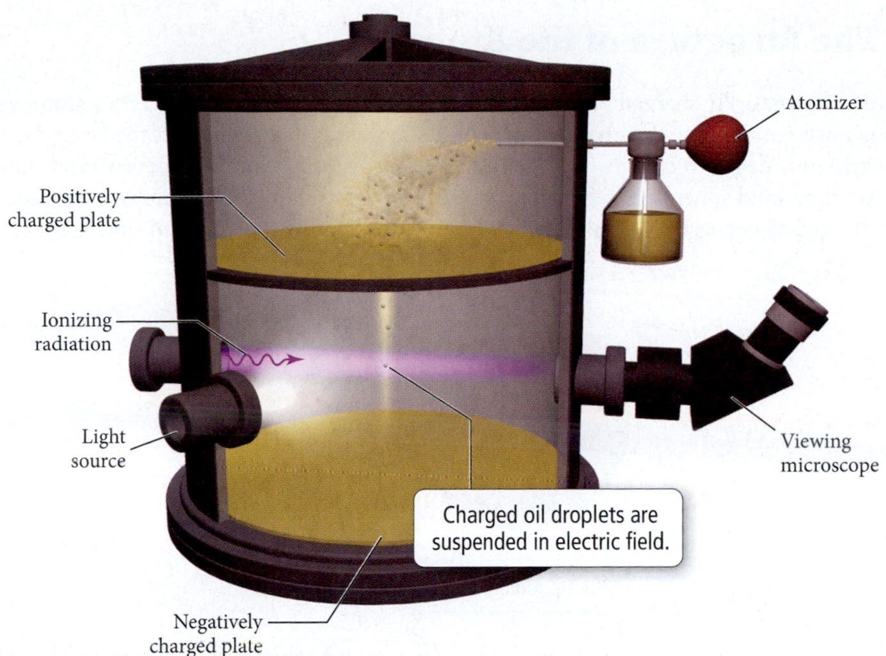

◀ **FIGURE 1.6 Millikan's Measurement of the Electron's Charge** Millikan calculated the charge on oil droplets falling in an electric field. He found that it was always a whole-number multiple of -1.60×10^{-19} C, the charge of a single electron.

whole) number of electrons, the charge of each drop must be a whole-number multiple of the electron's charge. Indeed, Millikan was correct; the measured charge on any drop was always a whole-number multiple of -1.60×10^{-19} C, the fundamental charge of a single electron.

With this number in hand, and knowing Thomson's mass-to-charge ratio for electrons, we can deduce the mass of an electron:

$$\text{Charge} \times \frac{\text{mass}}{\text{charge}} = \text{mass}$$

$$-1.60 \times 10^{-19}\ \text{C} \times \frac{\text{g}}{-1.76 \times 10^{8}\ \text{C}} = 9.10 \times 10^{-28}\ \text{g}$$

As Thomson had correctly determined, this mass is about 2000 times lighter than hydrogen, the lightest atom.

Why did scientists work so hard to measure the charge of the electron? Since the electron is a fundamental building block of matter, scientists want to know its properties, including its charge. The magnitude of the charge of the electron is of tremendous importance because it determines how strongly an atom holds its electrons. Imagine how matter would be different if electrons had a much smaller charge, so that atoms held them more loosely. Many atoms might not even be stable. On the other hand, imagine how matter would be different if electrons had a much greater charge, so that atoms held them more tightly. Since atoms form compounds by exchanging and sharing electrons (more on this in Chapter 6), there could be fewer compounds or maybe even none. Without the abundant diversity of compounds, life would not be possible. So, the magnitude of the charge of the electron—even though it may seem like an insignificantly small number—has great importance.

The Millikan Oil Drop Experiment

1.5 Cc Conceptual Connection

Suppose that one of Millikan's oil drops has a charge of -4.8×10^{-19} C. How many excess electrons does the drop contain?

1.7 The Structure of the Atom

The discovery of negatively charged particles within atoms raised a new question. Since atoms are charge-neutral, they must contain positive charge that neutralizes the negative charge of the electrons—but how do the positive and negative charges fit together? Are atoms just a jumble of even more fundamental particles? Are they solid spheres? Do they have some internal structure? J. J. Thomson proposed that the negatively charged electrons were small particles held within a positively charged sphere, as shown here.

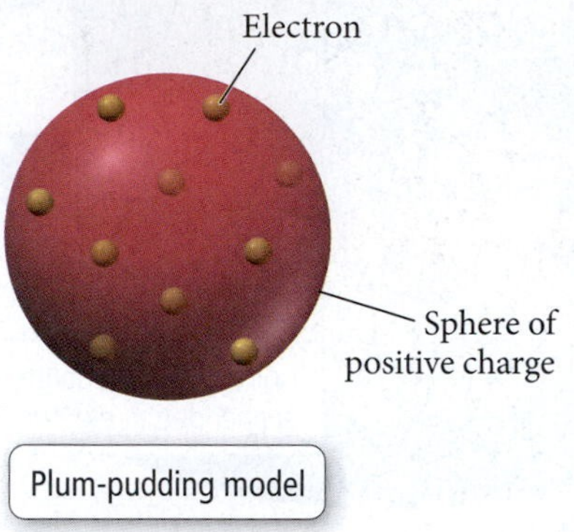

This model, the most popular of its time, became known as the plum-pudding model. The model suggested by Thomson, to those of us not familiar with plum pudding (a British dessert), was like a blueberry muffin, where the blueberries are the electrons and the muffin is the sphere of positive charge.

The discovery of **radioactivity**—the emission of small energetic particles from the core of certain unstable atoms—by scientists Henri Becquerel (1852–1908) and Marie Curie (1867–1934) at the end of the nineteenth century allowed researchers to experimentally probe the structure of the atom. At the time, scientists had identified three different types of radioactivity: alpha (α) particles, beta (β) particles, and gamma (γ) rays. We will discuss these and other types of radioactivity in more detail in Chapter 21. For now, just know that α particles are positively charged and that they are by far the most massive of the three.

Alpha particles are about 7000 times more massive than electrons.

In 1909, Ernest Rutherford (1871–1937) and his coworkers performed an experiment in an attempt to confirm Thomson's model (Rutherford had worked under Thomson and subscribed to his plum-pudding model). Instead, Rutherford's experiment, which employed α particles, proved Thomson wrong. In the experiment, positively charged α particles were directed at an ultrathin sheet of gold foil, as shown in Figure 1.7 ▼.

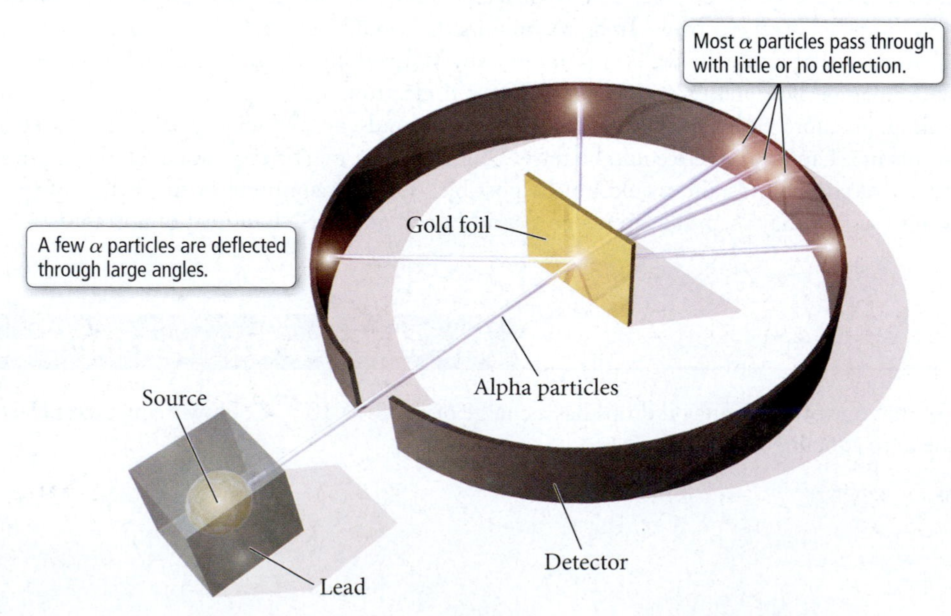

▶ **FIGURE 1.7 Rutherford's Gold Foil Experiment** Alpha particles were directed at a thin sheet of gold foil. Most of the particles passed through the foil, but a small fraction were deflected and a few even bounced backward.

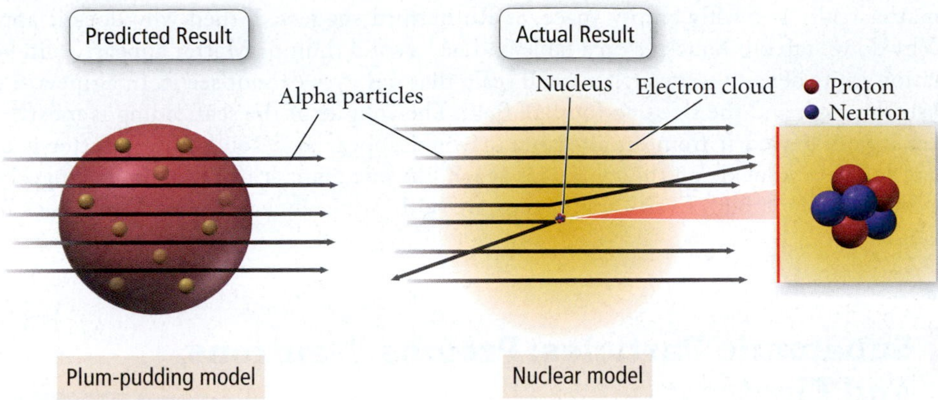

◄ **FIGURE 1.8 The Nuclear Atom** Rutherford's results could not be explained by the plum-pudding model. Instead, they suggested that the atom has a small, dense nucleus.

These particles were to act as probes of the gold atoms' structure. If the gold atoms were indeed like blueberry muffins or plum pudding—with their mass and charge spread throughout the entire volume of the atom—these speeding probes should pass right through the gold foil with minimum deflection.

When Rutherford and his coworkers performed the experiment, the results were not what he expected. A majority of the particles did pass directly through the foil, but some particles were deflected and some (approximately 1 in 20,000) even bounced back. The results puzzled Rutherford, who wrote that they were "about as credible as if you had fired a 15-inch shell at a piece of tissue paper and it came back and hit you." What sort of atomic structure could explain this odd behavior?

Rutherford created a new model—a modern version of which is shown in Figure 1.8 ▲ beside the plum-pudding model—to explain his results.

Rutherford realized that to account for the observed deflections, the mass and positive charge of an atom must be concentrated in a space much smaller than the size of the atom itself. He concluded that, in contrast to the plum-pudding model, matter must not be as uniform as it appears. It must contain large regions of empty space dotted with small regions of very dense matter. Building on this idea, he proposed the **nuclear theory** of the atom, with three basic parts:

1. Most of the atom's mass and all of its positive charge are contained in a small core called the **nucleus**.
2. Most of the volume of the atom is empty space, throughout which tiny, negatively charged electrons are dispersed.
3. There are as many negatively charged electrons outside the nucleus as there are positively charged particles (named **protons**) within the nucleus, so that the atom is electrically neutral.

Although Rutherford's model was highly successful, scientists realized that it was incomplete. For example, hydrogen atoms contain one proton, and helium atoms contain two, yet a hydrogen atom has only one-fourth the mass of a helium atom. Why? The helium atom must contain some additional mass. Subsequent work by Rutherford and one of his students, British scientist James Chadwick (1891–1974), demonstrated that the previously unaccounted for mass was due to **neutrons**, neutral particles within the nucleus. The mass of a neutron is similar to that of a proton, but a neutron has no electrical charge. The helium atom is four times as massive as the hydrogen atom because it contains two protons *and two neutrons* (while hydrogen contains only one proton and no neutrons).

The dense nucleus contains over 99.9% of the mass of the atom but occupies very little of its volume. For now, we can think of the electrons that surround the nucleus as analogous to the water droplets that make up a cloud—although their mass is relatively small, they are dispersed over a very large volume. Consequently, an atom, like a cloud, is mostly empty space.

Rutherford's nuclear theory was a success and is still valid today. The revolutionary part of this theory is the idea that matter—at its core—is much less uniform than it appears. If the nucleus of the atom were the size of the period at the end of this sentence, the average electron would be about 10 m away. Yet the period would contain nearly all of the atom's mass. Imagine what matter would be like if atomic structure were different. What if matter were composed of atomic nuclei piled on top of each other like marbles in a box? Such matter would be incredibly dense; a single grain of sand composed of solid atomic nuclei would have a mass of 5 million kg (or a weight of about 11 million pounds). Astronomers believe there are some objects in the universe composed of such matter—neutron stars.

If matter really is mostly empty space, as Rutherford suggested, then why does it appear so solid? Why do we tap our knuckles on a table and feel a solid thump? Matter appears solid because the variation in its density is on such a small scale that our eyes cannot see it. Imagine a scaffolding 100 stories high and the size of a football field. The volume of the scaffolding is mostly empty space. Yet if you viewed it from an airplane, it would appear as a solid mass. Matter is similar. When you tap your knuckle on the table, it is much like one giant scaffolding (your finger) crashing into another (the table). Even though they are both primarily empty space, one does not fall into the other.

1.8 Subatomic Particles: Protons, Neutrons, and Electrons

See Appendix I for a summary of the units used here.

All atoms are composed of the same subatomic particles: protons, neutrons, and electrons. Protons and neutrons, as we saw in Section 1.7, have nearly identical masses. In SI units, the mass of the proton is 1.67262×10^{-27} kg, and the mass of the neutron is 1.67493×10^{-27} kg. A more common unit to express these masses is the **atomic mass unit (amu)**, defined as 1/12 the mass of a carbon atom containing six protons and six neutrons. The mass of a proton or neutron is approximately 1 amu. Electrons, by contrast, have an almost negligible mass of 0.00091×10^{-27} kg or 0.00055 amu.

Recall that the proton and the electron both have electrical *charge*. We know from Millikan's oil drop experiment that the electron has a charge of -1.60×10^{-19} C. In atomic (or relative) units, the electron is assigned a charge of 1− and the proton is assigned a charge of 1+. *The charges of the proton and the electron are equal in magnitude but opposite in sign*, so that when the two particles are paired, the charges sum to zero. The neutron has no charge.

Most matter is charge-neutral (it has no overall charge) because protons and electrons are present in equal numbers. When matter does acquire charge imbalances, these imbalances usually equalize quickly, often in dramatic ways. For example, the shock you receive when touching a doorknob during dry weather is the equalization of a charge imbalance that develops as you walk across the carpet. Lightning is an equalization of charge imbalances that develop during electrical storms.

A sample of matter—even a tiny sample, such as a sand grain—composed of only protons or only electrons, would have extraordinary repulsive forces inherent within it, and would be unstable. Luckily, matter is not that way. Table 1.2 summarizes the properties of protons, neutrons, and electrons.

TABLE 1.2 Subatomic Particles

	Mass (kg)	Mass (amu)	Charge (relative)	Charge (C)
Proton	1.67262×10^{-27}	1.00727	1+	$+1.60218 \times 10^{-19}$
Neutron	1.67493×10^{-27}	1.00866	0	0
Electron	0.00091×10^{-27}	0.00055	1−	-1.60218×10^{-19}

Elements: Defined by Their Numbers of Protons

The number of protons in the nucleus of an atom determines the charge of the nucleus. For example, carbon has 6 protons and therefore a nuclear charge of 6+.

If all atoms are composed of the same subatomic particles, what makes the atoms of one element different from those of another? The answer is the *number* of these particles. The most important number to the *identity* of an atom is the number of protons in its nucleus. *The number of protons defines the element.* For example, an atom with 2 protons in its nucleus is a helium atom; an atom with 6 protons in its nucleus is a carbon atom (Figure 1.9 ▶); and an atom with 92 protons in its nucleus is a uranium atom. The number of protons in an atom's nucleus is its **atomic number** and is given the symbol *Z*. The atomic numbers of known elements range from 1 to 116 (although additional elements may still be discovered), as shown in the **periodic table** of the elements (Figure 1.10 ▶). In the periodic table, which we will describe in more detail in Chapter 4, the elements are arranged so that those with similar properties are in the same column.

The Number of Protons Defines the Element

Helium nucleus: two protons

Carbon nucleus: six protons

◀ **FIGURE 1.9 How Elements Differ** Each element is defined by a unique atomic number (*Z*), the number of protons in the nucleus of every atom of that element. The number of protons determines the charge of the nucleus.

The Periodic Table

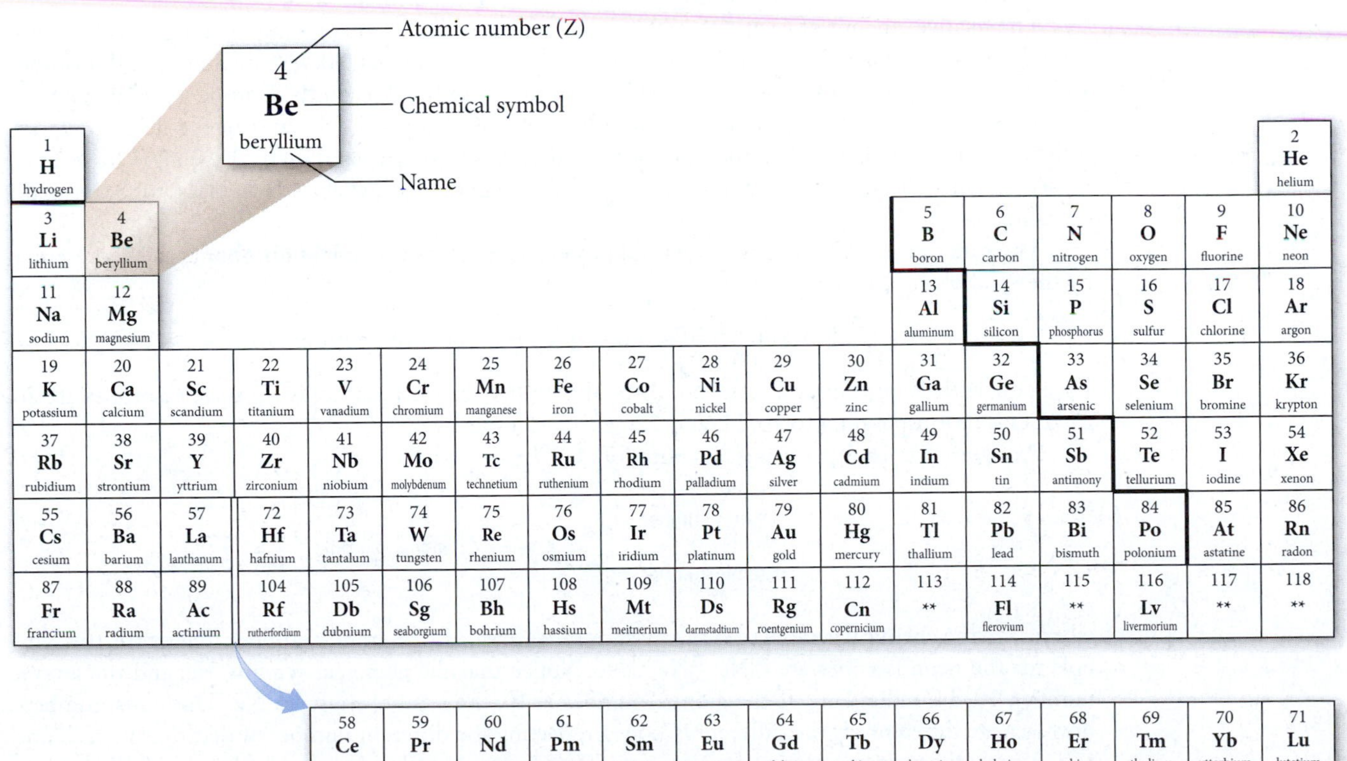

1	2	3	4	5	6	7	8	9	10	11	12	13	14	15	16	17	18
1 H hydrogen																	2 He helium
3 Li lithium	4 Be beryllium											5 B boron	6 C carbon	7 N nitrogen	8 O oxygen	9 F fluorine	10 Ne neon
11 Na sodium	12 Mg magnesium											13 Al aluminum	14 Si silicon	15 P phosphorus	16 S sulfur	17 Cl chlorine	18 Ar argon
19 K potassium	20 Ca calcium	21 Sc scandium	22 Ti titanium	23 V vanadium	24 Cr chromium	25 Mn manganese	26 Fe iron	27 Co cobalt	28 Ni nickel	29 Cu copper	30 Zn zinc	31 Ga gallium	32 Ge germanium	33 As arsenic	34 Se selenium	35 Br bromine	36 Kr krypton
37 Rb rubidium	38 Sr strontium	39 Y yttrium	40 Zr zirconium	41 Nb niobium	42 Mo molybdenum	43 Tc technetium	44 Ru ruthenium	45 Rh rhodium	46 Pd palladium	47 Ag silver	48 Cd cadmium	49 In indium	50 Sn tin	51 Sb antimony	52 Te tellurium	53 I iodine	54 Xe xenon
55 Cs cesium	56 Ba barium	57 La lanthanum	72 Hf hafnium	73 Ta tantalum	74 W tungsten	75 Re rhenium	76 Os osmium	77 Ir iridium	78 Pt platinum	79 Au gold	80 Hg mercury	81 Tl thallium	82 Pb lead	83 Bi bismuth	84 Po polonium	85 At astatine	86 Rn radon
87 Fr francium	88 Ra radium	89 Ac actinium	104 Rf rutherfordium	105 Db dubnium	106 Sg seaborgium	107 Bh bohrium	108 Hs hassium	109 Mt meitnerium	110 Ds darmstadtium	111 Rg roentgenium	112 Cn copernicium	113 **	114 Fl flerovium	115 **	116 Lv livermorium	117 **	118 **

58 Ce cerium	59 Pr praseodymium	60 Nd neodymium	61 Pm promethium	62 Sm samarium	63 Eu europium	64 Gd gadolinium	65 Tb terbium	66 Dy dysprosium	67 Ho holmium	68 Er erbium	69 Tm thulium	70 Yb ytterbium	71 Lu lutetium
90 Th thorium	91 Pa protactinium	92 U uranium	93 Np neptunium	94 Pu plutonium	95 Am americium	96 Cm curium	97 Bk berkelium	98 Cf californium	99 Es einsteinium	100 Fm fermium	101 Md mendelevium	102 No nobelium	103 Lr lawrencium

▲ **FIGURE 1.10 The Periodic Table**
Each element is represented by its symbol and atomic number. Elements in the same column have similar properties.

96
Cm
curium

▲ Element 96 is named curium, after Marie Curie, co-discoverer of radioactivity.

Each element, identified by its unique atomic number, is represented with a unique **chemical symbol**, a one- or two-letter abbreviation listed directly below its atomic number on the periodic table. The chemical symbol for helium is He; for carbon, it is C; and for uranium, it is U. The chemical symbol and the atomic number always go together. If the atomic number is 2, the chemical symbol *must be* He. If the atomic number is 6, the chemical symbol *must be* C. This is another way of saying that the number of protons defines the element.

Most chemical symbols are based on the English name of the element. For example, the symbol for sulfur is S; for oxygen, O; and for chlorine, Cl. Several of the oldest known elements, however, have symbols based on their Latin names. For example, the symbol for sodium is Na from the Latin *natrium*, and the symbol for tin is Sn from the Latin *stannum*. Early scientists often gave newly discovered elements names that reflect their properties. For example, argon originates from the Greek word *argos* meaning inactive, referring to argon's chemical inertness (it does not react with other elements). Chlorine originates from the Greek word *chloros* meaning pale green, referring to chlorine's pale green color. Other elements, including helium, selenium, and mercury, are named after figures from Greek or Roman mythology or astronomical bodies. Still others (such as europium, polonium, and berkelium) are named for the places where they were discovered or where their discoverers were born. More recently, elements have been named after scientists; for example, curium for Marie Curie, einsteinium for Albert Einstein, and rutherfordium for Ernest Rutherford.

Isotopes: When the Number of Neutrons Varies

All atoms of a given element have the same number of protons; however, they do not necessarily have the same number of neutrons. Since neutrons have nearly the same mass as protons (1 amu), this means that—contrary to what John Dalton originally proposed in his atomic theory—all atoms of a given element *do not* have the same mass. For example, all neon atoms contain 10 protons, but they may contain 10, 11, or 12 neutrons. All three types of neon atoms exist, and each has a slightly different mass. Atoms with the same number of protons but different numbers of neutrons are called **isotopes**. Some elements, such as beryllium (Be) and aluminum (Al), have only one naturally occurring isotope, while other elements, such as neon (Ne) and chlorine (Cl), have two or more.

The relative amount of each different isotope in a naturally occurring sample of a given element is roughly constant. For example, in any natural sample of neon atoms, 90.48% of them are the isotope with 10 neutrons, 0.27% are the isotope with 11 neutrons, and 9.25% are the isotope with 12 neutrons. These percentages are called the **natural abundance** of the isotopes. Each element has its own characteristic natural abundance of isotopes. However, advances in mass spectrometry have allowed accurate measurements that reveal small but significant variations in the natural abundance of isotopes for many elements.

The sum of the number of neutrons and protons in an atom is its **mass number** and is represented by the symbol A:

$$A = \text{number of protons } (p^+) + \text{number of neutrons } (n)$$

For neon, with 10 protons, the mass numbers of the three different naturally occurring isotopes are 20, 21, and 22, corresponding to 10, 11, and 12 neutrons, respectively.

We symbolize isotopes using this notation:

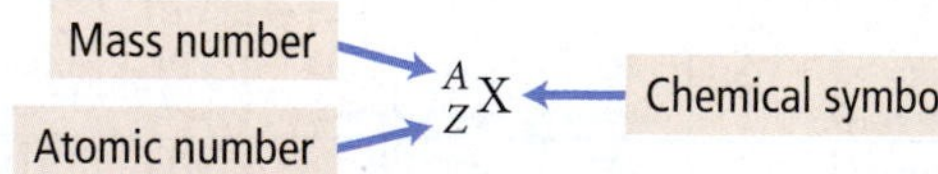

where X is the chemical symbol, A is the mass number, and Z is the atomic number. Therefore, the symbols for the neon isotopes are $^{20}_{10}Ne$ $^{21}_{10}Ne$ $^{22}_{10}Ne$. Notice that the chemical symbol, Ne, and the atomic number, 10, are redundant: If the atomic number is 10, the symbol must be Ne. The mass numbers, however, are different for the different isotopes, reflecting the different number of neutrons in each one.

A second common notation for isotopes is the chemical symbol (or chemical name) followed by a dash and the mass number of the isotope.

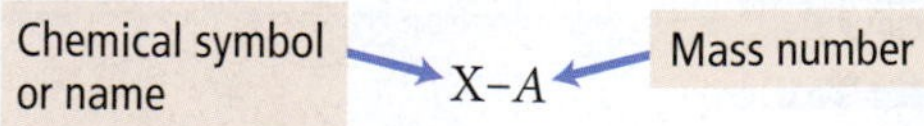

In this notation, the neon isotopes are:

Ne-20 Ne-21 Ne-22

neon-20 neon-21 neon-22

We can summarize what we have discussed about the neon isotopes in a table:

Symbol	Number of Protons	Number of Neutrons	*A* (Mass Number)	Natural Abundance (%)
Ne-20 or $^{20}_{10}\text{Ne}$	10	10	20	90.48
Ne-21 or $^{21}_{10}\text{Ne}$	10	11	21	0.27
Ne-22 or $^{22}_{10}\text{Ne}$	10	12	22	9.25

Notice that all isotopes of a given element have the same number of protons (otherwise they would be different elements). Notice also that the mass number is the *sum* of the number of protons and the number of neutrons. The number of neutrons in an isotope is therefore the difference between the mass number and the atomic number ($A - Z$). The different isotopes of an element generally exhibit the same chemical behavior—the three isotopes of neon, for example, all exhibit chemical inertness.

EXAMPLE 1.3
Atomic Numbers, Mass Numbers, and Isotope Symbols

(a) What are the atomic number (Z), mass number (A), and symbol of the chlorine isotope with 18 neutrons?

(b) How many protons, electrons, and neutrons are present in an atom of $^{52}_{24}\text{Cr}$?

SOLUTION

(a) Look up the atomic number (Z) for chlorine on the periodic table. The atomic number specifies the number of protons.	$Z = 17$, so chlorine has 17 protons.
The mass number (A) for an isotope is the sum of the number of protons and the number of neutrons.	A = number of protons + number of neutrons $= 17 + 18 = 35$
The symbol for an isotope is its two-letter abbreviation with the atomic number (Z) in the lower left corner and the mass number (A) in the upper left corner.	$^{35}_{17}\text{Cl}$
(b) For any isotope (in this case $^{52}_{24}\text{Cr}$) the number of protons is indicated by the atomic number located at the lower left. Since this is a neutral atom, the number of electrons equals the number of protons.	Number of protons $= Z = 24$ Number of electrons $= 24$ (neutral atom)
The number of neutrons is equal to the mass number (upper left) minus the atomic number (lower left).	Number of neutrons $= 52 - 24 = 28$

FOR PRACTICE 1.3

(a) What are the atomic number, mass number, and symbol for the carbon isotope with 7 neutrons?

(b) How many protons and neutrons are present in an atom of $^{39}_{19}\text{K}$?

Isotopes

1.6 Cc Conceptual Connection

Carbon has two naturally occurring isotopes: C-12 (natural abundance is 98.93%) and C-13 (natural abundance is 1.07%). Using circles to represent protons and squares to represent neutrons, draw the nucleus of each carbon isotope. How many C-13 atoms are present, on average, in a 10,000-atom sample of carbon?

Ions: Losing and Gaining Electrons

The number of electrons in a neutral atom is equal to the charge of its nucleus, which is determined by the number of protons in its nucleus (designated by its atomic number *Z*). During chemical changes, however, atoms can lose or gain electrons and become charged particles called **ions**. For example, neutral lithium (Li) atoms contain 3 protons and 3 electrons; however, in many chemical reactions lithium atoms lose one electron (e^-) to form Li^+ ions.

$$\text{Li} \longrightarrow \text{Li}^+ + 1\,\text{e}^-$$

The charge of an ion is indicated in the upper right corner of the chemical symbol. Since the Li^+ *ion* contains 3 protons and only 2 electrons, its charge is 1+ (ion charges are written as the magnitude first followed by the sign of the charge; for a charge of 1+, the 1 is usually dropped and the charge is written as simply +).

Ions can also be negatively charged. For example, neutral fluorine (F) atoms contain 9 protons and 9 electrons; however, in many chemical reactions fluorine atoms gain one electron to form F^- ions.

$$\text{F} + 1\,\text{e}^- \longrightarrow \text{F}^-$$

The F^- *ion* contains 9 protons and 10 electrons, resulting in a charge of 1− (written simply −). For many elements, such as lithium and fluorine, the ion is much more common than the neutral atom. Lithium and fluorine occur in nature mostly as ions.

Positively charged ions, such as Li^+, are **cations** and negatively charged ions, such as F^-, are **anions**. Ions behave quite differently than the atoms from which they are formed (because the structure of particles—including their charge—determines the properties of the matter they compose). Neutral sodium atoms, for example, are extremely unstable, reacting violently with most things they contact. Sodium cations (Na^+), by contrast, are relatively inert—we eat them all the time in sodium chloride (table salt). In ordinary matter, cations and anions always occur together so that matter is charge-neutral overall.

1.7 Cc Conceptual Connection

The Nuclear Atom, Isotopes, and Ions

In light of the nuclear model for the atom, which statement is true?

(a) For a given element, the size of an isotope with more neutrons is larger than one with fewer neutrons.

(b) For a given element, the size of an atom is the same for all of the element's isotopes.

1.9 Atomic Mass: The Average Mass of an Element's Atoms

Atomic mass is sometimes called *atomic weight* or *standard atomic weight.*

An important part of Dalton's atomic theory is that all atoms of a given element have the same mass. In Section 1.8, we learned that because of isotopes, the atoms of a given element often have different masses, so Dalton was not completely correct. We can, however, calculate an average mass—called the **atomic mass**—for each element.

The atomic mass of each element is listed directly beneath the element's symbol in the periodic table and represents the average mass of the isotopes that compose that element, *weighted according to the natural abundance of each isotope*. For example, the periodic table lists the atomic mass of chlorine as 35.45 amu. Naturally occurring chlorine consists of 75.77% chlorine-35 atoms (mass 34.97 amu) and 24.23% chlorine-37 atoms (mass 36.97 amu). We can calculate its atomic mass:

$$\text{Atomic mass} = 0.7577(34.97\text{ amu}) + 0.2423(36.97\text{ amu}) = 35.45\text{ amu}$$

Naturally occurring chlorine contains more chlorine-35 atoms than chlorine-37 atoms, so the weighted average mass of chlorine is closer to 35 amu than to 37 amu.

17
Cl
35.45
chlorine

In this book, we use the atomic masses recommended by the International Union of Pure and Applied Chemistry (IUPAC) for users seeking an atomic mass value for an unspecified sample. Detailed studies of the atomic masses of many samples, however, have shown that atomic masses are not constants of nature because the exact isotopic abundances in any given sample depend on the history of the sample.

We generally calculate the atomic mass with the equation:

$$\begin{aligned}\textbf{Atomic mass} &= \sum_n \textbf{(fraction of isotope } n\textbf{)} \times \textbf{(mass of isotope } n\textbf{)}\\ &= \textbf{(fraction of isotope 1} \times \textbf{mass of isotope 1)}\\ &\quad + \textbf{(fraction of isotope 2} \times \textbf{mass of isotope 2)}\\ &\quad + \textbf{(fraction of isotope 3} \times \textbf{mass of isotope 3)} + \ldots\end{aligned}$$

where the fractions of each isotope are the percent natural abundances converted to their decimal values. The concept of atomic mass is useful because it allows us to assign a characteristic mass to each element and, as we will see shortly, it allows us to quantify the number of atoms in a sample of that element. *The example that follows uses significant figure and rounding conventions discussed in Appendix II.*

When percentages are used in calculations, we convert them to their decimal value by dividing by 100.

EXAMPLE 1.4
Atomic Mass

Copper has two naturally occurring isotopes: Cu-63 with mass 62.9396 amu and a natural abundance of 69.17%, and Cu-65 with mass 64.9278 amu and a natural abundance of 30.83%. Calculate the atomic mass of copper.

SOLUTION

Convert the percent natural abundances into decimal form by dividing by 100.	$\text{Fraction Cu-63} = \dfrac{69.17}{100} = 0.6917$ $\text{Fraction Cu-65} = \dfrac{30.83}{100} = 0.3083$
Calculate the atomic mass using the equation given in the text. (See Appendix II for significant figure and rounding conventions.)	$\text{Atomic mass} = 0.6917(62.9396\text{ amu}) + 0.3083(64.9278\text{ amu})$ $= 43.5\underline{3}53\text{ amu} + 20.0\underline{1}72\text{ amu} = 63.5\underline{5}25 = 63.55\text{ amu}$ The magnitude of the answer makes sense given that approximately two-thirds of the atoms have a mass of nearly 63 amu and one-third have a mass of nearly 65. The weighted average should be closer to 63 than 65.

FOR PRACTICE 1.4

Magnesium has three naturally occurring isotopes with masses of 23.99 amu, 24.99 amu, and 25.98 amu and natural abundances of 78.99%, 10.00%, and 11.01%, respectively. Calculate the atomic mass of magnesium.

FOR MORE PRACTICE 1.4

Gallium has two naturally occurring isotopes: Ga-69 with a mass of 68.9256 amu and a natural abundance of 60.11%, and Ga-71. Use the atomic mass of gallium from the periodic table to find the mass of Ga-71.

1.8
Cc
Conceptual Connection

Recall from Conceptual Connection 1.6 that carbon has two naturally occurring isotopes: C-12 (natural abundance is 98.93%; mass is 12.0000 amu) and C-13 (natural abundance is 1.07%; mass is 13.0034 amu). Without doing any calculations, determine which mass is closest to the atomic mass of carbon.

(a) 12.00 amu **(b)** 12.50 amu **(c)** 13.00 amu

Mass Spectrometry: Measuring the Mass of Atoms and Molecules

The masses of atoms and the percent abundances of isotopes of elements are measured using **mass spectrometry**, a technique that separates particles according to their mass. In a mass spectrometer, such as the one in Figure 1.11 ▼, the sample (containing the atoms whose mass is to be measured) is injected into the instrument and vaporized. The vaporized atoms are ionized by an electron beam—the electrons in the beam collide with the atoms, removing electrons and creating positively charged ions. The ions are then accelerated into a magnetic field. When ions drift through a magnetic field, they experience a force that bends their trajectory. The amount of bending depends on the mass (and charge) of the ions—the trajectories of lighter ions are bent more than those of heavier ones (of the same charge).

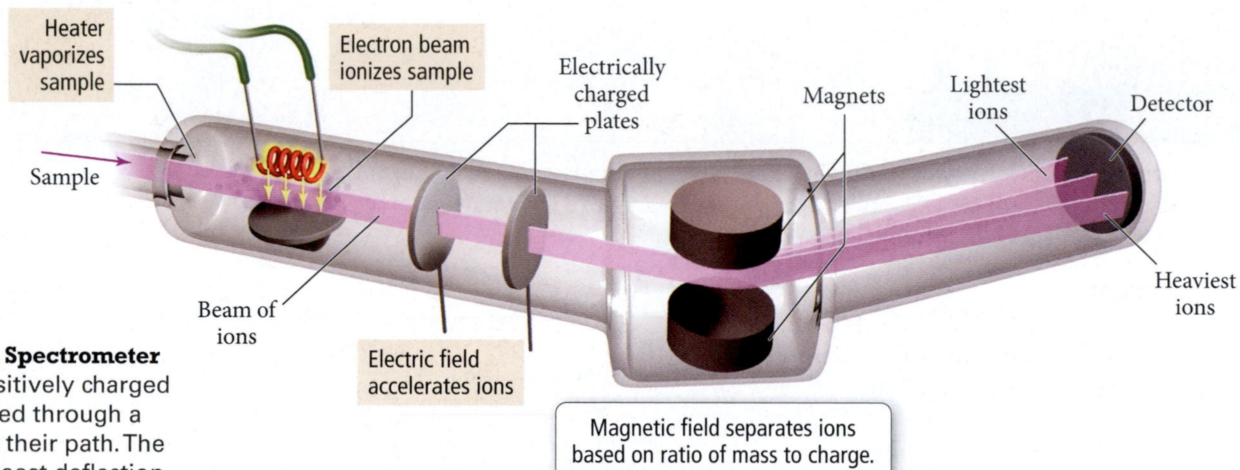

▶ **FIGURE 1.11 The Mass Spectrometer** Atoms are converted to positively charged ions, accelerated, and passed through a magnetic field that deflects their path. The heaviest ions undergo the least deflection.

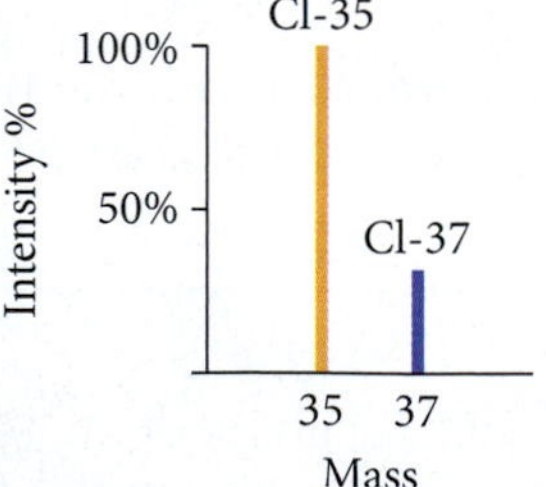

▲ **FIGURE 1.12 The Mass Spectrum of Chlorine** The position of each peak on the *x*-axis indicates the mass of the isotope. The intensity (or height) of the peak indicates the relative abundance of the isotope. The intensity of the highest peak is usually set to 100%, and the intensity of all other peaks is reported relative to the most intense one.

In the right side of the spectrometer shown in Figure 1.11, you can see three different paths, each corresponding to ions of different mass. Finally, the ions strike a detector and produce an electrical signal that is recorded. The result is the separation of the ions according to their mass, producing a mass spectrum such as the one in Figure 1.12 ◀. The *position* of each peak on the *x*-axis indicates the *mass of the isotope* that was ionized, and the *intensity* (indicated by the height of the peak) indicates the *relative abundance of that isotope*.

We can use the mass spectrum of an elemental sample to determine the atomic mass of that sample of the element. For example, consider the mass spectrum of a naturally occurring sample of silver:

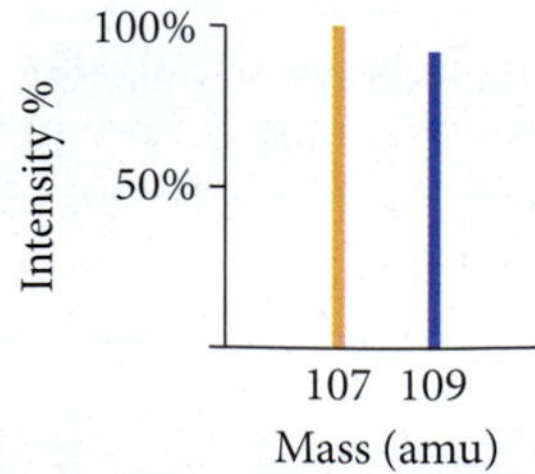

The two peaks correspond to the two naturally occurring isotopes of silver. We determine the percent abundance of each isotope from the intensity of each line. However, the *total* intensity must be *normalized*—it must be made to equal 100%. We accomplish this by dividing the intensity of each peak by the total intensity:

$$\text{Abundance of Ag-107} = \frac{100.0\%}{100.0\% + 92.90\%} \times 100\% = 51.84\%$$

$$\text{Abundance of Ag-109} = \frac{92.90\%}{100.0\% + 92.90\%} \times 100\% = 48.16\%$$

Then we calculate the atomic mass of silver as usual.

$$\begin{aligned}\text{Ag Atomic mass} &= 0.5184\,(106.905\text{ amu}) + 0.4816(108.904\text{ amu})\\ &= 55.4\underline{1}95\text{ amu} + 52.4\underline{4}82\text{ amu} = 107.8\underline{6}77 = 107.87\text{ amu}\end{aligned}$$

We can also use mass spectrometry on molecules. Because molecules often fragment (break apart) during ionization, the mass spectrum of a molecule usually contains many peaks representing the masses of different parts of the molecule, as well as a peak representing the mass of the molecule as a whole. The fragments that form upon ionization, and therefore the corresponding peaks that appear in the mass spectrum, are specific to the molecule, so that a mass spectrum is like a molecular fingerprint. Mass spectroscopy can be used to identify an unknown molecule and to determine how much of it is present in a particular sample.

Since the early 1990s, researchers have successfully applied mass spectrometry to biological molecules, including proteins (the workhorse molecules in cells) and nucleic acids (the molecules that carry genetic information). For a long time, these molecules could not be analyzed by mass spectrometry because they were difficult to vaporize and ionize without destroying them, but modern techniques have overcome this problem. A tumor, for example, can now be instantly analyzed by mass spectrometry to determine whether it contains specific proteins associated with cancer.

1.10 The Origins of Atoms and Elements

We have just discussed the elements and their isotopes. Where did these elements come from? The story of element formation is as old as the universe itself, and we have to go back to the very beginning to tell the story.

The birth of the universe is described by the Big Bang Theory, which asserts that the universe began as a hot, dense collection of matter and energy that expanded rapidly. As it expanded, it cooled, and within the first several hours, subatomic particles formed the first atomic nuclei: hydrogen and helium. These two elements were (and still are) the most abundant in the universe. As the universe continued expanding, some of the hydrogen and helium clumped together under the influence of gravity to form nebulae (clouds of gas) that eventually gave birth to stars and galaxies. These stars and galaxies became the nurseries where all other elements formed.

Stars are fueled by nuclear fusion, which we discuss in more detail in Chapter 21. Under the conditions within the core of a star, hydrogen nuclei can combine (or fuse) to form helium. Fusion gives off enormous quantities of energy, which is why stars emit so much heat and light. The fusion of hydrogen to helium can fuel a star for billions of years.

After it burns through large quantities of hydrogen, if a star is large enough, the helium that builds up in its core can in turn fuse to form carbon. The carbon then builds up in the core and (again, if the star is large enough) can fuse to form even heavier elements. The fusion process ends with iron, which has a highly stable nucleus. By the time iron is formed, however, the star is near the end of its existence and may enter a phase of expansion, transforming into a supernova. Within a supernova, which is in essence a large exploding star, a shower of neutrons allows the lighter elements (which formed during the lifetime of the star through the fusion processes just described) to capture extra neutrons. These neutrons can transform into protons (through processes that we discuss in Chapter 21) contributing ultimately to the formation of elements heavier than iron, all the way up to uranium. As the supernova continues to expand, the elements present within it are blown out into space, where they can incorporate into other nebulae and perhaps even eventually form planets that orbit stars like our own sun.

▲ Stars are born in nebulae such as the Eagle Nebula (also known as M16). This image was taken by the Hubble Space Telescope and shows a gaseous pillar in a star-forming region of the Eagle Nebula.

SELF-ASSESSMENT Quiz

1. Which statement is true about matter?
   a) Matter is particulate—it is composed of particles.
   b) The structure of the particles that compose matter determines the properties of matter.
   c) The particles that compose matter include atoms and molecules.
   d) All of the above statements are true.

2. This image represents a particulate view of a sample of matter. Classify the sample according to its composition.
   a) The sample is a pure element.
   b) The sample is a homogeneous mixture.
   c) The sample is a compound.
   d) The sample is a heterogeneous mixture.

3. A chemist mixes sodium with water and witnesses a violent reaction between the two substances. This is best classified as a(n):
   a) observation **b)** law
   c) hypothesis **d)** theory

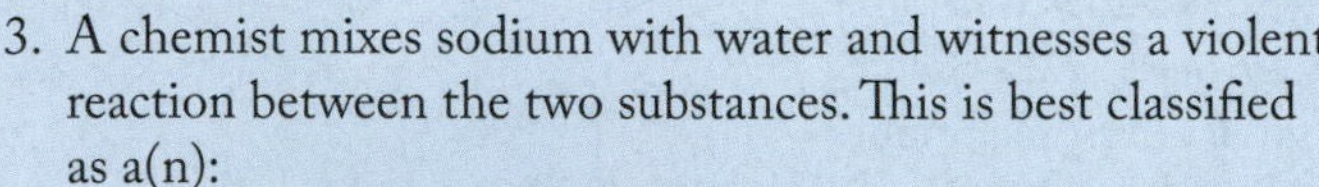

4. Two samples of a compound containing elements A and B are decomposed. The first sample produces 15 g A and 35 g B. The second sample produces 25 g of A and what mass of B?
   a) 11 g B **b)** 58 g B **c)** 21 g B **d)** 45 g B

5. A compound containing only carbon and hydrogen has a carbon-to-hydrogen mass ratio of 11.89. Which carbon-to-hydrogen mass ratio is possible for another compound composed only of carbon and hydrogen?
   a) 2.50 **b)** 3.97 **c)** 4.66 **d)** 7.89

6. Which concept was demonstrated by Rutherford's gold foil experiment?
   a) Atoms contain protons and neutrons.
   b) Matter is composed of atoms.
   c) Elements have isotopes.
   d) Atoms are mostly empty space.

7. A student re-creates Millikan's oil drop experiment and tabulates the relative charges of the oil drops in terms of a constant, α.

Drop #1	α
Drop #2	$\frac{3}{2}\alpha$
Drop #3	$\frac{5}{2}\alpha$
Drop #4	3α

   Which charge for the electron (in terms of α) is consistent with these data?
   a) $\frac{1}{2}\alpha$ **b)** α **c)** $\frac{3}{2}\alpha$ **d)** 2α

8. How many protons and neutrons are in the isotope Fe-58?
   a) 26 protons and 58 neutrons
   b) 32 protons and 26 neutrons
   c) 26 protons and 32 neutrons
   d) 58 protons and 58 neutrons

9. An isotope of an element contains 82 protons and 122 neutrons. What is the symbol for the isotope?
   a) $^{204}_{82}\text{Pb}$ **b)** $^{122}_{82}\text{Pb}$ **c)** $^{122}_{40}\text{Zr}$ **d)** $^{204}_{40}\text{Zr}$

10. How many electrons are in the Cr^{3+} ion?
    a) 24 electrons **b)** 27 electrons
    c) 3 electrons **d)** 21 electrons

11. A naturally occurring sample of an element contains only two isotopes. The first isotope has a mass of 68.9255 amu and a natural abundance of 60.11%. The second isotope has a mass of 70.9247 amu. Determine the atomic mass of the element.
    a) 70.13 amu **b)** 69.72 amu
    c) 84.06 amu **d)** 69.93 amu

12. Copper has an atomic mass of 63.55 amu and two naturally occurring isotopes with masses 62.94 amu and 64.93 amu. Which mass spectrum is most likely to correspond to a naturally occurring sample of copper?

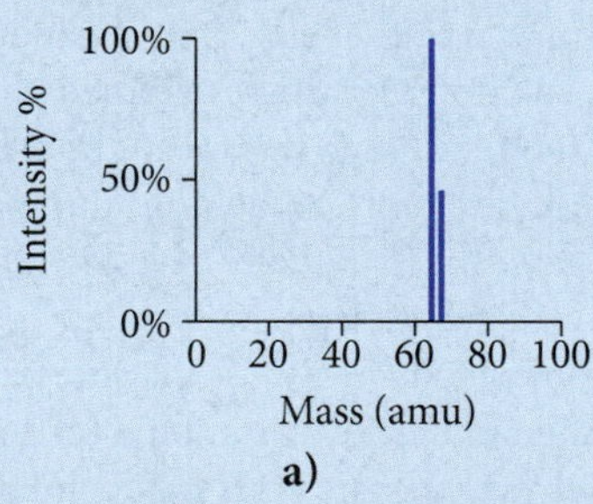

a)

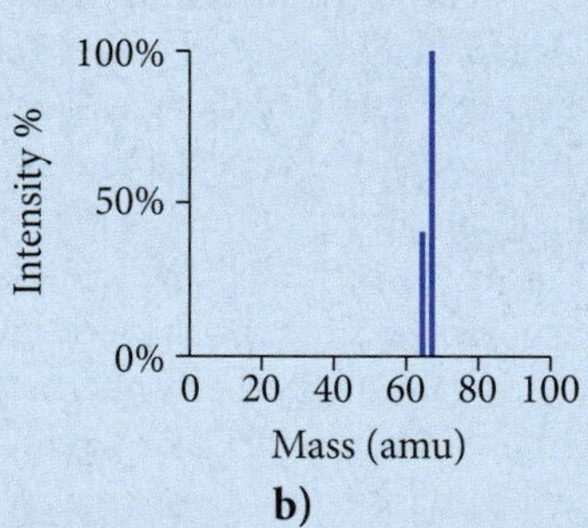

b)

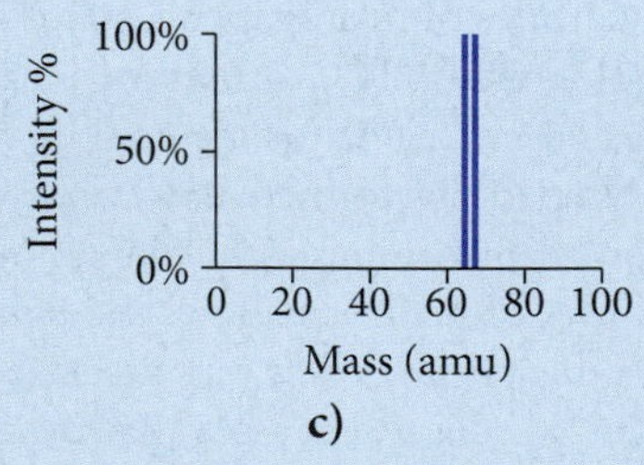

c)

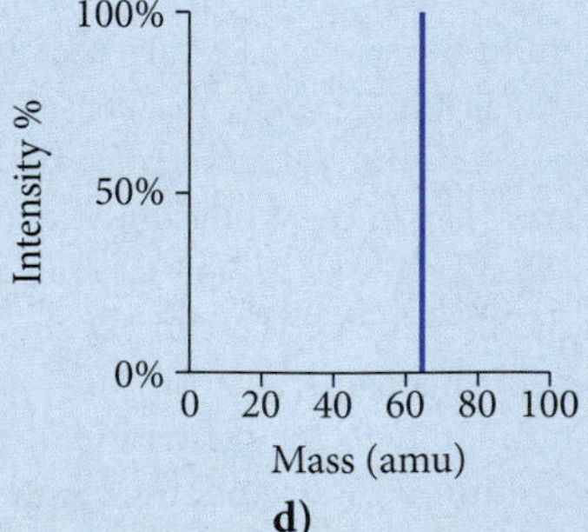

d)

Answers: 1:d; 2:c; 3:a; 4:b; 5:b; 6:d; 7:a; 8:c; 9:a; 10:d; 11:b; 12:a

CHAPTER SUMMARY 1

REVIEW

KEY LEARNING OUTCOMES

CHAPTER OBJECTIVES	ASSESSMENT
Classifying Matter by State and Composition (1.2)	• Exercises 33–40
Distinguishing between Laws and Theories (1.3)	• Exercises 41–42
Applying the Law of Definite Proportions (1.5)	• Example 1.1 • For Practice 1.1 • Exercises 47–50
Applying the Law of Multiple Proportions (1.5)	• Example 1.2 • For Practice 1.2 • Exercises 51–54
Working with Atomic Numbers, Mass Numbers, and Isotope Symbols (1.8)	• Example 1.3 • For Practice 1.3 • Exercises 63–70
Calculating Atomic Mass (1.9)	• Example 1.4 • For Practice 1.4 • Exercises 71, 72, 75, 76, 79, 80

KEY TERMS

Section 1.1
matter (3)
atom (4)
molecule (4)
chemistry (4)

Section 1.2
substance (4)
state (4)
composition (4)
solid (5)
liquid (5)
gas (5)
pure substance (6)
mixture (6)
element (6)
compound (6)
heterogeneous mixture (7)
homogeneous mixture (7)

Section 1.3
hypothesis (7)
experiment (7)
scientific law (7)
law of conservation of mass (7)
theory (8)
atomic theory (8)
unit (9)
metric system (9)
English system (9)
International System of Units (SI) (9)

Section 1.5
chemical reaction (10)
law of definite proportions (11)
law of multiple proportions (12)

Section 1.6
cathode ray (13)
cathode ray tube (13)
electrical charge (14)
electron (14)

Section 1.7
radioactivity (16)
nuclear theory (17)
nucleus (17)
proton (17)
neutron (17)

Section 1.8
atomic mass unit (amu) (18)
atomic number (*Z*) (18)
periodic table (18)
chemical symbol (20)
isotope (20)
natural abundance (20)
mass number (*A*) (20)
ion (22)
cation (22)
anion (22)

Section 1.9
atomic mass (22)
mass spectrometry (24)

KEY CONCEPTS

Matter Is Particulate (1.1)

- All matter is composed of particles.
- The structure of the particles that compose matter determines the properties of matter.
- Chemistry is the science that investigates the properties of matter by examining the atoms and molecules that compose matter it.

Classifying Matter Based on the Particles That Compose It (1.2)

- We classify matter according to its state (which depends on the relative positions of interactions between particles) or according to its composition (which depends on the type of particles).
- Matter has three common states: solid, liquid, and gas.
- Matter can be a pure substance (one type of particle) or a mixture (more than one type of particle).
- A pure substance can either be an element, which cannot be chemically broken down into simpler substances, or a compound, which is composed of two or more elements in fixed proportions.
- A mixture can be either homogeneous, with the same composition throughout, or heterogeneous, with different compositions in different regions.

The Scientific Approach to Knowledge (1.3)

- Science begins with the observation of the physical world. A number of related observations can often be summarized in a statement or generalization called a scientific law.

- A hypothesis is a tentative interpretation or explanation of observations. One or more well-established hypotheses may prompt the development of a scientific theory, a model for nature that explains the underlying reasons for observations and laws.
- Laws, hypotheses, and theories all give rise to predictions that can be tested by experiments, carefully controlled procedures designed to produce critical new observations. If scientists cannot confirm the predictions, they must modify or replace the law, hypothesis, or theory.

Atomic Theory (1.5)

- Each element is composed of indestructible particles called atoms.
- All atoms of a given element have the same mass and other properties.
- Atoms combine in simple, whole-number ratios to form compounds.
- Atoms of one element cannot change into atoms of another element. In a chemical reaction, atoms change the way that they are bound together with other atoms to form a new substance.

The Electron (1.6)

- J. J. Thomson discovered the electron in the late 1800s through experiments with cathode rays. He deduced that electrons are negatively charged, and he measured their charge-to-mass ratio.
- Robert Millikan measured the charge of the electron, which—in conjunction with Thomson's results—led to the calculation of the mass of an electron.

The Nuclear Atom (1.7)

- In 1909, Ernest Rutherford probed the inner structure of the atom by working with a form of radioactivity called alpha radiation and developed the nuclear theory of the atom.
- Nuclear theory states that the atom is mainly empty space, with most of its mass concentrated in a tiny region called the nucleus and most of its volume occupied by relatively light electrons.

Subatomic Particles (1.8)

- Atoms are composed of three fundamental particles: the proton (1 amu, +1 charge), the neutron (1 amu, 0 charge), and the electron (~0 amu, −1 charge).
- The number of protons in the nucleus of the atom is its atomic number (Z). The atomic number determines the charge of the nucleus and defines the element.
- The periodic table tabulates all known elements in order of increasing atomic number.
- The sum of the number of protons and neutrons is the mass number (A).
- Atoms of an element that have different numbers of neutrons (and therefore different mass numbers) are isotopes.
- Atoms that lose or gain electrons become charged and are called ions. Cations are positively charged and anions are negatively charged.

Atomic Mass (1.9)

- The atomic mass of an element, listed directly below its symbol in the periodic table, is a weighted average of the masses of the naturally occurring isotopes of the element.
- Atomic masses can be determined through mass spectrometry.

KEY EQUATIONS AND RELATIONSHIPS

Relationship between Mass Number (*A*), Number of Protons (p), and Number of Neutrons (n) (1.8)

$$A = \text{number of protons (p)} + \text{number of neutrons (n)}$$

Atomic Mass (1.9)

$$\text{Atomic mass} = \sum_n (\text{fraction of isotope } n) \times (\text{mass of isotope } n)$$

EXERCISES

REVIEW QUESTIONS

1. Explain this statement in your own words and give an example. *The properties of the substances around us depend on the structure of the particles that compose them.*
2. Explain the main goal of chemistry.
3. What are two different ways to classify matter?
4. How do solids, liquids, and gases differ?
5. Explain the difference between a pure substance and a mixture based on the composite particles of each.
6. Explain the difference between an element and a compound.
7. Explain the difference between a homogeneous and a heterogeneous mixture.
8. Describe the scientific approach to knowledge. How does it differ from other approaches?
9. Explain the differences between a hypothesis, a law, and a theory.
10. What observations did Antoine Lavoisier make? What law did he formulate?
11. What theory did John Dalton formulate?
12. What is wrong with the expression, "That is just a theory," if by theory the speaker is referring to a scientific theory?
13. Summarize the history of the atomic idea. How was Dalton able to convince others to accept an idea that had been controversial for 2000 years?
14. State and explain the law of conservation of mass.
15. State and explain the law of definite proportions.
16. State and explain the law of multiple proportions. How is the law of multiple proportions different from the law of definite proportions?
17. What are the main ideas in Dalton's atomic theory? How do they help explain the laws of conservation of mass, of constant composition, and of definite proportions?
18. How and by whom was the electron discovered? What basic properties of the electron were reported with its discovery?
19. Explain Millikan's oil drop experiment and how it led to the measurement of the electron's charge. Why is the magnitude of the charge of the electron so important?

20. Describe the plum-pudding model of the atom.
21. Describe Rutherford's gold foil experiment. How did the experiment prove that the plum-pudding model of the atom was wrong?
22. Describe Rutherford's nuclear model of the atom. What was revolutionary about his model?
23. If matter is mostly empty space, as suggested by Rutherford, then why does it appear so solid?
24. List the three subatomic particles that compose atoms and give the basic properties (mass and charge) of each.
25. What defines an element?
26. Explain the difference between Z (the atomic number) and A (the mass number).
27. Where do elements get their names?
28. What are isotopes? What is percent natural abundance of isotopes?
29. Describe the two different notations used to specify isotopes and give an example of each.
30. What is an ion? A cation? An anion?
31. What is atomic mass? How is it calculated?
32. Explain how a mass spectrometer works. What kind of information can be determined from a mass spectrum?

PROBLEMS BY TOPIC

Note: Answers to all odd-numbered Problems, numbered in blue, can be found in Appendix V. Exercises in the Problems by Topic section are paired, with each odd-numbered problem followed by a similar even-numbered problem. Exercises in the Cumulative Problems section are also paired, but more loosely. Challenge Problems and Conceptual Problems, because of their nature, are unpaired.

The Classification of Matter

33. Each shape represents a type of particle (such as an atom or molecule). Classify each image as a pure substance, homogeneous mixture, or heterogeneous mixture.

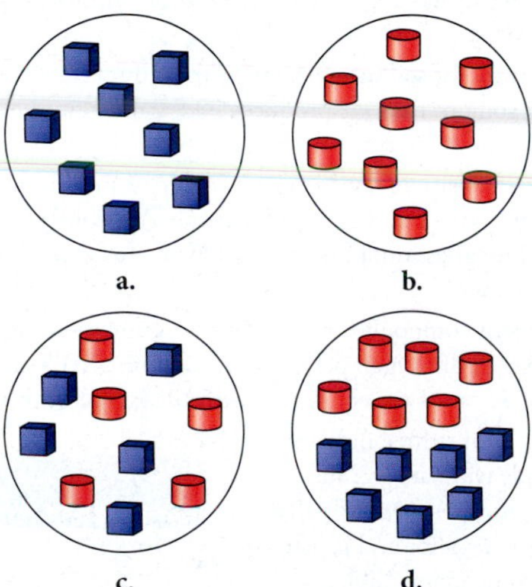

34. Using triangles to represent one type of atom and circles to represent another type of atom, draw one image to represent a mixture of the two atoms and draw another image to represent a compound composed of the two atoms.
35. Classify each substance as a pure substance or a mixture. If it is a pure substance, classify it as an element or a compound. If it is a mixture, classify it as homogeneous or heterogeneous.
    a. sweat **b.** carbon dioxide
    c. aluminum **d.** vegetable soup
36. Classify each substance as a pure substance or a mixture. If it is a pure substance, classify it as an element or a compound. If it is a mixture, classify it as homogeneous or heterogeneous.
    a. wine **b.** beef stew
    c. iron **d.** carbon monoxide
37. Complete the table.

Substance	Pure or mixture	Type
aluminum	pure	element
apple juice	_______	_______
hydrogen peroxide	_______	_______
chicken soup	_______	_______

38. Complete the table.

Substance	Pure or mixture	Type
water	pure	compound
coffee	_______	_______
ice	_______	_______
carbon	_______	_______

39. Determine whether each molecular diagram represents a pure substance or a mixture. If it represents a pure substance, classify the substance as an element or a compound. If it represents a mixture, classify the mixture as homogeneous or heterogeneous.

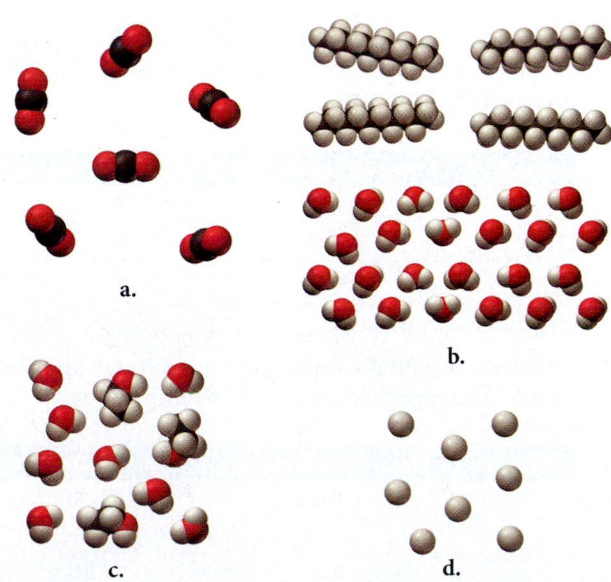

40. Determine whether each molecular diagram represents a pure substance or a mixture. If it represents a pure substance, classify the substance as an element or a compound. If it represents a mixture, classify the mixture as homogeneous or heterogeneous.

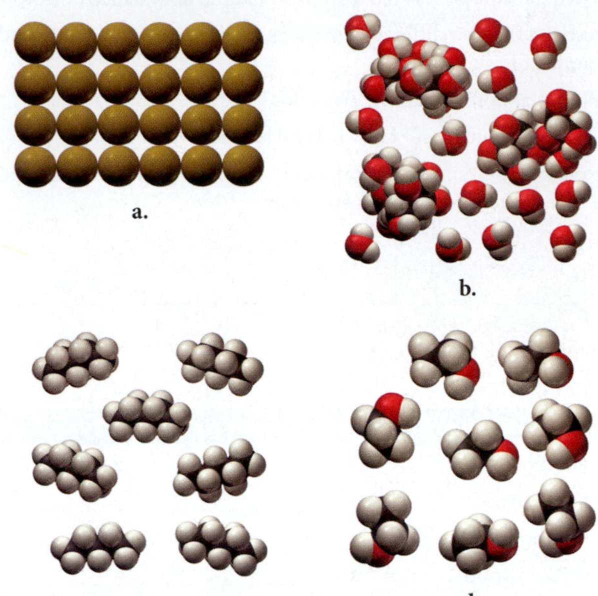

The Scientific Approach to Knowledge

41. Classify each statement as an observation, a law, or a theory.
   a. All matter is made of tiny, indestructible particles called atoms.
   b. When iron rusts in a closed container, the mass of the container and its contents do not change.
   c. In chemical reactions, matter is neither created nor destroyed.
   d. When a match burns, heat is released.

42. Classify each statement as an observation, a law, or a theory.
   a. Chlorine is a highly reactive gas.
   b. If elements are listed in order of increasing mass of their atoms, their chemical reactivity follows a repeating pattern.
   c. Neon is an inert (or nonreactive) gas.
   d. The reactivity of elements depends on the arrangement of their electrons.

43. A chemist decomposes several samples of carbon monoxide into carbon and oxygen and weighs the resultant elements. The results are shown in the table.

Sample	Mass of Carbon (g)	Mass of Oxygen (g)
1	6	8
2	12	16
3	18	24

   a. Describe any pattern you notice in these results.

   Next, the chemist decomposes several samples of hydrogen peroxide into hydrogen and oxygen. The results are shown in the table.

Sample	Mass of Hydrogen (g)	Mass of Oxygen (g)
1	0.5	8
2	1	16
3	1.5	24

   b. Describe any similarity you notice between these results and those for carbon monoxide in part a.
   c. Can you formulate a law from the observations in a and b?
   d. Can you formulate a hypothesis that might explain your law in c?

44. When astronomers observe distant galaxies, they can tell that most of them are moving away from one another. In addition, the more distant the galaxies, the more rapidly they are likely to be moving away from each other. Can you devise a hypothesis to explain these observations?

The Laws of Conservation of Mass, Definite Proportions, and Multiple Proportions

45. A hydrogen-filled balloon is ignited and 1.50 g of hydrogen reacts with 12.0 g of oxygen. How many grams of water vapor form? (Assume that water vapor is the only product.)

46. An automobile gasoline tank holds 21 kg of gasoline. When the gasoline burns, 84 kg of oxygen is consumed, and carbon dioxide and water are produced. What is the total combined mass of carbon dioxide and water that is produced?

47. Two samples of carbon tetrachloride are decomposed into their constituent elements. One sample produces 38.9 g of carbon and 448 g of chlorine, and the other sample produces 14.8 g of carbon and 134 g of chlorine. Are these results consistent with the law of definite proportions? Show why or why not.

48. Two samples of sodium chloride are decomposed into their constituent elements. One sample produces 6.98 g of sodium and 10.7 g of chlorine, and the other sample produces 11.2 g of sodium and 17.3 g of chlorine. Are these results consistent with the law of definite proportions? Explain your answer.

49. The mass ratio of sodium to fluorine in sodium fluoride is 1.21:1. A sample of sodium fluoride produces 28.8 g of sodium upon decomposition. How much fluorine (in grams) is formed?

50. Upon decomposition, one sample of magnesium fluoride produces 1.65 kg of magnesium and 2.57 kg of fluorine. A second sample produces 1.32 kg of magnesium. How much fluorine (in grams) does the second sample produce?

51. Two different compounds containing osmium and oxygen have the following masses of oxygen per gram of osmium: 0.168 and 0.3369 g. Show that these amounts are consistent with the law of multiple proportions.

52. Palladium forms three different compounds with sulfur. The mass of sulfur per gram of palladium in each compound is listed in the accompanying table. Show that these masses are consistent with the law of multiple proportions.

Compound	Grams S per Gram Pd
A	0.603
B	0.301
C	0.151

53. Sulfur and oxygen form both sulfur dioxide and sulfur trioxide. When samples of these are decomposed, the sulfur dioxide produces 3.49 g oxygen and 3.50 g sulfur, while the sulfur trioxide produces 6.75 g oxygen and 4.50 g sulfur. Calculate the mass of oxygen per gram of sulfur for each sample and show that these results are consistent with the law of multiple proportions.

54. Sulfur and fluorine form several different compounds including sulfur hexafluoride and sulfur tetrafluoride. Decomposition of a sample of sulfur hexafluoride produces 4.45 g of fluorine and 1.25 g of sulfur, while decomposition of a sample of sulfur tetrafluoride produces 4.43 g of fluorine and 1.87 g of sulfur. Calculate the mass of fluorine per gram of sulfur for each sample and show that these results are consistent with the law of multiple proportions.

Atomic Theory, Nuclear Theory, and Subatomic Particles

55. Which statements are *consistent* with Dalton's atomic theory as it was originally stated? Why?
    a. Sulfur and oxygen atoms have the same mass.
    b. All cobalt atoms are identical.
    c. Potassium and chlorine atoms combine in a 1:1 ratio to form potassium chloride.
    d. Lead atoms can be converted into gold.

56. Which statements are *inconsistent* with Dalton's atomic theory as it was originally stated? Why?
    a. All carbon atoms are identical.
    b. An oxygen atom combines with 1.5 hydrogen atoms to form a water molecule.
    c. Two oxygen atoms combine with a carbon atom to form a carbon dioxide molecule.
    d. The formation of a compound often involves the destruction of one or more atoms.

57. Which statements are *consistent* with Rutherford's nuclear theory as it was originally stated? Why?
    a. The volume of an atom is mostly empty space.
    b. The nucleus of an atom is small compared to the size of the atom.
    c. Neutral lithium atoms contain more neutrons than protons.
    d. Neutral lithium atoms contain more protons than electrons.

58. Which statements are *inconsistent* with Rutherford's nuclear theory as it was originally stated? Why?
    a. Since electrons are smaller than protons, and since a hydrogen atom contains only one proton and one electron, it must follow that the volume of a hydrogen atom is mostly due to the proton.
    b. A nitrogen atom has 7 protons in its nucleus and 7 electrons outside of its nucleus.
    c. A phosphorus atom has 15 protons in its nucleus and 150 electrons outside of its nucleus.
    d. The majority of the mass of a fluorine atom is due to its 9 electrons.

59. A chemist in an imaginary universe, where electrons have a different charge than they do in our universe, performs the Millikan oil drop experiment to measure the electron's charge. The charges of several drops are recorded here. What is the charge of the electron in this imaginary universe?

Drop #	Charge
A	-6.9×10^{-19} C
B	-9.2×10^{-19} C
C	-11.5×10^{-19} C
D	-4.6×10^{-19} C

60. Imagine a unit of charge called the zorg. A chemist performs the Millikan oil drop experiment and measures the charge of each drop in zorgs. Based on the results shown here, what is the charge of the electron in zorgs (z)? How many electrons are in each drop?

Drop #	Charge
A	-4.8×10^{-9} z
B	-9.6×10^{-9} z
C	-6.4×10^{-9} z
D	-12.8×10^{-9} z

61. Which statements about subatomic particles are true?
    a. If an atom has an equal number of protons and electrons, it will be charge-neutral.
    b. Electrons are attracted to protons.
    c. Electrons are much lighter than neutrons.
    d. Protons have twice the mass of neutrons.

62. Which statements about subatomic particles are false?
    a. Protons and electrons have charges of the same magnitude but opposite sign.
    b. Protons have about the same mass as neutrons.
    c. Some atoms don't have any protons.
    d. Protons and neutrons have charges of the same magnitude but opposite signs.

Isotopes and Ions

63. Write isotopic symbols in the form X–*A* (e.g., C-13) for each isotope.
    a. the silver isotope with 60 neutrons
    b. the silver isotope with 62 neutrons
    c. the uranium isotope with 146 neutrons
    d. the hydrogen isotope with 1 neutron

64. Write isotopic symbols in the form ${}^{A}_{Z}X$ for each isotope.
    a. the copper isotope with 34 neutrons
    b. the copper isotope with 36 neutrons
    c. the potassium isotope with 21 neutrons
    d. the argon isotope with 22 neutrons

65. Determine the number of protons and the number of neutrons in each isotope.
    a. ${}^{14}_{7}N$ b. ${}^{23}_{11}Na$ c. ${}^{222}_{86}Rn$ d. ${}^{208}_{82}Pb$

66. Determine the number of protons and the number of neutrons in each isotope.
    a. ${}^{40}_{19}K$ b. ${}^{226}_{88}Ra$ c. ${}^{99}_{43}Tc$ d. ${}^{33}_{15}P$

67. The amount of carbon-14 in ancient artifacts and fossils is often used to establish their age. Determine the number of protons and the number of neutrons in a carbon-14 isotope and write its symbol in the form ${}^{A}_{Z}X$.

68. Uranium-235 is used in nuclear fission. Determine the number of protons and the number of neutrons in uranium-235 and write its symbol in the form ${}^{A}_{Z}X$.

69. Determine the number of protons and the number of electrons in each ion.
    a. Ni^{2+} b. S^{2-} c. Br^{-} d. Cr^{3+}

70. Determine the number of protons and the number of electrons in each ion.
    a. Al^{3+} b. Se^{2-} c. Ga^{3+} d. Sr^{2+}

Atomic Mass and Mass Spectrometry

71. Gallium has two naturally occurring isotopes with the following masses and natural abundances:

Isotope	Mass (amu)	Abundance (%)
Ga-69	68.92558	60.108
Ga-71	70.92470	39.892

Sketch the mass spectrum of gallium.

72. Magnesium has three naturally occurring isotopes with the following masses and natural abundances:

Isotope	Mass (amu)	Abundance (%)
Mg-24	23.9850	78.99
Mg-25	24.9858	10.00
Mg-26	25.9826	11.01

Sketch the mass spectrum of magnesium.

73. The atomic mass of fluorine is 18.998 amu, and its mass spectrum shows a large peak at this mass. The atomic mass of chlorine is 35.45 amu, yet the mass spectrum of chlorine does not show a peak at this mass. Explain the difference.

74. The atomic mass of copper is 63.546 amu. Do any copper isotopes have a mass of 63.546 amu? Explain.

75. An element has two naturally occurring isotopes. Isotope 1 has a mass of 120.9038 amu and a relative abundance of 57.4%, and isotope 2 has a mass of 122.9042 amu. Find the atomic mass of this element and identify it.

76. An element has four naturally occurring isotopes with the masses and natural abundances given here. Find the atomic mass of the element and identify it.

Isotope	Mass (amu)	Abundance (%)
1	135.90714	0.19
2	137.90599	0.25
3	139.90543	88.43
4	141.90924	11.13

77. Bromine has two naturally occurring isotopes (Br-79 and Br-81) and has an atomic mass of 79.904 amu. The mass of Br-81 is 80.9163 amu, and its natural abundance is 49.31%. Calculate the mass and natural abundance of Br-79.

78. Silicon has three naturally occurring isotopes (Si-28, Si-29, and Si-30). The mass and natural abundance of Si-28 are 27.9769 amu and 92.2%, respectively. The mass and natural abundance of Si-29 are 28.9765 amu and 4.67%, respectively. Find the mass and natural abundance of Si-30.

79. Use the mass spectrum of europium shown here to determine the atomic mass of europium.

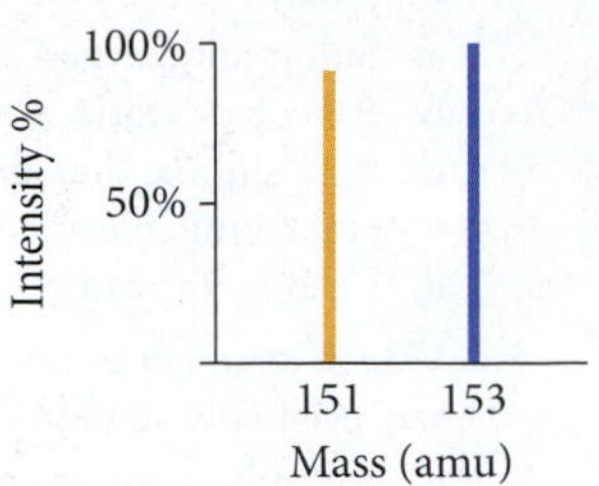

80. Use the mass spectrum of rubidium shown here to determine the atomic mass of rubidium.

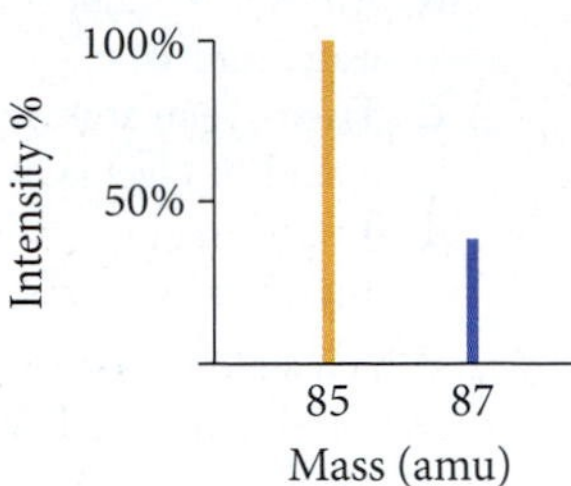

CUMULATIVE PROBLEMS

81. A 7.83-g sample of HCN contains 0.290 g of H and 4.06 g of N. Find the mass of carbon in a sample of HCN with a mass of 3.37 g.

82. The ratio of sulfur to oxygen by mass in SO_2 is 1.0:1.0.

a. Find the ratio of sulfur to oxygen by mass in SO_3.
b. Find the ratio of sulfur to oxygen by mass in S_2O.

83. Use the mass spectrum of lead shown here to estimate the atomic mass of lead. Estimate the mass and percent intensity values from the graph to three significant figures.

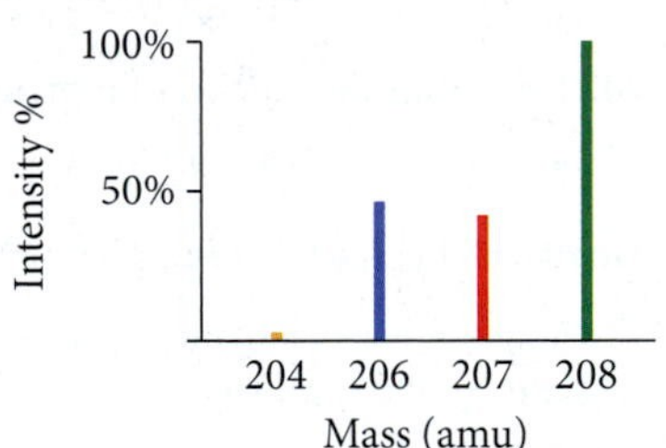
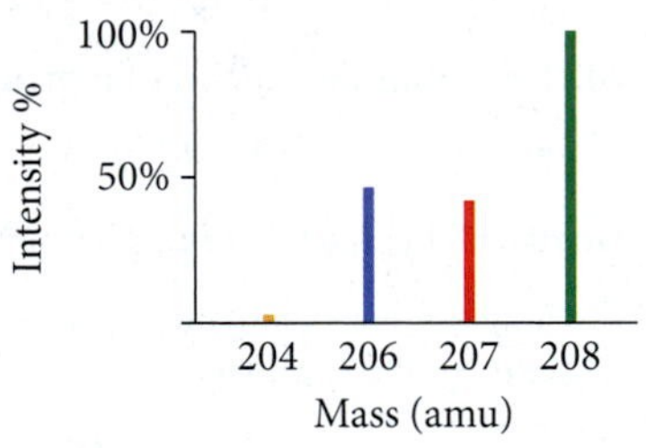

84. Use the mass spectrum of mercury shown here to estimate the atomic mass of mercury. Estimate the masses and percent intensity values from the graph to three significant figures.

85. Nuclei with the same number of *neutrons* but different mass numbers are called *isotones*. Write the symbols of four isotones of ^{236}Th.

86. Fill in the blanks to complete the table.

Symbol	Z	A	Number of p^+	Number of e^-	Number of n	Charge
Si	14	___	___	14	14	___
S^{2-}	___	32	___	___	___	2–
Cu^{2+}	___	___	___	___	34	2+
___	15	___	___	15	16	___

CHALLENGE PROBLEMS

87. Silver is composed of two naturally occurring isotopes: Ag-107 (51.839%) and Ag-109. The ratio of the masses of the two isotopes is 1.0187. What is the mass of Ag-107?

88. To the right is a representation of 50 atoms of a fictitious element called westmontium (Wt). The red spheres represent Wt-296, the blue spheres Wt-297, and the green spheres Wt-298.

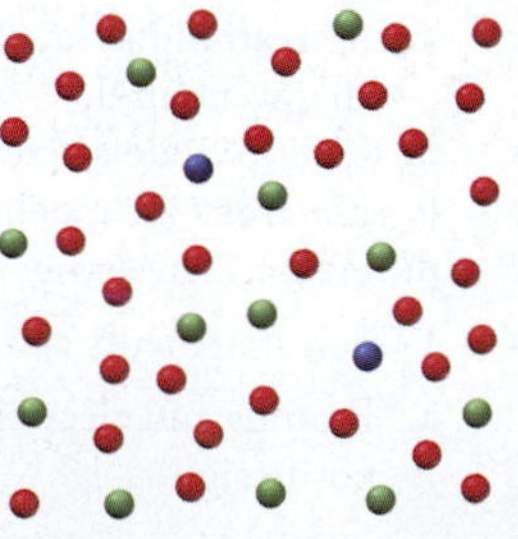

a. Assuming that the sample is statistically representative of a naturally occurring sample, calculate the percent natural abundance of each Wt isotope.

b. Draw the mass spectrum for a naturally occurring sample of Wt.

c. The mass of each Wt isotope is measured relative to C-12 and tabulated here. Use the mass of C-12 to convert each of the masses to amu and calculate the atomic mass of Wt.

Isotope	Mass
Wt-296	$24.6630 \times Mass(^{12}C)$
Wt-297	$24.7490 \times Mass(^{12}C)$
Wt-298	$24.8312 \times Mass(^{12}C)$

89. The ratio of oxygen to nitrogen by mass in NO_2 is 2.29. The ratio of fluorine to nitrogen by mass in NF_3 is 4.07. Find the ratio of oxygen to fluorine by mass in OF_2.

90. Naturally occurring cobalt consists of only one isotope, ^{59}Co, whose relative atomic mass is 58.9332. A synthetic radioactive isotope of cobalt, ^{60}Co, relative atomic mass 59.9338, is used in radiation therapy for cancer. A 1.5886-g sample of cobalt has an apparent "atomic mass" of 58.9901. Find the mass of ^{60}Co in this sample.

91. A 7.36-g sample of copper is contaminated with an additional 0.51 g of zinc. Suppose an atomic mass measurement is performed on this sample. What would be the apparent measured atomic mass?

92. The ratio of the mass of O to the mass of N in N_2O_3 is 12:7. Another binary compound of nitrogen has a ratio of O to N of 16:7. What is its formula? What is the ratio of O to N in the next member of this series of compounds?

93. Naturally occurring magnesium has an atomic mass of 24.312 and consists of three isotopes. The major isotope is ^{24}Mg, natural abundance 78.99%, relative atomic mass 23.98504. The next most abundant isotope is ^{26}Mg, relative atomic mass 25.98259. The third most abundant isotope is ^{25}Mg whose natural abundance is in the ratio of 0.9083 to that of ^{26}Mg. Find the relative atomic mass of ^{25}Mg.

CONCEPTUAL PROBLEMS

94. A volatile liquid (one that readily evaporates) is put into a jar, and the jar is then sealed. Does the mass of the sealed jar and its contents change upon the vaporization of the liquid?

95. The diagram to the right represents solid carbon dioxide, also known as dry ice.

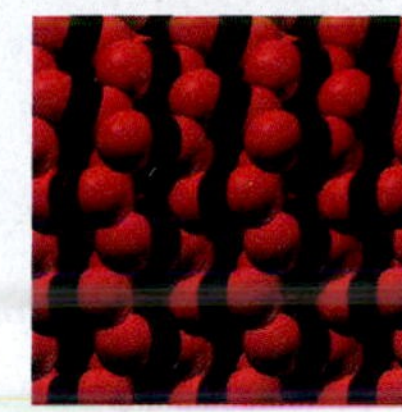

Which of the diagrams below best represents the dry ice after it has sublimed into a gas?

a.

b.

c.

96. Use triangles to represent atoms of element A and circles to represent atoms of element B. Draw an atomic level view of a homogeneous mixture of elements A and B. Draw an atomic view of the compound AB in a liquid state (molecules close together). Draw an atomic view of the compound AB after it has undergone a physical change (such as evaporation). Draw an atomic view of the compound after it has undergone a chemical change (such as decomposition of AB into A and B).

97. Identify each statement as being most like an observation, a law, or a theory.

a. All coastal areas experience two high tides and two low tides each day.

b. The tides in Earth's oceans are caused mainly by the gravitational attraction of the moon.

c. Yesterday, high tide in San Francisco Bay occurred at 2:43 A.M. and 3:07 P.M.

d. Tides are higher at the full moon and new moon than at other times of the month.

ANSWERS TO CONCEPTUAL CONNECTIONS

Cc 1.1

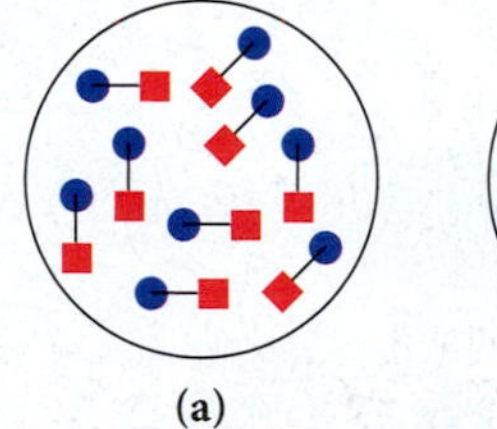

(a)

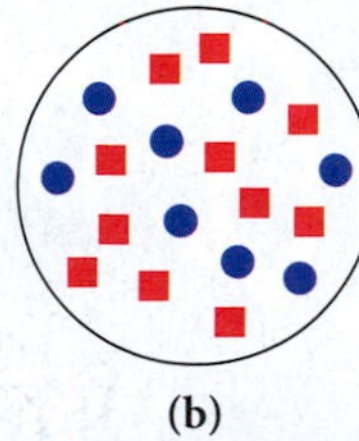

(b)

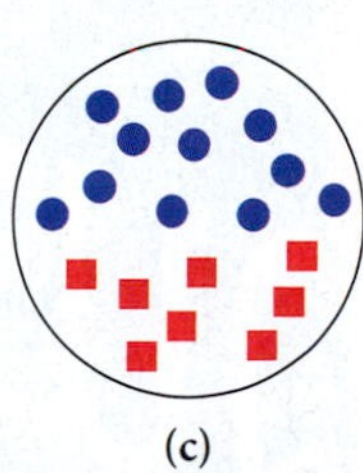

(c)

Cc 1.2 **(b)** A law only summarizes a series of related observations, while a theory gives the underlying reasons for them.

Cc 1.3 Most of the matter that composed the log reacts with oxygen molecules in the air. The products of the reaction (mostly carbon dioxide and water) are released as gases into the air.

Cc 1.4 The law of definite proportions applies to two or more samples of the *same compound* and states that the ratio of one element to the other is always the same. The law of multiple proportions applies to two *different compounds* containing the same two elements (A and B) and states that the masses of B that combine with 1 g of A are always related to each other as a small whole-number ratio.

Cc 1.5 The drop contains three excess electrons ($3 \times (-1.6 \times 10^{-19}\ C) = -4.8 \times 10^{-19}\ C$).

Cc 1.6

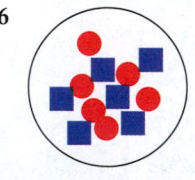

C-12 nucleus

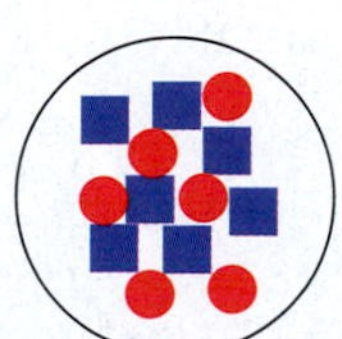

C-13 nucleus

A 10,000-atom sample of carbon, on average, contains 107 C-13 atoms.

Cc 1.7 **(b)** The number of neutrons in the nucleus of an atom does not affect the atom's size because the nucleus is miniscule compared to the atom itself.

Cc 1.8 **(a)** Since 98.93% of the atoms are C-12, we would expect the atomic mass to be very close to the mass of the C-12 isotope.

The most incomprehensible thing about the universe is that it is comprehensible.

—Albert Einstein (1879–1955)

CHAPTER 2

The $125 million Mars Climate Orbiter *was lost in the Martian atmosphere in 1999 because of a unit mix-up.*

Measurement, Problem Solving, and the Mole Concept

2.1 The Metric Mix-up: A $125 Million Unit Error 35

2.2 The Reliability of a Measurement 36

2.3 Density 38

2.4 Energy and Its Units 40

2.5 Converting between Units 43

2.6 Problem-Solving Strategies 45

2.7 Solving Problems Involving Equations 49

2.8 Atoms and the Mole: How Many Particles? 51

Key Learning Outcomes 56

QUANTIFICATION IS THE ASSIGNMENT of a number to some property of a substance or thing. For example, when we say that a pencil is 16 cm long, we assign a number to its length—we *quantify* how long it is. Quantification is among the most powerful tools in science. It requires the use of *units*, agreed upon quantities by which properties are quantified. We used the unit *centimeter* in quantifying the length of the pencil. People all over the world are in agreement about the length of a centimeter; therefore we can use that standard to specify the length of any object. In this chapter, we look closely at quantification and problem solving. Science would be much less powerful without these tools.

2.1 The Metric Mix-up: A $125 Million Unit Error

On December 11, 1998, NASA launched the Mars Climate Orbiter, which was to become the first weather satellite for a planet other than Earth. The Orbiter's mission was to monitor the Martian atmosphere and to serve as a communications relay for the Mars Polar Lander, a probe that was to follow the Orbiter and land on the planet surface three weeks later. Unfortunately, the mission ended in disaster. A unit mix-up caused the Orbiter to enter the Martian atmosphere at an altitude that was too low. Instead of settling into a stable orbit, the Orbiter likely disintegrated. The cost of the failed mission was estimated at $125 million.

There were hints of trouble several times during the Orbiter's nine-month cruise from Earth to Mars. Several adjustments made to its trajectory seemed to alter the course of the Orbiter less than expected. As the Orbiter neared the planet on September 8, 1999, discrepancies emerged about its trajectory. Some of the data indicated that the satellite was approaching Mars on a path that would place it too low in the Martian atmosphere. On September 15, engineers made the final adjustments that were supposed to put the Orbiter 226 km above the planet's surface. About a week later, as the Orbiter entered the atmosphere, communications were lost. The Orbiter had disappeared.

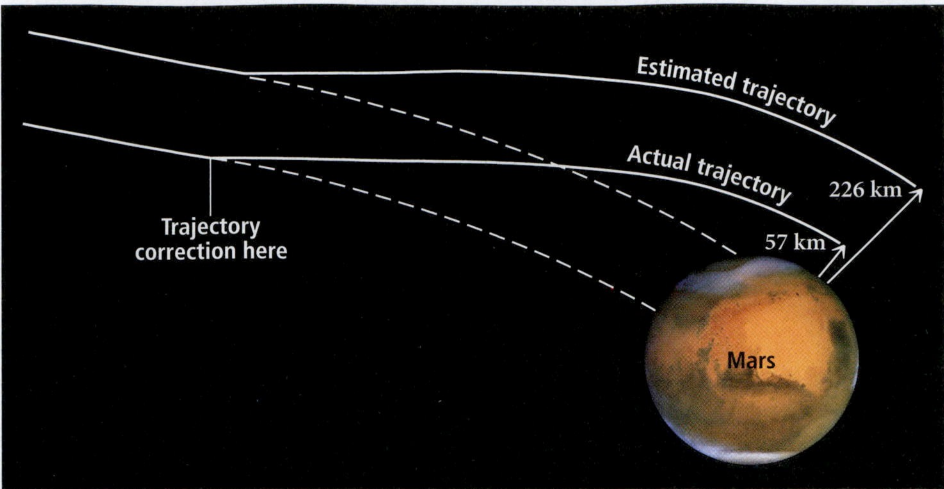

▶ **FIGURE 2.1 The Metric Mix-up** The top trajectory represents the expected Mars Climate Orbiter trajectory; the bottom trajectory represents the actual one.

Later investigations showed that the Orbiter had come within 57 km of the planet surface (Figure 2.1 ▲), an altitude that was too low for it to withstand (if a spacecraft enters a planet's atmosphere too close to the planet's surface, friction can cause the spacecraft to burn up). The on-board computers that controlled the trajectory corrections were programmed in metric units (newton•second), but the ground engineers entered the trajectory corrections in English units (pound•second). The English and the metric units are not equivalent (1 pound•second = 4.45 newton•second). The corrections that the ground engineers entered were 4.45 times too small and did not alter the trajectory enough to keep the Orbiter at a sufficiently high altitude. In chemistry as in space exploration, *units* (see Section 1.3) are critical. If we get them wrong, the consequences can be disastrous.

2.2 The Reliability of a Measurement

The reliability of a measurement depends on the instrument used to make the measurement. For example, a bathroom scale can reliably differentiate between 65 lbs and 75 lbs, but probably can't differentiate between 1.65 and 1.75 lbs. A more precise scale, such as the one a butcher uses to weigh meat, can differentiate between 1.65 and 1.75 lbs. The butcher shop scale is more precise than the bathroom scale. We must consider the reliability of measurements when reporting and manipulating them.

Reporting Measurements to Reflect Certainty

Scientists normally report measured quantities so that the number of reported digits reflects the certainty in the measurement: more digits, more certainty; fewer digits, less certainty.

For example, cosmologists report the age of the universe as 13.7 billion years. Measured values like this are usually written so that the uncertainty is in the last reported digit. (We assume the uncertainty to be ± 1 in the last digit unless otherwise indicated.) By reporting the age of the universe as 13.7 billion years, cosmologists mean that the uncertainty in the measurement is ± 0.1 billion years (or ± 100 million years). If the measurement was less certain, then the age would be reported differently. For example, reporting the age as 14 billion years would indicate that the uncertainty is ± 1 billion years. In general,

> **Scientific measurements are reported so that every digit is certain except the last, which is estimated.**

Consider the following reported number:

5.213
certain ↗ (5.21) ↖ estimated (3)

The first three digits are certain; the last digit is estimated.

The number of digits reported in a measurement depends on the measuring device. Consider weighing a sample on two different balances (Figure 2.2 ▶). These two balances have different levels of precision. The balance shown on top is accurate to the tenths place, so the uncertainty is ± 0.1 and the measurement should be reported as 10.5. The bottom balance is more precise, measuring to the ten-thousandths place, so the uncertainty is ± 0.0001 and the measurement should be reported as 10.4977 g. Many measuring instruments—such as laboratory glassware—are not digital. The measurement on these kinds of instruments must also be reported to reflect the instrument's precision. The usual procedure is to divide the space between the finest markings into ten and make that estimation the last digit reported. Example 2.1 demonstrates this procedure.

We must maintain the certainty reflected in the number of digits in a reported measurement throughout any calculation that involves the measurement. We can accomplish this by using the guidelines developed for significant figures, which are in Appendix II of this book. All calculations throughout the book adhere to the significant figure guidelines discussed in Appendix II.

Estimation in Weighing

(a)

Report as 10.5 g

(b)

Report as 10.4977 g

▲ **FIGURE 2.2 Precision in Weighing.** **(a)** This balance is precise to the tenths place. **(b)** This balance is precise to the ten-thousandths place.

EXAMPLE 2.1 Reporting the Correct Number of Digits

The graduated cylinder shown here has markings every 0.1 mL. Report the volume (which is read at the bottom of the meniscus) to the correct number of digits. (*Note*: The meniscus is the crescent-shaped surface at the top of a column of liquid.)

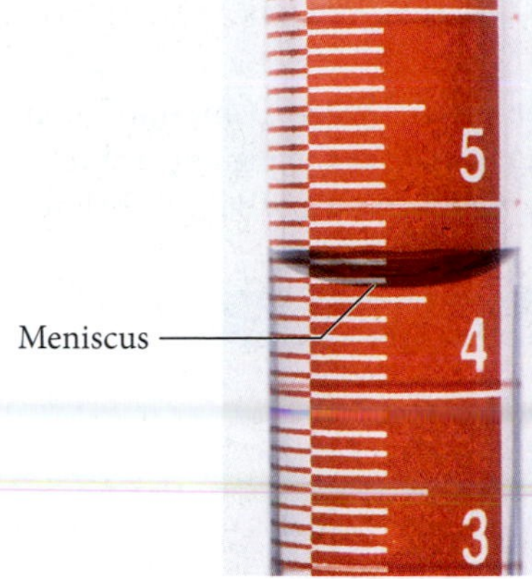

SOLUTION

Since the bottom of the meniscus is between the 4.5 and 4.6 mL markings, mentally divide the space between the markings into 10 equal spaces and estimate the next digit. In this case, the result is 4.57 mL.

What if you estimated a little differently and wrote 4.56 mL? In general, one unit difference in the last digit is acceptable because the last digit is estimated and different people might estimate it slightly differently. However, if you wrote 4.63 mL, you would have misreported the measurement.

FOR PRACTICE 2.1

Record the temperature on this thermometer to the correct number of digits.

Answers to For Practice and For More Practice problems are in Appendix VI.

Precision and Accuracy

Scientists often repeat measurements several times to increase their confidence in the result. We can distinguish between two different kinds of certainty—called accuracy and precision—associated with such measurements. **Accuracy** refers to how close the measured value is to the actual value. **Precision** refers to how close a series of measurements are to one another or how reproducible they

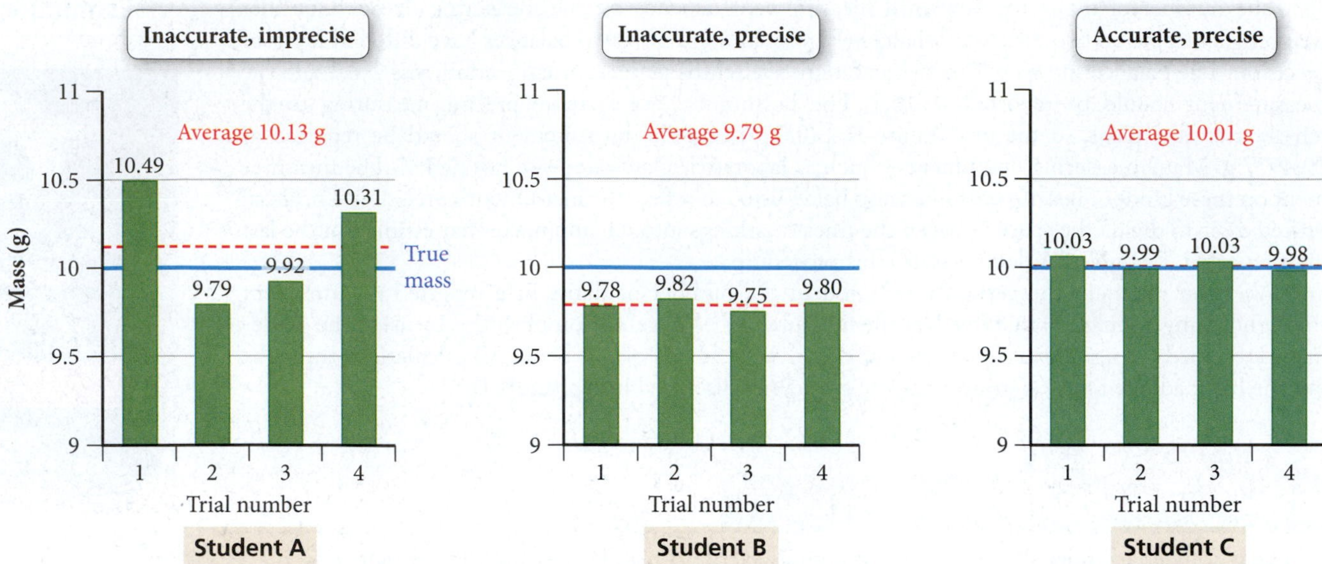

▲ **FIGURE 2.3 Precision and Accuracy** The results of three sets of measurements on the mass of a lead block. The blue horizontal line represents the true mass of the block (10.00 g). The red dashed line represents the average mass for each data set.

are. A series of measurements can be precise (close to one another in value and reproducible) but not accurate (not close to the true value). Consider the results of three students who repeatedly weighed a lead block known to have a true mass of 10.00 g tabulated below and displayed in Figure 2.3 ▲.

	Student A	Student B	Student C
Trial 1	10.49 g	9.78 g	10.03 g
Trial 2	9.79 g	9.82 g	9.99 g
Trial 3	9.92 g	9.75 g	10.03 g
Trial 4	10.31 g	9.80 g	9.98 g
Average	**10.13 g**	**9.79 g**	**10.01 g**

Measurements are precise if they are consistent with one another, but they are accurate only if they are close to the actual value.

- Student A's results are both inaccurate (not close to the true value) and imprecise (not consistent with one another). The inconsistency is the result of **random error**—error that has equal probability of being too high or too low. Almost all measurements have some degree of random error. Random error can, with enough trials, average itself out.
- Student B's results are precise (close to one another in value) but inaccurate. The inaccuracy is the result of **systematic error**—error that tends toward being either too high or too low. In contrast to random error, systematic error does not average out with repeated trials. For instance, if a balance is not properly calibrated, it may systematically read too high or too low.
- Student C's results display little systematic error or random error—they are both accurate and precise.

2.3 Density

An old riddle asks, "Which weighs more, a ton of bricks or a ton of feathers?" The answer, of course, is neither—they both weigh the same (1 ton). If you answered bricks, you confused weight with density. The **density (*d*)** of a substance is the ratio of its mass (*m*) to its volume (*V*):

$$\text{Density} = \frac{\text{mass}}{\text{volume}} \quad \text{or} \quad d = \frac{m}{V}$$

Density is a characteristic physical property of a substance. The density of a substance also depends on its temperature. Density is an example of an **intensive property**, one that is *independent* of the amount of the substance. The density of aluminum, for example, is the same whether we have a gram or a kilogram. We can use intensive properties to identify substances because these properties depend only on the type of substance, not on the amount of it. For example, in Table 2.1 we can see that pure gold has a density of 19.3 g/cm^3. One way to determine whether a substance is pure gold is to measure its density and compare it to 19.3 g/cm^3. Mass, in contrast, is an **extensive property**, one that depends on the amount of the substance. If we know only the mass of a sample of gold, that information alone will not allow us to identify it as gold.

The units of density are those of mass divided by volume. Although the SI derived unit for density is kg/m^3 (see Appendix I), we most often express the density of liquids and solids in g/cm^3 or g/mL. (Remember that cm^3 and mL are equivalent units: 1 cm^3 = 1 mL.) Aluminum is one of the least dense structural metals with a density of 2.7 g/cm^3, while platinum is one of the densest metals with a density of 21.4 g/cm^3.

We calculate the density of a substance by dividing the mass of a given amount of the substance by its volume. For example, suppose a small nugget we suspect to be gold has a mass of 22.5 g and a volume of 2.38 cm^3. To find its density, we divide the mass by the volume:

$$d = \frac{m}{V} = \frac{22.5\ \text{g}}{2.38\ \text{cm}^3} = 9.45\ \text{g/cm}^3$$

In this case, the density reveals that the nugget is not pure gold.

TABLE 2.1 The Density of Some Common Substances at 20 °C

Substance	Density (g/cm^3)
Charcoal (from oak)	0.57
Ethanol	0.789
Ice	0.917 (at 0 °C)
Water	1.00 (at 4 °C)
Sugar (sucrose)	1.58
Table salt (sodium chloride)	2.16
Glass	2.6
Aluminum	2.70
Titanium	4.51
Iron	7.86
Copper	8.96
Lead	11.4
Mercury	13.55
Gold	19.3
Platinum	21.4

EXAMPLE 2.2
Calculating Density

A man receives a ring from his fiancée, who tells him that it is made out of platinum. Before the wedding, he notices that the ring feels a little light for its size and decides to measure its density. He places the ring on a balance and finds that it has a mass of 3.15 g. He then finds that the ring displaces 0.233 cm^3 of water. Is the ring made of platinum? Assume that the measurements occurred at 20 °C. (*Note*: The volume of irregularly shaped objects is often measured by the displacement of water. In this method, the object is placed in water, and the change in volume of the water is measured. The increase in the total volume represents the volume of water *displaced* by the object and is equal to the volume of the object.)

Set up the problem by writing the important information that is *given* as well as the information that you are asked to *find*. In this case, you are to find the density of the ring and compare it to that of platinum.

Note: This standardized way of setting up problems is discussed in detail in Section 2.6.

GIVEN: $m = 3.15$ g

$V = 0.233$ cm^3

FIND: Density in g/cm^3

Next, write down the equation that defines density.

EQUATION: $d = \frac{m}{V}$

Solve the problem by substituting the correct values of mass and volume into the expression for density.

SOLUTION: $d = \frac{m}{V} = \frac{3.15\ \text{g}}{0.233\ \text{cm}^3} = 13.5\ \text{g/cm}^3$

The density of the ring is much too low to be platinum (platinum density is 21.4 g/cm^3). Therefore the ring is a fake.

FOR PRACTICE 2.2

The woman in Example 2.2 is shocked that the ring is fake and returns it. She buys a new ring that has a volume of 0.212 cm^3. If the new ring is indeed pure platinum, what is its mass?

FOR MORE PRACTICE 2.2

A metal cube has an edge length of 11.4 mm and a mass of 6.67 g. Calculate the density of the metal and refer to Table 2.1 to determine the likely identity of the metal.

2.1 Cc Conceptual Connection

Density

The density of copper decreases as temperature increases (as does the density of most substances). Which statement accurately describes the changes in a sample of copper when it is warmed from room temperature to 95 °C?

(a) The sample becomes lighter.

(b) The sample becomes heavier.

(c) The sample expands.

(d) The sample contracts.

2.4 Energy and Its Units

The two fundamental components of our universe are matter, which we discussed in Chapter 1, and energy, which we introduce briefly here. We first introduce the basic nature of energy, then we define its units, and lastly we discuss how we quantify changes in energy.

The Nature of Energy

The basic definition of **energy** is *the capacity to do work*. **Work** is defined as the action of a force through a distance. For instance, when you push a box across the floor or pedal your bicycle down the street, you do work.

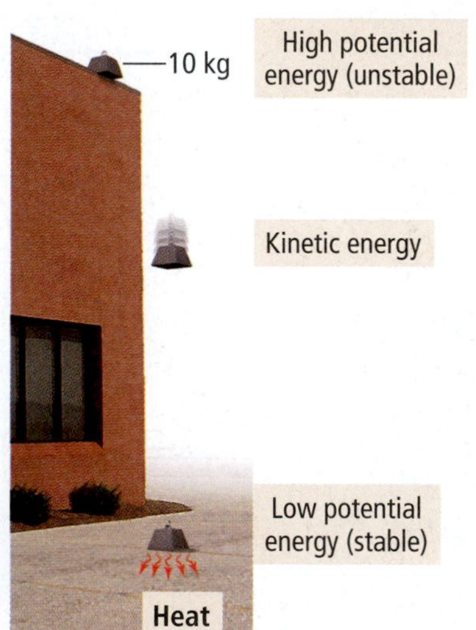

▲ **FIGURE 2.4 Energy Changes** Gravitational potential energy is converted into kinetic energy when the weight is released. The kinetic energy is converted mostly to thermal energy when the weight strikes the ground.

In Chapter 21 we will discuss how energy conservation is actually part of a more general law that allows for the interconvertibility of mass and energy.

The *total energy* of an object is a sum of its **kinetic energy**, the energy associated with its motion, and its **potential energy**, the energy associated with its position or composition. For example, a weight held several meters above the ground has potential energy due to its position within Earth's gravitational field (Figure 2.4 ◀). If you drop the weight, it accelerates, and its potential energy is converted to kinetic energy. When the weight hits the ground, its kinetic energy is converted primarily to **thermal energy**, the energy associated with the temperature of an object. Thermal energy is actually a type of kinetic energy because it arises from the motion of the individual atoms or molecules that make up an object. When the weight hits the ground, its kinetic energy is essentially transferred to the atoms and molecules that compose the ground, raising the temperature of the ground ever so slightly.

The first principle to note about the way energy changes as the weight falls to the ground is that *energy is neither created nor destroyed*. The potential energy of the weight becomes kinetic energy as the weight accelerates toward the ground. The kinetic energy then becomes thermal energy when the weight hits the ground. The total amount of thermal energy that is released through the process is exactly equal to the initial potential energy of the weight. The idea that energy is neither created nor destroyed is known as the **law of conservation of energy**. Although energy can change from one type into another and it can flow from one object to another, the *total quantity* of energy does not change—it remains constant.

The second principle to note about the raised weight and its fall is *the tendency of systems with high potential energy to change in a way that lowers their potential energy*. For this reason, objects or systems

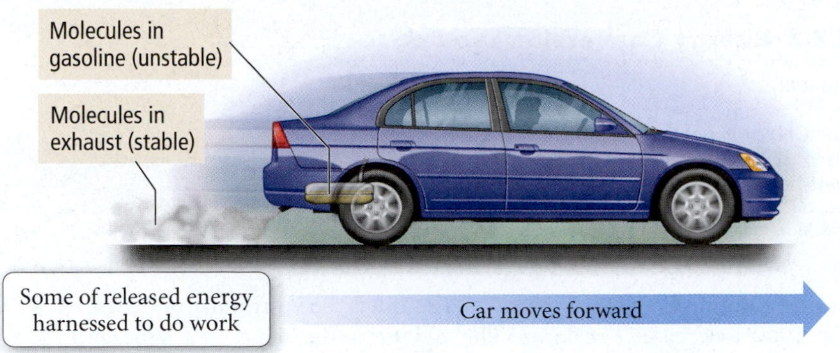

◀ **FIGURE 2.5 Using Chemical Energy to Do Work** Gasoline molecules have high potential energy. When they are burned, they form molecules with lower potential energy. The difference in potential energy can be harnessed to move the car.

with high potential energy tend to be *unstable*. The weight lifted several meters from the ground is unstable because it contains a significant amount of potential energy. Unless restrained, the weight will naturally fall, lowering its potential energy. We can harness some of the raised weight's potential energy to do work. For example, we can attach the weight to a rope that turns a paddle wheel or spins a drill as the weight falls. After the weight falls to the ground, it contains less potential energy—it has become more *stable*.

Some chemical substances are like a raised weight. For example, the molecules that compose gasoline have a relatively high potential energy—energy is concentrated in them just as energy is concentrated in a raised weight. The molecules in the gasoline tend to undergo chemical changes (specifically combustion) that lower their potential energy. As the energy of the molecules in gasoline is released, some of it can be harnessed to do work, such as moving a car forward (Figure 2.5 ▲). The molecules that result from the chemical change have less potential energy than the original molecules in gasoline and are more stable.

Chemical potential energy, such as the energy contained in the molecules that compose gasoline, arises primarily from electrostatic forces (see Section 1.6) between the electrically charged particles (protons and electrons) that compose atoms and molecules. Recall that molecules contain specific, usually complex, arrangements of protons and electrons. Some of these arrangements—such as the one within the molecules that compose gasoline—have a much higher potential energy than others. When gasoline undergoes combustion, the arrangement of these particles changes, creating molecules with lower potential energy and transferring energy (mostly in the form of heat) to the surroundings. In other words, the structure of a molecule—the way its protons and electrons are arranged—determines the potential energy of the molecule, which in turn determines its properties.

Energy Units

We can deduce the units of energy from the definition of kinetic energy. An object of mass m, moving at velocity v, has a kinetic energy KE given by:

$$\text{KE} = \frac{1}{2}mv^2$$

kg m/s

The SI unit of mass is the kg, and the unit of velocity is m/s. The SI unit of energy is kg · m^2/s^2, defined as the **joule (J)**, named after the English scientist James Joule (1818–1889).

SI units are covered in Appendix I.

$$1 \text{ kg} \frac{\text{m}^2}{\text{s}^2} = 1 \text{ J}$$

One joule is a relatively small amount of energy—for example, a 100-watt light bulb uses 3.6×10^5 J in 1 hour. Therefore, we often use the kilojoule (kJ) in our energy discussions and calculations (1 kJ = 1000 J). A second common unit of energy is the **calorie (cal)**, originally defined as the amount of energy required to raise the temperature of 1 g of water by 1 °C. The current definition is 1 cal = 4.184 J (exact);

The "calorie" referred to on all nutritional labels (regardless of the capitalization) is always the capital *C* Calorie.

TABLE 2.2 Energy Conversion Factors*

1 calorie (cal)	= 4.184 joules (J)
1 Calorie (Cal) or kilocalorie (kcal)	= 1000 cal = 4184 J
1 kilowatt-hour (kWh)	= 3.60×10^6 J

*All conversion factors in this table are exact.

3.6×10^5 J or 0.10 kWh used in 1 hour

▲ A watt (W) is 1 J/s, so a 100-W light bulb uses 100 J every second or 3.6×10^5 J every hour.

a calorie is a larger unit than a joule. A related energy unit is the nutritional, or uppercase "C" **Calorie (Cal)**, equivalent to 1000 lowercase "c" calories. The Calorie is the same as a kilocalorie (kcal): 1 Cal = 1 kcal = 1000 cal. Electricity bills typically are based on another, even larger, energy unit, the **kilowatt-hour (kWh)**: 1 kWh = 3.60×10^6 J. Electricity costs \$0.08–\$0.15 per kWh. Table 2.2 lists various energy units and their conversion factors. Table 2.3 shows the amount of energy required for various processes.

TABLE 2.3 Energy Uses in Various Units

Unit	Amount Required to Raise Temperature of 1 g of Water by 1 °C	Amount Required to Light 100-W Bulb for 1 Hour	Amount Used by Human Body in Running 1 Mile (Approximate)	Amount Used by Average U.S. Citizen in 1 Day
joule (J)	4.18	3.60×10^5	4.2×10^5	9.0×10^8
calorie (cal)	1.00	8.60×10^4	1.0×10^5	2.2×10^8
Calorie (Cal)	0.00100	86.0	100	2.2×10^5
kilowatt-hour (kWh)	1.16×10^{-6}	0.100	0.12	2.5×10^2

Quantifying Changes in Energy

Recall that energy is conserved in any process—it is neither created nor destroyed. However, energy can be transferred from one object or system to another. For example, when a raised weight is dropped, its potential energy is transferred to its surroundings as heat, or when gasoline is burned in an automobile engine, its potential energy is transferred to the kinetic energy associated with the motion of the car.

In this book, we always view the transfer of energy from the point of view of the *system* under observation. For example, if we drop a weight, we can define the weight as the system. As the weight drops, it loses energy to the *surroundings*. The *surroundings* include anything with which the system interacts. Notice that, for the falling weight, the system *loses* energy and the surroundings *gain* energy.

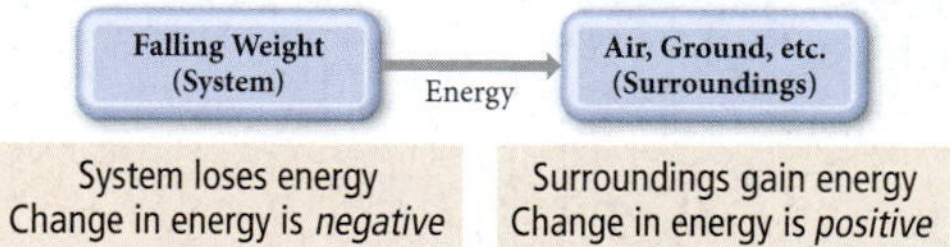
Falling Weight (System)
Energy
Air, Ground, etc. (Surroundings)
System loses energy
Change in energy is negative
Surroundings gain energy
Change in energy is positive

If we think of the energy of the system in the same way that we think of the dollar balance in a checking account, then the loss of energy from the system carries a negative sign, just like a withdrawal from a checking account carries a negative sign. Conversely, a gain in energy by the system carries a positive sign, just like a deposit in a checking account carries a positive sign.

Chemical processes almost always involve energy changes. For example, as we saw previously, when we burn gasoline, energy is given off. If we define the gasoline as the system, then the system loses energy (the change in energy is negative). Processes in which the system loses energy are **exothermic**, and the change in energy is negative. In contrast, processes in which the system gains energy are **endothermic**, and the change in energy is positive. For example, the chemical cold packs often used to ice athletic injuries contain substances within them that undergo an endothermic reaction when mixed. If we define the chemicals in the cold pack as the system, then the energy absorbed *by the system* cools down the muscle (which is acting as part of the surroundings) and helps prevent swelling.

Summarizing Energy and Its Units:

- Energy is the capacity to do work and is commonly measured in joules (J).
- Energy is conserved in any process but can be transferred between a system and its surroundings.
- An exothermic process involves the transfer of energy *from* the system *to* the surroundings and carries a negative sign (like a withdrawal from a checking account).
- An endothermic process involves the transfer of energy *to* the system *from* the surroundings and carries a positive sign (like a deposit into a checking account).

Energy Changes

2.2 Cc Conceptual Connection

Is the burning of natural gas in a stove exothermic or endothermic? What is the sign of the energy change?

2.5 Converting between Units

Knowing how to work with and manipulate units in calculations is central to solving chemical problems. In calculations, units can help us determine if an answer is correct. Using units as a guide to solving problems is called **dimensional analysis**. Always include units in your calculations; multiply, divide, and cancel them like any other algebraic quantity.

Consider converting 12.5 inches (in) to centimeters (cm). We know from the table in the inside back cover of this book that 1 in = 2.54 cm (exact), so we can use this quantity in the calculation:

$$12.5 \cancel{\text{in}} \times \frac{2.54 \text{ cm}}{1 \cancel{\text{in}}} = 31.8 \text{ cm}$$

The unit, in, cancels and we are left with cm as our final unit. The quantity $\frac{2.54 \text{ cm}}{1 \text{ in}}$ is a **conversion factor**—a fractional quantity with the units we are *converting from* on the bottom and the units we are *converting to* on the top. Conversion factors are constructed from any two equivalent quantities. In this example, 2.54 cm = 1 in, so we construct the conversion factor by dividing both sides of the equality by 1 in and canceling the units:

$$2.54 \text{ cm} = 1 \text{ in}$$

$$\frac{2.54 \text{ cm}}{1 \text{ in}} = \frac{1 \cancel{\text{in}}}{1 \cancel{\text{in}}}$$

$$\frac{2.54 \text{ cm}}{1 \text{ in}} = 1$$

Because the quantity $\frac{2.54 \text{ cm}}{1 \text{ in}}$ is equivalent to 1, multiplying by the conversion factor affects only the units, not the actual quantity. To convert the other way, from centimeters to inches, we must—using units as a guide—use a different form of the conversion factor. If we accidentally use the same form, we will get the wrong result, indicated by erroneous units. For example, suppose that we want to convert 31.8 cm to inches:

$$31.8 \text{ cm} \times \frac{2.54 \text{ cm}}{1 \text{ in}} = \frac{80.8 \text{ cm}^2}{\text{in}}$$

The units in the above answer (cm^2/in), as well as the value of the answer, are obviously wrong. We know that an inch is a larger unit than a centimeter; therefore the conversion of a value from cm to in should give us a smaller number, not a larger one. When we solve a problem, we always look at the final units. Are they the desired units? We always look at the magnitude of the numerical answer as well. Does it make sense? In this case, the mistake was the form of the conversion factor. It should have been inverted so that the units cancel as follows:

$$31.8 \cancel{\text{cm}} \times \frac{1 \text{ in}}{2.54 \cancel{\text{cm}}} = 12.5 \text{ in}$$

We can invert conversion factors because they are equal to 1 and the inverse of 1 is 1. Therefore,

$$\frac{2.54\ \text{cm}}{1\ \text{in}} = 1 = \frac{1\ \text{in}}{2.54\ \text{cm}}$$

Most unit conversion problems take the form:

$$\text{Information given} \times \text{conversion factor(s)} = \text{information sought}$$

$$\text{Given unit} \times \frac{\text{desired unit}}{\text{given unit}} = \text{desired unit}$$

In this book, we diagram problem solutions using a *conceptual plan*. A conceptual plan is a visual outline that helps us to see the general flow of the problem solution. For unit conversions, the conceptual plan focuses on units and the conversion from one unit to another. The conceptual plan for converting in to cm is:

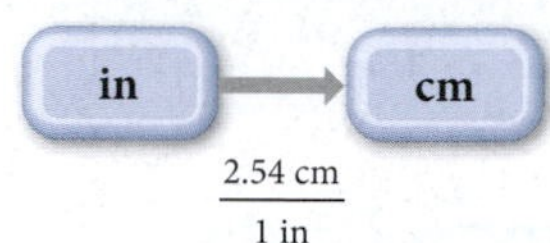

The conceptual plan for converting the other way, from cm to in, is just the reverse, with the reciprocal conversion factor:

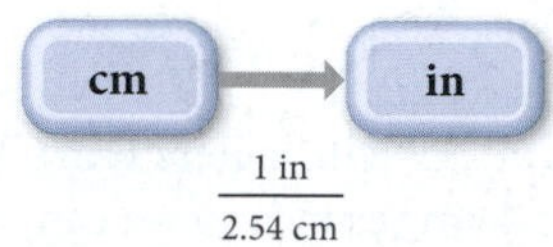

Each arrow in a conceptual plan for a unit conversion has an associated conversion factor with the units of the previous step in the denominator and the units of the following step in the numerator. In Section 2.6, we incorporate the idea of a conceptual plan into an overall approach to solving numerical chemical problems.

2.3
Cc
Conceptual Connection

The engineers involved in the Mars Climate Orbiter disaster entered the trajectory corrections in units of pound•second. Which conversion factor should they have multiplied their values by to convert them to the correct units of newton•second? (Recall from Section 2.1 that 1 pound•second = 4.45 newton•second.)

(a) $\frac{1\ \text{pound•second}}{4.45\ \text{newton•second}}$

(b) $\frac{4.45\ \text{newton•second}}{1\ \text{pound•second}}$

(c) $\frac{1\ \text{newton•second}}{4.45\ \text{pound•second}}$

(d) $\frac{4.45\ \text{pound•second}}{1\ \text{newton•second}}$

2.6 Problem-Solving Strategies

Problem solving is one of the most important skills you will acquire in this course. No one succeeds in chemistry—or in life, really—without the ability to solve problems. Although there is no simple formula you can apply to every chemistry problem, you can learn problem-solving strategies and begin to develop some chemical intuition. Many of the problems you will solve in this course are *unit conversion problems*, where you are given one or more quantities in some unit and asked to convert them into different units (see Section 2.5). Other problems require that you use *specific equations* to get to the information you are trying to find. In the sections that follow, you will find strategies to help you solve both of these types of problems. Of course, many problems contain both conversions and equations, requiring the combination of these strategies, and some problems require an altogether different approach.

In this book, we use a standard problem-solving procedure that you can adapt to many of the problems encountered in general chemistry and beyond. To solve any problem, you need to assess the information given in the problem and devise a way to get to the information asked for. In other words, you must:

- Identify the starting point (the ***given*** information).
- Identify the endpoint (what we must ***find***).
- Devise a way to get from the starting point to the endpoint using what is given as well as what you already know or can look up. (As we just discussed, we call this the ***conceptual plan***.)

In graphic form, we represent this progression as:

Given → Conceptual Plan → Find

One of the principal difficulties beginning students encounter when they try to solve problems in general chemistry is not knowing where to begin. While no problem-solving procedure is applicable to all problems, the following four-step procedure can be helpful in working through many of the numerical problems you encounter in this book:

1. **Sort.** Begin by sorting the information in the problem. *Given* information is the basic data provided by the problem—often one or more numbers with their associated units. *Find* indicates what the problem is asking you to find.
2. **Strategize.** This is usually the most challenging part of solving a problem. In this process, you must develop a *conceptual plan*—a series of steps that will get you from the given information to the information you are trying to find. You have already seen conceptual plans for simple unit conversion problems. Each arrow in a conceptual plan represents a computational step. On the left side of the arrow is the quantity you had before the step; on the right side of the arrow is the quantity you will have after the step; and below the arrow is the information you need to get from one to the other—the relationship between the quantities.

   Often such relationships take the form of conversion factors or equations. These may be given in the problem, in which case you will have written them down under "Given" in Step 1. Usually, however, you will need other information—which may include physical constants, formulas, or conversion factors—to help get you from what you are given to what you must find. This information comes from what you have learned or can look up in the chapter or in tables within the book.

   In some cases, you may get stuck at the strategize step. If you cannot figure out how to get from the given information to the information you are asked to find, you might try working backwards. For example, you can look at the units of the quantity you are trying to find and try to find conversion factors to get to the units of the given quantity. You may even try a combination of strategies; work forwards, backwards, or some of both. If you persist, you will develop a strategy to solve the problem.
3. **Solve.** This is the easiest part of solving a problem. Once you set up the problem properly and devise a conceptual plan, you follow the plan to solve the problem. Carry out any mathematical operations (paying attention to the rules for significant figures in calculations) and cancel units as needed.

Most problems can be solved in more than one way. The solutions in this book tend to be the most straightforward but certainly not the only way to solve the problem.

4. **Check.** This is the step beginning students most often overlook. Experienced problem solvers always ask, does this answer make sense? Are the units correct? Is the number of significant figures correct? When solving multistep problems, errors easily creep into the solution. You can catch most of these errors by simply checking the answer. For example, suppose you are calculating the number of atoms in a gold coin and end up with an answer of 1.1×10^{-6} atoms. Could the gold coin really be composed of one-millionth of one atom?

In Examples 2.3 and 2.4, we apply this problem-solving procedure to unit conversion problems. The procedure is summarized in the left column, and two examples of the procedure are provided in the middle and right columns. This three-column format is used in selected examples throughout this text. This format allows you to see how you can apply a particular procedure to two different problems. Work through one problem first (from top to bottom) and then apply the same procedure to the other problem. Recognizing the commonalities and differences between problems is a key part of developing problem-solving skills.

PROCEDURE FOR ▼ Solving Unit Conversion Problems	EXAMPLE 2.3 Unit Conversion	EXAMPLE 2.4 Unit Conversion
	Convert 1.76 yards to centimeters.	Convert 1.8 quarts to cubic centimeters.
SORT Begin by sorting the information in the problem into *given* and *find*.	**GIVEN:** 1.76 yd **FIND:** cm	**GIVEN:** 1.8 qt **FIND:** cm^3
STRATEGIZE Devise a *conceptual plan* for the problem. Begin with the *given* quantity and symbolize each conversion step with an arrow. Below each arrow, write the appropriate conversion factor for that step. Focus on the units. The conceptual plan should end at the *find* quantity and its units. In these examples, the other information you need consists of relationships between the various units as shown.	**CONCEPTUAL PLAN** 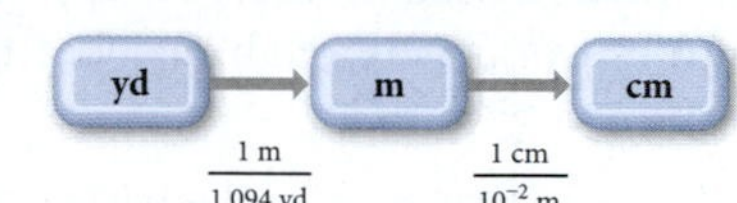 **RELATIONSHIPS USED** 1.094 yd = 1 m $1\ \text{cm} = 10^{-2}\ \text{cm}$ (These conversion factors are in the inside back cover of your book.)	**CONCEPTUAL PLAN** 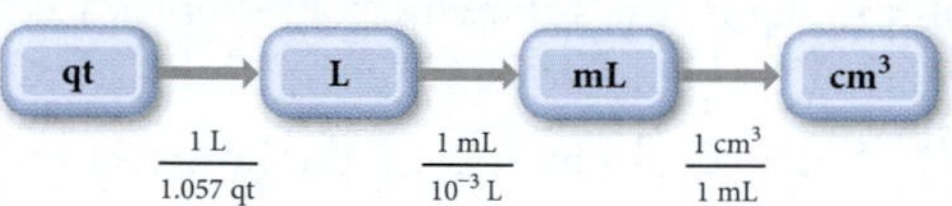 **RELATIONSHIPS USED** 1.057 qt = 1 L $1\ \text{mL} = 10^{-3}\ \text{L}$ $1\ \text{mL} = 1\ \text{cm}^3$ (These conversion factors are in the inside back cover of your book.)
SOLVE Follow the conceptual plan. Begin with the *given* quantity and its units. Multiply by the appropriate conversion factor(s), canceling units, to arrive at the *find* quantity. Round the answer to the correct number of significant figures following guidelines in Appendix II. Remember that exact conversion factors do not limit significant figures.	**SOLUTION** $1.76\ \cancel{\text{yd}} \times \dfrac{1\ \cancel{\text{m}}}{1.094\ \cancel{\text{yd}}} \times \dfrac{1\ \text{cm}}{10^{-2}\ \cancel{\text{m}}}$ $= 160.8775\ \text{cm}$ $160.8775\ \text{cm} = 161\ \text{cm}$	**SOLUTION** $1.8\ \cancel{\text{qt}} \times \dfrac{1\ \cancel{\text{L}}}{1.057\ \cancel{\text{qt}}} \times \dfrac{1\ \cancel{\text{mL}}}{10^{-3}\ \cancel{\text{L}}} \times \dfrac{1\ \text{cm}^3}{1\ \cancel{\text{mL}}}$ $= 1.70293 \times 10^3\ \text{cm}^3$ $1.70293 \times 10^3\ \text{cm}^3 = 1.7 \times 10^3\ \text{cm}^3$
CHECK Check your answer. Are the units correct? Does the answer make sense?	The units (cm) are correct. The magnitude of the answer (161) makes sense because a centimeter is a much smaller unit than a yard.	The units (cm^3) are correct. The magnitude of the answer (1700) makes sense because a cubic centimeter is a much smaller unit than a quart.
	FOR PRACTICE 2.3 Convert 288 cm to yards.	**FOR PRACTICE 2.4** Convert 9255 cm^3 to gallons.

Units Raised to a Power

When building conversion factors for units raised to a power, remember to raise both the number and the unit to the power. For example, to convert from in^2 to cm^2, you construct the conversion factor as follows:

$$2.54 \text{ cm} = 1 \text{ in}$$

$$(2.54 \text{ cm})^2 = (1 \text{ in})^2$$

$$(2.54)^2 \text{ cm}^2 = 1^2 \text{ in}^2$$

$$6.45 \text{ cm}^2 = 1 \text{ in}^2$$

$$\frac{6.45 \text{ cm}^2}{1 \text{ in}^2} = 1$$

Example 2.5 demonstrates how to use conversion factors involving units raised to a power.

EXAMPLE 2.5
Unit Conversions Involving Units Raised to a Power

Calculate the displacement (the total volume of the cylinders through which the pistons move) of a 5.70-L automobile engine in cubic inches.

SORT Sort the information in the problem into *given* and *find*.

GIVEN: 5.70 L
FIND: in^3

STRATEGIZE Write a conceptual plan. Begin with the given information and devise a path to the information that you are asked to find. Notice that for cubic units, you must cube the conversion factors.

CONCEPTUAL PLAN

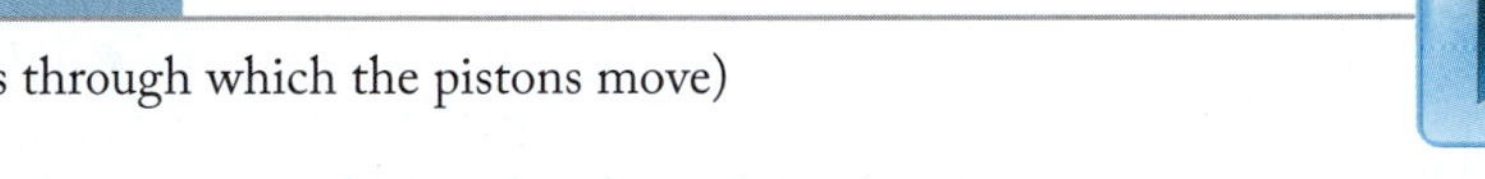

$$\frac{1 \text{ mL}}{10^{-3} \text{ L}} \qquad \frac{1 \text{ cm}^3}{1 \text{ mL}} \qquad \frac{(1 \text{ in})^3}{(2.54 \text{ cm})^3}$$

RELATIONSHIPS USED

$$1 \text{ mL} = 10^{-3} \text{ L}$$
$$1 \text{ mL} = 1 \text{ cm}^3$$
$$2.54 \text{ cm} = 1 \text{ in}$$

(These conversion factors are in the inside back cover of your book.)

SOLVE Follow the conceptual plan to solve the problem. Round the answer to three significant figures to reflect the three significant figures in the least precisely known quantity (5.70 L). These conversion factors are all exact and therefore do not limit the number of significant figures.

SOLUTION

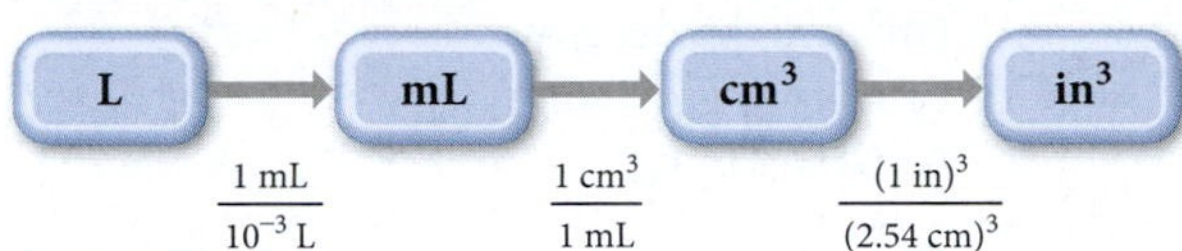

$$5.70 \text{ L} \times \frac{1 \text{ mL}}{10^{-3} \text{ L}} \times \frac{1 \text{ cm}^3}{1 \text{ mL}} \times \frac{(1 \text{ in})^3}{(2.54 \text{ cm})^3} = 347.835 \text{ in}^3$$

$$= 348 \text{ in}^3$$

CHECK The units of the answer are correct, and the magnitude makes sense. The unit cubic inches is smaller than liters, so the volume in cubic inches should be larger than the volume in liters.

FOR PRACTICE 2.5
How many cubic centimeters are there in 2.11 yd^3?

FOR MORE PRACTICE 2.5
A vineyard has 145 acres of Chardonnay grapes. A particular soil supplement requires 5.50 g for every square meter of vineyard. How many kilograms of the soil supplement are required for the entire vineyard? (1 km^2 = 247 acres)

EXAMPLE 2.6

Density as a Conversion Factor

The mass of fuel in a jet must be calculated before each flight to ensure that the jet is not too heavy to fly. A 747 is fueled with 173,231 L of jet fuel. If the density of the fuel is 0.768 g/cm^3, what is the mass of the fuel in kilograms?

SORT Begin by sorting the information in the problem into *given* and *find*.

GIVEN: fuel volume = 173,231 L
density of fuel = 0.768 g/cm^3
FIND: mass in kg

STRATEGIZE Draw the conceptual plan beginning with the given quantity, in this case the volume in liters (L). Your overall goal in this problem is to find the mass. You can convert between volume and mass using density (g/cm^3). However, you must first convert the volume to cm^3. Once you have converted the volume to cm^3, use the density to convert to g. Finally convert g to kg.

CONCEPTUAL PLAN

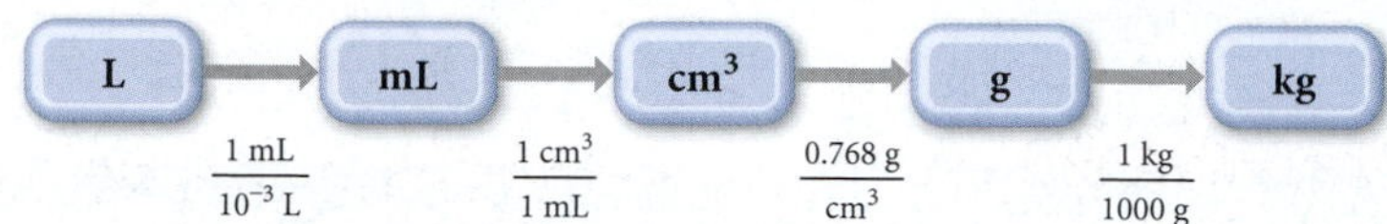

RELATIONSHIPS USED

$1\ \text{mL} = 10^{-3}\ \text{L}$
$1\ \text{mL} = 1\ \text{cm}^3$
$d = 0.768\ \text{g/cm}^3$
$1000\ \text{g} = 1\ \text{kg}$

(These conversion factors are in the inside back cover of your book.)

SOLVE Follow the conceptual plan to solve the problem. Round the answer to three significant figures to reflect the three significant figures in the density.

SOLUTION

$$173{,}231\ \cancel{\text{L}} \times \frac{1\ \cancel{\text{mL}}}{10^{-3}\ \cancel{\text{L}}} \times \frac{1\ \cancel{\text{cm}^3}}{1\ \cancel{\text{mL}}} \times \frac{0.768\ \cancel{\text{g}}}{1\ \cancel{\text{cm}^3}} \times \frac{1\ \text{kg}}{1000\ \cancel{\text{g}}}$$

$$= 1.33 \times 10^5\ \text{kg}$$

CHECK The units of the answer (kg) are correct. The magnitude makes sense because the mass (1.33×10^5 kg) is similar in magnitude to the given volume (173,231 L or 1.73231×10^5 L), as you would expect for a density close to one (0.768 g/cm^3).

FOR PRACTICE 2.6
Backpackers often use canisters of white gas to fuel a cooking stove's burner. If one canister contains 1.45 L of white gas and the density of the gas is 0.710 g/cm^3, what is the mass of the fuel in kilograms?

FOR MORE PRACTICE 2.6
A drop of gasoline has a mass of 22 mg and a density of 0.754 g/cm^3. What is its volume in cubic centimeters?

Order-of-Magnitude Estimations

Calculation is an integral part of chemical problem solving. But precise numerical calculation is not always necessary, or even possible. Sometimes data are only approximate; other times we do not need a high degree of precision—a rough estimate or a simplified "back of the envelope" calculation is enough. We can also use approximate calculations to get an initial feel for a problem, or as a quick check to see whether our solution is in the right ballpark.

One way to make such estimates is to simplify the numbers so that they can be manipulated easily. In the technique known as *order-of-magnitude estimation*, we focus only on the exponential part of numbers written in scientific notation, according to these guidelines:

- If the decimal part of the number is less than 5, we drop it. Thus, 4.36×10^5 becomes 10^5 and 2.7×10^{-3} becomes 10^{-3}.
- If the decimal part of the number is 5 or more, we round it up to 10 and rewrite the number as a power of 10. Thus, 5.982×10^7 becomes $10 \times 10^7 = 10^8$, and 6.1101×10^{-3} becomes $10 \times 10^{-3} = 10^{-2}$.

After we make these approximations, we are left with powers of 10, which are easier to multiply and divide. Of course our answer is only as reliable as the numbers used to get it, so we should not assume that the results of an order-of-magnitude calculation are accurate to more than an order of magnitude.

Suppose, for example, that we want to estimate the number of atoms that an immortal being could have counted in the 13.7 billion (1.37×10^{10}) years that the universe has been in existence, assuming a counting rate of 10 atoms per second. Since a year has 3.2×10^7 seconds, we can approximate the number of atoms counted as:

$$\underbrace{10^{10}\,\cancel{\text{years}}}_{\text{(number of years)}} \times \underbrace{10^{7}\,\frac{\cancel{\text{seconds}}}{\cancel{\text{year}}}}_{\text{(number of seconds per year)}} \times \underbrace{10^{1}\,\frac{\text{atoms}}{\cancel{\text{second}}}}_{\text{(number of atoms counted per second)}} \approx 10^{18}\ \text{atoms}$$

A million trillion atoms (10^{18}) may seem like a lot, but a speck of matter made up of a million trillion atoms is nearly impossible to see without a microscope.

In our general problem-solving procedure, the last step is to check whether the results seem reasonable. Order-of-magnitude estimations can often help us to catch mistakes that we may make in a detailed calculation, such as entering an incorrect exponent or sign into a calculator, or multiplying when we should have divided.

2.7 Solving Problems Involving Equations

We can solve problems involving equations in much the same way as problems involving conversions. Usually, in problems involving equations, we are asked to find one of the variables in the equation, given the others. For example, suppose we are given the mass (m) and volume (V) of a sample and asked to calculate its density. A conceptual plan shows how the *equation* takes us from the *given* quantities to the *find* quantity.

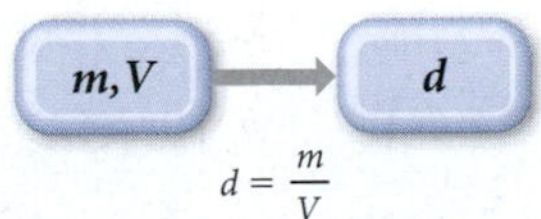

Here, instead of a conversion factor under the arrow, the conceptual plan has an equation. The equation shows the *relationship* between the quantities on the left of the arrow and the quantities on the right. Note that at this point, the equation need not be solved for the quantity on the right (although in this particular case it is). The procedure that follows, as well as Examples 2.7 and 2.8, guides you through solutions to problems involving equations. We again use the three-column format. Work through one problem from top to bottom and then apply the same general procedure to the second problem.

PROCEDURE FOR ▼ Solving Problems Involving Equations

EXAMPLE 2.7 Problems with Equations

Find the radius (r), in centimeters, of a spherical water droplet with a volume (V) of 0.058 cm^3. For a sphere, $V = (4/3)\pi r^3$.

EXAMPLE 2.8 Problems with Equations

Find the density (in g/cm^3) of a metal cylinder with a mass (m) of 8.3 g, a length (l) of 1.94 cm, and a radius (r) of 0.55 cm. For a cylinder, $V = \pi r^2 l$.

SORT Begin by sorting the information into *given* and *find*.

Example 2.7:

GIVEN: $V = 0.058\ \text{cm}^3$
FIND: r in cm

Example 2.8:

GIVEN: $m = 8.3\ \text{g}$
$l = 1.94\ \text{cm}$
$r = 0.55\ \text{cm}$
FIND: d in g/cm^3

STRATEGIZE Write a conceptual plan for the problem. Focus on the equation(s). The conceptual plan shows how the equation takes you from the *given* quantity (or quantities) to the *find* quantity. The conceptual plan may have several parts, involving other equations or required conversions. In these examples, you use the geometrical relationships given in the problem statements as well as the definition of density, $d = m/V$, which you learned in this chapter.

Example 2.7:

CONCEPTUAL PLAN

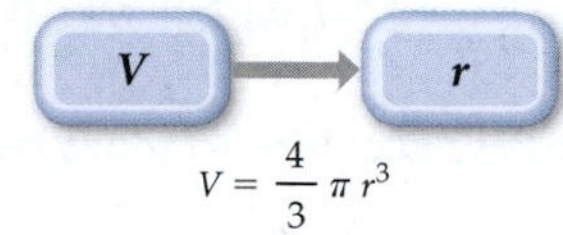

RELATIONSHIPS USED

$$V = \frac{4}{3}\pi r^3$$

Example 2.8:

CONCEPTUAL PLAN

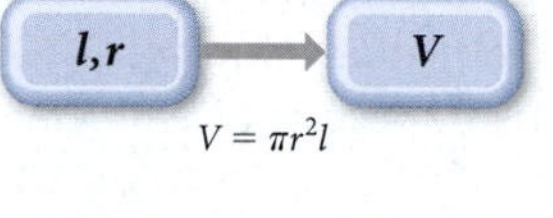

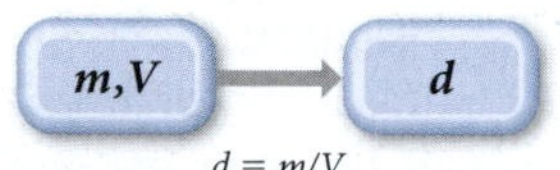

RELATIONSHIPS USED

$$V = \pi r^2 l$$

$$d = \frac{m}{V}$$

SOLVE Follow the conceptual plan. Solve the equation(s) for the *find* quantity (if it is not solved already). Gather each of the quantities that must go into the equation in the correct units. (Convert to the correct units if necessary.) Substitute the numerical values and their units into the equation(s) and calculate the answer.

Round the answer to the correct number of significant figures.

Example 2.7:

SOLUTION

$$V = \frac{4}{3}\pi r^3$$

$$r^3 = \frac{3}{4\pi}V$$

$$r = \left(\frac{3}{4\pi}V\right)^{1/3}$$

$$= \left(\frac{3}{4\pi}0.058\ \text{cm}^3\right)^{1/3}$$

$$= 0.24013\ \text{cm}$$

$$0.24013\ \text{cm} = 0.24\ \text{cm}$$

Example 2.8:

SOLUTION

$$V = \pi r^2 l$$

$$= \pi(0.55\ \text{cm})^2(1.94\ \text{cm})$$

$$= 1.8\underline{4}36\ \text{cm}^3$$

$$d = \frac{m}{V}$$

$$= \frac{8.3\ \text{g}}{1.8\underline{4}36\ \text{cm}^3} = 4.50195\ \text{g/cm}^3$$

$$4.50195\ \text{g/cm}^3 = 4.5\ \text{g/cm}^3$$

CHECK Check your answer. Are the units correct? Does the answer make sense?

Example 2.7: The units (cm) are correct, and the magnitude makes sense.

Example 2.8: The units (g/cm^3) are correct. The magnitude of the answer seems correct for one of the lighter metals (see Table 2.1).

FOR PRACTICE 2.7
Find the radius (r) of an aluminum cylinder that is 2.00 cm long and has a mass of 12.4 g. For a cylinder, $V = \pi r^2 l$.

FOR PRACTICE 2.8
Find the density, in g/cm^3, of a metal cube with a mass of 50.3 g and an edge length (l) of 2.65 cm. For a cube, $V = l^3$.

2.8 Atoms and the Mole: How Many Particles?

KEY CONCEPT VIDEO The mole concept

My seven-year-old sometimes asks, "How much eggs did the chickens lay?" or "How much pancakes do I get?" My wife immediately corrects him, "Do you mean how *many* eggs?" The difference between "how many" and "how much" depends on what you are specifying. If you are specifying something countable, such as eggs or pancakes, you say "how many." But if you are specifying something noncountable, such as water or milk, you say "how much."

Although samples of matter may seem noncountable—we normally say how *much* water—we know that *all matter is ultimately particulate and countable*. Even more importantly, when samples of matter interact with one another, they interact *particle by particle*. For example, when hydrogen and oxygen combine to form water, two hydrogen atoms combine with one oxygen atom to form one water molecule. Therefore, as chemists, we often ask of a sample of matter, not only *how much*, but also *how many*—how many particles does the sample contain?

The particles that compose matter are far too small to count by any ordinary means. Even if we could somehow count atoms, and counted them 24 hours a day for as long as we lived, as we saw in Section 2.6, we would barely begin to count the number of atoms in something as small as a sand grain. Therefore, if we want to know the number of atoms in anything of ordinary size, we must count them by weighing.

As an analogy, consider buying shrimp at your local fish market. Shrimp is normally sold by count, which indicates the number of shrimp per pound. For example, for 41–50 count shrimp there are between 41 and 50 shrimp per pound. The smaller the count, the larger the shrimp. Big tiger prawns have counts as low as 10–15, which means that each shrimp can weigh up to 1/10 of a pound. One advantage of categorizing shrimp this way is that we can count the shrimp by weighing them. For example, two pounds of 41–50 count shrimp contain between 82 and 100 shrimp. A similar concept exists for the particles that compose matter. We can determine the number of particles in a sample of matter from the mass of the sample.

The Mole: A Chemist's "Dozen"

When we count large numbers of objects, we use units such as a dozen (12 objects) or a gross (144 objects) to organize our counting and to keep our numbers more manageable. With atoms, quadrillions of which may be in a speck of dust, we need a much larger number for this purpose. The chemist's "dozen" is the **mole** (abbreviated mol). A mole is the *amount* of material containing 6.02214×10^{23} particles.

$$1 \text{ mol} = 6.02214 \times 10^{23} \text{ particles}$$

This number is **Avogadro's number**, named after Italian physicist Amedeo Avogadro (1776–1856), and is a convenient number to use when working with atoms, molecules, and ions. In this book, we usually round Avogadro's number to four significant figures or 6.022×10^{23}. Notice that the definition of the mole is an *amount* of a substance. We will often refer to the number of moles of substance as the *amount* of the substance.

The first thing to understand about the mole is that it can specify Avogadro's number of anything. For example, 1 mol of marbles corresponds to 6.022×10^{23} marbles, and 1 mol of sand grains corresponds to 6.022×10^{23} sand grains. *One mole of anything is 6.022×10^{23} units of that thing.* One mole of atoms, ions, or molecules, however, makes up objects of everyday sizes. Twenty-two copper pennies, for example, contain approximately 1 mol of copper atoms, and one tablespoon of water contains approximately 1 mol of water molecules.

The second, and more fundamental, thing to understand about the mole is how it gets its specific value.

The value of the mole is equal to the number of atoms in exactly 12 g of pure carbon-12 ($12 \text{ g C} = 1 \text{ mol C atoms} = 6.022 \times 10^{23}$ C atoms).

The definition of the mole gives us a relationship between mass (grams of carbon) and number of atoms (Avogadro's number). This relationship, as we will see shortly, allows us to count atoms by weighing them.

Twenty-two copper pennies contain approximately 1 mol of copper atoms.

▲ Before 1982, when they became almost all zinc with only a copper coating, pennies were mostly copper.

One tablespoon of water contains approximately 1 mole of water molecules.

▲ One tablespoon is approximately 15 mL; one mole of water occupies 18 mL.

Converting between Number of Moles and Number of Atoms

Converting between number of moles and number of atoms is similar to converting between dozens of eggs and number of eggs. For eggs, we use the conversion factor 1 dozen eggs = 12 eggs. For atoms we use the conversion factor 1 mol atoms = 6.022×10^{23} atoms. The conversion factors take the forms:

$$\frac{1 \text{ mol atoms}}{6.022 \times 10^{23} \text{ atoms}} \quad \text{or} \quad \frac{6.022 \times 10^{23} \text{ atoms}}{1 \text{ mol atoms}}$$

Example 2.9 demonstrates how to use these conversion factors in calculations.

EXAMPLE 2.9

Converting between Number of Moles and Number of Atoms

Calculate the number of copper atoms in 2.45 mol of copper.

SORT You are given the amount of copper in moles and asked to find the number of copper atoms.	**GIVEN:** 2.45 mol Cu **FIND:** Cu atoms
STRATEGIZE Convert between number of moles and number of atoms using Avogadro's number as a conversion factor.	**CONCEPTUAL PLAN** mol Cu → number of Cu atoms $\frac{6.022 \times 10^{23} \text{ Cu atoms}}{1 \text{ mol Cu}}$ **RELATIONSHIPS USED** $6.022 \times 10^{23} = 1$ mol (Avogadro's number)
SOLVE Follow the conceptual plan to solve the problem. Begin with 2.45 mol Cu and multiply by Avogadro's number to get to the number of Cu atoms.	**SOLUTION** $2.45 \text{ mol Cu} \times \frac{6.022 \times 10^{23} \text{ Cu atoms}}{1 \text{ mol Cu}}$ $= 1.48 \times 10^{24}$ Cu atoms

CHECK Since atoms are small, it makes sense that the answer is large. The given number of moles of copper is almost 2.5, so the number of atoms is almost 2.5 times Avogadro's number.

FOR PRACTICE 2.9

A pure silver ring contains 2.80×10^{22} silver atoms. How many moles of silver atoms does it contain?

Converting between Mass and Amount (Number of Moles)

To count atoms by weighing them, we need one other conversion factor—the mass of 1 mol of atoms. For the isotope carbon-12, we know that the mass of 1 mol of atoms is exactly 12 g, which is numerically equivalent to carbon-12's atomic mass in atomic mass units. Since the masses of all other elements are defined relative to carbon-12, the same relationship holds for all elements. The mass of 1 mol of atoms of an element is its **molar mass**.

An element's molar mass in grams per mole is numerically equal to the element's atomic mass in atomic mass units.

For example, copper has an atomic mass of 63.55 amu and a molar mass of 63.55 g/mol. One mole of copper atoms therefore has a mass of 63.55 g. Just as the count for shrimp depends on the size of the shrimp, the mass of 1 mol of atoms depends on the element (Figure 2.6 ▶): 1 mol of aluminum atoms (which are lighter than copper atoms) has a mass of 26.98 g; 1 mol of carbon atoms (which are even lighter than aluminum atoms) has a mass of 12.01 g; and 1 mol of helium atoms (lighter yet) has a mass of 4.003 g.

26.98 g aluminum = 1 mol aluminum = 6.022×10^{23} Al atoms

Al

12.01 g carbon = 1 mol carbon = 6.022×10^{23} C atoms

C

4.003 g helium = 1 mol helium = 6.022×10^{23} He atoms

He

The lighter the atom, the less mass in 1 mol of atoms.

The molar mass of any element is the conversion factor between the mass (in grams) of that element and the amount (in moles) of that element. For carbon:

$$12.01 \text{ g C} \quad \text{or} \quad \frac{12.01 \text{ g C}}{\text{mol C}} \quad \text{or} \quad \frac{1 \text{ mol C}}{12.01 \text{ g C}}$$

Example 2.10 demonstrates how to use these conversion factors.

▲ **FIGURE 2.6 Molar Mass** The two dishes contain the same number of objects (12), but the masses are different because peas are less massive than marbles. Similarly, a mole of light atoms has less mass than a mole of heavier atoms.

EXAMPLE 2.10
Converting between Mass and Amount (Number of Moles)

Calculate the amount of carbon (in moles) contained in a 0.0265-g pencil "lead." (Assume that the pencil "lead" is made of pure graphite, a form of carbon.)

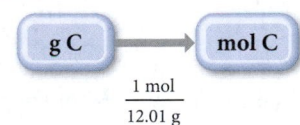

SORT You are given the mass of carbon and asked to find the amount of carbon in moles.	**GIVEN:** 0.0265 g C **FIND:** mol C
STRATEGIZE Convert between mass and amount (in moles) of an element using the molar mass of the element.	**CONCEPTUAL PLAN** g C → mol C $\frac{1 \text{ mol}}{12.01 \text{ g}}$ **RELATIONSHIPS USED** 12.01 g C = 1 mol C (carbon molar mass)
SOLVE Follow the conceptual plan to solve the problem.	**SOLUTION** $0.0265 \cancel{\text{g C}} \times \frac{1 \text{ mol C}}{12.01 \cancel{\text{g C}}} = 2.21 \times 10^{-3} \text{ mol C}$

CHECK The given mass of carbon is much less than the molar mass of carbon, so it makes sense that the answer (the amount in moles) is much less than 1 mol of carbon.

FOR PRACTICE 2.10
Calculate the amount of copper (in moles) in a 35.8-g pure copper sheet.

FOR MORE PRACTICE 2.10
Calculate the mass (in grams) of 0.473 mol of titanium.

We now have all the tools to count the number of atoms in a sample of an element by weighing it. First, we obtain the mass of the sample. Then we convert it to the amount in moles using the element's molar mass. Finally, we convert to number of atoms using Avogadro's number. The conceptual plan for these kinds of calculations is:

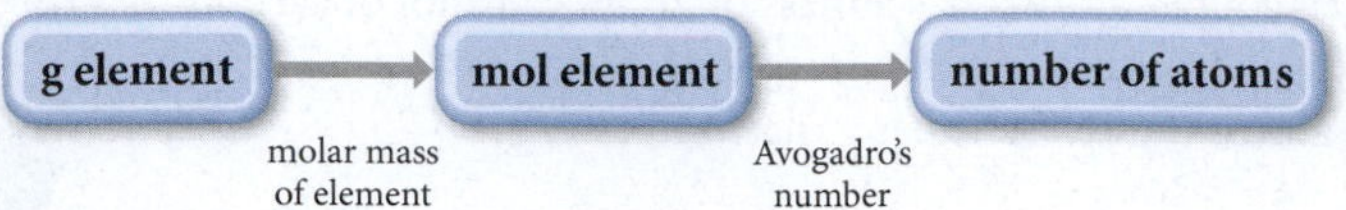

Examples 2.11 and 2.12 demonstrate these conversions.

EXAMPLE 2.11

The Mole Concept—Converting between Mass and Number of Atoms

How many copper atoms are in a copper penny with a mass of 3.10 g? (Assume that the penny is composed of pure copper.)

SORT You are given the mass of copper and asked to find the number of copper atoms.	**GIVEN:** 3.10 g Cu **FIND:** Cu atoms
STRATEGIZE Convert between the mass of an element in grams and the number of atoms of the element by first converting to moles (using the molar mass of the element) and then to number of atoms (using Avogadro's number).	**CONCEPTUAL PLAN** 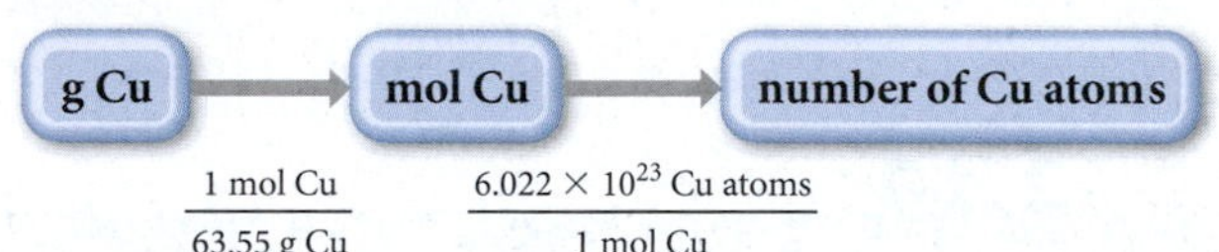 **RELATIONSHIPS USED** 63.55 g Cu = 1 mol Cu (molar mass of copper) $6.022 \times 10^{23} = 1$ mol (Avogadro's number)
SOLVE Follow the conceptual plan to solve the problem. Begin with 3.10 g Cu and multiply by the appropriate conversion factors to arrive at the number of Cu atoms.	**SOLUTION** $3.10 \text{ g Cu} \times \frac{1 \text{ mol Cu}}{63.55 \text{ g Cu}} \times \frac{6.022 \times 10^{23} \text{ Cu atoms}}{1 \text{ mol Cu}} = 2.94 \times 10^{22} \text{ Cu atoms}$

CHECK The answer (the number of copper atoms) is less than 6.022×10^{23} (one mole). This is consistent with the given mass of copper atoms, which is less than the molar mass of copper.

FOR PRACTICE 2.11
How many carbon atoms are there in a 1.3-carat diamond? Diamonds are a form of pure carbon. (1 carat = 0.20 g)

FOR MORE PRACTICE 2.11
Calculate the mass of 2.25×10^{22} tungsten atoms.

Notice that numbers with large exponents, such as 6.022×10^{23}, are almost unbelievably large. Twenty-two copper pennies contain 6.022×10^{23} or 1 mol of copper atoms, but 6.022×10^{23} pennies would cover the Earth's entire surface to a depth of 300 m. Even objects that are small by everyday standards occupy a huge space when we have a mole of them. For example, a grain of sand has a mass of less than 1 mg and a diameter of less than 0.1 mm, yet 1 mol of sand grains would cover the state of Texas to a depth of several feet. For every increase of 1 in the exponent of a number, the number increases by a factor of 10, so 10^{23} is incredibly large. Of course one mole has to be a large number if it is to have practical value, because atoms are so small.

EXAMPLE 2.12
The Mole Concept

An aluminum sphere contains 8.55×10^{22} aluminum atoms. What is the sphere's radius in centimeters? The density of aluminum is $2.70\ g/cm^3$.

SORT You are given the number of aluminum atoms in a sphere and the density of aluminum. You are asked to find the radius of the sphere.

GIVEN: 8.55×10^{22} Al atoms

$d = 2.70\ g/cm^3$

FIND: radius (r) of sphere

STRATEGIZE The heart of this problem is density, which relates mass to volume, and though you aren't given the mass directly, you are given the number of atoms, which you can use to find mass.

1. Convert from number of atoms to number of moles using Avogadro's number as a conversion factor.
2. Convert from number of moles to mass using molar mass as a conversion factor.
3. Convert from mass to volume (in cm^3) using density as a conversion factor.
4. Once you calculate the volume, find the radius from the volume using the formula for the volume of a sphere.

CONCEPTUAL PLAN

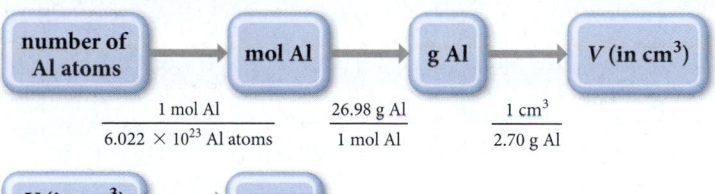

number of Al atoms → mol Al → g Al → V (in cm^3)

$$\frac{1\ \text{mol Al}}{6.022 \times 10^{23}\ \text{Al atoms}} \qquad \frac{26.98\ \text{g Al}}{1\ \text{mol Al}} \qquad \frac{1\ \text{cm}^3}{2.70\ \text{g Al}}$$

V (in cm^3) → r

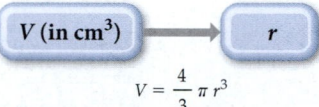

$$V = \frac{4}{3}\pi r^3$$

RELATIONSHIPS AND EQUATIONS USED

$6.022 \times 10^{23} = 1$ mol (Avogadro's number)

26.98 g Al $= 1$ mol Al (molar mass of aluminum)

$2.70\ g/cm^3$ (density of aluminum)

$V = \frac{4}{3}\pi r^3$ (volume of a sphere)

SOLVE Finally, follow the conceptual plan to solve the problem. Begin with 8.55×10^{22} Al atoms and multiply by the appropriate conversion factors to arrive at volume in cm^3.

Then solve the equation for the volume of a sphere for r and substitute the volume to calculate r.

SOLUTION

$$8.55 \times 10^{22}\ \text{Al atoms} \times \frac{1\ \text{mol Al}}{6.022 \times 10^{23}\ \text{Al atoms}} \times \frac{26.98\ \text{g Al}}{1\ \text{mol Al}} \times \frac{1\ \text{cm}^3}{2.70\ \text{g Al}} = 1.4\underline{1}87\ \text{cm}^3$$

$$V = \frac{4}{3}\pi r^3$$

$$r = \sqrt[3]{\frac{3V}{4\pi}} = \sqrt[3]{\frac{3(1.4\underline{1}87\ \text{cm}^3)}{4\pi}} = 0.697\ \text{cm}$$

CHECK The units of the answer (cm) are correct. The magnitude cannot be estimated accurately, but a radius of about one-half of a centimeter is reasonable for just over one-tenth of a mole of aluminum atoms.

FOR PRACTICE 2.12

A titanium cube contains 2.86×10^{23} atoms. What is the edge length of the cube? The density of titanium is $4.50\ g/cm^3$.

FOR MORE PRACTICE 2.12

Find the number of atoms in a copper rod with a length of 9.85 cm and a radius of 1.05 cm. The density of copper is $8.96\ g/cm^3$.

The Mole — 2.4 Cc Conceptual Connection

Without doing any calculations, determine which sample contains the most atoms.

(a) a 1-g sample of copper **(b)** a 1-g sample of carbon **(c)** a 10-g sample of uranium

SELF-ASSESSMENT Quiz

1. What is the mass of a 1.75-L sample of a liquid that has a density of 0.921 g/mL?
   a) 1.61×10^3 g **b)** 1.61×10^{-3} g
   c) 1.90×10^3 g **d)** 1.90×10^{-3} g

2. Convert 1,285 cm^2 to m^2.
   a) $1.285 \times 10^7\ m^2$ **b)** 12.85 m^2
   c) 0.1285 m^2 **d)** $1.285 \times 10^5\ m^2$

3. Three samples, each of a different substance, are weighed and their volume is measured. The results are tabulated. List the substances in order of decreasing density.

	Mass	Volume
Substance I	10.0 g	10.0 mL
Substance II	10.0 kg	12.0 L
Substance III	12.0 mg	10.0 μL

   a) III > II > I **b)** I > II > III
   c) III > I > II **d)** II > I > III

4. A solid metal sphere has a radius of 3.53 cm and a mass of 1.796 kg. What is the density of the metal in g/cm^3? The volume of sphere is $V = \frac{4}{3}\pi r^3$.
   a) 34.4 g/cm^3 **b)** 0.103 g/cm^3
   c) 121 g/cm^3 **d)** 9.75 g/cm^3

5. A German automobile's gas mileage is 22 km/L. Convert this quantity to miles per gallon.
   a) 9.4 mi/gal
   b) 1.3×10^2 mi/gal
   c) 52 mi/gal
   d) 3.6 mi/gal

6. A wooden block has a volume of 18.5 in^3. Express the volume of the cube in cm^3.
   a) 303 cm^3 **b)** 47.0 cm^3
   c) 1.13 cm^3 **d)** 7.28 cm^3

7. Which sample contains the greatest number of atoms?
   a) 14 g C **b)** 49 g Cr
   c) 102 g Ag **d)** 202 g Pb

8. A solid copper cube contains 4.3×10^{23} atoms. What is the edge length of the cube? The density of copper is 8.96 g/cm^3.
   a) 0.20 cm **b)** 1.7 cm
   c) 8.0 cm **d)** 6.4×10^3 cm

9. Determine the number of atoms in 1.85 mL of mercury. The density of mercury is 13.5 g/mL.
   a) 3.02×10^{27} atoms **b)** 4.11×10^{20} atoms
   c) 7.50×10^{22} atoms **d)** 1.50×10^{25} atoms

10. A 20.0-g sample of an element contains 4.95×10^{23} atoms. Identify the element.
    a) Cr **b)** O
    c) Mg **d)** Fe

Answers: 1:a; 2:c; 3:c; 4:d; 5:c; 6:a; 7:a; 8:b; 9:c; 10:c

CHAPTER SUMMARY 2

REVIEW

KEY LEARNING OUTCOMES

CHAPTER OBJECTIVES	ASSESSMENT
Reporting Scientific Measurements to the Correct Digit of Uncertainty (2.2)	• Example 2.1 For Practice 2.1 Exercises 15–18
Calculating the Density of a Substance (2.3)	• Example 2.2 For Practice 2.2 For More Practice 2.2 Exercises 19–26

Using Conversion Factors (2.5, 2.6)	• Examples 2.3, 2.4, 2.5, 2.6 For Practice 2.3, 2.4, 2.5, 2.6 For More Practice 2.5, 2.6 Exercises 27–46
Solving Problems Involving Equations (2.7)	• Examples 2.7, 2.8 For Practice 2.7, 2.8 Exercises 61, 62, 65, 66, 71, 72
Converting between Moles and Number of Atoms (2.8)	• Example 2.9 For Practice 2.9 Exercises 47, 48
Converting between Mass and Amount (Number of Moles) (2.8)	• Example 2.10 For Practice 2.10 For More Practice 2.10 Exercises 50, 51
Using the Mole Concept (2.8)	• Examples 2.11, 2.12 For Practice 2.11, 2.12 For More Practice 2.11, 2.12 Exercises 51–60

KEY TERMS

Section 2.2
accuracy (37)
precision (37)
random error (38)
systematic error (38)

Section 2.3
density (*d*) (38)
intensive property (39)
extensive property (39)

Section 2.4
energy (40)
work (40)
kinetic energy (40)
potential energy (40)
thermal energy (40)
law of conservation of energy (40)
joule (J) (41)
calorie (cal) (41)
Calorie (Cal) (42)
kilowatt-hour (kWh) (42)
exothermic (42)
endothermic (42)

Section 2.5
dimensional analysis (43)
conversion factor (43)

Section 2.8
mole (51)
Avogadro's number (51)
molar mass (52)

KEY CONCEPTS

The Reliability of a Measurement (2.2)

- Measurements usually involve the use of instruments, which have an inherent amount of uncertainty.
- In reported measurements, every digit is certain except the last, which is estimated. The precision of a measurement refers to its reproducibility.
- The accuracy of a measurement refers to how close a measurement is to the actual value of the quantity being measured.

Density (2.3)

- The density of a substance is the ratio of its mass to its volume.
- Density is an intensive property, which means it is independent of the amount of the substance.

Energy and Its Units (2.4)

- Energy is the capacity to do work and is often reported in units of joules. Systems with high potential energy tend to change in the direction of lower potential energy, releasing energy into the surroundings.
- In chemical and physical changes, matter often exchanges energy with its surroundings. In these exchanges, the total energy is always conserved; energy is neither created nor destroyed.
- A process in which a system transfers energy to the surroundings is exothermic, while a process in which a system gains energy from the surroundings is endothermic.

Converting between Units and Problem Solving (2.5, 2.6)

- Dimensional analysis—solving problems by using units as a guide—is useful in solving many chemical problems.
- An approach to solving many chemical problems involves four steps: sorting the information in the problem; strategizing about how to solve the problem; solving the problem; and checking the answer.

Atoms and the Mole (2.8)

- One mole of an element is the amount of that element that contains Avogadro's number (6.022×10^{23}) of atoms.
- Any sample of an element with a mass (in grams) that equals its atomic mass contains one mole of the element. For example, the atomic mass of carbon is 12.011 amu; therefore 12.011 g of carbon contains 1 mol of carbon atoms.

KEY EQUATIONS AND RELATIONSHIPS

Relationship between Density (*d*), Mass (*m*), and Volume (*V*) (2.3)

$$d = \frac{m}{V}$$

Avogadro's Number (2.8)

$$1 \text{ mol} = 6.0221421 \times 10^{23} \text{ particles}$$

EXERCISES

REVIEW QUESTIONS

1. Explain the relationship between the reliability of a measurement and the instrument used to make the measurement.
2. What is the significance of the number of digits reported in a measured quantity?
3. Explain the difference between precision and accuracy.
4. Explain the difference between random error and systematic error.
5. Explain the difference between density and mass.
6. Explain the difference between *intensive* and *extensive* properties.
7. What is energy? Explain the difference between kinetic energy and potential energy.
8. State the law of conservation of energy, and explain its significance.
9. What kind of energy is chemical energy? In what way is an elevated weight similar to a tank of gasoline?
10. Explain the difference between an exothermic process and an endothermic one.
11. What is dimensional analysis?
12. How should units be treated in calculations?
13. What is a mole? How is the mole concept useful in chemical calculations?
14. Why is the mass corresponding to a mole of one element different from the mass corresponding to a mole of another element?

PROBLEMS BY TOPIC

Note: Answers to all odd-numbered Problems, numbered in blue, can be found in Appendix V. Exercises in the Problems by Topic section are paired, with each odd-numbered problem followed by a similar even-numbered problem. Exercises in the Cumulative Problems section are also paired, but more loosely. Challenge Problems and Conceptual Problems, because of their nature, are unpaired.

The Reliability of a Measurement and Significant Figures

15. A ruler used to measure a penny has markings every 1 mm. Which measurement for the size of the penny is correctly reported for this ruler?

    a. 19.05 mm b. 19 mm c. 19.1 mm

16. A scale used to weigh produce at a market has markings every 0.1 kg. Which measurement for the mass of a dozen apples is correctly reported for this scale?

    a. 1.87 kg b. 1.9 kg c. 1.875 kg

17. Read each measurement to the correct number of significant figures. Laboratory glassware should always be read from the bottom of the meniscus.

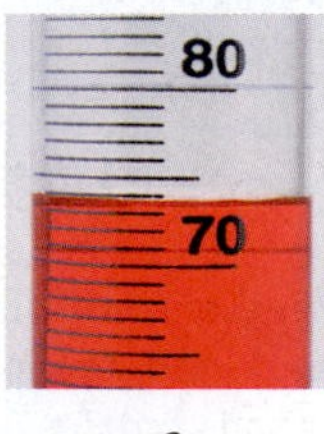

a.

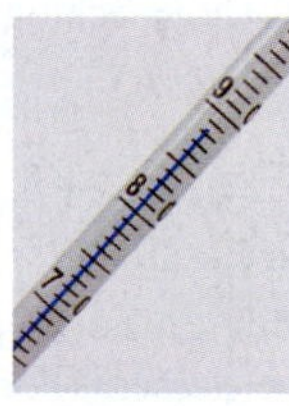
b.

c.

18. Read each measurement to the correct number of significant figures. Laboratory glassware should always be read from the bottom of the meniscus. Digital balances normally display mass to the correct number of significant figures for that particular balance.

a.

b.

c.

Density

19. A new penny has a mass of 2.49 g and a volume of 0.349 cm^3. Is the penny made of pure copper? Explain.
20. A titanium bicycle frame displaces 0.314 L of water and has a mass of 1.41 kg. What is the density of the titanium in g/cm^3?
21. Glycerol is a syrupy liquid used in cosmetics and soaps. A 3.25-L sample of pure glycerol has a mass of 4.10×10^3 g. What is the density of glycerol in g/cm^3?
22. An allegedly gold nugget is tested to determine its density. It is found to displace 19.3 mL of water and has a mass of 371 g. Could the nugget be made of gold?
23. Ethylene glycol (antifreeze) has a density of 1.11 g/cm^3.

    a. What is the mass in g of 417 mL of this liquid?
    b. What is the volume in L of 4.1 kg of this liquid?

24. Acetone (nail polish remover) has a density of 0.7857 g/cm^3.

    a. What is the mass, in g, of 28.56 mL of acetone?
    b. What is the volume, in mL, of 6.54 g of acetone?

25. A small airplane takes on 245 L of fuel. If the density of the fuel is 0.821 g/mL, what mass of fuel has the airplane taken on?
26. Human fat has a density of 0.918 g/cm^3. How much volume (in cm^3) is gained by a person who gains 10.0 lbs of pure fat?

Unit Conversions

27. Perform each unit conversion.

    a. 27.8 L to cm^3 b. 1898 mg to kg c. 198 km to cm

28. Perform each unit conversion.

    a. 28.9 nm to μm b. 1432 cm^3 to L c. 1211 Tm to Gm

29. Perform each unit conversion.

    a. 154 cm to in b. 3.14 kg to g
    c. 3.5 L to qt d. 109 mm to in

30. Perform each unit conversion.

    a. 1.4 in to mm b. 116 ft to cm
    c. 1845 kg to lb d. 815 yd to km

31. A runner wants to run 10.0 km. She knows that her running pace is 7.5 miles per hour. How many minutes must she run?

32. A cyclist rides at an average speed of 18 miles per hour. If she wants to bike 212 km, how long (in hours) must she ride?

33. A European automobile has a gas mileage of 17 km/L. What is the car's gas mileage in miles per gallon?

34. A gas can holds 5.0 gallons of gasoline. Express this quantity in cm^3.

35. A house has an area of 195 m^2. What is its area in:

a. km^2 **b.** dm^2 **c.** cm^2

36. A bedroom has a volume of 115 m^3. What is its volume in:

a. km^3 **b.** dm^3 **c.** cm^3

37. The average U.S. farm occupies 435 acres. How many square miles is this? (1 acre = 43,560 ft^2, 1 mile = 5280 ft)

38. Total U.S. farmland occupies 954 million acres. How many square miles is this? (1 acre = 43,560 ft^2, 1 mile = 5280 ft). Total U.S. land area is 3.537 million square miles. What percentage of U.S. land is farmland?

39. An acetaminophen suspension for infants contains 80 mg/0.80 mL suspension. The recommended dose is 15 mg/kg body weight. How many mL of this suspension should be given to an infant weighing 14 lbs? (Assume two significant figures.)

40. An ibuprofen suspension for infants contains 100 mg/5.0 mL suspension. The recommended dose is 10 mg/kg body weight. How many mL of this suspension should be given to an infant weighing 18 lbs? (Assume two significant figures.)

41. Convert between energy units.

a. 534 kWh to J **b.** 215 kJ to Cal
c. 567 Cal to J **d.** 2.85×10^3 J to cal

42. Convert between energy units.

a. 231 cal to kJ **b.** 132×10^4 kJ to kcal
c. 4.99×10^3 kJ to kWh **d.** 2.88×10^4 J to Cal

43. Suppose that a person eats 2387 Calories per day. Convert this amount of energy into each unit.

a. J **b.** kJ **c.** kWh

44. A particular frost-free refrigerator uses about 745 kWh of electrical energy per year. Express this amount of energy in each unit.

a. J **b.** kJ **c.** Cal

45. A household receives a $145 electricity bill. The cost of electricity is $0.120/kWh. How much energy, in joules, did the household use?

46. A 150-lb person burns about 2700 Calories to run a marathon. How much energy is burned in kJ? Assume two significant figures.

The Mole Concept

47. How many sulfur atoms are there in 5.52 mol of sulfur?

48. How many moles of aluminum do 3.7×10^{24} aluminum atoms represent?

49. What is the amount, in moles, of each elemental sample?

a. 11.8 g Ar **b.** 3.55 g Zn
c. 26.1 g Ta **d.** 0.211 g Li

50. What is the mass, in grams, of each elemental sample?

a. 2.3×10^{-3} mol Sb **b.** 0.0355 mol Ba
c. 43.9 mol Xe **d.** 1.3 mol W

51. How many silver atoms are there in 3.78 g of silver?

52. What is the mass of 4.91×10^{21} platinum atoms?

53. Calculate the number of atoms in each sample.

a. 5.18 g P **b.** 2.26 g Hg
c. 1.87 g Bi **d.** 0.082 g Sr

54. Calculate the number of atoms in each sample.

a. 14.955 g Cr **b.** 39.733 g S
c. 12.899 g Pt **d.** 97.552 g Sn

55. Calculate the mass, in grams, of each sample.

a. 1.1×10^{23} gold atoms **b.** 2.82×10^{22} helium atoms
c. 1.8×10^{23} lead atoms **d.** 7.9×10^{21} uranium atoms

56. Calculate the mass, in kg, of each sample.

a. 7.55×10^{26} cadmium atoms
b. 8.15×10^{27} nickel atoms
c. 1.22×10^{27} manganese atoms
d. 5.48×10^{29} lithium atoms

57. How many carbon atoms are there in a diamond (pure carbon) with a mass of 52 mg?

58. How many helium atoms are there in a helium blimp containing 536 kg of helium?

59. Calculate the average mass, in grams, of one platinum atom.

60. Using scanning tunneling microscopy, scientists at IBM wrote the initials of their company with 35 individual xenon atoms (as shown below). Calculate the total mass of these letters in grams.

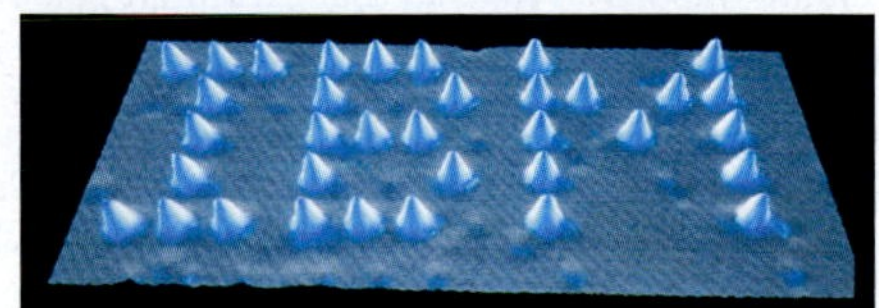

CUMULATIVE PROBLEMS

61. A thief uses a can of sand to replace a solid gold cylinder that sits on a weight-sensitive, alarmed pedestal. The can of sand and the gold cylinder have exactly the same dimensions (length = 22 cm and radius = 3.8 cm).

a. Calculate the mass of each cylinder (ignore the mass of the can itself). (density of gold = 19.3 g/cm^3, density of sand = 3.00 g/cm^3)
b. Did the thief set off the alarm? Explain.

62. The proton has a radius of approximately 1.0×10^{-13} cm and a mass of 1.7×10^{-24} g. Determine the density of a proton. For a sphere $V = (4/3)\pi r^3$.

63. The density of titanium is 4.51 g/cm^3. What is the volume (in cubic inches) of 3.5 lbs of titanium?

64. The density of iron is 7.86 g/cm^3. What is its density in pounds per cubic inch (lb/in^3)?

65. A steel cylinder has a length of 2.16 in, a radius of 0.22 in, and a mass of 41 g. What is the density of the steel in g/cm^3?

66. A solid aluminum sphere has a mass of 85 g. Use the density of aluminum to find the radius of the sphere in inches.

67. A backyard swimming pool holds 185 cu yd (yd^3) of water. What is the mass of the water in pounds?

68. An iceberg has a volume of 7655 cu ft. What is the mass of the ice (in kg) composing the iceberg (at 0 °C)?

69. The Toyota Prius, a hybrid electric vehicle, has a U.S. Environmental Protection Agency (EPA) gas mileage rating of 52 mi/gal in the city. How many kilometers can the Prius travel on 15 L of gasoline?

70. The Honda Insight, a hybrid electric vehicle, has an U.S. Environmental Protection Agency (EPA) gas mileage rating of 57 mi/gal in the city. How many kilometers can the Insight travel on the amount of gasoline that would fit in a soda can? The volume of a soda can is 355 mL.

71. The single proton that forms the nucleus of the hydrogen atom has a radius of approximately 1.0×10^{-13} cm. The hydrogen atom itself has a radius of approximately 52.9 pm. What fraction of the space within the atom is occupied by the nucleus?

72. A sample of gaseous neon atoms at atmospheric pressure and 0 °C contains 2.69×10^{22} atoms per liter. The atomic radius of neon is 69 pm. What fraction of the space is occupied by the atoms themselves? What does this reveal about the separation between atoms in the gaseous phase?

73. The diameter of a hydrogen atom is 212 pm. Find the length in kilometers of a row of 6.02×10^{23} hydrogen atoms. The diameter of a ping pong ball is 4.0 cm. Find the length in kilometers of a row of 6.02×10^{23} ping pong balls.

74. The world's record in the 100-m dash is 9.58 s, and in the 100-yd dash it is 9.07 s. Find the speed in mi/hr of the runners who set these records.

75. Table salt contains 39.33 g of sodium per 100 g of salt. The U.S. Food and Drug Administration (FDA) recommends that adults consume less than 2.40 g of sodium per day. A particular snack mix contains 1.25 g of salt per 100 g of the mix. What mass of the snack mix can an adult consume and not exceed the FDA limit?

76. Lead metal can be extracted from a mineral called galena, which contains 86.6% lead by mass. A particular ore contains 68.5% galena by mass. If the lead can be extracted with 92.5% efficiency, what mass of ore is required to make a lead sphere with a 5.00-cm radius?

77. A length of #8 copper wire (radius = 1.63 mm) has a mass of 24.0 kg and a resistance of 2.061 ohm per km (Ω/km). What is the overall resistance of the wire?

78. Rolls of aluminum foil are 304 mm wide and 0.016 mm thick. What maximum length of aluminum foil can be made from 1.10 kg of aluminum?

79. Liquid nitrogen has a density of 0.808 g/mL and boils at 77 K. Researchers often purchase liquid nitrogen in insulated 175-L tanks. The liquid vaporizes quickly to gaseous nitrogen (which has a density of 1.15 g/L at room temperature and atmospheric pressure) when the liquid is removed from the tank. Suppose that all 175 L of liquid nitrogen in a tank accidentally vaporized in a lab that measured 10.00 m × 10.00 m × 2.50 m. What maximum fraction of the air in the room could be displaced by the gaseous nitrogen?

80. Mercury is often used in thermometers. The mercury sits in a bulb on the bottom of the thermometer and rises up a thin capillary as the temperature rises. Suppose a mercury thermometer contains 3.380 g of mercury and has a capillary that is 0.200 mm in diameter. How far does the mercury rise in the capillary when the temperature changes from 0.0 °C to 25.0 °C? The density of mercury at these temperatures is 13.596 g/cm^3 and 13.534 g/cm^3, respectively.

81. Carbon-12 contains 6 protons and 6 neutrons. The radius of the nucleus is approximately 2.7 fm (femtometers), and the radius of the atom is approximately 70 pm (picometers). Calculate the volume of the nucleus and the volume of the atom. What percentage of the carbon atom's volume is occupied by the nucleus? (Assume two significant figures.)

82. A penny has a thickness of approximately 1.0 mm. If you stacked Avogadro's number of pennies one on top of the other on Earth's surface, how far would the stack extend (in km)? For comparison, the sun is about 150 million km from Earth and the nearest star, Proxima Centauri, is about 40 trillion km from Earth.

83. Consider the stack of pennies in problem 82. How much money (in dollars) would this represent? If this money were equally distributed among the world's population of 6.5 billion people, how much would each person receive? Would each person be a millionaire? A billionaire? A trillionaire?

84. The mass of an average blueberry is 0.75 g, and the mass of an automobile is 2.0×10^3 kg. How many automobiles have the same total combined mass as 1.0 mol blueberries?

85. A pure copper sphere has a radius of 0.935 in. How many copper atoms does it contain? The volume of a sphere is $(4/3)\pi r^3$, and the density of copper is 8.96 g/cm^3.

86. A pure titanium cube has an edge length of 2.78 in. How many titanium atoms does it contain? Titanium has a density of 4.50 g/cm^3.

87. A 67.2-g sample of a gold and palladium alloy contains 2.49×10^{23} atoms. What is the composition (by mass) of the alloy?

88. Common brass is a copper and zinc alloy containing 37.0% zinc by mass and having a density of 8.48 g/cm^3. A fitting composed of common brass has a total volume of 112.5 cm^3. How many atoms (copper and zinc) does the fitting contain?

89. The U.S. Environmental Protection Agency (EPA) sets limits on healthful levels of air pollutants. The maximum level that the EPA considers safe for lead air pollution is 1.5 μg/m^3. If your lungs were filled with air containing this level of lead, how many lead atoms would be in your lungs? (Assume a total lung volume of 5.50 L.)

90. Pure gold is usually too soft for jewelry, so it is often alloyed with other metals. How many gold atoms are in an 0.255-ounce 18 K gold bracelet? (18 K gold is 75% gold by mass.)

CHALLENGE PROBLEMS

91. In 1999, scientists discovered a new class of black holes with masses 100 to 10,000 times the mass of our sun that occupy less space than our moon. Suppose that one of these black holes has a mass of 1×10^3 suns and a radius equal to one-half the radius of our moon. What is the density of the black hole in g/cm^3? The radius of our sun is 7.0×10^5 km, and it has an average density of 1.4×10^3 kg/m^3. The diameter of the moon is 2.16×10^3 miles.

92. Polluted air can have carbon monoxide (CO) levels of 15.0 ppm. An average human inhales about 0.50 L of air per breath and takes about 20 breaths per minute. How many milligrams of carbon monoxide does the average person inhale in an 8-hour period in this level of carbon monoxide pollution? Assume that the carbon monoxide has a density of 1.2 g/L. (*Hint*: 15.0 ppm CO means 15.0 L CO per 10^6 L air.)

93. Nanotechnology, the field of building ultrasmall structures one atom at a time, has progressed in recent years. One potential application of nanotechnology is the construction of artificial cells. The simplest cells would probably mimic red blood cells, the body's oxygen transporters. Nanocontainers, perhaps constructed of carbon, could be pumped full of oxygen and injected into a person's bloodstream. If the person needed additional oxygen—due to a heart attack or for the purpose of space travel, for example—these containers could slowly release oxygen into the blood, allowing tissues that would otherwise die to remain alive. Suppose that the nanocontainers were cubic and had an edge length of 25 nanometers.

a. What is the volume of one nanocontainer? (Ignore the thickness of the nanocontainer's wall.)

b. Suppose that each nanocontainer could contain pure oxygen pressurized to a density of 85 g/L. How many grams of oxygen could be contained by each nanocontainer?

c. Air typically contains about 0.28 g of oxygen per liter. An average human inhales about 0.50 L of air per breath and takes about 20 breaths per minute. How many grams of oxygen does a human inhale per hour? (Assume two significant figures.)

d. What is the minimum number of nanocontainers that a person would need in his bloodstream to provide 1 hour's worth of oxygen?

e. What is the minimum volume occupied by the number of nanocontainers calculated in part d? Is such a volume feasible, given that total blood volume in an adult is about 5 L?

94. Determine the approximate percent increase in waist size that occurs when a 155-lb person gains 40.0 lbs of fat. Assume that the volume of the person can be modeled by a cylinder that is 4.0 ft tall. The average density of a human is about 1.0 g/cm^3, and the density of fat is 0.918 g/cm^3.

95. A box contains a mixture of small copper spheres and small lead spheres. The total volume of both metals is measured by the displacement of water to be 427 cm^3, and the total mass is 4.36 kg. What percentage of the spheres is copper?

96. In Section 2.8, it was stated that 1 mol of sand grains would cover the state of Texas to several feet. Estimate how many feet by assuming that the sand grains are roughly cube-shaped, each one with an edge length of 0.10 mm. Texas has a land area of 268,601 sq mi.

97. Use the concepts in this chapter to obtain an estimate for the number of atoms in the universe. Make the following assumptions: **(a)** All of the atoms in the universe are hydrogen atoms in stars. (This is not a ridiculous assumption because over three-fourths of the atoms in the universe are in fact hydrogen. Gas and dust between the stars represent only about 15% of the visible matter of our galaxy, and planets compose a far smaller fraction.) **(b)** The sun is a typical star composed of pure hydrogen with a density of 1.4 g/cm^3 and a radius of 7×10^8 m. **(c)** Each of the roughly 100 billion stars in the Milky Way galaxy contains the same number of atoms as our sun. **(d)** Each of the 10 billion galaxies in the visible universe contains the same number of atoms as our Milky Way galaxy.

CONCEPTUAL PROBLEMS

98. A cube has an edge length of 7 cm. If it is divided up into 1-cm cubes, how many 1-cm cubes are there?

99. Substance A has a density of 1.7 g/cm^3. Substance B has a density of 1.7 kg/m^3. Without doing any calculations, determine which substance is more dense.

100. For each box, examine the blocks attached to the balances. Based on their positions and sizes, determine which block is more dense (the dark block or the lighter-colored block), or if the relative densities cannot be determined. (Think carefully about the information being shown.)

a.

b.

c.

101. The mole is defined as the amount of a substance containing the same number of particles as exactly 12 g of C-12. The amu is defined as 1/12 of the mass of an atom of C-12. Why is it important that both of these definitions reference the same isotope? What would be the result, for example, of defining the mole with respect to C-12, but the amu with respect to Ne-20?

102. Without doing any calculations, determine which of the samples contains the greatest amount of the element in moles. Which contains the greatest mass of the element?

a. 55.0 g Cr **b.** 45.0 g Ti **c.** 60.0 g Zn

ANSWERS TO CONCEPTUAL CONNECTIONS

Cc 2.1 **(c)** The sample expands. However, because its mass remains constant while its volume increases, its density decreases.

Cc 2.2 Since burning natural gas gives off energy to the surroundings, the process is exothermic and the energy change is negative.

Cc 2.3 **(b)** When we multiply a value in pound•second by $\frac{4.45 \text{ newton•second}}{1 \text{ pound•second}}$, the pound•second cancels and we get newton•second.

Cc 2.4 **(b)** The carbon sample contains more atoms than the copper sample because carbon has a lower molar mass than copper. Carbon atoms are lighter than copper atoms, so a 1-g sample of carbon contains more atoms than a 1-g sample of copper. The carbon sample also contains more atoms than the uranium sample because, even though the uranium sample has 10 times the mass of the carbon sample, a uranium atom is more than 10 times as massive (238 g/mol for U versus 12 g/mol for carbon).

CHAPTER 3

"Anyone who is not shocked by quantum mechanics has not understood it."

—Niels Bohr (1885–1962)

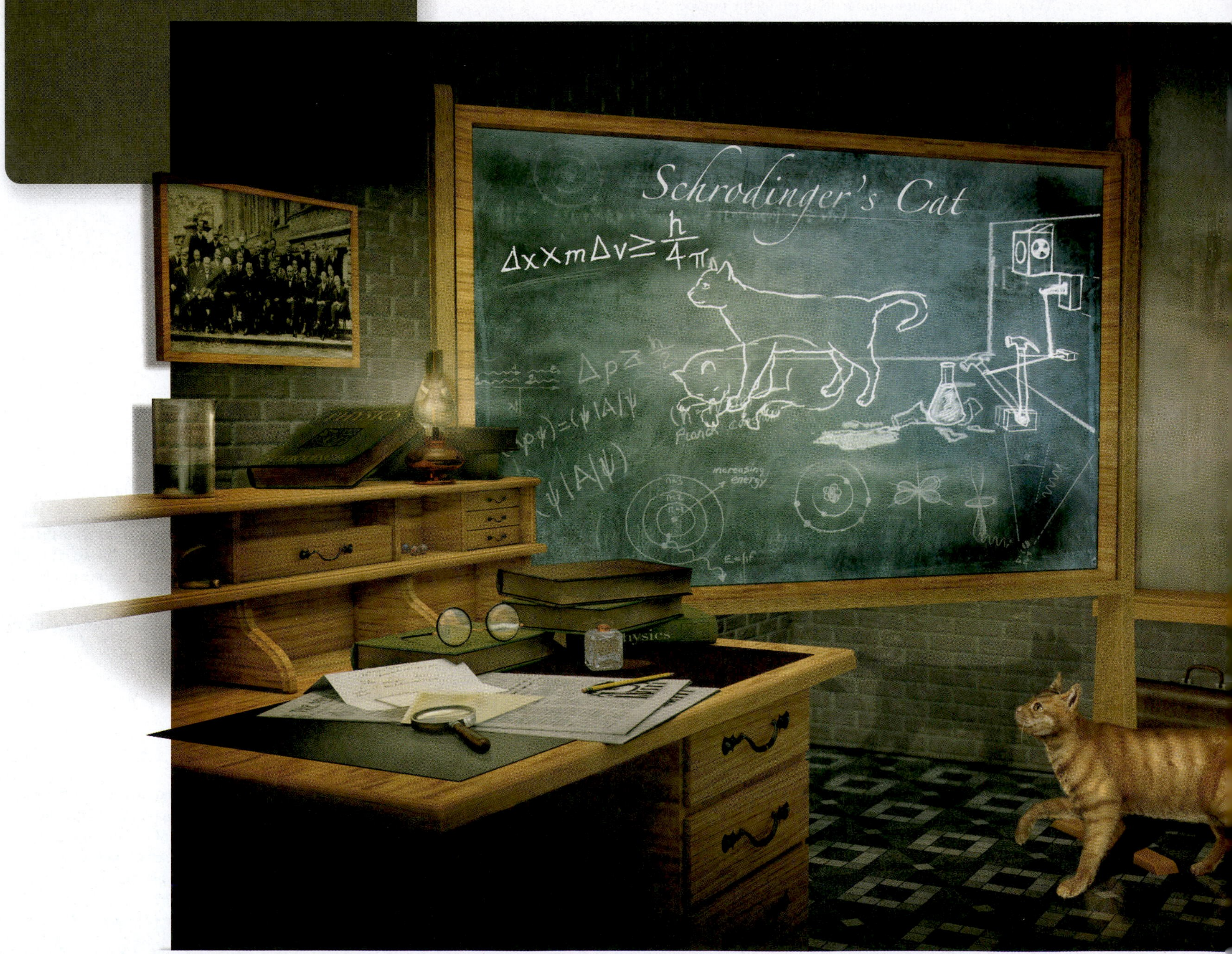

The thought experiment known as Schrödinger's cat is intended to show that the strangeness of the quantum world does not transfer to the macroscopic world.

The Quantum-Mechanical Model of the Atom

3.1 Schrödinger's Cat 63

3.2 The Nature of Light 64

3.3 Atomic Spectroscopy and the Bohr Model 73

3.4 The Wave Nature of Matter: The de Broglie Wavelength, the Uncertainty Principle, and Indeterminacy 77

3.5 Quantum Mechanics and the Atom 81

3.6 The Shapes of Atomic Orbitals 87

Key Learning Outcomes 93

THE EARLY TWENTIETH CENTURY revolutionized how we think about physical reality. Before that time, all descriptions of matter had been deterministic—the present completely determining the future. Quantum mechanics changed that. This new theory suggested that for subatomic particles—electrons, neutrons, and protons—the present does NOT completely determine the future. For example, if you shoot one electron down a path and measure where it lands, a second electron shot down the same path under the same conditions will most likely land in a different place! Several gifted scientists, such as Albert Einstein, Niels Bohr, Louis de Broglie, Max Planck, Werner Heisenberg, P. A. M. Dirac, and Erwin Schrödinger, developed quantum-mechanical theory. Their new theory, however, made even some of them uncomfortable. Bohr said, "Anyone who is not shocked by quantum mechanics has not understood it." Schrödinger wrote, "I don't like it, and I'm sorry I ever had anything to do with it." Albert Einstein disbelieved it stating, "God does not play dice with the universe." In fact, Einstein attempted to disprove quantum mechanics—without success—until he died. Today, quantum mechanics forms the foundation of chemistry—explaining the periodic table and the behavior of the elements in chemical bonding.

3.1 Schrödinger's Cat

Atoms and the particles that compose them are unimaginably small. Electrons have a mass of less than a trillionth of a trillionth of a gram, and a size so small that it is immeasurable. A single speck of dust contains more electrons than the number of people who have existed on Earth over all the centuries of time. Electrons are *small* in the absolute sense of the word—they are among the smallest particles that make up matter. Nonetheless, if we are to understand the main theme of this book—how the structure of the particles that compose matter determines the properties of matter—we must understand electrons. As we saw in

Section 1.7, most of the volume of the atom is occupied by electrons, so the size of an atom depends on its electrons. In fact, an atom's electrons determine many of its chemical and physical properties.

In the early twentieth century, scientists discovered that the *absolutely small* (or *quantum*) world of the electron behaves differently than the *large* (or *macroscopic*) world that we are used to observing. Chief among these differences is the idea that, when unobserved, *quantum particles like electrons can be in two different states at the same time*. For example, through a process called radioactive decay (see Chapter 21) an atom can emit small (that is, *absolutely* small) energetic particles from its nucleus. In the macroscopic world, something either emits an energetic particle or it doesn't. In the quantum world, however, the unobserved atom can be in a state in which it is doing both—emitting the particle and not emitting the particle—simultaneously. At first, this seems absurd. The absurdity resolves itself, however, upon observation. When we set out to measure the emitted particle, the act of measurement actually forces the atom into one state or other.

Early twentieth-century physicists struggled with this idea. Austrian physicist Erwin Schrödinger (1887–1961), in an attempt to demonstrate that this quantum strangeness could never transfer itself to the macroscopic world, published a paper in 1935 that contained a thought experiment about a cat, now known as Schrödinger's cat. In the thought experiment, the cat is put into a steel chamber that contains radioactive atoms such as the one described in the previous paragraph. The chamber is equipped with a mechanism that, upon the emission of an energetic particle by one of the radioactive atoms, causes a hammer to break a flask of hydrocyanic acid, a poison. If the flask breaks, the poison is released and the cat dies.

Now here comes the absurdity: If the steel chamber is closed, the whole system remains unobserved, and the radioactive atom is in a state in which it has emitted the particle and not emitted the particle (with equal probability). Therefore the cat is both dead and undead. Schrödinger put it this way: "[The steel chamber would have] *in it the living and dead cat (pardon the expression) mixed or smeared out in equal parts.*" When the chamber is opened, the act of observation forces the entire system into one state or the other: The cat is either dead or alive, not both. However, while unobserved, the cat is both dead and alive. The absurdity of the both dead and undead cat in Schrödinger's thought experiment was meant to demonstrate how quantum strangeness does not transfer to the macroscopic world.

In this chapter, we examine the **quantum-mechanical model** of the atom, a model that explains the strange behavior of electrons. In particular, we focus on how the model describes electrons as they exist within atoms, and later we shall see how those electrons determine the chemical and physical properties of elements.

3.2 The Nature of Light

Before we explore electrons and their behavior within the atom, we must understand some of the properties of light. As quantum-mechanical theory was developed, light was (surprisingly) found to have many characteristics in common with electrons. Chief among these characteristics is the *wave–particle duality* of light. Certain properties of light are best described by thinking of it as a wave, while other properties are best described by thinking of it as a particle. In this section, we first explore the wave behavior of light, and then its particle behavior. We then turn to electrons to see how they display the same wave–particle duality.

The Wave Nature of Light

Light is **electromagnetic radiation**, a type of energy embodied in oscillating electric and magnetic fields. A *magnetic field* is a region of space where a magnetic particle experiences a force (think of the space around a magnet). An *electric field* is a region of space where an electrically charged particle experiences a force. A proton, for example, has an electric field around it. If you bring another charged particle into that field, that particle experiences a force.

Electromagnetic radiation can be described as a wave composed of oscillating, mutually perpendicular electric and magnetic fields propagating through space, as shown in Figure 3.1 ◀. In a vacuum, these waves move at a constant speed of 3.00×10^8 m/s (186,000 mi/s)—fast enough to circle the Earth in one-seventh of a second. This great speed is the reason for the delay

▼ **FIGURE 3.1 Electromagnetic Radiation** Electromagnetic radiation can be described as a wave composed of oscillating electric and magnetic fields. The fields oscillate in perpendicular planes.

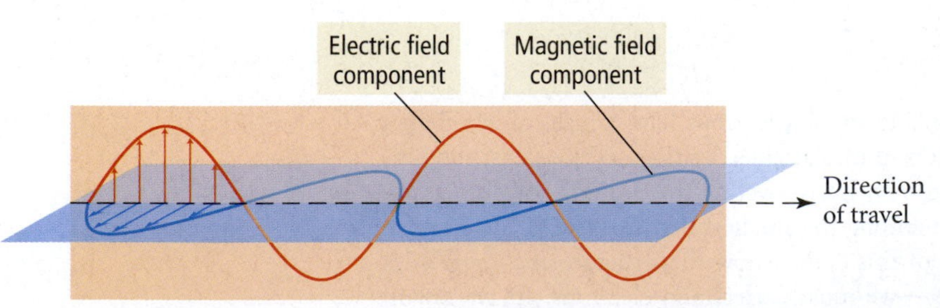

between the moment when you see a firework in the sky and the moment when you hear the sound of its explosion. The light from the exploding firework reaches your eye almost instantaneously. The sound, traveling much more slowly (340 m/s), takes longer. The same thing happens in a thunderstorm—you see the flash of lightning immediately, but the sound of thunder takes a few seconds to reach you. (The sound of thunder is delayed by five seconds for each mile between you and its origin.)

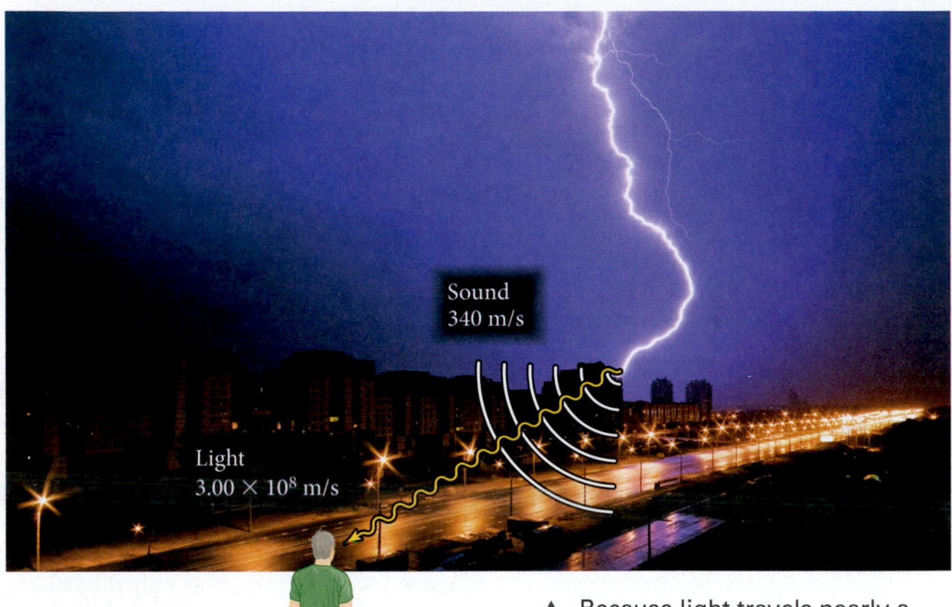

▲ Because light travels nearly a million times faster than sound, the flash of lightning reaches your eyes before the roll of thunder reaches your ears.

We can characterize a wave by its *amplitude* and its *wavelength*. In the graphical representation shown here, the **amplitude** of the wave is the vertical height of a crest (or depth of a trough). The amplitude of the electric and magnetic field waves in light determines the light's *intensity* or brightness—the greater the amplitude, the greater the intensity. The **wavelength** **(λ)** of the wave is the distance between adjacent crests (or any two analogous points) and is measured in units such as meters, micrometers, or nanometers.

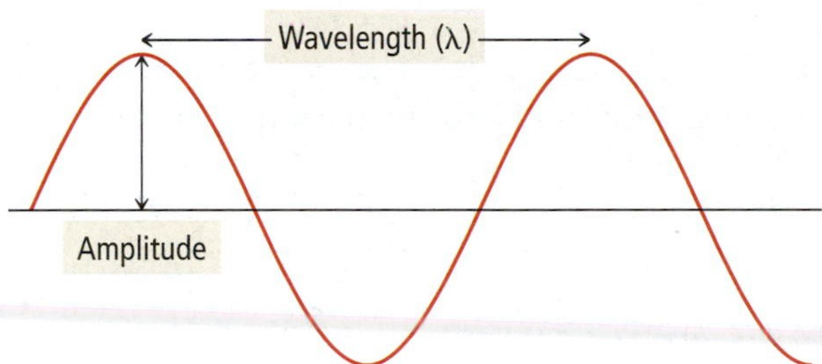

The symbol λ is the Greek letter lambda, pronounced "lamb-duh."

The wavelength of a light wave determines its color (Figure 3.2 ▼).

Like all waves, light is also characterized by its **frequency (ν)**, the number of cycles (or wave crests) that pass through a stationary point in a given period of time. The units of frequency are cycles per second (cycle/s) or simply s^{-1}. An equivalent unit of frequency is the hertz (Hz), defined as 1 cycle/s. The frequency of a wave is directly proportional to the speed at which the wave is traveling—the faster the wave, the more crests will pass a fixed location per unit time. Frequency is also *inversely*

The symbol ν is the Greek letter nu, pronounced "noo."

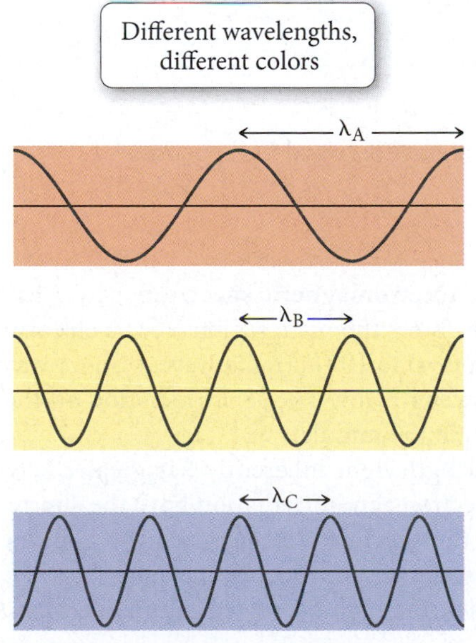

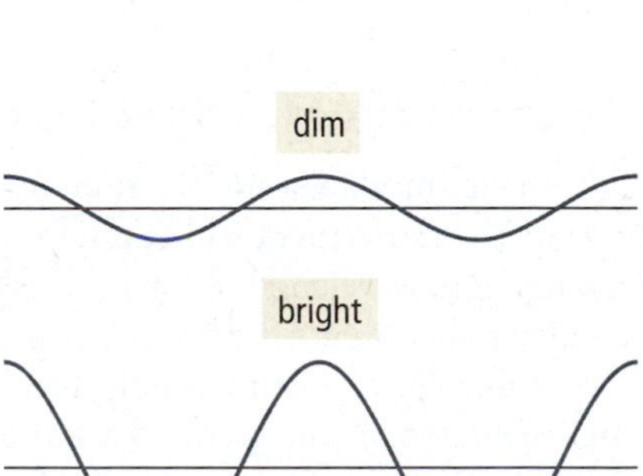

◀ **FIGURE 3.2 Wavelength and Amplitude** Wavelength and amplitude are independent properties. The wavelength of light determines its color. The amplitude, or intensity, determines its brightness.

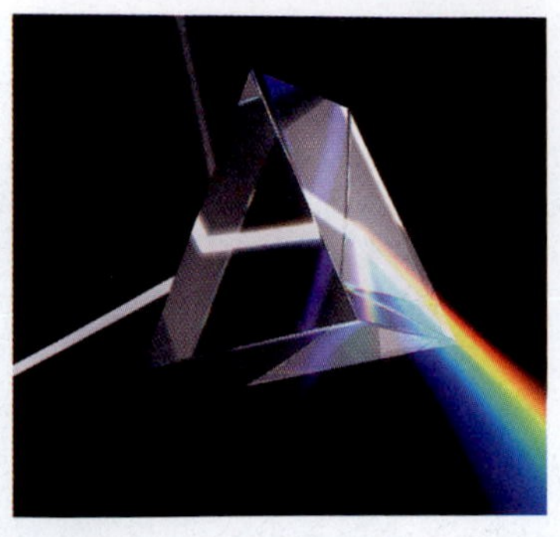

▲ **FIGURE 3.3 Components of White Light** We can pass white light through a prism and decompose it into its constituent colors, each with a different wavelength. The array of colors makes up the spectrum of visible light.

proportional to the wavelength (λ)—the farther apart the crests, the fewer will pass a fixed location per unit time. For light, therefore, we can write:

$$\nu = \frac{c}{\lambda} \quad [3.1]$$

where the speed of light, c, and the wavelength, λ, are both expressed in terms of the same unit of distance. Wavelength and frequency represent different ways of specifying the same information—if we know one, we can calculate the other.

The different colors in *visible light*—light that can be seen by the human eye—correspond to different wavelengths (or frequencies). White light, produced by the sun or by a light bulb, contains a spectrum of wavelengths and therefore a spectrum of colors. We see these colors—red, orange, yellow, green, blue, indigo, and violet—in a rainbow or when white light is passed through a prism (Figure 3.3 ◀). Red light, with a wavelength of about 750 nanometers (nm), has the longest wavelength of visible light; violet light, with a wavelength of about 400 nm, has the shortest. (Recall that nano means 10^{-9}.) The presence of a variety of wavelengths in white light is responsible for the way we perceive colors in objects. When a substance absorbs some colors while reflecting others, it appears colored. For example, a red shirt appears red because it reflects predominantly red light while absorbing most other colors (Figure 3.4 ◀). Our eyes see only the reflected light, making the shirt appear red.

▲ **FIGURE 3.4 The Color of an Object** A red shirt is red because it reflects predominantly red light while absorbing most other colors.

EXAMPLE 3.1

Wavelength and Frequency

Calculate the wavelength (in nm) of the red light emitted by a barcode scanner that has a frequency of $4.62 \times 10^{14}\ \text{s}^{-1}$.

SOLUTION

You are given the frequency of the light and asked to find its wavelength. Use Equation 3.1, which relates frequency to wavelength. You can convert the wavelength from meters to nanometers by using the conversion factor between the two (1 nm = 10^{-9} m).	$\nu = \frac{c}{\lambda}$ $\lambda = \frac{c}{\nu} = \frac{3.00 \times 10^8\ \text{m/}\cancel{\text{s}}}{4.62 \times 10^{14}/\cancel{\text{s}}}$ $= 6.49 \times 10^{-7}\ \text{m}$ $= 6.49 \times 10^{-7}\ \cancel{\text{m}} \times \frac{1\ \text{nm}}{10^{-9}\ \cancel{\text{m}}} = 649\ \text{mm}$

FOR PRACTICE 3.1

A laser dazzles the audience at a rock concert by emitting green light with a wavelength of 515 nm. Calculate the frequency of the light.

Answers to For Practice and For More Practice problems are found in Appendix VI.

The Electromagnetic Spectrum

Visible light makes up only a tiny portion of the entire **electromagnetic spectrum**, which includes all wavelengths of electromagnetic radiation. Figure 3.5 ▶ shows the main regions of the electromagnetic spectrum, ranging in wavelength from 10^{-15} m (gamma rays) to 10^5 m (radio waves). Short-wavelength, high-frequency radiation is on the right and long-wavelength, low-frequency radiation on the left. As you can see, visible light constitutes only a small region in the middle.

We will see later in this section that short-wavelength light inherently has greater energy than long-wavelength light. The most energetic forms of electromagnetic radiation have the shortest wavelengths. The form of electromagnetic radiation with the shortest wavelength is the **gamma (γ) ray**. Gamma rays are produced by the sun, other stars, and certain unstable atomic nuclei on Earth. Excessive exposure to gamma rays is dangerous to humans because the high energy of gamma rays can damage biological molecules.

We will discuss gamma rays in more detail in Chapter 21.

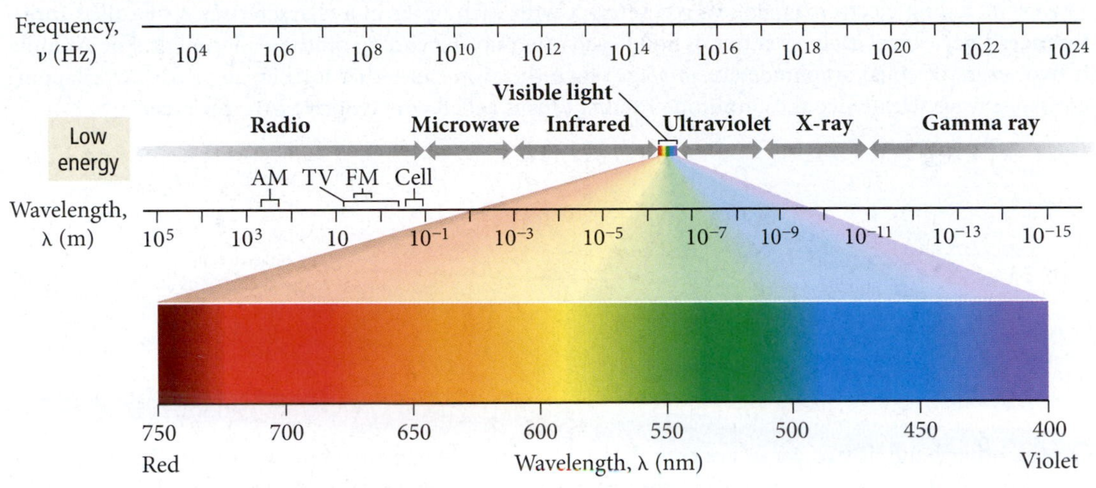

▲ **FIGURE 3.5 The Electromagnetic Spectrum** The right side of the spectrum consists of high-energy, high-frequency, short-wavelength radiation. The left side consists of low-energy, low-frequency, long-wavelength radiation.

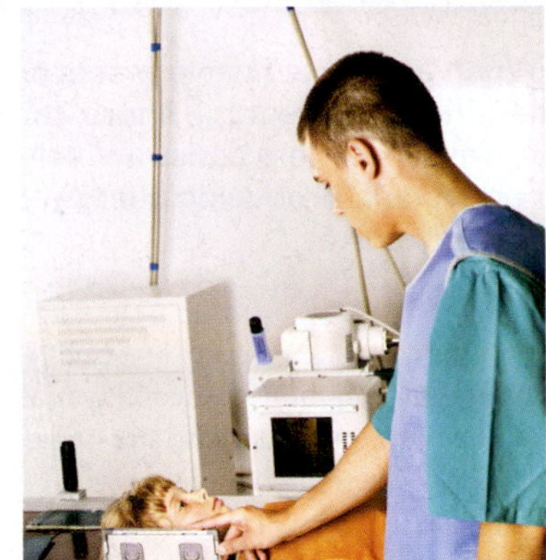

▲ To produce a medical X-ray, the patient is exposed to short-wavelength electromagnetic radiation that passes through the skin to create an image of bones and internal organs.

Next on the electromagnetic spectrum, with longer wavelengths than gamma rays, are **X-rays**, familiar to us from their medical use. X-rays pass through many substances that block visible light and are therefore used to image bones and internal organs. Like gamma rays, X-rays are energetic enough to damage biological molecules. While several annual exposures to X-rays are relatively harmless, too much exposure to X-rays increases cancer risk.

Sandwiched between X-rays and visible light in the electromagnetic spectrum is **ultraviolet (UV) radiation**, most familiar to us as the component of sunlight that produces a sunburn or suntan. While not as energetic as gamma rays or X-rays, ultraviolet light still carries enough energy to damage biological molecules. Excessive exposure to ultraviolet light increases the risk of skin cancer and cataracts and causes premature wrinkling of the skin.

Next on the spectrum is **visible light**, ranging from violet (shorter wavelength, higher energy) to red (longer wavelength, lower energy). Visible light—at low to moderate intensity—does not carry enough energy to damage biological molecules. It does, however, cause certain molecules in our eyes to change their shape, sending a signal to our brains that results in our ability to see.

Beyond visible light lies **infrared (IR) radiation**. The heat you feel when you place your hand near a hot object is infrared radiation. All warm objects, including human bodies, emit infrared light. Although infrared light is invisible to our eyes, infrared sensors can detect it and are often employed in night vision technology to help people "see" in the dark.

▲ Warm objects emit infrared light, which is invisible to the eye but can be captured on film or by detectors to produce an infrared photograph.

Beyond infrared light, at longer wavelengths still, are **microwaves**, used for radar and in microwave ovens. Although microwave radiation has longer wavelengths and therefore lower energies than visible or infrared light, it is efficiently absorbed by water and can therefore heat substances that contain water. The longest wavelengths are those of **radio waves**, which are used to transmit the signals responsible for AM and FM radio, cellular telephone, television, and other forms of communication.

3.1 Cc Conceptual Connection

Arrange the following types of electromagnetic radiation in order of (a) increasing frequency; and (b) increasing wavelength: visible, X-ray, infrared.

▲ When a reflected wave meets an incoming wave near the shore, the two waves interfere constructively for an instant, producing a large-amplitude spike.

Understanding interference in waves is critical to understanding the wave nature of the electron, as you will soon see.

Interference and Diffraction

Waves, including electromagnetic waves, interact with each other in a characteristic way called **interference:** They cancel each other out or build each other up, depending on their alignment. For example, if two waves of equal amplitude are *in phase* when they interact—that is, they align with overlapping crests—a wave with twice the amplitude results. This is called **constructive interference**.

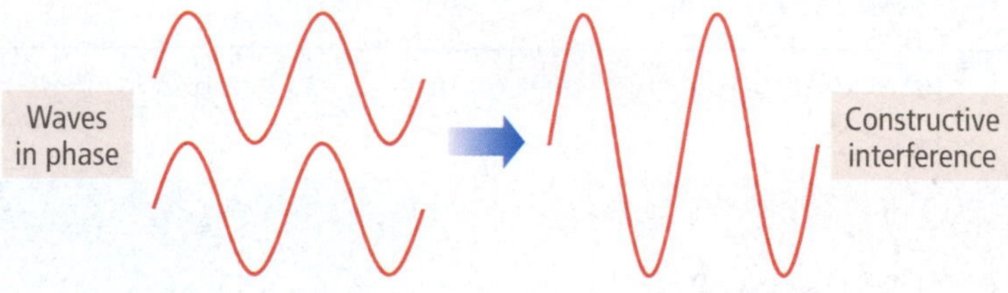

On the other hand, if two waves are completely *out of phase* when they interact—that is, they align so that the crest from one overlaps with the trough from the other—the waves cancel by **destructive interference**.

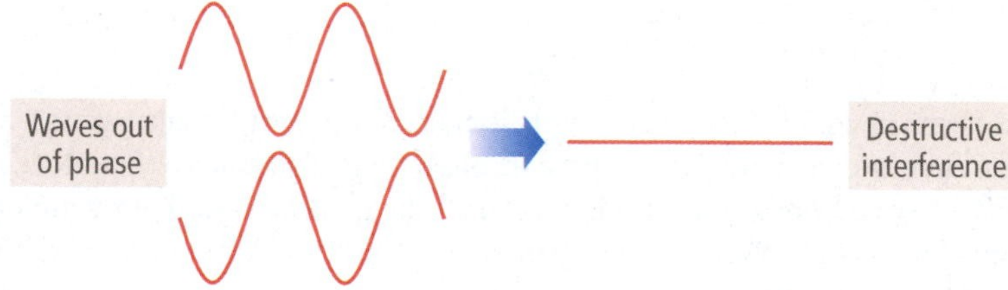

Waves also exhibit a characteristic behavior called **diffraction** (Figure 3.6 ▶). When a wave encounters an obstacle or a slit that is comparable in size to its wavelength, it bends (or *diffracts*) around it. The diffraction of light through two slits separated by a distance comparable to the wavelength of the light, coupled with interference, results in an *interference pattern* as shown in Figure 3.7 ▶. Each slit acts as a new wave source, and the two new waves interfere with each other. The resulting pattern is a series of bright and dark lines that can be viewed on a screen (or recorded on a film) placed a short distance behind the slits. At the center of the screen, the two waves travel equal distances and interfere constructively to produce a bright line. A small distance away from the center in either direction, the two waves travel slightly different distances, so that they are out of phase. At the point where the difference in distance is one-half of one wavelength, the interference is destructive and a dark line appears on the screen. Moving a bit further away from the center produces constructive interference again because the difference between the paths is one whole wavelength. The end result is the interference pattern. Notice that interference results from the ability of a wave to diffract through two slits—an inherent property of waves.

The term *classical*, as in classical electromagnetic theory or classical mechanics, refers to descriptions of matter and energy before the advent of quantum mechanics.

The Particle Nature of Light

Prior to the early 1900s, and especially after the discovery of the diffraction of light, light was thought to be purely a wave phenomenon. Its behavior was described adequately by classical electromagnetic theory, which treated the electric and magnetic fields that constitute light as waves propagating through space. However, a number of discoveries brought the classical view into question. Chief among these was the *photoelectric effect*.

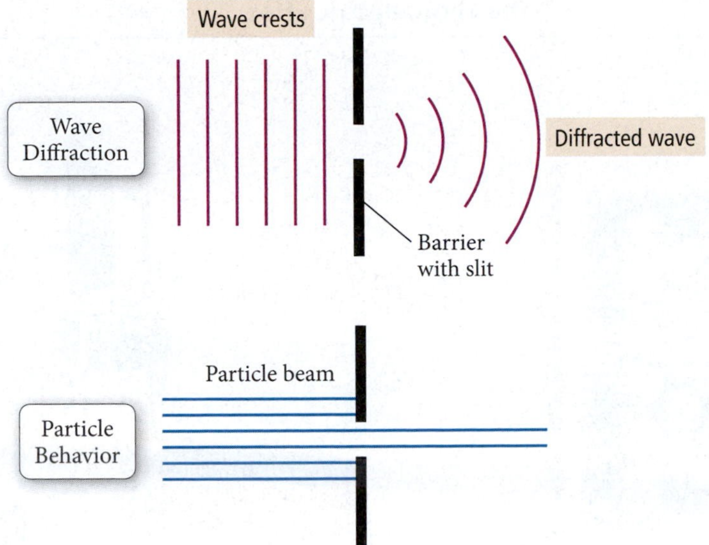

▲ **FIGURE 3.6 Diffraction** In this view from above, we can see how waves bend, or diffract, when they encounter an obstacle or slit with a size comparable to their wavelength. When a wave passes through a small opening, it spreads out. Particles, by contrast, do not diffract; they simply pass through the opening.

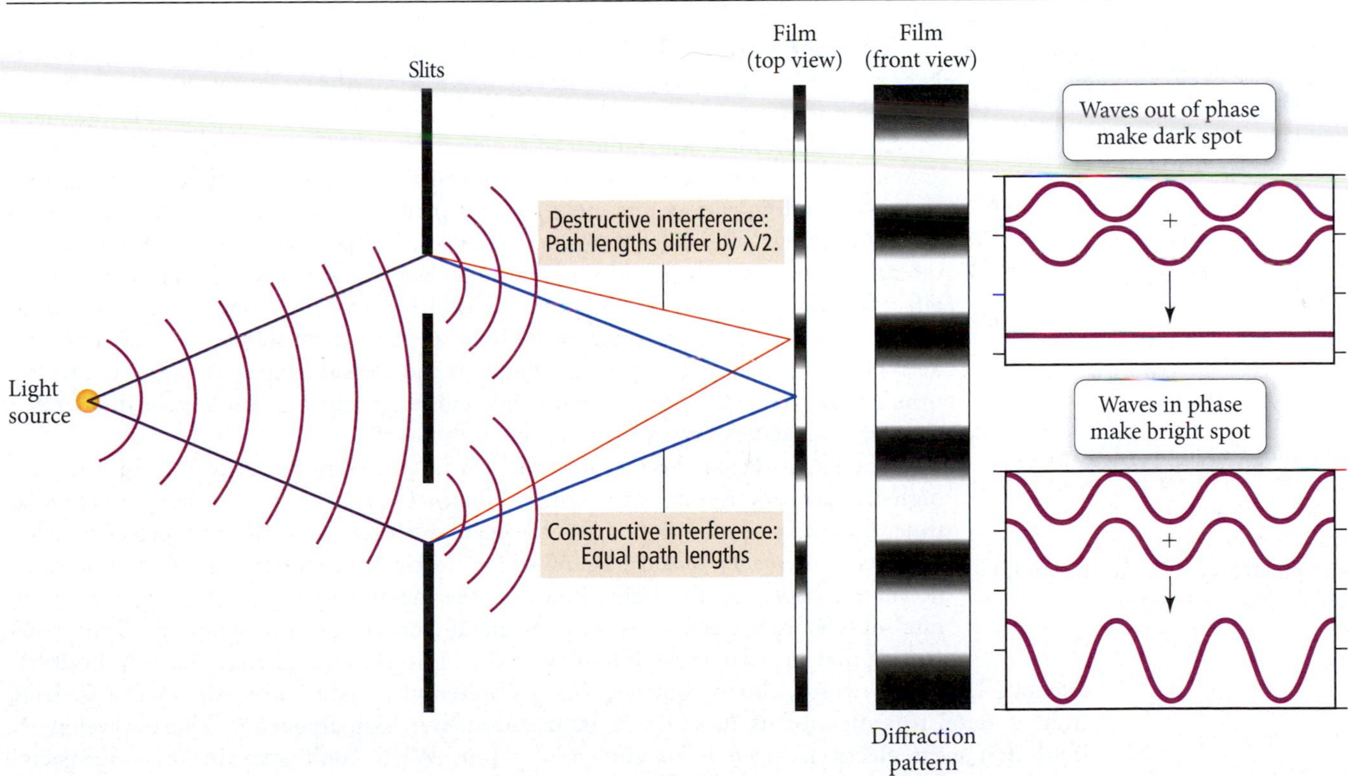

▲ **FIGURE 3.7 Interference from Two Slits** When a beam of light passes through two small slits, the two resulting waves interfere with each other. Whether the interference is constructive or destructive at any given point depends on the difference in the path lengths traveled by the waves. The resulting interference pattern appears as a series of bright and dark lines on a screen.

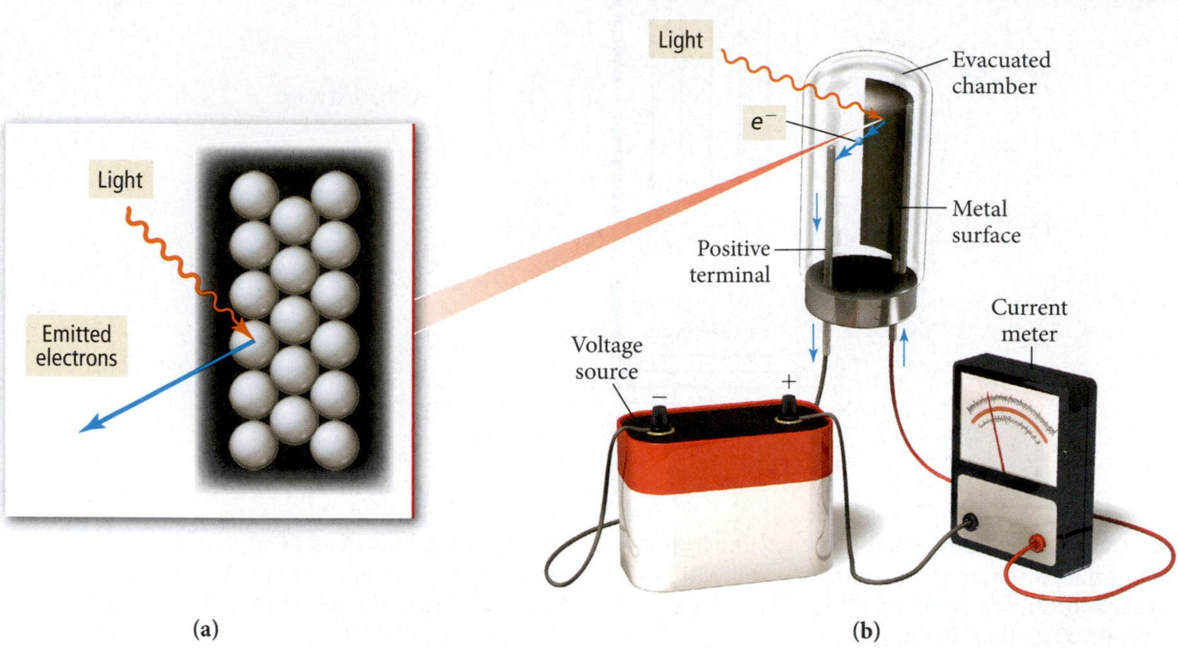

▲ **FIGURE 3.8 The Photoelectric Effect** **(a)** When sufficiently energetic light shines on a metal surface, the surface emits electrons. **(b)** The emitted electrons can be measured as an electrical current.

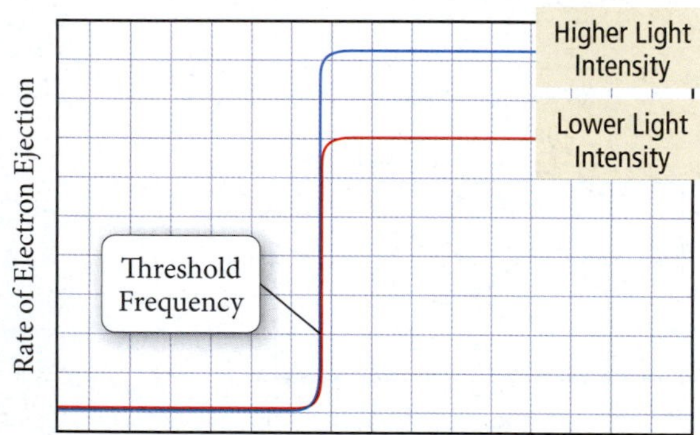

▲ **FIGURE 3.9 Threshold Frequency** A plot of the electron ejection rate versus frequency of light for the photoelectric effect. Electrons are only ejected when the energy of a photon exceeds the energy with which an electron is held to the metal. The frequency at which this occurs is the *threshold frequency*.

Einstein was not the first to suggest that energy was quantized. Max Planck used the idea in 1900 to account for certain characteristics of radiation from hot bodies. However, Planck did not suggest that light actually traveled in discrete packets.

The **photoelectric effect** is the observation that many metals emit electrons when light shines upon them, as shown in Figure 3.8 ▲. Classical electromagnetic theory attributed this effect to the transfer of energy from the light to an electron in the metal, which resulted in the dislodgment of the electron. If this explantion were correct, the amount of energy transferred from the light to the electron would have to exceed the electron's **binding energy**, the energy with which the electron is bound to the metal. Since the energy of a classical electromagnet wave depends only on its amplitude (or intensity), the rate at which electrons would leave the metal due to the photoelectric effect would depend only on the intensity of the light shining upon the surface (not on the wavelength). If the intensity of the light was low, there should be a *lag time* (or a delay) between the initial shining of the light and the subsequent emission of an electron. The lag time would be the minimum amount of time required for the dim light to transfer sufficient energy to the electron to dislodge it.

The experimental results, however, do not support the classical prediction. A high-frequency, low-intensity light produces electrons *without* the predicted lag time. Furthermore, the light used to dislodge electrons in the photoelectric effect exhibits a *threshold frequency*, below which no electrons are emitted from the metal, no matter how long the light shines on the metal. Figure 3.9 ◀ is a graph of the rate of electron ejection from the metal versus the frequency of light used. Notice that increasing the intensity of the light does not change the threshold frequency. In other words, low-frequency (long-wavelength) light *does not* eject electrons from a metal regardless of its intensity or its duration. But high-frequency (short-wavelength) light *does* eject electrons, even if its intensity is low. What could explain this unexpected behavior?

In 1905, Albert Einstein proposed a bold explanation for the photoelectric effect: *Light energy must come in packets*. According to Einstein, the amount of energy (E) in a light packet depends on its frequency (ν) according to the following equation:

$$E = h\nu \qquad [3.2]$$

where h, called *Planck's constant*, has the value $h = 6.626 \times 10^{-34}$ J · s. A *packet* of light is called a **photon** or a **quantum** of light. Since $\nu = c/\lambda$, the energy of a photon can also be expressed in terms of wavelength:

$$E = \frac{hc}{\lambda} \qquad [3.3]$$

The energy of a photon is directly proportional to its frequency.

The energy of a photon is inversely proportional to its wavelength.

Unlike classical electromagnetic theory, in which light was viewed purely as a wave whose intensity was *continuously variable*, Einstein suggested that light was *lumpy*. From this perspective, a beam of light is *not* a wave propagating through space, but a shower of particles (photons), each with energy $h\nu$.

EXAMPLE 3.2
Photon Energy

A nitrogen gas laser pulse with a wavelength of 337 nm contains 3.83 mJ of energy. How many photons does it contain?

SORT You are given the wavelength and total energy of a light pulse and asked to find the number of photons it contains.

GIVEN: $E_{pulse} = 3.83$ mJ
$\lambda = 337$ nm

FIND: number of photons

STRATEGIZE In the first part of the conceptual plan, calculate the energy of an individual photon from its wavelength.

CONCEPTUAL PLAN

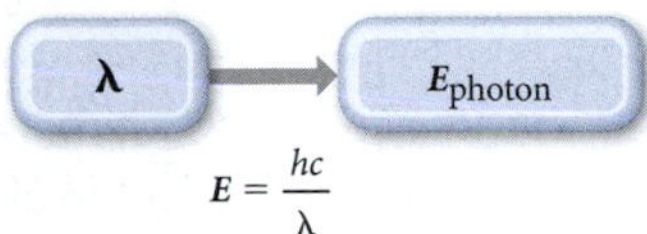

In the second part, divide the total energy of the pulse by the energy of a photon to get the number of photons in the pulse.

$$\frac{E_{pulse}}{E_{photon}} = \text{number of photons}$$

RELATIONSHIPS USED
$E = hc/\lambda$ (Equation 3.3)

SOLVE To execute the first part of the conceptual plan, convert the wavelength to meters and substitute it into Equation 3.3 to calculate the energy of a 337-nm photon.

SOLUTION

$$\lambda = 337\ \text{nm} \times \frac{10^{-9}\ \text{m}}{1\ \text{nm}} = 3.37 \times 10^{-7}\ \text{m}$$

$$E_{photon} = \frac{hc}{\lambda} = \frac{(6.626 \times 10^{-34}\ \text{J} \cdot \text{s})\left(3.00 \times 10^{8}\ \frac{\text{m}}{\text{s}}\right)}{3.37 \times 10^{-7}\ \text{m}}$$
$$= 5.8985 \times 10^{-19}\ \text{J}$$

To execute the second part of the conceptual plan, convert the energy of the pulse from mJ to J. Then divide the energy of the pulse by the energy of a photon to obtain the number of photons.

$$3.83\ \text{mJ} \times \frac{10^{-3}\ \text{J}}{1\ \text{mJ}} = 3.83 \times 10^{-3}\ \text{J}$$

$$\text{number of photons} = \frac{E_{pulse}}{E_{photon}} = \frac{3.83 \times 10^{-3}\ \text{J}}{5.8985 \times 10^{-19}\ \text{J}}$$
$$= 6.49 \times 10^{15}\ \text{photons}$$

CHECK The units of the answer, photons, are correct. The magnitude of the answer (10^{15}) is reasonable. Photons are small particles and any macroscopic collection should contain a large number of them.

FOR PRACTICE 3.2
A 100-watt light bulb radiates energy at a rate of 100 J/s. (The watt, a unit of power, or energy over time, is defined as 1 J/s.) If all of the light emitted has a wavelength of 525 nm, how many photons are emitted per second? (Assume three significant figures in this calculation.)

FOR MORE PRACTICE 3.2
The energy required to dislodge electrons from sodium metal via the photoelectric effect is 275 kJ/mol. What wavelength in nm of light has sufficient energy per photon to dislodge an electron from the surface of sodium?

EXAMPLE 3.3
Wavelength, Energy, and Frequency

Arrange these three types of electromagnetic radiation—visible light, X-rays, and microwaves—in order of increasing:

(a) wavelength **(b)** frequency **(c)** energy per photon

SOLUTION

Examine Figure 3.5 and note that X-rays have the shortest wavelength, followed by visible light and then microwaves.	**(a)** wavelength X-rays < visible < microwaves
Since frequency and wavelength are inversely proportional—the longer the wavelength, the shorter the frequency—the ordering with respect to frequency is the reverse of the order with respect to wavelength.	**(b)** frequency microwaves < visible < X-rays
Energy per photon decreases with increasing wavelength, but increases with increasing frequency; therefore the ordering with respect to energy per photon is the same as for frequency.	**(c)** energy per photon microwaves < visible < X-rays

FOR PRACTICE 3.3

Arrange these three colors of visible light—green, red, and blue—in order of increasing:

(a) wavelength **(b)** frequency **(c)** energy per photon

The symbol ϕ is the Greek letter phi, pronounced "fi."

Einstein's idea that light is *quantized* elegantly explains the photoelectric effect. The emission of electrons from the metal depends on whether or not a single photon has sufficient energy (as given by $h\nu$) to dislodge a single electron. For an electron bound to the metal with binding energy ϕ, the threshold frequency is reached when the energy of the photon is equal to ϕ.

Low-frequency light does not eject electrons because no single photon has the minimum energy necessary to dislodge the electron. We can draw an analogy between a photon ejecting an electron from a metal surface and a ball breaking a glass window. In this analogy, low-frequency photons are like ping-pong balls—a ping-pong ball thrown at a glass window does not break it (just as a low-frequency photon does not eject an electron). Increasing the *intensity* of low-frequency light is like increasing the number of ping-pong balls thrown at the window—doing so simply increases the number of low-energy photons but does not produce any single photon with sufficient energy. In contrast, increasing the *frequency* of the light, even at low intensity, *increases the energy of each photon*. In our analogy, a high-frequency photon is like a baseball—one baseball thrown at a glass window breaks it (just as a high-frequency photon dislodges an electron with no lag time).

Threshold frequency condition

$$h\nu = \phi$$

Energy of photon ($h\nu$); Binding energy of emitted electron (ϕ)

As the frequency of the light increases over the threshold frequency, the excess energy of the photon (beyond what is needed to dislodge the electron) transfers to the electron in the form of kinetic energy.

The kinetic energy (KE) of the ejected electron, therefore, is the difference between the energy of the photon ($h\nu$) and the binding energy of the electron, as given by the equation:

$$KE = h\nu - \phi$$

Although the quantization of light explains the photoelectric effect, the wave explanation of light continues to have explanatory power as well, depending on the circumstances of the particular observation. So the principle that slowly emerged (albeit with some measure of resistance) is what we now call the *wave–particle duality of light*. Sometimes light appears to behave like a wave, at other times like a particle. The behavior we observe depends on the particular experiment.

The Photoelectric Effect

3.2 Cc Conceptual Connection

We shine light of three different wavelengths—325 nm, 455 nm, and 632 nm—on a metal surface. The observations for each wavelength, labeled A, B, and C, are:

Observation A: No photoelectrons are observed.

Observation B: Photoelectrons with a kinetic energy of 155 kJ/mol are observed.

Observation C: Photoelectrons with a kinetic energy of 51 kJ/mol are observed.

Which observation corresponds to which wavelength of light?

3.3 Atomic Spectroscopy and the Bohr Model

The discovery of the particle nature of light began to break down the division that existed in nineteenth-century physics between electromagnetic radiation, which was thought of as a wave phenomenon, and the small particles (protons, neutrons, and electrons) that compose atoms, which were thought to follow Newton's laws of motion. Just as the photoelectric effect suggested the particle nature of light, so certain observations about atoms began to suggest a wave nature for particles. The most important of these observations came from *atomic spectroscopy*, the study of the electromagnetic radiation absorbed and emitted by atoms.

Atomic Spectra

When an atom absorbs energy—in the form of heat, light, or electricity—it often re-emits that energy as light. For example, a neon sign is composed of one or more glass tubes filled with neon gas. When an electric current is passed through the tube, the neon atoms absorb some of the electrical energy and re-emit it as the familiar red light of a neon sign. If the atoms in the tube are different (that is, not neon), they emit light of a different color. In other words, atoms of each element emit light of a characteristic color. Mercury atoms, for example, emit light that appears blue, helium atoms emit light that appears violet, and hydrogen atoms emit light that appears reddish (Figure 3.10 ▶).

Closer inspection of the light emitted by various atoms reveals that it contains several distinct wavelengths. We can separate the light emitted by a single element in a glass tube into its constituent wavelengths by passing it through

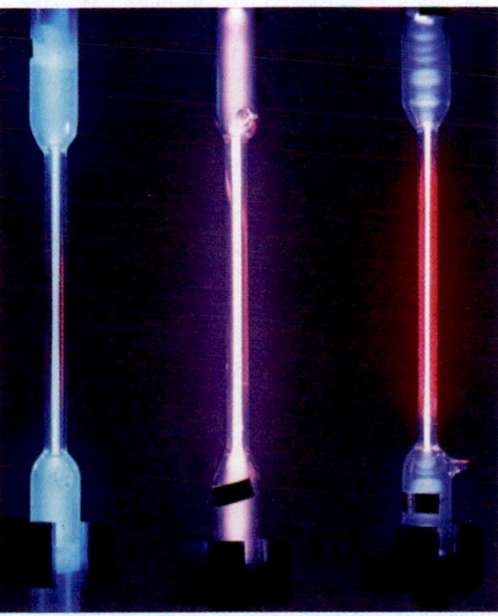

▲ **FIGURE 3.10 Mercury, Helium, and Hydrogen** Each element emits a characteristic color.

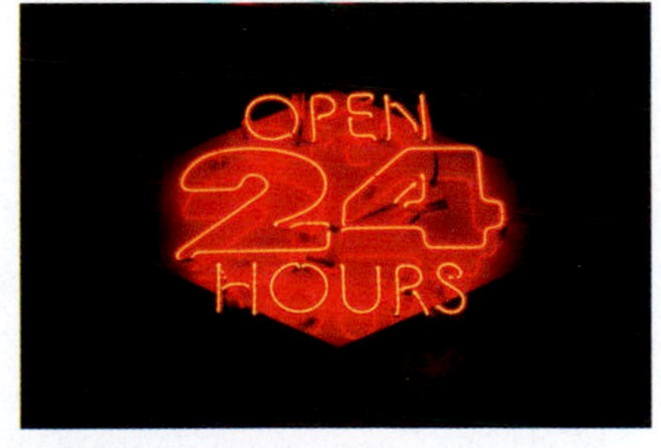

▲ The red light from a neon sign is emitted by neon atoms that have absorbed electrical energy, which they re-emit as visible radiation.

Remember that the color of visible light is determined by its wavelength.

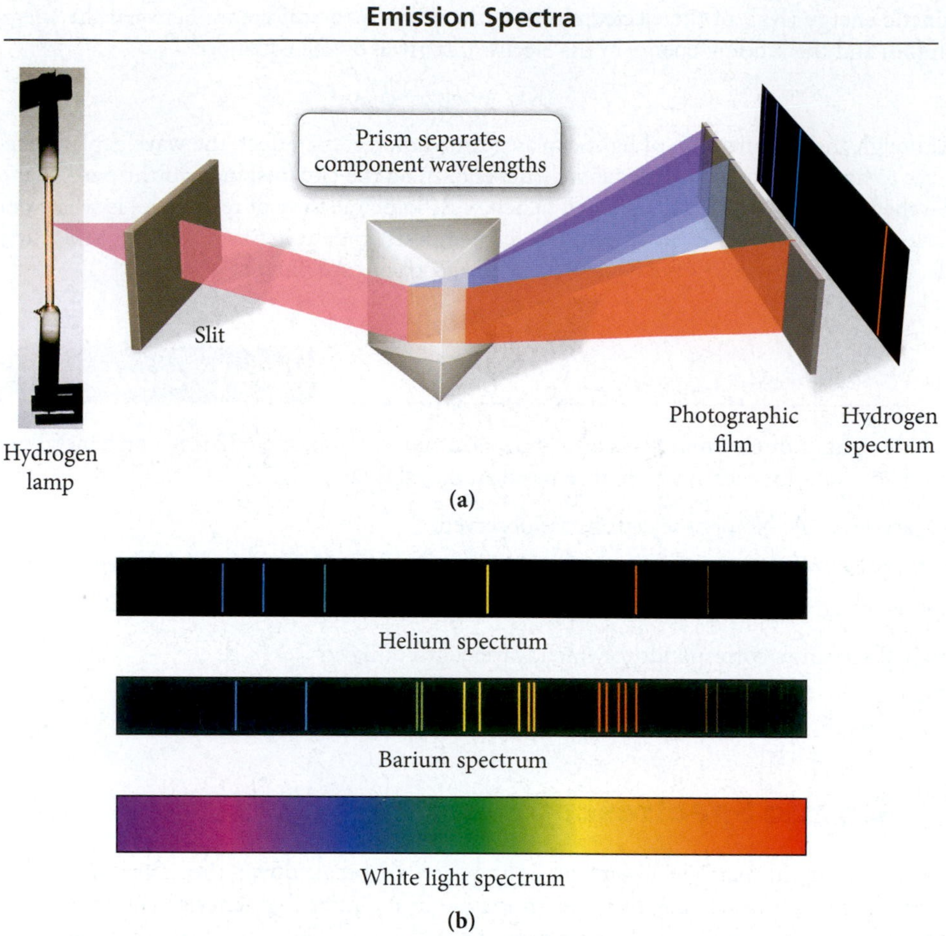

▶ **FIGURE 3.11 Emission Spectra** **(a)** The light emitted from a hydrogen, helium, or barium lamp consists of specific wavelengths, which can be separated by passing the light through a prism. **(b)** The resulting bright lines constitute an emission spectrum characteristic of the element that produced it.

a prism (just like we separate the white light from a light bulb), as shown in Figure 3.11a ▲. The result is a series of bright lines called an **emission spectrum**. The emission spectrum of a particular element is always the same—it consists of the same bright lines at the same characteristic wavelengths—and we can use it to identify the element. For example, light arriving from a distant star contains the emission spectra of the elements that compose the star. Analysis of the light allows us to identify the elements present in the star.

Notice the differences between the white light spectrum and the emission spectra of hydrogen, helium, and barium in Figure 3.11b. The white light spectrum is *continuous*, meaning that there are no sudden interruptions in the intensity of the light as a function of wavelength—the spectrum consists of light of all wavelengths. The emission spectra of hydrogen, helium, and barium, however, are not continuous—they consist of bright lines at specific wavelengths, with complete darkness in between. That is, only certain discrete wavelengths of light are present. Classical physics could not explain why these spectra consisted of discrete lines. In fact, according to classical physics, an atom composed of an electron orbiting a nucleus should emit a continuous white light spectrum. Even more problematic, the electron should lose energy as it emits the light and spiral into the nucleus. According to classical physics, an atom should not even be stable.

The Rydberg equation is $1/\lambda = R(1/m^2 - 1/n^2)$, where R is the Rydberg constant ($1.097 \times 10^7\ \text{m}^{-1}$), and m and n are integers.

Johannes Rydberg, a Swedish mathematician, analyzed many atomic spectra and developed an equation that predicts the wavelengths of the hydrogen emission spectrum. Although these predictions are accurate, Rydberg's equation (shown in the margin) gives little insight into *why* atomic spectra are discrete, *why* atoms are stable, or *why* his equation works.

The Bohr Model

The Danish physicist Niels Bohr (1885–1962) attempted to develop a model for the atom that explained atomic spectra. In his model, electrons travel around the nucleus in circular orbits (analogous to those of the planets around the sun). However, in contrast to planetary orbits—which can theoretically exist at any distance from the sun—Bohr's orbits exist only at specific, fixed

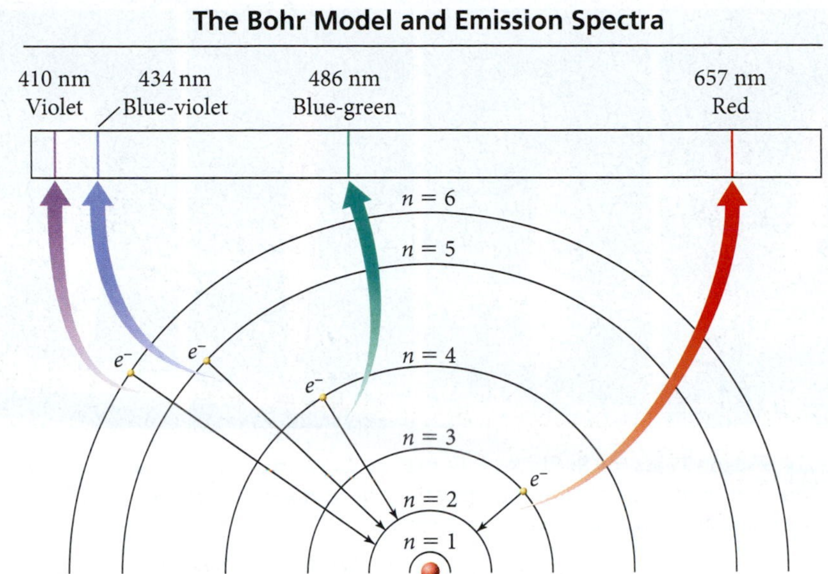

◀ **FIGURE 3.12 The Bohr Model and Emission Spectra for the Hydrogen Atom** According to the Bohr model, each spectral line is produced when an electron falls from one stable orbit, or stationary state, to another of lower energy.

distances from the nucleus. The energy of each Bohr orbit is also fixed, or quantized. Bohr called these orbits *stationary states* and suggested that, although they obey the laws of classical mechanics, they also possess "a peculiar, mechanically unexplainable, stability." We now know that the stationary states were really manifestations of the wave nature of the electron, which we will expand upon shortly. Bohr further proposed that, in contradiction to classical electromagnetic theory, no radiation is emitted by an electron orbiting the nucleus in a stationary state. It is only when an electron jumps, or makes a *transition*, from one stationary state to another that radiation is emitted or absorbed (Figure 3.12 ▲).

The transitions between stationary states in a hydrogen atom are quite unlike any transitions that we might be familiar with in the macroscopic world. The electron is *never* observed *between states;* it is observed only in one state or another. The emission spectrum of an atom consists of discrete lines because the stationary states exist only at specific, fixed energies. The energy of the photon emitted when an electron makes a transition from one stationary state to another is the energy difference between the two stationary states. Transitions between stationary states that are closer together, therefore, produce light of lower energy (longer wavelength) than transitions between stationary states that are farther apart.

In spite of its initial success in explaining the line spectrum of hydrogen, the Bohr model left many unanswered questions. It did, however, serve as an intermediate model between a classical view of the electron and a fully quantum-mechanical view, and therefore has great historical and conceptual importance. Nonetheless, it was ultimately replaced by a more complete quantum-mechanical theory that fully incorporated the wave nature of the electron. We examine this theory in Section 3.4.

Atomic Spectroscopy and the Identification of Elements

When you check out of the grocery store, a laser scanner reads the barcode on the items that you buy. Each item has a unique code that identifies the item and its price. Similarly, each element in the periodic table has a spectrum unlike that of any other element. Figure 3.13 ▼ shows the emission spectra of oxygen and neon. (In Figure 3.11, we saw the emission spectra of hydrogen, helium, and barium.) Each spectrum is unique and, as such, can be used to identify the substance.

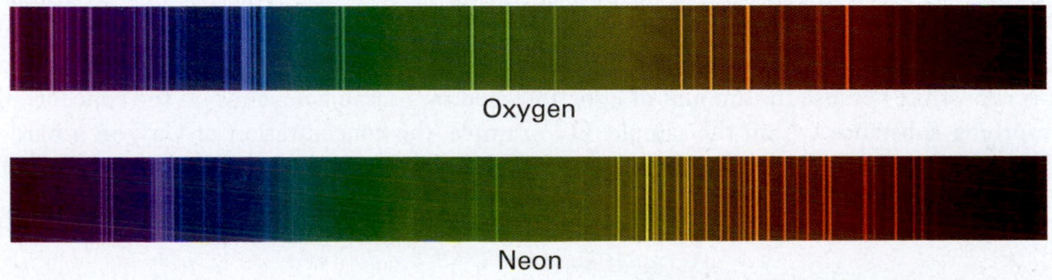

◀ **FIGURE 3.13 Emission Spectra of Oxygen and Neon** The emission spectrum of each element is unique, and we can use it to identify the element.

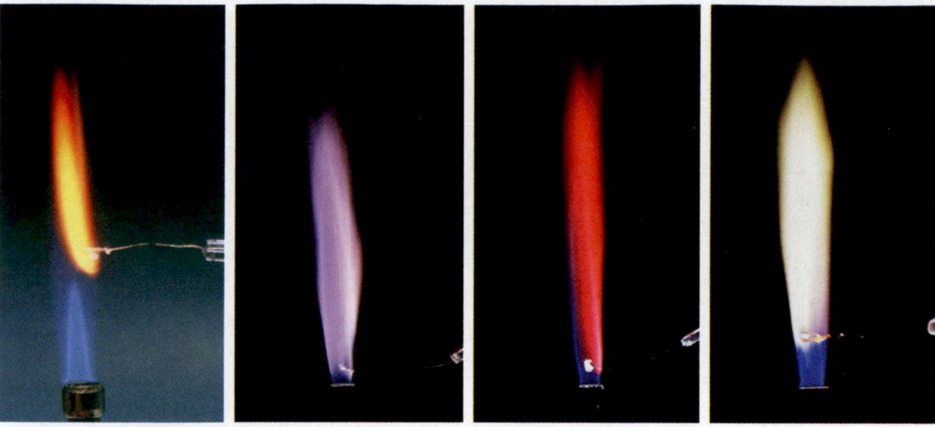

▲ **FIGURE 3.14 Flame Tests for Sodium, Potassium, Lithium, and Barium** We can identify elements by the characteristic color of the light they produce when heated. The colors derive from especially bright lines in their emission spectra.

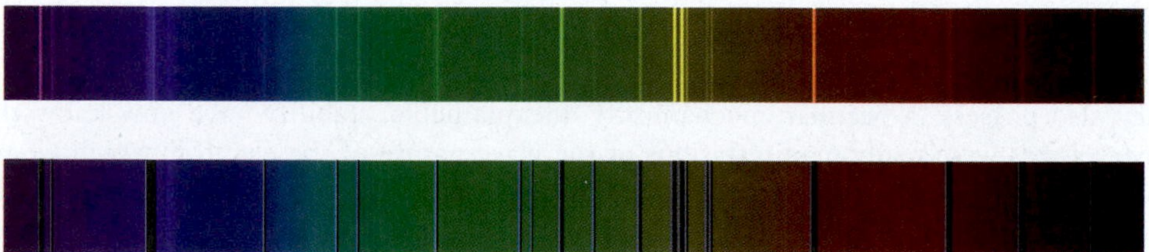

▲ **FIGURE 3.15 Emission and Absorption Spectrum of Mercury** Elements absorb light of the same wavelengths that they radiate when heated. When these wavelengths are subtracted from a beam of white light, the result is a pattern of dark lines corresponding to an absorption spectrum.

The presence of intense lines in the spectra of a number of metals is the basis for *flame tests*, simple tests used to identify elements in ionic compounds in the absence of a precise analysis of a compound's spectrum. For example, the emission spectrum of sodium features two closely spaced, bright yellow lines. When a crystal of a sodium salt (or a drop of a solution containing a sodium salt) is put into a flame, the flame glows bright yellow (Figure 3.14 ▲). As Figure 3.14 shows, other metals exhibit similarly characteristic colors in flame tests. Each color represents an especially bright spectral emission line (or a combination of two or more such lines). Similar emissions form the basis of the colors seen in fireworks.

▲ Fireworks typically contain the salts of such metals as sodium, calcium, strontium, barium, and copper. Emissions from these elements produce the brilliant colors of pyrotechnic displays.

Although the *emission* of light from elements is easier to detect, the *absorption* of light by elements is even more commonly used for purposes of identification. Whereas an emission spectrum consists of bright lines on a dark background, an **absorption spectrum** consists of dark lines on a bright background (Figure 3.15 ▲). An absorption spectrum is measured by passing white light through a sample and observing what wavelengths are *missing* due to absorption by the sample. Notice that, in the spectra of mercury in Figure 3.15, the absorption lines are at the same wavelengths as the emission lines. This is because the processes that produce them are mirror images. In emission, an electron makes a transition from a higher energy level to a lower energy one. In absorption, the transition is between the same two energy levels, but from the lower level to the higher one.

Absorption spectrometers, found in most chemistry laboratories, typically plot the intensity of absorption as a function of wavelength. Such plots are useful both for identifying substances (qualitative analysis) and for determining the concentration of substances (quantitative analysis). Quantitative analysis is possible because the amount of light absorbed by a sample depends on the concentration of the absorbing substance within the sample. For example, the concentration of Ca^{2+} in a hard water sample can be determined by measuring the quantity of light absorbed by the calcium ion at its characteristic wavelength.

3.4 The Wave Nature of Matter: The de Broglie Wavelength, the Uncertainty Principle, and Indeterminacy

KEY CONCEPT VIDEO
The wave nature of matter

The heart of the quantum-mechanical theory that replaced Bohr's model is the wave nature of the electron, first proposed by Louis de Broglie (1892–1987) in 1924 and later confirmed by experiments in 1927. It seemed incredible at the time, but electrons—which were then thought of only as particles and known to have mass—also have a wave nature. The wave nature of the electron is seen most clearly in its diffraction. If an electron beam is aimed at two closely spaced slits, and a series (or array) of detectors is arranged to detect the electrons after they pass through the slits, an interference pattern similar to that observed for light is recorded behind the slits (Figure 3.16a ▼). The detectors at the center of the array (midway between the two slits) detect a large number of electrons—exactly the opposite of what you would expect for particles (Figure 3.16b ▼). Moving outward from this center spot, the detectors alternately detect small numbers of electrons and then large numbers again and so on, forming an interference pattern characteristic of waves.

The first evidence of electron wave properties was provided by the Davisson-Germer experiment of 1927, in which electrons were observed to undergo diffraction by a metal crystal.

For interference to occur, the spacing of the slits has to be on the order of atomic dimensions.

Counter to what might be our initial intuition about electron interference, the interference pattern is *not caused by pairs of electrons interfering with each other, but rather by single electrons interfering with themselves*. If the electron source is turned down to a very low level, so that electrons come out only one at a time, *the interference pattern remains*. In other words, we can design an experiment in

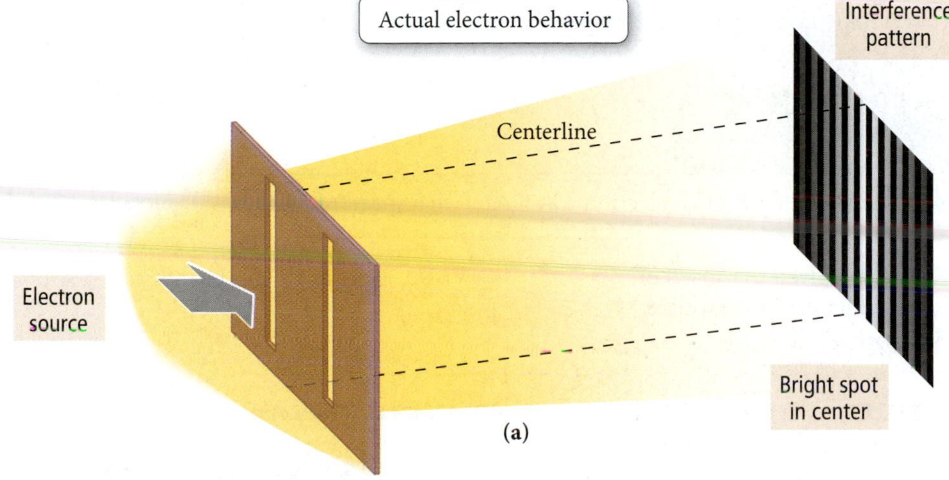

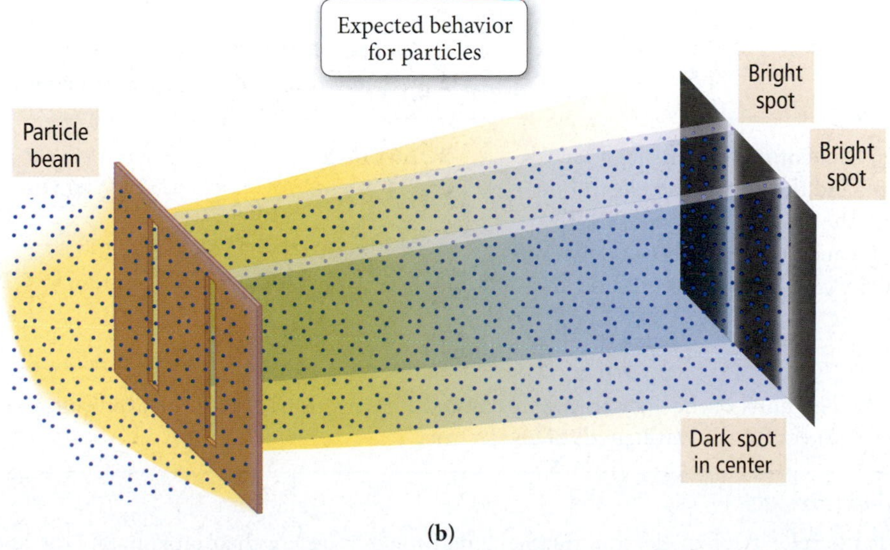

◀ **FIGURE 3.16 Electron Diffraction** When a beam of electrons goes through two closely spaced slits **(a)**, an interference pattern is created, as if the electrons were waves. By contrast, a beam of particles passing through two slits **(b)** produces two smaller beams of particles. Particle beams produce two bright stripes with darkness in between, but waves produce the brightest strip directly in the center of the screen.

which electrons come out of the source singly. We can then record where each electron strikes the detector after it has passed through the slits. If we individually record the positions of thousands of electrons over a long period of time, we find the same interference pattern shown in Figure 3.16a. This leads us to an important conclusion: *The wave nature of the electron is an inherent property of individual electrons.* In this case, the unobserved electron goes through both slits—it exists in two states simultaneously, just like Schrödinger's cat—and interferes with itself. As it turns out, this wave nature explains the existence of stationary states (in the Bohr model) and prevents the electrons in an atom from crashing into the nucleus as predicted by classical physics. We now turn to three important manifestations of the electron's wave nature: the de Broglie wavelength, the uncertainty principle, and indeterminacy.

The de Broglie Wavelength

As we have seen, a single electron traveling through space has a wave nature; its wavelength is related to its kinetic energy (the energy associated with its motion). The faster the electron is moving, the higher its kinetic energy and the shorter its wavelength. The wavelength (λ) of an electron of mass m moving at velocity v is given by the **de Broglie relation**:

The mass of an object (m) times its velocity (v) is its momentum. Therefore, the wavelength of an electron is inversely proportional to its momentum.

$$\lambda = \frac{h}{mv} \quad \text{de Broglie relation} \qquad [3.4]$$

where h is Planck's constant. *Notice that the velocity of a moving electron is related to its wavelength—knowing one is equivalent to knowing the other.*

EXAMPLE 3.4 De Broglie Wavelength

Calculate the wavelength of an electron traveling with a speed of 2.65×10^6 m/s.

SORT You are given the speed of an electron and asked to calculate its wavelength.

GIVEN: $v = 2.65 \times 10^6$ m/s

FIND: λ

STRATEGIZE The conceptual plan shows how the de Broglie relation relates the wavelength of an electron to its mass and velocity.

CONCEPTUAL PLAN

$v \rightarrow \lambda$

$\lambda = \frac{h}{mv}$

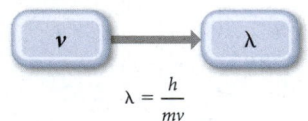

RELATIONSHIPS USED

$\lambda = h/mv$ (de Broglie relation, Equation 3.4)

SOLVE Substitute the velocity, Planck's constant, and the mass of an electron to calculate the electron's wavelength. To correctly cancel the units, break down the J in Planck's constant into its SI base units ($1\text{ J} = 1\text{ kg}\cdot\text{m}^2/\text{s}^2$).

SOLUTION

$$\lambda = \frac{h}{mv} = \frac{6.626 \times 10^{-34} \frac{\text{kg}\cdot\text{m}^2}{\text{s}^2}\text{s}}{(9.11 \times 10^{-31}\text{ kg})2.65 \times 10^6 \frac{\text{m}}{\text{s}}}$$

$$= 2.74 \times 10^{-10}\text{ m}$$

CHECK The units of the answer (m) are correct. The magnitude of the answer is very small, as we would expect for the wavelength of an electron.

FOR PRACTICE 3.4

What is the velocity of an electron that has a de Broglie wavelength approximately the length of a chemical bond? Assume this length to be 1.2×10^{-10} m.

The de Broglie Wavelength of Macroscopic Objects

3.3 Cc Conceptual Connection

Since quantum-mechanical theory is universal, it applies to all objects, regardless of size. Therefore, according to the de Broglie relation, a thrown baseball should also exhibit wave properties. Why don't we observe such properties at the ballpark?

The Uncertainty Principle

The wave nature of the electron is difficult to reconcile with its particle nature. How can a single entity behave as both a wave and a particle? We can begin to address this question by returning to the single-electron diffraction experiment. Specifically, we can ask the question: How does a single electron aimed at a double slit produce an interference pattern? We stated previously that the electron travels through both slits and interferes with itself. This idea is testable. We simply have to observe the single electron as it travels through both of the slits. If it travels through both slits simultaneously, our hypothesis is correct. But here is where nature gets tricky.

Any experiment designed to observe the electron as it travels through the slits results in the detection of an electron "particle" traveling through a single slit and no interference pattern. Recall from Section 3.1 that an *unobserved* electron can occupy two different states; however, the act of observation forces it into one state or the other. Similarly, the act of observing the electron as it travels through both slits forces it to go through only one slit. The electron diffraction experiment shown here is designed to observe which slit the electron travels through by using a laser beam placed directly behind the slits.

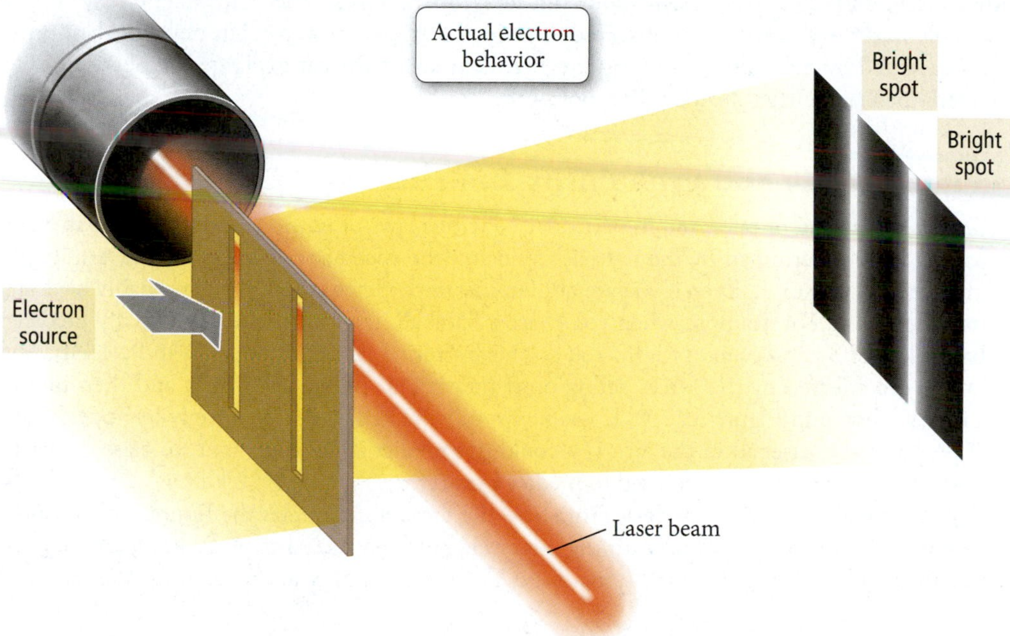

An electron that crosses the laser beam produces a tiny "flash" when a single photon is scattered at the point of crossing. If a flash shows up behind a particular slit, that indicates an electron is passing through that slit. When the experiment is performed, the flash always originates either from one slit *or* the other, but *never* from both at once. Furthermore, the interference pattern, which was present without the laser, is now absent. With the laser on, the electrons hit positions directly behind each slit, as if they were ordinary particles; their wave-like behavior is no longer manifested.

As it turns out, no matter how hard we try, or whatever method we set up, *we can never both see the interference pattern and simultaneously determine which hole the electron goes through*. It has never been done, and most scientists agree that it never will. In the words of P. A. M. Dirac,

> There is a limit to the fineness of our powers of observation and the smallness of the accompanying disturbance—a limit which is inherent in the nature of things and can never be surpassed by improved technique or increased skill on the part of the observer.

The single-electron diffraction experiment demonstrates that we cannot simultaneously observe both the wave nature and the particle nature of the electron. When we try to observe which hole the electron

▲ Werner Heisenberg (1901–1976)

Remember that velocity includes speed as well as direction of travel.

goes through (associated with the particle nature of the electron), we lose the interference pattern (associated with the wave nature of the electron). When we try to observe the interference pattern, we cannot determine which hole the electron goes through. The wave nature and particle nature of the electron are said to be **complementary properties**. Complementary properties exclude one another—the more we know about one, the less we know about the other. Which of two complementary properties we observe depends on the experiment we perform—in quantum mechanics, the observation of an event affects its outcome.

As we just saw in the de Broglie relation, the *velocity* of an electron is related to its *wave nature*. The *position* of an electron, however, is related to its *particle nature*. (Particles have well-defined positions, but waves do not.) Consequently, our inability to observe the electron simultaneously as both a particle and a wave means that *we cannot simultaneously measure its position and its velocity with infinite precision*. Werner Heisenberg formalized this idea with the equation:

$$\Delta x \times m\Delta \nu \geq \frac{h}{4\pi} \quad \text{Heisenberg's uncertainty principle} \qquad [3.5]$$

where Δx is the uncertainty in the position, $\Delta \nu$ is the uncertainty in the velocity, m is the mass of the particle, and h is Planck's constant. **Heisenberg's uncertainty principle** states that the product of Δx and $m\,\Delta \nu$ must be greater than or equal to a finite number ($h/4\pi$). In other words, the more accurately you know the position of an electron (the smaller Δx), the less accurately you can know its velocity (the bigger $\Delta \nu$) and vice versa. The complementarity of the wave nature and particle nature of the electron results in the complementarity of velocity and position.

Although Heisenberg's uncertainty principle may seem puzzling, it actually solves a great puzzle. Without the uncertainty principle, we are left with a paradox: How can something be *both* a particle and a wave? Saying that an object is both a particle and a wave is like saying that an object is both a circle and a square—a contradiction. Heisenberg solved the contradiction by introducing complementarity—an electron is observed as *either* a particle or a wave, but never both at once. This idea is captured by Schrödinger's thought experiment about the cat explained in Section 3.1: When observed, the cat is either dead or alive, not both.

Indeterminacy and Probability Distribution Maps

According to classical physics, and in particular Newton's laws of motion, particles move in a *trajectory* (or path) that is determined by the particle's velocity (the speed and direction of travel), its position, and the forces acting on it. Even if you are not familiar with Newton's laws, you probably have an intuitive sense of them. For example, when you chase a baseball in the outfield, you visually predict where the ball will land by observing its path. You do this by noting its initial position and velocity, watching how these are affected by the forces acting on it (gravity, air resistance, wind), and then inferring its trajectory, as shown in Figure 3.17 ▼. If you knew only the ball's velocity, or only its position (imagine a still photo of the baseball in the air), you could not predict its landing spot. In classical mechanics, both position and velocity are required to predict a trajectory.

Newton's laws of motion are **deterministic**—the present *determines* the future. This means that if two baseballs are hit consecutively with the same velocity from the same position under identical conditions, they will land in exactly the same place. The same is not true of electrons. We have just seen

The Classical Concept of Trajectory

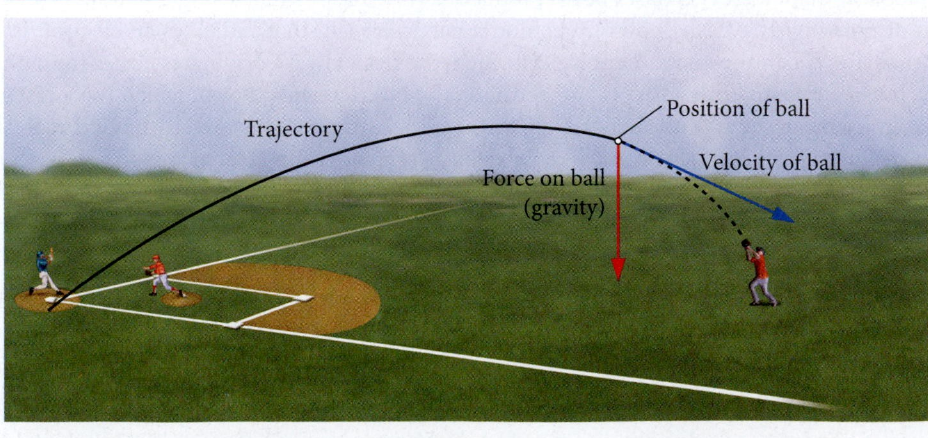

▶ **FIGURE 3.17 The Concept of Trajectory** In classical mechanics, the position and velocity of a particle determine its future trajectory, or path. Thus, an outfielder can catch a baseball by observing its position and velocity, allowing for the effects of forces acting on it, such as gravity, and estimating its trajectory. (For simplicity, air resistance and wind are not shown.)

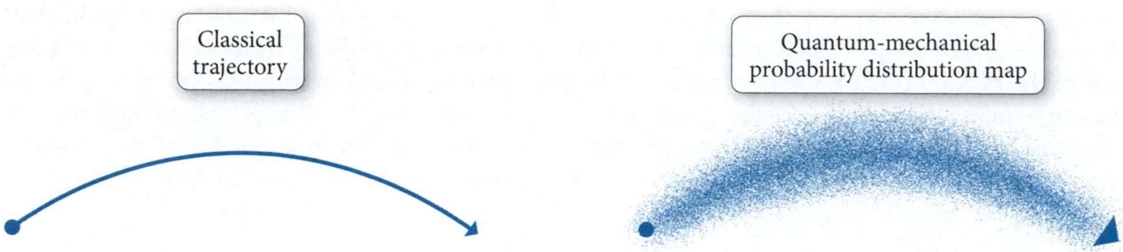

▲ **FIGURE 3.18 Trajectory versus Probability** In quantum mechanics, we cannot calculate deterministic trajectories. Instead, it is necessary to think in terms of probability maps: statistical pictures of where a quantum-mechanical particle, such as an electron, is most likely to be found. In this hypothetical map, darker shading indicates greater probability.

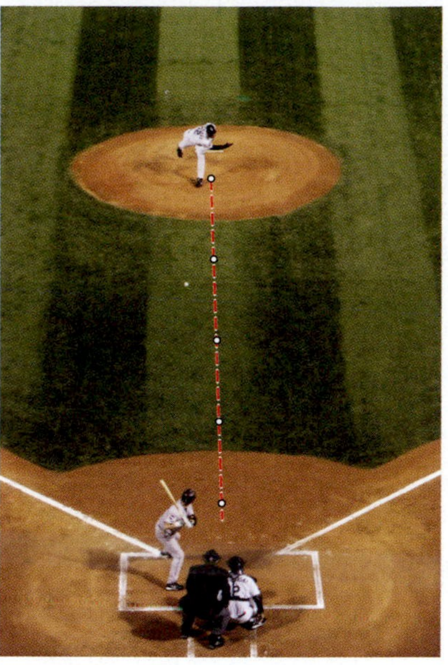

▲ **FIGURE 3.19 Trajectory of a Macroscopic Object** A baseball follows a well-defined trajectory from the hand of the pitcher to the mitt of the catcher.

that we cannot simultaneously know the position and velocity of an electron; therefore, we cannot know its trajectory. In quantum mechanics, trajectories are replaced with *probability distribution maps*, as shown in Figure 3.18 ▲. A probability distribution map is a statistical map that shows where an electron is likely to be found under a given set of conditions.

To understand the concept of a probability distribution map, let us return to baseball. Imagine a baseball thrown from the pitcher's mound to a catcher behind home plate (Figure 3.19 ▶.) The catcher can watch the baseball's path, predict exactly where it will cross home plate, and place his mitt in the correct place to catch it. As we have seen, the same predictions cannot be made for an electron. If an electron were thrown from the pitcher's mound to home plate, it would land in a different place every time, even if it were thrown in exactly the same way. This behavior is called **indeterminacy**. Unlike a baseball, whose future path is *determined* by its position and velocity when it leaves the pitcher's hand, the future path of an electron is indeterminate and can only be described statistically.

In the quantum-mechanical world of the electron, the catcher cannot know exactly where the electron will cross the plate for any given throw. However, if he were to record hundreds of identical electron throws, the catcher would observe a reproducible, *statistical pattern* of where the electron crosses the plate. He could even draw a map of the strike zone showing the probability of an electron crossing a certain area, as shown in Figure 3.20 ▼. This would be a probability distribution map. In the sections that follow, we discuss quantum-mechanical electron *orbitals*, which are essentially probability distribution maps for electrons as they exist within atoms.

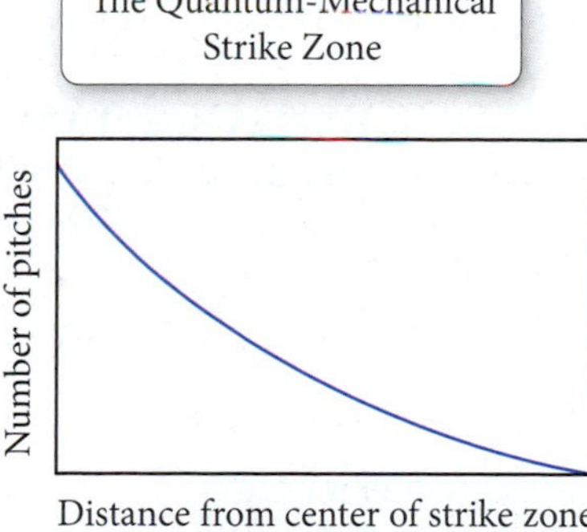

◀ **FIGURE 3.20 The Quantum-Mechanical Strike Zone** An electron does not have a well-defined trajectory. However, we can construct a probability distribution map to show the relative probability of it crossing home plate at different points.

3.5 Quantum Mechanics and the Atom

As we have seen, the position and velocity of the electron are complementary properties—if we know one accurately, the other becomes indeterminate. Since velocity is directly related to energy (recall that kinetic energy equals $\frac{1}{2}mv^2$), position and *energy* are also complementary properties—the more we know about one, the less we know about the other. Many of the properties of an element, however, depend on the energies of its electrons. In the following paragraphs, we describe the probability distribution maps for electron states in which the electron has well-defined energy, but not well-defined position. In other words, for each

These states are known as energy *eigenstates.*

of these states, we can specify the *energy* of the electron precisely, but not its location at a given instant. Instead, the electron's position is described in terms of an **orbital**, a probability distribution map showing where the electron is likely to be found. Since chemical bonding often involves the sharing of electrons between atoms (see Section 5.2), the spatial distribution of atomic electrons is important to bonding.

The mathematical derivation of energies and orbitals for electrons in atoms comes from solving the *Schrödinger* equation for the atom of interest. The general form of the Schrödinger equation is:

An operator is different from a normal algebraic entity. In general, an operator transforms a mathematical function into another mathematical function. For example, d/dx is an operator that means "take the derivative of." When d/dx operates on a function (such as x^2), it returns another function ($2x$).

$$H\psi = E\psi$$

The symbol H stands for the Hamiltonian operator, a set of mathematical operations that represents the total energy (kinetic and potential) of the electron within the atom. The symbol E is the actual energy of the electron. The symbol ψ is the **wave function**, a mathematical function that describes the wavelike nature of the electron. A plot of the wave function squared (ψ^2) represents an orbital, a probability density distribution map of the electron.

The symbol ψ is the Greek letter psi, pronounced "sigh."

Solutions to the Schrödinger Equation for the Hydrogen Atom

When the Schrödinger equation is solved, it yields many solutions—many possible wave functions. The wave functions themselves are fairly complicated mathematical functions, and we do not examine them in detail in this book. Instead, we will introduce graphical representations (or plots) of the orbitals that correspond to the wave functions. Each orbital is specified by three interrelated **quantum numbers:** n, the **principal quantum number**; l, the **angular momentum quantum number** (sometimes called the *azimuthal quantum number*); and m_l, the **magnetic quantum number**. These quantum numbers all have integer values, as had been hinted at by both the Rydberg equation and Bohr's model. A fourth quantum number, m_s, the **spin quantum number**, specifies the orientation of the spin of the electron. We examine each of these quantum numbers individually.

The Principal Quantum Number (n)

The principal quantum number is an integer that determines the overall size and energy of an orbital. Its possible values are $n = 1, 2, 3, \ldots$ and so on. For the hydrogen atom, the energy of an electron in an orbital with quantum number n is given by:

$$E_n = -2.18 \times 10^{-18}\ \text{J}\left(\frac{1}{n^2}\right) \qquad (n = 1, 2, 3, \ldots) \qquad [3.6]$$

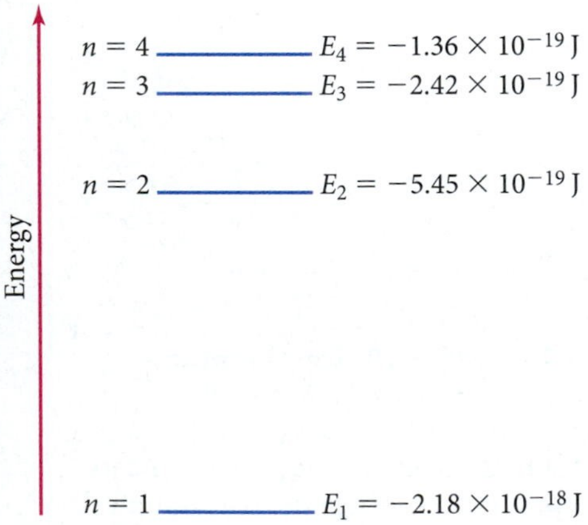

$n = 4$ — $E_4 = -1.36 \times 10^{-19}$ J
$n = 3$ — $E_3 = -2.42 \times 10^{-19}$ J
$n = 2$ — $E_2 = -5.45 \times 10^{-19}$ J
$n = 1$ — $E_1 = -2.18 \times 10^{-18}$ J
Energy

The energy is negative because the electron's energy is lowered (made more negative) by its interaction with the nucleus (see the description of Coulomb's law in Section 4.3). The constant, 2.18×10^{-18} J, is known as the Rydberg constant for hydrogen (R_H). Notice that orbitals with higher values of n have greater (less negative) energies, as shown in the energy level diagram at left. Notice also that, as n increases, the spacing between the energy levels becomes smaller.

The Angular Momentum Quantum Number (l)

The angular momentum quantum number is an integer that determines the shape of the orbital. We consider these shapes in Section 3.6. The possible values of l are $0, 1, 2, \ldots, (n - 1)$. In other words, for a given value of n, l can be any integer (including 0) up to $n - 1$. For example, if $n = 1$, then the only possible value of l is 0; if $n = 2$, the possible values of l are 0 and 1. In order to avoid confusion between n and l, values of l are often assigned letters as follows:

Value of l	Letter Designation
$l = 0$	*s*
$l = 1$	*p*
$l = 2$	*d*
$l = 3$	*f*

The values of l beyond 3 are designated with letters in alphabetical order so that $l = 4$ is designated *g*, $l = 5$ is designated *h*, and so on.

The Relationship between n and l

3.4 Cc Conceptual Connection

What values of l are possible for $n = 3$?

(a) 0 (or s) **(b)** 0 and 1 (or s and p)
(c) 0, 1, and 2 (or s, p, and d) **(d)** 0, 1, 2, and 3 (or s, p, d, and f)

The Magnetic Quantum Number (m_l) The magnetic quantum number is an integer that specifies the orientation of the orbital. We will consider these orientations in Section 3.6. The possible values of m_l are the integer values (including zero) ranging from $-l$ to $+l$. For example, if $l = 0$, then the only possible value of m_l is 0; if $l = 1$, the possible values of m_l are -1, 0, and $+1$; if $l = 2$, the possible values of m_l are $-2, -1, 0, +1, +2$, and so on.

The Relationship between l and m_l

3.5 Cc Conceptual Connection

What values of m_l are possible for $l = 2$?

(a) 0, 1, and 2 **(b)** 0 **(c)** -1, 0, and $+1$ **(d)** $-2, -1, 0, +1$, and $+2$

The Spin Quantum Number (m_s) The spin quantum number specifies the orientation of the *spin* of the electron. **Electron spin** is a fundamental property of an electron (like its negative charge). One electron does not have more or less spin than another—all electrons have the same amount of spin. The orientation of the electron's spin is quantized, with only two possibilities that we can call spin up ($m_s = +\frac{1}{2}$) and spin down ($m_s = -\frac{1}{2}$). The spin quantum becomes important in Section 4.3 when we begin to consider how electrons occupy orbitals. For now, we will focus on the first three quantum numbers.

The idea of a "spinning" electron is something of a metaphor. A more correct way to express the same idea is to say that an electron has inherent angular momentum.

Each specific combination of the first three quantum numbers (n, l, and m_l) specifies one atomic orbital. For example, the orbital with $n = 1$, $l = 0$, and $m_l = 0$ is known as the 1s orbital. The 1 in 1s is the value of n, and the s specifies that $l = 0$. There is only one 1s orbital in an atom, and its m_l value is zero. Orbitals with the same value of n are said to be in the same **principal level** (or **principal shell**). Orbitals with the same value of n and l are said to be in the same **sublevel** (or **subshell**). The diagram at right shows all of the orbitals, each represented by a small square, in the first three principal levels.

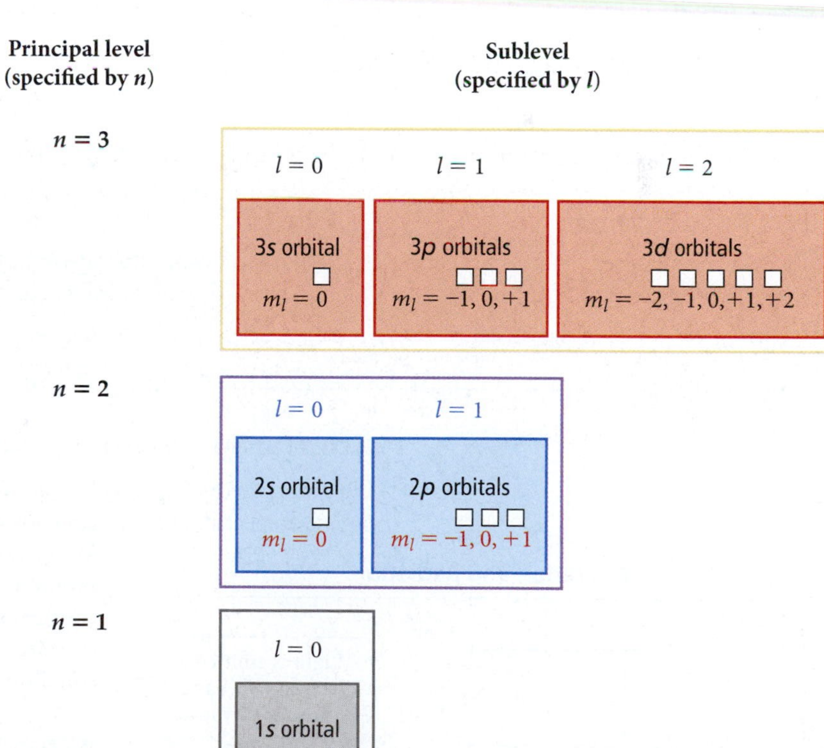

For example, the $n = 2$ level contains the $l = 0$ and $l = 1$ sublevels. Within the $n = 2$ level, the $l = 0$ sublevel—called the 2s sublevel—contains only one orbital (the 2s orbital), with $m_l = 0$. The $l = 1$ sublevel—called the 2p sublevel—contains three 2p orbitals, with $m_l = -1, 0, +1$.

In general, notice the following:

- The number of sublevels in any level is equal to n, the principal quantum number. Therefore, the $n = 1$ level has one sublevel, the $n = 2$ level has two sublevels, etc.
- The number of orbitals in any sublevel is equal to $2l + 1$. Therefore, the s sublevel ($l = 0$) has one orbital, the p sublevel ($l = 1$) has three orbitals, the d sublevel ($l = 2$) has five orbitals, etc.
- The number of orbitals in a level is equal to n^2. Therefore, the $n = 1$ level has one orbital, the $n = 2$ level has four orbitals, the $n = 3$ level has nine orbitals, etc.

EXAMPLE 3.5
Quantum Numbers I

What are the quantum numbers and names (for example, 2*s*, 2*p*) of the orbitals in the $n = 4$ principal level? How many $n = 4$ orbitals exist?

SOLUTION

First determine the possible values of l (from the given value of n). Then determine the possible values of m_l for each possible value of l. For a given value of n, the possible values of l are 0, 1, 2,..., $(n - 1)$.

$n = 4$; therefore $l = 0, 1, 2,$ and 3

For a given value of l, the possible values of m_l are the integer values including zero ranging from $-l$ to $+l$. The name of an orbital is its principal quantum number (n) followed by the letter corresponding to the value l. The total number of orbitals is given by n^2.

l	Possible m_l Values	Orbital name(s)
0	0	4*s* (1 orbital)
1	−1, 0, +1	4*p* (3 orbitals)
2	−2, −1, 0, +1, +2	4*d* (5 orbitals)
3	−3, −2, −1, 0, +1, +2, +3	4*f* (7 orbitals)

Total number of orbitals $= 4^2 = 16$

FOR PRACTICE 3.5

List the quantum numbers associated with all of the 5*d* orbitals. How many 5*d* orbitals exist?

EXAMPLE 3.6
Quantum Numbers II

These sets of quantum numbers are each supposed to specify an orbital. One set, however, is erroneous. Which one and why?

(a) $n = 3; l = 0; m_l = 0$ **(b)** $n = 2; l = 1; m_l = -1$ **(c)** $n = 1; l = 0; m_l = 0$ **(d)** $n = 4; l = 1; m_l = -2$

SOLUTION

Choice **(d)** is erroneous because, for $l = 1$, the possible values of m_l are only −1, 0, and +1.

FOR PRACTICE 3.6

Each set of quantum numbers is supposed to specify an orbital. However, each set contains one quantum number that is not allowed. Replace the quantum number that is not allowed with one that is allowed.

(a) $n = 3; l = 3; m_l = +2$ **(b)** $n = 2; l = 1; m_l = -2$ **(c)** $n = 1; l = 1; m_l = 0$

Atomic Spectroscopy Explained

Quantum theory explains the atomic spectra of atoms discussed in Section 3.3. Each wavelength in the emission spectrum of an atom corresponds to an electron *transition* between quantum-mechanical orbitals. When an atom absorbs energy, an electron in a lower energy orbital is *excited* or promoted to a higher energy orbital, as shown in Figure 3.21 ◀. In this new configuration, however, the atom is unstable, and the electron quickly falls back or *relaxes* to a lower energy orbital. As it does so, it releases a photon of light containing an amount of energy precisely equal to the energy difference

Excitation and Radiation
Energy
$n = 3$
$n = 2$
$n = 1$
Light is emitted as electron falls back to lower energy level.
Electron absorbs energy and is excited to unstable energy level.

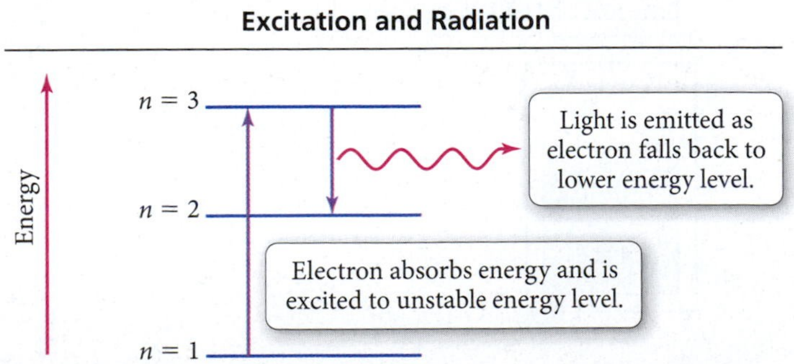

◀ **FIGURE 3.21 Excitation and Radiation** When an atom absorbs energy, an electron can be excited from an orbital in a lower energy level to an orbital in a higher energy level. The electron in this "excited state" is unstable, however, and relaxes to a lower energy level, releasing energy in the form of electromagnetic radiation.

between the two energy levels. We saw previously (see Equation 3.7) that the energy of an orbital with principal quantum number n is given by $E_n = -2.18 \times 10^{-18}\,\text{J}(1/n^2)$, where $n = 1, 2, 3, \ldots$. Therefore, the *difference* in energy between the two levels n_{initial} and n_{final} is given by $\Delta E = E_{\text{final}} - E_{\text{initial}}$. If we substitute the expression for E_n into the expression for ΔE, we get the following expression for the change in energy that occurs in an atom when an electron changes energy levels:

$$\Delta E = E_{\text{final}} - E_{\text{initial}}$$

$$= -2.18 \times 10^{-18}\,\text{J}\left(\frac{1}{n_f^2}\right) - \left[-2.18 \times 10^{-18}\,\text{J}\left(\frac{1}{n_i^2}\right)\right]$$

$$\Delta E = -2.18 \times 10^{-18}\,\text{J}\left(\frac{1}{n_f^2} - \frac{1}{n_i^2}\right) \quad [3.7]$$

The Rydberg equation, $1/\lambda = R(1/m^2 - 1/n^2)$, can be derived from the relationships just covered. We leave this derivation to an exercise (see Problem 3.96).

For example, suppose that an electron in a hydrogen atom relaxes from an orbital in the $n = 3$ level to an orbital in the $n = 2$ level. Then ΔE, the energy difference corresponding to the transition from $n = 3$ to $n = 2$, is determined as follows:

$$\Delta E_{\text{atom}} = E_2 - E_3$$

$$= -2.18 \times 10^{-18}\,\text{J}\left(\frac{1}{2^2}\right) - \left[-2.18 \times 10^{-18}\,\text{J}\left(\frac{1}{3^2}\right)\right]$$

$$= -2.18 \times 10^{-18}\,\text{J}\left(\frac{1}{2^2} - \frac{1}{3^2}\right)$$

$$= -3.03 \times 10^{-19}\,\text{J}$$

The energy carries a negative sign because the atom *emits* the energy as it relaxes from $n = 3$ to $n = 2$. Since energy must be conserved, the exact amount of energy emitted by the atom is carried away by the photon:

$$\Delta E_{\text{atom}} = -E_{\text{photon}}$$

This energy determines the frequency and wavelength of the photon. Since the wavelength of the photon is related to its energy as $E = hc/\lambda$, we calculate the wavelength of the photon as:

$$\lambda = \frac{hc}{E}$$

$$= \frac{(6.626 \times 10^{-34}\,\cancel{\text{J}} \cdot \cancel{\text{s}})(3.00 \times 10^{8}\,\text{m}/\cancel{\text{s}})}{3.03 \times 10^{-19}\,\cancel{\text{J}}}$$

$$= 6.56 \times 10^{-7}\,\text{m} \quad \text{or} \quad 656\,\text{nm}$$

Consequently, the light emitted by an excited hydrogen atom as it relaxes from an orbital in the $n = 3$ level to an orbital in the $n = 2$ level has a wavelength of 656 nm (red). Similarly, we can calculate the light emitted due to a transition from $n = 4$ to $n = 2$ to be 486 nm (green). Notice that transitions between orbitals that are further apart in energy produce light that is higher in energy, and therefore shorter in wavelength, than transitions between orbitals that are closer together. Figure 3.22 ▶ shows several of the transitions in the hydrogen atom and their corresponding wavelengths.

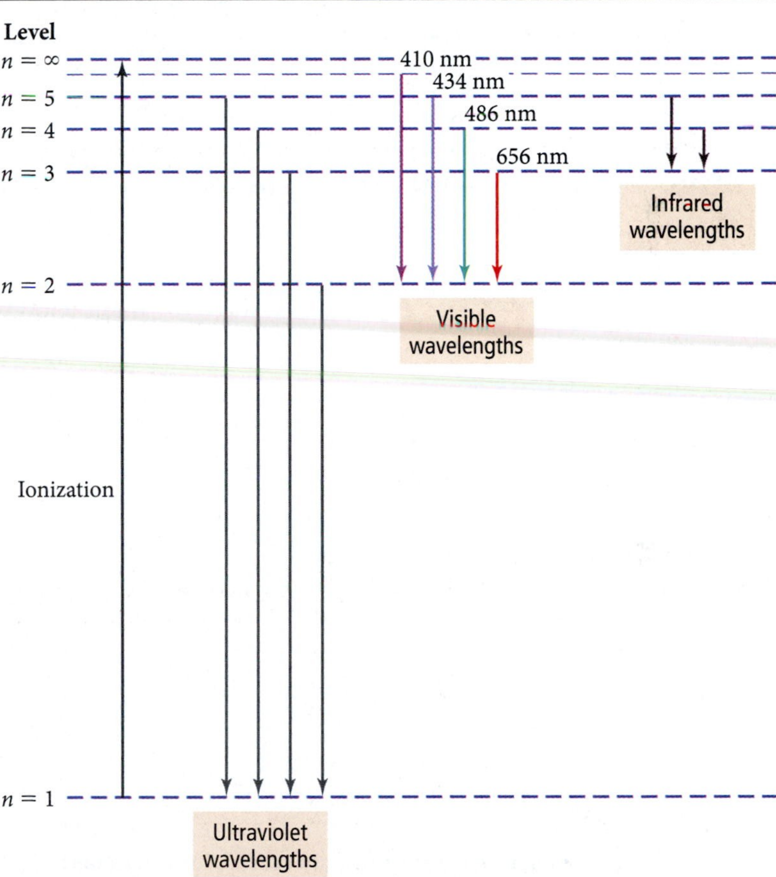

▲ **FIGURE 3.22 Hydrogen Energy Transitions and Radiation** An atomic energy level diagram for hydrogen, showing some possible electron transitions between levels and the corresponding wavelengths of emitted light.

Emission Spectra

3.6 Cc Conceptual Connection

Which transition results in emitted light with the shortest wavelength?

(a) $n = 5 \longrightarrow n = 4$ **(b)** $n = 4 \longrightarrow n = 3$ **(c)** $n = 3 \longrightarrow n = 2$

EXAMPLE 3.7

Wavelength of Light for a Transition in the Hydrogen Atom

Determine the wavelength of light emitted when an electron in a hydrogen atom makes a transition from an orbital in $n = 6$ to an orbital in $n = 5$.

SORT You are given the energy levels of an atomic transition and asked to find the wavelength of emitted light.

GIVEN: $n = 6 \longrightarrow n = 5$

FIND: λ

STRATEGIZE In the first part of the conceptual plan, calculate the energy of the electron in the $n = 6$ and $n = 5$ orbitals using Equation 3.6 and subtract to find ΔE_{atom}.

CONCEPTUAL PLAN

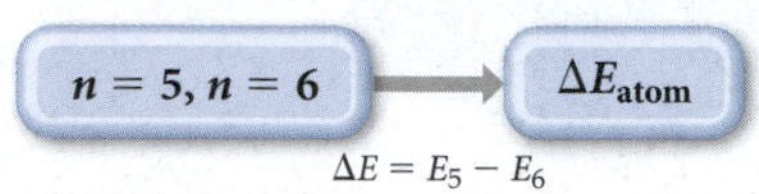

In the second part, find E_{photon} by taking the negative of ΔE_{atom}, and then calculating the wavelength corresponding to a photon of this energy using Equation 3.3. (The difference in sign between E_{photon} and ΔE_{atom} applies only to emission. *The energy of a photon must always be positive.*)

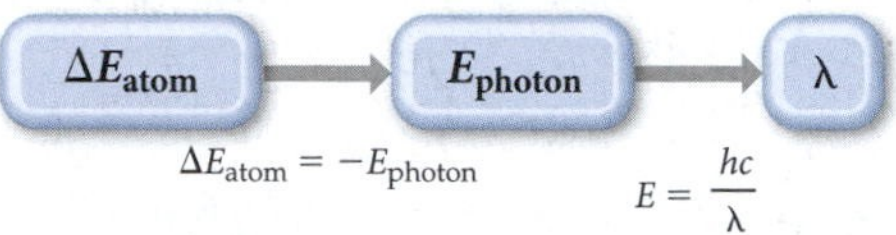

RELATIONSHIPS USED

$E_n = -2.18 \times 10^{-18}\,\text{J}(1/n^2)$

$E = hc/\lambda$

SOLVE Follow the conceptual plan. Begin by calculating ΔE_{atom}.

SOLUTION

$$\begin{aligned}\Delta E_{atom} &= E_5 - E_6 \\ &= -2.18 \times 10^{-18}\,\text{J}\left(\frac{1}{5^2}\right) - \left[-2.18 \times 10^{-18}\,\text{J}\left(\frac{1}{6^2}\right)\right] \\ &= -2.18 \times 10^{-18}\,\text{J}\left(\frac{1}{5^2} - \frac{1}{6^2}\right) \\ &= -2.66\underline{4}4 \times 10^{-20}\,\text{J}\end{aligned}$$

Calculate E_{photon} by changing the sign of ΔE_{atom}.

$$E_{photon} = -\Delta E_{atom} = +2.66\underline{4}4 \times 10^{-20}\,\text{J}$$

Solve the equation relating the energy of a photon to its wavelength for λ. Substitute the energy of the photon and calculate λ.

$$E = \frac{hc}{\lambda}$$

$$\begin{aligned}\lambda &= \frac{hc}{E} \\ &= \frac{(6.626 \times 10^{-34}\,\cancel{\text{J}} \cdot \cancel{\text{s}})(3.00 \times 10^8\,\text{m}/\cancel{\text{s}})}{2.66\underline{4}4 \times 10^{-20}\,\cancel{\text{J}}} \\ &= 7.46 \times 10^{-6}\text{ m or } 7460\text{ nm}\end{aligned}$$

CHECK The units of the answer (m) are correct for wavelength. The magnitude is reasonable because 10^{-6} m is in the infrared region of the electromagnetic spectrum. You know that transitions from $n = 3$ or $n = 4$ to $n = 2$ lie in the visible region, so it makes sense that a transition between levels of higher n value (which are energetically closer to one another) would result in light of longer wavelength.

FOR PRACTICE 3.7

Determine the wavelength of the light absorbed when an electron in a hydrogen atom makes a transition from an orbital in which $n = 2$ to an orbital in which $n = 7$.

FOR MORE PRACTICE 3.7

An electron in the $n = 6$ level of the hydrogen atom relaxes to a lower energy level, emitting light of $\lambda = 93.8$ nm. Find the principal level to which the electron relaxed.

3.6 The Shapes of Atomic Orbitals

The shapes of atomic orbitals are important because chemical bonds depend on the sharing of the electrons that occupy these orbitals. In one model of chemical bonding, for example, a bond consists of the overlap of atomic orbitals on adjacent atoms. The shapes of the overlapping orbitals determine the shape of the molecule. Although we limit ourselves in this chapter to the orbitals of the hydrogen atom, we will see in Chapter 4 that the orbitals of all atoms can be approximated as being hydrogen-like and therefore have very similar shapes to those of hydrogen.

The shape of an atomic orbital is determined primarily by l, the angular momentum quantum number. Recall that each value of l is assigned a letter that corresponds to particular orbitals. For example, the orbitals with $l = 0$ are called *s* orbitals; those with $l = 1$, *p* orbitals; those with $l = 2$, *d* orbitals, and so on. We now examine the shape of each of these orbitals.

s Orbitals ($l = 0$)

The lowest energy orbital is the spherically symmetrical 1*s* orbital shown in Figure 3.23a ▼. This image is actually a three-dimensional plot of the wave function squared (ψ^2), which represents **probability density**, the probability (per unit volume) of finding the electron at a point in space.

$$\psi^2 = \text{probability density} = \frac{\text{probability}}{\text{unit volume}}$$

The magnitude of ψ^2 in this plot is proportional to the density of the dots shown in the image. The high dot density near the nucleus (at the very center of the plot) indicates a higher probability density for the electron there. As you move away from the nucleus, the probability density decreases. Figure 3.23b ▼ shows a plot of probability density (ψ^2) versus r, the distance from the nucleus. The plot represents a slice through the three-dimensional plot of ψ^2 and illustrates how the probability density decreases as r increases.

We can understand probability density with the help of a thought experiment. Imagine an electron in the 1*s* orbital located within the volume surrounding the nucleus. Imagine taking a photograph of the electron every second for 10 or 15 minutes. In one photograph, the electron is very close to the nucleus, in another it is farther away, and so on. Each photo has a dot showing the electron's position relative to the nucleus when the photo was taken. Remember that you can never predict where the electron will be for any one photo. However, if you took hundreds of photos and superimposed all of them, you would have a plot similar to Figure 3.23a—a statistical representation of how likely the electron is to be found at each point.

The thought experiment we just engaged in can result in a possible misunderstanding: that the electron is moving around (like a moth near a flame) between photographs. However, in the quantum-mechanical model, that is not the case. Recall from Section 3.1 that the measurement affects the outcome of any quantum system. Between photographs, the location of the electron is uncertain—in a sense its location is spread out over the entire volume of the orbital. Only when the photograph is taken (that is, when a measurement of its location is made) does the location of the electron become localized to one spot. Between measurements, the electron has no single location.

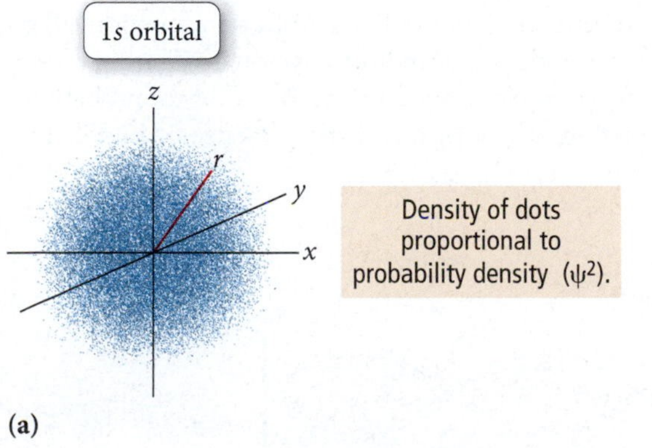
1s orbital
z
r
y
x
Density of dots proportional to probability density (ψ²).
(a)

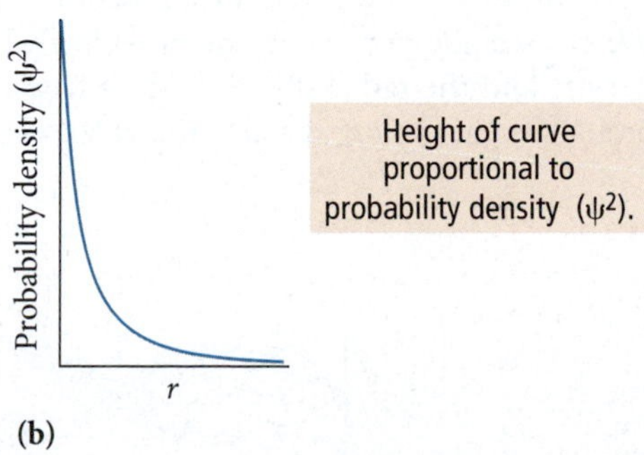
Probability density (ψ²)
r
Height of curve proportional to probability density (ψ²).
(b)

◀ **FIGURE 3.23 The 1*s* Orbital: Two Representations** In **(a)** the dot density is proportional to the electron probability density. In **(b)**, the height of the curve is proportional to the electron probability density. The *x*-axis is *r*, the distance from the nucleus.

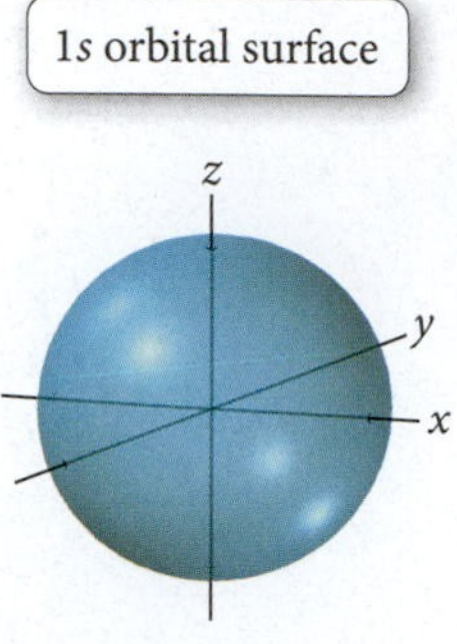

▲ **FIGURE 3.24 The 1s Orbital Surface** In this representation, the surface of the sphere encompasses the volume where the electron is found 90% of the time when the electron is in the 1s orbital.

An atomic orbital can also be represented by a geometrical shape that encompasses the volume where the electron is likely to be found most frequently—typically, 90% of the time. For example, the 1*s* orbital can be represented as the three-dimensional sphere shown in Figure 3.24 ◀. If we were to superimpose the dot-density representation of the 1*s* orbital on the shape representation, 90% of the dots would be within the sphere, meaning that when the electron is in the 1*s* orbital it has a 90% chance of being found within the sphere.

The plots we have just seen represent probability *density*. However, they are a bit misleading because they seem to imply that the electron is most likely to be found *at the nucleus*. To get a better idea of where the electron is most likely to be found, we can use a plot called the **radial distribution function**, shown in Figure 3.25 ◀ for the 1*s* orbital. The radial distribution function represents the *total probability of finding the electron within a thin spherical shell at a distance r from the nucleus.*

$$\text{Total radial probability (at a given } r) = \frac{\text{probability}}{\text{unit volume}} \times \text{volume of shell at } r$$

The radial distribution function represents, not probability density *at a point r*, but total probability *at a radius r*. In contrast to probability density, which has a maximum at the nucleus, the radial distribution function has a value of *zero* at the nucleus. It increases to a maximum at 52.9 pm and then decreases again with increasing *r*. 1 pm $= 10^{-12}$ m

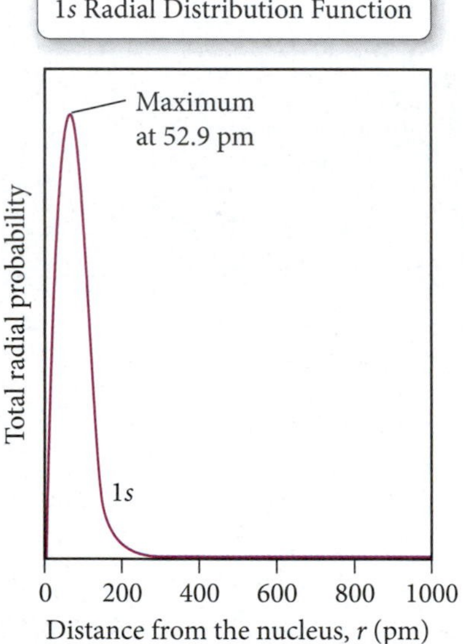

▲ **FIGURE 3.25 The Radial Distribution Function for the 1s Orbital** The curve shows the total probability of finding the electron within a thin shell at a distance *r* from the nucleus.

The shape of the radial distribution function is the result of multiplying together two functions with opposite trends in *r*:

1. The probability density function (ψ^2), which is the probability per unit volume, has a maximum at the nucleus and decreases with increasing *r*.
2. The volume of the thin shell, which is zero at the nucleus and increases with increasing *r*.

At the nucleus ($r = 0$), for example, the probability *density* is at a maximum; however, the volume of a thin spherical shell is zero, so the radial distribution function is zero. As *r* increases, the volume of the thin spherical shell increases. We can understand this by making an analogy to an onion. A spherical shell at a distance *r* from the nucleus is like a layer in an onion at a distance *r* from its center. If the layers of the onion all have the same thickness, then the volume of any one layer—think of this as the total amount of onion in the layer—is greater as *r* increases. Similarly, the volume of any one spherical shell in the radial distribution function increases with increasing distance from the nucleus, resulting in a greater total probability of finding the electron within that shell. Close to the nucleus, this increase in volume with increasing *r* outpaces the decrease in probability density, producing a maximum at 52.9 pm. Farther out, however, the density tapers off faster than the volume increases.

The maximum in the radial distribution function, 52.9 pm, turns out to be the very same radius that Bohr had predicted for the innermost orbit of the hydrogen atom. However, there is a significant conceptual difference between the two radii. In the Bohr model, every time you probe the atom (in its lowest energy state), you would find the electron at a radius of 52.9 pm. In the quantum-mechanical model, you would generally find the electron at various radii, with 52.9 pm having the greatest probability.

The probability densities and radial distribution functions for the 2*s* and 3*s* orbitals are shown in Figure 3.26 ▶. Like the 1*s* orbital, these orbitals are spherically symmetric. These orbitals are larger in size than the 1*s* orbital, and, unlike the 1*s* orbital, they contain at least one *node*. A **node** is a point where the wave function (ψ), and therefore the probability density (ψ^2) and radial distribution function, all go through zero. A node in a wave function is much like a node in a standing wave on a vibrating string. We can see nodes in an orbital most clearly by looking at a slice through the orbital. Plots of probability density and the radial distribution function as a function of *r* both reveal the presence of nodes. The probability of finding the electron at a node is zero.

▶ The nodes in quantum-mechanical atomic orbitals are three-dimensional analogs of the nodes on a vibrating string.

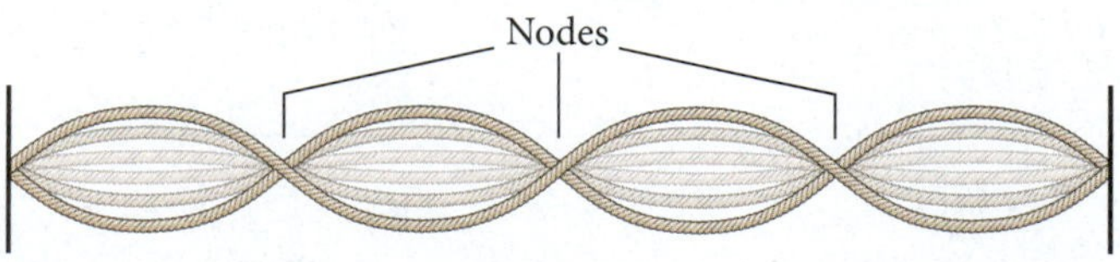

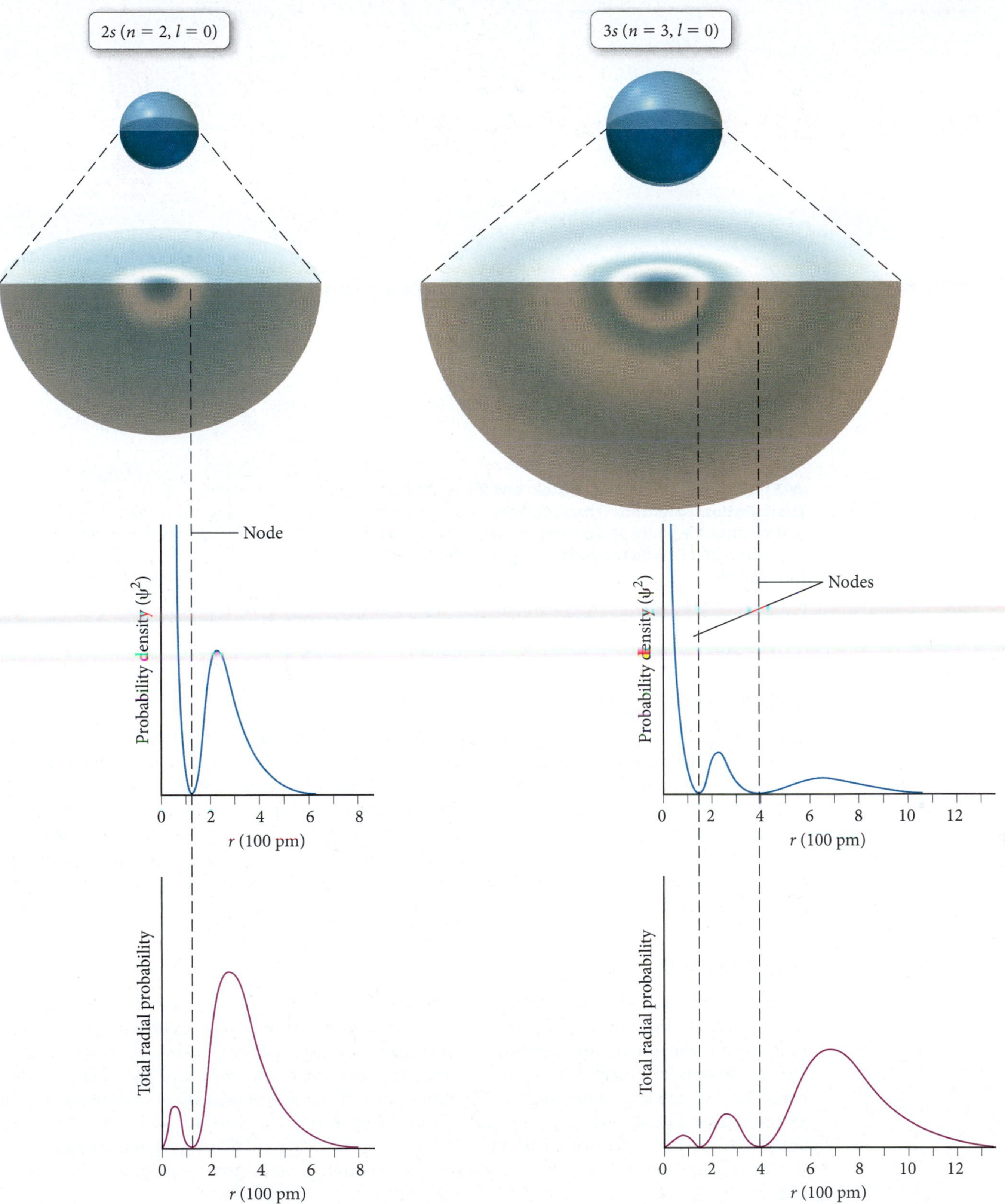

▲ **FIGURE 3.26 Probability Densities and Radial Distribution Functions for the 2*s* and 3*s* Orbitals**

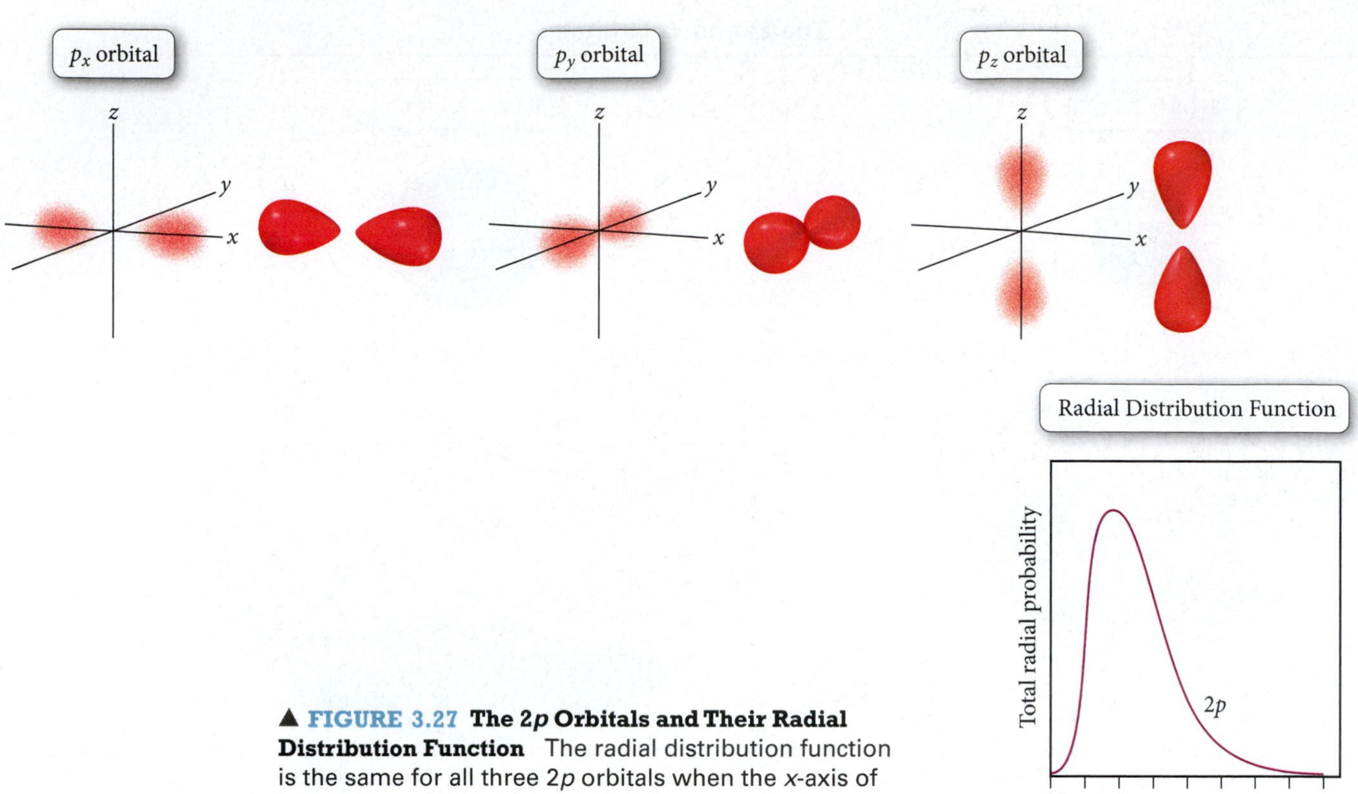

▲ **FIGURE 3.27 The 2*p* Orbitals and Their Radial Distribution Function** The radial distribution function is the same for all three 2*p* orbitals when the *x*-axis of the graph is taken as the axis containing the lobes of the orbital.

p Orbitals ($l = 1$)

Each principal level with $n = 2$ or greater contains three *p* orbitals ($m_l = -1, 0, +1$). The three 2*p* orbitals and their radial distribution functions are shown in Figure 3.27 ▲. The *p* orbitals are not spherically symmetric like the *s* orbitals, but have two *lobes* of electron density on either side of the nucleus and a node located at the nucleus. The three *p* orbitals differ only in their orientation and are orthogonal (mutually perpendicular) to one another. It is convenient to define an *x*-, *y*-, and *z*-axis system and then label each *p* orbital as p_x, p_y, and p_z. The 3*p*, 4*p*, 5*p*, and higher *p* orbitals are all similar in shape to the 2*p* orbitals, but they contain additional nodes (like the higher *s* orbitals) and are progressively larger in size.

d Orbitals ($l = 2$)

Each principal level with $n = 3$ or greater contains five *d* orbitals ($m_l = -2, -1, 0, +1, +2$). The five 3*d* orbitals are shown in Figure 3.28 ▶. Four of these orbitals have a cloverleaf shape, with four lobes of electron density around the nucleus and two perpendicular nodal planes. The d_{xy}, d_{xz}, and d_{yz} orbitals are oriented along the *xy*, *xz*, and *yz* planes, respectively, and their lobes are oriented *between* the corresponding axes. The four lobes of the $d_{x^2-y^2}$ orbital are oriented along the *x*- and *y*-axes. The d_{z^2}, etc., orbital is different in shape from the other four, having two lobes oriented along the *z*-axis and a donut-shaped ring along the *xy* plane. The 4*d*, 5*d*, 6*d*, orbitals are all similar in shape to the 3*d* orbitals, but they contain additional nodes and are progressively larger in size.

A nodal plane is a plane where the electron probability density is zero. For example, in the d_{xy} orbitals, the nodal planes lie in the *xz* and *yz* planes.

f Orbitals ($l = 3$)

Each principal level with $n = 4$ or greater contains seven *f* orbitals ($m_l = -3, -2, -1, 0, +1, +2, +3$), as shown in Figure 3.29 ▶. These *f* orbitals have more lobes and nodes than *d* orbitals.

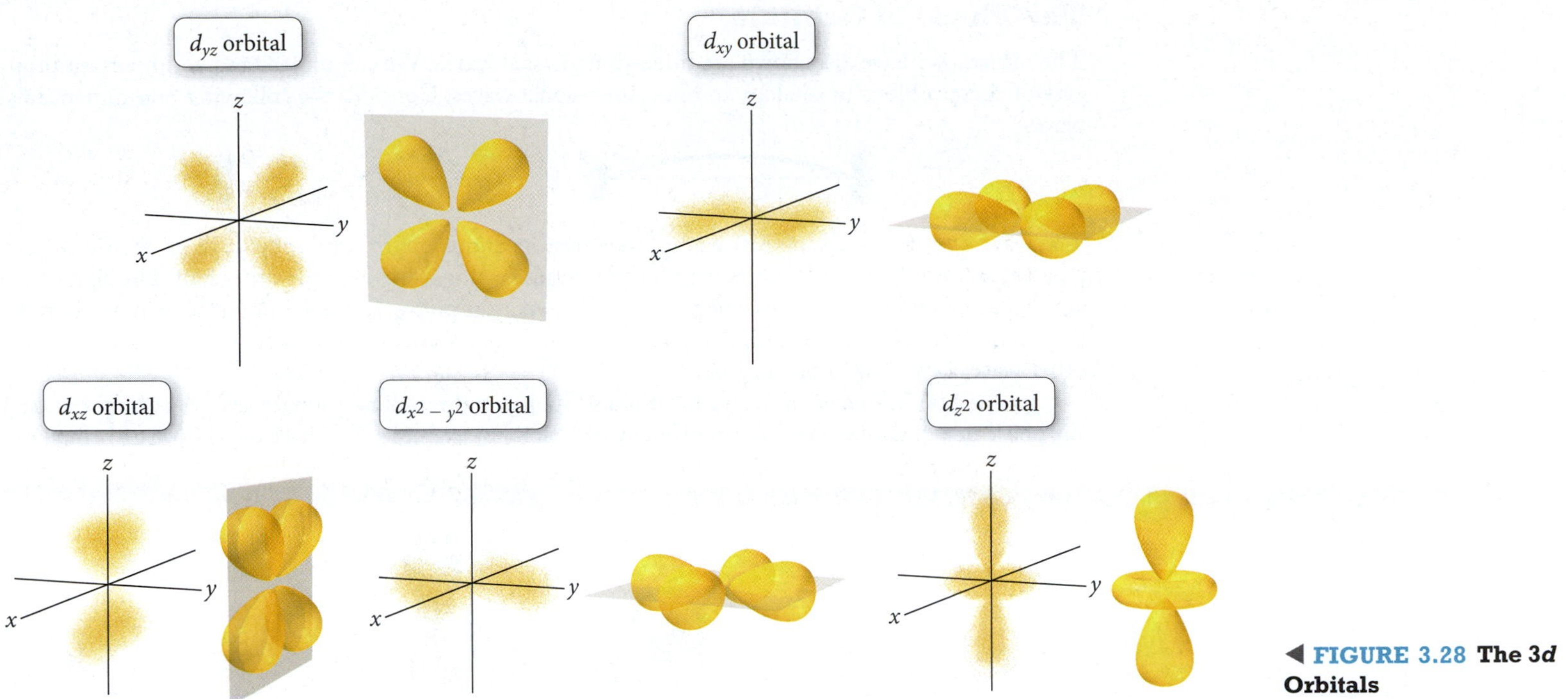

◀ **FIGURE 3.28 The 3*d* Orbitals**

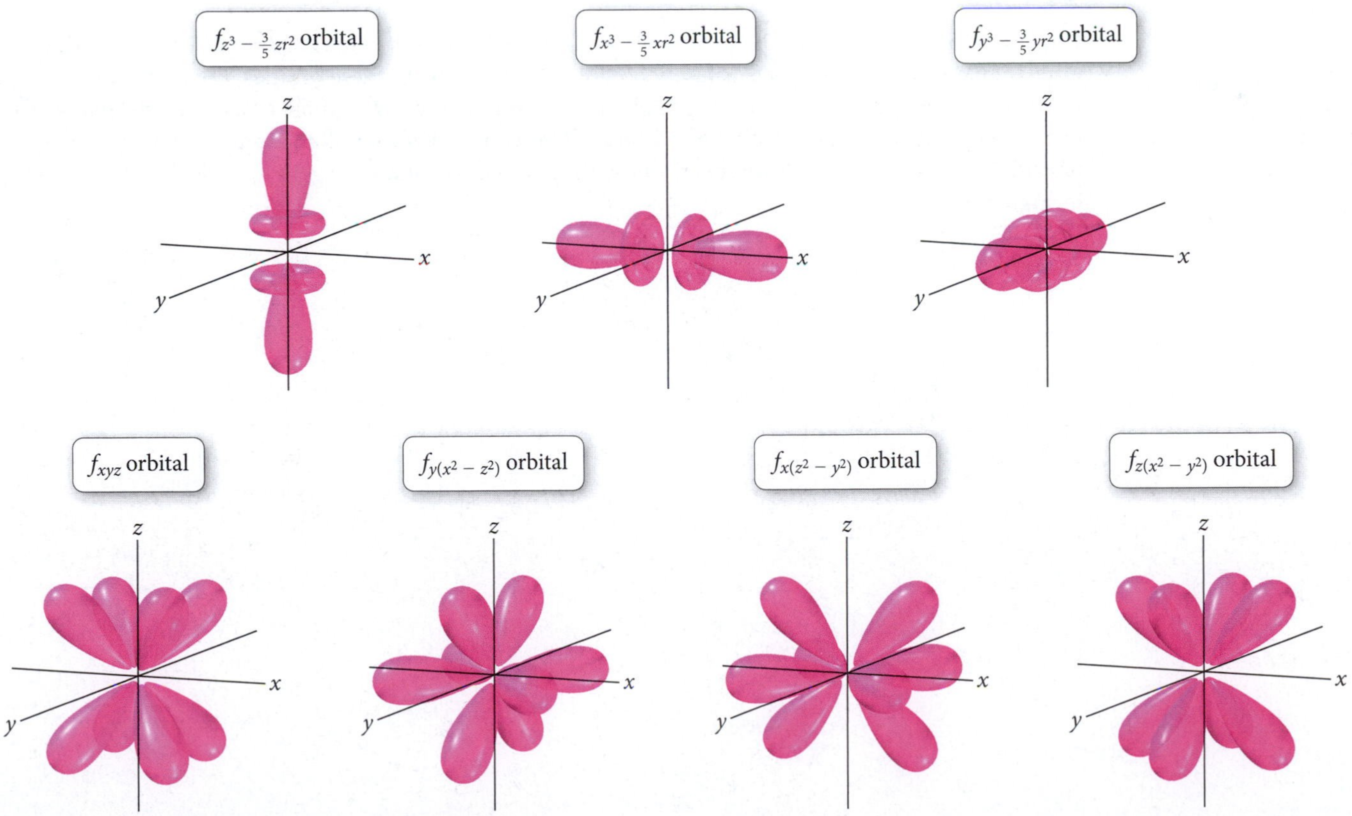

$f_{z^3 - \frac{3}{5}zr^2}$ orbital

$f_{x^3 - \frac{3}{5}xr^2}$ orbital

$f_{y^3 - \frac{3}{5}yr^2}$ orbital

f_{xyz} orbital

$f_{y(x^2 - z^2)}$ orbital

$f_{x(z^2 - y^2)}$ orbital

$f_{z(x^2 - y^2)}$ orbital

◀ **FIGURE 3.29 The 4*f* Orbitals**

The Phase of Orbitals

The orbitals we have just shown are three-dimensional waves. We can understand an important property of these orbitals by analogy to one-dimensional waves. Consider the following one-dimensional waves:

The wave on the left has a positive amplitude over its entire length, while the wave on the right has a positive amplitude over half of its length and a negative amplitude over the other half. The sign of the amplitude of a wave—positive or negative—is known as its **phase**. In these images, blue indicates positive phase and red indicates negative phase. The phase of a wave determines how it interferes with another wave, as we saw in Section 3.2.

Just as a one-dimensional wave has a phase, so does a three-dimensional wave. We often represent the phase of a quantum-mechanical orbital with color. For example, the phase of a 1*s* and 2*p* orbital can be represented as:

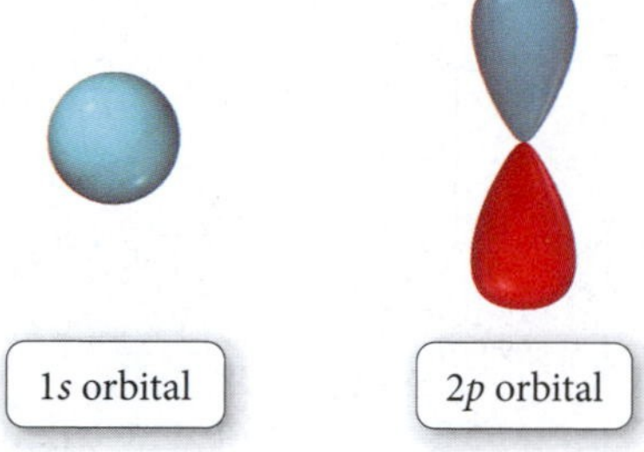

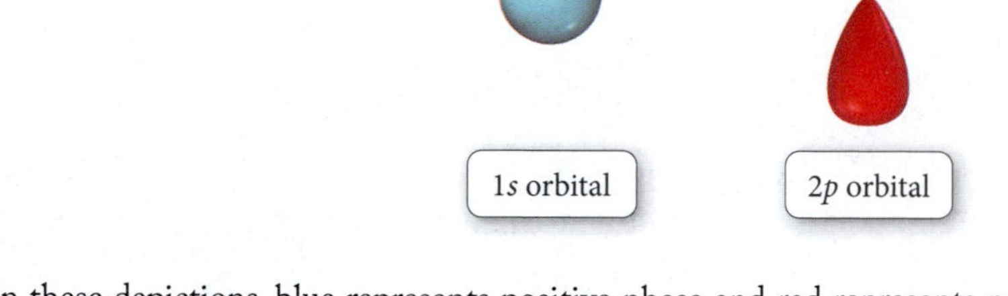

In these depictions, blue represents positive phase and red represents negative phase. The 1*s* orbital is all one phase, while the 2*p* orbital exhibits two different phases. The phase of quantum-mechanical orbitals is important in bonding, as we shall see in Chapter 7.

▲ **FIGURE 3.30 Why Atoms Are Depicted as Spherical** Atoms are depicted as spherical because all the orbitals together make up a spherical shape.

The Shape of Atoms

If some orbitals are shaped like dumbbells and three-dimensional cloverleafs, and if most of the volume of an atom is empty space diffusely occupied by electrons in these orbitals, then why do we often depict atoms as spheres? Atoms are usually drawn as spheres because most atoms contain many electrons occupying a number of different orbitals. Therefore, the shape of an atom is obtained by superimposing all of its orbitals. If we superimpose the *s*, *p*, and *d* orbitals, we get a spherical shape, as shown in Figure 3.30 ◀.

SELF-ASSESSMENT Quiz

1. Which wavelength of light has the highest frequency?
   a) 10 nm **b)** 10 mm **c)** 1 nm **d)** 1 mm

2. Which kind of electromagnetic radiation contains the greatest energy per photon?
   a) Microwaves **b)** Gamma rays
   c) X-rays **d)** Visible light

3. How much energy (in J) is contained in 1.00 mole of 552-nm photons?
   a) 3.60×10^{-19} J **b)** 2.17×10^{5} J
   c) 3.60×10^{-28} J **d)** 5.98×10^{-43} J

4. Light from three different lasers (A, B, and C), each with a different wavelength, is shined onto the same metal surface. Laser A produces no photoelectrons. Lasers B and C both produce photoelectrons, but the photoelectrons produced by laser B have a greater velocity than those produced by laser C. Arrange the lasers in order of increasing wavelength.
   a) $A < B < C$ **b)** $B < C < A$
   c) $C < B < A$ **d)** $A < C < B$

5. Calculate the frequency of an electron traveling at 1.85×10^{7} m/s.
   a) $1.31 \times 10^{-19}\ s^{-1}$ **b)** $1.18 \times 10^{-2}\ s^{-1}$
   c) $3.93 \times 10^{-11}\ s^{-1}$ **d)** $7.63 \times 10^{18}\ s^{-1}$

6. Which set of three quantum numbers *does not* specify an orbital in the hydrogen atom?
   a) $n = 2; l = 1; m_l = -1$ **b)** $n = 3; l = 3; m_l = -2$
   c) $n = 2; l = 0; m_l = 0$ **d)** $n = 3; l = 2; m_l = 2$

7. Calculate the wavelength of light emitted when an electron in the hydrogen makes a transition from an orbital with $n = 5$ to an orbital with $n = 3$.
   a) 1.28×10^{-6} m **b)** 6.04×10^{-7} m
   c) 2.28×10^{-6} m **d)** 1.55×10^{-19} m

8. Which electron transition produces light of the highest frequency in the hydrogen atom?
   a) $5p \longrightarrow 1s$ **b)** $4p \longrightarrow 1s$
   c) $3p \longrightarrow 1s$ **d)** $2p \longrightarrow 1s$

9. How much time (in seconds) does it take light to travel 1.00 billion km?
   a) 3.00×10^{17} s **b)** 3.33 s
   c) 3.33×10^{3} s **d)** 3.00×10^{20} s

10. Which figure represents a *d* orbital?
    a) 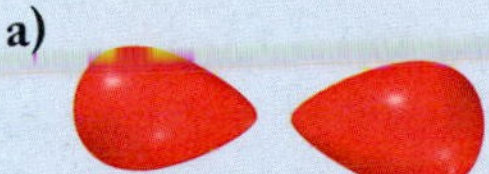
    b) 
    c)
    d) None of the above

Answers: 1:c; 2:b; 3:b; 4:b; 5:d; 6:b; 7:a; 8:a; 9:c; 10:b

CHAPTER SUMMARY 3

REVIEW

KEY LEARNING OUTCOMES

CHAPTER OBJECTIVES	ASSESSMENT
Calculating the Wavelength and Frequency of Light (3.2)	• Example 3.1 For Practice 3.1 Exercises 39, 40
Calculating the Energy of a Photon (3.2)	• Example 3.2 For Practice 3.2 For More Practice 3.2 Exercises 41–46
Relating Wavelength, Energy, and Frequency to the Electromagnetic Spectrum (3.2)	• Example 3.3 For Practice 3.3 Exercises 37, 38

Using the de Broglie Relation to Calculate Wavelength (3.4)	• Example 3.4 For Practice 3.4 Exercises 49–54
Relating Quantum Numbers to One Another and to Their Corresponding Orbitals (3.5)	• Examples 3.5, 3.6 For Practice 3.5, 3.6 Exercises 59–62
Relating the Wavelength of Light to Transitions in the Hydrogen Atom (3.5)	• Example 3.7 For Practice 3.7 For More Practice 3.7 Exercises 69–72

KEY TERMS

Section 3.1
quantum-mechanical model (64)

Section 3.2
electromagnetic radiation (64)
amplitude (65)
wavelength (λ) (65)
frequency (ν) (65)
electromagnetic spectrum (66)
gamma ray (66)
X-ray (67)
ultraviolet (UV) radiation (67)
visible light (67)
infrared (IR) radiation (67)
microwave (67)
radio wave (67)
interference (68)
constructive interference (68)
destructive interference (68)
diffraction (68)
photoelectric effect (70)
binding energy (70)
photon (quantum) (71)

Section 3.3
emission spectrum (74)
absorption spectrum (76)

Section 3.4
de Broglie relation (78)
complementary properties (80)
Heisenberg's uncertainty principle (80)
deterministic (80)
indeterminacy (81)

Section 3.5
orbital (82)
wave function (82)
quantum number (82)
principal quantum number (n) (82)
angular momentum quantum number (l) (82)
magnetic quantum number (m_l) (82)
spin quantum number (m_s) (82)
electron spin (83)
principal level (shell) (83)
sublevel (subshell) (83)

Section 3.6
probability density (87)
radial distribution function (88)
node (88)
phase (92)

KEY CONCEPTS

The Realm of Quantum Mechanics (3.1)

- The theory of quantum mechanics explains the behavior of absolutely small particles, such as electrons, in the atomic and subatomic realms.
- These particles behave differently than the sorts of particles we see in the macroscopic world.

The Nature of Light (3.2)

- Light is a type of electromagnetic radiation—a form of energy embodied in oscillating electric and magnetic fields that travels though space at 3.00×10^8 m/s.
- The wave nature of light is characterized by its wavelength—the distance between wave crests—and its ability to experience interference (constructive or destructive) and diffraction.
- The electromagnetic spectrum includes all wavelengths of electromagnetic radiation from gamma rays (high energy per photon, short wavelength) to radio waves (low energy per photon, long wavelength). Visible light is a tiny sliver in the middle of the electromagnetic spectrum.
- The particle nature of light is characterized by the specific quantity of energy carried in each photon.

Atomic Spectroscopy (3.3)

- Atomic spectroscopy is the study of the light absorbed and emitted by atoms when an electron makes a transition from one energy level to another.
- The wavelengths absorbed or emitted depend on the energy differences between the levels involved in the transition; large energy differences result in short wavelengths, and small energy differences result in long wavelengths.

The Wave Nature of Matter (3.4)

- Electrons have a wave nature with an associated wavelength, as quantified by the de Broglie relation.
- The wave nature and particle nature of matter are complementary—the more we know of one, the less we know of the other.
- The wave–particle duality of electrons is quantified in Heisenberg's uncertainty principle, which states that there is a limit to how well we can know both the position of an electron (associated with the electron's particle nature) and the velocity times the mass of an electron (associated with the electron's wave nature)—the more accurately one is measured, the greater the uncertainty in measurement of the other.
- The inability to simultaneously know both the position and the velocity of an electron results in indeterminacy, the inability to predict a trajectory for an electron. Consequently, electron behavior is described differently than the behavior of everyday-sized particles.
- The trajectory we normally associate with macroscopic objects is replaced, for electrons, with statistical descriptions that show, not the electron's path, but the region where it is most likely to be found.

The Quantum-Mechanical Model of the Atom (3.5, 3.6)

- The most common way to describe electrons in atoms according to quantum mechanics is to solve the Schrödinger equation for the energy states of the electrons within the atom. When the electron is in these states, its energy is well defined but its position is not. The position of an electron is described by a probability distribution map called an orbital.
- The solutions to the Schrödinger equation (including the energies and orbitals) are characterized by quantum numbers: n, l and m_l.
- The principal quantum number (n) determines the energy of the electron and the size of the orbital; the angular momentum quantum number (l) determines the shape of the orbital; the magnetic quantum number (m_l) determines the orientation of the orbital. A fourth quantum number, the spin quantum number (m_s), specifies the orientation of the spin of the electron.

KEY EQUATIONS AND RELATIONSHIPS

Relationship between Frequency (ν), Wavelength (λ), and the Speed of Light (*c*) (3.2)

$$\nu = \frac{c}{\lambda}$$

Relationship between Energy (*E*), Frequency (ν), Wavelength (λ), and Planck's Constant (*h*) (3.2)

$$E = h\nu$$

$$E = \frac{hc}{\lambda}$$

De Broglie Relation: Relationship between Wavelength (λ), Mass (*m*), and Velocity (ν) of a Particle (3.4)

$$\lambda = \frac{h}{m\nu}$$

Heisenberg's Uncertainty Principle: Relationship between a Particle's Uncertainty in Position (Δx) and Uncertainty in Velocity ($\Delta\nu$) (3.4)

$$\Delta x \times m\Delta\nu \geq \frac{h}{4\pi}$$

Energy of an Electron in an Orbital with Quantum Number *n* in a Hydrogen Atom (3.5)

$$E_n = -2.18 \times 10^{-18}\,\mathrm{J}\left(\frac{1}{n^2}\right) \quad (n = 1, 2, 3, \ldots)$$

Change in Energy That Occurs in an Atom When It Undergoes a Transition between Levels n_{initial} and n_{final} (3.5)

$$\Delta E = -2.18 \times 10^{-18}\,\mathrm{J}\left(\frac{1}{n_f^2} - \frac{1}{n_i^2}\right)$$

EXERCISES

REVIEW QUESTIONS

1. Why is the quantum-mechanical model of the atom important for understanding chemistry?

2. What is light? How fast does it travel in a vacuum?

3. Define the wavelength and amplitude of a wave.

4. Define the frequency of electromagnetic radiation. How is frequency related to wavelength?

5. What determines the color of light? Describe the difference between red light and blue light.

6. What determines the color of a colored object? Explain why grass appears green.

7. Give an approximate range of wavelengths for each type of electromagnetic radiation and summarize the characteristics and/or the uses of each.

a) gamma rays
b) X-rays
c) ultraviolet radiation
d) visible light
e) infrared radiation
f) microwave radiation
g) radio waves

8. Explain the wave behavior known as interference. Explain the difference between constructive and destructive interference.

9. Explain the wave behavior known as diffraction. Draw the diffraction pattern that occurs when light travels through two slits comparable in size and separation to the light's wavelength.

10. Describe the photoelectric effect. How did experimental observations of this phenomenon differ from the predictions of classical electromagnetic theory?

11. How did the photoelectric effect lead Einstein to propose that light is quantized?

12. What is a photon? How is the energy of a photon related to its wavelength? Its frequency?

13. What is an emission spectrum? How does an emission spectrum of a gas in a discharge tube differ from a white light spectrum?

14. Describe the Bohr model for the atom. How did the Bohr model account for the emission spectra of atoms?

15. Explain electron diffraction.

16. What is the de Broglie wavelength of an electron? What determines the value of the de Broglie wavelength for an electron?

17. What are complementary properties? How does electron diffraction demonstrate the complementarity of the wave nature and particle nature of the electron?

18. Explain Heisenberg's uncertainty principle. What paradox is at least partially solved by the uncertainty principle?

19. What is a trajectory? What kind of information do you need to predict the trajectory of a particle?

20. Why does the uncertainty principle make it impossible to predict a trajectory for the electron?

21. Newton's laws of motion are *deterministic*. Explain this statement.

22. An electron behaves in ways that are at least partially indeterminate. Explain this statement.

23. What is a probability distribution map?

24. For each solution to the Schrödinger equation, which quantity can be precisely specified: the electron's energy or its position? Explain.

25. What is a quantum-mechanical orbital?

26. What is the Schrödinger equation? What is a wave function? How is a wave function related to an orbital?

27. What are the possible values of the principal quantum number *n*? What does the principal quantum number determine?

28. What are the possible values of the angular momentum quantum number *l*? What does the angular momentum quantum number determine?

29. What are the possible values of the magnetic quantum number m_l? What does the magnetic quantum number determine?

30. List all the orbitals in each principal level. Specify the three quantum numbers for each orbital.

a) $n = 1$
b) $n = 2$
c) $n = 3$
d) $n = 4$

31. Explain the difference between a plot showing the probability density for an orbital and one showing the radial distribution function.

32. Sketch the general shapes of the *s*, *p*, and *d* orbitals.

33. List the four different sublevels. Given that only a maximum of two electrons can occupy an orbital, determine the maximum number of electrons that can exist in each sublevel.

34. Why are atoms usually portrayed as spheres when most orbitals are not spherically shaped?

PROBLEMS BY TOPIC

Electromagnetic Radiation

35. The distance from the sun to Earth is 1.496×10^8 km. How long does it take light to travel from the sun to Earth?

36. The nearest star to our sun is Proxima Centauri, at a distance of 4.3 light-years from the sun. A light-year is the distance that light travels in one year (365 days). How far away, in km, is Proxima Centauri from the sun?

37. List these types of electromagnetic radiation in order of (i) increasing wavelength and (ii) increasing energy per photon:

a. radio waves
b. microwaves
c. infrared radiation
d. ultraviolet radiation

38. List these types of electromagnetic radiation in order of (i) increasing frequency and (ii) decreasing energy per photon:

a. gamma rays
b. radio waves
c. microwaves
d. visible light

39. Calculate the frequency of each wavelength of electromagnetic radiation:

a. 632.8 nm (wavelength of red light from helium–neon laser)
b. 503 nm (wavelength of maximum solar radiation)
c. 0.052 nm (wavelength contained in medical X-rays)

40. Calculate the wavelength of each frequency of electromagnetic radiation:

a. 100.2 MHz (typical frequency for FM radio broadcasting)
b. 1070 kHz (typical frequency for AM radio broadcasting) (assume four significant figures)
c. 835.6 MHz (common frequency used for cell phone communication)

41. Calculate the energy of a photon of electromagnetic radiation at each of the wavelengths indicated in Problem 39.

42. Calculate the energy of a photon of electromagnetic radiation at each of the frequencies indicated in Problem 40.

43. A laser pulse with wavelength 532 nm contains 3.85 mJ of energy. How many photons are in the laser pulse?

44. A heat lamp produces 32.8 watts of power at a wavelength of 6.5 μm. How many photons are emitted per second? (1 watt = 1 J/s)

45. Determine the energy of 1 mol of photons for each kind of light. (Assume three significant figures.)

a. infrared radiation (1500 nm)
b. visible light (500 nm)
c. ultraviolet radiation (150 nm)

46. How much energy is contained in 1 mol of each?

a. X-ray photons with a wavelength of 0.135 nm
b. γ-ray photons with a wavelength of 2.15×10^{-5} nm

The Wave Nature of Matter and the Uncertainty Principle

47. Sketch the interference pattern that results from the diffraction of electrons passing through two closely spaced slits.

48. What happens to the interference pattern described in Problem 47 if the rate of electrons going through the slits is decreased to one electron per hour? What happens to the pattern if we try to determine which slit the electron goes through by using a laser placed directly behind the slits?

49. The resolution limit of a microscope is roughly equal to the wavelength of light used in producing the image. Electron microscopes use an electron beam (in place of photons) to produce much higher resolution images, about 0.20 nm in modern instruments. Assuming that the resolution of an electron microscope is equal to the de Broglie wavelength of the electrons used, to what speed must the electrons be accelerated to obtain a resolution of 0.20 nm?

50. The smallest atoms can themselves exhibit quantum-mechanical behavior. Calculate the de Broglie wavelength (in pm) of a hydrogen atom traveling 475 m/s.

51. What is the de Broglie wavelength of an electron traveling at 1.35×10^5 m/s?

52. A proton in a linear accelerator has a de Broglie wavelength of 122 pm. What is the speed of the proton?

53. Calculate the de Broglie wavelength of a 143-g baseball traveling at 95 mph. Why is the wave nature of matter not important for a baseball?

54. A 0.22-caliber handgun fires a 27-g bullet at a velocity of 765 m/s. Calculate the de Broglie wavelength of the bullet. Is the wave nature of matter significant for bullets?

55. An electron has an uncertainty in its position of 552 pm. What is the uncertainty in its velocity?

56. An electron traveling at 3.7×10^5 m/s has an uncertainty in its velocity of 1.88×10^5 m/s. What is the uncertainty in its position?

Orbitals and Quantum Numbers

57. Which electron is, on average, closer to the nucleus: an electron in a 2*s* orbital or an electron in a 3*s* orbital?

58. Which electron is, on average, further from the nucleus: an electron in a 3*p* orbital or an electron in a 4*p* orbital?

59. What are the possible values of l for each given value of n?

a. 1 b. 2 c. 3 d. 4

60. What are the possible values of m_l for each given value of l?

a. 0 b. 1 c. 2 d. 3

61. Which set of quantum numbers *cannot* occur together to specify an orbital?

a. $n = 2, l = 1, m_l = -1$
b. $n = 3, l = 2, m_l = 0$
c. $n = 3, l = 3, m_l = 2$
d. $n = 4, l = 3, m_l = 0$

62. Which combinations of n and l represent real orbitals, and which do not exist?

a. 1*s* b. 2*p* c. 4*s* d. 2*d*

63. Sketch the 1*s* and 2*p* orbitals. How do the 2*s* and 3*p* orbitals differ from the 1*s* and 2*p* orbitals?

64. Sketch the 3*d* orbitals. How do the 4*d* orbitals differ from the 3*d* orbitals?

Atomic Spectroscopy

65. An electron in a hydrogen atom is excited with electrical energy to an excited state with $n = 2$. The atom then emits a photon. What is the value of n for the electron following the emission?

66. Determine whether each transition in the hydrogen atom corresponds to absorption or emission of energy.

a. $n = 3 \longrightarrow n = 1$
b. $n = 2 \longrightarrow n = 4$
c. $n = 4 \longrightarrow n = 3$

67. According to the quantum-mechanical model for the hydrogen atom, which electron transition produces light with the longer wavelength: $2p \longrightarrow 1s$ or $3p \longrightarrow 1s$?

68. According to the quantum-mechanical model for the hydrogen atom, which electron transition produces light with the longer wavelength: $3p \longrightarrow 2s$ or $4p \longrightarrow 3p$?

69. Calculate the wavelength of the light emitted when an electron in a hydrogen atom makes each transition and indicate the region of the electromagnetic spectrum (infrared, visible, ultraviolet, etc.) where the light is found.

a. $n = 2 \longrightarrow n = 1$
b. $n = 3 \longrightarrow n = 1$
c. $n = 4 \longrightarrow n = 2$
d. $n = 5 \longrightarrow n = 2$

70. Calculate the frequency of the light emitted when an electron in a hydrogen atom makes each transition:

a. $n = 4 \longrightarrow n = 3$
b. $n = 5 \longrightarrow n = 1$
c. $n = 5 \longrightarrow n = 4$
d. $n = 6 \longrightarrow n = 5$

71. An electron in the $n = 7$ level of the hydrogen atom relaxes to a lower energy level, emitting light of 397 nm. What is the value of n for the level to which the electron relaxed?

72. An electron in a hydrogen atom relaxes to the $n = 4$ level, emitting light of 114 THz. What is the value of n for the level in which the electron originated?

CUMULATIVE PROBLEMS

73. Ultraviolet radiation and radiation of shorter wavelengths can damage biological molecules because they carry enough energy to break bonds within the molecules. A typical carbon–carbon bond requires 348 kJ/mol to break. What is the longest wavelength of radiation with enough energy to break carbon–carbon bonds?

74. The human eye contains a molecule called 11-*cis*-retinal that changes shape when struck with light of sufficient energy. The change in shape triggers a series of events that results in an electrical signal being sent to the brain. The minimum energy required to change the conformation of 11-*cis*-retinal within the eye is about 164 kJ/mol. Calculate the longest wavelength visible to the human eye.

75. An argon ion laser puts out 5.0 W of continuous power at a wavelength of 532 nm. The diameter of the laser beam is 5.5 mm. If the laser is pointed toward a pinhole with a diameter of 1.2 mm, how many photons will travel through the pinhole per second? Assume that the light intensity is equally distributed throughout the entire cross-sectional area of the beam. (1 W = 1 J/s)

76. A green leaf has a surface area of 2.50 cm^2. If solar radiation is 1000 W/m^2, how many photons strike the leaf every second? Assume three significant figures and an average wavelength of 504 nm for solar radiation.

77. In a technique used for surface analysis called auger electron spectroscopy (AES), electrons are accelerated toward a metal surface. These electrons cause the emissions of secondary electrons—called auger electrons—from the metal surface. The kinetic energy of the auger electrons depends on the composition of the surface. The presence of oxygen atoms on the surface results in auger electrons with a kinetic energy of approximately 506 eV. What is the de Broglie wavelength of one of these electrons?

$[\text{KE} = \frac{1}{2}mv^2; \text{ 1 electron volt (eV)} = 1.602 \times 10^{-19} \text{ J}]$

78. An X-ray photon of wavelength 0.989 nm strikes a surface. The emitted electron has a kinetic energy of 969 eV. What is the binding energy of the electron in kJ/mol?

$[\text{KE} = \frac{1}{2}mv^2; \text{ 1 electron volt (eV)} = 1.602 \times 10^{-19} \text{ J}]$

79. Ionization involves completely removing an electron from an atom. How much energy is required to ionize a hydrogen atom in its ground (or lowest energy) state? What wavelength of light contains enough energy in a single photon to ionize a hydrogen atom?

80. The energy required to ionize sodium is 496 kJ/mol. What minimum frequency of light is required to ionize sodium?

81. Suppose that in an alternate universe, the possible values of l are the integer values from 0 to n (instead of 0 to $n - 1$). Assuming no other differences between this imaginary universe and ours, how many orbitals would exist in each level?

a. $n = 1$
b. $n = 2$
c. $n = 3$

82. Suppose that, in an alternate universe, the possible values of m_l are the integer values including 0 ranging from $-l-1$ to $l+1$ (instead of simply $-l$ to $+l$). How many orbitals exist in each sublevel?

a. *s* sublevel
b. *p* sublevel
c. *d* sublevel

83. An atomic emission spectrum of hydrogen shows three wavelengths: 1875 nm, 1282 nm, and 1093 nm. Assign these wavelengths to transitions in the hydrogen atom.

84. An atomic emission spectrum of hydrogen shows three wavelengths: 121.5 nm, 102.6 nm, and 97.23 nm. Assign these wavelengths to transitions in the hydrogen atom.

85. The binding energy of electrons in a metal is 193 kJ/mol. Find the threshold frequency of the metal.

86. In order for a thermonuclear fusion reaction of two deuterons ($^2_1H^+$) to take place, the deuterons must collide with each deuteron traveling at 1×10^6 m/s. Find the wavelength of such a deuteron.

87. The speed of sound in air is 344 m/s at room temperature. The lowest frequency of a large organ pipe is 30 s^{-1} and the highest frequency of a piccolo is 1.5×10^4 s^{-1}. Determine the difference in wavelength between these two sounds.

88. The distance from Earth to the sun is 1.5×10^8 km. Find the number of crests in a light wave of frequency 1.0×10^{14} s^{-1} traveling from the sun to the Earth.

89. The iodine molecule can be photodissociated (broken apart with light) into iodine atoms in the gas phase with light of wavelengths shorter than about 792 nm. A glass tube contains 1.80×10^{17} iodine molecules. What minimum amount of light energy must be absorbed by the iodine in the tube to dissociate 15.0% of the molecules?

90. An ampule of napthalene in hexane contains 5.00×10^{-4} mol napthalene. The napthalene is excited with a flash of light and then emits 15.5 J of energy at an average wavelength of 349 nm. What percentage of the naphthalene molecules emitted a photon?

91. A laser produces 20.0 mW of red light. In 1.00 hr, the laser emits 2.29×10^{20} photons. What is the wavelength of the laser?

92. A particular laser consumes 150.0 Watts of electrical power and produces a stream of 1.33×10^{19} 1064 nm photons per second. What is the percent efficiency of the laser in converting electrical power to light?

CHALLENGE PROBLEMS

93. An electron confined to a one-dimensional box has energy levels given by the equation

$$E_n = n^2h^2/8\,mL^2$$

where n is a quantum number with possible values of 1, 2, 3,..., m is the mass of the particle, and L is the length of the box.

a. Calculate the energies of the $n = 1$, $n = 2$, and $n = 3$ levels for an electron in a box with a length of 155 pm.
b. Calculate the wavelength of light required to make a transition from $n = 1 \longrightarrow n = 2$ and from $n = 2 \longrightarrow n = 3$. In what region of the electromagnetic spectrum do these wavelengths lie?

94. The energy of a vibrating molecule is quantized much like the energy of an electron in the hydrogen atom. The energy levels of a vibrating molecule are given by the equation

$$E_n = \left(n + \frac{1}{2}\right)h\nu$$

where n is a quantum number with possible values of 1, 2,..., and ν is the frequency of vibration. The vibration frequency of HCl is approximately 8.85×10^{13} s^{-1}. What minimum energy is required to excite a vibration in HCl? What wavelength of light is required to excite this vibration?

95. The wave functions for the 1*s* and 2*s* orbitals are as follows:

$1s\ \psi = (1/\pi)^{1/2}\,(1/a_0^{3/2})\,\exp(-r/a_0)$

$2s\ \psi = (1/32\pi)^{1/2}\,(1/a_0^{3/2})(2 - r/a_0)\,\exp(-r/a_0)$

where a_0 is a constant ($a_0 = 53$ pm) and r is the distance from the nucleus. Use a spreadsheet to make a plot of each of these wave functions for values of r ranging from 0 pm to 200 pm. Describe the differences in the plots and identify the node in the 2*s* wave function.

96. Before quantum mechanics was developed, Johannes Rydberg developed an equation that predicted the wavelengths (λ) in the atomic spectrum of hydrogen:

$1/\lambda = R(1/m^2 - 1/n^2)$

In this equation R is a constant and m and n are integers. Use the quantum-mechanical model for the hydrogen atom to derive the Rydberg equation.

97. Find the velocity of an electron emitted by a metal whose threshold frequency is 2.25×10^{14} s^{-1} when it is exposed to visible light of wavelength 5.00×10^{-7} m.

98. Water is exposed to infrared radiation of wavelength 2.8×10^{-4} cm. Assume that all the radiation is absorbed and converted to heat. How many photons are required for the sample to absorb 16.72 J of heat?

99. The 2005 Nobel Prize in Physics was given, in part, to scientists who had made ultrashort pulses of light. These pulses are important in making measurements involving very short time periods. One challenge in making such pulses is the uncertainty principle, which can be stated with respect to energy and time as $\Delta E \cdot \Delta t \geq h/4\pi$. What is the energy uncertainty (ΔE) associated with a short pulse of laser light that lasts for only 5.0 femtoseconds (fs)? Suppose the low-energy end of the pulse had a wavelength of 722 nm. What is the wavelength of the high-energy end of the pulse that is limited only by the uncertainty principle?

100. A metal with a threshold frequency of 6.71×10^{14} s^{-1} emits an electron with a velocity of 6.95×10^5 m/s when radiation of 1.01×10^{15} s^{-1} strikes the metal. Calculate the mass of the electron.

101. Find the longest wavelength of a wave that can travel around in a circular orbit of radius 1.8 m.

102. The amount of heat to melt ice 0.333 is kJ/g. Find the number of photons of wavelength $= 6.42 \times 10^{-6}$ m that must be absorbed to melt 5.55×10^{-2} mol of ice.

CONCEPTUAL PROBLEMS

103. Explain the difference between the Bohr model for the hydrogen atom and the quantum-mechanical model. Is the Bohr model consistent with Heisenberg's uncertainty principle?

104. The light emitted from one of the following electronic transitions ($n = 4 \longrightarrow n = 3$ or $n = 3 \longrightarrow n = 2$) in the hydrogen atom causes the photoelectric effect in a particular metal while light from the other transition does not. Which transition causes the photoelectric effect and why?

105. Determine whether an interference pattern is observed on the other side of the slits in each experiment.

   a. An electron beam is aimed at two closely spaced slits. The beam is attenuated (made dimmer) to produce only 1 electron per minute.

   b. An electron beam is aimed at two closely spaced slits. A light beam is placed at each slit to determine when an electron goes through the slit.

   c. A high-intensity light beam is aimed at two closely spaced slits.

   d. A gun is fired at a solid wall containing two closely spaced slits. (Will the bullets that pass through the slits form an interference pattern on the other side of the solid wall?)

106. Which transition in the hydrogen atom results in emitted light with the longest wavelength?

   a. $n = 4 \longrightarrow n = 3$

   b. $n = 2 \longrightarrow n = 1$

   c. $n = 3 \longrightarrow n = 2$

ANSWERS TO CONCEPTUAL CONNECTIONS

Cc 3.1 **(a)** infrared < visible < X-ray **(b)** X-ray < visible < infrared

Cc 3.2 Observation A corresponds to 632 nm; observation B corresponds to 325 nm; and observation C corresponds to 455 nm. The shortest wavelength of light (highest energy per photon) corresponds to the photoelectrons with the greatest kinetic energy. The longest wavelength of light (lowest energy per photon) corresponds to the instance where no photoelectrons were observed.

Cc 3.3 Because of the baseball's large mass, its de Broglie wavelength is minuscule. (For a 150-g baseball, λ is on the order of 10^{-34} m.) This minuscule wavelength is insignificant compared to the size of the baseball itself, and therefore its effects are not observable.

Cc 3.4 **(c)** Since l can have a maximum value of $n - 1$, and since $n = 3$, then l can have a maximum value of 2.

Cc 3.5 **(d)** Since m_l can have the integer values (including 0) between $-l$ and $+l$, and since $l = 2$, the possible values of m_l are $-2, -1, 0, +1$, and $+2$.

Cc 3.6 **(c)** The energy difference between $n = 3$ and $n = 2$ is greatest because the energy differences get closer together with increasing n. The greater energy difference results in an emitted photon of greater energy and therefore shorter wavelength.

CHAPTER 4

"It is the function of science to discover the existence of a general reign of order in nature and to find the causes governing this order."

—Dmitri Mendeleev (1834–1907)

The majority of the material that composes most aircraft is aluminum.

Periodic Properties of the Elements

4.1 Aluminum: Low-Density Atoms Result in Low-Density Metal 101

4.2 Finding Patterns: The Periodic Law and the Periodic Table 102

4.3 Electron Configurations: How Electrons Occupy Orbitals 105

4.4 Electron Configurations, Valence Electrons, and the Periodic Table 112

4.5 How the Electron Configuration of an Element Relates to Its Properties 116

4.6 Periodic Trends in the Size of Atoms and Effective Nuclear Charge 119

4.7 Ions: Electron Configurations, Magnetic Properties, Ionic Radii, and Ionization Energy 124

4.8 Electron Affinities and Metallic Character 132

Key Learning Outcomes 137

GREAT ADVANCES IN SCIENCE occur not only when a scientist sees something new, but also when a scientist sees something everyone else has seen in a new way. That is what happened in 1869 when Dmitri Mendeleev, a Russian chemistry professor, saw a pattern in the properties of elements. Mendeleev's insight led to the development of the periodic table. Recall from Chapter 1 that theories explain the underlying reasons for observations. If we think of Mendeleev's periodic table as a compact way to summarize a large number of observations, then quantum mechanics is the theory that explains the underlying reasons. Quantum mechanics explains how electrons are arranged in an element's atoms, which in turn determines the element's properties. Because the periodic table is organized according to those properties, quantum mechanics elegantly accounts for Mendeleev's periodic table. In this chapter, we see a continuation of this book's theme—the properties of matter (in this case, the elements in the periodic table) are explained by the properties of the particles that compose them (in this case, atoms and their electrons).

4.1 Aluminum: Low-Density Atoms Result in Low-Density Metal

Look out the window of almost any airplane and you will see the large sheets of aluminum that compose the aircraft's wing. In fact, the majority of the plane is most likely made out of aluminum. Aluminum has several properties that make it suitable for airplane construction, but among the most important is its low density. Aluminum has a density of only 2.70 g/cm^3. For comparison, iron's density is 7.86 g/cm^3, and platinum's density is 21.4 g/cm^3. Why is the density of aluminum metal so low?

The density of an atom is not the only factor that determines the density of a solid. Other factors—such as how closely the atoms pack together—also contribute to determining the density.

The density of aluminum metal is low because the density of an aluminum atom is low. Few metal atoms have a lower mass-to-volume ratio than aluminum, and those that do can't be used in airplanes for other reasons (such as their high chemical reactivity). Although the *arrangements* of atoms in a solid must also be considered when evaluating the density of the solid, the mass-to-volume ratio of the composite atoms is a very important factor. For this reason, the densities of the elements generally follow a fairly well-defined trend: *The density of elements tends to increase as you move down a column in the periodic table.* For example, consider the densities of several of the elements in the column that includes aluminum in the periodic table:

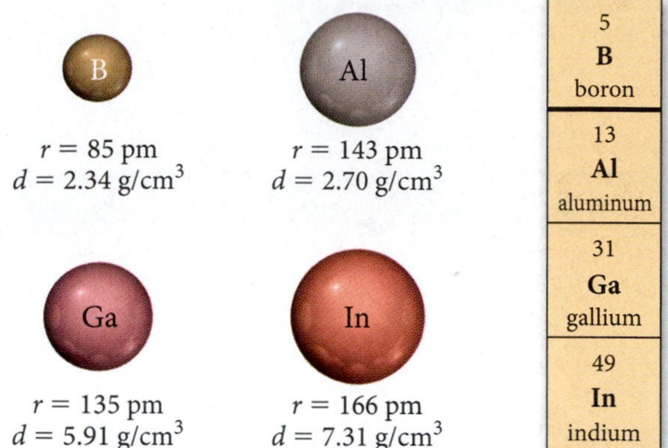

As you move down the column in the periodic table, the density of the elements increases even though the radius generally increases as well (with the exception of Ga whose radius decreases a bit). Why? *Because the mass of each successive atom increases even more than its volume does.* As you move down a column in the periodic table, the additional protons and neutrons add more mass to the atoms. This increase in mass is greater than the increase in volume, resulting in a higher denstity.

The densities of elements and the radii of their atoms are examples of *periodic properties*. A **periodic property** is one that is generally predictable based on an element's position within the periodic table. In this chapter, we examine several periodic properties of elements, including atomic radius, ionization energy, and electron affinity. As we do, we will see that these properties—as well as the overall arrangement of the periodic table—are explained by quantum-mechanical theory, which we first examined in Chapter 3. *Quantum-mechanical theory explains the electronic structure of atoms—this in turn determines the properties of those atoms.*

Notice again that *structure determines properties*. The arrangement of elements in the periodic table—originally based on similarities in the properties of the elements—reflects how electrons fill quantum-mechanical orbitals. Understanding the structure of atoms as explained by quantum mechanics allows us to predict the properties of elements from their position on the periodic table. If we need a metal with a high density, for example, we look toward the bottom of the periodic table. Platinum (as we saw previously) has density of 21.4 g/cm^3. It is among the densest metals and is found near the bottom of the periodic table. If we need a metal with a low density, we look toward the top of the periodic table. Aluminum is among the least dense metals and is found near the top of the periodic table.

4.2 Finding Patterns: The Periodic Law and the Periodic Table

Prior to the 1700s, the number of known elements was relatively small, consisting mostly of the metals used for coinage, jewelry, and weapons. From the early 1700s to the mid-1800s, however, chemists discovered over 50 new elements. The first attempt to organize these elements according to similarities in their properties was made by the German chemist Johann Döbereiner (1780–1849), who grouped elements into *triads*: A triad consisted of three elements with similar properties. For example, Döbereiner formed a triad out of barium, calcium, and strontium, three fairly reactive metals. About 50 years later, English chemist John Newlands (1837–1898) organized elements into *octaves*, in analogy to musical notes. When arranged this way, the properties of every eighth element were similar, much as every eighth note in the musical scale is similar. Newlands endured some ridicule for drawing an analogy between chemistry and music, including the derisive comments of one colleague who asked Newlands if he had ever tried ordering the elements according to the first letters of their names.

▲ Dmitri Mendeleev, a Russian chemistry professor who proposed the periodic law and arranged early versions of the periodic table, was honored on a Soviet postage stamp.

The modern periodic table is credited primarily to the Russian chemist Dmitri Mendeleev (1834–1907), even though a similar organization had been suggested by the German chemist Julius

The Periodic Law

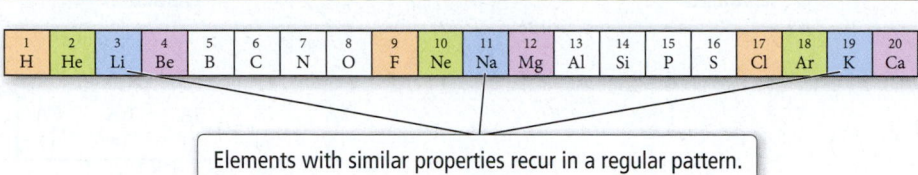

1 H	2 He	3 Li	4 Be	5 B	6 C	7 N	8 O	9 F	10 Ne	11 Na	12 Mg	13 Al	14 Si	15 P	16 S	17 Cl	18 Ar	19 K	20 Ca

Elements with similar properties recur in a regular pattern.

▲ **FIGURE 4.1 Recurring Properties** These elements are listed in order of increasing atomic number. Elements with similar properties are represented with the same color. Notice that the colors form a repeating pattern, much like musical notes form a repeating pattern on a piano keyboard.

A Simple Periodic Table

1 H							2 He
3 Li	4 Be	5 B	6 C	7 N	8 O	9 F	10 Ne
11 Na	12 Mg	13 Al	14 Si	15 P	16 S	17 Cl	18 Ar
19 K	20 Ca						

Elements with similar properties fall into columns.

▲ **FIGURE 4.2 Making a Periodic Table** We can arrange the elements in Figure 4.1 in a table where atomic number increases from left to right and elements with similar properties (as represented by the different colors) are aligned in columns.

Lothar Meyer (1830–1895). In 1869, Mendeleev noticed that certain groups of elements had similar properties. He also found that when he listed elements in order of increasing mass, these similar properties recurred in a periodic pattern (Figure 4.1 ▲). Mendeleev summarized these observations in the **periodic law**:

> **When the elements are arranged in order of increasing mass, certain sets of properties recur periodically.**

Mendeleev organized the known elements in a table consisting of a series of rows in which mass increases from left to right. He arranged the rows so that elements with similar properties fall in the same vertical columns (Figure 4.2 ▶).

Mendeleev's arrangement was a huge success, allowing him to predict the existence and properties of yet undiscovered elements such as eka-aluminum (later discovered and named gallium) and eka-silicon (later discovered and named germanium). The properties of these two elements are summarized in Figure 4.3 ▼.

However, Mendeleev did encounter some difficulties. For example, according to accepted values of atomic masses, tellurium (with higher mass) should come *after* iodine. But based on their properties, Mendeleev placed tellurium *before* iodine and suggested that the mass of tellurium was erroneous. The mass was correct; later work by the English physicist Henry Moseley (1887–1915) showed that listing elements according to *atomic number*, rather than atomic mass, resolved this problem and resulted in even better correlation with elemental properties. Mendeleev's original listing evolved into the modern

To be periodic means to exhibit a repeating pattern.

Eka means the one beyond or the next one in a family of elements. So, eka-silicon means the element beyond silicon in the same family as silicon.

Gallium (eka-aluminum)

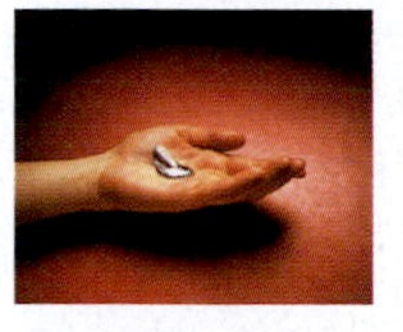

	Mendeleev's predicted properties	Actual properties
Atomic mass	About 68 amu	69.72 amu
Melting point	Low	29.8 °C
Density	5.9 g/cm^3	5.90 g/cm^3
Formula of oxide	X_2O_3	Ga_2O_3
Formula of chloride	XCl_3	$GaCl_3$

Germanium (eka-silicon)

	Mendeleev's predicted properties	Actual properties
Atomic mass	About 72 amu	72.64 amu
Density	5.5 g/cm^3	5.35 g/cm^3
Formula of oxide	XO_2	GeO_2
Formula of chloride	XCl_4	$GeCl_4$

▲ **FIGURE 4.3 Eka-aluminum and Eka-silicon** Mendeleev's arrangement of elements in the periodic table allowed him to predict the existence of these elements, now known as gallium and germanium, and to anticipate their properties.

Main-group elements | Transition elements | Main-group elements

Group number

Periods	1A 1	2A 2	3B 3	4B 4	5B 5	6B 6	7B 7	8B 8	8B 9	8B 10	1B 11	2B 12	3A 13	4A 14	5A 15	6A 16	7A 17	8A 18
1	1 **H**																	2 **He**
2	3 **Li**	4 **Be**											5 **B**	6 **C**	7 **N**	8 **O**	9 **F**	10 **Ne**
3	11 **Na**	12 **Mg**											13 **Al**	14 **Si**	15 **P**	16 **S**	17 **Cl**	18 **Ar**
4	19 **K**	20 **Ca**	21 **Sc**	22 **Ti**	23 **V**	24 **Cr**	25 **Mn**	26 **Fe**	27 **Co**	28 **Ni**	29 **Cu**	30 **Zn**	31 **Ga**	32 **Ge**	33 **As**	34 **Se**	35 **Br**	36 **Kr**
5	37 **Rb**	38 **Sr**	39 **Y**	40 **Zr**	41 **Nb**	42 **Mo**	43 **Tc**	44 **Ru**	45 **Rh**	46 **Pd**	47 **Ag**	48 **Cd**	49 **In**	50 **Sn**	51 **Sb**	52 **Te**	53 **I**	54 **Xe**
6	55 **Cs**	56 **Ba**	57 **La**	72 **Hf**	73 **Ta**	74 **W**	75 **Re**	76 **Os**	77 **Ir**	78 **Pt**	79 **Au**	80 **Hg**	81 **Tl**	82 **Pb**	83 **Bi**	84 **Po**	85 **At**	86 **Rn**
7	87 **Fr**	88 **Ra**	89 **Ac**	104 **Rf**	105 **Db**	106 **Sg**	107 **Bh**	108 **Hs**	109 **Mt**	110 **Ds**	111 **Rg**	112 **Cn**	113	114 **Fl**	115	116 **Lv**	117	118

58 **Ce** cerium	59 **Pr** praseodymium	60 **Nd** neodymium	61 **Pm** promethium	62 **Sm** samarium	63 **Eu** europium	64 **Gd** gadolinium	65 **Tb** terbium	66 **Dy** dysprosium	67 **Ho** holmium	68 **Er** erbium	69 **Tm** thulium	70 **Yb** ytterbium	71 **Lu** lutetium
90 **Th** thorium	91 **Pa** protactinium	92 **U** uranium	93 **Np** neptunium	94 **Pu** plutonium	95 **Am** americium	96 **Cm** curium	97 **Bk** berkelium	98 **Cf** californium	99 **Es** einsteinium	100 **Fm** fermium	101 **Md** mendelevium	102 **No** nobelium	103 **Lr** lawrencium

Inner transition elements

▲ **FIGURE 4.4 The Modern Periodic Table** The elements in the periodic table fall into columns. The two columns at the left and the six columns at the right constitute the main-group elements. The elements that constitute any one column are a *group* or *family*. The properties of main-group elements can generally be predicted from their position in the periodic table. The properties of the elements in the middle of the table, known as transition elements, and those at the bottom of the table, known as the inner transition elements, are less predictable based on their position within the table.

periodic table shown in Figure 4.4 ▲. In the modern table, elements are listed in order of increasing atomic number rather than increasing relative mass. The modern periodic table also contains more elements than Mendeleev's original table because more have been discovered since then.

We divide the periodic table, as shown in Figure 4.4, into **main-group elements**, whose properties tend to be largely predictable based on their position in the periodic table, and **transition elements** (or **transition metals**) and inner transition elements, whose properties tend to be less predictable based simply on their position in the periodic table. Main-group elements are in columns labeled with a number and the letter A. Transition elements are in columns labeled with a number and the letter B. An alternative numbering system does not use letters, but only the numbers 1–18. Both numbering systems are shown in most of the periodic tables in this book. Each column within the main-group regions of the periodic table is a **family** or **group** of elements. A family of elements has similar properties as observed by Mendeleev.

Notice the scientific approach in practice in the history of the periodic table. A number of related observations led to a scientific law—the periodic law. Mendeleev's table, an expression of the periodic law, has predictive power, as laws usually do. However, it does not explain *why* the properties of elements recur or *why* certain elements have similar properties. Quantum-mechanical theory explains the electronic structure of atoms, which in turn determines their properties. Since a family of elements has similar properties, we expect the electronic structure of their atoms to have similarities as well. We now turn to examining those similarities.

4.3 Electron Configurations: How Electrons Occupy Orbitals

As we saw in Chapter 3, electrons in atoms exist within orbitals. An **electron configuration** for an atom shows the particular orbitals that electrons occupy for that atom. For example, consider the **ground state**—or lowest energy state—electron configuration for a hydrogen atom:

H $1s^1$ ← Number of electrons in orbital

↑ Orbital

The electron configuration indicates that hydrogen's one electron is in the 1*s* orbital. Electrons generally occupy the lowest energy orbitals available. Since the 1*s* orbital is the lowest energy orbital in hydrogen (see Section 3.5), hydrogen's electron occupies that orbital. If we could write electron configurations for all the elements, we could see how the arrangements of the electrons within their atoms correlate with the element's chemical properties. However, the Schrödinger equation solutions (the atomic orbitals and their energies) that we described in Chapter 3 are for the hydrogen atom only. What do the atomic orbitals of *other atoms* look like? What are their relative energies?

The Schrödinger equation for multi-electron atoms includes terms to account for the interactions of the electrons with one another that make it too complicated to solve exactly. However, approximate solutions indicate that the orbitals in multi-electron atoms are hydrogen-like—they are similar to the *s*, *p*, *d*, and *f* orbitals that we examined in Chapter 3. In order to see how the electrons in multi-electron atoms occupy these hydrogen-like orbitals, we must examine two additional concepts: *the effects of electron spin*, a fundamental property of all electrons that affects the number of electrons allowed in any one orbital; and *sublevel energy splitting*, which determines the order of orbital filling within a level.

Electron Spin and the Pauli Exclusion Principle

We can represent the electron configuration of hydrogen ($1s^1$) in a slightly different way with an **orbital diagram**, which is similar to an electron configuration but symbolizes the electron as an arrow and the orbital as a box. The orbital diagram for a hydrogen atom is:

H [↿]
1*s*

In an orbital diagram, the direction of the arrow (pointing up or pointing down) represents the orientation of the *electron's spin*. Recall from Section 3.5 that the orientation of the electron's spin is quantized, with only two possibilities: spin up ($m_s = +\frac{1}{2}$) and spin down ($m_s = -\frac{1}{2}$). In an orbital diagram, we represent $m_s = +\frac{1}{2}$ with a half-arrow pointing up (↿) and $m_s = -\frac{1}{2}$ with a half-arrow pointing down (⇂). In a collection of hydrogen atoms, the electrons in about half of the atoms are spin up and the electrons in the other half are spin down. Since no additional electrons are present within the hydrogen atom, we conventionally represent the hydrogen atom electron configuration with its one electron as spin up.

Helium is the first element on the periodic table that contains two electrons. Its two electrons occupy the 1*s* orbital.

He $1s^2$

How do the spins of the two electrons in helium align relative to each other? The answer to this question is addressed by the **Pauli exclusion principle**, formulated by Wolfgang Pauli in 1925.

> **Pauli exclusion principle: No two electrons in an atom can have the same four quantum numbers.**

Since the two electrons occupying the same orbital have three identical quantum numbers (n, l, and m_l), they each must have a different spin quantum number. Since there are only two possible spin quantum numbers ($+\frac{1}{2}$ and $-\frac{1}{2}$), the Pauli exclusion principle implies that *each orbital can have a maximum of only two electrons, with opposing spins.* By applying the exclusion principle, we can write an electron configuration and orbital diagram for helium:

	Electron configuration	Orbital diagram
He	$1s^2$	⇅ (1s)

The following table shows the four quantum numbers for each of the two electrons in helium.

n	l	m_l	m_s
1	0	0	$+\frac{1}{2}$
1	0	0	$-\frac{1}{2}$

The two electrons have three quantum numbers in common (because they are in the same orbital), but each electron has a different spin quantum number (as indicated by the opposing half-arrows in the orbital diagram).

Sublevel Energy Splitting in Multi-electron Atoms

A major difference in the (approximate) solutions to the Schrödinger equation for multi-electron atoms compared to the solutions for the hydrogen atom is the energy ordering of the orbitals. In the hydrogen atom, the energy of an orbital depends only on n, the principal quantum number. For example, the $3s$, $3p$, and $3d$ orbitals (which are empty for hydrogen in its lowest energy state) all have the same energy—we say they are **degenerate**. The orbitals within a principal level of a *multi-electron atom*, in contrast, are not degenerate—their energy depends on the value of l. We say that the energies of the sublevels are *split*. In general, the lower the value of *l within a principal level*, the lower the energy (E) of the corresponding orbital. Thus, for a given value of n:

$$E(s\text{ orbital}) < E(p\text{ orbital}) < E(d\text{ orbital}) < E(f\text{ orbital})$$

To understand why the sublevels split in this way, we must examine three key concepts associated with the energy of an electron in the vicinity of a nucleus: (1) Coulomb's law, which describes the interactions between charged particles; (2) shielding, which describes how one electron can shield another electron from the full charge of the nucleus; and (3) penetration, which describes how one atomic orbital can overlap spatially with another, thus penetrating into a region that is close to the nucleus (and therefore less shielded from nuclear charge). We then examine how these concepts, together with the spatial distributions of electron probability for each orbital, result in the energy ordering presented above.

Coulomb's Law The attractions and repulsions between charged particles, first introduced in Section 1.6, are described by **Coulomb's law**, which states that the potential energy (E) of two charged particles depends on their charges (q_1 and q_2) and on their separation (r):

$$E = \frac{1}{4\pi\varepsilon_0}\frac{q_1 q_2}{r} \qquad [4.1]$$

In this equation, ε_0 is a constant ($\varepsilon_0 = 8.85 \times 10^{-12}\ \text{C}^2/\text{J}\cdot\text{m}$). The potential energy is positive for the interaction of charges with the same sign (plus × plus, or minus × minus), and negative for charges of the opposite sign (plus × minus, or minus × plus). The *magnitude* of the potential energy depends inversely on the separation between the charged particles. We can draw three important conclusions from Coulomb's law:

- The potential energy (E) associated with the interaction of like charges is positive, but decreases as the particles get *farther apart* (as r increases). Since systems tend toward lower potential energy (see Section 2.4), like charges that are close together have high potential energy and tend to move away from each other (toward lower potential energy). Like charges therefore repel one another (in much the same way that like poles of two magnets repel each other).

- The potential energy (E) associated with the interaction of unlike charges is negative and becomes more negative as the particles get closer together. Since systems tend toward lower potential energy, the interaction of unlike charges draws them closer together (toward lower potential energy). Unlike charges therefore attract one another (like the opposite poles of a magnet).
- The *magnitude* of the interaction between charged particles increases as the charges of the particles increases. Consequently, an electron with a charge of 1− is more strongly attracted to a nucleus with a charge of 2+ than it would be to a nucleus with a charge of 1+.

Coulomb's Law

According to Coulomb's law, what happens to the potential energy of two oppositely charged particles as they get closer together?

(a) Their potential energy decreases.

(b) Their potential energy increases.

(c) Their potential energy does not change.

Shielding For multi-electron atoms, any one electron experiences both the positive charge of the nucleus (which is attractive) and the negative charges of the other electrons (which are repulsive). We can think of the repulsion of one electron by other electrons as *screening* or **shielding** that electron from the full effects of the nuclear charge. For example, consider a lithium ion (Li^+). Because the lithium ion contains two electrons, its electron configuration is identical to that of helium:

$$Li^+ \quad 1s^2$$

Now imagine bringing a third electron toward the lithium ion. When the third electron is far from the nucleus, it experiences the 3+ charge of the nucleus through the *screen* or *shield* of the 2− charge of the two 1*s* electrons, as shown in Figure 4.5a ▼. We can think of the third electron as experiencing an **effective nuclear charge (Z_{eff})** of approximately 1+ (3+ from the nucleus and 2− from the electrons, for a net charge of 1+). The inner electrons in effect *shield* the outer electron from the full nuclear charge.

Penetration Now imagine allowing this third electron to come closer to the nucleus. As the electron *penetrates* the electron cloud of the 1*s* electrons, it begins to experience the 3+ charge of the nucleus more fully because it is less shielded by the intervening electrons. If the electron could somehow get closer to the nucleus than the 1*s* electrons, it would experience the full 3+ charge, as shown in Figure 4.5b ▼. In other words, as the outer electron undergoes **penetration** into the region occupied by the inner electrons, it experiences a greater nuclear charge and therefore (according to Coulomb's law) a lower energy.

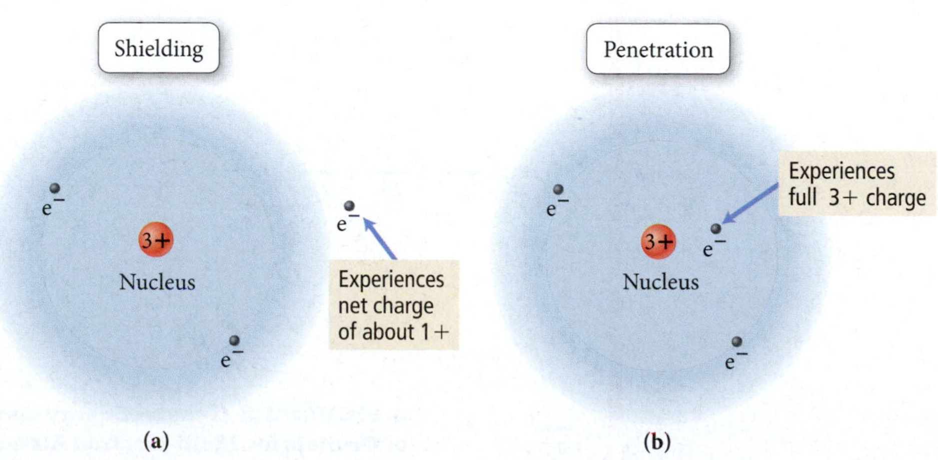

◀ **FIGURE 4.5 Shielding and Penetration** **(a)** An electron far from the nucleus is partly shielded by the electrons in the 1*s* orbital, reducing the effective net nuclear charge that it experiences. **(b)** An electron that penetrates the electron cloud of the 1*s* orbital experiences more of the nuclear charge.

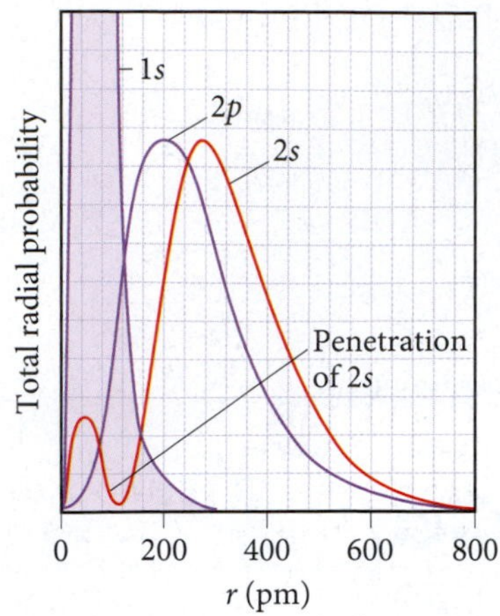

▲ **FIGURE 4.6 Radial Distribution Functions for the 1*s*, 2*s*, and 2*p* Orbitals**

Electron Spatial Distributions and Sublevel Splitting We have now examined all of the concepts we need to understand the energy splitting of the sublevels within a principal level. The splitting is a result of the spatial distributions of electrons within a sublevel. Recall from Section 3.6 that the radial distribution function for an atomic orbital shows the total probability of finding the electron within a thin spherical shell at a distance r from the nucleus. Figure 4.6 ◀ shows the radial distribution functions of the 2*s* and 2*p* orbitals superimposed on one another (the radial distribution function of the 1*s* orbital is also shown). Notice that, in general, an electron in a 2*p* orbital has a greater probability of being found closer to the nucleus than an electron in a 2*s* orbital. We might initially expect, therefore, that the 2*p* orbital would be lower in energy. However, exactly the opposite is true—the 2*s* orbital is actually lower in energy, *but only when the 1s orbital is occupied.* (When the 1*s* orbital is empty, the 2*s* and 2*p* orbitals are degenerate.) Why? The reason is the bump near $r = 0$, the nucleus for the 2*s* orbital. This bump represents a significant probability of the electron being found very close to the nucleus. Even more importantly, this area of the probability penetrates into the 1*s* orbital—it gets into the region where shielding by the 1*s* electrons is less effective. In contrast, most of the probability in the radial distribution function of the 2*p* orbital lies *outside* the radial distribution function of the 1*s* orbital. Consequently, almost all of the 2*p* orbital is shielded from nuclear charge by the 1*s* orbital. The 2*s* orbital—since it experiences more of the nuclear charge due to its greater *penetration*—is lower in energy than the 2*p* orbital. The situation is similar when we compare the 3*s*, 3*p*, and 3*d* orbitals. The *s* orbitals penetrate more fully than the *p* orbitals, which in turn penetrate more fully than the *d* orbitals, as shown in Figure 4.7 ▼.

Figure 4.8 ▼ shows the energy ordering of a number of orbitals in multi-electron atoms. Notice these features of Figure 4.8:

- Because of penetration, the sublevels of each principal level are *not* degenerate for multi-electron atoms.
- In the fourth and fifth principal levels, the effects of penetration become so important that the 4*s* orbital lies lower in energy than the 3*d* orbitals and the 5*s* orbital lies lower in energy than the 4*d* orbitals.
- The energy separations between one set of orbitals and the next become smaller for 4*s* orbitals and beyond, and the relative energy ordering of these orbitals can actually vary among elements. These variations result in irregularities in the electron configurations of the transition metals and their ions (as we shall see in Section 4.4).

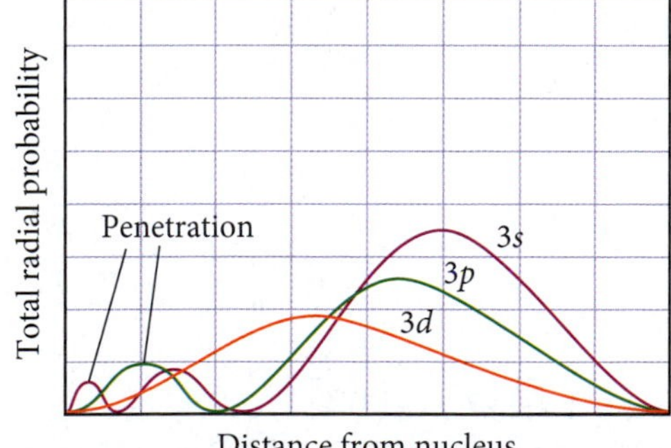

▲ **FIGURE 4.7 Radial Distribution Functions for the 3*s*, 3*p*, and 3*d* Orbitals** The 3*s* electrons penetrate most deeply into the inner orbitals, are least shielded, and experience the greatest effective nuclear charge. The 3*d* electrons penetrate least. This accounts for the energy ordering of the sublevels: $s < p < d$.

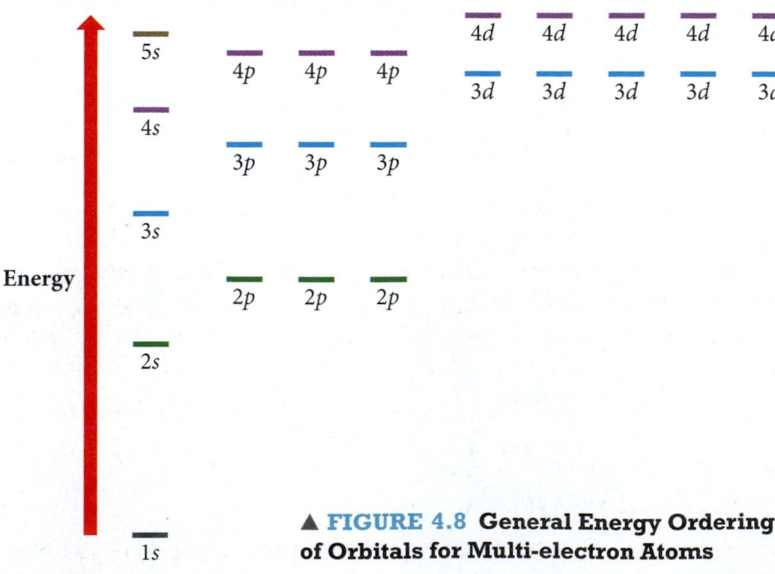

▲ **FIGURE 4.8 General Energy Ordering of Orbitals for Multi-electron Atoms**

Penetration and Shielding

4.2 Cc Conceptual Connection

Which statement is true?

(a) An orbital that penetrates into the region occupied by core electrons is more shielded from nuclear charge than an orbital that does not penetrate and will therefore have a higher energy.

(b) An orbital that penetrates into the region occupied by core electrons is less shielded from nuclear charge than an orbital that does not penetrate and will therefore have a higher energy.

(c) An orbital that penetrates into the region occupied by core electrons is less shielded from nuclear charge than an orbital that does not penetrate and will therefore have a lower energy.

(d) An orbital that penetrates into the region occupied by core electrons is more shielded from nuclear charge than an orbital that does not penetrate and will therefore have a lower energy.

Electron Configurations for Multi-electron Atoms

Now that we know the energy ordering of orbitals in multi-electron atoms, we can determine ground state electron configurations for the rest of the elements. Because electrons occupy the lowest energy orbitals available when the atom is in its ground state, and only two electrons (with opposing spins) are allowed in each orbital, we can systematically build up the electron configurations for the elements. This pattern of orbital filling is known as the **aufbau principle** (the German word *aufbau* means "build up"). For lithium, with three electrons, the electron configuration and orbital diagram are:

Unless otherwise specified, we use the term *electron configuration* to mean the ground state (or lowest energy) configuration.

Electron configuration — Orbital diagram

Li $1s^2 2s^1$

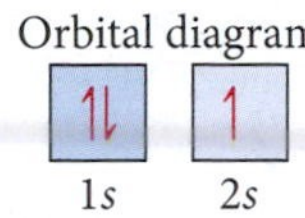

Remember that the number of electrons in a neutral atom is equal to its atomic number.

For carbon, which has six electrons, the electron configuration and orbital diagram are:

Electron configuration — Orbital diagram

C $1s^2 2s^2 2p^2$

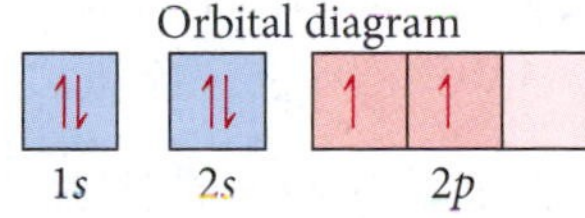

Notice that the $2p$ electrons occupy the p orbitals (of equal energy) singly, rather than pairing in one orbital. This way of filling orbitals follows **Hund's rule**, which states that *when filling degenerate orbitals, electrons fill them singly first, with parallel spins.* Hund's rule is a result of an atom's tendency to find the lowest energy state possible. When two electrons occupy separate orbitals of equal energy, the repulsive interaction between them is lower than when they occupy the same orbital because the electrons are spread out over a larger region of space. By convention we denote these parallel spins with half arrows pointing up.

Electrons with parallel spins have correlated motion that minimizes their mutual repulsion.

Summarizing Orbital Filling

- Electrons occupy orbitals so as to minimize the energy of the atom; therefore, lower energy orbitals fill before higher energy orbitals. Orbitals fill in the following order: $1s$ $2s$ $2p$ $3s$ $3p$ $4s$ $3d$ $4p$ $5s$ $4d$ $5p$ $6s$.
- Orbitals can hold no more than two electrons each. When two electrons occupy the same orbital, their spins are opposite. This is another way of expressing the Pauli exclusion principle (no two electrons in one atom can have the same four quantum numbers).
- When orbitals of identical energy are available, electrons first occupy these orbitals singly with parallel spins rather than in pairs (Hund's rule). Once the orbitals of equal energy are half-full, the electrons start to pair.

Consider the electron configurations and orbital diagrams for the elements with atomic numbers 3–10.

Symbol	Number of electrons	Electron configuration	Orbital diagram
Li	3	$1s^2 2s^1$	1s: ↑↓ 2s: ↑
Be	4	$1s^2 2s^2$	1s: ↑↓ 2s: ↑↓
B	5	$1s^2 2s^2 2p^1$	1s: ↑↓ 2s: ↑↓ 2p: ↑ _ _
C	6	$1s^2 2s^2 2p^2$	1s: ↑↓ 2s: ↑↓ 2p: ↑ ↑ _

Notice that, as a result of Hund's rule, the *p* orbitals fill with single electrons before the electrons pair.

Symbol	Number of electrons	Electron configuration	Orbital diagram
N	7	$1s^2 2s^2 2p^3$	1s: ↑↓ 2s: ↑↓ 2p: ↑ ↑ ↑
O	8	$1s^2 2s^2 2p^4$	1s: ↑↓ 2s: ↑↓ 2p: ↑↓ ↑ ↑
F	9	$1s^2 2s^2 2p^5$	1s: ↑↓ 2s: ↑↓ 2p: ↑↓ ↑↓ ↑
Ne	10	$1s^2 2s^2 2p^6$	1s: ↑↓ 2s: ↑↓ 2p: ↑↓ ↑↓ ↑↓

The electron configuration of neon represents the complete filling of the $n = 2$ principal level. When writing electron configurations for elements beyond neon, or beyond any other noble gas, we abbreviate the electron configuration of the previous noble gas—sometimes called the *inner electron configuration*—by the symbol for the noble gas in square brackets. For example, the electron configuration of sodium is:

$$\text{Na} \quad 1s^2 2s^2 2p^6 3s^1$$

We write this configuration more compactly by using [Ne] to represent the inner electrons:

$$\text{Na} \quad [\text{Ne}]\, 3s^1$$

[Ne] represents $1s^2\, 2s^2\, 2p^6$, the electron configuration for neon.

To write an electron configuration for an element, we first find its atomic number from the periodic table—this number equals the number of electrons. Then we use the order of filling to distribute the electrons in the appropriate orbitals. Remember that each orbital can hold a maximum of two electrons. Consequently,

- The *s* sublevel has only one orbital and can therefore hold only 2 electrons.
- The *p* sublevel has three orbitals and can hold 6 electrons.
- The *d* sublevel has five orbitals and can hold 10 electrons.
- The *f* sublevel has seven orbitals and can hold 14 electrons.

EXAMPLE 4.1
Electron Configurations

Write an electron configuration for each element.

(a) Mg **(b)** P **(c)** Br **(d)** Al

SOLUTION

(a) Mg Magnesium has 12 electrons. Distribute two electrons into the 1*s* orbital, two into the 2*s* orbital, six into the 2*p* orbitals, and two into the 3*s* orbital.	Mg $1s^2 2s^2 2p^6 3s^2$ or [Ne] $3s^2$
(b) P Phosphorus has 15 electrons. Distribute two electrons into the 1*s* orbital, two into the 2*s* orbital, six into the 2*p* orbitals, two into the 3*s* orbital, and three into the 3*p* orbitals.	P $1s^2 2s^2 2p^6 3s^2 3p^3$ or [Ne] $3s^2 3p^3$
(c) Br Bromine has 35 electrons. Distribute two electrons into the 1*s* orbital, two into the 2*s* orbital, six into the 2*p* orbitals, two into the 3*s* orbital, six into the 3*p* orbitals, two into the 4*s* orbital, ten into the 3*d* orbitals, and five into the 4*p* orbitals.	Br $1s^2 2s^2 2p^6 3s^2 3p^6 4s^2 3d^{10} 4p^5$ or [Ar] $4s^2 3d^{10} 4p^5$
(d) Al Aluminum has 13 electrons. Distribute two electrons into the 1*s* orbital, two into the 2*s* orbital, six into the 2*p* orbitals, two into the 3*s* orbital, and one into the 3*p* orbital.	Al $1s^2 2s^2 2p^6 3s^2 3p^1$ or [Ne] $3s^2 3p^1$

FOR PRACTICE 4.1

Write electron configurations for each element.

(a) Cl **(b)** Si **(c)** Sr **(d)** O

EXAMPLE 4.2
Writing Orbital Diagrams

Write the orbital diagram for sulfur and determine its number of unpaired electrons.

SOLUTION

Sulfur's atomic number is 16, so it has 16 electrons and the electron configuration $1s^2 2s^2 2p^6 3s^2 3p^4$. Draw a box for each orbital, putting the lowest energy orbital (1*s*) on the far left and proceeding to orbitals of higher energy to the right.	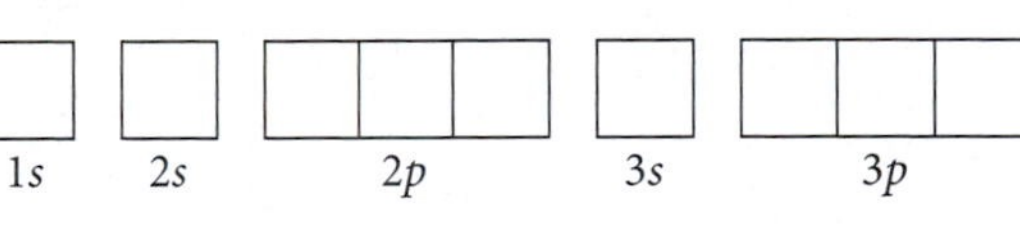
Distribute the 16 electrons into the boxes representing the orbitals, allowing a maximum of two electrons per orbital and remembering Hund's rule. You can see from the diagram that sulfur has two unpaired electrons.	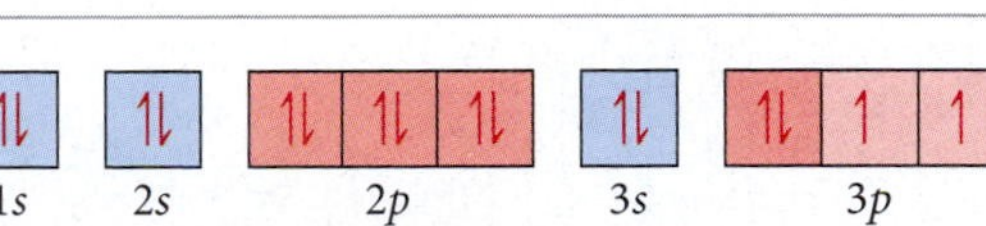 Two unpaired electrons

FOR PRACTICE 4.2

Write the orbital diagram for Ar and determine its number of unpaired electrons.

4.3 Cc Conceptual Connection

Electron Configurations and Quantum Numbers

What are the four quantum numbers for each of the two electrons in a 4*s* orbital?

4.4 Electron Configurations, Valence Electrons, and the Periodic Table

Recall from Section 4.2 that Mendeleev arranged the periodic table so that elements with similar chemical properties lie in the same column. We can begin to make the connection between an element's properties and its electron configuration by superimposing the electron configurations of the first 18 elements onto a partial periodic table, as shown in Figure 4.9 ▼. As we move to the right across a row (which is also called a period), the orbitals fill in the correct order. With each subsequent row, the highest principal quantum number increases by one. Notice that as we move down a column, *the number of electrons in the outermost principal energy level (highest n value) remains the same.* The key connection between the macroscopic world (an element's chemical properties) and the particulate world (an atom's electronic structure) lies in these outermost electrons.

An atom's **valence electrons** are the most important in chemical bonding. *For main-group elements, the valence electrons are those in the outermost principal energy level.* For transition elements, we also count the outermost *d* electrons among the valence electrons (even though they are not in an outermost principal energy level). The chemical properties of an element depend on its valence electrons, which are instrumental in bonding because they are held most loosely (and are therefore the easiest to lose or share). We can now see *why* the elements in a column of the periodic table have similar chemical properties: *They have the same number of valence electrons.*

We distinguish valence electrons from all the other electrons in an atom, which we call **core electrons**. The core electrons are those in *complete* principal energy levels and those in *complete d* and *f* sublevels. For example, silicon, with the electron configuration $1s^2 2s^2 2p^6 3s^2 3p^2$ has four valence electrons (those in the $n = 3$ principal level) and ten core electrons.

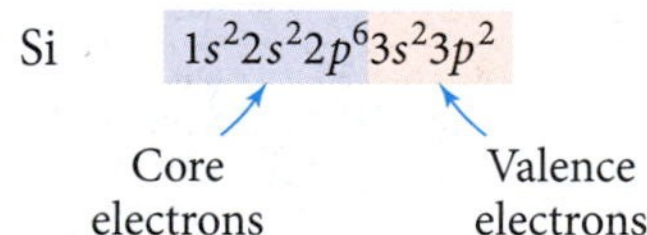

Outer Electron Configurations of Elements 1–18

1A	2A	3A	4A	5A	6A	7A	8A
1 **H** $1s^1$							2 **He** $1s^2$
3 **Li** $2s^1$	4 **Be** $2s^2$	5 **B** $2s^2 2p^1$	6 **C** $2s^2 2p^2$	7 **N** $2s^2 2p^3$	8 **O** $2s^2 2p^4$	9 **F** $2s^2 2p^5$	10 **Ne** $2s^2 2p^6$
11 **Na** $3s^1$	12 **Mg** $3s^2$	13 **Al** $3s^2 3p^1$	14 **Si** $3s^2 3p^2$	15 **P** $3s^2 3p^3$	16 **S** $3s^2 3p^4$	17 **Cl** $3s^2 3p^5$	18 **Ar** $3s^2 3p^6$

▲ **FIGURE 4.9 Outer Electron Configurations of the First 18 Elements in the Periodic Table**

EXAMPLE 4.3

Valence Electrons and Core Electrons

Write the electron configuration for Ge. Identify the valence electrons and the core electrons.

SOLUTION

To write the electron configuration for Ge, determine the total number of electrons from germanium's atomic number (32), and then distribute them into the appropriate orbitals.	Ge $1s^2 2s^2 2p^6 3s^2 3p^6 4s^2 3d^{10} 4p^2$
Because germanium is a main-group element, its valence electrons are those in the outermost principal energy level. For germanium, the $n = 1, 2$, and 3 principal levels are complete (or full), and the $n = 4$ principal level is outermost. Consequently, the $n = 4$ electrons are valence electrons and the rest are core electrons. *Note: In this book, we always write electron configurations with the orbitals in the* order of filling. *However, it is also common to write electron configurations in* the order of increasing principal quantum number. *The electron configuration of germanium written in order of increasing principal quantum number is* Ge $1s^2 2s^2 2p^6 3s^2 3p^6 3d^{10} 4s^2 4p^2$	4 valence electrons Ge $1s^2 2s^2 2p^6 3s^2 3p^6 4s^2 3d^{10} 4p^2$ 28 core electrons

FOR PRACTICE 4.3

Write an electron configuration for phosphorus. Identify the valence electrons and core electrons.

Orbital Blocks in the Periodic Table

A pattern similar to what we saw for the first 18 elements exists for the entire periodic table, as shown in Figure 4.10 ▼. Note that, because of the filling order of orbitals, the periodic table can be divided

Orbital Blocks of the Periodic Table

Legend: *s*-block elements; *p*-block elements; *d*-block elements; *f*-block elements

Periods / Groups	1 1A	2 2A	3 3B	4 4B	5 5B	6 6B	7 7B	8 8B	9 8B	10 8B	11 1B	12 2B	13 3A	14 4A	15 5A	16 6A	17 7A	18 8A
1	1 **H** $1s^1$																	2 **He** $1s^2$
2	3 **Li** $2s^1$	4 **Be** $2s^2$											5 **B** $2s^2 2p^1$	6 **C** $2s^2 2p^2$	7 **N** $2s^2 2p^3$	8 **O** $2s^2 2p^4$	9 **F** $2s^2 2p^5$	10 **Ne** $2s^2 2p^6$
3	11 **Na** $3s^1$	12 **Mg** $3s^2$											13 **Al** $3s^2 3p^1$	14 **Si** $3s^2 3p^2$	15 **P** $3s^2 3p^3$	16 **S** $3s^2 3p^4$	17 **Cl** $3s^2 3p^5$	18 **Ar** $3s^2 3p^6$
4	19 **K** $4s^1$	20 **Ca** $4s^2$	21 **Sc** $4s^2 3d^1$	22 **Ti** $4s^2 3d^2$	23 **V** $4s^2 3d^3$	24 **Cr** $4s^1 3d^5$	25 **Mn** $4s^2 3d^5$	26 **Fe** $4s^2 3d^6$	27 **Co** $4s^2 3d^7$	28 **Ni** $4s^2 3d^8$	29 **Cu** $4s^1 3d^{10}$	30 **Zn** $4s^2 3d^{10}$	31 **Ga** $4s^2 4p^1$	32 **Ge** $4s^2 4p^2$	33 **As** $4s^2 4p^3$	34 **Se** $4s^2 4p^4$	35 **Br** $4s^2 4p^5$	36 **Kr** $4s^2 4p^6$
5	37 **Rb** $5s^1$	38 **Sr** $5s^2$	39 **Y** $5s^2 4d^1$	40 **Zr** $5s^2 4d^2$	41 **Nb** $5s^1 4d^4$	42 **Mo** $5s^1 4d^5$	43 **Tc** $5s^2 4d^5$	44 **Ru** $5s^1 4d^7$	45 **Rh** $5s^1 4d^8$	46 **Pd** $4d^{10}$	47 **Ag** $5s^1 4d^{10}$	48 **Cd** $5s^2 4d^{10}$	49 **In** $5s^2 5p^1$	50 **Sn** $5s^2 5p^2$	51 **Sb** $5s^2 5p^3$	52 **Te** $5s^2 5p^4$	53 **I** $5s^2 5p^5$	54 **Xe** $5s^2 5p^6$
6	55 **Cs** $6s^1$	56 **Ba** $6s^2$	57 **La** $6s^2 5d^1$	72 **Hf** $6s^2 5d^2$	73 **Ta** $6s^2 5d^3$	74 **W** $6s^2 5d^4$	75 **Re** $6s^2 5d^5$	76 **Os** $6s^2 5d^6$	77 **Ir** $6s^2 5d^7$	78 **Pt** $6s^1 5d^9$	79 **Au** $6s^1 5d^{10}$	80 **Hg** $6s^2 5d^{10}$	81 **Tl** $6s^2 6p^1$	82 **Pb** $6s^2 6p^2$	83 **Bi** $6s^2 6p^3$	84 **Po** $6s^2 6p^4$	85 **At** $6s^2 6p^5$	86 **Rn** $6s^2 6p^6$
7	87 **Fr** $7s^1$	88 **Ra** $7s^2$	89 **Ac** $7s^2 6d^1$	104 **Rf** $7s^2 6d^2$	105 **Db** $7s^2 6d^3$	106 **Sg** $7s^2 6d^4$	107 **Bh**	108 **Hs**	109 **Mt**	110 **Ds**	111 **Rg**	112 **Cn**	113 **	114 **Fl**	115 **	116 **Lv**	117 **	118 **

Lanthanides	58 **Ce** $6s^2 4f^1 5d^1$	59 **Pr** $6s^2 4f^3$	60 **Nd** $6s^2 4f^4$	61 **Pm** $6s^2 4f^5$	62 **Sm** $6s^2 4f^6$	63 **Eu** $6s^2 4f^7$	64 **Gd** $6s^2 4f^7 5d^1$	65 **Tb** $6s^2 4f^9$	66 **Dy** $6s^2 4f^{10}$	67 **Ho** $6s^2 4f^{11}$	68 **Er** $6s^2 4f^{12}$	69 **Tm** $6s^2 4f^{13}$	70 **Yb** $6s^2 4f^{14}$	71 **Lu** $6s^2 4f^{14} 6d^1$
Actinides	90 **Th** $7s^2 6d^2$	91 **Pa** $7s^2 5f^2 6d^1$	92 **U** $7s^2 5f^3 6d^1$	93 **Np** $7s^2 5f^4 6d^1$	94 **Pu** $7s^2 5f^6$	95 **Am** $7s^2 5f^7$	96 **Cm** $7s^2 5f^7 6d^1$	97 **Bk** $7s^2 5f^9$	98 **Cf** $7s^2 5f^{10}$	99 **Es** $7s^2 5f^{11}$	100 **Fm** $7s^2 5f^{12}$	101 **Md** $7s^2 5f^{13}$	102 **No** $7s^2 5f^{14}$	103 **Lr** $7s^2 5f^{14} 6d^1$

▲ **FIGURE 4.10 The *s*, *p*, *d*, and *f* Blocks of the Periodic Table**

Helium is an exception. Even though it lies in the column with an outer electron configuration of $ns^2\ np^6$, its electron configuration is simply $1s^2$.

into blocks representing the filling of particular sublevels. The first two columns on the left side of the periodic table constitute the *s* block, with outer electron configurations of ns^1 (group 1A) and ns^2 (group 2A). The six columns on the right side of the periodic table constitute the *p* block, with outer electron configurations of $ns^2\ np^1$, $ns^2\ np^2$, $ns^2\ np^3$, $ns^2\ np^4$, $ns^2\ np^5$, and $ns^2\ np^6$. Together, the *s* and *p* blocks constitute the *main-group* elements. The *transition* elements consitute the *d* block, and the lanthanides and actinides (also called the inner transition elements) constitute the *f* block. (For compactness, the *f* block is normally printed below the *d* block instead of being embedded within it.)

Note also that *the number of columns in a block corresponds to the maximum number of electrons that can occupy the particular sublevel of that block.* The *s* block has 2 columns (corresponding to one *s* orbital holding a maximum of two electrons); the *p* block has 6 columns (corresponding to three *p* orbitals with two electrons each); the *d* block has 10 columns (corresponding to five *d* orbitals with two electrons each); and the *f* block has 14 columns (corresponding to seven *f* orbitals with two electrons each).

Except for helium, *the number of valence electrons for any main-group element is equal to its lettered group number.* For example, we can tell that chlorine has seven valence electrons because it is in group number 7A.

Lastly, note that, for main-group elements, *the row number in the periodic table is equal to the number (or n value) of the highest principal level.* For example, because chlorine is in row 3, its highest principal level is the $n = 3$ level.

Recall from Chapter 2 that main-group elements are those in the two far-left columns (groups 1A and 2A) and the six far-right columns (groups 3A–8A) of the periodic table.

Summarizing Periodic Table Organization

- The periodic table is divisible into four blocks corresponding to the filling of the four quantum sublevels (*s*, *p*, *d*, and *f*).
- The group number of a main-group element is equal to the number of valence electrons for that element.
- The row number of a main-group element is equal to the highest principal quantum number of that element.

Writing an Electron Configuration for an Element from Its Position in the Periodic Table

The organization of the periodic table allows us to write the electron configuration for any element based on its position in the periodic table. For example, suppose we want to write an electron configuration for Cl. The *inner electron configuration* of Cl is that of the noble gas that precedes it in the periodic table, Ne. So we can represent the inner electron configuration with [Ne]. We can obtain the *outer electron configuration*—the configuration of the electrons beyond the previous noble gas—by tracing the elements between Ne and Cl and assigning electrons to the appropriate orbitals, as shown here. Remember that the highest *n* value is indicated by the row number (3 for chlorine).

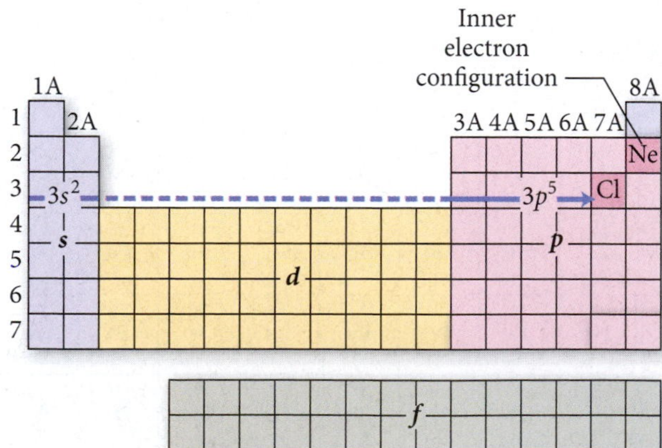

We begin with [Ne], then add in the two 3*s* electrons as we trace across the *s* block, followed by five 3*p* electrons as we trace across the *p* block to Cl, which is in the fifth column of the *p* block. The electron configuration is:

$$\text{Cl}\quad [\text{Ne}]\ 3s^2 3p^5$$

Notice that Cl is in column 7A and therefore has seven valence electrons and an outer electron configuration of $ns^2\ np^5$.

EXAMPLE 4.4

Writing Electron Configurations from the Periodic Table

Refer to the periodic table to write the electron configuration for selenium (Se).

SOLUTION

The atomic number of Se is 34. The noble gas that precedes Se in the periodic table is argon, so the inner electron configuration is [Ar]. Obtain the outer electron configuration by tracing the elements between Ar and Se and assigning electrons to the appropriate orbitals. Begin with [Ar]. Because Se is in row 4, add two 4*s* electrons as you trace across the *s* block (n = row number). Next, add ten 3*d* electrons as you trace across the *d* block (n = row number − 1). (See explanation below.) Lastly, add four 4*p* electrons as you trace across the *p* block to Se, which is in the fourth column of the *p* block (n = row number).

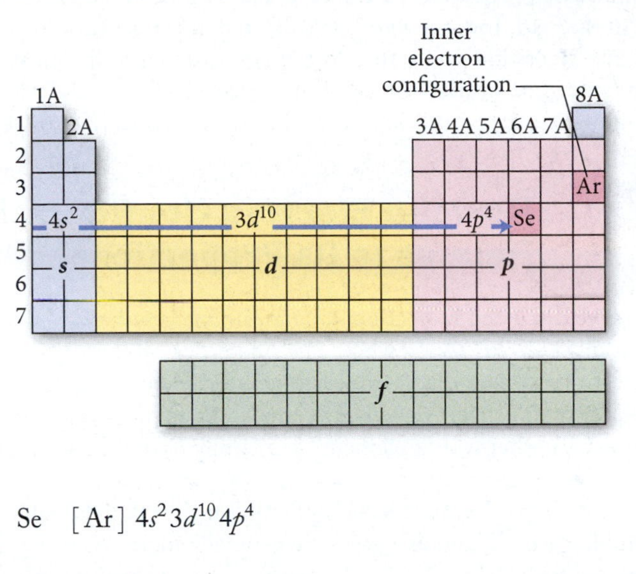

Se [Ar] $4s^2 3d^{10} 4p^4$

FOR PRACTICE 4.4

Refer to the periodic table to determine the electron configuration of bismuth (Bi).

FOR MORE PRACTICE 4.4

Refer to the periodic table to write the electron configuration for iodine (I).

The Transition and Inner Transition Elements

The electron configurations of the transition elements (*d* block) and inner transition elements (*f* block) exhibit trends that differ somewhat from those of the main-group elements. As we move to the right across a row in the *d* block, the *d* orbitals fill as shown here:

4	21 **Sc** $4s^23d^1$	22 **Ti** $4s^23d^2$	23 **V** $4s^23d^3$	24 **Cr** $4s^13d^5$	25 **Mn** $4s^23d^5$	26 **Fe** $4s^23d^6$	27 **Co** $4s^23d^7$	28 **Ni** $4s^23d^8$	29 **Cu** $4s^13d^{10}$	30 **Zn** $4s^23d^{10}$
5	39 **Y** $5s^24d^1$	40 **Zr** $5s^24d^2$	41 **Nb** $5s^14d^4$	42 **Mo** $5s^14d^5$	43 **Tc** $5s^24d^5$	44 **Ru** $5s^14d^7$	45 **Rh** $5s^14d^8$	46 **Pd** $4d^{10}$	47 **Ag** $5s^14d^{10}$	48 **Cd** $5s^24d^{10}$

Note that *the principal quantum number of the d orbitals that fill across each row in the transition series is equal to the row number minus one.* In the fourth row, the 3*d* orbitals fill, in the fifth row, the 4*d* orbitals fill, and so on. This happens because, as we discussed in Section 4.3, the 4*s* orbital is generally lower in energy than the 3*d* orbital (because the 4*s* orbital more efficiently penetrates into the region occupied by the core electrons). The result is that the 4*s* orbital fills before the 3*d* orbital, even though its principal quantum number ($n = 4$) is higher.

Keep in mind, however, that the 4*s* and the 3*d* orbitals are extremely close to each other in energy so that their relative energy ordering depends on the exact species under consideration; this causes some irregular behavior in the transition metals. For example, in the first transition series of the *d* block, the outer configuration is $4s^2 3d^x$ with two exceptions: Cr is $4s^1 3d^5$ and Cu is $4s^1 3d^{10}$. This behavior is related to the closely spaced $3d$ and $4s$ energy levels and the stability associated with a half-filled (as in Cr) or completely filled (as in Cu) sublevel. Actual electron configurations are determined experimentally (through spectroscopy) and do not always conform to the general pattern. Nonetheless, the patterns we have described allow us to accurately predict electron configurations for most of the elements in the periodic table.

As we move across the *f* block (the inner transition series), the *f* orbitals fill. For these elements, the principal quantum number of the *f* orbitals that fill across each row is the row number *minus two*. (In the sixth row, the $4f$ orbitals fill, and in the seventh row, the $5f$ orbitals fill.) In addition, within the inner transition series, the close energy spacing of the $5d$ and $4f$ orbitals sometimes causes an electron to enter a $5d$ orbital instead of the expected $4f$ orbital. For example, the electron configuration of gadolinium is [Xe] $6s^2 4f^7 5d^1$ (instead of the expected [Xe] $6s^2 4f^8$).

4.5 How the Electron Configuration of an Element Relates to Its Properties

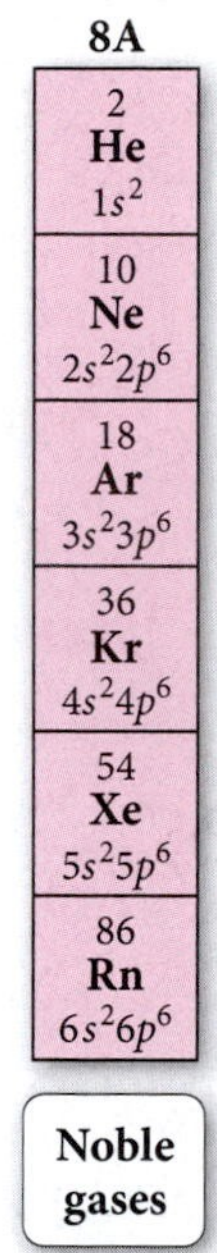

8A
2
He
$1s^2$
10
Ne
$2s^22p^6$
18
Ar
$3s^23p^6$
36
Kr
$4s^24p^6$
54
Xe
$5s^25p^6$
86
Rn
$6s^26p^6$
Noble gases

▲ The noble gases each have eight valence electrons except for helium, which has two. They have full outer quantum levels and are particularly stable and unreactive.

As we discussed in Section 4.4, *the chemical properties of elements are largely determined by the number of valence electrons they contain*. The properties of elements are periodic because the number of valence electrons is periodic. Mendeleev grouped elements into families (or columns) based on observations about their properties. We now know that elements in a family have the same number of valence electrons. In other words, elements in a family have similar properties because they have the same number of valence electrons.

Perhaps the most striking family in the periodic table is the column labeled 8A, known as the **noble gases**. The noble gases are generally inert—they are the most unreactive elements in the entire period table. Why? Notice that each noble gas has eight valence electrons (or two in the case of helium), and they all have full outer quantum levels. We do not cover the quantitative (or numerical) aspects of the quantum-mechanical model in this book, but calculations of the overall energy of the electrons within atoms with eight valence electrons (or two for helium) show that these atoms are particularly stable. *In other words, when a quantum level is completely full, the overall potential energy of the electrons that occupy that level is particularly low.*

Recall from Section 2.4 that systems with high potential energy tend to change in ways that lower their potential energy. Systems with low potential energy, on the other hand, tend not to change—they are stable. Because atoms with eight electrons (or two for helium) have particularly low potential energy, the noble gases are stable—they *cannot* lower their energy by reacting with other atoms or molecules.

We can explain a great deal of chemical behavior with the simple idea that *elements without a noble gas electron configuration react to attain a noble gas configuration*. This idea works particularly well for main-group elements. In this section, we first apply this idea to help differentiate between metals and nonmetals. We then apply the idea to understand the properties of several individual families of elements. Lastly, we apply the idea to the formation of ions.

Metals and Nonmetals

We can understand the broad chemical behavior of the elements by superimposing one of the most general properties of an element—whether it is a metal or nonmetal—with its outer electron configuration in the form of a periodic table (Figure 4.11 ▶). **Metals** lie on the lower left side and middle of the periodic table and share some common properties: They are good conductors of heat and electricity; they can be pounded into flat sheets (malleability); they can be drawn into wires (ductility); they are often shiny; and most importantly *they tend to lose electrons when they undergo chemical changes*. For example, sodium is among the most reactive metals. Its electron configuration is $1s^22s^22p^63s^1$. Notice that its electron configuration is one electron beyond the configuration of neon, a noble gas. Sodium can attain a noble gas electron configuration by losing that one valence electron—and that is exactly what it does. When we find sodium in nature, we most often find it as Na^+, which has the electron configuration of neon ($1s^22s^22p^6$). The other main-group metals in the periodic table behave similarly: They tend to lose their valence electrons in chemical changes to attain noble gas electron configurations. The transition metals also tend to lose electrons in their chemical changes, but they do not generally attain noble gas electron configurations.

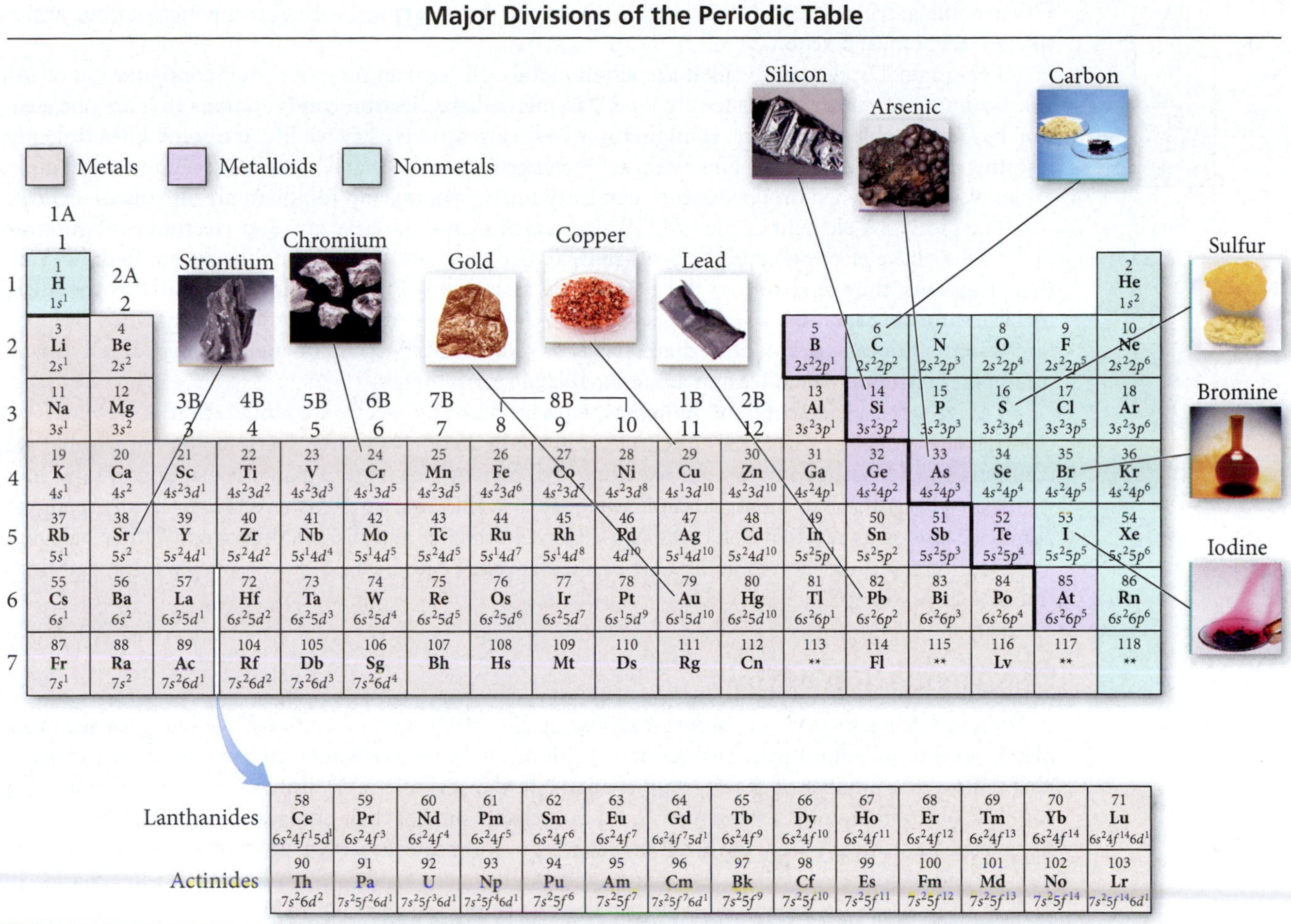

Major Divisions of the Periodic Table

Metals | Metalloids | Nonmetals

Silicon, Arsenic, Carbon, Strontium, Chromium, Gold, Copper, Lead, Sulfur, Bromine, Iodine

	1A 1	2A 2	3B 3	4B 4	5B 5	6B 6	7B 7	8B 8	8B 9	8B 10	1B 11	2B 12	13	14	15	16	17	18
1	1 H $1s^1$																	2 He $1s^2$
2	3 Li $2s^1$	4 Be $2s^2$											5 B $2s^22p^1$	6 C $2s^22p^2$	7 N $2s^22p^3$	8 O $2s^22p^4$	9 F $2s^22p^5$	10 Ne $2s^22p^6$
3	11 Na $3s^1$	12 Mg $3s^2$											13 Al $3s^23p^1$	14 Si $3s^23p^2$	15 P $3s^23p^3$	16 S $3s^23p^4$	17 Cl $3s^23p^5$	18 Ar $3s^23p^6$
4	19 K $4s^1$	20 Ca $4s^2$	21 Sc $4s^23d^1$	22 Ti $4s^23d^2$	23 V $4s^23d^3$	24 Cr $4s^13d^5$	25 Mn $4s^23d^5$	26 Fe $4s^23d^6$	27 Co $4s^23d^7$	28 Ni $4s^23d^8$	29 Cu $4s^13d^{10}$	30 Zn $4s^23d^{10}$	31 Ga $4s^24p^1$	32 Ge $4s^24p^2$	33 As $4s^24p^3$	34 Se $4s^24p^4$	35 Br $4s^24p^5$	36 Kr $4s^24p^6$
5	37 Rb $5s^1$	38 Sr $5s^2$	39 Y $5s^24d^1$	40 Zr $5s^24d^2$	41 Nb $5s^14d^4$	42 Mo $5s^14d^5$	43 Tc $5s^24d^5$	44 Ru $5s^14d^7$	45 Rh $5s^14d^8$	46 Pd $4d^{10}$	47 Ag $5s^14d^{10}$	48 Cd $5s^24d^{10}$	49 In $5s^25p^1$	50 Sn $5s^25p^2$	51 Sb $5s^25p^3$	52 Te $5s^25p^4$	53 I $5s^25p^5$	54 Xe $5s^25p^6$
6	55 Cs $6s^1$	56 Ba $6s^2$	57 La $6s^25d^1$	72 Hf $6s^25d^2$	73 Ta $6s^25d^3$	74 W $6s^25d^4$	75 Re $6s^25d^5$	76 Os $6s^25d^6$	77 Ir $6s^25d^7$	78 Pt $6s^15d^9$	79 Au $6s^15d^{10}$	80 Hg $6s^25d^{10}$	81 Tl $6s^26p^1$	82 Pb $6s^26p^2$	83 Bi $6s^26p^3$	84 Po $6s^26p^4$	85 At $6s^26p^5$	86 Rn $6s^26p^6$
7	87 Fr $7s^1$	88 Ra $7s^2$	89 Ac $7s^26d^1$	104 Rf $7s^26d^2$	105 Db $7s^26d^3$	106 Sg $7s^26d^4$	107 Bh	108 Hs	109 Mt	110 Ds	111 Rg	112 Cn	113 **	114 Fl	115 **	116 Lv	117 **	118 **

Lanthanides	58 Ce $6s^24f^15d^1$	59 Pr $6s^24f^3$	60 Nd $6s^24f^4$	61 Pm $6s^24f^5$	62 Sm $6s^24f^6$	63 Eu $6s^24f^7$	64 Gd $6s^24f^75d^1$	65 Tb $6s^24f^9$	66 Dy $6s^24f^{10}$	67 Ho $6s^24f^{11}$	68 Er $6s^24f^{12}$	69 Tm $6s^24f^{13}$	70 Yb $6s^24f^{14}$	71 Lu $6s^24f^{14}6d^1$
Actinides	90 Th $7s^26d^2$	91 Pa $7s^25f^26d^1$	92 U $7s^25f^36d^1$	93 Np $7s^25f^46d^1$	94 Pu $7s^25f^6$	95 Am $7s^25f^7$	96 Cm $7s^25f^76d^1$	97 Bk $7s^25f^9$	98 Cf $7s^25f^{10}$	99 Es $7s^25f^{11}$	100 Fm $7s^25f^{12}$	101 Md $7s^25f^{13}$	102 No $7s^25f^{14}$	103 Lr $7s^25f^{14}6d^1$

▲ **FIGURE 4.11 Metallic Behavior and Electron Configuration** The elements in the periodic table fall into these three broad classes. Notice the correlations.

Nonmetals lie on the upper right side of the periodic table. The division between metals and nonmetals is the zigzag diagonal line running from boron to astatine. Nonmetals have varied properties—some are solids at room temperature, others are liquids or gases—but as a whole they tend to be poor conductors of heat and electricity, and most importantly *they all tend to gain electrons when they undergo chemical changes*. Chlorine is among the most reactive nonmetals. Its electron configuration is $1s^22s^22p^6 3s^23p^5$. Notice that its electron configuration is one electron short of the configuration of argon, a noble gas. Chlorine can attain a noble gas electron configuration by gaining one electron—and that is exactly what it does. When we find chlorine in nature, we often find it as Cl^-, which has the electron configuration of argon ($1s^22s^22p^63s^23p^6$). The other nonmetals in the periodic table behave similarly: They tend to gain electrons in chemical changes to attain noble gas electron configurations.

Many of the elements that lie along the zigzag diagonal line that divides metals and nonmetals are **metalloids** and exhibit mixed properties. Several metalloids are classified as **semiconductors** because of their intermediate (and highly temperature-dependent) electrical conductivity. Our ability to change and control the conductivity of semiconductors makes them useful in the manufacture of the electronic chips and circuits central to computers, cellular telephones, and many other devices. Good examples of metalloids include silicon, arsenic, and antimony.

Metalloids are sometimes called semimetals.

Families of Elements

We have already seen that the group 8A elements, called the *noble gases*, are mostly unreactive. The most familiar noble gas is probably helium, used to fill buoyant balloons. Helium is chemically stable—it does not combine with other elements to form compounds—and is therefore safe to put into balloons.

2A

2A
4 **Be** $2s^2$
12 **Mg** $3s^2$
20 **Ca** $4s^2$
38 **Sr** $5s^2$
56 **Ba** $6s^2$
88 **Ra** $7s^2$

Alkaline earth metals

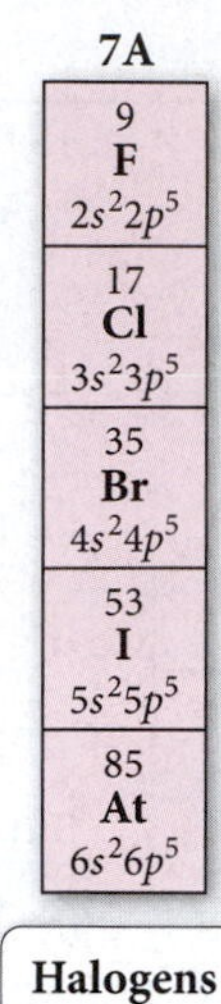

Other noble gases are neon (often used in electronic signs), argon (a small component of our atmosphere), krypton, and xenon.

The group 1A elements, called the **alkali metals**, all have an outer electron configuration of ns^1. Like sodium, a member of this family, the alkali metals have electron configurations that are one electron beyond a noble gas electron configuration. In their reactions, they readily, and sometimes violently, lose the ns^1 electron to form ions with a 1+ charge. A marble-sized piece of sodium, for example, explodes violently when dropped into water. Lithium, potassium, and rubidium are also alkali metals.

The group 2A elements, called the **alkaline earth metals**, all have an outer electron configuration of ns^2. They have electron configurations that are two electrons beyond a noble gas configuration. In their reactions, they tend to lose the two ns^2 electrons—though not quite as violently as the alkali metals—to form ions with a 2+ charge. Calcium, for example, reacts fairly vigorously when dropped into water but does not explode as dramatically as sodium. Magnesium (a common low-density structural metal), strontium, and barium are other alkaline earth metals.

The group 7A elements, the **halogens**, all have an outer electron configuration of ns^2np^5. Like chlorine, a member of this family, their electron configurations are one electron short of a noble gas configuration. Consequently, in their reactions with metals, they tend to gain one electron to form ions with a 1− charge. One of the most familiar halogens is chlorine, a greenish-yellow gas with a pungent odor. Because of its reactivity, chlorine is used as a sterilizing and disinfecting agent. Other halogens include bromine, a red-brown liquid that easily evaporates into a gas; iodine, a purple solid; and fluorine, a pale-yellow gas.

The Formation of Ions

In Section 1.8, we learned that atoms can lose or gain electrons to form ions. We have just seen that metals tend to form positively charged ions (cations) and nonmetals tend to form negatively charged ions (anions). A number of main-group elements in the periodic table always form ions with a noble gas electron configuration. Consequently, we can reliably predict their charges (Figure 4.12 ▼).

As we have already seen, the alkali metals tend to form cations with a 1+ charge, the alkaline earth metals tend to form ions with a 2+ charge, and the halogens tend to form ions with a 1− charge. In each of these cases, the ions have noble gas electron configurations. This is true of the rest of the ions in Figure 4.12. Nitrogen for example, has an electron configuration of $1s^22s^22p^3$. The N^{3-} ion has three additional electrons and an electron configuration of $1s^22s^22p^6$, which is the same as the configuration of neon, the nearest noble gas.

Notice that, for the main-group elements that form cations with predictable charge, the charge is equal to the group number. For main-group elements that form anions with predictable charge, the charge is equal to the group number minus eight. Transition elements may form various ions with different charges.

The tendency for many main-group elements to form ions with noble gas electron configurations *does not* mean that the process is in itself energetically favorable. In fact, forming cations always requires energy, and forming anions sometimes requires energy as well. However, the energy cost of forming a cation or anion with *a noble gas configuration* is often less than the energy payback that occurs when that cation or anion forms chemical bonds, as we shall see in Chapter 5.

Elements That Form Ions with Predictable Charges

	1A	2A	3B	4B	5B	6B	7B	8B	8B	8B	1B	2B	3A	4A	5A	6A	7A	8A
1	Li^+														N^{3-}	O^{2-}	F^-	
2	Na^+	Mg^{2+}											Al^{3+}			S^{2-}	Cl^-	
3	K^+	Ca^{2+}														Se^{2-}	Br^-	
4	Rb^+	Sr^{2+}														Te^{2-}	I^-	
5	Cs^+	Ba^{2+}																

▲ **FIGURE 4.12 Elements That Form Ions with Predictable Charges**

EXAMPLE 4.5
Predicting the Charge of Ions

Predict the charges of the monoatomic (single atom) ions formed by these main-group elements.

(a) Al **(b)** S

SOLUTION

(a) Aluminum is a main-group metal and tends to lose electrons to form a cation with the same electron configuration as the nearest noble gas. The electron configuration of aluminum is $1s^22s^22p^63s^23p^1$. The nearest noble gas is neon, which has an electron configuration of $1s^22s^22p^6$. Therefore, aluminum loses three electrons to form the cation Al^{3+}.

(b) Sulfur is a nonmetal and tends to gain electrons to form an anion with the same electron configuration as the nearest noble gas. The electron configuration of sulfur is $1s^22s^22p^63s^2 3p^4$. The nearest noble gas is argon, which has an electron configuration of $1s^22s^22p^63s^2 3p^6$. Therefore, sulfur gains two electrons to form the anion S^{2-}.

FOR PRACTICE 4.5

Predict the charges of the monoatomic ions formed by these main-group elements.

(a) N **(b)** Rb

4.6 Periodic Trends in the Size of Atoms and Effective Nuclear Charge

In previous chapters, we saw that the volume of an atom is taken up primarily by its electrons (Chapter 1) occupying quantum-mechanical orbitals (Chapter 3). We also saw that these orbitals do not have a definite boundary but represent only a statistical probability distribution for where the electron is found. So how do we define the size of an atom? One way to define atomic radii is to consider the distance between *nonbonding* atoms that are in direct contact. For example, krypton can be frozen into a solid in which the krypton atoms are touching each other but are not bonded together. The distance between the centers of adjacent krypton atoms—which can be determined from the solid's density—is then twice the radius of a krypton atom. An atomic radius determined in this way is called the **nonbonding atomic radius** or the **van der Waals radius**. The van der Waals radius represents the radius of an atom when it is not bonded to another atom.

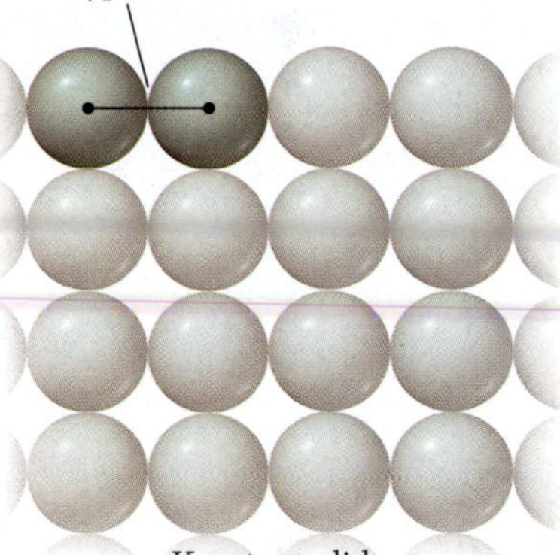

▲ The van der Waals radius of an atom is one-half the distance between adjacent nuclei in the atomic solid.

Another way to define the size of an atom, called the **bonding atomic radius** or **covalent radius**, is defined differently for nonmetals and metals, as follows:

Nonmetals: one-half the distance between two of the atoms bonded together

Metals: one-half the distance between two of the atoms next to each other in a crystal of the metal

For example, the distance between Br atoms in Br_2 is 228 pm; therefore, the Br covalent radius is assigned to be one-half of 228 pm, or 114 pm.

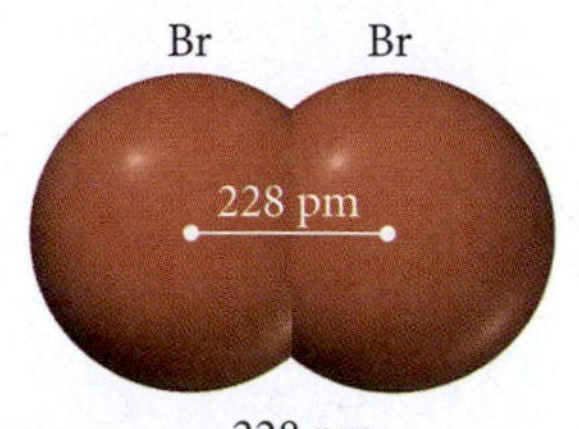

$$\text{Br radius} = \frac{228\text{ pm}}{2} = 114\text{ pm}$$

▲ The covalent radius of bromine is one-half the distance between two bonded bromine atoms.

Using this method, we can assign radii to all elements in the periodic table that form chemical bonds or form metallic crystals. A more general term, the **atomic radius**, refers to a set of average bonding radii determined from measurements on a large number of elements and compounds. The atomic radius represents the radius of an atom when it is bonded to another atom and is always smaller than the van der Waals radius. The approximate bond length of any two covalently bonded atoms is the sum of their atomic radii. For example, the approximate bond length for ICl is iodine's atomic radius (133 pm) plus chlorine's atomic radius (99 pm), for a bond length of 232 pm. (The actual experimentally measured bond length in ICl is 232.07 pm.)

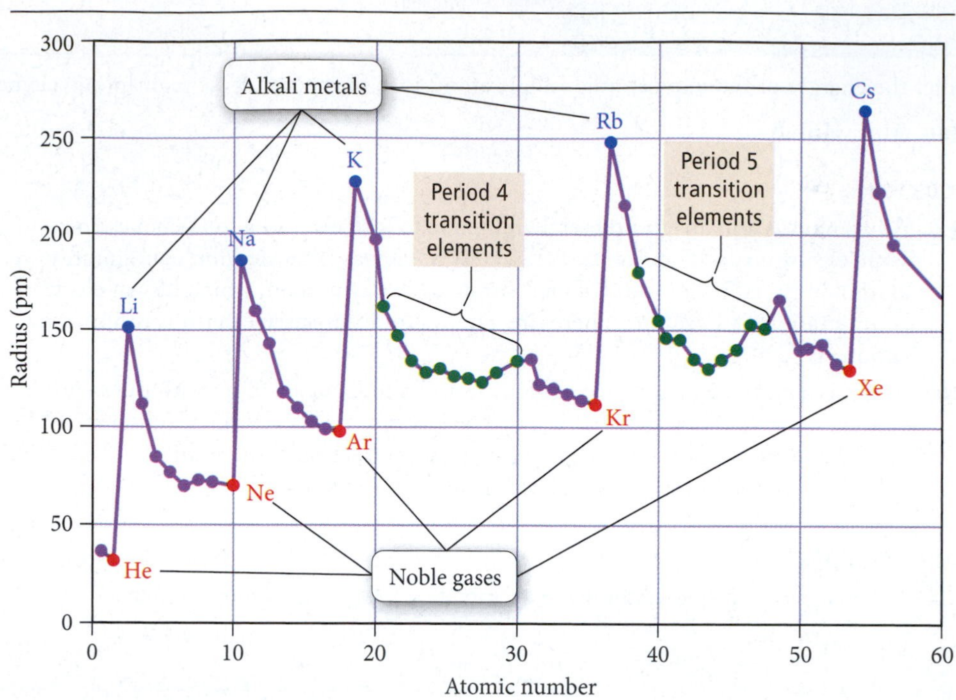

▶ **FIGURE 4.13 Atomic Radius versus Atomic Number** Notice the periodic trend in the atomic radius, starting at a peak with each alkali metal and falling to a minimum with each noble gas.

Figure 4.13 ▲ shows the atomic radius plotted as a function of atomic number for the first 57 elements in the periodic table. Notice the periodic trend in the radii. Atomic radii peak with each alkali metal. Figure 4.14 ▼ is a relief map of atomic radii for most of the elements in the periodic

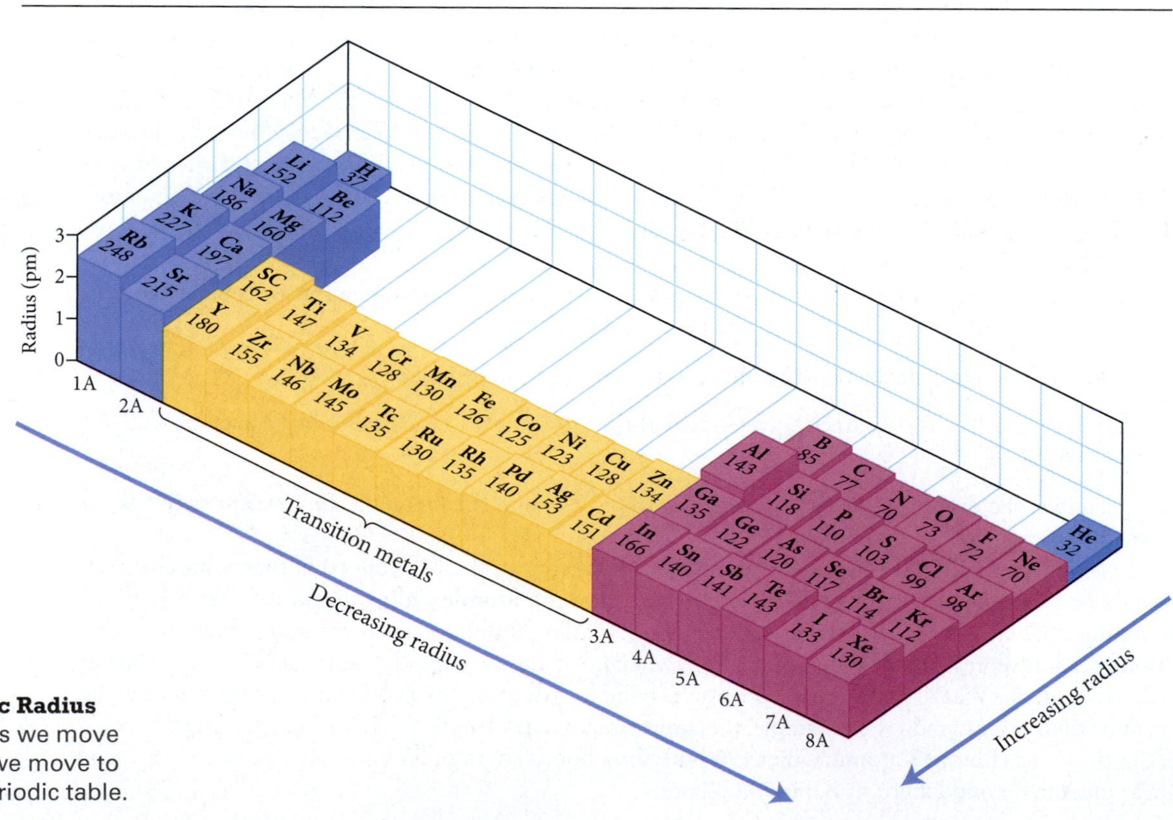

▶ **FIGURE 4.14 Trends in Atomic Radius** In general, atomic radii increase as we move down a column and decrease as we move to the right across a period in the periodic table.

table. The general trends in the atomic radii of main-group elements, which are the same as trends observed in van der Waals radii, are as follows:

1. As we move down a column (or family) in the periodic table, the atomic radius increases.
2. As we move to the right across a period (or row) in the periodic table, the atomic radius decreases.

KEY CONCEPT VIDEO
Periodic trends in the size of atoms and effective nuclear charge

We can understand the observed trend in radius as we move down a column based on the trends in the sizes of atomic orbitals. The atomic radius is largely determined by the valence electrons, the electrons farthest from the nucleus. As we move down a column in the periodic table, the highest principal quantum number (n) of the valence electrons increases. Consequently, the valence electrons occupy larger orbitals, resulting in larger atoms.

The observed trend in atomic radius as we move to the right across a row, however, is a bit more complex. To understand this trend, we now revisit some concepts from Section 4.3, including effective nuclear charge and shielding.

Effective Nuclear Charge

The trend in atomic radius as we move to the right across a row in the periodic table is determined by the inward pull of the nucleus on the electrons in the outermost principal energy level (highest n value). According to Coulomb's law, the attraction between a nucleus and an electron increases with increasing magnitude of nuclear charge. For example, compare the H atom to the He^+ ion:

$$\text{H} \quad 1s^1$$
$$\text{He}^+ \quad 1s^1$$

It takes 1312 kJ/mol of energy to remove the $1s$ electron from H, but 5251 kJ/mol of energy to remove it from He^+. Why? Although each electron is in a $1s$ orbital, the electron in the helium ion is attracted to the nucleus by a 2+ charge, while the electron in the hydrogen atom is attracted to the nucleus by only a 1+ charge. Therefore, the electron in the helium ion is held more tightly (it has lower potential energy according to Coulomb's law), making it more difficult to remove and making the helium ion smaller than the hydrogen atom.

As we saw in Section 4.3, any one electron in a multi-electron atom experiences both the positive charge of the nucleus (which is attractive) and the negative charges of the other electrons (which are repulsive). Consider again the outermost electron in the lithium atom:

$$\text{Li} \quad 1s^2 2s^1$$

As shown in Figure 4.15 ▶, even though the $2s$ orbital penetrates into the $1s$ orbital to some degree, the majority of the $2s$ orbital is outside of the $1s$ orbital. Therefore the electron in the $2s$ orbital is partially *screened* or *shielded* from the 3+ charge of the nucleus by the 2− charge of the $1s$ (or core) electrons, reducing the net charge experienced by the $2s$ electron.

Recall from Section 4.3 that we can define the average or net charge experienced by an electron as the *effective nuclear charge*. The effective nuclear charge experienced by a particular electron in an atom is the *actual nuclear charge (Z)* minus *the charge shielded by other electrons (S)*:

$$Z_{\text{eff}} = Z - S$$

Effective nuclear charge (Z_{eff}); Actual nuclear charge (Z); Charge screened by other electrons (S)

For lithium, we estimate that the two core electrons shield the valence electron from the nuclear charge with high efficiency (S is nearly 2). The effective nuclear charge experienced by lithium's valence electron is therefore slightly greater than 1+.

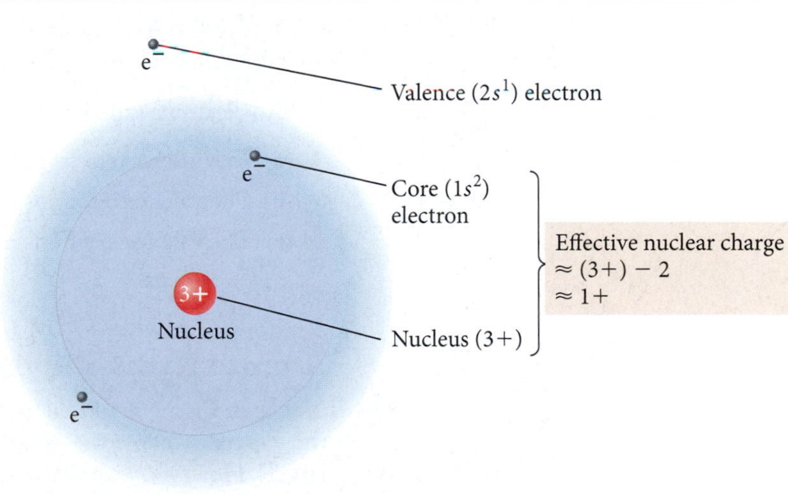

▲ **FIGURE 4.15 Screening and Effective Nuclear Charge** The valence electron in lithium experiences the 3+ charge of the nucleus through the screen of the 2− charge of the core electrons. The effective nuclear charge acting on the valence electron is approximately 1+.

Now consider the valence electrons in beryllium (Be), with atomic number 4. Its electron configuration is:

$$\text{Be} \quad 1s^2 2s^2$$

To estimate the effective nuclear charge experienced by the $2s$ electrons in beryllium, we must distinguish between two different types of shielding: (1) the shielding of the outermost electrons by the core electrons and (2) the shielding of the outermost electrons by *each other*. The key to understanding the trend in atomic radius is the difference between these two types of shielding.

> **Core electrons efficiently shield electrons in the outermost principal energy level from nuclear charge, but outermost electrons do not efficiently shield one another from nuclear charge.**

For example, the two outermost electrons in beryllium experience the 4+ charge of the nucleus through the shield of the two $1s$ core electrons without shielding each other from that charge very much. We estimate that the shielding (S) experienced by any one of the outermost electrons due to the core electrons is nearly 2, but that the shielding due to the other outermost electron is nearly zero. The effective nuclear charge experienced by beryllium's outermost electrons is therefore slightly greater than 2+.

The effective nuclear charge experienced by *beryllium's* outermost electrons is greater than that experienced by *lithium's* outermost electron. Consequently, beryllium's outermost electrons are held more tightly than lithium's, resulting in a smaller atomic radius for beryllium. The effective nuclear charge experienced by an atom's outermost electrons continues to become more positive as we move to the right across the rest of the second row in the periodic table, resulting in successively smaller atomic radii. The same trend is generally observed in all main-group elements.

Summarizing Atomic Radii for Main-Group Elements

- As we move down a column in the periodic table, the principal quantum number (n) of the electrons in the outermost principal energy level increases, resulting in larger orbitals and therefore larger atomic radii.
- As we move to the right across a row in the periodic table, the effective nuclear charge (Z_{eff}) experienced by the electrons in the outermost principal energy level increases, resulting in a stronger attraction between the outermost electrons and the nucleus, and smaller atomic radii.

4.4 Cc Conceptual Connection

Effective Nuclear Charge

Which electrons experience the greatest effective nuclear charge?

(a) The valence electrons in Mg

(b) The valence electrons in Al

(c) The valence electrons in S

Atomic Radii and the Transition Elements

Notice in Figure 4.14 that as *we move down the first two rows of a column* within the transition metals, the elements follow the same general trend in atomic radii as the main-group elements (the radii get larger). However, with the exception of the first couple of elements in each transition series, the atomic radii of the transition elements *do not* follow the same trend as the main-group elements as *we move to the right across a row*. Instead of decreasing in size, *the radii of transition elements stay roughly constant across each row*. Why? The difference is that, across a row of transition elements, the number of electrons in the outermost principal energy level (highest n value) is nearly constant (recall from Section 4.3, for example, that the $4s$ orbital fills before the $3d$). As another proton is added to the nucleus of each successive element, another electron is added as well, but the electron goes into an $n_{highest} - 1$ orbital. The number of outermost electrons stays constant and the electrons experience a roughly constant effective nuclear charge, keeping the radius approximately constant.

EXAMPLE 4.6
Atomic Size

On the basis of periodic trends, choose the larger atom in each pair (if possible). Explain your choices.

(a) N or F **(b)** C or Ge **(c)** N or Al **(d)** Al or Ge

SOLUTION

(a) N atoms are larger than F atoms because, as you trace the path between N and F on the periodic table, you move to the right within the same period. As you move to the right across a period, the effective nuclear charge experienced by the outermost electrons increases, resulting in a smaller radius.

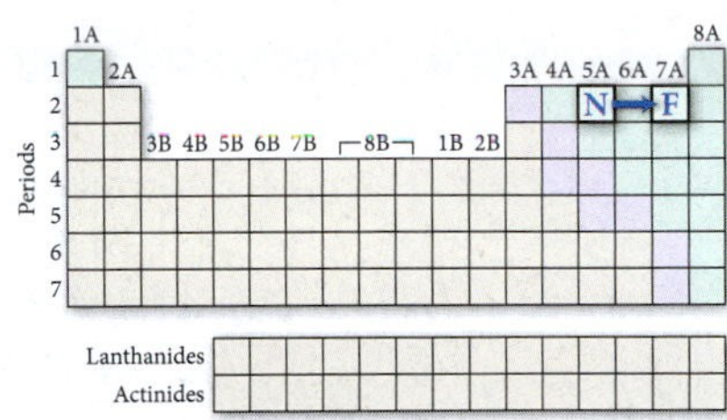

(b) Ge atoms are larger than C atoms because, as you trace the path between C and Ge on the periodic table, you move down a column. Atomic size increases as you move down a column because the outermost electrons occupy orbitals with a higher principal quantum number that are therefore larger, resulting in a larger atom.

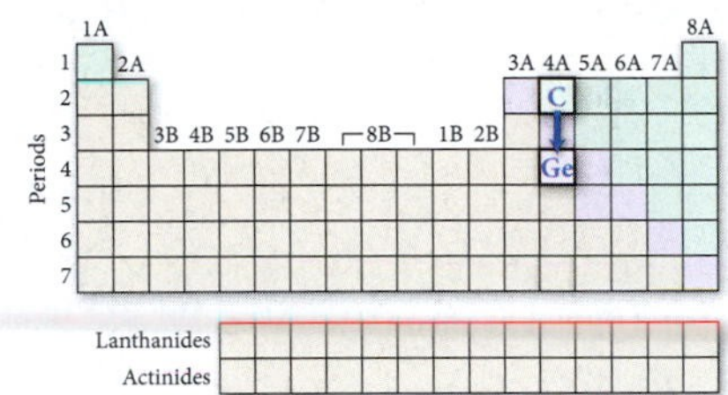

(c) Al atoms are larger than N atoms because, as you trace the path between N and Al on the periodic table, you move down a column (atomic size increases) and then to the left across a period (atomic size increases). These effects add together for an overall increase.

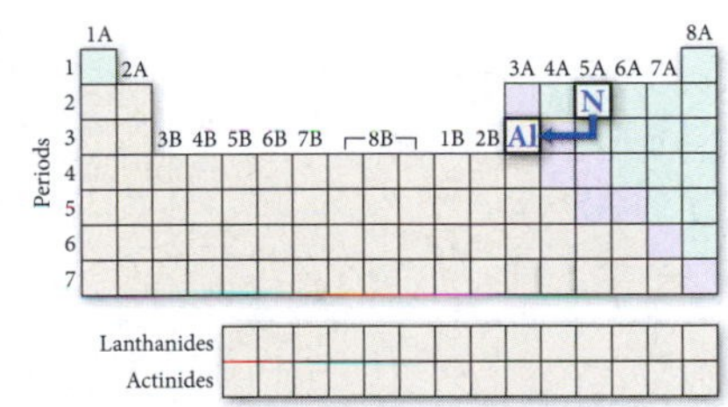

(d) Based on periodic trends alone, you cannot tell which atom is larger, because as you trace the path between Al and Ge you go to the right across a period (atomic size decreases) and then down a column (atomic size increases). These effects tend to counter each other, and it is not easy to tell which will predominate.

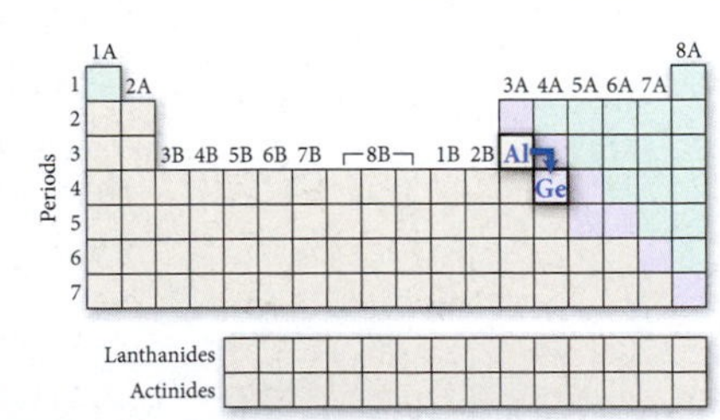

FOR PRACTICE 4.6

On the basis of periodic trends, choose the larger atom in each pair (if possible):

(a) Sn or I **(b)** Ge or Po **(c)** Cr or W **(d)** F or Se

FOR MORE PRACTICE 4.6

Arrange the elements in order of decreasing radius: S, Ca, F, Rb, Si.

4.7 Ions: Electron Configurations, Magnetic Properties, Ionic Radii, and Ionization Energy

Recall that ions are simply atoms (or groups of atoms) that have lost or gained electrons. In this section, we examine periodic trends in ionic electron configurations, magnetic properties, ionic radii, and ionization energies.

Electron Configurations and Magnetic Properties of Ions

As we saw in Section 4.5, we can deduce the electron configuration of a main-group monoatomic ion from the electron configuration of the neutral atom and the charge of the ion. For anions, we *add* the number of electrons indicated by the magnitude of the charge of the anion. For example, the electron configuration of fluorine (F) is $1s^2\,2s^2\,2p^5$ and that of the fluoride ion (F^-) is $1s^2 2s^2 2p^6$.

We determine the electron configuration of cations by *subtracting* the number of electrons indicated by the magnitude of the charge. For example, the electron configuration of lithium (Li) is $1s^2\,2s^1$ and that of the lithium ion (Li^+) is $1s^2\,2s^0$ (or simply $1s^2$). For main-group cations, we remove the required number of electrons in the reverse order of filling. However, for transition metal cations, the trend is different. When writing the electron configuration of a transition metal cation, we *remove the electrons in the highest n-value orbitals first, even if this does not correspond to the reverse order of filling*. For example, the electron configuration of vanadium is:

$$\text{V}\quad [\text{Ar}]\,4s^2\,3d^3$$

The V^{2+} ion, however, has the following electron configuration:

$$\text{V}^{2+}\quad [\text{Ar}]\,4s^0\,3d^3$$

In other words, for transition metal cations, the order in which electrons are removed upon ionization is *not* the reverse of the filling order. During filling, the $4s$ orbital normally fills before the $3d$ orbital. When a fourth period transition metal ionizes, however, it normally loses its $4s$ electrons before its $3d$ electrons. Why this unexpected behavior? The full answer to this question is beyond our scope, but the following two factors contribute to this phenomenon.

- As discussed previously, the ns and $(n - 1)d$ orbitals are extremely close in energy and, depending on the exact configuration, can vary in relative energy ordering.
- As the $(n - 1)d$ orbitals begin to fill in the first transition series, the increasing nuclear charge stabilizes the $(n - 1)d$ orbitals relative to the ns orbitals. This happens because the $(n - 1)d$ orbitals are not the outermost (or highest n) orbitals and are therefore not effectively shielded from the increasing nuclear charge by the ns orbitals.

The bottom-line experimental observation is that an $ns^0(n - 1)d^x$ configuration is lower in energy than an $ns^2(n - 1)d^{x-2}$ configuration for transition metal ions. Therefore, we remove the ns electrons before the $(n - 1)d$ electrons when writing electron configurations for transition metal ions.

The magnetic properties of transition metal ions support these assignments. Recall from Section 4.3 that an unpaired electron generates a magnetic field due to its spin. Consequently, an atom or ion that contains unpaired electrons is attracted to an external magnetic field, and we say that the atom or ion is **paramagnetic**. For example, consider the electron configuration of silver:

Ag [Kr] $5s^1 4d^{10}$

5s	4d				
↑	↑↓	↑↓	↑↓	↑↓	↑↓

Silver's unpaired $5s$ electron causes silver to be paramagnetic. In fact, an early demonstration of electron spin—called the Stern–Gerlach experiment—involved the interaction of a beam of silver atoms with a magnetic field. An atom or ion in which all electrons are paired is not attracted to an external magnetic field—it is instead slightly repelled—and we say that the atom or ion is **diamagnetic**. The zinc atom is diamagnetic.

Zn [Ar] $4s^2 3d^{10}$

4s	3d				
↑↓	↑↓	↑↓	↑↓	↑↓	↑↓

The magnetic properties of the zinc ion provide confirmation that the 4*s* electrons are indeed lost before 3*d* electrons in the ionization of zinc. If zinc lost two 3*d* electrons upon ionization, then the Zn^{2+} would become paramagnetic (because the two electrons would come out of two different filled *d* orbitals, leaving each of them with one unpaired electron). But the zinc ion, like the zinc atom, is diamagnetic because the 4*s* electrons are lost instead.

Zn^{2+} [Ar] $4s^0 3d^{10}$ 4*s*: [ ] 3*d*: [⇅][⇅][⇅][⇅][⇅]

Observations in other transition metals confirm that the *ns* electrons are lost before the $(n-1)d$ electrons upon ionization.

EXAMPLE 4.7
Electron Configurations and Magnetic Properties for Ions

Write the electron configuration and orbital diagram for each ion and determine whether each is diamagnetic or paramagnetic.

(a) Al^{3+} **(b)** S^{2-} **(c)** Fe^{3+}

SOLUTION

(a) Al^{3+}

Begin by writing the electron configuration of the neutral atom.

Al [Ne] $3s^2 3p^1$

Since this ion has a 3+ charge, remove three electrons to write the electron configuration of the ion. Write the orbital diagram by drawing half-arrows to represent each electron in boxes representing the orbitals. Because there are no unpaired electrons, Al^{3+} is diamagnetic.

Al^{3+} [Ne] or [He] $2s^2 2p^6$

Al^{3+} [He] 2*s*: [⇅] 2*p*: [⇅][⇅][⇅]

Diamagnetic

(b) S^{2-}

Begin by writing the electron configuration of the neutral atom.

S [Ne] $3s^2 3p^4$

Since this ion has a 2− charge, add two electrons to write the electron configuration of the ion. Write the orbital diagram by drawing half-arrows to represent each electron in boxes representing the orbitals. Because there are no unpaired electrons, S^{2-} is diamagnetic.

S^{2-} [Ne] $3s^2 3p^6$

S^{2-} [Ne] 3*s*: [⇅] 3*p*: [⇅][⇅][⇅]

Diamagnetic

(c) Fe^{3+}

Begin by writing the electron configuration of the neutral atom.

Fe [Ar] $4s^2 3d^6$

Since this ion has a 3+ charge, remove three electrons to write the electron configuration of the ion. Since it is a transition metal, remove the electrons from the 4*s* orbital before removing electrons from the 3*d* orbitals. Write the orbital diagram by drawing half-arrows to represent each electron in boxes representing the orbitals. There are unpaired electrons, so Fe^{3+} is paramagnetic.

Fe^{3+} [Ar] $4s^0 3d^5$

Fe^{3+} [Ar] 4*s*: [ ] 3*d*: [↿][↿][↿][↿][↿]

Paramagnetic

FOR PRACTICE 4.7

Write the electron configuration and orbital diagram for each ion and predict whether each will be paramagnetic or diamagnetic.

(a) Co^{2+} **(b)** N^{3-} **(c)** Ca^{2+}

Ionic Radii

What happens to the radius of an atom when it becomes a cation? An anion? Consider, for example, the difference between the Na atom and the Na^+ ion. Their electron configurations are:

$$\begin{array}{ll} \text{Na} & [\text{Ne}]\,3s^1 \\ \text{Na}^+ & [\text{Ne}]\,3s^0 \end{array}$$

The sodium atom has an outer 3*s* electron and a neon core. Since the 3*s* electron is the outermost electron, and since it is shielded from the nuclear charge by the core electrons, it contributes greatly to the size of the sodium atom. The sodium cation, having lost the outermost 3*s* electron, has only the neon core and carries a charge of 1+. Without the 3*s* electron, the sodium cation (ionic radius = 95 pm) is much smaller than the sodium atom (covalent radius = 186 pm). The trend is the same with all cations and their atoms, as shown in Figure 4.16 ▼.

Cations are much smaller than their corresponding neutral atoms.

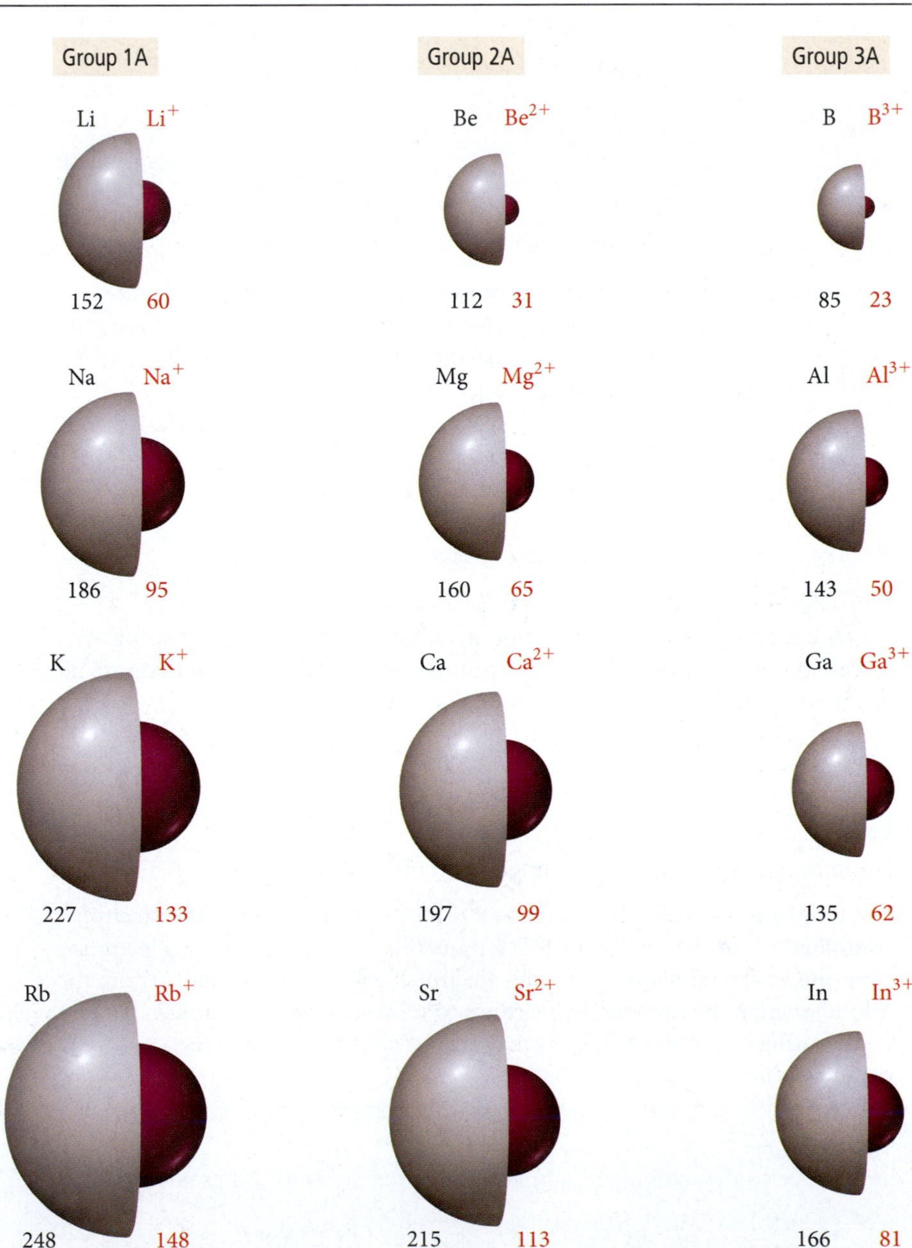

▶ **FIGURE 4.16 Sizes of Atoms and Their Cations** Atomic and ionic radii (pm) for the first three columns of main-group elements.

What about anions? Consider, for example, the difference between Cl and Cl^-. Their electron configurations are:

$$\text{Cl} \quad [\text{Ne}]\,3s^2 3p^5$$
$$\text{Cl}^- \quad [\text{Ne}]\,3s^2 3p^6$$

The chlorine anion has one additional outermost electron but no additional proton to increase the nuclear charge. The extra electron increases the repulsions among the outermost electrons, resulting in a chloride anion that is larger than the chlorine atom. The trend is the same with all anions and their atoms, as shown in Figure 4.17 ▶.

Anions are much larger than their corresponding neutral atoms.

We can observe an interesting trend in ionic size by examining the radii of an *isoelectronic* series of ions—ions with the same number of electrons. Consider the following ions and their radii:

S^{2-} (184 pm)	Cl^- (181 pm)	K^+ (133 pm)	Ca^{2+} (99 pm)
18 electrons	18 electrons	18 electrons	18 electrons
16 protons	17 protons	19 protons	20 protons

All of these ions have 18 electrons in exactly the same orbitals, but the radius of each ion gets successively smaller. Why? The reason is the progressively greater number of protons. The S^{2-} ion has 16 protons, and therefore a charge of 16+ pulling on 18 electrons. The Ca^{2+} ion, however, has 20 protons, and therefore a charge of 20+ pulling on the same 18 electrons. The result is a much smaller radius. In general, the greater the nuclear charge in atoms or ions with the same number of electrons, the smaller the atom or ion.

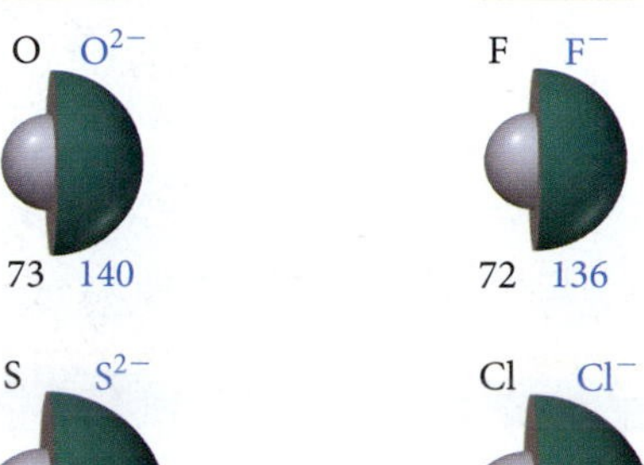

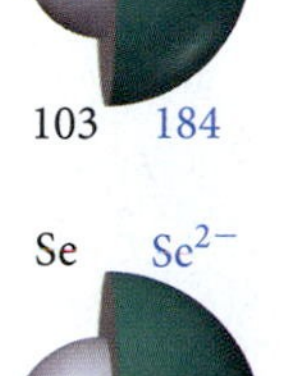

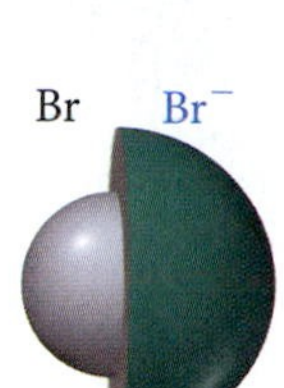

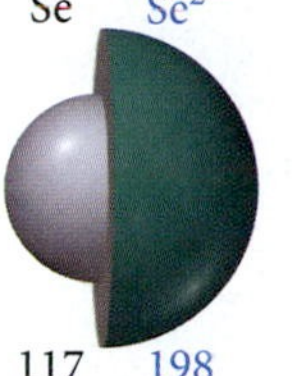

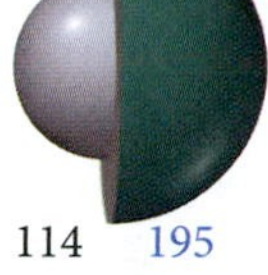

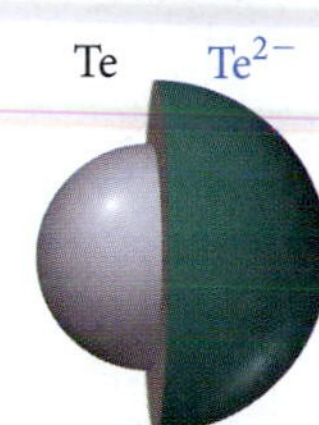

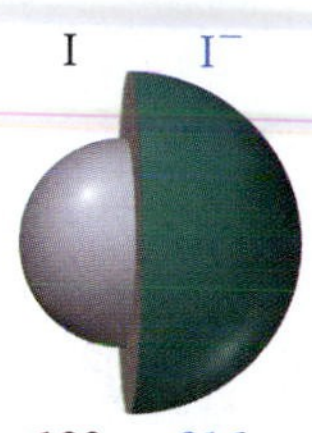

▶ **FIGURE 4.17 Sizes of Atoms and Their Anions** Atomic and ionic radii for groups 6A and 7A in the periodic table.

EXAMPLE 4.8

Ion Size

Choose the larger atom or ion from each pair.

(a) S or S^{2-} **(b)** Ca or Ca^{2+} **(c)** Br^- or Kr

SOLUTION

(a) The S^{2-} ion is larger than an S atom because the S^{2-} ion has the same number of protons as S, but two more electrons. The additional electron–electron repulsions cause the anion to be larger than the neutral atom.

(b) A Ca atom is larger than a Ca^{2+} ion because the Ca atom has an argon core and two 4*s* electrons. Since the 4*s* electrons are the outermost electrons, and since they are shielded from the nuclear charge by the core electrons, they contribute greatly to the size of the Ca atom. The Ca^{2+} cation, having lost the outermost 4*s* electrons, has only the argon core and carries a charge of 2^+, which makes it smaller than the Ca atom.

(c) A Br^- ion is larger than a Kr atom because, although they are isoelectronic, Br^- has one fewer proton than Kr, resulting in a smaller pull on the electrons and therefore a larger radius.

FOR PRACTICE 4.8

Choose the larger atom or ion from each pair.

(a) K or K^+ **(b)** F or F^- **(c)** Ca^{2+} or Cl^-

FOR MORE PRACTICE 4.8

Arrange the following in order of decreasing radius: Ca^{2+}, Ar, Cl^-.

4.5
Cc
Conceptual Connection

Ions, Isotopes, and Atomic Size

In the previous sections, we have seen how the number of electrons and the number of protons affect the size of an atom or ion. However, we have not considered how the number of neutrons affects the size of an atom. Why not? Would you expect isotopes—for example, C-12 and C-13—to have different atomic radii?

Ionization Energy

The **ionization energy (IE)** of an atom or ion is the energy required to remove an electron from the atom or ion in the gaseous state. Ionization energy is always positive because removing an electron always takes energy. (The process is endothermic, which, as discussed in Chapter 2, absorbs heat and therefore carries a positive sign.) The energy required to remove the first electron is called the *first ionization energy* (IE_1). For example, we represent the first ionization of sodium with the equation:

$$\text{Na}(g) \longrightarrow \text{Na}^+(g) + 1\,\text{e}^- \qquad \text{IE}_1 = 496\ \text{kJ/mol}$$

The energy required to remove the second electron is the *second ionization energy* (IE_2), the energy required to remove the third electron is the *third ionization energy* (IE_3), and so on. We represent the second ionization energy of sodium as:

$$\text{Na}^+(g) \longrightarrow \text{Na}^{2+}(g) + 1\,\text{e}^- \qquad \text{IE}_2 = 4560\ \text{kJ/mol}$$

Notice that the second ionization energy is not the energy required to remove *two* electrons from sodium (that quantity is the sum of IE_1 and IE_2), but rather the energy required to remove one electron from Na^+. We look at trends in IE_1 and IE_2 separately.

Trends in First Ionization Energy

The first ionization energies of the elements through Xe are shown in Figure 4.18 ▼. Notice the periodic trend in ionization energy, peaking at each noble gas and bottoming at each alkali metal. Based on

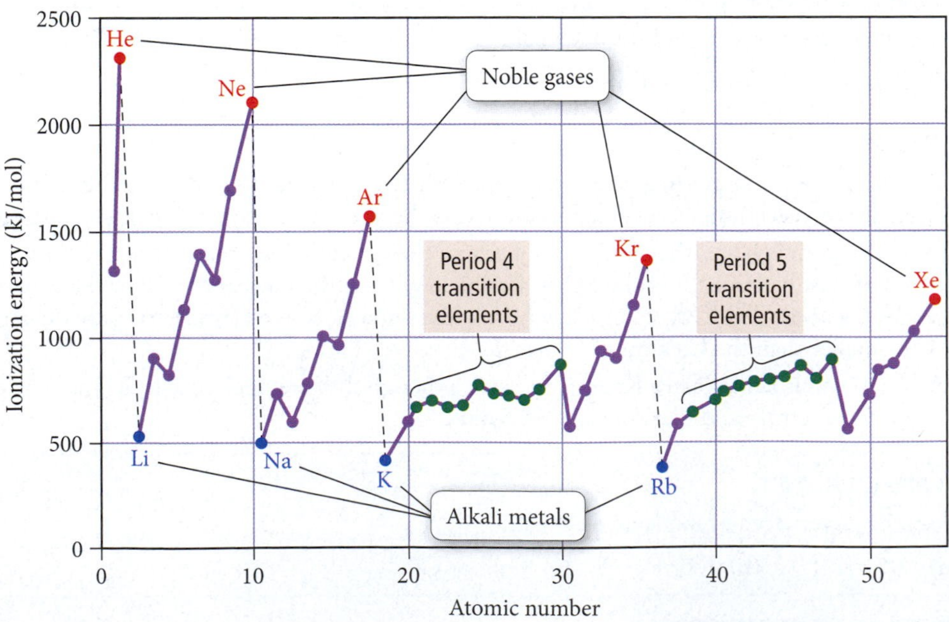

▶ **FIGURE 4.18 First Ionization Energy versus Atomic Number for the Elements through Xenon** Ionization starts at a minimum with each alkali metal and rises to a peak with each noble gas.

Trends in First Ionization Energy

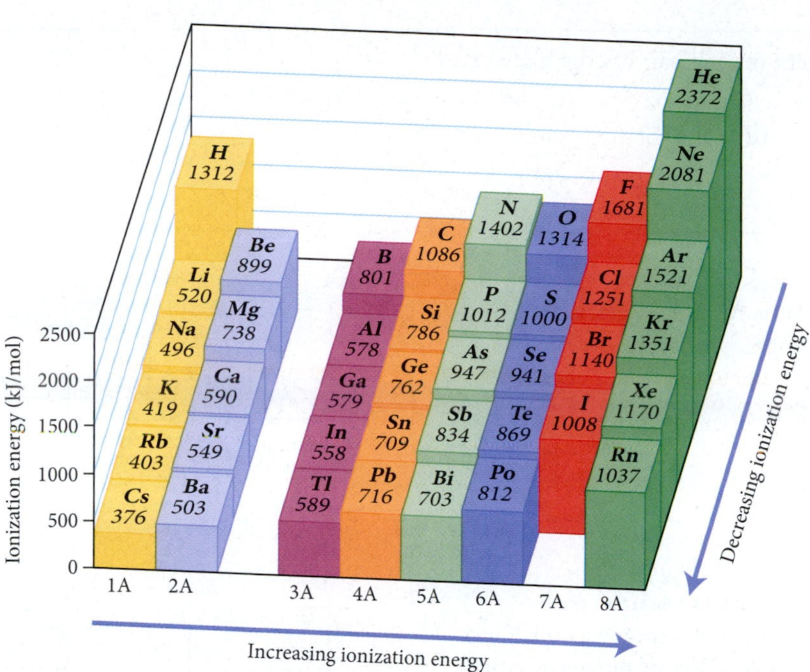

▲ **FIGURE 4.19 Trends in Ionization Energy** Ionization energy increases as we move to the right across a period and decreases as we move down a column in the periodic table.

what we have learned about electron configurations and effective nuclear charge, how can we account for the observed trend? As we have seen, the principal quantum number, n, increases as we move down a column. For a given sublevel, orbitals with higher principal quantum numbers are larger than orbitals with smaller principal quantum numbers. (For example, a 4s orbital is larger than a 3s orbital.) Consequently, electrons in the outermost principal level are farther away from the positively charged nucleus—and are therefore held less tightly—as we move down a column. This results in a lower ionization energy as we move down a column, as shown in Figure 4.19 ▲.

What about the trend as we move to the right across a row? For example, would it take more energy to remove an electron from Na or from Cl, two elements on either end of the third row in the periodic table? We know that Na has an outer electron configuration of $3s^1$ and Cl has an outer electron configuration of $3s^2\ 3p^5$. As discussed previously, the outermost electrons in chlorine experience a higher effective nuclear charge than the outermost electrons in sodium (which is why chlorine has a smaller atomic radius than sodium). Consequently, we would expect chlorine to have a higher ionization energy than sodium, which is indeed the case. We can make a similar argument for the other main-group elements: Ionization energy generally increases as we move to the right across a row in the periodic table, as shown in Figure 4.19.

Summarizing Ionization Energy for Main-Group Elements

- Ionization energy generally *decreases* as we move down a column (or family) in the periodic table because electrons in the outermost principal level are increasingly farther away from the positively charged nucleus and are therefore held less tightly.
- Ionization energy generally *increases* as we move to the right across a row (or period) in the periodic table because electrons in the outermost principal energy level generally experience a greater effective nuclear charge (Z_{eff}).

EXAMPLE 4.9

Ionization Energy

On the basis of periodic trends, determine which element in each pair has the higher first ionization energy (if possible).

(a) Al or S **(b)** As or Sb **(c)** N or Si **(d)** O or Cl

SOLUTION

(a) Al or S

S has a higher ionization energy than Al because, as you trace the path between Al and S on the periodic table, you move to the right within the same row. Ionization energy increases as you go to the right due to increasing effective nuclear charge.

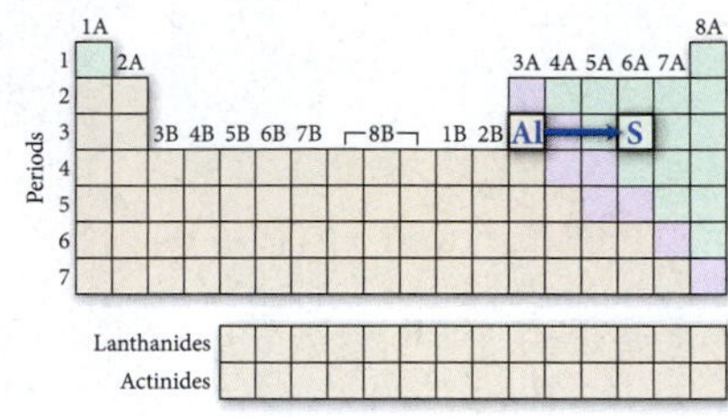

(b) As or Sb

As has a higher ionization energy than Sb because, as you trace the path between As and Sb on the periodic table, you move down a column. Ionization energy decreases as you go down a column as a result of the increasing size of orbitals with increasing *n*.

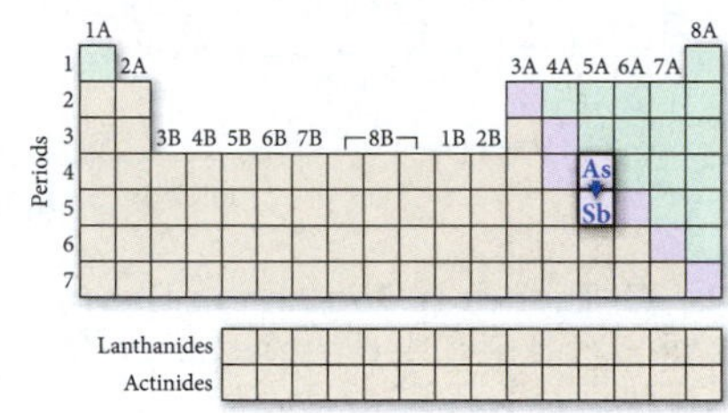

(c) N or Si

N has a higher ionization energy than Si because, as you trace the path between N and Si on the periodic table, you move down a column (ionization energy decreases due to increasing size of outermost orbitals) and then to the left across a row (ionization energy decreases due to decreasing effective nuclear charge). These effects sum together for an overall decrease.

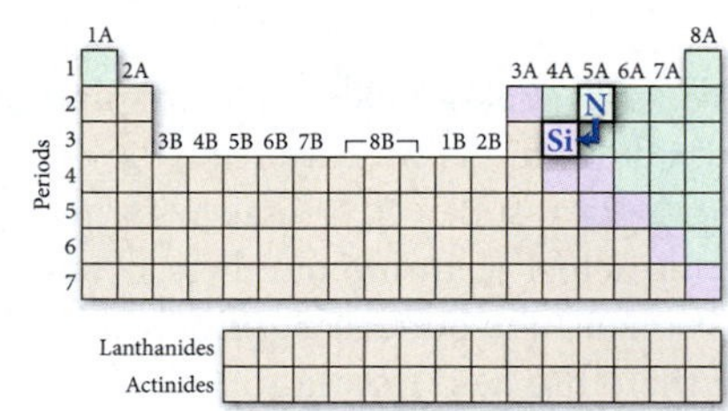

(d) O or Cl

Based on periodic trends alone, it is impossible to tell which has a higher ionization energy because, as you trace the path between O and Cl, you go to the right across a row (ionization energy increases) and then down a column (ionization energy decreases). These effects tend to counter each other, and it is not obvious which will dominate.

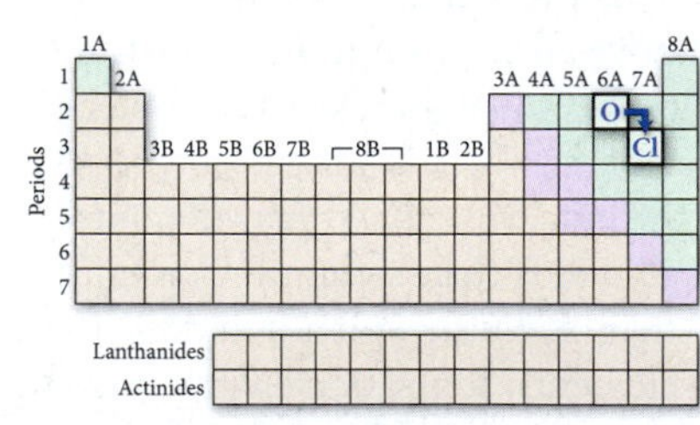

FOR PRACTICE 4.9

On the basis of periodic trends, determine the element in each pair with the higher first ionization energy (if possible).

(a) Sn or I

(b) Ca or Sr

(c) C or P

(d) F or S

FOR MORE PRACTICE 4.9

Arrange the following elements in order of decreasing first ionization energy: S, Ca, F, Rb, Si.

Exceptions to Trends in First Ionization Energy

If we carefully examine Figure 4.19, we can see some exceptions to the trends in first ionization energies. For example, boron has a smaller ionization energy than beryllium, even though it lies to the right of beryllium in the same row. This exception is caused by the change in going from the *s* block to the *p* block. Recall from Section 4.3 that the 2*p* orbital penetrates into the nuclear region *less than* the 2*s* orbital. Consequently, the 1*s* electrons shield the electron in the 2*p* orbital from nuclear charge more than they shield the electrons in the 2*s* orbital. The result is that the 2*p* orbitals are higher in energy, and therefore the electron is easier to remove (it has a lower ionization energy). Similar exceptions occur for aluminum and gallium, both directly below boron in group 3A.

Another exception occurs between nitrogen and oxygen. Although oxygen is to the right of nitrogen in the same row, it has a lower ionization energy. This exception is caused by the repulsion between electrons when they occupy the same orbital. Examine the electron configurations and orbital diagrams of nitrogen and oxygen shown here:

		1s	2s	2p		
N	$1s^22s^22p^3$	⇅	⇅	↑	↑	↑
O	$1s^22s^22p^4$	⇅	⇅	⇅	↑	↑

Nitrogen has three electrons in three *p* orbitals, while oxygen has four. In nitrogen, the 2*p* orbitals are half-filled (which makes the configuration particularly stable). Oxygen's fourth electron must pair with another electron, making it easier to remove (and less stable). Exceptions for similar reasons occur for S and Se, directly below oxygen in group 6A.

Trends in Second and Successive Ionization Energies

Notice the trends in the first, second, and third ionization energies of sodium (group 1A) and magnesium (group 2A), as shown here.

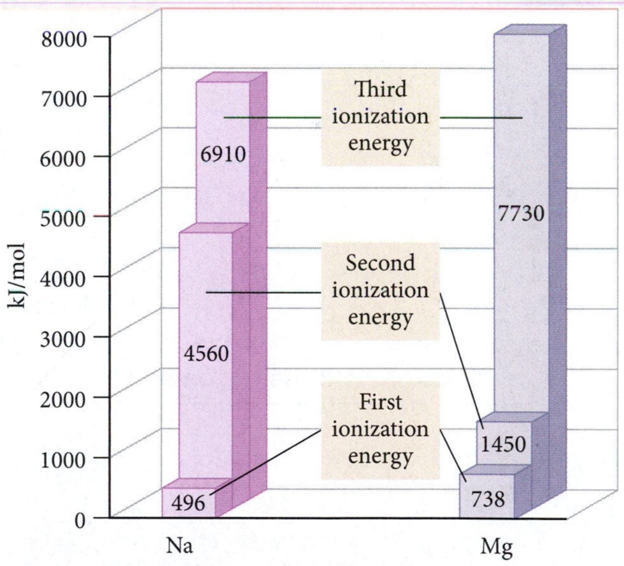

For sodium, there is a huge jump between the first and second ionization energies. For magnesium, the ionization energy roughly doubles from the first to the second, but then a huge jump occurs between the second and third ionization energies. What is the reason for these jumps?

We can understand these trends by examining the electron configurations of sodium and magnesium:

Na [Ne] $3s^1$
Mg [Ne] $3s^2$

The first ionization of sodium involves removing the valence electron in the 3*s* orbital. Recall that these valence electrons are held more loosely than the core electrons and that the resulting ion has a noble gas configuration, which is particularly stable. Consequently, the first ionization energy is fairly low. The second ionization of sodium, however, involves removing a core electron from an ion with a noble gas configuration. This requires a tremendous amount of energy, making the value of IE_2 very high.

As with sodium, the first ionization of magnesium involves removing a valence electron in the 3*s* orbital. This requires a bit more energy than the corresponding ionization of sodium because of the trends in Z_{eff} that we discussed earlier (Z_{eff} increases as we move to the right across a row). The second ionization of magnesium also involves removing an outer electron in the 3*s* orbital, but this time from an ion with a 1+ charge (instead of from a neutral atom). This requires roughly twice the energy as removing the electron from the neutral atom. The third ionization of magnesium is analogous to the second ionization of sodium—it requires removing a core electron from an ion with a noble gas configuration. This requires a tremendous amount of energy, making IE_3 very high.

As shown in Table 4.1, similar trends exist for the successive ionization energies of many elements. The ionization energy increases fairly uniformly with each successive removal of an outermost electron, but takes a large jump with the removal of the first core electron.

TABLE 4.1 Successive Values of Ionization Energies for the Elements Sodium through Argon (kJ/mol)

Element	IE_1	IE_2	IE_3	IE_4	IE_5	IE_6	IE_7
Na	496	4560					
Mg	738	1450	7730		Core electrons		
Al	578	1820	2750	11,600			
Si	786	1580	3230	4360	16,100		
P	1012	1900	2910	4960	6270	22,200	
S	1000	2250	3360	4560	7010	8500	27,100
Cl	1251	2300	3820	5160	6540	9460	11,000
Ar	1521	2670	3930	5770	7240	8780	12,000

4.6 Cc Conceptual Connection

Ionization Energies and Chemical Bonding

Based on what you just learned about ionization energies, explain why valence electrons are more important than core electrons in determining the reactivity and bonding in atoms.

4.8 Electron Affinities and Metallic Character

Electron affinity and metallic character also exhibit periodic trends. Electron affinity is a measure of how easily an atom accepts an additional electron and is crucial to chemical bonding because bonding involves the transfer or sharing of electrons. Metallic character is important because of the high proportion of metals in the periodic table and the large role they play in our lives. Of the roughly 110 elements, 87 are metals. We examine each of these periodic properties individually in this section.

Electron Affinity

The **electron affinity (EA)** of an atom or ion is the energy change associated with the gaining of an electron by the atom in the gaseous state. Electron affinity is usually—though not always—negative because an atom or ion usually releases energy when it gains an electron. (The process is exothermic, which, as discussed in Chapter 2, gives off heat and therefore carries a negative sign.) In other words, the coulombic attraction between the nucleus of an atom and the incoming electron usually results in the release of energy as the electron is gained. For example, we can represent the electron affinity of chlorine with the equation:

$$Cl(g) + 1\,e^- \longrightarrow Cl^-(g) \qquad EA = -349\ kJ/mol$$

Electron Affinities (kJ/mol)

1A	2A	3A	4A	5A	6A	7A	8A
H −73							He >0
Li −60	Be >0	B −27	C −122	N >0	O −141	F −328	Ne >0
Na −53	Mg >0	Al −43	Si −134	P −72	S −200	Cl −349	Ar >0
K −48	Ca −2	Ga −30	Ge −119	As −78	Se −195	Br −325	Kr >0
Rb −47	Sr −5	In −30	Sn −107	Sb −103	Te −190	I −295	Xe >0

▲ **FIGURE 4.20 Electron Affinities of Selected Main-Group Elements**

Figure 4.20 ◀ displays the electron affinities for a number of main-group elements. As you can see from this figure, the trends in electron affinity are not as regular as trends in other properties we have examined. For instance, we might expect electron affinities to become relatively more positive (so that the addition of an electron is less exothermic) as we move down a column because the electron is entering orbitals with successively higher principal quantum numbers and will therefore be farther from the nucleus. This trend applies to the group 1A metals but does not hold for the other columns in the periodic table.

A more regular trend in electron affinity, however, occurs as we move to the right across a row. Based on the periodic properties we have learned so far, would you expect more energy to be released when an electron is gained by Na or Cl? We know that Na has an outer electron configuration of $3s^1$ and Cl has an outer electron configuration of $3s^2\,3p^5$. Because adding an electron to chlorine gives it a noble gas configuration and adding an electron to sodium does not, and because the outermost electrons in chlorine experience a higher Z_{eff} than the outermost electrons in sodium, we would expect chlorine to have a more negative electron affinity—the process should be more exothermic for chlorine. This is in fact the case. For main-group elements, electron affinity generally becomes more negative (more exothermic) as we move to the right across a row in the periodic table. The halogens (group 7A) therefore have the most negative electron affinities. But exceptions do occur. For example, notice that nitrogen and the other group 5A elements do not follow the general trend. These elements have $ns^2\,np^3$ outer electron configurations. When an electron is added to this configuration, it must pair with another electron in an already occupied p orbital. The repulsion between two electrons occupying the same orbital causes the electron affinity to be more positive than for elements in the previous column.

Summarizing Electron Affinity for Main-Group Elements

- Most groups (columns) of the periodic table do not exhibit any definite trend in electron affinity. Among the group 1A metals, however, electron affinity becomes more positive as we move down the column (adding an electron becomes less exothermic).
- Electron affinity generally becomes more negative (adding an electron becomes more exothermic) as we move to the right across a period (row) in the periodic table.

Metallic Character

As we discussed in Section 4.5, metals are good conductors of heat and electricity; they can be pounded into flat sheets (malleability); they can be drawn into wires (ductility); they are often shiny; and they tend to lose electrons in chemical reactions. Nonmetals, in contrast, have more varied physical properties; some are solids at room temperature, others are gases, but in general nonmetals are typically poor conductors of heat and electricity, and they all tend to gain electrons in chemical reactions. As we move to the right across a row in the periodic table, ionization energy increases and electron affinity becomes more negative; therefore, elements on the left side of the periodic table are more likely to lose electrons than elements on the right side of the periodic table (which are more likely to gain them). The other properties associated with metals follow the same general trend (even though we do not quantify them here). Consequently, as shown in Figure 4.21 ▼:

As we move to the right across a row (or period) in the periodic table, metallic character decreases.

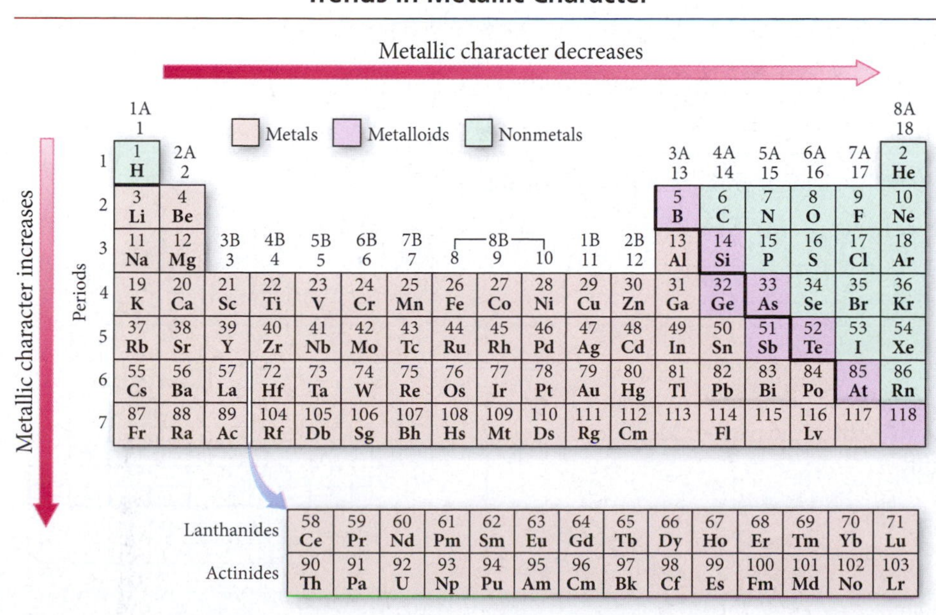

◀ **FIGURE 4.21 Trends in Metallic Character I** Metallic character decreases as we move to the right across a period and increases as we move down a column in the periodic table.

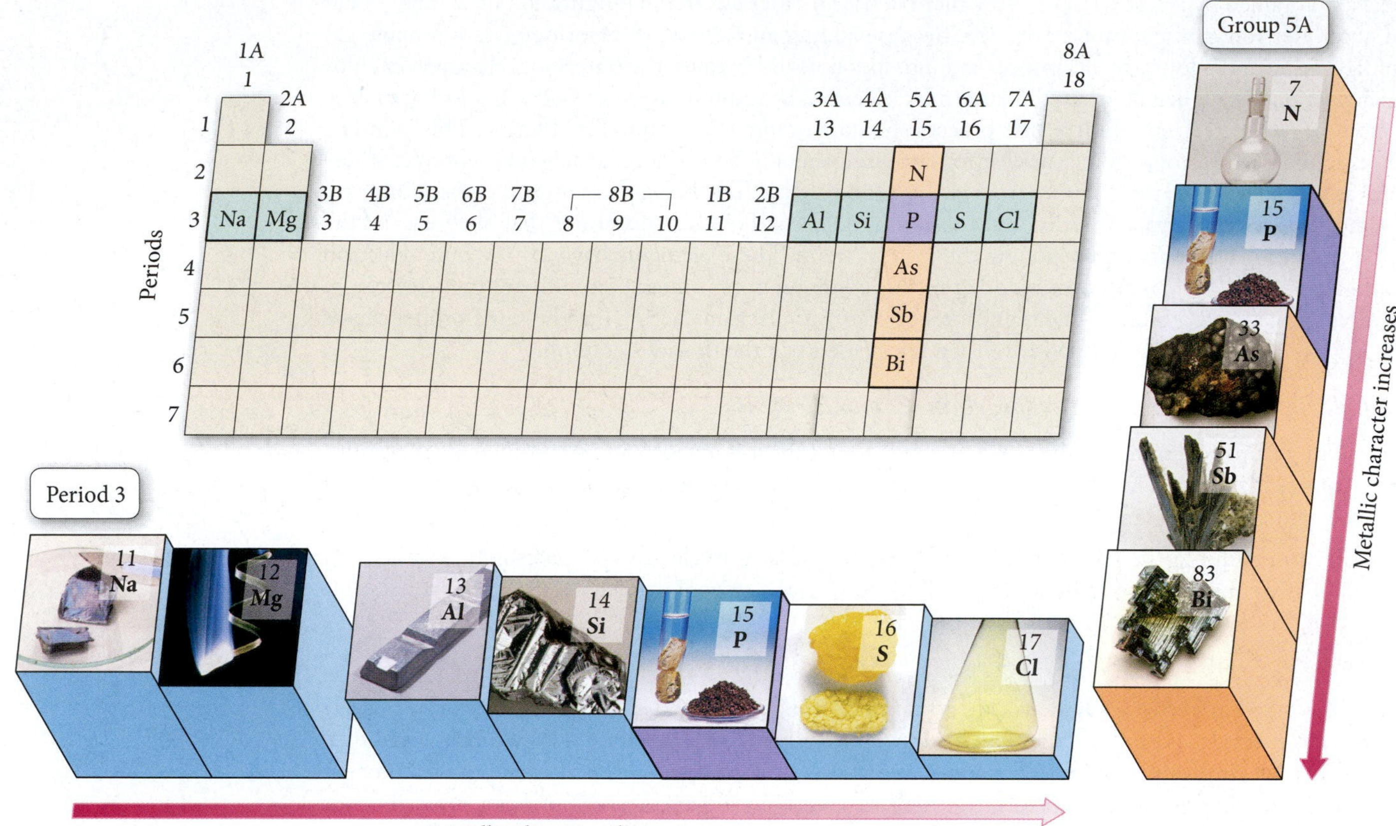

▲ **FIGURE 4.22 Trends in Metallic Character II** As we move down group 5A in the periodic table, metallic character increases. As we move across period 3, metallic character decreases.

As we move down a column in the periodic table, ionization energy decreases, making electrons more likely to be lost in chemical reactions. Consequently:

As we move down a column (or family) in the periodic table, metallic character increases.

These trends explain the overall distribution of metals and nonmetals in the periodic table first discussed in Section 4.5. Metals are found on the left side and toward the center and nonmetals on the upper right side. The change in chemical behavior from metallic to nonmetallic can be seen most clearly as we proceed to the right across period 3, or down along group 5A as shown in Figure 4.22 ▲.

EXAMPLE 4.10
Metallic Character

On the basis of periodic trends, choose the more metallic element from each pair (if possible).

(a) Sn or Te **(b)** P or Sb **(c)** Ge or In **(d)** S or Br

SOLUTION

(a) Sn or Te

Sn is more metallic than Te because, as you trace the path between Sn and Te on the periodic table, you move to the right within the same period. Metallic character decreases as you move to the right.

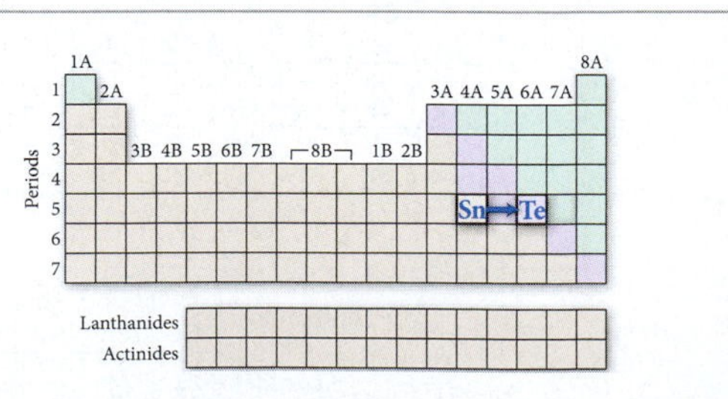

—*Continued on the next page*

Continued from the previous page—

(b) P or Sb

Sb is more metallic than P because, as you trace the path between P and Sb on the periodic table, you move down a column. Metallic character increases as you move down a column.

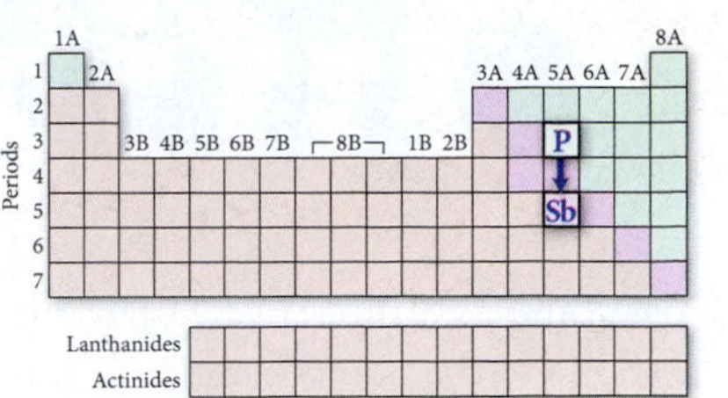

(c) Ge or In

In is more metallic than Ge because, as you trace the path between Ge and In on the periodic table, you move down a column (metallic character increases) and then to the left across a period (metallic character increases). These effects add together for an overall increase.

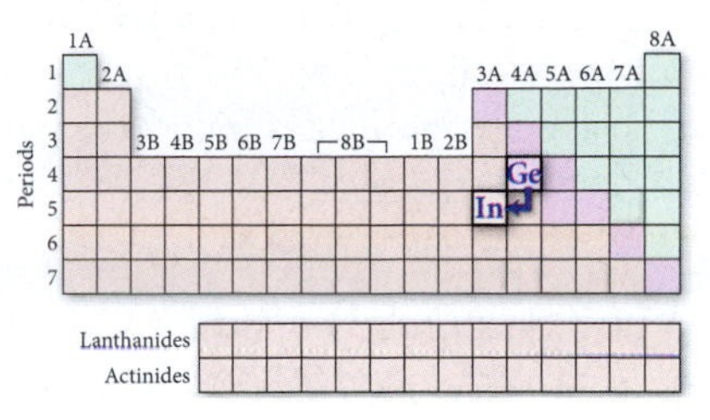

(d) S or Br

Based on periodic trends alone, we cannot tell which is more metallic because as you trace the path between S and Br, you go to the right across a period (metallic character decreases) and then down a column (metallic character increases). These effects tend to counter each other, and it is not obvious which will predominate.

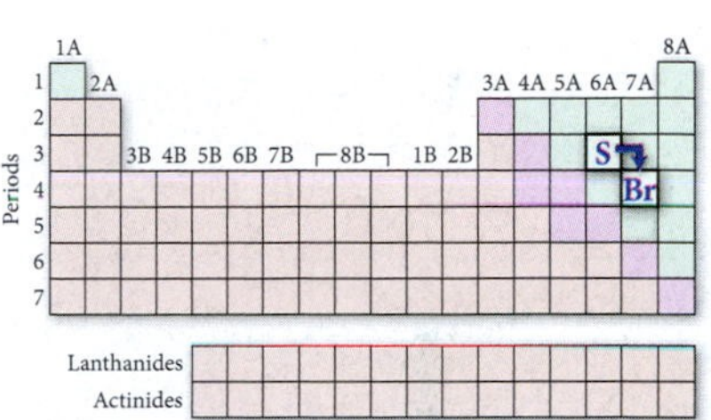

FOR PRACTICE 4.10

On the basis of periodic trends, choose the more metallic element from each pair (if possible).

(a) Ge or Sn **(b)** Ga or Sn **(c)** P or Bi **(d)** B or N

FOR MORE PRACTICE 4.10

Arrange the following elements in order of increasing metallic character: Si, Cl, Na, Rb.

Periodic Trends

4.7 Cc Conceptual Connection

Use the trends in ionization energy and electron affinity to explain why sodium chloride has the formula NaCl and not Na_2Cl or $NaCl_2$.

SELF-ASSESSMENT Quiz

1. According to Coulomb's law, if the separation between two particles of the same charge is doubled, the potential energy of the two particles ________________.
   a) becomes twice as high as it was before the distance separation
   b) becomes one-half as high as it was before the separation
   c) does not change
   d) becomes one-fourth as high as it was before the separation

2. Which electron in S is most shielded from nuclear charge?
   a) An electron in the 1*s* orbital.
   b) An electron in a 2*p* orbital.
   c) An electron in a 3*p* orbital.
   d) None of the above. (All of these electrons are equally shielded from nuclear charge.)

3. Choose the correct electron configuration for Se.
   a) $1s^22s^22p^63s^23p^4$
   b) $1s^22s^22p^63s^23p^64s^23d^{10}4p^4$
   c) $1s^22s^22p^63s^23p^64s^24p^4$
   d) $1s^22s^22p^63s^23p^64s^23d^4$

4. Choose the correct orbital diagram for vanadium.

   a) [Ar]

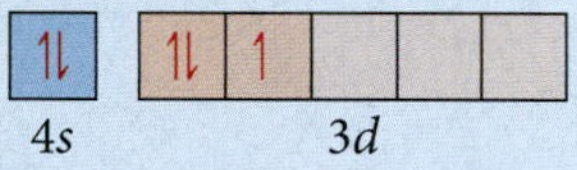

4s	3d				
↑↓	↑↓	↑			

   b) [Ar]

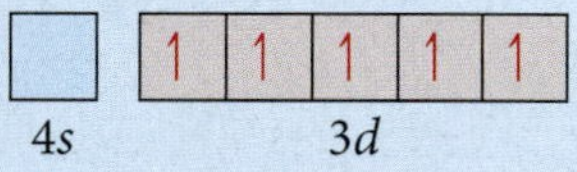

4s	3d				
	↑	↑	↑	↑	↑

   c) [Ar]

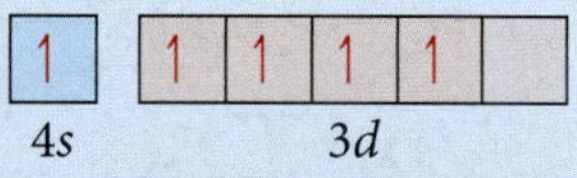

4s	3d				
↑	↑	↑	↑	↑	

   d) [Ar]

4s	3d				
↑↓	↑	↑	↑		

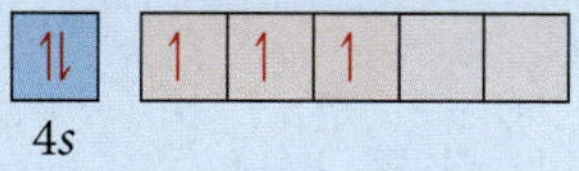

5. Which set of four quantum numbers corresponds to an electron in a 4*p* orbital?
   a) $n = 4, l = 1, m_l = 0, m_s = \frac{1}{2}$
   b) $n = 4, l = 3, m_l = 3, m_s = -\frac{1}{2}$
   c) $n = 4, l = 2, m_l = 0, m_s = \frac{1}{2}$
   d) $n = 4, l = 4, m_l = 3, m_s = -\frac{1}{2}$

6. Which element has the smallest atomic radius?
   a) C **b)** Si **c)** Be **d)** F

7. Which statement is true about electron shielding of nuclear charge?
   a) Outermost electrons efficiently shield one another from nuclear charge.
   b) Core electrons efficiently shield one another from nuclear charge.
   c) Outermost electrons efficiently shield core electrons from nuclear charge.
   d) Core electrons efficiently shield outermost electrons from nuclear charge.

8. Which statement is true about effective nuclear charge?
   a) Effective nuclear charge *decreases* as you move to the right across a row in the periodic table.
   b) Effective nuclear charge *increases* as you move to the right across a row in the periodic table.
   c) Effective nuclear charge remains relatively constant as you move to the right across a row in the periodic table.
   d) Effective nuclear charge *increases* then *decreases* at regular intervals as you move to the right across a row in the periodic table.

9. What is the electron configuration for Fe^{2+}?
   a) $[Ar]4s^23d^6$
   b) $[Ar]4s^23d^4$
   c) $[Ar]4s^03d^6$
   d) $[Ar]4s^23d^8$

10. Which species is diamagnetic?
    a) Cr^{2+} **b)** Zn **c)** Mn **d)** C

11. Arrange these atoms and ions in order of increasing radius: Cs^+, Xe, I^-.
    a) $I^- < Xe < Cs^+$ **b)** $Cs^+ < Xe < I^-$
    c) $Xe < Cs^+ < I^-$ **d)** $I^- < Cs^+ < Xe$

12. Arrange these elements in order of increasing first ionization energy: Cl, Sn, Si.
    a) Cl < Si < Sn **b)** Sn < Si < Cl
    c) Si < Cl < Sn **d)** Sn < Cl < Si

13. The ionization energies of an unknown third period element are shown here. Identify the element. $IE_1 = 786$ kJ/mol; $IE_2 = 1580$ kJ/mol; $IE_3 = 3230$ kJ/mol; $IE_4 = 4360$ kJ/mol; $IE_5 = 16{,}100$ kJ/mol
    a) Mg **b)** Al **c)** Si **d)** P

14. Identify the correct trends in metallic character.
    a) Metallic character *increases* as we move to the right across a row in the periodic table and *increases* as we move down a column.
    b) Metallic character *decreases* as we move to the right across a row in the periodic table and *increases* as we move down a column.

—Continued

Continued—

c) Metallic character *decreases* as we move to the right across a row in the periodic table and *decreases* as we move down a column.

d) Metallic character *increases* as we move to the right across a row in the periodic table and *decreases* as we move down a column.

15. For which element is the gaining of an electron most exothermic?
a) Li **b)** N **c)** F **d)** B

16. What is the charge of the ion most commonly formed by S?
a) 2+ **b)** + **c)** − **d)** 2−

Answers: 1:b; 2:c; 3:b; 4:d; 5:a; 6:d; 7:d; 8:b; 9:c; 10:b; 11:b; 12:b; 13:c; 14:b; 15:c; 16:d

CHAPTER SUMMARY 4

REVIEW

KEY LEARNING OUTCOMES

CHAPTER OBJECTIVES	ASSESSMENT
Writing Electron Configurations (4.3)	• Example 4.1 For Practice 4.1 Exercises 45, 46, 49, 50
Writing Orbital Diagrams (4.3)	• Example 4.2 For Practice 4.2 Exercises 47, 48
Valence Electrons and Core Electrons (4.4)	• Example 4.3 For Practice 4.3 Exercises 55–60
Electron Configurations from the Periodic Table (4.4)	• Example 4.4 For Practice 4.4 For More Practice 4.4 Exercises 49–52
Predicting the Charge of Ions (4.5)	• Example 4.5 For Practice 4.5 Exercises 63, 64
Using Periodic Trends to Predict Atomic Size (4.6)	• Example 4.6 For Practice 4.6 For More Practice 4.6 Exercises 71–74
Writing Electron Configurations for Ions (4.7)	• Example 4.7 For Practice 4.7 Exercises 63, 64, 75–78
Using Periodic Trends to Predict Ion Size (4.7)	• Example 4.8 For Practice 4.8 For More Practice 4.8 Exercises 79–82
Using Periodic Trends to Predict Relative Ionization Energies (4.7)	• Example 4.9 For Practice 4.9 For More Practice 4.9 Exercises 83–88
Predicting Metallic Character Based on Periodic Trends (4.8)	• Example 4.10 For Practice 4.10 For More Practice 4.10 Exercises 89–94

KEY TERMS

Section 4.1
periodic property (102)

Section 4.2
periodic law (103)
main-group elements (104)
transition elements (or transition metals) (104)
family (or group) of elements (104)

Section 4.3
electron configuration (105)
ground state (105)
orbital diagram (105)
Pauli exclusion principle (105)
degenerate (106)
Coulomb's law (106)
shielding (107)
effective nuclear charge (Z_{eff})(107)
penetration (107)
aufbau principle (109)
Hund's rule (109)

Section 4.4
valence electrons (112)
core electrons (112)

Section 4.5
noble gases (116)
metals (116)
nonmetals (117)
metalloids (117)
semiconductors (117)
alkali metals (118)
alkaline earth metals (118)
halogens (118)

Section 4.6
van der Waals radius (nonbonding atomic radius) (119)
covalent radius (bonding atomic radius) (119)
atomic radius (119)

Section 4.7
paramagnetic (124)
diamagnetic (124)
ionization energy (IE) (128)

Section 4.8
electron affinity (EA) (132)

KEY CONCEPTS

Periodic Properties and the Periodic Table (4.1, 4.2)

- The periodic table was primarily developed by Dmitri Mendeleev in the nineteenth century. Mendeleev arranged the elements in a table so that atomic mass increased from left to right in a row and elements with similar properties fell in the same columns.
- Periodic properties are predictable based on an element's position within the periodic table. Periodic properties include atomic radius, ionization energy, electron affinity, density, and metallic character.
- Quantum mechanics explains the periodic table by showing how electrons fill the quantum-mechanical orbitals within the atoms that compose the elements.

Electron Configurations (4.3)

- An electron configuration for an atom shows which quantum-mechanical orbitals the atom's electrons occupy. For example, the electron configuration of helium ($1s^2$) indicates that helium's two electrons exist within the $1s$ orbital.
- The order of filling quantum-mechanical orbitals in multi-electron atoms is: $1s\ 2s\ 2p\ 3s\ 3p\ 4s\ 3d\ 4p\ 5s\ 4d\ 5p\ 6s$.
- According to the Pauli exclusion principle, each orbital can hold a maximum of two electrons (with opposing spins).
- According to Hund's rule, orbitals of the same energy first fill singly with electrons with parallel spins before pairing.

Electron Configurations and the Periodic Table (4.4)

- An atom's outermost electrons (called valence electrons) are most important in determining the atom's properties.
- Because quantum-mechanical orbitals fill sequentially with increasing atomic number, we can infer the electron configuration of an element from its position in the periodic table.

Electron Configurations and the Properties of Elements (4.5)

- The most stable (or chemically unreactive) elements in the periodic table are the noble gases. These elements have completely full principal energy levels, which have particularly low potential energy compared to other possible electron configurations.
- Elements on the left side and in the center of the periodic table are metals and tend to lose electrons when they undergo chemical changes.
- Elements on the upper right side of the periodic table are nonmetals and tend to gain electrons when they undergo chemical changes.
- Elements with one or two valence electrons are among the most active metals, readily losing their valence electrons to attain noble gas configurations.
- Elements with six or seven valence electrons are among the most active nonmetals, readily gaining enough electrons to attain a noble gas configuration.
- Many main-group elements form ions with noble gas electron configurations.

Effective Nuclear Charge and Periodic Trends in Atomic Size (4.6)

- The size of an atom is largely determined by its outermost electrons. As we move down a column in the periodic table, the principal quantum number (n) of the outermost electrons increases, resulting in successively larger orbitals and therefore larger atomic radii.
- As we move across a row in the periodic table, atomic radii decrease because the effective nuclear charge—the net or average charge experienced by the atom's outermost electrons—increases.
- The atomic radii of the transition elements stay roughly constant as we move across each row because electrons are added to the $n_{highest} - 1$ orbitals while the number of highest n electrons stays roughly constant.

Ion Properties (4.7)

- We can determine the electron configuration of an ion by adding or subtracting the corresponding number of electrons to the electron configuration of the neutral atom.
- For main-group ions, the order of removing electrons is the same as the order in which they are added in building up the electron configuration.
- For transition metal atoms, ns electrons are removed before $(n-1)d$ electrons.
- The radius of a cation is much *smaller* than that of the corresponding atom, and the radius of an anion is much *larger* than that of the corresponding atom.
- The ionization energy—the energy required to remove an electron from an atom in the gaseous state—generally decreases as we move down a column in the periodic table and increases when we move to the right across a row.
- Successive ionization energies increase smoothly from one valence electron to the next, but the ionization energy increases dramatically for the first core electron.

Electron Affinities and Metallic Character (4.8)

- Electron affinity—the energy associated with an element in its gaseous state gaining an electron—does not show a general trend as we move down a column in the periodic table, but it generally becomes more negative (more exothermic) to the right across a row.
- Metallic character—the tendency to lose electrons in a chemical reaction—generally increases down a column in the periodic table and decreases to the right across a row.

KEY EQUATIONS AND RELATIONSHIPS

Order of Filling Quantum-Mechanical Orbitals (4.3)

$$1s\ 2s\ 2p\ 3s\ 3p\ 4s\ 3d\ 4p\ 5s\ 4d\ 5p\ 6s$$

EXERCISES

REVIEW QUESTIONS

1. What are periodic properties?
2. Use aluminum as an example to explain how density is a periodic property.
3. Explain the contributions of Johann Döbereiner and John Newlands to the organization of elements according to their properties.
4. Who is credited with arranging the periodic table? How were elements arranged in this table?
5. Explain the contributions of Meyer and Moseley to the periodic table.
6. The periodic table is a result of the periodic law. What observations led to the periodic law? What theory explains the underlying reasons for the periodic law?
7. What is an electron configuration? Provide an example.
8. What is Coulomb's law? Explain how the potential energy of two charged particles depends on the distance between the charged particles and on the magnitude and sign of their charges.
9. What is shielding? In an atom, which electrons tend to do the most shielding (core electrons or valence electrons)?
10. What is penetration? How does the penetration of an orbital into the region occupied by core electrons affect the energy of an electron in that orbital?
11. Why are the sublevels within a principal level split into different energies for multi-electron atoms but not for the hydrogen atom?
12. What is an orbital diagram? Provide an example.
13. Why is electron spin important when writing electron configurations? Explain in terms of the Pauli exclusion principle.
14. What are degenerate orbitals? According to Hund's rule, how are degenerate orbitals occupied?
15. List all orbitals from 1*s* through 5*s* according to increasing energy for multi-electron atoms.
16. What are valence electrons? Why are they important?
17. Copy this blank periodic table onto a sheet of paper and label each of the blocks within the table: *s* block, *p* block, *d* block, and *f* block.

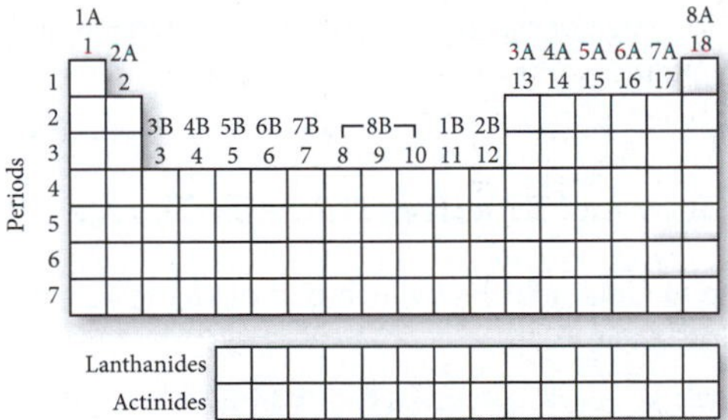

18. Explain why the *s* block in the periodic table has only two columns while the *p* block has six.
19. Explain why the rows in the periodic table become progressively longer as we move down the table. For example, the first row contains 2 elements, the second and third rows each contain 8 elements, and the fourth and fifth rows each contain 18 elements.
20. Explain the relationship between a main-group element's lettered group number (the number of the element's column) and its valence electrons.
21. Explain the relationship between an element's row number in the periodic table and the highest principal quantum number in the element's electron configuration. How does this relationship differ for main-group elements, transition elements, and inner transition elements?
22. Which of the transition elements in the first transition series have anomalous electron configurations?
23. Explain how to write the electron configuration for an element based on its position in the periodic table.
24. Explain the relationship between the properties of an element and the number of valence electrons that it contains.
25. List the number of valence electrons for each family in the periodic table, and explain the relationship between the number of valence electrons and the resulting chemistry of the elements in the family.
    a. alkali metals
    b. alkaline earth metals
    c. halogens
    d. oxygen family
26. Define atomic radius. For main-group elements, describe the observed trends in atomic radius as we:
    a. move across a period in the periodic table
    b. move down a column in the periodic table
27. What is effective nuclear charge? What is shielding?
28. When an alkali metal forms an ion, what is the charge of the ion? What is the charge of an alkaline earth metal ion?
29. When a halogen forms an ion, what is the charge of the ion? When the nonmetals in the oxygen family form an ion, what is the charge of the ion? What is the charge of the ion formed by N and Al?
30. Use the concepts of effective nuclear charge, shielding, and *n* value of the valence orbital to explain the trend in atomic radius as you move across a period in the periodic table.
31. For transition elements, describe the trends in atomic radius as we:
    a. move across a period in the periodic table
    b. move down a column in the periodic table

    Explain the reasons for the trends described in parts a and b.
32. How is the electron configuration of an anion different from that of the corresponding neutral atom? How is the electron configuration of a cation different?
33. Explain how to write an electron configuration for a transition metal cation. Is the order of electron removal upon ionization simply the reverse of electron addition upon filling? Why or why not?
34. Describe the relationship between:
    a. the radius of a cation and the radius of the atom from which it is formed
    b. the radius of an anion and the radius of the atom from which it is formed
35. What is ionization energy? What is the difference between first ionization energy and second ionization energy?
36. What is the general trend in ionization energy as we move down a column in the periodic table? As we move across a row?
37. What are the exceptions to the periodic trends in ionization energy? Why do they occur?
38. Examination of the first few successive ionization energies for a given element usually reveals a large jump between two ionization energies. For example, the successive ionization energies of magnesium show a large jump between IE_2 and IE_3. The successive ionization energies of aluminum show a large jump between IE_3 and IE_4. Explain why these jumps occur and how we might predict them.
39. What is electron affinity? What are the observed periodic trends in electron affinity?
40. What is metallic character? What are the observed periodic trends in metallic character?

PROBLEMS BY TOPIC

The Periodic Table

41. Write the name of each element and classify it as a metal, nonmetal, or metalloid.

 a. K b. Ba c. I d. O e. Sb

42. Write the symbol for each element and classify it as a metal, nonmetal, or metalloid.

 a. gold b. fluorine c. sodium
 d. tin e. argon

43. Determine whether each element is a main-group element.

 a. tellurium b. potassium
 c. vanadium d. manganese

44. Determine whether each element is a transition element.

 a. Cr b. Br c. Mo d. Cs

Electron Configurations

45. Write the full electron configuration for each element.

 a. Si b. O c. K d. Ne

46. Write the full electron configuration for each element.

 a. C b. P c. Ar d. Na

47. Write the full orbital diagram for each element.

 a. N b. F c. Mg d. Al

48. Write the full orbital diagram for each element.

 a. S b. Ca c. Ne d. He

49. Use the periodic table to write an electron configuration for each element. Represent core electrons with the symbol of the previous noble gas in brackets.

 a. P b. Ge c. Zr d. I

50. Use the periodic table to determine the element corresponding to each electron configuration.

 a. $[Ar]\ 4s^2 3d^{10} 4p^6$ b. $[Ar]\ 4s^2 3d^2$
 c. $[Kr]\ 5s^2 4d^{10} 5p^2$ d. $[Kr]\ 5s^2$

51. Use the periodic table to determine each quantity.

 a. the number of 2*s* electrons in Li
 b. the number of 3*d* electrons in Cu
 c. the number of 4*p* electrons in Br
 d. the number of 4*d* electrons in Zr

52. Use the periodic table to determine each quantity.

 a. the number of 3*s* electrons in Mg
 b. the number of 3*d* electrons in Cr
 c. the number of 4*d* electrons in Y
 d. the number of 6*p* electrons in Pb

53. Name an element in the fourth period (row) of the periodic table with:

 a. five valence electrons
 b. four 4*p* electrons
 c. three 3*d* electrons
 d. a complete outer shell

54. Name an element in the third period (row) of the periodic table with:

 a. three valence electrons
 b. four 3*p* electrons
 c. six 3*p* electrons
 d. two 3*s* electrons and no 3*p* electrons

Valence Electrons and Simple Chemical Behavior from the Periodic Table

55. Determine the number of valence electrons in each element.

 a. Ba b. Cs c. Ni d. S

56. Determine the number of valence electrons in each element. Which elements do you expect to lose electrons in chemical reactions? Which do you expect to gain electrons?

 a. Al b. Sn c. Br d. Se

57. Which outer electron configuration would you expect to correspond to a reactive metal? To a reactive nonmetal?

 a. ns^2 b. $ns^2 np^6$ c. $ns^2 np^5$ d. $ns^2 np^2$

58. Which outer electron configurations would you expect to correspond to a noble gas? To a metalloid?

 a. ns^2 b. $ns^2 np^6$ c. $ns^2 np^5$ d. $ns^2 np^2$

59. List the number of valence electrons for each element and classify each element as an alkali metal, alkaline earth metal, halogen, or noble gas.

 a. sodium b. iodine c. calcium
 d. barium e. krypton

60. List the number of valence electrons in each element and classify each element as an alkali metal, alkaline earth metal, halogen, or noble gas.

 a. F b. Sr c. K d. Ne e. At

61. Which pair of elements do you expect to be most similar? Why?

 a. N and Ni b. Mo and Sn c. Na and Mg
 d. Cl and F e. Si and P

62. Which pair of elements do you expect to be most similar? Why?

 a. nitrogen and oxygen
 b. titanium and gallium
 c. lithium and sodium
 d. germanium and arsenic
 e. argon and bromine

63. Predict the charge of the ion formed by each element and write the electron configuration of the ion.

 a. O b. K c. Al d. Rb

64. Predict the charge of the ion formed by each element and write the electron configuration of the ion.

 a. Mg b. N c. F d. Na

Coulomb's Law and Effective Nuclear Charge

65. According to Coulomb's law, which pair of charged particles has the lowest potential energy?

 a. a particle with a 1− charge separated by 150 pm from a particle with a 2+ charge
 b. a particle with a 1− charge separated by 150 pm from a particle with a 1+ charge
 c. a particle with a 1− charge separated by 100 pm from a particle with a 3+ charge

66. According to Coulomb's law, rank the interactions between charged particles from lowest potential energy to highest potential energy.

 a. a 1+ charge and a 1− charge separated by 100 pm
 b. a 2+ charge and a 1− charge separated by 100 pm
 c. a 1+ charge and a 1+ charge separated by 100 pm
 d. a 1+ charge and a 1− charge separated by 200 pm

67. Which electrons experience a greater effective nuclear charge: the valence electrons in beryllium or the valence electrons in nitrogen? Why?

68. Arrange the atoms according to decreasing effective nuclear charge experienced by their valence electrons: S, Mg, Al, Si.

69. If core electrons completely shielded valence electrons from nuclear charge (i.e., if each core electron reduced nuclear charge by one unit) and if valence electrons did not shield one another from nuclear charge at all, what would be the effective nuclear charge experienced by the valence electrons of each atom?

a. K **b.** Ca **c.** O **d.** C

70. In Section 4.6, we estimated the effective nuclear charge on beryllium's valence electrons to be slightly greater than 2+. What would a similar treatment predict for the effective nuclear charge on boron's valence electrons? Would you expect the effective nuclear charge to be different for boron's 2*s* electrons compared to its 2*p* electron? In what way? (*Hint:* Consider the shape of the 2*p* orbital compared to that of the 2*s* orbital.)

Atomic Radius

71. Choose the larger atom in each pair.

a. Al or In **b.** Si or N **c.** P or Pb **d.** C or F

72. Choose the larger atom in each pair.

a. Sn or Si **b.** Br or Ga **c.** Sn or Bi **d.** Se or Sn

73. Arrange these elements in order of increasing atomic radius: Ca, Rb, S, Si, Ge, F.

74. Arrange these elements in order of decreasing atomic radius: Cs, Sb, S, Pb, Se.

Ionic Electron Configurations, Ionic Radii, Magnetic Properties, and Ionization Energy

75. Write the electron configuration for each ion.

a. O^{2-} **b.** Br^- **c.** Sr^{2+} **d.** Co^{3+} **e.** Cu^{2+}

76. Write the electron configuration for each ion.

a. Cl^- **b.** P^{3-} **c.** K^+ **d.** Mo^{3+} **e.** V^{3+}

77. Write orbital diagrams for each ion and determine if the ion is diamagnetic or paramagnetic.

a. V^{5+} **b.** Cr^{3+} **c.** Ni^{2+} **d.** Fe^{3+}

78. Write orbital diagrams for each ion and determine if the ion is diamagnetic or paramagnetic.

a. Cd^{2+} **b.** Au^+ **c.** Mo^{3+} **d.** Zr^{2+}

79. Which is the larger species in each pair?

a. Li or Li^+ **b.** I^- or Cs^+ **c.** Cr or Cr^{3+} **d.** O or O^{2-}

80. Which is the larger species in each pair?

a. Sr or Sr^{2+} **b.** N or N^{3-} **c.** Ni or Ni^{2+} **d.** S^{2-} or Ca^{2+}

81. Arrange this isoelectronic series in order of decreasing radius: F^-, Ne, O^{2-}, Mg^{2+}, Na^+.

82. Arrange this isoelectronic series in order of increasing atomic radius: Se^{2-}, Kr, Sr^{2+}, Rb^+, Br^-.

83. Choose the element with the higher first ionization energy in each pair.

a. Br or Bi **b.** Na or Rb **c.** As or At **d.** P or Sn

84. Choose the element with the higher first ionization energy in each pair.

a. P or I **b.** Si or Cl **c.** P or Sb **d.** Ga or Ge

85. Arrange these elements in order of increasing first ionization energy: Si, F, In, N.

86. Arrange these elements in order of decreasing first ionization energy: Cl, S, Sn, Pb.

87. For each element, predict where the "jump" occurs for successive ionization energies. (For example, does the jump occur between the first and second ionization energies, the second and third, or the third and fourth?)

a. Be **b.** N **c.** O **d.** Li

88. Consider this set of successive ionization energies:

$IE_1 = 578$ kJ/mol $IE_2 = 1820$ kJ/mol
$IE_3 = 2750$ kJ/mol $IE_4 = 11{,}600$ kJ/mol

To which third period element do these ionization values belong?

Electron Affinities and Metallic Character

89. Choose the element with the more negative (more exothermic) electron affinity in each pair.

a. Na or Rb **b.** B or S **c.** C or N **d.** Li or F

90. Choose the element with the more negative (more exothermic) electron affinity in each pair.

a. Mg or S **b.** K or Cs **c.** Si or P **d.** Ga or Br

91. Choose the more metallic element in each pair.

a. Sr or Sb **b.** As or Bi **c.** Cl or O **d.** S or As

92. Choose the more metallic element in each pair.

a. Sb or Pb **b.** K or Ge **c.** Ge or Sb **d.** As or Sn

93. Arrange these elements in order of increasing metallic character: Fr, Sb, In, S, Ba, Se.

94. Arrange these elements in order of decreasing metallic character: Sr, N, Si, P, Ga, Al.

CUMULATIVE PROBLEMS

95. Bromine is a highly reactive liquid, whereas krypton is an inert gas. Explain the difference based on their electron configurations.

96. Potassium is a highly reactive metal, whereas argon is an inert gas. Explain the difference based on their electron configurations.

97. Both vanadium and its 3+ ion are paramagnetic. Use electron configurations to explain this statement.

98. Use electron configurations to explain why copper is paramagnetic while its 1+ ion is not.

99. Suppose you were trying to find a substitute for K^+ for some application. Where would you begin your search? What ions are most like K^+? For each ion you propose, explain the ways in which it is similar to K^+ and the ways it is different. Refer to periodic trends in your discussion.

100. Suppose you were trying to find a substitute for Na^+ some application. Where would you begin your search? What ions are most like Na^+? For each ion you propose, explain the ways in which it is similar to Na^+ and the ways it is different. Use periodic trends in your discussion.

101. Life on Earth evolved based on the element carbon. Based on periodic properties, what two or three elements would you expect to be most like carbon?

102. Which pair of elements would you expect to have the most similar atomic radii, and why?

a. Si and Ga **b.** Si and Ge **c.** Si and As

103. Consider these elements: N, Mg, O, F, Al.

a. Write the electron configuration for each element.
b. Arrange the elements in order of decreasing atomic radius.
c. Arrange the elements in order of increasing ionization energy.
d. Use the electron configurations in part a to explain the differences between your answers to parts b and c.

104. Consider these elements: P, Ca, Si, S, Ga.

a. Write the electron configuration for each element.
b. Arrange the elements in order of decreasing atomic radius.
c. Arrange the elements in order of increasing ionization energy.
d. Use the electron configurations in part a to explain the differences between your answers to parts b and c.

105. Explain why atomic radius decreases as we move to the right across a period for main-group elements but not for transition elements.

106. Explain why vanadium (radius = 134 pm) and copper (radius = 128 pm) have nearly identical atomic radii, even though the atomic number of copper is about 25% higher than that of vanadium. What would you predict about the relative densities of these two metals? Look up the densities in a reference book, periodic table, or on the Internet. Are your predictions correct?

107. The lightest noble gases, such as helium and neon, are completely inert—they do not form any chemical compounds whatsoever. The heavier noble gases, in contrast, do form a limited number of compounds. Explain this difference in terms of trends in fundamental periodic properties.

108. The lightest halogen is also the most chemically reactive, and reactivity generally decreases as we move down the column of halogens in the periodic table. Explain this trend in terms of periodic properties.

109. Write general outer electron configurations ($ns^x np^y$) for groups 6A and 7A in the periodic table. The electron affinity of each group 7A element is more negative than that of each corresponding group 6A element. Use the electron configurations to explain this observation.

110. The electron affinity of each group 5A element is more positive than that of each corresponding group 4A element. Use the outer electron configurations for these columns to suggest a reason for this behavior.

111. The elements with atomic numbers 35 and 53 have similar chemical properties. Based on their electronic configurations predict the atomic number of a heavier element that also should have these chemical properties.

112. Write the electronic configurations of the six cations that form from sulfur by the loss of one to six electrons. For those cations that have unpaired electrons, write orbital diagrams.

113. You have cracked a secret code that uses elemental symbols to spell words. The code uses numbers to designate the elemental symbols. Each number is the sum of the atomic number and the highest principal quantum number of the highest occupied orbital of the element whose symbol is to be used. Messages may be written forward or backward. Decode the following messages:

a. 10, 12, 58, 11, 7, 44, 63, 66
b. 9, 99, 30, 95, 19, 47, 79

114. The electron affinity of sodium is lower than that of lithium, while the electron affinity of chlorine is higher than that of fluorine. Suggest an explanation for this observation.

115. Use Coulomb's law to calculate the ionization energy in kJ/mol of an atom composed of a proton and an electron separated by 100.00 pm. What wavelength of light would have sufficient energy to ionize the atom?

116. The first ionization energy of sodium is 496 kJ/mol. Use Coulomb's law to estimate the average distance between the sodium nucleus and the 3*s* electron. How does this distance compare to the atomic radius of sodium? Explain the difference.

CHALLENGE PROBLEMS

117. Consider the densities and atomic radii of the noble gases at 25 °C:

Element	Atomic Radius (pm)	Density (g/L)
He	32	0.18
Ne	70	0.90
Ar	98	—
Kr	112	3.75
Xe	130	—
Rn	—	9.73

a. Estimate the densities of argon and xenon by interpolation from the data.
b. Provide an estimate of the density of the yet undiscovered element with atomic number 118 by extrapolation from the data.
c. Use the molar mass of neon to estimate the mass of a neon atom. Then use the atomic radius of neon to calculate the average density of a neon atom. How does this density compare to the density of neon gas? What does this comparison suggest about the nature of neon gas?
d. Use the densities and molar masses of krypton and neon to calculate the number of atoms of each element found in a volume of 1.0 L. Use these values to estimate the number of atoms that occur in 1.0 L of Ar. Now use the molar mass of argon to estimate the density of Ar. How does this estimate compare to that in part a?

118. As you have seen, the periodic table is a result of empirical observation (i.e., the periodic law), but quantum-mechanical theory explains *why* the table is so arranged. Suppose that, in another universe, quantum theory was such that there were one *s* orbital but only two *p* orbitals (instead of three) and only three *d* orbitals (instead of five). Draw out the first four periods of the periodic table in this alternative universe. Which elements would be the equivalent of the noble gases? Halogens? Alkali metals?

119. Consider the metals in the first transition series. Use periodic trends to predict a trend in density as you move to the right across the series.

120. Imagine a universe in which the value of m_s can be $+\frac{1}{2}$, 0, and $-\frac{1}{2}$. Assuming that all the other quantum numbers can take only the values possible in our world and that the Pauli exclusion principle applies, determine :

a. the new electronic configuration of neon
b. the atomic number of the element with a completed $n = 2$ shell
c. the number of unpaired electrons in fluorine

121. A carbon atom can absorb radiation of various wavelengths with resulting changes in its electronic configuration. Write orbital diagrams for the electronic configuration of carbon that would result from absorption of the three longest wavelengths of radiation that would change its electronic configuration.

122. Only trace amounts of the synthetic element darmstadtium, atomic number 110, have been obtained. The element is so highly unstable that no observations of its properties have been possible. Based on its position in the periodic table, propose three different reasonable valence electron configurations for this element.

123. What is the atomic number of the as yet undiscovered element in which the 8*s* and 8*p* electron energy levels fill? Predict the chemical behavior of this element.

124. The trend in second ionization energy for the elements from lithium to fluorine is not a regular one. Predict which of these elements has the highest second ionization energy and which has the lowest and explain. Of the elements N, O, and F, O has the highest and N the lowest second ionization energy. Explain.

125. Unlike the elements in groups 1A and 2A, those in group 3A do not show a regular decrease in first ionization energy in going down the column. Explain the irregularities.

126. Using the data in Figures 4.19 and 4.20, calculate ΔE (the change in energy) for the reaction

$$Na(g) + Cl(g) \longrightarrow Na^+(g) + Cl^-(g)$$

127. Even though adding two electrons to O or S forms an ion with a noble gas electron configuration, the second electron affinity of both of these elements is positive. Explain.

128. In Section 4.5 we discussed the metalloids, which form a diagonal band separating the metals from the nonmetals. There are other instances in which elements such as lithium and magnesium that are diagonal to each other have comparable metallic character. Suggest an explanation for this observation.

129. The heaviest known alkaline earth metal is radium, atomic number 88. Find the atomic numbers of the as yet undiscovered next two members of the series.

130. Predict the electronic configurations of the first two excited states (next higher energy states beyond the ground state) of Pd.

CONCEPTUAL PROBLEMS

131. Imagine that in another universe, atoms and elements are identical to ours, except that atoms with six valence electrons have particular stability (in contrast to our universe where atoms with eight valence electrons have particular stability). Give an example of an element in the alternative universe that corresponds to:

a. a noble gas **b.** a reactive nonmetal **c.** a reactive metal

132. The outermost valence electron in atom A experiences an effective nuclear charge of 2+ and is on average 225 pm from the nucleus. The outermost valence electron in atom B experiences an effective nuclear charge of 1+ and is on average 175 pm from the nucleus. Which atom (A or B) has the higher first ionization energy? Explain.

133. Determine whether each statement regarding penetration and shielding is true or false. (Assume that all lower energy orbitals are fully occupied.)

a. An electron in a 3*s* orbital is more shielded than an electron in a 2*s* orbital.

b. An electron in a 3*s* orbital penetrates into the region occupied by core electrons more than electrons in a 3*p* orbital.

c. An electron in an orbital that penetrates closer to the nucleus will always experience more shielding than an electron in an orbital that does not penetrate as far.

d. An electron in an orbital that penetrates close to the nucleus will tend to experience a higher effective nuclear charge than one that does not.

134. Give a combination of four quantum numbers that could be assigned to an electron occupying a 5*p* orbital. Do the same for an electron occupying a 6*d* orbital.

135. Use the trends in ionization energy and electron affinity to explain why calcium fluoride has the formula CaF_2 and not Ca_2F or CaF.

ANSWERS TO CONCEPTUAL CONNECTIONS

Cc 4.1 **(a)** Since the charges are opposite, the potential energy of the interaction is negative. As the charges get closer together, *r* becomes smaller and the potential energy decreases (it becomes more negative).

Cc 4.2 **(c)** Penetration results in less shielding from nuclear charge and therefore lower energy.

Cc 4.3 $n = 4, l = 0, m_l = 0, m_s = +\frac{1}{2}; n = 4, l = 0, m_l = 0, m_s = -\frac{1}{2}$

Cc 4.4 **(c)** Since Z_{eff} increases from left to right across a row in the periodic table, the valence electrons in S experience a greater effective nuclear charge than the valence electrons in Al or in Mg.

Cc 4.5 The isotopes of an element all have the same radii for two reasons: (1) neutrons are negligibly small compared to the size of an atom and therefore extra neutrons do not increase atomic size; and (2) neutrons have no charge and therefore do not attract electrons in the way that protons do.

Cc 4.6 As you can see from the successive ionization energies of any element, valence electrons are held most loosely and can therefore be transferred or shared most easily. Core electrons, on the other hand, are held tightly and are not easily transferred or shared. Consequently, valence electrons play a central role in chemical bonding.

Cc 4.7 The 3*s* electron in sodium has a relatively low ionization energy (496 kJ/mol) because it is a valence electron. The energetic cost for sodium to lose a second electron is extraordinarily high (4560 kJ/mol) because the next electron to be lost is a core electron (2*p*). Similarly, the electron affinity of chlorine to gain one electron (–349 kJ/mol) is highly exothermic since the added electron completes chlorine's valence shell. The gain of a second electron by the negatively charged chlorine anion is not so favorable. Therefore, we expect sodium and chlorine to combine in a 1:1 ratio.

> **"I have just returned from a short vacation for which the only books I took were a half-dozen detective stories and your 'Chemical Bond'. I found yours the most exciting of the lot."**
>
> **—Gilbert N. Lewis (1875–1946), in a letter to Linus Pauling in 1939**

When a balloon filled with H_2 and O_2 is ignited, the two elements react violently to form H_2O.

Molecules and Compounds

5.1 Hydrogen, Oxygen, and Water 145

5.2 Types of Chemical Bonds 146

5.3 Representing Compounds: Chemical Formulas and Molecular Models 148

5.4 The Lewis Model: Representing Valance Electrons with Dots 150

5.5 Ionic Bonding: The Lewis Model and Lattice Energies 152

5.6 Ionic Compounds: Formulas and Names 155

5.7 Covalent Bonding: Simple Lewis Structures 161

5.8 Molecular Compounds: Formulas and Names 163

5.9 Formula Mass and the Mole Concept for Compounds 165

5.10 Composition of Compounds 167

5.11 Determining a Chemical Formula from Experimental Data 172

5.12 Organic Compounds 177

Key Learning Outcomes 180

HOW MANY DIFFERENT SUBSTANCES exist? Our world contains about 91 different naturally existing elements, so there are at least 91 different substances. However, the world would be dull—not to mention lifeless—with only 91 different substances. Fortunately, elements combine with each other to form *compounds*. Just as combinations of only 26 letters in our English alphabet allow for an almost limitless number of words, each with its own specific meaning, so combinations of the 91 naturally occurring elements allow for an almost limitless number of compounds, each with its own specific properties. The great diversity of substances that we find in nature is a direct result of the ability of elements to form compounds. Life, for example, could not exist with just 91 different elements. It takes compounds, in all of their diversity, to make life possible.

5.1 Hydrogen, Oxygen, and Water

Hydrogen (H_2) is an explosive gas used as a fuel in rocket engines. Oxygen (O_2), also a gas, is a natural component of the air on Earth. Oxygen is not itself flammable but must be present for combustion (burning) to occur. Hydrogen and oxygen both have extremely low boiling points, as you can see from the table on the next page. When hydrogen and oxygen combine to form the compound water (H_2O), however, a dramatically different substance results.

Selected Properties	Hydrogen	Oxygen	Water
Boiling Point	−253 °C	−183 °C	100 °C
State at Room Temperature	Gas	Gas	Liquid
Flammability	Explosive	Necessary for combustion	Used to extinguish flame

First of all, water is a liquid rather than a gas at room temperature, and its boiling point is hundreds of degrees higher than the boiling points of hydrogen and oxygen. Second, instead of being flammable (like hydrogen gas) or supporting combustion (like oxygen gas), water actually extinguishes flames. Water is nothing like the hydrogen and oxygen from which it forms. The dramatic difference between the elements hydrogen and oxygen and the compound water is typical of the differences between elements and the compounds that they form. *When two or more elements combine to form a compound, an entirely new substance results.*

Although some of the substances that we encounter in everyday life are elements, most are compounds. As we discussed in Chapter 1, a compound is different from a mixture of elements. In a compound, elements combine in fixed, definite proportions; in a mixture, elements can mix in any proportions whatsoever. Consider the difference between a hydrogen–oxygen mixture and water as shown in Figure 5.1 ▼. A hydrogen–oxygen mixture can have any proportions of hydrogen and oxygen gas. Water, by contrast, is composed of water molecules that always contain two hydrogen atoms to every one oxygen atom. Water has a definite proportion of hydrogen to oxygen.

In this chapter we examine compounds. We discuss how to represent them, how to name them, and why they form. In the last part of the chapter, we learn how to quantify their elemental composition.

Mixtures and Compounds

Hydrogen and Oxygen Mixture
This can have any ratio of hydrogen to oxygen.

Water (A Compound)
Water molecules have a fixed ratio of hydrogen (two atoms) to oxygen (one atom).

▶ **FIGURE 5.1 Mixtures and Compounds** The balloon in this illustration contains a mixture of hydrogen gas and oxygen gas. The proportions of hydrogen and oxygen are variable. The glass contains water, a compound of hydrogen and oxygen. The ratio of hydrogen to oxygen in water is fixed: Water molecules always have two hydrogen atoms for each oxygen atom.

5.2 Types of Chemical Bonds

A **chemical bond** is the force that holds atoms together in a compound. We begin our discussion of chemical bonding by asking why bonds form in the first place. The answer to this question is not simple and involves not only quantum mechanics but also some thermodynamics that we do not cover until Chapter 19. Nonetheless, we can address an important *aspect* of the answer now: *Chemical bonds form because they lower the potential energy of the charged particles that compose atoms.*

Recall from Section 4.5 that when an outer principal quantum level is completely full, the overall potential energy of the electrons that occupy that level is particularly low. Recall also that only the noble gases have full outer principal quantum levels. Because the rest of the elements do *not* possess the stability of the noble gases, they form chemical bonds to become more stable (to lower the potential energy of the charged particles that compose them).

As we have already seen, atoms are composed of particles with positive charges (the protons in the nucleus) and negative charges (the electrons). When two atoms approach each other, the electrons of one atom are attracted to the nucleus of the other according to Coulomb's law (see Section 4.3) and vice versa. However, at the same time, the electrons of each atom repel the electrons of the other, and the nucleus of each atom repels the nucleus of the other. The result is a complex set of interactions among a large number of charged particles. If these interactions lead to an overall net reduction of potential energy between the charged particles, a chemical bond forms. Bonding theories help us to predict the circumstances under which bonds form and also the properties of the resultant molecules. We can broadly classify chemical bonds into two types—*ionic* and *covalent*—depending on the kind of atoms involved in the bonding (Figure 5.2 ▲).

▲ **FIGURE 5.2 Ionic and Covalent Bonding**

The bond that forms between a metal and a nonmetal is an **ionic bond**. Recall from Chapter 4 that metals have a tendency to lose electrons and that nonmetals have a tendency to gain them. Therefore, when a metal interacts with a nonmetal, it can transfer one or more of its electrons to the nonmetal. The metal atom then becomes a *cation* (a positively charged ion), and the nonmetal atom becomes an *anion* (a negatively charged ion) as shown in Figure 5.3 ▼. These oppositely charged ions

The Formation of an Ionic Compound

Sodium (a metal) loses an electron.

Chlorine (a nonmetal) gains an electron.

e^-

Neutral Na atom, 11e^-

Neutral Cl atom, 17e^-

Na$^+$ ion, 10e^-

Cl$^-$ ion, 18e^-

Sodium metal

anion

cation

Chlorine gas

Oppositely charged ions are held together by ionic bonds, forming a crystalline lattice.

Sodium chloride (table salt)

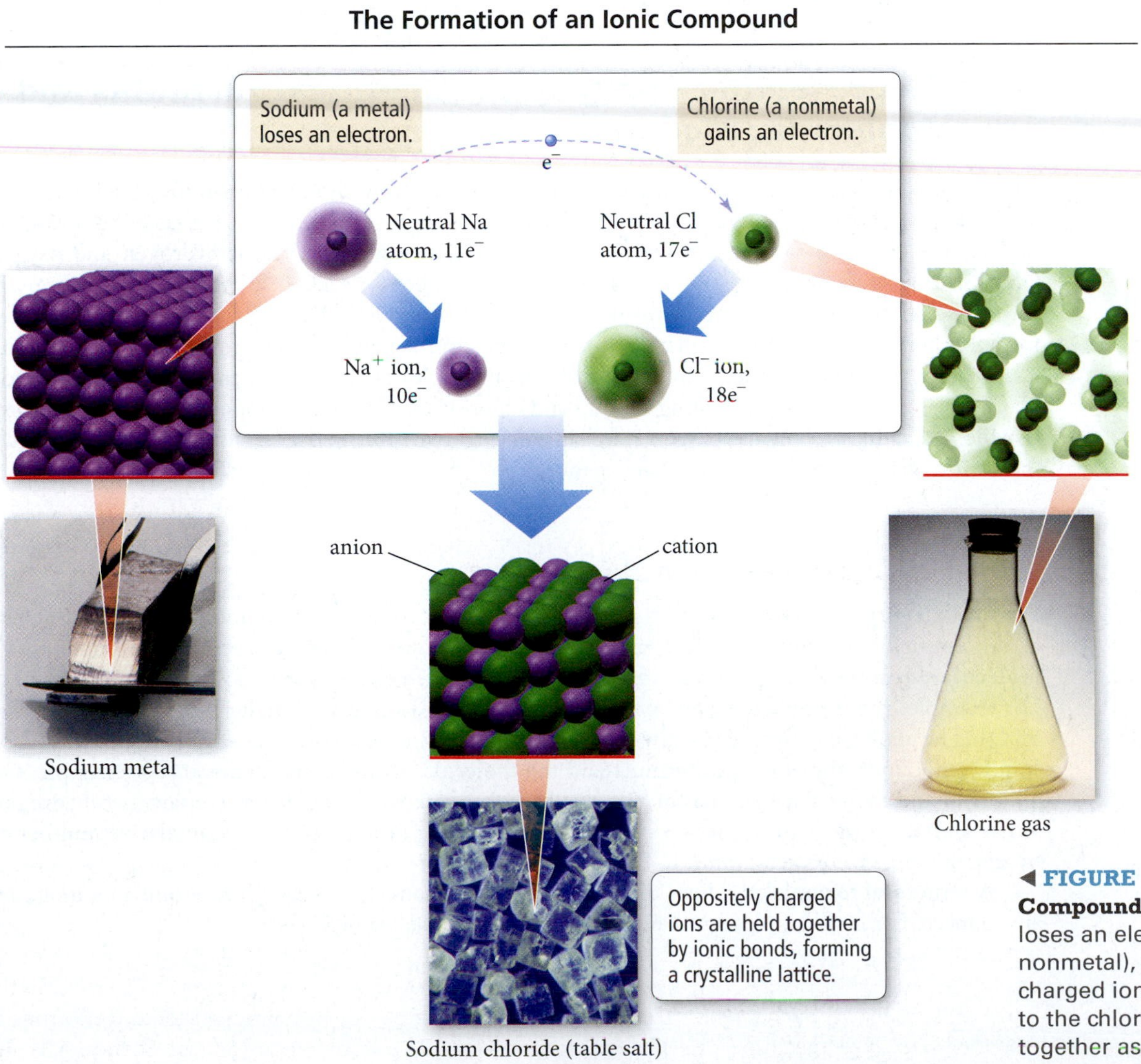

◀ **FIGURE 5.3 The Formation of an Ionic Compound** An atom of sodium (a metal) loses an electron to an atom of chlorine (a nonmetal), creating a pair of oppositely charged ions. The sodium cation is attracted to the chloride anion, and the two are held together as part of a crystalline lattice.

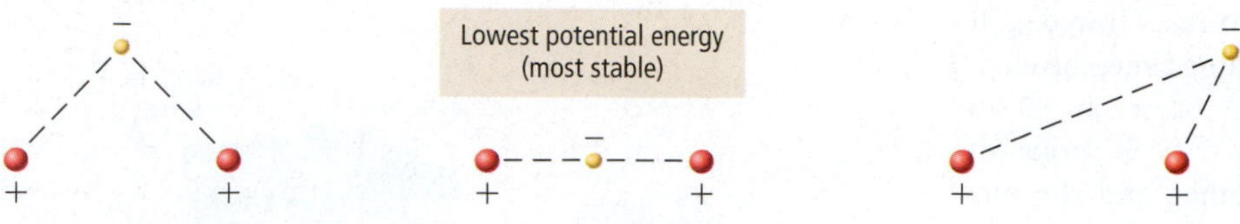

▶ **FIGURE 5.4 Possible Configurations of One Negatively Charged Particle and Two Positively Charged Ones**

attract one another according to Coulomb's law and form an **ionic compound,** which in the solid state is composed of a *lattice*—a regular three-dimensional array—of alternating cations and anions.

The bond that forms between two or more nonmetals is a **covalent bond**. In Chapter 4 we saw that nonmetals tend to have high ionization energies (their electrons are relatively difficult to remove). Therefore when a nonmetal bonds with another nonmetal, neither atom transfers electrons to the other. Instead, the two atoms *share* some electrons. The shared electrons interact with the nuclei of both of the bonding atoms, lowering their potential energy in accordance with Coulomb's law. Covalently bonded atoms form *molecules*, and the resulting compounds are called **molecular compounds**.

We can understand the stability of a covalent bond by considering the most stable arrangement (the one with the lowest potential energy) of two positively charged particles separated by a small distance and a negatively charged particle. As you can see in Figure 5.4 ▲, the arrangement in which the negatively charged particle lies *between* the two positively charged ones has the lowest potential energy because in this arrangement, the negatively charged particle interacts most strongly with *both of the positively charged ones*. In a sense, the negatively charged particle holds the two positively charged ones together. Similarly, shared electrons in a covalent chemical bond *hold* the bonding atoms together by attracting the positive charges of their nuclei.

5.3 Representing Compounds: Chemical Formulas and Molecular Models

The quickest and easiest way to represent a compound is with its **chemical formula**, which indicates the elements present in the compound and the relative number of atoms or ions of each. For example, H_2O is the chemical formula for water—it indicates that water consists of hydrogen and oxygen atoms in a two-to-one ratio. The formula contains the symbol for each element and a subscript indicating the relative number of atoms of the element. A subscript of 1 is typically omitted. Chemical formulas normally list the more metallic (or more positively charged) element first, followed by the less metallic (or more negatively charged) element. Examples of common chemical formulas include NaCl for sodium chloride, indicating sodium and chloride ions in a one-to-one ratio; CO_2 for carbon dioxide, indicating carbon and oxygen atoms in a one-to-two ratio; and CCl_4 for carbon tetrachloride, indicating carbon and chlorine in a one-to-four ratio.

Types of Chemical Formulas

We can categorize chemical formulas into three different types: empirical, molecular, and structural. An **empirical formula** gives the *relative* number of atoms of each element in a compound. A **molecular formula** gives the *actual* number of atoms of each element in a molecule of a compound. For example, the empirical formula for hydrogen peroxide is HO, but its molecular formula is H_2O_2. The molecular formula is always a whole-number multiple of the empirical formula. For some compounds, the empirical formula and the molecular formula are identical. For example, the empirical and molecular formula for water is H_2O because water molecules contain two hydrogen atoms and one oxygen atom, and no simpler whole-number ratio can express the relative number of hydrogen atoms to oxygen atoms.

A **structural formula** uses lines to represent covalent bonds and shows how atoms in a molecule are connected or bonded to each other. The structural formula for H_2O_2 is:

$$\mathrm{H{-}O{-}O{-}H}$$

Structural formulas may also be written to give a sense of the molecule's geometry. The structural formula for hydrogen peroxide can also be written this way:

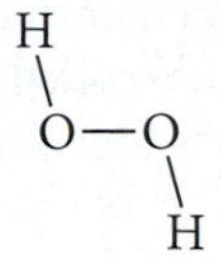

```
H
 \
  O—O
     \
      H
```

This version represents the approximate angles between bonds, giving a sense of the molecule's shape. Structural formulas can also depict the different types of bonds that occur within molecules. For example, consider the structural formula for carbon dioxide:

$$O{=}C{=}O$$

The two lines between each carbon and oxygen atom represent a double bond, which is generally stronger and shorter than a single bond (represented by a single line). A single bond corresponds to one shared electron pair, while a double bond corresponds to two shared electron pairs, as we will discuss in Section 5.7. The type of formula we use depends on how much we know about the compound and how much we want to communicate. A structural formula communicates the most information, while an empirical formula communicates the least.

Structural Formulas

5.1 Cc Conceptual Connection

Write the structural formula for water.

EXAMPLE 5.1

Molecular and Empirical Formulas

Write empirical formulas for the compounds represented by the molecular formulas.

(a) C_4H_8
(b) B_2H_6
(c) CCl_4

SOLUTION

To determine the empirical formula from a molecular formula, divide the subscripts by the greatest common factor (the largest number that divides exactly into all of the subscripts).

(a) For C_4H_8, the greatest common factor is 4. The empirical formula is therefore CH_2.
(b) For B_2H_6, the greatest common factor is 2. The empirical formula is therefore BH_3.
(c) For CCl_4, the only common factor is 1, so the empirical formula and the molecular formula are the same.

FOR PRACTICE 5.1

Write the empirical formula for the compounds represented by the molecular formulas.

(a) C_5H_{12}
(b) Hg_2Cl_2
(c) $C_2H_4O_2$

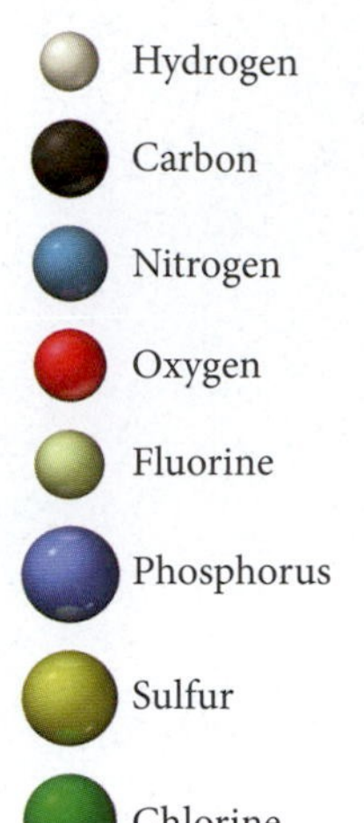

Molecular Models

A *molecular model* is a more accurate and complete way to specify a compound. A **ball-and-stick molecular model** represents atoms as balls and chemical bonds as sticks; how the two connect reflects a molecule's shape. The balls are typically color-coded to specific elements. For example, carbon is customarily black, hydrogen is white, nitrogen is blue, and oxygen is red. (For a complete list of colors of elements in the molecular models used in this book, see Appendix IV A.)

In a **space-filling molecular model**, atoms fill the space between each other to more closely represent our best estimates for how a molecule might appear if scaled to visible size. Consider the following ways to represent a molecule of methane, the main component of natural gas:

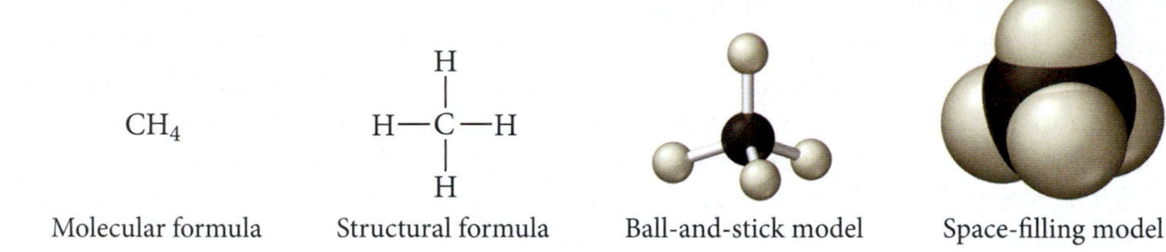

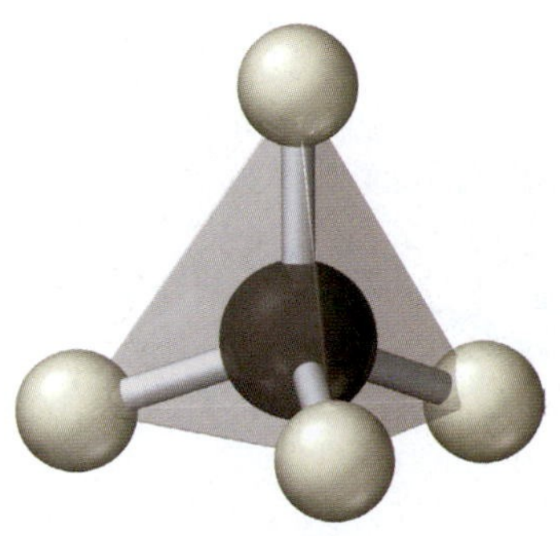

▲ A tetrahedron is a three-dimensional geometrical shape characterized by four equivalent triangular faces.

The molecular formula of methane indicates the number and type of each atom in the molecule: one carbon atom and four hydrogen atoms. The structural formula indicates how the atoms are connected: The carbon atom is bonded to the four hydrogen atoms. The ball-and-stick model clearly portrays the geometry of the molecule: The carbon atom sits in the center of a *tetrahedron* formed by the four hydrogen atoms. And finally, the space-filling model gives the best sense of the relative sizes of the atoms and how they merge together in bonding.

Throughout this book, you will see molecules represented in all of these ways. As you look at these representations, keep in mind what you learned in Chapter 1: The details about a molecule—the atoms that compose it, the lengths of the bonds between atoms, the angles of the bonds between atoms, and its overall shape—determine the properties of the substance that the molecule composes. Change any of these details and those properties change. Table 5.1 shows various compounds represented in the different ways we have just discussed.

Representing Molecules

Based on what you learned in Chapter 1 about atoms, what part of the atom do you think the spheres in the molecular models shown here represent? If you were to superimpose a nucleus on one of these spheres, how big would you draw it?

5.4 The Lewis Model: Representing Valence Electrons with Dots

KEY CONCEPT VIDEO
The Lewis model for chemical bonding

Bonding theories (or models) are central to chemistry because they explain how atoms bond together to form molecules. They explain why some combinations of atoms are stable and others are not. For example, bonding theories explain why table salt is NaCl and not $NaCl_2$ and why water is H_2O and not H_3O. Bonding theories also predict the shapes of molecules—a topic in the next chapter—which in turn determine many of the physical and chemical properties of compounds. The bonding model we introduce here is called the **Lewis model**, named after the American chemist G. N. Lewis (1875–1946). In the Lewis model, valence electrons are represented as dots, and we draw **Lewis electron-dot structures** (or simply **Lewis structures**) to depict molecules. These structures, which are fairly simple to draw, have tremendous predictive power. With minimal computation, we can use the Lewis model to predict whether a particular set of atoms will form a stable molecule and what that molecule might look like. Although we will also examine more advanced theories for

TABLE 5.1 Benzene, Acetylene, Glucose, and Ammonia

Name of Compound	Empirical Formula	Molecular Formula	Structural Formula	Ball-and-Stick Model	Space-Filling Model
Benzene	CH	C_6H_6			
Acetylene	CH	C_2H_2	H—C≡C—H		
Glucose	CH_2O	$C_6H_{12}O_6$			
Ammonia	NH_3	NH_3			

chemical bonding in Chapter 7, the Lewis model remains the simplest model for making quick, everyday predictions about most molecules.

The Lewis model focuses on valence electrons because chemical bonding involves the transfer or sharing of valence electrons between two or more atoms. Recall from Chapter 4 that, for main-group elements, the valence electrons are those in the outermost principal energy level. In a **Lewis symbol**, we represent the valence electrons of main-group elements as dots surrounding the abbreviation for the element. For example, the electron configuration of O is:

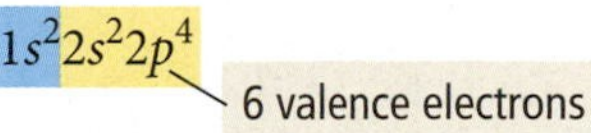

Remember, the number of valence electrons for any main group is equal to the group number of the element (except for helium, which is in group 8A but has only two valence electrons).

And its Lewis symbol is:

·Ö: — 6 dots representing valence electrons

Each dot represents a valence electron. The dots are placed around the element's symbol with a maximum of two dots per side. We draw the Lewis symbols for all of the period 2 elements in a similar way:

Li· ·Be· ·B· ·C· ·N: ·O: :F: :Ne:

While the exact location of dots is not critical, in this book we first place dots singly before pairing (except for helium, which always has two paired dots signifying its duet).

Lewis symbols provide a simple way to visualize the number of valence electrons in a main-group atom. Notice that atoms with eight valence electrons—which are particularly stable because they have a full outer principal level—are easily identified because they have eight dots, an **octet**.

Helium is an exception. Its electron configuration and Lewis symbol are:

$$1s^2 \quad \text{He:}$$

The Lewis symbol of helium contains only two dots (a **duet**). For helium, a duet represents a stable electron configuration because the $n = 1$ quantum level fills with only two electrons.

In the Lewis model, a chemical bond is the sharing or transfer of electrons to attain stable electron configurations for the bonding atoms. If electrons are transferred, as occurs between a metal and a nonmetal, the bond is an *ionic bond*. If the electrons are shared, as occurs between two nonmetals, the bond is a *covalent bond*. In either case, the bonding atoms obtain stable electron configurations; since the stable configuration is usually eight electrons in the outermost shell, this is known as the **octet rule**. When applying the Lewis model, we do not try to calculate the energies associated with the attractions and repulsions between electrons and nuclei on neighboring atoms. The energy changes that occur because of these interactions are central to chemical bonding (as we just discussed in Section 5.2), yet the Lewis model ignores them because calculating these energy changes is extremely complicated. Instead the Lewis model uses the simple octet rule, a practical approach which accurately predicts what we see in nature for a large number of compounds—hence the success and longevity of the Lewis model.

5.5 Ionic Bonding: The Lewis Model and Lattice Energies

As we have seen, ionic compounds are composed of cations (usually a metal) and anions (usually one or more nonmetals) bound together by ionic bonds. The basic unit of an ionic compound is the **formula unit**, the smallest, electrically neutral collection of ions. A formula unit is not a molecule—it does not usually exist as a discrete entity, but rather as part of a larger lattice. For example, table salt, an ionic compound with the formula unit NaCl, is composed of Na^+ and Cl^- ions in a one-to-one ratio. In table salt, Na^+ and Cl^- ions exist in a three-dimensional alternating array. Because ionic bonds are not directional, no one Na^+ ion pairs with a specific Cl^- ion. Rather, as we saw in Figure 5.2, any one Na^+ cation is surrounded by Cl^- anions and vice versa.

Some ionic compounds, such as K_2NaPO_4, contain more than one type of metal ion.

Although the Lewis model's strength is in modeling covalent bonding, we can also apply it to ionic bonding. To represent ionic bonding, we move electron dots from the Lewis symbol of the metal to the Lewis symbol of the nonmetal, so the metal becomes a cation and the nonmetal becomes an anion. This alone, however, does not account for the stability of ionic substances. To understand that stability, we must account for the formation of a crystalline lattice as a result of the attractions between the cations and anions. In this section of the chapter, we first look at the electron transfer and then examine the formation of the crystalline lattice.

Ionic Bonding and Electron Transfer

Consider potassium and chlorine, which have the following Lewis symbols:

$$\text{K}\cdot \qquad :\ddot{\underset{\cdot}{\text{Cl}}}:$$

When these atoms bond, potassium transfers its valence electron to chlorine:

$$\text{K}\cdot + :\ddot{\underset{\cdot}{\text{Cl}}}: \longrightarrow \text{K}^+ + [:\ddot{\underset{\cdot\cdot}{\text{Cl}}}:]^-$$

The transfer of the electron gives chlorine an octet (shown as eight dots around chlorine) and leaves potassium without any valence electrons but with an octet in the *previous* principal energy level (which is now its outermost level).

$$\text{K} \quad 1s^22s^22p^63s^23p^64s^1$$
$$\text{K}^+ \quad 1s^22s^22p^6\underbrace{3s^23p^6}_{\text{Octet in previous level}}4s^0$$

The potassium, having lost an electron, becomes positively charged (a cation), while the chlorine, which has gained an electron, becomes negatively charged (an anion). The Lewis symbol of an anion is usually written within brackets with the charge in the upper right-hand corner, outside the brackets. The positive and negative charges attract one another, resulting in the compound KCl.

We can use the Lewis model to predict the correct chemical formulas for ionic compounds. For the compound that forms between K and Cl, for example, the Lewis model predicts a ratio of one potassium cation to every one chloride anion, KCl. In nature, when we examine the compound formed between potassium and chlorine, we indeed find one potassium ion to every chloride ion. As another example, consider the ionic compound formed between sodium and sulfur. The Lewis symbols for sodium and sulfur are:

Na·   ·S̈:

Sodium must lose its one valence electron in order to have an octet (in the previous principal shell), while sulfur must gain two electrons to get an octet. Consequently, the compound that forms between sodium and sulfur requires two sodium atoms to every one sulfur atom—the formula is Na_2S. The two sodium atoms each lose their one valence electron while the sulfur atom gains two electrons and gets an octet. The Lewis model predicts that the correct chemical formula is Na_2S, exactly what we see in nature.

EXAMPLE 5.2

Using Lewis Symbols to Predict the Chemical Formula of an Ionic Compound

Use the Lewis model to predict the formula for the compound that forms between calcium and chlorine.

SOLUTION

Draw Lewis symbols for calcium and chlorine based on their number of valence electrons, obtained from their group number in the periodic table.	·Ca· :C̈l:
Calcium needs to lose its two valence electrons (to be left with an octet in its previous principal shell), while chlorine only needs to gain one electron to get an octet. Therefore, you must have two chlorine atoms for each calcium atom. The calcium atom loses its two electrons to form Ca^{2+}, and each chlorine atom gains an electron to form Cl^-. In this way, both calcium and chlorine attain octets.	Ca^{2+} 2[:C̈l:]$^-$
Finally, write the formula with subscripts to indicate the number of atoms.	$CaCl_2$

FOR PRACTICE 5.2
Use the Lewis model to predict the formula for the compound that forms between magnesium and nitrogen.

Lattice Energy: The Rest of the Story

The formation of an ionic compound from its constituent elements usually gives off quite a bit of energy as heat (the process is exothermic; see Section 2.4). For example, when one mole of sodium chloride forms from elemental sodium and chlorine, 411 kJ of heat is evolved in the following violent reaction:

$$Na(s) + {}^{1}\!/_{2}Cl_2(g) \longrightarrow NaCl(s)$$

Where does this energy come from? You might think that it comes solely from the tendency of metals to lose electrons and nonmetals to gain electrons—but it does not. In fact, the transfer of an electron from sodium to chlorine—by itself—actually *absorbs* energy. The first ionization energy of sodium is +496 kJ/mol, and the electron affinity of Cl is only −349 kJ/mol. (Recall from Section 2.4 that the positive sign indicates the absorption of energy and the negative sign indicates the emission of energy.) Based only on these energies, the reaction should absorb +147 kJ/mol. So why is the reaction so *exothermic*?

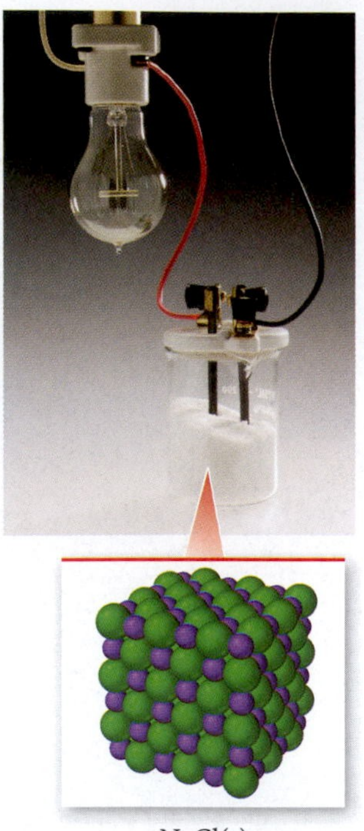

NaCl(s)

▲ Solid sodium chloride does not conduct electricity.

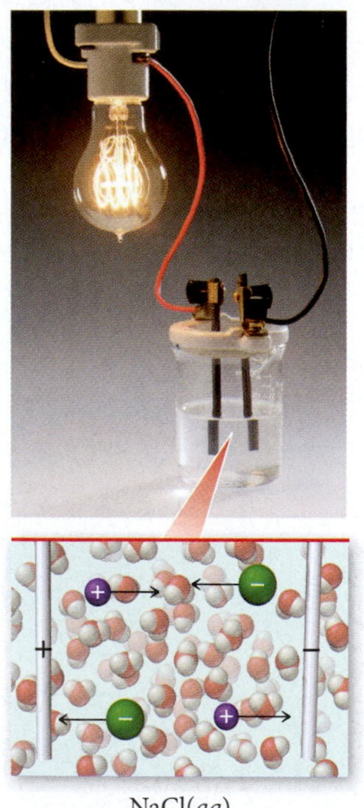

NaCl(aq)

▲ When sodium chloride dissolves in water, the resulting solution contains mobile ions that can create an electric current.

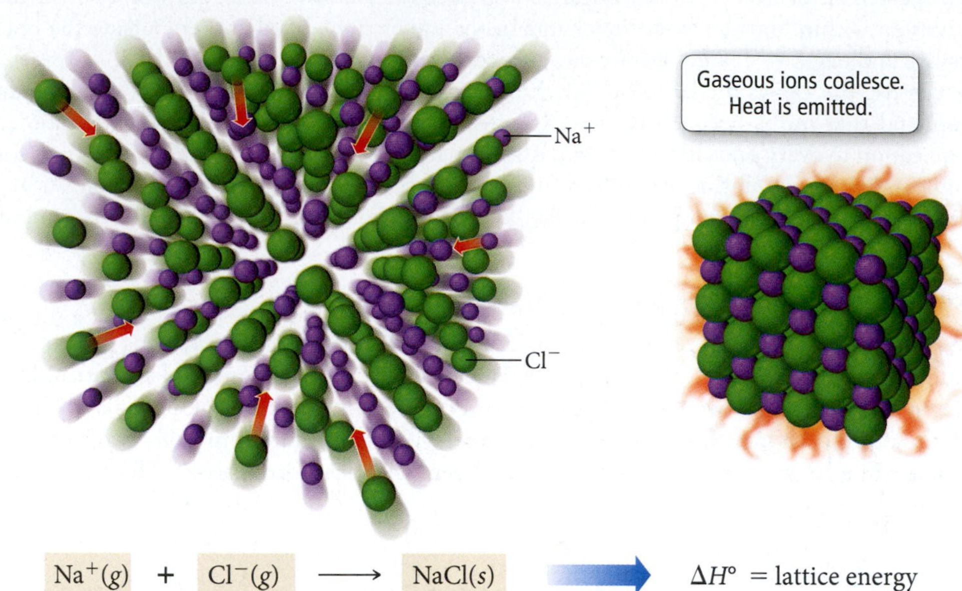

▲ **FIGURE 5.5 Lattice Energy** The lattice energy of an ionic compound is the energy associated with the formation of a crystalline lattice of the compound from the gaseous ions.

The answer lies in the **lattice energy**—the energy associated with the formation of a crystalline lattice of alternating cations and anions from the gaseous ions. Since the sodium ions are positively charged and the chlorine ions are negatively charged, the potential energy decreases—as described by Coulomb's law—when these ions come together to form a lattice. That energy is emitted as heat when the lattice forms, as illustrated in Figure 5.5 ▲. The exact value of the lattice energy, however, is not simple to determine because it involves a large number of interactions among many charged particles in a lattice. The easiest way to calculate lattice energy is with the *Born–Haber cycle*, which we will discuss in Section 10.11.

Ionic Bonding: Models and Reality

In this section, we applied the Lewis model to ionic bonding. The value of a model lies in how well it accounts for what we see in nature (through experiments). Does this ionic bonding model explain the properties of ionic compounds, including their high melting and boiling points, their tendency *not to conduct* electricity as solids, and their tendency *to conduct* electricity when dissolved in water?

We modeled an ionic solid as a lattice of individual ions held together by coulombic forces that are not directional (which means that, as you move away from the center of an ion, the forces are equally strong in all directions). To melt the solid, these forces must be overcome, which requires a significant amount of heat. Therefore, our model accounts for the high melting points of ionic solids. In the model, electrons are transferred from the metal to the nonmetal, but the transferred electrons remain localized on one atom. In other words, our model does not include any free electrons that might conduct electricity (the movement or flow of electrons in response to an electric potential, or voltage, is electrical current). In addition, the ions themselves are fixed in place; therefore, our model accounts for the nonconductivity of ionic solids. When our idealized ionic solid dissolves in water, however, the cations and anions dissociate, forming free ions in solution. These ions can move in response to electrical forces, creating an electrical current. Thus, our model predicts that solutions of ionic compounds conduct electricity (which in fact they do).

5.3 Cc Conceptual Connection

Melting Points of Ionic Solids

Use the ionic bonding model developed in this section to determine which has the higher melting point, NaCl or MgO. Explain your answer.

5.6 Ionic Compounds: Formulas and Names

In Section 5.5, we deduced the formula for a simple ionic compound from the Lewis symbols of its constituent atoms. Because ionic compounds must be charge-neutral, we can also deduce the formula from the charges of the ions. In this section, we examine how to write a formula for an ionic compound based on the charges of the constituent ions and how to systematically name ionic compounds. The process of naming compounds is called *nomenclature*.

Writing Formulas for Ionic Compounds

Because ionic compounds are charge-neutral and because many elements form only one type of ion with a predictable charge (see Figure 4.12), we can deduce the formulas for many ionic compounds from their constituent elements. For example, we know that the formula for the ionic compound composed of potassium and fluorine must be KF because, in compounds, K always forms 1+ cations and F always forms 1− anions. In order for the compound to be charge-neutral, it must contain one K^+ cation to every one F^- anion. The formula for the ionic compound composed of *magnesium* and fluorine, however, is MgF_2 because Mg always forms 2+ cations and F always forms 1− anions. In order for this compound to be charge-neutral, it must contain one Mg^{2+} cation for every two F^- anions.

Ionic Compound Formulas

- Ionic compounds always contain positive and negative ions.
- In a chemical formula, the sum of the charges of the positive ions (cations) must equal the sum of the charges of the negative ions (anions).
- The formula of an ionic compound reflects the smallest whole-number ratio of ions.

To write the formula for an ionic compound, follow the procedure in the left column in the following examples. Two examples of how to apply the procedure are provided in the center and right columns.

PROCEDURE FOR Writing Formulas for Ionic Compounds	EXAMPLE 5.3 Writing Formulas for Ionic Compounds	EXAMPLE 5.4 Writing Formulas for Ionic Compounds
	Write the formula for the ionic compound that forms between aluminum and oxygen.	Write the formula for the ionic compound that forms between calcium and oxygen.
1. Write the symbol for the metal cation and its charge followed by the symbol for the nonmetal anion and its charge. Determine charges from the element's group number in the periodic table (refer to Figure 4.12).	Al^{3+} O^{2-}	Ca^{2+} O^{2-}
2. Adjust the subscript on each cation and anion to balance the overall charge.	Al^{3+} O^{2-} ↓ Al_2O_3	Ca^{2+} O^{2-} ↓ CaO
3. Check that the sum of the charges of the cations equals the sum of the charges of the anions.	cations: 2(3+) = 6+ anions: 3(2−) = 6− The charges cancel.	cations: 2+ anions: 2− The charges cancel.
	FOR PRACTICE 5.3 Write the formula for the compound formed between potassium and sulfur.	**FOR PRACTICE 5.4** Write the formula for the compound formed between aluminum and nitrogen.

TABLE 5.2 Metals Whose Charge Is Invariant from One Compound to Another

Metal	Ion	Name	Group Number
Li	Li^+	Lithium	1A
Na	Na^+	Sodium	1A
K	K^+	Potassium	1A
Rb	Rb^+	Rubidium	1A
Cs	Cs^+	Cesium	1A
Be	Be^{2+}	Beryllium	2A
Mg	Mg^{2+}	Magnesium	2A
Ca	Ca^{2+}	Calcium	2A
Sr	Sr^{2+}	Strontium	2A
Ba	Ba^{2+}	Barium	2A
Al	Al^{3+}	Aluminum	3A
Zn	Zn^{2+}	Zinc	*
Sc	Sc^{3+}	Scandium	*
Ag**	Ag^+	Silver	*

*The charge of these metals cannot be inferred from their group number.

**Silver sometimes forms compounds with other charges, but these are rare.

Naming Ionic Compounds

Some ionic compounds—such as NaCl (table salt) and $NaHCO_3$ (baking soda)—have **common names**, which are nicknames of sorts learned by familiarity. Chemists have also developed **systematic names** for different types of compounds including ionic ones. Even if you are not familiar with a compound, you can determine its systematic name from its chemical formula. Conversely, you can deduce the formula of a compound from its systematic name.

The first step in naming an ionic compound is identifying it as one. Remember, *ionic compounds are usually composed of metals and nonmetals*; any time you see a metal and one or more nonmetals together in a chemical formula, assume that you have an ionic compound. We can categorize ionic compounds into two types, depending on the metal in the compound. The first type contains a metal whose charge is invariant from one compound to another. In other words, whenever the metal in this first type of compound forms an ion, the ion always has the same charge.

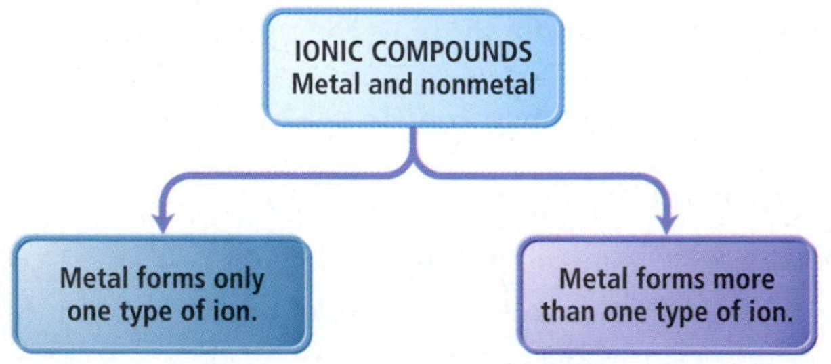

Since the charge of the metal in this first type of ionic compound is always the same, we don't need to specify its charge in the name of the compound. Sodium, for instance, has a 1+ charge in all of its compounds. Table 5.2 lists some examples of these types of metals; we can infer the charges of most of these metals from their group number in the periodic table. This table includes all of the metals in Figure 4.12 (which are all main-group metals) and three additional metals that are transition metals. The charges of the transition metals cannot be inferred from their group number.

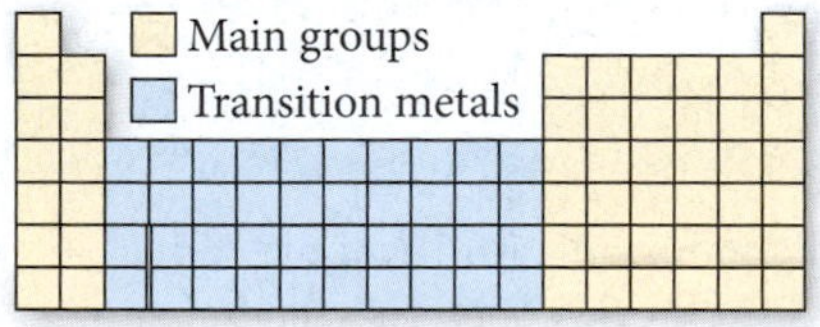

▲ **FIGURE 5.6 Transition Metals** Metals that can have different charges in different compounds are usually (but not always) transition metals.

The second type of ionic compound contains a metal with a charge that can differ in different compounds. In other words, the metal in this second type of ionic compound can form more than one kind of cation (depending on the compound); therefore, we must specify its charge for a given compound. Iron, for instance, forms a 2+ cation in some of its compounds and a 3+ cation in others. Metals of this type are usually *transition metals* (Figure 5.6 ◀). However, some transition metals, such as Zn and Ag, form cations with the same charge in all of their compounds (as shown in Table 5.2), and some main-group metals, such as Pb and Sn, form more than one type of cation.

Naming Binary Ionic Compounds Containing a Metal That Forms Only One Type of Cation

Binary compounds contain only two different elements. The names of binary ionic compounds take the form:

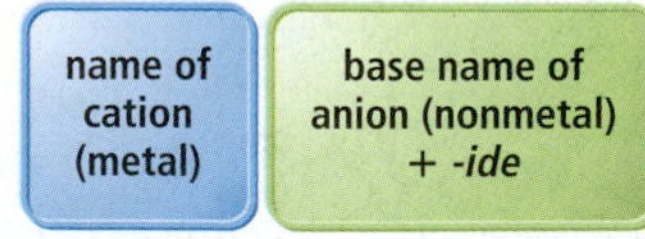

For example, the name for KCl consists of the name of the cation, *potassium*, followed by the base name of the anion, *chlor*, with the ending *-ide*. Its full name is *potassium chloride*.

KCl potassium chloride

The name for CaO consists of the name of the cation, *calcium*, followed by the base name of the anion, *ox*, with the ending *-ide*. Its full name is *calcium oxide*.

CaO calcium oxide

The base names for various nonmetals and their most common charges in ionic compounds are shown in Table 5.3.

TABLE 5.3 Some Common Monoatomic Anions

Nonmetal	Symbol for Ion	Base Name	Anion Name
Fluorine	F^-	fluor	Fluoride
Chlorine	Cl^-	chlor	Chloride
Bromine	Br^-	brom	Bromide
Iodine	I^-	iod	Iodide
Oxygen	O^{2-}	ox	Oxide
Sulfur	S^{2-}	sulf	Sulfide
Nitrogen	N^{3-}	nitr	Nitride
Phosphorus	P^{3-}	phosph	Phosphide

EXAMPLE 5.5

Naming Ionic Compounds Containing a Metal That Forms Only One Type of Cation

Name the compound $CaBr_2$.

SOLUTION

The cation is *calcium*. The anion is from bromine, which becomes *bromide*. The correct name is *calcium bromide*.

FOR PRACTICE 5.5

Name the compound Ag_3N.

FOR MORE PRACTICE 5.5

Write the formula for rubidium sulfide.

Naming Binary Ionic Compounds Containing a Metal That Forms More than One Kind of Cation

For these types of metals, the name of the cation is followed by a roman numeral (in parentheses), which indicates the charge of the metal in that particular compound. For example, we distinguish between Fe^{2+} and Fe^{3+} as follows:

Fe^{2+} iron(II)

Fe^{3+} iron(III)

Note that there is no space between the name of the cation and the parenthetical number indicating its charge.

The full names for compounds containing metals that form more than one kind of cation have the form:

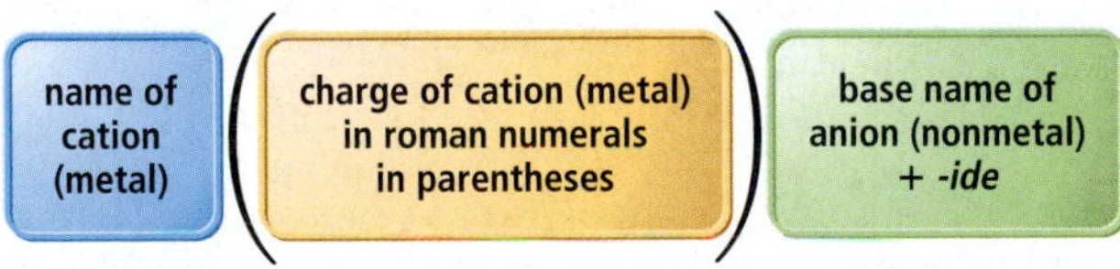

We can determine the charge of the metal cation by inference from the sum of the charges of the nonmetal anions—remember that the sum of all the charges in the compound must be zero. Table 5.4 (on the next page) shows some of the metals that form more than one cation and their most common charges. For example, in $CrBr_3$, the charge of chromium must be 3+ in order for the compound to be charge-neutral with three Br^- anions. The cation is named:

Cr^{3+} chromium(III)

The full name of the compound is:

$CrBr_3$ chromium(III) bromide

Similarly, in CuO, the charge of copper must be 2+ in order for the compound to be charge-neutral with one O^{2-} anion. The cation is therefore:

Cu^{2+} copper(II)

The full name of the compound is:

CuO copper(II) oxide

TABLE 5.4 Some Metals That Form Cations with Different Charges

Metal	Ion	Name	Older Name*
Chromium	Cr^{2+}	Chromium(II)	Chromous
	Cr^{3+}	Chromium(III)	Chromic
Iron	Fe^{2+}	Iron(II)	Ferrous
	Fe^{3+}	Iron(III)	Ferric
Cobalt	Co^{2+}	Cobalt(II)	Cobaltous
	Co^{3+}	Cobalt(III)	Cobaltic
Copper	Cu^{+}	Copper(I)	Cuprous
	Cu^{2+}	Copper(II)	Cupric
Tin	Sn^{2+}	Tin(II)	Stannous
	Sn^{4+}	Tin(IV)	Stannic
Mercury	Hg_2^{2+}	Mercury(I)	Mercurous
	Hg^{2+}	Mercury(II)	Mercuric
Lead	Pb^{2+}	Lead(II)	Plumbous
	Pb^{4+}	Lead(IV)	Plumbic

*An older naming system substitutes the names found in this column for the name of the metal and its charge. Under this system, chromium(II) oxide is named chromous oxide. In this system, the suffix *-ous* indicates the ion with the lesser charge and *-ic* indicates the ion with the greater charge. We will *not* use the older system in this text.

EXAMPLE 5.6

Naming Ionic Compounds Containing a Metal That Forms More than One Kind of Cation

Name the compound $PbCl_4$.

SOLUTION

The charge on Pb must be 4+ for the compound to be charge-neutral with 4 Cl^- anions. The name for $PbCl_4$ is the name of the cation, *lead*, followed by the charge of the cation in parentheses *(IV)*, and the base name of the anion, *chlor*, with the ending *-ide*. The full name is *lead(IV) chloride*.

$PbCl_4$ lead(IV) chloride

FOR PRACTICE 5.6

Name the compound FeS.

FOR MORE PRACTICE 5.6

Write the formula for ruthenium(IV) oxide.

▲ Polyatomic ions are common in household products such as bleach, which contains sodium hypochlorite (NaClO).

Naming Ionic Compounds Containing Polyatomic Ions

Many common ionic compounds contain ions that are themselves composed of a group of covalently bonded atoms with an overall charge. For example, the active ingredient in household bleach is sodium hypochlorite, which acts to chemically alter color-causing molecules in clothes (bleaching action) and to kill bacteria (disinfection). Hypochlorite is a **polyatomic ion**—an ion composed of two or more atoms—with the formula ClO^-. (Note that the charge on the hypochlorite ion is a property of the whole ion, not just the oxygen atom; this is true for all polyatomic ions.) The hypochlorite ion is often found as a unit in other compounds as well (such as $KClO$ and $Mg(ClO)_2$). Other common compounds that contain polyatomic ions include sodium bicarbonate ($NaHCO_3$), also known as baking soda, sodium nitrite ($NaNO_2$), an inhibitor of bacterial growth in packaged meats, and calcium carbonate ($CaCO_3$), the active ingredient in antacids such as Tums™.

Polyatomic Ions

Identify the polyatomic ion including its charge in each compound.

(a) KNO_2
(b) $CaSO_4$
(c) $Mg(NO_3)_2$

We name ionic compounds that contain a polyatomic ion in the same way that we name other ionic compounds, except that we use the name of the polyatomic ion whenever it occurs. Table 5.5 lists common polyatomic ions and their formulas. For example, $NaNO_2$ is named according to its cation, Na^+, *sodium*, and its polyatomic anion, NO_2^-, *nitrite*. Its full name is *sodium nitrite*.

$NaNO_2$ sodium nitrite

We name $FeSO_4$ according to its cation, *iron*, its charge *(II)*, and its polyatomic ion *sulfate*. Its full name is *iron(II) sulfate*.

$FeSO_4$ iron(II) sulfate

If the compound contains both a polyatomic cation and a polyatomic anion, we use the names of both polyatomic ions. For example, NH_4NO_3 is *ammonium nitrate*.

NH_4NO_3 ammonium nitrate

TABLE 5.5 Some Common Polyatomic Ions

Name	Formula	Name	Formula
Acetate	$C_2H_3O_2^-$	Hypochlorite	ClO^-
Carbonate	CO_3^{2-}	Chlorite	ClO_2^-
Hydrogen carbonate (or bicarbonate)	HCO_3^-	Chlorate	ClO_3^-
Hydroxide	OH^-	Perchlorate	ClO_4^-
Nitrite	NO_2^-	Permanganate	MnO_4^-
Nitrate	NO_3^-	Sulfite	SO_3^{2-}
Chromate	CrO_4^{2-}	Hydrogen sulfite (or bisulfite)	HSO_3^-
Dichromate	$Cr_2O_7^{2-}$	Sulfate	SO_4^{2-}
Phosphate	PO_4^{3-}	Hydrogen sulfate (or bisulfate)	HSO_4^-
Hydrogen phosphate	HPO_4^{2-}	Cyanide	CN^-
Dihydrogen phosphate	$H_2PO_4^-$	Peroxide	O_2^{2-}
Ammonium	NH_4^+		

You should be able to recognize polyatomic ions in a chemical formula, so become familiar with the ions listed in Table 5.5. Most common polyatomic ions are **oxyanions**, anions containing oxygen and another element. Notice that when a series of oxyanions contains different numbers of oxygen atoms, we name them systematically according to the number of oxygen atoms in the ion. If there are only two ions in the series, the one with more oxygen atoms has the ending *-ate* and the one with fewer oxygen atoms has the ending *-ite*. For example, NO_3^- is *nitrate* and NO_2^- is *nitrite*.

NO_3^- nitr*ate*

NO_2^- nitr*ite*

If there are more than two ions in the series, we use the prefixes *hypo-*, meaning *less than*, and *per*, meaning *more than*. So ClO^- is hypochlorite—less oxygen than chlorite, and ClO_4^- is perchlorate—more oxygen than chlorate.

Other halides (halogen ions) form similar series with similar names. Thus, IO_3^- is called iodate and BrO_3^- is called bromate.

ClO^- *hypo*chlor*ite*

ClO_2^- chlor*ite*

ClO_3^- chlor*ate*

ClO_4^- *per*chlor*ate*

EXAMPLE 5.7
Naming Ionic Compounds That Contain a Polyatomic Ion

Name the compound $Li_2Cr_2O_7$.

SOLUTION

The name for $Li_2Cr_2O_7$ is the name of the cation, *lithium*, followed by the name of the polyatomic ion, *dichromate*. Its full name is *lithium dichromate*.

$Li_2Cr_2O_7$ lithium dichromate

FOR PRACTICE 5.7

Name the compound $Sn(ClO_3)_2$.

FOR MORE PRACTICE 5.7

Write the formula for cobalt(II) phosphate.

$CoCl_2 \cdot 6H_2O$ $CoCl_2$

▲ **FIGURE 5.7 Hydrates** Cobalt(II) chloride hexahydrate is pink. Heating the compound removes the waters of hydration, leaving the blue anhydrous cobalt(II) chloride.

Hydrated Ionic Compounds

Some ionic compounds—called **hydrates**—contain a specific number of water molecules associated with each formula unit. For example, the formula for epsom salts is $MgSO_4 \cdot 7H_2O$, and its systematic name is magnesium sulfate heptahydrate. The seven H_2O molecules associated with the formula unit are *waters of hydration*. Waters of hydration can usually be removed by heating the compound. Figure 5.7 ◀ shows a sample of cobalt(II) chloride hexahydrate ($CoCl_2 \cdot 6H_2O$) before and after heating. The hydrate is pink and the anhydrous salt (the salt without any associated water molecules) is blue. We name hydrates like we name other ionic compounds, but we give them the additional name "*prefix*hydrate," where the *prefix* indicates the number of water molecules associated with each formula unit.

Other common hydrated ionic compounds and their names are as follows:

Common hydrate prefixes
hemi = 1/2
mono = 1
di = 2
tri = 3
tetra = 4
penta = 5
hexa = 6
hepta = 7
octa = 8

$CaSO_4 \cdot \frac{1}{2}H_2O$ calcium sulfate hemihydrate

$BaCl_2 \cdot 6H_2O$ barium chloride hexahydrate

$CuSO_4 \cdot 5H_2O$ copper(II) sulfate pentahydrate

5.7 Covalent Bonding: Simple Lewis Structures

In the Lewis model, we represent covalent bonding with a *Lewis structure*, which depicts neighboring atoms as sharing some (or all) of their valence electrons in order to attain octets (or duets for hydrogen).

Single Covalent Bonds

Consider hydrogen and oxygen, which have the following Lewis symbols:

H· ·Ö:

In water, these atoms share their unpaired valence electrons so that each hydrogen atom gets a duet and the oxygen atom gets an octet as represented with this Lewis structure:

H:Ö:H

The shared electrons—those that appear in the space between the two atoms—count toward the octets (or duets) of *both of the atoms*.

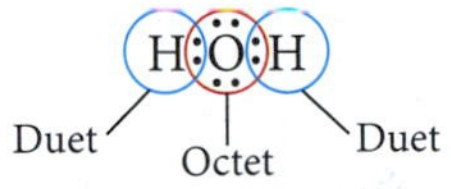

H:Ö:H

Duet Octet Duet

A shared pair of electrons is called a **bonding pair**, while a pair that is associated with only one atom—and therefore not involved in bonding—is called a **lone pair**. Lone pair electrons are also called **nonbonding electrons**.

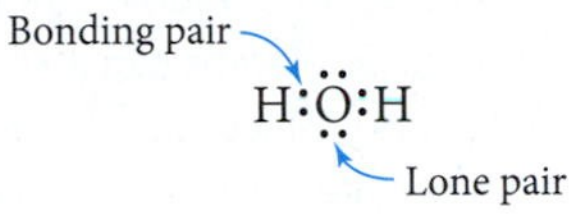

Bonding pair

H:Ö:H

Lone pair

We often represent a bonding pair of electrons by a dash to emphasize that the pair constitutes a chemical bond:

H—Ö—H

The Lewis model helps explain the tendency of some elements to form diatomic molecules. For example, consider the Lewis symbol for chlorine:

·C̈l:

If two Cl atoms pair together, they each get an octet:

:C̈l:C̈l: *or* :C̈l—C̈l:

Elemental chlorine does indeed exist as a diatomic molecule in nature, just as the Lewis model predicts. The same is true for the other halogens and several other elements as shown in Figure 5.8 ▼.

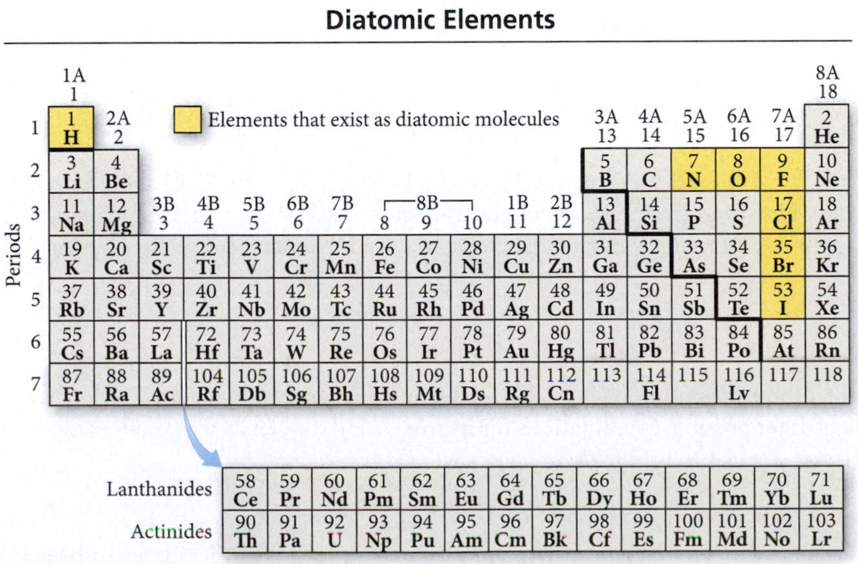

Diatomic Elements

Elements that exist as diatomic molecules

Periods	1A 1	2A 2	3B 3	4B 4	5B 5	6B 6	7B 7	8B 8	8B 9	8B 10	1B 11	2B 12	3A 13	4A 14	5A 15	6A 16	7A 17	8A 18
1	1 H																	2 He
2	3 Li	4 Be											5 B	6 C	7 N	8 O	9 F	10 Ne
3	11 Na	12 Mg											13 Al	14 Si	15 P	16 S	17 Cl	18 Ar
4	19 K	20 Ca	21 Sc	22 Ti	23 V	24 Cr	25 Mn	26 Fe	27 Co	28 Ni	29 Cu	30 Zn	31 Ga	32 Ge	33 As	34 Se	35 Br	36 Kr
5	37 Rb	38 Sr	39 Y	40 Zr	41 Nb	42 Mo	43 Tc	44 Ru	45 Rh	46 Pd	47 Ag	48 Cd	49 In	50 Sn	51 Sb	52 Te	53 I	54 Xe
6	55 Cs	56 Ba	57 La	72 Hf	73 Ta	74 W	75 Re	76 Os	77 Ir	78 Pt	79 Au	80 Hg	81 Tl	82 Pb	83 Bi	84 Po	85 At	86 Rn
7	87 Fr	88 Ra	89 Ac	104 Rf	105 Db	106 Sg	107 Bh	108 Hs	109 Mt	110 Ds	111 Rg	112 Cn	113	114 Fl	115	116 Lv	117	118

Lanthanides	58 Ce	59 Pr	60 Nd	61 Pm	62 Sm	63 Eu	64 Gd	65 Tb	66 Dy	67 Ho	68 Er	69 Tm	70 Yb	71 Lu
Actinides	90 Th	91 Pa	92 U	93 Np	94 Pu	95 Am	96 Cm	97 Bk	98 Cf	99 Es	100 Fm	101 Md	102 No	103 Lr

◀ **FIGURE 5.8 Diatomic Elements** The highlighted elements exist primarily as diatomic molecules.

Notice from Figure 5.8 that hydrogen also exists as a diatomic element. Similar to chlorine, the Lewis symbol for hydrogen has one unpaired electron.

H·

When two hydrogen atoms share their unpaired electron, each gets a duet, a stable configuration for hydrogen.

H:H *or* H—H

Again, the Lewis model helps us explain what we see in nature.

Double and Triple Covalent Bonds

In the Lewis model, two atoms may share more than one electron pair to get octets. For example, if we pair two oxygen atoms together, they share two electron pairs.

·Ö: + ·Ö:

↓

:Ö::Ö: *or* :Ö=Ö:

:Ö::Ö:

Octet Octet

One dash always stands for *two* electrons (a single bonding pair).

Each oxygen atom now has an octet because *the additional bonding pair counts toward the octet of both oxygen atoms.* When two atoms share two electron pairs, the resulting bond is a **double bond**. In general, double bonds are shorter and stronger than single bonds. The double bond that forms between two oxygen atoms explains why oxygen exists as a diatomic molecule. Atoms can also share three electron pairs. Consider the Lewis structure of another diatomic molecule, N_2. Since each N atom has five valence electrons, the Lewis structure for N_2 has 10 electrons. Both nitrogen atoms attain octets by sharing three electron pairs:

We will explore the characteristics of multiple bonds more fully in Section 7.3.

:N:::N: *or* :N≡N:

In this case, the bond is a **triple bond**. Triple bonds are even shorter and stronger than double bonds. When we examine nitrogen in nature, we find that it exists as a diatomic molecule with a very strong bond between the two nitrogen atoms. The bond is so strong that it is difficult to break, making N_2 a relatively unreactive molecule.

Covalent Bonding: Models and Reality

The Lewis model predicts the properties of molecular compounds in many ways. We have already seen how it explains the existence of several diatomic elements. The Lewis model also accounts for why particular combinations of atoms form molecules and others do not. For example, why is water H_2O and not H_3O? We can write a good Lewis structure for H_2O, but not for H_3O.

H—Ö—H (with lone pairs on O)

H—O—H with a third H bonded to O (O carries 5 nonbonding electrons)

Oxygen has nine electrons (one electron beyond an octet)

In this way, the Lewis model predicts that H_2O should be stable, while H_3O should not be, and that is in fact the case. However, if we remove an electron from H_3O, we get H_3O^+, which should be stable (according to the Lewis model) because, when we remove the extra electron, oxygen gets an octet.

[H—O—H with a third H bonded to O, lone pair on O]$^+$

This ion, called the hydronium ion, is in fact stable in aqueous solutions (see Section 9.7). The Lewis model predicts other possible combinations for hydrogen and oxygen as well. For example, we can write a Lewis structure for H_2O_2 as follows:

H—Ö—Ö—H

Indeed, H_2O_2, or hydrogen peroxide, exists and is often used as a disinfectant and a bleach.

The Lewis model also accounts for why covalent bonds are highly *directional*. The attraction between two covalently bonded atoms is due to the sharing of one or more electron pairs in the space between them. Thus, each bond links just one specific pair of atoms—*in contrast to ionic bonds, which are nondirectional and hold together an entire array of ions.* As a result, the fundamental units of covalently bonded compounds are individual molecules. These molecules can interact with one another in a number of different ways that we will cover in Chapter 12. However, in covalently bonded molecular compounds, the interactions *between* molecules (intermolecular forces) are generally much weaker than the bonding interactions within a molecule (intramolecular forces), as shown in Figure 5.9 ▶. When a molecular compound melts or boils, the molecules themselves remain intact—only the relatively weak interactions between molecules must be overcome. Consequently, molecular compounds tend to have lower melting and boiling points than ionic compounds.

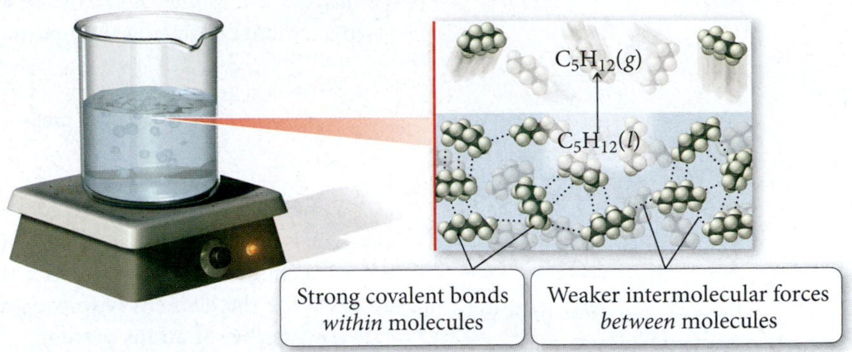

▲ **FIGURE 5.9 Intermolecular and Intramolecular Forces** The covalent bonds between atoms of a molecule are much stronger than the interactions between molecules. To bring a molecular substance to a boil, only the relatively weak intermolecular forces have to be overcome, so molecular compounds often have low boiling points.

Energy and the Octet Rule — 5.5 Cc Conceptual Connection

What is wrong with the following statement? *Atoms form bonds in order to satisfy the octet rule.*

Ionic and Molecular Compounds — 5.6 Cc Conceptual Connection

Which statement best summarizes the difference between ionic and molecular compounds?

(a) Molecular compounds contain highly directional covalent bonds, which results in the formation of molecules—discrete particles that do not covalently bond to each other. Ionic compounds contain nondirectional ionic bonds, which results (in the solid phase) in the formation of ionic lattices—extended networks of alternating cations and anions.

(b) Molecular compounds contain covalent bonds in which one of the atoms shares an electron with the other one, resulting in a new force that holds the atoms together in a covalent molecule. Ionic compounds contain ionic bonds in which one atom donates an electron to the other, resulting in a new force that holds the ions together in pairs (in the solid phase).

(c) The key difference between ionic and covalent compounds is the types of elements that compose them, not the way that the atoms bond together.

(d) A molecular compound is composed of covalently bonded molecules. An ionic compound is composed of ionically bonded molecules (in the solid phase).

5.8 Molecular Compounds: Formulas and Names

In contrast to an ionic compound, the formula for a molecular compound *cannot* always be determined from its constituent elements because the same combination of elements may form several different molecular compounds, each with a different formula. We just saw in Section 5.7 that hydrogen and oxygen can form both H_2O and H_2O_2 and each of these formulas is explained by its Lewis structure. Carbon and oxygen can form both CO and CO_2, and nitrogen and oxygen form NO, NO_2, N_2O, N_2O_3, N_2O_4, and N_2O_5. In Chapter 6, we will discuss how to draw Lewis structures for compounds such as these. For now, we focus on naming a molecular compound based on its formula and writing its formula based on its name.

Like ionic compounds, many molecular compounds have common names. For example, H_2O and NH_3 have the common names *water* and *ammonia*. However, the sheer number of existing molecular compounds—numbering in the millions—necessitates a systematic approach to naming them.

The first step in naming a molecular compound is identifying it as one. Remember, *molecular compounds are composed of two or more nonmetals.* In this section, we learn how to name binary (two-element) molecular compounds. Their names have the form:

prefix	name of 1st element	prefix	base name of 2nd element + *-ide*

When we write the name of a molecular compound, just as when we write its formula, the first element is the more metal-like one (toward the left and bottom of the periodic table). Generally, we write the name of the element with the smallest group number first. If the two elements lie in the same group, write the element with the greatest row number first. The prefixes given to each element indicate the number of atoms present:

These prefixes are the same ones we use when naming hydrates.

mono = 1	tri = 3	penta = 5	hepta = 7	nona = 9
di = 2	tetra = 4	hexa = 6	octa = 8	deca = 10

If there is only one atom of the *first element* in the formula, we normally omit the prefix *mono-*. For example, name NO_2 according to the first element, *nitrogen*, with no prefix because *mono-* is omitted for the first element, followed by the prefix *di*, to indicate two oxygen atoms, and the base name of the second element, *ox*, with the ending *-ide*. Its full name is *nitrogen dioxide.*

NO_2 nitrogen dioxide

We name the compound N_2O (sometimes called laughing gas) similarly except that we use the prefix *di-* before nitrogen to indicate two nitrogen atoms and the prefix *mono-* before oxide to indicate one oxygen atom. Its full name is *dinitrogen monoxide.*

N_2O dinitrogen monoxide

When a prefix ends with "o" and the base name begins with "o," the first "o" is often dropped. So mono-oxide becomes *monoxide*.

EXAMPLE 5.8
Naming Molecular Compounds

Name each compound.

(a) NI_3 **(b)** PCl_5 **(c)** P_4S_{10}

SOLUTION

(a) The name of the compound is the name of the first element, *nitrogen*, followed by the base name of the second element, *iod*, prefixed by *tri-* to indicate three and given the suffix *-ide.*

NI_3 nitrogen triiodide

(b) The name of the compound is the name of the first element, *phosphorus*, followed by the base name of the second element, *chlor*, prefixed by *penta-* to indicate five and given the suffix *-ide.*

PCl_5 phosphorus pentachloride

(c) The name of the compound is the name of the first element, *phosphorus*, prefixed by *tetra-* to indicate four, followed by the base name of the second element, *sulf*, prefixed by *deca* to indicate ten and given the suffix *-ide.*

P_4S_{10} tetraphosphorus decasulfide

FOR PRACTICE 5.8
Name the compound N_2O_5.

FOR MORE PRACTICE 5.8
Write the formula for phosphorus tribromide.

Nomenclature

The compound NCl_3 is nitrogen trichloride, but $AlCl_3$ is simply aluminum chloride. Why?

5.7 Cc Conceptual Connection

5.9 Formula Mass and the Mole Concept for Compounds

We have now seen how to name compounds and write formulas for them. It is also useful to be able to quantify the mass of a compound. In Chapter 1, we defined the average mass of an atom of an element as its *atomic mass*. Similarly, we now define the average mass of a molecule (or a formula unit) of a compound as its **formula mass**. (The terms *molecular mass* or *molecular weight* are synonymous with formula mass and are also common.) For any compound, the formula mass is the sum of the atomic masses of all the atoms in its chemical formula.

Formula mass = (**Number of atoms of 1st element in chemical formula** × **Atomic mass of 1st element**) + (**Number of atoms of 2nd element in chemical formula** × **Atomic mass of 2nd element**) + …

For example, the formula mass of carbon dioxide, CO_2, is:

Multiply by 2 because formula has 2 oxygen atoms.

$$\text{Formula mass} = 12.01\text{ amu} + 2(16.00\text{ amu}) = 44.01\text{ amu}$$

and that of sodium oxide, Na_2O, is:

Multiply by 2 because formula has 2 sodium atoms.

$$\text{Formula mass} = 2(22.99\text{ amu}) + 16.00\text{ amu} = 61.98\text{ amu}$$

EXAMPLE 5.9 Calculating Formula Mass

Calculate the formula mass of glucose, $C_6H_{12}O_6$.

SOLUTION

To find the formula mass, add the atomic masses of each atom in the chemical formula.

$$\begin{aligned}\text{Formula mass} &= 6 \times (\text{atomic mass C}) + 12 \times (\text{atomic mass H}) + 6 \times (\text{atomic mass O}) \\ &= 6(12.01\text{ amu}) + 12(1.008\text{ amu}) + 6(16.00\text{ amu}) \\ &= 180.16\text{ amu}\end{aligned}$$

FOR PRACTICE 5.9

Calculate the formula mass of calcium nitrate.

Molar Mass of a Compound

In Section 2.8, we saw that an element's molar mass—the mass in grams of one mole of its atoms—is numerically equivalent to its atomic mass. We use the molar mass in combination with Avogadro's number to determine the number of atoms in a given mass of the element. The same concept applies to compounds. The *molar mass of a compound*—the mass in grams of 1 mol of its molecules or formula units—is numerically equivalent to its formula mass. For example, we just calculated the formula mass of CO_2 to be 44.01 amu. The molar mass is, therefore:

$$CO_2\text{ molar mass} = 44.01\text{ g/mol}$$

Remember, ionic compounds do not contain individual molecules. In casual language, the smallest electrically neutral collection of ions is sometimes called a molecule but is more correctly called a formula unit.

Using Molar Mass to Count Molecules by Weighing

The molar mass of CO_2 is a conversion factor between mass (in grams) and amount (in moles) of CO_2. Suppose we want to find the number of CO_2 molecules in a sample of dry ice (solid CO_2) with a mass of 10.8 g. This calculation is analogous to Example 2.11, where we found the number of atoms in a sample of copper of a given mass. We begin with the mass of 10.8 g and use the molar mass to convert to the amount in moles. Then we use Avogadro's number to convert to number of molecules. The conceptual plan shows the progression:

Conceptual Plan

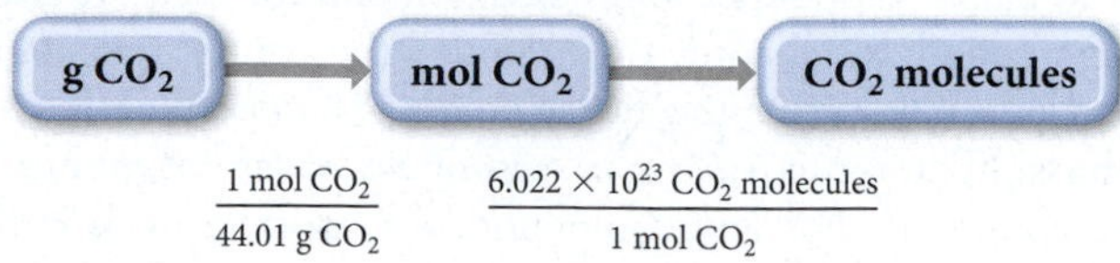

To solve the problem, we follow the conceptual plan, beginning with 10.8 g CO_2, converting to moles, and then to molecules.

Solution

$$10.8 \text{ g } CO_2 \times \frac{1 \text{ mol } CO_2}{44.01 \text{ g } CO_2} \times \frac{6.022 \times 10^{23} \text{ } CO_2 \text{ molecules}}{1 \text{ mol } CO_2} = 1.48 \times 10^{23} \text{ } CO_2 \text{ molecules}$$

EXAMPLE 5.10

The Mole Concept—Converting between Mass and Number of Molecules

An aspirin tablet contains 325 mg of acetylsalicylic acid ($C_9H_8O_4$). How many acetylsalicylic acid molecules does it contain?

SORT You are given the mass of acetylsalicylic acid and asked to find the number of molecules.

GIVEN: 325 mg $C_9H_8O_4$

FIND: number of $C_9H_8O_4$ molecules

STRATEGIZE First convert to grams and then to moles (using the molar mass of the compound) and then to number of molecules (using Avogadro's number). You will need both the molar mass of acetylsalicylic acid and Avogadro's number as conversion factors. You will also need the conversion factor between g and mg.

CONCEPTUAL PLAN

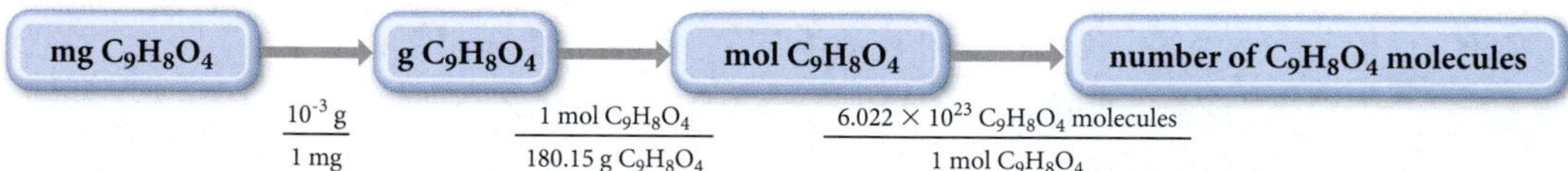

RELATIONSHIPS USED

$C_9H_8O_4$ molar mass $= 9(12.01) + 8(1.008) + 4(16.00)$
$= 180.15$ g/mol

$6.022 \times 10^{23} = 1$ mol

$1 \text{ mg} = 10^{-3} \text{ g}$

SOLVE Follow the conceptual plan to solve the problem.

SOLUTION

$$325 \text{ mg } C_9H_8O_4 \times \frac{10^{-3} \text{ g}}{1 \text{ mg}} \times \frac{1 \text{ mol } C_9H_8O_4}{180.15 \text{ g } C_9H_8O_4} \times \frac{6.022 \times 10^{23} \text{ } C_9H_8O_4 \text{ molecules}}{1 \text{ mol } C_9H_8O_4}$$

$$= 1.09 \times 10^{21} \text{ } C_9H_8O_4 \text{ molecules}$$

CHECK The units of the answer, $C_9H_8O_4$ molecules, are correct. The magnitude is smaller than Avogadro's number, as expected, since you have less than one molar mass of acetylsalicylic acid.

FOR PRACTICE 5.10

Find the number of ibuprofen molecules in a tablet containing 200.0 mg of ibuprofen ($C_{13}H_{18}O_2$).

FOR MORE PRACTICE 5.10

What is the mass of a sample of water containing 3.55×10^{22} H_2O molecules?

Molecular Models and the Size of Molecules

5.8 Cc Conceptual Connection

Throughout this book, we use space-filling molecular models to represent molecules. Which number is the best estimate for the scaling factor used in these models? In other words, by approximately what number would you have to multiply the radius of an actual oxygen atom to get the radius of the sphere used to represent the oxygen atom in the water molecule shown here?

(a) 10 **(b)** 10^4 **(c)** 10^8 **(d)** 10^{16}

5.10 Composition of Compounds

A chemical formula, in combination with the molar masses of its constituent elements, indicates the relative quantities of each element in a compound, which is extremely useful information. For example, about 35 years ago, scientists began to suspect that synthetic compounds known as chlorofluorocarbons (CFCs) were destroying ozone (O_3) in Earth's upper atmosphere. Upper atmospheric ozone is important because it acts as a shield, protecting life on Earth from the sun's harmful ultraviolet light.

CFCs are chemically inert compounds that were used primarily as refrigerants and industrial solvents. Over time CFCs have accumulated in the atmosphere. In the upper atmosphere, sunlight breaks bonds within CFCs, releasing chlorine atoms. The chlorine atoms then react with ozone, converting it into O_2. So the harmful part of CFCs is the chlorine atoms that they carry. How can we determine the mass of chlorine in a given mass of a CFC?

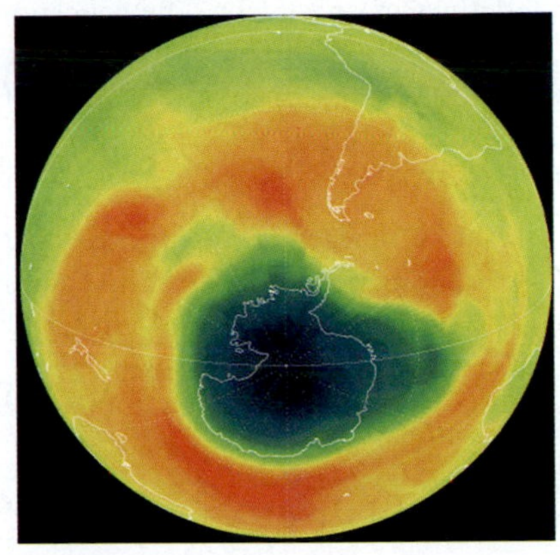

▲ The chlorine in chlorofluorocarbons caused the ozone hole over Antarctica. The dark blue color indicates depressed ozone levels.

One way to express how much of an element is in a given compound is to use the element's mass percent composition for that compound. The **mass percent composition** or **mass percent** of an element is that element's percentage of the compound's total mass. We can calculate the mass percent of element X in a compound from the chemical formula as follows:

$$\text{mass percent of element X} = \frac{\text{mass of element X in 1 mol of compound}}{\text{mass of 1 mol of the compound}} \times 100\%$$

Suppose, for example, that we want to calculate the mass percent composition of Cl in the chlorofluorocarbon CCl_2F_2. The mass percent Cl is given by:

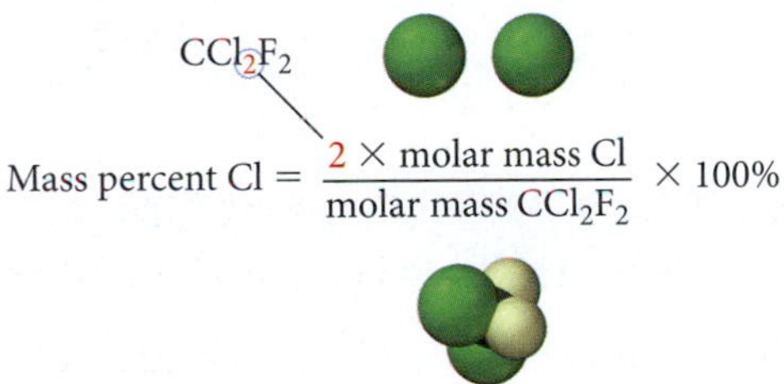

$$\text{Mass percent Cl} = \frac{2 \times \text{molar mass Cl}}{\text{molar mass } CCl_2F_2} \times 100\%$$

We multiply the molar mass of Cl by two because the chemical formula has a subscript of 2 for Cl, indicating that 1 mol of CCl_2F_2 contains 2 mol of Cl atoms. We calculate the molar mass of CCl_2F_2 as follows:

$$\begin{aligned}\text{Molar mass} &= 12.01 \text{ g/mol} + 2(35.45 \text{ g/mol}) + 2(19.00 \text{ g/mol}) \\ &= 120.91 \text{ g/mol}\end{aligned}$$

So the mass percent of Cl in CCl_2F_2 is:

$$\begin{aligned}\text{Mass percent Cl} &= \frac{2 \times \text{molar mass Cl}}{\text{molar mass } CCl_2F_2} \times 100\% \\ &= \frac{2 \times 35.45 \cancel{\text{g/mol}}}{120.91 \cancel{\text{g/mol}}} \times 100\% \\ &= 58.64\%\end{aligned}$$

EXAMPLE 5.11 Mass Percent Composition

Calculate the mass percent of Cl in Freon-112 ($C_2Cl_4F_2$), a CFC refrigerant.

SORT You are given the molecular formula of Freon-112 and asked to find the mass percent of Cl.

GIVEN: $C_2Cl_4F_2$

FIND: mass percent Cl

STRATEGIZE The molecular formula tells you that there are 4 mol of Cl in each mole of Freon-112. Find the mass percent composition from the chemical formula by using the equation that defines mass percent. The conceptual plan shows how to use the mass of Cl in 1 mol of $C_2Cl_4F_2$ and the molar mass of $C_2Cl_4F_2$ to find the mass percent of Cl.

CONCEPTUAL PLAN

$$\text{Mass \% Cl} = \frac{4 \times \text{molar mass Cl}}{\text{molar mass } C_2Cl_4F_2} \times 100\%$$

RELATIONSHIPS USED

$$\text{Mass percent of element X} = \frac{\text{mass of element X in 1 mol of compound}}{\text{mass of 1 mol of compound}} \times 100\%$$

SOLVE Calculate the necessary parts of the equation and substitute the values into the equation to find mass percent Cl.

SOLUTION

$$4 \times \text{molar mass Cl} = 4(35.45 \text{ g/mol}) = 141.8 \text{ g/mol}$$

$$\begin{aligned}\text{Molar mass } C_2Cl_4F_2 &= 2(12.01 \text{ g/mol}) + 4(35.45 \text{ g/mol}) + 2(19.00 \text{ g/mol}) \\ &= 24.02 \text{ g/mol} + 141.8 \text{ g/mol} + 38.00 \text{ g/mol} = 203.8 \text{ g/mol}\end{aligned}$$

$$\text{Mass \% Cl} = \frac{4 \times \text{molar mass Cl}}{\text{molar mass } C_2Cl_4F_2} \times 100\% = \frac{141.8 \cancel{\text{g/mol}}}{203.8 \cancel{\text{g/mol}}} \times 100\% = 69.58\%$$

CHECK The units of the answer (%) are correct, and the magnitude is reasonable because (a) it is between 0 and 100% and (b) chlorine is the heaviest atom in the molecule and there are four of them.

FOR PRACTICE 5.11
Acetic acid ($C_2H_4O_2$) is the active ingredient in vinegar. Calculate the mass percent composition of oxygen in acetic acid.

FOR MORE PRACTICE 5.14
Calculate the mass percent composition of sodium in sodium oxide.

5.9 Cc Conceptual Connection — Chemical Formula and Mass Percent Composition

Without doing any calculations, list the elements in C_6H_6O in order of decreasing mass percent composition.

$$C_6H_6O$$

Mass Percent Composition as a Conversion Factor

The mass percent composition of an element in a compound is a conversion factor between the mass of the element and the mass of the compound. For example, we saw that the mass percent composition of Cl in CCl_2F_2 is 58.64%. Since percent means *per hundred*, there are 58.64 g Cl *per hundred* grams CCl_2F_2, which can be expressed as the ratio:

$$58.64 \text{ g Cl} : 100 \text{ g } CCl_2F_2$$

or, in fractional form:

$$\frac{58.64 \text{ g Cl}}{100 \text{ g } CCl_2F_2} \quad \text{or} \quad \frac{100 \text{ g } CCl_2F_2}{58.64 \text{ g Cl}}$$

These ratios can function as conversion factors between grams of Cl and grams of CCl_2F_2. For example, to calculate the mass of Cl in 1.00 kg CCl_2F_2, we use the following conceptual plan:

Conceptual Plan

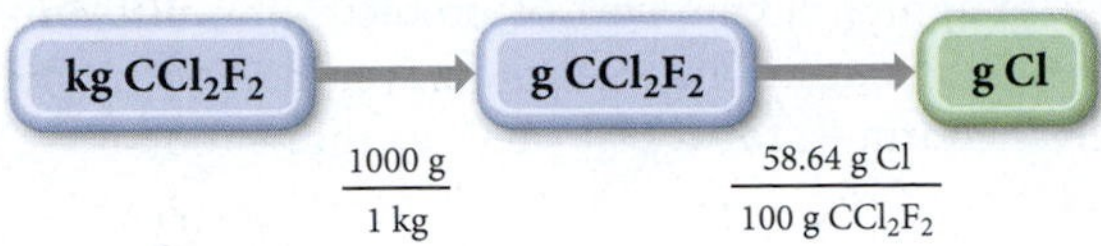

Notice that the mass percent composition acts as a conversion factor between grams of the compound and grams of the constituent element. To calculate grams Cl, we follow the conceptual plan:

Solution

$$1.00 \text{ kg } CCl_2F_2 \times \frac{1000 \text{ g}}{1 \text{ kg}} \times \frac{58.64 \text{ g Cl}}{100 \text{ g } CCl_2F_2} = 5.86 \times 10^2 \text{ g Cl}$$

EXAMPLE 5.12

Using Mass Percent Composition as a Conversion Factor

The U.S. Food and Drug Administration (FDA) recommends that a person consume less than 2.4 g of sodium per day. What mass of sodium chloride (in grams) can you consume and still be within the FDA guidelines? Sodium chloride is 39% sodium by mass.

SORT You are given a mass of sodium and the mass percent of sodium in sodium chloride. You are asked to find the mass of NaCl that contains the given mass of sodium.

GIVEN: 2.4 g Na
FIND: g NaCl

STRATEGIZE Convert between mass of a constituent element and mass of a compound by using mass percent composition as a conversion factor.

CONCEPTUAL PLAN

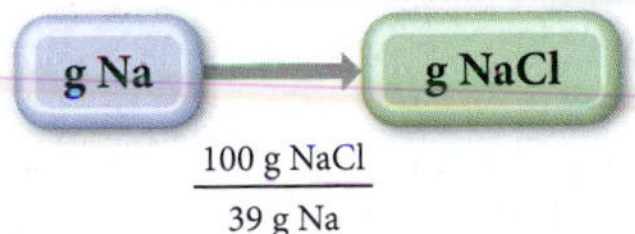

RELATIONSHIPS USED
39 g Na: 100 g NaCl

SOLVE Follow the conceptual plan to solve the problem.

SOLUTION

$$2.4 \text{ g Na} \times \frac{100 \text{ g NaCl}}{39 \text{ g Na}} = 6.2 \text{ g NaCl}$$

You can consume 6.2 g NaCl and still be within the FDA guidelines.

CHECK The units of the answer are correct. The magnitude seems reasonable because it is larger than the amount of sodium, as expected, because sodium is only one of the elements in NaCl.

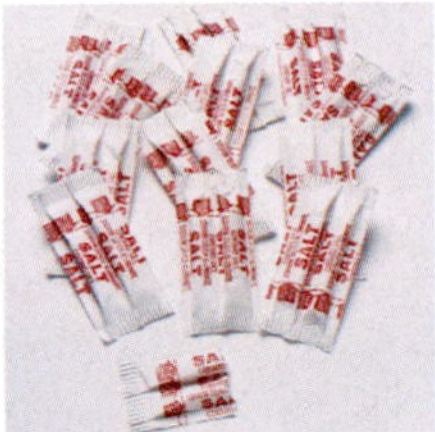

◀ 12.5 packets of salt contain 6.2 g of NaCl.

FOR PRACTICE 5.12
What mass (in grams) of iron(III) oxide contains 58.7 g of iron? Iron(III) oxide is 69.94% iron by mass.

FOR MORE PRACTICE 5.12
If someone consumes 22 g of sodium chloride per day, what mass (in grams) of sodium does that person consume? Sodium chloride is 39% sodium by mass.

Conversion Factors from Chemical Formulas

Mass percent composition is one way to understand how much chlorine is in a particular chlorofluorocarbon or, more generally, how much of a constituent element is present in a given mass of any compound. However, we can also approach this type of problem in a different way. Chemical formulas contain within them inherent relationships between atoms (or moles of atoms) and molecules (or moles of molecules). For example, the formula for CCl_2F_2 tells us that 1 mol of CCl_2F_2 contains 2 mol of Cl atoms. We write the ratio as:

$$1 \text{ mol } CCl_2F_2 : 2 \text{ mol Cl}$$

With ratios such as these—which come from the chemical formula—we can directly determine the amounts of the constituent elements present in a given amount of a compound without having to calculate mass percent composition. For example, we calculate the number of moles of Cl in 38.5 mol of CCl_2F_2 as follows:

Conceptual Plan

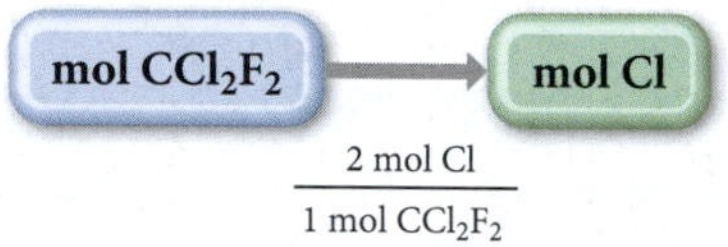

Solution

$$38.5 \text{ mol } CCl_2F_2 \times \frac{2 \text{ mol Cl}}{1 \text{ mol } CCl_2F_2} = 77.0 \text{ mol Cl}$$

As we have seen, however, we often want to know not the *amount in moles* of an element in a certain number of moles of compound, but the *mass in grams* (or other units) of a constituent element in a given *mass* of the compound. Suppose we want to know the mass (in grams) of Cl contained in 25.0 g CCl_2F_2. The relationship inherent in the chemical formula (2 mol Cl : 1 mol CCl_2F_2) applies to the amount in moles, not to mass. Therefore, we first convert the mass of CCl_2F_2 to moles CCl_2F_2. *Then* we use the conversion factor from the chemical formula to convert to moles Cl. Finally, we use the molar mass of Cl to convert to grams Cl.

Conceptual Plan

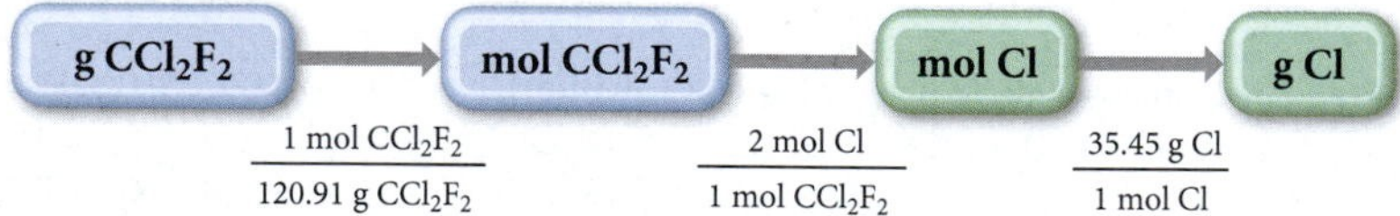

Solution

$$25.0 \text{ g } CCl_2F_2 \times \frac{1 \text{ mol } CCl_2F_2}{120.91 \text{ g } CCl_2F_2} \times \frac{2 \text{ mol Cl}}{1 \text{ mol } CCl_2F_2} \times \frac{35.45 \text{ g Cl}}{1 \text{ mol Cl}} = 14.7 \text{ g Cl}$$

Notice that we must convert from g CCl_2F_2 to mol CCl_2F_2 *before* we can use the chemical formula as a conversion factor. *Always remember that the chemical formula gives us a relationship between the amounts (in moles) of substances, not between the masses (in grams) of them.*

The general form for solving problems where we are asked to find the mass of an element present in a given mass of a compound is:

$$\text{Mass compound} \longrightarrow \text{moles compound} \longrightarrow \text{moles element} \longrightarrow \text{mass element}$$

We use the atomic or molar mass to convert between mass and moles, and we use relationships inherent in the chemical to convert between moles and moles.

EXAMPLE 5.13 Chemical Formulas as Conversion Factors

Hydrogen may potentially be used in the future to replace gasoline as a fuel. Most major automobile companies are developing vehicles that run on hydrogen. These cars are environmentally friendly because their only emission is water vapor. One way to obtain hydrogen for fuel is to use an emission-free energy source such as wind power to form elemental hydrogen from water. What mass of hydrogen (in grams) is contained in 1.00 gallon of water? (The density of water is 1.00 g/mL.)

SORT You are given a volume of water and asked to find the mass of hydrogen it contains. You are also given the density of water.

GIVEN: 1.00 gal H_2O

$d_{H_2O} = 1.00$ g/mL

FIND: g H

STRATEGIZE The first part of the conceptual plan shows how to convert the units of volume from gallons to liters and then to mL. It also shows how you can then use the density to convert mL to g.

The second part of the conceptual plan is the basic sequence of mass ⟶ moles ⟶ moles ⟶ mass. Convert between moles and mass using the appropriate molar masses, and convert from mol H_2O to mol H using the conversion factor derived from the molecular formula.

CONCEPTUAL PLAN

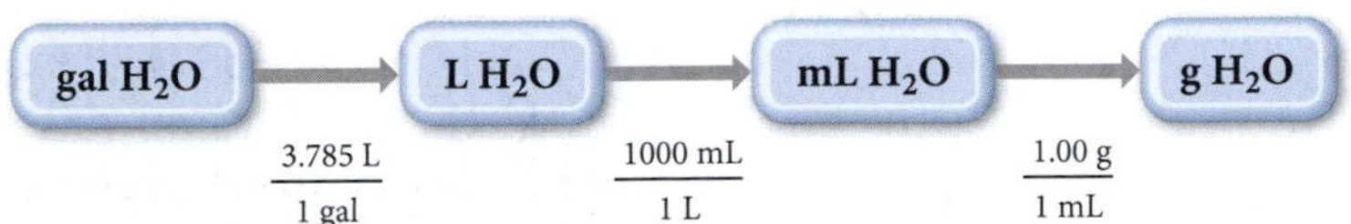

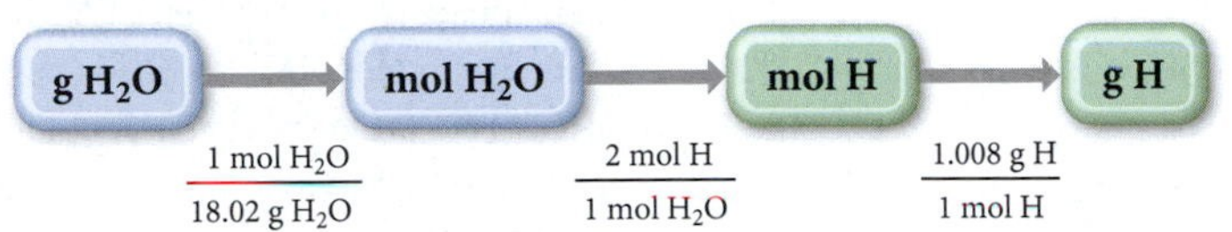

RELATIONSHIPS USED

3.785 L = 1 gal

1 mL = 10^{-3} L

1.00 g H_2O = 1 mL H_2O (density of H_2O)

Molar mass H_2O = 2(1.008) + 16.00 = 18.02 g/mol

2 mol H: 1 mol H_2O

1.008 g H = 1 mol H

SOLVE Follow the conceptual plan to solve the problem.

SOLUTION

$$1.00 \text{ gal } H_2O \times \frac{3.785 \text{ L}}{1 \text{ gal}} \times \frac{1 \text{ mL}}{10^{-3} \text{ L}} \times \frac{1.0 \text{ g}}{\text{mL}} = 3.78\underline{5} \times 10^3 \text{ g } H_2O$$

$$3.78\underline{5} \times 10^3 \text{ g } H_2O \times \frac{1 \text{ mol } H_2O}{18.02 \text{ g } H_2O} \times \frac{2 \text{ mol H}}{1 \text{ mol } H_2O} \times \frac{1.008 \text{ g H}}{1 \text{ mol H}} = 4.23 \times 10^2 \text{ g H}$$

CHECK The units of the answer (g H) are correct. Since a gallon of water is about 3.8 L, its mass is about 3.8 kg. H is a light atom, so its mass should be significantly less than 3.8 kg, as it is in the answer.

FOR PRACTICE 5.13

Determine the mass of oxygen in a 7.2-g sample of $Al_2(SO_4)_3$.

FOR MORE PRACTICE 5.13

Butane (C_4H_{10}) is the liquid fuel in lighters. How many grams of carbon are present within a lighter containing 7.25 mL of butane? (The density of liquid butane is 0.601 g/mL.)

5.10 Cc Conceptual Connection

Chemical Formulas and Elemental Composition

Which ratio can be correctly derived from the molecular formula for water (H_2O)? Explain your answer.

(a) 2 g H : 1 g H_2O
(b) 2 mL H : 1 mL H_2O
(c) 2 mol H : 1 mol H_2O

5.11 Determining a Chemical Formula from Experimental Data

In Section 5.10, we calculated mass percent composition from a chemical formula. Can we also do the reverse? Can we calculate a chemical formula from mass percent composition? This question is important because many laboratory analyses of compounds give the relative masses of each element present in the compound. For example, if we decompose water into hydrogen and oxygen in the laboratory, we can measure the masses of hydrogen and oxygen produced. Can we arrive at a chemical formula from this kind of data? The answer is a qualified yes. We can determine a chemical formula, but it is an empirical formula (not a molecular formula). To get a molecular formula, we need additional information, such as the molar mass of the compound.

Suppose we decompose a sample of water in the laboratory and find that it produces 0.857 g of hydrogen and 6.86 g of oxygen. How do we determine an empirical formula from these data? We know that an empirical formula represents a ratio of atoms or a ratio of moles of atoms, *not a ratio of masses*. So the first thing we must do is convert our data from mass (in grams) to amount (in moles). How many moles of each element are present in the sample? To convert to moles, we divide each mass by the molar mass of that element.

$$\text{Moles H} = 0.857\ \cancel{\text{g H}} \times \frac{1\ \text{mol H}}{1.01\ \cancel{\text{g H}}} = 0.849\ \text{mol H}$$

$$\text{Moles O} = 6.86\ \cancel{\text{g O}} \times \frac{1\ \text{mol O}}{16.00\ \cancel{\text{g O}}} = 0.429\ \text{mol O}$$

From these data, we know there are 0.849 mol H for every 0.429 mol O. We can now write a pseudoformula for water.

$$H_{0.849}O_{0.429}$$

To get the smallest whole-number subscripts in our formula, we divide all the subscripts by the smallest one, in this case 0.429.

$$H_{\frac{0.849}{0.429}}O_{\frac{0.429}{0.429}} = H_{1.98}O = H_2O$$

Our empirical formula for water, which also happens to be the molecular formula, is H_2O. You can use the following procedure to obtain the empirical formula of any compound from experimental data giving the relative masses of the constituent elements. The left column outlines the procedure, and the center and right columns contain two examples of how to apply the procedure.

PROCEDURE FOR Obtaining an Empirical Formula from Experimental Data

EXAMPLE 5.14 Obtaining an Empirical Formula from Experimental Data

A compound containing nitrogen and oxygen is decomposed in the laboratory and produces 24.5 g nitrogen and 70.0 g oxygen. Calculate the empirical formula of the compound.

EXAMPLE 5.15 Obtaining an Empirical Formula from Experimental Data

A laboratory analysis of aspirin determined the following mass percent composition:

C 60.00%
H 4.48%
O 35.52%

Find the empirical formula.

1. Write down (or calculate) as *given* the masses of each element present in a sample of the compound. If you are given mass percent composition, assume a 100-g sample and calculate the masses of each element from the given percentages.

Example 5.14: **GIVEN:** 24.5 g N, 70.0 g O
FIND: empirical formula

Example 5.15: **GIVEN:** In a 100-g sample: 60.00 g C, 4.48 g H, 35.52 g O
FIND: empirical formula

2. Convert each of the masses in Step 1 to moles by using the appropriate molar mass for each element as a conversion factor.

Example 5.14:

$$24.5 \text{ g N} \times \frac{1 \text{ mol N}}{14.01 \text{ g N}} = 1.75 \text{ mol N}$$

$$70.0 \text{ g O} \times \frac{1 \text{ mol O}}{16.00 \text{ g O}} = 4.38 \text{ mol O}$$

Example 5.15:

$$60.00 \text{ g C} \times \frac{1 \text{ mol C}}{12.01 \text{ g C}} = 4.996 \text{ mol C}$$

$$4.48 \text{ g H} \times \frac{1 \text{ mol H}}{1.008 \text{ g H}} = 4.44 \text{ mol H}$$

$$35.52 \text{ g O} \times \frac{1 \text{ mol O}}{16.00 \text{ g O}} = 2.220 \text{ mol O}$$

3. Write down a pseudoformula for the compound using the number of moles of each element (from Step 2) as subscripts.

Example 5.14: $N_{1.75}O_{4.38}$

Example 5.15: $C_{4.996}H_{4.44}O_{2.220}$

4. Divide all the subscripts in the formula by the smallest subscript.

Example 5.14: $N_{\frac{1.75}{1.75}}O_{\frac{4.38}{1.75}} \longrightarrow N_1O_{2.5}$

Example 5.15: $C_{\frac{4.996}{2.220}}H_{\frac{4.44}{2.220}}O_{\frac{2.220}{2.220}} \longrightarrow C_{2.25}H_2O_1$

5. If the subscripts are not whole numbers, multiply all the subscripts by a small whole number (see table) to get whole-number subscripts.

Fractional Subscript	Multiply by This
0.20	5
0.25	4
0.33	3
0.40	5
0.50	2
0.66	3
0.75	4
0.80	5

Example 5.14: $N_1O_{2.5} \times 2 \longrightarrow N_2O_5$
The correct empirical formula is N_2O_5.

Example 5.15: $C_{2.25}H_2O_1 \times 4 \longrightarrow C_9H_8O_4$
The correct empirical formula is $C_9H_8O_4$.

FOR PRACTICE 5.14
A sample of a compound is decomposed in the laboratory and produces 165 g carbon, 27.8 g hydrogen, and 220.2 g oxygen. Calculate the empirical formula of the compound.

FOR PRACTICE 5.15
Ibuprofen has the following mass percent composition:
C 75.69%, H 8.80%, O 15.51%.
What is the empirical formula of ibuprofen?

Calculating Molecular Formulas for Compounds

We can find the molecular formula of a compound from the empirical formula if we also know the molar mass of the compound. Recall from Section 5.3 that the molecular formula is always a whole-number multiple of the empirical formula.

$$\text{Molecular formula} = \text{empirical formula} \times n\text{, where } n = 1, 2, 3,...$$

Suppose we want to find the molecular formula for fructose (a sugar found in fruit) from its empirical formula, CH_2O, and its molar mass, 180.2 g/mol. We know that the molecular formula is a whole-number multiple of CH_2O.

$$\begin{aligned}\text{Molecular formula} &= (CH_2O) \times n\\ &= C_nH_{2n}O_n\end{aligned}$$

We also know that the molar mass is a whole-number multiple of the **empirical formula molar mass**, the sum of the masses of all the atoms in the empirical formula.

$$\text{Molar mass} = \text{empirical formula molar mass} \times n$$

For a particular compound, the value of n in both cases is the same. Therefore, we can find n by calculating the ratio of the molar mass to the empirical formula molar mass.

$$n = \frac{\text{molar mass}}{\text{empirical formula molar mass}}$$

For fructose, the empirical formula molar mass is:
empirical formula molar mass

$$= 12.01\text{ g/mol} + 2(1.01\text{ g/mol}) + 16.00\text{ g/mol} = 30.03\text{ g/mol}$$

Therefore, n is:

$$n = \frac{180.2\text{ g/mol}}{30.03\text{ g/mol}} = 6$$

We can then use this value of n to find the molecular formula.

$$\text{Molecular formula} = (CH_2O) \times 6 = C_6H_{12}O_6$$

EXAMPLE 5.16

Calculating a Molecular Formula from an Empirical Formula and Molar Mass

Butanedione—a main component responsible for the smell and taste of butter and cheese—contains the elements carbon, hydrogen, and oxygen. The empirical formula of butanedione is C_2H_3O and its molar mass is 86.09 g/mol. Find its molecular formula.

SORT You are given the empirical formula and molar mass of butanedione and asked to find the molecular formula.	**GIVEN:** Empirical formula $= C_2H_3O$ molar mass = 86.09 g/mol **FIND:** molecular formula
STRATEGIZE A molecular formula is always a whole-number multiple of the empirical formula. Divide the molar mass by the empirical formula mass to get the whole number.	Molecular formula = empirical formula $\times n$ $n = \frac{\text{molar mass}}{\text{empirical formula molar mass}}$
SOLVE Calculate the empirical formula mass. Divide the molar mass by the empirical formula mass to find n. Multiply the empirical formula by n to obtain the molecular formula.	Empirical formula molar mass $= 2(12.01\text{ g/mol}) + 3(1.008\text{ g/mol}) + 16.00\text{ g/mol} = 43.04\text{ g/mol}$ $n = \frac{\text{molar mass}}{\text{empirical formula mass}} = \frac{86.09\text{ g/mol}}{43.04\text{ g/mol}} = 2$ Molecular formula $= C_2H_3O \times 2 = C_4H_6O_2$

—Continued on the next page

Continued from the previous page—

CHECK Check the answer by calculating the molar mass of the formula as follows:

$4(12.01 \text{ g/mol}) + 6(1.008 \text{ g/mol}) + 2(16.00 \text{ g/mol}) = 86.09 \text{ g/mol}$

The calculated molar mass is in agreement with the given molar mass.

FOR PRACTICE 5.16

A compound has the empirical formula CH and a molar mass of 78.11 g/mol. What is its molecular formula?

FOR MORE PRACTICE 5.16

A compound with the percent composition shown here has a molar mass of 60.10 g/mol. Determine its molecular formula.

C, 39.97%
H, 13.41%
N, 46.62%

Combustion Analysis

In the previous section, we discussed how to determine the empirical formula of a compound from the relative masses of its constituent elements. Another common (and related) way to obtain empirical formulas for unknown compounds, especially those containing carbon and hydrogen, is **combustion analysis**. In combustion analysis, the unknown compound undergoes combustion (or burning) in the presence of pure oxygen, as shown in Figure 5.10 ▼. When the sample is burned, all of the carbon in the sample is converted to CO_2, and all of the hydrogen is converted to H_2O. The CO_2 and H_2O are then weighed. With these masses, we can use the numerical relationships between moles inherent in the formulas for CO_2 and H_2O (1 mol CO_2: 1 mol C and 1 mol H_2O: 2 mol H) to determine the amounts of C and H in the original sample. We can determine the amounts of any other elemental constituents, such as O, Cl, or N, by subtracting the original mass of the sample from the sum of the masses of C and H. Examples 5.17 and 5.18 show how to perform these calculations for a sample containing only C and H and for a sample containing C, H, and O.

Combustion is a type of *chemical reaction.* We discuss chemical reactions and their representation in Chapter 8.

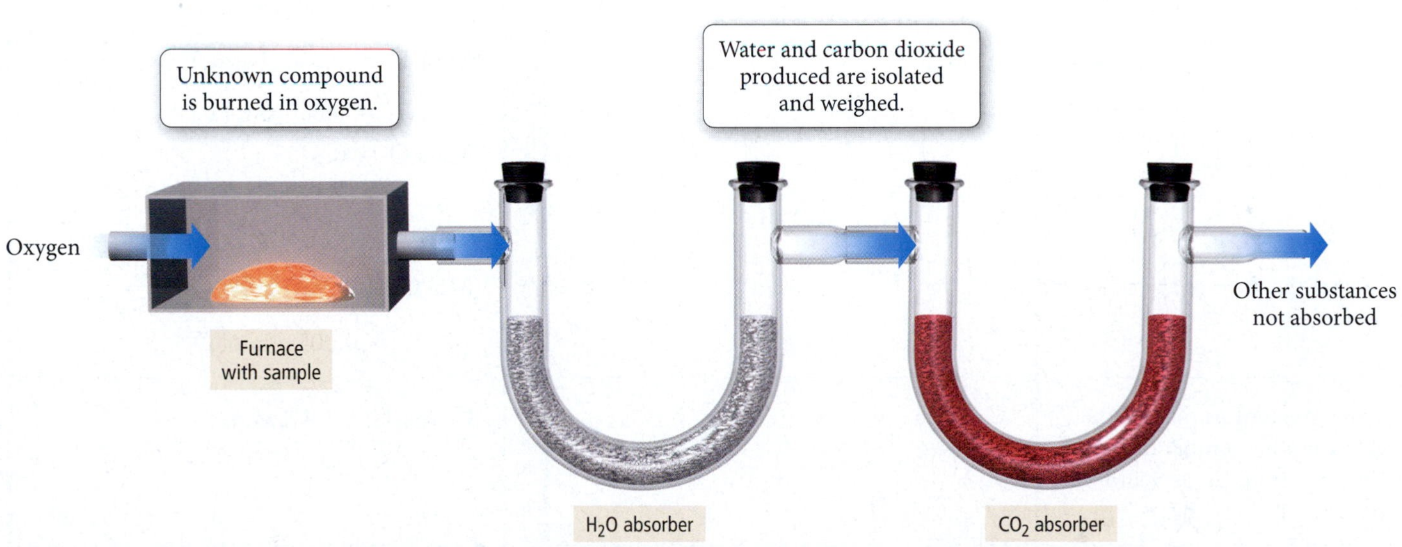
Combustion Analysis
Unknown compound is burned in oxygen.
Water and carbon dioxide produced are isolated and weighed.
Oxygen
Furnace with sample
Other substances not absorbed
H_2O absorber
CO_2 absorber

▲ **FIGURE 5.10 Combustion Analysis Apparatus** The sample to be analyzed is placed in a furnace and burned in oxygen. The water and carbon dioxide produced are absorbed into separate containers and weighed.

PROCEDURE FOR ▾ **Obtaining an Empirical Formula from Combustion Analysis**	EXAMPLE 5.17 **Determining an Empirical Formula from Combustion Analysis**	EXAMPLE 5.18 **Determining an Empirical Formula from Combustion Analysis**
	Upon combustion, a compound containing only carbon and hydrogen produces 1.83 g CO_2 and 0.901 g H_2O. Determine the empirical formula of the compound.	Upon combustion, a 0.8233-g sample of a compound containing only carbon, hydrogen, and oxygen produces 2.445 g CO_2 and 0.6003 g H_2O. Determine the empirical formula of the compound.
1. Write down as *given* the masses of each combustion product and the mass of the sample (if given).	**GIVEN:** 1.83 g CO_2, 0.901 g H_2O **FIND:** empirical formula	**GIVEN:** 0.8233-g sample, 2.445 g CO_2, 0.6003 g H_2O **FIND:** empirical formula
2. Convert the masses of CO_2 and H_2O from Step 1 to moles by using the appropriate molar mass for each compound as a conversion factor.	$1.83\ \text{g } CO_2 \times \frac{1\ \text{mol } CO_2}{44.01\ \text{g } CO_2}$ $= 0.0416\ \text{mol } CO_2$ $0.901\ \text{g } H_2O \times \frac{1\ \text{mol } H_2O}{18.02\ \text{g } H_2O}$ $= 0.0500\ \text{mol } H_2O$	$2.445\ \text{g } CO_2 \times \frac{1\ \text{mol } CO_2}{44.01\ \text{g } CO_2}$ $= 0.05556\ \text{mol } CO_2$ $0.6003\ \text{g } H_2O \times \frac{1\ \text{mol } H_2O}{18.01\ \text{g } H_2O}$ $= 0.03331\ \text{mol } H_2O$
3. Convert the moles of CO_2 and moles of H_2O from Step 2 to moles of C and moles of H using the conversion factors inherent in the chemical formulas of CO_2 and H_2O.	$0.0416\ \text{mol } CO_2 \times \frac{1\ \text{mol C}}{1\ \text{mol } CO_2}$ $= 0.0416\ \text{mol C}$ $0.0500\ \text{mol } H_2O \times \frac{2\ \text{mol H}}{1\ \text{mol } H_2O}$ $= 0.100\ \text{mol H}$	$0.05556\ \text{mol } CO_2 \times \frac{1\ \text{mol C}}{1\ \text{mol } CO_2}$ $= 0.05556\ \text{mol C}$ $0.03331\ \text{mol } H_2O \times \frac{2\ \text{mol H}}{1\ \text{mol } H_2O}$ $= 0.06662\ \text{mol H}$
4. If the compound contains an element other than C and H, find the mass of the other element by subtracting the sum of the masses of C and H (obtained in Step 3) from the mass of the sample. Finally, convert the mass of the other element to moles.	The sample contains no elements other than C and H, so proceed to next step.	$\text{Mass C} = 0.05556\ \text{mol C} \times \frac{12.01\ \text{g C}}{\text{mol C}}$ $= 0.6673\ \text{g C}$ $\text{Mass H} = 0.06662\ \text{mol H} \times \frac{1.008\ \text{g H}}{\text{mol H}}$ $= 0.06715\ \text{g H}$ $\text{Mass O} = 0.8233\ \text{g} - (0.6673\ \text{g} + 0.06715\ \text{g})$ $= 0.0889\ \text{g}$ $\text{Mol O} = 0.0889\ \text{g O} \times \frac{\text{mol O}}{16.00\ \text{g O}}$ $= 0.00556\ \text{mol O}$
5. Write down a pseudoformula for the compound using the number of moles of each element (from Steps 3 and 4) as subscripts.	$C_{0.0416}H_{0.100}$	$C_{0.05556}H_{0.06662}O_{0.00556}$
6. Divide all the subscripts in the formula by the smallest subscript. (Round all subscripts that are within 0.1 of a whole number.)	$C_{\frac{0.0416}{0.0416}}H_{\frac{0.100}{0.0416}} \longrightarrow C_1H_{2.4}$	$C_{\frac{0.05556}{0.00556}}H_{\frac{0.06662}{0.00556}}O_{\frac{0.00556}{0.00556}} \longrightarrow C_{10}H_{12}O_1$

—Continued on the next page

Continued from the previous page—

7. If the subscripts are not whole numbers, multiply all the subscripts by a small whole number to get whole-number subscripts.	$C_1H_{2.4} \times 5 \longrightarrow C_5H_{12}$ The correct empirical formula is C_5H_{12}.	The subscripts are whole numbers; no additional multiplication is needed. The correct empirical formula is $C_{10}H_{12}O$.
	FOR PRACTICE 5.17 Upon combustion, a compound containing only carbon and hydrogen produced 1.60 g CO_2 and 0.819 g H_2O. Find the empirical formula of the compound.	**FOR PRACTICE 5.18** Upon combustion, a 0.8009-g sample of a compound containing only carbon, hydrogen, and oxygen produced 1.6004 g CO_2 and 0.6551 g H_2O. Find the empirical formula of the compound.

5.12 Organic Compounds

In this chapter, we have examined chemical compounds. Early chemists divided compounds into two types: organic and inorganic. They designated organic compounds as those that originate from living things. Sugar—from sugarcane or the sugar beet—is a common example of an organic compound. Inorganic compounds, on the other hand, originate from the Earth. Salt—mined from the ground or from the ocean—is a common example of an inorganic compound.

Eighteenth-century chemists could synthesize inorganic compounds in the laboratory, but not organic compounds, so a clear difference existed between the two different types of compounds. Today, chemists can synthesize both organic and inorganic compounds, and even though organic chemistry is a subfield of chemistry, the differences between organic and inorganic compounds are now viewed as primarily organizational (not fundamental).

Organic compounds are common in everyday substances. Many smells—such as those in perfumes, spices, and foods—are caused by organic compounds. When we sprinkle cinnamon onto our French toast, some cinnamaldehyde—an organic compound present in cinnamon—evaporates into the air and we experience the unique smell of cinnamon. Organic compounds are the major components of living organisms. They are also the main components of most of our fuels, such as gasoline, oil, and natural gas, and they are the active ingredients in most pharmaceuticals, such as aspirin and ibuprofen.

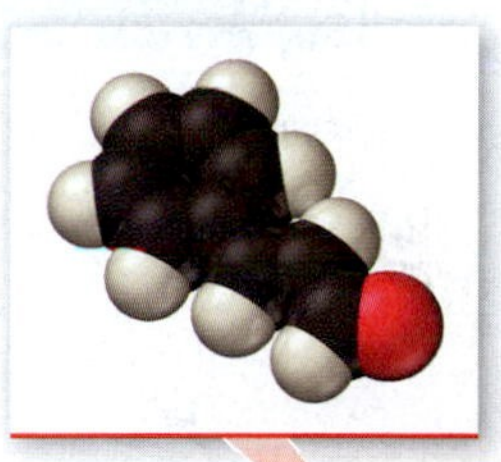

▲ The organic compound cinnamaldehyde is largely responsible for the taste and smell of cinnamon.

Organic compounds are composed of carbon and hydrogen and a few other elements, including nitrogen, oxygen, and sulfur. The key element in organic chemistry, however, is carbon. In its compounds, carbon always forms four bonds. The simplest organic compound is methane or CH_4.

Structural formula

```
    H
    |
H — C — H
    |
    H
```

Space-filling model

Methane, CH_4

The chemistry of carbon is unique and complex because carbon frequently bonds to itself to form chain, branched, and ring structures.

```
    H   H   H
    |   |   |
H — C — C — C — H
    |   |   |
    H   H   H
```

Propane (C_3H_8)

```
    H   H   H
    |   |   |
H — C — C — C — H
    |   |   |
    H   |   H
    H — C — H
        |
        H
```

Isobutane (C_4H_{10})

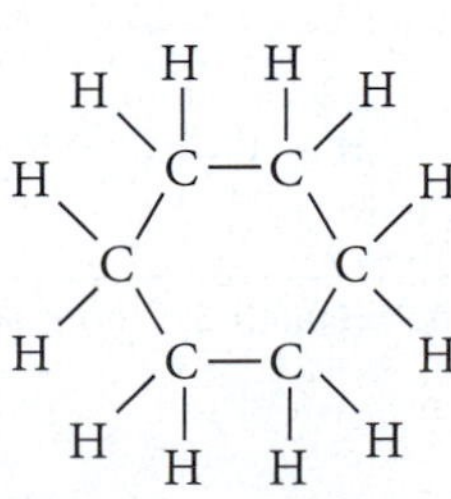

Cyclohexane (C_6H_{12})

Carbon can also form double bonds and triple bonds with itself and with other elements.

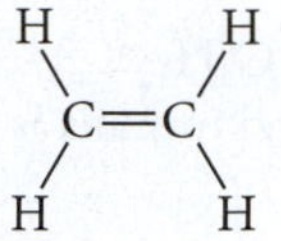

Ethene (C_2H_4) 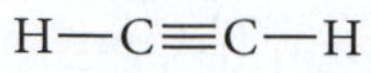 Ethyne (C_2H_2) 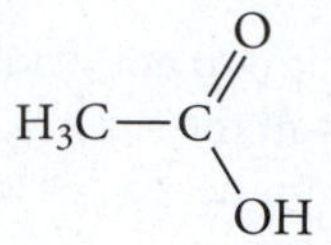 Acetic acid (CH_3COOH)

This versatility allows carbon to serve as the backbone of millions of different chemical compounds, which is why a general survey of organic chemistry is a year-long course. For now, all you really need to know is that the simplest organic compounds are called **hydrocarbons**, and they are composed of carbon and hydrogen. Hydrocarbons compose common fuels such as oil, gasoline, liquid propane gas, and natural gas. Table 5.6 lists some common hydrocarbons. You will recognize some of these compounds from your daily life, and we will often use organic compounds in examples throughout this book.

TABLE 5.6 Common Hydrocarbons

Name	Molecular Formula	Structural Formula	Space-filling Model	Common Uses
Methane	CH_4	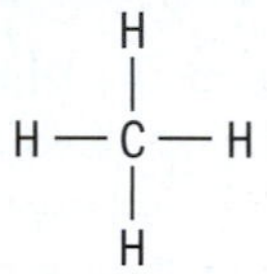		Primary component of natural gas
Propane	C_3H_8	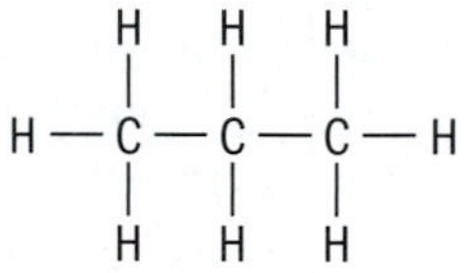	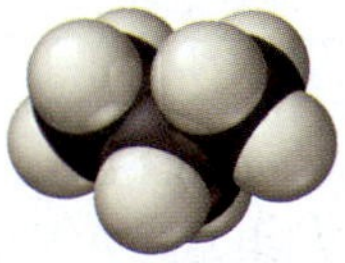	LP gas for grills and outdoor stoves
n-Butane*	C_4H_{10}	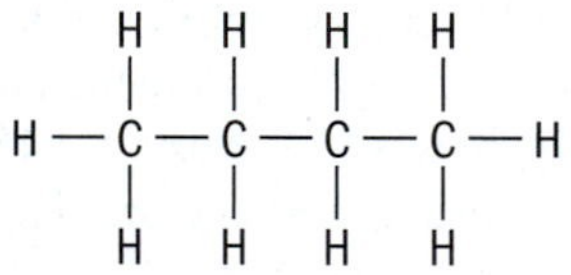		Common fuel for lighters
n-Pentane*	C_5H_{12}	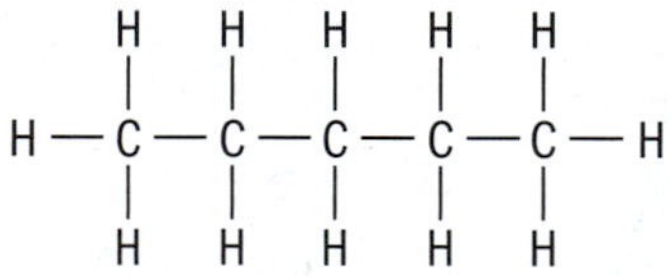		Component of gasoline
Ethene	C_2H_4	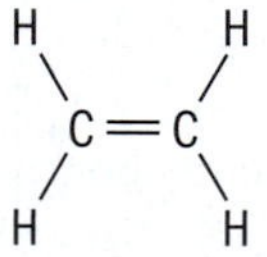	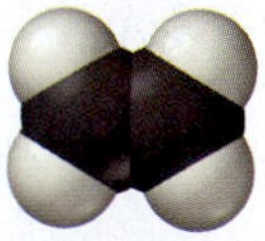	Ripening agent in fruit
Ethyne	C_2H_2	 H—C≡C—H		Fuel for welding torches

*The "*n*" in the names of these hydrocarbons stands for "normal," which means straight chain.

SELF-ASSESSMENT Quiz

1. What is the empirical formula of a compound with the molecular formula $C_{10}H_8$?
   a) C_5H_3
   b) C_2H_4
   c) C_5H_4
   d) CH

2. Which substance is an ionic compound?
   a) SrI_2
   b) N_2O_4
   c) He
   d) CCl_4

3. What is the correct formula for the compound formed between calcium and sulfur?
   a) CaS
   b) Ca_2S
   c) CaS_2
   d) CaS_3

4. Name the compound SrI_2.
   a) strontium iodide
   b) strontium diiodide
   c) strontium(II) iodide
   d) strontium(II) diiodide

5. What is the formula for manganese(IV) oxide?
   a) Mn_4O
   b) MnO_4
   c) Mn_2O
   d) MnO_2

6. Name the compound $Pb(C_2H_3O_2)_2$.
   a) lead(II) carbonate
   b) lead(II) acetate
   c) lead bicarbonate
   d) lead diacetate

7. Name the compound P_2I_4.
   a) phosphorus iodide
   b) phosphorus diiodide
   c) phosphorus(II) iodide
   d) diphosphorus tetraiodide

8. What is the correct Lewis symbol for S?
   a) :S:
   b) $:\ddot{S}\cdot$
   c) $:\ddot{\underset{..}{S}}:$
   d) :S

9. How many CH_2Cl_2 molecules are there in 25.0 g of CH_2Cl_2?
   a) 0.294 molecules
   b) 1.77×10^{23} molecules
   c) 1.28×10^{27} molecules
   d) 1.51×10^{25} molecules

10. List the elements in the compound CF_2Cl_2 in order of decreasing mass percent composition.
    a) $C > F > Cl$
    b) $F > Cl > C$
    c) $Cl > C > F$
    d) $Cl > F > C$

11. Determine the mass of potassium in 35.5 g of KBr.
    a) 17.4 g
    b) 0.298 g
    c) 11.7 g
    d) 32.9 g

12. A compound is 52.14% C, 13.13% H, and 34.73% O by mass. What is the empirical formula of the compound?
    a) $C_2H_8O_3$
    b) C_2H_6O
    c) C_4HO_3
    d) C_3HO_6

13. A compound has the empirical formula CH_2O and a formula mass of 120.10 amu. What is the molecular formula of the compound?
    a) CH_2O
    b) $C_2H_4O_2$
    c) $C_3H_6O_3$
    d) $C_4H_8O_4$

14. Combustion of 30.42 g of a compound containing only carbon, hydrogen, and oxygen produces 35.21 g CO_2 and 14.42 g H_2O. What is the empirical formula of the compound?
    a) $C_4H_8O_6$
    b) $C_2H_4O_3$
    c) $C_2H_2O_3$
    d) C_6HO_{12}

Answers: 1:c; 2:a; 3:a; 4:a; 5:d; 6:b; 7:d; 8:b; 9:b; 10:d; 11:c; 12:b; 13:d; 14:b

CHAPTER SUMMARY 5

REVIEW

KEY LEARNING OUTCOMES

CHAPTER OBJECTIVES	ASSESSMENT
Writing Molecular and Empirical Formulas (5.3)	• Example 5.1 For Practice 5.1 Exercises 31–36
Using Lewis Symbols to Predict the Chemical Formula of an Ionic Compound (5.5)	• Example 5.2 For Practice 5.2 Exercises 43, 44
Writing Formulas for Ionic Compounds (5.6)	• Examples 5.3, 5.4 For Practice 5.3, 5.4 Exercises 47–50
Naming Ionic Compounds (5.6)	• Examples 5.5, 5.6 For Practice 5.5, 5.6 For More Practice 5.5, 5.6 Exercises 51–54
Naming Ionic Compounds Containing Polyatomic Ions (5.6)	• Example 5.7 For Practice 5.7 For More Practice 5.7 Exercises 55, 56
Naming Molecular Compounds (5.8)	• Example 5.8 For Practice 5.8 For More Practice 5.8 Exercises 63–66
Calculating Formula Mass (5.9)	• Example 5.9 For Practice 5.9 Exercises 71, 72
Using Formula Mass to Count Molecules by Weighing (5.9)	• Example 5.10 For Practice 5.10 For More Practice 5.10 Exercises 77–80
Calculating Mass Percent Composition (5.10)	• Example 5.11 For Practice 5.11 For More Practice 5.11 Exercises 83–86
Using Mass Percent Composition as a Conversion Factor (5.10)	• Example 5.12 For Practice 5.12 For More Practice 5.12 Exercises 87–90
Using Chemical Formulas as Conversion Factors (5.10)	• Example 5.13 For Practice 5.13 For More Practice 5.13 Exercises 93–96
Obtaining an Empirical Formula from Experimental Data (5.11)	• Examples 5.14, 5.15 For Practice 5.14, 5.15 Exercises 97–104
Calculating a Molecular Formula from an Empirical Formula and Molar Mass (5.11)	• Example 5.16 For Practice 5.16 For More Practice 5.16 Exercises 105, 106
Determining an Empirical Formula from Combustion Analysis (5.11)	• Examples 5.17, 5.18 For Practice 5.17, 5.18 Exercises 107–110

KEY TERMS

Section 5.2
chemical bond (p. 146)
ionic bond (p. 147)
ionic compound (p. 148)
covalent bond (p. 148)
molecular compound (p. 148)

Section 5.3
chemical formula (p. 148)
empirical formula (p. 148)
molecular formula (p. 148)
structural formula (p. 148)
ball-and-stick molecular model (p. 150)
space-filling molecular model (p. 150)

Section 5.4
Lewis model (p. 150)
lewis electron-dot structure (Lewis structure) (p. 150)
Lewis symbol (p. 151)
octet (p. 152)
duet (p. 152)
octet rule (p. 152)

Section 5.5
formula unit (p. 152)
lattice energy (p. 154)

Section 5.6
common name (p. 156)
systematic name (p. 156)
binary compound (p. 156)
polyatomic ion (p. 158)
oxyanion (p. 160)
hydrate (p. 160)

Section 5.7
bonding pair (p. 161)
lone pair (nonbonding electrons) (p. 161)
double bond (p. 162)
triple bond (p. 162)

Section 5.9
formula mass (p. 165)

Section 5.10
mass percent composition (mass percent) (p. 167)

Section 5.11
empirical formula molar mass (p. 174)
combustion analysis (p. 175)

Section 5.12
organic compound (p. 177)
hydrocarbon (p. 178)

KEY CONCEPTS

Types of Chemical Bonds (5.2)

- Chemical bonds, the forces that hold atoms together in compounds, arise from the interactions between nuclei and electrons in atoms.
- In an ionic bond, one or more electrons are *transferred* from one atom to another, forming a cation (positively charged) and an anion (negatively charged). The two ions are then drawn together by the attraction between the opposite charges.
- In a covalent bond, one or more electrons are *shared* between two atoms. The atoms are held together by the attraction between their nuclei and the shared electrons.

Representing Molecules and Compounds (5.3)

- A compound is represented with a chemical formula, which indicates the elements present and the number of atoms of each.
- An empirical formula gives only the *relative* number of atoms, while a molecular formula gives the *actual* number of atoms present in the molecule.
- Structural formulas show how the atoms are bonded together, while molecular models portray the geometry of the molecule.

The Lewis Model (5.4)

- In the Lewis model, chemical bonds are formed when atoms transfer (ionic bonding) or share (covalent bonding) valence electrons to attain noble gas electron configurations.
- The Lewis model represents valence electrons as dots surrounding the symbol for an element. When two or more elements bond together, the dots are transferred or shared so that every atom gets eight dots, an octet (or two dots, a duet, in the case of hydrogen).

Ionic Bonding (5.5)

- In an ionic Lewis structure involving main-group metals, the metal transfers its valence electrons (dots) to the nonmetal.
- The formation of most ionic compounds is exothermic because of lattice energy, the energy released when metal cations and nonmetal anions coalesce to form the solid.

Covalent Bonding (5.7)

- In a covalent Lewis structure, neighboring atoms share valence electrons to attain octets (or duets).
- A single shared electron pair constitutes a single bond, while two or three shared pairs constitute double or triple bonds, respectively.

Naming Inorganic Ionic and Molecular Compounds (5.6, 5.8)

- A flowchart for naming simple inorganic compounds is provided at right.

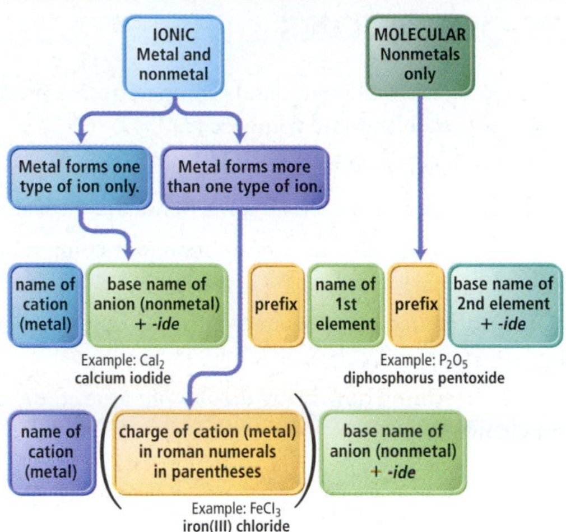

Formula Mass and Mole Concept for Compounds (5.9)

- The formula mass of a compound is the sum of the atomic masses of all the atoms in the chemical formula. Like the atomic masses of elements, the formula mass characterizes the average mass of a molecule (or a formula unit).
- The mass of one mole of a compound is its molar mass and equals its formula mass (in grams).

Chemical Composition (5.10, 5.11)

- The mass percent composition of a compound indicates each element's percentage of the total compound's mass. We can determine the mass percent composition from the compound's chemical formula and the molar masses of its elements.
- The chemical formula of a compound provides the relative number of atoms (or moles) of each element in a compound, and therefore we can use it to determine numerical relationships between moles of the compound and moles of its constituent elements.
- If we know the mass percent composition and molar mass of a compound, we can determine its empirical and molecular formulas.

Organic Compounds (5.12)

- Organic compounds are composed of carbon, hydrogen, and a few other elements such as nitrogen, oxygen, and sulfur.
- The simplest organic compounds are hydrocarbons, compounds composed of only carbon and hydrogen.

KEY EQUATIONS AND RELATIONSHIPS

Formula Mass (5.9)

$$\left(\begin{array}{c}\text{\# atoms of 1st element}\\ \text{in chemical formula}\end{array} \times \begin{array}{c}\text{atomic mass}\\ \text{of 1st element}\end{array}\right) + \left(\begin{array}{c}\text{\# atoms of 2nd element}\\ \text{in chemical formula}\end{array} \times \begin{array}{c}\text{atomic mass}\\ \text{of 2nd element}\end{array}\right) + \ldots$$

Mass Percent Composition (5.10)

$$\text{Mass \% of element X} = \frac{\text{mass of X in 1 mol compound}}{\text{mass of 1 mol compound}} \times 100\%$$

Empirical Formula Molar Mass (5.11)

$$\text{Molecular formula} = n \times (\text{empirical formula})$$

$$n = \frac{\text{molar mass}}{\text{empirical formula molar mass}}$$

EXERCISES

REVIEW QUESTIONS

1. How do the properties of compounds compare to the properties of the elements from which they are composed?
2. What is a chemical bond? Why do chemical bonds form?
3. Explain the difference between an ionic bond and a covalent bond.
4. List and describe the different ways to represent compounds. Why are there so many?
5. What is the difference between an empirical formula and a molecular formula?
6. How do you determine how many dots to put around the Lewis symbol of an element?
7. Describe the octet rule in the Lewis model.
8. According to the Lewis model, what is a chemical bond?
9. How can you use Lewis structures to determine the formula of ionic compounds? Give an example.
10. What is lattice energy?
11. Why is the formation of solid sodium chloride from solid sodium and gaseous chlorine exothermic, even though it takes more energy to form the Na^+ ion than the amount of energy released upon formation of Cl^-?
12. Explain how to write a formula for an ionic compound given the names of the metal and nonmetal (or polyatomic ion) in the compound.
13. Explain how to name binary ionic compounds. How do you name an ionic compound if it contains a polyatomic ion?
14. Why do the names of some ionic compounds include the charge of the metal ion while others do not?
15. Within a covalent Lewis structure, what is the difference between lone pair and bonding pair electrons?
16. In what ways are double and triple covalent bonds different from single covalent bonds?
17. How does the Lewis model for covalent bonding account for why certain combinations of atoms are stable while others are not?
18. How does the Lewis model for covalent bonding account for the relatively low melting and boiling points of molecular compounds (compared to ionic compounds)?
19. Explain how to name molecular inorganic compounds.
20. How many atoms are specified by each of these prefixes: mono, di, tri, tetra, penta, hexa?
21. What is the formula mass for a compound? Why is it useful?
22. Explain how the information in a chemical formula can be used to determine how much of a particular element is present in a given amount of a compound. Provide some examples of how this might be useful.
23. What is mass percent composition? Why is it useful?
24. Which kinds of conversion factors are inherent in chemical formulas? Provide an example.
25. Which kind of chemical formula can be obtained from experimental data showing the relative masses of the elements in a compound?
26. How can a molecular formula be obtained from an empirical formula? What additional information is required?
27. What is combustion analysis? What is it used for?
28. Which elements are normally present in organic compounds?

PROBLEMS BY TOPIC

Note: Answers to all odd-numbered Problems, numbered in blue, can be found in Appendix V. Exercises in the Problems by Topic section are paired, with each odd-numbered problem followed by a similar even-numbered problem. Exercises in the Cumulative Problems section are also paired, but somewhat more loosely. (Challenge Problems and Conceptual Problems, because of their nature, are unpaired.)

Types of Compounds and Chemical Formulas

29. Classify each compound as ionic or molecular.
    a. CO_2 **b.** $NiCl_2$ **c.** NaI **d.** PCl_3
30. Classify each compound as ionic or molecular.
    a. CF_2Cl_2 **b.** CCl_4 **c.** PtO_2 **d.** SO_3
31. Determine the empirical formula for the compound represented by each molecular formula.
    a. N_2O_4 **b.** C_5H_{12} **c.** C_4H_{10}
32. Determine the empirical formula for the compound represented by each molecular formula.
    a. C_2H_4 **b.** $C_6H_{12}O_6$ **c.** NH_3
33. Determine the number of each type of atom in each formula.
    a. $Mg_3(PO_4)_2$ **b.** $BaCl_2$ **c.** $Fe(NO_2)_2$ **d.** $Ca(OH)_2$
34. Determine the number of each type of atom in each formula.
    a. $Ca(NO_2)_2$ **b.** $CuSO_4$ **c.** $Al(NO_3)_3$ **d.** $Mg(HCO_3)_2$
35. Write a chemical formula for each molecular model. (See Appendix IV A for color codes.)

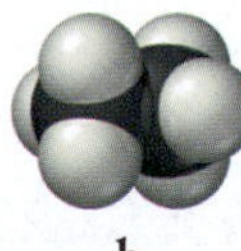
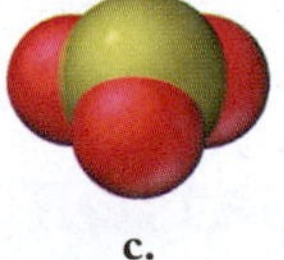

a. **b.** **c.**

36. Write a chemical formula for each molecular model. (See Appendix IV A for color codes.)

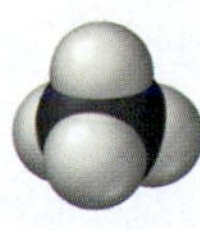

a. **b.** **c.**

Valence Electrons and Lewis Dot Structures

37. Write an electron configuration for N. Then write a Lewis symbol for N and show which electrons from the electron configuration are included in the Lewis symbol.
38. Write an electron configuration for Ne. Then write a Lewis symbol for Ne and show which electrons from the electron configuration are included in the Lewis symbol.
39. Write a Lewis symbol for each atom or ion.
    a. Al b. Na^+ c. Cl d. Cl^-
40. Write a Lewis symbol for each atom or ion.
    a. S^{2-} b. Mg c. Mg^{2+} d. P

Ionic Bonding and Lattice Energy

41. Write the Lewis symbols that represent the ions in each ionic compound.
    a. NaF b. CaO c. $SrBr_2$ d. K_2O
42. Write the Lewis symbols that represent the ions in each ionic compound.
    a. SrO b. Li_2S c. CaI_2 d. RbF
43. Use Lewis symbols to determine the formula for the compound that forms between each pair of elements.
    a. Sr and Se b. Ba and Cl c. Na and S d. Al and O
44. Use Lewis symbols to determine the formula for the compound that forms between each pair of elements.
    a. Ca and N b. Mg and I c. Ca and S d. Cs and F
45. The lattice energy of CsF is −744 kJ/mol, whereas that of BaO is −3029 kJ/mol. Explain this large difference in lattice energy.
46. Rubidium iodide has a lattice energy of −617 kJ/mol, while potassium bromide has a lattice energy of −671 kJ/mol. Why is the lattice energy of potassium bromide more exothermic than the lattice energy of rubidium iodide?

Formulas and Names for Ionic Compounds

47. Write a formula for the ionic compound that forms between each pair of elements.
    a. calcium and oxygen b. zinc and sulfur
    c. rubidium and bromine d. aluminum and oxygen
48. Write a formula for the ionic compound that forms between each pair of elements.
    a. silver and chlorine b. sodium and sulfur
    c. aluminum and sulfur d. potassium and chlorine
49. Write a formula for the compound that forms between calcium and each polyatomic ion.
    a. hydroxide b. chromate
    c. phosphate d. cyanide
50. Write a formula for the compound that forms between potassium and each polyatomic ion.
    a. carbonate b. phosphate
    c. hydrogen phosphate d. acetate
51. Name each ionic compound.
    a. Mg_3N_2 b. KF c. Na_2O
    d. Li_2S e. CsF f. KI
52. Name each ionic compound.
    a. $SnCl_4$ b. PbI_2 c. Fe_2O_3
    d. CuI_2 e. $HgBr_2$ f. $CrCl_2$
53. Name each ionic compound.
    a. SnO b. Cr_2S_3 c. RbI d. $BaBr_2$
54. Name each ionic compound.
    a. BaS b. $FeCl_3$ c. PbI_4 d. $SrBr_2$
55. Name each ionic compound containing a polyatomic ion.
    a. $CuNO_2$ b. $Mg(C_2H_3O_2)_2$
    c. $Ba(NO_3)_2$ d. $Pb(C_2H_3O_2)_2$
56. Name each ionic compound containing a polyatomic ion.
    a. $Ba(OH)_2$ b. NH_4I c. $NaBrO_4$ d. $Fe(OH)_3$
57. Write the formula for each ionic compound.
    a. sodium hydrogen sulfite
    b. lithium permanganate
    c. silver nitrate
    d. potassium sulfate
    e. rubidium hydrogen sulfate
    f. potassium hydrogen carbonate
58. Write the formula for each ionic compound.
    a. copper(II) chloride
    b. copper(I) iodate
    c. lead(II) chromate
    d. calcium fluoride
    e. potassium hydroxide
    f. iron(II) phosphate
59. Write the name from the formula or the formula from the name for each hydrated ionic compound.
    a. $CoSO_4 \cdot 7H_2O$
    b. iridium(III) bromide tetrahydrate
    c. $Mg(BrO_3)_2 \cdot 6H_2O$
    d. potassium carbonate dihydrate
60. Write the name from the formula or the formula from the name for each hydrated ionic compound.
    a. cobalt(II) phosphate octahydrate
    b. $BeCl_2 \cdot 2H_2O$
    c. chromium(III) phosphate trihydrate
    d. $LiNO_2 \cdot H_2O$

Simple Lewis Structures, Formulas, and Names for Molecular Compounds

61. Use covalent Lewis structures to explain why each element (or family of elements) occurs as diatomic molecules.
    a. hydrogen b. the halogens
    c. oxygen d. nitrogen
62. Use covalent Lewis structures to explain why the compound that forms between nitrogen and hydrogen has the formula NH_3. Show why NH_2 and NH_4 are not stable.
63. Name each molecular compound.
    a. CO b. NI_3 c. $SiCl_4$ d. N_4Se_4
64. Name each molecular compound.
    a. SO_3 b. SO_2 c. BrF_5 d. NO
65. Write a formula for each molecular compound.
    a. phosphorus trichloride
    b. chlorine monoxide
    c. disulfur tetrafluoride
    d. phosphorus pentafluoride

66. Write a formula for each molecular compound.

a. boron tribromide **b.** dichlorine monoxide
c. xenon tetrafluoride **d.** carbon tetrabromide

Naming Compounds (When the Type Is Not Specified)

67. Name each compound. (Refer to the nomenclature flowchart found in the Key Concepts section of the Chapter in Review.)

a. $SrCl_2$ **b.** SnO_2 **c.** P_2S_5

68. Name each compound. (Refer to the nomenclature flowchart found in the Key Concepts section of the Chapter in Review.)

a. B_2Cl_2 **b.** $BaCl_2$ **c.** $CrCl_3$

69. Name each compound. (Refer to the nomenclature flowchart found in the Key Concepts section of the Chapter in Review.)

a. $KClO_3$ **b.** I_2O_5 **c.** $PbSO_4$

70. Name each compound. (Refer to the nomenclature flowchart found in the Key Concepts section of the Chapter in Review.)

a. XeO_3 **b.** $KClO$ **c.** $CoSO_4$

Formula Mass and the Mole Concept for Compounds

71. Calculate the formula mass for each compound.

a. NO_2 **b.** C_4H_{10} **c.** $C_6H_{12}O_6$ **d.** $Cr(NO_3)_3$

72. Calculate the formula mass for each compound.

a. $MgBr_2$ **b.** HNO_2 **c.** CBr_4 **d.** $Ca(NO_3)_2$

73. Calculate the number of moles in each sample.

a. 72.5 g CCl_4 **b.** 12.4 g $C_{12}H_{22}O_{11}$
c. 25.2 kg C_2H_2 **d.** 12.3 g of dinitrogen monoxide

74. Calculate the mass of each sample.

a. 15.7 mol HNO_3
b. 1.04×10^{-3} mol H_2O_2
c. 72.1 mmol SO_2
d. 1.23 mol xenon difluoride

75. Determine the number of moles (of molecules or formula units) in each sample.

a. 25.5 g NO_2 **b.** 1.25 kg CO_2
c. 38.2 g KNO_3 **d.** 155.2 kg Na_2SO_4

76. Determine the number of moles (of molecules or formula units) in each sample.

a. 55.98 g CF_2Cl_2 **b.** 23.6 kg $Fe(NO_3)_2$
c. 0.1187 g C_8H_{18} **d.** 195 kg CaO

77. How many molecules are in each sample?

a. 6.5 g H_2O **b.** 389 g CBr_4
c. 22.1 g O_2 **d.** 19.3 g C_8H_{10}

78. How many molecules (or formula units) are in each sample?

a. 85.26 g CCl_4 **b.** 55.93 kg $NaHCO_3$
c. 119.78 g C_4H_{10} **d.** 4.59×10^5 g Na_3PO_4

79. Calculate the mass (in g) of each sample.

a. 5.94×10^{20} SO_3 molecules
b. 2.8×10^{22} H_2O molecules
c. 1 glucose molecule ($C_6H_{12}O_6$)

80. Calculate the mass (in g) of each sample.

a. 4.5×10^{25} O_3 molecules
b. 9.85×10^{19} CCl_2F_2 molecules
c. 1 water molecule

81. A sugar crystal contains approximately 1.8×10^{17} sucrose ($C_{12}H_{22}O_{11}$) molecules. What is its mass in mg?

82. A salt crystal has a mass of 0.12 mg. How many NaCl formula units does it contain?

Composition of Compounds

83. Calculate the mass percent composition of carbon in each carbon-containing compound.

a. CH_4 **b.** C_2H_6 **c.** C_2H_2 **d.** C_2H_5Cl

84. Calculate the mass percent composition of nitrogen in each nitrogen-containing compound.

a. N_2O **b.** NO **c.** NO_2 **d.** HNO_3

85. Most fertilizers consist of nitrogen-containing compounds such as NH_3, $CO(NH_2)_2$, NH_4NO_3, and $(NH_4)_2SO_4$. Plants use the nitrogen content in these compounds for protein synthesis. Calculate the mass percent composition of nitrogen in each of the fertilizers named above. Which fertilizer has the highest nitrogen content?

86. Iron in the earth is in the form of iron ore. Common ores include Fe_2O_3 (hematite), Fe_3O_4 (magnetite), and $FeCO_3$ (siderite). Calculate the mass percent composition of iron for each of these iron ores. Which ore has the highest iron content?

87. Copper(II) fluoride contains 37.42% F by mass. Calculate the mass of fluorine (in g) contained in 55.5 g of copper(II) fluoride.

88. Silver chloride, often used in silver plating, contains 75.27% Ag by mass. Calculate the mass of silver chloride required to plate 155 mg of pure silver.

89. The iodide ion is a dietary mineral essential to good nutrition. In countries where potassium iodide is added to salt, iodine deficiency (goiter) has been almost completely eliminated. The recommended daily allowance (RDA) for iodine is 150 μg/day. How much potassium iodide (76.45% I) must you consume to meet the RDA?

90. The American Dental Association recommends that an adult female should consume 3.0 mg of fluoride (F^-) per day to prevent tooth decay. If the fluoride is consumed in the form of sodium fluoride (45.24% F), what amount of sodium fluoride contains the recommended amount of fluoride?

91. Write a ratio showing the relationship between the molar amounts of each element for each compound.

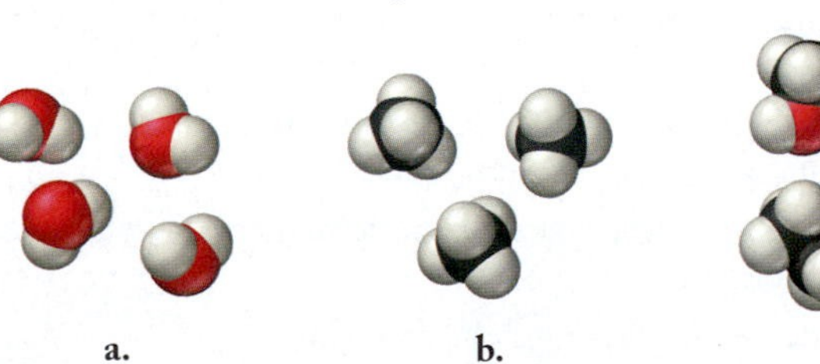

92. Write a ratio showing the relationship between the molar amounts of each element for each compound.

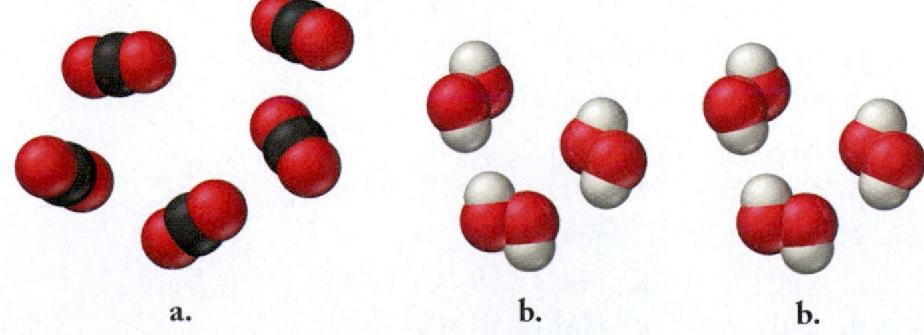

93. Determine the number of moles of hydrogen atoms in each sample.

a. 0.0885 mol C_4H_{10}
b. 1.3 mol CH_4
c. 2.4 mol C_6H_{12}
d. 1.87 mol C_8H_{18}

94. Determine the number of moles of oxygen atoms in each sample.

a. 4.88 mol H_2O_2
b. 2.15 mol N_2O
c. 0.0237 mol H_2CO_3
d. 24.1 mol CO_2

95. Calculate mass (in grams) of sodium in 8.5 g of each sodium-containing food additive.

a. NaCl (table salt)
b. Na_3PO_4 (sodium phosphate)
c. $NaC_7H_5O_2$ (sodium benzoate)
d. $Na_2C_6H_6O_7$ (sodium hydrogen citrate)

96. Calculate the mass (in kilograms) of chlorine in 25 kg of each chlorofluorocarbon (CFC).

a. CF_2Cl_2 b. $CFCl_3$ c. $C_2F_3Cl_3$ d. CF_3Cl

Chemical Formulas from Experimental Data

97. A chemist decomposes samples of several compounds; the masses of their constituent elements are shown below. Calculate the empirical formula for each compound.

a. 1.651 g Ag, 0.1224 g O
b. 0.672 g Co, 0.569 g As, 0.486 g O
c. 1.443 g Se, 5.841 g Br

98. A chemist decomposes samples of several compounds; the masses of their constituent elements are shown below. Calculate the empirical formula for each compound.

a. 1.245 g Ni, 5.381 g I
b. 2.677 g Ba, 3.115 g Br
c. 2.128 g Be, 7.557 g S, 15.107 g O

99. Calculate the empirical formula for each stimulant based on its elemental mass percent composition.

a. nicotine (found in tobacco leaves): C 74.03%, H 8.70%, N 17.27%
b. caffeine (found in coffee beans): C 49.48%, H 5.19%, N 28.85%, O 16.48%

100. Calculate the empirical formula for each natural flavor based on its elemental mass percent composition.

a. methyl butyrate (component of apple taste and smell): C 58.80%, H 9.87%, O 31.33%
b. vanillin (responsible for the taste and smell of vanilla): C 63.15%, H 5.30%, O 31.55%

101. The elemental mass percent composition of ibuprofen is 75.69% C, 8.80% H, and 15.51% O. Determine the empirical formula of ibuprofen.

102. The elemental mass percent composition of ascorbic acid (vitamin C) is 40.92% C, 4.58% H, and 54.50% O. Determine the empirical formula of ascorbic acid.

103. A 0.77-mg sample of nitrogen reacts with chlorine to form 6.61 mg of the chloride. Determine the empirical formula of nitrogen chloride.

104. A 45.2-mg sample of phosphorus reacts with selenium to form 131.6 mg of the selenide. Determine the empirical formula of phosphorus selenide.

105. The empirical formula and molar mass of several compounds are listed below. Find the molecular formula of each compound.

a. C_6H_7N, 186.24 g/mol
b. C_2HCl, 181.44 g/mol
c. $C_5H_{10}NS_2$, 296.54 g/mol

106. The molar mass and empirical formula of several compounds are listed below. Find the molecular formula of each compound.

a. C_4H_9, 114.22 g/mol
b. CCl, 284.77 g/mol
c. C_3H_2N, 312.29 g/mol

107. Combustion analysis of a hydrocarbon produced 33.01 g CO_2 and 13.51 g H_2O. Calculate the empirical formula of the hydrocarbon.

108. Combustion analysis of naphthalene, a hydrocarbon used in mothballs, produced 8.80 g CO_2 and 1.44 g H_2O. Calculate the empirical formula for naphthalene.

109. The foul odor of rancid butter is due largely to butyric acid, a compound containing carbon, hydrogen, and oxygen. Combustion analysis of a 4.30-g sample of butyric acid produces 8.59 g CO_2 and 3.52 g H_2O. Determine the empirical formula for butyric acid.

110. Tartaric acid is the white, powdery substance that coats tart candies such as Sour Patch Kids™. Combustion analysis of a 12.01-g sample of tartaric acid—which contains only carbon, hydrogen, and oxygen—produces 14.08 g CO_2 and 4.32 g H_2O. Determine the empirical formula for tartaric acid.

Organic Compounds

111. Classify each compound as organic or inorganic.

a. $CaCO_3$ b. C_4H_8 c. $C_4H_6O_6$ d. LiF

112. Classify each compound as organic or inorganic.

a. C_8H_{18} b. CH_3NH_2 c. CaO d. $FeCO_3$

113. Determine whether each compound is a hydrocarbon.

a. H_3C-CH_2OH
b. H_3C-CH_3
c. 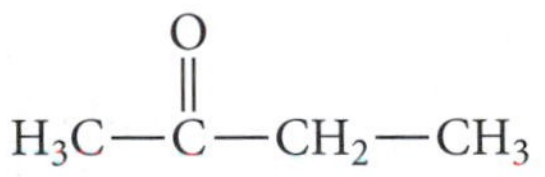 $H_3C-\overset{\overset{\displaystyle O}{\|}}{C}-CH_2-CH_3$
d. H_3C-NH_2

114. Determine whether each compound is a hydrocarbon.

a. 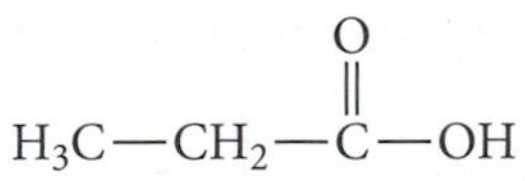 $H_3C-CH_2-\overset{\overset{\displaystyle O}{\|}}{C}-OH$
b. $H_3C-\overset{\overset{\displaystyle O}{\|}}{C}H$
c. $H_3C-\underset{\underset{\displaystyle CH_3}{|}}{\overset{\overset{\displaystyle CH_3}{|}}{C}}-CH_3$
d. $H_3C-CH_2-O-CH_3$

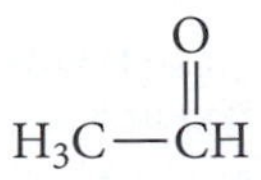

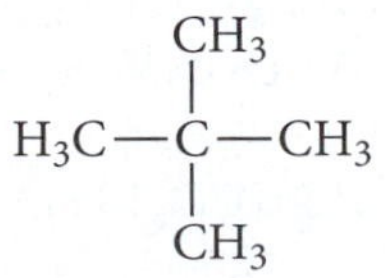

CUMULATIVE PROBLEMS

115. How many molecules of ethanol (C_2H_5OH) (the alcohol in alcoholic beverages) are present in 145 mL of ethanol? The density of ethanol is 0.789 g/cm^3.

116. A drop of water has a volume of approximately 0.05 mL. How many water molecules does it contain? The density of water is 1.0 g/cm^3.

117. Determine the chemical formula of each compound and use it to calculate the mass percent composition of each constituent element.

a. potassium chromate
b. lead(II) phosphate
c. cobalt(II) bromide

118. Determine the chemical formula of each compound and use it to calculate the mass percent composition of each constituent element.

a. phosphorus pentachloride
b. nitrogen triiodide
c. carbon dioxide

119. A Freon™ leak in the air conditioning system of an old car releases 25 g of CF_2Cl_2 per month. What mass of chlorine does this car emit into the atmosphere each year?

120. A Freon™ leak in the air conditioning system of a large building releases 12 kg of CHF_2Cl per month. If the leak is allowed to continue, how many kilograms of Cl are emitted into the atmosphere each year?

121. A metal (M) forms a compound with the formula MCl_3. If the compound contains 65.57% Cl by mass, what is the identity of the metal?

122. A metal (M) forms an oxide with the formula M_2O. If the oxide contains 16.99% O by mass, what is the identity of the metal?

123. Estradiol is a female sexual hormone that causes maturation and maintenance of the female reproductive system. Elemental analysis of estradiol gives the following mass percent composition: C 79.37%, H 8.88%, O 11.75%. The molar mass of estradiol is 272.37 g/mol. Find the molecular formula of estradiol.

124. Fructose is a common sugar found in fruit. Elemental analysis of fructose gives the following mass percent composition: C 40.00%, H 6.72%, O 53.28%. The molar mass of fructose is 180.16 g/mol. Find the molecular formula of fructose.

125. Combustion analysis of a 13.42-g sample of equilin (which contains only carbon, hydrogen, and oxygen) produces 39.61 g CO_2 and 9.01 g H_2O. The molar mass of equilin is 268.34 g/mol. Find its molecular formula.

126. Estrone, which contains only carbon, hydrogen, and oxygen, is a female sexual hormone that occurs in the urine of pregnant women. Combustion analysis of a 1.893-g sample of estrone produces 5.545 g of CO_2 and 1.388 g H_2O. The molar mass of estrone is 270.36 g/mol. Find its molecular formula.

127. Epsom salts is a hydrated ionic compound with the following formula: $MgSO_4 \cdot xH_2O$. A 4.93-g sample of Epsom salts was heated to drive off the water of hydration. The mass of the sample after complete dehydration was 2.41 g. Find the number of waters of hydration (x) in Epsom salts.

128. A hydrate of copper(II) chloride has the following formula: $CuCl_2 \cdot xH_2O$. The water in a 3.41-g sample of the hydrate is driven off by heating. The remaining sample has a mass of 2.69 g. Find the number of waters of hydration (x) in the hydrate.

129. A compound of molar mass 177 g/mol contains only carbon, hydrogen, bromine, and oxygen. Analysis reveals that the compound contains 8 times as much carbon as hydrogen by mass. Find the molecular formula.

130. Researchers obtain the following data from experiments to find the molecular formula of benzocaine, a local anesthetic, which contains only carbon, hydrogen, nitrogen, and oxygen. Complete combustion of a 3.54-g sample of benzocaine with excess O_2 forms 8.49 g of CO_2 and 2.14 g H_2O. Another sample of mass 2.35 g is found to contain 0.199 g of N. The molar mass of benzocaine is found to be 165 g/mol. Find the molecular formula of benzocaine.

131. Find the total number of atoms in a sample of cocaine hydrochloride, $C_{17}H_{22}ClNO_4$, of mass 23.5 mg.

132. Vanadium forms four different oxides in which the percent by mass of vanadium is respectively (a) 76%, (b) 68%, (c) 61%, and (d) 56%. Determine the formula and the name of each one of these oxides.

133. The chloride of an unknown metal is believed to have the formula MCl_3. A 2.395-g sample of the compound contains 3.606×10^{-2} mol Cl. Find the atomic mass of M.

134. Write the structural formulas of two different compounds that each has the molecular formula C_4H_{10}.

135. A chromium-containing compound has the formula $Fe_xCr_yO_4$ and is 28.59% oxygen by mass. Find x and y.

136. A phosphorus compound that contains 34.00% phosphorus by mass has the formula X_3P_2. Identify the element X.

137. A particular brand of beef jerky contains 0.0552% sodium nitrite by mass and is sold in an 8.00-oz bag. What mass of sodium does the sodium nitrite contribute to sodium content of the bag of beef jerky?

138. Phosphorus is obtained primarily from ores containing calcium phosphate. If a particular ore contains 57.8% calcium phosphate, what minimum mass of the ore must be processed to obtain 1.00 kg of phosphorus?

CHALLENGE PROBLEMS

139. A mixture of NaCl and NaBr has a mass of 2.00 g and contains 0.75 g of Na. What is the mass of NaBr in the mixture?

140. Three pure compounds form when 1.00-g samples of element X combine with, respectively, 0.472 g, 0.630 g, and 0.789 g of element Z. The first compound has the formula X_2X_3. Find the empirical formulas of the other two compounds.

141. A mixture of $CaCO_3$ and $(NH_4)_2CO_3$ is 61.9% CO_3 by mass. Find the mass percent of $CaCO_3$ in the mixture.

142. A mixture of 50.0 g of S and 1.00×10^2 g of Cl_2 reacts completely to form S_2Cl_2 and SCl_2. What mass of S_2Cl_2 forms?

143. Because of increasing evidence of damage to the ozone layer, chlorofluorocarbon (CFC) production was banned in 1996. However, there are about 100 million auto air conditioners in operation that still use CFC-12 (CF_2Cl_2). These air conditioners are recharged from stockpiled supplies of CFC-12. If each of the 100 million automobiles contains 1.1 kg of CFC-12 and leaks 25% of its CFC-12 into the atmosphere per year, how much chlorine, in kg, is added to the atmosphere each year due to auto air conditioners? (Assume two significant figures in your calculations.)

144. A particular coal contains 2.55% sulfur by mass. When the coal is burned, it produces SO_2 emissions, which combine with rainwater to produce sulfuric acid. Use the formula of sulfuric acid to calculate the mass percent of S in sulfuric acid. Then determine how much sulfuric acid (in metric tons) is produced by the combustion of 1.0 metric ton of this coal. (A metric ton is 1000 kg.)

145. Lead is found in Earth's crust as several different lead ores. Suppose a certain rock is 38.0% PbS (galena), 25.0% $PbCO_3$ (cerussite), and 17.4% $PbSO_4$ (anglesite). The remainder of the rock is composed of substances containing no lead. How much of this rock (in kg) must be processed to obtain 5.0 metric tons of lead? (A metric ton is 1000 kg.)

146. A 2.52-g sample of a compound containing only carbon, hydrogen, nitrogen, oxygen, and sulfur is burned in excess O to yield 4.23 g of CO_2 and 1.01 g of H_2O. Another sample of the same compound, of mass 4.14 g, yields 2.11 g of SO_3. A third sample, of mass 5.66 g, yields 2.27 g of HNO_3. Calculate the empirical formula of the compound.

147. A compound of molar mass 229 contains only carbon, hydrogen, iodine, and sulfur. Analysis shows that a sample of the compound contains 6 times as much carbon as hydrogen, by mass. Calculate the molecular formula of the compound.

148. The elements X and Y form a compound that is 40% X and 60% Y by mass. The atomic mass of X is twice that of Y. What is the empirical formula of the compound?

149. A compound of X and Y is $\frac{1}{3}$ X by mass. The atomic mass of element X is one-third the atomic mass of element Y. Find the empirical formula of the compound.

150. A mixture of carbon and sulfur has a mass of 9.0 g. Complete combustion with excess O_2 gives 23.3 g of a mixture of CO_2 and SO_2. Find the mass of sulfur in the original mixture.

CONCEPTUAL PROBLEMS

151. When molecules are represented by molecular models, what does each sphere represent? How big is the nucleus of an atom in comparison to the sphere used to represent an atom in a molecular model?

152. Without doing any calculations, determine which element in each of the compounds will have the highest mass percent composition.

a. CO
b. N_2O
c. $C_6H_{12}O_6$
d. NH_3

153. Explain the problem with this statement and correct it: "The chemical formula for ammonia (NH_3) indicates that ammonia contains three grams of hydrogen to each gram of nitrogen."

154. Without doing any calculations, arrange the elements in H_2SO_4 in order of decreasing mass percent composition.

ANSWERS TO CONCEPTUAL CONNECTIONS

Cc 5.1 H—O—H

Cc 5.2 The spheres represent the electron cloud of the atom. It would be nearly impossible to draw a nucleus to scale on any of the space-filling molecular models in this book—on this scale, the nucleus would be too small to see.

Cc 5.3 You would expect MgO to have the higher melting point because, in our bonding model, the magnesium and oxygen ions are held together in a crystalline lattice by charges of 2+ for magnesium and 2− for oxygen. In contrast, the NaCl lattice is held together by charges of 1+ for sodium and 1− for chlorine. According to Coloumb's law, as long as the spacing between the cation and the anion in the two compounds is not much different, the higher charges in MgO should result in lower potential energy (more stability), and therefore a higher melting point. The experimentally measured melting points of these compounds are 801 °C for NaCl and 2852 °C for MgO, in accordance with the model.

Cc 5.4 **(a)** NO_2^-; **(b)** SO_4^{2-}; **(c)** NO_3^-

Cc 5.5 The reasons that atoms form bonds are complex. One contributing factor is the lowering of their potential energy. The octet rule is just a handy way to predict the combinations of atoms that will have a lower potential energy when they bond together.

Cc 5.6 Choice **(a)** best describes the difference between ionic and molecular compounds. The **(b)** answer is incorrect because there are no "new" forces in bonding (just rearrangements that result in lower potential energy), and because ions do not group together in pairs in the solid phase. The **(c)** answer is incorrect because the main difference between ionic and molecular compounds is the way that the atoms bond. The **(d)** answer is incorrect because ionic compounds do not contain molecules.

Cc 5.7 This conceptual connection addresses one of the main errors you can make in nomenclature: the failure to correctly categorize the compound. Remember that you must first determine whether the compound is an ionic compound or a molecular compound, and then name it accordingly. NCl_3 is a molecular compound (two or more nonmetals), and therefore you must use prefixes to indicate the number of each type of atom—so NCl_3 is nitrogen trichloride. The compound $AlCl_3$, however, is an ionic compound (metal and nonmetal), and therefore does not require prefixes—so $AlCl_3$ is aluminum chloride.

Cc 5.8 **(c)** Atomic radii range in the hundreds of picometers, while the spheres in these models have radii of less than a centimeter. The scaling factor is therefore about 10^8 (100 million).

Cc 5.9 C > O > H. Since carbon and oxygen differ in atomic mass by only 4 amu, and since there are six carbon atoms in the formula, we can conclude that carbon constitutes the greatest fraction of the mass. Oxygen is next because its mass is 16 times that of hydrogen and there are only six hydrogen atoms to every one oxygen atom.

Cc 5.10 **(c)** The chemical formula for a compound gives relationships between *atoms* or *moles of atoms*. The chemical formula for water states that water molecules contain 2 H atoms to every 1 O atom or 2 mol H to every 1 mol H_2O. This *does not* imply a two-to-one relationship between *masses* of hydrogen and oxygen because these atoms have different masses. It also does not imply a two-to-one relationship between volumes.

CHAPTER 6

A geometrical and mechanical basis of the physical science cannot be contructed until we know the forms, sizes, and positions of the molecules of substances.

—George Gore (1826–1908)

Morphinan (a morphine analog) binding to an opiod receptor (based on research done by Kobilka and co-workers at Stanford University). Morphine is derived from the sap of the opium poppy.

Chemical Bonding I

Drawing Lewis Structures and Determining Molecular Shapes

6.1 Morphine: A Molecular Imposter 189

6.2 Electronegativity and Bond Polarity 190

6.3 Writing Lewis Structures for Molecular Compounds and Polyatomic Ions 194

6.4 Resonance and Formal Charge 196

6.5 Exceptions to the Octet Rule: Odd-Electron Species, Incomplete Octets, and Expanded Octets 201

6.6 Bond Energies and Bond Lengths 204

6.7 VSEPR Theory: The Five Basic Shapes 207

6.8 VSEPR Theory: The Effect of Lone Pairs 211

6.9 VSEPR Theory: Predicting Molecular Geometries 215

6.10 Molecular Shape and Polarity 219

Key Learning Outcomes 225

CHEMICAL BONDING IS AT THE HEART of chemistry. In this book, we examine three different theories for chemical bonding. Recall from Section 5.4 that bonding theories explain why atoms bond together to form molecules and predict many of the properties (such as the shape) of molecules. Therefore, bonding theories play an important role in helping us to see the relationship between the structure of a molecule and its properties. The first and simplest bonding theory is the Lewis model, which we introduced in Chapter 5 and expand upon in this chapter. With just a few dots, dashes, and chemical symbols, the Lewis model can help us to understand and predict a myriad of chemical observations. The Lewis model, combined with a theory called *valence shell electron pair repulsion theory (VSEPR)*, allows us to predict the shapes of molecules. The other two bonding theories are valence bond theory and molecular orbital theory, which we will cover in Chapter 7.

6.1 Morphine: A Molecular Imposter

Morphine—a drug named after Morpheus, the Greek god of dreams—is the silver bullet in the human arsenal against pain. Morphine is often prescribed after surgery to aid recovery or to alleviate the severe pain associated with illnesses such as cancer. It is also prescribed to patients who have chronic pain toward the end of their lives. For these patients, prescribed morphine provides relief from an otherwise tortuous existence.

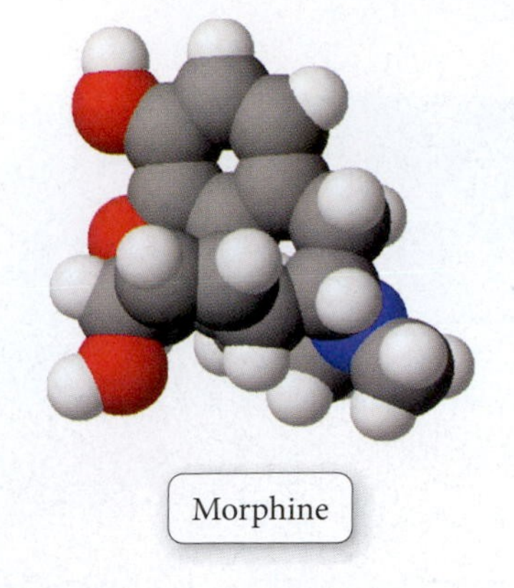

Endogenous means produced within the organism.

Proteins are among the most important biological molecules and serve many functions in living organisms.

Morphine is a natural product derived from the sap of the opium poppy. The effects of opium sap have been known for thousands of years, but morphine itself was not isolated from opium until the early 1800s. Morphine acts by binding to receptors (called opioid receptors) that exist within nerve cells. When morphine binds to an opioid receptor, the transmission of nerve signals is altered, resulting in less pain, sedation, and feelings of euphoria and tranquility.

Why do humans (and other mammals) have receptors within their nerve cells that bind to molecules derived from the sap of a plant? Researchers long suspected that these receptors must also bind other molecules as well; otherwise, why would the receptors exist? In the 1970s, researchers discovered some of these molecules, known as endorphins (short for **endo**genous mor**phine**). Endorphins are the body's natural painkillers. Our bodies naturally produce endorphins during periods of pain such as childbirth or intense exercise. Endorphins are at least partially responsible for the so-called runner's high, a feeling of well-being that often follows an athlete's intense workout.

Morphine binds to opioid receptors because it fits into a special pocket (called the active site) on the opioid receptor protein (just as a key fits into a lock) that normally binds endorphins. Certain parts of the morphine molecule have a similar enough shape to endorphins that they fit the lock (even though they are not the original key). In other words, morphine is a *molecular imposter*, mimicking the action of endorphins because of similarities in shape.

The lock-and-key fit between the active site of a protein and a particular molecule is important, not only to our perception of pain, but to many other biological functions as well. Immune response, the sense of smell, the sense of taste, and many types of drug action depend on shape-specific interactions between molecules and proteins. In fact, our ability to determine the shapes of key biological molecules is largely responsible for the revolution that has occurred in biology over the last 50 years.

In this chapter, we look at ways to predict and account for the shapes of molecules. The molecules we examine are much smaller than the protein molecules we just discussed, but the same principles apply to both. The simple model we examine to account for molecular shape is valence shell electron pair repulsion (VSEPR) theory, and in this chapter we use it in conjunction with the Lewis model to make important predictions about the shapes of molecules.

6.2 Electronegativity and Bond Polarity

In Chapter 5, we introduced the Lewis model for chemical bonding. In the Lewis model, we represent valence electrons with dots and draw Lewis structures that show how atoms attain octets by sharing electrons with other atoms. The shared electrons are covalent chemical bonds. However, we know from quantum mechanics (Chapter 3) that representing electrons with dots, as we do in the Lewis model, is a drastic oversimplification. This does not invalidate the Lewis model—which is an extremely useful theory—but we must recognize and compensate for its inherent limitations. One limitation of representing electrons as dots, and covalent bonds as two dots shared between two atoms, is that the shared electrons always appear to be *equally* shared. Such is not the case. For example, consider the Lewis structure of hydrogen fluoride.

$$\mathrm{H{:}\ddot{\underset{..}{F}}{:}}$$

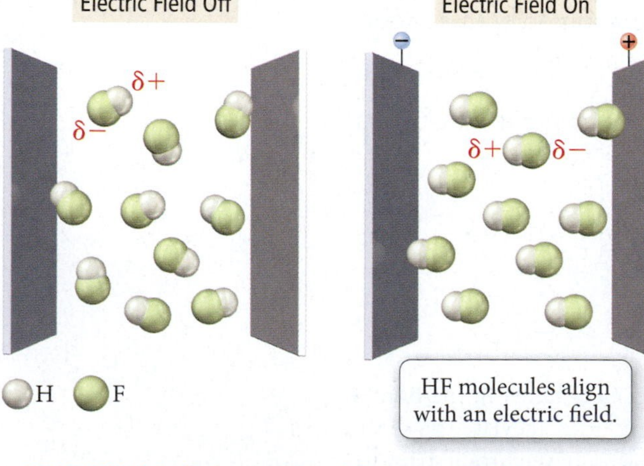

▲ **FIGURE 6.1 Orientation of Gaseous Hydrogen Fluoride in an Electric Field** Because one side of the HF molecule has a slight positive charge and the other side a slight negative charge, the molecules align themselves with an external electric field.

The two shared electron dots sitting between the H and the F atoms appear to be equally shared between hydrogen and fluorine. However, based on laboratory measurements, we know they are not. When HF is put in an electric field, the molecules orient as shown in Figure 6.1 ◀. From this observation, we know that the hydrogen side of the molecule must have a slight positive charge and the fluorine side of the molecule must have a slight negative charge. We represent this partial separation of charge as follows:

$$\overset{\longleftrightarrow}{\mathrm{H{-}F}} \quad or \quad \overset{\delta^+}{\mathrm{H}}{-}\overset{\delta^-}{\mathrm{F}}$$

The red arrow above, with a positive sign on the tail, indicates that the left side of the molecule has a partial positive charge and that the right side of the molecule (the side the arrow is pointing *toward*) has a partial negative charge. Similarly, the red δ+ (delta plus) on the right H—F structure represents a partial positive charge, and the red δ− (delta minus) represents a partial negative charge. Does this make the bond ionic? No. In an ionic bond, the electron is essentially *transferred* from one atom to another. In HF, the electron is *unequally shared*. In other words, even though the Lewis structure of HF portrays the bonding electrons as residing *between* the two atoms, in reality the electron density is greater on the fluorine atom than on the hydrogen atom (Figure 6.2 ▶). The bond is said to be *polar*—having a positive pole and a negative pole. A **polar covalent bond** is intermediate in nature between a pure covalent bond and an ionic bond. In fact, the categories of pure covalent and ionic are really two extremes within a broad continuum. Most covalent bonds between dissimilar atoms are actually *polar covalent*, somewhere between the two extremes.

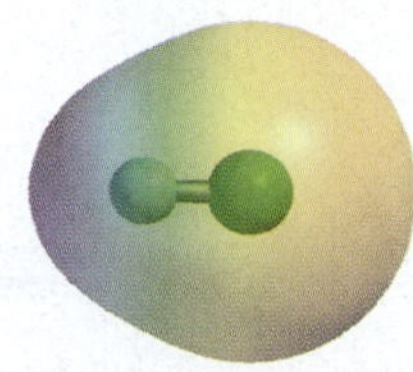

▲ **FIGURE 6.2 Electrostatic Potential Map for the HF Molecule** Red indicates electron-rich regions and blue indicates electron-poor regions.

Electronegativity

The ability of an atom to attract electrons to itself in a chemical bond (which results in polar and ionic bonds) is called **electronegativity**. We say that fluorine is more *electronegative* than hydrogen because it takes a greater share of the electron density in HF.

Electronegativity was quantified by the American chemist Linus Pauling in his classic book, *The Nature of the Chemical Bond*. Pauling compared the bond energy—the energy required to break a bond—of a heteronuclear diatomic molecule such as HF with the bond energies of its homonuclear counterparts, in this case H_2 and F_2. The bond energies of H_2 and F_2 are 436 kJ/mol and 155 kJ/mol, respectively. Pauling reasoned that if the HF bond were purely covalent—that is, if the electrons were shared equally—the bond energy of HF should simply be an average of the bond energies of H_2 and F_2, which would be 296 kJ/mol. However, the bond energy of HF is experimentally measured to be 565 kJ/mol. Pauling suggested that the additional bond energy was due to the *ionic character* of the bond. Based on many such comparisons of bond energies, and by arbitrarily assigning an electronegativity of 4.0 to fluorine (the most electronegative element on the periodic table), Pauling developed the electronegativity values shown in Figure 6.3 ▼.

We cover the concept of bond energy in more detail in Section 6.6.

Pauling's "average" bond energy was actually calculated a little bit differently than the normal average shown here. He took the square root of the product of the bond energies of the homonuclear counterparts as the "average."

For main-group elements, notice the following periodic trends in electronegativity in Figure 6.3:

- Electronegativity generally increases across a period in the periodic table.
- Electronegativity generally decreases down a column in the periodic table.
- Fluorine is the most electronegative element.
- Francium is the least electronegative element (sometimes called the most *electropositive*).

Trends in Electronegativity

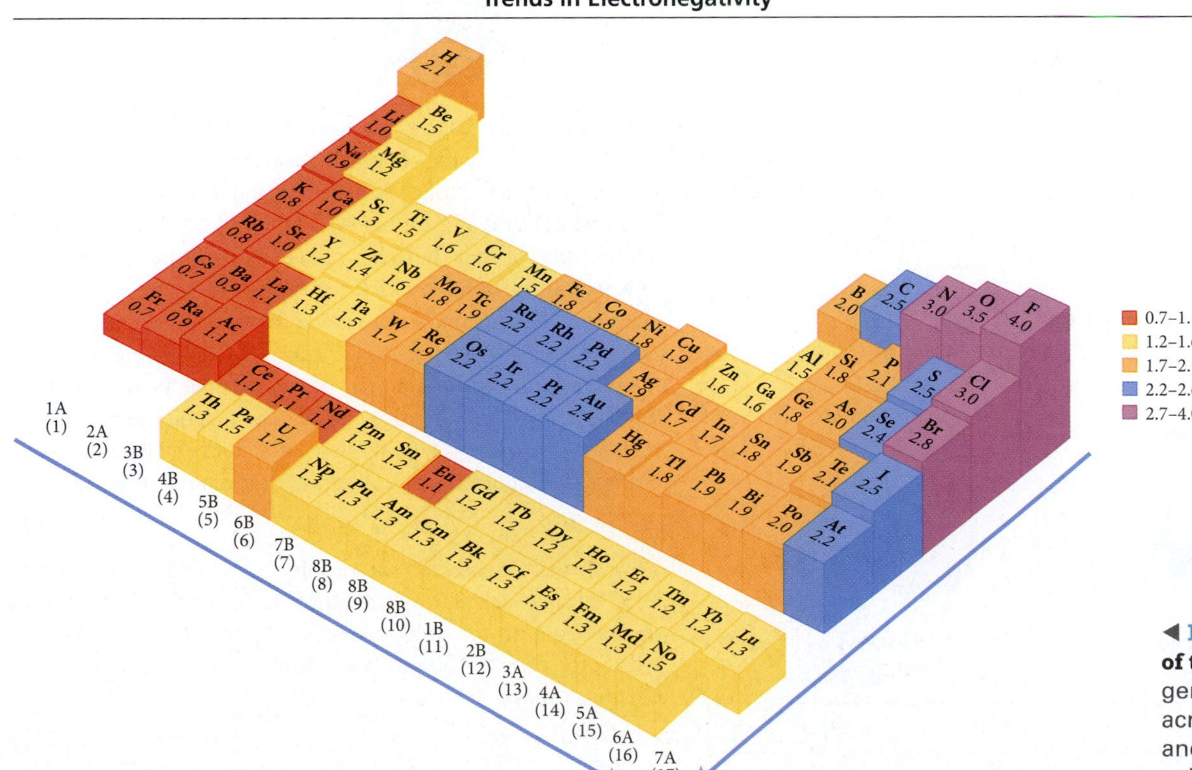

◀ **FIGURE 6.3 Electronegativities of the Elements** Electronegativity generally increases as we move across a row in the periodic table and decreases as we move down a column.

The periodic trends in electronegativity are consistent with other periodic trends we have seen. In general, electronegativity is inversely related to atomic size—the larger the atom, the less ability it has to attract electrons to itself in a chemical bond.

6.1 Cc Conceptual Connection

Periodic Trends in Electronegativity

Arrange these elements in order of decreasing electronegativity: P, Na, N, Al.

Bond Polarity, Dipole Moment, and Percent Ionic Character

The degree of polarity in a chemical bond depends on the electronegativity difference (sometimes abbreviated ΔEN) between the two bonding atoms. The greater the electronegativity difference, the more polar the bond. If two atoms with identical electronegativities form a covalent bond, they share the electrons equally, and the bond is purely covalent or *nonpolar*. For example, the chlorine molecule (shown at left), composed of two chlorine atoms (which necessarily have identical electronegativities), has a covalent bond in which electrons are evenly shared.

If there is a large electronegativity difference between the two atoms in a bond, such as normally occurs between a metal and a nonmetal, the electron from the metal is almost completely transferred to the nonmetal, and the bond is ionic. For example, sodium and chlorine form an ionic bond.

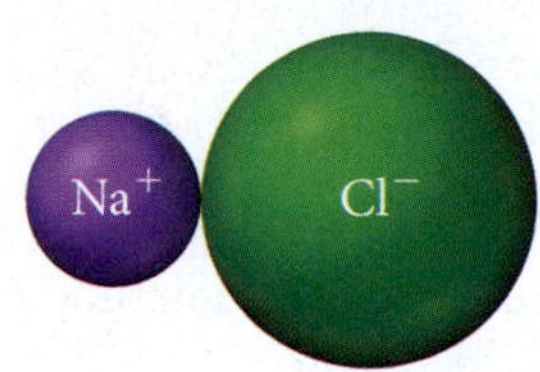

If there is an intermediate electronegativity difference between the two atoms, such as between two different nonmetals, then the bond is polar covalent. For example, HCl has a polar covalent bond.

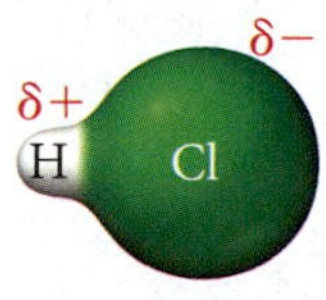

TABLE 6.1 The Effect of Electronegativity Difference on Bond Type

Electronegativity Difference (ΔEN)	Bond Type	Example
Small (0–0.4)	Covalent	Cl_2
Intermediate (0.4–2.0)	Polar covalent	HCl
Large (2.0+)	Ionic	NaCl

While all attempts to divide the bond polarity continuum into specific regions are necessarily arbitrary, it is helpful to classify bonds as covalent, polar covalent, and ionic, based on the electronegativity difference between the bonding atoms as shown in Table 6.1 and Figure 6.4 ◀. The bond between C and H lies at the border between covalent and polar covalent; however, this bond—which is very important in organic chemistry—is normally considered covalent (nonpolar).

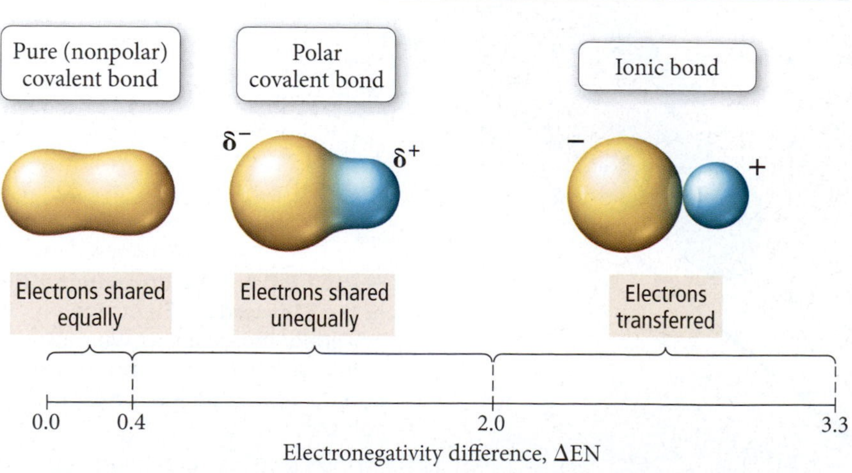

▲ **FIGURE 6.4 Electronegativity Difference (ΔEN) and Bond Type**

We quantify the polarity of a bond by the size of its dipole moment. A **dipole moment (μ)** occurs anytime there is a separation of positive and negative charge. The magnitude of a dipole moment created by separating two particles of equal but opposite charges of magnitude q by a distance r is given by the equation:

$$\mu = qr \qquad [6.1]$$

We can get a sense for the dipole moment of a completely ionic bond by calculating the dipole moment that results from separating a proton and an electron ($q = 1.6 \times 10^{-19}$ C) by a distance of $r = 130$ pm (the approximate length of a short chemical bond).

$$\begin{aligned} \mu &= qr \\ &= (1.6 \times 10^{-19}\ \text{C})(130 \times 10^{-12}\ \text{m}) \\ &= 2.1 \times 10^{-29}\ \text{C} \cdot \text{m} \\ &= 6.2\ \text{D} \end{aligned}$$

The debye (D) is the unit commonly used for reporting dipole moments ($1 \text{ D} = 3.34 \times 10^{-30} \text{ C}\cdot\text{m}$). Based on this calculation, we would expect the dipole moment of completely ionic bonds with bond lengths close to 130 pm to be about 6 D. The smaller the magnitude of the charge separation, and the smaller the distance between the charges, the smaller the dipole moment. Table 6.2 lists the dipole moments of several molecules along with the electronegativity differences of their atoms.

We cover bond lengths in more detail in Section 6.6.

TABLE 6.2 Dipole Moments of Several Molecules in the Gas Phase

Molecule	ΔEN	Dipole Moment (D)
Cl_2	0	0
ClF	1.0	0.88
HF	1.9	1.82
LiF	3.0	6.33

By comparing the *actual* dipole moment of a bond to what the dipole moment would be if the electron were completely transferred from one atom to the other, we can get a sense of the degree to which the electron is transferred (or the degree to which the bond is ionic). The **percent ionic character** is the ratio of a bond's actual dipole moment to the dipole moment it would have if the electron were completely transferred from one atom to the other, multiplied by 100%.

$$\text{Percent ionic character} = \frac{\text{measured dipole moment of bond}}{\text{dipole moment if electron were completely transferred}} \times 100\%$$

For example, suppose a diatomic molecule with a bond length of 130 pm has a dipole moment of 3.5 D. We previously calculated that separating a proton and an electron by 130 pm results in a dipole moment of 6.2 D. Therefore, the percent ionic character of the bond is 56%.

$$\begin{aligned}\text{Percent ionic character} &= \frac{3.5 \cancel{\text{D}}}{6.2 \cancel{\text{D}}} \times 100\% \\ &= 56\%\end{aligned}$$

A bond in which an electron is completely transferred from one atom to another would have 100% ionic character (although even the most ionic bonds do not reach this ideal). Figure 6.5 ▶ shows the percent ionic character of a number of diatomic gaseous molecules plotted against the electronegativity difference between the bonding atoms. As expected, the percent ionic character generally increases as the electronegativity difference increases. However, as we can see, no bond is 100% ionic. In general, bonds with greater than 50% ionic character are referred to as ionic bonds.

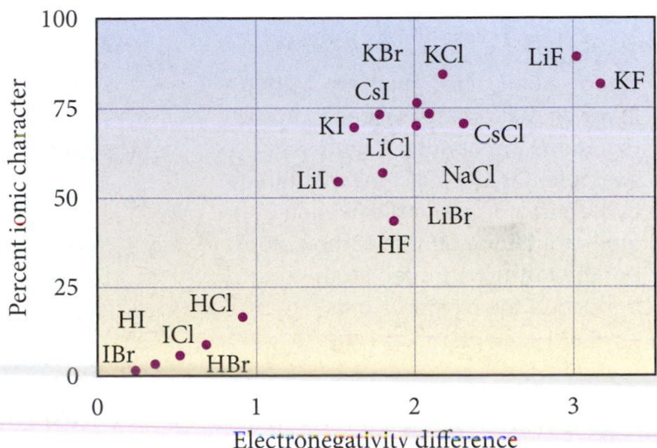

▲ **FIGURE 6.5 Percent Ionic Character versus Electronegativity Difference for Some Compounds**

EXAMPLE 6.1

Classifying Bonds as Pure Covalent, Polar Covalent, or Ionic

Determine if the bond formed between each pair of atoms is covalent, polar covalent, or ionic.

(a) Sr and F **(b)** N and Cl **(c)** N and O

SOLUTION

(a) In Figure 6.3, find the electronegativity of Sr (1.0) and of F (4.0). The electronegativity difference (ΔEN) is $\Delta\text{EN} = 4.0 - 1.0 = 3.0$. According to Table 6.1, this bond is ionic.

(b) In Figure 6.3, find the electronegativity of N (3.0) and of Cl (3.0). The electronegativity difference (ΔEN) is $\Delta\text{EN} = 3.0 - 3.0 = 0$. According to Table 6.1, this bond is covalent.

(c) In Figure 6.3, find the electronegativity of N (3.0) and of O (3.5). The electronegativity difference (ΔEN) is $\Delta\text{EN} = 3.5 - 3.0 = 0.5$. According to Table 6.1, this bond is polar covalent.

FOR PRACTICE 6.1

Determine if the bond formed between each pair of atoms is pure covalent, polar covalent, or ionic.

(a) I and I **(b)** Cs and Br **(c)** P and O

6.2 Cc Conceptual Connection

Percent Ionic Character

The HCl(*g*) molecule has a bond length of 127 pm and a dipole moment of 1.08 D. Without doing detailed calculations, determine the best estimate for its percent ionic character.

(a) 5%
(b) 15%
(c) 50%
(d) 80%

6.3 Writing Lewis Structures for Molecular Compounds and Polyatomic Ions

In Chapter 5, we saw how the Lewis model can be used to describe ionic and covalent bonds, and in Section 6.2 we saw how some covalent bonds could be polar. We now turn to the basic sequence of steps involved in writing Lewis structures for given combinations of atoms.

Often, chemical formulas are written in a way that provides clues to how the atoms are bonded together. For example, CH_3OH indicates that three hydrogen atoms and the oxygen atom are bonded to the carbon atom, but the fourth hydrogen atom is bonded to the oxygen atom.

Writing Lewis Structures for Molecular Compounds

To write a Lewis structure for a molecular compound, follow these steps:

1. **Write the correct skeletal structure for the molecule.** The Lewis structure of a molecule must have the atoms in the correct positions. For example, you could not write a Lewis structure for water if you started with the hydrogen atoms next to each other and the oxygen atom at the end (HHO). In nature, oxygen is the central atom and the hydrogen atoms are *terminal* (at the ends). The correct skeletal structure is HOH. The only way to determine the skeletal structure of a molecule with absolute certainty is to examine its structure experimentally. However, you can write likely skeletal structures by remembering two guidelines. First, *hydrogen atoms are always terminal.* Hydrogen does not ordinarily occur as a central atom because central atoms must form at least two bonds, and hydrogen, which has only a single valence electron to share and requires only a duet, can form just one. Second, *put the more electronegative elements in terminal positions* and the less electronegative elements (other than hydrogen) in the central position. Later in this chapter, you will learn how to distinguish between competing skeletal structures by applying the concept of formal charge.

There are a few exceptions to this rule, such as diborane (B_2H_6), which contains *bridging hydrogens*, but these are rare and cannot be adequately addressed by the Lewis model.

2. **Calculate the total number of electrons for the Lewis structure by summing the valence electrons of each atom in the molecule.** Remember that the number of valence electrons for any main-group element is equal to its group number in the periodic table. *If you are writing a Lewis structure for a polyatomic ion, you must consider the charge of the ion when calculating the total number of electrons.* Add one electron for each negative charge and subtract one electron for each positive charge. Don't worry about which electron comes from which atom—only the total number is important.

Sometimes distributing all the remaining electrons to the central atom results in more than an octet. This is called an expanded octet and is covered in Section 6.5.

3. **Distribute the electrons among the atoms, giving octets (or duets in the case of hydrogen) to as many atoms as possible.** Begin by placing two electrons between every two atoms. These represent the minimum number of bonding electrons. Then distribute the remaining electrons as lone pairs, first to terminal atoms, and then to the central atom, giving octets (or duets for hydrogen) to as many atoms as possible.

4. **If any atoms lack an octet, form double or triple bonds as necessary to give them octets.** Do this by moving lone electron pairs from terminal atoms into the bonding region with the central atom.

5. **Check** Count the number of electrons in the Lewis structure. The total should equal the number from step 2.

The left column in the examples that follow contains an abbreviated version of the procedure for writing Lewis structures; the center and right columns contain two examples of applying the procedure.

PROCEDURE FOR **Writing Lewis Structures for Covalent Compounds**	EXAMPLE 6.2 **Writing Lewis Structures**	EXAMPLE 6.3 **Writing Lewis Structures**
	Write the Lewis structure for CO_2. 	Write the Lewis structure for NH_3.
1. Write the correct skeletal structure for the molecule.	**SOLUTION** Because carbon is the less electronegative atom, put it in the central position. O C O	**SOLUTION** Since hydrogen is always terminal, put nitrogen in the central position. H N H / H
2. Calculate the total number of electrons for the Lewis structure by summing the valence electrons of each atom in the molecule.	Total number of electrons for Lewis structure = $\begin{pmatrix}\text{number of}\\ \text{valence}\\ e^- \text{ for C}\end{pmatrix} + 2\begin{pmatrix}\text{number of}\\ \text{valence}\\ e^- \text{ for O}\end{pmatrix}$ $= 4 + 2(6) = 16$	Total number of electrons for Lewis structure = $\begin{pmatrix}\text{number of}\\ \text{valence}\\ e^- \text{ for N}\end{pmatrix} + 3\begin{pmatrix}\text{number of}\\ \text{valence}\\ e^- \text{ for H}\end{pmatrix}$ $= 5 + 3(1) = 8$
3. Distribute the electrons among the atoms, giving octets (or duets for hydrogen) to as many atoms as possible. Begin with the bonding electrons and then proceed to lone pairs on terminal atoms and finally to lone pairs on the central atom.	Bonding electrons are first. O:C:O (4 of 16 electrons used) Lone pairs on terminal atoms are next. :Ö:C:Ö: (16 of 16 electrons used)	Bonding electrons are first. H:N:H / H (6 of 8 electrons used) Lone pairs on terminal atoms are next, but none are needed on hydrogen. Lone pairs on central atom are last. H—N̈—H / | / H (8 of 8 electrons used)
4. If any atom lacks an octet, form double or triple bonds as necessary to give them octets.	Since carbon lacks an octet, move lone pairs from the oxygen atoms to bonding regions to form double bonds. :Ö:C:Ö: → :Ö=C=Ö:	Since all of the atoms have octets (or duets for hydrogen), the Lewis structure for NH_3 is complete as shown above.
5. **Check** Count the number of electrons in the Lewis structure. The total should equal the number from step 2.	The Lewis structure has 16 electrons as calculated in step 2.	The Lewis structure has 8 electrons as calculated in step 2.
	FOR PRACTICE 6.2 Write the Lewis structure for CO.	**FOR PRACTICE 6.3** Write the Lewis structure for H_2CO.

Writing Lewis Structures for Polyatomic Ions

Write Lewis structures for polyatomic ions by following the same procedure, but pay special attention to the charge of the ion when calculating the number of electrons for the Lewis structure. Add one electron for each negative charge and subtract one electron for each positive charge. Write the Lewis structure for a polyatomic ion within brackets with the charge of the ion in the upper right-hand corner, outside the bracket.

EXAMPLE 6.4

Writing Lewis Structures for Polyatomic Ions

Write the Lewis structure for the NH_4^+ ion.

SOLUTION

Begin by writing the skeletal structure. Since hydrogen is always terminal, put the nitrogen atom in the central position.	H H N H H
Calculate the total number of electrons for the Lewis structure by summing the number of valence electrons for each atom and subtracting 1 for the 1+ charge.	Total number of electrons for Lewis structure = (number of valence e^- in N) + (number of valence e^- in H) − 1 = 5 + 4(1) − 1 = 8 Subtract 1 e^- to account for 1+ charge of ion.
Place two bonding electrons between every two atoms. Since all of the atoms have complete octets, no double bonds are necessary.	H H:N:H H (8 of 8 electrons used)
Write the Lewis structure in brackets with the charge of the ion in the upper right-hand corner.	[H—N(—H)(—H)—H]$^+$

FOR PRACTICE 6.4

Write the Lewis structure for the hypochlorite ion, ClO^-.

6.4 Resonance and Formal Charge

We need two additional concepts to write the best possible Lewis structures for a large number of compounds. The concepts are *resonance*, used when two or more valid Lewis structures can be drawn for the same compound, and *formal charge*, an electron bookkeeping system that allows us to discriminate between alternative Lewis structures.

Resonance

For some molecules, we can write more than one valid Lewis structure. For example, consider writing a Lewis structure for O_3. The following two Lewis structures, with the double bond on alternate sides, are equally correct:

$$:\ddot{O}=\ddot{O}-\ddot{\underset{..}{O}}: \qquad :\ddot{\underset{..}{O}}-\ddot{O}=\ddot{O}:$$

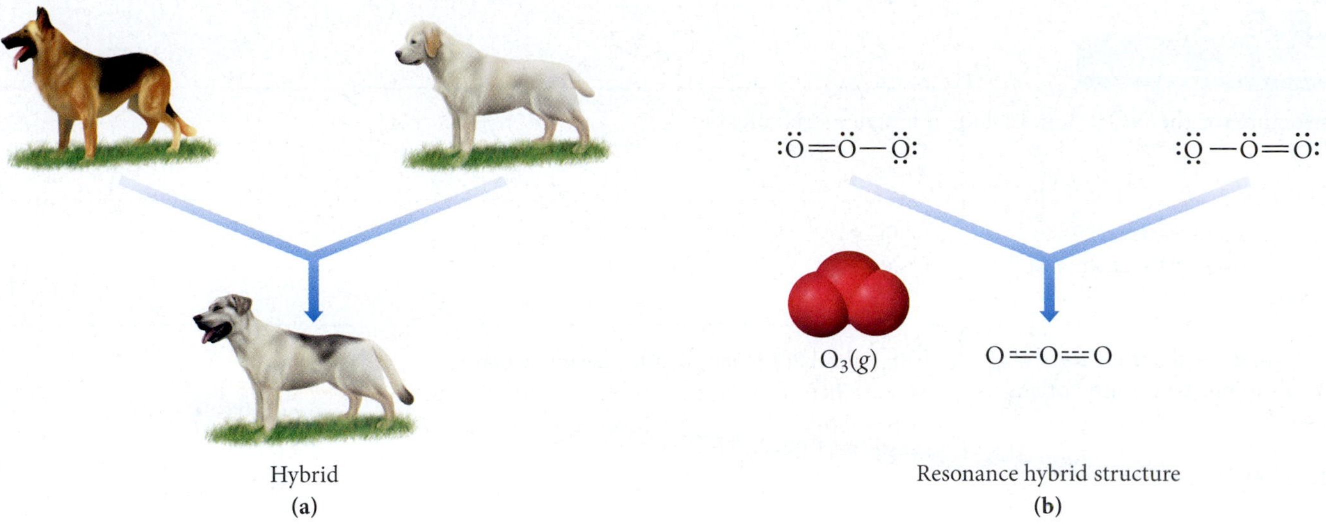

▲ **FIGURE 6.6 Hybridization** Just as the offspring of two different dog breeds is a hybrid that is intermediate between the two breeds **(a)**, the structure of a resonance hybrid is intermediate between that of the contributing resonance structures **(b)**.

In cases such as this—where we can write two or more valid Lewis structures for the same molecule—we find that, in nature, the molecule exists as an *average* of the two Lewis structures. Both of the Lewis structures for O_3 predict that O_3 contains two different bonds (one double bond and one single bond). However, when we experimentally examine the structure of O_3, we find that the bonds in the O_3 molecule are equivalent and that each bond is intermediate in strength and length between a double bond and single bond. We account for this by representing the molecule with both structures, called *resonance structures*, with a double-headed arrow between them.

$$:\ddot{O}=\ddot{O}-\underset{\cdot\cdot}{\ddot{O}}: \longleftrightarrow :\underset{\cdot\cdot}{\ddot{O}}-\ddot{O}=\ddot{O}:$$

A **resonance structure** is one of two or more Lewis structures that have the same skeletal formula (the atoms are in the same locations), but different electron arrangements. The actual structure of the molecule is intermediate between the two (or more) resonance structures and is called a **resonance hybrid**. The term *hybrid* comes from breeding and means the offspring of two animals or plants of different varieties or breeds. If we breed a Labrador retriever with a German shepherd, we get a *hybrid* that is intermediate between the two breeds (Figure 6.6a ▲). Similarly, the actual structure of a resonance hybrid is intermediate between the two resonance structures (Figure 6.6b ▲). The only structure that actually exists is the hybrid structure. The individual resonance structures do not exist and are merely a convenient way to describe the actual structure. Notice that the actual structure of ozone has two equivalent bonds and a bent geometry (we discuss molecular geometries in more detail later in this chapter).

The concept of resonance is an adaptation of the Lewis model that helps account for the complexity of actual molecules. In the Lewis model, electrons are *localized* either on one atom (lone pair) or between atoms (bonding pair). However, in nature, the electrons in molecules are often *delocalized* over several atoms or bonds. The delocalization of electrons lowers their energy; it stabilizes them (for reasons that are beyond the scope of this book). Resonance depicts two or more structures with the electrons in different places in an attempt to more accurately reflect the delocalization of electrons. In the real hybrid structure—an average between the resonance structures—the electrons are more spread out (or delocalized) than in any of the resonance structures. The resulting stabilization of the electrons (that is, the lowering of their potential energy due to delocalization) is sometimes called *resonance stabilization*. Resonance stabilization makes an important contribution to the stability of many molecules.

EXAMPLE 6.5
Writing Resonance Structures

Write the Lewis structure for the NO_3^- ion. Include resonance structures.

SOLUTION

Begin by writing the skeletal structure. Since nitrogen is the least electronegative atom, put it in the central position.	O O N O
Calculate the total number of electrons for the Lewis structure by summing the number of valence electrons for each atom and adding 1 for the 1− charge.	Total number of electrons for Lewis structure = (number of valence e^- in N) + 3 (number of valence e^- in O) + 1 = 5 + 3(6) + 1 = 24 Add 1 e^- to account for 1− charge of ion.
Place two bonding electrons between each pair of atoms.	 O O:N:O (6 of 24 electrons used)
Distribute the remaining electrons, first to terminal atoms. There are not enough electrons to complete the octet on the central atom.	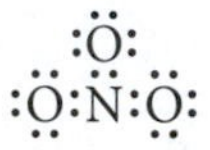 :Ö: :Ö:N:Ö: (24 of 24 electrons used)
Form a double bond by moving a lone pair from one of the oxygen atoms into the bonding region with nitrogen. Enclose the structure in brackets and include the charge.	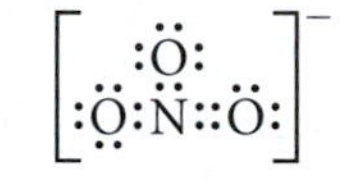 [:Ö:N::Ö: with :Ö: above N]$^-$ *or* [:Ö—N=Ö: with :Ö: above N]$^-$ 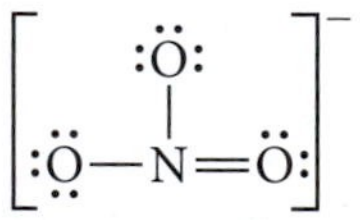
Since the double bond can form equally well with any of the three oxygen atoms, write all three structures as resonance structures. (The actual space-filling model of NO_3^- is shown here for comparison. Note that all three bonds are equal in length.)	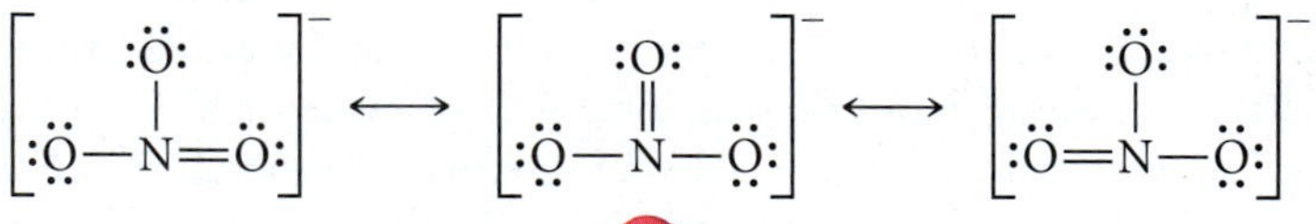  [:Ö—N=Ö: with :Ö: single-bonded above]$^-$ ⟷ [:Ö—N—Ö: with :O: double-bonded above]$^-$ ⟷ [:O=N—Ö: with :Ö: single-bonded above]$^-$ NO_3^-

FOR PRACTICE 6.5

Write the Lewis structure for the NO_2^- ion. Include resonance structures.

In the examples of resonance hybrids that we have examined so far, the contributing structures have been equivalent (or equally valid) Lewis structures. In these cases, the true structure is an equally weighted average of the resonance structures. In some cases, however, we can write resonance structures that are not equivalent. For reasons we cover in the material that follows—such as formal charge, for example—one possible resonance structure may be somewhat better than another. In such cases, the true structure is still an average of the resonance structures, but the better resonance structure contributes more to the true structure. In other words, multiple nonequivalent resonance structures may be weighted differently in their contributions to the true overall structure of a molecule (see Example 6.6).

Formal Charge

Formal charge is a fictitious charge assigned to each atom in a Lewis structure that helps us to distinguish among competing Lewis structures. The **formal charge** of an atom in a Lewis structure is *the charge it would have if all bonding electrons were shared equally between the bonded atoms.* In other words, formal charge is the calculated charge for an atom if we completely ignore the effects of electronegativity. For example, we know that because fluorine is more electronegative than hydrogen, HF has a dipole moment. The hydrogen atom has a slight positive charge, and the fluorine atom has a slight negative charge. However, the *formal charges* of hydrogen and fluorine in HF (the calculated charges if we ignore their differences in electronegativity) are both zero.

H:F: (fluorine with three lone pairs)

Formal charge = 0 (H)   Formal charge = 0 (F)

We can calculate the formal charge on any atom as the difference between the number of valence electrons in the atom and the number of electrons that it "owns" in a Lewis structure. An atom in a Lewis structure can be thought of as "owning" all of its nonbonding electrons and one-half of its bonding electrons.

$$\text{Formal charge} = \text{number of valence electrons} - \left(\text{number of nonbonding electrons} + \tfrac{1}{2}\text{ number of bonding electrons}\right)$$

So the formal charge of hydrogen in HF is 0:

$$\text{Formal charge} = 1 - \left[0 + \tfrac{1}{2}(2)\right] = 0$$

Number of valence electrons for H (→ 1)

Number of electrons that H "owns" in the Lewis structure (→ $[0 + \tfrac{1}{2}(2)]$)

Similarly, we calculate the formal charge of fluorine in HF as 0:

$$\text{Formal charge} = 7 - \left[6 + \tfrac{1}{2}(2)\right] = 0$$

Number of valence electrons for F (→ 7)

Number of electrons that F "owns" in the Lewis structure (→ $[6 + \tfrac{1}{2}(2)]$)

The concept of formal charge is useful because it can help us distinguish between competing skeletal structures or competing resonance structures. In general, these four rules apply:

1. The sum of all formal charges in a neutral molecule must be zero.
2. The sum of all formal charges in an ion must equal the charge of the ion.
3. Small (or zero) formal charges on individual atoms are better than large ones.
4. When formal charge cannot be avoided, negative formal charge should reside on the most electronegative atom.

Let's apply the concept of formal charge to distinguish between possible skeletal structures for the molecule formed by H, C, and N. The three atoms can bond with C in the center (HCN) or N in the center (HNC). The table below shows the two possible structures and the corresponding formal charges.

	Structure A			Structure B		
	H	—C≡	N:	H	—N≡	C:
number of valence e^-	1	4	5	1	5	4
− number of nonbonding e^-	−0	−0	−2	−0	−0	−2
$-\frac{1}{2}$ (number of bonding e^-)	$-\frac{1}{2}(2)$	$-\frac{1}{2}(8)$	$-\frac{1}{2}(6)$	$-\frac{1}{2}(2)$	$-\frac{1}{2}(8)$	$-\frac{1}{2}(6)$
Formal charge	**0**	**0**	**0**	**0**	**+1**	**−1**

The sum of the formal charges for each of these structures is zero (as it always must be for neutral molecules). However, Structure B has formal charges on both the N atom and the C atom, while Structure A has no formal charges on any atom. Furthermore, in Structure B, the negative formal charge is not on

Both HCN and HNC exist, but—as we predicted by assigning formal charges—HCN is more stable than HNC.

the most electronegative element (nitrogen is more electronegative than carbon). Consequently, Structure A is the better Lewis structure. Since atoms in the middle of a molecule tend to have more bonding electrons and fewer nonbonding electrons, they also tend to have more positive formal charges. Consequently, the best skeletal structure usually has the least electronegative atom in the central position, as we learned in step 1 of our procedure for writing Lewis structures in Section 6.3.

EXAMPLE 6.6
Assigning Formal Charges

Assign formal charges to each atom in the resonance forms of the cyanate ion (OCN^-). Which resonance form is likely to contribute most to the correct structure of OCN^-?

A	B	C
$[:\ddot{\underset{..}{O}}-C\equiv N:]^-$	$[:\ddot{O}=C=\ddot{N}:]^-$	$[:O\equiv C-\ddot{\underset{..}{N}}:]^-$

SOLUTION

Calculate the formal charge on each atom by finding the number of valence electrons and subtracting the number of nonbonding electrons and one-half the number of bonding electrons.

	A $[:\ddot{\underset{..}{O}}-C\equiv N:]^-$			B $[:\ddot{O}=C=\ddot{N}:]^-$			C $[:O\equiv C-\ddot{\underset{..}{N}}:]^-$		
Number of valence e^-	6	4	5	6	4	5	6	4	5
− number of nonbonding e^-	−6	−0	−2	−4	−0	−4	−2	−0	−6
$-\frac{1}{2}$(number of bonding e^-)	−1	−4	−3	−2	−4	−2	−3	−4	−1
Formal charge	−1	**0**	**0**	**0**	**0**	−1	+1	**0**	−2

The sum of all formal charges for each structure is −1, as it should be for a 1− ion. Structures A and B have the least amount of formal charge and are therefore to be preferred over Structure C. Structure A is preferable to B because it has the negative formal charge on the more electronegative atom. You would therefore expect Structure A to make the biggest contribution to the resonance forms of the cyanate ion.

FOR PRACTICE 6.6

Assign formal charges to each atom in the resonance forms of N_2O. Which resonance form is likely to contribute most to the correct structure of N_2O?

A	B	C
$:\ddot{N}=N=\ddot{O}:$	$:N\equiv N-\ddot{\underset{..}{O}}:$	$:\ddot{\underset{..}{N}}-N\equiv O:$

FOR MORE PRACTICE 6.6

Assign formal charges to each of the atoms in the nitrate ion (NO_3^-). The Lewis structure for the nitrate ion is shown in Example 6.5.

EXAMPLE 6.7
Drawing Resonance Structures and Assigning Formal Charge for Organic Compounds

Draw the Lewis structure (including resonance structures) for nitromethane (CH_3NO_2). For each resonance structure, assign formal charges to all atoms that have formal charge.

SOLUTION

Begin by writing the skeletal structure. For organic compounds, the condensed structural formula (in this case CH_3NO_2) indicates how the atoms are connected.

```
   H  O
H  C  N  O
   H
```

—*Continued on the next page*

Continued from the previous page—

Calculate the total number of electrons for the Lewis structure by summing the number of valence electrons for each atom.	Total number of electrons for Lewis structure $= \#$ valence e^- in C $+ 3(\#$ valence e^- in H$) + \#$ valence e^- in N $+ 2(\#$ valence e^- in O$)$ $= 4 + 3(1) + 5 + 2(6)$ $= 24$
Place a dash between each pair of atoms to indicate a bond. Each dash counts for two electrons.	H O | | H—C—N—O | H (12 of 24 electrons used)
Distribute the remaining electrons, first to terminal atoms, then to interior atoms.	H :Ö: | | H—C—N—Ö: | H (24 of 24 electrons used)
If there are not enough electrons to complete the octets on the interior atoms, form double bonds by moving lone pair electrons from terminal atoms into the bonding region with interior atoms.	H :Ö | ‖ H—C—N—Ö: | H
Draw any necessary resonance structures by moving only electron dots. (In this case, you can form a double bond between the nitrogen atom and the other oxygen atom.)	H :Ö: H :Ö | | | ‖ H—C—N=Ö: ⟷ H—C—N—Ö: | | H H
Assign formal charges (FC) to each atom. FC $= \#$ valence $e^- - (\#$ nonbonding $e^- + \frac{1}{2}\#$ bonding $e^-)$	H :Ö:$^{-1}$ H :Ö | |$^{+1}$ | ‖$^{+1}$ H—C—N=Ö: ⟷ H—C—N—Ö:$^{-1}$ | | H H Carbon, hydrogen, and the doubly bonded oxygen atoms have no formal charge. Nitrogen has a $+1$ formal charge $[5 - \frac{1}{2}(8)]$ and the singly bonded oxygen atom in each resonance structure has a -1 formal charge $[6 - (6 + \frac{1}{2}(2))]$.

FOR PRACTICE 6.7

Draw the Lewis structure (including resonance structures) for diazomethane (CH_2N_2). For each resonance structure, assign formal charges to all atoms that have formal charge.

6.5 Exceptions to the Octet Rule: Odd-Electron Species, Incomplete Octets, and Expanded Octets

The octet rule in the Lewis model has some exceptions, which we examine in this section of the chapter. They include (1) *odd-electron species,* molecules or ions with an odd number of electrons; (2) *incomplete octets,* molecules or ions with *fewer than eight electrons* around an atom; and (3) *expanded octets,* molecules or ions with *more than eight electrons* around an atom.

Odd-Electron Species

Molecules and ions with an odd number of electrons in their Lewis structures are called **free radicals** (or simply *radicals*). For example, nitrogen monoxide—a pollutant found in motor vehicle exhaust—has 11 electrons. If we try to write a Lewis structure for nitrogen monoxide, we can't achieve octets for both atoms.

The unpaired electron in nitrogen monoxide is put on the nitrogen rather than the oxygen in order to minimize formal charges.

:Ṅ::Ö: *or* :Ṅ=Ö:

The nitrogen atom does not have an octet, so this Lewis structure does not satisfy the octet rule. Yet, nitrogen monoxide exists, especially in polluted air. Why? As with any simple theory, the Lewis model is not sophisticated enough to handle every single case. We can't write good Lewis structures for free radicals; nevertheless some of these molecules exist in nature. Perhaps it is a testament to the Lewis model, however, that *relatively few* such molecules exist and that, in general, they tend to be somewhat unstable and reactive. NO, for example, reacts with oxygen in the air to form NO_2, another odd-electron molecule represented with the following 17-electron resonance structures:

:Ö=Ṅ—Ö̤: ⟷ :Ö̤—Ṅ=Ö:

In turn, NO_2 reacts with water to form nitric acid (a component of acid rain) and also reacts with other atmospheric pollutants to form peroxyacetylnitrate (PAN), an active component of photochemical smog. For free radicals, such as NO and NO_2, we simply write the best Lewis structure that we can.

Odd-Electron Species

Which molecule would you expect to be a free radical?

(a) CO **(b)** CO_2 **(c)** N_2O **(d)** ClO

Incomplete Octets

Another significant exception to the octet rule involves those elements that tend to form *incomplete octets.* The most important of these is boron, which forms compounds with only six electrons around B, rather than eight. For example, BF_3 and BH_3 lack an octet for B.

Beryllium compounds, such as BeH_2, also have incomplete octets.

```
   :F̤:          H
:F̤:B:F̤:       H:B:H
```

You might be wondering why we don't just form double bonds to increase the number of electrons around B. For BH_3, of course, we can't, because there are no additional electrons to move into the bonding region. For BF_3, however, we could attempt to give B an octet by moving a lone pair from an F atom into the bonding region with B.

```
       F̈:
       ‖
  :F̤—B—F̤:
```

This Lewis structure has octets for all atoms, including boron. However, when we assign formal charges to this structure, we get a negative formal charge on B and a positive formal charge on F.

```
      +1 F̈:
         ‖
  0 :F̤—B—F̤: 0
         −1
```

The positive formal charge on fluorine—the most electronegative element in the periodic table—makes this an unfavorable structure. This leaves us with some questions. Do we complete the octet on B at the expense of giving fluorine a positive formal charge? Or do we leave B without an octet in order to avoid the positive formal charge on fluorine? The answers to these kinds of questions are not always clear because we are pushing the limits of the Lewis model. In the case of boron, we usually accept the incomplete octet as the better Lewis structure. However, doing so does not rule out the possibility that the doubly bonded Lewis structure might be a minor contributing resonance structure. The ultimate answers to these kinds of questions must be determined from experiments. Experimental measurements of the B—F bond length in BF_3 suggest that the bond may be slightly shorter than expected for a single B—F bond, indicating that it may indeed have a small amount of double-bond character.

BF_3 can complete its octet in another way—via a chemical reaction. The Lewis model predicts that BF_3 might react in ways that would complete its octet, and indeed it does. For example, BF_3 reacts with NH_3 as follows:

$$:\ddot{F}-\underset{:\ddot{F}:}{\overset{}{B}}-\ddot{F}: + H-\underset{H}{\ddot{N}}-H \longrightarrow :\ddot{F}-\underset{:\ddot{F}:}{\overset{:\ddot{F}:}{B}}-\underset{H}{\overset{H}{N}}-H$$

When nitrogen bonds to boron, the nitrogen atom provides both of the electrons. This kind of bond is called a *coordinate covalent bond*, which we will discuss in Chapter 23.

The product has complete octets for all atoms in the structure.

Expanded Octets

Elements in the third row of the periodic table and beyond often exhibit *expanded octets* of up to 12 (and occasionally 14) electrons. Consider the Lewis structures of arsenic pentafluoride and sulfur hexafluoride:

$$AsF_5 \qquad SF_6$$

In AsF_5 arsenic has an expanded octet of 10 electrons, and in SF_6 sulfur has an expanded octet of 12 electrons. Both of these compounds exist and are stable. Ten- and twelve-electron expanded octets are common in third-period elements and beyond because the *d* orbitals in these elements are energetically accessible (they are not much higher in energy than the orbitals occupied by the valence electrons) and can accommodate the extra electrons (see Section 4.3). Expanded octets *never* occur in second-period elements.

In some Lewis structures, we must decide whether or not to expand an octet in order to lower formal charge. For example, consider the Lewis structure of H_2SO_4:

$$\underset{-1}{:\ddot{O}}-\underset{\underset{-1}{:\ddot{O}:}}{\overset{\overset{H}{|}\;:\ddot{O}:}{\overset{+2}{S}}}-\ddot{O}-H$$

Notice that both of the oxygen atoms have a -1 formal charge and that sulfur has a $+2$ formal charge. While this amount of formal charge is acceptable, especially since the negative formal charge resides on the more electronegative atom, it is possible to eliminate the formal charge by expanding the octet on sulfur:

$$\underset{0}{:\ddot{O}}=\underset{\underset{0}{:\ddot{O}}}{\overset{\overset{H}{|}\;:\ddot{O}:}{\overset{0}{S}}}-\ddot{O}-H$$

Which of these two Lewis structures for H_2SO_4 is better? Again, the answer is not straightforward. Experiments show that the sulfur–oxygen bond lengths in the two sulfur–oxygen bonds without the hydrogen atoms are shorter than expected for sulfur–oxygen single bonds, indicating that the double-bonded Lewis structure plays an important role in describing the bonding in H_2SO_4. In general, we expand octets in third-row (or beyond) elements in order to lower formal charge. However, we should *never* expand the octets of second-row elements. Second-row elements do not have energetically accessible *d* orbitals and therefore never exhibit expanded octets.

EXAMPLE 6.8

Writing Lewis Structures for Compounds Having Expanded Octets

Write the Lewis structure for XeF_2.

SOLUTION

Begin by writing the skeletal structure. Since xenon is the less electronegative atom, put it in the central position.	F Xe F
Calculate the total number of electrons for the Lewis structure by summing the number of valence electrons for each atom.	Total number of electrons for Lewis structure = (number of valence e^- in Xe) + 2(number of valence e^- in F) = 8 + 2(7) = 22
Place two bonding electrons between the atoms of each pair of atoms.	F:Xe:F (4 of 22 electrons used)
Distribute the remaining electrons to give octets to as many atoms as possible, beginning with terminal atoms and finishing with the central atom. Arrange additional electrons around the central atom, giving it an expanded octet of up to 12 electrons.	:F̤̈:Xe:F̤̈: (16 of 22 electrons used) :F̤̈:Ẍ̤e:F̤̈: or :F̤̈—Ẍ̤e—F̤̈: (22 of 22 electrons used)

FOR PRACTICE 6.8

Write the Lewis structure for XeF_4.

FOR MORE PRACTICE 6.8

Write the Lewis structure for H_3PO_4. If necessary, expand the octet on any appropriate atoms to lower formal charge.

6.4 Cc Conceptual Connection

Expanded Octets

Which molecule could have an expanded octet?

(a) H_2CO_3
(b) H_3PO_4
(c) HNO_2

6.6 Bond Energies and Bond Lengths

In the Lewis model, a bond is a shared electron pair; when we draw Lewis structures for molecular compounds, all bonds appear identical. However, from experiments we know that they are not identical—they can vary both in their strength (how strong the bond is) and their length. In this section, we examine the concepts of bond energy and bond length for a number of commonly encountered bonds. In Chapter 10, we will learn how to use these bond energies to calculate energy changes occurring in chemical reactions.

Bond Energy

The **bond energy** of a chemical bond is the energy required to break 1 mole of the bond in the gas phase. For example, the bond energy of the Cl—Cl bond in Cl_2 is 243 kJ/mol. Bond energies are positive because energy must be put into a molecule to break a bond (the process is endothermic, which, as discussed in Chapter 2, absorbs heat and carries a positive sign).

Bond energy is also called bond enthalpy or bond dissociation energy.

$$Cl_2(g) \longrightarrow 2\,Cl(g) \quad \text{Bond energy} = 243\text{ kJ}$$

The bond energy of HCl is 431 kJ/mol.

$$HCl(g) \longrightarrow H(g) + Cl(g) \quad \text{Bond energy} = 431\text{ kJ}$$

We say that the HCl bond is *stronger* than the Cl_2 bond because it requires more energy to break it. In general, compounds with stronger bonds tend to be more chemically stable, and therefore less chemically reactive, than compounds with weaker bonds. The triple bond in N_2 has a bond energy of 946 kJ/mol.

$$N_2(g) \longrightarrow N(g) + N(g) \quad \text{Bond energy} = 946\text{ kJ}$$

It is a very strong and stable bond, which explains nitrogen's relative inertness.

The bond energy of a particular bond in a polyatomic molecule is a little more difficult to determine because a particular type of bond can have different bond energies in different molecules. For example, consider the C—H bond. In CH_4, the energy required to break one C—H bond is 438 kJ/mol.

$$H_3C{-}H(g) \longrightarrow H_3C(g) + H(g) \quad \text{Bond energy} = 438\text{ kJ}$$

However, the energy required to break a C—H bond in other molecules varies slightly, as shown here:

$$F_3C{-}H(g) \longrightarrow F_3C(g) + H(g) \quad \text{Bond energy} = 446\text{ kJ}$$

$$Br_3C{-}H(g) \longrightarrow Br_3C(g) + H(g) \quad \text{Bond energy} = 402\text{ kJ}$$

$$Cl_3C{-}H(g) \longrightarrow Cl_3C(g) + H(g) \quad \text{Bond energy} = 401\text{ kJ}$$

We can calculate an *average bond energy* for a chemical bond, which is an average of the bond energies for that bond in a large number of compounds. For the limited number of compounds we just listed, we calculate an average C—H bond energy of 422 kJ/mol. Table 6.3 lists average bond energies for a number of common chemical bonds averaged over a large number of compounds. Notice that the C—H bond energy listed is 414 kJ/mol, which is not too different from the value we calculated from our limited number of compounds. Notice also that bond energies depend not only on the kind of atoms involved in the bond, but also on the type of bond: single, double, or triple. In general, for a given pair of atoms, triple bonds are stronger than double bonds, which are, in turn, stronger than single bonds. For example, for carbon–carbon bonds, notice the increase in bond energy in going from a single to a double and then to a triple bond.

TABLE 6.3 Average Bond Energies

Bond	Bond Energy (kJ/mol)	Bond	Bond Energy (kJ/mol)	Bond	Bond Energy (kJ/mol)
H—H	436	C—C	347	N≡N	946
H—C	414	C=C	611	O—O	142
H—N	389	C≡C	837	O=O	498
H—O	464	C—O	360	F—F	159
H—F	565	C=O	736*	Cl—Cl	243
H—Cl	431	C—Cl	339	Br—Br	193
H—Br	364	N—N	163	I—I	151
H—I	297	N=N	418		

*799 in CO_2

Bond Length

Just as we can tabulate average bond energies, which represent the average energy of a bond between two particular atoms in a large number of compounds, we can tabulate average bond lengths (Table 6.4). The average **bond length** represents the average length of a bond between two particular atoms in a large number of compounds. Like bond energies, bond lengths depend not only on the kind of atoms involved in the bond, but also on the type of bond: single, double, or triple. In general, for a particular pair of atoms, triple bonds are shorter than double bonds, which are in turn shorter than single bonds. For example, consider the bond lengths (shown here with bond energies, repeated from earlier in this section) of carbon–carbon triple, double, and single bonds:

Bond	Bond Length (pm)	Bond Energy (kJ/mol)
C≡C	120 pm	837 kJ/mol
C=C	134 pm	611 kJ/mol
C—C	154 pm	347 kJ/mol

Notice that, as the bond gets longer, it also becomes weaker. This relationship between the length of a bond and the strength of a bond does not necessarily hold true for all bonds. Consider the following series of nitrogen–halogen single bonds:

Bond	Bond Length (pm)	Bond Energy (kJ/mol)
N—F	139	272
N—Cl	191	200
N—Br	214	243
N—I	222	159

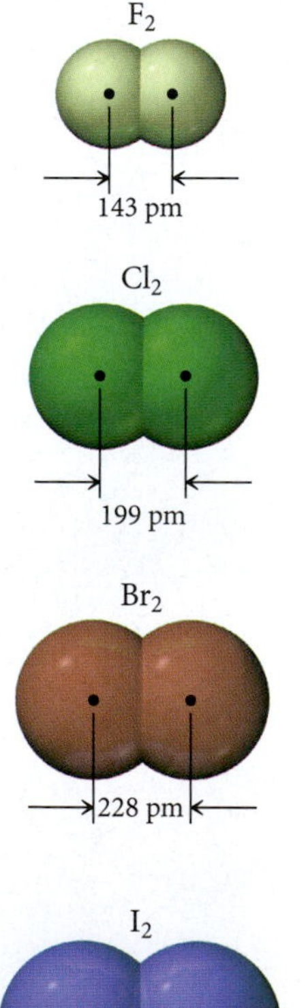

▲ Bond lengths in the diatomic halogen molecules.

Although the bonds generally get weaker as they get longer, the trend is not a smooth one.

In this chapter, we look at ways to predict and account for the shapes of molecules. The molecules we examine are much smaller than the molecules we discussed in Section 6.1, but the same principles apply to both. The simple model we examine to account for molecular shape is called *valence shell electron pair repulsion* (VSEPR) theory, and we will use it in conjunction with the Lewis model. In Chapter 7, we will explore two additional bonding theories: valence bond theory and molecular orbital theory. These bonding theories are more complex, but also more powerful, than the Lewis model. They predict and account for molecular shape as well as other properties of molecules.

TABLE 6.4 Average Bond Lengths

Bond	Bond Length (pm)	Bond	Bond Length (pm)	Bond	Bond Length (pm)
H—H	74	C—C	154	N≡N	110
H—C	110	C=C	134	O—O	145
H—N	100	C≡C	120	O=O	121
H—O	97	C—O	143	F—F	143
H—F	92	C=O	120	Cl—Cl	199
H—Cl	127	C—Cl	178	Br—Br	228
H—Br	141	N—N	145	I—I	266
H—I	161	N=N	123		

6.7 VSEPR Theory: The Five Basic Shapes

Valence shell electron pair repulsion (VSEPR) theory is based on the simple idea that **electron groups**—which we define as lone pairs, single bonds, multiple bonds, and even single electrons—repel one another through coulombic forces. The electron groups, of course, are also attracted to the nucleus (otherwise the molecule would fall apart), but VSEPR theory focuses on the repulsions. According to VSEPR theory, the repulsions between electron groups on *interior atoms* of a molecule determine the geometry of the molecule (Figure 6.7 ▶). The preferred geometry of a molecule is the one in which the electron groups have the maximum separation (and therefore the minimum energy) possible. Consequently, for molecules having just one interior atom (the central atom) molecular geometry depends on (a) the number of electron groups around the central atom and (b) how many of those electron groups are bonding groups and how many are lone pairs. In this section, we first look at the molecular geometries associated with two to six electron groups around the central atom when all of those groups are bonding groups (single or multiple bonds). The resulting geometries constitute the five basic shapes of molecules. We will then consider how these basic shapes are modified if one or more of the electron groups are lone pairs.

▲ **FIGURE 6.7 Repulsion between Electron Groups** The basic idea of VSEPR theory is that repulsions between electron groups determine molecular geometry.

Two Electron Groups: Linear Geometry

KEY CONCEPT VIDEO VSEPR theory

Consider the Lewis structure of $BeCl_2$, which has two electron groups (two single bonds) about the central atom:

$$:\ddot{\underset{..}{Cl}}:Be:\ddot{\underset{..}{Cl}}:$$

Beryllium often forms incomplete octets, as it does in this structure.

According to VSEPR theory, the geometry of $BeCl_2$ is determined by the repulsion between these two electron groups, which maximize their separation by assuming a 180° bond angle or a **linear geometry**. Experimental measurements of the geometry of $BeCl_2$ indicate that the molecule is indeed linear, as predicted by the theory.

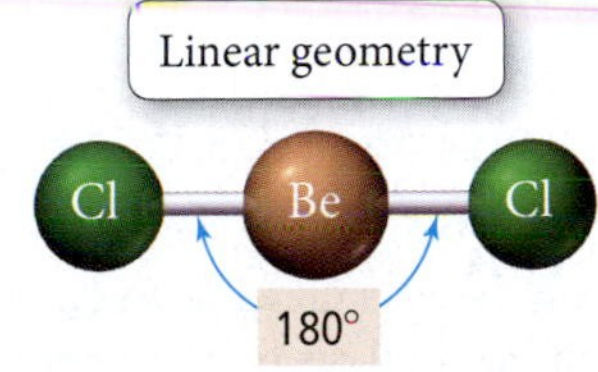

Molecules that form only two single bonds, with no lone pairs, are rare because they do not follow the octet rule. However, the same geometry is observed in all molecules that have two electron groups (and no lone pairs). Consider the Lewis structure of CO_2, which has two electron groups (the double bonds) around the central carbon atom:

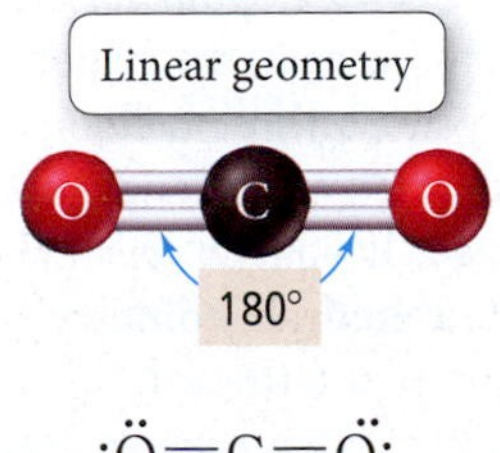

$$:\ddot{O}=C=\ddot{O}:$$

According to VSEPR theory, the two double bonds repel each other (just as the two single bonds in $BeCl_2$ repel each other), resulting in a linear geometry for CO_2. Experimental observations confirm that CO_2 is indeed a linear molecule.

A double bond counts as one electron group.

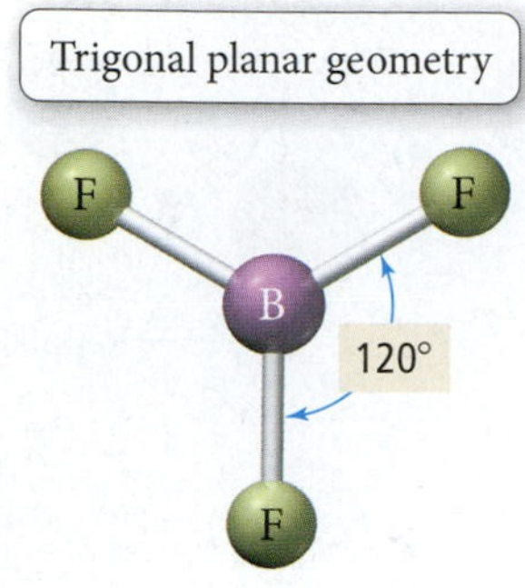

Three Electron Groups: Trigonal Planar Geometry

The Lewis structure of BF_3 (another molecule with an incomplete octet) has three electron groups around the central atom:

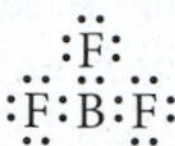

These three electron groups can maximize their separation by assuming 120° bond angles in a plane—a **trigonal planar geometry**. Experimental observations of the structure of BF_3 again confirm the predictions of VSEPR theory.

Another molecule with three electron groups, formaldehyde, has one double bond and two single bonds around the central atom:

```
   :O:
    ‖
H—C—H
```

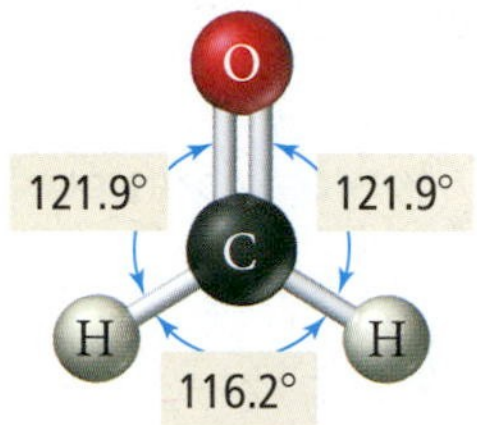

Since formaldehyde has three electron groups around the central atom, we initially predict that the bond angles should also be 120°. However, experimental observations show that the HCO bond angles are 121.9° and that the HCH bond angle is 116.2°. These bond angles are close to the idealized 120° that we originally predicted, but the HCO bond angles are slightly greater than the HCH bond angle because the double bond contains more electron density than the single bond and therefore exerts a slightly greater repulsion on the single bonds. In general, *different types of electron groups exert slightly different repulsions*—the resulting bond angles reflect these differences.

6.5 Cc Conceptual Connection

Electron Groups and Molecular Geometry

In determining electron geometry, why do we consider only the electron groups on the central atom? Why don't we consider electron groups on terminal atoms?

Four Electron Groups: Tetrahedral Geometry

The VSEPR geometries of molecules with two or three electron groups around the central atom are two-dimensional and therefore can easily be visualized and represented on paper. For molecules with four or more electron groups around the central atom, the geometries are three-dimensional and are therefore more difficult to imagine and draw. One common way to help visualize these basic shapes is by analogy to balloons tied together. In this analogy, each electron group around a central atom is like a balloon tied to a central point. The bulkiness of the balloons causes them to spread out as much as possible, much as the repulsion between electron groups causes them to position themselves as far apart as possible. For example, if you tie two balloons together, they assume a roughly linear arrangement, as shown in Figure 6.8a ◀, analogous to the linear geometry of $BeCl_2$ that we just examined. Keep in mind that the balloons do not represent atoms, but *electron groups*. Similarly, if you tie three balloons together—in analogy to three electron groups—they assume a trigonal planar geometry, as shown in Figure 6.8b ◀, much like the BF_3 molecule. If you tie *four* balloons together, however, they assume a three-dimensional **tetrahedral geometry** with 109.5° angles between the balloons. That is, the balloons point toward the vertices of a *tetrahedron*—a geometrical shape with four identical faces, each an equilateral triangle, as shown here:

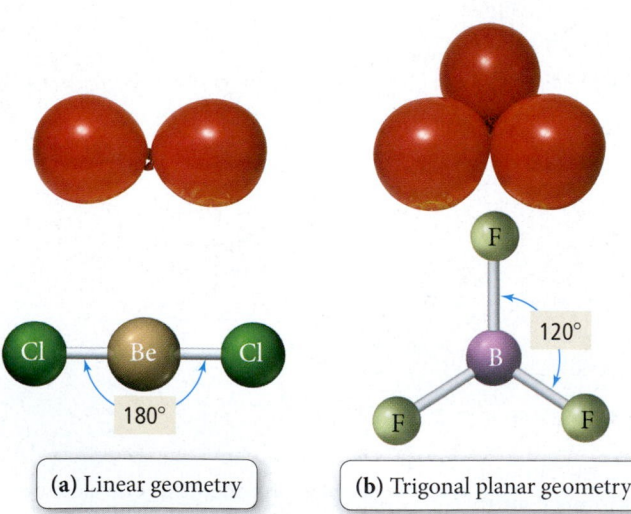

Cl Be Cl 180° — **(a)** Linear geometry

F B F F 120° — **(b)** Trigonal planar geometry

▲ **FIGURE 6.8 Representing Electron Geometry with Balloons** **(a)** The bulkiness of balloons causes them to assume a linear arrangement when two of them are tied together. Similarly, the repulsion between two electron groups produces a linear geometry. **(b)** Like three balloons tied together, three electron groups adopt a trigonal planar geometry.

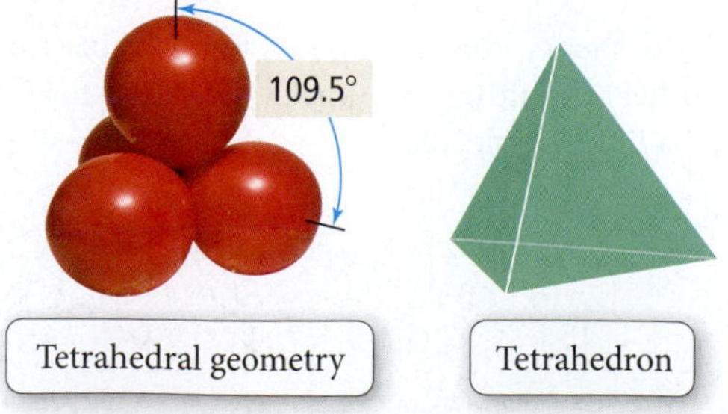

Methane is an example of a molecule with four electron groups around the central atom:

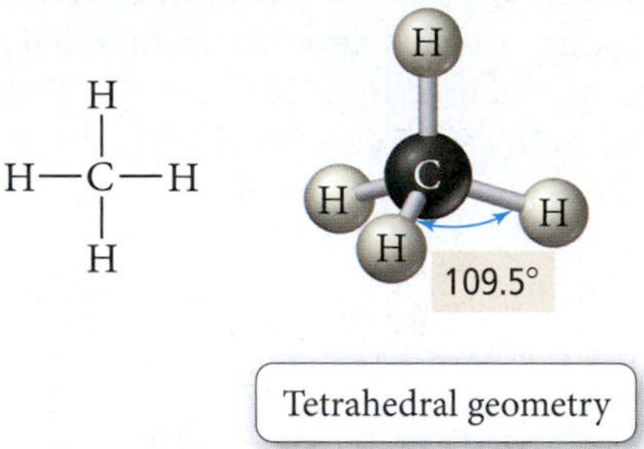

Tetrahedral geometry

For four electron groups, the tetrahedron is the three-dimensional shape that allows the maximum separation among the groups. The repulsions among the four electron groups in the C—H bonds cause the molecule to assume the tetrahedral shape. When we write the Lewis structure of CH_4 on paper, it may seem that the molecule should be square planar, with bond angles of 90°. However, in three dimensions, the electron groups can get farther away from each other by forming the tetrahedral geometry, as illustrated in our balloon analogy.

Molecular Geometry

6.6 Cc Conceptual Connection

What is the geometry of the HCN molecule? The Lewis structure of HCN is H—C≡N:

(a) Linear **(b)** Trigonal planar **(c)** Tetrahedral

Five Electron Groups: Trigonal Bipyramidal Geometry

Five electron groups around a central atom assume a **trigonal bipyramidal geometry**, like five balloons tied together. In this structure, three of the groups lie in a single plane, as in the trigonal planar configuration, while the other two are positioned above and below this plane. The angles in the trigonal bipyramidal structure are not all the same. The angles between the *equatorial positions* (the three bonds in the trigonal plane) are 120°, while the angle between the *axial positions* (the two bonds on either side of the trigonal plane) and the trigonal plane is 90°. As an example of a molecule with five electron groups around the central atom, consider PCl_5:

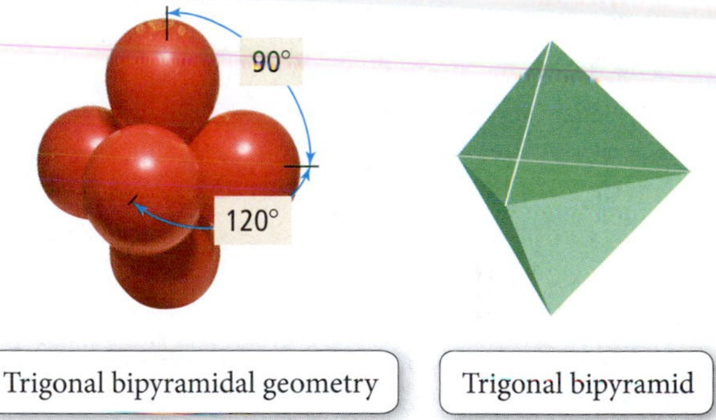

Trigonal bipyramidal geometry

Trigonal bipyramid

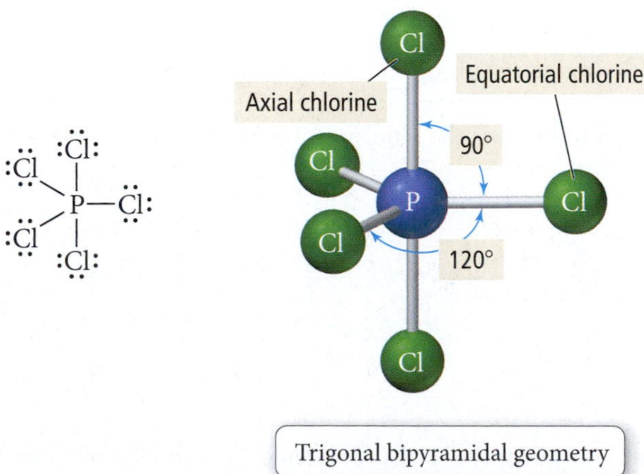

Trigonal bipyramidal geometry

The three equatorial chlorine atoms are separated by 120° bond angles, and the two axial chlorine atoms are separated from the equatorial atoms by 90° bond angles.

90°

90°

Octahedral geometry

Octahedron

Six Electron Groups: Octahedral Geometry

Six electron groups around a central atom assume an **octahedral geometry**, like six balloons tied together. In this structure—named after the eight-sided geometrical shape called the octahedron—four of the groups lie in a single plane, with a fifth group above the plane and another below it. The angles in this geometry are all 90°. As an example of a molecule with six electron groups around the central atom, consider SF_6:

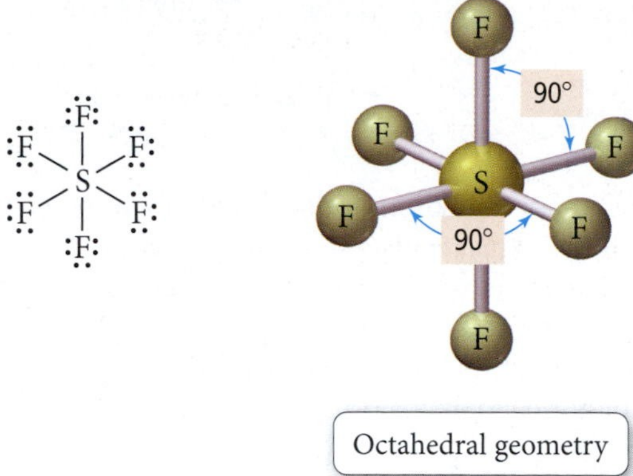

Octahedral geometry

The structure of this molecule is highly symmetrical; all six bonds are equivalent.

EXAMPLE 6.9

VSEPR Theory and the Basic Shapes

Determine the molecular geometry of NO_3^-.

SOLUTION

The molecular geometry of NO_3^- is determined by the number of electron groups around the central atom (N). Begin by drawing the Lewis structure of NO_3^-.	NO_3^- has $5 + 3(6) + 1 = 24$ valence electrons. The Lewis structure is as follows: [:Ö—N—Ö: with N=O:]$^-$ ⟷ [Ö=N—Ö: with N—:Ö:]$^-$ ⟷ [:Ö—N=Ö with N—:Ö:]$^-$ The hybrid structure is intermediate between these three and has three equivalent bonds.
Use any one of the resonance structures to determine the number of electron groups around the central atom.	[:Ö—N—Ö: with N=O:]$^-$ The nitrogen atom has three electron groups.
Based on the number of electron groups, determine the geometry that minimizes the repulsions between the groups.	The electron geometry that minimizes the repulsions between three electron groups is trigonal planar. 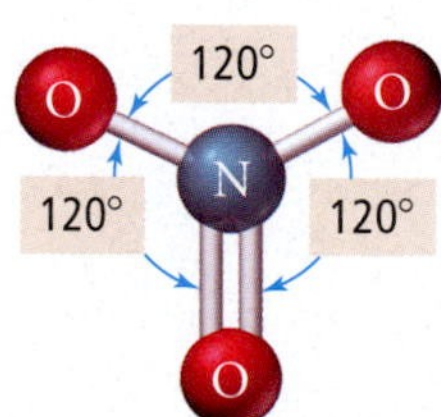 Since the three bonds are equivalent (because of the resonance structures), they each exert the same repulsion on the other two and the molecule has three equal bond angles of 120°.

FOR PRACTICE 6.9

Determine the molecular geometry of CCl_4.

6.8 VSEPR Theory: The Effect of Lone Pairs

Each of the examples we looked at in Section 6.7 had only bonding electron groups around the central atom. What happens in molecules that have lone pairs around the central atom as well? The lone pairs also repel other electron groups, as we see in the examples that follow.

Four Electron Groups with Lone Pairs

Consider the Lewis structure of ammonia:

H
|
H—N—H
··

The central nitrogen atom has four electron groups (one lone pair and three bonding pairs) that repel each other. If we do not distinguish between bonding electron groups and lone pairs, we find that the **electron geometry**—the geometrical arrangement of the *electron groups*—is still tetrahedral, as we expect for four electron groups. However, the **molecular geometry**—the geometrical arrangement of the atoms—is **trigonal pyramidal**, as shown here:

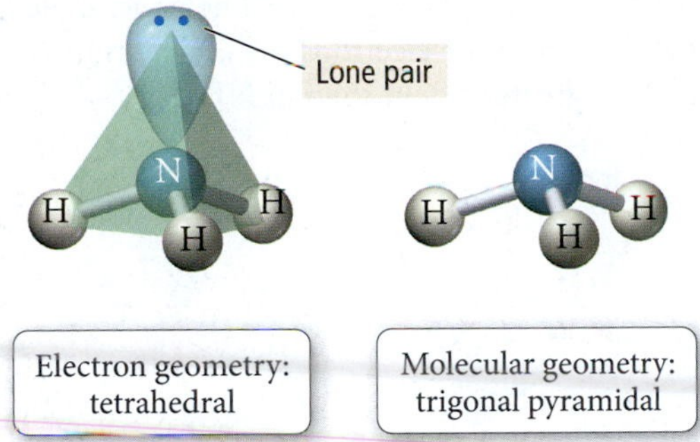

Notice that although the electron geometry and the molecular geometry are different, *the electron geometry is relevant to the molecular geometry*. The lone pair exerts its influence on the bonding pairs.

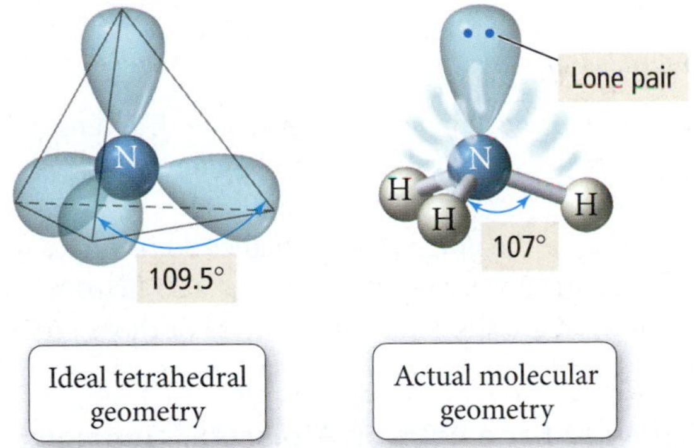

As we noted previously, different kinds of electron groups generally result in different amounts of repulsion. Lone pair electrons generally exert slightly greater repulsions than bonding electrons. If all four electron groups in NH_3 exerted equal repulsions on one another, the bond angles in the molecule would all be the ideal tetrahedral angle, 109.5°. However, the actual angle between N—H bonds in ammonia is slightly smaller, 107°. A lone electron pair is more spread out in space than a bonding electron pair because a lone pair is attracted to only one nucleus while a bonding pair is attracted to two (Figure 6.9 ▶). The lone pair occupies more of the angular space around a nucleus, exerting a greater repulsive force on neighboring electrons and compressing the N—H bond angles.

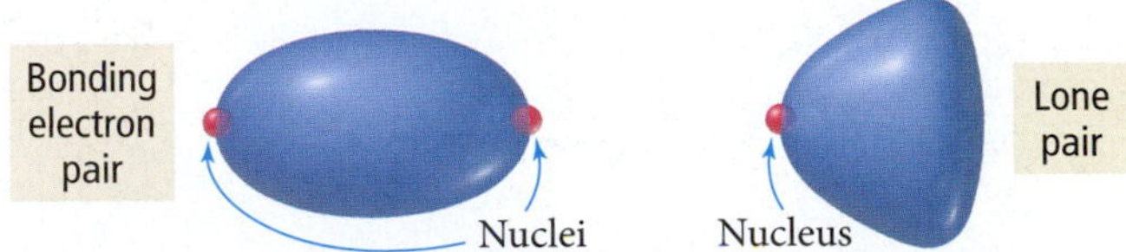

▲ **FIGURE 6.9 Nonbonding versus Bonding Electron Pairs** A lone electron pair occupies more angular space than a bonding pair.

We see a similar effect in water. The Lewis structure of water has two bonding pairs and two lone pairs:

$$\mathrm{H{-}\ddot{\underset{..}{O}}{-}H}$$

Since it has four electron groups, its *electron geometry* is tetrahedral (like that of ammonia), but its *molecular geometry* is **bent**.

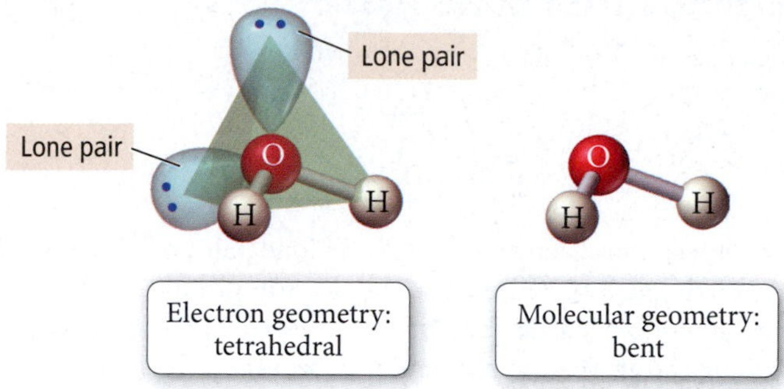

As in NH_3, the bond angles in H_2O are smaller (104.5°) than the ideal tetrahedral bond angles because of the greater repulsion exerted by the lone pair electrons. The bond angle in H_2O is even smaller than in NH_3 because H_2O has *two* lone pairs of electrons on the central oxygen atom. These lone pairs compress the H_2O bond angle to a greater extent than in NH_3:

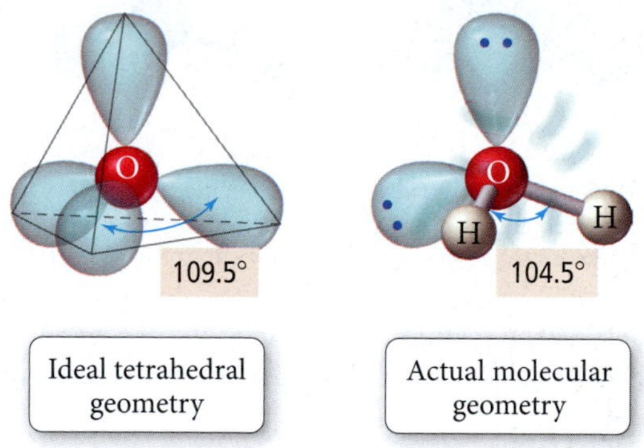

In general, electron group repulsions vary as follows:

Lone pair—lone pair > Lone pair—bonding pair > Bonding pair—bonding pair

Most repulsive — Least repulsive

We see the effects of this ordering in the progressively smaller bond angles of CH_4, NH_3, and H_2O, as shown in Figure 6.10 ▼. The relative ordering of repulsions also helps to determine the geometry of molecules with five and six electron groups when one or more of those groups are lone pairs.

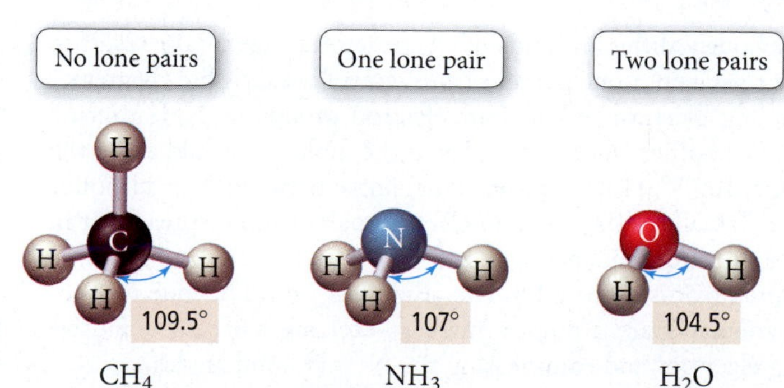

▶ **FIGURE 6.10 The Effect of Lone Pairs on Molecular Geometry** The bond angles get progressively smaller as the number of lone pairs on the central atom increases from zero in CH_4 to one in NH_3 to two in H_2O.

Five Electron Groups with Lone Pairs

Consider the Lewis structure of SF_4:

The central sulfur atom has five electron groups (one lone pair and four bonding pairs). The *electron geometry*, due to the five electron groups, is trigonal bipyramidal. In determining the molecular geometry, notice that the lone pair could occupy either an equatorial position or an axial position within the trigonal bipyramidal electron geometry. Which position is more favorable? To answer this question, we must consider that, as we have just seen, lone pair–bonding pair repulsions are greater than bonding pair–bonding pair repulsions. Therefore, the lone pair occupies the position that minimizes its interaction with the bonding pairs. If the lone pair were in an axial position, it would have three 90° interactions with bonding pairs. In an equatorial position, however, it has only two 90° interactions. Consequently, the lone pair occupies an equatorial position. The resulting molecular geometry is called **seesaw** because it resembles a seesaw (or teeter-totter).

The seesaw molecular geometry is sometimes called an *irregular tetrahedron.*

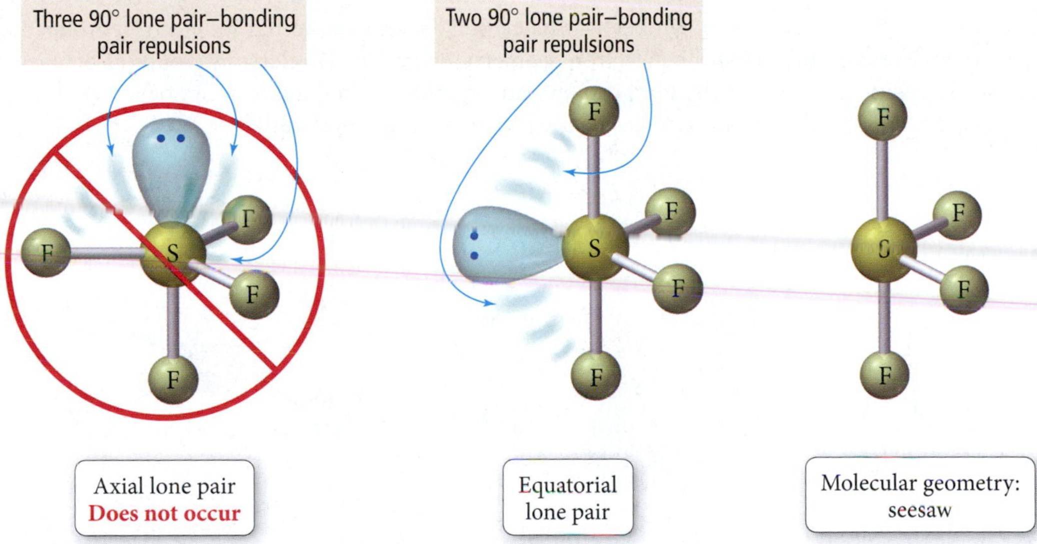

When two of the five electron groups around the central atom are lone pairs, as in BrF_3, the lone pairs occupy two of the three equatorial positions—again minimizing 90° interactions with bonding pairs and also avoiding a lone pair–lone pair 90° repulsion. The resulting molecular geometry is **T-shaped**.

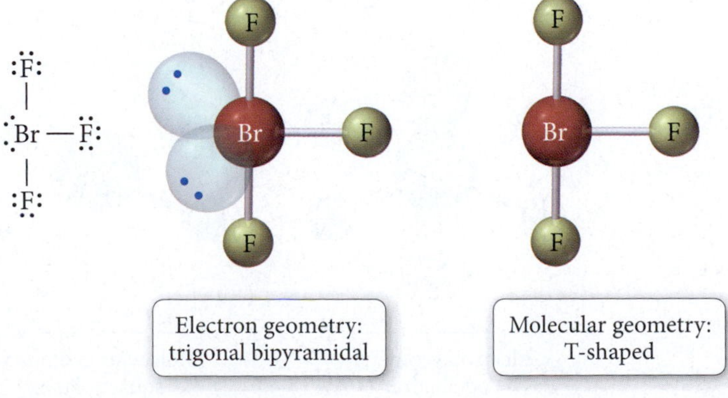

When three of the five electron groups around the central atom are lone pairs, as in XeF_2, the lone pairs occupy all three of the equatorial positions and the resulting molecular geometry is linear.

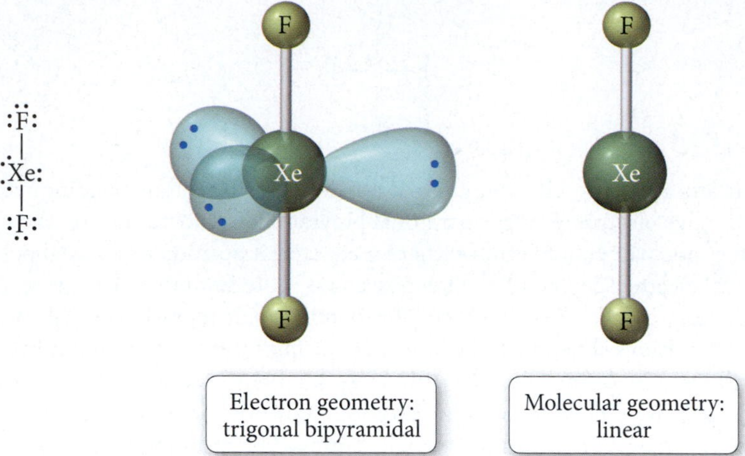

Six Electron Groups with Lone Pairs

The Lewis structure of BrF_5 is shown below. The central bromine atom has six electron groups (one lone pair and five bonding pairs). The electron geometry, due to the six electron groups, is octahedral. Since all six positions in the octahedral geometry are equivalent, the lone pair can be situated in any one of these positions. The resulting molecular geometry is **square pyramidal**.

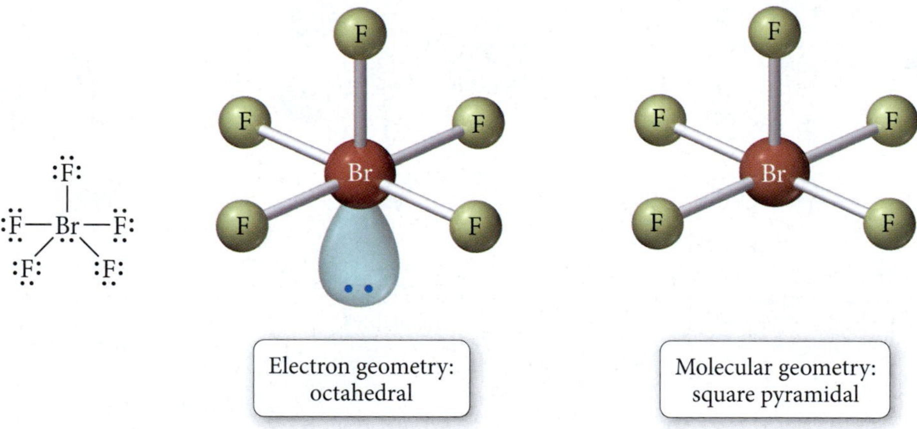

When two of the six electron groups around the central atom are lone pairs, as in XeF_4, the lone pairs occupy positions across from one another (to minimize lone pair–lone pair repulsions), and the resulting molecular geometry is **square planar**.

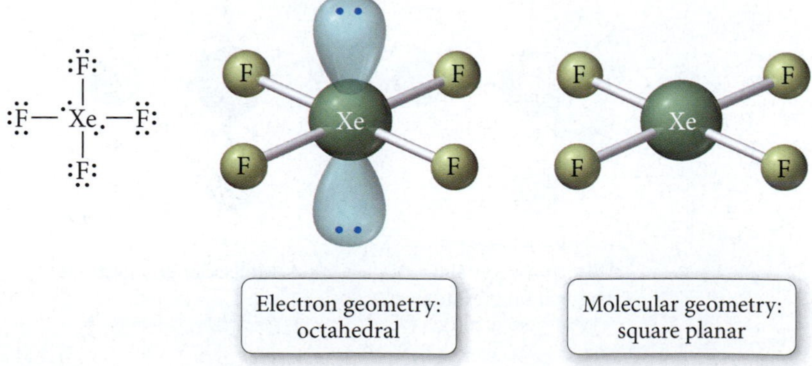

Lone Pair Electrons and Molecular Geometry

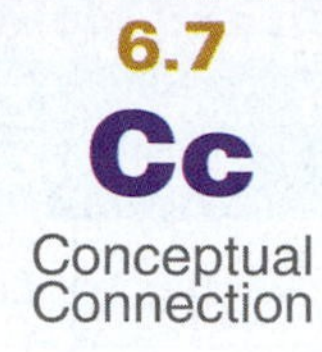

Suppose that a molecule with six electron groups were confined to two dimensions and therefore had a hexagonal planar electron geometry. If two of the six groups were lone pairs, where would they be located in the figure shown here?

(a) positions 1 and 2
(b) positions 1 and 3
(c) positions 1 and 4

Summarizing VSEPR Theory:

- The geometry of a molecule is determined by the number of electron groups on the central atom (or on each interior atom, if there is more than one).
- The number of electron groups is determined from the Lewis structure of the molecule. If the Lewis structure contains resonance structures, we can use any one of the resonance structures to determine the number of electron groups.
- Each of the following counts as a single electron group: a lone pair, a single bond, a double bond, a triple bond, or a single electron.
- The geometry of the electron groups is determined by their repulsions as summarized in Table 6.5 on the next page. In general, electron group repulsions vary as follows:

  Lone pair—lone pair > Lone pair—bonding pair > Bonding pair—bonding pair

- Bond angles can vary from the idealized angles because double and triple bonds occupy more space than single bonds (they are bulkier even though they are shorter), and lone pairs occupy more space than bonding groups. The presence of lone pairs usually makes bond angles smaller than the ideal angle for the particular geometry.

Molecular Geometry and Electron Group Repulsions

6.8
Cc
Conceptual Connection

Which statement is *always* true according to VSEPR theory?

(a) The shape of a molecule is determined only by repulsions among bonding electron groups.
(b) The shape of a molecule is determined only by repulsions among nonbonding electron groups.
(c) The shape of a molecule is determined by the polarity of its bonds.
(d) The shape of a molecule is determined by repulsions among all electron groups on the central atom (or interior atoms, if there is more than one).

6.9 VSEPR Theory: Predicting Molecular Geometries

To determine the geometry of a molecule, follow the procedure presented in Examples 6.10 and 6.11. As in other examples, we provide the steps in the left column and provide two examples of applying the steps in the center and right columns.

TABLE 6.5 Electron and Molecular Geometries

Electron Groups*	Bonding Groups	Lone Pairs	Electron Geometry	Molecular Geometry	Approximate Bond Angles	Example
2	2	0	Linear	Linear	180°	:Ö=C=Ö:
3	3	0	Trigonal planar	Trigonal planar	120°	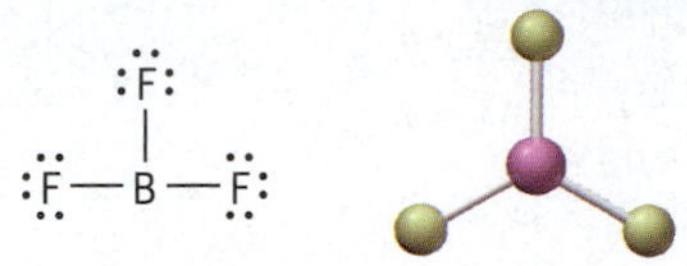 BF_3
3	2	1	Trigonal planar	Bent	<120°	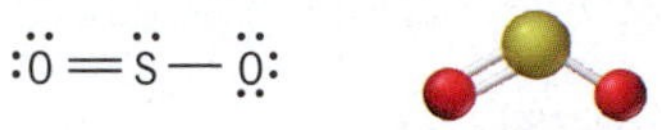 :Ö=S̈—Ö:
4	4	0	Tetrahedral	Tetrahedral	109.5°	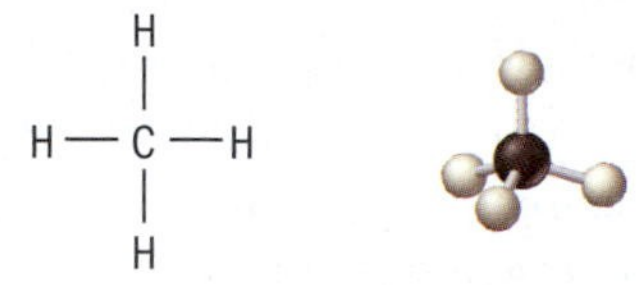 CH_4
4	3	1	Tetrahedral	Trigonal pyramidal	<109.5°	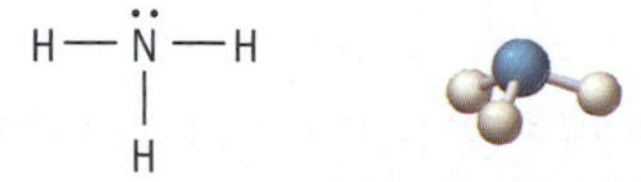 NH_3
4	2	2	Tetrahedral	Bent	<109.5°	 H—Ö—H
5	5	0	Trigonal bipyramidal	Trigonal bipyramidal	120° (equatorial) 90° (axial)	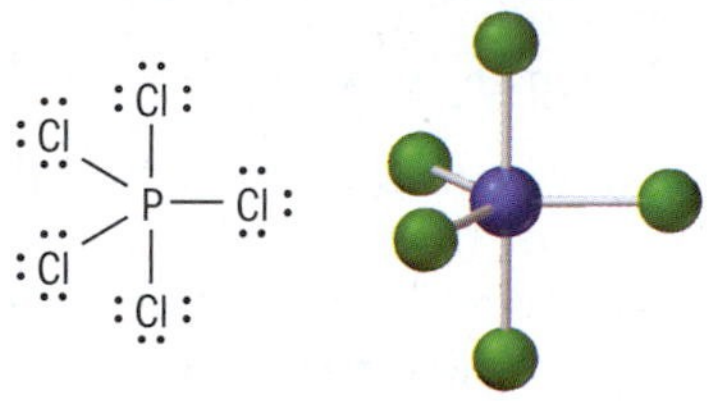 PCl_5
5	4	1	Trigonal bipyramidal	Seesaw	<120° (equatorial) <90° (axial)	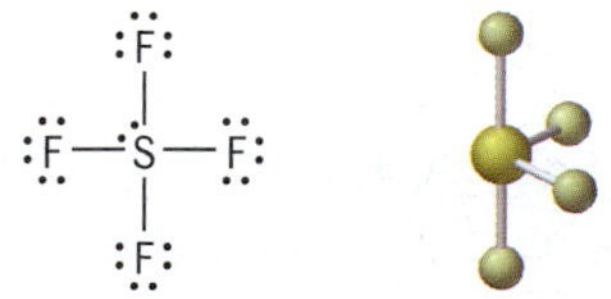 SF_4
5	3	2	Trigonal bipyramidal	T-shaped	<90°	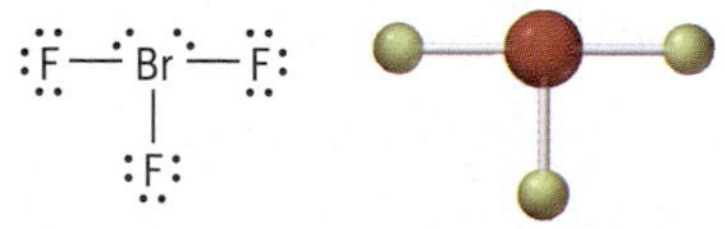 BrF_3
5	2	3	Trigonal bipyramidal	Linear	180°	 :F̤—Xe—F̤:
6	6	0	Octahedral	Octahedral	90°	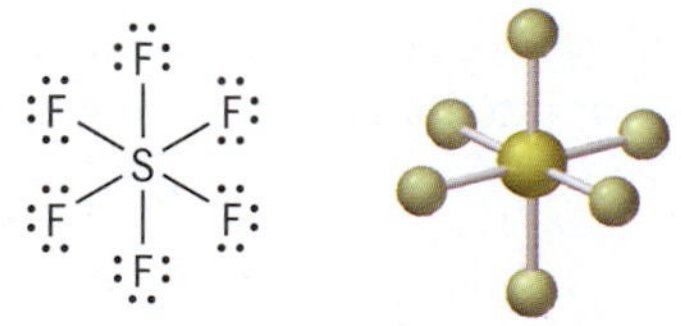 SF_6
6	5	1	Octahedral	Square pyramidal	<90°	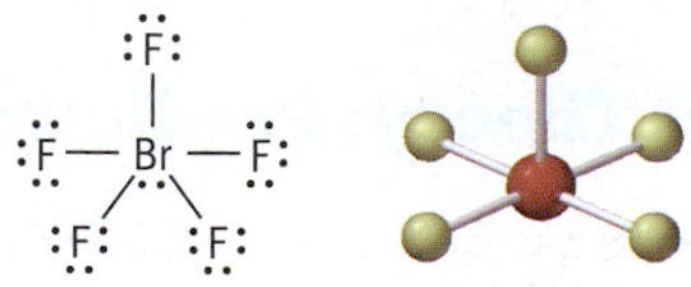 BrF_5
6	4	2	Octahedral	Square planar	90°	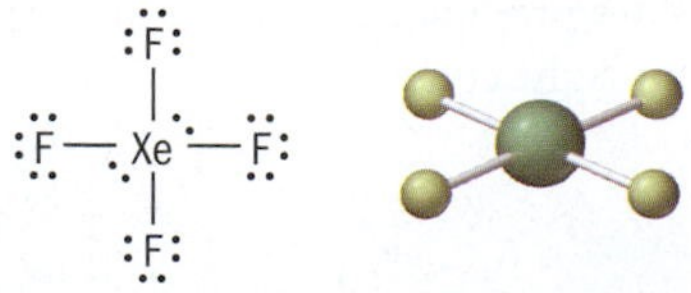 XeF_4

*Count only electron groups around the central atom. Each of the following is considered one electron group: a lone pair, a single bond, a double bond, a triple bond, or a single electron.

PROCEDURE FOR ▼ Predicting Molecular Geometries

EXAMPLE 6.10 Predicting Molecular Geometries

Predict the geometry and bond angles of PCl_3.

EXAMPLE 6.11 Predicting Molecular Geometries

Predict the geometry and bond angles of ICl_4^-.

---

1. ***Draw the Lewis structure for the molecule.***

Example 6.10: PCl_3 has 26 valence electrons.

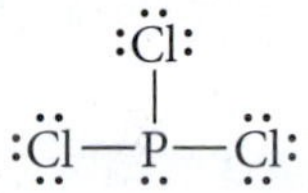

```
       :C̈l:
        |
  :C̈l—P—C̈l:
   ¨   ¨   ¨
```

Example 6.11: ICl_4^- has 36 valence electrons.

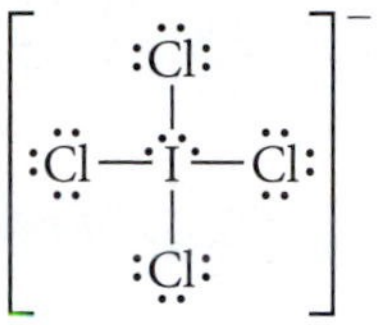

```
  ┌     :C̈l:     ┐⁻
  |      |       |
  | :C̈l—Ï—C̈l:   |
  |      |       |
  └     :C̈l:     ┘
```

2. ***Determine the total number of electron groups around the central atom.*** Lone pairs, single bonds, double bonds, triple bonds, and single electrons each count as one group.

Example 6.10: The central atom (P) has four electron groups.

Example 6.11: The central atom (I) has six electron groups.

3. ***Determine the number of bonding groups and the number of lone pairs around the central atom.*** These should sum to your result from step 2. Bonding groups include single bonds, double bonds, and triple bonds.

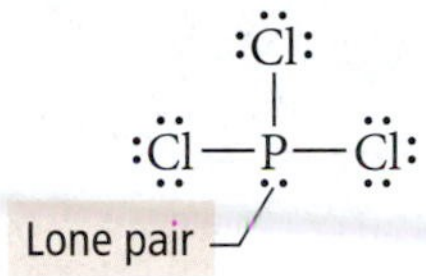

Example 6.10:

```
       :C̈l:
        |
  :C̈l—P—C̈l:
        ¨  ← Lone pair
```

Three of the four electron groups around P are bonding groups and one is a lone pair.

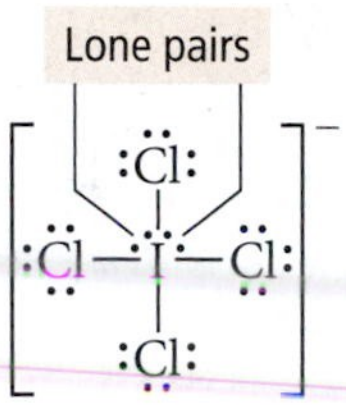

Example 6.11:

```
        Lone pairs
  ┌     :C̈l:     ┐⁻
  |      |       |
  | :C̈l—Ï—C̈l:   |
  |      |       |
  └     :C̈l:     ┘
```

Four of the six electron groups around I are bonding groups, and two are lone pairs.

4. ***Refer to Table 6.5 to determine the electron geometry and molecular geometry.*** If no lone pairs are present around the central atom, the bond angles will be that of the ideal geometry. If lone pairs are present, the bond angles may be smaller than the ideal geometry.

Example 6.10: The electron geometry is tetrahedral (four electron groups) and the molecular geometry—the shape of the molecule—is *trigonal pyramidal* (three bonding groups and one lone pair). Because of the presence of a lone pair, the bond angles are less than 109.5°.

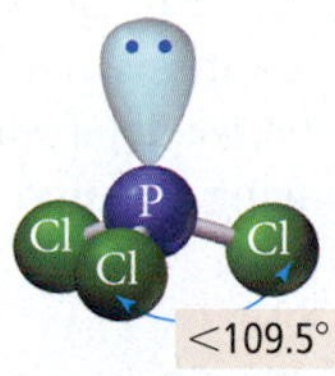

Trigonal pyramidal

Example 6.11: The electron geometry is octahedral (six electron groups) and the molecular geometry—the shape of the molecule—is *square planar* (four bonding groups and two lone pairs). Even though lone pairs are present, the bond angles are 90° because the lone pairs are symmetrically arranged and do not compress the I—Cl bond angles.

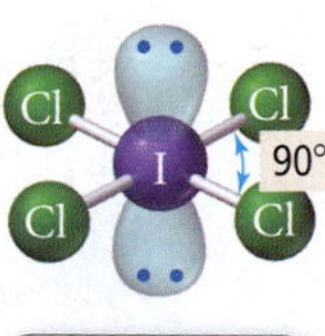

Square planar

FOR PRACTICE 6.10
Predict the molecular geometry and bond angle of ClNO.

FOR PRACTICE 6.11
Predict the molecular geometry of I_3^-.

Representing Molecular Geometries on Paper

Since molecular geometries are three-dimensional, they are often difficult to represent on two-dimensional paper. Many chemists use the notation shown here for bonds to indicate three-dimensional structures on two-dimensional paper.

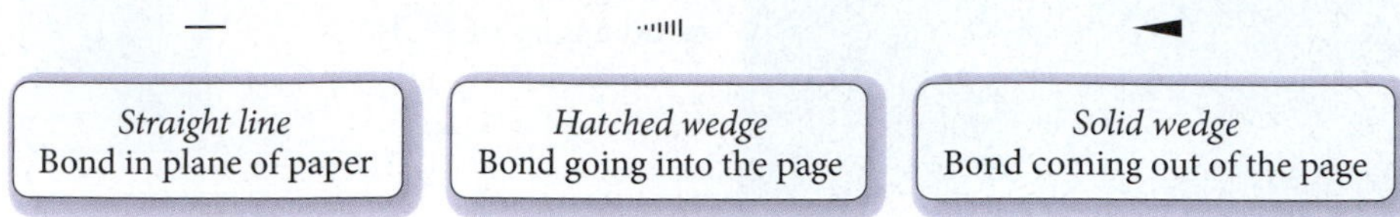

Some examples of the molecular geometries used in this book are shown here using this notation.

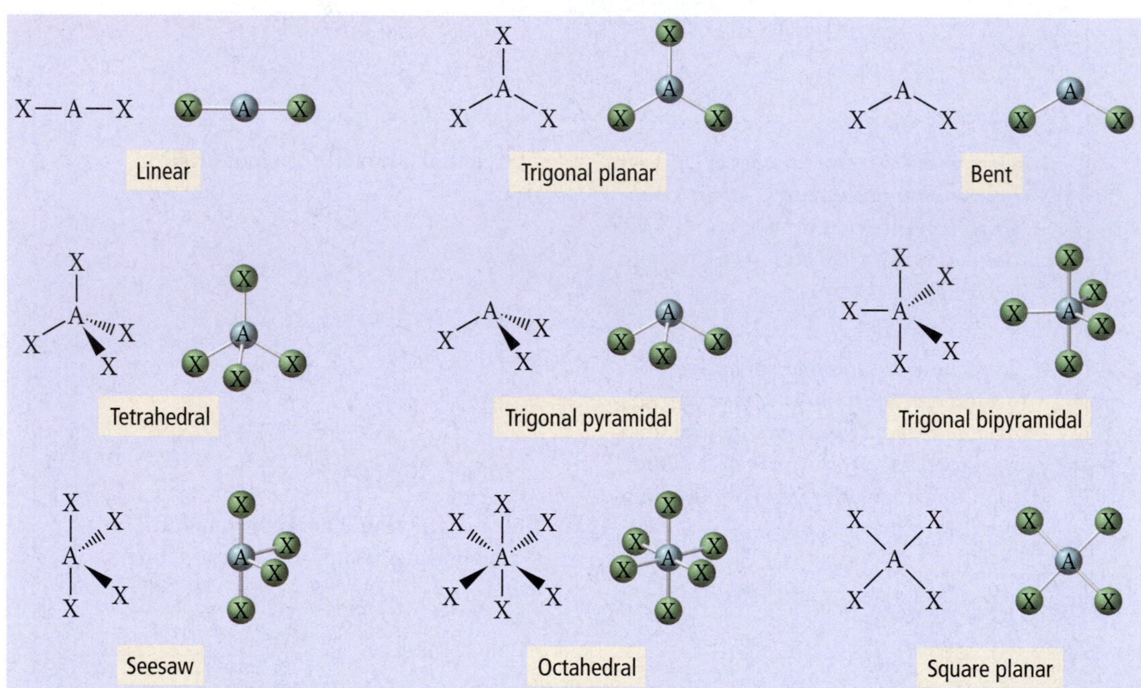

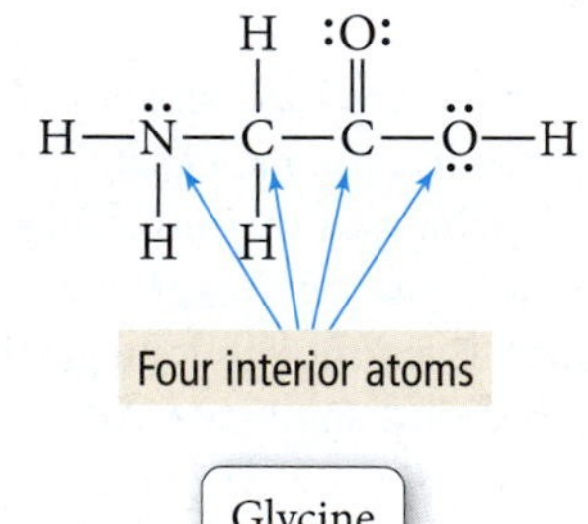

Glycine

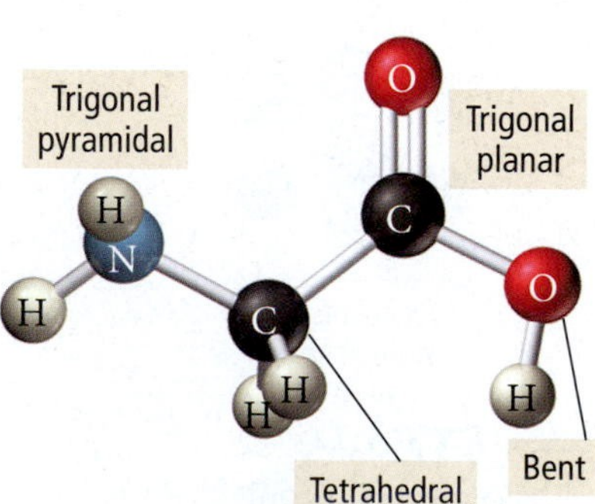

Ball-and-Stick Model of Glycine

Predicting the Shapes of Larger Molecules

Larger molecules may have two or more *interior* atoms. When predicting the shapes of these molecules, we apply the principles we just covered to each interior atom. Consider glycine, an amino acid found in many proteins (such as those involved in opioid receptors discussed in Section 6.1). Glycine, shown at left, contains four interior atoms: one nitrogen atom, two carbon atoms, and an oxygen atom. To determine the shape of glycine, we determine the geometry about each interior atom as shown in the following table and accompanying ball-and-stick model shown at left.

Atom	Number of Electron Groups	Number of Lone Pairs	Molecular Geometry
Nitrogen	4	1	Trigonal pyramidal
Leftmost carbon	4	0	Tetrahedral
Rightmost carbon	3	0	Trigonal planar
Oxygen	4	2	Bent

EXAMPLE 6.12

Predicting the Shape of Larger Molecules

Predict the geometry about each interior atom in methanol (CH_3OH) and make a sketch of the molecule.

SOLUTION

Begin by drawing the Lewis structure of CH_3OH. CH_3OH contains two interior atoms: one carbon atom and one oxygen atom. To determine the shape of methanol, determine the geometry about each interior atom.

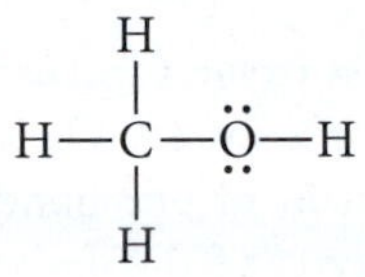

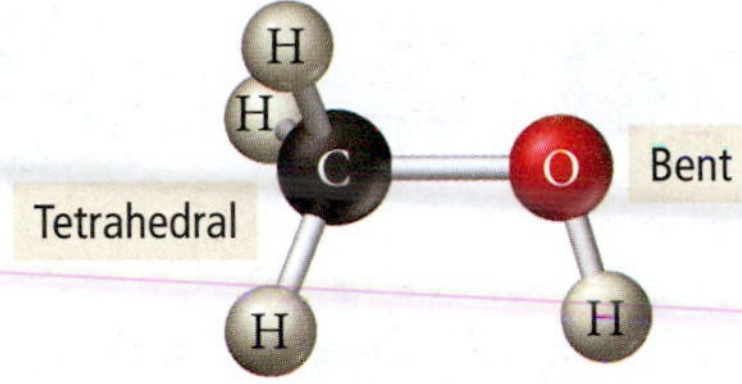

Atom	Number of Electron Groups	Number of Lone Pairs	Molecular Geometry
Carbon	4	0	Tetrahedral
Oxygen	4	2	Bent

Using the geometries of each of these, draw a three-dimensional sketch of the molecule as shown here.

FOR PRACTICE 6.12

Predict the geometry about each interior atom in acetic acid:

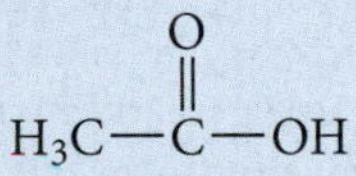

$$H_3C—\overset{\overset{\displaystyle O}{\|}}{C}—OH$$

and make a sketch of the molecule.

6.10 Molecular Shape and Polarity

In Section 6.2, we discussed polar bonds. Entire molecules can also be polar, depending on their shape and the nature of their bonds. For example, if a diatomic molecule has a polar bond, the molecule as a whole will be polar.

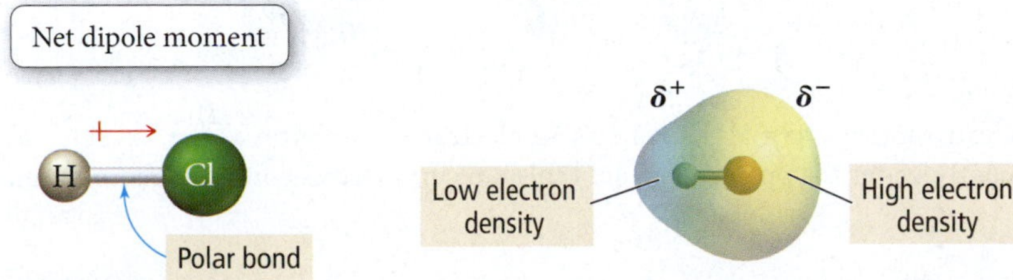

In the figure shown here the image to the right is an electrostatic potential map of HCl. In these maps, red areas indicate electron-rich regions in the molecule and the blue areas indicate electron-poor regions. Yellow indicates moderate electron density. Notice that the region around the more electronegative atom (chlorine) is more electron rich than the region around the hydrogen atom. Thus the molecule itself is polar. If the bond in a diatomic molecule is *nonpolar*, the molecule as a whole will be *nonpolar*.

In polyatomic molecules, the presence of polar bonds may or may not result in a polar molecule, depending on the molecular geometry. If the molecular geometry is such that the dipole moments of individual polar bonds sum together to a net dipole moment, then the molecule is polar. But if the molecular geometry is such that the dipole moments of the individual polar bonds cancel each other (that is, sum to zero), then the molecule is nonpolar. It all depends on the geometry of the molecule. Consider carbon dioxide:

$$:\ddot{O}=C=\ddot{O}:$$

Each C═O bond in CO_2 is polar because oxygen and carbon have significantly different electronegativities (3.5 and 2.5, respectively). However, since CO_2 is a linear molecule, the polar bonds directly oppose one another and the dipole moment of one bond exactly opposes the dipole moment of the other. The two dipole moments sum to zero and the *molecule* is nonpolar. Dipole moments cancel each other because they are *vector quantities*; they have both a magnitude and a direction. Think of each polar bond as a vector, pointing in the direction of the more electronegative atom. The length of the vector is proportional to the electronegativity difference between the bonding atoms. In CO_2, we have two identical vectors pointing in exactly opposite directions—the vectors sum to zero, much as +1 and −1 sum to zero:

See the next section entitled Vector Addition for instructions about adding vectors.

No net dipole moment

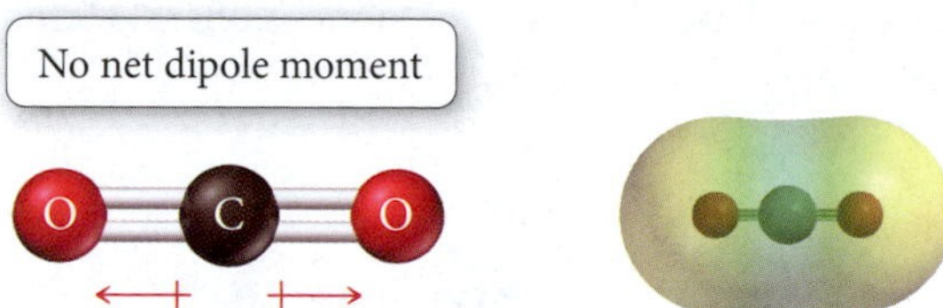

Notice that the electrostatic potential map of CO_2 shows regions of moderately high electron density (yellow with slight red) positioned symmetrically on either end of the molecule with a region of low electron density (blue) located in the middle.

In contrast, consider water:

$$H-\ddot{\underset{\cdot\cdot}{O}}-H$$

The O—H bonds in water are also polar; oxygen and hydrogen have electronegativities of 3.5 and 2.1, respectively. However, the water molecule is not linear but bent, so the two dipole moments do not sum to zero. If we imagine each bond as a vector pointing toward oxygen (the more electronegative atom), we see that, because of the angle between the vectors, they do not cancel, but sum to an overall vector or a net dipole moment (shown by the dashed arrow).

Net dipole moment

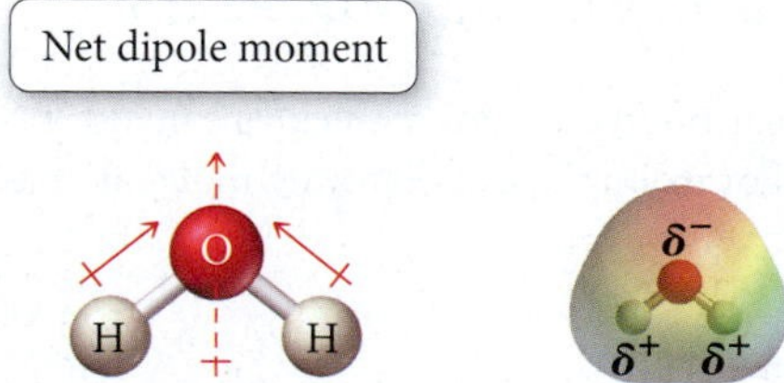

Water's electrostatic potential map shows an electron-rich region at the oxygen end of the molecule. Consequently, water is a polar molecule. Table 6.6 summarizes common geometries and molecular polarity.

Summarizing Determining Molecular Shape and Polarity:

- ***Draw the Lewis structure for the molecule and determine its molecular geometry.***
- ***Determine if the molecule contains polar bonds.*** A bond is polar if the two bonding atoms have sufficiently different electronegativities (see Figure 6.4). If the molecule contains polar bonds, superimpose a vector, pointing toward the more electronegative atom, on each bond. Make the length of the vector proportional to the electronegativity difference between the bonding atoms.
- ***Determine if the polar bonds add together to form a net dipole moment.*** Sum the vectors corresponding to the polar bonds together. If the vectors sum to zero, the molecule is nonpolar. If the vectors sum to a net vector, the molecule is polar.

Vector Addition

As discussed previously, we can determine whether a molecule is polar by summing the vectors associated with the dipole moments of all the polar bonds in the molecule. If the vectors sum to zero, the molecule will be nonpolar. If they sum to a net vector, the molecule will be polar. Here, we demonstrate how to add vectors together in one dimension and in two or more dimensions.

One Dimension To add two vectors that lie on the same line, assign one direction as positive. Vectors pointing in that direction have positive magnitudes. Consider vectors pointing in the opposite direction to have negative magnitudes. Then sum the vectors (always remembering to include their signs), as shown in Examples 1–3.

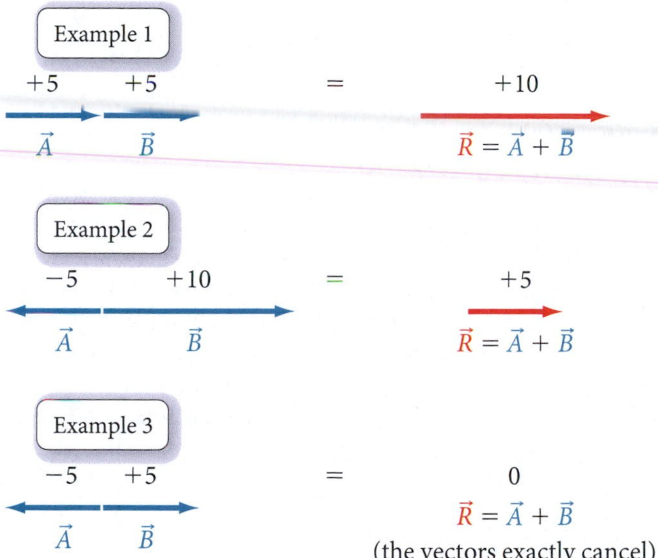

Two or More Dimensions To add two vectors, draw a parallelogram in which the two vectors form two adjacent sides. Draw the other two sides of the parallelogram parallel to and the same length as the two original vectors. Draw the resultant vector beginning at the origin and extending to the far corner of the parallelogram as shown in Examples 4 and 5.

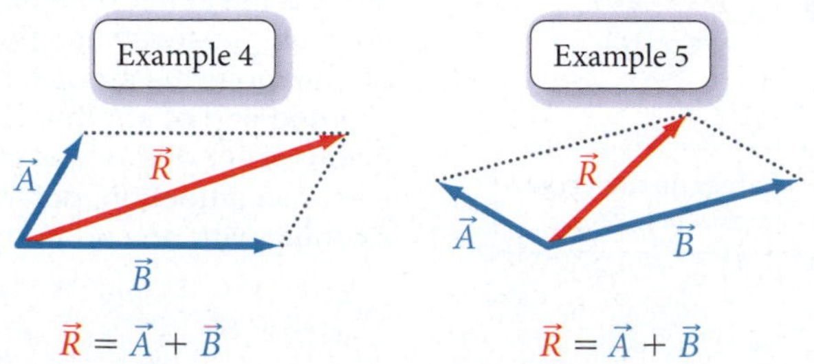

TABLE 6.6 Common Cases of Adding Dipole Moments to Determine whether a Molecule Is Polar

Linear

Nonpolar

The dipole moments of two identical polar bonds pointing in opposite directions will cancel. The molecule is nonpolar.

Bent

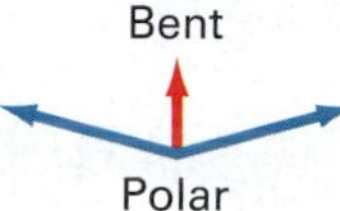

Polar

The dipole moments of two polar bonds with an angle of less than 180° between them will not cancel. The resultant dipole moment vector is shown in red. The molecule is polar.

Trigonal planar

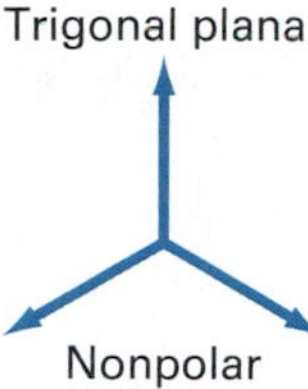

Nonpolar

The dipole moments of three identical polar bonds at 120° from each other will cancel. The molecule is nonpolar.

Tetrahedral

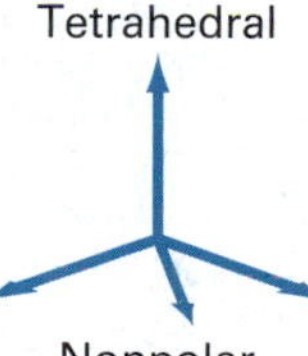

Nonpolar

The dipole moments of four identical polar bonds in a tetrahedral arrangement (109.5° from each other) will cancel. The molecule is nonpolar.

Trigonal pyramidal

Polar

The dipole moments of three polar bonds in a trigonal pyramidal arrangement (109.5° from each other) will not cancel. The resultant dipole moment vector is shown in red. The molecule is polar.

Note: In all cases in which the dipoles of two or more polar bonds cancel, the bonds are assumed to be identical. If one or more of the bonds are different from the other(s), the dipoles will not cancel and the molecule will be polar.

To add three or more vectors, add two of them together first, and then add the third vector to the result as shown in Examples 6 and 7.

Oil is nonpolar.
Water is polar.

▲ Oil and water do not mix because water molecules are polar and the molecules that compose oil are nonpolar.

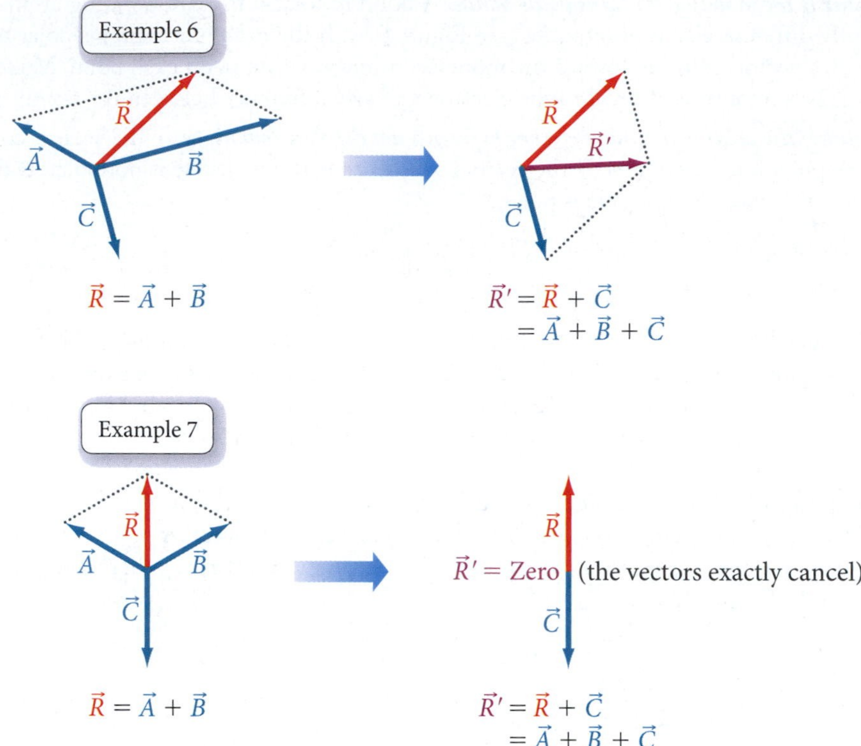

Example 6
$\vec{R} = \vec{A} + \vec{B}$
$\vec{R}' = \vec{R} + \vec{C}$
$= \vec{A} + \vec{B} + \vec{C}$
Example 7
$\vec{R} = \vec{A} + \vec{B}$
$\vec{R}'$ = Zero (the vectors exactly cancel)
$\vec{R}' = \vec{R} + \vec{C}$
$= \vec{A} + \vec{B} + \vec{C}$

The ability to predict and examine a molecule's polarity is a key connection between the structure of a molecule and its properties, the theme of this book. Water and oil do not mix, for example, because water molecules are polar and the molecules that compose oil are generally nonpolar. Polar molecules interact strongly with other polar molecules because the positive end of one molecule is attracted to the negative end of another, just as the south pole of a magnet is attracted to the north pole of another magnet (Figure 6.11 ▼). A mixture of polar and nonpolar molecules is similar to a mixture of small magnetic particles and nonmagnetic ones. The magnetic particles (which are like polar molecules) clump together, excluding the nonmagnetic particles (which are like nonpolar molecules) and separating into distinct regions.

▲ A mixture of polar and nonpolar molecules is analogous to a mixture of magnetic marbles (opaque) and nonmagnetic marbles (transparent). As with the magnetic marbles, mutual attraction causes polar molecules to clump together, excluding the nonpolar molecules.

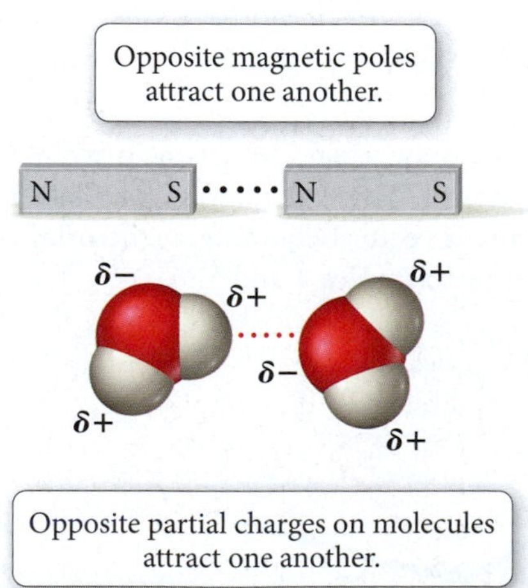

Opposite magnetic poles attract one another.
N S N S
δ− δ+ δ+ δ− δ+ δ+
Opposite partial charges on molecules attract one another.

◀ **FIGURE 6.11 Interaction of Polar Molecules** The north pole of one magnet attracts the south pole of another magnet. In an analogous way, the positively charged end of one molecule attracts the negatively charged end of another (although the forces involved are different). As a result of this electrical attraction, polar molecules interact strongly with one another.

EXAMPLE 6.13

Determining If a Molecule Is Polar

Determine if NH_3 is polar.

SOLUTION ***Draw the Lewis structure for the molecule and determine its molecular geometry.***	H | H—N—H The Lewis structure has three bonding groups and one lone pair about the central atom. Therefore the molecular geometry is trigonal pyramidal.
Determine if the molecule contains polar bonds. Sketch the molecule and superimpose a vector for each polar bond. The relative length of each vector should be proportional to the electronegativity difference between the atoms forming each bond. The vector should point in the direction of the more electronegative atom.	The electronegativities of nitrogen and hydrogen are 3.0 and 2.1, respectively. Therefore the bonds are polar. 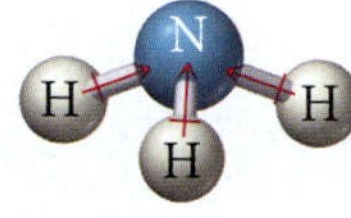
Determine if the polar bonds add together to form a net dipole moment. Examine the symmetry of the vectors (representing dipole moments) and determine if they cancel each other or sum to a net dipole moment.	The three dipole moments sum to a net dipole moment. The molecule is polar. 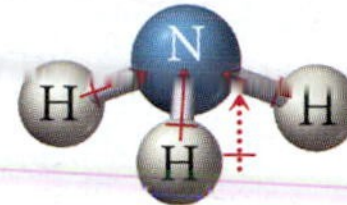

FOR PRACTICE 6.13

Determine if CF_4 is polar.

SELF-ASSESSMENT Quiz

1. Which set of elements is arranged in order of increasing electronegativity?
   a) O < S < As < Ge **b)** Ge < As < S < O
   c) S < O < As < Ge **d)** As < O < Ge < S

2. Which compound is likely to have an incomplete octet?
   a) NH_3 **b)** SO_3 **c)** N_2O **d)** BH_3

3. Which pair of atoms forms the most polar bond?
   a) N and O **b)** C and O **c)** C and F **d)** N and F

4. Which pair of atoms forms a nonpolar covalent bond?
   a) C and S **b)** C and O **c)** B and O **d)** Na and Cl

5. Which is the correct Lewis structure for nitrogen trifluoride?

   a) 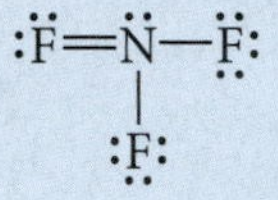
   ```
   :F̤̈=N̈—F̤̈:
        |
       :F̤̈:
   ```
   b) 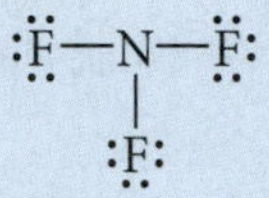
   ```
   :F̤̈—N—F̤̈:
        |
       :F̤̈:
   ```
   c) 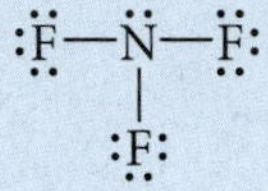
   ```
   :F̤̈—N̈—F̤̈:
        |
       :F̤̈:
   ```
   d) 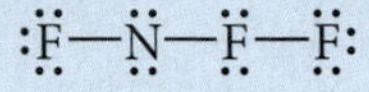
   ```
   :F̤̈—N̤̈—F̤̈—F̤̈:
   ```

6. Determine the correct Lewis structure for CO_3^{2-}.

   a) 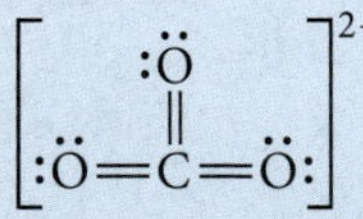
   ```
   [     :Ö      ]2−
   [      ||     ]
   [ :Ö=C=Ö:     ]
   ```
   b)
   ```
   [     :Ö      ]2−       [    :Ö:     ]2−       [    :Ö:     ]2−
   [      ||     ]   ←→   [     |      ]   ←→   [     |      ]
   [ :Ö̤—C—Ö̤:     ]         [ :Ö=C—Ö̤:   ]         [ :Ö̤—C=Ö:   ]
   ```
   c)
   ```
   [    :Ö̤:     ]2−
   [     |      ]
   [ :Ö̤—C̤—Ö̤:   ]
   ```
   d) 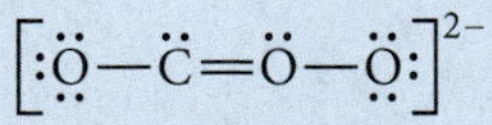
   ```
   [:Ö̤—C̈=Ö—Ö̤:]2−
   ```

7. What is the formal charge of nitrogen in this structure?

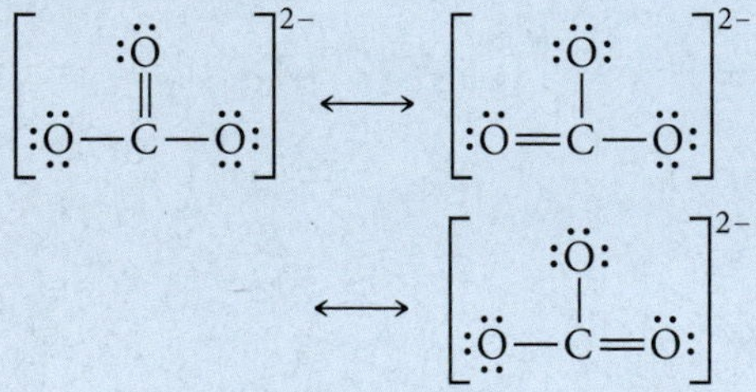

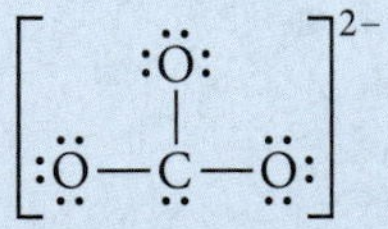

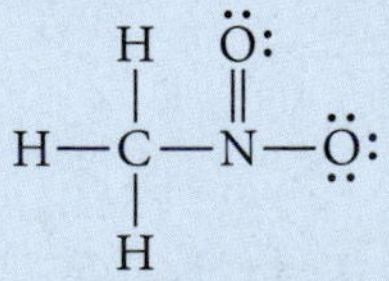

   ```
      H   Ö:
      |   ||
   H—C—N—Ö̤:
      |
      H
   ```

   a) +1 **b)** +2 **c)** −1 **d)** −2

8. Which of the following structures is the best resonance structure for the acetate ion shown here?

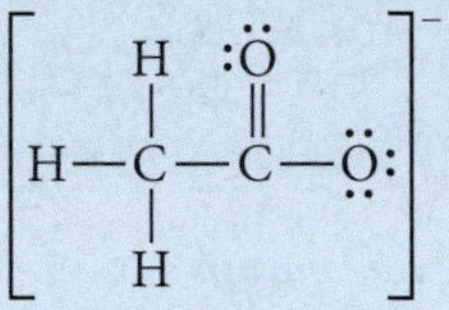

   ```
   [   H  :Ö      ]−
   [   |   ||     ]
   [ H—C—C—Ö̤:    ]
   [   |          ]
   [   H          ]
   ```

   a)
   ```
   [   H              ]−
   [   |              ]
   [ H—C—C̤—Ö̤—Ö̤:     ]
   [   |              ]
   [   H              ]
   ```
   b)
   ```
   [   H  :Ö:     ]−
   [   |   |      ]
   [ H—C=C—Ö̤:    ]
   [   |          ]
   [   H          ]
   ```
   c)
   ```
   [  :Ö—H        ]−
   [   |          ]
   [ H—C—C̤—Ö̤:    ]
   [   |          ]
   [   H          ]
   ```
   d) 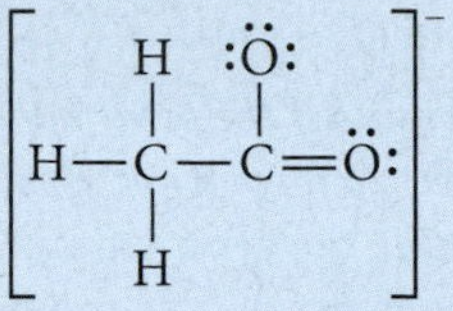
   ```
   [   H  :Ö:     ]−
   [   |   |      ]
   [ H—C—C=Ö:    ]
   [   |          ]
   [   H          ]
   ```

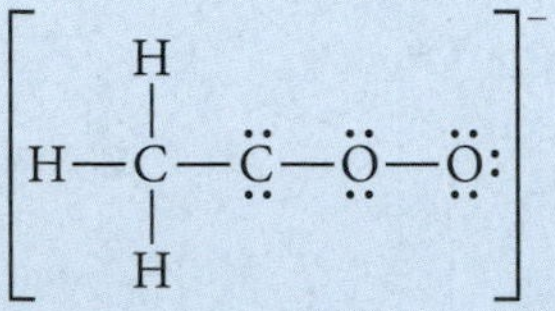

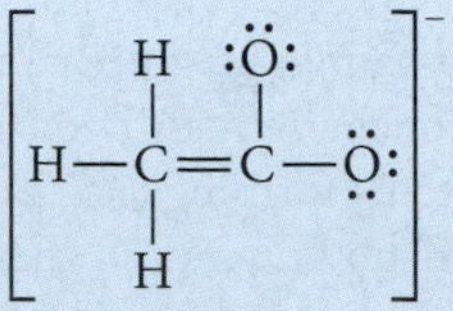

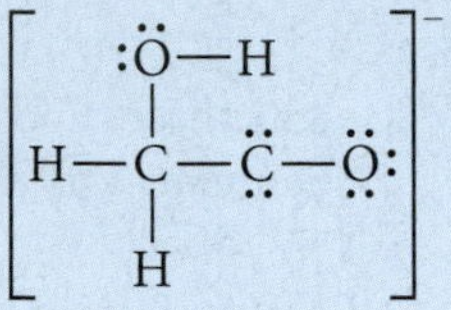

9. Use formal charge to determine the best Lewis structure for CH_3SOCH_3.

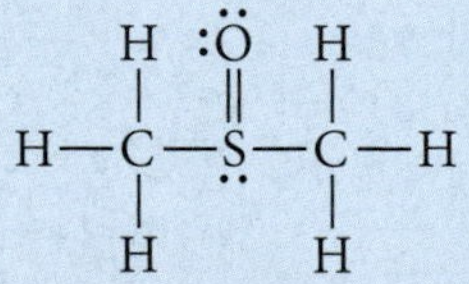

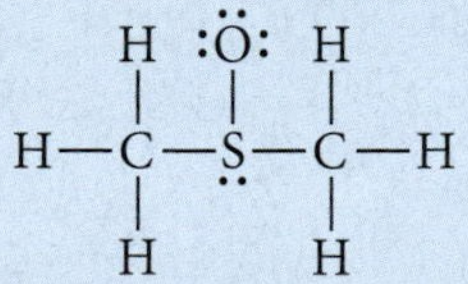

   a) 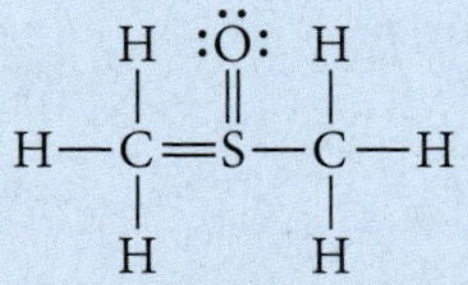
   ```
     H  :Ö:  H
     |   ||  |
   H—C=S—C—H
     |       |
     H       H
   ```
   b)
   ```
     H  :Ö   H
     |   ||  |
   H—C—S̤—C—H
     |       |
     H       H
   ```
   c)
   ```
     H  :Ö:  H
     |   |   |
   H—C—S̤—C—H
     |       |
     H       H
   ```
   d) 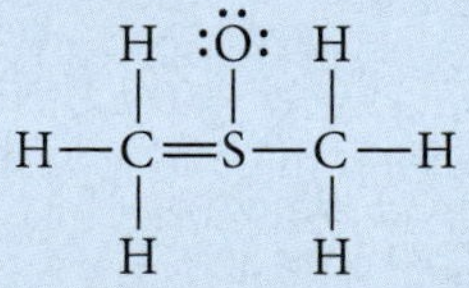
   ```
     H  :Ö:  H
     |   |   |
   H—C=S—C—H
     |       |
     H       H
   ```

10. Determine the molecular geometry of CBr_4.
    a) Linear **b)** Trigonal planar
    c) Tetrahedral **d)** Trigonal pyramidal

11. Determine the molecular geometry of SeF_4.
    a) Tetrahedral **b)** Trigonal bipyramidal
    c) T-shaped **d)** Seesaw

12. Predict the relative bond angles in BF_3 and SO_2.
    a) BF_3 bond angles > SO_2 bond angle
    b) SO_2 bond angle > BF_3 bond angles
    c) BF_3 bond angles = SO_2 bond angle
    d) Relative bond angles cannot be predicted

—*Continued*

Continued—

13. Predict the molecular geometry about N in the molecule CH_3NHCH_3.
    a) Linear **b)** Trigonal planar
    c) Trigonal pyramidal **d)** Bent

14. Which molecule is polar?
    a) SF_2 **b)** BH_3 **c)** PF_5 **d)** CS_2

Answers: 1:b; 2:d; 3:c; 4:a; 5:c; 6:b; 7:a; 8:d; 9:b; 10:c; 11:d; 12:a; 13:c; 14:a

CHAPTER SUMMARY 6

REVIEW

KEY LEARNING OUTCOMES

CHAPTER OBJECTIVES	ASSESSMENT
Classifying Bonds: Pure Covalent, Polar Covalent, or Ionic (6.2)	• Example 6.1 For Practice 6.1 Exercises 23, 24
Writing Lewis Structures for Covalent Compounds (6.3)	• Examples 6.2, 6.3 For Practice 6.2, 6.3 Exercises 27–30
Writing Lewis Structures for Polyatomic Ions (6.3)	• Example 6.4 For Practice 6.4 Exercises 31–34
Writing Resonance Lewis Structures (6.4)	• Example 6.5 For Practice 6.5 Exercises 35, 36
Assigning Formal Charges to Assess Competing Resonance Structures (6.4)	• Example 6.6 For Practice 6.6 For More Practice 6.6 Exercises 37–40
Drawing Resonance Structures and Assigning Formal Charge for Organic Compounds (6.4)	• Example 6.7 For Practice 6.7 Exercises 41–44
Writing Lewis Structures for Compounds Having Expanded Octets (6.5)	• Example 6.8 For Practice 6.8 For More Practice 6.8 Exercises 47–50
Using VSEPR Theory to Predict the Basic Shapes of Molecules (6.7)	• Example 6.9 For Practice 6.9 Exercises 53, 54
Predicting Molecular Geometries Using VSEPR Theory and the Effects of Lone Pairs (6.8)	• Examples 6.10, 6.11 For Practice 6.10, 6.11 Exercises 55–58
Predicting the Shapes of Larger Molecules (6.8)	• Example 6.12 For Practice 6.12 Exercises 63, 64, 67, 68
Using Molecular Shape to Determine Polarity of a Molecule (6.10)	• Example 6.13 For Practice 6.13 Exercises 71–74

KEY TERMS

Section 6.2
polar covalent bond (191)
electronegativity (191)
dipole moment (μ) (192)
percent ionic character (193)

Section 6.4
resonance structures (197)
resonance hybrid (197)
formal charge (199)

Section 6.5
free radical (202)

Section 6.6
bond energy (205)
bond length (206)

Section 6.7
valence shell electron pair repulsion (VSEPR) theory (207)
electron groups (207)
linear geometry (207)
trigonal planar geometry (208)
tetrahedral geometry (208)
trigonal bipyramidal geometry (209)
octahedral geometry (210)

Section 6.8
electron geometry (211)
molecular geometry (211)
trigonal pyramidal geometry (211)
bent geometry (212)
seesaw geometry (213)
T-shaped geometry (213)
square pyramidal geometry (214)
square planar geometry (214)

KEY CONCEPTS

Morphine: A Molecular Imposter (6.1)

- The properties of molecules are directly related to their shapes.
- Many biological functions, such as drug action and the immune response, are determined by the shapes of molecules.

Electronegativity and Bond Polarity (6.2)

- The shared electrons in a covalent bond are not always *equally* shared; when two dissimilar nonmetals form a covalent bond, the electron density is greater on the more electronegative element. The result is a polar bond, with one element carrying a partial positive charge and the other a partial negative charge.
- Electronegativity—the ability of an atom to attract electrons to itself in chemical bonding—increases as we move across a period to the right in the periodic table and decreases as we move down a column.
- Elements with very dissimilar electronegativities form ionic bonds; those with very similar electronegativities form nonpolar covalent bonds; and those with intermediate electronegativity differences form polar covalent bonds. The degree of polarity of a bond depends on the electronegativity difference between the bonding atoms.

Drawing Lewis Structures of Molecular Compounds and Polyatomic Ions (6.3)

- Follow the procedure in Section 6.3 to draw Lewis structures for compounds.
- Electrons must be added (anions) or removed (cations) to account for the charge of a polyatomic ion.

Resonance and Formal Charge (6.4)

- Some molecules are best represented not by a single Lewis structure, but by two or more resonance structures. The actual structure of these molecules is a resonance hybrid: a combination or average of the contributing structures.
- The formal charge of an atom in a Lewis structure is the charge the atom would have if all bonding electrons were shared equally between bonding atoms.
- In general, the best Lewis structures will have the fewest atoms with formal charge, and any negative formal charge will be on the most electronegative atom.

Exceptions to the Octet Rule (6.5)

- Although the octet rule is normally used in drawing Lewis structures, some exceptions occur.
- These exceptions include odd-electron species, which necessarily have Lewis structures with only seven electrons around an atom. Such molecules, called free radicals, tend to be unstable and chemically reactive.
- Other exceptions to the octet rule include molecules with incomplete octets—usually totaling just six electrons (especially important in compounds containing boron)—and molecules with expanded octets—usually 10 or 12 electrons (which can occur in compounds containing elements from the third row of the periodic table and below). Expanded octets never occur in second-period elements.

Bond Energies and Bond Lengths (6.6)

- The bond energy of a chemical bond is the energy required to break 1 mole of the bond in the gas phase. Average bond energies for a number of different bonds are tabulated.
- Average bond lengths are also tabulated. In general, triple bonds are shorter and stronger than double bonds, which are in turn shorter and stronger than single bonds.

VSEPR Theory: Predicting Molecular Shape (6.7–6.9)

- In VSEPR theory, molecular geometries are determined by the repulsions between electron groups on the central atom. An electron group can be a single bond, double bond, triple bond, lone pair, or even a single electron.
- The five basic molecular shapes are linear (two electron groups), trigonal planar (three electron groups), tetrahedral (four electron groups), trigonal bipyramidal (five electron groups), and octahedral (six electron groups).
- When lone pairs are present on the central atom, the *electron* geometry is still one of the five basic shapes, but one or more positions are occupied by lone pairs. The *molecular* geometry is therefore different from the electron geometry. Lone pairs are positioned so as to minimize repulsions with other lone pairs and with bonding pairs.
- To determine the geometry of a molecule, follow the procedure presented in Section 6.9.

Molecular Shape and Polarity (6.10)

- The polarity of a polyatomic molecule containing polar bonds depends on its geometry. If the dipole moments of the polar bonds are aligned in such a way that they cancel one another, the molecule will not be polar. If they are aligned in such a way as to sum together, the molecule will be polar.
- Highly symmetric molecules tend to be nonpolar, while asymmetric molecules containing polar bonds tend to be polar. The polarity of a molecule dramatically affects its properties.

KEY EQUATIONS AND RELATIONSHIPS

Dipole Moment (μ): Separation of Two Particles of Equal but Opposite Charges of Magnitude *q* by a Distance *r* (6.2)

$$\mu = qr$$

Percent Ionic Character (6.2)

$$\text{Percent ionic character} = \frac{\text{measured dipole moment of bond}}{\text{dipole moment if electron were completely transferred}} \times 100\%$$

Formal Charge (6.4)

$$\text{Formal charge} = \text{number of valence electrons} - \left(\text{number of nonbonding electrons} + \tfrac{1}{2}\,\text{number of shared electrons}\right)$$

EXERCISES

REVIEW QUESTIONS

1. What is electronegativity? What are the periodic trends in electronegativity?
2. Explain the difference between a pure covalent bond, a polar covalent bond, and an ionic bond.

3. What is meant by the percent ionic character of a bond? Do any bonds have 100% ionic character?
4. What is a dipole moment?
5. What is the magnitude of the dipole moment formed by separating a proton and an electron by 100 pm? 200 pm?
6. What is the basic procedure for writing a covalent Lewis structure?
7. How do you determine the number of electrons that go into the Lewis structure of a molecule? A polyatomic ion?
8. What are resonance structures? What is a resonance hybrid?
9. Do resonance structures always contribute equally to the overall structure of a molecule? Explain.
10. What is formal charge? How is formal charge calculated? How is it helpful?
11. Why does the octet rule have exceptions? Give the three major categories of exceptions and an example of each.
12. Which elements can have expanded octets? Which elements cannot have expanded octets?
13. What is bond energy?
14. Give some examples of some typical bond lengths. Which factors influence bond lengths?
15. Why is molecular geometry important? Cite some examples.
16. According to VSEPR theory, what determines the geometry of a molecule?
17. Name and draw the five basic electron geometries, and state the number of electron groups corresponding to each. What constitutes an *electron group*?
18. Explain the difference between electron geometry and molecular geometry. Under what circumstances are they not the same?
19. List the correct electron and molecular geometries that correspond to each set of electron groups around the central atom of a molecule.
    a. four electron groups overall; three bonding groups and one lone pair
    b. four electron groups overall; two bonding groups and two lone pairs
    c. five electron groups overall; four bonding groups and one lone pair
    d. five electron groups overall; three bonding groups and two lone pairs
    e. five electron groups overall; two bonding groups and three lone pairs
    f. six electron groups overall; five bonding groups and one lone pair
    g. six electron groups overall; four bonding groups and two lone pairs
20. How do you apply VSEPR theory to predict the shape of a molecule with more than one interior atom?
21. How do you determine if a molecule is polar?
22. Why is polarity a key connection between the structure of a molecule and its properties?

PROBLEMS BY TOPIC

Electronegativity and Bond Polarity

23. Determine if a bond between each pair of atoms would be pure covalent, polar covalent, or ionic.
    a. Br and Br b. C and Cl c. C and S d. Sr and O
24. Determine if a bond between each pair of atoms would be pure covalent, polar covalent, or ionic.
    a. C and N b. N and S c. K and F d. N and N
25. Draw the Lewis structure for CO with an arrow representing the dipole moment. Use Figure 6.5 to estimate the percent ionic character of the CO bond.
26. Draw the Lewis structure for BrF with an arrow representing the dipole moment. Use Figure 6.5 to estimate the percent ionic character of the BrF bond.

Writing Lewis Structures, Resonance, and Formal Charge

27. Write the Lewis structure for each molecule.
    a. PH_3 b. SCl_2 c. HI d. CH_4
28. Write the Lewis structure for each molecule.
    a. NF_3 b. HBr c. SBr_2 d. CCl_4
29. Write the Lewis structure for each molecule.
    a. SF_2 b. SiH_4
    c. HCOOH (both O bonded to C) d. CH_3SH (C and S central)
30. Write the Lewis structure for each molecule.
    a. CH_2O b. C_2Cl_4
31. Write the Lewis structure for each molecule or ion.
    a. CI_4 b. N_2O c. SiH_4 d. Cl_2CO
32. Write the Lewis structure for each molecule or ion.
    a. H_3COH b. OH^- c. BrO^- d. O_2^{2-}
33. Write the Lewis structure for each molecule or ion.
    a. N_2H_2 b. N_2H_4 c. C_2H_2 d. C_2H_4
34. Write the Lewis structure for each molecule or ion.
    a. H_3COCH_3 b. CN^- c. NO_2^- d. ClO^-
35. Write a Lewis structure that obeys the octet rule for each molecule or ion. Include resonance structures if necessary and assign formal charges to each atom.
    a. SeO_2 b. CO_3^{2-} c. ClO^- d. NO_2^-
36. Write a Lewis structure that obeys the octet rule for each ion. Include resonance structures if necessary and assign formal charges to each atom.
    a. ClO_3^- b. ClO_4^- c. NO_3^- d. NH_4^+
37. Use formal charge to determine which Lewis structure is better.

```
   H           H
   |           |
H—C=S:      H—S=C:
```

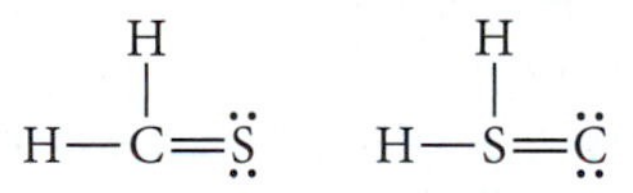

38. Use formal charge to determine which Lewis structure is better.

```
    H              H
    |              |
H—S—C—H        H—C—S—H
    |              |
    H              H
```

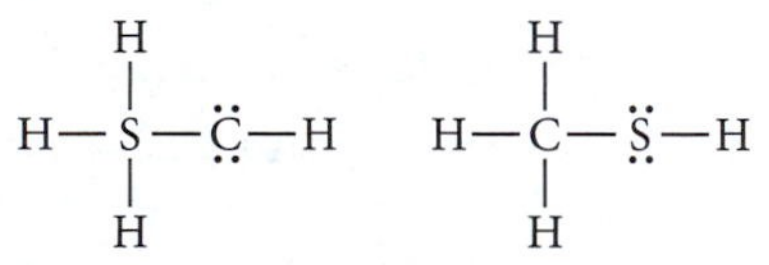

39. How important is this resonance structure to the overall structure of carbon dioxide? Explain.

:O≡C—O:

40. In N_2O, nitrogen is the central atom and the oxygen atom is terminal. In OF_2, however, oxygen is the central atom. Use formal charges to explain why.
41. Draw the Lewis structure (including resonance structures) for the acetate ion (CH_3COO^-). For each resonance structure, assign formal charges to all atoms that have formal charge.
42. Draw the Lewis structure (including resonance structures) for methyl azide (CH_3N_3). For each resonance structure, assign formal charges to all atoms that have formal charge.

43. What are the formal charges of the atoms shown in red?

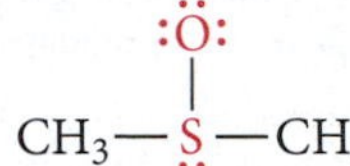

44. What are the formal charges of the atoms shown in red?

:Ö:
|
CH_3—S—CH_3

Odd-Electron Species, Incomplete Octets, and Expanded Octets

45. Write the Lewis structure for each molecule (octet rule not followed).
   a. BCl_3 b. NO_2 c. BH_3

46. Write the Lewis structure for each molecule (octet rule not followed).
   a. BBr_3 b. NO c. ClO_2

47. Write the Lewis structure for each ion. Include resonance structures if necessary and assign formal charges to all atoms. If necessary, expand the octet on the central atom to lower formal charge.
   a. PO_4^{3-} b. CN^- c. SO_3^{2-} d. ClO_2^-

48. Write Lewis structures for each molecule or ion. Include resonance structures if necessary and assign formal charges to all atoms. If you need to, expand the octet on the central atom to lower formal charge.
   a. SO_4^{2-} b. HSO_4^- c. SO_3 d. BrO_2^-

49. Write Lewis structures for each molecule or ion. Use expanded octets as necessary.
   a. PF_5 b. I_3^- c. SF_4 d. GeF_4

50. Write Lewis structures for each molecule or ion. Use expanded octets as necessary.
   a. ClF_5 b. AsF_6^- c. Cl_3PO d. IF_5

Bond Energies and Bond Lengths

51. Order these compounds in order of increasing carbon–carbon bond *strength* and in order of decreasing carbon–carbon bond *length*: HCCH, H_2CCH_2, H_3CCH_3.

52. Which of these compounds has the stronger nitrogen–nitrogen bond? The shorter nitrogen–nitrogen bond?

   H_2NNH_2, HNNH

VSEPR Theory and Molecular Geometry

53. A molecule with the formula AB_3 has a trigonal pyramidal geometry. How many electron groups are on the central atom (A)?

54. A molecule with the formula AB_3 has a trigonal planar geometry. How many electron groups are on the central atom?

55. For each molecular geometry shown here, list the number of total electron groups, the number of bonding groups, and the number of lone pairs on the central atom.

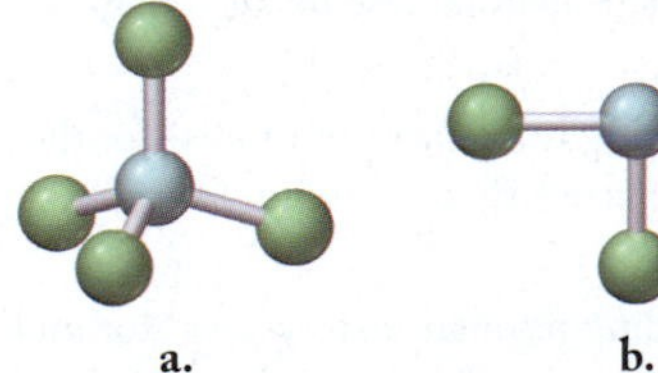
a.

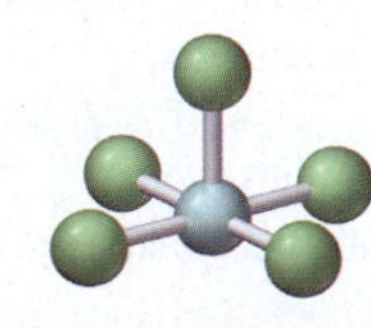
b.

c.

56. For each molecular geometry shown here, list the number of total electron groups, the number of bonding groups, and the number of lone pairs on the central atom.

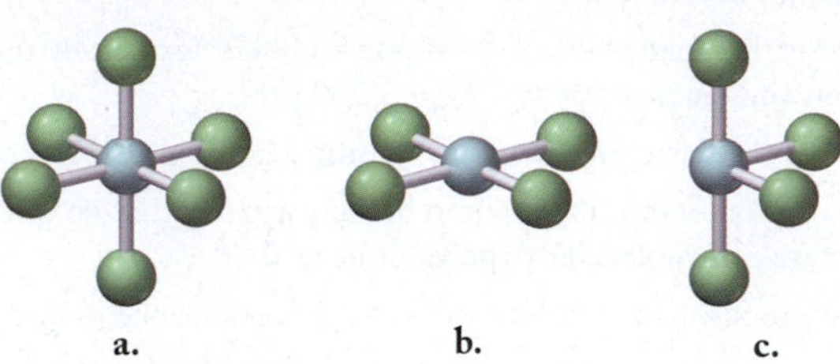
a. b. c.

57. Determine the electron geometry, molecular geometry, and idealized bond angles for each molecule. In which cases do you expect deviations from the idealized bond angle?
   a. PF_3 b. SBr_2 c. $CHCl_3$ d. CS_2

58. Determine the electron geometry, molecular geometry, and idealized bond angles for each molecule. In which cases do you expect deviations from the idealized bond angle?
   a. CF_4 b. NF_3 c. OF_2 d. H_2S

59. Which species has the smaller bond angle, H_3O^+ or H_2O? Explain.

60. Which species has the smaller bond angle, ClO_4^- or ClO_3^-? Explain.

61. Determine the molecular geometry and draw each molecule or ion using the bond conventions shown in the "Representing Molecular Geometries on Paper" section of this chapter (see Section 6.9).
   a. SF_4 b. ClF_3 c. IF_2^- d. IBr_4^-

62. Determine the molecular geometry and draw each molecule or ion, using the bond conventions shown in the "Representing Molecular Geometries on Paper" section of this chapter (see Section 6.9).
   a. BrF_5 b. SCl_6 c. PF_5 d. IF_4^+

63. Determine the molecular geometry about each interior atom and draw each molecule. (Skeletal structure is indicated in parentheses.)
   a. C_2H_2 (HCCH)
   b. C_2H_4 (H_2CCH_2)
   c. C_2H_6 (H_3CCH_3)

64. Determine the molecular geometry about each interior atom and draw each molecule. (Skeletal structure is indicated in parentheses.)
   a. N_2
   b. N_2H_2 (HNNH)
   c. N_2H_4 (H_2NNH_2)

65. Each ball-and-stick model shows the electron and molecular geometry of a generic molecule. Explain what is wrong with each molecular geometry and provide the correct molecular geometry, given the number of lone pairs and bonding groups on the central atom.

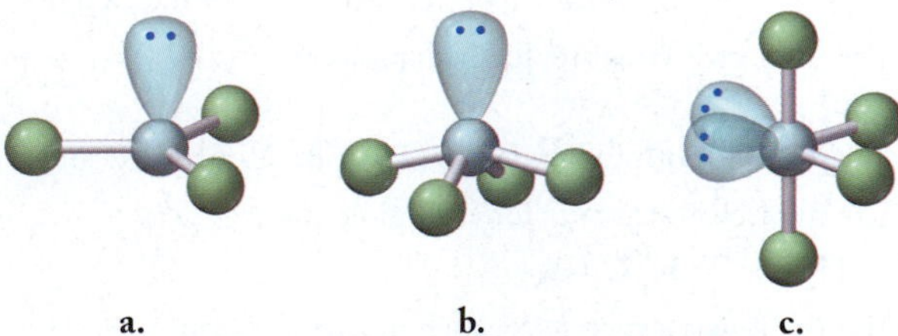
a. b. c.

66. Each ball-and-stick model shows the electron and molecular geometry of a generic molecule. Explain what is wrong with each molecular geometry and provide the correct molecular geometry, given the number of lone pairs and bonding groups on the central atom.

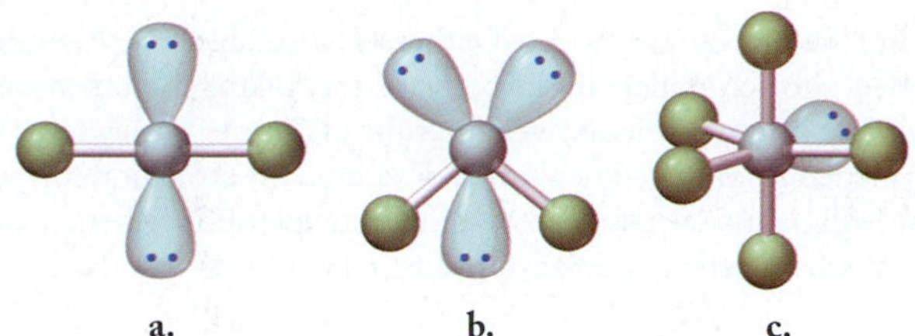

67. Determine the geometry about each interior atom in each molecule and draw the molecule. (Skeletal structure is indicated in parentheses.)

a. CH_3OH (H_3COH) **b.** CH_3OCH_3 (H_3COCH_3)
c. H_2O_2 (HOOH)

68. Determine the geometry about each interior atom in each molecule and draw the molecule. (Skeletal structure is indicated in parentheses.)

a. CH_3NH_2 (H_3CNH_2)
b. $CH_3CO_2CH_3$ ($H_3CCOOCH_3$ both O atoms attached to second C)
c. NH_2CO_2H (H_2NCOOH both O atoms attached to C)

Molecular Shape and Polarity

69. Explain why CO_2 and CCl_4 are both nonpolar even though they contain polar bonds.

70. CH_3F is a polar molecule, even though the tetrahedral geometry often leads to nonpolar molecules. Explain.

71. Determine whether each molecule in Exercise 57 is polar or nonpolar.

72. Determine whether each molecule in Exercise 58 is polar or nonpolar.

73. Determine whether each molecule or ion is polar or nonpolar.

a. ClO_3^- **b.** SCl_2 **c.** SCl_4 **d.** $BrCl_5$

74. Determine whether each molecule is polar or nonpolar.

a. $SiCl_4$ **b.** CF_2Cl_2 **c.** SeF_6 **d.** IF_5

CUMULATIVE PROBLEMS

75. Each compound contains both ionic and covalent bonds. Write ionic Lewis structures for each, including the covalent structure for the ion in brackets. Write resonance structures if necessary.

a. $BaCO_3$ **b.** $Ca(OH)_2$ **c.** KNO_3 **d.** LiIO

76. Each compound contains both ionic and covalent bonds. Write ionic Lewis structures for each, including the covalent structure for the ion in brackets. Write resonance structures if necessary.

a. $RbIO_2$ **b.** NH_4Cl **c.** KOH **d.** $Sr(CN)_2$

77. Carbon ring structures are common in organic chemistry. Draw a Lewis structure for each carbon ring structure, including any necessary resonance structures.

a. C_4H_8 **b.** C_4H_4 **c.** C_6H_{12} **d.** C_6H_6

78. Amino acids are the building blocks of proteins. The simplest amino acid is glycine (H_2NCH_2COOH). Draw a Lewis structure for glycine. (*Hint*: The central atoms in the skeletal structure are nitrogen bonded to carbon, which is bonded to another carbon. The two oxygen atoms are bonded directly to the rightmost carbon atom.)

79. Formic acid is partly responsible for the sting of ant bites. By mass, formic acid is 26.10% C, 4.38% H, and 69.52% O. The molar mass of formic acid is 46.02 g/mol. Find the molecular formula of formic acid and draw its Lewis structure.

80. Diazomethane is a highly poisonous, explosive compound because it readily evolves N_2. Diazomethane has the following composition by mass: 28.57% C; 4.80% H; and 66.64% N. The molar mass of diazomethane is 42.04 g/mol. Find the molecular formula of diazomethane, draw its Lewis structure, and assign formal charges to each atom. Why is diazomethane not very stable? Explain.

81. Draw the Lewis structure for nitric acid (the hydrogen atom is attached to one of the oxygen atoms). Include all three resonance structures by alternating the double bond among the three oxygen atoms. Use formal charge to determine which of the resonance structures is most important to the structure of nitric acid.

82. Phosgene (Cl_2CO) is a poisonous gas that was used as a chemical weapon during World War I. It is a potential agent for chemical terrorism today. Draw the Lewis structure of phosgene. Include all three resonance forms by alternating the double bond among the three terminal atoms. Which resonance structure is the best?

83. The cyanate ion (OCN^-) and the fulminate ion (CNO^-) share the same three atoms, but have vastly different properties. The cyanate ion is stable, while the fulminate ion is unstable and forms explosive compounds. The resonance structures of the cyanate ion were explored in Example 6.6. Draw Lewis structures for the fulminate ion—including possible resonance forms—and use formal charge to explain why the fulminate ion is less stable (and therefore more reactive) than the cyanate ion.

84. Draw the Lewis structure for each organic compound from its condensed structural formula.

a. C_3H_8 **b.** CH_3OCH_3 **c.** CH_3COCH_3
d. CH_3COOH **e.** CH_3CHO

85. Draw the Lewis structure for each organic compound from its condensed structural formula.

a. C_2H_4 **b.** CH_3NH_2 **c.** HCHO
d. CH_3CH_2OH **e.** HCOOH

86. Use Lewis structures to explain why Br_3^- and I_3^- are stable, while F_3^- is not.

87. Draw the Lewis structure for $HCSNH_2$. (The carbon and nitrogen atoms are bonded together, and the sulfur atom is bonded to the carbon atom.) Label each bond in the molecule as polar or nonpolar.

88. Draw the Lewis structure for urea, H_2NCONH_2, one of the compounds responsible for the smell of urine. (The central carbon atom is bonded to both nitrogen atoms and to the oxygen atom.) Does urea contain polar bonds? Which bond in urea is most polar?

89. Some theories of aging suggest that free radicals cause certain diseases and perhaps aging in general. As you know from the Lewis model, such molecules are not chemically stable and will quickly react with other molecules. According to some theories, free radicals may attack molecules within the cell, such as DNA, changing them and causing cancer or other diseases. Free radicals may also attack molecules on the surfaces of cells, making them appear foreign to the body's immune system. The immune system then attacks the cells and destroys them, weakening the body. Draw Lewis structures for the free radicals implicated in this theory of aging, which are given here.

a. O_2^- **b.** O^- **c.** OH
d. CH_3OO (unpaired electron on terminal oxygen)

90. Free radicals are important in many environmentally significant reactions. For example, photochemical smog—smog that results from the action of sunlight on air pollutants—forms in part by these two steps:

$$NO_2 \xrightarrow{\text{UV light}} NO + O$$

$$O + O_2 \longrightarrow O_3$$

The product of this reaction, ozone, is a pollutant in the lower atmosphere. (Upper atmospheric ozone is a natural part of the atmosphere that protects life on Earth from ultraviolet light.) Ozone is an eye and lung irritant and also accelerates the weathering of rubber products. Rewrite the above reactions using the Lewis structure of each reactant and product. Identify the free radicals.

91. A compound composed of only carbon and hydrogen is 7.743% hydrogen by mass. Draw a Lewis structure for the compound.

92. A compound composed of only carbon and chlorine is 85.5% chlorine by mass. Draw a Lewis structure for the compound.

93. Amino acids are biological compounds that link together to form proteins, the workhorse molecules in living organisms. The skeletal structures of several simple amino acids are shown here. For each skeletal structure, complete the Lewis structure, determine the geometry about each interior atom, and draw the molecule, using the bond conventions of Section 6.9.

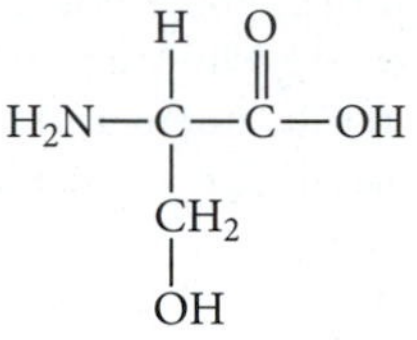

a. serine

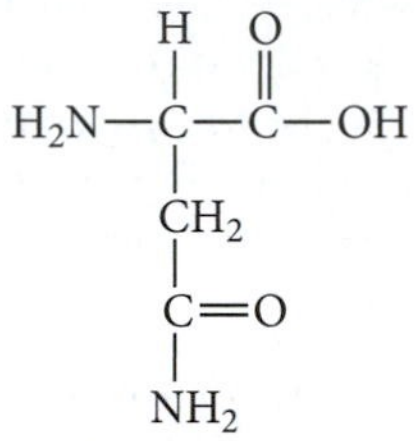

b. asparagine

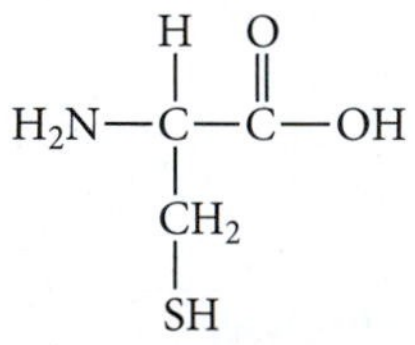

c. cysteine

94. The genetic code is based on four different bases with the structures shown here. Assign a geometry to each interior atom in these four bases.

a. cytosine **b.** adenine **c.** thymine **d.** guanine

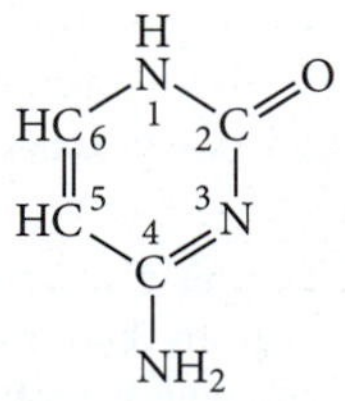

a.

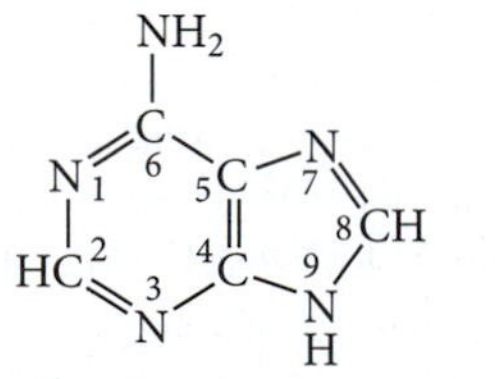

b.

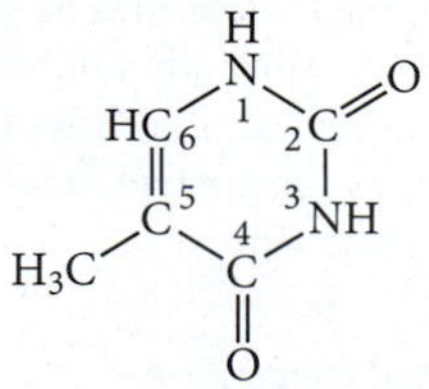

c)

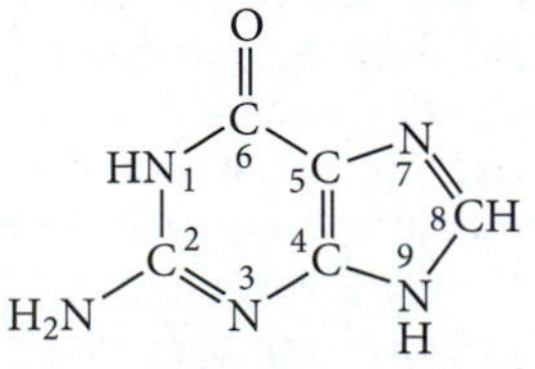

d.

95. Most vitamins can be classified either as fat soluble, which results in their tendency to accumulate in the body (so that taking too much can be harmful), or water soluble, which results in their tendency to be quickly eliminated from the body in urine. Examine the structural formulas and space-filling models of these vitamins and determine whether each one is fat soluble (mostly nonpolar) or water soluble (mostly polar).

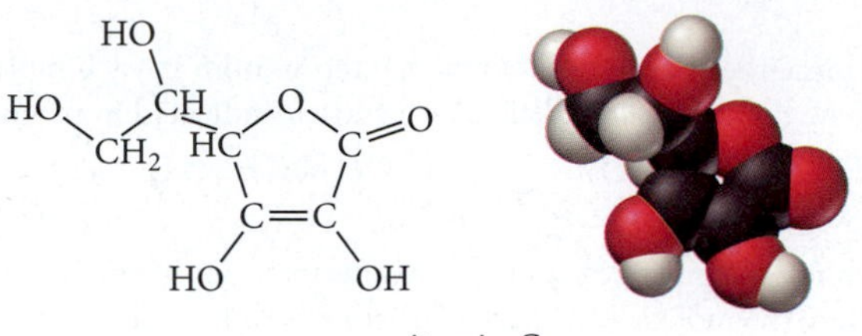

a. vitamin C

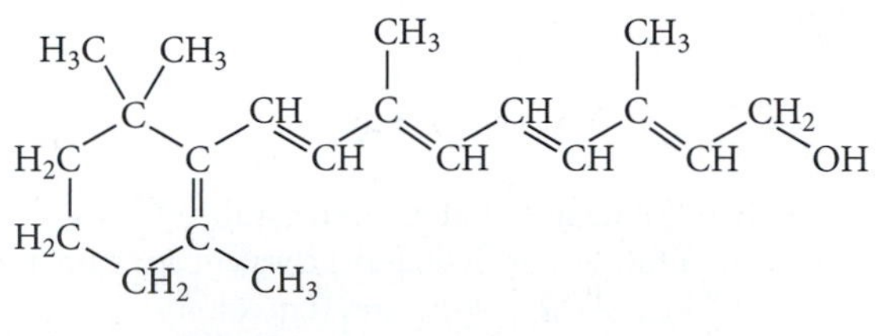

b. vitamin A

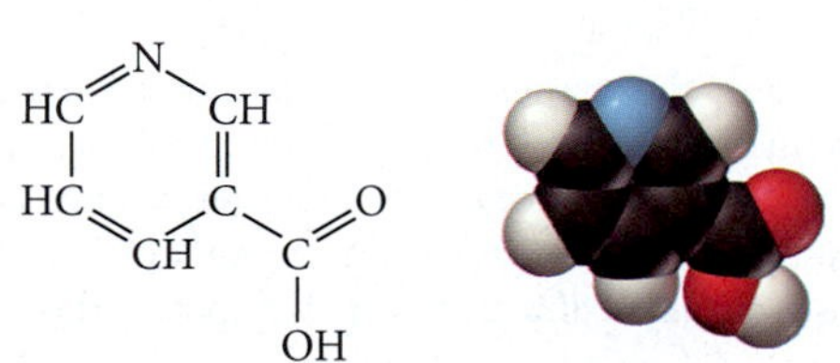

c. niacin (vitamin B_3)

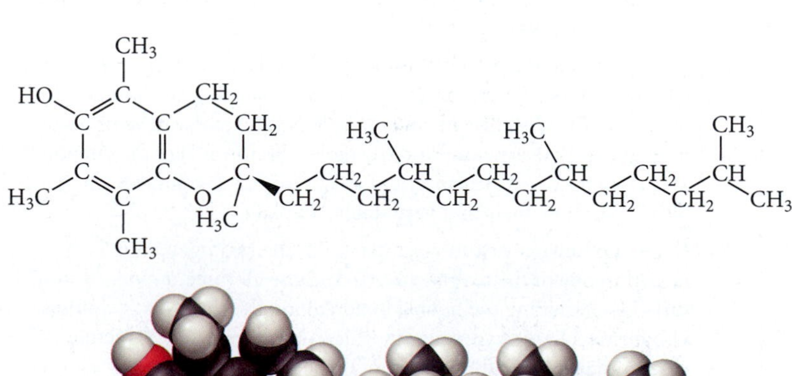

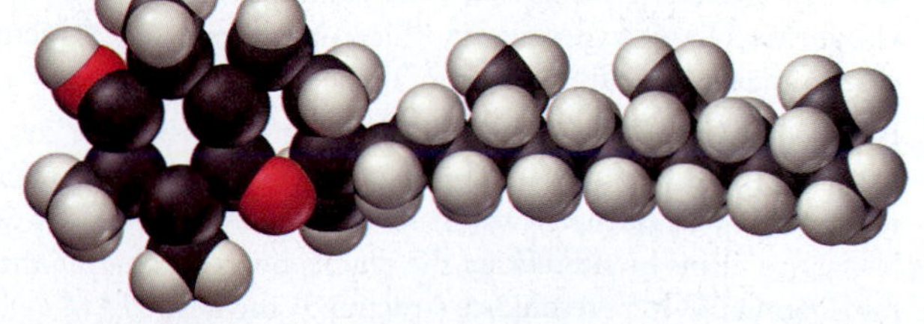

d. vitamin E

96. Water alone does not easily remove grease from dishes or hands because grease is nonpolar and water is polar. The addition of soap to water, however, allows the grease to dissolve. Study the structure of sodium stearate (a soap) and describe how it works.

$$CH_3(CH_2)_{16}C(=O)-O^-Na^+$$

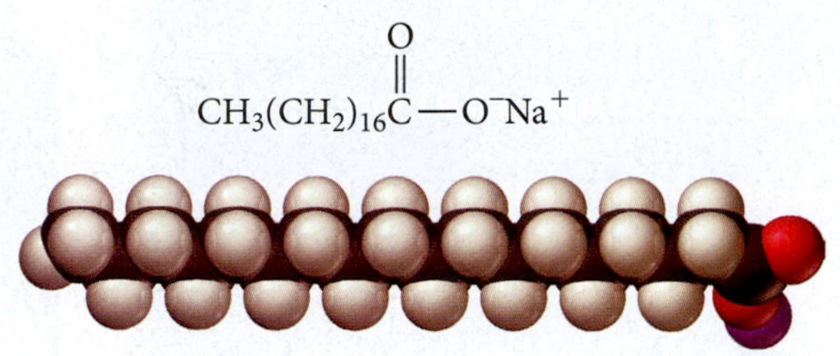

CHALLENGE PROBLEMS

97. The azide ion, N_3^-, is a symmetrical ion and all of its contributing resonance structures have formal charges. Draw three important contributing structures for this ion.

98. A 0.167-g sample of an unknown compound contains 0.00278 mol of the compound. Elemental analysis of the compound gives the following percentages by mass: 40.00% C; 6.71% H; 53.29% O. Determine the molecular formula, molar mass, and Lewis structure of the unknown compound.

99. Use the dipole moments of HF and HCl (given at the end of the problem) together with the percent ionic character of each bond (Figure 6.5) to estimate the bond length in each molecule. How well does your estimated bond length agree with the bond length in Table 6.4?

$$\text{HCl } \mu = 1.08 \text{ D}$$
$$\text{HF } \mu = 1.82 \text{ D}$$

100. One form of phosphorus exists as P_4 molecules. Each P_4 molecule has four equivalent P atoms, no double or triple bonds, and no expanded octets. Draw the Lewis structure for P_4.

101. A compound has the formula C_8H_8 and does not contain any double or triple bonds. All the carbon atoms are chemically identical, as are all the hydrogen atoms. Draw the Lewis structure for this molecule.

102. The species NO_2, NO_2^+, and NO_2^-, in which N is the central atom, have very different bond angles. Predict what these bond angles might be with respect to the ideal angles and justify your prediction.

103. The bond angles increase steadily in the series PF_3, PCl_3, PBr_3, and PI_3. After consulting the data on atomic radii in Chapter 4, provide an explanation for this observation.

104. Draw the Lewis structure for acetamide (CH_3CONH_2), an organic compound, and determine the geometry about each interior atom. Experiments show that the geometry about the nitrogen atom in acetamide is nearly planar. Which resonance structure can account for the planar geometry about the nitrogen atom?

105. Use VSEPR to predict the geometry (including bond angles) about each interior atom of methyl azide (CH_3N_3) and draw the molecule. Would you expect the bond angle between the two interior nitrogen atoms to be the same or different? Would you expect the two nitrogen–nitrogen bond lengths to be the same or different?

CONCEPTUAL PROBLEMS

106. In the very first chapter of this book, we described the scientific approach and put a special emphasis on scientific models or theories. In this chapter, we looked carefully at the Lewis model of chemical bonding. Why is this theory successful? What are some of the limitations of the theory?

107. Which statement best captures the fundamental idea behind VSEPR theory? Explain what is wrong with the statements you do not choose.

a. The angle between two or more bonds is determined primarily by the repulsions between the electrons within those bonds and other (lone pair) electrons on the central atom of a molecule. Each of these electron groups (bonding electrons or lone pair electrons) lowers its potential energy by maximizing its separation from other electron groups, thus determining the geometry of the molecule.

b. The angle between two or more bonds is determined primarily by the repulsions between the electrons within those bonds. Each of these bonding electrons lowers its potential energy by maximizing its separation from other electron groups, thus determining the geometry of the molecule.

c. The geometry of a molecule is determined by the shapes of the overlapping orbitals that form the chemical bonds. Therefore, to determine the geometry of a molecule, you must determine the shapes of the orbitals involved in bonding.

108. Suppose that a molecule has four bonding groups and one lone pair on the central atom. Suppose further that the molecule is confined to two dimensions (this is a purely hypothetical assumption for the sake of understanding the principles behind VSEPR theory). Draw the molecule and estimate the bond angles.

ANSWERS TO CONCEPTUAL CONNECTIONS

Cc 6.1 N > P > Al > Na

Cc 6.2 **(b)** The dipole moment of the HCl bond is about 1 D and the bond length is 127 pm. Previously you calculated the dipole moment for a 130-pm bond that is 100% ionic to be about 6.2 D. You can therefore estimate the bond's ionic character as $\frac{1}{6} \times 100\%$, which is closest to 15%.

Cc 6.3 **(d)** ClO because the sum of its valence electrons is odd.

Cc 6.4 **(b)** The only molecule in this group that could have an expanded octet is H_3PO_4 because phosphorus is a third-period element. Expanded octets *never* occur in second-period elements such as carbon and nitrogen.

Cc 6.5 The geometry of a molecule is determined by how the terminal atoms are arranged around the central atom, which is in turn determined by how the electron groups are arranged around the central atom. The electron groups on the terminal atoms do not affect this arrangement.

Cc 6.6 **(a)** HCN has two electron groups (the single bond and the triple bond) resulting in a linear geometry.

Cc 6.7 **(c)** Positions 1 and 4 would put the greatest distance between the lone pairs and minimize lone pair–lone pair repulsions.

Cc 6.8 **(d)** All electron groups on the central atom (or interior atoms, if there is more than one) determine the shape of a molecule according to VSEPR theory.

CHAPTER 7

"It is structure that we look for whenever we try to understand anything. All science is built upon this search...."

—Linus Pauling (1901–1994)

Chemical Bonding II

Valence Bond Theory and Molecular Orbital Theory

7.1 Oxygen: A Magnetic Liquid 233

7.2 Valence Bond Theory: Orbital Overlap as a Chemical Bond 234

7.3 Valence Bond Theory: Hybridization of Atomic Orbitals 236

7.4 Molecular Orbital Theory: Electron Delocalization 248

7.5 Molecular Orbital Theory: Polyatomic Molecules 259

7.6 Bonding in Metals and Semiconductors 261

Key Learning Outcomes 264

IN CHAPTER 6, WE EXAMINED a simple model for chemical bonding called the Lewis model. When we combine the Lewis model with valence shell electron pair repulsion (VSEPR), we can predict the shapes of molecules. However, in spite of the success of the Lewis model and VSEPR, we know that electrons do not act like dots. Instead, electrons exist within quantum-mechanical orbitals. In this chapter, we examine two additional bonding theories—valence bond theory and molecular orbital theory—that treat electrons quantum mechanically. These theories are more sophisticated than the Lewis model and also highly quantitative—they numerically predict quantities such as bond angles, bond strengths, and bond lengths. In other words, these theories make accurate predictions about the structure of molecules. As Linus Pauling points out in this chapter's opening quote, understanding structure is central to understanding anything. In chemistry, that is especially true.

7.1 Oxygen: A Magnetic Liquid

Do you remember the first time you played with a magnet? As a child, I was in awe of the invisible force that allowed my magnet to attract objects at a distance, or even through other substances. I once had a pair of magnets so strong that they interacted with one another on opposite sides of my palm. A magnetic substance contains unpaired electrons—it is *paramagnetic* (see Section 4.7). The spin of an unpaired electron generates a tiny magnetic field, much like the circular motion of electrical current running through a coil in an electromagnet generates a magnetic field. When a substance containing unpaired electrons is brought near an external magnetic field, the orientation of the electrons' spin aligns, resulting in a magnetic interaction between the substance and the magnet. This occurs, for example, when you bring a paper clip close to a permanent magnet. The paper clip is attracted to the magnet by the interaction between the permanent magnetic field of the magnet and the magnetic field generated by the spin of unpaired electrons in the paper clip.

Most common liquids, such as water or gasoline, are not magnetic—they are not attracted to a magnet. But there is one liquid that is magnetic: liquid oxygen, which forms when you cool gaseous oxygen below −183 °C (see chapter opening image). What causes liquid oxygen to be magnetic? We already know that oxygen is a diatomic molecule with the following Lewis structure:

$$:\ddot{O}::\ddot{O}:$$

Notice that, in this structure, the electrons are all paired. Yet oxygen is magnetic and *must therefore contain unpaired electrons*. The Lewis model for chemical bonding, while very useful, fails to predict the correct magnetic properties for oxygen. In this chapter, we explore more advanced theories for chemical bonding. First we explore valence bond theory and then molecular orbital theory. We will see that molecular orbital theory—the most sophisticated and accurate bonding theory—correctly predicts the magnetic properties of oxygen. When we apply molecular orbital theory to the O_2 molecule, it shows that O_2 contains two unpaired electrons. These unpaired electrons are responsible for the magnetic behavior of oxygen. We will also examine bonding models that apply to metals and semiconductors.

7.2 Valence Bond Theory: Orbital Overlap as a Chemical Bond

Valence bond theory is an application of a general quantum-mechanical approximation method called *perturbation theory*. In perturbation theory, a complex system (such as a molecule) is viewed as a simpler system (such as two atoms) that is slightly altered or perturbed by some additional force or interaction (such as the interaction between the two atoms).

In the Lewis model, a chemical bond is a shared electron pair. In **valence bond theory**, a chemical bond is the overlap between two half-filled **atomic orbitals (AOs)**. Unlike Lewis theory, which represents valence electrons as dots, valence bond theory treats valence electrons as residing in quantum-mechanical atomic orbitals. In some cases, these orbitals are simply the standard *s*, *p*, *d*, and *f* atomic orbitals that we learned about in Chapter 3. In other cases, these orbitals are *hybridized atomic orbitals*, a kind of blend or combination of two or more standard atomic orbitals.

When two atoms approach each other, the electrons and nucleus of one atom interact with the electrons and nucleus of the other atom. In valence bond theory, we calculate the effect of these interactions on the energies of the valence electrons in the atomic orbitals. If the energy of the system is lowered because of the interactions, then a chemical bond forms. If the energy of the system is raised by the interactions, then a chemical bond does not form.

The interaction energy is usually calculated as a function of the internuclear distance between the two bonding atoms. Figure 7.1 ▼ shows the calculated interaction energy between two hydrogen

KEY CONCEPT VIDEO Valence Bond theory

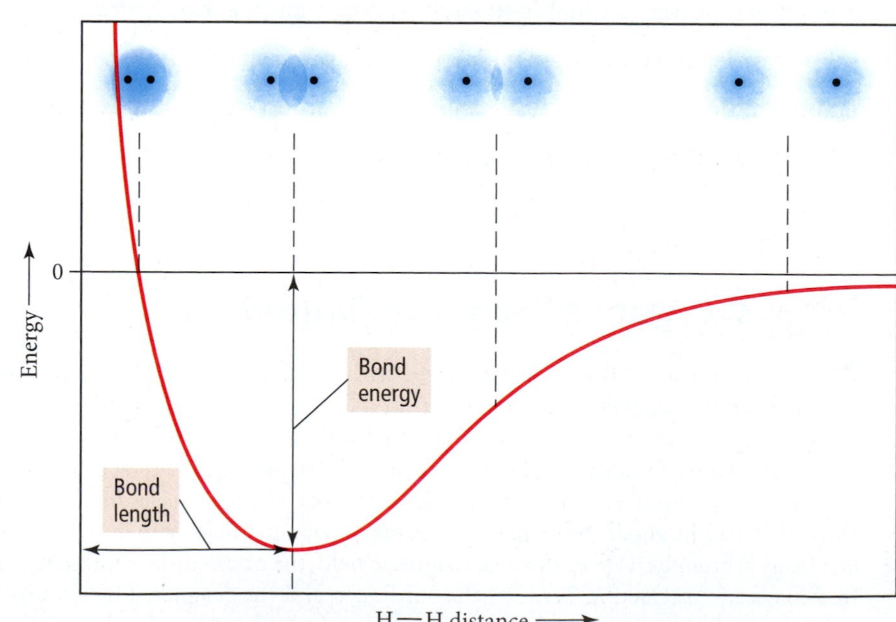

▶ **FIGURE 7.1 Interaction Energy Diagram for H_2** The potential energy of two hydrogen atoms is lowest when they are separated by a distance that allows their 1*s* orbitals a substantial degree of overlap without too much repulsion between their nuclei. This distance, at which the system is most stable, is the bond length of the H_2 molecule.

atoms as a function of the distance between them. The y-axis of the graph is the potential energy that results from the interactions between the charged particles in the two atoms. The x-axis is the separation (or internuclear distance) between the two atoms. As we can see from the graph, when the atoms are far apart (right side of the graph), the interaction energy is nearly zero because the two atoms do not interact to any significant extent. As the atoms get closer, the interaction energy becomes negative. This is a net stabilization that attracts one hydrogen atom to the other. If the atoms get too close, however, the interaction energy begins to rise, primarily because of the mutual repulsion of the two positively charged nuclei. The most stable point on the curve occurs at the minimum of the interaction energy—this is the equilibrium bond length. At this distance, the two atomic 1*s* orbitals have a significant amount of overlap, and the electrons spend time in the internuclear region where they can interact with both nuclei. The value of the interaction energy at the equilibrium bond distance is the bond energy.

When we apply valence bond theory to calculate the interaction energies for a number of atoms and their corresponding molecules, we arrive at the following general observation: *The interaction energy is usually negative (or stabilizing) when the interacting atomic orbitals contain a total of two electrons that can spin-pair (orient with opposing spins).* Most commonly, the two electrons come from two half-filled orbitals, but in some cases, the two electrons come from one filled orbital overlapping with a completely empty orbital (this is called a coordinate covalent bond, and we will cover it in more detail in Chapter 23). In other words, when two atoms with half-filled orbitals approach each other, the half-filled orbitals *overlap*—parts of the orbitals occupy the same space—and the electrons occupying them align with opposite spins. This results in a net energy stabilization that constitutes a covalent chemical bond. The resulting geometry of the molecule emerges from the geometry of the overlapping orbitals.

When *completely filled* orbitals overlap, the interaction energy is positive (or destabilizing) and no bond forms.

Summarizing Valence Bond Theory:

- The valence electrons of the atoms in a molecule reside in quantum-mechanical atomic orbitals. The orbitals can be the standard *s*, *p*, *d*, and *f* orbitals, or they may be hybrid combinations of these.
- A chemical bond results from the overlap of two half-filled orbitals and spin-pairing of the two valence electrons (or less commonly the overlap of a completely filled orbital with an empty orbital).
- The geometry of the overlapping orbitals determines the shape of the molecule.

We can apply the general concepts of valence bond theory to explain bonding in hydrogen sulfide, H_2S. The valence electron configurations of the atoms in the molecule are as follows:

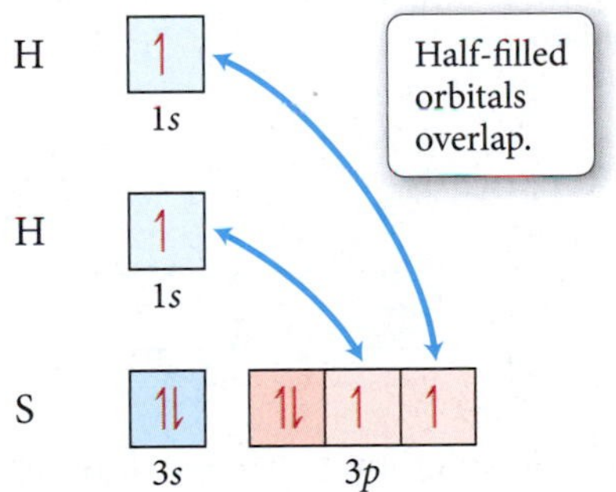

The hydrogen atoms each have one half-filled orbital, and the sulfur atom has two half-filled orbitals. The half-filled orbitals on each hydrogen atom overlap with the two half-filled orbitals on the sulfur atom, forming two chemical bonds.

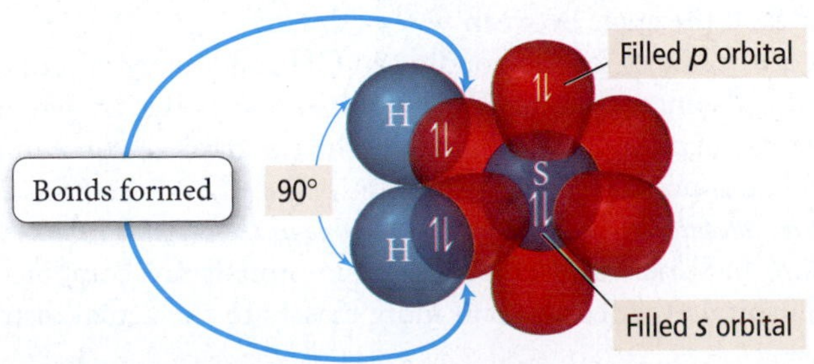

To illustrate the spin-pairing of the electrons in the overlapping orbitals, we superimpose a half-arrow for each electron in each half-filled orbital and show that, within a bond, the electrons are spin-paired (one half-arrow is pointing up, and the other is pointing down). We also superimpose paired half-arrows in the filled sulfur *s* and *p* orbitals to represent the lone pair electrons in those orbitals. (Since those orbitals are full, they are not involved in bonding.)

A quantitative calculation of H_2S using valence bond theory yields bond energies, bond lengths, and bond angles. In our qualitative treatment, we simply show how orbital overlap leads to bonding and make a rough sketch of the molecule based on the overlapping orbitals. Notice that, because the overlapping orbitals on the central atom (sulfur) are *p* orbitals and because *p* orbitals are oriented at 90° to one another, the predicted bond angle is 90°. The actual bond angle in H_2S is 92°. In the case of H_2S, a simple valence bond treatment matches well with the experimentally measured bond angle (in contrast to VSEPR theory, which predicts a bond angle of a bit less than 109.5°).

7.1 Cc Conceptual Connection

What Is a Chemical Bond, Part I?

The answer to the question, *what is a chemical bond*, depends on the bonding model. Answer these three questions.

(a) What is a covalent chemical bond according to the Lewis model?

(b) What is a covalent chemical bond according to valence bond theory?

(c) Why are the answers different?

7.3 Valence Bond Theory: Hybridization of Atomic Orbitals

Although the overlap of half-filled *standard* atomic orbitals adequately explains the bonding in H_2S, it cannot adequately explain the bonding in many other molecules. For example, suppose we try to explain the bonding between hydrogen and carbon using the same approach. The valence electron configurations of H and C are shown here:

H 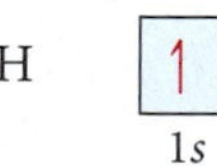

C 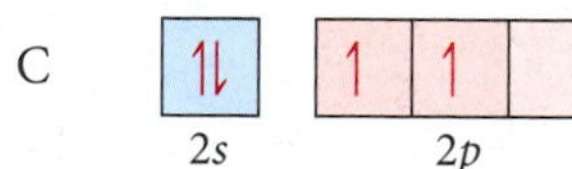

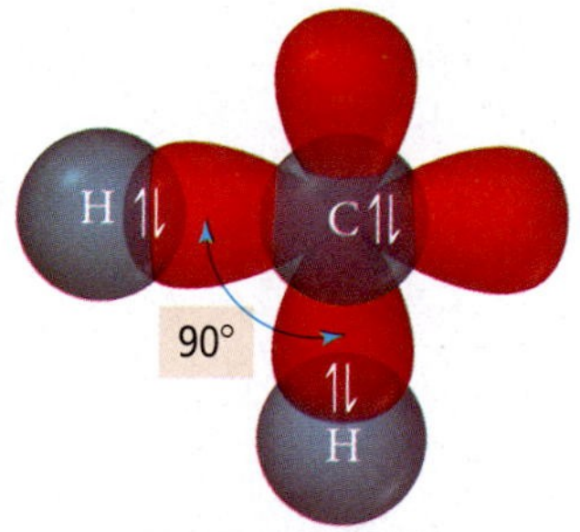

Theoretical prediction

Observed reality

Carbon has only two half-filled orbitals and should therefore form only two bonds with two hydrogen atoms. We would predict that carbon and hydrogen would form a molecule with the formula CH_2 and with a bond angle of 90° (corresponding to the angle between any two *p* orbitals).

However, experiments show that the stable compound formed from carbon and hydrogen is CH_4 (methane), with bond angles of 109.5°. The experimental reality is different from our simple theoretical prediction in two ways. First, carbon forms bonds to four hydrogen atoms, not two. Second, the bond angles are much larger than the angle between two *p* orbitals.

Valence bond theory accounts for the bonding in CH_4 and many other polyatomic molecules by incorporating an additional concept called *orbital hybridization*. So far, we have assumed that the overlapping orbitals that form chemical bonds are simply the standard *s*, *p*, or *d* atomic orbitals, but this is an oversimplification. The concept of hybridization in valence bond theory is a step toward recognizing that *the orbitals in a molecule are not necessarily the same as the orbitals in an atom.* **Hybridization** is a mathematical procedure in which the standard atomic orbitals are combined to form new atomic orbitals called **hybrid orbitals** that correspond more closely to the actual distribution of electrons in

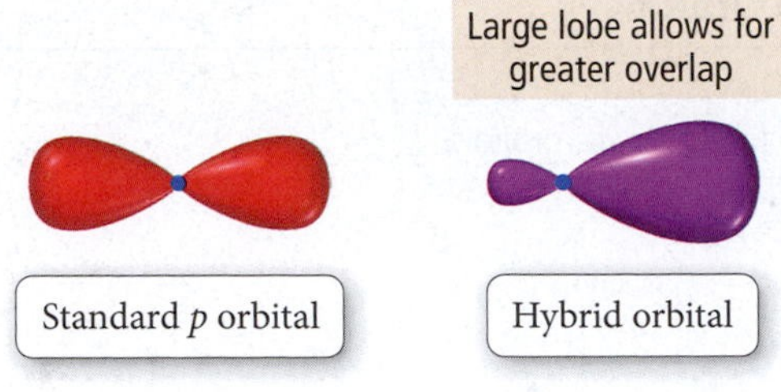

◀ **FIGURE 7.2 Comparison of Standard Atomic Orbital and Hybrid Orbital** A hybrid orbital has a large lobe that allows for better overlap with other orbitals.

chemically *bonded* atoms. Hybrid orbitals are still localized on individual atoms, but they have different shapes and energies from those of standard atomic orbitals.

Why does valence bond theory propose that electrons in some molecules occupy hybrid orbitals instead of the standard atomic orbitals? To answer this question, remember that, according to valence bond theory, a chemical bond is the overlap of two orbitals that together contain two electrons. The overlap of orbitals lowers the potential energy of the electrons in those orbitals. *The greater the overlap, the lower the energy, and the stronger the bond.* Hybrid orbitals allow greater overlap because the electron density in a hybrid orbital is concentrated along a single directional lobe as shown in Figure 7.2 ▲. This concentration of electron density in a single direction allows for greater overlap between orbitals. In other words, hybrid orbitals *minimize* the energy of the molecule by *maximizing* the orbital overlap in a bond.

Hybridization, however, is not a free lunch—in most cases it actually costs some energy. So hybridization occurs only to the degree that the energy payback through bond formation is large. In general, therefore, the more bonds that an atom forms, the greater the tendency of its orbitals to hybridize. Central or interior atoms, which form the most bonds, have the greatest tendency to hybridize. Terminal atoms, which form the fewest bonds, have the least tendency to hybridize. *In this book, we focus on the hybridization of interior atoms and assume that all terminal atoms—those bonding to only one other atom—are unhybridized.* Hybridization is particularly important in carbon, which tends to form four bonds in its compounds and therefore always hybridizes.

As we saw in Chapter 6, the word *hybrid* comes from breeding. A *hybrid* is an offspring of two animals or plants of different standard breeds. Similarly, a hybrid orbital is a product of mixing two or more standard atomic orbitals.

Although we cannot examine the procedure for obtaining hybrid orbitals in mathematical detail here, we can make the following general statements regarding hybridization:

- The *number of standard atomic orbitals* added together always equals the *number of hybrid orbitals* formed. The total number of orbitals is conserved.
- The *particular combinations* of standard atomic orbitals added together determine the *shapes and energies* of the hybrid orbitals formed.
- The *particular type of hybridization that occurs* is the one that yields the *lowest overall energy for the molecule*. Since actual energy calculations are beyond the scope of this book, we use electron geometries as determined by VSEPR theory to predict the type of hybridization.

In a more nuanced treatment, hybridization is not an all-or-nothing process—it can occur to varying degrees that are not always easy to predict. We saw earlier, for example, that sulfur does not hybridize very much in forming H_2S.

sp^3 Hybridization

We can account for the tetrahedral geometry in CH_4 by the hybridization of the one $2s$ orbital and the three $2p$ orbitals on the carbon atom. The four new orbitals that result, called sp^3 hybrids, are shown in the following energy diagram:

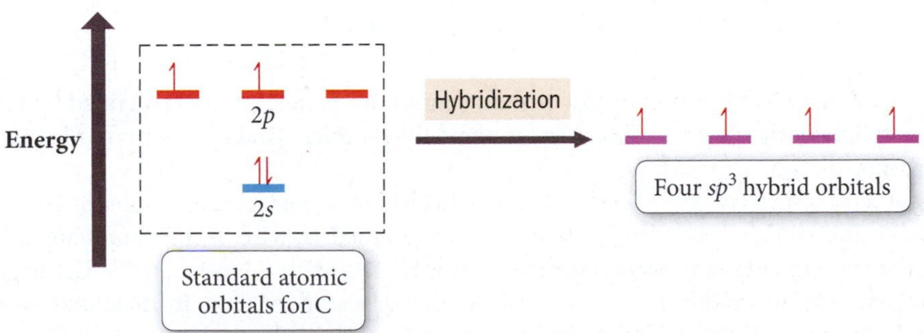

The notation sp^3 indicates that the hybrid orbitals are mixtures of one s orbital and three p orbitals. Notice that the hybrid orbitals all have the same energy—they are degenerate. The shapes of the sp^3

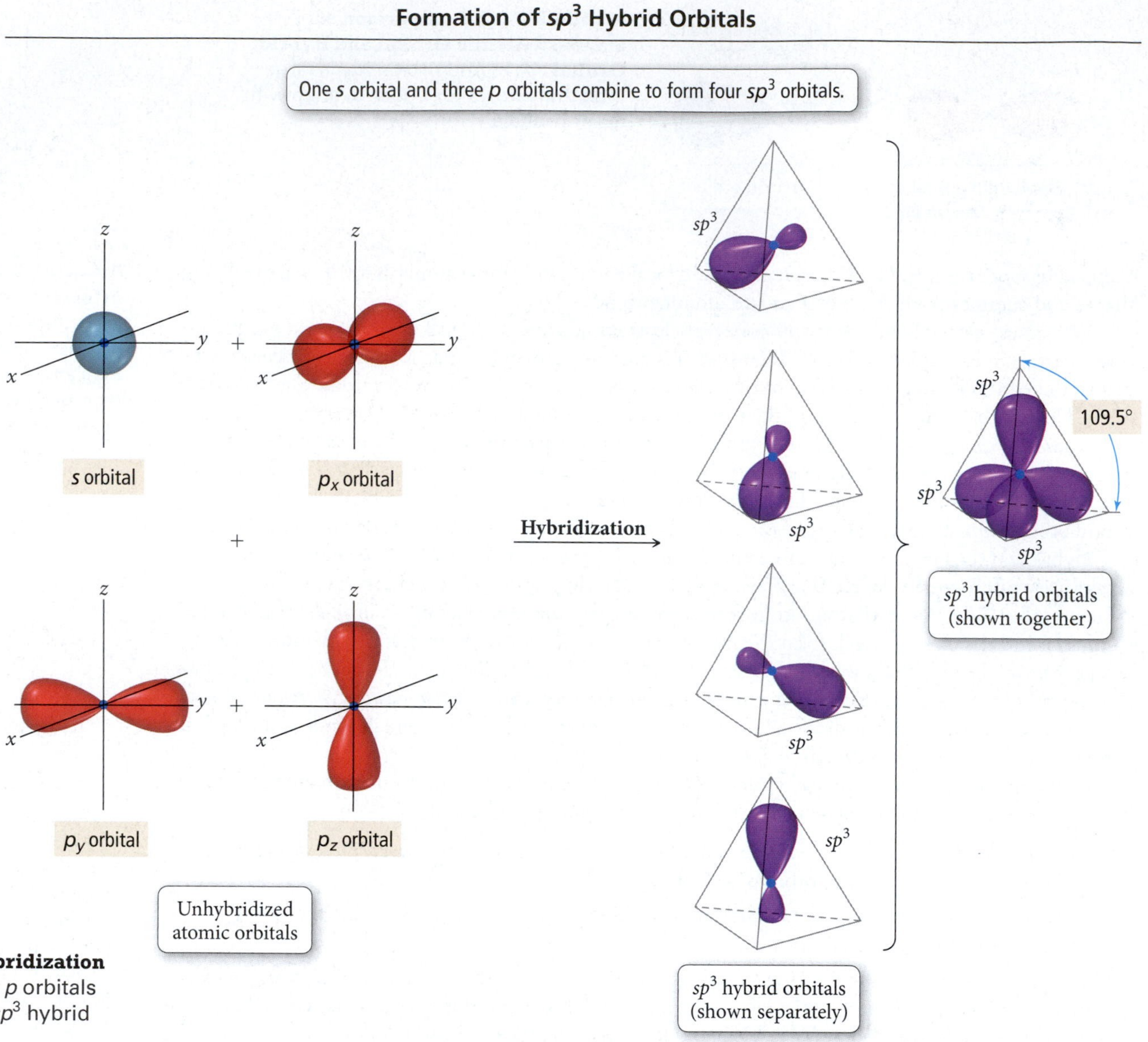

Formation of sp^3 Hybrid Orbitals

One s orbital and three p orbitals combine to form four sp^3 orbitals.

s orbital + p_x orbital + p_y orbital + p_z orbital

Unhybridized atomic orbitals

Hybridization →

sp^3 hybrid orbitals (shown separately)

109.5°

sp^3 hybrid orbitals (shown together)

▶ **FIGURE 7.3** ***sp*³ Hybridization** One s orbital and three p orbitals combine to form four sp^3 hybrid orbitals.

hybrid orbitals are shown in Figure 7.3 ▲. The four hybrid orbitals are arranged in a tetrahedral geometry with 109.5° angles between them.

We can write an orbital diagram for carbon using these hybrid orbitals:

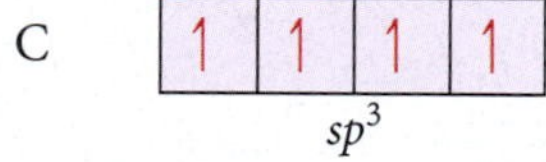

Carbon's four valence electrons occupy the orbitals singly with parallel spins as dictated by Hund's rule. With this electron configuration, carbon has four half-filled orbitals and can form four bonds with four hydrogen atoms.

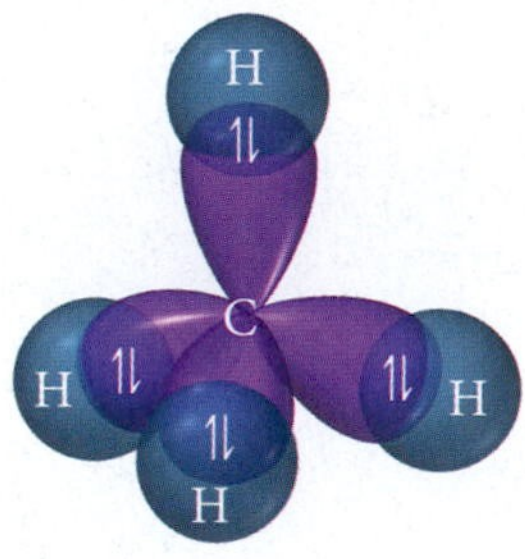

The geometry of the *overlapping orbitals* (the hybrids) is tetrahedral, with angles of 109.5° between the orbitals, so the *resulting geometry of the molecule* is tetrahedral, with 109.5° bond angles, in agreement with the experimentally measured geometry of CH_4 and with the predicted VSEPR geometry.

Hybridized orbitals readily form chemical bonds because they tend to maximize overlap with other orbitals. However, if the central atom of a molecule contains lone pairs, hybrid orbitals can also accommodate them. For example, the nitrogen orbitals in ammonia are sp^3 hybrids. Three of the hybrids are involved in bonding with three hydrogen atoms, but the fourth hybrid contains a lone pair. The presence of the lone pair lowers the tendency of nitrogen's orbitals to hybridize. (Recall that the tendency to hybridize increases with the number of bonds formed.) Therefore the bond angle in NH_3 is 107°, a bit closer to the unhybridized *p* orbital bond angle of 90°.

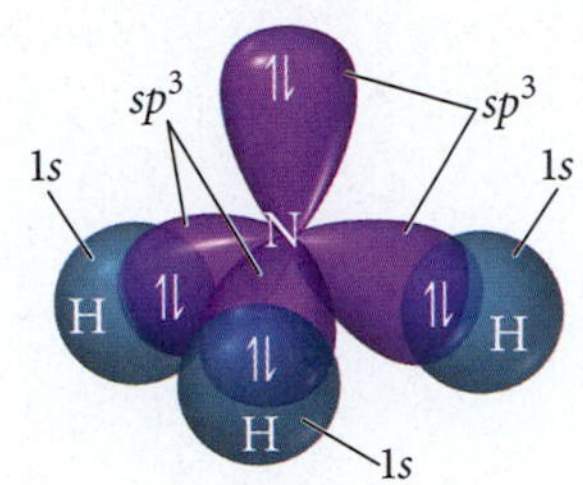

*sp*² Hybridization and Double Bonds

Hybridization of one *s* and two *p* orbitals results in three *sp*² hybrids and one leftover unhybridized *p* orbital.

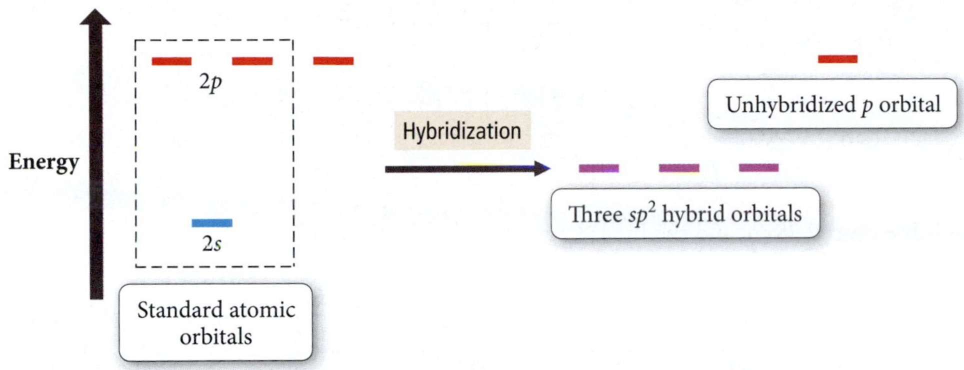

In valence bond theory, the particular hybridization scheme (*sp*² versus *sp*³, for example) for a given molecule is determined computationally, which is beyond our scope. In this book, we determine the particular hybridization scheme from the VSEPR geometry of the molecule, as shown later in this section.

The notation *sp*² indicates that the hybrids are mixtures of one *s* orbital and two *p* orbitals. The shapes of the *sp*² hybrid orbitals are shown in Figure 7.4 ▼. Notice that the three hybrid orbitals have a trigonal planar geometry with 120° angles between them. The unhybridized *p* orbital is oriented perpendicular to the three hybridized orbitals.

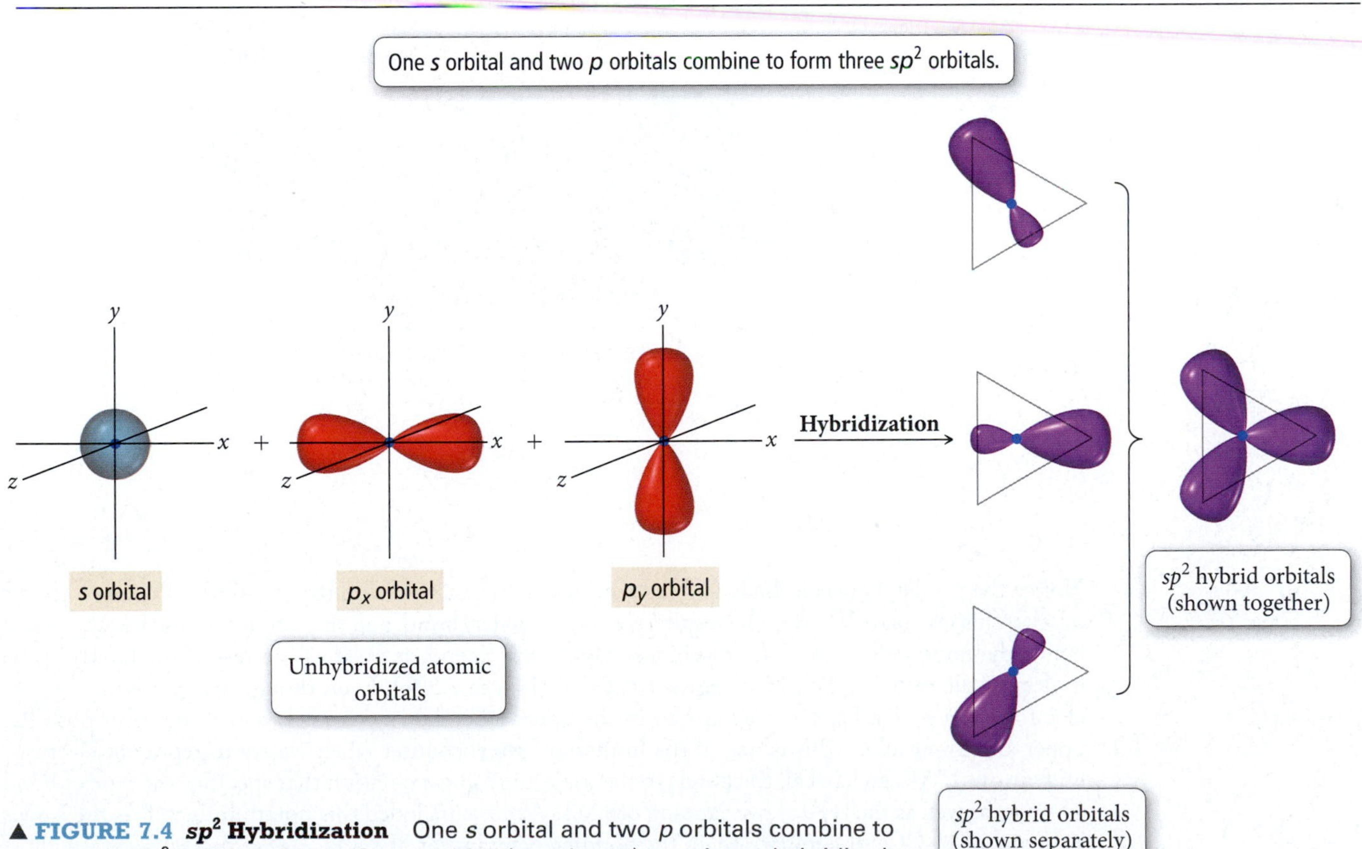

▲ **FIGURE 7.4** ***sp*² Hybridization** One *s* orbital and two *p* orbitals combine to form three *sp*² hybrid orbitals. One *p* orbital (not shown) remains unhybridized.

As an example of a molecule with *sp*^{2} hybrid orbitals, consider H_2CO. The unhybridized valence electron configurations of each of the atoms are as follows:

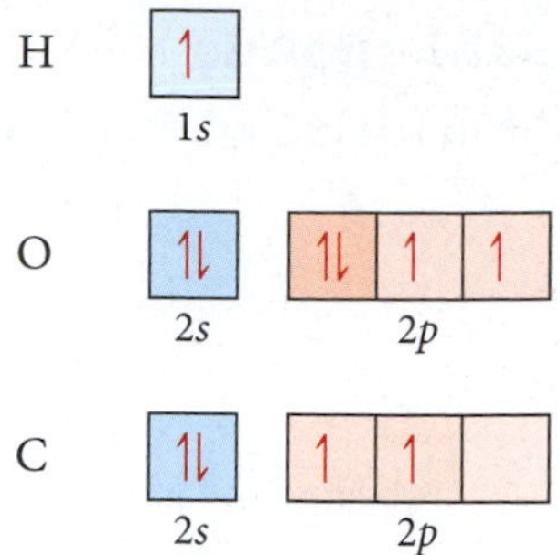

Carbon is the central atom, and the hybridization of its orbitals is *sp*^{2}:

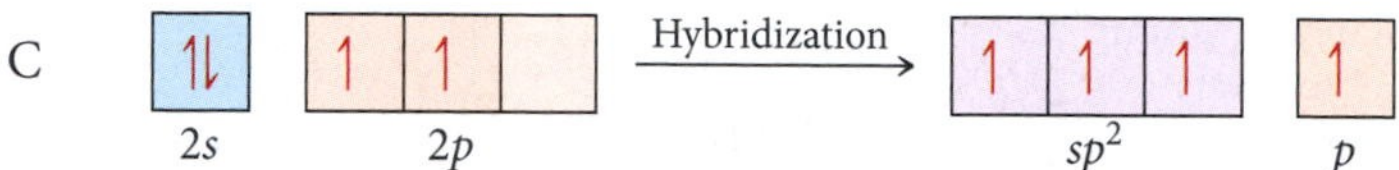

Each of the *sp*^{2} orbitals is half-filled. The remaining electron occupies the leftover *p* orbital, even though it is slightly higher in energy. We can now see that the carbon atom has four half-filled orbitals and can therefore form four bonds: two with two hydrogen atoms and two (a double bond) with the oxygen atom. We draw the molecule and the overlapping orbitals as follows:

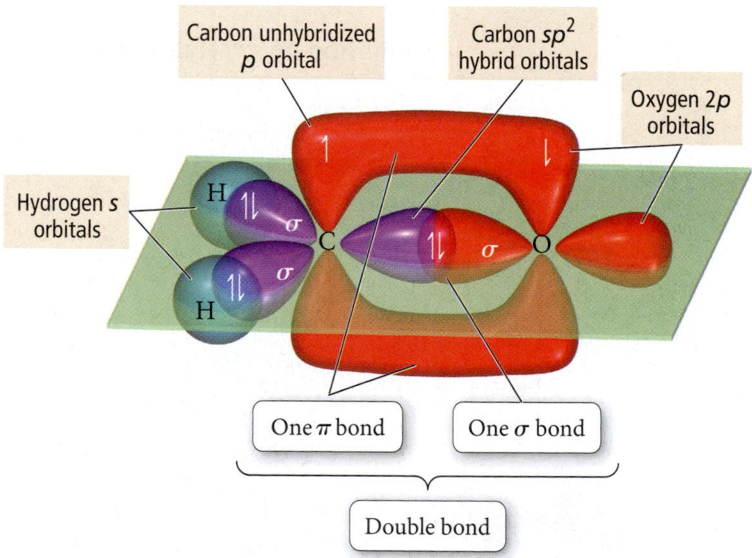

Notice the overlap between the half-filled *p* orbitals on the carbon and oxygen atoms. When *p* orbitals overlap this way (side by side), the resulting bond is a **pi (π) bond**, and the electron density is above and below the internuclear axis. When orbitals overlap end to end, as in all of the rest of the bonds in the molecule, the resulting bond is a **sigma (σ) bond** (Figure 7.5 ▶). Even though we represent the two electrons in a π bond as two half arrows in the upper lobe, they are actually spread out over both the upper and lower lobes (this is one of the limitations we encounter when we try to represent electrons with arrows). We can label all the bonds in the molecule using a notation that specifies the type of bond (σ or π) as well as the type of overlapping orbitals. We have included this notation, as well as the Lewis structure of H_2CO for comparison, in the bonding diagram for H_2CO:

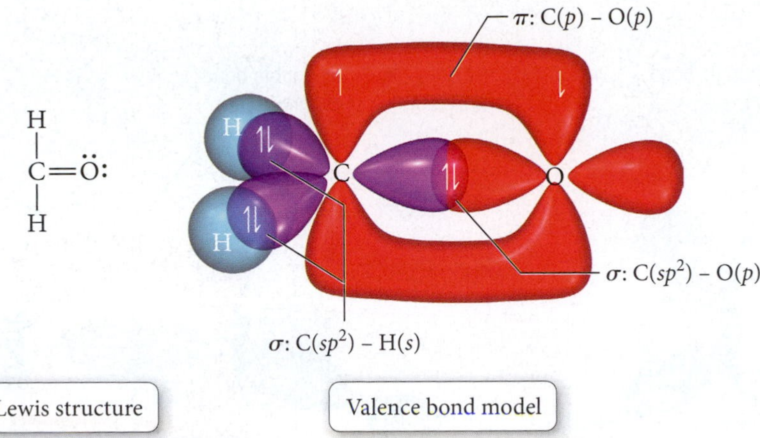

In this image, nonbonding electrons have been omitted (for clarity).

Notice the correspondence between the valence bond model and the Lewis structure. In both cases, the central carbon atom is forming four bonds: two single bonds and one double bond. However, valence bond theory gives us more insight into the bonds. The double bond between carbon and oxygen according to valence bond theory consists of two different kinds of bonds—one σ and one π—while in the Lewis model the two bonds within the double bond appear identical. *A double bond in the Lewis model always corresponds to one σ and one π bond in valence bond theory.* In general, π bonds are weaker than σ bonds because the side-to-side orbital overlap tends to be less efficient than the end-to-end orbital overlap. Consequently, the π bond in a double bond is generally easier to break than the σ bond.

One—and only one—σ bond forms between any two atoms. Additional bonds must be π bonds.

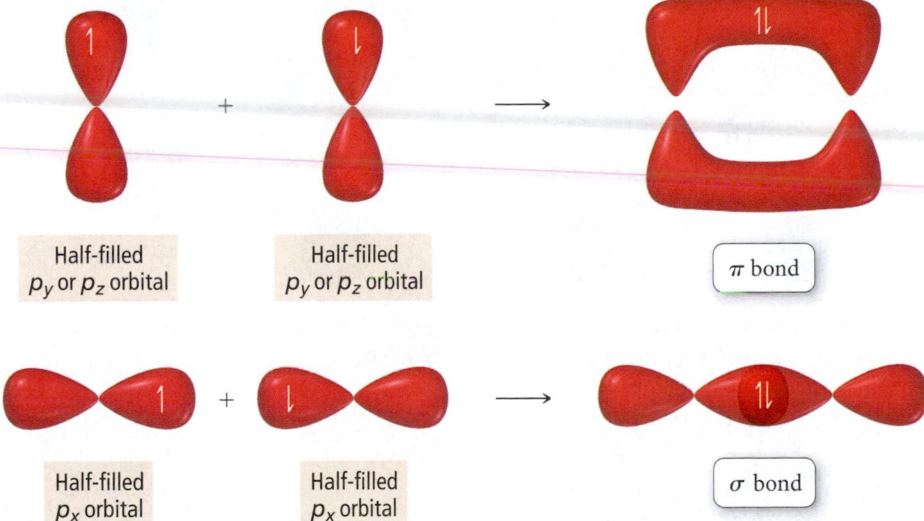

▲ **FIGURE 7.5 Sigma and Pi Bonding** When orbitals overlap side by side, the result is a pi (π) bond. When orbitals overlap end to end, they form a sigma (σ) bond. Two atoms can form only one sigma bond. A single bond is a sigma bond; a double bond consists of a sigma bond and a pi bond; a triple bond consists of a sigma bond and two pi bonds.

Valence bond theory allows us to see why the rotation about a double bond is severely restricted. Because of the side-by-side overlap of the *p* orbitals, the π bond must essentially break for rotation to occur. Valence bond theory shows us the types of orbitals involved in the bonding and their shapes. In H_2CO, the sp^2 hybrid orbitals on the central atom are trigonal planar with 120° angles between them, so the resulting predicted geometry of the molecule is trigonal planar with 120° bond angles. The experimentally measured bond angles in H_2CO, as discussed previously, are 121.9° for the HCO bond and 116.2° for the HCH bond angle, close to the predicted values.

Although rotation about a double bond is highly restricted, rotation about a single bond is relatively unrestricted. Consider, for example, the structures of two chlorinated hydrocarbons, 1,2-dichloroethane and 1,2-dichloroethene:

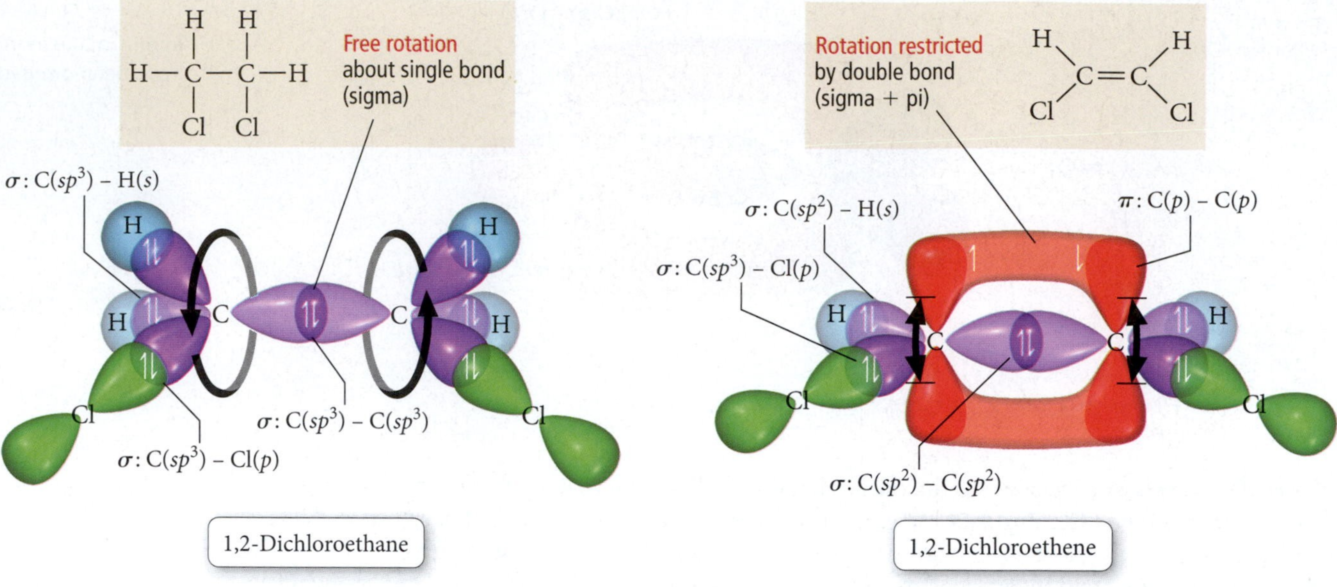

1,2-Dichloroethane

1,2-Dichloroethene

The hybridization of the carbon atoms in 1,2-dichloroethane is sp^3, resulting in relatively free rotation about the sigma single bond. Consequently, there is no difference between the following two structures at room temperature because they quickly interconvert:

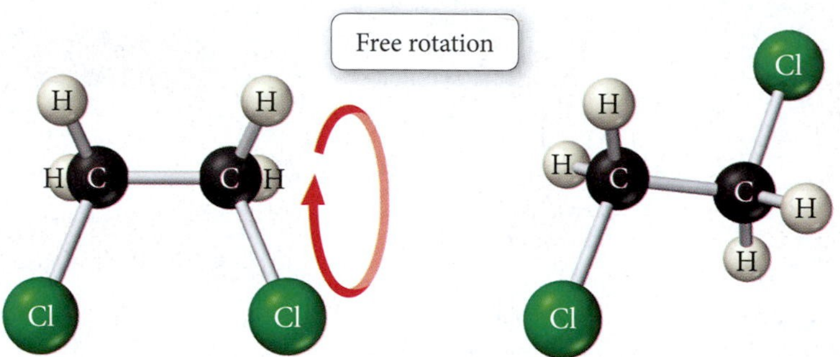

In contrast, rotation about the double bond (sigma + pi) in 1,2-dichloroethene is restricted, so that, at room temperature, 1,2-dichloroethene exists in these two forms:

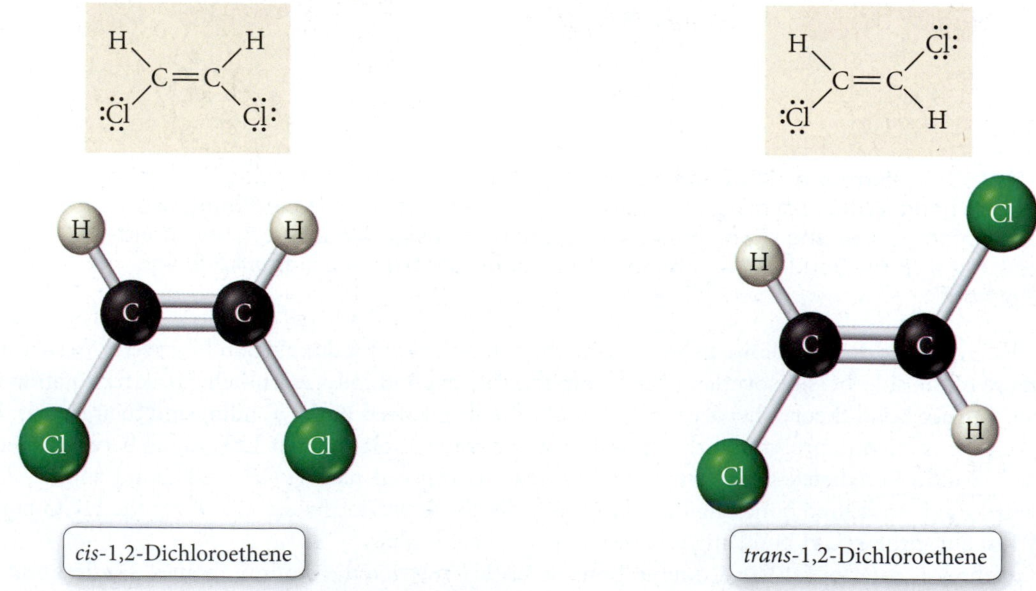

cis-1,2-Dichloroethene

trans-1,2-Dichloroethene

We distinguish between them with the designations *cis* (meaning "same side") and *trans* (meaning "opposite sides"). These two forms of 1,2-dichloroethene have different structures and therefore different properties. For example, the *cis* form boils at 60.3 °C and the *trans* form boils at 47.2 °C. Notice that *structure* (in this case the nature of a double bond and the arrangement of the chlorine atoms) *affects properties* (boiling point). Compounds such as the two forms of 1,2-dichloroethene, with the same molecular formula but different structures or different spatial arrangement of atoms, are called **isomers**. Nature can—and does—make different compounds out of the same atoms by arranging the atoms in different ways. Isomerism is common throughout chemistry and is especially important in organic chemistry, as we will discuss in Chapter 22.

Single and Double Bonds

In Section 6.6 we learned that double bonds are stronger and shorter than single bonds. For example, a C—C single bond has an average bond energy of 347 kJ/mole, while a C═C double bond has an average bond energy of 611 kJ/mole. Use valence bond theory to explain why a double bond is *not* simply twice as strong as a single bond.

sp Hybridization and Triple Bonds

Hybridization of one *s* and one *p* orbital results in two *sp* hybrid orbitals and two leftover unhybridized *p* orbitals.

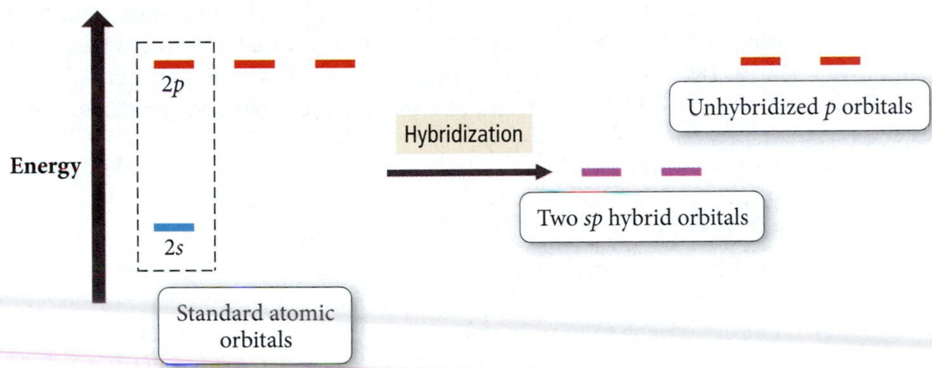

The shapes of the *sp* hybrid orbitals are shown in Figure 7.6 ▼. Notice that the two *sp* hybrid orbitals are arranged in a linear geometry with a 180° angle between them. The unhybridized *p* orbitals are oriented in the plane that is perpendicular to the hybridized *sp* orbitals.

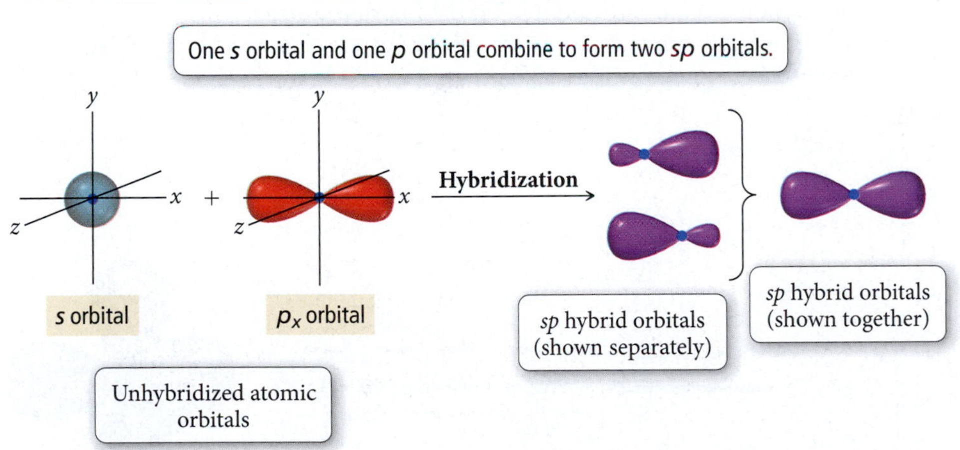

◀ **FIGURE 7.6** ***sp* Hybridization** One *s* orbital and one *p* orbital combine to form two *sp* hybrid orbitals. Two *p* orbitals (not shown) remain unhybridized.

The acetylene molecule, HC≡CH, has *sp* hybrid orbitals. The four valence electrons of carbon can distribute themselves among the two *sp* hybrid orbitals and the two *p* orbitals:

H [↿] 1*s*

C [⥮] 2*s* [↿][↿][ ] 2*p* —Hybridization→ [↿][↿] *sp* [↿][↿] 2*p*

Each carbon atom then has four half-filled orbitals and can form four bonds: one with a hydrogen atom and three (a triple bond) with the other carbon atom. We draw the molecule and the overlapping orbitals as follows:

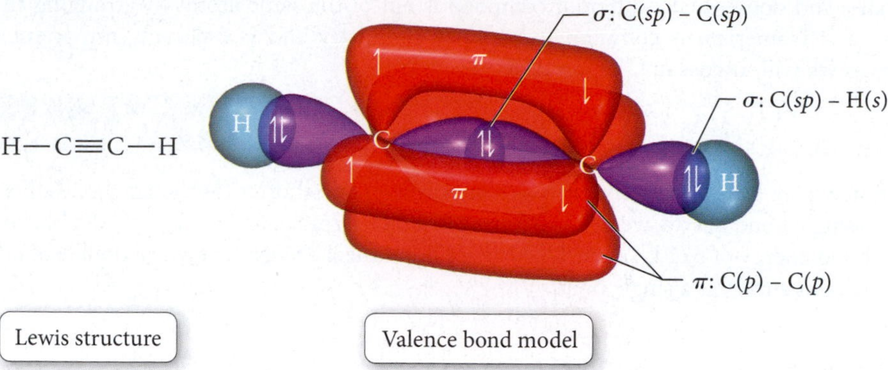

Notice that the triple bond between the two carbon atoms consists of two π bonds (overlapping p orbitals) and one σ bond (overlapping sp orbitals). The sp orbitals on the carbon atoms are linear with 180° between them, so the resulting geometry of the molecule is linear with 180° bond angles, in agreement with the experimentally measured geometry of HC≡CH, and also in agreement with the prediction of VSEPR theory.

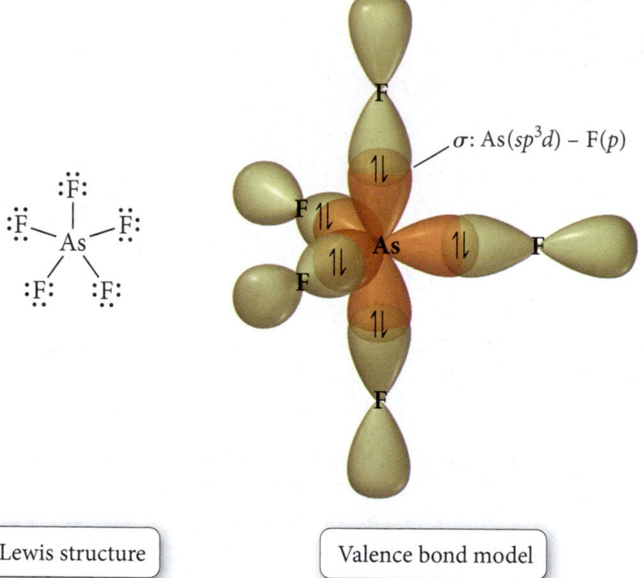

sp^3d and sp^3d^2 Hybridization

Recall that, according to the Lewis model, elements occurring in the third row of the periodic table (or below) can exhibit expanded octets (see Section 6.5). The equivalent concept in valence bond theory is hybridization involving the d orbitals. For third-period elements, the 3d orbitals are involved in hybridization because their energies are close to the energies of the 3s and 3p orbitals. The hybridization of one s orbital, three p orbitals, and one d orbital results in sp^3d hybrid orbitals, as shown in Figure 7.7(a) ▼. The five sp^3d hybrid orbitals have a trigonal bipyramidal arrangement, as shown in Figure 7.7(b) ▼. Arsenic pentafluoride, AsF_5 (shown in the margin), is an example of sp^3d hybridization. The arsenic atom bonds to five fluorine atoms by overlap between the sp^3d hybrid orbitals on arsenic and p orbitals on the fluorine atoms.

The sp^3d orbitals on the arsenic atom are trigonal bipyramidal, so the molecular geometry is trigonal bipyramidal.

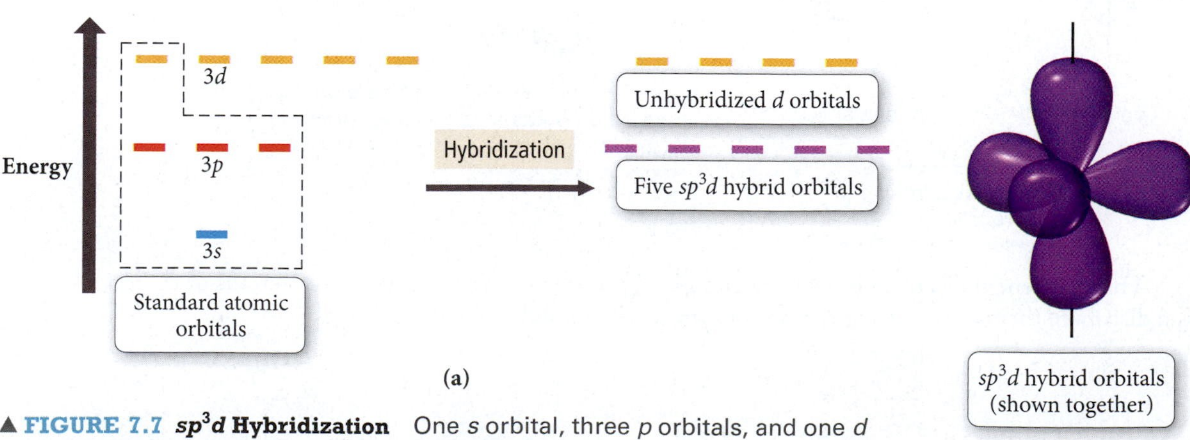

▲ **FIGURE 7.7** ***sp^3d* Hybridization** One s orbital, three p orbitals, and one d orbital combine to form five sp^3d hybrid orbitals.

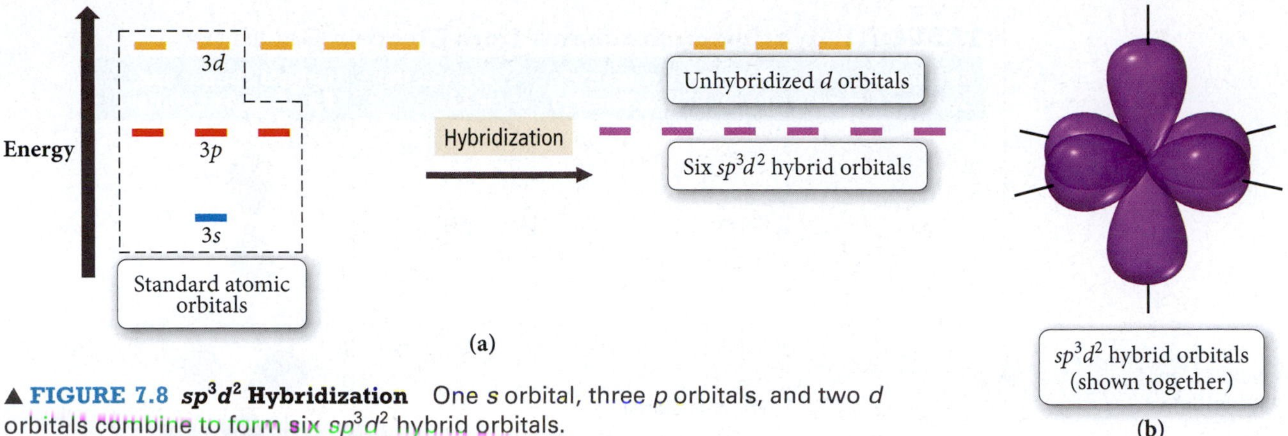

▲ **FIGURE 7.8 sp^3d^2 Hybridization** One *s* orbital, three *p* orbitals, and two *d* orbitals combine to form six sp^3d^2 hybrid orbitals.

The hybridization of one *s* orbital, three *p* orbitals, and *two d* orbitals results in sp^3d^2 hybrid orbitals, as shown in Figure 7.8(a) ▲. The six sp^3d^2 hybrid orbitals have an octahedral geometry, shown in Figure 7.8(b) ▲. Sulfur hexafluoride, SF_6, is an example of sp^3d^2 hybridization. The sulfur atom bonds to six fluorine atoms by overlap between the sp^3d^2 hybrid orbitals on sulfur and *p* orbitals on the fluorine atoms.

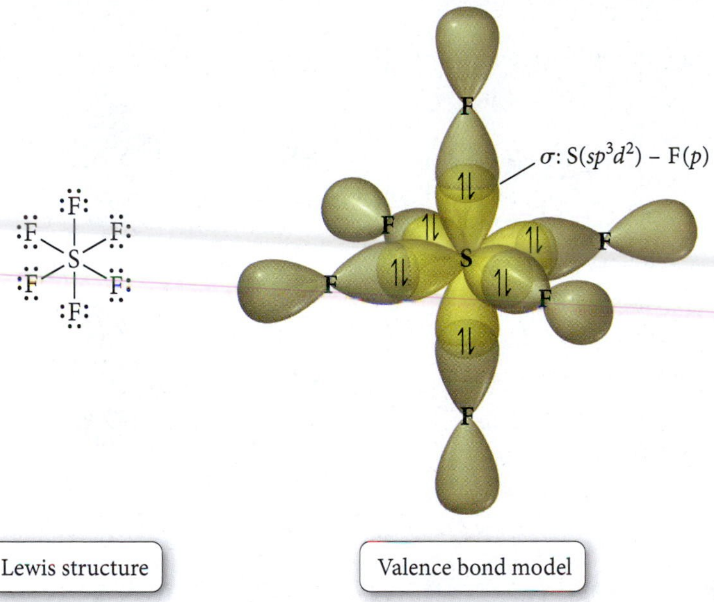

The sp^3d^2 orbitals on the sulfur atom are octahedral, so the molecular geometry is octahedral, again in agreement with VSEPR theory and with the experimentally observed geometry.

Writing Hybridization and Bonding Schemes

We have now studied examples of the five main types of atomic orbital hybridization. *But how do we know which hybridization scheme best describes the orbitals of a specific atom in a specific molecule?* In computational valence bond theory, the energy of the molecule is calculated using a computer; the degree of hybridization as well as the type of hybridization are varied to find the combination that gives the molecule the lowest overall energy. For our purposes, we can assign a hybridization scheme from the electron geometry—determined using VSEPR theory—of the central atom (or interior atoms) of the molecule. The five VSEPR electron geometries and the corresponding hybridization schemes are shown in Table 7.1. For example, if the electron geometry of the central atom is tetrahedral, then the hybridization is sp^3, and if the electron geometry is octahedral, then the hybridization is sp^3d^2, and so on. Although this method of determining the hybridization scheme is not 100% accurate (for example, it predicts that H_2S should be sp^3 when in fact H_2S is largely unhybridized), it is the best we can do without more complex computer-based calculations.

TABLE 7.1 Hybridization Scheme from Electron Geometry

Number of Electron Groups	Electron Geometry (from VSEPR Theory)	Hybridization Scheme	Orbital Shapes and Relative Orientation
2	Linear	*sp*	
3	Trigonal planar	sp^2	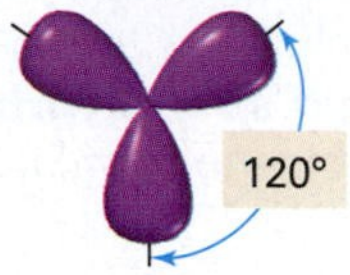 120°
4	Tetrahedral	sp^3	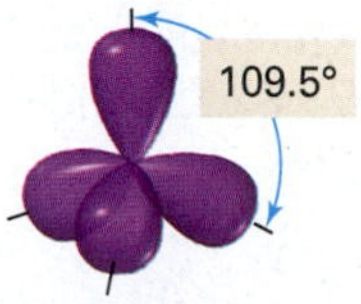 109.5°
5	Trigonal bipyramidal	sp^3d	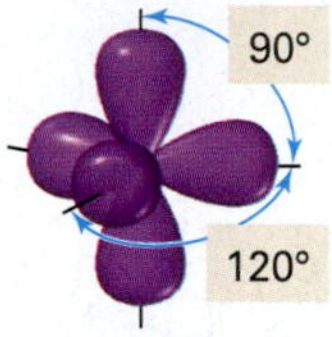 90°, 120°
6	Octahedral	sp^3d^2	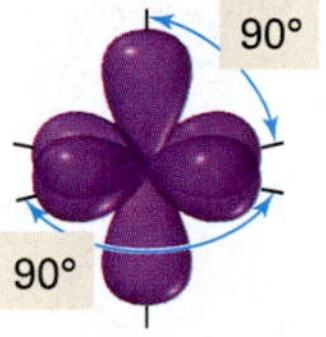 90°, 90°

As defined in the previous chapter, an *electron group* is a lone pair, bonding pair, or multiple bond.

7.3 Cc Conceptual Connection

Hybridization

What is the hybridization of C in CO_2?

(a) *sp*
(b) sp^2
(c) sp^3

We are now ready to put the Lewis model and valence bond theory together to describe bonding in molecules. In the procedure and examples that follow, you learn how to write a *hybridization and bonding scheme* for a molecule. This involves drawing the Lewis structure for the molecule, determining its geometry using VSEPR theory, determining the hybridization of the interior atoms, drawing the molecule with its overlapping orbitals, and labeling each bond with the σ and π notation followed by the type of overlapping orbitals. This procedure involves virtually everything you have learned about bonding so far in this chapter and Chapter 6. The procedure for writing a hybridization and bonding scheme is shown in the left column, with two examples of how to apply the procedure in the columns to the right.

PROCEDURE FOR ▼ Hybridization and Bonding Scheme

1. Draw the Lewis structure for the molecule.
2. Apply VSEPR theory to predict the electron geometry about the central atom (or interior atoms).
3. Select the correct hybridization for the central atom (or interior atoms) based on the electron geometry (refer to Table 7.1).
4. Sketch the molecule, beginning with the central atom and its orbitals. Show overlap with the appropriate orbitals on the terminal atoms.
5. Label all bonds using the σ or π notation followed by the type of overlapping orbitals.

EXAMPLE 7.1 Hybridization and Bonding Scheme

Write a hybridization and bonding scheme for bromine trifluoride, BrF_3.

SOLUTION

BrF_3 has 28 valence electrons and the following Lewis structure:

```
 :F:
  |
 :Br—F:
  |
 :F:
```

The bromine atom has five electron groups and therefore has a trigonal bipyramidal electron geometry.

A trigonal bipyramidal electron geometry corresponds to sp^3d hybridization.

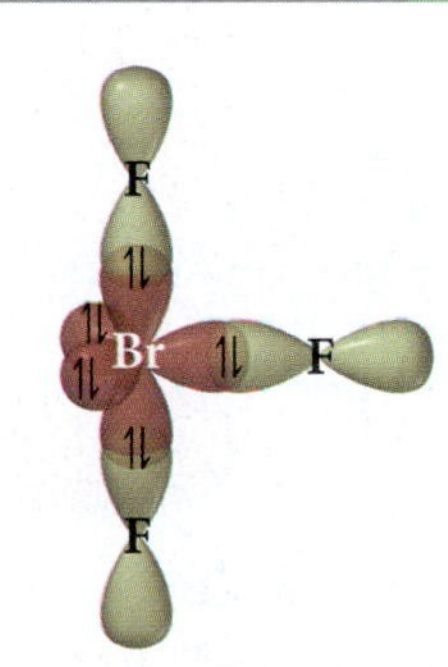

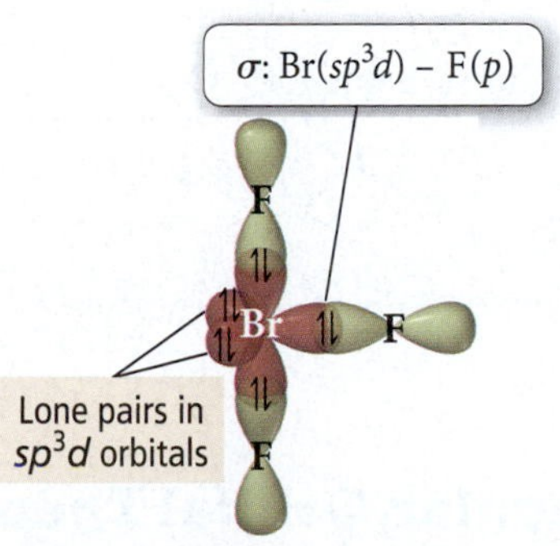

FOR PRACTICE 7.1

Write a hybridization and bonding scheme for XeF_4.

EXAMPLE 7.2 Hybridization and Bonding Scheme

Write a hybridization and bonding scheme for acetaldehyde.

```
        O
        ‖
H3C—C—H
```

SOLUTION

Acetaldehyde has 18 valence electrons and the following Lewis structure:

```
    H  :O:
    |   ‖
H—C—C—H
    |
    H
```

The leftmost carbon atom has four electron groups and a tetrahedral electron geometry. The rightmost carbon atom has three electron groups and trigonal planar geometry.

The leftmost carbon atom is sp^3 hybridized, and the rightmost carbon atom is sp^2 hybridized.

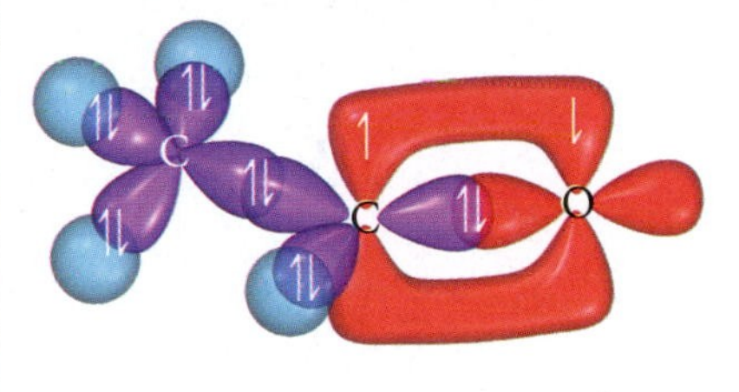

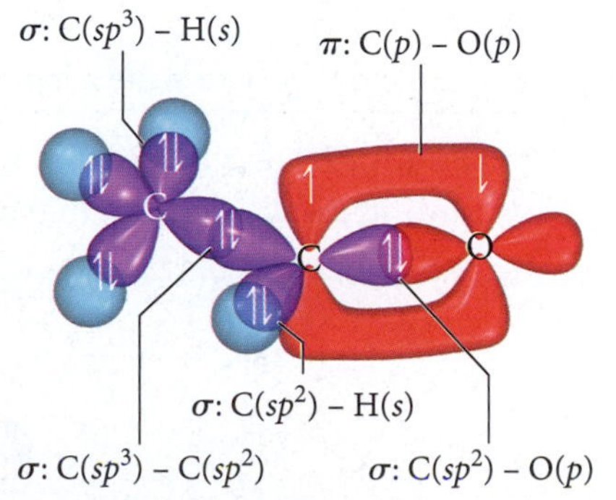

FOR PRACTICE 7.2

Write a hybridization and bonding scheme for HCN.

EXAMPLE 7.3

Hybridization and Bonding Scheme

Apply valence bond theory to write a hybridization and bonding scheme for ethene, $H_2C{=}CH_2$.

SOLUTION

1. Draw the Lewis structure for the molecule.	H H | | H—C=C—H
2. Apply VSEPR theory to predict the electron geometry about the central atom (or interior atoms).	The molecule has two interior atoms. Since each atom has three electron groups (one double bond and two single bonds), the electron geometry about each atom is trigonal planar.
3. Select the correct hybridization for the central atom (or interior atoms) based on the electron geometry (refer to Table 7.1).	A trigonal planar geometry corresponds to sp^2 hybridization.
4. Sketch the molecule, beginning with the central atom and its orbitals. Show overlap with the appropriate orbitals on the terminal atoms.	H H C C H H
5. Label all bonds using the σ or π notation followed by the type of overlapping orbitals.	π: C(p) – C(p) σ: C(sp^2) – C(sp^2) σ: C(sp^2) – H(s)

FOR PRACTICE 7.3
Apply valence bond theory to write a hybridization and bonding scheme for CO_2.

FOR MORE PRACTICE 7.3
What is the hybridization of the central iodine atom in I_3^-?

7.4 Molecular Orbital Theory: Electron Delocalization

Valence bond theory can account for much of what we observe in chemical bonding—such as the rigidity of a double bond—but it also has limitations. In valence bond theory, we treat electrons as if they reside in the quantum-mechanical orbitals that we calculated *for atoms*. This is a significant oversimplification that we partially compensate for by hybridization. Nevertheless, we can do better.

In Chapter 3, we learned that the mathematical derivation of energies and orbitals for electrons *in atoms* comes from solving the Schrödinger equation for the atom of interest. For a molecule, we can theoretically do the same thing. The resulting orbitals would be the actual **molecular orbitals (MOs)** of the molecule as a whole (in contrast to valence bond theory, in which the orbitals are the atomic orbitals (AOs) of individual atoms). As it turns out, however, solving the Schrödinger equation exactly for even the simplest molecules is impossible without making some approximations.

In **molecular orbital (MO) theory**, we do not actually solve the Schrödinger equation for a molecule directly. Instead, we use a trial function, an "educated guess" as to what the solution might be. In other words, rather than mathematically solving the Schrödinger equation, which would give us a mathematical function describing an orbital, we start with a trial mathematical function for the orbital. We then test the trial function to see how well it "works."

Molecular orbital theory is a specific application of a more general quantum-mechanical approximation technique called the variational principle. In this technique, the energy of a trial function within the Schrödinger equation is minimized.

We can understand this process by analogy to solving an algebraic equation. Suppose we want to know what x is in the equation $4x + 5 = 70$ without actually solving the equation. For an easy equation like this one, we might first estimate that $x = 16$. We can then determine how well our estimate works by substituting $x = 16$ into the equation. If the estimate does not work, we can try again until we find the right value of x. (In this case, we can quickly see that x must be a little more than 16.)

In molecular orbital theory, the estimating procedure is analogous. However, we need one more concept to get at the heart of molecular orbital theory. In order to determine how well a trial function for an orbital "works" in molecular orbital theory, we calculate its energy. No matter how good our trial function, *we can never do better than nature at minimizing the energy of the orbital*. In other words, we can devise any trial function for an orbital in a molecule and calculate its energy. The energy we calculate for the devised orbital is always greater than or (at best) equal to the energy of the actual orbital.

We calculate the energy of an estimated orbital by substituting it into the Schrödinger equation and solving for the energy.

How does this help us? The best possible orbital is therefore the one with the minimum energy. In modern molecular orbital theory, computer programs are designed to try many different variations of a guessed orbital and compare the energies of each one. The variation with the lowest energy is the best approximation for the actual molecular orbital.

Linear Combination of Atomic Orbitals (LCAO)

The simplest trial functions that work reasonably well in molecular orbital theory turn out to be linear combinations of atomic orbitals, or LCAOs. An LCAO molecular orbital is a *weighted linear sum—analogous to a weighted average—of the valence atomic orbitals* of the atoms in the molecule. At first glance, this concept might seem very similar to that of hybridization in valence bond theory. However, in valence bond theory, hybrid orbitals are weighted linear sums of the valence atomic orbitals of a *particular atom*, and the hybrid orbitals remain *localized* on that atom. In molecular orbital theory, the molecular orbitals are weighted linear sums of the valence atomic orbitals of *all the atoms* in a molecule, and many of the molecular orbitals are *delocalized* over the entire molecule.

When molecular orbitals are calculated mathematically, it is actually the *wave functions* corresponding to the orbitals that are combined.

Consider the H_2 molecule. One of the molecular orbitals for H_2 is simply an equally weighted sum of the 1s orbital from one atom and the 1s orbital from the other. We can represent this pictorially and energetically as follows:

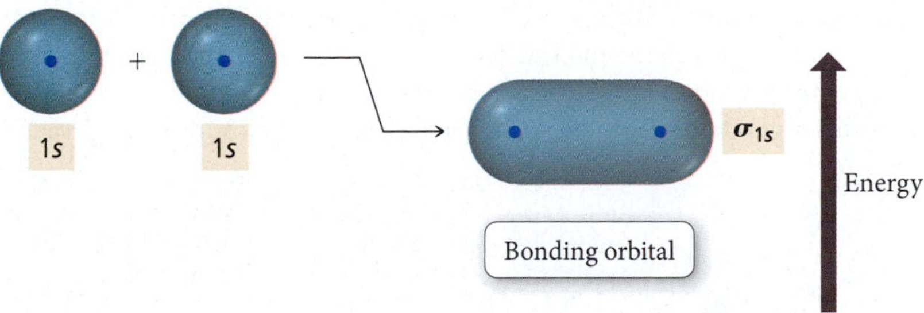

The name of this molecular orbital is σ_{1s}. The σ comes from the shape of the orbital, which looks like a σ bond in valence bond theory, and the 1s comes from its formation by a linear sum of 1s orbitals. The σ_{1s} orbital is lower in energy than either of the two 1s atomic orbitals from which it was formed. For this reason, we call this orbital a **bonding orbital**. When electrons occupy bonding molecular orbitals, the energy of the electrons is lower than it would be if they were occupying atomic orbitals.

We can think of a molecular orbital in a molecule in much the same way that we think about an atomic orbital in an atom. Electrons will seek the lowest energy molecular orbital available, but just as an atom has more than one atomic orbital (and some may be empty), a molecule has more than one molecular orbital (and some may be empty). The next molecular orbital of H_2 is approximated by summing the 1*s* orbital on one hydrogen atom with the *negative* (opposite phase) of the 1*s* orbital on the other hydrogen atom.

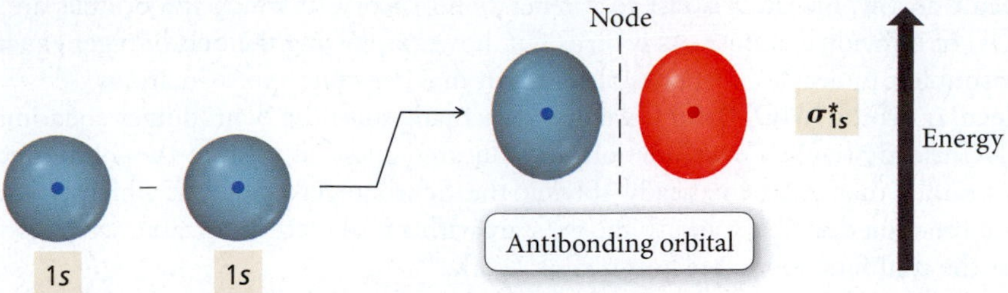

The different phases of the orbitals result in *destructive* interference between them. The resulting molecular orbital therefore has a node between the two atoms. The different colors (red and blue) on either side of the node represent the different phases of the orbital (see Section 3.6). The name of this molecular orbital is σ^*_{1s}. The star indicates that this orbital is an **antibonding orbital**. Electrons in antibonding orbitals have higher energies than they had in their respective atomic orbitals and therefore tend to raise the energy of the system (relative to the unbonded atoms).

In general, when two atomic orbitals are added together to form molecular orbitals, one of the resultant molecular orbitals will be lower in energy (the bonding orbital) than the atomic orbitals and the other will be higher in energy (the antibonding orbital). Remember that electrons in orbitals behave like waves. The bonding molecular orbital arises out of constructive interference between the atomic orbitals because both orbitals have the same phase. The antibonding orbital arises out of destructive interference between the atomic orbitals because *subtracting* one from the other means the two interacting orbitals have opposite phases (Figure 7.9 ▼).

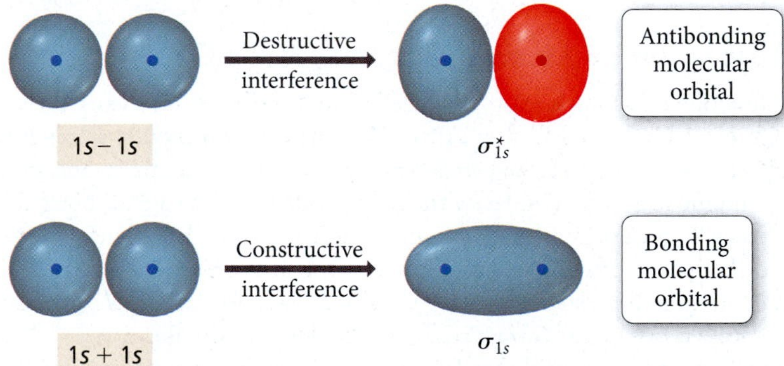

▶ **FIGURE 7.9 Formation of Bonding and Antibonding Orbitals** Constructive interference between two atomic orbitals gives rise to a molecular orbital that is lower in energy than the atomic orbitals. This is the bonding orbital. Destructive interference between two atomic orbitals gives rise to a molecular orbital that is higher in energy than the atomic orbitals. This is the antibonding orbital.

For this reason, the bonding orbital has an *increased* electron density in the internuclear region while the antibonding orbital has a *node* in the internuclear region. The greater electron density in the internuclear region for a bonding orbital *lowers its energy* compared to the orbitals in nonbonded atoms. The diminished electron density in the internuclear region for an antibonding orbital (due to the node) *increases its energy* compared to the orbitals of nonbonded atoms.

We put all of this together in the molecular orbital (MO) diagram for H_2:

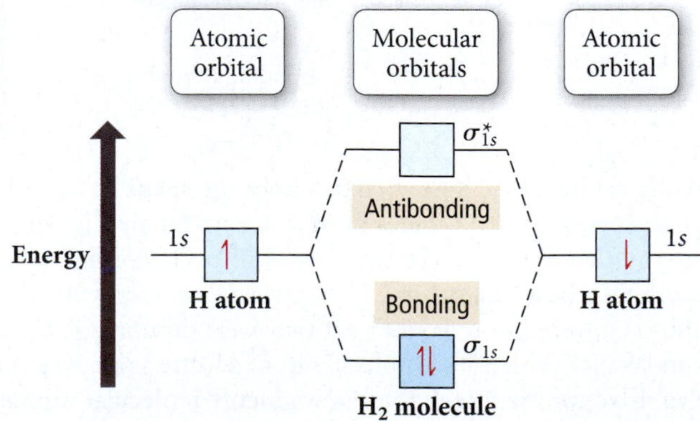

The **molecular orbital (MO) diagram** shows the atomic orbitals of the atoms, the molecular orbitals of the molecule, and their relative energies (higher on the diagram corresponds to higher energy). Notice that two hydrogen atoms can lower their overall energy by forming H_2 because the electrons can move from higher energy atomic orbitals into the lower energy σ_{1s} bonding molecular orbital. In molecular orbital theory, we define the **bond order** of a diatomic molecule such as H_2 as follows:

$$\text{Bond order} = \frac{(\text{number of electrons in bonding MOs}) - (\text{number of electrons in antibonding MOs})}{2}$$

For H_2, the bond order is 1.

$$H_2 \text{ bond order} = \frac{2-0}{2} = 1$$

A positive bond order means that there are more electrons in bonding molecular orbitals than in antibonding molecular orbitals. The electrons therefore have lower energy than they had in the orbitals of the isolated atoms, and a chemical bond forms. In general, the higher the bond order, the stronger the bond. A negative or zero bond order indicates that a bond will *not* form between the atoms. For example, consider the MO diagram for He_2:

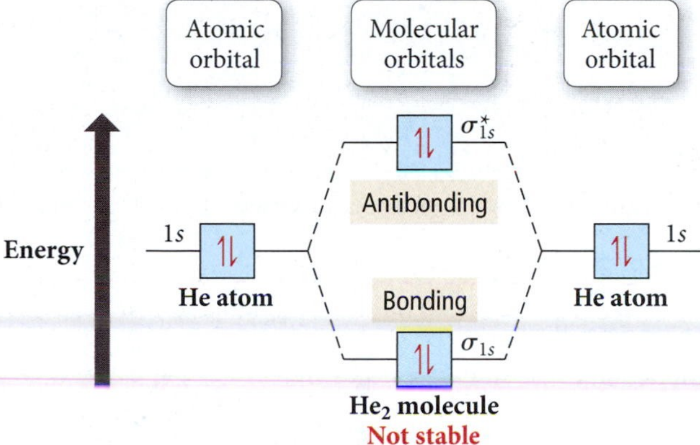

Notice that the two additional electrons must go into the higher energy antibonding orbital. There is no net stabilization by joining two helium atoms to form a helium molecule, as indicated by the bond order:

$$He_2 \text{ bond order} = \frac{2-2}{2} = 0$$

So according to MO theory, He_2 should not exist as a stable molecule, and indeed it does not. Another interesting case is the helium–helium ion, He_2^+, which we can represent with the following MO diagram:

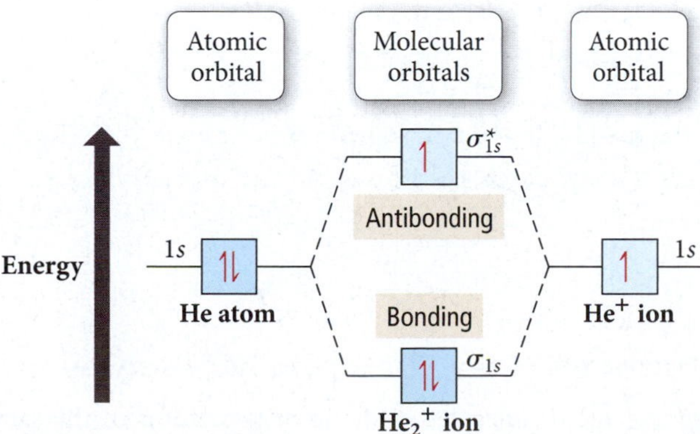

The bond order is $\frac{1}{2}$, indicating that He_2^+ should exist, and indeed it does.

Summarizing LCAO–MO Theory:

- We can approximate molecular orbitals (MOs) as a linear combination of atomic orbitals (AOs). The total number of MOs formed from a particular set of AOs always equals the number of AOs in the set.
- When two AOs combine to form two MOs, one MO is lower in energy (the bonding MO) and the other is higher in energy (the antibonding MO).
- When assigning the electrons of a molecule to MOs, we fill the lowest energy MOs first with a maximum of two spin-paired electrons per orbital.
- When assigning electrons to two MOs of the same energy, we follow Hund's rule—fill the orbitals singly first, with parallel spins, before pairing.
- The bond order in a diatomic molecule is the number of electrons in bonding MOs minus the number in antibonding MOs divided by two. Stable bonds require a positive bond order (more electrons in bonding MOs than in antibonding MOs).

Notice the power of the molecular orbital approach. Every electron that enters a bonding molecular orbital stabilizes the molecule or polyatomic ion, and every electron that enters an antibonding molecular orbital destabilizes it. The emphasis on electron pairs has been removed. One electron in a bonding molecular orbital stabilizes half as much as two, so a bond order of one-half is nothing mysterious.

EXAMPLE 7.4
Bond Order

Apply molecular orbital theory to predict the bond order in H_2^-. Is the H_2^- bond a stronger or weaker bond than the H_2 bond?

SOLUTION

The H_2^- ion has three electrons. Assign the three electrons to the molecular orbitals, filling lower energy orbitals first and proceeding to higher energy orbitals.

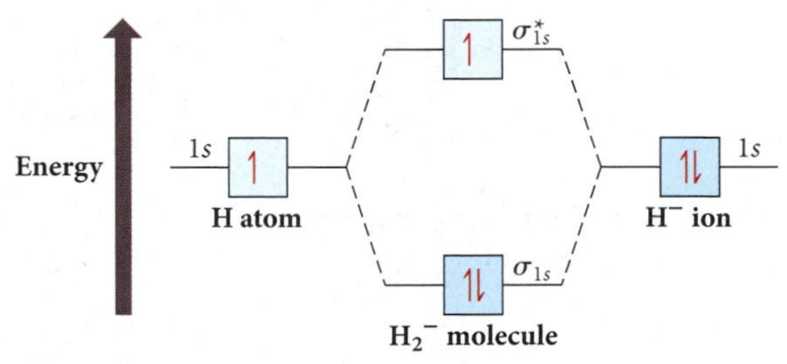

Calculate the bond order by subtracting the number of electrons in antibonding orbitals from the number in bonding orbitals and dividing the result by two.

$$H_2^- \text{ bond order} = \frac{2-1}{2} = +\tfrac{1}{2}$$

Since the bond order is positive, H_2^- should be stable. However, the bond order of H_2^- is lower than the bond order of H_2 (which is 1); therefore, the bond in H_2^- is weaker than in H_2.

FOR PRACTICE 7.4

Apply molecular orbital theory to predict the bond order in H_2^+. Is the H_2^+ bond a stronger or weaker bond than the H_2 bond?

The core electrons can be ignored (as they are in other models for bonding), because these electrons do not contribute significantly to chemical bonding.

Second-Period Homonuclear Diatomic Molecules

Homonuclear diatomic molecules (molecules made up of two atoms of the same kind) formed from second-period elements have between 2 and 16 valence electrons. To explain bonding in these molecules, we must consider the next set of higher energy molecular orbitals, which we can approximate by linear combinations of the valence atomic orbitals of the period 2 elements.

We begin with Li_2. Even though lithium is normally a metal, we can use MO theory to predict whether or not the Li_2 molecule should exist in the gas phase. We approximate the molecular orbitals in Li_2 as linear combinations of the $2s$ atomic orbitals. The resulting molecular orbitals look much like those of the H_2 molecule. The MO diagram for Li_2 therefore looks a lot like the MO diagram for H_2:

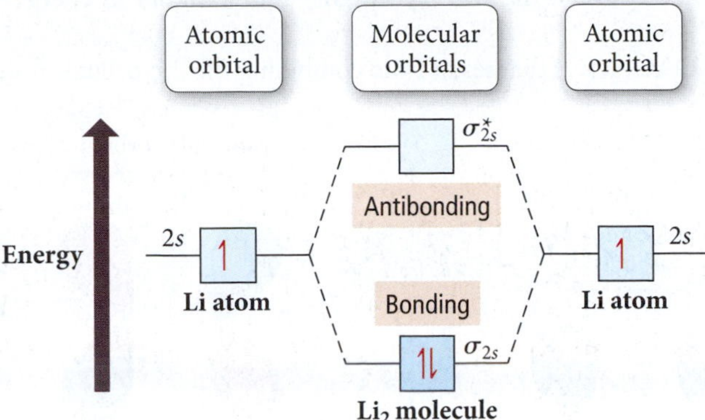

The two valence electrons of Li_2 occupy a bonding molecular orbital. We would predict that the Li_2 molecule is stable with a bond order of 1. Experiments confirm this prediction. In contrast, consider the MO diagram for Be_2:

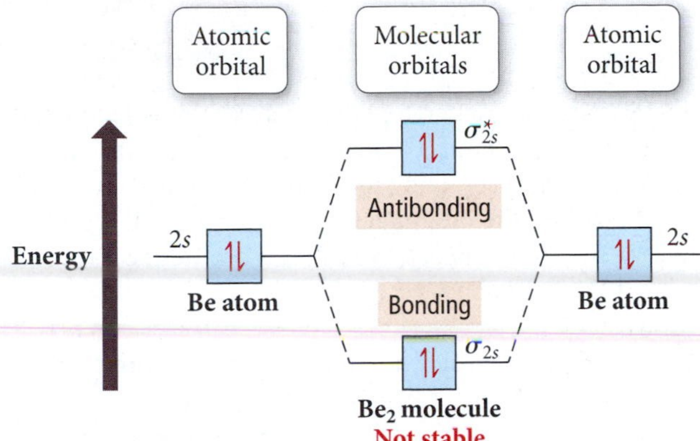

The four valence electrons of Be_2 occupy one bonding MO and one antibonding MO. The bond order is 0, and we predict that Be_2 should not be stable. This is again consistent with experimental findings.

The next homonuclear molecule composed of second row elements is B_2, which has six total valence electrons to accommodate. We can approximate the next higher energy molecular orbitals for B_2 and the rest of the period 2 diatomic molecules as linear combinations of the $2p$ orbitals taken pairwise. Because the three $2p$ orbitals orient along three orthogonal axes, we must assign similar axes to the molecule. In this book, we assign the internuclear axis to be the x direction. The LCAO–MOs that result from combining the $2p_x$ orbitals—the ones that lie along the internuclear axis—from each atom are represented pictorially as follows:

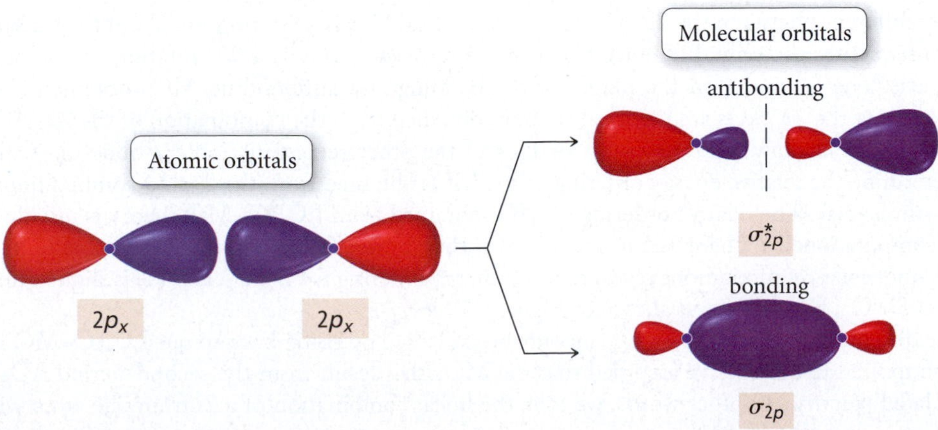

The bonding MO in this pair looks something like candy in a wrapper, with increased electron density in the internuclear region due to constructive interference between the two $2p$ atomic orbitals. It has the characteristic σ shape (it is cylindrically symmetrical about the bond axis) and is therefore called the σ_{2p} bonding orbital. The antibonding orbital, called σ^*_{2p}, has a node between the two nuclei (due to destructive interference between the two $2p$ orbitals) and is higher in energy than either of the $2p_x$ orbitals.

We represent the LCAO–MOs that result from combining the $2p_z$ orbitals from each atom as follows:

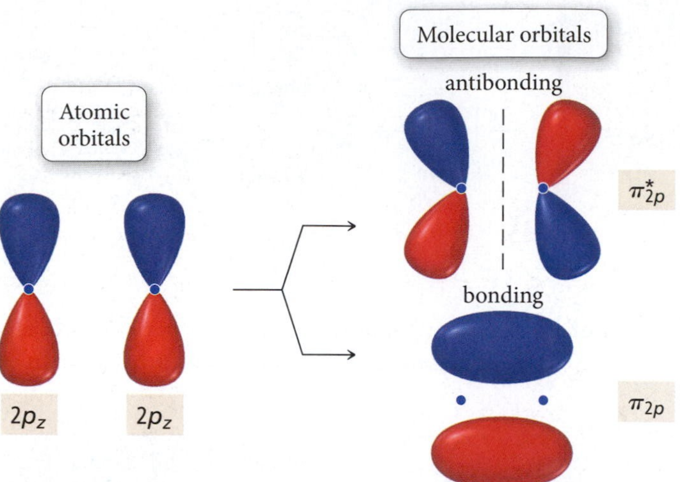

Notice that in this case the p orbitals are added together in a side-by-side orientation (in contrast to the $2p_x$ orbitals which were oriented end to end). The resultant molecular orbitals consequently have a different shape. The electron density in the bonding molecular orbital is above and below the internuclear axis with a nodal plane that includes the internuclear axis. This orbital resembles the electron density distribution of a π bond in valence bond theory. We call this orbital the π_{2p} orbital. The corresponding antibonding orbital has an additional node *between* the nuclei (perpendicular to the internuclear axis) and is called the π^*_{2p} orbital.

We represent the LCAO–MOs that result from combining the $2p_y$ orbitals from each atom as follows:

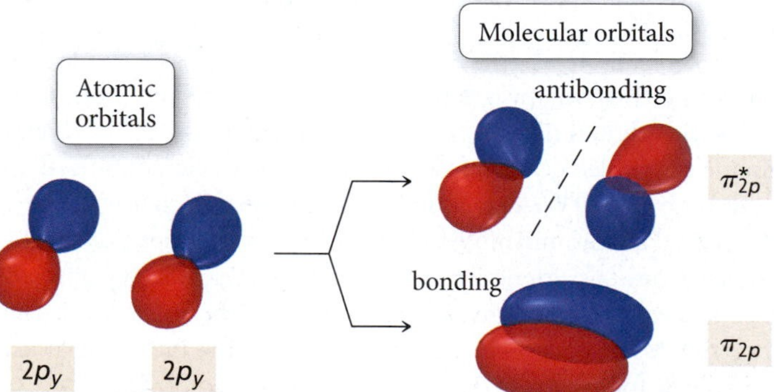

The only difference between the $2p_y$ and the $2p_z$ atomic orbitals is a 90° rotation about the internuclear axis. Consequently, the only difference between the resulting MOs is a 90° rotation about the internuclear axis. The energies and the names of the bonding and antibonding MOs obtained from the combination of the $2p_y$ AOs are identical to those obtained from the combination of the $2p_z$ AOs.

Before we can draw MO diagrams for B_2 and the other second-period diatomic molecules, we must determine the relative energy ordering of the MOs obtained from the $2p$ AO combinations. This is not a simple task. The relative ordering of MOs obtained from LCAO–MO theory is usually determined computationally. There is no single order that works for all molecules. For second-period diatomic molecules, computations reveal that the energy ordering for B_2, C_2, and N_2 is slightly different from that for O_2, F_2, and Ne_2 as shown in Figure 7.10 ▶.

The difference in energy ordering can only be explained by going back to our LCAO–MO model. In our simplified treatment, we assumed that the MOs that result from the second-period AOs could be calculated pairwise. In other words, we took the linear combination of a $2s$ from one atom with the

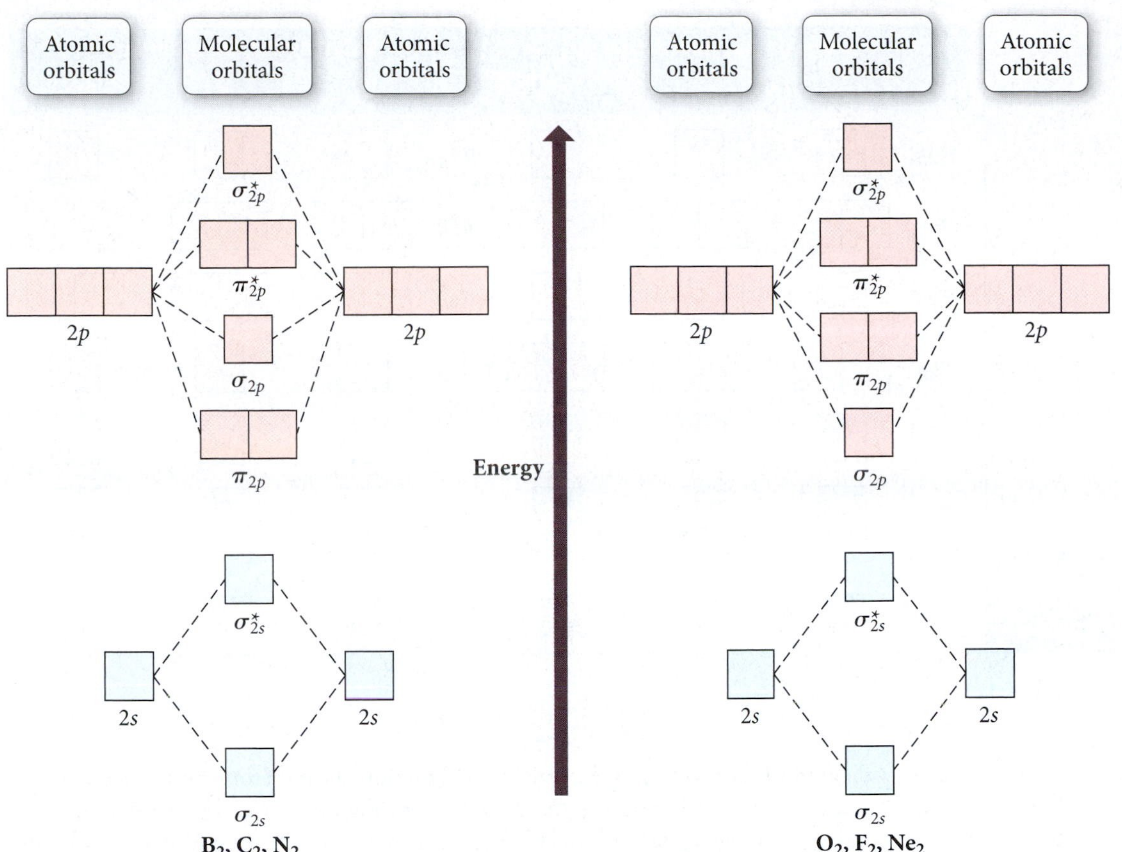

◀ **FIGURE 7.10 Molecular Orbital Energy Ordering** Molecular orbital energy diagrams for second-period diatomic molecules show that the energy ordering of the π_{2p} and σ_{2p} molecular orbitals can vary.

2*s* from another, a 2p_x from one atom with a 2p_x from the other, and so on. However, in a more comprehensive treatment, the MOs are formed from linear combinations that include all of the AOs that are relatively close to each other in energy and of the correct symmetry. Specifically, in a more detailed treatment, the two 2*s* orbitals and the two 2p_x orbitals should all be combined to form a total of four molecular orbitals. The extent to which you include this type of mixing affects the energy levels of the corresponding MOs, as shown in Figure 7.11 ▼. The bottom line is that *s*–*p* mixing is significant in B_2, C_2, and N_2 but not in O_2, F_2, and Ne_2. The result is a different energy ordering, depending on the specific molecule.

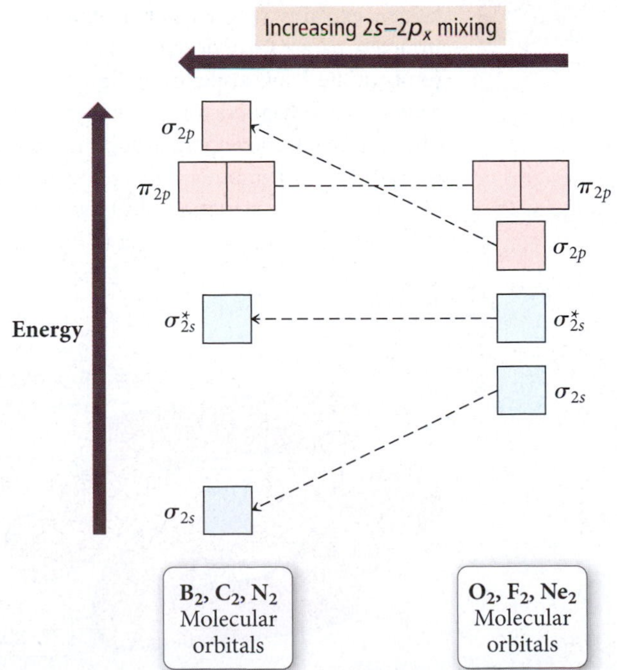

◀ **FIGURE 7.11 The Effects of 2*s*–2*p* Mixing** The degree of mixing between two orbitals decreases with increasing energy difference between them. Mixing of the 2*s* and 2p_x orbitals is therefore greater in B_2, C_2, and N_2 than in O_2, F_2, and Ne_2 because in B, C, and N the energy levels of the atomic orbitals are more closely spaced than in O, F, and Ne. This mixing produces a change in energy ordering for the π_{2p} and σ_{2p} molecular orbitals.

▶ **FIGURE 7.12 Molecular Orbital Energy Diagrams for Second-Period *p*-Block Homonuclear Diatomic Molecules**

	Large $2s$–$2p_x$ interaction				Small $2s$–$2p_x$ interaction		
	B_2	C_2	N_2		O_2	F_2	Ne_2
σ^*_{2p}				σ^*_{2p}			↑↓
π^*_{2p}				π^*_{2p}	↑ ↑	↑↓ ↑↓	↑↓ ↑↓
σ_{2p}			↑↓	π_{2p}	↑↓ ↑↓	↑↓ ↑↓	↑↓ ↑↓
π_{2p}	↑ ↑	↑↓ ↑↓	↑↓ ↑↓	σ_{2p}	↑↓	↑↓	↑↓
σ^*_{2s}	↑↓	↑↓	↑↓	σ^*_{2s}	↑↓	↑↓	↑↓
σ_{2s}	↑↓	↑↓	↑↓	σ_{2s}	↑↓	↑↓	↑↓
Bond order	1	2	3		2	1	0
Bond energy (kJ/mol)	290	620	946		498	159	—
Bond length (pm)	159	131	110		121	143	—

The MO energy diagrams for the rest of the second-period homonuclear diatomic molecules, as well as their bond orders, bond energies, and bond lengths, are shown in Figure 7.12 ▲. Notice that as bond order increases, the bond gets stronger (greater bond energy) and shorter (smaller bond length). For B_2, with six electrons, the bond order is 1. For C_2, the bond order is 2, and for N_2, the bond order reaches a maximum with a value of 3. Recall that the Lewis structure for N_2 has a triple bond, so both the Lewis model and molecular orbital theory predict a strong bond for N_2, which is experimentally observed.

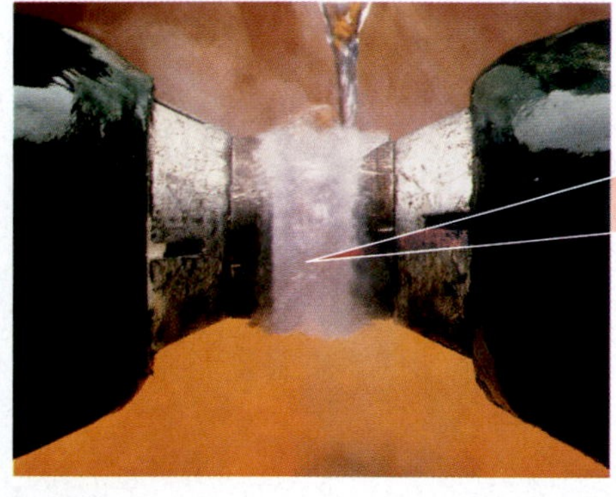

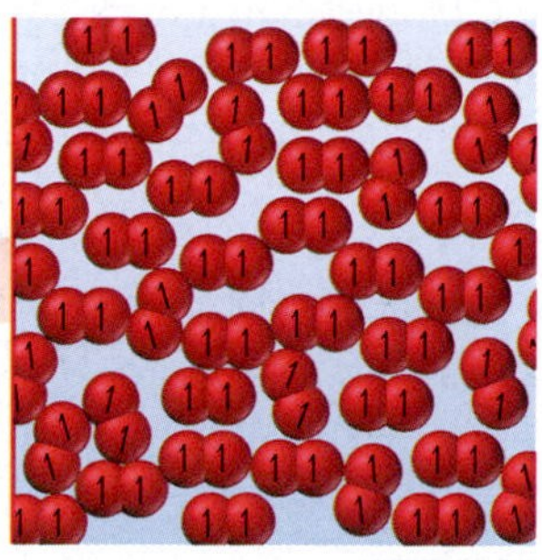

▲ Liquid oxygen can be suspended between the poles of a magnet because it is paramagnetic. It contains unpaired electrons (depicted here in the inset) that generate tiny magnetic fields, which align with and interact with the external field.

In O_2, the two additional electrons occupy antibonding orbitals and the bond order is 2. These two electrons are unpaired—they occupy the π^*_{2p} orbitals *singly with parallel spins*, as indicated by Hund's rule. The presence of unpaired electrons in the molecular orbital diagram of oxygen explains why oxygen is *paramagnetic* (see Section 7.1)—it is attracted to a magnetic field. The paramagnetism of oxygen can be demonstrated by suspending liquid oxygen between the poles of a magnet. This magnetic property is the direct result of *unpaired electrons*, whose spin and movement around the nucleus (more accurately known as orbital angular momentum) generate tiny magnetic fields.

In the Lewis structure of O_2, as well as in the valence bond model of O_2, all of the electrons seem to be paired:

The $2s$ orbital on each O atom contains two electrons, but for clarity neither the s orbitals nor the electrons that occupy them are shown.

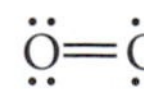

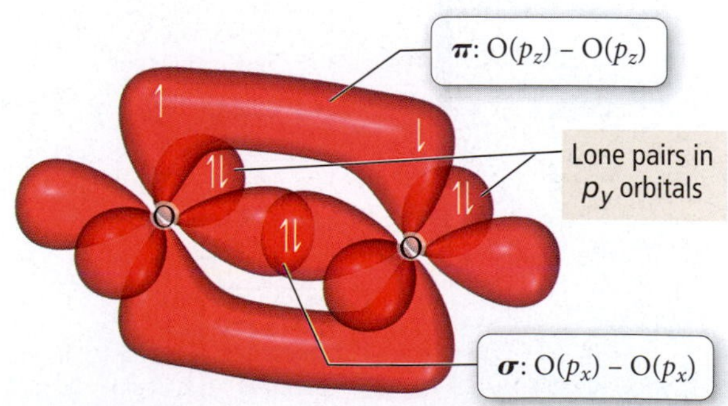

In the MO diagram for O_2, in contrast, we can account for the unpaired electrons. Molecular orbital theory is the more powerful theory in that it can account for the paramagnetism of O_2—it gives us a picture of bonding that more closely corresponds to what we see in experiments. Continuing along the second-period homonuclear diatomic molecules, we see that F_2 has a bond order of 1 and Ne_2 has a bond order of 0, again consistent with experiment since F_2 exists and Ne_2 does not.

EXAMPLE 7.5

Molecular Orbital Theory

Draw an MO energy diagram and determine the bond order for the N_2^- ion. Do you expect the bond in the N_2^- ion to be stronger or weaker than the bond in the N_2 molecule? Is N_2^- diamagnetic or paramagnetic?

SOLUTION

Write an energy-level diagram for the molecular orbitals in N_2^-. Use the energy ordering for N_2.	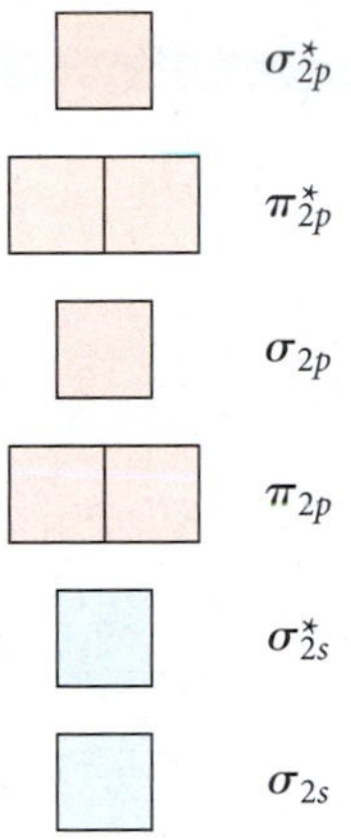    
The N_2^- ion has 11 valence electrons (5 for each nitrogen atom plus 1 for the negative charge). Assign the electrons to the molecular orbitals beginning with the lowest energy orbitals and following Hund's rule.	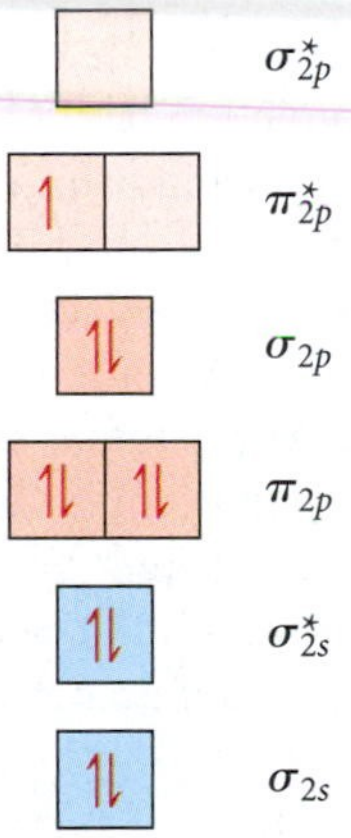 
Calculate the bond order by subtracting the number of electrons in antibonding orbitals from the number in bonding orbitals and dividing the result by two.	N_2^- bond order $= \frac{8 - 3}{2} = +2.5$

The bond order is 2.5, which is a lower bond order than in the N_2 molecule (bond order = 3); therefore, the bond is weaker. The MO diagram shows that the N_2^- ion has one unpaired electron and is therefore paramagnetic.

FOR PRACTICE 7.5

Draw an MO energy diagram and determine the bond order for the N_2^+ ion. Do you expect the bond in the N_2^+ ion to be stronger or weaker than the bond in the N_2 molecule? Is N_2^+ diamagnetic or paramagnetic?

FOR MORE PRACTICE 7.5

Apply molecular orbital theory to determine the bond order of Ne_2.

Second-Period Heteronuclear Diatomic Molecules

We can also apply molecular orbital theory to heteronuclear diatomic molecules (two different atoms). For example, we can draw an MO diagram for NO as follows:

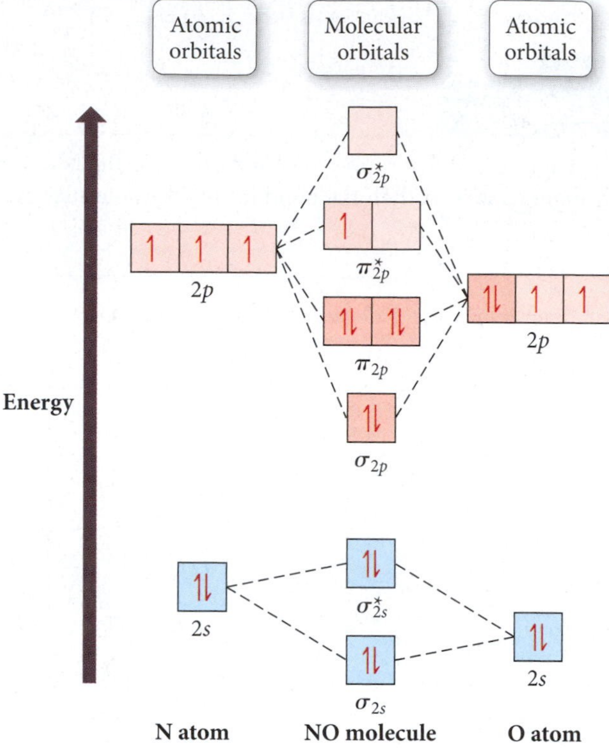

A given orbital will have lower energy in a more electronegative atom. For this reason, electronegative atoms have the ability to attract electrons to themselves.

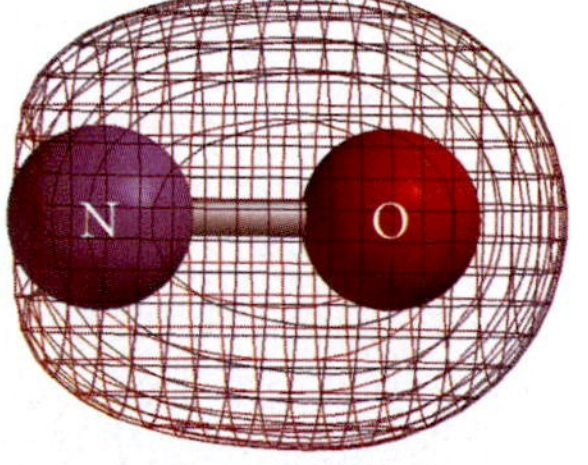

▲ **FIGURE 7.13 Shape of σ_{2s} Bonding Orbital in NO** The molecular orbital shows more electron density at the oxygen end of the molecule because the atomic orbitals of oxygen, the more electronegative element, are lower in energy than those of nitrogen. The atomic orbitals of oxygen therefore contribute more to the bonding molecular orbital.

Oxygen is more electronegative than nitrogen, so its atomic orbitals are lower in energy than nitrogen's atomic orbitals. When two atomic orbitals are identical and of equal energy, the weighting of each orbital in forming a molecular orbital is identical. However, when two atomic orbitals are different, the weighting of each orbital in forming a molecular orbital may be different. More specifically, when we approximate a molecular orbital as a linear combination of atomic orbitals of different energies, the lower energy atomic orbital makes a greater contribution to the bonding molecular orbital and the higher energy atomic orbital makes a greater contribution to the antibonding molecular orbital. For example, notice that the σ_{2s} bonding orbital is closer in energy to the oxygen $2s$ orbital than to the nitrogen $2s$ orbital. We can also see this unequal weighting in the shape of the resultant molecular orbital, in which the electron density is concentrated on the oxygen atom, as shown in Figure 7.13 ◀.

As another example of a heteronuclear diatomic molecule, consider the MO diagram for HF:

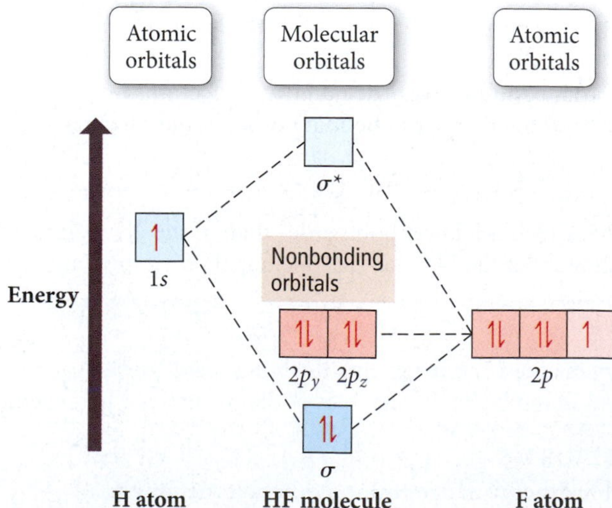

Fluorine is so electronegative that all of its atomic orbitals are lower in energy than hydrogen's atomic orbitals. In fact, fluorine's $2s$ orbital is so low in energy compared to hydrogen's $1s$ orbital that it does not contribute appreciably to the molecular orbitals. The molecular orbitals in HF are approximated by the linear combination of the fluorine $2p_x$ orbital and the hydrogen $1s$ orbital. The other $2p$ orbitals remain localized on the fluorine and appear in the energy diagram as **nonbonding orbitals**. The electrons in these nonbonding orbitals remain localized on the fluorine atom.

EXAMPLE 7.6

Molecular Orbital Theory for Heteronuclear Diatomic Molecules and Ions

Apply molecular orbital theory to determine the bond order of the CN^- ion. Is the ion paramagnetic or diamagnetic?

SOLUTION

Determine the number of valence electrons in the molecule or ion.	Number of valence electrons $= 4$ (from C) $+ 5$ (from N) $+ 1$(from negative charge) $= 10$
Write an energy-level diagram using Figure 7.12 as a guide. Fill the orbitals beginning with the lowest energy orbital and progressing upward until all electrons have been assigned to an orbital. Remember to allow no more than two electrons (with paired spins) per orbital and to fill degenerate orbitals with single electrons (with parallel spins) before pairing.	σ^*_{2p}: (empty) π^*_{2p}: (empty) (empty) σ_{2p}: ⇅ π_{2p}: ⇅ ⇅ σ^*_{2s}: ⇅ σ_{2s}: ⇅
Calculate the bond order using the appropriate formula: $\text{Bond order} = \frac{(\text{number of e}^- \text{ in bonding MOs}) - (\text{number of e}^- \text{ in antibonding MOs})}{2}$	CN^- bond order $= \frac{8-2}{2} = +3$
If the MO energy diagram has unpaired electrons, the molecule or ion is paramagnetic. If the electrons are all paired, the molecule or ion is diamagnetic.	Since the MO diagram has no unpaired electrons, the ion is diamagnetic.

FOR PRACTICE 7.6

Apply molecular orbital theory to determine the bond order of NO. (Use the energy ordering of N_2.) Is the molecule paramagnetic or diamagnetic?

7.5 Molecular Orbital Theory: Polyatomic Molecules

With the aid of computers, we can apply molecular orbital theory to polyatomic molecules and ions, yielding results that correlate very well with experimental measurements. These applications are beyond the scope of this text. However, the delocalizaton of electrons over an entire molecule is an important contribution of molecular orbital theory to our basic understanding of chemical bonding. For example, consider the Lewis structure and valence bond diagram of ozone:

:Ö—O=Ö ⟷ Ö=O—Ö:

Lewis structure

Valence bond model

In the Lewis model, we use resonance forms to represent the two equivalent bonds. In valence bond theory, it appears that the two oxygen–oxygen bonds should be different. In molecular orbital theory, however, the π molecular orbitals in ozone are formed from a linear combination of the three oxygen $2p$ orbitals and are delocalized over the entire molecule. The lowest energy π bonding molecular orbital is shown here:

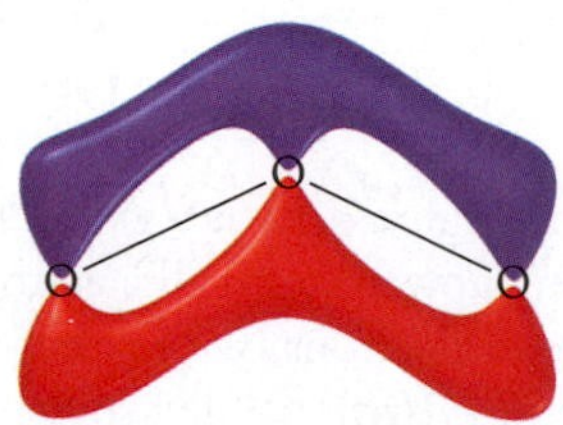

When we examine ozone in nature, we indeed find two equivalent bonds. A similar situation occurs with benzene (C_6H_6). In the Lewis model, we represent the structure with two resonance forms:

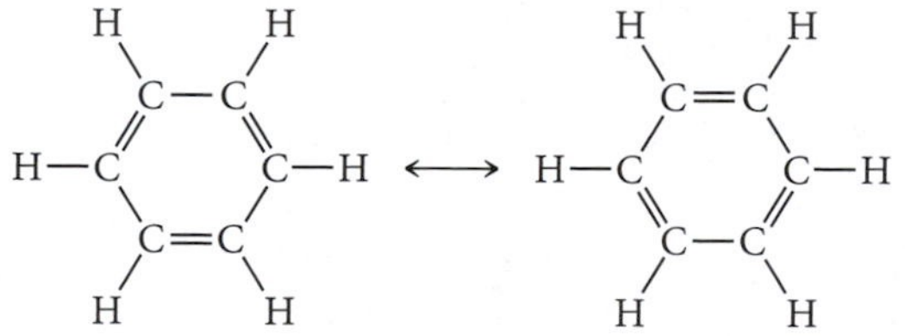

H—C₆H₄ ring (C—C, C=C alternating) ⟷ H—C₆H₄ ring (C=C, C—C alternating)

In molecular orbital theory, the π molecular orbitals in benzene are formed from a linear combination of the six carbon $2p$ orbitals and are delocalized over the entire molecule. The lowest energy π bonding molecular orbital is shown here:

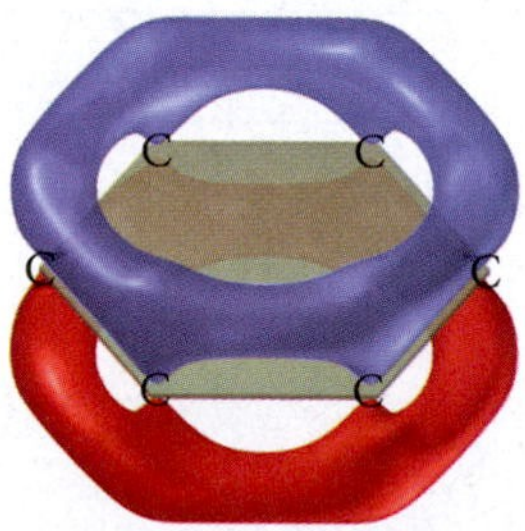

Benzene is in fact a highly symmetric molecule with six identical carbon–carbon bonds. The best picture of the π electrons in benzene is one in which the electrons occupy roughly circular-shaped orbitals above and below the plane of the molecule in accordance with the molecular orbital theory approach.

7.4
Cc
Conceptual Connection

What Is a Chemical Bond, Part II?

We have learned that the Lewis model portrays a chemical bond as the transfer or sharing of electrons represented as dots. Valence bond theory portrays a chemical bond as the overlap of two half-filled atomic orbitals. What is a chemical bond according to molecular orbital theory?

7.6 Bonding in Metals and Semiconductors

So far, we have developed models for bonding between a metal and a nonmetal (ionic bonding) and for bonding between two nonmetals (covalent bonding). We have seen how these models account for and predict the properties of ionic and molecular compounds. In this section, we explore two theories that help us to understand bonding in crystalline solids. The first and simplest theory is the **electron sea model**, which adequately explains many of the properties of metallic solids. The second, and more sophisticated model, is **band theory**, which is particularly useful in understanding semiconductors.

▲ **FIGURE 7.14 The Electron Sea Model for Sodium** In this model of metallic bonding, Na^+ ions are immersed in a "sea" of electrons.

Bonding in Metals: The Electron Sea Model

As we have already discussed, metals have a tendency to lose electrons, which means that they have relatively low ionization energies. When metal atoms bond together to form a solid, each metal atom donates one or more electrons to an *electron sea*. For example, we can think of sodium metal as an array of positively charged Na^+ ions immersed in a sea of negatively charged electrons (e^-), as shown in Figure 7.14 ▶. Each sodium atom donates its one valence electron to the "sea" and becomes a sodium ion. The sodium cations are held together by their attraction to the sea of electrons.

Although this model is simple, it accounts for many of the properties of metals. For example, metals conduct electricity because—in contrast to ionic solids where electrons are localized on an ion—the electrons in a metal are free to move. The movement or flow of electrons in response to an electric potential (or voltage) is an electric current. Metals are also excellent conductors of heat, again because of the highly mobile electrons, which help to disperse thermal energy throughout the metal.

The electron sea model also accounts for the *malleability* of metals (their capacity to be pounded into sheets) and the *ductility* of metals (their capacity to be drawn into wires). Since there are no localized or specific "bonds" in a metal, we can deform it relatively easily by forcing the metal ions to slide past one another. The electron sea easily accommodates deformations by flowing into the new shape.

▲ Copper can easily be drawn into fine strands like those used in household electrical cords.

Semiconductors and Band Theory

We now turn to a model for bonding in solids that is both more sophisticated and more broadly applicable—it applies to both metallic and covalently bonded crystalline solids. The model is *band theory,* and it grows out of molecular orbital theory, discussed in Section 7.4.

Recall that in molecular orbital theory, we combine the atomic orbitals of the atoms within a molecule to form molecular orbitals. These molecular orbitals are not localized on individual atoms, but *delocalized over the entire molecule*. Similarly, in band theory, we combine the atomic orbitals of the atoms within a solid crystal to form orbitals that are not localized on individual atoms, but delocalized over the entire *crystal*. In some sense then, the crystal is like a very large molecule, and its valence electrons occupy the molecular orbitals formed from the atomic orbitals of each atom in the crystal.

Consider a series of molecules constructed from individual lithium atoms. The energy levels of the atomic orbitals and resulting molecular orbitals for Li, Li_2, Li_3, Li_4, and Li_N (where N is a large number on the order of 10^{23}) are shown in Figure 7.15 ▶. The lithium atom has a single electron in a single 2*s* atomic orbital. The Li_2 molecule contains two electrons and two molecular orbitals. The electrons occupy the lower energy bonding orbital—the higher energy, or antibonding, molecular orbital is empty. The Li_4 molecule contains four electrons and four molecular orbitals. The electrons occupy the two bonding molecular orbitals—the two antibonding orbitals are completely empty.

The Li_N molecule contains N electrons and N molecular orbitals. However, because there are so many molecular orbitals, the energy spacings between them are infinitesimally small; they are no longer discrete energy levels, but rather form a *band* of energy levels. One-half of the orbitals in the band ($N/2$) are bonding molecular orbitals and (at 0 K) contain the N valence electrons. The other $N/2$ molecular orbitals are antibonding and (at 0 K) are completely empty. If the atoms composing a solid have *p* orbitals available, then the same process leads to another band of orbitals at higher energies.

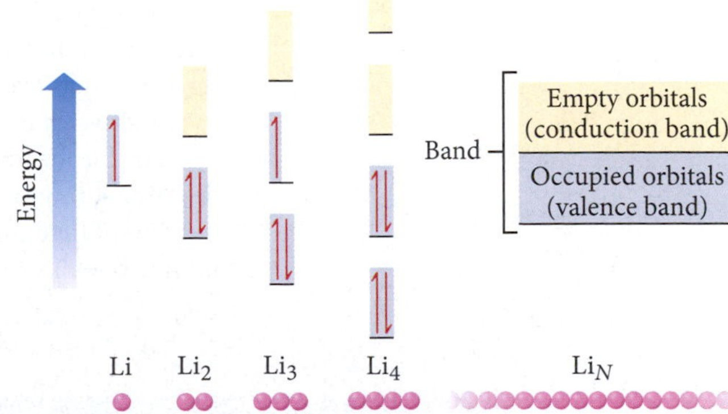

▲ **FIGURE 7.15 Energy Levels of Molecular Orbitals in Lithium Molecules** When many Li atoms are present, the energy levels of the molecular orbitals are so closely spaced that they fuse to form a band. Half of the orbitals are bonding orbitals and contain valence electrons; the other half are antibonding orbitals and are empty.

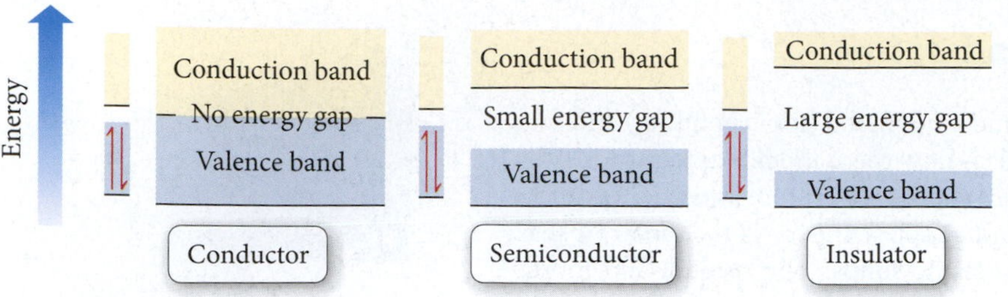

▲ **FIGURE 7.16 Band Gap** In a conductor, there is no energy gap between the valence band and the conduction band. In semiconductors there is a small energy gap, and in insulators there is a large energy gap.

In band theory, electrons become mobile when they make a transition from the highest occupied molecular orbital into higher energy empty molecular orbitals. For this reason, we call the occupied molecular orbitals the *valence band* and the unoccupied orbitals the *conduction band.* In lithium metal, the highest occupied molecular orbital lies in the middle of a band of orbitals, and the energy difference between it and the next higher energy orbital is infinitesimally small. Therefore, above 0 K, electrons can easily make the transition from the valence band to the conduction band. Because electrons in the conduction band are mobile, lithium, like all metals, is a good electrical conductor. Mobile electrons in the conduction band are also responsible for the thermal conductivity of metals. When a metal is heated, electrons are excited to higher energy molecular orbitals. These electrons can then quickly transport the thermal energy throughout the crystal lattice.

In metals, as we have just seen, the valence band and conduction band are energetically continuous—the energy difference between the top of the valence band and the bottom of the conduction band is infinitesimally small. In semiconductors and insulators, however, an energy gap, called the **band gap**, exists between the valence band and conduction band as shown in Figure 7.16 ▲. In insulators, the band gap is large, and electrons are not promoted into the conduction band at ordinary temperatures, resulting in no electrical conductivity. In semiconductors, the band gap is small, allowing some electrons to be promoted at ordinary temperatures and resulting in limited conductivity. However, the conductivity of semiconductors can be increased in a controlled way by adding minute amounts of other substances, called *dopants*, to the semiconductor.

Doping: Controlling the Conductivity of Semiconductors

Doped semiconductors contain minute amounts of impurities that result in additional electrons in the conduction band or electron "holes" in the valence band. For example, silicon is a group 4A semiconductor. Its valence electrons just fill its valence band. The band gap in silicon is large enough that only a few electrons are promoted into the conduction band at room temperature; therefore silicon is a poor electrical conductor. However, silicon can be doped with phosphorus, a group 5A element with five valence electrons, to increase its conductivity. The phosphorus atoms are incorporated into the silicon crystal structure, and each phosphorous atom brings with it one additional electron. Since the valence band is completely full, the additional electrons must go into the conduction band. These electrons are then mobile and can conduct electrical current. This type of semiconductor is called an **n-type semiconductor** because the charge carriers are negatively charged electrons in the conduction band. Silicon can also be doped with a group 3A element, such as gallium, which has only three valence electrons. When gallium is incorporated into the silicon crystal structure, it results in electron "holes," or empty molecular orbitals, in the valence band. The presence of holes also allows for the movement of electrical current because electrons in the valence band can move between holes. In this way, the holes move in the opposite direction as the electrons. This type of semiconductor is called a **p-type semiconductor** because each hole acts as a positive charge.

At the heart of most modern electronic devices are silicon chips containing millions of **p–n junctions**, tiny spots that are p-type on one side and n-type on the other. These junctions can serve a number of functions, including acting as **diodes** (circuit elements that allow the flow of electrical current in only one direction) or amplifiers (elements that amplify a small electrical current into a larger one).

7.5 Cc Conceptual Connection

Semiconductors

A semiconductor is composed of silicon doped with antimony. Is the semiconductor n-type or p-type?

SELF-ASSESSMENT Quiz

1. Determine the hybridization about O in CH_3OH.
   a) *sp*
   b) sp^2
   c) sp^3
   d) sp^3d

2. Determine the hybridization about C in H_2CO.
   a) *sp*
   b) sp^2
   c) sp^3
   d) sp^3d

3. According to valence bond theory, which kind of orbitals overlap to form the P—Cl bonds in PCl_5?
   a) $P(sp^3)-Cl(p)$
   b) $P(sp^3d)-Cl(s)$
   c) $P(sp^3)-Cl(s)$
   d) $P(sp^3d)-Cl(p)$

4. Use molecular orbital theory to determine the bond order in C_2.
   a) 0
   b) 1
   c) 2
   d) 3

5. Use molecular orbital theory to predict which species has the strongest bond.
   a) N_2
   b) N_2^-
   c) N_2^+
   d) All bonds are equivalent according to molecular orbital theory.

6. Use molecular orbital theory to determine which molecule is diamagnetic.
   a) CO
   b) B_2
   c) O_2
   d) None of the above (all are paramagnetic)

7. Which hybridization scheme occurs about nitrogen when nitrogen forms a double bond?
   a) *sp*
   b) sp^2
   c) sp^3
   d) sp^3d

8. What is the hybridization about a central atom that has five total electron groups, with three of those being bonding groups and two being lone pairs?
   a) *sp* **b)** sp^2
   c) sp^3 **d)** sp^3d

9. Determine the correct hybridizaton (from left to right) about each interior atom in CH_3CH_2OH.
   a) 1st C sp^3; 2nd C sp^2; O *sp*
   b) 1st C sp^3; 2nd C sp^3; O sp^3
   c) 1st C *sp*; 2nd C sp^2; O sp^3
   d) 1st C sp^3; 2nd C sp^2; O sp^2

10. The central atom in a molecule has a bent molecular geometry. Determine the hybridization of the orbitals in the atom.
    a) *sp*
    b) sp^2
    c) sp^3
    d) Unable to determine from information given.

11. Which type of orbitals overlap to form the sigma bond between C and N in $H-C\equiv N:$?
    a) $C(sp^3)-N(p)$
    b) $C(sp^2)-N(p)$
    c) $C(sp)-N(p)$
    d) $C(sp^3)-N(sp^3)$

12. With which element would you dope silicon to produce a p-type semiconductor?
    a) Sb
    b) As
    c) Ge
    d) Ga

Answers: 1:c; 2:b; 3:d; 4:c; 5:a; 6:a; 7:b; 8:d; 9:b; 10:d; 11:c; 12:d

CHAPTER 7 SUMMARY

REVIEW

KEY LEARNING OUTCOMES

CHAPTER OBJECTIVES	ASSESSMENT
Writing Hybridization and Bonding Schemes Using Valence Bond Theory (7.3)	• Examples 7.1, 7.2, 7.3 For Practice 7.1, 7.2, 7.3 For More Practice 7.3 Exercises 39–44
Drawing Molecular Orbital Energy Diagrams and Predicting Bond Order in a Homonuclear Diatomic Molecule (7.4)	• Examples 7.4, 7.5 For Practice 7.4, 7.5 For More Practice 7.5 Exercises 47–58
Drawing Molecular Orbital Energy Diagrams and Predicting Magnetic Properties in a Heteronuclear Diatomic Molecule (7.4)	• Examples 7.6 For Practice 7.6 Exercises 59, 60
Predicting Whether a Semiconductor is n-type or p-type (7.6)	• Conceptual Connection 7.5 Exercises 65, 66

KEY TERMS

Section 7.2
valence bond theory (234)
atomic orbitals (AO) (234)

Section 7.3
hybridization (236)
hybrid orbitals (236)
pi (π) bond (240)
sigma (σ) bond (240)
isomer (243)

Section 7.4
molecular orbital (MO) (249)
molecular orbital (MO) theory (249)
bonding orbital (249)
antibonding orbital (250)
molecular orbital (MO) diagram (251)
bond order (251)
nonbonding orbitals (259)

Section 7.6
electron sea model (261)
band theory (261)
band gap (262)
n-type semiconductor (262)
p-type semiconductor (262)
p–n junctions (262)
diodes (262)

KEY CONCEPTS

Valence Bond Theory (7.2, 7.3)

- In contrast to the Lewis model, in which a covalent chemical bond is the sharing of electrons represented by dots, in valence bond theory a chemical bond is the overlap of half-filled atomic orbitals (or in some cases the overlap between a completely filled orbital and an empty one).
- The overlapping orbitals may be the standard atomic orbitals, such as 1*s* or 2*p* or they may be hybridized atomic orbitals, which are mathematical combinations of the standard orbitals on a single atom. The basic hybridized orbitals are sp, sp^2, sp^3, sp^3d, and sp^3d^2.
- The geometry of the molecule is determined by the geometry of the overlapping orbitals.
- In valence bond theory, we distinguish between two types of bonds, σ (sigma) and π (pi). In a σ bond, the orbital overlap occurs in the region that lies directly between the two bonding atoms. In a π bond, formed from the side-by-side overlap of *p* orbitals, the overlap occurs above and below the region that lies directly between the two bonding atoms.
- Rotation about a σ bond is relatively free, while rotation about a π bond is restricted.
- In our treatment of valence bond theory, we use the molecular geometry determined by VSEPR theory to determine the correct hybridization scheme.

Molecular Orbital Theory (7.4, 7.5)

- The simplest molecular orbitals are linear combinations of atomic orbitals (LCAOs), weighted averages of the atomic orbitals of the different atoms in the molecule.
- When two atomic orbitals combine to form molecular orbitals, they form one molecular orbital of lower energy (the bonding orbital) and one of higher energy (the antibonding orbital).
- A set of molecular orbitals is filled in much the same way as a set of atomic orbitals.
- The stability of the molecule and the strength of the bond depend on the number of electrons in bonding orbitals compared to the number in antibonding orbitals.

Bonding in Metals and Semiconductors (7.6)

- According to the electron sea model, each metal atom in a solid metal donates one or more electrons to an *electron sea.* The metal cations are then held together by their attraction to the sea of electrons.
- According to band theory, the atomic orbitals of each atom in a solid combine and delocalize over the entire crystal solid to create energy bands. The valence electrons occupy the energy band known as the valence band. In semiconductors and insulators, a band gap exists between the valence band and the higher energy band, which is called the conduction band.

KEY EQUATIONS AND RELATIONSHIPS

Bond Order of a Diatomic Molecule (7.4)

$$\text{Bond order} = \frac{(\text{number of electrons in bonding MOs}) - (\text{number of electrons in antibonding MOs})}{2}$$

EXERCISES

REVIEW QUESTIONS

1. Why do we use other bonding theories in addition to the Lewis model?

2. What is a chemical bond according to valence bond theory?

3. In valence bond theory, what determines the geometry of a molecule?

4. In valence bond theory, the interaction energy between the electrons and nucleus of one atom with the electrons and nucleus of another atom is usually negative (stabilizing) when ____________.

5. What is hybridization? Why is hybridization necessary in valence bond theory?

6. How does hybridization of the atomic orbitals in the central atom of a molecule help lower the overall energy of the molecule?

7. How is the *number* of hybrid orbitals related to the number of standard atomic orbitals that are hybridized?

8. Sketch each hybrid orbital.

a. sp **b.** sp^2 **c.** sp^3 **d.** sp^3d **e.** sp^3d^2

9. In the Lewis model, the two bonds in a double bond look identical. However, valence bond theory shows that they are not. Describe a double bond according to valence bond theory. Explain why rotation is restricted about a double bond, but not about a single bond.

10. Name the hybridization scheme that corresponds to each electron geometry.

a. linear **b.** trigonal planar **c.** tetrahedral
d. trigonal bipyramidal **e.** octahedral

11. What is a chemical bond according to molecular orbital theory?

12. Explain the difference between hybrid atomic orbitals in valence bond theory and LCAO molecular orbitals in molecular orbital theory.

13. What is a bonding molecular orbital?

14. What is an antibonding molecular orbital?

15. What is the role of wave interference in determining whether a molecular orbital is bonding or antibonding?

16. In molecular orbital theory, what is bond order? Why is it important?

17. How is the number of molecular orbitals approximated by a linear combination of atomic orbitals related to the number of atomic orbitals used in the approximation?

18. Sketch each molecular orbital.

a. σ_{2s} **b.** σ^*_{2s} **c.** σ_{2p} **d.** σ^*_{2p} **e.** π_{2p} **f.** π^*_{2p}

19. Draw an energy diagram for the molecular orbitals of period 2 diatomic molecules. Show the difference in ordering for B_2, C_2, and N_2 compared to O_2, F_2, and Ne_2.

20. Why does the energy ordering of the molecular orbitals of the second-period diatomic molecules change in going from N_2 to O_2?

21. Explain the difference between a paramagnetic species and a diamagnetic one.

22. When applying molecular orbital theory to heteronuclear diatomic molecules, the atomic orbitals used may be of different energies. If two atomic orbitals of different energies make two molecular orbitals, how are the energies of the molecular orbitals related to the energies of the atomic orbitals? How is the shape of the resultant molecular orbitals related to the shapes of the atomic orbitals?

23. In molecular orbital theory, what is a nonbonding orbital?

24. Write a short paragraph describing chemical bonding according to the Lewis model, valence bond theory, and molecular orbital theory. Indicate how the theories differ in their description of a chemical bond and indicate the strengths and weaknesses of each theory. Which theory is correct?

25. Describe the electron sea model for bonding in metals.

26. How does the electron sea model explain the conductivity of metals? The malleability and ductility of metals?

27. In band theory of bonding for solids, what is a *band*? What is the difference between the *valence band* and the *conduction band*?

28. In band theory of bonding for solids, what is a band gap? How does the band gap differ in metals, semiconductors, and insulators?

29. Explain how doping can increase the conductivity of a semiconductor.

30. What is the difference between an n-type semiconductor and a p-type semiconductor?

PROBLEMS BY TOPIC

Valence Bond Theory

31. The valence electron configurations of several atoms are shown below. How many bonds can each atom make without hybridization?

a. Be $2s^2$ **b.** P $3s^23p^3$ **c.** F $2s^22p^5$

32. The valence electron configurations of several atoms are shown below. How many bonds can each atom make without hybridization?

a. B $2s^22p^1$ **b.** N $2s^22p^3$ **c.** O $2s^22p^4$

33. Draw orbital diagrams (boxes with arrows in them) to represent the electron configurations—without hybridization—for all the atoms in PH_3. Circle the electrons involved in bonding. Draw a three-dimensional sketch of the molecule and show orbital overlap. What bond angle do you expect from the unhybridized orbitals? How well does valence bond theory agree with the experimentally measured bond angle of 93.3°?

34. Draw orbital diagrams (boxes with arrows in them) to represent the electron configurations—without hybridization—for all the atoms in SF_2. Circle the electrons involved in bonding. Draw a three-dimensional sketch of the molecule and show orbital overlap. What bond angle do you expect from the unhybridized orbitals? How well does valence bond theory agree with the experimentally measured bond angle of 98.2°?

35. Draw orbital diagrams (boxes with arrows in them) to represent the electron configuration of carbon before and after sp^3 hybridization.

36. Draw orbital diagrams (boxes with arrows in them) to represent the electron configurations of carbon before and after sp hybridization.

37. Which hybridization scheme allows the formation of at least one π bond?

sp^3, sp^2, sp^3d^2

38. Which hybridization scheme allows the central atom to form more than four bonds?

sp^3, sp^3d, sp^2

39. Write a hybridization and bonding scheme for each molecule. Sketch the molecule, including overlapping orbitals, and label all bonds using the notation shown in Examples 7.1 and 7.2.

a. CCl_4 **b.** NH_3 **c.** OF_2 **d.** CO_2

40. Write a hybridization and bonding scheme for each molecule. Sketch the molecule, including overlapping orbitals, and label all bonds using the notation shown in Examples 7.1 and 7.2.

a. CH_2Br_2 **b.** SO_2 **c.** NF_3 **d.** BF_3

41. Write a hybridization and bonding scheme for each molecule or ion. Sketch the structure, including overlapping orbitals, and label all bonds using the notation shown in Examples 7.1 and 7.2.

a. $COCl_2$ (carbon is the central atom) **b.** BrF_5 **c.** XeF_2 **d.** I_3^-

42. Write a hybridization and bonding scheme for each molecule or ion. Sketch the structure, including overlapping orbitals, and label all bonds using the notation shown in Examples 7.1 and 7.2.

a. SO_3^{2-} **b.** PF_6^- **c.** BrF_3 **d.** HCN

43. Write a hybridization and bonding scheme for each molecule that contains more than one interior atom. Indicate the hybridization about each interior atom. Sketch the structure, including overlapping orbitals, and label all bonds using the notation shown in Examples 7.1 and 7.2.

a. N_2H_2 (skeletal structure HNNH)
b. N_2H_4 (skeletal structure H_2NNH_2)
c. CH_3NH_2 (skeletal structure H_3CNH_2)

44. Write a hybridization and bonding scheme for each molecule that contains more than one interior atom. Indicate the hybridization about each interior atom. Sketch the structure, including overlapping orbitals, and label all bonds using the notation shown in Examples 7.1 and 7.2.

a. C_2H_2 (skeletal structure HCCH)
b. C_2H_4 (skeletal structure H_2CCH_2)
c. C_2H_6 (skeletal structure H_3CCH_3)

45. Consider the structure of the amino acid alanine. Indicate the hybridization about each interior atom.

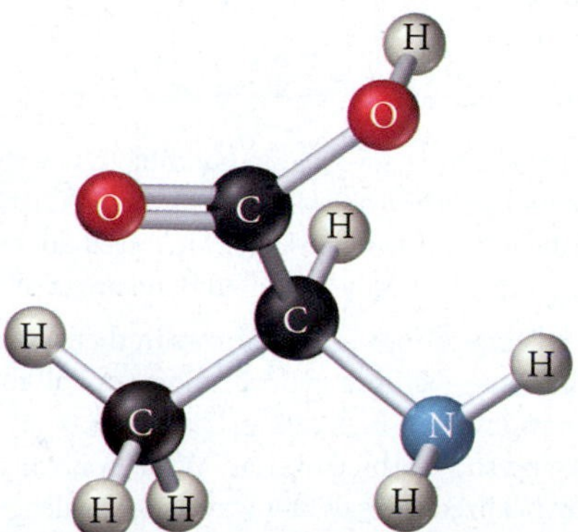

46. Consider the structure of the amino acid aspartic acid. Indicate the hybridization about each interior atom.

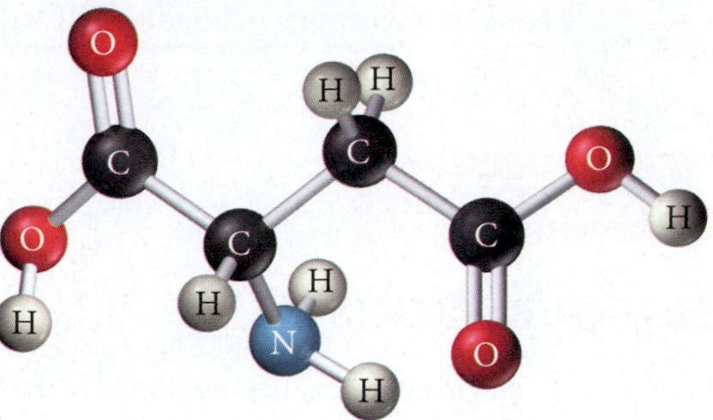

Molecular Orbital Theory

47. Sketch the bonding molecular orbital that results from the linear combination of two 1*s* orbitals. Indicate the region where interference occurs and state the kind of interference (constructive or destructive).

48. Sketch the antibonding molecular orbital that results from the linear combination of two 1*s* orbitals. Indicate the region where interference occurs and state the kind of interference (constructive or destructive).

49. Draw an MO energy diagram and predict the bond order of Be_2^+ and Be_2^-. Do you expect these molecules to exist in the gas phase?

50. Draw an MO energy diagram and predict the bond order of Li_2^+ and Li_2^-. Do you expect these molecules to exist in the gas phase?

51. Sketch the bonding and antibonding molecular orbitals that result from linear combinations of the $2p_x$ atomic orbitals in a homonuclear diatomic molecule. (The $2p_x$ orbitals are those whose lobes are oriented along the bonding axis.)

52. Sketch the bonding and antibonding molecular orbitals that result from linear combinations of the $2p_z$ atomic orbitals in a homonuclear diatomic molecule. (The $2p_z$ orbitals are those whose lobes are oriented perpendicular to the bonding axis.) How do these molecular orbitals differ from those obtained from linear combinations of the $2p_y$ atomic orbitals? (The $2p_y$ orbitals are also oriented perpendicular to the bonding axis, but also perpendicular to the $2p_z$ orbitals.)

53. Using the molecular orbital energy ordering for second-period homonuclear diatomic molecules in which the π_{2p} orbitals lie at *lower* energy than the σ_{2p}, draw MO energy diagrams and predict the bond order in a molecule or ion with each number of total valence electrons. Will the molecule or ion be diamagnetic or paramagnetic?

a. 4 **b.** 6 **c.** 8 **d.** 9

54. Using the molecular orbital energy ordering for second-period homonuclear diatomic molecules in which the π_{2p} orbitals lie at *higher* energy than the σ_{2p}, draw MO energy diagrams and predict the bond order in a molecule or ion with each number of total valence electrons. Will the molecule or ion be diamagnetic or paramagnetic?

a. 10 **b.** 12 **c.** 13 **d.** 14

55. Apply molecular orbital theory to predict if each molecule or ion exists in a relatively stable form.

a. H_2^{2-} **b.** Ne_2 **c.** He_2^{2+} **d.** F_2^{2-}

56. Apply molecular orbital theory to predict if each molecule or ion exists in a relatively stable form.

a. C_2^{2+} **b.** Li_2 **c.** Be_2^{2+} **d.** Li_2^{2-}

57. According to MO theory, which molecule or ion has the highest bond order? Highest bond energy? Shortest bond length?

C_2, C_2^+, C_2^-

58. According to MO theory, which molecule or ion has the highest bond order? Highest bond energy? Shortest bond length?
O_2, C_2^-, O_2^{2-}

59. Draw an MO energy diagram for CO. (Use the energy ordering of O_2.) Predict the bond order and make a sketch of the lowest energy bonding molecular orbital.

60. Draw an MO energy diagram for HCl. Predict the bond order and make a sketch of the lowest energy bonding molecular orbital.

Electron Sea Model and Band Theory

61. How many electrons are in the electron sea of a 5.68 g sample of magnesium metal?

62. How many electrons are in the electron sea of a 28.5 g sample of potassium metal?

63. Which solid would you expect to have little or no band gap?

a. Zn(*s*) **b.** Si(*s*) **c.** As(*s*)

64. How many molecular orbitals are present in the valence band of a sodium crystal with a mass of 5.45 g?

65. Indicate if each solid forms an n-type or a p-type semiconductor.

a. germanium doped with gallium
b. silicon doped with arsenic

66. Indicate if each solid forms an n-type or a p-type semiconductor.

a. silicon doped with gallium
b. germanium doped with antimony

CUMULATIVE PROBLEMS

67. For each compound, draw the Lewis structure, determine the geometry using VSEPR theory, determine whether the molecule is polar, identify the hybridization of all interior atoms, and make a sketch of the molecule, according to valence bond theory, showing orbital overlap.

a. COF_2 (carbon is the central atom)
b. S_2Cl_2 (ClSSCl)
c. SF_4

68. For each compound, draw the Lewis structure, determine the geometry using VSEPR theory, determine whether the molecule is polar, identify the hybridization of all interior atoms, and make a sketch of the molecule, according to valence bond theory, showing orbital overlap.

a. IF_5
b. CH_2CHCH_3
c. CH_3SH

69. Amino acids are biological compounds that link together to form proteins, the workhorse molecules in living organisms. The skeletal structures of several simple amino acids are shown here. For each skeletal structure, determine the hybridization about each interior atom.

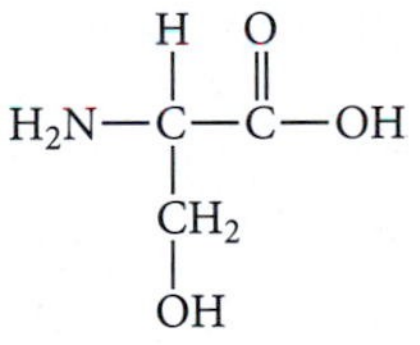

a. serine 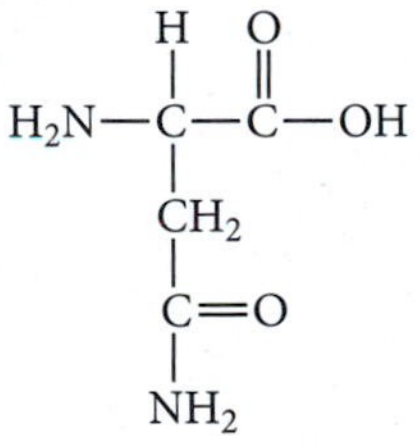 b. asparagine 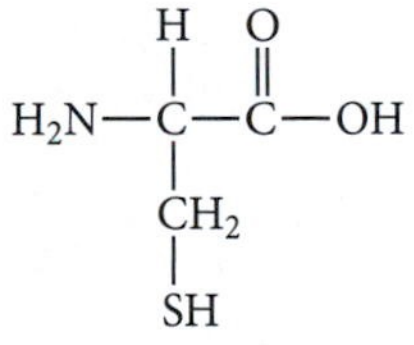 c. cysteine

70. The genetic code is based on four different bases with the structures shown here. Determine the hybridization in each interior atom in these four bases.

a. cytosine **b.** adenine **c.** thymine **d.** guanine

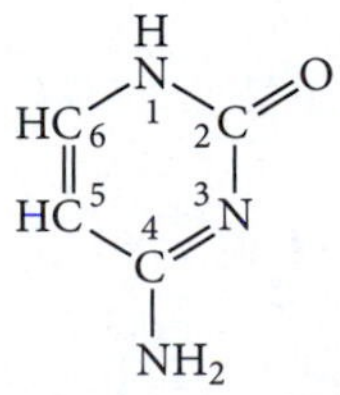

a. 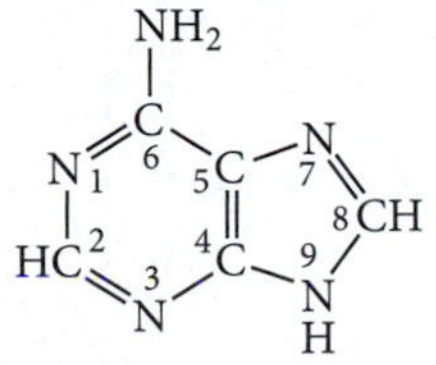 b. 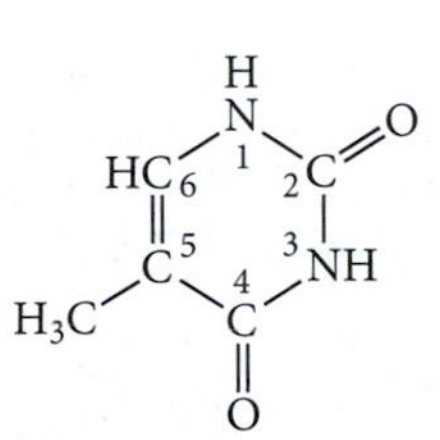 c. 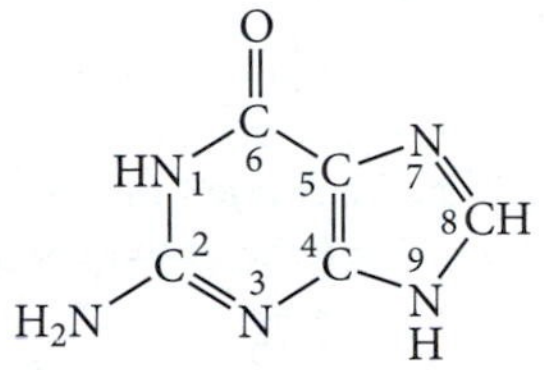 d.

71. The structure of caffeine, present in coffee and many soft drinks, is shown here. How many pi bonds are present in caffeine? How many sigma bonds? Insert the lone pairs in the molecule. Which kinds of orbitals do the lone pairs occupy?

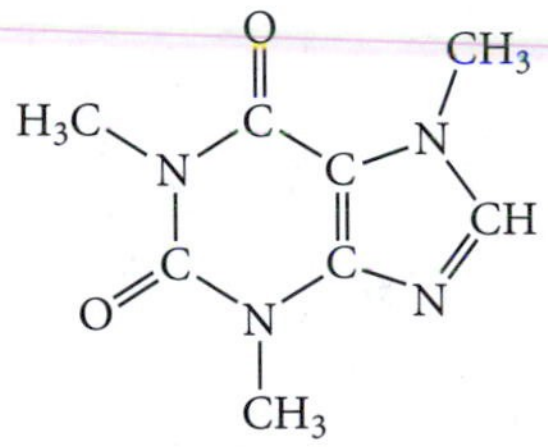

72. The structure of acetylsalicylic acid (aspirin) is shown here. How many pi bonds are present in acetylsalicylic acid? How many sigma bonds? Which parts of the molecule are free to rotate? Which parts are rigid?

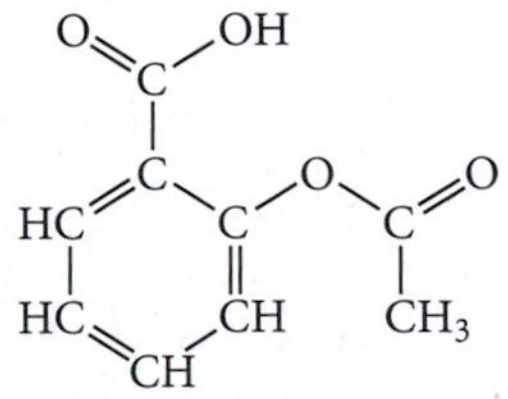

73. Draw a molecular orbital energy diagram for ClF. (Assume that the σ_p orbitals are lower in energy than the π orbitals.) What is the bond order in ClF?

74. Draw Lewis structures and MO diagrams for CN^+, CN, and CN^-. According to the Lewis model, which species is most stable? According to MO theory, which species is most stable? Do the two theories agree?

75. Bromine can form compounds or ions with any number of fluorine atoms from one to five. Write the formulas of all five of these species, assign a hybridization, and describe their electron and molecular geometry.

76. The compound C_3H_4 has two double bonds. Describe its bonding and geometry, using a valence bond approach.

77. How many hybrid orbitals do we use to describe each molecule?
   a. N_2O_5
   b. C_2H_5NO (4 C—H bonds and one O—H bond)
   c. BrCN (no formal charges)

78. Indicate which orbitals overlap to form the σ bonds in each compound.
   a. $BeBr_2$ b. $HgCl_2$ c. ICN

CHALLENGE PROBLEMS

79. In VSEPR theory, which uses the Lewis model to determine molecular geometry, the trend of decreasing bond angle in CH_4, NH_3, and H_2O is accounted for by the greater repulsion of lone pair electrons compared to bonding pair electrons. How is this trend accounted for in valence bond theory?

80. The results of a molecular orbital calculation for H_2O are shown here. Examine each of the orbitals and classify them as bonding, antibonding, or nonbonding. Assign the correct number of electrons to the energy diagram. According to this energy diagram, is H_2O stable? Explain.

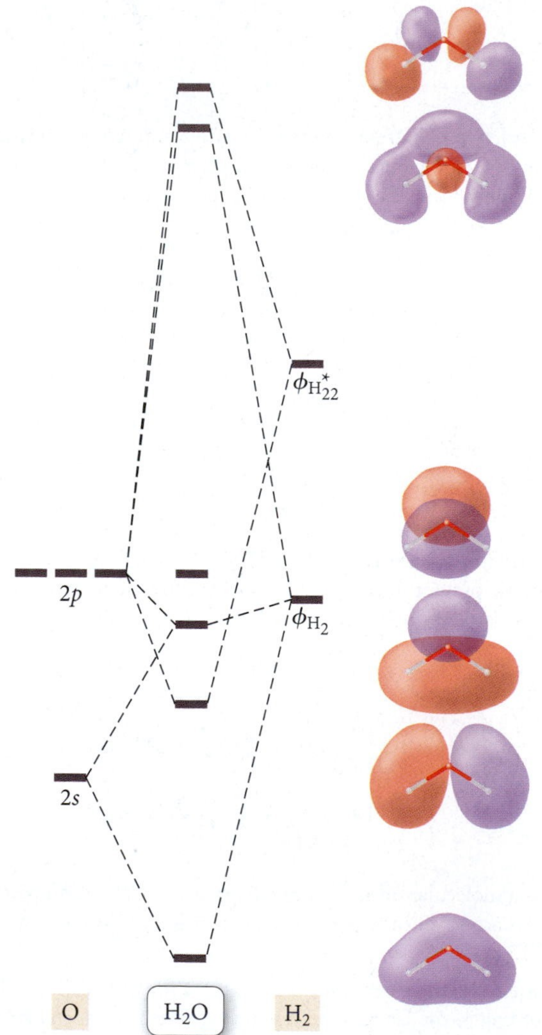

81. The results of a molecular orbital calculation for NH_3 are shown here. Examine each of the orbitals and classify them as bonding, antibonding, or nonbonding. Assign the correct number of electrons to the energy diagram. According to this energy diagram, is NH_3 stable? Explain.

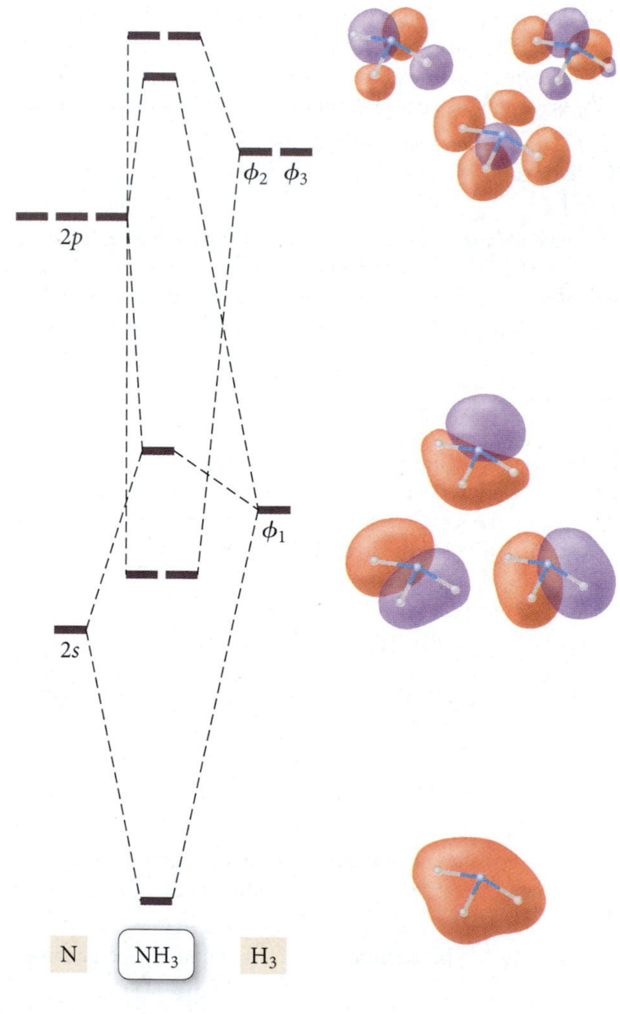

82. *cis*-2-Butene isomerizes (changes its structure) to *trans*-2-butene via the reaction:

```
 H  H H  H               H  H
  \ /  \ /                \ /
H—C     C—H            H—C      H
   \   /                   \   /
   C=C          ⟶          C=C
   /  \                    /  \
  H    H                  H    C—H
                               / \
                              H   H
```

a. If isomerization requires breaking the pi bond, what minimum energy is required for isomerization in J/mol? In J/molecule?

b. If the energy for isomerization comes from light, what minimum frequency of light is required? In what portion of the electromagnetic spectrum does this frequency lie?

83. The ion CH_5^+ can form under very special high-energy conditions in the vapor phase in a mass spectrometer. Propose a hybridization for the carbon atom and predict the geometry.

84. Neither the VSEPR model nor the hybridization model is able to account for the experimental observation that the F—Ba—F bond angle in gaseous BaF_2 is 108° rather than the predicted 180°. Suggest some possible explanations for this observation.

CONCEPTUAL PROBLEMS

85. How does each of the three major bonding theories (the Lewis model, valence bond theory, and molecular orbital theory) define a single chemical bond? A double bond? A triple bond? How are these definitions similar? How are they different?

86. The most stable forms of the nonmetals in groups 4A, 5A, and 6A of the first period are molecules with multiple bonds. Beginning with the second period, the most stable forms of the nonmetals of these groups are molecules without multiple bonds. Propose an explanation for this observation based on valence bond theory.

87. Metals are among the best conductors of heat. Use the electron sea model to explain why metals conduct heat so well.

88. Consider the bond energies of three iodine halides:

Bond	Bond Energy
Br—Cl	218 kJ/mol
Br—Br	193 kJ/mol
I—Br	175 kJ/mol

How might you use valence bond theory to help explain this trend?

ANSWERS TO CONCEPTUAL CONNECTIONS

Cc 7.1 **(a)** In the Lewis model, a covalent chemical bond is the sharing of electrons (represented by dots). **(b)** In valence bond theory, a covalent chemical bond is the overlap of half-filled atomic orbitals. **(c)** The answers are different because the Lewis model and valence bond theory are different models for chemical bonding. They both make useful and often similar predictions, but the assumptions of each model are different, and so are their respective descriptions of a chemical bond.

Cc 7.2 According to valence bond theory, a double bond is actually composed of two different kinds of bonds, one σ and one π. The orbital overlap in the π bond is side to side between two *p* orbitals and consequently different from the end-to-end overlap in a σ bond. Because the bonds are different types, the bond energy of the double bond is not just the bond energy of the single bond doubled.

Cc 7.3 **(a)** Because carbon has two electron groups in CO_2 (the two double bonds), the geometry is linear and the hybridization is *sp*.

Cc 7.4 In molecular orbital theory, atoms join together (or bond) when the electrons in the atoms can lower their energy by occupying the molecular orbitals of the resultant molecule. Unlike the Lewis model or valence bond theory, the chemical "bonds" in MO theory are not localized between atoms, but spread throughout the entire molecule.

Cc 7.5 The semiconductor is n-type because antimony contains one more valence electron than silicon. The extra valence electrons enter the conduction band and conduct electricity.

CHAPTER 8

> **"I feel sorry for people who don't understand anything about chemistry. They are missing an important source of happiness."**
>
> **—Linus Pauling (1901–1994)**

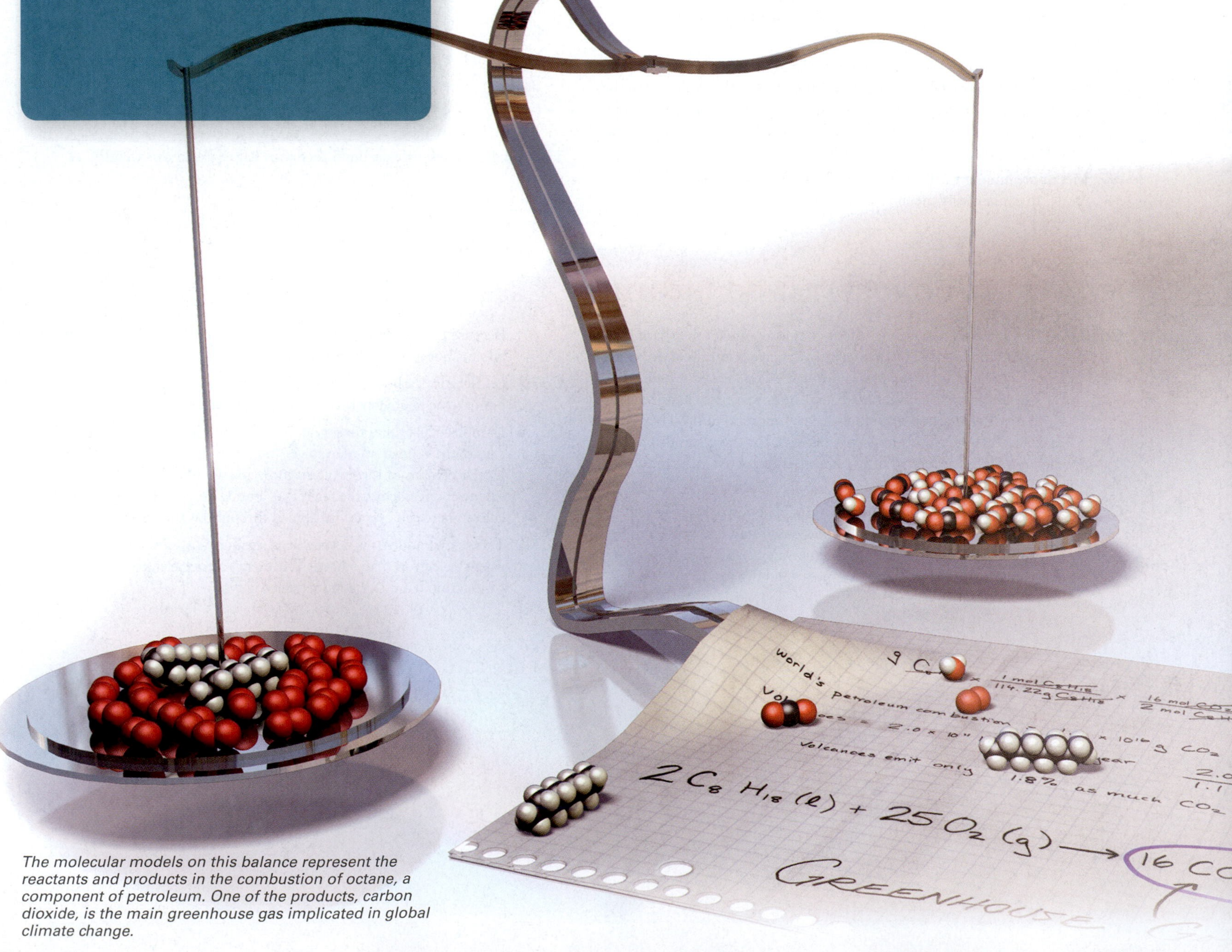

The molecular models on this balance represent the reactants and products in the combustion of octane, a component of petroleum. One of the products, carbon dioxide, is the main greenhouse gas implicated in global climate change.

Chemical Reactions and Chemical Quantities

8.1 Climate Change and the Combustion of Fossil Fuels 271

8.2 Chemical Change 273

8.3 Writing and Balancing Chemical Equations 274

8.4 Reaction Stoichiometry: How Much Carbon Dioxide? 279

8.5 Limiting Reactant, Theoretical Yield, and Percent Yield 283

8.6 Three Examples of Chemical Reactions: Combustion, Alkali Metals, and Halogens 289

Key Learning Outcomes 292

WE HAVE SPENT THE LAST THREE CHAPTERS examining compounds and bonding within compounds. We now turn to the process that can create or transform compounds: *chemical reactions*. As we have seen, matter is composed of particles. Those particles can be atoms, ions, or molecules. The particles that compose matter are in constant motion (above 0 K), vibrating, jostling, and colliding with one another. In some cases, a collision between particles leads to a remarkable change. The electrons in one particle are drawn to the nuclei in the other particle and vice versa. If the conditions are right, a chemical reaction occurs and the particles are transformed. In this chapter, we examine chemical reactions. We learn how to write equations to represent reactions, and we look closely at chemical *stoichiometry*—the numerical relationships between the relative amounts of particles in a chemical reaction.

8.1 Climate Change and the Combustion of Fossil Fuels

The temperature outside my office today is a cool 48 °F, lower than normal for this time of year on the California coast. However, today's "chill" pales in comparison with how cold it would be without the presence of *greenhouse gases* in the atmosphere. These gases act like the glass of a greenhouse, allowing sunlight to enter the atmosphere and warm Earth's surface, but preventing some of the heat generated by

▶ **FIGURE 8.1 The Greenhouse Effect** Greenhouse gases in the atmosphere act as a one-way filter. They allow visible light to pass through and warm Earth's surface, but they prevent heat energy from radiating back out into space.

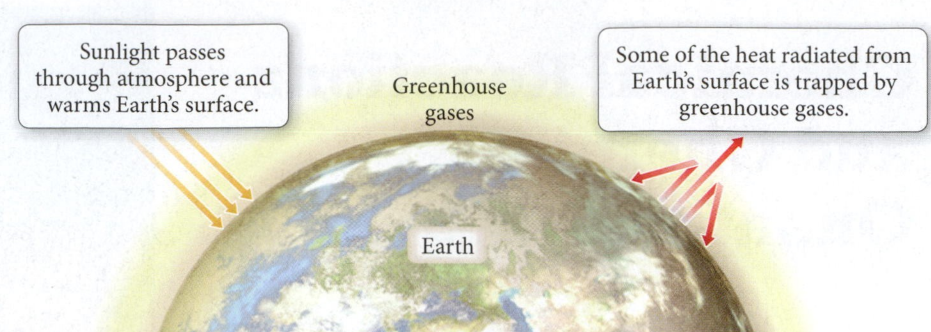

the sunlight from escaping, as shown in Figure 8.1 ▲. The balance between incoming and outgoing energy from the sun determines Earth's average temperature.

If the greenhouse gases in the atmosphere were not present, more heat energy would escape, and Earth's average temperature would be about 60 °F colder. The temperature outside of my office today would be below 0 °F, and even the sunniest U.S. cities would most likely be covered with snow. However, if the concentration of greenhouse gases in the atmosphere were to increase, Earth's average temperature would rise.

In recent years scientists have become increasingly concerned because the quantity of atmospheric carbon dioxide (CO_2)—Earth's most significant greenhouse gas—is rising. More CO_2 enhances the atmosphere's ability to hold heat and is believed to lead to *global warming*, an increase in Earth's average temperature. Since 1860, atmospheric CO_2 levels have risen by 38% (Figure 8.2 ▼), and Earth's average temperature has risen by 0.7 °C (about 1.2 °F), as shown in Figure 8.3 ▼.

Most scientists assert that the primary cause of rising atmospheric CO_2 concentration is the burning of fossil fuels (natural gas, petroleum, and coal), which provide about 82% of our society's energy. The burning of fossil fuels is a *chemical reaction*, the subject of this chapter. Some people, however, have suggested that fossil fuel combustion does not significantly contribute to global warming and climate change. They argue that the amount of carbon dioxide emitted into the atmosphere by natural sources, such as volcanic eruptions, far exceeds that from fossil fuel combustion. Which group is right? We can judge the validity of the naysayers' argument by examining the combustion reaction that forms carbon

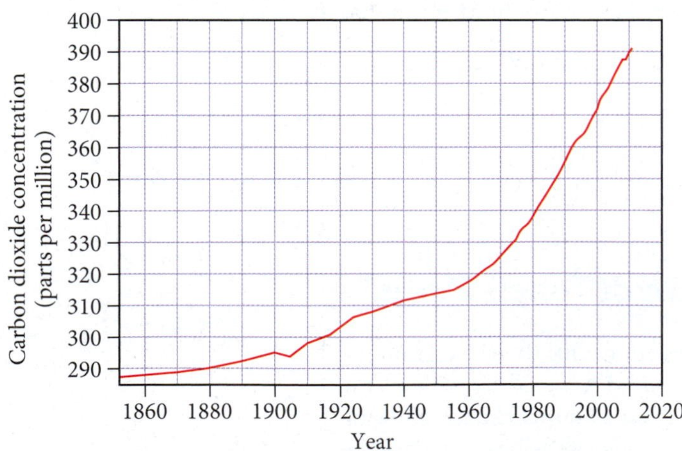

▲ **FIGURE 8.2 Carbon Dioxide Concentrations in the Atmosphere** The rise in carbon dioxide levels is due largely to fossil fuel combustion.

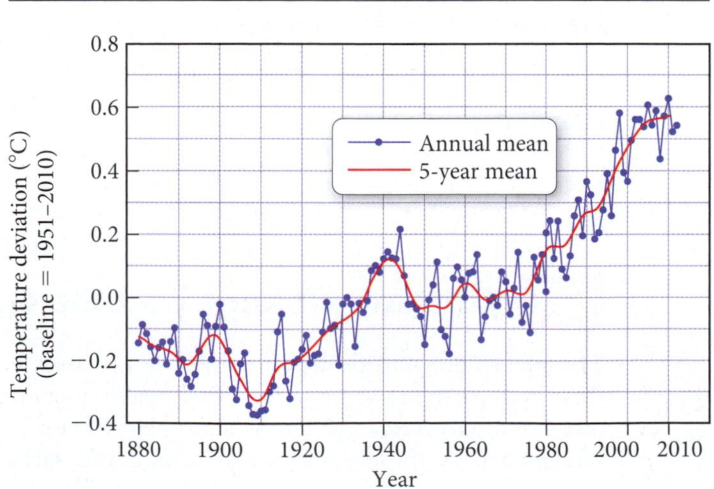

▲ **FIGURE 8.3 Global Temperature** Average temperatures worldwide have risen by about 0.7 °C since 1880.

dioxide. Governments keep records on the amount of fossil fuel that is burned. By understanding the combustion reaction by which fossil fuels burn, we can calculate the amount of carbon dioxide that is formed. We can then compare that amount to the amount released by volcanic eruptions. In this chapter, we first look at chemical changes in general, and then turn to examining this debate.

8.2 Chemical Change

The process by which a fossil fuel such as gasoline burns in the presence of oxygen is a *chemical change*. In a **chemical change**, atoms rearrange, transforming the original substances (in this case gasoline) into different substances (in this case water and carbon dioxide). Other chemical changes include the rusting of iron (Figure 8.4 ▼), the burning of sugar in a hot pan, and the transformation of oxygen and carbon dioxide into glucose that occurs within plants during photosynthesis.

Matter can also undergo **physical changes**, which—in contrast to chemical changes—do not change its composition. The atoms or molecules that compose a substance *do not change* their identity during a physical change. For example, when water boils, it changes from the liquid state to the gas state, but the molecules composing the water remain intact (Figure 8.5 ▼).

A physical change results in a different form of the same substance, while a chemical change results in a completely different substance.

Chemical and physical changes are manifestations of chemical and physical properties. A **chemical property** is a property that a substance displays only by changing its composition via a chemical change, while a **physical property** is a property that a substance displays without changing its composition. The flammability of gasoline is a chemical property—gasoline changes its composition when it burns, turning into completely new substances (primarily carbon dioxide and water as we have already seen). The smell of gasoline, by contrast, is a physical property—gasoline does not change its composition when it exhibits its odor. Chemical properties include corrosiveness, flammability, acidity, toxicity, and other such characteristics. Physical properties include odor, taste, color, appearance, melting point, boiling point, and density.

The differences between chemical and physical changes are not always apparent. Only chemical examination can confirm whether any particular change is chemical or physical. In many cases, however,

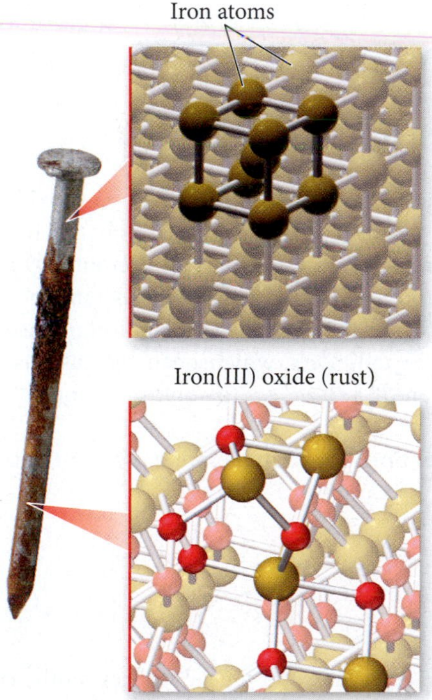

▲ **FIGURE 8.4 Rusting, a Chemical Change** When iron rusts, the iron atoms exchange electrons with oxygen atoms and combine with them to form a different chemical substance, the compound iron(III) oxide. Rusting is therefore a chemical change.

▲ **FIGURE 8.5 Boiling, a Physical Change** When water boils, it turns into a gas but does not alter its chemical identity—the water molecules are the same in both the liquid and gaseous states. Boiling is thus a physical change, and the boiling point of water is a physical property.

we can identify chemical and physical changes based on what we know about the changes. Chemical changes are often evidenced by temperature or color changes. Changes in the state of matter, such as melting or boiling, or changes in the physical condition of matter, such as those that result from cutting or crushing, are typically physical changes.

8.1 Cc Conceptual Connection

Chemical Change

Which change is a chemical change?

(a) the evaporation of rubbing alcohol
(b) the burning of lamp oil
(c) the forming of frost on a cold night

8.2 Cc Conceptual Connection

Physical Change

The diagram on the right represents liquid water molecules in a pan.

Which of the three diagrams that follow best represents the water molecules after they have been vaporized by the boiling of liquid water?

(a)

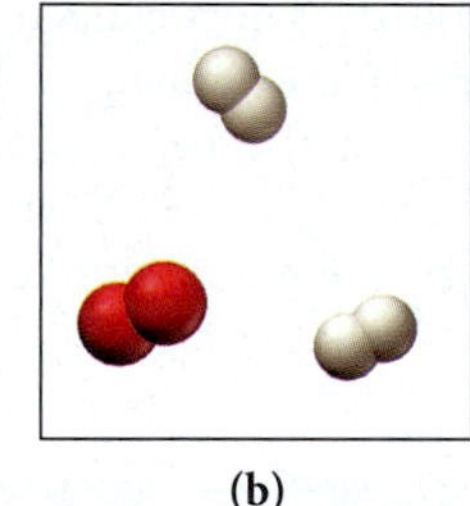

(b)

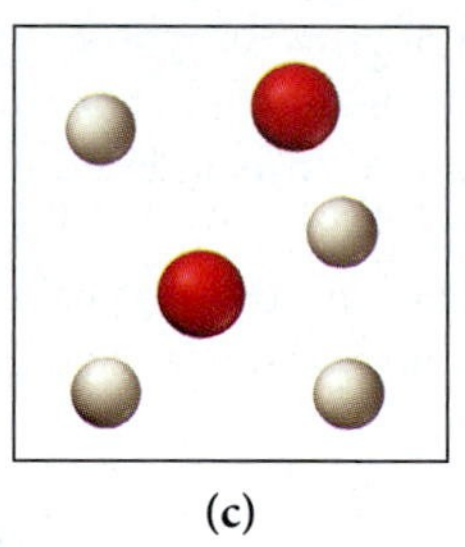

(c)

8.3 Writing and Balancing Chemical Equations

Chemical changes occur via **chemical reactions**. For example, fossil fuels form carbon dioxide via combustion reactions. A *combustion reaction* (see Section 8.6) is a particular type of chemical reaction in which a substance combines with oxygen to form one or more oxygen-containing compounds. Combustion reactions also emit heat. The heat produced in fossil fuel combustion reactions supplies much of our society's energy needs. For example, the heat from the combustion of gasoline expands the gaseous combustion products in a car engine's cylinders, which push the pistons and propel the car. We use the heat released by the combustion of *natural gas* to cook food and to heat our homes.

We represent a chemical reaction with a **chemical equation**. For example, we represent the combustion of natural gas with the equation:

$$\underbrace{CH_4 + O_2}_{\text{reactants}} \longrightarrow \underbrace{CO_2 + H_2O}_{\text{products}}$$

The substances on the left side of the equation are the **reactants**, and the substances on the right side are the **products**. We often specify the state of each reactant or product in parentheses next to the formula as shown here:

$$CH_4(g) + O_2(g) \longrightarrow CO_2(g) + H_2O(g)$$

The (*g*) indicates that these substances are gases in the reaction. Table 8.1 summarizes the common states of reactants and products and their symbols used in chemical equations.

TABLE 8.1 States of Reactants and Products in Chemical Equations

Abbreviation	State
(*g*)	Gas
(*l*)	Liquid
(*s*)	Solid
(*aq*)	Aqueous (water solution)

The equation just presented for the combustion of natural gas is not complete, however.

$$CH_4(g) + O_2(g) \longrightarrow CO_2(g) + H_2O(g)$$

2 O atoms — 2 O atoms + 1 O atom = 3 O atoms

The left side of the equation has two oxygen atoms, while the right side has three. The reaction as written, therefore, violates the law of conservation of mass because an oxygen atom formed out of nothing. Notice also that the left side has four hydrogen atoms, while the right side has only two.

$$CH_4(g) + O_2(g) \longrightarrow CO_2(g) + H_2O(g)$$

4 H atoms — 2 H atoms

Two hydrogen atoms have vanished, again violating mass conservation. To correct these problems—that is, to write an equation that more closely represents *what actually happens*—we must *balance* the equation. We need to change the coefficients (the numbers *in front of* the chemical formulas), not the subscripts (the numbers within the chemical formulas), to ensure that the number of each type of atom on the left side of the equation is equal to the number on the right side. New atoms do not form during a reaction, nor do atoms vanish—matter must be conserved.

When we add coefficients to the reactants and products to balance an equation, we change the number of molecules in the equation but not the *kind of* molecules. To balance the equation for the combustion of methane, we put the coefficient 2 before O_2 in the reactants, and the coefficient 2 before H_2O in the products.

The reason that we cannot change the subscripts when balancing a chemical equation is that changing the subscripts changes the substance itself, while changing the coefficients changes the number of molecules of the substance. For example, 2 H_2O is simply two water molecules, but H_2O_2 is hydrogen peroxide, a drastically different compound.

$$CH_4(g) + 2\,O_2(g) \longrightarrow CO_2(g) + 2\,H_2O(g)$$

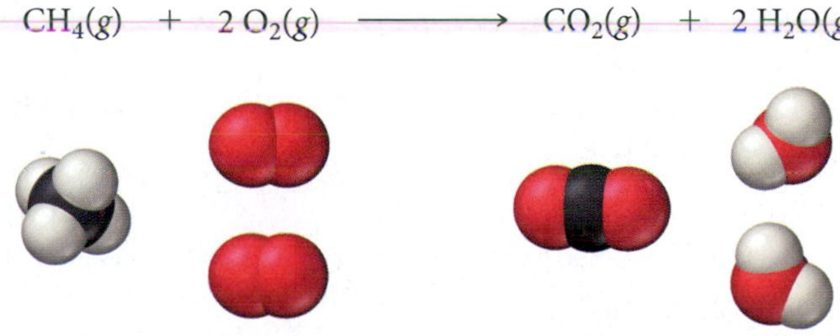

The equation is now balanced because the numbers of each type of atom on either side of the equation are equal. The balanced equation tells us that one CH_4 molecule reacts with two O_2 molecules to form one CO_2 molecule and two H_2O molecules. We can verify that the equation is balanced by summing the number of each type of atom on each side of the equation.

$$CH_4(g) + 2\,O_2(g) \longrightarrow CO_2(g) + 2\,H_2O(g)$$

Reactants	Products
1 C atom (1 × CH_4)	1 C atom (1 × CO_2)
4 H atoms (1 × CH_4)	4 H atoms (2 × H_2O)
4 O atoms (2 × O_2)	4 O atoms (1 × CO_2 + 2 × H_2O)

The number of each type of atom on both sides of the equation is now equal—we have balanced the equation.

We can balance many chemical equations simply by trial and error. However, some guidelines are useful. For example, balancing the atoms in the most complex substances first and the atoms in the simplest substances (such as pure elements) last often makes the process shorter. The following illustrations of how to balance chemical equations are presented in a three-column format. The general guidelines are shown on the left, with two examples of how to apply them on the right. This procedure is meant only as a flexible guide, not a rigid set of steps.

PROCEDURE FOR Balancing Chemical Equations	EXAMPLE 8.1 Balancing Chemical Equations	EXAMPLE 8.2 Balancing Chemical Equations
	Write a balanced equation for the reaction between solid cobalt(III) oxide and solid carbon to produce solid cobalt and carbon dioxide gas.	Write a balanced equation for the combustion of gaseous butane (C_4H_{10}), a fuel used in portable stoves and grills, in which it combines with gaseous oxygen to form gaseous carbon dioxide and gaseous water.
1. Write a skeletal (unbalanced) equation by writing chemical formulas for each of the reactants and products. Review Sections 5.6 and 5.8 for nomenclature rules. (If a skeletal equation is provided, go to Step 2.)	$Co_2O_3(s) + C(s) \longrightarrow Co(s) + CO_2(g)$	$C_4H_{10}(g) + O_2(g) \longrightarrow CO_2(g) + H_2O(g)$
2. Balance atoms that occur in more complex substances first. Always balance atoms in compounds before atoms in pure elements.	**Begin with O:** $Co_2O_3(s) + C(s) \longrightarrow Co(s) + CO_2(g)$ 3 O atoms ⟶ 2 O atoms To balance O, put a 2 before $Co_2O_3(s)$ and a 3 before $CO_2(g)$. $2\,Co_2O_3(s) + C(s) \longrightarrow Co(s) + 3\,CO_2(g)$ 6 O atoms ⟶ 6 O atoms	**Begin with C:** $C_4H_{10}(g) + O_2(g) \longrightarrow CO_2(g) + H_2O(g)$ 4 C atoms ⟶ 1 C atom To balance C, put a 4 before $CO_2(g)$. $C_4H_{10}(g) + O_2(g) \longrightarrow 4\,CO_2(g) + H_2O(g)$ 4 C atoms ⟶ 4 C atoms **Balance H:** $C_4H_{10}(g) + O_2(g) \longrightarrow 4\,CO_2(g) + H_2O(g)$ 10 H atoms ⟶ 2 H atoms To balance H, put a 5 before $H_2O(g)$: $C_4H_{10}(g) + O_2(g) \longrightarrow 4\,CO_2(g) + 5\,H_2O(g)$ 10 H atoms ⟶ 10 H atoms
3. Balance atoms that occur as free elements on either side of the equation last. Always balance free elements by adjusting the coefficient on the free element.	**Balance Co:** $2\,Co_2O_3(s) + C(s) \longrightarrow Co(s) + 3\,CO_2(g)$ 4 Co atoms ⟶ 1 Co atom To balance Co, put a 4 before $Co(s)$. $2\,Co_2O_3(s) + C(s) \longrightarrow 4\,Co(s) + 3\,CO_2(g)$ 4 Co atoms ⟶ 4 Co atoms	**Balance O:** $C_4H_{10}(g) + O_2(g) \longrightarrow 4\,CO_2(g) + 5\,H_2O(g)$ 2 O atoms ⟶ 8 O + 5 O = 13 O atoms To balance O, put a 13/2 before $O_2(g)$: $C_4H_{10}(g) + 13/2\,O_2(g) \longrightarrow 4\,CO_2(g) + 5\,H_2O(g)$ 13 O atoms ⟶ 13 O atoms

—Continued on the next page

Continued from the previous page—

	Balance C: $2\ Co_2O_3(s) + C(s) \longrightarrow 4\ Co(s) + 3\ CO_2(g)$ 1 C atom ⟶ 3 C atoms To balance C, put a 3 before C(s). $2\ Co_2O_3(s) + 3\ C(s) \longrightarrow 4\ Co(s) + 3\ CO_2(g)$	
4. If the balanced equation contains coefficient fractions, clear these by multiplying the entire equation by the denominator of the fraction.	This step is not necessary in this example. Proceed to Step 5.	$[C_4H_{10}(g) + 13/2\ O_2(g) \longrightarrow 4\ CO_2(g) + 5\ H_2O(g)] \times 2$ $2\ C_4H_{10}(g) + 13\ O_2(g) \longrightarrow 8\ CO_2(g) + 10\ H_2O(g)$
5. Check to make certain the equation is balanced by summing the total number of each type of atom on both sides of the equation.	$2\ Co_2O_3(s) + 3\ C(s) \longrightarrow 4\ Co(s) + 3\ CO_2(g)$ Left / Right: 4 Co atoms / 4 Co atoms 6 O atoms / 6 O atoms 3 C atoms / 3 C atoms The equation is balanced.	$2\ C_4H_{10}(g) + 13\ O_2(g) \longrightarrow 8\ CO_2(g) + 10\ H_2O(g)$ Left / Right: 8 C atoms / 8 C atoms 20 H atoms / 20 H atoms 26 O atoms / 26 O atoms The equation is balanced.

Left	Right
4 Co atoms	4 Co atoms
6 O atoms	6 O atoms
3 C atoms	3 C atoms

Left	Right
8 C atoms	8 C atoms
20 H atoms	20 H atoms
26 O atoms	26 O atoms

FOR PRACTICE 8.1
Write a balanced equation for the reaction between solid silicon dioxide and solid carbon to produce solid silicon carbide (SiC) and carbon monoxide gas.

FOR PRACTICE 8.2
Write a balanced equation for the combustion of gaseous ethane (C_2H_6), a minority component of natural gas, in which it combines with gaseous oxygen to form gaseous carbon dioxide and gaseous water.

Balanced Chemical Equations

8.3 Cc Conceptual Connection

Which quantity or quantities must always be the same on both sides of a chemical equation?

(a) the number of atoms of each kind
(b) the number of molecules of each kind
(c) the number of moles of each kind of molecule
(d) the sum of the masses of all substances involved

EXAMPLE 8.3

Balancing Chemical Equations Containing Ionic Compounds with Polyatomic Ions

Write a balanced equation for the reaction between aqueous strontium chloride and aqueous lithium phosphate to form solid strontium phosphate and aqueous lithium chloride.

SOLUTION

1. Write a skeletal equation by writing chemical formulas for each of the reactants and products. Review Sections 5.6 and 5.8 for naming rules. (If a skeletal equation is provided, proceed to step 2.)

$SrCl_2(aq) + Li_3PO_4(aq) \longrightarrow Sr_3(PO_4)_2(s) + LiCl(aq)$

2. Balance metal ions (cations) first. If a polyatomic cation exists on both sides of the equation, balance it as a unit.

Begin with Sr^{2+}:

$SrCl_2(aq) + Li_3PO_4(aq) \longrightarrow Sr_3(PO_4)_2(s) + LiCl(aq)$

1 Sr^{2+} ion $\longrightarrow$ 3 Sr^{2+} ions

To balance Sr^{2+}, put a 3 before $SrCl_2(aq)$.

$3\ SrCl_2(aq) + Li_3PO_4(aq) \longrightarrow Sr_3(PO_4)_2(s) + LiCl(aq)$

3 Sr^{2+} ions $\longrightarrow$ 3 Sr^{2+} ions

Balance Li^+:

$3\ SrCl_2(aq) + Li_3PO_4(aq) \longrightarrow Sr_3(PO_4)_2(s) + LiCl(aq)$

3 Li^+ ions $\longrightarrow$ 1 Li^+ ion

To balance Li^+, put a 3 before $LiCl(aq)$.

$3\ SrCl_2(aq) + Li_3PO_4(aq) \longrightarrow Sr_3(PO_4)_2(s) + 3\ LiCl(aq)$

3 Li^+ ions $\longrightarrow$ 3 Li^+ ions

3. Balance nonmetal ions (anions) second. If a polyatomic anion exists on both sides of the equation, balance it as a unit.

Balance PO_4^{3-}:

$3\ SrCl_2(aq) + Li_3PO_4(aq) \longrightarrow Sr_3(PO_4)_2(s) + 3\ LiCl(aq)$

1 PO_4^{3-} ion $\longrightarrow$ 2 PO_4^{3-} ions

To balance PO_4^{3-}, put a 2 before $Li_3PO_4(aq)$.

$3\ SrCl_2(aq) + 2\ Li_3PO_4(aq) \longrightarrow Sr_3(PO_4)_2(s) + 3\ LiCl(aq)$

2 PO_4^{3-} ions $\longrightarrow$ 2 PO_4^{3-} ions

Balance Cl^-:

$3\ SrCl_2(aq) + 2\ Li_3PO_4(aq) \longrightarrow Sr_3(PO_4)_2(s) + 3\ LiCl(aq)$

6 Cl^- ions $\longrightarrow$ 1 Cl^- ion

To balance Cl^-, replace the 3 before $LiCl(aq)$ with a 6. This also corrects the balance for Li^+, which was thrown off in the previous step.

$3\ SrCl_2(aq) + 2\ Li_3PO_4(aq) \longrightarrow Sr_3(PO_4)_2(s) + 6\ LiCl(aq)$

6 Cl^- ions $\longrightarrow$ 6 Cl^- ions

4. Check to make certain the equation is balanced by summing the total number of each type of ion on both sides of the equation.

$3\ SrCl_2(aq) + 2\ Li_3PO_4(aq) \longrightarrow Sr_3(PO_4)_2(s) + 6\ LiCl(aq)$

Left	Right
3 Sr^{2+} ions	3 Sr^{2+} ions
6 Li^+ ions	6 Li^+ ions
2 PO_4^{3-} ions	2 PO_4^{3-} ions
6 Cl^- ions	6 Cl^- ions

The equation is balanced.

FOR PRACTICE 8.3

Write a balanced equation for the reaction between aqueous lead(II) nitrate and aqueous potassium chloride to form solid lead(II) chloride and aqueous potassium nitrate.

8.4 Reaction Stoichiometry: How Much Carbon Dioxide?

Now that we have examined how to write balanced chemical equations, we can return to the question posed in Section 8.1: How much carbon dioxide is produced from the world's combustion of fossil fuels (and how does that compare to the amount produced by volcanoes)? The balanced chemical equations for fossil fuel combustion reactions provide the relationship between the amount of fossil fuel burned and the amount of carbon dioxide emitted. In this discussion, we use octane (a component of gasoline) as a representative fossil fuel. The balanced equation for the combustion of octane is:

$$2\, C_8H_{18}(l) + 25\, O_2(g) \longrightarrow 16\, CO_2(g) + 18\, H_2O(g)$$

The balanced equation shows that 16 CO_2 molecules are produced for every 2 molecules of octane burned. We can extend this numerical relationship between molecules to the amounts in moles as follows:

The coefficients in a chemical equation specify the relative amounts in moles of each of the substances involved in the reaction.

In other words, from the equation, we know that 16 *moles* of CO_2 are produced for every 2 *moles* of octane burned. The numerical relationships between chemical amounts in a balanced chemical equation are called reaction **stoichiometry**. Stoichiometry allows us to predict the amounts of products that will form in a chemical reaction based on the amounts of reactants that react. Stoichiometry also allows us to determine the amount of reactants necessary to form a given amount of product. These calculations are central to chemistry, allowing chemists to plan and carry out chemical reactions to obtain products in the desired quantities.

Stoichiometry is pronounced stoy-kee-AHM-e-tree.

Making Pizza: The Relationships among Ingredients

The concepts of stoichiometry are similar to those in a cooking recipe. Calculating the amount of carbon dioxide produced by the combustion of a given amount of a fossil fuel is analogous to calculating the number of pizzas that can be made from a given amount of cheese. For example, suppose we use the following pizza recipe:

$$1 \text{ crust} + 5 \text{ ounces tomato sauce} + 2 \text{ cups cheese} \longrightarrow 1 \text{ pizza}$$

The recipe contains the numerical relationships between the pizza ingredients. It says that if we have 2 cups of cheese—and enough of everything else—we can make 1 pizza. We can write this relationship as a ratio between the cheese and the pizza:

$$2 \text{ cups cheese} : 1 \text{ pizza}$$

What if we have 6 cups of cheese? Assuming that we have enough of everything else, we can use the above ratio as a conversion factor to calculate the number of pizzas.

$$6 \text{ cups cheese} \times \frac{1 \text{ pizza}}{2 \text{ cups cheese}} = 3 \text{ pizzas}$$

Six cups of cheese are sufficient to make 3 pizzas. The pizza recipe contains numerical ratios between other ingredients as well, including the following:

$$1 \text{ crust} : 1 \text{ pizza}$$

$$5 \text{ ounces tomato sauce} : 1 \text{ pizza}$$

Making Molecules: Mole-to-Mole Conversions

In a balanced chemical equation, we have a "recipe" for how reactants combine to form products. From our balanced equation for the combustion of octane, for example, we can write the following stoichiometric ratio:

$$2 \text{ mol } C_8H_{18} : 16 \text{ mol } CO_2$$

We can use this ratio to determine how many moles of CO_2 form when a given number of moles of C_8H_{18} burns. Suppose that we burn 22.0 moles of C_8H_{18}; how many moles of CO_2 form? We use the ratio from the balanced chemical equation in the same way that we used the ratio from the pizza recipe. The ratio acts as a conversion factor between the amount in moles of the reactant (C_8H_{18}) and the amount in moles of the product (CO_2).

$$22.0\ \cancel{\text{mol } C_8H_{18}} \times \frac{16\ \text{mol } CO_2}{2\ \cancel{\text{mol } C_8H_{18}}} = 176\ \text{mol } CO_2$$

The combustion of 22 moles of C_8H_{18} adds 176 moles of CO_2 to the atmosphere.

Making Molecules: Mass-to-Mass Conversions

According to the U.S. Department of Energy, the world burned 3.2×10^{10} barrels of petroleum in 2012, the equivalent of approximately 3.6×10^{15} g of gasoline. We can estimate the mass of CO_2 emitted into the atmosphere from burning this much gasoline using the combustion of 3.6×10^{15} g octane as the representative reaction. This calculation is similar to the one we just did, except that we are now given the *mass* of octane instead of the *amount* of octane in moles. Consequently, we must first convert the mass (in grams) to the amount (in moles). The general conceptual plan for calculations in which we are given the mass of a reactant or product in a chemical reaction and asked to find the mass of a different reactant or product takes this form:

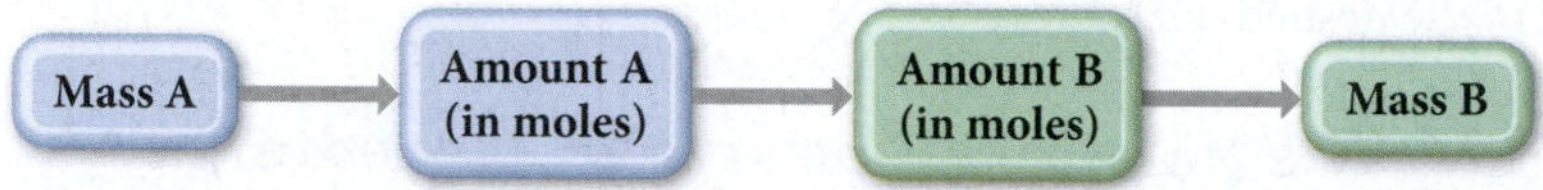

where A and B are two different substances involved in the reaction. We use the molar mass of A to convert from the mass of A to the amount of A (in moles). We use the appropriate ratio from the balanced chemical equation to convert from the amount of A (in moles) to the amount of B (in moles). And finally, we use the molar mass of B to convert from the amount of B (in moles) to the mass of B. To calculate the mass of CO_2 emitted upon the combustion of 3.6×10^{15} g of octane, we use the following conceptual plan:

Conceptual Plan

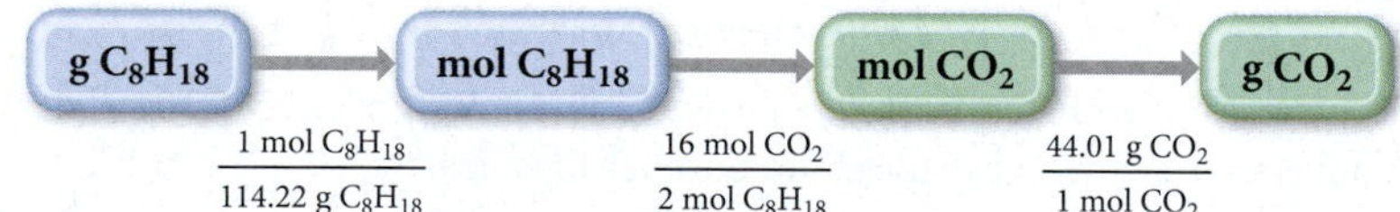

Relationships Used

molar mass $C_8H_{18} = 114.22$ g/mol

2 mol C_8H_{18} : 16 mol CO_2 (from chemical equation)

molar mass $CO_2 = 44.01$ g/mol

Solution

We follow the conceptual plan to solve the problem, beginning with g C_8H_{18} and canceling units to arrive at g CO_2.

$$3.6 \times 10^{15}\ \text{g } C_8H_{18} \times \frac{1\ \cancel{\text{mol } C_8H_{18}}}{114.22\ \cancel{\text{g } C_8H_{18}}} \times \frac{16\ \cancel{\text{mol } CO_2}}{2\ \cancel{\text{mol } C_8H_{18}}} \times \frac{44.01\ \text{g } CO_2}{1\ \cancel{\text{mol } CO_2}} = 1.1 \times 10^{16}\ \text{g } CO_2$$

The world's petroleum combustion produces 1.1×10^{16} g CO_2 (1.1×10^{13} kg) per year. In comparison, volcanoes produce about 2×10^{11} kg CO_2 per year.* In other words, volcanoes emit only $\frac{2.0 \times 10^{11}\text{ kg}}{1.1 \times 10^{13}\text{ kg}} \times 100\% = 1.8\%$ as much CO_2 per year as petroleum combustion. The argument that volcanoes emit more carbon dioxide than fossil fuel combustion is clearly mistaken. Examples 8.4 and 8.5 provide additional practice with stoichiometric calculations.

The percentage of CO_2 emitted by volcanoes relative to all fossil fuels is even less than 1.8% because the combustion of coal and natural gas also emits CO_2.

EXAMPLE 8.4

Stoichiometry

In photosynthesis, plants convert carbon dioxide and water into glucose ($C_6H_{12}O_6$) according to the reaction:

$$6\,CO_2(g) + 6\,H_2O(l) \xrightarrow{\text{sunlight}} 6\,O_2(g) + C_6H_{12}O_6(aq)$$

Suppose you determine that a particular plant consumes 37.8 g of CO_2 in one week. Assuming that there is more than enough water present to react with all of the CO_2, what mass of glucose (in grams) can the plant synthesize from the CO_2?

SORT The problem gives the mass of carbon dioxide and asks you to find the mass of glucose that the plant can produce.

GIVEN: 37.8 g CO_2

FIND: g $C_6H_{12}O_6$

STRATEGIZE The conceptual plan follows the general pattern of mass A $\longrightarrow$ amount A (in moles) $\longrightarrow$ amount B (in moles) $\longrightarrow$ mass B. From the chemical equation, deduce the relationship between moles of carbon dioxide and moles of glucose. Use the molar masses to convert between grams and moles.

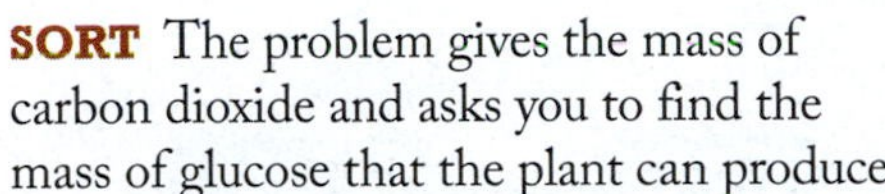

CONCEPTUAL PLAN

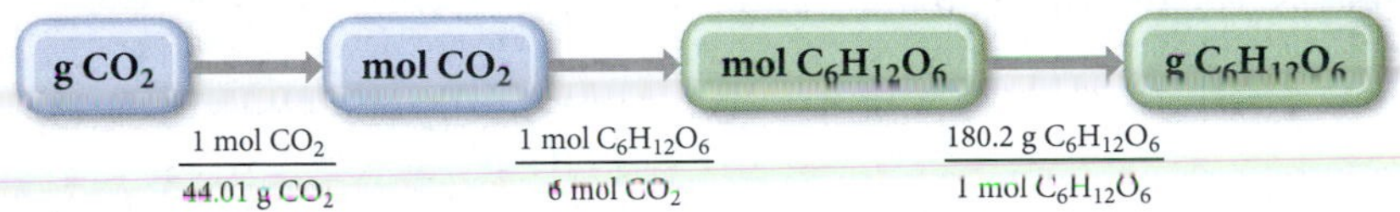

RELATIONSHIPS USED

molar mass CO_2 = 44.01 g/mol

6 mol CO_2 : 1 mol $C_6H_{12}O_6$ (from chemical equation)

molar mass $C_6H_{12}O_6$ = 180.2 g/mol

SOLVE Follow the conceptual plan to solve the problem. Begin with g CO_2 and use the conversion factors to arrive at g $C_6H_{12}O_6$.

SOLUTION

$$37.8\text{ g }CO_2 \times \frac{1\text{ mol }CO_2}{44.01\text{ g }CO_2} \times \frac{1\text{ mol }C_6H_{12}O_6}{6\text{ mol }CO_2} \times \frac{180.2\text{ g }C_6H_{12}O_6}{1\text{ mol }C_6H_{12}O_6} = 25.8\text{ g }C_6H_{12}O_6$$

CHECK The units of the answer are correct. The magnitude of the answer (25.8 g) is less than the initial mass of CO_2 (37.8 g). This is reasonable because each carbon in CO_2 has two oxygen atoms associated with it, while in $C_6H_{12}O_6$ each carbon has only one oxygen atom associated with it and two hydrogen atoms, which are much lighter than oxygen. Therefore the mass of glucose the plant produces should be less than the mass of carbon dioxide for this reaction.

FOR PRACTICE 8.4

Magnesium hydroxide, the active ingredient in milk of magnesia, neutralizes stomach acid, primarily HCl, according to the reaction:

$$Mg(OH)_2(aq) + 2\,HCl(aq) \longrightarrow 2\,H_2O(l) + MgCl_2(aq)$$

What mass of HCl, in grams, is neutralized by a dose of milk of magnesia containing 3.26 g $Mg(OH)_2$?

*Gerlach, T. M., Present-day CO_2 emissions from volcanoes; *Eos, Transactions, American Geophysical Union*, Vol. 72, No. 23, June 4, 1991, pp. 249 and 254–255.

EXAMPLE 8.5 Stoichiometry

Sulfuric acid (H_2SO_4) is a component of acid rain that forms when SO_2, a pollutant, reacts with oxygen and water according to the simplified reaction:

$$2\,SO_2(g) + O_2(g) + 2\,H_2O(l) \longrightarrow 2\,H_2SO_4(aq)$$

The generation of the electricity used by a medium-sized home produces about 25 kg of SO_2 per year. Assuming that there is more than enough O_2 and H_2O, what mass of H_2SO_4, in kg, can form from this much SO_2?

SORT The problem gives the mass of sulfur dioxide and asks you to find the mass of sulfuric acid.

GIVEN: 25 kg SO_2

FIND: kg H_2SO_4

STRATEGIZE The conceptual plan follows the standard format of mass $\longrightarrow$ amount (in moles) $\longrightarrow$ amount (in moles) $\longrightarrow$ mass. Because the original quantity of SO_2 is given in kg, you must first convert to grams. You can deduce the relationship between moles of sulfur dioxide and moles of sulfuric acid from the balanced chemical equation. Because the final quantity is requested in kg, you convert to kg at the end.

CONCEPTUAL PLAN

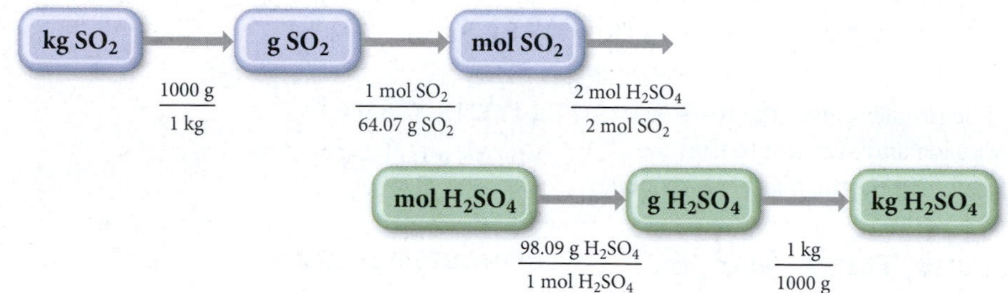

RELATIONSHIPS USED

1 kg = 1000 g

molar mass SO_2 = 64.07 g/mol

2 mol SO_2 : 2 mol H_2SO_4 (from chemical equation)

molar mass H_2SO_4 = 98.09 g/mol

SOLVE Follow the conceptual plan to solve the problem. Begin with the given amount of SO_2 in kilograms and use the conversion factors to arrive at kg H_2SO_4.

SOLUTION

$$25\ \text{kg SO}_2 \times \frac{1000\ \text{g}}{1\ \text{kg}} \times \frac{1\ \text{mol SO}_2}{64.07\ \text{g SO}_2} \times \frac{2\ \text{mol H}_2\text{SO}_4}{2\ \text{mol SO}_2} \times \frac{98.09\ \text{g H}_2\text{SO}_4}{1\ \text{mol H}_2\text{SO}_4} \times \frac{1\ \text{kg}}{1000\ \text{g}} = 38\ \text{kg H}_2\text{SO}_4$$

CHECK The units of the final answer are correct. The magnitude of the final answer (38 kg H_2SO_4) is larger than the given amount of SO_2 (25 kg). This is reasonable because in the reaction each SO_2 molecule "gains weight" by reacting with O_2 and H_2O.

FOR PRACTICE 8.5

Another component of acid rain is nitric acid, which forms when NO_2, also a pollutant, reacts with oxygen and water according to the simplified equation:

$$4\,NO_2(g) + O_2(g) + 2\,H_2O(l) \longrightarrow 4\,HNO_3(aq)$$

The generation of the electricity used by a medium-sized home produces about 16 kg of NO_2 per year. Assuming that there is adequate O_2 and H_2O, what mass of HNO_3, in kg, can form from this amount of NO_2 pollutant?

8.4 Cc Conceptual Connection — Stoichiometry

Under certain conditions, sodium can react with oxygen to form sodium oxide according to the reaction:

$$4\,Na(s) + O_2(g) \longrightarrow 2\,Na_2O(s)$$

A flask contains the amount of oxygen represented by the diagram on the right.

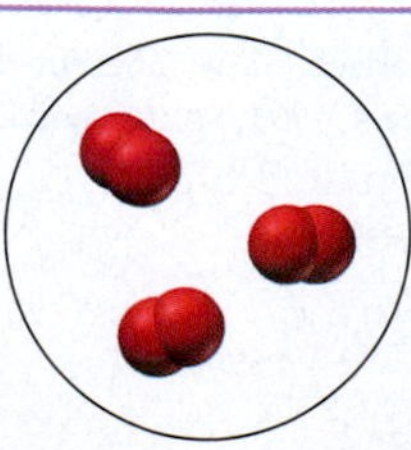

—*Continued on the next page*

Continued from the previous page—

Which image best represents the amount of sodium required to completely react with all of the oxygen in the flask according to the above equation?

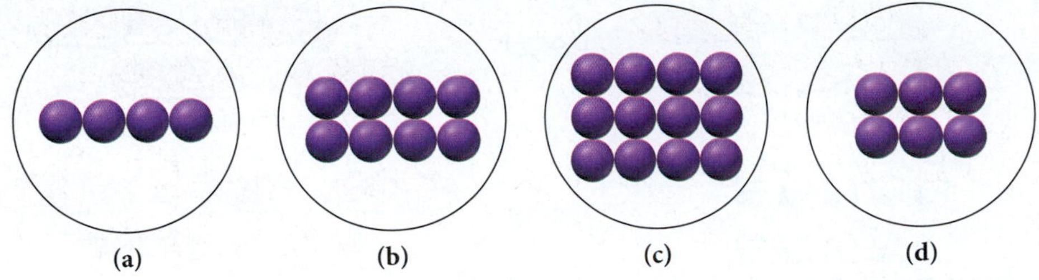

Stoichiometry II

8.5 Cc Conceptual Connection

Consider the generic chemical equation A + 3B $\longrightarrow$ 2C. Let circles represent molecules of A, squares represent molecules of B, and triangles represent molecules of C. The diagram at right represents the amount of B available for reaction. Draw similar diagrams showing: (a) the amount of A necessary to completely react with B; and (b) the amount of C that forms if B completely reacts.

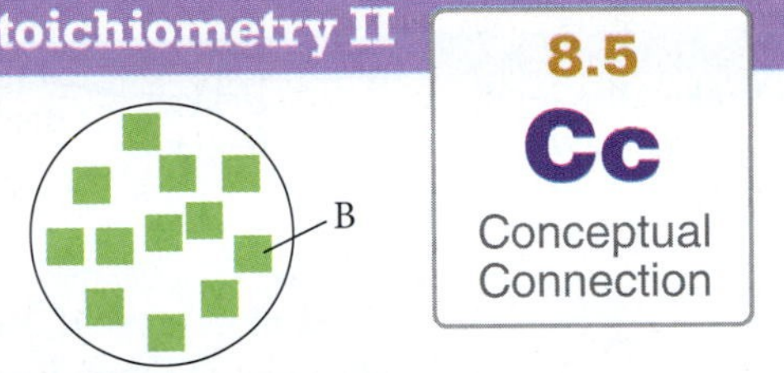

8.5 Limiting Reactant, Theoretical Yield, and Percent Yield

We return to our pizza analogy to understand three more important concepts in reaction stoichiometry: *limiting reactant, theoretical yield,* and *percent yield.* Recall our pizza recipe from Section 8.4:

$$1 \text{ crust} + 5 \text{ ounces tomato sauce} + 2 \text{ cups cheese} \longrightarrow 1 \text{ pizza}$$

Suppose that we have 4 crusts, 10 cups of cheese, and 15 ounces of tomato sauce. How many pizzas can we make?

We have enough crusts to make:

$$4 \text{ crusts} \times \frac{1 \text{ pizza}}{1 \text{ crust}} = 4 \text{ pizzas}$$

We have enough cheese to make:

$$10 \text{ cups cheese} \times \frac{1 \text{ pizza}}{2 \text{ cups cheese}} = 5 \text{ pizzas}$$

We have enough tomato sauce to make:

$$15 \text{ ounces tomato sauce} \times \frac{1 \text{ pizza}}{5 \text{ ounces tomato sauce}} = 3 \text{ pizzas}$$

Limiting reactant — Smallest number of pizzas

We have enough crusts for 4 pizzas, enough cheese for 5 pizzas, but enough tomato sauce for only 3 pizzas. Consequently, unless we get more ingredients, we can make only 3 pizzas. The tomato sauce *limits* how many pizzas we can make. If the pizza recipe were a chemical reaction, the tomato sauce would be the **limiting reactant**, the reactant that limits the amount of product in a chemical reaction. Notice that the limiting reactant is the reactant that makes *the least amount of product.* The reactants

The term *limiting reagent* is sometimes used in place of *limiting reactant.*

that *do not* limit the amount of product—such as the crusts and the cheese in this example—are said to be *in excess*. If this were a chemical reaction, 3 pizzas would be the **theoretical yield**, the amount of product that can be made in a chemical reaction based on the amount of limiting reactant.

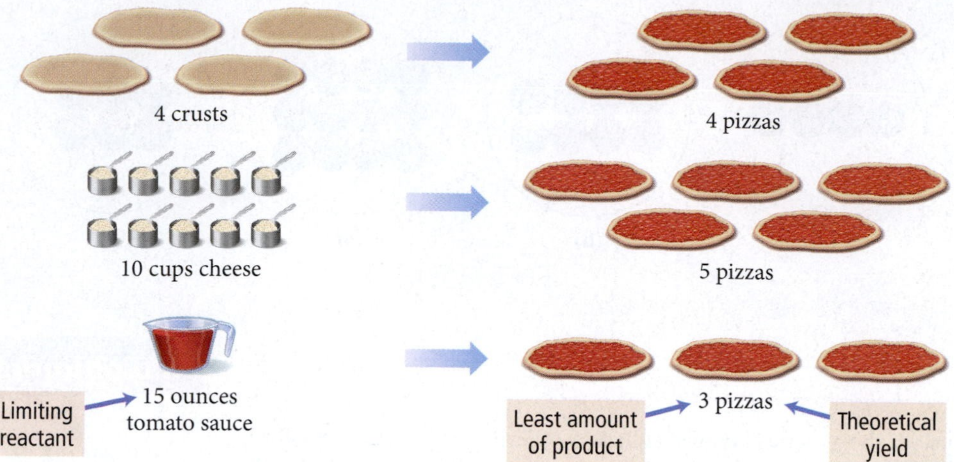

▶The ingredient that makes the least amount of pizza determines how many pizzas we can make.

Let us carry this analogy one step further. Suppose we go on to cook our pizzas and accidentally burn one of them. Even though we theoretically have enough ingredients for 3 pizzas, we end up with only 2. If this were a chemical reaction, the 2 pizzas would be our **actual yield**, the amount of product actually produced by a chemical reaction. (The actual yield is always equal to or less than the theoretical yield because a small amount of product is usually lost to other reactions or does not form during a reaction.) Finally, we calculate our **percent yield**, the percentage of the theoretical yield that was actually attained, as the ratio of the actual yield to the theoretical yield:

Actual yield

$$\% \text{ yield} = \frac{2 \text{ pizzas}}{3 \text{ pizzas}} \times 100\% = 67\%$$

Theoretical yield

Since one of our pizzas burned, we obtained only 67% of our theoretical yield.

Summarizing Limiting Reactant and Yield:

- **The limiting reactant** (or **limiting reagent**) is the reactant that is completely consumed in a chemical reaction and limits the amount of product.
- **The reactant in excess** is any reactant that occurs in a quantity greater than is required to completely react with the limiting reactant.
- **The theoretical yield** is the amount of product that can be made in a chemical reaction based on the amount of limiting reactant.
- **The actual yield** is the amount of product actually produced by a chemical reaction.
- **The percent yield** is calculated as $\frac{\text{actual yield}}{\text{theoretical yield}} \times 100\%$.

We can apply these concepts to a chemical reaction. Recall from Section 8.3 our balanced equation for the combustion of methane:

$$CH_4(g) + 2\,O_2(g) \longrightarrow CO_2(g) + 2\,H_2O(l)$$

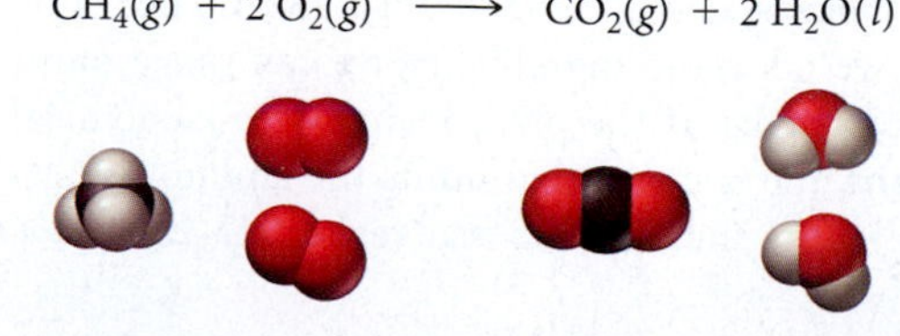

If we start out with 5 CH_4 molecules and 8 O_2 molecules, what is our limiting reactant? What is our theoretical yield of carbon dioxide molecules? First, we calculate the number of CO_2 molecules that can be made from 5 CH_4 molecules:

$$5\ \cancel{CH_4} \times \frac{1\ CO_2}{1\ \cancel{CH_4}} = 5\ CO_2$$

We then calculate the number of CO_2 molecules that can be made from 8 O_2 molecules:

$$8\ \cancel{O_2} \times \frac{1\ CO_2}{2\ \cancel{O_2}} = 4\ CO_2$$

Limiting reactant

Least amount of product

Theoretical yield

We have enough CH_4 to make 5 CO_2 molecules and enough O_2 to make 4 CO_2 molecules; therefore O_2 is the limiting reactant, and 4 CO_2 molecules is the theoretical yield. The CH_4 is in excess.

An alternative way to calculate the limiting reactant (which we mention here but do not use in this book) is to pick any reactant and determine how much of the *other reactant* is necessary to completely react with it. For the reaction we just examined, we have 5 CH_4 molecules and 8 O_2 molecules. Let's pick the 5 CH_4 molecules and determine how many O_2 molecules are necessary to completely react with them:

$$5\ \cancel{CH_4} \times \frac{2\,O_2}{1\ \cancel{CH_4}} = 10\ O_2$$

Since we need 10 O_2 molecules to completely react with the 5 CH_4 molecules, and since we have only 8 O_2 molecules, we know that the O_2 is the limiting reactant. We can apply the same method by comparing the amounts of reactants in moles.

Limiting Reactant and Theoretical Yield

Nitrogen and hydrogen gas react to form ammonia according to the reaction:

$$N_2(g) + 3\ H_2(g) \longrightarrow 2\ NH_3(g)$$

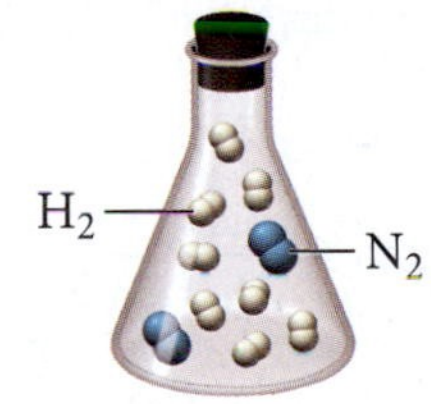

If a flask contains a mixture of reactants represented by the image on the right, which image best represents the mixture in the flask after the reactants have reacted as completely as possible? Which is the limiting reactant? Which reactant is in excess?

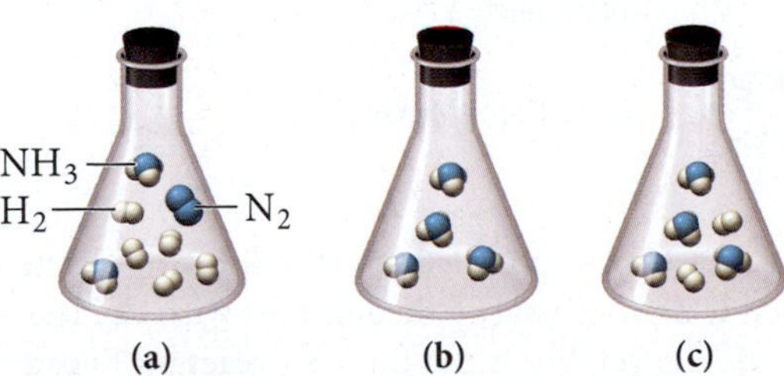

When working in the laboratory, we normally measure the initial quantities of reactants in grams, not in number of molecules. To find the limiting reactant and theoretical yield from initial masses, we must first convert the masses to amounts in moles. Consider the reaction:

$$2\ Mg(s) + O_2(g) \longrightarrow 2\ MgO(s)$$

A reaction mixture contains 42.5 g Mg and 33.8 g O_2; what are the limiting reactant and theoretical yield? To solve this problem, we must determine which of the reactants makes the least amount of product.

Conceptual Plan

We can find the limiting reactant by calculating how much product can be made from each reactant. However, we are given the initial quantities in grams, and stoichiometric relationships are between moles, so we must first convert to moles. We then convert from moles of the reactant to moles of product. The reactant that makes the *least amount of product* is the limiting reactant. The conceptual plan is:

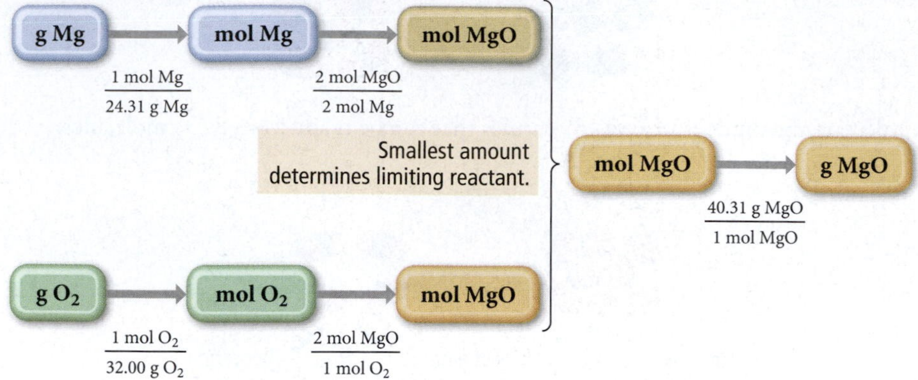

In this conceptual plan, we compare the number of moles of MgO made by each reactant and convert only the smaller amount to grams. (Alternatively, we can convert both quantities to grams and determine the limiting reactant based on the mass of the product.)

Relationships Used

molar mass Mg = 24.31 g Mg

molar mass O_2 = 32.00 g O_2

2 mol Mg : 2 mol MgO (from chemical equation)

1 mol O_2 : 2 mol MgO (from chemical equation)

molar mass MgO = 40.31 g MgO

Solution

Beginning with the masses of each reactant, we follow the conceptual plan to calculate how much product can be made from each.

$$42.5\ \text{g Mg} \times \frac{1\ \text{mol Mg}}{24.31\ \text{g Mg}} \times \frac{2\ \text{mol MgO}}{2\ \text{mol Mg}} = 1.7\underline{4}83\ \text{mol MgO}$$

Limiting reactant

Least amount of product

$$33.8\ \text{g } O_2 \times \frac{1\ \text{mol } O_2}{32.00\ \text{g } O_2} \times \frac{2\ \text{mol MgO}}{1\ \text{mol } O_2} = 2.1\underline{1}25\ \text{mol MgO}$$

$$1.7\underline{4}83\ \text{mol MgO} \times \frac{40.31\ \text{g MgO}}{1\ \text{mol MgO}} = 70.5\ \text{g MgO}$$

Because Mg makes the least amount of product, it is the limiting reactant, and O_2 is in excess. Notice that the limiting reactant is not necessarily the reactant with the least mass. In this case, the mass of O_2 is less than the mass of Mg, yet Mg is the limiting reactant because it makes the least amount of MgO. The theoretical yield is 70.5 g of MgO, the mass of product possible based on the limiting reactant.

Suppose that after the synthesis, the actual yield of MgO is 55.9 g. What is the percent yield? We calculate the percent yield as follows:

$$\%\ \text{yield} = \frac{\text{actual yield}}{\text{theoretical yield}} \times 100\% = \frac{55.9\ \text{g}}{70.5\ \text{g}} \times 100\% = 79.3\%$$

EXAMPLE 8.6 Limiting Reactant and Theoretical Yield

Ammonia, NH_3, can be synthesized by the reaction:

$$2\,NO(g) + 5\,H_2(g) \longrightarrow 2\,NH_3(g) + 2\,H_2O(g)$$

Starting with 86.3 g NO and 25.6 g H_2, find the theoretical yield of ammonia in grams.

SORT You are given the mass of each reactant in grams and asked to find the theoretical yield of a product.

GIVEN: 86.3 g NO, 25.6 g H_2

FIND: theoretical yield of $NH_3(g)$

STRATEGIZE Determine which reactant makes the least amount of product by converting from grams of each reactant to moles of the reactant to moles of the product. Use molar masses to convert between grams and moles and use the stoichiometric relationships (deduced from the balanced chemical equation) to convert between moles of reactant and moles of product. Remember that the reactant that makes *the least amount of product* is the limiting reactant. Convert the number of moles of product obtained using the limiting reactant to grams of product.

CONCEPTUAL PLAN

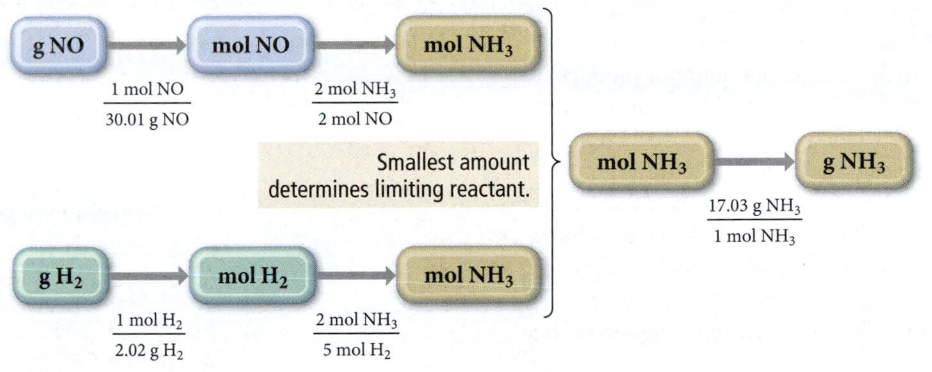

RELATIONSHIPS USED

molar mass NO = 30.01 g/mol

molar mass H_2 = 2.02 g/mol

2 mol NO : 2 mol NH_3 (from chemical equation)

5 mol H_2 : 2 mol NH_3 (from chemical equation)

molar mass NH_3 = 17.03 g/mol

SOLVE Beginning with the given mass of each reactant, calculate the amount of product that can be made in moles. Convert the amount of product made by the limiting reactant to grams—this is the theoretical yield.

SOLUTION

$$86.3 \text{ g NO} \times \frac{1 \text{ mol NO}}{30.01 \text{ g NO}} \times \frac{2 \text{ mol } NH_3}{2 \text{ mol NO}} = 2.8\underline{7}57 \text{ mol } NH_3$$

Limiting reactant — Least amount of product

$$2.8\underline{7}57 \text{ mol } NH_3 \times \frac{17.03 \text{ g } NH_3}{\text{mol } NH_3} = 49.0 \text{ g } NH_3$$

$$25.6 \text{ g } H_2 \times \frac{1 \text{ mol } H_2}{2.02 \text{ g } H_2} \times \frac{2 \text{ mol } NH_3}{5 \text{ mol } H_2} = 5.0\underline{6}93 \text{ mol } NH_3$$

Since NO makes the least amount of product, it is the limiting reactant, and the theoretical yield of ammonia is 49.0 g.

CHECK The units of the answer (g NH_3) are correct. The magnitude (49.0 g) seems reasonable given that 86.3 g NO is the limiting reactant. NO contains one oxygen atom per nitrogen atom, and NH_3 contains three hydrogen atoms per nitrogen atom. Three hydrogen atoms have less mass than one oxygen atom, so it is reasonable that the mass of NH_3 obtained is less than the mass of NO.

FOR PRACTICE 8.6

Ammonia can also be synthesized by the reaction:

$$3\,H_2(g) + N_2(g) \longrightarrow 2\,NH_3(g)$$

What is the theoretical yield of ammonia, in kg, that we can synthesize from 5.22 kg of H_2 and 31.5 kg of N_2?

EXAMPLE 8.7
Limiting Reactant and Theoretical Yield

We can obtain titanium metal from its oxide according to the following balanced equation:

$$TiO_2(s) + 2\,C(s) \longrightarrow Ti(s) + 2\,CO(g)$$

When 28.6 kg of C reacts with 88.2 kg of TiO_2, 42.8 kg of Ti is produced. Find the limiting reactant, theoretical yield (in kg), and percent yield.

SORT You are given the mass of each reactant and the mass of product formed. You are asked to find the limiting reactant, theoretical yield, and percent yield.

GIVEN: 28.6 kg C, 88.2 kg TiO_2, 42.8 kg Ti produced

FIND: limiting reactant, theoretical yield, % yield

STRATEGIZE Determine which of the reactants makes the least amount of product by converting from kilograms of each reactant to moles of product. Convert between grams and moles using molar mass. Convert between moles of reactant and moles of product using the stoichiometric relationships derived from the balanced chemical equation. Remember that the reactant that makes the *least amount of product* is the limiting reactant.

CONCEPTUAL PLAN

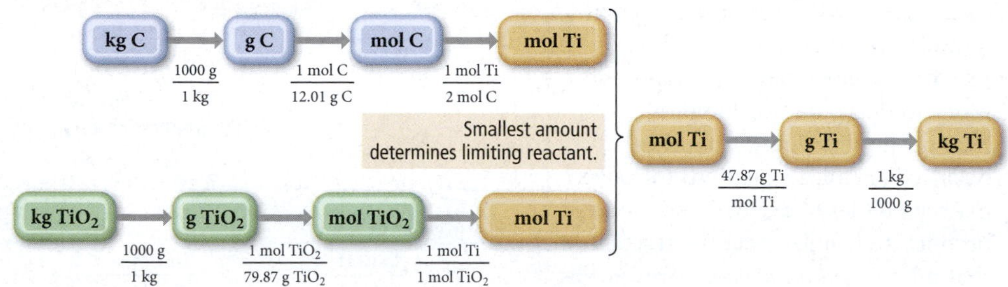

Determine the theoretical yield (in kg) by converting the number of moles of product obtained with the limiting reactant to kilograms of product.

RELATIONSHIPS USED

1000 g = 1 kg

molar mass of C = 12.01 g/mol

molar mass of TiO_2 = 79.87 g/mol

1 mol TiO_2 : 1 mol Ti (from chemical equation)

2 mol C : 1 mol Ti (from chemical equation)

molar mass of Ti = 47.87 g/mol

SOLVE Beginning with the actual amount of each reactant, calculate the amount of product that can be made in moles. Convert the amount of product made by the limiting reactant to kilograms—this is the theoretical yield.

SOLUTION

$$\mathbf{28.6\ \cancel{kg\ C}} \times \frac{1000\ \cancel{g}}{1\ \cancel{kg}} \times \frac{1\ \cancel{mol\ C}}{12.01\ \cancel{g\ C}} \times \frac{1\ mol\ Ti}{2\ \cancel{mol\ C}} = \mathbf{1.1\underline{9}07 \times 10^3\ mol\ Ti}$$

Limiting reactant

Least amount of product

$$\mathbf{88.2\ \cancel{kg\ TiO_2}} \times \frac{1000\ \cancel{g}}{1\ \cancel{kg}} \times \frac{1\ \cancel{mol\ TiO_2}}{79.87\ \cancel{g\ TiO_2}} \times \frac{1\ \cancel{mol\ Ti}}{1\ \cancel{mol\ TiO_2}} = \mathbf{1.1\underline{0}43 \times 10^3\ mol\ Ti}$$

$$\mathbf{1.1\underline{0}43 \times 10^3\ \cancel{mol\ Ti}} \times \frac{47.87\ \cancel{g\ Ti}}{1\ \cancel{mol\ Ti}} \times \frac{1\ kg}{1000\ \cancel{g}} = \mathbf{52.9\ kg\ Ti}$$

Calculate the percent yield by dividing the actual yield (42.8 kg Ti) by the theoretical yield.

Since TiO_2 makes the least amount of product, it is the limiting reactant, and 52.9 kg Ti is the theoretical yield.

$$\%\ \text{yield} = \frac{\text{actual yield}}{\text{theoretical yield}} \times 100\% = \frac{42.8\ g}{52.9\ g} \times 100\% = 80.9\%$$

CHECK The theoretical yield has the correct units (kg Ti) and has a reasonable magnitude compared to the mass of TiO_2. Because Ti has a lower molar mass than TiO_2, the amount of Ti made from TiO_2 should have a lower mass. The percent yield is reasonable (under 100% as it should be).

FOR PRACTICE 8.7

Mining companies use this reaction to obtain iron from iron ore:

$$Fe_2O_3(s) + 3\,CO(g) \longrightarrow 2\,Fe(s) + 3\,CO_2(g)$$

The reaction of 167 g Fe_2O_3 with 85.8 g CO produces 72.3 g Fe. Determine the limiting reactant, theoretical yield, and percent yield.

Reactant in Excess

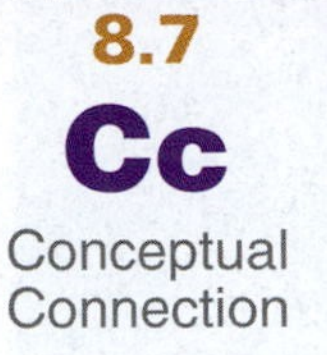

Nitrogen dioxide reacts with water to form nitric acid and nitrogen monoxide according to the equation:

$$3\,NO_2(g) + H_2O(l) \longrightarrow 2\,HNO_3(l) + NO(g)$$

Suppose that 5 mol NO_2 and 1 mol H_2O combine and react completely. How many moles of the reactant in excess are present after the reaction has completed?

8.6 Three Examples of Chemical Reactions: Combustion, Alkali Metals, and Halogens

In this section, we examine three types of reactions. The first is combustion reactions, which we encountered in Section 8.1. The second is the reactions of the alkali metals. As we discussed in Chapter 4, alkali metals have low first ionization energies and are among the most active metals. Their reactions are good examples of the types of reactions that many metals undergo. The third type of reactions involves the halogens. Halogens have among the most negative (most exothermic) electron affinities and are therefore among the most active nonmetals.

Combustion Reactions

A **combustion reaction** involves the reaction of a substance with O_2 to form one or more oxygen-containing compounds, often including water. Combustion reactions also emit heat. For example, as you saw earlier in this chapter, natural gas (CH_4) reacts with oxygen to form carbon dioxide and water:

$$CH_4(g) + 2\,O_2(g) \longrightarrow CO_2(g) + 2\,H_2O(g)$$

Ethanol, the alcohol in alcoholic beverages, also reacts with oxygen in a combustion reaction to form carbon dioxide and water:

$$C_2H_5OH(l) + 3\,O_2(g) \longrightarrow 2\,CO_2(g) + 3\,H_2O(g)$$

Compounds containing carbon and hydrogen—or carbon, hydrogen, and oxygen—always form carbon dioxide and water upon complete combustion. Other combustion reactions include the reaction of carbon with oxygen to form carbon dioxide:

$$C(s) + O_2(g) \longrightarrow CO_2(g)$$

and the reaction of hydrogen with oxygen to form water:

$$2\,H_2(g) + O_2(g) \longrightarrow 2\,H_2O(g)$$

We can write chemical equations for most combustion reactions by noticing the pattern of reactivity. Any carbon in a combustion reaction reacts with oxygen to produce carbon dioxide, and any hydrogen reacts with oxygen to form water.

EXAMPLE 8.8

Writing Equations for Combustion Reactions

Write a balanced equation for the combustion of liquid methyl alcohol (CH_3OH).

SOLUTION

Begin by writing an unbalanced equation showing the reaction of CH_3OH with O_2 to form CO_2 and H_2O.	$CH_3OH(l) + O_2(g) \longrightarrow CO_2(g) + H_2O(g)$
Balance the equation using the guidelines from Section 8.3.	$\mathbf{2}\,CH_3OH(l) + \mathbf{3}\,O_2(g) \longrightarrow \mathbf{2}\,CO_2(g) + \mathbf{4}\,H_2O(g)$

FOR PRACTICE 8.8

Write a balanced equation for the complete combustion of liquid C_2H_5SH.

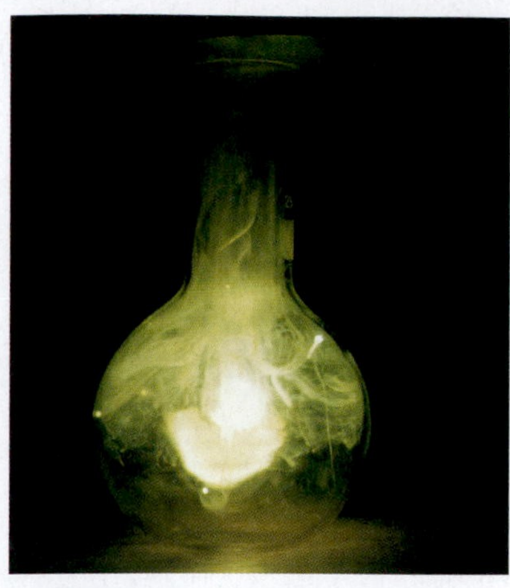

▲ **FIGURE 8.6 Reaction of Sodium and Chlorine to Form Sodium Chloride**

Alkali Metal Reactions

The alkali metals (group 1A) have ns^1 outer electron configurations. The single valence electron that keeps these metals from having noble gas configurations is easily removed (the metals have low ionization energies), making these elements the most active metals in the periodic table. The reactions of the alkali metals with nonmetals are vigorous. For example, the alkali metals (M) react with halogens (X) according to the reaction:

$$2\,M + X_2 \longrightarrow 2\,MX$$

The reaction of sodium and chlorine to form sodium chloride is typical:

$$2\,Na(s) + Cl_2(g) \longrightarrow 2\,NaCl(s)$$

This reaction emits heat and sparks as it occurs (Figure 8.6 ◀). Each successive alkali metal reacts even more vigorously with chlorine. The alkali metals also react with water to form the dissolved alkali metal ion, the hydroxide ion, and hydrogen gas:

$$2\,M(s) + 2\,H_2O(l) \longrightarrow 2\,M^+(aq) + 2\,OH^-(aq) + H_2(g)$$

The reaction is highly exothermic and can be explosive because the heat from the reaction can ignite the hydrogen gas. The reaction becomes more explosive as we move down the column from one metal to the next, as shown in Figure 8.7 ▼.*

Reactions of the Alkali Metals with Water

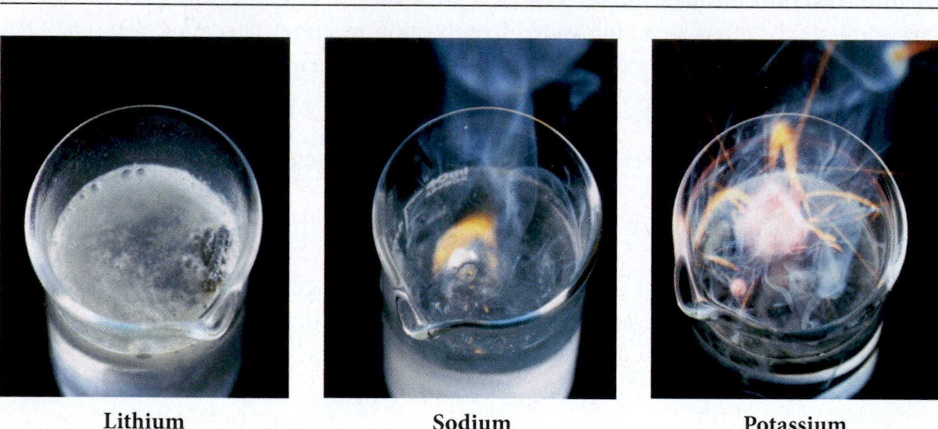

▶ **FIGURE 8.7 Reactions of the Alkali Metals with Water** The reactions become progressively more vigorous as we move down the group.

*The rate of the alkali metal reaction with water, and therefore its vigor, is enhanced by the successively lower melting points of the alkali metals as we move down the column. The low melting points of the heavier metals allow the emitted heat to actually melt the metal, increasing the reaction rate.

Halogen Reactions

The halogens (group 7A) have ns^2np^5 outer electron configurations. The one electron needed to attain a noble gas configuration is easily acquired (the halogens have highly negative electron affinities), making these elements among the most active nonmetals in the periodic table. The halogens all react with many metals to form *metal halides* according to the equation:

$$2\,M + n\,X_2 \longrightarrow 2\,MX_n$$

where M is the metal, X is the halogen, and MX_n is the metal halide. For example, chlorine reacts with iron according to the equation:

$$2\,Fe(s) + 3\,Cl_2(g) \longrightarrow 2\,FeCl_3(s)$$

Since metals tend to lose electrons and the halogens tend to gain them, the metal halides—like all compounds that form between metals and nonmetals—contain ionic bonds.

The halogens also react with hydrogen to form *hydrogen halides* according to the equation:

$$H_2(g) + X_2 \longrightarrow 2\,HX(g)$$

The hydrogen halides—like all compounds that form between two nonmetals—contain covalent bonds. As we will see in Chapter 9, all of the hydrogen halides form acidic solutions when combined with water.

The halogens also react with each other to form *interhalogen compounds*. For example, bromine reacts with fluorine according to the equation:

$$Br_2(l) + F_2(g) \longrightarrow 2\,BrF(g)$$

Again, like all compounds that form between two nonmetals, the interhalogen compounds contain covalent bonds.

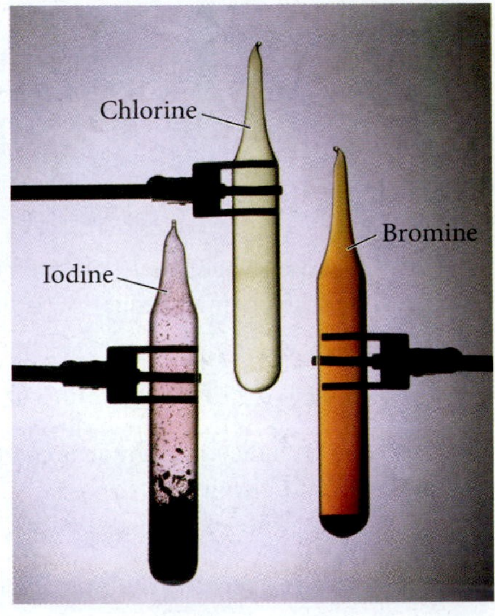

▲ Three Halogens

EXAMPLE 8.9
Alkali Metal and Halogen Reactions

Write a balanced chemical equation for each reaction.

(a) the reaction between potassium metal and bromine gas
(b) the reaction between rubidium metal and liquid water
(c) the reaction between gaseous chlorine and solid iodine

SOLUTION

(a) Alkali metals react with halogens to form metal halides. Write the formulas for the reactants and the metal halide product (making sure to write the correct ionic chemical formula for the metal halide, as outlined in Section 5.6), and then balance the equation.	$2\,K(s) + Br_2(g) \longrightarrow 2\,KBr(s)$
(b) Alkali metals react with water to form the dissolved metal ion, the hydroxide ion, and hydrogen gas. Write the skeletal equation including each of these and then balance it.	$2\,Rb(s) + 2\,H_2O(l) \longrightarrow 2\,Rb^+(aq) + 2\,OH^-(aq) + H_2(g)$
(c) Halogens react with each other to form interhalogen compounds. Write the skeletal equation with each of the halogens as the reactants and the interhalogen compound as the product and balance the equation.	$Cl_2(g) + I_2(s) \longrightarrow 2\,ICl(g)$

FOR PRACTICE 8.9

Write a balanced chemical equation for each reaction.

(a) the reaction between aluminum metal and chlorine gas
(b) the reaction between lithium metal and liquid water
(c) the reaction between gaseous hydrogen and liquid bromine

SELF-ASSESSMENT Quiz

1. What are the correct coefficients (reading from left to right) when the chemical equation is balanced?
$$\_PCl_3(l) + \_H_2O(l) \longrightarrow \_H_3PO_3(aq) + \_HCl(aq)$$
a) 1, 3, 1, 3 **b)** 1, 2, 1, 1 **c)** 1, 3, 2, 1 **d)** 3, 6, 1, 9

2. Which change is a physical change?
a) Wood burning **b)** Iron rusting
c) Dynamite exploding **d)** Gasoline evaporating

3. Which property of rubbing alcohol is a chemical property?
a) Its density (0.786 g/cm^3) **b)** Its flammability
c) Its boiling point (82.5 °C) **d)** Its melting point (−89 °C)

4. For the reaction shown here, 3.5 mol A is mixed with 5.9 mol B and 2.2 mol C. What is the limiting reactant?
$$3\,A + 2\,B + C \longrightarrow 2\,D$$
a) A **b)** B **c)** C **d)** D

5. Manganese(IV) oxide reacts with aluminum to form elemental manganese and aluminum oxide.
$$3\,MnO_2 + 4\,Al \longrightarrow 3\,Mn + 2\,Al_2O_3$$
What mass of Al is required to completely react with 25.0 g MnO_2?
a) 7.76 g Al **b)** 5.82 g Al **c)** 33.3 g Al **d)** 10.3 g Al

6. Sodium and chlorine react to form sodium chloride.
$$2\,Na(s) + Cl_2(g) \longrightarrow 2NaCl(s)$$
What is the theoretical yield of sodium chloride for the reaction of 55.0 g Na with 67.2 g Cl_2?
a) 1.40×10^2 g NaCl **b)** 111 g NaCl
c) 55.4 g NaCl **d)** 222 g NaCl

7. Sulfur and fluorine react to form sulfur hexafluoride.
$$S(s) + 3\,F_2(g) \longrightarrow SF_6(g)$$
If 50.0 g S reacts as completely as possible with 105.0 g $F_2(g)$, what mass of the excess reactant remains?
a) 20.5 g S **b)** 45.7 g F_2 **c)** 15.0 g S **d)** 36.3 g F_2

8. A reaction has a theoretical yield of 45.8 g. When the reaction is carried out, 37.2 g of the product forms. What is the percent yield?
a) 55.1% **b)** 44.8% **c)** 123% **d)** 81.2%

9. Identify the correct balanced equation for the combustion of propane (C_3H_8).
a) $C_3H_8(g) \longrightarrow 4\,H_2(g) + 3\,C(s)$
b) $C_3H_8(g) + 5\,O_2(g) \longrightarrow 4\,H_2O(g) + 3\,CO_2(g)$
c) $C_3H_8(g) + 3\,O_2(g) \longrightarrow 4\,H_2O(g) + 3\,CO_2(g)$
d) $2\,C_3H_8(g) + 9\,O_2(g) \longrightarrow 6\,H_2CO_3(g) + 2\,H_2(g)$

10. Solid potassium chlorate ($KClO_3$) decomposes into potassium chloride and oxygen gas when heated. How many moles of oxygen form when 55.8 g completely decomposes?
a) 0.455 mol O_2 **b)** 0.304 mol O_2
c) 83.7 mol O_2 **d)** 0.683 mol O_2

Answers: 1:a; 2:d; 3:b; 4:a; 5:d; 6:b; 7:a; 8:d; 9:b; 10:d

CHAPTER SUMMARY 8

REVIEW

KEY LEARNING OUTCOMES

CHAPTER OBJECTIVES	ASSESSMENT
Balancing Chemical Equations (8.3)	• Examples 8.1, 8.2, 8.3 For Practice 8.1, 8.2, 8.3 Exercises 23–34
Calculations Involving the Stoichiometry of a Reaction (8.4)	• Examples 8.4, 8.5 For Practice 8.4, 8.5 Exercises 35–44

Determining the Limiting Reactant and Calculating Theoretical and Percent Yield (8.5)	• Examples 8.6, 8.7	For Practice 8.6, 8.7	Exercises 45–60
Writing Equations for Combustion Reactions (8.6)	• Example 8.8	For Practice 8.8	Exercises 61, 62
Writing Reactions for Alkali Metal and Halogen Reactions (8.6)	• Example 8.9	For Practice 8.9	Exercises 63–68

KEY TERMS

Section 8.2
chemical change (273)
physical change (273)
chemical property (273)
physical property (273)

Section 8.3
chemical reactions (274)
chemical equation (274)
reactants (274)
products (274)

Section 8.4
stoichiometry (279)

Section 8.5
limiting reactant (283)
theoretical yield (284)
actual yield (284)
percent yield (284)

Section 8.6
combustion reaction (289)

KEY CONCEPTS

Climate Change and the Combustion of Fossil Fuels (8.1)

- Greenhouse gases warm Earth by trapping some of the sunlight that penetrates Earth's atmosphere. Increases in atmospheric carbon dioxide levels (a major greenhouse gas) has led to global warming.
- The largest source of atmospheric carbon dioxide is the burning of fossil fuels. This can be verified by reaction stoichiometry.

Chemical Change (8.2)

- Changes in matter in which composition changes are chemical changes. Changes in matter in which composition does not change are physical changes.
- We can classify the properties of matter into two types: physical and chemical. Matter displays its physical properties without changing its composition.

Writing and Balancing Chemical Equations (8.3)

- In chemistry, we represent chemical reactions with chemical equations. The substances on the left-hand side of a chemical equation are the reactants, and the substances on the right-hand side are the products.
- Chemical equations are balanced when the number of each type of atom on the left side of the equation is equal to the number on the right side.

Reaction Stoichiometry (8.4)

- Reaction stoichiometry refers to the numerical relationships between the reactants and products in a balanced chemical equation.
- Reaction stoichiometry allows us to predict, for example, the amount of product that can form from a given amount of reactant, or how much of one reactant is required to react with a given amount of another.

Limiting Reactant, Theoretical Yield, and Percent Yield (8.5)

- When a chemical reaction actually occurs, the reactants are usually not present in the exact stoichiometric ratios specified by the balanced chemical equation. The limiting reactant is the one that is available in the smallest stoichiometric quantity—it is completely consumed in the reaction, and it limits the amount of product that can be made.
- Any reactant that does not limit the amount of product is said to be in excess.
- The amount of product that can be made from the limiting reactant is the theoretical yield.
- The actual yield—always equal to or less than the theoretical yield—is the amount of product that is actually made when the reaction is carried out.
- The percentage of the theoretical yield that is actually produced is the percent yield.

Combustion, Alkali Metal, and Halogen Reactions (8.6)

- In a combustion reaction a substance reacts with oxygen—emitting heat and forming one or more oxygen-containing products. The alkali metals react with nonmetals, losing electrons in the process.
- The halogens react with many metals to form metal halides. They also react with hydrogen to form hydrogen halides and with one another to form interhalogen compounds.

KEY EQUATIONS AND RELATIONSHIPS

Mass-to-Mass Conversion: Stoichiometry (8.4)

$$\text{mass A} \longrightarrow \text{amount A (in moles)} \longrightarrow \text{amount B (in moles)} \longrightarrow \text{mass B}$$

Percent Yield (8.5)

$$\%\text{ yield} = \frac{\text{actual yield}}{\text{theoretical yield}} \times 100\%$$

EXERCISES

REVIEW QUESTIONS

1. What is the greenhouse effect?
2. Why are scientists concerned about increases in atmospheric carbon dioxide? What is the source of the increase?
3. What is the difference between a physical change and a chemical change? List some examples of each.
4. What is the difference between a physical property and a chemical property?
5. What is a balanced chemical equation?
6. Why must chemical equations be balanced?

7. What is reaction stoichiometry? What is the significance of the coefficients in a balanced chemical equation?
8. In a chemical reaction, what is the limiting reactant? What do we mean when we say a reactant is in excess?
9. In a chemical reaction, what is the theoretical yield? The percent yield?
10. We typically calculate the percent yield using the actual yield and theoretical yield in units of mass (g or kg). Would the percent yield be different if the actual yield and theoretical yield were in units of amount (moles)?
11. Where does our society get the majority of its energy?
12. What is a combustion reaction? Why are they important? Give an example.
13. Write a general equation for the reaction of an alkali metal with:
   a) a halogen b) water
14. Write a general equation for the reaction of a halogen with:
   a) a metal b) hydrogen c) another halogen

PROBLEMS BY TOPIC

Chemical and Physical Changes

15. Classify each change as physical or chemical.
   a. Natural gas burns in a stove.
   b. The liquid propane in a gas grill evaporates because the valve was left open.
   c. The liquid propane in a gas grill burns in a flame.
   d. A bicycle frame rusts on repeated exposure to air and water.
16. Classify each change as physical or chemical.
   a. Sugar burns when heated on a skillet.
   b. Sugar dissolves in water.
   c. A platinum ring becomes dull because of continued abrasion.
   d. A silver surface becomes tarnished after exposure to air for a long period of time.
17. Based on the molecular diagram, classify each change as physical or chemical.

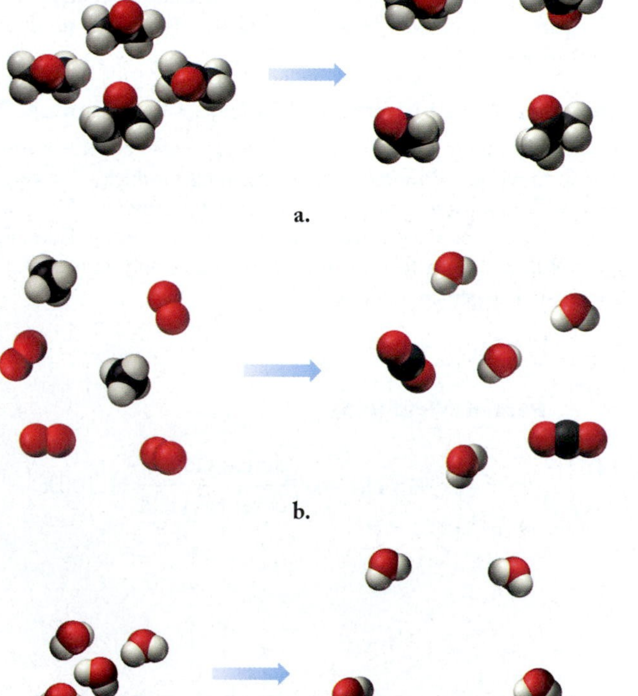

a.

b.

c.

18. Based on the molecular diagram, classify each change as physical or chemical.

a.

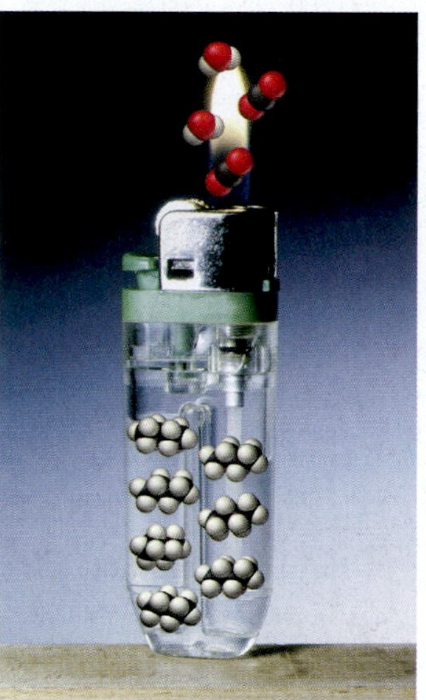

b.

c.

19. Classify each of the listed properties of isopropyl alcohol (also known as rubbing alcohol) as physical or chemical.
    a. colorless
    b. flammable
    c. liquid at room temperature
    d. density = 0.79 g/mL
    e. mixes with water

20. Classify each of the listed properties of ozone (a pollutant in the lower atmosphere, but part of a protective shield against UV light in the upper atmosphere) as physical or chemical.
    a. bluish color
    b. pungent odor
    c. very reactive
    d. decomposes on exposure to ultraviolet light
    e. gas at room temperature

21. Classify each property as physical or chemical.
    a. the tendency of ethyl alcohol to burn
    b. the shine of silver
    c. the odor of paint thinner
    d. the flammability of propane gas

22. Classify each property as physical or chemical.
    a. the boiling point of ethyl alcohol
    b. the temperature at which dry ice evaporates
    c. the tendency of iron to rust
    d. the color of gold

Writing and Balancing Chemical Equations

23. Sulfuric acid (H_2SO_4) is a component of acid rain that forms when gaseous sulfur dioxide pollutant reacts with gaseous oxygen and liquid water to form aqueous sulfuric acid. Write a balanced chemical equation for this reaction. (Note: This is a simplified representation of this reaction.)

24. Nitric acid (HNO_3) is a component of acid rain that forms when gaseous nitrogen dioxide pollutant reacts with gaseous oxygen and liquid water to form aqueous nitric acid. Write a balanced chemical equation for this reaction. (Note: This is a simplified representation of this reaction.)

25. In a popular classroom demonstration, solid sodium is added to liquid water and reacts to produce hydrogen gas and aqueous sodium hydroxide. Write a balanced chemical equation for this reaction.

26. When iron rusts, solid iron reacts with gaseous oxygen to form solid iron(III) oxide. Write a balanced chemical equation for this reaction.

27. Write a balanced chemical equation for the fermentation of sucrose ($C_{12}H_{22}O_{11}$) by yeasts in which the aqueous sugar reacts with liquid water to form aqueous ethyl alcohol (C_2H_5OH) and carbon dioxide gas.

28. Write a balanced equation for the photosynthesis reaction in which gaseous carbon dioxide and liquid water react in the presence of chlorophyll to produce aqueous glucose ($C_6H_{12}O_6$) and oxygen gas.

29. Write a balanced chemical equation for each reaction.
    a. Solid lead(II) sulfide reacts with aqueous hydrobromic acid (HBr) to form solid lead(II) bromide and dihydrogen monosulfide gas.
    b. Gaseous carbon monoxide reacts with hydrogen gas to form gaseous methane (CH_4) and liquid water.
    c. Aqueous hydrochloric acid (HCl) reacts with solid manganese(IV) oxide to form aqueous manganese(II) chloride, liquid water, and chlorine gas.
    d. Liquid pentane (C_5H_{12}) reacts with gaseous oxygen to form gaseous carbon dioxide and liquid water.

30. Write a balanced chemical equation for each reaction.
    a. Solid copper reacts with solid sulfur to form solid copper(I) sulfide.
    b. Solid iron(III) oxide reacts with hydrogen gas to form solid iron and liquid water.
    c. Sulfur dioxide gas reacts with oxygen gas to form sulfur trioxide gas.
    d. Gaseous ammonia (NH_3) reacts with gaseous oxygen to form gaseous nitrogen monoxide and gaseous water.

31. Write a balanced chemical equation for the reaction of aqueous sodium carbonate with aqueous copper(II) chloride to form solid copper(II) carbonate and aqueous sodium chloride.

32. Write a balanced chemical equation for the reaction of aqueous potassium hydroxide with aqueous iron(III) chloride to form solid iron(III) hydroxide and aqueous potassium chloride.

33. Balance each chemical equation.
    a. $CO_2(g) + CaSiO_3(s) + H_2O(l) \longrightarrow SiO_2(s) + Ca(HCO_3)_2(aq)$
    b. $Co(NO_3)_3(aq) + (NH_4)_2S(aq) \longrightarrow Co_2S_3(s) + NH_4NO_3(aq)$
    c. $Cu_2O(s) + C(s) \longrightarrow Cu(s) + CO(g)$
    d. $H_2(g) + Cl_2(g) \longrightarrow HCl(g)$

34. Balance each chemical equation.
    a. $Na_2S(aq) + Cu(NO_3)_2(aq) \longrightarrow NaNO_3(aq) + CuS(s)$
    b. $N_2H_4(l) \longrightarrow NH_3(g) + N_2(g)$
    c. $HCl(aq) + O_2(g) \longrightarrow H_2O(l) + Cl_2(g)$
    d. $FeS(s) + HCl(aq) \longrightarrow FeCl_2(aq) + H_2S(g)$

Reaction Stoichiometry

35. Consider the unbalanced equation for the combustion of hexane:

$$C_6H_{14}(g) + O_2(g) \longrightarrow CO_2(g) + H_2O(g)$$

Balance the equation and determine how many moles of O_2 are required to react completely with 7.2 moles C_6H_{14}.

36. Consider the unbalanced equation for the neutralization of acetic acid:

$$HC_2H_3O_2(aq) + Ba(OH)_2(aq) \longrightarrow H_2O(l) + Ba(C_2H_3O_2)_2(aq)$$

Balance the equation and determine how many moles of $Ba(OH)_2$ are required to completely neutralize 0.461 mole of $HC_2H_3O_2$.

37. Calculate how many moles of NO_2 form when each quantity of reactant completely reacts.

$$2\,N_2O_5(g) \longrightarrow 4\,NO_2(g) + O_2(g)$$

    a. 2.5 mol N_2O_5
    b. 6.8 mol N_2O_5
    c. 15.2 g N_2O_5
    d. 2.87 kg N_2O_5

38. Calculate how many moles of NH_3 form when each quantity of reactant completely reacts.

$$3\,N_2H_4(l) \longrightarrow 4\,NH_3(g) + N_2(g)$$

    a. 2.6 mol N_2H_4
    b. 3.55 mol N_2H_4
    c. 65.3 g N_2H_4
    d. 4.88 kg N_2H_4

39. Consider the balanced equation:

$$SiO_2(s) + 3\,C(s) \longrightarrow SiC(s) + 2\,CO(g)$$

Complete the table showing the appropriate number of moles of reactants and products. If the number of moles of a *reactant* is provided, fill in the required amount of the other reactant, as well as the moles of each product that forms. If the number of moles of a *product* is provided, fill in the required amount of each reactant to make that amount of product, as well as the amount of the other product that is made.

Mol SiO_2	Mol C	Mol SiC	Mol CO
3	____	____	____
____	6	____	____
____	____	____	10
2.8	____	____	____
____	1.55	____	____

40. Consider the balanced equation:

$$2\,N_2H_4(g) + N_2O_4(g) \longrightarrow 3\,N_2(g) + 4\,H_2O(g)$$

Complete the table showing the appropriate number of moles of reactants and products. If the number of moles of a *reactant* is provided, fill in the required amount of the other reactant, as well as the moles of each product that forms. If the number of moles of a *product* is provided, fill in the required amount of each reactant to make that amount of product, as well as the amount of the other product that is made.

Mol N_2H_4	Mol N_2O_4	Mol N_2	Mol H_2O
2	____	____	____
____	5	____	____
____	____	____	10
2.5	____	____	____
____	4.2	____	____
____	____	11.8	____

41. Hydrobromic acid (HBr) dissolves solid iron according to the reaction:

$$Fe(s) + 2\,HBr(aq) \longrightarrow FeBr_2(aq) + H_2(g)$$

What mass of HBr (in g) do you need to dissolve a 3.2-g pure iron bar on a padlock? What mass of H_2 can the complete reaction of the iron bar produce?

42. Sulfuric acid (H_2SO_4) dissolves aluminum metal according to the reaction:

$$2\,Al(s) + 3\,H_2SO_4(aq) \longrightarrow Al_2(SO_4)_3(aq) + 3\,H_2(g)$$

Suppose you want to dissolve an aluminum block with a mass of 15.2 g. What minimum mass of H_2SO_4 (in g) do you need? What mass of H_2 gas (in g) can the complete reaction of the aluminum block produce?

43. For each of the reactions, calculate the mass (in grams) of the product that forms when 3.67 g of the underlined reactant completely reacts. Assume that there is more than enough of the other reactant.

a. $\underline{Ba(s)} + Cl_2(g) \longrightarrow BaCl_2(s)$
b. $\underline{CaO(s)} + CO_2(g) \longrightarrow CaCO_3(s)$
c. $2\,\underline{Mg(s)} + O_2(g) \longrightarrow 2\,MgO(s)$
d. $4\,\underline{Al(s)} + 3\,O_2(g) \longrightarrow 2\,Al_2O_3(s)$

44. For each of the reactions, calculate the mass (in grams) of the product that forms when 15.39 g of the underlined reactant completely reacts. Assume that there is more than enough of the other reactant.

a. $2\,K(s) + \underline{Cl_2(g)} \longrightarrow 2\,KCl(s)$
b. $2\,K(s) + \underline{Br_2(l)} \longrightarrow 2\,KBr(s)$
c. $4\,Cr(s) + 3\,\underline{O_2(g)} \longrightarrow 2\,Cr_2O_3(s)$
d. $2\,\underline{Sr(s)} + O_2(g) \longrightarrow 2\,SrO(s)$

Limiting Reactant, Theoretical Yield, and Percent Yield

45. For the reaction below, determine the limiting reactant for each of the initial amounts of reactants.

$$2\,Na(s) + Br_2(g) \longrightarrow 2\,NaBr(s)$$

a. 2 mol Na, 2 mol Br_2
b. 1.8 mol Na, 1.4 mol Br_2
c. 2.5 mol Na, 1 mol Br_2
d. 12.6 mol Na, 6.9 mol Br_2

46. Find the limiting reactant for each initial amount of reactants.

$$4\,Al(s) + 3\,O_2(g) \longrightarrow 2\,Al_2O_3(s)$$

a. 1 mol Al, 1 mol O_2
b. 4 mol Al, 2.6 mol O_2
c. 16 mol Al, 13 mol O_2
d. 7.4 mol Al, 6.5 mol O_2

47. Consider the reaction:

$$4\,HCl(g) + O_2(g) \longrightarrow 2\,H_2O(g) + 2\,Cl_2(g)$$

Each molecular diagram represents an initial mixture of the reactants. How many molecules of Cl_2 would form from the reaction mixture that produces the greatest amount of products?

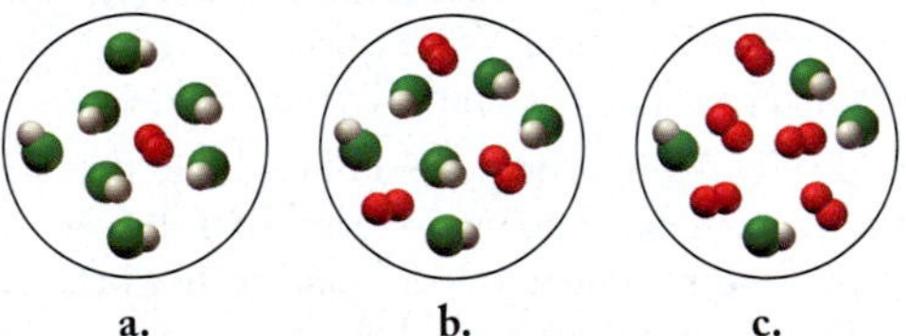

48. Consider the reaction:

$$2\,CH_3OH(g) + 3\,O_2(g) \longrightarrow 2\,CO_2(g) + 4\,H_2O(g)$$

Each of the molecular diagrams represents an initial mixture of the reactants. How many CO_2 molecules would form from the reaction mixture that produces the greatest amount of products?

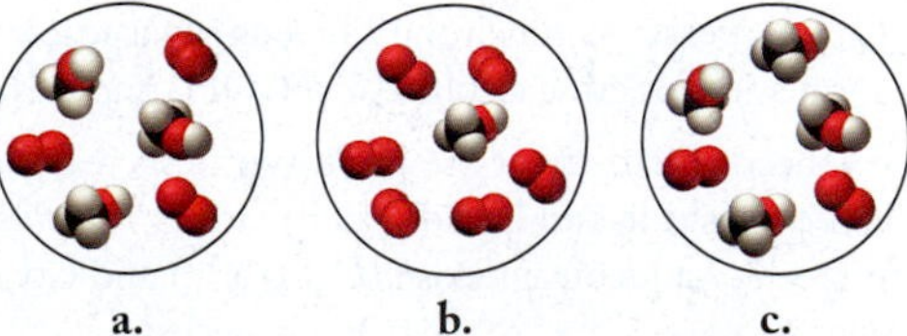

49. Calculate the theoretical yield of the product (in moles) for each initial amount of reactants.

$$Ti(s) + 2\,Cl_2(g) \longrightarrow TiCl_4(s)$$

a. 4 mol Ti, 4 mol Cl_2
b. 7 mol Ti, 17 mol Cl_2
c. 12.4 mol Ti, 18.8 mol Cl_2

50. Calculate the theoretical yield of product (in moles) for each initial amount of reactants.

$$2\,Mn(s) + 2\,O_2(g) \longrightarrow 2\,MnO_2(s)$$

a. 3 mol Mn, 3 mol O_2
b. 4 mol Mn, 7 mol O_2
c. 27.5 mol Mn, 43.8 mol O_2

51. Zinc sulfide reacts with oxygen according to the reaction:

$$2\,ZnS(s) + 3\,O_2(g) \longrightarrow 2\,ZnO(s) + 2\,SO_2(g)$$

A reaction mixture initially contains 4.2 mol ZnS and 6.8 mol O_2. Once the reaction has occurred as completely as possible, what amount (in moles) of the excess reactant remains?

52. Iron(II) sulfide reacts with hydrochloric acid according to the reaction:

$$FeS(s) + 2\,HCl(aq) \longrightarrow FeCl_2(s) + H_2S(g)$$

A reaction mixture initially contains 0.223 mol FeS and 0.652 mol HCl. Once the reaction has occurred as completely as possible, what amount (in moles) of the excess reactant is left?

53. For the reaction shown, calculate the theoretical yield of product (in grams) for each initial amount of reactants.

$$2\,Al(s) + 3\,Cl_2(g) \longrightarrow 2\,AlCl_3(s)$$

a. 2.0 g Al, 2.0 g Cl_2
b. 7.5 g Al, 24.8 g Cl_2
c. 0.235 g Al, 1.15 g Cl_2

54. For the reaction shown, calculate the theoretical yield of the product (in grams) for each initial amount of reactants.

$$Ti(s) + 2\,F_2(g) \longrightarrow TiF_4(s)$$

a. 5.0 g Ti, 5.0 g F_2
b. 2.4 g Ti, 1.6 g F_2
c. 0.233 g Ti, 0.288 g F_2

55. Iron(III) oxide reacts with carbon monoxide according to the equation:

$$Fe_2O_3(s) + 3\,CO(g) \longrightarrow 2\,Fe(s) + 3\,CO_2(g)$$

A reaction mixture initially contains 22.55 g Fe_2O_3 and 14.78 g CO. Once the reaction has occurred as completely as possible, what mass (in g) of the excess reactant remains?

56. Elemental phosphorus reacts with chlorine gas according to the equation:

$$P_4(s) + 6\,Cl_2(g) \longrightarrow 4\,PCl_3(l)$$

A reaction mixture initially contains 45.69 g P_4 and 131.3 g Cl_2. Once the reaction has occurred as completely as possible, what mass (in g) of the excess reactant remains?

57. Lead(II) ions can be removed from solution with KCl according to the reaction:

$$Pb^{2+}(aq) + 2\,KCl(aq) \longrightarrow PbCl_2(s) + 2\,K^+(aq)$$

When 28.5 g KCl is added to a solution containing 25.7 g Pb^{2+}, a $PbCl_2(s)$ forms. The solid is filtered and dried and found to have a mass of 29.4 g. Determine the limiting reactant, theoretical yield of $PbCl_2$, and percent yield for the reaction.

58. Magnesium oxide can be made by heating magnesium metal in the presence of oxygen. The balanced equation for the reaction is:

$$2\,Mg(s) + O_2(g) \longrightarrow 2\,MgO(s)$$

When 10.1 g of Mg reacts with 10.5 g O_2, 11.9 g MgO forms. Determine the limiting reactant, theoretical yield, and percent yield for the reaction.

59. Urea (CH_4N_2O) is a common fertilizer that is synthesized by the reaction of ammonia (NH_3) with carbon dioxide:

$$2\,NH_3(aq) + CO_2(aq) \longrightarrow CH_4N_2O(aq) + H_2O(l)$$

In an industrial synthesis of urea, a chemist combines 136.4 kg of ammonia with 211.4 kg of carbon dioxide and obtains 168.4 kg of urea. Determine the limiting reactant, theoretical yield of urea, and percent yield for the reaction.

60. Many computer chips are manufactured from silicon, which occurs in nature as SiO_2. When SiO_2 is heated to melting, it reacts with solid carbon to form liquid silicon and carbon monoxide gas. In an industrial preparation of silicon, 155.8 kg of SiO_2 reacts with 78.3 kg of carbon to produce 66.1 kg of silicon. Determine the limiting reactant, theoretical yield, and percent yield for the reaction.

Combustion, Alakali Metal, and Halogen Reactions

61. Complete and balance each combustion reaction equation.

a. $S(s) + O_2(g) \longrightarrow$
b. $C_3H_6(g) + O_2(g) \longrightarrow$
c. $Ca(s) + O_2(g) \longrightarrow$
d. $C_5H_{12}S(l) + O_2(g) \longrightarrow$

62. Complete and balance each combustion reaction equation.

a. $C_4H_6(g) + O_2(g) \longrightarrow$
b. $C(s) + O_2(g) \longrightarrow$
c. $CS_2(s) + O_2(g) \longrightarrow$
d. $C_3H_8O(l) + O_2(g) \longrightarrow$

63. Write a balanced chemical equation for the reaction of solid strontium with iodine gas.

64. Based on the ionization energies of the alkali metals (see Section 4.7), which alkali metal would you expect to undergo the most exothermic reaction with chlorine gas? Write a balanced chemical equation for the reaction.

65. Write a balanced chemical equation for the reaction of solid lithium with liquid water.

66. Write a balanced chemical equation for the reaction of solid potassium with liquid water.

67. Write a balanced equation for the reaction of hydrogen gas with bromine gas.

68. Write a balanced equation for the reaction of chlorine gas with fluorine gas.

CUMULATIVE PROBLEMS

69. The combustion of gasoline produces carbon dioxide and water. Assume gasoline to be pure octane (C_8H_{18}) and calculate the mass (in kg) of carbon dioxide that is added to the atmosphere per 1.0 kg of octane burned. (*Hint*: Begin by writing a balanced equation for the combustion reaction.)

70. Many home barbeques are fueled with propane gas (C_3H_8). What mass of carbon dioxide (in kg) forms upon the complete combustion of 18.9 L of propane (approximate contents of one 5-gallon tank)? Assume that the density of the liquid propane in the tank is 0.621 g/mL. (*Hint*: Begin by writing a balanced equation for the combustion reaction.)

71. Aspirin can be made in the laboratory by reacting acetic anhydride ($C_4H_6O_3$) with salicylic acid ($C_7H_6O_3$) to form aspirin ($C_9H_8O_4$) and acetic acid ($C_2H_4O_2$). The balanced equation is:

$$C_4H_6O_3 + C_7H_6O_3 \longrightarrow C_9H_8O_4 + C_2H_4O_2$$

In a laboratory synthesis, a student begins with 3.00 mL of acetic anhydride (density = 1.08 g/mL) and 1.25 g of salicylic acid. Once the reaction is complete, the student collects 1.22 g of aspirin. Determine the limiting reactant, theoretical yield of aspirin, and percent yield for the reaction.

72. The combustion of liquid ethanol (C_2H_5OH) produces carbon dioxide and water. After 4.62 mL of ethanol (density = 0.789 g/mL) burns in the presence of 15.55 g of oxygen gas, 3.72 mL of water (density = 1.00 g/mL) is collected. Determine the limiting reactant, theoretical yield of H_2O, and percent yield for the reaction. (*Hint*: Write a balanced equation for the combustion of ethanol.)

73. A loud classroom demonstration involves igniting a hydrogen-filled balloon. The hydrogen within the balloon reacts explosively with oxygen in the air to form water. If the balloon is filled with a mixture of hydrogen and oxygen, the explosion is even louder than if the balloon is filled only with hydrogen—the intensity of the explosion depends on the relative amounts of oxygen and hydrogen within the balloon. Look at the molecular views representing different amounts of hydrogen and oxygen in four different balloons. Based on the balanced chemical equation, which balloon will make the loudest explosion?

a.

b.

c.

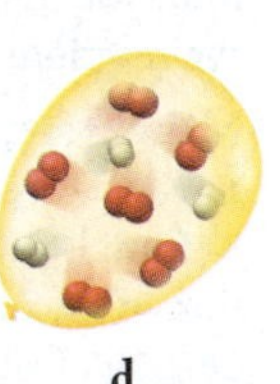
d.

74. The nitrogen in sodium nitrate and in ammonium sulfate is available to plants as fertilizer. Which is the more economical source of nitrogen, a fertilizer containing 30.0% sodium nitrate by weight and costing \$9.00 per 100 lb or one containing 20.0% ammonium sulfate by weight and costing \$8.10 per 100 lb?

75. The reaction of NH_3 and O_2 forms NO and water. The NO can be used to convert P_4 to P_4O_6, forming N_2 in the process. The P_4O_6 can be treated with water to form H_3PO_3, which forms PH_3 and H_3PO_4 when heated. Find the mass of PH_3 that forms from the reaction of 1.00 g NH_3.

76. An important reaction that takes place in a blast furnace during the production of iron is the formation of iron metal and CO_2 from Fe_2O_3 and CO. Determine the mass of Fe_2O_3 required to form 910 kg of iron. Determine the amount of CO_2 that forms in this process.

77. A liquid fuel mixture contains 30.35% hexane (C_6H_{14}) and 15.85% heptane (C_7H_{16}). The remainder is octane (C_8H_{18}). What maximum mass of carbon dioxide is produced by the complete combustion of 10.0 kg of this fuel mixture?

78. Titanium occurs in the magnetic mineral ilmenite ($FeTiO_3$), which is often found mixed with sand. The ilmenite can be separated from the sand with magnets. The titanium can then be extracted from the ilmenite by the following set of reactions:

$$FeTiO_3(s) + 3\,Cl_2(g) + 3\,C(s) \longrightarrow 3\,CO(g) + FeCl_2(s) + TiCl_4(g)$$

$$TiCl_4(g) + 2\,Mg(s) \longrightarrow 2\,MgCl_2(l) + Ti(s)$$

Suppose that an ilmenite–sand mixture contains 22.8% ilmenite by mass and that the first reaction is carried out with a 90.8% yield. If the second reaction is carried out with an 85.9% yield, what mass of titanium can we obtain from 1.00 kg of the ilmenite–sand mixture?

CHALLENGE PROBLEMS

79. A mixture of C_3H_8 and C_2H_2 has a mass of 2.0 g. It is burned in excess O_2 to form a mixture of water and carbon dioxide that contains 1.5 times as many moles of CO_2 as of water. Find the mass of C_2H_2 in the original mixture.

80. A mixture of 20.6 g of P and 79.4 g Cl_2 reacts completely to form PCl_3 and PCl_5, which are the only products. Determine the mass of PCl_3 that forms.

81. Lead poisoning is a serious condition resulting from the ingestion of lead in food, water, or other environmental sources. It affects the central nervous system, leading to a variety of symptoms such as distractibility, lethargy, and loss of motor coordination. Lead poisoning is treated with chelating agents, substances that bind to metal ions, allowing them to be eliminated in the urine. A modern chelating agent used for this purpose is succimer ($C_4H_6O_4S_2$). Suppose you are trying to determine the appropriate dose for succimer treatment of lead poisoning. What minimum mass of succimer (in mg) is needed to bind all of the lead in a patient's bloodstream? Assume that patient blood lead levels are 45 μg/dL, that total blood volume is 5.0 L, and that one mole of succimer binds one mole of lead.

82. A particular kind of emergency breathing apparatus—often placed in mines, caves, or other places where oxygen might become depleted or where the air might become poisoned—works via the following chemical reaction:

$$4\,KO_2(s) + 2\,CO_2(g) \longrightarrow 2\,K_2CO_3(s) + 3\,O_2(g)$$

Notice that the reaction produces O_2, which can be breathed, and absorbs CO_2, a product of respiration. Suppose you work for a company interested in producing a self-rescue breathing apparatus (based on the above reaction) that would allow the user to survive for 10 minutes in an emergency situation. What are the important chemical considerations in designing such a unit? Estimate how much KO_2 would be required for the apparatus. (Find any necessary additional information—such as human breathing rates—from appropriate sources. Assume that normal air is 20% oxygen.)

83. Metallic aluminum reacts with MnO_2 at elevated temperatures to form manganese metal and aluminum oxide. A mixture of the two reactants is 67.2% mole percent Al. Determine the theoretical yield (in grams) of manganese from the reaction of 250 g of this mixture.

84. Hydrolysis of the compound B_5H_9 forms boric acid, H_3BO_3. Fusion of boric acid with sodium oxide forms a borate salt, $Na_2B_4O_7$. Without writing complete equations, find the mass (in grams) of B_5H_9 required to form 151 g of the borate salt by this reaction sequence.

CONCEPTUAL PROBLEMS

85. Consider the reaction:

$$4\ K(s) + O_2(g) \longrightarrow 2\ K_2O(s)$$

The molar mass of K is 39.09 g/mol and that of O_2 is 32.00 g/mol. Without doing any calculations, choose the conditions under which potassium is the limiting reactant and explain your reasoning.

a. 170 g K, 31 g O_2 **b.** 16 g K, 2.5 g O_2
c. 165 kg K, 28 kg O_2 **d.** 1.5 g K, 0.38 g O_2

86. Consider the reaction:

$$2\ NO(g) + 5\ H_2(g) \longrightarrow 2\ NH_3(g) + 2\ H_2O(g)$$

A reaction mixture initially contains 5 moles of NO and 10 moles of H_2. Without doing any calculations, determine which set of amounts best represents the mixture after the reactants have reacted as completely as possible. Explain your reasoning.

a. 1 mol NO, 0 mol H_2, 4 mol NH_3, 4 mol H_2O
b. 0 mol NO, 1 mol H_2, 5 mol NH_3, 5 mol H_2O
c. 3 mol NO, 5 mol H_2, 2 mol NH_3, 2 mol H_2O
d. 0 mol NO, 0 mol H_2, 4 mol NH_3, 4 mol H_2O

87. Consider the reaction:

$$2\ N_2H_4(g) + N_2O_4(g) \longrightarrow 3\ N_2(g) + 4\ H_2O(g)$$

Consider also this representation of an initial mixture of N_2H_4 and N_2O_4:

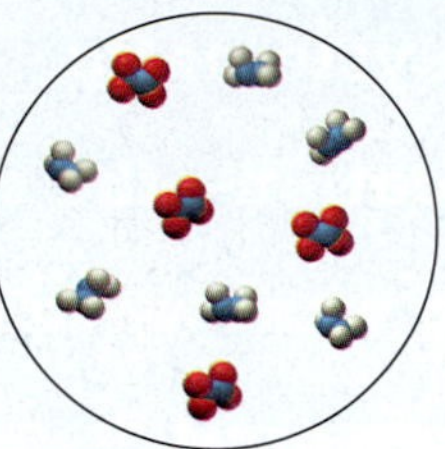

Which diagram best represents the reaction mixture after the reactants have reacted as completely as possible?

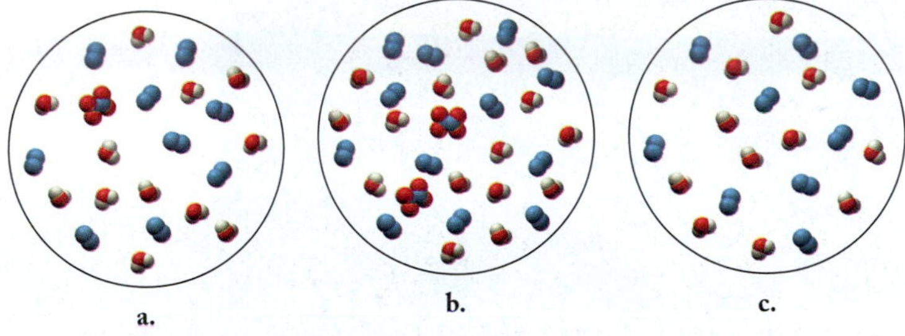

ANSWERS TO CONCEPTUAL CONNECTIONS

Cc 8.1 **(b)** The burning of lamp oil is like the burning of gasoline. The lamp oil is transformed into other substances (primarily carbon dioxide and water). The evaporation of rubbing alcohol and the forming of frost are both changes of state and therefore physical changes.

Cc 8.2 View **(a)** best represents the water after vaporization. Vaporization is a physical change, so the molecules remain the same before and after the change.

Cc 8.3 Both **(a)** and **(d)** are correct. When the number of atoms of each type is balanced, the sum of the masses of the substances involved will be the same on both sides of the equation. Since molecules change during a chemical reaction, their number is not the same on both sides, nor is the number of moles necessarily the same.

Cc 8.4 **(c)** Because each O_2 molecule reacts with 4 Na atoms, 12 Na atoms are required to react with 3 O_2 molecules.

Cc 8.5

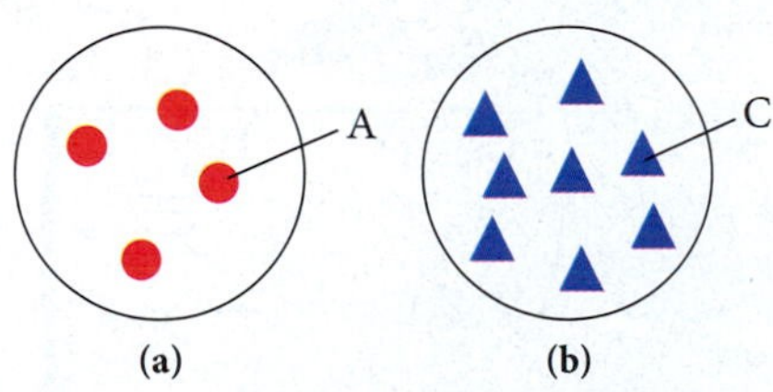

Cc 8.6 **(c)** Nitrogen is the limiting reactant, and there is enough nitrogen to make 4 NH_3 molecules. Hydrogen is in excess, and 2 hydrogen molecules remain after the reactants have reacted as completely as possible.

Cc 8.7 The limiting reactant is the H_2O, which is completely consumed. The 1 mol of H_2O requires 3 mol of NO_2 to completely react; therefore 2 mol NO_2 remain after the reaction is complete.

CHAPTER 9

> **"Science may be described as the art of systematic oversimplification—the art of discerning what we may with advantage omit."**
>
> **—Karl Popper (1902–1994)**

The spherical olive is a creation of the style of cuisine known as molecular gastronomy.

Introduction to Solutions and Aqueous Reactions

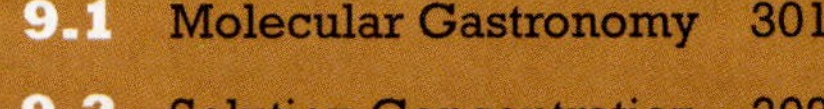

9.1 Molecular Gastronomy 301

9.2 Solution Concentration 302

9.3 Solution Stoichiometry 307

9.4 Types of Aqueous Solutions and Solubility 308

9.5 Precipitation Reactions 313

9.6 Representing Aqueous Reactions: Molecular, Ionic, and Complete Ionic Equations 318

9.7 Acid–Base Reactions 319

9.8 Gas-Evolution Reactions 327

9.9 Oxidation–Reduction Reactions 328

Key Learning Outcomes 335

YOU HAVE NOW LEARNED how to represent a chemical reaction, the process by which substances transform to other substances. In this chapter, we turn our attention to chemical reactions that occur with water as the medium. You have probably witnessed many of these types of reactions in your daily life because they are so common. Have you ever mixed baking soda with vinegar and observed the subsequent bubbling, or noticed the hard water deposits that form on plumbing fixtures? These reactions—and many others, including those that occur within the watery environment of living cells—are aqueous chemical reactions, the subject of this chapter.

9.1 Molecular Gastronomy

On July 30, 2011, one of the most famous restaurants in the world—which boasted over one million reservation requests per year for a mere 8000 available spots—shut down. (It is scheduled to reopen in 2014 as a culinary institute.) The restaurant, located on the coast north of Barcelona, Spain, was called elBulli, and it made the combination of chemistry and cooking famous. The style of cuisine practiced by the chefs at elBulli, which relies heavily on chemical processes, is often called *molecular gastronomy*.

A common chemical reaction in molecular gastronomy is precipitation. In a *precipitation reaction*, two *solutions*—homogeneous mixtures often containing a solid dissolved in a liquid—are mixed. Upon mixing, a solid forms. For example, when you mix a solution of lead (II) nitrate with postassium iodide, a brilliant yellow solid forms (see Section 9.4). The solid is lead(II) iodide.

In molecular gastronomy, chefs use a similar precipitation reaction—called spherification—to encapsulate liquids. One of the most famous creations at elBulli is the spherical olive (Figure 9.1 ▶). To make a spherical olive, chefs take the olive juice from real olives and mix it

▲ **FIGURE 9.1 The Spherical Olive** The spherical olive is made by precipitating an encapsulating layer around olive juice.

with a calcium salt (such as calcium chloride), which dissolves in the olive juice. They then carefully pour the olive juice into a bath of sodium alginate. Sodium alginate is a sodium salt that dissolves into water, resulting in the presence of alginate ions. When the calcium ions in the olive juice encounter the alginate ions in the bath, a precipitation reaction occurs. In this case, the precipitation reaction forms an encapsulating sphere around the olive juice. The result is a spherical, edible "olive" that pops in the mouth and releases its juice.

In this chapter, we explore solutions, focusing especially on *aqueous* solutions (solutions in which one component is water). The olive juice and calcium chloride mixture just discussed is an example of an aqueous solution. Other common aqueous solutions include seawater, vinegar, and the watery environment within biological cells. We will also explore the chemical reactions that occur within solutions, such as precipitation reactions, which have many common applications in addition to encapsulating the spherical olive.

9.2 Solution Concentration

A homogeneous mixture of two substances—such as salt and water—is a **solution**. The major component of the mixture is the **solvent**, and the minor component is the **solute**. An **aqueous solution** is one in which water acts as the solvent. In this section, we examine how to quantify the concentration of a solution (the amount of solute relative to solvent), how to carry out calculations involving the concentration, and how to calculate the effects of diluting a solution.

Quantifying Solution Concentration

The amount of solute in a solution is variable. For example, we can add just a little salt to water to make a **dilute solution**, one that contains a small amount of solute relative to the solvent, or we can add a lot of salt to water to make a **concentrated solution**, one that contains a large amount of solute relative to the solvent (Figure 9.2 ▼). A common way to express solution concentration is **molarity (M)**, the amount of solute (in moles) divided by the volume of solution (in liters).

$$\text{Molarity (M)} = \frac{\text{amount of solute (in mol)}}{\text{volume of solution (in L)}}$$

Notice that molarity is a ratio of the amount of solute per liter of *solution*, not per liter of solvent. To make an aqueous solution of a specified molarity, we usually put the solute into a flask and then add water to reach the desired volume of solution. For example, to make 1 L of a 1 M NaCl solution, we add 1 mol of NaCl to a flask and then add enough water to make 1 L of solution (Figure 9.3 ▼).

▼ **FIGURE 9.2 Concentrated and Dilute Solutions** A concentrated solution contains a relatively large amount of solute relative to solvent. A dilute solution contains a relatively small amount of solute relative to solvent.

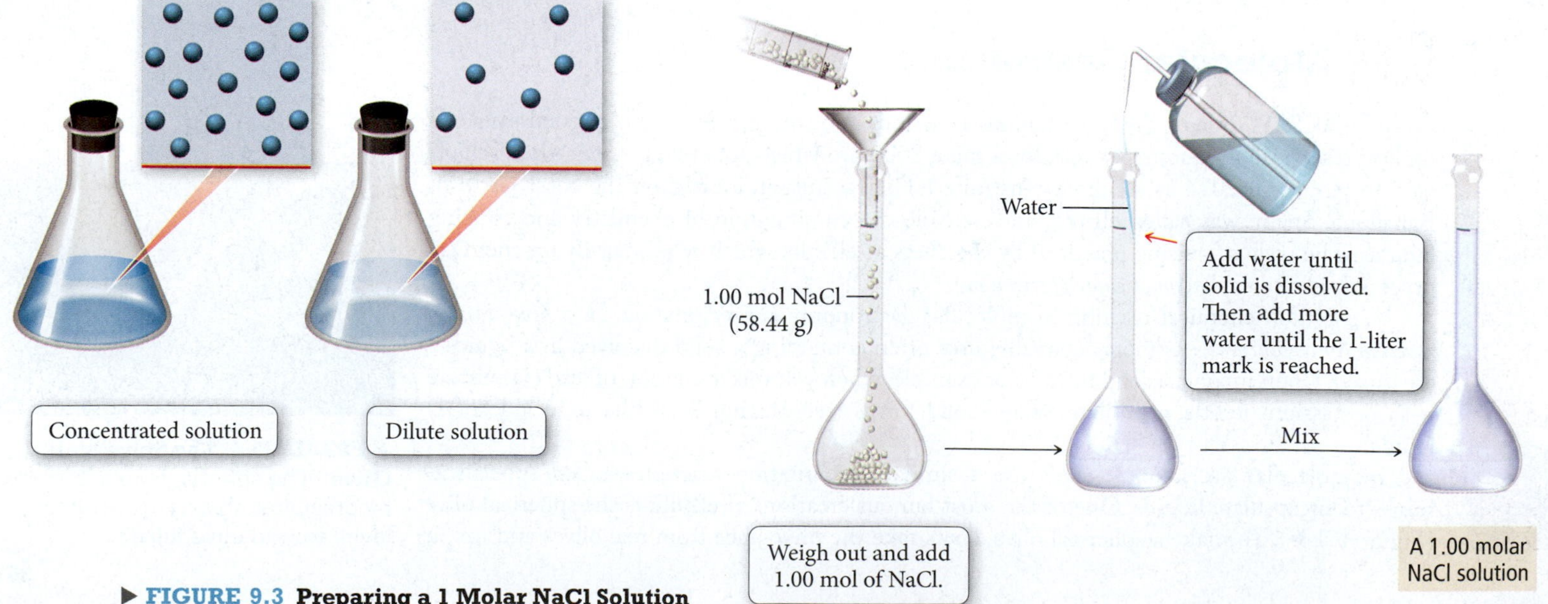
Concentrated and Dilute Solutions
Concentrated solution
Dilute solution
Preparing a Solution of Specified Concentration
1.00 mol NaCl
(58.44 g)
Weigh out and add
1.00 mol of NaCl.
Water
Add water until
solid is dissolved.
Then add more
water until the 1-liter
mark is reached.
Mix
A 1.00 molar
NaCl solution

▶ **FIGURE 9.3 Preparing a 1 Molar NaCl Solution**

We *do not* combine 1 mol of NaCl with 1 L of water because the resulting solution would have a total volume exceeding 1 L and therefore a molarity of less than 1 M. To calculate molarity, we divide the amount of the solute in moles by the volume of the solution (solute *and* solvent) in liters, as shown in Example 9.1.

EXAMPLE 9.1
Calculating Solution Concentration

If you dissolve 25.5 g KBr in enough water to make 1.75 L of solution, what is the molarity of the solution?

SORT You are given the mass of KBr and the volume of a solution and asked to find its molarity.

GIVEN: 25.5 g KBr, 1.75 L of solution

FIND: molarity (M)

STRATEGIZE When formulating the conceptual plan, think about the definition of molarity: the amount of solute *in moles* per *liter of solution.*

You are given the mass of KBr, so first use the molar mass of KBr to convert from g KBr to mol KBr.

CONCEPTUAL PLAN

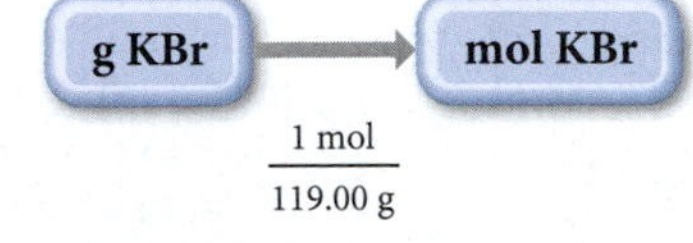

$$\frac{1\text{ mol}}{119.00\text{ g}}$$

Then use the number of moles of KBr and liters of solution to find the molarity.

mol KBr, L solution → Molarity

$$\text{Molarity (M)} = \frac{\text{amount of solute (in mol)}}{\text{volume of solution (in L)}}$$

RELATIONSHIPS USED

molar mass of KBr = 119.00 g/mol

SOLVE Follow the conceptual plan. Begin with g KBr and convert to mol KBr; then use mol KBr and L solution to calculate molarity.

SOLUTION

$$25.5\text{ g KBr} \times \frac{1\text{ mol KBr}}{119.00\text{ g KBr}} = 0.21\underline{4}29\text{ mol KBr}$$

$$\text{molarity (M)} = \frac{\text{amount of solute (in mol)}}{\text{volume of solution (in L)}}$$

$$= \frac{0.21\underline{4}29\text{ mol KBr}}{1.75\text{ L solution}}$$

$$= 0.122\text{ M}$$

CHECK The units of the answer (M) are correct. The magnitude is reasonable since common solutions range in concentration from 0 to about 18 M. Concentrations significantly above 18 M are suspect and should be double-checked.

FOR PRACTICE 9.1

Calculate the molarity of a solution made by adding 45.4 g of $NaNO_3$ to a flask and dissolving it with water to create a total volume of 2.50 L.

FOR MORE PRACTICE 9.1

What mass of KBr (in grams) do you need to make 250.0 mL of a 1.50 M KBr solution?

Using Molarity in Calculations

We can use the molarity of a solution as a conversion factor between moles of the solute and liters of the solution. For example, a 0.500 M NaCl solution contains 0.500 mol NaCl for every liter of solution.

$$\frac{0.500\text{ mol NaCl}}{\text{L solution}}$$ *converts*

This conversion factor converts from L solution to mol NaCl. If we want to go the other way, we invert the conversion factor.

$$\frac{\text{L solution}}{0.500 \text{ mol NaCl}} \quad \textit{converts} \quad \text{mol NaCl} \rightarrow \text{L solution}$$

Example 9.2 illustrates how to use molarity in this way.

EXAMPLE 9.2 Using Molarity in Calculations

How many liters of a 0.125 M NaOH solution contain 0.255 mol of NaOH?

SORT You are given the concentration of an NaOH solution. You are asked to find the volume of the solution that contains a given amount (in moles) of NaOH.

GIVEN: 0.125 M NaOH solution, 0.255 mol NaOH
FIND: volume of NaOH solution (in L)

STRATEGIZE The conceptual plan begins with mol NaOH and shows the conversion to L of solution using molarity as a conversion factor.

CONCEPTUAL PLAN

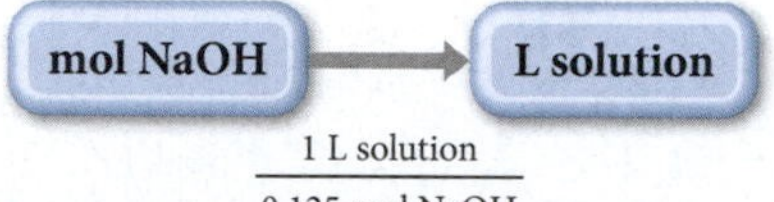

RELATIONSHIPS USED

$$0.125 \text{ M NaOH} = \frac{0.125 \text{ mol NaOH}}{1 \text{ L solution}}$$

SOLVE Follow the conceptual plan. Begin with mol NaOH and convert to L solution.

SOLUTION

$$0.255 \cancel{\text{mol NaOH}} \times \frac{1 \text{ L solution}}{0.125 \cancel{\text{mol NaOH}}} = 2.04 \text{ L solution}$$

CHECK The units of the answer (L) are correct. The magnitude is reasonable because the solution contains 0.125 mol per liter. Therefore, roughly 2 L contains the given amount of moles (0.255 mol).

FOR PRACTICE 9.2
How many grams of sucrose ($C_{12}H_{22}O_{11}$) are in 1.55 L of 0.758 M sucrose solution?

FOR MORE PRACTICE 9.2
How many mL of a 0.155 M KCl solution contains 2.55 g KCl?

9.1 Cc Conceptual Connection — Solutions

If we dissolve 25 grams of salt in 251 grams of water, what is the mass of the resulting solution?

(a) 251 g **(b)** 276 g **(c)** 226 g

When diluting acids, always add the concentrated acid to the water. Never add water to concentrated acid solutions, as the heat generated may cause the concentrated acid to splatter and burn your skin.

Solution Dilution

To save space in storerooms, laboratories often store solutions in concentrated forms called **stock solutions**. For example, hydrochloric acid is frequently stored as a 12 M stock solution. However, many lab procedures call for much less concentrated hydrochloric acid solutions, so we must dilute the stock solution to the required concentration. How do we know how much of the stock solution to use? When you dilute a solution by adding more solvent, the number of moles of solute do not change; the same number of moles are simply contained in a greater volume, which changes the concentration. Therefore, the most direct way to solve dilution problems is to use the following dilution equation:

$$M_1V_1 = M_2V_2 \qquad [9.1]$$

where M_1 and V_1 are the molarity and volume of the initial concentrated solution, and M_2 and V_2 are the molarity and volume of the final diluted solution. This equation works because the molarity multiplied by the volume gives the number of moles of solute, which does not change upon dilution.

$$M_1V_1 = M_2V_2$$
$$\text{mol}_1 = \text{mol}_2$$

For example, suppose a procedure for spherification (see Section 9.1) calls for 3.00 L of a 0.500 M $CaCl_2$ solution. How can we prepare this solution from a 10.0 M stock solution? We solve Equation 9.1 for V_1, the volume of the stock solution required for the dilution, and then substitute in the correct values to calculate it.

$$\begin{aligned} M_1V_1 &= M_2V_2 \\ V_1 &= \frac{M_2V_2}{M_1} \\ &= \frac{0.500\ \cancel{\text{mol/L}} \times 3.00\ \text{L}}{10.0\ \cancel{\text{mol/L}}} \\ &= 0.150\ \text{L} \end{aligned}$$

Consequently, we make the solution by diluting 0.150 L of the stock solution to a total volume of 3.00 L (V_2). The resulting solution will be 0.500 M $CaCl_2$ (Figure 9.4 ▼).

Diluting a Solution

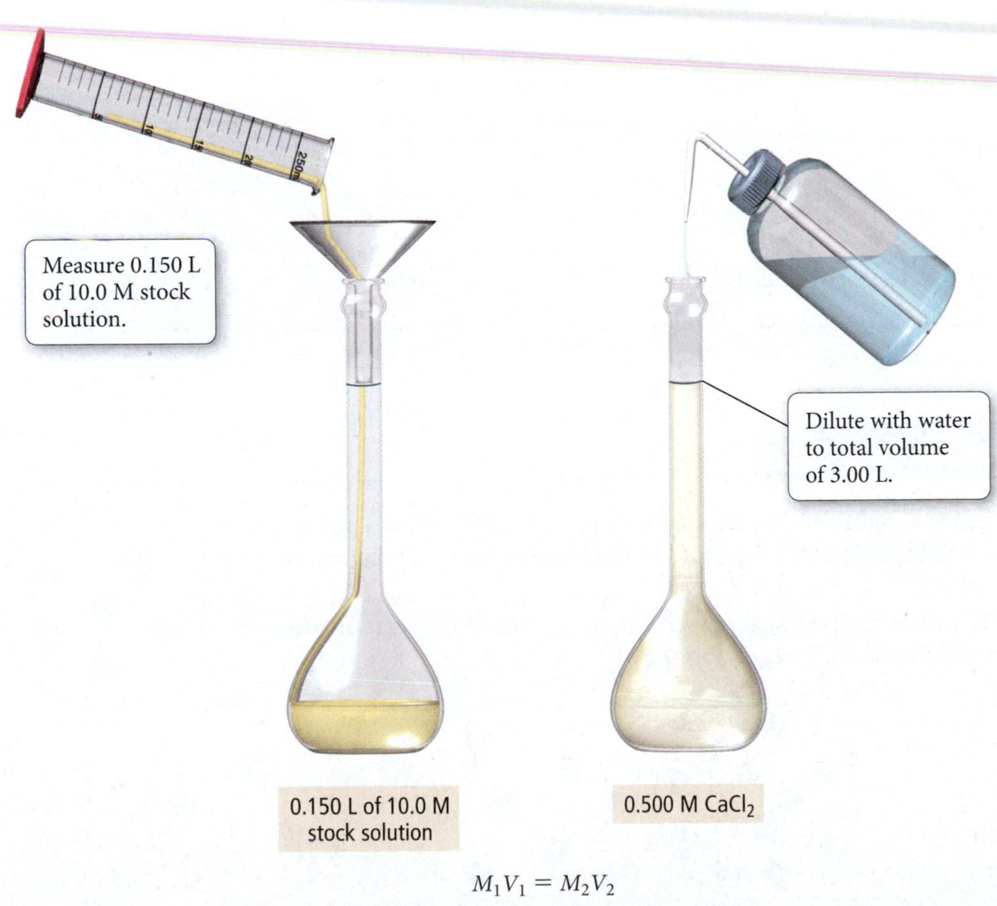
Measure 0.150 L of 10.0 M stock solution.
Dilute with water to total volume of 3.00 L.
0.150 L of 10.0 M stock solution
0.500 M CaCl₂

$$M_1V_1 = M_2V_2$$
$$\frac{10.0\ \text{mol}}{\cancel{\text{L}}} \times 0.150\ \cancel{\text{L}} = \frac{0.500\ \text{mol}}{\cancel{\text{L}}} \times 3.00\ \cancel{\text{L}}$$
$$1.50\ \text{mol} = 1.50\ \text{mol}$$

◀ **FIGURE 9.4 Preparing 3.00 L of 0.500 M $CaCl_2$ from a 10.0 M Stock Solution**

EXAMPLE 9.3 Solution Dilution

To what volume should you dilute 0.200 L of a 15.0 M NaOH solution to obtain a 3.00 M NaOH solution?

SORT You are given the initial volume, initial concentration, and final concentration of a solution, and you need to determine the final volume.

GIVEN: $V_1 = 0.200$ L
$M_1 = 15.0$ M
$M_2 = 3.00$ M

FIND: V_2

STRATEGIZE Equation 9.1 relates the initial and final volumes and concentrations for solution dilution problems. You are asked to find V_2. The other quantities (V_1, M_1, and M_2) are all given in the problem.

CONCEPTUAL PLAN

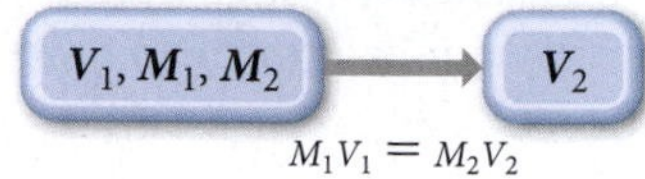

$M_1V_1 = M_2V_2$

RELATIONSHIPS USED

$M_1V_1 = M_2V_2$

SOLVE Begin with the solution dilution equation and solve it for V_2.

Substitute in the required quantities and calculate V_2.

Make the solution by diluting 0.200 L of the stock solution to a total volume of 1.00 L (V_2). The resulting solution has a concentration of 3.00 M.

SOLUTION

$$M_1V_1 = M_2V_2$$

$$V_2 = \frac{M_1V_1}{M_2}$$

$$= \frac{15.0 \text{ mol/L} \times 0.200 \text{ L}}{3.00 \text{ mol/L}}$$

$$= 1.00 \text{ L}$$

CHECK The final units (L) are correct. The magnitude of the answer is reasonable because the solution is diluted from 15.0 M to 3.00 M, a factor of five. Therefore the volume should increase by a factor of five.

FOR PRACTICE 9.3
To what volume (in mL) should you dilute 100.0 mL of a 5.00 M $CaCl_2$ solution to obtain a 0.750 M $CaCl_2$ solution?

FOR MORE PRACTICE 9.3
What volume of a 6.00 M $NaNO_3$ solution should you use to make 0.525 L of a 1.20 M $NaNO_3$ solution?

9.2 Cc Conceptual Connection — Solution Dilution

The image at right represents a small volume within 500 mL of aqueous ethanol (CH_3CH_2OH) solution. (The water molecules have been omitted for clarity.)

Which image best represents the same volume of the solution after we add an additional 500 mL of water?

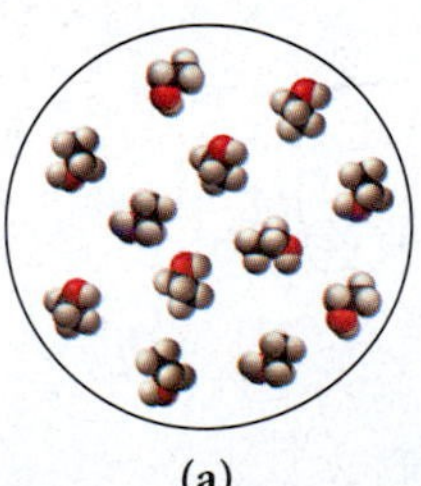
(a)

(b)

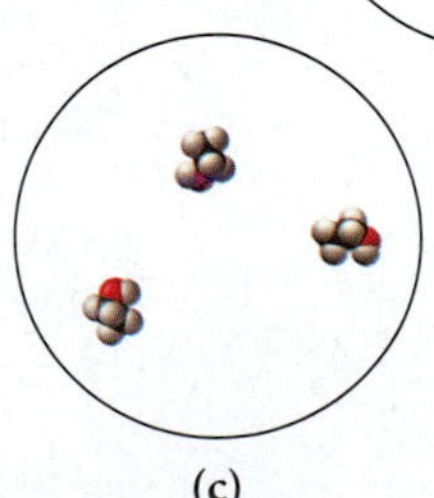
(c)

9.3 Solution Stoichiometry

In Section 8.4 we discussed how to use the coefficients in chemical equations as conversion factors between the amounts of reactants (in moles) and the amounts of products (in moles). In aqueous reactions, quantities of reactants and products are often specified in terms of volumes and concentrations. We can use the volume and concentration of a reactant or product to calculate its amount in moles. We can then use the stoichiometric coefficients in the chemical equation to convert to the amount in moles of another reactant or product. The general conceptual plan for these kinds of calculations begins with the volume of a reactant or product:

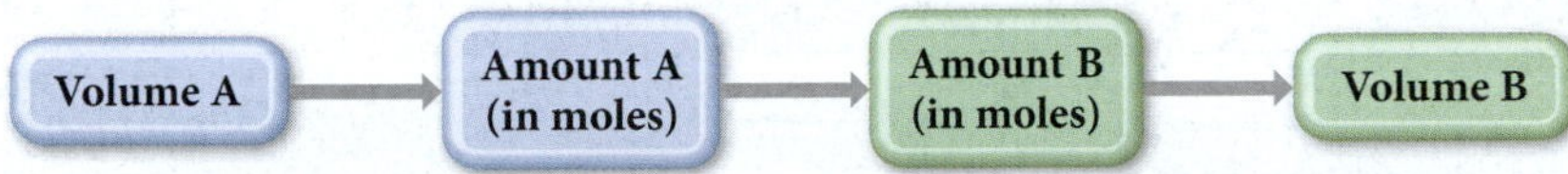

We make the conversions between solution volumes and amounts in moles of solute using the molarities of the solutions. We make the conversions between amounts in moles of A and B using the stoichiometric coefficients from the balanced chemical equation. Example 9.4 demonstrates solution stoichiometry.

EXAMPLE 9.4 Solution Stoichiometry

What volume (in L) of 0.150 M KCl solution will completely react with 0.150 L of a 0.175 M $Pb(NO_3)_2$ solution according to this balanced chemical equation?

$$2\,KCl(aq) + Pb(NO_3)_2(aq) \longrightarrow PbCl_2(s) + 2\,KNO_3(aq)$$

SORT You are given the volume and concentration of a $Pb(NO_3)_2$ solution. You are asked to find the volume of KCl solution (of a given concentration) required to react with it.

GIVEN: 0.150 L of $Pb(NO_3)_2$ solution, 0.175 M $Pb(NO_3)_2$ solution, 0.150 M KCl solution

FIND: volume KCl solution (in L)

STRATEGIZE The conceptual plan has the form: volume A $\longrightarrow$ amount A (in moles) $\longrightarrow$ amount B (in moles) $\longrightarrow$ volume B. Use the molar concentrations of the KCl and $Pb(NO_3)_2$ solutions as conversion factors between the number of moles of reactants in these solutions and their volumes. Use the stoichiometric coefficients from the balanced equation to convert between number of moles of $Pb(NO_3)_2$ and number of moles of KCl.

CONCEPTUAL PLAN

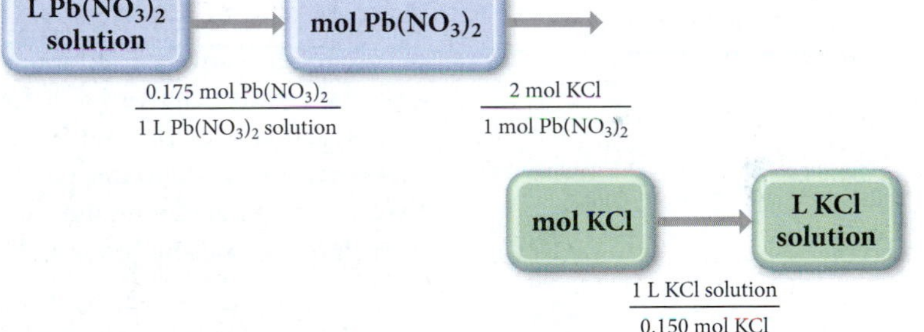

RELATIONSHIPS USED

$$M\,[Pb(NO_3)_2] = \frac{0.175\text{ mol }Pb(NO_3)_2}{1\text{ L }Pb(NO_3)_2\text{ solution}}$$

2 mol KCl : 1 mol $Pb(NO_3)_2$ (from chemical equation)

$$M\,[KCl] = \frac{0.150\text{ mol KCl}}{1\text{ L KCl solution}}$$

SOLVE Begin with L $Pb(NO_3)_2$ solution and follow the conceptual plan to arrive at L KCl solution.

SOLUTION

$$0.150\text{ L }\cancel{Pb(NO_3)_2\text{ solution}} \times \frac{0.175\ \cancel{\text{mol }Pb(NO_3)_2}}{1\text{ L }\cancel{Pb(NO_3)_2\text{ solution}}} \times \frac{2\ \cancel{\text{mol KCl}}}{1\ \cancel{\text{mol }Pb(NO_3)_2}} \times \frac{1\text{ L KCl solution}}{0.150\ \cancel{\text{mol KCl}}} = 0.350\text{ L KCl solution}$$

—*Continued on the next page*

Continued from the previous page—

CHECK The final units (L KCl solution) are correct. The magnitude (0.350 L) is reasonable because the reaction stoichiometry requires 2 mol of KCl per mole of $Pb(NO_3)_2$. Since the concentrations of the two solutions are not very different (0.150 M compared to 0.175 M), the volume of KCl required is roughly two times the 0.150 L of $Pb(NO_3)_2$ given in the problem.

FOR PRACTICE 9.4

What volume (in mL) of a 0.150 M HNO_3 solution will completely react with 35.7 mL of a 0.108 M Na_2CO_3 solution according to this balanced chemical equation?

$$Na_2CO_3(aq) + 2\,HNO_3(aq) \longrightarrow 2\,NaNO_3(aq) + CO_2(g) + H_2O(l)$$

FOR MORE PRACTICE 9.4

In the reaction in For Practice 9.4, what mass (in grams) of carbon dioxide forms?

9.4 Types of Aqueous Solutions and Solubility

Consider two familiar aqueous solutions: salt water and sugar water. Salt water is a homogeneous mixture of NaCl and H_2O, and sugar water is a homogeneous mixture of $C_{12}H_{22}O_{11}$ and H_2O. You may have made these solutions yourself by adding table salt or sugar to water. As you stir either of these two substances into the water, the substance seems to disappear. However, you know that the original substance is still present because you can taste saltiness or sweetness in the water. How do solids such as salt and sugar dissolve in water?

When a solid is put into a liquid solvent, the attractive forces that hold the solid together (the solute–solute interactions) compete with the attractive forces between the solvent molecules and the particles that compose the solid (the solvent–solute interactions), as shown in Figure 9.5 ◀. For example, when we add sodium chloride to water, there is a competition between the attraction of Na^+ cations and Cl^- anions to each other (due to their opposite charges) and the attraction of Na^+ and Cl^- to water molecules. The attraction of Na^+ and Cl^- to water is based on the *polar nature* of the water molecule (see Section 6.10). The oxygen atom in water is electron-rich, giving it a partial negative charge (δ^-), as shown in Figure 9.6 ▼. The hydrogen atoms, in contrast, are electron-poor, giving them a partial positive charge (δ^+). As a result, the positively charged sodium ions are strongly attracted to the oxygen side of the water molecule, and the negatively charged chloride ions are attracted to the hydrogen side of the water molecule, as shown in Figure 9.7 ▼. In the case of NaCl, the attraction between the separated ions and the water molecules overcomes the attraction of sodium and chloride ions to each other, and the sodium chloride dissolves in the water (Figure 9.8 ▶).

Solute and Solvent Interactions

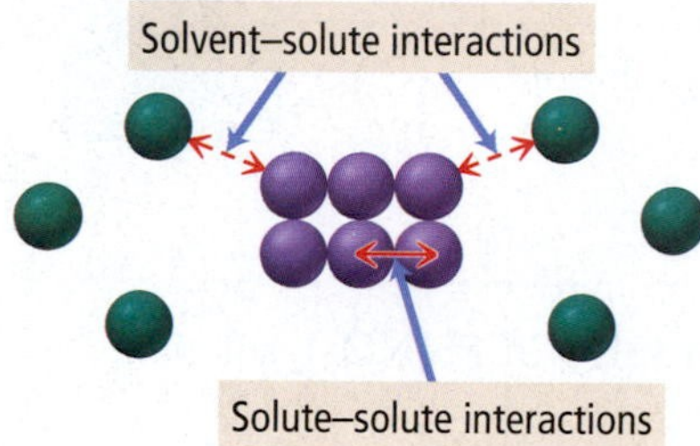

▲ **FIGURE 9.5 Solute and Solvent Interactions** When a solid is put into a solvent, the interactions among solvent and solute particles compete with the interactions among the solute particles themselves.

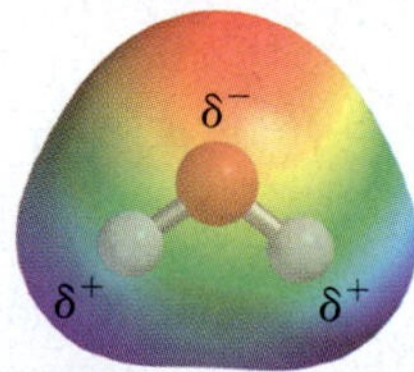

▲ **FIGURE 9.6 Electrostatic Potential Map of a Water Molecule** An uneven distribution of electrons within the water molecule causes the oxygen side of the water molecule to have a partial negative charge and the hydrogen side to have a partial positive charge.

Interactions in a Sodium Chloride Solution

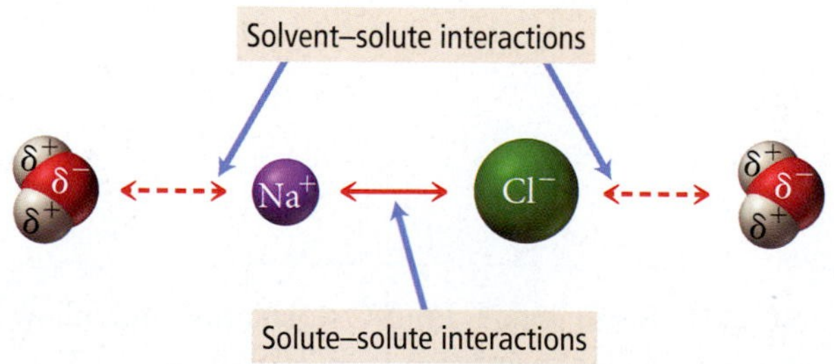

▲ **FIGURE 9.7 Solute and Solvent Interactions in a Sodium Chloride Solution** When sodium chloride is put into water, the attraction of Na^+ and Cl^- ions to water molecules competes with the attraction among the oppositely charged ions themselves.

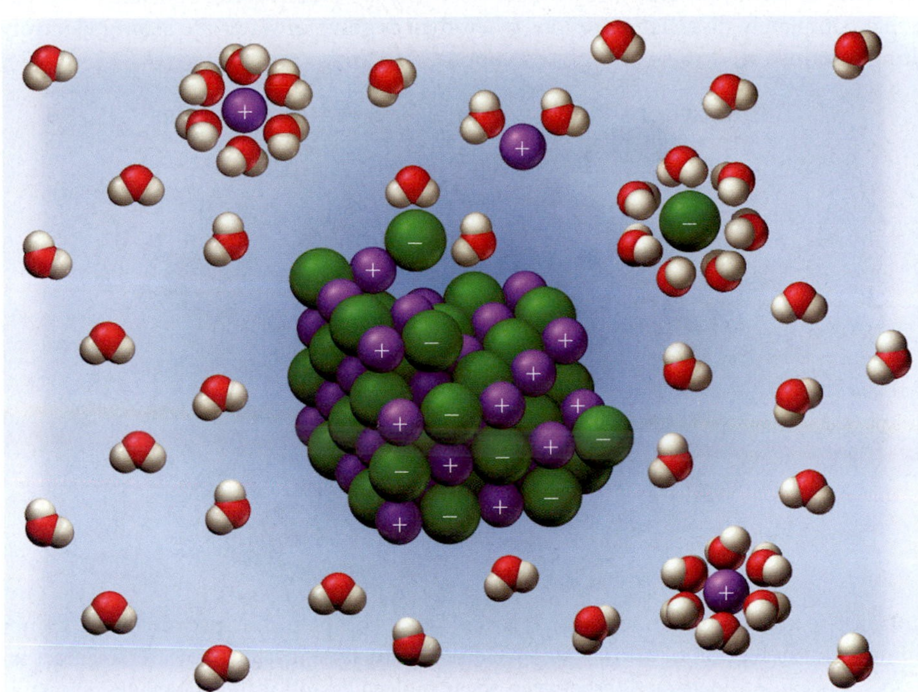

◀ **FIGURE 9.8 Sodium Chloride Dissolving in Water** The attraction between water molecules and the ions of sodium chloride causes NaCl to dissolve in the water.

Electrolyte and Nonelectrolyte Solutions

As Figure 9.9 ▼ illustrates, a salt solution conducts electricity while a sugar solution does not. The difference between the way that salt (an ionic compound) and sugar (a molecular compound) dissolve in water illustrates a fundamental difference between types of solutions. Ionic compounds such as the sodium chloride in the previous example (or the calcium chloride used for spherification in molecular gastronomy discussed in Section 9.1), dissociate into their component ions when they dissolve in water. An NaCl solution, represented as $NaCl(aq)$, does not contain any NaCl units, but rather dissolved Na^+ ions and Cl^- ions.

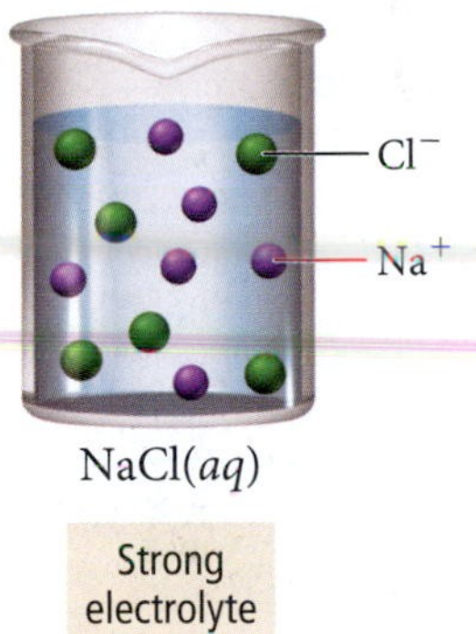

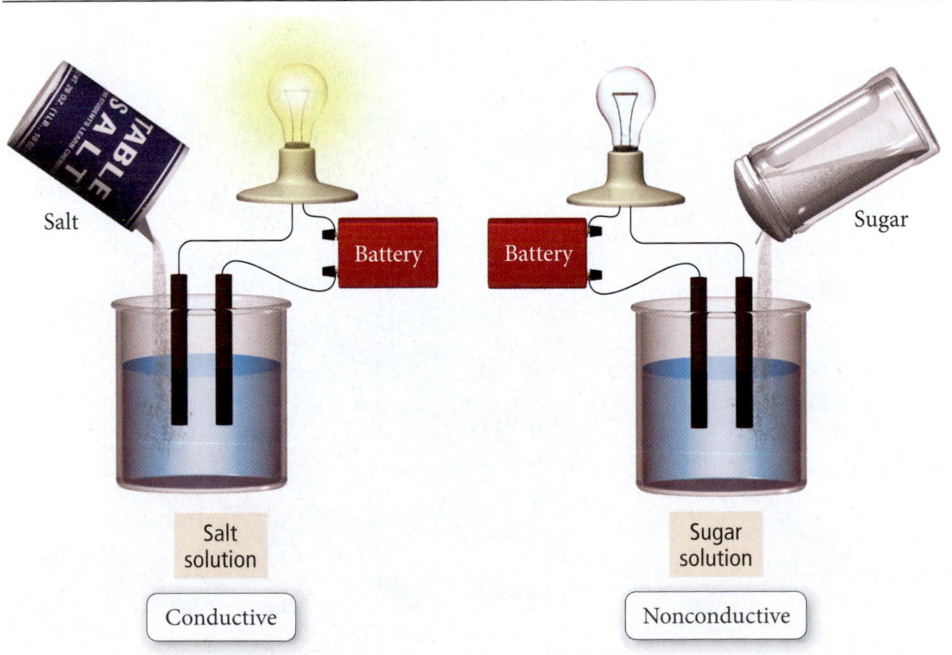

◀ **FIGURE 9.9 Electrolyte and Nonelectrolyte Solutions** A solution of salt (an electrolyte) conducts electrical current. A solution of sugar (a nonelectrolyte) does not.

The dissolved ions act as charge carriers, allowing the solution to conduct electricity. Substances that dissolve in water to form solutions that conduct electricity are **electrolytes**. Substances such as sodium chloride that completely dissociate into ions when they dissolve in water are **strong electrolytes**, and the resulting solutions are strong electrolyte solutions.

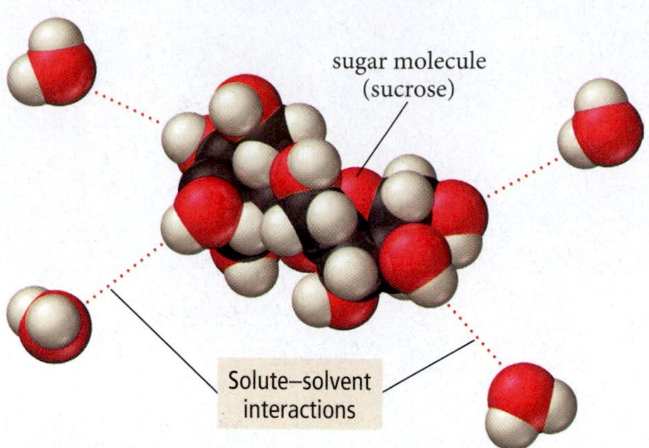

▶ **FIGURE 9.10 Sugar and Water Interactions** Partial charges on sugar molecules and water molecules (discussed more fully in Chapter 14) result in attractions between the sugar molecules and water molecules.

In contrast to sodium chloride, sugar is a molecular compound. Most molecular compounds—with the important exception of acids, which we discuss shortly—dissolve in water as intact molecules. Sugar dissolves because the attraction between sugar molecules and water molecules, shown in Figure 9.10 ▲, overcomes the attraction of sugar molecules to each other (Figure 9.11 ▼). So unlike a sodium chloride solution (which is composed of dissociated ions), a sugar solution is composed of intact $C_{12}H_{22}O_{11}$ molecules homogeneously mixed with the water molecules.

Compounds such as sugar that do not dissociate into ions when dissolved in water are **nonelectrolytes**, and the resulting solutions—called *nonelectrolyte solutions*—do not conduct electricity.

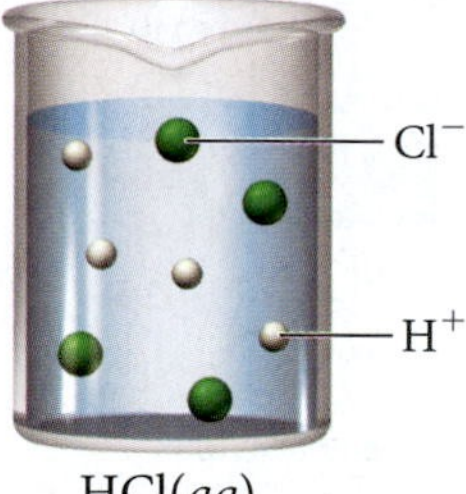

Acids are molecular compounds that ionize to form H^+ ions when they dissolve in water. Hydrochloric acid (HCl), for example, ionizes into H^+ and Cl^- when it dissolves in water. HCl is an example of a **strong acid**, one that completely ionizes in solution. Since strong acids completely ionize in

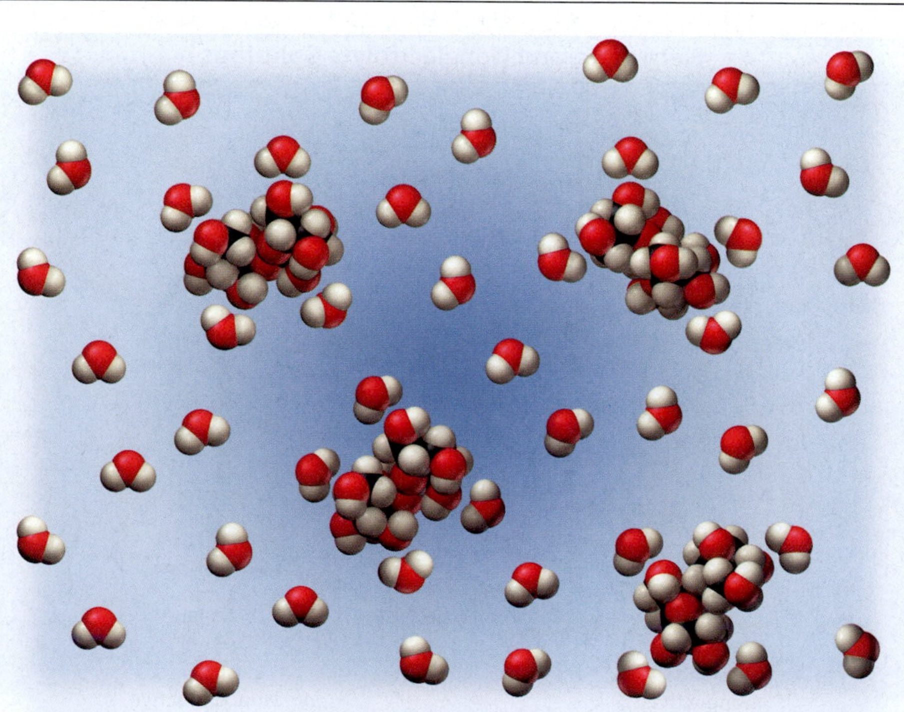

▶ **FIGURE 9.11 A Sugar Solution** Sugar dissolves because the attractions between sugar molecules and water molecules, which both contain a distribution of electrons that results in partial positive and partial negative charges, overcome the attractions among sugar molecules to each other.

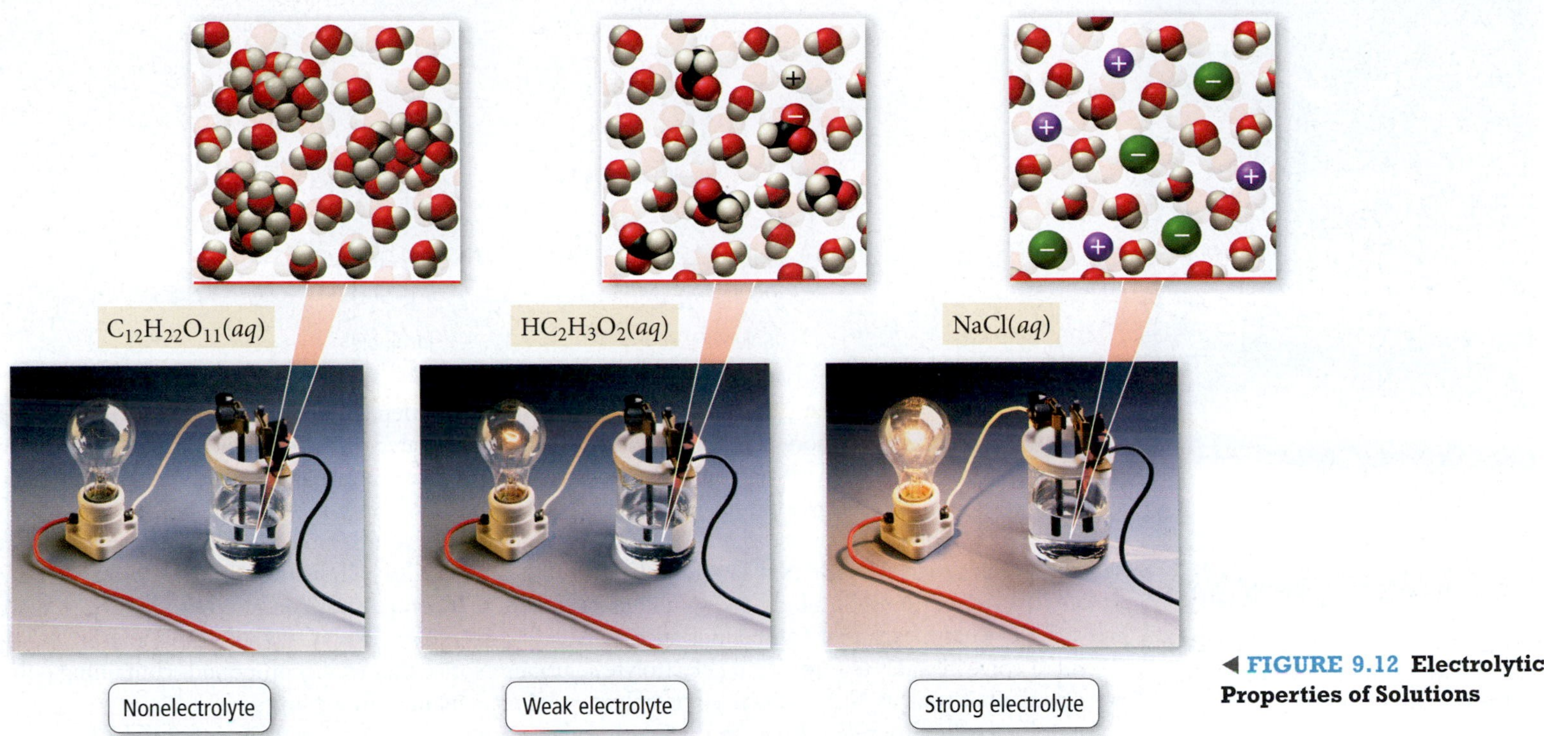

◀ **FIGURE 9.12 Electrolytic Properties of Solutions**

solution, they are also strong electrolytes. We represent the complete ionization of a strong acid with a single reaction arrow between the acid and its ionized form:

$$HCl(aq) \longrightarrow H^+(aq) + Cl^-(aq)$$

Many acids are **weak acids**; they do not completely ionize in water. For example, acetic acid ($HC_2H_3O_2$), the acid in vinegar, is a weak acid. A solution of a weak acid is composed mostly of the nonionized acid—only a small percentage of the acid molecules ionize. We represent the partial ionization of a weak acid with opposing half arrows between the reactants and products:

$$HC_2H_3O_2(aq) \rightleftharpoons H^+(aq) + C_2H_3O_2^-(aq)$$

Weak acids are classified as **weak electrolytes**, and the resulting solutions—called *weak electrolyte solutions*—conduct electricity only weakly. Figure 9.12 ▲ summarizes the electrolytic properties of solutions.

Unlike soluble ionic compounds, which contain ions and therefore *dissociate* in water, acids are molecular compounds that *ionize* in water.

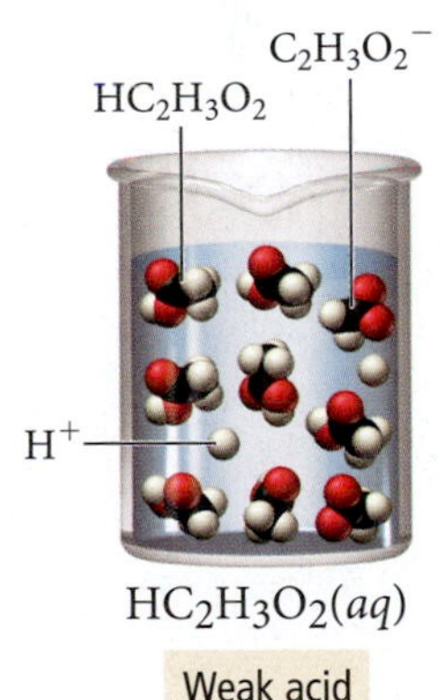

The Solubility of Ionic Compounds

As we have just discussed, when an ionic compound dissolves in water, the resulting solution contains, not the intact ionic compound itself, but its component ions dissolved in water. However, not all ionic compounds dissolve in water. If we add AgCl to water, for example, it remains solid and appears as a white powder at the bottom of the water.

In general, a compound is termed **soluble** if it dissolves in water and **insoluble** if it does not. However, these classifications are a bit of an oversimplification. In reality, solubility is a continuum and even "insoluble" compounds dissolve to some extent, though usually orders of magnitude less than soluble compounds. However, this oversimplification is useful in allowing us to systematically categorize a large number of compounds. (See Karl Popper's quote at the beginning of this chapter.)

As an example, consider silver nitrate, which is soluble. If we mix solid $AgNO_3$ with water, it dissolves and forms a strong electrolyte solution. Silver chloride, on the other hand, is almost

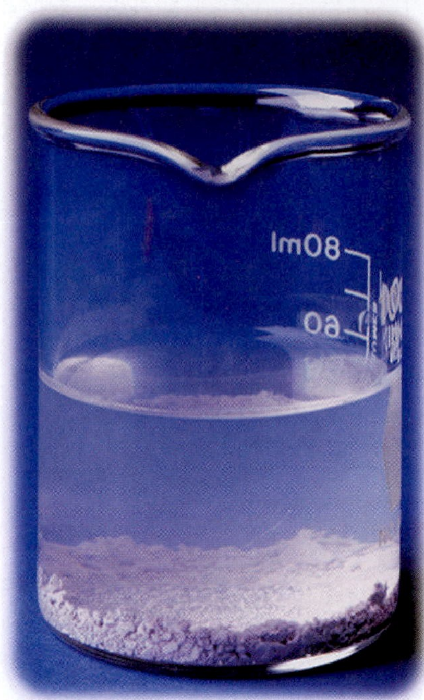
▲ AgCl does not dissolve in water; it remains as a white powder at the bottom of the beaker.

completely insoluble. If we mix solid AgCl with water, virtually all of it remains as a solid within the liquid water.

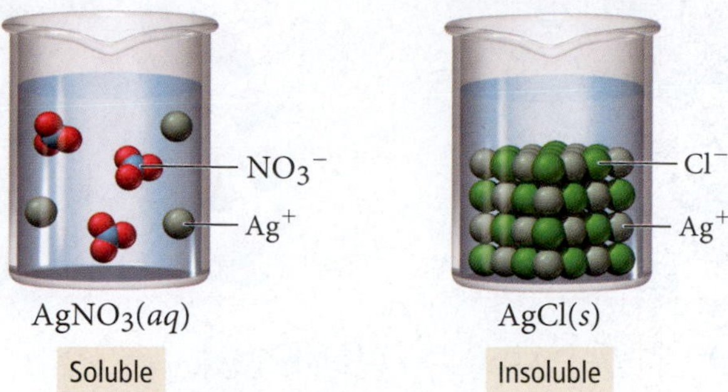

Whether a particular compound is soluble or insoluble depends on several factors. In Section 14.3, we will examine more closely the energy changes associated with solution formation. For now, however, we can follow a set of empirical rules that chemists have inferred from observations on many ionic compounds. Table 9.1 summarizes these *solubility rules*.

The solubility rules state that compounds containing the sodium ion are soluble. That means that compounds such as NaBr, $NaNO_3$, Na_2SO_4, NaOH, and Na_2CO_3 all dissolve in water to form strong electrolyte solutions. Similarly, the solubility rules state that compounds containing the NO_3^- ion are soluble. That means that compounds such as $AgNO_3$, $Pb(NO_3)_2$, $NaNO_3$, $Ca(NO_3)_2$, and $Sr(NO_3)_2$ all dissolve in water to form strong electrolyte solutions. Notice that when compounds containing polyatomic ions such as NO_3^- dissolve, the polyatomic ions remain as intact units.

The solubility rules also state that, with some exceptions, compounds containing the CO_3^{2-} ion are insoluble. Therefore, compounds such as $CuCO_3$, $CaCO_3$, $SrCO_3$, and $FeCO_3$ do not dissolve in water. Note that the solubility rules contain many exceptions. For example, compounds containing CO_3^{2-} *are* soluble when paired with Li^+, Na^+, K^+, or NH_4^+. Thus Li_2CO_3, Na_2CO_3, K_2CO_3, and $(NH_4)_2CO_3$ are all soluble.

TABLE 9.1 Solubility Rules for Ionic Compounds in Water

Compounds Containing the Following Ions Are Generally Soluble	Exceptions
Li^+, Na^+, K^+, and NH_4^+	None
NO_3^- and $C_2H_3O_2^-$	None
Cl^-, Br^-, and I^-	When these ions pair with Ag^+, Hg_2^{2+} or Pb^{2+}, the resulting compounds are insoluble.
SO_4^{2-}	When SO_4^{2-} pairs with Sr^{2+}, Ba^{2+}, Pb^{2+}, Ag^+, or Ca^{2+}, the resulting compound is insoluble.
Compounds Containing the Following Ions Are Generally Insoluble	**Exceptions**
OH^- and S^{2-}	When these ions pair with Li^+, Na^+, K^+, or NH_4^+, the resulting compounds are soluble.
	When S^{2-} pairs with Ca^{2+}, Sr^{2+}, or Ba^{2+}, the resulting compound is soluble.
	When OH^- pairs with Ca^{2+}, Sr^{2+}, or Ba^{2+}, the resulting compound is slightly soluble.
CO_3^{2-} and PO_4^{3-}	When these ions pair with Li^+, Na^+, K^+, or NH_4^+, the resulting compounds are soluble.

EXAMPLE 9.5
Predicting Ionic Compound Solubility

Predict whether each compound is soluble or insoluble.

(a) $PbCl_2$ **(b)** $CuCl_2$ **(c)** $Ca(NO_3)_2$ **(d)** $BaSO_4$

SOLUTION

(a) Insoluble. Compounds containing Cl^- are normally soluble, but Pb^{2+} is an exception.
(b) Soluble. Compounds containing Cl^- are normally soluble, and Cu^{2+} is not an exception.
(c) Soluble. Compounds containing NO_3^- are always soluble.
(d) Insoluble. Compounds containing SO_4^{2-} are normally soluble, but Ba^{2+} is an exception.

FOR PRACTICE 9.5
Predict whether each compound is soluble or insoluble.

(a) NiS **(b)** $Mg_3(PO_4)_2$ **(c)** Li_2CO_3 **(d)** NH_4Cl

9.5 Precipitation Reactions

We discussed an example of a **precipitation reaction**—a reaction in which a solid forms upon the mixing of two solutions—in Section 9.1. You have probably seen another precipitation reaction if you have taken a bath in hard water. Hard water contains dissolved ions such as Ca^{2+} and Mg^{2+} that form a **precipitate** when they react with ions in soap. This precipitate is a gray curd that may appear as "bathtub ring" after you drain the tub. Hard water is particularly troublesome when washing clothes. Consider how your white shirt would look covered with the gray curd from the bathtub. Most laundry detergents include substances designed to remove Ca^{2+} and Mg^{2+} from the laundry mixture. The most common substance used for this purpose is sodium carbonate, which dissolves in water to form sodium cations (Na^+) and carbonate (CO_3^{2-}) anions:

▲ The reaction of ions in hard water with soap produces a gray curd you can see after you drain the bathwater.

$$Na_2CO_3(aq) \longrightarrow 2\,Na^+(aq) + CO_3^{2-}(aq)$$

Sodium carbonate is soluble, but calcium carbonate and magnesium carbonate are not (see the solubility rules in Table 9.1). Consequently, the carbonate anions react with dissolved Mg^{2+} and Ca^{2+} ions in hard water to form solids that *precipitate* from (or come out of) solution:

$$Mg^{2+}(aq) + CO_3^{2-}(aq) \longrightarrow MgCO_3(s)$$
$$Ca^{2+}(aq) + CO_3^{2-}(aq) \longrightarrow CaCO_3(s)$$

The precipitation of these ions prevents their reaction with the soap, eliminating curd and preventing white shirts from turning gray.

The reactions between CO_3^{2-} and Mg^{2+} and Ca^{2+} are also examples of precipitation reactions. Precipitation reactions are common in chemistry. Potassium iodide and lead(II) nitrate, which each form colorless, strong electrolyte solutions when dissolved in water, form a brilliant yellow precipitate when combined (Figure 9.13 ▶). We describe this precipitation reaction with the following chemical equation:

$$2\,KI(aq) + Pb(NO_3)_2(aq) \longrightarrow PbI_2(s) + 2\,KNO_3(aq)$$

Precipitation reactions do not always occur when two aqueous solutions are mixed. For example, if we combine solutions of KI(*aq*) and NaCl(*aq*), nothing happens (Figure 9.14 ▶):

KEY CONCEPT VIDEO Reactions in solution

$$KI(aq) + NaCl(aq) \longrightarrow \text{NO REACTION}$$

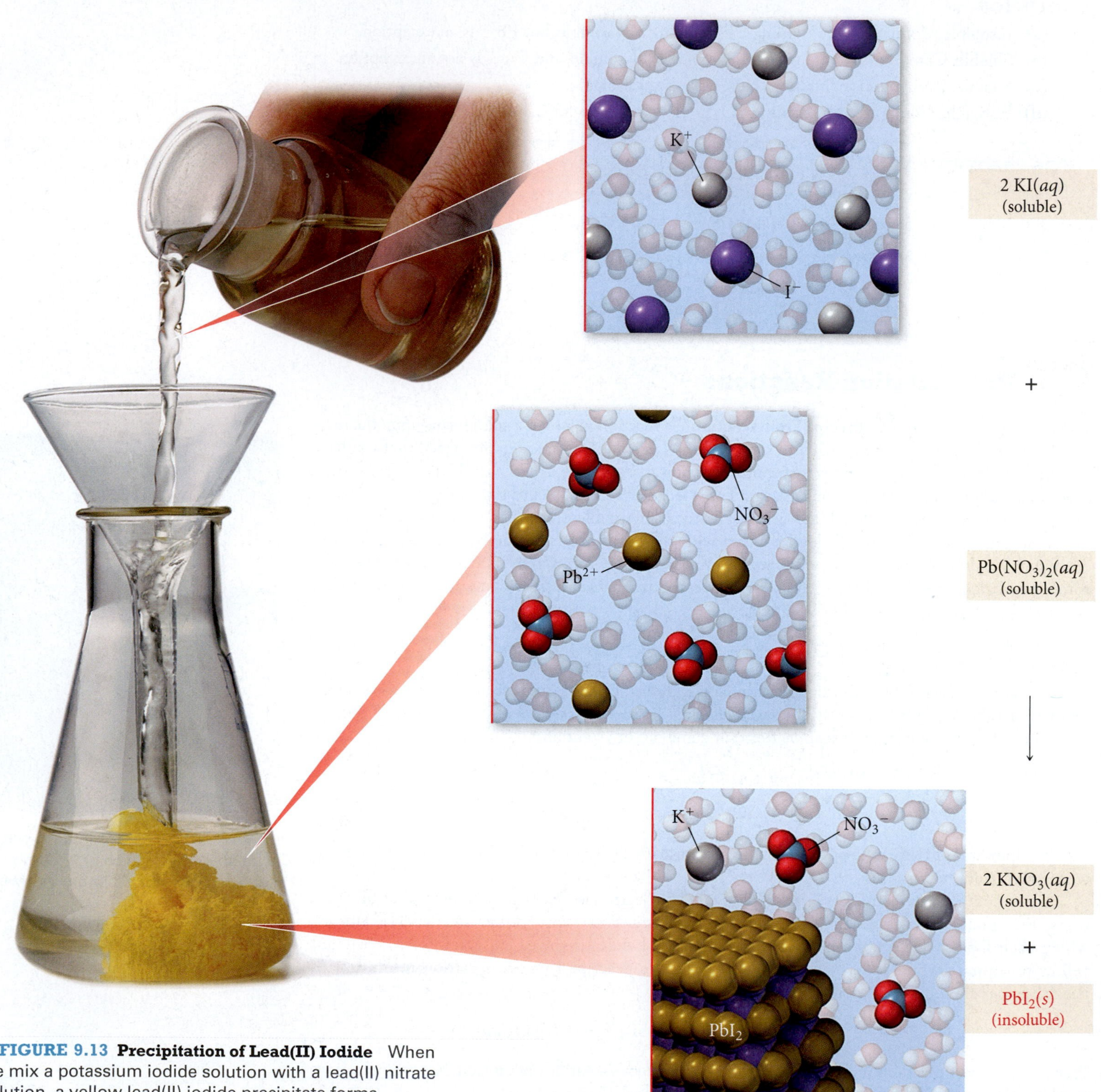

▲ **FIGURE 9.13 Precipitation of Lead(II) Iodide** When we mix a potassium iodide solution with a lead(II) nitrate solution, a yellow lead(II) iodide precipitate forms.

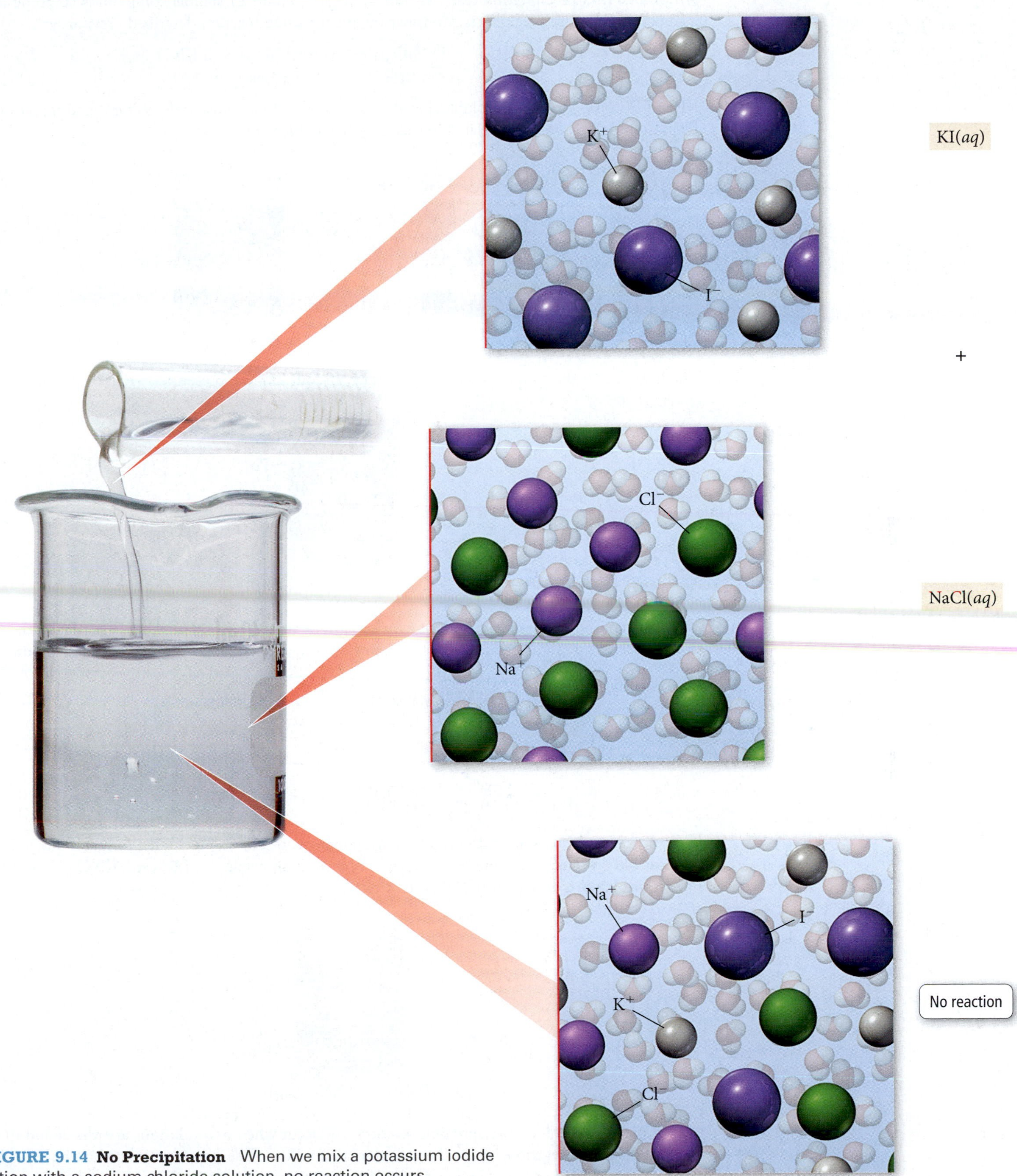

▲ **FIGURE 9.14 No Precipitation** When we mix a potassium iodide solution with a sodium chloride solution, no reaction occurs.

The key to predicting precipitation reactions is understanding that *only insoluble compounds form precipitates.* In a precipitation reaction, two solutions containing soluble compounds combine and an insoluble compound precipitates. Consider the precipitation reaction described previously:

$$\underset{\text{soluble}}{2\,KI(aq)} + \underset{\text{soluble}}{Pb(NO_3)_2(aq)} \longrightarrow \underset{\text{insoluble}}{PbI_2(s)} + \underset{\text{soluble}}{2\,KNO_3(aq)}$$

KI and $Pb(NO_3)_2$ are both soluble, but the precipitate, PbI_2, is insoluble. Before mixing, KI(*aq*) and $Pb(NO_3)_2(aq)$ are both dissociated in their respective solutions:

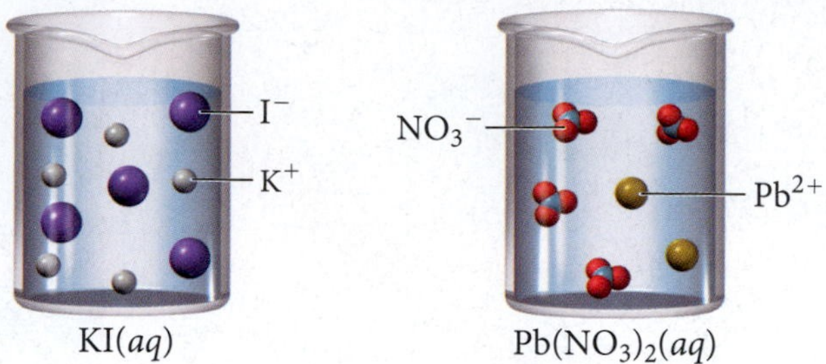

The instant that the solutions come into contact, all four ions are present:

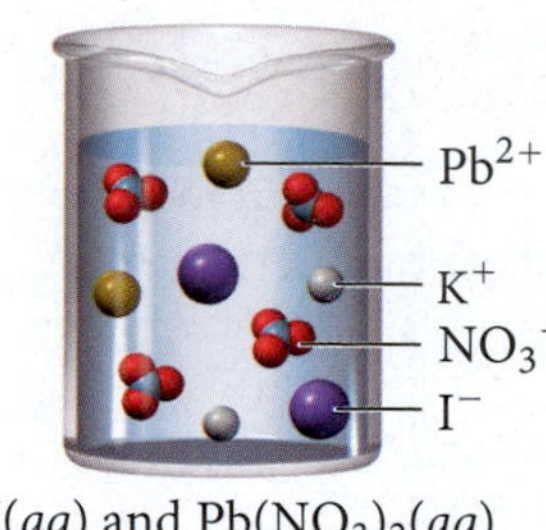

KI(*aq*) and $Pb(NO_3)_2(aq)$

Now, new compounds—one or both of which might be insoluble—are possible. Specifically, the cation from either compound can pair with the anion from the other to form possibly insoluble products:

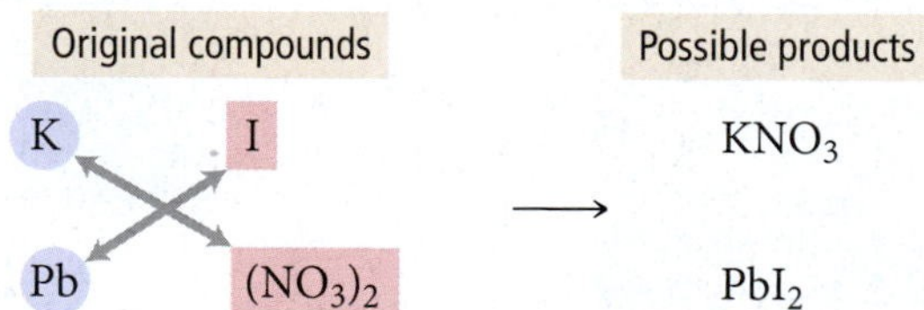

If the possible products are both soluble, no reaction occurs and no precipitate forms. If one or both of the possible products are insoluble, a precipitation reaction occurs. In this case, KNO_3 is soluble, but PbI_2 is insoluble. Consequently, PbI_2 precipitates.

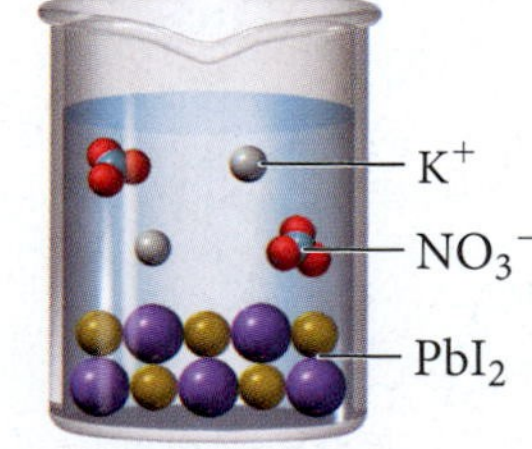

$PbI_2(s)$ and $KNO_3(aq)$

To predict whether a precipitation reaction will occur when two solutions are mixed and to write an equation for the reaction, we use the procedure that follows. The steps are outlined in the left column, and two examples illustrating how to apply the procedure are shown in the center and right columns.

PROCEDURE FOR **Writing Equations for Precipitation Reactions**	**EXAMPLE 9.6** **Writing Equations for Precipitation Reactions**	**EXAMPLE 9.7** **Writing Equations for Precipitation Reactions**
	Write an equation for the precipitation reaction that occurs (if any) when you mix solutions of potassium carbonate and nickel(II) chloride.	Write an equation for the precipitation reaction that occurs (if any) when you mix solutions of sodium nitrate and lithium sulfate.
1. Write the formulas of the two compounds being mixed as reactants in a chemical equation.	$K_2CO_3(aq) + NiCl_2(aq) \longrightarrow$	$NaNO_3(aq) + Li_2SO_4(aq) \longrightarrow$
2. Below the equation, write the formulas of the products that could form from the reactants. Obtain these by combining the cation from each reactant with the anion from the other. Make sure to write correct formulas for these ionic compounds, as described in Section 5.6.	$K_2CO_3(aq) + NiCl_2(aq) \longrightarrow$ Possible products KCl $NiCO_3$	$NaNO_3(aq) + Li_2SO_4(aq) \longrightarrow$ Possible products $LiNO_3$ Na_2SO_4
3. Refer to the solubility rules in Table 9.1 to determine whether any of the possible products are insoluble.	KCl is soluble. (Compounds containing Cl^- are usually soluble, and K^+ is not an exception.) $NiCO_3$ is insoluble. (Compounds containing CO_3^{2-} are usually insoluble, and Ni^{2+} is not an exception.)	$LiNO_3$ is soluble. (Compounds containing NO_3^- are soluble, and Li^+ is not an exception.) Na_2SO_4 is soluble. (Compounds containing SO_4^{2-} are generally soluble, and Na^+ is not an exception.)
4. If all of the possible products are soluble, there will be no precipitate. Write NO REACTION after the arrow.	Since this example has an insoluble product, we proceed to the next step.	Since this example has no insoluble product, there is no reaction. $NaNO_3(aq) + Li_2SO_4(aq) \longrightarrow$ NO REACTION
5. If any of the possible products are insoluble, write their formulas as the products of the reaction using (*s*) to indicate solid. Write any soluble products with (*aq*) to indicate aqueous.	$K_2CO_3(aq) + NiCl_2(aq) \longrightarrow NiCO_3(s) + KCl(aq)$	
6. Balance the equation. Remember to adjust only coefficients, not subscripts.	$K_2CO_3(aq) + NiCl_2(aq) \longrightarrow NiCO_3(s) + 2KCl(aq)$	
	FOR PRACTICE 9.6 Write an equation for the precipitation reaction that occurs (if any) when you mix solutions of ammonium chloride and iron(III) nitrate.	**FOR PRACTICE 9.7** Write an equation for the precipitation reaction that occurs (if any) when you mix solutions of sodium hydroxide and copper(II) bromide.

9.3
Cc
Conceptual Connection

Precipitation Reactions

Consider the generic ionic compounds with the formulas AX and BY and the following solubility rules: AX soluble; BY soluble; AY soluble; BX insoluble. Let circles represent A^+ ions; squares represent B^+ ions; triangles represent X^- ions; and diamonds represent Y^- ions. Solutions of the two compounds (AX and BY) are represented as follows:

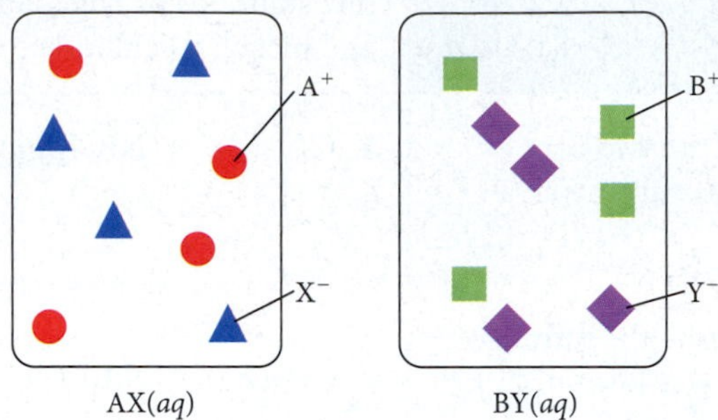

Draw a molecular-level representation showing the result of mixing these two solutions (AX and BY) and write an equation to represent the reaction.

9.6 Representing Aqueous Reactions: Molecular, Ionic, and Complete Ionic Equations

Consider the following equation for a precipitation reaction:

$$Pb(NO_3)_2(aq) + 2\,KCl(aq) \longrightarrow PbCl_2(s) + 2\,KNO_3(aq)$$

This equation is a **molecular equation**, an equation that shows the complete neutral formulas for each compound in the reaction as if they existed as molecules. In actual solutions of soluble ionic compounds, dissolved substances are present as ions. We can write equations for reactions occurring in aqueous solution in a way that better shows the dissociated nature of dissolved ionic compounds. For example, we can rewrite the above equation as:

$$Pb^{2+}(aq) + 2NO_3^-(aq) + 2K^+(aq) + 2Cl^-(aq) \longrightarrow PbCl_2(s) + 2\,K^+(aq) + 2NO_3^-(aq)$$

Equations such as this, which list all of the ions present as either reactants or products in a chemical reaction, are **complete ionic equations**.

Notice that in the complete ionic equation, some of the ions in solution appear unchanged on both sides of the equation. These ions are called **spectator ions** because they do not participate in the reaction.

$$Pb^{2+}(aq) + \cancel{2NO_3^-}(aq) + \cancel{2K^+}(aq) + 2\,Cl^-(aq) \longrightarrow PbCl_2(s) + \cancel{2K^+}(aq) + \cancel{2NO_3^-}(aq)$$

Spectator ions

To simplify the equation and to show more clearly what is happening, we can omit spectator ions:

$$Pb^{2+}(aq) + 2Cl^-(aq) \longrightarrow PbCl_2(s)$$

Equations that show only the species that actually change during the reaction are **net ionic equations**.

As another example, consider the reaction between HCl(*aq*) and KOH(*aq*):

$$HCl(aq) + KOH(aq) \longrightarrow H_2O(l) + KCl(aq)$$

Since HCl, KOH, and KCl all exist in solution primarily as independent ions, the complete ionic equation is:

$$H^+(aq) + Cl^-(aq) + K^+(aq) + OH^-(aq) \longrightarrow H_2O(l) + K^+(aq) + Cl^-(aq)$$

To write the net ionic equation, we remove the spectator ions, those that are unchanged on both sides of the equation:

$$H^+(aq) + Cl^-(aq) + K^+(aq) + OH^-(aq) \longrightarrow H_2O(l) + K^+(aq) + Cl^-(aq)$$

Spectator ions

The net ionic equation is $H^+(aq) + OH^-(aq) \longrightarrow H_2O(l)$.

Summarizing Aqueous Equations

- A **molecular equation** is a chemical equation showing the complete, neutral formulas for every compound in a reaction.
- A **complete ionic equation** is a chemical equation showing all of the species as they are actually present in solution.
- A **net ionic equation** is an equation showing only the species that actually change during the reaction.

EXAMPLE 9.8

Writing Complete Ionic and Net Ionic Equations

Consider the following precipitation reaction occurring in aqueous solution:

$$3\ SrCl_2(aq) + 2\ Li_3PO_4(aq) \longrightarrow Sr_3(PO_4)_2(s) + 6\ LiCl(aq)$$

Write the complete ionic equation and net ionic equation for this reaction.

SOLUTION

Write the complete ionic equation by separating aqueous ionic compounds into their constituent ions. The $Sr_3(PO_4)_2(s)$, precipitating as a solid, remains as one unit.	**Complete ionic equation:** $3\ Sr^{2+}(aq) + 6\ Cl^-(aq) + 6\ Li^+(aq) + 2\ PO_4^{3-}(aq) \longrightarrow Sr_3(PO_4)_2(s) + 6\ Li^+(aq) + 6\ Cl^-(aq)$
Write the net ionic equation by eliminating the spectator ions, those that do not change from one side of the reaction to the other.	**Net ionic equation:** $3\ Sr^{2+}(aq) + 2\ PO_4^{3-}(aq) \longrightarrow Sr_3(PO_4)_2(s)$

FOR PRACTICE 9.8

Consider the following reaction occurring in aqueous solution:

$$2\ HI(aq) + Ba(OH)_2(aq) \longrightarrow 2\ H_2O(l) + BaI_2(aq)$$

Write the complete ionic equation and net ionic equation for this reaction.

FOR MORE PRACTICE 9.8

Write complete ionic and net ionic equations for the following reaction occurring in aqueous solution:

$$2\ AgNO_3(aq) + MgCl_2(aq) \longrightarrow 2\ AgCl(s) + Mg(NO_3)_2(aq)$$

9.7 Acid–Base Reactions

An important class of reactions that occurs in aqueous solution is the acid–base reaction. In an **acid–base reaction** (also called a **neutralization reaction**), an acid reacts with a base and the two neutralize each other, producing water (or in some cases a weak electrolyte). In this section we discuss how to define acids and bases, how to name acids, and how to write equations for the reactions between acids and bases.

We discuss acids and bases in more detail in Chapter 17.

Properties of Acids and Bases

As we saw in Section 9.4, we can define *acids* as molecular compounds that release hydrogen ions (H^+) when dissolved in water. For example, HCl(*aq*) is an acid because it produces H^+ ions in solution:

$$HCl(aq) \longrightarrow H^+(aq) + Cl^-(aq)$$

This definition of an acid based on acid behavior is the **Arrhenius definition**.

An H^+ ion is a proton. In solution, bare protons normally associate with water molecules to form **hydronium ions** (Figure 9.15 ◀):

$$H^+(aq) + H_2O(l) \longrightarrow H_3O^+(aq)$$

H^+ and H_3O^+ are used interchangeably because, even though H^+ associates with water to form hydronium, it is the H^+ part that reacts with other substances such as bases.

Chemists use $H^+(aq)$ and $H_3O^+(aq)$ interchangeably to mean the same thing—a hydronium ion. The chemical equation for the ionization of HCl and other acids is often written to show the association of the proton with a water molecule to form the hydronium ion:

$$HCl(aq) + H_2O(l) \longrightarrow H_3O^+(aq) + Cl^-(aq)$$

Some acids—called **polyprotic acids**—contain more than one ionizable proton and release them sequentially. For example, sulfuric acid, H_2SO_4, is a **diprotic acid**. It is strong in its first ionizable proton, but weak in its second:

$$H_2SO_4(aq) \longrightarrow H^+(aq) + HSO_4^-(aq)$$

$$HSO_4^-(aq) \rightleftharpoons H^+(aq) + SO_4^{2-}(aq)$$

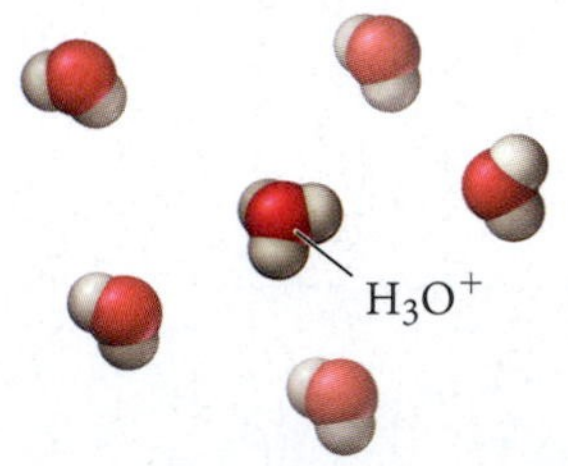

▲ **FIGURE 9.15 The Hydronium Ion** Protons normally associate with water molecules in solution to form H_3O^+ ions, which in turn interact with other water molecules.

Acids are characterized by their sour taste and their ability to dissolve many metals. For example, hydrochloric acid is present in stomach fluids; its sour taste becomes painfully obvious during vomiting. Hydrochloric acid dissolves some metals. For example, if you put a strip of zinc into a test tube of hydrochloric acid, it slowly dissolves as the $H^+(aq)$ ions convert the zinc metal into $Zn^{2+}(aq)$ cations (Figure 9.16 ▼). Acids are present in foods such as lemons and limes and are used in household products such as toilet bowl cleaners and Lime-Away®.

▲ Many fruits are acidic and have the characteristically sour taste of acids.

According to the Arrhenius definition, **bases** produce OH^- in solution. Sodium hydroxide (NaOH) is a base because it produces OH^- ions in solution:

$$NaOH(aq) \longrightarrow Na^+(aq) + OH^-(aq)$$

In analogy to diprotic acids, some bases, such as $Sr(OH)_2$, produce two moles of OH^- per mole of the base.

$$Sr(OH)_2(aq) \longrightarrow Sr^{2+}(aq) + 2OH^-(aq)$$

Acids Dissolve Many Metals

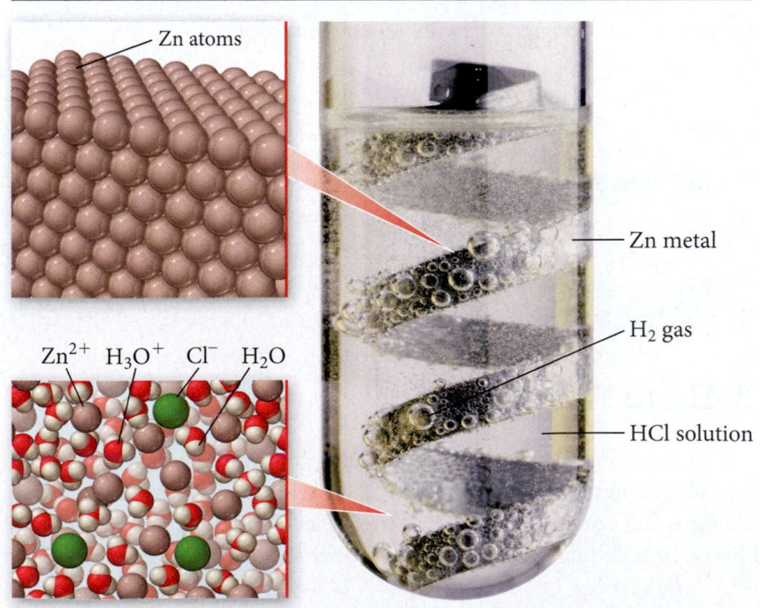

► **FIGURE 9.16 Hydrochloric Acid Dissolving Zinc Metal** The zinc atoms are ionized to zinc ions, which dissolve in the water. The HCl forms H_2 gas, which is responsible for the bubbles you can see in the test tube.

TABLE 9.2 Some Common Acids and Bases

Name of Acid	Formula	Name of Base	Formula
Hydrochloric acid	HCl	Sodium hydroxide	$NaOH$
Hydrobromic acid	HBr	Lithium hydroxide	$LiOH$
Hydroiodic acid	HI	Potassium hydroxide	KOH
Nitric acid	HNO_3	Calcium hydroxide	$Ca(OH)_2$
Sulfuric acid	H_2SO_4	Barium hydroxide	$Ba(OH)_2$
Perchloric acid	$HClO_4$	Ammonia*	NH_3 (weak base)
Acetic acid	$HC_2H_3O_2$ (weak acid)		
Hydrofluoric acid	HF (weak acid)		

*Ammonia does not contain OH^-, but it produces OH^- in a reaction with water that occurs only to a small extent: $NH_3(aq) + H_2O(l) \rightleftharpoons NH_4^+(aq) + OH^-(aq)$.

▲ Many common household products are bases.

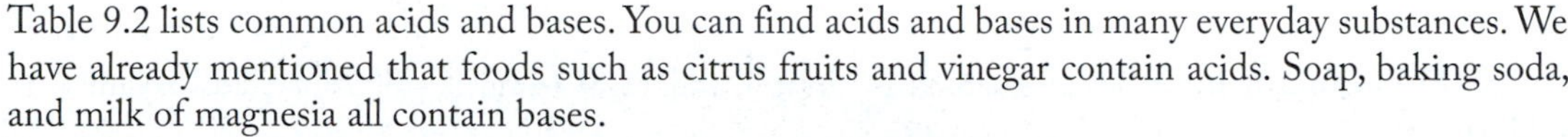

Table 9.2 lists common acids and bases. You can find acids and bases in many everyday substances. We have already mentioned that foods such as citrus fruits and vinegar contain acids. Soap, baking soda, and milk of magnesia all contain bases.

We can categorize acids into two types: binary acids and oxyacids.

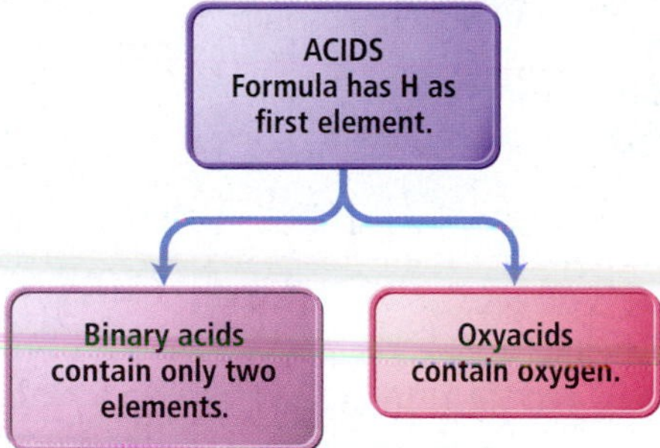

Naming Binary Acids **Binary acids** are composed of hydrogen and a nonmetal. Names for binary acids have the form:

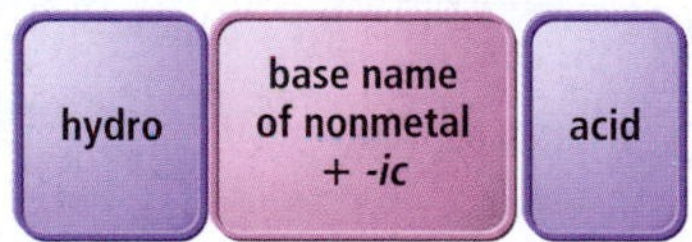

For example, $HCl(aq)$ is hydro*chlor*ic acid and $HBr(aq)$ is hydro*brom*ic acid.

$HCl(aq)$ hydrochloric acid $\quad$ $HBr(aq)$ hydrobromic acid

EXAMPLE 9.9

Naming Binary Acids

Name $HI(aq)$.

SOLUTION

The base name of I is *iod* so $HI(aq)$ is hydroiodic acid.

$HI(aq)$ hydroiodic acid

FOR PRACTICE 9.9

Name $HF(aq)$.

Naming Oxyacids

Oxyacids contain hydrogen and an oxyanion (an anion containing a nonmetal and oxygen). The common oxyanions are listed in the table of polyatomic ions in Chapter 5 (Table 5.5). For example, $HNO_3(aq)$ contains the nitrate (NO_3^-) ion, $H_2SO_3(aq)$ contains the sulfite (SO_3^{2-}) ion, and $H_2SO_4(aq)$ contains the sulfate (SO_4^{2-}) ion. Oxyacids are a combination of one or more H^+ ions with an oxyanion (see Table 5.5). The number of H^+ ions depends on the charge of the oxyanion; the formula is always charge-neutral. The names of oxyacids depend on the ending of the oxyanion and take the following forms:

oxyanions ending with *-ate* — base name of oxyanion + *-ic* | acid

oxyanions ending with *-ite* — base name of oxyanion + *-ous* | acid

So $HNO_3(aq)$ is nitric acid (oxyanion is nitrate), and $H_2SO_3(aq)$ is sulfurous acid (oxyanion is sulfite).

$HNO_3(aq)$ nitric acid $H_2SO_3(aq)$ sulfurous acid

EXAMPLE 9.10
Naming Oxyacids

Name $HC_2H_3O_2(aq)$.

SOLUTION

The oxyanion is acetate, which ends in *-ate*; therefore, the name of the acid is *acetic acid.*

$HC_2H_3O_2(aq)$ acetic acid

FOR PRACTICE 9.10

Name $HNO_2(aq)$.

FOR MORE PRACTICE 9.10

Write the formula for perchloric acid.

Acid–Base Reactions

Our stomachs contain hydrochloric acid ($HCl(aq)$), which acts in the digestion of food. Certain foods or stress, however, can increase the stomach's acidity to uncomfortable levels, causing acid stomach or heartburn. Antacids are over-the-counter medicines that work by reacting with and neutralizing stomach acid. Antacids employ different bases as neutralizing agents. Milk of magnesia, for example, contains $Mg(OH)_2$ and Mylanta® contains $Al(OH)_3$. All antacids, regardless of the base they employ, neutralize stomach acid and relieve heartburn through *acid–base reactions.*

When we mix an acid and a base, the $H^+(aq)$ from the acid—whether it is weak or strong—combines with the $OH^-(aq)$ from the base to form $H_2O(l)$ (Figure 9.17 ▶). Consider the reaction between hydrochloric acid and sodium hydroxide:

The word *salt* in this sense applies to any ionic compound and is therefore more general than the common usage, which refers only to table salt (NaCl).

$$\underset{\text{Acid}}{HCl(aq)} + \underset{\text{Base}}{NaOH(aq)} \longrightarrow \underset{\text{Water}}{H_2O(l)} + \underset{\text{Salt}}{NaCl(aq)}$$

Acid–Base Reaction

$$HCl(aq) + NaOH(aq) \longrightarrow H_2O(l) + NaCl(aq)$$

The reaction between hydrochloric acid and sodium hydroxide forms water and a salt, sodium chloride, which remains dissolved in the solution.

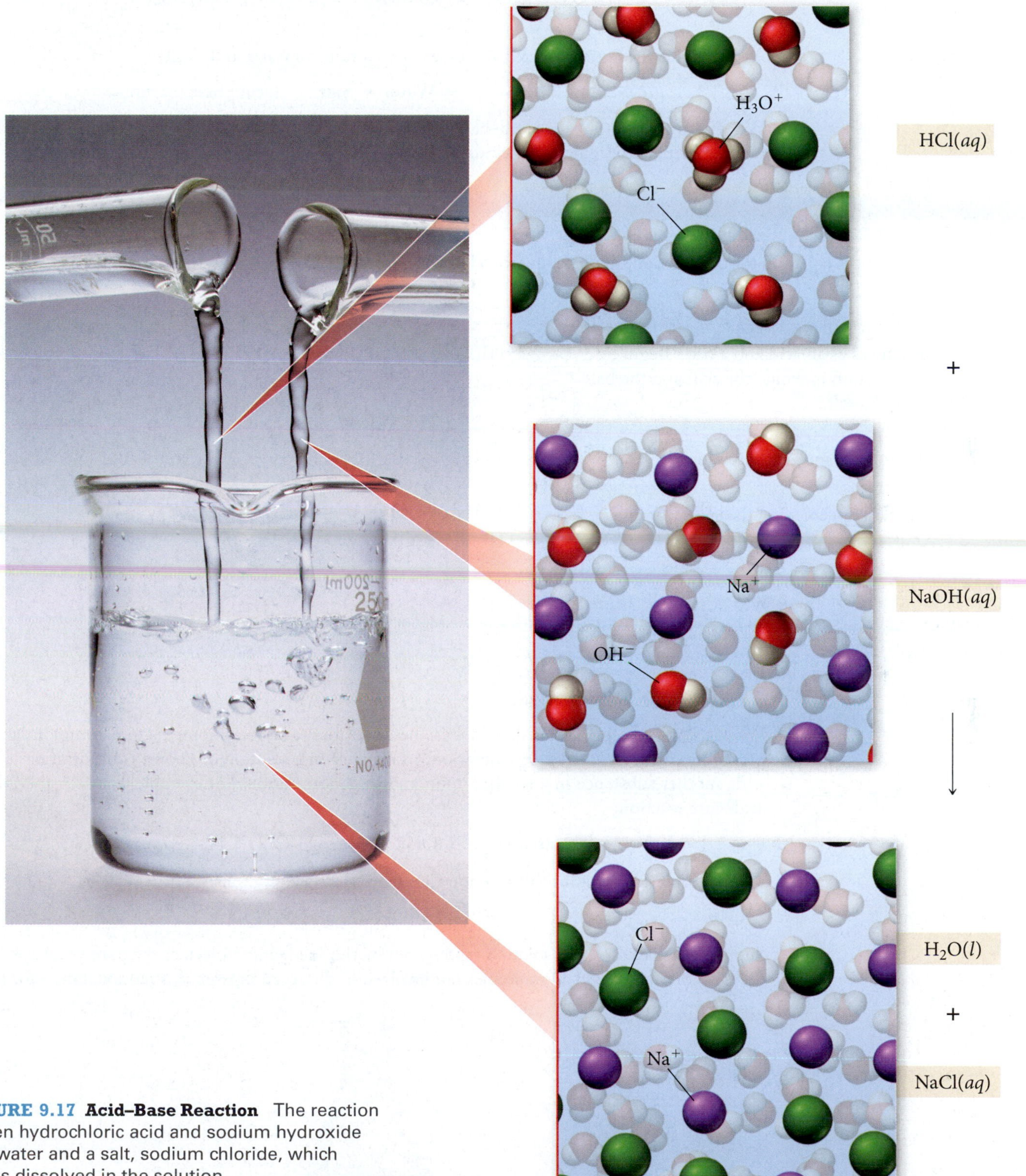

▲ **FIGURE 9.17 Acid–Base Reaction** The reaction between hydrochloric acid and sodium hydroxide forms water and a salt, sodium chloride, which remains dissolved in the solution.

Acid–base reactions generally form water and an ionic compound—called a **salt**—that usually remains dissolved in the solution. The net ionic equation for many acid–base reactions is:

$$H^+(aq) + OH^-(aq) \longrightarrow H_2O(l)$$

Another example of an acid–base reaction is the reaction between sulfuric acid and potassium hydroxide:

$$\underset{\text{acid}}{H_2SO_4(aq)} + \underset{\text{base}}{2\,KOH(aq)} \longrightarrow \underset{\text{water}}{2\,H_2O(l)} + \underset{\text{salt}}{K_2SO_4(aq)}$$

Again, notice the pattern of acid and base reacting to form water and a salt.

$$\textbf{Acid} + \textbf{Base} \longrightarrow \textbf{Water} + \textbf{Salt} \qquad \text{(acid–base reactions)}$$

When writing equations for acid–base reactions, write the formula of the salt using the procedure for writing formulas of ionic compounds demonstrated in Section 5.6.

EXAMPLE 9.11
Writing Equations for Acid–Base Reactions

Write a molecular and net ionic equation for the reaction between aqueous HI and aqueous $Ba(OH)_2$.

SOLUTION

First identify these substances as an acid and a base. Begin by writing the unbalanced equation in which the acid and the base combine to form water and a salt.	$\underset{\text{acid}}{HI(aq)} + \underset{\text{base}}{Ba(OH)_2(aq)} \longrightarrow \underset{\text{water}}{H_2O(l)} + \underset{\text{salt}}{BaI_2(aq)}$
Balance the equation; this is the molecular equation.	$2\,HI(aq) + Ba(OH)_2(aq) \longrightarrow 2\,H_2O(l) + BaI_2(aq)$
Write the net ionic equation by removing the spectator ions.	$2\,H^+(aq) + 2\,OH^-(aq) \longrightarrow 2\,H_2O(l)$ or simply $H^+(aq) + OH^-(aq) \longrightarrow H_2O(l)$

FOR PRACTICE 9.11
Write a molecular and a net ionic equation for the reaction that occurs between aqueous H_2SO_4 and aqueous LiOH.

Acid–Base Titrations

We can apply the principles of acid–base neutralization and stoichiometry to a common laboratory procedure called a *titration*. In a **titration**, a substance in a solution of known concentration is reacted with another substance in a solution of unknown concentration. For example, consider the following acid–base reaction:

$$HCl(aq) + NaOH(aq) \longrightarrow H_2O(l) + NaCl(aq)$$

The net ionic equation for this reaction eliminates the spectator ions:

$$H^+(aq) + OH^-(aq) \longrightarrow H_2O(l)$$

Suppose we have an HCl solution represented by the following molecular diagram (we have omitted the Cl^- ions and the H_2O molecules not involved in the reaction from this representation for clarity):

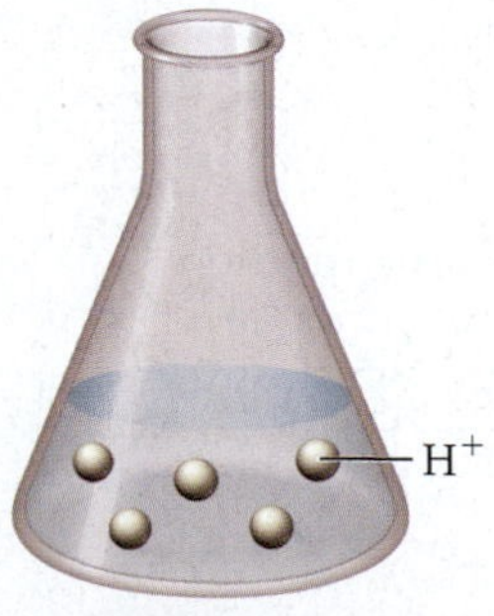

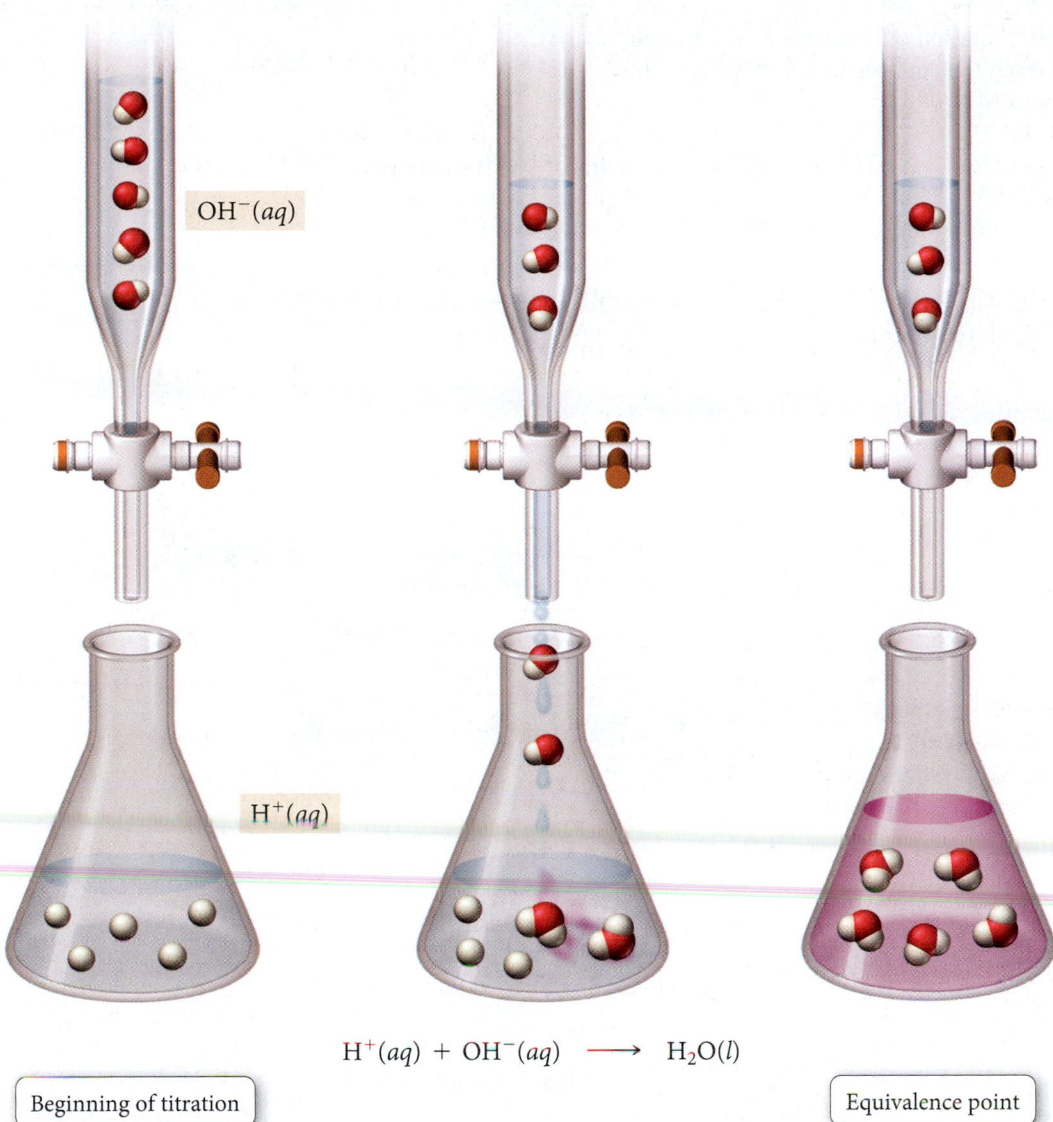

◀ **FIGURE 9.18 Acid–Base Titration**

In titrating this sample, we slowly add a solution of known OH^- concentration, as shown in the molecular diagrams in Figure 9.18 ▲. As the OH^- is added, it reacts with and neutralizes the H^+, forming water. At the **equivalence point**—the point in the titration when the number of moles of OH^- equals the number of moles of H^+ in solution—the titration is complete. The equivalence point is typically signaled by an **indicator**, a dye whose color depends on the acidity or basicity of the solution (Figure 9.19 ▶).

We cover acid–base titrations and indicators in more detail in Chapter 17. In most laboratory titrations, the concentration of one of the reactant solutions is unknown, and the concentration of the other is precisely known. By carefully measuring the volume of each solution required to reach the equivalence point, we can determine the concentration of the unknown solution, as demonstrated in Example 9.12.

Indicator in Titration

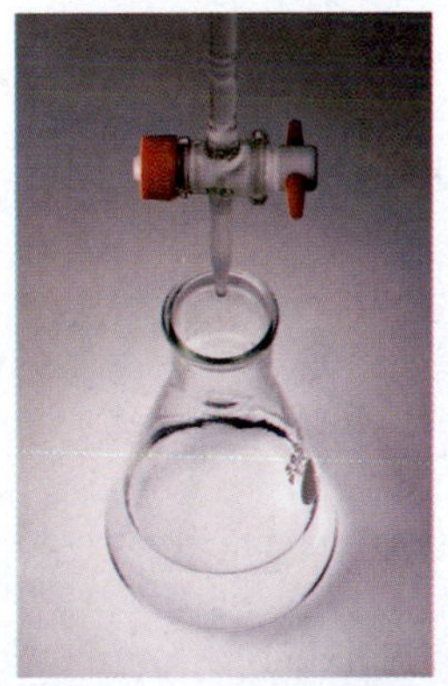

▲ **FIGURE 9.19 Titration** In this titration, NaOH is added to a dilute HCl solution. When the NaOH and HCl reach stoichiometric proportions (the equivalence point), the phenolphthalein indicator changes color to pink.

EXAMPLE 9.12
Acid–Base Titration

The titration of a 10.00-mL sample of an HCl solution of unknown concentration requires 12.54 mL of a 0.100 M NaOH solution to reach the equivalence point. What is the concentration of the unknown HCl solution in M?

SORT You are given the volume and concentration of NaOH solution required to titrate a given volume of HCl solution. You are asked to find the concentration of the HCl solution.

GIVEN: 12.54 mL of NaOH solution, 0.100 M NaOH solution, 10.00 mL of HCl solution

FIND: concentration of HCl solution

STRATEGIZE Since this problem involves an acid–base neutralization reaction between HCl and NaOH, you start by writing the balanced equation.

$$HCl(aq) + NaOH(aq) \longrightarrow H_2O(l) + NaCl(aq)$$

The first part of the conceptual plan has the form: volume A $\longrightarrow$ moles A $\longrightarrow$ moles B. The concentration of the NaOH solution is a conversion factor between moles and volume of NaOH. The balanced equation provides the relationship between number of moles of NaOH and number of moles of HCl.

CONCEPTUAL PLAN

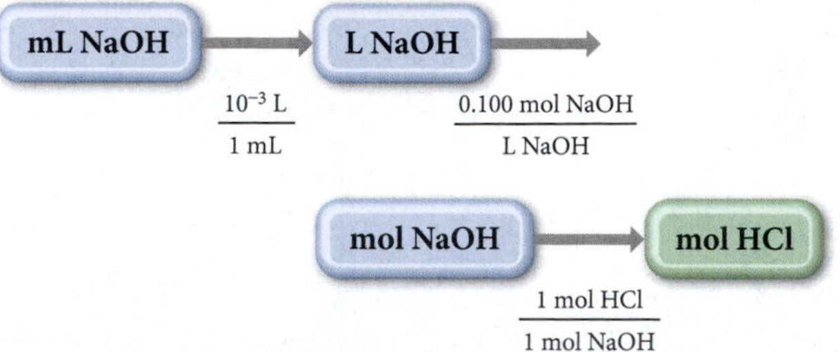

In the second part of the conceptual plan, use the number of moles of HCl (from the first part) and the volume of HCl solution (given) to calculate the molarity of the HCl solution.

mol HCl, L HCl solution $\longrightarrow$ molarity

$$M = \frac{mol}{L}$$

RELATIONSHIPS USED

$10^{-3}\ L = 1\ mL$

$$M\ (NaOH) = \frac{0.100\ mol\ NaOH}{L\ NaOH}$$

1 mol HCl : 1 mol NaOH

$$Molarity\ (M) = \frac{moles\ of\ solute\ (mol)}{volume\ of\ solution\ (L)}$$

SOLVE First determine the number of moles of HCl in the unknown solution.

SOLUTION

$$12.54\ \cancel{mL}\ \cancel{NaOH} \times \frac{10^{-3}\ \cancel{L}}{1\ \cancel{mL}} \times \frac{0.100\ \cancel{mol\ NaOH}}{\cancel{L}\ \cancel{NaOH}} \times \frac{1\ mol\ HCl}{1\ \cancel{mol\ NaOH}} = 1.25 \times 10^{-3}\ mol\ HCl$$

Next divide the number of moles of HCl by the volume of the HCl solution in L. 10.0 mL is equivalent to 0.010 L.

$$Molarity = \frac{1.25 \times 10^{-3}\ mol\ HCl}{0.01000\ L} = 0.125\ M\ HCl$$

CHECK The units of the answer (M HCl) are correct. The magnitude of the answer (0.125 M) is reasonable because it is similar to the molarity of the NaOH solution, as expected from the reaction stoichiometry (1 mol HCl reacts with 1 mol NaOH) and the similar volumes of NaOH and HCl.

FOR PRACTICE 9.12
The titration of a 20.0-mL sample of an H_2SO_4 solution of unknown concentration requires 22.87 mL of a 0.158 M KOH solution to reach the equivalence point. What is the concentration of the unknown H_2SO_4 solution?

FOR MORE PRACTICE 9.12
What volume (in mL) of 0.200 M NaOH do we need to titrate 35.00 mL of 0.140 M HBr to the equivalence point?

9.8 Gas-Evolution Reactions

In a **gas-evolution reaction**, a gas forms, resulting in bubbling. As in precipitation reactions (see Section 9.5), the reactions occur when the anion from one reactant combines with the cation of the other. Many gas-evolution reactions are also acid–base reactions. Some gas-evolution reactions form a gaseous product directly when the cation of one reactant combines with the anion of the other. For example, when sulfuric acid reacts with lithium sulfide, dihydrogen sulfide gas forms:

$$H_2SO_4(aq) + Li_2S(aq) \longrightarrow \underset{\text{gas}}{H_2S(g)} + Li_2SO_4(aq)$$

Other gas-evolution reactions form an intermediate product that then decomposes (breaks down into simpler substances) to form a gas. For example, when aqueous hydrochloric acid is mixed with aqueous sodium bicarbonate, the following reaction occurs (Figure 9.20 ▼):

$$HCl(aq) + NaHCO_3(aq) \longrightarrow \underset{\text{intermediate product}}{H_2CO_3(aq)} + NaCl(aq) \longrightarrow H_2O(l) + \underset{\text{gas}}{CO_2(g)} + NaCl(aq)$$

▲ Gas-evolution reactions, such as the reaction of hydrochloric acid with limestone ($CaCO_3$), often produce CO_2; bubbling occurs as the gas is released.

Gas-Evolution Reaction

$$NaHCO_3(aq) + HCl(aq) \longrightarrow H_2O(l) + NaCl(aq) + CO_2(g)$$

When aqueous sodium bicarbonate is mixed with aqueous hydrochloric acid, gaseous CO_2 bubbles are the result of the reaction.

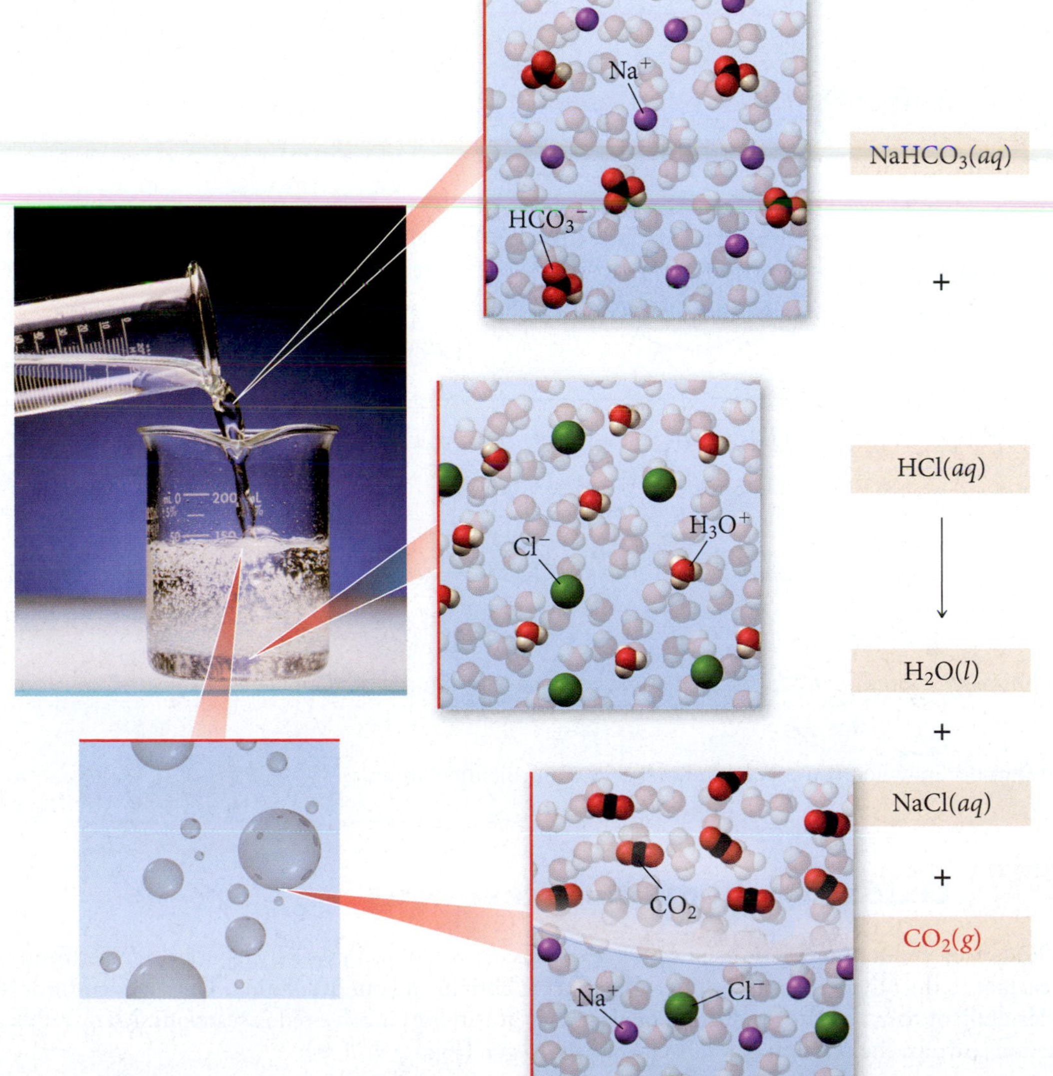

◀ **FIGURE 9.20 Gas-Evolution Reaction** When we mix aqueous hydrochloric acid with aqueous sodium bicarbonate, gaseous CO_2 bubbles out of the reaction mixture.

TABLE 9.3 Types of Compounds That Undergo Gas-Evolution Reactions

Reactant Type	Intermediate Product	Gas Evolved	Example
Sulfides	None	H_2S	$2\ HCl(aq) + K_2S(aq) \longrightarrow H_2S(g) + 2\ KCl(aq)$
Carbonates and bicarbonates	H_2CO_3	CO_2	$2\ HCl(aq) + K_2CO_3(aq) \longrightarrow H_2O(l) + CO_2(g) + 2\ KCl(aq)$
Sulfites and bisulfites	H_2SO_3	SO_2	$2\ HCl(aq) + K_2SO_3(aq) \longrightarrow H_2O + SO_2(g) + 2\ KCl(aq)$
Ammonium	NH_4OH	NH_3	$NH_4Cl(aq) + KOH(aq) \longrightarrow H_2O(l) + NH_3(g) + KCl(aq)$

The intermediate product, H_2CO_3, is not stable and decomposes into H_2O and gaseous CO_2. Other important gas-evolution reactions form either H_2SO_3 or NH_4OH as intermediate products:

The intermediate product NH_4OH provides a convenient way to think about this reaction, but the extent to which it actually forms is debatable.

$$HCl(aq) + NaHSO_3(aq) \longrightarrow \underbrace{H_2SO_3(aq)}_{\text{intermediate product}} + NaCl(aq) \longrightarrow H_2O(l) + \underbrace{SO_2(g)}_{\text{gas}} + NaCl(aq)$$

$$NH_4Cl(aq) + NaOH(aq) \longrightarrow \underbrace{NH_4OH(aq)}_{\text{intermediate product}} + NaCl(aq) \longrightarrow H_2O(l) + \underbrace{NH_3(g)}_{\text{gas}} + NaCl(aq)$$

Table 9.3 lists the main types of compounds that form gases in aqueous reactions, as well as the gases that form.

EXAMPLE 9.13
Writing Equations for Gas-Evolution Reactions

Write a molecular equation for the gas-evolution reaction that occurs when you mix aqueous nitric acid and aqueous sodium carbonate.

Begin by writing an unbalanced equation in which the cation of each reactant combines with the anion of the other.	$HNO_3(aq) + Na_2CO_3(aq) \longrightarrow H_2CO_3(aq) + NaNO_3(aq)$
You must then recognize that $H_2CO_3(aq)$ decomposes into $H_2O(l)$ and $CO_2(g)$ and write these products into the equation.	$HNO_3(aq) + Na_2CO_3(aq) \longrightarrow H_2O(l) + CO_2(g) + NaNO_3(aq)$
Finally, balance the equation.	$2\ HNO_3(aq) + Na_2CO_3(aq) \longrightarrow H_2O(l) + CO_2(g) + 2\ NaNO_3(aq)$

FOR PRACTICE 9.13
Write a molecular equation for the gas-evolution reaction that occurs when you mix aqueous hydrobromic acid and aqueous potassium sulfite.

FOR MORE PRACTICE 9.13
Write a net ionic equation for the reaction that occurs when you mix hydroiodic acid with calcium sulfide.

9.9 Oxidation–Reduction Reactions

Oxidation–reduction reactions are covered in more detail in Chapter 20.

Oxidation–reduction reactions or **redox reactions** are reactions in which electrons transfer from one reactant to the other. These types of reactions occur both in and out of solution. The rusting of iron, the bleaching of hair, and the production of electricity in batteries involve redox reactions. Many redox reactions involve the reaction of a substance with oxygen (Figure 9.21 ▶):

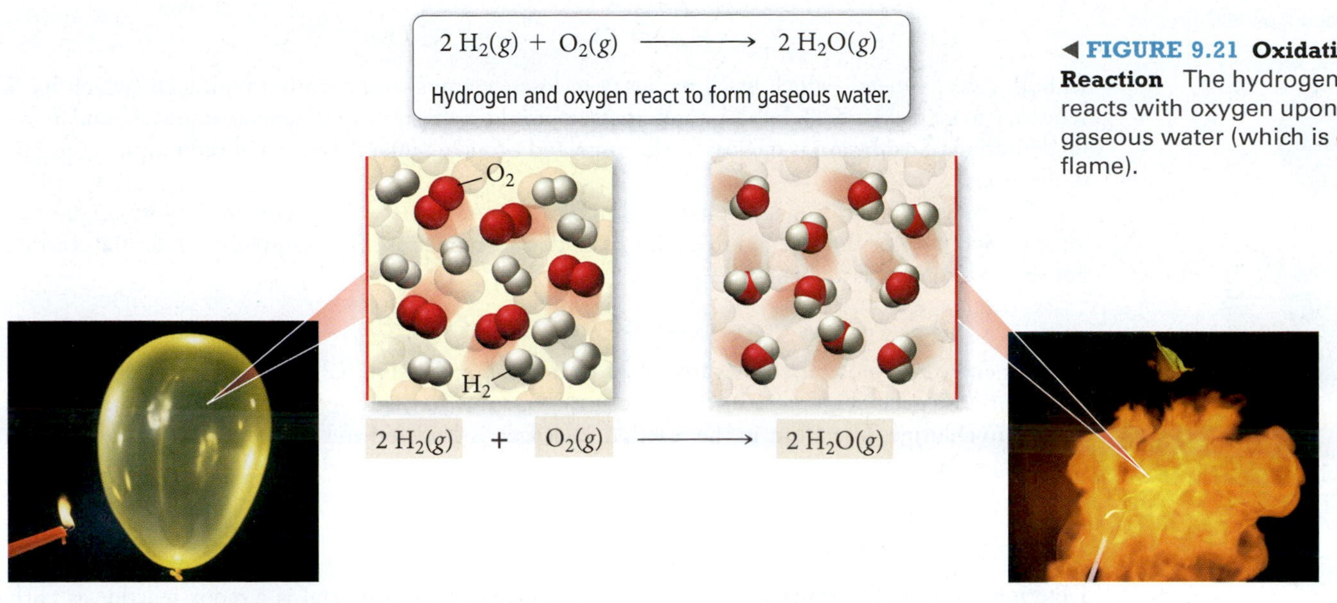

◀ **FIGURE 9.21 Oxidation–Reduction Reaction** The hydrogen in the balloon reacts with oxygen upon ignition to form gaseous water (which is dispersed in the flame).

$$4\ Fe(s) + 3\ O_2(g) \longrightarrow 2\ Fe_2O_3(s) \quad \text{(rusting of iron)}$$
$$2\ C_8H_{18}(l) + 25\ O_2(g) \longrightarrow 16\ CO_2(g) + 18\ H_2O(g) \quad \text{(combustion of octane)}$$
$$2\ H_2(g) + O_2(g) \longrightarrow 2H_2O(g) \quad \text{(combustion of hydrogen)}$$

Combustion reactions, first covered in Section 8.6, are a type of redox reaction. However, redox reactions need not involve oxygen. Consider, for example, the reaction between sodium and chlorine to form sodium chloride (NaCl), depicted in Figure 9.22 ▼:

$$2\ Na(s) + Cl_2(g) \longrightarrow 2\ NaCl(s)$$

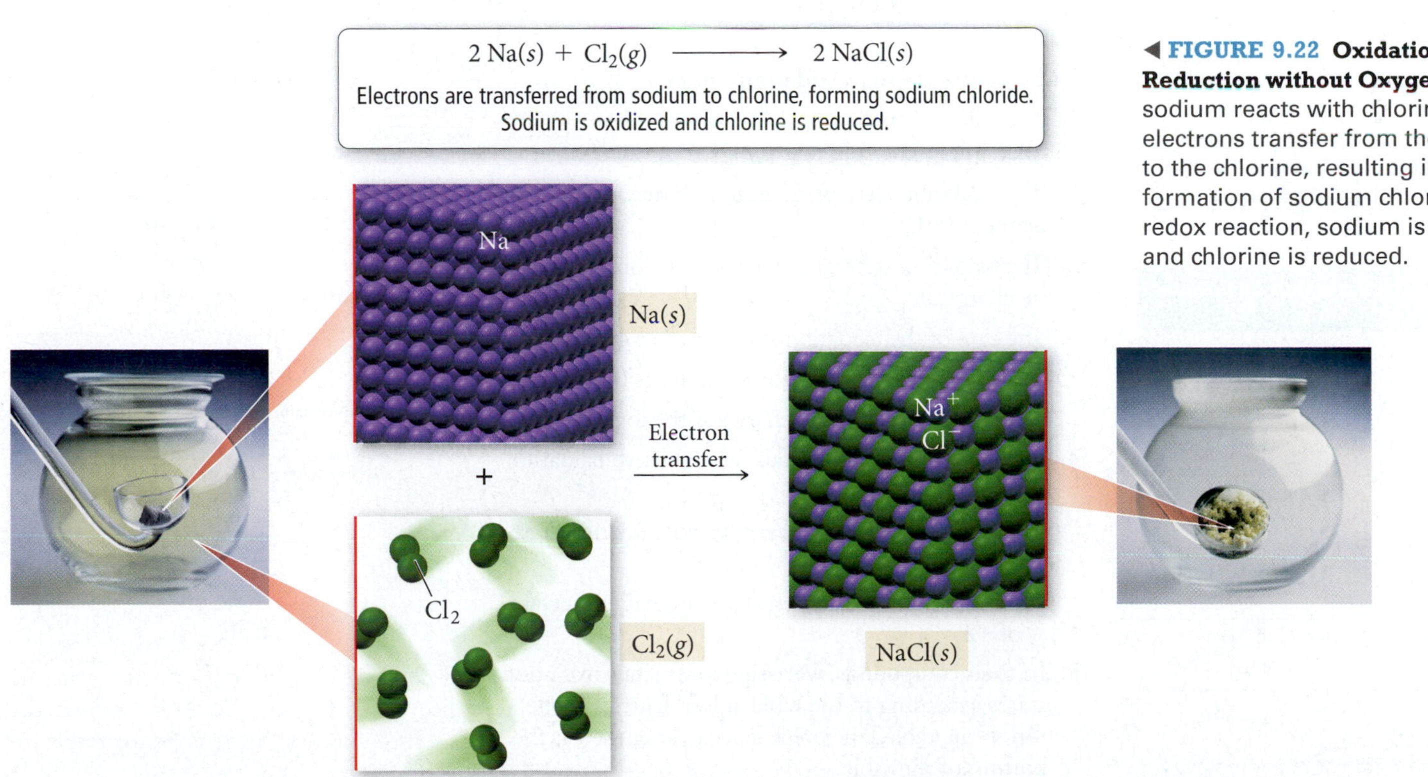

◀ **FIGURE 9.22 Oxidation–Reduction without Oxygen** When sodium reacts with chlorine, electrons transfer from the sodium to the chlorine, resulting in the formation of sodium chloride. In this redox reaction, sodium is oxidized and chlorine is reduced.

Helpful Mnemonic: O I L R I G—
Oxidation **I**s **L**oss; **R**eduction **I**s **G**ain.

This reaction is similar to the reaction between sodium and oxygen, which forms sodium oxide (as one possible product):

$$4\,\mathrm{Na}(s) + \mathrm{O_2}(g) \longrightarrow 2\,\mathrm{Na_2O}(s)$$

In both cases, a metal (which has a tendency to lose electrons) reacts with a nonmetal (which has a tendency to gain electrons). In both cases, metal atoms lose electrons to nonmetal atoms. A fundamental definition of **oxidation** is the loss of electrons, and a fundamental definition of **reduction** is the gain of electrons.

The transfer of electrons does not need to be a *complete* transfer (as occurs in the formation of an ionic compound) for the reaction to qualify as oxidation–reduction. For example, consider the reaction between hydrogen gas and chlorine gas:

$$\mathrm{H_2}(g) + \mathrm{Cl_2}(g) \longrightarrow 2\,\mathrm{HCl}(g)$$

Because chorine is more electronegative than hydrogen (see Section 6.2) the bond between hydrogen and chlorine is polar (Figure 9.23 ◀). Notice that in the reaction, hydrogen has lost some of its electron density to chlorine. Therefore, in the reaction, hydrogen is oxidized and chlorine is reduced—this is a redox reaction.

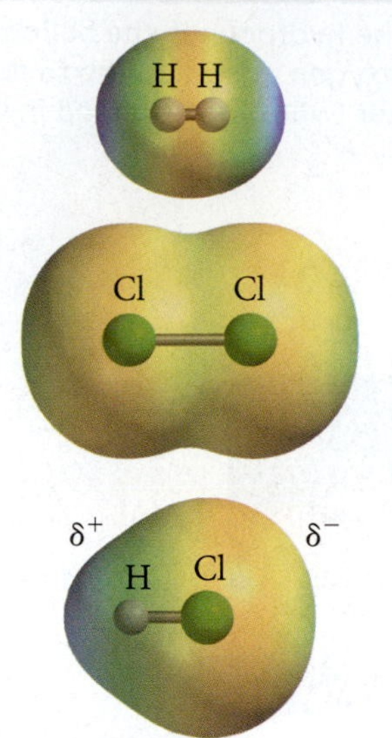
H H
Cl Cl
δ+
δ−
H Cl
Hydrogen loses electron density (oxidation), and chlorine gains electron density (reduction).

▲ **FIGURE 9.23 Electrostatic Potential Maps Showing Electron Transfer** When hydrogen bonds to chlorine, the electrons are unevenly shared, resulting in an increase of electron density (reduction) for chlorine and a decrease in electron density (oxidation) for hydrogen.

Oxidation States

Determining whether or not a reaction between a metal and a nonmetal is a redox reaction is fairly straightforward because of ion formation. But how do we identify redox reactions that occur between nonmetals? Chemists have devised a scheme to track electrons before and after a chemical reaction. In this scheme—which is like bookkeeping for electrons—we assign each shared electron to the most electronegative atom. Then we give a number, called the **oxidation state** or **oxidation number**, to each atom based on the electron assignments. In other words, *the oxidation number of an atom in a compound is the "charge" it would have if all shared electrons were assigned to the more electronegative atom.*

For example, consider HCl. Since chlorine is more electronegative than hydrogen, we assign the two shared electrons in the bond to chlorine; then H (which lost an electron in our assignment) has an oxidation state of +1, and Cl (which gained one electron in our assignment) has an oxidation state of −1. Notice that, in contrast to ionic charges, which are usually written with the sign of the charge *after* the magnitude (1+ and 1−, for example), we write oxidation states with the sign of the charge *before* the magnitude (+1 and −1, for example). We use the following rules to assign oxidation states to atoms in elements and compounds:

Do not confuse oxidation state with ionic charge. Unlike ionic charge—which is a real property of an ion—the oxidation state of an atom is merely a theoretical (but useful) construct.

Nonmetal	Oxidation State	Example
Fluorine	−1	MgF_2 (−1 ox state)
Hydrogen	+1	H_2O (+1 ox state)
Oxygen	−2	CO_2 (−2 ox state)
Group 7A	−1	CCl_4 (−1 ox state)
Group 6A	−2	H_2S (−2 ox state)
Group 5A	−3	NH_3 (−3 ox state)

Rules for Assigning Oxidation States	Examples
(These rules are hierarchical. If any two rules conflict, follow the rule that is higher on the list.)	
1. The oxidation state of an atom in a free element is 0.	Cu (0 ox state); Cl_2 (0 ox state)
2. The oxidation state of a monoatomic ion is equal to its charge.	Ca^{2+} (+2 ox state); Cl^- (−1 ox state)
3. The sum of the oxidation states of all atoms in:	
• a neutral molecule or formula unit is 0.	H_2O: 2(H ox state) + 1(O ox state) = 0
• an ion is equal to the charge of the ion.	NO_3^-: 1(N ox state) + 3(O ox state) = −1
4. In their compounds, metals have positive oxidation states.	
• Group 1A metals *always* have an oxidation state of +1.	NaCl (+1 ox state)
• Group 2A metals *always* have an oxidation state of +2.	CaF_2 (+2 ox state)
5. In their compounds, we assign nonmetals oxidation states according to the table at left. Entries at the top of the table take precedence over entries at the bottom of the table.	

When assigning oxidation states, these points also apply:

- The oxidation state of any given element generally depends on what other elements are present in the compound. (The exceptions are the group 1A and 2A metals, which are *always* +1 and +2, respectively.)
- Rule 3 must always be followed. Therefore, when following the hierarchy shown in rule 5, we give priority to the element(s) highest on the list and then assign the oxidation state of the element lowest on the list using rule 3.
- When assigning oxidation states to elements that are not covered by rules 4 and 5 (such as carbon), we use rule 3 to deduce their oxidation state once all other oxidation states have been assigned.

EXAMPLE 9.14
Assigning Oxidation States

Assign an oxidation state to each atom in each element, ion, or compound.

(a) Cl_2 **(b)** Na^+ **(c)** KF **(d)** CO_2 **(e)** SO_4^{2-} **(f)** K_2O_2

SOLUTION

Because Cl_2 is a free element, the oxidation state of both Cl atoms is 0 (rule 1).	**(a)** Cl_2 ClCl 0 0
Because Na^+ is a monoatomic ion, the oxidation state of the Na^+ ion is +1 (rule 2).	**(b)** Na^+ Na^+ +1
The oxidation state of K is +1 (rule 4). The oxidation state of F is −1 (rule 5). Because this is a neutral compound, the sum of the oxidation states is 0.	**(c)** KF KF +1 −1 sum: +1 −1=0
The oxidation state of oxygen is −2 (rule 5). You must deduce the oxidation state of carbon using rule 3, which says that the sum of the oxidation states of all the atoms must be 0.	**(d)** CO_2 (C ox state) + 2(O ox state) = 0 (C ox state) + 2(−2) = 0 C ox state = +4 CO_2 +4 −2 sum: +4 +2(−2)=0
The oxidation state of oxygen is −2 (rule 5). You would ordinarily expect the oxidation state of S to be −2 (rule 5). However, if that were the case, the sum of the oxidation states would not equal the charge of the ion. Because O is higher on the list than S, it takes priority and you calculate the oxidation state of sulfur by setting the *sum* of all of the oxidation states equal to −2 (the charge of the ion).	**(e)** SO_4^{2-} (S ox state) + 4(O ox state) = −2 (S ox state) + 4(−2) = −2 S ox state = +6 SO_4^{2-} +6 −2 sum: +6 +4(−2)=−2
The oxidation state of potassium is +1 (rule 4). You would ordinarily expect the oxidation state of O to be −2 (rule 5), but rule 4 takes priority, and you need to deduce the oxidation state of O by setting the sum of all of the oxidation states equal to 0.	**(f)** K_2O_2 2(K ox state) + 2(O ox state) = 0 2(+1) + 2(O ox state) = 0 O ox state = −1 K_2O_2 +1 −1 sum: 2(+1) +2(−1)=0

FOR PRACTICE 9.14

Assign an oxidation state to each atom in each element, ion, or compound.

(a) Cr **(b)** Cr^{3+} **(c)** CCl_4 **(d)** $SrBr_2$ **(e)** SO_3 **(f)** NO_3^-

In most cases, oxidation states are positive or negative integers; however, on occasion an atom within a compound can have a fractional oxidation state. Consider KO_2. We assign the oxidation states as follows:

$$\underset{+1\;-\frac{1}{2}}{KO_2}$$

$$\text{sum: } +1+2\left(-\frac{1}{2}\right)=0$$

In KO_2, oxygen has a $-\frac{1}{2}$ oxidation state. Although this seems unusual, it is accepted because oxidation states are merely an imposed electron bookkeeping scheme, not an actual physical quantity.

9.4
Cc
Conceptual Connection

Oxidation Numbers in Polyatomic Ions

Which statement best describes the *difference* between the *charge* of a polyatomic ion and the *oxidation states* of its constituent atoms? (For example, the charge of NO_3^- is 1−, and the oxidation states of its atoms are +5 for the nitrogen atom and −2 for each oxygen atom.)

(a) The charge of a polyatomic ion is a property of the entire ion, while the oxidation states are assigned to each individual atom.

(b) The oxidation state of the ion is the same as its charge.

(c) The charge of a polyatomic ion is not a real physical property, while the oxidation states of atoms are actual physical properties.

We discuss how to balance redox reactions in Section 20.2.

Identifying Redox Reactions

We can use oxidation states to identify redox reactions, even between nonmetals. For example, is the following reaction between carbon and sulfur a redox reaction?

$$C + 2\,S \longrightarrow CS_2$$

If so, what element is oxidized? What element is reduced? We use the oxidation state rules to assign oxidation states to all elements on both sides of the equation:

$$\underset{0}{C} + \underset{0}{2\,S} \longrightarrow \underset{+4\;-2}{CS_2}$$

Oxidation states: 0, 0 → +4 −2 (Reduction: S 0 → −2; Oxidation: C 0 → +4)

Carbon changes from an oxidation state of 0 to an oxidation state of +4. In terms of our electron bookkeeping scheme (the assigned oxidation state), carbon *loses electrons* and is *oxidized.* Sulfur changes from an oxidation state of 0 to an oxidation state of −2. In terms of our electron bookkeeping scheme, sulfur *gains electrons* and is *reduced.* In terms of oxidation states, oxidation and reduction are defined as follows:

Remember that a reduction is a *reduction* in oxidation state.

- Oxidation: An increase in oxidation state
- Reduction: A decrease in oxidation state

EXAMPLE 9.15

Using Oxidation States to Identify Oxidation and Reduction

Use oxidation states to identify the element that is oxidized and the element that is reduced in the redox reaction.

$$Mg(s) + 2\,H_2O(l) \longrightarrow Mg(OH)_2(aq) + H_2(g)$$

SOLUTION

Begin by assigning an oxidation state to each atom in the reaction.

$$Mg(s) + 2\,H_2O(l) \longrightarrow Mg(OH)_2(aq) + H_2(g)$$

Oxidation states: 0 (Mg); +1 −2 (H_2O); +2 −2 +1 ($Mg(OH)_2$); 0 (H_2)

Oxidation (Mg: 0 → +2); Reduction (H: +1 → 0)

Since Mg increased in the oxidation state, it was oxidized. Since H decreased in the oxidation state, it was reduced.

FOR PRACTICE 9.15

Use oxidation states to identify the element that is oxidized and the element that is reduced in the redox reaction.

$$Sn(s) + 4\,HNO_3(aq) \longrightarrow SnO_2(s) + 4\,NO_2(g) + 2\,H_2O(g)$$

FOR MORE PRACTICE 9.15

Determine whether or not each reaction is a redox reaction. If the reaction is a redox reaction, identify which element is oxidized and which is reduced.

(a) $Hg_2(NO_3)_2(aq) + 2\,KBr(aq) \longrightarrow Hg_2Br_2(s) + 2\,KNO_3(aq)$

(b) $4\,Al(s) + 3\,O_2(g) \longrightarrow 2\,Al_2O_3(s)$

(c) $CaO(s) + CO_2(g) \longrightarrow CaCO_3(s)$

Notice that *oxidation and reduction must occur together.* If one substance loses electrons (oxidation), then another substance must gain electrons (reduction). A substance that causes the oxidation of another substance is an **oxidizing agent**. Oxygen, for example, is an excellent oxidizing agent because it causes the oxidation of many substances. In a redox reaction, *the oxidizing agent is always reduced.* A substance that causes the reduction of another substance is a **reducing agent**. Hydrogen, for example, as well as the group 1A and group 2A metals (because of their tendency to lose electrons) are excellent reducing agents. In a redox reaction, *the reducing agent is always oxidized.*

In Section 20.2, we will further discuss redox reactions, including how to balance them. For now, you need to be able to identify redox reactions, as well as oxidizing and reducing agents, according to the following guidelines.

Redox reactions:

- Any reaction in which there is a change in the oxidation states of atoms in going from reactants to products.

In a redox reaction:

- The oxidizing agent oxidizes another substance (and is itself reduced).
- The reducing agent reduces another substance (and is itself oxidized).

EXAMPLE 9.16

Identifying Redox Reactions, Oxidizing Agents, and Reducing Agents

Determine whether or not each reaction is an oxidation–reduction reaction. For each oxidation–reduction reaction, identify the oxidizing agent and the reducing agent.

(a) $2\,Mg(s) + O_2(g) \longrightarrow 2\,MgO(s)$

(b) $2\,HBr(aq) + Ca(OH)_2(aq) \longrightarrow 2\,H_2O(l) + CaBr_2(aq)$

(c) $Zn(s) + Fe^{2+}(aq) \longrightarrow Zn^{2+}(aq) + Fe(s)$

SOLUTION

This is a redox reaction because magnesium increases in oxidation number (oxidation) and oxygen decreases in oxidation number (reduction).	**(a)** $2\,Mg(s) + O_2(g) \longrightarrow 2\,MgO(s)$ Oxidation numbers: Mg 0, O 0 → Mg +2, O −2 Reduction; Oxidation Oxidizing agent: O_2 Reducing agent: Mg
This is not a redox reaction because none of the atoms undergoes a change in oxidation number.	**(b)** $2\,HBr(aq) + Ca(OH)_2(aq) \longrightarrow 2H_2O(l) + CaBr_2(aq)$ Oxidation numbers: +1 −1; +2 −2 +1; +1 −2; +2 −1
This is a redox reaction because zinc increases in oxidation number (oxidation) and iron decreases in oxidation number (reduction).	**(c)** $Zn(s) + Fe^{2+}(aq) \longrightarrow Zn^{2+}(aq) + Fe(s)$ Oxidation numbers: 0; +2 → +2; 0 Reduction; Oxidation Oxidizing agent: Fe^{2+} Reducing agent: Zn

FOR PRACTICE 9.16

Determine whether or not each reaction is a redox reaction. For all redox reactions, identify the oxidizing agent and the reducing agent.

(a) $2\,Li(s) + Cl_2(g) \longrightarrow 2\,LiCl(s)$

(b) $2\,Al(s) + 3\,Sn^{2+}(aq) \longrightarrow 2\,Al^{3+}(aq) + 3\,Sn(s)$

(c) $Pb(NO_3)_2(aq) + 2\,LiCl(aq) \longrightarrow PbCl_2(s) + 2\,LiNO_3(aq)$

(d) $C(s) + O_2(g) \longrightarrow CO_2(g)$

9.5 Cc Conceptual Connection — Oxidation and Reduction

Which statement is true?

(a) A redox reaction involves *either* the transfer of an electron *or* a change in the oxidation state of an element.

(b) If any of the reactants or products in a reaction contains oxygen, the reaction is a redox reaction.

(c) In a reaction, oxidation can occur independently of reduction.

(d) In a redox reaction, any increase in the oxidation state of a reactant must be accompanied by a decrease in the oxidation state of a reactant.

SELF-ASSESSMENT Quiz

1. What is the molarity of a solution containing 55.8 g of $MgCl_2$ dissolved in 1.00 L of solution?
   a) 55.8 M **b)** 1.71 M **c)** 0.586 M **d)** 0.558 M

2. What mass (in grams) of $Mg(NO_3)_2$ is present in 145 mL of a 0.150 M solution of $Mg(NO_3)_2$?
   a) 3.23 g **b)** 0.022 g **c)** 1.88 g **d)** 143 g

3. What volume of a 1.50 M HCl solution should you use to prepare 2.00 L of a 0.100 M HCl solution?
   a) 0.300 L **b)** 0.133 L **c)** 30.0 L **d)** 2.00 L

4. Potassium iodide reacts with lead(II) nitrate in this precipitation reaction:

   $$2\ KI(aq) + Pb(NO_3)_2(aq) \longrightarrow 2\ KNO_3(aq) + PbI_2(s)$$

   What minimum volume of 0.200 M potassium iodide solution is required to completely precipitate all of the lead in 155.0 mL of a 0.112 M lead(II) nitrate solution?
   a) 348 mL **b)** 86.8 mL **c)** 174 mL **d)** 43.4 mL

5. Which solution forms a precipitate when mixed with a solution of aqueous Na_2CO_3?
   a) $KNO_3(aq)$ **b)** $NaBr(aq)$ **c)** $NH_4Cl(aq)$ **d)** $CuCl_2(aq)$

6. What is the net ionic equation for the reaction that occurs when you mix aqueous solutions of KOH and $SrCl_2$?
   a) $K^+(aq) + Cl^-(aq) \longrightarrow KCl(s)$
   b) $Sr^{2+}(aq) + 2\ OH^-(aq) \longrightarrow Sr(OH)_2(s)$
   c) $H^+(aq) + OH^-(aq) \longrightarrow H_2O(l)$
   d) None of the above because no reaction occurs.

7. What is the net ionic equation for the reaction that occurs when you mix aqueous solutions of KOH and HNO_3?
   a) $K^+(aq) + NO_3^-(aq) \longrightarrow KNO_3(s)$
   b) $NO_3^-(aq) + OH^-(aq) \longrightarrow NO_3OH(s)$
   c) $H^+(aq) + OH^-(aq) \longrightarrow H_2O(l)$
   d) None of the above because no reaction occurs.

8. What is the net ionic equation for the reaction that occurs when you mix aqueous solutions of $KHCO_3$ and $HC_2H_3O_2$?
   a) $K^+(aq) + C_2H_3O_2^-(aq) \longrightarrow KC_2H_3O_2(s)$
   b) $HCO_3^-(aq) + HC_2H_3O_2(aq) \longrightarrow H_2O(l) + CO_2(g) + C_2H_3O_2^-(aq)$
   c) $H^+(aq) + OH^-(aq) \longrightarrow H_2O(l)$
   d) None of the above because no reaction occurs.

9. What is the oxidation state of carbon in CO_3^{2-}?
   a) +4 **b)** +3 **c)** −3 **d)** −2

10. Sodium reacts with water according to the reaction:

    $$2\ Na(s) + 2\ H_2O(l) \longrightarrow 2\ NaOH(aq) + H_2(g)$$

    Identify the oxidizing agent.
    a) $Na(s)$ **b)** $H_2O(l)$
    c) $NaOH(aq)$ **d)** $H_2(aq)$

11. Name the compound $HNO_2(aq)$.
    a) Hydrogen nitrogen dioxide
    b) Hydrogen nitrate
    c) Nitric acid
    d) Nitrous acid

Answers: 1:c; 2:a; 3:b; 4:c; 5:d; 6:b; 7:c; 8:b; 9:a; 10:b; 11:d

CHAPTER SUMMARY 9

REVIEW

KEY LEARNING OUTCOMES

CHAPTER OBJECTIVES	ASSESSMENT
Calculating and Using Molarity as a Conversion Factor (9.2)	• Examples 9.1, 9.2 For Practice 9.1, 9.2 For More Practice 9.1, 9.2 Exercises 21–28
Determining Solution Dilutions (9.2)	• Example 9.3 For Practice 9.3 For More Practice 9.3 Exercises 29–32

Using Solution Stoichiometry to Find Volumes and Amounts (9.3)	• Example 9.4 For Practice 9.4 For More Practice 9.4 Exercises 33–38
Predicting Compound Solubility (9.4)	• Example 9.5 For Practice 9.5 Exercises 39–42
Writing Equations for Precipitation Reactions (9.5)	• Examples 9.6, 9.7 For Practice 9.6, 9.7 Exercises 43–46
Writing Complete Ionic and Net Ionic Equations (9.6)	• Example 9.8 For Practice 9.8 For More Practice 9.8 Exercises 47–50
Naming Acids (9.7)	• Examples 9.9, 9.10 For Practice 9.9, 9.10 For More Practice 9.10 Exercises 51–54
Writing Equations for Acid–Base Reactions (9.7)	• Example 9.11 For Practice 9.11 Exercises 55–58
Calculations Involving Acid–Base Titrations (9.7)	• Example 9.12 For Practice 9.12 For More Practice 9.12 Exercises 59, 60
Writing Equations for Gas-Evolution Reactions (9.8)	• Example 9.13 For Practice 9.13 For More Practice 9.13 Exercises 61–64
Assigning Oxidation States (9.9)	• Example 9.14 For Practice 9.14 Exercises 65–68
Identifying Redox Reactions, Oxidizing Agents, and Reducing Agents Using Oxidation States (9.9)	• Examples 9.15, 9.16 For Practice 9.15, 9.16 For More Practice 9.15 Exercises 69, 70

KEY TERMS

Section 9.2
solution (302)
solvent (302)
solute (302)
aqueous solution (302)
dilute solution (302)
concentrated solution (302)
molarity (M) (302)
stock solution (304)

Section 9.4
electrolyte (310)
strong electrolyte (310)
nonelectrolyte (310)
acid (310)
strong acid (310)
weak acid (311)
weak electrolyte (311)
soluble (311)
insoluble (311)

Section 9.5
precipitation reaction (313)
precipitate (313)

Section 9.6
molecular equation (318)
complete ionic equation (318)
spectator ion (318)
net ionic equation (318)

Section 9.7
acid–base reaction (neutralization reaction) (319)
Arrhenius definition (320)
hydronium ion (320)
polyprotic acid (320)
diprotic acid (320)
base (320)
binary acid (321)
oxyacid (322)
salt (324)
titration (324)
equivalence point (325)
indicator (325)

Section 9.8
gas-evolution reaction (327)

Section 9.9
oxidation–reduction (redox) reaction (328)
oxidation (330)
reduction (330)
oxidation state (oxidation number) (330)
oxidizing agent (333)
reducing agent (333)

KEY CONCEPTS

Solution Concentration (9.2)

- An aqueous solution is a homogeneous mixture of water (the solvent) with another substance (the solute).
- We often express the concentration of a solution in molarity, the number of moles of solute per liter of solution.

Solution Stoichiometry (9.3)

- We can use the molarities and volumes of reactant solutions to predict the amount of product that will form in an aqueous reaction or the amount of one reactant needed to react with a given amount of another reactant.

Aqueous Solutions and Precipitation Reactions (9.4, 9.5)

- Solutes that completely dissociate (or completely ionize in the case of the acids) to ions in solution are strong electrolytes and are good conductors of electricity. Water soluble ionic compounds, strong acids and strong bases are strong electrolytes.
- Solutes that only partially dissociate (or partially ionize) are weak electrolytes. Weak acids are weak electrolytes.
- Solutes that do not dissociate (or ionize) are nonelectrolytes.
- A substance that dissolves in water to form a solution is soluble.
- The solubility rules are an empirical set of guidelines that help predict the solubilities of ionic compounds; these rules are especially useful when determining whether or not a precipitate will form.

- In a precipitation reaction, we mix two aqueous solutions and a solid—a precipitate—forms.

Equations for Aqueous Reactions (9.6)

- We can represent an aqueous reaction with a molecular equation, which shows the complete neutral formula for each compound in the reaction.
- We can also represent an aqueous reaction with a complete ionic equation, which shows the dissociated nature of the aqueous ionic compounds.
- A third representation of an aqueous reaction is the net ionic equation, in which the spectator ions—those that do not change in the course of the reaction—are left out of the equation.

Acid–Base Reactions (9.7)

- In an acid–base reaction, an acid, a substance that produces H^+ in solution, reacts with a base, a substance that produces OH^- in solution, and the two neutralize each other, producing water (or in some cases a weak electrolyte).
- An acid–base titration is a laboratory procedure in which a reaction is carried to its equivalence point—the point at which the reactants are in exact stoichiometric proportions; titrations are useful in determining the concentrations of unknown solutions.

Gas-Evolution Reactions (9.8)

- In gas-evolution reactions, two aqueous solutions combine and a gas is produced.

Oxidation–Reduction Reactions (9.9)

- In oxidation–reduction reactions, one substance transfers electrons to another substance.
- The substance that loses electrons is oxidized, and the substance that gains them is reduced.
- An oxidation state is a fictitious charge given to each atom in a redox reaction by assigning all shared electrons to the atom with the greater attraction for those electrons. Oxidation states are an imposed electronic bookkeeping scheme, not an actual physical state.
- The oxidation state of an atom increases upon oxidation and decreases upon reduction.

KEY EQUATIONS AND RELATIONSHIPS

Molarity (M): Solution Concentration (9.2)

$$M = \frac{\text{amount of solute (in mol)}}{\text{volume of solution (in L)}}$$

Solution Dilution (9.2)

$$M_1V_1 = M_2V_2$$

Solution Stoichiometry (9.3)

$$\text{volume A} \longrightarrow \text{amount A (in moles)} \longrightarrow \text{amount B (in moles)} \longrightarrow \text{volume B}$$

EXERCISES

REVIEW QUESTIONS

1. What is an aqueous solution? What is the difference between the solute and the solvent?
2. What is molarity? How is it useful?
3. Explain how a strong electrolyte, a weak electrolyte, and a nonelectrolyte differ.
4. What is an acid? Explain the difference between a strong acid and a weak acid.
5. What does it mean for a compound to be soluble? Insoluble?
6. What are the solubility rules? How are they useful?
7. Which cations and anions form compounds that are usually soluble? What are the exceptions? Which anions form compounds that are mostly insoluble? What are the exceptions?
8. What is a precipitation reaction? Give an example.
9. How can you predict whether a precipitation reaction will occur upon mixing two aqueous solutions?
10. Explain how a molecular equation, a complete ionic equation, and a net ionic equation differ.
11. What is the Arrhenius definition of a base?
12. Explain how to name binary acids and oxyacids.
13. What is an acid–base reaction? Provide an example.
14. Explain the principles behind an acid–base titration. What is an indicator?
15. What is a gas-evolution reaction? Provide an example.
16. Which reactant types give rise to gas-evolution reactions?
17. What is an oxidation–reduction reaction? Provide an example.
18. What are oxidation states? How can oxidation states be used to identify redox reactions?
19. What happens to a substance when it becomes oxidized? Reduced?
20. In a redox reaction, which reactant is the oxidizing agent? The reducing agent?

PROBLEMS BY TOPIC

Solution Concentration and Solution Stoichiometry

21. Calculate the molarity of each solution.
   a. 3.25 mol of LiCl in 2.78 L solution
   b. 28.33 g $C_6H_{12}O_6$ in 1.28 L of solution
   c. 32.4 mg NaCl in 122.4 mL of solution
22. Calculate the molarity of each solution.
   a. 0.38 mol of $LiNO_3$ in 6.14 L of solution
   b. 72.8 g C_2H_6O in 2.34 L of solution
   c. 12.87 mg KI in 112.4 mL of solution
23. What is the molarity of NO_3^- in each solution?
   a. 0.150 M KNO_3
   b. 0.150 M $Ca(NO_3)_2$
   c. 0.150 M $Al(NO_3)_3$
24. What is the molarity of Cl^- in each solution?
   a. 0.200 M NaCl b. 0.150 M $SrCl_2$ c. 0.100 M $AlCl_3$
25. How many moles of KCl are contained in each solution?
   a. 0.556 L of a 2.3 M KCl solution
   b. 1.8 L of a 0.85 M KCl solution
   c. 114 mL of a 1.85 M KCl solution
26. What volume of 0.200 M ethanol solution contains each of the following amounts?
   a. 0.45 mol ethanol
   b. 1.22 mol ethanol
   c. 1.2×10^{-2} mol ethanol
27. A laboratory procedure calls for making 400.0 mL of a 1.1 M $NaNO_3$ solution. What mass of $NaNO_3$ (in g) do you need?
28. A chemist wants to make 5.5 L of a 0.300 M $CaCl_2$ solution. What mass of $CaCl_2$ (in g) should the chemist use?
29. If 123 mL of a 1.1 M glucose solution is diluted to 500.0 mL, what is the molarity of the diluted solution?
30. If 3.5 L of a 4.8 M $SrCl_2$ solution is diluted to 45 L, what is the molarity of the diluted solution?
31. To what volume should you dilute 50.0 mL of a 12 M stock HNO_3 solution to obtain a 0.100 M HNO_3 solution?
32. To what volume should you dilute 25 mL of a 10.0 M H_2SO_4 solution to obtain a 0.150 M H_2SO_4 solution?
33. Consider the precipitation reaction:

$$2\,Na_3PO_4(aq) + 3\,CuCl_2(aq) \longrightarrow Cu_3(PO_4)_2(s) + 6\,NaCl(aq)$$

   What volume of 0.175 M Na_3PO_4 solution is necessary to completely react with 95.4 mL of 0.102 M $CuCl_2$?
34. Consider the reaction:

$$Li_2S(aq) + Co(NO_3)_2(aq) \longrightarrow 2\,LiNO_3(aq) + CoS(s)$$

   What volume of 0.150 M Li_2S solution is required to completely react with 125 mL of 0.150 M $Co(NO_3)_2$?
35. What is the minimum amount of 6.0 M H_2SO_4 necessary to produce 25.0 g of $H_2(g)$ according to the reaction between aluminum and sulfuric acid?

$$2\,Al(s) + 3\,H_2SO_4(aq) \longrightarrow Al_2(SO_4)_3(aq) + 3H_2(g)$$

36. What molarity of $ZnCl_2$ forms when 25.0 g of zinc completely reacts with $CuCl_2$ according to the following reaction? Assume a final volume of 275 mL.

$$Zn(s) + CuCl_2(aq) \longrightarrow ZnCl_2(aq) + Cu(s)$$

37. You mix a 25.0 mL sample of a 1.20 M potassium chloride solution with 15.0 mL of a 0.900 M barium nitrate solution, and this precipitation reaction occurs:

$$2\,KCl(aq) + Ba(NO_3)_2(aq) \longrightarrow BaCl_2(s) + 2\,KNO_3(aq)$$

   You collect and dry the solid $BaCl_2$ and find it has a mass of 2.45 g. Determine the limiting reactant, the theoretical yield, and the percent yield.
38. You mix a 55.0 mL sample of a 0.102 M potassium sulfate solution with 35.0 mL of a 0.114 M lead acetate solution, and this precipitation reaction occurs:

$$K_2SO_4(aq) + Pb(C_2H_3O_2)_2(aq) \longrightarrow 2\,KC_2H_3O_2(aq) + PbSO_4(s).$$

   You collect and dry the solid $PbSO_4$ and find it has a mass of 1.01 g. Determine the the limiting reactant, the theoretical yield, and the percent yield.

Types of Aqueous Solutions and Solubility

39. For each compound (all water soluble), would you expect the resulting aqueous solution to conduct electrical current?
   a. CsCl b. CH_3OH c. $Ca(NO_2)_2$ d. $C_6H_{12}O_6$
40. Classify each compound as a strong electrolyte or nonelectrolyte.
   a. $MgBr_2$ b. $C_{12}H_{22}O_{11}$ c. Na_2CO_3 d. KOH
41. Determine whether each compound is soluble or insoluble. If the compound is soluble, list the ions present in solution.
   a. $AgNO_3$ b. $Pb(C_2H_3O_2)_2$
   c. KNO_3 d. $(NH_4)_2S$
42. Determine whether each compound is soluble or insoluble. For the soluble compounds, list the ions present in solution.
   a. AgI b. $Cu_3(PO_4)_2$
   c. $CoCO_3$ d. K_3PO_4

Precipitation Reactions

43. Complete and balance each equation. If no reaction occurs, write NO REACTION.
   a. $LiI(aq) + BaS(aq) \longrightarrow$
   b. $KCl(aq) + CaS(aq) \longrightarrow$
   c. $CrBr_2(aq) + Na_2CO_3(aq) \longrightarrow$
   d. $NaOH(aq) + FeCl_3(aq) \longrightarrow$
44. Complete and balance each equation. If no reaction occurs, write NO REACTION.
   a. $NaNO_3(aq) + KCl(aq) \longrightarrow$
   b. $NaCl(aq) + Hg_2(C_2H_3O_2)_2(aq) \longrightarrow$
   c. $(NH_4)_2SO_4(aq) + SrCl_2(aq) \longrightarrow$
   d. $NH_4Cl(aq) + AgNO_3(aq) \longrightarrow$
45. Write a molecular equation for the precipitation reaction that occurs (if any) when each pair of aqueous solutions is mixed. If no reaction occurs, write NO REACTION.
   a. potassium carbonate and lead(II) nitrate
   b. lithium sulfate and lead(II) acetate
   c. copper(II) nitrate and magnesium sulfide
   d. strontium nitrate and potassium iodide
46. Write a molecular equation for the precipitation reaction that occurs (if any) when each pair of aqueous solutions is mixed. If no reaction occurs, write NO REACTION.
   a. sodium chloride and lead(II) acetate
   b. potassium sulfate and strontium iodide
   c. cesium chloride and calcium sulfide
   d. chromium(III) nitrate and sodium phosphate

Ionic and Net Ionic Equations

47. Write balanced complete ionic and net ionic equations for each reaction.

a. $HCl(aq) + LiOH(aq) \longrightarrow H_2O(l) + LiCl(aq)$
b. $MgS(aq) + CuCl_2(aq) \longrightarrow CuS(s) + MgCl_2(aq)$
c. $NaOH(aq) + HNO_3(aq) \longrightarrow H_2O(l) + NaNO_3(aq)$
d. $Na_3PO_4(aq) + NiCl_2(aq) \longrightarrow Ni_3(PO_4)_2(s) + NaCl(aq)$

48. Write balanced complete ionic and net ionic equations for each reaction.

a. $K_2SO_4(aq) + CaI_2(aq) \longrightarrow CaSO_4(s) + KI(aq)$
b. $NH_4Cl(aq) + NaOH(aq) \longrightarrow H_2O(l) + NH_3(g) + NaCl(aq)$
c. $AgNO_3(aq) + NaCl(aq) \longrightarrow AgCl(s) + NaNO_3(aq)$
d. $HC_2H_3O_2(aq) + K_2CO_3(aq) \longrightarrow H_2O(l) + CO_2(g) + KC_2H_3O_2(aq)$

49. Mercury ions (Hg_2^{2+}) can be removed from solution by precipitation with Cl^-. Suppose that a solution contains aqueous $Hg_2(NO_3)_2$. Write complete ionic and net ionic equations to show the reaction of aqueous $Hg_2(NO_3)_2$ with aqueous sodium chloride to form solid Hg_2Cl_2 and aqueous sodium nitrate.

50. Lead ions can be removed from solution by precipitation with sulfate ions. Suppose that a solution contains lead(II) nitrate. Write complete ionic and net ionic equations to show the reaction of aqueous lead(II) nitrate with aqueous potassium sulfate to form solid lead(II) sulfate and aqueous potassium nitrate.

Naming Acids

51. Name each acid.

a. $HI(aq)$ b. $HNO_3(aq)$ c. $H_2CO_3(aq)$

52. Name each acid.

a. $HCl(aq)$ b. $HClO_2(aq)$ c. $H_2SO_4(aq)$

53. Provide the formula for each acid.

a. hydrofluoric acid
b. hydrobromic acid
c. sulfurous acid

54. Provide the formula for each acid.

a. phosphoric acid
b. hydrocyanic acid
c. chlorous acid

Acid–Base Reactions

55. Write balanced molecular and net ionic equations for the reaction between hydrobromic acid and potassium hydroxide.

56. Write balanced molecular and net ionic equations for the reaction between nitric acid and calcium hydroxide.

57. Complete and balance each acid–base equation.

a. $H_2SO_4(aq) + Ca(OH)_2(aq) \longrightarrow$
b. $HClO_4(aq) + KOH(aq) \longrightarrow$
c. $H_2SO_4(aq) + NaOH(aq) \longrightarrow$

58. Complete and balance each acid–base equation.

a. $HI(aq) + LiOH(aq) \longrightarrow$
b. $HC_2H_3O_2(aq) + Ca(OH)_2(aq) \longrightarrow$
c. $HCl(aq) + Ba(OH)_2(aq) \longrightarrow$

59. A 25.00-mL sample of an unknown $HClO_4$ solution requires titration with 22.62 mL of 0.2000 M NaOH to reach the equivalence point. What is the concentration of the unknown $HClO_4$ solution? The neutralization reaction is:

$$HClO_4(aq) + NaOH(aq) \longrightarrow H_2O(l) + NaClO_4(aq)$$

60. A 30.00-mL sample of an unknown H_3PO_4 solution is titrated with a 0.100 M NaOH solution. The equivalence point is reached when 26.38 mL of NaOH solution is added. What is the concentration of the unknown H_3PO_4 solution? The neutralization reaction is:

$$H_3PO_4(aq) + 3\,NaOH(aq) \longrightarrow 3\,H_2O(l) + Na_3PO_4(aq)$$

Gas-Evolution Reactions

61. Complete and balance each gas-evolution equation.

a. $HBr(aq) + NiS(s) \longrightarrow$
b. $NH_4I(aq) + NaOH(aq) \longrightarrow$
c. $HBr(aq) + Na_2S(aq) \longrightarrow$

62. Complete and balance each gas-evolution equation.

a. $HCl(aq) + KHCO_3(aq) \longrightarrow$
b. $HC_2H_3O_2(aq) + NaHSO_3(aq) \longrightarrow$
c. $(NH_4)_2SO_4(aq) + Ca(OH)_2(aq) \longrightarrow$

63. Write a balanced equation for the reaction between perchloric acid and lithium carbonate.

64. Write a balanced equation for the reaction between nitric acid and sodium sulfite.

Oxidation and Reduction

65. Assign oxidation states to each atom in each element, ion, or compound.

a. Ag b. Ag^+ c. CaF_2
d. H_2S e. CO_3^{2-} f. CrO_4^{2-}

66. Assign oxidation states to each atom in each element, ion, or compound.

a. Cl_2 b. Fe^{3+} c. $CuCl_2$
d. CH_4 e. $Cr_2O_7^{2-}$ f. HSO_4^-

67. What is the oxidation state of Cr in each compound?

a. CrO b. CrO_3 c. Cr_2O_3

68. What is the oxidation state of Cl in each ion?

a. ClO^- b. ClO_2^- c. ClO_3^- d. ClO_4^-

69. Determine whether or not each reaction is a redox reaction. For each redox reaction, identify the oxidizing agent and the reducing agent.

a. $4\,Li(s) + O_2(g) \longrightarrow 2\,Li_2O(s)$
b. $Mg(s) + Fe^{2+}(aq) \longrightarrow Mg^{2+}(aq) + Fe(s)$
c. $Pb(NO_3)_2(aq) + Na_2SO_4(aq) \longrightarrow PbSO_4(s) + 2\,NaNO_3(aq)$
d. $HBr(aq) + KOH(aq) \longrightarrow H_2O(l) + KBr(aq)$

70. Determine whether or not each reaction is a redox reaction. For each redox reaction, identify the oxidizing agent and the reducing agent.

a. $Al(s) + 3\,Ag^+(aq) \longrightarrow Al^{3+}(aq) + 3\,Ag(s)$
b. $SO_3(g) + H_2O(l) \longrightarrow H_2SO_4(aq)$
c. $Ba(s) + Cl_2(g) \longrightarrow BaCl_2(s)$
d. $Mg(s) + Br_2(l) \longrightarrow MgBr_2(s)$

CUMULATIVE PROBLEMS

71. The density of a 20.0% by mass ethylene glycol ($C_2H_6O_2$) solution in water is 1.03 g/mL. Find the molarity of the solution.

72. Find the percent by mass of sodium chloride in a 1.35 M NaCl solution. The density of the solution is 1.05 g/mL.

73. People often use sodium bicarbonate as an antacid to neutralize excess hydrochloric acid in an upset stomach. What mass of hydrochloric acid (in grams) can 2.5 g of sodium bicarbonate neutralize? (*Hint*: Begin by writing a balanced equation for the reaction between aqueous sodium bicarbonate and aqueous hydrochloric acid.)

74. Toilet bowl cleaners often contain hydrochloric acid, which dissolves the calcium carbonate deposits that accumulate within a toilet bowl. What mass of calcium carbonate (in grams) can 3.8 g of HCl dissolve? (*Hint*: Begin by writing a balanced equation for the reaction between hydrochloric acid and calcium carbonate.)

75. A hydrochloric acid solution will neutralize a sodium hydroxide solution. Consider these molecular views of one beaker of HCl and four beakers of NaOH. Which NaOH beaker will just neutralize the HCl beaker? Begin by writing a balanced chemical equation for the neutralization reaction.

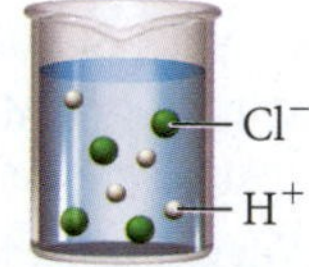

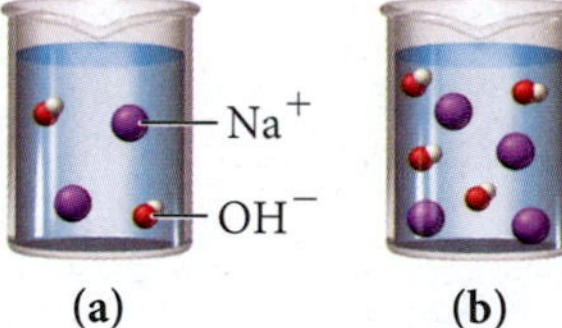

76. These two beakers represent solutions of HCl and NaOH. Draw a third beaker showing the ions that remain after the reaction has gone to completion.

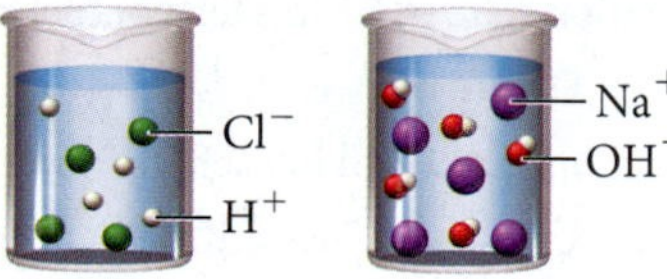

77. Predict the products and write a balanced molecular equation for each reaction. If no reaction occurs, write NO REACTION.
   a. $HCl(aq) + Hg_2(NO_3)_2(aq) \longrightarrow$
   b. $KHSO_3(aq) + HNO_3(aq) \longrightarrow$
   c. aqueous ammonium chloride and aqueous lead(II) nitrate
   d. aqueous ammonium chloride and aqueous calcium hydroxide

78. Predict the products and write a balanced molecular equation for each reaction. If no reaction occurs, write NO REACTION.
   a. $H_2SO_4(aq) + HNO_3(aq) \longrightarrow$
   b. $Cr(NO_3)_3(aq) + LiOH(aq) \longrightarrow$
   c. liquid pentanol ($C_5H_{12}O$) and gaseous oxygen
   d. aqueous strontium sulfide and aqueous copper(II) sulfate

79. Hard water often contains dissolved Ca^{2+} and Mg^{2+} ions. One way to soften water is to add phosphates. The phosphate ion forms insoluble precipitates with calcium and magnesium ions, removing them from solution. A solution is 0.050 M in calcium chloride and 0.085 M in magnesium nitrate. What mass of sodium phosphate would have to be added to 1.5 L of this solution to completely eliminate the hard water ions? Assume a complete reaction.

80. An acid solution is 0.100 M in HCl and 0.200 M in H_2SO_4. What volume of a 0.150 M KOH solution would completely neutralize all the acid in 500.0 mL of this solution?

81. Find the mass of barium metal (in grams) that must react with O_2 to produce enough barium oxide to prepare 1.0 L of a 0.10 M solution of OH^-. (*Hint*: Barium metal reacts with oxygen to form BaO; BaO reacts with water to form $Ba(OH)_2$.)

82. A solution contains Cr^{3+} ion and Mg^{2+} ion. The addition of 1.00 L of 1.51 M NaF solution causes the complete precipitation of these ions as $CrF_3(s)$ and $MgF_2(s)$. The total mass of the precipitate is 49.6 g. Find the mass of Cr^{3+} in the original solution.

83. Find the volume of 0.110 M hydrochloric acid necessary to react completely with 1.52 g $Al(OH)_3$.

84. Find the volume of 0.150 M sulfuric acid necessary to react completely with 75.3 g sodium hydroxide.

85. Treatment of gold metal with BrF_3 and KF produces Br_2 and $KAuF_4$, a salt of gold. Identify the oxidizing agent and the reducing agent in this reaction. What mass of the gold salt forms when a 73.5-g mixture of equal masses of all three reactants is prepared?

86. We prepare a solution by mixing 0.10 L of 0.12 M sodium chloride with 0.23 L of a 0.18 M $MgCl_2$ solution. What volume of a 0.20 M silver nitrate solution do we need to precipitate all the Cl^- ion in the solution as AgCl?

CHALLENGE PROBLEMS

87. A solution contains Ag^+ and Hg^{2+} ions. The addition of 0.100 L of 1.22 M NaI solution is just enough to precipitate all the ions as AgI and HgI_2. The total mass of the precipitate is 28.1 g. Find the mass of AgI in the precipitate.

88. The water in lakes that have been acidified by acid rain (HNO_3 and H_2SO_4) can be neutralized by a process called liming, in which limestone ($CaCO_3$) is added to the acidified water. What mass of limestone (in kg) will completely neutralize a 15.2 billion-liter lake that is 1.8×10^{-5} M in H_2SO_4 and 8.7×10^{-6} M in HNO_3?

89. Recall from Section 9.5 that sodium carbonate is often added to laundry detergents to soften hard water and make the detergent more effective. Suppose that a particular detergent mixture is designed to soften hard water that is 3.5×10^{-3} M in Ca^{2+} and 1.1×10^{-3} M in Mg^{2+} and that the average capacity of a washing machine is 19.5 gallons of water. If 0.65 kg detergent is required per load of laundry, what percentage (by mass) of the detergent should be sodium carbonate in order to completely precipitate all of the calcium and magnesium ions in an average load of laundry water?

90. A solution contains one or more of the following ions: Ag^+, Ca^{2+}, and Cu^{2+}. When you add sodium chloride to the solution, no precipitate forms. When you add sodium sulfate to the solution, a white precipitate forms. You filter off the precipitate and add sodium carbonate to the remaining solution, producing another precipitate. Which ions were present in the original solution? Write net ionic equations for the formation of each of the precipitates observed.

91. A solution contains one or more of the following ions: Hg_2^{2+}, Ba^{2+}, and Fe^{2+}. When potassium chloride is added to the solution, a precipitate forms. The precipitate is filtered off, and potassium sulfate is added to the remaining solution, producing no precipitate. When potassium carbonate is added to the remaining solution, a precipitate forms. Which ions were present in the original solution? Write net ionic equations for the formation of each of the precipitates observed.

CONCEPTUAL PROBLEMS

92. The circle shown here represents 1.0 liter of a solution with a solute concentration of 1 M:

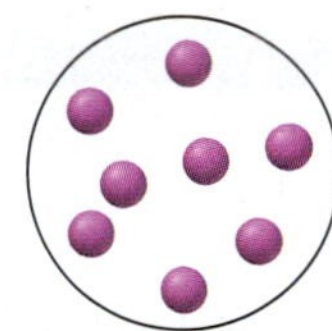

Explain what you would add (the amount of solute or volume of solvent) to the solution to obtain a solution represented by each diagram:

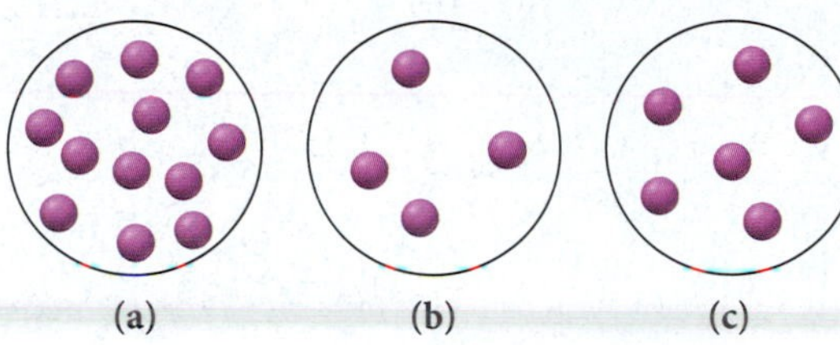

93. Consider the generic ionic compounds with the formulas A_2X and BY_2 and the following solubility rules: A_2X soluble; BY_2 soluble; AY insoluble; BX soluble. Let circles represent A^+ ions, squares represent B^{2+} ions, triangles represent X^{2-} ions, and diamonds represent Y^- ions. Solutions of the two compounds (A_2X and BY_2) can be represented as follows:

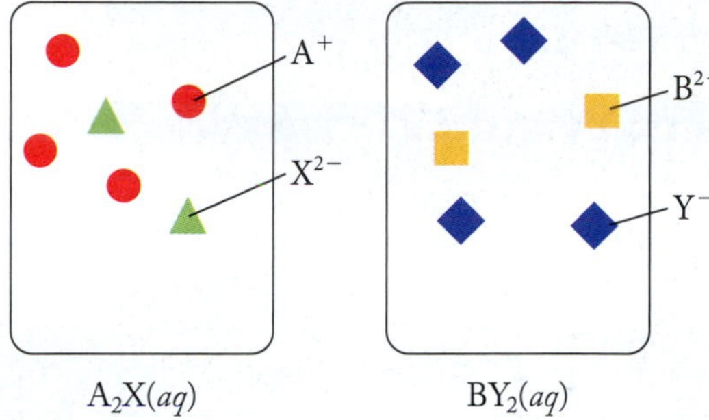

Draw a molecular-level representation showing the result of mixing the two solutions (A_2X and BY_2) and write an equation to represent the reaction.

94. If you dissolve 27 g of sugar into 155 mL of water, what can you conclude about the mass and volume of the resulting solution? Assume a density of 1.00 g/mL for the water.

a. mass = 155 g; volume = 155 mL
b. mass = 182 g; volume = 155 mL
c. mass = 182 g; volume > 155 mL
d. mass = 155 g; volume > 155 mL

95. Explain the difference between the charge of an ion, such as a charge of 2– for an O^{2-} ion, and the oxidation state of an atom, such as the −4 oxidation state of carbon in CH_4.

ANSWERS TO CONCEPTUAL CONNECTIONS

Cc 9.1 **(b)** The mass of a solution is equal to the mass of the solute plus the mass of the solvent. Although the solute seems to disappear, it really does not, and its mass becomes part of the mass of the solution, in accordance with the law of mass conservation.

Cc 9.2 **(c)** Because the volume has doubled, the concentration is halved, so the same volume should contain half as many solute molecules.

Cc 9.3

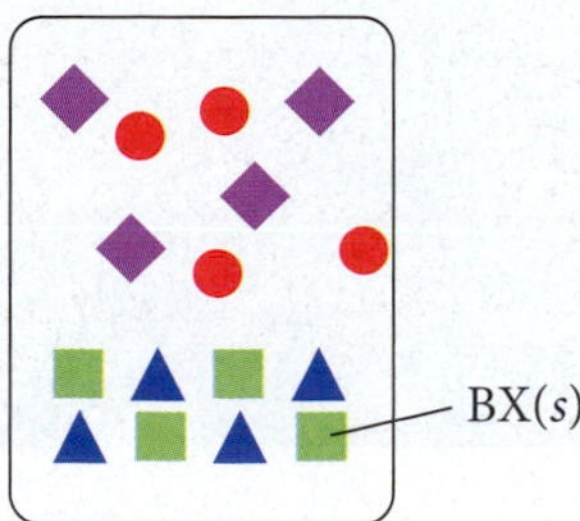

$$AX(aq) + BY(aq) \longrightarrow BX(s) + AY(aq)$$

Cc 9.4 **(a)** The charge of a polyatomic ion is the charge associated with the ion *as a whole*. The oxidation states of the individual atoms must sum to the charge of the ion, but they are assigned to *the individual atoms themselves*.

(b) is incorrect because oxidation state and charge *are not identical*, even though the charge of a *monoatomic* ion is equal to its oxidation state.

(c) is incorrect because charge *is* a physical property of ions. Conversely, the oxidation states of atoms are *not* real physical properties, but an imposed electron bookkeeping scheme.

Cc 9.5 **(d)** Since oxidation and reduction must occur together, an increase in the oxidation state of a reactant is always accompanied by a decrease in the oxidation state of a reactant.

CHAPTER 10

"There is a fact, or if you wish, a law, governing all natural phenomena that are known to date. There is no exception to this law—it is exact as far as we know. The law is called the conservation of energy. It states that there is a certain quantity, which we call energy, that does not change in the manifold changes which nature undergoes."

—Richard P. Feynman (1918–1988)

A twenty-dollar bill bursts into flames, but is not consumed.

Thermochemistry

10.1 On Fire, But Not Consumed 343

10.2 The Nature of Energy: Key Definitions 344

10.3 The First Law of Thermodynamics: There Is No Free Lunch 346

10.4 Quantifying Heat and Work 349

10.5 Measuring ΔE for Chemical Reactions: Constant-Volume Calorimetry 355

10.6 Enthalpy: The Heat Evolved in a Chemical Reaction at Constant Pressure 358

10.7 Measuring ΔH for Chemical Reactions: Constant-Pressure Calorimetry 362

10.8 Relationships Involving ΔH_{rxn} 364

10.9 Determining Enthalpies of Reaction from Bond Energies 367

10.10 Determining Enthalpies of Reaction from Standard Enthalpies of Formation 370

10.11 Lattice Energies for Ionic Compounds 375

Key Learning Outcomes 380

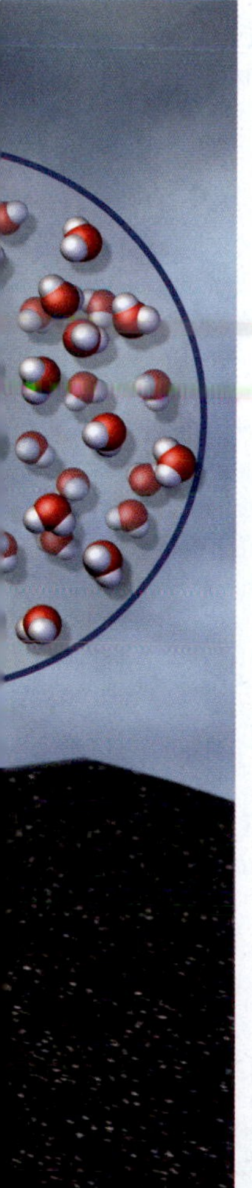

WE HAVE SPENT NEARLY THE FIRST HALF of this book examining one of the two major components of our universe—matter. We now turn our attention to the other major component—energy. As far as we know, matter and energy—which can be interchanged but not destroyed—make up the physical universe. Unlike matter, energy is not something we can touch or hold in our hand, but we experience it in many ways. The warmth of sunlight, the feel of wind on our faces, and the force that presses us back when a car accelerates are all manifestations of energy and its interconversions. And of course energy is critical to society and to the world. The standard of living around the globe is strongly correlated with the access to and use of energy resources. Most of those resources, as we shall see, are chemical, and we can understand their advantages as well as their drawbacks in terms of chemistry.

10.1 On Fire, But Not Consumed

Have you ever seen a flammable object burst into flames but not be consumed? One of my favorite classroom demonstrations involves soaking a twenty-dollar bill in water for several minutes, then dipping it into ethanol and lighting it on fire. The bill erupts in flames, but after a few seconds the flames diminish and the bill is left intact.

The bill is not destroyed because two processes—one that emits heat and one that absorbs heat—happen simultaneously. The first process is the combustion of ethanol, which is an exothermic chemical reaction (a reaction that emits heat; see Section 2.4). If I had dipped the bill only in ethanol, the burning ethanol would raise the temperature of the bill to its combustion temperature and it would burn to ashes. However, because I initially soaked the bill in water, another process was simultaneously occurring: the evaporation of water. The evaporation of water is endothermic and absorbs the heat emitted by the combustion of ethanol. As a result, the two processes cancel each other out: The temperature of the bill does not change much, and the bill is unharmed.

The demonstration of the burning bill involves many of the principles of **thermochemistry**, the study of the relationships between chemistry and energy. In this chapter, we look at how chemical reactions *exchange* energy with their surroundings and how we quantify the magnitude of those exchanges. These kinds of calculations are important, not only for classroom demonstrations involving burning money, but also to many other important processes, such as the production of energy, that drive the world economy.

10.2 The Nature of Energy: Key Definitions

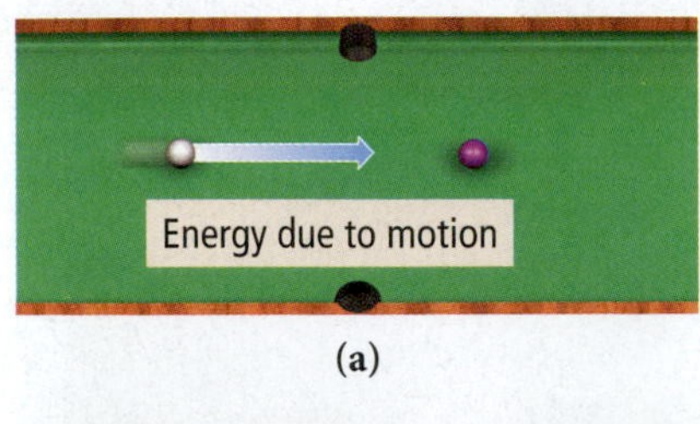

(a)

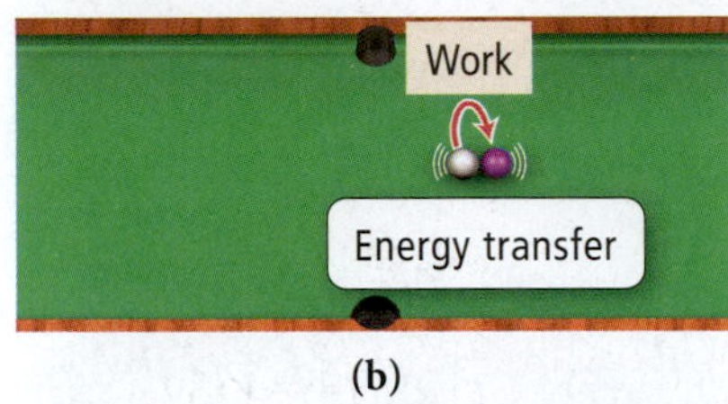

(b)

(c)

▲ **(a)** A rolling billiard ball has energy due to its motion. **(b)** When the ball collides with a second ball it does work, transferring energy to the second ball. **(c)** The second ball now has energy as it rolls away from the collision.

Recall from our discussion of energy in Section 2.4 that **energy** is the capacity to do work and **work** is the result of a force acting through a distance. When you push a box across the floor, you have done work. Consider another example of work: a billiard ball rolling across a billiard table and colliding straight on with a second, stationary billiard ball as shown in the margin. The rolling ball has *energy* due to its motion. When it collides with another ball it does *work*, resulting in the *transfer* of energy from one ball to the other. The second billiard ball absorbs the energy and begins to roll across the table.

Energy can also be transferred through **heat**, the flow of energy caused by a temperature difference. For example, if you hold a cup of coffee in your hand, energy is transferred, in the form of heat, from the hot coffee to your cooler hand. Think of *energy* as something that an object or set of objects possesses. Think of *heat* and *work* as ways that objects or sets of objects *exchange* energy.

The energy contained in a rolling billiard ball is an example of **kinetic energy**, the energy associated with the *motion* of an object. The energy contained in a hot cup of coffee is **thermal energy**, the energy associated with the *temperature* of an object. Thermal energy is actually a type of kinetic energy because it arises from the motions of atoms or molecules within a substance.

If you raise a billiard ball off the table, you increase its **potential energy**, the energy associated with the *position* or *composition* of an object. The potential energy of the billiard ball, for example, is a result of its position in Earth's gravitational field. Raising the ball off the table, against Earth's gravitational pull, gives it more potential energy. Another example of potential energy is the energy contained in a compressed spring. When you compress a spring, you push against the forces that tend to maintain the spring's uncompressed shape, storing energy as potential energy. **Chemical energy**, the energy associated with the relative positions of electrons and nuclei in atoms and molecules, is also a form of potential energy. Some chemical compounds, such as the methane in natural gas or the ethanol on the burning twenty-dollar bill, also contain potential energy, and chemical reactions can release that potential energy. Figure 10.1 ▼ summarizes these different kinds of energy.

As we discussed in Section 2.4, the SI unit of energy is the $kg \cdot m^2/s^2$, defined as the **joule (J)** and named after the English scientist James Joule (1818–1889). A second common unit of energy is the **calorie (cal)**, originally defined as the amount of energy required to raise the temperature of 1 g of water by 1 °C. The current definition is 1 cal = 4.184 J (exact); a calorie is a larger unit than a joule. A related energy unit is the nutritional, or uppercase "C" **Calorie (Cal)**, equivalent to 1000 lowercase "c" calories. The Calorie is the same as a kilocalorie (kcal): 1 Cal = 1 kcal = 1000 cal.

Energy: Capacity to Do Work

Kinetic Energy: Due to Motion

Potential Energy: Due to Position or Composition

Thermal Energy: Associated with Temperature

Chemical Energy: Associated with Positions of Electrons and Nuclei

▲ **FIGURE 10.1 The Different Manifestations of Energy**

The **law of conservation of energy** states that *energy can be neither created nor destroyed*. However, energy can be transferred from one object to another, and it can assume different forms. For example, if you drop a raised billiard ball, some of its potential energy becomes kinetic energy as

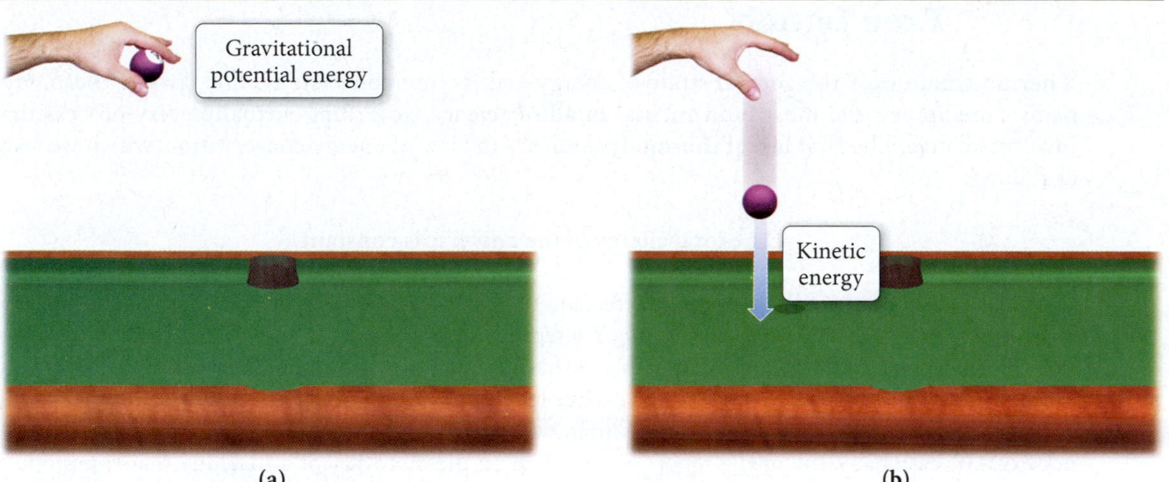

◀ **FIGURE 10.2 Energy Transformation: Potential and Kinetic Energy I** **(a)** A billiard ball held above the ground has gravitational potential energy. **(b)** When the ball is released, the potential energy is transformed into kinetic energy, the energy of motion.

the ball falls toward the table, as shown in Figure 10.2 ▲. If you release a compressed spring, the potential energy becomes kinetic energy as the spring expands outward, as shown in Figure 10.3 ▼. When you burn natural gas in an oven, the chemical energy of the natural gas and oxygen becomes thermal energy that increases the temperature of the air in the oven.

When natural gas is burned, it combines with oxygen to form carbon dioxide and water, which have lower potential energy than the natural gas and the oxygen. The change in potential energy is the source of heat upon burning.

A good way to understand and track energy changes is to define the **system** under investigation. For example, the system may be the chemicals in a beaker, or it may be the ethanol burning on the twenty-dollar bill. The system's **surroundings** are everything with which the system can exchange energy. If we define the chemicals in a beaker as the system, the surroundings may include the water that the chemicals are dissolved in (for aqueous solutions), the beaker itself, the lab bench on which the beaker sits, the air in the room, and so on. If we define the ethanol on the burning bill as the system, then surroundings include the bill itself, the water on the bill, and the air that surrounds the burning bill.

In an energy exchange, energy is transferred between the system and the surroundings, as shown in Figure 10.4 ▶. If the system loses energy, the surroundings gain the same exact amount of energy, and vice versa. When the ethanol on the bill burns, it transfers energy to the water in the bill, which absorbs the energy as it evaporates.

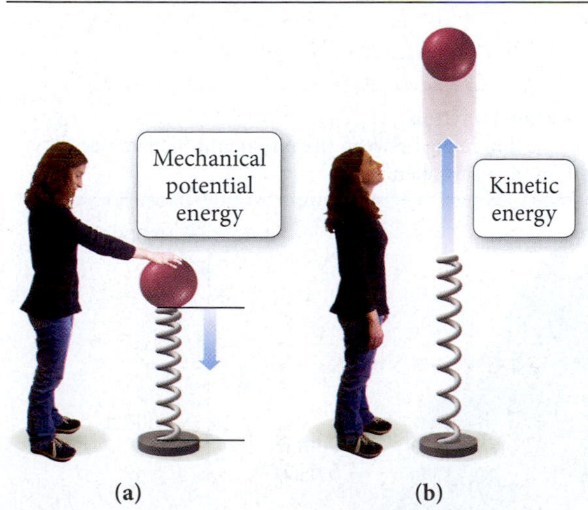

▲ **FIGURE 10.3 Energy Transformation: Potential and Kinetic Energy II** **(a)** A compressed spring has potential energy. **(b)** When the spring is released, the potential energy is transformed into kinetic energy.

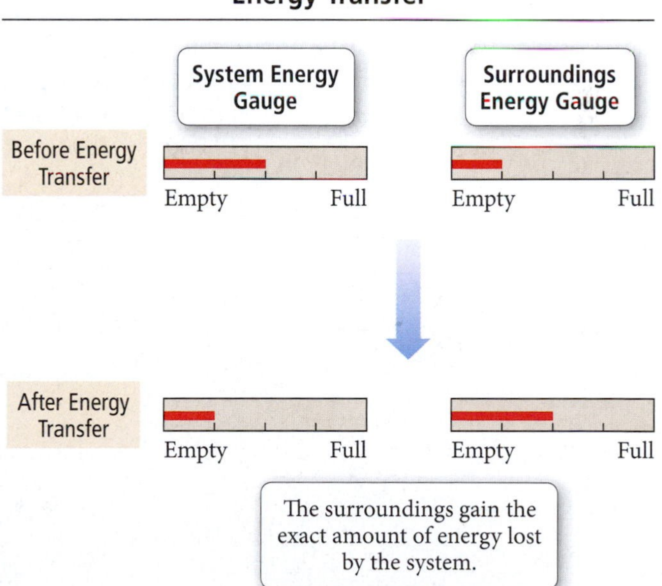

▲ **FIGURE 10.4 Energy Transfer** If a system and surroundings had energy gauges (which would measure energy content in the way a fuel gauge measures fuel content), an energy transfer in which the system transfers energy to the surroundings would result in a decrease in the energy content of the system and an increase in the energy content of the surroundings. The total amount of energy, however, must be conserved.

10.3 The First Law of Thermodynamics: There Is No Free Lunch

Einstein showed that it is mass–energy that is conserved; one can be converted into the other. This equivalence becomes important in nuclear reactions, discussed in Chapter 21. In ordinary chemical reactions, however, the interconversion of mass and energy is not a significant factor, and we can regard mass and energy as independently conserved.

Thermodynamics is the general study of energy and its interconversions. The laws of thermodynamics are among the most fundamental in all of science, governing virtually every process that involves change. The **first law of thermodynamics** is the law of energy conservation, which we state as follows:

The total energy of the universe is constant.

In other words, because energy is neither created nor destroyed, and because the universe does not exchange energy with anything else, its energy content does not change. The first law has many implications: The most important one is that, with energy, we cannot get something for nothing. The best you can do with energy is break even—there is no free lunch. According to the first law, a device that would continually produce energy with no energy input, sometimes known as a *perpetual motion machine*, cannot exist. Occasionally, the media report or speculate on the discovery of a machine that can produce energy without the need for energy input. For example, you may have heard claims about an electric car that recharges itself while driving, or a new motor that can create additional usable electricity as well as the electricity to power itself. Although some hybrid (electric and gasoline-powered) vehicles can capture energy from braking and use that energy to recharge their batteries, they cannot run indefinitely without additional fuel. As for a motor that powers an external load as well as itself—no such thing exists. Our society has a continual need for energy, and as our present energy resources dwindle, new energy sources will be required. However, those sources, whatever they may be, will follow the first law of thermodynamics—energy must be conserved.

The **internal energy (*E*)** of a system is *the sum of the kinetic and potential energies of all of the particles that compose the system*. Internal energy is a **state function**, which means that its value depends *only on the state of the system*, not on how the system arrived at that state. The state of a chemical system is specified by parameters such as temperature, pressure, concentration, and physical state (solid, liquid, or gas). Consider the mountain-climbing analogy depicted in Figure 10.5 ▼. The elevation at any point during a mountain climb is analogous to a state function. For example, when we reach 10,000 ft, our elevation is 10,000 ft, no matter how we got there. The distance we traveled to get there, by contrast, is not a state function; we could have climbed the mountain by any number of routes, each requiring us to cover a different distance.

Because state functions depend only on the state of the system, the value of a *change* in a state function is always the difference between its final and initial values. If we start climbing a mountain at an elevation of 3000 ft and reach the summit at 10,000 ft, then our elevation change is 7000 ft (10,000 ft − 3000 ft), regardless of what path we took.

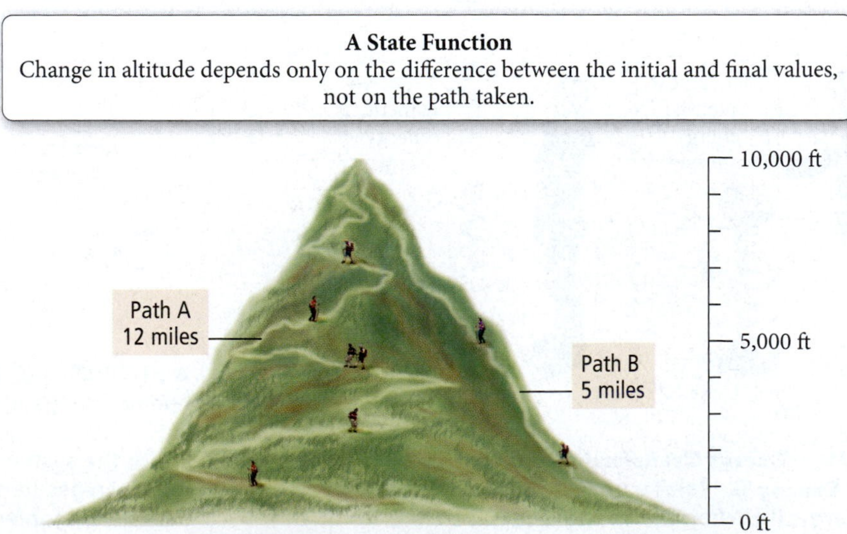

▶ **FIGURE 10.5 Altitude as a State Function** The change in altitude during a climb depends only on the difference between the final and initial altitudes, not on the route traveled.

Like an altitude change, we determine an internal energy change (ΔE) by the difference in internal energy between the final and initial states.

$$\Delta E = E_{\text{final}} - E_{\text{initial}}$$

In a chemical system, the reactants constitute the initial state and the products constitute the final state. So ΔE is the difference in internal energy between the products and the reactants.

$$\Delta E = E_{\text{products}} - E_{\text{reactants}} \qquad [10.1]$$

For example, consider the reaction between carbon and oxygen to form carbon dioxide:

$$C(s) + O_2(g) \longrightarrow CO_2(g)$$

Just as we can portray the changes that occur when climbing a mountain with an *altitude* diagram which depicts the *altitude* before and after the climb (see Figure 10.5), we can portray the energy changes that occur during a reaction with an *energy* diagram, which compares the *internal energy* of the reactants and the products.

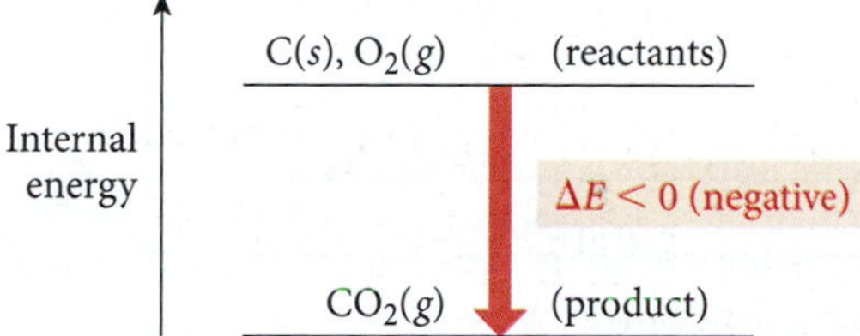

The vertical axis of the diagram is internal *energy*, which increases as we move up on the diagram. For this reaction, the reactants are *higher* on the diagram than the products because they have higher internal energy. As the reaction occurs, the reactants become products, which have lower internal energy. Therefore, the reaction gives off energy, and ΔE (that is, $E_{\text{products}} - E_{\text{reactants}}$) is *negative.*

Where does the energy lost by the reactants (as they transform to products) go? If we define the thermodynamic *system* as the reactants and products of the reaction, then energy flows *out of the system* and *into the surroundings.*

System → Energy flow → Surroundings

$\Delta E_{\text{sys}} < 0$ (negative) $\quad \Delta E_{\text{surr}} > 0$ (positive)

According to the first law, energy must be conserved. Therefore, the amount of energy lost by the system must exactly equal the amount gained by the surroundings:

$$\Delta E_{\text{sys}} = -\Delta E_{\text{surr}} \qquad [10.2]$$

Now, suppose the reaction is reversed:

$$CO_2(g) \longrightarrow C(s) + O_2(g)$$

The energy-level diagram is nearly identical, with one important difference: $CO_2(g)$ is now the reactant, and $C(s)$ and $O_2(g)$ are the products. Instead of decreasing in energy as the reaction occurs, the system increases in energy.

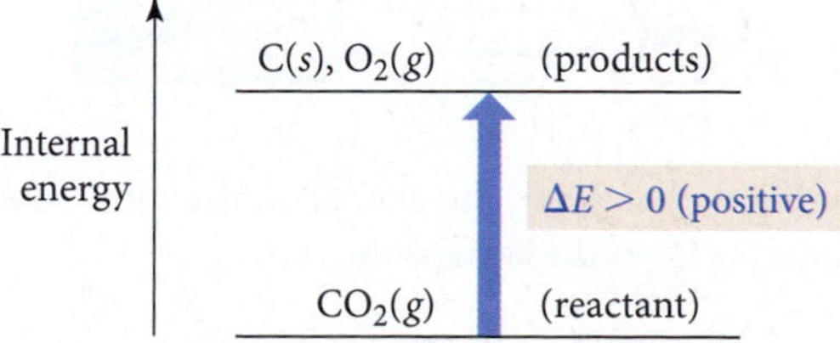

In this reversed reaction ΔE is *positive* and energy flows *into the system* and *out of the surroundings.*

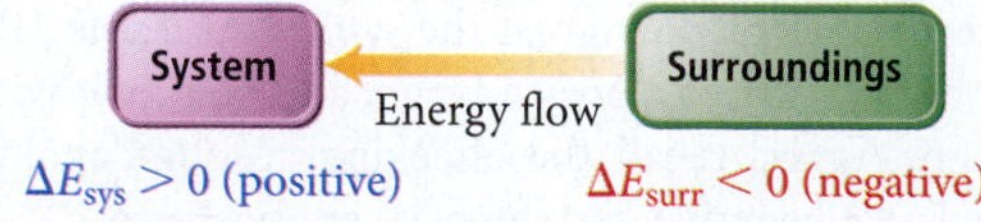

Summarizing Energy Flow:

- If the reactants have a higher internal energy than the products, ΔE_{sys} is negative and energy flows out of the system into the surroundings.
- If the reactants have a lower internal energy than the products, ΔE_{sys} is positive and energy flows into the system from the surroundings.

We can think of the internal energy of the system in the same way we think about the balance in a checking account. Energy flowing *out of* the system is like a withdrawal and therefore carries a negative sign. Energy flowing *into* the system is like a deposit and carries a positive sign.

10.1 Cc Conceptual Connection

System and Surroundings

Consider these fictitious internal energy gauges for a chemical system and its surroundings.

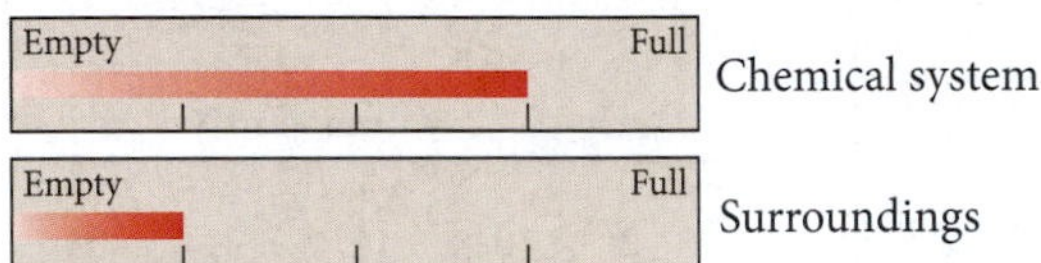

Which diagram best represents the energy gauges for the same system and surroundings after an energy exchange in which ΔE_{sys} is negative?

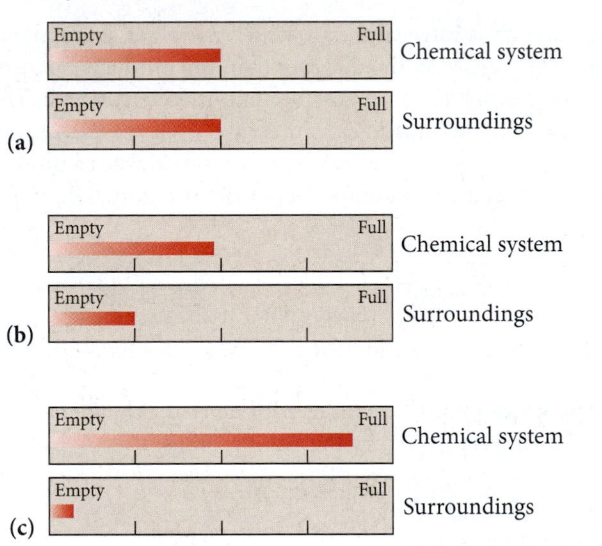

As we saw earlier, a system can exchange energy with its surroundings through *heat* and *work*:

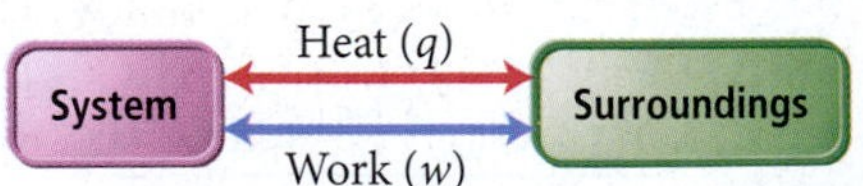

According to the first law of thermodynamics, the change in the internal energy of the system (ΔE) is the sum of the heat transferred (q) and the work done (w).

$$\Delta E = q + w \qquad [10.3]$$

In the above equation, and from this point forward, we follow the standard convention that ΔE (with no subscript) refers to the internal energy change of the *system*. As Table 10.1 illustrates, energy entering the system through heat or work carries a positive sign, and energy leaving the system through heat or work carries a negative sign. Again, recall the checking account analogy. The system is like the checking account—withdrawals are negative and deposits are positive.

TABLE 10.1 Sign Conventions for q, w, and ΔE

q (heat)	+ system *gains* thermal energy	− system *loses* thermal energy
w (work)	+ work done *on* the system	− work done *by* the system
ΔE (change in internal energy)	+ energy flows *into* the system	− energy flows *out* of the system

Heat and Work

Identify each energy exchange as heat or work and determine whether the sign of heat or work (relative to the system) is positive or negative.

(a) An ice cube melts and cools the surrounding beverage. (The ice cube is the system.)

(b) A metal cylinder is rolled up a ramp. (The metal cylinder is the system.)

(c) Steam condenses on skin, causing a burn. (The condensing steam is the system.)

EXAMPLE 10.1
Internal Energy, Heat, and Work

The firing of a potato cannon provides a good example of the heat and work associated with a chemical reaction. In the potato cannon, a potato is stuffed into a long cylinder that is capped on one end and open at the other. Some kind of fuel is introduced under the potato at the capped end—usually through a small hole—and ignited. The potato shoots out of the cannon, sometimes flying hundreds of feet, and the cannon emits heat to the surroundings. If the burning of the fuel performs 855 J of work on the potato and produces 1422 J of heat, what is ΔE for the burning of the fuel? (*Note:* A potato cannon can be dangerous and should not be constructed without proper training and experience.)

SOLUTION

To solve the problem, substitute the values of q and w into the equation for ΔE. Because work is done by the system on the surroundings, w is negative. Similarly, because heat is released by the system to the surroundings, q is also negative.

$$\begin{aligned}\Delta E &= q + w \\ &= -1422\text{ J} - 855\text{ J} \\ &= -2277\text{ J}\end{aligned}$$

FOR PRACTICE 10.1
A cylinder and piston assembly (defined as the system) is warmed by an external flame. The contents of the cylinder expand, doing work on the surroundings by pushing the piston outward against the external pressure. If the system absorbs 559 J of heat and does 488 J of work during the expansion, what is the value of ΔE?

10.4 Quantifying Heat and Work

In the previous section, we calculated ΔE based on *given values of q (heat) and w (work)*. We now turn to *calculating q and w* based on changes in temperature and volume.

Heat

Recall from Section 10.2 that *heat* is the exchange of thermal energy between a system and its surroundings caused by a temperature difference. Notice the distinction we make between heat and temperature. Temperature is a *measure* of the thermal energy within a sample of matter. Heat is the *transfer* of thermal energy. Thermal energy always flows from hot to cold—from matter at higher temperatures to matter at lower temperatures. For example, a hot cup of coffee transfers thermal energy—as heat—to the lower temperature surroundings as it cools down.

The reason for this one-way transfer (from hot to cold) is related to the second law of thermodynamics, which we will discuss in Chapter 19.

TABLE 10.2 Specific Heat Capacities of Some Common Substances

Substance	Specific Heat Capacity, C_s (J/g · °C)*
Elements	
Lead	0.128
Gold	0.128
Silver	0.235
Copper	0.385
Iron	0.449
Aluminum	0.903
Compounds	
Ethanol	2.42
Water	4.18
Materials	
Glass (Pyrex)	0.75
Granite	0.79
Sand	0.84

*At 298 K.

▲ The high heat capacity of the water surrounding San Francisco results in relatively cool summer temperatures.

Imagine a world where the cooler surroundings actually became colder as they transferred thermal energy to the hot coffee, which became hotter. Such a scenario is impossible because the spontaneous transfer of heat from hot to cold is a fundamental principle of our universe—no exception has ever been observed. The thermal energy in the molecules that compose the hot coffee distributes itself to the molecules in the surroundings. The heat transfer from the coffee to the surroundings stops when the two reach the same temperature, a condition called **thermal equilibrium**. At thermal equilibrium, there is no additional net transfer of heat.

Temperature Changes and Heat Capacity When a system absorbs heat (q), its temperature changes by ΔT, where $\Delta T = T_{\text{final}} - T_{\text{initial}}$.

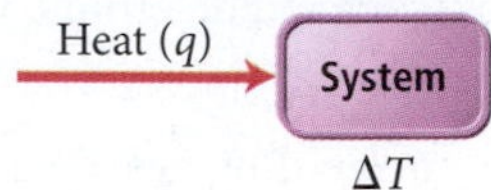

Experiments show that the heat absorbed by a system and its corresponding temperature change are directly proportional: $q \propto \Delta T$. The constant of proportionality between q and ΔT is the system's *heat capacity* (C), a measure of how much heat a system must absorb to undergo a specified change in temperature.

$$q = C \times \Delta T \qquad [10.4]$$

(Heat capacity)

Notice that the higher the heat capacity of a system, the smaller the change in temperature for a given amount of absorbed heat. We define the **heat capacity (C)** of a system as the quantity of heat required to change its temperature by 1 °C. As we can see by solving Equation 10.4 for heat capacity, the units of heat capacity are those of heat (typically J) divided by those of temperature (typically °C).

$$C = \frac{q}{\Delta T} = \frac{\text{J}}{°\text{C}}$$

In order to understand two important concepts related to heat capacity, consider a steel saucepan on a kitchen flame. The saucepan's temperature rises rapidly as it absorbs heat from the flame. However, if we add some water to the saucepan, the temperature rises more slowly. Why? The first reason is that, when we add the water, the same amount of heat must now warm more matter, so the temperature rises more slowly. In other words, heat capacity is an extensive property—*it depends on the amount of matter being heated* (see Section 2.3). The second (and more fundamental) reason is that *water is more resistant to temperature change than steel*—water has an intrinsically higher capacity to absorb heat without undergoing a large temperature change. The measure of the *intrinsic capacity* of a substance to absorb heat is its **specific heat capacity (C_s)**, the amount of heat required to raise the temperature of *1 gram* of the substance by 1 °C. The units of specific heat capacity (also called *specific heat*) are J/g · °C. Table 10.2 lists the values of the specific heat capacity for several substances. Heat capacity is also sometimes reported as **molar heat capacity**, the amount of heat required to raise the temperature of *1 mole* of a substance by 1 °C. The units of molar heat capacity are J/mol · °C. *Specific* heat capacity and *molar* heat capacity are intensive properties—they depend on the *kind* of substance being heated, not on the amount.

Notice that water has the highest specific heat capacity of all the substances in Table 10.2—changing the temperature of water requires a lot of heat. If you have ever experienced the drop in temperature that occurs when traveling from an inland region to the coast during the summer, you have experienced the effects of water's high specific heat capacity. On a summer's day in California, for example, the temperature difference between Sacramento (an inland city) and San Francisco (a coastal city) can be 18 °C (30 °F)—San Francisco enjoys a cool 20 °C (68 °F), while Sacramento bakes at nearly 38 °C (100 °F). Yet the intensity of sunlight falling on these two cities is the same. Why the large temperature difference? San Francisco sits on a peninsula, surrounded by the water of the Pacific Ocean. Water, with its high heat capacity, absorbs much of the sun's heat without undergoing a large increase in temperature, keeping San Francisco cool. Sacramento, by contrast, is about 160 km (100 mi) inland. The land surrounding Sacramento, with its low heat capacity, undergoes a large increase in temperature as it absorbs a similar amount of heat.

Similarly, only two U.S. states have never recorded a temperature above 100 °F. One of them is obvious: Alaska. It is too far north to get that hot. The other one, however, may come as a surprise. It is Hawaii. The high heat capacity of the water that surrounds the only island state moderates the temperature, preventing Hawaii from ever getting too hot.

We can use the specific heat capacity of a substance to quantify the relationship between the amount of heat added to a given amount of the substance and the corresponding temperature increase. The equation that relates these quantities is:

$$q = m \times C_s \times \Delta T \quad [10.5]$$

Heat (J) → q; Mass (g) → m; Specific heat capacity J/g · °C → C_s; Temperature change (°C) → ΔT

where q is the amount of heat in J, m is the mass of the substance in g, C_s is the specific heat capacity in J/g·°C, and ΔT is the temperature change in °C. Example 10.2 demonstrates the use of this equation.

ΔT in °C is equal to ΔT in K, but not equal to ΔT in °F (see Appendix I).

EXAMPLE 10.2
Temperature Changes and Heat Capacity

You find a penny (minted before 1982, when pennies were made of nearly pure copper) in the snow. How much heat is absorbed by the penny as it warms from the temperature of the snow, which is −8.0 °C, to the temperature of your body, 37.0 °C? Assume the penny is pure copper and has a mass of 3.10 g.

SORT You are given the mass of copper as well as its initial and final temperature. You are asked to find the heat required for the given temperature change.

GIVEN: $m = 3.10$ g copper

$T_i = -8.0$ °C

$T_f = 37.0$ °C

FIND: q

STRATEGIZE The equation $q = m \times C_s \times \Delta T$ gives the relationship between the amount of heat (q) and the temperature change (ΔT).

CONCEPTUAL PLAN

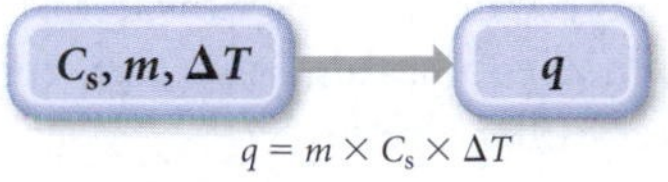

$q = m \times C_s \times \Delta T$

RELATIONSHIPS USED

$q = m \times C_s \times \Delta T$ (Equation 10.5)

$C_s = 0.385$ J/g·°C (Table 10.2)

SOLVE Gather the necessary quantities for the equation in the correct units and substitute these into the equation to calculate q.

SOLUTION

$\Delta T = T_f - T_i = 37.0\ °C - (-8.0\ °C) = 45.0\ °C$

$$q = m \times C_s \times \Delta T$$

$$= 3.10\ \cancel{g} \times 0.385 \frac{J}{\cancel{g} \cdot \cancel{°C}} \times 45.0\ \cancel{°C} = 53.7\ J$$

CHECK The units (J) are correct for heat. The sign of q is *positive,* as it should be because the penny *absorbed* heat from the surroundings.

FOR PRACTICE 10.2

To determine whether a shiny gold-colored rock is actually gold, a chemistry student decides to measure its heat capacity. She first weighs the rock and finds it has a mass of 4.7 g. She then finds that upon absorption of 57.2 J of heat, the temperature of the rock rises from 25 °C to 57 °C. Find the specific heat capacity of the substance composing the rock and determine whether the value is consistent with the rock being pure gold.

FOR MORE PRACTICE 10.2

A 55.0-g aluminum block initially at 27.5 °C absorbs 725 J of heat. What is the final temperature of the aluminum?

10.3
Cc
Conceptual Connection

The Heat Capacity of Water

Suppose you are cold-weather camping and decide to heat some objects to bring into your sleeping bag for added warmth. You place a large water jug and a rock of equal mass near the fire. Over time, both the rock and the water jug warm to about 38 °C (100 °F). You can bring only one into your sleeping bag. Which one should you choose to keep you warmer? Why?

Thermal Energy Transfer As we noted earlier, when two substances of different temperature are combined, thermal energy flows as heat from the hotter substance to the cooler one. If we assume that the two substances are thermally isolated from everything else, then the heat lost by one substance exactly equals the heat gained by the other (according to the law of energy conservation). If we define one substance as the system and the other as the surroundings, we can quantify the heat exchange as:

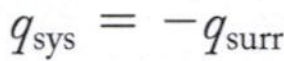

$$q_{sys} = -q_{surr}$$

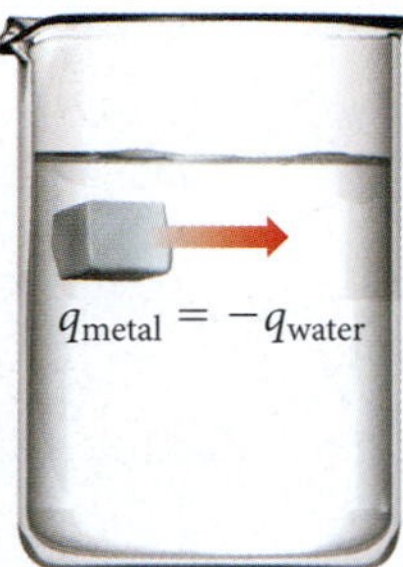

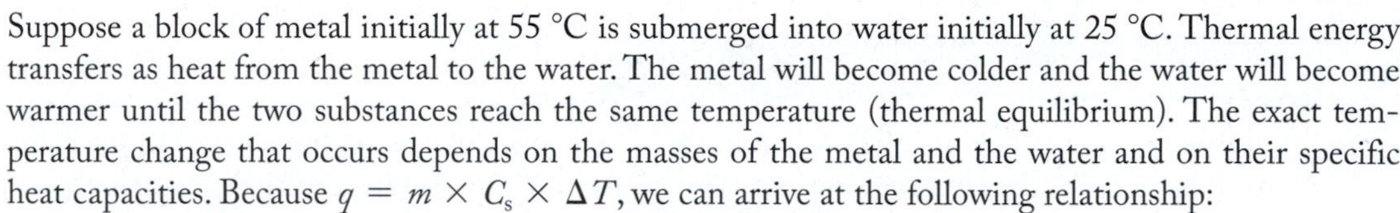

Suppose a block of metal initially at 55 °C is submerged into water initially at 25 °C. Thermal energy transfers as heat from the metal to the water. The metal will become colder and the water will become warmer until the two substances reach the same temperature (thermal equilibrium). The exact temperature change that occurs depends on the masses of the metal and the water and on their specific heat capacities. Because $q = m \times C_s \times \Delta T$, we can arrive at the following relationship:

$$q_{metal} = -q_{water}$$

$$m_{metal} \times C_{s,\,metal} \times \Delta T_{metal} = -m_{water} \times C_{s,\,water} \times \Delta T_{water}$$

Example 10.3 demonstrates how to work with thermal energy transfer.

EXAMPLE 10.3
Thermal Energy Transfer

A 32.5-g cube of aluminum initially at 45.8 °C is submerged into 105.3 g of water at 15.4 °C. What is the final temperature of both substances at thermal equilibrium? (Assume that the aluminum and the water are thermally isolated from everything else.)

SORT You are given the masses of aluminum and water and their initial temperatures. You are asked to find the final temperature.

GIVEN: $m_{Al} = 32.5\text{ g}$ $\quad m_{H_2O} = 105.3\text{ g}$
$T_{i,\,Al} = 45.8\text{ °C};\ T_{i,\,H_2O} = 15.4\text{ °C}$

FIND: T_f

STRATEGIZE The heat lost by the aluminum (q_{Al}) equals the heat gained by the water (q_{H_2O}).

Use the relationship beween q and ΔT and the given variables to find a relationship between ΔT_{Al} and ΔT_{H_2O}.

CONCEPTUAL PLAN
$q_{Al} = -q_{H_2O}$

$m_{Al}, C_{s,\,Al}, m_{H_2O}\ C_{s,\,H_2O}$ → $\Delta T_{Al} = \text{constant} \times \Delta T_{H_2O}$

$m_{Al} \times C_{s,\,Al} \times \Delta T_{Al} = -m_{H_2O} \times C_{s,\,H_2O} \times \Delta T_{H_2O}$

Use the relationship between ΔT_{Al} and ΔT_{H_2O} (that you just found) along with the initial temperatures of the aluminum and the water to determine the final temperature. Note that at thermal equilibrium, the final temperature of the aluminum and the water is the same, that is, $T_{f,\,Al} = T_{f,\,H_2O} = T_f$.

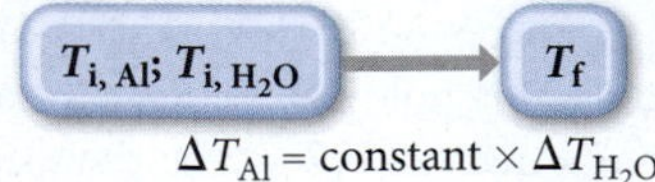

RELATIONSHIPS USED
$C_{s,\,H_2O} = 4.18\text{ J/g}\cdot\text{°C};\ C_{s,\,Al} = 0.903\text{ J/g}\cdot\text{°C}$ (Table 10.2)
$q = m \times C_s \times \Delta T$ (Equation 10.5)

—*Continued on the next page*

Continued from the previous page—

SOLVE Write the equation for the relationship between the heat lost by the aluminum (q_{Al}) and the heat gained by the water (q_{H_2O}) and substitute $q = m \times C_s \times \Delta T$ for each substance.	**SOLUTION** $q_{Al} = -q_{H_2O}$ $m_{Al} \times C_{s,\,Al} \times \Delta T_{Al} = -m_{H_2O} \times C_{s,\,H_2O} \times \Delta T_{H_2O}$
Substitute the values of m (given) and C_s (from Table 10.2) for each substance and solve the equation for ΔT_{Al}. (Alternatively, you can solve the equation for ΔT_{H_2O}.)	$32.5\ \text{g} \times \frac{0.903\ \text{J}}{\text{g} \cdot {}^\circ\text{C}} \cdot \Delta T_{Al} = -105.3\ \text{g} \times \frac{4.18\ \text{J}}{\text{g} \cdot {}^\circ\text{C}} \cdot \Delta T_{H_2O}$ $29.348 \cdot \Delta T_{Al} = -440.15 \cdot \Delta T_{H_2O}$ $\Delta T_{Al} = -14.998 \cdot \Delta T_{H_2O}$
Substitute the initial temperatures of aluminum and water into the relationship from the previous step and solve the expression for the final temperature (T_f). Remember that the final temperature for both substances is the same.	$T_f - T_{i,\,Al} = -14.998(T_f - T_{i,\,H_2O})$ $T_f = -14.998 \cdot T_f + 14.998 \cdot T_{i,\,H_2O} + T_{i,\,Al}$ $T_f + 14.998 \cdot T_f = 14.998 \cdot T_{i,\,H_2O} + T_{i,\,Al}$ $15.998 \cdot T_f = 14.998 \cdot T_{i,\,H_2O} + T_{i,\,Al}$ $T_f = \frac{14.998 \cdot T_{i,\,H_2O} + T_{i,\,Al}}{15.998} = \frac{14.998 \cdot 15.4\ {}^\circ\text{C} + 45.8\ {}^\circ\text{C}}{15.998}$ $= 17.3\ {}^\circ\text{C}$

CHECK The units °C are correct. The final temperature of the mixture is closer to the initial temperature of the *water* than the *aluminum*. This makes sense for two reasons: (1) Water has a higher specific heat capacity than aluminum; and (2) there is more water than aluminum. Because the aluminum loses the same amount of heat that the water gains, the greater mass and specific heat capacity of the water make the temperature change in the water *less than* the temperature change in the aluminum.

FOR PRACTICE 10.3

A block of copper of unknown mass has an initial temperature of 65.4 °C. The copper is immersed in a beaker containing 95.7 g of water at 22.7 °C. When the two substances reach thermal equilibrium, the final temperature is 24.2 °C. What is the mass of the copper block?

Thermal Energy Transfer

10.4 Cc Conceptual Connection

Substances A and B, initially at different temperatures, come in contact with each other and reach thermal equilibrium. The mass of substance A is twice the mass of substance B. The specific heat capacity of substance B is twice the specific heat capacity of substance A. Which statement is true about the final temperature of the two substances once thermal equilibrium is reached?

(a) The final temperature is closer to the initial temperature of substance A than substance B.

(b) The final temperature is closer to the initial temperature of substance B than substance A.

(c) The final temperature is exactly midway between the initial temperatures of substance A and B.

Work: Pressure–Volume Work

Recall that energy transfer between a system and its surroundings can occur via heat (q) or work (w). We just discussed how to calculate the *heat* associated with an observed *temperature* change. We now turn to calculating the *work* associated with an observed *volume* change. Although a chemical reaction can do several different types of work, for now we limit our discussion to **pressure–volume work**. We have already defined work as a force acting through a distance. Pressure–volume work occurs when the force is caused by a volume change against an external pressure. **Pressure** is the force that pushes against the cylinder divided by the area of the cylinder. For example, pressure–volume work occurs in the cylinder of an automobile engine. The combustion of gasoline causes gases within the cylinders to expand, pushing the piston outward (against an external pressure) and ultimately moving the wheels of the car.

See Section 11.2 for a detailed description of pressure.

We can derive an equation for the value of pressure–volume work from the definition of work as a force (F) acting through a distance (D).

$$w = F \times D \qquad [10.6]$$

When the volume of a cylinder increases (Figure 10.6 ▼), it pushes against an external force. That external force is pressure (P), which is defined as force (F) divided by area (A):

The force in this equation must be a constant force.

$$P = \frac{F}{A} \quad \text{or} \quad F = P \times A$$

If we substitute this expression for force into the definition of work given in Equation 10.6, we arrive at:

$$\begin{aligned} w &= F \times D \\ &= P \times A \times D \end{aligned}$$

The distance through which the force acts is the change in the height of the piston as it moves during the expansion (Δh). Substituting Δh for D, we get:

$$w = P \times A \times \Delta h$$

Because the volume of a cylinder is the area of its base times its height, $A \times \Delta h$ is actually the change in volume (ΔV) that occurs during the expansion. Thus, the expression for work becomes:

$$w = P\Delta V$$

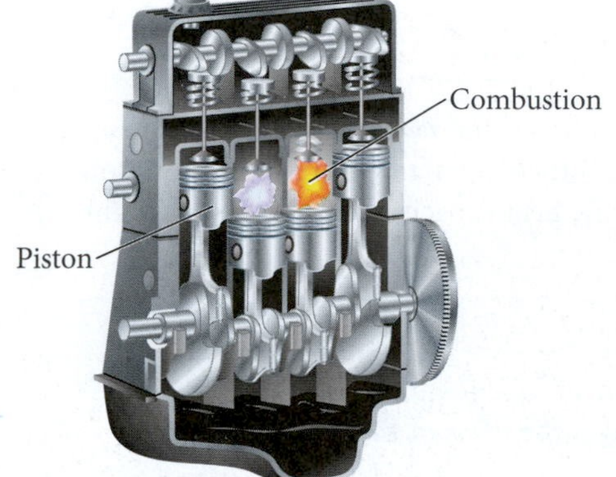

▲ The combustion of gasoline within an engine's cylinders does pressure–volume work that results in the motion of the car.

Still missing from the equation is the *sign* of the work done by the expanding gases. As the volume of the cylinder increases, work is done *on* the surroundings *by* the system, so w should be negative. However, upon expansion, V_2 (the final volume) is greater than V_1 (the initial volume) so ΔV is positive. In order for w to be negative for a positive expansion, we need to add a negative sign to our equation. In other words, w and ΔV must be opposite in sign:

$$w = -P\Delta V \qquad [10.7]$$

So the work caused by an expansion of volume is the negative of the pressure that the volume expands against multiplied by the change in volume that occurs during the expansion. The units of the work we obtain by using this equation depend on the units of pressure and volume. Most commonly, the units of pressure are atmospheres (atm) (see Section 11.2) and those of volume are liters (L), so the resulting units are L · atm. To convert between L · atm and J, use the conversion factor 101.3 J = 1 L · atm.

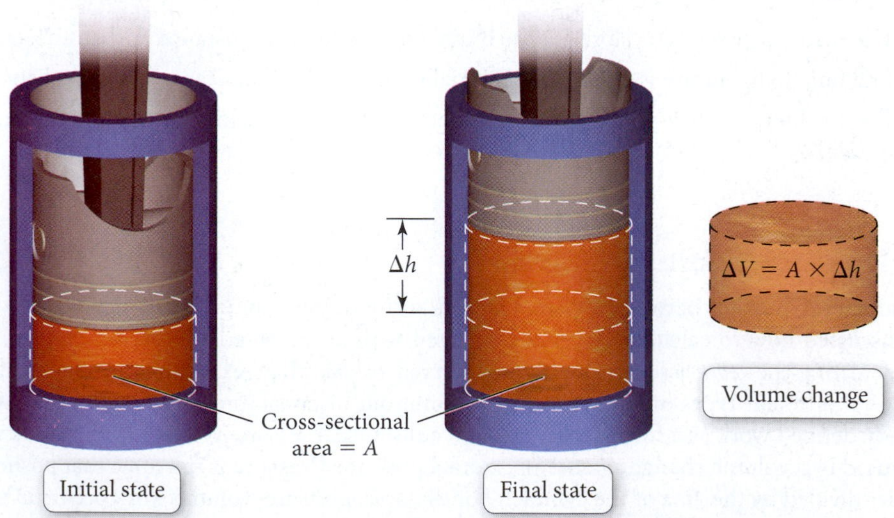

▲ **FIGURE 10.6 Piston Moving within a Cylinder against an External Pressure**

EXAMPLE 10.4

Pressure–Volume Work

To inflate a balloon you must do pressure–volume work on the surroundings. If you inflate a balloon from a volume of 0.100 L to 1.85 L against an external pressure of 1.00 atm, how much work is done (in joules)?

SORT You know the initial and final volumes of the balloon and the pressure against which it expands. The balloon and its contents are the system.	**GIVEN:** $V_1 = 0.100$ L, $V_2 = 1.85$ L, $P = 1.00$ atm **FIND:** w
STRATEGIZE The equation $w = -P\,\Delta V$ specifies the amount of work done during a volume change against an external pressure.	**CONCEPTUAL PLAN** 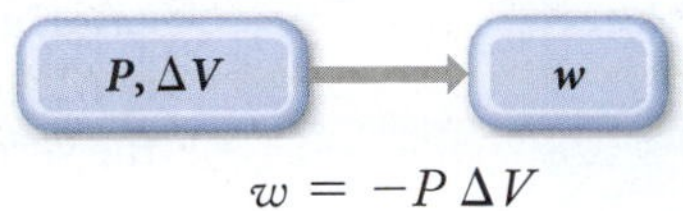 $w = -P\,\Delta V$
SOLVE To solve the problem, calculate the value of ΔV and substitute it, together with P, into Equation 10.7.	**SOLUTION** $\Delta V = V_2 - V_1$ $= 1.85\text{ L} - 0.100\text{ L}$ $= 1.75\text{ L}$ $w = -P\,\Delta V$ $= -1.00\text{ atm} \times 1.75\text{ L}$ $= -1.75\text{ L}\cdot\text{atm}$
Convert the units of the answer (L · atm) to joules using 101.3 J = 1 L · atm.	$-1.75\ \cancel{\text{L}\cdot\text{atm}} \times \dfrac{101.3\text{ J}}{1\ \cancel{\text{L}\cdot\text{atm}}} = -177\text{ J}$

CHECK The units (J) are correct for work. The sign of the work is negative, as it should be for an expansion: Work is done on the surroundings by the expanding balloon.

FOR PRACTICE 10.4

A cylinder equipped with a piston expands against an external pressure of 1.58 atm. If the initial volume is 0.485 L and the final volume is 1.245 L, how much work (in J) is done?

FOR MORE PRACTICE 10.4

When fuel is burned in a cylinder equipped with a piston, the volume expands from 0.255 L to 1.45 L against an external pressure of 1.02 atm. In addition, 875 J is emitted as heat. What is ΔE for the burning of the fuel?

10.5 Measuring ΔE for Chemical Reactions: Constant-Volume Calorimetry

We now have a complete picture of how a system exchanges energy with its surroundings via heat and pressure–volume work:

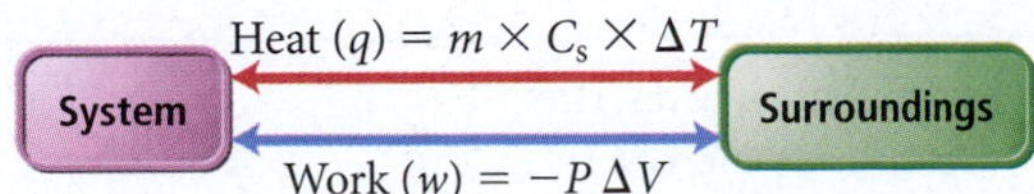

Recall from Section 10.3 that the change in internal energy that occurs during a chemical reaction (ΔE) is a measure of *all of the energy* (heat and work) exchanged with the surroundings ($\Delta E = q + w$). Therefore, we can measure the changes in temperature (to calculate heat) and the changes in volume (to calculate work) that occur during a chemical reaction, and then sum them together to calculate ΔE. However,

The Bomb Calorimeter

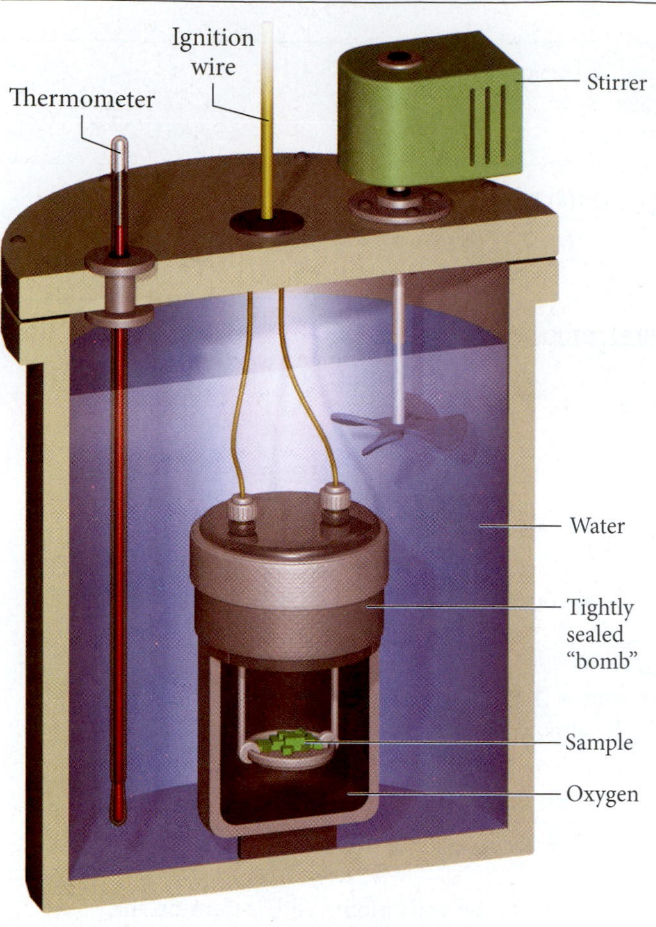

▲ **FIGURE 10.7 The Bomb Calorimeter** A bomb calorimeter measures changes in internal energy for combustion reactions.

The heat capacity of the calorimeter, C_{cal}, has units of energy over temperature; its value accounts for all of the heat absorbed by all of the components within the calorimeter (including the water).

an easier way to obtain the value of ΔE for a chemical reaction is to force all of the energy change associated with a reaction to manifest itself as heat rather than work. We can then measure the temperature change caused by the heat flow.

Recall that $\Delta E = q + w$ and that $w = -P\,\Delta V$. If a reaction is carried out at constant volume, then $\Delta V = 0$ and $w = 0$. The heat evolved (or given off), called the *heat at constant volume* (q_v), is then equal to ΔE_{rxn}.

$$\Delta E_{rxn} = q_v + w \quad \text{(Equals zero at constant volume)}$$
$$\Delta E_{rxn} = q_v \qquad [10.8]$$

We can measure the heat evolved in a chemical reaction using *calorimetry*. In **calorimetry**, we measure the thermal energy the reaction (defined as the system) and the surroundings exchange by observing the change in temperature of the surroundings.

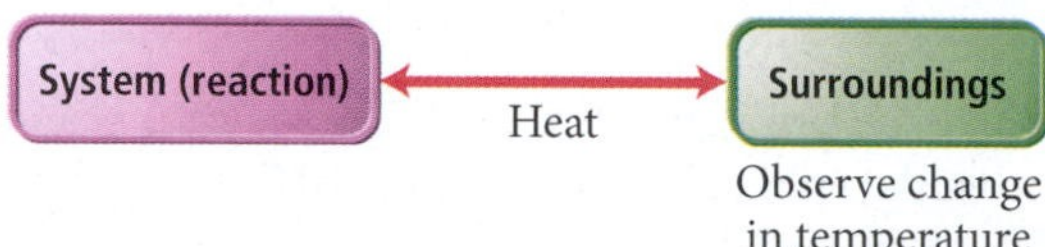

The magnitude of the temperature change in the surroundings depends on the magnitude of ΔE for the reaction and on the heat capacity of the surroundings.

Figure 10.7 ◀ shows a **bomb calorimeter**, a piece of equipment designed to measure ΔE for combustion reactions. In a bomb calorimeter, the reaction occurs in a sealed container called a *bomb*. This ensures that the reaction occurs at constant volume. To use a bomb calorimeter, we put the sample to be burned (of known mass) into a cup equipped with an ignition wire. We then seal the cup into the bomb, which is filled with oxygen gas, and place the bomb in a water-filled, insulated container. The container is equipped with a stirrer and a thermometer. Finally, we ignite the sample with a wire coil, and monitor the temperature with the thermometer. The temperature change (ΔT) is related to the heat absorbed by the entire calorimeter assembly (q_{cal}) by the equation:

$$q_{cal} = C_{cal} \times \Delta T \qquad [10.9]$$

where C_{cal} is the heat capacity of the entire calorimeter assembly (which is usually determined in a separate measurement involving the burning of a substance that gives off a known amount of heat). If no heat escapes from the calorimeter, the amount of heat *gained by* the calorimeter exactly equals that *released by* the reaction (the two are equal in magnitude but opposite in sign):

$$q_{cal} = -q_{rxn} \qquad [10.10]$$

Because the reaction occurs under conditions of constant volume, $q_{rxn} = q_v = \Delta E_{rxn}$. This measured quantity is the change in the internal energy of the reaction for the specific amount of reactant burned. To get ΔE_{rxn} per mole of a particular reactant—a more general quantity—we divide by the number of moles that actually react, as demonstrated in Example 10.5.

EXAMPLE 10.5
Measuring ΔE_{rxn} in a Bomb Calorimeter

When 1.010 g of sucrose ($C_{12}H_{22}O_{11}$) undergoes combustion in a bomb calorimeter, the temperature rises from 24.92 °C to 28.33 °C. Find ΔE_{rxn} for the combustion of sucrose in kJ/mol sucrose. The heat capacity of the bomb calorimeter, determined in a separate experiment, is 4.90 kJ/°C. (You can ignore the heat capacity of the small sample of sucrose because it is negligible compared to the heat capacity of the calorimeter.)

—Continued on the next page

Continued from the previous page—

SORT You are given the mass of sucrose, the heat capacity of the calorimeter, and the initial and final temperatures. You are asked to find the change in internal energy for the reaction.

GIVEN: 1.010 g $C_{12}H_{22}O_{11}$, $T_i = 24.92\ °C$, $T_f = 28.33\ °C$, $C_{cal} = 4.90\ kJ/°C$

FIND: ΔE_{rxn}

STRATEGIZE The conceptual plan has three parts. In the first part, use the temperature change and the heat capacity of the calorimeter to find q_{cal}.

CONCEPTUAL PLAN

$C_{cal}, \Delta T \rightarrow q_{cal}$

$q_{cal} = C_{cal} \times \Delta T$

In the second part, use q_{cal} to determine q_{rxn} (which just involves changing the sign). Because the bomb calorimeter ensures constant volume, q_{rxn} is equivalent to ΔE_{rxn} for the amount of sucrose burned.

$q_{cal} \rightarrow q_{rxn}$

$q_{rxn} = -q_{cal}$

In the third part, divide q_{rxn} by the number of moles of sucrose to find ΔE_{rxn} per mole of sucrose.

$$\Delta E_{rxn} = \frac{q_{rxn}}{\text{mol } C_{12}H_{22}O_{11}}$$

RELATIONSHIPS USED

$q_{cal} = C_{cal} \times \Delta T = -q_{rxn}$

molar mass $C_{12}H_{22}O_{11} = 342.3\ g/mol$

SOLVE Gather the necessary quantities in the correct units and substitute these into the equation to calculate q_{cal}.

SOLUTION

$$\Delta T = T_f - T_i$$

$$= 28.33\ °C - 24.92\ °C = 3.41\ °C$$

$$q_{cal} = C_{cal} \times \Delta T$$

$$q_{cal} = 4.90\ \frac{kJ}{°C} \times 3.41\ °C = 16.7\ kJ$$

Find q_{rxn} by taking the negative of q_{cal}.

$$q_{rxn} = -q_{cal} = -16.7\ kJ$$

Find ΔE_{rxn} per mole of sucrose by dividing q_{rxn} by the number of moles of sucrose (calculated from the given mass of sucrose and its molar mass).

$$\Delta E_{rxn} = \frac{q_{rxn}}{\text{mol } C_{12}H_{22}O_{11}}$$

$$= \frac{-16.7\ kJ}{1.010\ g\ C_{12}H_{22}O_{11} \times \frac{1\ mol\ C_{12}H_{22}O_{11}}{342.3\ g\ C_{12}H_{22}O_{11}}}$$

$$= -5.66 \times 10^3\ kJ/mol\ C_{12}H_{22}O_{11}$$

CHECK The units of the answer (kJ) are correct for a change in internal energy. The sign of ΔE_{rxn} is negative, as it should be for a combustion reaction that gives off energy.

FOR PRACTICE 10.5

When 1.550 g of liquid hexane (C_6H_{14}) undergoes combustion in a bomb calorimeter, the temperature rises from 25.87 °C to 38.13 °C. Find ΔE_{rxn} for the reaction in kJ/mol hexane. The heat capacity of the bomb calorimeter, determined in a separate experiment, is 5.73 kJ/°C.

FOR MORE PRACTICE 10.5

The combustion of toluene has a ΔE_{rxn} of -3.91×10^3 kJ/mol. When 1.55 g of toluene (C_7H_8) undergoes combustion in a bomb calorimeter, the temperature rises from 23.12 °C to 37.57 °C. Find the heat capacity of the bomb calorimeter.

10.6 Enthalpy: The Heat Evolved in a Chemical Reaction at Constant Pressure

In Section 10.5, we saw that when a chemical reaction occurs in a sealed container under conditions of constant volume, the energy evolves only as heat. However, when a chemical reaction occurs open to the atmosphere under conditions of constant pressure—for example, a reaction occurring in an open beaker or the burning of natural gas on a stove—the energy can evolve as both heat and work. As we have also seen, ΔE_{rxn} is a measure of the *total energy change* (both heat and work) that occurs during a reaction. However, in many cases, we are interested only in the heat exchanged, not the work done. For example, when we burn natural gas on a stove to cook food, we do not really care how much work the combustion reaction does on the atmosphere by expanding against it—we just want to know how much heat is given off to cook the food. Under conditions of constant pressure, a thermodynamic quantity called *enthalpy* represents exactly this.

We define the **enthalpy (*H*)** of a system as the sum of its internal energy and the product of its pressure and volume:

$$H = E + PV \qquad [10.11]$$

Because internal energy, pressure, and volume are all state functions, enthalpy is also a state function. The *change in enthalpy* (ΔH) for any process occurring under constant pressure is given by the expression:

$$\Delta H = \Delta E + P\Delta V \qquad [10.12]$$

To better understand this expression, we can interpret the two terms on the right with the help of relationships already familiar to us. We saw previously that $\Delta E = q + w$. If we represent the heat at constant pressure as q_p, then the change in internal energy at constant pressure is $\Delta E = q_p + w$. In addition, from our definition of pressure–volume work, we know that $P\Delta V = -w$. Substituting these expressions into the expression for ΔH gives us:

$$\begin{aligned} \Delta H &= \Delta E + P\Delta V \\ &= (q_p + w) + P\Delta V \qquad [10.13] \\ &= q_p + w - w \\ \Delta H &= q_p \end{aligned}$$

We can see that ΔH is equal to q_p, the heat at constant pressure.

Conceptually (and often numerically), ΔH and ΔE are similar: They both represent changes in a state function for the system. However, ΔE is a measure of *all of the energy* (heat and work) exchanged with the surroundings, while ΔH is a measure of only the heat exchanged under conditions of constant pressure. For chemical reactions that do not exchange much work with the surroundings—that is, those that do not cause a large change in reaction volume as they occur—ΔH and ΔE are nearly identical in value. For chemical reactions that produce or consume large amounts of gas, and therefore result in large volume changes, ΔH and ΔE will be slightly different in value.

10.5 Cc Conceptual Connection

The Difference between ΔH and ΔE

Lighters are usually fueled by butane (C_4H_{10}). When 1 mole of butane burns at constant pressure, it produces 2658 kJ of heat and does 3 kJ of work. What are the values of ΔH and ΔE for the combustion of one mole of butane?

The signs of ΔH and ΔE follow the same conventions. A positive ΔH indicates that heat flows into the system as the reaction occurs. A chemical reaction with a positive ΔH, called an **endothermic reaction**, absorbs heat from its surroundings. A chemical cold pack used to ice athletic injuries is a good example of an endothermic reaction. When a barrier separating the reactants in a chemical cold pack is broken, the substances mix, react, and absorb heat from the surroundings. The surroundings—including, say, your bruised wrist—become *colder* because they *lose* energy as the cold pack absorbs it.

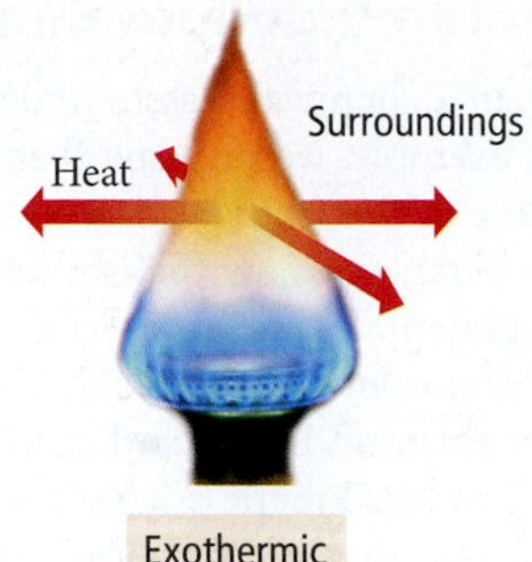

▲ The reaction that occurs in a chemical cold pack is endothermic—it absorbs energy from the surroundings. The combustion of natural gas is an exothermic reaction—it releases energy to the surroundings.

A chemical reaction with a negative ΔH, called an **exothermic reaction**, gives off heat to its surroundings. The reaction that occurs when ethanol burns (discussed in Section 10.1) is a good example of an exothermic reaction. As the reaction occurs, heat is given off into the surroundings making them warmer. The burning of natural gas is another example of an exothermic reaction. As the gas burns, it gives off energy, raising the temperature of its surroundings.

***Summarizing Enthalpy*:**

- The value of ΔH for a chemical reaction is the amount of heat absorbed or evolved in the reaction under conditions of constant pressure.
- An endothermic reaction has a *positive* ΔH and absorbs heat from the surroundings. An endothermic reaction feels cold to the touch (if it is occurring near room temperature).
- An exothermic reaction has a *negative* ΔH and gives off heat to the surroundings. An exothermic reaction feels warm to the touch (if it is occurring near room temperature).

EXAMPLE 10.6

Exothermic and Endothermic Processes

Identify each process as endothermic or exothermic and indicate the sign of ΔH.

(a) sweat evaporating from skin
(b) water freezing in a freezer
(c) wood burning in a fire

SOLUTION

(a) Sweat evaporating from skin cools the skin and is therefore endothermic, with a positive ΔH. The skin must supply heat to the perspiration in order for it to continue to evaporate.
(b) Water freezing in a freezer releases heat and is therefore exothermic, with a negative ΔH. The refrigeration system in the freezer must remove this heat for the water to continue to freeze.
(c) Wood burning in a fire releases heat and is therefore exothermic, with a negative ΔH.

FOR PRACTICE 10.6

Identify each process as endothermic or exothermic and indicate the sign of ΔH.

(a) an ice cube melting
(b) nail polish remover quickly evaporating after it is accidentally spilled on the skin
(c) gasoline burning within the cylinder of an automobile engine

Exothermic and Endothermic Processes: A Particulate View

When a chemical system undergoes a change in enthalpy, where does the energy come from or go to? For example, we just saw that an exothermic chemical reaction gives off *thermal energy*—what is the source of that energy?

First, we know that the emitted thermal energy *does not* come from the original thermal energy of the system. Recall from Section 10.2 that the thermal energy of a system is the total kinetic energy of the atoms and molecules that compose the system. This kinetic energy *cannot* be the source of the energy given off in an exothermic reaction because, if the atoms and molecules that compose the system were to lose kinetic energy, their temperature would necessarily fall—the system would get colder. Yet, we know that in exothermic reactions, the temperature of the system and the surroundings rises. So there must be some other source of energy.

Recall from Section 10.3 that the internal energy of a chemical system is the sum of its kinetic energy and its *potential energy*. This potential energy is the source in an exothermic chemical reaction. Under normal circumstances, chemical potential energy (or simply chemical energy) arises primarily from the electrostatic forces between the protons and electrons that compose the atoms and molecules within the system. In an exothermic reaction, some bonds break and new ones form, and the nuclei and electrons reorganize into an arrangement with lower potential energy. As the molecules rearrange, their potential energy converts into thermal energy, the heat emitted in the reaction. In an endothermic reaction, the opposite happens: As some bonds break and others form, the nuclei and electrons reorganize into an arrangement with higher potential energy, absorbing thermal energy in the process.

10.6 Cc Conceptual Connection

Exothermic and Endothermic Reactions

If an endothermic reaction absorbs heat, why does it feel cold to the touch?

Stoichiometry Involving ΔH: Thermochemical Equations

The enthalpy change for a chemical reaction, abbreviated ΔH_{rxn}, is also called the **enthalpy of reaction** or **heat of reaction** and is an extensive property, one that depends on the amount of material undergoing the reaction. In other words, the amount of heat generated or absorbed when a chemical reaction occurs depends on the *amounts* of reactants that actually react. We usually specify ΔH_{rxn} in combination with the balanced chemical equation for the reaction. *The magnitude of* ΔH_{rxn} *is for the stoichiometric amounts of reactants and products for the reaction as written.* For example, the balanced equation and ΔH_{rxn} for the combustion of propane (the main component of LP gas) are as follows:

Throughout this book, combustion reactions such as this one are written using gaseous water as the product because, when we actually burn a hydrocarbon such as propane, water is formed in the gaseous state. Some books, however, use liquid water as the product, which gives a moderately different value for ΔH_{rxn}.

$$C_3H_8(g) + 5\,O_2(g) \longrightarrow 3\,CO_2(g) + 4\,H_2O(g) \qquad \Delta H_{rxn} = -2044\text{ kJ}$$

This means that when 1 mol of C_3H_8 reacts with 5 mol of O_2 to form 3 moles of CO_2 and 4 mol of H_2O, 2044 kJ of heat is emitted. We can write these relationships in the same way that we expressed stoichiometric relationships in Chapter 8 as ratios between two quantities. For example, for the reactants, we write:

$$1\text{ mol } C_3H_8 : -2044\text{ kJ} \qquad \text{or} \qquad 5\text{ mol } O_2 : -2044\text{ kJ}$$

These ratios indicate that 2044 kJ of heat evolves when 1 mol of C_3H_8 and 5 mol of O_2 completely react. We can use these ratios to construct conversion factors between amounts of reactants or products and the quantity of heat emitted (for exothermic reactions) or absorbed (for endothermic reactions).

To find out how much heat is emitted upon the combustion of a certain mass in grams of C_3H_8, we use the following conceptual plan:

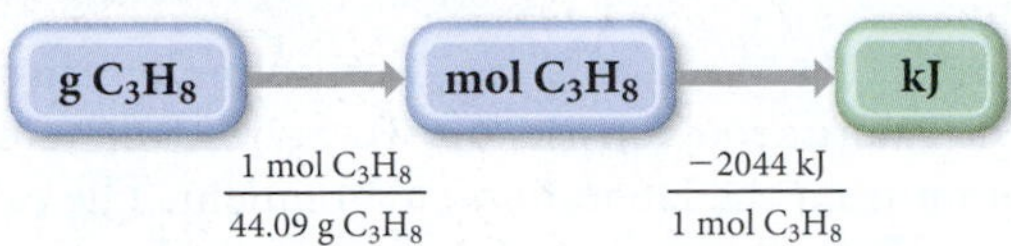

We use the molar mass to convert between grams and moles, and the stoichiometric relationship between moles of C_3H_8 and the heat of reaction to convert between moles and kilojoules, as demonstrated in Example 10.7.

EXAMPLE 10.7

Stoichiometry Involving ΔH

An LP gas tank in a home barbeque contains 13.2 kg of propane, C_3H_8. Calculate the heat (in kJ) associated with the complete combustion of all of the propane in the tank.

$$C_3H_8(g) + 5\,O_2(g) \longrightarrow 3\,CO_2(g) + 4\,H_2O(g) \qquad \Delta H_{rxn} = -2044 \text{ kJ}$$

SORT You are given the mass of propane and asked to find the heat evolved in its combustion.

GIVEN: 13.2 kg C_3H_8

FIND: q

STRATEGIZE Starting with kg C_3H_8, convert to g C_3H_8 and then use the molar mass of C_3H_8 to find the number of moles. Next, use the stoichiometric relationship between mol C_3H_8 and kJ to determine the heat evolved.

CONCEPTUAL PLAN

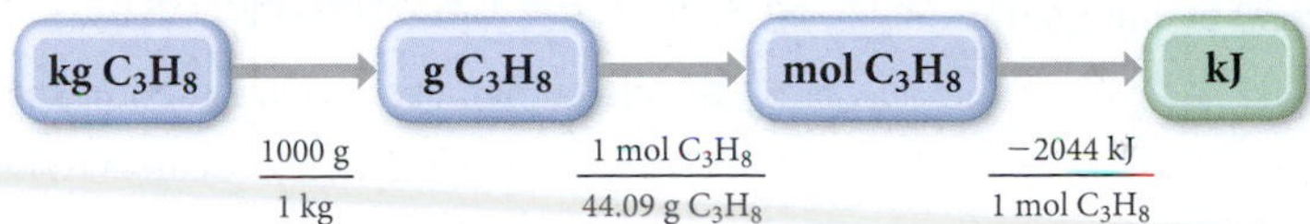

RELATIONSHIPS USED

1000 g = 1 kg

molar mass C_3H_8 = 44.09 g/mol

1 mol C_3H_8 : −2044 kJ (from balanced equation)

SOLVE Follow the conceptual plan to solve the problem. Begin with 13.2 kg C_3H_8 and multiply by the appropriate conversion factors to arrive at kJ.

SOLUTION

$$13.2 \text{ kg } C_3H_8 \times \frac{1000 \text{ g}}{1 \text{ kg}} \times \frac{1 \text{ mol } C_3H_8}{44.09 \text{ g } C_3H_8} \times \frac{-2044 \text{ kJ}}{1 \text{ mol } C_3H_8}$$

$$= -6.12 \times 10^5 \text{ kJ}$$

CHECK The units of the answer (kJ) are correct for energy. The answer is negative, as it should be for heat evolved by the reaction.

FOR PRACTICE 10.7

Ammonia reacts with oxygen according to the equation:

$$4\,NH_3(g) + 5\,O_2(g) \longrightarrow 4\,NO(g) + 6\,H_2O(g) \qquad \Delta H_{rxn} = -906 \text{ kJ}$$

Calculate the heat (in kJ) associated with the complete reaction of 155 g of NH_3.

FOR MORE PRACTICE 10.7

What mass of butane in grams is necessary to produce 1.5×10^3 kJ of heat? What mass of CO_2 is produced?

$$C_4H_{10}(g) + \frac{13}{2}O_2(g) \longrightarrow 4\,CO_2(g) + 5\,H_2O(g) \qquad \Delta H_{rxn} = -2658 \text{ kJ}$$

The Coffee-Cup Calorimeter

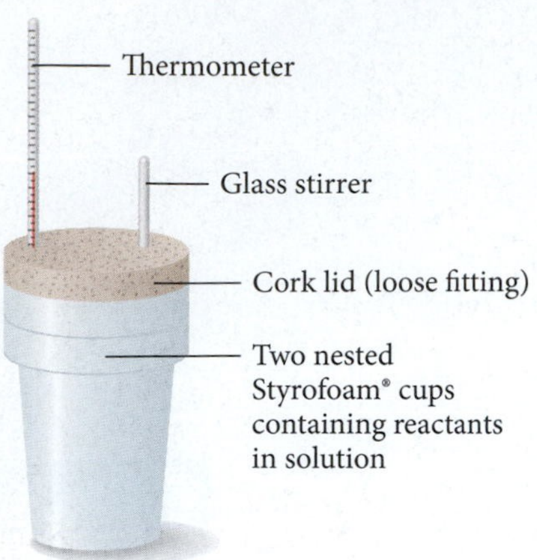

▲ **FIGURE 10.8 The Coffee-Cup Calorimeter** A coffee-cup calorimeter measures enthalpy changes for chemical reactions in solution.

The equation, $q_{rxn} = -q_{soln}$, assumes that no heat is lost to the calorimeter itself. If heat absorbed by the calorimeter is accounted for, the equation becomes $q_{rxn} = -(q_{soln} + q_{cal})$.

10.7 Measuring ΔH for Chemical Reactions: Constant-Pressure Calorimetry

For many aqueous reactions, we can measure ΔH_{rxn} fairly simply using a **coffee-cup calorimeter** shown in Figure 10.8 ◄. The calorimeter consists of two Styrofoam® coffee cups, one inserted into the other, to provide insulation from the laboratory environment. The calorimeter is equipped with a thermometer and a stirrer. The reaction occurs in a specifically measured quantity of solution within the calorimeter, so that the mass of the solution is known. During the reaction, the heat evolved (or absorbed) causes a temperature change in the solution, which the thermometer measures. If we know the specific heat capacity of the solution, normally assumed to be that of water, we can calculate q_{soln}, the heat absorbed by or lost from the solution (which is acting as the surroundings) using the equation:

$$q_{soln} = m_{soln} \times C_{s,\,soln} \times \Delta T$$

The insulated calorimeter prevents heat from escaping, so we assume that the heat gained by the solution equals that lost by the reaction (or vice versa):

$$q_{rxn} = -q_{soln}$$

Because the reaction happens under conditions of constant pressure (open to the atmosphere), $q_{rxn} = q_p = \Delta H_{rxn}$ (Equation 10.13). This measured quantity is the heat of reaction for the specific amount (which is measured ahead of time) of reactants that reacted. To get ΔH_{rxn} per mole of a particular reactant—a more general quantity—we divide by the number of moles that actually reacted, as demonstrated in Example 10.8.

Summarizing Calorimetry:

- Bomb calorimetry occurs at constant *volume* and measures ΔE for a reaction.
- Coffee-cup calorimetry occurs at constant *pressure* and measures ΔH for a reaction.

EXAMPLE 10.8
Measuring ΔH_{rxn} in a Coffee-Cup Calorimeter

Magnesium metal reacts with hydrochloric acid according to the balanced equation:

$$Mg(s) + 2\,HCl(aq) \longrightarrow MgCl_2(aq) + H_2(g)$$

In an experiment to determine the enthalpy change for this reaction, you combine 0.158 g of Mg metal with enough HCl to make 100.0 mL of solution in a coffee-cup calorimeter. The HCl is sufficiently concentrated so that the Mg completely reacts. The temperature of the solution rises from 25.6 °C to 32.8 °C as a result of the reaction. Find ΔH_{rxn} for the reaction as written. Use 1.00 g/mL as the density of the solution and $C_{s,\,soln} = 4.18\ J/g \cdot {}^\circ C$ as the specific heat capacity of the solution.

SORT You are given the mass of magnesium, the volume of solution, the initial and final temperatures, the density of the solution, and the heat capacity of the solution. You are asked to find the change in enthalpy for the reaction.	**GIVEN:** 0.158 g Mg 100.0 mL soln $T_i = 25.6\ {}^\circ C$ $T_f = 32.8\ {}^\circ C$ $d_{soln} = 1.00\ g/mL$ $C_{s,\,soln} = 4.18\ J/g \cdot {}^\circ C$ **FIND:** ΔH_{rxn}

—Continued on the next page

Continued from the previous page—

STRATEGIZE The conceptual plan has three parts. In the first part, use the temperature change and the other given quantities, together with the equation $q = m \times C_s \times \Delta T$, to find q_{soln}.

CONCEPTUAL PLAN

$C_{s,\,soln}, m_{soln}, \Delta T \rightarrow q_{soln}$

$q = m \times C_s \times \Delta T$

In the second part, use q_{soln} to get q_{rxn} (which simply involves changing the sign). Because the pressure is constant, q_{rxn} is equivalent to ΔH_{rxn} for the amount of magnesium that reacts.

$q_{soln} \rightarrow q_{rxn}$

$q_{rxn} = -q_{soln}$

In the third part, divide q_{rxn} by the number of moles of magnesium to get ΔH_{rxn} per mole of magnesium.

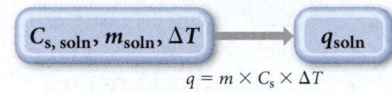

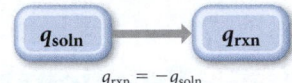

$$\Delta H_{rxn} = \frac{q_{rxn}}{\text{mol Mg}}$$

RELATIONSHIPS USED

$q = m \times C_s \times \Delta T$

$q_{rxn} = -q_{soln}$

SOLVE Gather the necessary quantities in the correct units for the equation $q = m \times C_s \times \Delta T$ and substitute into the equation to calculate q_{soln}. Notice that the sign of q_{soln} is positive, meaning that the solution absorbed heat from the reaction.

SOLUTION

$C_{s,\,soln} = 4.18\ \text{J/g} \cdot ^\circ\text{C}$

$$m_{soln} = 100.0\ \cancel{\text{mL soln}} \times \frac{1.00\ \text{g}}{1\ \cancel{\text{mL soln}}} = 1.00 \times 10^2\ \text{g}$$

$$\Delta T = T_f - T_i = 32.8\ ^\circ\text{C} - 25.6\ ^\circ\text{C} = 7.2\ ^\circ\text{C}$$

$$q_{soln} = m_{soln} \times C_{s,\,soln} \times \Delta T = 1.00 \times 10^2\ \cancel{\text{g}} \times 4.18 \frac{\text{J}}{\cancel{\text{g} \cdot ^\circ\text{C}}} \times 7.2\ \cancel{^\circ\text{C}} = 3.0 \times 10^3\ \text{J}$$

Find q_{rxn} by taking the negative of q_{soln}. Notice that q_{rxn} is negative, as expected for an exothermic reaction.

$$q_{rxn} = -q_{soln} = -3.0 \times 10^3\ \text{J}$$

Finally, find ΔH_{rxn} per mole of magnesium by dividing q_{rxn} by the number of moles of magnesium that reacts. Find the number of moles of magnesium from the given mass of magnesium and its molar mass.

$$\Delta H_{rxn} = \frac{q_{rxn}}{\text{mol Mg}} = \frac{-3.0 \times 10^3\ \text{J}}{0.158\ \cancel{\text{g Mg}} \times \dfrac{1\ \text{mol Mg}}{24.31\ \cancel{\text{g Mg}}}} = -4.6 \times 10^5\ \text{J/mol Mg}$$

Because the stoichiometric coefficient for magnesium in the balanced chemical equation is 1, the calculated value represents ΔH_{rxn} for the reaction as written.

$$Mg(s) + 2\,HCl(aq) \longrightarrow MgCl_2(aq) + H_2(g)$$

$$\Delta H_{rxn} = -4.6 \times 10^5\ \text{J}$$

CHECK The units of the answer (J) are correct for the change in enthalpy of a reaction. The sign is negative, as you would expect for an exothermic reaction.

FOR PRACTICE 10.8

The addition of hydrochloric acid to a silver nitrate solution precipitates silver chloride according to the reaction:

$$AgNO_3(aq) + HCl(aq) \longrightarrow AgCl(s) + HNO_3(aq)$$

When you combine 50.0 mL of 0.100 M $AgNO_3$ with 50.0 mL of 0.100 M HCl in a coffee-cup calorimeter, the temperature changes from 23.40 °C to 24.21 °C. Calculate ΔH_{rxn} for the reaction as written. Use 1.00 g/mL as the density of the solution and $C = 4.18\ \text{J/g} \cdot ^\circ\text{C}$ as the specific heat capacity.

10.7 Cc Conceptual Connection

Constant-Pressure versus Constant-Volume Calorimetry

The same reaction, with exactly the same amount of reactant, is conducted in a bomb calorimeter and in a coffee-cup calorimeter. In one measurement, $q_{rxn} = -12.5$ kJ and in the other $q_{rxn} = -11.8$ kJ. Which value was obtained in the bomb calorimeter? (Assume that the reaction has a positive ΔV in the coffee-cup calorimeter.)

10.8 Relationships Involving ΔH_{rxn}

The change in enthalpy for a reaction is always associated with a *particular* reaction. If we change the reaction in well-defined ways, then ΔH_{rxn} also changes in well-defined ways. We now turn our attention to three quantitative relationships between a chemical equation and ΔH_{rxn}.

1. **If a chemical equation is multiplied by some factor, then ΔH_{rxn} is also multiplied by the same factor.**

Recall from Section 10.6 that ΔH_{rxn} is an extensive property; it depends on the quantity of reactants undergoing reaction. Recall also that ΔH_{rxn} is usually reported for a reaction involving stoichiometric amounts of reactants. For example, for a reaction $A + 2\,B \longrightarrow C$, we typically report ΔH_{rxn} as the amount of heat emitted or absorbed when 1 mol A reacts with 2 mol B to form 1 mol C. Therefore, if a chemical equation is multiplied by a factor, then ΔH_{rxn} is also multiplied by the same factor. For example,

$$A + 2\,B \longrightarrow C \qquad \Delta H_1$$

$$2\,A + 4\,B \longrightarrow 2\,C \qquad \Delta H_2 = 2 \times \Delta H_1$$

2. **If a chemical equation is reversed, then ΔH_{rxn} changes sign.**

Recall from Section 10.6 that ΔH_{rxn} is a state function, which means that its value depends only on the initial and final states of the system.

$$\Delta H = H_{final} - H_{initial}$$

When a reaction is reversed, the final state becomes the initial state and vice versa. Consequently, ΔH_{rxn} changes sign. For example,

$$A + 2\,B \longrightarrow C \qquad \Delta H_1$$

$$C \longrightarrow A + 2B \qquad \Delta H_2 = -\Delta H_1$$

3. **If a chemical equation can be expressed as the sum of a series of steps, then ΔH_{rxn} for the overall equation is the sum of the heats of reactions for each step.**

This last relationship, known as **Hess's law**, follows from the enthalpy of reaction being a state function. Because ΔH_{rxn} is dependent only on the initial and final states, and not on the pathway the reaction follows, then ΔH obtained from summing the individual steps that lead to an overall reaction must be the same as ΔH for that overall reaction. For example,

$$A + 2\,B \longrightarrow \cancel{C} \qquad \Delta H_1$$

$$\cancel{C} \longrightarrow 2\,D \qquad \Delta H_2$$

$$\overline{A + 2\,B \longrightarrow 2\,D \qquad \Delta H_3 = \Delta H_1 + \Delta H_2}$$

Hess's Law
The change in enthalpy for a stepwise process is the sum of the enthalpy changes of the steps.

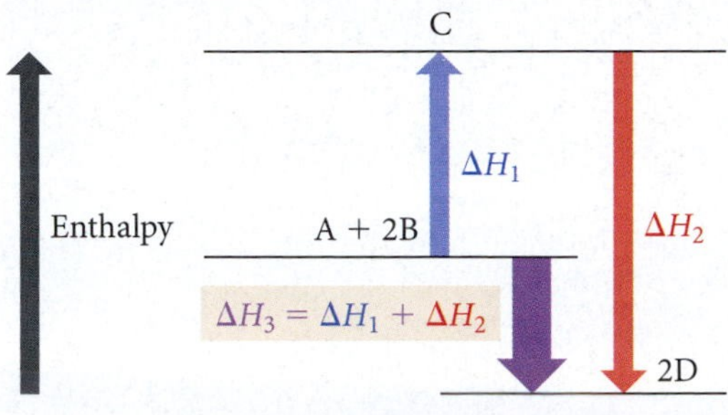

◀ **FIGURE 10.9 Hess's Law** The change in enthalpy for a stepwise process is the sum of the enthalpy changes of the steps.

We illustrate Hess's law with the energy level diagram shown in Figure 10.9 ▲.

These three quantitative relationships make it possible to determine ΔH for a reaction without directly measuring it in the laboratory. (For some reactions, direct measurement can be difficult.) If we can find related reactions (with known ΔH's) that sum to the reaction of interest, we can find ΔH for the reaction of interest. For example, the reaction between C(*s*) and $H_2O(g)$ is an industrially important method of generating hydrogen gas.

$$C(s) + H_2O(g) \longrightarrow CO(g) + H_2(g) \qquad \Delta H_{rxn} = ?$$

We can find ΔH_{rxn} from the following reactions with known ΔH's:

$$C(s) + O_2(g) \longrightarrow CO_2(g) \qquad \Delta H = -393.5 \text{ kJ}$$
$$2\,CO(g) + O_2(g) \longrightarrow 2\,CO_2(g) \qquad \Delta H = -566.0 \text{ kJ}$$
$$2\,H_2(g) + O_2(g) \longrightarrow 2\,H_2O(g) \qquad \Delta H = -483.6 \text{ kJ}$$

We just have to determine how to sum these reactions to get the overall reaction of interest. We do this by manipulating the reactions with known ΔH's in such a way as to get the reactants of interest on the left, the products of interest on the right, and other species to cancel.

Because the first reaction has C(*s*) as a reactant, and the reaction of interest also has C(*s*) as a reactant, we write the first reaction unchanged.

$$C(s) + O_2(g) \longrightarrow CO_2(g) \qquad \Delta H = -393.5 \text{ kJ}$$

The second reaction has 2 mol of CO(*g*) as a reactant. However, the reaction of interest has 1 mol of CO(*g*) as a product. Therefore, we reverse the second reaction, change the sign of ΔH, and multiply the reaction and ΔH by $\frac{1}{2}$.

$$\tfrac{1}{2} \times [2\,CO_2(g) \longrightarrow 2\,CO(g) + O_2(g)]$$
$$\Delta H = \tfrac{1}{2} \times (+566.0 \text{ kJ})$$

The third reaction has $H_2(g)$ as a reactant. In the reaction of interest, however, $H_2(g)$ is a product. Therefore, we reverse the equation and change the sign of ΔH. In addition, to obtain coefficients that match the reaction of interest, and to cancel O_2, we must multiply the reaction and ΔH by $\frac{1}{2}$.

$$\tfrac{1}{2} \times [2\,H_2O(g) \longrightarrow 2\,H_2(g) + O_2(g)]$$
$$\Delta H = \tfrac{1}{2} \times (+483.6 \text{ kJ})$$

Lastly, we rewrite the three reactions after multiplying through by the indicated factors and show how they sum to the reaction of interest. ΔH for the reaction of interest is then just the sum of the ΔH's for the steps.

$$\begin{array}{llr} C(s) + \cancel{O_2(g)} \longrightarrow \cancel{CO_2(g)} & & \Delta H = -393.5 \text{ kJ} \\ \cancel{CO_2(g)} \longrightarrow CO(g) + \cancel{\tfrac{1}{2}O_2(g)} & & \Delta H = +283.0 \text{ kJ} \\ H_2O(g) \longrightarrow H_2(g) + \cancel{\tfrac{1}{2}O_2(g)} & & \Delta H = +241.8 \text{ kJ} \\ \hline C(s) + H_2O(g) \longrightarrow CO(g) + H_2(g) & & \Delta H_{rxn} = +131.3 \text{ kJ} \end{array}$$

EXAMPLE 10.9

Hess's Law

Find ΔH_{rxn} for the reaction:

$$3\,C(s) + 4\,H_2(g) \longrightarrow C_3H_8(g)$$

Use these reactions with known ΔH's:

$$C_3H_8(g) + 5\,O_2(g) \longrightarrow 3\,CO_2(g) + 4\,H_2O(g) \qquad \Delta H = -2043\text{ kJ}$$
$$C(s) + O_2(g) \longrightarrow CO_2(g) \qquad \Delta H = -393.5\text{ kJ}$$
$$2\,H_2(g) + O_2(g) \longrightarrow 2\,H_2O(g) \qquad \Delta H = -483.6\text{ kJ}$$

SOLUTION

To solve this and other Hess's law problems, you manipulate the reactions with known ΔH's in such a way as to get the reactants of interest on the left, the products of interest on the right, and other species to cancel.

The first reaction has C_3H_8 as a reactant, and the reaction of interest has C_3H_8 as a product, so you can reverse the first reaction and change the sign of ΔH.	$3\,CO_2(g) + 4\,H_2O(g) \longrightarrow C_3H_8(g) + 5\,O_2(g) \qquad \Delta H = +2043\text{ kJ}$
The second reaction has C as a reactant and CO_2 as a product, as required in the reaction of interest. However, the coefficient for C is 1, and in the reaction of interest, the coefficient for C is 3. You need to multiply this equation and its ΔH by 3.	$3 \times [C(s) + O_2(g) \longrightarrow CO_2(g)] \qquad \Delta H = 3 \times (-393.5\text{ kJ})$
The third reaction has $H_2(g)$ as a reactant, as required. However, the coefficient for H_2 is 2, and in the reaction of interest, the coefficient for H_2 is 4. Multiply this reaction and its ΔH by 2.	$2 \times [2\,H_2(g) + O_2(g) \longrightarrow 2\,H_2O(g)] \qquad \Delta H = 2 \times (-483.6\text{ kJ})$
Lastly, rewrite the three reactions after multiplying by the indicated factors and show how they sum to the reaction of interest. ΔH for the reaction of interest is the sum of the ΔH's for the steps.	$\cancel{3CO_2(g)} + \cancel{4H_2O(g)} \longrightarrow C_3H_8(g) + \cancel{5O_2(g)} \quad \Delta H = +2043\text{ kJ}$ $3C(s) + \cancel{3O_2(g)} \longrightarrow \cancel{3CO_2(g)} \quad \Delta H = -1181\text{ kJ}$ $4H_2(g) + \cancel{2O_2(g)} \longrightarrow \cancel{4H_2O(g)} \quad \Delta H = -967.2\text{ kJ}$ $3C(s) + 4H_2(g) \longrightarrow C_3H_8(g) \quad \Delta H_{rxn} = -105\text{ kJ}$

FOR PRACTICE 10.9

Find ΔH_{rxn} for the reaction:

$$N_2O(g) + NO_2(g) \longrightarrow 3\,NO(g)$$

Use these reactions with known ΔH's:

$$2\,NO(g) + O_2(g) \longrightarrow 2\,NO_2(g) \qquad \Delta H = -113.1\text{ kJ}$$
$$N_2(g) + O_2(g) \longrightarrow 2\,NO(g) \qquad \Delta H = +182.6\text{ kJ}$$
$$2\,N_2O(g) \longrightarrow 2\,N_2(g) + O_2(g) \qquad \Delta H = -163.2\text{ kJ}$$

FOR MORE PRACTICE 10.9

Find ΔH_{rxn} for the reaction:

$$3\,H_2(g) + O_3(g) \longrightarrow 3\,H_2O(g)$$

Use these reactions with known ΔH's:

$$2\,H_2(g) + O_2(g) \longrightarrow 2\,H_2O(g) \qquad \Delta H = -483.6\text{ kJ}$$
$$3\,O_2(g) \longrightarrow 2\,O_3(g) \qquad \Delta H = +285.4\text{ kJ}$$

10.9 Determining Enthalpies of Reaction from Bond Energies

In Section 10.7, we saw that we can *measure* ΔH for a chemical reaction with calorimetry. In this section we turn to *estimating* ΔH for a chemical reaction from bond energies.

We saw in Section 10.6 that the energy changes associated with chemical reactions correspond to potential energy changes in the particles that compose atoms and molecules. In other words, during a chemical reaction, the *structure* of the particles that compose matter changes. That structure change causes a potential energy change that results in an energy exchange with the surroundings.

We can estimate the magnitude and sign of the energy exchange associated with a particular chemical reaction by focusing on the bonds that are broken and formed during the reaction. Recall from Section 6.6 that, based on experimental measurements, we can assign energies to specific bonds within a molecule (see Table 10.3). These bond energies correspond to the amount of energy *necessary to break* the particular chemical bond, but they also correspond to the amount of energy *emitted* when the bond is formed. Therefore, by adding the energies associated with the bonds that break in a reaction (remember that endothermic processes such as bond breaking carry a positive sign), and subtracting the energies associated with the bonds that form (remember that exothermic processes such as bond formation carry a negative sign), we can estimate the overall ΔH for a reaction.

For example, consider the reaction between methane and chlorine:

$$H_3C{-}H(g) + Cl{-}Cl(g) \longrightarrow H_3C{-}Cl(g) + H{-}Cl(g)$$

We can imagine this reaction occurring by the breaking of a C—H bond and a Cl—Cl bond and the forming of a C—Cl bond and an H—Cl bond. So we can calculate the overall enthalpy change as a sum

TABLE 10.3 Average Bond Energies

Bond	Bond Energy (kJ/mol)	Bond	Bond Energy (kJ/mol)	Bond	Bond Energy (kJ/mol)
H—H	436	N—N	163	Br—F	237
H—C	414	N=N	418	Br—Cl	218
H—N	389	N≡N	946	Br—Br	193
H—O	464	N—O	222	I—Cl	208
H—S	368	N=O	590	I—Br	175
H—F	565	N—F	272	I—I	151
H—Cl	431	N—Cl	200	Si—H	323
H—Br	364	N—Br	243	Si—Si	226
H—I	297	N—I	159	Si—C	301
C—C	347	O—O	142	S—O	265
C=C	611	O=O	498	Si=O	368
C≡C	837	O—F	190	S=O	523
C—N	305	O—Cl	203	Si—Cl	464
C=N	615	O—I	234	S=S	418
C≡N	891	F—F	159	S—F	327
C—O	360	Cl—F	253	S—Cl	253
C=O	736*	Cl—Cl	243	S—Br	218
C≡O	1072			S—S	266
C—Cl	339				

*799 in CO_2

▶ **FIGURE 10.10 Estimating ΔH_{rxn} from Bond Energies** We can approximate the enthalpy change of a reaction by summing the enthalpy changes involved in breaking old bonds and forming new ones.

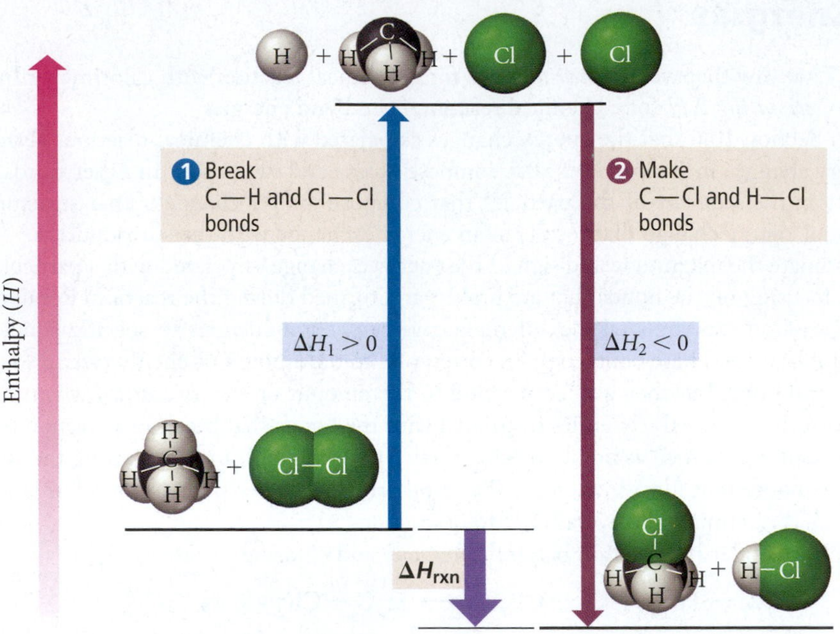

of the enthalpy changes associated with breaking the required bonds in the reactants and forming the required bonds in the products, as shown in Figure 10.10 ▲.

$$H_3C—H(g) + Cl—Cl(g) \longrightarrow H_3C—Cl(g) + H—Cl(g)$$

Bonds Broken		Bonds Formed	
C—H break	+414 kJ	C—Cl form	−339 kJ
Cl—Cl break	+243 kJ	H—Cl form	−431 kJ
Sum (Σ) *ΔH's bonds broken*: +657 kJ		*Sum* (Σ) *ΔH's bonds formed*: −770 kJ	

$$\Delta H_{rxn} = \Sigma(\Delta H\text{'s bonds broken}) + \Sigma(\Delta H\text{'s bonds formed})$$
$$= +657 \text{ kJ} - 770 \text{ kJ} = -113 \text{ kJ}$$

Using this method, we determine that $\Delta H_{rxn} = -113$ kJ. The actual $\Delta H^\circ_{rxn} = -101$ kJ, which is fairly close to the value we obtained from average bond energies. Bond energies don't give exact values for the change in enthalpy of a reaction because bond energies are average values obtained from measurements on many different molecules. Nonethless, we can estimate ΔH_{rxn} from average bond energies fairly accurately by summing the ΔH's for the bonds broken and bonds formed. Remember that ΔH is positive for breaking bonds and negative for forming bonds.

$$\Delta H_{rxn} = \underbrace{\Sigma(\Delta H\text{'s bonds broken})}_{\text{Positive}} + \underbrace{\Sigma(\Delta H\text{'s bonds formed})}_{\text{Negative}}$$

As we can see from the above equation:

- A reaction is *exothermic* when weak bonds break and strong bonds form.
- A reaction is *endothermic* when strong bonds break and weak bonds form.

Scientists sometimes say that "energy is stored in chemical bonds or in a chemical compound," which may make it sound as if breaking the bonds in the compound releases energy. For example, in biology we often hear that energy is stored in glucose or in ATP. However, *breaking a chemical bond always requires energy.* When we say that energy is stored in a compound, or that a compound is energy rich, we mean that the compound can undergo a reaction in which weak bonds break and strong bonds form, thereby releasing energy in the overall process. *Bond formation always releases energy.*

Bond Energies and ΔH_{rxn}

10.8 Cc Conceptual Connection

The reaction between hydrogen and oxygen to form water is highly exothermic. Which statement is true of the energies of the bonds that break and form during the reaction?

(a) The energy required to break the required bonds is greater than the energy released when the new bonds form.

(b) The energy required to break the required bonds is less than the energy released when the new bonds form.

(c) The energy required to break the required bonds is about the same as the energy released when the new bonds form.

EXAMPLE 10.10

Calculating ΔH_{rxn} from Bond Energies

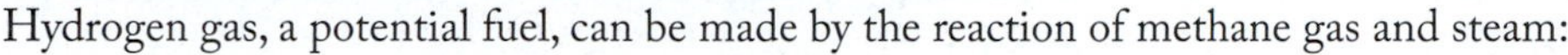

Hydrogen gas, a potential fuel, can be made by the reaction of methane gas and steam:

$$CH_4(g) + 2\,H_2O(g) \longrightarrow 4\,H_2(g) + CO_2(g)$$

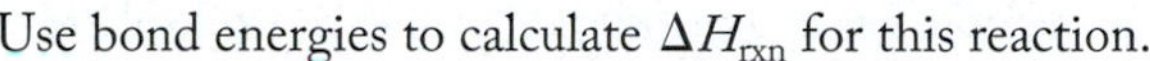

Use bond energies to calculate ΔH_{rxn} for this reaction.

SOLUTION

Begin by rewriting the reaction using the Lewis structures of the molecules involved.	H_4C (H—C—H with H above and below C) + 2 H—Ö—H ⟶ 4 H—H + Ö=C=Ö
Determine which bonds are broken in the reaction and sum the bond energies of these. (Find bond energies in Table 10.3.)	H_4C (H—C—H with H above and below C) + 2 H—Ö—H $\sum(\Delta H\text{'s bonds broken})$ $= 4(C—H) + 4(O—H) = 4(414\text{ kJ}) + 4(464\text{ kJ}) = 3512\text{ kJ}$
Determine which bonds are formed in the reaction and sum the negatives of their bond energies.	4 H—H + Ö=C=Ö $\sum(\Delta H\text{'s bonds formed})$ $= -4(H—H) - 2(C=O) = -4(436\text{ kJ}) - 2(799\text{ kJ}) = -3342\text{ kJ}$
Find ΔH_{rxn} by summing the results of the previous two steps.	$\Delta H_{rxn} = \Sigma(\Delta H\text{'s bonds broken}) + \Sigma(\Delta H\text{'s bonds formed})$ $= 3512 - 3342 = 1.70 \times 10^2\text{ kJ}$

FOR PRACTICE 10.10

Another potential future fuel is methanol (CH_3OH). Write a balanced equation for the combustion of gaseous methanol and use bond energies to calculate the enthalpy of combustion of methanol in kJ/mol.

FOR MORE PRACTICE 10.10

Use bond energies to calculate ΔH_{rxn} for this reaction: $N_2(g) + 3\,H_2(g) \longrightarrow 2\,NH_3(g)$.

10.10 Determining Enthalpies of Reaction from Standard Enthalpies of Formation

We have examined how to measure ΔH for a chemical reaction directly through calorimetry and how to infer ΔH for a reaction from related reactions through Hess's law. We have also seen how to estimate ΔH for a reaction from bond energies. We now turn to a fourth and more convenient way to determine ΔH for a large number of chemical reactions: from tabulated *standard enthalpies of formation*.

Standard States and Standard Enthalpy Changes

Recall that ΔH is the *change* in enthalpy for a chemical reaction—the difference in enthalpy between the products and the reactants. Because we are interested in changes in enthalpy (and not in absolute values of enthalpy itself), we are free to define the *zero* of enthalpy as conveniently as possible. Returning to our mountain-climbing analogy from Section 10.3, a change in altitude (like a change in enthalpy) is an absolute quantity. Altitude itself (like enthalpy), however, is a relative quantity, defined relative to some standard (such as sea level in the case of altitude). We define a similar, albeit slightly more complex, standard for enthalpy. This standard has three parts: the **standard state**, the **standard enthalpy change** (ΔH°), and the **standard enthalpy of formation** (ΔH_f°).

The standard state was changed in 1997 to a pressure of 1 bar, which is very close to 1 atm (1 atm = 1.013 bar). Both standards are now in common use.

1. **Standard State**
   - *For a Gas:* The standard state for a gas is the pure gas at a pressure of exactly 1 atmosphere.
   - *For a Liquid or Solid:* The standard state for a liquid or solid is the pure substance in its most stable form at a pressure of 1 atm and at the temperature of interest (often taken to be 25 °C).
   - *For a Substance in Solution:* The standard state for a substance in solution is a concentration of exactly 1 M.
2. **Standard Enthalpy Change (ΔH°)**
   - The change in enthalpy for a process when all reactants and products are in their standard states. The degree sign indicates standard states.
3. **Standard Enthalpy of Formation (ΔH_f°)**
   - *For a Pure Compound:* The change in enthalpy when 1 mole of the compound forms from its constituent elements in their standard states.
   - *For a Pure Element in Its Standard State:* $\Delta H_f^\circ = 0$.

The standard enthalpy of formation is also called the **standard heat of formation**.

Assigning the value of zero to the standard enthalpy of formation for an element in its standard state is the equivalent of assigning an altitude of zero to sea level. Once we assume sea level is zero, we can measure all subsequent changes in altitude relative to sea level. Similarly, we can measure all changes in enthalpy relative to those of pure elements in their standard states. For example, consider the standard enthalpy of formation of methane gas at 25 °C:

The carbon in this equation must be graphite (the most stable form of carbon at 1 atm and 25 °C).

$$C(s, \text{graphite}) + 2\,H_2(g) \longrightarrow CH_4(g) \qquad \Delta H_f^\circ = -74.6\ \text{kJ/mol}$$

For methane, as with most compounds, ΔH_f° is negative. Continuing our analogy, if we think of pure elements in their standard states as being at *sea level*, then most compounds lie *below sea level*. We always write the chemical equation for the enthalpy of formation of a compound to form 1 mole of the compound, so ΔH_f° has the units of kJ/mol. Table 10.4 lists ΔH_f° values for some selected compounds. A more complete list can be found in Appendix IVB.

EXAMPLE 10.11

Standard Enthalpies of Formation

Write equations for the formation of **(a)** $MgCO_3(s)$ and **(b)** $C_6H_{12}O_6(s)$ from their respective elements in their standard states. Include the value of ΔH_f° for each equation.

—*Continued on the next page*

Continued from the previous page—

SOLUTION

(a) $MgCO_3(s)$

Write the equation with the constituent elements in $MgCO_3$ in their standard states as the reactants and 1 mol of $MgCO_3$ as the product.

$$Mg(s) + C(s, \text{graphite}) + O_2(g) \longrightarrow MgCO_3(s)$$

Balance the equation and look up ΔH_f° in Appendix IVB. (Use fractional coefficients so that the product of the reaction is 1 mol of $MgCO_3$.)

$$Mg(s) + C(s, \text{graphite}) + \frac{3}{2}O_2(g) \longrightarrow MgCO_3(s)$$

$$\Delta H_f^\circ = -1095.8 \text{ kJ/mol}$$

(b) $C_6H_{12}O_6(s)$

Write the equation with the constituent elements in $C_6H_{12}O_6$ in their standard states as the reactants and 1 mol of $C_6H_{12}O_6$ as the product.

$$C(s, \text{graphite}) + H_2(g) + O_2(g) \longrightarrow C_6H_{12}O_6(s)$$

Balance the equation and look up ΔH_f° in Appendix IVB.

$$6\,C(s, \text{graphite}) + 6\,H_2(g) + 3\,O_2(g) \longrightarrow C_6H_{12}O_6(s)$$

$$\Delta H_f^\circ = -1273.3 \text{ kJ/mol}$$

FOR PRACTICE 10.11

Write equations for the formation of **(a)** $NaCl(s)$ and **(b)** $Pb(NO_3)_2(s)$ from their respective elements in their standard states. Include the value of ΔH_f° for each equation.

TABLE 10.4 Standard Enthalpies (or Heats) of Formation, ΔH_f°, at 298 K

Formula	ΔH_f° (kJ/mol)	Formula	ΔH_f° (kJ/mol)	Formula	ΔH_f° (kJ/mol)
Bromine		$C_3H_8O(l$, isopropanol)	−318.1	***Oxygen***	
$Br(g)$	111.9	$C_6H_6(l)$	49.1	$O_2(g)$	0
$Br_2(l)$	0	$C_6H_{12}O_6(s$, glucose)	−1273.3	$O_3(g)$	142.7
$HBr(g)$	−36.3	$C_{12}H_{22}O_{11}(s$, sucrose)	−2226.1	$H_2O(g)$	−241.8
Calcium		***Chlorine***		$H_2O(l)$	−285.8
$Ca(s)$	0	$Cl(g)$	121.3	***Silver***	
$CaO(s)$	−634.9	$Cl_2(g)$	0	$Ag(s)$	0
$CaCO_3(s)$	−1207.6	$HCl(g)$	−92.3	$AgCl(s)$	−127.0
Carbon		***Fluorine***		***Sodium***	
$C(s$, graphite)	0	$F(g)$	79.38	$Na(s)$	0
$C(s$, diamond)	1.88	$F_2(g)$	0	$Na(g)$	107.5
$CO(g)$	−110.5	$HF(g)$	−273.3	$NaCl(s)$	−411.2
$CO_2(g)$	−393.5	***Hydrogen***		$Na_2CO_3(s)$	−1130.7
$CH_4(g)$	−74.6	$H(g)$	218.0	$NaHCO_3(s)$	−950.8
$CH_3OH(l)$	−238.6	$H_2(g)$	0	***Sulfur***	
$C_2H_2(g)$	227.4	***Nitrogen***		$S_8(s$, rhombic)	0
$C_2H_4(g)$	52.4	$N_2(g)$	0	$S_8(s$, monoclinic)	0.3
$C_2H_6(g)$	−84.68	$NH_3(g)$	−45.9	$SO_2(g)$	−296.8
$C_2H_5OH(l)$	−277.6	$NH_4NO_3(s)$	−365.6	$SO_3(g)$	−395.7
$C_3H_8(g)$	−103.85	$NO(g)$	91.3	$H_2SO_4(l)$	−814.0
$C_3H_6O(l$, acetone)	−248.4	$N_2O(g)$	81.6		

Calculating the Standard Enthalpy Change for a Reaction

We have just seen that the standard enthalpy of formation corresponds to the *formation* of a compound from its constituent elements in their standard states:

$$\text{elements} \longrightarrow \text{compound} \qquad \Delta H_f^\circ$$

Therefore, the *negative* of the standard enthalpy of formation corresponds to the *decomposition* of a compound into its constituent elements in their standard states:

$$\text{compound} \longrightarrow \text{elements} \qquad -\Delta H_f^\circ$$

We can use these two concepts—the decomposition of a compound into its elements and the formation of a compound from its elements—to calculate the enthalpy change of any reaction by mentally taking the reactants through two steps. In the first step we *decompose the reactants* into their constituent elements in their standard states; in the second step we *form the products* from the constituent elements in their standard states.

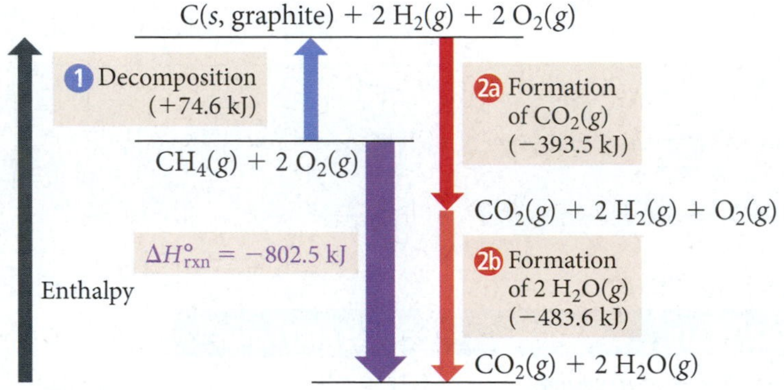

▲ **FIGURE 10.11 Calculating the Enthalpy Change for the Combustion of Methane**

$$\begin{array}{lll} \text{reactants} \longrightarrow \text{elements} & \Delta H_1 = -\sum \Delta H_f^\circ(\text{reactants}) \\ \text{elements} \longrightarrow \text{products} & \Delta H_2 = +\sum \Delta H_f^\circ(\text{products}) \\ \hline \text{reactants} \longrightarrow \text{products} & \Delta H_{rxn}^\circ = \Delta H_1 + \Delta H_2 \end{array}$$

In these equations, Σ means "the sum of" so that ΔH_1 is the sum of the negatives of the heats of formation of the reactants and ΔH_2 is the sum of the heats of formation of the products.

We can demonstrate this procedure by calculating the standard enthalpy change (ΔH_{rxn}°) for the combustion of methane.

$$CH_4(g) + 2\,O_2(g) \longrightarrow CO_2(g) + 2\,H_2O(g) \qquad \Delta H_{rxn}^\circ = ?$$

The enthalpy changes associated with the decomposition of the reactants and the formation of the products are shown in Figure 10.11 ◄. The first step (1) is the decomposition of 1 mol of methane into its constituent elements in their standard states. We can obtain the change in enthalpy for this step by reversing the enthalpy of formation equation for methane and changing the sign of ΔH_f°.

$$(1)\ CH_4(g) \longrightarrow C(s, \text{graphite}) + 2\,H_2(g) \qquad -\Delta H_f^\circ = +74.6\ \text{kJ/mol}$$

The second step, the formation of the products from their constituent elements, has two parts: (2a) the formation of 1 mol CO_2 and (2b) the formation of 2 mol H_2O. Because part (2b) forms 2 mol H_2O, we multiply the ΔH_f° for that step by 2.

$$(2a)\ C(s, \text{graphite}) + O_2(g) \longrightarrow CO_2(g) \qquad \Delta H_f^\circ = -393.5\ \text{kJ/mol}$$

$$(2b)\ 2 \times [H_2(g) + \tfrac{1}{2}O_2(g) \longrightarrow H_2O(g)] \qquad 2 \times \Delta H_f^\circ = 2 \times (-241.8\ \text{kJ/mol})$$

As we know from Hess's law, the enthalpy of reaction for the overall reaction is the sum of the enthalpies of reaction of the individual steps.

$$\begin{array}{lr} (1)\ CH_4(g) \longrightarrow \cancel{C(s, \text{graphite})} + \cancel{2\,H_2(g)} & -\Delta H_f^\circ = +74.6\ \text{kJ/mol} \\ (2a)\ \cancel{C(s, \text{graphite})} + O_2(g) \longrightarrow CO_2(g) & \Delta H_f^\circ = -393.5\ \text{kJ/mol} \\ (2b)\ \cancel{2\,H_2(g)} + O_2(g) \longrightarrow 2\,H_2O(g) & 2 \times \Delta H_f^\circ = -483.6\ \text{kJ/mol} \\ \hline CH_4(g) + 2\,O_2(g) \longrightarrow CO_2(g) + 2\,H_2O(g) & \Delta H_{rxn}^\circ = -802.5\ \text{kJ/mol} \end{array}$$

We can streamline and generalize this process as follows:

> **To calculate ΔH_{rxn}°, subtract the enthalpies of formation of the reactants multiplied by their stoichiometric coefficients from the enthalpies of formation of the products multiplied by their stoichiometric coefficients.**

In the form of an equation,

$$\Delta H^\circ_{rxn} = \sum n_p \Delta H^\circ_f \text{ (products)} - \sum n_r \Delta H^\circ_f \text{ (reactants)} \quad [10.14]$$

In this equation, n_p represents the stoichiometric coefficients of the products, n_r represents the stoichiometric coefficients of the reactants, and ΔH°_f represents the standard enthalpies of formation. Keep in mind when using this equation that elements in their standard states have $\Delta H^\circ_f = 0$. Examples 10.12 and 10.13 demonstrate this process.

EXAMPLE 10.12

ΔH°_{rxn} and Standard Enthalpies of Formation

Use the standard enthalpies of formation to determine ΔH°_{rxn} for the reaction.

$$4\,NH_3(g) + 5\,O_2(g) \longrightarrow 4\,NO(g) + 6\,H_2O(g)$$

SORT You are given the balanced equation and asked to find the enthalpy of reaction.

GIVEN: $4\,NH_3(g) + 5\,O_2(g) \longrightarrow 4\,NO(g) + 6\,H_2O(g)$

FIND: ΔH°_{rxn}

STRATEGIZE To calculate ΔH°_{rxn} from standard enthalpies of formation, subtract the heats of formation of the reactants multiplied by their stoichiometric coefficients from the heats of formation of the products multiplied by their stoichiometric coefficients.

CONCEPTUAL PLAN

$$\Delta H^\circ_{rxn} = \Sigma\, n_p\, \Delta H^\circ_f \text{ (products)} - \Sigma\, n_r \Delta H^\circ_f \text{ (reactants)}$$

SOLVE Begin by looking up (in Appendix IVB) the standard enthalpy of formation for each reactant and product. Remember that the standard enthalpy of formation of pure elements in their standard state is zero. Calculate ΔH°_{rxn} by substituting into the equation.

SOLUTION

Reactant or product	ΔH°_f (kJ/mol, from Appendix IVB)
$NH_3(g)$	−45.9
$O_2(g)$	0.0
$NO(g)$	+91.3
$H_2O(g)$	−241.8

$$\begin{aligned}\Delta H^\circ_{rxn} &= \Sigma\, n_p \Delta H^\circ_f \text{ (products)} - \Sigma\, n_r \Delta H^\circ_f \text{ (reactants)}\\ &= [4(\Delta H^\circ_{f,\,NO(g)}) + 6(\Delta H^\circ_{f,\,H_2O(g)})] - [4(\Delta H^\circ_{f,\,NH_3(g)}) + 5(\Delta H^\circ_{f,\,O_2(g)})]\\ &= [4(+91.3\text{ kJ}) + 6(-241.8\text{ kJ})] - [4(-45.9\text{ kJ}) + 5(0.0\text{ kJ})]\\ &= -1085.6\text{ kJ} - (-183.6\text{ kJ})\\ &= -902.0\text{ kJ}\end{aligned}$$

CHECK The units of the answer (kJ) are correct. The answer is negative, which means that the reaction is exothermic.

FOR PRACTICE 10.12

The thermite reaction, in which powdered aluminum reacts with iron(III) oxide, is highly exothermic.

$$2\,Al(s) + Fe_2O_3(s) \longrightarrow Al_2O_3(s) + 2\,Fe(s)$$

Use standard enthalpies of formation to find ΔH°_{rxn} for the thermite reaction.

▶The reaction of powdered aluminum with iron(III) oxide, known as the thermite reaction, releases a large amount of heat.

EXAMPLE 10.13

ΔH°_{rxn} and Standard Enthalpies of Formation

A city of 100,000 people uses approximately 1.0×10^{11} kJ of energy per day. Suppose all of the city's energy comes from the combustion of liquid octane (C_8H_{18}) to form gaseous water and gaseous carbon dioxide. Use standard enthalpies of formation to calculate ΔH°_{rxn} for the combustion of octane and determine how many kilograms of octane are necessary to provide this amount of energy.

SORT You are given the amount of energy used and asked to find the mass of octane required to produce the energy.

GIVEN: 1.0×10^{11} kJ

FIND: kg C_8H_{18}

STRATEGIZE The conceptual plan has three parts. In the first part, write a balanced equation for the combustion of octane.

CONCEPTUAL PLAN

1. Write balanced equation.

In the second part, calculate ΔH°_{rxn} from the ΔH°_f's of the reactants and products.

2.

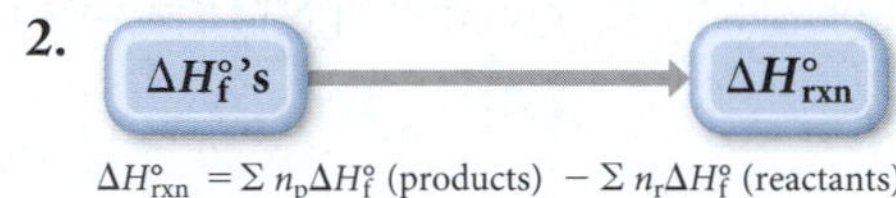

$\Delta H^\circ_{rxn} = \Sigma n_p \Delta H^\circ_f \text{ (products)} - \Sigma n_r \Delta H^\circ_f \text{ (reactants)}$

In the third part, convert from kilojoules of energy to moles of octane using the conversion factor found in Step 2, and then convert from moles of octane to mass of octane using the molar mass.

3.

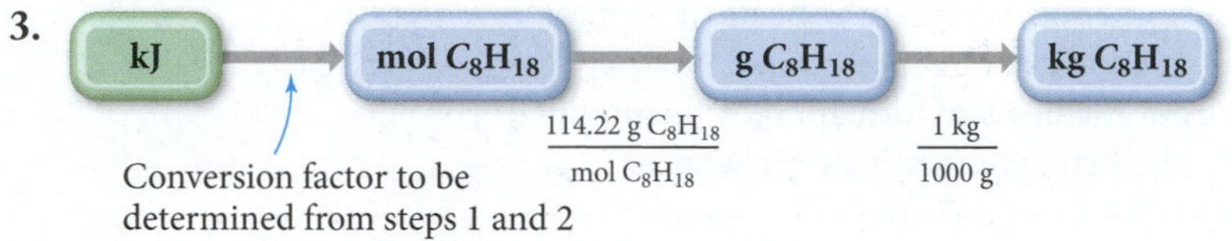

RELATIONSHIPS USED

molar mass C_8H_{18} = 114.22 g/mol

1 kg = 1000 g

SOLVE Begin by writing the balanced equation for the combustion of octane. For convenience, do not clear the $\frac{25}{2}$ fraction in order to keep the coefficient on octane as 1.

SOLUTION STEP 1

$$C_8H_{18}(l) + \frac{25}{2} O_2(g) \rightarrow 8\,CO_2(g) + 9\,H_2O(g)$$

Look up (in Appendix IVB) the standard enthalpy of formation for each reactant and product and calculate ΔH°_{rxn}.

SOLUTION STEP 2

Reactant or product	ΔH°_f (kJ/mol, from Appendix IVB)
$C_8H_{18}(l)$	−250.1
$O_2(g)$	0.0
$CO_2(g)$	−393.5
$H_2O(g)$	−241.8

$$\begin{aligned}\Delta H^\circ_{rxn} &= \Sigma n_p \Delta H^\circ_f \text{ (products)} - \Sigma n_r \Delta H^\circ_f \text{ (reactants)}\\ &= [8(\Delta H^\circ_{f,\,CO_2(g)}) + 9(\Delta H^\circ_{f,\,H_2O(g)})] - [1(\Delta H^\circ_{f,\,C_8H_{18}(l)}) + +\tfrac{25}{2}(\Delta H^\circ_{f,\,O_2(g)})]\\ &= [8(-393.5\text{ kJ}) + 9(-241.8\text{ kJ})] - [1(-250.1\text{ kJ}) + \tfrac{25}{2}(0.0\text{ kJ})]\\ &= -5324.2\text{ kJ} - (-250.1\text{ kJ})\\ &= -5074.1\text{ kJ}\end{aligned}$$

—Continued on the next page

Continued from the previous page—

From Steps 1 and 2 build a conversion factor between mol C_8H_{18} and kJ.	**SOLUTION STEP 3** $1 \text{ mol } C_8H_{18} : -5074.1 \text{ kJ}$
Follow Step 3 of the conceptual plan. Begin with -1.0×10^{11} kJ (because the city uses this much energy, the reaction must emit this amount, and therefore the sign is negative) and follow the steps to determine kg octane.	$-1.0 \times 10^{11} \text{ kJ} \times \frac{1 \text{ mol } C_8H_{18}}{-5074.1 \text{ kJ}} \times \frac{114.22 \text{ g } C_8H_{18}}{1 \text{ mol } C_8H_{18}} \times \frac{1 \text{ kg}}{1000 \text{ g}} = 2.3 \times 10^6 \text{ kg } C_8H_{18}$

CHECK The units of the answer (kg C_8H_{18}) are correct. The answer is positive, as it should be for mass. The magnitude is fairly large, as you would expect because this amount of octane is supposed to provide the energy for an entire city.

FOR PRACTICE 10.13
Chemical hand warmers popular with skiers and snowboarders produce heat when they are removed from their airtight plastic wrappers. They utilize the oxidation of iron to form iron(III) oxide according to the reaction: $4\,Fe(s) + 3\,O_2(g) \longrightarrow 2\,Fe_2O_3(s)$. Calculate ΔH°_{rxn} for this reaction and calculate how much heat is produced from a hand warmer containing 15.0 g of iron powder.

10.11 Lattice Energies for Ionic Compounds

As we discussed in Section 5.5, the formation of an ionic compound from its constituent elements is usually quite exothermic because of lattice energy—the energy associated with the formation of a crystalline lattice of alternating cations and anions from the gaseous ions (see Figure 5.5). In this section of the chapter we examine how to calculate lattice energy for an ionic compound and two trends in lattice energy.

Calculating Lattice Energy: The Born–Haber Cycle

The Born–Haber cycle is a hypothetical series of steps that represents the formation of an ionic compound from its constituent elements. The steps are chosen so that the change in enthalpy of each of the steps is known except for that of the last step, which is the lattice energy. The change in enthalpy for the overall process is also known. Using Hess's law (see Section 10.8), we can determine the enthalpy change for the unknown last step, the lattice energy.

Consider the formation of NaCl from its constituent elements in their standard states. The enthalpy change for the overall reaction is the standard enthalpy of formation of NaCl(*s*).

$$Na(s) + \tfrac{1}{2}Cl_2(g) \longrightarrow NaCl(s) \qquad \Delta H^\circ_f = -411 \text{ kJ/mol}$$

Now consider the following set of steps—the Born–Haber cycle—from which NaCl(*s*) can also be made from Na(*s*) and $Cl_2(g)$:

- Step 1 is the formation of gaseous sodium from solid sodium.

$$Na(s) \longrightarrow Na(g) \qquad \Delta H^\circ_{step\ 1} \text{ (sublimation energy of Na)} = +108 \text{ kJ}$$

The sublimation energy is the energy required to convert one mole of substance from a solid to gas.

- Step 2 is the formation of a chlorine atom from a chlorine molecule.

$$\tfrac{1}{2}Cl_2(g) \longrightarrow Cl(g) \qquad \Delta H^\circ_{step\ 2} \text{ (bond energy of } Cl_2 \times \tfrac{1}{2}) = +122 \text{ kJ}$$

- Step 3 is the ionization of gaseous sodium. The enthalpy change for this step is the ionization energy of sodium.

$$Na(g) \longrightarrow Na^+(g) + e^- \qquad \Delta H^\circ_{step\ 3} \text{ (ionization energy of Na)} = +496 \text{ kJ}$$

- Step 4 is the addition of an electron to gaseous chlorine. The enthalpy change for this step is the electron affinity of chlorine.

$$Cl(g) + e^- \longrightarrow Cl^-(g) \qquad \Delta H^\circ_{step\ 4}\ (\text{electron affinity of Cl}) = -349\ \text{kJ}$$

- Step 5 is the formation of the crystalline solid from the gaseous ions. The enthalpy change for this step is the lattice energy, the unknown quantity.

$$Na^+(g) + Cl^-(g) \longrightarrow NaCl(s) \qquad \Delta H^\circ_{step\ 5} = \Delta H^\circ_{lattice} = ?$$

Figure 10.12 ▼ illustrates the entire Born–Haber cycle for NaCl.

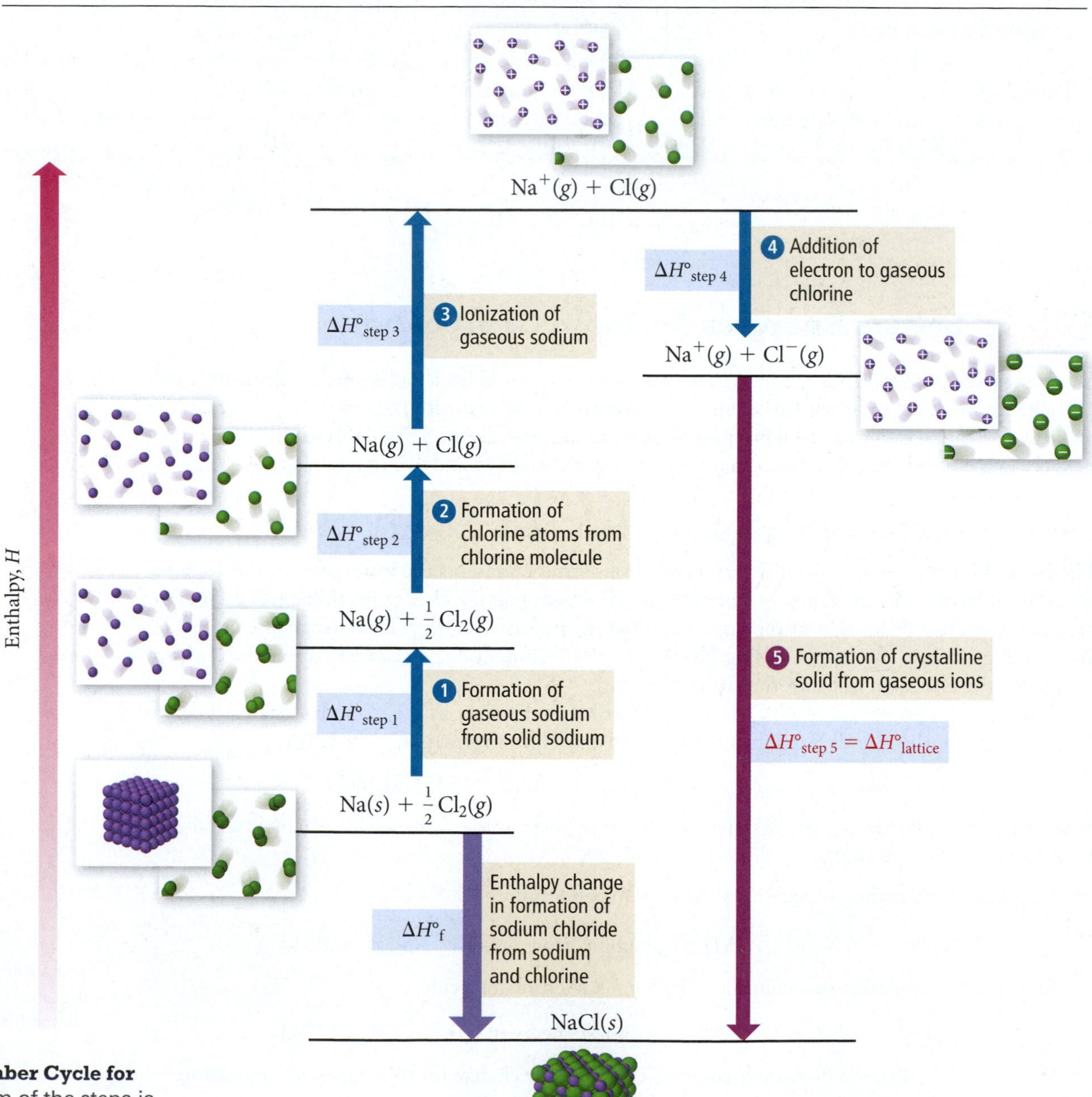

▶ **FIGURE 10.12 Born–Haber Cycle for Sodium Chloride** The sum of the steps is the formation of NaCl from elemental Na and Cl. The enthalpy change of the last step (5) is the lattice energy.

Because the overall reaction obtained by summing the steps in the Born–Haber cycle is equivalent to the formation of NaCl from its constituent elements, we use Hess's law to set the overall enthalpy of formation for NaCl(s) equal to the sum of the steps in the Born–Haber cycle:

Lattice energy ↓

$$\Delta H^\circ_f = \Delta H^\circ_{\text{step 1}} + \Delta H^\circ_{\text{step 2}} + \Delta H^\circ_{\text{step 3}} + \Delta H^\circ_{\text{step 4}} + \Delta H^\circ_{\text{step 5}}$$

We then solve this equation for $\Delta H^\circ_{\text{step 5}}$, which is $\Delta H^\circ_{\text{lattice}}$, and substitute the appropriate values to calculate the lattice energy.

$$\begin{aligned}\Delta H^\circ_{\text{lattice}} = \Delta H^\circ_{\text{step 5}} &= \Delta H^\circ_f - (\Delta H^\circ_{\text{step1}} + \Delta H^\circ_{\text{step 2}} + \Delta H^\circ_{\text{step 3}} + \Delta H^\circ_{\text{step 4}})\\ &= -411\text{ kJ} - (+108\text{ kJ} + 122\text{ kJ} + 496\text{ kJ} - 349\text{ kJ})\\ &= -788\text{ kJ}\end{aligned}$$

The value of the lattice energy is a large negative number. The formation of the crystalline NaCl lattice from sodium cations and chloride anions is highly exothermic and more than compensates for the endothermicity of the electron transfer process. In other words, the formation of ionic compounds is not exothermic because sodium "wants" to lose electrons and chlorine "wants" to gain them; rather, it is exothermic because of the large amount of heat released when sodium and chlorine ions coalesce to form a crystalline lattice.

Trends in Lattice Energies: Ion Size

Consider the lattice energies of the following alkali metal chlorides:

Metal Chloride	Lattice Energy (kJ/mol)
LiCl	–834
NaCl	–788
KCl	–701
CsCl	–657

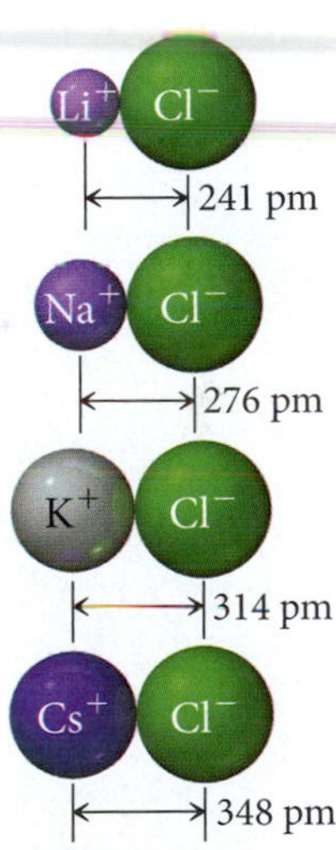

▲ Bond lengths of the group 1A metal chlorides.

Why does the magnitude of the lattice energy decrease as we move down the column? We know from the periodic trends discussed in Chapter 4 that ionic radius increases as we move down a column in the periodic table (see Section 4.7). We also know, from the discussion of Coulomb's law in Section 4.3, that the potential energy of oppositely charged ions becomes less negative (or more positive) as the distance between the ions increases. As the size of the alkali metal ions increases down the column, so does the distance between the metal cations and the chloride anions. The magnitude of the lattice energy of the chlorides decreases accordingly, making the formation of the chlorides less exothermic. In other words, *as the ionic radii increase as we move down the column, the ions cannot get as close to each other and therefore do not release as much energy when the lattice forms.*

Trends in Lattice Energies: Ion Charge

Consider the lattice energies of the following two compounds:

Compound	Lattice Energy (kJ/mol)
NaF	–910
CaO	–3414
Na^+ F^- 231 pm	Ca^{2+} O^{2-} 239 pm

Why is the magnitude of the lattice energy of CaO so much greater than the lattice energy of NaF? Na^+ has a radius of 95 pm and F^- has a radius of 136 pm, resulting in a distance between ions of 231 pm. Ca^{2+} has a radius of 99 pm and O^{2-} has a radius of 140 pm, resulting in a distance between ions of 239 pm. Even though the separation between the calcium and oxygen is slightly greater (which would tend to lower the lattice energy), the lattice energy for CaO is almost four times *greater*. The explanation lies in the charges of the ions. Recall from Coulomb's law that the magnitude of the potential energy of two interacting charges depends not only on the distance between the charges, but also on the product of the charges:

$$E = \frac{1}{4\pi\varepsilon_0}\frac{q_1 q_2}{r}$$

For NaF, E is proportional to (1+)(1−)=1−, while for CaO, E is proportional to (2+)(2−)=4−, so the relative stabilization for CaO relative to NaF is roughly four times greater, as observed in the lattice energy.

Summarizing Trends in Lattice Energies:

- *Lattice energies become less exothermic (less negative) with increasing ionic radius.*
- *Lattice energies become more exothermic (more negative) with increasing magnitude of ionic charge.*

EXAMPLE 10.14 Predicting Relative Lattice Energies

Arrange these ionic compounds in order of increasing *magnitude* of lattice energy: CaO, KBr, KCl, SrO.

SOLUTION

KBr and KCl should have lattice energies of smaller magnitude than CaO and SrO because of their lower ionic charges (1+, 1− compared to 2+, 2−). When you compare KBr and KCl, you expect KBr to have a lattice energy of lower magnitude due to the larger ionic radius of the bromide ion relative to the chloride ion. Between CaO and SrO, you expect SrO to have a lattice energy of lower magnitude due to the larger ionic radius of the strontium ion relative to the calcium ion.

Order of increasing *magnitude* of lattice energy:

KBr < KCl < SrO < CaO

Actual lattice energy values:

Compound	Lattice Energy (kJ/mol)
KBr	−671
KCl	−701
SrO	−3217
CaO	−3414

FOR PRACTICE 10.14

Arrange these ionic compounds in order of increasing magnitude of lattice energy: LiBr, KI, and CaO.

FOR MORE PRACTICE 10.14

Which compound has a higher magnitude lattice energy: NaCl or $MgCl_2$?

SELF-ASSESSMENT Quiz

1. A chemical system produces 155 kJ of heat and does 22 kJ of work. What is ΔE for the *surroundings*?
   a) 177 kJ **b)** −177 kJ
   c) 133 kJ **d)** −133 kJ

2. Which sample is most likely to undergo the smallest change in temperature upon the absorption of 100 kJ of heat?
   a) 15 g water **b)** 15 g lead
   c) 50 g water **d)** 50 g lead

3. How much heat must a 15.0 g sample of water absorb to raise its temperature from 25.0 °C to 55.0 °C? (For water, $C_s = 4.18$ J/g·°C.)
   a) 1.57 kJ **b)** 1.88 kJ
   c) 3.45 kJ **d)** 107 J

4. A 12.5 g sample of granite initially at 82.0 °C is immersed into 25.0 g of water that is initially at 22.0 °C. What is the final temperature of both substances when they reach thermal equilibrium? (For water, $C_s = 4.18$ J/g · °C and for granite, $C_s = 0.790$ J/g·°C.)
   a) 52.0 °C **b)** 1.55×10^3 °C
   c) 15.7 °C **d)** 27.2 °C

5. A cylinder with a moving piston expands from an initial volume of 0.250 L against an external pressure of 2.00 atm. The expansion does 288 J of work on the surroundings. What is the final volume of the cylinder?
   a) 1.42 L **b)** 1.17 L
   c) 144 L **d)** 1.67 L

6. When a 3.80 g sample of liquid octane (C_8H_{18}) burns in a bomb calorimeter, the temperature of the calorimeter rises by 27.3 °C. The heat capacity of the calorimeter, measured in a separate experiment, is 6.18 kJ/°C. What is ΔE for the combustion for octane?
   a) -5.07×10^3 kJ/mol
   b) 5.07×10^3 kJ/mol
   c) -44.4×10^3 kJ/mol
   d) -16.7×10^3 kJ/mol

7. Hydrogen gas reacts with oxygen to form water.
   $$2\ H_2(g) + O_2(g) \longrightarrow 2\ H_2O(g) \qquad \Delta H = -483.5\ \text{kJ}$$
   Determine the minimum mass of hydrogen gas required to produce 226 kJ of heat.
   a) 8.63 g **b)** 1.88 g
   c) 0.942 g **d)** 0.935 g

8. Manganese reacts with hydrochloric acid to produce manganese(II) chloride and hydrogen gas.
   $$Mn(s) + 2\ HCl(aq) \longrightarrow MnCl_2(aq) + H_2(g)$$
   When 0.625 g Mn is combined with enough hydrochloric acid to make 100.0 mL of solution in a coffee-cup calorimeter, all of the Mn reacts, raising the temperature of the solution from 23.5 °C to 28.8 °C. Find ΔH_{rxn} for the reaction as written. (Assume that the specific heat capacity of the solution is 4.18 J/g °C and that the density of the solution is 1.00 g/mL.)
   a) −195 kJ **b)** −3.54 kJ
   c) −1.22 kJ **d)** −2.21 kJ

9. Consider the generic reactions:
   $$A \longrightarrow 2\ B \quad \Delta H_1$$
   $$A \longrightarrow 3\ C \quad \Delta H_2$$
   What is ΔH for the related generic reaction $2\ B \longrightarrow 3\ C$?
   a) $\Delta H_1 + \Delta H_2$ **b)** $\Delta H_1 - \Delta H_2$
   c) $\Delta H_2 - \Delta H_1$ **d)** $2 \times (\Delta H_1 + \Delta H_2)$

10. Use standard enthalpies of formation to determine ΔH°_{rxn} for the reaction.
    $$Fe_2O_3(s) + 3\ CO(g) \longrightarrow 2\ Fe(s) + 3\ CO_2(g)$$
    a) −541.2 kJ **b)** −2336 kJ
    c) 541.2 kJ **d)** −24.8 kJ

11. Two substances, A and B, which are of equal mass but at different temperatures, come into thermal contact. The specific heat capacity of substance A is twice the specific heat capacity of substance B. Which statement is true of the temperature of the two substances when they reach thermal equilibrium? (Assume no other heat loss other than the thermal transfer between the substances.)
    a) The final temperature of both substances is closer to the initial temperature of substance A than to the initial temperature of substance B.
    b) The final temperature of both substances is closer to the initial temperature of substance B than to the initial temperature of substance A.
    c) The final temperature of both substances is exactly midway between the initial temperatures of substance A and substance B.
    d) The final temperature of substance B is greater than the final temperature of substance A.

12. The standard enthalpy of formation for glucose ($C_6H_{12}O_6(s)$) is −1273.3 kJ/mol. What is the correct formation equation corresponding to this ΔH°_f?
    a) $6\ C(s, \textit{graphite}) + 6\ H_2O(g) \longrightarrow C_6H_{12}O_6(s, \text{glucose})$
    b) $6\ C(s, \textit{graphite}) + 6\ H_2O(l) \longrightarrow C_6H_{12}O_6(s, \text{glucose})$
    c) $6\ C(s, \textit{graphite}) + 6\ H_2(l) + 3\ O_2(l) \longrightarrow C_6H_{12}O_6(s, \text{glucose})$
    d) $6\ C(s, \textit{graphite}) + 6\ H_2(g) + 3\ O_2(g) \longrightarrow C_6H_{12}O_6(s, \text{glucose})$

—*Continued*

Continued—

13. Natural gas burns in air to form carbon dioxide and water, releasing heat.

$$CH_4(g) + O_2(g) \longrightarrow CO_2(g) + H_2O(g) \qquad \Delta H^\circ_{rxn} = -802.3\ kJ$$

What minimum mass of CH_4 is required to heat 55 g of water by 25 °C? (Assume 100% heating efficiency.)

a) 0.115 g **b)** 2.25×10^3 g
c) 115 g **d)** 8.70 g

14. Which set of compounds is arranged in order of increasing magnitude of lattice energy?

a) CsI < NaCl < MgS **b)** NaCl < CsI < MgS
c) MgS < NaCl < CsI **d)** CsI < MgS < NaCl

15. Use bond energies to determine ΔH_{rxn} for the reaction between ethanol and hydrogen chloride.

$$CH_3CH_2OH(g) + HCl(g) \longrightarrow CH_3CH_2Cl(g) + H_2O(g)$$

a) −1549 kJ **b)** 1549 kJ
c) −12 kJ **d)** 12 kJ

Answers: 1:a; 2:c; 3:b; 4:d; 5:d; 6:a; 7:b; 8:a; 9:c; 10:d; 11:a; 12:d; 13:a; 14:a; 15:c

CHAPTER SUMMARY 10

REVIEW

KEY LEARNING OUTCOMES

CHAPTER OBJECTIVES	ASSESSMENT
Calculating Changes in Internal Energy from Heat and Work (10.3)	• Example 10.1 For Practice 10.1 Exercises 33–36
Finding Heat from Temperature Changes (10.4)	• Example 10.2 For Practice 10.2 For More Practice 10.2 Exercises 39, 40
Determining Quantities in Thermal Energy Transfer (10.4)	• Example 10.3 For Practice 10.3 Exercises 57–62
Calculating Work from Volume Changes (10.4)	• Example 10.4 For Practice 10.4 For More Practice 10.4 Exercises 43, 44
Using Bomb Calorimetry to Calculate ΔE_{rxn} (10.5)	• Example 10.5 For Practice 10.5 For More Practice 10.5 Exercises 65, 66
Distinquishing between Endothermic and Exothermic Processes (10.6)	• Example 10.6 For Practice 10.6 Exercises 49, 50
Determining Heat from ΔH and Stoichiometry (10.6)	• Example 10.7 For Practice 10.7 For More Practice 10.7 Exercises 51–56
Finding ΔH_{rxn} Using Calorimetry (10.7)	• Example 10.8 For Practice 10.8 Exercises 67, 68
Finding ΔH_{rxn} Using Hess's Law (10.8)	• Example 10.9 For Practice 10.9 For More Practice 10.9 Exercises 71–74
Finding ΔH_{rxn} Using Bond Energies (10.9)	• Example 10.10 For Practice 10.10 For More Practice 10.10 Exercises 75–78
Finding ΔH°_{rxn} Using Standard Enthalpies of Formation (10.10)	• Examples 10.11, 10.12, 10.13 For Practice 10.11, 10.12, 10.13 Exercises 79–86
Predicting Relative Lattice Energies (10.11)	• Example 10.14 For Practice 10.14 For More Practice 10.14 Exercises 89-92

KEY TERMS

Section 10.1
thermochemistry (344)

Section 10.2
energy (344)
work (344)
heat (344)
kinetic energy (344)
thermal energy (344)
potential energy (344)
chemical energy (344)
joule (J) (344)
calorie (cal) (344)
Calorie (Cal) (344)
law of conservation of energy (344)
system (345)
surroundings (345)

Section 10.3
thermodynamics (346)
first law of thermodynamics (346)
internal energy (E) (346)
state function (346)

Section 10.4
thermal equilibrium (350)
heat capacity (C) (350)
specific heat capacity (C_s) (350)
molar heat capacity (350)
pressure–volume work (353)
pressure (353)

Section 10.5
calorimetry (356)
bomb calorimeter (356)

Section 10.6
enthalpy (H) (358)
endothermic reaction (358)
exothermic reaction (359)
enthalpy (heat) of reaction (ΔH_{rxn}) (360)

Section 10.7
coffee-cup calorimeter (362)

Section 10.8
Hess's law (364)

Section 10.10
standard state (370)
standard enthalpy change ($\Delta H°$) (370)
standard enthalpy of formation(ΔH_f°) (370)
standard heat of formation (370)

Section 10.11
Born–Haber cycle (375)

KEY CONCEPTS

The Nature of Energy and Thermodynamics (10.1–10.3)

- Energy, which is measured in the SI unit of joules (J), is the capacity to do work.
- Work is the result of a force acting through a distance.
- Many different kinds of energy exist, including kinetic energy, thermal energy, potential energy, and chemical energy, a type of potential energy associated with the relative positions of electrons and nuclei in atoms and molecules.
- The first law of thermodynamics states that energy can be converted from one form to another, but the total amount of energy is always conserved.
- The internal energy (E) of a system is the sum of all of its kinetic and potential energy. Internal energy is a state function, which means that it depends only on the state of the system and not on the pathway by which it got to that state.
- A chemical system exchanges energy with its surroundings through heat (the transfer of thermal energy caused by a temperature difference) or work. The total change in internal energy is the sum of these two quantities.
- The change in internal energy (ΔE) that occurs during a chemical reaction is the sum of the heat (q) exchanged and the work (w) done: $\Delta E = q + w$.

Heat and Work (10.4)

- We quantify heat with the equation $q = m \times C_s \times \Delta T$. In this expression, C_s is the specific heat capacity, the amount of heat required to change the temperature of 1 g of the substance by 1 °C. Compared to most substances, water has a very high heat capacity—it takes a lot of heat to change its temperature.
- The type of work most characteristic of chemical reactions is pressure–volume work, which occurs when a gas expands against an external pressure. Pressure–volume work can be quantified with the equation $w = -P\Delta V$.

Enthalpy (10.6)

- The heat evolved in a chemical reaction occurring at constant pressure is the change in enthalpy (ΔH) for the reaction. Like internal energy, enthalpy is a state function.
- An endothermic reaction has a positive enthalpy of reaction; an exothermic reaction has a negative enthalpy of reaction.
- The enthalpy of reaction can be used to determine stoichiometrically the heat evolved when a specific amount of reactant reacts.

Calorimetry (10.5, 10.7)

- Calorimetry is a method of measuring ΔE or ΔH for a reaction.
- In bomb calorimetry, the reaction is carried out under conditions of *constant volume*, so the energy is released only as heat ($\Delta E = q_v$). The temperature change of the calorimeter can therefore be used to calculate ΔE for the reaction.
- When a reaction takes place at *constant pressure*, energy may be released both as heat and as work.
- In coffee-cup calorimetry, a reaction is carried out under atmospheric pressure in a solution, so $q = \Delta H$. The temperature change of the solution is then used to calculate ΔH for the reaction.

Calculating ΔH_{rxn} (10.8–10.10)

- We can calculate the enthalpy of reaction (ΔH_{rxn}) from known thermochemical data using the following relationships: (a) when a reaction is multiplied by a factor, ΔH_{rxn} is multiplied by the same factor; (b) when a reaction is reversed, ΔH_{rxn} changes sign; and (c) if a chemical reaction can be expressed as a sum of two or more steps, ΔH_{rxn} is the sum of the ΔH's for the individual steps (Hess's law). We can use these relationships together to determine the enthalpy change of an unknown reaction from reactions with known enthalpy changes.
- We can also calculate the enthalpy of reaction (ΔH_{rxn}) from the bond energies of the reactants and products. The ΔH_{rxn} is the sum of the ΔH's for bonds broken (which are positive) plus the sum of the sum of the ΔH's for the bond formed (which are negative).
- A third way to calculate ΔH_{rxn} from known thermochemical data involves using tabulated standard enthalpies of formation for the reactants and products of the reaction. These are usually tabulated for substances in their standard states, and the enthalpy of reaction is called the standard enthalpy of reaction (ΔH°_{rxn}). We can obtain ΔH°_{rxn} by subtracting the sum of the enthalpies of formation of the reactants multiplied by their stoichiometric coefficients from the sum of the enthalpies of formation of the products multiplied by their stoichiometric coefficients.

Lattice Energy (10.11)

- The formation of most ionic compounds is exothermic because of lattice energy, the energy released when metal cations and nonmetal anions coalesce to form the solid. The smaller the radius of the ions and the greater their charge, the more exothermic the lattice energy.

KEY EQUATIONS AND RELATIONSHIPS

Change in Internal Energy (ΔE) of a Chemical System (10.3)

$$\Delta E = E_{\text{products}} - E_{\text{reactants}}$$

Energy Flow between System and Surroundings (10.3)

$$\Delta E_{\text{system}} = -\Delta E_{\text{surroundings}}$$

Relationship between Internal Energy (ΔE), Heat (q), and Work (w) (10.3)

$$\Delta E = q + w$$

Relationship between Heat (q), Temperature (T), and Heat Capacity (C) (10.4)

$$q = C \times \Delta T$$

Relationship between Heat (q), Mass (m), Temperature (T), and Specific Heat Capacity of a Substance (C_s) (10.4)

$$q = m \times C_s \times \Delta T$$

Relationship between Work (w), Force (F), and Distance (D) (10.4)

$$w = F \times D$$

Relationship between Work (w), Pressure (P), and Change in Volume (ΔV) (10.4)

$$w = -P\Delta V$$

Change in Internal Energy (ΔE) of System at Constant Volume (10.5)

$$\Delta E = q_v$$

Heat of a Bomb Calorimeter (q_{cal}) (10.5)

$$q_{\text{cal}} = C_{\text{cal}} \times \Delta T$$

Heat Exchange between a Calorimeter and a Reaction (10.5)

$$q_{\text{cal}} = -q_{\text{rxn}}$$

Relationship between Enthalpy (ΔH), Internal Energy (ΔE), Pressure (P), and Volume (V) (10.6)

$$\Delta H = \Delta E + P\Delta V$$

$$\Delta H = q_p$$

Enthalpy Change of a Reaction (ΔH_{rxn}): Relationship of Bond Energies (10.9)

$$\Delta H_{\text{rxn}} = \Sigma(\Delta H\text{'s bonds broken}) + \Sigma(\Delta H\text{'s bonds formed})$$

Relationship between Enthalpy of a Reaction ($\Delta H^\circ_{\text{rxn}}$) and the Heats of Formation (ΔH°_f) (10.10)

$$\Delta H^\circ_{\text{rxn}} = \sum n_p \Delta H^\circ_f(\text{products}) - \sum n_r \Delta H^\circ_f(\text{reactants})$$

EXERCISES

REVIEW QUESTIONS

1. What is thermochemistry? Why is it important?
2. What is energy? What is work? List some examples of each.
3. What is kinetic energy? What is potential energy? List some examples of each.
4. What is the law of conservation of energy? How does it relate to energy exchanges between a thermodynamic system and its surroundings?
5. A friend claims to have constructed a machine that creates electricity, but requires no energy input. Explain why you should be suspicious of your friend's claim.
6. What is a state function? List some examples of state functions.
7. What is internal energy? Is internal energy a state function?
8. If energy flows out of a chemical system and into the surroundings, what is the sign of ΔE_{system}?
9. If the internal energy of the products of a reaction is higher than the internal energy of the reactants, what is the sign of ΔE for the reaction? In which direction does energy flow?
10. What is heat? Explain the difference between heat and temperature.
11. How is the change in internal energy of a system related to heat and work?
12. Explain how the sum of heat and work can be a state function, even though heat and work are themselves not state functions.
13. What is heat capacity? Explain the difference between heat capacity and specific heat capacity.
14. Explain how the high specific heat capacity of water can affect the weather in coastal regions.
15. If two objects, A and B, of different temperature come into direct contact, what is the relationship between the heat lost by one object and the heat gained by the other? What is the relationship between the temperature changes of the two objects? (Assume that the two objects do not lose any heat to anything else.)
16. What is pressure–volume work? How is it calculated?
17. What is calorimetry? Explain the difference between a coffee-cup calorimeter and a bomb calorimeter. What is each designed to measure?
18. What is the change in enthalpy (ΔH) for a chemical reaction? How is ΔH different from ΔE?
19. Explain the difference between an exothermic and an endothermic reaction. Give the sign of ΔH for each type of reaction.
20. From a molecular viewpoint, where does the energy emitted in an exothermic chemical reaction come from? Why does the reaction mixture undergo an increase in temperature even though energy is emitted?
21. From a molecular viewpoint, where does the energy absorbed in an endothermic chemical reaction go? Why does the reaction mixture undergo a decrease in temperature even though energy is absorbed?

22. Is the change in enthalpy for a reaction an extensive property? Explain the relationship between ΔH for a reaction and the amounts of reactants and products that undergo reaction.

23. Explain how the value of ΔH for a reaction changes upon:
   a. multiplying the reaction by a factor
   b. reversing the reaction

   Why do these relationships hold?

24. What is Hess's law? Why is it useful?

25. What is a standard state? What is the standard enthalpy change for a reaction?

26. How can bond energies be used to estimate ΔH for a reaction?

27. Explain the difference between exothermic and endothermic reactions in terms of the relative strengths of the bonds that are broken and the bonds that are formed.

28. What is the standard enthalpy of formation for a compound? For a pure element in its standard state?

29. How do you calculate ΔH°_{rxn} from tabulated standard enthalpies of formation?

30. What is lattice energy? How does lattice energy depend on ion size? On ion charge?

PROBLEMS BY TOPIC

Internal Energy, Heat, and Work

31. Which statement is true of the internal energy of a system and its surroundings during an energy exchange with a negative ΔE_{sys}?
   a. The internal energy of the system increases, and the internal energy of the surroundings decreases.
   b. The internal energy of both the system and the surroundings increases.
   c. The internal energy of both the system and the surroundings decreases.
   d. The internal energy of the system decreases, and the internal energy of the surroundings increases.

32. During an energy exchange, a chemical system absorbs energy from its surroundings. What is the sign of ΔE_{sys} for this process? Explain.

33. Identify each energy exchange as primarily heat or work and determine the sign of ΔE (positive or negative) for the system.
   a. Sweat evaporates from skin, cooling the skin. (The evaporating sweat is the system.)
   b. A balloon expands against an external pressure. (The contents of the balloon are the system.)
   c. An aqueous chemical reaction mixture is warmed with an external flame. (The reaction mixture is the system.)

34. Identify each energy exchange as primarily heat or work and determine the sign of ΔE (positive or negative) for the system.
   a. A rolling billiard ball collides with another billiard ball. The first billiard ball (defined as the system) stops rolling after the collision.
   b. A book is dropped to the floor. (The book is the system.)
   c. A father pushes his daughter on a swing. (The daughter and the swing are the system.)

35. A system releases 622 kJ of heat and does 105 kJ of work on the surroundings. What is the change in internal energy of the system?

36. A system absorbs 196 kJ of heat, and the surroundings do 117 kJ of work on the system. What is the change in internal energy of the system?

37. The gas in a piston (defined as the system) warms and absorbs 655 J of heat. The expansion performs 344 J of work on the surroundings. What is the change in internal energy for the system?

38. The air in an inflated balloon (defined as the system) warms over a toaster and absorbs 115 J of heat. As it expands, it does 77 kJ of work. What is the change in internal energy for the system?

Heat, Heat Capacity, and Work

39. A person packs two identical coolers for a picnic, placing twenty-four 12-ounce soft drinks and 5 pounds of ice in each. However, the drinks put into cooler A were refrigerated for several hours before they were packed in the cooler, while the drinks put into cooler B were at room temperature. When the picnickers open the two coolers 3 hours later, most of the ice in cooler A is still present, while nearly all of the ice in cooler B has melted. Explain this difference.

40. A kilogram of aluminum metal and a kilogram of water are each warmed to 75 °C and placed in two identical insulated containers. One hour later, the two containers are opened, and the temperature of each substance is measured. The aluminum has cooled to 35 °C, while the water has cooled only to 66 °C. Explain this difference.

41. How much heat is required to warm 1.50 L of water from 25.0 °C to 100.0 °C? (Assume a density of 1.0 g/mL for the water.)

42. How much heat is required to warm 1.50 kg of sand from 25.0 °C to 100.0 °C?

43. Suppose that 25 g of each substance is initially at 27.0 °C. What is the final temperature of each substance upon absorbing 2.35 kJ of heat?
   a. gold
   b. silver
   c. aluminum
   d. water

44. An unknown mass of each substance, initially at 23.0 °C, absorbs 1.95×10^3 J of heat. The final temperature is recorded as indicated. Find the mass of each substance.
   a. Pyrex glass ($T_f = 55.4$ °C)
   b. sand ($T_f = 62.1$ °C)
   c. ethanol ($T_f = 44.2$ °C)
   d. water ($T_f = 32.4$ °C)

45. How much work (in J) is required to expand the volume of a pump from 0.0 L to 2.5 L against an external pressure of 1.1 atm?

46. The average human lung expands by about 0.50 L during each breath. If this expansion occurs against an external pressure of 1.0 atm, how much work (in J) is done during the expansion?

47. The air within a piston equipped with a cylinder absorbs 565 J of heat and expands from an initial volume of 0.10 L to a final volume of 0.85 L against an external pressure of 1.0 atm. What is the change in internal energy of the air within the piston?

48. A gas is compressed from an initial volume of 5.55 L to a final volume of 1.22 L by an external pressure of 1.00 atm. During the compression the gas releases 124 J of heat. What is the change in internal energy of the gas?

Enthalpy and Thermochemical Stoichiometry

49. When 1 mol of a fuel burns at constant pressure, it produces 3452 kJ of heat and does 11 kJ of work. What are ΔE and ΔH for the combustion of the fuel?

50. The change in internal energy for the combustion of 1.0 mol of octane at a pressure of 1.0 atm is 5084.3 kJ. If the change in enthalpy is 5074.1 kJ, how much work is done during the combustion?

51. Is each process exothermic or endothermic? Indicate the sign of ΔH.

a. natural gas burning on a stove
b. isopropyl alcohol evaporating from skin
c. water condensing from steam

52. Is each process exothermic or endothermic? Indicate the sign of ΔH.

a. dry ice evaporating
b. a sparkler burning
c. the reaction that occurs in a chemical cold pack used to ice athletic injuries

53. Consider the thermochemical equation for the combustion of acetone (C_3H_6O), the main ingredient in nail polish remover.

$$C_3H_6O(l) + 4\,O_2(g) \longrightarrow 3\,CO_2(g) + 3\,H_2O(g) \qquad \Delta H^\circ_{rxn} = -1790\text{ kJ}$$

If a bottle of nail polish remover contains 177 mL of acetone, how much heat is released by its complete combustion? The density of acetone is 0.788 g/mL.

54. What mass of natural gas (CH_4) must burn to emit 267 kJ of heat?

$$CH_4(g) + 2\,O_2(g) \longrightarrow CO_2(g) + 2\,H_2O(g) \qquad \Delta H^\circ_{rxn} = -802.3\text{ kJ}$$

55. Nitromethane (CH_3NO_2) burns in air to produce significant amounts of heat.

$$2\,CH_3NO_2(l) + \tfrac{3}{2}\,O_2(g) \longrightarrow 2\,CO_2(g) + 3\,H_2O(l) + N_2(g) \qquad \Delta H^\circ_{rxn} = -1418\text{ kJ}$$

How much heat is produced by the complete reaction of 5.56 kg of nitromethane?

56. Titanium reacts with iodine to form titanium(III) iodide, emitting heat.

$$2\,Ti(s) + 3\,I_2(g) \longrightarrow 2\,TiI_3(s) \qquad \Delta H^\circ_{rxn} = -839\text{ kJ}$$

Determine the masses of titanium and iodine that react if 1.55×10^3 kJ of heat is emitted by the reaction.

57. The propane fuel (C_3H_8) used in gas barbeques burns according to a thermochemical equation.

$$C_3H_8(g) + 5\,O_2(g) \longrightarrow 3\,CO_2(g) + 4\,H_2O(g) \qquad \Delta H^\circ_{rxn} = -2217\text{ kJ}$$

If a pork roast must absorb 1.6×10^3 kJ to fully cook, and if only 10% of the heat produced by the barbeque is actually absorbed by the roast, what mass of CO_2 is emitted into the atmosphere during the grilling of the pork roast?

58. Charcoal is primarily carbon. Determine the mass of CO_2 produced by burning enough carbon (in the form of charcoal) to produce 5.00×10^2 kJ of heat.

$$C(s) + O_2(g) \longrightarrow CO_2(g) \qquad \Delta H^\circ_{rxn} = -393.5\text{ kJ}$$

Thermal Energy Transfer

59. We submerge a silver block, initially at 58.5 °C, into 100.0 g of water at 24.8 °C, in an insulated container. The final temperature of the mixture upon reaching thermal equilibrium is 26.2 °C. What is the mass of the silver block?

60. We submerge a 32.5-g iron rod, initially at 22.7 °C, into an unknown mass of water at 63.2 °C, in an insulated container. The final temperature of the mixture upon reaching thermal equilibrium is 59.5 °C. What is the mass of the water?

61. We submerge a 31.1-g wafer of pure gold initially at 69.3 °C into 64.2 g of water at 27.8 °C in an insulated container. What is the final temperature of both substances at thermal equilibrium?

62. We submerge a 2.85-g lead weight, initially at 10.3 °C, in 7.55 g of water at 52.3 °C in an insulated container. What is the final temperature of both substances at thermal equilibrium?

63. Two substances, A and B, initially at different temperatures, come into contact and reach thermal equilibrium. The mass of substance A is 6.15 g, and its initial temperature is 20.5 °C. The mass of substance B is 25.2 g, and its initial temperature is 52.7 °C. The final temperature of both substances at thermal equilibrium is 46.7 °C. If the specific heat capacity of substance B is 1.17 J/g·°C, what is the specific heat capacity of substance A?

64. A 2.74-g sample of a substance suspected of being pure gold is warmed to 72.1 °C and submerged into 15.2 g of water initially at 24.7 °C. The final temperature of the mixture is 26.3 °C. What is the heat capacity of the unknown substance? Could the substance be pure gold?

Calorimetry

65. Exactly 1.5 g of a fuel burns under conditions of constant pressure and then again under conditions of constant volume. In measurement A the reaction produces 25.9 kJ of heat, and in measurement B the reaction produces 23.3 kJ of heat. Which measurement (A or B) corresponds to conditions of constant pressure? Which one corresponds to conditions of constant volume? Explain.

66. In order to obtain the largest possible amount of heat from a chemical reaction in which there is a large increase in the number of moles of gas, should you carry out the reaction under conditions of constant volume or constant pressure? Explain.

67. When 0.514 g of biphenyl ($C_{12}H_{10}$) undergoes combustion in a bomb calorimeter, the temperature rises from 25.8 °C to 29.4 °C. Find ΔE_{rxn} for the combustion of biphenyl in kJ/mol biphenyl. The heat capacity of the bomb calorimeter, determined in a separate experiment, is 5.86 kJ/°C.

68. Mothballs are composed primarily of the hydrocarbon naphthalene ($C_{10}H_8$). When 1.025 g of naphthalene burns in a bomb calorimeter, the temperature rises from 24.25 °C to 32.33 °C. Find ΔE_{rxn} for the combustion of naphthalene. The heat capacity of the calorimeter, determined in a separate experiment, is 5.11 kJ/°C.

69. Zinc metal reacts with hydrochloric acid according to the balanced equation:

$$Zn(s) + 2\,HCl(aq) \longrightarrow ZnCl_2(aq) + H_2(g)$$

When 0.103 g of Zn(*s*) is combined with enough HCl to make 50.0 mL of solution in a coffee-cup calorimeter, all of the zinc reacts, raising the temperature of the solution from 22.5 °C to 23.7 °C. Find ΔH_{rxn} for this reaction as written. (Use 1.0 g/mL for the density of the solution and 4.18 J/g·°C as the specific heat capacity.)

70. Instant cold packs used to ice athletic injuries on the field contain ammonium nitrate and water separated by a thin plastic divider. When the divider is broken, the ammonium nitrate dissolves according to the endothermic reaction:

$$NH_4NO_3(s) \longrightarrow NH_4^+(aq) + NO_3^-(aq)$$

In order to measure the enthalpy change for this reaction, 1.25 g of NH_4NO_3 is dissolved in enough water to make 25.0 mL of solution. The initial temperature is 25.8 °C, and the final temperature (after the solid dissolves) is 21.9 °C. Calculate the change in enthalpy for the reaction in kJ. (Use 1.0 g/mL as the density of the solution and 4.18 J/g·°C as the specific heat capacity.)

Quantitative Relationships Involving ΔH and Hess's Law

71. For each generic reaction, determine the value of ΔH_2 in terms of ΔH_1.

a. $A + B \longrightarrow 2\,C$ ΔH_1
   $2\,C \longrightarrow A + B$ $\Delta H_2 = ?$

b. $A + \frac{1}{2}B \longrightarrow C$ ΔH_1
   $2\,A + B \longrightarrow 2\,C$ $\Delta H_2 = ?$

c. $A \longrightarrow B + 2\,C$ ΔH_1
   $\frac{1}{2}B + C \longrightarrow \frac{1}{2}A$ $\Delta H_2 = ?$

72. Consider the generic reaction:

$$A + 2\,B \longrightarrow C + 3\,D \qquad \Delta H = 155\text{ kJ}$$

Determine the value of ΔH for each related reaction:

a. $3\,A + 6\,B \longrightarrow 3\,C + 9\,D$
b. $C + 3\,D \longrightarrow A + 2\,B$
c. $\frac{1}{2}C + \frac{3}{2}D \longrightarrow \frac{1}{2}A + B$

73. Calculate ΔH_{rxn} for the reaction.

$$Fe_2O_3(s) + 3\,CO(g) \longrightarrow 2\,Fe(s) + 3\,CO_2(g)$$

Use the following reactions and given ΔH's:

$2\,Fe(s) + \frac{3}{2}O_2(g) \longrightarrow Fe_2O_3(s)$ $\Delta H = -824.2\text{ kJ}$
$CO(g) + \frac{1}{2}O_2(g) \longrightarrow CO_2(g)$ $\Delta H = -282.7\text{ kJ}$

74. Calculate ΔH_{rxn} for the reaction.

$$CaO(s) + CO_2(g) \longrightarrow CaCO_3(s)$$

Use the following reactions and given ΔH's:

$Ca(s) + CO_2(g) + \frac{1}{2}O_2(g) \longrightarrow CaCO_3(s)$ $\Delta H = -812.8\text{ kJ}$
$2\,Ca(s) + O_2(g) \longrightarrow 2\,CaO(s)$ $\Delta H = -1269.8\text{ kJ}$

75. Calculate ΔH_{rxn} for the reaction.

$5\,C(s) + 6\,H_2(g) \longrightarrow C_5H_{12}(l)$

Use the following reactions and given ΔH's:

$C_5H_{12}(l) + 8\,O_2(g) \longrightarrow 5\,CO_2(g) + 6\,H_2O(g)$ $\Delta H = -3244.8\text{ kJ}$
$C(s) + O_2(g) \longrightarrow CO_2(g)$ $\Delta H = -393.5\text{ kJ}$
$2\,H_2(g) + O_2(g) \longrightarrow 2\,H_2O(g)$ $\Delta H = -483.5\text{ kJ}$

76. Calculate ΔH_{rxn} for the reaction.

$CH_4(g) + 4\,Cl_2(g) \longrightarrow CCl_4(g) + 4\,HCl(g)$

Use the following reactions and given ΔH's:

$C(s) + 2\,H_2(g) \longrightarrow CH_4(g)$ $\Delta H = -74.6\text{ kJ}$
$C(s) + 2\,Cl_2(g) \longrightarrow CCl_4(g)$ $\Delta H = -95.7\text{ kJ}$
$H_2(g) + Cl_2(g) \longrightarrow 2\,HCl(g)$ $\Delta H = -92.3\text{ kJ}$

Using Bond Energies to Calculate ΔH_{rxn}

77. Hydrogenation reactions are used to add hydrogen across double bonds in hydrocarbons and other organic compounds. Use average bond energies to calculate ΔH_{rxn} for the hydrogenation reaction.

$$H_2C{=}CH_2(g) + H_2(g) \longrightarrow H_3C{-}CH_3(g)$$

78. Ethanol is a possible fuel. Use average bond energies to calculate ΔH_{rxn} for the combustion of ethanol.

$$CH_3CH_2OH(g) + 3\,O_2(g) \longrightarrow 2\,CO_2(g) + 3\,H_2O(g)$$

79. Hydrogen, a potential future fuel, can be produced from carbon (from coal) and steam by this reaction:

$$C(s) + 2\,H_2O(g) \longrightarrow 2\,H_2(g) + CO_2(g)$$

Use average bond energies to calculate ΔH_{rxn} for the reaction.

80. Hydroxyl radicals react with and eliminate many atmospheric pollutants. However, the hydroxyl radical does not clean up everything. For example, chlorofluorocarbons—which destroy stratospheric ozone—are not attacked by the hydroxyl radical. Consider the hypothetical reaction by which the hydroxyl radical might react with a chlorofluorocarbon:

$$OH(g) + CF_2Cl_2(g) \longrightarrow HOF(g) + CFCl_2(g)$$

Use bond energies to explain why this reaction is improbable.

Enthalpies of Formation and ΔH

81. Write an equation for the formation of each compound from its elements in their standard states, and find ΔH_f° for each from Appendix IVB.

a. $NH_3(g)$ b. $CO_2(g)$ c. $Fe_2O_3(s)$ d. $CH_4(g)$

82. Write an equation for the formation of each compound from its elements in their standard states, and find ΔH_{rxn}° for each from Appendix IVB.

a. $NO_2(g)$ b. $MgCO_3(s)$
c. $C_2H_4(g)$ d. $CH_3OH(l)$

83. Hydrazine (N_2H_4) is a fuel used by some spacecraft. It is normally oxidized by N_2O_4 according to the equation:

$N_2H_4(l) + N_2O_4(g) \longrightarrow 2\,N_2O(g) + 2\,H_2O(g)$

Calculate ΔH_{rxn}° for this reaction using standard enthalpies of formation.

84. Pentane (C_5H_{12}) is a component of gasoline that burns according to the following balanced equation:

$$C_5H_{12}(l) + 8\,O_2(g) \longrightarrow 5\,CO_2(g) + 6\,H_2O(g)$$

Calculate ΔH°_{rxn} for this reaction using standard enthalpies of formation. (The standard enthalpy of formation of liquid pentane is −146.8 kJ/mol.)

85. Use standard enthalpies of formation to calculate ΔH°_{rxn} for each reaction.

a. $C_2H_4(g) + H_2(g) \longrightarrow C_2H_6(g)$
b. $CO(g) + H_2O(g) \longrightarrow H_2(g) + CO_2(g)$
c. $3\,NO_2(g) + H_2O(l) \longrightarrow 2\,HNO_3(aq) + NO(g)$
d. $Cr_2O_3(s) + 3\,CO(g) \longrightarrow 2\,Cr(s) + 3\,CO_2(g)$

86. Use standard enthalpies of formation to calculate ΔH°_{rxn} for each reaction.

a. $2\,H_2S(g) + 3\,O_2(g) \longrightarrow 2\,H_2O(l) + 2\,SO_2(g)$
b. $SO_2(g) + \frac{1}{2}\,O_2(g) \longrightarrow SO_3(g)$
c. $C(s) + H_2O(g) \longrightarrow CO(g) + H_2(g)$
d. $N_2O_4(g) + 4\,H_2(g) \longrightarrow N_2(g) + 4\,H_2O(g)$

87. During photosynthesis, plants use energy from sunlight to form glucose ($C_6H_{12}O_6$) and oxygen from carbon dioxide and water. Write a balanced equation for photosynthesis and calculate ΔH°_{rxn}.

88. Ethanol can be made from the fermentation of crops and has been used as a fuel additive to gasoline. Write a balanced equation for the combustion of ethanol and calculate ΔH°_{rxn}.

89. Top fuel dragsters and funny cars burn nitromethane as fuel according to the balanced combustion equation:

$$2\,CH_3NO_2(l) + \tfrac{3}{2}\,O_2(g) \longrightarrow 2\,CO_2(g) + 3\,H_2O(l) + N_2(g) \qquad \Delta H^\circ_{rxn} = -1418\text{ kJ}$$

The enthalpy of combustion for nitromethane is −709.2 kJ/mol. Calculate the standard enthalpy of formation (ΔH°_f) for nitromethane.

90. The explosive nitroglycerin ($C_3H_5N_3O_9$) decomposes rapidly upon ignition or sudden impact according to the balanced equation:

$$4\,C_3H_5N_3O_9(l) \longrightarrow 12\,CO_2(g) + 10\,H_2O(g) + 6\,N_2(g) + O_2(g) \qquad \Delta H^\circ_{rxn} = -5678\text{ kJ}$$

Calculate the standard enthalpy of formation (ΔH°_f) for nitroglycerin.

Lattice Energies

91. Explain the trend in the lattice energies (shown here) of the alkaline earth metal oxides.

Metal Oxide	Lattice Energy (kJ/mol)
MgO	−3795
CaO	−3414
SrO	−3217
BaO	−3029

92. Rubidium iodide has a lattice energy of −617 kJ/mol, while potassium bromide has a lattice energy of −671 kJ/mol. Why is the lattice energy of potassium bromide more exothermic than the lattice energy of rubidium iodide?

93. The lattice energy of CsF is −744 kJ/mol whereas that of BaO is −3029 kJ/mol. Explain this large difference in lattice energy.

94. Arrange these compounds in order of increasing magnitude of lattice energy: KCl, SrO, RbBr, CaO.

95. Use the Born–Haber cycle and data from Appendix IVB and Chapters 4 and 10 to calculate the lattice energy of KCl. (ΔH_{sub} for potassium is 89.0 kJ/mol.)

96. Use the Born–Haber cycle and data from Appendix IVB and Table 10.3 to calculate the lattice energy of CaO. (ΔH_{sub} for calcium is 178 kJ/mol; IE_1 and IE_2 for calcium are 590 kJ/mol and 1145 kJ/mol, respectively; EA_1 and EA_2 for O are −141 kJ/mol and 744 kJ/mol, respectively.)

CUMULATIVE PROBLEMS

97. The kinetic energy of a rolling billiard ball is given by $KE = \frac{1}{2}mv^2$. Suppose a 0.17-kg billiard ball is rolling down a pool table with an initial speed of 4.5 m/s. As it travels, it loses some of its energy as heat. The ball slows down to 3.8 m/s and then collides head-on with a second billiard ball of equal mass. The first billiard ball completely stops, and the second one rolls away with a velocity of 3.8 m/s. Assume the first billiard ball is the system and calculate w, q, and ΔE for the process.

98. A 100-W light bulb is placed in a cylinder equipped with a moveable piston. The light bulb is turned on for 0.015 hour, and the assembly expands from an initial volume of 0.85 L to a final volume of 5.88 L against an external pressure of 1.0 atm. Use the wattage of the light bulb and the time it is on to calculate ΔE in joules (assume that the cylinder and light bulb assembly is the system and assume two significant figures). Calculate w and q.

99. Evaporating sweat cools the body because evaporation is an endothermic process:

$$H_2O(l) \longrightarrow H_2O(g) \qquad \Delta H^\circ_{rxn} = +44.01\text{ kJ}$$

Estimate the mass of water that must evaporate from the skin to cool the body by 0.50 °C. Assume a body mass of 95 kg and assume that the specific heat capacity of the body is 4.0 J/g · °C.

100. LP gas burns according to the exothermic reaction:

$$C_3H_8(g) + 5\,O_2(g) \longrightarrow 3\,CO_2(g) + 4\,H_2O(g) \qquad \Delta H^\circ_{rxn} = -2044\text{ kJ}$$

What mass of LP gas is necessary to heat 1.5 L of water from room temperature (25.0 °C) to boiling (100.0 °C)? Assume that during heating, 15% of the heat emitted by the LP gas combustion goes to heat the water. The rest is lost as heat to the surroundings.

101. Use standard enthalpies of formation to calculate the standard change in enthalpy for the melting of ice. (The ΔH°_f for $H_2O(s)$ is −291.8 kJ/mol.) Use this value to calculate the mass of ice required to cool 355 mL of a beverage from room temperature (25.0 °C) to 0.0 °C. Assume that the specific heat capacity and density of the beverage are the same as those of water.

102. Dry ice is solid carbon dioxide. Instead of melting, solid carbon dioxide sublimes according to the equation:

$$CO_2(s) \longrightarrow CO_2(g)$$

When dry ice is added to warm water, heat from the water causes the dry ice to sublime more quickly. The evaporating carbon dioxide produces a dense fog often used to create special effects. In a simple dry ice fog machine, dry ice is added to warm water in a Styrofoam cooler. The dry ice produces fog until it evaporates away, or until the water gets too cold to sublime the dry ice quickly enough. A small Styrofoam cooler holds 15.0 L of water heated to 85 °C. Use standard enthalpies of formation to calculate the change in enthalpy for dry ice sublimation, and calculate the mass of dry ice that should be added to the water so that the dry ice completely sublimes away when the water reaches 25 °C. Assume no heat loss to the surroundings. (The ΔH_f° for $CO_2(s)$ is −427.4 kJ/mol.)

▲ When carbon dioxide sublimes, the gaseous CO_2 is cold enough to cause water vapor in the air to condense, forming fog.

103. A 25.5-g aluminum block is warmed to 65.4 °C and plunged into an insulated beaker containing 55.2 g water initially at 22.2 °C. The aluminum and the water are allowed to come to thermal equilibrium. Assuming that no heat is lost, what is the final temperature of the water and aluminum?

104. We mix 50.0 mL of ethanol (density = 0.789 g/mL) initially at 7.0 °C with 50.0 mL of water (density = 1.0 g/mL) initially at 28.4 °C in an insulated beaker. Assuming that no heat is lost, what is the final temperature of the mixture?

105. Palmitic acid ($C_{16}H_{32}O_2$) is a dietary fat found in beef and butter. The caloric content of palmitic acid is typical of fats in general. Write a balanced equation for the complete combustion of palmitic acid and calculate the standard enthalpy of combustion. What is the caloric content of palmitic acid in Cal/g? Do the same calculation for table sugar (sucrose, $C_{12}H_{22}O_{11}$). Which dietary substance (sugar or fat) contains more Calories per gram? The standard enthalpy of formation of palmitic acid is −208 kJ/mol, and that of sucrose is −2226.1 kJ/mol. (Use $H_2O(l)$ in the balanced chemical equations because the metabolism of these compounds produces liquid water.)

106. Hydrogen and methanol have both been proposed as alternatives to hydrocarbon fuels. Write balanced reactions for the complete combustion of hydrogen and methanol and use standard enthalpies of formation to calculate the amount of heat released per kilogram of the fuel. Which fuel contains the most energy in the least mass? How does the energy of these fuels compare to that of octane (C_8H_{18})?

107. One tablespoon of peanut butter has a mass of 16 g. It is combusted in a calorimeter whose heat capacity is 120.0 kJ/°C. The temperature of the calorimeter rises from 22.2 °C to 25.4 °C. Find the food caloric content of peanut butter.

108. A mixture of 2.0 mol of $H_2(g)$ and 1.0 mol of $O_2(g)$ is placed in a sealed evacuated container made of a perfect insulating material at 25 °C. The mixture is ignited with a spark, and it reacts to form liquid water. Determine the temperature of the water.

109. A 20.0-L volume of an ideal gas in a cylinder with a piston is at a pressure of 3.0 atm. Enough weight is suddenly removed from the piston to lower the external pressure to 1.5 atm. The gas then expands at constant temperature until its pressure is 1.5 atm. Find ΔE, ΔH, q, and w for this change in state.

110. When we burn 10.00 g of phosphorus in $O_2(g)$ to form $P_4O_{10}(s)$, we generate enough heat to raise the temperature of 2950 g of water from 18.0 °C to 38.0 °C. Calculate the enthalpy of formation of $P_4O_{10}(s)$ under these conditions.

111. The ΔH for the oxidation of S in the gas phase to SO_3 is −204 kJ/mol, and for the oxidation of SO_2 to SO_3 it is 89.5 kJ/mol. Find the enthalpy of formation of SO_2 under these conditions.

112. The ΔH_f° of $TiI_3(s)$ is −328 kJ/mol, and the ΔH° for the reaction $2\,Ti(s) + 3\,I_2(g) \longrightarrow 2\,TiI_3(s)$ is −839 kJ. Calculate the ΔH of sublimation (the state transition from solid to gas) of $I_2(s)$, which is a solid at 25 °C.

113. A copper cube measuring 1.55 cm on edge and an aluminum cube measuring 1.62 cm on edge are both heated to 55.0 °C and submerged in 100.0 mL of water at 22.2 °C. What is the final temperature of the water when equilibrium is reached? (Assume a density of 0.998 g/mL for water.)

114. A pure gold ring and pure silver ring have a total mass of 14.9 g. We heat the two rings to 62.0 °C and drop them into 15.0 mL of water at 23.5 °C. When equilibrium is reached, the temperature of the water is 25.0 °C. What is the mass of each ring? (Assume a density of 0.998 g/mL for water.)

115. The reaction of $Fe_2O_3(s)$ with $Al(s)$ to form $Al_2O_3(s)$ and $Fe(s)$ is called the thermite reaction and is highly exothermic. What role does lattice energy play in the exothermicity of the reaction?

116. NaCl has a lattice energy −787 kJ/mol. Consider a hypothetical salt XY. X^{3+} has the same radius of Na^+, and Y^{3-} has the same radius as Cl^-. Estimate the lattice energy of XY.

117. If hydrogen were used as a fuel, it could be burned according to this reaction:

$$H_2(g) + \tfrac{1}{2}\,O_2(g) \longrightarrow H_2O(g)$$

Use average bond energies to calculate ΔH_{rxn} for this reaction and also for the combustion of methane (CH_4). Which fuel yields more energy per mole? Per gram?

118. Calculate ΔH_{rxn} for the combustion of octane (C_8H_{18}), a component of gasoline, by using average bond energies, and then calculate it using enthalpies of formation from Appendix IVB. What is the percent difference between your results? Which result would you expect to be more accurate?

119. The heat of atomization is the heat required to convert a molecule in the gas phase into its constituent atoms in the gas phase. The heat of atomization is used to calculate average bond energies. Without using any tabulated bond energies, calculate the average C—Cl bond energy from the following data: The heat of atomization of CH_4 is 1660 kJ/mol, and that of CH_2Cl_2 is 1495 kJ/mol.

120. Calculate the heat of atomization (see previous problem) of C_2H_3Cl, using the average bond energies in Table 10.3.

CHALLENGE PROBLEMS

121. A typical frostless refrigerator uses 655 kWh of energy per year in the form of electricity. Suppose that all of this electricity is generated at a power plant that burns coal containing 3.2% sulfur by mass and that all of the sulfur is emitted as SO_2 when the coal is burned. If all of the SO_2 goes on to react with rainwater to form H_2SO_4, what mass of H_2SO_4 does the annual operation of the refrigerator produce? (*Hint:* Assume that the remaining percentage of the coal is carbon and begin by calculating ΔH°_{rxn} for the combustion of carbon.)

122. A large sport utility vehicle has a mass of 2.5×10^3 kg. Calculate the mass of CO_2 emitted into the atmosphere upon accelerating the SUV from 0.0 mph to 65.0 mph. Assume that the required energy comes from the combustion of octane with 30% efficiency. (*Hint:* Use $KE = \frac{1}{2}mv^2$ to calculate the kinetic energy required for the acceleration.)

123. Combustion of natural gas (primarily methane) occurs in most household heaters. The heat given off in this reaction is used to raise the temperature of the air in the house. Assuming that all the energy given off in the reaction goes to heating up only the air in the house, determine the mass of methane required to heat the air in a house by 10.0 °C. Assume that the house dimensions are 30.0 m × 30.0 m × 3.0 m; specific heat capacity of air is 30 J/K · mol; and 1.00 mol of air occupies 22.4 L for all temperatures concerned.

124. When backpacking in the wilderness, hikers often boil water to sterilize it for drinking. Suppose that you are planning a backpacking trip and will need to boil 35 L of water for your group. What volume of fuel should you bring? Assume that the fuel has an average formula of C_7H_{16}; 15% of the heat generated from combustion goes to heat the water (the rest is lost to the surroundings); the density of the fuel is 0.78 g/mL; the initial temperature of the water is 25.0 °C; and the standard enthalpy of formation of C_7H_{16} is −224.4 kJ/mol.

125. An ice cube of mass 9.0 g is added to a cup of coffee. The coffee's initial temperature is 90.0 °C and the cup contains 120.0 g of liquid. Assume the specific heat capacity of the coffee is the same as that of water. The heat of fusion of ice (the heat associated with ice melting) is 6.0 kJ/mol. Find the temperature of the coffee after the ice melts.

126. Find ΔH, ΔE, q, and w for the freezing of water at −10.0 °C. The specific heat capacity of ice is 2.04 J/g · °C and its heat of fusion (the quantity of heat associated with melting) is −332 J/g.

127. The heat of vaporization of water at 373 K is 40.7 kJ/mol. Find q, w, ΔE, and ΔH for the evaporation of 454 g of water at this temperature at 1 atm.

128. Find ΔH for the combustion of ethanol (C_2H_6O) to carbon dioxide and liquid water from the following data. The heat capacity of the bomb calorimeter is 34.65 kJ/K, and the combustion of 1.765 g of ethanol raises the temperature of the calorimeter from 294.33 K to 295.84 K.

129. The main component of acid rain (H_2SO_4) forms from SO_2, a pollutant in the atmosphere, via these steps:

$$SO_2 + OH\cdot \longrightarrow HSO_3\cdot$$
$$HSO_3\cdot + O_2 \longrightarrow SO_3 + HOO\cdot$$
$$SO_3 + H_2O \longrightarrow H_2SO_4$$

Draw the Lewis structure for each of the species in these steps and use bond energies and Hess's law to estimate ΔH_{rxn} for the overall process. (Use 265 kJ/mol for the S—O single-bond energy.)

130. Use average bond energies together with the standard enthalpy of formation of C(g) (718.4 kJ/mol) to estimate the standard enthalpy of formation of gaseous benzene, $C_6H_6(g)$. (Remember that average bond energies apply to the gas phase only.) Compare the value you obtain using average bond energies to the actual standard enthalpy of formation of gaseous benzene, 82.9 kJ/mol. What does the difference between these two values tell you about the stability of benzene?

131. The standard heat of formation of $CaBr_2$ is −675 kJ/mol. The first ionization energy of Ca is 590 kJ/mol, and its second ionization energy is 1145 kJ/mol. The heat of sublimation of Ca [$Ca(s) \longrightarrow Ca(g)$] is 178 kJ/mol. The bond energy of Br_2 is 193 kJ/mol, the heat of vaporization of $Br_2(l)$ is 31 kJ/mol, and the electron affinity of Br is −325 kJ/mol. Calculate the lattice energy of $CaBr_2$.

132. The standard heat of formation of $PI_3(s)$ is −24.7 kJ/mol, and the PI bond energy in this molecule is 184 kJ/mol. The standard heat of formation of P(g) is 334 kJ/mol, and that of $I_2(g)$ is 62 kJ/mol. The I_2 bond energy is 151 kJ/mol. Calculate the heat of sublimation of PI_3[$PI_3(s) \longrightarrow PI_3(g)$].

CONCEPTUAL PROBLEMS

133. Which statement is true of the internal energy of the system and its surroundings following a process in which $\Delta E_{sys} = +65$ kJ? Explain.

   a. The system and the surroundings both lose 65 kJ of energy.
   b. The system and the surroundings both gain 65 kJ of energy.
   c. The system loses 65 kJ of energy, and the surroundings gain 65 kJ of energy.
   d. The system gains 65 kJ of energy, and the surroundings lose 65 kJ of energy.

134. Which expression describes the heat emitted in a chemical reaction when the reaction is carried out at constant pressure? Explain.

   a. $\Delta E - w$
   b. ΔE
   c. $\Delta E - q$

135. Two identical refrigerators are plugged in for the first time. Refrigerator A is empty (except for air), and refrigerator B is filled with jugs of water. The compressors of both refrigerators immediately turn on and begin cooling the interiors of the refrigerators. After two hours, the compressor of refrigerator A turns off while the compressor of refrigerator B continues to run. The next day, the compressor of refrigerator A can be heard turning on and off every few minutes, while the compressor of refrigerator B turns off and on every hour or so (and stays on longer each time). Explain these observations.

136. A 1-kg cylinder of aluminum and a 1-kg jug of water, both at room temperature, are put into a refrigerator. After one hour, the temperature of each object is measured. One of the objects is much cooler than the other. Which one is cooler and why?

137. Two substances A and B, initially at different temperatures, are thermally isolated from their surroundings and allowed to come into thermal contact. The mass of substance A is twice the mass of substance B, but the specific heat capacity of substance B is four times the specific heat capacity of substance A. Which substance will undergo a larger change in temperature?

138. When 1 mol of a gas burns at constant pressure, it produces 2418 J of heat and does 5 J of work. Determine ΔE, ΔH, q, and w for the process.

139. In an exothermic reaction, the reactants lose energy and the reaction feels hot to the touch. Explain why the reaction feels hot even though the reactants are losing energy. Where does the energy come from?

140. Which statement is true of a reaction in which ΔV is positive? Explain.

a. $\Delta H = \Delta E$
b. $\Delta H > \Delta E$
c. $\Delta H < \Delta E$

141. Which statement is true of an endothermic reaction?

a. Strong bonds break and weak bonds form.
b. Weak bonds break and strong bonds form.
c. The bonds that break and those that form are of approximately the same strength.

142. When a firecracker explodes, energy is obviously released. The compounds in the firecracker can be viewed as being "energy rich." What does this mean? Explain the source of the energy in terms of chemical bonds.

ANSWERS TO CONCEPTUAL CONNECTIONS

Cc 10.1 The correct answer is **(a)**. When ΔE_{sys} is negative, energy flows out of the system and into the surroundings. The energy increase in the surroundings must exactly match the decrease in the system.

Cc 10.2 **(a)** heat, sign is positive **(b)** work, sign is positive **(c)** heat, sign is negative

Cc 10.3 Bring the water; it has the higher heat capacity and will therefore release more heat as it cools (because it absorbed more heat when it was heated to 38 °C).

Cc 10.4 **(c)** The specific heat capacity of substance B is twice that of A, but because the mass of B is half that of A, the quantity $m \times C_s$ is identical for both substances so that the final temperature is exactly midway between the two initial temperatures.

Cc 10.5 ΔH represents only the heat exchanged; therefore $\Delta H = -2658$ kJ. ΔE represents the heat *and work* exchanged; therefore $\Delta E = -2661$ kJ. The signs of both ΔH and ΔE are negative because heat and work are flowing out of the system and into the surroundings. Notice that the values of ΔH and ΔE are similar in magnitude, as is the case in many chemical reactions.

Cc 10.6 An endothermic reaction feels cold to the touch because the reaction (acting here as the system) absorbs heat from the surroundings. When you touch the vessel in which the reaction occurs, you, being part of the surroundings, lose heat to the system (the reaction), which makes you feel cold. The heat absorbed by the reaction (from your body, in this case) does not contribute to increasing its temperature, but rather becomes potential energy.

Cc 10.7 The value of q_{rxn} with the greater magnitude (−12.5 kJ) must have come from the bomb calorimeter. Recall that $\Delta E_{rxn} = q_{rxn} + w_{rxn}$. In a bomb calorimeter, the energy change that occurs in the course of the reaction all takes the form of heat (q). In a coffee-cup calorimeter, the amount of energy released as heat may be smaller because some of the energy may be used to do work (w).

Cc 10.8 **(b)** In a highly exothermic reaction, the energy required to break bonds is less than the energy released when the new bonds form, resulting in a net release of energy.

CHAPTER 11

"So many of the properties of matter, especially when in the gaseous form, can be deduced from the hypothesis that their minute parts are in rapid motion, the velocity increasing with the temperature, that the precise nature of this motion becomes a subject of rational curiosity."

—James Clerk Maxwell (1831–1879)

Space suit helmet design prototype

PV=nRT

A pressurized suit protected Felix Baumgartner from the vacuum of space during his record-breaking skydive. Any significant damage to the suit carried the risk of uncontrolled decompression, which would likely result in Baumgartner's death.

Gases

11.1 Supersonic Skydiving and the Risk of Decompression 391

11.2 Pressure: The Result of Particle Collisions 392

11.3 The Simple Gas Laws: Boyle's Law, Charles's Law, and Avogadro's Law 395

11.4 The Ideal Gas Law 401

11.5 Applications of the Ideal Gas Law: Molar Volume, Density, and Molar Mass of a Gas 404

11.6 Mixtures of Gases and Partial Pressures 407

11.7 A Particulate Model for Gases: Kinetic Molecular Theory 414

11.8 Temperature and Molecular Velocities 417

11.9 Mean Free Path, Diffusion, and Effusion of Gases 420

11.10 Gases in Chemical Reactions: Stoichiometry Revisited 422

11.11 Real Gases: The Effects of Size and Intermolecular Forces 425

Key Learning Outcomes 430

WE CAN SURVIVE FOR WEEKS without food, days without water, but only minutes without air. Fortunately, we live at the bottom of a vast ocean of air, held to Earth by gravity. The air around us is matter in the gaseous state. The behavior of gases can be explained (and in fact predicted) by a model called kinetic molecular theory. The core of this model is that *gases are composed of particles in constant motion.* Here, we again see the main theme of this book played out: *Matter is particulate, and its behavior can be understood in terms of particles.* In this chapter, we first look at some observations that led to laws that govern the behavior of gases. We then turn to the kinetic molecular theory, the model that explains those laws in terms of particles. Notice the progression of the scientific approach to knowledge in this sequence: (1) scientists made observations on the properties of gases; (2) those observations led to scientific laws (that summarized the observed behavior); and (3) the observations and laws led to the development of a theory—the kinetic molecular theory—that models the nature of a gas.

11.1 Supersonic Skydiving and the Risk of Decompression

On October 14, 2012, just after midday in New Mexico, Austrian daredevil Felix Baumgartner stepped into the dark void of space 24 miles (38.6 km) above the Earth's surface. Baumgartner's 20-minute journey

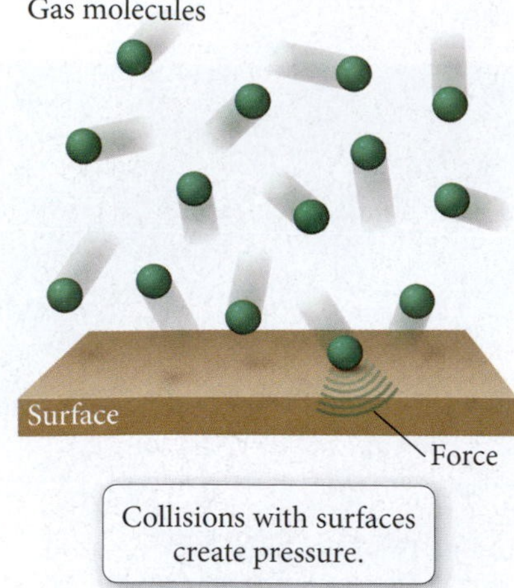

▲ **FIGURE 11.1 Gas Pressure** Pressure is the force per unit area exerted by gas molecules colliding with the surfaces around them.

back to the desert floor broke the sound barrier and shattered the previous skydiving record of 19.5 miles (31.4 km).

When Baumgartner stepped into space, he was protected from the surrounding vacuum by a pressurized suit. The suit contained air at a pressure similar to that found on the surface of Earth. **Pressure** is the force exerted per unit area by gas particles as they strike the surfaces around them (Figure 11.1 ◀). Just as a ball exerts a force when it bounces against a wall, so a gaseous atom or molecule exerts a force when it collides with a surface. The result of many of these collisions is pressure—a constant force on the surfaces exposed to any gas. The total pressure exerted by a gas depends on several factors, including the concentration of gas particles in the sample; the lower the concentration, the lower the pressure. At 24 miles (38.6 km) above the Earth's surface, the concentration of gas particles is much lower than at sea level; consequently, the pressure is very low. Without the pressurized suit, Baumgartner could not survive the space-like conditions.

Among the risks that Baumgartner faced while in space was uncontrolled decompression. Any significant damage to the suit would cause the air within the suit to escape, resulting in a large pressure drop. While the effects of a large pressure drop are sometimes exaggerated—for example, one urban myth erroneously claims that a person can explode—they are the nonetheless lethal. For example, if a pressure drop had occurred fast enough, a large pressure difference would have quickly developed between the air in Baumgartner's lungs and the surrounding vacuum. The pressure difference would have caused his lungs to expand too much, resulting in severe lung damage. Fortunately, Baumgartner's suit worked just as it was designed to do, and he plunged safely back to Earth without incident.

11.2 Pressure: The Result of Particle Collisions

As we just discussed, particles in a gas collide with each other and with the surfaces around them. Each collision exerts only a small force, but when the forces of the many particles are summed, they quickly add up. The result of the constant collisions between the atoms or molecules in a gas and the surfaces around them is *pressure*. Because of pressure, we can drink from straws, inflate basketballs, and breathe. Variation in pressure in Earth's atmosphere creates wind, and changes in pressure help us to predict weather (Figure 11.2 ▼). Pressure is all around us and inside of us. The pressure that a gas sample exerts is the *force* that results from the collisions of gas particles divided by the *area* of the surface with which they collide.

$$\text{Pressure} = \frac{\text{force}}{\text{area}} = \frac{F}{A} \qquad [11.1]$$

▶ **FIGURE 11.2 Pressure and Weather** Pressure variations in Earth's atmosphere create wind and weather. The H in this map indicates a region of high pressure, usually associated with clear weather. The L indicates a region of low pressure, often associated with unstable weather.

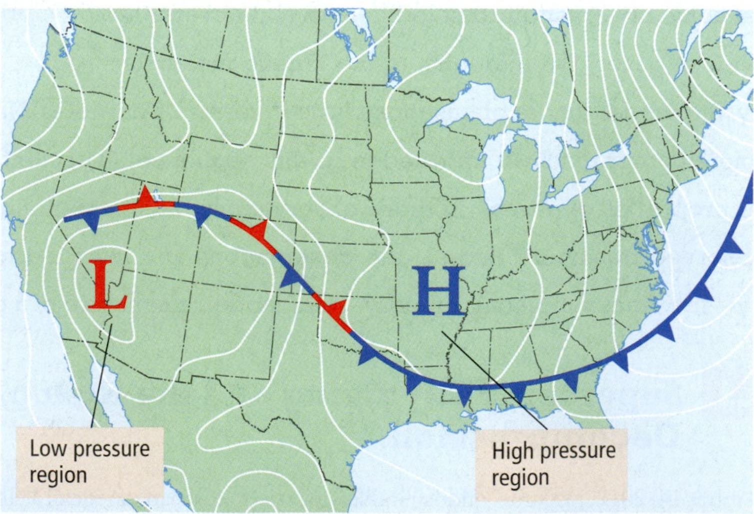

The pressure exerted by a gas sample, therefore, depends on the number of gas particles in a given volume—the fewer the gas particles, the lower the force per unit area and the lower the pressure (Figure 11.3 ▶). Because the number of gas particles in a given volume generally decreases with increasing altitude, *pressure decreases with increasing altitude.* Above 30,000 ft (about 5.6 miles or 9 km), for example, where most commercial airplanes fly, the pressure is so low that a person could pass out due to a lack of oxygen. For this reason, most airplane cabins are artificially pressurized. At 24 miles, the altitude from which Baumgartner jumped, the pressure is less than 1% of the pressure at sea level.

You may sometimes feel the effect of a drop in pressure as a brief pain in your ears. This pain arises within the air-containing cavities in your ear (Figure 11.4 ▼). When you ascend in a plane or hike up a mountain, the external pressure (the pressure that surrounds you) drops, while the pressure within your ear cavities (the internal pressure) remains the same. This creates an imbalance—the greater internal pressure forces your eardrum to bulge outward, causing pain. With time, and with the help of a yawn or two, the excess air within your ear's cavities escapes, equalizing the internal and external pressure and relieving the pain.

Pressure and Density

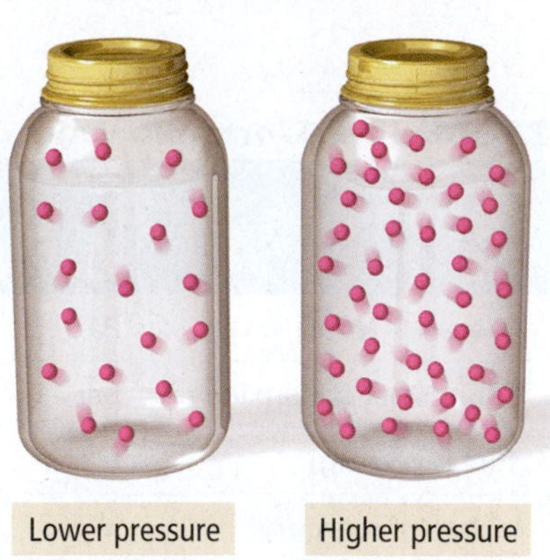

▲ **FIGURE 11.3 Pressure and Particle Density** A low density of gas particles results in low pressure. A high density of gas particles results in high pressure.

Pressure Imbalance

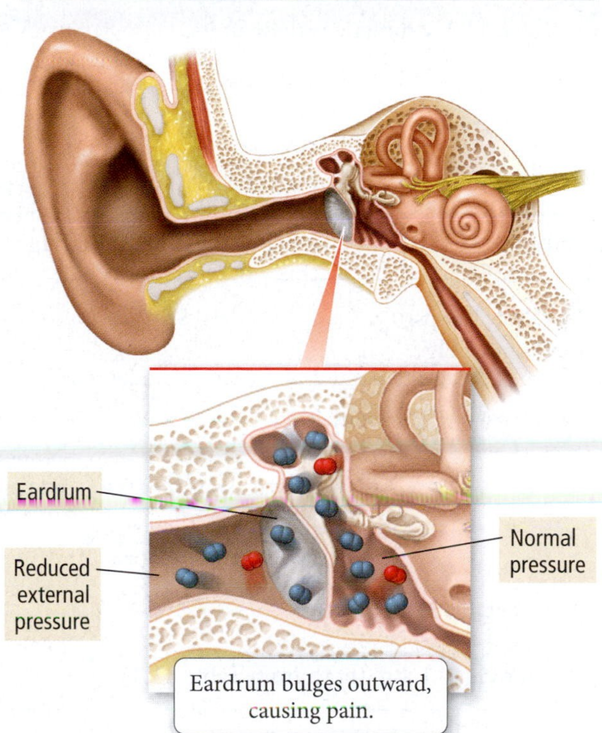

▲ **FIGURE 11.4 Pressure Imbalance** The discomfort you may feel in your ears upon ascending a mountain is caused by a pressure imbalance between the cavities in your ears and the outside air.

Pressure Units

We measure pressure in several different units. A common unit of pressure, the **millimeter of mercury (mmHg)**, originates from how pressure is measured with a **barometer** (Figure 11.5 ▶). A barometer is an evacuated glass tube, the tip of which is submerged in a pool of mercury. Atmospheric pressure on the liquid's surface forces the liquid mercury upward into the evacuated tube. Because mercury is so dense (13.5 times more dense than water), atmospheric pressure can support a column of Hg that is only about 0.760 m or 760 mm (about 30 in) tall. (By contrast, atmospheric pressure can support a column of water that is about 10.3 m tall.) This makes a column of mercury a convenient way to measure pressure.

In a barometer, when the atmospheric pressure rises, the height of the mercury column rises as well. Similarly, when atmospheric pressure falls, the height of the column falls. The unit *millimeter of mercury* is often called a **torr**, after the Italian physicist Evangelista Torricelli (1608–1647) who invented the barometer.

$$1 \text{ mmHg} = 1 \text{ torr}$$

The Mercury Barometer

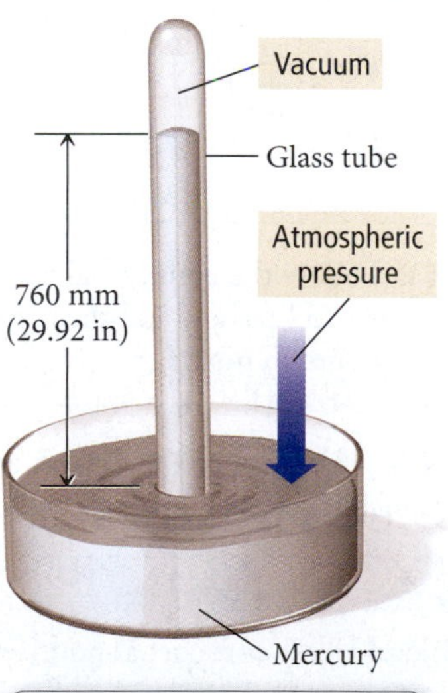

▲ **FIGURE 11.5 The Mercury Barometer** Average atmospheric pressure at sea level can support a column of mercury 760 mm in height.

A second unit of pressure is the **atmosphere (atm)**, the average pressure at sea level. One atmosphere of pressure pushes a column of mercury to a height of 760 mm; 1 atm and 760 mmHg are equal.

$$1 \text{ atm} = 760 \text{ mmHg}$$

A fully inflated mountain bike tire has a pressure of about 6 atm, and the pressure at the top of Mount Everest is about 0.31 atm.

The SI unit of pressure is the **pascal (Pa)**, defined as 1 newton (N) per square meter.

$$1 \text{ Pa} = 1 \text{ N/m}^2$$

The pascal is a much smaller unit of pressure than the atmosphere.

$$1 \text{ atm} = 101{,}325 \text{ Pa}$$

Other common units of pressure include the bar, inches of mercury (in Hg), and pounds per square inch (psi).

$$1 \text{ atm} = 1.013 \text{ bar} \qquad 1 \text{ atm} = 29.92 \text{ in Hg} \qquad 1 \text{ atm} = 14.7 \text{ psi}$$

Table 11.1 summarizes these units.

TABLE 11.1 Common Units of Pressure

Unit	Abbreviation	Average Air Pressure at Sea Level
Pascal (1 N/m^2)	Pa	101,325 Pa
Pounds per square inch	psi	14.7 psi
Torr (1 mmHg)	torr	760 torr (exact)
Inches of mercury	in Hg	29.92 in Hg
Bar	bar	1.013 atm
Atmosphere	atm	1.00 atm

EXAMPLE 11.1 Converting between Pressure Units

A cyclist inflates her high-performance road bicycle tire to a total pressure of 132 psi. What is this pressure in mmHg?

SORT The problem gives a pressure in psi and asks you to convert the units to mmHg.

GIVEN: 132 psi
FIND: mmHg

STRATEGIZE Table 11.1 does not have a direct conversion factor between psi and mmHg, but it does provide relationships between both of these units and atmospheres, so you can convert to atm as an intermediate step.

CONCEPTUAL PLAN

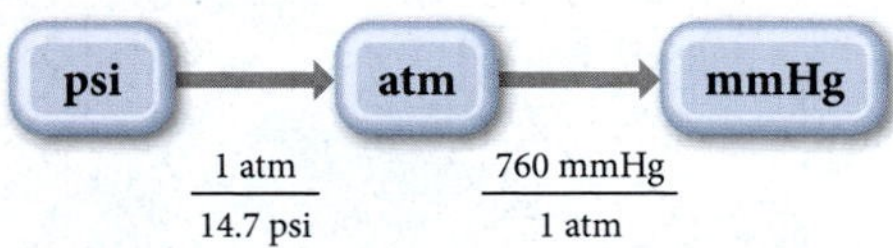

RELATIONSHIPS USED
1 atm = 14.7 psi
760 mmHg = 1 atm (both from Table 11.1)

SOLVE Follow the conceptual plan to solve the problem. Begin with 132 psi and use the conversion factors to arrive at the pressure in mmHg.

SOLUTION

$$132 \text{ psi} \times \frac{1 \text{ atm}}{14.7 \text{ psi}} \times \frac{760 \text{ mmHg}}{1 \text{ atm}} = 6.82 \times 10^3 \text{ mmHg}$$

CHECK The units of the answer are correct. The magnitude of the answer, 6.82×10^3 mmHg, is greater than the given pressure in psi. This is reasonable because mmHg is a much smaller unit than psi.

FOR PRACTICE 11.1
Your local weather report announces that the barometric pressure is 30.44 in Hg. Convert this pressure to psi.

FOR MORE PRACTICE 11.1
Convert a pressure of 23.8 in Hg to kPa.

The Manometer: A Way to Measure Pressure in the Laboratory

We can measure the pressure of a gas sample in the laboratory with a **manometer**. A manometer is a U-shaped tube containing a dense liquid, usually mercury. In a manometer such as the one shown in

Figure 11.6 ▶, one end of the tube is open to atmospheric pressure and the other is attached to a flask containing a gas sample. If the pressure of the gas sample is exactly equal to atmospheric pressure, then the mercury levels on both sides of the tube are the same. If the pressure of the sample is *greater than* atmospheric pressure, the mercury level on the left side of the tube is *higher than* the level on the right. If the pressure of the sample is *less than* atmospheric pressure, the mercury level on the left side is *lower than* the level on the right. This type of manometer always measures the pressure of the gas sample relative to atmospheric pressure. The difference in height between the two levels is equal to the difference between the sample's pressure and atmospheric pressure. To accurately calculate the absolute pressure of the sample, we also need a barometer to measure atmospheric pressure (which can vary from day to day).

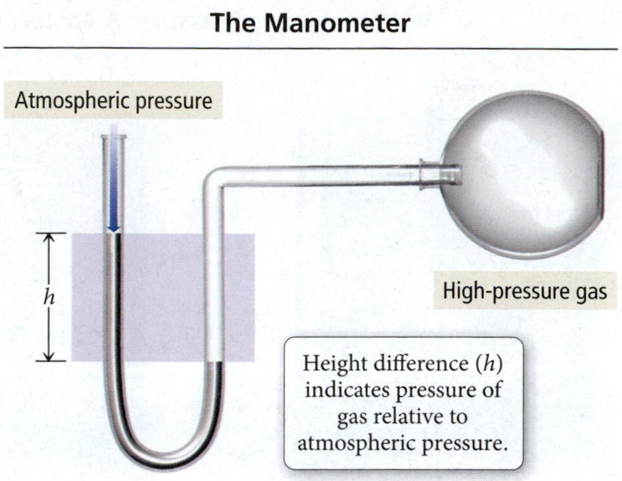

The Manometer

Atmospheric pressure

h

High-pressure gas

Height difference (*h*) indicates pressure of gas relative to atmospheric pressure.

▲ **FIGURE 11.6 The Manometer** A manometer measures the pressure exerted by a sample of gas.

11.3 The Simple Gas Laws: Boyle's Law, Charles's Law, and Avogadro's Law

In this section, we broaden our discussion of gases to include the four basic properties of a gas sample: pressure (P), volume (V), temperature (T), and amount in moles (n). These properties are interrelated—when one changes, it affects the others. The *simple gas laws* describe the relationships between pairs of these properties. For example, one simple gas law describes how *volume* varies with *pressure* at constant temperature and amount of gas; another law describes how volume varies with *temperature* at constant pressure and amount of gas.

Boyle's Law: Volume and Pressure

In the early 1660s, the pioneering English scientist Robert Boyle (1627–1691) and his assistant Robert Hooke (1635–1703) used a J-tube (Figure 11.7 ▼) to measure the volume of a sample of gas at different pressures. They trapped a sample of air in the J-tube and added mercury to increase the pressure on the gas. Boyle and Hook observed an *inverse relationship* between volume and pressure—an increase in one causes a decrease in the other—as shown in Figure 11.8 ▶. This relationship is now known as **Boyle's law**.

$$\text{Boyle's law:} \quad V \propto \frac{1}{P} \quad (\text{constant } T \text{ and } n)$$

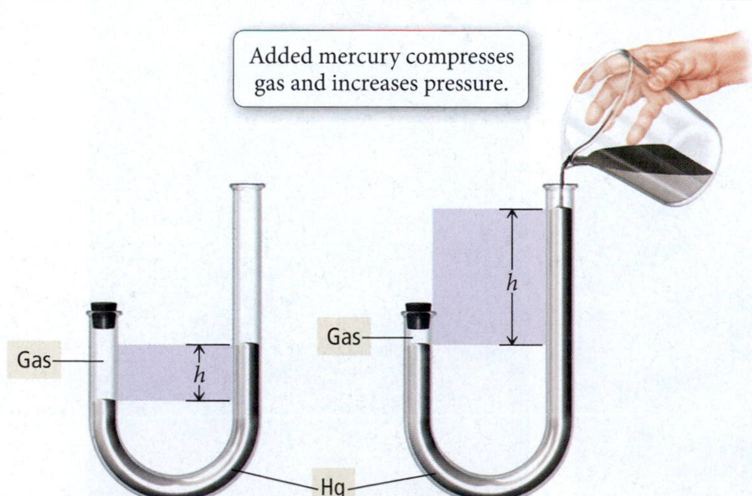

▲ **FIGURE 11.7 The J-Tube** In a J-tube, a column of mercury traps a sample of gas. We can increase the pressure on the gas by increasing the height (*h*) of mercury in the column.

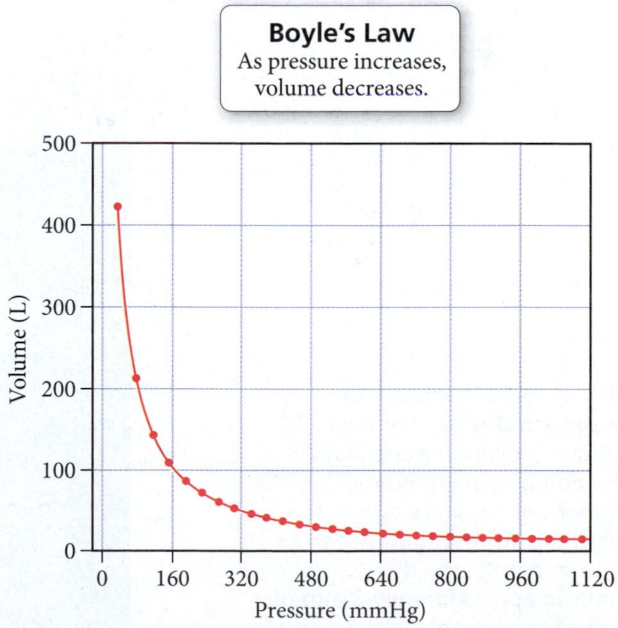

Boyle's Law
As pressure increases, volume decreases.

Volume (L): 0, 100, 200, 300, 400, 500
Pressure (mmHg): 0, 160, 320, 480, 640, 800, 960, 1120

▲ **FIGURE 11.8 Volume versus Pressure** A plot of the volume of a gas sample—as measured in a J-tube—versus pressure. The plot shows that volume and pressure are inversely related.

Volume versus Pressure: A Molecular View

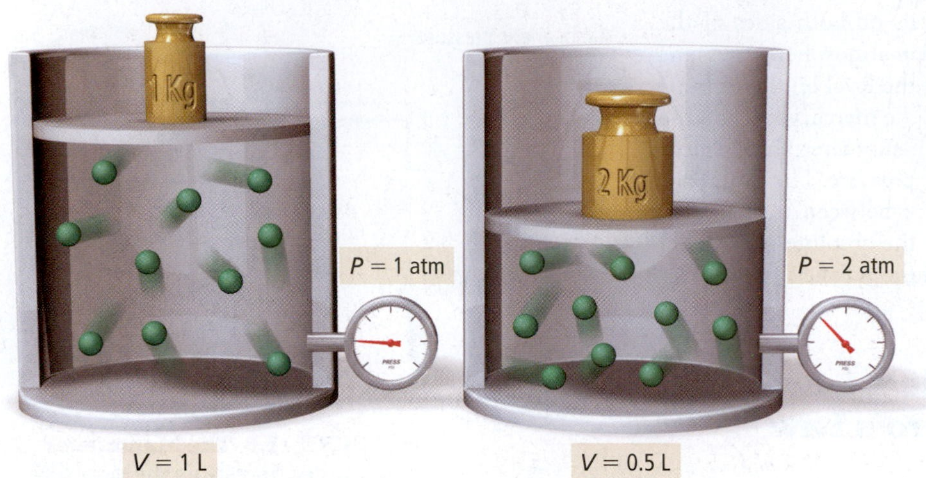

▲ **FIGURE 11.9 Molecular Interpretation of Boyle's Law** As the volume of a gas sample decreases, gas molecules collide with surrounding surfaces more frequently, resulting in greater pressure.

Boyle's law follows from the idea that pressure results from the collisions of the gas particles with the walls of their container (even though this was not known in Boyle's time). If the volume of a gas sample is decreased, the same number of gas particles is crowded into a smaller volume, resulting in more collisions with the walls and therefore an increase in the pressure (Figure 11.9 ◀).

Scuba divers learn about Boyle's law during certification because it explains why a diver should never ascend toward the surface without continuous breathing. For every 10 m of depth that a diver descends in water, she experiences an additional 1 atm of pressure due to the weight of the water above her (Figure 11.10 ▼). The pressure regulator used in scuba diving delivers air into the diver's lungs at a pressure that matches the external pressure; otherwise the diver could not inhale the air. For example, when a diver is 20 m below the surface, the regulator delivers air at a pressure of 3 atm to match the 3 atm of pressure around the diver (1 atm due to normal atmospheric pressure and 2 additional atmospheres due to the weight of the water at 20 m).

Suppose that a diver inhaled a lungful of air at a pressure of 3 atm and swam quickly to the surface (where the pressure is 1 atm) while holding her breath. What would happen to the volume of air in her lungs? The pressure decreases by a factor of 3 so that the volume of the air in her lungs increases by a factor of 3—a dangerous situation similar to that of the uncontrolled decompression we discussed in Section 11.1. For the scuba diver, the volume increase would prevent her from holding her breath all the way to the surface—the air would force itself out of her mouth, but probably not before the expanded air damaged her lungs, possibly killing her. Consequently, the most important rule in diving is *never hold your breath*. To avoid such catastrophic results, divers must ascend slowly and breathe continuously, allowing the regulator to bring the air pressure in their lungs back to 1 atm by the time they reach the surface.

Boyle's law assumes constant temperature and constant amount of gas.

We can use Boyle's law to calculate the volume of a gas following a pressure change or the pressure of a gas following a volume change *as long as the temperature and the amount of gas remain constant*. For these types of calculations, we write Boyle's law in a slightly different way:

If two quantities are proportional, then one is equal to the other multiplied by a constant.

$$\text{Since } V \propto \frac{1}{P}, \quad \text{then} \quad V = (\text{constant}) \times \frac{1}{P} \quad \text{or} \quad V = \frac{(\text{constant})}{P}$$

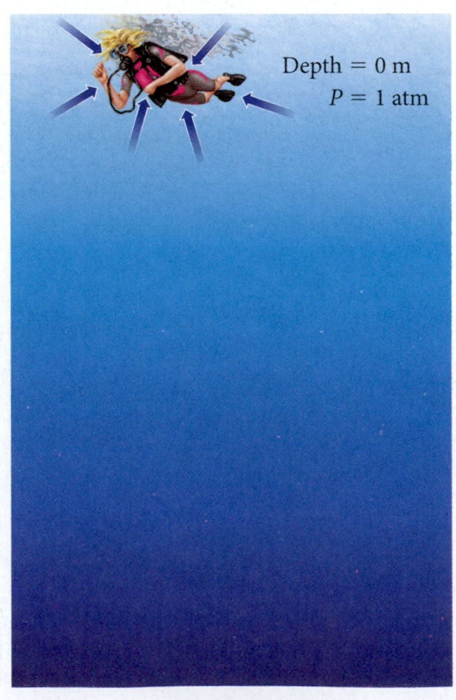

▶ **FIGURE 11.10 Increase in Pressure with Depth** For every 10 m of depth, a diver experiences approximately one additional atmosphere of pressure due to the weight of the surrounding water. At 20 m, for example, the diver experiences approximately 3 atm of pressure (1 atm of normal atmospheric pressure plus an additional 2 atm due to the weight of the water).

If we multiply both sides by P, we get:

$$PV = \text{constant}$$

This relationship indicates that if the pressure increases, the volume decreases, but the product $P \times V$ always equals the same constant. For two different sets of conditions, we can say that:

$$P_1 V_1 = \text{constant} = P_2 V_2$$

or

$$P_1 V_1 = P_2 V_2 \quad [11.2]$$

where P_1 and V_1 are the initial pressure and volume of the gas and P_2 and V_2 are the final volume and pressure.

EXAMPLE 11.2

Boyle's Law

As you breathe, you inhale by increasing your lung volume. A woman has an initial lung volume of 2.75 L, which is filled with air at an atmospheric pressure of 1.02 atm. If she increases her lung volume to 3.25 L without inhaling any additional air, what is the pressure in her lungs?

To solve the problem, first solve Boyle's law (Equation 11.2) for P_2 and then substitute the given quantities to calculate P_2.

SOLUTION

$$P_1 V_1 = P_2 V_2$$

$$P_2 = \frac{V_1}{V_2} P_1 = \frac{2.75\ \cancel{\text{L}}}{3.25\ \cancel{\text{L}}} 1.02\ \text{atm} = 0.863\ \text{atm}$$

FOR PRACTICE 11.2

A snorkeler takes a syringe filled with 16 mL of air from the surface, where the pressure is 1.0 atm, to an unknown depth. The volume of the air in the syringe at this depth is 7.5 mL. What is the pressure at this depth? If the pressure increases by 1 atm for every additional 10 m of depth, how deep is the snorkeler?

Charles's Law: Volume and Temperature

Suppose we keep the pressure of a gas sample constant and measure its volume at a number of different temperatures. Figure 11.11 ▶ shows the results of several such measurements. From the plot we can see a relationship between volume and temperature: The volume of a gas increases with increasing temperature. Closer examination of the plot reveals more—volume and temperature are *linearly related*. If two variables are linearly related, plotting one against the other produces a straight line.

Another interesting feature emerges if we extend or *extrapolate* the line in the plot in Figure 11.11 backwards from the lowest measured temperature. The dotted extrapolated line shows that the gas should have a zero volume at −273.15 °C. A temperature of −273.15 °C corresponds to 0 K (zero on the Kelvin scale), the coldest possible temperature (see Appendix I). The extrapolated

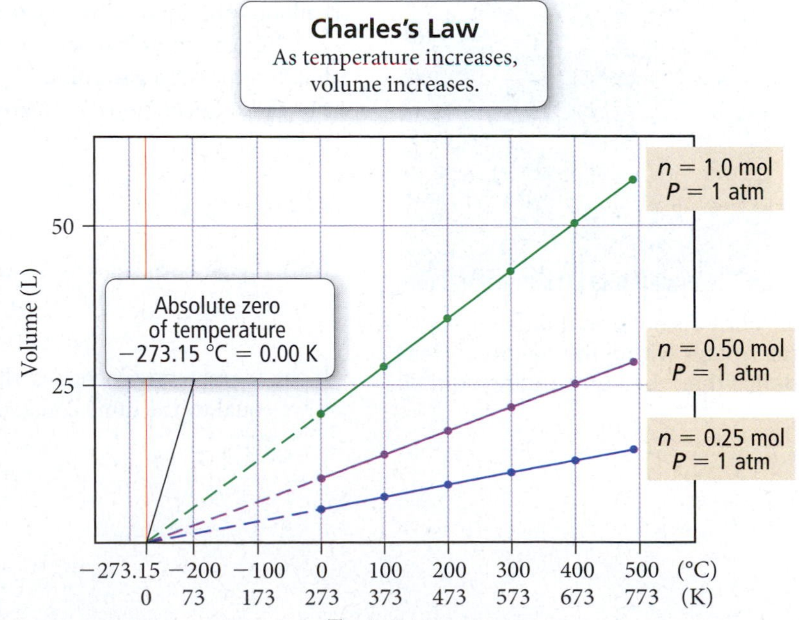

▶ **FIGURE 11.11 Volume versus Temperature** The volume of a fixed amount of gas at a constant pressure increases linearly with increasing temperature in kelvins. (The dotted extrapolated lines cannot be measured experimentally because all gases condense into liquids before −273.15 °C is reached.)

Volume versus Temperature: A Molecular View

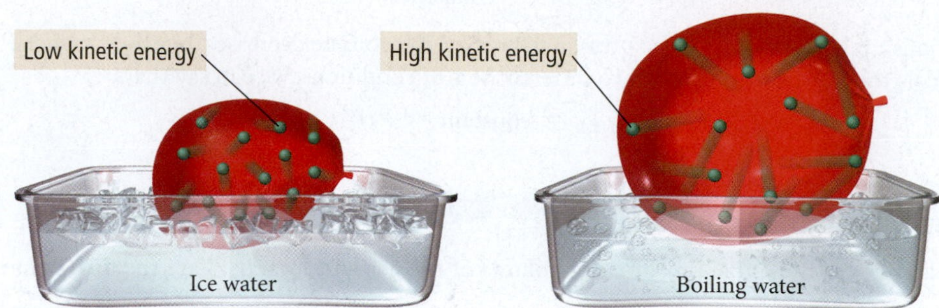

▲ **FIGURE 11.12 Particulate Interpretation of Charles's Law** If we move a balloon from an ice water bath to a boiling water bath, its volume expands as the gas particles within the balloon move faster (due to the increased temperature) and collectively occupy more space.

line indicates that below −273.15 °C, the gas would have a negative volume, which is physically impossible. For this reason, we refer to 0 K as *absolute zero*—colder temperatures do not exist.

The first person to carefully quantify the relationship between the volume of a gas and its temperature was J. A. C. Charles (1746–1823), a French mathematician and physicist. Charles was interested in gases and was among the first people to ascend in a hydrogen-filled balloon. The direct proportionality between volume and temperature is named **Charles's law** after him.

Charles's law assumes constant pressure and constant amount of gas.

$$\text{Charles's law: } V \propto T \qquad (\text{constant } P \text{ and } n)$$

When the temperature of a gas sample increases, the gas particles move faster; collisions with the walls are more frequent, and the force exerted with each collision is greater. The only way for the pressure (the force per unit area) to remain constant is for the gas to occupy a larger volume, so that collisions become less frequent and occur over a larger area (Figure 11.12 ▲).

Charles's law explains why the second floor of a house is usually warmer than the ground floor. According to Charles's law, when air is heated, its volume increases, resulting in a lower density. The warm, less dense air tends to rise in a room filled with colder, denser air. Similarly, Charles's law explains why a hot-air balloon can take flight. The gas that fills a hot-air balloon is warmed with a burner, increasing its volume and lowering its density, and causing it to float in the colder, denser surrounding air.

▲ A hot-air balloon floats because the hot air within the balloon is less dense than the surrounding cold air.

We can experience Charles's law directly by holding a partially inflated balloon over a warm toaster. As the air in the balloon warms, we can feel the balloon expanding. Alternatively, we can put an inflated balloon into liquid nitrogen and watch it become smaller as it cools.

We can use Charles's law to calculate the volume of a gas following a temperature change or the temperature of a gas following a volume change *as long as the pressure and the amount of gas are constant.* For these calculations, we rearrange Charles's law as follows:

$$\text{Since } V \propto T, \quad \text{then } V = \text{constant} \times T$$

If we divide both sides by T, we get:

$$V/T = \text{constant}$$

If the temperature increases, the volume increases in direct proportion so that the quotient, V/T, is always equal to the same constant. So, for two different measurements, we can say that:

$$V_1/T_1 = \text{constant} = V_2/T_2$$

or

$$\frac{V_1}{T_1} = \frac{V_2}{T_2} \qquad [11.3]$$

where V_1 and T_1 are the initial volume and temperature of the gas and V_2 and T_2 are the final volume and temperature. *We must always express the temperatures in kelvins (K)*, because, as shown in Figure 11.11, the volume of a gas is directly proportional to its absolute temperature, not its temperature in °C. For example, doubling the temperature of a gas sample from 1 °C to 2 °C does not double its volume, but doubling the temperature from 200 K to 400 K does.

EXAMPLE 11.3

Charles's Law

A sample of gas has a volume of 2.80 L at an unknown temperature. When you submerge the sample in ice water at $T = 0.00$ °C, its volume decreases to 2.57 L. What was its initial temperature (in K and in °C)?

	SOLUTION
To solve the problem, first solve Charles's law for T_1.	$\frac{V_1}{T_1} = \frac{V_2}{T_2}$ $T_1 = \frac{V_1}{V_2} T_2$
Before you substitute the numerical values to calculate T_1, convert the temperature to kelvins (K). *Remember, you must always work gas law problems with Kelvin temperatures.*	$T_2(\text{K}) = 0.00 + 273.15 = 273.15$ K
Substitute T_2 and the other given quantities to calculate T_1.	$T_1 = \frac{V_1}{V_2} T_2$ $= \frac{2.80 \cancel{\text{L}}}{2.57 \cancel{\text{L}}} 273.15 \text{ K}$ $= 29\underline{7}.6\text{K}$
Calculate T_1 in °C by subtracting 273 from the value in kelvins.	$T_1(°\text{C}) = 29\underline{7}.6 - 273.15 = 24$ °C

FOR PRACTICE 11.3

A gas in a cylinder with a moveable piston has an initial volume of 88.2 mL. If you heat the gas from 35 °C to 155 °C, what is its final volume (in mL)?

11.1 Cc Conceptual Connection

Boyle's Law and Charles's Law

The pressure exerted on a sample of a fixed amount of gas is doubled at constant temperature, and then the temperature of the gas in kelvins is doubled at constant pressure. What is the final volume of the gas?

(a) The final volume of the gas is twice the initial volume.

(b) The final volume of the gas is four times the initial volume.

(c) The final volume of the gas is one-half the initial volume.

(d) The final volume of the gas is one-fourth the initial volume.

(e) The final volume of the gas is the same as the initial volume.

Avogadro's law assumes constant temperature and constant pressure and is independent of the nature of the gas.

Avogadro's Law: Volume and Amount (in Moles)

So far, we have discussed the relationships between volume and pressure, and volume and temperature, but we have considered only a constant amount of a gas. What happens when the amount of gas changes? The volume of a gas sample (at constant temperature and pressure) as a function of the amount of gas (in moles) in the sample is shown in Figure 11.13 ◀. We can see that the relationship between volume and amount is linear. As we might expect, extrapolation to zero moles shows zero volume. This relationship, first stated formally by Amadeo Avogadro (1776–1856), is **Avogadro's law**:

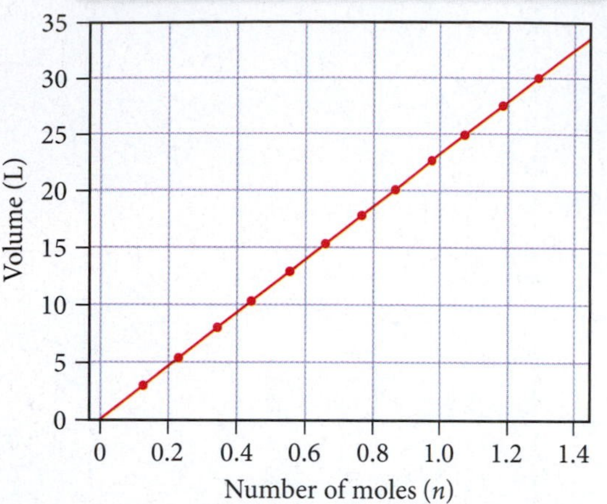

▲ **FIGURE 11.13 Volume versus Number of Moles** The volume of a gas sample increases linearly with the number of moles of gas in the sample.

$$\text{Avogadro's law:} \quad V \propto n \quad (\text{constant } T \text{ and } P)$$

When the amount of gas in a sample increases at constant temperature and pressure, its volume increases in direct proportion because the greater number of gas particles fills more space.

We experience Avogadro's law when we inflate a balloon. With each exhaled breath, we add more gas particles to the inside of the balloon, increasing its volume. We can use Avogadro's law to calculate the volume of a gas following a change in the amount of the gas *as long as the pressure and temperature of the gas are constant*. For these types of calculations, we express Avogadro's law as:

$$\frac{V_1}{n_1} = \frac{V_2}{n_2} \qquad [11.4]$$

where V_1 and n_1 are the initial volume and number of moles of the gas and V_2 and n_2 are the final volume and number of moles. In calculations, we use Avogadro's law in a manner similar to the other gas laws, as Example 11.4 demonstrates.

EXAMPLE 11.4 Avogadro's Law

A male athlete in a kinesiology research study has a lung volume of 6.15 L during a deep inhalation. At this volume, his lungs contain 0.254 moles of air. During exhalation, his lung volume decreases to 2.55 L. How many moles of gas does the athlete exhale during exhalation? Assume constant temperature and pressure.

SOLUTION

To solve the problem, first solve Avogadro's law for the number of moles of gas left in the athlete's lungs after exhalation, n_2.

$$\frac{V_1}{n_1} = \frac{V_2}{n_2}$$

$$n_2 = \frac{V_2}{V_1} n_1$$

Then substitute the given quantities to calculate n_2.

$$= \frac{2.55 \cancel{L}}{6.15 \cancel{L}} 0.254 \text{ mol}$$

$$= 0.105 \text{ mol}$$

Because the lungs initially contained 0.254 mol of air, you calculate the amount of air exhaled by subtracting the result from 0.254 mol.

$$\text{moles exhaled} = 0.254 \text{ mol} - 0.105 \text{ mol} = 0.149 \text{ mol}$$

FOR PRACTICE 11.4

A chemical reaction occurring in a cylinder equipped with a moveable piston produces 0.621 mol of a gaseous product. If the cylinder contains 0.120 mol of gas before the reaction and has an initial volume of 2.18 L, what is its volume after the reaction? (Assume constant pressure and temperature and that the initial amount of gas completely reacts.)

11.4 The Ideal Gas Law

The relationships that we discussed in Section 11.3 can be combined into a single law that encompasses all of the simple gas laws. So far, we have shown that:

$$V \propto \frac{1}{P} \quad \text{(Boyle's law)}$$

$$V \propto T \quad \text{(Charles's law)}$$

$$V \propto n \quad \text{(Avogadro's law)}$$

Combining these three expressions, we get:

$$V \propto \frac{nT}{P}$$

The volume of a gas is directly proportional to the number of moles of gas and to the temperature of the gas, but is inversely proportional to the pressure of the gas. We can replace the proportionality sign with an equals sign by incorporating R, a proportionality constant called the *ideal gas constant*:

$$V = \frac{RnT}{P}$$

Rearranging, we get:

$$PV = nRT \quad [11.5]$$

As we shall see in Section 11.11, the particles that compose an ideal gas have two properties: (1) negligible intermolecular forces and (2) low densities (the space between the particles is large compared to the size of the particles themselves).

This equation is the **ideal gas law**, and a hypothetical gas that exactly follows this law is an **ideal gas**. The value of R, the **ideal gas constant**, is the same for all gases and has the value:

$$R = 0.08206 \frac{\text{L} \cdot \text{atm}}{\text{mol} \cdot \text{K}}$$

L = liters
atm = atmospheres
mol = moles
K = kelvins

The ideal gas law contains within it the simple gas laws that we have discussed. For example, recall that Boyle's law states that $V \propto 1/P$ when the amount of gas (n) and the temperature of the gas (T) are kept constant. We can rearrange the ideal gas law as follows:

The ideal gas law contains the simple gas laws within it.

$$PV = nRT$$

First, we divide both sides by P.

$$V = \frac{nRT}{P}$$

Then we put the variables that are constant, along with R, in parentheses.

$$V = (nRT)\frac{1}{P}$$

Because n and T are constant in this case, and R is always a constant, we can write:

$$V \propto (\text{constant}) \times \frac{1}{P}$$

which means that $V \propto 1/P$.

The ideal gas law also shows how other pairs of variables are related. For example, from Charles's law we know that $V \propto T$ at constant pressure and constant number of moles. But what if we heat a sample of gas at constant *volume* and constant number of moles? This question applies to the warning labels on aerosol cans such as hair spray or deodorants. These labels warn against excessive heating or incineration of the cans, even after the contents are used up. Why? An "empty" aerosol can is not really empty but contains a fixed amount of gas trapped in a fixed volume. What would happen if we were to

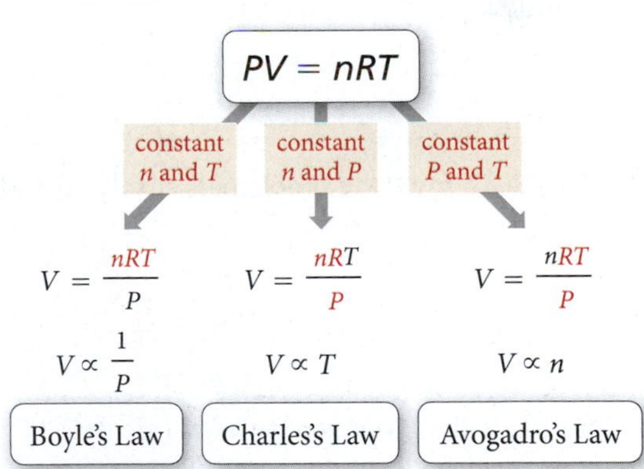

▲ The ideal gas law contains the simple gas laws within it.

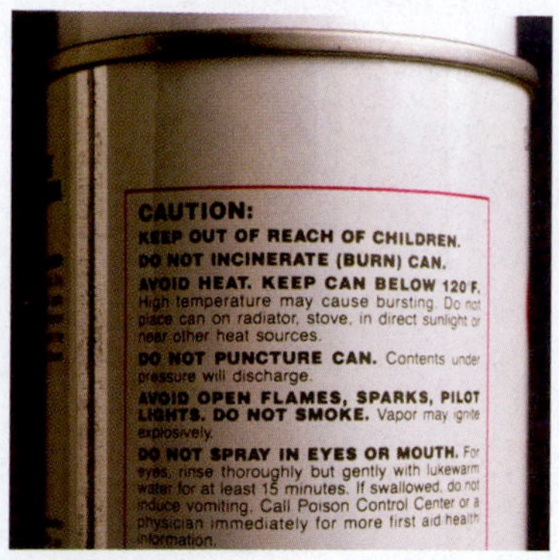

▲ The labels on most aerosol cans warn against incineration. Because the volume of the can is constant, an increase in temperature causes an increase in pressure and possibly an explosion.

heat the can? We can rearrange the ideal gas law to clearly see the relationship between pressure and temperature at constant volume and constant number of moles:

$$PV = nRT$$

$$P = \frac{nRT}{V} = \left(\frac{nR}{V}\right)T$$

We divide both sides by V.

Because n and V are constant and R is always a constant:

$$P = (\text{constant}) \times T$$

This relationship between pressure and temperature is known as *Gay-Lussac's law*. As the temperature of a fixed amount of gas in a fixed volume increases, the pressure increases. In an aerosol can, this pressure increase can blow the can apart, which is why aerosol cans should not be heated or incinerated. They might explode.

We can use the ideal gas law to determine the value of any one of the four variables (P, V, n, or T) given the other three. To do so, we must express each of the quantities in the ideal gas law in the units within R:

- pressure (P) in atm, volume (V) in L, moles (n) in mol, temperature (T) in K

EXAMPLE 11.5
Ideal Gas Law I

Calculate the volume occupied by 0.845 mol of nitrogen gas at a pressure of 1.37 atm and a temperature of 315 K.

SORT The problem gives you the number of moles of nitrogen gas, the pressure, and the temperature. You are asked to find the volume.

GIVEN: $n = 0.845\text{ mol}$, $P = 1.37\text{ atm}$, $T = 315\text{ K}$

FIND: V

STRATEGIZE You are given three of the four variables (P, T, and n) in the ideal gas law and asked to find the fourth (V). The conceptual plan shows how the ideal gas law provides the relationship between the known quantities and the unknown quantity.

CONCEPTUAL PLAN

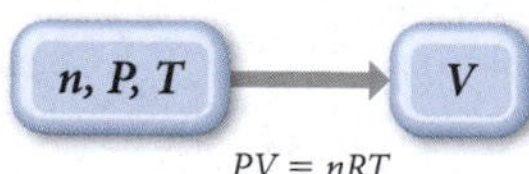

RELATIONSHIP USED

$PV = nRT$ (ideal gas law)

SOLVE To solve the problem, first solve the ideal gas law for V.

SOLUTION

$$PV = nRT$$

$$V = \frac{nRT}{P}$$

Substitute the given quantities to calculate V.

$$V = \frac{0.845\text{ mol} \times 0.08206\frac{\text{L}\cdot\text{atm}}{\text{mol}\cdot\text{K}} \times 315\text{ K}}{1.37\text{ atm}}$$

$$= 15.9\text{ L}$$

CHECK The units of the answer are correct. The magnitude of the answer (15.9 L) makes sense because, as you will see in the next section, one mole of an ideal gas under standard temperature and pressure (273 K and 1 atm) occupies 22.4 L. Although this is not standard temperature and pressure, the conditions are close enough for a ballpark check of the answer. This gas sample contains 0.845 mol, so a volume of 15.9 L is reasonable.

FOR PRACTICE 11.5

An 8.50-L tire contains 0.552 mol of gas at a temperature of 305 K. What is the pressure (in atm and psi) of the gas in the tire?

EXAMPLE 11.6
Ideal Gas Law II

Calculate the number of moles of gas in a 3.24-L basketball inflated to a *total pressure* of 24.3 psi at 25 °C. (*Note:* The *total pressure* is not the same as the pressure read on the type of pressure gauge used for checking a car or bicycle tire. That pressure, called the *gauge pressure*, is the *difference* between the total pressure and atmospheric pressure. In this case, if atmospheric pressure is 14.7 psi, the gauge pressure would be 9.6 psi. However, for calculations involving the ideal gas law, you must use the *total pressure* of 24.3 psi.)

SORT The problem gives you the pressure, the volume, and the temperature. You need to find the number of moles of gas.

GIVEN: $P = 24.3\text{ psi}, V = 3.24\text{ L}, T(°\text{C}) = 25\,°\text{C}$

FIND: n

STRATEGIZE The conceptual plan shows how the ideal gas law provides the relationship between the given quantities and the quantity to be found.

CONCEPTUAL PLAN

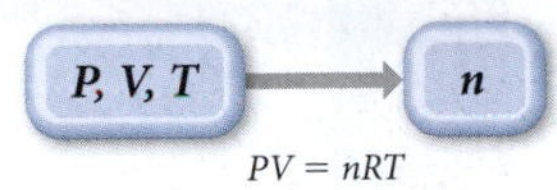

$PV = nRT$

RELATIONSHIP USED

$PV = nRT$ (ideal gas law)

SOLVE To solve the problem, first solve the ideal gas law for n.

SOLUTION

$$PV = nRT$$

$$n = \frac{PV}{RT}$$

Before substituting into the equation, convert P and T into the correct units.

$$P = 24.3\ \cancel{\text{psi}} \times \frac{1\text{ atm}}{14.7\ \cancel{\text{psi}}} = 1.65\underline{3}1\text{ atm}$$

(Because rounding the intermediate answer would result in a slightly different final answer, we mark the least significant digit in the intermediate answer, but don't round until the end.)

$$T(\text{K}) = 25 + 273 = 298\text{ K}$$

Substitute into the equation and calculate n.

$$n = \frac{1.65\underline{3}1\ \cancel{\text{atm}} \times 3.24\ \cancel{\text{L}}}{0.08206\ \frac{\cancel{\text{L}}\cdot\cancel{\text{atm}}}{\text{mol}\cdot\cancel{\text{K}}} \times 298\ \cancel{\text{K}}} = 0.219\text{ mol}$$

CHECK The units of the answer are correct. The magnitude of the answer (0.219 mol) makes sense because, as you will see in the next section, one mole of an ideal gas under standard temperature and pressure (273 K and 1 atm) occupies 22.4 L. At a pressure that is 65% higher than standard pressure, the volume of 1 mol of gas would be proportionally lower. This gas sample occupies 3.24 L, so the answer of 0.219 mol is reasonable.

FOR PRACTICE 11.6

What volume does 0.556 mol of gas occupy at a pressure of 715 mmHg and a temperature of 58 °C?

FOR MORE PRACTICE 11.6

Determine the pressure in mmHg of a 0.133-g sample of helium gas in a 648-mL container at a temperature of 32 °C.

11.5 Applications of the Ideal Gas Law: Molar Volume, Density, and Molar Mass of a Gas

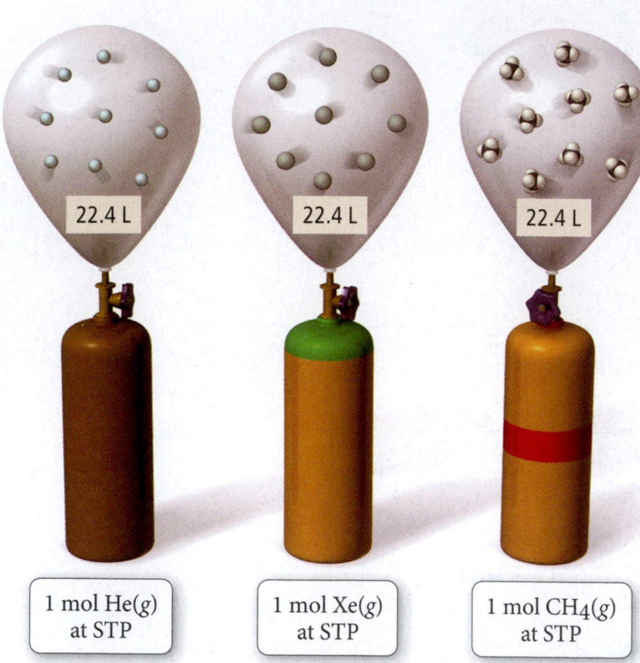

▲ One mole of any gas occupies approximately 22.4 L at standard temperature (273 K) and pressure (1.0 atm).

We just examined how we can use the ideal gas law to calculate one of the variables (P, V, T, or n) given the other three. We now turn to three other applications of the ideal gas law: molar volume, density, and molar mass.

Molar Volume at Standard Temperature and Pressure

The volume occupied by one mole of a substance is its **molar volume**. For gases, we often specify the molar volume under conditions known as **standard temperature** ($T = 0$ °C or 273 K) **and pressure** ($P = 1.00$ atm), abbreviated as **STP**. Using the ideal gas law, we can determine that the molar volume of an ideal gas at STP is:

$$V = \frac{nRT}{P}$$

$$= \frac{1.00\ \cancel{\text{mol}} \times 0.08206 \dfrac{\text{L} \cdot \cancel{\text{atm}}}{\cancel{\text{mol}} \cdot \cancel{\text{K}}} \times 273\ \cancel{\text{K}}}{1.00\ \cancel{\text{atm}}}$$

$$= 22.4\ \text{L}$$

The molar volume of an ideal gas at STP is useful because—as we saw in the *Check* steps of Examples 11.5 and 11.6—it gives us a way to approximate the volume of an ideal gas under conditions that are close to STP.

11.2 Cc Conceptual Connection — Molar Volume

Assuming ideal behavior, which of these gas samples has the greatest volume at STP?

(a) 1 g of H_2 **(b)** 1 g of O_2 **(c)** 1 g of Ar

Density of a Gas

Because one mole of an ideal gas occupies 22.4 L under standard temperature and pressure, we can readily calculate the density of an ideal gas under these conditions. Density is mass/volume, and because the mass of one mole of a gas is simply its molar mass, the *density of a gas* is:

$$\text{Density} = \frac{\text{molar mass}}{\text{molar volume}}$$

We can calculate the density of a gas at STP by using 22.4 L as the molar volume. For example, the densities of helium and nitrogen gas at STP are:

$$d_{\text{He}} = \frac{4.00\ \text{g/mol}}{22.4\ \text{L/mol}} = 0.179\ \text{g/L} \quad d_{\text{N}_2} = \frac{28.02\ \text{g/mol}}{22.4\ \text{L/mol}} = 1.25\ \text{g/L}$$

Notice that *the density of a gas is directly proportional to its molar mass.* The greater the molar mass of a gas, the more dense the gas. For this reason, a gas with a molar mass lower than that of air tends to rise in air. For example, both helium and hydrogen gas (molar masses of 4.00 and 2.01 g/mol, respectively) have molar masses that are lower than the average molar mass of air (approximately 28.8 g/mol). Therefore a balloon filled with either helium or hydrogen gas floats in air.

The primary components of air are nitrogen (about four-fifths) and oxygen (about one-fifth). We discuss the detailed composition of air in Section 11.6.

We can calculate the density of a gas more generally (under any conditions) by using the ideal gas law. To do so, we arrange the ideal gas law as:

$$PV = nRT$$

$$\frac{n}{V} = \frac{P}{RT}$$

Because the left-hand side of this equation has units of moles/liter, it represents the *molar* density. We can obtain the density in grams/liter from the molar density by multiplying by the molar mass ($\mathcal{M}$).

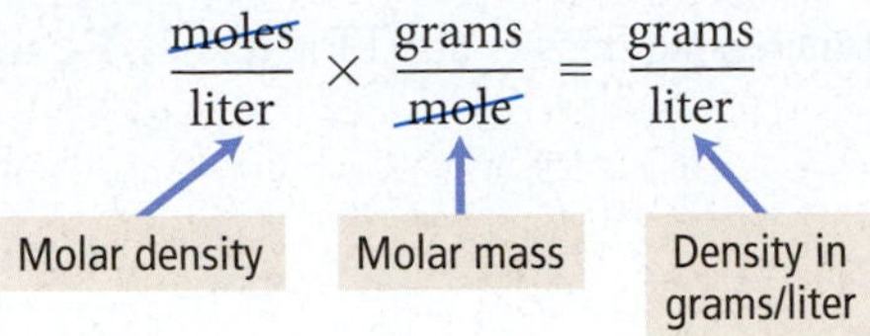

Density

$\frac{n}{V} \mathcal{M} = \frac{P\mathcal{M}}{RT}$

Molar density Molar mass

$d = \frac{P\mathcal{M}}{RT}$

Therefore,

$$d = \frac{P\mathcal{M}}{RT} \qquad [11.6]$$

Notice that, as expected, density increases with increasing molar mass. Notice also that as we discussed in Section 11.3, density decreases with increasing temperature.

EXAMPLE 11.7
Density of a Gas

Calculate the density of nitrogen gas at 125 °C and a pressure of 755 mmHg.

SORT The problem gives you the temperature and pressure of a gas and asks you to find its density. The problem states that the gas is nitrogen.

GIVEN: $T(°C) = 125\ °C, P = 755$ mmHg
FIND: d

STRATEGIZE Equation 11.6 provides the relationship between the density of a gas and its temperature, pressure, and molar mass. The temperature and pressure are given. You can calculate the molar mass from the formula of the gas, N_2.

CONCEPTUAL PLAN

$P, T, \mathcal{M} \rightarrow d$

$d = \frac{P\mathcal{M}}{RT}$

RELATIONSHIPS USED

$d = \frac{P\mathcal{M}}{RT}$ (density of a gas)

Molar mass N_2 = 28.02 g/mol

SOLVE To solve the problem, gather each of the required quantities in the correct units. Convert the temperature to kelvins and the pressure to atmospheres.

SOLUTION

$T(K) = 125 + 273 = 398$ K

$P = 755\ \text{mmHg} \times \frac{1\ \text{atm}}{760\ \text{mmHg}} = 0.99\underline{3}42$ atm

Substitute the quantities into the equation to calculate density.

$$d = \frac{P\mathcal{M}}{RT} = \frac{0.99\underline{3}42\ \text{atm}\left(28.02\frac{\text{g}}{\text{mol}}\right)}{0.08206\frac{\text{L}\cdot\text{atm}}{\text{mol}\cdot\text{K}}(398\ \text{K})} = 0.852\ \text{g/L}$$

CHECK The units of the answer are correct. The magnitude of the answer (0.852 g/L) makes sense because earlier you calculated the density of nitrogen gas at STP as 1.25 g/L. The temperature is higher than standard temperature, so it follows that the density is lower.

FOR PRACTICE 11.7
Calculate the density of xenon gas at a pressure of 742 mmHg and a temperature of 45 °C.

FOR MORE PRACTICE 11.7
A gas has a density of 1.43 g/L at a temperature of 23 °C and a pressure of 0.789 atm. Calculate its molar mass.

11.3 Cc Conceptual Connection

Density of a Gas

Arrange these gases in order of increasing density at STP: Ne, Cl_2, F_2, and O_2.

Molar Mass of a Gas

We can use the ideal gas law in combination with mass measurements to calculate the molar mass of an unknown gas. First we measure the mass and volume of an unknown gas under conditions of known pressure and temperature. Then, we determine the amount of the gas in moles from the ideal gas law. Finally, we calculate the molar mass by dividing the mass (in grams) by the amount (in moles) as shown in Example 11.8.

EXAMPLE 11.8 Molar Mass of a Gas

A sample of gas has a mass of 0.311 g. Its volume is 0.225 L at a temperature of 55 °C and a pressure of 886 mmHg. Find its molar mass.

SORT The problem gives you the mass of a gas sample, along with its volume, temperature, and pressure. You are asked to find the molar mass.

GIVEN: $m = 0.311$ g, $V = 0.225$ L, $T(°C) = 55$ °C, $P = 886$ mmHg

FIND: molar mass (g/mol)

STRATEGIZE The conceptual plan has two parts. In the first part, use the ideal gas law to find the number of moles of gas.

CONCEPTUAL PLAN

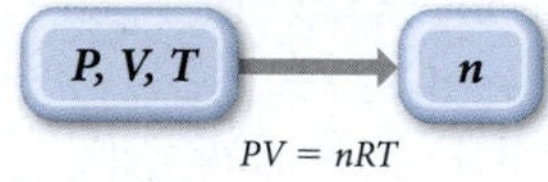

In the second part, use the definition of molar mass to find the molar mass.

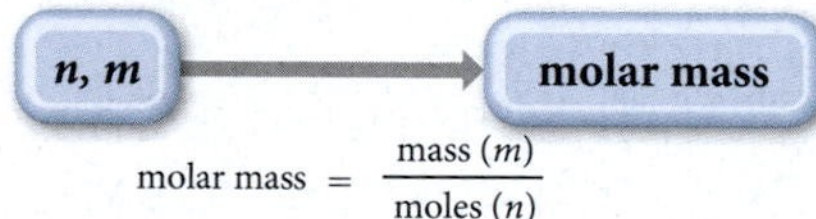

RELATIONSHIPS USED

$PV = nRT$

$$\text{Molar mass} = \frac{\text{mass } (m)}{\text{moles } (n)}$$

SOLVE To find the number of moles, first solve the ideal gas law for n.

SOLUTION

$$PV = nRT$$

$$n = \frac{PV}{RT}$$

Before substituting into the equation for n, convert the pressure to atm and the temperature to K.

$$P = 886 \text{ mmHg} \times \frac{1 \text{ atm}}{760 \text{ mmHg}} = 1.1\underline{6}58 \text{ atm}$$

$$T(\text{K}) = 55 + 273 = 328 \text{ K}$$

Substitute into the equation and calculate n, the number of moles.

$$n = \frac{1.1\underline{6}58 \text{ atm} \times 0.225 \text{ L}}{0.08206 \frac{\text{L} \cdot \text{atm}}{\text{mol} \cdot \text{K}} \times 328 \text{ K}} = 9.7\underline{4}54 \times 10^{-3} \text{ mol}$$

Finally, use the number of moles (n) and the given mass (m) to find the molar mass.

$$\text{molar mass} = \frac{\text{mass } (m)}{\text{moles } (n)}$$

$$= \frac{0.311 \text{ g}}{9.7\underline{4}54 \times 10^{-3} \text{ mol}} = 31.9 \text{ g/mol}$$

—Continued on the next page

Continued from the previous page—

CHECK The units of the answer are correct. The magnitude of the answer (31.9 g/mol) is a reasonable number for a molar mass. If your answer is some very small number (such as any number smaller than 1) or a very large number, you solved the problem incorrectly. Most gases have molar masses between one and several hundred grams per mole.

FOR PRACTICE 11.8

A sample of gas has a mass of 827 mg. Its volume is 0.270 L at a temperature of 88 °C and a pressure of 975 mmHg. Find its molar mass.

11.6 Mixtures of Gases and Partial Pressures

Many gas samples are not pure; they are mixtures of gases. Dry air, for example, is a mixture containing nitrogen, oxygen, argon, carbon dioxide, and a few other gases in trace amounts (Table 11.2).

Because the molecules in an ideal gas do not interact (as we discuss further in Section 11.7), each of the components in an ideal gas mixture acts independently of the others. For example, the nitrogen molecules in air exert a certain pressure—78% of the total pressure—that is independent of the other gases in the mixture. Likewise, the oxygen molecules in air exert a certain pressure—21% of the total pressure—that is also independent of the other gases in the mixture. The pressure due to any individual component in a gas mixture is its **partial pressure (P_n)**. We can calculate partial pressure from the ideal gas law by assuming that each gas component acts independently.

$$P_n = n_n \frac{RT}{V}$$

For a multicomponent gas mixture, we calculate the partial pressure of each component from the ideal gas law and the number of moles of that component (n_n) as follows:

$$P_a = n_a \frac{RT}{V}; \quad P_b = n_b \frac{RT}{V}; \quad P_c = n_c \frac{RT}{V}; \cdots \qquad [11.7]$$

The sum of the partial pressures of the components in a gas mixture equals the total pressure:

$$P_{total} = P_a + P_b + P_c + \cdots \qquad [11.8]$$

where P_{total} is the total pressure and $P_a, P_b, P_c, \ldots$, are the partial pressures of the components. This relationship is known as **Dalton's law of partial pressures**.

Combining Equations 11.7 and 11.8, we get:

$$\begin{aligned} P_{total} &= P_a + P_b + P_c + \cdots \\ &= n_a \frac{RT}{V} + n_b \frac{RT}{V} + n_c \frac{RT}{V} + \cdots \qquad [11.9] \\ &= (n_a + n_b + n_c + \cdots) \frac{RT}{V} \\ &= (n_{total}) \frac{RT}{V} \end{aligned}$$

The total number of moles in the mixture, when substituted into the ideal gas law, indicates the total pressure of the sample.

If we divide Equation 11.7 by Equation 11.9, we get:

$$\frac{P_a}{P_{total}} = \frac{n_a(\cancel{RT/V})}{n_{total}(\cancel{RT/V})} = \frac{n_a}{n_{total}} \qquad [11.10]$$

TABLE 11.2 Composition of Dry Air

Gas	Percent by Volume (%)
Nitrogen (N_2)	78
Oxygen (O_2)	21
Argon (Ar)	0.9
Carbon dioxide (CO_2)	0.04

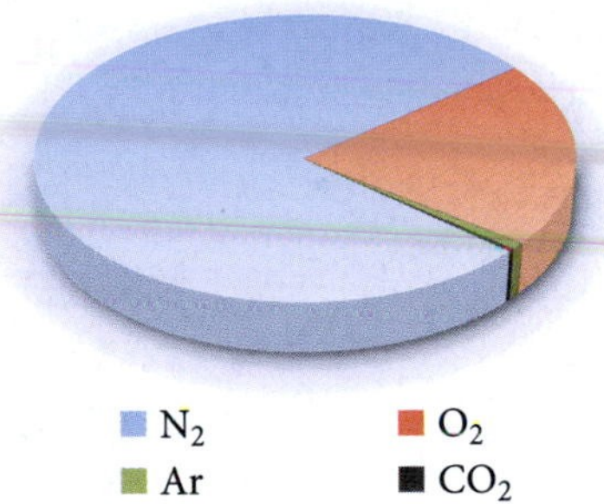

The quantity n_a/n_{total}, the number of moles of a component in a mixture divided by the total number of moles in the mixture, is the **mole fraction (χ_a)**.

$$\chi_a = \frac{n_a}{n_{total}} \qquad [11.11]$$

Rearranging Equation 11.10 and substituting the definition of mole fraction (Equation 11.11) gives us:

$$\frac{P_a}{P_{total}} = \frac{n_a}{n_{total}}$$

$$P_a = \frac{n_a}{n_{total}} P_{total} = \chi_a P_{total}$$

or more simply:

$$P_a = \chi_a P_{total} \qquad [11.12]$$

The partial pressure of a component in a gaseous mixture is its mole fraction multiplied by the total pressure. For gases, the mole fraction of a component is equivalent to its percent by volume divided by 100%. Therefore, based on Table 11.2, we calculate the partial pressure of nitrogen (P_{N_2}) in air at 1.00 atm to be:

$$P_{N_2} = 0.78 \times 1.00 \text{ atm}$$
$$= 0.78 \text{ atm}$$

Likewise, the partial pressure of oxygen in air at 1.00 atm is 0.21 atm, and the partial pressure of argon in air is 0.01 atm. Applying Dalton's law of partial pressures to air at 1.00 atm:

For these purposes, we ignore the contribution of the CO_2 and other trace gases because they are so small.

$$P_{total} = P_{N_2} + P_{O_2} + P_{Ar}$$
$$P_{total} = 0.78 \text{ atm} + 0.21 \text{ atm} + 0.01 \text{ atm}$$
$$= 1.00 \text{ atm}$$

11.4 Cc Conceptual Connection

Partial Pressures

A gas mixture contains an equal number of moles of He and Ne. The total pressure of the mixture is 3.0 atm. What are the partial pressures of He and Ne?

EXAMPLE 11.9

Total Pressure and Partial Pressures

A 1.00-L mixture of helium, neon, and argon has a total pressure of 662 mmHg at 298 K. If the partial pressure of helium is 341 mmHg and the partial pressure of neon is 112 mmHg, what mass of argon is present in the mixture?

SORT The problem gives you partial pressures for two of the three components in a gas mixture, along with the total pressure, the volume, and the temperature, and asks you to find the mass of the third component.	**GIVEN:** $P_{He} = 341$ mmHg, $P_{Ne} = 112$ mmHg, $P_{total} = 662$ mmHg, $V = 1.00$ L, $T = 298$ K **FIND:** m_{Ar}

—Continued on the next page

Continued from the previous page—

STRATEGIZE You can find the mass of argon from the number of moles of argon, which you can calculate from the partial pressure of argon and the ideal gas law. Begin by using Dalton's law to determine the partial pressure of argon.

Then use the partial pressure of argon together with the volume of the sample and the temperature to find the number of moles of argon.

Finally, use the molar mass of argon to calculate the mass of argon from the number of moles of argon.

CONCEPTUAL PLAN

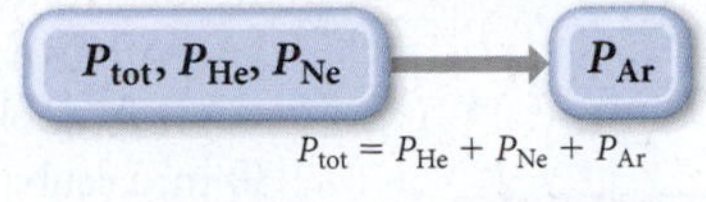

$P_{tot} = P_{He} + P_{Ne} + P_{Ar}$

$P_{Ar}, V, T \rightarrow n_{Ar}$

$PV = nRT$

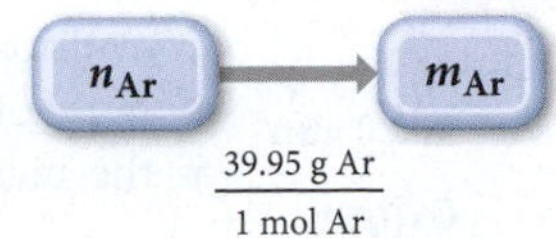

$\dfrac{39.95 \text{ g Ar}}{1 \text{ mol Ar}}$

RELATIONSHIPS USED

$P_{total} = P_{He} + P_{Ne} + P_{Ar}$ (Dalton's law)

$PV = nRT$ (ideal gas law)

molar mass Ar = 39.95 g/mol

SOLVE Follow the conceptual plan. To find the partial pressure of argon, solve the equation for P_{Ar} and substitute the values of the other partial pressures to calculate P_{Ar}.

SOLUTION

$$P_{total} = P_{He} + P_{Ne} + P_{Ar}$$

$$\begin{aligned} P_{Ar} &= P_{total} - P_{He} - P_{Ne} \\ &= 662 \text{ mmHg} - 341 \text{ mmHg} - 112 \text{ mmHg} \\ &= 209 \text{ mmHg} \end{aligned}$$

Convert the partial pressure from mmHg to atm and use it in the ideal gas law to calculate the amount of argon in moles.

$$209 \cancel{\text{mmHg}} \times \frac{1 \text{ atm}}{760 \cancel{\text{mmHg}}} = 0.275 \text{ atm}$$

$$n = \frac{PV}{RT} = \frac{0.275 \cancel{\text{atm}}(1.00 \cancel{\text{L}})}{0.08206 \dfrac{\cancel{\text{L}} \cdot \cancel{\text{atm}}}{\text{mol} \cdot \cancel{\text{K}}}(298 \cancel{\text{K}})} = 1.1\underline{2}5 \times 10^{-2} \text{ mol Ar}$$

Use the molar mass of argon to convert from amount of argon in moles to mass of argon.

$$1.1\underline{2}5 \times 10^{-2} \cancel{\text{mol Ar}} \times \frac{39.95 \text{ g Ar}}{1 \cancel{\text{mol Ar}}} = 0.449 \text{ g Ar}$$

CHECK The units of the answer are correct. The magnitude of the answer makes sense because the volume is 1.0 L, which at STP would contain about 1/22 mol. Because the partial pressure of argon in the mixture is about 1/3 of the total pressure, you can roughly estimate about 1/66 of one molar mass of argon, which is fairly close to your answer.

FOR PRACTICE 11.9

A sample of hydrogen gas is mixed with water vapor. The mixture has a total pressure of 755 torr, and the water vapor has a partial pressure of 24 torr. What amount (in moles) of hydrogen gas is contained in 1.55 L of this mixture at 298 K?

Deep-Sea Diving and Partial Pressures

Our lungs have evolved to breathe oxygen at a partial pressure of $P_{O_2} = 0.21$ atm. If the total air pressure decreases—when a person climbs a mountain, for example—the partial pressure of oxygen also decreases. On top of Mount Everest, where the total pressure is 0.311 atm, the partial pressure of oxygen is only 0.065 atm. Low oxygen levels produce a physiological condition called **hypoxia** or oxygen

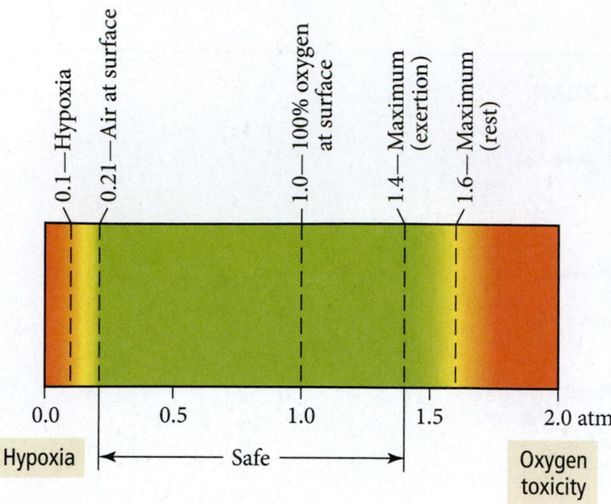

▲ **FIGURE 11.14 Oxygen Partial Pressure Limits** The partial pressure of oxygen in air at sea level is 0.21 atm. Partial pressures of oxygen below 0.1 atm and above 1.4 atm are dangerous to humans.

starvation (Figure 11.14 ◀). Mild hypoxia causes dizziness, headache, and shortness of breath. Severe hypoxia, which occurs when (P_{O_2}) drops below 0.1 atm, may result in unconsciousness or even death. For this reason, climbers hoping to make the summit of Mount Everest, usually carry oxygen to breathe.

While not as dangerous as a lack of oxygen, too much oxygen can also cause physiological problems. Recall from Section 11.3 that scuba divers breathe pressurized air. At 30 m, a scuba diver breathes air at a total pressure of 4.0 atm, which means P_{O_2} is about 0.84 atm. This elevated partial pressure of oxygen raises the density of oxygen molecules in the lungs, resulting in a higher concentration of oxygen in body tissues. When P_{O_2} increases beyond 1.4 atm, the increased oxygen concentration in body tissues causes a condition called **oxygen toxicity** that results in muscle twitching, tunnel vision, and convulsions. Divers who venture too deep without proper precautions have drowned because of oxygen toxicity. A second problem associated with breathing pressurized air is the increase of nitrogen in the lungs. At 30 m, a scuba diver breathes nitrogen at $P_{N_2} = 3.12$ atm, which increases the nitrogen concentration in body tissues and fluids. When P_{N_2} increases beyond about 4 atm, a condition called **nitrogen narcosis** or *rapture of the deep* results. Divers describe the effects of this condition as similar to being inebriated or drunk. A diver breathing compressed air at 60 m feels as if he has consumed too much wine.

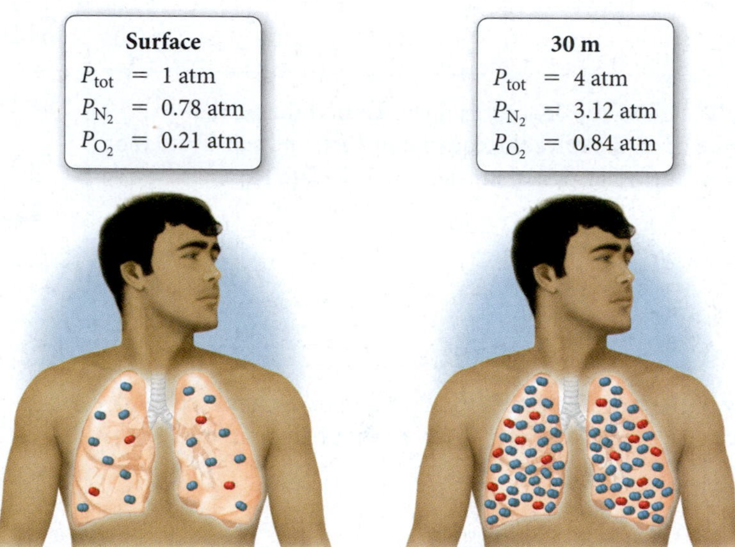

▶ When a diver breathes compressed air, the abnormally high partial pressure of oxygen in the lungs leads to an elevated concentration of oxygen in body tissues.

To avoid oxygen toxicity and nitrogen narcosis, deep-sea divers—those who descend beyond 50 m—breathe specialized mixtures of gases. One common mixture is heliox, a mixture of helium and oxygen. These mixtures usually contain a smaller percentage of oxygen than is typically found in air, thereby lowering the risk of oxygen toxicity. Heliox also contains helium instead of nitrogen, eliminating the risk of nitrogen narcosis.

EXAMPLE 11.10

Partial Pressures and Mole Fractions

A 12.5-L scuba diving tank contains a helium-oxygen (heliox) mixture of 24.2 g of He and 4.32 g of O_2 at 298 K. Calculate the mole fraction and partial pressure of each component in the mixture and the total pressure of the mixture.

SORT The problem gives the masses of two gases in a mixture and the volume and temperature of the mixture. You are to find the mole fraction and partial pressure of each component, as well as the total pressure.	**GIVEN:** $m_{He} = 24.2$ g, $m_{O_2} = 4.32$ g, $V = 12.5$ L, $T = 298$ K **FIND:** $\chi_{He}, \chi_{O_2}, P_{He}, P_{O_2}, P_{total}$

—Continued on the next page

Continued from the previous page—

STRATEGIZE The conceptual plan has several parts. To calculate the mole fraction of each component, you must first find the number of moles of each component. In the first part of the conceptual plan, convert the masses to moles using the molar masses.

CONCEPTUAL PLAN

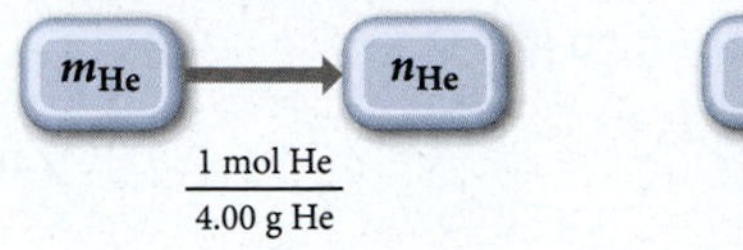

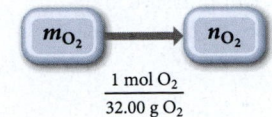

$\frac{1 \text{ mol He}}{4.00 \text{ g He}}$ $\quad$ $\frac{1 \text{ mol } O_2}{32.00 \text{ g } O_2}$

In the second part, calculate the mole fraction of each component using the mole fraction definition (Equation 11.11).

$$\chi_{He} = \frac{n_{He}}{n_{He} + n_{O_2}}; \chi_{O_2} = \frac{n_{O_2}}{n_{He} + n_{O_2}}$$

To calculate *partial pressures* calculate the *total pressure* and then use the mole fractions from the previous calculation to determine the partial pressures. Calculate the total pressure from the sum of the moles of both components. (Alternatively, you can calculate the partial pressures of the components individually, using the number of moles of each component and adding them to obtain the total pressure.)

$$P_{total} = \frac{(n_{He} + n_{O_2})RT}{V}$$

$$P_{He} = \chi_{He}P_{total}; \quad P_{O_2} = \chi_{O_2}P_{total}$$

Use the mole fractions of each component and the total pressure to calculate the partial pressure of each component.

RELATIONSHIPS USED

$\chi_a = n_a/n_{total}$ (mole fraction definition)

$P_{total}V = n_{total}RT$ (ideal gas law)

$P_a = \chi_a P_{total}$

SOLVE Follow the plan to solve the problem. Begin by converting each of the masses to amounts in moles.

SOLUTION

$$24.2 \text{ g He} \times \frac{1 \text{ mol He}}{4.00 \text{ g He}} = 6.05 \text{ mol He}$$

$$4.32 \text{ g } O_2 \times \frac{1 \text{ mol } O_2}{32.00 \text{ g } O_2} = 0.135 \text{ mol } O_2$$

Calculate each of the mole fractions.

$$\chi_{He} = \frac{n_{He}}{n_{He} + n_{O_2}} = \frac{6.05}{6.05 + 0.135} = 0.97\underline{8}17$$

$$\chi_{O_2} = \frac{n_{O_2}}{n_{He} + n_{O_2}} = \frac{0.135}{6.05 + 0.135} = 0.021\underline{8}27$$

Calculate the total pressure.

$$P_{total} = \frac{(n_{He} + n_{O_2})RT}{V}$$

$$= \frac{(6.05 \text{ mol} + 0.135 \text{ mol})\left(0.08206 \frac{\text{L} \cdot \text{atm}}{\text{mol} \cdot \text{K}}\right)(298 \text{ K})}{12.5 \text{ L}}$$

$$= 12.\underline{0}99 \text{ atm}$$

Finally, calculate the partial pressure of each component.

$$P_{He} = \chi_{He}P_{total} = 0.97\underline{8}17 \times 12.\underline{0}99 \text{ atm}$$

$$= 11.8 \text{ atm}$$

$$P_{O_2} = \chi_{O_2}P_{total} = 0.021\underline{8}27 \times 12.\underline{0}99 \text{ atm}$$

$$= 0.264 \text{ atm}$$

CHECK The units of the answers are correct, and the magnitudes are reasonable.

FOR PRACTICE 11.10

A diver breathes a heliox mixture with an oxygen mole fraction of 0.050. What must the total pressure be for the partial pressure of oxygen to be 0.21 atm?

Collecting Gases over Water

We will discuss vapor pressure in detail in Chapter 12.

When the desired product of a chemical reaction is a gas, we can collect the gas by the displacement of water. For example, suppose we use the reaction of zinc with hydrochloric acid as a source of hydrogen gas.

$$Zn(s) + 2HCl(aq) \longrightarrow ZnCl_2(aq) + H_2(g)$$

Appendix IV includes a more complete table of the vapor pressure of water versus temperature.

To collect the gas, we can set up an apparatus like the one shown in Figure 11.15 ▼. As the hydrogen gas forms, it bubbles through the water and gathers in the collection flask. The hydrogen gas collected in this way is not pure. It is mixed with water vapor because some water molecules evaporate and mix with the hydrogen molecules.

The partial pressure of water in the mixture, which we call its **vapor pressure**, depends on temperature (Table 11.3). Vapor pressure increases with increasing temperature because higher temperatures cause more water molecules to evaporate.

TABLE 11.3 Vapor Pressure of Water versus Temperature

Temperature (°C)	Pressure (mmHg)	Temperature (°C)	Pressure (mmHg)
0	4.58	55	118.2
5	6.54	60	149.6
10	9.21	65	187.5
15	12.79	70	233.7
20	17.55	75	289.1
25	23.78	80	355.1
30	31.86	85	433.6
35	42.23	90	525.8
40	55.40	95	633.9
45	71.97	100	760.0
50	92.6		

Suppose we collect the hydrogen gas over water at a total pressure of 758.2 mmHg and a temperature of 25 °C. What is the partial pressure of the hydrogen gas? We know that the total pressure is 758.2 mmHg and that the partial pressure of water is 23.78 mmHg (its vapor pressure at 25 °C).

$$P_{\text{total}} = P_{H_2} + P_{H_2O}$$

$$758.2 \text{ mmHg} = P_{H_2} + 23.78 \text{ mmHg}$$

Therefore:

$$P_{H_2} = 758.2 \text{ mmHg} - 23.78 \text{ mmHg}$$
$$= 734.4 \text{ mmHg}$$

The partial pressure of the hydrogen in the mixture is 734.4 mmHg.

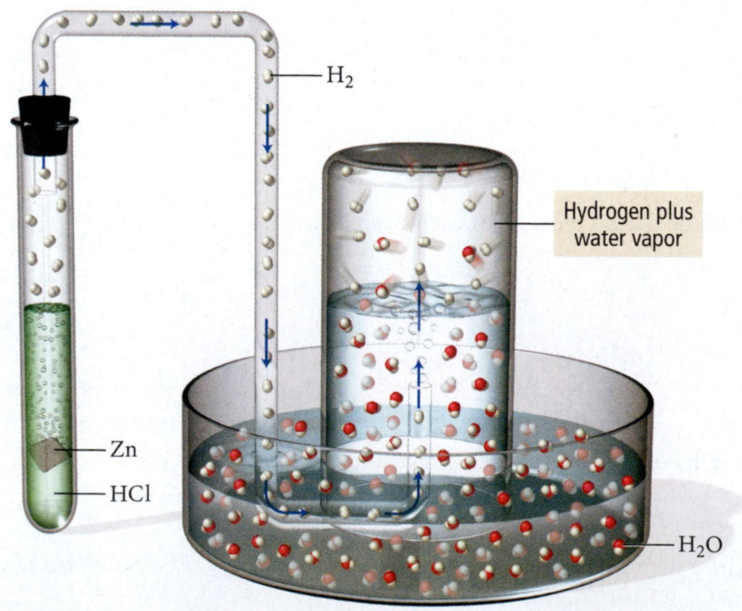

▶ **FIGURE 11.15 Collecting a Gas over Water** When we collect the gaseous product of a chemical reaction over water, the product molecules (in this case H_2) are mixed with water molecules. The pressure of water in the final mixture is equal to the vapor pressure of water at the temperature at which the gas is collected. The partial pressure of the product is the total pressure minus the partial pressure of water.

EXAMPLE 11.11 Collecting Gases over Water

In order to determine the rate of photosynthesis, the oxygen gas emitted by an aquatic plant is collected over water at a temperature of 293 K and a total pressure of 755.2 mmHg. Over a specific time period, a total of 1.02 L of gas is collected. What mass of oxygen gas (in grams) is formed?

SORT The problem gives the volume of gas collected over water as well as the temperature and the pressure. You are asked to find the mass in grams of oxygen that forms.

GIVEN: $V = 1.02$ L, $P_{total} = 755.2$ mmHg, $T = 293$ K

FIND: g O_2

STRATEGIZE You can find the mass of oxygen from moles of oxygen, which you can get from the ideal gas law and the partial pressure of oxygen. Find the partial pressure of oxygen by subtracting the partial pressure of water at 293 K (20 °C) from the total pressure.

CONCEPTUAL PLAN

$$P_{O_2} = P_{total} - P_{H_2O}\ (20\ °C)$$

Use the ideal gas law to determine the number of moles of oxygen from its partial pressure, volume, and temperature.

$P_{O_2}, V, T \rightarrow n_{O_2}$

$P_{O_2}V = n_{O_2}RT$

Then use the molar mass of oxygen to convert the number of moles to grams.

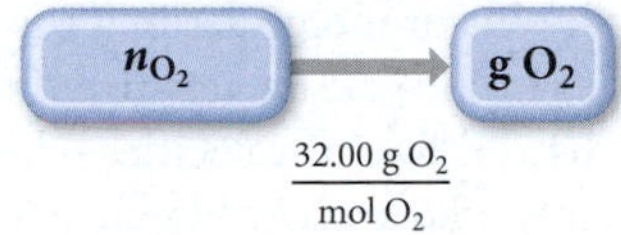

RELATIONSHIPS USED

$$P_{total} = P_a + P_b + P_c + \cdots \text{ (Dalton's law)}$$

$$PV = nRT \text{ (ideal gas law)}$$

SOLVE Follow the conceptual plan to solve the problem. Begin by calculating the partial pressure of oxygen in the oxygen/water mixture. You can find the partial pressure of water at 20 °C in Table 11.3.

SOLUTION

$$P_{O_2} = P_{total} - P_{H_2O}\ (20\ °C)$$
$$= 755.2 \text{ mmHg} - 17.55 \text{ mmHg} = 737.\underline{6}5 \text{ mmHg}$$

Next, solve the ideal gas law for number of moles.

$$n_{O_2} = \frac{P_{O_2}V}{RT}$$

Before substituting into the ideal gas law, convert the partial pressure of oxygen from mmHg to atm.

$$737.\underline{6}5 \text{ mmHg} \times \frac{1 \text{ atm}}{760 \text{ mmHg}} = 0.970\underline{5}9 \text{ atm}$$

Substitute into the ideal gas law to find the number of moles of oxygen.

$$n_{O_2} = \frac{P_{O_2}V}{RT} = \frac{0.970\underline{5}9 \text{ atm } (1.02 \text{ L})}{0.08206 \frac{\text{L} \cdot \text{atm}}{\text{mol} \cdot \text{K}} (293 \text{ K})} = 4.1\underline{1}75 \times 10^{-2} \text{ mol}$$

Finally, use the molar mass of oxygen to convert to grams of oxygen.

$$4.1\underline{1}75 \times 10^{-2} \text{ mol } O_2 \times \frac{32.00 \text{ g } O_2}{1 \text{ mol } O_2} = 1.32 \text{ g } O_2$$

CHECK The answer is in the correct units. You can quickly check the magnitude of the answer by using molar volume. Under STP one liter is about 1/22 of one mole. Therefore the answer should be about 1/22 the molar mass of oxygen ($1/22 \times 32 = 1.45$). The magnitude of the answer seems reasonable.

FOR PRACTICE 11.11

A common way to make hydrogen gas in the laboratory is to place a metal such as zinc in hydrochloric acid. The hydrochloric acid reacts with the metal to produce hydrogen gas, which is then collected over water. Suppose a student carries out this reaction and collects a total of 154.4 mL of gas at a pressure of 742 mmHg and a temperature of 25 °C. What mass of hydrogen gas (in mg) does the student collect?

Kinetic Molecular Theory

▲ **FIGURE 11.16 A Model for Gas Behavior** In the kinetic molecular theory of gases, a gas sample is modeled as a collection of particles in constant straight-line motion. The size of each particle is negligibly small, and their collisions are elastic.

KEY CONCEPT VIDEO
Kinetic Molecular Theory

11.7 A Particulate Model for Gases: Kinetic Molecular Theory

In Chapter 1, we discussed how the scientific approach proceeds from observations to laws and eventually to theories. Remember that laws summarize behavior—for example, Charles's law summarizes *how* the volume of a gas depends on temperature—while theories give the underlying reasons for the behavior. A theory of gas behavior explains, for example, *why* the volume of a gas increases with increasing temperature.

The simplest model for the behavior of gases is the **kinetic molecular theory**. In this theory, a gas is modeled as a collection of particles (either molecules or atoms, depending on the gas) in constant motion (Figure 11.16 ◀). A single particle moves in a straight line until it collides with another particle (or with the wall of the container). The basic postulates (or assumptions) of kinetic molecular theory are listed here.

1. **The size of a particle is negligibly small.** Kinetic molecular theory assumes that the particles themselves occupy no volume, even though they have mass. This postulate is justified because, under normal pressures, the space between atoms or molecules in a gas is very large compared to the size of the atoms or molecule themselves. For example, in a sample of argon gas at STP, atoms occupy only about 0.01% of the volume, and the average distance from one argon atom to another is 3.3 nm. In comparison, the atomic radius of argon is 97 pm (0.097 nm). If an argon atom were the size of a golf ball, its nearest neighbor would be, on average, just over 4 ft away at STP.
2. **The average kinetic energy of a particle is proportional to the temperature in kelvins.** The motion of atoms or molecules in a gas is due to thermal energy, which distributes itself among the particles in the gas. At any given moment, some particles are moving faster than others—there is a distribution of velocities—but the higher the temperature, the faster the overall motion, and the greater the average kinetic energy. Notice that *kinetic energy* ($\frac{1}{2}mv^2$)—not *velocity*—is proportional to temperature. The atoms in a sample of helium and a sample of argon at the same temperature have the same average *kinetic energy*, but not the same average *velocity*. Because the helium atoms are lighter, they must move faster to have the same kinetic energy as argon atoms.
3. **The collision of one particle with another (or with the walls of its container) is completely elastic.** This means that when two particles collide, they may *exchange energy*, but there is no overall *loss of energy*. Any kinetic energy lost by one particle is completely gained by the other. In other words, the particles have no "stickiness," and they are not deformed by the collision. An encounter between two particles in kinetic molecular theory is more like the collision between two billiard balls than the collision between two lumps of clay (Figure 11.17 ▼). Between collisions, the particles do not exert any forces on one another.

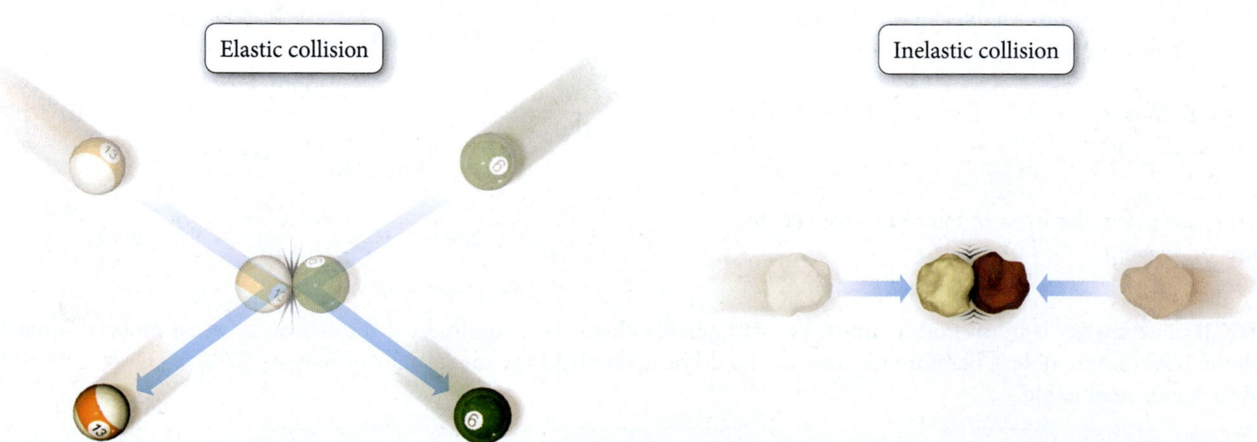

▲ **FIGURE 11.17 Elastic versus Inelastic Collisions** When two billiard balls collide, the collision is elastic—the total kinetic energy of the colliding bodies is the same before and after the collision. When two lumps of clay collide, the collision is inelastic—the kinetic energy of the colliding bodies dissipates in the form of heat during the collision.

Kinetic Molecular Theory, Pressure, and the Simple Gas Laws

If we start with the postulates of kinetic molecular theory, we can mathematically derive the ideal gas law (as we demonstrate later). In other words, the ideal gas law follows directly from kinetic molecular theory, which gives us confidence that the assumptions of the theory are valid, at least under conditions in which the ideal gas law applies. Let's see how the concept of pressure as well as each of the gas laws we have examined in this chapter follow conceptually from kinetic molecular theory.

The Nature of Pressure In Section 11.2, we define pressure as force divided by area:

$$P = \frac{F}{A}$$

According to kinetic molecular theory, a gas is a collection of particles in constant motion. The motion results in collisions between the particles and the surfaces around them. As each particle collides with a surface, it exerts a force upon that surface. Many particles in a gas sample exerting forces on the surfaces around them result in constant pressure.

The force (F) associated with an individual collision is given by $F = ma$, where m is the mass of the particle and a is its acceleration as it changes its direction of travel due to the collision.

Boyle's Law Boyle's law states that, for a constant number of particles at constant temperature, the volume of a gas is inversely proportional to its pressure. According to kinetic molecular theory, if you decrease the volume of a gas, you force the gas particles to occupy a smaller space. As long as the temperature remains the same, the number of collisions with the surrounding surfaces (per unit surface area) must necessarily increase, resulting in a greater pressure.

Charles's Law Charles's law states that, for a constant number of particles at constant pressure, the volume of a gas is proportional to its temperature. According to kinetic molecular theory, when we increase the temperature of a gas, the average kinetic energy, and therefore the average speed, of the particles increases. Because this greater kinetic energy results in more frequent collisions and more force per collision, the pressure of the gas increases if its volume is held constant (Gay-Lussac's law). The only way for the pressure to remain constant is for the volume to increase. The greater volume spreads the collisions out over a greater surface area, so that the pressure (defined as force per unit area) does not change.

Avogadro's Law Avogadro's law states that, at constant temperature and pressure, the volume of a gas is proportional to the number of particles. According to kinetic molecular theory, when we increase the number of particles in a gas sample, the number of collisions with the surrounding surfaces increases. The greater number of collisions results in a greater overall force on surrounding surfaces; the only way for the pressure to remain constant is for the volume to increase so that the number of particles per unit volume (and thus the number of collisions) remains constant.

Dalton's Law Dalton's law states that the total pressure of a gas mixture is the sum of the partial pressures of its components. In other words, according to Dalton's law, the components in a gas mixture act identically to, and independently of, one another. According to kinetic molecular theory, the particles have negligible size and they do not interact. Consequently, the only property that would distinguish one type of particle from another is its mass. However, even particles of different masses have the same average kinetic energy at a given temperature, so they exert the same force upon collision with a surface. Consequently, adding components to a gas mixture—even different *kinds* of gases—has the same effect as simply adding more particles. The partial pressures of all the components sum to the overall pressure.

11.5 Cc Conceptual Connection

Kinetic Molecular Theory I

Draw a depiction of a gas sample containing equal molar amounts of argon and xenon as described by kinetic molecular theory. Use red dots to represent argon atoms and blue dots to represent xenon atoms. Draw each atom with a "tail" that represents its velocity relative to the others in the mixture.

Calculating Gas Pressure: A Molecular View

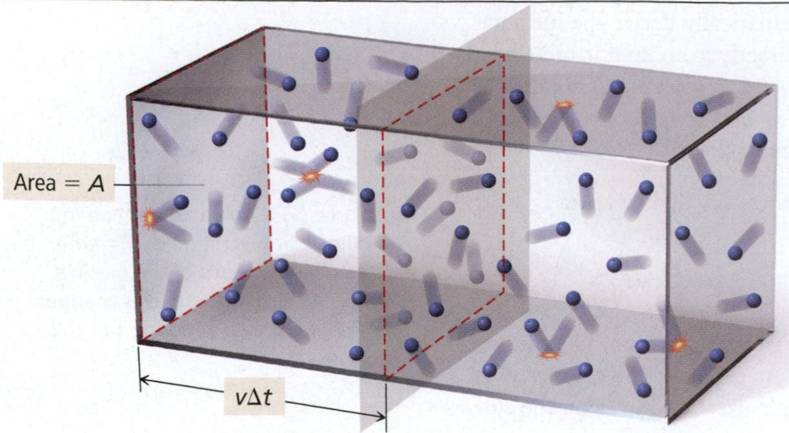

▲ **FIGURE 11.18 The Pressure on the Wall of a Container** We can calculate the pressure on the wall of a container by determining the total force due to collisions of the particles with the wall.

Kinetic Molecular Theory and the Ideal Gas Law

We have just seen how each of the simple gas laws conceptually follows from kinetic molecular theory. We can also *derive* the ideal gas law from the postulates of kinetic molecular theory. In other words, the kinetic molecular theory is a quantitative model that *implies* $PV = nRT$. Let's explore this derivation.

The pressure on a wall of a container (Figure 11.18 ◀) occupied by particles in constant motion is the total force on the wall (due to the collisions) divided by the area of the wall.

$$P = \frac{F_{total}}{A} \qquad [11.13]$$

According to Newton's second law, the force (F) associated with an individual collision is given by $F = ma$, where m is the mass of the particle and a is its acceleration as it changes its direction of travel due to the collision. The acceleration for each collision is the change in velocity (Δv) divided by the time interval (Δt), so the force imparted for each collision is:

$$F_{collision} = m\frac{\Delta v}{\Delta t} \qquad [11.14]$$

If a particle collides elastically with the wall, it bounces off the wall with no loss of energy. For a straight-line collision, the change in velocity is $2v$ (the particle's velocity was v before the collision and $-v$ after the collision; therefore, the change is $2v$). The force per collision is given by:

$$F_{collision} = m\frac{2v}{\Delta t} \qquad [11.15]$$

The total number of collisions in the time interval Δt on a wall of surface area A is proportional to the number of particles that can reach the wall in this time interval—in other words, all particles within a distance of $v\,\Delta t$ of the wall. These particles occupy a volume given by $v\,\Delta t \times A$, and their total number is equal to this volume multiplied by the density of particles in the container (n/V).

$$\text{Number of collisions} \propto \text{number of particles within } v\,\Delta t$$

$$\propto \underbrace{v\,\Delta t \times A}_{\text{Volume}} \times \underbrace{\frac{n}{V}}_{\text{Density of particles}} \qquad [11.16]$$

The *total force* on the wall is equal to the force per collision multiplied by the number of collisions.

$$F_{total} = F_{collision} \times \text{number of collisions}$$

$$\propto m\frac{2v}{\cancel{\Delta t}} \times v\cancel{\Delta t} \times A \times \frac{n}{V} \qquad [11.17]$$

$$\propto mv^2 \times A \times \frac{n}{V}$$

The pressure on the wall is equal to the total force divided by the surface area of the wall.

$$P = \frac{F_{total}}{A}$$

$$\propto \frac{mv^2 \times \cancel{A} \times \frac{n}{V}}{\cancel{A}} \qquad [11.18]$$

$$P \propto mv^2 \times \frac{n}{V}$$

Notice that Equation 11.18 contains within it Boyle's law ($P \propto 1/V$) and Avogadro's law ($V \propto n$). We can get the complete ideal gas law from postulate 2 of the kinetic molecular theory, which states that the average kinetic energy ($\frac{1}{2}mv^2$) is proportional to the temperature in kelvins (T).

$$mv^2 \propto T \quad [11.19]$$

Combining Equations 11.18 and 11.19, we get:

$$P \propto \frac{T \times n}{V} \quad [11.20]$$

$$PV \propto nT$$

The proportionality can be replaced by an equals sign if we provide the correct constant, R.

$$PV = nRT \quad [11.21]$$

In other words, the kinetic molecular theory (a model for how gases behave) predicts behavior that is consistent with our observations and measurements of gases—the theory agrees with the experiments. Recall from Chapter 1 that a scientific theory is the most powerful kind of scientific knowledge. In the kinetic molecular theory, we have a model for what a gas is like. Although the model is not perfect—indeed, it breaks down under certain conditions, as we shall see later in this chapter—it predicts a great deal about the behavior of gases. Therefore, the model is a good approximation of what a gas is actually like. A careful examination of the conditions under which the model breaks down (see Section 11.11) gives us even more insight into the behavior of gases.

11.8 Temperature and Molecular Velocities

According to kinetic molecular theory, particles of different masses have the same average kinetic energy at a given temperature. The kinetic energy of a particle depends on its mass and velocity according to the equation:

$$KE = \frac{1}{2}mv^2$$

The only way for particles of different masses to have the same kinetic energy is for them to have different velocities as Conceptual Connection 11.5 demonstrates.

In a gas mixture at a given temperature, lighter particles travel faster (on average) than heavier ones.

In kinetic molecular theory, we define the root mean square velocity (u_{rms}) of a particle as:

$$u_{rms} = \sqrt{\overline{u^2}} \quad [11.22]$$

where $\overline{u^2}$ is the average of the squares of the particle velocities. Even though the root mean square velocity of a collection of particles is not identical to the average velocity, the two are close in value and conceptually similar. Root mean square velocity is a special *type* of average. The average kinetic energy of one mole of gas particles is given by:

$$KE_{avg} = \frac{1}{2}N_A m\overline{u^2} \quad [11.23]$$

where N_A is Avogadro's number.

Postulate 2 of the kinetic molecular theory states that the average kinetic energy is proportional to the temperature in kelvins. The constant of proportionality in this relationship is (3/2) R.

$$KE_{avg} = (3/2)RT \quad [11.24]$$

The (3/2) R proportionality constant comes from a derivation that is beyond the current scope of this textbook.

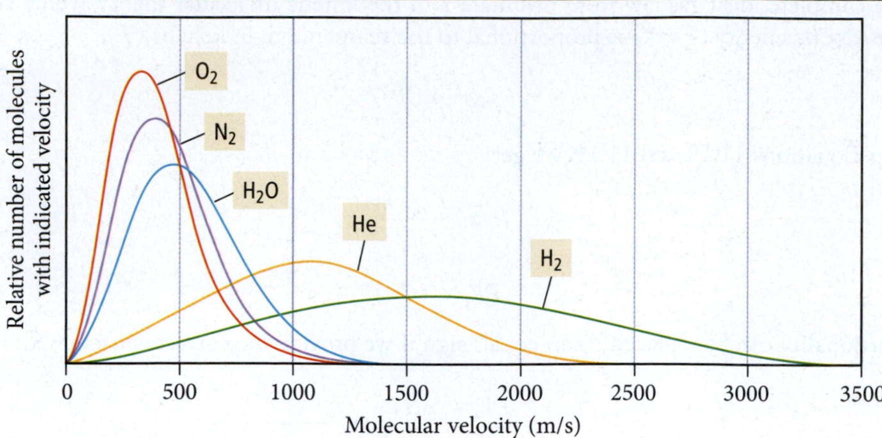

▲ **FIGURE 11.19 Velocity Distribution for Several Gases at 25 °C** At a given temperature, there is a distribution of velocities among the particles in a sample of gas. The exact shape and peak of the distribution vary with the molar mass of the gas.

where R is the gas constant, but in different units ($R = 8.314$ J/mol · K) than those we use in the ideal gas law. If we combine Equations 11.23 and 11.24, and solve for $\overline{u^2}$, we get:

The joule (J) is a unit of energy discussed in more detail in Section 10.2. $\left(1\text{ J} = 1\text{ kg}\,\frac{\text{m}^2}{\text{s}^2}\right)$

$$(1/2)N_A m\overline{u^2} = (3/2)RT$$

$$\overline{u^2} = \frac{(3/2)RT}{(1/2)N_A m} = \frac{3RT}{N_A m}$$

Taking the square root of both sides we get:

$$\sqrt{\overline{u^2}} = u_{\text{rms}} = \sqrt{\frac{3RT}{N_A m}} \qquad [11.25]$$

In Equation 11.25, m is the mass of a particle in kg and N_A is Avogadro's number. The product $N_A m$, then, is the molar mass in kg/mol. If we call this quantity $\mathcal{M}$, the expression for mean square velocity as a function of temperature is the following important result:

$$u_{\text{rms}} = \sqrt{\frac{3RT}{\mathcal{M}}} \qquad [11.26]$$

The root mean square velocity of a collection of gas particles is proportional to the square root of the temperature in kelvins and inversely proportional to the square root of the molar mass of the particles (which because of the units of R is in kilograms per mole). The root mean square velocity of nitrogen molecules at 25 °C, for example, is 515 m/s (1152 mi/h). The root mean square velocity of hydrogen molecules at room temperature is 1920 m/s (4295 mi/h). Notice that the lighter molecules move much faster at a given temperature.

The root mean square velocity, as we have seen, is a kind of average velocity. Some particles are moving faster, and some are moving slower than this average. The velocities of all the particles in a gas sample form distributions such as those shown in Figure 11.19 ▲. We can see from these distributions that some particles are indeed traveling at the root mean square velocity. However, many particles are traveling faster and many slower than the root mean square velocity. For lighter particles, such as helium and hydrogen, the velocity distribution is shifted toward higher velocities and the curve becomes broader, indicating a wider range of velocities. Figure 11.20 ◀ is the velocity distribution for nitrogen at different temperatures. As the temperature increases, the root mean square velocity increases and the distribution becomes broader.

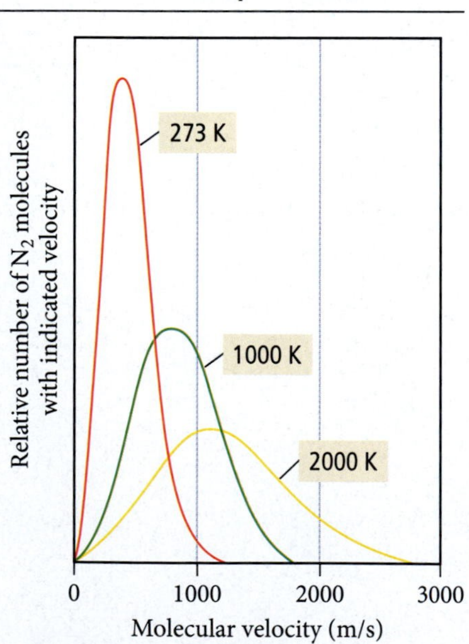

▲ **FIGURE 11.20 Velocity Distribution for Nitrogen at Several Temperatures** As the temperature of a gas sample increases, the velocity distribution of the molecules shifts toward higher velocity and becomes less sharply peaked.

EXAMPLE 11.12 Root Mean Square Velocity

Calculate the root mean square velocity of oxygen molecules at 25 °C.

SORT The problem describes the kind of molecule and its temperature and asks you to find the root mean square velocity.

GIVEN: $O_2, t = 25\ °C$

FIND: u_{rms}

STRATEGIZE The conceptual plan for this problem illustrates how to use the molar mass of oxygen and the temperature (in kelvins) with the equation that defines the root mean square velocity to calculate root mean square velocity.

CONCEPTUAL PLAN

$\mathcal{M}, T \rightarrow u_{rms}$

$$u_{rms} = \sqrt{\frac{3RT}{\mathcal{M}}}$$

RELATIONSHIP USED

$$u_{rms} = \sqrt{\frac{3RT}{\mathcal{M}}} \text{ (Equation 11.26)}$$

SOLVE Gather the required quantities in the correct units. Note that molar mass must be in kg/mol.

SOLUTION

$$T = 25 + 273 = 298\ \text{K}$$

$$\mathcal{M} = \frac{32.00\ \text{g O}_2}{1\ \text{mol O}_2} \times \frac{1\ \text{kg}}{1000\ \text{g}} = \frac{32.00 \times 10^{-3}\ \text{kg O}_2}{1\ \text{mol O}_2}$$

Substitute the quantities into the equation to calculate root mean square velocity. Note that $1\ \text{J} = 1\ \text{kg} \cdot \text{m}^2/\text{s}^2$.

$$u_{rms} = \sqrt{\frac{3RT}{\mathcal{M}}} = \sqrt{\frac{3\left(8.314\ \frac{\text{J}}{\text{mol} \cdot \text{K}}\right)(298\ \text{K})}{\frac{32.00 \times 10^{-3}\ \text{kg O}_2}{1\ \text{mol O}_2}}}$$

$$= \sqrt{2.32 \times 10^5\ \frac{\text{J}}{\text{kg}}} = \sqrt{2.32 \times 10^5\ \frac{\frac{\text{kg} \cdot \text{m}^2}{\text{s}^2}}{\text{kg}}} = 482\ \text{m/s}$$

CHECK The units of the answer (m/s) are correct. The magnitude of the answer is reasonable because oxygen is slightly heavier than nitrogen and should therefore have a slightly lower root mean square velocity at the same temperature. Recall that earlier we stated that the root mean square velocity of nitrogen is 515 m/s at 25 °C.

FOR PRACTICE 11.12

Calculate the root mean square velocity of gaseous xenon atoms at 25 °C.

11.6 Cc Conceptual Connection — Kinetic Molecular Theory II

Which sample of an ideal gas has the greatest pressure? Assume that the mass of each particle is proportional to its size and that all the gas samples are at the same temperature.

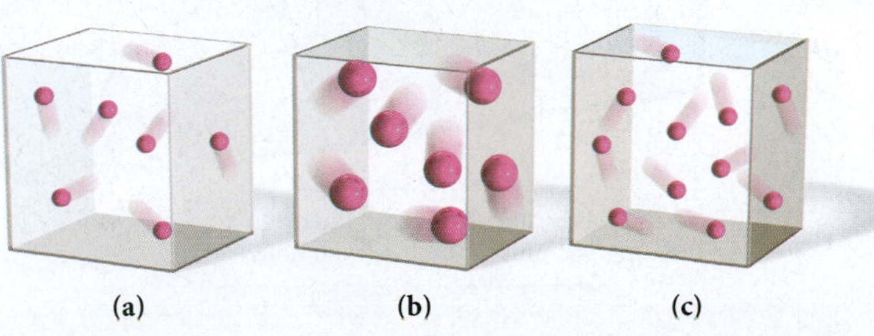

(a) (b) (c)

11.9 Mean Free Path, Diffusion, and Effusion of Gases

We saw in Section 11.8 that the root mean square velocity of gas molecules at room temperature is in the range of hundreds of meters per second. However, suppose that your roommate just put on too much perfume in the bathroom only 2 m away. Why does it take a minute or two before you can smell the fragrance? Although most molecules in a perfume bottle have higher molar masses than nitrogen, their velocities are still hundreds of meters per second, so why the delay? The answer is that, even though gaseous particles travel at tremendous speeds, they also travel in haphazard paths (Figure 11.21 ◀). To a perfume molecule, the path from the perfume bottle in the bathroom to your nose 2 m away is like a bargain hunter's path through a busy shopping mall during a clearance sale. The molecule travels only a short distance before it collides with another molecule, changes direction, only to collide again, and so on. In fact, at room temperature and atmospheric pressure, a molecule in the air experiences several billion collisions per second. The average distance that a molecule travels between collisions is its **mean free path**. At room temperature and atmospheric pressure, the mean free path of a nitrogen molecule, which has a molecular diameter of 300 pm (four times the covalent radius), is 93 nm, or about 310 molecular diameters. If a nitrogen molecule were the size of a golf ball, it would travel about 40 ft between collisions. Mean free path increases with *decreasing* pressure. Under conditions of ultrahigh vacuum (10^{-10} torr), the mean free path of a nitrogen molecule is hundreds of kilometers.

Typical Gas Molecule Path

The average distance between collisions is the mean free path.

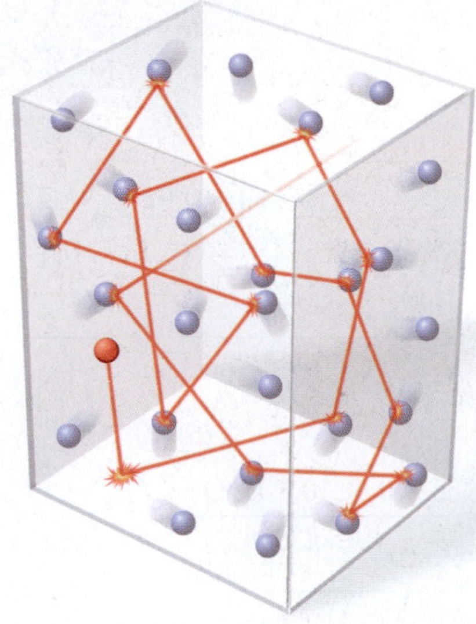

▲ **FIGURE 11.21 Mean Free Path** A molecule in a volume of gas follows a haphazard path, involving many collisions with other molecules.

The process by which gas molecules spread out in response to a concentration gradient is **diffusion**, and even though the particles undergo many collisions, the root mean square velocity still influences the rate of diffusion. Heavier molecules diffuse more slowly than lighter ones, so the first molecules you smell from a perfume mixture (in a room with no air currents) are the lighter ones.

In a ventilated room, air currents greatly enhance the transport of gas molecules.

A process related to diffusion is **effusion**, the process by which a gas escapes from a container into a vacuum through a small hole (Figure 11.22 ▼). The rate of effusion is also related to root mean square velocity—heavier molecules effuse more slowly than lighter ones. The rate of effusion—the amount of gas that effuses in a given time—is inversely proportional to the square root of the molar mass of the gas as follows:

$$\text{rate} \propto \frac{1}{\sqrt{\mathcal{M}}}$$

Effusion

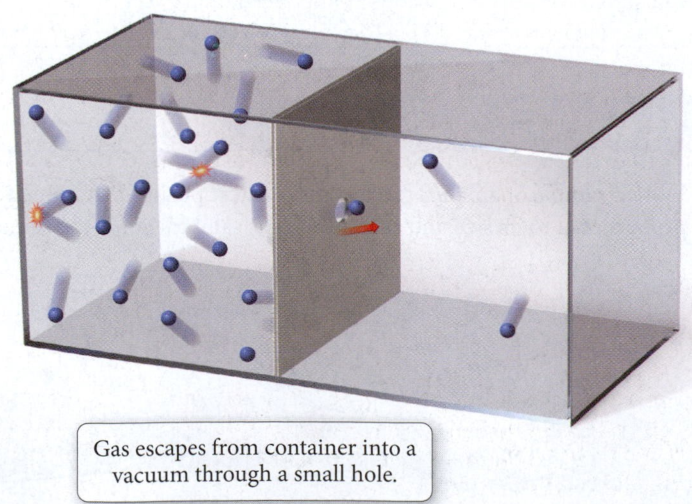

Gas escapes from container into a vacuum through a small hole.

▶ **FIGURE 11.22 Effusion** Effusion is the escape of a gas from a container into a vacuum through a small hole.

The ratio of effusion rates of two different gases is given by **Graham's law of effusion**, named after Thomas Graham (1805–1869):

$$\frac{\text{rate}_A}{\text{rate}_B} = \sqrt{\frac{\mathcal{M}_B}{\mathcal{M}_A}} \qquad [11.27]$$

In this expression, rate_A and rate_B are the effusion rates of gases A and B and $\mathcal{M}_A$ and $\mathcal{M}_B$ are their molar masses.

Graham's law explains, in part, why helium balloons only float for a day or so. Because helium has such a low molar mass, it escapes from the balloon quite quickly. A balloon filled with air, by contrast, remains inflated longer because the gas particles within it have a higher average molar mass.

EXAMPLE 11.13

Graham's Law of Effusion

An unknown gas effuses at a rate that is 0.462 times that of nitrogen gas (at the same temperature). Calculate the molar mass of the unknown gas in g/mol.

SORT The problem gives you the ratio of effusion rates for the unknown gas and nitrogen and asks you to find the molar mass of the unknown gas.

GIVEN: $\frac{\text{Rate}_{\text{unk}}}{\text{Rate}_{N_2}} = 0.462$

FIND: $\mathcal{M}_{\text{unk}}$

STRATEGIZE The conceptual plan uses Graham's law of effusion. You are given the ratio of rates, and you know the molar mass of the nitrogen. You can use Graham's law to determine the molar mass of the unknown gas.

CONCEPTUAL PLAN

$\frac{\text{Rate}_{\text{unk}}}{\text{Rate}_{N_2}}, \mathcal{M}_{N_2} \rightarrow \mathcal{M}_{unk}$

$$\frac{\text{Rate}_{\text{unk}}}{\text{Rate}_{N_2}} = \sqrt{\frac{\mathcal{M}_{N_2}}{\mathcal{M}_{\text{unk}}}}$$

RELATIONSHIP USED

$$\frac{\text{rate}_A}{\text{rate}_B} = \sqrt{\frac{\mathcal{M}_B}{\mathcal{M}_A}} \text{ (Graham's law)}$$

SOLVE Solve the equation for $\mathcal{M}_{\text{unk}}$, substitute the correct values, and calculate.

SOLUTION

$$\frac{\text{rate}_{\text{unk}}}{\text{rate}_{N_2}} = \sqrt{\frac{\mathcal{M}_{N_2}}{\mathcal{M}_{\text{unk}}}}$$

$$\mathcal{M}_{\text{unk}} = \frac{\mathcal{M}_{N_2}}{\left(\frac{\text{rate}_{\text{unk}}}{\text{rate}_{N_2}}\right)^2}$$

$$= \frac{28.02 \text{ g/mol}}{(0.462)^2} = 131 \text{ g/mol}$$

CHECK The units of the answer are correct. The magnitude of the answer seems reasonable for the molar mass of a gas. In fact, from the answer, you can even conclude that the gas is probably xenon, which has a molar mass of 131.29 g/mol.

FOR PRACTICE 11.13

Find the ratio of effusion rates of hydrogen gas and krypton gas.

11.10 Gases in Chemical Reactions: Stoichiometry Revisited

In Chapter 8, we discussed how to use the coefficients in chemical equations as conversion factors between number of moles of reactants and number of moles of products in a chemical reaction. We can use these conversion factors to determine, for example, the mass of product obtained in a chemical reaction based on a given mass of reactant, or the mass of one reactant needed to react completely with a given mass of another reactant. The general conceptual plan for these kinds of calculations is:

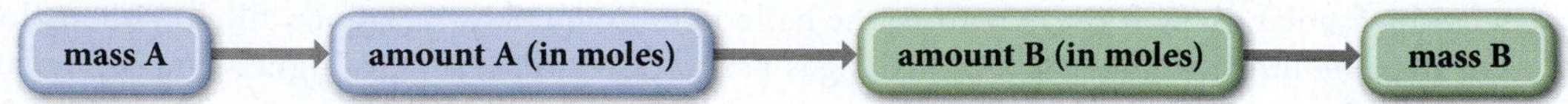

where A and B are two different substances involved in the reaction and the conversion factor between amounts (in moles) of each comes from the stoichiometric coefficients in the balanced chemical equation.

In reactions involving *gaseous* reactant or products, we often specify the quantity of a gas in terms of its volume at a given temperature and pressure. As we have seen, stoichiometry involves relationships between amounts in moles. For stoichiometric calculations involving gases, we can use the ideal gas law to determine the amounts in moles from the volumes, or to determine the volumes from the amounts in moles.

The pressures here could also be partial pressures.

$$n = \frac{PV}{RT} \qquad V = \frac{nRT}{P}$$

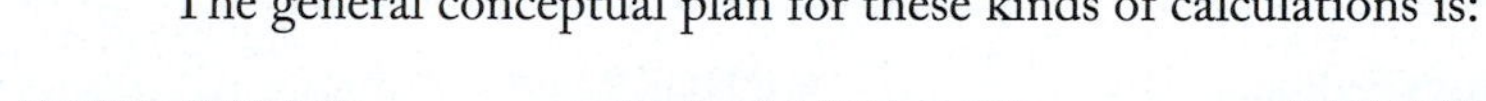

The general conceptual plan for these kinds of calculations is:

Examples 11.14 and 11.15 demonstrate this kind of calculation.

EXAMPLE 11.14 Gases in Chemical Reactions

Methanol (CH_3OH) can be synthesized by the reaction:

$$CO(g) + 2\,H_2(g) \longrightarrow CH_3OH(g)$$

What volume (in liters) of hydrogen gas, at a temperature of 355 K and a pressure of 738 mmHg, do we need to synthesize 35.7 g of methanol?

SORT The problem gives the mass of methanol, the product of a chemical reaction. You are asked to find the required volume of one of the reactants (hydrogen gas) at a specified temperature and pressure.	**GIVEN:** 35.7 g CH_3OH, $T = 355$ K, $P = 738$ mmHg **FIND:** V_{H_2}
STRATEGIZE You can calculate the required volume of hydrogen gas from the number of moles of hydrogen gas, which you can obtain from the number of moles of methanol via the stoichiometry of the reaction. First, find the number of moles of methanol from its mass by using the molar mass.	**CONCEPTUAL PLAN** 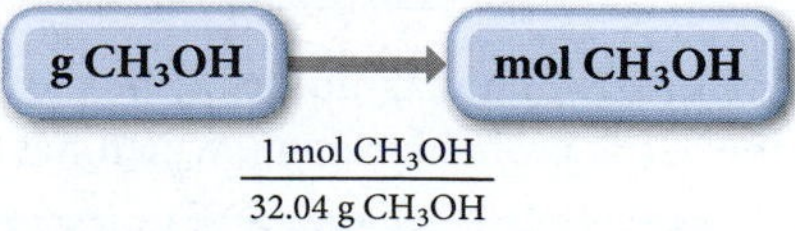

—*Continued on the next page*

Continued from the previous page—

Then use the stoichiometric relationship from the balanced chemical equation to find the number of moles of hydrogen you need to form that quantity of methanol.

Finally, substitute the number of moles of hydrogen together with the pressure and temperature into the ideal gas law to find the volume of hydrogen.

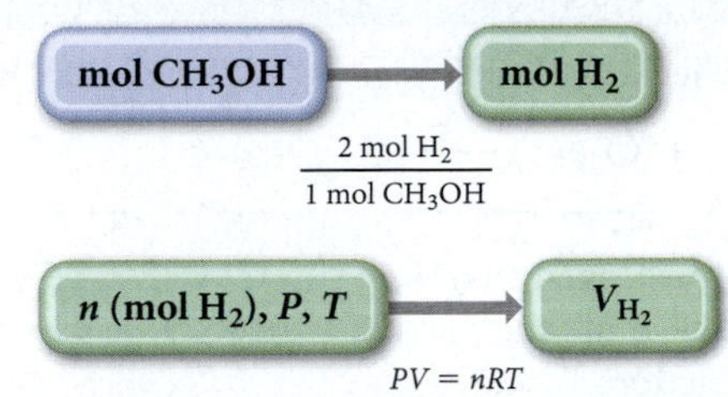

RELATIONSHIPS USED

$PV = nRT$ (ideal gas law)

2 mol H_2 : 1 mol CH_3OH (from balanced chemical equation)

molar mass CH_3OH = 32.04 g/mol

SOLVE Follow the conceptual plan to solve the problem. Begin by using the mass of methanol to determine the number of moles of methanol.

Next, convert the number of moles of methanol to moles of hydrogen.

Finally, use the ideal gas law to find the volume of hydrogen. Before substituting into the equation, you need to convert the pressure to atmospheres.

SOLUTION

$$35.7 \text{ g } CH_3OH \times \frac{1 \text{ mol } CH_3OH}{32.04 \text{ g } CH_3OH} = 1.1142 \text{ mol } CH_3OH$$

$$1.1142 \text{ mol } CH_3OH \times \frac{2 \text{ mol } H_2}{1 \text{ mol } CH_3OH} = 2.2284 \text{ mol } H_2$$

$$V_{H_2} = \frac{n_{H_2}RT}{P}$$

$$P = 738 \text{ mmHg} \times \frac{1 \text{ atm}}{760 \text{ mmHg}} = 0.97105 \text{ atm}$$

$$V_{H_2} = \frac{(2.2284 \text{ mol})\left(0.08206 \frac{\text{L} \cdot \text{atm}}{\text{mol} \cdot \text{K}}\right)(355 \text{ K})}{0.97105 \text{ atm}}$$

$$= 66.9 \text{ L}$$

CHECK The units of the answer are correct. The magnitude of the answer (66.9 L) seems reasonable. You are given slightly more than one molar mass of methanol, which is therefore slightly more than one mole of methanol. From the equation you can see that you need 2 mol hydrogen to make 1 mol methanol, so the answer must be slightly greater than 2 mol hydrogen. Under standard temperature and pressure, slightly more than 2 mol hydrogen occupies slightly more than 2×22.4 L = 44.8 L. At a temperature greater than standard temperature, the volume would be even greater; therefore, this answer is reasonable.

FOR PRACTICE 11.14

In the following reaction, 4.58 L of O_2 was formed at P = 745 mmHg and T = 308 K. How many grams of Ag_2O decomposed?

$$2\,Ag_2O(s) \longrightarrow 4\,Ag(s) + O_2(g)$$

FOR MORE PRACTICE 11.14

In the reaction in For Practice 11.14, what mass of $Ag_2O(s)$ (in grams) is required to form 388 mL of oxygen gas at P = 734 mmHg and 25.0 °C?

Molar Volume and Stoichiometry

In Section 11.5, we saw that, under standard temperature and pressure, 1 mol of an ideal gas occupies 22.4 L. Consequently, if a reaction occurs at or near standard temperature and pressure, we can use 1 mol = 22.4 L as a conversion factor in stoichiometric calculations, as we demonstrate in Example 11.15.

EXAMPLE 11.15

Using Molar Volume in Gas Stoichiometric Calculations

How many grams of water form when 1.24 L of H_2 gas at STP completely reacts with O_2?

$$2\,H_2(g) + O_2(g) \longrightarrow 2\,H_2O(g)$$

SORT You are given the volume of hydrogen gas (a reactant) at STP and asked to determine the mass of water that forms upon complete reaction.

GIVEN: 1.24 L H_2
FIND: g H_2O

STRATEGIZE Because the reaction occurs under standard temperature and pressure, you can convert directly from the volume (in L) of hydrogen gas to the amount in moles. Then use the stoichiometric relationship from the balanced equation to find the number of moles of water formed. Finally, use the molar mass of water to obtain the mass of water formed.

CONCEPTUAL PLAN

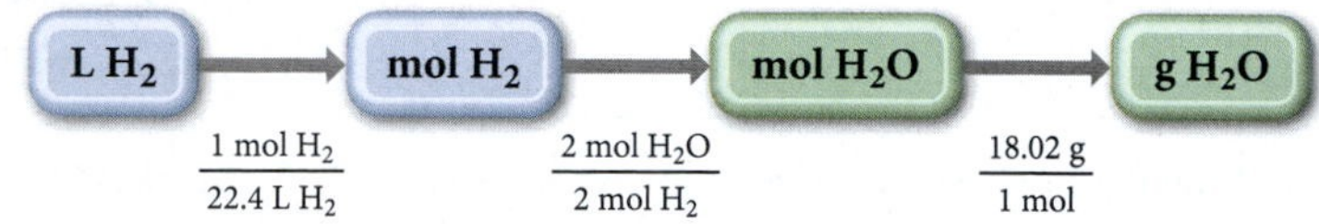

RELATIONSHIPS USED

1 mol = 22.4 L (at STP)

2 mol H_2 : 2 mol H_2O (from balanced equation)

molar mass H_2O = 18.02 g/mol

SOLVE Follow the conceptual plan to solve the problem.

$$1.24\ \text{L H}_2 \times \frac{1\ \text{mol H}_2}{22.4\ \text{L H}_2} \times \frac{2\ \text{mol H}_2\text{O}}{2\ \text{mol H}_2} \times \frac{18.02\ \text{g H}_2\text{O}}{1\ \text{mol H}_2\text{O}} = 0.998\ \text{g H}_2\text{O}$$

CHECK The units of the answer are correct. The magnitude of the answer (0.998 g) is about 1/18 of the molar mass of water, roughly equivalent to the approximately 1/22 of a mole of hydrogen gas given, as you would expect for the 1:1 stoichiometric relationship between number of moles of hydrogen and number of moles of water.

FOR PRACTICE 11.15

How many liters of oxygen (at STP) are required to form 10.5 g of H_2O?

$$2\,H_2(g) + O_2(g) \longrightarrow 2\,H_2O(g)$$

11.7 Cc Conceptual Connection — Pressure and Number of Moles

Nitrogen and hydrogen react to form ammonia according to the following equation:

$$N_2(g) + 3\,H_2(g) \rightleftharpoons 2\,NH_3(g)$$

Consider these representations of the initial mixture of reactants and the resulting mixture after the reaction has been allowed to react for some time.

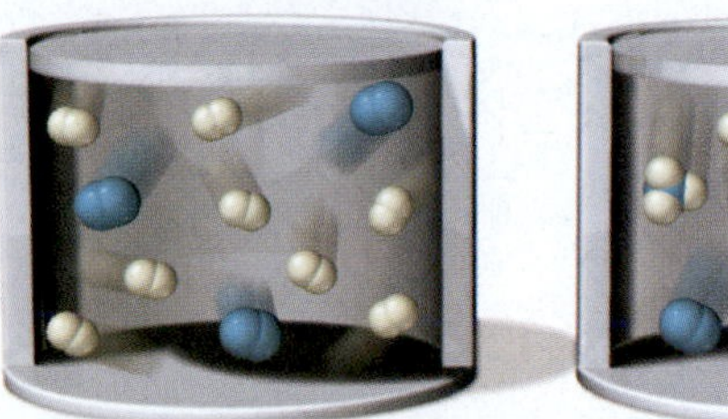

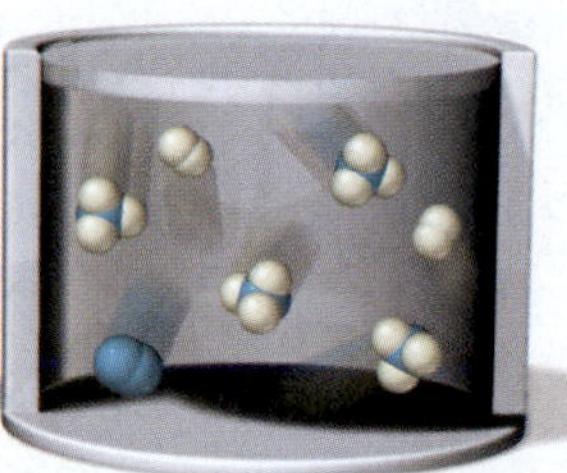

If the volume is kept constant, and nothing is added to the reaction mixture, what happens to the total pressure during the course of the reaction?

(a) The pressure increases. **(b)** The pressure decreases. **(c)** The pressure does not change.

11.11 Real Gases: The Effects of Size and Intermolecular Forces

One mole of an ideal gas has a volume of 22.41 L at STP. Figure 11.23 ▶ shows the molar volume of several real gases at STP. As you can see, most of these gases have a volume that is very close to 22.41 L, meaning that they act very nearly as ideal gases. Gases behave ideally when both of the following are true: (a) the volume of the gas particles is small compared to the space between them; and (b) the forces between the gas particles are not significant. At STP, these assumptions are valid for most common gases. However, these assumptions break down at higher pressures or lower temperatures.

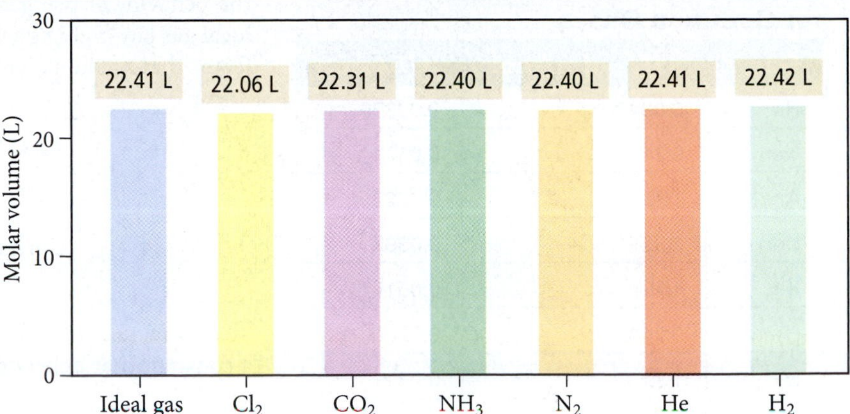

▲ **FIGURE 11.23 Molar Volumes of Real Gases** The molar volumes of several gases at STP are close to 22.414 L, indicating that their departures from ideal behavior are small.

The Effect of the Finite Volume of Gas Particles

The finite volume of gas particles—that is, their actual *size*—becomes important at high pressure because the volume of the particles themselves occupies a significant portion of the total gas volume (Figure 11.24 ▶). We can see the effect of particle volume by comparing the molar volume of argon to the molar volume of an ideal gas as a function of pressure at 500 K as shown in Figure 11.25 ▼. At low pressures, the molar volume of argon is nearly identical to that of an ideal gas. But as the pressure increases, the molar volume of argon becomes *greater than* that of an ideal gas. At the higher pressures, the argon atoms themselves occupy a significant portion of the gas volume, making the actual volume greater than that predicted by the ideal gas law.

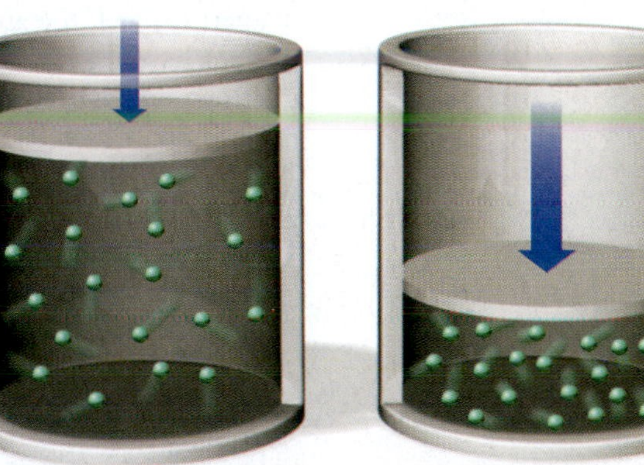

▲ **FIGURE 11.24 Particle Volume and Ideal Behavior** As a gas is compressed the gas particles themselves begin to occupy a significant portion of the total gas volume, leading to deviations from ideal behavior.

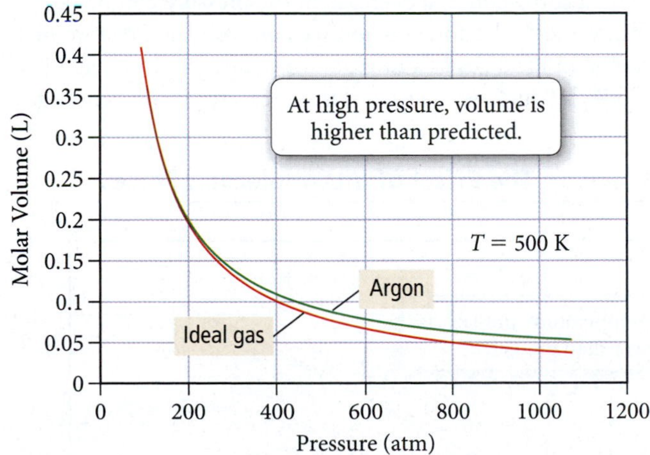

▲ **FIGURE 11.25 The Effect of Particle Volume** At high pressures, 1 mol of argon occupies a larger volume than 1 mol of an ideal gas because of the volume of the argon atoms themselves. (This example was chosen to minimize the effects of intermolecular forces, which are very small in argon at 500 K, thereby isolating the effect of particle volume.)

TABLE 11.4 Van der Waals Constants for Common Gases

Gas	a ($L^2 \cdot atm/mol^2$)	*b* (L/mol)
He	0.0342	0.02370
Ne	0.211	0.0171
Ar	1.35	0.0322
Kr	2.32	0.0398
Xe	4.19	0.0511
H_2	0.244	0.0266
N_2	1.39	0.0391
O_2	1.36	0.0318
Cl_2	6.49	0.0562
H_2O	5.46	0.0305
CH_4	2.25	0.0428
CO_2	3.59	0.0427
CCl_4	20.4	0.1383

In 1873, Johannes van der Waals (1837–1923) modified the ideal gas equation to fit the behavior of real gases. From the graph for argon in Figure 11.25 we can see that the ideal gas law predicts a volume that is too small. Van der Waals suggested a small correction factor that accounts for the volume of the gas particles themselves.

Ideal behavior $$V = \frac{nRT}{P}$$

Corrected for volume of gas particles $$V = \frac{nRT}{P} + nb \quad [11.28]$$

The correction adds the quantity nb to the volume, where n is the number of moles and b is a constant that depends on the gas (Table 11.4). We can rearrange the corrected equation as follows:

$$(V - nb) = \frac{nRT}{P} \quad [11.29]$$

The Effect of Intermolecular Forces

Intermolecular forces, which we will discuss in more detail in Chapter 12, are attractions between the atoms or molecules that compose any substance. These attractions are typically small in gases and therefore do not matter much at low pressure because the molecules are too far apart to "feel" the attractions. They also do not matter much at high temperatures because the molecules have a lot of kinetic energy and when two particles with high kinetic energies collide, a weak attraction between them does not affect the collision much. At lower temperatures, however, the collisions occur with less kinetic energy, and weak attractions can affect the collisions. We can understand this difference with an analogy to billiard balls. Imagine two billiard balls that are coated with a substance that makes them slightly sticky. If they collide when moving at high velocities, the stickiness does not have much of an effect—the balls bounce off one another as if the sticky substance is not even there. However, if the two billiard balls collide when moving very slowly (say, barely rolling), the sticky substance has an effect—the billiard balls might even stick together and not bounce off one another.

The effect of these weak attractions between particles is a decrease in the number of collisions with the surfaces of the container, and a corresponding decrease in the pressure compared to that of an ideal gas. We can see the effect of intermolecular forces when we compare the pressure of 1.0 mol of xenon gas to the pressure of 1.0 mol of an ideal gas as a function of temperature and at a fixed volume of 1.0 L, as shown in Figure 11.26 ▼. At high temperature, the pressure of the xenon gas is nearly

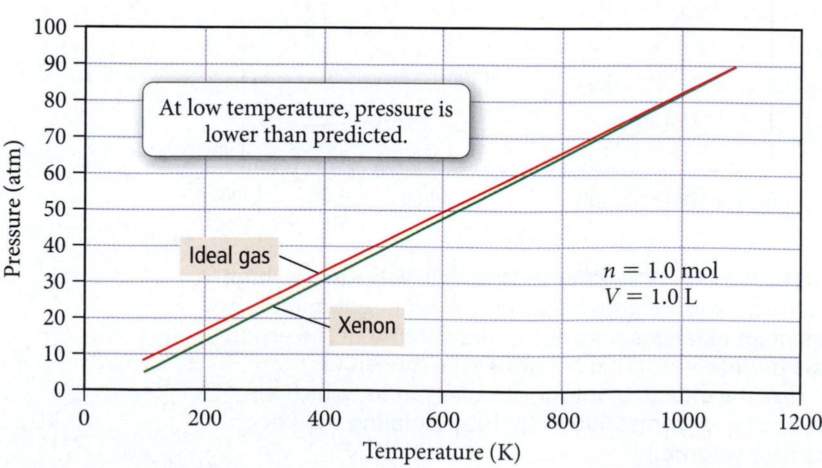

▶ **FIGURE 11.26 The Effect of Intermolecular Forces** At low temperatures, the pressure of xenon is less than an ideal gas exerts because interactions among xenon molecules reduce the number of collisions with the walls of the container.

identical to that of an ideal gas. But at lower temperatures, the pressure of xenon is *less than* that of an ideal gas. At the lower temperatures, the xenon atoms spend more time interacting with each other and less time colliding with the walls, making the actual pressure less than that predicted by the ideal gas law.

From the graph for xenon shown in Figure 11.26 we can see that the ideal gas law predicts a pressure that is too large at low temperatures. Van der Waals suggested a small correction factor that accounts for the intermolecular forces between gas particles.

Ideal behavior $$P = \frac{nRT}{V}$$

Corrected for intermolecular forces $$P = \frac{nRT}{V} - a\left(\frac{n}{V}\right)^2 \quad [11.30]$$

The correction subtracts the quantity $a(n/V)^2$ from the pressure, where n is the number of moles, V is the volume, and a is a constant that depends on the gas (see Table 11.4). Notice that the correction factor increases as n/V (the number of moles of particles per unit volume) increases because a greater concentration of particles makes it more likely that they will interact with one another. We can rearrange the corrected equation as:

$$P + a\left(\frac{n}{V}\right)^2 = \frac{nRT}{V} \quad [11.31]$$

Van der Waals Equation

We can combine the effects of particle volume (Equation 11.29) and particle intermolecular forces (Equation 11.31) into one equation that describes nonideal gas behavior.

$$\left[P + a\left(\frac{n}{V}\right)^2\right] \times [V - nb] = nRT$$

Correction for intermolecular forces (the $a(n/V)^2$ term); Correction for particle volume (the nb term)

This equation is the **van der Waals equation**, and we can use it to calculate the properties of a gas under nonideal conditions.

Real Gases

We can see the combined effects of particle volume and intermolecular forces by examining a plot of PV/RT versus P for 1 mol of a number of real gases (Figure 11.27 ▶). For an ideal gas, $PV/RT = n$, the number of moles of gas. Therefore, for 1 mol of an ideal gas, PV/RT is equal to 1, as shown in the plot. For real gases, PV/RT deviates from 1, but the deviations are not uniform. For example, water displays a large negative deviation from PV/RT because, for water, the effect of intermolecular forces on lowering the pressure (relative to an ideal gas) is far greater than the effect of particle

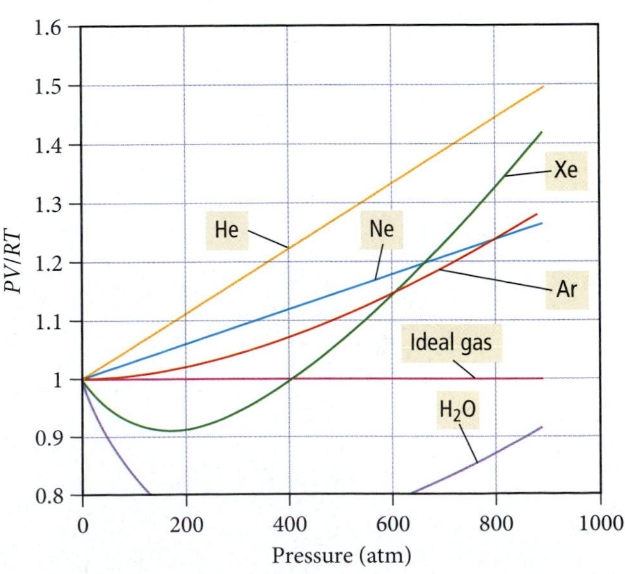

▶ **FIGURE 11.27 Real versus Ideal Behavior** For 1 mol of an ideal gas, *PV/RT* is equal to 1. The combined effects of the volume of gas particles and the interactions among them cause each real gas to deviate from ideal behavior in a slightly different way. These curves were calculated at a temperature of 500 K.

size on increasing the volume. We can see in Table 11.4 that water has a high value of a, the constant that corrects for intermolecular forces, but a moderate value of b, the constant that corrects for particle size. Therefore PV/RT for water is lower than predicted from the ideal gas law.

By contrast, consider the behavior of helium, which displays a positive deviation from the ideal behavior. Helium has very weak intermolecular forces, and their effect on lowering the pressure (relative to ideal gas) is small compared to the effect of particle size on increasing the volume. Therefore PV/RT is greater than predicted from the ideal gas law for helium.

11.8 Cc Conceptual Connection — Real Gases

This graph shows PV/RT for carbon dioxide at three different temperatures. Rank the curves in order of increasing temperature.

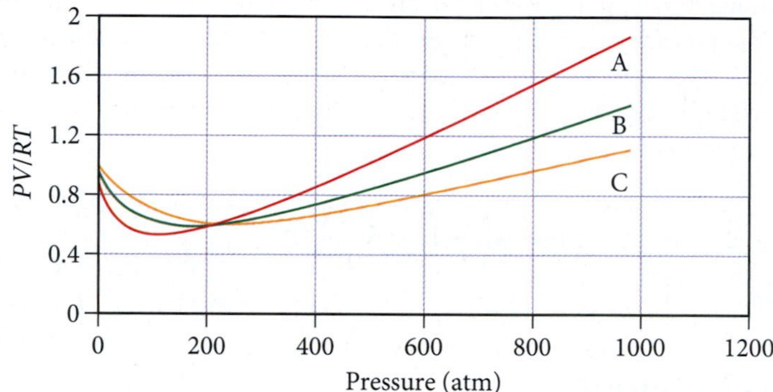

SELF-ASSESSMENT Quiz

1. A gas sample has an initial pressure of 547 mmHg and an initial volume of 0.500 L. What is the pressure (in atm) if we decrease the volume of the sample to 225 mL? (Assume constant temperature and constant number of moles of gas.)
   a) 1.60×10^{-3} atm **b)** 1.60 atm
   c) 0.324 atm **d)** 1.22 atm

2. A gas sample has a volume of 178 mL at 0.00 °C. The temperature is raised (at constant pressure) until the volume reaches 211 mL. What is the temperature of the gas sample in °C at this volume?
   a) 0.00 °C **b)** 324 °C
   c) −43 °C **d)** 51 °C

3. What is the pressure of 1.78 g of nitrogen gas confined to a volume of 0.118 L at 25 °C?
   a) 13.2 atm **b)** 369 atm
   c) 1.10 atm **d)** 26.3 atm

4. What is the density of a sample of argon gas at 55 °C and 765 mmHg?
   a) 2.99 g/L **b)** 1.13×10^3 g/L
   c) 1.49 g/L **d)** 8.91 g/L

5. Which gas sample has the greatest volume at STP?
   a) 10.0 g Ar
   b) 10.0 g Kr
   c) 10.0 g Xe
   d) None of the above. (They all have the same volume.)

6. A 1.25 g gas sample occupies 663 mL at 25 °C and 1.00 atm. What is the molar mass of the gas?
   a) 0.258 g/mol **b)** 0.0461 g/mol
   c) 3.87 g/mol **d)** 46.1 g/mol

7. A 255 mL gas sample contains argon and nitrogen at a temperature of 65 °C. The total pressure of the sample is 725 mmHg, and the partial pressure of argon is 231 mmHg. What mass of nitrogen is present in the sample?
   a) 0.324 g nitrogen **b)** 0.167 g nitrogen
   c) 0.0837 g nitrogen **d)** 0.870 g nitrogen

8. A gas mixture in a 1.55 L at 298 K container contains 10.0 g of Ne and 10.0 g of Ar. Calculate the partial pressure (in atm) of Ne and Ar in the container.
   a) $P_{Ne} = 10.5$ atm, $P_{Ar} = 5.29$ atm
   b) $P_{Ne} = 5.83$ atm, $P_{Ar} = 2.95$ atm
   c) $P_{Ne} = 5.88$ atm, $P_{Ar} = 5.88$ atm
   d) $P_{Ne} = 7.82$ atm, $P_{Ar} = 3.95$ atm

9. A gas sample at STP contains 1.15 g oxygen and 1.55 g nitrogen. What is the volume of the gas sample?
   a) 1.26 L **b)** 2.04 L
   c) 4.08 L **d)** 61.0 L

10. Aluminum reacts with chlorine gas to form aluminum chloride.

$$2\ Al(s) + 3\ Cl_2(g) \longrightarrow 2\ AlCl_3(s)$$

    What minimum volume of chlorine gas (at 298 K and 225 mm Hg) is required to completely react with 7.85 g of aluminum?
    a) 36.0 L **b)** 24.0 L
    c) 0.0474 L **d)** 16.0 L

11. Calculate the root mean square velocity of $I_2(g)$ at 373 K.
    a) 19.0 m/s **b)** 191 m/s
    c) 6.05 m/s **d)** 99.1 m/s

12. Which gas has the greatest kinetic energy at STP?
    a) He **b)** Ne
    c) Ar
    d) None of the above. (All have the same kinetic energy.)

13. A sample of Xe takes 75 seconds to effuse out of a container. An unknown gas takes 37 seconds to effuse out of the identical container under identical conditions. What is the most likely identity of the unknown gas?
    a) He **b)** O_2
    c) Br_2 **d)** Kr

14. Consider the generic reaction: $2\ A(g) + B(g) \longrightarrow 2\ C(g)$. If a flask initially contains 1.0 atm of A and 1.0 atm of B, what is the pressure in the flask if the reaction proceeds to completion? (Assume constant volume and temperature.)
    a) 1.0 atm **b)** 1.5 atm
    c) 2.0 atm **d)** 3.0 atm

15. Rank the gases Ar, N_2, CH_4, and C_2H_6 in order of increasing density at STP.
    a) $CH_4 < C_2H_6 < N_2 < Ar$
    b) $CH_4 < N_2 < Ar < C_2H_6$
    c) $Ar < C_2H_6 < N_2 < CH_4$
    d) $CH_4 < N_2 < C_2H_6 < Ar$

Answers: 1:b; 2:d; 3:a; 4:c; 5:a; 6:d; 7:b; 8:d; 9:b; 10:a; 11:b; 12:d; 13:b; 14:b; 15:d

CHAPTER SUMMARY 11

REVIEW

KEY LEARNING OUTCOMES

CHAPTER OBJECTIVES	ASSESSMENT
Converting between Pressure Units (11.2)	• Example 11.1 For Practice 11.1 For More Practice 11.1 Exercises 25–28
Relating Volume and Pressure: Boyle's Law (11.3)	• Example 11.2 For Practice 11.2 Exercises 31, 32
Relating Volume and Temperature: Charles's Law (11.3)	• Example 11.3 For Practice 11.3 Exercises 33, 34
Relating Volume and Moles: Avogadro's Law (11.3)	• Example 11.4 For Practice 11.4 Exercises 35, 36
Determining *P, V, n,* or *T* Using the Ideal Gas Law (11.4)	• Examples 11.5, 11.6 For Practice 11.5, 11.6 For More Practice 11.6 Exercises 37–46, 51, 52
Relating the Density of a Gas to Its Molar Mass (11.5)	• Example 11.7 For Practice 11.7 For More Practice 11.7 Exercises 55, 56
Calculating the Molar Mass of a Gas with the Ideal Gas Law (11.5)	• Example 11.8 For Practice 11.8 Exercises 57–60
Calculating Total Pressure, Partial Pressures, and Mole Fractions of Gases in a Mixture (11.6)	• Examples 11.9, 11.10, 11.11 For Practice 11.9, 11.10, 11.11 Exercises 61, 62, 65, 67, 68, 70
Calculating the Root Mean Square Velocity of a Gas (11.8)	• Example 11.12 For Practice 11.12 Exercises 73, 74
Calculating the Effusion Rate or the Ratio of Effusion Rates of Two Gases (11.9)	• Example 11.13 For Practice 11.13 Exercises 75–78
Relating the Amounts of Reactants and Products in Gaseous Reactions: Stoichiometry (11.10)	• Examples 11.14, 11.15 For Practice 11.14, 11.15 For More Practice 11.14 Exercises 81–87

KEY TERMS

Section 11.1
pressure (392)

Section 11.2
millimeter of mercury (mmHg) (393)
barometer (393)
torr (393)
atmosphere (atm) (394)
pascal (Pa) (394)
manometer (394)

Section 11.3
Boyle's law (395)
Charles's law (398)
Avogadro's law (400)

Section 11.4
ideal gas law (401)
ideal gas (401)
ideal gas constant (401)

Section 11.5
molar volume (404)
standard temperature and pressure (STP) (404)

Section 11.6
partial pressure (P_n) (407)
Dalton's law of partial pressures (407)
mole fraction (χ_a) (408)
hypoxia (409)
oxygen toxicity (410)
nitrogen narcosis (410)
vapor pressure (412)

Section 11.7
kinetic molecular theory (414)

Section 11.9
mean free path (420)
diffusion (420)
effusion (420)
Graham's law of effusion (421)

Section 11.11
van der Waals equation (427)

KEY CONCEPTS

Pressure (11.1, 11.2)

- Gas pressure is the force per unit area that results from gas particles colliding with the surfaces around them. We use a variety of units to measure pressure, including mmHg, torr, Pa, psi, in Hg, and atm.

The Simple Gas Laws (11.3)

- The simple gas laws express relationships between pairs of variables when other variables are held constant. These variables correspond to the four basic properties of a gas sample: pressure (P), volume (V), temperature (T), and amount in moles (n).
- Boyle's law states that the volume of a gas is inversely proportional to its pressure.
- Charles's law states that the volume of a gas is directly proportional to its temperature.
- Avogadro's law states that the volume of a gas is directly proportional to the amount (in moles).

The Ideal Gas Law (11.4, 11.5)

- The ideal gas law, $PV = nRT$, relates the relationship among all four gas variables and contains the simple gas laws within it.
- We can use the ideal gas law to find one of the four variables if we know the other three. We can use it to calculate the molar volume of an ideal gas, which is 22.4 L at STP, and to calculate the density and molar mass of a gas.

Mixtures of Gases and Partial Pressures (11.6)

- In a mixture of gases, each gas acts independently of the others so that any overall property of the mixture is the sum of the properties of the individual components.
- The pressure due to any individual component is its partial pressure.

Kinetic Molecular Theory (11.7–11.9)

- Kinetic molecular theory is a quantitative model for gases. The theory has three main assumptions: (1) the gas particles are negligibly small; (2) the average kinetic energy of a gas particle is proportional to the temperature in kelvins; and (3) the collision of one gas particle with another is completely elastic (the particles do not stick together). The gas laws all follow from the kinetic molecular theory.
- We can use kinetic molecular theory to derive the expression for the root mean square velocity of gas particles. This velocity is inversely proportional to the molar mass of the gas, and therefore—at a given temperature—smaller gas particles are (on average) moving more quickly than larger ones.
- The kinetic molecular theory also allows us to predict the mean free path of a gas particle (the distance it travels between collisions) and relative rates of diffusion or effusion.

Gas Stoichiometry (11.10)

- In reactions involving gaseous reactants and products, we often report quantities in volumes at specified pressures and temperatures. We can convert these quantities to amounts (in moles) using the ideal gas law. Then we can use the stoichiometric coefficients from the balanced equation to determine the stoichiometric amounts of other reactants or products.
- The general form for these types of calculations is: volume A $\longrightarrow$ amount A (in moles) $\longrightarrow$ amount B (in moles) $\longrightarrow$ quantity of B (in desired units).
- In cases where the reaction is carried out at STP, we can use the molar volume at STP (22.4 L = 1 mol) to convert between volume in liters and amount in moles.

Real Gases (11.11)

- Real gases, unlike ideal gases, do not always fit the assumptions of kinetic molecular theory.
- These assumptions tend to break down at high pressures, where the volume is higher than predicted for an ideal gas because the particles are no longer negligibly small compared to the space between them.
- The assumptions also break down at low temperatures where the pressure is lower than predicted because the attraction between molecules combined with low kinetic energies causes partially inelastic collisions.
- We can use the van der Waals equation to predict gas properties under nonideal conditions.

KEY EQUATIONS AND RELATIONSHIPS

Relationship between Pressure (*P*), Force (*F*), and Area (*A*) (11.2)

$$P = \frac{F}{A}$$

Boyle's Law: Relationship between Pressure (*P*) and Volume (*V*) (11.3)

$$V \propto \frac{1}{P}$$

$$P_1V_1 = P_2V_2$$

Charles's Law: Relationship between Volume (*V*) and Temperature (*T*) (11.3)

$$V \propto T \quad (\text{in K})$$

$$\frac{V_1}{T_1} = \frac{V_2}{T_2}$$

Avogadro's Law: Relationship between Volume (*V*) and Amount in Moles (*n*) (11.3)

$$V \propto n$$

$$\frac{V_1}{n_1} = \frac{V_2}{n_2}$$

Ideal Gas Law: Relationship between Volume (*V*), Pressure (*P*), Temperature (*T*), and Amount (*n*) (11.4)

$$PV = nRT$$

Density of a Gas (11.5)

$$d = \frac{P\mathcal{M}}{RT}$$

Dalton's Law: Relationship between Partial Pressures P_n in Mixture of Gases and Total Pressure (P_{total}) (11.6)

$$P_{total} = P_a + P_b + P_c + \cdots$$

$$P_a = \frac{n_aRT}{V} \quad P_b = \frac{n_bRT}{V} \quad P_c = \frac{n_cRT}{V}$$

Mole Fraction (χ_a) (11.6)

$$\chi_a = \frac{n_a}{n_{total}}$$

$$P_a = \chi_a P_{total}$$

Average Kinetic Energy (KE_{avg}) (11.8)

$$KE_{avg} = \frac{3}{2}RT$$

Relationship between Root Mean Square Velocity (u_{rms}) and Temperature (*T*) (11.8)

$$u_{rms} = \sqrt{\frac{3RT}{\mathcal{M}}}$$

Relationship of Effusion Rates of Two Different Gases (11.9)

$$\frac{\text{rate A}}{\text{rate B}} = \sqrt{\frac{\mathcal{M}_B}{\mathcal{M}_A}}$$

Van der Waals Equation: The Effects of Volume and Intermolecular Forces on Nonideal Gas Behavior (11.11)

$$[P + a(n/V)^2] \times (V - nb) = nRT$$

EXERCISES

REVIEW QUESTIONS

1. What is pressure? What causes pressure?
2. How does pressure change as a function of altitude?
3. Explain the risks associated with uncontrolled decompression.
4. What are the common units of pressure? List them in order of smallest to largest unit.
5. What is a manometer? How does it measure the pressure of a sample of gas?
6. Summarize each of the simple gas laws (Boyle's law, Charles's law, and Avogadro's law). For each law, explain the relationship between the two variables it addresses and also state which variables must be kept constant.
7. Explain the source of ear pain experienced due to a rapid change in altitude.
8. Why must scuba divers never hold their breath as they ascend to the surface?
9. Why is the second story of a house usually warmer than the ground story?
10. Explain why hot-air balloons float above the ground.
11. What is the ideal gas law? Why is it useful?
12. Explain how the ideal gas law contains within it the simple gas laws (show an example).
13. Define molar volume and give its value for a gas at STP.
14. How does the density of a gas depend on temperature? Pressure? How does it depend on the molar mass of the gas?
15. What is partial pressure? What is the relationship between the partial pressures of each gas in a sample and the total pressure of gas in the sample?
16. Why do deep-sea divers breathe a mixture of helium and oxygen?
17. When a gas is collected over water, is the gas pure? Why or why not? How can the partial pressure of the collected gas be determined?
18. What are the basic postulates of kinetic molecular theory? How does the concept of pressure follow from kinetic molecular theory?
19. Explain how Boyle's law, Charles's law, Avogadro's law, and Dalton's law all follow from kinetic molecular theory.
20. How is the kinetic energy of a gas related to temperature? How is the root mean square velocity of a gas related to its molar mass?
21. Describe how perfume molecules travel from the bottle to your nose. What is mean free path?
22. Explain the difference between diffusion and effusion. How is the effusion rate of a gas related to its molar mass?
23. If a reaction occurs in the gas phase at STP, the mass of a product can be determined from the volumes of reactants. Explain.
24. Deviations from the ideal gas law are often observed at high pressure and low temperature. Explain this in light of kinetic molecular theory.

PROBLEMS BY TOPIC

Converting between Pressure Units

25. The pressure in Denver, Colorado (elevation 5280 ft or 1600 m), averages about 24.9 in Hg. Convert this pressure to

    a. atm b. mmHg
    c. psi d. Pa

26. The pressure on top of Mount Everest (29,029 ft or 8848 m) averages about 235 mmHg. Convert this pressure to

    a. torr b. psi
    c. in Hg d. atm

27. The North American record for highest recorded barometric pressure is 31.85 in Hg, set in 1989 in Northway, Alaska. Convert this pressure to:

    a. mmHg b. atm
    c. torr d. kPa (kilopascals)

28. The world record for lowest pressure (at sea level) was 652.5 mmHg recorded inside Typhoon Tip on October 12, 1979, in the western Pacific Ocean. Convert this pressure to:

    a. torr b. atm
    c. in Hg d. psi

29. If the barometric pressure is 762.4 mmHg, what is the pressure of the gas sample shown in each illustration?

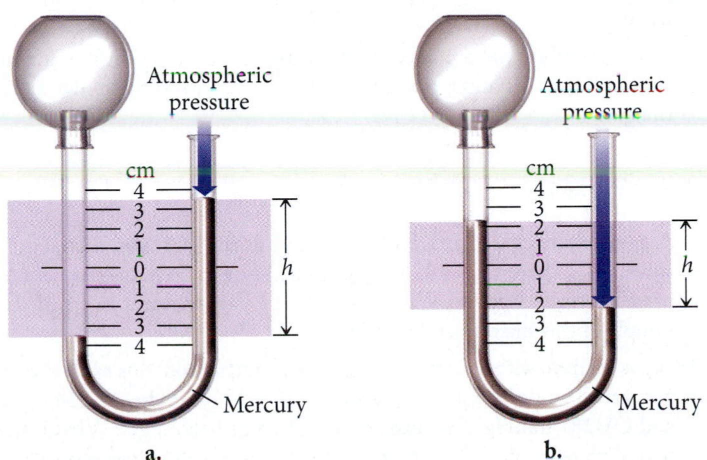

30. If the barometric pressure is 751.5 mmHg, what is the pressure of the gas sample shown in each illustration?

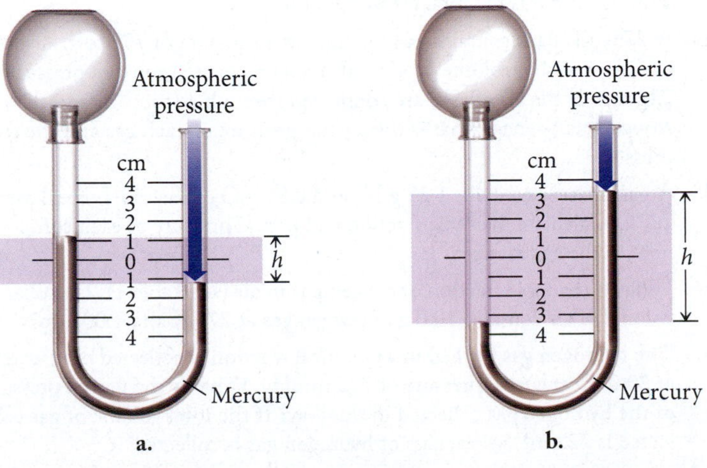

Simple Gas Laws

31. A sample of gas has an initial volume of 5.6 L at a pressure of 735 mmHg. If the volume of the gas is increased to 9.4 L, what is its pressure?

32. A sample of gas has an initial volume of 13.9 L at a pressure of 1.22 atm. If the sample is compressed to a volume of 10.3 L, what is its pressure?

33. A 48.3-mL sample of gas in a cylinder is warmed from 22 °C to 87 °C. What is its volume at the final temperature?

34. A syringe containing 1.55 mL of oxygen gas is cooled from 95.3 °C to 0.0 °C. What is the final volume of oxygen gas?

35. A balloon contains 0.158 mol of gas and has a volume of 2.46 L. If we add 0.113 mol of gas to the balloon (at the same temperature and pressure), what is its final volume?

36. A cylinder with a moveable piston contains 0.553 mol of gas and has a volume of 253 mL. What is its volume if we add 0.365 mol of gas to the cylinder? (Assume constant temperature and pressure.)

Ideal Gas Law

37. What volume does 0.118 mol of helium gas at a pressure of 0.97 atm and a temperature of 305 K occupy? Would the volume be different if the gas was argon (under the same conditions)?

38. What volume does 12.5 g of argon gas at a pressure of 1.05 atm and a temperature of 322 K occupy? Would the volume be different if the sample were 12.5 g of helium (under identical conditions)?

39. What is the pressure in a 10.0-L cylinder filled with 0.448 mol of nitrogen gas at a temperature of 315 K?

40. What is the pressure in a 15.0-L cylinder filled with 32.7 g of oxygen gas at a temperature of 302 K?

41. A cylinder contains 28.5 L of oxygen gas at a pressure of 1.8 atm and a temperature of 298 K. How much gas (in moles) is in the cylinder?

42. What is the temperature of 0.52 mol of gas at a pressure of 1.3 atm and a volume of 11.8 L?

43. An automobile tire has a maximum rating of 38.0 psi (gauge pressure). The tire is inflated (while cold) to a volume of 11.8 L and a gauge pressure of 36.0 psi at a temperature of 12.0 °C. When the car is driven on a hot day, the tire warms to 65.0 °C and its volume expands to 12.2 L. Does the pressure in the tire exceed its maximum rating? (*Note*: The *gauge pressure* is the *difference* between the total pressure and atmospheric pressure. In this case, assume that atmospheric pressure is 14.7 psi.)

44. A weather balloon is inflated to a volume of 28.5 L at a pressure of 748 mmHg and a temperature of 28.0 °C. The balloon rises in the atmosphere to an altitude of approximately 25,000 feet, where the pressure is 385 mmHg and the temperature is −15.0 °C. Assuming the balloon can freely expand, calculate the volume of the balloon at this altitude.

45. A piece of dry ice (solid carbon dioxide) with a mass of 28.8 g sublimes (converts from solid to gas) into a large balloon. Assuming that all of the carbon dioxide ends up in the balloon, what is the volume of the balloon at a temperature of 22 °C and a pressure of 742 mmHg?

46. A 1.0-L container of liquid nitrogen is kept in a closet measuring 1.0 m by 1.0 m by 2.0 m. Assuming that the container is completely full, that the temperature is 25.0 °C, and that the atmospheric pressure is 1.0 atm, calculate the percent (by volume) of air that is displaced if all of the liquid nitrogen evaporates. (Liquid nitrogen has a density of 0.807 g/mL.)

47. A wine-dispensing system uses argon canisters to pressurize and preserve wine in the bottle. An argon canister for the system has a volume of 55.0 mL and contains 26.0 g of argon. Assuming ideal gas behavior, what is the pressure in the canister at 295 K? When the argon is released from the canister, it expands to fill the wine bottle. How many 750.0-mL wine bottles can be purged with the argon in the canister at a pressure of 1.20 atm and a temperature of 295 K?

48. Pressurized carbon dioxide inflators can be used to inflate a bicycle tire in the event of a flat. These inflators use metal cartridges that contain 16.0 g of carbon dioxide. At 298 K, to what gauge pressure (in psi) can the carbon dioxide in the cartridge inflate a 3.45-L mountain bike tire? (*Note*: The *gauge pressure* is the *difference* between the total pressure and atmospheric pressure. In this case, assume that atmospheric pressure is 14.7 psi.)

49. Which gas sample illustrated here has the greatest pressure? Assume that all the samples are at the same temperature. Explain.

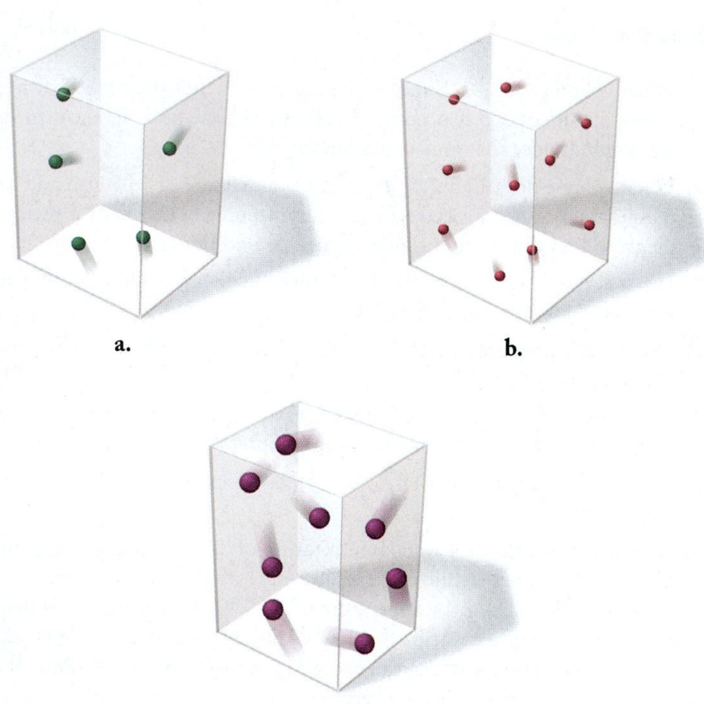

50. This picture represents a sample of gas at a pressure of 1 atm, a volume of 1 L, and a temperature of 25 °C. Draw a similar picture showing what would happen to the sample if the volume were reduced to 0.5 L and the temperature were increased to 250 °C. What would happen to the pressure?

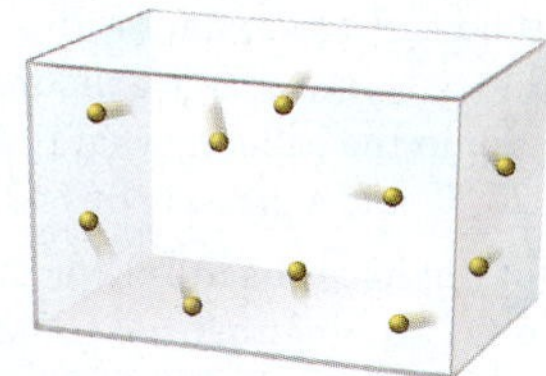

51. Aerosol cans carry clear warnings against incineration because of the high pressures that can develop if they are heated. Suppose that a can contains a residual amount of gas at a pressure of 755 mmHg and a temperature of 25 °C. What would the pressure be if the can were heated to 1155 °C?

52. A sample of nitrogen gas in a 1.75-L container exerts a pressure of 1.35 atm at 25 °C. What is the pressure if the volume of the container is maintained constant and the temperature is raised to 355 °C?

Molar Volume, Density, and Molar Mass of a Gas

53. Use the molar volume of a gas at STP to determine the volume (in L) occupied by 33.6 g of neon at STP.

54. Use the molar volume of a gas at STP to calculate the density (in g/L) of nitrogen gas at STP.

55. What is the density (in g/L) of hydrogen gas at 20.0 °C and a pressure of 1655 psi?

56. A sample of N_2O gas has a density of 2.85 g/L at 298 K. What is the pressure of the gas (in mmHg)?

57. A 248-mL gas sample has a mass of 0.433 g at a pressure of 745 mmHg and a temperature of 28 °C. What is the molar mass of the gas?

58. A 113-mL gas sample has a mass of 0.171 g at a pressure of 721 mmHg and a temperature of 32 °C. What is the molar mass of the gas?

59. A sample of gas has a mass of 38.8 mg. Its volume is 224 mL at a temperature of 55 °C and a pressure of 886 torr. Find the molar mass of the gas.

60. A sample of gas has a mass of 0.555 g. Its volume is 117 mL at a temperature of 85 °C and a pressure of 753 mmHg. Find the molar mass of the gas.

Partial Pressure

61. A gas mixture contains each of these gases at the indicated partial pressures: N_2, 215 torr; O_2, 102 torr; and He, 117 torr. What is the total pressure of the mixture? What mass of each gas is present in a 1.35-L sample of this mixture at 25.0 °C?

62. A gas mixture with a total pressure of 745 mmHg contains each of these gases at the indicated partial pressures: CO_2, 125 mmHg; Ar, 214 mmHg; and O_2, 187 mmHg. The mixture also contains helium gas. What is the partial pressure of the helium gas? What mass of helium gas is present in a 12.0-L sample of this mixture at 273 K?

63. We add a 1.20-g sample of dry ice to a 755-mL flask containing nitrogen gas at a temperature of 25.0 °C and a pressure of 725 mmHg. The dry ice sublimes (converts from solid to gas) and the mixture returns to 25.0 °C. What is the total pressure in the flask?

64. A 275-mL flask contains pure helium at a pressure of 752 torr. A second flask with a volume of 475 mL contains pure argon at a pressure of 722 torr. If the two flasks are connected through a stopcock and the stopcock is opened, what is the partial pressure of each gas and the total pressure?

65. A gas mixture contains 1.25 g N_2 and 0.85 g O_2 in a 1.55-L container at 18 °C. Calculate the mole fraction and partial pressure of each component in the gas mixture.

66. What is the mole fraction of oxygen gas in air (see Table 11.2)? What volume of air contains 10.0 g of oxygen gas at 273 K and 1.00 atm?

67. The hydrogen gas formed in a chemical reaction is collected over water at 30.0 °C at a total pressure of 732 mmHg. What is the partial pressure of the hydrogen gas collected in this way? If the total volume of gas collected is 722 mL, what mass of hydrogen gas is collected?

68. The air in a bicycle tire is bubbled through water and collected at 25 °C. If the total volume of gas collected is 5.45 L at a temperature of 25 °C and a pressure of 745 torr, how many moles of gas were in the bicycle tire?

69. The zinc within a copper-plated penny will dissolve in hydrochloric acid if the copper coating is filed down in several spots (so that the hydrochloric acid can get to the zinc). The reaction between the acid and the zinc is $2\ H^+(aq) + Zn(s) \longrightarrow H_2(g) + Zn^{2+}(aq)$. When the zinc in a certain penny dissolves, the total volume of gas collected over water at 25 °C is 0.951 L at a total pressure of 748 mmHg. What mass of hydrogen gas is collected?

70. A heliox deep-sea diving mixture contains 2.0 g of oxygen to every 98.0 g of helium. What is the partial pressure of oxygen when this mixture is delivered at a total pressure of 8.5 atm?

Kinetic Molecular Theory

71. Consider a 1.0-L sample of helium gas and a 1.0-L sample of argon gas, both at room temperature and atmospheric pressure.
   a. Do the atoms in the helium sample have the same *average kinetic energy* as the atoms in the argon sample?
   b. Do the atoms in the helium sample have the same *average velocity* as the atoms in the argon sample?
   c. Do the argon atoms, because they are more massive, exert a greater pressure on the walls of the container? Explain.
   d. Which gas sample has the faster rate of effusion?

72. A flask at room temperature contains exactly equal amounts (in moles) of nitrogen and xenon.
   a. Which of the two gases exerts the greater partial pressure?
   b. The molecules or atoms of which gas will have the greater average velocity?
   c. The molecules of which gas will have the greater average kinetic energy?
   d. If a small hole were opened in the flask, which gas would effuse more quickly?

73. Calculate the root mean square velocity and kinetic energy of F_2, Cl_2, and Br_2 at 298 K. Rank the three halogens with respect to their rate of effusion.

74. Calculate the root mean square velocity and kinetic energy of CO, CO_2, and SO_3 at 298 K. Which gas has the greatest velocity? The greatest kinetic energy? The greatest effusion rate?

75. We obtain uranium-235 from U-238 by fluorinating the uranium to form UF_6 (which is a gas) and then taking advantage of the different rates of effusion and diffusion for compounds containing the two isotopes. Calculate the ratio of effusion rates for $^{238}UF_6$ and $^{235}UF_6$. The atomic mass of U-235 is 235.054 amu, and that of U-238 is 238.051 amu.

76. Calculate the ratio of effusion rates for Ar and Kr.

77. A sample of neon effuses from a container in 76 seconds. The same amount of an unknown noble gas requires 155 seconds. Identify the gas.

78. A sample of N_2O effuses from a container in 42 seconds. How long will it take the same amount of gaseous I_2 to effuse from the same container under identical conditions?

79. This graph shows the distribution of molecular velocities for two different molecules (A and B) at the same temperature. Which molecule has the higher molar mass? Which molecule has the higher rate of effusion?

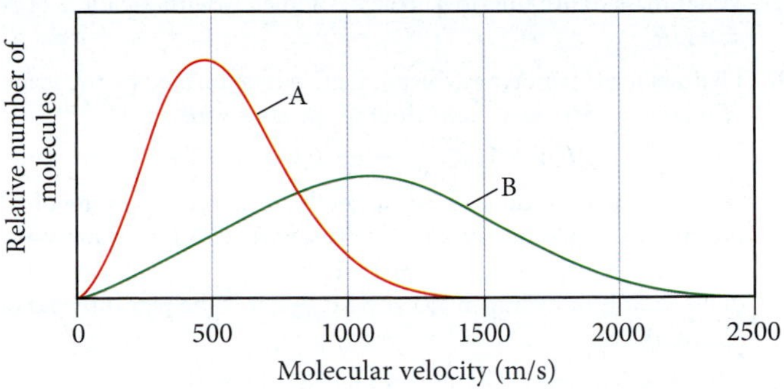

80. This graph shows the distribution of molecular velocities for the same molecule at two different temperatures (T_1 and T_2). Which temperature is greater? Explain.

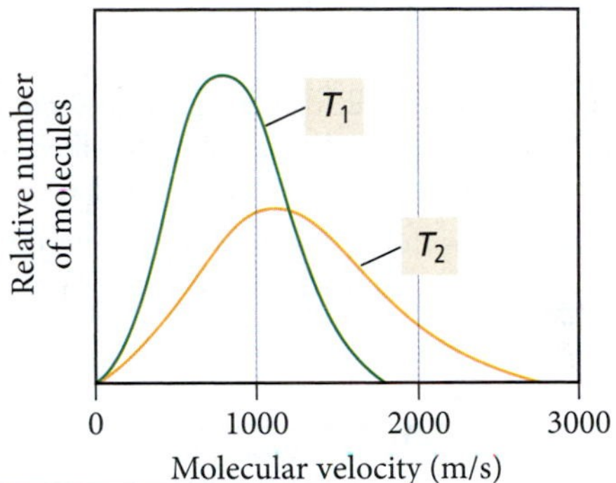

Reaction Stoichiometry Involving Gases

81. Consider the chemical reaction:

$$C(s) + H_2O(g) \longrightarrow CO(g) + H_2(g)$$

How many liters of hydrogen gas are formed from the complete reaction of 15.7 g C? Assume that the hydrogen gas is collected at a pressure of 1.0 atm and a temperature of 355 K.

82. Consider the chemical reaction:

$$2\ H_2O(l) \longrightarrow 2\ H_2(g) + O_2(g)$$

What mass of H_2O is required to form 1.4 L of O_2 at a temperature of 315 K and a pressure of 0.957 atm?

83. CH_3OH can be synthesized by the reaction:

$$CO(g) + 2\ H_2(g) \longrightarrow CH_3OH(g)$$

What volume of H_2 gas (in L), at 748 mmHg and 86 °C, is required to synthesize 25.8 g CH_3OH? How many liters of CO gas, measured under the same conditions, are required?

84. Oxygen gas reacts with powdered aluminum according to the reaction:

$$4\ Al(s) + 3\ O_2(g) \longrightarrow 2\ Al_2O_3(s)$$

What volume of O_2 gas (in L), measured at 782 mmHg and 25 °C, completely reacts with 53.2 g Al?

85. Automobile air bags inflate following serious impacts, which trigger the chemical reaction:

$$2\ NaN_3(s) \longrightarrow 2\ Na(s) + 3\ N_2(g)$$

If an automobile air bag has a volume of 11.8 L, what mass of NaN_3 (in g) is required to fully inflate the air bag upon impact? Assume STP conditions.

86. Lithium reacts with nitrogen gas according to the reaction:

$$6\ Li(s) + N_2(g) \longrightarrow 2\ Li_3N(s)$$

What mass of lithium (in g) reacts completely with 58.5 mL of N_2 gas at STP?

87. Hydrogen gas (a potential future fuel) can be formed by the reaction of methane with water according to the equation:

$$CH_4(g) + H_2O(g) \longrightarrow CO(g) + 3\,H_2(g)$$

In a particular reaction, 25.5 L of methane gas (at a pressure of 732 torr and a temperature of 25 °C) mixes with 22.8 L of water vapor (at a pressure of 702 torr and a temperature of 125 °C). The reaction produces 26.2 L of hydrogen gas at STP. What is the percent yield of the reaction?

88. Ozone is depleted in the stratosphere by chlorine from CF_3Cl according to this set of equations:

$$CF_3Cl + \text{UV light} \longrightarrow CF_3 + Cl$$
$$Cl + O_3 \longrightarrow ClO + O_2$$
$$O_3 + \text{UV light} \longrightarrow O_2 + O$$
$$ClO + O \longrightarrow Cl + O_2$$

What total volume of ozone at a pressure of 25.0 mmHg and a temperature of 225 K is destroyed when all of the chlorine from 15.0 g of CF_3Cl goes through ten cycles of the above reactions?

89. Chlorine gas reacts with fluorine gas to form chlorine trifluoride.

$$Cl_2(g) + 3\,F_2(g) \longrightarrow 2\,ClF_3(g)$$

A 2.00-L reaction vessel, initially at 298 K, contains chlorine gas at a partial pressure of 337 mmHg and fluorine gas at a partial pressure of 729 mmHg. Identify the limiting reactant and determine the theoretical yield of ClF_3 in grams.

90. Carbon monoxide gas reacts with hydrogen gas to form methanol.

$$CO(g) + 2\,H_2(g) \longrightarrow CH_3OH(g)$$

A 1.50-L reaction vessel, initially at 305 K, contains carbon monoxide gas at a partial pressure of 232 mmHg and hydrogen gas at a partial pressure of 397 mmHg. Identify the limiting reactant and determine the theoretical yield of methanol in grams.

Real Gases

91. Which postulate of the kinetic molecular theory breaks down under conditions of high pressure? Explain.

92. Which postulate of the kinetic molecular theory breaks down under conditions of low temperature? Explain.

93. Use the van der Waals equation and the ideal gas equation to calculate the volume of 1.000 mol of neon at a pressure of 500.0 atm and a temperature of 355.0 K. Explain why the two values are different. (*Hint*: One way to solve the van der Waals equation for V is to use successive approximations. Use the ideal gas law to get a preliminary estimate for V.)

CUMULATIVE PROBLEMS

94. Use the van der Waals equation and the ideal gas equation to calculate the pressure exerted by 1.000 mol of Cl_2 in a volume of 5.000 L at a temperature of 273.0 K. Explain why the two values are different.

95. Pennies that are currently being minted are composed of zinc coated with copper. A student determines the mass of a penny to be 2.482 g and then makes several scratches in the copper coating (to expose the underlying zinc). The student puts the scratched penny in hydrochloric acid, where the following reaction occurs between the zinc and the HCl (the copper remains undissolved):

$$Zn(s) + 2\,HCl(aq) \longrightarrow H_2(g) + ZnCl_2(aq)$$

The student collects the hydrogen produced over water at 25 °C. The collected gas occupies a volume of 0.899 L at a total pressure of 791 mmHg. Calculate the percent zinc (by mass) in the penny. (Assume that all the Zn in the penny dissolves.)

96. A 2.85-g sample of an unknown chlorofluorocarbon decomposes and produces 564 mL of chlorine gas at a pressure of 752 mmHg and a temperature of 298 K. What is the percent chlorine (by mass) in the unknown chlorofluorocarbon?

97. The mass of an evacuated 255-mL flask is 143.187 g. The mass of the flask filled with 267 torr of an unknown gas at 25 °C is 143.289 g. Calculate the molar mass of the unknown gas.

98. A 118-mL flask is evacuated and found to have a mass of 97.129 g. When the flask is filled with 768 torr of helium gas at 35 °C, it has a mass of 97.171 g. Is the helium gas pure?

99. A gaseous hydrogen and carbon containing compound is decomposed and found to contain 82.66% carbon and 17.34% hydrogen by mass. The mass of 158 mL of the gas, measured at 556 mmHg and 25 °C, is 0.275 g. What is the molecular formula of the compound?

100. A gaseous hydrogen-and-carbon containing compound is decomposed and found to contain 85.63% C and 14.37% H by mass. The mass of 258 mL of the gas, measured at STP, is 0.646 g. What is the molecular formula of the compound?

101. Consider the reaction:

$$2\,NiO(s) \longrightarrow 2\,Ni(s) + O_2(g)$$

If O_2 is collected over water at 40.0 °C and a total pressure of 745 mmHg, what volume of gas is collected for the complete reaction of 24.78 g of NiO?

102. Consider the reaction:

$$2\,Ag_2O(s) \longrightarrow 4\,Ag(s) + O_2(g)$$

If this reaction produces 15.8 g of Ag(s), what total volume of gas can be collected over water at a temperature of 25 °C and a total pressure of 752 mmHg?

103. When hydrochloric acid is poured over potassium sulfide, 42.9 mL of hydrogen sulfide gas is produced at a pressure of 752 torr and 25.8 °C. Write an equation for the gas-evolution reaction and determine how much potassium sulfide (in grams) reacted.

104. Consider the reaction:

$$2\,SO_2(g) + O_2(g) \longrightarrow 2\,SO_3(g)$$

a. If 285.5 mL of SO_2 reacts with 158.9 mL of O_2 (both measured at 315 K and 50.0 mmHg), what is the limiting reactant and the theoretical yield of SO_3?

b. If 187.2 mL of SO_3 is collected (measured at 315 K and 50.0 mmHg), what is the percent yield for the reaction?

105. Ammonium carbonate decomposes upon heating according to the balanced equation:

$$(NH_4)_2CO_3(s) \longrightarrow 2\,NH_3(g) + CO_2(g) + H_2O(g)$$

Calculate the total volume of gas produced at 22 °C and 1.02 atm by the complete decomposition of 11.83 g of ammonium carbonate.

106. Ammonium nitrate decomposes explosively upon heating according to the balanced equation:

$$2\ NH_4NO_3(s) \longrightarrow 2\ N_2(g) + O_2(g) + 4\ H_2O(g)$$

Calculate the total volume of gas (at 125 °C and 748 mmHg) produced by the complete decomposition of 1.55 kg of ammonium nitrate.

107. Olympic cyclists fill their tires with helium to make them lighter. Calculate the mass of air in an air-filled tire and the mass of helium in a helium-filled tire. What is the mass difference between the two? Assume that the volume of the tire is 855 mL, that it is filled to a total pressure of 125 psi, and that the temperature is 25 °C. Also, assume an average molar mass for air of 28.8 g/mol.

108. In a common classroom demonstration, a balloon is filled with air and submerged in liquid nitrogen. The balloon contracts as the gases within the balloon cool. Suppose a balloon initially contains 2.95 L of air at a temperature of 25.0 °C and a pressure of 0.998 atm. Calculate the expected volume of the balloon upon cooling to −196 °C (the boiling point of liquid nitrogen). When the demonstration is carried out, the actual volume of the balloon decreases to 0.61 L. How does the observed volume of the balloon compare to your calculated value? Explain the difference.

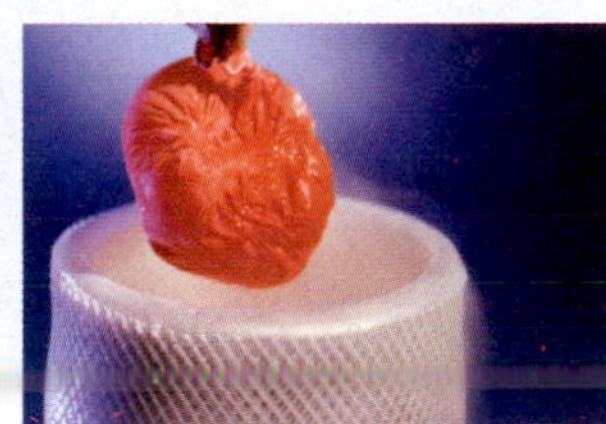

109. Gaseous ammonia is injected into the exhaust stream of a coal-burning power plant to reduce the pollutant NO to N_2 according to the reaction:

$$4\ NH_3(g) + 4\ NO(g) + O_2(g) \longrightarrow 4\ N_2(g) + 6\ H_2O(g)$$

Suppose that the exhaust stream of a power plant has a flow rate of 335 L/s at a temperature of 955 K, and that the exhaust contains a partial pressure of NO of 22.4 torr. What should be the flow rate of ammonia delivered at 755 torr and 298 K into the stream to react completely with the NO if the ammonia is 65.2% pure (by volume)?

110. The emission of NO_2 by fossil fuel combustion can be prevented by injecting gaseous urea into the combustion mixture. The urea reduces NO (which oxidizes in air to form NO_2) according to the reaction:

$$2\ CO(NH_2)_2(g) + 4\ NO(g) + O_2(g) \longrightarrow 4\ N_2(g) + 2\ CO_2(g) + 4\ H_2O(g)$$

Suppose that the exhaust stream of an automobile has a flow rate of 2.55 L/s at 655 K and contains a partial pressure of NO of 12.4 torr. What total mass of urea is necessary to react completely with the NO formed during 8.0 hours of driving?

111. An ordinary gasoline can measuring 30.0 cm by 20.0 cm by 15.0 cm is evacuated with a vacuum pump. Assuming that virtually all of the air can be removed from inside the can, and that atmospheric pressure is 14.7 psi, what is the total force (in pounds) on the surface of the can? Do you think that the can will withstand the force?

112. Twenty-five milliliters of liquid nitrogen (density = 0.807 g/mL) is poured into a cylindrical container with a radius of 10.0 cm and a length of 20.0 cm. The container initially contains only air at a pressure of 760.0 mmHg (atmospheric pressure) and a temperature of 298 K. If the liquid nitrogen completely vaporizes, what is the total force (in lb) on the interior of the container at 298 K?

113. A 160.0-L helium tank contains pure helium at a pressure of 1855 psi and a temperature of 298 K. How many 3.5-L helium balloons will the helium in the tank fill? (Assume an atmospheric pressure of 1.0 atm and a temperature of 298 K.)

114. An 11.5-mL sample of liquid butane (density = 0.573 g/mL) is evaporated in an otherwise empty container at a temperature of 28.5 °C. The pressure in the container following evaporation is 892 torr. What is the volume of the container?

115. A scuba diver creates a spherical bubble with a radius of 2.5 cm at a depth of 30.0 m where the total pressure (including atmospheric pressure) is 4.00 atm. What is the radius of the bubble when it reaches the surface of the water? (Assume that the atmospheric pressure is 1.00 atm and the temperature is 298 K.)

116. A particular balloon can be stretched to a maximum surface area of 1257 cm^2. The balloon is filled with 3.0 L of helium gas at a pressure of 755 torr and a temperature of 298 K. The balloon is then allowed to rise in the atmosphere. If the atmospheric temperature is 273 K, at what pressure will the balloon burst? (Assume the balloon is a sphere.)

117. A catalytic converter in an automobile uses a palladium or platinum catalyst (a substance that increases the rate of a reaction without being consumed by the reaction) to convert carbon monoxide gas to carbon dioxide according to the reaction:

$$2\ CO(g) + O_2(g) \longrightarrow 2\ CO_2(g)$$

A chemist researching the effectiveness of a new catalyst combines a 2.0 : 1.0 mole ratio mixture of carbon monoxide and oxygen gas (respectively) over the catalyst in a 2.45-L flask at a total pressure of 745 torr and a temperature of 552 °C. When the reaction is complete, the pressure in the flask has dropped to 552 torr. What percentage of the carbon monoxide was converted to carbon dioxide?

118. A quantity of N_2 occupies a volume of 1.0 L at 300 K and 1.0 atm. The gas expands to a volume of 3.0 L as the result of a change in both temperature and pressure. Find the density of the gas at these new conditions.

119. A mixture of $CO(g)$ and $O_2(g)$ in a 1.0-L container at 1.0×10^3 K has a total pressure of 2.2 atm. After some time the total pressure falls to 1.9 atm as the result of the formation of CO_2. Determine the mass (in grams) of CO_2 that forms.

120. The radius of a xenon atom is 1.3×10^{-8} cm. A 100-mL flask is filled with Xe at a pressure of 1.0 atm and a temperature of 273 K. Calculate the fraction of the volume that is occupied by Xe atoms. (*Hint:* The atoms are spheres.)

121. A natural gas storage tank is a cylinder with a moveable top whose volume can change only as its height changes. Its radius remains fixed. The height of the cylinder is 22.6 m on a day when the temperature is 22 °C. The next day the height of the cylinder increases to 23.8 m when the gas expands because of a heat wave. Find the temperature on the second day, assuming that the pressure and amount of gas in the storage tank have not changed.

122. A mixture of 8.0 g CH_4 and 8.0 g Xe is placed in a container, and the total pressure is found to be 0.44 atm. Determine the partial pressure of CH_4.

123. A steel container of volume 0.35 L can withstand pressures up to 88 atm before exploding. What mass of helium can be stored in this container at 299 K?

124. Binary compounds of alkali metals and hydrogen react with water to liberate $H_2(g)$. The H_2 from the reaction of a sample of NaH with an excess of water fills a volume of 0.490 L above the water. The temperature of the gas is 35 °C, and the total pressure is 758 mmHg. Determine the mass of H_2 liberated and the mass of NaH that reacted.

125. In a given diffusion apparatus, 15.0 mL of HBr gas diffused in 1.0 min. In the same apparatus and under the same conditions, 20.3 mL of an unknown gas diffused in 1.0 min. The unknown gas is a hydrocarbon. Find its molecular formula.

126. A sample of $N_2O_3(g)$ has a pressure of 0.017 atm. The temperature (in K) is doubled, and the N_2O_3 undergoes complete decomposition to $NO_2(g)$ and $NO(g)$. Find the total pressure of the mixture of gases assuming constant volume and no additional temperature change.

127. When 0.583 g of neon is added to an 800 cm^3 bulb containing a sample of argon, the total pressure of the gases is 1.17 atm at a temperature of 295 K. Find the mass of the argon in the bulb.

128. A gas mixture composed of helium and argon has a density of 0.670 g/L at 755 mmHg and 298 K. What is the composition of the mixture by volume?

129. A gas mixture contains 75.2% nitrogen and 24.8% krypton by mass. What is the partial pressure of krypton in the mixture if the total pressure is 745 mmHg?

CHALLENGE PROBLEMS

130. A 10-L container is filled with 0.10 mol of $H_2(g)$ and heated to 3000 K, causing some of the $H_2(g)$ to decompose into $H(g)$. The total pressure is 3.0 atm. Find the partial pressure of the $H(g)$ that forms from H_2 at this temperature. (Assume two significant figures for the temperature.)

131. A mixture of $NH_3(g)$ and $N_2H_4(g)$ is placed in a sealed container at 300 K. The total pressure is 0.50 atm. The container is heated to 1200 K, at which time both substances decompose completely according to the equations: $2\ NH_3(g) \longrightarrow N_2(g) + 3\ H_2(g)$; $N_2H_4(g) \longrightarrow N_2(g) + 2\ H_2(g)$. After decomposition is complete, the total pressure at 1200 K is 4.5 atm. Find the percent of $N_2H_4(g)$ in the original mixture. (Assume two significant figures for the temperature.)

132. A quantity of CO gas occupies a volume of 0.48 L at 1.0 atm and 275 K. The pressure of the gas is lowered and its temperature is raised until its volume is 1.3 L. Determine the density of the CO under the new conditions.

133. When $CO_2(g)$ is put in a sealed container at 701 K and a pressure of 10.0 atm and is heated to 1401 K, the pressure rises to 22.5 atm. Some of the CO_2 decomposes to CO and O_2. Calculate the mole percent of CO_2 that decomposes.

134. The world burns approximately 3.5×10^{12} kg of fossil fuel per year. Use the combustion of octane as the representative reaction and determine the mass of carbon dioxide (the most significant greenhouse gas) formed per year. The current concentration of carbon dioxide in the atmosphere is approximately 394 ppm (by volume). By what percentage does the concentration increase each year due to fossil fuel combustion? Approximate the average properties of the entire atmosphere by assuming that the atmosphere extends from sea level to 15 km and that it has an average pressure of 381 torr and average temperature of 275 K. Assume Earth is a perfect sphere with a radius of 6371 km.

135. The atmosphere slowly oxidizes hydrocarbons in a number of steps that eventually convert the hydrocarbon into carbon dioxide and water. The overall reaction of a number of such steps for methane gas is:

$$CH_4(g) + 5\ O_2(g) + 5\ NO(g) \longrightarrow CO_2(g) + H_2O(g) + 5\ NO_2(g) + 2\ OH(g)$$

Suppose that an atmospheric chemist combines 155 mL of methane at STP, 885 mL of oxygen at STP, and 55.5 mL of NO at STP in a 2.0-L flask. The flask stands for several weeks at 275 K. If the reaction reaches 90.0% of completion (90.0% of the limiting reactant is consumed), what is the partial pressure of each of the reactants and products in the flask at 275 K? What is the total pressure in the flask?

136. Two identical balloons are filled to the same volume, one with air and one with helium. The next day, the volume of the air-filled balloon has decreased by 5.0%. By what percent has the volume of the helium-filled balloon decreased? (Assume that the air is four-fifths nitrogen and one-fifth oxygen, and that the temperature did not change.)

137. A mixture of $CH_4(g)$ and $C_2H_6(g)$ has a total pressure of 0.53 atm. Just enough $O_2(g)$ is added to the mixture to bring about its complete combustion to $CO_2(g)$ and $H_2O(g)$. The total pressure of the two product gases is found to be 2.2 atm. Assuming constant volume and temperature, find the mole fraction of CH_4 in the mixture.

138. A sample of $C_2H_2(g)$ has a pressure of 7.8 kPa. After some time, a portion of it reacts to form $C_6H_6(g)$. The total pressure of the mixture of gases is then 3.9 kPa. Assume the volume and the temperature do not change. What fraction of $C_2H_2(g)$ has undergone reaction?

CONCEPTUAL PROBLEMS

139. When the driver of an automobile applies the brakes, the passengers are pushed toward the front of the car, but a helium balloon is pushed toward the back of the car. Upon forward acceleration, the passengers are pushed toward the back of the car, but the helium balloon is pushed toward the front of the car. Why?

140. Suppose that a liquid is 10 times denser than water. If you were to sip this liquid at sea level using a straw, what is the maximum length your straw can be?

141. The generic reaction occurs in a closed container:

$$A(g) + 2\ B(g) \longrightarrow 2\ C(g)$$

A reaction mixture initially contains 1.5 L of A and 2.0 L of B. Assuming that the volume and temperature of the reaction mixture remain constant, what is the percent change in pressure if the reaction goes to completion?

142. One mole of nitrogen and one mole of neon are combined in a closed container at STP. How big is the container?

143. Exactly equal amounts (in moles) of gas A and gas B are combined in a 1-L container at room temperature. Gas B has a molar mass that is twice that of gas A. Which statement is true for the mixture of gases and why?

a. The molecules of gas B have greater kinetic energy than those of gas A.
b. Gas B has a greater partial pressure than gas A.
c. The molecules of gas B have a greater average velocity than those of gas A.
d. Gas B makes a greater contribution to the average density of the mixture than gas A.

144. Which gas would you expect to deviate most from ideal behavior under conditions of low temperature: F_2, Cl_2, or Br_2? Explain.

145. The volume of a sample of a fixed amount of gas is decreased from 2.0 L to 1.0 L. The temperature of the gas in kelvins is then doubled. What is the final pressure of the gas in terms of the initial pressure?

146. Which gas sample has the greatest volume at STP?

a. 10.0 g Kr
b. 10.0 g Xe
c. 10.0 g He

147. Draw a depiction of a gas sample, as described by kinetic molecular theory, containing equal molar amounts of helium, neon, and krypton. Use different color dots to represent each element. Give each atom a "tail" to represent its velocity relative to the others in the mixture.

ANSWERS TO CONCEPTUAL CONNECTIONS

Cc 11.1 **(e)** The final volume of the gas is the same as the initial volume because doubling the pressure *decreases* the volume by a factor of 2, but doubling the temperature *increases* the volume by a factor of 2. The two changes in volume are equal in magnitude but opposite in sign, resulting in a final volume that is equal to the initial volume.

Cc 11.2 **(a)** Because 1 g of H_2 contains the greatest number of moles (due to H_2 having the lowest molar mass of the listed gases), and one mole of *any* ideal gas occupies the same volume, the H_2 occupies the greatest volume.

Cc 11.3 $Ne < O_2 < F_2 < Cl_2$; Because each gas occupies the same volume at STP (assuming ideal behavior), the densities increase with increasing molar mass.

Cc 11.4 $P_{He} = 1.5\,atm$; $P_{Ne} = 1.5\,atm$. Because the number of moles of each gas are equal, the mole fraction of each gas is 0.50 and the partial pressure of each gas is simply $0.50 \times P_{tot}$.

Cc 11.5 Although the velocity "tails" of each atom will have different lengths, the average length of the tails on the argon atoms in your drawing should be longer than the average length of the tails on the xenon atoms. Because the argon atoms are lighter, they must on average move faster than the xenon atoms to have the same kinetic energy.

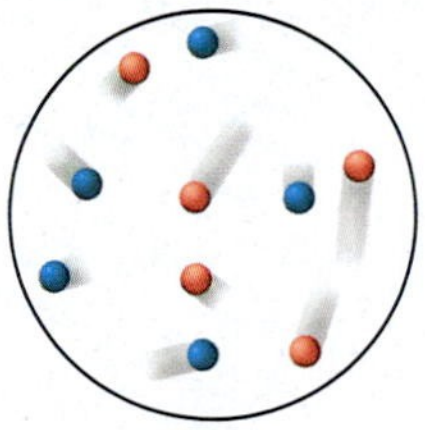

Cc 11.6 **(c)** Because the temperature and the volume are both constant, the ideal gas law tells us that the pressure depends solely on the number of particles. Sample **(c)** has the greatest number of particles per unit volume, and therefore has the greatest pressure. The pressures of samples **(a)** and **(b)** at a given temperature are identical. Even though the particles in **(b)** are more massive than those in **(a)**, they have the same average kinetic energy at a given temperature. The particles in **(b)** move more slowly than those in **(a)**, and therefore exert the same pressure as the particles in **(a)**.

Cc 11.7 **(b)** Because the total number of gas molecules decreases, the total pressure—the sum of all the partial pressures—must also decrease.

Cc 11.8 A < B < C. Curve A is the lowest temperature curve because it deviates the most from ideality. The tendency for the intermolecular forces in carbon dioxide to lower the pressure (relative to that of an ideal gas) is greatest at low temperature (because the molecules are moving more slowly and are therefore less able to overcome the intermolecular forces). As a result, the curve that dips the lowest must correspond to the lowest temperature.

"It's a wild dance floor there at the molecular level."

—Roald Hoffmann (1937–)

CHAPTER 12

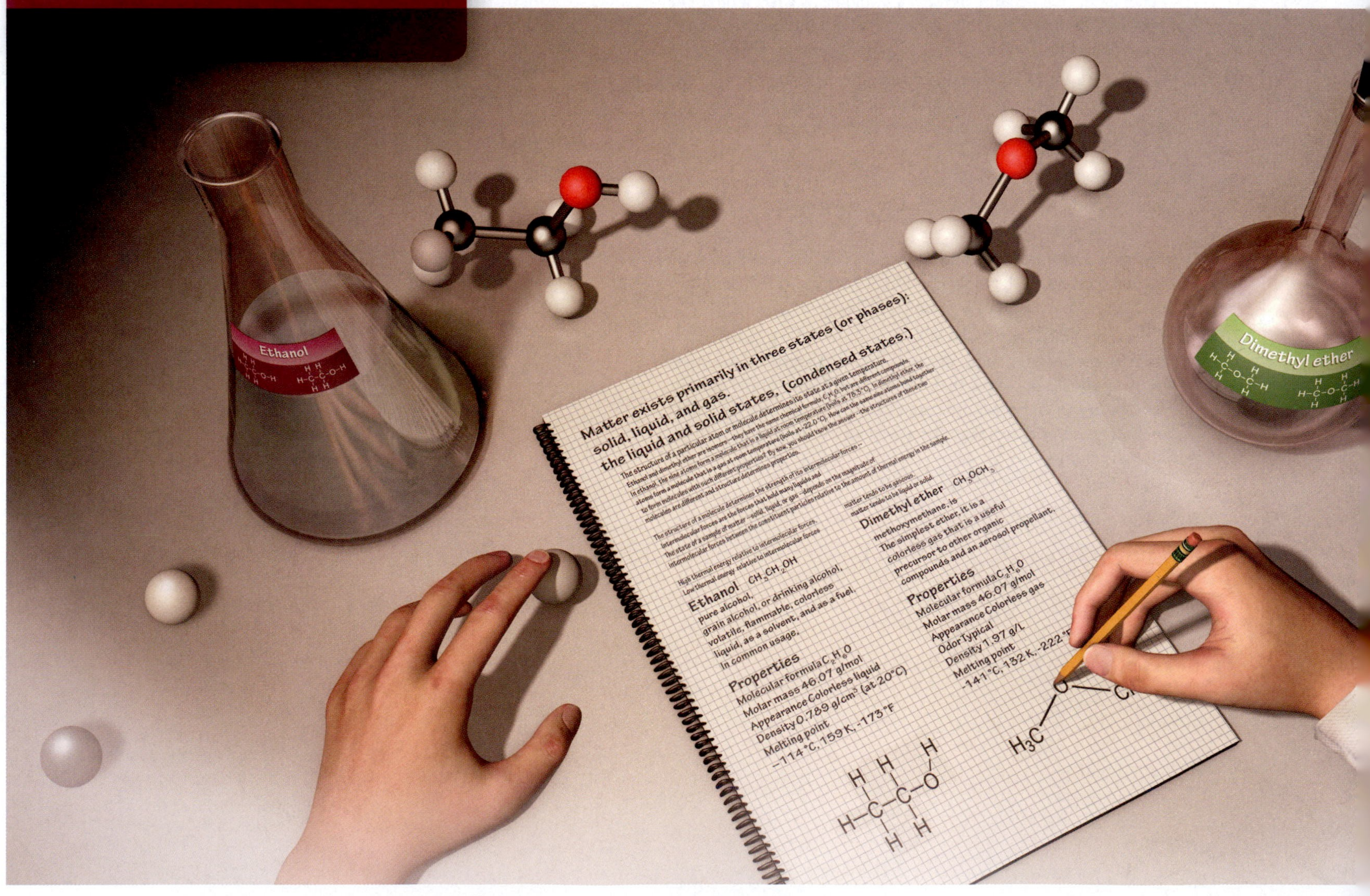

Ethanol and dimethyl ether are isomers—they have the same chemical formula, C_2H_6O but different structures. In ethanol, the nine atoms form a molecule that is a liquid at room temperature. In dimethyl ether, however, the same 9 atoms atoms form a molecule that is a gas at room temperature.

Liquids, Solids, and Intermolecular Forces

12.1 Structure Determines Properties 441

12.2 Solids, Liquids, and Gases: A Molecular Comparison 442

12.3 Intermolecular Forces: The Forces That Hold Condensed States Together 445

12.4 Intermolecular Forces in Action: Surface Tension, Viscosity, and Capillary Action 454

12.5 Vaporization and Vapor Pressure 456

12.6 Sublimation and Fusion 466

12.7 Heating Curve for Water 468

12.8 Water: An Extraordinary Substance 470

Key Learning Outcomes 473

RECALL FROM CHAPTER 1 that matter exists primarily in three states (or phases): solid, liquid, and gas. In Chapter 11, we examined the gas state. In this chapter and the next we turn to the liquid and solid states, known collectively as the *condensed* states. The liquid and solid states are more similar to each other than they are to the gas state. In the gas state, the constituent particles—atoms or molecules—are separated by large distances and do not interact with each other very much. In the condensed states, the constituent particles are close together and exert moderate to strong attractive forces on one another. Whether a substance is a solid, liquid, or gas depends on the structure of the particles that compose the substance. Remember the theme we have emphasized since Chapter 1 of this book: The properties of matter are determined by the properties of the particles that compose it. In this chapter, we will see how the structure of a particular atom or molecule determines its state at a given temperature.

12.1 Structure Determines Properties

Ethanol and dimethyl ether are isomers—they have the same chemical formula, C_2H_6O but are different compounds. In ethanol, the nine atoms form a molecule that is a liquid at room temperature (boils at 78.3 °C). In dimethyl ether, the atoms form a molecule that is a gas at room temperature (boils at −22.0 °C). How can the same nine atoms bond together to form molecules with such different properties? By now, you should know the answer—the structures of these two molecules are different, and *structure determines properties*.

Name	Formula	Space-Filling Model	Structural Formula	bp (°C)
Ethanol	C_2H_6O		CH_3CH_2OH	78.3
Dimethyl Ether	C_2H_6O		CH_3OCH_3	−22.0

In this chapter, we examine how the structure of a molecule determines the strength of its *intermolecular forces*—the attractive forces that exist *between* all molecules and atoms. Intermolecular forces are the forces that hold many liquids and solids—such as ethanol or water—together. In fact, intermolecular forces are responsible for the very existence of condensed states (in molecular compounds). The state of a sample of matter—solid, liquid, or gas—depends on the magnitude of intermolecular forces between the constituent particles relative to the amount of thermal energy in the sample. Recall from Chapter 10 that the molecules and atoms composing matter are in constant random motion that increases with increasing temperature. The energy associated with this motion is *thermal energy.* When thermal energy is high relative to intermolecular forces, matter tends to be gaseous. When thermal energy is low relative to intermolecular forces, matter tends to be liquid or solid.

12.2 Solids, Liquids, and Gases: A Molecular Comparison

We are all familiar with solids, liquids, and gases. Water, gasoline, rubbing alcohol, and nail polish remover are common liquids. Ice, dry ice, and diamond are familiar solids. Oxygen, nitrogen, and helium are common gases. To begin to understand the differences between the three common states of matter, examine Table 12.1, which shows the density and molar volume of water in its three different states,

TABLE 12.1 The Three States of Water

Phase	Temperature (°C)	Density (g/cm^3, at 1 atm)	Molar Volume	Molecular View
Gas (steam)	100	5.90×10^{-4}	30.5 L	
Liquid (water)	20	0.998	18.0 mL	
Solid (ice)	0	0.917	19.6 mL	

along with molecular representations of each state. Notice that the densities of the solid and liquid states are much greater than the density of the gas state. Notice also that the solid and liquid states are more similar in density and molar volume to one another than they are to the gas state. The molecular representations show the reason for these differences. The molecules in liquid water and ice are in close contact with one another—essentially touching—while those in gaseous water are separated by large distances. The molecular representation of gaseous water in Table 12.1 is actually out of proportion—the water molecules in the figure should be much farther apart for their size. (Only a fraction of a molecule could be included in the table if it were drawn to scale.) From the molar volumes, we know that 18.0 mL of liquid water (slightly more than a tablespoon) occupies 30.5 L when converted to gas at 100 °C (at atmospheric pressure). The low density of gaseous water is a direct result of this large separation between molecules.

Notice also that, for water, the solid is slightly less dense than the liquid. This is *atypical* behavior. Most solids are slightly more dense than their corresponding liquids because the molecules move closer together upon freezing. As we will discuss in Section 12.8, ice is less dense than liquid water because the unique crystal structure of ice results in water molecules moving slightly farther apart upon freezing.

A major difference between liquids and solids is the freedom of movement of the constituent molecules or atoms. Even though the atoms or molecules in a liquid are in close contact, thermal energy partially overcomes the attractions between them, allowing them to move around one another. This is not the case in solids; the atoms or molecules in a solid are virtually locked in their positions, only vibrating back and forth about a fixed point. Table 12.2 summarizes the properties of liquids and solids, as well as the properties of gases for comparison.

TABLE 12.2 Properties of the States of Matter

State	Density	Shape	Volume	Strength of Intermolecular Forces (Relative to Thermal Energy)
Gas	Low	Indefinite	Indefinite	Weak
Liquid	High	Indefinite	Definite	Moderate
Solid	High	Definite	Definite	Strong

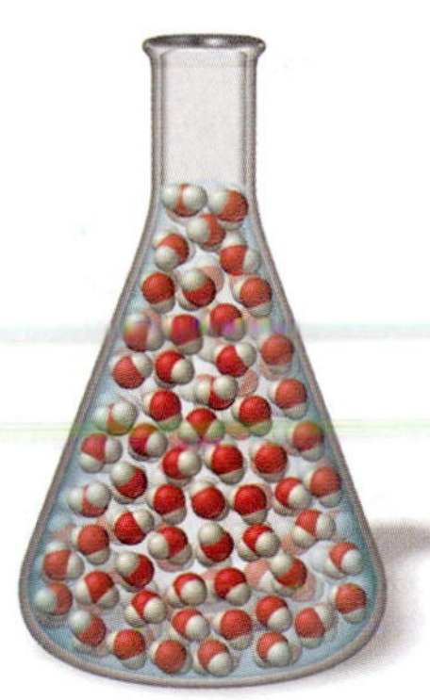

▲ **FIGURE 12.1 Liquids Assume the Shapes of Their Containers** When we pour water into a flask, it assumes the shape of the flask because water molecules are free to flow.

Liquids assume the shape of their containers because the atoms or molecules that compose liquids are free to flow (or move around one another). When we pour water into a beaker, the water flows and assumes the shape of the beaker (Figure 12.1 ▶). Liquids are not easily compressed because the molecules or atoms that compose them are already in close contact—they cannot be pushed much closer together. The molecules in a gas, by contrast, have a great deal of space between them and are easily forced into a smaller volume by an increase in external pressure (Figure 12.2 ▼).

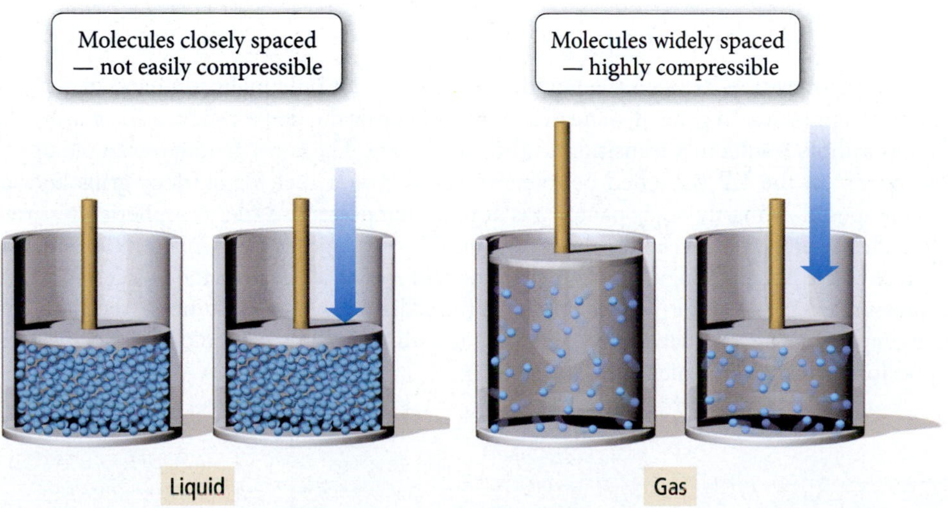

◀ **FIGURE 12.2 Gases Are Compressible** Molecules in a liquid are closely spaced and are not easily compressed. Molecules in a gas have a great deal of space between them, making gases compressible.

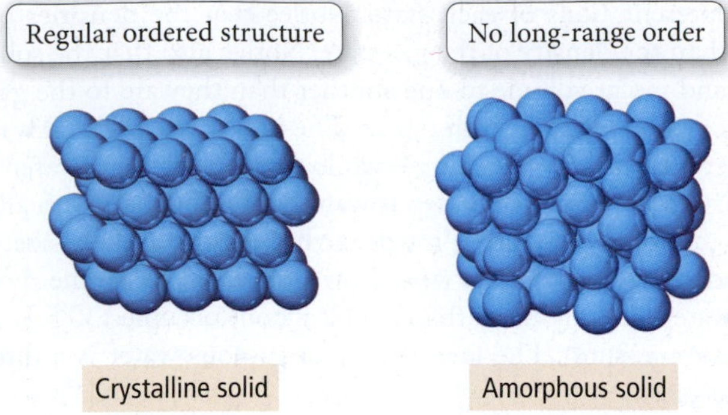

▶ **FIGURE 12.3 Crystalline and Amorphous Solids** In a crystalline solid, the arrangement of the particles displays long-range order. In an amorphous solid, the arrangement of the particles has no long-range order.

According to some definitions, an amorphous solid is a unique state, different from the normal solid state because it lacks any long-range order.

The definite shape of solids is due to the relative immobility of their atoms or molecules. Like liquids, solids have a definite volume and generally cannot be compressed because the molecules or atoms composing them are already in close contact. Solids may be **crystalline**, in which case the atoms or molecules that compose them are arranged in a well-ordered three-dimensional array, or they may be **amorphous**, in which case the atoms or molecules that compose them have no long-range order (Figure 12.3 ▲).

Changes between States

We can transform one state of matter to another by changing the temperature, pressure, or both. For example, we can convert solid ice to liquid water by heating, and liquid water to solid ice by cooling. The following diagram shows the three states of matter and the changes in conditions that commonly induce transitions between the states.

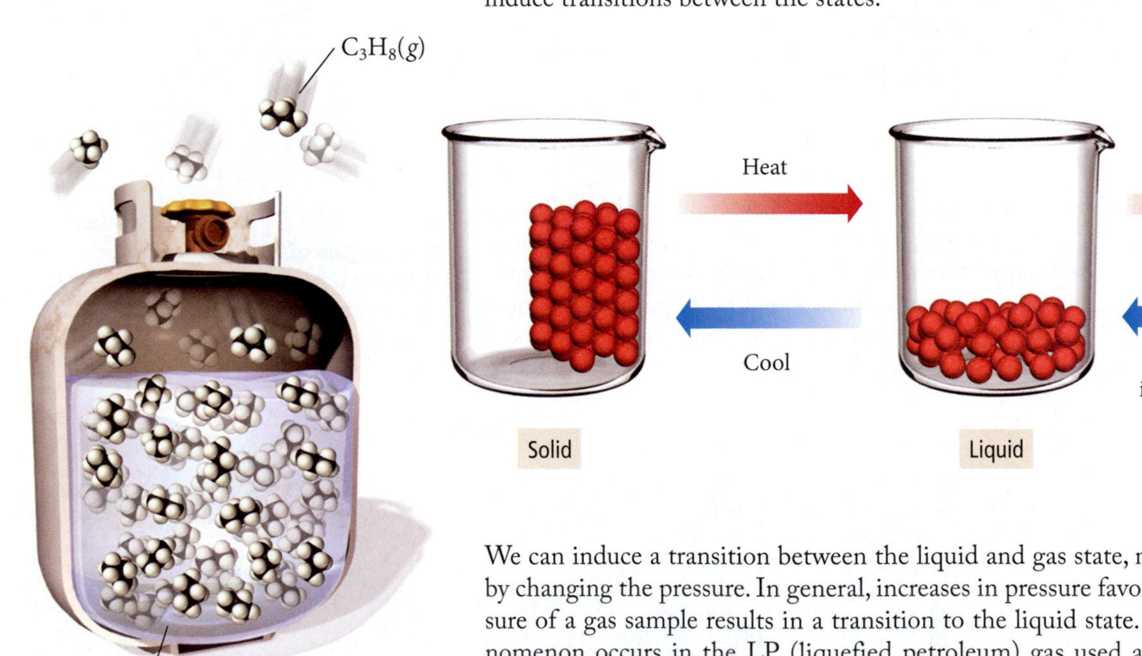

▲The propane in an LP gas tank is in the liquid state. When you open the tank, some propane vaporizes and escapes as a gas.

We can induce a transition between the liquid and gas state, not only by heating and cooling, but also by changing the pressure. In general, increases in pressure favor the denser state, so increasing the pressure of a gas sample results in a transition to the liquid state. The most familiar example of this phenomenon occurs in the LP (liquefied petroleum) gas used as a fuel for outdoor grills and lanterns. LP gas is composed primarily of propane, a gas at room temperature and atmospheric pressure. However, it liquefies at pressures exceeding about 2.7 atm. The propane you buy in a tank is under pressure and therefore in the liquid form. When you open the tank, some of the propane escapes as a gas, lowering the pressure in the tank for a brief moment. Immediately, however, some of the liquid propane evaporates, replacing the gas that escaped. Storing gases like propane as liquids is efficient because, in their liquid form, they occupy much less space.

State Changes **12.1 Cc Conceptual Connection**

This molecular diagram shows a sample of liquid water.

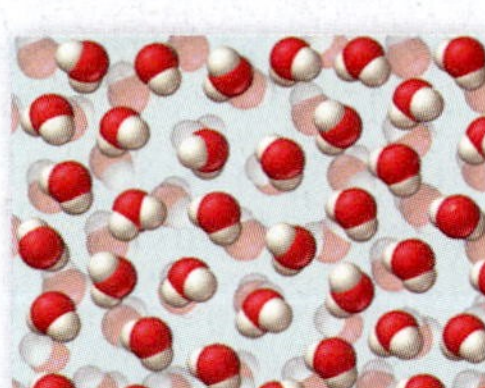

Which of these diagrams best depicts the vapor emitted from a pot of boiling water?

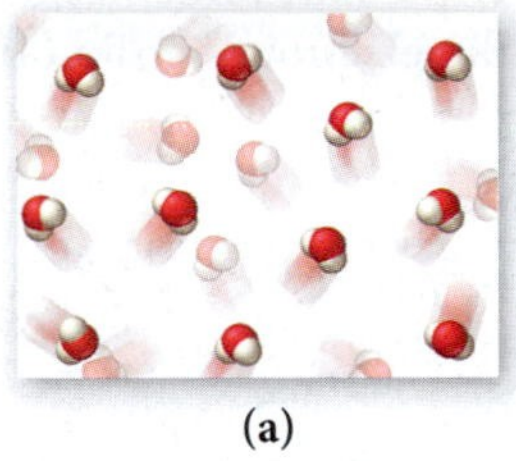
(a)

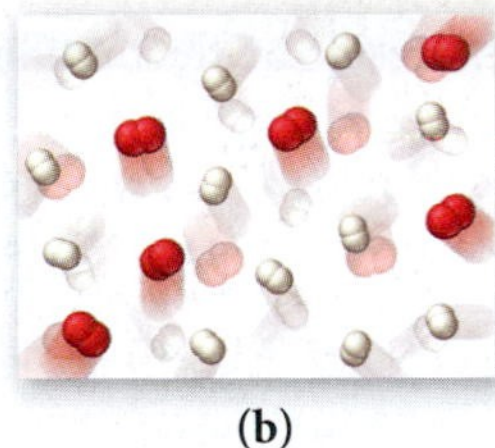
(b)

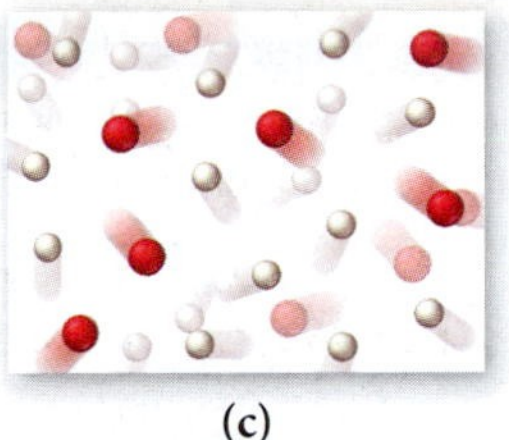
(c)

12.3 Intermolecular Forces: The Forces That Hold Condensed States Together

The structure of the particles that compose a substance determines the strength of the intermolecular forces that hold the substance together, which in turn determine whether the substance is a solid, liquid, or gas at a given temperature. At room temperature, moderate to strong intermolecular forces tend to result in liquids and solids (high melting and boiling points), and weak intermolecular forces tend to result in gases (low melting and boiling points).

Intermolecular forces originate from the interactions between charges, partial charges, and temporary charges on molecules (or atoms and ions), much as bonding forces originate from interactions between charged particles in atoms. Recall from Section 4.3 that according to Coulomb's law, the potential energy (E) of two oppositely charged particles (with charges q_1 and q_2) decreases (becomes more negative) with increasing magnitude of charge and with decreasing separation (r).

$$E = \frac{1}{4\pi\varepsilon_0}\frac{q_1 q_2}{r} \quad \text{(When } q_1 \text{ and } q_2 \text{ are opposite in sign, } E \text{ is negative.)}$$

Therefore, as we have seen, protons and electrons are attracted to each other because their potential energy decreases as they get closer together. Similarly, molecules with partial or temporary charges are attracted to each other because *their* potential energy decreases as they get closer together. However, intermolecular forces, even the strongest ones, are generally *much weaker* than bonding forces.

The reason intermolecular forces are relatively weak compared to bonding forces is also related to Coulomb's law. Bonding forces are the result of large charges (the charges on protons and electrons) interacting at very close distances. Intermolecular forces are the result of smaller charges (as we shall see in the following discussion) interacting at greater distances. For example, consider the interaction between two water molecules in liquid water:

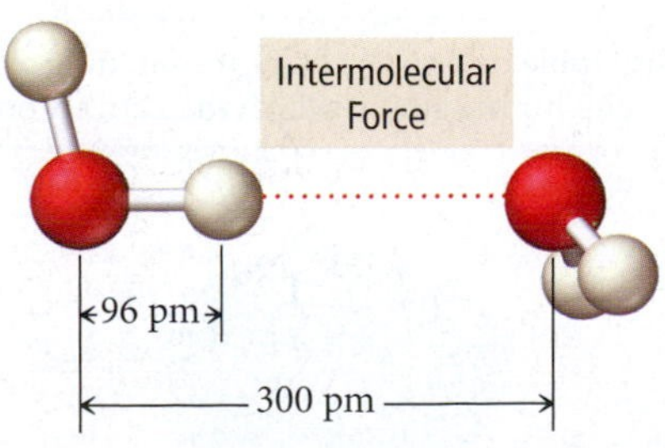

The length of an O—H bond in liquid water is 96 pm; however, the average distance between water molecules in liquid water is about 300 pm. The larger distances between molecules, as well as the smaller charges involved (the partial charges on the hydrogen and oxygen atoms), result in weaker forces. To break the O—H bonds in water, we have to heat the water to thousands of degrees Celsius. However, to completely overcome the intermolecular forces *between* water molecules, we have to heat water only to its boiling point, 100 °C (at sea level).

In this section we examine several different types of intermolecular forces, including dispersion forces, dipole–dipole forces, hydrogen bonding, and ion–dipole forces. The first three of these can potentially occur in all substances; the last one occurs only in mixtures.

Dispersion Force

The nature of dispersion forces was first recognized by Fritz W. London (1900–1954), a German-American physicist.

The intermolecular force present in all molecules and atoms is the **dispersion force** (also called the London force). Dispersion forces are the result of fluctuations in the electron distribution within molecules or atoms. Because all atoms and molecules have electrons, they all exhibit dispersion forces. The electrons in an atom or molecule may, *at any one instant,* be unevenly distributed. Imagine a frame-by-frame movie of a helium atom in which each "frame" captures the position of the helium atom's two electrons.

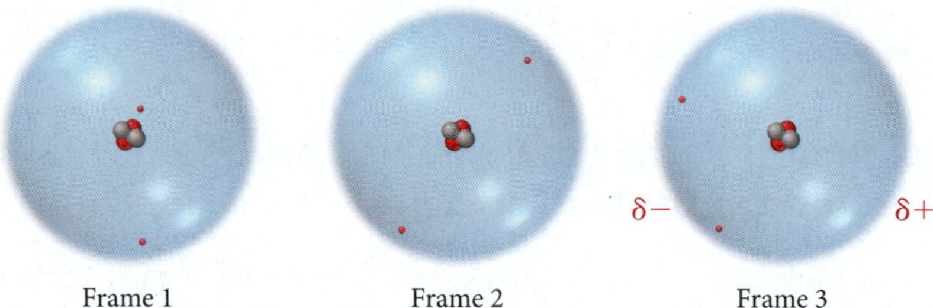

In any one frame, the electrons may not be symmetrically arranged around the nucleus. In frame 3, for example, helium's two electrons are on the left side of the helium atom. At that instant, the left side of the atom will have a slightly negative charge (δ−). The right side of the atom, which temporarily has no electrons, will have a slightly positive charge (δ+) because of the charge of the nucleus. This fleeting charge separation is called an *instantaneous dipole* or a *temporary dipole.* As shown in Figure 12.4 ▼, an instantaneous dipole on one helium atom induces an instantaneous dipole on its neighboring atoms because the positive end of the instantaneous dipole attracts electrons in the neighboring atoms. The neighboring atoms then attract one another—the positive end of one instantaneous dipole attracting the negative end of another. This attraction is the dispersion force.

To polarize means to form a dipole moment (see Section 6.2).

The *magnitude* of the dispersion force depends on how easily the electrons in the atom or molecule move or *polarize* in response to an instantaneous dipole, which in turn depends on the size (or volume) of the electron cloud. A larger electron cloud results in a greater dispersion force because the electrons are held less tightly by the nucleus and therefore polarize more easily. If all other variables are constant, the dispersion force increases with increasing molar mass because molecules or atoms of higher molar mass generally have more electrons dispersed over a greater volume. We can see evidence for the increase in dispersion force with increasing molar mass by examining the boiling points of the noble gases, listed in Table 12.3. Boiling points generally increase with increasing strength of intermolecular

Dispersion Force

An instantaneous dipole on any one helium atom induces instantaneous dipoles on neighboring atoms, which then attract one another.

▶ **FIGURE 12.4 Dispersion Interactions** The temporary dipole in one helium atom induces a temporary dipole in its neighbor. The resulting attraction between the positive and negative charges creates the dispersion force.

forces because more thermal energy is required to separate the particles from the liquid state into the gas state when those particles are more strongly attracted to one another. Notice in Table 12.3 that as the molar masses and electron cloud volumes of the noble gases increase, the greater dispersion forces result in higher boiling points.

Molar mass alone, however, does not determine the magnitude of the dispersion force. Compare the molar masses and boiling points of *n*-pentane and neopentane:

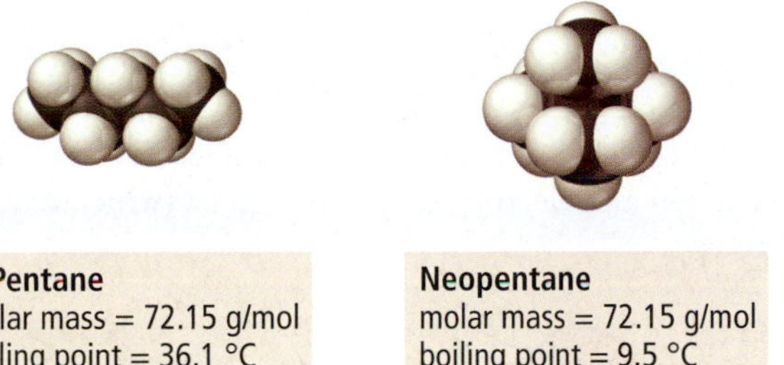

_n_-Pentane
molar mass = 72.15 g/mol
boiling point = 36.1 °C

Neopentane
molar mass = 72.15 g/mol
boiling point = 9.5 °C

TABLE 12.3 Boiling Points of the Noble Gases

Noble Gas	Molar Mass (g/mol)	Boiling Point (K)
He	4.00	4.2
Ne	20.18	27
Ar	39.95	87
Kr	83.80	120
Xe	131.30	165

These molecules have identical molar masses (they are isomers), but *n*-pentane has a higher boiling point than neopentane. Why? Because, like ethanol and dimethyl ether, which we discussed in Section 12.1, the two molecules have different structures. The *n*-pentane molecules are long and can interact with one another along their entire length, as shown in Figure 12.5(a) ▼. In contrast, the bulky, round shape of neopentane molecules results in a smaller area of interaction between neighboring molecules, as shown in Figure 12.5(b) ▼. The result is a lower boiling point for neopentane.

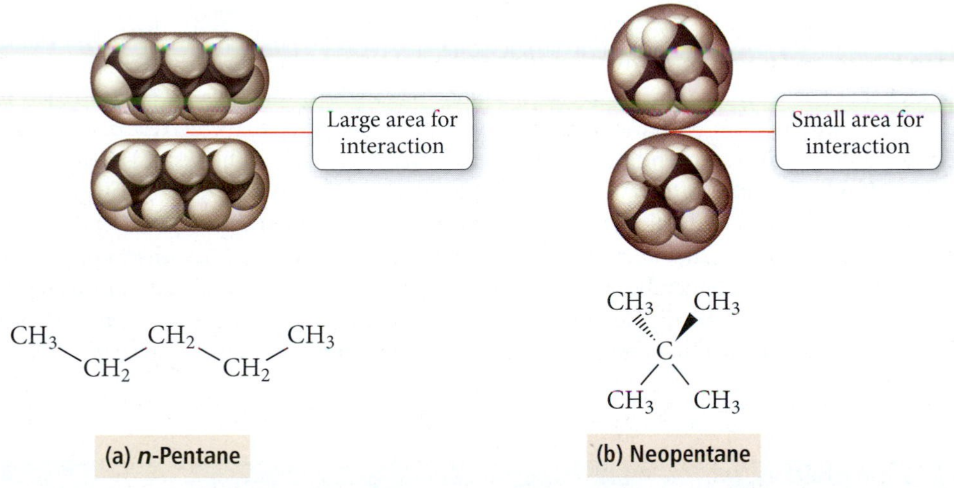

◀ **FIGURE 12.5 Dispersion Force and Molecular Shape** **(a)** The straight shape of *n*-pentane molecules allows them to interact with one another along the entire length of the molecules. **(b)** The nearly spherical shape of neopentane molecules allows for only a small area of interaction. Thus, dispersion forces are weaker in neopentane than in *n*-pentane, resulting in a lower boiling point.

Although molecular shape and other factors must always be considered in determining the magnitude of dispersion forces, molar mass can act as a guide when comparing dispersion forces within a family of similar elements or compounds, as shown in Figure 12.6 ▼.

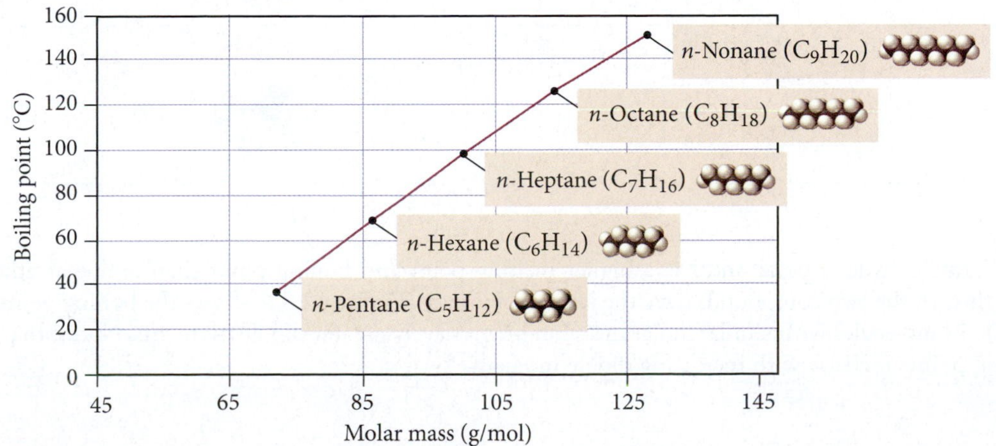

◀ **FIGURE 12.6 Boiling Points of the _n_-Alkanes** The boiling points of the *n*-alkanes rise with increasing molar mass and the consequent stronger dispersion forces.

12.2 Cc Conceptual Connection

Dispersion Forces

Which halogen has the highest boiling point?

(a) Cl_2

(b) Br_2

(c) I_2

See Section 6.10 to review how to determine if a molecule is polar.

Dipole–Dipole Force

The **dipole–dipole force** exists in all molecules that are polar. Polar molecules have electron-rich regions (which have a partial negative charge) and electron-deficient regions (which have a partial positive charge). For example, consider acetone:

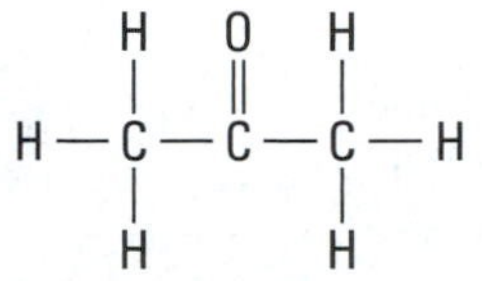

```
    H   O   H
    |   ‖   |
H — C — C — C — H
    |       |
    H       H
```

Structural formula

Space-filling model

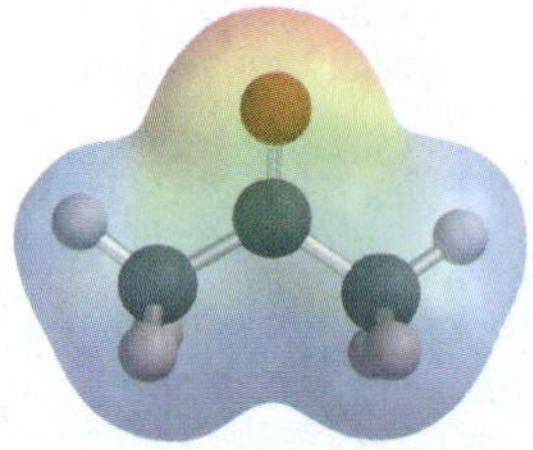

Electrostatic potential map

The image on the right is an electrostatic potential map of acetone; these kinds of maps were first introduced in Chapter 6 (see Section 6.10). Recall that the red areas indicate electron-rich regions in the molecule and that the blue areas indicate electron-poor regions. Acetone has an electron-rich region surrounding the oxygen atom (because oxygen is more electronegative than the rest of the molecule) and electron-poorer regions surrounding the carbon and hydrogen atoms. The result is that acetone has a **permanent dipole** that can interact with other acetone molecules, as shown in Figure 12.7 ▶. The positive end of one permanent dipole attracts the negative end of another; this attraction is the dipole–dipole force. Polar molecules, therefore, have higher melting and boiling points than nonpolar molecules of similar molar mass. Remember that all molecules (including polar ones) have dispersion forces. *In addition,* polar molecules have dipole–dipole forces. This additional attractive force raises their melting and boiling points relative to nonpolar molecules of similar molar mass. For example, consider formaldehyde and ethane.

Name	Formula	Molar Mass (amu)	Structure	bp (°C)	mp (°C)
Formaldehyde	CH_2O	30.03	O ‖ H—C—H	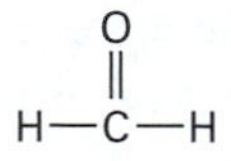  −19.5	−92
Ethane	C_2H_6	30.07	H H / H—C—C—H / H H	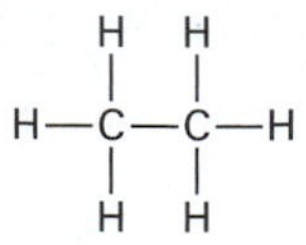  −88	−172

Formaldehyde is polar and has a higher melting point and boiling point than nonpolar ethane, even though the two compounds have the same molar mass. Figure 12.8 ▶ shows the boiling points of a series of molecules with similar molar mass but progressively greater dipole moments. Notice that the boiling points increase with increasing dipole moment.

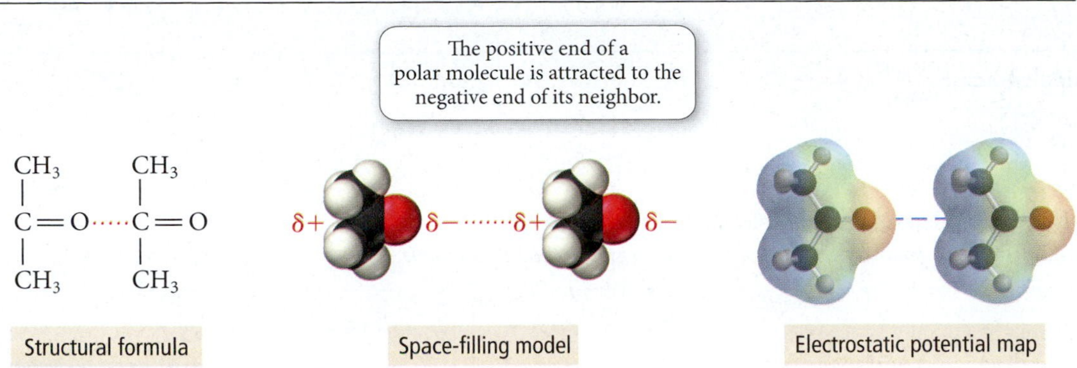

▲ **FIGURE 12.7 Dipole–Dipole Interaction** Molecules with permanent dipoles, such as acetone, are attracted to one another via dipole–dipole interactions.

The polarity of molecules is also important in determining the **miscibility**—the ability to mix without separating into two states—of liquids. In general, polar liquids are miscible with other polar liquids but are not miscible with nonpolar liquids. For example, water, a polar liquid, is not miscible with pentane (C_5H_{12}), a nonpolar liquid (Figure 12.9 ▶). Similarly, water and oil (also nonpolar) do not mix. Consequently, oily hands or oily stains on clothes cannot be cleaned with plain water.

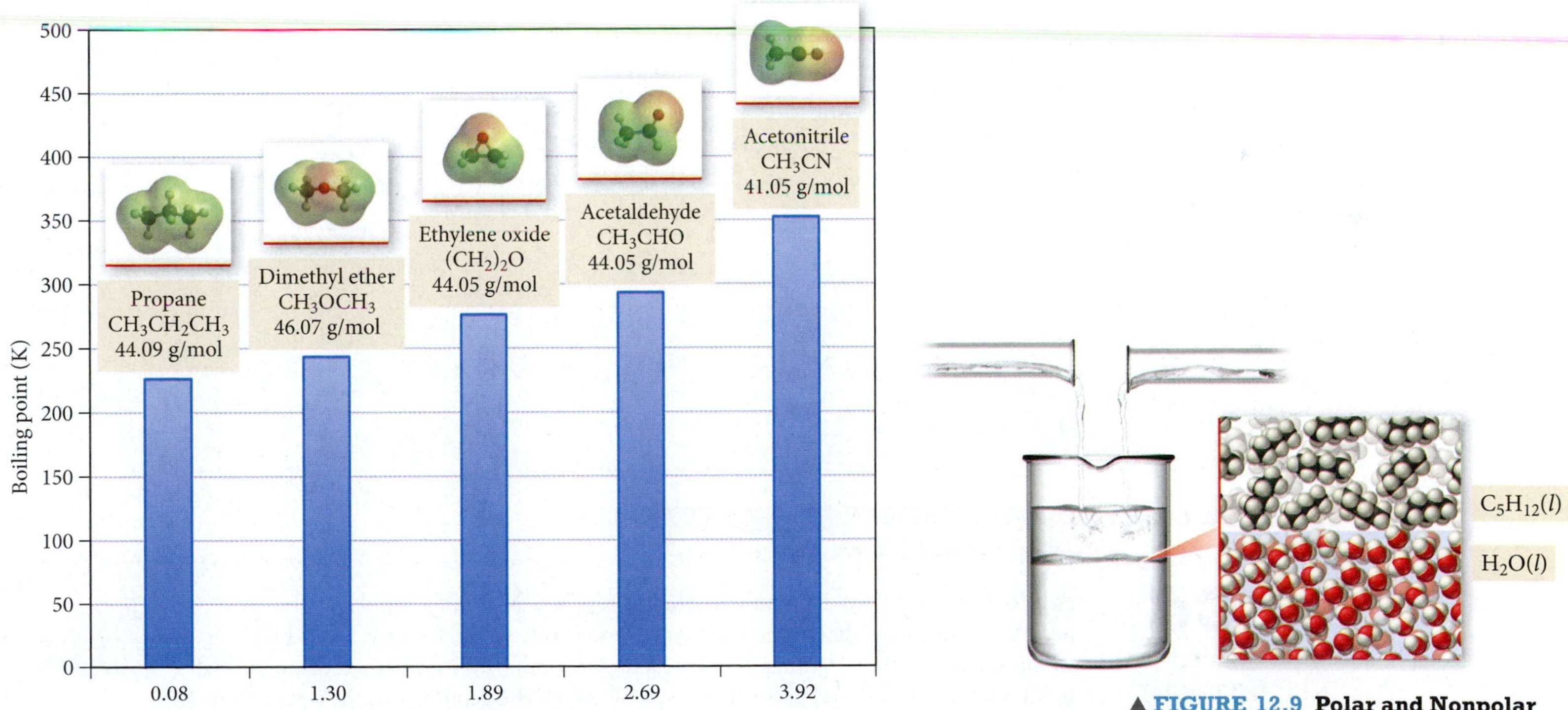

▲ **FIGURE 12.8 Dipole Moment and Boiling Point** The molecules shown here all have similar molar masses but different dipole moments. The boiling points increase with increasing dipole moment.

▲ **FIGURE 12.9 Polar and Nonpolar Compounds** Water and pentane do not mix because water molecules are polar and pentane molecules are nonpolar.

EXAMPLE 12.1

Dipole–Dipole Forces

Which of these molecules have dipole–dipole forces?

(a) CO_2 **(b)** CH_2Cl_2 **(c)** CH_4

SOLUTION

A molecule has dipole–dipole forces if it is polar. To determine if a molecule is polar, (1) *determine if the molecule contains polar bonds* and (2) *determine if the polar bonds add together to form a net dipole moment.*

(a) CO_2

1. Because the electronegativity of carbon is 2.5 and that of oxygen is 3.5 (Figure 6.3), CO_2 has polar bonds.

2. The geometry of CO_2 is linear. Consequently, the dipoles of the polar bonds cancel, so the molecule is *nonpolar* and does not have dipole–dipole forces.

O=C=O

No dipole forces present

(b) CH_2Cl_2

1. The electronegativity of C is 2.5, that of H is 2.1, and that of Cl is 3.0. Consequently, CH_2Cl_2 has two polar bonds (C—Cl) and two bonds that are nearly nonpolar (C—H).

2. The geometry of CH_2Cl_2 is tetrahedral. Because the (C—Cl) bonds and the (C—H) bonds are different, their dipoles do not cancel but sum to a net dipole moment. The molecule is polar and has dipole–dipole forces.

CH_2Cl_2

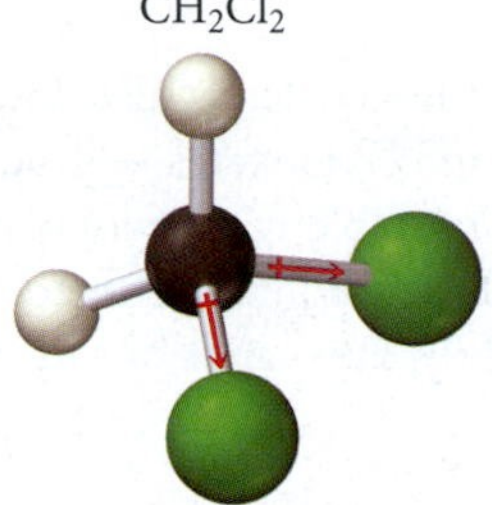

Dipole forces present

(c) CH_4

1. Because the electronegativity of C is 2.5 and that of hydrogen is 2.1, the (C—H) bonds are nearly nonpolar.

2. In addition, because the geometry of the molecule is tetrahedral, any slight polarities that the bonds might have will cancel. CH_4 is therefore nonpolar and does not have dipole–dipole forces.

CH_4

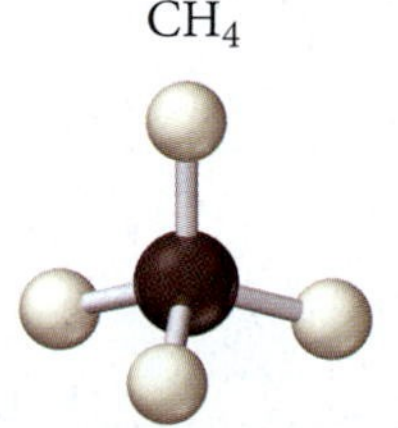

No dipole forces present

FOR PRACTICE 12.1

Which molecules have dipole–dipole forces?

(a) CI_4 **(b)** CH_3Cl **(c)** HCl

12.3 Cc Conceptual Connection

Dipole–Dipole Interaction

An electrostatic potential map for acetonitrile (CH_3CN), which is polar, is shown here. From this map, determine the geometry for how two acetonitrile molecules would interact with each other. Draw structural formulas, using the three-dimensional bond notation introduced in Section 6.9, to illustrate the geometry of the interaction.

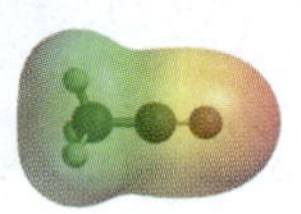

Hydrogen Bonding

Polar molecules containing hydrogen atoms bonded directly to small electronegative atoms—most importantly fluorine, oxygen, or nitrogen—exhibit an intermolecular force called **hydrogen bonding**.

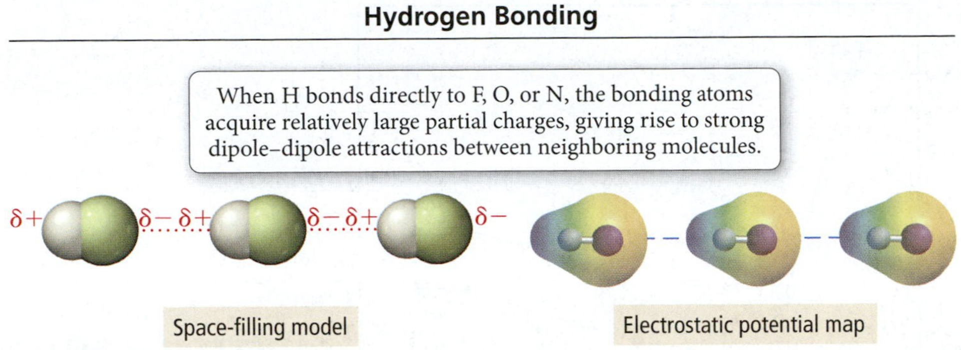

Hydrogen Bonding

When H bonds directly to F, O, or N, the bonding atoms acquire relatively large partial charges, giving rise to strong dipole–dipole attractions between neighboring molecules.

δ+ δ− δ+ δ− δ+ δ−

Space-filling model

Electrostatic potential map

◀ **FIGURE 12.10 Hydrogen Bonding in HF** The hydrogen of one HF molecule, with its partial positive charge, is attracted to the fluorine of its neighbor, with its partial negative charge. This dipole–dipole interaction is an example of a hydrogen bond.

HF, NH_3, and H_2O, for example, all undergo hydrogen bonding. The hydrogen bond is a sort of *super* dipole–dipole force. The large electronegativity difference between hydrogen and these electronegative elements causes the hydrogen atom to have a fairly large partial positive charge (δ+) within the bond, while the F, O, or N atom has a fairly large partial negative charge (δ−). In addition, since these atoms are all quite small, the H atom on one molecule can approach the F, O, or N atom on an adjacent molecule very closely. The result is a strong attraction between the H atom on one molecule and the F, O, or N on its neighbor—an attraction called a **hydrogen bond**. For example, in HF, the hydrogen atom in one molecule is strongly attracted to the fluorine atom on a neighboring molecule (Figure 12.10 ▲). The electrostatic potential maps in Figure 12.10 show the large differences in electron density that result in unusually large partial charges.

Hydrogen bonds should not be confused with chemical bonds. Chemical bonds occur *between individual atoms within a molecule,* whereas hydrogen bonds—like dispersion forces and dipole–dipole forces—are intermolecular forces that occur *between molecules.* A typical hydrogen bond is only 2–5% as strong as a typical covalent chemical bond. Hydrogen bonds are, however, the strongest of the three *intermolecular* forces we have discussed so far. Substances composed of molecules that form hydrogen bonds often have higher melting and boiling points than substances composed of molecules that do not form hydrogen bonds. Hydrogen bonding is responsible for the differences between ethanol and dimethyl ether, which we first examined in Section 12.1.

Hydrogen Bonding in Ethanol

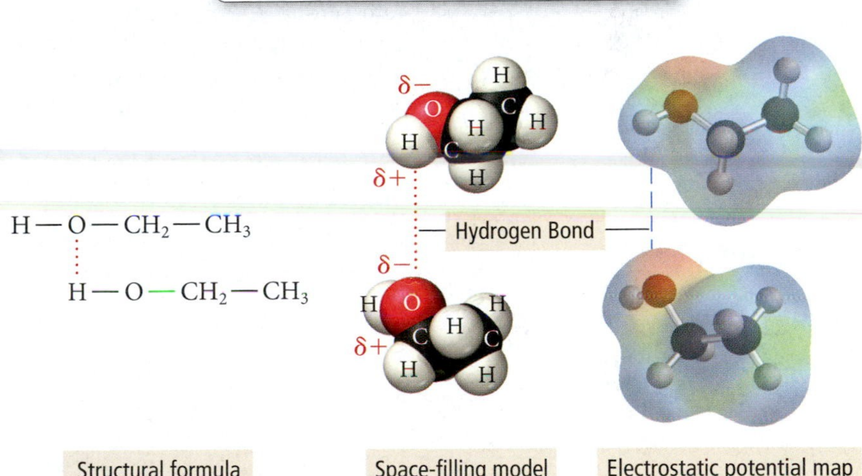

▲ **FIGURE 12.11 Hydrogen Bonding in Ethanol** The left side shows the structural formula, the center shows the space-filling models, and the right side shows the electrostatic potential maps.

Because ethanol contains hydrogen bonded directly to oxygen, ethanol molecules form hydrogen bonds with each other as shown in Figure 12.11 ▶. The hydrogen that is directly bonded to oxygen in an individual ethanol molecule is also strongly attracted to the oxygen on neighboring molecules. This strong attraction makes the boiling point of ethanol 78.3 °C. Consequently, ethanol is a liquid at room temperature. In contrast, dimethyl ether has an identical molar mass to ethanol but does not exhibit hydrogen bonding because in the dimethyl ether molecule the oxygen atom is not bonded directly to hydrogen; this results in lower boiling and melting points, and dimethyl ether is a gas at room temperature.

Name	Formula	Molar Mass (amu)	Structure	bp (°C)	mp (°C)
Ethanol	C_2H_6O	 46.07	CH_3CH_2OH	78.3	−114.1
Dimethyl Ether	C_2H_6O	 46.07	CH_3OCH_3	−22.0	−141

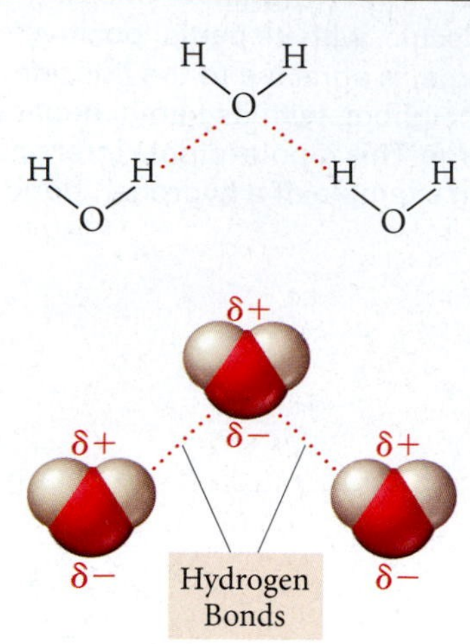

▲ **FIGURE 12.12 Hydrogen Bonding in Water**

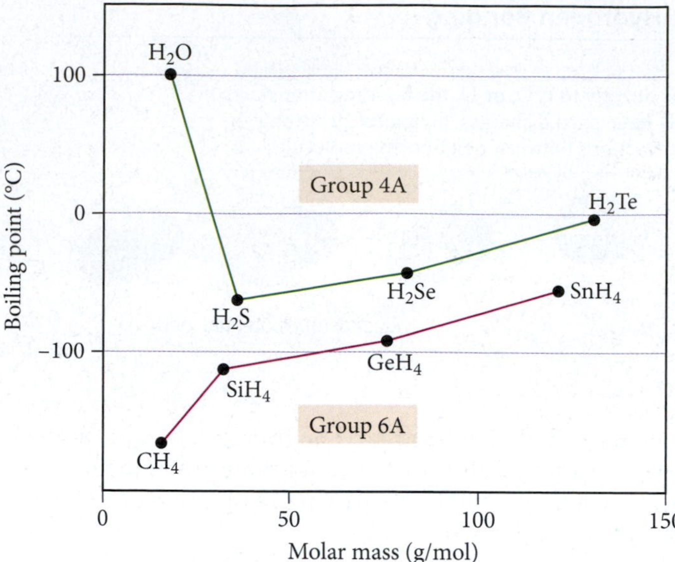

◀ **FIGURE 12.13 Boiling Points of Group 4A and 6A Compounds** Because of hydrogen bonding, the boiling point of water is anomalous compared to the boiling points of other hydrogen-containing compounds.

Water is another good example of a molecule with hydrogen bonding (Figure 12.12 ◀). Figure 12.13 ▲ shows the boiling points of the simple hydrogen compounds of the Group 4A and Group 6A elements. In general, boiling points increase with increasing molar mass, as expected, based on increasing dispersion forces. However, because of hydrogen bonding, the boiling point of water (100 °C) is much higher than expected based on its molar mass (18.0 g/mol). Without hydrogen bonding, all the water on our planet would be gaseous.

EXAMPLE 12.2
Hydrogen Bonding

One of these compounds is a liquid at room temperature. Which one and why?

Formaldehyde	Fluoromethane	Hydrogen peroxide
$H-C(=O)-H$	$H-CH_2-F$	$H-O-O-H$

SOLUTION

The three compounds have similar molar masses.

Formaldehyde	30.03 g/mol
Fluoromethane	34.03 g/mol
Hydrogen peroxide	34.02 g/mol

Therefore the strengths of their dispersion forces are similar. All three compounds are also polar, so they have dipole–dipole forces. Hydrogen peroxide, however, is the only one of these compounds that also contains H bonded directly to F, O, or N. Therefore, it also has hydrogen bonding and is likely to have the highest boiling point of the three. Because the example states that only one of the compounds is a liquid, you can safely assume that hydrogen peroxide is the liquid. Note that, although fluoromethane *contains* both H and F, H is not *directly bonded* to F, so fluoromethane does not have hydrogen bonding as an intermolecular force. Similarly, formaldehyde *contains* both H and O, but H is not *directly bonded* to O, so formaldehyde does not have hydrogen bonding either.

FOR PRACTICE 12.2

Which compound has the higher boiling point, HF or HCl? Why?

Ion–Dipole Force

The **ion–dipole force** occurs when an ionic compound is mixed with a polar compound; it is especially important in aqueous solutions of ionic compounds. For example, when sodium chloride is mixed with water, the sodium and chloride ions interact with water molecules via ion–dipole forces, as shown in Figure 12.14 ▶. The positive sodium ions interact with the negative poles of water molecules, while the negative chloride ions interact with the positive poles. Ion–dipole forces are the strongest types of intermolecular forces discussed in this section and are responsible for the ability of ionic substances to form solutions with water. We will discuss aqueous solutions more thoroughly in Chapter 14.

Ion–Dipole Forces

The positively charged end of a polar molecule such as H_2O is attracted to negative ions, and the negatively charged end of the molecule is attracted to positive ions.

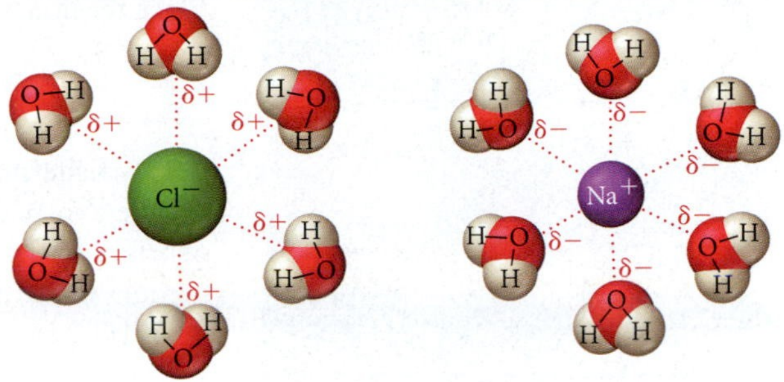

▲ **FIGURE 12.14 Ion–Dipole Forces** Ion–dipole forces exist between Na^+ and the negative ends of H_2O molecules and between Cl^- and the positive ends of H_2O molecules.

Summarizing Intermolecular Forces (as shown in Table 12.4):

- Dispersion forces are present in all molecules and atoms and increase with increasing molar mass. These forces are always weak in small molecules but can be significant in molecules with high molar masses.
- Dipole–dipole forces are present in polar molecules.
- Hydrogen bonds, the strongest of the intermolecular forces that can occur in pure substances (second only to ion–dipole forces in general), are present in molecules containing hydrogen bonded directly to fluorine, oxygen, or nitrogen.
- Ion–dipole forces are present in mixtures of ionic compounds and polar compounds. These forces are very strong and are especially important in aqueous solutions of ionic compounds.

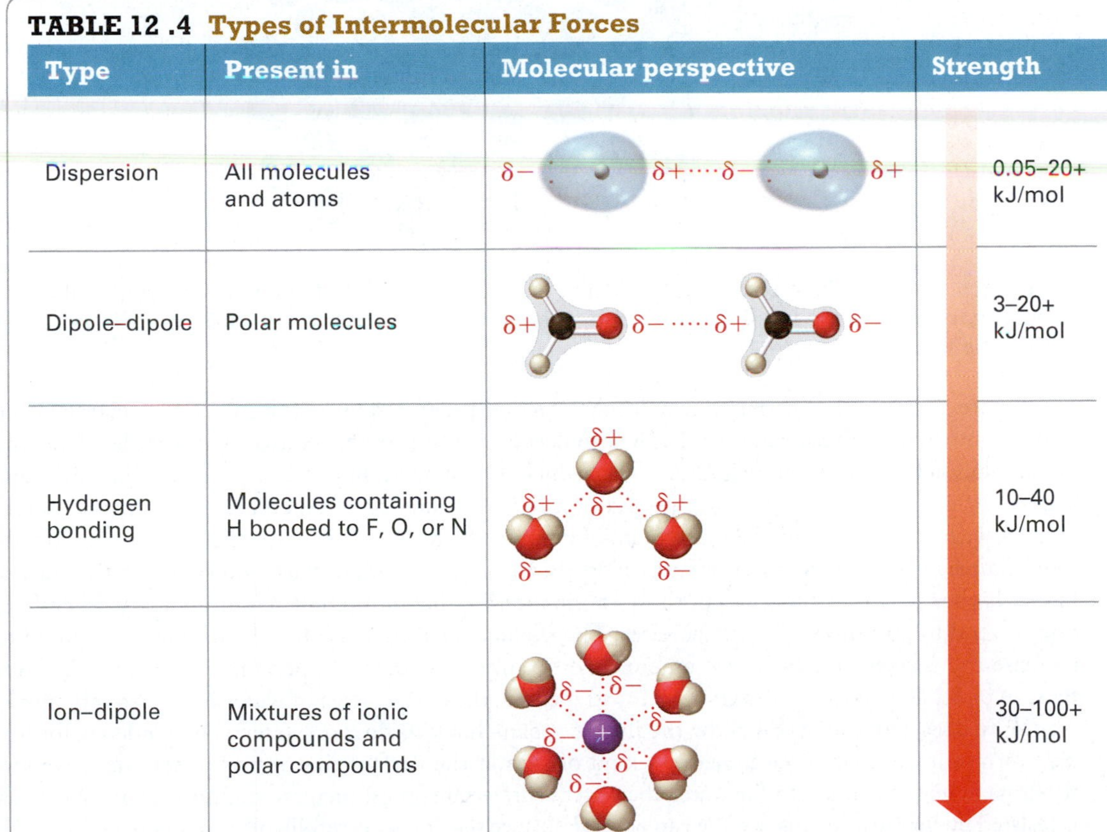

TABLE 12.4 Types of Intermolecular Forces

Type	Present in	Molecular perspective	Strength
Dispersion	All molecules and atoms	δ− δ+····δ− δ+	0.05–20+ kJ/mol
Dipole–dipole	Polar molecules	δ+ δ−·····δ+ δ−	3–20+ kJ/mol
Hydrogen bonding	Molecules containing H bonded to F, O, or N	δ+ δ+ δ− δ+ δ− δ−	10–40 kJ/mol
Ion–dipole	Mixtures of ionic compounds and polar compounds	+ δ− δ− δ− δ− δ− δ−	30–100+ kJ/mol

Intermolecular Forces and Boiling Point

12.4 Cc Conceptual Connection

Which substance has the highest boiling point?

(a) CH_3OH **(b)** CO **(c)** N_2

▲ A trout fly can float on water because of surface tension.

12.4 Intermolecular Forces in Action: Surface Tension, Viscosity, and Capillary Action

The most important manifestation of intermolecular forces is the existence of molecular liquids and solids. In liquids, we also observe several other manifestations of intermolecular forces including surface tension, viscosity, and capillary action.

Surface Tension

A fly fisherman delicately casts a small fishing fly (a metal hook with a few feathers and strings attached to make it look like an insect) onto the surface of a moving stream. The fly floats on the surface of the water—even though the metal composing the hook is denser than water—and attracts trout. Why? The hook floats because of *surface tension,* the tendency of liquids to minimize their surface area.

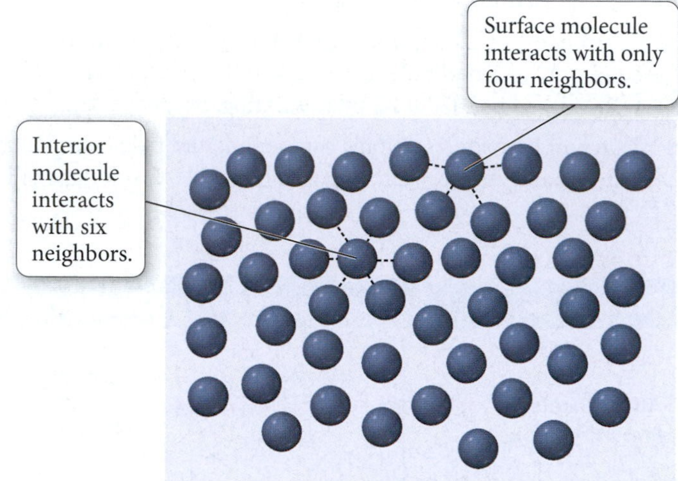
Surface molecule interacts with only four neighbors.
Interior molecule interacts with six neighbors.

▲ **FIGURE 12.15 Surface Tension** Molecules at the liquid surface have a higher potential energy than those in the interior. As a result, a liquid tends to minimize its surface area; its surface behaves like a "skin."

Recall from Section 12.3 that the interactions between molecules lower their potential energy in much the same way that the interaction between protons and electrons lowers their potential energy, in accordance with Coulomb's law.

Figure 12.15 ▲ depicts the intermolecular forces experienced by a molecule at the surface of the liquid compared to those experienced by a molecule in the interior. Notice that a molecule at the surface has relatively fewer neighbors with which to interact, and it is therefore inherently less stable—it has higher potential energy—than those in the interior. (Remember that attractive interactions with other molecules lower potential energy.) In order to increase the surface area of the liquid, molecules from the interior must move to the surface, and, because molecules at the surface have a higher potential energy than those in the interior, this movement requires energy. Therefore, liquids tend to minimize their surface area. The **surface tension** of a liquid is the energy required to increase the surface area by a unit amount. For example, at room temperature, water has a surface tension of $72.8\ \text{mJ/m}^2$—it takes 72.8 mJ to increase the surface area of water by one square meter.

Why does surface tension allow the fly fisherman's hook to float on water? The tendency for liquids to minimize their surface creates a kind of skin at the surface that resists penetration. For the fisherman's hook to sink into the water, the water's surface area must increase slightly—an increase that is resisted by the surface tension. We can observe surface tension by carefully placing a paper clip on the surface of water (Figure 12.16 ◀). The paper clip, even though it is denser than water, floats on the surface of the water. A slight tap on the clip provides the energy necessary to overcome the surface tension and causes the clip to sink.

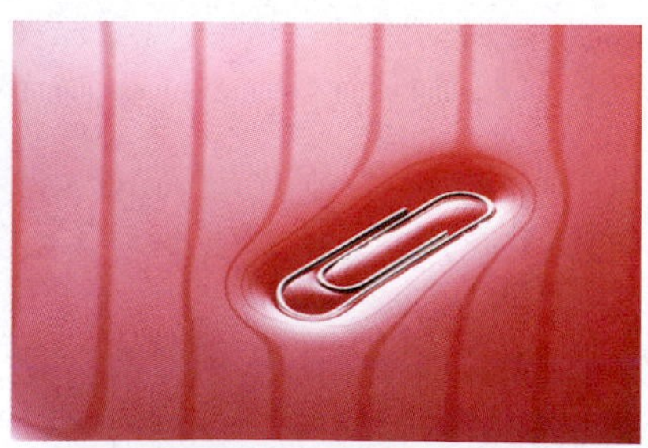

▲ **FIGURE 12.16 Surface Tension in Action** A paper clip floats on water because of surface tension.

Surface tension decreases as intermolecular forces decrease. You can't float a paper clip on benzene, for example, because the dispersion forces among the molecules composing benzene are significantly weaker than the hydrogen bonds among water molecules. The surface tension of benzene is only $28\ \text{mJ/m}^2$—just 40% that of water.

Surface tension is also the reason that small water droplets (those not large enough to be distorted by gravity) form nearly perfect spheres. On the space shuttle, the complete absence of gravity allows even large samples of water to form nearly perfect spheres (Figure 12.17 ▶). Why? Just as gravity pulls the matter of a planet or star inward to form a sphere, so intermolecular forces among water molecules pull the water into a sphere. A sphere is the geometrical shape with the smallest surface area to volume ratio; the formation of a sphere minimizes the number of molecules at the surface, thus minimizing the potential energy of the system.

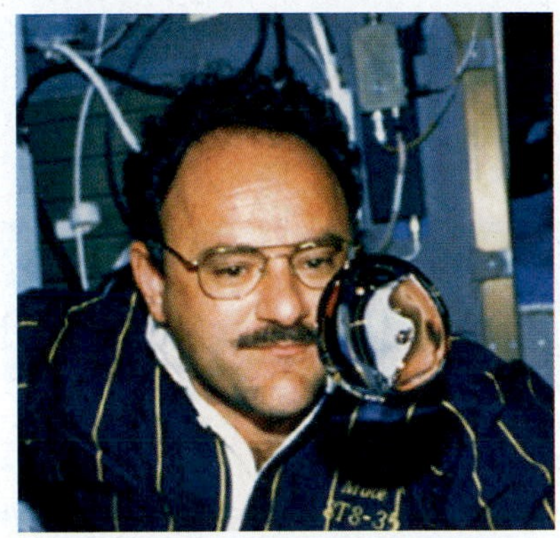

▲ **FIGURE 12.17 Spherical Water** On the space shuttle in orbit, under weightless conditions, collections of water molecules coalesce into nearly perfect spheres held together by intermolecular forces between molecules.

Viscosity

Another manifestation of intermolecular forces is **viscosity**, the resistance of a liquid to flow. Motor oil, for example, is more viscous than gasoline, and maple syrup is more viscous than water. Viscosity is measured in a unit called the poise (P), defined as 1 g/cm · s. The viscosity of water at room temperature is approximately one centipoise (cP). Viscosity is greater in substances with stronger intermolecular forces because if molecules are more strongly attracted to each other, they do not flow around each other as freely. Viscosity also depends on molecular shape, increasing in longer molecules that can interact over a greater area and possibly become entangled. Table 12.5 lists the viscosity of several hydrocarbons. Notice the increase in viscosity with increasing molar mass (and therefore increasing magnitude of dispersion forces) and with increasing length (and therefore increasing potential for molecular entanglement).

Viscosity also depends on temperature because thermal energy partially overcomes the intermolecular forces, allowing molecules to flow past each other more easily. Table 12.6 lists the viscosity of water as a function of temperature. Nearly all liquids become less viscous as temperature increases.

TABLE 12.5 Viscosity of Several Hydrocarbons at 20 °C

Hydrocarbon	Molar Mass (g/mol)	Formula	Viscosity (cP)
n-Pentane	72.15	$CH_3CH_2CH_2CH_2CH_3$	0.240
n-Hexane	86.17	$CH_3CH_2CH_2CH_2CH_2CH_3$	0.326
n-Heptane	100.2	$CH_3CH_2CH_2CH_2CH_2CH_2CH_3$	0.409
n-Octane	114.2	$CH_3CH_2CH_2CH_2CH_2CH_2CH_2CH_3$	0.542
n-Nonane	128.3	$CH_3CH_2CH_2CH_2CH_2CH_2CH_2CH_2CH_3$	0.711

TABLE 12.6 Viscosity of Liquid Water at Several Temperatures

Temperature (°C)	Viscosity (cP)
20	1.002
40	0.653
60	0.467
80	0.355
100	0.282

Capillary Action

Medical technicians take advantage of **capillary action**—the ability of a liquid to flow against gravity up a narrow tube—when taking a blood sample. The technician pokes the patient's finger with a pin, squeezes some blood out of the puncture, and collects the blood with a thin tube. When the tube's tip comes into contact with the blood, the blood is drawn into the tube by capillary action. The same force helps trees and plants draw water from the soil.

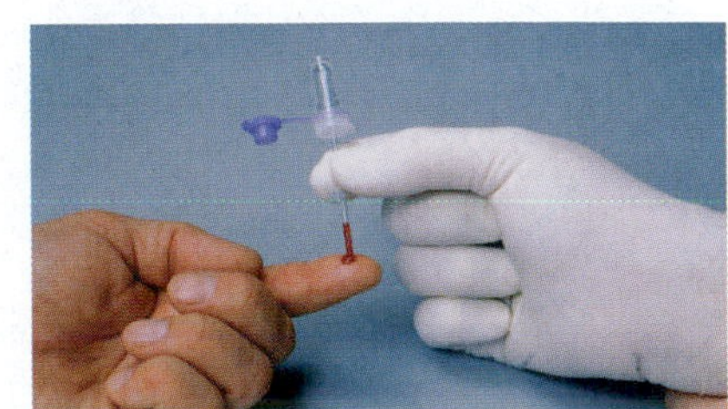

▲ Blood is drawn into a capillary tube by capillary action.

Capillary action results from a combination of two forces: the attraction between molecules in a liquid, called *cohesive forces,* and the attraction between these molecules and the surface of the tube, called *adhesive forces.* The adhesive forces cause the liquid to spread out over the surface of the tube, while the cohesive forces cause the liquid to stay together. If the adhesive forces are greater than the cohesive forces (as is the case for water in a thin glass tube), the attraction to the surface draws the liquid up the tube and the cohesive forces pull along those molecules not in direct contact with the tube walls

▲ **FIGURE 12.18 Capillary Action** The attraction of water molecules to the glass surface draws the liquid around the edge of the tube up the walls. The water in the rest of the column is pulled along by the attraction of water molecules to one another. As can be seen in this figure, the narrower the tube, the higher the liquid will rise.

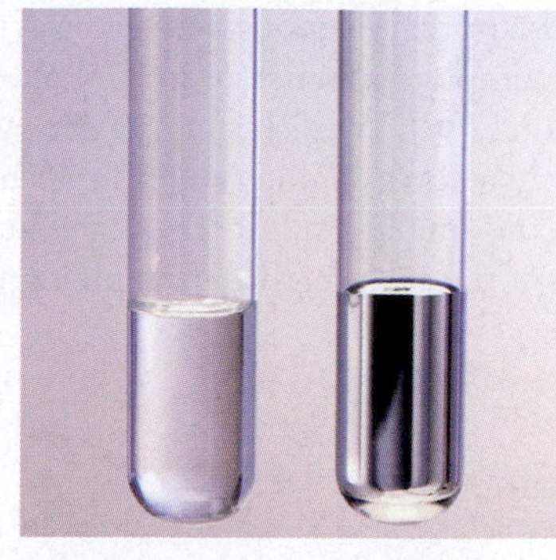

▲ **FIGURE 12.19 Meniscuses of Water and Mercury** The meniscus of water (on the left) is concave because water molecules are more strongly attracted to the glass wall than to one another. The meniscus of mercury is convex (on the right) because mercury atoms are more strongly attracted to one another than to the glass walls.

(Figure 12.18 ◀). The water rises up the tube until the force of gravity balances the capillary action—the thinner the tube, the higher the rise. If the adhesive forces are smaller than the cohesive forces (as is the case for liquid mercury), the liquid does not rise up the tube at all (and in fact will drop to a level below the level of the surrounding liquid).

We can see the result of the differences in the relative magnitudes of cohesive and adhesive forces by comparing the meniscus of water to the meniscus of mercury (Figure 12.19 ◀). (The meniscus is the curved shape of a liquid surface within a tube.) The meniscus of water is concave (rounded inward) because the *adhesive forces* are greater than the cohesive forces, causing the edges of the water to creep up the sides of the tube a bit, forming the familiar cupped shape. The meniscus of mercury is convex (rounded outward) because the *cohesive forces*—due to metallic bonding between the atoms—are greater than the adhesive forces. The mercury atoms crowd toward the interior of the liquid to maximize their interactions with each other, resulting in the upward bulge at the center of the surface.

12.5 Vaporization and Vapor Pressure

We now turn our attention to vaporization, the process by *which thermal energy can overcome intermolecular forces and produce a state change from liquid to gas.* We first introduce the process of vaporization itself, then we discuss the energetics of vaporization, and finally the concepts of vapor pressure, dynamic equilibrium, and critical point. Vaporization is a common occurrence that we experience every day and even depend on to maintain proper body temperature.

The Process of Vaporization

Imagine water molecules at room temperature in a beaker that is open to the atmosphere (Figure 12.20 ◀). The molecules are in constant motion due to thermal energy. If we could actually see the molecules at the surface, we would witness what Roald Hoffmann described as a "wild dance floor" (see the chapter-opening quote) because of all the vibrating, jostling, and molecular movement. *The higher the temperature, the greater the average energy of the collection of molecules.* However, at any one time, some molecules have more thermal energy than the average and some have less.

The distributions of thermal energies for the molecules in a sample of water at two different temperatures are shown in Figure 12.21 ▼. The molecules at the high end of the distribution curve have

▲ **FIGURE 12.20 Vaporization of Water** Some molecules in an open beaker have enough kinetic energy to vaporize from the surface of the liquid.

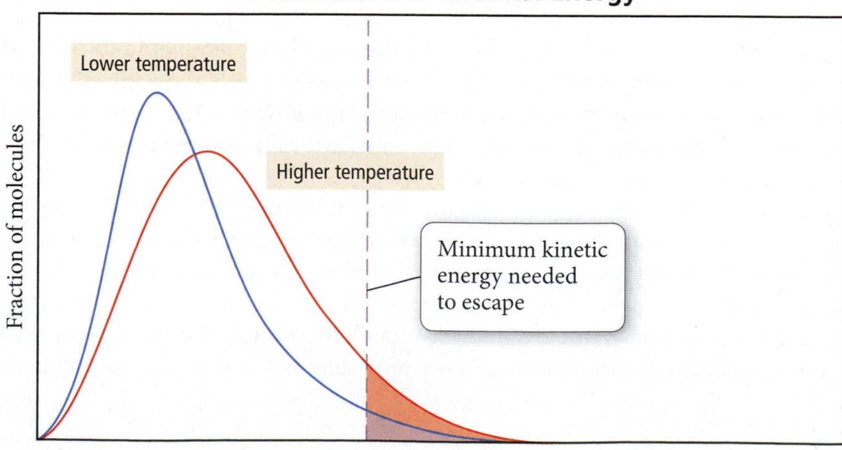

▶ **FIGURE 12.21 Distribution of Thermal Energy** The thermal energies of the molecules in a liquid are distributed over a range. The peak energy increases with increasing temperature.

enough energy to break free from the surface—where molecules are held less tightly than in the interior due to fewer neighbor–neighbor hydrogen bonds—and into the gas state. This transition, from liquid to gas, is called **vaporization**. Some of the water molecules in the gas state, at the low end of the energy distribution curve for the gaseous molecules, plunge back into the water and are captured by intermolecular forces. This transition, from gas to liquid, is the opposite of vaporization and is called **condensation**.

Although both evaporation and condensation occur in a beaker open to the atmosphere, under normal conditions, evaporation takes place at a greater rate because most of the newly evaporated molecules escape into the surrounding atmosphere and never come back. The result is a noticeable decrease in the water level within the beaker over time (usually several days).

What happens if we increase the temperature of the water within the beaker? Because of the shift in the energy distribution to higher energies (see Figure 12.21), more molecules now have enough energy to break free and evaporate, so vaporization occurs more quickly. What happens if we spill the water on the table or floor? The same amount of water is now spread over a wider area, resulting in more molecules at the surface of the liquid. Because molecules at the surface have the greatest tendency to evaporate—because they are held less tightly—vaporization also occurs more quickly in this case. You probably know from experience that water in a beaker or glass may take many days to evaporate completely, while the same amount of water spilled on a table or floor typically evaporates within a few hours (depending on the conditions).

What happens if the liquid in the beaker is not water, but some other substance with weaker intermolecular forces, such as acetone (the main component in nail polish remover)? The weaker intermolecular forces allow more molecules to evaporate at a given temperature, increasing the rate of vaporization. We call liquids that vaporize easily **volatile**, and those that do not vaporize easily **nonvolatile**. Acetone is more volatile than water. Motor oil is virtually nonvolatile at room temperature.

Summarizing the Process of Vaporization:

- The rate of vaporization increases with increasing temperature.
- The rate of vaporization increases with increasing surface area.
- The rate of vaporization increases with decreasing strength of intermolecular forces.

The Energetics of Vaporization

To understand the energetics of vaporization, consider again a beaker of water from the molecular point of view, except now imagine that the beaker is thermally insulated so that heat from the surroundings cannot enter the beaker. What happens to the temperature of the water left in the beaker as molecules evaporate? To answer this question, think about the energy distribution curve again (see Figure 12.21). The molecules that leave the beaker are the ones at the high end of the energy curve—the most energetic. If no additional heat enters the beaker, the average energy of the entire collection of molecules decreases—much as the class average on an exam goes down if we eliminate the highest-scoring students. Vaporization is an *endothermic* process; it takes energy to vaporize the molecules in a liquid. Another way to understand the endothermicity of vaporization is to remember that vaporization requires overcoming the intermolecular forces that hold liquids together. Because energy is needed to pull the molecules away from one another, the process is endothermic.

See Chapter 10 to review endothermic and exothermic processes.

Our bodies use the endothermic nature of vaporization for cooling. When we overheat, we sweat and our skin becomes covered with liquid water. As this water evaporates, it absorbs heat from the body, cooling the skin. A fan makes us feel cooler because it blows newly vaporized water away from our skin, allowing more sweat to vaporize and causing even more cooling. High humidity, on the other hand, slows down the net rate of evaporation, preventing cooling. When the air already contains large amounts of water vapor, our sweat evaporates more slowly, making the body's cooling system less efficient.

Condensation, the opposite of vaporization, is exothermic—heat is released when a gas condenses to a liquid. If you have ever accidentally put your hand above a steaming kettle or opened a bag of microwaved popcorn too soon, you may have experienced a *steam burn.* As the steam condenses to a liquid on your skin, it releases a lot of heat, causing the burn. The condensation of water vapor is also the reason that winter overnight temperatures in coastal regions, which tend to have water vapor in the air, do not get as low as in deserts, which tend to have dry air. As the air temperature in a coastal area drops, water condenses out of the air, releasing heat and preventing the temperature from dropping further. In deserts, the air contains almost no moisture to condense, so the temperature drop is more extreme.

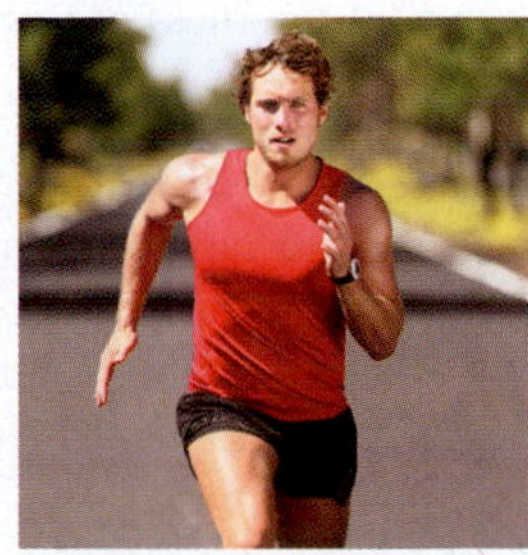

▲ When we sweat, water evaporates from the skin. Since evaporation is endothermic, the result is a cooling effect.

The amount of heat required to vaporize one mole of a liquid to gas is its **heat (or enthalpy) of vaporization (ΔH_{vap})**. The heat of vaporization of water at its normal boiling point of 100 °C is +40.7 kJ/mol.

$$H_2O(l) \longrightarrow H_2O(g) \quad \Delta H_{vap} = +40.7 \text{ kJ/mol}$$

The sign conventions of ΔH were introduced in Chapter 10.

The heat of vaporization is always positive because the process is endothermic—energy must be absorbed to vaporize a substance. The heat of vaporization is somewhat temperature-dependent. For example, at 25 °C the heat of vaporization of water is +44.0 kJ/mol, slightly more than at 100 °C because the water contains less thermal energy at 25 °C. Table 12.7 lists the heats of vaporization of several liquids at their boiling points and at 25 °C.

When a substance condenses from a gas to a liquid, the same amount of heat is involved, but the heat is emitted rather than absorbed.

$$H_2O(g) \longrightarrow H_2O(l) \quad \Delta H = -\Delta H_{vap} = -40.7 \text{ kJ (at 100 °C)}$$

When one mole of water condenses, it releases 40.7 kJ of heat. The sign of ΔH in this case is negative because the process is exothermic.

We can use the heat of vaporization of a liquid to calculate the amount of energy required to vaporize a given mass of the liquid (or the amount of heat given off by the condensation of a given mass of liquid), using concepts similar to those covered in Section 10.6. The heat of vaporization is like a conversion factor between the number of moles of a liquid and the amount of heat required to vaporize it (or the amount of heat emitted when it condenses), as demonstrated in Example 12.3.

TABLE 12.7 Heats of Vaporization of Several Liquids at Their Boiling Points and at 25 °C

Liquid	Chemical Formula	Normal Boiling Point (°C)	ΔH_{vap}(kJ/mol) at Boiling Point	ΔH_{vap}(kJ/mol) at 25 °C
Water	H_2O	100	40.7	44.0
Rubbing alcohol (isopropyl alcohol)	C_3H_8O	82.3	39.9	45.4
Acetone	C_3H_6O	56.1	29.1	31.0
Diethyl ether	$C_4H_{10}O$	34.6	26.5	27.1

EXAMPLE 12.3 Using the Heat of Vaporization in Calculations

Calculate the mass of water (in g) that can be vaporized at its boiling point with 155 kJ of heat.

SORT You are given a certain amount of heat in kilojoules and asked to find the mass of water that can be vaporized.

GIVEN: 155 kJ
FIND: g H_2O

STRATEGIZE The heat of vaporization gives the relationship between heat absorbed and moles of water vaporized. Begin with the given amount of heat (in kJ) and convert to moles of water that can be vaporized. Then use the molar mass as a conversion factor to convert from moles of water to mass of water.

CONCEPTUAL PLAN

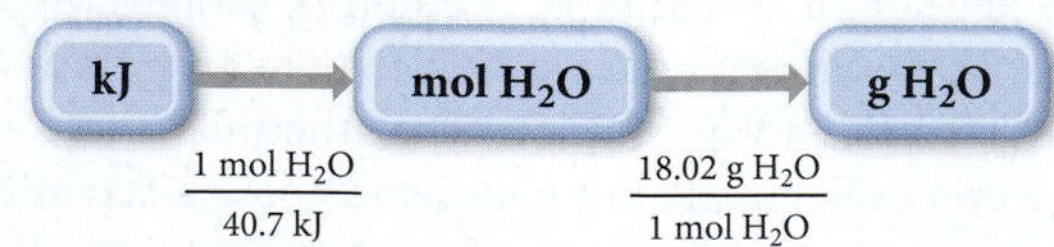

RELATIONSHIPS USED

$\Delta H_{vap} = 40.7$ kJ/mol (at 100 °C)

18.02 g H_2O = 1 mol H_2O

—*Continued on the next page*

Continued from the previous page—

SOLVE Follow the conceptual plan to solve the problem.	**SOLUTION** $155\ \text{kJ} \times \frac{1\ \text{mol H}_2\text{O}}{40.7\ \text{kJ}} \times \frac{18.02\ \text{g H}_2\text{O}}{1\ \text{mol H}_2\text{O}} = 68.6\ \text{g H}_2\text{O}$

FOR PRACTICE 12.3
Calculate the amount of heat (in kJ) required to vaporize 2.58 kg of water at its boiling point.

FOR MORE PRACTICE 12.3
Suppose that 0.48 g of water at 25 °C condenses on the surface of a 55 g block of aluminum that is initially at 25 °C. If the heat released during condensation goes only toward heating the metal, what is the final temperature (in °C) of the metal block? (The specific heat capacity of aluminum is 0.903 J/g °C).

Vapor Pressure and Dynamic Equilibrium

We have already seen that if a container of water is left uncovered at room temperature, the water slowly evaporates away. But what happens if the container is sealed? Imagine a sealed evacuated flask—one from which the air has been removed—containing liquid water, as shown in Figure 12.22 ▼. Initially, the water molecules evaporate, as they did in the open beaker. However, because of the seal, the

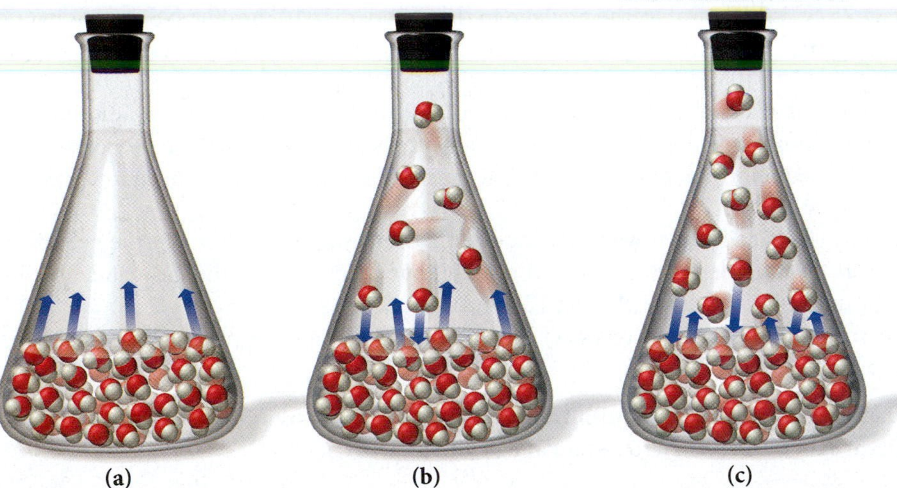

▲ **FIGURE 12.22 Vaporization in a Sealed Flask** **(a)** When water is in a sealed container, water molecules begin to vaporize. **(b)** As water molecules build up in the gas state, they begin to recondense into the liquid. **(c)** When the rate of evaporation equals the rate of condensation, dynamic equilibrium is reached.

evaporated molecules cannot escape into the atmosphere. As water molecules enter the gas state, some start condensing back into the liquid. As the concentration (or partial pressure) of gaseous water molecules increases, the rate of condensation also increases. However, as long as the water remains at a constant temperature, the rate of evaporation remains constant. Eventually the rate of condensation and the rate of vaporization become equal—in other words, **dynamic equilibrium** has been reached (Figure 12.23 ►). Condensation and vaporization continue at equal rates, and the concentration of water vapor above the liquid is constant.

The pressure of a gas in dynamic equilibrium with its liquid is its **vapor pressure**. The vapor pressure of a particular liquid depends on the intermolecular forces present in the liquid and the temperature. Weak intermolecular forces result in volatile substances with high vapor pressures because the

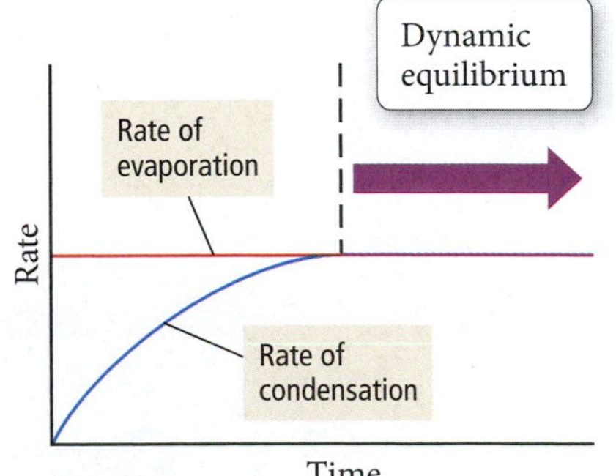

▲ **FIGURE 12.23 Dynamic Equilibrium** Dynamic equilibrium occurs when the rate of condensation is equal to the rate of evaporation.

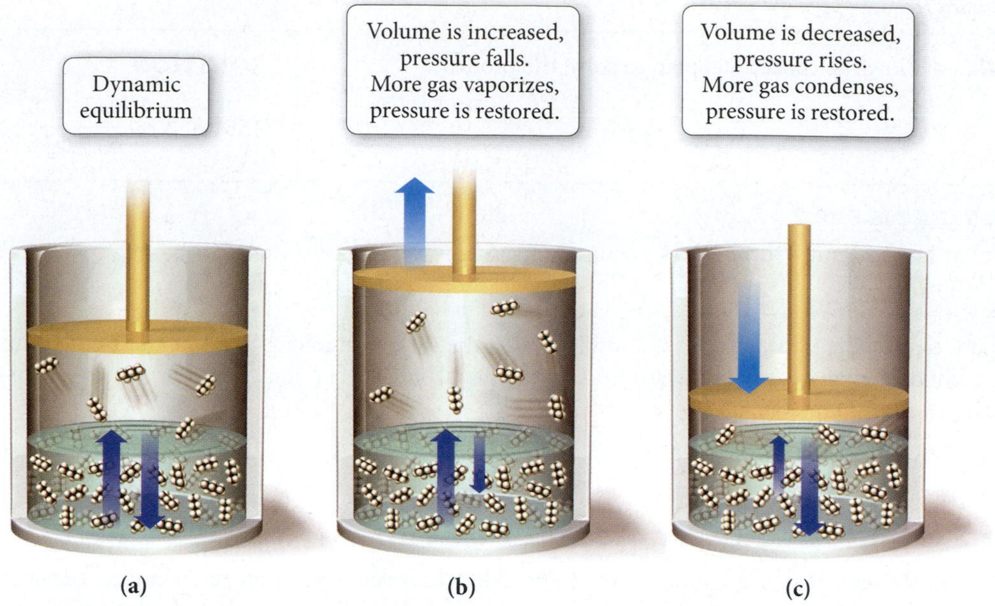

▶ **FIGURE 12.24 Dynamic Equilibrium in *n*-Pentane** **(a)** Liquid *n*-pentane is in dynamic equilibrium with its vapor. **(b)** When the volume is increased, the pressure drops and some liquid is converted to gas to bring the pressure back up. **(c)** When the volume is decreased, the pressure increases and some gas is converted to liquid to bring the pressure back down.

Dynamic equilibrium
Volume is increased, pressure falls. More gas vaporizes, pressure is restored.
Volume is decreased, pressure rises. More gas condenses, pressure is restored.
(a)
(b)
(c)

intermolecular forces are easily overcome by thermal energy. Strong intermolecular forces result in nonvolatile substances with low vapor pressures.

A liquid in dynamic equilibrium with its vapor is a balanced system that tends to return to equilibrium if disturbed. For example, consider a sample of *n*-pentane (a component of gasoline) at 25 °C in a cylinder equipped with a moveable piston (Figure 12.24(a) ▲). The cylinder contains no other gases except *n*-pentane vapor in dynamic equilibrium with the liquid. Because the vapor pressure of *n*-pentane at 25 °C is 510 mmHg, the pressure in the cylinder is 510 mmHg. Now, what happens when we move the piston upward to expand the volume within the cylinder? Initially, the pressure in the cylinder drops below 510 mmHg, in accordance with Boyle's law. Then, however, more liquid vaporizes until equilibrium is reached once again (Figure 12.24(b)). If we expand the volume of the cylinder again, the same thing happens—the pressure initially drops and more *n*-pentane vaporizes to bring the system back into equilibrium. Further expansion causes the same result *as long as any liquid n-pentane remains in the cylinder*.

Boyle's law is discussed in Section 11.3.

Conversely, what happens if we lower the piston, decreasing the volume in the cylinder? Initially, the pressure in the cylinder rises above 510 mmHg, but then some of the gas condenses into liquid until equilibrium is reached again (Figure 12.24(c)).

We describe the tendency of a system in dynamic equilibrium to return to equilibrium with the following general statement:

> **When a system in dynamic equilibrium is disturbed, the system responds so as to minimize the disturbance and return to a state of equilibrium.**

If the pressure above a liquid–vapor system in equilibrium decreases, some of the liquid evaporates, restoring the equilibrium pressure. If the pressure increases, some of the vapor condenses, bringing the pressure back down to the equilibrium pressure. This basic principle—*Le Châtelier's principle*—is applicable to any chemical system in equilibrium, as we shall see in Chapter 16.

12.5 Cc Conceptual Connection

Vapor Pressure

What happens to the vapor pressure of a substance when its surface area is increased at constant temperature?

(a) The vapor pressure increases.

(b) The vapor pressure remains the same.

(c) The vapor pressure decreases.

Temperature Dependence of Vapor Pressure and Boiling Point

When the temperature of a liquid increases, its vapor pressure rises because the higher thermal energy increases the number of molecules that have enough energy to vaporize (see Figure 12.21). Because of the shape of the thermal energy distribution curve, a small change in temperature makes a large difference in the number of molecules that have enough energy to vaporize, which results in a large increase in vapor pressure. For example, the vapor pressure of water at 25 °C is 23.3 torr, while at 60 °C the vapor pressure is 149.4 torr. Figure 12.25 ▶ shows the vapor pressure of water and several other liquids as a function of temperature.

The **boiling point** of a liquid is *the temperature at which the liquid's vapor pressure equals the external pressure.* When a liquid reaches its boiling point, the thermal energy is enough for molecules in the interior of the liquid (not just those at the surface) to break free of their neighbors and enter the gas state (Figure 12.26 ▼). The bubbles in boiling water are pockets of gaseous water that have formed within the liquid water. The bubbles float to the surface and leave as gaseous water or steam.

The **normal boiling point** of a liquid is *the temperature at which its vapor pressure equals 1 atm.* The normal boiling point of pure water is 100 °C. However, at a lower pressure, water boils at a lower temperature. In Denver, Colorado, where the altitude is around 1600 m (5200 ft) above sea level, for example, the average atmospheric pressure is about 83% of what it is at sea level, and water boils at approximately 94 °C. For this reason, it takes slightly longer to cook food in boiling water in Denver than in San Francisco, California (which is at sea level). Table 12.8 shows the boiling point of water at several locations of varied altitudes.

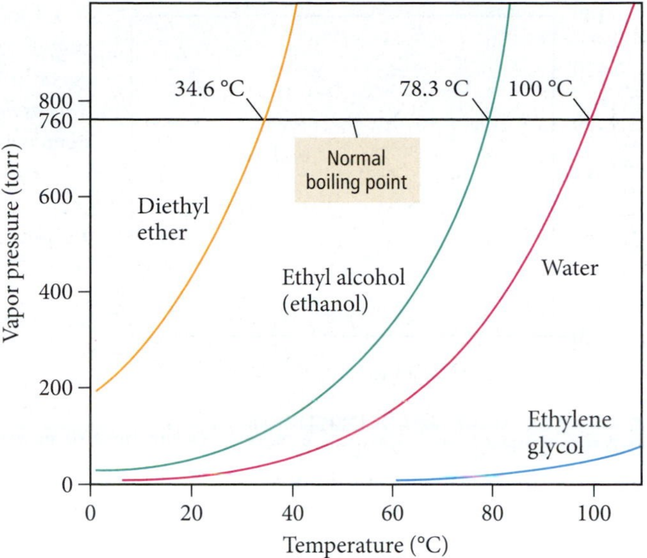

▲ **FIGURE 12.25 Vapor Pressure of Several Liquids at Different Temperatures** At higher temperatures, more molecules have enough thermal energy to escape into the gas state, so vapor pressure increases with increasing temperature.

TABLE 12.8 Boiling Points of Water at Several Locations of Varied Altitudes

Location	Elevation (ft)	Approximate Pressure (atm)*	Approximate Boiling Point of Water (°C)
Mount Everest, Tibet (highest mountain peak on Earth)	29,035	0.32	78
Mount McKinley (Denali), Alaska (highest mountain peak in North America)	20,320	0.46	83
Mount Whitney, California (highest mountain peak in 48 contiguous United States)	14,495	0.60	87
Denver, Colorado (mile high city)	5,280	0.83	94
Boston, Massachusetts (sea level)	20	1.0	100

*The atmospheric pressure in each of these locations is subject to weather conditions and can vary significantly from these values.

◀ **FIGURE 12.26 Boiling** A liquid boils when thermal energy is high enough to cause molecules in the interior of the liquid to become gaseous, forming bubbles that rise to the surface. Sometimes you see bubbles begin to form in hot water below 100 °C. These bubbles are dissolved air—not gaseous water—leaving the liquid.

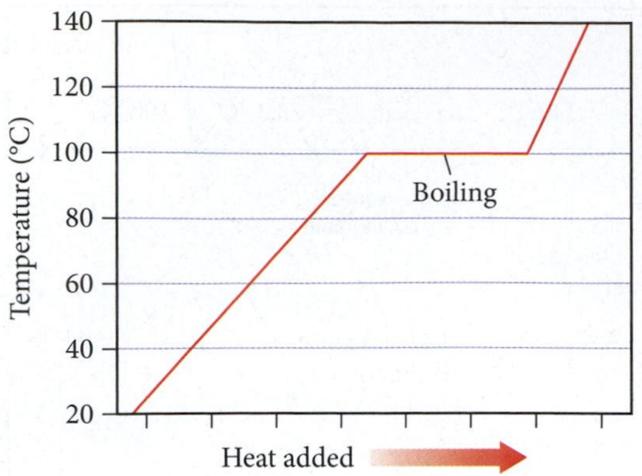

▲ **FIGURE 12.27 The Temperature During Boiling** The temperature of water during boiling remains at 100 °C.

Once the boiling point of a liquid is reached, additional heating only causes more rapid boiling; it does not raise the temperature of the liquid above its boiling point, as shown in the *heating curve* in Figure 12.27 ◀. Therefore, at 1 atm, boiling water always has a temperature of 100 °C. *As long as liquid water is present, its temperature cannot rise above its boiling point.* After all the water has been converted to steam, the temperature of the steam can continue to rise beyond 100 °C.

Now, let's return our attention to Figure 12.25 (on the previous page). As we can see from the graph, the vapor pressure of a liquid increases with increasing temperature. However, *the relationship is not linear.* In other words, doubling the temperature results in more than a doubling of the vapor pressure. The relationship between vapor pressure and temperature is exponential, and we express it as:

$$P_{vap} = \beta \exp\left(\frac{-\Delta H_{vap}}{RT}\right) \qquad [12.1]$$

In this expression P_{vap} is the vapor pressure, β is a constant that depends on the gas, ΔH_{vap} is the heat of vaporization, R is the gas constant (8.314 J/mol·K), and T is the temperature in kelvins. We can rearrange Equation 12.1 by taking the natural logarithm of both sides:

$$\ln P_{vap} = \ln\left[\beta \exp\left(\frac{-\Delta H_{vap}}{RT}\right)\right] \qquad [12.2]$$

Because $\ln AB = \ln A + \ln B$, we can rearrange the right side of Equation 12.2:

$$\ln P_{vap} = \ln \beta + \ln\left[\exp\left(\frac{-\Delta H_{vap}}{RT}\right)\right] \qquad [12.3]$$

Because $\ln e^x = x$ (see Appendix IIIB), we can simplify Equation 12.3:

$$\ln P_{vap} = \ln \beta + \frac{-\Delta H_{vap}}{RT} \qquad [12.4]$$

A slight additional rearrangement gives the important result:

$$\ln P_{vap} = \frac{-\Delta H_{vap}}{R}\left(\frac{1}{T}\right) + \ln \beta \qquad \text{Clausius–Clapeyron equation}$$

$$y = m(x) + b \qquad \text{Equation for a straight line}$$

When we use the Clausius–Clapeyron equation in this way, we ignore the relatively small temperature dependence of ΔH_{vap}.

Notice the parallel relationship between the **Clausius–Clapeyron equation** and the equation for a straight line. Just as a plot of y versus x yields a straight line with slope m and intercept b, so a plot of $\ln P_{vap}$ (equivalent to y) versus $1/T$ (equivalent to x) gives a straight line with slope $-\Delta H_{vap}/R$ (equivalent to m) and y-intercept $\ln \beta$ (equivalent to b), as shown in Figure 12.28 ▼. The Clausius–Clapeyron equation gives a linear relationship—not between the vapor pressure and the temperature (which have an exponential relationship)—but between the *natural log* of the vapor pressure and the *inverse* of temperature.

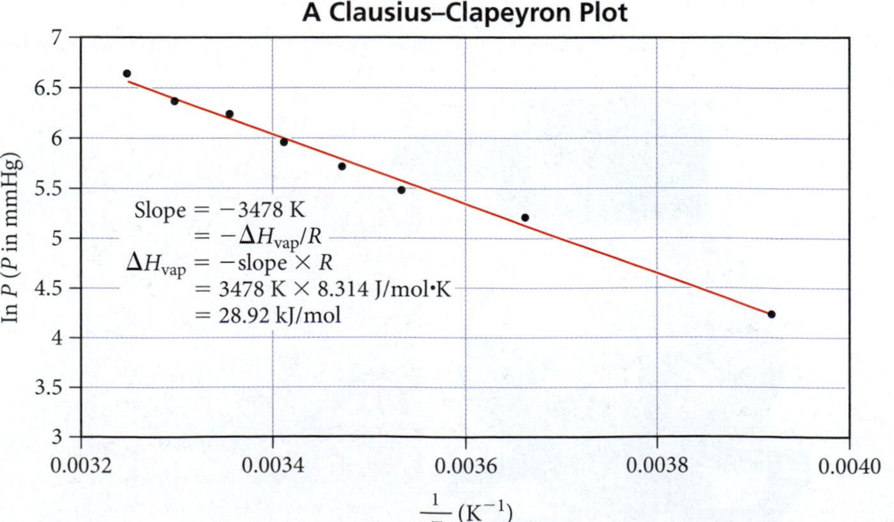

▶ **FIGURE 12.28 A Clausius–Clapeyron Plot for Diethyl Ether ($CH_3CH_2OCH_2CH_3$)** A plot of the natural log of the vapor pressure versus the inverse of the temperature in K yields a straight line with slope $-\Delta H_{vap}/R$.

This is a common technique in the analysis of chemical data. If two variables are not linearly related, it is often convenient to find ways to graph *functions of those variables* that are linearly related.

The Clausius–Clapeyron equation leads to a convenient way to measure the heat of vaporization in the laboratory. We measure the vapor pressure of a liquid as a function of temperature and create a plot of the natural log of the vapor pressure versus the inverse of the temperature. We can then determine the slope of the line to find the heat of vaporization, as demonstrated in Example 12.4.

EXAMPLE 12.4

Using the Clausius–Clapeyron Equation to Determine Heat of Vaporization from Experimental Measurements of Vapor Pressure

The vapor pressure of dichloromethane is measured as a function of temperature, and the following results are obtained:

Temperature (K)	Vapor Pressure (torr)
200	0.8
220	4.5
240	21
260	71
280	197
300	391

Determine the heat of vaporization of dichloromethane.

SOLUTION

To find the heat of vaporization, use an Excel™ spreadsheet or a graphing calculator to make a plot of the natural log of vapor pressure (ln P) as a function of the inverse of the temperature in kelvins ($1/T$). Fit the points to a line and determine the slope of the line. The slope of the best-fitting line is −3773 K. Because the slope equals $-\Delta H_{vap}/R$, we find the heat of vaporization as follows:

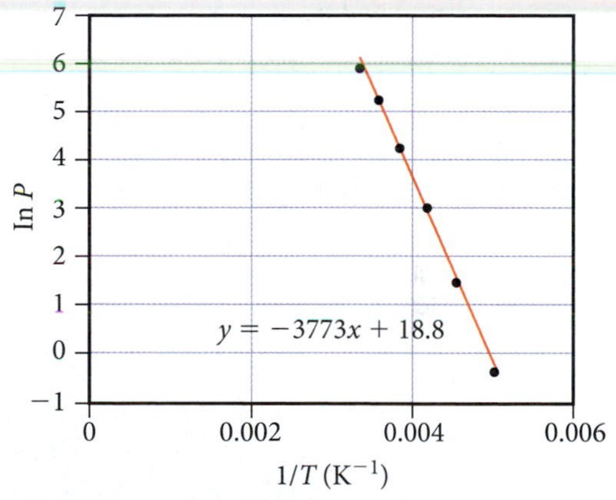

$$\begin{aligned} \text{slope} &= -\Delta H_{vap}/R \\ \Delta H_{vap} &= -\text{slope} \times R \\ &= -(-3773\ \text{K})(8.314\ \text{J/mol} \cdot \text{K}) \\ &= 3.14 \times 10^4\ \text{J/mol} \\ &= 31.4\ \text{kJ/mol} \end{aligned}$$

FOR PRACTICE 12.4

The vapor pressure of carbon tetrachloride is measured as a function of the temperature, and the following results are obtained:

Temperature (K)	Vapor Pressure (torr)
255	11.3
265	21.0
275	36.8
285	61.5
295	99.0
300	123.8

Determine the heat of vaporization of carbon tetrachloride.

The Clausius–Clapeyron equation can also be expressed in a two-point form that we can use with just two measurements of vapor pressure and temperature to determine the heat of vaporization.

$$\ln \frac{P_2}{P_1} = \frac{-\Delta H_{vap}}{R}\left(\frac{1}{T_2} - \frac{1}{T_1}\right)$$ Clausius–Clapeyron equation (two-point form)

The two-point method is generally inferior to plotting multiple points because fewer data points result in more chance for error.

We can use this form of the equation to predict the vapor pressure of a liquid at any temperature if we know the enthalpy of vaporization and the normal boiling point (or the vapor pressure at some other temperature), as demonstrated in Example 12.5.

EXAMPLE 12.5

Using the Two-Point Form of the Clausius–Clapeyron Equation to Predict the Vapor Pressure at a Given Temperature

Methanol has a normal boiling point of 64.6 °C and a heat of vaporization (ΔH_{vap}) of 35.2 kJ/mol. What is the vapor pressure of methanol at 12.0 °C?

SORT You are given the normal boiling point of methanol (the temperature at which the vapor pressure is 760 mmHg) and the heat of vaporization. You are asked to find the vapor pressure at a specified temperature that is also given.

GIVEN: $T_1(°C) = 64.6\ °C$

$P_1 = 760\ \text{torr}$

$\Delta H_{vap} = 35.2\ \text{kJ/mol}$

$T_2(°C) = 12.0\ °C$

FIND: P_2

STRATEGIZE The conceptual plan is essentially the Clausius–Clapeyron equation, which relates the given and find quantities.

CONCEPTUAL PLAN

$$\ln \frac{P_2}{P_1} = \frac{-\Delta H_{vap}}{R}\left(\frac{1}{T_2} - \frac{1}{T_1}\right)$$

(Clausius–Clapeyron equation, two-point form)

SOLVE First, convert T_1 and T_2 from °C to K.

SOLUTION

$$\begin{aligned} T_1(K) &= T_1(°C) + 273.15 \\ &= 64.6 + 273.15 \\ &= 337.8\ K \end{aligned}$$

$$\begin{aligned} T_2(K) &= T_2(°C) + 273.15 \\ &= 12.0 + 273.15 \\ &= 285.2\ K \end{aligned}$$

Substitute the required values into the Clausius–Clapeyron equation and solve for P_2. Use the heat of vaporization in J/mol for the correct canceling of units with J in R.

$$\ln \frac{P_2}{P_1} = \frac{-\Delta H_{vap}}{R}\left(\frac{1}{T_2} - \frac{1}{T_1}\right)$$

$$\begin{aligned} \ln \frac{P_2}{P_1} &= \frac{-35.2 \times 10^3 \frac{J}{mol}}{8.314 \frac{J}{mol \cdot K}}\left(\frac{1}{285.2\ K} - \frac{1}{337.8\ K}\right) \\ &= -2.31 \end{aligned}$$

$$\frac{P_2}{P_1} = e^{-2.31}$$

$$\begin{aligned} P_2 &= P_1(e^{-2.31}) \\ &= 760\ \text{torr}(0.0993) \\ &= 75.4\ \text{torr} \end{aligned}$$

—*Continued on the next page*

Continued from the previous page—

CHECK The units of the answer are correct. The magnitude of the answer makes sense because vapor pressure should be significantly lower at the lower temperature.

FOR PRACTICE 12.5

Propane has a normal boiling point of −4.20 °C and a heat of vaporization (ΔH_{vap}) of 19.04 kJ/mol. What is the vapor pressure of propane at 25.0 °C?

The Critical Point: The Transition to an Unusual State of Matter

We have considered the vaporization of a liquid in a container open to the atmosphere with and without heating, and the vaporization of a liquid in a *sealed* container without heating. We now examine the vaporization of a liquid in a *sealed* container *during heating*. Consider liquid *n*-pentane in equilibrium with its vapor in a sealed container initially at 25 °C. At this temperature, the vapor pressure of *n*-pentane is 0.67 atm. What happens if we heat the liquid? As the temperature rises, more *n*-pentane vaporizes and the pressure within the container increases. At 100 °C, the pressure is 5.5 atm, and at 190 °C the pressure is 29 atm. As the temperature and pressure increase, more and more gaseous *n*-pentane is forced into the same amount of space, and the density of the *gas* gets higher and higher. At the same time, the increasing temperature causes the density of the *liquid* to become lower and lower. At 197 °C, the meniscus between the liquid and gaseous *n*-pentane disappears and the gas and liquid states commingle to form a *supercritical fluid* (Figure 12.29 ▼). For any substance, the *temperature* at which this transition occurs is the **critical temperature** (T_c). The liquid cannot exist (regardless of pressure) above this temperature. The *pressure* at which this transition occurs is the **critical pressure** (P_c).

Researchers are interested in supercritical fluids because of their unique properties. A supercritical fluid has properties of both liquids and gases—it is in some sense intermediate between the two. Supercritical fluids can act as good solvents, selectively dissolving a number of compounds. For example, supercritical carbon dioxide is used as a solvent to extract caffeine from coffee beans. The caffeine dissolves in the supercritical carbon dioxide, but other substances—such as those responsible for the flavor of coffee—do not dissolve. Consequently, the caffeine is removed without substantially altering the coffee's flavor. The supercritical carbon dioxide is easily removed from a mixture by lowering the pressure below the critical pressure, at which point the carbon dioxide evaporates, leaving no residue.

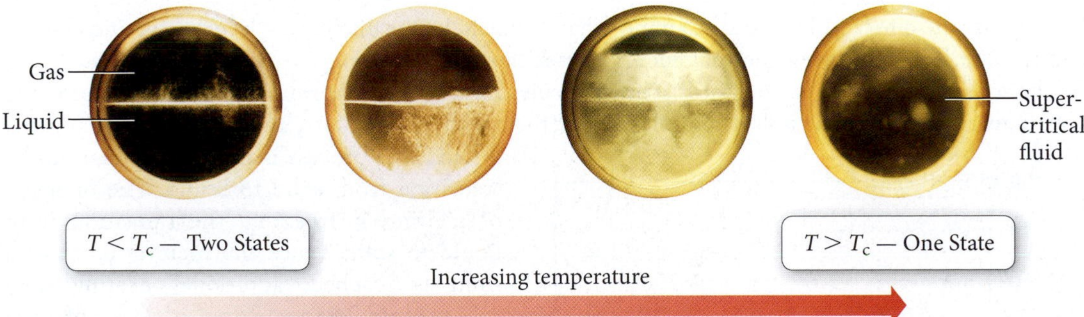

▲ **FIGURE 12.29 Critical Point Transition** As *n*-pentane is heated in a sealed container, it undergoes a transition to a supercritical fluid. At the critical point, the meniscus separating the liquid and gas disappears, and the fluid becomes supercritical—neither a liquid nor a gas.

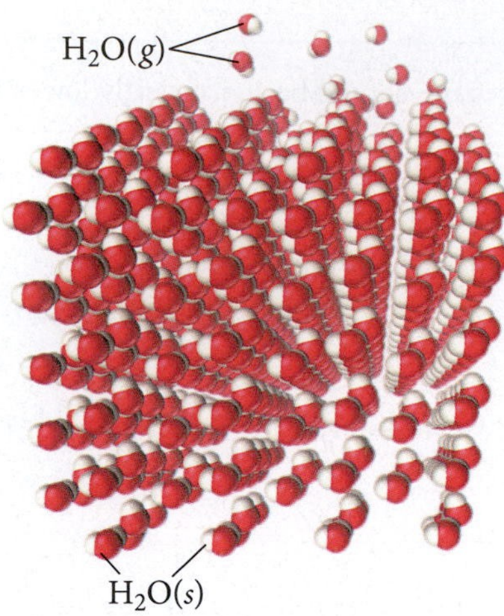

▲ **FIGURE 12.30 The Sublimation of Ice** The water molecules at the surface of an ice cube can sublime directly into the gas state.

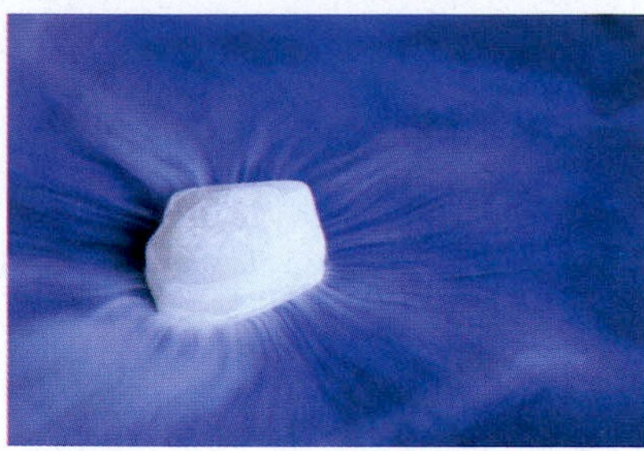

▲ Dry ice (solid CO_2) sublimes but does not melt at atmospheric pressure.

12.6 Sublimation and Fusion

In Section 12.5, we examined a beaker of liquid water at room temperature from the molecular viewpoint. Now, let's examine a block of ice at −10 °C from the same molecular perspective, paying close attention to two common processes: sublimation and fusion.

Sublimation

Even though a block of ice is solid, the water molecules have thermal energy, which causes each one to vibrate about a fixed point. The motion is much less vigorous than in a liquid, but it is significant nonetheless. As in liquids, at any instant some molecules in the block of ice have more thermal energy than the average and some have less. The molecules with high enough thermal energy can break free from the ice surface—where, as in liquids, molecules are held less tightly than in the interior due to fewer neighbor–neighbor interactions—and go directly into the gas state (Figure 12.30 ◀). This process is **sublimation**, the transition from solid to gas. Some of the water molecules in the gas state (those at the low end of the energy distribution curve for the gaseous molecules) collide with the surface of the ice and are captured by the intermolecular forces with other molecules. This process—the opposite of sublimation—is **deposition**, the transition from gas to solid. As is the case with liquids, the pressure of a gas in dynamic equilibrium with its solid is the vapor pressure of the solid.

Although both sublimation and deposition occur on the surface of an ice block open to the atmosphere at −10 °C, sublimation usually occurs at a greater rate because most of the newly sublimed molecules escape into the surrounding atmosphere and never come back. The result is a noticeable decrease in the size of the ice block over time (even though the temperature is below the melting point).

If you live in a cold climate, you may have noticed the disappearance of ice and snow from the ground even though the temperature remains below 0 °C. Similarly, ice cubes left in the freezer for a long time slowly shrink, even though the freezer is always below 0 °C. In both cases, the ice is *subliming*, turning directly into water vapor. Ice also sublimes out of frozen foods. You may have noticed, for example, the gradual growth of ice crystals on the *inside* of airtight plastic food-storage bags in a freezer. The ice crystals are composed of water that has sublimed out of the food and redeposited on the surface of the bag or on the surface of the food. For this reason, food that remains frozen for too long becomes dried out. Such dehydration can be avoided to some degree by freezing foods to colder temperatures, a process called deep-freezing. The colder temperature lowers the vapor pressure of ice and preserves the food longer. Freezer burn on meats is another common manifestation of sublimation. When you improperly store meat (for example, in a container that is not airtight) sublimation continues unabated. The result is the dehydration of the surface of the meat, which becomes discolored and loses flavor and texture.

A substance commonly associated with sublimation is solid carbon dioxide or dry ice, which does not melt under atmospheric pressure no matter what the temperature is. However, at −78 °C the CO_2 molecules have enough energy to leave the surface of the dry ice and become gaseous through sublimation.

Fusion

The term *fusion* is used for melting because if we heat several crystals of a solid, they *fuse* into a continuous liquid upon melting.

Let's return to our ice block and examine what happens at the molecular level as we increase its temperature. The increasing thermal energy causes the water molecules to vibrate faster and faster. At the **melting point** (0 °C for water), the molecules have enough thermal energy to overcome the intermolecular forces that hold the molecules at their stationary points, and the solid turns into a liquid. This process is **melting** or **fusion**, the transition from solid to liquid. The opposite of melting is **freezing**, the transition from liquid to solid. Once the melting point of a solid is reached, additional heating only causes more rapid melting; it does not raise the temperature of the solid above its melting point (Figure 12.31 ◀). Only after all of the ice has melted does additional heating raise the temperature of the liquid water past 0 °C. A mixture of water *and* ice always has a temperature of 0 °C (at 1 atm pressure).

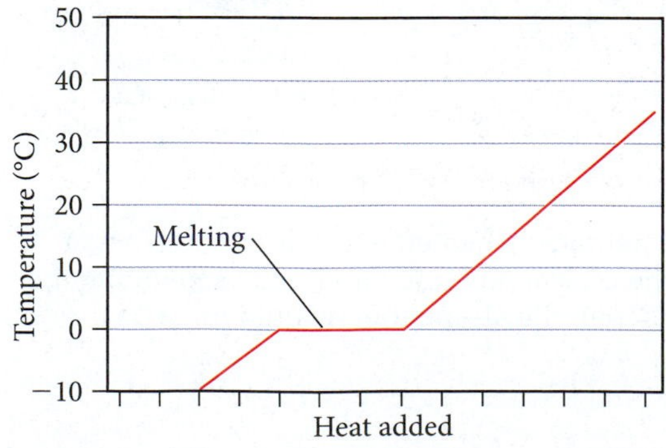

▶ **FIGURE 12.31 Temperature during Melting** The temperature of water during melting remains at 0.0 °C as long as both solid and liquid water remain.

Energetics of Melting and Freezing

The most common way to cool a beverage quickly is to drop several ice cubes into it. As the ice melts, the drink cools because melting is endothermic—the melting ice absorbs heat from the liquid. The amount of heat required to melt 1 mol of a solid is the **heat of fusion (ΔH_{fus})**. The heat of fusion for water is 6.02 kJ/mol.

$$H_2O(s) \longrightarrow H_2O(l) \qquad \Delta H_{fus} = 6.02 \text{ kJ/mol}$$

The heat of fusion is positive because melting is endothermic.

Freezing, the opposite of melting, is exothermic—heat is released when a liquid freezes into a solid. For example, as water in the freezer turns into ice, it releases heat, which must be removed by the refrigeration system of the freezer. If the refrigeration system did not remove the heat, the water would not completely freeze into ice. The heat released as the water began to freeze would warm the freezer, preventing further freezing. The change in enthalpy for freezing has the same magnitude as the heat of fusion but the opposite sign.

$$H_2O(l) \longrightarrow H_2O(s) \qquad \Delta H = -\Delta H_{fus} = -6.02 \text{ kJ/mol}$$

Different substances have different heats of fusion, as shown in Table 12.9.

TABLE 12.9 Heats of Fusion of Several Substances

Liquid	Chemical Formula	Melting Point (°C)	ΔH_{fus}(kJ/mol)
Water	H_2O	0.00	6.02
Rubbing alcohol (isopropyl alcohol)	C_3H_8O	–89.5	5.37
Acetone	C_3H_6O	–94.8	5.69
Diethyl ether	$C_3H_{10}O$	–116.3	7.27

In general, the heat of fusion for a substance is significantly less than its heat of vaporization, as shown in Figure 12.32 ▼. We have already seen that the solid and liquid states are closer to each other in many ways than they are to the gas state. It takes less energy to melt 1 mol of ice into liquid than it does to vaporize 1 mol of liquid water into gas because vaporization requires complete separation of molecules from one another, so the intermolecular forces must be completely overcome. Melting, however, requires that intermolecular forces be only partially overcome, allowing molecules to move around one another while still remaining in contact.

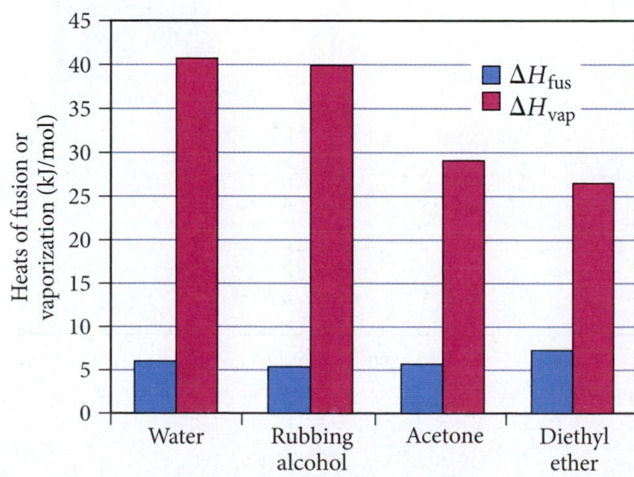

◀ **FIGURE 12.32 Heat of Fusion and Heat of Vaporization** Typical heats of fusion are significantly less than heats of vaporization.

12.7 Heating Curve for Water

We can combine and build on the concepts from Sections 12.5 and 12.6 by examining the *heating curve* for 1.00 mol of water at 1.00 atm pressure shown in Figure 12.33 ▼. The y-axis of the heating curve represents the temperature of the water sample. The x-axis represents the amount of heat added (in kilojoules) during heating. In the diagram, we divide the process into five segments: (1) ice warming; (2) ice melting into liquid water; (3) liquid water warming; (4) liquid water vaporizing into steam; and (5) steam warming.

In two of the segments in Figure 12.33 (2 and 4) the temperature is constant as heat is added because the added heat goes into producing the transition between states, not into increasing the temperature. The two states are in equilibrium during the transition and the temperature remains constant. The amount of heat required to achieve the state change is given by $q = n\Delta H$.

In the other three segments (1, 3, and 5), temperature increases linearly. These segments represent the heating of a single state in which the deposited heat raises the temperature in accordance with the substance's heat capacity ($q = mC_s\,\Delta T$). Let's examine each of these segments individually.

Segment 1 In Segment 1, solid ice is warmed from −25 °C to 0 °C. Since no transition between states occurs here, the amount of heat required to heat the solid ice is given by $q = mC_s\,\Delta T$ (see

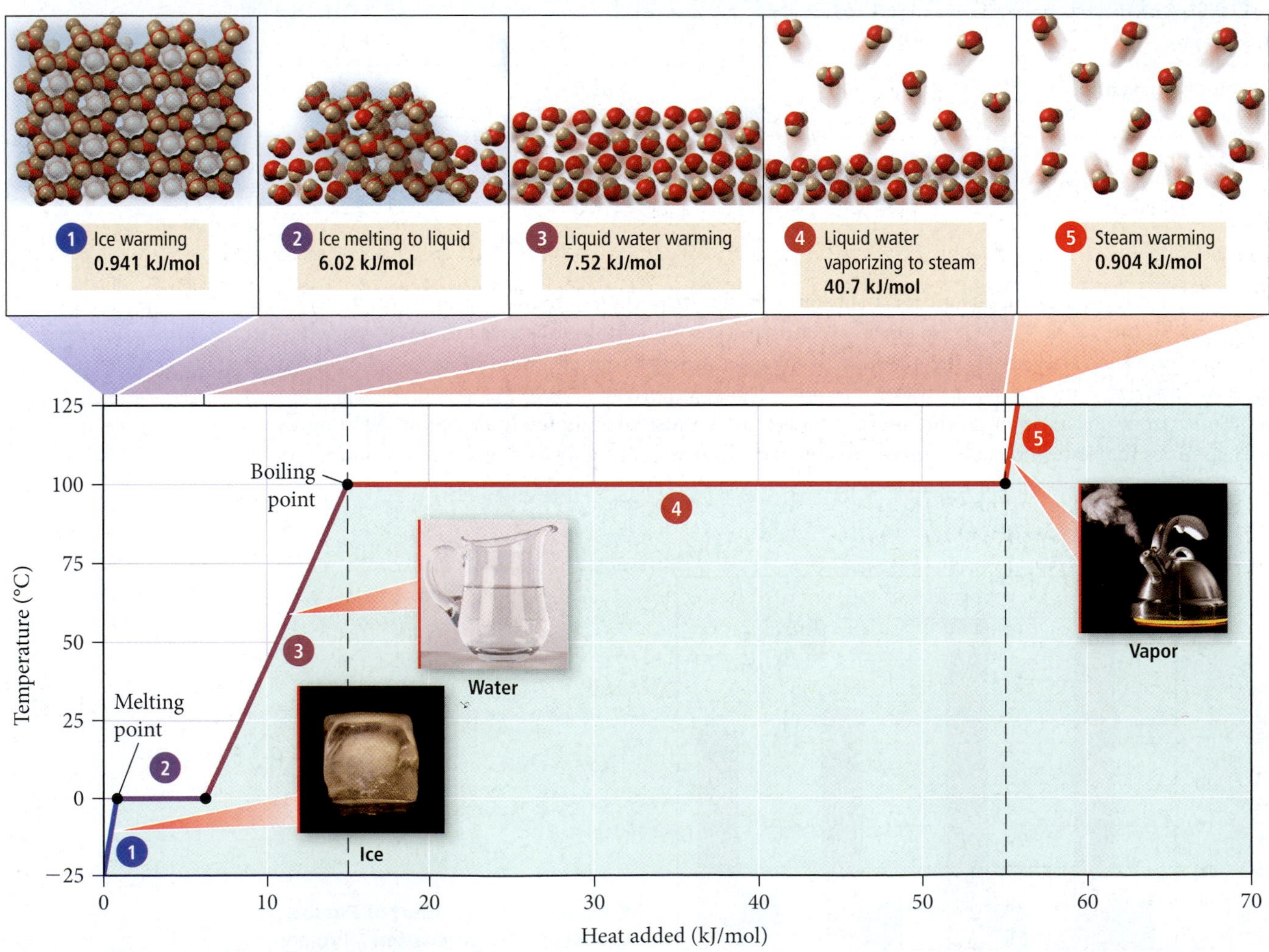

▲ **FIGURE 12.33 Heating Curve for Water**

Section 10.4), where C_s is the specific heat capacity of ice $C_{s,\,ice} = 2.09\ J/g\cdot{}^\circ C$). For 1.00 mol of water (18.0 g), we calculate the amount of heat as follows:

$$
\begin{aligned}
q &= mC_{s,\,ice}\ \Delta T \\
&= 18.0\ \cancel{g}\left(2.09\frac{J}{\cancel{g}\cdot\cancel{^\circ C}}\right)[0.0\ \cancel{^\circ C} - (-25.0\ \cancel{^\circ C})] \\
&= 941\ J = 0.941\ kJ
\end{aligned}
$$

So in Segment 1, 0.941 kJ of heat is added to the ice, warming it from −25° C to 0 °C.

Segment 2 In Segment 2, the added heat does not change the temperature of the ice and water mixture because the heat is absorbed by the transition from solid to liquid. The amount of heat required to convert the ice to liquid water is given by $q = n\ \Delta H_{fus}$, where n is the number of moles of water and ΔH_{fus} is the heat of fusion (see Section 12.6):

$$
\begin{aligned}
q &= n\ \Delta H_{fus} \\
&= 1.00\ \cancel{mol}\left(\frac{6.02\ kJ}{\cancel{mol}}\right) \\
&= 6.02\ kJ
\end{aligned}
$$

In Segment 2, 6.02 kJ is added to the ice, melting it into liquid water. Notice that the temperature does not change during melting. The liquid and solid coexist at 0 °C as the melting occurs.

Segment 3 In Segment 3, the liquid water is warmed from 0 °C to 100 °C. Because no transition between states occurs here, the amount of heat required to heat the liquid water is given by $q = mC_s\,\Delta T$, as in segment 1. However, now we use the heat capacity of liquid water (not ice) for the calculation. For 1.00 mol of water (18.0 g), we calculate the amount of heat as follows:

$$
\begin{aligned}
q &= mC_{s,\,liq}\ \Delta T \\
&= 18.0\ \cancel{g}\left(4.18\frac{J}{\cancel{g}\cdot\cancel{^\circ C}}\right)(100.0\ \cancel{^\circ C} - 0.0\ \cancel{^\circ C}) \\
&= 7.52 \times 10^3\ J = 7.52\ kJ
\end{aligned}
$$

In Segment 3, 7.52 kJ of heat is added to the liquid water, warming it from 0 °C to 100 °C.

Segment 4 In Segment 4, the water undergoes a second transition between states, this time from liquid to gas. The amount of heat required to convert the liquid to gas is given by $q = n\ \Delta H_{vap}$, where n is the number of moles and ΔH_{vap} is the heat of vaporization (see Section 12.5):

$$
\begin{aligned}
q &= n\ \Delta H_{vap} \\
&= 1.00\ \cancel{mol}\left(\frac{40.7\ kJ}{\cancel{mol}}\right) \\
&= 40.7\ kJ
\end{aligned}
$$

Thus, in Segment 4, 40.7 kJ is added to the water, vaporizing it into steam. Notice that the temperature does not change during boiling. The liquid and gas coexist at 100 °C as the boiling occurs.

Segment 5 In Segment 5, the steam is warmed from 100 °C to 125 °C. Because no transition between states occurs here, the amount of heat required to heat the steam is given by $q = mC_s\,\Delta T$ (as in segments 1 and 3) except that we use the heat capacity of steam (not water or ice) (2.01 J/g · °C):

$$
\begin{aligned}
q &= mC_{s,\,steam}\ \Delta T \\
&= 18.0\ \cancel{g}\left(2.01\frac{J}{\cancel{g}\cdot\cancel{^\circ C}}\right)(125.0\ \cancel{^\circ C} - 100.0\ \cancel{^\circ C}) \\
&= 904 = 0.904\ kJ
\end{aligned}
$$

In Segment 5, 0.904 kJ of heat is added to the steam, warming it from 100 °C to 125 °C.

12.6 Cc Conceptual Connection **Cooling of Water with Ice**

You just saw that the heat capacity of ice is $C_{s,\,ice} = 2.09\ J/g \cdot °C$ and that the heat of fusion of ice is 6.02 kJ/mol. When a small ice cube at −10 °C is put into a cup of water at room temperature, which of the following plays a greater role in cooling the liquid water: the warming of the ice from −10 °C to 0 °C, or the melting of the ice?

12.8 Water: An Extraordinary Substance

▲The *Mars Curiosity Rover* has found evidence that liquid water (which could possibly sustain life) once existed on the surface of Mars.

Water is the most common and important liquid on Earth. It fills our oceans, lakes, and streams. In its solid form, it caps our mountains, and in its gaseous form, it humidifies our air. We drink water, we sweat water, and we excrete bodily wastes dissolved in water. Indeed, the majority of our body mass *is* water. Life is impossible without water, and in most places on Earth where liquid water exists, life exists. Recent evidence for the past existence of water on Mars has fueled hopes of finding evidence of past life there. And though it may not be obvious to us (because we take water for granted), this familiar substance has many remarkable properties.

Among liquids, water is unique. It has a low molar mass (18.02 g/mol), yet it is a liquid at room temperature. Other main-group hydrides have higher molar masses but lower boiling points, as shown in Figure 12.34 ▼. No other substance of similar molar mass (except HF) comes close to being a liquid at room temperature. We can understand water's high boiling point (in spite of its low molar mass) by examining its molecular structure. The bent geometry of the water molecule and the highly polar nature of the O—H bonds result in a molecule with a significant dipole moment. Water's two O—H bonds (hydrogen directly bonded to oxygen) allow a water molecule to form strong hydrogen bonds with four other water molecules (Figure 12.35 ▼), resulting in a relatively high boiling point. Water's high polarity also allows it to dissolve many other polar and ionic compounds and even a number of nonpolar gases such as oxygen and carbon dioxide (by inducing a dipole moment in their molecules). Consequently, water is the main solvent within living organisms, transporting nutrients and other important compounds throughout the body. Water is the main solvent in our environment as well, allowing aquatic animals, for example, to survive by breathing dissolved oxygen and allowing aquatic plants to survive by using dissolved carbon dioxide for photosynthesis.

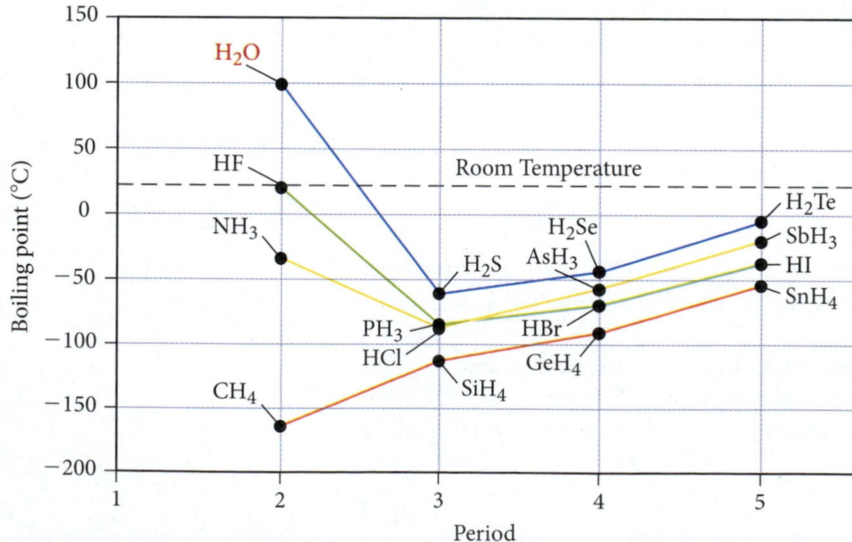

▲ **FIGURE 12.34 Boiling Points of Main-Group Hydrides** Water is the only common main-group hydride that is a liquid at room temperature.

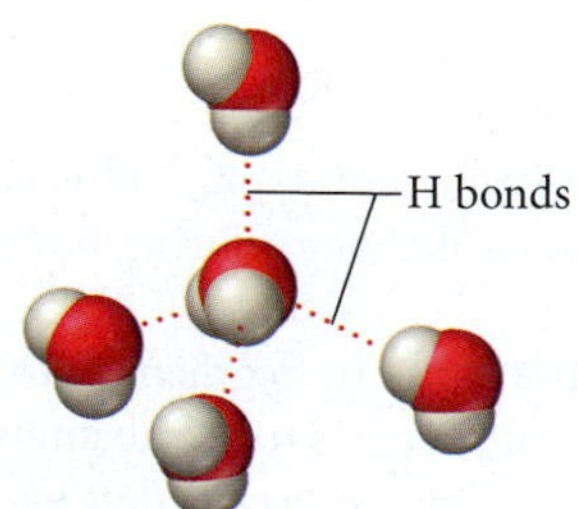

▲ **FIGURE 12.35 Hydrogen Bonding in Water** A water molecule can form four strong hydrogen bonds with four other water molecules.

We already saw in Section 10.4 that water has an exceptionally high specific heat capacity, which has a moderating effect on the climate of coastal cities. In some cities, such as San Francisco, for example, the daily fluctuation in temperature can be less than 10 °C. This same moderating effect occurs over the entire planet, two-thirds of which is covered by water. Without water, the daily temperature fluctuations on Earth might be more like those on Mars, where temperature fluctuations of 63 °C (113 °F) occur between early morning and midday. Imagine awakening to below-freezing temperatures, only to bake at summer desert temperatures in the afternoon! The presence of water on Earth and water's uniquely high specific heat capacity are largely responsible for our planet's much smaller daily fluctuations.

▲ When lettuce freezes, the water within its cells expands, rupturing them.

The way that water freezes is also unique. Unlike other substances, which contract upon freezing, water expands upon freezing. Consequently, ice is less dense than liquid water, which is why ice floats. This seemingly trivial property has significant consequences. The frozen layer of ice at the surface of a winter lake insulates the water in the lake from further freezing. If this ice layer sank, it would kill bottom-dwelling aquatic life and possibly allow the lake to freeze solid, eliminating virtually all life in the lake.

The expansion of water upon freezing, however, is one reason that most organisms do not survive freezing. When the water within a cell freezes, it expands and often ruptures the cell, just as water freezing within a pipe bursts the pipe. Many foods, especially those with high water content, do not survive freezing very well either. Have you ever tried to freeze your own vegetables? If you put lettuce or spinach in the freezer, it is limp and damaged upon thawing. The frozen-food industry gets around this problem by *flash freezing* vegetables and other foods. In this process, foods are frozen nearly instantaneously, which prevents water molecules from settling into their preferred crystalline structure. Consequently, the water does not expand very much, and the food remains largely undamaged.

SELF-ASSESSMENT Quiz

1. Which state of matter is compressible?
   a) gas
   b) liquid
   c) solid
   d) none of the above

2. Liquid nitrogen boils at 77 K. This image depicts a sample of liquid nitrogen.

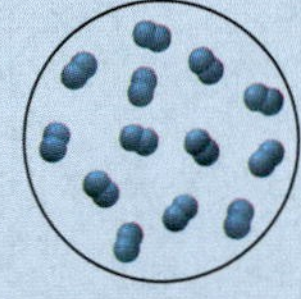

Which image best depicts the nitrogen after it has boiled?

a)

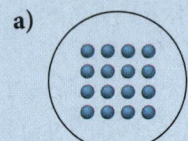

b)

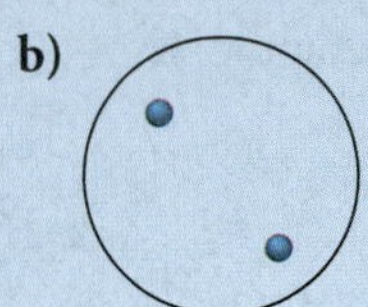

c)

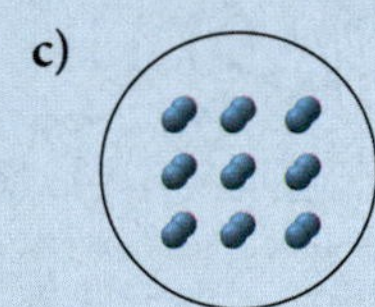

d)

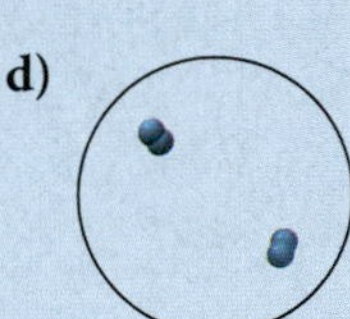

3. Taking intermolecular forces into account, which halogen would you expect to have the highest boiling point?
   a) F_2 **b)** Cl_2
   c) Br_2 **d)** I_2

4. Which substance experiences dipole–dipole forces?
   a) CCl_4 **b)** NF_3
   c) CS_2 **d)** SO_3

5. Which substance is a liquid at room temperature?
   a) CH_3OH **b)** CF_4
   c) SiH_4 **d)** CO_2

6. Which property of a liquid increases with increasing temperature?
   a) surface tension
   b) viscosity
   c) vapor pressure
   d) none of the above

7. Determine the amount of heat (in kJ) required to vaporize 1.55 kg of water at its boiling point. For water, $\Delta H_{vap} = 40.7$ kJ/mol (at 100 °C).
   a) 3.50×10^3 kJ
   b) 1.14×10^6 kJ
   c) 2.11 kJ
   d) 686 kJ

8. The vapor pressure of a substance is measured over a range of temperatures. A plot of the natural log of the vapor pressure versus the inverse of the temperature (in Kelvin) produces a straight line with a slope of -3.46×10^3 K. What is the enthalpy of vaporization of the substance?
   a) 2.40×10^{-3} kJ/mol
   b) 28.8 kJ/mol
   c) 0.416 kJ/mol
   d) 3.22 kJ/mol

9. Acetic acid has a normal boiling point of 118 °C and a ΔH_{vap} of 23.4 kJ/mol. What is the vapor pressure (in mmHg) of acetic acid at 25 °C?
   a) 2.92×10^{-39} mmHg
   b) 7.16×10^3 mmHg
   c) 758 mmHg
   d) 80.6 mmHg

10. A mixture containing 21.4 g of ice (at exactly 0.00 °C) and 75.3 g of water (at 55.3 °C) is placed in an insulated container. Assuming no heat is lost to the surroundings, what is the final temperature of the mixture?
    a) 22.5 °C
    b) 25.4 °C
    c) 32.6 °C
    d) 41.9 °C

11. Which process releases the greatest amount of heat?
    a) the condensation of 10 g of gaseous water
    b) the freezing of 10 g of liquid water
    c) the boiling of 10 g of liquid water
    d) the melting of 10 g of ice

Answers: 1:a; 2:d; 3:d; 4:b; 5:a; 6:c; 7:a; 8:b; 9:d; 10:b; 11:a

CHAPTER 12 SUMMARY

REVIEW

KEY LEARNING OUTCOMES

CHAPTER OBJECTIVES	ASSESSMENT
Determining Whether a Molecule Has Dipole–Dipole Forces (12.3)	• Example 12.1 For Practice 12.1 Exercises 33–36
Determining Whether a Molecule Displays Hydrogen Bonding (12.3)	• Example 12.2 For Practice 12.2 Exercises 33–36
Using the Heat of Vaporization in Calculations (12.5)	• Example 12.3 For Practice 12.3 For More Practice 12.3 Exercises 55–58
Using the Clausius–Clapeyron Equation (12.5)	• Examples 12.4, 12.5 For Practice 12.4, 12.5 Exercises 59–62

KEY TERMS

Section 12.2
crystalline (444)
amorphous (444)

Section 12.3
dispersion force (446)
dipole–dipole force (448)
permanent dipole (448)
miscibility (449)
hydrogen bonding (450)
hydrogen bond (451)
ion–dipole force (453)

Section 12.4
surface tension (454)
viscosity (455)
capillary action (455)

Section 12.5
vaporization (457)
condensation (457)
volatile (457)
nonvolatile (457)
heat (or enthalpy) of vaporization (ΔH_{vap}) (458)
dynamic equilibrium (459)
vapor pressure (459)
boiling point (461)
normal boiling point (461)
Clausius–Clapeyron equation (462)
critical temperature (T_c) (465)
critical pressure (P_c) (465)

Section 12.6
sublimation (466)
deposition (466)
melting point (466)
melting (fusion) (466)
freezing (466)
heat of fusion (ΔH_{fus}) (467)

KEY CONCEPTS

Solids, Liquids, and Intermolecular Forces (12.1, 12.2, 12.3)

- The forces that hold molecules or atoms together in a liquid or solid are intermolecular forces. The strength of the intermolecular forces in a substance determines its state.
- Dispersion forces are present in all elements and compounds; they arise from the fluctuations in electron distribution within atoms and molecules. Dispersion forces are the weakest intermolecular forces, but they are significant in molecules with high molar masses.
- Dipole–dipole forces, generally stronger than dispersion forces, are present in all polar molecules.
- Hydrogen bonding occurs in polar molecules that contain hydrogen atoms bonded directly to fluorine, oxygen, or nitrogen. These are the strongest intermolecular forces.
- Ion–dipole forces occur when ionic compounds are mixed with polar compounds; they are especially important in aqueous solutions.

Surface Tension, Viscosity, and Capillary Action (12.4)

- Surface tension results from the tendency of liquids to minimize their surface area in order to maximize the interactions between their constituent particles, thus lowering potential energy. Surface tension causes water droplets to form spheres and allows insects and paper clips to "float" on the surface of water.
- Viscosity is the resistance of a liquid to flow. Viscosity increases with increasing strength of intermolecular forces and decreases with increasing temperature.
- Capillary action is the ability of a liquid to flow against gravity up a narrow tube. It is the result of adhesive forces, the attraction between the molecules and the surface of the tube, and cohesive forces, the attraction between the molecules in the liquid.

Vaporization and Vapor Pressure (12.5, 12.7)

- Vaporization, the transition from liquid to gas, occurs when thermal energy overcomes the intermolecular forces present in a liquid. The opposite process is condensation. Vaporization is endothermic, and condensation is exothermic.

- The rate of vaporization increases with increasing temperature, increasing surface area, and decreasing strength of intermolecular forces.
- The heat of vaporization (ΔH_{vap}) is the heat required to vaporize one mole of a liquid.
- In a sealed container, a solution and its vapor come into dynamic equilibrium, at which point the rate of vaporization equals the rate of condensation. The pressure of a gas that is in dynamic equilibrium with its liquid is its vapor pressure.
- The vapor pressure of a substance increases with increasing temperature and with decreasing strength of its intermolecular forces.
- The boiling point of a liquid is the temperature at which its vapor pressure equals the external pressure.
- The Clausius–Clapeyron equation expresses the relationship between the vapor pressure of a substance and its temperature, and we can use it to calculate the heat of vaporization from experimental measurements.
- When a liquid is heated in a sealed container, it eventually forms a supercritical fluid, which has properties intermediate between a liquid and a gas. This occurs at critical temperature and critical pressure.

Fusion and Sublimation (12.6, 12.7)

- Sublimation is the transition from solid to gas. The opposite process is deposition.
- Fusion, or melting, is the transition from solid to liquid. The opposite process is freezing.
- The heat of fusion (ΔH_{fus}) is the amount of heat required to melt one mole of a solid. Fusion is endothermic.
- The heat of fusion is generally less than the heat of vaporization because intermolecular forces do not have to be completely overcome for melting to occur.

The Uniqueness of Water (12.8)

- Water is a liquid at room temperature despite its low molar mass. Water forms strong hydrogen bonds and therefore has a high boiling point.
- The polarity of water enables it to dissolve many polar and ionic compounds and even nonpolar gases.
- Water expands upon freezing, so ice is less dense than liquid water. Water is critical both to the existence of life and to human health.

KEY EQUATIONS AND RELATIONSHIPS

Clausius–Clapeyron Equation: Relationship between Vapor Pressure (P_{vap}), the Heat of Vaporization (ΔH_{vap}), and Temperature (T) (12.5)

$$\ln P_{vap} = \frac{-\Delta H_{vap}}{RT} + \ln \beta \text{ } (\beta \text{ is a constant})$$

$$\ln \frac{P_2}{P_1} = \frac{-\Delta H_{vap}}{R}\left(\frac{1}{T_2} - \frac{1}{T_1}\right)$$

EXERCISES

REVIEW QUESTIONS

1. Why do ethanol and dimethyl ether have such different properties even though they have the same chemical formula?
2. Why are intermolecular forces important?
3. What are the key properties of liquids (in contrast to gases and solids)?
4. What are the key properties of solids (in contrast to liquids and gases)?
5. What is the fundamental difference between an amorphous solid and a crystalline solid?
6. Which factors cause transitions between the solid and liquid state? The liquid and gas state?
7. Describe the relationship between the state of a substance, its temperature, and the strength of its intermolecular forces.
8. From which kinds of interactions do intermolecular forces originate?
9. Why are intermolecular forces generally much weaker than bonding forces?
10. What is the dispersion force? What does the magnitude of the dispersion force depend on? How can we predict the magnitude of the dispersion force for closely related elements or compounds?
11. What is the dipole–dipole force? How can we predict the presence of dipole–dipole forces in a compound?
12. How is the miscibility of two liquids related to their polarity?
13. What is hydrogen bonding? How can we predict the presence of hydrogen bonding in a compound?
14. What is the ion–dipole force? Why is it important?
15. What is surface tension? How does surface tension result from intermolecular forces? How is it related to the strength of intermolecular forces?

16. What is viscosity? How does viscosity depend on intermolecular forces? What other factors affect viscosity?
17. What is capillary action? How does it depend on the relative strengths of adhesive and cohesive forces?
18. Explain what happens during the processes of vaporization and condensation. Why does the rate of vaporization increase with increasing temperature and surface area?
19. Why is vaporization endothermic? Why is condensation exothermic?
20. How is the volatility of a substance related to the intermolecular forces present within the substance?
21. What is the heat of vaporization for a liquid, and why is it useful?
22. Explain the process of dynamic equilibrium. How is dynamic equilibrium related to vapor pressure?
23. What happens to a system in dynamic equilibrium when it is disturbed in some way?
24. How is vapor pressure related to temperature? What happens to the vapor pressure of a substance when the temperature is increased? Decreased?
25. Define the terms *boiling point* and *normal boiling point*.
26. What is the Clausius–Clapeyron equation, and why is it important?
27. Explain what happens to a substance when it is heated in a closed container to its critical temperature.
28. What is sublimation? Cite a common example of sublimation.
29. What is fusion? Is fusion exothermic or endothermic? Why?
30. What is the heat of fusion, and why is it important?
31. Examine the heating curve for water in Section 12.7 (Figure 12.33). Explain why the curve has two segments in which heat is added to the water but the temperature does not rise.
32. Examine the heating curve for water in Section 12.7 (Figure 12.33). Explain the significance of the slopes of each of the three rising segments. Why are the slopes different?

PROBLEMS BY TOPIC

Intermolecular Forces

33. Determine the kinds of intermolecular forces that are present in each element or compound.
    a. N_2
    b. NH_3
    c. CO
    d. CCl_4
34. Determine the kinds of intermolecular forces that are present in each element or compound.
    a. Kr
    b. NCl_3
    c. SiH_4
    d. HF
35. Determine the kinds of intermolecular forces that are present in each element or compound.
    a. HCl
    b. H_2O
    c. Br_2
    d. He
36. Determine the kinds of intermolecular forces that are present in each element or compound.
    a. PH_3
    b. HBr
    c. CH_3OH
    d. I_2
37. Arrange these compounds in order of increasing boiling point. Explain your reasoning.
    a. CH_4
    b. CH_3CH_3
    c. CH_3CH_2Cl
    d. CH_3CH_2OH
38. Arrange these compounds in order of increasing boiling point. Explain your reasoning.
    a. H_2S
    b. H_2Se
    c. H_2O
39. Pick the compound with the highest boiling point in each pair. Explain your reasoning.
    a. CH_3OH or CH_3SH
    b. CH_3OCH_3 or CH_3CH_2OH
    c. CH_4 or CH_3CH_3
40. Pick the compound with the highest boiling point in each pair. Explain your reasoning.
    a. NH_3 or CH_4
    b. CS_2 or CO_2
    c. CO_2 or NO_2
41. In each pair of compounds, pick the one with the higher vapor pressure at a given temperature. Explain your reasoning.
    a. Br_2 or I_2
    b. H_2S or H_2O
    c. NH_3 or PH_3
42. In each pair of compounds, pick the one with the higher vapor pressure at a given temperature. Explain your reasoning.
    a. CH_4 or CH_3Cl
    b. $CH_3CH_2CH_2OH$ or CH_3OH
    c. CH_3OH or H_2CO
43. Determine if each pair of compounds forms a homogeneous solution when combined. For those that form homogeneous solutions, indicate the type of forces that are involved.
    a. CCl_4 and H_2O
    b. KCl and H_2O
    c. Br_2 and CCl_4
    d. CH_3CH_2OH and H_2O

44. Determine if each pair of compounds forms a homogeneous solution when combined. For those that form homogeneous solutions, indicate the type of forces that are involved.

a. $CH_3CH_2CH_2CH_2CH_3$ and $CH_3CH_2CH_2CH_2CH_2CH_3$
b. CBr_4 and H_2O
c. $LiNO_3$ and H_2O
d. CH_3OH and $CH_3CH_2CH_2CH_2CH_3$

Surface Tension, Viscosity, and Capillary Action

45. Which compound would you expect to have greater surface tension: acetone $[(CH_3)_2CO]$ or water (H_2O)? Explain.

46. Water (a) "wets" some surfaces and beads up on others. Mercury (b), in contrast, beads up on almost all surfaces. Explain this difference.

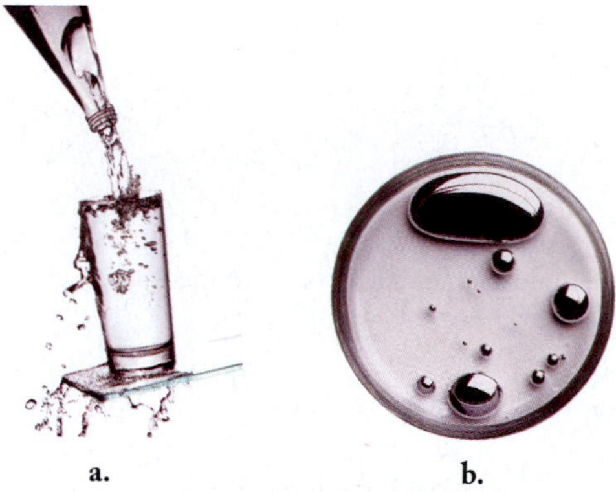

a. b.

47. The structures of two isomers of heptane are shown here. Which of these two compounds would you expect to have the greater viscosity?

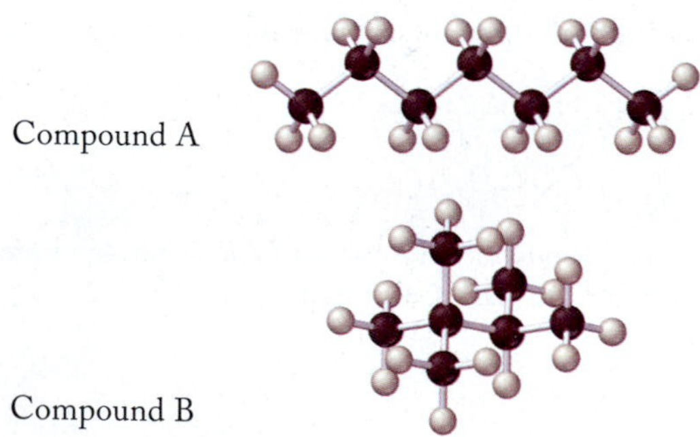

48. The viscosity of a multigrade motor oil (such as one rated 10W-40) is less temperature dependent than the viscosities of most substances. These oils contain polymers (long molecules composed of repeating units) that coil up at low temperatures, but unwind at higher temperatures. Explain how the addition of these polymers to the motor oil might make the viscosity less temperature dependent than a normal liquid.

49. Water in a glass tube that contains grease or oil residue displays a flat meniscus (the tube on the left in the accompanying photo), whereas water in a clean glass tube displays a concave meniscus (the tube on the right). Explain this observation.

50. When a thin glass tube is put into water, the water rises 1.4 cm. When the same tube is put into hexane, the hexane rises only 0.4 cm. Explain.

Vaporization and Vapor Pressure

51. Which evaporates more quickly: 55 mL of water in a beaker with a diameter of 4.5 cm, or 55 mL of water in a dish with a diameter of 12 cm? Is the vapor pressure of the water different in the two containers? Explain.

52. Which evaporates more quickly: 55 mL of water (H_2O) in a beaker or 55 mL of acetone $[(CH_3)_2CO]$ in an identical beaker under identical conditions? Is the vapor pressure of the two substances different? Explain.

53. Spilling room-temperature water over your skin on a hot day cools you down. Spilling room-temperature vegetable oil over your skin on a hot day does not. Explain the difference.

54. Why is the heat of vaporization of water greater at room temperature than it is at its boiling point?

55. The human body obtains 915 kJ of energy from a candy bar. If this energy were used to vaporize water at 100.0 °C, how much water (in liters) could be vaporized? (Assume the density of water is 1.00 g/mL.)

56. A 100.0 mL sample of water is heated to its boiling point. How much heat (in kJ) is required to vaporize it? (Assume a density of 1.00 g/mL.)

57. Suppose that 0.95 g of water condenses on a 75.0 g block of iron that is initially at 22 °C. If the heat released during condensation is used only to warm the iron block, what is the final temperature (in °C) of the iron block? (Assume a constant enthalpy of vaporization for water of 44.0 kJ/mol.)

58. Suppose that 1.15 g of rubbing alcohol (C_3H_8O) evaporates from a 65.0 g aluminum block. If the aluminum block is initially at 25 °C, what is the final temperature of the block after the evaporation of the alcohol? Assume that the heat required for the vaporization of the alcohol comes only from the aluminum block and that the alcohol vaporizes at 25 °C.

59. This table displays the vapor pressure of ammonia at several different temperatures. Use the data to determine the heat of vaporization and normal boiling point of ammonia.

Temperature (K)	Pressure (torr)
200	65.3
210	134.3
220	255.7
230	456.0
235	597.0

60. This table displays the vapor pressure of nitrogen at several different temperatures. Use the data to determine the heat of vaporization and normal boiling point of nitrogen.

Temperature (K)	Pressure (torr)
65	130.5
70	289.5
75	570.8
80	1028
85	1718

61. Ethanol has a heat of vaporization of 38.56 kJ/mol and a normal boiling point of 78.4 °C. What is the vapor pressure of ethanol at 15 °C?

62. Benzene has a heat of vaporization of 30.72 kJ/mol and a normal boiling point of 80.1 °C. At what temperature does benzene boil when the external pressure is 445 torr?

Sublimation and Fusion

63. How much energy is released when 65.8 g of water freezes?

64. Calculate the amount of heat required to completely sublime 50.0 g of solid dry ice (CO_2) at its sublimation temperature. The heat of sublimation for carbon dioxide is 32.3 kJ/mol.

65. An 8.5 g ice cube is placed into 255 g of water. Calculate the temperature change in the water upon the complete melting of the ice. Assume that all of the energy required to melt the ice comes from the water.

66. How much ice (in grams) would have to melt to lower the temperature of 352 mL of water from 25 °C to 5 °C? (Assume the density of water is 1.0 g/ml.)

67. How much heat (in kJ) is required to warm 10.0 g of ice, initially at −10.0 °C, to steam at 110.0 °C? The heat capacity of ice is 2.09 J/g·°C, and that of steam is 2.01 J/g·°C.

68. How much heat (in kJ) is evolved in converting 1.00 mol of steam at 145 °C to ice at −50 °C? The heat capacity of steam is 2.01 J/g·°C, and that of ice is 2.09 J/g·°C.

The Uniqueness of Water

69. Water has a high boiling point given its relatively low molar mass. Explain.

70. Water is a good solvent for many substances. What is the molecular basis for this property, and why is it significant?

71. Explain the role water plays in moderating Earth's climate.

72. How is the density of solid water compared to that of liquid water atypical among substances? Why is this significant?

CUMULATIVE PROBLEMS

73. Explain the observed trend in the melting points of the hydrogen halides.

HI	−50.8 °C
HBr	−88.5 °C
HCl	−114.8 °C
HF	−83.1 °C

74. Explain the observed trend in the boiling points of these compounds.

H_2Te	−2 °C
H_2Se	−41.5 °C
H_2S	−60.7 °C
H_2O	−100 °C

75. The vapor pressure of water at 25 °C is 23.76 torr. If 1.25 g of water is enclosed in a 1.5 L container, is any liquid present? If so, what is the mass of the liquid?

76. The vapor pressure of CCl_3F at 300 K is 856 torr. If 11.5 g of CCl_3F is enclosed in a 1.0 L container, is any liquid present? If so, what is the mass of the liquid?

77. Four ice cubes at exactly 0 °C with a total mass of 53.5 g are combined with 115 g of water at 75 °C in an insulated container. If no heat is lost to the surroundings, what is the final temperature of the mixture?

78. A sample of steam with a mass of 0.552 g at a temperature of 100 °C condenses into an insulated container holding 4.25 g of water at 5.0 °C. Assuming that no heat is lost to the surroundings, what is the final temperature of the mixture?

79. Draw a heating curve (such as the one in Figure 12.33) for 1 mole of methanol beginning at 170 K and ending at 350 K. Assume that the values given here are constant over the relevant temperature ranges.

Melting point	176 K
Boiling point	338 K
ΔH_{fus}	2.2 kJ/mol
ΔH_{vap}	35.2 kJ/mol
$C_{s,\,solid}$	105 J/mol·K
$C_{s,\,liquid}$	81.3 J/mol·K
$C_{s,\,gas}$	48 J/mol·K

80. Draw a heating curve (such as the one in Figure 12.33) for 1 mole of benzene beginning at 0 °C and ending at 100 °C. Assume that the values given here are constant over the relevant temperature ranges.

Melting point	5.4 °C
Boiling point	90.1 °C
ΔH_{fus}	9.9 kJ/mol
ΔH_{vap}	30.7 kJ/mol
$C_{s,\ solid}$	118 J/mol·K
$C_{s,\ liquid}$	135 J/mol·K
$C_{s,\ gas}$	104 J/mol·K

81. Air conditioners not only cool air, but dry it as well. A room in a home measures $6.0\ m \times 10.0\ m \times 2.2\ m$. If the outdoor temperature is 30 °C and the vapor pressure of water in the air is 85% of the vapor pressure of water at this temperature, what mass of water must be removed from the air each time the volume of air in the room cycles through the air conditioner? The vapor pressure for water at 30 °C is 31.8 torr.

82. A sealed flask contains 0.55 g of water at 28 °C. The vapor pressure of water at this temperature is 28.36 mmHg. What is the minimum volume of the flask in order that there is no liquid water present in the flask?

CHALLENGE PROBLEMS

83. Two liquids, A and B, have vapor pressures at a given temperature of 24 mmHg and 36 mmHg, respectively. We prepare solutions of A and B at a given temperature and measure the total pressures above the solutions. We obtain these data:

Solution	Amt A (mol)	Amt B (mol)	*P* (mmHg)
1	1	1	30
2	2	1	28
3	1	2	32
4	1	3	33

Predict the total pressure above a solution of 5 mol A and 1 mol B.

84. Butane (C_4H_{10}) has a heat of vaporization of 22.44 kJ/mol and a normal boiling point of −0.4 °C. A 250.0 mL sealed flask contains 0.55 g of butane at −22 °C. How much liquid butane is present? If the butane is warmed to 25 °C, how much liquid butane is present?

85. Liquid nitrogen can be used as a cryogenic substance to obtain low temperatures. Under atmospheric pressure, liquid nitrogen boils at 77 K, allowing low temperatures to be reached. However, if the nitrogen is placed in a sealed, insulated container connected to a vacuum pump, even lower temperatures can be reached. Why? If the vacuum pump has sufficient capacity and is left on for an extended period of time, the liquid nitrogen starts to freeze. Explain.

86. Given that the heat of fusion of water is −6.02 kJ/mol, the heat capacity of $H_2O(l)$ is 75.2 J/mol·K, and the heat capacity of $H_2O(s)$ is 37.7 J/mol·K, calculate the heat of fusion of water at −10 °C.

87. The heat of combustion of CH_4 is 890.4 kJ/mol, and the heat capacity of H_2O is 75.2 J/mol·K. Find the volume of methane measured at 298 K and 1.00 atm required to convert 1.00 L of water at 298 K to water vapor at 373 K.

88. Three 1.0 L flasks, maintained at 308 K, are connected to each other with stopcocks. Initially the stopcocks are closed. One of the flasks contains 1.0 atm of N_2; the second, 2.0 g of H_2O; and the third, 0.50 g of ethanol, C_2H_6O. The vapor pressure of H_2O at 308 K is 42 mmHg, and that of ethanol is 102 mmHg. When the stopcocks are opened and the contents mix freely, what is the pressure?

CONCEPTUAL PROBLEMS

89. One prediction of global warming is the melting of global ice, which may result in coastal flooding. A criticism of this prediction is that the melting of icebergs does not increase ocean levels any more than the melting of ice in a glass of water increases the level of liquid in the glass. Is this a valid criticism? Does the melting of an ice cube in a cup of water raise the level of the liquid in the cup? Why or why not? In response to this criticism, scientists have asserted that they are not worried about melting icebergs, but rather the melting of ice sheets that sit on the continent of Antarctica. Would the melting of this ice increase ocean levels? Why or why not?

90. The rate of vaporization depends on the surface area of the liquid. However, the vapor pressure of a liquid does not depend on the surface area. Explain.

91. Substance A has a smaller heat of vaporization than substance B. Which of the two substances undergoes a larger change in vapor pressure for a given change in temperature?

92. A substance has a heat of vaporization of ΔH_{vap} and heat of fusion of ΔH_{fus}. Express the heat of sublimation in terms of ΔH_{vap} and ΔH_{fus}.

93. Examine the heating curve for water in Section 12.7 (Figure 12.33). If heat is added to the water at a constant rate, which of the three segments in which temperature is rising will have the least steep slope? Why?

94. A root cellar is an underground chamber used to store fruits, vegetables, and sometimes meats. In extreme cold, farmers put large vats of water into the root cellar to prevent fruits and vegetables from freezing. Explain why this works.

95. Suggest an explanation for the observation that the heat of fusion of a substance is always smaller than its heat of vaporization.

96. Refer to Figure 12.33 to answer each question.

a. A sample of steam begins on the line segment labeled 5 on the graph. Is heat absorbed or released in moving from the line segment labeled 5 to the line segment labeled 3? What is the sign of q for this change?

b. In moving from left to right along the line segment labeled 2 on the graph, heat is absorbed, but the temperature remains constant. Where does the heat go?

c. How would the graph change if it were for another substance (other than water)?

97. The following image is an electrostatic potential map for ethylene oxide, $(CH_2)_2O$, a polar molecule. Use the electrostatic potential map to predict the geometry for how one ethylene oxide molecule interacts with another. Draw structural formulas, using the three-dimensional bond notation introduced in Section 6.9, to show the geometry of the interaction.

ANSWERS TO CONCEPTUAL CONNECTIONS

Cc 12.1 **(a)** When water boils, it simply changes state from liquid to gas. Water molecules do not decompose during boiling.

Cc 12.2 **(c)** I_2 has the highest boiling point because it has the highest molar mass. Because the halogens are all similar in other ways, you would expect I_2 to have the greatest dispersion forces and therefore the highest boiling point (and in fact it does).

Cc 12.3

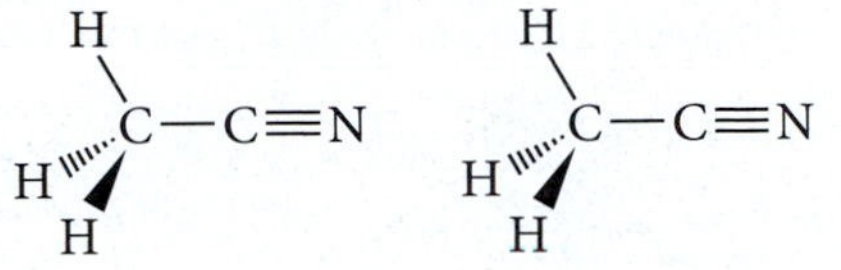

```
  H                    H
   \                    \
    C—C≡N                C—C≡N
 H''' ◀               H''' ◀
   H                    H
```

Cc 12.4 **(a)** CH_3OH. The compounds all have similar molar masses, so the dispersion forces are similar in all three. CO is polar, but because CH_3OH contains H directly bonded to O, it has hydrogen bonding, resulting in the highest boiling point.

Cc 12.5 **(b)** Although the *rate of vaporization* increases with increasing surface area, the *vapor pressure* of a liquid is independent of surface area. An increase in surface area increases both the rate of vaporization and the rate of condensation—the effects exactly cancel, and the vapor pressure does not change.

Cc 12.6 The warming of the ice from −10 °C to 0 °C absorbs only 20.9 J/g of ice. The melting of the ice, however, absorbs about 334 J/g of ice. (You can obtain this value by dividing the heat of fusion of water by its molar mass.) Therefore, the melting of the ice produces a larger temperature decrease in the water than the warming of the ice.

CHAPTER 13

"Should we not suppose that in the formation of a crystal, the particles are not only established in rows and columns set in regular figures, but also by means of some polar property have turned identical sides in identical directions?"

—Isaac Newton (1642–1727)

A form of glacial movement, called basal sliding, occurs in part because the great mass of the glacier causes the ice at the bottom of the glacier to melt and turn into liquid water.

Phase Diagrams and Crystalline Solids

13.1 Sliding Glaciers 481

13.2 Phase Diagrams 482

13.3 Crystalline Solids: Determining Their Structure by X-Ray Crystallography 485

13.4 Crystalline Solids: Unit Cells and Basic Structures 487

13.5 Crystalline Solids: The Fundamental Types 495

13.6 The Structures of Ionic Solids 497

13.7 Network Covalent Atomic Solids: Carbon and Silicates 498

Key Learning Outcomes 503

In Chapter 12, we discussed state transitions such as fusion (melting), vaporization, and sublimation. In this chapter, we learn about phase diagrams, which are maps of the states of a substance as a function of temperature and pressure. Phase diagrams are useful because they allow us to predict the state of a substance under a given set of conditions. They also allow us to predict whether a state change will occur when the conditions are changed. For example, from the phase diagram for water, we can predict that water will boil at room temperature if we lower the external pressure to about 20 torr. In this chapter, we also focus on one of the states in the phase diagram—the solid state. Specifically, we examine the crystalline structures of solids. The symmetric three-dimensional shapes that we observe in solid crystals—such as the hexagonal shape of snowflakes or the cubic shape of salt grains—result from the well-ordered repeating patterns of the particles that form these crystals. In other words, the properties of solids (in this case their shape) are determined by the structure of the particles that compose them.

13.1 Sliding Glaciers

Glaciers are beautiful and important bodies of ice that, because of their great mass, flow slowly under the influence of gravity. Glaciers have cut dramatic valleys, such as the Yosemite Valley in California, and also hold much of our planet's freshwater supply. Glaciers move because their great mass, and the corresponding high pressure, cause changes in the properties of ice. An important kind of glacial movement—called basal sliding—occurs in part because solid water (ice) has the unique property of being less dense than its liquid form, as we saw in Chapter 12 (see Section 12.8).

In a glacier, most of the water is in its solid phase. However, the ice at the bottom of the glacier is under intense pressure (tens to hundreds of atmospheres, depending on the thickness of the glacier). Extremely high pressures such as this can—depending on the exact conditions—cause the ice to melt, even below 0 °C. This happens because pressure favors the denser state. Water molecules in liquid water are closer together than they are in solid ice, and the effect of pressure on ice is to push the water molecules closer together forcing them to become a liquid. This liquid layer at the bottom of the glacier acts as a lubricant, allowing the glacier to slide down mountains at rates that are much faster—meters per day instead of centimeters per day—than would occur without the liquid water.

In this chapter, we examine *phase diagrams,* which map the phase of a substance as a function of temperature and pressure. The phase diagrams of most substances show that an increase in pressure favors the solid state (because for most substances, the solid state is denser than the liquid state). However, the phase diagram for water illustrates the exact opposite trend: An increase in pressure in water favors the liquid state. Phase diagrams also show how changes in pressure affect boiling points. For every substance, an increase in pressure raises its boiling point because, for every substance, the liquid state is denser than the gaseous state. Therefore an increase in pressure causes any substance to remain a liquid at a higher temperature.

13.2 Phase Diagrams

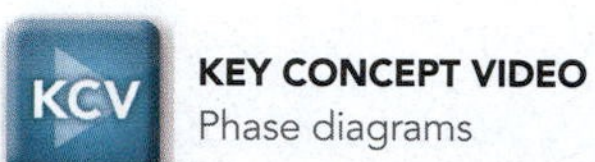

A **phase diagram** is a map of the state or *phase* of a substance as a function of pressure (on the *y*-axis) and temperature (on the *x*-axis). Let's first examine the major features of a phase diagram. Once we are familiar with these features, we can turn to navigating within a phase diagram, and finally examine and compare the phase diagrams of selected substances.

The Major Features of a Phase Diagram

We can become familiar with the major features of a phase diagram by examining the phase diagram for water as an example (Figure 13.1 ▼). The *y*-axis displays the pressure in torr, and the *x*-axis displays the temperature in degrees Celsius. We categorize the main features of the phase diagram as regions, lines, and points.

Regions *Any of the three main regions—solid, liquid, and gas—in the phase diagram represents conditions where that particular state is stable.* For example, under any of the temperatures and pressures within the liquid region in the phase diagram of water, the liquid is the stable state. Notice that the point 25 °C and 760 torr falls within the liquid region, as we know from everyday experience. In general, low temperature and high pressure favor the solid state, high temperature and low pressure favor the gas

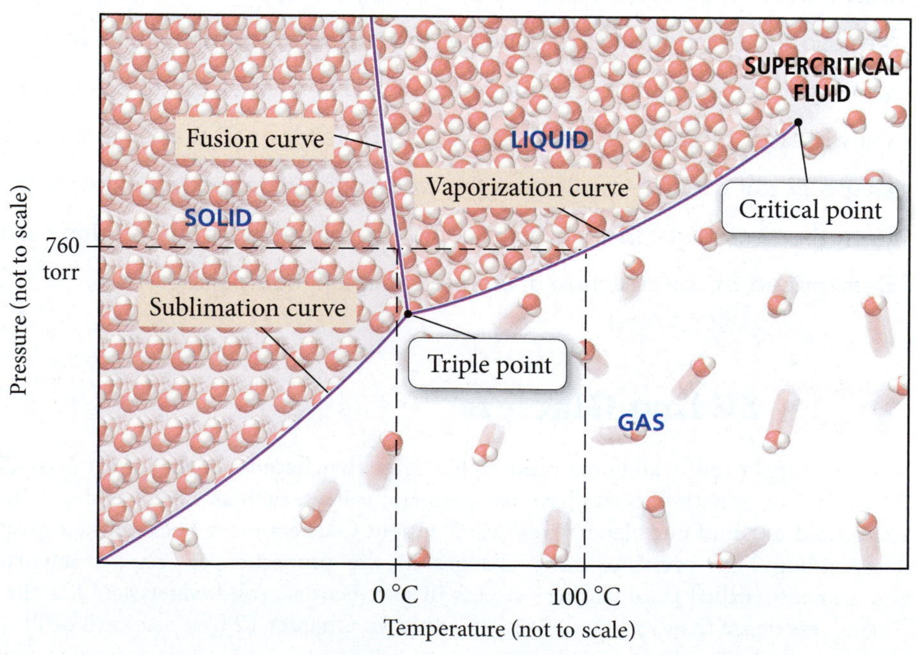

▶ **FIGURE 13.1 Phase Diagram for Water**

state, and intermediate conditions favor the liquid state. A sample of matter that is not in the state indicated by its phase diagram for a given set of conditions converts to that state when those conditions are imposed. For example, steam that is cooled to room temperature at 1 atm condenses to liquid.

Lines *Each of the lines (or curves) in the phase diagram represents a set of temperatures and pressures at which the substance is in equilibrium between the two states on either side of the line.* In the phase diagram for water, consider the curved line beginning just beyond 0 °C that separates the liquid from the gas. This line is the *vaporization curve* (also called the vapor pressure curve) for water that we examined in Section 12.5. At any of the temperatures and pressures that fall along this line, the liquid and gas states of water are equally stable and in equilibrium. For example, at 100 °C and 760 torr pressure, water and its vapor are in equilibrium—they are equally stable and coexist. The other two major lines in a phase diagram are the *sublimation curve* (separating the solid and the gas) and the *fusion curve* (separating the solid and the liquid).

The Triple Point *The* ***triple point*** *in a phase diagram represents the unique set of conditions at which the three states are equally stable and in equilibrium.* In the phase diagram for water, the triple point occurs at 0.0098 °C and 4.58 torr. Under these unique conditions (and only under these conditions), the solid, liquid, and gas states of water are equally stable and coexist in equilibrium.

The triple point of a substance such as water can be reproduced anywhere to calibrate a thermometer or pressure gauge with a known temperature and pressure.

The Critical Point As we discussed in Section 12.5, at the critical temperature and pressure, the liquid and gas states coalesce into a *supercritical fluid. The* ***critical point*** *in a phase diagram represents the temperature and pressure above which a supercritical fluid exists.*

Navigation within a Phase Diagram

We represent changes in the temperature or pressure of a sample of water as movement within the phase diagram. For example, suppose we heat a block of ice initially at 1.0 atm and −25 °C. We represent the change in temperature at constant pressure as movement along the horizontal line marked A in Figure 13.2 ▼. As the temperature rises, we move to the right along the line. At the fusion curve, the temperature stops rising and melting occurs until the solid ice is completely converted to liquid water. Crossing the fusion curve requires the complete transition from solid to liquid. Once the ice has completely melted, the temperature of the liquid water begins to rise until the vaporization curve is reached. At this point, the temperature again stops rising, and boiling occurs until all the liquid is converted to gas.

We represent a change in pressure with a vertical line on the phase diagram. For example, suppose we lower the pressure above a sample of water initially at 1.0 atm and 25 °C. We represent the change in pressure at constant temperature as movement along the line marked B in Figure 13.2. As the pressure drops, we move down the line and approach the vaporization curve. At the vaporization curve, the

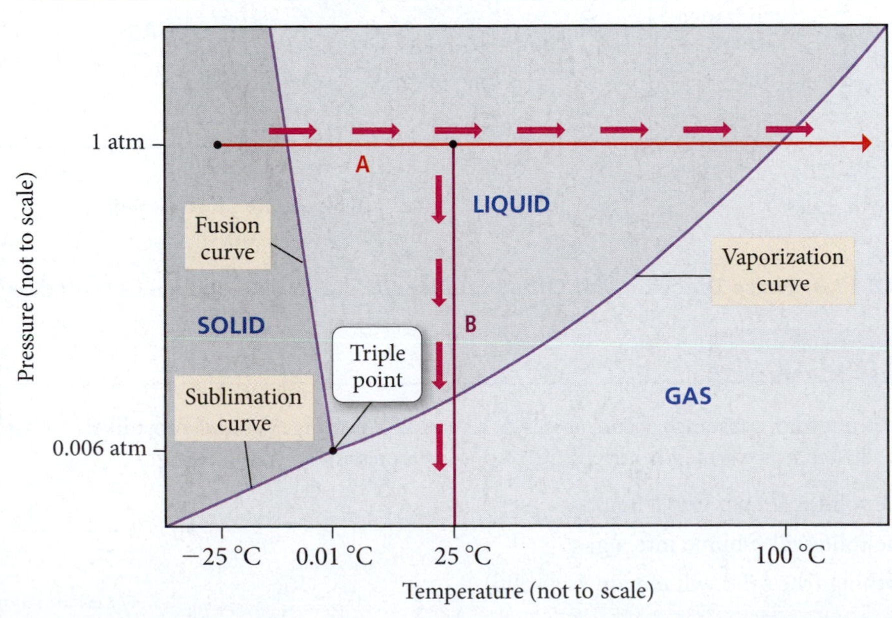

◀ **FIGURE 13.2 Navigation on the Phase Diagram for Water**

pressure stops dropping, and vaporization occurs until the liquid is completely converted to vapor. Crossing the vaporization curve requires the complete transition from liquid to gas. Only after all the liquid has vaporized does the pressure continue to drop. Notice that, for water, the fusion curve (the line separating the solid and the liquid) slopes to the left. This means that, as the pressure increases, the liquid state is favored over the solid. This behavior is unique to water and is one of the reasons responsible for glacial sliding, as we saw in Section 13.1.

The Phase Diagrams of Other Substances

Examine the phase diagrams of iodine and carbon dioxide, shown in Figure 13.3 ▼. The phase diagrams are similar to that of water in most of their general features, but some significant differences exist.

The fusion curves for both carbon dioxide and iodine have a positive slope—as the temperature increases, the pressure also increases—in contrast to the fusion curve for water, which has a negative slope. As we saw in Section 13.1, the behavior of water is atypical. The fusion curve within the phase diagrams for most substances has a positive slope because increasing pressure favors the denser state, which for most substances is the solid state. For example, suppose we increase the pressure on a sample of iodine from 1 atm to 100 atm at 184 °C. This change is represented by line A in Figure 13.3(a) ▼. Notice that this change crosses the fusion curve, converting the liquid into a solid. In contrast, a pressure increase from 1 atm to 100 atm at −0.1 °C in water causes a state transition from solid to liquid. Unlike most substances, the liquid state of water is actually denser than the solid state.

Both water and iodine have stable solid, liquid, and gaseous states at a pressure of 1 atm. Notice, however, that carbon dioxide has no stable liquid state at a pressure of 1 atm. If we increase the temperature of a block of solid carbon dioxide (dry ice) at 1 atm, as indicated by line B in Figure 13.3(b) ▼, we cross the sublimation curve at −78.5 °C. At this temperature, the solid sublimes to a gas. This is one reason that dry ice is useful; it does not melt into a liquid at atmospheric pressure. Carbon dioxide forms a liquid only above pressures of 5.1 atm.

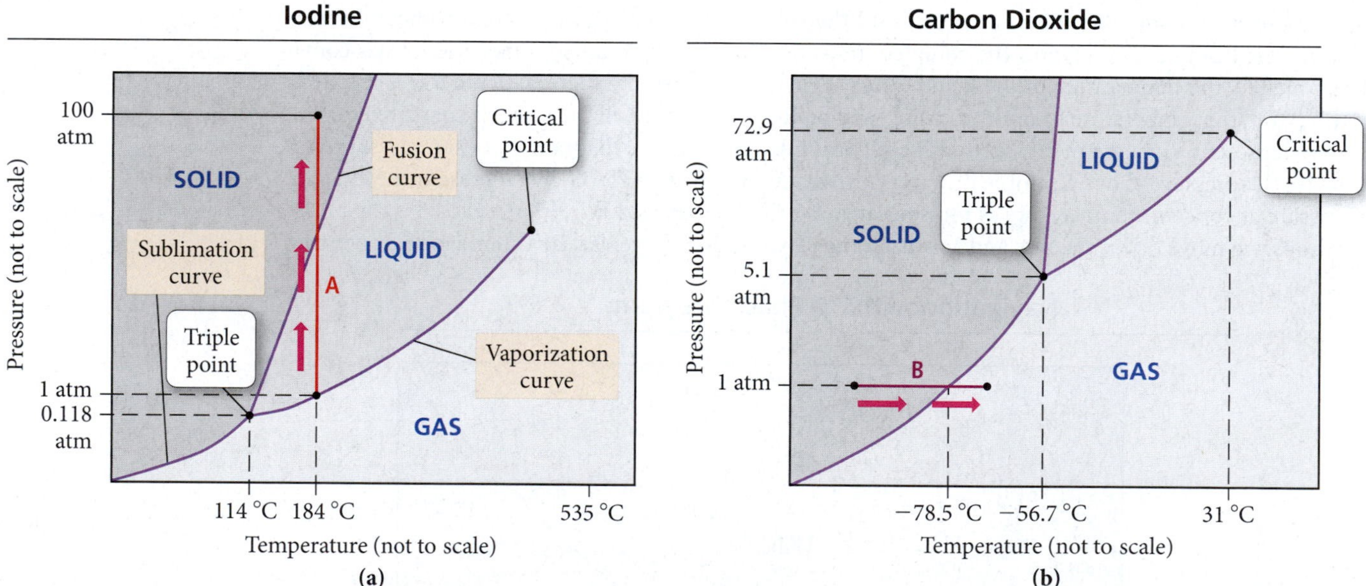

▲ **FIGURE 13.3 Phase Diagrams for Other Substances** **(a)** Iodine and **(b)** Carbon dioxide

13.1 Cc Conceptual Connection

Phase Diagrams

The triple point for a substance occurs at −24.5 °C and 225 mmHg. What is most likely to happen when this substance is warmed from −35 °C to 0 °C at a pressure of 220 mmHg?

(a) The solid will melt into a liquid.

(b) The solid will sublime into a gas.

(c) Nothing (the solid will remain as a solid).

EXAMPLE 13.1
Navigation within a Phase Diagram

What state transitions occur when a sample of carbon dioxide at −60.0 °C at 10.0 atm is warmed to 30.0 °C and 10.0 atm?

SOLUTION

To solve this problem, draw a horizontal line on the phase diagram in Figure 13.3(b) beginning at −60.0 °C at 10.0 atm and ending at 30.0 °C and 10.0 atm.

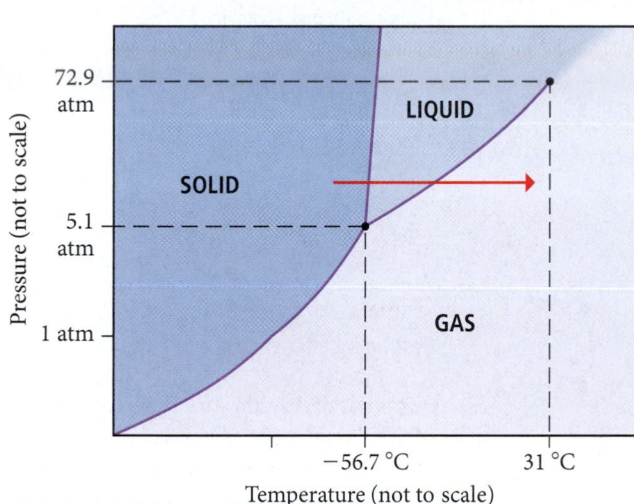

Since the line begins in the solid region of the phase diagram, the sample is initially a solid. As the sample warms, it crosses the fusion curve and becomes a liquid. Continued warming causes it to cross the vaporization curve and become a gas. So the state transitions that occur are solid to liquid and liquid to gas.

FOR PRACTICE 13.1

What state transitions occur in carbon dioxide if you begin with a sample of carbon dioxide at −60.0 °C at 0.50 atm and warm the sample to 30.0 °C and 0.50 atm?

▲ The well-defined angles and smooth faces of crystalline solids reflect the underlying order of the atoms composing them.

13.3 Crystalline Solids: Determining Their Structure by X-Ray Crystallography

We now turn to one of the states in phase diagrams—the solid state. Recall that crystalline solids are composed of atoms or molecules arranged in structures with long-range order (see Section 12.2). If you have ever visited the mineral section of a natural history museum and seen crystals with smooth faces and well-defined angles between them, or if you have carefully observed the hexagonal shapes of snowflakes, you have witnessed some of the effects of the underlying order in crystalline solids. The often beautiful geometric shapes that we see on the macroscopic scale are the result of specific structural patterns on the molecular and atomic scales. But how do we study these patterns? How do we look into the atomic and molecular world to determine the arrangement of the atoms and measure the distances between them? In this section, we examine **X-ray diffraction**, a powerful laboratory technique that enables us to do exactly that.

In Section 3.2 we saw that electromagnetic (or light) waves interact with each other in a characteristic way called *interference:* Waves can cancel each other out or reinforce each other, depending on the alignment of their crests and troughs. *Constructive interference* occurs when two waves interact with their crests and troughs in alignment. *Destructive interference* occurs when two waves interact with the crests of one

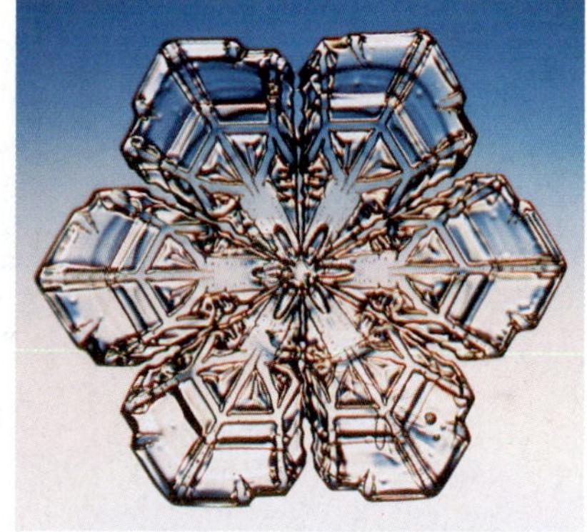

▲ The hexagonal shape of a snowflake derives from the hexagonal arrangement of water molecules in crystalline ice.

aligning with the troughs of the other. Recall also that when light encounters two slits separated by a distance comparable to the wavelength of the light, constructive and destructive interference between the resulting beams produces a characteristic *interference pattern,* consisting of alternating bright and dark lines.

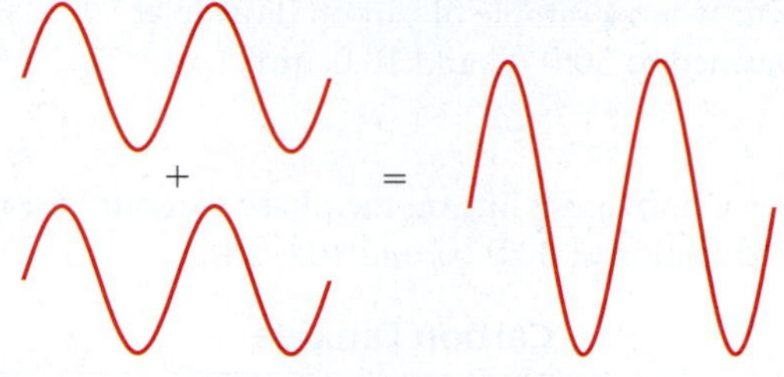

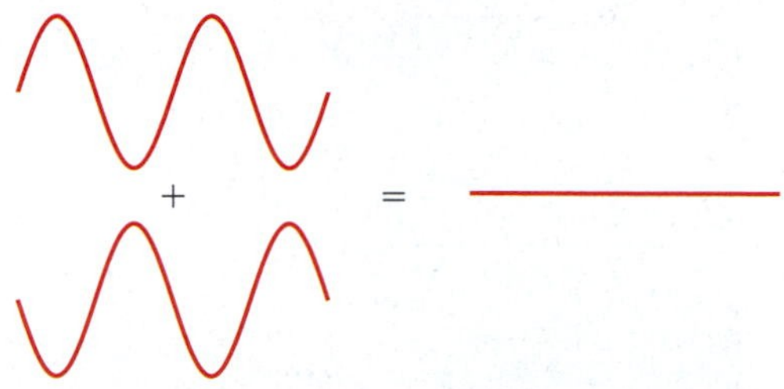

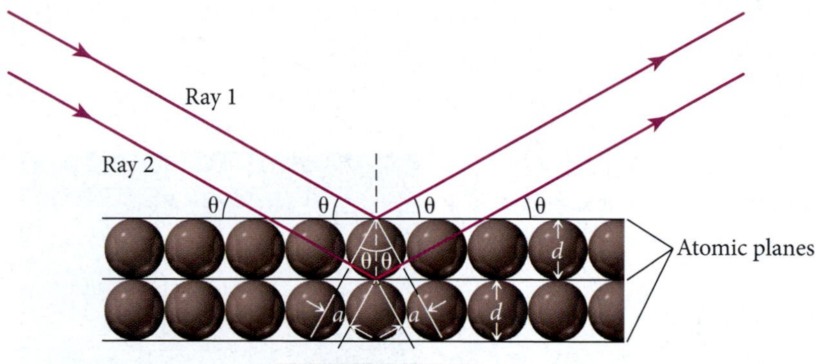

▲ **FIGURE 13.4 Diffraction from a Crystal** When X-rays strike parallel planes of atoms in a crystal, constructive interference occurs if the difference in path length between beams reflected from adjacent planes is an integral number of wavelengths.

Atoms within crystal structures have spaces between them on the order of 10^2 pm, so light of similar wavelength (which happens to fall in the X-ray region of the electromagnetic spectrum) forms interference patterns or *diffraction patterns* when it interacts with those atoms in the crystals. The exact pattern of diffraction reveals the spacing between planes of atoms. Consider two planes of atoms within a crystalline lattice separated by a distance d, as shown in Figure 13.4 ◀. If two rays of light with wavelength λ that are initially in phase (that is, the crests of one wave are aligned with the crests of the other) diffract from the two planes, the diffracted rays may interfere with each other constructively or destructively, depending on the difference between the path lengths traveled by each ray. If the difference between the two path lengths ($2a$) is an integral number (n) of wavelengths, then the interference will be constructive:

$$n\lambda = 2a \quad \text{(criterion for constructive interference)} \qquad [13.1]$$

Using trigonometry, we can see that the angle of reflection (θ) is related to the distance a and the separation between layers (d) by the following relation:

$$\sin\theta = \frac{a}{d} \qquad [13.2]$$

Rearranging, we get:

$$a = d\sin\theta \qquad [13.3]$$

By substituting Equation 13.1 into Equation 13.3, we arrive at the following important relationship:

$$n\lambda = 2d\sin\theta \qquad \text{Bragg's law}$$

This equation is known as *Bragg's law.* For a given wavelength of light incident on atoms arranged in layers, we can measure the angle that produces constructive interference (which appears as a bright spot on the X-ray diffraction pattern) and then calculate d, the distance between the atomic layers:

$$d = \frac{n\lambda}{2\sin\theta} \qquad [13.4]$$

In a modern X-ray diffractometer (Figure 13.5 ▶), the diffraction pattern from a crystal is collected and analyzed by a computer. By rotating the crystal and collecting the resulting diffraction patterns at different angles, the distances between various crystalline planes can be measured, eventually yielding the entire crystalline structure. This process is called X-ray crystallography. Researchers use X-ray crystallography to determine not only the structures of simple atomic lattices, but also the structures of proteins, DNA, and other biologically important molecules. For example, the famous X-ray diffraction photograph shown at right in the margin, obtained by Rosalind Franklin and Maurice Wilkins, helped Watson and Crick determine the double-helical structure of DNA. Researchers also used X-ray diffraction to determine the structure of HIV protease, a protein critical to the reproduction of HIV and the development of AIDS. That structure was then used to design drug molecules that would inhibit the action of HIV protease, thus halting the advance of the disease.

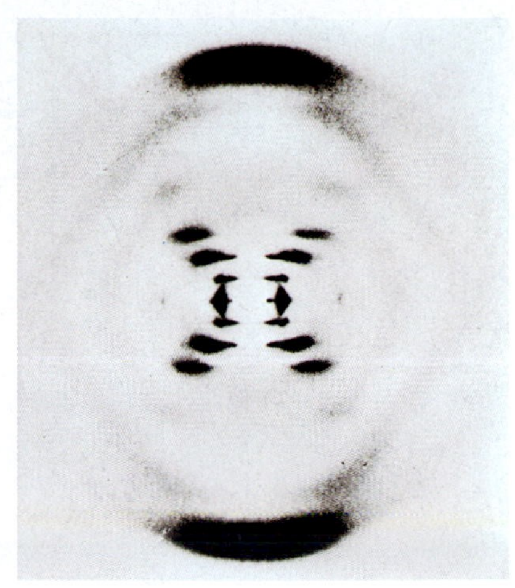

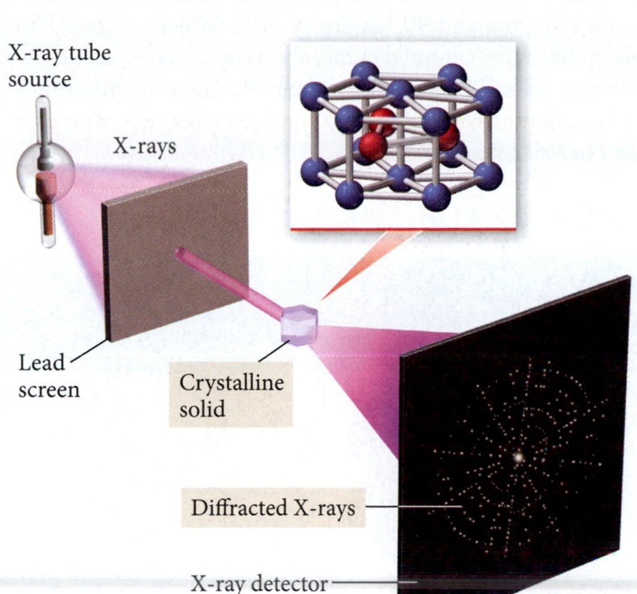

◀ **FIGURE 13.5 X-Ray Diffraction Analysis** In X-ray crystallography, an X-ray beam is passed through a sample, which is rotated to allow diffraction from different crystalline planes. The resulting patterns, representing constructive interference from various planes, are analyzed to determine crystalline structure.

EXAMPLE 13.2
Using Bragg's Law

When an X-ray beam of $\lambda = 154$ pm is incident on the surface of an iron crystal, it produces a maximum reflection at an angle of $\theta = 32.6°$. Assuming $n = 1$, calculate the separation between layers of iron atoms in the crystal.

SOLUTION

To solve this problem, use Bragg's law in the form of Equation 13.4. The distance, d, is the separation between layers in the crystal.

$$d = \frac{n\lambda}{2\ \sin\theta}$$

$$= \frac{154\ \text{pm}}{2\ \sin(32.6°)}$$

$$= 143\ \text{pm}$$

FOR PRACTICE 13.2

The spacing between layers of molybdenum atoms is 157 pm. Calculate the angle at which 154 pm X-rays produce a maximum reflection for $n = 1$.

13.4 Crystalline Solids: Unit Cells and Basic Structures

X-ray crystallography allows us to determine the regular arrangements of atoms within a crystalline solid. This arrangement is called the **crystalline lattice**. The crystalline lattice of any solid is nature's way of aggregating the particles to minimize their energy.

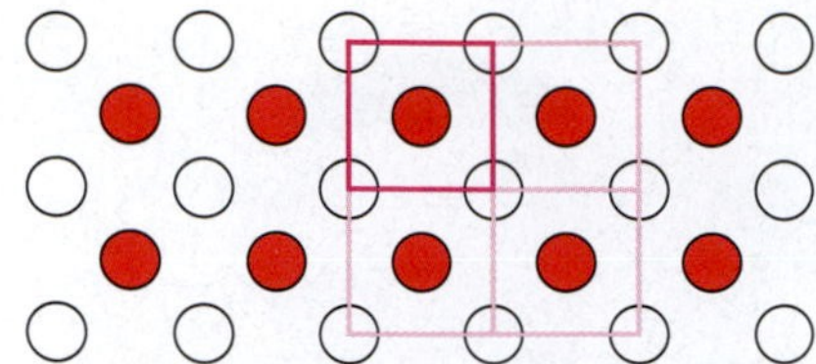

The Unit Cell

We can represent the crystalline lattice with a small collection of atoms, ions, or molecules called the **unit cell**. When the unit cell is repeated over and over—like the tiles of a floor or the pattern in a wallpaper design, but in three dimensions—the entire lattice is reproduced. For example, consider the two-dimensional crystalline lattice shown at left. The unit cell for this lattice is the purple square. Each circle represents a *lattice point,* a point in space occupied by an atom, ion, or molecule. Repeating the unit cell pattern throughout the two-dimensional space generates the entire lattice.

Many different unit cells exist, and we often classify unit cells by their symmetry. In this book, we focus primarily on *cubic unit cells* (although we look at one hexagonal unit cell). Cubic unit cells are characterized by equal edge lengths and 90° angles at their corners. Figure 13.6 ▼ presents the three cubic unit cells—simple cubic, body-centered cubic, and face-centered cubic—and some of their basic characteristics. There are two colors in this figure to help illustrate the different positions of the atoms; the colors *do not* represent different *kinds* of atoms. For these unit cells, *each atom in any one structure is identical to the other atoms in that structure.*

Cubic Cell Name	Atoms per Unit Cell	Structure	Coordination Number	Edge Length in Terms of *r*	Packing Efficiency (fraction of volume occupied)
Simple Cubic	1		6	$2r$	52%
Body-Centered Cubic	2		8	$\frac{4r}{\sqrt{3}}$	68%
Face-Centered Cubic	4		12	$2\sqrt{2}r$	74%

▲ **FIGURE 13.6 The Cubic Crystalline Lattices** The different colors used for the atoms in this figure are for clarity only. All atoms within each structure are identical.

Unit cells, such as the cubic ones shown here, are customarily portrayed with "whole" atoms, even though only a part of the whole atom may actually be in the unit cell.

Simple cubic

r l

$l = 2r$

▲ In the simple cubic lattice, the atoms touch along each edge so that the edge length is $2r$.

The **simple cubic** unit cell (Figure 13.7 ▶) consists of a cube with one atom at each corner. The atoms touch along each edge of the cube, so the edge length is twice the radius of the atoms ($l = 2r$). Even though it may seem like the unit cell contains eight atoms, it actually contains only one. Each corner atom is shared by eight other unit cells. In other words, any one unit cell actually contains only one-eighth of each of the eight atoms at its corners, for a total of only one atom per unit cell.

A characteristic feature of any unit cell is the **coordination number**, the number of atoms with which each atom is in *direct contact.* The coordination number is the number of atoms with which a particular atom can strongly interact. The simple cubic unit cell has a coordination number of 6; any one atom touches only six others, as Figure 13.7 illustrates. A quantity closely related to the coordination number is the **packing efficiency**, the percentage of the volume of the unit cell occupied by the spheres. The higher the coordination number, the greater the packing efficiency. The simple cubic unit cell has a packing efficiency of 52%—the simple cubic unit cell contains a lot of empty space.

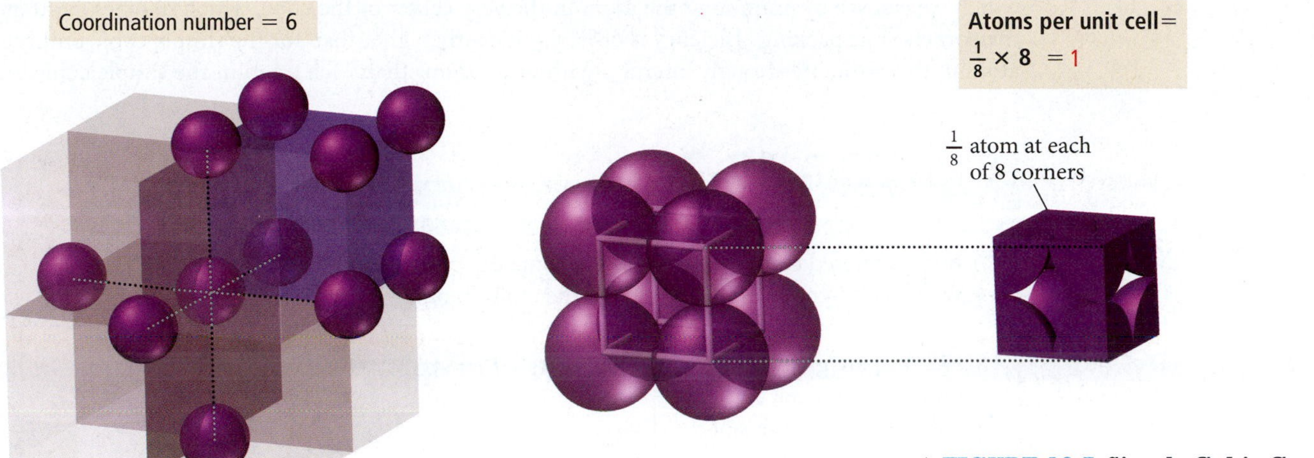

▲ **FIGURE 13.7 Simple Cubic Crystal Structure**

The **body-centered cubic** unit cell (Figure 13.8 ▼) consists of a cube with one atom at each corner and one atom (of the same kind) in the very center of the cube. Note that in the body-centered unit cell, the atoms *do not* touch along each edge of the cube, but instead along the diagonal line that runs from one corner, through the middle of the cube, to the opposite corner. The edge length in terms of the atomic radius is therefore $l = 4r/\sqrt{3}$ as shown in this diagram:

Body-centered cubic

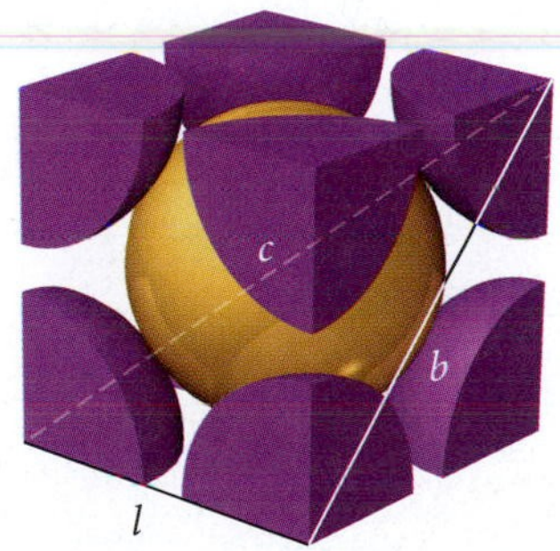

$$c^2 = b^2 + l^2 \qquad b^2 = l^2 + l^2$$
$$c = 4r \qquad b^2 = 2l^2$$
$$(4r)^2 = 2l^2 + l^2$$
$$(4r)^2 = 3l^2$$
$$l^2 = \frac{(4r)^2}{3}$$
$$l = \frac{4r}{\sqrt{3}}$$

▶ In the body-centered cubic lattice, the atoms touch only along the cube diagonal. The edge length is $4r/\sqrt{3}$.

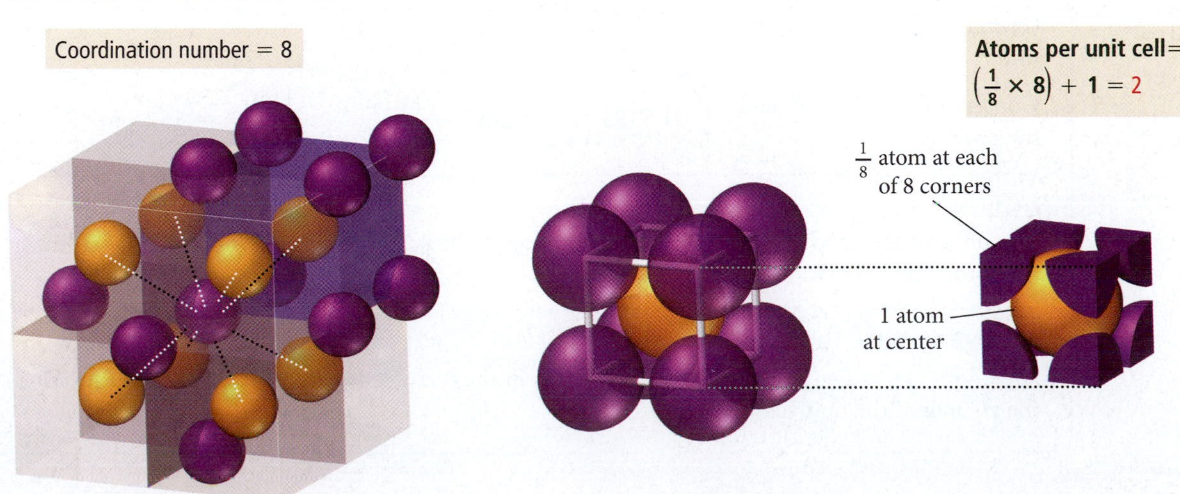

◀ **FIGURE 13.8 Body-Centered Cubic Crystal Structure** The different colors used for the atoms in this figure are for clarity only. All atoms within the structure are identical.

The body-centered unit cell contains two atoms per unit cell because the center atom is not shared with any other neighboring cells. The coordination number of the body-centered cubic unit cell is 8, which we can see by examining the atom in the very center of the cube, which touches eight atoms at the corners. The packing efficiency is 68%, significantly higher than for the simple cubic unit cell. Each atom in this structure strongly interacts with more atoms than each atom in the simple cubic unit cell.

EXAMPLE 13.3

Relating Unit Cell Volume, Edge Length, and Atomic Radius

A body-centered cubic unit cell has a volume of 4.32×10^{-23} cm^3. Find the radius of the atom in pm.

SORT You are given the volume of a unit cell and asked to find the radius of the atom.

GIVEN: $V = 4.32 \times 10^{-23}\ \text{cm}^3$
FIND: r (in pm)

STRATEGIZE Use the volume to find the edge length of the unit cell.

Then use the edge length to find the radius of the atom.

CONCEPTUAL PLAN

$$V \xrightarrow{V = l^3} l$$

$$l \xrightarrow{l = \frac{4r}{\sqrt{3}}} r$$

RELATIONSHIPS USED

$V = l^3$ (Volume of a cube)

$l = \dfrac{4r}{\sqrt{3}}$ (Edge length of body-centered cubic unit cell)

SOLVE

Solve the equation for the volume of a cube for l and substitute in the given value for V to find l.

SOLUTION

$$V = l^3$$

$$l = \sqrt[3]{V} = \sqrt[3]{4.32 \times 10^{-23}\text{cm}^3} = 3.50\underline{8}8 \times 10^{-8}\ \text{cm}$$

Solve the equation for the edge length of a body-centered cubic unit cell for r and substitute in the value of l (from the previous step) to find r.

$$l = \frac{4r}{\sqrt{3}}$$

$$r = \frac{\sqrt{3}\,l}{4} = \frac{\sqrt{3}(3.50\underline{8}8 \times 10^{-8}\text{cm})}{4} = 1.5\underline{1}93 \times 10^{-8}\ \text{cm}$$

Convert r from cm to m and then to pm.

$$1.5\underline{1}93 \times 10^{-8}\ \cancel{\text{cm}} \times \frac{0.01\ \cancel{\text{m}}}{1\ \cancel{\text{cm}}} \times \frac{1\ \text{pm}}{10^{-12}\ \cancel{\text{m}}} = 152\ \text{pm}$$

CHECK

The units of the answer (pm) are correct. The magnitude is also reasonable since atomic radii range roughly from 50 to 200 pm.

FOR PRACTICE 13.3

An atom has a radius of 138 pm and crystallizes in the body-centered cubic unit cell. What is the volume of the unit cell in cm^3?

Face-Centered Cubic Unit Cell

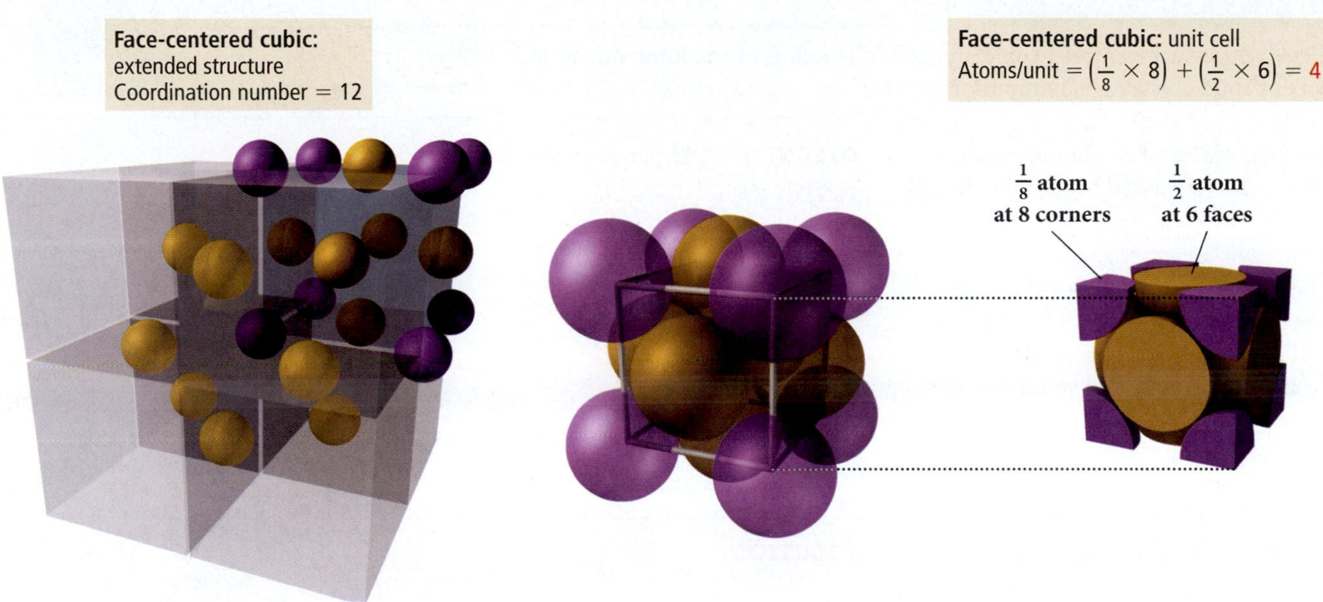

▲ **FIGURE 13.9 Face-Centered Cubic Crystal Structure** The different colors used on the atoms in this figure are for clarity only. All atoms within the structure are identical.

The **face-centered cubic** unit cell (Figure 13.9 ▲) is a cube with one atom at each corner and one atom (of the same kind) in the center of each cube face. Note that in the face-centered unit cell (like the body-centered unit cell), the atoms *do not* touch along each edge of the cube. Instead, the atoms touch *along the diagonal face.* The edge length in terms of the atomic radius is therefore $l = 2\sqrt{2}r$, as shown in this figure:

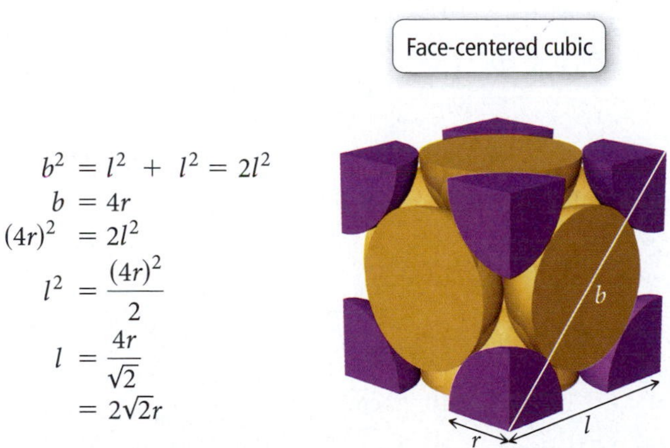

▲ In the face-centered cubic lattice, the atoms touch along a face diagonal. The edge length is $2\sqrt{2}r$.

The face-centered unit cell contains four atoms per unit cell because the center atoms on each of the six faces are shared between two unit cells. There are $\frac{1}{2} \times 6 = 3$ face-centered atoms plus $\frac{1}{8} \times 8 = 1$ corner atom, for a total of four atoms per unit cell. The coordination number of the face-centered cubic unit cell is 12, and its packing efficiency is 74%. In this structure, any one atom strongly interacts with more atoms than in either the simple cubic unit cell or the body-centered cubic unit cell.

EXAMPLE 13.4

Relating Density to Crystal Structure

Aluminum crystallizes with a face-centered cubic unit cell. The radius of an aluminum atom is 143 pm. Calculate the density of solid crystalline aluminum in g/cm^3.

SORT You are given the radius of an aluminum atom and its crystal structure. You are asked to find the density of solid aluminum.

GIVEN: $r = 143$ pm, face-centered cubic

FIND: d

STRATEGIZE The conceptual plan is based on the definition of density.
Because the unit cell has the physical properties of the entire crystal, you can find the mass and volume of the unit cell and use these to calculate its density.

CONCEPTUAL PLAN

$d = m/V$

$m =$ mass of unit cell

$=$ number of atoms in unit cell $\times$ mass of each atom

$V =$ volume of unit cell

$= (\text{edge length})^3$

SOLVE Begin by finding the mass of the unit cell. Determine the mass of an aluminum atom from its molar mass. Because the face-centered cubic unit cell contains four atoms per unit cell, you multiply the mass of aluminum by 4 to get the mass of a unit cell.

SOLUTION

$$m(\text{Al atom}) = 26.98\frac{\text{g}}{\text{mol}} \times \frac{1\ \text{mol}}{6.022 \times 10^{23}\ \text{atoms}}$$

$$= 4.480 \times 10^{-23}\ \frac{\text{g}}{\text{atom}}$$

$$m(\text{unit cell}) = 4\ \text{atoms}\left(4.480 \times 10^{-23}\ \frac{\text{g}}{\text{atom}}\right)$$

$$= 1.792 \times 10^{-22}\ \text{g}$$

Next, calculate the edge length (l) of the unit cell (in m) from the atomic radius of aluminum. For the face-centered cubic structure, $l = 2\sqrt{2}r$.

$$l = 2\sqrt{2}r$$

$$= 2\sqrt{2}(143\ \text{pm})$$

$$= 2\sqrt{2}(143 \times 10^{-12}\ \text{m})$$

$$= 4.045 \times 10^{-10}\ \text{m}$$

Calculate the volume of the unit cell (in cm) by converting the edge length to cm and cubing the edge length. (Use centimeters because you will want to report the density in units of g/cm^3.)

$$V = l^3$$

$$= \left(4.045 \times 10^{-10}\ \text{m} \times \frac{1\ \text{cm}}{10^{-2}\ \text{m}}\right)^3$$

$$= 6.618 \times 10^{-23}\ \text{cm}^3$$

Finally, calculate the density by dividing the mass of the unit cell by the volume of the unit cell.

$$d = \frac{m}{V} = \frac{1.792 \times 10^{-22}\ \text{g}}{6.618 \times 10^{-23}\ \text{cm}^3}$$

$$= 2.71\ \text{g/cm}^3$$

CHECK
The units of the answer are correct. The magnitude of the answer is reasonable because the density is greater than 1 g/cm^3 (as you would expect for metals), but still not too high (because aluminum is a low-density metal).

FOR PRACTICE 13.4
Chromium crystallizes with a body-centered cubic unit cell. The radius of a chromium atom is 125 pm. Calculate the density of solid crystalline chromium in g/cm^3.

Closest-Packed Structures

Another way to envision crystal structures, especially useful in metals where bonds are not usually directional, is to think of the atoms as stacking in layers, much as fruit is stacked at the grocery store. For example, we can envision the simple cubic structure as one layer of atoms arranged in a square pattern with the next layer stacking directly over the first, so that the atoms in one layer align exactly on top of the atoms in the layer beneath it, as shown here:

As we saw previously, this simple cubic unit crystal structure has a great deal of empty space—only 52% of the volume is occupied by the spheres, and the coordination number is 6.

More space-efficient packing can be achieved by aligning neighboring rows of atoms in a pattern with one row offset from the next by one-half a sphere, as shown here:

In this way, the atoms pack more closely to each other in any one layer. We can further increase the packing efficiency by placing the next layer *not directly on top of the first,* but again offset so that any one atom actually sits in the indentation formed by three atoms in the layer beneath it, as shown here:

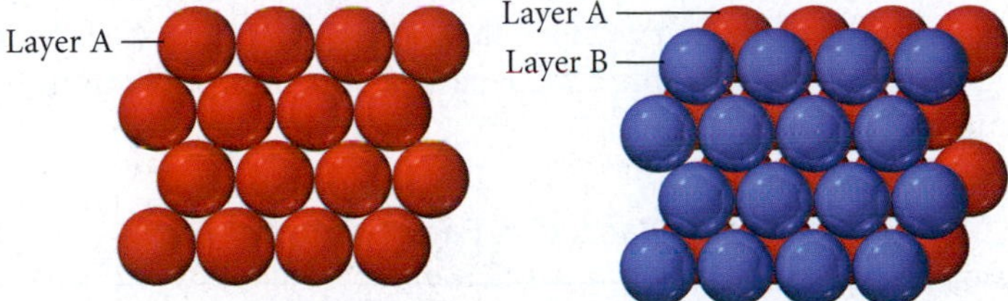

This kind of packing leads to two different crystal structures called *closest-packed structures,* both of which have a packing efficiency of 74% and a coordination number of 12. In the first of these two closest-packed structures—called **hexagonal closest packing**—the third layer of atoms aligns exactly on top of the first, as shown here:

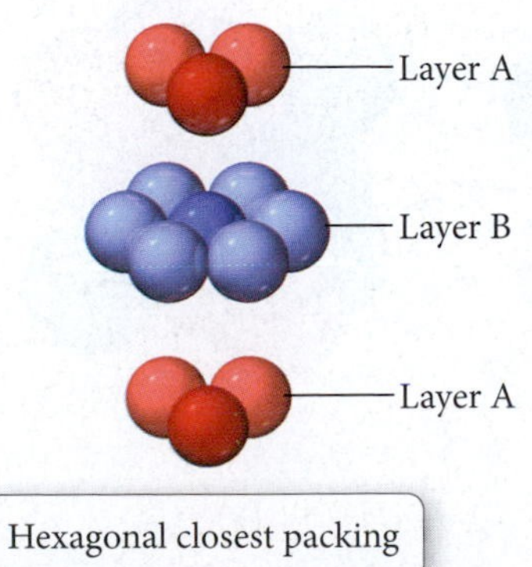

▶ **FIGURE 13.10 Hexagonal Closest-Packing Crystal Structure** The unit cell is outlined in bold.

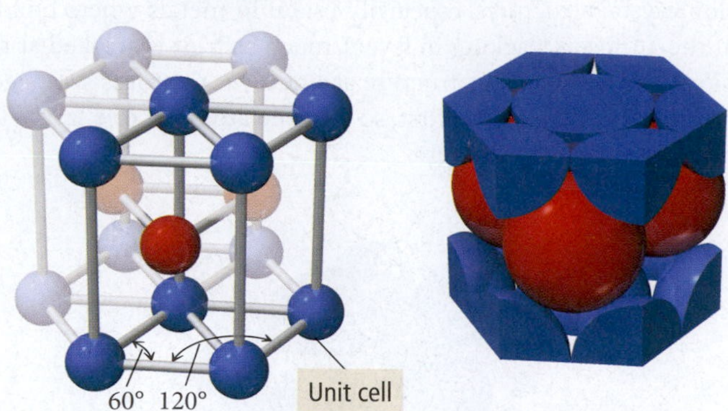

The pattern from one layer to the next is ABAB, with the third layer aligning exactly above the first. Notice that the central atom in layer B of this structure is touching six atoms in its own layer, three atoms in the layer above it, and three atoms in the layer below, for a coordination number of 12. The unit cell for this crystal structure is not a cubic unit cell, but a hexagonal one, as shown in Figure 13.10 ▲.

In the second of the two closest-packed structures—called **cubic closest packing**—the third layer of atoms is offset from the first, as shown here:

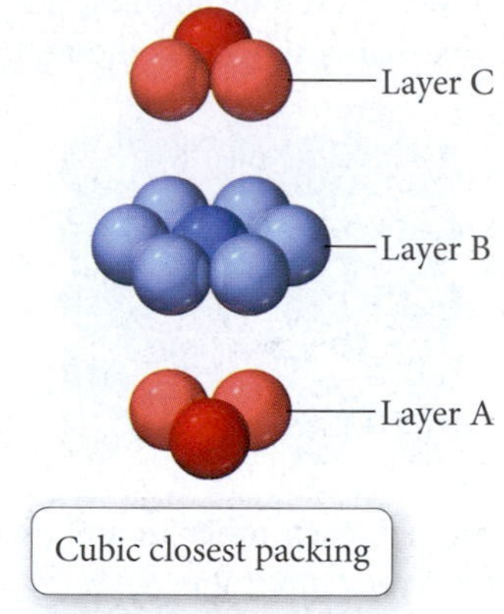

The pattern from one layer to the next is ABCABC, with every fourth layer aligning with the first. Although not simple to visualize, the unit cell for cubic closest packing is the face-centered cubic unit cell, as shown in Figure 13.11 ▼. The cubic closest-packed structure is identical to the face-centered cubic unit cell structure.

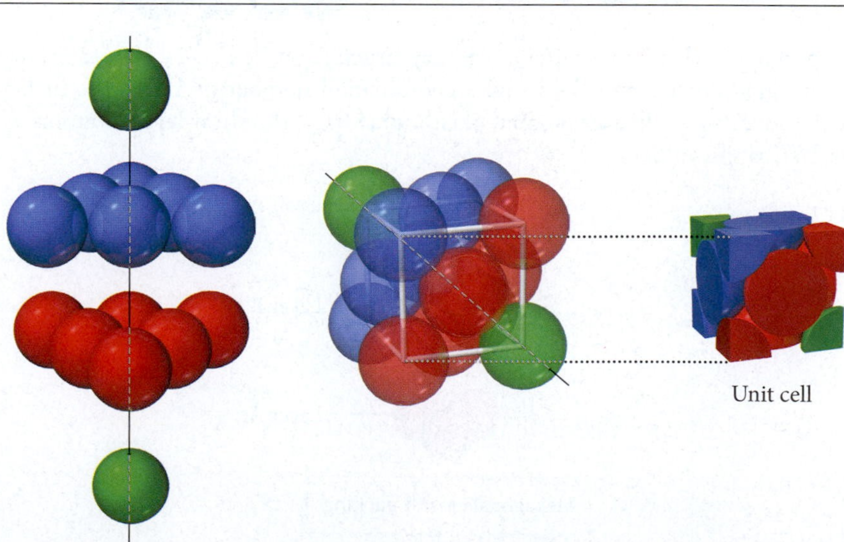

▶ **FIGURE 13.11 Cubic Closest-Packing Crystal Structure** The unit cell of the cubic closest-packed structure is face-centered cubic.

13.5 Crystalline Solids: The Fundamental Types

As we discussed in Section 12.2, solids may be crystalline (comprising a well-ordered array of atoms or molecules) or amorphous (having no long-range order). We classify crystalline solids into three categories—molecular, ionic, and atomic—based on the individual units that compose the solid. Atomic solids can themselves be classified into three categories—nonbonding, metallic, and network covalent—depending on the types of interactions between atoms within the solid. Figure 13.12 ▼ shows the different categories of crystalline solids.

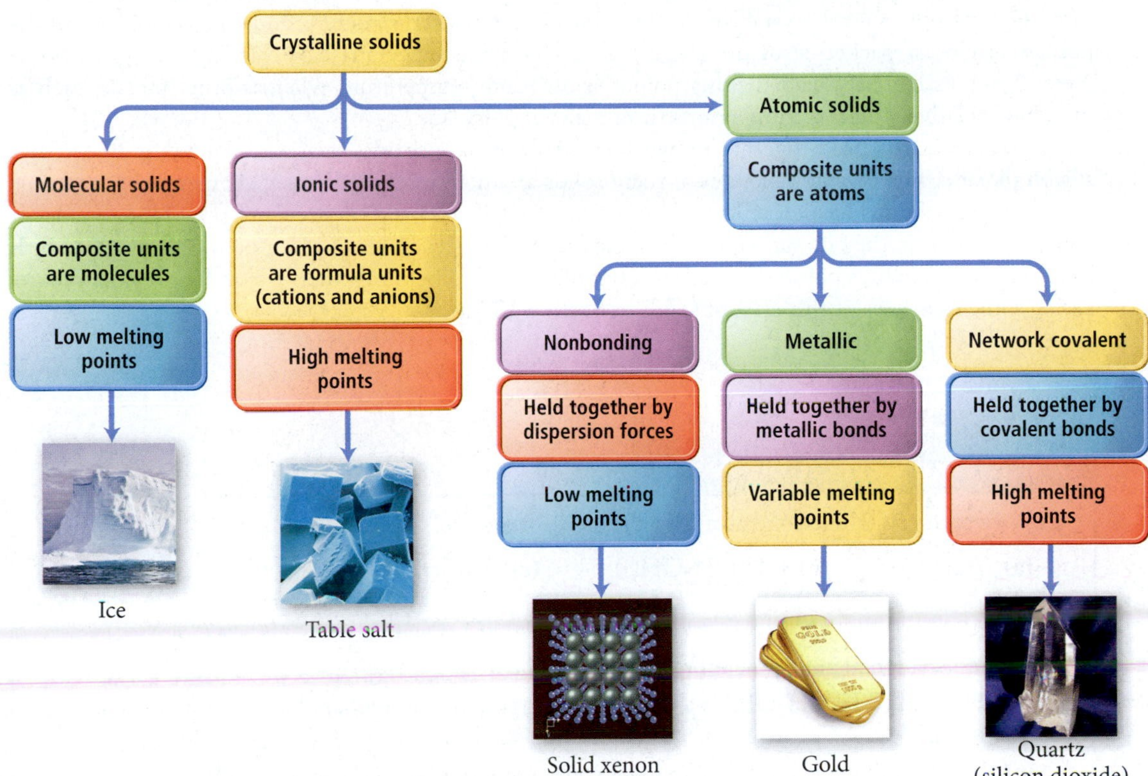

◀ **FIGURE 13.12 Types of Crystalline Solids**

Molecular Solids

Molecular solids are solids whose composite units are *molecules*. The lattice sites in a crystalline molecular solid are therefore occupied by molecules. Ice (solid H_2O) and dry ice (solid CO_2) are examples of molecular solids. Molecular solids are held together by the kinds of intermolecular forces—dispersion forces, dipole–dipole forces, and hydrogen bonding—discussed in Chapter 12. Molecular solids as a whole tend to have low to moderately low melting points. However, strong intermolecular forces (such as the hydrogen bonds in water) increase the melting points of some molecular solids.

Ionic Solids

The composite units of **ionic solids** are ions. Table salt (NaCl) and calcium fluoride (CaF_2) are examples of ionic solids. The forces holding ionic solids together are strong coulombic forces (or ionic bonds), and because these forces are much stronger than intermolecular forces, ionic solids tend to have much higher melting points than molecular solids. For example, sodium chloride melts at 801 °C, while carbon disulfide (CS_2)—a molecular solid with a higher molar mass—melts at −110 °C. We examine the structure of ionic solids in Section 13.6.

Atomic Solids

In **atomic solids** the composite units are individual atoms. Solid xenon (Xe), iron (Fe), and silicon dioxide (SiO_2) are examples of atomic solids. We classify atomic solids into three categories—*nonbonding atomic solids*, *metallic atomic solids*, and *network covalent atomic solids*—each type is held together by a different kind of force.

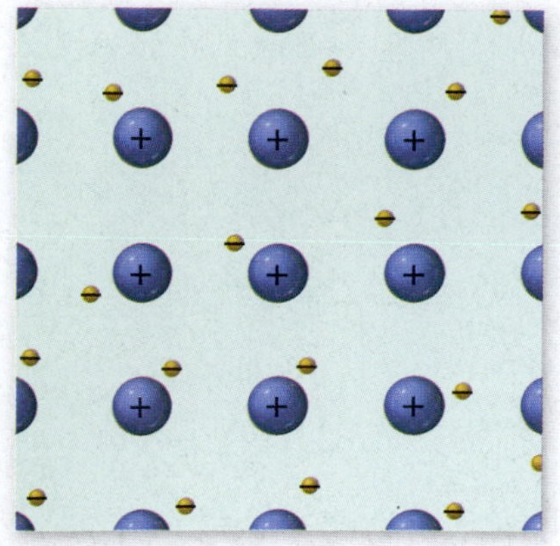

▲ **FIGURE 13.13 The Electron Sea Model** In the electron sea model for metals, the metal cations exist in a "sea" of electrons.

Nonbonding atomic solids are held together by relatively weak dispersion forces. In order to maximize these interactions, nonbonding atomic solids form closest-packed structures, maximizing their coordination numbers and minimizing the distance between composite units. Nonbonding atomic solids have very low melting points that increase uniformly with molar mass. The only nonbonding atomic solids are noble gases in their solid form. Argon, for example, has a melting point of −189 °C, and xenon has a melting point of −112 °C.

Metallic atomic solids, such as iron or gold, are held together by *metallic bonds,* which in the simplest model are represented by the interaction of metal cations with the sea of electrons that surrounds them, as described in Section 7.6 (Figure 13.13 ◀).

Because metallic bonds are not directional, many metals tend to form closest-packed crystal structures. For example, nickel crystallizes in the cubic closest-packed structure, and zinc crystallizes in the hexagonal closest-packed structure (Figure 13.14 ▼). Metallic bonds have varying strengths. Some metals, such as mercury, have melting points below room temperature, whereas other metals, such as iron, have relatively high melting points (iron melts at 1809 °C).

Covalent bonds hold together **network covalent atomic solids**, such as diamond, graphite, and silicon dioxide. The crystal structures of these solids are more restricted by the geometrical constraints of the covalent bonds (which tend to be more directional than intermolecular forces, ionic bonds, or metallic bonds), so they *do not* tend to form closest-packed structures. Network covalent atomic solids have very high melting points because the crystalline solid is held together by covalent bonds. We examine some examples of this class of solids in Section 13.7.

EXAMPLE 13.5

Classifying Crystalline Solids

Classify each crystalline solid as molecular, ionic, or atomic.

(a) Au(*s*) **(b)** $CH_3CH_2OH(s)$ **(c)** $CaF_2(s)$

SOLUTION

(a) Au is a metal and is therefore an atomic solid.

(b) Ethanol (CH_3CH_2OH) is a molecular compound. Solid ethanol is therefore a molecular solid.

(c) Calcium fluoride (CaF_2) is an ionic compound, so $CaF_2(s)$ is an ionic solid.

FOR PRACTICE 13.5

What type of atomic solid is Au(*s*)?

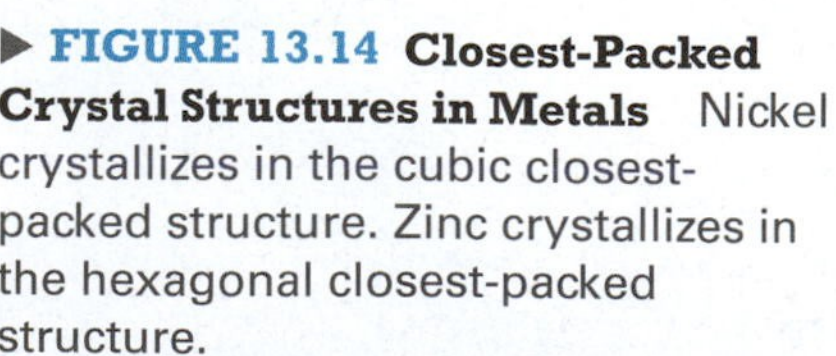

▶ **FIGURE 13.14 Closest-Packed Crystal Structures in Metals** Nickel crystallizes in the cubic closest-packed structure. Zinc crystallizes in the hexagonal closest-packed structure.

Crystalline Solid Types and Melting Points

13.2 Cc Conceptual Connection

Which solid would you expect to have the highest melting point?

(a) $MgO(s)$ **(b)** $I_2(s)$ **(c)** $Kr(s)$

13.6 The Structures of Ionic Solids

Many ionic solids have crystalline structures that are closely related to unit cells that we examined in Section 13.4. However, because ionic compounds necessarily contain both cations and anions, their structures must accommodate the two different types of ions. In an ionic solid, the cations and anions attract one another. The coordination number of the unit cell represents the number of close cation–anion interactions. Because these interactions lower potential energy, the crystal structure of a particular ionic compound is the structure that maximizes the coordination number while accommodating both charge neutrality (each unit cell must be charge neutral) and the different sizes of the cations and anions that compose the particular compound. In general, the more similar the radii of the cation and the anion, the higher the coordination number.

Cesium chloride (CsCl) is a good example of an ionic compound with cations and anions of similar size (Cs^+ radius = 167 pm; Cl^- radius = 181 pm). In the cesium chloride structure, the chloride ions occupy the lattice sites of a simple cubic cell and one cesium ion lies in the very center of the cell, as shown in Figure 13.15 ▶. (In this and subsequent figures of ionic crystal structures, the different colored spheres represent different ions.) The coordination number for cesium chloride is 8, meaning that each cesium ion is in direct contact with eight chloride ions (and vice versa). The cesium chloride unit cell contains one chloride anion ($8 \times \frac{1}{8} = 1$) and one cesium cation for a ratio of Cs to Cl of 1:1, as the formula for the compound indicates. (Note that complete chloride ions are shown in Figure 13.15, even though only $\frac{1}{8}$ of each ion is in the unit cell.) Calcium sulfide (CaS) has the same structure as cesium chloride.

Cesium chloride (CsCl)

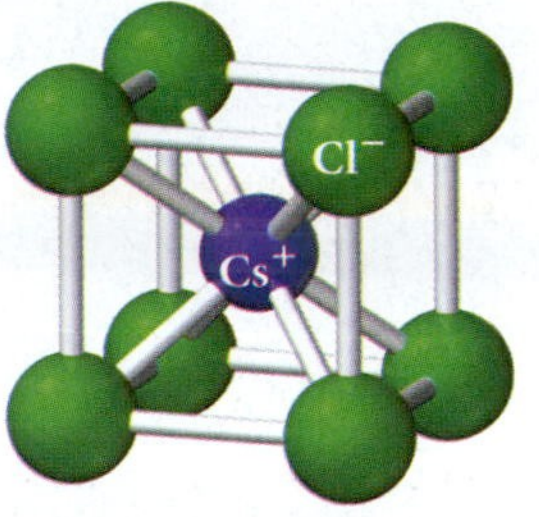

▲ **FIGURE 13.15 Cesium Chloride Unit Cell** The different colored spheres in this figure represent the different ions in the compound.

The crystal structure of sodium chloride must accommodate the more disproportionate sizes of Na^+(radius = 97 pm) and Cl^-(radius = 181 pm). If ion size were the only consideration, the larger chloride anion could theoretically fit many of the smaller sodium cations around it, but charge neutrality requires that each sodium cation be surrounded by an equal number of chloride anions. Therefore, the coordination number is limited by the number of chloride anions that can fit around the relatively small sodium cation. The structure that minimizes the energy is shown in Figure 13.16 ▼ and has a coordination number of 6 (each chloride anion is surrounded by six sodium cations and vice versa). We can visualize this structure, called the *rock salt* structure, as chloride anions occupying the lattice sites of a face-centered cubic structure, with the smaller sodium cations occupying the holes between the anions. (Alternatively, we can visualize this structure as the *sodium cations* occupying the lattice sites of a face-centered cubic structure with the *larger chloride anions* occupying the spaces between the cations.) Each unit cell contains four chloride anions $[(8 \times \frac{1}{8} = 1) + (6 \times \frac{1}{2}) = 4]$ and four sodium cations $[(12 \times \frac{1}{4}) + 1 = 4]$, resulting in a ratio of 1:1, as the formula of the compound specifies. Other compounds exhibiting the sodium chloride structure include LiF, KCl, KBr, AgCl, MgO, and CaO.

Sodium chloride (NaCl)

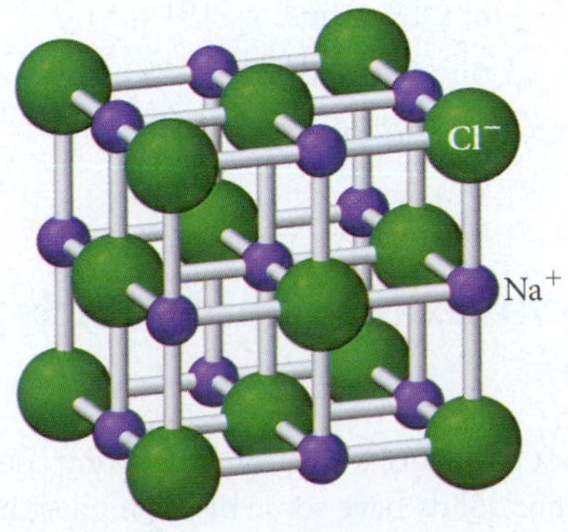

◀ **FIGURE 13.16 Sodium Chloride Unit Cell** The different colored spheres in this figure represent the different ions in the compound.

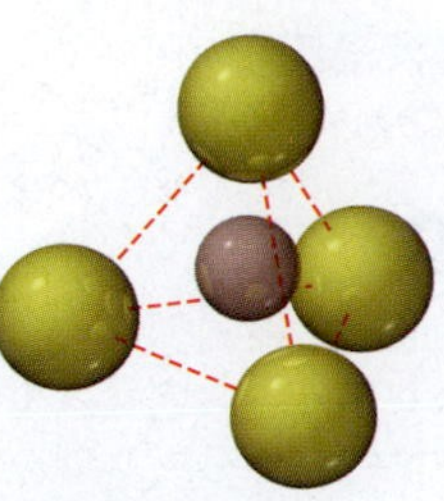

▲ A tetrahedral hole

When there is a greater disproportion between the sizes of the cations and anions in a compound, a coordination number of even 6 is physically impossible. For example, in ZnS (Zn^{2+} radius = 74 pm; S^{2-} radius = 184 pm) the crystal structure, shown in Figure 13.17 ▼, has a coordination number of only 4. We can visualize this structure, called the *zinc blende* structure, as sulfide anions occupying the lattice sites of a face-centered cubic structure, with the smaller zinc cations occupying four of the eight tetrahedral holes located directly beneath each corner atom. A tetrahedral hole is the empty space that lies in the center of a tetrahedral arrangement of four atoms, as shown in the figure. Each unit cell contains four sulfide anions $[(8 \times \frac{1}{8}) + (6 \times \frac{1}{2} = 4)]$ and four zinc cations (each of the four zinc cations is completely contained within the unit cell), resulting in a ratio of 1:1, just as the formula of the compound indicates. Other compounds that exhibit the zinc blende structure include CuCl, AgI, and CdS.

When the ratio of cations to anions is not 1:1, the crystal structure must accommodate the unequal number of cations and anions. Many compounds that contain a cation-to-anion ratio of 1:2 adopt the *fluorite* (CaF_2) *structure* shown in Figure 13.18 ▼. We can visualize this structure as calcium cations occupying the lattice sites of a face-centered cubic structure, with the larger fluoride anions occupying all eight of the tetrahedral holes located directly beneath each corner atom. Each unit cell contains four calcium cations $[(8 \times \frac{1}{8}) + (6 \times \frac{1}{2}) = 4]$ and eight fluoride anions (each of the eight fluoride anions is completely contained within the unit cell), resulting in a cation-to-anion ratio of 1:2, just as in the formula of the compound. Other compounds exhibiting the fluorite structure include PbF_2, SrF_2, and $BaCl_2$. Compounds with a cation-to-anion ratio of 2:1 often exhibit the *antifluorite structure*, in which the anions occupy the lattice sites of a face-centered cubic structure and the cations occupy the tetrahedral holes beneath each corner atom.

Zinc blende (ZnS)

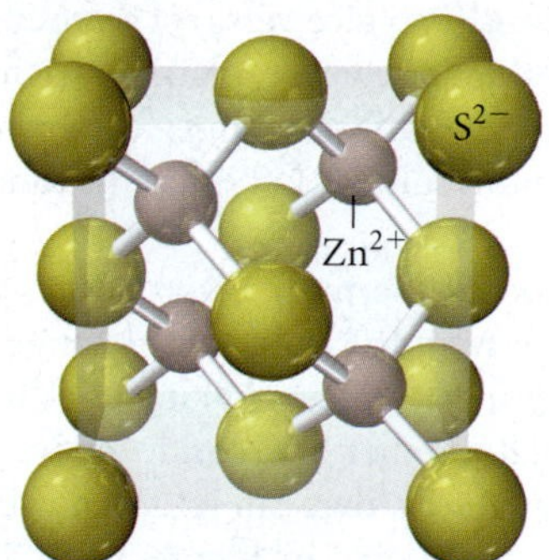

▲ **FIGURE 13.17 Zinc Sulfide (Zinc Blende) Unit Cell** The different colored spheres in this figure represent the different ions in the compound.

Calcium fluoride (CaF_2)

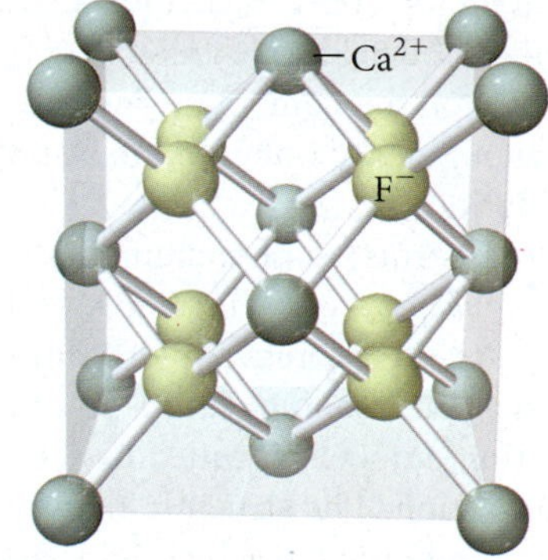

▲ **FIGURE 13.18 Calcium Fluoride Unit Cell** The different colored spheres in this figure represent the different ions in the compound.

13.3 Cc Conceptual Connection

Ionic Crystalline Solid Unit Cells

Which compound is most likely to crystallize in the zinc blende structure?

(a) RbCl (Rb^+ radius = 148 pm; Cl^- radius = 181 pm)
(b) $MgCl_2$ (Mg^{2+} radius = 65 pm; Cl^- radius = 181 pm)
(c) CuI (Cu^+ radius = 96 pm; I^- = 216 pm)

13.7 Network Covalent Atomic Solids: Carbon and Silicates

As we saw in Section 13.5, network covalent atomic solids are composed of atoms held together by covalent bonds. Network covalent atomic solids have some of the highest melting points of all substances. In this section, we examine two different families of network covalent atomic solids: carbon and silicates.

Carbon

Elemental carbon exists in several different forms. Two well-known naturally occurring crystalline forms of carbon are **graphite** and **diamond**. Graphite's structure, shown in Figure 13.19(a) ▼, consists of flat sheets of carbon atoms covalently bonded together as interconnected hexagonal rings. The bond length between carbon atoms *within a sheet* is 142 pm. However, the forces *between* sheets are much different. There are no covalent bonds between sheets, only relatively weak dispersion force, and the separation between sheets is 341 pm. Consequently, the sheets slide past each other relatively easily, which explains the slippery feel of graphite and its extensive use as a lubricant. The electrons in the extended pi bonding network within a sheet make graphite a good electrical conductor in the direction of the plane of the sheets. Because of its relative stability and electrical properties, graphite is used for electrodes in electrochemical applications and for heating elements in furnaces.

The density of graphite is $2.2\,g/cm^3$. Under high pressure, the carbon atoms in graphite rearrange to form diamond, which has a higher density of $3.5\,g/cm^3$. Diamond forms naturally when carbon is exposed to high pressures deep underground. Through movements in Earth's crust, diamond rises toward the surface. Most diamond is found in Africa, mainly in the Congo region and in South Africa. The first synthetic diamonds were produced in the 1940s, using pressures of 50,000 atm and a temperature of 1600 °C.

The diamond structure consists of carbon atoms covalently bonded to four other carbon atoms at the corners of a tetrahedron (Figure 13.19(b) ▼). This structure extends throughout the entire crystal, so that a diamond crystal can be thought of as a giant molecule, held together by these covalent bonds. Because covalent bonds are very strong, diamond has a very high melting point (it is estimated to melt at about 3800 °C). The electrons in diamond are confined to the covalent bonds and are not free to flow. Therefore, diamond does not conduct electricity.

Diamond is very hard and is an excellent conductor of heat. Consequently, the largest use of diamonds is for abrasives and cutting tools. Small diamonds are used at the cutting edge of tools, making the edges much harder and giving them a longer life. Natural diamonds are valued as gems for their brilliance and relative inertness.

◀ **FIGURE 13.19 Network Covalent Atomic Solids** **(a)** In graphite, carbon atoms are arranged in sheets. Within each sheet, the atoms are covalently bonded to one another by a network of sigma and pi bonds. Neighboring sheets are held together by dispersion forces. **(b)** In diamond, each carbon atom forms four covalent bonds to four other carbon atoms in a tetrahedral geometry.

Phase Changes and Pressure

13.4 Cc Conceptual Connection

Why do high pressures favor the formation of diamond from graphite?

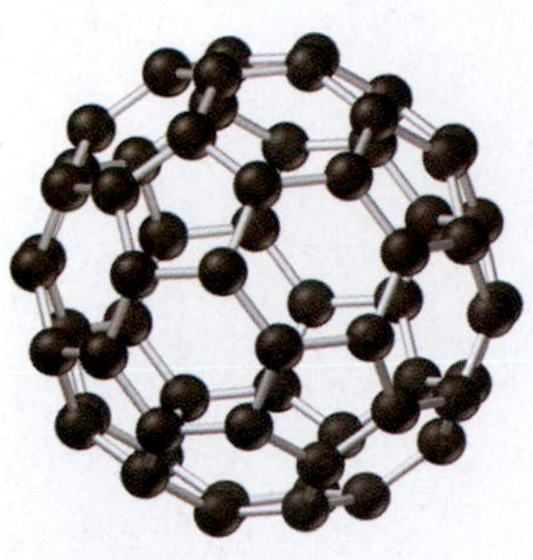

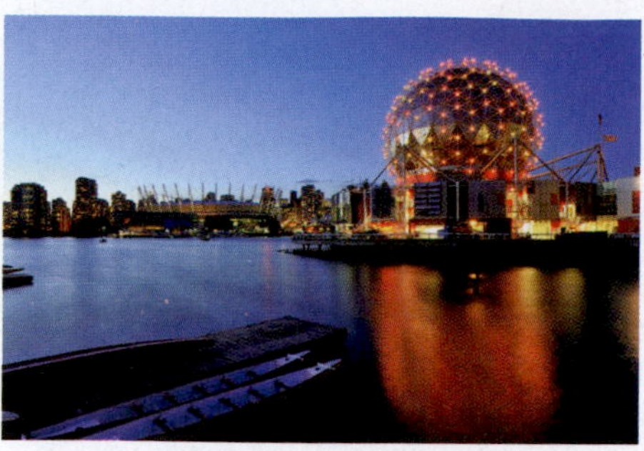

▲ **FIGURE 13.20 C_{60} and a Geodesic Dome** The C_{60} structure resembles Buckminster Fuller's geodesic dome.

In the 1980s, researchers discovered a new form of carbon when they aimed a powerful laser at a graphite surface. This new form of carbon occurs as soccer-ball-shaped clusters of 60 carbon atoms (C_{60}). The atoms form five- and six-membered carbon rings wrapped into a 20-sided icosahedral structure (Figure 13.20 ◀). The compound was named *buckminsterfullerene*, honoring R. Buckminster Fuller, a twentieth-century engineer and architect who advocated the construction of buildings using a structurally strong geodesic dome shape that he patented.

Researchers have since identified carbon clusters similar to C_{60} containing from 36 to over 100 carbon atoms. As a class, these carbon clusters are called **fullerenes** and nicknamed *buckyballs*. At room temperature, fullerenes are black solids—the individual clusters are held to one another by dispersion forces. Fullerenes are somewhat soluble in nonpolar solvents, and the different fullerenes form solutions of different colors.

Researchers have also synthesized long carbon structures called **nanotubes**, which consist of sheets of interconnected C_6 rings that assume the shape of a cylinder (like a roll of chicken wire). The first nanotubes discovered consisted of tubes with double walls of C_6 rings with closed ends. The ends of the tubes can be opened when they are heated under the proper conditions. Today, two general types of nanotubes can be produced: (1) single-walled nanotubes (SWNT) that have one layer of interconnected C_6 rings forming the walls, and (2) multiwalled nanotubes (MWNT) that have concentric layers of interconnected C_6 rings forming the walls. In addition, researchers have been able to form *nanoribbons* by slicing open nanotubes. These types of nanotubes are shown in Figure 13.21 ▼.

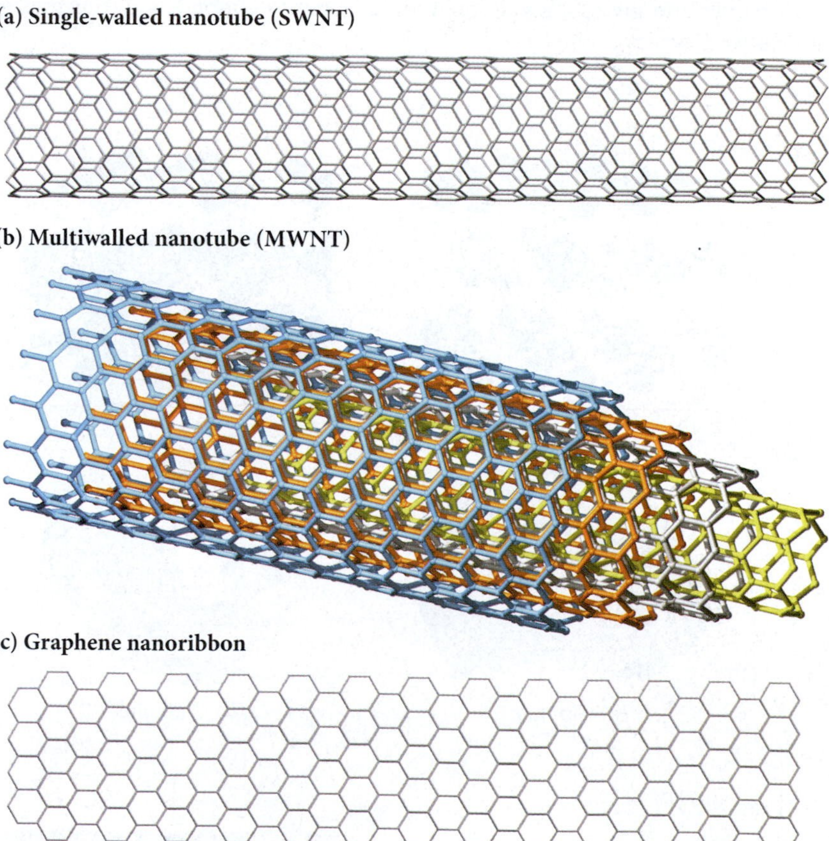

▲ **FIGURE 13.21 Novel Carbon Structures** **(a)** A single-walled nanotube, **(b)** a multiwallled nanotube; **(c)** a graphene nanoribbon

Nanotubes are 100 times stronger than steel and only one-sixteenth as dense. Consequently, we use carbon nanotubes commercially for lightweight applications that require strength, such as golf clubs and bicycle frames. When the nanotubes are lined up parallel to one another, a bundle of the tubes forms a "wire" with very low electrical resistance. These tiny wires raise the possibility of making incredibly small electronic devices.

Silicates

The **silicates** (extended arrays of silicon and oxygen) are the most common network covalent atomic solids. Geologists estimate that 90% of Earth's crust is composed of silicates. Silicon and oxygen form a network covalent structure in which a silicon atom bonds to four oxygen atoms, forming a tetrahedral shape with the silicon atom in the middle and the four oxygen atoms at the corners of the tetrahedron (Figure 13.22 ▼). In this structure, the silicon atom bonds to each oxygen atom with a single covalent sigma bond. In contrast to carbon, which often bonds to oxygen with a double bond (one sigma and one pi bond), silicon forms only a single bond with oxygen because the silicon atom is too large to allow substantial overlap between the *p* orbitals on the two atoms. The silicon atom in this structure, by bonding to four oxygen atoms, obtains a complete octet. However, each oxygen atom is one electron short of an octet. Therefore, each O atom forms a second covalent bond to a different Si atom, forming the three-dimensional structure of **quartz**. Quartz has a formula unit of SiO_2 and is generally called **silica**. In silica, each Si atom is in a tetrahedron surrounded by four O atoms, and each O atom acts as a bridge connecting the corners of two tetrahedrons (Figure 13.23 ▼). Silica melts when heated above 1500 °C. After melting, if cooled quickly, silica does not crystallize back into the quartz structure. Instead, the Si atoms and O atoms form a randomly ordered amorphous structure called a glass. Common glass is amorphous SiO_2.

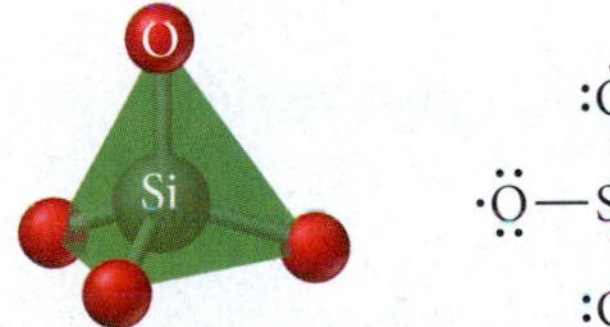

:Ö:
·Ö—Si—Ö·
:Ö:

▲ **FIGURE 13.22 SiO_4 Tetrahedron** In a SiO_4 tetrahedron, silicon occupies the center of the tetrahedron, and one oxygen atom occupies each corner.

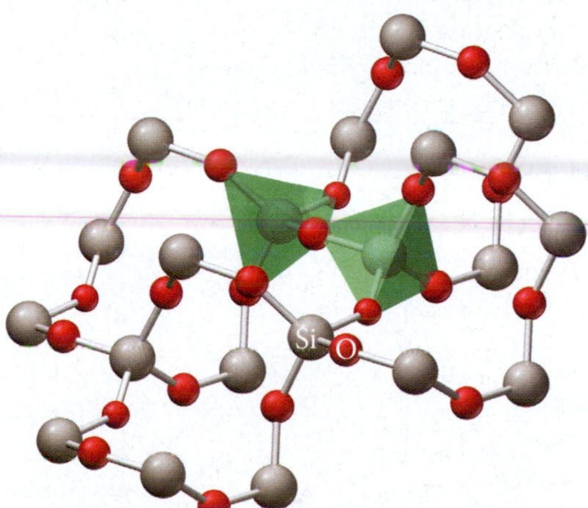

▲ **FIGURE 13.23 Structure of Quartz** In the quartz structure, each Si atom is in a tetrahedron surrounded by four O atoms, and each O atom is a bridge connecting the corners of two tetrahedra.

SELF-ASSESSMENT Quiz

1. Determine what state this substance is in at 1 atm and 298 K by referring to its phase diagram.

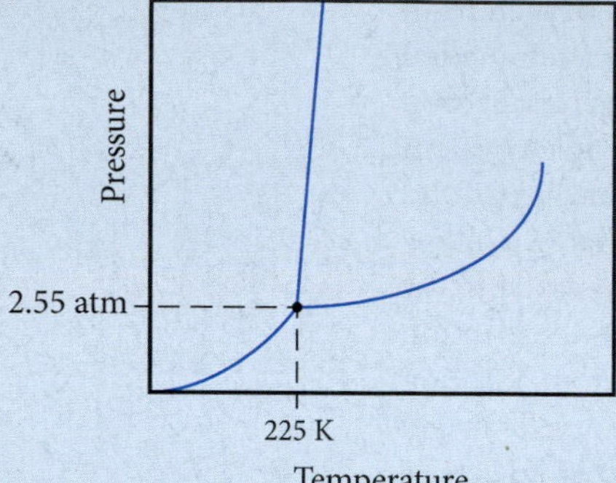

   a) Solid
   b) Liquid
   c) Gas
   d) All three states in equilibrium

2. A sample of the substance in this phase diagram is initially at 175 °C and 925 mmHg. Which phase transition occurs when the pressure is decreased to 760 mmHg at constant temperature?

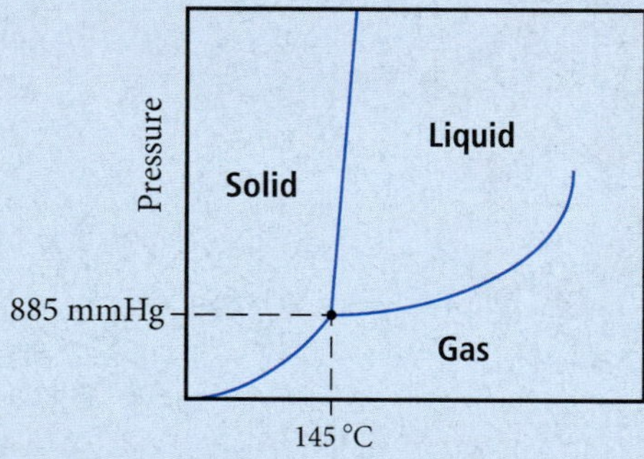

   a) Solid to liquid
   b) Liquid to gas
   c) Solid to gas
   d) Liquid to solid

3. An X-ray beam of $\lambda = 71.07$ pm is incident on the surface of an atomic crystal. The maximum reflection occurs at an angle of 39.8 °. Calculate the separation between layers of atoms in the crystal (assume $n = 1$).
   a) 90.1 pm
   b) 41.2 pm
   c) 111 pm
   d) 55.5 pm

4. A crystalline solid has a body-centered cubic structure. If the radius of the atoms is 172 pm, what is the edge length of the unit cell?
   a) 397 pm
   b) 344 pm
   c) 486 pm
   d) 688 pm

5. What is the coordination number (the number of atoms each atom touches) in a face-centered cubic unit cell?
   a) 4
   b) 6
   c) 8
   d) 12

6. How many atoms are in a body-centered cubic unit cell?
   a) 1
   b) 2
   c) 4
   d) 5

7. Rhodium crystallizes in a face-centered cubic unit cell. The radius of a rhodium atom is 135 pm. Determine the density of rhodium in g/cm^3.
   a) 3.07 g/cm^3
   b) 12.4 g/cm^3
   c) 278 g/cm^3
   d) 0.337 g/cm^3

8. Which type of solid is dry ice (solid carbon dioxide)?
   a) Ionic
   b) Molecular
   c) Atomic
   d) None of the above

9. Which crystalline solid would you expect to have the highest melting point?
   a) Au(*s*)
   b) H_2O(*s*)
   c) MgS(*s*)
   d) Xe(*s*)

10. Which compound is most likely to crystallize in the fluorite structure?
   a) $BaCl_2$
   b) MgO
   c) K_2S
   d) LiF

Answers: 1:c; 2:b; 3:d; 4:a; 5:d; 6:b; 7:b; 8:b; 9:c; 10:a

CHAPTER SUMMARY 13

REVIEW

KEY LEARNING OUTCOMES

CHAPTER OBJECTIVES	ASSESSMENT
Navigating Within a Phase Diagram (13.2)	• Example 13.1 For Practice 13.1 Exercises 19, 20, 23, 24
Using Bragg's Law in X-Ray Diffraction Calculations (13.3)	• Example 13.2 For Practice 13.2 Exercises 25, 26
Relating Unit Cell Volume, Edge Length, and Atomic Radius (13.4)	• Example 13.3 For Practice 13.3 Exercises 29–32
Relating Density to Crystal Structure (13.4)	• Examples 13.3, 13.4 For Practice 13.3, 13.4 Exercises 29–34
Classifying Crystalline Solids (13.5)	• Example 13.5 For Practice 13.5 Exercises 37–42

KEY TERMS

Section 13.2
phase diagram (482)
triple point (483)
critical point (483)

Section 13.3
X-ray diffraction (485)

Section 13.4
crystalline lattice (487)
unit cell (488)
simple cubic (488)
coordination number (488)
packing efficiency (488)
body-centered cubic (488)
face-centered cubic (491)
hexagonal closest packing (493)
cubic closest packing (494)

Section 13.5
molecular solids (495)
ionic solids (495)
atomic solids (495)
nonbonding atomic solids (496)
metallic atomic solids (496)
network covalent atomic solids (496)

Section 13.7
graphite (499)
diamond (499)
fullerenes (500)
nanotubes (500)
silicates (501)
quartz (501)
silica (501)

KEY CONCEPTS

Phase Diagrams (13.2)

- A phase diagram is a map of the states of a substance as a function of its pressure (y-axis) and temperature (x-axis).
- The regions in a phase diagram represent conditions under which a single stable state (solid, liquid, gas) exists.
- The lines in a phase diagram represent conditions under which two states are in equilibrium.
- The triple point represents the conditions under which all three states coexist.
- The critical point is the temperature and pressure above which a supercritical fluid exists.

Crystalline Structures (13.3–13.4)

- X-ray crystallography uses the diffraction pattern of X-rays to determine the crystal structure of solids.
- The crystal lattice is represented by a unit cell, a structure that reproduces the entire lattice when repeated in three dimensions.
- Three basic cubic unit cells are the simple cubic, the body-centered cubic, and the face-centered cubic.
- Some crystal lattices can also be depicted as closest-packed structures, including the hexagonal closest-packing structure (not cubic) and the cubic closest-packing structure (which has a face-centered cubic unit cell).

Types of Crystalline Solids (13.5)

- The basic types of crystal solids are molecular, ionic, and atomic solids. We divide atomic solids into three different types: nonbonding, metallic, and covalent.

Structure of Ionic Solids (13.6)

- Ionic solids have structures that accommodate both cations and anions.
- Common cubic structures for ionic compounds include the cesium chloride structure, the rock salt structure, the zinc blende structure, the fluorite structure, and the antifluorite structure.

Network Covalent Atomic Solids (13.7)

- Carbon forms the network covalent atomic solids graphite and diamond.
- SiO_2 forms the network covalent atomic solid quartz.

KEY EQUATIONS AND RELATIONSHIPS

Bragg's Law: Relationship between Light Wavelength (λ), Angle of Reflection (θ), and Distance (d) between the Atomic Layers (13.3)

$$n\lambda = 2d \sin \theta \ (n = \text{integer})$$

EXERCISES

REVIEW QUESTIONS

1. What is a phase diagram?
2. Draw a generic phase diagram and label its important features.
3. What is the significance of crossing a line in a phase diagram?
4. Explain the basic principles involved in X-ray crystallography.
5. What is Bragg's law, and how is it used in X-ray crystallography?
6. What is a crystalline lattice? How is the lattice represented with the unit cell?
7. Sketch each unit cell: simple cubic, body-centered cubic, and face-centered cubic.
8. For each of the cubic cells in the previous problem, give the coordination number, edge length in terms of r, and number of atoms per unit cell.
9. What is the difference between hexagonal closest packing and cubic closest packing? What are the unit cells for each of these structures?
10. What are the three basic types of solids and the composite units of each? What types of forces hold each type of solid together?
11. In an ionic compound, how are the relative sizes of the cation and anion related to the coordination number of the crystal structure?
12. Show how the cesium chloride, sodium chloride, and zinc blende unit cells each contain a cation-to-anion ratio of 1:1.
13. Show how the fluorite structure accommodates a cation-to-anion ratio of 1:2.
14. List the three basic subtypes of atomic solids. What kinds of forces hold each of these subtypes together?
15. Describe the main structural differences between graphite and diamond. How do these structural differences affect the properties of each substance?
16. Describe the structures of fullerenes, nanotubes, and nanoribbons.
17. Why does silicon form only single bonds with oxygen, but carbon, which is in the same family as silicon, forms double bonds with oxygen in many compounds?
18. What is the difference between SiO_2 that is cooled slowly and SiO_2 that is cooled quickly?

PROBLEMS BY TOPIC

Phase Diagrams

19. Identify the states present at points a through g in the phase diagram shown here.

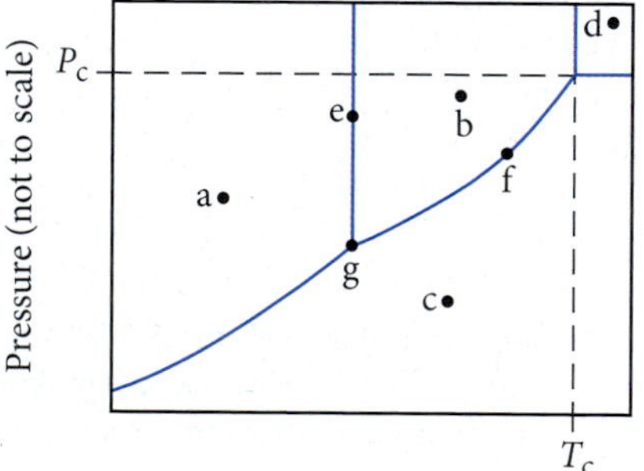

20. Consider the phase diagram for iodine shown here.
    a. What is the normal boiling point for iodine?
    b. What is the melting point for iodine at 1 atm?
    c. What state is present at room temperature and normal atmospheric pressure?
    d. What state is present at 186 °C and 1.0 atm?

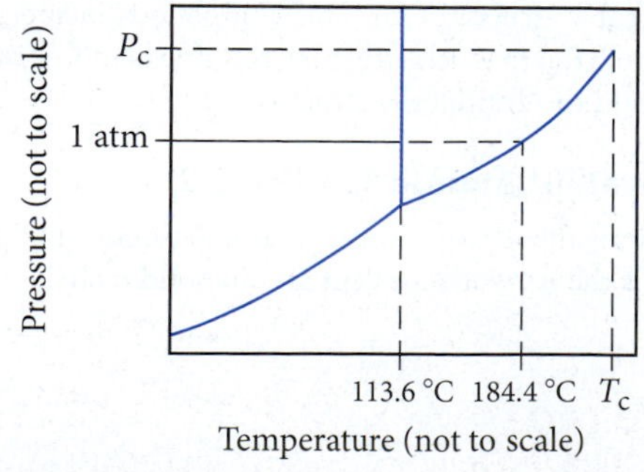

21. Nitrogen has a normal boiling point of 77.3 K and a melting point (at 1 atm) of 63.1 K. Its critical temperature is 126.2 K, and its critical pressure is 2.55×10^4 torr. It has a triple point at 63.1 K and 94.0 torr. Sketch the phase diagram for nitrogen. Does nitrogen have a stable liquid state at 1 atm?
22. Argon has a normal boiling point of 87.2 K and a melting point (at 1 atm) of 84.1 K. Its critical temperature is 150.8 K, and its critical pressure is 48.3 atm. It has a triple point at 83.7 K and 0.68 atm. Sketch the phase diagram for argon. Which has the greater density, solid argon or liquid argon?
23. Consider the phase diagram for sulfur shown here. The rhombic and monoclinic states are two solid states with different structures.
    a. Below what pressure does solid sulfur sublime?
    b. Which of the two solid states of sulfur is most dense?

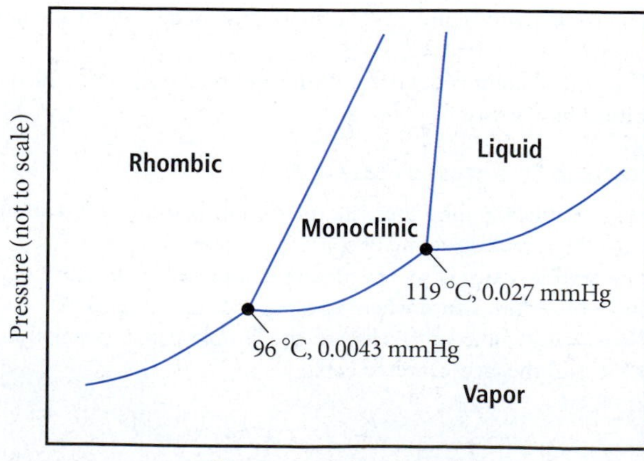

24. The high-pressure phase diagram of ice is shown here. Notice that, under high pressure, ice can exist in several different solid forms. What three forms of ice are present at the triple point marked O? What is the density of ice II compared to ice I (the familiar form of ice)? Would ice III sink or float in liquid water?

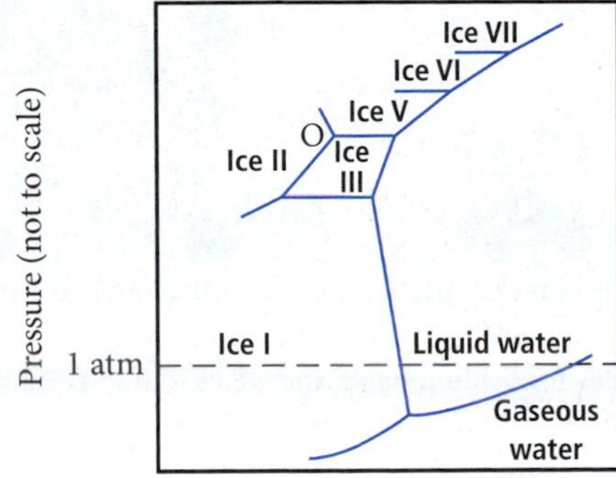

X-Ray Crystallography

25. An X-ray beam with $\lambda = 154$ incident on the surface of a crystal produces a maximum reflection at an angle of $\theta = 28.3°$. Assuming $n = 1$, calculate the separation between layers of atoms in the crystal.

26. An X-ray beam of unknown wavelength is diffracted from a NaCl surface. If the interplanar distance in the crystal is 286 pm and the angle of maximum reflection is found to be 7.23°, what is the wavelength of the X-ray beam? (Assume $n = 1$.)

Crystalline Structures and Unit Cells

27. Determine the number of atoms per unit cell for each metal.

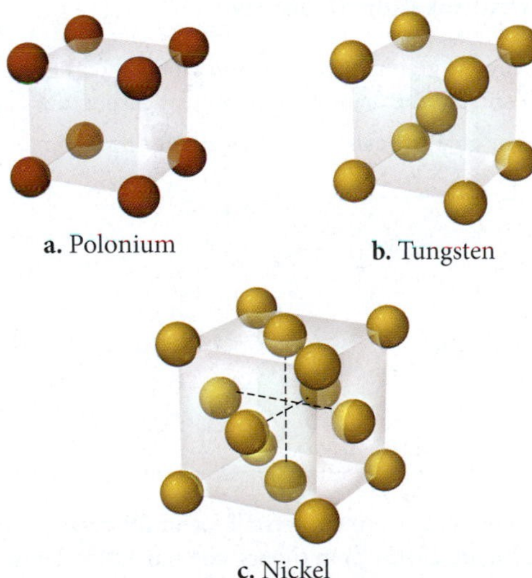

a. Polonium **b.** Tungsten **c.** Nickel

28. Determine the coordination number for each structure.

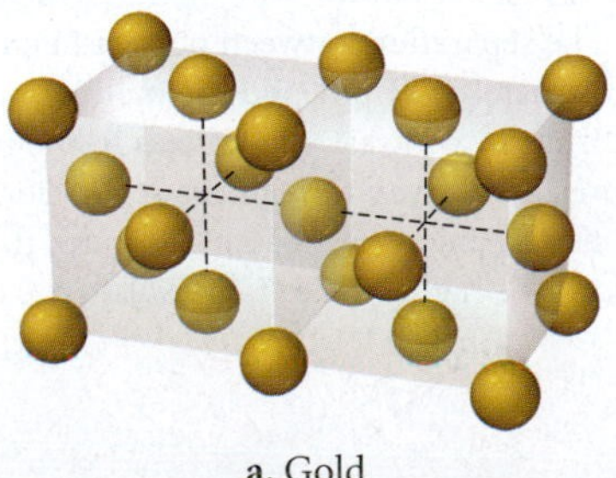

a. Gold

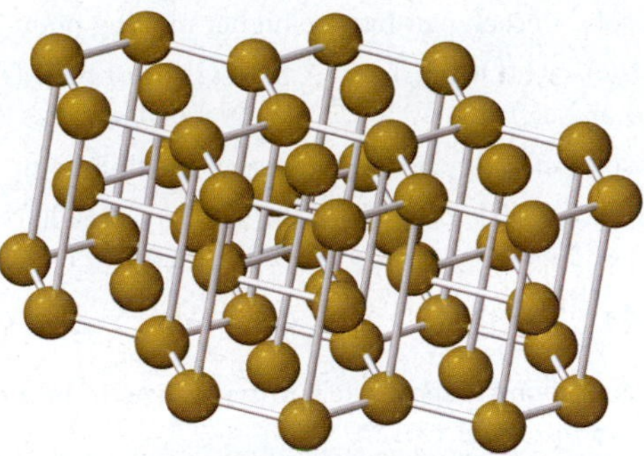

b. Ruthenium

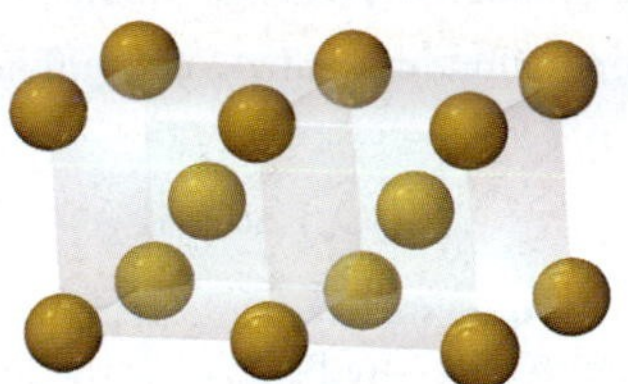

c. Chromium

29. Platinum crystallizes with the face-centered cubic unit cell. The radius of a platinum atom is 139 pm. Calculate the edge length of the unit cell and the density of platinum in g/cm^3.

30. Molybdenum crystallizes with the body-centered unit cell. The radius of a molybdenum atom is 136 pm. Calculate the edge length of the unit cell and the density of molybdenum.

31. A body-centered cubic unit cell has a volume of 2.62×10^{-23} cm^3. Find the radius of the atom in pm.

32. An atom has a radius of 142 pm and crystallizes in the face-centered cubic unit cell. What is the volume of the unit cell in cm^3?

33. Rhodium has a density of 12.41 g/cm^3 and crystallizes with the face-centered cubic unit cell. Calculate the radius of a rhodium atom.

34. Barium has a density of 3.59 g/cm^3 and crystallizes with the body-centered cubic unit cell. Calculate the radius of a barium atom.

35. Polonium crystallizes with a simple cubic structure. It has a density of 9.3 g/cm^3, a radius of 167 pm, and a molar mass of 209 g/mol. Use these data to estimate Avogadro's number (the number of atoms in one mole).

36. Palladium crystallizes with a face-centered cubic structure. It has a density of 12.0 g/cm^3, a radius of 138 pm, and a molar mass of 106.42 g/mol. Use these data to estimate Avogadro's number.

Types of Crystalline Solids

37. Identify each solid as molecular, ionic, or atomic.
    a. $Ar(s)$ **b.** $H_2O(s)$
    c. $K_2O(s)$ **d.** $Fe(s)$

38. Identify each solid as molecular, ionic, or atomic.
    a. $CaCl_2(s)$ **b.** $CO_2(s)$
    c. $Ni(s)$ **d.** $I_2(s)$

39. Which solid has the highest melting point? Why?
    $Ar(s)$, $CCl_4(s)$, $LiCl(s)$, $CH_3OH(s)$

40. Which solid has the highest melting point? Why?
    $C(s, \text{diamond})$, $Kr(s)$, $NaCl(s)$, $H_2O(s)$

41. Which solid in each pair has the higher melting point and why?
 a. $TiO_2(s)$ or HOOH(s)
 b. $CCl_4(s)$ or $SiCl_4(s)$
 c. Kr(s) or Xe(s)
 d. NaCl(s) or CaO(s)

42. Which solid in each pair has the higher melting point and why?
 a. Fe(s) or $CCl_4(s)$
 b. KCl(s) or HCl(s)
 c. Ti(s) or Ne(s)
 d. $H_2O(s)$ or $H_2S(s)$

Ionic Solids

43. Which compound is most likely to crystallize in the zinc blende structure shown in Figure 13.17?
 CuI(s) (Cu^+ radius = 74 pm; I^- radius = 216 pm)
 KCl(s) (K^+ radius = 133 pm; Cl^- radius = 181 pm)
 $MgCl_2(s)$ (Mg^{2+} radius = 65 pm; Cl^- radius = 181 pm)

44. One of these compounds crystallizes in the fluorite structure. Which one and why?

Rb_2O, CaS, BaF_2

45. An oxide of titanium crystallizes with the unit cell shown here (titanium = gray; oxygen = red). What is the formula of the oxide?

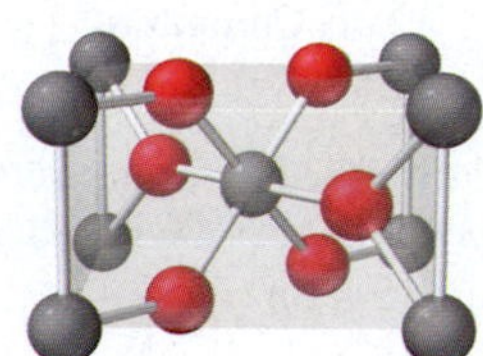

46. An oxide of rhenium crystallizes with the unit cell shown here (rhenium = gray; oxygen = red). What is the formula of the oxide?

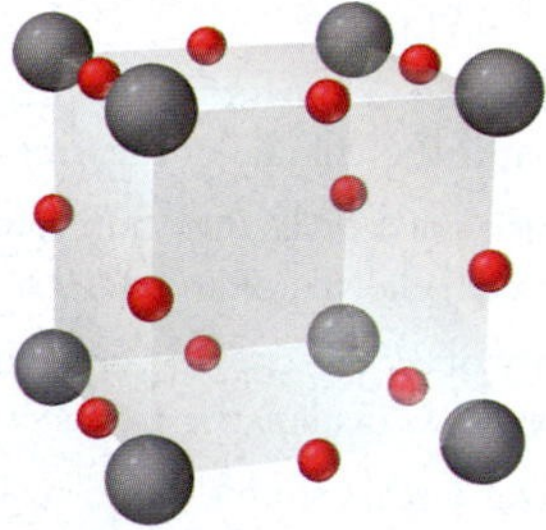

47. The unit cells for cesium chloride and barium chloride are shown here. Show that the ratio of cations to anions in each unit cell corresponds to the ratio of cations to anions in the formula of each compound.

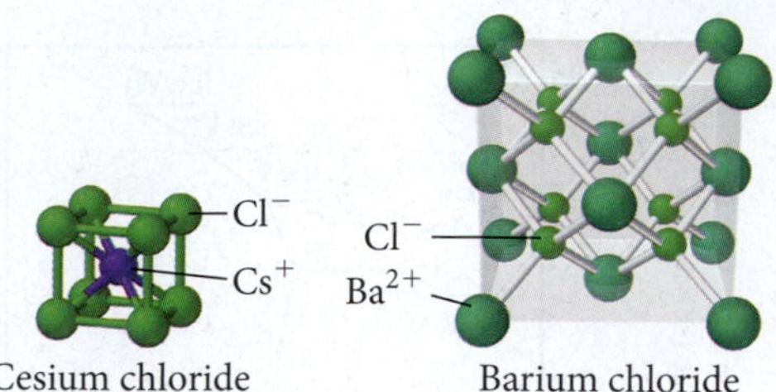

Cesium chloride Barium chloride

48. The unit cells for lithium oxide and silver iodide are shown here. Show that the ratio of cations to anions in each unit cell corresponds to the ratio of cations to anions in the formula of each compound.

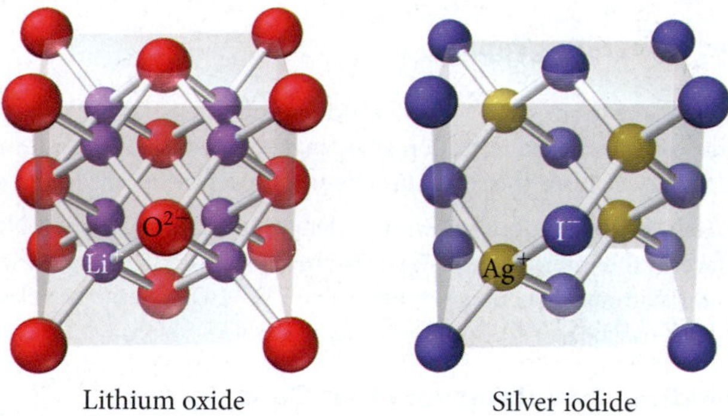

Lithium oxide Silver iodide

Network Covalent Atomic Solids

49. Explain why the graphite structure of carbon allows graphite to be used as a lubricant, but the diamond structure of carbon does not.

50. Explain why graphite can conduct electricity, but diamond does not.

CUMULATIVE PROBLEMS

51. Examine the phase diagram for iodine shown in Figure 13.3(a). What state transitions occur as we uniformly increase the pressure on a gaseous sample of iodine from 0.010 atm at 185 °C to 100 atm at 185 °C? Make a graph, analogous to the heating curve for water shown in Figure 12.33. Plot pressure versus time during the pressure increase.

52. Carbon tetrachloride displays a triple point at 249.0 K and a melting point (at 1 atm) of 250.3 K. Which state of carbon tetrachloride is more dense, the solid or the liquid? Explain.

53. Silver iodide crystallizes in the zinc blende structure shown in Figure 13.17. The separation between nearest neighbor cations and anions is approximately 325 pm, and the melting point is 558 °C. Cesium chloride, by contrast, crystallizes in the cesium chloride structure shown in Figure 13.15. Even though the separation between nearest neighbor cations and anions is greater (348 pm), the melting point of cesium chloride is higher (645 °C). Explain.

54. Copper iodide crystallizes in the zinc blende structure shown in Figure 13.17. The separation between nearest neighbor cations and anions is approximately 311 pm, and the melting point is 606 °C. Potassium chloride, by contrast, crystallizes in the rock salt structure shown in Figure 13.16. Even though the separation between nearest neighbor cations and anions is greater (319 pm), the melting point of potassium chloride is higher (776 °C). Explain.

55. Consider the face-centered cubic structure shown here.

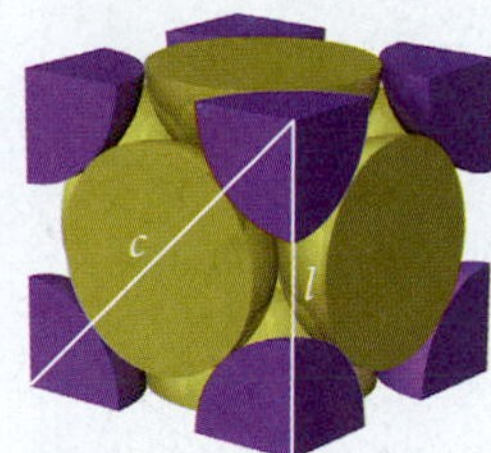

a. What is the length of the line (labeled c) that runs diagonally across one of the faces of the cube in terms of r (the atomic radius)?

b. Use the answer to part (a) and the Pythagorean theorem to derive the expression for the edge length (l) in terms of r.

56. Consider the body-centered cubic structure shown here.

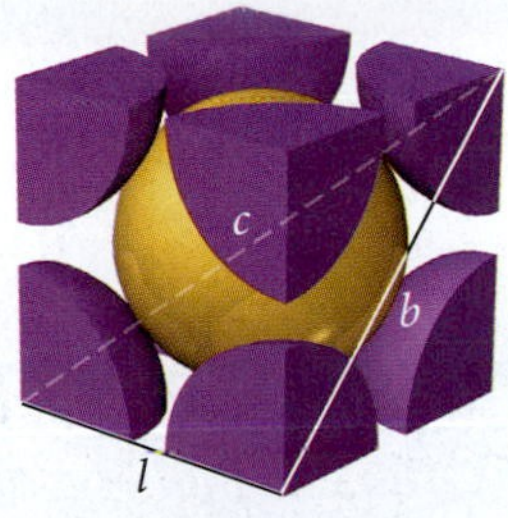

a. What is the length of the line (labeled c) that runs from one corner of the cube diagonally through the center of the cube to the other corner in terms of r (the atomic radius)?

b. Use the Pythagorean theorem to derive an expression for the length of the line (labeled b) that runs diagonally across one of the faces of the cube in terms of the edge length (l).

c. Use the answer to parts (a) and (b) along with the Pythagorean theorem to derive the expression for the edge length (l) in terms of r.

57. The unit cell in a crystal of diamond belongs to a crystal system different from any we discussed in this chapter. The volume of a unit cell of diamond is 0.0454 nm^3, and the density of diamond is 3.52 g/cm^3. Find the number of carbon atoms in a unit cell of diamond.

58. The density of an unknown metal is 12.3 g/cm^3, and its atomic radius is 0.134 nm. It has a face-centered cubic lattice. Find the atomic mass of this metal.

59. Based on the phase diagram of CO_2 shown in Figure 13.3(b), describe the state changes that occur when the temperature of CO_2 is increased from 190 K to 350 K at a constant pressure of (a) 1 atm, (b) 5.1 atm, (c) 10 atm, (d) 100 atm.

60. Consider a planet where the pressure of the atmosphere at sea level is 2500 mmHg. Does water behave in a way that can sustain life on the planet?

61. An unknown metal is found to have a density of 7.8748 g/cm^3 and to crystallize in a body-centered cubic lattice. The edge of the unit cell is 0.28664 nm. Calculate the atomic mass of the metal.

62. When spheres of radius r are packed in a body-centered cubic arrangement, they occupy 68.0% of the available volume. Use the fraction of occupied volume to calculate the value of a, the length of the edge of the cube, in terms of r.

CHALLENGE PROBLEMS

63. Potassium chloride crystallizes in the rock salt structure. Estimate the density of potassium chloride using the ionic radii provided in Chapter 4.

64. Calculate the fraction of empty space in cubic closest packing to five significant figures.

65. A tetrahedral site in a closest-packed lattice is formed by four spheres at the corners of a regular tetrahedron. This is equivalent to placing the spheres at alternate corners of a cube. In such a closest-packed arrangement the spheres are in contact, and if the spheres have a radius r, the diagonal of the face of the cube is $2r$. The tetrahedral hole is inside the middle of the cube. Find the length of the body diagonal of this cube and then find the radius of the tetrahedral hole.

CONCEPTUAL PROBLEMS

66. The density of a substance is greater in its solid state than in its liquid state. If the triple point in the phase diagram of the substance is below 1.0 atm, which must be at a lower temperature, the triple point or the normal melting point?

67. A substance has a triple point at a temperature of 17 °C and a pressure of 3.2 atmospheres. In what states can the substance exist on the surface of Earth at sea level (open to the atmosphere)?

ANSWERS TO CONCEPTUAL CONNECTIONS

Cc 13.1 **(b)** The solid will sublime into a gas. Since the pressure is below the triple point, the liquid state is not stable.

Cc 13.2 **(a)** MgO is an ionic solid, which means that the forces holding the solid together are strong ionic bonds. The other two substances are $I_2(s)$, a molecular solid, and $Kr(s)$, a nonbonding atomic solid. These two solids are held together by dispersion forces and will therefore have much lower melting points.

Cc 13.3 **(c)** The zinc blende structure occurs in ionic compounds with a one-to-one cation-to-anion ratio in which the cation is much smaller than the anion.

Cc 13.4 An increase in pressure favors the denser phase, in this case diamond.

"One molecule of nonsaline substance (held in the solvent) dissolved in 100 molecules of any volatile liquid decreases the vapor pressure of this liquid by a nearly constant fraction, nearly 0.0105."

—François-Marie Raoult (1830–1901)

CHAPTER 14

The wood frog protects its cells by flooding them with glucose, which acts as an antifreeze.

Solutions

14.1 Antifreeze in Frogs 509
14.2 Types of Solutions and Solubility 510
14.3 Energetics of Solution Formation 514
14.4 Solution Equilibrium and Factors Affecting Solubility 518
14.5 Expressing Solution Concentration 522
14.6 Colligative Properties: Vapor Pressure Lowering, Freezing Point Depression, Boiling Point Elevation, and Osmotic Pressure 528
14.7 Colligative Properties of Strong Electrolyte Solutions 539
Key Learning Outcomes 544

WE LEARNED IN Chapter 1 that matter often exists in the form of a mixture, two or more different types of particles mixed together. In this chapter, we focus on homogeneous mixtures, also known as solutions. Solutions are mixtures in which atoms, ions, and molecules intermingle on the molecular and atomic scale. Common examples of solutions include ocean water, gasoline, and air. In this chapter we answer the question: Why do solutions form? We also discuss how the properties of solutions differ from the properties of the pure substances that compose them. As you read this chapter, keep in mind the large number of solutions that surround you at every moment, including those that exist within your own body.

14.1 Antifreeze in Frogs

The wood frog *(Rana sylvatica)* looks like most other frogs. It is a few inches long and has characteristic greenish-brown skin. At first glance, the wood frog seems quite unremarkable. But a wood frog survives cold winters in a remarkable way—it partially freezes. In its partially frozen state, a wood frog has no heartbeat, no blood circulation, no breath, and no brain activity. Within 1–2 hours of thawing, however, these vital functions restart, and the frog hops off to find food. How does the wood frog do this?

Most cold-blooded animals cannot survive freezing temperatures because the water within their cells freezes. As we learned in Section 12.8, when water freezes, it expands. If the water in cells freezes and expands, it irreversibly damages cells. To prevent this, the wood frog, before hibernating, secretes a large

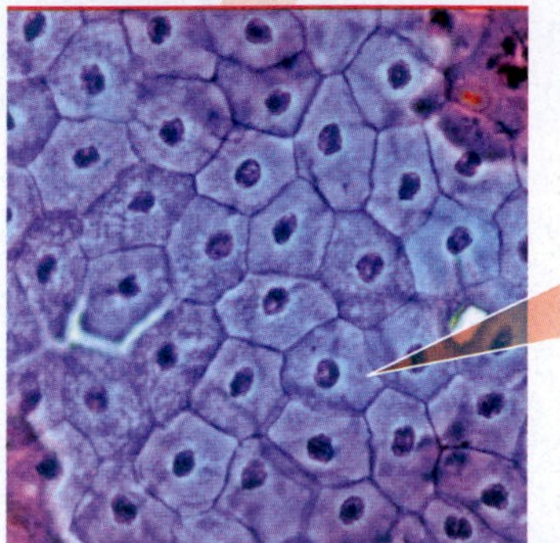

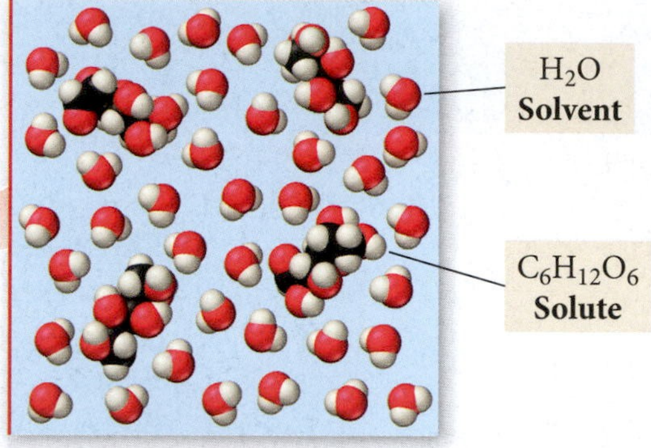

▼ **FIGURE 14.1 A Glucose Solution in a Wood Frog's Cells** In a glucose solution, glucose ($C_6H_{12}O_6$) is the solute and water is the solvent.

amount of glucose into its bloodstream. The glucose then incorporates into the frog's cells resulting in cells that are filled with a concentrated glucose solution. As we shall see in this chapter, concentrated solutions have a lower freezing point than the corresponding pure liquid. When the temperature drops below freezing, extracellular body fluids, such as those in the frog's abdominal cavity, freeze solid. Fluids within the frog's cells, however, remain liquid because of the high glucose concentration. In other words, the concentrated glucose solution within the frog's cells acts as antifreeze, preventing the water within the cells from freezing and allowing the frog to survive.

The concentrated glucose and water mixture within the wood frog's cells is an example of a **solution**, a homogeneous mixture of two or more substances or components (Figure 14.1 ◀). The majority component in a solution is typically called the **solvent**, and the minority component is the **solute**. In a glucose solution, water is the solvent and glucose is the solute. Solutions form in part because of the intermolecular forces we discussed in Chapter 12. In most solutions, the particles of the solute interact with the particles of the solvent through intermolecular forces.

14.2 Types of Solutions and Solubility

A solution may be composed of a solid and a liquid (such as the salt and water that are the primary components of seawater), but it may also be composed of a gas and a liquid, two different liquids, or other combinations (see Table 14.1). In **aqueous solutions**, water is the solvent, and a solid, liquid, or gas is the solute. For example, sugar water and salt water are both aqueous solutions. Similarly, ethyl alcohol—the alcohol in alcoholic beverages—readily mixes with water to form a solution, and carbon dioxide dissolves in water to form the aqueous solution that we know as club soda.

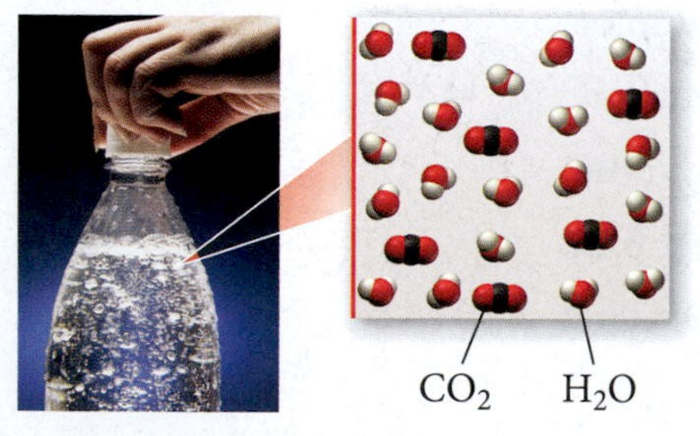

▲ Club soda is a solution of carbon dioxide and water.

You probably know from experience that a particular solvent, such as water, does not dissolve all possible solutes. For example, you cannot clean your greasy hands with just water because the water does not dissolve the grease. However, another solvent, such as paint thinner, can easily dissolve the grease. The grease is *insoluble* in water but *soluble* in the paint thinner. The **solubility** of a substance is the amount of the substance that will dissolve in a given amount of solvent. The solubility of sodium chloride in water at 25 °C is 36 g NaCl per 100 g water, while the solubility of grease in water is nearly zero. The solubility of one substance in another depends both on nature's tendency toward mixing and on the types of intermolecular forces that we discussed in Chapter 12.

The general solubilities of a number of ionic compounds are described by the solubility rules in Section 4.5.

TABLE 14.1 Common Types of Solutions

Solution Phase	Solute Phase	Solvent Phase	Example
Gaseous solution	Gas	Gas	Air (mainly oxygen and nitrogen)
Liquid solution	Gas	Liquid	Club soda (CO_2 and water)
	Liquid	Liquid	Vodka (ethanol and water)
	Solid	Liquid	Seawater (salt and water)
Solid solution	Solid	Solid	Brass (copper and zinc) and other alloys

Nature's Tendency toward Mixing: Entropy

So far in this book, we have seen that many physical systems tend toward lower *potential energy*. For example, two particles with opposite charges (such as a proton and an electron or a cation and an anion) move toward each other because their potential energy decreases as their separation decreases according to Coulomb's law (see Section 4.3). The formation of a solution, however, *does not necessarily* lower the potential energy of its constituent particles. The clearest example of this phenomenon is the formation of a homogeneous mixture (a *solution*) of two ideal gases. Suppose that we enclose neon and argon in a container with a removable barrier between them, as shown in Figure 14.2(a) ▶. As soon as we remove the barrier, the neon and argon mix together to form a solution, as shown in Figure 14.2(b) ▶. *Why*?

At low pressures and moderate temperatures, both neon and argon behave as ideal gases—they do not interact with each other in any way (that is, there are no significant intermolecular forces between their constituent particles). When the barrier is removed, the two gases mix, but their potential energy remains unchanged. In other words, *we cannot think of the mixing of two ideal gases as lowering their potential energy*. Rather, the tendency to mix is related to a concept called *entropy*.

Entropy is a measure of *energy randomization* or *energy dispersal* in a system. Recall that a gas at any temperature above 0 K has kinetic energy due to the motion of its atoms. When neon and argon are confined to their individual compartments, their kinetic energies are also confined to those compartments. When the barrier between the compartments is removed, each gas—along with its kinetic energy—becomes *spread out* or *dispersed* over a larger volume. Thus, the mixture of the two gases has greater energy dispersal, or greater *entropy*, than the separated components.

The pervasive tendency for energy to spread out, or disperse, whenever it is not restrained from doing so is the reason that two ideal gases mix. Another common example of the tendency toward energy dispersal is the transfer of thermal energy from hot to cold. If we heat one end of an iron rod, the thermal energy deposited at the end of the rod will spontaneously spread along the entire length of the rod. In similarity to the mixing of two ideal gases—where the kinetic energy of the particles becomes dispersed over a larger volume as the particles become dispersed—the thermal energy in the rod, initially concentrated in relatively fewer particles, becomes dispersed as it is distributed over a larger number of particles. The tendency for energy to disperse is why thermal energy flows from the hot end of the rod to the cold one, and not the other way around. Imagine a metal rod that became spontaneously hotter on one end and ice cold on the other. This does not happen because energy does not spontaneously concentrate itself. In Chapter 19, we will see that the dispersal of energy is actually the fundamental criterion that ultimately determines the spontaneity of any process.

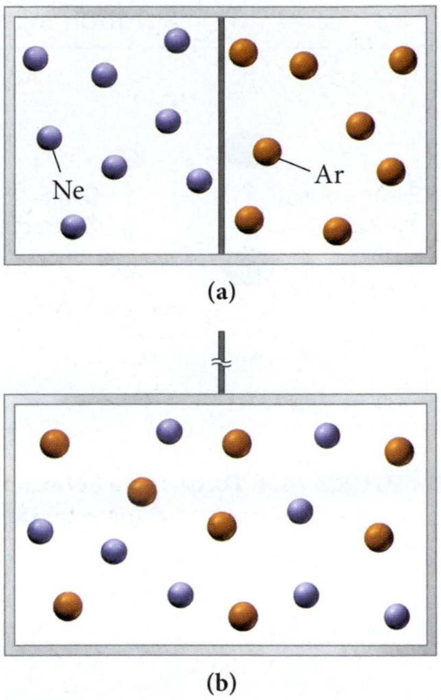

▲ **FIGURE 14.2 Spontaneous Mixing of Two Ideal Gases** **(a)** Neon and argon are separated by a barrier. **(b)** When the barrier is removed, the two gases spontaneously mix to form a uniform solution.

The Effect of Intermolecular Forces

We have just seen that, in the absence of intermolecular forces, two substances spontaneously mix to form a homogeneous solution. We know from Chapter 12, however, that solids and liquids exhibit a number of different types of intermolecular forces, including dispersion forces, dipole–dipole forces, hydrogen bonding, and ion–dipole forces (Figure 14.3)▼. These forces, and the possible localized

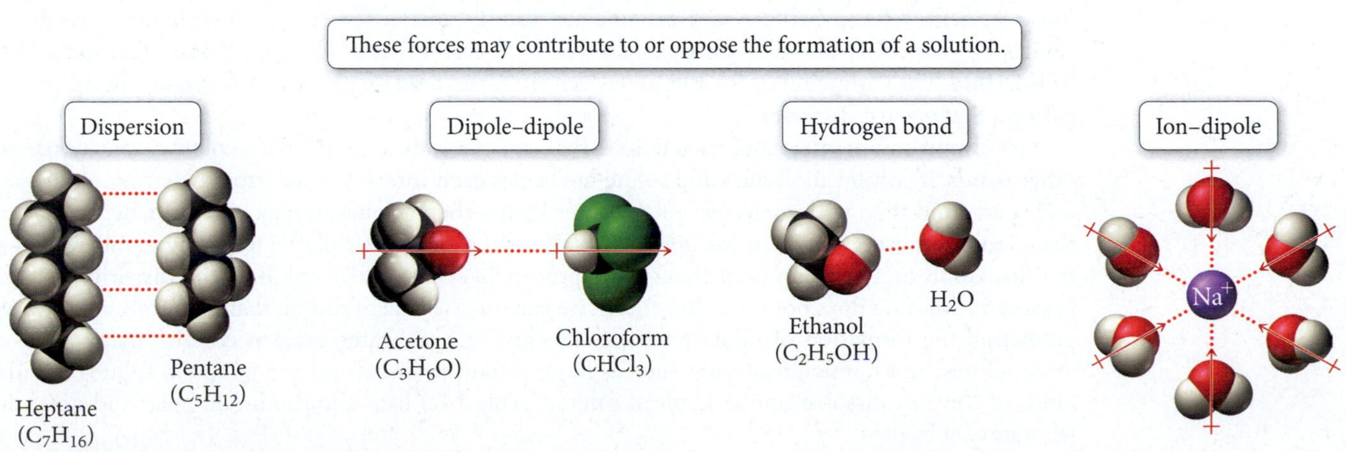

▲ **FIGURE 14.3 Intermolecular Forces Involved in Solutions**

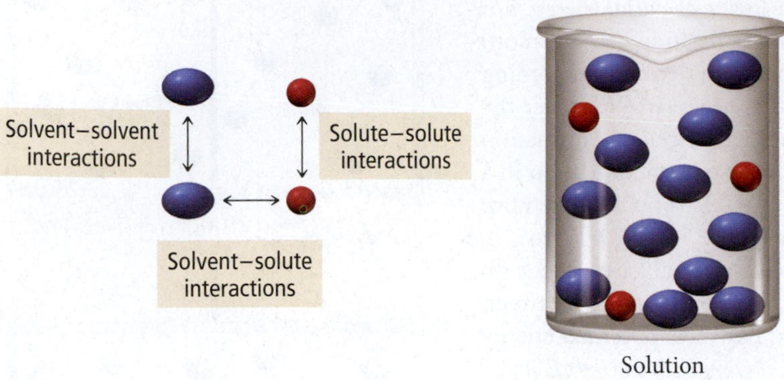

▲ **FIGURE 14.4 Forces in a Solution** The relative strengths of these three interactions determine whether a solution will form.

structures that result from them, may promote the formation of a solution or prevent it, depending on the nature of the forces in the particular combination of solute and solvent.

Intermolecular forces exist between (a) the solvent and solute particles, (b) the solvent particles themselves, and (c) the solute particles themselves, as shown in Figure 14.4 ◀.

Solvent–solute interactions:	The interactions between a solvent particle and a solute particle.
Solvent–solvent interactions:	The interactions between a solvent particle and another solvent particle.
Solute–solute interactions:	The interactions between a solute particle and another solute particle.

TABLE 14.2 Relative Interactions and Solution Formation

Solvent–solute interactions	>	Solvent–solvent and solute–solute interactions	Solution generally forms*
Solvent–solute interactions	=	Solvent–solvent and solute–solute interactions	Solution generally forms*
Solvent–solute interactions	<	Solvent–solvent and solute–solute interactions	Solution may or may not form, depending on relative disparity

*In some cases, especially solutions involving water, solvent structural changes that occur during solvation result in *decreases in entropy*, which may also prevent solution formation.

TABLE 14.3 Common Laboratory Solvents

Common Polar Solvents	Common Nonpolar Solvents
Water (H_2O)	Hexane (C_6H_{14})
Acetone (CH_3COCH_3)	Diethyl ether ($CH_3CH_2OCH_2CH_3$)*
Methanol (CH_3OH)	Toluene (C_7H_8)
Ethanol (CH_3CH_2OH)	Carbon tetrachloride (CCl_4)

*Diethyl ether has a small dipole moment and can be considered intermediate between polar and nonpolar.

As shown in Table 14.2, a solution generally forms if the solvent–solute interactions are comparable to, or stronger than, the solvent–solvent interactions and the solute–solute interactions. For example, consider mixing the hydrocarbons pentane (C_5H_{12}) and heptane (C_7H_{16}). The intermolecular forces present within both pentane and heptane are dispersion forces. Similarly, the intermolecular forces present *between* heptane and pentane are also dispersion forces. All three interactions are of similar magnitude, so the two substances are soluble in each other in all proportions—they are said to be **miscible**. The formation of the solution is driven by the tendency toward mixing, or toward greater entropy, that we just discussed.

If solvent–solute interactions are weaker than solvent–solvent and solute–solute interactions—in other words, if solvent molecules and solute molecules each interact more strongly with molecules of their own kind than with molecules of the other kind—then a solution may still form, depending on the relative disparities between the interactions. If the disparity is small, the tendency to mix results in the formation of a solution even though the process is energetically uphill. If the disparity is large, however, a solution does not form. In general, we can use the rule of thumb that *like dissolves like* when predicting the formation of solutions. Polar solvents, such as water, tend to dissolve many polar or ionic solutes, and nonpolar solvents, such as hexane, tend to dissolve many nonpolar solutes. Similar kinds of solvents dissolve similar kinds of solutes. Table 14.3 lists some common polar and nonpolar laboratory solvents.

EXAMPLE 14.1
Solubility

Vitamins are categorized as either fat soluble or water soluble. Water-soluble vitamins dissolve in body fluids and are easily eliminated in the urine, so there is little danger of overconsumption. Fat-soluble vitamins, on the other hand, can accumulate in the body's fatty deposits. Overconsumption of a fat-soluble vitamin can be dangerous to your health. Examine the structure of each vitamin shown here and classify it as either fat soluble or water soluble.

(a) Vitamin C

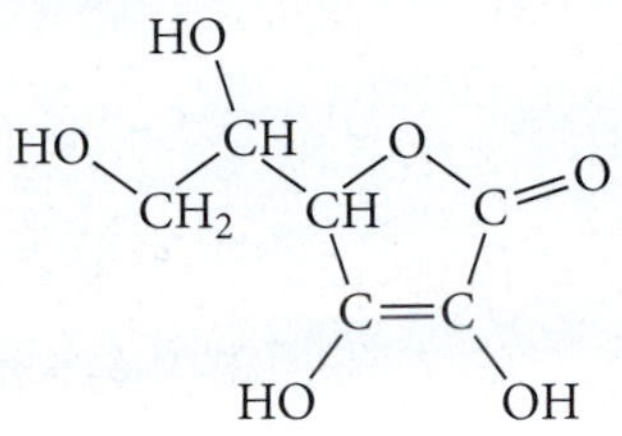

(b) Vitamin K_3

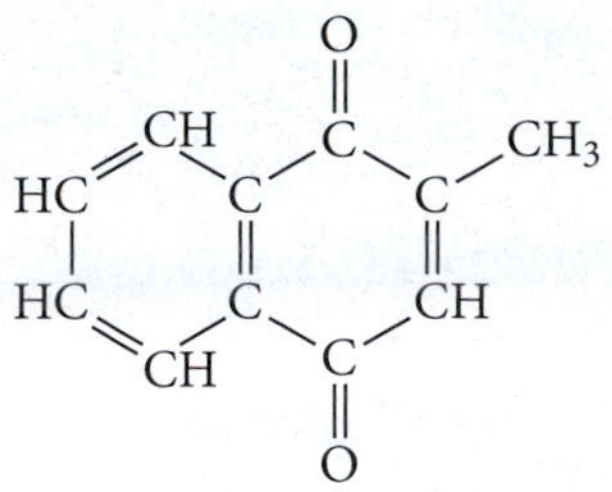

(c) Vitamin A

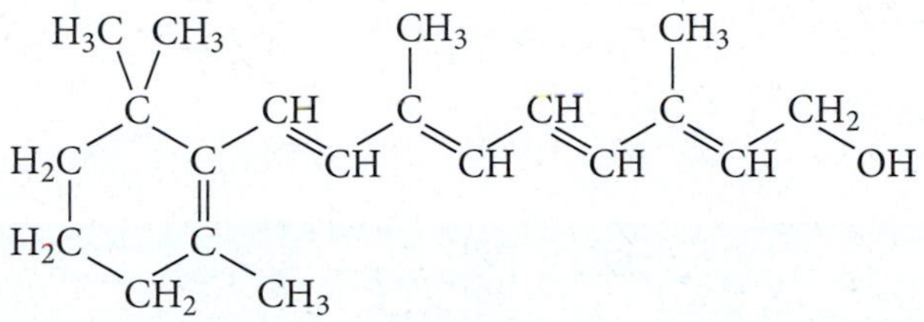

(d) Vitamin B_5

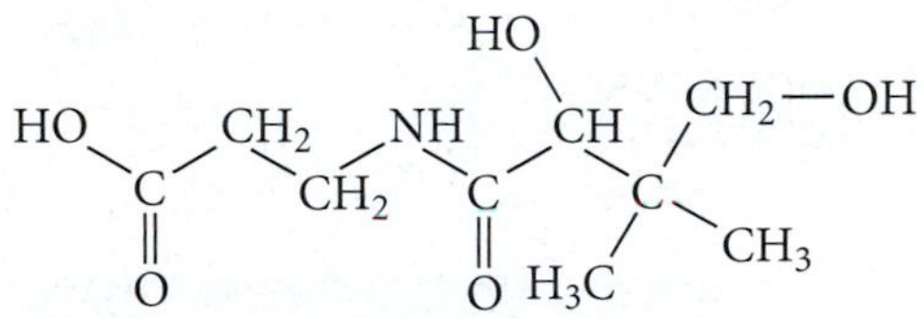

SOLUTION

(a) The four —OH bonds in vitamin C make it highly polar and allow it to hydrogen bond with water. Vitamin C is water soluble.

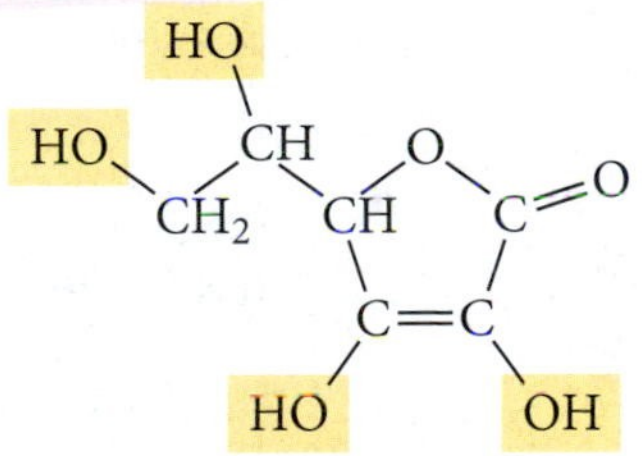

(b) The C—C bonds in vitamin K_3 are nonpolar and the C—H bonds are nearly so. The C=O bonds are polar, but the bond dipoles oppose and largely cancel each other, so the molecule is dominated by the nonpolar bonds. Vitamin K_3 is fat soluble.

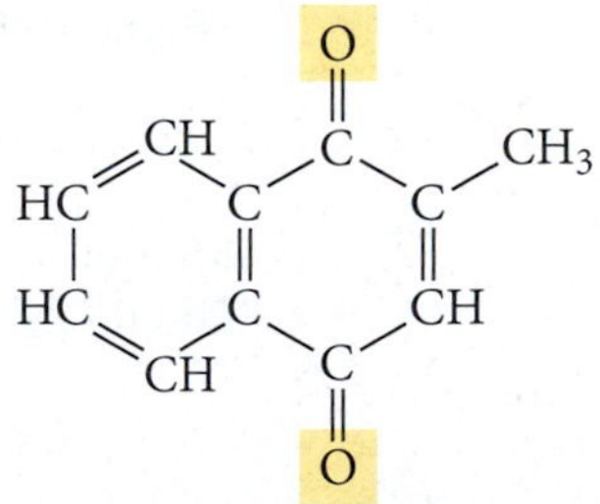

(c) The C—C bonds in vitamin A are nonpolar, and the C—H bonds are nearly so. The one polar —OH bond may increase vitamin A's water solubility slightly, but overall it is nonpolar and therefore fat soluble.

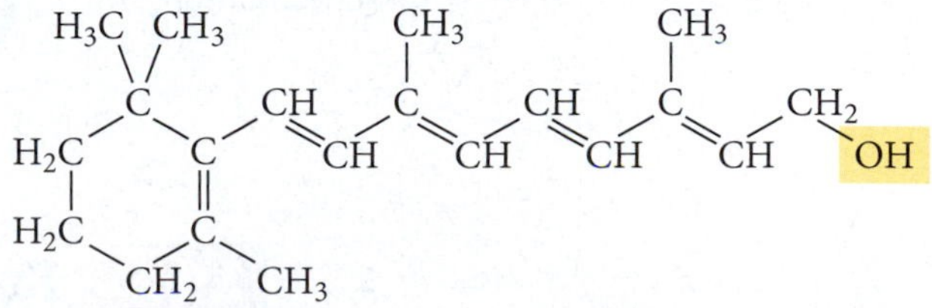

—*Continued on the next page*

Continued from the previous page—

(d) The three —OH bonds and one —NH bond in vitamin B_5 make it highly polar and allow it to hydrogen bond with water. Vitamin B_5 is water soluble.

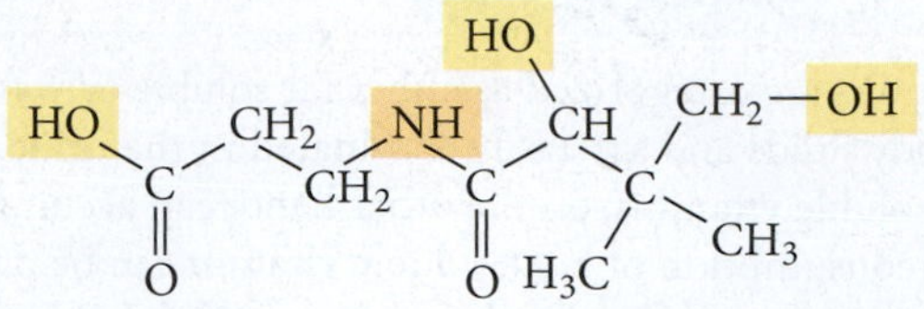

FOR PRACTICE 14.1

Determine whether each compound is soluble in hexane.

(a) water (H_2O)
(b) propane ($CH_3CH_2CH_3$)
(c) ammonia (NH_3)
(d) hydrogen chloride (HCl)

14.1 Cc Conceptual Connection

Solubility

Examine the table listing the solubilities of several alcohols in water and in hexane. Explain the observed trend in terms of intermolecular forces.

Alcohol	Space-Filling Model	Solubility in H_2O (mol alcohol/100 g H_2O)	Solubility in Hexane (C_6H_{14}) (mol alcohol/100 g C_6H_{14})
Methanol (CH_3OH)		Miscible	0.12
Ethanol (CH_3CH_2OH)		Miscible	Miscible
Propanol ($CH_3CH_2CH_2OH$)		Miscible	Miscible
Butanol ($CH_3CH_2CH_2CH_2OH$)		0.11	Miscible
Pentanol ($CH_3CH_2CH_2CH_2CH_2OH$)		0.030	Miscible

14.3 Energetics of Solution Formation

In Chapter 10, we examined the energy changes associated with chemical reactions. Similar energy changes can occur when a solution forms, depending on the magnitude of the interactions between the solute and solvent particles. For example, when we dissolve sodium hydroxide in water, heat is evolved—the solution process is *exothermic*. In contrast, when we dissolve ammonium nitrate (NH_4NO_3) in water, heat is absorbed—the solution process is *endothermic*. Other solutions, such as sodium chloride in water, barely absorb or evolve heat upon formation. What causes these different behaviors?

Energy Changes during Solution Formation

We can understand the energy changes associated with solution formation by envisioning the process as occurring in three steps, each with an associated change in enthalpy:

1. Separating the solute into its constituent particles.

   This step is always endothermic (positive ΔH) because energy is required to overcome the forces that hold the solute particles together.

2. Separating the solvent particles from each other to make room for the solute particles.

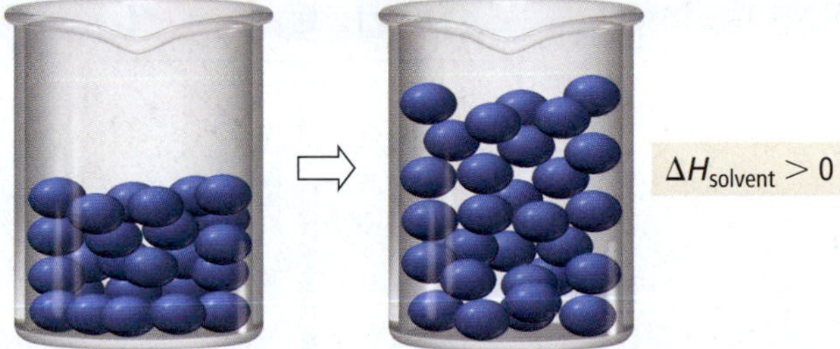

   This step is also endothermic because energy is required to overcome the intermolecular forces among the solvent particles.

3. Mixing the solute particles with the solvent particles.

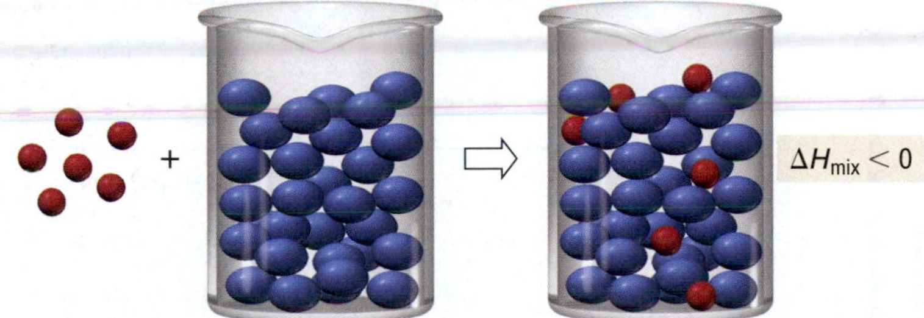

   This step is exothermic because energy is released as the solute particles interact (through intermolecular forces) with the solvent particles.

According to Hess's law, the overall enthalpy change upon solution formation, called the **enthalpy of solution (ΔH_{soln})**, is the sum of the changes in enthalpy for each step:

$$\Delta H_{soln} = \underset{\text{endothermic }(+)}{\Delta H_{solute}} + \underset{\text{endothermic }(+)}{\Delta H_{solvent}} + \underset{\text{exothermic }(-)}{\Delta H_{mix}}$$

Since the first two terms are endothermic (positive ΔH) and the third term is exothermic (negative ΔH), the overall sign of ΔH_{soln} depends on the magnitudes of the individual terms, as shown in Figure 14.5 ▶ (on the next page).

1. *If the sum of the endothermic terms is approximately equal in magnitude to the exothermic term, then ΔH_{soln} is about zero.* The increasing entropy upon mixing drives the formation of a solution, while the overall energy of the system remains nearly constant.
2. *If the sum of the endothermic terms is smaller in magnitude than the exothermic term, then ΔH_{soln} is negative and the solution process is exothermic.* In this case, both the tendency toward lower energy and the tendency toward greater entropy drive the formation of a solution.
3. *If the sum of the endothermic terms is greater in magnitude than the exothermic term, then ΔH_{soln} is positive and the solution process is endothermic.* In this case, as long as ΔH_{soln} is not too large, the tendency toward greater entropy still drives the formation of a solution. If, on the other hand, ΔH_{soln} is too large, a solution does not form.

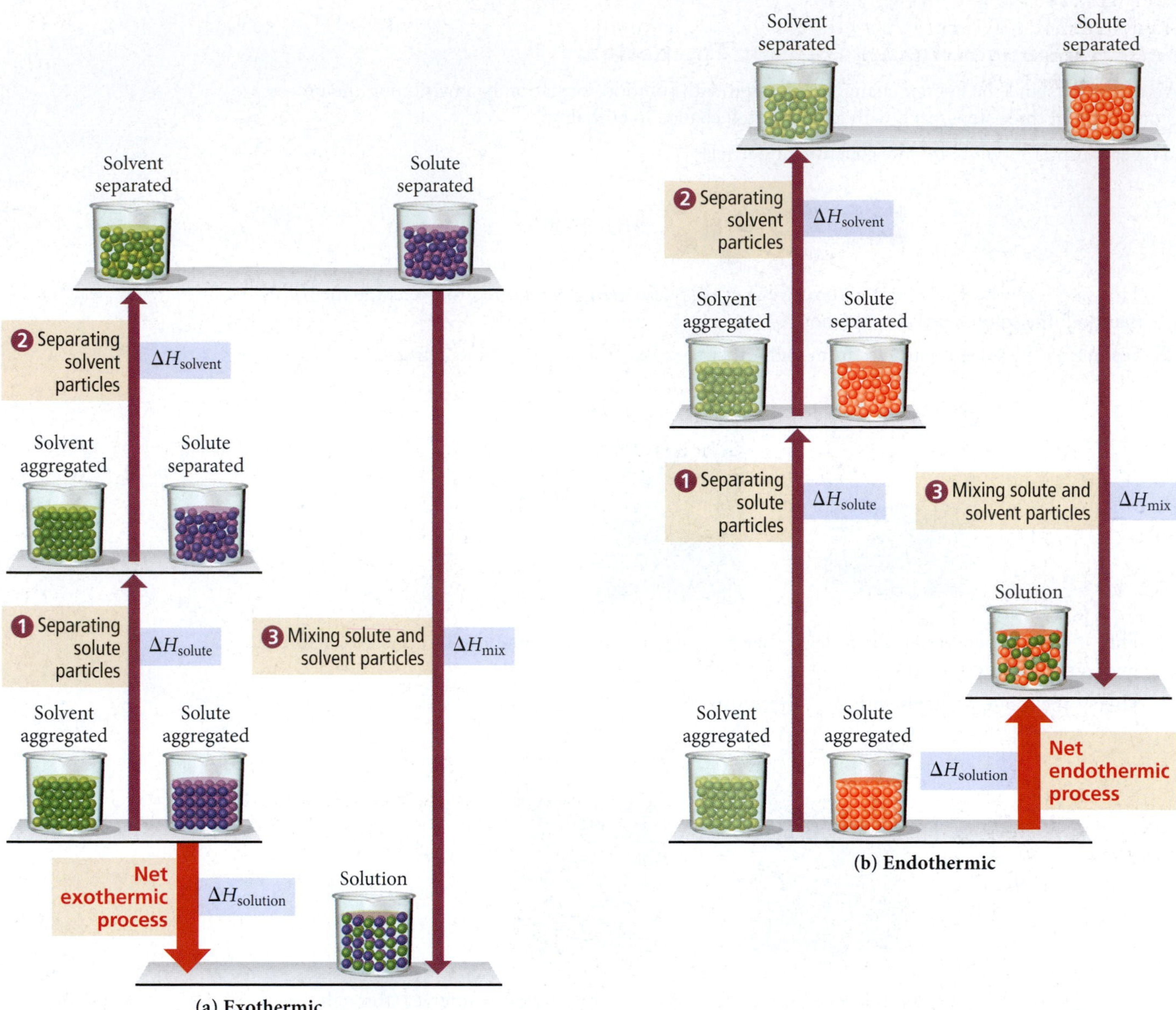

▲ **FIGURE 14.5 Energetics of the Solution Process** **(a)** When ΔH_{mix} is greater in magnitude than the sum of ΔH_{solute} and $\Delta H_{solvent}$, the heat of solution is negative (exothermic). **(b)** When ΔH_{mix} is smaller in magnitude than the sum of ΔH_{solute} and $\Delta H_{solvent}$, the heat of solution is positive (endothermic).

Aqueous Solutions and Heats of Hydration

Many common solutions, such as seawater, contain an ionic compound dissolved in water. For these aqueous solutions, we combine $\Delta H_{solvent}$ and ΔH_{mix} into a single term called the **heat of hydration** **($\Delta H_{hydration}$)** (Figure 14.6 ▶). The heat of hydration is the enthalpy change that occurs when 1 mol of the gaseous solute ions is dissolved in water. Because the ion–dipole interactions that occur between a dissolved ion and the surrounding water molecules (Figure 14.7 ▶) are much stronger than the hydrogen bonds in water, $\Delta H_{hydration}$ is always largely negative (exothermic) for ionic compounds. Using the heat of hydration, we can write the enthalpy of solution as a sum of just two terms, one endothermic and one exothermic:

$$\Delta H_{soln} = \Delta H_{solute} + \underbrace{\Delta H_{solvent} + \Delta H_{mix}}$$

$$\Delta H_{soln} = \underset{\substack{\text{endothermic} \\ \text{(positive)}}}{\Delta H_{solute}} + \underset{\substack{\text{exothermic} \\ \text{(negative)}}}{\Delta H_{hydration}}$$

For ionic compounds, ΔH_{solute}, the energy required to separate the solute into its constituent particles, is the negative of the solute's lattice energy ($\Delta H_{solute} = -\Delta H_{lattice}$), discussed in Section 10.11. For ionic aqueous solutions, then, the overall enthalpy of solution depends on the relative magnitudes of ΔH_{solute} and $\Delta H_{hydration}$, with three possible scenarios (in each case we refer to the *magnitude or absolute value* of ΔH):

1. $|\Delta H_{solute}| < |\Delta H_{hydration}|$ The amount of energy required to separate the solute into its constituent ions is less than the energy given off when the ions are hydrated. ΔH_{soln} is therefore negative, and the solution process is exothermic. Solutes with negative enthalpies of solution include lithium bromide and potassium hydroxide. When these solutes dissolve in water, the resulting solutions feel warm to the touch.

$$LiBr(s) \xrightarrow{H_2O} Li^+(aq) + Br^-(aq) \qquad \Delta H_{soln} = -48.78 \text{ kJ/mol}$$

$$KOH(s) \xrightarrow{H_2O} K^+(aq) + OH^-(aq) \qquad \Delta H_{soln} = -57.56 \text{ kJ/mol}$$

2. $|\Delta H_{solute}| > |\Delta H_{hydration}|$ The amount of energy required to separate the solute into its constituent ions is greater than the energy given off when the ions are hydrated. ΔH_{soln} is therefore positive, and the solution process is endothermic (if a solution forms at all). Solutes that form aqueous solutions with positive enthalpies of solution include ammonium nitrate and silver nitrate. When these solutes dissolve in water, the resulting solutions feel cool to the touch.

$$NH_4NO_3(s) \xrightarrow{H_2O} NH_4^+(aq) + NO_3^-(aq) \qquad \Delta H_{soln} = +25.67 \text{ kJ/mol}$$

$$AgNO_3(s) \xrightarrow{H_2O} Ag^+(aq) + NO_3^-(aq) \qquad \Delta H_{soln} = +36.91 \text{ kJ/mol}$$

3. $|\Delta H_{solute}| \approx |\Delta H_{hydration}|$ The amount of energy required to separate the solute into its constituent ions is roughly equal to the energy given off when the ions are hydrated. ΔH_{soln} is therefore approximately zero, and the solution process is neither appreciably exothermic nor appreciably endothermic. Solutes with enthalpies of solution near zero include sodium chloride and sodium fluoride. When these solutes dissolve in water, the resulting solutions do not undergo a noticeable change in temperature.

$$NaCl(s) \xrightarrow{H_2O} Na^+(aq) + Cl^-(aq) \qquad \Delta H_{soln} = +3.88 \text{ kJ/mol}$$

$$NaF(s) \xrightarrow{H_2O} Na^+(aq) + F^-(aq) \qquad \Delta H_{soln} = +0.91 \text{ kJ/mol}$$

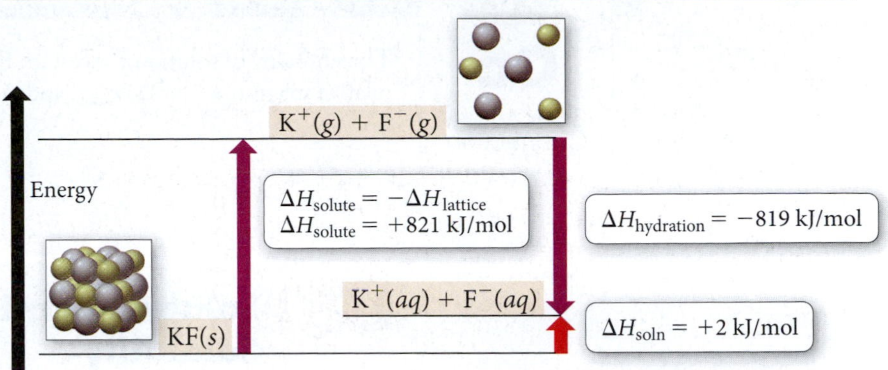

▲ **FIGURE 14.6 Heat of Hydration and Heat of Solution** The heat of hydration is the heat emitted when 1 mol of gaseous solute ions is dissolved in water. The sum of the negative of the lattice energy (which is ΔH_{solute}) and the heat of hydration is the heat of solution.

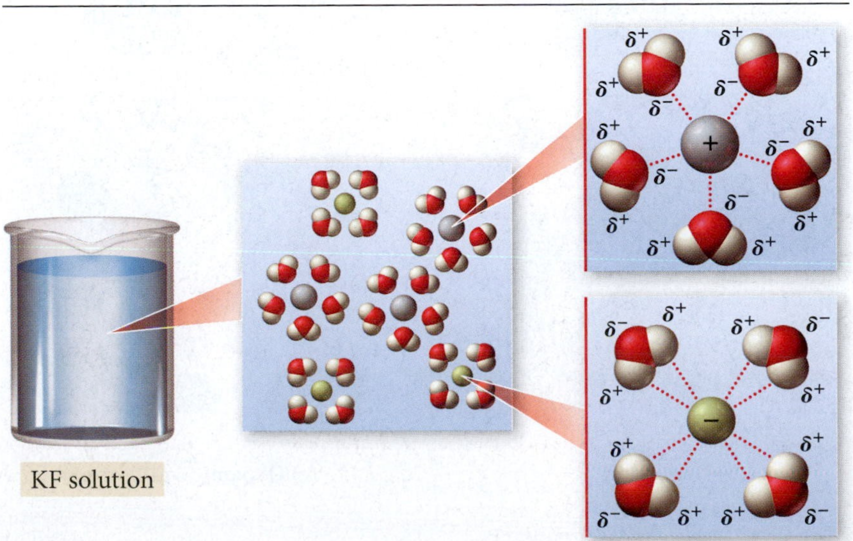

◀ **FIGURE 14.7 Ion–Dipole Interactions** Ion–dipole interactions such as those between potassium ions, fluoride ions, and water molecules cause the heat of hydration to be largely negative (exothermic) for ionic compounds.

14.2 Cc Conceptual Connection

Energetics of Aqueous Solution Formation

The enthalpy of solution for cesium fluoride is −36.8 kJ/mol. What can you conclude about the relative magnitudes of ΔH_{solute} and $\Delta H_{\text{hydration}}$?

14.4 Solution Equilibrium and Factors Affecting Solubility

The dissolution of a solute in a solvent is an equilibrium process similar to the equilibrium process associated with a state change (discussed in Chapter 12). Imagine, from a particulate viewpoint, the dissolution of a solid solute such as sodium chloride in a liquid solvent such as water (Figure 14.8 ▼). Initially, water molecules rapidly solvate sodium cations and chloride anions, resulting in a noticeable decrease in the amount of solid sodium chloride in the water. Over time, however, the concentration of dissolved sodium chloride in the solution increases. This dissolved sodium chloride begins to recrystallize as solid sodium chloride. Initially, the rate of dissolution far exceeds the rate of recrystallization, but

▼ **FIGURE 14.8 Dissolution of NaCl** **(a)** Initial addition of NaCl, **(b)** Dissolution, **(c)** Dynamic equilibrium.

Solution Equilibrium

$NaCl(s)$

When sodium chloride is first added to water, sodium and chloride ions begin to dissolve into the water.

$NaCl(s) \longrightarrow Na^+(aq) + Cl^-(aq)$

As the solution becomes more concentrated, some of the sodium and chloride ions can begin to recrystallize as solid sodium chloride.

$NaCl(s) \rightleftharpoons Na^+(aq) + Cl^-(aq)$

When the rate of dissolution equals the rate of recrystallization, dynamic equilibrium has been reached.

(a) Initial

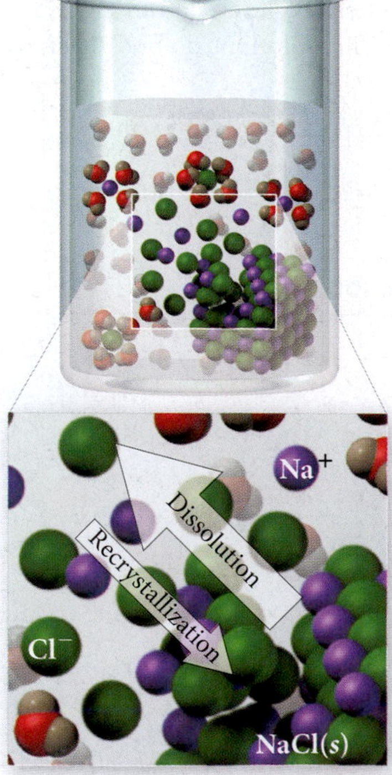

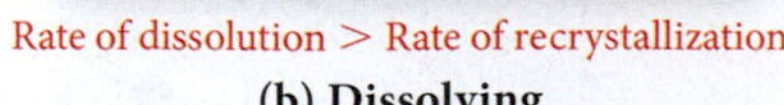

Rate of dissolution > Rate of recrystallization

(b) Dissolving

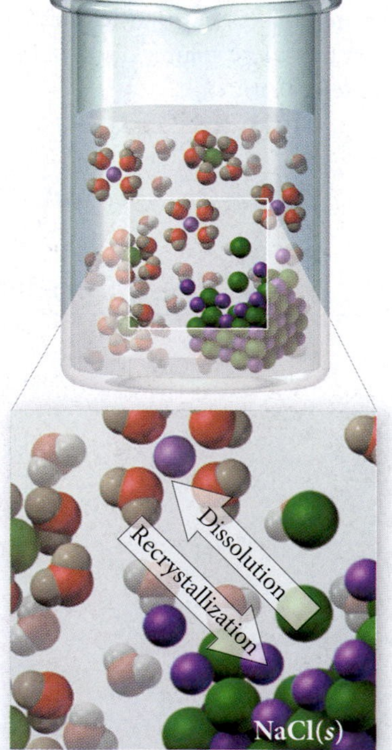

Rate of dissolution = Rate of recrystallization

(c) Dynamic equilibrium

◀ **FIGURE 14.9 Precipitation from a Supersaturated Solution** When a small piece of solid sodium acetate is added to a supersaturated sodium acetate solution, the excess solid precipitates out of the solution.

as the concentration of dissolved sodium chloride increases, the rate of recrystallization also increases. Eventually, the rates of dissolution and recrystallization become equal—**dynamic equilibrium** has been reached.

$$NaCl(s) \rightleftharpoons Na^+(aq) + Cl^-(aq)$$

A solution in which the dissolved solute is in dynamic equilibrium with the solid (undissolved) solute is a **saturated solution**. *If we add additional solute to a saturated solution, it will not dissolve.* A solution containing less than the equilibrium amount of solute is an **unsaturated solution**. *If we add additional solute to an unsaturated solution, it will dissolve.*

Under certain circumstances, a **supersaturated solution**—one containing more than the equilibrium amount of solute—may form. Supersaturated solutions are unstable, and the excess solute normally precipitates out of the solution. However, in some cases, if left undisturbed, a supersaturated solution can exist for an extended period of time. For example, in a common classroom demonstration a tiny piece of solid sodium acetate is added to a supersaturated solution of sodium acetate. This triggers the precipitation of the solute, which crystallizes out of solution in a dramatic and often beautiful way (Figure 14.9 ▲).

The Effect of Temperature on the Solubility of Solids

The solubility of solids in water can be highly dependent on temperature. Have you ever noticed how much more sugar you can dissolve in hot tea than in cold tea? Although exceptions exist, *the solubility of most solids in water increases with increasing temperature*, as shown in Figure 14.10 ▼. For example, the solubility of potassium nitrate (KNO_3) at room temperature is about 37 g KNO_3 per 100 g of water. At 50 °C, the solubility rises to 88 g KNO_3 per 100 g of water.

In the case of sugar dissolving in water, the higher temperature increases both *how fast* the sugar dissolves and *how much* sugar dissolves.

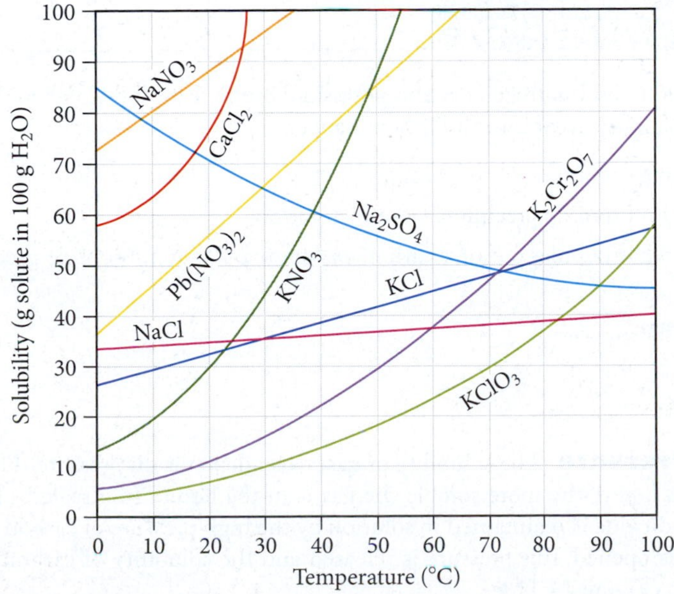

◀ **FIGURE 14.10 Solubility and Temperature** The solubility of most solids increases with increasing temperature.

▲ Rock candy is formed by the recrystallization of sugar.

A common way to purify a solid is **recrystallization**. In this technique, enough solid is added to water (or some other solvent) to create a saturated solution at an elevated temperature. As the solution cools, it becomes supersaturated and the excess solid precipitates out of solution. If the solution cools slowly, the solid forms crystals as it comes out of solution. The crystalline structure tends to reject impurities, resulting in a purer solid.

You can use the temperature dependence of the solubility of solids to make rock candy. Prepare a saturated sucrose (table sugar) solution at an elevated temperature, and allow a string or stick to dangle into the solution for several days. As the solution cools and the solvent evaporates, the solution becomes supersaturated and sugar crystals grow on the string or stick. After several days, beautiful edible crystals or "rocks" of sugar cover the string.

Factors Affecting the Solubility of Gases in Water

Solutions of gases dissolved in water are common. Club soda, for example, is a solution of carbon dioxide and water, and most liquids exposed to air contain dissolved gases. Fish depend on the oxygen dissolved in lake or ocean water for life, and blood contains dissolved nitrogen, oxygen, and carbon dioxide. Even tap water contains dissolved gases. The solubility of a gas in a liquid is affected by both temperature and pressure.

▲ Warm soda pop bubbles more than cold soda pop because carbon dioxide is less soluble in the warm solution.

The Effect of Temperature We can observe the effect of temperature on the solubility of a gas in water when we heat ordinary tap water on a stove. Before the water reaches its boiling point, small bubbles develop in the water. These bubbles are the dissolved air (mostly nitrogen and oxygen) coming out of solution. (Once the water boils, the bubbling becomes more vigorous—these larger bubbles are composed of water vapor.) The dissolved air comes out of the heated solution because—unlike solids, whose solubility generally increases with increasing temperature—*the solubility of gases in liquids decreases with increasing temperature.*

The inverse relationship between gas solubility and temperature is the reason that warm soda pop bubbles more than cold soda pop when you open it and warm beer goes flat faster than cold beer. More carbon dioxide comes out of these solutions at room temperature than at a lower temperature because the gas is less soluble at room temperature. The decreasing solubility of gases with increasing temperature is also the reason that fish don't bite much in a warm lake. The warm temperature results in a lower oxygen concentration. With lower oxygen levels, the fish become lethargic and do not strike as aggressively at the lure or bait you cast their way.

14.3 Cc Conceptual Connection

Solubility and Temperature

A solution is saturated in both nitrogen gas and potassium bromide at 75 °C. When the solution is cooled to room temperature, what is most likely to happen?

(a) Some nitrogen gas bubbles out of solution.

(b) Some potassium bromide precipitates out of solution.

(c) Some nitrogen gas bubbles out of solution, *and* some potassium bromide precipitates out of solution.

(d) Nothing happens.

The Effect of Pressure The solubility of gases also depends on pressure. The higher the pressure of a gas above a liquid, the more soluble the gas is in the liquid. In a sealed can of soda pop, for example, the carbon dioxide is maintained in solution by the high pressure of carbon dioxide within the can. When the can is opened, this pressure is released and the solubility of carbon dioxide decreases, resulting in bubbling (Figure 14.11 ▶).

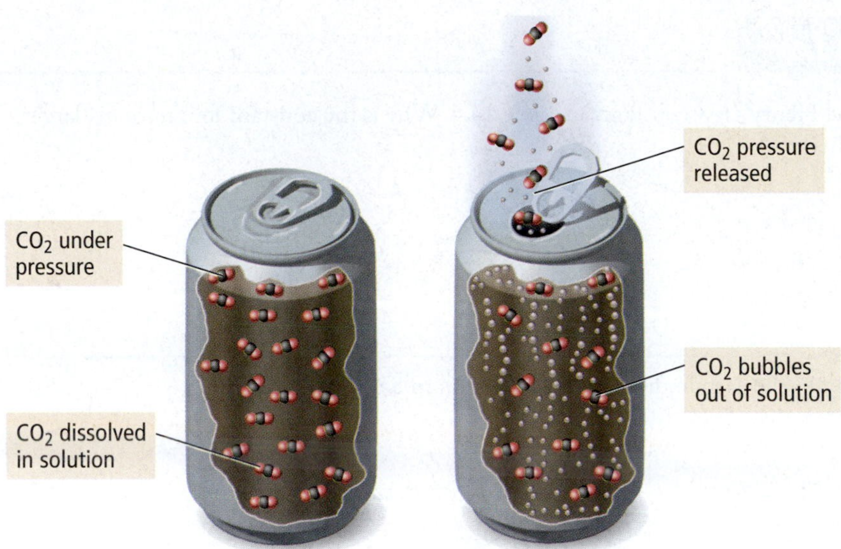

◀ **FIGURE 14.11 Soda Fizz** The bubbling that occurs when a can of soda is opened results from the reduced pressure of carbon dioxide over the liquid. At lower pressure, the carbon dioxide is less soluble and bubbles out of solution.

We can better understand the increased solubility of a gas in a liquid by considering cylinders containing water and carbon dioxide gas.

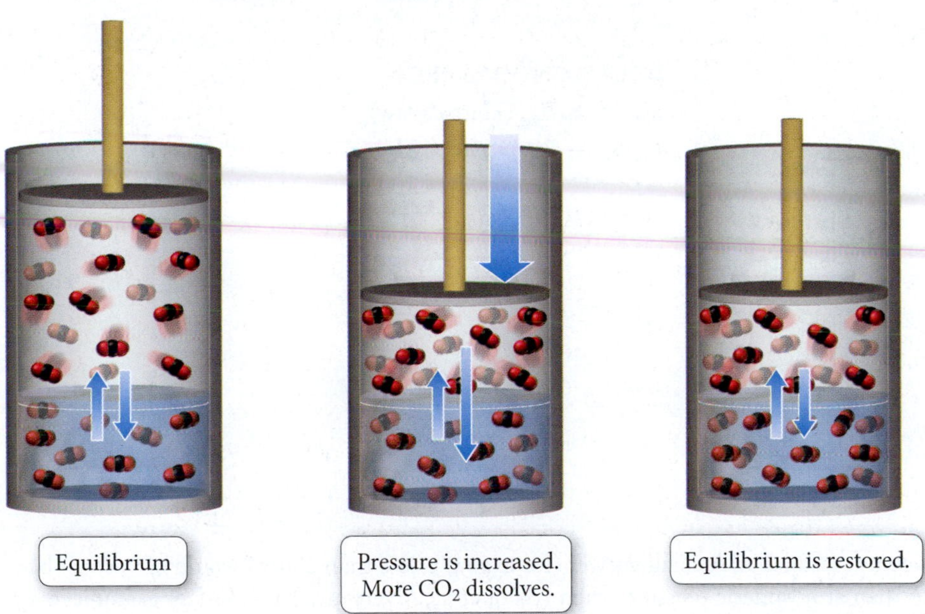

The first cylinder represents an equilibrium between gaseous and dissolved carbon dioxide—the rate of carbon dioxide molecules entering solution exactly equals the rate of molecules leaving the solution. Now imagine we decrease the volume, as shown in the second cylinder. The pressure of carbon dioxide increases, causing the rate of molecules entering the solution (represented by the longer arrow on the right) to rise. The number of molecules in solution increases until equilibrium is established again, as shown in the third cylinder. However, the amount of carbon dioxide in solution is now greater.

We can quantify the solubility of gases with increasing pressure with **Henry's law**:

$$S_{gas} = k_H P_{gas}$$

where S_{gas} is the solubility of the gas (usually in M), k_H is a constant of proportionality (called the *Henry's law constant*) that depends on the specific solute and solvent and also on temperature, and P_{gas} is the partial pressure of the gas (usually in atm). The equation shows that the solubility of a gas in a liquid is directly proportional to the pressure of the gas above the liquid. Table 14.4 lists the Henry's law constants for several common gases.

TABLE 14.4 Henry's Law Constants for Several Gases in Water at 25 °C

Gas	k_H (M/atm)
O_2	1.3×10^{-3}
N_2	6.1×10^{-4}
CO_2	3.4×10^{-2}
NH_3	5.8×10^{1}
He	3.7×10^{-4}

14.4 Cc Conceptual Connection

Henry's Law

Examine the Henry's law constants in Table 14.4. Why is the constant for ammonia larger than the others?

EXAMPLE 14.2 Henry's Law

What pressure of carbon dioxide is required to maintain the carbon dioxide concentration in a bottle of club soda at 0.12 M at 25 °C?

SORT You are given the desired solubility of carbon dioxide and asked to find the pressure required to achieve this solubility.

GIVEN: $S_{CO_2} = 0.12\ M$

FIND: P_{CO_2}

STRATEGIZE Use Henry's law to find the required pressure from the solubility. Use the Henry's law constant for carbon dioxide listed in Table 14.4.

CONCEPTUAL PLAN

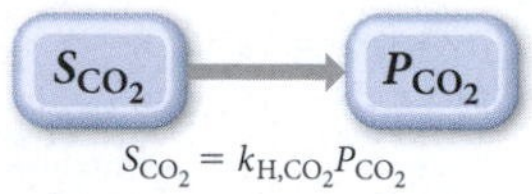

$S_{CO_2} = k_{H,CO_2} P_{CO_2}$

RELATIONSHIPS USED

$S_{gas} = k_H P_{gas}$ (Henry's law)

$k_{H,CO_2} = 3.4 \times 10^{-2}\ M/atm$ (from Table 14.4)

SOLVE Solve the Henry's law equation for P_{CO_2} and substitute the other quantities to calculate it.

SOLUTION

$$S_{CO_2} = k_{H,CO_2} P_{CO_2}$$

$$P_{CO_2} = \frac{S_{CO_2}}{k_{H,CO_2}} = \frac{0.12\ \cancel{M}}{3.4 \times 10^{-2}\ \frac{\cancel{M}}{atm}} = 3.5\ atm$$

CHECK The answer is in the correct units and seems reasonable. A small answer (for example, less than 1 atm) would be suspect because you know that the soda is under a pressure greater than atmospheric pressure when you open it. A very large answer (for example, over 100 atm) would be suspect because an ordinary can or bottle probably could not sustain such high pressures without bursting.

FOR PRACTICE 14.2

Determine the solubility of oxygen in water at 25 °C exposed to air at 1.0 atm. Assume a partial pressure for oxygen of 0.21 atm.

14.5 Expressing Solution Concentration

The amount of solute in a solution is an important property of the solution. A **dilute solution** contains small quantities of solute relative to the amount of solvent. A **concentrated solution** contains large quantities of solute relative to the amount of solvent. Common ways of reporting solution concentration include molarity, molality, parts by mass, parts by volume, mole fraction, and mole percent, as summarized in Table 14.5. We have seen two of these units before: molarity in Section 9.2, and mole fraction in Section 11.6. In this section of the chapter, we review the terms we have already covered and introduce the new ones.

TABLE 14.5 Solution Concentration Terms

Unit	Definition	Units
Molarity (M)	$\frac{\text{amount solute (in mol)}}{\text{volume solution (in L)}}$	$\frac{\text{mol}}{\text{L}}$
Molality (*m*)	$\frac{\text{amount solute (in mol)}}{\text{mass solvent (in kg)}}$	$\frac{\text{mol}}{\text{kg}}$
Mole fraction (χ)	$\frac{\text{amount solute (in mol)}}{\text{total amount of solute and solvent (in mol)}}$	None
Mole percent (mol %)	$\frac{\text{amount solute (in mol)}}{\text{total amount of solute and solvent (in mol)}} \times 100\%$	%
Parts by mass	$\frac{\text{mass solute}}{\text{mass solution}} \times \text{multiplication factor}$	
Percent by mass (%)	Multiplication factor = 100	%
Parts per million by mass (ppm)	Multiplication factor = 10^6	ppm
Parts per billion by mass (ppb)	Multiplication factor = 10^9	ppb
Parts by volume (%, ppm, ppb)	$\frac{\text{volume solute}}{\text{volume solution}} \times \text{multiplication factor}^*$	

*Multiplication factors for parts by volume are identical to those for parts by mass.

Molarity

The **molarity (M)** of a solution is the amount of solute (in moles) divided by the volume of solution (in liters).

$$\text{Molarity (M)} = \frac{\text{amount solute (in mol)}}{\text{volume solution (in L)}}$$

Note that molarity is moles of solute per liter of *solution*, not per liter of solvent. To make a solution of a specified molarity, we usually put the solute into a flask and then add water (or another solvent) to the desired volume of solution, as shown in Figure 14.12 ▼. Molarity is a convenient unit to use when making, diluting, and transferring solutions because it specifies the amount of solute per unit of solution.

Weigh out
1.00 mol NaCl
(58.44 g).
Add water until
solid is dissolved.
Then add additional
water until the 1 L
mark is reached.
Mix
A 1.00 molar NaCl solution

◀ **FIGURE 14.12 Preparing a Solution of Known Concentration** To make a 1 M NaCl solution, we add 1 mol of the solid to a flask and dilute with water to make 1 L of solution.

Molarity depends on volume, and because volume varies with temperature, molarity also varies with temperature. For example, a 1 M aqueous solution at room temperature is slightly less than 1 M at an elevated temperature because the volume of the solution is greater at the elevated temperature.

Molality

A concentration unit that is independent of temperature is **molality (*m*)**, the amount of solute (in moles) divided by the mass of solvent (in kilograms).

Molality is abbreviated with a lowercase italic *m*, while molarity is abbreviated with a capital M.

$$\text{Molality } (m) = \frac{\text{amount solute (in mol)}}{\text{mass solvent (in kg)}}$$

Notice that we define molality with respect to kilograms *solvent*, not kilograms solution. Molality is particularly useful when we need to compare concentrations over a range of different temperatures.

Parts by Mass and Parts by Volume

It is often convenient to report a concentration as a ratio of masses. A **parts by mass** concentration is the ratio of the mass of the solute to the mass of the solution, all multiplied by a multiplication factor.

$$\frac{\text{Mass solute}}{\text{Mass solution}} \times \text{multiplication factor}$$

The particular parts by mass unit we use, which determines the size of the multiplication factor, depends on the concentration of the solution. For example, for **percent by mass** the multiplication factor is 100.

$$\text{Percent by mass} = \frac{\text{mass solute}}{\text{mass solution}} \times 100$$

Percent means *per hundred*; a solution with a concentration of 14% by mass contains 14 g of solute per 100 g of solution.

For more dilute solutions, we can use **parts per million (ppm)**, which has a multiplication factor of 10^6, or **parts per billion (ppb)**, which has a multiplication factor of 10^9.

For dilute aqueous solutions near room temperature, the units of ppm are equivalent to milligrams solute/per liter of solution. This is because the density of a dilute aqueous solution near room temperature is 1.0 g/mL, so that 1 L has a mass of 1000 g.

$$\text{ppm} = \frac{\text{mass solute}}{\text{mass solution}} \times 10^6$$

$$\text{ppb} = \frac{\text{mass solute}}{\text{mass solution}} \times 10^9$$

A solution with a concentration of 15 ppm by mass, for example, contains 15 g of solute per 10^6 g of solution.

Sometimes, we report concentrations as a ratio of volumes, especially for solutions in which both the solute and solvent are liquids. A **parts by volume** concentration is usually the ratio of the volume of the solute to the volume of the solution, all multiplied by a multiplication factor.

$$\frac{\text{Volume solute}}{\text{Volume solution}} \times \text{multiplication factor}$$

The multiplication factors are identical to those just described for parts by mass concentrations. For example, a 22% ethanol solution by volume contains 22 mL of ethanol for every 100 mL of solution.

We can use the parts by mass (or parts by volume) concentration of a solution as a conversion factor between mass (or volume) of the solute and mass (or volume) of the solution. For example, for a solution containing 3.5% sodium chloride by mass, we write the following conversion factor:

$$\frac{3.5 \text{ g NaCl}}{100 \text{ g solution}} \quad \text{converts} \quad \text{g solution} \longrightarrow \text{g NaCl}$$

This conversion factor converts from grams solution to grams NaCl. To convert the other way, we invert the conversion factor:

$$\frac{100 \text{ g solution}}{3.5 \text{ g NaCl}} \quad \text{converts} \quad \text{g NaCl} \rightarrow \text{g solution}$$

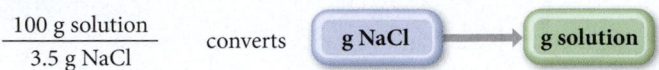

EXAMPLE 14.3
Using Parts by Mass in Calculations

What volume (in mL) of a soft drink that is 10.5% sucrose ($C_{12}H_{22}O_{11}$) by mass contains 78.5 g of sucrose? (The density of the solution is 1.04 g/mL.)

SORT You are given a mass of sucrose and the concentration and density of a sucrose solution, and you are asked to find the volume of solution containing the given mass.

GIVEN: 78.5 g $C_{12}H_{22}O_{11}$
10.5% $C_{12}H_{22}O_{11}$ by mass
density = 1.04 g/mL

FIND: mL

STRATEGIZE Begin with the mass of sucrose in grams. Use the mass percent concentration of the solution (written as a ratio, as shown under relationships used) to find the number of grams of solution containing this quantity of sucrose. Then use the density of the solution to convert grams to milliliters of solution.

CONCEPTUAL PLAN

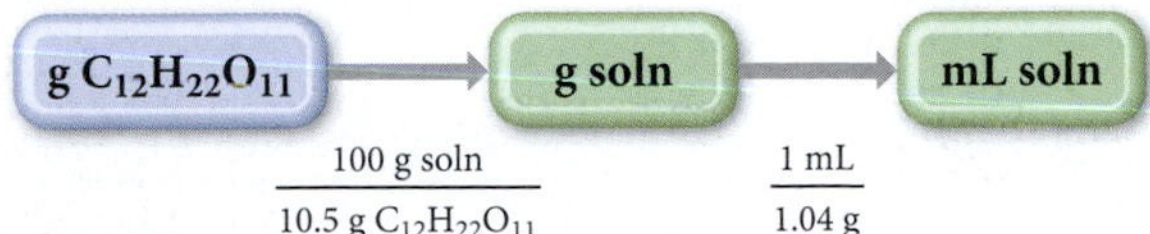

RELATIONSHIPS USED

$\frac{10.5 \text{ g } C_{12}H_{22}O_{11}}{100 \text{ g soln}}$ (percent by mass written as ratio)

$\frac{1 \text{ mL}}{1.04 \text{ g}}$ (given density of the solution)

SOLVE Begin with 78.5 g $C_{12}H_{22}O_{11}$ and multiply by the conversion factors to arrive at the volume of solution.

SOLUTION

$$78.5 \text{ g } C_{12}H_{22}O_{11} \times \frac{100 \text{ g soln}}{10.5 \text{ g } C_{12}H_{22}O_{11}} \times \frac{1 \text{ mL}}{1.04 \text{ g}} = 719 \text{ mL soln}$$

CHECK The units of the answer are correct. The magnitude seems correct because the solution is approximately 10% sucrose by mass. As the density of the solution is approximately 1 g/mL, the volume containing 78.5 g sucrose should be roughly 10 times larger, as calculated ($719 \approx 10 \times 78.5$).

FOR PRACTICE 14.3
What mass of sucrose ($C_{12}H_{22}O_{11}$), in g, is in 355 mL (12 fl oz) of a soft drink that is 11.5% sucrose by mass? (Assume a density of 1.04 g/mL.)

FOR MORE PRACTICE 14.3
A water sample contains the pollutant chlorobenzene with a concentration of 15 ppb (by mass). What volume of this water contains 5.00×10^2 mg of chlorobenzene? (Assume a density of 1.00 g/mL.)

Mole Fraction and Mole Percent

For some applications, especially those in which the ratio of solute to solvent can vary widely, the most useful way to express concentration is the amount of solute (in moles) divided by the total amount of solute and solvent (in moles). This ratio is the **mole fraction (χ_{solute})**.

$$\chi_{\text{solute}} = \frac{\text{amount solute (in mol)}}{\text{total amount of solute and solvent (in mol)}} = \frac{n_{\text{solute}}}{n_{\text{solute}} + n_{\text{solvent}}}$$

The mole fraction can also be defined for the solvent:

$$\chi_{\text{solvent}} = \frac{n_{\text{solvent}}}{n_{\text{solute}} + n_{\text{solvent}}}$$

Also in common use is the **mole percent (mol %)**, which is the mole fraction × one hundred percent.

$$\text{mol } \% = \chi_{\text{solute}} \times 100\%$$

EXAMPLE 14.4 Calculating Concentrations

You prepare a solution by dissolving 17.2 g of ethylene glycol ($C_2H_6O_2$) in 0.500 kg of water. The final volume of the solution is 515 mL. For this solution, calculate the concentration in each unit.

(a) molarity **(b)** molality **(c)** percent by mass
(d) mole fraction **(e)** mole percent

SOLUTION

(a) To calculate molarity, first find the amount of ethylene glycol in moles from the mass and molar mass.	$\text{mol } C_2H_6O_2 = 17.2 \text{ g } C_2H_6O_2 \times \frac{1 \text{ mol g } C_2H_6O_2}{62.07 \text{ g } C_2H_6O_2} = 0.2771 \text{ mol } C_2H_6O_2$
Divide the amount in moles by the volume of the solution in liters.	$\text{Molarity (M)} = \frac{\text{amount solute (in mol)}}{\text{volume solution (in L)}} = \frac{0.2771 \text{ mol } C_2H_6O_2}{0.515 \text{ L solution}} = 0.538 \text{ M}$
(b) To calculate molality, use the amount of ethylene glycol in moles from (a), and divide by the mass of the water in kilograms.	$\text{Molality } (m) = \frac{\text{amount solute (in mol)}}{\text{mass solvent (in kg)}} = \frac{0.2771 \text{ mol } C_2H_6O_2}{0.500 \text{ kg } H_2O} = 0.554\ m$
(c) To calculate percent by mass, divide the mass of the solute by the sum of the masses of the solute and solvent and multiply the ratio by 100%.	$\text{Percent by mass} = \frac{\text{mass solute}}{\text{mass solution}} \times 100\% = \frac{17.2 \text{ g}}{17.2 \text{ g} + 5.00 \times 10^2 \text{ g}} \times 100\% = 3.33\%$
(d) To calculate mole fraction, first determine the amount of water in moles from the mass of water and its molar mass.	$\text{mol } H_2O = 5.00 \times 10^2 \text{ g } H_2O \times \frac{1 \text{ mol } H_2O}{18.02 \text{ g } H_2O} = 27.75 \text{ mol } H_2O$
Divide the amount of ethylene glycol in moles from (a) by the total number of moles.	$\chi_{\text{solute}} = \frac{n_{\text{solute}}}{n_{\text{solute}} + n_{\text{solvent}}} = \frac{0.2771 \text{ mol}}{0.2771 \text{ mol} + 27.75 \text{ mol}} = 9.89 \times 10^{-3}$
(e) To calculate mole percent, multiply the mole fraction from (d) by 100%.	$\text{mol } \% = \chi_{\text{solute}} \times 100\% = 0.989\%$

FOR PRACTICE 14.4

You prepare a solution by dissolving 50.4 g sucrose ($C_{12}H_{22}O_{11}$) in 0.332 kg of water. The final volume of the solution is 355 mL. Calculate the concentration of the solution in each unit.

(a) molarity **(b)** molality **(c)** percent by mass
(d) mole fraction **(e)** mole percent

EXAMPLE 14.5 Converting between Concentration Units

What is the molarity of a 6.56% by mass glucose ($C_6H_{12}O_6$) solution? (The density of the solution is 1.03 g/mL.)

SORT You are given the concentration of a glucose solution in percent by mass and the density of the solution. Find the concentration of the solution in molarity.

GIVEN: 6.56% $C_6H_{12}O_6$
density = 1.03 g/mL

FIND: M

STRATEGIZE Begin with the mass percent concentration of the solution written as a ratio, and separate the numerator from the denominator. Convert the numerator from g $C_6H_{12}O_6$ to mol $C_6H_{12}O_6$. Convert the denominator from g soln to mL of solution and then to L solution. Divide the numerator (now in mol) by the denominator (now in L) to obtain molarity.

CONCEPTUAL PLAN

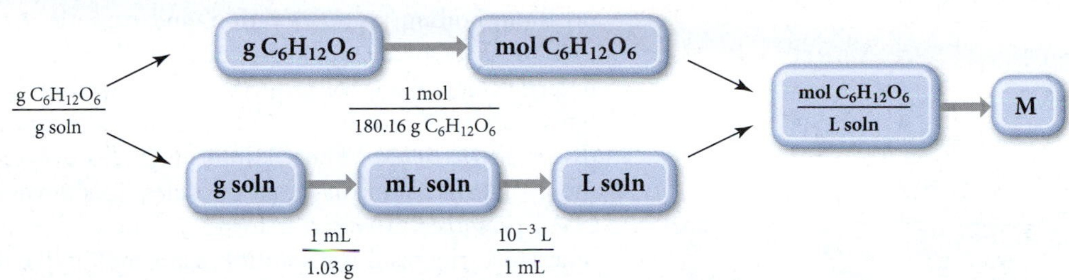

RELATIONSHIPS USED

$$\frac{6.56 \text{ g } C_6H_{12}O_6}{100 \text{ g soln}} \text{ (percent by mass written as ratio)}$$

$$\frac{1 \text{ mol}}{180.16 \text{ g } C_6H_{12}O_6} \text{ (from molar mass of glucose)}$$

$$\frac{1 \text{ mL}}{1.03 \text{ g}} \text{ (from given density of the solution)}$$

SOLVE Begin with the numerator (6.56 g $C_6H_{12}O_6$) and use the molar mass to convert to mol $C_6H_{12}O_6$.

SOLUTION

$$6.56 \text{ g } C_6H_{12}O_6 \times \frac{1 \text{ mol } C_6H_{12}O_6}{180.16 \text{ g } C_6H_{12}O_6} = 0.036412 \text{ mol } C_6H_{12}O_6$$

Convert the denominator (100 g solution) into mL of solution (using the density) and then to L of solution.

$$100 \text{ g soln} \times \frac{1 \text{ mL}}{1.03 \text{ g}} \times \frac{10^{-3} \text{ L}}{\text{mL}} = 0.097087 \text{ L soln}$$

Finally, divide mol $C_6H_{12}O_6$ by L solution to arrive at molarity.

$$\frac{0.036412 \text{ mol } C_6H_{12}O_6}{0.097087 \text{ L soln}} = 0.375 \text{ M } C_6H_{12}O_6$$

CHECK The units of the answer are correct. The magnitude seems correct. Very high molarities (especially those above 25 M) should immediately appear suspect. One liter of water contains about 55 moles of water molecules, so molarities higher than 55 M are physically impossible.

FOR PRACTICE 14.5
What is the molarity of a 10.5% by mass glucose ($C_6H_{12}O_6$) solution? (The density of the solution is 1.03 g/mL.)

FOR MORE PRACTICE 14.5
What is the molality of a 10.5% by mass glucose ($C_6H_{12}O_6$) solution? (The density of the solution is 1.03 g/mL.)

▲ In winter, salt is applied to roads so that the ice will melt at lower temperatures.

14.6 Colligative Properties: Vapor Pressure Lowering, Freezing Point Depression, Boiling Point Elevation, and Osmotic Pressure

Have you ever wondered why salt is added to the ice in an ice-cream maker? Or why salt is scattered on icy roads in cold climates? Salt lowers the temperature at which a saltwater solution freezes. A salt and water solution is liquid even below 0 °C. When salt is added to ice in the ice-cream maker, an ice/water/salt mixture forms that can reach a temperature of about −10 °C, at which point the cream freezes. On the winter road, the salt allows the ice to melt when the ambient temperature is below freezing.

The depression of the freezing point of water by salt is an example of a **colligative property**, a property that depends on the number of particles dissolved in solution, not on the type of particle. In this section, we examine four colligative properties: vapor pressure lowering, freezing point depression, boiling point elevation, and osmotic pressure. Because these properties depend on the *number* of dissolved particles, nonelectrolytes are treated slightly differently than electrolytes when determining colligative properties. (See Section 9.4 for a review of the difference between electrolytes and nonelectrolytes.) When 1 mol of a nonelectrolyte dissolves in water, it forms 1 mol of dissolved particles. When 1 mol of an electrolyte dissolves in water, however, it normally forms more than 1 mol of dissolved particles (as shown in Figure 14.13 ◀). For example, when 1 mol of NaCl dissolves in water, it forms 1 mol of dissolved Na^+ ions and 1 mol of dissolved Cl^- ions. Therefore, the resulting solution has 2 mol of dissolved particles. The colligative properties of electrolyte solutions reflect this higher concentration of dissolved particles. In this section we examine colligative properties of nonelectrolyte solutions; we then expand the concept to include electrolyte solutions in Section 14.7.

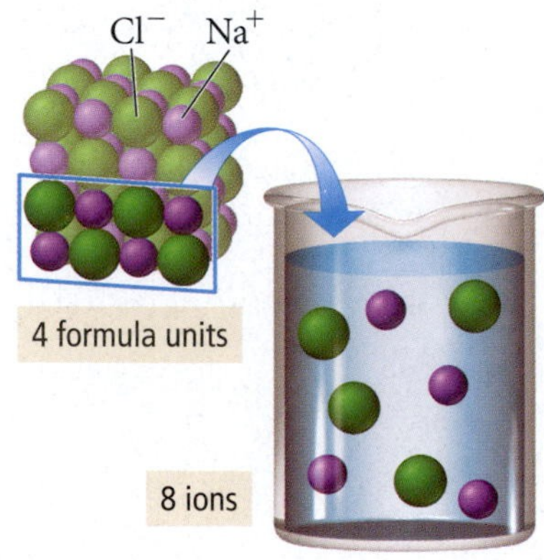

▲ **FIGURE 14.13** When sodium chloride dissolves in water, each mole of NaCl produces 2 mol of particles: 1 mol of Na^+ cations and 1 mol of Cl^- anions.

Vapor Pressure Lowering

Recall from Section 12.5 that the vapor pressure of a liquid is the pressure of the gas above the liquid when the two are in dynamic equilibrium (that is, when the rate of vaporization equals the rate of condensation). What is the effect of a nonvolatile nonelectrolyte solute on the vapor pressure of the liquid into which it dissolves? The basic answer to this question is that *the vapor pressure of the solution is lower than the vapor pressure of the pure solvent*.

The vapor pressure of a solution is lower than that of the pure solvent because of nature's tendency toward mixing (toward greater entropy) that we discussed in Section 14.2. We can see a dramatic demonstration of this tendency by placing a concentrated solution of a nonvolatile solute and a beaker of the pure solvent in a sealed container. Over time, the level of the pure solvent will drop, and the level of the solution will rise as molecules vaporize out of the pure solvent and condense into the solution. Why? The reason is nature's tendency to mix. If a pure solvent and concentrated solution are combined in a beaker, they naturally form a mixture in which the concentrated solution becomes less concentrated than it was initially. Similarly, if a pure solvent and concentrated solution are combined in a sealed container—even though they are in separate beakers—the two mix so that the concentrated solution becomes less concentrated.

KEY CONCEPT VIDEO Colligative properties

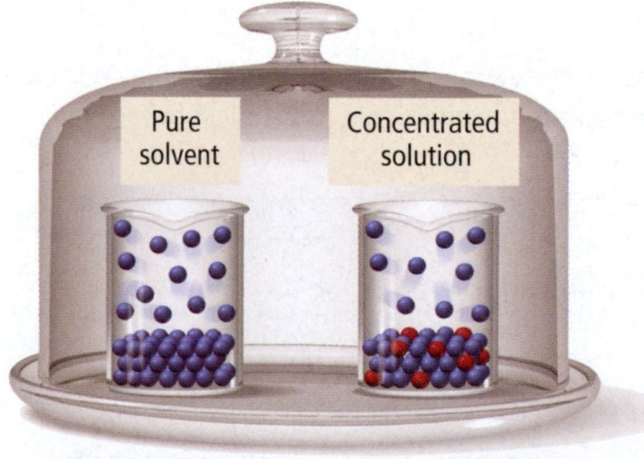

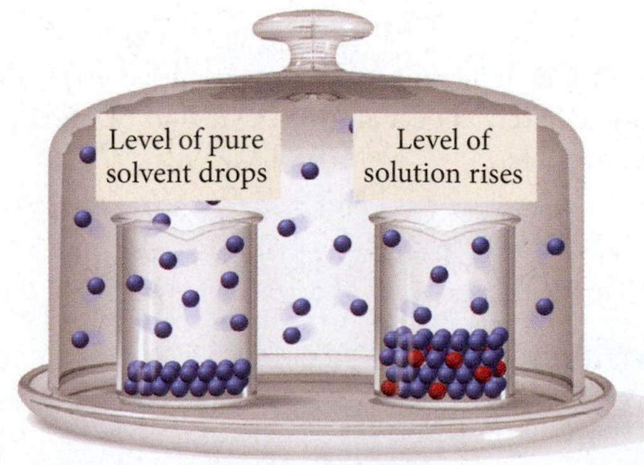

The net transfer of solvent from the beaker containing pure solvent to the beaker containing the solution indicates that the vapor pressure of the solution is lower than that of the pure solvent. As solvent molecules vaporize, the vapor pressure in the sealed container rises. Before dynamic equilibrium can be attained, however, the pressure in the sealed container exceeds the vapor pressure of the solution, causing molecules to condense into the solution (the beaker on the right). Molecules constantly vaporize from the pure solvent (the beaker on the left), but the solvent's vapor pressure is never reached because molecules are constantly condensing into the solution. The result is a continuous transfer of solvent molecules from the pure solvent to the solution.

We can quantify the vapor pressure of a solution with **Raoult's law**.

$$P_{\text{solution}} = \chi_{\text{solvent}} P^{\circ}_{\text{solvent}}$$

In this equation, P_{solution} is the vapor pressure of the solution, χ_{solvent} is the mole fraction of the solvent, and $P^{\circ}_{\text{solvent}}$ is the vapor pressure of the pure solvent at the same temperature. For example, suppose a water sample at 25 °C contains 0.90 mol of water and 0.10 mol of a nonvolatile solute such as sucrose. The pure water has a vapor pressure of 23.8 torr. We calculate the vapor pressure of the solution by substituting into Raoult's law:

$$\begin{aligned} P_{\text{solution}} &= \chi_{H_2O} P^{\circ}_{H_2O} \\ &= 0.90(23.8 \text{ torr}) \\ &= 21.4 \text{ torr} \end{aligned}$$

Notice that the vapor pressure of the solution is directly proportional to the amount of the solvent in the solution. Because the solvent particles compose 90% of all of the particles in the solution, the vapor pressure of the solution is 90% of the vapor pressure of the pure solvent.

EXAMPLE 14.6

Calculating the Vapor Pressure of a Solution Containing a Nonvolatile Nonelectrolyte Solute

Calculate the vapor pressure at 25 °C of a solution containing 99.5 g sucrose ($C_{12}H_{22}O_{11}$) and 300.0 mL water. The vapor pressure of pure water at 25 °C is 23.8 torr. Assume the density of water to be 1.00 g/mL.

SORT You are given the mass of sucrose and volume of water in a solution. You are also given the vapor pressure and density of pure water and asked to find the vapor pressure of the solution.

GIVEN: 99.5 g $C_{12}H_{22}O_{11}$

300.0 mL H_2O

$P^{\circ}_{H_2O} = 23.8$ torr at 25 °C

$d_{H_2O} = 1.00$ g/ml

FIND: P_{solution}

STRATEGIZE Raoult's law relates the vapor pressure of a solution to the mole fraction of the solvent and the vapor pressure of the pure solvent. Begin by calculating the amount in moles of sucrose and water.

Calculate the mole fraction of the solvent from the calculated amounts of solute and solvent.

Then use Raoult's law to calculate the vapor pressure of the solution.

CONCEPTUAL PLAN

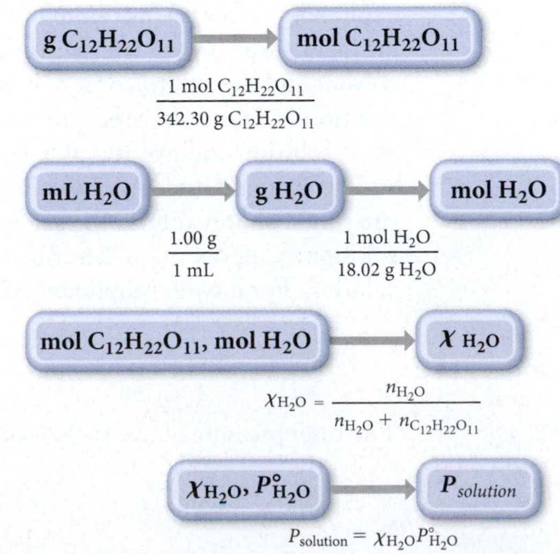

—*Continued on the next page*

Continued from the previous page—

SOLVE Calculate the number of moles of each solution component.	**SOLUTION** $99.5\ \text{g}\ C_{12}H_{22}O_{11} \times \frac{1\ \text{mol}\ C_{12}H_{22}O_{11}}{342.30\ \text{g}\ C_{12}H_{22}O_{11}} = 0.2907\ \text{mol}\ C_{12}H_{22}O_{11}$ $300.0\ \text{mL}\ H_2O \times \frac{1.00\ \text{g}}{1\ \text{mL}} \times \frac{1\ \text{mol}\ H_2O}{18.02\ \text{g}\ H_2O} = 16.65\ \text{mol}\ H_2O$
Use the number of moles of each component to calculate the mole fraction of the solvent (H_2O).	$\chi_{H_2O} = \frac{n_{H_2O}}{n_{C_{12}H_{22}O_{11}} + n_{H_2O}}$ $= \frac{16.65\ \text{mol}}{0.2907\ \text{mol} + 16.65\ \text{mol}}$ $= 0.9828$
Use the mole fraction of water and the vapor pressure of pure water to calculate the vapor pressure of the solution.	$P_{\text{solution}} = \chi_{H_2O} P^{\circ}_{H_2O}$ $= 0.9828\ (23.8\ \text{torr})$ $= 23.4\ \text{torr}$

CHECK The units of the answer are correct. The magnitude of the answer seems right because the calculated vapor pressure of the solution is just below that of the pure liquid, as you would expect for a solution with a large mole fraction of solvent.

FOR PRACTICE 14.6

Calculate the vapor pressure at 25 °C of a solution containing 55.3 g ethylene glycol ($HOCH_2CH_2OH$) and 285.2 g water. The vapor pressure of pure water at 25 °C is 23.8 torr.

FOR MORE PRACTICE 14.6

A solution containing ethylene glycol and water has a vapor pressure of 7.88 torr at 10 °C. Pure water has a vapor pressure of 9.21 torr at 10 °C. What is the mole fraction of ethylene glycol in the solution?

Vapor Pressures of Solutions Containing a Volatile (Nonelectrolyte) Solute

Some solutions contain not only a volatile solvent, but also a volatile *solute*. In this case, *both* the solvent and the solute contribute to the overall vapor pressure of the solution. A solution like this may be an **ideal solution** (in which case its behavior follows Raoult's law at all concentrations for both the solvent and the solute), or it may be nonideal (in which case it does not follow Raoult's law). An ideal solution is similar in concept to an ideal gas. Just as an ideal gas follows the ideal gas law exactly, an ideal solution follows Raoult's law exactly. In an ideal solution, the solute–solvent interactions are similar in magnitude to the solute–solute and solvent–solvent interactions. In this type of solution, the solute simply dilutes the solvent and ideal behavior is observed. The vapor pressure of each of the solution components is described by Raoult's law throughout the entire composition range of the solution. For a two-component solution containing liquids A and B, we can write:

Over a complete range of composition of a solution, it no longer makes sense to designate a solvent and solute, so we simply label the two components A and B.

$$P_A = \chi_A P^{\circ}_A$$
$$P_B = \chi_B P^{\circ}_B$$

The total pressure above such a solution is the sum of the partial pressures of the components:

$$P_{\text{tot}} = P_A + P_B$$

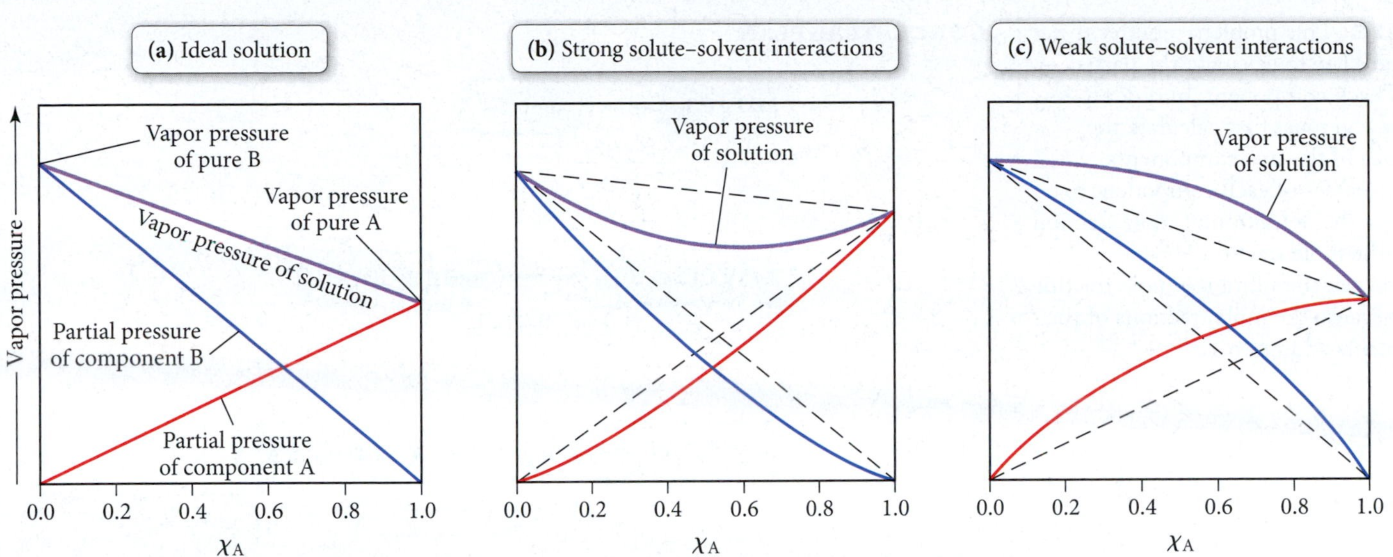

▲ **FIGURE 14.14 Behavior of Ideal and Nonideal Solutions** **(a)** An ideal solution follows Raoult's law for both components. **(b)** A solution with particularly strong solute–solvent interactions displays negative deviations from Raoult's law. **(c)** A solution with particularly weak solute–solvent interactions displays positive deviations from Raoult's law. (The dashed lines in parts b and c represent ideal behavior.)

Figure 14.14(a) ▲ is a plot of vapor pressure versus solution composition for an ideal two-component solution.

In a nonideal solution, the solute–solvent interactions are either stronger or weaker than the solvent–solvent interactions. If the solute–solvent interactions are stronger, then the solute tends to prevent the solvent from vaporizing as readily as it would otherwise. If the solution is sufficiently dilute, then the effect is small and Raoult's law works as an approximation. However, if the solution is not dilute, the effect is significant and the vapor pressure of the solution is *less than* that predicted by Raoult's law, as shown in Figure 14.14(b) ▲.

If, on the other hand, the solute–solvent interactions are weaker than solvent–solvent interactions, then the solute tends to allow more vaporization than would occur with just the solvent. If the solution is not dilute, the effect is significant and the vapor pressure of the solution is *greater than* predicted by Raoult's law, as shown in Figure 14.14(c) ▲.

EXAMPLE 14.7
Calculating the Vapor Pressure of a Two-Component Solution

A solution contains 3.95 g of carbon disulfide (CS_2) and 2.43 g of acetone (CH_3COCH_3). The vapor pressures at 35 °C of pure carbon disulfide and pure acetone are 515 torr and 332 torr, respectively. Assuming ideal behavior, calculate the vapor pressures of each of the components and the total vapor pressure above the solution. The experimentally measured total vapor pressure of the solution at 35 °C is 645 torr. Is the solution ideal? If not, what can you say about the relative strength of carbon disulfide–acetone interactions compared to the acetone–acetone and carbon disulfide–carbon disulfide interactions?

SORT You are given the masses and vapor pressures of carbon disulfide and acetone and are asked to find the vapor pressures of each component in the mixture and the total pressure assuming ideal behavior.

GIVEN: 3.95 g CS_2
2.34 g CH_3COCH_3
$P^{\circ}_{CS_2} = 515$ torr (at 35 °C)
$P^{\circ}_{CH_3COCH_3} = 332$ torr (at 35 °C)
$P_{tot}(\text{exp}) = 645$ torr (at 35 °C)

FIND: P_{CS_2}, $P_{CH_3COCH_3}$, $P_{tot}(\text{ideal})$

—Continued on the next page

Continued from the previous page—

STRATEGIZE This problem requires the use of Raoult's law to calculate the partial pressures of each component. In order to use Raoult's law, you must first calculate the mole fractions of the two components. Convert the masses of each component to moles and use the definition of mole fraction to calculate the mole fraction of carbon disulfide. You can then find the mole fraction of acetone because the mole fractions of the two components add up to 1.	**CONCEPTUAL PLAN** $3.95 \text{ g } CS_2 \rightarrow \text{mol } CS_2$ $\frac{1 \text{ mol } CS_2}{76.15 \text{ g } CS_2}$ $2.43 \text{ g } CH_3COCH_3 \rightarrow \text{mol } CH_3COCH_3$ $\frac{1 \text{ mol } CH_3COCH_3}{58.08 \text{ g } CH_3COCH_3}$ $\text{mol } CS_2, \text{ mol } CH_3OCH_3 \rightarrow \chi_{CS_2}, \chi_{CH_3COCH_3}$ $\chi_{CS_2} = \frac{n_{CS_2}}{n_{CS_2} + n_{CH_3COCH_3}}$
Use the mole fraction of each component along with Raoult's law to calculate the partial pressure of each component. The total pressure is the sum of the partial pressures.	$P_{CS_2} = \chi_{CS_2} P^{\circ}_{CS_2}$ $P_{CH_3COCH_3} = \chi_{CH_3COCH_3} P^{\circ}_{CH_3COCH_3}$ $P_{tot} = P_{CS_2} + P_{CH_3COCH_3}$ **RELATIONSHIPS USED** $\chi_A = \frac{n_A}{n_A + n_B}$ (mole fraction definition) $P_A = \chi_A P^{\circ}_A$ (Raoult's law)
SOLVE Begin by converting the mass of each component to the amounts in moles.	**SOLUTION** $3.95 \text{ g } CS_2 \times \frac{1 \text{ mol } CS_2}{76.15 \text{ g } CS_2} = 0.05187 \text{ mol } CS_2$ $2.43 \text{ g } CH_3COCH_3 \times \frac{1 \text{ mol } CH_3COCH_3}{58.08 \text{ g } CH_3COCH_3} = 0.04184 \text{ mol } CH_3COCH_3$
Calculate the mole fraction of carbon disulfide.	$\chi_{CS_2} = \frac{n_{CS_2}}{n_{CS_2} + n_{CH_3COCH_3}}$ $= \frac{0.05187 \text{ mol}}{0.05187 \text{ mol} + 0.04184 \text{ mol}}$ $= 0.5535$
Calculate the mole fraction of acetone by subtracting the mole fraction of carbon disulfide from 1.	$\chi_{CH_3COCH_3} = 1 - 0.5535$ $= 0.4465$

—Continued on the next page

Continued from the previous page—

Calculate the partial pressures of carbon disulfide and acetone by using Raoult's law and the given values of the vapor pressures of the pure substances.	$P_{CS_2} = \chi_{CS_2} P^{\circ}_{CS_2}$ $= 0.55\underline{3}5\ (515\ \text{torr})$ $= 285\ \text{torr}$ $P_{CH_3COCH_3} = \chi_{CH_3COCH_3} P^{\circ}_{CH_3COCH_3}$ $= 0.44\underline{6}5\ (332\ \text{torr})$ $= 148\ \text{torr}$
Calculate the total pressure by summing the partial pressures.	$P_{tot}(\text{ideal}) = 285\ \text{torr} + 148\ \text{torr}$ $= 433\ \text{torr}$
Lastly, compare the calculated total pressure for the ideal case to the experimentally measured total pressure. Since the experimentally measured pressure is greater than the calculated pressure, you can conclude that the interactions between the two components are weaker than the interactions between the components themselves.	$P_{tot}(\text{exp}) = 645\ \text{torr}$ $P_{tot}(\text{exp}) > P_{tot}(\text{ideal})$ The solution is not ideal and shows positive deviations from Raoult's law. Therefore, carbon disulfide–acetone interactions must be weaker than acetone–acetone and carbon disulfide–carbon disulfide interactions.

CHECK The units of the answer (torr) are correct. The magnitude seems reasonable given the partial pressures of the pure substances.

FOR PRACTICE 14.7

A solution of benzene (C_6H_6) and toluene (C_7H_8) is 25.0% benzene by mass. The vapor pressures of pure benzene and pure toluene at 25 °C are 94.2 torr and 28.4 torr, respectively. Assuming ideal behavior, calculate the following:

(a) The vapor pressure of each of the solution components in the mixture.
(b) The total pressure above the solution.
(c) The composition of the vapor in mass percent.

Why is the composition of the vapor different from the composition of the solution?

Raoult's Law

14.5 Cc Conceptual Connection

A solution contains equal amounts (in moles) of liquid components A and B. The vapor pressure of pure A is 100 mmHg and that of pure B is 200 mmHg. The experimentally measured vapor pressure of the solution is 120 mmHg. What are the relative strengths of the solute–solute, solute–solvent, and solvent–solvent interactions in this solution?

Freezing Point Depression and Boiling Point Elevation

Vapor pressure lowering occurs at all temperatures. We can see the effect of vapor pressure lowering over a range of temperatures by comparing the phase diagrams for a pure solvent and for a solution containing a nonvolatile solute:

▶ A nonvolatile solute lowers the vapor pressure of a solution, resulting in a lower freezing point and a higher boiling point.

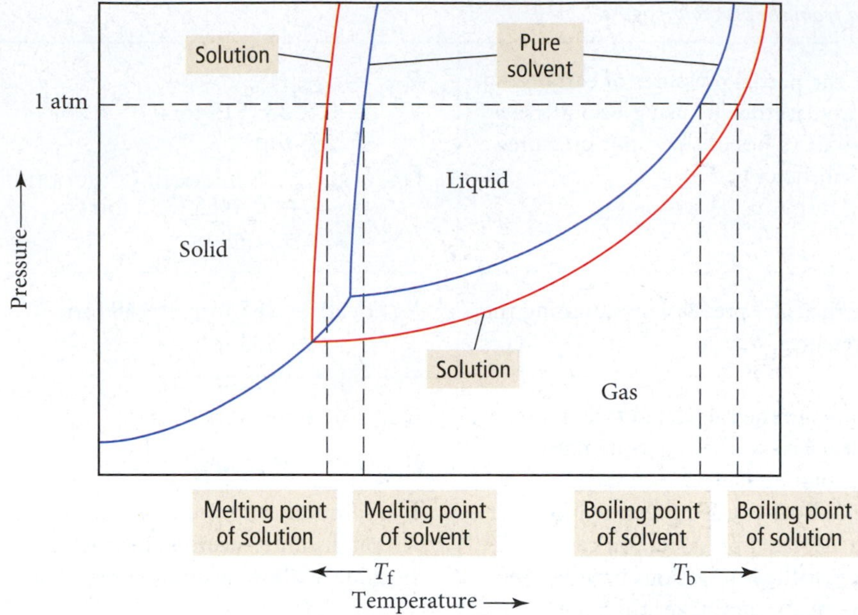

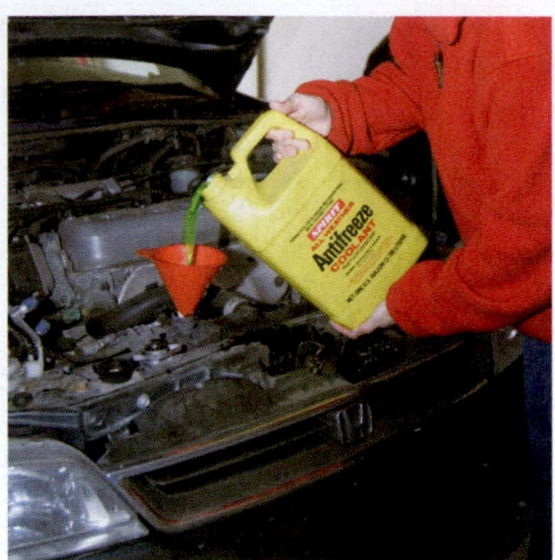

▲ Antifreeze is an aqueous solution of ethylene glycol or propylene glycol. The solution has a lower freezing point and higher boiling point than pure water.

Notice that the vapor pressure for the solution is shifted downward compared to that of the pure solvent. Consequently, the vapor pressure curve intersects the solid–gas curve at a lower temperature. The net effect is that the solution has a *lower melting point* and a *higher boiling point* than the pure solvent. These effects are called **freezing point depression** and **boiling point elevation**, both of which are colligative properties (like vapor pressure lowering).

The freezing point of a solution containing a nonvolatile solute is lower than the freezing point of the pure solvent. For example, antifreeze, used to prevent the freezing of engine blocks in cold climates, is an aqueous solution of ethylene glycol ($C_2H_6O_2$). The more concentrated the solution, the lower the freezing point becomes.

The amount that the freezing point decreases is given by the following equation:

$$\Delta T_f = m \times K_f$$

where

- ΔT_f is the change in temperature of the freezing point in Celsius degrees (relative to the freezing point of the pure solvent), usually reported as a positive number;
- m is the molality of the solution in moles solute per kilogram solvent; and
- K_f is the freezing point depression constant for the solvent.

For water,

$$K_f = 1.86\ °\mathrm{C}/m$$

When an aqueous solution containing a dissolved solid solute freezes slowly, the ice that forms does not normally contain much of the solute. For example, when ice forms in ocean water, the ice is not salt water, but freshwater. As the ice forms, the crystal structure of the ice tends to exclude the solute particles. You can verify this yourself by partially freezing a saltwater solution in the freezer. Take out the newly formed ice, rinse it several times, and taste it. Compare its taste to the taste of the original solution. The ice is much less salty.

Table 14.6 provides freezing point depression and boiling point elevation constants for several liquids. Calculating the freezing point of a solution involves substituting into the freezing point depression equation, as Example 14.8 demonstrates.

TABLE 14.6 Freezing Point Depression and Boiling Point Elevation Constants for Several Liquid Solvents

Solvent	Normal Freezing Point (°C)	K_f (°C/m)	Normal Boiling Point (°C)	K_b (°C/m)
Benzene (C_6H_6)	5.5	5.12	80.1	2.53
Carbon tetrachloride (CCl_4)	−22.9	29.9	76.7	5.03
Chloroform ($CHCl_3$)	−63.5	4.70	61.2	3.63
Ethanol (C_2H_5OH)	−114.1	1.99	78.3	1.22
Diethyl ether ($C_4H_{10}O$)	−116.3	1.79	34.6	2.02
Water (H_2O)	0.00	1.86	100.0	0.512

EXAMPLE 14.8
Freezing Point Depression

What is the freezing point of a 1.7 m aqueous ethylene glycol solution?

SORT You are given the molality of a solution and asked to find its freezing point.

GIVEN: 1.7 m solution

FIND: freezing point (from ΔT_f)

STRATEGIZE To solve this problem, use the freezing point depression equation.

CONCEPTUAL PLAN

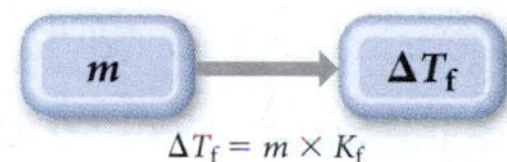

$\Delta T_f = m \times K_f$

SOLVE Substitute into the equation to calculate ΔT_f. The actual freezing point is the freezing point of pure water (0.00 °C) $-\Delta T_f$.

SOLUTION

$$\Delta T_f = m \times K_f$$
$$= 1.7\ \cancel{m} \times 1.86\ °C/\cancel{m}$$
$$= 3.2\ °C$$

$$\text{Freezing point} = 0.00\ °C - 3.2\ °C$$
$$= -3.2\ °C$$

CHECK The units of the answer are correct. The magnitude seems about right. The expected range for freezing points of an aqueous solution is anywhere from −10 °C to just below 0 °C. Any answers out of this range would be suspect.

FOR PRACTICE 14.8
Calculate the freezing point of a 2.6 m aqueous sucrose solution.

The boiling point of a solution containing a nonvolatile solute is higher than the boiling point of the pure solvent. In automobiles, antifreeze not only prevents the freezing of water within engine blocks in cold climates, it also prevents the boiling of water within engine blocks in hot climates. The amount that the boiling point rises in solutions is given by the equation:

$$\Delta T_b = m \times K_b$$

where

- ΔT_b is the change in temperature of the boiling point in Celsius degrees (relative to the boiling point of the pure solvent);
- m is the molality of the solution in moles solute per kilogram solvent; and
- K_b is the boiling point elevation constant for the solvent.

For water,

$$K_b = 0.512\ °C/m$$

Calculating the boiling point of a solution involves substituting into the boiling point elevation equation, as Example 14.9 demonstrates.

14.6 Cc Conceptual Connection

Boiling Point Elevation

Solution A is a 1.0 M solution with a nonionic solute and water as the solvent. Solution B is a 1.0 M solution with the same nonionic solute as solution A and ethanol as the solvent. Which solution has the greater increase in its boiling point relative to its pure solvent?

EXAMPLE 14.9
Boiling Point Elevation

What mass of ethylene glycol ($C_2H_6O_2$), in grams, must you add to 1.0 kg of water to produce a solution that boils at 105.0 °C? (Since pure water boils at 100.0 °C, the change in boiling point is 5.0 °C.)

SORT You are given the desired boiling point of an ethylene glycol solution containing 1.0 kg of water and asked to determine the mass of ethylene glycol needed to achieve the boiling point.

GIVEN: $\Delta T_b = 5.0\ °C$, 1.0 kg H_2O

FIND: g $C_2H_6O_2$

STRATEGIZE To solve this problem, use the boiling point elevation equation to calculate the desired molality of the solution from ΔT_b.

CONCEPTUAL PLAN

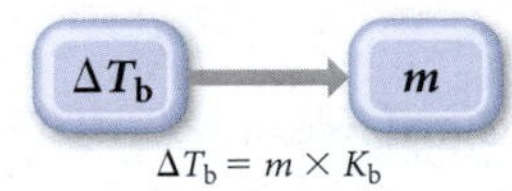

Then use that molality to determine how many moles of ethylene glycol are needed per kilogram of water. Finally, calculate the molar mass of ethylene glycol and use it to convert from moles of ethylene glycol to mass of ethylene glycol.

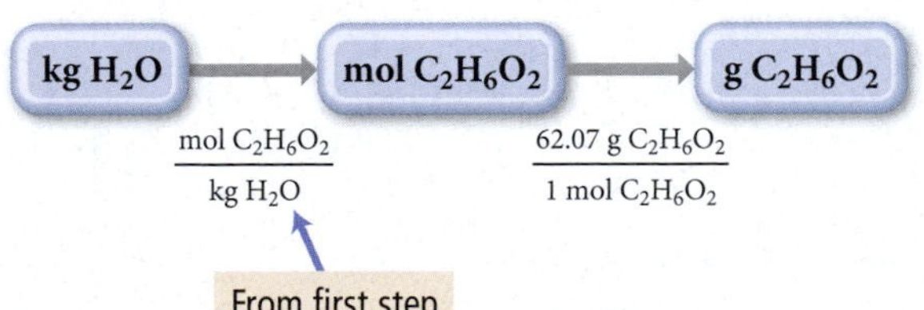

RELATIONSHIPS USED

$C_2H_6O_2$ molar mass = 62.07 g/mol

$\Delta T_b = m \times K_b$ (boiling point elevation)

SOLVE Begin by solving the boiling point elevation equation for molality and substituting the required quantities to calculate m.

SOLUTION

$$\Delta T_b = m \times K_b$$

$$m = \frac{\Delta T_b}{K_b} = \frac{5.0\ \cancel{°C}}{0.512\ \frac{\cancel{°C}}{m}} = 9.7\underline{7}m$$

$$1.0\ \cancel{kg\ H_2O} \times \frac{9.7\underline{7}\ \cancel{mol\ C_2H_6O_2}}{\cancel{kg\ H_2O}} \times \frac{62.07\ g\ C_2H_6O_2}{1\ \cancel{mol\ C_2H_6O_2}} = 6.1 \times 10^2\ g\ C_2H_6O_2$$

CHECK The units of the answer are correct. The magnitude might seem a little high initially, but the boiling point elevation constant is so small that a lot of solute is required to raise the boiling point by a small amount.

FOR PRACTICE 14.9

Calculate the boiling point of a 3.60 m aqueous sucrose solution.

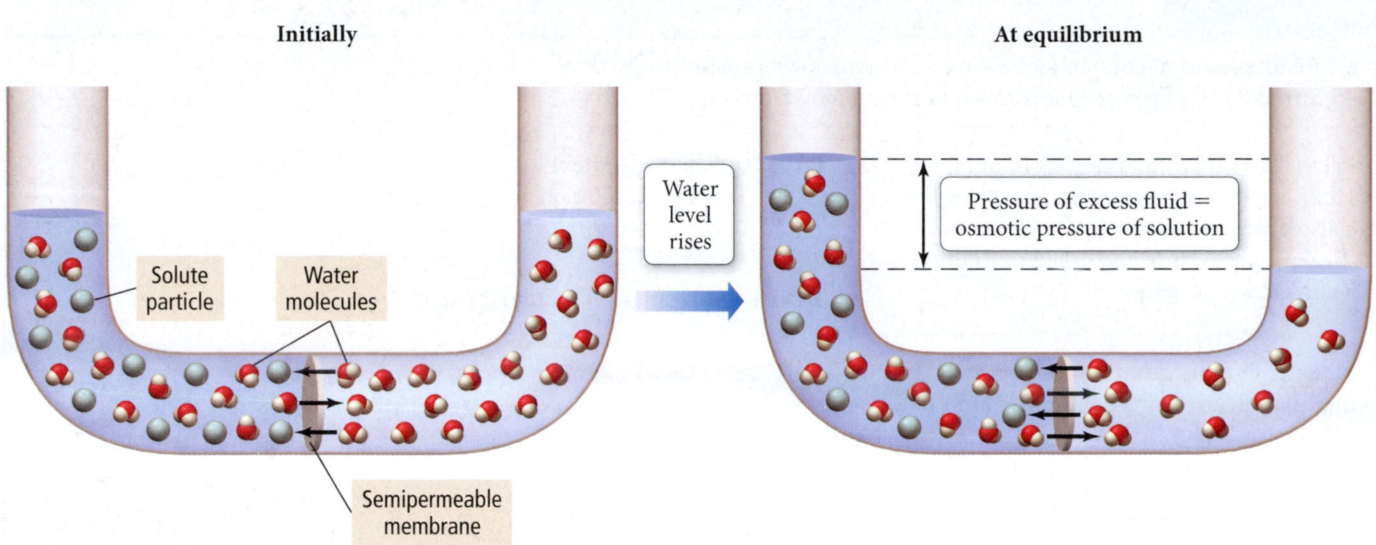

▲ **FIGURE 14.15 An Osmosis Cell** In an osmosis cell, water flows from the pure-water side of the cell through the semipermeable membrane to the saltwater side.

Osmotic Pressure

Osmosis is the flow of solvent from a solution of lower solute concentration to one of higher solute concentration. Concentrated solutions draw solvent from more dilute solutions because of nature's tendency to mix. Figure 14.15 ▲ illustrates an osmosis cell. The left side of the cell contains a concentrated saltwater solution, and the right side of the cell contains pure water. A **semipermeable membrane**—a membrane that selectively allows some substances to pass through but not others—separates the two halves of the cell. Water flows by osmosis from the pure-water side of the cell through the semipermeable membrane and into the saltwater side. Over time, the water level on the left side of the cell rises, while the water level on the right side of the cell falls. Notice the similarity between osmosis and vapor pressure lowering. In both cases, as a solution becomes concentrated, it develops a tendency to draw pure solvent to itself. In the case of vapor pressure lowering, the pure solvent is drawn from the gas state. In the case of osmosis, the pure solvent is drawn from the liquid state. In both cases, the solution becomes more dilute as it draws the pure solvent to itself—nature's tendency to mix is powerful.

If external pressure is applied to the solution in the left side of the cell, this process can be opposed and even stopped. The pressure required to stop the osmotic flow, called the **osmotic pressure**, is given by the following equation:

$$\Pi = MRT$$

where M is the molarity of the solution, T is the temperature (in Kelvin), and R is the ideal gas constant ($0.08206 \text{ L} \cdot \text{atm/mol} \cdot \text{K}$).

Osmosis is the reason that you should never drink seawater—or any concentrated solution. Seawater draws water *out of the body* as it passes through the stomach and intestines, resulting in diarrhea and further dehydration (Figure 14.16 ▶). We can think of seawater as a *thirsty solution*—one that draws more water to itself. Consequently, seawater should never be consumed as drinking water.

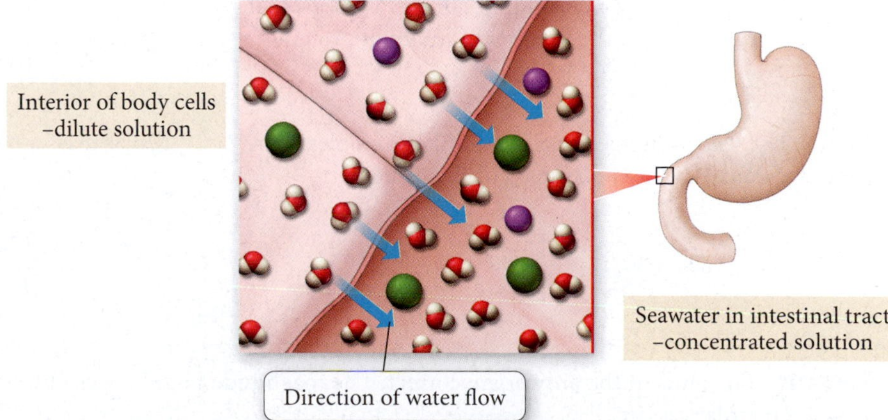

▲ **FIGURE 14.16 The Effect of Drinking Seawater** Seawater is a more concentrated solution than the fluids in body cells. As a result, when seawater flows through the digestive tract, it draws water out of the surrounding tissues.

EXAMPLE 14.10
Osmotic Pressure

The osmotic pressure of a solution containing 5.87 mg of an unknown protein per 10.0 mL of solution is 2.45 torr at 25 °C. Find the molar mass of the unknown protein.

SORT The problem states that a solution of an unknown protein contains 5.87 mg of the protein per 10.0 mL of solution. You are also given the osmotic pressure of the solution at a particular temperature and asked to find the molar mass of the unknown protein.

GIVEN: 5.87 mg protein
10.0 mL solution
$\Pi = 2.45$ torr
$T = 25$ °C

FIND: molar mass of protein (g/mol)

STRATEGIZE Step 1: Use the given osmotic pressure and temperature to find the molarity of the protein solution.

CONCEPTUAL PLAN

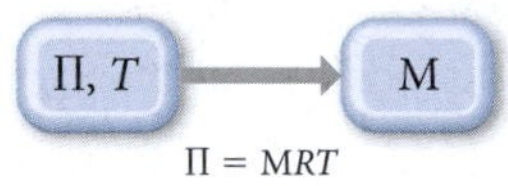

$\Pi = MRT$

Step 2: Use the molarity calculated in Step 1 to find the number of moles of protein in 10 mL of solution.

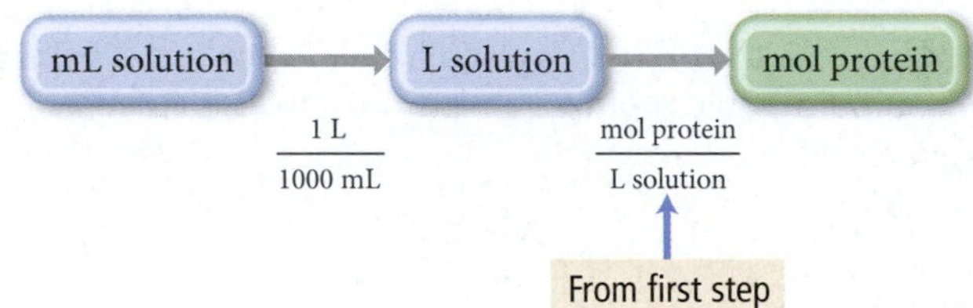

$\frac{1\text{ L}}{1000\text{ mL}}$ $\frac{\text{mol protein}}{\text{L solution}}$

From first step

Step 3: Finally, use the number of moles of the protein calculated in Step 2 and the given mass of the protein in 10.0 mL of solution to find the molar mass.

$$\text{Molar mass} = \frac{\text{mass protein}}{\text{moles protein}}$$

RELATIONSHIPS USED $\Pi = MRT$ (osmotic pressure equation)

SOLVE
Step 1: Begin by solving the osmotic pressure equation for molarity and substituting in the required quantities in the correct units to calculate M.

SOLUTION

$$\Pi = MRT$$

$$M = \frac{\Pi}{RT} = \frac{2.45\ \text{torr} \times \frac{1\ \text{atm}}{760\ \text{torr}}}{0.08206\frac{\text{L}\cdot\text{atm}}{\text{mol}\cdot\text{K}}(298\ \text{K})}$$

$$= 1.318 \times 10^{-4}\ \text{M}$$

Step 2: Begin with the given volume, convert to liters, then use the molarity to find the number of moles of protein.

$$10.0\ \text{mL} \times \frac{1\ \text{L}}{1000\ \text{mL}} \times \frac{1.318 \times 10^{-4}\ \text{mol}}{\text{L}} = 1.318 \times 10^{-6}\ \text{mol}$$

Step 3: Use the given mass and the number of moles from Step 2 to calculate the molar mass of the protein.

$$\text{Molar mass} = \frac{\text{mass protein}}{\text{moles protein}}$$

$$= \frac{5.87 \times 10^{-3}\ \text{g}}{1.318 \times 10^{-6}\ \text{mol}}$$

$$= 4.45 \times 10^{3}\ \text{g/mol}$$

CHECK The units of the answer are correct. The magnitude might seem a little high initially, but proteins are large molecules and therefore have high molar masses.

FOR PRACTICE 14.10
Calculate the osmotic pressure (in atm) of a solution containing 1.50 g ethylene glycol ($C_2H_6O_2$) in 50.0 mL of solution at 25 °C.

14.7 Colligative Properties of Strong Electrolyte Solutions

At the beginning of Section 14.6, we saw that colligative properties depend on the number of dissolved particles and that we therefore must treat electrolytes slightly differently than nonelectrolytes when determining colligative properties. For example, the freezing point depression of a 0.10 *m* sucrose solution is $\Delta T_f = 0.186\ ^\circ C$. However, the freezing point depression of a 0.10 *m* sodium chloride solution is nearly twice this large. Why? Because 1 mol of sodium chloride dissociates into nearly 2 mol of ions in solution. The ratio of moles of particles in solution to moles of formula units dissolved is called the **van't Hoff factor (*i*)**.

$$i = \frac{\text{moles of particles in solution}}{\text{moles of fomula units dissolved}}$$

Because 1 mol of NaCl produces 2 mol of particles in solution, we expect the van't Hoff factor for NaCl to be exactly 2. In reality, this expected factor only occurs in very dilute solutions. For example, the van't Hoff factor for a 0.10 *m* NaCl solution is 1.87, and that for a 0.010 *m* NaCl solution is 1.94. The van't Hoff factor approaches the expected value at infinite dilution (as the concentration approaches zero). Table 14.7 lists the actual and expected van't Hoff factors for a number of solutes.

The reason that van't Hoff factors do not exactly equal expected values is that some ions effectively pair in solution. We expect the dissociation of an ionic compound to be complete in solution. In reality, however, the dissociation is not complete—at any given moment, some cations are pairing with anions (Figure 14.17 ▼), slightly reducing the number of particles in solution.

To calculate freezing point depression, boiling point elevation, and osmotic pressure of ionic solutions, we use the van't Hoff factor in each equation as follows:

$$\Delta T_f = im \times K_f \text{ (freezing point depression)}$$

$$\Delta T_b = im \times K_b \text{ (boiling point elevation)}$$

$$\Pi = iMRT \text{ (osmotic pressure)}$$

TABLE 14.7 Van't Hoff Factors at 0.05 *m* Concentration in Aqueous Solution

Solute	*i* Expected	*i* Measured
Nonelectrolyte	1	1
NaCl	2	1.9
$MgSO_4$	2	1.3
$MgCl_2$	3	2.7
K_2SO_4	3	2.6
$FeCl_3$	4	3.4

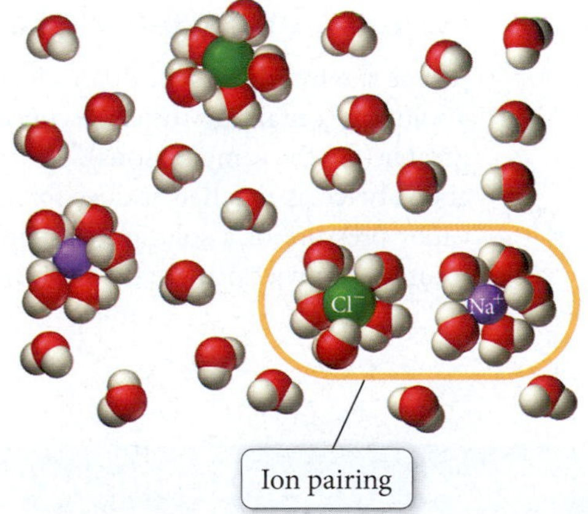

◀ **FIGURE 14.17 Ion Pairing** Hydrated anions and cations may get close enough together to effectively pair, lowering the concentration of particles more than would be expected.

Colligative Properties — 14.7 Cc Conceptual Connection

Which aqueous solution has the highest boiling point?

(a) 0.50 M $C_{12}H_{22}O_{11}$
(b) 0.50 M NaCl
(c) 0.50 M $MgCl_2$

EXAMPLE 14.11
Van't Hoff Factor and Freezing Point Depression

The freezing point of an aqueous 0.050 *m* $CaCl_2$ solution is −0.27 °C. What is the van't Hoff factor (i) for $CaCl_2$ at this concentration? How does it compare to the expected value of i?

SORT You are given the molality of a solution and its freezing point. You are asked to find the value of i—the van't Hoff factor—and compare it to the expected value.

GIVEN: 0.050 *m* $CaCl_2$ solution, $\Delta T_f = 0.27\ °C$
FIND: i

STRATEGIZE To solve this problem, use the freezing point depression equation including the van't Hoff factor.

CONCEPTUAL PLAN
$\Delta T_f = im \times K_f$

SOLVE Solve the freezing point depression equation for i, substituting in the given quantities to calculate its value.
The expected value of i for $CaCl_2$ is 3 because calcium chloride forms 3 mol of ions for each mole of calcium chloride that dissolves. The experimental value is slightly less than 3, probably because of ion pairing.

SOLUTION

$$\Delta T_f = im \times K_f$$

$$i = \frac{\Delta T_f}{m \times K_f} = \frac{0.27\ \cancel{°C}}{0.050\ \cancel{m} \times \frac{1.86\ \cancel{°C}}{\cancel{m}}} = 2.9$$

CHECK The answer has no units, as expected, as i is a ratio. The magnitude is about right because it is close to the value you would expect upon complete dissociation of $CaCl_2$.

FOR PRACTICE 14.11
Calculate the freezing point of an aqueous 0.10 *m* $FeCl_3$ solution using a van't Hoff factor of 3.2.

Strong Electrolytes and Vapor Pressure

Just as the freezing point depression of a solution containing an electrolyte solute is greater than that of a solution containing the same concentration of a nonelectrolyte solute, so the vapor pressure lowering is greater (for the same reasons). The vapor pressure for a sodium chloride solution, for example, decreases about twice as much as it does for a nonelectrolyte solution of the same concentration. To calculate the vapor pressure of a solution containing an ionic solute, we need to account for the dissociation of the solute when we calculate the mole fraction of the solvent, as demonstrated in Example 14.12.

EXAMPLE 14.12
Calculating the Vapor Pressure of a Solution Containing an Ionic Solute

A solution contains 0.102 mol $Ca(NO_3)_2$ and 0.927 mol H_2O. Calculate the vapor pressure of the solution at 55 °C. The vapor pressure of pure water at 55 °C is 118.1 torr. (Assume that the solute completely dissociates.)

SORT You are given the number of moles of each component of a solution and asked to find the vapor pressure of the solution. You are also given the vapor pressure of pure water at the appropriate temperature.

GIVEN: 0.102 mol $Ca(NO_3)_2$
0.927 mol H_2O
$P^{\circ}_{H_2O} = 118.1$ torr (at 55 °C)
FIND: $P_{solution}$

—*Continued on the next page*

Continued from the previous page—

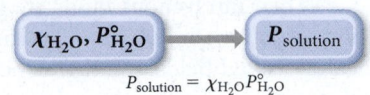

STRATEGIZE To solve this problem, use Raoult's law as you did in Example 14.6. Calculate χ_{solvent} from the given amounts of solute and solvent.	**CONCEPTUAL PLAN** $\chi_{H_2O}, P^\circ_{H_2O} \rightarrow P_{\text{solution}}$ $P_{\text{solution}} = \chi_{H_2O} P^\circ_{H_2O}$
SOLVE The key to solving this problem is understanding the dissociation of calcium nitrate. Write an equation showing the dissociation. Since 1 mol of calcium nitrate dissociates into 3 mol of dissolved particles, you must multiply the number of moles of calcium nitrate by 3 when you calculate the mole fraction.	**SOLUTION** $Ca(NO_3)_2(s) \longrightarrow Ca^{2+}(aq) + 2\,NO_3^-(aq)$ $\chi_{H_2O} = \dfrac{n_{H_2O}}{3 \times n_{Ca(NO_3)_2} + n_{H_2O}}$ $= \dfrac{0.927\ \text{mol}}{3(0.102)\ \text{mol} + 0.927\ \text{mol}}$ $= 0.7518$
Use the mole fraction of water and the vapor pressure of pure water to calculate the vapor pressure of the solution.	$P_{\text{solution}} = \chi_{H_2O} P^\circ_{H_2O}$ $= 0.7518(118.1\ \text{torr})$ $= 88.8\ \text{torr}$

CHECK The units of the answer are correct. The magnitude also seems right because the calculated vapor pressure of the solution is significantly less than that of the pure solvent, as you would expect for a solution with a significant amount of solute.

FOR PRACTICE 14.12
A solution contains 0.115 mol H_2O and an unknown number of moles of sodium chloride. The vapor pressure of the solution at 30 °C is 25.7 torr. The vapor pressure of pure water at 30 °C is 31.8 torr. Calculate the number of moles of sodium chloride in the solution.

Colligative Properties and Medical Solutions

Doctors and other health care workers often administer solutions to patients. The osmotic pressure of these solutions is controlled to ensure the desired effect on the patient. Intravenous (IV) solutions—those that are administered directly into a patient's veins—must have osmotic pressures equal to those of body fluids. These solutions are called *isosmotic* (or *isotonic*) (Figure 14.18(a) ▼). Solutions having osmotic pressures greater than those of body fluids are *hyperosmotic*. These solutions take water out of cells and tissues. When a human cell is placed in a hyperosmotic solution, it tends to shrivel as it loses water to the surrounding solution (Figure 14.18(b) ▼). Solutions having osmotic pressures less than

Isosmotic solution

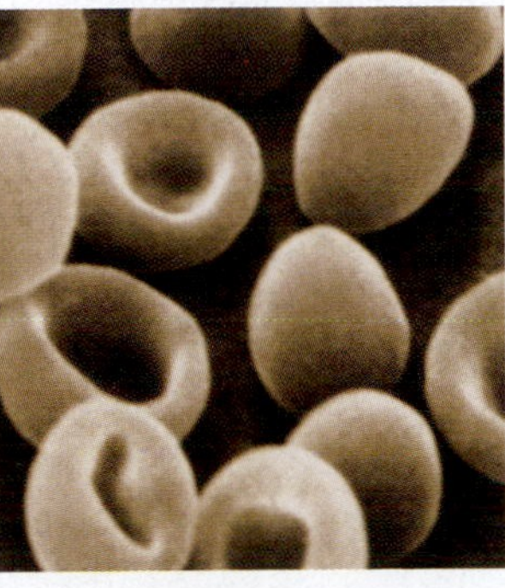

(a) Normal red blood cells

Hyperosmotic solution

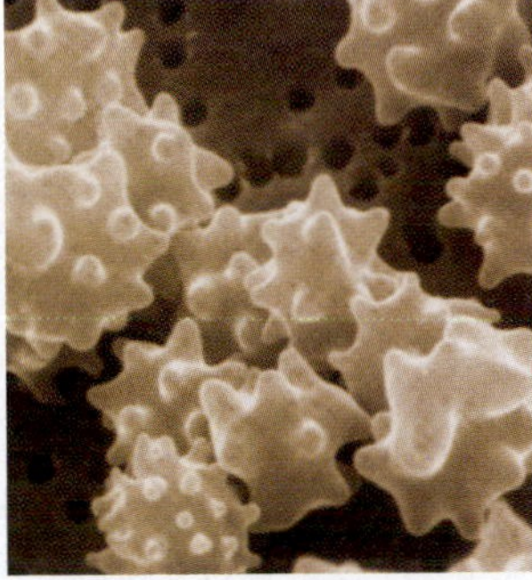

(b) Shriveled red blood cells

Hyposmotic solution

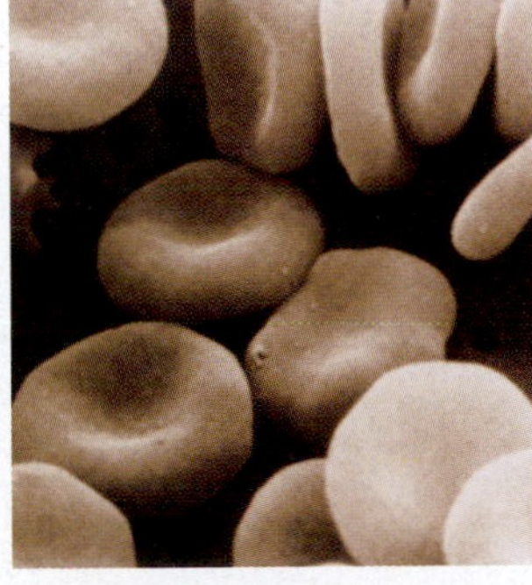

(c) Swollen red blood cells

◀ **FIGURE 14.18 Red Blood Cells and Osmosis** **(a)** In an isosmotic solution, red blood cells have the normal shape shown here. In a hyperosmotic solution **(b)**, they lose water and shrivel. In a hyposmotic solution **(c)**, they swell up and may burst as water flows into the cell.

those of body fluids are *hyposmotic*. These solutions pump water into cells. When a human cell is placed in a hyposmotic solution—such as pure water, for example—water enters the cell, sometimes causing it to burst (Figure 14.18(c) ▲).

When intravenous fluids are given in a hospital, the majority of the fluid is usually an isosmotic saline solution—a solution containing 0.9 g NaCl per 100 mL of solution. In medicine and in other health-related fields, solution concentrations are often reported in units that indicate the mass of the solute per given volume of solution. Also common is percent mass to volume—which is the mass of the solute in grams divided by the volume of the solution in milliliters times 100%. In these units, the concentration of an isotonic saline solution is 0.9% mass/volume.

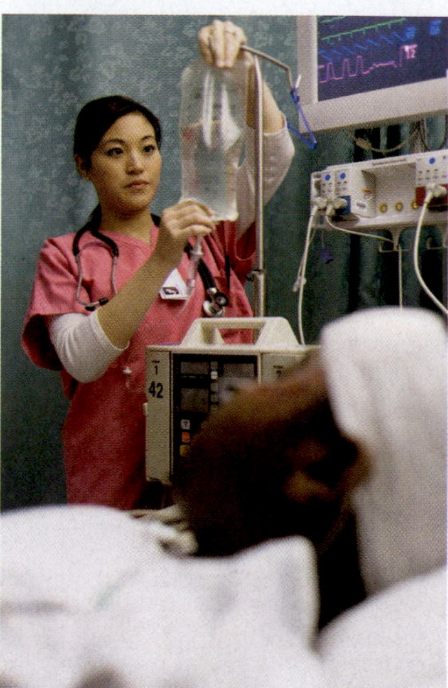

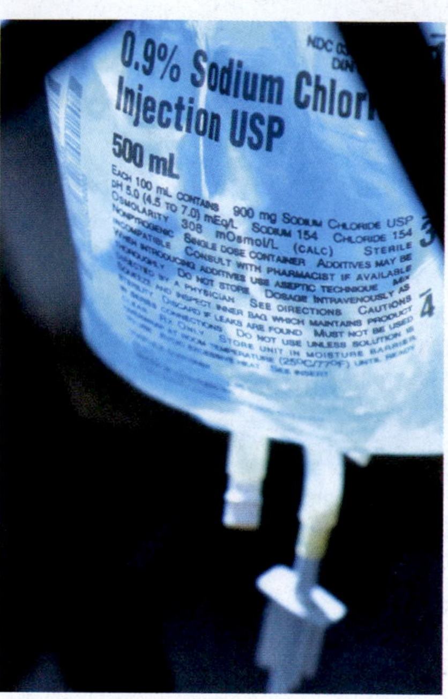

▲ Fluids used for intravenous transfusion must be isosmotic with bodily fluids—that is, they must have the same osmotic pressure.

SELF-ASSESSMENT Quiz

1. Which compound is most soluble in octane (C_8H_{18})?
   a) CH_3OH
   b) CBr_4
   c) H_2O
   d) NH_3

2. An aqueous solution is saturated in both potassium chlorate and carbon dioxide gas at room temperature. What happens when the solution is warmed to 85 °C?
   a) Potassium chlorate precipitates out of solution.
   b) Carbon dioxide bubbles out of solution.
   c) Potassium chlorate precipitates out of solution and carbon dioxide bubbles out of solution.
   d) Nothing happens; all of the potassium chlorate and the carbon dioxide remain dissolved in solution.

3. A 500.0 mL sample of pure water is allowed to come to equilibrium with pure oxygen gas at a pressure of 755 mmHg. What mass of oxygen gas dissolves in the water? (The Henry's law constant for oxygen gas at 25 °C is 1.3×10^{-3} M/atm.)
   a) 15.7 g
   b) 6.5×10^{-3} g
   c) 0.041 g
   d) 0.021 g

4. A potassium bromide solution is 7.55% potassium bromide by mass, and its density is 1.03 g/mL. What mass of potassium bromide is contained in 35.8 mL of the solution?
   a) 2.78 g
   b) 2.70 g
   c) 4.88 g
   d) 2.62 g

5. A solution contains 22.4 g glucose ($C_6H_{12}O_6$) dissolved in 0.500 L of water. What is the molality of the solution? (Assume a density of 1.00 g/mL for water.)
   a) 0.238 *m*
   b) 44.8 *m*
   c) 0.249 *m*
   d) 4.03 *m*

6. A sodium nitrate solution is 12.5% $NaNO_3$ by mass and has a density of 1.02 g/mL. Calculate the molarity of the solution.
   a) 1.44 M
   b) 12.8 M
   c) 6.67 M
   d) 1.50 M

7. What is the vapor pressure of an aqueous ethylene glycol ($C_2H_6O_2$) solution that is 14.8% $C_2H_6O_2$ by mass? The vapor pressure of pure water at 25 °C is 23.8 torr.
   a) 3.52 torr
   b) 22.7 torr
   c) 1.14 torr
   d) 20.3 torr

8. A solution contains a mixture of two volatile substances A and B. The mole fraction of substance A is 0.35. At 32 °C the vapor pressure of pure A is 87 mmHg and the vapor pressure of pure B is 122 mmHg. What is the total vapor pressure of the solution at this temperature?
   a) 110 mmHg
   b) 209 mmHg
   c) 99.3 mmHg
   d) 73.2 mmHg

9. What mass of glucose ($C_6H_{12}O_6$) should you dissolve in 10.0 kg of water to obtain a solution with a freezing point of −4.2 °C?
   a) 0.023 kg
   b) 4.1 kg
   c) 0.41 kg
   d) 14.1 kg

10. Which aqueous solution has the highest boiling point?
    a) 1.25 M $C_6H_{12}O_6$
    b) 1.25 M KNO_3
    c) 1.25 M $Ca(NO_3)_2$
    d) None of the above (they all have the same boiling point)

11. The osmotic pressure of a solution containing 22.7 mg of an unknown protein in 50.0 mL of solution is 2.88 mmHg at 25 °C. Determine the molar mass of the protein.
    a) 246 g/mol
    b) 3.85 g/mol
    c) 2.93×10^3 g/mol
    d) 147 g/mol

12. The enthalpy of solution for NaOH is −44.46 kJ/mol. What can you conclude about the relative magnitudes of the absolute values of ΔH_{solute} and $\Delta H_{hydration}$, where ΔH_{solute} is the heat associated with separating the solute particles and $\Delta H_{hydration}$ is the heat associated with dissolving the solute particles in water?
    a) $|\Delta H_{solute}| > |\Delta H_{hydration}|$
    b) $|\Delta H_{solute}| < |\Delta H_{hydration}|$
    c) $|\Delta H_{solute}| = |\Delta H_{hydration}|$
    d) None of the above (nothing can be concluded about the relative magnitudes)

13. A 2.4 *m* aqueous solution of an ionic compound with the formula MX_2 has a boiling point of 103.4 °C. Calculate the van't Hoff factor (i) for MX_2 at this concentration.
    a) 2.8
    b) 83
    c) 0.73
    d) 1.0

14. A solution is an equimolar mixture of two volatile components A and B. Pure A has a vapor pressure of 50 torr, and pure B has a vapor pressure of 100 torr. The vapor pressure of the mixture is 85 torr. What can you conclude about the relative strengths of the intermolecular forces between particles of A and B (relative to those between particles of A and those between particles of B)?
    a) The intermolecular forces between particles A and B are *weaker* than those between particles of A and those between particles of B.
    b) The intermolecular forces between particles A and B are *stronger* than those between particles of A and those between particles of B.
    c) The intermolecular forces between particles A and B are *the same as* those between particles of A and those between particles of B.
    d) Nothing can be concluded about the relative strengths of intermolecular forces from this observation.

15. An aqueous solution is in equilibrium with a gaseous mixture containing an equal number of moles of oxygen, nitrogen, and helium. Rank the relative concentrations of each gas in the aqueous solution from highest to lowest.
    a) $[O_2] > [N_2] > [He]$
    b) $[He] > [N_2] > [O_2]$
    c) $[N_2] > [He] > [O_2]$
    d) $[N_2] > [O_2] > [He]$

Answers: 1:b; 2:b; 3:d; 4:a; 5:c; 6:d; 7:b; 8:a; 9:b; 10:c; 11:c; 12:b; 13:a; 14:a; 15:a

CHAPTER SUMMARY 14

REVIEW

KEY LEARNING OUTCOMES

CHAPTER OBJECTIVES	ASSESSMENT
Determining Whether a Solute Is Soluble in a Solvent (14.2)	• Example 14.1 For Practice 14.1 Exercises 25–30
Using Henry's Law to Predict the Solubility of Gases with Increasing Pressure (14.4)	• Example 14.2 For Practice 14.2 Exercises 41–46
Calculating Concentrations of Solutions (14.5)	• Examples 14.3, 14.4 For Practice 14.3, 14.4 For More Practice 14.3 Exercises 47, 48, 59, 60
Converting between Concentration Units (14.5)	• Example 14.5 For Practice 14.5 Exercises 61–64
Determining the Vapor Pressure of a Solution Containing a Nonelectrolyte and Nonvolatile Solute (14.6)	• Example 14.6 For Practice 14.6 For More Practice 14.6 Exercises 67, 68
Determining the Vapor Pressure of a Two-Component Solution (14.6)	• Example 14.7 For Practice 14.7 Exercises 69–72
Calculating Freezing Point Depression (14.6)	• Example 14.8 For Practice 14.8 Exercises 73–76, 83–85

Calculating Boiling Point Elevation (14.6)	• Example 14.9	For Practice 14.9	Exercises 73, 74, 83, 84, 86, 87
Determining the Osmotic Pressure (14.6)	• Example 14.10	For Practice 14.10	Exercises 79–82
Determining and Using the van't Hoff Factor (14.7)	• Example 14.11	For Practice 14.11	Exercises 83–85, 87–92
Determining the Vapor Pressure of a Solution Containing an Ionic Solute (14.7)	• Example 14.12	For Practice 14.12	Exercises 93, 94

KEY TERMS

Section 14.1
solution (510)
solvent (510)
solute (510)

Section 14.2
aqueous solution (510)
solubility (510)
entropy (511)
miscible (512)

Section 14.3
enthalpy of solution (ΔH_{soln}) (515)
heat of hydration ($\Delta H_{hydration}$) (516)

Section 14.4
dynamic equilibrium (519)
saturated solution (519)
unsaturated solution (519)
supersaturated solution (519)
recrystallization (520)
Henry's law (521)

Section 14.5
dilute solution (522)
concentrated solution (522)
molarity (M) (523)
molality (m) (524)
parts by mass (524)
percent by mass (524)
parts per million (ppm) (524)
parts per billion (ppb) (524)
parts by volume (524)
mole fraction (χ_{solute}) (525)
mole percent (mol %) (526)

Section 14.6
colligative property (528)
Raoult's law (529)
ideal solution (530)
freezing point depression (534)
boiling point elevation (534)
osmosis (537)
semipermeable membrane (537)
osmotic pressure (537)

Section 14.7
van't Hoff factor (i) (539)

KEY CONCEPTS

Solutions (14.1, 14.2)

- A solution is a homogeneous mixture of two or more substances. In a solution, the majority component is the solvent and the minority component is the solute.
- The tendency toward greater entropy (or greater energy dispersal) is the driving force for solution formation.
- In aqueous solutions, water is a solvent, and a solid, liquid, or gas is the solute.

Solubility and Energetics of Solution Formation (14.2, 14.3)

- The solubility of a substance is the amount of the substance that will dissolve in a given amount of solvent. The solubility of one substance in another depends on the types of intermolecular forces that exist *between* the substances as well as *within* each substance.
- We determine the overall enthalpy change upon solution formation by adding the enthalpy changes for the three steps of solution formation: (1) separation of the solute particles, (2) separation of the solvent particles, and (3) mixing of the solute and solvent particles. The first two steps are both endothermic, while the last is exothermic.
- In aqueous solutions of an ionic compound, we combine the change in enthalpy for Steps 2 and 3 as the heat of hydration ($\Delta H_{hydration}$), which is always negative.

Solution Equilibrium (14.4)

- Dynamic equilibrium in a solution occurs when the rates of dissolution and recrystallization in a solution are equal. A solution in this state is saturated. Solutions containing less than or more than the equilibrium amount of solute are unsaturated or supersaturated, respectively.
- The solubility of most solids in water increases with increasing temperature.
- The solubility of gases in liquids generally decreases with increasing temperature, but it increases with increasing pressure.

Concentration Units (14.5)

- Common units used to express solution concentration are molarity (M), molality (m), mole fraction (χ), mole percent (mol %), percent (%) by mass or volume, parts per million (ppm) by mass or volume, and parts per billion (ppb) by mass or volume. These units are summarized in Table 14.5.

Vapor Pressure Lowering, Freezing Point Depression, Boiling Point Elevation, and Osmosis (14.6, 14.7)

- The presence of a nonvolatile solute in a liquid results in a lower vapor pressure of the solution relative to the vapor pressure of the pure liquid. This lower vapor pressure is predicted by Raoult's law for an ideal solution.

- If the solute–solvent interactions are particularly strong, the actual vapor pressure of the solution is lower than that predicted by Raoult's law.
- If the solute–solvent interactions are particularly weak, the actual vapor pressure of the solution is higher than that predicted by Raoult's law.
- The addition of a nonvolatile solute to a liquid results in a solution that has a lower freezing point and a higher boiling point than the pure solvent.
- The flow of solvent from a solution of lower concentration to a solution of higher concentration is osmosis.
- Vapor pressure lowering, freezing point depression, boiling point elevation, and osmosis are colligative properties and depend only on the number of solute particles added, not the type of solute particles.
- Electrolyte solutes have a greater effect on these colligative properties than the corresponding amount of a nonelectrolyte solute as specified by the van't Hoff factor.

KEY EQUATIONS AND RELATIONSHIPS

Henry's Law: Solubility of Gases with Increasing Pressure (14.4)

$$S_{gas} = k_H P_{gas} \text{ is Henry's law constant}$$

Molarity (M) of a Solution (14.5)

$$(M) = \frac{\text{amount solute (in mol)}}{\text{volume solution (in L)}}$$

Molality (*m*) of a Solution (14.5)

$$(m) = \frac{\text{amount solute (in mol)}}{\text{mass solvent (in kg)}}$$

Concentration of a Solution in Parts by Mass and Parts by Volume (14.5)

$$\text{Percent by mass} = \frac{\text{mass solute} \times 100\%}{\text{mass solution}}$$

$$\text{Parts per million (ppm)} = \frac{\text{mass solute} \times 10^6}{\text{mass solution}}$$

$$\text{Parts per billion (ppb)} = \frac{\text{mass solute} \times 10^9}{\text{mass solution}}$$

$$\text{Parts by volume} = \frac{\text{volume solute} \times \text{multiplication factor}}{\text{volume solution}}$$

Concentration of a Solution in Mole Fraction (χ) and Mole Percent (14.5)

$$\chi_{solute} = \frac{n_{solute}}{n_{solute} + n_{solvent}}$$

$$\text{Mol\%} = \chi \times 100\%$$

Raoult's Law: Relationship between the Vapor Pressure of a Solution ($P_{solution}$) the Mole Fraction of the Solvent ($\chi_{solvent}$) and the Vapor Pressure of the Pure Solvent ($P^{\circ}_{solvent}$) (14.6)

$$P_{solution} = \chi_{solvent} P^{\circ}_{solvent}$$

The Vapor Pressure of a Solution Containing Two Volatile Components (14.6)

$$P_A = \chi_A P^{\circ}_A$$

$$P_B = \chi_B P^{\circ}_B$$

$$P_{tot} = P_A + P_B$$

Relationship between Freezing Point Depression (ΔT_f), Molality (*m*), and Freezing Point Depression Constant K_f (14.6)

$$\Delta T_f = m \times K_f$$

Relationship between Boiling Point Elevation (ΔT_b), Molality (*m*), and Boiling Point Elevation Constant (K_f) (14.6)

$$\Delta T_b = m \times K_b$$

Relationship between Osmotic Pressure (Π), Molarity (M), the Ideal Gas Constant (*R*), and Temperature (*T*, in K) (14.6)

$$\Pi = MRT \; (R = 0.08206 \text{ L} \cdot \text{atm/mol} \cdot \text{K})$$

van't Hoff Factor (*i*): Ratio of Moles of Particles in Solution to Moles of Formula Units Dissolved (14.7)

$$i = \frac{\text{moles of particles in solution}}{\text{moles of formula units dissolved}}$$

EXERCISES

REVIEW QUESTIONS

1. What is a solution? What are the solute and solvent?
2. What does it mean when we say that a substance is soluble in another substance? Which units do we use to report solubility?
3. Why do two ideal gases thoroughly mix when combined? What drives the mixing?
4. What is entropy? What role does entropy play in the formation of solutions?
5. What kinds of intermolecular forces are involved in solution formation?
6. Explain how the relative strengths of solute–solute interactions, solvent–solvent interactions, and solvent–solute interactions affect solution formation.

7. What does the statement *like dissolves like* mean with respect to solution formation?
8. List the three steps involved in evaluating the enthalpy changes associated with solution formation.
9. What is the heat of hydration ($\Delta H_{hydration}$)? How does the enthalpy of solution depend on the relative magnitudes of ΔH_{solute} and $\Delta H_{hydration}$?
10. Explain dynamic equilibrium with respect to solution formation. What is a saturated solution? An unsaturated solution? A supersaturated solution?
11. How does temperature affect the solubility of a solid in a liquid? How is this temperature dependence exploited to purify solids through recrystallization?
12. How does temperature affect the solubility of a gas in a liquid ? How does this temperature dependence affect the amount of oxygen available for fish and other aquatic animals?
13. How does pressure affect the solubility of a gas in a liquid ? How does this pressure dependence account for the bubbling that occurs upon opening a can of soda?
14. What is Henry's law? For what kinds of calculations is Henry's law useful?
15. What are the common units for expressing solution concentration?
16. How are parts by mass and parts by volume used in calculations?
17. What is the effect of a nonvolatile solute on the vapor pressure of a liquid? Why is the vapor pressure of a solution different from the vapor pressure of the pure liquid solvent?
18. What is Raoult's law? For what kind of calculations is Raoult's law useful?
19. Explain the difference between an ideal and a nonideal solution.
20. What is the effect on vapor pressure of a solution with particularly *strong* solute–solvent interactions? With particularly *weak* solute–solvent interactions?
21. Explain why the lower vapor pressure for a solution containing a nonvolatile solute results in a higher boiling point and lower melting point compared to the pure solvent.
22. What are colligative properties?
23. What is osmosis? What is osmotic pressure?
24. Explain the significance of the van't Hoff factor (i) and its role in determining the colligative properties of solutions containing ionic solutes.

PROBLEMS BY TOPIC

Solubility

25. Pick an appropriate solvent from Table 14.3 to dissolve each substance. State the kind of intermolecular forces that would occur between the solute and solvent in each case.
    a. motor oil (nonpolar)
    b. ethanol (polar, contains an OH group)
    c. lard (nonpolar)
    d. potassium chloride (ionic)
26. Pick an appropriate solvent from Table 14.3 to dissolve each substance. State the kind of intermolecular forces that would occur between the solute and solvent in each case.
    a. isopropyl alcohol (polar, contains an OH group)
    b. sodium chloride (ionic)
    c. vegetable oil (nonpolar)
    d. sodium nitrate (ionic)
27. Which molecule would you expect to be more soluble in water, $CH_3CH_2CH_2CH_2OH$ or $HOCH_2CH_2CH_2OH$?
28. Which molecule would you expect to be more soluble in water, CCl_4 or CH_2Cl_2?
29. For each compound, would you expect greater solubility in water or in hexane? Indicate the kinds of intermolecular forces that occur between the solute and the solvent in which the molecule is most soluble.
    a. glucose 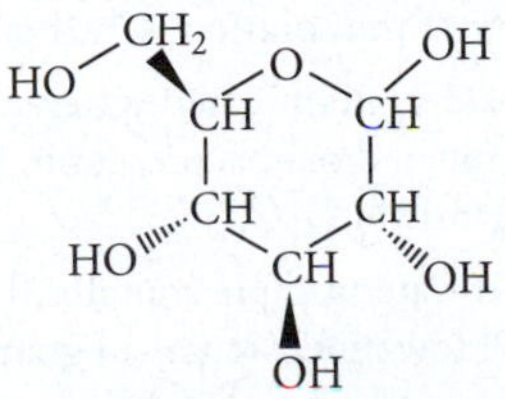
    b. naphthalene 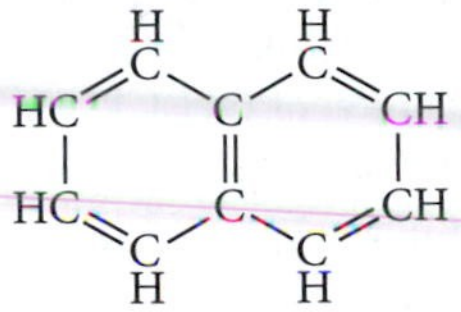
    c. dimethyl ether 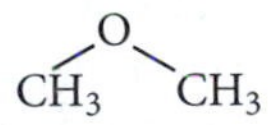
    d. alanine (an amino acid) 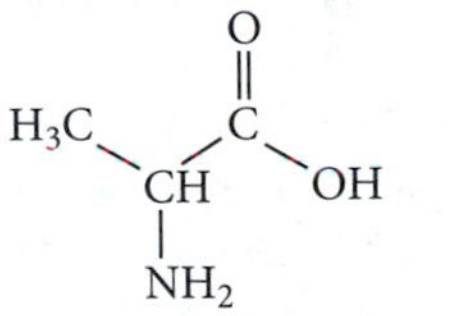
30. For each compound, would you expect greater solubility in water or in hexane? Indicate the kinds of intermolecular forces that would occur between the solute and the solvent in which the molecule is most soluble.
    a. toluene

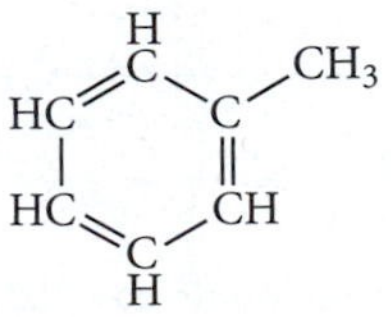

b. sucrose (table sugar)

c. isobutene

d. ethylene glycol

Energetics of Solution Formation

31. When ammonium chloride (NH_4Cl) is dissolved in water, the solution becomes colder.
    - **a.** Is the dissolution of ammonium chloride endothermic or exothermic?
    - **b.** What can you conclude about the relative magnitudes of the lattice energy of ammonium chloride and its heat of hydration?
    - **c.** Sketch a qualitative energy diagram similar to Figure 14.6 for the dissolution of NH_4Cl.
    - **d.** Why does the solution form? What drives the process?

32. When lithium iodide (LiI) is dissolved in water, the solution becomes hotter.
    - **a.** Is the dissolution of lithium iodide endothermic or exothermic?
    - **b.** What can you conclude about the relative magnitudes of the lattice energy of lithium iodide and its heat of hydration?
    - **c.** Sketch a qualitative energy diagram similar to Figure 14.6 for the dissolution of LiI.
    - **d.** Why does the solution form? What drives the process?

33. Silver nitrate has a lattice energy of −820 kJ/mol and a heat of solution of −22.6 kJ/mol. Calculate the heat of hydration for silver nitrate.

34. Use the given data to calculate the heats of hydration of lithium chloride and sodium chloride. Which of the two cations, lithium or sodium, has stronger ion–dipole interactions with water? Why?

Compound	Lattice Energy (kJ/mol)	ΔH_{soln} (kJ/mol)
LiCl	−834	−37.0
NaCl	−769	+3.88

35. Lithium iodide has a lattice energy of -7.3×10^2 kJ/mol and a heat of hydration of −793 kJ/mol. Find the heat of solution for lithium iodide and determine how much heat is evolved or absorbed when 15.0 g of lithium iodide completely dissolves in water.

36. Potassium nitrate has a lattice energy of −163.8 kcal/mol and a heat of hydration of −155.5 kcal/mol. How much potassium nitrate has to dissolve in water to absorb 1.00×10^2 kJ of heat?

Solution Equilibrium and Factors Affecting Solubility

37. A solution contains 25 g of NaCl per 100.0 g of water at 25 °C. Is the solution unsaturated, saturated, or supersaturated? (Refer to Figure 14.10.)

38. A solution contains 32 g of KNO_3 per 100.0 g of water at 25 °C. Is the solution unsaturated, saturated, or supersaturated? (Refer to Figure 14.10.)

39. A KNO_3 solution containing 45 g of KNO_3 per 100.0 g of water is cooled from 40 °C to 0 °C. What happens during cooling? (Refer to Figure 14.10.)

40. A KCl solution containing 42 g of KCl per 100.0 g of water is cooled from 60 °C to 0 °C. What happens during cooling? (Refer to Figure 14.10.)

41. Some laboratory procedures involving oxygen-sensitive reactants or products call for using water that has been boiled (and then cooled). Explain.

42. A person preparing a fish tank fills the tank with water that has been boiled (and then cooled). When the person puts fish into the tank, they die. Explain.

43. Scuba divers breathing air at increased pressure can suffer from nitrogen narcosis—a condition resembling drunkenness—when the partial pressure of nitrogen exceeds about 4 atm. What property of gas/water solutions causes this to happen? How can a diver reverse this effect?

44. Scuba divers breathing air at increased pressure can suffer from oxygen toxicity—too much oxygen in their bloodstream—when the partial pressure of oxygen exceeds about 1.4 atm. What happens to the amount of oxygen in a diver's bloodstream when he or she breathes oxygen at elevated pressures? How can this be reversed?

45. Calculate the mass of nitrogen dissolved at room temperature in an 80.0 L home aquarium. Assume a total pressure of 1.0 atm and a mole fraction for nitrogen of 0.78.

46. Use Henry's law to determine the molar solubility of helium at a pressure of 1.0 atm and 25 °C.

Concentrations of Solutions

47. An aqueous NaCl solution is made using 112 g of NaCl diluted to a total solution volume of 1.00 L. Calculate the molarity, molality, and mass percent of the solution. (Assume a density of 1.08 g/mL for the solution.)

48. An aqueous KNO_3 solution is made using 72.5 g of KNO_3 diluted to a total solution volume of 2.00 L. Calculate the molarity, molality, and mass percent of the solution. (Assume a density of 1.05 g/mL for the solution.)

49. To what volume should you dilute 50.0 mL of a 5.00 M KI solution so that 25.0 mL of the diluted solution contains 3.05 g of KI?

50. To what volume should you dilute 125 mL of an 8.00 M $CuCl_2$ solution so that 50.0 mL of the diluted solution contains 4.67 g $CuCl_2$?

51. Silver nitrate solutions are used to plate silver onto other metals. What is the maximum amount of silver (in grams) that can be plated out of 4.8 L of an $AgNO_3$ solution containing 3.4% Ag by mass? Assume that the density of the solution is 1.01 g/mL.

52. A dioxin-contaminated water source contains 0.085% dioxin by mass. How much dioxin is present in 2.5 L of this water? Assume a density of 1.00 g/mL.

53. A hard water sample contains 0.0085% Ca by mass (in the form of Ca^{2+} ions). How much water (in grams) contains 1.2 g of Ca? (1.2 g of Ca is the recommended daily allowance of calcium for adults between 19 and 24 years old.)

54. Lead is a toxic metal that affects the central nervous system. A Pb-contaminated water sample contains 0.0011% Pb by mass. How much of the water (in mL) contains 150 mg of Pb? (Assume a density of 1.0 g/mL.)

55. You can purchase nitric acid in a concentrated form that is 70.3% HNO_3 by mass and has a density of 1.41 g/mL. How can you prepare 1.15 L of 0.100 M HNO_3 from the concentrated solution?

56. You can purchase hydrochloric acid in a concentrated form that is 37.0% HCl by mass and that has a density of 1.20 g/mL. How can you prepare 2.85 L of 0.500 M HCl from the concentrated solution?

57. Describe how to prepare each solution from the dry solute and the solvent.

a. 1.00×10^2 mL of 0.500 M KCl
b. 1.00×10^2 g of 0.500 *m* KCl
c. 1.00×10^2 g of 5.0% KCl solution by mass

58. Describe how to prepare each solution from the dry solute and the solvent.

a. 125 mL of 0.100 M $NaNO_3$
b. 125 g of 0.100 *m* $NaNO_3$
c. 125 g of 1.0 $NaNO_3$ solution by mass

59. A solution is prepared by dissolving 28.4 g of glucose ($C_6H_{12}O_6$) in 355 g of water. The final volume of the solution is 378 mL. For this solution, calculate the concentration in each unit.

a. molarity
b. molality
c. percent by mass
d. mole fraction
e. mole percent

60. A solution is prepared by dissolving 20.2 mL of methanol (CH_3OH) in 100.0 mL of water at 25 °C. The final volume of the solution is 118 mL. The densities of methanol and water at this temperature are 0.782 g/mL and 1.00 g/mL, respectively. For this solution, calculate the concentration in each unit.

a. molarity
b. molality
c. percent by mass
d. mole fraction
e. mole percent

61. Household hydrogen peroxide is an aqueous solution containing 3.0% hydrogen peroxide by mass. What is the molarity of this solution? (Assume a density of 1.01 g/mL.)

62. One brand of laundry bleach is an aqueous solution containing 4.55% sodium hypochlorite (NaOCl) by mass. What is the molarity of this solution? (Assume a density of 1.02 g/mL.)

63. An aqueous solution contains 36% HCl by mass. Calculate the molality and mole fraction of the solution.

64. An aqueous solution contains 5.0% NaCl by mass. Calculate the molality and mole fraction of the solution.

Vapor Pressure of Solutions

65. A beaker contains 100.0 mL of pure water. A second beaker contains 100.0 mL of seawater. The two beakers are left side by side on a lab bench for one week. At the end of the week, the liquid level in both beakers has decreased. However, the level has decreased more in one of the beakers than in the other. Which one and why?

66. Which solution has the highest vapor pressure?

a. 20.0 g of glucose ($C_6H_{12}O_6$) in 100.0 mL of water
b. 20.0 g of sucrose ($C_{12}H_{22}O_{11}$) in 100.0 mL of water
c. 10.0 g of potassium acetate $KC_2H_3O_2$ in 100.0 mL of water

67. Calculate the vapor pressure of a solution containing 24.5 g of glycerin ($C_3H_8O_3$) in 135 mL of water at 30.0 °C. The vapor pressure of pure water at this temperature is 31.8 torr. Assume that glycerin is not volatile and dissolves molecularly (that is, it is not ionic), and use a density of 1.00 g/mL for the water.

68. A solution contains naphthalene ($C_{10}H_8$) dissolved in hexane (C_6H_{14}) at a concentration of 12.35% naphthalene by mass. Calculate the vapor pressure at 25 °C of hexane above the solution. The vapor pressure of pure hexane at 25 °C is 151 torr.

69. A solution contains 50.0 g of heptane (C_7H_{16}) and 50.0 g of octane (C_8H_{18}) at 25 °C. The vapor pressures of pure heptane and pure octane at 25 °C are 45.8 torr and 10.9 torr, respectively. Assuming ideal behavior, answer each question.

a. What is the vapor pressure of each of the solution components in the mixture?
b. What is the total pressure above the solution?
c. What is the composition of the vapor in mass percent?
d. Why is the composition of the vapor different from the composition of the solution?

70. A solution contains a mixture of pentane and hexane at room temperature. The solution has a vapor pressure of 258 torr. Pure pentane and hexane have vapor pressures of 425 torr and 151 torr, respectively, at room temperature. What is the mole fraction composition of the mixture? (Assume ideal behavior.)

71. A solution contains 4.08 g of chloroform ($CHCl_3$) and 9.29 g of acetone (CH_3COCH_3). The vapor pressures at 35 °C of pure chloroform and pure acetone are 295 torr and 332 torr, respectively. Assuming ideal behavior, calculate the vapor pressures of each of the components and the total vapor pressure above the solution. The experimentally measured total vapor pressure of the solution at 35 °C is 312 torr. Is the solution ideal? If not, what can you say about the relative strength of chloroform–acetone interactions compared to the acetone–acetone and chloroform–chloroform interactions?

72. A solution of methanol and water has a mole fraction of water of 0.312 and a total vapor pressure of 211 torr at 39.9 °C. The vapor pressures of pure methanol and pure water at this temperature are 256 torr and 55.3 torr, respectively. Is the solution ideal? If not, what can you say about the relative strengths of the solute–solvent interactions compared to the solute–solute and solvent–solvent interactions?

Freezing Point Depression, Boiling Point Elevation, and Osmosis

73. A glucose solution contains 55.8 g of glucose ($C_6H_{12}O_6$) in 455 g of water. Determine the freezing point and boiling point of the solution. (Assume a density of 1.00 g/mL for water.)

74. An ethylene glycol solution contains 21.2 g of ethylene glycol ($C_2H_6O_2$) in 85.4 mL of water. Determine the freezing point and boiling point of the solution. (Assume a density of 1.00 g/mL for water.)

75. Calculate the freezing point and melting point of a solution containing 10.0 g of naphthalene ($C_{10}H_8$) in 100.0 mL of benzene. Benzene has a density of 0.877 g/cm^3.

76. Calculate the freezing point and melting point of a solution containing 7.55 g of ethylene glycol ($C_2H_6O_2$) in 85.7 mL of ethanol. Ethanol has a density of 0.789 g/cm^3.

77. An aqueous solution containing 17.5 g of an unknown molecular (nonelectrolyte) compound in 100.0 g of water has a freezing point of −1.8 °C. Calculate the molar mass of the unknown compound.

78. An aqueous solution containing 35.9 g of an unknown molecular (nonelectrolyte) compound in 150.0 g of water has a freezing point of −1.3 °C. Calculate the molar mass of the unknown compound.

79. Calculate the osmotic pressure of a solution containing 24.6 g of glycerin ($C_3H_8O_3$) in 250.0 mL of solution at 298 K.

80. What mass of sucrose ($C_{12}H_{22}O_{11}$) would you combine with 5.00×10^2 g of water to make a solution with an osmotic pressure of 8.55 atm at 298 K? (Assume a density of 1.0 g/mL for the solution.)

81. A solution containing 27.55 mg of an unknown protein per 25.0 mL solution was found to have an osmotic pressure of 3.22 torr at 25 °C. What is the molar mass of the protein?

82. Calculate the osmotic pressure of a solution containing 18.75 mg of hemoglobin in 15.0 mL of solution at 25 °C. The molar mass of hemoglobin is 6.5×10^4 g/mol.

83. Calculate the freezing point and boiling point of each aqueous solution, assuming complete dissociation of the solute.

a. 0.100 *m* K_2S
b. 21.5 g of $CuCl_2$ in 4.50×10^2 g water
c. 5.5% $NaNO_3$ by mass (in water)

84. Calculate the freezing point and boiling point in each solution, assuming complete dissociation of the solute.

a. 10.5 g $FeCl_3$ in 1.50×10^2 g water
b. 3.5% KCl by mass (in water)
c. 0.150 *m* MgF_2

85. What mass of salt (NaCl) should you add to 1.00 L of water in an ice-cream maker to make a solution that freezes at −10.0 °C? Assume complete dissociation of the NaCl and a density of 1.00 g/mL for water.

86. Determine the required concentration (in percent by mass) for an aqueous ethylene glycol ($C_2H_6O_2$) solution to have a boiling point of 104.0 °C.

87. Use the van't Hoff factors in Table 14.7 to calculate each colligative property.

a. the melting point of a 0.100 *m* iron(III) chloride solution
b. the osmotic pressure of a 0.085 M potassium sulfate solution at 298 K
c. the boiling point of a 1.22% by mass magnesium chloride solution

88. Referring to the van't Hoff factors in Table 14.7, calculate the mass of solute required to make each aqueous solution.

a. a sodium chloride solution containing 1.50×10^2 g of water that has a melting point of −1.0 °C
b. 2.50×10^2 mL of a magnesium sulfate solution that has an osmotic pressure of 3.82 atm at 298 K
c. an iron(III) chloride solution containing 2.50×10^2 g of water that has a boiling point of 102 °C

89. A 1.2 *m* aqueous solution of an ionic compound with the formula MX_2 has a boiling point of 101.4 °C. Calculate the van't Hoff factor (i) for MX_2 at this concentration.

90. A 0.95 *m* aqueous solution of an ionic compound with the formula MX has a freezing point of −3.0 °C. Calculate the van't Hoff factor (i) for MX at this concentration.

91. A 0.100 M ionic solution has an osmotic pressure of 8.3 atm at 25 °C. Calculate the van't Hoff factor (i) for this solution.

92. A solution contains 8.92 g of KBr in 500.0 mL of solution and has an osmotic pressure of 6.97 atm at 25 °C. Calculate the van't Hoff factor (i) for KBr at this concentration.

93. Calculate the vapor pressure at 25 °C of an aqueous solution that is 5.50% NaCl by mass. (Assume complete dissociation of the solute.)

94. An aqueous $CaCl_2$ solution has a vapor pressure of 81.6 mmHg at 50 °C. The vapor pressure of pure water at this temperature is 92.6 mmHg. What is the concentration of $CaCl_2$ in mass percent? (Assume complete dissociation of the solute.)

CUMULATIVE PROBLEMS

95. The solubility of carbon tetrachloride (CCl_4) in water at 25 °C is 1.2 g/L. The solubility of chloroform ($CHCl_3$) at the same temperature is 10.1 g/L. Why is chloroform almost ten times more soluble in water than carbon tetrachloride?

96. The solubility of phenol in water at 25 °C is 8.7 g/L. The solubility of naphthol at the same temperature is only 0.074 g/L. Examine the structures of phenol and naphthol shown here and explain why phenol is so much more soluble than naphthol.

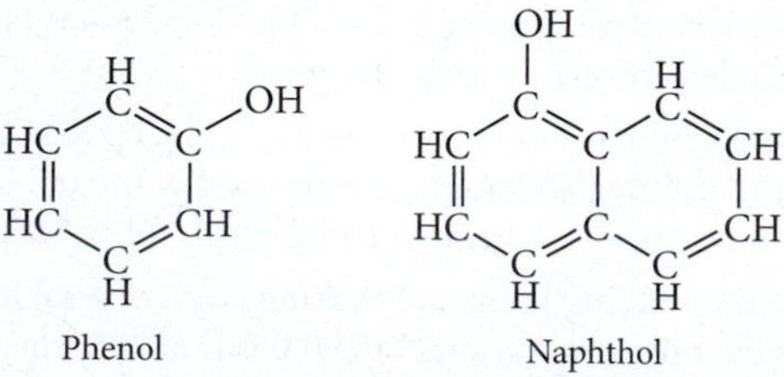

97. Potassium perchlorate ($KClO_4$) has a lattice energy of −599 kJ/mol and a heat of hydration of −548 kJ/mol. Find the heat of solution for potassium perchlorate and determine the temperature change that occurs when 10.0 g of potassium perchlorate is dissolved with enough water to make 100.0 mL of solution. (Assume a heat capacity of 4.05 J/g·°C for the solution and a density of 1.05 g/mL.)

98. Sodium hydroxide (NaOH) has a lattice energy of −887 kJ/mol and a heat of hydration of −932 kJ/mol. How much solution could be heated to boiling by the heat evolved by the dissolution of 25.0 g of NaOH? (For the solution, assume a heat capacity of 4.0 J/g·°C, an initial temperature of 25.0 °C, a boiling point of 100.0 °C, and a density of 1.05 g/mL.)

99. A saturated solution forms when 0.0537 L of argon, at a pressure of 1.0 atm and temperature of 25 °C, is dissolved in 1.0 L of water. Calculate the Henry's law constant for argon.

100. A gas has a Henry's law constant of 0.112 M/atm. What total volume of solution is needed to completely dissolve 1.65 L of the gas at a pressure of 725 torr and a temperature of 25 °C?

101. The Safe Drinking Water Act (SDWA) sets a limit for mercury—a toxin to the central nervous system—at 0.0020 ppm by mass. Water suppliers must periodically test their water to ensure that mercury levels do not exceed this limit. Suppose water becomes contaminated with mercury at twice the legal limit (0.0040 ppm). How much of this water would a person have to consume to ingest 50.0 mg of mercury?

102. Water softeners often replace calcium ions in hard water with sodium ions. Because sodium compounds are soluble, the presence of sodium ions in water does not cause the white, scaly residues caused by calcium ions. However, calcium is more beneficial to human health than sodium because calcium is a necessary part of the human diet, while high levels of sodium intake are linked to increases in blood pressure. The U.S. Food and Drug Administration (FDA) recommends that adults ingest less than 2.4 g of sodium per day. How many liters of softened water, containing a sodium concentration of 0.050% sodium by mass, does a person have to consume to exceed the FDA recommendation? (Assume a water density of 1.0 g/mL.)

103. An aqueous solution contains 12.5% NaCl by mass. What mass of water (in grams) is contained in 2.5 L of the vapor above this solution at 55 °C? The vapor pressure of pure water at 55 °C is 118 torr. (Assume complete dissociation of NaCl.)

104. The vapor above an aqueous solution contains 19.5 mg water per liter at 25 °C. Assuming ideal behavior, what is the concentration of the solute within the solution in mole percent?

105. What is the freezing point of an aqueous solution that boils at 106.5 °C?

106. What is the boiling point of an aqueous solution that has a vapor pressure of 20.5 torr at 25 °C? (Assume a nonvolatile solute.)

107. An isotonic solution contains 0.90% NaCl mass to volume. Calculate the percent mass to volume for isotonic solutions containing each solute at 25 °C. Assume a van't Hoff factor of 1.9 for all *ionic* solutes.

a. KCl
b. NaBr
c. glucose ($C_6H_{12}O_6$)

108. Magnesium citrate, $Mg_3(C_6H_5O_7)_2$, belongs to a class of laxatives called *hyperosmotics*, which cause rapid emptying of the bowel. When a person consumes a concentrated solution of magnesium citrate, it passes through the intestines, drawing water and promoting diarrhea, usually within 6 hours. Calculate the osmotic pressure of a magnesium citrate laxative solution containing 28.5 g of magnesium citrate in 235 mL of solution at 37 °C (approximate body temperature). Assume complete dissociation of the ionic compound.

109. A solution is prepared from 4.5701 g of magnesium chloride and 43.238 g of water. The vapor pressure of water above this solution is 0.3624 atm at 348.0 K. The vapor pressure of pure water at this temperature is 0.3804 atm. Find the value of the van't Hoff factor (i) for magnesium chloride in this solution.

110. When HNO_2 dissolves in water, it partially dissociates according to the equation $HNO_2(aq) \rightleftharpoons H^+(aq) + NO_2^-(aq)$. A solution contains 7.050 g of HNO_2 in 1.000 kg of water. Its freezing point is −0.2929 °C. Calculate the fraction of HNO_2 that has dissociated.

111. A solution of a nonvolatile solute in water has a boiling point of 375.3 K. Calculate the vapor pressure of water above this solution at 338 K. The vapor pressure of pure water at this temperature is 0.2467 atm.

112. The density of a 0.438 M solution of potassium chromate (K_2CrO_4) at 298 K is 1.063 g/mL. Calculate the vapor pressure of water above the solution. The vapor pressure of pure water at this temperature is 0.0313 atm. (Assume complete dissociation of the solute.)

113. The vapor pressure of carbon tetrachloride, CCl_4, is 0.354 atm, and the vapor pressure of chloroform, $CHCl_3$, is 0.526 atm at 316 K. A solution is prepared from equal masses of these two compounds at this temperature. Calculate the mole fraction of the chloroform in the vapor above the solution. If the vapor above the original solution is condensed and isolated into a separate flask, what would the vapor pressure of chloroform be above this new solution?

114. Distillation is a method of purification based on successive separations and recondensations of vapor above a solution. Use the result of the previous problem to calculate the mole fraction of chloroform in the vapor above a solution obtained by three successive separations and condensations of the vapors above the original solution of carbon tetrachloride and chloroform. Show how this result supports the use of distillation as a separation method.

115. A solution of 49.0% H_2SO_4 by mass has a density of 1.39 g/cm^3 at 293 K. A 25.0 cm^3 sample of this solution is mixed with enough water to increase the volume of the solution to 99.8 cm^3. Find the molarity of sulfuric acid in this solution.

116. Find the mass of urea (CH_4N_2O) needed to prepare 50.0 g of a solution in water in which the mole fraction of urea is 0.0770.

117. A solution contains 10.05 g of unknown compound dissolved in 50.0 mL of water. (Assume a density of 1.00 g/mL for water.) The freezing point of the solution is −3.16 °C. The mass percent composition of the compound is 60.97% C, 11.94% H, and the rest is O. What is the molecular formula of the compound?

118. The osmotic pressure of a solution containing 2.10 g of an unknown compound dissolved in 175.0 mL of solution at 25 °C is 1.93 atm. The combustion of 24.02 g of the unknown compound produced 28.16 g CO_2 and 8.64 g H_2O. What is the molecular formula of the compound (which contains only carbon, hydrogen, and oxygen)?

119. A 100.0 mL aqueous sodium chloride solution is 13.5% NaCl by mass and has a density of 1.12 g/mL. What would you add (solute or solvent) and what mass of it to make the boiling point of the solution 104.4 °C? (Use $i = 1.8$ for NaCl.)

120. A 50.0 mL solution is initially 1.55% $MgCl_2$ by mass and has a density of 1.05 g/mL. What is the freezing point of the solution after you add an additional 1.35 g $MgCl_2$? (Use $i = 2.5$ for $MgCl_2$.)

CHALLENGE PROBLEMS

121. The small bubbles that form on the bottom of a water pot that is being heated (before boiling) are due to dissolved air coming out of solution. Use Henry's law and the solubilities given to calculate the total volume of nitrogen and oxygen gas that should bubble out of 1.5 L of water upon warming from 25 °C to 50 °C. Assume that the water is initially saturated with nitrogen and oxygen gas at 25 °C and a total pressure of 1.0 atm. Assume that the gas bubbles out at a temperature of 50 °C. The solubility of oxygen gas at 50 °C is 27.8 mg/L at an oxygen pressure of 1.00 atm. The solubility of nitrogen gas at 50 °C is 14.6 mg/L at a nitrogen pressure of 1.00 atm. Assume that the air above the water contains an oxygen partial pressure of 0.21 atm and a nitrogen partial pressure of 0.78 atm.

122. The vapor above a mixture of pentane and hexane at room temperature contains 35.5% pentane by mass. What is the mass percent composition of the solution? Pure pentane and hexane have vapor pressures of 425 torr and 151 torr, respectively, at room temperature.

123. A 1.10 g sample contains only glucose ($C_6H_{12}O_6$) and sucrose ($C_{12}H_{22}O_{11}$). When the sample is dissolved in water to a total solution volume of 25.0 mL, the osmotic pressure of the solution is 3.78 atm at 298 K. What is the mass percent composition of glucose and sucrose in the sample?

124. A solution is prepared by mixing 631 mL of methanol with 501 mL of water. The molarity of methanol in the resulting solution is 14.29 M. The density of methanol at this temperature is 0.792 g/mL. Calculate the difference in volume between this solution and the total volume of water and methanol that were mixed to prepare the solution.

125. Two alcohols, isopropyl alcohol and propyl alcohol, have the same molecular formula, C_3H_8O. A solution of the two that is two-thirds by mass isopropyl alcohol has a vapor pressure of 0.110 atm at 313 K. A solution that is one-third by mass isopropyl alcohol has a vapor pressure of 0.089 atm at 313 K. Calculate the vapor pressure of each pure alcohol at this temperature. Explain the difference given that the formula of propyl alcohol is $CH_3CH_2CH_2OH$ and that of isopropyl alcohol is $(CH_3)_2CHOH$.

126. A metal, M, of atomic mass 96 amu reacts with fluorine to form a salt that can be represented as MF_x. In order to determine x and therefore the formula of the salt, a boiling point elevation experiment is performed. A 9.18 g sample of the salt is dissolved in 100.0 g of water and the boiling point of the solution is found to be 374.38 K. Find the formula of the salt. (Assume complete dissociation of the salt in solution.)

127. Sulfuric acid in water dissociates completely into H^+ and HSO_4^- ions. The HSO_4^- ion dissociates to a limited extent into H^+ and SO_4^-. The freezing point of a 0.1000 m solution of sulfuric acid in water is 272.76 K. Calculate the molality of SO_4^- in the solution, assuming ideal solution behavior.

128. A solution of 75.0 g of benzene (C_6H_6) and 75.0 g of toluene (C_7H_8) has a total vapor pressure of 80.9 mmHg at 303 K. Another solution of 100.0 g benzene and 50.0 g toluene has a total vapor pressure of 93.9 mmHg at this temperature. Find the vapor pressure of pure benzene and pure toluene at 303 K. (Assume ideal solutions.)

129. A solution is prepared by dissolving 11.60 g of a mixture of sodium carbonate and sodium bicarbonate in 1.00 L of water. A 300.0 cm^3 sample of the solution is treated with excess HNO_3 and boiled to remove all the dissolved gas. A total of 0.940 L of dry CO_2 is collected at 298 K and 0.972 atm. Find the molarity of the carbonate and bicarbonate in the solution.

CONCEPTUAL PROBLEMS

130. Substance A is a nonpolar liquid and has only dispersion forces among its constituent particles. Substance B is also a nonpolar liquid and has about the same magnitude of dispersion forces among its constituent particles as substance A. When substance A and B are combined, they spontaneously mix.

a. Why do the two substances mix?
b. Predict the sign and magnitude of ΔH_{soln}.
c. Determine the signs and relative magnitudes of ΔH_{solute}, $\Delta H_{solvent}$, and ΔH_{mix}.

131. A power plant built on a river uses river water as a coolant. The water is warmed as it is used in heat exchangers within the plant. Should the warm water be immediately cycled back into the river? Why or why not?

132. The vapor pressure of a 1 M ionic solution is different from the vapor pressure of a 1 M nonelectrolyte solution. In both cases, the solute is nonvolatile. Which set of diagrams best represents the differences between the two solutions and their vapors?

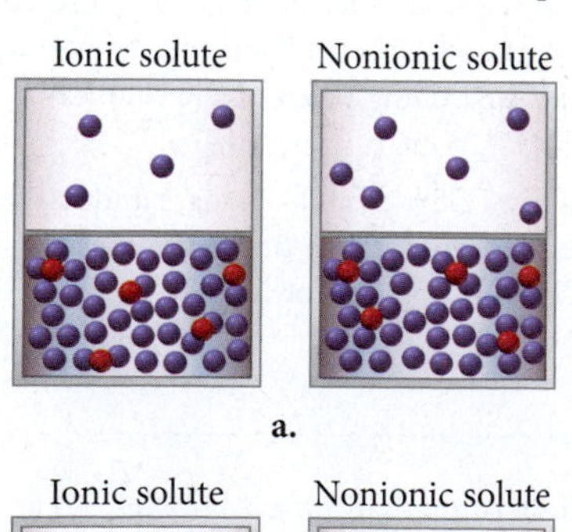

Solvent particles

Solute particles

a.

Ionic solute Nonionic solute

b.

Ionic solute Nonionic solute

c.

Ionic solute Nonionic solute

d.

133. If each substance listed here costs the same amount per kilogram, which would be most cost-effective as a way to lower the freezing point of water? (Assume complete dissociation for all ionic compounds.) Explain.

a. $HOCH_2CH_2OH$
b. NaCl
c. KCl
d. $MgCl_2$
e. $SrCl_2$

134. A helium balloon inflated on one day falls to the ground the next day. The volume of the balloon decreases somewhat overnight but not by enough to explain why it no longer floats. (If you inflate a new balloon with helium to the same size as the balloon that fell to the ground, the newly inflated balloon floats.) Explain.

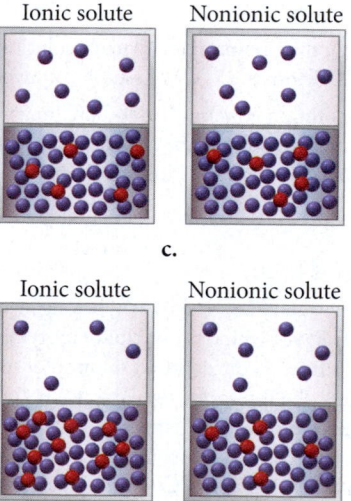

ANSWERS TO CONCEPTUAL CONNECTIONS

Cc 14.1 The first alcohol on the list is methanol, which is highly polar and forms hydrogen bonds with water. It is miscible in water and has only limited solubility in hexane, which is nonpolar. However, as the carbon chain gets longer in the series of alcohols, the OH group becomes less important relative to the growing nonpolar carbon chain. Therefore, the alcohols become progressively less soluble in water and more soluble in hexane. This table demonstrates the rule of thumb *like dissolves like.* Methanol is like water and therefore dissolves in water. It is unlike hexane and therefore has limited solubility in hexane. As you move down the list, the alcohols become increasingly like hexane and increasingly unlike water and therefore become increasingly soluble in hexane and increasingly insoluble in water.

Cc 14.2 You can conclude that $|\Delta H_{solute}| < |\Delta H_{hydration}|$. Because ΔH_{soln} is negative, the absolute value of the negative term ($\Delta H_{hydration}$) must be greater than the absolute value of the positive term (ΔH_{solute}).

Cc 14.3 **(b)** Some potassium bromide precipitates out of solution. The solubility of most solids decreases with decreasing temperature. However, the solubility of gases increases with decreasing temperature. Therefore, the nitrogen becomes more soluble and will not bubble out of solution.

Cc 14.4 Ammonia is the only compound on the list that is polar, so we would expect its solubility in water to be greater than those of the other gases (which are all nonpolar).

Cc 14.5 The solute–solvent interactions must be stronger than the solute–solute and solvent–solvent interactions. The stronger interactions lower the vapor pressure from the expected ideal value of 150 mmHg.

Cc 14.6 Solution B because K_b for ethanol is greater than K_b for water.

Cc 14.7 **(c)** The 0.50 M $MgCl_2$ solution has the highest boiling point because it has the highest concentration of particles. We expect 1 mol of $MgCl_2$ to form 3 mol of particles in solution (although it effectively forms slightly fewer).

CHAPTER 15

"Nobody, I suppose, could devote many years to the study of chemical kinetics without being deeply conscious of the fascination of time and change: this is something that goes outside science into poetry. . . ."

—Sir Cyril N. Hinshelwood (1897–1967)

Pouring ice water on a lizard slows it down, making it easier to catch.

Chemical Kinetics

15.1 Catching Lizards 555
15.2 Rates of Reaction and the Particulate Nature of Matter 556
15.3 Defining and Measuring the Rate of a Chemical Reaction 557
15.4 The Rate Law: The Effect of Concentration on Reaction Rate 563
15.5 The Integrated Rate Law: The Dependence of Concentration on Time 568
15.6 The Effect of Temperature on Reaction Rate 576
15.7 Reaction Mechanisms 583
15.8 Catalysis 588
Key Learning Outcomes 595

IN THE PASSAGE QUOTED in this chapter opener, Oxford chemistry professor Sir Cyril Hinshelwood calls attention to an aspect of chemistry often overlooked by the casual observer—the mystery of change with time. Since the opening chapter of this book, you have learned that the goal of chemistry is to understand the macroscopic world by examining the structure of the particles that compose it. In this chapter, we focus on understanding how these particles change with time, an area of study called chemical kinetics. The particulate world is anything but static. Thermal energy produces constant molecular motion, causing molecules to repeatedly collide with one another. In a tiny fraction of these collisions, something extraordinary happens—the electrons on one molecule or atom are attracted to the nuclei of another molecule or atom. Some bonds weaken and new bonds form—a chemical reaction occurs. Chemical kinetics is the study of how these kinds of changes occur in time.

15.1 Catching Lizards

The children who live in my neighborhood (including my own kids) have a unique way of catching lizards. Armed with cups of ice water, they chase one of these cold-blooded reptiles into a corner, take aim, and pour the cold water directly onto the lizard's body. The lizard's body temperature drops, and it becomes virtually immobilized—easy prey for little hands. The kids scoop up the lizard and place it in a tub filled with sand and leaves. They then watch as the lizard warms back up and becomes active again. They usually release the lizard back into the yard within hours. I guess you could call them catch-and-release lizard hunters.

Unlike mammals, which actively regulate their body temperature through metabolic activity, lizards are *ectotherms*—their body temperature depends on their surroundings. When splashed with cold water, a lizard's body simply gets colder. The drop in body temperature immobilizes the lizard because its movement depends on chemical reactions that occur within its muscles, and the *rates* of those reactions—how fast they occur—are highly sensitive to temperature. In other words, when the temperature drops, the reactions that produce movement in the lizard occur more slowly; therefore, the movement itself slows down. When reptiles get cold, they become lethargic, unable to move very quickly. For this reason, reptiles try to maintain their body temperature within a narrow range by moving between sun and shade.

The ability to understand and control reaction rates is important not only to reptile movement but to many other phenomena as well. For example, a successful rocket launch depends on the rate at which fuel burns—too quickly and the rocket can explode, too slowly and it will not leave the ground. Chemists must always consider reaction rates when synthesizing compounds. No matter how stable a compound might be, its synthesis is impossible if the rate at which it forms is too slow. As we have seen with reptiles, reaction rates are important to life. In fact, the human body's ability to switch a specific reaction on or off at a specific time is achieved largely by controlling the rate of that reaction through the use of enzymes (biological molecules that we explore more fully in Section 15.8). Knowledge of reaction rates is not only practically important—giving us the ability to control how fast a reaction occurs—but also theoretically important. As you will see in Section 15.7, the rate of a reaction can tell us much about how the reaction occurs on the molecular scale.

15.2 Rates of Reaction and the Particulate Nature of Matter

We have seen throughout this book that matter is composed of particles (atoms, ions, and molecules). The simplest way to begin to understand the factors that influence a reaction rate is to think of a chemical reaction as the result of a collision between these particles, which is the basis of *the collision model* (which we cover in more detail in Section 15.6). For example, consider the following simple generic reaction occurring in the gaseous state:

$$\text{A—A} + \text{B} \longrightarrow \text{A—B} + \text{A}$$

According to the collision model, the reaction occurs as a result of a collision between A-A particles and B particles.

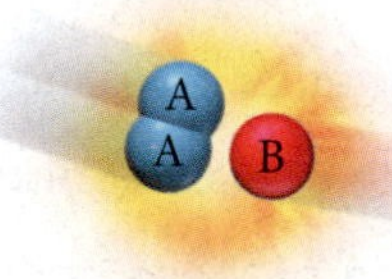

The rate at which the reaction occurs—that is, how many particles react per unit time—depends on several factors: (a) the concentration of the reacting particles; (b) the temperature; and (c) the structure and relative orientation of the reacting particles. We examine each of these individually.

The Concentration of the Reactant Particles

We saw in Chapter 11 that we can model a gas as a collection of particles in constant motion. The particles frequently collide with one another and with the walls of their container. The greater the number of particles in a given volume—that is, the greater their *concentration*—the greater the number of collisions per unit time. Since a chemical reaction requires a collision between particles, the rate of the reaction will depend on the concentration of the particles.

The first person to accurately measure this effect was Ludwig Wilhelmy. In 1850, he measured how fast sucrose, upon treatment with acid, hydrolyzed (or broke apart) into glucose and fructose. This reaction occurred over several hours, and Wilhelmy was able to show how the rate depended on the initial amount of sugar present—the greater the initial amount, the faster the initial rate. We more thoroughly examine the relationship between reaction rate and reactant concentration in Section 15.4.

The Temperature of the Reactant Mixture

We saw in Section 15.1 that chemical reactions generally occur faster with increasing temperature. We can understand this behavior based on particle collisions. Let's return to our simple generic reaction:

$$\text{A—A} + \text{B} \longrightarrow \text{A—B} + \text{A}$$

When B reacts with A-A, the A-A bond breaks and a new bond forms between A and B. *The initial breaking of the A-A bond, however, requires energy*, and even though the forming of A-B gives off energy, some energy, called the *activation energy*, is required to get the reaction started. The energy needed to begin to break the A-A bond can come from the kinetic energy of the colliding particles. In other words, the reaction between A-A and B can occur if there is enough energy in the collision to begin to break the A-A bond.

Recall from Section 11.7 that the average kinetic energy of a gas increases with increasing temperature. At low temperatures, very few of the particle collisions occur with enough energy to begin to break the A-A bond, so the reaction rate is slow. At high temperatures, many particle collisions occur with enough energy to begin to break the A-A bond, and the reaction rate is fast. Therefore the reaction rate increases with increasing temperature. We more thoroughly examine the relationship between reaction rate and temperature in Section 15.6.

The Structure and Orientation of the Colliding Particles

The rates of chemical reactions also depend on the *structure* and relative *orientation* of the colliding particles. We can understand this behavior by comparing two simple reactions:

$$\text{A—A} + \text{B} \longrightarrow \text{A—B} + \text{A}$$
$$\text{A—X} + \text{B} \longrightarrow \text{A—B} + \text{X}$$

In the first reaction, B can collide with A-A from either side and result in a reaction. In the second reaction, however, B must collide with A-X only from one side (the A side). If B collides with A-X on the X side, the reaction cannot occur because B must form a bond with A, not X. As a result, in the second reaction many of the collisions that occur, even if they occur with enough kinetic energy, will not result in a reaction because the orientation of the colliding particles is not correct. In this way, the structure of the reactant particles influences the rate of the reaction. All other factors being equal, we expect the first reaction to occur with a faster rate than the second one. We examine examples of the structure dependence of reaction rates in Section 15.6.

Particle Collisions and Reaction Rates

15.1
Cc
Conceptual Connection

According to the collision model, why does increasing the temperature increase the rate of a reaction?

(a) Increasing the temperature causes more of the collisions to occur with the correct particle orientation for the reaction to occur.

(b) Increasing the temperature increases the number of collisions that can occur with enough energy for the reaction to occur.

(c) Increasing the temperature causes more particles to occupy the same amount of space which therefore increases the reaction rate.

15.3 Defining and Measuring the Rate of a Chemical Reaction

The rate of a chemical reaction is a measure of how fast the reaction occurs, as shown in Figure 15.1 ▶. If a chemical reaction has a fast rate, a large fraction of molecules reacts to form products in a given period of time. If a chemical reaction has a slow rate, a relatively small fraction of molecules reacts to form products in a given period of time.

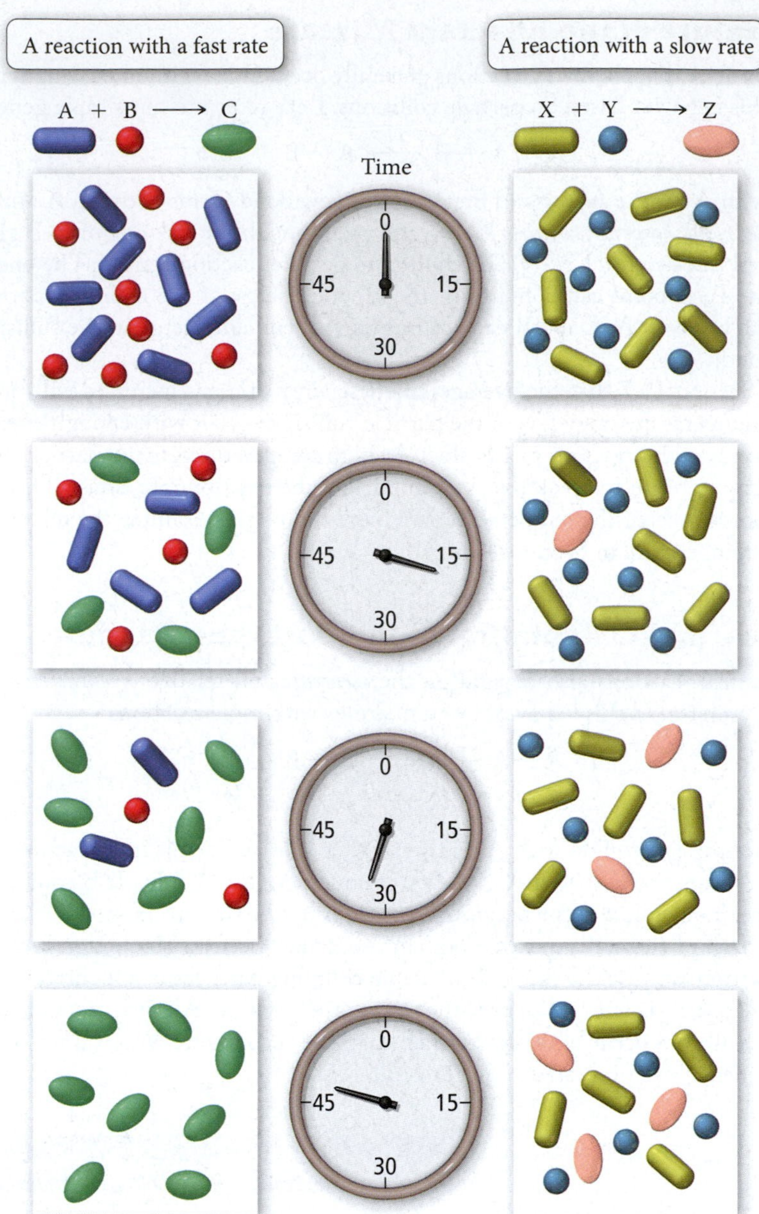

▶ **FIGURE 15.1 The Rate of a Chemical Reaction**

Defining Reaction Rate

When we measure how fast something occurs, or more specifically the *rate* at which it occurs, we usually express the measurement as a change in some quantity per unit of time. For example, we measure the speed of a car—the rate at which it travels—in *miles per hour*, and we might measure how fast people lose weight in *pounds per week*. We report these rates in units that represent the change in what we are measuring (distance or weight) divided by the change in time.

$$\text{Speed} = \frac{\text{change in distance}}{\text{change in time}} = \frac{\Delta x}{\Delta t} \qquad \text{Weight loss} = \frac{\text{change in weight}}{\text{change in time}} = \frac{\Delta\ \text{weight}}{\Delta t}$$

Similarly, the rate of a chemical reaction is measured as a change in the amounts of reactants or products (usually in concentration units) divided by the change in time. For example, consider the gas-phase reaction between $H_2(g)$ and $I_2(g)$:

$$H_2(g) + I_2(g) \longrightarrow 2\ HI(g)$$

We can define the rate of this reaction in the time interval t_1 to t_2 as follows:

$$\text{Rate} = -\frac{\Delta[H_2]}{\Delta t} = -\frac{[H_2]_{t_2} - [H_2]_{t_1}}{t_2 - t_1} \qquad [15.1]$$

Recall that [A] means the concentration of A in M (mol/L).

In this expression, $[H_2]_{t_2}$ is the hydrogen concentration at time t_2 and $[H_2]_{t_1}$ is the hydrogen concentration at time t_1. We define the reaction rate as *the negative* of the change in concentration of a reactant divided by the change in time. The negative sign is usually part of the definition when we define the reaction rate in terms of a reactant because reactant concentrations decrease as a reaction proceeds; therefore, *the change in the concentration of a reactant is negative*. The negative sign thus makes the overall *rate* positive. (By convention, reaction rates are reported as positive quantities.)

We can also define the reaction rate in terms of the other reactant as follows:

$$\text{Rate} = -\frac{\Delta[I_2]}{\Delta t} \qquad [15.2]$$

Because 1 mol of H_2 reacts with 1 mol of I_2, the rates are defined in the same way. We can also define the rate with respect to the *product* of the reaction as follows:

$$\text{Rate} = +\frac{1}{2}\frac{\Delta[HI]}{\Delta t} \qquad [15.3]$$

Because product concentrations *increase* as the reaction proceeds, the change in concentration of a product is positive. Therefore, when we define the rate with respect to a product, we do not include a negative sign in the definition—the rate is naturally positive. The factor of $\frac{1}{2}$ in this definition is related to the stoichiometry of the reaction. In order to have a single rate for the entire reaction, the definition of the rate with respect to each reactant and product must reflect the stoichiometric coefficients of the reaction. For this particular reaction, 2 mol of HI are produced from 1 mol of H_2 and 1 mol of I_2:

The concentration of HI increases at twice the rate that the concentration of H_2 or I_2 decreases. In other words, if 100 I_2 molecules react per second, then 200 HI molecules form per second. In order for the overall rate to have the same value when defined with respect to any of the reactants or products, the change in HI concentration must be multiplied by a factor of one-half.

Consider the graph shown in Figure 15.2 ▶, which represents the changes in concentration for H_2 (one of the reactants) and HI (the product) versus time. Let's examine several features of this graph individually.

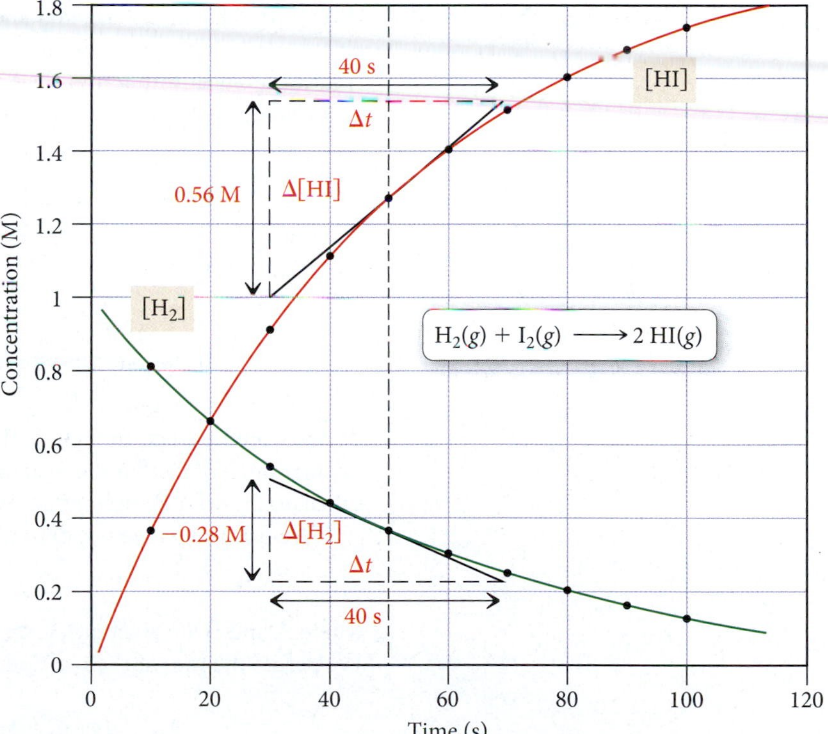

▲ **FIGURE 15.2 Reactant and Product Concentrations as a Function of Time** The graph shows the concentration of one of the reactants (H_2) and the product (HI) as a function of time. The other reactant (I_2) is omitted for clarity.

Change in Reactant and Product Concentrations The reactant concentration, as expected, *decreases* with time because reactants are consumed in a reaction. The product concentration *increases* with time because products are formed in a reaction. The increase in HI concentration occurs at exactly twice the rate of the decrease in H_2 concentration because of the stoichiometry of the reaction: 2 mol of HI form for every 1 mol of H_2 consumed.

The Average Rate of the Reaction We can calculate the average rate of the reaction for any time interval using Equation 15.1 for H_2. The table on the following page lists the H_2 concentration ($[H_2]$) at various times, the change in H_2 concentration for each interval ($\Delta[H_2]$), the change in time for each interval (Δt), and the rate for each interval ($-\Delta[H_2]/\Delta t$). The rate is the average rate within the given time interval. For example, the average rate of the reaction in the time interval between

10 and 20 seconds is 0.0149 M/s, while the average rate in the time interval between 20 and 30 seconds is 0.0121 M/s. Notice that the average rate *decreases* as the reaction progresses. In other words, the reaction slows down as it proceeds, because for most reactions, the rate depends on the concentrations of the reactants. As the reactants transform to products, their concentrations decrease, and the reaction slows down.

Time (s)	$[H_2]$ (M)	$\Delta[H_2]$ (M)	Δt (s)	Rate $= -\Delta[H_2]/\Delta t$ (M/s)
0.000	1.000			
		−0.181	10.000	0.0181
10.000	0.819			
		−0.149	10.000	0.0149
20.000	0.670			
		−0.121	10.000	0.0121
30.000	0.549			
		−0.100	10.000	0.0100
40.000	0.449			
		−0.081	10.000	0.0081
50.000	0.368			
		−0.067	10.000	0.0067
60.000	0.301			
		−0.054	10.000	0.0054
70.000	0.247			
		−0.045	10.000	0.0045
80.000	0.202			
		−0.037	10.000	0.0037
90.000	0.165			
		−0.030	10.000	0.0030
100.000	0.135			

The Instantaneous Rate of the Reaction The instantaneous rate of the reaction is the rate at any one point in time. We can determine the instantaneous rate by calculating the slope of the tangent to the curve at the point of interest. In Figure 15.2, we have drawn the tangent lines for both $[H_2]$ and [HI] at 50 seconds and have labeled the changes in H_2 concentration ($\Delta[H_2]$) and the changes in time (Δt) corresponding to the tangent line. We calculate the instantaneous rate at 50 seconds by substituting into the expression for rate.

Using $[H_2]$:

$$\text{Instantaneous rate (at 50 s)} = -\frac{\Delta[H_2]}{\Delta t} = -\frac{-0.28\text{ M}}{40\text{ s}} = 0.0070\text{ M/s}$$

Using [HI]:

$$\text{Instantaneous rate (at 50 s)} = +\frac{1}{2}\frac{\Delta[\text{HI}]}{\Delta t} = +\frac{1}{2}\frac{0.56\text{ M}}{40\text{ s}} = 0.0070\text{ M/s}$$

As we would expect, the rate is the same whether we use one of the reactants or the product for the calculation. Notice that the instantaneous rate at 50 seconds (0.0070 M/s) is between the average rates calculated for the 10-second intervals just before and just after 50 seconds.

We can generalize our definition of reaction rate for the generic reaction:

$$a\text{A} + b\text{B} \longrightarrow c\text{C} + d\text{D} \qquad [15.4]$$

where A and B are reactants, C and D are products, and *a, b, c,* and *d* are the stoichiometric coefficients. We define the rate of the reaction as follows:

$$\text{Rate} = -\frac{1}{a}\frac{\Delta[\text{A}]}{\Delta t} = -\frac{1}{b}\frac{\Delta[\text{B}]}{\Delta t} = +\frac{1}{c}\frac{\Delta[\text{C}]}{\Delta t} = +\frac{1}{d}\frac{\Delta[\text{D}]}{\Delta t} \qquad [15.5]$$

From the definition, we can see that knowing the rate of change in the concentration of any one reactant or product at a point in time allows us to determine the rate of change in the concentration of any other reactant or product at that point in time (from the balanced equation). *However, predicting the rate at some future time is not possible from just the balanced equation.*

EXAMPLE 15.1
Expressing Reaction Rates

Consider this balanced chemical equation:

$$H_2O_2(aq) + 3\,I^-(aq) + 2\,H^+(aq) \longrightarrow I_3^-(aq) + 2\,H_2O(l)$$

In the first 10.0 seconds of the reaction, the concentration of I^- drops from 1.000 M to 0.868 M.

(a) Calculate the average rate of this reaction in this time interval.

(b) Determine the rate of change in the concentration of H^+ (that is, $\Delta[H^+]/\Delta t$) during this time interval.

SOLUTION

(a) Use Equation 15.5 to calculate the average rate of the reaction. Since the stoichiometric coefficient is 3, the multiplication factor is 1/3.	$\text{Rate} = -\frac{1}{3}\frac{\Delta[I^-]}{\Delta t}$ $= -\frac{1}{3}\frac{[I^-]_{\text{final}} - [I^-]_{\text{initial}}}{t_{\text{final}} - t_{\text{initial}}}$ $= -\frac{1}{3}\frac{(0.868\text{ M} - 1.000\text{ M})}{(10.0\text{ s} - 0.00\text{ s})}$ $= 4.40 \times 10^{-3}\text{ M/s}$
(b) Use Equation 15.5 again to determine the relationship between the rate of the reaction and $\Delta[H^+]/\Delta t$. Since the stoichiometric coefficient is 2, the multiplication factor is 1/2. After solving for $\Delta[H^+]/\Delta t$, substitute the calculated rate from part (a) and calculate $\Delta[H^+]/\Delta t$.	$\text{Rate} = -\frac{1}{2}\frac{\Delta[H^+]}{\Delta t}$ $\frac{\Delta[H^+]}{\Delta t} = -2(\text{rate})$ $= -2(4.40 \times 10^{-3}\text{ M/s})$ $= -8.80 \times 10^{-3}\text{ M/s}$

FOR PRACTICE 15.1

For the reaction shown in Example 15.1, predict the rate of change in concentration of H_2O_2 ($\Delta[H_2O_2]/\Delta t$) and I_3^- ($\Delta[I_3^-]/\Delta t$) during this time interval.

Reaction Rates

15.2 Cc Conceptual Connection

For the reaction $A + 2B \longrightarrow C$ under a given set of conditions, the initial rate is 0.100 M/s. What is $\Delta[B]/\Delta t$ under the same conditions?

(a) −0.0500 M/s
(b) −0.100 M/s
(c) −0.200 M/s
(d) +0.200 M/s

Measuring Reaction Rates

In order to study the kinetics of a reaction, we must have an experimental way to measure the concentration of at least one of the reactants or products as a function of time. For example, Ludwig Wilhelmy, whose experiment on the rate of the conversion of sucrose to glucose and fructose we discussed briefly in Section 15.2, took advantage of sucrose's ability to rotate polarized light. (Polarized light is

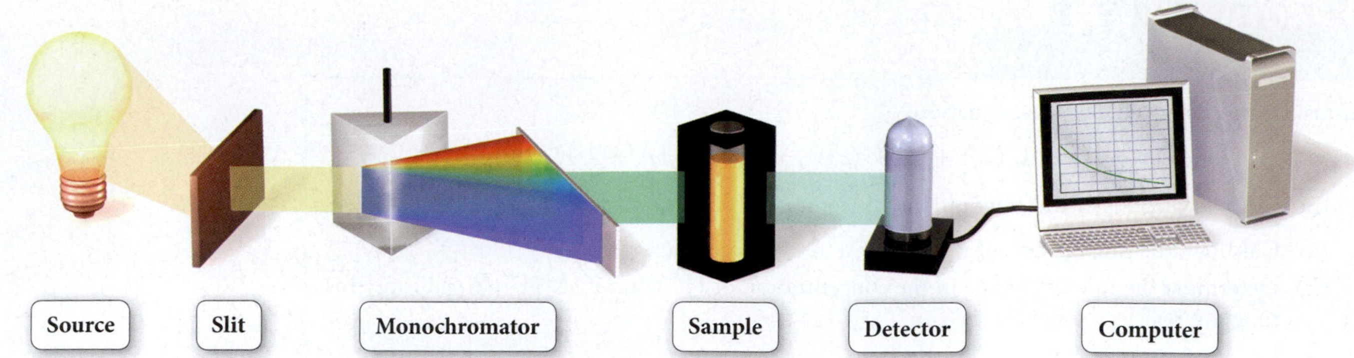

▲ **FIGURE 15.3 The Spectrometer** In a spectrometer, light of a specific wavelength is passed through the sample and the intensity of the transmitted light—which depends on how much light is absorbed by the sample—is measured and recorded.

light with an electric field oriented along one plane.) When a beam of polarized light is passed through a sucrose solution, the polarization of the light is rotated clockwise. In contrast, the combined products of the reaction (glucose and fructose) rotate polarized light counterclockwise. By measuring the degree of polarization of light passing through a reacting solution—a technique known as polarimetry—Wilhelmy was able to determine the relative concentrations of the reactants and products as a function of time.

Perhaps the most common way to study the kinetics of a reaction is through spectroscopy (see Section 3.3). For example, we can follow the reaction of H_2 and I_2 to form HI spectroscopically because I_2 is violet and H_2 and HI are colorless. As I_2 reacts with H_2 to form HI, the violet color of the reaction mixture fades. We can monitor the fading color with a spectrometer, a device that passes light through a sample and measures how strongly the light is absorbed (Figure 15.3 ▲). If the sample contains the reacting mixture, the intensity of the light absorption will decrease as the reaction proceeds, providing a direct measure of the concentration of I_2 as a function of time. Because light travels so fast and because experimental techniques can produce very short pulses of light, spectroscopy can be used to measure reactions that happen on time scales as short as several femtoseconds.

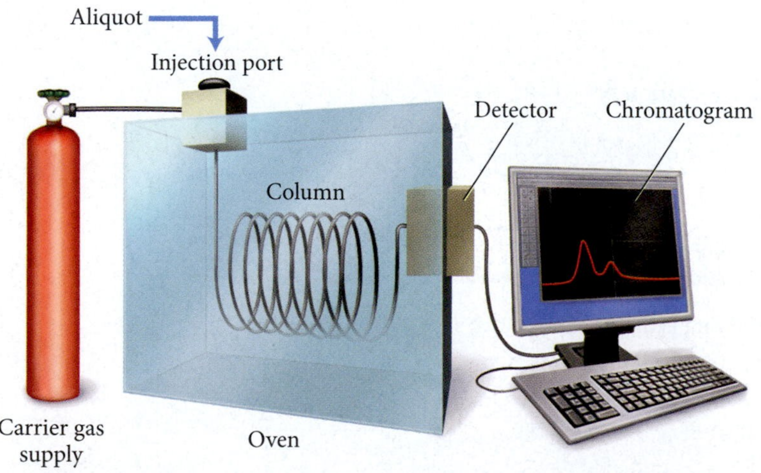

▲ **FIGURE 15.4 The Gas Chromatograph** In a gas chromatograph (GC), a sample of the reaction mixture, or aliquot, is injected into a specially constructed column. Because of their characteristic physical and chemical properties, different components of the mixture pass through the column at different rates and thus exit at different times. As each component leaves the column, it is identified electronically and a chromatogram is recorded. The area under each peak in the chromatogram is proportional to the amount of one particular component in the sample mixture.

We can monitor reactions in which the number of moles of gaseous reactants and products changes as the reaction proceeds by measuring changes in pressure. Consider the reaction in which dinitrogen monoxide reacts to form nitrogen and oxygen gas:

$$2\ N_2O(g) \longrightarrow 2\ N_2(g) + O_2(g)$$

For every 2 mol of N_2O that reacts, the reaction vessel will contain one additional mole of gas. As the reaction proceeds and the amount of gas increases, the pressure steadily rises. We can use the rise in pressure to determine the relative concentrations of reactants and products as a function of time.

The three techniques mentioned here—polarimetry, spectroscopy, and pressure measurement—can all be used to monitor a reaction as it occurs in a reaction vessel. Some reactions occur slowly enough that samples, or *aliquots,* can be periodically withdrawn from the reaction vessel and analyzed to determine the progress of the reaction. Instrumental techniques such as gas chromatography (Figure 15.4 ◀) or mass spectrometry (see Section 1.9), as well as wet chemical techniques such as titration, can be used to measure the relative amounts of reactants or products in the aliquot. By taking aliquots at regular time intervals, we can determine the relative amounts of reactants and products as a function of time.

15.4 The Rate Law: The Effect of Concentration on Reaction Rate

As we saw in Section 15.2, the rate of a reaction often depends on the concentration of one or more of the reactants. For simplicity, let's consider a reaction in which a single reactant, A, decomposes into products:

$$\text{A} \longrightarrow \text{products}$$

As long as the rate of the reverse reaction (in which the products return to reactants) is negligibly slow, we can express the relationship between the rate of the reaction and the concentration of the reactant—called the **rate law**—as follows:

$$\text{Rate} = k[\text{A}]^n \qquad [15.6]$$

where k is a constant of proportionality called the **rate constant** and n is the **reaction order**. The value of n (usually an integer) determines how the rate depends on the concentration of the reactant.

- If $n = 0$, the reaction is *zero order* and the rate is independent of the concentration of A.
- If $n = 1$, the reaction is *first order* and the rate is directly proportional to the concentration of A.
- If $n = 2$, the reaction is *second order* and the rate is proportional to the square of the concentration of A.

By definition, $[\text{A}]^0 = 1$, so the rate is equal to k regardless of [A].

Although other orders are possible, including noninteger (or fractional) orders, we limit our current discussion to three orders.

Figure 15.5 ▼ shows three plots illustrating how the *concentration of A changes with time* for the three common reaction orders with identical numerical values for the rate constant (k) and identical initial concentrations. Figure 15.6 ▼ has three plots showing the *rate of the reaction* (the slope of the lines in Figure 15.5) *as a function of the reactant concentration* for each reaction order.

Zero-Order Reaction In a zero-order reaction, the rate of the reaction is independent of the concentration of the reactant.

$$\text{Rate} = k[\text{A}]^0 = k \qquad [15.7]$$

Consequently, for a zero-order reaction, the concentration of the reactant decreases linearly with time, as shown in Figure 15.5. The slope of the line is constant, indicating a constant rate. The rate is constant because the reaction does not slow down as the concentration of A decreases. The graph in Figure 15.6

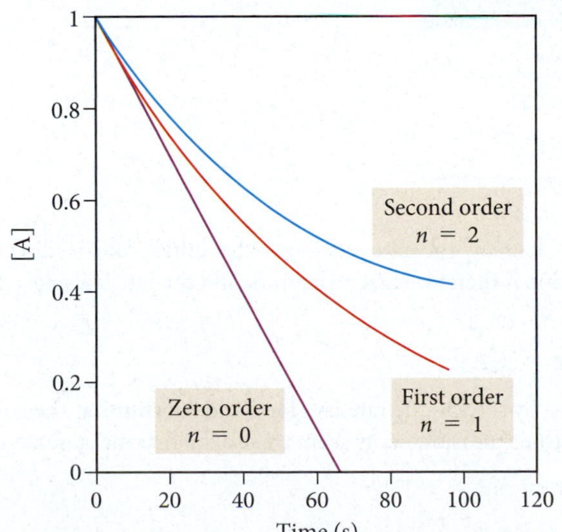

▲ **FIGURE 15.5 Reactant Molar Concentration as a Function of Time for Different Reaction Orders**

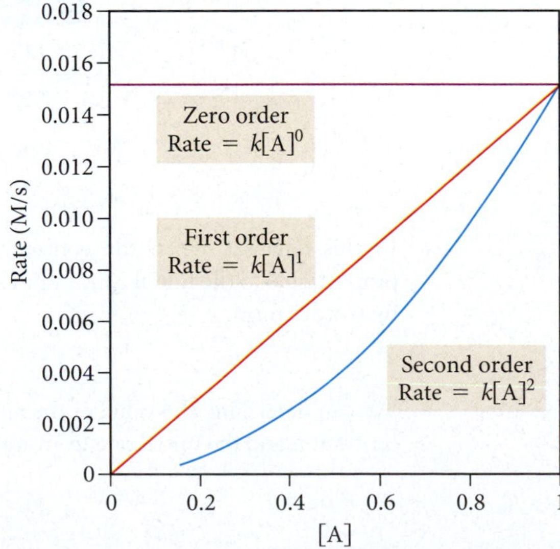

▲ **FIGURE 15.6 Reaction Rate as a Function of Reactant Molar Concentration for Different Reaction Orders**

shows that the rate of a zero-order reaction is the same at any concentration of A. Zero-order reactions occur under conditions where the amount of reactant actually *available for reaction* is unaffected by changes in the *overall quantity of reactant.* For example, sublimation is normally zero order because only molecules at the surface can sublime, and their concentration does not change as the amount of subliming substance decreases (Figure 15.7 ◀).

Sublimation Is Zero Order

When one layer of particles sublimes, the next layer is exposed. The number of particles available to sublime remains constant.

▲ **FIGURE 15.7 Sublimation** When a layer of particles sublimes, another identical layer is just below it. Consequently, the number of particles available to sublime at any one time does not change with the total number of particles in the sample, and the process is zero order.

First-Order Reaction In a first-order reaction, the rate of the reaction is directly proportional to the concentration of the reactant.

$$\text{Rate} = k[\text{A}]^1 \qquad [15.8]$$

For a first-order reaction, the rate slows down as the reaction proceeds because the concentration of the reactant decreases. You can see this in Figure 15.5—the slope of the curve (the rate) becomes less steep (slower) with time. Figure 15.6 shows the rate as a function of the concentration of A. Notice the linear relationship—the rate is directly proportional to the concentration.

Second-Order Reaction In a second-order reaction, the rate of the reaction is proportional to the square of the concentration of the reactant.

$$\text{Rate} = k[\text{A}]^2 \qquad [15.9]$$

Consequently, for a second-order reaction, the rate is even more sensitive to the reactant concentration. You can see this in Figure 15.5—the slope of the curve (the rate) flattens out more quickly than it does for a first-order reaction. Figure 15.6 shows the rate as a function of the concentration of A. Notice the quadratic relationship—the rate is proportional to the square of the concentration.

Determining the Order of a Reaction

The order of a reaction can be determined only by experiment. A common way to determine reaction order is the *method of initial rates.* In this method, the initial rate—the rate for a short period of time at the beginning of the reaction—is measured by running the reaction several times with different initial reactant concentrations to determine the effect of the concentration on the rate. For example, let's return to our simple reaction in which a single reactant, A, decomposes into products:

$$\text{A} \longrightarrow \text{products}$$

In an experiment, the initial rate is measured at several different initial concentrations with the following results:

[A] (M)	Initial Rate (M/s)
0.10	0.015
0.20	0.030
0.40	0.060

In this data set, when the concentration of A doubles, the rate doubles—the initial rate is directly proportional to the initial concentration. The reaction is therefore first order in A, and the rate law takes the first-order form:

$$\text{Rate} = k[\text{A}]^1$$

We can determine the value of the rate constant, k, by solving the rate law for k and substituting the concentration and the initial rate from any one of the three measurements (here we use the first measurement).

$$\text{Rate} = k[\text{A}]^1$$

$$k = \frac{\text{rate}}{[\text{A}]} = \frac{0.015\ \text{M/s}}{0.10\ \text{M}} = 0.15\ \text{s}^{-1}$$

Notice that the rate constant for a first-order reaction has units of s^{-1}.

The following two data sets show how measured initial rates are different for zero-order and for second-order reactions having the same initial rate at [A] = 0.10 M:

Zero Order ($n = 0$)

[A] (M)	Initial Rate (M/s)
0.10	0.015
0.20	0.015
0.40	0.015

Second Order ($n = 2$)

[A] (M)	Initial Rate (M/s)
0.10	0.015
0.20	0.060
0.40	0.240

For a zero-order reaction, the initial rate is independent of the reactant concentration—the rate is the same at all measured initial concentrations. For a second-order reaction, the initial rate quadruples for a doubling of the reactant concentration—the relationship between concentration and rate is quadratic. If we are unsure about how the initial rate is changing with the initial reactant concentration, or if the numbers are not as obvious as they are in these examples, we can substitute any two initial concentrations and the corresponding initial rates into a ratio of the rate laws to determine the order (n):

$$\frac{\text{rate } 2}{\text{rate } 1} = \frac{k[\text{A}]_2^n}{k[\text{A}]_1^n}$$

For example, we can substitute the last two measurements in the data set just given for the second-order reaction as follows:

Remember that $\frac{x^n}{y^n} = \left(\frac{x}{y}\right)^n$.

$$\frac{0.240 \text{ M/s}}{0.060 \text{ M/s}} = \frac{k(0.40 \text{ M})^n}{k(0.20 \text{ M})^n}$$

$$4.0 = \left(\frac{0.40}{0.20}\right)^n = 2^n$$

$$\log 4.0 = \log (2^n)$$

$$= n \log 2$$

$$n = \frac{\log 4}{\log 2}$$

$$= 2$$

Remember that $\log x^n = n \log x$.

The rate constants for zero- and second-order reactions have different units than for first-order reactions. The rate constant for a zero-order reaction has units of $\text{M} \cdot \text{s}^{-1}$, and the rate constant for a second-order reaction has units of $\text{M}^{-1} \cdot \text{s}^{-1}$.

Order of Reaction

15.3 Cc Conceptual Connection

The reaction A $\longrightarrow$ B has been experimentally determined to be second order. The initial rate is 0.0100 M/s at an initial concentration of A of 0.100 M. What is the initial rate at [A] = 0.500 M?

(a) 0.00200 M/s **(b)** 0.0100 M/s **(c)** 0.0500 M/s **(d)** 0.250 M/s

Reaction Order for Multiple Reactants

So far, we have considered a simple reaction with only one reactant. How is the rate law defined for reactions with more than one reactant? Consider the generic reaction:

$$a\text{A} + b\text{B} \longrightarrow c\text{C} + d\text{D}$$

As long as the reverse reaction is negligibly slow, the rate law is proportional to the concentration of [A] raised to the m power multiplied by the concentration of [B] raised to the n power:

$$\text{Rate} = k[\text{A}]^m[\text{B}]^n \qquad [15.10]$$

where m is the reaction order with respect to A and n is the reaction order with respect to B. The **overall order** is the sum of the exponents ($m + n$). For example, the reaction between hydrogen and iodine has been experimentally determined to be first order with respect to hydrogen, first order with respect to iodine, and thus second order overall.

$$H_2(g) + I_2(g) \longrightarrow 2\ HI(g) \qquad \text{Rate} = k[H_2]^1[I_2]^1$$

Similarly, the reaction between hydrogen and nitrogen monoxide has been experimentally determined to be first order with respect to hydrogen, second order with respect to nitrogen monoxide, and thus third order overall.

$$2\ H_2(g) + 2\ NO(g) \longrightarrow N_2(g) + 2\ H_2O(g) \quad \text{Rate} = k[H_2]^1[NO]^2$$

As we have already noted, the rate law for any reaction must always be determined by experiment, often by the method of initial rates described previously. There is no way to simply look at a chemical equation and determine the rate law for the reaction. When the reaction has two or more reactants, the concentration of each reactant is usually varied independently of the others to determine the dependence of the rate on the concentration of that reactant. Example 15.2 shows how to use the method of initial rates to determine the order of a reaction with multiple reactants.

EXAMPLE 15.2
Determining the Order and Rate Constant of a Reaction

Consider the reaction between nitrogen dioxide and carbon monoxide:

$$NO_2(g) + CO(g) \longrightarrow NO(g) + CO_2(g)$$

The initial rate of the reaction is measured at several different concentrations of the reactants with the results shown at right. From the data, determine:

(a) the rate law for the reaction

(b) the rate constant (k) for the reaction

Experiment Number	$[NO_2]$ (M)	[CO] (M)	Initial Rate (M/s)
1	0.10	0.10	0.0021
2	0.20	0.10	0.0082
3	0.20	0.20	0.0083
4	0.40	0.10	0.033

SOLUTION

(a) Begin by examining how the rate changes for each change in concentration. Between the first two experiments, the concentration of NO_2 doubles, the concentration of CO stays constant, and the rate quadruples, suggesting that the reaction is second order in NO_2.

Between the second and third experiments, the concentration of NO_2 stays constant, the concentration of CO doubles, and the rate remains constant (the small change in the least significant figure is simply experimental error), suggesting that the reaction is zero order in CO.

Between the third and fourth experiments, the concentration of NO_2 again doubles and the concentration of CO halves, yet the rate quadruples again, confirming that the reaction is second order in NO_2 and zero order in CO.

Write the overall rate expression.

$[NO_2]$ (M)	[CO] (M)	Initial Rate (M/s)
0.10 M	0.10 M	0.0021
↓ × 2	↓ constant	↓ × 4
0.20 M	0.10 M	0.0082 M
↓ constant	↓ × 2	↓ × 1
0.20 M	0.20 M	0.0083 M
↓ × 2	↓ × $\frac{1}{2}$	↓ × 4
0.40 M	0.10 M	0.033 M

$$\text{Rate} = k[NO_2]^2[CO]^0 = k[NO_2]^2$$

(a) ALTERNATIVE APPROACH

If you can't easily see the relationships between the changes in concentrations and the changes in initial rates, you can determine the reaction order for any reactant by substituting any two initial rates and the corresponding initial concentrations into a ratio of the rate laws to determine the order (n).

$$\frac{\text{rate } 2}{\text{rate } 1} = \frac{k[A]_2^n}{k[A]_1^n}$$

For NO_2 use the first and second concentrations and rates (because $[NO_2]$ changes here, but [CO] is constant).

For NO_2 $$\frac{\text{rate } 2}{\text{rate } 1} = \frac{k[NO_2]_2^n}{k[NO_2]_1^n}$$

—*Continued on the next page*

Continued from the previous page—

Substitute the rates and concentrations into the expression for the ratio of the rate constants.

$$\frac{0.082\ \cancel{M/s}}{0.021\ \cancel{M/s}} = \frac{\cancel{k}(0.20\ \cancel{M})^n}{\cancel{k}(0.10\ \cancel{M})^n}$$

$$3.9 = \left(\frac{0.20}{0.10}\right)^n = 2^n$$

Take the log of both sides of the equation and solve for n.

$$\log 3.9 = \log 2^n$$
$$= n \log 2$$
$$n = \frac{\log 3.9}{\log 2}$$
$$= 1.96 = 2$$

For CO, use the second and third concentrations and rates (because [CO] changes here, but $[NO_2]$ is constant).

For CO $$\frac{\text{rate } 3}{\text{rate } 2} = \frac{k[CO]_3^n}{k[CO]_2^n}$$

Substitute the rates and concentrations into the expression for the ratio of the rate constants.

$$\frac{0.083\ \cancel{M/s}}{0.082\ \cancel{M/s}} = \frac{\cancel{k}(0.20\ \cancel{M})^n}{\cancel{k}(0.10\ \cancel{M})^n}$$

$$1.01 = \left(\frac{0.20}{0.10}\right)^n = 2^n$$

Take the log of both sides of the equation and solve for n.

$$\log 1.01 = \log 2^n$$
$$= n \log 2$$
$$n = \frac{\log 1.01}{\log 2}$$
$$= 0.01 = 0.0$$

Write the overall rate expression from the orders of each reactant.

$$\text{Rate} = k[NO_2]^2[CO]^0 = k[NO_2]^2$$

(b) To determine the rate constant for the reaction, solve the rate law for k and substitute the concentration and the initial rate from any one of the four measurements. In this case, we use the first measurement.

$$\text{Rate} = k[NO_2]^2$$

$$k = \frac{\text{rate}}{[NO_2]^2} = \frac{0.0021\ M/s}{(0.10\ M)^2} = 0.21\ M^{-1} \cdot s^{-1}$$

FOR PRACTICE 15.2

Consider the equation:

$$CHCl_3(g) + Cl_2(g) \longrightarrow CCl_4(g) + HCl(g)$$

The initial rate of reaction is measured at several different concentrations of the reactants with the following results:

$[CHCl_3]$ (M)	$[Cl_2]$ (M)	Initial Rate (M/s)
0.010	0.010	0.0035
0.020	0.010	0.0069
0.020	0.020	0.0098
0.040	0.040	0.027

From the data, determine:

(a) the rate law for the reaction **(b)** the rate constant (k) for the reaction

15.4 Cc Conceptual Connection

Rate and Concentration

This reaction was experimentally determined to be first order with respect to O_2 and second order with respect to NO.

$$O_2(g) + 2\ NO(g) \longrightarrow 2\ NO_2(g)$$

These diagrams represent reaction mixtures in which the number of each type of molecule represents its relative initial concentration. Which mixture has the fastest initial rate?

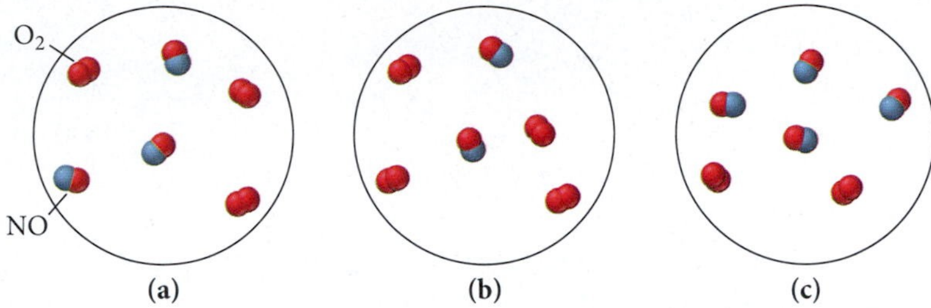

15.5 The Integrated Rate Law: The Dependence of Concentration on Time

The rate laws we have examined so far show the relationship between *the rate of a reaction and the concentration of a reactant.* But we often want to know the relationship between *the concentration of a reactant and time.* For example, the presence of chlorofluorocarbons (CFCs) in the atmosphere threatens the ozone layer. One of the reasons that CFCs pose such a significant threat is that the reactions that consume them are so slow (see Table 15.1). Legislation has significantly reduced CFC emissions, but even if we were to completely stop adding CFCs to the atmosphere, their concentration would decrease only very slowly. Nonetheless, we would like to know how their concentration changes with time. How much will be left in 20 years? In 50 years?

TABLE 15.1 Atmospheric Lifetimes of Several CFCs

CFC Name	Structure	Atmospheric Lifetime*
CFC-11 (CCl_3F) Trichlorofluoromethane		45 years
CFC-12 (CCl_2F_2) Dichlorodifluoromethane		100 years
CFC-113 ($C_2F_3Cl_3$) 1,1,2-Trichloro-1,2,2-trifluoroethane		85 years
CFC-114 ($C_2F_4Cl_2$) 1,2-Dichlorotetrafluoroethane		300 years
CFC-115 (C_2F_5Cl) Monochloropentafluoroethane		1700 years

*Data taken from EPA site (under section 602 of Clean Air Act).

Integrated Rate Laws

The **integrated rate law** for a chemical reaction is a relationship between the concentrations of the reactants and time. For simplicity, we return to a single reactant decomposing into products:

$$\text{A} \longrightarrow \text{products}$$

The integrated rate law for this reaction depends on the order of the reaction; let's examine each of the common reaction orders individually.

First-Order Integrated Rate Law If our simple reaction is first order, the rate is directly proportional to the concentration of A:

$$\text{Rate} = k[\text{A}]$$

Because Rate $= -\Delta[\text{A}]/\Delta t$, we can write:

$$-\frac{\Delta[\text{A}]}{\Delta t} = k[\text{A}] \qquad [15.11]$$

In this form, the rate law is also known as the *differential rate law*.

Although we do not show the steps here, we can use calculus (see End-of-Chapter Exercise 114) to integrate the differential rate law and obtain the first-order *integrated rate law*:

$$\ln[\text{A}]_t = -kt + \ln[\text{A}]_0 \qquad [15.12]$$

or

$$\ln\frac{[\text{A}]_t}{[\text{A}]_0} = -kt \qquad [15.13]$$

$$\ln[\text{A}]_t = -kt + \ln[\text{A}]_0$$
$$\ln[\text{A}]_t - \ln[\text{A}]_0 = -kt$$
$$\ln\frac{[\text{A}]_t}{[\text{A}]_0} = -kt$$

Remember that $\ln \text{A} - \ln \text{B} = \ln\left(\frac{\text{A}}{\text{B}}\right)$.

where $[\text{A}]_t$ is the concentration of A at any time t, k is the rate constant, and $[\text{A}]_0$ is the initial concentration of A. These two forms of the equation are equivalent, as shown in the margin.

Notice that the integrated rate law shown in Equation 15.12 has the form of an equation for a straight line:

$$\ln[\text{A}]_t = -kt + \ln[\text{A}]_0$$

$$y = mx + b$$

For a first-order reaction, a plot of the natural log of the reactant concentration as a function of time yields a straight line with a slope of $-k$ and a y-intercept of $\ln[\text{A}]_0$, as shown in Figure 15.8 ▼. (Note that the slope is negative but that the rate constant is always positive.)

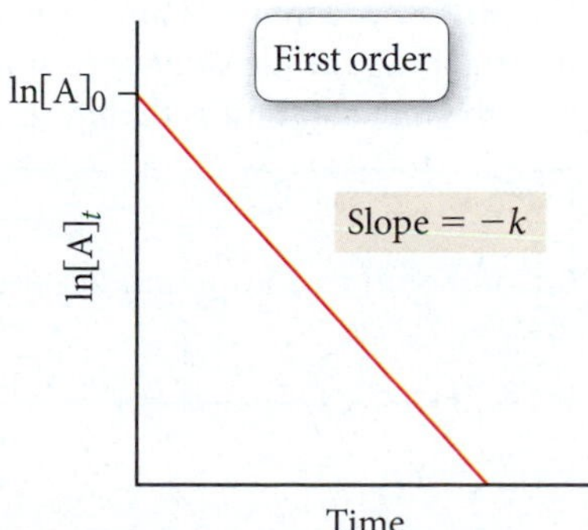

◀ **FIGURE 15.8 First-Order Integrated Rate Law** For a first-order reaction, a plot of the natural log of the reactant concentration as a function of time yields a straight line. The slope of the line is equal to $-k$ and the y-intercept is $\ln[\text{A}]_0$.

EXAMPLE 15.3

The First-Order Integrated Rate Law: Using Graphical Analysis of Reaction Data

Consider the equation for the decomposition of SO_2Cl_2:

$$SO_2Cl_2(g) \longrightarrow SO_2(g) + Cl_2(g)$$

The concentration of SO_2Cl_2 is monitored at a fixed temperature as a function of time during the decomposition reaction, and the following data are tabulated:

Time (s)	$[SO_2Cl_2]$ (M)	Time (s)	$[SO_2Cl_2]$ (M)
0	0.100	800	0.0793
100	0.0971	900	0.0770
200	0.0944	1000	0.0748
300	0.0917	1100	0.0727
400	0.0890	1200	0.0706
500	0.0865	1300	0.0686
600	0.0840	1400	0.0666
700	0.0816	1500	0.0647

Show that the reaction is first order, and determine the rate constant for the reaction.

SOLUTION

In order to show that the reaction is first order, prepare a graph of $\ln [SO_2Cl_2]$ versus time as shown.

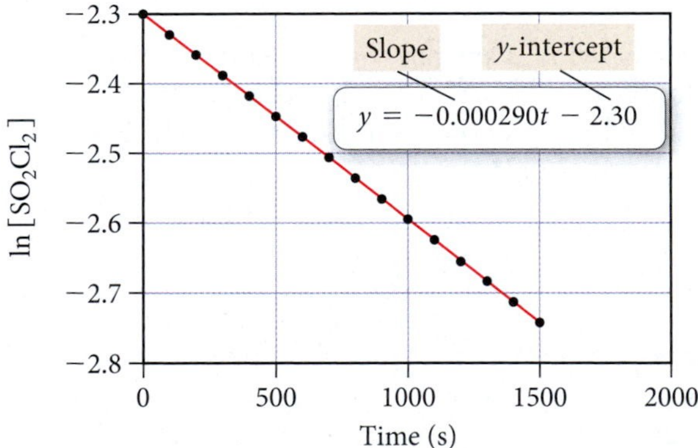

The plot is linear, confirming that the reaction is indeed first order. To obtain the rate constant, fit the data to a line. The slope of the line will be equal to $-k$. Since the slope of the best-fitting line (which is most easily determined on a graphing calculator or with spreadsheet software such as Microsoft Excel) is $-2.90 \times 10^{-4}\,s^{-1}$, the rate constant is therefore $+2.90 \times 10^{-4}\,s^{-1}$.

FOR PRACTICE 15.3

Use the graph and the best-fitting line in Example 15.3 to predict the concentration of SO_2Cl_2 at 1900 s.

EXAMPLE 15.4

The First-Order Integrated Rate Law: Determining the Concentration of a Reactant at a Given Time

In Example 15.3, you determined that the decomposition of SO_2Cl_2 (under the given reaction conditions) is first order and has a rate constant of $+2.90 \times 10^{-4}\ s^{-1}$. If the reaction is carried out at the same temperature, and the initial concentration of SO_2Cl_2 is 0.0225 M, what is the concentration of SO_2Cl_2 after 865 s?

SORT You are given the rate constant of a first-order reaction and the initial concentration of the reactant and asked to find the concentration at 865 seconds.	**GIVEN:** $k = +2.90 \times 10^{-4}\ s^{-1}$ $[SO_2Cl_2]_0 = 0.0225\ M$ **FIND:** $[SO_2Cl_2]$ at $t = 865\ s$
STRATEGIZE Refer to the first-order integrated rate law to determine the information you are asked to find.	**EQUATION** $\ln[A]_t = -kt + \ln[A]_0$
SOLVE Substitute the rate constant, the initial concentration, and the time into the integrated rate law. Solve the integrated rate law for the concentration of $[SO_2Cl_2]_t$.	**SOLUTION** $\ln[SO_2Cl_2]_t = -kt + \ln[SO_2Cl_2]_0$ $\ln[SO_2Cl_2]_t = -(2.90 \times 10^{-4}\ \cancel{s^{-1}})865\ \cancel{s} + \ln[0.0225]$ $\ln[SO_2Cl_2]_t = -0.251 - 3.79$ $[SO_2Cl_2]_t = e^{-4.04}$ $= 0.0175\ M$

CHECK The concentration is smaller than the original concentration as expected. If the concentration were larger than the initial concentration, this would indicate a mistake in the signs of one of the quantities on the right-hand side of the equation.

FOR PRACTICE 15.4

Cyclopropane rearranges to form propene in the gas phase.

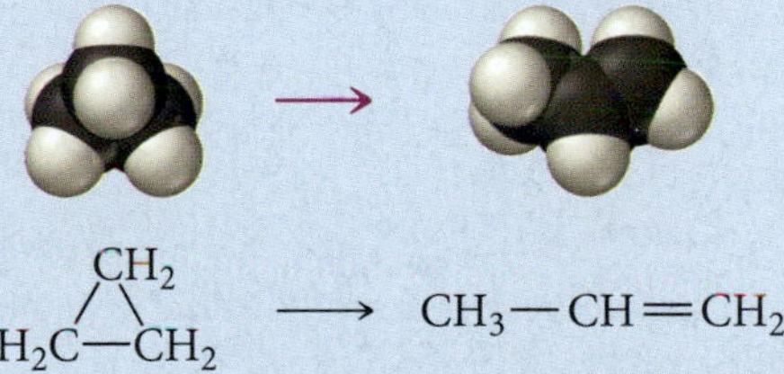

The reaction is first order in cyclopropane and has a measured rate constant of $3.36 \times 10^{-5}\ s^{-1}$ at 720 K. If the initial cyclopropane concentration is 0.0445 M, what is the cyclopropane concentration after 235.0 minutes?

Second-Order Integrated Rate Law If our simple reaction ($A \longrightarrow$ products) is second order, the rate is proportional to the square of the concentration of A:

$$\text{Rate} = k[A]^2$$

Because Rate $= -\Delta[A]/\Delta t$, we can write:

$$-\frac{\Delta[A]}{\Delta t} = k[A]^2 \qquad [15.14]$$

Again, although we do not show the steps here, this differential rate law can be integrated to obtain the *second-order integrated rate law*:

$$\frac{1}{[A]_t} = kt + \frac{1}{[A]_0} \qquad [15.15]$$

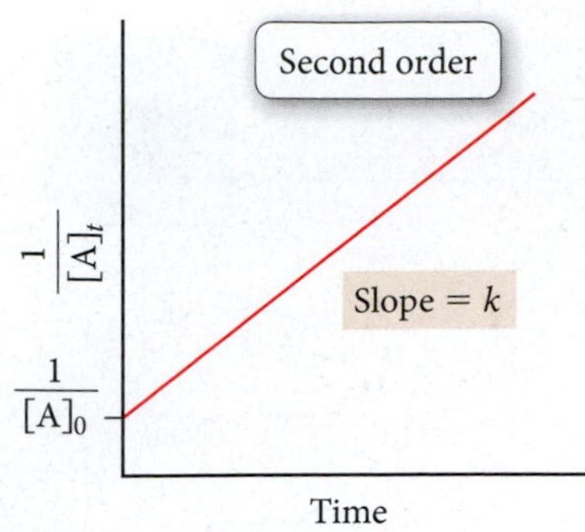

▲ **FIGURE 15.9 Second-Order Integrated Rate Law** For a second-order reaction, a plot of the inverse of the reactant concentration as a function of time yields a straight line. The slope of the line is equal to k and the y-intercept is $\ln[A]_0$.

Time (s)	$[NO_2]$ (M)
0	0.01000
50	0.00887
100	0.00797
150	0.00723
200	0.00662
250	0.00611
300	0.00567
350	0.00528
400	0.00495
450	0.00466
500	0.00440
550	0.00416
600	0.00395
650	0.00376
700	0.00359
750	0.00343
800	0.00329
850	0.00316
900	0.00303
950	0.00292
1000	0.00282

The second-order integrated rate law is also in the form of an equation for a straight line:

$$\frac{1}{[A]_t} = kt + \frac{1}{[A]_0}$$

$$y = mx + b$$

However, you must plot the inverse of the concentration of the reactant as a function of time. The plot yields a straight line with a slope of k and an intercept of $1/[A]_0$ as shown in Figure 15.9 ◀.

EXAMPLE 15.5

The Second-Order Integrated Rate Law: Using Graphical Analysis of Reaction Data

Consider the equation for the decomposition of NO_2:

$$NO_2(g) \longrightarrow NO(g) + O(g)$$

The concentration of NO_2 is monitored at a fixed temperature as a function of time during the decomposition reaction, and the data are tabulated in the table in the margin. Show by graphical analysis that the reaction is not first order and that it is second order. Determine the rate constant for the reaction.

SOLUTION

In order to show that the reaction is not first order, prepare a graph of $\ln[NO_2]$ versus time as shown.

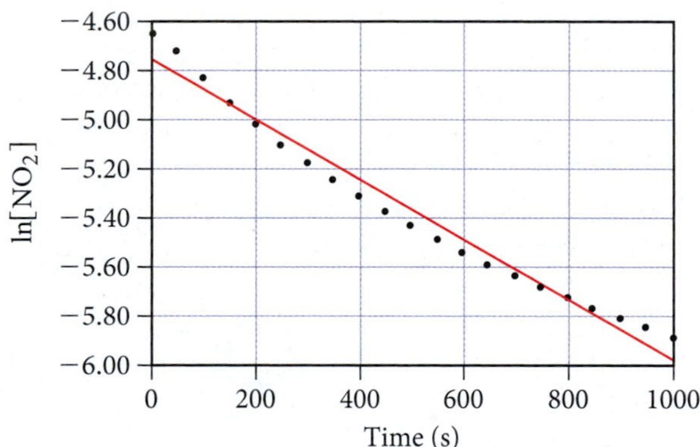

The plot is *not* linear (the straight line does not fit the data points), confirming that the reaction is not first order. In order to show that the reaction is second order, prepare a graph of $1/[NO_2]$ versus time as shown.

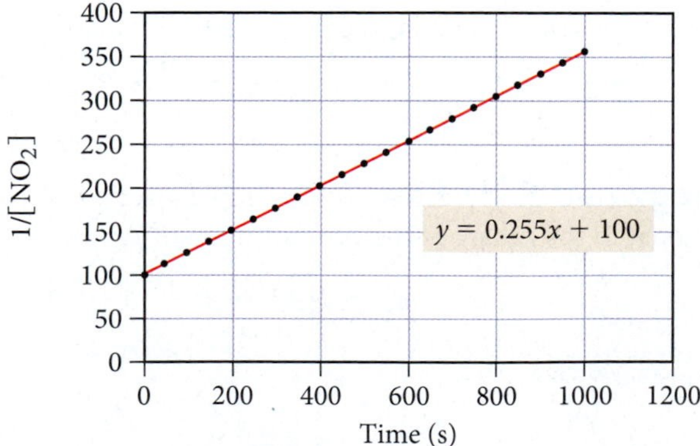

This graph is linear (the data points fit well to a straight line), confirming that the reaction is indeed second order. To obtain the rate constant, determine the slope of the best-fitting line. The slope is $0.255\ M^{-1} \cdot s^{-1}$; therefore, the rate constant is $0.255\ M^{-1} \cdot s^{-1}$.

FOR PRACTICE 15.5

Use the graph and the best-fitting line in Example 15.5 to predict the concentration of NO_2 at 2000 s.

Zero-Order Integrated Rate Law If our simple reaction is zero order, the rate is proportional to a constant:

$$\text{Rate} = k[\text{A}]^0 = k$$

Since Rate $= -\Delta[\text{A}]/\Delta t$, we can write:

$$-\frac{\Delta[\text{A}]}{\Delta t} = k \quad [15.16]$$

We can integrate this differential rate law to obtain the *zero-order integrated rate law*:

$$[\text{A}]_t = -kt + [\text{A}]_0 \quad [15.17]$$

The zero-order integrated rate law in Equation 15.17 is also in the form of an equation for a straight line. A plot of the concentration of the reactant as a function of time yields a straight line with a slope of $-k$ and an intercept of $[\text{A}]_0$, as shown in Figure 15.10 ▶.

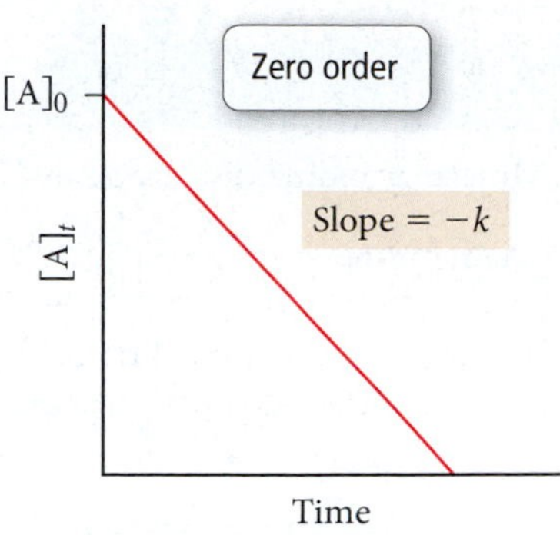

▲ **FIGURE 15.10 Zero-Order Integrated Rate Law** For a zero-order reaction, a plot of the reactant concentration as a function of time yields a straight line. The slope of the line is equal to $-k$ and the y-intercept is $[\text{A}]_0$.

The Half-Life of a Reaction

The **half-life ($t_{1/2}$)** of a reaction is the time required for the concentration of a reactant to fall to one-half of its initial value. For example, if a reaction has a half-life of 100 seconds, and if the initial concentration of the reactant is 1.0 M, the concentration will fall to 0.50 M in 100 s. The half-life expression—which defines the dependence of half-life on the rate constant and the initial concentration—is different for different reaction orders.

First-Order Reaction Half-Life From the definition of half-life, and from the integrated rate law, we can derive an expression for the half-life. For a first-order reaction, the integrated rate law is:

$$\ln\frac{[\text{A}]_t}{[\text{A}]_0} = -kt$$

At a time equal to the half-life ($t = t_{1/2}$), the concentration is exactly half of the initial concentration: ($[\text{A}]_t = \frac{1}{2}[\text{A}]_0$). Therefore, when $t = t_{1/2}$ we can write the following expression:

$$\ln\frac{\frac{1}{2}[\text{A}]_0}{[\text{A}]_0} = \ln\frac{1}{2} = -kt_{1/2} \quad [15.18]$$

Solving for $t_{1/2}$, and substituting –0.693 for $\ln\frac{1}{2}$, we arrive at the expression for the half-life of a first-order reaction:

$$t_{1/2} = \frac{0.693}{k} \quad [15.19]$$

Notice that, for a first-order reaction, $t_{1/2}$ *is independent of the initial concentration*. For example, if $t_{1/2}$ is 100 s, and if the initial concentration is 1.0 M, the concentration falls to 0.50 M in 100 s, to 0.25 M in another 100 s, to 0.125 M in another 100 s, and so on (Figure 15.11 ▶). Even though the concentration is changing as the reaction proceeds, the half-life (how long it takes for the concentration to halve) is constant. A constant half-life is unique to first-order reactions, making the concept of half-life particularly useful for first-order reactions.

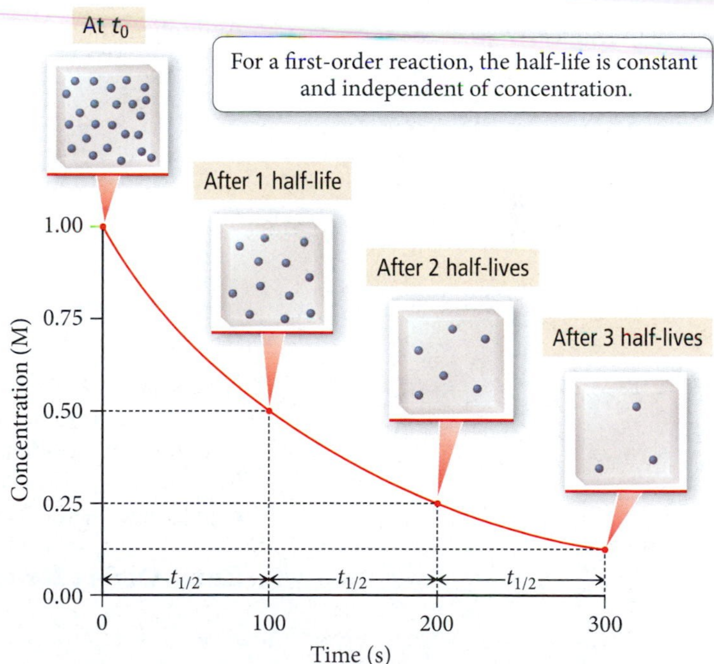

▲ **FIGURE 15.11 Half-Life: Concentration versus Time for a First-Order Reaction** For this reaction, the concentration decreases by one-half every 100 seconds ($t_{1/2} = 100$ s). The blue spheres represent reactant molecules (the products are omitted for clarity).

EXAMPLE 15.6
Half-Life

Molecular iodine dissociates at 625 K with a first-order rate constant of $0.271\ s^{-1}$. What is the half-life of this reaction?

SOLUTION

Because the reaction is first order, the half-life is given by Equation 15.19. Substitute the value of k into the expression and calculate $t_{1/2}$.	$t_{1/2} = \frac{0.693}{k}$ $= \frac{0.693}{0.271/s} = 2.56\ s$

FOR PRACTICE 15.6

A first-order reaction has a half-life of 26.4 seconds. How long does it take for the concentration of the reactant in the reaction to fall to one-eighth of its initial value?

Second-Order Reaction Half-Life For a second-order reaction, the integrated rate law is:

$$\frac{1}{[A]_t} = kt + \frac{1}{[A]_0}$$

At a time equal to the half-life ($t = t_{1/2}$), the concentration is exactly one-half of the initial concentration ($[A]_t = \frac{1}{2}[A]_0$). We can therefore write the following expression at $t = t_{1/2}$:

$$\frac{1}{\frac{1}{2}[A]_0} = kt_{1/2} + \frac{1}{[A]_0} \qquad [15.20]$$

And solve for $t_{1/2}$:

$$kt_{1/2} = \frac{1}{\frac{1}{2}[A]_0} - \frac{1}{[A]_0}$$

$$kt_{1/2} = \frac{2}{[A]_0} - \frac{1}{[A]_0}$$

$$t_{1/2} = \frac{1}{k[A]_0} \qquad [15.21]$$

Notice that, for a second-order reaction, the half-life depends on the initial concentration. So if the initial concentration of a reactant in a second-order reaction is 1.0 M, and the half-life is 100 s, the concentration falls to 0.50 M in 100 s. However, the time it takes for the concentration to fall to 0.25 M is now *longer than 100 s* because the initial concentration has decreased. Thus, the half-life continues to get longer as the concentration decreases.

Zero-Order Reaction Half-Life For a zero-order reaction, the integrated rate law is:

$$[A]_t = -kt + [A]_0$$

Making the substitutions ($t = t_{1/2}$; $[A]_t = \frac{1}{2}[A]_0$), we can write the expression at $t = t_{1/2}$:

$$\frac{1}{2}[A]_0 = -kt_{1/2} + [A]_0 \qquad [15.22]$$

We then solve for $t_{1/2}$:

$$t_{1/2} = \frac{[A]_0}{2k} \qquad [15.23]$$

Notice that, for a zero-order reaction, the half-life also depends on the initial concentration; however, unlike in the second-order case, the two are directly proportional—the half-life gets *shorter* as the concentration decreases.

Summarizing Basic Kinetic Relationships (see Table 15.2):

- The reaction order and rate law must be determined experimentally.
- The rate law relates the *rate* of the reaction to the *concentration* of the reactant(s).
- The integrated rate law (which is mathematically derived from the rate law) relates the *concentration* of the reactant(s) to *time*.
- The half-life is the time it takes for the concentration of a reactant to fall to one-half of its initial value.
- The half-life of a first-order reaction is independent of the initial concentration.
- The half-lives of zero-order and second-order reactions depend on the initial concentration.

TABLE 15.2 Rate Law Summary Table

Order	Rate Law	Units of *k*	Integrated Rate Law	Straight-Line Plot	Half-Life Expression
0	Rate $= k[A]^0$	$M \cdot s^{-1}$	$[A]_t = -kt + [A]_0$	[A] vs. Time *t*; y-intercept = $[A]_0$; Slope = $-k$	$t_{1/2} = \frac{[A]_0}{2k} = \frac{1}{k}\frac{[A]_0}{2}$
1	Rate $= k[A]^1$	s^{-1}	$\ln[A]_t = -kt + \ln[A]_0$ $\ln\frac{[A]_t}{[A]_0} = -kt$	ln[A] vs. Time *t*; y-intercept = $\ln[A]_0$; Slope = $-k$	$t_{1/2} = \frac{0.693}{k} = \frac{1}{k}(0.693)$
2	Rate $= k[A]^2$	$M^{-1} \cdot s^{-1}$	$\frac{1}{[A]_t} = kt + \frac{1}{[A]_0}$	1/[A] vs. Time *t*; Slope = k; y-intercept = $1/[A]_0$	$t_{1/2} = \frac{1}{k[A]_0} = \frac{1}{k}\frac{1}{[A]_0}$

Rate Law and Integrated Rate Law

15.5 Cc Conceptual Connection

A decomposition reaction, with a rate that is observed to slow down as the reaction proceeds, has a half-life that depends on the initial concentration of the reactant. Which statement is most likely true for this reaction?

(a) A plot of the natural log of the concentration of the reactant as a function of time is linear.

(b) The half-life of the reaction increases as the initial concentration increases.

(c) A doubling of the initial concentration of the reactant results in a quadrupling of the rate.

15.6 The Effect of Temperature on Reaction Rate

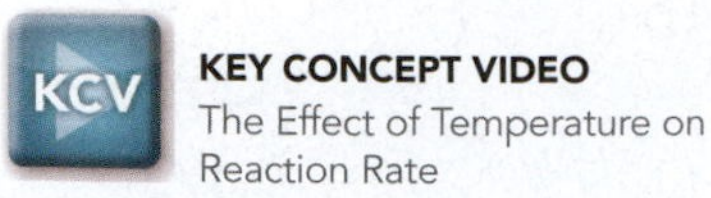

KEY CONCEPT VIDEO
The Effect of Temperature on Reaction Rate

As we saw in Sections 15.1 and 15.2, the rates of chemical reactions are, in general, highly sensitive to temperature. Around room temperature, for example, a 10 °C increase in temperature increases the rate of a typical biological reaction by two or three times. Recall that the rate law for a reaction is Rate $= k[A]^n$. *The temperature dependence of the reaction rate is contained in the rate constant,* k (which is actually a constant only when the temperature remains constant). An increase in temperature generally results in an increase in k, which results in a faster rate.

The Arrhenius Equation

In 1889, Swedish chemist Svante Arrhenius wrote a paper quantifying the temperature dependence of the rate constant. The modern form of the **Arrhenius equation** shows the relationship between the rate constant (k) and the temperature in kelvin (T):

$$k = Ae^{\frac{-E_a}{RT}} \qquad [15.24]$$

Activation energy (E_a); Frequency factor (A); Exponential factor ($e^{-E_a/RT}$)

In this equation, R is the gas constant (8.314 J/mol · K), A is a constant called the *frequency factor* (or the *pre-exponential factor*), and E_a is the *activation energy* (or *activation barrier*), which we briefly examined in Section 15.2.

The **activation energy** E_a is an energy barrier or hump that must be surmounted for the reactants to be transformed into products (Figure 15.12 ▼). We examine the frequency factor more closely in the next section of the chapter; for now, we can think of the **frequency factor (**A**)** as the number of times that the reactants approach the activation barrier per unit time.

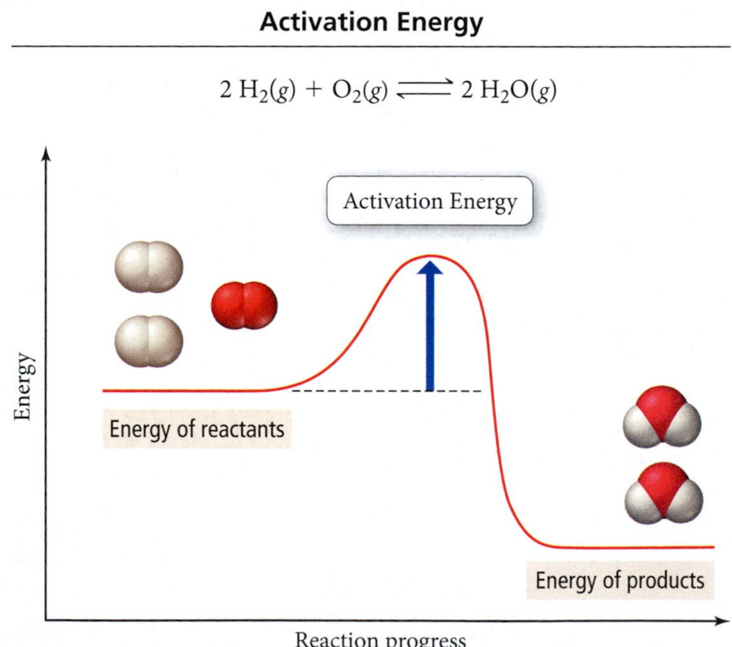

▶ **FIGURE 15.12 The Activation Energy Barrier** Even though the reaction is energetically favorable (the energy of the products is lower than that of the reactants), the reactants must overcome the energy barrier in order to react.

To understand each of these quantities better, we can look more closely at the simple reaction in which CH_3NC (methyl isonitrile) rearranges to form CH_3CN (acetonitrile):

$$CH_3{-}N{\equiv}C \longrightarrow CH_3{-}C{\equiv}N$$

Let's examine the physical meaning of the activation energy, frequency factor, and exponential factor for this reaction.

The Activation Energy Figure 15.13 ▶ illustrates the energy of a molecule as the reaction proceeds. The *x*-axis represents the progress of the reaction from left (reactant) to right (product). To get from the reactant to the product, the molecule must go through a high-energy intermediate state called the **activated complex** or **transition state**. Even though the overall reaction is energetically downhill (exothermic), it must first go uphill to reach the activated complex because energy is required to initially weaken the $H_3C—N$ bond and allow the NC group to begin to rotate:

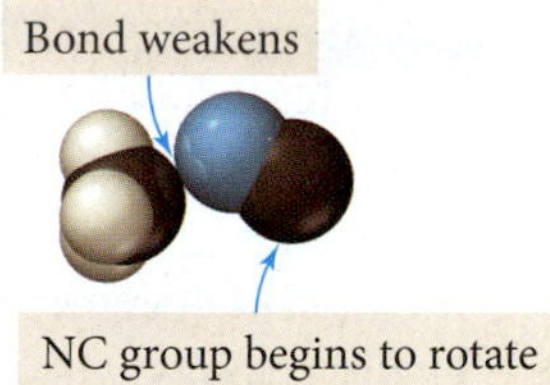

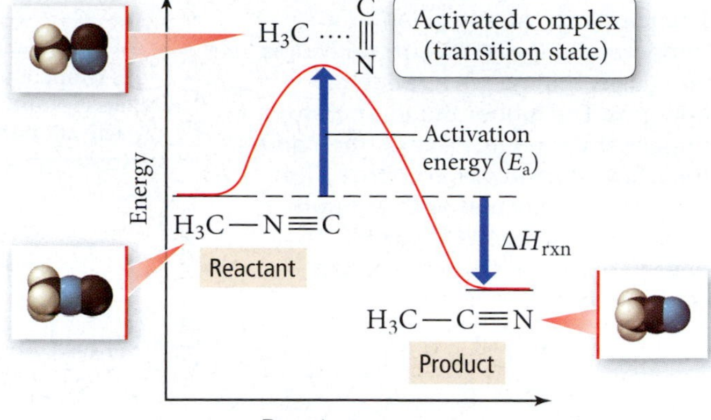

▲ **FIGURE 15.13 The Activated Complex** The reaction pathway includes a transition state—the activated complex—that has a higher energy than either the reactant or the product.

The energy required to reach the activated complex is the *activation energy. In general, the higher the activation energy, the slower the reaction rate (at a given temperature).*

The Frequency Factor Recall that the frequency factor represents the number of approaches to the activation barrier per unit time. Any time that it begins to rotate, the NC group approaches the activation barrier. For this reaction, the frequency factor represents the rate at which the NC part of the molecule wags (vibrates side to side). With each wag, the reactant approaches the activation barrier. However, approaching the activation barrier is not equivalent to surmounting it. Most of the approaches do not have enough total energy to make it over the activation barrier.

The Exponential Factor The **exponential factor** is a number between 0 and 1 that represents the fraction of molecules that have enough energy to make it over the activation barrier on a given approach. The exponential factor is the fraction of approaches that are actually successful and result in the product. For example, if the frequency factor is 10^9/s and the exponential factor is 10^{-7} at a certain temperature, then the overall rate constant at that temperature is $10^9/\text{s} \times 10^{-7} = 10^2/\text{s}$. In this case, the CN group is wagging at a rate of 10^9/s. With each wag, the activation barrier is approached. However, for a given wag only 1 in 10^7 molecules has sufficient energy to actually make it over the activation barrier.

The exponential factor depends on both the temperature (T) and the activation energy (E_a) of the reaction.

$$\text{Exponential factor} = e^{-E_a/RT}$$

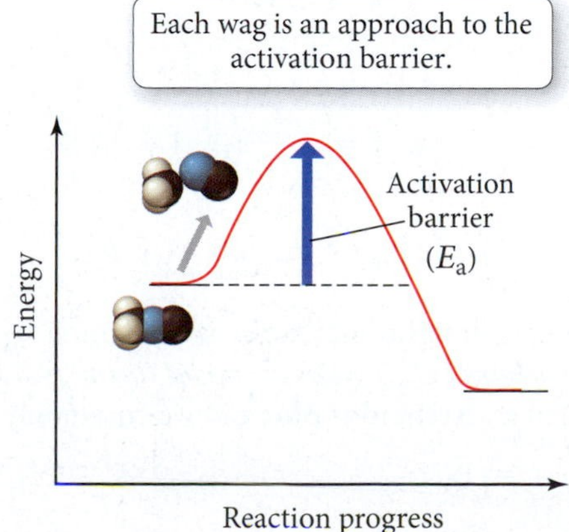

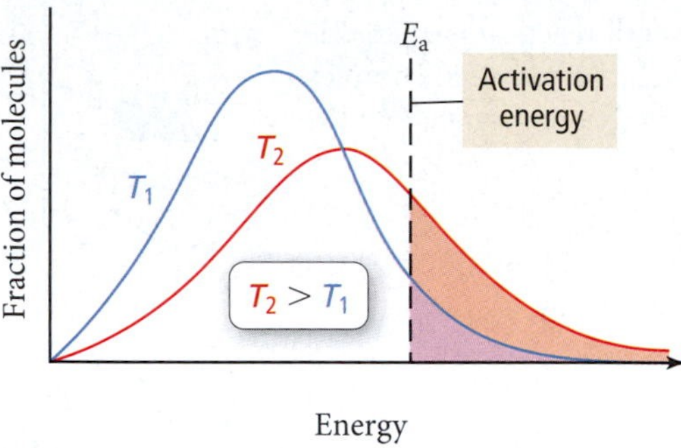

▶ **FIGURE 15.14 Thermal Energy Distribution** At any given temperature, the atoms or molecules in a gas sample have a range of energies. The higher the temperature, the wider the energy distribution and the greater the average energy. The fraction of molecules with enough energy to surmount the activation energy barrier and react (shaded regions) increases sharply as the temperature rises.

A low activation energy and a high temperature make the negative exponent small, so the exponential factor approaches one. For example, if the activation energy is zero, the exponent is zero, and the exponential factor is exactly one ($e^{-0} = 1$): every approach to the activation barrier is successful. By contrast, a large activation energy and a low temperature make the exponent a very large negative number, so the exponential factor becomes very small. For example, as the temperature approaches 0 K, the exponent approaches an infinitely large number, and the exponential factor approaches zero ($e^{-\infty} = 0$).

As the temperature increases, the number of molecules having enough thermal energy to surmount the activation barrier increases. At any given temperature, a sample of molecules has a distribution of energies, as shown in Figure 15.14 ◀. Under common circumstances, only a small fraction of the molecules has enough energy to make it over the activation barrier. Because of the shape of the energy distribution curve, however, a small change in temperature results in a large difference in the number of molecules having enough energy to surmount the activation barrier. This explains the sensitivity of reaction rates to temperature.

Summarizing Temperature and Reaction Rate:

- The frequency factor is the number of times that the reactants approach the activation barrier per unit time.
- The exponential factor is the fraction of the approaches that are successful in surmounting the activation barrier and forming products.
- The exponential factor increases with increasing temperature but decreases with increasing activation energy.

Arrhenius Plots: Experimental Measurements of the Frequency Factor and the Activation Energy

The frequency factor and activation energy are important quantities in understanding the kinetics of any reaction. To see how we measure these factors in the laboratory, consider again Equation 15.24: $k = Ae^{-E_a/RT}$. Taking the natural log of both sides of this equation, we get the following result:

Remember that ln (AB) = ln A + ln B.

$$\ln k = \ln (Ae^{-E_a/RT})$$

$$\ln k = \ln A + \ln e^{-E_a/RT}$$

Remember that ln $e^x = x$.

$$\ln k = \ln A - \frac{E_a}{RT} \qquad [15.25]$$

In an Arrhenius analysis, the pre-exponential factor (A) is assumed to be independent of temperature. Although the pre-exponential factor does depend on temperature to some degree, its temperature dependence is much less than that of the exponential factor and is often ignored.

$$\ln k = -\frac{E_a}{R}\left(\frac{1}{T}\right) + \ln A$$

$$y = \quad mx \quad + b \qquad [15.26]$$

Equation 15.26 is in the form of a straight line. *A plot of the natural log of the rate constant (ln k) versus the inverse of the temperature in kelvins (1/T) yields a straight line with a slope of* $-E_a/R$ *and a y-intercept of ln A.* This type of plot is called an **Arrhenius plot** and we commonly use it in the analysis of kinetic data, as shown in Example 15.7.

EXAMPLE 15.7

Using an Arrhenius Plot to Determine Kinetic Parameters

The decomposition of ozone shown here is important to many atmospheric reactions.

$$O_3(g) \longrightarrow O_2(g) + O(g)$$

A study of the kinetics of the reaction results in the following data:

Temperature (K)	Rate Constant ($M^{-1} \cdot s^{-1}$)	Temperature (K)	Rate Constant ($M^{-1} \cdot s^{-1}$)
600	3.37×10^3	1300	7.83×10^7
700	4.85×10^4	1400	1.45×10^8
800	3.58×10^5	1500	2.46×10^8
900	1.70×10^6	1600	3.93×10^8
1000	5.90×10^6	1700	5.93×10^8
1100	1.63×10^7	1800	8.55×10^8
1200	3.81×10^7	1900	1.19×10^9

Determine the value of the frequency factor and activation energy for the reaction.

SOLUTION

To determine the frequency factor and activation energy, prepare a graph of the natural log of the rate constant ($\ln k$) versus the inverse of the temperature ($1/T$).

The plot is linear, as expected for Arrhenius behavior. The line that fits best has a slope of -1.12×10^4 K and a y-intercept of 26.8. Calculate the activation energy from the slope by setting the slope equal to $-E_a/R$ and solving for E_a:

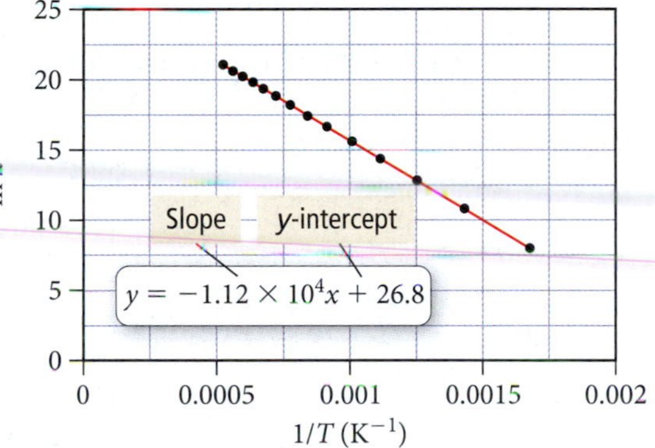

$$-1.12 \times 10^4 \text{ K} = \frac{-E_a}{R}$$

$$E_a = 1.12 \times 10^4 \text{ K}\left(8.314 \frac{\text{J}}{\text{mol} \cdot \text{K}}\right)$$

$$= 9.31 \times 10^4 \text{ J/mol}$$

$$= 93.1 \text{ kJ/mol}$$

Calculate the frequency factor (A) by setting the intercept equal to $\ln A$:

$$26.8 = \ln A$$

$$A = e^{26.8}$$

$$= 4.36 \times 10^{11}$$

Because the rate constants are measured in units of $M^{-1} \cdot s^{-1}$, the frequency factor is in the same units. Consequently, we can conclude that the reaction has an activation energy of 93.1 kJ/mol and a frequency factor of $4.36 \times 10^{11}\ M^{-1} \cdot s^{-1}$.

FOR PRACTICE 15.7

For the decomposition of ozone reaction in Example 15.7, use the results of the Arrhenius analysis to predict the rate constant at 298 K.

In some cases, when either data are limited or plotting capabilities are absent, we can calculate the activation energy if we know the rate constant at just two different temperatures. We can apply the Arrhenius expression in Equation 15.26 to the two different temperatures as follows:

$$\ln k_2 = -\frac{E_a}{R}\left(\frac{1}{T_2}\right) + \ln A \qquad \ln k_1 = -\frac{E_a}{R}\left(\frac{1}{T_1}\right) + \ln A$$

We can then subtract $\ln k_1$ from $\ln k_2$:

$$\ln k_2 - \ln k_1 = \left[-\frac{E_a}{R}\left(\frac{1}{T_2}\right) + \ln A\right] - \left[-\frac{E_a}{R}\left(\frac{1}{T_1}\right) + \ln A\right]$$

Rearranging, we get the two-point form of the Arrhenius equation:

$$\ln\frac{k_2}{k_1} = \frac{E_a}{R}\left(\frac{1}{T_1} - \frac{1}{T_2}\right) \qquad [15.27]$$

Example 15.8 demonstrates how to use this equation to calculate the activation energy from experimental measurements of the rate constant at two different temperatures.

EXAMPLE 15.8

Using the Two-Point Form of the Arrhenius Equation

Consider the reaction between nitrogen dioxide and carbon monoxide.

$$NO_2(g) + CO(g) \longrightarrow NO(g) + CO_2(g)$$

The rate constant at 701 K is measured as $2.57\ M^{-1} \cdot s^{-1}$ and that at 895 K is measured as $567\ M^{-1} \cdot s^{-1}$. Find the activation energy for the reaction in kJ/mol.

SORT You are given the rate constant of a reaction at two different temperatures and asked to find the activation energy.

GIVEN: $T_1 = 701\ K,\ k_1 = 2.57\ M^{-1} \cdot s^{-1}$

$T_2 = 895\ K,\ k_2 = 567\ M^{-1} \cdot s^{-1}$

FIND: E_a

STRATEGIZE Use the two-point form of the Arrhenius equation, which relates the activation energy to the given information and R (a constant).

EQUATION $\ln\frac{k_2}{k_1} = \frac{E_a}{R}\left(\frac{1}{T_1} - \frac{1}{T_2}\right)$

SOLVE Substitute the two rate constants and the two temperatures into the equation.

SOLUTION

$$\ln\frac{567\ \cancel{M^{-1} \cdot s^{-1}}}{2.57\ \cancel{M^{-1} \cdot s^{-1}}} = \frac{E_a}{R}\left(\frac{1}{701\ K} - \frac{1}{895\ K}\right)$$

$$5.40 = \frac{E_a}{R}\left(\frac{3.09 \times 10^{-4}}{K}\right)$$

Solve the equation for E_a, the activation energy, and convert to kJ/mol.

$$E_a = 5.40\left(\frac{K}{3.09 \times 10^{-4}}\right)R$$

$$= 5.40\left(\frac{\cancel{K}}{3.09 \times 10^{-4}}\right)8.314\ \frac{J}{mol \cdot \cancel{K}}$$

$$= 1.45 \times 10^5\ kJ/mol$$

$$= 145\ kJ/mol$$

CHECK The magnitude of the answer is reasonable. Activation energies for most reactions range from tens to hundreds of kilojoules per mole.

FOR PRACTICE 15.8

Use the results from Example 15.8 and the given rate constant of the reaction at either of the two temperatures to predict the rate constant for this reaction at 525 K.

Temperature Dependence of Reaction Rate

15.6 Cc Conceptual Connection

Reaction A and reaction B have identical frequency factors. However, reaction B has a higher activation energy than reaction A. Which reaction has a greater rate constant at room temperature?

The Collision Model: A Closer Look at the Frequency Factor

We saw previously that the frequency factor in the Arrhenius equation represents the number of approaches to the activation barrier per unit time. Let's refine that idea for a reaction involving two gas-phase reactants:

$$A(g) + B(g) \longrightarrow \text{products}$$

Recall from Section 15.2, that in the **collision model**, a chemical reaction occurs after a sufficiently energetic collision between two reactant molecules (Figure 15.15 ▶). In collision theory, therefore, each approach to the activation barrier is a collision between the reactant molecules. Consequently, the value of the frequency factor should simply be the number of collisions that occur per second. However, the frequency factors of most (though not all) gas-phase chemical reactions tend to be smaller than the number of collisions that occur per second. As we saw in Section 15.2, not all sufficiently energetic collisions lead to products because the *orientations* of some of the collisions may not be adequate for the reaction to occur.

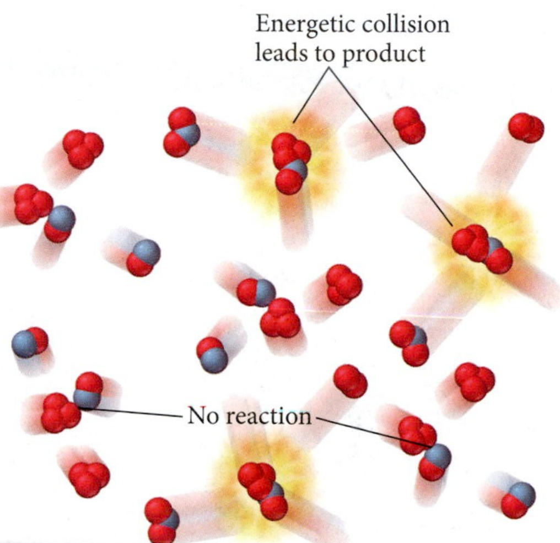

▲ **FIGURE 15.15 The Collision Model** In the collision model, two molecules react after a sufficiently energetic collision with the correct orientation brings the reacting groups together.

In the collision model, we can separate the frequency factor into two distinct parts, as shown in the following equations:

$$k = Ae^{\frac{-E_a}{RT}}$$

$$= pze^{\frac{-E_a}{RT}}$$

Orientation factor (p) Collision frequency (z)

where p is the **orientation factor** and z is the **collision frequency**. The collision frequency is the number of collisions that occur per unit time, which we can calculate for a gas-phase reaction from the pressure of the gases and the temperature of the reaction mixture. Under typical gaseous conditions, a single molecule undergoes on the order of 10^9 collisions every second. The orientation factor is a number, usually between 0 and 1, which represents the fraction of collisions with an orientation that allows the reaction to occur.

The orientation factor is a measure of how specific the orientation of the colliding molecules must be. A large orientation factor (near 1) indicates that the colliding molecules can have virtually any orientation and the reaction will still occur. A small orientation factor indicates that the colliding molecules must have a highly specific orientation for the reaction to occur. As an example, consider the reaction represented by the following equation:

$$NOCl(g) + NOCl(g) \longrightarrow 2\,NO(g) + Cl_2(g)$$

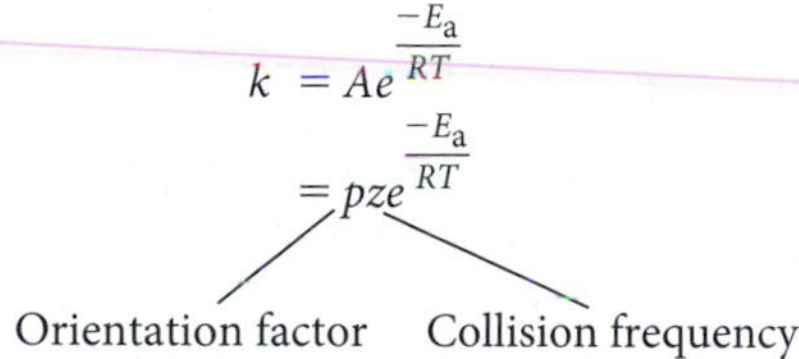

In order for the reaction to occur, two NOCl molecules must collide with sufficient energy. However, not all collisions with sufficient energy will lead to products, because the reactant molecules must also be properly oriented. Consider three possible collision orientations of the reactant molecules shown below. The first two collisions, even if they occur with sufficient energy, will not result in a reaction, because the reactant molecules are not oriented in a way that allows the chlorine atoms to bond. In other words, if two molecules are to react with each other, they must collide in such a way that allows the necessary bonds to break and form. For the reaction of NOCl(g), the orientation factor is $p = 0.16$. This means that only 16 out of 100 sufficiently energetic collisions are actually successful in forming the products.

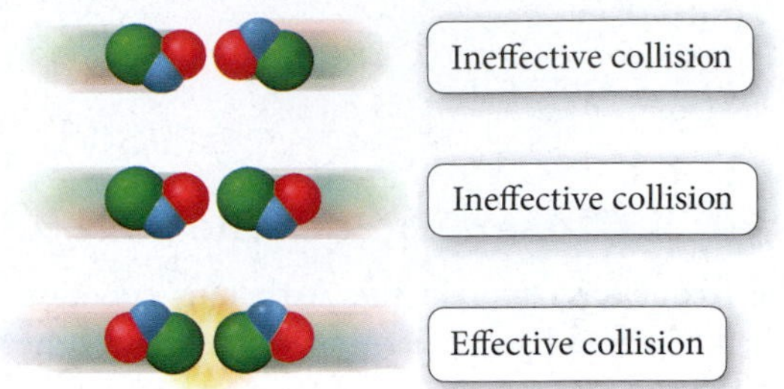

Some reactions have orientation factors that are much smaller than one. Consider the reaction between hydrogen and ethene:

$$H_2(g) + CH_2{=}CH_2(g) \longrightarrow CH_3{-}CH_3(g)$$

The orientation factor for this reaction is 1.7×10^{-6}, which means that fewer than two out of each million sufficiently energetic collisions actually form products. The small orientation factor indicates that the orientation requirements for this reaction are very stringent—the molecules must be aligned in a *very specific way* for the reaction to occur. Notice that, as we have seen many times, *structure affects properties*. In this case, the structure of the reactants affects how fast the reaction occurs.

Reactions between *individual atoms* usually have orientation factors of approximately 1 because atoms are spherically symmetric and thus any orientation can lead to the formation of products. A few reactions have orientation factors greater than one. Consider the reaction between potassium and bromine:

$$K(g) + Br_2(g) \longrightarrow KBr(g) + Br(g)$$

This reaction has an orientation factor of $p = 4.8$. In other words, there are more reactions than collisions—the reactants do not even have to collide to react! Apparently, through a process dubbed *the harpoon mechanism*, a potassium atom can actually transfer an electron to a bromine molecule without a collision. The resulting positive charge on the potassium and the negative charge on the bromine cause the two species to attract each other and form a bond. The potassium atom essentially *harpoons* a passing bromine molecule with an electron and *reels it in* through the coulombic attraction between unlike charges.

We can picture a sample of reactive gases as a frenzy of collisions between the reacting atoms or molecules. At normal temperatures, the vast majority of these collisions do not have sufficient energy to overcome the activation barrier, and the atoms or molecules simply bounce off one another. Of the collisions having sufficient energy to overcome the activation barrier, most do not have the proper orientation for the reaction to occur (for the majority of common reactions). When two molecules with sufficient energy *and* the correct orientation collide, something extraordinary happens. The electrons on one of the atoms or molecules are attracted to the nuclei of the other; some bonds begin to weaken while other bonds begin to form and, if all goes well, the reactants go through the transition state and are transformed into the products. This is how a chemical reaction occurs.

15.7 Cc Conceptual Connection

Collision Theory

Which reaction would you expect to have the smallest orientation factor?

(a) $H(g) + I(g) \longrightarrow HI(g)$
(b) $H_2(g) + I_2(g) \longrightarrow 2\ HI(g)$
(c) $HCl(g) + HCl(g) \longrightarrow H_2(g) + Cl_2(g)$

15.7 Reaction Mechanisms

Most chemical reactions do not occur in a single step, but through several steps. When we write a chemical equation to represent a chemical reaction, *we usually represent the overall reaction, not the series of individual steps by which the reaction occurs*. Consider the reaction in which hydrogen gas reacts with iodine monochloride:

$$H_2(g) + 2\ ICl(g) \longrightarrow 2\ HCl(g) + I_2(g)$$

The overall equation shows only the substances present at the beginning of the reaction and the substances formed by the reaction—it does not show the intermediate steps. The **reaction mechanism** is the series of individual chemical steps by which an overall chemical reaction occurs. For example, the proposed mechanism for the reaction between hydrogen and iodine monochloride contains two steps:

Step 1 $H_2(g) + ICl(g) \longrightarrow HI(g) + HCl(g)$

Step 2 $HI(g) + ICl(g) \longrightarrow HCl(g) + I_2(g)$

In the first step, an H_2 molecule collides with an ICl molecule and forms an HI molecule and an HCl molecule. In the second step, the HI molecule formed in the first step collides with a second ICl molecule to form another HCl molecule and an I_2 molecule. Each step in a reaction mechanism is an **elementary step**. Elementary steps cannot be broken down into simpler steps—they occur as they are written (they represent the exact species that are colliding in the reaction).

An elementary step represents an actual interaction between the reactant molecules in the step. An overall reaction equation shows only the starting substances and the ending substances, not the path between them.

In a valid reaction mechanism, the individual steps in the mechanism add to the overall reaction. For example, the mechanism just shown sums to the overall reaction as shown here:

$$\begin{array}{l} H_2(g) + ICl(g) \longrightarrow \cancel{HI(g)} + HCl(g) \\ \underline{\cancel{HI(g)} + ICl(g) \longrightarrow HCl(g) + I_2(g)} \\ H_2(g) + 2\ ICl(g) \longrightarrow 2\ HCl(g) + I_2(g) \end{array}$$

Notice that the HI molecule appears in the reaction mechanism but not in the overall reaction equation. We call species such as HI *reaction intermediates*. A **reaction intermediate** forms in one elementary step and is consumed in another. The reaction mechanism is a complete, detailed description of the reaction at the molecular level—it specifies the individual collisions and reactions that result in the overall reaction. As such, reaction mechanisms are highly sought-after pieces of chemical knowledge.

How do we determine reaction mechanisms? Recall from the opening section of this chapter that chemical kinetics are not only practically important (because they allow us to control the rate of a particular reaction), but also theoretically important because they can help us determine the mechanism of the reaction. We can piece together possible reaction mechanisms by measuring the kinetics of the overall reaction and working backward to write a mechanism consistent with the measured kinetics.

Rate Laws for Elementary Steps

We characterize elementary steps by **molecularity**, the number of reactant particles involved in the step. The most common molecularities are unimolecular and bimolecular.

$A \longrightarrow$ products **Unimolecular**

$A + A \longrightarrow$ products **Bimolecular**

$A + B \longrightarrow$ products **Bimolecular**

Elementary steps in which three reactant particles collide, called **termolecular** steps, are very rare because the probability of three particles simultaneously colliding is small.

Although we cannot deduce the rate law for an *overall chemical reaction* from the balanced chemical equation, we can deduce the rate law for an *elementary step* from its equation. Since we know that an elementary step occurs through the collision of the reactant particles, the rate is proportional to the product of the concentrations of those particles. For example, the rate for the bimolecular elementary step in which A reacts with B is proportional to the concentration of A multiplied by the concentration of B.

$$A + B \longrightarrow \text{products} \qquad \text{Rate} = k[A][B]$$

Similarly, the rate law for the bimolecular step in which A reacts with A is proportional to the square of the concentration of A.

$$\text{A} + \text{A} \longrightarrow \text{products} \qquad \text{Rate} = k[\text{A}]^2$$

Table 15.3 summarizes the rate laws for the common elementary steps, as well as those for the rare termolecular step. Notice that the molecularity of the elementary step is equal to the overall order of the step.

TABLE 15.3 Rate Laws for Elementary Steps

Elementary Step	Molecularity	Rate Law
$\text{A} \longrightarrow \text{products}$	1	$\text{Rate} = k[\text{A}]$
$\text{A} + \text{A} \longrightarrow \text{products}$	2	$\text{Rate} = k[\text{A}]^2$
$\text{A} + \text{B} \longrightarrow \text{products}$	2	$\text{Rate} = k[\text{A}][\text{B}]$
$\text{A} + \text{A} + \text{A} \longrightarrow \text{products}$	3 (rare)	$\text{Rate} = k[\text{A}]^3$
$\text{A} + \text{A} + \text{B} \longrightarrow \text{products}$	3 (rare)	$\text{Rate} = k[\text{A}]^2[\text{B}]$
$\text{A} + \text{B} + \text{C} \longrightarrow \text{products}$	3 (rare)	$\text{Rate} = k[\text{A}][\text{B}][\text{C}]$

Rate-Determining Steps and Overall Reaction Rate Laws

In most chemical reactions, one of the elementary steps—called the **rate-determining step**—is much slower than the others. The rate-determining step (also called the rate-limiting step) in a chemical reaction is analogous to the narrowest section on a freeway. If a section of a freeway narrows from four lanes to two lanes, for even a short distance, the rate at which cars travel along the freeway is limited by the rate at which they can travel through the narrow section (even though the rate might be much faster along the four-lane section). Similarly, the rate-determining step in a reaction mechanism limits the overall rate of the reaction (even though the other steps occur much faster) and therefore determines *the rate law for the overall reaction.*

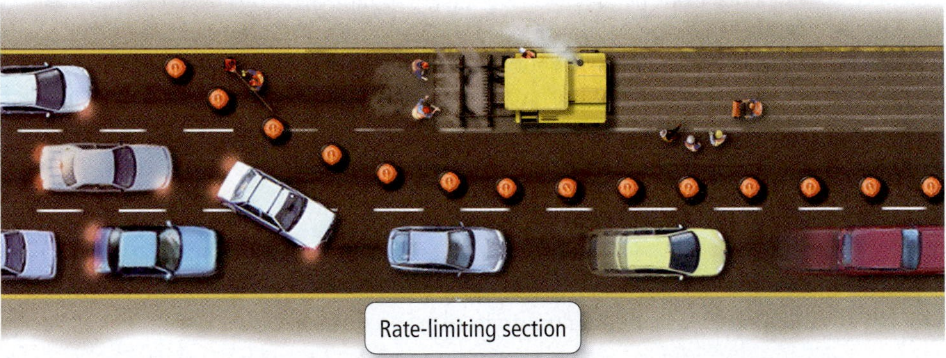

▶The rate-determining step in a reaction mechanism limits the overall rate of the reaction just as the narrowest section of a highway limits the rate at which traffic can pass.

As an example, consider the reaction between nitrogen dioxide gas and carbon monoxide gas.

$$\text{NO}_2(g) + \text{CO}(g) \longrightarrow \text{NO}(g) + \text{CO}_2(g)$$

The experimentally determined rate law for this reaction is $\text{Rate} = k[\text{NO}_2]^2$. We can see from this rate law that the reaction must not be a single-step reaction—otherwise the rate law would be $\text{Rate} = k[\text{NO}_2][\text{CO}]$. A possible mechanism for this reaction involves two steps.

$$\text{NO}_2(g) + \text{NO}_2(g) \longrightarrow \text{NO}_3(g) + \text{NO}(g) \qquad \text{Slow}$$
$$\text{NO}_3(g) + \text{CO}(g) \longrightarrow \text{NO}_2(g) + \text{CO}_2(g) \qquad \text{Fast}$$

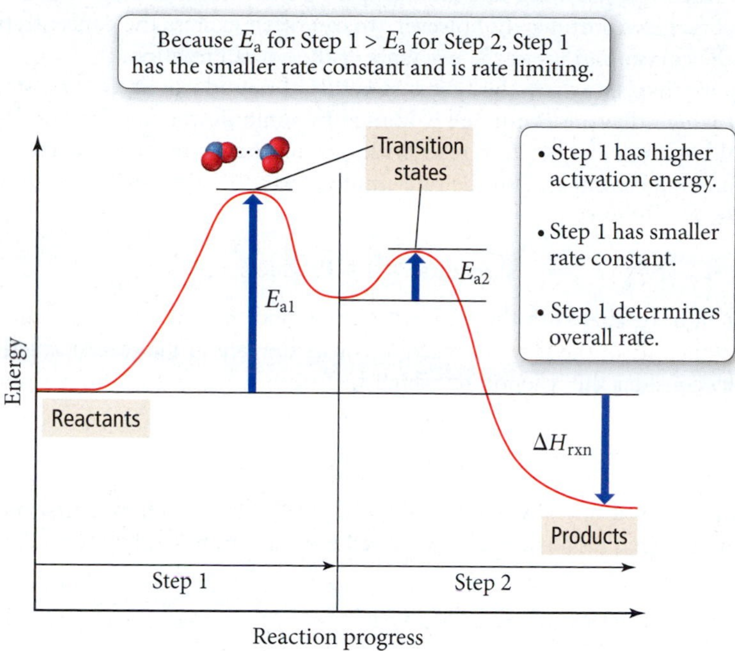
Energy Diagram for a Two-Step Mechanism
Because E_a for Step 1 > E_a for Step 2, Step 1 has the smaller rate constant and is rate limiting.
Transition states
• Step 1 has higher activation energy.
• Step 1 has smaller rate constant.
• Step 1 determines overall rate.
E_{a2}
E_{a1}
Energy
Reactants
ΔH_{rxn}
Products
Step 1
Step 2
Reaction progress

◀ **FIGURE 15.16 Energy Diagram for a Two-Step Mechanism**

Figure 15.16 ▲ shows the energy diagram for this mechanism. The first step has a much larger activation energy than the second step. This greater activation energy results in a much smaller rate constant for the first step compared to the second step. The first step determines the overall rate of the reaction, and the predicted *overall* rate law is therefore Rate $= k[NO_2]^2$, which is consistent with the observed experimental rate law.

For a proposed reaction mechanism, such as the one shown here for NO_2 and CO, to be *valid* (mechanisms can only be validated, not proven), two conditions must be met.

1. **The elementary steps in the mechanism must sum to the overall reaction.**
2. **The rate law predicted by the mechanism must be consistent with the experimentally observed rate law.**

We have already seen that the rate law predicted by the earlier mechanism is consistent with the experimentally observed rate law. We can check whether the elementary steps sum to the overall reaction by adding them together.

$$\begin{array}{lll} NO_2(g) + \cancel{NO_2(g)} \longrightarrow \cancel{NO_3(g)} + NO(g) & & \text{Slow} \\ \cancel{NO_3(g)} + CO(g) \longrightarrow \cancel{NO_2(g)} + CO_2(g) & & \text{Fast} \\ \hline NO_2(g) + CO(g) \longrightarrow NO(g) + CO_2(g) & & \text{Overall} \end{array}$$

The mechanism fulfills both of the given requirements and is therefore valid. A valid mechanism is not a *proven* mechanism (because other mechanisms may also fulfill both of the given requirements). We can only say that a given mechanism is consistent with kinetic observations of the reaction and therefore possible. Other types of data—such as the experimental evidence for a proposed intermediate—are necessary to further strengthen the validity of a proposed mechanism.

Mechanisms with a Fast Initial Step

When the proposed mechanism for a reaction has a slow initial step—like the one shown for the reaction between NO_2 and CO—the rate law predicted by the mechanism normally contains only reactants involved in the overall reaction. However, when a mechanism begins with a fast initial step, some other subsequent step in the mechanism is the rate-limiting step. In these cases, the rate law predicted by the rate-limiting step may contain reaction intermediates. Since reaction intermediates do not appear in the overall reaction equation, a rate law containing intermediates cannot generally correspond to the experimental rate law. Fortunately, however, we can often express the concentration of intermediates in terms of the concentrations of the reactants of the overall reaction.

In a multistep mechanism where the first step is fast, the products of the first step may build up because the rate at which they are consumed is limited by some slower step further down the line. As those products build up, they can begin to react with one another to re-form the reactants. As long as the first step is fast enough compared to the rate-limiting step, the first-step reaction will reach equilibrium. We indicate equilibrium as follows:

$$\text{Reactants} \underset{k_{-1}}{\overset{k_1}{\rightleftharpoons}} \text{Products}$$

The double arrows indicate that both the forward reaction and the reverse reaction occur. If equilibrium is reached, then the rate of the forward reaction equals the rate of the reverse reaction.

As an example, consider the reaction by which hydrogen reacts with nitrogen monoxide to form water and nitrogen gas:

$$2\ H_2(g) + 2\ NO(g) \longrightarrow 2\ H_2O(g) + N_2(g)$$

The experimentally observed rate law is Rate $= k[H_2][NO]^2$. The reaction is first order in hydrogen and second order in nitrogen monoxide. The proposed mechanism has a slow second step.

$$\begin{array}{rcll} 2\ NO(g) & \underset{k_{-1}}{\overset{k_1}{\rightleftharpoons}} & \cancel{N_2O_2(g)} & \text{Fast} \\ H_2(g) + \cancel{N_2O_2(g)} & \xrightarrow[k_2]{} & H_2O(g) + \cancel{N_2O(g)} & \text{Slow (rate limiting)} \\ \cancel{N_2O(g)} + H_2(g) & \xrightarrow[k_3]{} & N_2(g) + H_2O(g) & \text{Fast} \\ \hline 2\ H_2(g) + 2\ NO(g) & \longrightarrow & 2\ H_2O(g) + N_2(g) & \text{Overall} \end{array}$$

To determine whether the mechanism is valid, we must determine whether the two conditions described previously are met. As you can see, the steps do indeed sum to the overall reaction, so the first condition is met.

The second condition is that the rate law predicted by the mechanism must be consistent with the experimentally observed rate law. Because the second step is rate limiting, we write the following expression for the rate law:

$$\text{Rate} = k_2[H_2][N_2O_2] \qquad [15.28]$$

This rate law contains an intermediate (N_2O_2) and can therefore not be consistent with the experimentally observed rate law (which does not contain intermediates). Because of the equilibrium in the first step, however, *we can express the concentration of the intermediate in terms of the reactants of the overall equation.* The first step reaches equilibrium, so the rate of the forward reaction in the first step equals the rate of the reverse reaction:

$$\text{Rate (forward)} = \text{Rate (reverse)}$$

The rate of the forward reaction is given by the rate law:

$$\text{Rate} = k_1[NO]^2$$

The rate of the reverse reaction is given by the rate law:

$$\text{Rate} = k_{-1}[N_2O_2]$$

Since these two rates are equal at equilibrium, we write the expression:

$$k_1[NO]^2 = k_{-1}[N_2O_2]$$

Rearranging, we get:

$$[N_2O_2] = \frac{k_1}{k_{-1}}[NO]^2$$

We can now substitute this expression into Equation 15.28, the rate law obtained from the slow step.

$$\begin{aligned}\text{Rate} &= k_2[H_2][N_2O_2] \\ &= k_2[H_2]\frac{k_1}{k_{-1}}[NO]^2 \\ &= \frac{k_2k_1}{k_{-1}}[H_2][NO]^2\end{aligned}$$

If we combine the individual rate constants into one overall rate constant, we get the predicted rate law.

$$\text{Rate} = k[H_2][NO]^2 \qquad [15.29]$$

Since this rate law is consistent with the experimentally observed rate law, the second condition is met and the proposed mechanism is valid.

EXAMPLE 15.9

Reaction Mechanisms

Ozone naturally decomposes to oxygen by this reaction:

$$2\ O_3(g) \longrightarrow 3\ O_2(g)$$

The experimentally observed rate law for this reaction is:

$$\text{Rate} = k[O_3]^2[O_2]^{-1}$$

Show that this proposed mechanism is consistent with the experimentally observed rate law:

$$O_3(g) \underset{k_{-1}}{\overset{k_1}{\rightleftharpoons}} O_2(g) + O(g) \quad \text{Fast}$$

$$O_3(g) + O(g) \xrightarrow[k_2]{} 2\ O_2(g) \quad \text{Slow}$$

SOLUTION

To determine whether the mechanism is valid, you must first determine whether the steps sum to the overall reaction. The steps do indeed sum to the overall reaction, so the first condition is met.	$O_3(g) \underset{k_{-1}}{\overset{k_1}{\rightleftharpoons}} O_2(g) + \cancel{O}(g)$ $O_3(g) + \cancel{O}(g) \xrightarrow[k_2]{} 2\ O_2(g)$ $2\ O_3(g) \longrightarrow 3\ O_2(g)$
The second condition is that the rate law predicted by the mechanism must be consistent with the experimentally observed rate law. Because the second step is rate limiting, write the rate law based on the second step.	$\text{Rate} = k_2[O_3][O]$
Because the rate law contains an intermediate (O), you must express the concentration of the intermediate in terms of the concentrations of the reactants of the overall reaction. To do this, set the rates of the forward reaction and the reverse reaction of the first step equal to each other. Solve the expression from the previous step for [O], the concentration of the intermediate.	Rate (forward) = Rate (backward) $k_1[O_3] = k_{-1}[O_2][O]$ $[O] = \dfrac{k_1[O_3]}{k_{-1}[O_2]}$

—*Continued on the next page*

—Continued from the previous page

Finally, substitute [O] into the rate law predicted by the slow step, and combine the rate constants into one overall rate constant, k.	$\text{Rate} = k_2[O_3][O]$ $= k_2[O_3]\dfrac{k_1[O_3]}{k_{-1}[O_2]}$ $= k_2\dfrac{k_1[O_3]^2}{k_{-1}[O_2]}$ $= k[O_3]^2[O_2]^{-1}$

CHECK Because the two steps in the proposed mechanism sum to the overall reaction, and because the rate law obtained from the proposed mechanism is consistent with the experimentally observed rate law, the proposed mechanism is valid. The −1 reaction order with respect to $[O_2]$ indicates that the rate slows down as the concentration of oxygen increases—oxygen inhibits, or slows down, the reaction.

FOR PRACTICE 15.9

Predict the overall reaction and rate law that results from the following two-step mechanism:

$$2\,A \longrightarrow A_2 \quad \text{Slow}$$
$$A_2 + B \longrightarrow A_2B \quad \text{Fast}$$

15.8 Catalysis

Throughout this chapter, we have discussed ways to control the rates of chemical reactions. We can speed up the rate of a reaction by increasing the concentration of the reactants or by increasing the temperature. However, these approaches are not always feasible. There are limits to how concentrated we can make a reaction mixture, and increases in temperature may allow unwanted reactions—such as the decomposition of a reactant—to occur.

Alternatively, reaction rates can be increased by using a **catalyst**, a substance that increases the rate of a chemical reaction but is not consumed by the reaction. A catalyst works by providing an alternative mechanism for the reaction—one in which the rate-determining step has a lower activation energy. For example, consider the noncatalytic destruction of ozone in the upper atmosphere, which happens according to this reaction:

$$O_3(g) + O(g) \longrightarrow 2\,O_2(g)$$

In this reaction, an ozone molecule collides with an oxygen atom to form two oxygen molecules in a single elementary step. The reason that Earth has a protective ozone layer in the upper atmosphere is that the activation energy for this reaction is fairly high and the reaction, therefore, proceeds at a fairly slow rate; the ozone layer does not rapidly decompose into O_2. However, the addition of Cl atoms (which come from the photodissociation of human-made chlorofluorocarbons) to the upper atmosphere makes available another pathway by which O_3 can be destroyed. The first step in this pathway—called the catalytic destruction of ozone—is the reaction of Cl with O_3 to form ClO and O_2:

Photodissociation means *light-induced* dissociation. The energy from a photon of light can break chemical bonds and therefore dissociate, or break apart, a molecule.

$$Cl + O_3 \longrightarrow ClO + O_2$$

This is followed by a second step in which ClO reacts with O, regenerating Cl:

$$ClO + O \longrightarrow Cl + O_2$$

If we add the two reactions, the overall reaction is identical to the noncatalytic reaction:

$$\cancel{Cl} + O_3 \longrightarrow \cancel{ClO} + O_2$$
$$\cancel{ClO} + O \longrightarrow \cancel{Cl} + O_2$$
$$\overline{O_3 + O \longrightarrow 2\,O_2}$$

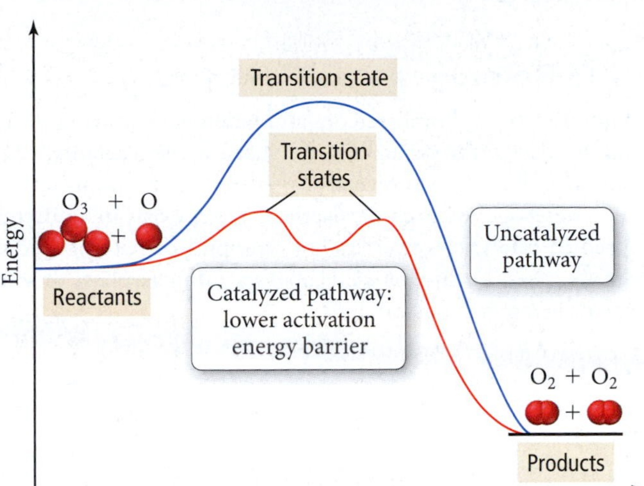

◀ **FIGURE 15.17 Catalyzed and Uncatalyzed Decomposition of Ozone** In the catalytic destruction of ozone (red), the activation barrier for the rate-limiting step is much lower than in the uncatalyzed process (blue).

However, the activation energy for the rate-limiting step in this pathway is much smaller than for the first, uncatalyzed pathway (as shown in Figure 15.17 ▲); therefore the reaction occurs at a much faster rate. Note that the Cl is not consumed in the overall reaction—this is characteristic of a catalyst.

In the case of the catalytic destruction of ozone, the catalyst speeds up a reaction that we *do not* want to happen. Most of the time, however, catalysts are used to speed up reactions that we *do* want to happen. For example, most cars have a catalytic converter in their exhaust systems. The catalytic converter contains solid catalysts, such as platinum, rhodium, or palladium, dispersed on an underlying high-surface-area ceramic structure. These catalysts convert exhaust pollutants such as nitrogen monoxide and carbon monoxide into less harmful substances.

$$2\,NO(g) + 2\,CO(g) \xrightarrow[\text{catalyst}]{} N_2(g) + 2\,CO_2(g)$$

The catalytic converter also promotes the complete combustion of any fuel fragments present in the exhaust.

$$CH_3CH_2CH_3(g) + 5\,O_2(g) \xrightarrow[\text{catalyst}]{} 3\,CO_2(g) + 4\,H_2O(g)$$

Fuel fragment ($CH_3CH_2CH_3$)

Fuel fragments in exhaust are harmful because they lead to the formation of ozone. Although ozone is a natural part of our *upper* atmosphere that protects us from excess exposure to ultraviolet light, it is a pollutant in the *lower* atmosphere, interfering with cardiovascular function and acting as an eye and

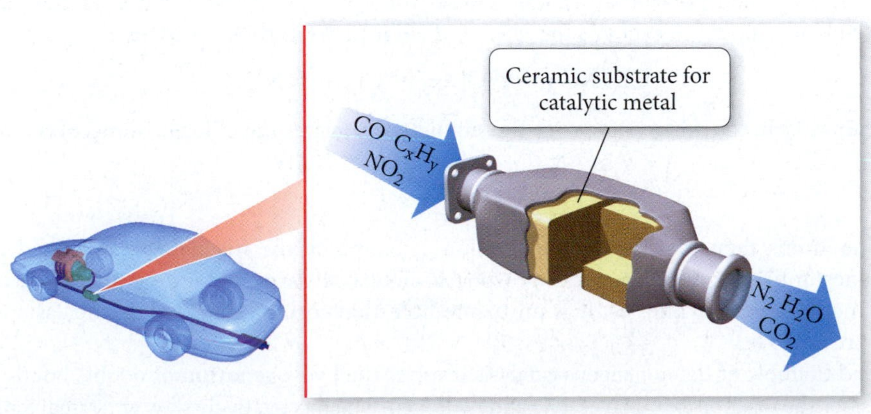

◀ The catalytic converter in the exhaust system of a car helps eliminate pollutants in the exhaust.

TABLE 15.4 Change in Pollutant Levels

Pollutant	Change 1980–2010
NO_2	−52%
O_3	−28%
CO	−82%

Source: U.S. Environmental Protection Agency, Our Nation's Air: Status and Trends through 2010.

lung irritant. The use of catalytic converters in motor vehicles has resulted in lower levels of these pollutants over most U.S. cities in the last 30 years even though the number of cars on the roadways has dramatically increased (see Table 15.4).

Homogeneous and Heterogeneous Catalysis

We categorize catalysis into two types: homogeneous and heterogeneous (Figure 15.18 ▼). In **homogeneous catalysis**, the catalyst exists in the same phase (or state) as the reactants. The catalytic destruction of ozone by Cl is an example of homogeneous catalysis—the chlorine atoms exist in the gas phase with the gas-phase reactants. In **heterogeneous catalysis**, the catalyst exists in a different phase than the reactants. The catalysts used in catalytic converters are examples of heterogeneous catalysts—they are solids while the reactants are gases. The use of solid catalysts with gas-phase or solution-phase reactants is the most common type of heterogeneous catalysis.

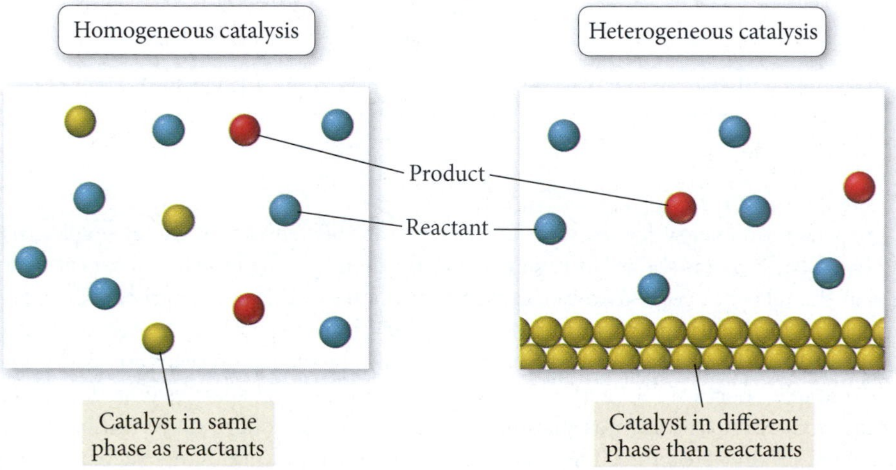

▲ **FIGURE 15.18 Homogeneous and Heterogeneous Catalysis** A homogeneous catalyst exists in the same phase as the reactants. A heterogeneous catalyst exists in a different phase than the reactants. Often a heterogeneous catalyst provides a solid surface on which the reaction can take place.

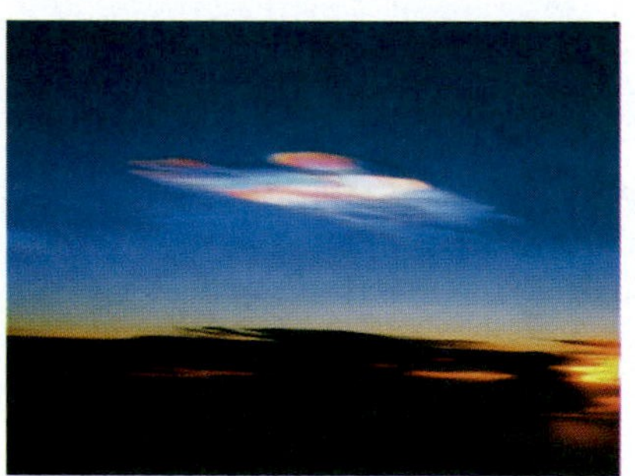

▲ Polar stratospheric clouds contain ice particles that catalyze reactions by which chlorine is released from its atmospheric chemical reservoirs.

Research has shown that heterogeneous catalysis is most likely responsible for the annual formation of an ozone hole over Antarctica. After the discovery of the Antarctic ozone hole in 1985, scientists wondered why there was such a dramatic drop in ozone over Antarctica but not over the rest of the planet. After all, the chlorine from chlorofluorocarbons that catalyzes ozone destruction is evenly distributed throughout the entire atmosphere.

As it turns out, most of the chlorine that enters the atmosphere from chlorofluorocarbons gets bound up in chlorine reservoirs, substances such as $ClONO_2$ that hold chlorine and prevent it from catalyzing ozone destruction. The unique conditions over Antarctica—especially the cold isolated air mass that exists during the long dark winter—result in clouds that contain solid ice particles. These unique clouds are called polar stratospheric clouds (or PSCs), and the surfaces of the ice particles within these clouds appear to catalyze the release of chlorine from their reservoirs:

$$ClONO_2 + HCl \xrightarrow[\text{PSCs}]{} Cl_2 + HNO_3$$

When the sun rises in the Antarctic spring, the sunlight dissociates the chlorine molecules into chlorine atoms:

$$Cl_2 \xrightarrow[\text{light}]{} 2\ Cl$$

The chlorine atoms then catalyze the destruction of ozone by the mechanism discussed previously. This continues until the sun melts the stratospheric clouds, allowing chlorine atoms to be reincorporated into their reservoirs. The result is an ozone hole that forms every spring and lasts about 6–8 weeks (Figure 15.19 ▶).

A second example of heterogeneous catalysis involves the **hydrogenation** of double bonds within alkenes. Consider the reaction between ethene and hydrogen, which is relatively slow at normal temperatures.

$$H_2C{=}CH_2(g) + H_2(g) \longrightarrow H_3C{-}CH_3(g) \quad \text{Slow at room temperature}$$

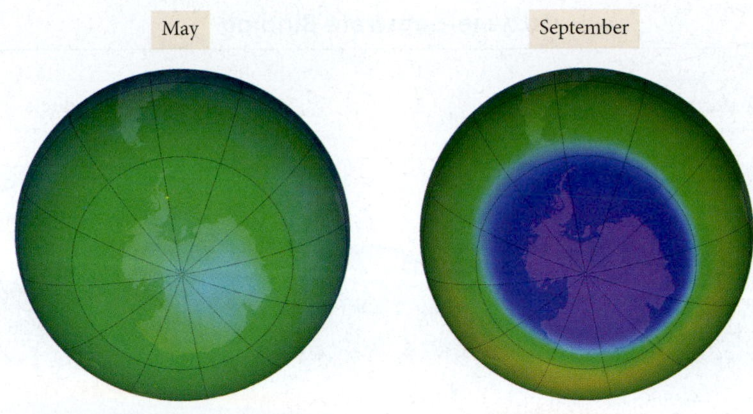

◀ **FIGURE 15.19 Ozone Depletion in the Antarctic Spring** The concentration of ozone over Antarctica drops sharply during the months of September and October due to the catalyzed destruction of ozone by chlorine. The image at the left shows the average ozone levels in May 2011 while the image at the right shows the average levels from September 2011. (The lowest ozone levels are represented in purple.) *Source:* NASA Ozone Watch, OMI instrument (KNMI/NASA) onboard the Aura satellite.

However, in the presence of finely divided platinum, palladium, or nickel, the reaction happens rapidly. The catalysis occurs by the four-step process depicted in Figure 15.20▼.

1. Adsorption: the reactants are adsorbed onto the metal surface.
2. Diffusion: the reactants diffuse on the surface until they approach each other.
3. Reaction: the reactants react to form the products.
4. Desorption: the products desorb from the surface into the gas phase.

The large activation energy of the hydrogenation reaction—due primarily to the strength of the hydrogen–hydrogen bond in H_2 which must be broken—is greatly lowered when the reactants adsorb onto the surface.

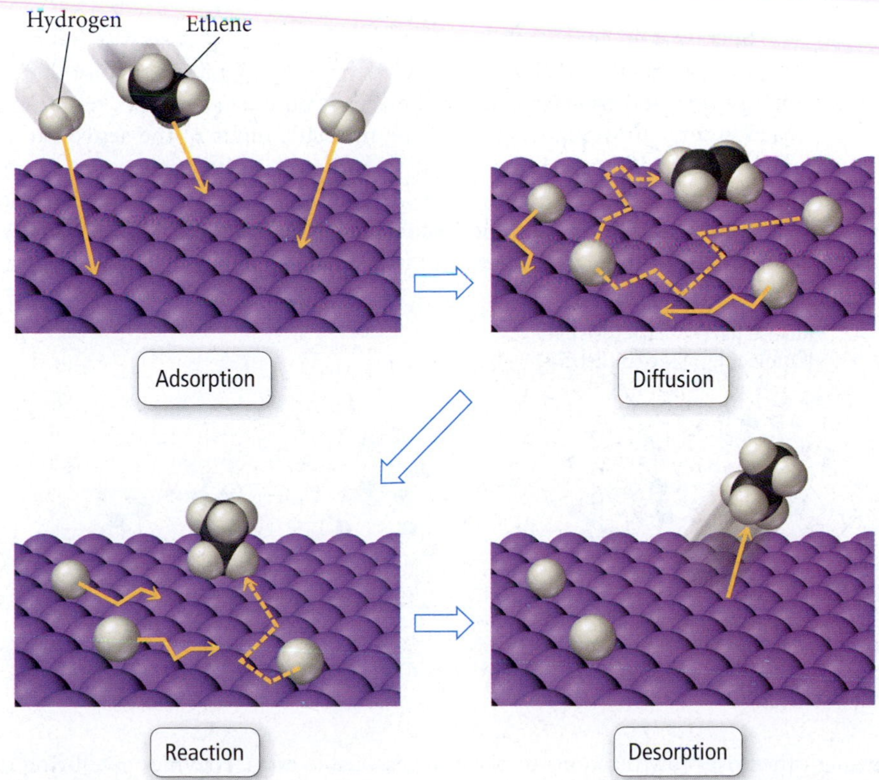

◀ **FIGURE 15.20 Catalytic Hydrogenation of Ethene**

Enzymes: Biological Catalysts

We find perhaps the best example of chemical catalysis in living organisms. Most of the thousands of reactions that must occur for an organism to survive are too slow at normal temperatures. So living organisms rely on **enzymes,** biological catalysts that increase the rates of biochemical reactions. Enzymes

The strategies used to speed up chemical reactions in the laboratory—high temperatures, high pressures, strongly acidic or alkaline conditions—are not available to living organisms, since those strategies would be fatal to cells.

▶ **FIGURE 15.21 Enzyme–Substrate Binding** A substrate (or reactant) fits into the active site of an enzyme much as a key fits into a lock. It is held in place by intermolecular forces and forms an enzyme–substrate complex. (Sometimes temporary covalent bonding may also be involved.) After the reaction occurs, the products are released from the active site.

Enzyme–Substrate Binding

Substrate
Active site
Enzyme
Enzyme–substrate complex
Products

are usually large protein molecules with complex three-dimensional structures. Within each enzyme's structure is a specific area called the **active site**. The properties and shape of the active site are just right to bind the reactant molecule, usually called the **substrate**. The substrate fits into the active site in a manner that is analogous to a key fitting into a lock (Figure 15.21▲). When the substrate binds to the active site of the enzyme—through intermolecular forces such as hydrogen bonding and dispersion forces, or even covalent bonds—the activation energy of the reaction is greatly lowered, allowing the reaction to occur at a much faster rate. The general mechanism by which an enzyme (E) binds a substrate (S) and then reacts to form the products (P) is:

$$\mathrm{E + S \rightleftharpoons ES} \qquad \text{Fast}$$

$$\mathrm{ES \longrightarrow E + P} \qquad \text{Slow, rate limiting}$$

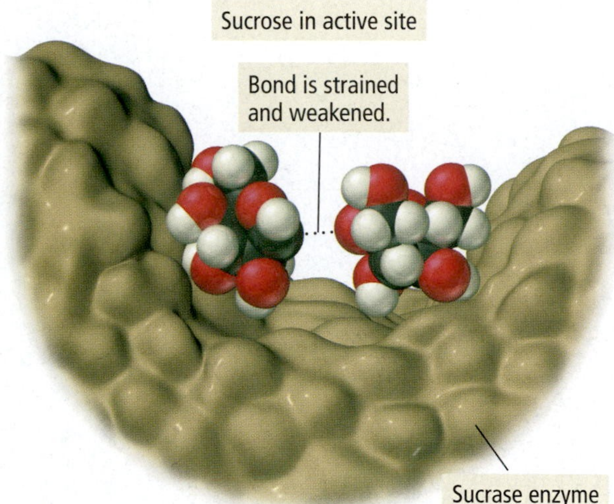

▲ **FIGURE 15.22 An Enzyme-Catalyzed Reaction** Sucrase catalyzes the conversion of sucrose into glucose and fructose by weakening the bond that joins the two rings.

Sucrase is an enzyme that catalyzes the breaking up of sucrose (table sugar) into glucose and fructose within the body. At body temperature, sucrose does not break into glucose and fructose because the activation energy is high, resulting in a slow reaction rate. However, when a sucrose molecule binds to the active site within sucrase, the bond between the glucose and fructose units weakens because glucose is forced into a geometry that stresses the bond (Figure 15.22◀). Weakening of this bond lowers the activation energy for the reaction, increasing the reaction rate. The reaction can then proceed toward equilibrium—which favors the products—at a much lower temperature.

▶ Sucrose breaks up into glucose and fructose during digestion.

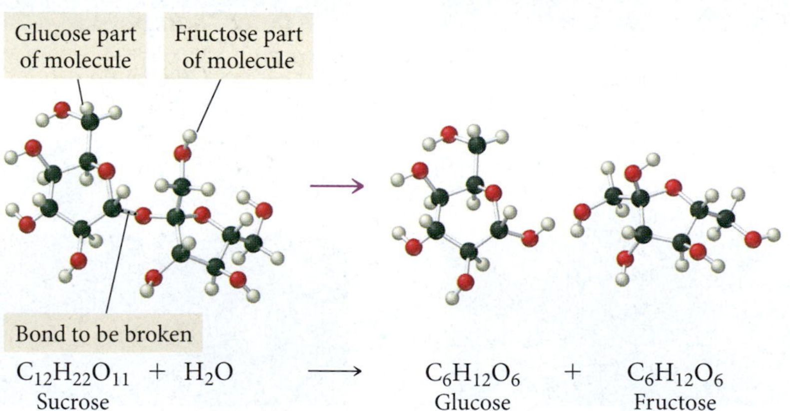

$$\underset{\text{Sucrose}}{C_{12}H_{22}O_{11}} + H_2O \longrightarrow \underset{\text{Glucose}}{C_6H_{12}O_6} + \underset{\text{Fructose}}{C_6H_{12}O_6}$$

By allowing otherwise slow reactions to occur at reasonable rates, enzymes give living organisms tremendous control over which reactions occur and when they occur. Enzymes are extremely specific (each enzyme catalyzes only a single reaction) and efficient, speeding up reaction rates by factors of as much as a billion. To turn a particular reaction on, a living organism produces or activates the correct enzyme to catalyze that reaction. Because organisms are so dependent on the reactions enzymes catalyze, many substances that inhibit the action of enzymes are highly toxic. Locking up a single enzyme molecule can stop the reaction of billions of substrates, much as one motorist stalled at a tollbooth can paralyze an entire highway full of cars.

SELF-ASSESSMENT Quiz

1. This graph shows the concentration of the reactant A in the reaction A $\longrightarrow$ B. Determine the average rate of the reaction between 0 and 10 seconds.

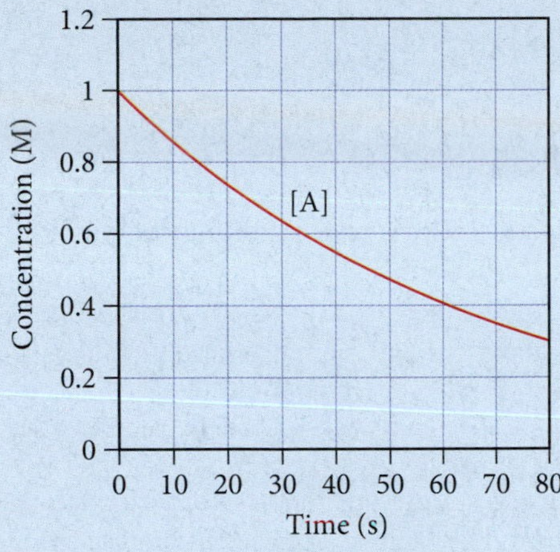

a) 0.07 M/s
b) 0.007 M/s
c) 0.86 M/s
d) 0.014 M/s

2. Dinitrogen monoxide decomposes into nitrogen and oxygen when heated. The initial rate of the reaction is 0.022 M/s. What is the initial rate of change of the concentration of N_2O (that is, $\Delta[N_2O]/\Delta t$)?

$$2\ N_2O(g) \longrightarrow 2\ N_2(g) + O_2(g)$$

a) −0.022 M/s
b) −0.011 M/s
c) −0.044 M/s
d) +0.022 M/s

3. This plot shows the rate of the decomposition of SO_2Cl_2 into SO_2 and Cl_2 as a function of the concentration of SO_2Cl_2. What is the order of the reaction?

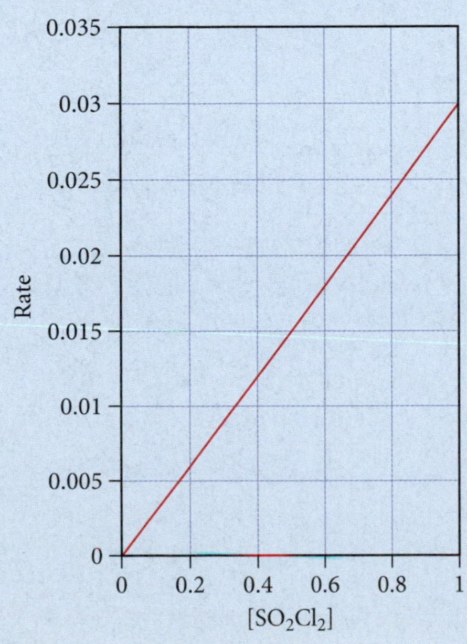

a) first order
b) second order
c) zero order
d) Order cannot be determined without more information.

4. For the reaction 2A + B $\longrightarrow$ C, the initial rate is measured at several different reactant concentrations. From the tabulated data, determine the rate law for the reaction.

[A] (M)	[B] (M)	Initial Rate (M/s)
0.05	0.05	0.035
0.10	0.05	0.070
0.20	0.10	0.56

a) Rate $= k[A][B]$
b) Rate $= k[A]^2[B]$
c) Rate $= k[A][B]^2$
d) Rate $= k[A]^2[B]^2$

5. What is the rate constant for the reaction in Question 4?
a) $2.8 \times 10^2\ M^{-2} \cdot s^{-1}$
b) $14\ M^{-2} \cdot s^{-1}$
c) $1.4 \times 10^2\ M^{-2} \cdot s^{-1}$
d) $1.4 \times 10^3\ M^{-2} \cdot s^{-1}$

6. The decomposition of Br_2 is followed as a function of time; two different plots of the data are shown here. Determine the order and rate constant for the reaction.

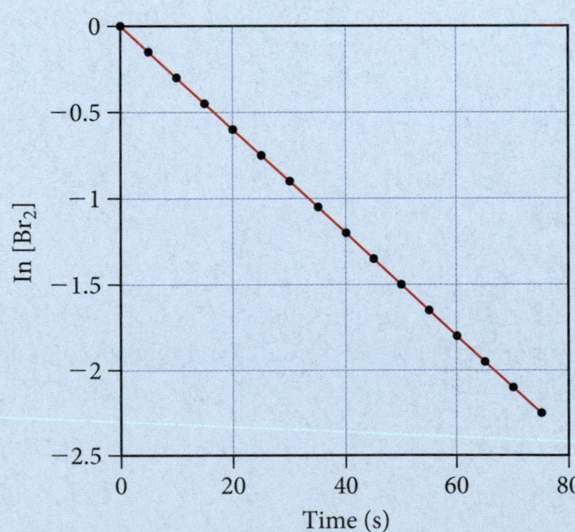

—*Continued*

Continued—

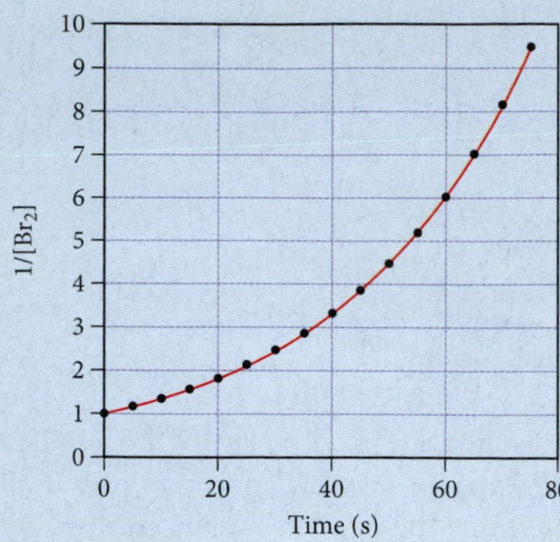

a) first order; 0.030 s^{-1}
b) first order; 33.3 s^{-1}
c) second order; 0.045 $M^{-1} \cdot s^{-1}$
d) second order; 22.2 $M^{-1} \cdot s^{-1}$

7. The reaction X $\longrightarrow$ products is second order in X and has a rate constant of 0.035 $M^{-1}s^{-1}$. If a reaction mixture is initially 0.45 M in X, what is the concentration of X after 155 seconds?
a) 7.6 M
b) 2.0×10^{-3} M
c) 0.13 M
d) 0.00 M

8. A decomposition reaction has a half-life that does not depend on the initial concentration of the reactant. What is the order of the reaction?
a) zero order
b) first order
c) second order
d) Order cannot be determined without more information.

9. The rate constant of a reaction is measured at different temperatures. A plot of the natural log of the rate constant as a function of the inverse of the temperature (in kelvins) yields a straight line with a slope of -8.55×10^3 K^{-1}. What is the activation energy (E_a) for the reaction?
a) −71 kJ
b) 71 kJ
c) 1.0 kJ
d) −1.0 kJ

10. The rate constant for a reaction at 25.0 °C is 0.010 s^{-1} and its activation energy is 35.8 kJ/mol. Find the rate constant at 50.0 °C.
a) 0.021 s^{-1}
b) 0.010 s^{-1}
c) 0.0033 s^{-1}
d) 0.031 s^{-1}

11. The mechanism shown here is proposed for the gas-phase reaction, $2\ N_2O_5 \longrightarrow 2\ NO_2 + O_2$. What rate law does the mechanism predict?

$$N_2O_5 \underset{k_{-1}}{\overset{k_1}{\rightleftharpoons}} NO_2 + NO_3 \quad \text{Fast}$$

$$NO_2 + NO_3 \longrightarrow NO_2 + O_2 + NO \quad \text{Slow}$$

$$NO + N_2O_5 \longrightarrow 3\ NO_2 \quad \text{Fast}$$

a) Rate $= k[N_2O_5]$
b) Rate $= k[N_2O_5]^2$
c) Rate $= k[N_2O_5]^0$
d) Rate $= k[NO_2][NO_3]$

12. Which statement is true regarding the function of a catalyst in a chemical reaction?
a) A catalyst increases the rate of a reaction.
b) A catalyst provides an alternate mechanism for the reaction.
c) A catalyst is not consumed by the reaction.
d) All of the above are true.

13. These images represent the first-order reaction A $\longrightarrow$ B initially and at some later time. The rate law for the reaction is Rate = 0.010 s^{-1} [A]. How much time has passed between the two images?

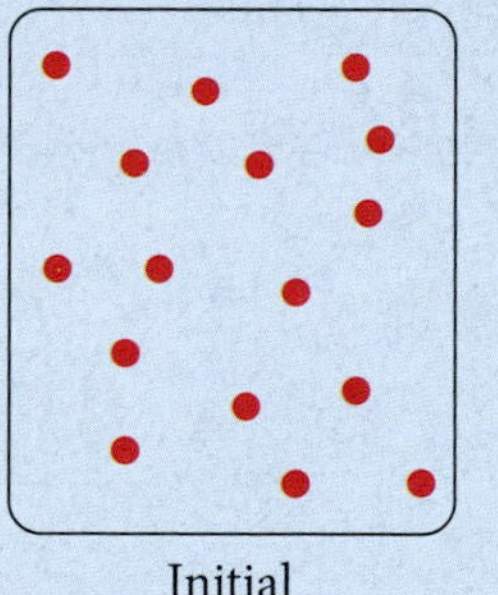
Initial

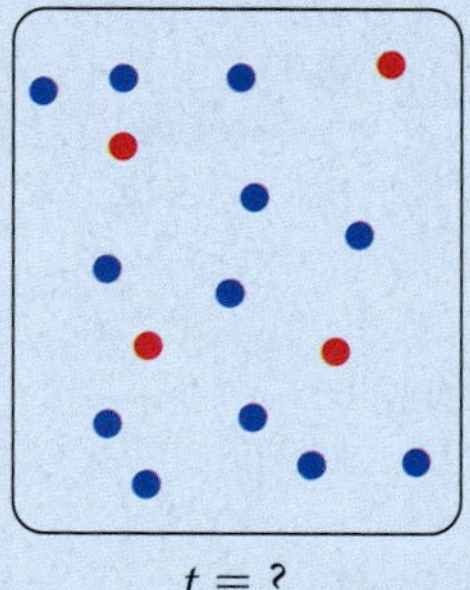
t = ?

A ●
B ●

a) 69 s
b) 139 s
c) 60 s
d) 12.5 s

14. Pick the single-step reaction which, according to collision theory, is likely to have the smallest orientation factor.
a) $H + H \longrightarrow H_2$
b) $I + HI \longrightarrow I_2 + H$
c) $H_2 + H_2C{=}CH_2 \longrightarrow H_3C{-}CH_3$
d) All of these reactions have the same orientation factor.

—*Continued*

Continued—

15. Carbon monoxide and chlorine gas react to form phosgene ($COCl_2$) according to the equation:

$$CO(g) + Cl_2(g) \longrightarrow COCl_2(g)$$

The rate law for the reaction is Rate $= k[Cl_2]^{3/2}[CO]$. Which mixture of chlorine gas and carbon monoxide gas has the fastest initial rate?

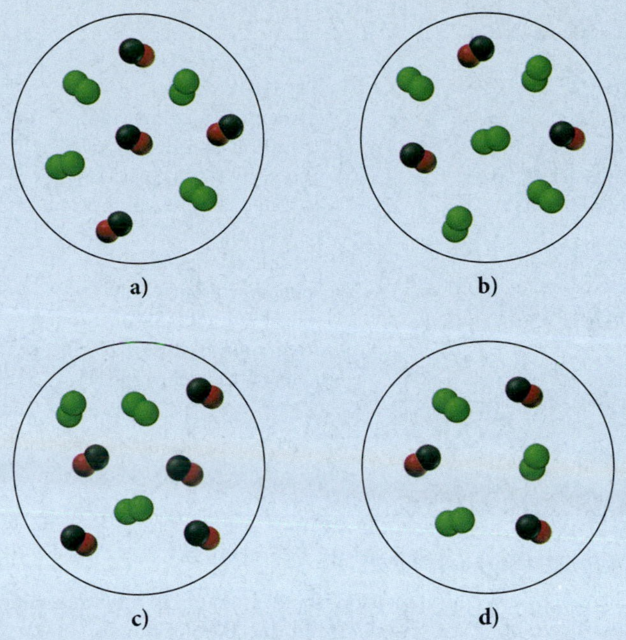

Answers: 1:d; 2:c; 3:a; 4:c; 5:a; 6:a; 7:c; 8:b; 9:b; 10:d; 11:a; 12:d; 13:b; 14:c; 15:b

CHAPTER SUMMARY 15

REVIEW

KEY LEARNING OUTCOMES

CHAPTER OBJECTIVES	ASSESSMENT
Expressing Reaction Rates (15.3)	• Example 15.1 For Practice 15.1 Exercises 27–36
Determining the Order, Rate Law, and Rate Constant of a Reaction (15.4)	• Example 15.2 For Practice 15.2 Exercises 43–46
Using Graphical Analysis of Reaction Data to Determine Reaction Order and Rate Constants (15.5)	• Examples 15.3, 15.5 For Practice 15.3, 15.5 Exercises 49–54
Determining the Concentration of a Reactant at a Given Time (15.5)	• Example 15.4 For Practice 15.4 Exercises 53–56
Working with the Half-Life of a Reaction (15.5)	• Example 15.6 For Practice 15.6 Exercises 55–58
Using the Arrhenius Equation to Determine Kinetic Parameters (15.6)	• Examples 15.7, 15.8 For Practice 15.7, 15.8 Exercises 61–72
Determining whether a Reaction Mechanism Is Valid (15.7)	• Example 15.9 For Practice 15.9 Exercises 75–78

KEY TERMS

Section 15.4
rate law (563)
rate constant (k) (563)
reaction order (n) (563)
overall order (565)

Section 15.5
integrated rate law (569)
half-life ($t_{1/2}$) (573)

Section 15.6
Arrhenius equation (576)
activation energy (E_a) (576)
frequency factor (A) (576)
activated complex (transition state) (577)
exponential factor (577)
Arrhenius plot (578)
collision model (581)
orientation factor (581)
collision frequency (581)

Section 15.7
reaction mechanism (583)
elementary step (583)
reaction intermediates (583)
molecularity (583)
unimolecular (583)
bimolecular (583)
termolecular (583)
rate-determining step (584)

Section 15.8
catalyst (588)
homogeneous catalysis (590)
heterogeneous catalysis (590)
hydrogenation (590)
enzyme (591)
active site (592)
substrate (592)

KEY CONCEPTS

Reaction Rates (15.1–15.3)

- The rate of a chemical reaction is a measure of how fast a reaction occurs. The rate reflects the change in the concentration of a reactant or product per unit time and is usually reported in units of M/s.
- Reaction rates generally depend on the concentration of the reactants, the temperature at which the reaction occurs, and the structure of the reactants.

Reaction Rate Laws and Orders (15.4)

- The rate of a first-order reaction is directly proportional to the concentration of the reactant, the rate of a second-order reaction is proportional to the square of the concentration of the reactant, and the rate of a zero-order reaction is independent of the concentration of the reactant.
- For a reaction with more than one reactant, the order with respect to each reactant is, in general, independent of the order with respect to other reactants. The rate law shows the relationship between the rate and the concentrations of each reactant and must be determined experimentally.

Integrated Rate Laws and Half-Life (15.5)

- The rate law for a reaction describes the relationship between the rate of the reaction and the concentrations of the reactants.
- The integrated rate law for a reaction describes the relationship between the concentration of a reactant and time.
- The integrated rate law for a zero-order reaction shows that the concentration of the reactant varies linearly with time. For a first-order reaction, the *natural log* of the concentration of the reactant varies linearly with time, and for a second-order reaction, the *inverse* of the concentration of the reactant varies linearly with time.
- The half-life of a reaction can be derived from the integrated rate law and represents the time required for the concentration of a reactant to fall to one-half of its initial value. The half-life of a first-order reaction is *independent* of initial concentration of the reactant. The half-life of a zero-order or second-order reaction *depends* on the initial concentration of reactant.

The Effect of Temperature on Reaction Rate (15.6)

- The rate constant of a reaction generally depends on temperature and can be expressed by the Arrhenius equation, which consists of a frequency factor and an exponential factor.
- The frequency factor represents the number of times that the reactants approach the activation barrier per unit time. The exponential factor is the fraction of approaches that are successful in surmounting the activation barrier and forming products.
- The exponential factor depends on both the temperature and the activation energy, a barrier that the reactants must overcome to become products. The exponential factor increases with increasing temperature but decreases with increasing activation energy.
- We can determine the frequency factor and activation energy for a reaction by measuring the rate constant at different temperatures and constructing an Arrhenius plot.
- For reactions in the gas phase, Arrhenius behavior can be described with the collision model. In this model, reactions occur as a result of sufficiently energetic collisions. The colliding molecules must be oriented in such a way that the reaction can occur. The frequency factor contains two terms: p, which represents the fraction of collisions that have the proper orientation, and z, which represents the number of collisions per unit time.

Reaction Mechanisms (15.7)

- Most chemical reactions occur in several steps. The series of individual steps by which a reaction occurs is the reaction mechanism.
- In order for a proposed reaction mechanism to be valid, it must fulfill two conditions: (a) the steps must sum to the overall reaction, and (b) the mechanism must predict the experimentally observed rate law.
- For mechanisms with a slow initial step, we derive the rate law from the slow step.
- For mechanisms with a fast initial step, we first write the rate law based on the slow step but then assume that the fast steps reach equilibrium, so we can write concentrations of intermediates in terms of the reactants.

Catalysis (15.8)

- A catalyst is a substance that increases the rate of a chemical reaction by providing an alternative mechanism that has a lower activation energy for the rate-determining step.
- Catalysts can be homogeneous or heterogeneous. A homogeneous catalyst exists in the same phase as the reactants and forms a homogeneous mixture with them. A heterogeneous catalyst generally exists in a different phase than the reactants.
- Enzymes are biological catalysts capable of increasing the rate of specific biochemical reactions by many orders of magnitude.

KEY EQUATIONS AND RELATIONSHIPS

The Rate of Reaction (15.3)

For a reaction $aA + bB \longrightarrow cC + dD$, the rate is defined as

$$\text{Rate} = -\frac{1}{a}\frac{\Delta[A]}{\Delta t} = -\frac{1}{b}\frac{\Delta[B]}{\Delta t} = +\frac{1}{c}\frac{\Delta[C]}{\Delta t} = +\frac{1}{d}\frac{\Delta[D]}{\Delta t}$$

The Rate Law (15.4)

$\text{Rate} = k[A]^n$ (single reactant)

$\text{Rate} = k[A]^m[B]^n$ (multiple reactant)

Integrated Rate Laws and Half-Life (15.5)

Reaction Order	Integrated Rate Law	Units of k	Half-Life Expression
0	$[A]_t = -kt + [A]_0$	$M \cdot s^{-1}$	$t_{1/2} = \frac{[A]_0}{2k}$
1	$\ln [A]_t = -kt + \ln [A]_0$	s^{-1}	$t_{1/2} = \frac{0.693}{k}$
2	$\frac{1}{[A]_t} = kt + \frac{1}{[A]_0}$	$M^{-1} \cdot s^{-1}$	$t_{1/2} = \frac{1}{k[A]_0}$

Arrhenius Equation (15.6)

$$k = Ae^{-E_a/RT}$$

$$\ln k = -\frac{E_a}{R}\left(\frac{1}{T}\right) + \ln A \quad \text{(linearized form)}$$

$$\ln\frac{k_2}{k_1} = \frac{E_a}{R}\left(\frac{1}{T_1} - \frac{1}{T_2}\right) \quad \text{(two-point form)}$$

$$k = pze^{-E_a/RT} \quad \text{(collision theory)}$$

Rate Laws for Elementary Steps (15.7)

Elementary Step	Molecularity	Rate Law
$A \longrightarrow$ products	1	$\text{Rate} = k[A]$
$A + A \longrightarrow$ products	2	$\text{Rate} = k[A]^2$
$A + B \longrightarrow$ products	2	$\text{Rate} = k[A][B]$
$A + A + A \longrightarrow$ products	3 (rare)	$\text{Rate} = k[A]^3$
$A + A + B \longrightarrow$ products	3 (rare)	$\text{Rate} = k[A]^2[B]$
$A + B + C \longrightarrow$ products	3 (rare)	$\text{Rate} = k[A][B][C]$

EXERCISES

REVIEW QUESTIONS

1. Explain why lizards become sluggish in cold weather. How is this phenomenon related to chemistry?
2. Why are reaction rates important (both practically and theoretically)?
3. Using the idea that reactions occur as a result of collisions between particles, explain why reaction rates depend on the concentration of the reactants.
4. Using the idea that reactions occur as a result of collisions between particles, explain why reaction rates depend on the temperature of the reaction mixture.
5. What units are typically used to express the rate of a reaction?
6. Why is the reaction rate for reactants defined as the *negative* of the change in reactant concentration with respect to time, whereas for products it is defined as the change in reactant concentration with respect to time (with a positive sign)?
7. Explain the difference between the average rate of reaction and the instantaneous rate of reaction.
8. Consider a simple reaction in which a reactant A forms products:

   $$A \longrightarrow \text{products}$$

   What is the rate law if the reaction is zero order with respect to A? First order? Second order? For each case, explain how a doubling of the concentration of A would affect the rate of reaction.
9. How is the order of a reaction generally determined?
10. For a reaction with multiple reactants, how is the overall order of the reaction defined?
11. Explain the difference between the rate law for a reaction and the integrated rate law for a reaction. What relationship does each kind of rate law express?
12. Write integrated rate laws for zero-order, first-order, and second-order reactions of the form $A \longrightarrow$ products.
13. What does the term *half-life* mean? Write the expressions for the half-lives of zero-order, first-order, and second-order reactions.
14. How do reaction rates typically depend on temperature? What part of the rate law is temperature dependent?
15. Explain the meaning of each term within the Arrhenius equation: activation energy, frequency factor, and exponential factor. Use these terms and the Arrhenius equation to explain why small changes in temperature can result in large changes in reaction rates.
16. What is an Arrhenius plot? Explain the significance of the slope and intercept of an Arrhenius plot.
17. Explain the meaning of the orientation factor in the colllision model.
18. Explain the difference between a normal chemical equation for a chemical reaction and the mechanism of that reaction.
19. In a reaction mechanism, what is an elementary step? Write down the three most common elementary steps and the corresponding rate law for each one.
20. What are the two requirements for a proposed mechanism to be valid for a given reaction?
21. What is an intermediate within a reaction mechanism?
22. What is a catalyst? How does a catalyst increase the rate of a chemical reaction?
23. Explain the difference between homogeneous catalysis and heterogeneous catalysis.
24. What are the four basic steps involved in heterogeneous catalysis?
25. What are enzymes? What is the active site of an enzyme? What is a substrate?
26. What is the general two-step mechanism by which most enzymes work?

PROBLEMS BY TOPIC

Reaction Rates

27. Consider the reaction.

$$2\ HBr(g) \longrightarrow H_2(g) + Br_2(g)$$

a. Express the rate of the reaction in terms of the change in concentration of each of the reactants and products.

b. In the first 25.0 s of this reaction, the concentration of HBr drops from 0.600 M to 0.512 M. Calculate the average rate of the reaction during this time interval.

c. If the volume of the reaction vessel in part b is 1.50 L, what amount of Br_2 (in moles) forms during the first 15.0 s of the reaction?

28. Consider the reaction.

$$2\ N_2O(g) \longrightarrow 2\ N_2(g) + O_2(g)$$

a. Express the rate of the reaction in terms of the change in concentration of each of the reactants and products.

b. In the first 15.0 s of the reaction, 0.015 mol of O_2 is produced in a reaction vessel with a volume of 0.500 L. What is the average rate of the reaction during this time interval?

c. Predict the rate of change in the concentration of N_2O during this time interval. In other words, what is $\Delta[N_2O]/\Delta t$?

29. For the reaction $2\ A(g) + B(g) \longrightarrow 3\ C(g)$,

a. determine the expression for the rate of the reaction in terms of the change in concentration of each of the reactants and products.

b. when A is decreasing at a rate of 0.100 M/s, how fast is B decreasing? How fast is C increasing?

30. For the reaction $A(g) + \frac{1}{2}B(g) \longrightarrow 2\ C(g)$,

a. determine the expression for the rate of the reaction in terms of the change in concentration of each of the reactants and products.

b. when C is increasing at a rate of 0.0025 M/s, how fast is B decreasing? How fast is A decreasing?

31. Consider the reaction.

$$Cl_2(g) + 3\ F_2(g) \longrightarrow 2\ ClF_3(g)$$

Complete the table.

$\Delta[Cl_2]/\Delta t$	$\Delta[F_2]/\Delta t$	$\Delta[ClF_3]/\Delta t$	Rate
−0.012 M/s			

32. Consider the reaction.

$$8\ H_2S(g) + 4\ O_2(g) \longrightarrow 8\ H_2O(g) + S_8(g)$$

Complete the table.

$\Delta[H_2S]/\Delta t$	$\Delta[O_2]/\Delta t$	$\Delta[H_2O]/\Delta t$	$\Delta[S_8]/\Delta t$	Rate
−0.080 M/s				

33. Consider the reaction:

$$C_4H_8(g) \longrightarrow 2\ C_2H_4(g)$$

The tabulated data were collected for the concentration of C_4H_8 as a function of time.

Time (s)	C_4H_8 (M)
0	1.000
10	0.913
20	0.835
30	0.763
40	0.697
50	0.637

a. What is the average rate of the reaction between 0 and 10 s? Between 40 and 50 s?

b. What is the rate of formation of C_2H_4 between 20 and 30 s?

34. Consider the reaction.

$$NO_2(g) \longrightarrow NO(g) + \tfrac{1}{2}O_2(g)$$

The tabulated data were collected for the concentration of NO_2 as a function of time.

Time (s)	$[NO_2]$ (M)
0	1.000
10	0.951
20	0.904
30	0.860
40	0.818
50	0.778
60	0.740
70	0.704
80	0.670
90	0.637
100	0.606

a. What is the average rate of the reaction between 10 and 20 s? Between 50 and 60 s?

b. What is the rate of formation of O_2 between 50 and 60 s?

35. Consider the reaction.

$$H_2(g) + Br_2(g) \longrightarrow 2\ HBr(g)$$

The graph shows the concentration of Br_2 as a function of time.

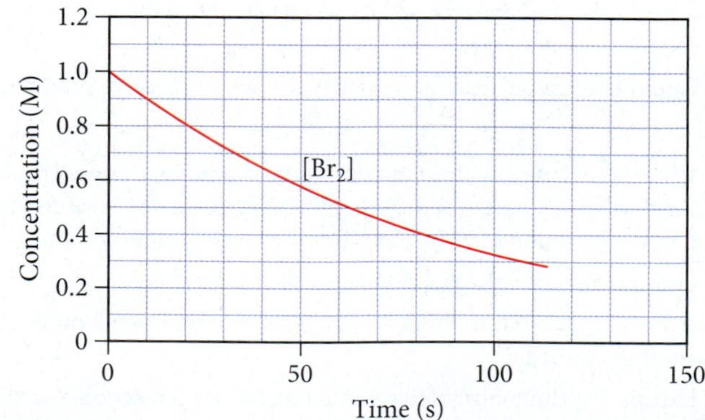

a. Use the graph to calculate each quantity.
   (i) the average rate of the reaction between 0 and 25 s
   (ii) the instantaneous rate of the reaction at 25 s
   (iii) the instantaneous rate of formation of HBr at 50 s

b. Make a rough sketch of a curve representing the concentration of HBr as a function of time. Assume that the initial concentration of HBr is zero.

36. Consider the reaction.

$$2\ H_2O_2(aq) \longrightarrow 2\ H_2O(l) + O_2(g)$$

The graph shows the concentration of H_2O_2 as a function of time.

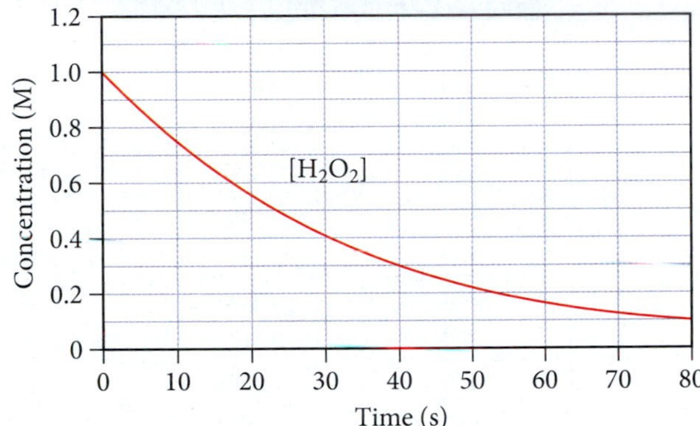

a. Use the graph to calculate each quantity.
   (i) the average rate of the reaction between 10 and 20 s
   (ii) the instantaneous rate of the reaction at 30 s
   (iii) the instantaneous rate of formation of O_2 at 50 s

b. If the initial volume of the H_2O_2 is 1.5 L, what total amount of O_2 (in moles) is formed in the first 50 s of reaction?

The Rate Law and Reaction Orders

37. This graph shows a plot of the rate of a reaction versus the concentration of the reactant A for the reaction A $\longrightarrow$ products.

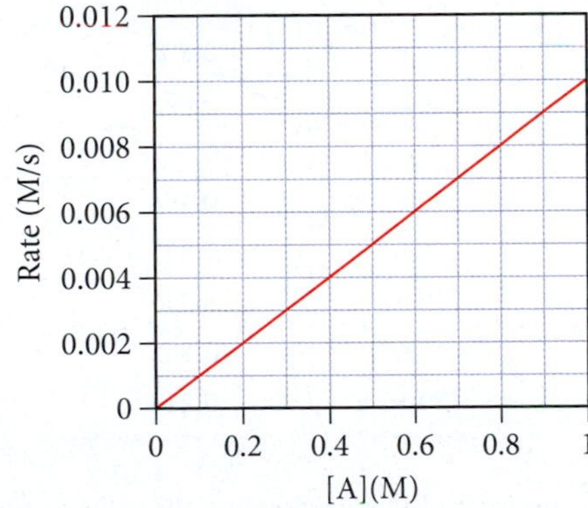

a. What is the order of the reaction with respect to A?
b. Make a rough sketch of a plot of [A] versus *time*.
c. Write a rate law for the reaction including an estimate for the value of k.

38. This graph shows a plot of the rate of a reaction versus the concentration of the reactant.

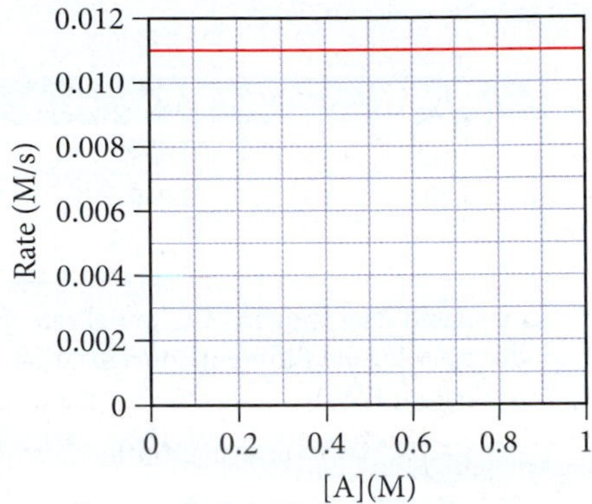

a. What is the order of the reaction with respect to A?
b. Make a rough sketch of a plot of [A] versus *time*.
c. Write a rate law for the reaction including the value of k.

39. What are the units of k for each type of reaction?

a. first-order reaction
b. second-order reaction
c. zero-order reaction

40. This reaction is first order in N_2O_5:

$$N_2O_5(g) \longrightarrow NO_3(g) + NO_2(g)$$

The rate constant for the reaction at a certain temperature is 0.053/s.

a. Calculate the rate of the reaction when $[N_2O_5] = 0.055$ M.
b. What is the rate of the reaction at the concentration indicated in part a if the reaction is second order? Zero order? (Assume the same *numerical* value for the rate constant with the appropriate units.)

41. A reaction in which A, B, and C react to form products is first order in A, second order in B, and zero order in C.

a. Write a rate law for the reaction.
b. What is the overall order of the reaction?
c. By what factor does the reaction rate change if [A] is doubled (and the other reactant concentrations are held constant)?
d. By what factor does the reaction rate change if [B] is doubled (and the other reactant concentrations are held constant)?
e. By what factor does the reaction rate change if [C] is doubled (and the other reactant concentrations are held constant)?
f. By what factor does the reaction rate change if the concentrations of all three reactants are doubled?

42. A reaction in which A, B, and C react to form products is zero order in A, one-half order in B, and second order in C.

a. Write a rate law for the reaction.
b. What is the overall order of the reaction?
c. By what factor does the reaction rate change if [A] is doubled (and the other reactant concentrations are held constant)?
d. By what factor does the reaction rate change if [B] is doubled (and the other reactant concentrations are held constant)?
e. By what factor does the reaction rate change if [C] is doubled (and the other reactant concentrations are held constant)?
f. By what factor does the reaction rate change if the concentrations of all three reactants are doubled?

43. Consider the tabulated data showing the initial rate of a reaction (A ⟶ products) at several different concentrations of A. What is the order of the reaction? Write a rate law for the reaction including the value of the rate constant, *k*.

[A] (M)	Initial Rate (M/s)
0.100	0.053
0.200	0.210
0.300	0.473

44. Consider the tabulated data showing the initial rate of a reaction (A ⟶ products) at several different concentrations of A. What is the order of the reaction? Write a rate law for the reaction including the value of the rate constant, *k*.

[A] (M)	Initial Rate (M/s)
0.15	0.008
0.30	0.016
0.60	0.032

45. The tabulated data were collected for this reaction:

$$2\ NO_2(g) + F_2(g) \longrightarrow 2\ NO_2F(g)$$

$[NO_2]$ (M)	$[F_2]$ (M)	Initial Rate (M/s)
0.100	0.100	0.026
0.200	0.100	0.051
0.200	0.200	0.103
0.400	0.400	0.411

Write an expression for the reaction rate law and calculate the value of the rate constant, *k*. What is the overall order of the reaction?

46. The tabulated data were collected for this reaction:

$$CH_3Cl(g) + 3\ Cl_2(g) \longrightarrow CCl_4(g) + 3\ HCl(g)$$

$[CH_3Cl]$ (M)	$[Cl_2]$ (M)	Initial Rate (M/s)
0.050	0.050	0.014
0.100	0.050	0.029
0.100	0.100	0.041
0.200	0.200	0.115

Write an expression for the reaction rate law and calculate the value of the rate constant, *k*. What is the overall order of the reaction?

The Integrated Rate Law and Half-Life

47. Indicate the order of reaction consistent with each observation.
   - **a.** A plot of the concentration of the reactant versus time yields a straight line.
   - **b.** The reaction has a half-life that is independent of initial concentration.
   - **c.** A plot of the inverse of the concentration versus time yields a straight line.

48. Indicate the order of reaction consistent with each observation.
   - **a.** The half-life of the reaction gets shorter as the initial concentration is increased.
   - **b.** A plot of the natural log of the concentration of the reactant versus time yields a straight line.
   - **c.** The half-life of the reaction gets longer as the initial concentration is increased.

49. The tabulated data show the concentration of AB versus time for this reaction:

$$AB(g) \longrightarrow A(g) + B(g)$$

Time (s)	[AB] (M)
0	0.950
50	0.459
100	0.302
150	0.225
200	0.180
250	0.149
300	0.128
350	0.112
400	0.0994
450	0.0894
500	0.0812

Determine the order of the reaction and the value of the rate constant. Predict the concentration of AB at 25 s.

50. The tabulated data show the concentration of N_2O_5 versus time for this reaction:

$$N_2O_5(g) \longrightarrow NO_3(g) + NO_2(g)$$

Time (s)	$[N_2O_5]$ (M)
0	1.000
25	0.822
50	0.677
75	0.557
100	0.458
125	0.377
150	0.310
175	0.255
200	0.210

Determine the order of the reaction and the value of the rate constant. Predict the concentration of N_2O_5 at 250 s.

51. The tabulated data show the concentration of cyclobutane (C_4H_8) versus time for this reaction:

$$C_4H_8 \longrightarrow 2\ C_2H_4(g)$$

Time (s)	$[C_4H_8]$ (M)
0	1.000
10	0.894
20	0.799
30	0.714
40	0.638
50	0.571
60	0.510
70	0.456
80	0.408
90	0.364
100	0.326

Determine the order of the reaction and the value of the rate constant. What is the rate of reaction when $[C_4H_8] = 0.25$ M?

52. A reaction in which A $\longrightarrow$ products is monitored as a function of time. The results are tabulated here.

Time (s)	[A] (M)
0	1.000
25	0.914
50	0.829
75	0.744
100	0.659
125	0.573
150	0.488
175	0.403
200	0.318

Determine the order of the reaction and the value of the rate constant. What is the rate of reaction when [A] = 0.10 M?

53. This reaction was monitored as a function of time:

$$A \longrightarrow B + C$$

A plot of ln[A] versus time yields a straight line with slope −0.0045/s.

a. What is the value of the rate constant (k) for this reaction at this temperature?
b. Write the rate law for the reaction.
c. What is the half-life?
d. If the initial concentration of A is 0.250 M, what is the concentration after 225 s?

54. This reaction was monitored as a function of time:

$$AB \longrightarrow A + B$$

A plot of 1/[AB] versus time yields a straight line with slope $-0.055/\text{M}\cdot\text{s}$.

a. What is the value of the rate constant (k) for this reaction at this temperature?
b. Write the rate law for the reaction.
c. What is the half-life when the initial concentration is 0.55 M?
d. If the initial concentration of AB is 0.250 M, and the reaction mixture initially contains no products, what are the concentrations of A and B after 75 s?

55. The decomposition of SO_2Cl_2 is first order in SO_2Cl_2 and has a rate constant of $1.42 \times 10^{-4}\ \text{s}^{-1}$ at a certain temperature.

a. What is the half-life for this reaction?
b. How long will it take for the concentration of SO_2Cl_2 to decrease to 25% of its initial concentration?
c. If the initial concentration of SO_2Cl_2 is 1.00 M, how long will it take for the concentration to decrease to 0.78 M?
d. If the initial concentration of SO_2Cl_2 is 0.150 M, what is the concentration of SO_2Cl_2 after 2.00×10^2 s? After 5.00×10^2 s?

56. The decomposition of XY is second order in XY and has a rate constant of $7.02 \times 10^{-3}\ \text{M}^{-1}\cdot\text{s}^{-1}$ at a certain temperature.

a. What is the half-life for this reaction at an initial concentration of 0.100 M?
b. How long will it take for the concentration of XY to decrease to 12.5% of its initial concentration when the initial concentration is 0.100 M? When the initial concentration is 0.200 M?
c. If the initial concentration of XY is 0.150 M, how long will it take for the concentration to decrease to 0.062 M?
d. If the initial concentration of XY is 0.050 M, what is the concentration of XY after 5.0×10^1 s? After 5.50×10^2 s?

57. The half-life for the radioactive decay of U-238 is 4.5 billion years and is independent of initial concentration. How long will it take for 10% of the U-238 atoms in a sample of U-238 to decay? If a sample of U-238 initially contained 1.5×10^{18} atoms when the universe was formed 13.8 billion years ago, how many U-238 atoms does it contain today?

58. The half-life for the radioactive decay of C-14 is 5730 years and is independent of the initial concentration. How long does it take for 25% of the C-14 atoms in a sample of C-14 to decay? If a sample of C-14 initially contains 1.5 mmol of C-14, how many millimoles are left after 2255 years?

The Effect of Temperature and the Collision Model

59. The diagram shows the energy of a reaction as the reaction progresses. Label each blank box in the diagram.

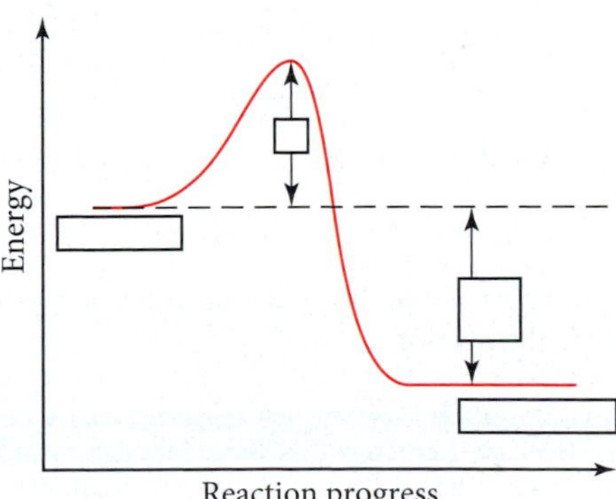

a. reactants
b. products
c. activation energy (E_a)
d. enthalpy of reaction (ΔH_{rxn})

60. A chemical reaction is endothermic and has an activation energy that is twice the value of the enthalpy change of the reaction. Draw a diagram depicting the energy of the reaction as it progresses. Label the position of the reactants and products and indicate the activation energy and enthalpy of reaction.

61. The activation energy of a reaction is 56.8 kJ/mol and the frequency factor is 1.5×10^{11}/s. Calculate the rate constant of the reaction at 25 °C.

62. The rate constant of a reaction at 32 °C is 0.055/s. If the frequency factor is 1.2×10^{13}/s, what is the activation barrier?

63. The rate constant (k) for a reaction is measured as a function of temperature. A plot of ln k versus $1/T$ (in K) is linear and has a slope of −7445 K. Calculate the activation energy for the reaction.

64. The rate constant (k) for a reaction is measured as a function of temperature. A plot of ln k versus $1/T$ (in K) is linear and has a slope of -1.01×10^4 K. Calculate the activation energy for the reaction.

65. The tabulated data shown here were collected for the first-order reaction:

$$N_2O(g) \longrightarrow N_2(g) + O(g)$$

Use an Arrhenius plot to determine the activation barrier and frequency factor for the reaction.

Temperature (K)	Rate Constant (s^{-1})
800	3.24×10^{-5}
900	0.00214
1000	0.0614
1100	0.955

66. The tabulated data show the rate constant of a reaction measured at several different temperatures. Use an Arrhenius plot to determine the activation barrier and frequency factor for the reaction.

Temperature (K)	Rate Constant (s^{-1})
300	0.0134
310	0.0407
320	0.114
330	0.303
340	0.757

67. The tabulated data were collected for the second-order reaction:

$$Cl(g) + H_2(g) \longrightarrow HCl(g) + H(g)$$

Use an Arrhenius plot to determine the activation barrier and frequency factor for the reaction.

Temperature (K)	Rate Constant (L/mol · s)
90	0.00357
100	0.0773
110	0.956
120	7.781

68. The tabulated data show the rate constant of a reaction measured at several different temperatures. Use an Arrhenius plot to determine the activation barrier and frequency factor for the reaction.

Temperature (K)	Rate Constant (s^{-1})
310	0.00434
320	0.0140
330	0.0421
340	0.118
350	0.316

69. A reaction has a rate constant of 0.0117/s at 400.0 K and 0.689/s at 450.0 K.

a. Determine the activation barrier for the reaction.
b. What is the value of the rate constant at 425 K?

70. A reaction has a rate constant of 0.000122/s at 27 °C and 0.228/s at 77 °C.

a. Determine the activation barrier for the reaction.
b. What is the value of the rate constant at 17 °C?

71. If a temperature increase from 10.0 °C to 20.0 °C doubles the rate constant for a reaction, what is the value of the activation barrier for the reaction?

72. If a temperature increase from 20.0 °C to 35.0 °C triples the rate constant for a reaction, what is the value of the activation barrier for the reaction?

73. Consider these two gas-phase reactions:

a. $AA(g) + BB(g) \longrightarrow 2AB(g)$
b. $AB(g) + CD(g) \longrightarrow AC(g) + BD(g)$

If the reactions have identical activation barriers and are carried out under the same conditions, which one would you expect to have the faster rate?

74. Which of these two reactions would you expect to have the smaller orientation factor? Explain.

a. $O(g) + N_2(g) \longrightarrow NO(g) + N(g)$
b. $NO(g) + Cl_2(g) \longrightarrow NOCl(g) + Cl(g)$

Reaction Mechanisms

75. Consider this overall reaction, which is experimentally observed to be second order in AB and zero order in C.

$$AB + C \longrightarrow A + BC$$

Is the following mechanism valid for this reaction?

$$AB + AB \xrightarrow[k_1]{} AB_2 + A \quad \text{Slow}$$

$$AB_2 + C \xrightarrow[k_2]{} AB + BC \quad \text{Fast}$$

76. Consider this overall reaction, which is experimentally observed to be second order in X and first order in Y:

$$X + Y \longrightarrow XY$$

a. Does the reaction occur in a single step in which X and Y collide?
b. Is this two-step mechanism valid?

$$2\,X \underset{k_2}{\overset{k_1}{\rightleftharpoons}} X_2 \quad \text{Fast}$$

$$X_2 + Y \xrightarrow[k_3]{} XY + X \quad \text{Slow}$$

77. Consider this three-step mechanism for a reaction.

$$Cl_2(g) \underset{k_2}{\overset{k_1}{\rightleftharpoons}} 2\ Cl(g) \quad \text{Fast}$$
$$Cl(g) + CHCl_3(g) \xrightarrow{k_3} HCl(g) + CCl_3(g) \quad \text{Slow}$$
$$Cl(g) + CCl_3(g) \xrightarrow{k_4} CCl_4(g) \quad \text{Fast}$$

a. What is the overall reaction?
b. Identify the intermediates in the mechanism.
c. What is the predicted rate law?

78. Consider this two-step mechanism for a reaction.

$$NO_2(g) + Cl_2(g) \xrightarrow{k_1} ClNO_2(g) + Cl(g) \quad \text{Slow}$$
$$NO_2(g) + Cl(g) \xrightarrow{k_2} ClNO_2(g) \quad \text{Fast}$$

a. What is the overall reaction?
b. Identify the intermediates in the mechanism.
c. What is the predicted rate law?

Catalysis

79. Many heterogeneous catalysts are deposited on high surface-area supports. Why?

80. Suppose that the reaction A $\longrightarrow$ products is exothermic and has an activation barrier of 75 kJ/mol. Sketch an energy diagram showing the energy of the reaction as a function of the progress of the reaction. Draw a second energy curve showing the effect of a catalyst.

81. Suppose that a catalyst lowers the activation barrier of a reaction from 125 kJ/mol to 55 kJ/mol. By what factor would you expect the reaction rate to increase at 25 °C? (Assume that the frequency factors for the catalyzed and uncatalyzed reactions are identical.)

82. The activation barrier for the hydrolysis of sucrose into glucose and fructose is 108 kJ/mol. If an enzyme increases the rate of the hydrolysis reaction by a factor of 1 million, how much lower must the activation barrier be when sucrose is in the active site of the enzyme? (Assume that the frequency factors for the catalyzed and uncatalyzed reactions are identical and a temperature of 25 °C.)

CUMULATIVE PROBLEMS

83. The tabulated data were collected for this reaction at 500 °C:

$$CH_3CN(g) \longrightarrow CH_3NC(g)$$

Time (h)	$[CH_3CN]$ (M)
0.0	1.000
5.0	0.794
10.0	0.631
15.0	0.501
20.0	0.398
25.0	0.316

a. Determine the order of the reaction and the value of the rate constant at this temperature.
b. What is the half-life for this reaction (at the initial concentration)?
c. How long will it take for 90% of the CH_3CN to convert to CH_3NC?

84. The tabulated data were collected for this reaction at a certain temperature:

$$X_2Y \longrightarrow 2\,X + Y$$

Time (h)	$[X_2Y]$ (M)
0.0	0.100
1.0	0.0856
2.0	0.0748
3.0	0.0664
4.0	0.0598
5.0	0.0543

a. Determine the order of the reaction and the value of the rate constant at this temperature.
b. What is the half-life for this reaction (at the initial concentration)?
c. What is the concentration of X after 10.0 hours?

85. Consider the reaction.

$$A + B + C \longrightarrow D$$

The rate law for this reaction is:

$$\text{Rate} = k\frac{[A][C]^2}{[B]^{1/2}}$$

Suppose the rate of the reaction at certain initial concentrations of A, B, and C is 0.0115 M/s. What is the rate of the reaction if the concentrations of A and C are doubled and the concentration of B is tripled?

86. Consider the reaction.

$$2\ O_3(g) \longrightarrow 3\ O_2(g)$$

The rate law for this reaction is:

$$\text{Rate} = k\frac{[O_3]^2}{[O_2]}$$

Suppose that a 1.0 L reaction vessel initially contains 1.0 mol of O_3 and 1.0 mol of O_2. What fraction of the O_3 has reacted when the rate falls to one-half of its initial value?

87. At 700 K acetaldehyde decomposes in the gas phase to methane and carbon monoxide. The reaction is:

$$CH_3CHO(g) \longrightarrow CH_4(g) + CO(g)$$

A sample of CH_3CHO is heated to 700 K, and the pressure is measured as 0.22 atm before any reaction takes place. The kinetics of the reaction are followed by measurements of total pressure, and these data are obtained:

t(s)	0	1000	3000	7000
P_{Total}(atm)	0.22	0.24	0.27	0.31

Determine the rate law, the rate constant, and the total pressure after 2.00×10^4 s.

88. At 400 K oxalic acid decomposes according to the reaction:

$$H_2C_2O_4(g) \longrightarrow CO_2(g) + HCOOH(g)$$

In three separate experiments, the initial pressure of oxalic acid and the final total pressure after 20,000 s are measured.

Experiment	1	2	3
$P_{H_2C_2O_4}$ at $t = 0$	65.8	92.1	111
P_{Total} at $t = 20{,}000$ s	94.6	132	160

Find the rate law of the reaction and its specific rate constant.

89. Dinitrogen pentoxide decomposes in the gas phase to form nitrogen dioxide and oxygen gas. The reaction is first order in dinitrogen pentoxide and has a half-life of 2.81 h at 25 °C. If a 1.5 L reaction vessel initially contains 745 torr of N_2O_5 at 25 °C, what partial pressure of O_2 is present in the vessel after 215 minutes?

90. Cyclopropane (C_3H_6) reacts to form propene (C_3H_6) in the gas phase. The reaction is first order in cyclopropane and has a rate constant of 5.87×10^{-4}/s at 485 °C. If a 2.5 L reaction vessel initially contains 722 torr of cyclopropane at 485 °C, how long will it take for the partial pressure of cyclopropane to drop to below 1.00×10^2 torr?

91. Iodine atoms combine to form I_2 in liquid hexane solvent with a rate constant of 1.5×10^{10} L/mol·s. The reaction is second order in I. Since the reaction occurs so quickly, the only way to study the reaction is to create iodine atoms almost instantaneously, usually by photochemical decomposition of I_2. Suppose a flash of light creates an initial [I] concentration of 0.0100 M. How long will it take for 95% of the newly created iodine atoms to recombine to form I_2?

92. The hydrolysis of sucrose ($C_{12}H_{22}O_{11}$) into glucose and fructose in acidic water has a rate constant of $1.8 \times 10^{-4}\ s^{-1}$ at 25 °C. Assuming the reaction is first order in sucrose, determine the mass of sucrose that is hydrolyzed when 2.55 L of a 0.150 M sucrose solution is allowed to react for 195 minutes.

93. The reaction $AB(aq) \longrightarrow A(g) + B(g)$ is second order in AB and has a rate constant of $0.0118\ M^{-1} \cdot s^{-1}$ at 25.0 °C. A reaction vessel initially contains 250.0 mL of 0.100 M AB that is allowed to react to form the gaseous product. The product is collected over water at 25.0 °C. How much time is required to produce 200.0 mL of the products at a barometric pressure of 755.1 mmHg? (The vapor pressure of water at this temperature is 23.8 mmHg.)

94. The reaction $2\ H_2O_2(aq) \longrightarrow 2\ H_2O(l) + O_2(g)$ is first order in H_2O_2 and under certain conditions has a rate constant of $0.00752\ s^{-1}$ at 20.0 °C. A reaction vessel initially contains 150.0 mL of 30.0% H_2O_2 by mass solution (the density of the solution is 1.11 g/mL). The gaseous oxygen is collected over water at 20.0 °C as it forms. What volume of O_2 forms in 85.0 seconds at a barometric pressure of 742.5 mmHg? (The vapor pressure of water at this temperature is 17.5 mmHg.)

95. Consider this energy diagram.

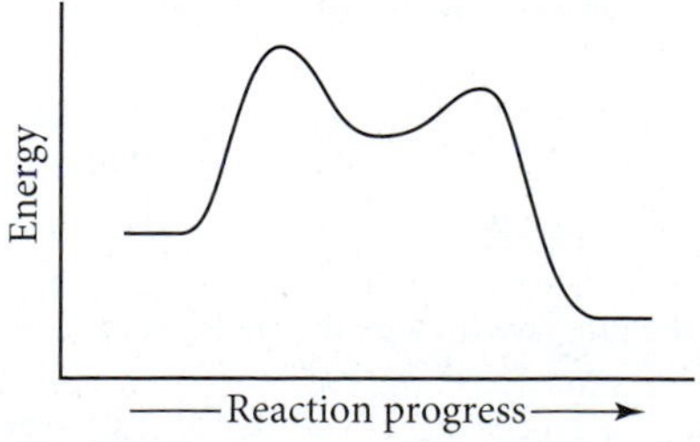

a. How many elementary steps are involved in this reaction?

b. Label the reactants, products, and intermediates.

c. Which step is rate limiting?

d. Is the overall reaction endothermic or exothermic?

96. Consider the reaction in which HCl adds across the double bond of ethene.

$$HCl + H_2C{=}CH_2 \longrightarrow H_3C{-}CH_2Cl$$

The following mechanism, with the accompanying energy diagram, has been suggested for this reaction:

Step 1 $HCl + H_2C{=}CH_2 \longrightarrow H_3C{=}CH_2^+ + Cl^-$

Step 2 $H_3C{=}CH_2^+ + Cl^- \longrightarrow H_3C{-}CH_2Cl$

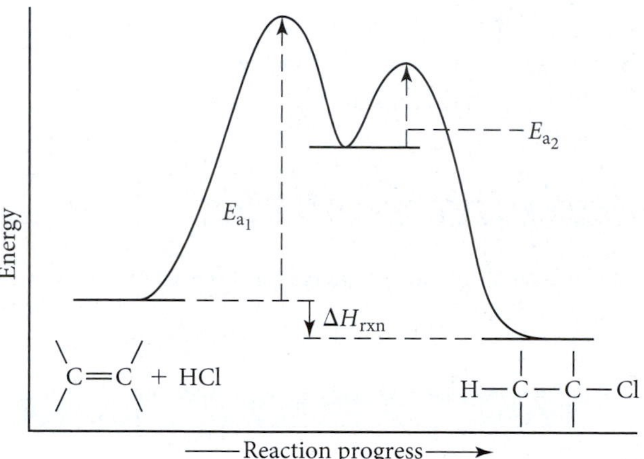

a. Based on the energy diagram, determine which step is rate limiting.

b. What is the expected order of the reaction based on the proposed mechanism?

c. Is the overall reaction exothermic or endothermic?

97. The desorption of a single molecular layer of *n*-butane from a single crystal of aluminum oxide is found to be first order with a rate constant of 0.128/s at 150 K.

a. What is the half-life of the desorption reaction?

b. If the surface is initially completely covered with *n*-butane at 150 K, how long will it take for 25% of the molecules to desorb? For 50% to desorb?

c. If the surface is initially completely covered, what fraction will remain covered after 10 s? After 20 s?

98. The evaporation of a 120 nm film of *n*-pentane from a single crystal of aluminum oxide is zero order with a rate constant of 1.92×10^{13} molecules/cm^2·s at 120 K.

a. If the initial surface coverage is 8.9×10^{16} molecules/cm^2, how long will it take for one-half of the film to evaporate?

b. What fraction of the film is left after 10 s? Assume the same initial coverage as in part a.

99. The kinetics of this reaction were studied as a function of temperature. (The reaction is first order in each reactant and second order overall.)

$$C_2H_5Br(aq) + OH^-(aq) \longrightarrow C_2H_5OH(l) + Br^-(aq)$$

Temperature (°C)	k(L/mol · s)
25	8.81×10^{-5}
35	0.000285
45	0.000854
55	0.00239
65	0.00633

a. Determine the activation energy and frequency factor for the reaction.

b. Determine the rate constant at 15 °C.

c. If a reaction mixture is 0.155 M in C_2H_5Br and 0.250 M in OH^-, what is the initial rate of the reaction at 75 °C?

100. The reaction $2\ N_2O_5 \longrightarrow 2\ N_2O_4 + O_2$ takes place at around room temperature in solvents such as CCl_4. The rate constant at 293 K is $2.35 \times 10^{-4}\ s^{-1}$, and at 303 K the rate constant is $9.15 \times 10^{-4}\ s^{-1}$. Calculate the frequency factor for the reaction.

101. This reaction has an activation energy of zero in the gas phase.

$$CH_3 + CH_3 \longrightarrow C_2H_6$$

a. Would you expect the rate of this reaction to change very much with temperature?

b. Why might the activation energy be zero?

c. What other types of reactions would you expect to have little or no activation energy?

102. Consider the two reactions.

$$O + N_2 \longrightarrow NO + N \quad E_a = 315\ kJ/mol$$

$$Cl + H_2 \longrightarrow HCl + H \quad E_a = 23\ kJ/mol$$

a. Why is the activation barrier for the first reaction so much higher than that for the second?

b. The frequency factors for these two reactions are very close to each other in value. Assuming that they are the same, calculate the ratio of the reaction rate constants for these two reactions at 25 °C.

103. Anthropologists can estimate the age of a bone or other sample of organic matter by its carbon-14 content. The carbon-14 in a living organism is constant until the organism dies, after which carbon-14 decays with first-order kinetics and a half-life of 5730 years. Suppose a bone from an ancient human contains 19.5% of the C-14 found in living organisms. How old is the bone?

104. Geologists can estimate the age of rocks by their uranium-238 content. The uranium is incorporated in the rock as it hardens and then decays with first-order kinetics and a half-life of 4.5 billion years. A rock contains 83.2% of the amount of uranium-238 that it contained when it was formed. (The amount that the rock contained when it was formed can be deduced from the presence of the decay products of U-238.) How old is the rock?

105. Consider the gas-phase reaction.

$$H_2(g) + I_2(g) \longrightarrow 2\ HI(g)$$

The reaction was experimentally determined to be first order in H_2 and first order in I_2. Consider the proposed mechanisms.

Proposed mechanism I:

$$H_2(g) + I_2(g) \longrightarrow 2\ HI(g) \quad \text{Single step}$$

Proposed mechanism II:

$$I_2(g) \underset{k_{-1}}{\overset{k_1}{\rightleftharpoons}} 2\ I(g) \quad \text{Fast}$$

$$H_2(g) + 2\ I(g) \xrightarrow{k_2} 2\ HI(g) \quad \text{Slow}$$

a. Show that both of the proposed mechanisms are valid.

b. What kind of experimental evidence might lead you to favor mechanism II over mechanism I?

106. Consider the reaction.

$$2\ NH_3(aq) + OCl^-(aq) \longrightarrow N_2H_4(aq) + H_2O(l) + Cl^-(aq)$$

This three-step mechanism is proposed.

$$NH_3(aq) + OCl^-(aq) \underset{k_2}{\overset{k_1}{\rightleftharpoons}} NH_2Cl(aq) + OH^-(aq) \quad \text{Fast}$$

$$NH_2Cl(aq) + NH_3(aq) \xrightarrow{k_3} N_2H_5^+(aq) + Cl^-(aq) \quad \text{Slow}$$

$$N_2H_5^+(aq) + OH^-(aq) \xrightarrow{k_4} N_2H_4(aq) + H_2O(l) \quad \text{Fast}$$

a. Show that the mechanism sums to the overall reaction.

b. What is the rate law predicted by this mechanism?

107. The proposed mechanism for the formation of hydrogen bromide can be written in a simplified form as:

$$Br_2(g) \underset{k_{-1}}{\overset{k_1}{\rightleftharpoons}} 2\ Br(g) \quad \text{Fast}$$

$$Br(g) + H_2(g) \xrightarrow{k_2} HBr(g) + H(g) \quad \text{Slow}$$

$$H(g) + Br_2(g) \xrightarrow{k_3} HBr(g) + Br(g) \quad \text{Fast}$$

What rate law corresponds to this mechanism?

108. A proposed mechanism for the formation of hydrogen iodide can be written in simplified form as:

$$I_2 \underset{k_{-1}}{\overset{k_1}{\rightleftharpoons}} 2\ I \quad \text{Fast}$$

$$I + H_2 \underset{k_{-2}}{\overset{k_2}{\rightleftharpoons}} H_2I \quad \text{Fast}$$

$$H_2I + I \xrightarrow{k_3} 2\ HI \quad \text{Slow}$$

What rate law corresponds to this mechanism?

109. A certain substance X decomposes. Fifty percent of X remains after 100 minutes. How much X remains after 200 minutes if the reaction order with respect to X is (a) zero order, (b) first order, (c) second order?

110. The half-life for radioactive decay (a first-order process) of plutonium-239 is 24,000 years. How many years does it take for one mole of this radioactive material to decay until just one atom remains?

111. The energy of activation for the decomposition of 2 mol of HI to H_2 and I_2 in the gas phase is 185 kJ. The heat of formation of HI(g) from $H_2(g)$ and $I_2(g)$ is −5.68 kJ/mol. Find the energy of activation for the reaction of 1 mol of H_2 and 1 mol of I_2 to form 2 mol of HI in the gas phase.

112. Ethyl chloride vapor decomposes by the first-order reaction:

$$C_2H_5Cl \longrightarrow C_2H_4 + HCl$$

The activation energy is 249 kJ/mol, and the frequency factor is $1.6 \times 10^{14}\ s^{-1}$. Find the value of the specific rate constant at 710 K. What fraction of the ethyl chloride decomposes in 15 minutes at this temperature? Find the temperature at which the rate of the reaction would be twice as fast.

CHALLENGE PROBLEMS

113. In this chapter we have seen a number of reactions in which a single reactant forms products. For example, consider the following first-order reaction:

$$CH_3NC(g) \longrightarrow CH_3CN(g)$$

However, we also learned that gas-phase reactions occur through collisions.

a. One possible explanation is that two molecules of CH_3NC collide with each other and form two molecules of the product in a single elementary step. If that is the case, what reaction order would you expect?

b. Another possibility is that the reaction occurs through more than one step. For example, a possible mechanism involves one step in which the two CH_3NC molecules collide, resulting in the "activation" of one of them. In a second step, the activated molecule goes on to form the product. Write down this mechanism and determine which step must be rate determining in order for the kinetics of the reaction to be first order. Show explicitly how the mechanism predicts first-order kinetics.

114. The first-order *integrated* rate law for a reaction A $\longrightarrow$ products is derived from the rate law using calculus.

$$\text{Rate} = k[\text{A}] \quad \text{(first-order rate law)}$$

$$\text{Rate} = \frac{d[\text{A}]}{dt}$$

$$\frac{d[\text{A}]}{dt} = -k[\text{A}]$$

The equation just given is a first-order, separable differential equation that can be solved by separating the variables and integrating:

$$\frac{d[\text{A}]}{[\text{A}]} = -kdt$$

$$\int_{[\text{A}]_0}^{[\text{A}]} \frac{d[\text{A}]}{[\text{A}]} = -\int_0^t kdt$$

In the integral just given, $[\text{A}]_0$ is the initial concentration of A. We then evaluate the integral:

$$[\ln [\text{A}]]_{[\text{A}]_0}^{[\text{A}]} = -k[t]_0^t$$

$$\ln [\text{A}] - \ln [\text{A}]_0 = -kt$$

$$\ln [\text{A}] = -kt + \ln [\text{A}]_0 \quad \text{(integrated rate law)}$$

a. Use a procedure similar to the one just shown to derive an integrated rate law for a reaction A $\longrightarrow$ products, which is one-half-order in the concentration of A (that is, Rate $= k[\text{A}]^{1/2}$).

b. Use the result from part a to derive an expression for the half-life of a one-half-order reaction.

115. The previous exercise shows how the first-order integrated rate law is derived from the first-order differential rate law. Begin with the second-order differential rate law and derive the second-order integrated rate law.

116. The rate constant for the first-order decomposition of $N_2O_5(g)$ to $NO_2(g)$ and $O_2(g)$ is $7.48 \times 10^{-3}\ s^{-1}$ at a given temperature.

a. Find the length of time required for the total pressure in a system containing N_2O_5 at an initial pressure of 0.100 atm to rise to 0.145 atm.

b. Find the length of time required for the total pressure in a system containing N_2O_5 at an initial pressure of 0.100 atm to rise to 0.200 atm.

c. Find the total pressure after 100 s of reaction.

117. Phosgene (Cl_2CO), a poison gas used in World War I, is formed by the reaction of Cl_2 and CO. The proposed mechanism for the reaction is:

$$Cl_2 \rightleftharpoons 2\ Cl \quad \text{(fast, equilibrium)}$$

$$Cl + CO \rightleftharpoons ClCO \quad \text{(fast, equilibrium)}$$

$$ClCO + Cl_2 \longrightarrow Cl_2CO + Cl \quad \text{(slow)}$$

What rate law is consistent with this mechanism?

118. The rate of decomposition of $N_2O_3(g)$ to $NO_2(g)$ and $NO(g)$ is monitored by measuring $[NO_2]$ at different times. The following tabulated data are obtained.

$[NO_2]$(mol/L)	0	0.193	0.316	0.427	0.784
t(s)	0	884	1610	2460	50,000

The reaction follows a first-order rate law. Calculate the rate constant. Assume that after 50,000 s all the $N_2O_3(g)$ had decomposed.

119. At 473 K, for the elementary reaction $2\ NOCl(g) \underset{k_{-1}}{\overset{k_1}{\rightleftharpoons}} 2\ NO(g) + Cl_2(g)$

$$k_1 = 7.8 \times 10^{-2}\ \text{L/mol s and}$$

$$k_{-1} = 4.7 \times 10^2\ \text{L}^2/\text{mol}^2\ \text{s}$$

A sample of NOCl is placed in a container and heated to 473 K. When the system comes to equilibrium, [NOCl] is found to be 0.12 mol/L. What are the concentrations of NO and Cl_2?

CONCEPTUAL PROBLEMS

120. Consider the reaction.

$$CHCl_3(g) + Cl_2(g) \longrightarrow CCl_4(g) + HCl(g)$$

The reaction is first order in $CHCl_3$ and one-half order in Cl_2. Which reaction mixture would you expect to have the fastest initial rate?

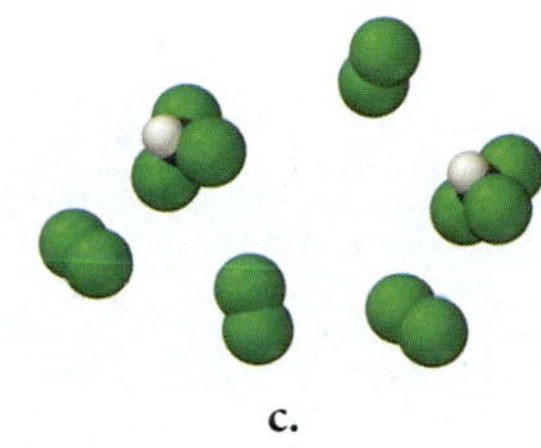

121. The accompanying graph shows the concentration of a reactant as a function of time for two different reactions. One of the reactions is first order and the other is second order. Which of the two reactions is first order? Second order? How would you change each plot to make it linear?

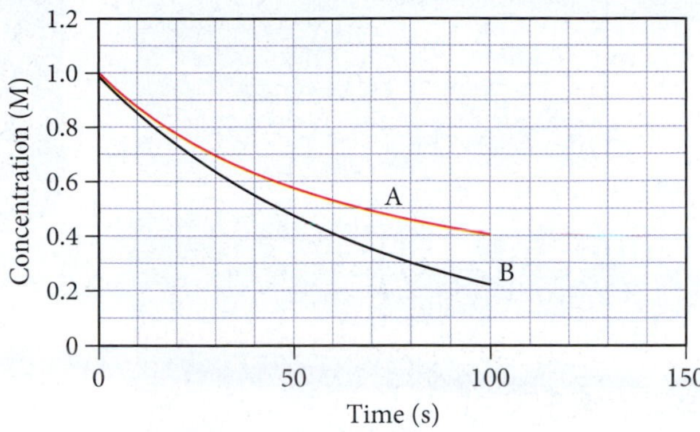

122. A particular reaction, A $\longrightarrow$ products, has a rate that slows down as the reaction proceeds. The half-life of the reaction is found to depend on the initial concentration of A. Determine whether each statement is likely to be true or false for this reaction.

a. A doubling of the concentration of A doubles the rate of the reaction.

b. A plot of 1/[A] versus time is linear.

c. The half-life of the reaction gets longer as the initial concentration of A increases.

d. A plot of the concentration of A versus time has a constant slope.

ANSWERS TO CONCEPTUAL CONNECTIONS

Cc 15.1 **(b)** Increasing the temperature increases the number of collisions that can occur with enough energy for the reaction to occur.

Cc 15.2 **(c)** The rate at which B changes is twice the rate of the reaction because its coefficient is 2, and it is negative because B is a reactant.

Cc 15.3 **(d)** Because the reaction is second order, increasing the concentration of A by a factor of 5 causes the rate to increase by 5^2 or 25.

Cc 15.4 **(c)** All three mixtures have the same total number of molecules, but mixture **(c)** has the greatest number of NO molecules. Since the reaction is second order in NO and only first order in O_2, mixture **(c)** has the fastest initial rate.

Cc 15.5 **(c)** The reaction is most likely second order because its rate depends on the concentration (therefore it cannot be zero order), and its half-life depends on the initial concentration (therefore it cannot be first order). For a second-order reaction, a doubling of the initial concentration results in the quadrupling of the rate.

Cc 15.6 Reaction A has a faster rate because it has a lower activation energy; therefore, the exponential factor is larger at a given temperature, making the rate constant larger. (With a larger rate constant and the same initial concentration, the rate will be faster).

Cc 15.7 **(c)** Since the reactants in part **(a)** are atoms, the orientation factor should be about one. The reactants in parts **(b)** and **(c)** are both molecules, so we expect orientation factors of less than one. Since the reactants in **(b)** are symmetrical, we would not expect the collision to have as specific an orientation requirement as in **(c)**, where the reactants are asymmetrical and must therefore collide in such way that a hydrogen atom is in close proximity to another hydrogen atom. Therefore, we expect **(c)** to have the smallest orientation factor.

CHAPTER 16

"Every system in chemical equilibrium, under the influence of a change of any one of the factors of equilibrium, undergoes a transformation . . . [that produces a change] . . . in the opposite direction of the factor in question."

—Henri Le Châtelier (1850–1936)

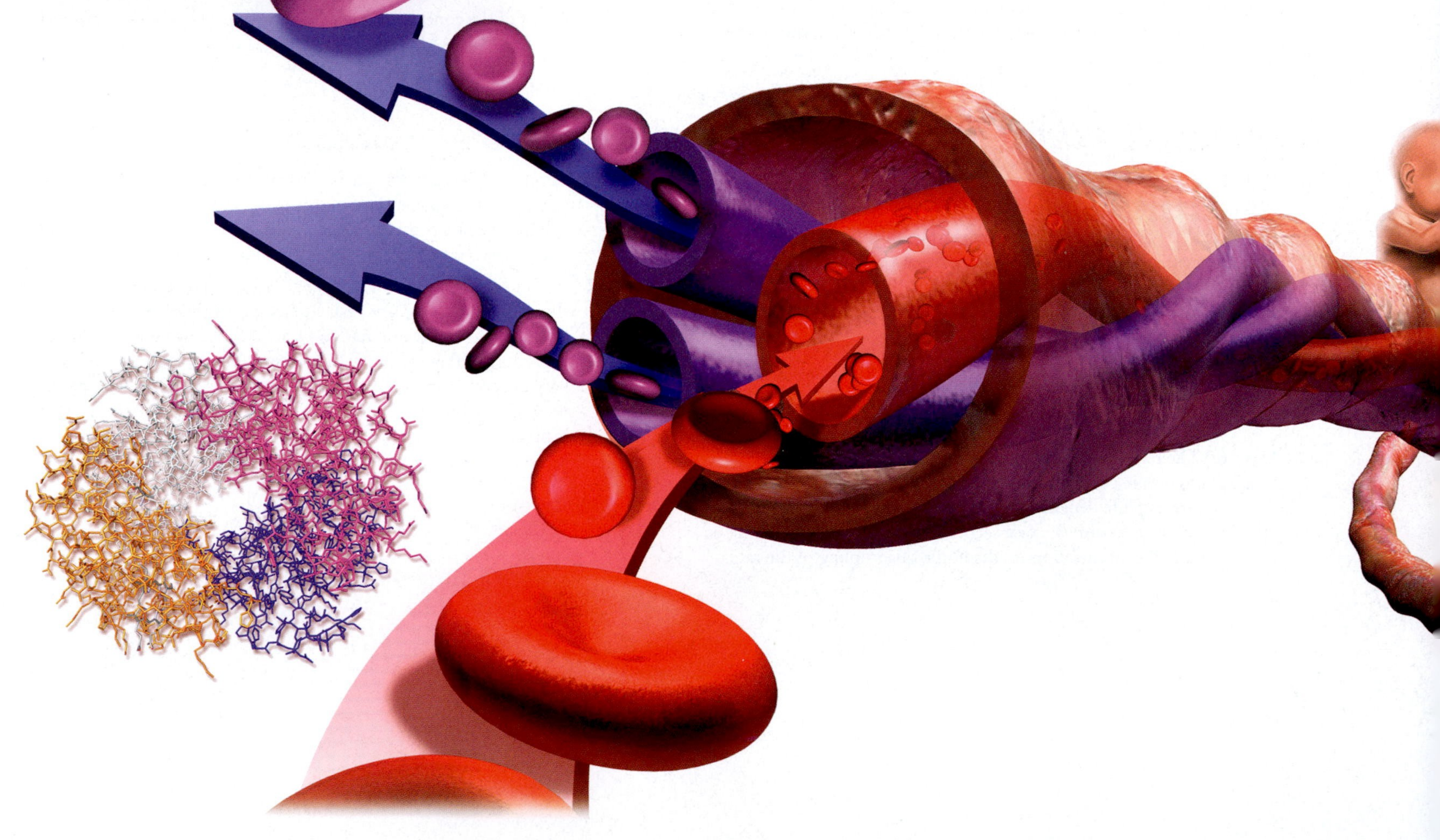

A developing fetus gets oxygen from the mother's blood because the reaction between oxygen and fetal hemoglobin has a larger equilibrium constant than the reaction between oxygen and maternal hemoglobin.

Chemical Equilibrium

16.1 Fetal Hemoglobin and Equilibrium 609

16.2 The Concept of Dynamic Equilibrium 611

16.3 The Equilibrium Constant (*K*) 612

16.4 Expressing the Equilibrium Constant in Terms of Pressure 617

16.5 Heterogeneous Equilibria: Reactions Involving Solids and Liquids 620

16.6 Calculating the Equilibrium Constant from Measured Equilibrium Concentrations 621

16.7 The Reaction Quotient: Predicting the Direction of Change 623

16.8 Finding Equilibrium Concentrations 626

16.9 Le Châtelier's Principle: How a System at Equilibrium Responds to Disturbances 636

Key Learning Outcomes 645

IN CHAPTER 15, we examined *how fast* a chemical reaction occurs. In this chapter we examine *how far* a chemical reaction goes. The *speed* of a chemical reaction is determined by kinetics, whereas the *extent* of a chemical reaction is determined by thermodynamics. In this chapter, we focus on describing and quantifying how far a chemical reaction goes based on an experimentally measurable quantity called *the equilibrium constant*. A reaction with a large equilibrium constant proceeds nearly to completion—nearly all the reactants react to form products. A reaction with a small equilibrium constant barely proceeds at all—nearly all the reactants remain as reactants, hardly forming any products. In this chapter we simply accept the equilibrium constant as an experimentally measurable quantity and learn how to use it to predict and quantify the extent of a reaction. In Chapter 19, we will explore the reasons underlying the magnitude of equilibrium constants.

16.1 Fetal Hemoglobin and Equilibrium

Have you ever wondered how a baby in the womb gets oxygen? Unlike you and me, a fetus does not breathe air. Yet, like you and me, a fetus needs oxygen. Where does that oxygen come from? After we are born, we inhale air into our lungs and that air diffuses into capillaries, where it comes into contact with our blood. Within our red blood cells, a protein called hemoglobin (Hb) reacts with oxygen according to the chemical equation:

$$\mathrm{Hb} + \mathrm{O_2} \rightleftharpoons \mathrm{HbO_2}$$

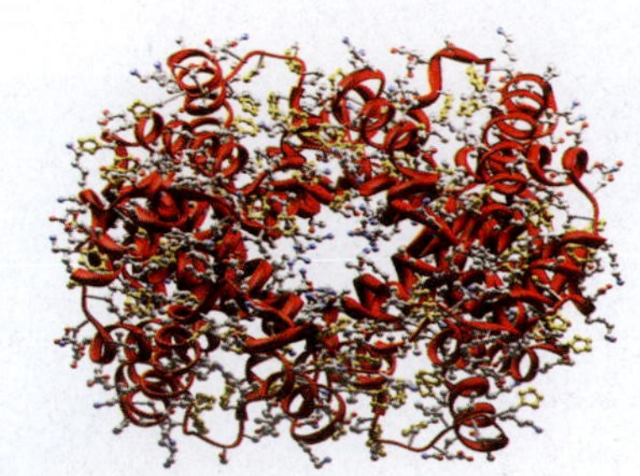

▲ Hemoglobin is the oxygen-carrying protein in red blood cells.

The double arrows in this equation indicate that the reaction can occur in both the forward and reverse directions and can reach chemical *equilibrium*. We first encountered this term in Chapter 12 (see Section 12.5), and we define it more carefully in the next section of this chapter. For now, understand that the concentrations of the reactants and products in a reaction at equilibrium are described by the *equilibrium constant*, K. A large value of K means that the reaction lies far to the right at equilibrium—a high concentration of products and a low concentration of reactants. A small value of K means that the reaction lies far to the left at equilibrium—a high concentration of reactants and a low concentration of products. In other words, the value of K is a measure of how far a reaction proceeds—the larger the value of K, the more the reaction proceeds toward the products.

The equilibrium constant for the reaction between hemoglobin and oxygen is such that hemoglobin efficiently binds oxygen at typical lung oxygen concentrations, but it can also release oxygen under the appropriate conditions. Any system at equilibrium, including the hemoglobin–oxygen system, responds to changes in ways that maintain equilibrium. If any of the concentrations of the reactants or products change, the reaction shifts to counteract that change. For the hemoglobin system, as blood flows through the lungs where oxygen concentrations are high, the reaction shifts to the right—hemoglobin binds oxygen.

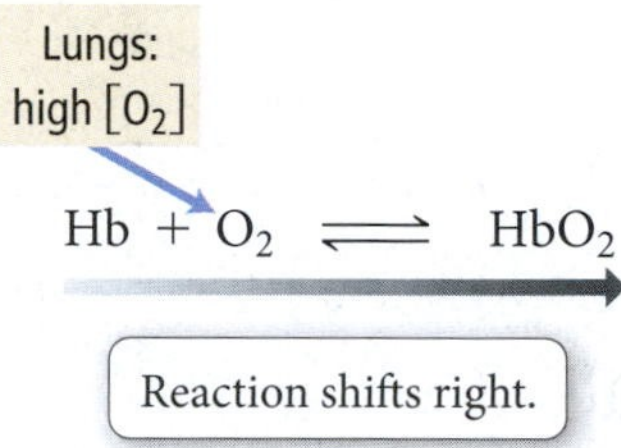

As blood flows out of the lungs and into muscles and organs where oxygen concentrations have been depleted (because muscles and organs use oxygen), the reaction shifts to the left—hemoglobin releases oxygen.

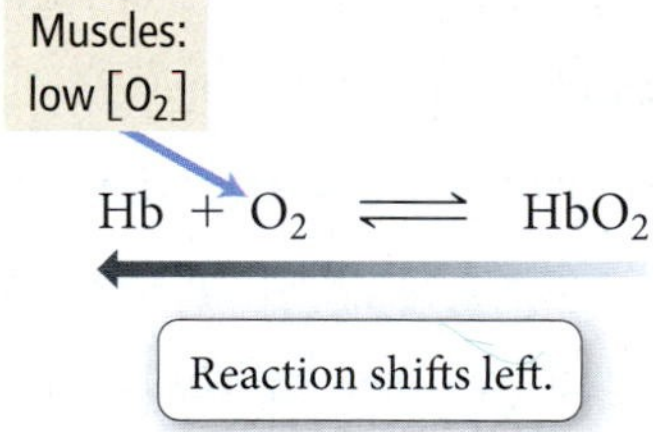

In other words, in order to maintain equilibrium, *hemoglobin binds oxygen when the surrounding oxygen concentration is high, but it releases oxygen when the surrounding oxygen concentration is low*. In this way, hemoglobin transports oxygen from the lungs to all parts of the body that use oxygen.

A fetus has its own circulatory system. The mother's blood never flows into the fetus's body, and the fetus cannot get any air in the womb. How, then, does the fetus get oxygen? The answer lies in the properties of fetal hemoglobin (HbF), which is slightly different from adult hemoglobin. Like adult hemoglobin, fetal hemoglobin is in equilibrium with oxygen:

$$HbF + O_2 \rightleftharpoons HbFO_2$$

However, the equilibrium constant for fetal hemoglobin is larger than the equilibrium constant for adult hemoglobin, meaning that the reaction tends to go farther in the direction of the product. Consequently, fetal hemoglobin loads oxygen at a lower oxygen concentration than does adult hemoglobin. In the placenta, fetal blood flows in close proximity to maternal blood. Although the two never mix, because of the different equilibrium constants, the maternal hemoglobin releases oxygen that the fetal hemoglobin then binds and carries into its own circulatory system (Figure 16.1 ▶). Nature has evolved a chemical system through which the mother's hemoglobin can in effect *hand off* oxygen to the hemoglobin of the fetus.

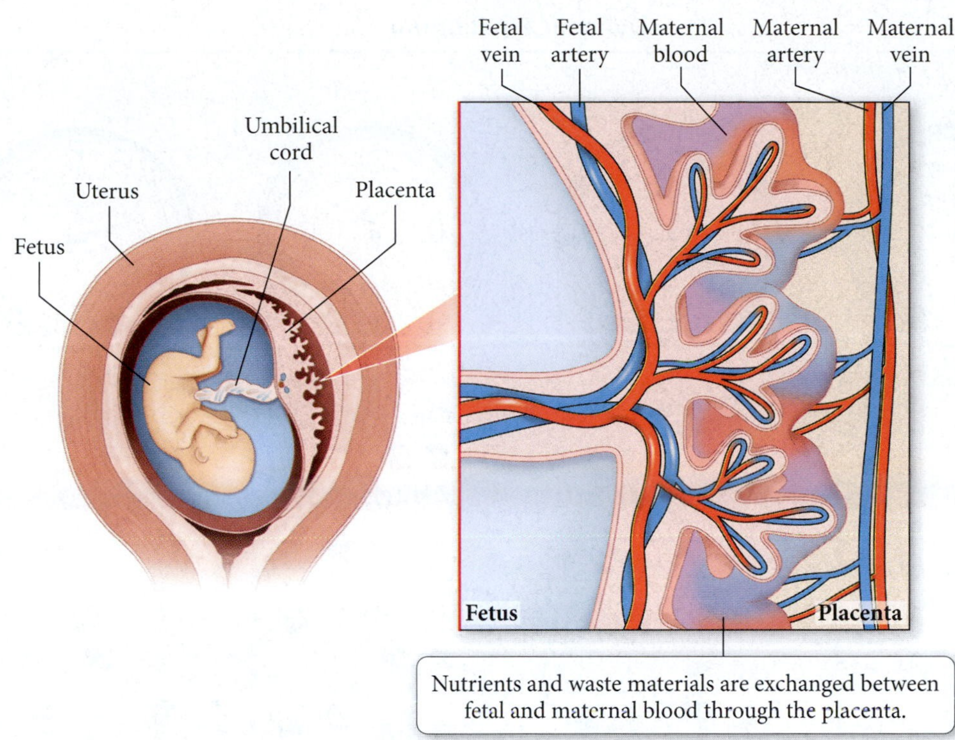

◀ **FIGURE 16.1 Oxygen Exchange between the Maternal and Fetal Circulation** In the placenta, the blood of the fetus comes into close proximity with that of the mother, although the two do not mix directly. Because the reaction of fetal hemoglobin with oxygen has a larger equilibrium constant than the reaction of maternal hemoglobin with oxygen, the fetus receives oxygen from the mother's blood.

16.2 The Concept of Dynamic Equilibrium

Recall from Chapter 15 that reaction rates generally increase with increasing concentration of the reactants (unless the reaction order is zero) and decrease with decreasing concentration of the reactants. With this in mind, consider the reaction between hydrogen and iodine:

$$H_2(g) + I_2(g) \rightleftharpoons 2HI(g)$$

In this reaction, H_2 and I_2 react to form 2 HI molecules, but the 2 HI molecules can also react to re-form H_2 and I_2. A reaction such as this one—that can proceed in both the forward and reverse directions—is said to be **reversible**. Suppose we begin with only H_2 and I_2 in a container (Figure 16.2(a) ▶). What happens? Initially, H_2 and I_2 begin to react to form HI (Figure 16.2(b)). However, as H_2 and I_2 react, their concentrations decrease, which in turn *decreases the rate of the forward reaction*. At the same time, HI begins to form. As the concentration of HI increases, the reverse reaction begins to occur at a faster and faster rate. Eventually, the rate of the reverse reaction (which has been increasing) equals the rate of the forward reaction (which has been decreasing). At that point, **dynamic equilibrium** is reached (Figure 16.2(c, d)).

Nearly all chemical reactions are at least theoretically reversible. In many cases, however, the reversibility is so small that it can be ignored.

Dynamic equilibrium for a chemical reaction is the condition in which the rate of the forward reaction equals the rate of the reverse reaction.

Dynamic equilibrium is "dynamic" because the forward and reverse reactions are still occurring; however, they are occurring at the same rate. When dynamic equilibrium is reached, the concentrations of H_2, I_2, and HI no longer change (as long as the temperature is constant). The concentrations remain constant because the reactants and products form at the same rate that they are depleted. Note that although the concentrations of reactants and products no longer change at equilibrium, *the concentrations of reactants and products are not equal to one another* at equilibrium. Some reactions reach equilibrium only after most of the reactants have formed products; others reach equilibrium when only a small fraction of the reactants have formed products. It depends on the reaction.

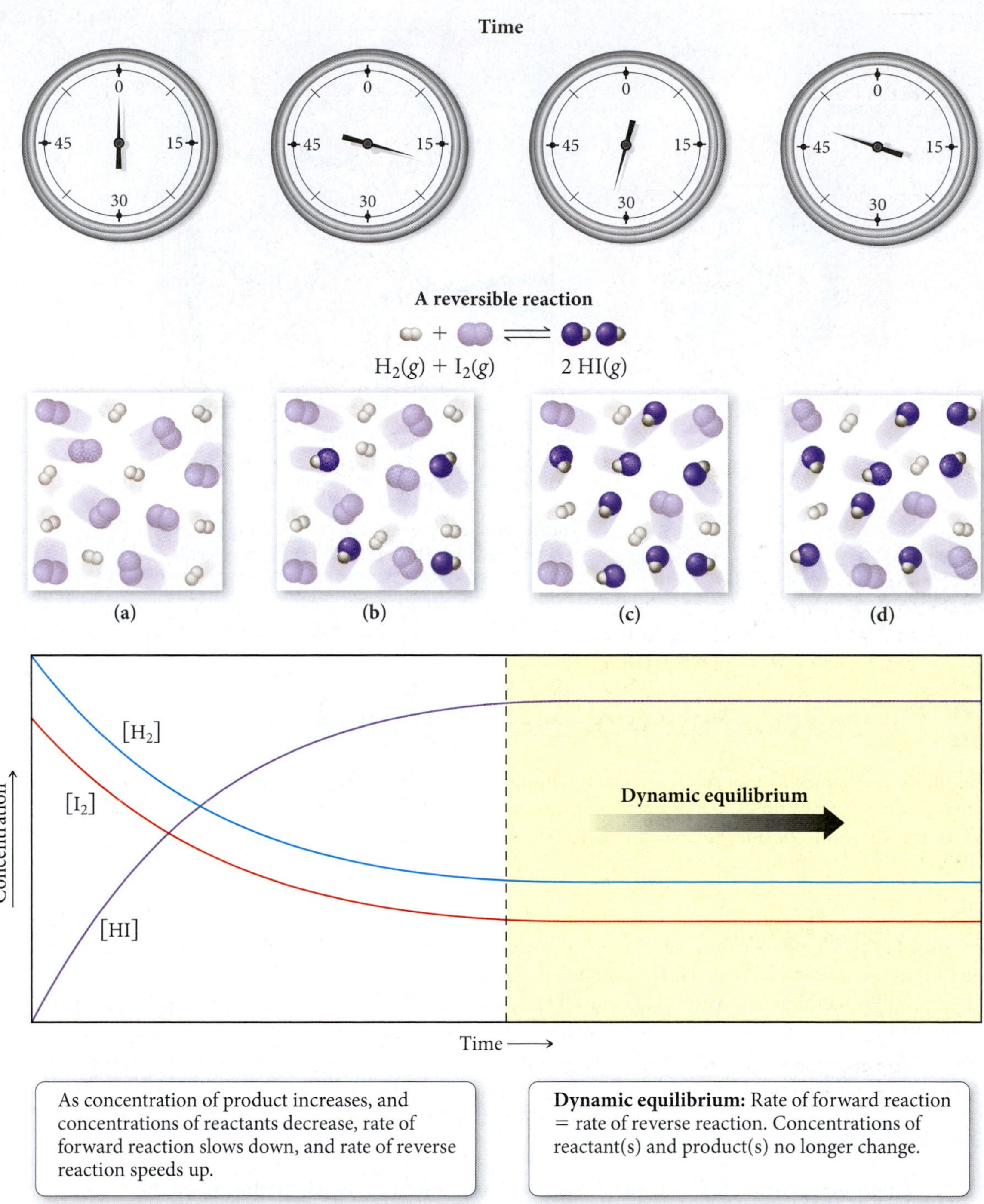
Dynamic Equilibrium
Time
0
15
30
45
A reversible reaction
$H_2(g) + I_2(g) \rightleftharpoons 2\,HI(g)$
(a)
(b)
(c)
(d)
[H_2]
[I_2]
[HI]
Concentration
Time
Dynamic equilibrium
As concentration of product increases, and concentrations of reactants decrease, rate of forward reaction slows down, and rate of reverse reaction speeds up.
Dynamic equilibrium: Rate of forward reaction = rate of reverse reaction. Concentrations of reactant(s) and product(s) no longer change.

▶ **FIGURE 16.2 Dynamic Equilibrium** Equilibrium is reached in a chemical reaction when the concentrations of the reactants and products no longer change. The molecular images depict the progress of the reaction $H_2(g) + I_2(g) \rightleftharpoons 2\,HI(g)$. The graph shows the concentrations of H_2, I_2, and HI as a function of time. When equilibrium is reached, both the forward and reverse reactions continue, but at equal rates, so the concentrations of the reactants and products remain constant.

16.3 The Equilibrium Constant (K)

We have just seen that the *concentrations of reactants and products* are not equal at equilibrium—rather, the *rates of the forward and reverse reactions* are equal. So what about the concentrations? The concentrations, as you can see by reexamining Figure 16.2, become constant; they don't change once equilibrium is reached (as long as the temperature is constant). We can quantify the relative concentrations of reactants and products at equilibrium with a quantity called the *equilibrium constant (K)*. Consider an equation for a generic chemical reaction:

$$aA + bB \rightleftharpoons cC + dD$$

where A and B are reactants, C and D are products, and a, b, c, and d are the respective stoichiometric coefficients in the chemical equation. We define the **equilibrium constant (*K*)** for the reaction as the ratio—*at equilibrium*—of the concentrations of the products raised to their stoichiometric coefficients divided by the concentrations of the reactants raised to their stoichiometric coefficients.

We distinguish between the equilibrium constant (*K*) and the Kelvin unit of temperature (K) by italicizing the equilibrium constant.

Law of Mass Action

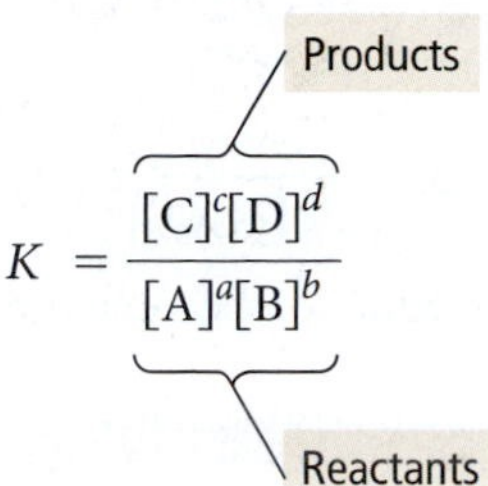

$$K = \frac{[C]^c[D]^d}{[A]^a[B]^b}$$

In this notation, [A] represents the molar concentration of A. The equilibrium constant quantifies the relative concentrations of reactants and products *at equilibrium*. The relationship between the balanced chemical equation and the expression of the equilibrium constant is known as the **law of mass action**.

Why is *this* particular ratio of concentrations at equilibrium—and not some other ratio—defined as the equilibrium constant? Because this particular ratio always equals the same number at equilibrium (at constant temperature), regardless of the initial concentrations of the reactants and products. For example, Table 16.1 shows several different equilibrium concentrations of H_2, I_2, and HI, each from a different set of initial concentrations. Notice that the ratio defined by the law of mass action is always the same, regardless of the initial concentrations. Whether we start with only reactants or only products, the reaction reaches equilibrium at concentrations in which the equilibrium constant is the same. No matter what the initial concentrations are, the reaction always goes in a direction that ensures that the equilibrium concentrations—when substituted into the equilibrium expression—give the same constant, *K* (at constant temperature).

TABLE 16.1 Initial and Equilibrium Concentrations for the Reaction $H_2(g) + I_2(g) \rightleftharpoons 2\,HI(g)$ at 445 °C

Initial Concentrations			Equilibrium Concentrations			Equilibrium Constant as Defined by the Law of Mass Action
$[H_2]$	$[I_2]$	[HI]	$[H_2]$	$[I_2]$	[HI]	$K = \frac{[HI]^2}{[H_2][I_2]}$
0.50	0.50	0.0	0.11	0.11	0.78	$\frac{(0.78)^2}{(0.11)(0.11)} = 50$
0.0	0.0	0.50	0.055	0.055	0.39	$\frac{(0.39)^2}{(0.055)(0.055)} = 50$
0.50	0.50	0.50	0.165	0.165	1.17	$\frac{(1.17)^2}{(0.165)(0.165)} = 50$
1.0	0.50	0.0	0.53	0.033	0.934	$\frac{(0.934)^2}{(0.53)(0.033)} = 50$
0.50	1.0	0.0	0.033	0.53	0.934	$\frac{(0.934)^2}{(0.033)(0.53)} = 50$

Expressing Equilibrium Constants for Chemical Reactions

To express an equilibrium constant for a chemical reaction, we examine the balanced chemical equation and apply the law of mass action. For example, suppose we want to express the equilibrium constant for the reaction:

$$2\,N_2O_5(g) \rightleftharpoons 4\,NO_2(g) + O_2(g)$$

The equilibrium constant is $[NO_2]$ raised to the fourth power multiplied by $[O_2]$ raised to the first power divided by $[N_2O_5]$ raised to the second power:

$$K = \frac{[NO_2]^4[O_2]}{[N_2O_5]^2}$$

Notice that the *coefficients* in the chemical equation become the *exponents* in the expression of the equilibrium constant.

EXAMPLE 16.1 Expressing Equilibrium Constants for Chemical Equations

Express the equilibrium constant for the chemical equation:

$$CH_3OH(g) \rightleftharpoons CO(g) + 2\,H_2(g)$$

SOLUTION

The equilibrium constant is the equilibrium concentrations of the products raised to their stoichiometric coefficients divided by the equilibrium concentrations of the reactants raised to their stoichiometric coefficients.	$K = \frac{[CO][H_2]^2}{[CH_3OH]}$

FOR PRACTICE 16.1

Express the equilibrium constant for the combustion of propane as shown by the balanced chemical equation:

$$C_3H_8(g) + 5\,O_2(g) \rightleftharpoons 3\,CO_2(g) + 4\,H_2O(g)$$

The Significance of the Equilibrium Constant

We now know how to express the equilibrium constant, but what does it mean? What, for example, does a large equilibrium constant ($K \gg 1$) imply about a reaction? A large equilibrium constant indicates that the numerator (which specifies the amounts of products at equilibrium) is larger than the denominator (which specifies the amounts of reactants at equilibrium). Therefore, when the equilibrium constant is large, the forward reaction is favored. For example, consider the reaction:

$$H_2(g) + Br_2(g) \rightleftharpoons 2\,HBr(g) \qquad K = 1.9 \times 10^{19} \text{ (at 25 °C)}$$

The equilibrium constant is large, indicating that the equilibrium point for the reaction lies far to the right—high concentrations of products, low concentrations of reactants (Figure 16.3 ◀). Remember that the equilibrium constant says nothing about *how fast* a reaction reaches equilibrium, only *how far* the reaction has proceeded once equilibrium is reached. A reaction with a large equilibrium constant may be kinetically very slow and take a long time to reach equilibrium.

Conversely, what does a *small* equilibrium constant ($K \ll 1$) mean? It indicates that the reverse reaction is favored and that there will be more reactants than products when equilibrium is reached. For example, consider the reaction:

$$N_2(g) + O_2(g) \rightleftharpoons 2\,NO(g) \qquad K = 4.1 \times 10^{-31} \text{ (at 25 °C)}$$

$$H_2(g) + Br_2(g) \rightleftharpoons 2\,HBr(g)$$

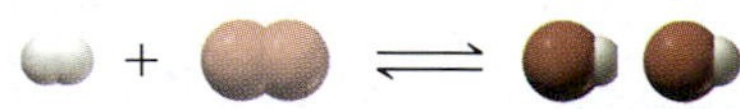

$$K = \frac{[HBr]^2}{[H_2][Br_2]} = \text{large number}$$

▲ **FIGURE 16.3 The Meaning of a Large Equilibrium Constant** If the equilibrium constant for a reaction is large, the equilibrium point of the reaction lies far to the right—the concentration of products is large and the concentration of reactants is small.

$$N_2(g) + O_2(g) \rightleftharpoons 2\,NO(g)$$

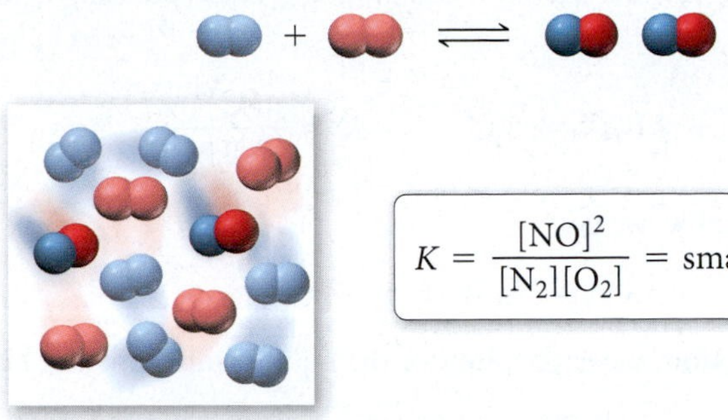

$$K = \frac{[NO]^2}{[N_2][O_2]} = \text{small number}$$

◀ **FIGURE 16.4 The Meaning of a Small Equilibrium Constant** If the equilibrium constant for a reaction is small, the equilibrium point of the reaction lies far to the left—the concentration of products is small and the concentration of reactants is large.

The equilibrium constant is very small, indicating that the equilibrium point for the reaction lies far to the left—high concentrations of reactants, low concentrations of products (Figure 16.4 ▲). This is fortunate because N_2 and O_2 are the main components of air. If this equilibrium constant were large, much of the N_2 and O_2 in air would react to form NO, a toxic gas.

Summarizing the Significance of the Equilibrium Constant:

- $K \ll 1$ Reverse reaction is favored; forward reaction does not proceed very far.
- $K \approx 1$ Neither direction is favored; forward reaction proceeds about halfway.
- $K \gg 1$ Forward reaction is favored; forward reaction proceeds essentially to completion.

Equilibrium Constants

16.1 Cc Conceptual Connection

The equilibrium constant for the reaction $A(g) \rightleftharpoons B(g)$ is 10. A reaction mixture initially contains $[A] = 1.1$ M and $[B] = 0.0$ M. Which statement about this reaction is true at equilibrium?

(a) The reaction mixture contains $[A] = 1.0$ M and $[B] = 0.1$ M.

(b) The reaction mixture contains $[A] = 0.1$ M and $[B] = 1.0$ M.

(c) The reaction mixture contains equal concentrations of A and B.

Relationships between the Equilibrium Constant and the Chemical Equation

If a chemical equation is modified in some way, the equilibrium constant for the equation changes because of the modification. The three modifications we discuss here are common.

1. **If we reverse the equation, we invert the equilibrium constant.** For example, consider this equilibrium equation:

$$A + 2\,B \rightleftharpoons 3\,C$$

The expression for the equilibrium constant of this reaction is:

$$K_{\text{forward}} = \frac{[C]^3}{[A][B]^2}$$

If we reverse the equation

$$3\,C \rightleftharpoons A + 2\,B$$

then, according to the law of mass action, the expression for the equilibrium constant becomes:

$$K_{\text{reverse}} = \frac{[A][B]^2}{[C]^3} = \frac{1}{K_{\text{forward}}}$$

2. **If we multiply the coefficients in the equation by a factor, we raise the equilibrium constant to the same factor.** Consider again this chemical equation and corresponding expression for the equilibrium constant:

$$A + 2\,B \rightleftharpoons 3\,C \qquad K = \frac{[C]^3}{[A][B]^2}$$

If we multiply the equation by n, we get:

$$n\,A + 2n\,B \rightleftharpoons 3n\,C$$

Applying the law of mass action, the expression for the equilibrium constant becomes:

$$K' = \frac{[C]^{3n}}{[A]^n[B]^{2n}} = \left(\frac{[C]^3}{[A][B]^2}\right)^n = K^n$$

If n is a fractional quantity, we raise K to the same fractional quantity.

Remember that $(X^a)^b = X^{ab}$.

3. **If we add two or more individual chemical equations to obtain an overall equation, we multiply the corresponding equilibrium constants by each other to obtain the overall equilibrium constant.** Consider these two chemical equations and their corresponding equilibrium constant expressions:

$$A \rightleftharpoons 2\,B \qquad K_1 = \frac{[B]^2}{[A]}$$

$$2\,B \rightleftharpoons 3\,C \qquad K_2 = \frac{[C]^3}{[B]^2}$$

The two equations sum as follows:

$$A \rightleftharpoons 2\,B$$
$$2\,B \rightleftharpoons 3\,C$$
$$A \rightleftharpoons 3\,C$$

According to the law of mass action, the equilibrium constant for this overall equation is then:

$$K_{\text{overall}} = \frac{[C]^3}{[A]}$$

Notice that K_{overall} is the product of K_1 and K_2:

$$\begin{aligned} K_{\text{overall}} &= K_1 \times K_2 \\ &= \frac{[B]^2}{[A]} \times \frac{[C]^3}{[B]^2} \\ &= \frac{[C]^3}{[A]} \end{aligned}$$

16.2 Cc Conceptual Connection

The Equilibrium Constant and the Chemical Equation

The reaction $A(g) \rightleftharpoons 2\,B(g)$ has an equilibrium constant of $K = 0.010$. What is the equilibrium constant for the reaction $B(g) \rightleftharpoons \frac{1}{2}A(g)$?

(a) 1
(b) 10
(c) 100
(d) 0.0010

EXAMPLE 16.2

Manipulating the Equilibrium Constant to Reflect Changes in the Chemical Equation

Consider the chemical equation and equilibrium constant for the synthesis of ammonia at 25 °C:

$$N_2(g) + 3\,H_2(g) \rightleftharpoons 2\,NH_3(g) \qquad K = 5.6 \times 10^5$$

Calculate the equilibrium constant for the following reaction at 25 °C:

$$NH_3(g) \rightleftharpoons \tfrac{1}{2}N_2(g) + \tfrac{3}{2}H_2(g) \qquad K' = ?$$

SOLUTION

You want to manipulate the given reaction and value of K to obtain the desired reaction and value of K. Note that the given reaction is the reverse of the desired reaction, and its coefficients are twice those of the desired reaction.

Begin by reversing the given reaction and taking the inverse of the value of K.	$N_2(g) + 3\,H_2(g) \rightleftharpoons 2\,NH_3(g) \quad K = 5.6 \times 10^5$ $2\,NH_3(g) \rightleftharpoons N_2(g) + 3\,H_2(g) \quad K_{reverse} = \dfrac{1}{5.6 \times 10^5}$
Next, multiply the reaction by $\frac{1}{2}$ and raise the equilibrium constant to the $\frac{1}{2}$ power.	$NH_3(g) \rightleftharpoons \frac{1}{2}N_2(g) + \frac{3}{2}H_2(g)$ $K' = K_{reverse}^{1/2} = \left(\dfrac{1}{5.6 \times 10^5}\right)^{1/2}$
Calculate the value of K'.	$K' = 1.3 \times 10^{-3}$

FOR PRACTICE 16.2

Consider the following chemical equation and equilibrium constant at 25 °C:

$$2\,COF_2(g) \rightleftharpoons CO_2(g) + CF_4(g) \qquad K = 2.2 \times 10^6$$

Calculate the equilibrium constant for the following reaction at 25 °C:

$$2\,CO_2(g) + 2\,CF_4(g) \rightleftharpoons 4\,COF_2(g) \qquad K' = ?$$

FOR MORE PRACTICE 16.2

Predict the equilibrium constant for the first reaction given the equilibrium constants for the second and third reactions:

$$CO_2(g) + 3\,H_2(g) \rightleftharpoons CH_3OH(g) + H_2O(g) \qquad K_1 = ?$$
$$CO(g) + H_2O(g) \rightleftharpoons CO_2(g) + H_2(g) \qquad K_2 = 1.0 \times 10^5$$
$$CO(g) + 2\,H_2(g) \rightleftharpoons CH_3OH(g) \qquad K_3 = 1.4 \times 10^7$$

16.4 Expressing the Equilibrium Constant in Terms of Pressure

So far, we have expressed the equilibrium constant only in terms of the *concentrations* of the reactants and products. For gaseous reactions, the partial pressure of a particular gas is proportional to its concentration. Therefore, we can also express the equilibrium constant in terms of the *partial pressures* of the reactants and products. Consider the gaseous reaction:

$$2\,SO_3(g) \rightleftharpoons 2\,SO_2(g) + O_2(g)$$

From this point on, we designate K_c as the equilibrium constant with respect to concentration in molarity. For the reaction just given, we can express K_c using the law of mass action:

$$K_c = \frac{[SO_2]^2[O_2]}{[SO_3]^2}$$

We now designate K_p as the equilibrium constant with respect to partial pressures in atmospheres. *The expression for K_p takes the form of the expression for K_c, except that we use the partial pressure of each gas in place of its concentration.* For the SO_3 reaction, we write K_p as:

$$K_p = \frac{(P_{SO_2})^2 P_{O_2}}{(P_{SO_3})^2}$$

where P_A is the partial pressure of gas A in units of atmospheres.

Because the partial pressure of a gas in atmospheres is not the same as its concentration in molarity, the value of K_p for a reaction is not necessarily equal to the value of K_c. However, as long as the gases are behaving ideally, we can derive a relationship between the two constants. The concentration of an ideal gas A is the number of moles of A (n_A) divided by its volume (V) in liters:

$$[A] = \frac{n_A}{V}$$

From the ideal gas law, we can relate the quantity n_A/V to the partial pressure of A as follows:

$$P_A V = n_A RT$$

$$P_A = \frac{n_A}{V} RT$$

Since $[A] = n_A/V$, we can write:

$$P_A = [A]RT \qquad \text{or} \qquad [A] = \frac{P_A}{RT} \qquad [16.1]$$

Now consider the following general equilibrium chemical equation:

$$aA + bB \rightleftharpoons cC + dD$$

According to the law of mass action, we write K_c as follows:

$$K_c = \frac{[C]^c[D]^d}{[A]^a[B]^b}$$

Substituting $[X] = P_X/RT$ for each concentration term, we get:

$$K_c = \frac{\left(\frac{P_C}{RT}\right)^c\left(\frac{P_D}{RT}\right)^d}{\left(\frac{P_A}{RT}\right)^a\left(\frac{P_B}{RT}\right)^b} = \frac{P_C^c P_D^d\left(\frac{1}{RT}\right)^{c+d}}{P_A^a P_B^b\left(\frac{1}{RT}\right)^{a+b}} = \frac{P_C^c P_D^d}{P_A^a P_B^b}\left(\frac{1}{RT}\right)^{c+d-(a+b)}$$

$$= K_p\left(\frac{1}{RT}\right)^{c+d-(a+b)}$$

Rearranging,

$$K_p = K_c(RT)^{c+d-(a+b)}$$

Finally, if we let $\Delta n = c + d - (a + b)$, which is the sum of the stoichiometric coefficients of the gaseous products minus the sum of the stoichiometric coefficients of the gaseous reactants, we get the following general result:

$$K_p = K_c(RT)^{\Delta n} \qquad [16.2]$$

In the equation $K_p = K_c(RT)^{\Delta n}$, the quantity Δn represents the difference between the number of moles of gaseous products and gaseous reactants.

Notice that if the total number of moles of gas is the same after the reaction as before, $\Delta n = 0$, and K_p is equal to K_c.

EXAMPLE 16.3

Relating K_p and K_c

Nitrogen monoxide, a pollutant in automobile exhaust, is oxidized to nitrogen dioxide in the atmosphere according to the equation:

$$2\,NO(g) + O_2(g) \rightleftharpoons 2\,NO_2(g) \qquad K_p = 2.2 \times 10^{12} \text{ at } 25\,°C$$

Find K_c for this reaction.

SORT You are given K_p for the reaction and asked to find K_c.

GIVEN: $K_p = 2.2 \times 10^{12}$
FIND: K_c

STRATEGIZE Use Equation 16.2 to relate K_p and K_c.

EQUATION $K_p = K_c(RT)^{\Delta n}$

SOLVE Solve the equation for K_c.

SOLUTION

$$K_c = \frac{K_p}{(RT)^{\Delta n}}$$

Calculate Δn.

$$\Delta n = 2 - 3 = -1$$

Substitute the required quantities to calculate K_c. The temperature must be in kelvins. The units are dropped when reporting K_c as described below.

$$K_c = \frac{2.2 \times 10^{12}}{\left(0.08206 \dfrac{\text{L} \cdot \text{atm}}{\text{mol} \cdot \text{K}} \times 298\ \text{K}\right)^{-1}}$$

$$= 5.4 \times 10^{13}$$

CHECK The most straightforward way to check this answer is to substitute it back into Equation 16.2 and confirm that you get the original value for K_p.

$$K_p = K_c(RT)^{\Delta n}$$

$$= 5.4 \times 10^{13}\left(0.08206 \frac{\text{L} \cdot \text{atm}}{\text{mol} \cdot \text{K}} \times 298\ \text{K}\right)^{-1}$$

$$= 2.2 \times 10^{12}$$

FOR PRACTICE 16.3

Consider the following reaction and corresponding value of K_c:

$$H_2(g) + I_2(g) \rightleftharpoons 2\,HI(g) \qquad K_c = 6.2 \times 10^2 \text{ at } 25\,°C$$

What is the value of K_p at this temperature?

Units of K

Throughout this book, we express concentrations and partial pressures within the equilibrium constant expression in units of molarity and atmospheres, respectively. When expressing the value of the equilibrium constant, however, we have not included the units. Formally, the values of concentration or partial pressure that we substitute into the equilibrium constant expression are ratios of the concentration or pressure to a reference concentration (exactly 1 M) or a reference pressure (exactly 1 atm). For example, within the equilibrium constant expression, a pressure of 1.5 atm becomes:

$$\frac{1.5\ \text{atm}}{1\ \text{atm}} = 1.5$$

Similarly, a concentration of 1.5 M becomes:

$$\frac{1.5\ \text{M}}{1\ \text{M}} = 1.5$$

As long as concentration units are expressed in molarity for K_c and pressure units are expressed in atmospheres for K_p, we skip this formality and enter the quantities directly into the equilibrium expression, dropping their corresponding units.

16.3 Cc Conceptual Connection

The Relationship between K_p and K_c

Under which circumstances are K_p and K_c equal for the reaction $a\text{A}(g) + b\text{B}(g) \rightleftharpoons c\text{C}(g) + d\text{D}(g)$?

(a) When $a + b = c + d$.

(b) When the reaction is reversible.

(c) When the equilibrium constant is small.

16.5 Heterogeneous Equilibria: Reactions Involving Solids and Liquids

Many chemical reactions involve pure solids or pure liquids as reactants or products. Consider, for example, the reaction:

$$2\,\text{CO}(g) \rightleftharpoons \text{CO}_2(g) + \text{C}(s)$$

We might expect the expression for the equilibrium constant to be:

$$K_c = \frac{[\text{CO}_2][\text{C}]}{[\text{CO}]^2} \quad \text{(incorrect)}$$

However, since carbon is a solid, its concentration is constant (if you double the amount of carbon, its *concentration* remains the same). The concentration of a solid does not change because a solid does not expand to fill its container. Its concentration, therefore, depends only on its density, which is constant as long as *some* solid is present (Figure 16.5 ▼). Consequently, pure solids—those reactants or products labeled in the chemical equation with an (s)—are not included in the equilibrium expression (because their constant value is incorporated into the value of K). The correct equilibrium expression for this reaction is therefore:

$$K_c = \frac{[\text{CO}_2]}{[\text{CO}]^2}$$

Similarly, the concentration of a pure liquid does not change. So, pure liquids—reactants or products labeled in the chemical equation with an (l)—are also excluded from the equilibrium expression. For example, consider the equilibrium expression for the reaction between carbon dioxide and water:

$$\text{CO}_2(g) + \text{H}_2\text{O}(\ell) \rightleftharpoons \text{H}^+(aq) + \text{HCO}_3^-(aq)$$

A Heterogeneous Equilibrium

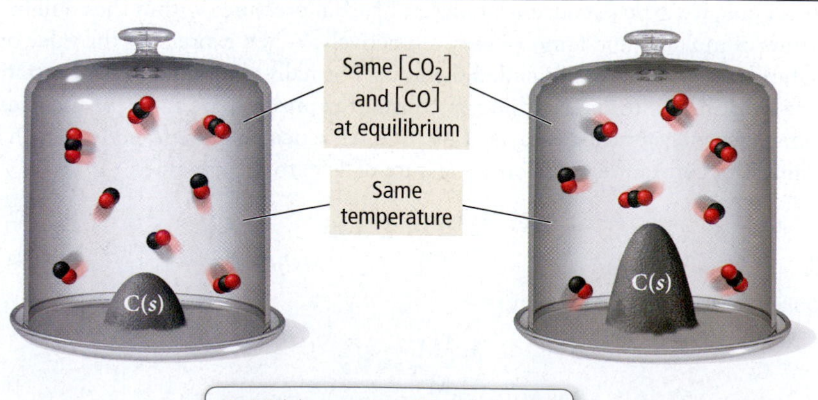

▶ **FIGURE 16.5 Heterogeneous Equilibrium** The concentration of solid carbon (the number of atoms per unit volume) is constant as long as some solid carbon is present. The same is true for pure liquids. For this reason, the concentrations of solids and pure liquids are not included in equilibrium constant expressions.

Since $H_2O(\ell)$ is pure liquid, we omit it from the equilibrium expression:

$$K_c = \frac{[H^+][HCO_3^-]}{[CO_2]}$$

EXAMPLE 16.4
Writing Equilibrium Expressions for Reactions Involving a Solid or a Liquid

Write an expression for the equilibrium constant (K_c) for the chemical equation.

$$CaCO_3(s) \rightleftharpoons CaO(s) + CO_2(g)$$

SOLUTION

Since $CaCO_3(s)$ and $CaO(s)$ are both solids, omit them from the equilibrium expression.	$K_c = [CO_2]$

FOR PRACTICE 16.4

Write an equilibrium expression (K_c) for the equation.

$$4\,HCl(g) + O_2(g) \rightleftharpoons 2\,H_2O(\ell) + 2\,Cl_2(g)$$

Heterogeneous Equilibria, K_p and K_c

16.4 Cc Conceptual Connection

For which reaction does $K_p = K_c$?

(a) $2\,Na_2O_2(s) + 2\,CO_2(g) \rightleftharpoons 2\,Na_2CO_3(s) + O_2(g)$

(b) $Fe_2O_3(s) + 3\,CO(g) \rightleftharpoons 2\,Fe(s) + 3\,CO_2(g)$

(c) $NH_4NO_3(s) \rightleftharpoons N_2O(g) + 2\,H_2O(g)$

16.6 Calculating the Equilibrium Constant from Measured Equilibrium Concentrations

The most direct way to obtain an experimental value for the equilibrium constant of a reaction is to measure the concentrations of the reactants and products in a reaction mixture at equilibrium. Consider again the reaction between hydrogen and iodine to form hydrogen iodide:

$$H_2(g) + I_2(g) \rightleftharpoons 2\,HI(g)$$

We saw in Table 16.1 (see Section 16.3) that the measured concentrations of the reactants and products, when substituted into the expression for K, always equal a constant (the equilibrium constant) at a constant temperature. For example, one set of measurements at 445 °C results in equilibrium concentrations of $[H_2] = 0.11$ M, $[I_2] = 0.11$ M, and $[HI] = 0.78$ M. What is the value of the equilibrium constant at this temperature? We can write the expression for K_c from the balanced equation and substitute the equilibrium concentrations to obtain the value of K_c:

Because equilibrium constants depend on temperature, many equilibrium problems state the temperature even though it has no formal part in the calculation.

$$K_c = \frac{[HI]^2}{[H_2][I_2]} = \frac{(0.78)^2}{(0.11)(0.11)} = 5.0 \times 10^1$$

The concentrations within K_c should always be written in moles per liter (M); however, as noted in Section 16.4, we do not normally include the units when expressing the value of the equilibrium constant, so K_c is unitless.

We just calculated the equilibrium constant from values of the equilibrium concentrations of all the reactants and products. In most cases, however, we need only know the initial concentrations of the reactant(s) and the equilibrium concentration of any *one* reactant or product. We can deduce the other equilibrium concentrations from the stoichiometry of the reaction. For example, consider the simple reaction:

$$A(g) \rightleftharpoons 2\,B(g)$$

Suppose that we have a reaction mixture in which the initial concentration of A is 1.00 M and the initial concentration of B is 0.00 M. When equilibrium is reached, the concentration of A is 0.75 M. Since [A] has changed by −0.25 M, we can deduce (based on the stoichiometry) that [B] must have changed by 2 × (+0.25 M) or +0.50 M. We summarize the initial conditions, the changes, and the equilibrium conditions in the following table:

	[A]	[B]
Initial	1.00	0.00
Change	−0.25	+2(0.25)
Equilibrium	0.75	0.50

The last row in an ICE table is the sum of the two rows above it.

We refer to this type of table as an ICE table (I = initial, C = change, E = equilibrium). To calculate the equilibrium constant, we use the balanced equation to write an expression for the equilibrium constant and then substitute the equilibrium concentrations from the ICE table:

$$K = \frac{[B]^2}{[A]} = \frac{(0.50)^2}{(0.75)} = 0.33$$

In Examples 16.5 and 16.6, the general procedure for solving these kinds of equilibrium problems is in the left column and two worked examples exemplifying the procedure are in the center and right columns.

PROCEDURE FOR ▼

Finding Equilibrium Constants from Experimental Concentration Measurements

To solve these types of problems, follow the given procedure.

1. Using the balanced equation as a guide, prepare an ICE table showing the known initial concentrations and equilibrium concentrations of the reactants and products. Leave space in the middle of the table for determining the changes in concentration that occur during the reaction. If initial concentrations of some reactants or products are not given, you may assume they are zero.

EXAMPLE 16.5

Finding Equilibrium Constants from Experimental Concentration Measurements

Consider the following reaction:

$$CO(g) + 2\,H_2(g) \rightleftharpoons CH_3OH(g)$$

A reaction mixture at 780 °C initially contains [CO] = 0.500 M and $[H_2]$ = 1.00 M. At equilibrium, the CO concentration is 0.15 M. What is the value of the equilibrium constant?

$$CO(g) + 2\,H_2(g) \rightleftharpoons CH_3OH(g)$$

	[CO]	$[H_2]$	$[CH_3OH]$
Initial	0.500	1.00	0.00
Change			
Equil	0.15		

EXAMPLE 16.6

Finding Equilibrium Constants from Experimental Concentration Measurements

Consider the following reaction:

$$2\,CH_4(g) \rightleftharpoons C_2H_2(g) + 3\,H_2(g)$$

A reaction mixture at 1700 °C initially contains $[CH_4]$ = 0.115 M. At equilibrium, the mixture contains $[C_2H_2]$ = 0.035 M. What is the value of the equilibrium constant?

$$2\,CH_4(g) \rightleftharpoons C_2H_2(g) + 3\,H_2(g)$$

	$[CH_4]$	$[C_2H_2]$	$[H_2]$
Initial	0.115	0.00	0.00
Change			
Equil		0.035	

—*Continued on the next page*

Continued from the previous page—

2. For the reactant or product whose concentration is known both initially and at equilibrium, calculate the change in concentration that occurs.

$$CO(g) + 2\,H_2(g) \rightleftharpoons CH_3OH(g)$$

	$[CO]$	$[H_2]$	$[CH_3OH]$
Initial	0.500	1.00	0.00
Change	−0.35		
Equil	0.15		

$$2\,CH_4(g) \rightleftharpoons C_2H_2(g) + 3\,H_2(g)$$

	$[CH_4]$	$[C_2H_2]$	$[H_2]$
Initial	0.115	0.00	0.00
Change		+0.035	
Equil		0.035	

3. Use the change calculated in step 2 and the stoichiometric relationships from the balanced chemical equation to determine the changes in concentration of all other reactants and products. Since reactants are consumed during the reaction, the changes in their concentrations are negative. Since products are formed, the changes in their concentrations are positive.

$$CO(g) + 2\,H_2(g) \rightleftharpoons CH_3OH(g)$$

	$[CO]$	$[H_2]$	$[CH_3OH]$
Initial	0.500	1.00	0.00
Change	−0.35	−2(0.35)	+0.35
Equil	0.15		

$$2\,CH_4(g) \rightleftharpoons C_2H_2(g) + 3\,H_2(g)$$

	$[CH_4]$	$[C_2H_2]$	$[H_2]$
Initial	0.115	0.00	0.00
Change	−2(0.035)	+0.035	+3(0.035)
Equil		0.035	

4. Sum each column for each reactant and product to determine the equilibrium concentrations.

	$[CO]$	$[H_2]$	$[CH_3OH]$
Initial	0.500	1.00	0.00
Change	−0.35	−0.70	+0.35
Equil	0.15	0.30	0.35

	$[CH_4]$	$[C_2H_2]$	$[H_2]$
Initial	0.115	0.00	0.00
Change	−0.070	+0.035	+0.105
Equil	0.045	0.035	0.105

5. Use the balanced equation to write an expression for the equilibrium constant and substitute the equilibrium concentrations to calculate *K*.

$$K_c = \frac{[CH_3OH]}{[CO][H_2]^2} = \frac{0.35}{(0.15)(0.30)^2} = 26$$

$$K_c = \frac{[C_2H_2][H_2]^3}{[CH_4]^2} = \frac{(0.035)(0.105)^3}{(0.045)^2} = 0.020$$

FOR PRACTICE 16.5

The reaction in Example 16.5 between CO and H_2 is carried out at a different temperature with initial concentrations of $[CO] = 0.27$ M and $[H_2] = 0.49$ M. At equilibrium, the concentration of CH_3OH is 0.11 M. Find the equilibrium constant at this temperature.

FOR PRACTICE 16.6

The reaction of CH_4 in Example 16.6 is carried out at a different temperature with an initial concentration of $[CH_4] = 0.087$ M. At equilibrium, the concentration of H_2 is 0.012 M. Find the equilibrium constant at this temperature.

16.7 The Reaction Quotient: Predicting the Direction of Change

When the reactants of a chemical reaction mix, they generally react to form products—we say that the reaction proceeds to the right (toward the products). The amount of products formed when equilibrium is reached depends on the magnitude of the equilibrium constant, as we have seen. However, what if a reaction mixture not at equilibrium contains both reactants *and products*? Can we predict the direction of change for such a mixture?

To gauge the progress of a reaction relative to equilibrium, we use a quantity called the *reaction quotient.* The definition of the reaction quotient takes the same form as the definition of the equilibrium constant, except that the reaction need not be at equilibrium. So, for the general reaction:

$$a\text{A} + b\text{B} \rightleftharpoons c\text{C} + d\text{D}$$

we define the **reaction quotient (Q_c)** as the ratio—at any point in the reaction—of the concentrations of the products raised to their stoichiometric coefficients divided by the concentrations of the reactants raised to their stoichiometric coefficients. For gases with amounts measured in atmospheres, the reaction quotient uses the partial pressures in place of concentrations and is called Q_p:

$$Q_c = \frac{[\text{C}]^c[\text{D}]^d}{[\text{A}]^a[\text{B}]^b} \qquad Q_p = \frac{P_\text{C}^c P_\text{D}^d}{P_\text{A}^a P_\text{B}^b}$$

The difference between the reaction quotient and the equilibrium constant is that, at a given temperature, the equilibrium constant has only one value and it specifies the relative amounts of reactants and products *at equilibrium.* The reaction quotient, by contrast, depends on the current state of the reaction and has many different values as the reaction proceeds. For example, in a reaction mixture containing only reactants, the reaction quotient is zero ($Q_c = 0$):

$$Q_c = \frac{[0]^c[0]^d}{[\text{A}]^a[\text{B}]^b} = 0$$

In a reaction mixture containing only products, the reaction quotient is infinite ($Q_c = \infty$):

$$Q_c = \frac{[\text{C}]^c[\text{D}]^d}{[0]^a[0]^b} = \infty$$

In a reaction mixture containing both reactants and products, each at a concentration of 1 M, the reaction quotient is one ($Q_c = 1$):

$$Q_c = \frac{[1]^c[1]^d}{[1]^a[1]^b} = 1$$

The reaction quotient is useful because *the value of Q relative to K is a measure of the progress of the reaction toward equilibrium. At equilibrium, the reaction quotient is equal to the equilibrium constant.* Figure 16.6 ▶ shows a plot of Q as a function of the concentrations of A and B for the simple reaction $\text{A}(g) \rightleftharpoons \text{B}(g)$, which has an equilibrium constant of $K = 1.45$. The following points are representative of three possible conditions:

Q	K	Predicted Direction of Reaction
0.55	1.45	To the right (toward products)
2.55	1.45	To the left (toward reactants)
1.45	1.45	No change (at equilibrium)

For the first set of values in the table, Q is less than K and must therefore get larger as the reaction proceeds toward equilibrium. Q becomes larger as the reactant concentration decreases and the product concentration increases—the reaction proceeds to the right. For the second set of values, Q is greater than K and must therefore get smaller as the reaction proceeds toward equilibrium. Q gets smaller as the reactant concentration increases and the product concentration decreases—the reaction proceeds to the left. In the third set of values, $Q = K$, implying that the reaction is at equilibrium—the reaction will not proceed in either direction.

Summarizing Direction of Change Predictions:

The reaction quotient (Q) relative to the equilibrium constant (K) is a measure of the progress of a reaction toward equilibrium.

- $Q < K$ Reaction goes to the right (toward products).
- $Q > K$ Reaction goes to the left (toward reactants).
- $Q = K$ Reaction is at equilibrium.

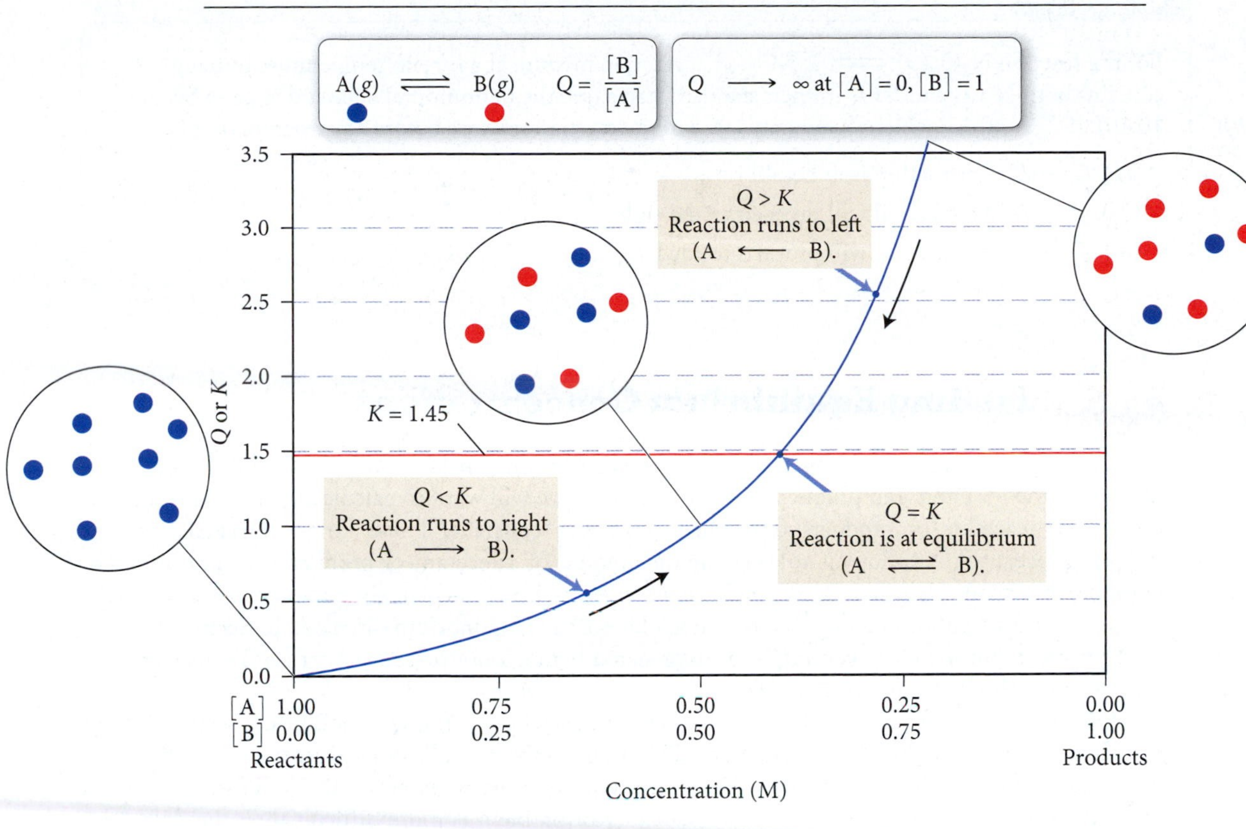

◀ **FIGURE 16.6** ***Q*, *K*, and the Direction of a Reaction** The graph shows a plot of *Q* as a function of the concentrations of the reactants and products in a simple reaction A ⇌ B, in which $K = 1.45$ and the sum of the reactant and product concentrations is 1 M. The far left of the graph represents pure reactant, and the far right represents pure product. The midpoint of the graph represents an equal mixture of A and B. When *Q* is less than *K*, the reaction moves in the forward direction (A ⟶ B). When *Q* is greater than *K*, the reaction moves in the reverse direction (A ⟵ B). When *Q* is equal to *K*, the reaction is at equilibrium.

EXAMPLE 16.7

Predicting the Direction of a Reaction by Comparing *Q* and *K*

Consider the reaction and its equilibrium constant.

$$I_2(g) + Cl_2(g) \rightleftharpoons 2\,ICl(g) \qquad K_p = 81.9$$

A reaction mixture contains $P_{I_2} = 0.114$ atm, $P_{Cl_2} = 0.102$ atm, and $P_{ICl} = 0.355$ atm. Is the reaction mixture at equilibrium? If not, in which direction will the reaction proceed?

SOLUTION

To determine the progress of the reaction relative to the equilibrium state, first calculate *Q*.	$Q_p = \left(\frac{P_{ICl}^2}{P_{I_2}P_{Cl_2}}\right) = \frac{(0.355)^2}{(0.114)(0.102)} = 10.8$
Compare *Q* to *K*.	$Q_p = 10.8;\ K_p = 81.9$ Since $Q_p < K_p$, the reaction is not at equilibrium and will proceed to the right.

FOR PRACTICE 16.7

Consider the reaction and its equilibrium constant.

$$N_2O_4(g) \rightleftharpoons 2\,NO_2(g) \qquad K_c = 5.85 \times 10^{-3} \text{ (at some temperature)}$$

A reaction mixture contains $[NO_2] = 0.0255$ M and $[N_2O_4] = 0.0331$ M. Calculate Q_c and determine the direction in which the reaction will proceed.

16.5 Cc Conceptual Connection

Q and *K*

For the reaction $N_2O_4(g) \rightleftharpoons 2\,NO_2(g)$, a reaction mixture at a certain temperature initially contains both N_2O_4 and NO_2 in their standard states (see the definition of standard state in Section 10.10). If $K_p = 0.15$, which statement is true of the reaction mixture before any reaction occurs?

(a) $Q = K$; the reaction is at equilibrium.
(b) $Q < K$; the reaction will proceed to the right.
(c) $Q > K$; the reaction will proceed to the left.

16.8 Finding Equilibrium Concentrations

In Section 16.6, we discussed how to calculate an equilibrium constant from the equilibrium concentrations of the reactants and products. Just as commonly we will want to calculate equilibrium concentrations of reactants or products from the equilibrium constant. These kinds of calculations are important because they allow us to calculate the amount of a reactant or product at equilibrium. For example, in a synthesis reaction, we might want to know how much of the product forms when the reaction reaches equilibrium. Or for the hemoglobin–oxygen equilibrium discussed in Section 16.1, we might want to know the concentration of oxygenated hemoglobin present under certain oxygen concentrations within the lungs or muscles.

We can divide these types of problems into two categories: (1) finding equilibrium concentrations when we know the equilibrium constant and all but one of the equilibrium concentrations of the reactants and products, and (2) finding equilibrium concentrations when we know the equilibrium constant and only initial concentrations. The second category of problem is more difficult than the first. Let's examine each separately.

Finding Equilibrium Concentrations from the Equilibrium Constant and All but One of the Equilibrium Concentrations of the Reactants and Products

We can use the equilibrium constant to calculate the equilibrium concentration of one of the reactants or products, given the equilibrium concentrations of the others. To solve this type of problem, we can follow our general problem-solving procedure as demonstrated in Example 16.8.

EXAMPLE 16.8

Finding Equilibrium Concentrations When You Know the Equilibrium Constant and All but One of the Equilibrium Concentrations of the Reactants and Products

Consider the following reaction:

$$2\,COF_2(g) \rightleftharpoons CO_2(g) + CF_4(g) \qquad K_c = 2.00 \text{ at } 1000\,^\circ C$$

In an equilibrium mixture, the concentration of COF_2 is 0.255 M and the concentration of CF_4 is 0.118 M. What is the equilibrium concentration of CO_2?

SORT You are given the equilibrium constant of a chemical reaction, together with the equilibrium concentrations of the reactant and one product. You are asked to find the equilibrium concentration of the other product.	**GIVEN:** $[COF_2] = 0.255$ M $[CF_4] = 0.118$ M $K_c = 2.00$ **FIND:** $[CO_2]$

—Continued on the next page

Continued from the previous page—

STRATEGIZE You can calculate the concentration of the product using the given quantities and the expression for K_c.

CONCEPTUAL PLAN

$[COF_2], [CF_4], K_c \rightarrow [CO_2]$

$$K_c = \frac{[CO_2][CF_4]}{[COF_2]^2}$$

SOLVE Solve the equilibrium expression for $[CO_2]$ and substitute in the appropriate values to calculate it.

SOLUTION

$$[CO_2] = K_c\frac{[COF_2]^2}{[CF_4]}$$

$$[CO_2] = 2.00\left(\frac{(0.255)^2}{0.118}\right) = 1.10\text{ M}$$

CHECK Check your answer by mentally substituting the given values of $[COF_2]$ and $[CF_4]$ as well as your calculated value for CO_2 back into the equilibrium expression.

$$K_c = \frac{[CO_2][CF_4]}{[COF_2]^2}$$

$[CO_2]$ was found to be roughly equal to 1. $[COF_2]^2 \approx 0.06$ and $[CF_4] \approx 0.12$. Therefore K_c is approximately 2, as given in the problem.

FOR PRACTICE 16.8
Diatomic iodine $[I_2]$ decomposes at high temperature to form I atoms according to the reaction:

$$I_2(g) \rightleftharpoons 2\,I(g) \qquad K_c = 0.011 \text{ at } 1200\text{ °C}$$

In an equilibrium mixture, the concentration of I_2 is 0.10 M. What is the equilibrium concentration of I?

Finding Equilibrium Concentrations from the Equilibrium Constant and Initial Concentrations or Pressures

More commonly, we know the equilibrium constant and only the initial concentrations of reactants and need to find the *equilibrium concentrations* of the reactants or products. These kinds of problems are generally more involved than those we just examined and require a specific procedure to solve them. The procedure has some similarities to the one used in Examples 16.5 and 16.6 in that we set up an ICE table showing the initial conditions, the changes, and the equilibrium conditions. However, unlike Examples 16.5 and 16.6, here the changes in concentration are not known and are represented with the variable x. For example, consider again the simple reaction:

$$A(g) \rightleftharpoons 2\,B(g)$$

Suppose that, as before (see Section 16.6), we have a reaction mixture in which the initial concentration of A is 1.0 M and the initial concentration of B is 0.00 M. However, now we know the equilibrium constant, $K = 0.33$, and want to find the equilibrium concentrations. We know that since $Q = 0$, the reaction proceeds to the right (toward the products). We set up the ICE table with the given initial concentrations and *represent the unknown change in [A] with the variable* x as follows:

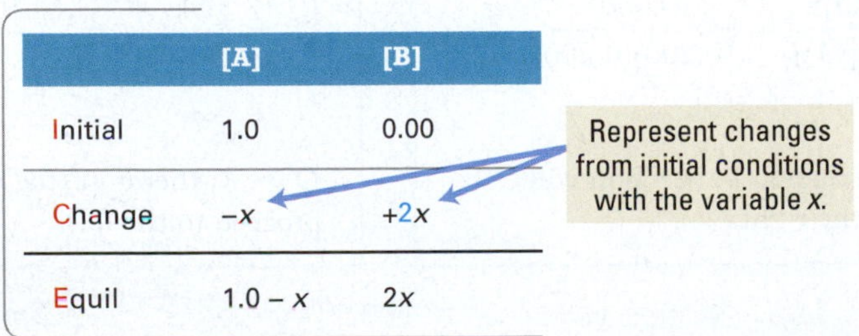

	[A]	[B]
Initial	1.0	0.00
Change	$-x$	$+2x$
Equil	$1.0 - x$	$2x$

Represent changes from initial conditions with the variable x.

Notice that, due to the stoichiometry of the reaction, the change in [B] must be $+2x$. As before, each *equilibrium* concentration is the sum of the two entries above it in the ICE table. In order to find the equilibrium concentrations of A and B, we must find the value of the variable x. Since we know the value of the equilibrium constant, we can use the equilibrium constant expression to set up an equation in which x is the only variable:

$$K_c = \frac{[B]^2}{[A]} = \frac{(2x)^2}{1.0 - x} = 0.33$$

or more simply:

$$\frac{4x^2}{1.0 - x} = 0.33$$

This equation is a *quadratic* equation—it contains the variable x raised to the second power. In general, we can solve quadratic equations with the quadratic formula, which we introduce in Example 16.10. If the quadratic equation is a perfect square, however, we can solve it by simpler means, as shown in Example 16.9. For these examples, we give the general procedure in the left column and apply the procedure to the two different example problems in the center and right columns. Later in this section, we see that quadratic equations can often be simplified by making some approximations based on our chemical knowledge.

PROCEDURE FOR Finding Equilibrium Concentrations from Initial Concentrations and the Equilibrium Constant

To solve these types of problems, follow the given procedure.

EXAMPLE 16.9 Finding Equilibrium Concentrations from Initial Concentrations and the Equilibrium Constant

Consider the reaction:

$$N_2(g) + O_2(g) \rightleftharpoons 2\,NO(g) \qquad K_c = 0.10 \text{ (at 2000 °C)}$$

A reaction mixture at 2000 °C initially contains $[N_2] = 0.200$ M and $[O_2] = 0.200$ M. Find the equilibrium concentrations of the reactants and product at this temperature.

EXAMPLE 16.10 Finding Equilibrium Concentrations from Initial Concentrations and the Equilibrium Constant

Consider the reaction:

$$N_2O_4(g) \rightleftharpoons 2\,NO_2(g) \qquad K_c = 0.36 \text{ (at 100 °C)}$$

A reaction mixture at 100 °C initially contains $[NO_2] = 0.100$ M. Find the equilibrium concentrations of NO_2 and N_2O_4 at this temperature.

1. Using the balanced equation as a guide, prepare a table showing the known initial concentrations of the reactants and products. Leave room in the table for the changes in concentrations and for the equilibrium concentrations.

Example 16.9:

$$N_2(g) + O_2(g) \rightleftharpoons 2\,NO(g)$$

	$[N_2]$	$[O_2]$	[NO]
Initial	0.200	0.200	0.00
Change			
Equil			

Example 16.10:

$$N_2O_4(g) \rightleftharpoons 2\,NO_2(g)$$

	$[N_2O_4]$	$[NO_2]$
Initial	0.00	0.100
Change		
Equil		

2. Use the initial concentrations to calculate the reaction quotient (Q) for the initial concentrations. Compare Q to K to predict the direction in which the reaction will proceed.

Example 16.9:

$$Q_c = \frac{[NO]^2}{[N_2][O_2]} = \frac{(0.000)^2}{(0.200)(0.200)} = 0$$

$Q < K$; therefore, the reaction will proceed to the right.

Example 16.10:

$$Q_c = \frac{[NO_2]^2}{[N_2O_4]} = \frac{(0.100)^2}{0.00} = \infty$$

$Q > K$; therefore, the reaction will proceed to the left.

—*Continued on the next page*

Continued from the previous page—

3. Represent the change in the concentration of one of the reactants or products with the variable x. Define the changes in the concentrations of the other reactants or products in terms of x. It is usually most convenient to let x represent the change in concentration of the reactant or product with the smallest stoichiometric coefficient.

$$N_2(g) + O_2(g) \rightleftharpoons 2\,NO(g)$$

	$[N_2]$	$[O_2]$	$[NO]$
Initial	0.200	0.200	0.00
Change	$-x$	$-x$	$+2x$
Equil			

$$N_2O_4(g) \rightleftharpoons 2\,NO_2(g)$$

	$[N_2O_4]$	$[NO_2]$
Initial	0.00	0.100
Change	$+x$	$-2x$
Equil		

4. Sum each column for each reactant and each product to determine the equilibrium concentrations in terms of the initial concentrations and the variable x.

$$N_2(g) + O_2(g) \rightleftharpoons 2\,NO(g)$$

	$[N_2]$	$[O_2]$	$[NO]$
Initial	0.200	0.200	0.00
Change	$-x$	$-x$	$+2x$
Equil	$0.200-x$	$0.200-x$	$2x$

$$N_2O_4(g) \rightleftharpoons 2\,NO_2(g)$$

	$[N_2O_4]$	$[NO_2]$
Initial	0.00	0.100
Change	$+x$	$-2x$
Equil	x	$0.100-2x$

5. Substitute the expressions for the equilibrium concentrations (from Step 4) into the expression for the equilibrium constant. Using the given value of the equilibrium constant, solve the expression for the variable x. In some cases, such as Example 16.9, you can take the square root of both sides of the expression to solve for x. In other cases, such as Example 16.10, you must solve a quadratic equation to find x.

Remember the quadratic formula:

$$ax^2 + bx + c = 0$$

$$x = \frac{-b \pm \sqrt{b^2 - 4ac}}{2a}$$

$$K_c = \frac{[NO]^2}{[N_2][O_2]}$$

$$= \frac{(2x)^2}{(0.200 - x)(0.200 - x)}$$

$$0.10 = \frac{(2x)^2}{(0.200 - x)^2}$$

$$\sqrt{0.10} = \frac{2x}{0.200 - x}$$

$$\sqrt{0.10}\,(0.200 - x) = 2x$$

$$\sqrt{0.10}\,(0.200) - \sqrt{0.10}x = 2x$$

$$0.063 = 2x + \sqrt{0.10}x$$

$$0.063 = 2.3x$$

$$x = 0.027$$

$$K_c = \frac{[NO_2]^2}{[N_2O_4]}$$

$$= \frac{(0.100 - 2x)^2}{x}$$

$$0.36 = \frac{0.0100 - 0.400x + 4x^2}{x}$$

$$0.36x = 0.0100 - 0.400x + 4x^2$$

$$4x^2 - 0.76x + 0.0100 = 0 \ (\textit{quadratic})$$

$$x = \frac{-b \pm \sqrt{b^2 - 4ac}}{2a}$$

$$= \frac{-(-0.76) \pm \sqrt{(-0.76)^2 - 4(4)(0.0100)}}{2(4)}$$

$$= \frac{0.76 \pm 0.65}{8}$$

$$x = 0.176 \quad \text{or} \quad x = 0.014$$

6. Substitute x into the expressions for the equilibrium concentrations of the reactants and products (from Step 4) and calculate the concentrations. In cases where you solved a quadratic and have two values for x, choose the value for x that gives a physically realistic answer. For example, reject the value of x that results in any negative concentrations.

$$[NO_2] = 0.200 - 0.027$$

$$= 0.173\text{ M}$$

$$[O_2] = 0.200 - 0.027$$

$$= 0.173\text{ M}$$

$$[NO] = 2(0.027)$$

$$= 0.054\text{ M}$$

We reject the root $x = 0.176$ because it gives a negative concentration for NO_2. Using $x = 0.014$, we get the following concentrations:

$$[NO_2] = 0.100 - 2x$$

$$= 0.100 - 2(0.014)$$

$$= 0.072\text{ M}$$

$$[N_2O_4] = x$$

$$= 0.014\text{ M}$$

—Continued on the next page

Continued from the previous page—

7. Check your answer by substituting the calculated equilibrium values into the equilibrium expression. The calculated value of K should match the given value of K. Note that rounding errors could cause a difference in the least significant digit when comparing values of the equilibrium constant.

$$K_c = \frac{[NO]^2}{[N_2][O_2]} = \frac{(0.054)^2}{(0.173)(0.173)} = 0.097$$

Since the calculated value of K_c matches the given value (to within one digit in the least significant figure), the answer is valid.

$$K_c = \frac{[NO_2]^2}{[N_2O_4]} = \frac{(0.072)^2}{0.014} = 0.37$$

Since the calculated value of K_c matches the given value (to within one digit in the least significant figure), the answer is valid.

FOR PRACTICE 16.9

The reaction in Example 16.9 is carried out at a different temperature at which $K_c = 0.055$. This time the reaction mixture starts with only the product, $[NO] = 0.0100$ M, and no reactants. Find the equilibrium concentrations of N_2, O_2, and NO at equilibrium.

FOR PRACTICE 16.10

The reaction in Example 16.10 is carried out at the same temperature, but this time the reaction mixture initially contains only the reactant, $[N_2O_4] = 0.0250$ M, and no NO_2. Find the equilibrium concentrations of N_2O_4 and NO_2.

When the initial conditions are given in terms of partial pressures (instead of concentrations) and the equilibrium constant is given as K_p instead of K_c, use the same procedure, but substitute partial pressures for concentrations, as shown in Example 16.11.

EXAMPLE 16.11

Finding Equilibrium Partial Pressures When You Are Given the Equilibrium Constant and Initial Partial Pressures

Consider the reaction:

$$I_2(g) + Cl_2(g) \rightleftharpoons 2\,ICl(g) \qquad K_p = 81.9 \text{ (at 25 °C)}$$

A reaction mixture at 25 °C initially contains $P_{I_2} = 0.100$ atm, $P_{Cl_2} = 0.100$ atm, and $P_{ICl} = 0.100$ atm. Find the equilibrium partial pressures of I_2, Cl_2, and ICl at this temperature.

SOLUTION

Follow the procedure used in Examples 16.5 and 16.6 (using partial pressures in place of concentrations) to solve the problem.

1. Using the balanced equation as a guide, prepare a table showing the known initial partial pressures of the reactants and products.

$$I_2(g) + Cl_2(g) \rightleftharpoons 2\,ICl(g)$$

	P_{I_2} (atm)	P_{Cl_2} (atm)	P_{ICl} (atm)
Initial	0.100	0.100	0.100
Change			
Equil			

—Continued on the next page

Continued from the previous page—

2. Use the initial partial pressures to calculate the reaction quotient (Q). Compare Q to K to predict the direction in which the reaction will proceed.	$Q_p = \frac{(P_{ICl})^2}{P_{I_2}P_{Cl_2}} = \frac{(0.100)^2}{(0.100)(0.100)} = 1$ $K_p = 81.9$ (given) $Q < K$; therefore, the reaction will proceed to the right.
3. Represent the change in the partial pressure of one of the reactants or products with the variable x. Define the changes in the partial pressures of the other reactants or products in terms of x.	$I_2(g) + Cl_2(g) \rightleftharpoons 2\, ICl(g)$ (see table below)
4. Sum each column for each reactant and product to determine the equilibrium partial pressures in terms of the initial partial pressures and the variable x.	$I_2(g) + Cl_2(g) \rightleftharpoons 2\, ICl(g)$ (see table below)
5. Substitute the expressions for the equilibrium partial pressures (from Step 4) into the expression for the equilibrium constant. Use the given value of the equilibrium constant to solve the expression for the variable x.	$K_p = \frac{(P_{ICl})^2}{P_{I_2}P_{Cl_2}} = \frac{(0.100 + 2x)^2}{(0.100 - x)(0.100 - x)}$ $81.9 = \frac{(0.100 + 2x)^2}{(0.100 - x)^2}$ (perfect square) $\sqrt{81.9} = \frac{(0.100 + 2x)}{(0.100 - x)}$ $\sqrt{81.9}\,(0.100 - x) = 0.100 + 2x$ $\sqrt{81.9}\,(0.100) - \sqrt{81.9}\,x = 0.100 + 2x$ $\sqrt{81.9}\,(0.100) - 0.100 = 2x + \sqrt{81.9}\,x$ $0.805 = 11.05x$ $x = 0.0729$
6. Substitute x into the expressions for the equilibrium partial pressures of the reactants and products (from Step 4) and calculate the partial pressures.	$P_{I_2} = 0.100 - 0.0729 = 0.027$ atm $P_{Cl_2} = 0.100 - 0.0729 = 0.027$ atm $P_{ICl} = 0.100 + 2(0.0729) = 0.246$ atm
7. Check your answer by substituting the calculated equilibrium partial pressures into the equilibrium expression. The calculated value of K should match the given value of K.	$K_p = \frac{(P_{ICl})^2}{P_{I_2}P_{Cl_2}} = \frac{(0.246)^2}{(0.027)(0.027)} = 83$ Since the calculated value of K_p matches the given value (within the uncertainty indicated by the significant figures), the answer is valid.

Step 3 table:

	P_{I_2} (atm)	P_{Cl_2} (atm)	P_{ICl} (atm)
Initial	0.100	0.100	0.100
Change	$-x$	$-x$	$+2x$
Equil			

Step 4 table:

	P_{I_2} (atm)	P_{Cl_2} (atm)	P_{ICl} (atm)
Initial	0.100	0.100	0.100
Change	$-x$	$-x$	$+2x$
Equil	$0.100 - x$	$0.100 - x$	$0.100 + 2x$

FOR PRACTICE 16.11

The reaction between I_2 and Cl_2 in Example 16.11 is carried out at the same temperature, but with these initial partial pressures: $P_{I_2} = 0.150$ atm, $P_{Cl_2} = 0.150$ atm, $P_{ICl} = 0.00$ atm. Find the equilibrium partial pressures of all three substances.

Simplifying Approximations in Working Equilibrium Problems

For some equilibrium problems of the type shown in Examples 16.9, 16.10, and 16.11, we can make an approximation that simplifies the calculations without any significant loss of accuracy. For example, if the equilibrium constant is relatively small, the reaction will not proceed very far to the right. Therefore, if the initial reactant concentration is relatively large, we can make the assumption that x is small relative to the initial concentration of reactant. To see how this approximation works, consider again the simple reaction $A \rightleftharpoons 2\,B$. Suppose that, as before, we have a reaction mixture in which the initial concentration of A is 1.0 M and the initial concentration of B is 0.0 M and that we want to find the equilibrium concentrations. However, suppose that in this case the equilibrium constant is much smaller, say $K_c = 3.3 \times 10^{-5}$. The ICE table is identical to the one we set up previously:

	[A]	[B]
Initial	1.0	0.0
Change	$-x$	$+2x$
Equil	$1.0 - x$	$2x$

Except for the value of K_c, we end up with the exact quadratic equation that we had before:

$$K_c = \frac{[B]^2}{[A]} = \frac{(2x)^2}{1.0 - x} = 3.3 \times 10^{-5}$$

or more simply:

$$\frac{4x^2}{1.0 - x} = 3.3 \times 10^{-5}$$

We can rearrange this quadratic equation into standard form and solve it using the quadratic formula. But because K_c is small, the reaction will not proceed very far toward products and, therefore, x will also be small. If x is much smaller than 1.0, then $(1.0 - x)$ (the quantity in the denominator) can be approximated by (1.0):

$$\frac{4x^2}{(1.0 - \cancel{x})} = 3.3 \times 10^{-5}$$

This approximation greatly simplifies the equation, which we can then solve for x as follows:

$$\frac{4x^2}{1.0} = 3.3 \times 10^{-5}$$

$$4x^2 = 3.3 \times 10^{-5}$$

$$x = \sqrt{\frac{3.3 \times 10^{-5}}{4}} = 0.0029$$

We can check the validity of this approximation by comparing the calculated value of x to the number it was subtracted from. The ratio of x to the number it is subtracted from should be less than 0.05 (or 5%) for the approximation to be valid. In this case, x was subtracted from 1.0, and therefore the ratio of the value of x to 1.0 is calculated as follows:

$$\frac{0.0029}{1.0} \times 100\% = 0.29\%$$

The approximation is therefore valid. In Examples 16.12 and 16.13, we treat two nearly identical problems—the only difference is the initial concentration of the reactant. In Example 16.12, the initial concentration of the reactant is relatively large, the equilibrium constant is small, and the *x is small* approximation works well. In Example 16.13, however, the initial concentration of the reactant is much smaller, and even though the equilibrium constant is the same, the *x is small* approximation does not work (because the initial concentration is also small). In cases such as this, we have a couple of options to solve the problem. We can either solve the equation exactly (using the quadratic formula, for example) or use the *method of successive approximations*, which is introduced in Example 16.13. In this method, we essentially solve for x as if it were small, and then substitute the value obtained back into the equation (where x was initially neglected) to solve for x again. This can be repeated until the calculated value of x stops changing with each iteration, an indication that we have arrived at an acceptable value for x.

Note that the *x is small* approximation does not imply that *x is zero.* If that were the case, the reactant and product concentrations would not change from their initial values. The *x is small* approximation just means that when x is added or subtracted to another number, it does not change that number by very much. For example, we can calculate the value of the difference $1.0 - x$ when $x = 3.0 \times 10^{-4}$:

$$1.0 - x = 1.0 - 3.0 \times 10^{-4} = 0.9\underline{99}7 = 1.0$$

Since the value of 1.0 is known only to two significant figures, subtracting the small x does not change the value at all. This situation is similar to weighing yourself on a bathroom scale with and without a penny in your pocket. Unless your scale is unusually precise, removing the penny from your pocket does not change the reading on the scale. This does not imply that the penny is weightless, only that its weight is small when compared to your body weight. You can neglect the weight of the penny in reading your weight with no detectable loss in accuracy.

PROCEDURE FOR

Finding Equilibrium Concentrations from Initial Concentrations in Cases with a Small Equilibrium Constant

To solve these types of problems, follow the given procedure.

EXAMPLE 16.12

Finding Equilibrium Concentrations from Initial Concentrations in Cases with a Small Equilibrium Constant

Consider the reaction for the decomposition of hydrogen disulfide:

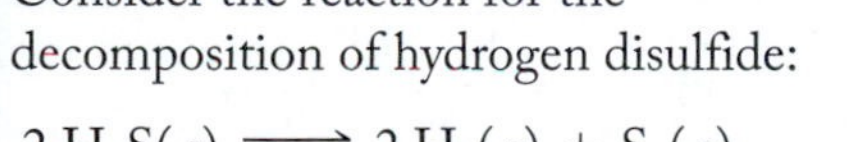

$$2\,H_2S(g) \rightleftharpoons 2\,H_2(g) + S_2(g) \qquad K_c = 1.67 \times 10^{-7} \text{ at } 800\ °C$$

A 0.500 L reaction vessel initially contains 0.0125 mol of H_2S at 800 °C. Find the equilibrium concentrations of H_2 and S_2.

EXAMPLE 16.13

Finding Equilibrium Concentrations from Initial Concentrations in Cases with a Small Equilibrium Constant

Consider the reaction for the decomposition of hydrogen disulfide:

$$2\,H_2S(g) \rightleftharpoons 2\,H_2(g) + S_2(g) \qquad K_c = 1.67 \times 10^{-7} \text{ at } 800\ °C$$

A 0.500 L reaction vessel initially contains 1.25×10^{-4} mol of H_2S at 800 °C. Find the equilibrium concentrations of H_2 and S_2.

1. Using the balanced equation as a guide, prepare a table showing the known initial concentrations of the reactants and products. (In these examples, you must first calculate the concentration of H_2S from the given number of moles and volume.)

Example 16.12:

$$[H_2S] = \frac{0.0125 \text{ mol}}{0.500 \text{ L}} = 0.0250 \text{ M}$$

$$2\,H_2S(g) \rightleftharpoons 2\,H_2(g) + S_2(g)$$

	$[H_2S]$	$[H_2]$	$[S_2]$
Initial	0.0250	0.00	0.00
Change			
Equil			

Example 16.13:

$$[H_2S] = \frac{1.25 \times 10^{-4} \text{ mol}}{0.500 \text{ L}} = 2.50 \times 10^{-4} \text{ M}$$

$$2\,H_2S(g) \rightleftharpoons 2\,H_2(g) + S_2(g)$$

	$[H_2S]$	$[S_2]$	$[S_2]$
Initial	2.50×10^{-4}	0.00	0.00
Change			
Equil			

—*Continued on the next page*

Continued from the previous page—

2. Use the initial concentrations to calculate the reaction quotient (Q). Compare Q to K to predict the direction in which the reaction will proceed.

By inspection, $Q_c = 0$; the reaction will proceed to the right.

By inspection, $Q_c = 0$; the reaction will proceed to the right.

3. Represent the change in the concentration of one of the reactants or products with the variable x. Define the changes in the concentrations of the other reactants or products with respect to x.

$$2\,H_2S(g) \rightleftharpoons 2H_2(g) + S_2(g)$$

	$[H_2S]$	$[H_2]$	$[S_2]$
Initial	0.0250	0.00	0.00
Change	$-2x$	$+2x$	$+x$
Equil			

$$2\,H_2S(g) \rightleftharpoons 2\,H_2(g) + S_2(g)$$

	$[H_2S]$	$[H_2]$	$[S_2]$
Initial	2.50×10^{-4}	0.00	0.00
Change	$-2x$	$+2x$	$+x$
Equil			

4. Sum each column for each reactant and product to determine the equilibrium concentrations in terms of the initial concentrations and the variable x.

$$2\,H_2S(g) \rightleftharpoons 2\,H_2(g) + S_2(g)$$

	$[H_2S]$	$[H_2]$	$[S_2]$
Initial	0.0250	0.00	0.00
Change	$-2x$	$+2x$	$+x$
Equil	$0.0250 - 2x$	$2x$	x

$$2\,H_2S(g) \rightleftharpoons 2\,H_2(g) + S_2(g)$$

	$[H_2S]$	$[H_2]$	$[S_2]$
Initial	2.50×10^{-4}	0.00	0.00
Change	$-2x$	$+2x$	$+x$
Equil	$2.50 \times 10^{-4} - 2x$	$2x$	x

5. Substitute the expressions for the equilibrium concentrations (from Step 4) into the expression for the equilibrium constant. Use the given value of the equilibrium constant to solve the resulting equation for the variable x. In this case, the resulting equation is cubic in x. Although cubic equations can be solved, the solutions are not usually simple. However, since the equilibrium constant is small, you know that the reaction does not proceed very far to the right. Therefore, x will be a small number and can be dropped from any quantities in which it is added to or subtracted from another number (as long as the number itself is not too small).

$$K_c = \frac{[H_2]^2[S_2]}{[H_2S]^2} = \frac{(2x)^2x}{(0.0250-2x)^2}$$

$$1.67 \times 10^{-7} = \frac{4x^3}{(0.0250 - 2x)^2}$$

x is small.

$$1.67 \times 10^{-7} = \frac{4x^3}{(0.0250 - \cancel{2x})^2}$$

$$1.67 \times 10^{-7} = \frac{4x^3}{6.25 \times 10^{-4}}$$

$$6.25 \times 10^{-4}(1.67 \times 10^{-7}) = 4x^3$$

$$x^3 = \frac{6.25 \times 10^{-4}(1.67 \times 10^{-7})}{4}$$

$$x = 2.97 \times 10^{-4}$$

$$K_c = \frac{[H_2]^2[S_2]}{[H_2S]^2} = \frac{(2x)^2x}{(2.50 \times 10^{-4} - 2x)^2}$$

$$1.67 \times 10^{-7} = \frac{4x^3}{(2.50 \times 10^{-4}-2x)^2}$$

x is small.

$$1.67 \times 10^{-7} = \frac{4x^3}{(2.50 \times 10^{-4} - \cancel{2x})^2}$$

$$1.67 \times 10^{-7} = \frac{4x^3}{6.25 \times 10^{-8}}$$

$$6.25 \times 10^{-8}(1.67 \times 10^{-7}) = 4x^3$$

$$x^3 = \frac{6.25 \times 10^{-8}(1.67 \times 10^{-7})}{4}$$

$$x = 1.38 \times 10^{-5}$$

Check whether your approximation was valid by comparing the calculated value of x to the number it was added to or subtracted from. The ratio of the two numbers should be less than 0.05 (or 5%) for the approximation to be valid. If approximation is not valid, proceed to Step 5a.

Checking the *x is small* approximation:

$$\frac{2.97 \times 10^{-4}}{0.0250} \times 100\% = 1.19\%$$

The *x is small* approximation is valid. Proceed to Step 6.

Checking the *x is small* approximation:

$$\frac{1.38 \times 10^{-5}}{2.50 \times 10^{-4}} \times 100\% = 5.52\%$$

The approximation does not satisfy the <5% rule (although it is close).

—Continued on the next page

Continued from the previous page—

	Example 16.12	Example 16.13
5a. If the approximation is not valid, you can either solve the equation exactly (by hand or with your calculator), or use the method of successive approximations. In Example 16.13, use the method of successive approximations.		$1.67 \times 10^{-7} = \frac{4x^3}{(2.50 \times 10^{-4} - 2x)^2}$ $x = 1.38 \times 10^{-5}$ $1.67 \times 10^{-7} = \frac{4x^3}{(2.50 \times 10^{-4} - 2.76 \times 10^{-5})^2}$ $x = 1.27 \times 10^{-5}$
Substitute the value obtained for x in Step 5 back into the original cubic equation, but only at the exact spot where x was assumed to be negligible, and then solve the equation for x again. Continue this procedure until the value of x obtained from solving the equation is the same as the one that is substituted into the equation.		If you substitute this value of x back into the cubic equation and solve it, you get $x = 1.28 \times 10^{-5}$, which is nearly identical to 1.27×10^{-5}. Therefore, you have arrived at the best approximation for x.
6. Substitute x into the expressions for the equilibrium concentrations of the reactants and products (from Step 4) and calculate the concentrations.	$[H_2S] = 0.0250 - 2(2.97 \times 10^{-4}) = 0.0244$ M $[H_2] = 2(2.97 \times 10^{-4}) = 5.94 \times 10^{-4}$ M $[S_2] = 2.97 \times 10^{-4}$ M	$[H_2S] = 2.50 \times 10^{-4} - 2(1.28 \times 10^{-5}) = 2.24 \times 10^{-4}$ M $[H_2] = 2(1.28 \times 10^{-5}) = 2.56 \times 10^{-5}$ M $[S_2] = 1.28 \times 10^{-5}$ M
7. Check your answer by substituting the calculated equilibrium values into the equilibrium expression. The calculated value of K should match the given value of K. Note that the approximation method and rounding errors could cause a difference of up to about 5% when comparing values of the equilibrium constant.	$K_c = \frac{(5.94 \times 10^{-4})^2(2.97 \times 10^{-4})}{(0.0244)^2} = 1.76 \times 10^{-7}$ The calculated value of K is close enough to the given value when you consider the uncertainty introduced by the approximation. Therefore the answer is valid.	$K_c = \frac{(2.56 \times 10^{-5})^2(1.28 \times 10^{-5})}{(2.24 \times 10^{-4})^2} = 1.67 \times 10^{-7}$ The calculated value of K is equal to the given value. Therefore the answer is valid.
	FOR PRACTICE 16.12 The reaction in Example 16.12 is carried out at the same temperature with the following initial concentrations: $[H_2S] = 0.100$ M, $[H_2] = 0.100$ M, and $[S_2] = 0.000$ M. Find the equilibrium concentration of $[S_2]$.	**FOR PRACTICE 16.13** The reaction in Example 16.13 is carried out at the same temperature with the following initial concentrations: $[H_2S] = 1.00 \times 10^{-4}$ M, $[H_2] = 0.00$ M, and $[S_2] = 0.000$ M. Find the equilibrium concentration of $[S_2]$.

The *x is small* Approximation

16.6 Cc Conceptual Connection

For the generic reaction, $A(g) \rightleftharpoons B(g)$, consider each value of K and initial concentration of A. For which set will the *x is small* approximation most likely apply?

(a) $K = 1.0 \times 10^{-5}$; [A] = 0.250 M

(b) $K = 1.0 \times 10^{-2}$; [A] = 0.250 M

(c) $K = 1.0 \times 10^{-5}$; [A] = 0.00250 M

(d) $K = 1.0 \times 10^{-2}$; [A] = 0.00250 M

KCV
KEY CONCEPT VIDEO
Le Châtelier's Principle

16.9 Le Châtelier's Principle: How a System at Equilibrium Responds to Disturbances

Le Châtelier is pronounced "Le-sha-te-lyay."

We have seen that a chemical system not in equilibrium tends to progress toward equilibrium and that the relative concentrations of the reactants and products at equilibrium are characterized by the equilibrium constant, *K*. What happens, however, when a chemical system already at equilibrium is disturbed? **Le Châtelier's principle** states that the chemical system responds to minimize the disturbance.

> **Le Châtelier's principle: When a chemical system at equilibrium is disturbed, the system shifts in a direction that minimizes the disturbance.**

In other words, a system at equilibrium tends to maintain that equilibrium—it bounces back when disturbed.

The Effect of a Concentration Change on Equilibrium

Consider the following reaction in chemical equilibrium:

$$N_2O_4(g) \rightleftharpoons 2\,NO_2(g)$$

Suppose we disturb the equilibrium by adding NO_2 to the equilibrium mixture (Figure 16.7 ▼). In other words, we increase the concentration of NO_2, the product. What happens? According to Le Châtelier's principle, the system shifts in a direction to minimize the disturbance. The reaction goes to the left (it proceeds in the reverse direction), consuming some of the added NO_2 and thus bringing its concentration back down, as shown graphically in Figure 16.8(a) ▶.

Add NO_2.

$$N_2O_4(g) \rightleftharpoons 2\,NO_2(g)$$

Reaction shifts left.

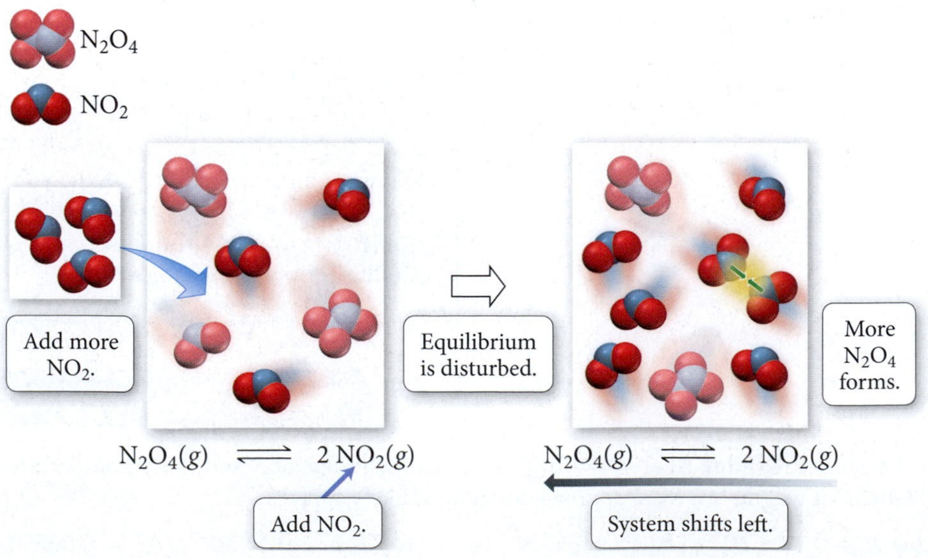
Le Châtelier's Principle: Changing Concentration
N2O4
NO2
Add more NO2.
Equilibrium is disturbed.
More N2O4 forms.
N2O4(g) ⇌ 2 NO2(g)
Add NO2.
N2O4(g) ⇌ 2 NO2(g)
System shifts left.

▶ **FIGURE 16.7 Le Châtelier's Principle: The Effect of a Concentration Change** Adding NO_2 causes the reaction to shift left, consuming some of the added NO_2 and forming more N_2O_4.

The reaction shifts to the left because the value of Q increases when we add a product to the reaction mixture. For example, suppose we double the conentration of NO_2.

- Before doubling the concentration of NO_2: $Q_1 = K = \dfrac{[NO_2]^2}{[N_2O_4]}$
- Immediately after doubling the concentration of NO_2:

$$Q_2 = \frac{(2[NO_2])^2}{[N_2O_4]} = 4 \times \frac{[NO_2]^2}{[N_2O_4]} = 4 \times Q_1 > K.$$

- Since $Q > K$, the reaction shifts to left to reestablish equilibrium.

What happens, however, if we add extra N_2O_4 (the reactant), increasing its concentration? In this case, the reaction shifts to the right, consuming some of the added N_2O_4 and bringing *its* concentration back down, as shown graphically in Figure 16.8(b) ▼.

Add N_2O_4.

$$N_2O_4(g) \rightleftharpoons 2\,NO_2(g)$$

Reaction shifts right.

The reaction shifts to the right in this case because the value of Q changes. For example, suppose we double the concentration of N_2O_4.

- Before doubling the concentration of N_2O_4: $Q_1 = K = \dfrac{[NO_2]^2}{[N_2O_4]}$
- Immediately after doubling the concentration of N_2O_4:

$$Q_2 = \frac{[NO_2]^2}{2[N_2O_4]} = \frac{1}{2} \times \frac{[NO_2]^2}{[N_2O_4]} = \frac{1}{2} \times Q_1 < K.$$

- Since $Q < K$, the reaction shifts to the right to reestablish equilibrium.

In both of these cases, the system shifts in a direction that minimizes the disturbance. Lowering the concentration of a reactant (which makes $Q > K$) causes the system to shift in the direction of the reactants to minimize the disturbance. Lowering the concentration of a product (which makes $Q < K$) causes the system to shift in the direction of products.

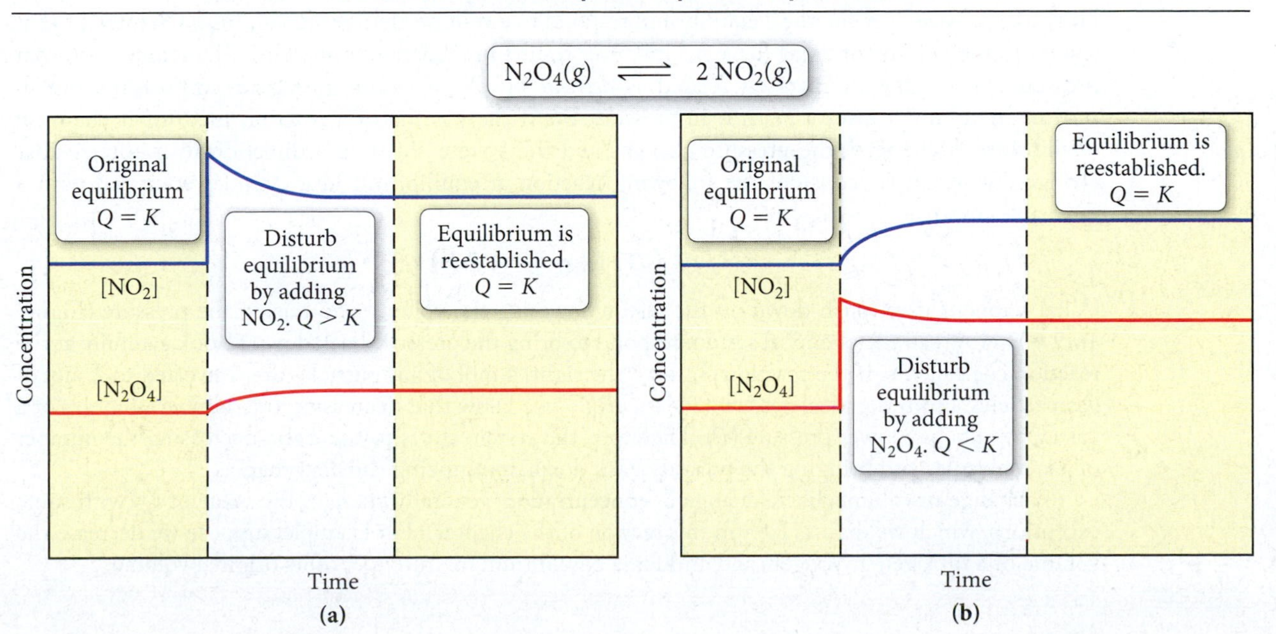

◀ **FIGURE 16.8** **Le Châtelier's Principle: Changing Concentration** These two graphs each show the concentrations of NO_2 and N_2O_4 for the reaction $N_2O_4(g) \longrightarrow 2NO_2(g)$ as a function of time in three distinct stages of the reaction: initially at equilibrium (left), upon disturbance of the equilibrium by addition of more NO_2 **(a)** or N_2O_4 **(b)** to the reaction mixture (center), and upon reestablishment of equilibrium (right).

Summarizing the Effect of a Concentration Change on Equilibrium:
If a chemical system is at equilibrium:

- *Increasing* the concentration of one or more of the *reactants* (which makes $Q < K$) causes the reaction to *shift to the right* (in the direction of the products).
- *Increasing* the concentration of one or more of the *products* (which makes $Q > K$) causes the reaction to *shift to the left* (in the direction of the reactants).
- *Decreasing* the concentration of one or more of the *reactants* (which makes $Q > K$) causes the reaction to *shift to the left* (in the direction of the reactants).
- *Decreasing* the concentration of one or more of the *products* (which makes $Q < K$) causes the reaction to *shift to the right* (in the direction of the products).

EXAMPLE 16.14
The Effect of a Concentration Change on Equilibrium

Consider the following reaction at equilibrium:

$$CaCO_3(s) \rightleftharpoons CaO(s) + CO_2(g)$$

What is the effect of adding CO_2 to the reaction mixture? What is the effect of adding $CaCO_3$?

SOLUTION

Adding CO_2 increases the concentration of CO_2 and causes the reaction to shift to the left. Adding additional $CaCO_3$, however, does *not* increase the concentration of $CaCO_3$ because $CaCO_3$ is a solid and therefore has a constant concentration. Thus, adding $CaCO_3$ has no effect on the position of the equilibrium. (Note that, as we saw in Section 16.5, solids are not included in the equilibrium expression.)

FOR PRACTICE 16.14

Consider the following reaction in chemical equilibrium:

$$2\,BrNO(g) \rightleftharpoons 2\,NO(g) + Br_2(g)$$

What is the effect of adding Br_2 to the reaction mixture? What is the effect of adding BrNO?

The Effect of a Volume (or Pressure) Change on Equilibrium

In considering the effect of a change in volume, we are assuming that the change in volume is carried out at constant temperature.

How does a system in chemical equilibrium respond to a volume change? Recall from Chapter 11 that changing the volume of a gas (or a gas mixture) results in a change in pressure. Remember also that pressure and volume are inversely related: A *decrease* in volume causes an *increase* in pressure, and an *increase* in volume causes a *decrease* in pressure. So, if the volume of a reaction mixture at chemical equilibrium is changed, the pressure changes and the system shifts in a direction to minimize that change. For example, consider the following reaction at equilibrium in a cylinder equipped with a moveable piston:

$$N_2(g) + 3\,H_2(g) \rightleftharpoons 2\,NH_3(g)$$

What happens if we push down on the piston, lowering the volume and raising the pressure (Figure 16.9 ▶)? How can the chemical system respond to bring the pressure back down? Look carefully at the reaction coefficients. If the reaction shifts to the right, 4 mol of gas particles are converted to 2 mol of gas particles. From the ideal gas law ($PV = nRT$), we know that decreasing the number of moles of a gas (n) results in a lower pressure (P). Therefore, the system shifts to the right, decreasing the number of gas molecules and bringing the pressure back down, minimizing the disturbance.

A change in volume, like a change in concentration, generally changes the value of Q (with some exceptions, which we discuss later in this section of the chapter). For example, suppose we decrease the volume of a nitrogen, hydrogen, and ammonia equilibrium mixture to $\frac{1}{2}$ of its original volume.

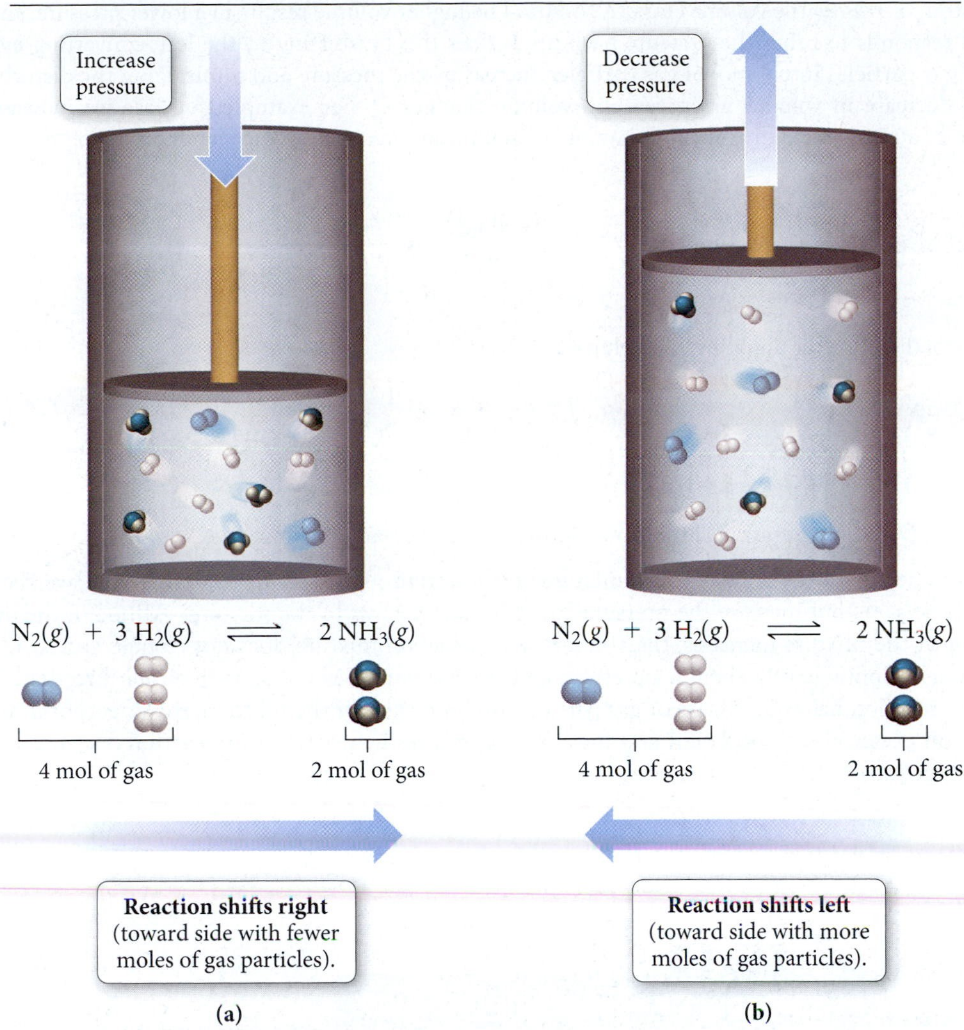

▲ **FIGURE 16.9 Le Châtelier's Principle: The Effect of a Pressure Change** **(a)** Decreasing the volume increases the pressure, causing the reaction to shift to the right (fewer moles of gas, lower pressure). **(b)** Increasing the volume reduces the pressure, causing the reaction to shift to the left (more moles of gas, higher pressure).

- Before halving the volume: $Q_1 = K = \dfrac{[NH_3]^2}{[N_2][H_2]^3} = \dfrac{\left(\dfrac{n_{NH_3}}{V}\right)^2}{\left(\dfrac{n_{N_2}}{V}\right)\left(\dfrac{n_{H_2}}{V}\right)^3}$

We can express the concentration of any substance as the number of moles of the substance divided by the volume: $[A] = n_a/V$.

- Immediately after halving the volume:

$$Q_2 = \frac{\left(\dfrac{n_{NH_3}}{\frac{1}{2}V}\right)^2}{\left(\dfrac{n_{N_2}}{\frac{1}{2}V}\right)\left(\dfrac{n_{H_2}}{\frac{1}{2}V}\right)^3} = \frac{4\left(\dfrac{n_{NH_3}}{V}\right)^2}{2\left(\dfrac{n_{N_2}}{V}\right)8\left(\dfrac{n_{H_2}}{V}\right)^3} = \frac{1}{4} \times \frac{\left(\dfrac{n_{NH_3}}{V}\right)^2}{\left(\dfrac{n_{N_2}}{V}\right)\left(\dfrac{n_{H_2}}{V}\right)^3} = \frac{1}{4} \times Q_1 < K.$$

- Since $Q < K$, the reaction shifts to the right—toward the direction of fewer gas moles—to reestablish equilibrium.

Consider the same reaction mixture at equilibrium again. What happens if, this time, we pull *up* on the piston, *increasing* the volume (Figure 16.9b)? The higher volume results in a lower pressure, and the system responds to bring the pressure back up. It does this by shifting to the left, converting every 2 mol of gas particles into 4 mol of gas particles, increasing the pressure and minimizing the disturbance. Like a decrease in volume, an increase in volume changes Q. For example, suppose we increase the volume of a nitrogen, hydrogen, and ammonia equilibrium mixture to twice its original volume.

- Before doubling the volume: $Q_1 = K = \frac{[NH_3]^2}{[N_2][H_2]^3} = \frac{\left(\frac{n_{NH_3}}{V}\right)^2}{\left(\frac{n_{N_2}}{V}\right)\left(\frac{n_{H_2}}{V}\right)^3}$
- Immediately after doubling the volume:

$$Q_2 = \frac{\left(\frac{n_{NH_3}}{2V}\right)^2}{\left(\frac{n_{N_2}}{2V}\right)\left(\frac{n_{H_2}}{2V}\right)^3} = \frac{\frac{1}{4}\left(\frac{n_{NH_3}}{V}\right)^2}{\frac{1}{2}\left(\frac{n_{N_2}}{V}\right)\frac{1}{8}\left(\frac{n_{H_2}}{V}\right)^3} = 4 \times \frac{\left(\frac{n_{NH_3}}{V}\right)^2}{\left(\frac{n_{N_2}}{V}\right)\left(\frac{n_{H_2}}{V}\right)^3} = 4 \times Q_1 > K.$$

- Since $Q > K$, the reaction shifts to the left to reestablish equilibrium.

Consider again the same reaction mixture at equilibrium. What happens if, this time, we keep the volume the same but increase the pressure *by adding an inert gas* to the mixture? Although the overall pressure of the mixture increases, the volume of the reaction mixture does not change and so Q does not change. Consequently, there is no effect, and the reaction does not shift in either direction. Similarly, if a reaction has equal moles of gas particles on both sides of the reaction, the effects of a volume change on Q cancel each other out and there is no effect on the reaction (the reaction does not shift in either direction).

EXAMPLE 16.15
The Effect of a Volume Change on Equilibrium

Consider the following reaction at chemical equilibrium:

$$2\ KClO_3(s) \rightleftharpoons 2\ KCl(s) + 3\ O_2(g)$$

What is the effect of decreasing the volume of the reaction mixture? Increasing the volume of the reaction mixture? Adding an inert gas at constant volume?

SOLUTION

The chemical equation has 3 mol of gas on the right and zero mol of gas on the left. Decreasing the volume of the reaction mixture increases the pressure and causes the reaction to shift to the left (toward the side with fewer moles of gas particles). Increasing the volume of the reaction mixture decreases the pressure and causes the reaction to shift to the right (toward the side with more moles of gas particles.) Adding an inert gas has no effect.

FOR PRACTICE 16.15

Consider the following reaction at chemical equilibrium:

$$2\ SO_2(g) + O_2(g) \rightleftharpoons 2\ SO_3(g)$$

What is the effect of decreasing the volume of the reaction mixture? Increasing the volume of the reaction mixture?

Summarizing the Effect of Volume Change on Equilibrium:
If a chemical system is at equilibrium:

- *Decreasing* the volume causes the reaction to shift in the direction that has *the fewer moles of gas particles.*
- *Increasing* the volume causes the reaction to shift in the direction that has *the greater number of moles of gas particles.*
- Adding an inert gas to the mixture at a fixed volume has no effect on the equilibrium.
- When a reaction has an equal number of moles of gas on both sides of the chemical equation, a change in volume produces no effect on the equilibrium.

The Effect of a Temperature Change on Equilibrium

When a system at equilibrium is disturbed by a change in concentration or a change in volume, the equilibrium shifts to counter the change, but *the equilibrium constant does not change.* In other words, changes in volume or concentration generally change Q, not K, and the system responds by shifting so that Q becomes equal to K. However, a change in temperature changes the actual value of the equilibrium constant. Nonetheless, we can use Le Châtelier's principle to predict the effects of a temperature change. If we increase the temperature of a reaction mixture at equilibrium, the reaction shifts in the direction that tends to decrease the temperature and vice versa. Recall from Chapter 10 that an exothermic reaction (negative ΔH) emits heat:

$$\text{Exothermic reaction: } A + B \rightleftharpoons C + D + \text{heat}$$

We can think of heat as a product in an exothermic reaction. In an endothermic reaction (positive ΔH), the reaction absorbs heat.

$$\text{Endothermic reaction: } A + B + \text{heat} \rightleftharpoons C + D$$

We can think of heat as a reactant in an endothermic reaction.

At constant pressure, raising the temperature of an *exothermic* reaction—think of this as adding heat—is similar to adding more product, causing the reaction to shift left. For example, the reaction of nitrogen with hydrogen to form ammonia is exothermic:

Add heat

$$N_2(g) + 3\,H_2(g) \rightleftharpoons 2\,NH_3(g) + \text{heat}$$

Reaction shifts left.
Smaller K

Raising the temperature of an equilibrium mixture of these three gases causes the reaction to shift left, absorbing some of the added heat and forming less products and more reactants. Note that, unlike adding NH_3 to the reaction mixture (which does *not* change the value of the equilibrium constant), *changing the temperature does change the value of the equilibrium constant.* The new equilibrium mixture will have more reactants and fewer products and therefore a smaller value of K.

Conversely, lowering the temperature causes the reaction to shift right, releasing heat and producing more products because the value of K has increased:

Remove heat

$$N_2(g) + 3\,H_2(g) \rightleftharpoons 2NH_3(g) + \text{heat}$$

Reaction shifts right.
Larger K

In contrast, for an *endothermic* reaction, raising the temperature (adding heat) causes the reaction to shift right to absorb the added heat. For example, the following reaction is endothermic:

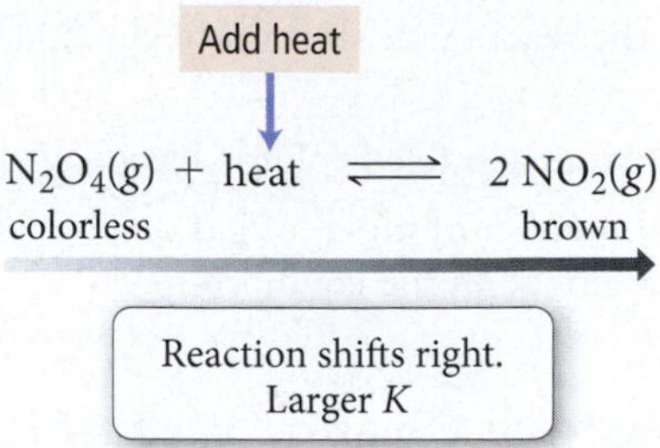

Raising the temperature of an equilibrium mixture of these two gases causes the reaction to shift right, absorbing some of the added heat and producing more products because the value of *K* has increased. Since N_2O_4 is colorless and NO_2 is brown, the effects of changing the temperature of this reaction are easily seen (Figure 16.10 ▶). In contrast, lowering the temperature (removing heat) of a reaction mixture of these two gases causes the reaction to shift left, releasing heat, forming less products, and lowering the value of *K*:

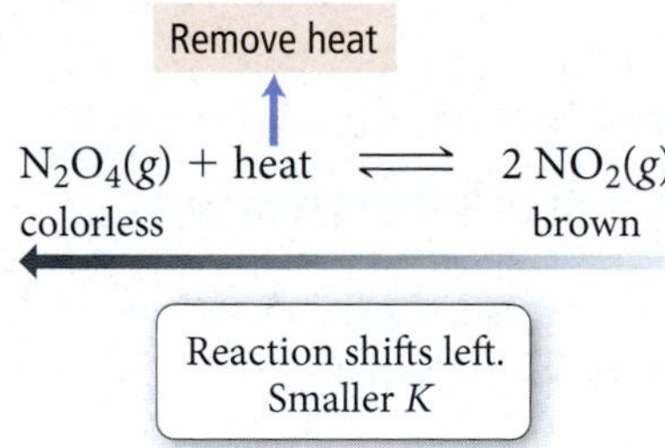

We can quantify the relationship between the value of the equilibrium constant and the temperature using an equation derived from thermodynamic considerations. We will cover this equation in Section 19.9.

Summarizing the Effect of a Temperature Change on Equilibrium:

In an *exothermic* chemical reaction, heat is a product.

- *Increasing* the temperature causes an exothermic reaction to *shift left* (in the direction of the reactants); the value of the equilibrium constant decreases.
- *Decreasing* the temperature causes an exothermic reaction to *shift right* (in the direction of the products); the value of the equilibrium constant increases.

In an *endothermic* chemical reaction, heat is a reactant.

- *Increasing* the temperature causes an endothermic reaction to *shift right* (in the direction of the products); the equilibrium constant increases.
- *Decreasing* the temperature causes an endothermic reaction to *shift left* (in the direction of the reactants); the equilibrium constant decreases.

Adding heat favors the endothermic direction. Removing heat favors the exothermic direction.

Le Châtelier's Principle: Changing Temperature

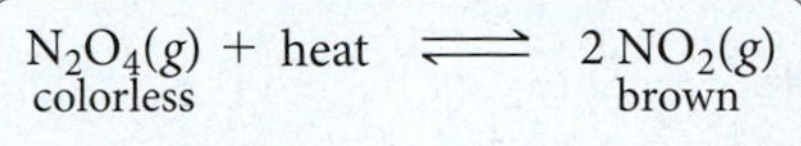

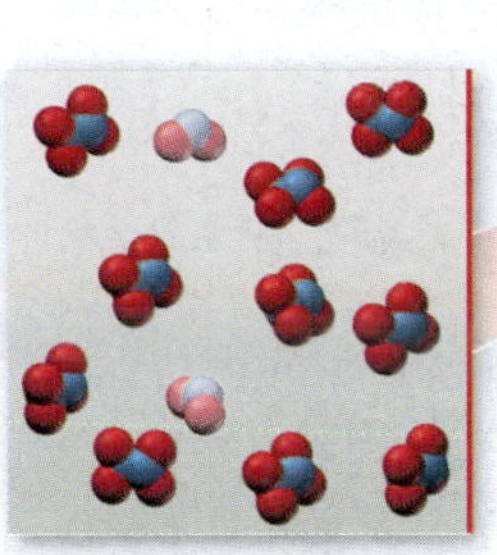

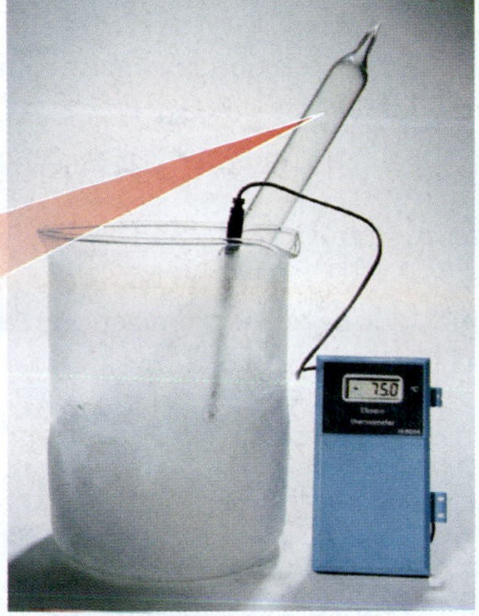

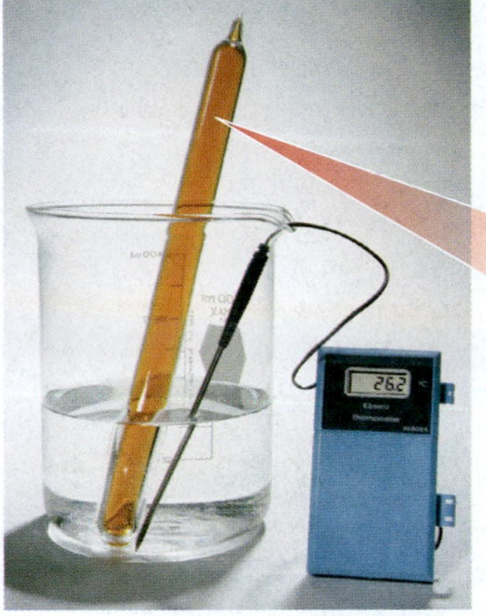

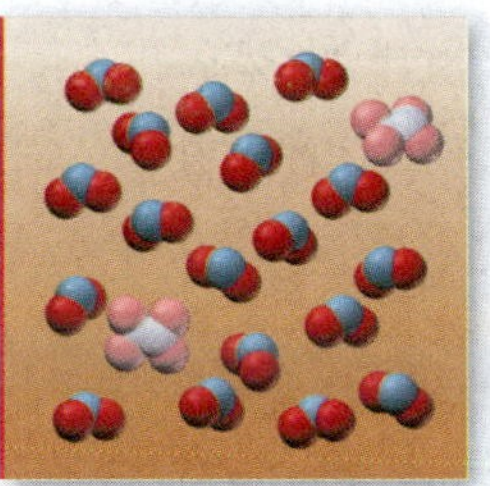

▲ **FIGURE 16.10 Le Châtelier's Principle: The Effect of a Temperature Change** Because the reaction is endothermic, raising the temperature causes a shift to the right, toward the formation of brown NO_2.

EXAMPLE 16.16

The Effect of a Temperature Change on Equilibrium

The following reaction is endothermic:

$$CaCO_3(s) \rightleftharpoons CaO(s) + CO_2(g)$$

What is the effect of increasing the temperature of the reaction mixture? Decreasing the temperature?

SOLUTION

Since the reaction is endothermic, we can think of heat as a reactant:

$$\text{heat} + CaCO_3(s) \rightleftharpoons CaO(s) + CO_2(g)$$

Raising the temperature is equivalent to adding a reactant, causing the reaction to shift to the right. Lowering the temperature is equivalent to removing a reactant, causing the reaction to shift to the left.

FOR PRACTICE 16.16

The following reaction is exothermic:

$$2\,SO_2(g) + O_2(g) \rightleftharpoons 2\,SO_3(g)$$

What is the effect of increasing the temperature of the reaction mixture? Decreasing the temperature?

SELF-ASSESSMENT Quiz

1. What is the correct expression for the equilibrium constant (K_c) for the reaction between carbon and hydrogen gas to form methane shown here?

$$C(s) + 2\,H_2(g) \rightleftharpoons CH_4(g)$$

a) $K_c = \frac{[CH_4]}{[H_2]}$ **b)** $K_c = \frac{[CH_4]}{[C][H_2]}$

c) $K_c = \frac{[CH_4]}{[C][H_2]^2}$ **d)** $K_c = \frac{[CH_4]}{[H_2]^2}$

2. The equilibrium constant for the reaction shown here is $K_c = 1.0 \times 10^3$. A reaction mixture at equilibrium contains $[A] = 1.0 \times 10^{-3}$ M. What is the concentration of B in the mixture?

$$A(g) \rightleftharpoons B(g)$$

a) 1.0×10^{-3} M **b)** 1.0 M

c) 2.0 M **d)** 1.0×10^3 M

3. Use the data below to find the equilibrium constant (K_c) for the reaction $A(g) \rightleftharpoons 2\,B(g) + C(g)$.

$$A(g) \rightleftharpoons 2\,X(g) + C(g) \quad K_c = 1.55$$
$$B(g) \rightleftharpoons X(g) \quad K_c = 25.2$$

a) 984 **b)** 26.8

c) 6.10×10^{-4} **d)** 2.44×10^{-3}

4. The reaction shown here has a $K_p = 4.5 \times 10^2$ at 825 K. Find K_c for the reaction at this temperature.

$$CH_4(g) + CO_2(g) \rightleftharpoons 2\,CO(g) + 2\,H_2(g)$$

a) 0.098 **b)** 2.1×10^6

c) 6.6 **d)** 4.5×10^{-2}

5. Consider the reaction between NO and Cl_2 to form NOCl.

$$2\,NO(g) + Cl_2(g) \rightleftharpoons 2\,NOCl(g)$$

A reaction mixture at a certain temperature initially contains only $[NO] = 0.50$ M and $[Cl_2] = 0.50$ M. After the reaction comes to equilibrium, the concentration of NOCl is 0.30 M. Find the value of the equilibrium constant (K_c) at this temperature.

a) 11 **b)** 4.3 **c)** 6.4 **d)** 0.22

6. For the reaction $2\,A(g) \rightleftharpoons B(g)$, the equilibrium constant is $K_p = 0.76$. A reaction mixture initially contains 2.0 atm of each gas ($P_A = 2.0$ atm and $P_B = 2.0$ atm). Which statement is true of the reaction mixture?

a) The reaction mixture is at equilibrium.

b) The reaction mixture will proceed toward products.

c) The reaction mixture will proceed toward reactants.

d) It is not possible to determine from the information given the future direction of the reaction mixture.

7. Consider the reaction between iodine gas and chlorine gas to form iodine monochloride.

$$I_2(g) + Cl_2(g) \rightleftharpoons 2\,ICl(g) \quad K_p = 81.9 \text{ (at 298 K)}$$

A reaction mixture at 298 K initially contains $P_{I_2} = 0.25$ atm and $P_{Cl_2} = 0.25$ atm. What is the partial pressure of iodine monochloride when the reaction reaches equilibrium?

a) 0.17 atm **b)** 0.64 atm **c)** 0.41 atm **d)** 2.3 atm

8. Consider the reaction of A to form B.

$$2\,A(g) \rightleftharpoons B(g) \quad K_c = 1.8 \times 10^{-5} \text{ (at 298 K)}$$

A reaction mixture at 298 K initially contains $[A] = 0.50$ M. What is the concentration of B when the reaction reaches equilibrium?

a) 9.0×10^{-6} M **b)** 0.060 M

c) 0.030 M **d)** 4.5×10^{-6} M

9. The decomposition of NH_4HS is endothermic:

$$NH_4HS(s) \rightleftharpoons NH_3(g) + H_2S(g)$$

Which change to an equilibrium mixture of this reaction results in the formation of more H_2S?

a) a decrease in the volume of the reaction vessel (at constant temperature)

b) an increase in the amount of NH_4HS in the reaction vessel

c) an increase in temperature

d) all of the above

10. The solid XY decomposes into gaseous X and Y:

$$XY(s) \rightleftharpoons X(g) + Y(g) \quad K_p = 4.1 \text{ (at 0 °C)}$$

If the reaction is carried out in a 22.4 L container, which initial amounts of X and Y result in the formation of solid XY?

a) 5 mol X; 0.5 mol Y **b)** 2.0 mol X; 2.0 mol Y

c) 1 mol X; 1 mol Y **d)** none of the above

11. What is the effect of adding helium gas (at constant volume) to an equilibrium mixture of this reaction:

$$CO(g) + Cl_2(g) \rightleftharpoons COCl_2(g)$$

a) The reaction shifts toward the products.

b) The reaction shifts toward the reactants.

c) The reaction does not shift in either direction.

d) The reaction slows down.

12. The reaction $X_2(g) \rightleftharpoons 2\,X(g)$ occurs in a closed reaction vessel at constant volume and temperature. Initially, the vessel contains only X_2 at a pressure of 1.55 atm. After the reaction reaches equilibrium, the total pressure is 2.85 atm. What is the value of the equilibrium constant, K_p, for the reaction?

a) 27 **b)** 10 **c)** 5.2 **d)** 32

Answers: 1:d; 2:b; 3:d; 4:a; 5:c; 6:b; 7:c; 8:d; 9:c; 10:d; 11:c; 12:a

CHAPTER SUMMARY 16

REVIEW

KEY LEARNING OUTCOMES

CHAPTER OBJECTIVES	ASSESSMENT
Expressing Equilibrium Constants for Chemical Equations (16.3)	• Example 16.1 For Practice 16.1 Exercises 21, 22
Manipulating the Equilibrium Constant to Reflect Changes in the Chemical Equation (16.3)	• Example 16.2 For Practice 16.2 For More Practice 16.2 Exercises 27–30
Relating K_p and K_c (16.4)	• Example 16.3 For Practice 16.3 Exercises 31, 32
Writing Equilibrium Expressions for Reactions Involving a Solid or a Liquid (16.5)	• Example 16.4 For Practice 16.4 Exercises 33, 34
Finding Equilibrium Constants from Experimental Concentration Measurements (16.6)	• Examples 16.5, 16.6 For Practice 16.5, 16.6 Exercises 35, 36, 43, 44
Predicting the Direction of a Reaction by Comparing Q and K (16.7)	• Example 16.7 For Practice 16.7 Exercises 47–50
Calculating Equilibrium Concentrations from the Equilibrium Constant and One or More Equilibrium Concentrations (16.8)	• Example 16.8 For Practice 16.8 Exercises 37–46
Finding Equilibrium Concentrations from Initial Concentrations and the Equilibrium Constant (16.8)	• Examples 16.9, 16.10 For Practice 16.9, 16.10 Exercises 53–58
Calculating Equilibrium Partial Pressures from the Equilibrium Constant and Initial Partial Pressures (16.8)	• Example 16.11 For Practice 16.11 Exercises 59, 60
Finding Equilibrium Concentrations from Initial Concentrations in Cases with a Small Equilibrium Constant (16.8)	• Examples 16.12, 16.13 For Practice 16.12, 16.13 Exercises 61, 62
Determining the Effect of a Concentration Change on Equilibrium (16.9)	• Example 16.14 For Practice 16.14 Exercises 63–66
Determining the Effect of a Volume Change on Equilibrium (16.9)	• Example 16.15 For Practice 16.15 Exercises 67, 68
Determining the Effect of a Temperature Change on Equilibrium (16.9)	• Example 16.16 For Practice 16.16 Exercises 69, 70

KEY TERMS

Section 16.2
reversible (611)
dynamic equilibrium (611)

Section 16.3
equilibrium constant (K) (613)
law of mass action (613)

Section 16.7
reaction quotient (Q) (624)

Section 16.9
Le Châtelier's principle (636)

KEY CONCEPTS

The Equilibrium Constant (16.1)

- The equilibrium constant, K expresses the relative concentrations of the reactants and the products at equilibrium..
- The equilibrium constant measures how far a reaction proceeds toward products: a large K (much greater than 1) indicates a high concentration of products at equilibrium, and a small K (much less than 1) indicates a low concentration of products at equilibrium.

Dynamic Equilibrium (16.2)

- Most chemical reactions are reversible; they can proceed in either the forward or the reverse direction.
- When a chemical reaction is in dynamic equilibrium, the rate of the forward reaction equals the rate of the reverse reaction, so the net concentrations of reactants and products do not change. However, this does *not* imply that the concentrations of the reactants and the products are equal at equilibrium.

The Equilibrium Constant Expression (16.3)

- The equilibrium constant expression is given by the law of mass action and is equal to the concentrations of the products, raised to their stoichiometric coefficients, divided by the concentrations of the reactants, raised to their stoichiometric coefficients.
- When the equation for a chemical reaction is reversed, multiplied, or added to another equation, K must be modified accordingly.

The Equilibrium Constant, K (16.4)

- The equilibrium constant can be expressed in terms of concentrations (K_c) or in terms of partial pressures (K_p). These two constants are related by Equation 16.2. Concentration must always be expressed in units of molarity for K_c. Partial pressures must always be expressed in units of atmospheres for K_p.

States of Matter and the Equilibrium Constant (16.5)

- The equilibrium constant expression contains only partial pressures or concentrations of reactants and products that exist as gases or solutes dissolved in solution. Pure liquids and solids are not included in the expression for the equilibrium constant.

Calculating K (16.6)

- We can calculate the equilibrium constant from equilibrium concentrations or partial pressures by substituting measured values into the expression for the equilibrium constant (as obtained from the law of mass action).
- In most cases, we can calculate the equilibrium concentrations of the reactants and products—and therefore the value of the equilibrium constant—from the initial concentrations of the reactants and products and the equilibrium concentration of *just one* reactant or product.

The Reaction Quotient, Q (16.7)

- The reaction quotient, Q, is the ratio of the concentrations (or partial pressures) of products raised to their stoichiometric coefficients to the concentrations of reactants raised to their stoichiometric coefficients *at any point in the reaction.*
- Like K, Q can be expressed in terms of concentrations (Q_c) or partial pressures (Q_p).
- At equilibrium, Q is equal to K; therefore, we can determine the direction in which a reaction proceeds by comparing Q to K. If $Q < K$, the reaction moves in the direction of the products; if $Q > K$, the reaction moves in the reverse direction.

Finding Equilibrium Concentrations (16.8)

- There are two general types of problems in which K is given and one (or more) equilibrium concentrations can be found:
  1. K and all but one equilibrium concentration are given.
  2. K and *only* initial concentrations are given.
- We solve the first type by rearranging the law of mass action and substituting the given values.
- We solve the second type by creating an ICE table and using a variable x to represent the change in concentration.

Le Châtelier's Principle (16.9)

- When a system at equilibrium is disturbed—by a change in the amount of a reactant or product, a change in volume, or a change in temperature—the system shifts in the direction that minimizes the disturbance.

KEY EQUATIONS AND RELATIONSHIPS

Expression for the Equilibrium Constant, K_c (16.3)

$$aA + bB \rightleftharpoons cC + dD$$

$$K = \frac{[C]^c[D]^d}{[A]^a[B]^b} \quad \text{(equilibrium concentrations only)}$$

Relationship between the Equilibrium Constant and the Chemical Equation (16.3)

1. If you reverse the equation, invert the equilibrium constant.
2. If you multiply the coefficients in the equation by a factor, raise the equilibrium constant to the same factor.
3. If you add two or more individual chemical equations to obtain an overall equation, multiply the corresponding equilibrium constants by each other to obtain the overall equilibrium constant.

Expression for the Equilibrium Constant, K_p (16.4)

$$aA + bB \rightleftharpoons cC + dD$$

$$K_p = \frac{P_C^c P_D^d}{P_A^a P_B^b} \quad \text{(equilibrium partial pressures only)}$$

Relationship between the Equilibrium Constants, K_c and K_p (16.4)

$$K_p = K_c(RT)^{\Delta n}$$

The Reaction Quotient, Q_c (16.7)

$$aA + bB \rightleftharpoons cC + dD$$

$$Q_c = \frac{[C]^c[D]^d}{[A]^a[B]^b} \quad \text{(concentration at any point in the reaction)}$$

The Reaction Quotient, Q_p (16.7)

$$aA + bB \rightleftharpoons cC + dD$$

$$Q_p = \frac{P_C^c P_D^d}{P_A^a P_B^b} \quad \text{(partial pressures at any point in the reaction)}$$

Relationship of Q to the Direction of the Reaction (16.7)

$Q < K$ Reaction goes to the right.

$Q > K$ Reaction goes to the left.

$Q = K$ Reaction is at equilibrium.

EXERCISES

REVIEW QUESTIONS

1. How does a developing fetus get oxygen in the womb?

2. What is dynamic equilibrium? Why is it called *dynamic*?

3. Give the general expression for the equilibrium constant of the following generic reaction:

$$aA + bB \rightleftharpoons cC + dD$$

4. What is the significance of the equilibrium constant? What does a large equilibrium constant tell us about a reaction? A small one?

5. What happens to the value of the equilibrium constant for a reaction if the reaction equation is reversed? Multiplied by a constant?

6. If two reactions sum to an overall reaction, and the equilibrium constants for the two reactions are K_1 and K_2, what is the equilibrium constant for the overall reaction?

7. Explain the difference between K_c and K_p. For a given reaction, how are the two constants related?

8. What units should you use when expressing concentrations or partial pressures in the equilibrium constant? What are the units of K_p and K_c? Explain.

9. Why do we omit the concentrations of solids and liquids from equilibrium expressions?

10. Does the value of the equilibrium constant depend on the initial concentrations of the reactants and products? Do the equilibrium concentrations of the reactants and products depend on their initial concentrations? Explain.

11. Explain how you might deduce the equilibrium constant for a reaction in which you know the initial concentrations of the reactants and products and the equilibrium concentration of only one reactant or product.

12. What is the definition of the reaction quotient (Q) for a reaction? What does Q measure?

13. What is the value of Q when each reactant and product is in its standard state? (See Section 10.10 for the definition of standard states.)

14. In what direction does a reaction proceed for each condition: (a) $Q < K$; (b) $Q > K$; and (c) $Q = K$?

15. Many equilibrium calculations involve finding the equilibrium concentrations of reactants and products given their initial concentrations and the equilibrium constant. Outline the general procedure used in solving these kinds of problems.

16. In equilibrium problems involving equilibrium constants that are small relative to the initial concentrations of reactants, we can often assume that the quantity x (which represents how far the reaction proceeds toward products) is small. When this assumption is made, we can ignore the quantity x when it is subtracted from a large number but not when it is multiplied by a large number. In other words, $2.5 - x \approx 2.5$, but $2.5x \neq 2.5$. Explain why we can ignore a small x in the first case but not in the second.

17. What happens to a chemical system at equilibrium when equilibrium is disturbed?

18. What is the effect of a change in concentration of a reactant or product on a chemical reaction initially at equilibrium?

19. What is the effect of a change in volume on a chemical reaction (that includes gaseous reactants or products) initially at equilibrium?

20. What is the effect of a temperature change on a chemical reaction initially at equilibrium? How does the effect differ for an exothermic reaction compared to an endothermic one?

PROBLEMS BY TOPIC

Equilibrium and the Equilibrium Constant Expression

21. Write an expression for the equilibrium constant of each chemical equation.

a. $SbCl_5(g) \rightleftharpoons SbCl_3(g) + Cl_2(g)$

b. $2\ BrNO(g) \rightleftharpoons 2\ NO(g) + Br_2(g)$

c. $CH_4(g) + 2\ H_2S(g) \rightleftharpoons CS_2(g) + 4\ H_2(g)$

d. $2\ CO(g) + O_2(g) \rightleftharpoons 2\ CO_2(g)$

22. Find and fix each mistake in the equilibrium constant expressions.

a. $2\ H_2S(g) \rightleftharpoons 2\ H_2(g) + S_2(g)\ K_c = \dfrac{[H_2][S_2]}{[H_2S]}$

b. $CO(g) + Cl_2(g) \rightleftharpoons COCl_2(g)\ K_c = \dfrac{[CO][Cl_2]}{[COCl_2]}$

23. When this reaction comes to equilibrium, will the concentrations of the reactants or products be greater? Does the answer to this question depend on the initial concentrations of the reactants and products?

$$A(g) + B(g) \rightleftharpoons 2\ C(g) \quad K_c = 1.4 \times 10^{-5}$$

24. Ethene (C_2H_4) can be halogenated by this reaction:

$$C_2H_4(g) + X_2(g) \rightleftharpoons C_2H_4X_2(g)$$

where X_2 can be Cl_2 (green), Br_2 (brown), or I_2 (purple). Examine the three figures representing equilibrium concentrations in this reaction at the same temperature for the three different halogens. Rank the equilibrium constants for the three reactions from largest to smallest.

25. H_2 and I_2 are combined in a flask and allowed to react according to the reaction:

$$H_2(g) + I_2(g) \rightleftharpoons 2\,HI(g)$$

Examine the figures (sequential in time) and answer the questions:

a. Which figure represents the point at which equilibrium is reached?
b. How would the series of figures change in the presence of a catalyst?
c. Would there be different amounts of reactants and products in the final figure (vi) in the presence of a catalyst?

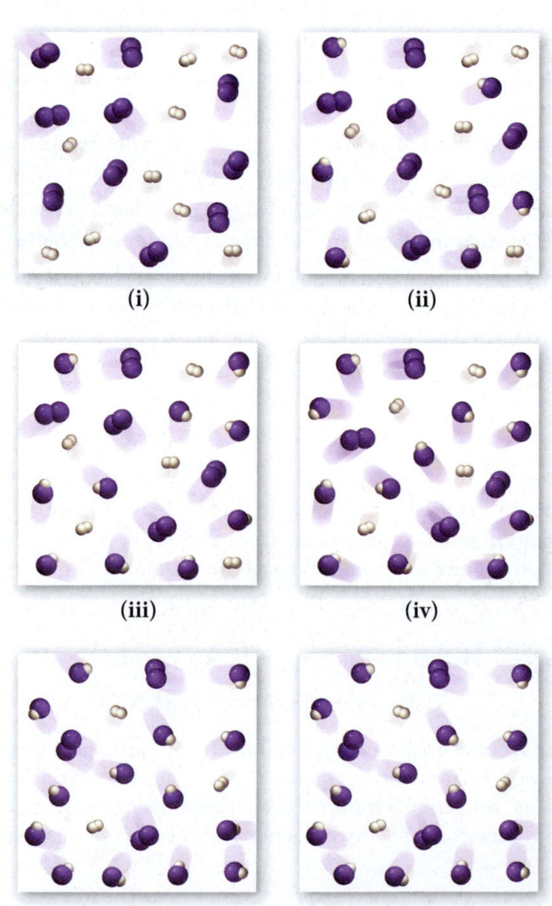

26. A chemist trying to synthesize a particular compound attempts two different synthesis reactions. The equilibrium constants for the two reactions are 23.3 and 2.2×10^4 at room temperature. However, upon carrying out both reactions for 15 minutes, the chemist finds that the reaction with the smaller equilibrium constant produces more of the desired product. Explain how this might be possible.

27. This reaction has an equilibrium constant of $K_p = 2.26 \times 10^4$ at 298 K.

$$CO(g) + 2\,H_2(g) \rightleftharpoons CH_3OH(g)$$

Calculate K_p for each reaction and predict whether reactants or products will be favored at equilibrium.

a. $CH_3OH(g) \rightleftharpoons CO(g) + 2\,H_2(g)$
b. $\frac{1}{2}CO(g) + H_2(g) \rightleftharpoons \frac{1}{2}CH_3OH(g)$
c. $2\,CH_3OH(g) \rightleftharpoons 2\,CO(g) + 4\,H_2(g)$

28. This reaction has an equilibrium constant of $K_p = 2.2 \times 10^6$ at 298 K.

$$2\,COF_2(g) \rightleftharpoons CO_2(g) + CF_4(g)$$

Calculate K_p for each reaction and predict whether reactants or products will be favored at equilibrium.

a. $COF_2(g) \rightleftharpoons \frac{1}{2}CO_2(g) + \frac{1}{2}CF_4(g)$
b. $6\,COF_2(g) \rightleftharpoons 3\,CO_2(g) + 3\,CF_4(g)$
c. $2\,CO_2(g) + 2\,CF_4(g) \rightleftharpoons 4\,COF_2(g)$

29. Consider the reactions and their respective equilibrium constants.

$$NO(g) + \tfrac{1}{2}Br_2(g) \rightleftharpoons NOBr(g) \qquad K_p = 5.3$$
$$2\,NO(g) \rightleftharpoons N_2(g) + O_2(g) \qquad K_p = 2.1 \times 10^{30}$$

Use these reactions and their equilibrium constants to predict the equilibrium constant for the following reaction:

$$N_2(g) + O_2(g) + Br_2(g) \rightleftharpoons 2\,NOBr(g)$$

30. Use the following reactions and their equilibrium constants to predict the equilibrium constant for this reaction: $2\,A(s) \rightleftharpoons 3\,D(g)$.

$$A(s) \rightleftharpoons \tfrac{1}{2}B(g) + C(g) \qquad K_1 = 0.0334$$
$$3\,D(g) \rightleftharpoons B(g) + 2\,C(g) \qquad K_2 = 2.35$$

K_p, K_c, and Heterogeneous Equilibria

31. Calculate K_c for each reaction.

a. $I_2(g) \rightleftharpoons 2\,I(g) \quad K_p = 6.26 \times 10^{-22}$ (at 298 K)
b. $CH_4(g) + H_2O(g) \rightleftharpoons CO(g) + 3\,H_2(g)$
$K_p = 7.7 \times 10^{24}$ (at 298 K)
c. $I_2(g) + Cl_2(g) \rightleftharpoons 2\,ICl(g) \quad K_p = 81.9$ (at 298 K)

32. Calculate K_p for each reaction.

a. $N_2O_4(g) \rightleftharpoons 2\,NO_2(g) \quad K_c = 5.9 \times 10^{-3}$ (at 298 K)
b. $N_2(g) + 3\,H_2(g) \rightleftharpoons 2\,NH_3(g) \quad K_c = 3.7 \times 10^{8}$ (at 298 K)
c. $N_2(g) + O_2(g) \rightleftharpoons 2\,NO(g) \quad K_c = 4.10 \times 10^{-31}$ (at 298 K)

33. Write an equilibrium expression for each chemical equation involving one or more solid or liquid reactants or products.

a. $CO_3^{2-}(aq) + H_2O(l) \rightleftharpoons HCO_3^-(aq) + OH^-(aq)$
b. $2\,KClO_3(s) \rightleftharpoons 2\,KCl(s) + 3\,O_2(g)$
c. $HF(aq) + H_2O(l) \rightleftharpoons H_3O^+(aq) + F^-(aq)$
d. $NH_3(aq) + H_2O(l) \rightleftharpoons NH_4^+(aq) + OH^-(aq)$

34. Find and fix the mistake in the equilibrium expression.

$$PCl_5(g) \rightleftharpoons PCl_3(l) + Cl_2(g)$$

$$K_c = \frac{[PCl_3][Cl_2]}{[PCl_5]}$$

Relating the Equilibrium Constant to Equilibrium Concentrations and Equilibrium Partial Pressures

35. Consider the reaction:

$$CO(g) + 2\,H_2(g) \rightleftharpoons CH_3OH(g)$$

An equilibrium mixture of this reaction at a certain temperature has $[CO] = 0.105$ M, $[H_2] = 0.114$ M, and $[CH_3OH] = 0.185$ M. What is the value of the equilibrium constant (K_c) at this temperature?

36. Consider the reaction:

$$NH_4HS(s) \rightleftharpoons NH_3(g) + H_2S(g)$$

An equilibrium mixture of this reaction at a certain temperature has $[NH_3] = 0.278$ M and $[H_2S] = 0.355$ M. What is the value of the equilibrium constant (K_c) at this temperature?

37. Consider the reaction:

$$N_2(g) + 3\,H_2(g) \rightleftharpoons 2\,NH_3(g)$$

Complete the table. Assume that all concentrations are equilibrium concentrations in M.

T (K)	$[N_2]$	$[H_2]$	$[NH_3]$	K_c
500	0.115	0.105	0.439	____
575	0.110	____	0.128	9.6
775	0.120	0.140	____	0.0584

38. Consider the reaction:

$$H_2(g) + I_2(g) \rightleftharpoons 2\,HI(g)$$

Complete the table. Assume that all concentrations are equilibrium concentrations in M.

T (°C)	$[H_2]$	$[I_2]$	$[HI]$	K_c
25	0.0355	0.0388	0.922	____
340	____	0.0455	0.387	9.6
445	0.0485	0.0468	____	50.2

39. Consider the reaction:

$$2\,NO(g) + Br_2(g) \rightleftharpoons 2\,NOBr(g) \qquad K_p = 28.4 \text{ at } 298\text{ K}$$

In a reaction mixture at equilibrium, the partial pressure of NO is 108 torr and that of Br_2 is 126 torr. What is the partial pressure of NOBr in this mixture?

40. Consider the reaction:

$$SO_2Cl_2(g) \rightleftharpoons SO_2(g) + Cl_2(g) \qquad K_p = 2.91 \times 10^3 \text{ at } 298\text{ K}$$

In a reaction at equilibrium, the partial pressure of SO_2 is 137 torr and that of Cl_2 is 285 torr. What is the partial pressure of SO_2Cl_2 in this mixture?

41. For the reaction $A(g) \rightleftharpoons 2\,B(g)$, a reaction vessel initially contains only A at a pressure of $P_A = 1.32$ atm. At equilibrium, $P_A = 0.25$ atm. Calculate the value of K_p. (Assume no changes in volume or temperature.)

42. For the reaction $2\,A(g) \rightleftharpoons B(g) + 2\,C(g)$, a reaction vessel initially contains only A at a pressure of $P_A = 225$ mmHg. At equilibrium, $P_A = 55$ mmHg. Calculate the value of K_p. (Assume no changes in volume or temperature.)

43. Consider the reaction:

$$Fe^{3+}(aq) + SCN^-(aq) \rightleftharpoons FeSCN^{2+}(aq)$$

A solution is made containing an initial $[Fe^{3+}]$ of 1.0×10^{-3} M and an initial $[SCN^-]$ of 8.0×10^{-4} M. At equilibrium, $[FeSCN^{2+}] = 1.7 \times 10^{-4}$ M. Calculate the value of the equilibrium constant (K_c).

44. Consider the reaction:

$$SO_2Cl_2(g) \rightleftharpoons SO_2(g) + Cl_2(g)$$

A reaction mixture is made containing an initial $[SO_2Cl_2]$ of 0.020 M. At equilibrium, $[Cl_2] = 1.2 \times 10^{-2}$ M. Calculate the value of the equilibrium constant (K_c).

45. Consider the reaction:

$$H_2(g) \rightleftharpoons I_2(g) + 2\,HI(g)$$

A reaction mixture in a 3.67 L flask at a certain temperature initially contains 0.763 g H_2 and 96.9 g I_2. At equilibrium, the flask contains 90.4 g HI. Calculate the equilibrium constant (K_c) for the reaction at this temperature.

46. Consider the reaction:

$$CO(g) + 2\,H_2(g) \rightleftharpoons CH_3OH(g)$$

A reaction mixture in a 5.19 L flask at a certain temperature contains 26.9 g CO and 2.34 g H_2. At equilibrium, the flask contains 8.65 g CH_3OH. Calculate the equilibrium constant (K_c) for the reaction at this temperature.

The Reaction Quotient and Reaction Direction

47. Consider the reaction:

$$NH_4HS(s) \rightleftharpoons NH_3(g) + H_2S(g)$$

At a certain temperature, $K_c = 8.5 \times 10^{-3}$. A reaction mixture at this temperature containing solid NH_4HS has $[NH_3] = 0.166$ M and $[H_2S] = 0.166$ M. Will more of the solid form, or will some of the existing solid decompose as equilibrium is reached?

48. Consider the reaction:

$$2\,H_2S(g) \rightleftharpoons 2\,H_2(g) + S_2(g) \qquad K_p = 2.4 \times 10^{-4} \text{ at } 1073\text{ K}$$

A reaction mixture contains 0.112 atm of H_2, 0.055 atm of S_2, and 0.445 atm of H_2S. Is the reaction mixture at equilibrium? If not, in what direction will the reaction proceed?

49. Silver sulfate dissolves in water according to the reaction:

$$Ag_2SO_4(s) \rightleftharpoons 2\,Ag^+(aq) + SO_4^{2-}(aq) \qquad K_c = 1.1 \times 10^{-5} \text{ at } 298\text{ K}$$

A 1.5 L solution contains 6.55 g of dissolved silver sulfate. If additional solid silver sulfate is added to the solution, will it dissolve?

50. Nitrogen dioxide reacts with itself according to the reaction:

$$2NO_2(g) \rightleftharpoons N_2O_4(g) \qquad K_p = 6.7 \text{ at } 298\text{ K}$$

A 2.25 L container contains 0.055 mol of NO_2 and 0.082 mol of N_2O_4 at 298 K. Is the reaction at equilibrium? If not, in what direction will the reaction proceed?

Finding Equilibrium Concentrations from Initial Concentrations and the Equilibrium Constant

51. Consider the reaction and the associated equilibrium constant:

$$aA(g) \rightleftharpoons bB(g) \quad K_c = 4.0$$

Find the equilibrium concentrations of A and B for each value of *a* and *b*. Assume that the initial concentration of A in each case is 1.0 M and that no B is present at the beginning of the reaction.

a. $a = 1; b = 1$
b. $a = 2; b = 2$
c. $a = 1; b = 2$

52. Consider the reaction and the associated equilibrium constant:

$$aA(g) + bB(g) \rightleftharpoons cC(g) \quad K_c = 5.0$$

Find the equilibrium concentrations of A, B, and C for each value of *a*, *b*, and *c*. Assume that the initial concentrations of A and B are each 1.0 M and that no product is present at the beginning of the reaction.

a. $a = 1; b = 1; c = 2$
b. $a = 1; b = 1; c = 1$
c. $a = 2; b = 1; c = 1$ (set up equation for *x*; don't solve)

53. For the reaction, $K_c = 0.513$ at 500 K.

$$N_2O_4(g) \rightleftharpoons 2\ NO_2(g)$$

If a reaction vessel initially contains an N_2O_4 concentration of 0.0500 M at 500 K, what are the equilibrium concentrations of N_2O_4 and NO_2 at 500 K?

54. For the reaction, $K_c = 255$ at 1000 K.

$$CO(g) + Cl_2(g) \rightleftharpoons COCl_2(g)$$

If a reaction mixture initially contains a CO concentration of 0.1500 M and a Cl_2 concentration of 0.175 M at 1000 K, what are the equilibrium concentrations of CO, Cl_2, and $COCl_2$ at 1000 K?

55. Consider the reaction:

$$NiO(s) + CO(g) \rightleftharpoons Ni(s) + CO_2(g) \quad K_c = 4.0 \times 10^3 \text{ at } 1500 \text{ K}$$

If a mixture of solid nickel (II) oxide and 0.20 M carbon monoxide comes to equilibrium at 1500 K, what is the equilibrium concentration of CO_2?

56. Consider the reaction:

$$CO(g) + H_2O(g) \rightleftharpoons CO_2(g) + H_2(g) \quad K_c = 102 \text{ at } 500 \text{ K}$$

If a reaction mixture initially contains 0.110 M CO and 0.110 M H_2O, what is the equilibrium concentration of each of the reactants and products?

57. Consider the reaction:

$$HC_2H_3O_2(aq) + H_2O(l) \rightleftharpoons H_3O^+(aq) + C_2H_3O_2^-(aq) \quad K_c = 1.8 \times 10^{-5} \text{ at } 25\ °C$$

If a solution initially contains 0.210 M $HC_2H_3O_2$, what is the equilibrium concentration of H_3O^+ at 25 °C?

58. Consider the reaction:

$$SO_2Cl_2(g) \rightleftharpoons SO_2(g) + Cl_2(g) \quad K_c = 2.99 \times 10^{-7} \text{ at } 227\ °C$$

If a reaction mixture initially contains 0.175 M SO_2Cl_2, what is the equilibrium concentration of Cl_2 at 227 °C?

59. Consider the reaction:

$$Br_2(g) + Cl_2(g) \rightleftharpoons 2\ BrCl(g) \quad K_p = 1.11 \times 10^{-4} \text{ at } 150 \text{ K}$$

A reaction mixture initially contains a Br_2 partial pressure of 755 torr and a Cl_2 partial pressure of 735 torr at 150 K. Calculate the equilibrium partial pressure of BrCl.

60. Consider the reaction:

$$CO(g) + H_2O(g) \rightleftharpoons CO_2(g) + H_2(g) \quad K_p = 0.0611 \text{ at } 2000 \text{ K}$$

A reaction mixture initially contains a CO partial pressure of 1344 torr and a H_2O partial pressure of 1766 torr at 2000 K. Calculate the equilibrium partial pressures of each of the products.

61. Consider the reaction:

$$A(g) \rightleftharpoons B(g) + C(g)$$

Find the equilibrium concentrations of A, B, and C for each value of K_c. Assume that the initial concentration of A in each case is 1.0 M and that the reaction mixture initially contains no products. Make any appropriate simplifying assumptions.

a. $K_c = 1.0$
b. $K_c = 0.0010$
c. $K_c = 1.0 \times 10^{-5}$

62. Consider the reaction:

$$A(g) \rightleftharpoons 2\ B(g)$$

Find the equilibrium partial pressures of A and B for each value of *K*. Assume that the initial partial pressure of B in each case is 1.0 atm and that the initial partial pressure of A is 0.0 atm. Make any appropriate simplifying assumptions.

a. $K_c = 1.0$
b. $K_c = 1.0 \times 10^{-4}$
c. $K_c = 1.0 \times 10^5$

Le Châtelier's Principle

63. Consider this reaction at equilibrium:

$$CO(g) + Cl_2(g) \rightleftharpoons COCl_2(g)$$

Predict whether the reaction will shift left, shift right, or remain unchanged after each disturbance:

a. $COCl_2$ is added to the reaction mixture.
b. Cl_2 is added to the reaction mixture.
c. $COCl_2$ is removed from the reaction mixture.

64. Consider this reaction at equilibrium:

$$2\ BrNO(g) \rightleftharpoons 2\ NO(g) + Br_2(g)$$

Predict whether the reaction will shift left, shift right, or remain unchanged after each disturbance.

a. NO is added to the reaction mixture.
b. BrNO is added to the reaction mixture.
c. Br_2 is removed from the reaction mixture.

65. Consider this reaction at equilibrium:

$$2\ KClO_3(s) \rightleftharpoons 2\ KCl(s) + 3\ O_2(g)$$

Predict whether the reaction will shift left, shift right, or remain unchanged after each disturbance.

a. O_2 is removed from the reaction mixture.
b. KCl is added to the reaction mixture.
c. $KClO_3$ is added to the reaction mixture.
d. O_2 is added to the reaction mixture.

66. Consider this reaction at equilibrium:

$$C(s) + H_2O(g) \rightleftharpoons CO(g) + H_2(g)$$

Predict whether the reaction will shift left, shift right, or remain unchanged after each disturbance.

a. C is added to the reaction mixture.
b. H_2O is condensed and removed from the reaction mixture.
c. CO is added to the reaction mixture.
d. H_2 is removed from the reaction mixture.

67. Each reaction is allowed to come to equilibrium, and then the volume is changed as indicated. Predict the effect (shift right, shift left, or no effect) of the indicated volume change.

a. $I_2(g) \rightleftharpoons 2\,I(g)$ (volume is increased)
b. $2\,H_2S(g) \rightleftharpoons 2\,H_2(g) + S_2(g)$ (volume is decreased)
c. $I_2(g) + Cl_2(g) \rightleftharpoons 2\,ICl(g)$ (volume is decreased)

68. Each reaction is allowed to come to equilibrium, and then the volume is changed as indicated. Predict the effect (shift right, shift left, or no effect) of the indicated volume change.

a. $CO(g) + H_2O(g) \rightleftharpoons CO_2(g) + H_2(g)$ (volume is decreased)
b. $PCl_3(g) + Cl_2(g) \rightleftharpoons PCl_5(g)$ (volume is increased)
c. $CaCO_3(s) \rightleftharpoons CaO(s) + CO_2(g)$ (volume is increased)

69. This reaction is endothermic.

$$C(s) + CO_2(g) \rightleftharpoons 2\,CO(g)$$

Predict the effect (shift right, shift left, or no effect) of increasing and decreasing the reaction temperature. How does the value of the equilibrium constant depend on temperature?

70. This reaction is exothermic.

$$C_6H_{12}O_6(s) + 6\,O_2(g) \rightleftharpoons 6\,CO_2(g) + 6\,H_2O(g)$$

Predict the effect (shift right, shift left, or no effect) of increasing and decreasing the reaction temperature. How does the value of the equilibrium constant depend on temperature?

71. Coal, which is primarily carbon, can be converted to natural gas, primarily CH_4, by the exothermic reaction:

$$C(s) + 2\,H_2(g) \rightleftharpoons CH_4(g)$$

Which disturbance favors CH_4 at equilibrium?

a. adding more C to the reaction mixture
b. adding more H_2 to the reaction mixture
c. raising the temperature of the reaction mixture
d. lowering the volume of the reaction mixture
e. adding a catalyst to the reaction mixture
f. adding neon gas to the reaction mixture

72. Coal can be used to generate hydrogen gas (a potential fuel) by the endothermic reaction:

$$C(s) + H_2O(g) \rightleftharpoons CO(g) + H_2(g)$$

If this reaction mixture is at equilibrium, predict whether each disturbance will result in the formation of additional hydrogen gas, the formation of less hydrogen gas, or have no effect on the quantity of hydrogen gas.

a. adding C to the reaction mixture
b. adding H_2O to the reaction mixture
c. raising the temperature of the reaction mixture
d. increasing the volume of the reaction mixture
e. adding a catalyst to the reaction mixture
f. adding an inert gas to the reaction mixture

CUMULATIVE PROBLEMS

73. Carbon monoxide replaces oxygen in oxygenated hemoglobin according to the reaction:

$$HbO_2(aq) + CO(aq) \rightleftharpoons HbCO(aq) + O_2(aq)$$

a. Use the reactions and associated equilibrium constants at body temperature to find the equilibrium constant for the reaction just shown.

$$Hb(aq) + O_2(aq) \rightleftharpoons HbO_2(aq) \quad K_c = 1.8$$
$$Hb(aq) + CO(aq) \rightleftharpoons HbCO(aq) \quad K_c = 306$$

b. Suppose that an air mixture becomes polluted with carbon monoxide at a level of 0.10%. Assuming the air contains 20.0% oxygen and that the oxygen and carbon monoxide ratios that dissolve in the blood are identical to the ratios in the air, what is the ratio of HbCO to HbO_2 in the bloodstream? Comment on the toxicity of carbon monoxide.

74. Nitrogen monoxide is a pollutant in the lower atmosphere that irritates the eyes and lungs and leads to the formation of acid rain. Nitrogen monoxide forms naturally in the atmosphere according to the endothermic reaction:

$$N_2(g) + O_2(g) \rightleftharpoons 2\,NO(g) \quad K_p = 4.1 \times 10^{-31} \text{ at } 298\text{ K}$$

Use the ideal gas law to calculate the concentrations of nitrogen and oxygen present in air at a pressure of 1.0 atm and a temperature of 298 K. Assume that nitrogen composes 78% of air by volume and that oxygen composes 21% of air. Find the "natural" equilibrium concentration of NO in air in units of molecules/cm^3. How would you expect this concentration to change in an automobile engine in which combustion is occurring?

75. The reaction $CO_2(g) + C(s) \rightleftharpoons 2\,CO(g)$ has $K_p = 5.78$ at 1200 K.

a. Calculate the total pressure at equilibrium when 4.45 g of CO_2 is introduced into a 10.0 L container and heated to 1200 K in the presence of 2.00 g of graphite.
b. Repeat the calculation of part a in the presence of 0.50 g of graphite.

76. A mixture of water and graphite is heated to 600 K. When the system comes to equilibrium, it contains 0.13 mol of H_2, 0.13 mol of CO, 0.43 mol of H_2O, and some graphite. Some O_2 is added to the system, and a spark is applied so that the H_2 reacts completely with the O_2. Find the amount of CO in the flask when the system returns to equilibrium.

77. At 650 K, the reaction $MgCO_3(s) \rightleftharpoons MgO(s) + CO_2(g)$ has $K_p = 0.026$. A 10.0 L container at 650 K has 1.0 g of MgO(s) and CO_2 at P = 0.0260 atm. The container is then compressed to a volume of 0.100 L. Find the mass of $MgCO_3$ that is formed.

78. A system at equilibrium contains $I_2(g)$ at a pressure of 0.21 atm and $I(g)$ at a pressure of 0.23 atm. The system is then compressed to half its volume. Find the pressure of each gas when the system returns to equilibrium.

79. Consider the exothermic reaction:

$$C_2H_4(g) + Cl_2(g) \rightleftharpoons C_2H_4Cl_2(g)$$

If you were trying to maximize the amount of $C_2H_4Cl_2$ produced, which tactic might you try? Assume that the reaction mixture reaches equilibrium.

a. increasing the reaction volume
b. removing $C_2H_4Cl_2$ from the reaction mixture as it forms
c. lowering the reaction temperature
d. adding Cl_2

80. Consider the endothermic reaction:

$$C_2H_4(g) + I_2(g) \rightleftharpoons C_2H_4I_2(g)$$

If you were trying to maximize the amount of $C_2H_4I_2$ produced, which tactic might you try? Assume that the reaction mixture reaches equilibrium.

a. decreasing the reaction volume
b. removing I_2 from the reaction mixture
c. raising the reaction temperature
d. adding C_2H_4 to the reaction mixture

81. Consider the reaction:

$$H_2(g) + I_2(g) \rightleftharpoons 2\,HI(g)$$

A reaction mixture at equilibrium at 175 K contains $P_{H_2} = 0.958$ atm, $P_{I_2} = 0.877$ atm, and $P_{HI} = 0.020$ atm. A second reaction mixture, also at 175 K, contains $P_{H_2} = P_{I_2} = 0.621$ atm, and $P_{HI} = 0.101$ atm. Is the second reaction at equilibrium? If not, what will be the partial pressure of HI when the reaction reaches equilibrium at 175 K?

82. Consider the reaction:

$$2\,H_2S(g) + SO_2(g) \rightleftharpoons 3\,S(s) + 2\,H_2O(g)$$

A reaction mixture initially containing 0.500 M H_2S and 0.500 M SO_2 contains 0.0011 M H_2O at a certain temperature. A second reaction mixture at the same temperature initially contains $[H_2S] = 0.250$ M and $[SO_2] = 0.325$ M. Calculate the equilibrium concentration of H_2O in the second mixture at this temperature.

83. Ammonia can be synthesized according to the reaction:

$$N_2(g) + 3\,H_2(g) \rightleftharpoons 2\,NH_3(g) \quad K_p = 5.3 \times 10^{-5} \text{ at } 725\text{ K}$$

A 200.0 L reaction container initially contains 1.27 kg of N_2 and 0.310 kg of H_2 at 725 K. Assuming ideal gas behavior, calculate the mass of NH_3 (in g) present in the reaction mixture at equilibrium. What is the percent yield of the reaction under these conditions?

84. Hydrogen can be extracted from natural gas according to the reaction:

$$CH_4(g) + CO_2(g) \rightleftharpoons 2\,CO(g) + 2\,H_2(g) \quad K_p = 45 \times 10^2 \text{ at } 825\text{ K}$$

An 85.0 L reaction container initially contains 22.3 kg of CH_4 and 55.4 kg of CO_2 at 825 K. Assuming ideal gas behavior, calculate the mass of H_2 (in g) present in the reaction mixture at equilibrium. What is the percent yield of the reaction under these conditions?

85. The system described by the reaction: $CO(g) + Cl_2(g) \rightleftharpoons COCl_2(g)$ is at equilibrium at a given temperature when $P_{CO} = 0.30$ atm, $P_{Cl_2} = 0.10$ atm, and $P_{COCl_2} = 0.60$ atm. An additional pressure of $Cl_2(g) = 0.40$ atm is added. Find the pressure of CO when the system returns to equilibrium.

86. A reaction vessel at 27 °C contains a mixture of SO_2 ($P = 3.00$ atm) and O_2 ($P = 1.00$ atm). When a catalyst is added, this reaction takes place: $2\,SO_2(g) + O_2(g) \rightleftharpoons 2\,SO_3(g)$.

At equilibrium, the total pressure is 3.75 atm. Find the value of K_c.

87. At 70 K, CCl_4 decomposes to carbon and chlorine. The K_p for the decomposition is 0.76. Find the starting pressure of CCl_4 at this temperature that will produce a total pressure of 1.0 atm at equilibrium.

88. The equilibrium constant for the reaction $SO_2(g) + NO_2(g) \rightleftharpoons SO_3(g) + NO(g)$ is 3.0. Find the amount of NO_2 that must be added to 2.4 mol of SO_2 in order to form 1.2 mol of SO_3 at equilibrium.

89. A sample of $CaCO_3(s)$ is introduced into a sealed container of volume 0.654 L and heated to 1000 K until equilibrium is reached. The K_p for the reaction $CaCO_3(s) \rightleftharpoons CaO(s) + CO_2(g)$ is 3.9×10^{-2} at this temperature. Calculate the mass of $CaO(s)$ that is present at equilibrium.

90. An equilibrium mixture contains N_2O_4, ($P = 0.28$ atm) and NO_2 ($P = 1.1$ atm) at 350 K. The volume of the container is doubled at constant temperature. Calculate the equilibrium pressures of the two gases when the system reaches a new equilibrium.

91. Carbon monoxide and chlorine gas react to form phosgene:

$$CO(g) + Cl_2(g) \rightleftharpoons COCl_2(g) \quad K_p = 3.10 \text{ at } 700\text{ K}$$

If a reaction mixture initially contains 215 torr of CO and 245 torr of Cl_2, what is the mole fraction of $COCl_2$ when equilibrium is reached?

92. Solid carbon can react with gaseous water to form carbon monoxide gas and hydrogen gas. The equilibrium constant for the reaction at 700.0 K is $K_p = 1.60 \times 10^{-3}$. If a 1.55 L reaction vessel initially contains 145 torr of water at 700.0 K in contact with excess solid carbon, find the percent by mass of hydrogen gas of the gaseous reaction mixture at equilibrium.

CHALLENGE PROBLEMS

93. Consider the reaction:

$$2\,NO(g) + O_2(g) \rightleftharpoons 2\,NO_2(g)$$

a. A reaction mixture at 175 K initially contains 522 torr of NO and 421 torr of O_2. At equilibrium, the total pressure in the reaction mixture is 748 torr. Calculate K_p at this temperature.
b. A second reaction mixture at 175 K initially contains 255 torr of NO and 185 torr of O_2. What is the equilibrium partial pressure of NO_2 in this mixture?

94. Consider the reaction:

$$2\,SO_2(g) + O_2(g) \rightleftharpoons 2\,SO_3(g) \quad K_p = 0.355 \text{ at } 950\text{ K}$$

A 2.75 L reaction vessel at 950 K initially contains 0.100 mol of SO_2 and 0.100 mol of O_2. Calculate the total pressure (in atmospheres) in the reaction vessel when equilibrium is reached.

95. Nitrogen monoxide reacts with chlorine gas according to the reaction:

$$2\,NO(g) + Cl_2(g) \rightleftharpoons 2\,NOCl(g) \quad K_p = 0.27 \text{ at } 700\text{ K}$$

A reaction mixture initially contains equal partial pressures of NO and Cl_2. At equilibrium, the partial pressure of NOCl is 115 torr. What were the initial partial pressures of NO and Cl_2?

96. At a given temperature, a system containing $O_2(g)$ and some oxides of nitrogen are described by these reactions:

$$2\,NO(g) + O_2(g) \rightleftharpoons 2\,NO_2(g) \quad K_p = 10^4$$
$$2\,NO_2(g) \rightleftharpoons N_2O_4(g) \quad K_p = 0.10$$

A pressure of 1 atm of $N_2O_4(g)$ is placed in a container at this temperature. Predict which, if any, component (other than N_2O_4) will be present at a pressure greater than 0.2 atm at equilibrium.

97. A sample of pure NO_2 is heated to 337 °C, at which temperature it partially dissociates according to the equation:

$$2\ NO_2(g) \rightleftharpoons 2\ NO(g) + O_2(g)$$

At equilibrium the density of the gas mixture is 0.520 g/L at 0.750 atm. Calculate K_c for the reaction.

98. When $N_2O_5(g)$ is heated, it dissociates into $N_2O_3(g)$ and $O_2(g)$ according to the reaction:

$$N_2O_5(g) \rightleftharpoons N_2O_3(g) + O_2(g)$$

$K_c = 7.75$ at a given temperature

The $N_2O_3(g)$ dissociates to give $N_2O(g)$ and $O_2(g)$ according to the reaction:

$$N_2O_3(g) \rightleftharpoons N_2O(g) + O_2(g)$$

$K_c = 4.00$ at the same temperature

When 4.00 mol of $N_2O_5(g)$ is heated in a 1.00 L reaction vessel to this temperature, the concentration of $O_2(g)$ at equilibrium is 4.50 mol/L. Find the concentrations of all the other species in the equilibrium system.

99. A sample of SO_3 is introduced into an evacuated sealed container and heated to 600 K. The following equilibrium is established:

$$2\ SO_3(g) \rightleftharpoons 2\ SO_2(g) + O_2(g)$$

The total pressure in the system is 3.0 atm, and the mole fraction of O_2 is 0.12. Find K_p.

CONCEPTUAL PROBLEMS

100. A reaction $A(g) \rightleftharpoons B(g)$ has an equilibrium constant of 1.0×10^{-4}. For which of the initial reaction mixtures is the *x is small* approximation most likely to apply?

a. [A] = 0.0010 M; [B] = 0.00 M
b. [A] = 0.00 M; [B] = 0.10 M
c. [A] = 0.10 M; [B] = 0.10 M
d. [A] = 0.10 M; [B] = 0.00 M

101. The reaction $A(g) \rightleftharpoons 2\ B(g)$ has an equilibrium constant of $K_c = 1.0$ at a given temperature. If a reaction vessel contains equal initial amounts (in moles) of A and B, will the direction in which the reaction proceeds depend on the volume of the reaction vessel? Explain.

102. A particular reaction has an equilibrium constant of $K_p = 0.50$. A reaction mixture is prepared in which all the reactants and products are in their standard states. In which direction will the reaction proceed?

103. Consider the reaction:

$$aA(g) \rightleftharpoons bB(g)$$

Each of the entries in the table represents equilibrium partial pressures of A and B under different initial conditions. What are the values of *a* and *b* in the reaction?

P_A (atm)	P_B (atm)
4.0	2.0
2.0	1.4
1.0	1.0
0.50	0.71
0.25	0.50

104. Consider the simple one-step reaction:

$$A(g) \rightleftharpoons B(g)$$

Since the reaction occurs in a single step, the forward reaction has a rate of k_{for}[A] and the reverse reaction has a rate of k_{rev}[B]. What happens to the rate of the forward reaction when we increase the concentration of A? How does this explain the reason behind Le Châtelier's principle?

ANSWERS TO CONCEPTUAL CONNECTIONS

Cc 16.1 **(b)** The reaction mixture will contain [A] = 0.1 M and [B] = 1.0 M so that [B]/[A] = 10.

Cc 16.2 **(b)** The reaction is reversed and divided through by two. Therefore, you invert the equilibrium constant and take the square root of the result. $K = (1/0.010)^{1/2} = 10$.

Cc 16.3 **(a)** When $a + b = c + d$, the quantity Δn is zero so that $K_p = K_c(RT)^0$. Since $(RT)^0$ is equal to 1, $K_p = K_c$.

Cc 16.4 **(b)** Since Δn for gaseous reactants and products is zero, K_p equals K_c.

Cc 16.5 **(c)** Because N_2O_4 and NO_2 are both in their standard states, they each have a partial pressure of 1.0 atm. Consequently, $Q_p = 1$. Since $K_p = 0.15$, $Q_p > K_p$, and the reaction proceeds to the left.

Cc 16.6 **(a)** The *x is small* approximation is most likely to apply to a reaction with a small equilibrium constant and an initial concentration of reactant that is not too small. The bigger the equilibrium constant and the smaller the initial concentration of reactant, the less likely that the *x is small* approximation will apply.

CHAPTER 17

We frequently define an acid or a base as a substance whose aqueous solution gives, respectively, a higher concentration of hydrogen ion or of hydroxide ion than that furnished by pure water. This is a very one-sided definition.

—Gilbert N. Lewis (1875–1946)

"Don't worry Mask, this base will neutralize that acid."

Batman neutralizes acid with base (a strategy he learned in his chemistry class).

Acids and Bases

17.1 Batman's Basic Blunder 655

17.2 The Nature of Acids and Bases 656

17.3 Definitions of Acids and Bases 658

17.4 Acid Strength and Molecular Structure 661

17.5 Acid Strength and the Acid Ionization Constant (K_a) 663

17.6 Autoionization of Water and pH 666

17.7 Finding the $[H_3O^+]$ and pH of Strong and Weak Acid Solutions 670

17.8 Finding the $[OH^-]$ and pH of Strong and Weak Base Solutions 680

17.9 The Acid–Base Properties of Ions and Salts 684

17.10 Polyprotic Acids 691

17.11 Lewis Acids and Bases 696

Key Learning Outcomes 699

IN THIS CHAPTER, we apply the equilibrium concepts introduced in the previous chapter to acid–base phenomena. Acids are common in many foods, such as limes, lemons, and vinegar, and in a number of consumer products, such as toilet cleaners and batteries. Bases are less common in foods but are key ingredients in consumer products such as drain openers and antacids. We will examine three different models for acid–base behavior, all of which define that behavior differently. In spite of their differences, the three models coexist, and each is useful at explaining a particular range of acid–base phenomena. We also examine how to calculate the acidity or basicity of solutions and define a useful scale, called the pH scale, to quantify acidity and basicity. These types of calculations often involve solving the kind of equilibrium problems that we explored in Chapter 16.

17.1 Batman's Basic Blunder

In an episode of the Batman comic book series (*War Crimes 4: Judgement at Gotham*), the Joker attacks another villain named Black Mask (who is masquerading as Batman) with an acid-filled dart. The acid begins to burn into Black Mask's head, but the real Batman arrives on the scene and showers Black Mask with a base, saving him. Batman tells Black Mask that he will be fine because the base neutralizes the acid.

Did Batman take the right course of action in counteracting the acid dart? Presumably the acid used by the Joker was fairly concentrated; otherwise it would not have begun to burn into Black Mask's head. It is true that a concentrated base would neutralize the acid in the dart. However, the concentrated base would likely cause further problems for two reasons. First, the neutralization reaction between a concentrated acid and a concentrated base is highly exothermic, so Batman's treatment is likely to have produced a significant amount of heat, leading to potential burns for Black Mask. Second, concentrated bases are themselves caustic to the skin, so Batman's base has the potential to cause further additional burns as well. The standard treatment for spilling concentrated acid on the skin is to rinse with large amounts of water for an extended period of time. The water dilutes the acid and washes it away. Black Mask would have been better off with a shower.

In this chapter, we examine acid and base behavior. Acids and bases are not only important to our health (as we have just seen), but are also found in many household products, foods, medicines, and in nearly every chemistry laboratory. Acid–base chemistry is central to much of biochemistry and molecular biology. The building blocks of proteins, for example, are acids (called amino acids), and the molecules that carry the genetic code in DNA are bases.

17.2 The Nature of Acids and Bases

Litmus paper contains certain dyes that change color in the presence of acids and bases.

Acids have the following general properties: a sour taste; the ability to dissolve many metals; the ability to turn blue litmus paper red; and the ability to neutralize bases. Table 17.1 lists some common acids.

You can find hydrochloric acid in most chemistry laboratories. In industry, it is used to clean metals, to prepare and process some foods, and to refine metal ores. Hydrochloric acid is also the main component of stomach acid, which is why heartburn results in a sour taste (the sour taste occurs when HCl backs up out of the stomach and into the throat and mouth).

For a review of acid naming, see Section 9.7.

HCl

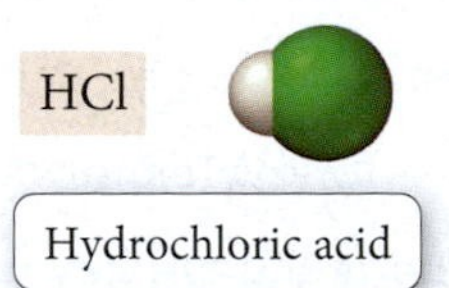

Hydrochloric acid

The formula for acetic acid can also be written as CH_3COOH.

TABLE 17.1 Common Acids

Name	Occurrence/Uses
Hydrochloric acid (HCl)	Metal cleaning; food preparation; ore refining; primary component of stomach acid
Sulfuric acid (H_2SO_4)	Fertilizer and explosives manufacturing; dye and glue production; automobile batteries; electroplating of copper
Nitric acid (HNO_3)	Fertilizer and explosives manufacturing; dye and glue production
Acetic acid ($HC_2H_3O_2$)	Plastic and rubber manufacturing; food preservative; active component of vinegar
Citric acid ($H_3C_6H_5O_7$)	Present in citrus fruits such as lemons and limes; used to adjust pH in foods and beverages
Carbonic acid (H_2CO_3)	Found in carbonated beverages due to the reaction of carbon dioxide with water
Hydrofluoric acid (HF)	Metal cleaning; glass frosting and etching
Phosphoric acid (H_3PO_4)	Fertilizer manufacturing; biological buffering; preservative in beverages

Sulfuric acid and nitric acid are also common in the laboratory. They are important in the manufacture of fertilizers, explosives, dyes, and glues. Sulfuric acid, produced in larger quantities than any other industrial chemical, is present in most automobile batteries.

$$\mathrm{H{-}O{-}\overset{\overset{\displaystyle O}{\|}}{\underset{\underset{\displaystyle \underset{|}{O}\atop{\displaystyle H}}{|}}{S}}{=}O}$$

H_2SO_4

Sulfuric acid

$$\mathrm{H{-}O{-}\overset{\overset{\displaystyle O}{\|}}{N}{-}O}$$

HNO_3

Nitric acid

You can probably find acetic acid in your home—it is the active component of vinegar. It is also produced in improperly stored wines. The word "vinegar" originates from the French words *vin aigre*, which means sour wine. Wine experts consider the presence of vinegar in wines a serious fault because it makes the wine taste like salad dressing.

$$\mathrm{H{-}O{-}\overset{\overset{\displaystyle O}{\|}}{C}{-}\overset{\overset{\displaystyle H}{|}}{\underset{\underset{\displaystyle H}{|}}{C}}{-}H}$$

$HC_2H_3O_2$

Acetic acid

Acetic acid is a **carboxylic acid**, an acid that contains the following grouping of atoms:

$$\mathrm{H{-}O{-}\overset{\overset{\displaystyle O}{\|}}{C}{-}}$$

Carboxylic acid group

Carboxylic acids are often found in substances derived from living organisms. Other examples of carboxylic acids are citric acid, the main acid in lemons and limes, and malic acid, found in apples, grapes, and wine.

◄ Citrus fruits, apples, and grapes all contain acids.

▲ Acetic acid makes vinegar taste sour.

▲ Many common household products and remedies contain bases.

TABLE 17.2 Common Bases

Name	Occurrence/Uses
Sodium hydroxide (NaOH)	Petroleum processing; soap and plastic manufacturing
Potassium hydroxide (KOH)	Cotton processing; electroplating; soap production; batteries
Sodium bicarbonate ($NaHCO_3$)	Antacid; active ingredient of baking soda; source of CO_2
Sodium carbonate (Na_2CO_3)	Glass and soap manufacturing; general cleanser; water softener
Ammonia (NH_3)	Detergent; fertilizer and explosives manufacturing; synthetic fiber production

Bases have the following general properties: a bitter taste; a slippery feel; the ability to turn red litmus paper blue; and the ability to neutralize acids. Because of their bitterness, bases are less common in foods than are acids. Our aversion to the taste of bases is probably an evolutionary adaptation to warn us against **alkaloids**, organic bases found in plants that are often poisonous. (For example, the active component of hemlock—the poisonous plant that killed the Greek philosopher Socrates—is the alkaloid coniine.) Nonetheless, some foods, such as coffee and chocolate (especially dark chocolate), contain bitter flavors. Many people enjoy the bitterness, but only after acquiring the taste over time.

Coffee is acidic overall, but bases present in coffee—such as caffeine—and other compounds impart a bitter flavor.

Bases feel slippery because they react with oils on the skin to form soap-like substances. Some household cleaning solutions, such as ammonia, are basic and have the characteristic slippery feel of a base. Bases turn red litmus paper blue; in the laboratory, litmus paper is routinely used to test the basicity of solutions.

Table 17.2 lists some common bases. You can find sodium hydroxide and potassium hydroxide in most chemistry laboratories. They are used in petroleum and cotton processing, as well as in soap and plastic manufacturing. Sodium hydroxide is the active ingredient in products such as Drāno® that unclog drains. In many homes, you can find sodium bicarbonate in the medicine cabinet (it is an active ingredient in some antacids) as well as in the kitchen (labeled as baking soda).

17.3 Definitions of Acids and Bases

What are the main characteristics of the molecules and ions that exhibit acid and base behavior? In this section, we examine two different definitions of acid and bases, and in the next section we will see how the structure of a molecule determines its degree of acidity according to these definitions. The two definitions we examine here are the Arrhenius definition and the Brønsted–Lowry definition (later in the chapter we examine a third one called the Lewis definition). Why three definitions, and which one is correct? As Lewis himself noted in the opening quote of this chapter, each definition has limits, and no single definition is "correct." Rather, each definition is useful in a given instance.

The Arrhenius Definition

In the 1880s, Swedish chemist Svante Arrhenius proposed the following molecular definitions of acids and bases:

Acid—A substance that produces H^+ ions in aqueous solution

Base—A substance that produces OH^- ions in aqueous solution

According to the **Arrhenius definition**, HCl is an acid because it produces H^+ ions in solution (Figure 17.1 ◀).

$$HCl(aq) \longrightarrow H^+(aq) + Cl^-(aq)$$

Hydrogen monochloride (HCl) is a covalent compound and does not contain ions. However, in water it *ionizes* completely to form $H^+(aq)$ ions and $Cl^-(aq)$ ions. The H^+ ions are highly reactive. In aqueous solution, the ions bond to water to form H_3O^+.

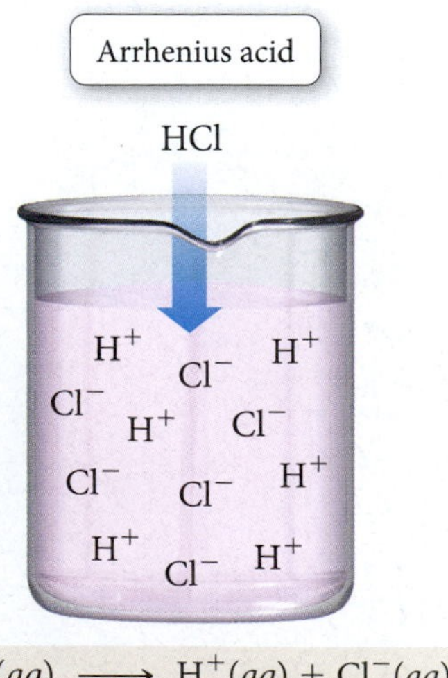

▲ **FIGURE 17.1 Arrhenius Acid** An Arrhenius acid produces H^+ ions in solution.

$$H^+ + H{:}\ddot{\underset{..}{O}}{:}H \longrightarrow \left[H{:}\underset{..}{\overset{H}{\ddot{O}}}{:}H\right]^+$$

The H_3O^+ ion is called the **hydronium ion**. In water, H^+ ions *always* associate with H_2O molecules to form hydronium ions and other associated species with the general formula $H(H_2O)_n{}^+$. For example, an H^+ ion can associate with two water molecules to form $H(H_2O)_2{}^+$, with three to form $H(H_2O)_3{}^+$, and so on. Chemists often use $H^+(aq)$ and $H_3O^+(aq)$ interchangeably, to mean the same thing—an H^+ ion that has been solvated (or dissolved) in water.

Recall from Section 9.4 that the strength of an electrolyte (a substance, such as an acid, that forms ions in solution) depends on the extent of the electrolyte's dissociation into its component ions in solution. A *strong electrolyte* completely dissociates into ions in solution, whereas a *weak electrolyte* only partially dissociates. We define strong and weak acids accordingly. A **strong acid** completely ionizes in solution, whereas a **weak acid** only partially ionizes. We represent the ionization of a strong acid with a single arrow and that of a weak acid with an equilibrium arrow.

Single arrow indicates complete ionizaton

$$HA(aq) \longrightarrow H^+(aq) + A^-(aq)$$

Strong acid

Double arrow indicates partial ionizaton

$$HA(aq) \rightleftharpoons H^+(aq) + A^-(aq)$$

Weak acid

In this notation, HA represents a generic acid with the anion A^-.

According to the Arrhenius definition, NaOH is a base because it produces OH^- ions in solution (Figure 17.2 ▶).

$$NaOH(aq) \longrightarrow Na^+(aq) + OH^-(aq)$$

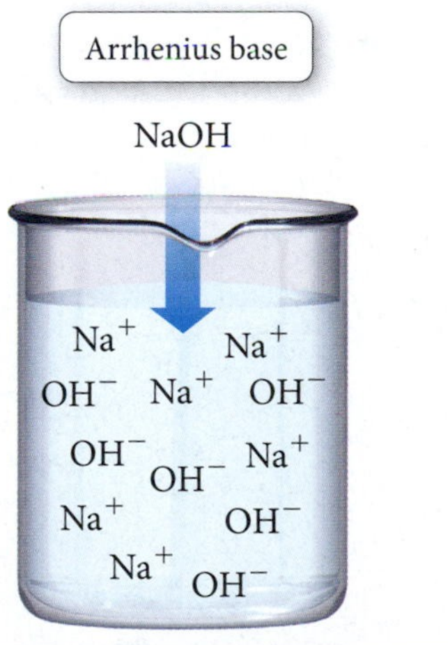

$$NaOH(aq) \longrightarrow Na^+(aq) + OH^-(aq)$$

▲ **FIGURE 17.2 Arrhenius Base** An Arrhenius base produces OH^- ions in solution.

NaOH is an ionic compound and therefore contains Na^+ and OH^- ions. When NaOH is added to water, it *dissociates* or breaks apart into its component ions. NaOH is an example of a **strong base**, one that completely dissociates in solution (analogous to a strong acid). A **weak base** is analogous to a weak acid. Unlike strong bases that contain OH^- and *dissociate* in water, the most common weak bases produce OH^- by accepting a proton from water and ionizing water to form OH^- according to the general equation:

$$B(aq) + H_2O(l) \rightleftharpoons BH^+(aq) + OH^-(aq)$$

In this equation, B is a generic symbol for a weak base. Ammonia, for example, ionizes water as follows:

$$NH_3(aq) + H_2O(l) \rightleftharpoons NH_4{}^+(aq) + OH^-(aq)$$

We examine weak bases more thoroughly in Section 17.8.

Under the Arrhenius definition, acids and bases combine to form water, neutralizing each other in the process:

$$H^+(aq) + OH^-(aq) \longrightarrow H_2O(l)$$

The Brønsted–Lowry Definition

A second, more widely applicable definition of acids and bases, called the **Brønsted–Lowry definition**, was introduced in 1923. This definition focuses on the *transfer of H^+ ions* in an acid–base reaction. Since an H^+ ion is a proton—a hydrogen atom without its electron—this definition focuses on the idea of a proton donor and a proton acceptor.

Acid—Proton (H^+ ion) *donor*

Base—Proton (H^+ ion) *acceptor*

According to this definition, HCl is an acid because, in solution, it donates a proton to water:

$$HCl(aq) + H_2O(l) \longrightarrow H_3O^+(aq) + Cl^-(aq)$$

This definition clearly describes what happens to the H^+ ion from an acid—it associates with a water molecule to form H_3O^+ (a hydronium ion). The Brønsted–Lowry definition also applies nicely to bases (such as NH_3) that do not inherently contain OH^- ions but still produce OH^- ions in solution. According to the Brønsted–Lowry definition, NH_3 is a base because it accepts a proton from water:

All Arrhenius acids and bases are acids and bases under the Brønsted–Lowry definition. However, some Brønsted–Lowry acids and bases cannot be classified as Arrhenius acids and bases.

$$NH_3(aq) + H_2O(l) \rightleftharpoons NH_4^+(aq) + OH^-(aq)$$

According to the Brønsted–Lowry definition, acids (proton donors) and bases (proton acceptors) always occur together. In the reaction between HCl and H_2O, HCl is the proton donor (acid) and H_2O is the proton acceptor (base).

$$\underset{\substack{\text{acid} \\ \text{(proton donor)}}}{HCl(aq)} + \underset{\substack{\text{base} \\ \text{(proton acceptor)}}}{H_2O(l)} \longrightarrow H_3O^+(aq) + Cl^-(aq)$$

In the reaction between NH_3 and H_2O, H_2O is the proton donor (acid) and NH_3 is the proton acceptor (base).

$$\underset{\substack{\text{base} \\ \text{(proton acceptor)}}}{NH_3(aq)} + \underset{\substack{\text{acid} \\ \text{(proton donor)}}}{H_2O(l)} \rightleftharpoons NH_4^+(aq) + OH^-(aq)$$

According to the Brønsted–Lowry definition, some substances—such as water in the previous two equations—can act as acids *or* bases. Substances that can act as acids or bases are **amphoteric**. Notice what happens when we reverse an equation representing Brønsted–Lowry acid–base behavior:

$$\underset{\substack{\text{acid} \\ \text{(proton donor)}}}{NH_4^+(aq)} + \underset{\substack{\text{base} \\ \text{(proton acceptor)}}}{OH^-(aq)} \rightleftharpoons NH_3(aq) + H_2O(l)$$

In this reaction, NH_4^+ is the proton donor (acid) and OH^- is the proton acceptor (base). The substance that was the base (NH_3) has become the acid (NH_4^+) and vice versa. NH_4^+ and NH_3 are often referred to as a **conjugate acid–base pair**, two substances related to each other by the transfer of a proton (Figure 17.3 ▼). A **conjugate acid** is any base to which a proton has been added, and a **conjugate base** is any acid from which a proton has been removed. Going back to the original forward reaction, we can identify the conjugate acid–base pairs:

$$\underset{\text{Base}}{NH_3(aq)} + \underset{\text{Acid}}{H_2O(l)} \rightleftharpoons \underset{\text{Conjugate acid}}{NH_4^+(aq)} + \underset{\text{Conjugate base}}{OH^-(aq)}$$

Summarizing the Brønsted–Lowry Definition of an Acid–Base Reaction:

- A base accepts a proton and becomes a conjugate acid.
- An acid donates a proton and becomes a conjugate base.

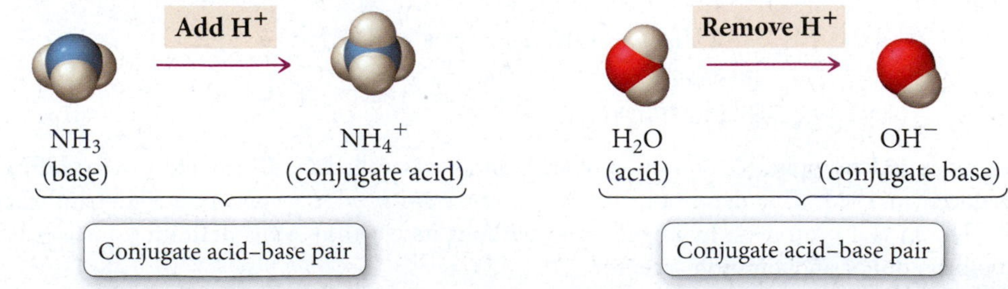

▶ **FIGURE 17.3 Conjugate Acid–Base Pairs** A conjugate acid–base pair consists of two substances related to each other by the transfer of a proton.

EXAMPLE 17.1

Identifying Brønsted–Lowry Acids and Bases and Their Conjugates

In each reaction, identify the Brønsted–Lowry acid, the Brønsted–Lowry base, the conjugate acid, and the conjugate base.

(a) $H_2SO_4(aq) + H_2O(l) \longrightarrow HSO_4^-(aq) + H_3O^+(aq)$

(b) $HCO_3^-(aq) + H_2O(l) \rightleftharpoons H_2CO_3(aq) + OH^-(aq)$

SOLUTION

(a) Because H_2SO_4 donates a proton to H_2O in this reaction, it is the acid (proton donor). After H_2SO_4 donates the proton, it becomes HSO_4^-, the conjugate base. Because H_2O accepts a proton, it is the base (proton acceptor). After H_2O accepts the proton, it becomes H_3O^+, the conjugate acid.

$H_2SO_4(aq) + H_2O(l) \longrightarrow HSO_4^-(aq) + H_3O^+(aq)$

$H_2SO_4(aq)$ (Acid) + $H_2O(l)$ (Base) $\longrightarrow$ $HSO_4^-(aq)$ (Conjugate base) + $H_3O^+(aq)$ (Conjugate acid)

(b) Because H_2O donates a proton to HCO_3^- in this reaction, it is the acid (proton donor). After H_2O donates the proton, it becomes OH^-, the conjugate base. Because HCO_3^- accepts a proton, it is the base (proton acceptor). After HCO_3^- accepts the proton, it becomes H_2CO_3, the conjugate acid.

$HCO_3^-(aq) + H_2O(l) \rightleftharpoons H_2CO_3(aq) + OH^-(aq)$

$HCO_3^-(aq)$ (Base) + $H_2O(l)$ (Acid) $\rightleftharpoons$ $H_2CO_3(aq)$ (Conjugate acid) + $OH^-(aq)$ (Conjugate base)

FOR PRACTICE 17.1

In each reaction, identify the Brønsted–Lowry acid, the Brønsted–Lowry base, the conjugate acid, and the conjugate base.

(a) $C_5H_5N(aq) + H_2O(l) \rightleftharpoons C_5H_5NH^+(aq) + OH^-(aq)$

(b) $HNO_3(aq) + H_2O(l) \rightleftharpoons H_3O^+(aq) + NO_3^-(aq)$

Conjugate Acid–Base Pairs

17.1 Cc Conceptual Connection

Which pair is not a conjugate acid–base pair?

(a) $(CH_3)_3N$; $(CH_3)_3NH^+$ **(b)** H_2SO_4; H_2SO_3 **(c)** HNO_2; NO_2^-

17.4 Acid Strength and Molecular Structure

We have learned that a Brønsted–Lowry acid is a proton (H^+) donor. Now we explore why some hydrogen-containing molecules act as proton donors while others do not. In other words, we want to explore *how the structure of a molecule affects its acidity.* Why is H_2S acidic while CH_4 is not? Or why is HF a weak acid while HCl is a strong acid? We divide our discussion about these issues into two categories: binary acids (those containing hydrogen and only one other element) and oxyacids (those containing hydrogen bonded to an oxygen atom that is bonded to another element).

Binary Acids

Consider the bond between a hydrogen atom and some other generic element (which we will call Y):

$$H—Y$$

The factors affecting the ease with which this hydrogen is donated (and therefore acidic) are the *polarity* of the bond and the *strength* of the bond.

Bond Polarity Using the notation introduced in Chapter 6, the H—Y bond must be polarized with the hydrogen atom as the positive pole in order for HY to be acidic:

$$\delta^{+}\text{H—Y}^{\delta-}$$

This requirement makes physical sense because the hydrogen atom must be lost as a positively charged ion (H^+). A partial positive charge on the hydrogen atom facilitates its loss. Consider the following three bonds and their corresponding dipole moments:

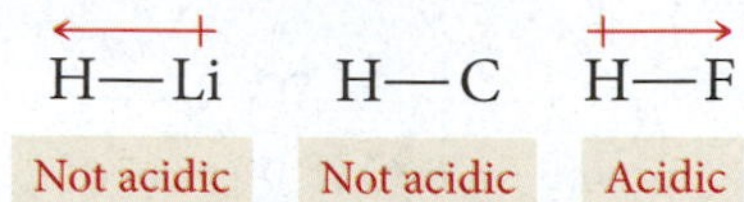

LiH is ionic with *the negative charge on the hydrogen atom*; therefore LiH is not acidic. The C—H bond is virtually nonpolar because the electronegativities of carbon and hydrogen are similar; therefore C—H is not acidic. In contrast, the H—F bond is polar with the positive charge on the hydrogen atom. HF is an acid. This is because the partial positive charge on the hydrogen atom makes it easier for the hydrogen to be lost as an H^+ ion.

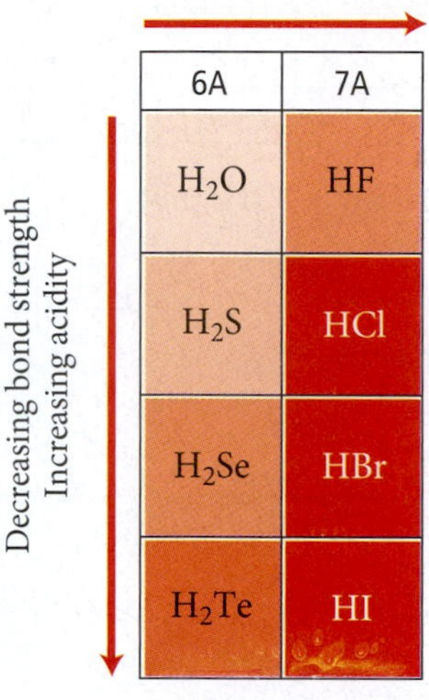

▲ **FIGURE 17.4 Acidity of the Group 6A and 7A Hydrides** From left to right, the hydrides become more acidic because the H—Y bond becomes more polar. From top to bottom, these hydrides become more acidic because the H—Y bond becomes weaker.

Bond Strength The strength of the H—Y bond also affects the strength of the corresponding acid. As you might expect, the stronger the bond, the weaker the acid. The more tightly the hydrogen atom is held, the less likely it is to come off. We can see the effect of bond strength by comparing the bond strengths and acidities of the hydrogen halides.

Acid	Bond Energy (kJ/mol)	Acid Strength
H—F	565	Weak
H—Cl	431	Strong
H—Br	364	Strong

HCl and HBr have weaker bonds and are both strong acids. HF, however, has a stronger bond and is therefore a weak acid, despite the greater bond polarity of HF. Other factors also effect acid strength, but they are beyond our scope.

The Combined Effect of Bond Polarity and Bond Strength We can see the combined effect of bond polarity and bond strength by examining the trends in acidity of the group 6A and 7A hydrides illustrated in Figure 17.4 ◀. The hydrides become more acidic from left to right as the H—Y bond becomes more polar. The hydrides also become more acidic from top to bottom as the H—Y bond becomes weaker.

Oxyacids

Oxyacids are sometimes called oxoacids.

Oxyacids contain a hydrogen atom bonded to an oxygen atom. The oxygen atom is in turn bonded to another atom (which we will call Y):

$$\text{H—O—Y—}$$

Y may or may not be bonded to additional atoms. The factors affecting the ease with which the hydrogen in an oxyacid will be donated (and therefore be acidic) are the *electronegativity of the element* Y and the *number of oxygen atoms attached to the element* Y.

The Electronegativity of Y The more electronegative the element Y is, the more it weakens and polarizes the H—O bond and the more acidic the oxyacid is. We can see this effect by comparing the electronegativity of Y and the relative strengths of the following weak oxyacids:

Acid	Electronegativity of Y	Acid Strength
H—O—I	2.5	Weakest
H—O—Br	2.8	Weaker
H—O—Cl	3.0	Weak

Chlorine is the most electronegative of the three elements, and the corresponding acid is the least weak of these three weak acids.

The Number of Oxygen Atoms Bonded to Y Oxyacids may contain additional oxygen atoms bonded to the element Y. Because these additional oxygen atoms are electronegative, they draw electron density away from Y, which in turn draws electron density away from the H—O bond, further weakening and polarizing it, and leading to increasing acidity. We can see this effect by comparing the relative strengths of the following series of acids.

Acid	Structure	Acid Strength
$HClO_4$	H—O—Cl(=O)(=O)=O	Strong
$HClO_3$	H—O—Cl(=O)=O	Weak
$HClO$	H—O—Cl	Very Weak

The greater the number of oxygen atoms bonded to Y, the stronger the acid. On this basis we would predict that H_2SO_4 is a stronger acid than H_2SO_3 and that HNO_3 is stronger than HNO_2. In both cases, our predictions are correct. Both H_2SO_4 and HNO_3 are strong acids, while H_2SO_3 and HNO_2 are weak acids.

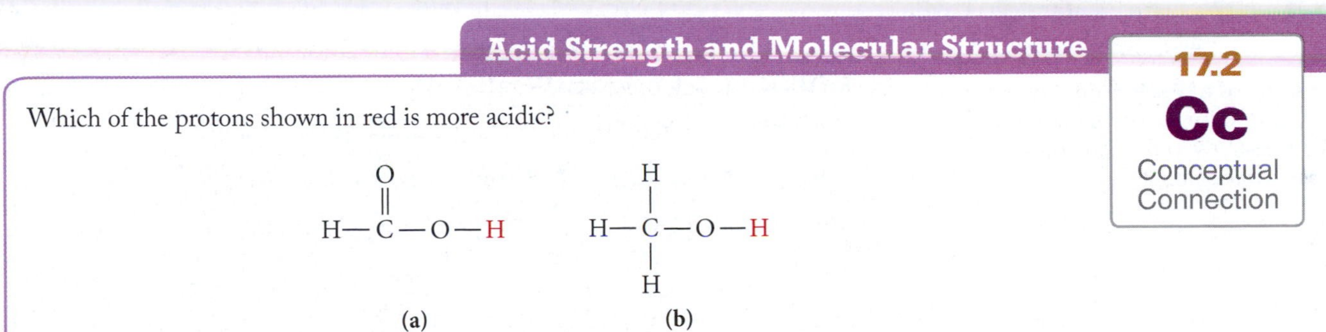

Acid Strength and Molecular Structure
17.2
Cc
Conceptual Connection
Which of the protons shown in red is more acidic?
O
H—C—O—H
(a)
H
H—C—O—H
H
(b)

17.5 Acid Strength and the Acid Ionization Constant (K_a)

We have seen that, depending on the structure of an acid, it can be either strong (in which case it completely ionizes in solution) or weak (in which case it only partially ionizes in solution). We can think of acid strength in terms of the equilibrium concepts we discussed in Chapter 16. For the generic acid HA, we can write the following ionization equation:

$$HA(aq) + H_2O(l) \rightleftharpoons H_3O^+(aq) + A^-(aq)$$

If the equilibrium lies far to the right, the acid is strong. If the equilibrium lies to the left, the acid is weak. The range of acid strength is continuous, but for most purposes, the categories of strong and weak are useful.

Strong Acids

Hydrochloric acid (HCl) is an example of a strong acid. An HCl solution contains virtually no intact HCl; the HCl has essentially all ionized to form $H_3O^+(aq)$ and $Cl^-(aq)$ (Figure 17.5 ▶). A 1.0 M HCl solution has an H_3O^+ concentration of 1.0 M.

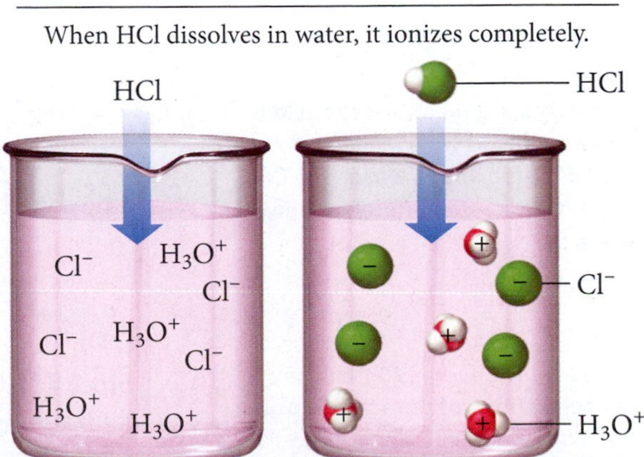

▲ **FIGURE 17.5 Ionization of a Strong Acid** When HCl dissolves in water, it completely ionizes to form H_3O^+ and Cl^-. The solution contains virtually no intact HCl.

Abbreviating the concentration of H_3O^+ as $[H_3O^+]$, we say that a 1.0 M HCl solution has $[H_3O^+] = 1.0$ M.

Table 17.3 lists the six important strong acids. The first five acids in the table are **monoprotic acids**, acids containing only one ionizable proton. Sulfuric acid is an example of a **diprotic acid**, an acid containing two ionizable protons.

An ionizable proton is one that ionizes in solution. We discuss polyprotic acids in more detail in Section 17.10.

TABLE 17.3 Strong Acids

Hydrochloric acid (HCl)	Nitric acid (HNO_3)
Hydrobromic acid (HBr)	Perchloric acid ($HClO_4$)
Hydriodic acid (HI)	Sulfuric acid (H_2SO_4) (*diprotic*)

A Weak Acid

When HF dissolves in water, only a fraction of the molecules ionize.

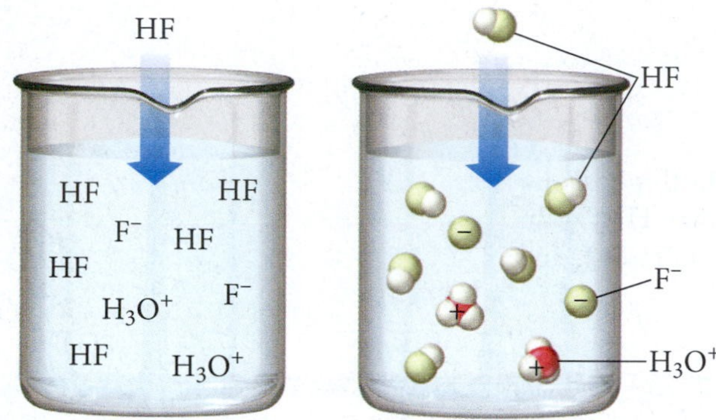

▲ **FIGURE 17.6 Ionization of a Weak Acid** When HF dissolves in water, only a fraction of the dissolved molecules ionize to form H_3O^+ and F^-. The solution contains many intact HF molecules.

Weak Acids

In contrast to HCl, HF is a weak acid, one that does not completely ionize in solution. As we saw in Section 17.4, the strength of the HF bond relative to the HCl bond makes HF less likely to ionize in solution than HCl. Therefore, an HF solution contains a large number of intact (or un-ionized) HF molecules; it also contains some $H_3O^+(aq)$ and $F^-(aq)$ (Figure 17.6 ◀). In other words, a 1.0 M HF solution has $[H_3O^+]$ that is much less than 1.0 M because only some of the HF molecules ionize to form H_3O^+. Table 17.4 lists some other common weak acids.

Notice that two of the weak acids in Table 17.4 are diprotic, meaning that they have two ionizable protons, and one is **triprotic** (three ionizable protons). We discuss polyprotic acids in more detail in Section 17.10.

TABLE 17.4 Common Weak Acids

Hydrofluoric acid (HF)	Sulfurous acid (H_2SO_3) (*diprotic*)
Acetic acid ($HC_2H_3O_2$)	Carbonic acid (H_2CO_3) (*diprotic*)
Formic acid ($HCHO_2$)	Phosphoric acid (H_3PO_4) (*triprotic*)

We can also write the formulas for acetic acid and formic acid as CH_3COOH and HCOOH, respectively, to indicate that in these compounds the only H that ionizes is the one attached to an oxygen atom.

The Acid Ionization Constant (K_a)

We quantify the relative strength of a weak acid with the **acid ionization constant (K_a)**, which is the equilibrium constant for the ionization reaction of the weak acid. Based on the law of mass action (see Section 16.3) for the two equivalent reactions:

$$HA(aq) + H_2O(l) \rightleftharpoons H_3O^+(aq) + A^-(aq)$$

$$HA(aq) \rightleftharpoons H^+(aq) + A^-(aq)$$

The equilibrium constant is:

$$K_a = \frac{[H_3O^+][A^-]}{[HA]} = \frac{[H^+][A^-]}{[HA]}$$

Since $[H_3O^+]$ is equivalent to $[H^+]$, both forms of the above expression are equal. Although the ionization constants for all weak acids are relatively small (otherwise the acid would not be a weak acid), they do vary in magnitude. The smaller the constant, the less the acid ionizes, and the weaker the acid. Table 17.5 lists the acid ionization constants for a number of common weak acids in order of decreasing acid strength.

The terms *strong* and *weak* acids are often confused with the terms *concentrated* and *dilute* acids. Can you articulate the difference between these terms?

Recall from Chapter 16 that the concentrations of pure solids or pure liquids are not included in the expression for K_c. Therefore, $H_2O(l)$ is not included in the expression for K_a.

TABLE 17.5 Acid Ionization Constants (K_a) for Some Monoprotic Weak Acids at 25 °C

Acid	Formula	Structural Formula	Ionization Reaction	K_a
Chlorous acid	$HClO_2$	H—O—Cl=O	$HClO_2(aq) + H_2O(l) \rightleftharpoons H_3O^+(aq) + ClO_2^-(aq)$	1.1×10^{-2}
Nitrous acid	HNO_2	H—O—N=O	$HNO_2(aq) + H_2O(l) \rightleftharpoons H_3O^+(aq) + NO_2^-(aq)$	4.6×10^{-4}
Hydrofluoric acid	HF	H—F	$HF(aq) + H_2O(l) \rightleftharpoons H_3O^+(aq) + F^-(aq)$	3.5×10^{-4}
Formic acid	$HCHO_2$	H—O—C(=O)—H	$HCHO_2(aq) + H_2O(l) \rightleftharpoons H_3O^+(aq) + CHO_2^-(aq)$	1.8×10^{-4}
Benzoic acid	$HC_7H_5O_2$	H—O—C(=O)—C_6H_5	$HC_7H_5O_2(aq) + H_2O(l) \rightleftharpoons H_3O^+(aq) + C_7H_5O_2^-(aq)$	6.5×10^{-5}
Acetic acid	$HC_2H_3O_2$	H—O—C(=O)—CH_3	$HC_2H_3O_2(aq) + H_2O(l) \rightleftharpoons H_3O^+(aq) + C_2H_3O_2^-(aq)$	1.8×10^{-5}
Hypochlorous acid	HClO	H—O—Cl	$HClO(aq) + H_2O(l) \rightleftharpoons H_3O^+(aq) + ClO^-(aq)$	2.9×10^{-8}
Hydrocyanic acid	HCN	H—C≡N	$HCN(aq) + H_2O(l) \rightleftharpoons H_3O^+(aq) + CN^-(aq)$	4.9×10^{-10}
Phenol	HC_6H_5O	HO—C_6H_5	$HC_6H_5O(aq) + H_2O(l) \rightleftharpoons H_3O^+(aq) + C_6H_5O^-(aq)$	1.3×10^{-10}

Relative Strengths of Weak Acids

17.3 Cc Conceptual Connection

Consider these two acids and their K_a values.

$$\text{HF} \quad K_a = 3.5 \times 10^{-4}$$

$$\text{HClO} \quad K_a = 2.9 \times 10^{-8}$$

Which acid is stronger?

17.6 Autoionization of Water and pH

We saw previously that water acts as a base when it reacts with HCl and as an acid when it reacts with NH_3:

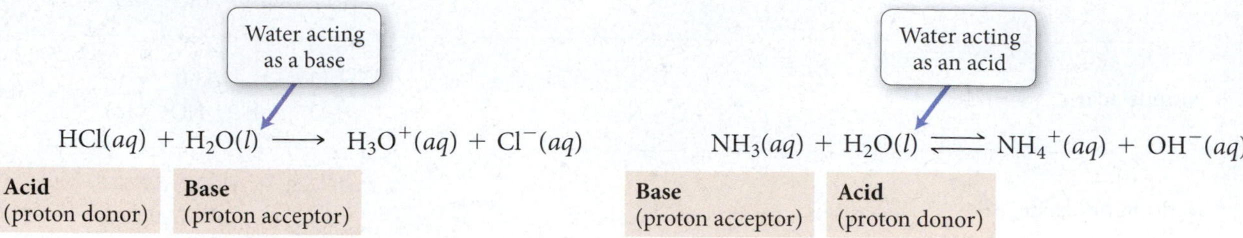

Water is *amphoteric*; it can act as either an acid or a base. Even when pure, water acts as an acid and a base with itself, a process called **autoionization**:

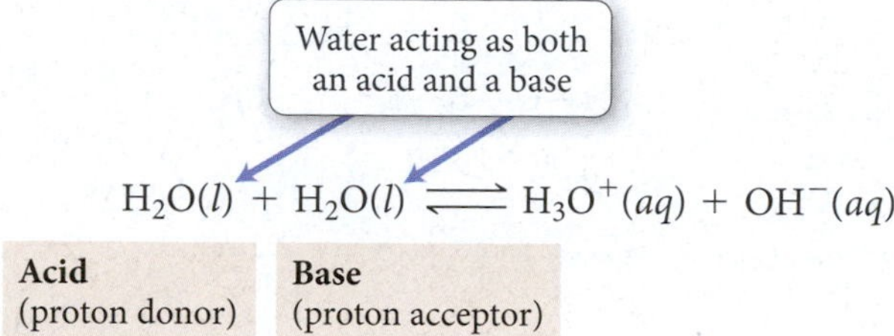

We can also write the autoionization reaction as:

$$H_2O(l) \rightleftharpoons H^+(aq) + OH^-(aq)$$

The equilibrium constant for this reaction is the product of the concentration of the two ions:

$$K_w = [H_3O^+][OH^-] = [H^+][OH^-]$$

This equilibrium constant is the **ion product constant for water** (K_w) (sometimes called the *dissociation constant for water*). At 25 °C, $K_w = 1.0 \times 10^{-14}$. In pure water, since H_2O is the only source of these ions, the concentrations of H_3O^+ and OH^- are equal, and the solution is **neutral**. Since the concentrations are equal, we can easily calculate them from K_w.

$$[H_3O^+] = [OH^-] = \sqrt{K_w} = 1.0 \times 10^{-7} \quad \text{(in pure water at 25 °C)}$$

As you can see, in pure water, the concentrations of H_3O^+ and OH^- are *very small* (1.0×10^{-7} M) at room temperature.

An **acidic solution** contains an acid that creates additional H_3O^+ ions, causing $[H_3O^+]$ to increase. However, the *ion product constant still applies*:

$$[H_3O^+][OH^-] = K_w = 1.0 \times 10^{-14}$$

The concentration of H_3O^+ times the concentration of OH^- is always 1.0×10^{-14} at 25 °C. If $[H_3O^+]$ increases, then $[OH^-]$ must decrease for the ion product constant to remain 1.0×10^{-14}. For example, if $[H_3O^+] = 1.0 \times 10^{-3}$, we can find $[OH^-]$ by solving the ion product constant expression for $[OH^-]$:

$$(1.0 \times 10^{-3})[OH^-] = 1.0 \times 10^{-14}$$

$$[OH^-] = \frac{1.0 \times 10^{-14}}{1.0 \times 10^{-3}} = 1.0 \times 10^{-11}\ \text{M}$$

In an acidic solution $[H_3O^+] > [OH^-]$.

A **basic solution** contains a base that creates additional OH^- ions, causing $[OH^-]$ to increase and $[H_3O^+]$ to decrease, but again the *ion product constant still applies.* Suppose $[OH^-] = 1.0 \times 10^{-2}$; we can find $[H_3O^+]$ by solving the ion product constant expression for $[H_3O^+]$:

$$[H_3O^+](1.0 \times 10^{-2}) = 1.0 \times 10^{-14}$$

$$[H_3O^+] = \frac{1.0 \times 10^{-14}}{1.0 \times 10^{-2}} = 1.0 \times 10^{-12}\ M$$

In a basic solution $[OH^-] > [H_3O^+]$.

Notice that changing $[H_3O^+]$ in an aqueous solution produces an inverse change in $[OH^-]$ and vice versa.

Summarizing K_w:

- A *neutral solution* contains $[H_3O^+] = [OH^-] = 1.0 \times 10^{-7}$ M (at 25 °C).
- An *acidic solution* contains $[H_3O^+] > [OH^-]$.
- A *basic solution* contains $[OH^-] > [H_3O^+]$.
- In *all aqueous solutions* both H_3O^+ and OH^- are present and $[H_3O^+][OH^-] = K_w = 1.0 \times 10^{-14}$ (at 25 °C).

EXAMPLE 17.2

Using K_w in Calculations

Calculate $[OH^-]$ at 25 °C for each solution and determine if the solution is acidic, basic, or neutral.

(a) $[H_3O^+] = 7.5 \times 10^{-5}$ M

(b) $[H_3O^+] = 1.5 \times 10^{-9}$ M

(c) $[H_3O^+] = 1.0 \times 10^{-7}$ M

SOLUTION

(a) To find $[OH^-]$ use the ion product constant. Substitute the given value for $[H_3O^+]$ and solve the equation for $[OH^-]$. Since $[H_3O^+] > [OH^-]$, the solution is acidic.	$[H_3O^+][OH^-] = K_w = 1.0 \times 10^{-14}$ $(7.5 \times 10^{-5})[OH^-] = 1.0 \times 10^{-14}$ $[OH^-] = \frac{1.0 \times 10^{-14}}{7.5 \times 10^{-5}} = 1.3 \times 10^{-10}$ M Acidic solution
(b) Substitute the given value for $[H_3O^+]$ and solve the acid ionization equation for $[OH^-]$. Since $[H_3O^+] < [OH^-]$, the solution is basic.	$(1.5 \times 10^{-9})[OH^-] = 1.0 \times 10^{-14}$ $[OH^-] = \frac{1.0 \times 10^{-14}}{1.5 \times 10^{-9}} = 6.7 \times 10^{-6}$ M Basic solution
(c) Substitute the given value for $[H_3O^+]$ and solve the acid ionization equation for $[OH^-]$. Since $[H_3O^+] = 1.0 \times 10^{-7}$ and $[OH^-] = 1.0 \times 10^{-7}$, the solution is neutral.	$(1.0 \times 10^{-7})[OH^-] = 1.0 \times 10^{-14}$ $[OH^-] = \frac{1.0 \times 10^{-14}}{1.0 \times 10^{-7}} = 1.0 \times 10^{-7}$ M Neutral solution

FOR PRACTICE 17.2

Calculate $[H_3O^+]$ at 25 °C for each solution and determine if the solution is acidic, basic, or neutral.

(a) $[OH^-] = 1.5 \times 10^{-2}$ M **(b)** $[OH^-] = 1.0 \times 10^{-7}$ M **(c)** $[OH^-] = 8.2 \times 10^{-10}$ M

The log of a number is the exponent to which 10 must be raised to obtain that number. Thus, $\log 10^1 = 1$; $\log 10^2 = 2$; $\log 10^{-1} = -1$; $\log 10^{-2} = -2$, etc. (see Appendix III).

Specifying the Acidity or Basicity of a Solution: The pH Scale

The pH scale is a compact way to specify the acidity of a solution. We define **pH** as the negative of the log of the hydronium ion concentration:

$$pH = -\log[H_3O^+]$$

A solution with $[H_3O^+] = 1.0 \times 10^{-3}$ M (acidic) has a pH of:

$$\begin{aligned} pH &= -\log[H_3O^+] \\ &= -\log(1.0 \times 10^{-3}) \\ &= -(-3.00) \\ &= 3.00 \end{aligned}$$

Notice that we report the pH to two *decimal places* here. This is because only the numbers to the right of the decimal point are significant in a logarithm. Because our original value for the concentration had two significant figures, the log of that number has two decimal places.

2 significant digits → 2 decimal places

$$\log 1.0 \times 10^{-3} = 3.00$$

If the original number had three significant digits, we would report the log to three decimal places.

3 significant digits → 3 decimal places

$$\log 1.00 \times 10^{-3} = 3.000$$

A solution with $[H_3O^+] = 1.0 \times 10^{-7}$ M (neutral) has a pH of:

$$\begin{aligned} pH &= -\log[H_3O^+] \\ &= -\log(1.0 \times 10^{-7}) \\ &= -(-7.00) \\ &= 7.00 \end{aligned}$$

In general, at 25 °C:

- $pH < 7$ The solution is *acidic*.
- $pH > 7$ The solution is *basic*.
- $pH = 7$ The solution is *neutral*.

TABLE 17.6 The pH of Some Common Substances

Substance	pH
Gastric juice (human stomach)	1.0–3.0
Limes	1.8–2.0
Lemons	2.2–2.4
Soft drinks	2.0–4.0
Plums	2.8–3.0
Wines	2.8–3.8
Apples	2.9–3.3
Peaches	3.4–3.6
Cherries	3.2–4.0
Beers	4.0–5.0
Rainwater (unpolluted)	5.6
Human blood	7.3–7.4
Egg whites	7.6–8.0
Milk of magnesia	10.5
Household ammonia	10.5–11.5
4% NaOH solution	14

Table 17.6 lists the pH of some common substances. As we discussed in Section 17.2, many foods, especially fruits, are acidic and have low pH values. Relatively few foods are basic. The foods with the lowest pH values are limes and lemons, and they are among the sourest. Because the pH scale is a *logarithmic scale*, a change of 1 pH unit corresponds to a tenfold change in H_3O^+ concentration (Figure 17.7 ▼). For example, a lime with a pH of 2.0 is 10 times more acidic than a plum with a pH of 3.0 and 100 times more acidic than a cherry with a pH of 4.0.

Concentrated acid solutions can have a negative pH. For example, if $[H_3O^+] = 2.0$ M, the pH is -0.30.

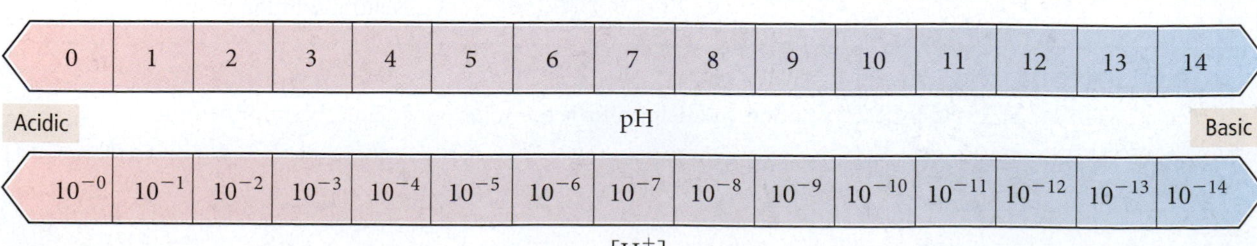

▶ **FIGURE 17.7 The pH Scale** An increase of 1 on the pH scale corresponds to a factor of 10 decrease in $[H_3O^+]$.

EXAMPLE 17.3
Calculating pH from $[H_3O^+]$ or $[OH^-]$

Calculate the pH of each solution at 25 °C and indicate whether the solution is acidic or basic.

(a) $[H_3O^+] = 1.8 \times 10^{-4}$ M **(b)** $[OH^-] = 1.3 \times 10^{-2}$ M

SOLUTION

(a) To calculate pH, substitute the given $[H_3O^+]$ into the pH equation. Since pH < 7, this solution is acidic.	$\text{pH} = -\log[H_3O^+]$ $= -\log(1.8 \times 10^{-4})$ $= -(-3.74)$ $= 3.74$ (acidic)
(b) First use K_w to find $[H_3O^+]$ from $[OH^-]$.	$[H_3O^+][OH^-] = K_w = 1.0 \times 10^{-14}$ $[H_3O^+](1.3 \times 10^{-2}) = 1.0 \times 10^{-14}$ $[H_3O^+] = \frac{1.0 \times 10^{-14}}{1.3 \times 10^{-2}} = 7.7 \times 10^{-13}$ M
Then substitute $[H_3O^+]$ into the pH expression to find pH. Since pH > 7, this solution is basic.	$\text{pH} = -\log[H_3O^+]$ $= -\log(7.7 \times 10^{-13})$ $= -(-12.11)$ $= 12.11$ (basic)

FOR PRACTICE 17.3
Calculate the pH of each solution and indicate whether the solution is acidic or basic.

(a) $[H_3O^+] = 9.5 \times 10^{-9}$ M **(b)** $[OH^-] = 7.1 \times 10^{-3}$ M

EXAMPLE 17.4
Calculating $[H_3O^+]$ from pH

Calculate $[H_3O^+]$ for a solution with a pH of 4.80.

SOLUTION

To determine the $[H_3O^+]$ from pH, start with the equation that defines pH. Substitute the given value of pH and then solve for $[H_3O^+]$. Because the given pH value was reported to two decimal places, the $[H_3O^+]$ is written to two significant figures. (Remember that $10^{\log x} = x$; see Appendix III. Some calculators use an inv log key to represent this function.)	$\text{pH} = -\log[H_3O^+]$ $4.80 = -\log[H_3O^+]$ $-4.80 = \log[H_3O^+]$ $10^{-4.80} = 10^{\log[H_3O^+]}$ $10^{-4.80} = [H_3O^+]$ $[H_3O^+] = 1.6 \times 10^{-5}$ M

FOR PRACTICE 17.4
Calculate $[H_3O^+]$ for a solution with a pH of 8.37.

pOH and Other p Scales

The pOH scale is analogous to the pH scale, but we define pOH with respect to $[OH^-]$ instead of $[H_3O^+]$.

$$\text{pOH} = -\log[OH^-]$$

▶ **FIGURE 17.8 pH and pOH**

0.0	1.0	2.0	3.0	4.0	5.0	6.0	7.0	8.0	9.0	10.0	11.0	12.0	13.0	14.0

Acidic pH Basic

14.0	13.0	12.0	11.0	10.0	9.0	8.0	7.0	6.0	5.0	4.0	3.0	2.0	1.0	0.0

pOH

A solution with an $[OH^-]$ of 1.0×10^{-3} M (basic) has a pOH of 3.00. On the pOH scale, a pOH less than 7 is basic and a pOH greater than 7 is acidic. A pOH of 7 is neutral (Figure 17.8 ▲). We can derive a relationship between pH and pOH at 25 °C from the expression for K_w:

$$[H_3O^+][OH^-] = 1.0 \times 10^{-14}$$

Taking the log of both sides, we get:

$$\begin{aligned} \log([H_3O^+][OH^-]) &= \log(1.0 \times 10^{-14}) \\ \log[H_3O^+] + \log[OH^-] &= -14.00 \\ -\log[H_3O^+] - \log[OH^-] &= 14.00 \\ \text{pH} + \text{pOH} &= 14.00 \end{aligned}$$

The sum of pH and pOH is always equal to 14.00 at 25 °C. Therefore, a solution with a pH of 3 has a pOH of 11.

Another common p scale is the pK_a scale defined as:

Notice that p is the mathematical operator −log; thus, pX = −log X.

$$pK_a = -\log K_a$$

The pK_a of a weak acid is another way to quantify strength. The smaller the pK_a, the stronger the acid. For example, chlorous acid, with a K_a of 1.1×10^{-2}, has a pK_a of 1.96 and formic acid, with a K_a of 1.8×10^{-4}, has a pK_a of 3.74.

17.7 Finding the $[H_3O^+]$ and pH of Strong and Weak Acid Solutions

In a solution containing a strong or weak acid, there are two potential sources of H_3O^+: the ionization of the acid itself and the autoionization of water. If we let HA be a strong or weak acid, the ionization reactions are:

$$HA(aq) + H_2O(l) \rightleftharpoons H_3O^+(aq) + A^-(aq) \qquad \text{Strong or Weak Acid}$$

$$H_2O(l) + H_2O(l) \rightleftharpoons H_3O^+(aq) + OH^-(aq) \qquad K_w = 1.0 \times 10^{-14}$$

Except in extremely dilute acid solutions, the autoionization of water contributes a negligibly small amount of H_3O^+ compared to the ionization of the strong or weak acid. Recall from Section 17.6 that autoionization in pure water produces an H_3O^+ concentration of 1.0×10^{-7} M. In a strong or weak acid solution, the additional H_3O^+ from the acid causes the autoionization of water equilibrium to shift left (as described by Le Châtelier's principle). Consequently, in most strong or weak acid solutions, the autoionization of water produces even less H_3O^+ than in pure water and can be ignored. Therefore we can focus exclusively on the amount of H_3O^+ produced by the acid.

Strong Acids

The only exceptions are extremely dilute ($<10^{-5}$ M) strong acid solutions.

Because strong acids, by definition, completely ionize in solution, and because we can (in nearly all cases) ignore the contribution of the autoionization of water, *the concentration of* H_3O^+ *in a strong acid solution is equal to the concentration of the strong acid.* For example, a 0.10 M HCl solution has an H_3O^+ concentration of 0.10 M and a pH of 1.00:

$$0.10 \text{ M HCl} \Rightarrow [H_3O^+] = 0.10 \text{ M} \Rightarrow \text{pH} = -\log(0.10) = 1.00$$

Weak Acids

Determining the pH of a weak acid solution is more complicated because the concentration of H_3O^+ is *not equal* to the concentration of the weak acid. For example, if we make solutions of 0.10 M HCl (a strong acid) and 0.10 M acetic acid (a weak acid) in the laboratory and measure the pH of each, we get the following results:

$$0.10 \text{ M HCl} \qquad \text{pH} = 1.00$$
$$0.10 \text{ M } HC_2H_3O_2 \qquad \text{pH} = 2.87$$

The pH of the acetic acid solution is higher (it is less acidic) because acetic acid only partially ionizes. Calculating the $[H_3O^+]$ formed by the ionization of a weak acid requires solving an equilibrium problem similar to those in Chapter 16. Consider, for example, a 0.10 M solution of the generic weak acid HA with an acid ionization constant K_a. Since we can ignore the contribution of the autoionization of water, we only have to determine the concentration of H_3O^+ formed by the following equilibrium:

$$HA(aq) + H_2O(l) \rightleftharpoons H_3O^+(aq) + A^-(aq) \qquad K_a$$

We can summarize the initial conditions, the changes, and the equilibrium conditions in the following ICE table:

	[HA]	$[H_3O^+]$	$[A^-]$
Initial	0.10	≈0.00	0.00
Change	$-x$	$+x$	$+x$
Equilibrium	$0.10 - x$	x	x

ICE tables were first introduced in Section 16.6. The reactant $H_2O(l)$ is a pure liquid and is therefore not included in either the equilibrium constant expression or the ICE table (see Section 16.5).

The initial H_3O^+ concentration is listed as *approximately* zero because of the negligibly small contribution of H_3O^+ due to the autoionization of water (discussed previously). The variable x represents the amount of HA that ionizes. As discussed in Chapter 16, each *equilibrium* concentration is the sum of the two entries above it in the ICE table. In order to find the equilibrium concentration of H_3O^+, we must find the value of the variable x. We can use the equilibrium expression to set up an equation in which x is the only variable:

$$K_a = \frac{[H_3O^+][A^-]}{[HA]}$$
$$= \frac{x^2}{0.10 - x}$$

As is often the case with equilibrium problems, we arrive at a quadratic equation in x, which we can solve using the quadratic formula. However, in many cases we can apply the *x is small* approximation (first discussed in Section 16.8). In Examples 17.5 and 17.6, we examine the general procedure for solving weak acid equilibrium problems. In both of these examples, the *x is small* approximation works well. In Example 17.7, we solve a problem in which the *x is small* approximation does not work. In such cases, we can solve the quadratic equation explicitly, or we can apply the method of successive approximations (also discussed in Section 16.8). Finally, in Example 17.8, we work a problem in which we find the equilibrium constant of a weak acid from its pH.

PROCEDURE FOR

Finding the pH (or $[H_3O^+]$) of a Weak Acid Solution

To solve these types of problems, follow the procedure outlined below.

EXAMPLE 17.5 Finding the $[H_3O^+]$ of a Weak Acid Solution

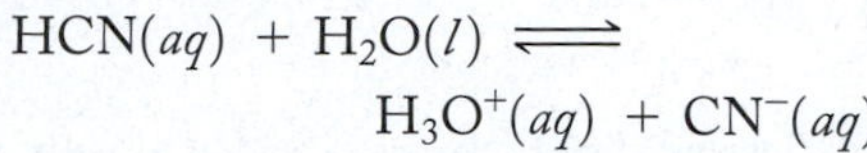

Find the $[H_3O^+]$ of a 0.100 M HCN solution.

EXAMPLE 17.6 Finding the pH of a Weak Acid Solution

Find the pH of a 0.200 M HNO_2 solution.

1. Write the balanced equation for the ionization of the acid and use it as a guide to prepare an ICE table showing the given concentration of the weak acid as its initial concentration. Leave room in the table for the changes in concentrations and for the equilibrium concentrations.

(Note that the $[H_3O^+]$ is listed as approximately zero because the autoionization of water produces a negligibly small amount of H_3O^+.)

Example 17.5:

$$HCN(aq) + H_2O(l) \rightleftharpoons H_3O^+(aq) + CN^-(aq)$$

	[HCN]	$[H_3O^+]$	$[CN^-]$
Initial	0.100	≈0.00	0.00
Change			
Equil			

Example 17.6:

$$HNO_2(aq) + H_2O(l) \rightleftharpoons H_3O^+(aq) + NO_2^-(aq)$$

	$[HNO_2]$	$[H_3O^+]$	$[NO_2^-]$
Initial	0.200	≈0.00	0.00
Change			
Equil			

2. Represent the change in the concentration of H_3O^+ with the variable x. Define the changes in the concentrations of the other reactants and products in terms of x. Always keep in mind the stoichiometry of the reaction.

Example 17.5:

$$HCN(aq) + H_2O(l) \rightleftharpoons H_3O^+(aq) + CN^-(aq)$$

	[HCN]	$[H_3O^+]$	$[CN^-]$
Initial	0.100	≈0.00	0.00
Change	$-x$	$+x$	$+x$
Equil			

Example 17.6:

$$HNO_2(aq) + H_2O(l) \rightleftharpoons H_3O^+(aq) + NO_2^-(aq)$$

	$[HNO_2]$	$[H_3O^+]$	$[NO_2^-]$
Initial	0.200	≈0.00	0.00
Change	$-x$	$+x$	$+x$
Equil			

3. Sum each column to determine the equilibrium concentrations in terms of the initial concentrations and the variable x.

Example 17.5:

$$HCN(aq) + H_2O(l) \rightleftharpoons H_3O^+(aq) + CN^-(aq)$$

	[HCN]	$[H_3O^+]$	$[CN^-]$
Initial	0.100	≈0.00	0.00
Change	$-x$	$+x$	$+x$
Equil	$0.100 - x$	x	x

Example 17.6:

$$HNO_2(aq) + H_2O(l) \rightleftharpoons H_3O^+(aq) + NO_2^-(aq)$$

	$[HNO_2]$	$[H_3O^+]$	$[NO_2^-]$
Initial	0.200	≈0.00	0.00
Change	$-x$	$+x$	$+x$
Equil	$0.200 - x$	x	x

4. Substitute the expressions for the equilibrium concentrations (from step 3) into the expression for the acid ionization constant (K_a).

In many cases, you can make the approximation that x *is small* (as discussed in Section 16.8). **Substitute the value of the acid ionization constant (from Table 17.5) into the K_a expression and solve for x.**

Example 17.5:

$$K_a = \frac{[H_3O^+][CN^-]}{[HCN]}$$

$$= \frac{x^2}{0.100 - \cancel{x}} \quad (x \text{ is small})$$

$$4.9 \times 10^{-10} = \frac{x^2}{0.100}$$

$$\sqrt{4.9 \times 10^{-10}} = \sqrt{\frac{x^2}{0.100}}$$

$$x = \sqrt{(0.100)(4.9 \times 10^{-10})}$$

$$= 7.0 \times 10^{-6}$$

Example 17.6:

$$K_a = \frac{[H_3O^+][NO_2^-]}{[HNO_2]}$$

$$= \frac{x^2}{0.200 - \cancel{x}} \quad (x \text{ is small})$$

$$4.6 \times 10^{-4} = \frac{x^2}{0.200}$$

$$\sqrt{4.6 \times 10^{-4}} = \sqrt{\frac{x^2}{0.200}}$$

$$x = \sqrt{(0.200)(4.6 \times 10^{-4})}$$

$$= 9.6 \times 10^{-3}$$

—*Continued on the next page*

Continued from the previous page—

Confirm that the *x is small* approximation is valid by calculating the ratio of x to the number it was subtracted from in the approximation. The ratio should be less than 0.05 (or 5%).	$\frac{7.0 \times 10^{-6}}{0.100} \times 100\% = 7.0 \times 10^{-3}\%$ Therefore the approximation is valid.	$\frac{9.6 \times 10^{-3}}{0.200} \times 100\% = 4.8\%$ Therefore the approximation is valid (but barely so).
5. Determine the $[H_3O^+]$ concentration from the calculated value of x and calculate the pH if necessary.	$[H_3O^+] = 7.0 \times 10^{-6}$ M (pH was not asked for in this problem.)	$[H_3O^+] = 9.6 \times 10^{-3}$ M $pH = -\log[H_3O^+] = -\log(9.6 \times 10^{-3}) = 2.02$
6. Check your answer by substituting the calculated equilibrium values into the acid ionization expression. The calculated value of K_a should match the given value of K_a. Note that rounding errors and the *x is small* approximation could result in a difference in the least significant digit when comparing values of K_a.	$K_a = \frac{[H_3O^+][CN^-]}{[HCN]} = \frac{(7.0 \times 10^{-6})^2}{0.100} = 4.9 \times 10^{-10}$ Since the calculated value of K_a matches the given value, the answer is valid.	$K_a = \frac{[H_3O^+][NO_2^-]}{[HNO_2]} = \frac{(9.6 \times 10^{-3})^2}{0.200} = 4.6 \times 10^{-4}$ Since the calculated value of K_a matches the given value, the answer is valid.
	FOR PRACTICE 17.5 Find the H_3O^+ concentration of a 0.250 M hydrofluoric acid solution.	**FOR PRACTICE 17.6** Find the pH of a 0.0150 M acetic acid solution.

EXAMPLE 17.7
Finding the pH of a Weak Acid Solution in Cases Where the *x is small* Approximation Does Not Work

Find the pH of a 0.100 M $HClO_2$ solution.

SOLUTION

1. Write the balanced equation for the ionization of the acid and use it as a guide to prepare an ICE table showing the given concentration of the weak acid as its initial concentration.

(Note that the H_3O^+ concentration is listed as approximately zero. Although a little H_3O^+ is present from the autoionization of water, this amount is negligibly small compared to the amount of H_3O^+ produced by the acid.)

$$HClO_2(aq) + H_2O(l) \rightleftharpoons H_3O^+(aq) + ClO_2^-(aq)$$

	$[HClO_2]$	$[H_3O^+]$	$[ClO_2^-]$
Initial	0.100	≈0.00	0.00
Change			
Equil			

2. Represent the change in $[H_3O^+]$ with the variable x. Define the changes in the concentrations of the other reactants and products in terms of x.

$$HClO_2(aq) + H_2O(l) \rightleftharpoons H_3O^+(aq) + ClO_2^-(aq)$$

	$[HClO_2]$	$[H_3O^+]$	$[ClO_2^-]$
Initial	0.100	≈0.00	0.00
Change	$-x$	$+x$	$+x$
Equil			

—Continued on the next page

Continued from the previous page—

3. Sum each column to determine the equilibrium concentrations in terms of the initial concentrations and the variable x.	$HClO_2(aq) + H_2O(l) \rightleftharpoons H_3O^+(aq) + ClO_2^-(aq)$ Initial: $[HClO_2]$ 0.100; $[H_3O^+]$ ≈0.00; $[ClO_2^-]$ 0.00 Change: $-x$; $+x$; $+x$ Equil: $0.100-x$; x; x
4. Substitute the expressions for the equilibrium concentrations (from step 3) into the expression for the acid ionization constant (K_a). Make the *x is small* approximation and substitute the value of the acid ionization constant (from Table 17.5) into the K_a expression. Solve for x. Check to see if the *x is small* approximation is valid by calculating the ratio of x to the number it was subtracted from in the approximation. The ratio should be less than 0.05 (or 5%).	$K_a = \frac{[H_3O^+][ClO_2^-]}{[HClO_2]}$ $= \frac{x^2}{0.100 - \cancel{x}}$ (*x is small*) $0.011 = \frac{x^2}{0.100}$ $\sqrt{0.011} = \sqrt{\frac{x^2}{0.100}}$ $x = \sqrt{(0.100)(0.011)}$ $= 0.033$ $\frac{0.033}{0.100} \times 100\% = 33\%$ Therefore, the *x is small* approximation is *not valid.*
4a. If the *x is small* approximation is not valid, solve the quadratic equation explicitly or use the method of successive approximations to find x. In this case, we solve the quadratic equation.	$0.011 = \frac{x^2}{0.100 - x}$ $0.011(0.100 - x) = x^2$ $0.0011 - 0.011x = x^2$ $x^2 + 0.011x - 0.0011 = 0$ $x = \frac{-b \pm \sqrt{b^2 - 4ac}}{2a}$ $= \frac{-(0.011) \pm \sqrt{(0.011)^2 - 4(1)(-0.0011)}}{2(1)}$ $= \frac{-0.011 \pm 0.067\underline{2}}{2}$ $x = -0.039$ *or* $x = 0.028$ Since x represents the concentration of H_3O^+, and since concentrations cannot be negative, we reject the negative root. $x = 0.028$
5. Determine the H_3O^+ concentration from the calculated value of x and calculate the pH.	$[H_3O^+] = 0.028$ M $pH = -\log[H_3O^+]$ $= -\log 0.028$ $= 1.55$

—Continued on the next page

Continued from the previous page—

6. Check your answer by substituting the calculated equilibrium values into the acid ionization expression. The calculated value of K_a should match the given value of K_a. Note that rounding errors could result in a difference in the least significant digit when comparing values of K_a.	$K_a = \frac{[H_3O^+][ClO_2^-]}{[HClO_2]} = \frac{0.028^2}{0.100 - 0.028}$ $= 0.011$ Since the calculated value of K_a matches the given value, the answer is valid.

FOR PRACTICE 17.7

Find the pH of a 0.010 M HNO_2 solution.

EXAMPLE 17.8
Finding the Equilibrium Constant from pH

A 0.100 M weak acid (HA) solution has a pH of 4.25. Find K_a for the acid.

SOLUTION

Use the given pH to find the equilibrium concentration of $[H_3O^+]$. Then write the balanced equation for the ionization of the acid and use it as a guide to prepare an ICE table showing all known concentrations.

$$\begin{aligned} pH &= -\log[H_3O^+] \\ 4.25 &= -\log[H_3O^+] \\ [H_3O^+] &= 5.6 \times 10^{-5}\,M \end{aligned}$$

$$HA(aq) + H_2O(l) \rightleftharpoons H_3O^+(aq) + A^-(aq)$$

	[HA]	$[H_3O^+]$	$[A^-]$
Initial	0.100	≈0.00	0.00
Change			
Equil		5.6×10^{-5}	

Use the equilibrium concentration of H_3O^+ and the stoichiometry of the reaction to predict the changes and equilibrium concentration for all species. For most weak acids, the initial and equilibrium concentrations of the weak acid (HA) are equal because the amount that ionizes is usually very small compared to the initial concentration.

$$HA(aq) + H_2O(l) \rightleftharpoons H_3O^+(aq) + A^-(aq)$$

	[HA]	$[H_3O^+]$	$[A^-]$
Initial	0.100	≈0.00	0.00
Change	-5.6×10^{-5}	$+5.6 \times 10^{-5}$	$+5.6 \times 10^{-5}$
Equil	$(0.100 - 5.6 \times 10^{-5}) \approx 0.100$	5.6×10^{-5}	5.6×10^{-5}

Substitute the equilibrium concentrations into the expression for K_a and calculate its value.

$$\begin{aligned} K_a &= \frac{[H_3O^+][A^-]}{[HA]} \\ &= \frac{(5.6 \times 10^{-5})(5.6 \times 10^{-5})}{0.100} \\ &= 3.1 \times 10^{-8} \end{aligned}$$

FOR PRACTICE 17.8

A 0.175 M weak acid solution has a pH of 3.25. Find K_a for the acid.

17.4
Cc
Conceptual Connection

The *x is small* Approximation

The initial concentration and the K_a values of several weak acid (HA) solutions are listed here. For which of these is the *x is small* approximation *least* likely to work in finding the pH of the solution?

(a) initial $[HA] = 0.100$ M; $K_a = 1.0 \times 10^{-5}$
(b) initial $[HA] = 1.00$ M; $K_a = 1.0 \times 10^{-6}$
(c) initial $[HA] = 0.0100$ M; $K_a = 1.0 \times 10^{-3}$
(d) initial $[HA] = 1.0$ M; $K_a = 1.5 \times 10^{-3}$

17.5
Cc
Conceptual Connection

Strong and Weak Acids

Which solution is most acidic (that is, which one has the lowest pH)?

(a) 0.10 M HCl
(b) 0.10 M HF
(c) 0.20 M HF

Percent Ionization of a Weak Acid

We can quantify the ionization of a weak acid according to the percentage of acid molecules that actually ionize. We define the **percent ionization** of a weak acid as the ratio of the ionized acid concentration to the initial acid concentration, multiplied by 100%:

$$\text{Percent ionization} = \frac{\text{concentration of ionized acid}}{\text{initial concentration of acid}} \times 100\% = \frac{[H_3O^+]_{\text{equil}}}{[HA]_{\text{init}}} \times 100\%$$

Because the concentration of ionized acid is equal to the H_3O^+ concentration at equilibrium (for a monoprotic acid), we can use $[H_3O^+]_{\text{equil}}$ and $[HA]_{\text{init}}$ in the formula to calculate the percent ionization. For instance, in Example 17.6, we found that a 0.200 M HNO_2 solution contains 9.6×10^{-3} M H_3O^+. The 0.200 M HNO_2 solution therefore has the following percent ionization:

$$\%\ \text{ionization} = \frac{[H_3O^+]_{\text{equil}}}{[HA]_{\text{init}}} \times 100\% = \frac{9.6 \times 10^{-3}\ \cancel{M}}{0.200\ \cancel{M}} \times 100\% = 4.8\%$$

As we can see, the percent ionization is relatively small. In this case, even though HNO_2 has the second largest K_a in Table 17.5, less than five molecules out of one hundred ionize. For most other weak acids (with smaller K_a values) the percent ionization is even less.

In Example 17.9, we calculate the percent ionization of a more concentrated HNO_2 solution. In the example, notice that the calculated H_3O^+ concentration is much greater (as we would expect for a more concentrated solution), but the *percent ionization* is actually smaller.

EXAMPLE 17.9

Finding the Percent Ionization of a Weak Acid

Find the percent ionization of a 2.5 M HNO_2 solution.

SOLUTION

To find the percent ionization, you must find the equilibrium concentration of H_3O^+. Follow the procedure in Example 17.5, shown in condensed form here.

$$HNO_2(aq) + H_2O(l) \rightleftharpoons H_3O^+(aq) + NO_2^-(aq)$$

	$[HNO_2]$	$[H_3O^+]$	$[NO_2^-]$
Initial	2.5	≈0.00	0.00
Change	−x	+x	+x
Equil	2.5 − x	x	x

$$K_a = \frac{[H_3O^+][NO_2^-]}{[HNO_2]} = \frac{x^2}{2.5 - \cancel{x}} \quad (x \text{ is small})$$

$$4.6 \times 10^{-4} = \frac{x^2}{2.5}$$

$$x = 0.034$$

Therefore, $[H_3O^+] = 0.034$ M.

Use the definition of percent ionization to calculate it. (Since the percent ionization is less than 5%, the *x is small* approximation is valid.)

$$\% \text{ ionization} = \frac{[H_3O^+]_{\text{equil}}}{[HA]_{\text{init}}} \times 100\%$$

$$= \frac{0.034\ \cancel{M}}{2.5\ \cancel{M}} \times 100\%$$

$$= 1.4\%$$

FOR PRACTICE 17.9

Find the percent ionization of a 0.250 M $HC_2H_3O_2$ solution.

We can summarize the results of Examples 17.6 and 17.9:

$[HNO_2]$	$[H_3O^+]$	Percent Ionization
0.200	0.0096	4.8%
2.5	0.034	1.4%

The trend you can see in the above table applies to all weak acids.

- **The *equilibrium H_3O^+ concentration* of a weak acid *increases* with increasing initial concentration of the acid.**
- **The *percent ionization* of a weak acid *decreases* with increasing concentration of the acid.**

In other words, as the concentration of a weak acid solution increases, the concentration of the hydronium ion also increases, but the increase is not linear. The H_3O^+ concentration increases more slowly than the concentration of the acid because as the acid concentration increases, a smaller fraction of weak acid molecules ionizes.

We can understand this behavior by analogy with Le Châtelier's principle. Consider the following weak acid ionization equilibrium:

$$\underset{\text{1 mol dissolved particles}}{HA(aq)} \rightleftharpoons \underset{\text{2 mol dissolved particles}}{H^+(aq) + A^-(aq)}$$

If we dilute a weak acid solution initially at equilibrium, the system (according to Le Châtelier's principle) responds to minimize the disturbance. The equilibrium shifts to the right because the right side of the equation contains more particles in solution (2 mol versus 1 mol) than the left side. If the system shifts to the right, the percent ionization is greater in the more dilute solution, which is what we observe.

17.6
Cc
Conceptual Connection

Percent Ionization

Which of these weak acid solutions has the greatest percent ionization? Which solution has the lowest (most acidic) pH?

(a) 0.100 M $HC_2H_3O_2$
(b) 0.500 M $HC_2H_3O_2$
(c) 0.0100 M $HC_2H_3O_2$

Mixtures of Acids

Finding the pH of a mixture of acids may seem difficult at first. However, in many cases, the relative strengths of the acids in the mixture allow us to neglect the weaker acid and focus only on the stronger one. Here, we consider two possible acid mixtures: a strong acid with a weak acid and a weak acid with another weak acid.

A Strong Acid and a Weak Acid Consider a mixture that is 0.10 M in HCl and 0.10 M in $HCHO_2$. There are three sources of H_3O^+ ions: the strong acid (HCl), the weak acid ($HCHO_2$), and the autoionization of water.

$$HCl(aq) + H_2O(l) \longrightarrow H_3O^+(aq) + Cl^-(aq) \quad \text{Strong}$$

$$HCHO_2(aq) + H_2O(l) \rightleftharpoons H_3O^+(aq) + CHO_2^-(aq) \quad K_a = 1.8 \times 10^{-4}$$

$$H_2O(l) + H_2O(l) \rightleftharpoons H_3O^+(aq) + OH^-(aq) \quad K_w = 1.0 \times 10^{-14}$$

Since HCl is strong, we know that it completely ionizes to produce a significant concentration of H_3O^+(0.10 M). The H_3O^+ formed by HCl then *suppresses* the formation of additional H_3O^+ formed by the ionization of $HCHO_2$ or the autoionization of water. In other words, according to Le Châtelier's principle, the formation of H_3O^+ by the strong acid causes the weak acid to ionize even less than it would in the absence of the strong acid. To see this clearly, we can calculate $[H_3O^+]$ and $[CHO_2^-]$ in this solution.

In an initial estimate of $[H_3O^+]$, we can neglect the contribution of $HCHO_2$ and H_2O. The concentration of H_3O^+ is then equal to the initial concentration of HCl.

$$[H_3O^+] = [HCl] = 0.10 \text{ M}$$

To find $[CHO_2^-]$ we must solve an equilibrium problem. However, the initial concentration of H_3O^+ in this case is not negligible (as it has been in all the other weak acid equilibrium problems that we have worked so far) because HCl has formed a significant amount of H_3O^+. The concentration of H_3O^+ formed by HCl becomes the *initial* concentration of H_3O^+ in the ICE table for $HCHO_2$, as shown here:

$$HCHO_2(aq) + H_2O(l) \rightleftharpoons H_3O^+(aq) + CHO_2^-(aq)$$

	$[HCHO_2]$	$[H_3O^+]$	$[CHO_2^-]$
Initial	0.10	0.10	0.00
Change	$-x$	$+x$	$+x$
Equil	$0.10 - x$	$0.10 + x$	x

We then use the equilibrium expression to set up an equation in which x is the only variable:

$$K_a = \frac{[H_3O^+][CHO_2^-]}{[HCHO_2]} = \frac{(0.10 + x)x}{0.10 - x}$$

Since the equilibrium constant is small relative to the initial concentration of the acid, we can make the *x is small* approximation:

$$K_a = \frac{(0.10 + \cancel{x})x}{0.10 - \cancel{x}}$$

$$1.8 \times 10^{-4} = \frac{(0.10)x}{0.10}$$

$$x = 1.8 \times 10^{-4}$$

Checking the *x is small* approximation:

$$\frac{1.8 \times 10^{-4}}{0.10} \times 100\% = 0.18\%$$

We find that the approximation is valid. Therefore, $[CHO_2^-] = 1.8 \times 10^{-4}$ M. We can now see that we can completely ignore the ionization of the weak acid ($HCHO_2$) in calculating $[H_3O^+]$ for the mixture. The contribution to the concentration of H_3O^+ by the weak acid must necessarily be equal to the concentration of CHO_2^- that we just calculated (because of the stoichiometry of the ionization reaction). Therefore, we have the following contributions to $[H_3O^+]$:

HCl contributes 0.10 M.

$HCHO_2$ contributes 1.8×10^{-4} M or 0.00018 M.

Total $[H_3O^+] = 0.10\text{ M} + 0.00018\text{ M} = 0.10\text{ M}$.

As we can see, because the significant figure rules for addition limit the answer to two decimal places, the amount of H_3O^+ contributed by $HCHO_2$ is completely negligible. The amount of H_3O^+ contributed by the autoionization of water is even smaller and therefore similarly negligible.

A Mixture of Two Weak Acids When we mix two weak acids, we again have three potential sources of H_3O^+ to consider: each of the two weak acids and the autoionization of water. However, if the K_a values of the two weak acids are sufficiently different in magnitude (if they differ by more than a factor of several hundred), then as long as the concentrations of the two acids are similar in magnitude (or the concentration of the stronger one is greater than that of the weaker), we can assume that the weaker acid will not make a significant contribution to the concentration of H_3O^+. We make this assumption for the same reason that we made a similar assumption for a mixture of a strong acid and a weak one: The stronger acid suppresses the ionization of the weaker one, in accordance with Le Châtelier's principle. Example 17.10 shows how to calculate the concentration of H_3O^+ in a mixture of two weak acids.

EXAMPLE 17.10 Mixtures of Weak Acids

Find the pH of a mixture that is 0.150 M in HF and 0.100 M in HClO.

SOLUTION

The three possible sources of H_3O^+ ions are HF, HClO, and H_2O. Write the ionization equations for the three sources and their corresponding equilibrium constants. Because the equilibrium constant for the ionization of HF is about 12,000 times larger than that for the ionization of HClO, the contribution of HF to $[H_3O^+]$ is by far the greatest. You can therefore only calculate the $[H_3O^+]$ formed by HF and neglect the other two potential sources of H_3O^+.

$$HF(aq) + H_2O(l) \rightleftharpoons H_3O^+(aq) + F^-(aq) \qquad K_a = 3.5 \times 10^{-4}$$

$$HClO(aq) + H_2O(l) \rightleftharpoons H_3O^+(aq) + ClO^-(aq) \qquad K_a = 2.9 \times 10^{-8}$$

$$H_2O(l) + H_2O(l) \rightleftharpoons H_3O^+(aq) + OH^-(aq) \qquad K_w = 1.0 \times 10^{-14}$$

Write the balanced equation for the ionization of HF and use it as a guide to prepare an ICE table.

$$HF(aq) + H_2O(l) \rightleftharpoons H_3O^+(aq) + F^-(aq)$$

	[HF]	$[H_3O^+]$	$[F^-]$
Initial	0.150	≈0.00	0.00
Change	$-x$	$+x$	$+x$
Equil	$0.150 - x$	x	x

—Continued on the next page

Continued from the previous page—

Substitute the expressions for the equilibrium concentrations into the expression for the acid ionization constant (K_a). Because the equilibrium constant is small relative to the initial concentration of HF, you can make the *x is small* approximation. Substitute the value of the acid ionization constant (from Table 17.5) into the K_a expression and solve for *x*.	$K_a = \frac{[H_3O^+][F^-]}{[HF]} = \frac{x^2}{0.150 - \cancel{x}}$ (*x is small*) $3.5 \times 10^{-4} = \frac{x^2}{0.150}$ $\sqrt{(0.150)(3.5 \times 10^{-4})} = \sqrt{x^2}$ $x = 7.2 \times 10^{-3}$
Confirm that the *x is small* approximation is valid by calculating the ratio of *x* to the number it was subtracted from in the approximation. The ratio should be less than 0.05 (or 5%).	$\frac{7.2 \times 10^{-3}}{0.150} \times 100\% = 4.8\%$ Therefore the approximation is valid (though barely so).
Determine the H_3O^+ concentration from the calculated value of *x* and find the pH.	$[H_3O^+] = 7.2 \times 10^{-3}$ M $pH = -\log(7.2 \times 10^{-3}) = 2.14$

FOR PRACTICE 17.10

Find the ClO^- concentration of the above mixture of HF and HClO.

17.7 Cc Conceptual Connection

Judging Relative pH

Which solution is most acidic (that is, has the lowest pH)?

(a) 1.0 M HCl
(b) 2.0 M HF
(c) A solution that is 1.0 M in HF and 1.0 M in HClO

17.8 Finding the [OH⁻] and pH of Strong and Weak Base Solutions

Like acids, bases can be strong or weak. Most strong bases are ionic compounds containing the OH^- ion. Most weak bases act as a base by accepting a proton from water. We examine each separately.

Strong Bases

Just as we define a strong acid as one that completely ionizes in solution, analogously we define a *strong base* as one that completely dissociates in solution. NaOH, for example, is a strong base:

$$NaOH(aq) \longrightarrow Na^+(aq) + OH^-(aq)$$

An NaOH solution contains no intact NaOH—it has all dissociated to form $Na^+(aq)$ and $OH^-(aq)$ (Figure 17.9 ◀). In other words, a 1.0 M NaOH solution has $[OH^-] = 1.0$ M and $[Na^+] = 1.0$ M. Table 17.7 lists the common strong bases.

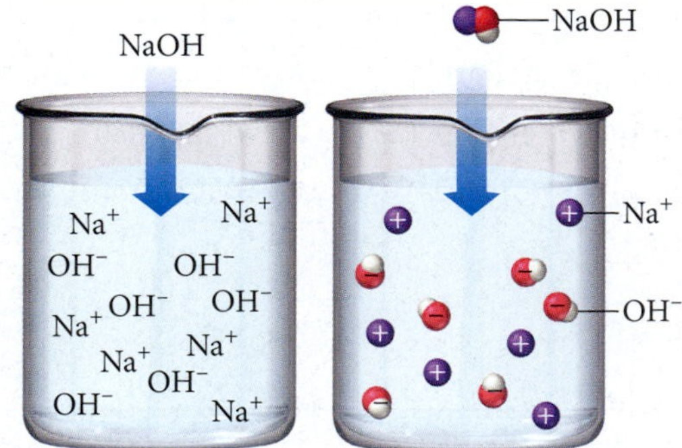

▲ **FIGURE 17.9 Ionization of a Strong Base** When NaOH dissolves in water, it dissociates completely into Na^+ and OH^-. The solution contains virtually no intact NaOH.

TABLE 17.7 Common Strong Bases

Lithium hydroxide (LiOH)	Strontium hydroxide [$Sr(OH)_2$]
Sodium hydroxide (NaOH)	Calcium hydroxide [$Ca(OH)_2$]
Potassium hydroxide (KOH)	Barium hydroxide [$Ba(OH)_2$]

As Table 17.7 illustrates, most strong bases are group 1A or group 2A metal hydroxides. The group 1A metal hydroxides are highly soluble in water and can form concentrated base solutions. The group 2A metal hydroxides, however, are only slightly soluble, a useful property for some applications. Notice that the general formula for the group 2A metal hydroxides is $M(OH)_2$. When they dissolve in water, they produce 2 mol of OH^- per mole of the base. For example, $Sr(OH)_2$ dissociates as follows:

$$Sr(OH)_2(aq) \longrightarrow Sr^{2+}(aq) + 2\,OH^-(aq)$$

Unlike diprotic acids, which ionize in two steps, bases containing two OH^- ions dissociate in a single step.

Weak Bases

Recall that a *weak base* is analogous to a weak acid. Unlike strong bases that contain OH^- and *dissociate* in water, the most common weak bases produce OH^- by accepting a proton from water, ionizing water to form OH^- according to the general equation:

$$B(aq) + H_2O(l) \rightleftharpoons BH^+(aq) + OH^-(aq)$$

In this equation, B is a generic symbol for a weak base. Ammonia, for example, ionizes water as follows:

$$NH_3(aq) + H_2O(l) \rightleftharpoons NH_4^+(aq) + OH^-(aq)$$

The double arrow indicates that the ionization is not complete. An NH_3 solution contains mostly NH_3 with only some NH_4^+ and OH^- (Figure 17.10 ▶). A 1.0 M NH_3 solution will have $[OH^-] < 1.0$ M.

We quantify the extent of ionization of a weak base with the **base ionization constant,** K_b. For the general reaction in which a weak base ionizes water, we define K_b as follows:

$$B(aq) + H_2O(l) \rightleftharpoons BH^+(aq) + OH^-(aq) \qquad K_b = \frac{[BH^+][OH^-]}{[B]}$$

By analogy with K_a, the smaller the value of K_b, the weaker the base. Table 17.8 lists some common weak bases, their ionization reactions, and values for K_b. The "p" scale can also be applied to K_b, so that $pK_b = -\log K_b$.

All but two of the weak bases listed in Table 17.8 are either ammonia or *amines*, which we can think of as ammonia with one or more hydrocarbon groups substituted for one or

A Weak Base

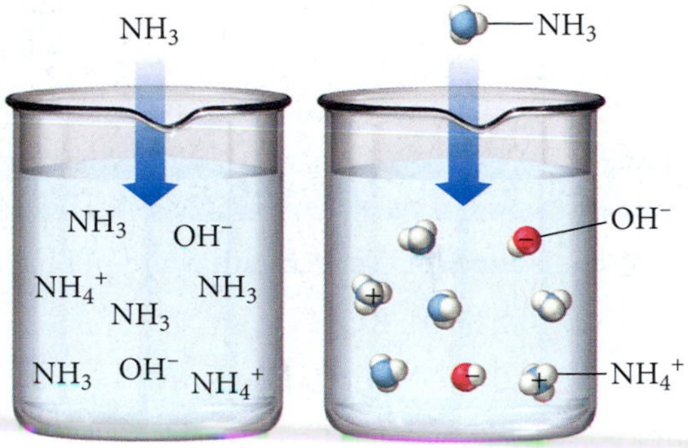

▲ **FIGURE 17.10 Ionization of a Weak Base** When NH_3 dissolves in water, it partially ionizes water to form NH_4^+ and OH^-. Most of the NH_3 molecules in solution remain as NH_3.

TABLE 17.8 Some Common Weak Bases

Weak Base	Ionization Reaction	K_b
Carbonate ion (CO_3^{2-})*	$CO_3^{2-}(aq) + H_2O(l) \rightleftharpoons HCO_3^-(aq) + OH^-(aq)$	1.8×10^{-4}
Methylamine (CH_3NH_2)	$CH_3NH_2(aq) + H_2O(l) \rightleftharpoons CH_3NH_3^+(aq) + OH^-(aq)$	4.4×10^{-4}
Ethylamine ($C_2H_5NH_2$)	$C_2H_5NH_2(aq) + H_2O(l) \rightleftharpoons C_2H_5NH_3^+(aq) + OH^-(aq)$	5.6×10^{-4}
Ammonia (NH_3)	$NH_3(aq) + H_2O(l) \rightleftharpoons NH_4^+(aq) + OH^-(aq)$	1.76×10^{-5}
Bicarbonate ion (HCO_3^-)* (or hydrogen carbonate)	$HCO_3^-(aq) + H_2O(l) \rightleftharpoons H_2CO_3(aq) + OH^-(aq)$	2.3×10^{-8}
Pyridine (C_5H_5N)	$C_5H_5N(aq) + H_2O(l) \rightleftharpoons C_5H_5NH^+(aq) + OH^-(aq)$	1.7×10^{-9}
Aniline ($C_6H_5NH_2$)	$C_6H_5NH_2(aq) + H_2O(l) \rightleftharpoons C_6H_5NH_3^+(aq) + OH^-(aq)$	3.9×10^{-10}

*The carbonate and bicarbonate ions must occur with a positively charged ion such as Na^+ that serves to balance the charge but does not have any part in the ionization reaction. For example, it is the bicarbonate ion that makes sodium bicarbonate ($NaHCO_3$) basic. We look more closely at ionic bases in Section 17.9.

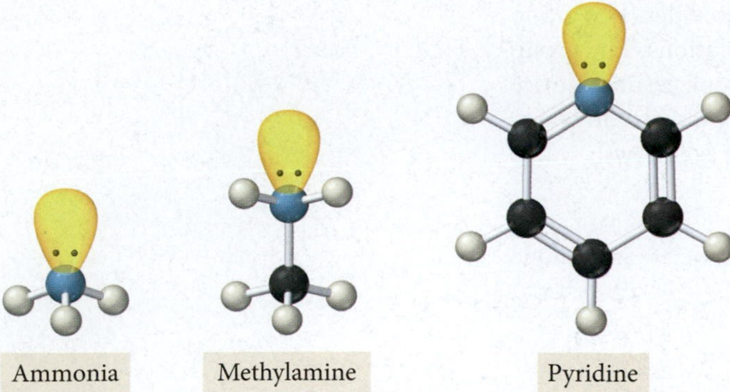

▲ **FIGURE 17.11 Lone Pairs in Weak Bases** Many weak bases have a nitrogen atom with a lone pair that acts as the proton acceptor.

more hydrogen atoms. All of these bases have a nitrogen atom with a lone pair (Figure 17.11 ◀). This lone pair acts as the proton acceptor that makes the substance a base, as shown in the reactions for ammonia and methylamine:

$$H-\ddot{N}H_2(aq) + H-\ddot{\underset{\cdot\cdot}{O}}-H(l) \rightleftharpoons H-NH_3^{+}(aq) + :\ddot{\underset{\cdot\cdot}{O}}^{-}-H(aq)$$

$$H_3C-\ddot{N}H_2(aq) + H-\ddot{\underset{\cdot\cdot}{O}}-H(l) \rightleftharpoons H_3C-NH_3^{+}(aq) + :\ddot{\underset{\cdot\cdot}{O}}^{-}-H(aq)$$

Finding the $[OH^-]$ and pH of Basic Solutions

Finding the $[OH^-]$ and pH of a strong base solution is relatively straightforward, as shown in Example 17.11. As we did in calculating the $[H_3O^+]$ in strong acid solutions, we can neglect the contribution of the autoionization of water to the $[OH^-]$ and focus solely on the strong base itself.

EXAMPLE 17.11
Finding the $[OH^-]$ and pH of a Strong Base Solution

What is the OH^- concentration and pH in each solution?

(a) 0.225 M KOH

(b) 0.0015 M $Sr(OH)_2$

SOLUTION

(a) Since KOH is a strong base, it completely dissociates into K^+ and OH^- in solution. The concentration of OH^- will therefore be the same as the given concentration of KOH.	$KOH(aq) \longrightarrow K^+(aq) + OH^-(aq)$ $[OH^-] = 0.225$ M
Use this concentration and K_w to find $[H_3O^+]$.	$[H_3O^+][OH^-] = K_w = 1.00 \times 10^{-14}$ $[H_3O^+](0.225) = 1.00 \times 10^{-14}$ $[H_3O^+] = 4.44 \times 10^{-14}$ M
Then substitute $[H_3O^+]$ into the pH expression to find the pH.	$pH = -\log[H_3O^+]$ $= -\log(4.44 \times 10^{-14})$ $= 13.352$
(b) Since $Sr(OH)_2$ is a strong base, it completely dissociates into 1 mol of Sr^{2+} and 2 mol of OH^- in solution. The concentration of OH^- will therefore be twice the given concentration of $Sr(OH)_2$.	$Sr(OH)_2(aq) \longrightarrow Sr^{2+}(aq) + 2\,OH^-(aq)$ $[OH^-] = 2(0.0015)$ M $= 0.0030$ M
Use this concentration and K_w to find $[H_3O^+]$.	$[H_3O^+][OH^-] = K_w = 1.0 \times 10^{-14}$ $[H_3O^+](0.0030) = 1.0 \times 10^{-14}$ $[H_3O^+] = 3.3 \times 10^{-12}$ M
Substitute $[H_3O^+]$ into the pH expression to find the pH.	$pH = -\log[H_3O^+]$ $= -\log(3.3 \times 10^{-12})$ $= 11.48$

FOR PRACTICE 17.11

Find the $[OH^-]$ and pH of a 0.010 M $Ba(OH)_2$ solution.

Finding the $[OH^-]$ and pH of a *weak base* solution is analogous to finding the $[H_3O^+]$ and pH of a weak acid. Similarly, we can neglect the contribution of the autoionization of water to the $[OH^-]$ and focus solely on the weak base itself. We find the contribution of the weak base by preparing an ICE table showing the relevant concentrations of all species and then use the base ionization constant expression to find the $[OH^-]$. Example 17.12 demonstrates how to find the $[OH^-]$ and pH of a weak base solution.

EXAMPLE 17.12

Finding the [OH^-] and pH of a Weak Base Solution

Find the $[OH^-]$ and pH of a 0.100 M NH_3 solution.

SOLUTION

1. Write the balanced equation for the ionization of water by the base and use it as a guide to prepare an ICE table showing the given concentration of the weak base as its initial concentration. Leave room in the table for the changes in concentrations and for the equilibrium concentrations. (Note that you should list the OH^- concentration as approximately zero. Although a little OH^- is present from the autoionization of water, this amount is negligibly small compared to the amount of OH^- formed by the base.)

$$NH_3(aq) + H_2O(l) \rightleftharpoons NH_4^+(aq) + OH^-(aq)$$

	$[NH_3]$	$[NH_4^+]$	$[OH^-]$
Initial	0.100	0.00	≈0.00
Change			
Equil			

2. Represent the change in the concentration of OH^- with the variable x. Define the changes in the concentrations of the other reactants and products in terms of x.

$$NH_3(aq) + H_2O(l) \rightleftharpoons NH_4^+(aq) + OH^-(aq)$$

	$[NH_3]$	$[NH_4^+]$	$[OH^-]$
Initial	0.100	0.00	≈0.00
Change	$-x$	$+x$	$+x$
Equil			

3. Sum each column to determine the equilibrium concentrations in terms of the initial concentrations and the variable x.

$$NH_3(aq) + H_2O(l) \rightleftharpoons NH_4^+(aq) + OH^-(aq)$$

	$[NH_3]$	$[NH_4^+]$	$[OH^-]$
Initial	0.100	0.00	≈0.00
Change	$-x$	$+x$	$+x$
Equil	$0.100 - x$	x	x

4. Substitute the expressions for the equilibrium concentrations (from step 3) into the expression for the base ionization constant.

$$K_b = \frac{[NH_4^+][OH^-]}{[NH_3]}$$

In many cases, you can make the approximation that *x is small*.

$$= \frac{x^2}{0.100 - \cancel{x}} \quad (x \text{ is small})$$

Substitute the value of the base ionization constant (from Table 17.8) into the K_b expression and solve for x.

$$1.76 \times 10^{-5} = \frac{x^2}{0.100}$$

$$\sqrt{1.76 \times 10^{-5}} = \sqrt{\frac{x^2}{0.100}}$$

$$x = \sqrt{(0.100)(1.76 \times 10^{-5})}$$

$$= 1.33 \times 10^{-3}$$

Confirm that the *x is small* approximation is valid by calculating the ratio of x to the number it was subtracted from in the approximation. The ratio should be less than 0.05 (or 5%).

$$\frac{1.33 \times 10^{-3}}{0.100} \times 100\% = 1.33\%$$

Therefore the approximation is valid.

—*Continued on the next page*

Continued from the previous page—

5. Determine the OH^- concentration from the calculated value of x. Use the expression for K_w to find $[H_3O^+]$.	$[OH^-] = 1.33 \times 10^{-3}$ M $[H_3O^+][OH^-] = K_w = 1.00 \times 10^{-14}$ $[H_3O^+](1.33 \times 10^{-3}) = 1.00 \times 10^{-14}$ $[H_3O^+] = 7.52 \times 10^{-12}$ M
Substitute $[H_3O^+]$ into the pH equation to find pH.	$pH = -\log[H_3O^+]$ $= -\log(7.52 \times 10^{-12})$ $= 11.124$

FOR PRACTICE 17.12
Find the $[OH^-]$ and pH of a 0.33 M methylamine solution.

17.9 The Acid–Base Properties of Ions and Salts

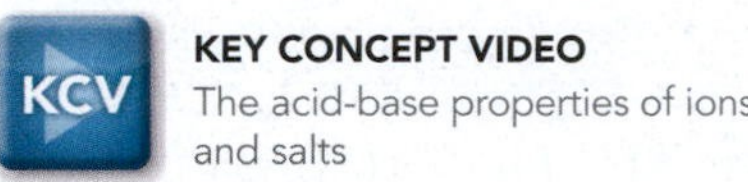

KEY CONCEPT VIDEO
The acid-base properties of ions and salts

We have already seen that some ions act as bases. For example, the bicarbonate ion acts as a base according to the following equation:

$$HCO_3^-(aq) + H_2O(l) \rightleftharpoons H_2CO_3(aq) + OH^-(aq)$$

The bicarbonate ion, like any ion, does not exist by itself; in order to be charge neutral, it must pair with a counter ion (in this case a cation) to form an ionic compound, called a *salt*. For example, the sodium salt of bicarbonate is sodium bicarbonate. Like all soluble salts, sodium bicarbonate dissociates in solution to form a sodium cation and bicarbonate anion:

$$NaHCO_3(s) \longrightarrow Na^+(aq) + HCO_3^-(aq)$$

The sodium ion has neither acidic nor basic properties (it does not ionize water), as we will see shortly. The bicarbonate ion, by contrast, acts as a weak base, ionizing water as just shown to form a basic solution. Consequently, the pH of a sodium bicarbonate solution is above 7 (the solution is basic). In this section, we consider some of the acid–base properties of salts and the ions they contain. Some salts are pH-neutral when put into water, others are acidic, and still others are basic, depending on their constituent anions and cations. In general, anions tend to form either *basic* or neutral solutions, while cations tend to form either *acidic* or neutral solutions.

Anions as Weak Bases

We can think of any anion as the conjugate base of an acid. Consider the following anions and their corresponding acids:

This anion	is the conjugate base of	this acid
Cl^-		HCl
F^-		HF
NO_3^-		HNO_3
$C_2H_3O_2^-$		$HC_2H_3O_2$

In general, the anion A^- is the conjugate base of the acid HA. Since every anion can be regarded as the conjugate base of an acid, every anion itself can potentially act as a base. However, *not every anion does act as a base*—it depends on the strength of the corresponding acid. In general:

- An anion that is the conjugate base of a *weak acid* is itself a *weak base.*
- An anion that is the conjugate base of a *strong acid* is pH-*neutral* (forms solutions that are neither acidic nor basic).

For example, the Cl^- anion is the conjugate base of HCl, a strong acid. Therefore, the Cl^- anion is pH-neutral (neither acidic nor basic). The F^- anion, however, is the conjugate base of HF, a weak acid. Therefore, the F^- ion is itself a weak base and ionizes water according to the reaction:

$$F^-(aq) + H_2O(l) \rightleftharpoons OH^-(aq) + HF(aq)$$

We can understand why the conjugate base of a weak acid is basic by asking why an acid is weak to begin with. Hydrofluoric acid is a weak acid because, as we saw in Section 17.4, the HF bond is particularly strong. Therefore, the following reaction lies to the left:

$$HF(aq) + H_2O(l) \rightleftharpoons H_3O^+(aq) + F^-(aq)$$

The strength of the HF bond causes the F^- ion to have significant affinity for H^+ ions. Consequently, when F^- is put into water, its affinity for H^+ ions allows it to remove H^+ ions from water molecules, thus acting as a weak base. In general, as shown in Figure 17.12 ▶, the weaker the acid, the stronger the conjugate base. In contrast, the conjugate base of a strong acid, such as Cl^-, does not act as a base because the HCl bond is weaker than the HF bond reaction; as a result, this reaction lies far to the right:

$$HCl(aq) + H_2O(l) \longrightarrow H_3O^+(aq) + Cl^-(aq)$$

The Cl^- ion has a relatively lower affinity for H^+ ions. Consequently, when Cl^- is put into water, it does not remove H^+ ions from water molecules.

Acid Strength	Acid	Base	Base Strength
Strong	HCl	Cl^-	Neutral
Strong	H_2SO_4	HSO_4^-	Neutral
Strong	HNO_3	NO_3^-	Neutral
Strong	H_3O^+	H_2O	Weak
Weak	HSO_4^-	SO_4^{2-}	Weak
Weak	H_2SO_3	HSO_3^-	Weak
Weak	H_3PO_4	$H_2PO_4^-$	Weak
Weak	HF	F^-	Weak
Weak	$HC_2H_3O_2$	$C_2H_3O_2^-$	Weak
Weak	H_2CO_3	HCO_3^-	Weak
Weak	H_2S	HS^-	Weak
Weak	HSO_3^-	SO_3^{2-}	Weak
Weak	$H_2PO_4^-$	HPO_4^{2-}	Weak
Weak	HCN	CN^-	Weak
Weak	NH_4^+	NH_3	Weak
Weak	HCO_3^-	CO_3^{2-}	Weak
Weak	HPO_4^{2-}	PO_4^{3-}	Weak
Weak	H_2O	OH^-	Strong
Negligible	HS^-	S^{2-}	Strong
Negligible	OH^-	O^{2-}	Strong

(Acid Strength increases upward; Base Strength increases downward.)

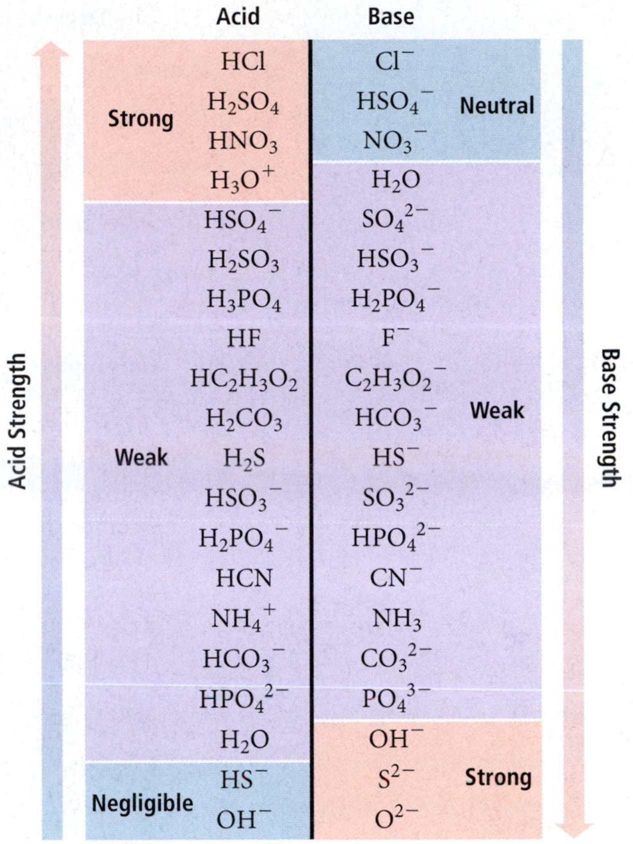

▲ **FIGURE 17.12 Strength of Conjugate Acid–Base Pairs** The stronger an acid, the weaker is its conjugate base.

EXAMPLE 17.13

Determining Whether an Anion Is Basic or pH-Neutral

Classify each anion as a weak base or pH-neutral.

(a) NO_3^- **(b)** NO_2^- **(c)** $C_2H_3O_2^-$

SOLUTION

(a) From Table 17.3, you can see that NO_3^- is the conjugate base of a *strong* acid (HNO_3) and and is therefore pH-neutral.

(b) From Table 17.5 (or from its absence in Table 17.3), you know that NO_2^- is the conjugate base of a weak acid (HNO_2) and is therefore a weak base.

(c) From Table 17.5 (or from its absence in Table 17.3), you know that $C_2H_3O_2^-$ is the conjugate base of a weak acid ($HC_2H_3O_2$) and is therefore a weak base.

FOR PRACTICE 17.13

Classify each anion as a weak base or pH-neutral.

(a) CHO_2^- **(b)** ClO_4^-

We can determine the pH of a solution containing an anion that acts as a weak base in a manner similar to how we determine the pH of any weak base solution. However, we need to know K_b for the anion acting as a base, which we can readily determine from K_a of the corresponding acid. Recall from Section 17.5 the expression for K_a for a generic acid HA:

$$HA(aq) + H_2O(l) \rightleftharpoons H_3O^+(aq) + A^-(aq)$$

$$K_a = \frac{[H_3O^+][A^-]}{[HA]}$$

Similarly, the expression for K_b for the conjugate base (A^-) is:

$$A^-(aq) + H_2O(l) \rightleftharpoons OH^-(aq) + HA(aq)$$

$$K_b = \frac{[OH^-][HA]}{[A^-]}$$

If we multiply the expressions for K_a and K_b we get K_w:

$$K_a \times K_b = \frac{[H_3O^+]\cancel{[A^-]}}{\cancel{[HA]}}\frac{[OH^-]\cancel{[HA]}}{\cancel{[A^-]}} = [H_3O^+][OH^-] = K_w$$

Or simply,

$$K_a \times K_b = K_w$$

The product of K_a for an acid and K_b for its conjugate base is K_w (1.0×10^{-14} at 25 °C). Consequently, we can find K_b for an anion acting as a base from the value of K_a for the corresponding acid. For example, for acetic acid ($HC_2H_3O_2$), $K_a = 1.8 \times 10^{-5}$. We calculate K_b for the conjugate base ($C_2H_3O_2^-$) by substituting into the equation:

$$K_a \times K_b = K_w$$

$$K_b = \frac{K_w}{K_a} = \frac{1.0 \times 10^{-14}}{1.8 \times 10^{-5}} = 5.6 \times 10^{-10}$$

Knowing K_b, we can find the pH of a solution containing an anion acting as a base, as demonstrated in Example 17.14.

EXAMPLE 17.14

Finding the pH of a Solution Containing an Anion Acting as a Base

Find the pH of a 0.100 M $NaCHO_2$ solution. The salt completely dissociates into $Na^+(aq)$ and $CHO_2^-(aq)$, and the Na^+ ion has no acid or base properties.

SOLUTION

1. Since the Na^+ ion does not have any acidic or basic properties, you can ignore it. Write the balanced equation for the ionization of water by the basic anion and use it as a guide to prepare an ICE table showing the given concentration of the weak base as its initial concentration.

$$CHO_2^-(aq) + H_2O(l) \rightleftharpoons HCHO_2(aq) + OH^-(aq)$$

	$[CHO_2^-]$	$[HCHO_2]$	$[OH^-]$
Initial	0.100	0.00	≈0.00
Change			
Equil			

2. Represent the change in the concentration of OH^- with the variable x. Define the changes in the concentrations of the other reactants and products in terms of x.

$$CHO_2^-(aq) + H_2O(l) \rightleftharpoons HCHO_2(aq) + OH^-(aq)$$

	$[CHO_2^-]$	$[HCHO_2]$	$[OH^-]$
Initial	0.100	0.00	≈0.00
Change	$-x$	$+x$	$+x$
Equil			

—Continued on the next page

Continued from the previous page—

3. Sum each column to determine the equilibrium concentrations in terms of the initial concentrations and the variable x.	$CHO_2^-(aq) + H_2O(l) \rightleftharpoons HCHO_2(aq) + OH^-(aq)$

	$[CHO_2^-]$	$[HCHO_2]$	$[OH^-]$
Initial	0.100	0.00	≈0.00
Change	−x	+x	+x
Equil	0.100−x	x	x

4. Find K_b from K_a (for the conjugate acid from Table 17.5).

$$K_a \times K_b = K_w$$

$$K_b = \frac{K_w}{K_a} = \frac{1.0 \times 10^{-14}}{1.8 \times 10^{-4}} = 5.6 \times 10^{-11}$$

Substitute the expressions for the equilibrium concentrations (from step 3) into the expression for K_b. In many cases, you can make the approximation that *x is small.*

$$K_b = \frac{[HCHO_2][OH^-]}{[CHO_2^-]}$$

$$= \frac{x^2}{0.100 - \cancel{x}} \quad (x \text{ is small})$$

Substitute the value of K_b into the K_b expression and solve for x.

$$5.6 \times 10^{-11} = \frac{x^2}{0.100}$$

$$x = 2.4 \times 10^{-6}$$

Confirm that the *x is small* approximation is valid by calculating the ratio of x to the number it was subtracted from in the approximation. The ratio should be less than 0.05 (or 5%).

$$\frac{2.4 \times 10^{-6}}{0.100} \times 100\% = 0.0024\%$$

Therefore the approximation is valid.

5. Determine the OH^- concentration from the calculated value of x.

$$[OH^-] = 2.4 \times 10^{-6}\ M$$

Use the expression for K_w to find $[H_3O^+]$.

$$[H_3O^+][OH^-] = K_w = 1.0 \times 10^{-14}$$

$$[H_3O^+](2.4 \times 10^{-6}) = 1.0 \times 10^{-14}$$

$$[H_3O^+] = 4.2 \times 10^{-9}\ M$$

Substitute $[H_3O^+]$ into the pH equation to find pH.

$$pH = -\log[H_3O^+]$$

$$= -\log(4.2 \times 10^{-9})$$

$$= 8.38$$

FOR PRACTICE 17.14

Find the pH of a 0.250 M $NaC_2H_3O_2$ solution.

We can also express the relationship between K_a and K_b in terms of pK_a and pK_b. By taking the log of both sides of $K_a \times K_b = K_w$, we get:

$$\log(K_a \times K_b) = \log K_w$$

$$\log K_a + \log K_b = \log K_w$$

Because $K_w = 10^{-14}$, we can rearrange the equation to get:

$$\log K_a + \log K_b = \log 10^{-14} = -14$$

Rearranging further:

$$-\log K_a - \log K_b = 14$$

Because $-\log K = pK$, we get:

$$pK_a + pK_b = 14$$

Cations as Weak Acids

In contrast to anions, which in some cases act as weak bases, cations can in some cases act as *weak acids*. We can generally divide cations into three categories: cations that are the counterions of strong bases; cations that are the conjugate acids of *weak* bases; and cations that are small, highly-charged metals. We examine each individually.

Cations That Are the Counterions of Strong Bases Strong bases such as NaOH or $Ca(OH)_2$ generally contain hydroxide ions and a counterion. In solution, a strong base completely dissociates to form $OH^-(aq)$ and the solvated (in solution) counterion. Although these counterions interact with water molecules via ion–dipole forces, they do not ionize water and they do not contribute to the acidity or basicity of the solution. In general *cations that are the counterions of strong bases are themselves pH-neutral* (they form solutions that are neither acidic nor basic). For example, Na^+, K^+, and Ca^{2+} are the counterions of the strong bases NaOH, KOH, and $Ca(OH)_2$ and are therefore themselves pH-neutral.

Cations That Are the Conjugate Acids of Weak Bases A cation can be formed from any nonionic weak base by adding a proton (H^+) to its formula. The cation is the conjugate acid of the base. Consider the following cations and their corresponding weak bases:

This cation	is the conjugate acid of	this weak base
NH_4^+		NH_3
$C_2H_5NH_3^+$		$C_2H_5NH_2$
$CH_3NH_3^+$		CH_3NH_2

Any of these cations, with the general formula BH^+, acts as a weak acid according to the equation:

$$BH^+(aq) + H_2O(l) \rightleftharpoons H_3O^+(aq) + B(aq)$$

In general, *a cation that is the conjugate acid of a weak base is a weak acid.*

We can calculate the pH of a solution containing the conjugate acid of a weak base just like that of any other weakly acidic solution. However, the value of K_a for the acid must be derived from K_b using the previously derived relationship: $K_a \times K_b = K_w$.

Cations That Are Small, Highly Charged Metals *Small, highly-charged metal cations such as Al^{3+} and Fe^{3+} form weakly acidic solutions.* For example, when Al^{3+} is dissolved in water, it becomes hydrated according to the equation:

$$Al^{3+}(aq) + 6\,H_2O(l) \longrightarrow Al(H_2O)_6^{\ 3+}(aq)$$

The hydrated form of the ion then acts as a Brønsted–Lowry acid:

$$Al(H_2O)_6^{3+}(aq) \quad + \quad H_2O(l) \rightleftharpoons Al(H_2O)_5(OH)^{2+}(aq) + H_3O^+(aq)$$

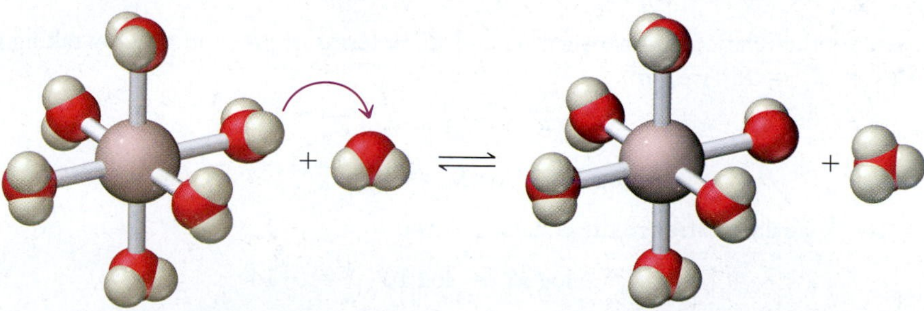

Neither the alkali metal cations nor the alkaline earth metal cations ionize water in this way, but the cations of many other metals do. The smaller and more highly-charged the cation, the more acidic its behavior.

EXAMPLE 17.15 Determining Whether a Cation Is Acidic or pH-Neutral

Classify each cation as a weak acid or pH-neutral.

(a) $C_5H_5NH^+$ **(b)** Ca^{2+} **(c)** Cr^{3+}

SOLUTION

(a) The $C_5H_5NH^+$ cation is the conjugate acid of a weak base and is therefore a weak acid.

(b) The Ca^{2+} cation is the counterion of a strong base and is therefore pH-neutral (neither acidic nor basic).

(c) The Cr^{3+} cation is a small, highly charged metal cation and is therefore a weak acid.

FOR PRACTICE 17.15

Classify each cation as a weak acid or pH-neutral.

(a) Li^+ **(b)** $CH_3NH_3^+$ **(c)** Fe^{3+}

Classifying Salt Solutions as Acidic, Basic, or Neutral

Since salts contain both a cation and an anion, they can form acidic, basic, or neutral solutions when dissolved in water. The pH of the solution depends on the specific cation and anion involved. We examine the four possibilities individually.

1. **Salts in which neither the cation nor the anion acts as an acid or a base form pH-neutral solutions.** A salt in which the cation is the counterion of a strong base and in which the anion is the conjugate base of a strong acid forms a *neutral* solution. Some salts in this category include:

$NaCl$ — sodium chloride; $Ca(NO_3)_2$ — calcium nitrate; KBr — potassium bromide

Cations are pH-neutral — Anions are conjugates bases of strong acids

2. **Salts in which the cation does not act as an acid and the anion acts as a base form basic solutions.** A salt in which the cation is the counterion of a strong base and in which the anion is the conjugate base of a *weak* acid forms a *basic* solution. Salts in this category include:

NaF — sodium fluoride; $Ca(C_2H_3O_2)_2$ — calcium acetate; KNO_2 — potassium nitrite

Cations are pH-neutral — Anions are conjugates bases of weak acids

3. **Salts in which the cation acts as an acid and the anion does not act as a base form acidic solutions.** A salt in which the cation is either the conjugate acid of a weak base or a small, highly-charged metal ion and in which the anion is the conjugate base of a *strong* acid forms an *acidic* solution. Salts in this category include:

$FeCl_3$ — iron(III) chloride; $Al(NO_3)_3$ — aluminum nitrate; NH_4Br — ammonium bromide

Cations are conjugate acids of *weak* bases or small, highly-charged metal ions — Anions are conjugates bases of *strong* acids

4. **Salts in which the cation acts as an acid and the anion acts as a base form solutions in which the pH depends on the relative strengths of the acid and the base.** A salt in which the cation is either the conjugate acid of a weak base or a small, highly-charged metal ion and in which the anion is the conjugate base of a *weak* acid forms a solution in which the pH depends on the relative strengths of the acid and base. Salts in this category include:

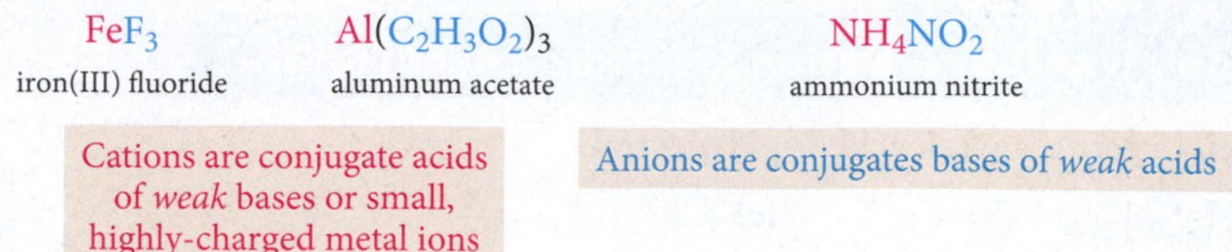

We can determine the overall pH of a solution containing one of these salts by comparing the K_a of the acid to the K_b of the base—the ion with the higher value of K dominates and determines whether the solution will be acidic or basic, as shown in part (e) of Example 17.16. Table 17.9 summarizes these possibilities.

TABLE 17.9 pH of Salt Solutions

		ANION	
		Conjugate base of strong acid	**Conjugate base of weak acid**
CATION	**Conjugate acid of weak base**	*Acidic*	*Depends on relative strengths*
	Small, highly-charged metal ion	*Acidic*	*Depends on relative strengths*
	Counterion of strong base	*Neutral*	*Basic*

EXAMPLE 17.16
Determining the Overall Acidity or Basicity of Salt Solutions

Determine if the solution formed by each salt is acidic, basic, or neutral.

(a) $SrCl_2$

(b) $AlBr_3$

(c) $CH_3NH_3NO_3$

(d) $NaCHO_2$

(e) NH_4F

SOLUTION

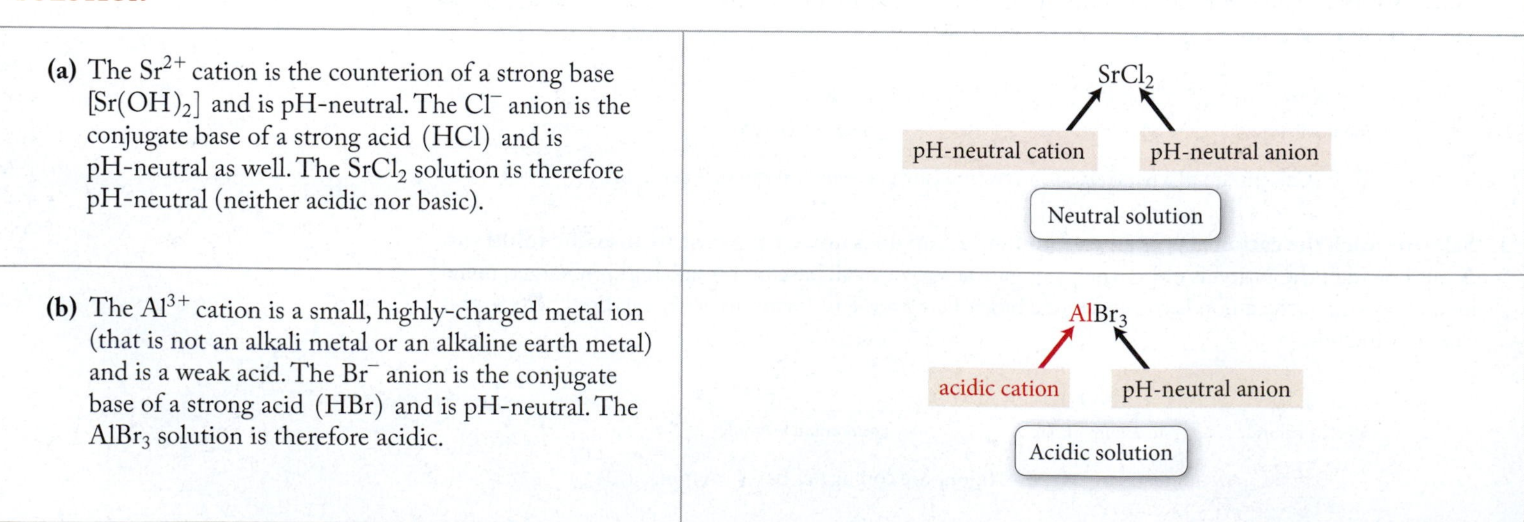

(a) The Sr^{2+} cation is the counterion of a strong base $[Sr(OH)_2]$ and is pH-neutral. The Cl^- anion is the conjugate base of a strong acid (HCl) and is pH-neutral as well. The $SrCl_2$ solution is therefore pH-neutral (neither acidic nor basic).

(b) The Al^{3+} cation is a small, highly-charged metal ion (that is not an alkali metal or an alkaline earth metal) and is a weak acid. The Br^- anion is the conjugate base of a strong acid (HBr) and is pH-neutral. The $AlBr_3$ solution is therefore acidic.

—Continued on the next page

Continued from the previous page—

(c) The $CH_3NH_3^+$ ion is the conjugate acid of a weak base (CH_3NH_2) and is acidic. The NO_3^- anion is the conjugate base of a strong acid (HNO_3) and is pH-neutral. The $CH_3NH_3NO_3$ solution is therefore acidic.

$CH_3NH_3NO_3$

acidic cation — pH-neutral anion

Acidic solution

(d) The Na^+ cation is the counterion of a strong base and is pH-neutral. The CHO_2^- anion is the conjugate base of a weak acid and is basic. The $NaCHO_2$ solution is therefore basic.

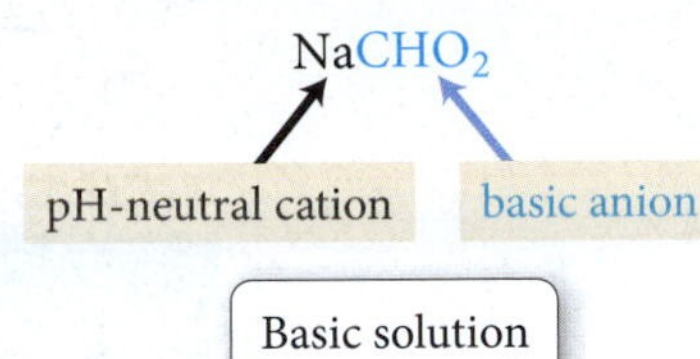

(e) The NH_4^+ ion is the conjugate acid of a weak base (NH_3) and is acidic. The F^- ion is the conjugate base of a weak acid and is basic. To determine the overall acidity or basicity of the solution, compare the values of K_a for the acidic cation and K_b for the basic anion. Obtain each value of K from the conjugate by using $K_a \times K_b = K_w$.

NH_4F

acidic cation — basic anion

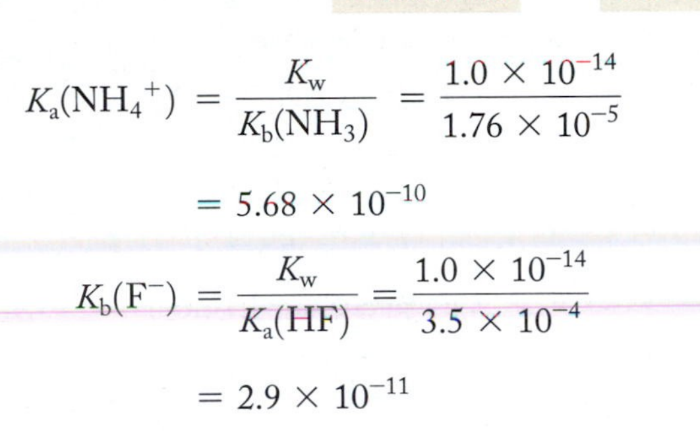

$$K_a(NH_4^+) = \frac{K_w}{K_b(NH_3)} = \frac{1.0 \times 10^{-14}}{1.76 \times 10^{-5}}$$

$$= 5.68 \times 10^{-10}$$

$$K_b(F^-) = \frac{K_w}{K_a(HF)} = \frac{1.0 \times 10^{-14}}{3.5 \times 10^{-4}}$$

$$= 2.9 \times 10^{-11}$$

$K_a > K_b$ — Acidic solution

Since K_a is greater than K_b the solution is acidic.

FOR PRACTICE 17.16

Determine if the solution formed by each salt is acidic, basic, or neutral.

(a) $NaHCO_3$

(b) $CH_3CH_2NH_3Cl$

(c) KNO_3

(d) $Fe(NO_3)_3$

17.10 Polyprotic Acids

In Section 17.5, we stated that some acids, called polyprotic acids, contain two or more ionizable protons. Recall that sulfurous acid (H_2SO_3) is a diprotic acid containing two ionizable protons and that phosphoric acid (H_3PO_4) is a triprotic acid containing three ionizable protons. Typically, a **polyprotic acid** ionizes in successive steps, each with its own K_a. For example, sulfurous acid ionizes in two steps:

$$H_2SO_3(aq) \rightleftharpoons H^+(aq) + HSO_3^-(aq) \qquad K_{a_1} = 1.6 \times 10^{-2}$$

$$HSO_3^-(aq) \rightleftharpoons H^+(aq) + SO_3^{2-}(aq) \qquad K_{a_2} = 6.4 \times 10^{-8}$$

K_{a_1} is the acid ionization constant for the first step, and K_{a_2} is the acid ionization constant for the second step. Notice that K_{a_2} is smaller than K_{a_1}. This is true for all polyprotic acids and makes sense because the first proton separates from a neutral molecule, while the second must separate from an anion. The negatively charged anion holds the positively charged proton more tightly, making the proton more difficult to remove and resulting in a smaller value of K_a. Table 17.10 lists some common polyprotic acids and their acid ionization constants. Notice that in all cases, the value of K_a for each step becomes successively smaller. The value of K_{a_1} for sulfuric acid is listed as strong because sulfuric acid is strong in the first step and weak in the second.

TABLE 17.10 Common Polyprotic Acids and Ionization Constants

Name (Formula)	Structure	Space-filling model	K_{a_1}	K_{a_2}	K_{a_3}
Sulfuric Acid (H_2SO_4)	$O{=}S({=}O)(OH)$—OH		Strong	1.2×10^{-2}	
Oxalic Acid ($H_2C_2O_4$)	HO—C(=O)—C(=O)—OH		6.0×10^{-2}	6.1×10^{-5}	
Sulfurous Acid (H_2SO_3)	HO—S(=O)—OH		1.6×10^{-2}	6.4×10^{-8}	
Phosphoric Acid (H_3PO_4)	HO—P(=O)(OH)—OH		7.5×10^{-3}	6.2×10^{-8}	4.2×10^{-13}
Citric Acid ($H_3C_6H_5O_7$)	HO—C(=O)—CH_2—C(OH)(C(=O)OH)—CH_2—C(=O)—OH		7.4×10^{-4}	1.7×10^{-5}	4.0×10^{-7}
Ascorbic Acid ($H_2C_6H_6O_6$)	HO—CH_2—CH(OH)—HC ring: O, C=O, C=C, HO, OH		8.0×10^{-5}	1.6×10^{-12}	
Carbonic Acid (H_2CO_3)	HO—C(=O)—OH		4.3×10^{-7}	5.6×10^{-11}	

Finding the pH of Polyprotic Acid Solutions

Finding the pH of a polyprotic acid solution is less difficult than we might first imagine because, for most polyprotic acids, K_{a_1} is much larger than K_{a_2} (or K_{a_3} for triprotic acids). Therefore the amount of H_3O^+ formed by the first ionization step is much larger than that formed by the second or third ionization step. In addition, the formation of H_3O^+ in the first step inhibits the formation of additional H_3O^+ in the second step (because of Le Châtelier's principle). Consequently, we treat most polyprotic acid solutions as if the first step were the only one that contributes to the H_3O^+ concentration, as shown in Example 17.17. A major exception is a dilute solution of sulfuric acid, which we examine in Example 17.18.

EXAMPLE 17.17 Finding the pH of a Polyprotic Acid Solution

Find the pH of a 0.100 M ascorbic acid ($H_2C_6H_6O_6$) solution.

SOLUTION

To find the pH, you must find the equilibrium concentration of H_3O^+. Treat the problem as a weak acid pH problem with a single ionizable proton. The second proton contributes a negligible amount to the concentration of H_3O^+ and can be ignored. Follow the procedure from Example 17.6, shown in condensed form here. Use K_{a_1} for ascorbic acid from Table 17.10.

$$H_2C_6H_6O_6(aq) + H_2O(l) \rightleftharpoons H_3O^+(aq) + HC_6H_6O_6^-(aq)$$

	$[H_2C_6H_6O_6]$	$[H_3O^+]$	$[HC_6H_6O_6^-]$
Initial	0.100	≈0.00	0.000
Change	$-x$	$+x$	$+x$
Equil	$0.100 - x$	x	x

$$K_{a_1} = \frac{[H_3O^+][HC_6H_6O_6^-]}{[H_2C_6H_6O_6]}$$

$$= \frac{x^2}{0.100 - \cancel{x}} \quad (x \text{ is small})$$

$$8.0 \times 10^{-5} = \frac{x^2}{0.100}$$

Confirm that the *x is small* approximation is valid by calculating the ratio of x to the number it was subtracted from in the approximation. The ratio should be less than 0.05 (or 5%).

$$x = 2.8 \times 10^{-3}$$

$$\frac{2.8 \times 10^{-3}}{0.100} \times 100\% = 2.8\%$$

The approximation is valid. Therefore,

$$[H_3O^+] = 2.8 \times 10^{-3}\ M$$

Calculate the pH from H_3O^+ concentration.

$$pH = -\log(2.8 \times 10^{-3}) = 2.55$$

FOR PRACTICE 17.17

Find the pH of a 0.050 M H_2CO_3 solution.

EXAMPLE 17.18 Finding the pH of a Dilute H_2SO_4 Solution

Find the pH of a 0.0100 M sulfuric acid (H_2SO_4) solution.

SOLUTION

Sulfuric acid is strong in its first ionization step and weak in its second. Begin by writing the equations for the two steps. As the concentration of an H_2SO_4 solution becomes smaller, the second ionization step becomes more significant because the percent ionization increases (as discussed in Section 17.7). Therefore, for a concentration of 0.0100 M, you can't neglect the H_3O^+ contribution from the second step, as you can for other polyprotic acids. You must calculate the H_3O^+ contributions from both steps.

$$H_2SO_4(aq) + H_2O(l) \longrightarrow H_3O^+(aq) + HSO_4^-(aq) \quad \text{Strong}$$

$$HSO_4^-(aq) + H_2O(l) \rightleftharpoons H_3O^+(aq) + SO_4^{2-}(aq) \quad K_{a_2} = 0.012$$

The $[H_3O^+]$ that results from the first ionization step is 0.0100 M (because the first step is strong). To determine the $[H_3O^+]$ formed by the second step, prepare an ICE table for the second step in which the initial concentration of H_3O^+ is 0.0100 M. The initial concentration of HSO_4^- must also be 0.0100 M (due to the stoichiometry of the ionization reaction).

	$[HSO_4^-]$	$[H_3O^+]$	$[SO_4^{2-}]$
Initial	0.0100	≈ 0.0100	0.000
Change	$-x$	$+x$	$+x$
Equil	$0.0100 - x$	$0.0100 + x$	x

Substitute the expressions for the equilibrium concentrations (from the table in the previous step) into the expression for K_{a_2}. In this case, you cannot make the *x is small* approximation because the equilibrium constant (0.012) *is not small* relative to the initial concentration (0.0100).

$$K_{a_2} = \frac{[H_3O^+][SO_4^{2-}]}{[HSO_4^-]} = \frac{(0.0100 + x)x}{0.0100 - x}$$

Substitute the value of K_{a_2} and multiply out the expression to arrive at the standard quadratic form.

$$0.012 = \frac{0.0100x + x^2}{0.0100 - x}$$

$$0.012(0.0100 - x) = 0.0100x + x^2$$

$$0.00012 - 0.012x = 0.0100x + x^2$$

$$x^2 + 0.022x - 0.00012 = 0$$

Solve the quadratic equation using the quadratic formula.

$$x = \frac{-b \pm \sqrt{b^2 - 4ac}}{2a} = \frac{-(0.022) \pm \sqrt{(0.022)^2 - 4(1)(-0.00012)}}{2(1)} = \frac{-0.022 \pm 0.031}{2}$$

$x = -0.027$ or $x = 0.0045$

Since x represents a concentration, and since concentrations cannot be negative, we reject the negative root.

$x = 0.0045$

Determine the H_3O^+ concentration from the calculated value of x and calculate the pH. Notice that the second step produces almost half as much H_3O^+ as the first step—an amount that must not be neglected. This will always be the case with dilute H_2SO_4 solutions.

$$[H_3O^+] = 0.0100 + x = 0.0100 + 0.0045 = 0.0145 \text{ M}$$

—*Continued on the next page*

Continued from the previous page—

	$pH = -\log[H_3O^+]$ $= -\log(0.0145)$ $= 1.84$

FOR PRACTICE 17.18

Find the pH and $[SO_4^{\ 2-}]$ of a 0.0075 M sulfuric acid solution.

Finding the Concentration of the Anions for a Weak Diprotic Acid Solution

In some cases, we may want to know the concentrations of the anions formed by a polyprotic acid. Consider the following generic polyprotic acid H_2X and its ionization steps:

$$H_2X(aq) + H_2O(l) \rightleftharpoons H_3O^+(aq) + HX^-(aq) \qquad K_{a_1}$$

$$HX^-(aq) + H_2O(l) \rightleftharpoons H_3O^+(aq) + X^{2-}(aq) \qquad K_{a_2}$$

In Examples 17.17 and 17.18, we illustrated how to find the concentration of H_3O^+ for such a solution, which is equal to the concentration of HX^-. What if instead we needed to find the concentration of X^{2-}? To find the concentration of X^{2-}, we use the concentration of HX^- and H_3O^+ (from the first ionization step) as the initial concentrations for the second ionization step. We then solve a second equilibrium problem using the second ionization equation and K_{a_2}, as demonstrated in Example 17.19.

EXAMPLE 17.19

Finding the Concentration of the Anions for a Weak Diprotic Acid Solution

Find the $[C_6H_6O_6^{\ 2-}]$ of the 0.100 M ascorbic acid $(H_2C_6H_6O_6)$ solution in Example 17.17.

SOLUTION

To find the $[C_6H_6O_6^{\ 2-}]$, use the concentrations of $[HC_6H_6O_6^{\ -}]$ and H_3O^+ produced by the first ionization step (as calculated in Example 17.17) as the initial concentrations for the second step. Because of the 1:1 stoichiometry, $[HC_6H_6O_6^{\ -}] = [H_3O^+]$. Then solve an equilibrium problem for the second step similar to that of Example 17.6, shown in condensed form here. Use K_{a_2} for ascorbic acid from Table 17.10.

$$HC_6H_6O_6^{\ -}(aq) + H_2O(l) \rightleftharpoons H_3O^+(aq) + C_6H_6O_6^{\ 2-}(aq)$$

	$[HC_6H_6O_6^{\ -}]$	$[H_3O^+]$	$[C_6H_6O_6^{\ 2-}]$
Initial	2.8×10^{-3}	2.8×10^{-3}	0.000
Change	$-x$	$+x$	$+x$
Equil	$2.8 \times 10^{-3} - x$	$2.8 \times 10^{-3} + x$	x

$$K_{a_2} = \frac{[H_3O^+][C_6H_6O_6^{\ 2-}]}{[HC_6H_6O_6^{\ -}]}$$

$$= \frac{(2.8 \times 10^{-3} + \cancel{x})x}{2.8 \times 10^{-3} - \cancel{x}} \quad (x \text{ is small})$$

$$= \frac{\cancel{(2.8 \times 10^{-3})}x}{\cancel{2.8 \times 10^{-3}}}$$

$$x = K_{a_2} = 1.6 \times 10^{-12}$$

Since x is much smaller than 2.8×10^{-3}, the *x is small* approximation is valid. Therefore, $[C_6H_6O_6^{\ 2-}] = 1.6 \times 10^{-12}$ M.

FOR PRACTICE 17.19

Find the $[CO_3^{\ 2-}]$ of the 0.050 M carbonic acid (H_2CO_3) solution in For Practice 17.17.

Notice from the results of Example 17.19 that the concentration of X^{2-} for a weak diprotic acid H_2X is equal to K_{a_2}. This general result applies to all diprotic acids in which the *x is small* approximation is valid. Notice also that the concentration of H_3O^+ produced by the second ionization step of a diprotic acid is very small compared to the concentration produced by the first step, as shown in Figure 17.13 ▼.

Dissociation of a Polyprotic Acid

$$H_2C_6H_6O_6(aq) + H_2O(l) \rightleftharpoons H_3O^+(aq) + HC_6H_6O_6^-(aq)$$

$$[H_3O^+] = 2.8 \times 10^{-3}\ M$$

$$HC_6H_6O_6^-(aq) + H_2O(l) \rightleftharpoons H_3O^+(aq) + C_6H_6O_6^{2-}(aq)$$

$$[H_3O^+] = 1.6 \times 10^{-12}\ M$$

$$\text{Total } [H_3O^+] = 2.8 \times 10^{-3}\ M + 1.6 \times 10^{-12}\ M = 2.8 \times 10^{-3}\ M$$

0.100 M $H_2C_6H_6O_6$

▶ **FIGURE 17.13 Dissociation of a Polyprotic Acid** A 0.100 M $H_2C_6H_6O_6$ solution contains an H_3O^+ concentration of 2.8×10^{-3} M from the first step. The amount of H_3O^+ contributed by the second step is only 1.6×10^{-12} M, which is insignificant compared to the amount produced by the first step.

17.11 Lewis Acids and Bases

We began our definitions of acids and bases with the Arrhenius model. We then saw how the Brønsted–Lowry model, by introducing the concept of a proton donor and proton acceptor, expanded the range of substances that we consider acids and bases. We now introduce a third model which further broadens the range of substances that we can consider acids. This third model is the *Lewis model*, named after G. N. Lewis, the American chemist who devised the electron-dot representation of chemical bonding (Section 5.4). While the Brønsted–Lowry model focuses on the transfer of a proton, the Lewis model focuses on the donation of an electron pair. Consider the simple acid–base reaction between the H^+ ion and NH_3, shown here with Lewis structures:

$$H^+ + :NH_3 \longrightarrow [H:NH_3]^+$$

Brønsted–Lowry model focuses on the proton

Lewis model focuses on the electron pair

According to the Brønsted–Lowry model, the ammonia accepts a proton, thus acting as a base. According to the Lewis model, the ammonia acts as a base by *donating an electron pair*. The general definitions of acids and bases according to the Lewis model focus on the electron pair.

Lewis acid: Electron pair acceptor

Lewis base: Electron pair donor

According to the Lewis definition, H^+ in the above reaction is acting as an acid because it is accepting an electron pair from NH_3. NH_3 is acting as a Lewis base because it is donating an electron pair to H^+.

Although the Lewis model does not significantly expand the substances that can be considered bases—because all proton acceptors must have an electron pair to bind the proton—it does significantly expand the substances that can be considered acids. According to the Lewis model, a substance doesn't even need to contain hydrogen to be an acid. For example, consider the gas-phase reaction between boron trifluoride and ammonia shown here:

$$\underset{\text{Lewis acid}}{BF_3} + \underset{\text{Lewis base}}{:NH_3} \longrightarrow \underset{\text{adduct}}{F_3B:NH_3}$$

Boron trifluoride has an empty orbital that can accept the electron pair from ammonia and form the product (the product of a Lewis acid–base reaction is sometimes called an *adduct*). The above reaction demonstrates an important property of Lewis acids:

A Lewis acid has an empty orbital (or can rearrange electrons to create an empty orbital) that can accept an electron pair.

Consequently, the Lewis definition subsumes a whole new class of acids. Next we examine a few examples.

Molecules That Act as Lewis Acids

Since molecules with incomplete octets have empty orbitals, they can serve as Lewis acids. For example, both $AlCl_3$ and BCl_3 have incomplete octets:

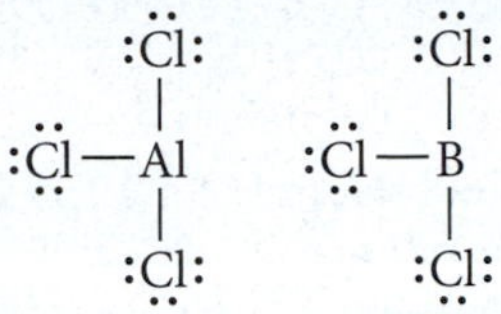

$$AlCl_3 \qquad BCl_3$$

These both act as Lewis acids, as shown in the following reactions:

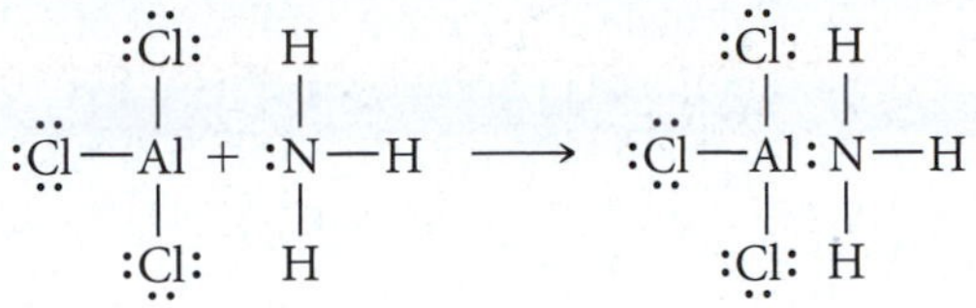

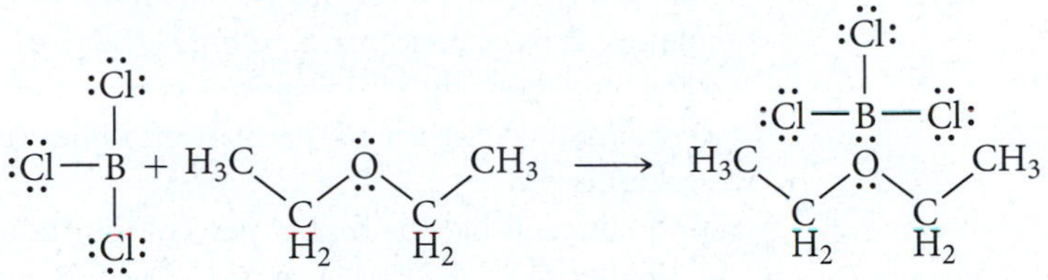

$$AlCl_3 + :NH_3 \longrightarrow Cl_3Al{:}NH_3$$

$$BCl_3 + H_3C{-}CH_2{-}\ddot{O}{-}CH_2{-}CH_3 \longrightarrow Cl_3B{-}\ddot{O}(CH_2CH_3)_2$$

Some molecules that may not initially contain empty orbitals can rearrange their electrons to act as Lewis acids. Consider the reaction between carbon dioxide and water:

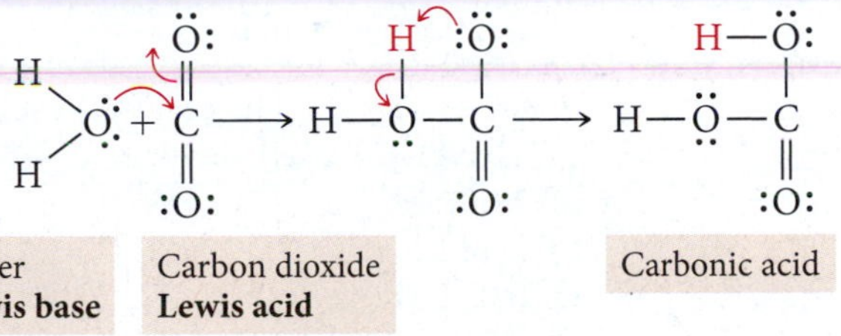

$$H_2\ddot{O}: + O{=}C{=}O \longrightarrow H{-}\ddot{O}(H){-}C({-}\ddot{O}:^{-}){=}O \longrightarrow H{-}\ddot{O}{-}C({-}\ddot{O}{-}H){=}O$$

Water
Lewis base

Carbon dioxide
Lewis acid

Carbonic acid

The electrons in the double bond on carbon move to the terminal oxygen atom, allowing carbon dioxide to act as a Lewis acid by accepting an electron pair from water. The molecule then undergoes a rearrangement in which the hydrogen atom shown in red bonds with the terminal oxygen atom instead of the internal one.

Cations That Act as Lewis Acids

Some cations, since they are positively charged and have lost some electrons, have empty orbitals that allow them to also act as Lewis acids. Consider the hydration process of the Al^{3+} ion discussed in Section 17.9 shown here:

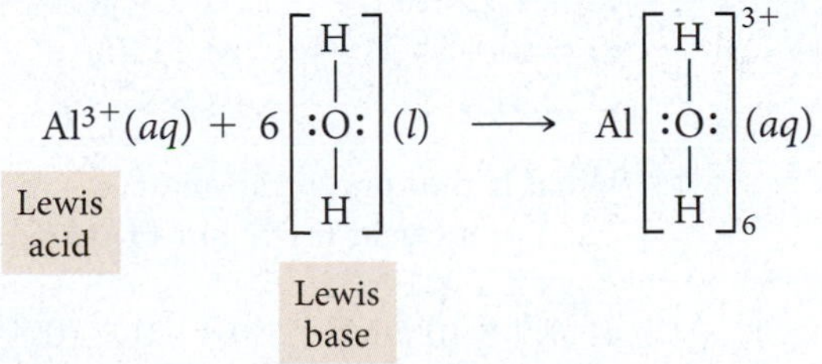

$$Al^{3+}(aq) + 6\left[H{-}\ddot{O}{-}H\right](l) \longrightarrow Al\left[:\!O(H)H\right]_6^{3+}(aq)$$

Lewis
acid

Lewis
base

The aluminum ion acts as a Lewis acid, accepting lone pairs from six water molecules to form the hydrated ion. Many other small, highly-charged metal ions also act as Lewis acids in this way.

SELF-ASSESSMENT Quiz

1. Identify the conjugate base in the reaction shown here.
$$HClO_2(aq) + H_2O(l) \rightleftharpoons H_3O^+(aq) + ClO_2^-(aq)$$
a) $HClO_2$ **b)** H_2O **c)** H_3O^+ **d)** ClO_2^-

2. Which pair is a Brønsted–Lowry conjugate acid–base pair?
a) NH_3; NH_4^+
b) H_3O^+; OH^-
c) HCl; HBr
d) ClO_4^-; ClO_3^-

3. Which acid has the largest K_a: $HClO_2(aq)$, $HBrO_2(aq)$, or $HIO_2(aq)$?
a) $HClO_2(aq)$
b) $HBrO_2(aq)$
c) $HIO_2(aq)$
d) All three acids have the same K_a.

4. Consider the given acid ionization constants and identify the strongest conjugate base.

Acid	K_a
$HNO_2(aq)$	4.6×10^{-4}
$HCHO_2(aq)$	1.8×10^{-4}
$HClO(aq)$	2.9×10^{-8}
$HCN(aq)$	4.9×10^{-10}

a) $NO_2^-(aq)$
b) $CHO_2^-(aq)$
c) $ClO^-(aq)$
d) $CN^-(aq)$

5. What is the OH^- concentration in an aqueous solution at 25 °C in which $[H_3O^+] = 1.9 \times 10^{-9}$ M?
a) 1.9×10^{-9} M
b) 5.3×10^{-6} M
c) $5.3 \times 10^{+6}$ M
d) 1.9×10^{-23} M

6. An $HNO_3(aq)$ solution has a pH of 1.75. What is the molar concentration of the $HNO_3(aq)$ solution?
a) 1.75 M **b)** 5.6×10^{-13} M
c) 56 M **d)** 0.018 M

7. Find the pH of a 0.350 M aqueous benzoic acid solution. For benzoic acid, $K_a = 6.5 \times 10^{-5}$.
a) 4.64 **b)** 4.19
c) 2.32 **d)** 11.68

8. Find the pH of a 0.155 M $HClO_2(aq)$ solution. For $HClO_2$, $K_a = 0.011$.
a) 0.92 **b)** 1.44
c) 1.39 **d)** 0.69

9. Calculate the percent ionization of 1.45 M aqueous acetic acid solution. For acetic acid, $K_a = 1.8 \times 10^{-5}$.
a) 0.35% **b)** 0.0018%
c) 0.29% **d)** 0.0051%

10. Consider two aqueous solutions of nitrous acid (HNO_2). Solution A has a concentration of $[HNO_2] = 0.55$ M and solution B has a concentration of $[HNO_2] = 1.25$ M. Which statement about the two solutions is true?
a) Solution A has the higher percent ionization and the higher pH.
b) Solution B has the higher percent ionization and the higher pH.
c) Solution A has the higher percent ionization and solution B has the higher pH.
d) Solution B has the higher percent ionization and solution A has the higher pH.

11. What is the $[OH^-]$ in a 0.200 M solution of ethylamine ($C_2H_5NH_2$)? For ethylamine, $K_b = 5.6 \times 10^{-4}$.
a) 11.52 M
b) 2.48 M
c) 0.033 M
d) 0.011 M

12. Which ion will be basic in aqueous solution?
a) Br^- **b)** NO_3^-
c) HSO_4^- **d)** SO_3^{2-}

13. Which compound will form an acidic solution when dissolved in water?
a) NH_4Cl **b)** NaCl
c) KNO_2 **d)** $Ca(NO_3)_2$

14. Find the pH of 0.175 M NaCN solution. For HCN, $K_a = 4.9 \times 10^{-10}$.
a) 5.03 **b)** 11.28
c) 2.31 **d)** 8.97

15. What is the concentration of X^{2-} in a 0.150 M solution of the diprotic acid H_2X? For H_2X, $K_{a1} = 4.5 \times 10^{-6}$ and $K_{a2} = 1.2 \times 10^{-11}$.
a) 9.9×10^{-8} M **b)** 2.0×10^{-9} M
c) 8.2×10^{-4} M **d)** 1.2×10^{-11} M

Answers: 1:d; 2:a; 3:a; 4:d; 5:b; 6:d; 7:c; 8:b; 9:a; 10:a; 11:d; 12:d; 13:a; 14:b; 15:d

CHAPTER SUMMARY 17

REVIEW

KEY LEARNING OUTCOMES

CHAPTER OBJECTIVES	ASSESSMENT
Identifying Brønsted–Lowry Acids and Bases and Their Conjugates (17.3)	• Example 17.1 For Practice 17.1 Exercises 33, 34
Determining Relative Acid Strength from Molecular Structure (17.4)	• Exercises 39–42
Using K_w in Calculations (17.6)	• Example 17.2 For Practice 17.2 Exercises 49, 50
Calculating pH from $[H_3O^+]$ or $[OH^-]$ (17.6)	• Examples 17.3, 17.4 For Practice 17.3, 17.4 Exercises 51–54
Finding the pH of a Weak Acid Solution (17.7)	• Examples 17.5, 17.6, 17.7 For Practice 17.5, 17.6, 17.7 Exercises 65–70
Finding the Acid Ionization Constant from pH (17.7)	• Example 17.8 For Practice 17.8 Exercises 71, 72
Finding the Percent Ionization of a Weak Acid (17.7)	• Example 17.9 For Practice 17.9 Exercises 73–76
Mixtures of Weak Acids (17.7)	• Example 17.10 For Practice 17.10 Exercises 81, 82
Finding the $[OH^-]$ and pH of a Strong Base Solution (17.8)	• Example 17.11 For Practice 17.11 Exercises 83, 84
Finding the $[OH^-]$ and pH of a Weak Base Solution (17.8)	• Example 17.12 For Practice 17.12 Exercises 91, 92
Determining Whether an Anion Is Basic or Neutral (17.9)	• Example 17.13 For Practice 17.13 Exercises 97, 98
Determining the pH of a Solution Containing an Anion Acting as a Base (17.9)	• Example 17.14 For Practice 17.14 Exercises 99, 100
Determining Whether a Cation Is Acidic or Neutral (17.9)	• Example 17.15 For Practice 17.15 Exercises 101, 102
Determining the Overall Acidity or Basicity of Salt Solutions (17.9)	• Example 17.16 For Practice 17.16 Exercises 103, 104
Finding the pH of a Polyprotic Acid Solution (17.10)	• Example 17.17 For Practice 17.17 Exercises 115, 116
Finding the $[H_3O^+]$ in Dilute H_2SO_4 Solutions (17.10)	• Example 17.18 For Practice 17.18 Exercise 119
Finding the Concentration of the Anions for a Weak Diprotic Acid Solution (17.10)	• Example 17.19 For Practice 17.19 Exercises 117, 118

KEY TERMS

Section 17.2
carboxylic acid (657)
alkaloid (658)

Section 17.3
Arrhenius definitions (of acids and bases) (658)
hydronium ion (659)
strong acid (659)
weak acid (659)
strong base (659)
weak base (659)
Brønsted–Lowry definitions (of acids and bases) (659)
amphoteric (660)
conjugate acid–base pair (660)
conjugate acid (660)
conjugate base (660)

Section 17.5
monoprotic acid (664)
diprotic acid (664)
triprotic acid (664)
acid ionization constant (K_a) (664)

Section 17.6
autoionization (666)
ion product constant for water (K_w) (666)
neutral (666)
acidic solution (666)
basic solution (667)
pH (668)

Section 17.7
percent ionization (676)

Section 17.8
base ionization constant (K_b) (681)

Section 17.10
polyprotic acid (691)

Section 17.11
Lewis acid (696)
Lewis base (696)

KEY CONCEPTS

The Nature of Acids and Bases (17.2)

- Acids generally taste sour, dissolve metals, turn blue litmus paper red, and neutralize bases. Common acids are hydrochloric, sulfuric, nitric, and carboxylic acids.
- Bases generally taste bitter, feel slippery, turn red litmus paper blue, and neutralize acids. Common bases are sodium hydroxide, sodium bicarbonate, and potassium hydroxide.

Definitions of Acids and Bases (17.3)

- The Arrhenius definition of acids and bases states that in an aqueous solution, an acid produces hydrogen ions and a base produces hydroxide ions.
- The Brønsted–Lowry definition of acids and bases states that an acid is a proton (hydrogen ion) donor and a base is a proton acceptor. According to the Brønsted–Lowry definition, two substances related by the transfer of a proton are a conjugate acid–base pair.

Acid Strength and Molecular Structure (17.4)

- For binary acids, acid strength decreases with increasing bond energy and increases with increasing bond polarity.
- For oxyacids, acid strength increases with the electronegativity of the atoms bonded to the oxygen atom and also increases with the number of oxygen atoms in the molecule.

Acid Strength and the Acid Dissociation Constant (K_a) (17.5)

- In a solution, a strong acid completely ionizes, but a weak acid only partially ionizes.
- The extent of dissociation of a weak acid is quantified by the acid dissociation constant, K_a, which is the equilibrium constant for the ionization of the weak acid.

Autoionization of Water and pH (17.6)

- In an acidic solution, the concentration of hydrogen ions is always greater than the concentration of hydroxide ions. $[H_3O^+]$ multiplied by $[OH^-]$ is always constant at a constant temperature.
- There are two types of logarithmic acid–base scales: pH and pOH. At 25 °C, the sum of the pH and pOH is always 14.

Finding the $[H_3O^+]$ and pH of Strong and Weak Acid Solutions (17.7)

- In a strong acid solution, the hydrogen ion concentration equals the initial concentration of the acid.
- In a weak acid solution, the hydrogen ion concentration—which can be determined by solving an equilibrium problem—is lower than the initial acid concentration.
- The percent ionization of weak acids decreases as the acid (and hydrogen ion) concentration increases.
- In mixtures of two acids with large K_a differences, the concentration of hydrogen ions can usually be determined by considering only the stronger of the two acids.

Finding the $[OH^-]$ and pH of Strong and Weak Base Solutions (17.8)

- A strong base dissociates completely; a weak base does not.
- Most weak bases produce hydroxide ions through the ionization of water. The base ionization constant, K_b, indicates the extent of ionization.

The Acid–Base Properties of Ions and Salts (17.9)

- An anion is a weak base if it is the conjugate base of a weak acid; it is neutral if it is the conjugate base of a strong acid.
- A cation is a weak acid if it is the conjugate acid of a weak base; it is neutral if it is the conjugate acid of a strong base.
- To calculate the pH of an acidic cation or basic anion, we determine K_a or K_b from the equation $K_a \times K_b = K_w$.

Polyprotic Acids (17.10)

- Polyprotic acids contain two or more ionizable protons.
- Generally, polyprotic acids ionize in successive steps, with the value of K_a becoming smaller for each step.
- In many cases, we can determine the $[H_3O^+]$ of a polyprotic acid solution by considering only the first ionization step; then, the concentration of the acid anion formed in the second ionization step is equivalent to the value of K_{a_2}.

Lewis Acids and Bases (17.11)

- A third model of acids and bases, the Lewis model, defines a base as an electron pair donor and an acid as an electron pair acceptor; therefore, according to this definition, an acid does not have to contain hydrogen. According to this definition, an acid can be a compound with an empty orbital—or one that will rearrange to make an empty orbital—or a cation.

KEY EQUATIONS AND RELATIONSHIPS

Note: In all of these equations $[H^+]$ is interchangeable with $[H_3O^+]$.

Expression for the Acid Ionization Constant, K_a (17.5)

$$K_a = \frac{[H_3O^+][A^-]}{[HA]}$$

The Ion Product Constant for Water, K_w (17.6)

$$K_w = [H_3O^+][OH^-] = 1.0 \times 10^{-14} \text{ (at 25 °C)}$$

Expression for the pH Scale (17.6)

$$\text{pH} = -\log[H_3O^+]$$

Expression for the pOH Scale (17.6)

$$\text{pOH} = -\log[OH^-]$$

Relationship between pH and pOH (17.6)

$$\text{pH} + \text{pOH} = 14.00$$

Expression for the pK_a Scale (17.6)

$$pK_a = -\log K_a$$

Expression for Percent Ionization (17.7)

$$\text{Percent ionization} = \frac{\text{concentration of ionized acid}}{\text{initial concentration of acid}} \times 100\%$$

$$= \frac{[H_3O^+]_{\text{equil}}}{[HA]_{\text{init}}} \times 100\%$$

Relationship between K_a, K_b, and K_w (17.9)

$$K_a \times K_b = K_w$$

EXERCISES

REVIEW QUESTIONS

1. In the opening section of this chapter, we see that in the Batman comic book series, Batman treats a strong acid burn with strong base. What is problematic about this treatment?
2. What are the general physical and chemical properties of acids? Of bases?
3. What is a carboxylic acid? Give an example.
4. What is the Arrhenius definition of an acid? Of a base?
5. What is a hydronium ion? Does H^+ exist in solution by itself?
6. What is the Brønsted–Lowry definition of an acid? Of a base?
7. Why is there more than one definition of acid–base behavior? Which definition is the right one?
8. Describe amphoteric behavior and give an example.
9. What is a conjugate acid–base pair? Provide an example.
10. Explain the difference between a strong acid and a weak acid and list one example of each.
11. For a binary acid, H—Y, which factors affect the relative ease with which the acid ionizes?
12. Which factors affect the relative acidity of an oxyacid?
13. What are diprotic and triprotic acids? List an example of each.
14. Define the acid ionization constant and explain its significance.
15. Write an equation for the autoionization of water and an expression for the ion product constant for water (K_w). What is the value of K_w at 25 °C?
16. What happens to the $[OH^-]$ of a solution when the $[H_3O^+]$ is increased? Decreased?
17. Define pH. What pH range is considered acidic? Basic? Neutral?
18. Define pOH. What pOH range is considered acidic? Basic? Neutral?
19. In most solutions containing a strong or weak acid, the autoionization of water can be neglected when calculating $[H_3O^+]$. Explain why this statement is valid.
20. When calculating $[H_3O^+]$ for weak acid solutions, we can often use the *x is small* approximation. Explain the nature of this approximation and why it is valid.
21. What is the percent ionization of an acid? Explain what happens to the percent ionization of a weak acid as a function of the concentration of the weak acid solution.
22. In calculating $[H_3O^+]$ for a mixture of a strong acid and weak acid, the weak acid can often be neglected. Explain why this statement is valid.
23. Write a generic equation showing how a weak base ionizes water.
24. How can you determine if an anion will act as a weak base? Write a generic equation showing the reaction by which an anion, A^-, acts as a weak base.
25. What is the relationship between the acid ionization constant for a weak acid (K_a) and the base ionization constant for its conjugate base (K_b)?
26. What kinds of cations act as weak acids? List some examples.
27. When calculating the $[H_3O^+]$ for a polyprotic acid, the second ionization step can often be neglected. Explain why this statement is valid.
28. For a weak diprotic acid H_2X, what is the relationship between $[X^{2-}]$ and K_{a_2}? Under what conditions does this relationship exist?
29. What is the Lewis definition of an acid? Of a base?
30. What is a general characteristic of a Lewis acid? Of a Lewis base?

PROBLEMS BY TOPIC

The Nature and Definitions of Acids and Base

31. Identify each substance as an acid or a base and write a chemical equation showing how it is an acid or a base according to the Arrhenius definition.
    - **a.** $HNO_3(aq)$
    - **b.** $NH_4^+(aq)$
    - **c.** $KOH(aq)$
    - **d.** $HC_2H_3O_2(aq)$
32. Identify each substance as an acid or a base and write a chemical equation showing how it is an acid or a base according to the Arrhenius definition.
    - **a.** $NaOH(aq)$
    - **b.** $H_2SO_4(aq)$
    - **c.** $HBr(aq)$
    - **d.** $Sr(OH)_2(aq)$
33. In each reaction, identify the Brønsted–Lowry acid, the Brønsted–Lowry base, the conjugate acid, and the conjugate base.
    - **a.** $H_2CO_3(aq) + H_2O(l) \rightleftharpoons H_3O^+(aq) + HCO_3^-(aq)$
    - **b.** $NH_3(aq) + H_2O(l) \rightleftharpoons NH_4^+(aq) + OH^-(aq)$
    - **c.** $HNO_3(aq) + H_2O(l) \longrightarrow H_3O^+(aq) + NO_3^-(aq)$
    - **d.** $C_5H_5N(aq) + H_2O(l) \rightleftharpoons C_5H_5NH^+(aq) + OH^-(aq)$
34. In each reaction, identify the Brønsted–Lowry acid, the Brønsted–Lowry base, the conjugate acid, and the conjugate base.
    - **a.** $HI(aq) + H_2O(l) \longrightarrow H_3O^+(aq) + I^-(aq)$
    - **b.** $CH_3NH_2(aq) + H_2O(l) \rightleftharpoons CH_3NH_3^+(aq) + OH^-(aq)$
    - **c.** $CO_3^{2-}(aq) + H_2O(l) \rightleftharpoons HCO_3^-(aq) + OH^-(aq)$
    - **d.** $HBr(aq) + H_2O(l) \longrightarrow H_3O^+(aq) + Br^-(aq)$
35. Write the formula for the conjugate base of each acid.
    - **a.** HCl
    - **b.** H_2SO_3
    - **c.** $HCHO_2$
    - **d.** HF
36. Write the formula for the conjugate acid of each base.
    - **a.** NH_3
    - **b.** ClO_4^-
    - **c.** HSO_4^-
    - **d.** CO_3^{2-}
37. Both H_2O and $H_2PO_4^-$ are amphoteric. Write an equation to show how each substance can act as an acid, and another equation to show how each can act as a base.
38. Both HCO_3^- and HS^- are amphoteric. Write an equation to show how each substance can act as an acid, and another equation to show how each can act as a base.

Acid Strength and Molecular Structure

39. Based on their molecular structure, pick the stronger acid from each pair of binary acids. Explain your choice.
    a. HF and HCl
    b. H_2O or HF
    c. H_2Se or H_2S
40. Based on molecular structure, arrange the binary compounds in order of increasing acid strength. Explain your choice.

$$H_2Te, HI, H_2S, NaH$$

41. Based on their molecular structure, pick the stronger acid from each pair of oxyacids. Explain your choice.
    a. H_2SO_4 or H_2SO_3
    b. $HClO_2$ or HClO
    c. HClO or HBrO
    d. CCl_3COOH or CH_3COOH
42. Based on molecular structure, arrange the oxyacids in order of increasing acid strength. Explain your choice.

$$HClO_3, HIO_3, HBrO_3$$

43. Which is a stronger base, S^{2-} or Se^{2-}? Explain.
44. Which is a stronger base, PO_4^{3-} or AsO_4^{3-}? Explain.

Acid Strength and K_a

45. Classify each acid as strong or weak. If the acid is weak, write an expression for the acid ionization constant (K_a).
    a. HNO_3
    b. HCl
    c. HBr
    d. H_2SO_3
46. Classify each acid as strong or weak. If the acid is weak, write an expression for the acid ionization constant (K_a).
    a. HF
    b. $HCHO_2$
    c. H_2SO_4
    d. H_2CO_3
47. The three diagrams represent three different solutions of the binary acid HA. Water molecules have been omitted for clarity, and hydronium ions (H_3O^+) are represented by hydrogen ions (H^+). Rank the acids in order of decreasing acid strength.

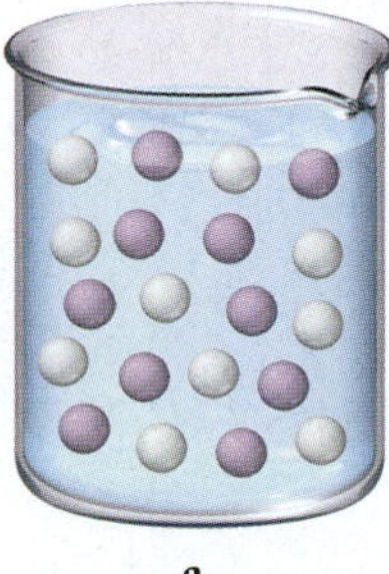
a.

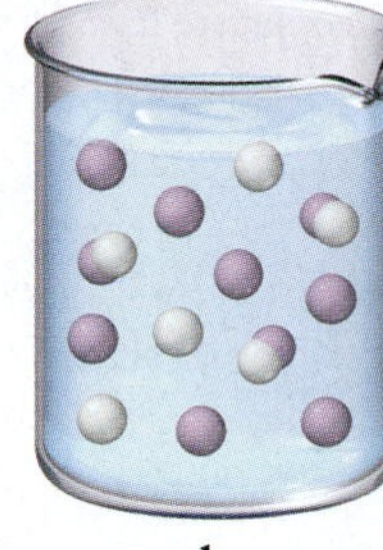
b.

c.

48. Rank the solutions in order of decreasing $[H_3O^+]$: 0.10 M HCl; 0.10 M HF; 0.10 M HClO; 0.10 M HC_6H_5O.

Autoionization of Water and pH

49. Calculate $[OH^-]$ in each aqueous solution at 25 °C, and classify the solution as acidic or basic.
    a. $[H_3O^+] = 1.2 \times 10^{-8}$ M
    b. $[H_3O^+] = 8.5 \times 10^{-5}$ M
    c. $[H_3O^+] = 3.5 \times 10^{-2}$ M
50. Calculate $[H_3O^+]$ in each aqueous solution at 25 °C, and classify each solution as acidic or basic.
    a. $[OH^-] = 1.1 \times 10^{-9}$ M
    b. $[OH^-] = 2.9 \times 10^{-2}$ M
    c. $[OH^-] = 6.9 \times 10^{-12}$ M
51. Calculate the pH and pOH of each solution.
    a. $[H_3O^+] = 1.7 \times 10^{-8}$ M
    b. $[H_3O^+] = 1.0 \times 10^{-7}$ M
    c. $[H_3O^+] = 2.2 \times 10^{-6}$ M
52. Calculate $[H_3O^+]$ and $[OH^-]$ for each solution.
    a. pH = 8.55
    b. pH = 11.23
    c. pH = 2.87
53. Complete the table. (All solutions are at 25 °C.)

$[H_3O^+]$	$[OH^-]$	pH	Acidic or Basic
____	____	3.15	____
3.7×10^{-9} M	____	____	____
____	____	11.1	____
____	1.6×10^{-11} M	____	____

54. Complete the table. (All solutions are at 25 °C.)

$[H_3O^+]$	$[OH^-]$	pH	Acidic or Basic
3.5×10^{-3} M	____	____	____
____	3.8×10^{-7} M	____	____
1.8×10^{-9} M	____	____	____
____	____	7.15	____

55. Like all equilibrium constants, the value of K_w depends on temperature. At body temperature (37 °C), $K_w = 2.4 \times 10^{-14}$. What is the $[H_3O^+]$ and pH of pure water at body temperature?
56. The value of K_w increases with increasing temperature. Is the autoionization of water endothermic or exothermic?
57. Calculate the pH of each acid solution. Explain how the pH values you calculate demonstrate that the pH of an acid solution should carry as many digits to the right of the decimal place as the number of significant figures in the concentration of the solution.

$$[H_3O^+] = 0.044 \text{ M}$$
$$[H_3O^+] = 0.045 \text{ M}$$
$$[H_3O^+] = 0.046 \text{ M}$$

58. Find the concentration of H_3O^+, to the correct number of significant figures, in a solution with each pH. Describe how your calculations show the relationship between the number of digits to the right of the decimal place in pH and the number of significant figures in concentration.

pH = 2.50
pH = 2.51
pH = 2.52

Acid Solutions

59. For each strong acid solution, determine $[H_3O^+]$, $[OH^-]$, and pH.
    a. 0.25 M HCl
    b. 0.015 M HNO_3
    c. a solution that is 0.052 M in HBr and 0.020 M in HNO_3
    d. a solution that is 0.655% HNO_3 by mass (Assume a density of 1.01 g/mL for the solution.)
60. Determine the pH of each solution.
    a. 0.048 M HI
    b. 0.0895 M $HClO_4$
    c. a solution that is 0.045 M in $HClO_4$ and 0.048 M in HCl
    d. a solution that is 1.09% HCl by mass (Assume a density of 1.01 g/mL for the solution.)
61. What mass of HI should be present in 0.250 L of solution to obtain a solution with each pH value?
    a. pH = 1.25
    b. pH = 1.75
    c. pH = 2.85
62. What mass of $HClO_4$ should be present in 0.500 L of solution to obtain a solution with each pH value?
    a. pH = 2.50
    b. pH = 1.50
    c. pH = 0.50
63. What is the pH of a solution in which 224 mL of HCl(*g*), measured at 27.2 °C and 1.02 atm, is dissolved in 1.5 L of aqueous solution?
64. What volume of a concentrated HCl solution, which is 36.0% HCl by mass and has a density of 1.179 g/mL, should you use to make 5.00 L of an HCl solution with a pH of 1.8?
65. Determine the $[H_3O^+]$ and pH of a 0.100 M solution of benzoic acid.
66. Determine the $[H_3O^+]$ and pH of a 0.200 M solution of formic acid.
67. Determine the pH of an HNO_2 solution of each concentration. In which cases can you *not* make the assumption that *x is small*?
    a. 0.500 M
    b. 0.100 M
    c. 0.0100 M
68. Determine the pH of an HF solution of each concentration. In which cases can you *not* make the assumption that *x is small*?
    a. 0.250 M
    b. 0.0500 M
    c. 0.0250 M
69. If 15.0 mL of glacial acetic acid (pure $HC_2H_3O_2$) is diluted to 1.50 L with water, what is the pH of the resulting solution? The density of glacial acetic acid is 1.05 g/mL.
70. Calculate the pH of a formic acid solution that contains 1.35% formic acid by mass. (Assume a density of 1.01 g/mL for the solution.)
71. A 0.185 M solution of a weak acid (HA) has a pH of 2.95. Calculate the acid ionization constant (K_a) for the acid.
72. A 0.115 M solution of a weak acid (HA) has a pH of 3.29. Calculate the acid ionization constant (K_a) for the acid.
73. Determine the percent ionization of a 0.125 M HCN solution.
74. Determine the percent ionization of a 0.225 M solution of benzoic acid.
75. Calculate the percent ionization of an acetic acid solution having the given concentrations.
    a. 1.00 M
    b. 0.500 M
    c. 0.100 M
    d. 0.0500 M
76. Calculate the percent ionization of a formic acid solution having the given concentrations.
    a. 1.00 M
    b. 0.500 M
    c. 0.100 M
    d. 0.0500 M
77. A 0.148 M solution of a monoprotic acid has a percent ionization of 1.55%. Determine the acid ionization constant (K_a) for the acid.
78. A 0.085 M solution of a monoprotic acid has a percent ionization of 0.59%. Determine the acid ionization constant (K_a) for the acid.
79. Find the pH and percent ionization of each HF solution.
    a. 0.250 M HF
    b. 0.100 M HF
    c. 0.050 M HF
80. Find the pH and percent ionization of a 0.100 M solution of a weak monoprotic acid having the given K_a values.
    a. $K_a = 1.0 \times 10^{-5}$
    b. $K_a = 1.0 \times 10^{-3}$
    c. $K_a = 1.0 \times 10^{-1}$
81. Find the pH of each mixture of acids.
    a. 0.115 M in HBr and 0.125 M in $HCHO_2$
    b. 0.150 M in HNO_2 and 0.085 M in HNO_3
    c. 0.185 M in $HCHO_2$ and 0.225 M in $HC_2H_3O_2$
    d. 0.050 M in acetic acid and 0.050 M in hydrocyanic acid
82. Find the pH of each mixture of acids.
    a. 0.075 M in HNO_3 and 0.175 M in $HC_7H_5O_2$
    b. 0.020 M in HBr and 0.015 M in $HClO_4$
    c. 0.095 M in HF and 0.225 M in HC_6H_5O
    d. 0.100 M in formic acid and 0.050 M in hypochlorous acid

Base Solutions

83. For each strong base solution, determine $[OH^-]$, $[H_3O^+]$, pH, and pOH.
    a. 0.15 M NaOH
    b. 1.5×10^{-3} M $Ca(OH)_2$
    c. 4.8×10^{-4} M $Sr(OH)_2$
    d. 8.7×10^{-5} M KOH
84. For each strong base solution, determine $[OH^-]$, $[H_3O^+]$, pH, and pOH.
    a. 8.77×10^{-3} M LiOH
    b. 0.0112 M $Ba(OH)_2$
    c. 1.9×10^{-4} M KOH
    d. 5.0×10^{-4} M $Ca(OH)_2$
85. Determine the pH of a solution that is 3.85% KOH by mass. Assume that the solution has density of 1.01 g/mL.

86. Determine the pH of a solution that is 1.55% NaOH by mass. Assume that the solution has density of 1.01 g/mL.

87. What volume of 0.855 M KOH solution do you need to make 3.55 L of a solution with pH of 12.4?

88. What volume of a 15.0% by mass NaOH solution, which has a density of 1.116 g/mL, should you use to make 5.00 L of an NaOH solution with a pH of 10.8?

89. Write equations showing how each weak base ionizes water to form OH^-. Also write the corresponding expression for K_b.

a. NH_3
b. HCO_3^-
c. CH_3NH_2

90. Write equations showing how each weak base ionizes water to form OH^-. Also write the corresponding expression for K_b.

a. CO_3^{2-}
b. $C_6H_5NH_2$
c. $C_2H_5NH_2$

91. Determine the $[OH^-]$, pH, and pOH of a 0.15 M ammonia solution.

92. Determine the $[OH^-]$, pH, and pOH of a solution that is 0.125 M in CO_3^{2-}.

93. Caffeine ($C_8H_{10}N_4O_2$) is a weak base with a pK_b of 10.4. Calculate the pH of a solution containing a caffeine concentration of 455 mg/L.

94. Amphetamine ($C_9H_{13}N$) is a weak base with a pK_b of 4.2. Calculate the pH of a solution containing an amphetamine concentration of 225 mg/L.

95. Morphine is a weak base. A 0.150 M solution of morphine has a pH of 10.5. What is K_b for morphine?

96. A 0.135 M solution of a weak base has a pH of 11.23. Determine K_b for the base.

Acid–Base Properties of Ions and Salts

97. Determine if each anion acts as a weak base in solution. For the anions that are basic, write an equation that shows how the anion acts as a base.

a. Br^-
b. ClO^-
c. CN^-
d. Cl^-

98. Determine whether each anion is basic or neutral. For the anions that are basic, write an equation that shows how the anion acts as a base.

a. $C_7H_5O_2^-$
b. I^-
c. NO_3^-
d. F^-

99. Determine the $[OH^-]$ and pH of a solution that is 0.140 M in F^-.

100. Determine the $[OH^-]$ and pH of a solution that is 0.250 M in HCO_3^-.

101. Determine whether each cation is acidic or pH-neutral. For the cations that are acidic, write an equation that shows how the cation acts as an acid.

a. NH_4^+
b. Na^+
c. Co^{3+}
d. $CH_3NH_3^+$

102. Determine whether each cation is acidic or pH-neutral. For the cations that are acidic, write an equation that shows how the cation acts as an acid.

a. Sr^{2+}
b. Mn^{3+}
c. $C_5H_5NH^+$
d. Li^+

103. Determine if each salt will form a solution that is acidic, basic, or pH-neutral.

a. $FeCl_3$
b. NaF
c. $CaBr_2$
d. NH_4Br
e. $C_6H_5NH_3NO_2$

104. Determine if each salt will form a solution that is acidic, basic, or pH-neutral.

a. $Al(NO_3)_3$
b. $C_2H_5NH_3NO_3$
c. K_2CO_3
d. RbI
e. NH_4ClO

105. Arrange the solutions in order of increasing acidity.

NaCl, NH_4Cl, $NaHCO_3$, NH_4ClO_2, NaOH

106. Arrange the solutions in order of increasing basicity.

CH_3NH_3Br, KOH, KBr, KCN, $C_5H_5NHNO_2$

107. Determine the pH of each solution.

a. 0.10 M NH_4Cl
b. 0.10 M $NaC_2H_3O_2$
c. 0.10 NaCl

108. Determine the pH of each solution.

a. 0.20 M $KCHO_2$
b. 0.20 M CH_3NH_3I
c. 0.20 M KI

109. Calculate the concentration of all species in a 0.15 M KF solution.

110. Calculate the concentration of all species in a 0.225 M $C_6H_5NH_3Cl$ solution.

111. Pick the stronger base from each pair.

a. F^- or Cl^-
b. NO_2^- or NO_3^-
c. F^- or ClO^-

112. Pick the stronger base from each pair.

a. ClO_4^- or ClO_2^-
b. Cl^- or H_2O
c. CN^- or ClO^-

Polyprotic Acids

113. Write chemical equations and corresponding equilibrium expressions for each of the three ionization steps of phosphoric acid.

114. Write chemical equations and corresponding equilibrium expressions for each of the two ionization steps of carbonic acid.

115. Calculate the $[H_3O^+]$ and pH of each polyprotic acid solution.

a. 0.350 M H_3PO_4
b. 0.350 M $H_2C_2O_4$

116. Calculate the $[H_3O^+]$ and pH of each polyprotic acid solution.

a. 0.125 M H_2CO_3
b. 0.125 M $H_3C_6H_5O_7$

117. Calculate the concentration of each species in a 0.500 M solution of H_2SO_3.

118. Calculate the concentration of each species in a 0.155 M solution of H_2CO_3.

119. Calculate the $[H_3O^+]$ and pH of each H_2SO_4 solution. At approximately what concentration does the *x is small* approximation break down?

a. 0.50 M
b. 0.10 M
c. 0.050 M

120. Consider a 0.10 M solution of a weak polyprotic acid (H_2A) with the possible values of K_{a_1} and K_{a_2} given here.

a. $K_{a_1} = 1.0 \times 10^{-4}$; $K_{a_2} = 5.0 \times 10^{-5}$
b. $K_{a_1} = 1.0 \times 10^{-4}$; $K_{a_2} = 1.0 \times 10^{-5}$
c. $K_{a_1} = 1.0 \times 10^{-4}$; $K_{a_2} = 1.0 \times 10^{-6}$

Calculate the contributions to $[H_3O^+]$ from each ionization step. At what point can the contribution of the second step be neglected?

Lewis Acids and Bases

121. Classify each species as a Lewis acid or a Lewis base.

a. Fe^{3+}
b. BH_3
c. NH_3
d. F^-

122. Classify each species as a Lewis acid or a Lewis base.

a. $BeCl_2$
b. OH^-
c. $B(OH)_3$
d. CN^-

123. Identify the Lewis acid and Lewis base among the reactants in each equation.

a. $Fe^{3+}(aq) + 6H_2O(l) \rightleftharpoons Fe(H_2O)_6{}^{3+}(aq)$
b. $Zn^{2+}(aq) + 4NH_3(aq) \rightleftharpoons Zn(NH_3)_4{}^{2+}(aq)$
c. $(CH_3)_3N(g) + BF_3(g) \rightleftharpoons (CH_3)_3NBF_3(s)$

124. Identify the Lewis acid and Lewis base among the reactants in each equation.

a. $Ag^+(aq) + 2NH_3(aq) \rightleftharpoons Ag(NH_3)_2{}^+(aq)$
b. $AlBr_3 + NH_3 \rightleftharpoons H_3NAlBr_3$
c. $F^-(aq) + BF_3(aq) \rightleftharpoons BF_4{}^-(aq)$

CUMULATIVE PROBLEMS

125. Based on these molecular views, determine whether each pictured acid is weak or strong.

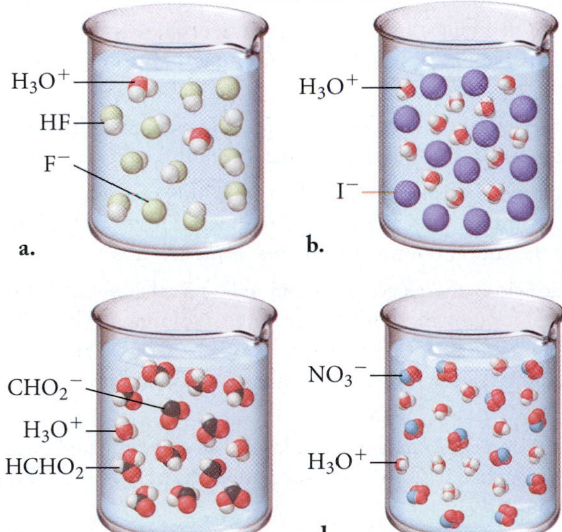

126. Based on these molecular views, determine whether each pictured base is weak or strong.

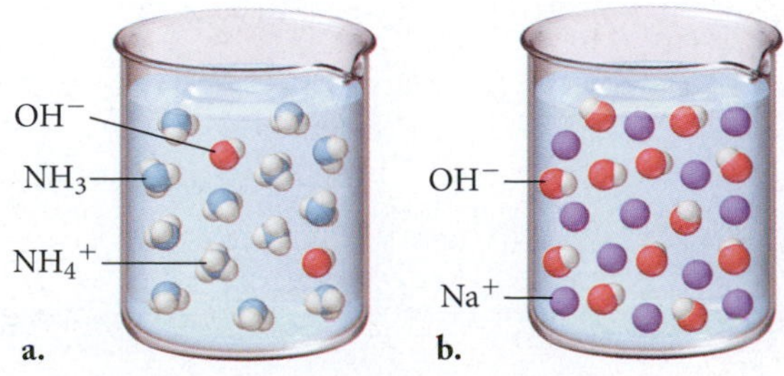

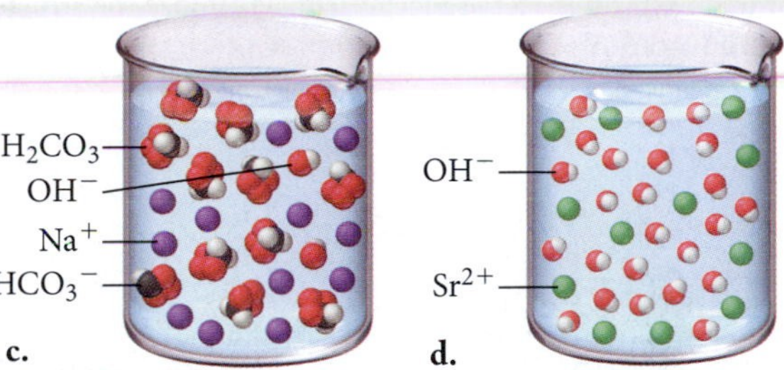

127. The binding of oxygen by hemoglobin in the blood involves the equilibrium reaction:

$$HbH^+(aq) + O_2(aq) \rightleftharpoons HbO_2(aq) + H^+(aq)$$

In this equation, Hb is hemoglobin. The pH of normal human blood is highly controlled within a range of 7.35 to 7.45. Given the above equilibrium, why is this important? What would happen to the oxygen-carrying capacity of hemoglobin if blood became too acidic (a dangerous condition known as acidosis)?

128. Carbon dioxide dissolves in water according to the equations:

$$CO_2(g) + H_2O(l) \rightleftharpoons H_2CO_3(aq)$$
$$H_2CO_3(aq) + H_2O(l) \rightleftharpoons HCO_3{}^-(aq) + H_3O^+(aq)$$

Carbon dioxide levels in the atmosphere have increased about 20% over the last century. Given that Earth's oceans are exposed to atmospheric carbon dioxide, what effect might the increased CO_2 have on the pH of the world's oceans? What effect might this change have on the limestone structures (primarily $CaCO_3$) of coral reefs and marine shells?

129. People often take milk of magnesia to reduce the discomfort associated with acid stomach or heartburn. The recommended dose is 1 teaspoon, which contains 4.00×10^2 mg of $Mg(OH)_2$. What volume of an HCl solution with a pH of 1.3 can be neutralized by one dose of milk of magnesia? If the stomach contains 2.00×10^2 mL of pH 1.3 solution, will all the acid be neutralized? If not, what fraction is neutralized?

130. Lakes that have been acidified by acid rain (which is caused by air pollutants) can be neutralized by liming, the addition of limestone $(CaCO_3)$. How much limestone (in kg) is required to completely neutralize a 4.3 billion liter lake with a pH of 5.5?

Liming a lake.

131. Acid rain over the Great Lakes has a pH of about 4.5. Calculate the $[H_3O^+]$ of this rain and compare that value to the $[H_3O^+]$ of rain over the West Coast that has a pH of 5.4. How many times more concentrated is the acid in rain over the Great Lakes?

132. White wines tend to be more acidic than red wines. Find the $[H_3O^+]$ in a Sauvignon Blanc with a pH of 3.23 and a Cabernet Sauvignon with a pH of 3.64. How many times more acidic is the Sauvignon Blanc?

133. Common aspirin is acetylsalicylic acid, which has the structure shown here and a pK_a of 3.5.

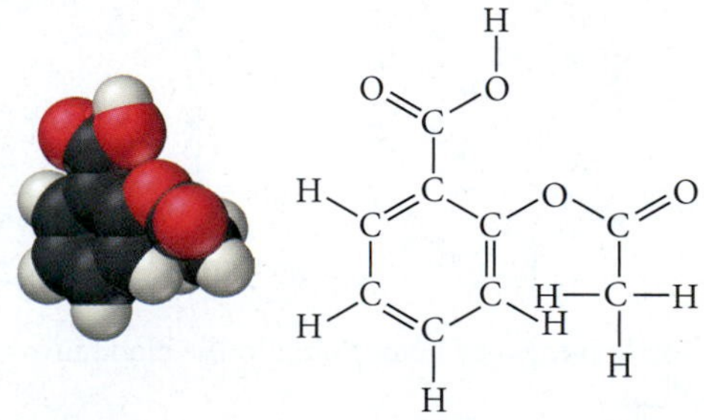

Calculate the pH of a solution in which one normal adult dose of aspirin (6.5×10^2 mg) is dissolved in 8.0 ounces of water.

134. The AIDS drug zalcitabine (also known as ddC) is a weak base with a pK_b of 9.8. What percentage of the base is protonated (in the form BH^+) in an aqueous zalcitabine solution containing 565 mg/L?

135. Determine the pH of each solution.

a. 0.0100 M $HClO_4$
b. 0.115 M $HClO_2$
c. 0.045 M $Sr(OH)_2$
d. 0.0852 M KCN
e. 0.155 M NH_4Cl

136. Determine the pH of each solution.

a. 0.0650 M HNO_3
b. 0.150 M HNO_2
c. 0.0195 M KOH
d. 0.245 M CH_3NH_3I
e. 0.318 M KC_6H_5O

137. Determine the pH of each two-component solution.

a. 0.0550 M in HI and 0.00850 M in HF
b. 0.112 M in NaCl and 0.0953 M in KF
c. 0.132 M in NH_4Cl and 0.150 M HNO_3
d. 0.0887 M in sodium benzoate and 0.225 M in potassium bromide
e. 0.0450 M in HCl and 0.0225 M in HNO_3

138. Determine the pH of each two-component solution.

a. 0.050 M KOH and 0.015 M $Ba(OH)_2$
b. 0.265 M NH_4NO_3 and 0.102 M HCN
c. 0.075 M RbOH and 0.100 M $NaHCO_3$
d. 0.088 M $HClO_4$ and 0.022 M KOH
e. 0.115 M NaClO and 0.0500 M KI

139. Write net ionic equations for the reactions that take place when aqueous solutions of each pair of substances is mixed.

a. sodium cyanide and nitric acid
b. ammonium chloride and sodium hydroxide
c. sodium cyanide and ammonium bromide
d. potassium hydrogen sulfate and lithium acetate
e. sodium hypochlorite and ammonia

140. Morphine has the formula $C_{17}H_{19}NO_3$. It is a base and accepts one proton per molecule. It is isolated from opium. A 0.682 g sample of opium is found to require 8.92 mL of a 0.0116 M solution of sulfuric acid for neutralization. Assuming that morphine is the only acid or base present in opium, calculate the percent morphine in the sample of opium.

141. The pH of a 1.00 M solution of urea, a weak organic base, is 7.050. Calculate the K_a of protonated urea (i.e., the conjugate acid of urea).

142. A solution is prepared by dissolving 0.10 mol of acetic acid and 0.10 mol of ammonium chloride in enough water to make 1.0 L of solution. Find the concentration of ammonia in the solution.

143. Lactic acid is a weak acid found in milk. Its calcium salt is a source of calcium for growing animals. A saturated solution of this salt, which we can represent as $Ca(Lact)_2$, has a $[Ca^{2+}] = 0.26$ M and a pH $= 8.40$. Assuming the salt is completely dissociated, calculate the K_a of lactic acid.

144. A solution of 0.23 mol of the chloride salt of protonated quinine (QH^+), a weak organic base, in 1.0 L of solution has pH $= 4.58$. Find the K_b of quinine (Q).

CHALLENGE PROBLEMS

145. A student mistakenly calculates the pH of a 1.0×10^{-7} M HI solution to be 7.0. Explain why the student is incorrect and calculate the correct pH.

146. When 2.55 g of an unknown weak acid (HA) with a molar mass of 85.0 g/mol is dissolved in 250.0 g of water, the freezing point of the resulting solution is −0.257 °C. Calculate K_a for the unknown weak acid.

147. Calculate the pH of a solution that is 0.00115 M in HCl and 0.0100 M in $HClO_2$.

148. To what volume should you dilute 1 L of a solution of a weak acid HA to reduce the $[H^+]$ to one-half of that in the original solution?

149. HA, a weak acid, with $K_a = 1.0 \times 10^{-8}$, also forms the ion HA_2^-. The reaction is $HA(aq) + A^-(aq) \rightleftharpoons HA_2^-(aq)$ and its $K_a = 4.0$. Calculate the $[H^+]$, $[A^-]$, and $[HA_2^-]$ in a 1.0 M solution of HA.

150. Basicity in the gas phase can be defined as the proton affinity of the base, for example, $CH_3NH_2(g) + H^+(g) \rightleftharpoons CH_3NH_3^+(g)$. In the gas phase, $(CH_3)_3N$ is more basic than CH_3NH_2 while in solution the reverse is true. Explain this observation.

151. Calculate the pH of a solution prepared from 0.200 mol of NH_4CN and enough water to make 1.00 L of solution.

152. To 1.0 L of a 0.30 M solution of $HClO_2$ is added 0.20 mol of NaF. Calculate the $[HClO_2]$ at equilibrium.

153. A mixture of Na_2CO_3 and $NaHCO_3$ has a mass of 82.2 g. It is dissolved in 1.00 L of water, and the pH is found to be 9.95. Determine the mass of $NaHCO_3$ in the mixture.

154. A mixture of NaCN and $NaHSO_4$ consists of a total of 0.60 mol. When the mixture is dissolved in 1.0 L of water and comes to equilibrium, the pH is found to be 9.9. Determine the amount of NaCN in the mixture.

CONCEPTUAL PROBLEMS

155. Without doing any calculations, determine which solution in each pair is most acidic.

a. 0.0100 M in HCl and 0.0100 M in KOH
b. 0.0100 M in HF and 0.0100 in KBr
c. 0.0100 M in NH_4Cl and 0.0100 M in CH_3NH_3Br
d. 0.100 M in NaCN and 0.100 M in $CaCl_2$

156. Without doing any calculations, determine which solution in each pair is most basic.

a. 0.100 M in NaClO and 0.100 M in NaF
b. 0.0100 M in KCl and 0.0100 M in $KClO_2$
c. 0.0100 M in HNO_3 and 0.0100 M in NaOH
d. 0.0100 M in NH_4Cl and 0.0100 M in HCN

157. Rank the acids in order of increasing acid strength.

CH_3COOH $\quad CH_2ClCOOH$ $\quad CHCl_2COOH$ $\quad CCl_3COOH$

ANSWERS TO CONCEPTUAL CONNECTIONS

Cc 17.1 **(b)** H_2SO_4 and H_2SO_3 are both acids; this is not a conjugate acid–base pair.

Cc 17.2 **(a)** Because the carbon atom in **(a)** is bonded to another oxygen atom which draws electron density away from the O—H bond (weakening and polarizing it), and the carbon atom in **(b)** is bonded only to other hydrogen atoms, the proton in structure **(a)** is more acidic.

Cc 17.3 The larger ionization constant indicates that HF is stronger.

Cc 17.4 **(c)** The validity of the *x is small* approximation depends on both the value of the equilibrium constant and the initial concentration—the closer that these are to one another, the less likely the approximation will be valid.

Cc 17.5 **(a)** A weak acid solution will usually be less than 5% dissociated. Since HCl is a strong acid, the 0.10 M solution is much more acidic than either a weak acid with the same concentration or even a weak acid that is twice as concentrated.

Cc 17.6 Solution **(c)** has the greatest percent ionization because percent ionization increases with decreasing weak acid concentration. Solution **(b)** has the lowest pH because the equilibrium H_3O^+ concentration increases with increasing weak acid concentration.

Cc 17.7 **(a)** A weak acid solution will usually be less than 5% dissociated. Therefore, because HCl is the only strong acid, the 1.0 M solution is much more acidic than either a weak acid that is twice as concentrated or a combination of two weak acids with the same concentrations.

CHAPTER 18

"In the strictly scientific sense of the word, insolubility does not exist, and even those substances characterized by the most obstinate resistance to the solvent action of water may properly be designated as extraordinarily difficult of solution, not as insoluble."

—Otto N. Witt (1853–1915)

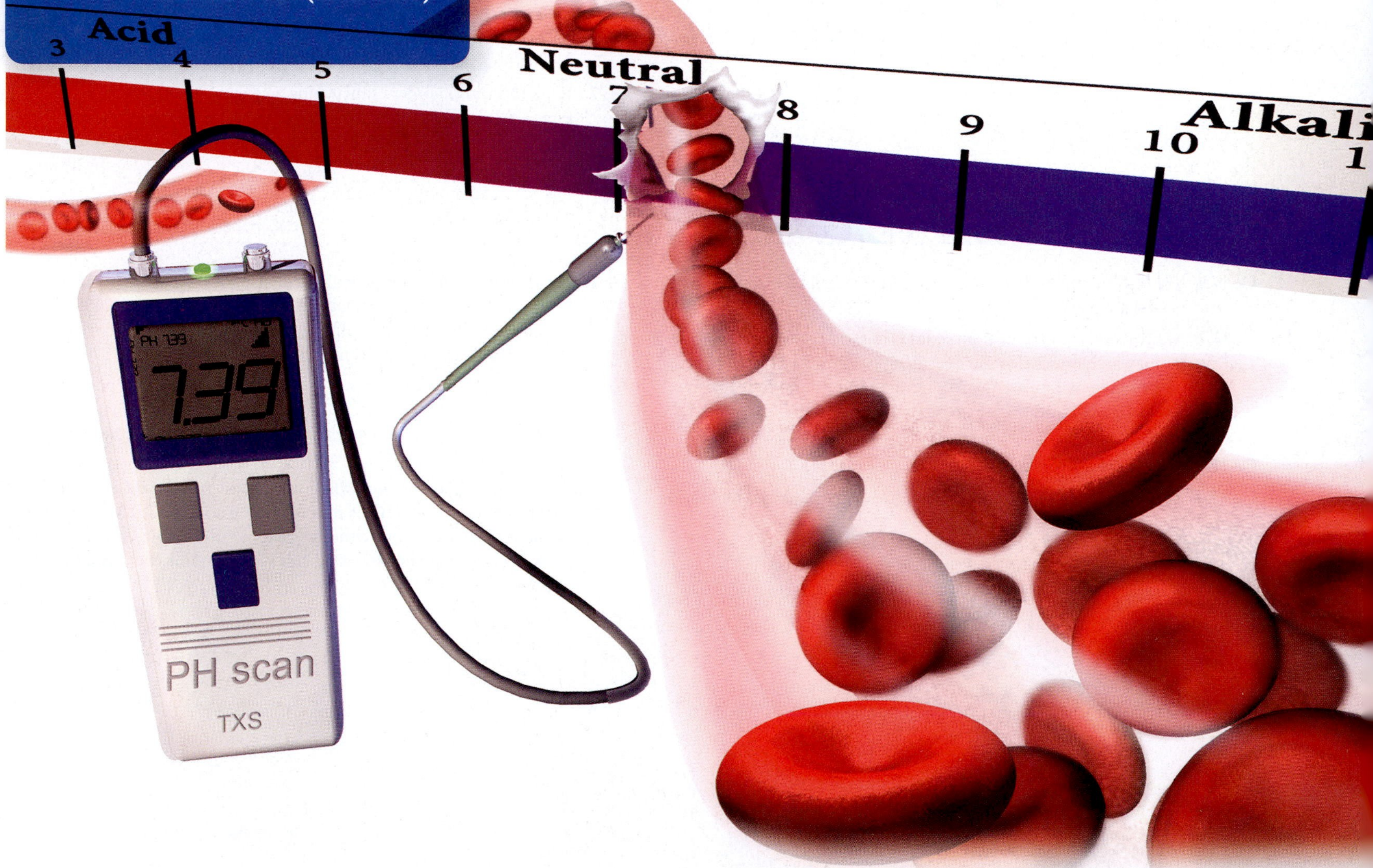

Human blood is held at nearly constant pH by the action of buffers, a main topic of this chapter.

Aqueous Ionic Equilibrium

18.1 The Danger of Antifreeze 709

18.2 Buffers: Solutions That Resist pH Change 710

18.3 Buffer Effectiveness: Buffer Range and Buffer Capacity 722

18.4 Titrations and pH Curves 725

18.5 Solubility Equilibria and the Solubility-Product Constant 739

18.6 Precipitation 745

18.7 Complex Ion Equilibria 748

Key Learning Outcomes 755

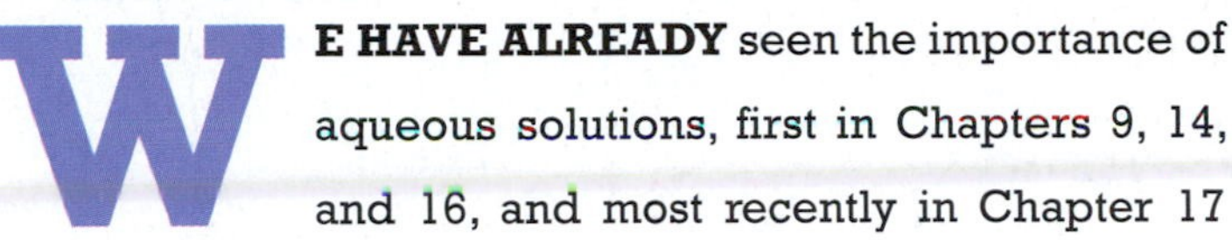

WE HAVE ALREADY seen the importance of aqueous solutions, first in Chapters 9, 14, and 16, and most recently in Chapter 17 on acids and bases. We now turn our attention to two additional topics involving aqueous solutions: buffers (solutions that resist pH change) and solubility equilibria (the extent to which slightly soluble ionic compounds dissolve in water). Buffers are tremendously important in biology because nearly all physiological processes must occur within a narrow pH range. Solubility equilibria are related to the solubility rules that we discussed in Chapter 9. In this chapter, we find a more complicated picture: solids that we considered insoluble under the simple "solubility rules" are actually better described as being only very slightly soluble, as the chapter-opening quotation from Otto Witt suggests. Solubility equilibria are important in predicting not only solubility, but also precipitation reactions that might occur when aqueous solutions are mixed.

18.1 The Danger of Antifreeze

Every year, thousands of dogs and cats die from consuming a common household product: antifreeze that was improperly stored or that leaked out of a car radiator. Most types of antifreeze used in cars are aqueous solutions of ethylene glycol:

Some brands of antifreeze use propyleneglycol, which is less toxic than ethylene glycol.

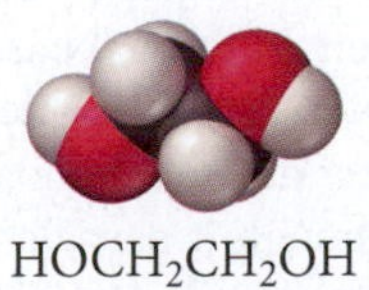

$HOCH_2CH_2OH$

Ethylene glycol has a somewhat sweet taste that can attract curious dogs and cats—and sometimes even young children, who are also vulnerable to this toxic compound. The first stage of ethylene glycol poisoning is a state resembling drunkenness. Since the compound is an alcohol, it affects the brain much as an alcoholic beverage would. Once ethylene glycol starts to be metabolized, however, a second and more deadly stage commences.

In the liver, ethylene glycol is oxidized to glycolic acid ($HOCH_2COOH$), which enters the bloodstream. The acidity of blood is critically important and tightly regulated because many proteins only function in a narrow pH range. In human blood, for example, pH is held between 7.36 and 7.42. This nearly constant blood pH is maintained by *buffers*. We discuss buffers in more depth later in this chapter, but for now know that a buffer is a chemical system that resists pH changes by neutralizing added acid or base. An important buffer in blood is a mixture of carbonic acid (H_2CO_3) and the bicarbonate ion (HCO_3^-). The carbonic acid neutralizes added base:

$$H_2CO_3(aq) + \underset{\text{added base}}{OH^-(aq)} \longrightarrow H_2O(l) + HCO_3^-(aq)$$

The bicarbonate ion neutralizes added acid:

$$HCO_3^-(aq) + \underset{\text{added acid}}{H^+(aq)} \longrightarrow H_2CO_3(aq)$$

In this way, the carbonic acid and bicarbonate ion buffering system keeps blood pH nearly constant.

When the glycolic acid generated by antifreeze consumption first enters the bloodstream, the acid's tendency to lower blood pH is countered by the buffering action of the bicarbonate ion. However, if the quantities of consumed antifreeze are large enough, the glycolic acid overwhelms the capacity of the buffer (we discuss buffer capacity in Section 18.3), causing blood pH to drop to dangerously low levels.

Low blood pH results in *acidosis*, a condition in which the acid affects the equilibrium between hemoglobin (Hb) and oxygen:

Excess H^+

$$HbH^+(aq) + O_2(g) \rightleftharpoons HbO_2(aq) + H^+(aq)$$

Shift left

The excess acid causes the equilibrium to shift to the left, reducing the blood's ability to carry oxygen. At this point, the cat or dog may begin hyperventilating in an effort to overcome the acidic blood's lowered oxygen-carrying capacity. If no treatment is administered, the animal will eventually go into a coma and die.

One treatment for ethylene glycol poisoning is the administration of ethyl alcohol (the alcohol found in alcoholic beverages). The two molecules are similar enough in structure that the liver enzyme that catalyzes the metabolism of ethylene glycol also acts on ethyl alcohol, but the enzyme has a higher affinity for ethyl alcohol than for ethylene glycol. Consequently, the enzyme preferentially metabolizes ethyl alcohol, allowing the unmetabolized ethylene glycol to escape through the urine. If administered early, this treatment can save the life of a dog or cat that has consumed ethylene glycol.

18.2 Buffers: Solutions That Resist pH Change

KEY CONCEPT VIDEO Buffers

Most solutions significantly change pH when an acid or base is added to them. As we have just learned, however, a **buffer** resists pH change by neutralizing added acid or added base. A buffer contains either:

1. significant amounts of both a weak acid and its conjugate base or
2. significant amounts of both a weak base and its conjugate acid.

For example, the buffer in blood is composed of carbonic acid (H_2CO_3) and its conjugate base, the bicarbonate ion (HCO_3^-). When additional base is added to a buffer, the weak acid reacts with the base, neutralizing it. When additional acid is added to a buffer, the conjugate base reacts with the acid, neutralizing it. In this way, a buffer can maintain a nearly constant pH.

A weak acid by itself, even though it partially ionizes to form some of its conjugate base, does not contain sufficient base to be a buffer. Similarly, a weak base by itself, even though it partially ionizes water to form some of its conjugate acid, does not contain sufficient acid to be a buffer. *A buffer must contain*

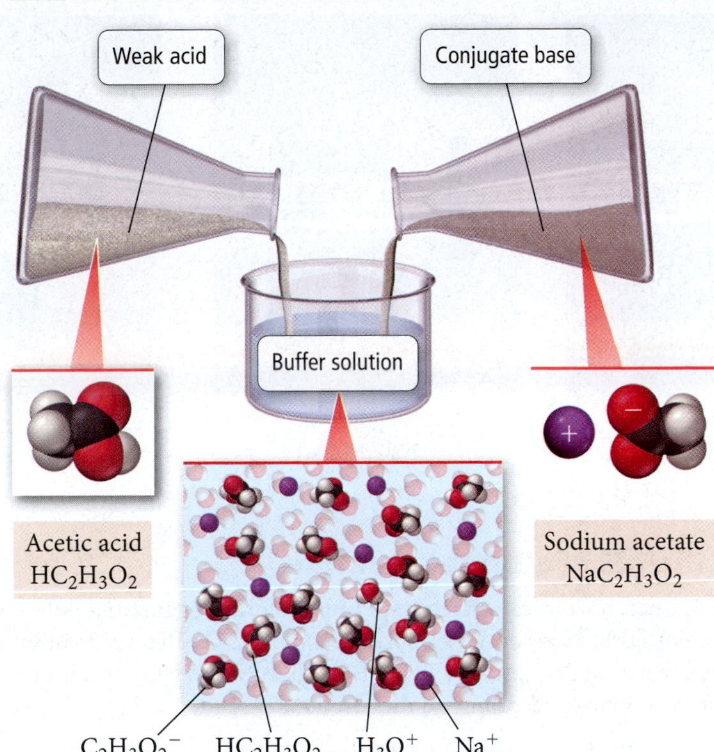

◀ **FIGURE 18.1 A Buffer Solution** A buffer typically consists of a weak acid (which can neutralize added base) and its conjugate base (which can neutralize added acid).

significant amounts of both a weak acid and its conjugate base (or vice versa). Consider the simple buffer made by dissolving acetic acid ($HC_2H_3O_2$) and sodium acetate ($NaC_2H_3O_2$) in water (Figure 18.1 ▲).

$C_2H_3O_2^-$ is the conjugate base of $HC_2H_3O_2$.

Suppose that we add a strong base, such as NaOH, to this solution. The acetic acid neutralizes the base:

$$NaOH(aq) + HC_2H_3O_2(aq) \longrightarrow H_2O(l) + NaC_2H_3O_2(aq)$$

As long as the amount of added NaOH is less than the amount of $HC_2H_3O_2$ in solution, the buffer neutralizes the added NaOH and the resulting pH change is small. Suppose, on the other hand, that we add a strong acid, such as HCl, to the solution. In this case, the conjugate base, $NaC_2H_3O_2$, neutralizes the added HCl:

$$HCl(aq) + NaC_2H_3O_2(aq) \longrightarrow HC_2H_3O_2(aq) + NaCl(aq)$$

As long as the amount of added HCl is less than the amount of $NaC_2H_3O_2$ in solution, the buffer neutralizes the added HCl and the resulting pH change is small.

Summarizing Buffer Characteristics:

- Buffers resist pH change.
- A buffer contains significant amounts of either (1) a weak acid and its conjugate base, or (2) a weak base and its conjugate acid.
- The weak acid neutralizes added base.
- The weak base neutralizes added acid.

Buffers **18.1 Cc Conceptual Connection**

Which solution is a buffer?

(a) a solution that is 0.100 M in HNO_2 and 0.100 M in HCl
(b) a solution that is 0.100 M in HNO_3 and 0.100 M in $NaNO_3$
(c) a solution that is 0.100 M in HNO_2 and 0.100 M in NaCl
(d) a solution that is 0.100 M in HNO_2 and 0.100 M in $NaNO_2$

▶ **FIGURE 18.2 The Common Ion Effect** The pH of a 0.100 M acetic acid solution is 2.90. The pH of a 0.100 M sodium acetate solution is 8.90. The pH of a solution that is 0.100 M in acetic acid and 0.100 M in sodium acetate is 4.70. (The solution contains an indicator that causes the color differences.)

pH = 2.90

0.100 M $HC_2H_3O_2$

pH = 8.90

0.100 M $NaC_2H_3O_2$

pH = 4.70

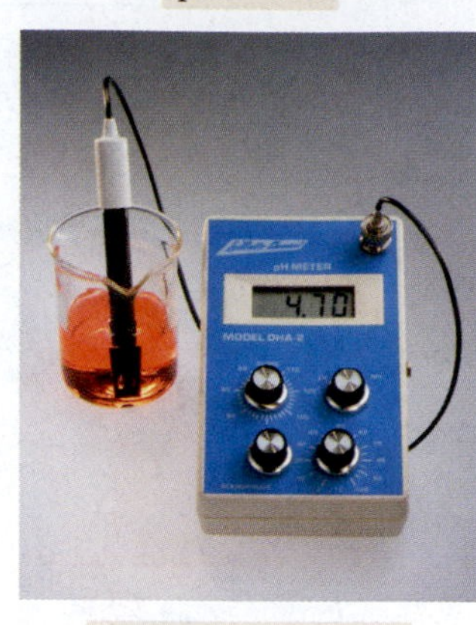

0.100 M $HC_2H_3O_2$
0.100 M $NaC_2H_3O_2$

Calculating the pH of a Buffer Solution

In Chapter 17, we learned how to calculate the pH of a solution containing either a weak acid or its conjugate base, but not both. How do we calculate the pH of a buffer—a solution containing both? Consider a solution that initially contains $HC_2H_3O_2$ and $NaC_2H_3O_2$, each at a concentration of 0.100 M. The acetic acid ionizes according to the reaction:

Le Châtelier's principle is discussed in Section 16.9.

$$HC_2H_3O_2(aq) + H_2O(l) \longrightarrow H_3O^+(aq) + C_2H_3O_2^-(aq)$$

Initial concentration: 0.100 M ($HC_2H_3O_2$), 0.100 M ($C_2H_3O_2^-$)

However, the ionization of $HC_2H_3O_2$ in this solution is suppressed compared to its ionization in a solution that does not initially contain any $C_2H_3O_2^-$. Why? To answer this question, consider an aqueous solution of only $HC_2H_3O_2$ at equilibrium. What happens to the position of the equilibrium if we add $C_2H_3O_2^-$ to the solution? According to Le Châtelier's principle, the reaction shifts to the left, so that less of the $HC_2H_3O_2$ is ionizied. In other words, the presence of the $C_2H_3O_2^-$ ion causes the acid to ionize even less than it normally would (Figure 18.2 ▲), resulting in a less acidic solution (higher pH). This effect is known as the **common ion effect**, so named because the solution contains two substances ($HC_2H_3O_2$ and $NaC_2H_3O_2$) that share a common ion ($C_2H_3O_2^-$). To determine the pH of a buffer solution containing common ions, we work an equilibrium problem in which the initial concentrations include both the acid and its conjugate base, as demonstrated in Example 18.1.

EXAMPLE 18.1
Calculating the pH of a Buffer Solution

Calculate the pH of a buffer solution that is 0.100 M in $HC_2H_3O_2$ and 0.100 M in $NaC_2H_3O_2$.

SOLUTION

1. Write the balanced equation for the ionization of the acid and use it as a guide to prepare an ICE table showing the given concentrations of the acid and its conjugate base as the initial concentrations. Leave room in the table for the changes in concentrations and for the equilibrium concentrations.

$$HC_2H_3O_2(aq) + H_2O(l) \rightleftharpoons H_3O^+(aq) + C_2H_3O_2^-(aq)$$

	$[HC_2H_3O_2]$	$[H_3O^+]$	$[C_2H_3O_2^-]$
Initial	0.100	≈0.00	0.100
Change			
Equil			

—*Continued on the next page*

Continued from the previous page—

2. Represent the change in the concentration of H_3O^+ with the variable x. Express the changes in the concentrations of the other reactants and products in terms of x.

$$HC_2H_3O_2(aq) + H_2O(l) \rightleftharpoons H_3O^+(aq) + C_2H_3O_2^-(aq)$$

	$[HC_2H_3O_2]$	$[H_3O^+]$	$[C_2H_3O_2^-]$
Initial	0.100	≈0.00	0.100
Change	$-x$	$+x$	$+x$
Equil			

3. Sum each column to determine the equilibrium concentrations in terms of the initial concentrations and the variable x.

$$HC_2H_3O_2(aq) + H_2O(l) \rightleftharpoons H_3O^+(aq) + C_2H_3O_2^-(aq)$$

	$[HC_2H_3O_2]$	$[H_3O^+]$	$[C_2H_3O_2^-]$
Initial	0.100	≈0.00	0.100
Change	$-x$	$+x$	$+x$
Equil	$0.100 - x$	x	$0.100 + x$

4. Substitute the expressions for the equilibrium concentrations (from Step 3) into the expression for the acid ionization constant.

In most cases, you can make the approximation that *x is small.* (See Secti 16.8 and 17.7 to review the *x is small* approximation.)

$$K_a = \frac{[H_3O^+][C_2H_3O^-]}{[HC_2H_3O_2]}$$

$$= \frac{x(0.100 + \cancel{x})}{0.100 - \cancel{x}} \quad (x \text{ is small})$$

Substitute the value of the acid ionization constant (from Table 17.5) into the K_a expression and solve for x.

$$1.8 \times 10^{-5} = \frac{x(\cancel{0.100})}{\cancel{0.100}}$$

$$x = 1.8 \times 10^{-5}$$

Confirm that *x is small* by calculating the ratio of x and the number it was subtracted from in the approximation. The ratio should be less than 0.05 (or 5%).

$$\frac{1.8 \times 10^{-5}}{0.100} \times 100\% = 0.018\%$$

Therefore the approximation is valid.

5. Determine the H_3O^+ concentration from the calculated value of x and substitute into the pH equation to find pH.

$$[H_3O^+] = x = 1.8 \times 10^{-5}\ M$$

$$pH = -\log[H_3O^+] = -\log(1.8 \times 10^{-5}) = 4.74$$

FOR PRACTICE 18.1

Calculate the pH of a buffer solution that is 0.200 M in $HC_2H_3O_2$ and 0.100 M in $NaC_2H_3O_2$.

FOR MORE PRACTICE 18.1

Calculate the pH of the buffer that results from mixing 60.0 mL of 0.250 M $HCHO_2$ and 15.0 mL of 0.500 M $NaCHO_2$.

The Henderson–Hasselbalch Equation

We can derive an equation that relates the pH of a buffer solution to the initial concentration of the buffer components, thus simplifying the calculation of the pH of a buffer solution. Consider a buffer containing the generic weak acid HA and its conjugate base A^-. The acid ionizes as follows:

$$HA(aq) + H_2O(l) \rightleftharpoons H_3O^+(aq) + A^-(aq)$$

We derive an expression for the concentration of H_3O^+ from the acid ionization equilibrium expression by solving the expression for $[H_3O^+]$:

$$K_a = \frac{[H_3O^+][A^-]}{[HA]}$$

$$[H_3O^+] = K_a \frac{[HA]}{[A^-]} \quad [18.1]$$

Recall that the variable x in a weak acid equilibrium problem represents the change in the initial acid concentration. The *x is small* approximation is valid because so little of the weak acid ionizes compared to its initial concentration.

If we make the same *x is small* approximation that we make for weak acid or weak base equilibrium problems, *we can consider the equilibrium concentrations of HA and* A^- *to be essentially identical to the initial concentrations of HA and* A^- (see Step 4 of Example 18.1). Therefore, to determine $[H_3O^+]$ for any buffer solution, we multiply K_a by the ratio of the concentrations of the acid and the conjugate base. To find the $[H_3O^+]$ of the buffer in Example 18.1 (a solution that is 0.100 M in $HC_2H_3O_2$ and 0.100 M in $NaC_2H_3O_2$), we substitute the concentrations of $HC_2H_3O_2$ and $C_2H_3O_2^-$ into Equation 18.1:

$$[H_3O^+] = K_a \frac{[HC_2H_3O_2]}{[C_2H_3O_2^-]}$$

$$= K_a \frac{0.100}{0.100}$$

$$= K_a$$

In this buffer solution, as in any solution in which the acid and conjugate base concentrations are equal, $[H_3O^+]$ is equal to K_a.

We can derive an equation for the pH of a buffer by taking the negative logarithm of both sides of Equation 18.1:

Recall that log AB = log A + log B, so −log AB = −log A − log B.

$$[H_3O^+] = K_a \frac{[HA]}{[A^-]}$$

$$-\log[H_3O^+] = -\log\left(K_a \frac{[HA]}{[A^-]}\right)$$

$$-\log[H_3O^+] = -\log K_a - \log \frac{[HA]}{[A^-]} \quad [18.2]$$

We can rearrange Equation 18.2 to get:

Recall that log (A/B) = −log (B/A), so −log (A/B) = log (B/A).

$$-\log[H_3O^+] = -\log K_a + \log \frac{[A^-]}{[HA]}$$

Since pH $= -\log[H_3O^+]$ and since $pK_a = -\log K_a$, we obtain the result:

$$\text{pH} = pK_a + \log \frac{[A^-]}{[HA]}$$

Since A^- is a weak base and HA is a weak acid, we can generalize the equation:

Note that, as expected, the pH of a buffer increases with an increase in the amount of base relative to the amount of acid.

$$\text{pH} = pK_a + \log \frac{[\text{base}]}{[\text{acid}]} \quad [18.3]$$

where the base is the conjugate base of the acid or the acid is the conjugate acid of the base. This equation, known as the **Henderson–Hasselbalch equation**, allows us to quickly calculate the pH of a buffer solution from the initial concentrations of the buffer components *as long as the x is small approximation is valid.* In Example 18.2, we demonstrate two ways to find the pH of a buffer: In the left column we solve a common ion effect equilibrium problem using a method similar to the one we used in Example 18.1; in the right column we use the Henderson–Hasselbalch equation.

EXAMPLE 18.2

Calculating the pH of a Buffer Solution as an Equilibrium Problem and with the Henderson–Hasselbalch Equation

Calculate the pH of a buffer solution that is 0.050 M in benzoic acid ($HC_7H_5O_2$) and 0.150 M in sodium benzoate ($NaC_7H_5O_2$). For benzoic acid, $K_a = 6.5 \times 10^{-5}$.

SOLUTION

Equilibrium Approach

Write the balanced equation for the ionization of the acid and use it as a guide to prepare an ICE table.

$$HC_7H_5O_2(aq) + H_2O(l) \rightleftharpoons H_3O^+(aq) + C_7H_5O_2^-(aq)$$

	$[HC_7H_5O_2]$	$[H_3O^+]$	$[C_7H_5O_2^-]$
Initial	0.050	≈0.00	0.150
Change	$-x$	$+x$	$+x$
Equil	$0.050 - x$	x	$0.150 - x$

Substitute the expressions for the equilibrium concentrations into the expression for the acid ionization constant. Make the *x is small* approximation and solve for x.

$$K_a = \frac{[H_3O^+][C_7H_5O_2^-]}{[HC_7H_5O_2]}$$

$$= \frac{x(0.150 + x)}{0.050 - x} \quad (x \text{ is small})$$

$$6.5 \times 10^{-5} = \frac{x(0.150)}{0.050}$$

$$x = 2.2 \times 10^{-5}$$

Since $[H_3O^+] = x$, you calculate pH as follows:

$$pH = -\log[H_3O^+]$$
$$= -\log(2.2 \times 10^{-5})$$
$$= 4.66$$

Confirm that the *x is small* approximation is valid by calculating the ratio of x to the smallest number it was subtracted from in the approximation. The ratio should be less than 0.05 (or 5%). (See Sections 16.8 and 17.7 to review the *x is small* approximation.)

$$\frac{2.2 \times 10^{-5}}{0.050} \times 100\% = 0.044\%$$

The approximation is valid.

Henderson–Hasselbalch Approach

To find the pH of this solution, determine which component is the acid and which is the base and substitute their concentrations into the Henderson–Hasselbalch equation to calculate pH.

$HC_7H_5O_2$ is the acid and $NaC_7H_5O_2$ is the base. Therefore, you calculate the pH as follows:

$$pH = pK_a + \log\frac{[\text{base}]}{[\text{acid}]}$$

$$= -\log(6.5 \times 10^{-5}) + \log\frac{0.150}{0.050}$$

$$= 4.1\underline{8}7 + 0.4\underline{7}7 = 4.66$$

Confirm that the *x is small* approximation is valid by calculating the $[H_3O^+]$ from the pH. Since $[H_3O^+]$ is formed by ionization of the acid, the calculated $[H_3O^+]$ has to be less than 0.05 (or 5%) of the initial concentration of the acid in order for the *x is small* approximation to be valid.

$$pH = 4.66 = -\log[H_3O^+]$$

$$[H_3O^+] = 10^{-4.66} = 2.2 \times 10^{-5} \text{ M}$$

$$\frac{2.2 \times 10^{-5}}{0.050} \times 100\% = 0.044\%$$

The approximation is valid.

FOR PRACTICE 18.2

Calculate the pH of a buffer solution that is 0.250 M in HCN and 0.170 M in KCN. For HCN, $K_a = 4.9 \times 10^{-10}$ ($pK_a = 9.31$). Use both the equilibrium approach and the Henderson–Hasselbalch approach.

How do you decide whether to use the equilibrium approach or the Henderson–Hasselbalch equation when calculating the pH of buffer solutions? The answer depends on the specific problem.

In cases where you can make the *x is small* approximation, the Henderson–Hasselbalch equation is adequate. However, as you can see from Example 18.2, checking the *x is small* approximation is not as convenient with the Henderson–Hasselbalch equation (because the approximation is implicit). Thus, the equilibrium approach, though lengthier, gives you a better sense of the important quantities in the problem and the nature of the approximation.

When first working buffer problems, use the equilibrium approach until you get a good sense for when the *x is small* approximation is adequate. Then, you can switch to the more streamlined approach in cases where the approximation applies (and only in those cases). In general, remember that the *x is small* approximation applies to problems in which both of the following are true: (a) the initial concentrations of acids (and/or bases) are not too dilute; and (b) the equilibrium constant is fairly small. Although the exact values depend on the details of the problem, for many buffer problems this means that the initial concentrations of acids and conjugate bases should be at least 10^2–10^3 times greater than the equilibrium constant (depending on the required accuracy).

18.2 Cc Conceptual Connection

pH of Buffer Solutions

A buffer contains the weak acid HA and its conjugate base A^-. The weak acid has a pK_a of 4.82, and the buffer has a pH of 4.25. Which statement is true of the relative concentrations of the weak acid and conjugate base in the buffer?

(a) $[HA] > [A^-]$

(b) $[HA] < [A^-]$

(c) $[HA] = [A^-]$

Which buffer component would you add to change the pH of the buffer to 4.72?

Calculating pH Changes in a Buffer Solution

When we add acid or base to a buffer, the buffer resists a pH change. Nonetheless, the pH does change by a small amount. Calculating the pH change requires breaking up the problem into two parts:

1. **The stoichiometry calculation** in which we calculate how the addition changes the relative amounts of acid and conjugate base.
2. **The equilibrium calculation** in which we calculate the pH based on the new amounts of acid and conjugate base.

We demonstrate this calculation with a 1.0 L buffer solution that is 0.100 M in the generic acid HA and 0.100 M in its conjugate base A^-. Since the concentrations of the weak acid and the conjugate base are equal, the pH of the buffer is equal to pK_a. Let's calculate the pH of the solution after we add 0.025 mol of strong acid (H^+) (assuming that the change in volume from adding the acid is negligible).

The Stoichiometry Calculation As the added acid is neutralized, it converts a stoichiometric amount of the base into its conjugate acid through the neutralization reaction (Figure 18.3a):

$$\underset{\text{added acid}}{H^+(aq)} + \underset{\text{weak base in buffer}}{A^-(aq)} \longrightarrow HA(aq)$$

Neutralizing 0.025 mol of the strong acid (H^+) requires 0.025 mol of the weak base (A^-). Consequently, the amount of A^- *decreases* by 0.025 mol and the amount of HA *increases* by 0.025 mol (because of the 1:1:1 stoichiometry of the neutralization reaction). We can track these changes in tabular form as follows:

It is best to work with amounts in moles instead of concentrations when tracking these changes, as explained later.

	$H^+(aq)$ +	$A^-(aq)$ ⟶	$HA(aq)$
Before addition	≈0.00 mol	0.100 mol	0.100 mol
Addition	+0.025 mol	—	—
After addition	≈0.00 mol	0.075 mol	0.125 mol

Notice that this table is not an ICE table. This table simply tracks the stoichiometric changes that occur during the neutralization of the added acid. We write ≈ 0.00 mol for the amount of H^+ because the amount is so small compared to the amounts of A^- and HA. (Remember that weak acids ionize only to a small extent and that the presence of the common ion further suppresses the ionization.) The amount of H^+, of course, is not *exactly* zero, as we can see by completing the equilibrium part of the calculation.

The Equilibrium Calculation We have just seen that adding a small amount of acid to a buffer is equivalent to changing the initial concentrations of the acid and conjugate base present in the buffer (in this case, since the volume is 1.0 L, [HA] increased from 0.100M to 0.125 M and $[A^-]$ decreased from 0.100 M to 0.075 M). Knowing these new initial concentrations, we can calculate the new pH in the same way that we calculate the pH of any buffer: either by working a full equilibrium problem or by using the Henderson–Hasselbalch equation (see Examples 18.1 and 18.2). In this case, we work the full equilibrium problem. We begin by writing the balanced equation for the ionization of the acid and using it as a guide to prepare an ICE table. The initial concentrations for the ICE table are those that we calculated in the stoichiometry part of the calculation:

$$HA(aq) + H_2O(l) \rightleftharpoons H_3O^+(aq) + A^-(aq)$$

	[HA]	$[H_3O^+]$	$[A^-]$
Initial	0.125	≈ 0.00	0.075
Change	$-x$	$+x$	$+x$
Equil	$0.125 - x$	x	$0.075 + x$

From stoichiometry calculation

We then substitute the expressions for the equilibrium concentrations into the expression for the acid ionization constant. As long as K_a is sufficiently small relative to the initial concentrations, we can make the *x is small* approximation and solve for *x*, which is equal to $[H_3O^+]$.

$$K_a = \frac{[H_3O^+][A^-]}{[HA]}$$

$$= \frac{x(0.075 + \cancel{x})}{0.125 - \cancel{x}} \quad (x \text{ is small})$$

$$K_a = \frac{x(0.075)}{0.125}$$

$$x = [H_3O^+] = K_a \frac{0.125}{0.075}$$

Once we calculate $[H_3O^+]$, we can calculate the pH with the equation $pH = -\log[H_3O^+]$.

Notice that, since the expression for *x* contains a *ratio* of concentrations $[HA]/[A^-]$, we can substitute the *amounts of acid and base in moles* in place of concentration because, in a single buffer solution, the volume is the same for both the acid and the base. Therefore, the volumes cancel:

$$[HA]/[A^-] = \frac{\dfrac{n_{HA}}{\cancel{V}}}{\dfrac{n_{A^-}}{\cancel{V}}} = n_{HA}/n_{A^-}$$

The effect of adding a small amount of strong base to the buffer is exactly the opposite of adding acid. The added base converts a stoichiometric amount of the acid into its conjugate base through the neutralization reaction (Figure 18.3b ▶):

$$\underset{\text{added base}}{OH^-(aq)} + \underset{\text{weak acid in buffer}}{HA(aq)} \longrightarrow H_2O(l) + A^-(aq)$$

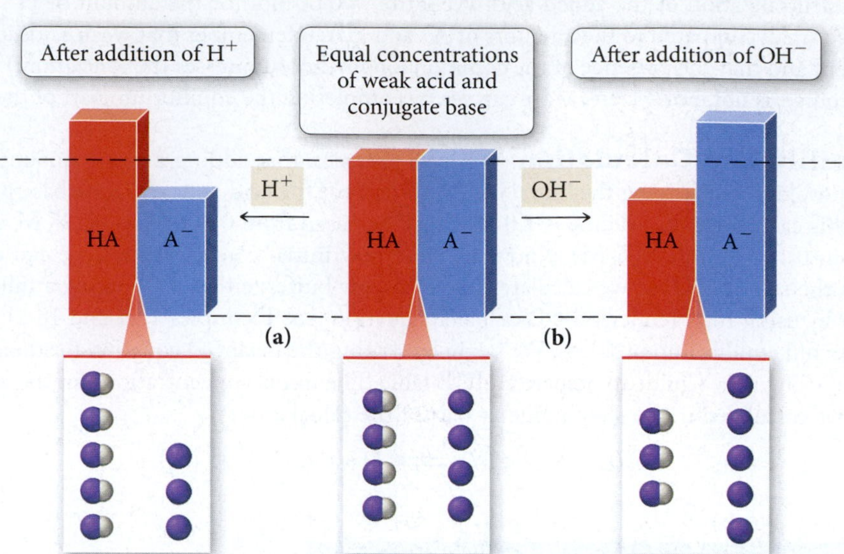

▲ **FIGURE 18.3 Buffering Action** **(a)** When an acid is added to a buffer, a stoichiometric amount of the weak base is converted to the conjugate acid. **(b)** When a base is added to a buffer, a stoichiometric amount of the weak acid is converted to the conjugate base.

If we add 0.025 mol of OH^-, the amount of A^- goes *up* by 0.025 mol and the amount of HA goes *down* by 0.025 mol as shown in the following table:

	$OH^-(aq)$	+	$HA(aq)$	$\longrightarrow$	$H_2O(l)$	+	$A^-(aq)$
Before addition	≈0.00 mol		0.100 mol				0.100 mol
Addition	+0.025 mol		—				—
After addition	≈0.00 mol		0.075 mol				0.125 mol

When you calculate the pH of a buffer after adding small amounts of acid or base, remember the following:

- Adding a small amount of strong acid to a buffer converts a stoichiometric amount of the base to the conjugate acid and decreases the pH of the buffer (adding acid decreases pH just as we would expect).
- Adding a small amount of strong base to a buffer converts a stoichiometric amount of the acid to the conjugate base and increases the pH of the buffer (adding base increases the pH just as we would expect).

The easiest way to remember these changes is relatively simple: adding acid creates more acid; adding base creates more base.

Example 18.3 and the For Practice Problems that follow it involve calculating pH changes in a buffer solution after small amounts of strong acid or strong base are added. As we have seen, these problems generally have two parts:

- Part I. Stoichiometry—use the stoichiometry of the neutralization equation to calculate the changes in the amounts (in moles) of the buffer components upon addition of the acid or base.
- Part II. Equilibrium—use the new amounts of buffer components to work an equilibrium problem to find pH. (For most buffers, this can also be done with the Henderson–Hasselbalch equation.)

EXAMPLE 18.3

Calculating the pH Change in a Buffer Solution after the Addition of a Small Amount of Strong Acid or Base

A 1.0 L buffer solution contains 0.100 mol $HC_2H_3O_2$ and 0.100 mol $NaC_2H_3O_2$. The value of K_a for $HC_2H_3O_2$ is 1.8×10^{-5}. Because the initial amounts of acid and conjugate base are equal, the pH of the buffer is equal to $pK_a = -\log(1.8 \times 10^{-5}) = 4.74$. Calculate the new pH after adding 0.010 mol of solid NaOH to the buffer. For comparison, calculate the pH after adding 0.010 mol of solid NaOH to 1.0 L of pure water. (Ignore any small changes in volume that might occur upon addition of the base.)

SOLUTION

Part I: Stoichiometry. The addition of the base converts a stoichiometric amount of acid to the conjugate base (adding base creates more base). Write an equation showing the neutralization reaction and then set up a table to track the changes.

	$OH^-(aq)$ +	$HC_2H_3O_2(aq)$ ⟶	$H_2O(l)$ +	$C_2H_3O_2^-(aq)$
Before addition	≈0.00 mol	0.100 mol		0.100 mol
Addition	0.010 mol	—		—
After addition	≈0.00 mol	0.090 mol		0.110 mol

Part II: Equilibrium. Write the balanced equation for the ionization of the acid and use it as a guide to prepare an ICE table. Use the amounts of acid and conjugate base from part I as the initial amounts of acid and conjugate base in the ICE table.

$$HC_2H_3O_2(aq) + H_2O(l) \rightleftharpoons H_3O^+(aq) + C_2H_3O_2^-(aq)$$

	$[HC_2H_3O_2]$	$[H_3O^+]$	$[C_2H_3O_2^-]$
Initial	0.090	≈0.00	0.110
Change	$-x$	$+x$	$+x$
Equil	$0.090 - x$	x	$0.110 + x$

Substitute the expressions for the equilibrium concentrations of acid and conjugate base into the expression for the acid ionization constant. Make the *x is small* approximation and solve for *x*. Calculate the pH from the value of *x*, which is equal to $[H_3O^+]$.

$$K_a = \frac{[H_3O^+][C_2H_3O_2^-]}{[HC_2H_3O_2]}$$

$$= \frac{x(0.110 + \cancel{x})}{0.090 - \cancel{x}} \quad (x \text{ is small})$$

$$1.8 \times 10^{-5} = \frac{x(0.110)}{0.090}$$

$$x = [H_3O^+] = 1.4\underline{7} \times 10^{-5}\ \text{M}$$

$$\text{pH} = -\log[H_3O^+]$$

$$= -\log(1.4\underline{7} \times 10^{-5})$$

$$= 4.83$$

Confirm the validity of the *x is small* approximation by calculating the ratio of *x* to the smallest number it was subtracted from in the approximation. The ratio should be less than 0.05 (or 5%).

$$\frac{1.4\underline{7} \times 10^{-5}}{0.090} \times 100\% = 0.016\%$$

The approximation is valid.

Part II: Equilibrium Alternative (using the Henderson–Hasselbalch equation). As long as the *x is small* approximation is valid, you can substitute the quantities of acid and conjugate base after the addition (from part I) into the Henderson–Hasselbalch equation and calculate the new pH.

$$\text{pH} = pK_a + \log\frac{[\text{base}]}{[\text{acid}]}$$

$$= -\log(1.8 \times 10^{-5}) + \log\frac{0.110}{0.090}$$

$$= 4.74 + 0.087$$

$$= 4.83$$

—*Continued on the next page*

Continued from the previous page—

The pH of 1.0 L of water after adding 0.010 mol of NaOH is calculated from the $[OH^-]$. For a strong base, $[OH^-]$ is simply the number of moles of OH^- divided by the number of liters of solution.	$[OH^-] = \frac{0.010 \text{ mol}}{1.0 \text{ L}} = 0.010 \text{ M}$ $pOH = -\log[OH^-] = -\log(0.010)$ $= 2.00$ $pOH + pH = 14.00$ $pH = 14.00 - pOH$ $= 14.00 - 2.00$ $= 12.00$

CHECK Notice that the buffer solution changed from pH = 4.74 to pH = 4.83 upon addition of the base (a small fraction of a single pH unit). In contrast, the pure water changed from pH = 7.00 to pH = 12.00, five whole pH units (a factor of 10^5). Notice also that even the buffer solution got slightly more basic upon addition of a base, as you would expect. To check your answer, always make sure the pH goes in the direction you expect: adding base should make the solution more basic (higher pH); adding acid should make the solution more acidic (lower pH).

FOR PRACTICE 18.3
Calculate the pH of the solution in Example 18.3 upon addition of 0.015 mol of NaOH to the original buffer.

FOR MORE PRACTICE 18.3
Calculate the pH of the solution in Example 18.3 upon addition of 10.0 mL of 1.00 M HCl to the original buffer in Example 18.3.

18.3 Cc Conceptual Connection

Adding Acid or Base to a Buffer

A buffer contains equal amounts of a weak acid and its conjugate base and has a pH of 5.25. Which is a reasonable value of buffer pH after the addition of a small amount of acid?

(a) 4.15
(b) 5.15
(c) 5.35
(d) 6.35

Buffers Containing a Base and Its Conjugate Acid

So far, we have seen examples of buffers composed of an acid and its conjugate base (where the conjugate base is an ion). A buffer can also be composed of a base and its conjugate acid (where the conjugate acid is an ion). For example, a solution containing significant amounts of both NH_3 and NH_4Cl acts as a buffer (Figure 18.4 ▶). The NH_3 is a weak base that neutralizes small amounts of added acid, and the NH_4^+ ion is the conjugate acid that neutralizes small amounts of added base. We calculate the pH of a solution like this in the same way that we calculated the pH of a buffer containing a weak acid and its conjugate base. When using the Henderson–Hasselbalch equation, however, we must first calculate pK_a for the conjugate acid of the weak base. Recall from Section 17.9 that for a conjugate acid–base pair, $K_a \times K_b = K_w$ and $pK_a + pK_b = 14$. Consequently, we can find pK_a of the conjugate acid by subtracting pK_b of the weak base from 14. Example 18.4 illustrates the procedure for calculating the pH of a buffer composed of a weak base and its conjugate acid.

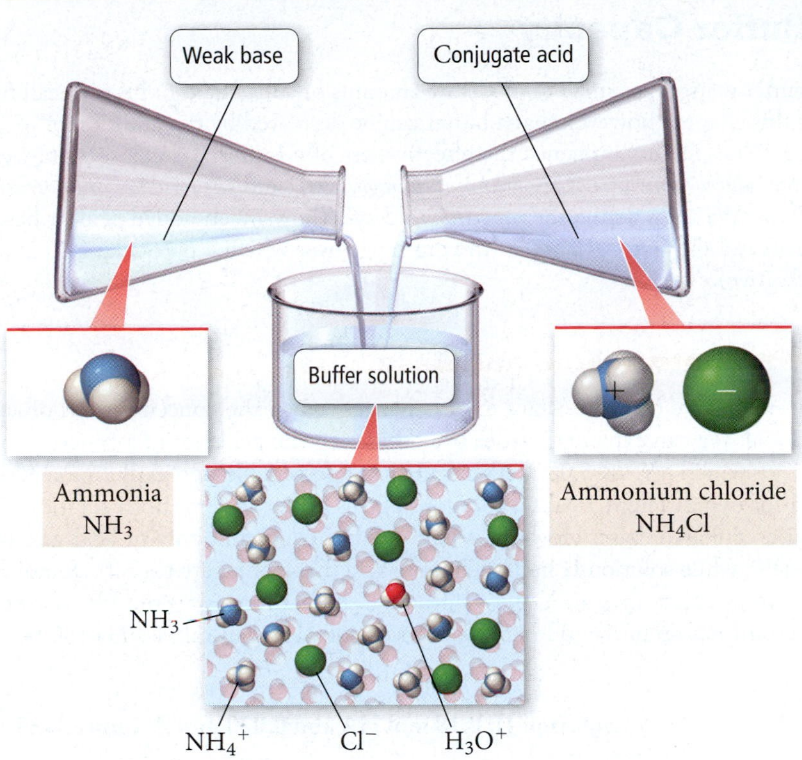

FIGURE 18.4 Buffer Containing a Base A buffer can also consist of a weak base and its conjugate acid.

EXAMPLE 18.4

Using the Henderson–Hasselbalch Equation to Calculate the pH of a Buffer Solution Composed of a Weak Base and Its Conjugate Acid

Use the Henderson–Hasselbalch equation to calculate the pH of a buffer solution that is 0.50 M in NH_3 and 0.20 M in NH_4Cl. For ammonia, $pK_b = 4.75$.

SOLUTION

Since K_b for (1.8×10^{-5}) is much smaller than the initial concentrations in this problem, you can use the Henderson–Hasselbalch equation to calculate the pH of the buffer. First calculate pK_a from pK_b.

$$\begin{aligned} pK_a + pK_b &= 14 \\ pK_a &= 14 - pK_b \\ &= 14 - 4.75 \\ &= 9.25 \end{aligned}$$

Then substitute the given quantities into the Henderson–Hasselbalch equation and calculate pH.

$$\begin{aligned} pH &= pK_a + \log\frac{[\text{base}]}{[\text{acid}]} \\ &= 9.25 + \log\frac{0.50}{0.20} \\ &= 9.25 + 0.40 \\ &= 9.65 \end{aligned}$$

FOR PRACTICE 18.4

Calculate the pH of 1.0 L of the solution in Example 18.4 upon addition of 0.010 mol of solid NaOH to the original buffer solution.

FOR MORE PRACTICE 18.4

Calculate the pH of 1.0 L of the solution in Example 18.4 upon addition of 30.0 mL of 1.0 M HCl to the original buffer solution.

18.3 Buffer Effectiveness: Buffer Range and Buffer Capacity

An effective buffer neutralizes small to moderate amounts of added acid or base. Recall from the opening section of this chapter, however, that a buffer can be destroyed by the addition of too much acid or too much base. What factors influence the effectiveness of a buffer? In this section, we examine two such factors: *the relative amounts of the acid and conjugate base* and *the absolute concentrations of the acid and conjugate base*. We then define the *capacity of a buffer* (how much added acid or base it can effectively neutralize) and the *range of a buffer* (the pH range over which a particular acid and its conjugate base can be effective).

Relative Amounts of Acid and Base

A buffer is most effective (most resistant to pH changes) when the concentrations of acid and conjugate base are equal. We can explore this idea by considering the behavior of a generic buffer composed of HA and A^- for which $pK_a = 5.00$. Let's calculate the percent change in pH upon addition of 0.010 mol of NaOH for two different 1.0-liter solutions of this buffer system. Both solutions have 0.20 mol of *total* acid and conjugate base. However, solution I has equal amounts of acid and conjugate base (0.10 mol of each), while solution II has much more acid than conjugate base (0.18 mol HA and 0.020 mol A^-). We can calculate the initial pH values of each solution using the Henderson–Hasselbalch equation. Solution I has an initial pH of 5.00, and solution II has an initial pH of 4.05.

Solution I: 0.10 mol HA and 0.10 mol A^-; initial pH = 5.00

	$OH^-(aq)$ +	$HA(aq)$ ⟶	$H_2O(l)$ +	$A^-(aq)$
Before addition	≈0.00 mol	0.100 mol		0.100 mol
Addition	0.010 mol	—		—
After addition	≈0.00 mol	0.090 mol		0.110 mol

$$\begin{aligned} pH &= pK_a + \log\frac{[\text{base}]}{[\text{acid}]} \\ &= 5.00 + \log\frac{0.110}{0.090} \\ &= 5.09 \end{aligned}$$

$$\begin{aligned} \%\ \text{change} &= \frac{5.09 - 5.00}{5.00} \times 100\% \\ &= 1.8\% \end{aligned}$$

Solution II: 0.18 mol HA and 0.020 mol A^-; initial pH = 4.05

	$OH^-(aq)$ +	$HA(aq)$ ⟶	$H_2O(l)$ +	$A^-(aq)$
Before addition	≈0.00 mol	0.18 mol		0.020 mol
Addition	0.010 mol	—		—
After addition	≈0.00 mol	0.17 mol		0.030 mol

$$\begin{aligned} pH &= pK_a + \log\frac{[\text{base}]}{[\text{acid}]} \\ &= 5.00 + \log\frac{0.030}{0.17} \\ &= 4.25 \end{aligned}$$

$$\begin{aligned} \%\ \text{change} &= \frac{4.25 - 4.05}{4.05} \times 100\% \\ &= 4.9\% \end{aligned}$$

We can see from the calculations that the buffer with equal amounts of acid and conjugate base is more resistant to pH change and is therefore the more effective buffer. A buffer becomes less effective as the difference in the relative amounts of acid and conjugate base increases. As a guideline, we say that an effective buffer must have a [base]/[acid] ratio in the range of 0.10 to 10. *In order for a buffer to be reasonably effective, the relative concentrations of acid and conjugate base should not differ by more than a factor of 10.*

Absolute Concentrations of the Acid and Conjugate Base

A buffer is most effective (most resistant to pH changes) when the concentrations of acid and conjugate base are high. Let's explore this idea by again considering a generic buffer composed of HA and A^- and a pK_a of 5.00 and calculating the percent change in pH upon addition of 0.010 mol of NaOH to two 1.0-liter solutions of this buffer system. In this case, both the acid and the base in solution I are ten times more concentrated than the acid and base in solution II. Both solutions have equal relative amounts of acid and conjugate base and therefore have the same initial pH of 5.00.

Solution I: 0.50 mol HA and 0.50 mol A^-; initial pH = 5.00

	$OH^-(aq)$	+ $HA(aq)$	$\longrightarrow H_2O(l)$	+ $A^-(aq)$
Before addition	≈0.00 mol	0.50 mol		0.50 mol
Addition	0.010 mol	—		—
After addition	≈0.00 mol	0.49 mol		0.51 mol

$$\text{pH} = \text{p}K_a + \log\frac{[\text{base}]}{[\text{acid}]}$$
$$= 5.00 + \log\frac{0.51}{0.49}$$
$$= 5.02$$
$$\%\ \text{change} = \frac{5.02 - 5.00}{5.00} \times 100\%$$
$$= 0.4\%$$

Solution II: 0.050 mol HA and 0.050 mol A^-; initial pH = 5.00

	$OH^-(aq)$	+ $HA(aq)$	$\longrightarrow H_2O(l)$	+ $A^-(aq)$
Before addition	≈0.00 mol	0.050 mol		0.050 mol
Addition	0.010 mol	—		—
After addition	≈0.00 mol	0.040 mol		0.060 mol

$$\text{pH} = \text{p}K_a + \log\frac{[\text{base}]}{[\text{acid}]}$$
$$= 5.00 + \log\frac{0.060}{0.040}$$
$$= 5.18$$
$$\%\ \text{change} = \frac{5.18 - 5.00}{5.00} \times 100\%$$
$$= 3.6\%$$

As this calculation shows, the buffer with greater amounts of acid and conjugate base is more resistant to pH changes and therefore the more effective buffer. The more dilute the buffer components, the less effective the buffer.

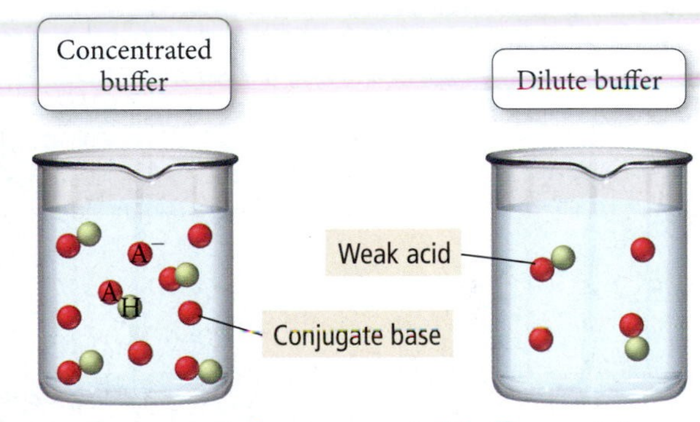

▲ A concentrated buffer contains more of the weak acid and its conjugate base than a weak buffer does. It can therefore neutralize more added acid or added base.

Buffer Range

In light of the guideline that the relative concentrations of acid and conjugate base should not differ by more than a factor of 10 in order for a buffer to be reasonably effective, we can calculate the pH range over which a particular acid and its conjugate base make an effective buffer. Since the pH of a buffer is given by the Henderson–Hasselbalch equation, we can calculate the outermost points of the effective range as follows:

***Lowest pH* for effective buffer occurs when the base is one-tenth as concentrated as the acid.**

$$\text{pH} = \text{p}K_a + \log\frac{[\text{base}]}{[\text{acid}]}$$
$$= \text{p}K_a + \log 0.10$$
$$= \text{p}K_a - 1$$

***Highest pH* for effective buffer occurs when the base is ten times as concentrated as the acid.**

$$\text{pH} = \text{p}K_a + \log\frac{[\text{base}]}{[\text{acid}]}$$
$$= \text{p}K_a + \log 10$$
$$= \text{p}K_a + 1$$

The effective range for a buffering system is one pH unit on either side of pK_a. For example, we can use a weak acid with a $\text{p}K_a$ of 5.0 (and its conjugate base) to prepare a buffer in the range of 4.0–6.0. We can adjust the relative amounts of acid and conjugate base to achieve any pH within this range. As we noted earlier, however, the buffer is most effective at pH 5.0 because the buffer components are exactly equal at that pH. Example 18.5 demonstrates how to pick an acid/conjugate base system for a buffer and how to calculate the relative amounts of acid and conjugate base required for a desired pH.

EXAMPLE 18.5

Preparing a Buffer

Which acid would you choose to combine with its sodium salt to make a solution buffered at pH 4.25? For the best choice, calculate the ratio of the conjugate base to the acid required to attain the desired pH.

chlorous acid ($HClO_2$)	$pK_a = 1.95$	**formic acid** ($HCHO_2$)	$pK_a = 3.74$
nitrous acid (HNO_2)	$pK_a = 3.34$	**hypochlorous acid** ($HClO$)	$pK_a = 7.54$

SOLUTION

The best choice is formic acid because its pK_a lies closest to the desired pH. You can calculate the required ratio of conjugate base (CHO_2^-) to acid ($HCHO_2$) by using the Henderson–Hasselbalch equation as follows:

$$\text{pH} = \text{p}K_a + \log\frac{[\text{base}]}{[\text{acid}]}$$

$$4.25 = 3.74 + \log\frac{[\text{base}]}{[\text{acid}]}$$

$$\log\frac{[\text{base}]}{[\text{acid}]} = 4.25 - 3.74$$

$$= 0.51$$

$$\frac{[\text{base}]}{[\text{acid}]} = 10^{0.51}$$

$$= 3.24$$

FOR PRACTICE 18.5

Which acid in Example 18.5 would you choose to create a buffer with pH = 7.35? If you have 500.0 mL of a 0.10 M solution of the acid, what mass of the corresponding sodium salt of the conjugate base do you need to make the buffer?

Buffer Capacity

Buffer capacity is the amount of acid or base that we can add to a buffer without causing a large change in pH. Given what we just learned about the absolute concentrations of acid and conjugate base in an effective buffer, we can conclude that the *buffer capacity increases with increasing absolute concentrations of the buffer components*. The more concentrated the weak acid and conjugate base that compose the buffer, the higher the buffer capacity. In addition, *overall buffer capacity increases as the relative concentrations of the buffer components become more similar to each other*. As the ratio of the buffer components gets closer to 1, the *overall* capacity of the buffer (the ability to neutralize added acid *and* added base) becomes greater. In some cases, however, a buffer that must neutralize primarily added acid (or primarily added base) can be overweighted in one of the buffer components.

18.4 Cc Conceptual Connection

Buffer Capacity

A 1.0 L buffer solution is 0.10 M in HF and 0.050 M in NaF. Which action destroys the buffer?

(a) adding 0.050 mol of HCl
(b) adding 0.050 mol of NaOH
(c) adding 0.050 mol of NaF
(d) none of the above

18.4 Titrations and pH Curves

We first examined acid–base titrations in Section 9.7. In a typical **acid–base titration**, a basic (or acidic) solution of unknown concentration reacts with an acidic (or basic) solution of known concentration. The known solution is slowly added to the unknown one while the pH is monitored with either a pH meter or an **indicator** (a substance whose color depends on the pH). As the acid and base combine, they neutralize each other. At the **equivalence point**—the point in the titration when the number of moles of base is stoichiometrically equal to the number of moles of acid—the titration is complete. When the equivalence point is reached, neither reactant is in excess and the number of moles of the reactants are related by the reaction stoichiometry (Figure 18.5 ▼).

The equivalence point is so named because the number of moles of acid and base are stoichiometrically equivalent at this point.

In this section, we examine acid–base titrations more closely, concentrating on the pH changes that occur during the titration. A plot of the pH of the solution during a titration is a *titration curve* or *pH curve.* Figure 18.6 ▼ is a pH curve for the titration of HCl with NaOH. Before any base is added to the solution, the pH is low (as expected for a solution of HCl). As the NaOH is added, the solution becomes less acidic because the NaOH begins to neutralize the HCl. The point of inflection in the middle of the curve is the equivalence point. Notice that the pH changes very quickly near the

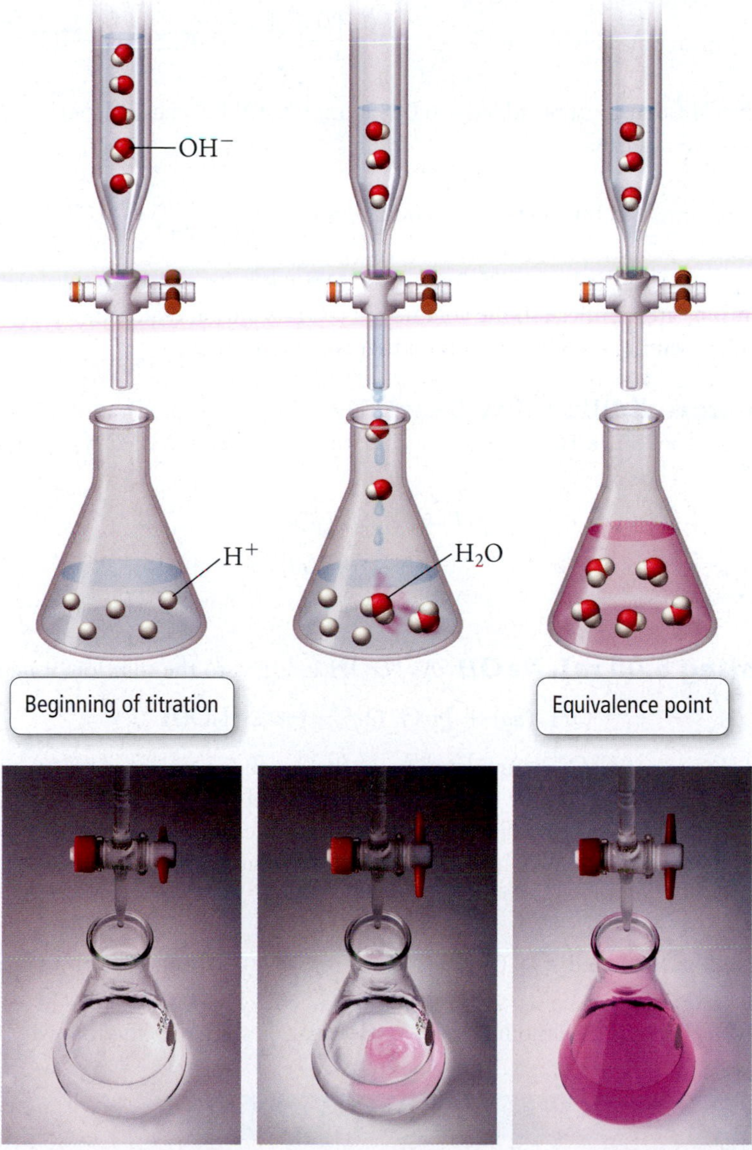

▲ **FIGURE 18.5 Acid–Base Titration** As OH^- is added in a titration, it neutralizes the H^+, forming water. At the equivalence point, the titration is complete. (The solution in the flask contains the indicator phenolthalein, which produces the color change at or near the equivlance point.)

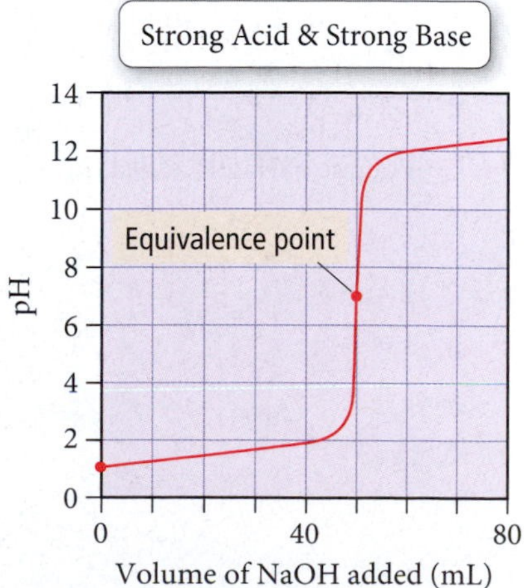

▲ **FIGURE 18.6 Titration Curve: Strong Acid + Strong Base** This curve represents the titration of 50.0 mL of 0.100 M HCl with 0.100 M NaOH.

equivalence point (small amounts of added base cause large changes in pH). Beyond the equivalence point, the solution is basic because the HCl has been completely neutralized and excess base is being added to the solution. The exact shape of the pH curve depends on several factors, including the strength of the acid or base being titrated. Let's consider several combinations individually.

The Titration of a Strong Acid with a Strong Base

Consider the titration of 25.0 mL of 0.100 M HCl with 0.100 M NaOH. We begin by calculating the volume of base required to reach the equivalence point, and then we determine the pH at several points during the titration.

Volume of NaOH Required to Reach the Equivalence Point During the titration, the added sodium hydroxide neutralizes the hydrochloric acid:

$$\text{HCl}(aq) + \text{NaOH}(aq) \longrightarrow \text{H}_2\text{O}(l) + \text{NaCl}(aq)$$

The equivalence point is reached when the number of moles of base added equals the number of moles of acid initially in solution. We calculate the amount of acid initially in solution from its volume and its concentration:

$$\text{Initial mol HCl} = 0.0250\ \cancel{\text{L}} \times \frac{0.100\ \text{mol}}{1\ \cancel{\text{L}}} = 0.00250\ \text{mol HCl}$$

The amount of NaOH that must be added is 0.00250 mol NaOH. We calculate the volume of NaOH required from its concentration:

$$\text{Volume NaOH solution} = 0.00250\ \cancel{\text{mol}} \times \frac{1\ \text{L}}{0.100\ \cancel{\text{mol}}} = 0.0250\ \text{L}$$

The equivalence point is reached when 25.0 mL of NaOH has been added. In this case, the concentrations of both solutions are identical, so the volume of NaOH solution required to reach the equivalence point is equal to the volume of the HCl solution that is being titrated.

Initial pH (Before Adding Any Base) The initial pH of the solution is simply the pH of a 0.100 M HCl solution. Since HCl is a strong acid, the concentration of H_3O^+ is also 0.100 M and the pH is 1.00:

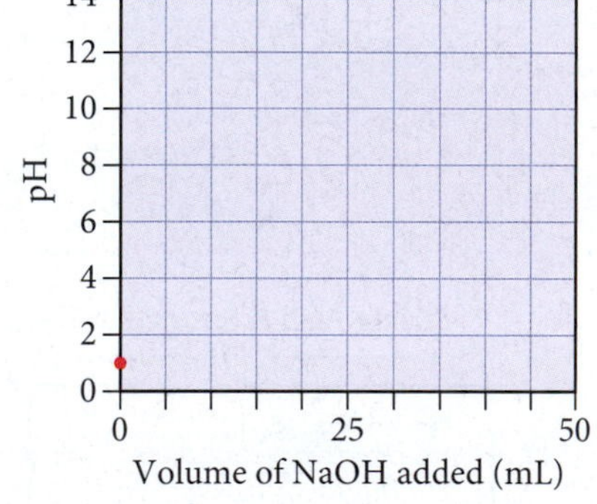

$$\begin{aligned} \text{pH} &= -\log[\text{H}_3\text{O}^+] \\ &= -\log(0.100) \\ &= 1.00 \end{aligned}$$

pH After Adding 5.00 mL NaOH As NaOH is added to the solution, it neutralizes H_3O^+:

$$\text{OH}^-(aq) + \text{H}_3\text{O}^+(aq) \longrightarrow 2\ \text{H}_2\text{O}(l)$$

We calculate the amount of H_3O^+ at any given point (before the equivalence point) by using the reaction stoichiometry—1 mol of NaOH neutralizes 1 mol of H_3O^+. The initial number of moles of H_3O^+ (as we just calculated) is 0.00250 mol. We calculate the number of moles of NaOH added at 5.00 mL by multiplying the added volume (in L) by the concentration of the NaOH solution:

$$\text{mol NaOH added} = 0.00500\ \cancel{\text{L}} \times \frac{0.100\ \text{mol}}{1\ \cancel{\text{L}}} = 0.000500\ \text{mol NaOH}$$

The addition of OH^- causes the amount of H^+ to decrease as shown in the following table:

	$\text{OH}^-(aq)$ +	$\text{H}_3\text{O}^+(aq) \longrightarrow 2\ \text{H}_2\text{O}(l)$
Before addition	≈0.00 mol	0.00250 mol
Addition	0.000500 mol	—
After addition	≈0.00 mol	0.00200 mol

We calculate the H_3O^+ concentration by dividing the number of moles of H_3O^+ remaining by the *total volume* (initial volume plus added volume):

$$[H_3O^+] = \frac{0.00200 \text{ mol } H_3O^+}{0.0250 \text{ L} + 0.00500 \text{ L}} = 0.0667 \text{ M}$$

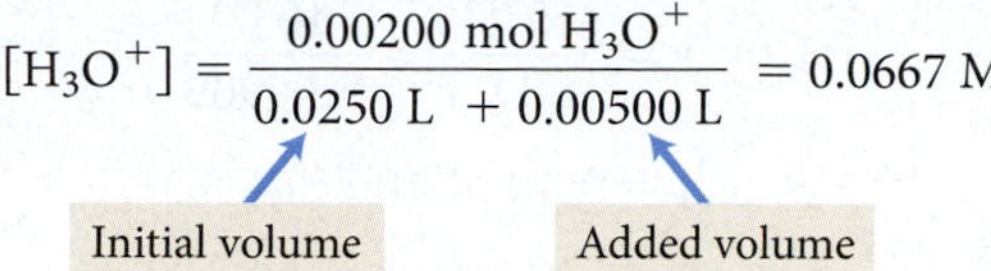

The pH is therefore 1.18.

$$\begin{aligned} pH &= -\log 0.0667 \\ &= 1.18 \end{aligned}$$

pH's After Adding 10.0, 15.0, and 20.0 mL NaOH As more NaOH is added, it further neutralizes the H_3O^+ in the solution. We calculate the pH at each of these points in the same way that we calculated the pH at the 5.00 mL point. The results are tabulated as follows:

Volume (mL)	pH
10.0	1.37
15.0	1.60
20.0	1.95

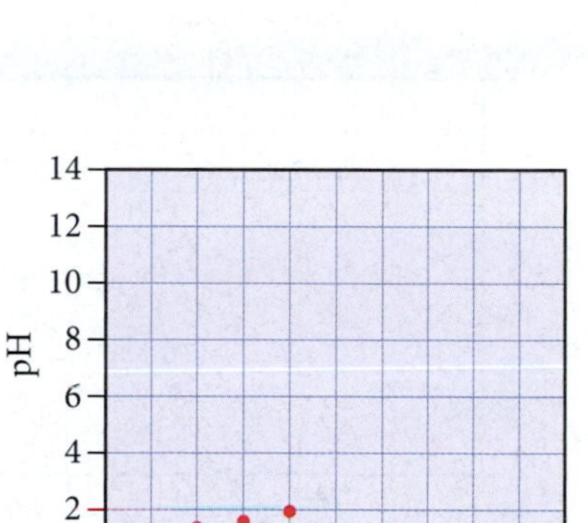

pH After Adding 25.0 mL NaOH (Equivalence Point) The pH at the equivalence point of a strong acid–strong base titration is always 7.00 (at 25 °C). At the equivalence point, the strong base has completely neutralized the strong acid. The only source of hydronium ions then is the ionization of water. The $[H_3O^+]$ at 25 °C from the ionization of water is 1.00×10^{-7} M, and the pH is 7.00.

pH After Adding 30.00 mL NaOH As NaOH is added beyond the equivalence point, it becomes the excess reagent. We calculate the amount of OH^- at any given point (past the equivalence point) by subtracting the initial amount of H_3O^+ from the amount of OH^- added. The number of moles of OH^- added at 30.00 mL is:

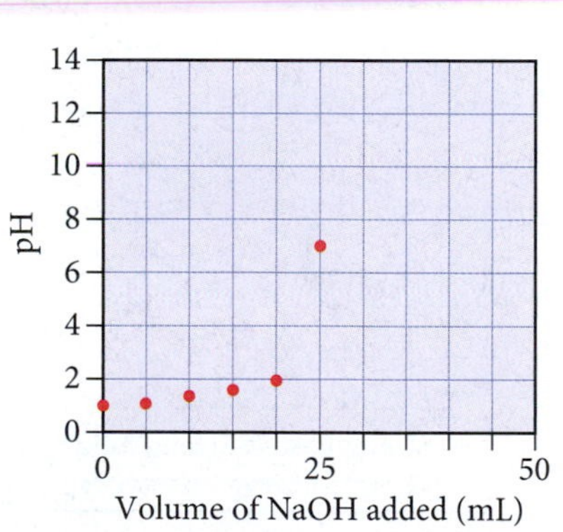

$$\text{mol } OH^- \text{ added} = 0.0300 \cancel{L} \times \frac{0.100 \text{ mol}}{1 \cancel{L}} = 0.00300 \text{ mol } OH^-$$

The number of moles of OH^- remaining after neutralization is shown in the following table:

	$OH^-(aq)$ +	$H_3O^+(aq)$ $\longrightarrow$	$2 H_2O(l)$
Before addition	≈0.00 mol	0.00250 mol	
Addition	0.00300 mol	—	
After addition	0.00050 mol	0.00 mol	

We calculate the OH^- concentration by dividing the number of moles of OH^- remaining by the *total volume* (initial volume plus added volume):

$$[OH^-] = \frac{0.000500 \text{ mol } OH^-}{0.0250 \text{ L} + 0.0300 \text{ L}} = 0.00909 \text{ M}$$

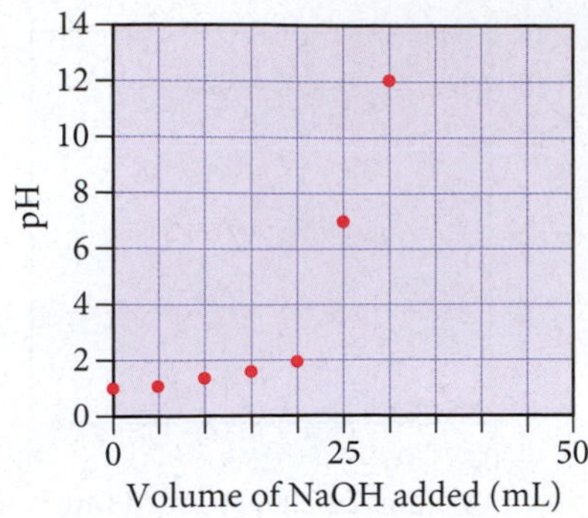

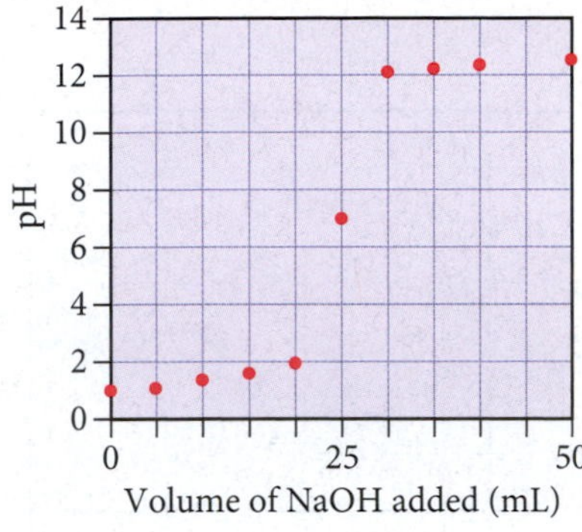

We can then calculate the $[H_3O^+]$ and pH:

$$[H_3O^+][OH^-] = 10^{-14}$$

$$[H_3O^+] = \frac{10^{-14}}{[OH^-]} = \frac{10^{-14}}{0.00909}$$

$$= 1.10 \times 10^{-12}\ M$$

$$pH = -\log(1.10 \times 10^{-12})$$

$$= 11.96$$

pH's After Adding 35.0, 40.0, and 50.0 mL NaOH As more NaOH is added, it further increases the basicity of the solution. We calculate the pH at each of these points in the same way that we calculated the pH at the 30.00 mL point. We tabulate the results as follows:

Volume (mL)	pH
35.0	12.22
40.0	12.36
50.0	12.52

The Overall pH Curve The overall pH curve for the titration of a strong acid with a strong base has the characteristic S-shape we just plotted. The overall curve is as follows:

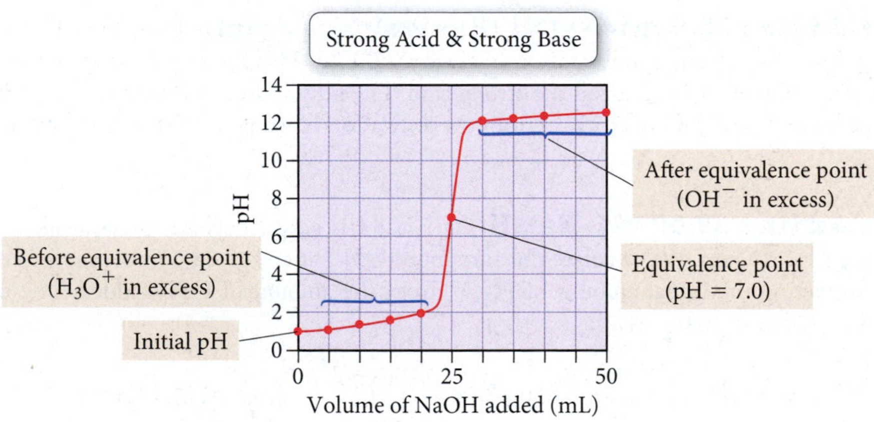

Summarizing the Titration of a Strong Acid with a Strong Base:

- The initial pH is simply the pH of the strong acid solution to be titrated.
- Before the equivalence point, H_3O^+ is in excess. Calculate the $[H_3O^+]$ by subtracting the number of moles of added OH^- from the initial number of moles of H_3O^+ and dividing by the *total* volume.
- At the equivalence point, neither reactant is in excess and the pH = 7.00.
- Beyond the equivalence point, OH^- is in excess. Calculate the $[OH^-]$ by subtracting the initial number of moles of H_3O^+ from the number of moles of added OH^- and dividing by the *total* volume.

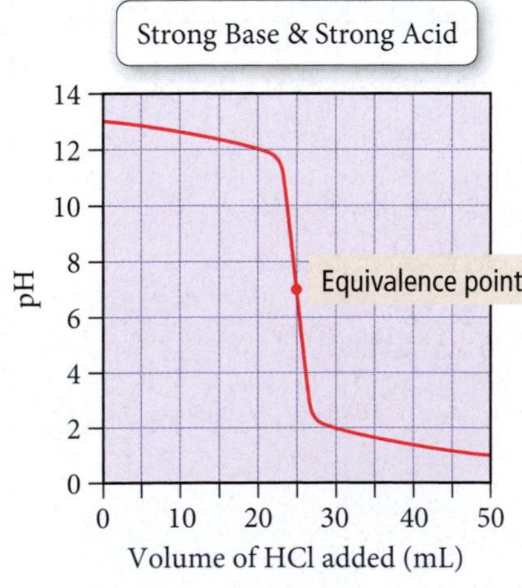

▲ **FIGURE 18.7 Titration Curve: Strong Base + Strong Acid** This curve represents the titration of 25.0 mL of 0.100 M NaOH with 0.100 M HCl.

The pH curve for the titration of a strong base with a strong acid is shown in Figure 18.7 ◄. Calculating the points along this curve is very similar to calculating the points along the pH curve for the titration of a strong acid with a strong base. The primary difference is that the curve starts basic and then becomes acidic after the equivalence point (instead of vice versa).

EXAMPLE 18.6

Strong Base–Strong Acid Titration pH Curve

A 50.0 mL sample of 0.200 M sodium hydroxide is titrated with 0.200 M nitric acid. Calculate the pH:

(a) after adding 30.00 mL of HNO_3

(b) at the equivalence point

SOLUTION

(a) Begin by calculating the initial amount of NaOH (in moles) from the volume and molarity of the NaOH solution. Because NaOH is a strong base, it dissociates completely, so the amount of OH^- is equal to the amount of NaOH.

$$\text{moles NaOH} = 0.0500\ \cancel{L} \times \frac{0.200\ \text{mol}}{1\ \cancel{L}} = 0.0100\ \text{mol}$$

$$\text{moles } OH^- = 0.0100\ \text{mol}$$

Calculate the amount of HNO_3 (in moles) added at 30.0 mL from the molarity of the HNO_3 solution.

$$\text{mol } HNO_3 \text{ added} = 0.0300\ \text{L} \times \frac{0.200\ \text{mol}}{1\ \text{L}} = 0.00600\ \text{mol } HNO_3$$

As HNO_3 is added to the solution, it neutralizes some of the OH^-. Calculate the number of moles of OH^- remaining by setting up a table based on the neutralization reaction that shows the amount of OH^- before the addition, the amount of H_3O^+ added, and the amounts left after the addition.

	$OH^-(aq)$	+ $H_3O^+(aq)$	$\longrightarrow$ $2\ H_2O(l)$
Before addition	0.0100 mol	≈0.00 mol	
Addition	—	0.00600 mol	
After addition	0.0040 mol	0.00 mol	

Calculate the OH^- concentration by dividing the amount of OH^- remaining by the *total volume* (initial volume plus added volume).

Calculate the pOH from $[OH^-]$.

$$[OH^-] = \frac{0.0040\ \text{mol}}{0.0500\ \text{L} + 0.0300\ \text{L}} = 0.0500\ \text{M}$$

$$\text{pOH} = -\log(0.0500) = 1.30$$

Calculate the pH from the pOH using the equation $\text{pH} + \text{pOH} = 14$.

$$\text{pH} = 14 - \text{pOH} = 14 - 1.30 = 12.70$$

(b) At the equivalence point, the strong base has completely neutralized the strong acid. The $[H_3O^+]$ at 25 °C from the ionization of water is 1.00×10^{-7} M and the pH is therefore 7.00.

$$\text{pH} = 7.00$$

FOR PRACTICE 18.6

Calculate the pH in the titration in Example 18.6 after the addition of 60.0 mL of 0.200 M HNO_3.

Titration Equivalence Point

18.5 Cc Conceptual Connection

You want to titrate the acid shown in the flask with a strong base. Which mark on the burette next to the flask indicates the amount of base you will need to add to reach the equivalence point?

(a) A
(b) B
(c) C
(d) D

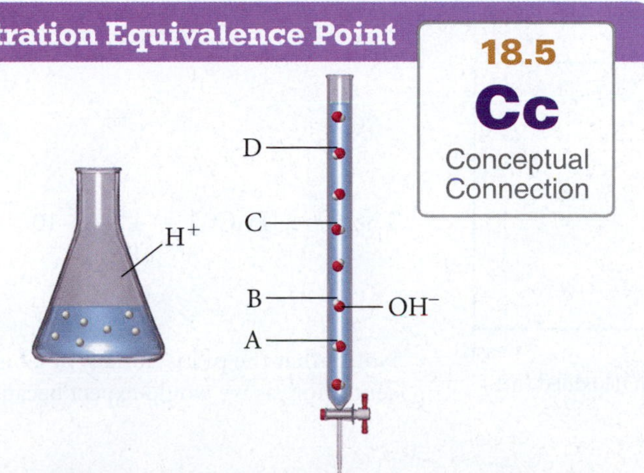

The Titration of a Weak Acid with a Strong Base

Let's consider the titration of 25.0 mL of 0.100 M $HCHO_2$ with 0.100 M NaOH.

$$NaOH(aq) + HCHO_2(aq) \longrightarrow H_2O(l) + NaCHO_2(aq)$$

The concentrations and the volumes here are identical to those in our previous titration, in which we calculated the pH curve for the titration of a *strong* acid with a strong base. The only difference is that $HCHO_2$ is a *weak* acid rather than a strong one. We begin by calculating the volume required to reach the equivalence point of the titration.

Volume of NaOH Required to Reach the Equivalence Point From the stoichiometry of the equation, we can see that the equivalence point occurs when the amount (in moles) of added base equals the amount (in moles) of acid initially in solution.

$$\text{Initial mol } HCHO_2 = 0.0250 \cancel{\text{L}} \times \frac{0.100 \text{ mol}}{1 \cancel{\text{L}}} = 0.00250 \text{ mol } HCHO_2$$

The amount of NaOH that must be added is 0.00250 mol NaOH. The volume of NaOH required is therefore:

$$\text{Volume NaOH solution} = 0.00250 \cancel{\text{mol}} \times \frac{1 \text{ L}}{0.100 \cancel{\text{mol}}} = 0.0250 \text{ L NaOH solution}$$

The equivalence point occurs when 25.0 mL of base has been added. Notice that the volume of NaOH required to reach the equivalence point for this weak acid is identical to that required for a strong acid. *The volume at the equivalence point in an acid–base titration does not depend on whether the acid being titrated is a strong acid or a weak acid; it depends only on the amount (in moles) of acid present in solution before the titration begins, the stoichiometry of the reaction, and the concentration of the added base.*

Initial pH (Before Adding Any Base) The initial pH of the solution is the pH of a 0.100 M $HCHO_2$ solution. Since $HCHO_2$ is a weak acid, we calculate the concentration of H_3O^+ and the pH by doing an equilibrium problem for the ionization of $HCHO_2$. The procedure for solving weak acid ionization problems is demonstrated in Examples 17.5 and 17.6. We show a highly condensed calculation here (K_a for $HCHO_2$ is 1.8×10^{-4}).

$$HCHO_2(aq) + H_2O(l) \rightleftharpoons H_3O^+(aq) + CHO_2^-(aq)$$

	$[HCHO_2]$	$[H_3O^+]$	$[HCO_2^-]$
Initial	0.100	≈0.00	0.00
Change	$-x$	$+x$	$+x$
Equil	$0.100 - x$	x	x

$$K_a = \frac{[H_3O^+][CHO_2^-]}{[HCHO_2]}$$

$$= \frac{x^2}{0.100 - \cancel{x}} \quad (x \text{ is small})$$

$$1.8 \times 10^{-4} = \frac{x^2}{0.100}$$

$$x = 4.24 \times 10^{-3}$$

Therefore, $[H_3O^+] = 4.24 \times 10^{-3}$ M.

$$pH = -\log(4.24 \times 10^{-3}) = 2.37$$

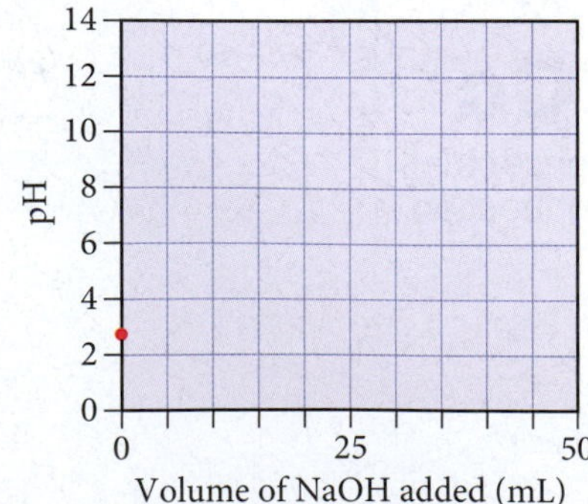

Notice that the pH is initially at a higher value (less acidic) than it is for a strong acid of the same concentration, as we would expect because the acid is weak.

pH After Adding 5.00 mL NaOH When we tritrate a *weak acid* with a strong base, the added NaOH *converts a stoichiometric amount of the acid into its conjugate base*. As we calculated previously, 5.00 mL of the 0.100 M NaOH solution contains 0.000500 mol OH^-. When the 0.000500 mol OH^- is added to the weak acid solution, the OH^- reacts stoichiometrically with $HCHO_2$ causing the amount of $HCHO_2$ to *decrease* by 0.000500 mol and the amount of CHO_2^- to *increase* by 0.000500 mol. This is very similar to what happens when we add strong base to a buffer, and it is summarized in the following table:

	$OH^-(aq)$ +	$HCHO_2(aq)$ ⟶	$H_2O(l)$ +	$CHO_2^-(aq)$
Before addition	≈0.00 mol	0.00250 mol	—	0.00 mol
Addition	0.000500 mol	—	—	—
After addition	≈0.00 mol	0.00200 mol	—	0.000500 mol

Notice that, after the addition, the solution contains significant amounts of both an acid ($HCHO_2$) and its conjugate base (CHO_2^-)—*the solution is now a buffer*. To calculate the pH of a buffer (when the *x is small* approximation applies as it does here), we can use the Henderson–Hasselbalch equation and pK_a for $HCHO_2$ (which is 3.74):

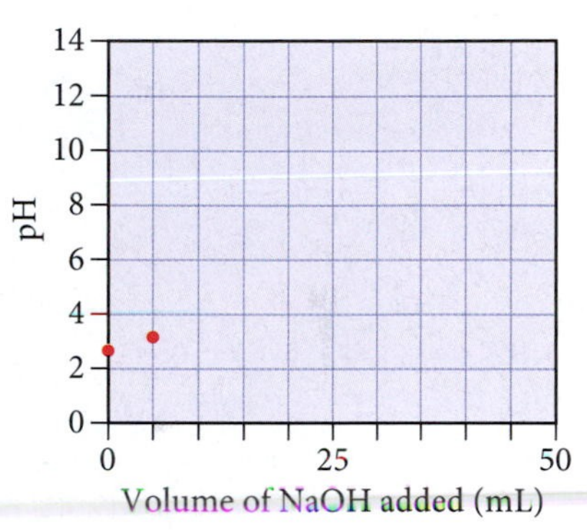

$$\begin{aligned} pH &= pK_a + \log \frac{[\text{base}]}{[\text{acid}]} \\ &= 3.74 + \log \frac{0.000500}{0.00200} \\ &= 3.74 - 0.60 \\ &= 3.14 \end{aligned}$$

pH's After Adding 10.0, 12.5, 15.0, and 20.0 mL NaOH As more NaOH is added, it converts more $HCHO_2$ into CHO_2^-. We calculate the relative amounts of $HCHO_2$ and CHO_2^- at each of these volumes using the reaction stoichiometry and then calculate the pH of the resulting buffer using the Henderson–Hasselbalch equation (as we did for the pH at 5.00 mL). The amounts of $HCHO_2$ and CHO_2^- (after addition of the OH^-) at each volume and the corresponding pH's are tabulated as follows:

Volume (mL)	mol $HCHO_2$	mol CHO_2^-	pH
10.0	0.00150	0.00100	3.56
12.5	0.00125	0.00125	3.74
15.0	0.00100	0.00150	3.92
20.0	0.00050	0.00200	4.34

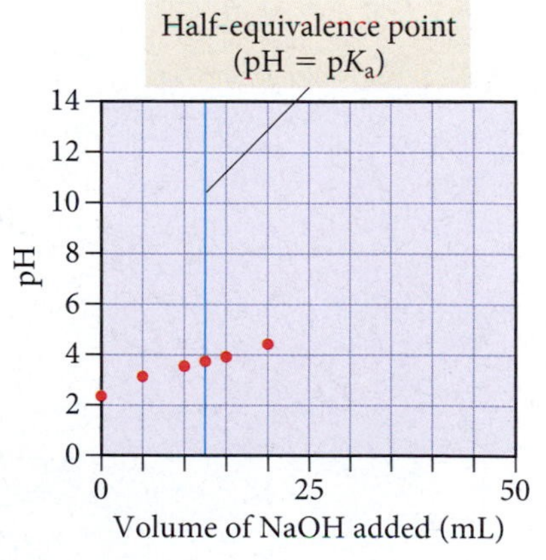

As the titration proceeds, more of the $HCHO_2$ is converted to the conjugate base (CHO_2^-). Notice that an added NaOH volume of 12.5 mL corresponds to one-half of the equivalence point. At this volume, one-half of the initial amount of $HCHO_2$ has been converted to CHO_2^-, resulting in *equal amounts of weak acid and conjugate base*. For any buffer in which the amounts of weak acid and conjugate base are equal, $pH = pK_a$:

$$pH = pK_a + \log \frac{[\text{base}]}{[\text{acid}]}$$

If [base] = [acid], then [base]/[acid] = 1.

$$\begin{aligned} pH &= pK_a + \log 1 \\ &= pK_a + 0 \\ &= pK_a \end{aligned}$$

Since $pH = pK_a$ halfway to the equivalence point, we can use titrations to determine the pK_a of an acid.

pH After Adding 25.0 mL NaOH (Equivalence Point) At the equivalence point, 0.00250 mol of OH^- have been added and therefore all of the $HCHO_2$ has been converted into its conjugate base (CHO_2^-) as summarized in the following table:

	$OH^-(aq)$ +	$HCHO_2(aq)$ ⟶	$H_2O(l)$ +	$CHO_2^-(aq)$
Before addition	≈0.00 mol	0.00250 mol	—	0.00 mol
Addition	0.00250 mol	—	—	—
After addition	≈0.00 mol	0.00 mol	—	0.00250 mol

The solution is no longer a buffer (it no longer contains significant amounts of both a weak acid and its conjugate base). Instead, the solution contains an ion (CHO_2^-) acting as a weak base. We demonstrated how to calculate the pH of solutions such as this in Section 17.9 (see Example 17.14) by solving an equilibrium problem involving the ionization of water by the weak base (CHO_2^-):

$$CHO_2^-(aq) + H_2O(l) \rightleftharpoons HCHO_2(aq) + OH^-(aq)$$

We calculate the initial concentration of CHO_2^- for the equilibrium problem by dividing the number of moles of CHO_2^- (0.00250 mol) by the *total* volume at the equivalence point (initial volume plus added volume):

Moles CHO_2^- at equivalence point

$$[CHO_2^-] = \frac{0.00250 \text{ mol}}{0.0250 \text{ L} + 0.0250 \text{ L}} = 0.0500 \text{ M}$$

Initial volume

Added volume at equivalence point

We then proceed to solve the equilibrium problem as shown in condensed form here:

$$CHO_2^-(aq) + H_2O(l) \rightleftharpoons HCHO_2(aq) + OH^-(aq)$$

	$[CHO_2^-]$	$[HCHO_2]$	$[OH^-]$
Initial	0.0500	0.00	≈0.00
Change	$-x$	$+x$	$+x$
Equil	$0.0500-x$	x	x

Before substituting into the expression for K_b, we find the value of K_b from K_a for formic acid ($K_a = 1.8 \times 10^{-4}$) and K_w:

$$K_a \times K_b = K_w$$

$$K_b = \frac{K_w}{K_a} = \frac{1.0 \times 10^{-14}}{1.8 \times 10^{-4}} = 5.6 \times 10^{-11}$$

Then we substitute the equilibrium concentrations from the previous table into the expression for K_b:

$$K_b = \frac{[HCHO_2][OH^-]}{[CHO_2^-]}$$

$$= \frac{x^2}{0.0500 - \cancel{x}} \quad (x \text{ is small})$$

$$5.6 \times 10^{-11} = \frac{x^2}{0.0500}$$

$$x = 1.7 \times 10^{-6}$$

Remember that x represents the concentration of the hydroxide ion. We calculate $[H_3O^+]$ and pH:

$$[OH^-] = 1.7 \times 10^{-6}\ M$$
$$[H_3O^+][OH^-] = K_w = 1.0 \times 10^{-14}$$
$$[H_3O^+](1.7 \times 10^{-6}) = 1.0 \times 10^{-14}$$
$$[H_3O^+] = 5.9 \times 10^{-9}\ M$$
$$pH = -\log[H_3O^+]$$
$$= -\log(5.9 \times 10^{-9})$$
$$= 8.23$$

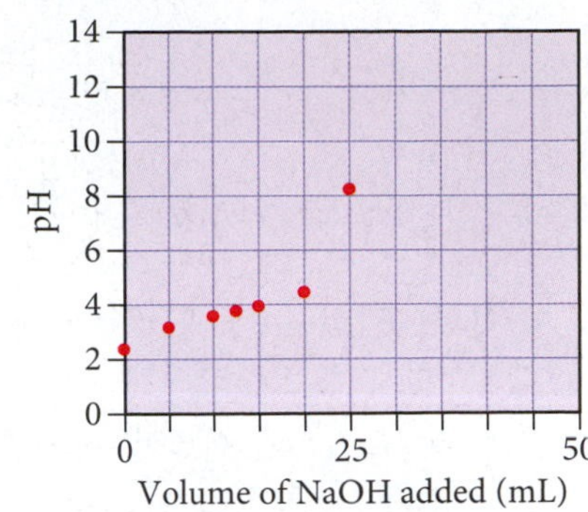

Notice that the pH at the equivalence point is *not* neutral but basic. *The titration of a weak acid by a strong base always has a basic equivalence point* because, at the equivalence point, all of the acid has been converted into its conjugate base, resulting in a weakly basic solution.

pH After Adding 30.00 mL NaOH At this point in the titration, 0.00300 mol of OH^- have been added. NaOH has thus become the excess reagent as shown in the following table:

	$OH^-(aq)$ +	$HCHO_2(aq)$ ⟶	$H_2O(l)$ +	$CHO_2^-(aq)$
Before addition	≈0.00 mol	0.00250 mol	—	0.00 mol
Addition	0.00300 mol	—	—	—
After addition	0.00050 mol	≈0.00 mol	—	0.00250 mol

The solution is now a mixture of a strong base (NaOH) and a weak base (CHO_2^-). The strong base completely overwhelms the weak base, and we can calculate the pH by considering the strong base alone (as we did for the titration of a strong acid and a strong base). We calculate the OH^- concentration by dividing the amount of OH^- remaining by the *total volume* (initial volume plus added volume):

$$[OH^-] = \frac{0.00050\ \text{mol } OH^-}{0.0250\ L + 0.0300\ L} = 0.0091\ M$$

We then calculate the $[H_3O^+]$ and pH:

$$[H_3O^+][OH^-] = 1.0 \times 10^{-14}$$
$$[H_3O^+] = \frac{1.0 \times 10^{-14}}{[OH^-]} = \frac{1.0 \times 10^{-14}}{0.0091} = 1.10 \times 10^{-12}\ M$$
$$pH = -\log(1.10 \times 10^{-12})$$
$$= 11.96$$

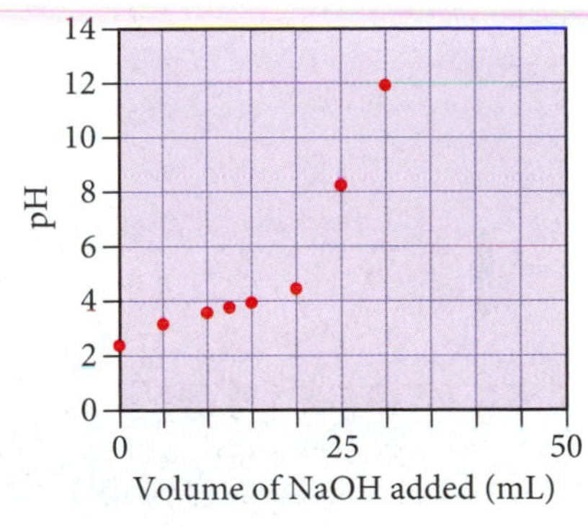

pH's After Adding 35.0, 40.0, and 50.0 mL NaOH As more NaOH is added, the basicity of the solution increases further. We calculate the pH at each of these volumes in the same way we calculated the pH at 30.00 mL of added NaOH. The results are tabulated as follows:

Volume (mL)	pH
35.0	12.22
40.0	12.36
50.0	12.52

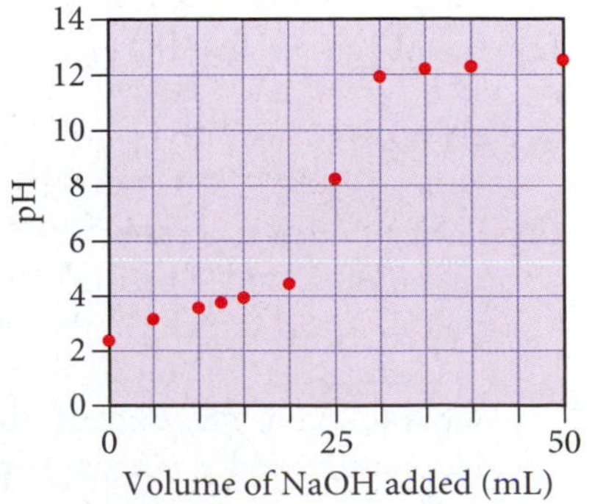

The Overall pH Curve The overall pH curve for the titration of a weak acid with a strong base has a characteristic S-shape similar to that for the titration of a strong acid with a strong base. The primary difference is that the equivalence point pH is basic (not neutral). Notice that calculating the pH in different regions throughout the titration involves working different kinds of acid–base problems, all of which we have encountered before.

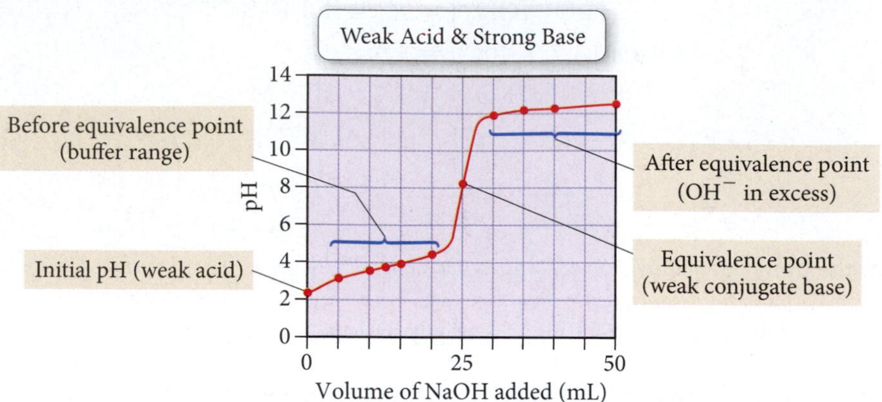

Summarizing Titration of a Weak Acid with a Strong Base:

- The initial pH is that of the weak acid solution to be titrated. Calculate the pH by working an equilibrium problem (similar to Examples 17.5 and 17.6) using the concentration of the weak acid as the initial concentration.
- Between the initial pH and the equivalence point, the solution becomes a buffer. Use the reaction stoichiometry to calculate the amounts of each buffer component and then use the Henderson–Hasselbalch equation to calculate the pH (as in Example 18.3).
- Halfway to the equivalence point, the buffer components are exactly equal and $pH = pK_a$.
- At the equivalence point, the acid has all been converted into its conjugate base. Calculate the pH by working an equilibrium problem for the ionization of water by the ion acting as a weak base (similar to Example 17.14). (Calculate the concentration of the ion acting as a weak base by dividing the number of moles of the ion by the total volume at the equivalence point.)
- Beyond the equivalence point, OH^- is in excess. Ignore the weak base and calculate the $[OH^-]$ by subtracting the initial number of moles of the weak acid from the number of moles of added OH^- and dividing by the *total* volume.

EXAMPLE 18.7

Weak Acid–Strong Base Titration pH Curve

A 40.0 mL sample of 0.100 M HNO_2 is titrated with 0.200 M KOH. Calculate:

(a) the volume required to reach the equivalence point

(b) the pH after adding 5.00 mL of KOH

(c) the pH at one-half the equivalence point

SOLUTION

(a) The equivalence point occurs when the amount (in moles) of added base equals the amount (in moles) of acid initially in the solution. Begin by calculating the amount (in moles) of acid initially in the solution. The amount (in moles) of KOH that must be added is equal to the amount of the weak acid.	$\text{mol } HNO_2 = 0.0400 \cancel{\text{L}} \times \frac{0.100 \text{ mol}}{\cancel{\text{L}}} = 4.00 \times 10^{-3} \text{ mol}$ $\text{mol KOH required} = 4.00 \times 10^{-3} \text{ mol}$
Calculate the volume of KOH required from the number of moles of KOH and the molarity.	$\text{volume KOH solution} = 4.00 \times 10^{-3} \cancel{\text{mol}} \times \frac{1 \text{ L}}{0.200 \cancel{\text{mol}}} = 0.0200 \text{ L KOH solution} = 20.0 \text{ mL KOH solution}$

Continued from the previous page—

(b) Use the concentration of the KOH solution to calculate the amount (in moles) of OH^- in 5.00 mL of the solution.

$$\text{mol } OH^- = 5.00 \times 10^{-3} \text{ L} \times \frac{0.200 \text{ mol}}{1 \text{ L}} = 1.00 \times 10^{-3} \text{ mol } OH^-$$

Prepare a table showing the amounts of HNO_2 and NO_2^- before and after the addition of 5.00 mL KOH. The addition of the KOH stoichiometrically reduces the concentration of HNO_2 and increases the concentration of NO_2^-.

	$OH^-(aq)$ +	$HNO_2(aq)$ ⟶	$H_2O(l)$ +	$NO_2^-(aq)$
Before addition	≈0.00 mol	4.00×10^{-3} mol	—	0.00 mol
Addition	1.00×10^{-3} mol	—	—	—
After addition	≈0.00 mol	3.00×10^{-3} mol	—	1.00×10^{-3} mol

Since the solution now contains significant amounts of a weak acid and its conjugate base, use the Henderson–Hasselbalch equation and pK_a for HNO_2 (which is 3.34) to calculate the pH of the solution.

$$\begin{aligned} pH &= pK_a + \log\frac{[\text{base}]}{[\text{acid}]} \\ &= 3.34 + \log\frac{1.00 \times 10^{-3}}{3.00 \times 10^{-3}} \\ &= 3.34 - 0.48 = 2.86 \end{aligned}$$

(c) At one-half the equivalence point, the amount of added base is exactly one-half the initial amount of acid. The base converts exactly half of the HNO_2 into NO_2^-, resulting in equal amounts of the weak acid and its conjugate base. The pH is therefore equal to pK_a.

	$OH^-(aq)$ +	$HNO_2(aq)$ ⟶	$H_2O(l)$ +	$NO_2^-(aq)$
Before addition	≈0.00 mol	4.00×10^{-3} mol	—	0.00 mol
Addition	2.00×10^{-3} mol	—	—	—
After addition	≈0.00 mol	2.00×10^{-3} mol	—	2.00×10^{-3} mol

$$\begin{aligned} pH &= pK_a + \log\frac{[\text{base}]}{[\text{acid}]} \\ &= 3.34 + \log\frac{2.00 \times 10^{-3}}{2.00 \times 10^{-3}} \\ &= 3.34 + 0 = 3.34 \end{aligned}$$

FOR PRACTICE 18.7

Determine the pH at the equivalence point for the titration of HNO_2 and KOH in Example 18.7.

The Titration of a Weak Base with a Strong Acid

Figure 18.8 ▶ is the pH curve for the titration of a weak base with a strong acid. Calculating the points along this curve is very similar to calculating the points along the pH curve for the titration of a weak acid with a strong base (which we just did). The main differences are that the curve starts basic and has an acidic equivalence point. We calculate the pH in the buffer region using $pH = pK_a + \log\frac{[\text{base}]}{[\text{acid}]}$ where the pK_a corresponds to the conjugate acid of the base being titrated.

The Half-Equivalence Point

18.6 Cc Conceptual Connection

What is the pH at the half-equivalence point in the titration of a weak base with a strong acid? The pK_b of the weak base is 8.75.

(a) 8.75 **(b)** 7.0 **(c)** 5.25 **(d)** 4.37

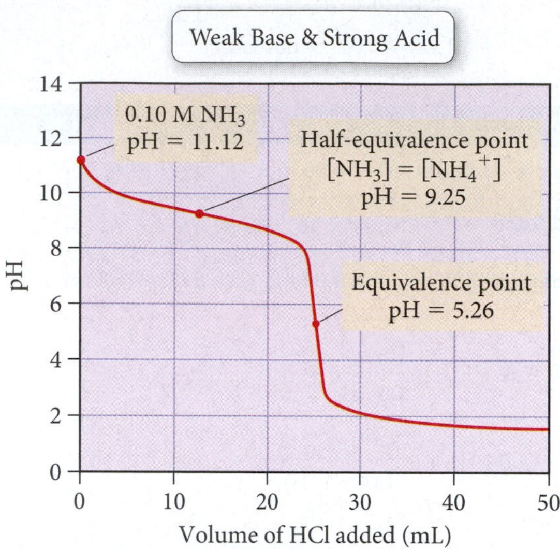

▲ **FIGURE 18.8 Titration Curve: Weak Base with Strong Acid** This curve represents the titration of 25.0 mL of 0.100 M NH_3 with 0.100 M HCl.

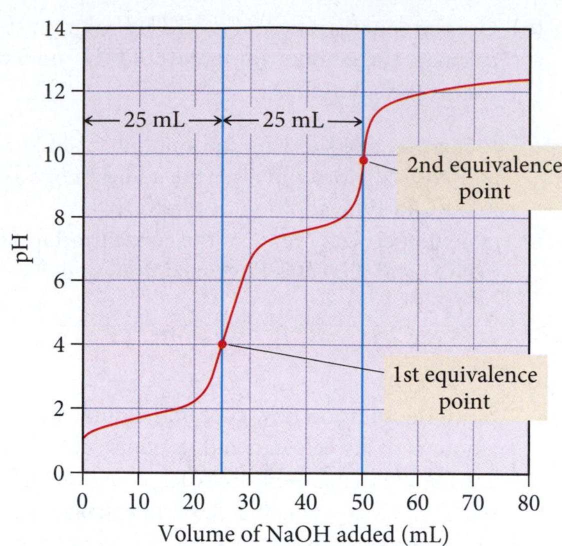

▲ **FIGURE 18.9 Titration Curve: Diprotic Acid with Strong Base** This curve represents the titration of 25.0 mL of 0.100 M H_2SO_3 with 0.100 M NaOH.

The Titration of a Polyprotic Acid

When a diprotic acid is titrated with a strong base, if K_{a_1} and K_{a_2} are sufficiently different, the pH curve will have two equivalence points. For example, Figure 18.9 ▲ shows the pH curve for the titration of sulfurous acid (H_2SO_3) with sodium hydroxide. Recall from Section 17.10 that sulfurous acid ionizes in two steps as follows:

$$H_2SO_3(aq) \rightleftharpoons H^+(aq) + HSO_3^-(aq) \quad K_{a_1} = 1.6 \times 10^{-2}$$

$$HSO_3^-(aq) \rightleftharpoons H^+(aq) + SO_3^{2-}(aq) \quad K_{a_2} = 6.4 \times 10^{-8}$$

The first equivalence point in the titration curve represents the titration of the first proton while the second equivalence point represents the titration of the second proton. Notice that the volume required to reach the first equivalence point is identical to the volume required to the reach the second one because the number of moles of H_2SO_3 in the first step determines the number of moles of HSO_3^- in the second step.

18.7 Cc Conceptual Connection

Acid–Base Titrations

Consider these three titrations:

(i) the titration of 25.0 mL of a 0.100 M monoprotic weak acid with 0.100 M NaOH
(ii) the titration of 25.0 mL of a 0.100 M diprotic weak acid with 0.100 M NaOH
(iii) the titration of 25.0 mL of a 0.100 M strong acid with 0.100 M NaOH

Which statement is most likely to be true?

(a) All three titrations have the same initial pH.
(b) All three titrations have the same pH at their first equivalence point.
(c) All three titrations require the same volume of NaOH to reach their first equivalence point.

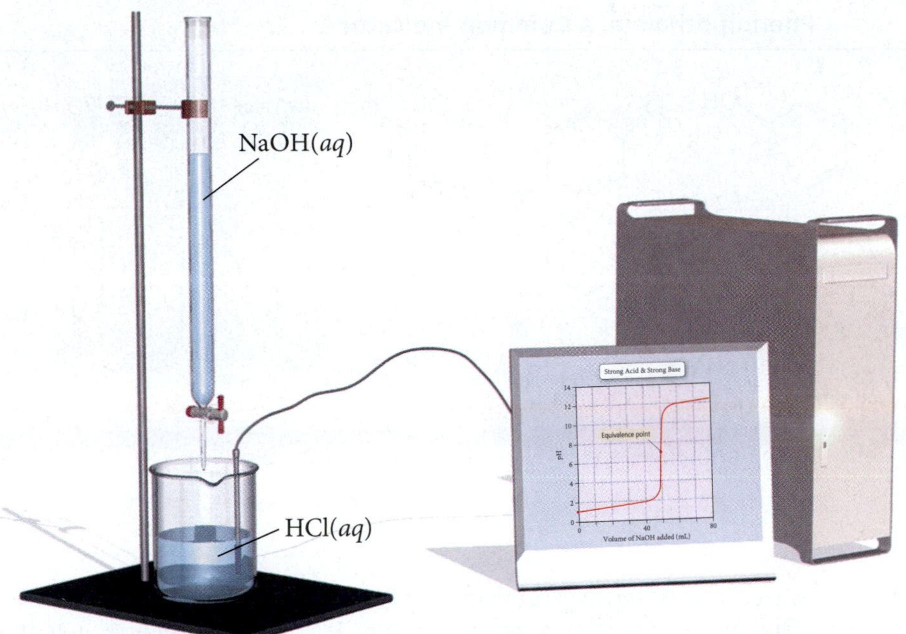

◀ **FIGURE 18.10 Monitoring pH during a Titration** A pH meter monitors pH during titration. The inflection point in the resulting pH curve signifies the equivalence point.

Indicators: pH-Dependent Colors

We can monitor the pH of a titration with either a pH meter or an indicator. The direct monitoring of pH with a meter yields data like the pH curves we have examined previously, allowing determination of the equivalence point from the pH curve itself, as shown in Figure 18.10 ▲. With an indicator, we rely on the point where the indicator changes color—called the **endpoint**—to determine the equivalence point, as shown in Figure 18.11 ▼. With the correct indicator, the endpoint of the titration (indicated by the color change) occurs at the equivalence point (when the amount of acid equals the amount of base).

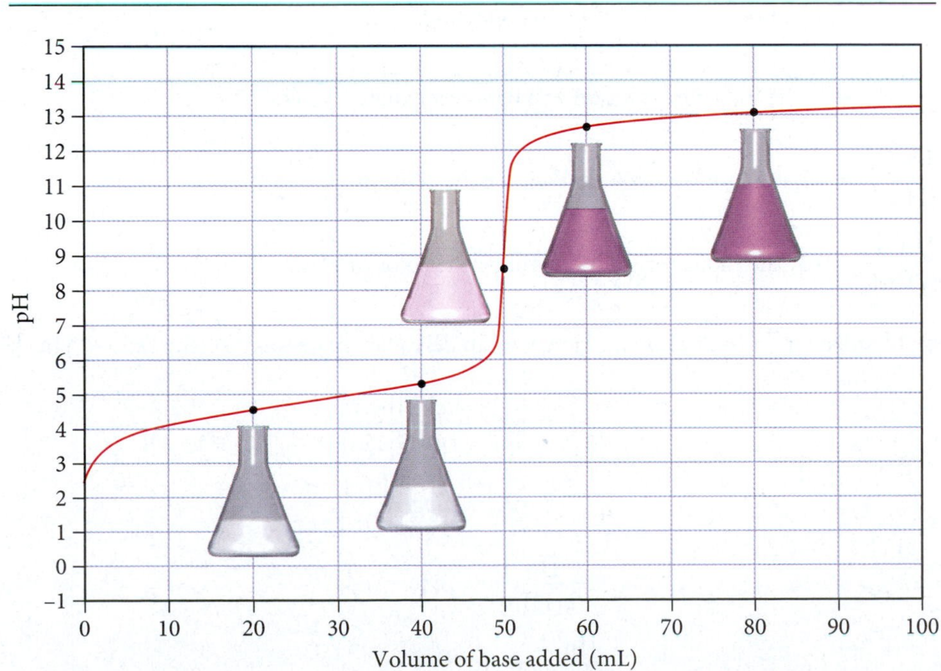

◀ **FIGURE 18.11 Monitoring Color Change during a Titration** Titration of 50.0 mL of 0.100 M $HC_2H_3O_2$ with 0.100 M NaOH. The endpoint of a titration is signaled by a color change in an appropriate indicator (in this case, phenolphthalein).

Phenolphthalein, a Common Indicator

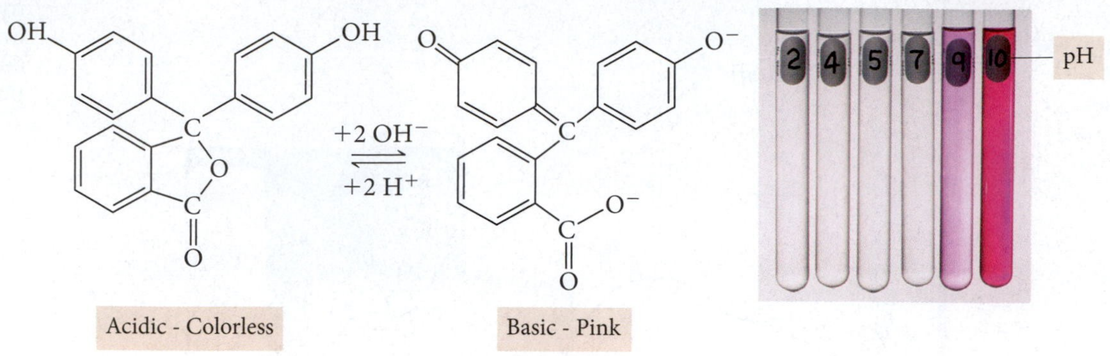

Acidic - Colorless

+2 OH^- ⇌ +2 H^+

Basic - Pink

pH: 2, 4, 5, 7, 9, 10

▲ **FIGURE 18.12 Phenolphthalein** Phenolphthalein, a weakly acidic compound, is colorless. Its conjugate base is pink.

An indicator is itself a weak organic acid that is a different color than its conjugate base. For example, phenolphthalein (whose structure is shown in Figure 18.12 ▲) is a common indicator whose acid form is colorless and conjugate base form is pink. If we let HIn represent the acid form of a generic indicator and In^- the conjugate base form, we have the following equilibrium:

$$\underset{\text{color 1}}{\mathrm{HIn}(aq)} + \mathrm{H_2O}(l) \rightleftharpoons \mathrm{H_3O^+}(aq) + \underset{\text{color 2}}{\mathrm{In^-}(aq)}$$

Because its color is intense, only a small amount of indicator is required—an amount that will not affect the pH of the solution or the equivalence point of the neutralization reaction. When the $[H_3O^+]$ changes during the titration, the equilibrium shifts in response. At low pH, the $[H_3O^+]$ is high and the equilibrium lies far to the left, resulting in a solution of color 1. As the titration proceeds, the $[H_3O^+]$ decreases, shifting the equilibrium to the right. Since the pH change is large near the equivalence point of the titration, there is a large change in $[H_3O^+]$ near the equivalence point. Provided that the correct indicator is chosen, there will also be a correspondingly significant change in color. For the titration of a strong acid with a strong base, one drop of the base near the endpoint is usually enough to change the indicator from color 1 to color 2.

The color of a solution containing an indicator depends on the relative concentrations of HIn and In^-. As a useful guideline, we can assume the following:

If $\dfrac{[\mathrm{In^-}]}{[\mathrm{HIn}]} = 1$, the indicator solution will be intermediate in color.

If $\dfrac{[\mathrm{In^-}]}{[\mathrm{HIn}]} > 10$, the indicator solution will be the color of In^-.

If $\dfrac{[\mathrm{In^-}]}{[\mathrm{HIn}]} < 0.1$, the indicator solution will be the color of HIn.

From the Henderson–Hasselbalch equation, we can derive an expression for the ratio of $[In^-]/[HIn]$:

$$\mathrm{pH} = \mathrm{p}K_a + \log\frac{[\text{base}]}{[\text{acid}]}$$

$$= \mathrm{p}K_a + \log\frac{[\mathrm{In^-}]}{[\mathrm{HIn}]}$$

$$\log\frac{[\mathrm{In^-}]}{[\mathrm{HIn}]} = \mathrm{pH} - \mathrm{p}K_a$$

$$\frac{[\mathrm{In^-}]}{[\mathrm{HIn}]} = 10^{(\mathrm{pH}-\mathrm{p}K_a)}$$

Consider the following three pH values relative to pK_a and the corresponding colors of the indicator solution:

pH (relative to pK_a)	$[In^-]/[HIn]$ ratio	Color of Indicator Solution
$pH = pK_a$	$\frac{[In^-]}{[HIn]} = 10^0 = 1$	Intermediate Color
$pH = pK_a + 1$	$\frac{[In^-]}{[HIn]} = 10^1 = 10$	Color of In^-
$pH = pK_a - 1$	$\frac{[In^-]}{[HIn]} = 10^{-1} = 0.10$	Color of HIn

When the pH of the solution equals the pK_a of the indicator, the solution will have an intermediate color. When the pH is 1 unit (or more) above pK_a, the indicator will be the color of In^-, and when the pH is 1 unit (or more) below pK_a, the indicator will be the color of HIn. As we can see, the indicator changes color within a range of two pH units centered at pK_a (Figure 18.13 ▶). Table 18.1 shows various indicators and their colors as a function of pH.

Indicator Color Change: Methyl Red

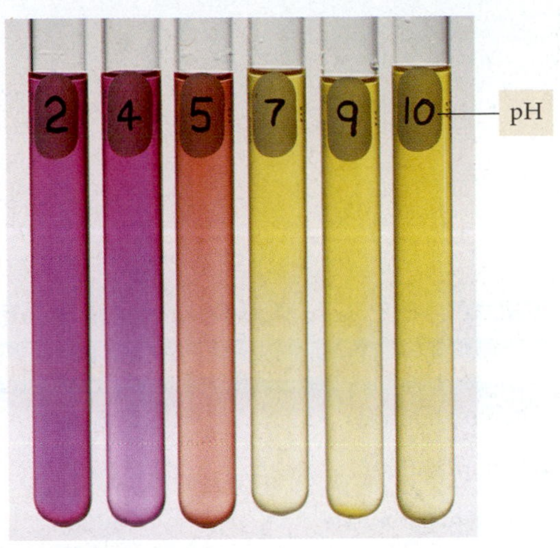

▲ **FIGURE 18.13 Indicator Color Change** An indicator (in this case, methyl red) generally changes color within a range of two pH units. (The pH for each solution is marked on its test tube.)

TABLE 18.1 Ranges of Color Changes for Several Acid–Base Indicators

	pH												
	0	1	2	3	4	5	6	7	8	9	10	11	12
Crystal Violet													
Thymol Blue													
Erythrosin B													
2,4-Dinitrophenol													
Bromphenol Blue													
Bromocresol Green													
Methyl Red													
Eriochrome* Black T													
Bromocresol Purple													
Alizarin													
Bromthymol Blue													
Phenol Red													
m-Nitrophenol													
o-Cresolphthalein													
Phenolphthalein													
Thymolphthalein													
Alizarin yellow R													

*Trademark of CIBA GEIGY CORP.

18.5 Solubility Equilibria and the Solubility-Product Constant

Recall from Chapter 9 that a compound is considered *soluble* if it dissolves in water and *insoluble* if it does not. Recall also that the *solubility rules* in Table 9.1 allow us to classify ionic compounds simply as soluble or insoluble. Now we have the tools to examine *degrees* of solubility.

We can better understand the solubility of an ionic compound by applying the concept of equilibrium to the process of dissolution. For example, we can represent the dissolution of calcium fluoride in water as an equilibrium:

$$CaF_2(s) \rightleftharpoons Ca^{2+}(aq) + 2\,F^-(aq)$$

The equilibrium constant for a chemical equation representing the dissolution of an ionic compound is the **solubility-product constant (K_{sp})**. For CaF_2, the expression of the solubility-product constant is:

$$K_{sp} = [Ca^{2+}][F^-]^2$$

Notice that, as we discussed in Section 16.5, solids are omitted from the equilibrium expression because the concentration of a solid is constant (it is determined by its density and does not change when more solid is added).

The value of K_{sp} is a measure of the solubility of a compound. Table 18.2 lists the values of K_{sp} for a number of ionic compounds. Appendix IV C includes a more complete list.

TABLE 18.2 Selected Solubility-Product Constants (K_{sp}) at 25 °C

Compound	Formula	K_{sp}	Compound	Formula	K_{sp}
Barium fluoride	BaF_2	2.45×10^{-5}	Lead(II) chloride	$PbCl_2$	1.17×10^{-5}
Barium sulfate	$BaSO_4$	1.07×10^{-10}	Lead(II) bromide	$PbBr_2$	4.67×10^{-6}
Calcium carbonate	$CaCO_3$	4.96×10^{-9}	Lead(II) sulfate	$PbSO_4$	1.82×10^{-8}
Calcium fluoride	CaF_2	1.46×10^{-10}	Lead(II) sulfide*	PbS	9.04×10^{-29}
Calcium hydroxide	$Ca(OH)_2$	4.68×10^{-6}	Magnesium carbonate	$MgCO_3$	6.82×10^{-6}
Calcium sulfate	$CaSO_4$	7.10×10^{-5}	Magnesium hydroxide	$Mg(OH)_2$	2.06×10^{-13}
Copper(II) sulfide*	CuS	1.27×10^{-36}	Silver chloride	$AgCl$	1.77×10^{-10}
Iron(II) carbonate	$FeCO_3$	3.07×10^{-11}	Silver chromate	Ag_2CrO_4	1.12×10^{-12}
Iron(II) hydroxide	$Fe(OH)_2$	4.87×10^{-17}	Silver bromide	$AgBr$	5.36×10^{-13}
Iron(II) sulfide*	FeS	3.72×10^{-19}	Silver iodide	AgI	8.51×10^{-17}

*Sulfide equilibrium is of the type: $MS(s) + H_2O(l) \rightleftharpoons M^{2+}(aq) + HS^-(aq) + OH^-(aq)$.

K_{sp} and Molar Solubility

Recall from Section 14.2 that the *solubility* of a compound is the quantity of the compound that dissolves in a certain amount of liquid. The **molar solubility** is the solubility in units of moles per liter (mol/L). We can calculate the molar solubility of a compound directly from K_{sp}. Consider silver chloride:

$$AgCl(s) \rightleftharpoons Ag^+(aq) + Cl^-(aq) \quad K_{sp} = 1.77 \times 10^{-10}$$

Notice that K_{sp} is *not* the molar solubility but the solubility-product constant. The solubility-product constant has only one value at a given temperature. The solubility, however, can have different values in different kinds of solutions. For example, due to the common ion effect, the solubility of AgCl in pure water is different from its solubility in an NaCl solution, even though the solubility-product constant is the same for both solutions. Notice also that the solubility of AgCl is directly related (by the reaction stoichiometry) to the amount of Ag^+ or Cl^- present in solution when equilibrium is reached. Consequently, determining molar solubility from K_{sp} involves solving an equilibrium problem. For AgCl, we set up an ICE table for the dissolution of AgCl into its ions in pure water:

$$AgCl(s) \rightleftharpoons Ag^+(aq) + Cl^-(aq)$$

	$[Ag^+]$	$[Cl^-]$
Initial	0.00	0.00
Change	$+S$	$+S$
Equil	S	S

We let S represent the concentration of AgCl that dissolves (which is the molar solubility) and then represent the concentrations of the ions formed in terms of S. In this case, for every 1 mol of AgCl that dissolves, 1 mol of Ag^+ and 1 mol of Cl^- are produced. Therefore, the concentrations of Ag^+ and Cl^- present in solution are equal to S. Substituting the equilibrium concentrations of Ag^+ and Cl^- into the expression for the solubility-product constant, we get:

Alternatively, we can use the variable x in place of S, as we have in other equilibrium calculations.

$$K_{sp} = [Ag^+][Cl^-] = S \times S = S^2$$

Therefore,

$$S = \sqrt{K_{sp}} = \sqrt{1.77 \times 10^{-10}} = 1.33 \times 10^{-5}\ M$$

The molar solubility of AgCl is 1.33×10^{-5} mol per liter.

EXAMPLE 18.8
Calculating Molar Solubility from K_{sp}

Calculate the molar solubility of $PbCl_2$ in pure water.

SOLUTION

Begin by writing the reaction by which solid $PbCl_2$ dissolves into its constituent aqueous ions and write the corresponding expression for K_{sp}.

$$PbCl_2(s) \rightleftharpoons Pb^{2+}(aq) + 2\,Cl^-(aq)$$
$$K_{sp} = [Pb^{2+}][Cl^-]^2$$

Refer to the stoichiometry of the reaction and prepare an ICE table, showing the equilibrium concentrations of Pb^{2+} and Cl^- relative to S, the amount of $PbCl_2$ that dissolves.

$$PbCl_2(s) \rightleftharpoons Pb^{2+}(aq) + 2\,Cl^-(aq)$$

	$[Pb^{2+}]$	$[Cl^-]$
Initial	0.00	0.00
Change	$+S$	$+2S$
Equil	S	$2S$

Substitute the equilibrium expressions for $[Pb^{2+}]$ and $[Cl^-]$ from the previous step into the expression for K_{sp}.

$$K_{sp} = [Pb^{2+}][Cl^-]^2 = S(2S)^2 = 4S^3$$

Solve for S and substitute the numerical value of K_{sp} (from Table 18.2) to calculate S.

$$S = \sqrt[3]{\frac{K_{sp}}{4}}$$

$$S = \sqrt[3]{\frac{1.17 \times 10^{-5}}{4}} = 1.43 \times 10^{-2}\ M$$

FOR PRACTICE 18.8

Calculate the molar solubility of $Fe(OH)_2$ in pure water.

EXAMPLE 18.9
Calculating K_{sp} from Molar Solubility

The molar solubility of Ag_2SO_4 in pure water is 1.2×10^{-5} M. Calculate K_{sp}.

SOLUTION

Begin by writing the reaction by which solid Ag_2SO_4 dissolves into its constituent aqueous ions; then write the corresponding expression for K_{sp}.

$$Ag_2SO_4(s) \rightleftharpoons 2\,Ag^+(aq) + SO_4^{2-}(aq)$$
$$K_{sp} = [Ag^+]^2[SO_4^{2-}]$$

Use an ICE table to define $[Ag^+]$ and $[SO_4^{2-}]$ in terms of S, the amount of Ag_2SO_4 that dissolves.

$$Ag_2SO_4(s) \rightleftharpoons 2\,Ag^+(aq) + SO_4^{2-}(aq)$$

	$[Ag^+]$	$[SO_4^{2-}]$
Initial	0.00	0.00
Change	$+2S$	$+S$
Equil	$2S$	S

Substitute the expressions for $[Ag^+]$ and $[SO_4^{2-}]$ from the previous step into the expression for K_{sp}. Substitute the given value *of the molar solubility for S and calculate* K_{sp}.

$$\begin{aligned} K_{sp} &= [Ag^+]^2[SO_4^{2-}] \\ &= (2S)^2S \\ &= 4S^3 \\ &= 4(1.2 \times 10^{-5})^3 \\ &= 6.9 \times 10^{-15} \end{aligned}$$

FOR PRACTICE 18.9

The molar solubility of AgBr in pure water is 7.3×10^{-7} M. Calculate K_{sp}.

K_{sp} and Relative Solubility

As we have just seen, molar solubility and K_{sp} are related, and we can calculate each from the other; however, we cannot generally use the K_{sp} values of two different compounds to directly compare their relative solubilities. For example, consider the following compounds, their K_{sp} values, and their molar solubilities:

Compound	K_{sp}	Solubility
$Mg(OH)_2$	2.06×10^{-13}	3.72×10^{-5} M
$FeCO_3$	3.07×10^{-11}	5.54×10^{-6} M

Magnesium hydroxide has a smaller K_{sp} than iron(II) carbonate, but a higher molar solubility. Why? The relationship between K_{sp} and molar solubility depends on the stoichiometry of the dissociation reaction. Consequently, we can only make a direct comparison of K_{sp} values for different compounds if the compounds have the same dissociation stoichiometry. Consider the following compounds with the same dissociation stoichiometry, their K_{sp} values, and their molar solubilities:

Compound	K_{sp}	Solubility
$Mg(OH)_2$	2.06×10^{-13}	3.72×10^{-5} M
CaF_2	1.46×10^{-10}	3.32×10^{-4} M

In this case, magnesium hydroxide and calcium fluoride have the same dissociation stoichiometry (1 mol of each compound produces 3 mol of dissolved ions); therefore, we can directly compare the K_{sp} values as a measure of relative solubility.

The Effect of a Common Ion on Solubility

How is the solubility of an ionic compound affected when the compound is dissolved in a solution that already contains one of its ions? For example, what is the solubility of CaF_2 in a solution that is 0.100 M in NaF? We can determine the change in solubility by considering the common ion effect, which we first encountered in Section 18.2. We represent the dissociation of CaF_2 in a 0.100 M NaF solution as shown at right.

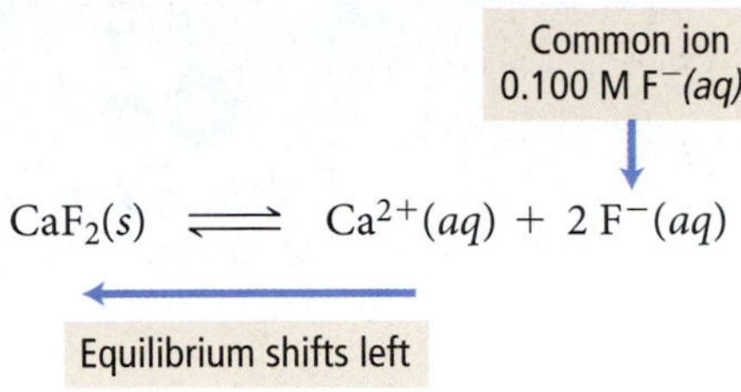

In accordance with Le Châtelier's principle, the presence of the F^- ion in solution causes the equilibrium to shift to the left (compared to its position in pure water), which means that less CaF_2 dissolves—that is, its solubility is decreased.

In general, the solubility of an ionic compound is lower in a solution containing a common ion than in pure water.

We can calculate the exact value of the solubility by working an equilibrium problem in which the concentration of the common ion is accounted for in the initial conditions, as shown in Example 18.10.

EXAMPLE 18.10
Calculating Molar Solubility in the Presence of a Common Ion

What is the molar solubility of CaF_2 in a solution containing 0.100 M NaF?

SOLUTION

Begin by writing the reaction by which solid CaF_2 dissolves into its constituent aqueous ions. Write the corresponding expression for K_{sp}.

$$CaF_2(s) \rightleftharpoons Ca^{2+}(aq) + 2\,F^-(aq)$$
$$K_{sp} = [Ca^{2+}][F^-]^2$$

Use the stoichiometry of the reaction to prepare an ICE table showing the initial concentration of the common ion. Fill in the equilibrium concentrations of Ca^{2+} and F^- relative to S, the amount of CaF_2 that dissolves.

$$CaF_2(s) \rightleftharpoons Ca^{2+}(aq) + 2\,F^-(aq)$$

	$[Ca^{2+}]$	$[F^-]$
Initial	0.00	0.100
Change	$+S$	$+2S$
Equil	S	$0.100 + 2S$

Substitute the equilibrium expressions for $[Ca^{2+}]$ and $[F^-]$ from the previous step into the expression for K_{sp}. Since K_{sp} is small, you can make the approximation that $2S$ is much less than 0.100 and will therefore be insignificant when added to 0.100 (this is similar to the *x is small* approximation in equilibrium problems).

$$K_{sp} = [Ca^{2+}][F^-]^2$$
$$= S(0.100 + \cancel{2S})^2 \quad (S \textit{ is small})$$
$$= S(0.100)^2$$

Solve for S and substitute the numerical value of K_{sp} (from Table 18.2) to calculate S.

Note that the calculated value of S is indeed small compared to 0.100; your approximation is valid.

$$K_{sp} = S(0.100)^2$$
$$S = \frac{K_{sp}}{0.0100} = \frac{1.46 \times 10^{-10}}{0.0100} = 1.46 \times 10^{-8}\ \text{M}$$

For comparison, the molar solubility of CaF_2 in pure water is 3.32×10^{-4} M, which means CaF_2 is over 20,000 times more soluble in water than in the NaF solution. (Confirm this for yourself by calculating its solubility in pure water from the value of K_{sp}).

FOR PRACTICE 18.10

Calculate the molar solubility of CaF_2 in a solution containing 0.250 M $Ca(NO_3)_2$.

18.8
Cc
Conceptual Connection

Common Ion Effect

In which solution is $BaSO_4$ most soluble?

(a) a solution that is 0.10 M in $Ba(NO_3)_2$
(b) a solution that is 0.10 M in Na_2SO_4
(c) a solution that is 0.10 M in $NaNO_3$

The Effect of pH on Solubility

The pH of a solution can affect the solubility of a compound in that solution. For example, consider the dissociation of $Mg(OH)_2$, the active ingredient in milk of magnesia:

$$Mg(OH)_2(s) \rightleftharpoons Mg^{2+}(aq) + 2\,OH^-(aq)$$

The solubility of this compound is highly dependent on the pH of the solution into which it dissolves. If the pH is high, then the concentration of OH^- in the solution is high. In accordance with the common ion effect, this shifts the equilibrium to the left, lowering the solubility:

High $[OH^-]$

$$Mg(OH)_2(s) \rightleftharpoons Mg^{2+}(aq) + 2\,OH^-(aq)$$

Equilibrium shifts left

If the pH is low, then the concentration of $H_3O^+(aq)$ in the solution is high. As the $Mg(OH)_2$ dissolves, these H_3O^+ ions neutralize the newly dissolved OH^- ions, driving the reaction to the right:

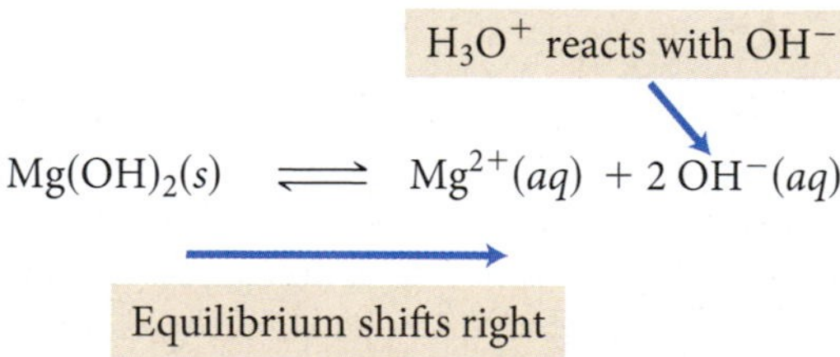

Consequently, the solubility of $Mg(OH)_2$ in an acidic solution is higher than in a pH-neutral or basic solution.

In general, the solubility of an ionic compound with a strongly basic or weakly basic anion increases with increasing acidity (decreasing pH).

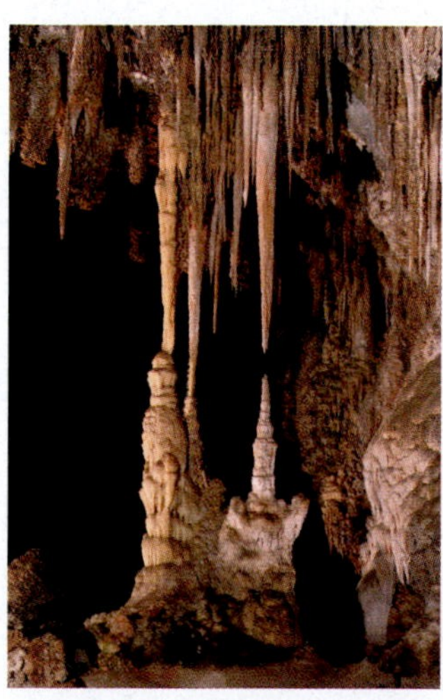

▲ Stalactites (which hang from the ceiling) and stalagmites (which grow up from the ground) form as calcium carbonate precipitates out of the water evaporating in underground caves.

Common basic anions include OH^-, S^{2-}, and CO_3^{2-}. Therefore, hydroxides, sulfides, and carbonates are more soluble in acidic water than in pure water. Since rainwater is naturally acidic due to dissolved carbon dioxide, it can dissolve rocks high in limestone ($CaCO_3$) as it flows through the ground, sometimes resulting in huge underground caverns such as those at Carlsbad Caverns National Park in New Mexico. Dripping water saturated in $CaCO_3$ within the cave creates the dramatic mineral formations known as stalagmites and stalactites.

EXAMPLE 18.11
The Effect of pH on Solubility

Determine whether each compound is more soluble in an acidic solution than in a neutral solution.

(a) BaF_2 **(b)** AgI **(c)** $Ca(OH)_2$

SOLUTION

(a) BaF_2 is more soluble in acidic solution because the F^- ion is a weak base. (F^- is the conjugate base of the weak acid HF and is therefore a weak base.)

(b) The solubility of AgI is not greater in acidic solution because the I^- is *not* a base. (I^- is the conjugate base of the *strong* acid HI and is therefore pH-neutral.)

(c) $Ca(OH)_2$ is more soluble in acidic solution because the OH^- ion is a strong base.

FOR PRACTICE 18.11

Which compound, $FeCO_3$ or $PbBr_2$, is more soluble in acid than in base? Why?

18.6 Precipitation

In Chapter 9, we learned that a precipitation reaction can occur upon the mixing of two solutions containing ionic compounds when one of the possible cross products—the combination of a cation from one solution and the anion from the other—is insoluble. In this chapter, however, we have seen that the terms *soluble* and *insoluble* are extremes in a continuous range of solubility—many compounds are slightly soluble and even those that we categorized as insoluble in Chapter 9 actually have some limited degree of solubility (they have very small solubility-product constants).

We can better understand precipitation reactions by revisiting a concept from Chapter 16—the reaction quotient (Q). The reaction quotient for the reaction by which an ionic compound dissolves is the product of the concentrations of the ionic components raised to their stoichiometric coefficients. For example, consider the reaction by which CaF_2 dissolves:

$$CaF_2(s) \rightleftharpoons Ca^{2+}(aq) + 2\,F^-(aq)$$

The reaction quotient for this reaction is:

$$Q = [Ca^{2+}][F^-]^2$$

The difference between Q and K_{sp} is that K_{sp} is the value of this product *at equilibrium only*, whereas Q is the value of the product under any conditions. We can therefore use the value of Q to compare a solution containing any concentrations of the component ions to a solution that is at equilibrium.

Consider a solution of calcium fluoride in which Q is less than K_{sp}. Recall from Chapter 16 that if Q is less than K_{sp}, the reaction will proceed to the right (toward products). Consequently, if the solution contains any solid CaF_2, the CaF_2 will continue to dissolve. If all of the solid has already dissolved, the solution will simply remain as it is, containing less than the equilibrium amount of the dissolved ions. Such a solution is an *unsaturated solution.* If more solid is added to an unsaturated solution, it will dissolve, as long as Q remains less than K_{sp}.

Now consider a solution in which Q is exactly equal to K_{sp}. In this case, the reaction is at equilibrium and will not make progress in either direction. Such a solution most likely contains at least a small amount of the solid in equilibrium with its component ions. However, the amount of solid may be too small to be visible. Such a solution is a *saturated solution.*

Finally, consider a solution in which Q is greater than K_{sp}. In this case, the reaction will proceed to the left (toward the reactants) and solid calcium fluoride will form from the dissolved calcium and fluoride ions. In other words, the solid normally precipitates out of a solution in which Q is greater than K_{sp}.

▲ **FIGURE 18.14 Precipitation from a Supersaturated Solution** The excess solute in a supersaturated solution of sodium acetate precipitates out if a small sodium acetate crystal is added.

Under certain circumstances, however, Q can remain greater than K_{sp} for an unlimited period of time. Such a solution, called a *supersaturated solution,* is unstable and will form a precipitate when sufficiently disturbed. Figure 18.14 ◄ shows a supersaturated solution of sodium acetate. When a small seed crystal of solid sodium acetate is dropped into the solution, it triggers the precipitation reaction.

Summarizing the Relationship of Q and K_{sp} in Solutions Containing an Ionic Compound:

- If $Q < K_{sp}$, the solution is unsaturated and more of the solid ionic compound can dissolve in the solution.
- If $Q = K_{sp}$, the solution is saturated. The solution is holding the equilibrium amount of the dissolved ions and additional solid does not dissolve in the solution.
- If $Q > K_{sp}$, the solution is supersaturated. Under most circumstances, the excess solid precipitates out of a supersaturated solution.

We can use Q to predict whether a precipitation reaction will occur upon the mixing of two solutions containing dissolved ionic compounds. For example, consider mixing a silver nitrate solution with a potassium iodide solution to form a mixture that is 0.010 M in $AgNO_3$ and 0.015 M in KI. Will a precipitate form in the newly mixed solution? From Chapter 9 we know that one of the cross products, KNO_3, is soluble and will therefore not precipitate. The other cross product, AgI, *may* precipitate if the concentrations of Ag^+ and I^- are high enough in the newly mixed solution: We can compare Q to K_{sp} to determine whether a precipitate will form. For AgI, $K_{sp} = 8.51 \times 10^{-17}$. For the newly mixed solution, $[Ag^+] = 0.010$ M and $[I^-] = 0.015$ M. We calculate Q as follows:

$$Q = [Ag^+][I^-] = (0.010)(0.015) = 1.5 \times 10^{-4}$$

The value of Q is much greater than K_{sp}; therefore, AgI should precipitate out of the newly mixed solution.

EXAMPLE 18.12

Predicting Precipitation Reactions by Comparing Q and K_{sp}

A solution containing lead(II) nitrate is mixed with one containing sodium bromide to form a solution that is 0.0150 M in $Pb(NO_3)_2$ and 0.00350 M in NaBr. Does a precipitate form in the newly mixed solution?

SOLUTION

First, determine the possible cross products and their K_{sp} values (Table 18.2). Any cross products that are soluble will *not* precipitate (see Table 9.1).	Possible cross products: $NaNO_3$ soluble $PbBr_2$ $K_{sp} = 4.67 \times 10^{-6}$
Calculate Q and compare it to K_{sp}. A precipitate only forms if $Q > K_{sp}$.	$Q = [Pb^{2+}][Br^-]^2$ $= (0.0150)(0.00350)^2$ $= 1.84 \times 10^{-7}$ $Q < K_{sp}$; therefore no precipitate forms.

FOR PRACTICE 18.12

If the original solutions in Example 18.12 are concentrated through evaporation and mixed again to form a solution that is 0.0600 M in $Pb(NO_3)_2$ and 0.0158 M in NaBr, does a precipitate form in the newly mixed solution?

Selective Precipitation

A solution may contain several different dissolved metal cations that we can separate by **selective precipitation**, a process involving the addition of a reagent that forms a precipitate with one of the dissolved cations but not the others. For example, seawater contains dissolved magnesium and calcium cations with the concentrations $[Mg^{2+}] = 0.059$ M and $[Ca^{2+}] = 0.011$ M. We can separate these ions by adding a reagent that precipitates one of the ions but not the other. From Table 18.2, we find that $Mg(OH)_2$ has a K_{sp} of 2.06×10^{-13} and that $Ca(OH)_2$ has a K_{sp} of 4.68×10^{-6}, indicating that the hydroxide ion forms a precipitate with magnesium at a much lower concentration than it does with calcium. Consequently, a soluble hydroxide—such as KOH or NaOH—is a good choice for the precipitating reagent. When we add an appropriate amount of KOH or NaOH to seawater, the hydroxide ion causes the precipitation of $Mg(OH)_2$ (the compound with the lowest K_{sp}) but not $Ca(OH)_2$. Calculations for this selective precipitation are shown in Examples 18.13 and 18.14. In these calculations, we compare Q to K_{sp} to determine the concentration that triggers precipitation.

For selective precipitation to work, K_{sp} values must differ by at least a factor of 10^3.

EXAMPLE 18.13

Finding the Minimum Required Reagent Concentration for Selective Precipitation

The magnesium and calcium ions present in seawater ($[Mg^{2+}] = 0.059$ M and $[Ca^{2+}] = 0.011$ M) can be separated by selective precipitation with KOH. What minimum $[OH^-]$ triggers the precipitation of the Mg^{2+} ion?

SOLUTION

The precipitation commences when the value of Q for the precipitating compound just equals the value of K_{sp}. Set the expression for Q for magnesium hydroxide equal to the value of K_{sp}, and solve for $[OH^-]$. This is the concentration above which $Mg(OH)_2$ precipitates.

$$Q = [Mg^{2+}][OH^-]^2$$
$$= (0.059)[OH^-]^2$$

When $Q = K_{sp}$,

$$(0.059)[OH^-]^2 = K_{sp} = 2.06 \times 10^{-13}$$

$$[OH^-]^2 = \frac{2.06 \times 10^{-13}}{0.059}$$

$$[OH^-] = 1.9 \times 10^{-6}\ \text{M}$$

FOR PRACTICE 18.13

If the concentration of Mg^{2+} in the solution in Example 18.13 were 0.025 M, what minimum $[OH^-]$ triggers precipitation of the Mg^{2+} ion?

EXAMPLE 18.14

Finding the Concentrations of Ions Left in Solution after Selective Precipitation

You add potassium hydroxide to the solution in Example 18.13. When the $[OH^-]$ reaches 1.9×10^{-6} M (as you just calculated), magnesium hydroxide begins to precipitate out of solution. As you continue to add KOH, the magnesium hydroxide continues to precipitate. At some point, the $[OH^-]$ becomes high enough to begin to precipitate the calcium ions as well. What is the concentration of Mg^{2+} when Ca^{2+} begins to precipitate?

SOLUTION

First, calculate the OH^- concentration at which Ca^{2+} begins to precipitate by writing the expression for Q for calcium hydroxide and substituting the concentration of Ca^{2+} from Example 18.13.

$$Q = [Ca^{2+}][OH^-]^2$$
$$= (0.011)[OH^-]^2$$

—Continued on the next page

Continued from the previous page—

Set the expression for Q equal to the value of K_{sp} for calcium hydroxide and solve for $[OH^-]$. $Ca(OH)_2$ precipitates above this concentration.	When $Q = K_{sp}$, $(0.011)[OH^-]^2 = K_{sp} = 4.68 \times 10^{-6}$ $[OH^-]^2 = \frac{4.68 \times 10^{-6}}{0.011}$ $[OH^-] = 2.\underline{0}6 \times 10^{-2}$ M
Find the concentration of Mg^{2+} when OH^- reaches the concentration you just calculated by writing the expression for Q *for magnesium hydroxide* and substituting the concentration of OH^- that you just calculated. Then set the expression for Q equal to the value of K_{sp} for magnesium hydroxide and solve for $[Mg^{2+}]$. This is the concentration of Mg^{2+} that remains when $Ca(OH)_2$ begins to precipitate.	$Q = [Mg^{2+}][OH^-]^2$ $= [Mg^{2+}](2.\underline{0}6 \times 10^{-2})^2$ When $Q = K_{sp}$, $[Mg^{2+}](2.\underline{0}6 \times 10^{-2})^2 = K_{sp} = 2.06 \times 10^{-13}$ $[Mg^{2+}] = \frac{2.06 \times 10^{-13}}{(2.\underline{0}6 \times 10^{-2})^2}$ $[Mg^{2+}] = 4.9 \times 10^{-10}$ M

As you can see from the results, the selective precipitation worked very well. The concentration of Mg^{2+} dropped from 0.059 M to 4.9×10^{-10} M before any calcium began to precipitate, which means that 99.99% of the magnesium separated out of the solution.

FOR PRACTICE 18.14

A solution is 0.085 M in Pb^{2+} and 0.025 M in Ag^+. (a) If selective precipitation is to be achieved using NaCl, what minimum concentration of NaCl do you need to begin to precipitate the ion that precipitates first? (b) What is the concentration of each ion left in solution at the point when the second ion begins to precipitate?

18.7 Complex Ion Equilibria

We have discussed several different types of equilibria so far, including acid–base equilibria and solubility equilibria. We now turn to equilibria of another type, which primarily involve transition metal ions in solution. Transition metal ions tend to be good electron acceptors (good Lewis acids). In aqueous solutions, water molecules can act as electron donors (Lewis bases) to hydrate transition metal ions. For example, silver ions are hydrated by water in solution to form $Ag(H_2O)_2^+(aq)$. Chemists often write $Ag^+(aq)$ as a shorthand notation for the hydrated silver ion, but the bare ion does not really exist by itself in solution.

Species such as $Ag(H_2O)_2^{2+}$ are known as *complex ions*. A **complex ion** contains a central metal ion bound to one or more *ligands*. A **ligand** is a neutral molecule or ion that acts as a Lewis base with the central metal ion. In $Ag(H_2O)_2^+$, water is the ligand. If we put a stronger Lewis base into a solution containing $Ag(H_2O)_2^+$, the stronger Lewis base displaces the water in the complex ion. For example, ammonia reacts with $Ag(H_2O)_2^+$ according to the following reaction:

We cover complex ions in more detail in Chapter 23. Here, we focus on the equilibria associated with their formation.

$$Ag(H_2O)_2^+(aq) + 2\,NH_3(aq) \rightleftharpoons Ag(NH_3)_2^+(aq) + 2\,H_2O(l)$$

For simplicity, we often leave water out of the equation:

$$Ag^+(aq) + 2\,NH_3(aq) \rightleftharpoons Ag(NH_3)_2^+(aq) \quad K_f = 1.7 \times 10^7$$

The equilibrium constant associated with the reaction for the formation of a complex ion, such as the one just shown, is the **formation constant (K_f)**. We determine the expression for K_f by the law of mass action, like any equilibrium constant. For $Ag(NH_3)_2^+$, the expression for K_f is:

$$K_f = \frac{[Ag(NH_3)_2{}^+]}{[Ag^+][NH_3]^2}$$

Notice that the value of K_f for $Ag(NH_3)_2{}^+$ is large, indicating that the formation of the complex ion is highly favored. Table 18.3 lists the formation constants for some common complex ions. Note that, in general, values of K_f are very large, indicating that the formation of complex ions is highly favored in each case. Example 18.15 illustrates the use K_f in calculations.

TABLE 18.3 Formation Constants of Selected Complex Ions in Water at 25 °C

Complex Ion	K_f	Complex Ion	K_f
$Ag(CN)_2{}^-$	1×10^{21}	$Cu(NH_3)_4{}^{2+}$	1.7×10^{13}
$Ag(NH_3)_2{}^+$	1.7×10^{7}	$Fe(CN)_6{}^{4-}$	1.5×10^{35}
$Ag(S_2O_3)_2{}^{3-}$	2.8×10^{13}	$Fe(CN)_6{}^{3-}$	2×10^{43}
$AlF_6{}^{3-}$	7×10^{19}	$Hg(CN)_4{}^{2-}$	1.8×10^{41}
$Al(OH)_4{}^-$	3×10^{33}	$HgCl_4{}^{2-}$	1.1×10^{16}
$CdBr_4{}^{2-}$	5.5×10^{3}	$HgI_4{}^{2-}$	2×10^{30}
$CdI_4{}^{2-}$	2×10^{6}	$Ni(NH_3)_6{}^{2+}$	2.0×10^{8}
$Cd(CN)_4{}^{2-}$	3×10^{18}	$Pb(OH)_3{}^-$	8×10^{13}
$Co(NH_3)_6{}^{3+}$	2.3×10^{33}	$Sn(OH)_3{}^-$	3×10^{25}
$Co(OH)_4{}^{2-}$	5×10^{9}	$Zn(CN)_4{}^{2-}$	2.1×10^{19}
$Co(SCN)_4{}^{2-}$	1×10^{3}	$Zn(NH_3)_4{}^{2+}$	2.8×10^{9}
$Cr(OH)_4{}^-$	8.0×10^{29}	$Zn(OH)_4{}^{2-}$	2×10^{15}
$Cu(CN)_4{}^{2-}$	1.0×10^{25}		

EXAMPLE 18.15

Complex Ion Equilibria

You mix a 200.0 mL sample of a solution that is 1.5×10^{-3} M in $Cu(NO_3)_2$ with a 250.0 mL sample of a solution that is 0.20 M in NH_3. After the solution reaches equilibrium, what concentration of $Cu^{2+}(aq)$ remains?

SOLUTION

Write the balanced equation for the complex ion equilibrium that occurs and look up the value of K_f in Table 18.3. Since this is an equilibrium problem, you have to create an ICE table, which requires the initial concentrations of Cu^{2+} and NH_3 in the combined solution. Calculate those concentrations from the given values.

$$Cu^{2+}(aq) + 4\,NH_3(aq) \rightleftharpoons Cu(NH_3)_4{}^{2+}(aq)$$

$$K_f = 1.7 \times 10^{13}$$

$$[Cu^{2+}]_{initial} = \frac{0.200\,\cancel{L} \times \dfrac{1.5 \times 10^{-3}\,mol}{\cancel{L}}}{0.200\,L + 0.250\,L} = 6.7 \times 10^{-4}\,M$$

$$[NH_3]_{initial} = \frac{0.250\,\cancel{L} \times \dfrac{0.20\,mol}{1\,\cancel{L}}}{0.200\,L + 0.250\,L} = 0.11\,M$$

—Continued on the next page

Continued from the previous page—

Construct an ICE table for the reaction and write down the initial concentrations of each species.

$$Cu^{2+}(aq) + 4\,NH_3(aq) \rightleftharpoons Cu(NH_3)_4^{\,2+}(aq)$$

	$[Cu^{2+}]$	$[NH_3]$	$[Cu(NH_3)_4^{\,2+}]$
Initial	6.7×10^{-4}	0.11	0.0
Change			
Equil			

Since the equilibrium constant is large and the concentration of ammonia is much larger than the concentration of Cu^{2+}, you can assume that the reaction will be driven to the right so that most of the Cu^{2+} is consumed. Unlike in previous ICE tables, where you let x represent the change in concentration in going to equilibrium, here you let x represent the small amount of Cu^{2+} that remains when equilibrium is reached.

$$Cu^{2+}(aq) + 4\,NH_3(aq) \rightleftharpoons Cu(NH_3)_4^{\,2+}(aq)$$

	$[Cu^{2+}]$	$[NH_3]$	$[Cu(NH_3)_4^{\,2+}]$
Initial	6.7×10^{-4}	0.11	0.0
Change	$\approx(-6.7 \times 10^{-4})$	$\approx 4(-6.7 \times 10^{-4})$	$\approx(+6.7 \times 10^{-4})$
Equil	x	0.11	6.7×10^{-4}

Substitute the expressions for the equilibrium concentrations into the expression for K_f and solve for x.

$$K_f = \frac{[Cu(NH_3)_4^{\,2+}]}{[Cu^{2+}][NH_3]^4}$$

$$= \frac{6.7 \times 10^{-4}}{x(0.11)^4}$$

$$x = \frac{6.7 \times 10^{-4}}{K_f(0.11)^4}$$

$$= \frac{6.7 \times 10^{-4}}{1.7 \times 10^{13}(0.11)^4}$$

$$= 2.7 \times 10^{-13}$$

Confirm that x is indeed small compared to the initial concentration of the metal cation.

The remaining Cu^{2+} is very small because the formation constant is very large.

Since $x = 2.7 \times 10^{-13} \ll 6.7 \times 10^{-4}$, the approximation is valid. The remaining $[Cu^{2+}] = 2.7 \times 10^{-13}$ M.

FOR PRACTICE 18.15

You mix a 125.0 mL sample of a solution that is 0.0117 M in $NiCl_2$ with a 175.0 mL sample of a solution that is 0.250 M in NH_3. After the solution reaches equilibrium, what concentration of $Ni^{2+}(aq)$ remains?

The Effect of Complex Ion Equilibria on Solubility

Recall from Section 18.5 that the solubility of an ionic compound with a basic anion increases with increasing acidity because the acid reacts with the anion and drives the reaction to the right. Similarly, *the solubility of an ionic compound containing a metal cation that forms complex ions increases in the presence of Lewis bases that form complex ions with the cation.* The most common Lewis bases that increase the solubility of metal cations are NH_3, CN^-, and OH^-. For example, silver chloride is only slightly soluble in pure water:

$$AgCl(s) \rightleftharpoons Ag^+(aq) + Cl^-(aq) \qquad K_{sp} = 1.77 \times 10^{-10}$$

However, adding ammonia increases its solubility dramatically because, as we saw previously in this section, the ammonia forms a complex ion with the silver cations:

$$Ag^+(aq) + 2\,NH_3(aq) \rightleftharpoons Ag(NH_3)_2^{\,+}(aq) \qquad K_f = 1.7 \times 10^7$$

The large value of K_f significantly lowers the concentration of $Ag^+(aq)$ in solution and therefore drives the dissolution of $AgCl(s)$. The two previous reactions can be added together:

$$\begin{array}{ll} AgCl(s) \rightleftharpoons \cancel{Ag^+}(aq) + Cl^-(aq) & K_{sp} = 1.77 \times 10^{-10} \\ \cancel{Ag^+}(aq) + 2\,NH_3(aq) \rightleftharpoons Ag(NH_3)_2^+(aq) & K_f = 1.7 \times 10^7 \\ \hline AgCl(s) + 2\,NH_3(aq) \rightleftharpoons Ag(NH_3)_2^+(aq) + Cl^-(aq) & K = K_{sp} \times K_f = 3.0 \times 10^{-3} \end{array}$$

As we learned in Section 16.3, the equilibrium constant for a reaction that is the sum of two other reactions is the product of the equilibrium constants for the two other reactions. Adding ammonia changes the equilibrium constant for the dissolution of $AgCl(s)$ by a factor of $3.0 \times 10^{-3}/1.77 \times 10^{-10} = 1.7 \times 10^7$ (17 million), which makes the otherwise relatively insoluble $AgCl(s)$ quite soluble, as shown in Figure 18.15 ▼.

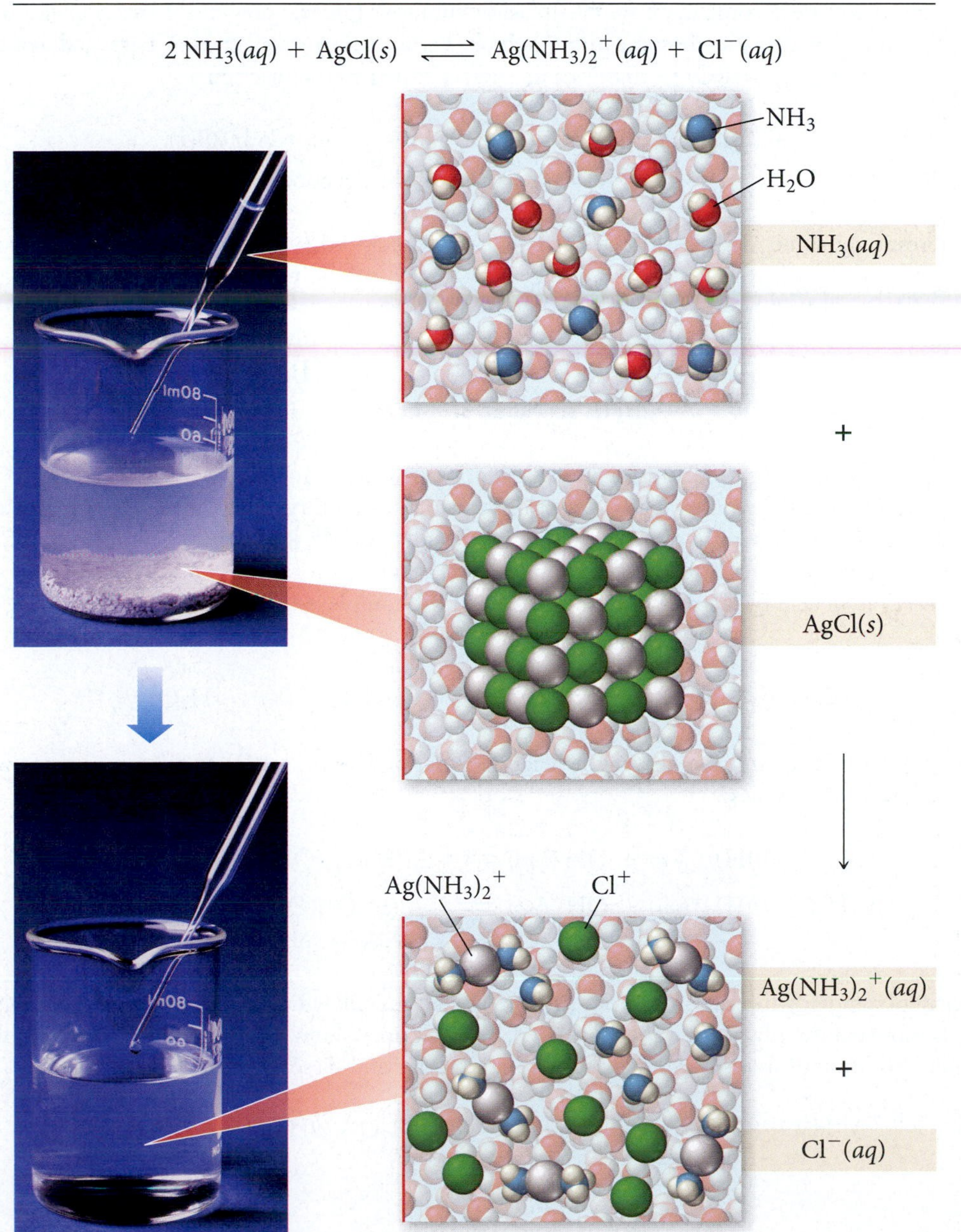

◀ **FIGURE 18.15 Complex Ion Formation** Normally insoluble AgCl is made soluble by the addition of NH_3, which forms a complex ion with Ag^+ and dissolves the AgCl.

18.9
Cc
Conceptual Connection

Solubility and Complex Ion Equilibria

Which compound, when added to water, is most likely to increase the solubility of CuS?

(a) NaCl
(b) KNO_3
(c) NaCN
(d) $MgBr_2$

The Solubility of Amphoteric Metal Hydroxides

Many metal hydroxides are insoluble or only very slightly soluble in pH-neutral water. For example, $Al(OH)_3$ has $K_{sp} = 2 \times 10^{-32}$, which means that if we put $Al(OH)_3$ in water, the vast majority of it will settle to the bottom as an undissolved solid. All metal hydroxides, however, have a basic anion (OH^-) and therefore become more soluble in acidic solutions (see the previous subsection and Section 18.5). The metal hydroxides become more soluble because they can act as a base and react with $H_3O^+(aq)$. For example, $Al(OH)_3$ dissolves in acid according to the reaction:

$$Al(OH)_3(s) + 3\,H_3O^+(aq) \longrightarrow Al^{3+}(aq) + 6\,H_2O(l)$$

$Al(OH)_3$ acts as a base in this reaction.

Recall from Section 17.3 that a substance that can act as either an acid or a base is said to be amphoteric.

Interestingly, some metal hydroxides can also act as acids—they are *amphoteric*. The ability of an amphoteric metal hydroxide to act as an acid increases its solubility in basic solution. For example, $Al(OH)_3(s)$ dissolves in basic solution according to the reaction:

$$Al(OH)_3(s) + OH^-(aq) \longrightarrow Al(OH)_4{}^-(aq)$$

$Al(OH)_3$ acts as an acid in this reaction.

$Al(OH)_3$ is soluble at high pH and soluble at low pH but *insoluble* in a pH-neutral solution.

We can observe the whole range of the pH-dependent solubility behavior of Al^{3+} by considering a hydrated aluminum ion in solution, beginning at an acidic pH. We know from Section 17.9 that Al^{3+} in solution is inherently acidic because it complexes with water to form $Al(H_2O)_6{}^{3+}(aq)$. The complex ion then acts as an acid by losing a proton from one of the complexed water molecules according to the reaction:

$$Al(H_2O)_6{}^{3+}(aq) + H_2O(l) \rightleftharpoons Al(H_2O)_5(OH)^{2+}(aq) + H_3O^+(aq)$$

Addition of base to the solution drives the reaction to the right and continues to remove protons from complexed water molecules:

$$Al(H_2O)_5(OH)^{2+}(aq) + OH^-(aq) \rightleftharpoons Al(H_2O)_4(OH)_2{}^+(aq) + H_2O(l)$$

$$Al(H_2O)_4(OH)_2{}^+(aq) + OH^-(aq) \rightleftharpoons \underbrace{Al(H_2O)_3(OH)_3(s)}_{\text{equivalent to } Al(OH)_3(s)} + H_2O(l)$$

The result of removing three protons from $Al(H_2O)_6{}^{3+}$ is the solid white precipitate $Al(H_2O)_3(OH)_3(s)$, which is more commonly written as $Al(OH)_3(s)$. The solution is now pH-neutral, and the hydroxide is insoluble. Addition of more OH^- makes the solution basic and dissolves the solid precipitate:

$$Al(H_2O)_3(OH)_3(s) + OH^-(aq) \rightleftharpoons Al(H_2O)_2(OH)_4{}^-(aq) + H_2O(l)$$

As the solution goes from acidic to neutral to basic, the solubility of Al^{3+} changes accordingly, as illustrated in Figure 18.16 ▼.

The extent to which a metal hydroxide dissolves in both acid and base depends on the degree to which it is amphoteric. Cations that form amphoteric hydroxides include Al^{3+}, Cr^{3+}, Zn^{2+}, Pb^{2+}, and Sn^{2+}. Other metal hydroxides, such as those of Ca^{2+}, Fe^{2+}, and Fe^{3+}, are not amphoteric—they become soluble in acidic solutions but not in basic ones.

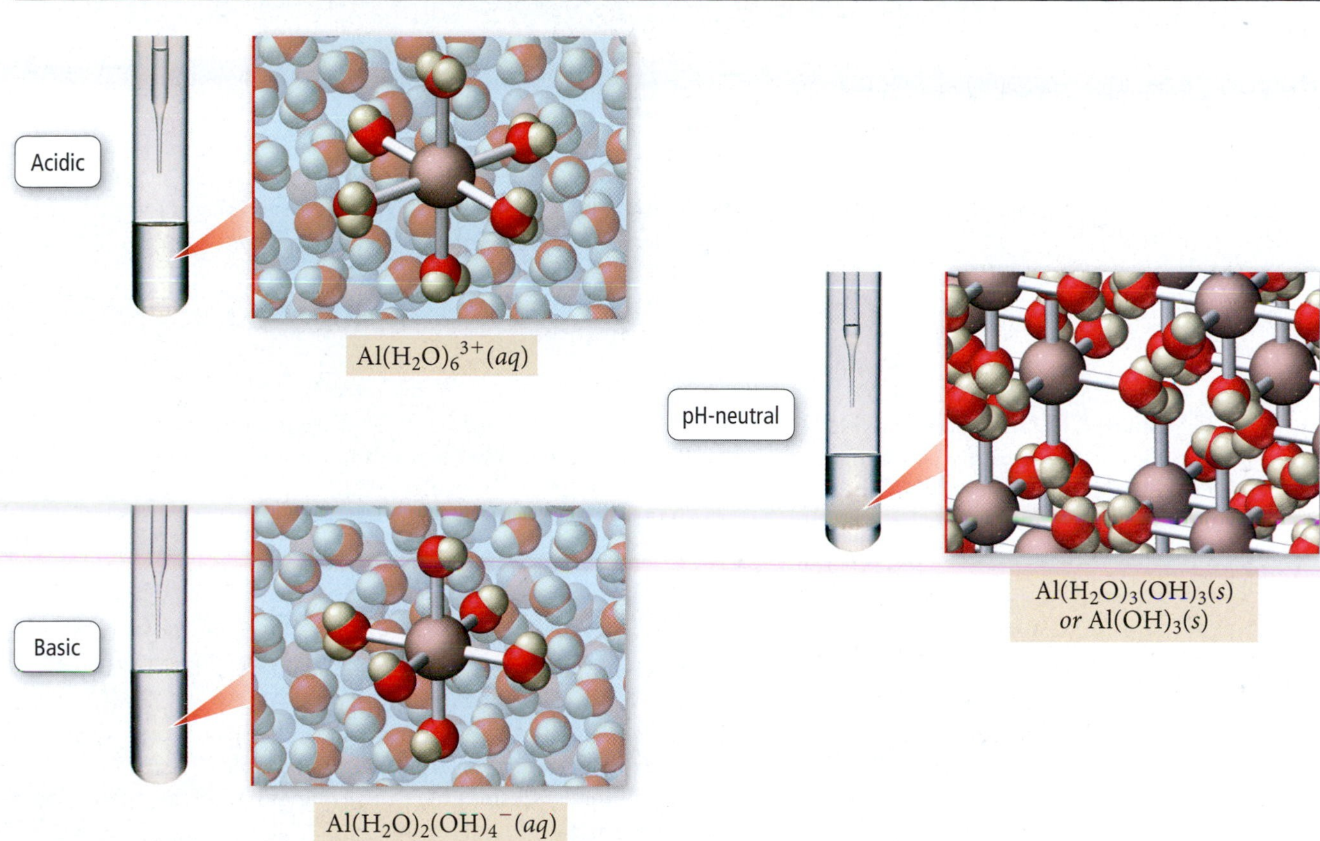

▲ **FIGURE 18.16 Solubility of an Amphoteric Hydroxide** Because aluminum hydroxide is amphoteric, its solubility is pH-dependent. At low pH, the formation of $Al(H_2O)_6{}^{3+}$ drives the dissolution. At neutral pH, insoluble $Al(OH)_3$ precipitates out of solution. At high pH, the formation of $Al(H_2O)_2(OH)_4{}^{-}$ drives the dissolution.

SELF-ASSESSMENT Quiz

1. A buffer is 0.100 M in NH_4Cl and 0.100 M in NH_3. When a small amount of hydrobromic acid is added to this buffer, which buffer component neutralizes the added acid?
   a) NH_4^+
   b) Cl^-
   c) NH_3
   d) None of the above (hydrobromic acid will not be neutralized by this buffer).

2. What is the pH of a buffer that is 0.120 M in formic acid ($HCHO_2$) and 0.080 M in potassium formate ($KCHO_2$)? For formic acid, $K_a = 1.8 \times 10^{-4}$.
   a) 2.33
   b) 3.57
   c) 3.74
   d) 3.91

3. A buffer with a pH of 9.85 contains CH_3NH_2 and CH_3NH_3Cl in water. What can you conclude about the relative concentrations of CH_3NH_2 and CH_3NH_3Cl in this buffer? For CH_3NH_2, $pK_b = 3.36$.
   a) $CH_3NH_2 > CH_3NH_3Cl$
   b) $CH_3NH_2 < CH_3NH_3Cl$
   c) $CH_3NH_2 = CH_3NH_3Cl$
   d) Nothing can be concluded about the relative concentrations of CH_3NH_2 and CH_3NH_3Cl.

4. A 500.0 mL buffer solution is 0.10 M in benzoic acid and 0.10 M in sodium benzoate and has an initial pH of 4.19. What is the pH of the buffer upon the addition of 0.010 mol of NaOH?
   a) 1.70
   b) 4.01
   c) 4.29
   d) 4.37

5. Consider a buffer composed of the weak acid HA and its conjugate base A^-. Which pair of concentrations results in the most effective buffer?
   a) 0.10 M HA; 0.10 M A^-
   b) 0.50 M HA; 0.50 M A^-
   c) 0.90 M HA; 0.10 M A^-
   d) 0.10 M HA; 0.90 M A^-

6. Which combination is the best choice to use to prepare a buffer with a pH of 9.0?
   a) NH_3; NH_4Cl (pK_b for NH_3 is 4.75)
   b) C_5H_5N; C_5H_5NHCl (pK_b for C_5H_5N is 8.76)
   c) HNO_2; $NaNO_2$(pK_a for HNO_2 is 3.33)
   d) $HCHO_2$; $NaCHO_2$ (pK_a for $HCHO_2$ is 3.74)

7. A 25.0 mL sample of an unknown HBr solution is titrated with 0.100 M NaOH. The equivalence point is reached upon the addition of 18.88 mL of the base. What is the concentration of the HBr solution?
   a) 0.0755 M
   b) 0.0376 M
   c) 0.100 M
   d) 0.00188 M

8. A 10.0 mL sample of 0.200 M hydrocyanic acid (HCN) is titrated with 0.0998 M NaOH. What is the pH at the equivalence point? For hydrocyanic acid, $pK_a = 9.31$.
   a) 7.00
   b) 8.76
   c) 9.31
   d) 11.07

9. A 20.0 mL sample of 0.150 M ethylamine is titrated with 0.0981 M HCl. What is the pH after the addition of 5.0 mL of HCl? For ethylamine, $pK_b = 3.25$.
   a) 10.75
   b) 11.04
   c) 2.96
   d) 11.46

10. Three 15.0 mL acid samples—0.10 M HA, 0.10 M HB, and 0.10 M H_2C —are all titrated with 0.100 M NaOH. If HA is a weak acid, HB is a strong acid, and H_2C is a diprotic acid, which statement is true of all three titrations?
    a) All three titrations have the same pH at the first equivalence point.
    b) All three titrations have the same initial pH.
    c) All three titrations have the same final pH.
    d) All three titrations require the same volume of NaOH to reach the first equivalence point.

11. A weak unknown monoprotic acid is titrated with a strong base. The titration curve is shown below. What is K_a for the unknown acid?

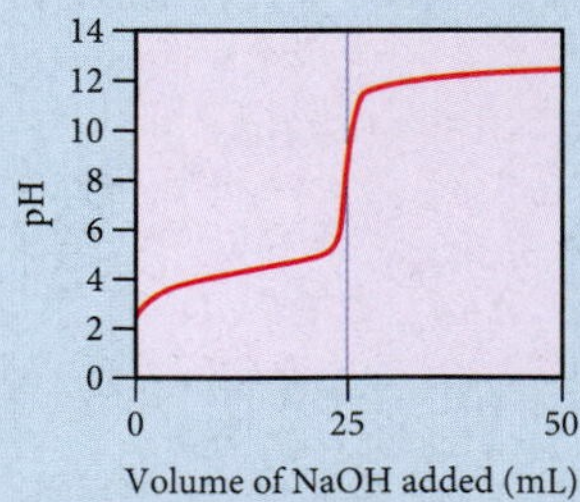

    a) 2.5×10^{-3}
    b) 3.2×10^{-5}
    c) 3.2×10^{-7}
    d) 2.5×10^{-9}

—*Continued*

Continued—

12. Calculate the molar solubility of lead(II) bromide ($PbBr_2$). For lead(II) bromide, $K_{sp} = 4.67 \times 10^{-6}$.
    a) 0.00153 M
    b) 0.0105 M
    c) 0.0167 M
    d) 0.0211 M

13. Calculate the molar solubility of magnesium fluoride (MgF_2) in a solution that is 0.250 M in NaF. For magnesium fluoride, $K_{sp} = 5.16 \times 10^{-11}$.
    a) 2.35×10^{-4} M
    b) 2.06×10^{-10} M
    c) 2.87×10^{-5} M
    d) 8.26×10^{-10} M

14. A solution is 0.025 M in Pb^{2+}. What minimum concentration of Cl^- is required to begin to precipitate $PbCl_2$? For $PbCl_2$, $K_{sp} = 1.17 \times 10^{-5}$.
    a) 1.17×10^{-5} M
    b) 0.0108 M
    c) 0.0216 M
    d) 5.41×10^{-4} M

15. Which compound is more soluble in an acidic solution than in a neutral solution?
    a) $PbBr_2$
    b) CuCl
    c) AgI
    d) BaF_2

Answers: 1:c; 2:b; 3:b; 4:d; 5:b; 6:a; 7:a; 8:d; 9:d; 10:d; 11:b; 12:b; 13:d; 14:c; 15:d

CHAPTER SUMMARY 18

REVIEW

KEY LEARNING OUTCOMES

CHAPTER OBJECTIVES	ASSESSMENT		
Calculating the pH of a Buffer Solution (18.2)	• Example 18.1 Exercises 27, 28, 31, 32	For Practice 18.1	For More Practice 18.1
Using the Henderson–Hasselbalch Equation to Calculate the pH of a Buffer Solution (18.2)	• Example 18.2	For Practice 18.2	Exercises 35–40
Calculating the pH Change in a Buffer Solution after the Addition of a Small Amount of Strong Acid or Base (18.2)	• Example 18.3 Exercises 45–48	For Practice 18.3	For More Practice 18.3
Using the Henderson–Hasselbalch Equation to Calculate the pH of a Buffer Solution Composed of a Weak Base and Its Conjugate Acid (18.2)	• Example 18.4 Exercises 35–40	For Practice 18.4	For More Practice 18.4
Determining Buffer Range (18.3)	• Example 18.5	For Practice 18.5	Exercises 55–56
Determining pH's for a Strong Acid–Strong Base Titration (18.4)	• Example 18.6	For Practice 18.6	Exercises 65–68
Determining pH's for a Weak Acid–Strong Base Titration (18.4)	• Example 18.7	For Practice 18.7	Exercises 63–64, 69–70, 73, 75–78
Calculating Molar Solubility from K_{sp} (18.5)	• Example 18.8	For Practice 18.8	Exercises 85–86
Calculating K_{sp} from Molar Solubility (18.5)	• Example 18.9	For Practice 18.9	Exercises 87–88, 90, 92
Calculating Molar Solubility in the Presence of a Common Ion (18.5)	• Example 18.10	For Practice 18.10	Exercises 93–94
Determining the Effect of pH on Solubility (18.5)	• Example 18.11	For Practice 18.11	Exercises 95–98

Predicting Precipitation Reactions by Comparing Q and K_{sp} (18.6)	• Example 18.12	For Practice 18.12	Exercises 99–102
Finding the Minimum Required Reagent Concentration for Selective Precipitation (18.6)	• Example 18.13	For Practice 18.13	Exercises 103–104
Finding the Concentrations of Ions Left in Solution after Selective Precipitation (18.6)	• Example 18.14	For Practice 18.14	Exercises 105–106
Working with Complex Ion Equilibria (18.7)	• Example 18.15	For Practice 18.15	Exercises 107–110

KEY TERMS

Section 18.2
buffer (710)
common ion effect (712)
Henderson–Hasselbalch equation (714)

Section 18.3
buffer capacity (724)

Section 18.4
acid–base titration (725)
indicator (725)
equivalence point (725)
endpoint (737)

Section 18.5
solubility-product constant (K_{sp}) (740)
molar solubility (740)

Section 18.6
selective precipitation (747)

Section 18.7
complex ion (748)
ligand (748)
formation constant (K_f) (749)

KEY CONCEPTS

The Dangers of Antifreeze (18.1)

- Although buffers closely regulate the pH of mammalian blood, the capacity of these buffers to neutralize can be overwhelmed.
- Ethylene glycol, the main component of antifreeze, is metabolized by the liver into glycolic acid. The resulting acidity can exceed the buffering capacity of blood and cause acidosis, a serious condition that results in oxygen deprivation.

Buffers: Solutions That Resist pH Change (18.2)

- A buffer contains significant amounts of both a weak acid and its conjugate base (or a weak base and its conjugate acid), enabling the buffer to neutralize added acid or added base.
- Adding a small amount of acid to a buffer converts a stoichiometric amount of base to the conjugate acid. Adding a small amount of base to a buffer converts a stoichiometric amount of the acid to the conjugate base.
- We can determine the pH of a buffer solution by solving an equilibrium problem, focusing on the common ion effect, or using the Henderson–Hasselbalch equation.

Buffer Range and Buffer Capacity (18.3)

- A buffer works best when the amounts of acid and conjugate base it contains are large and approximately equal.
- If the relative amounts of acid and base in a buffer differ by more than a factor of 10, the ability of the buffer to neutralize added acid and added base diminishes. The maximum pH range at which a buffer is effective is one pH unit on either side of the acid's pK_a.

Titrations and pH Curves (18.4)

- A titration curve is a graph of the change in pH versus the volume of acid or base that is added during a titration.
- This chapter examines three types of titration curves, representing three types of acid–base reactions: a strong acid with a strong base, a weak acid with a strong base (or vice versa), and a polyprotic acid with a base.
- The equivalence point of a titration can be made visible by an indicator, a compound that changes color over a specific pH range.

Solubility Equilibria and the Solubility-Product Constant (18.5)

- The solubility-product constant (K_{sp}) is an equilibrium constant for the dissolution of an ionic compound in water.
- We can determine the molar solubility of an ionic compound from K_{sp} and vice versa. Although the value of K_{sp} is constant at a given temperature, the solubility of an ionic substance can depend on other factors such as the presence of common ions and the pH of the solution.

Precipitation (18.6)

- We can compare the magnitude of K_{sp} to the reaction quotient, Q, in order to determine the relative saturation of a solution.
- Substances with cations that have sufficiently different values of K_{sp} can be separated by selective precipitation, in which an added reagent forms a precipitate with one of the dissolved cations but not others.

Complex Ion Equilibria (18.7)

- A complex ion contains a central metal ion bound to two or more ligands.
- The equilibrium constant for the formation of a complex ion is called a formation constant and is usually quite large.
- The solubility of an ionic compound containing a metal cation that forms complex ions increases in the presence of Lewis bases that form a complex ion with the cation because the formation of the complex ion drives the dissolution reaction to the right.
- All metal hydroxides become more soluble in the presence of acids, but amphoteric metal hydroxides also become more soluble in the presence of bases.

KEY EQUATIONS AND RELATIONSHIPS

The Henderson–Hasselbalch Equation (18.2)

$$pH = pK_a + \log\frac{[base]}{[acid]}$$

Effective Buffer Range (18.3)

$$pH\ range = pK_a \pm 1$$

The Relation between Q and K_{sp} (18.6)

If $Q < K_{sp}$, the solution is unsaturated. More of the solid ionic compound can dissolve in the solution.

If $Q = K_{sp}$, the solution is saturated. The solution is holding the equilibrium amount of the dissolved ions, and additional solid will not dissolve in the solution.

If $Q > K_{sp}$, the solution is supersaturated. Under most circumstances, the solid will precipitate out of a supersaturated solution.

EXERCISES

REVIEW QUESTIONS

1. What is the pH range of human blood? How is human blood maintained in this pH range?
2. What is a buffer? How does a buffer work? How does it neutralize added acid? Added base?
3. What is the common ion effect?
4. What is the Henderson–Hasselbalch equation, and why is it useful?
5. What is the pH of a buffer when the concentrations of both buffer components (the weak acid and its conjugate base) are equal? What happens to the pH when the buffer contains more of the weak acid than the conjugate base? More of the conjugate base than the weak acid?
6. Suppose that a buffer contains equal amounts of a weak acid and its conjugate base. What happens to the relative amounts of the weak acid and conjugate base when a small amount of strong acid is added to the buffer? What happens when a small amount of strong base is added?
7. How do you use the Henderson–Hasselbalch equation to calculate the pH of a buffer containing a base and its conjugate acid? Specifically, how do you determine the correct value for pK_a?
8. What factors influence the effectiveness of a buffer? What are the characteristics of an effective buffer?
9. What is the effective pH range of a buffer (relative to the pK_a of the weak acid component)?
10. Describe acid–base titration. What is the equivalence point?
11. The pH at the equivalence point of the titration of a strong acid with a strong base is 7.0. However, the pH at the equivalence point of the titration of a *weak* acid with a strong base is above 7.0. Explain.
12. The volume required to reach the equivalence point of an acid–base titration depends on the volume and concentration of the acid or base to be titrated and on the concentration of the acid or base used to do the titration. It does not, however, depend on the strength or weakness of the acid or base being titrated. Explain.
13. In the titration of a strong acid with a strong base, how do you calculate these quantities?
    a) initial pH
    b) pH before the equivalence point
    c) pH at the equivalence point
    d) pH beyond the equivalence point
14. In the titration of a weak acid with a strong base, how do you calculate these quantities?
    a) initial pH
    b) pH before the equivalence point
    c) pH at one-half the equivalence point
    d) pH at the equivalence point
    e) pH beyond the equivalence point
15. The titration of a diprotic acid with sufficiently different pK_a's displays two equivalence points. Why?
16. In the titration of a polyprotic acid, the volume required to reach the first equivalence point is identical to the volume required to reach the second one. Why?
17. What is the difference between the endpoint and the equivalence point in a titration?
18. What is an indicator? How can an indicator signal the equivalence point of a titration?
19. What is the solubility-product constant? Write a general expression for the solubility constant of a compound with the general formula A_mX_n.
20. What is molar solubility? How do you obtain the molar solubility of a compound from K_{sp}?
21. How does a common ion affect the solubility of a compound? More specifically, how is the solubility of a compound with the general formula AX different in a solution containing one of the common ions (A^+ or X^-) than it is in pure water? Explain.
22. How is the solubility of an ionic compound with a basic anion affected by pH? Explain.
23. For a given solution containing an ionic compound, what is the relationship between Q, K_{sp}, and the relative saturation of the solution?
24. What is selective precipitation? Under which conditions does selective precipitation occur?

PROBLEMS BY TOPIC

The Common Ion Effect and Buffers

25. In which of these solutions does HNO_2 ionize less than it does in pure water?

a. 0.10 M NaCl
b. 0.10 M KNO_3
c. 0.10 M NaOH
d. 0.10 M $NaNO_2$

26. A formic acid solution has a pH of 3.25. Which of these substances raises the pH of the solution upon addition? Explain your answer.

a. HCl
b. NaBr
c. $NaCHO_2$
d. KCl

27. Solve an equilibrium problem (using an ICE table) to calculate the pH of each solution.

a. a solution that is 0.20 M in $HCHO_2$ and 0.15 M in $NaCHO_2$
b. a solution that is 0.16 M in NH_3 and 0.22 M in NH_4Cl

28. Solve an equilibrium problem (using an ICE table) to calculate the pH of each solution.

a. a solution that is 0.195 M in $HC_2H_3O_2$ and 0.125 M in $KC_2H_3O_2$
b. a solution that is 0.255 M in CH_3NH_2 and 0.135 M in CH_3NH_3Br

29. Calculate the percent ionization of a 0.15 M benzoic acid solution in pure water and in a solution containing 0.10 M sodium benzoate. Why does the percent ionization differ significantly in the two solutions?

30. Calculate the percent ionization of a 0.13 M formic acid solution in pure water and also in a solution containing 0.11 M potassium formate. Explain the difference in percent ionization in the two solutions.

31. Solve an equilibrium problem (using an ICE table) to calculate the pH of each solution. (K_a(HF) = 3.5×10^{-4})

a. 0.15 M HF
b. 0.15 M NaF
c. a mixture that is 0.15 M in HF and 0.15 M in NaF

32. Solve an equilibrium problem (using an ICE table) to calculate the pH of each solution.

a. 0.18 M CH_3NH_2
b. 0.18 M CH_3NH_3Cl
c. a mixture that is 0.18 M in CH_3NH_2 and 0.18 M in CH_3NH_3Cl

33. A buffer contains significant amounts of acetic acid and sodium acetate. Write equations that demonstrate how this buffer neutralizes added acid and added base.

34. A buffer contains significant amounts of ammonia and ammonium chloride. Write equations that demonstrate how this buffer neutralizes added acid and added base.

35. Use the Henderson–Hasselbalch equation to calculate the pH of each solution in Problem 27.

36. Use the Henderson–Hasselbalch equation to calculate the pH of each solution in Problem 28.

37. Use the Henderson–Hasselbalch equation to calculate the pH of each solution.

a. a solution that is 0.135 M in HClO and 0.155 M in KClO
b. a solution that contains 1.05% $C_2H_5NH_2$ by mass and 1.10% $C_2H_5NH_3Br$ by mass
c. a solution that contains 10.0 g of $HC_2H_3O_2$ and 10.0 g of $NaC_2H_3O_2$ in 150.0 mL of solution

38. Use the Henderson–Hasselbalch equation to calculate the pH of each solution.

a. a solution that is 0.145 M in propanoic acid and 0.115 M in potassium propanoate
b. a solution that contains 0.785% C_5H_5N by mass and 0.985% C_5H_5NHCl by mass
c. a solution that contains 15.0 g of HF and 25.0 g of NaF in 125 mL of solution

39. Calculate the pH of the solution that results from each mixture.

a. 50.0 mL of 0.15 M $HCHO_2$ with 75.0 mL of 0.13 M $NaCHO_2$
b. 125.0 mL of 0.10 M NH_3 with 250.0 mL of 0.10 M NH_4Cl

40. Calculate the pH of the solution that results from each mixture.

a. 150.0 mL of 0.25 M HF with 225.0 mL of 0.30 M NaF
b. 175.0 mL of 0.10 M $C_2H_5NH_2$ with 275.0 mL of 0.20 M $C_2H_5NH_3Cl$

41. Calculate the ratio of NaF to HF required to create a buffer with pH = 4.00.

42. Calculate the ratio of CH_3NH_2 to CH_3NH_3Cl concentration required to create a buffer with pH = 10.24.

43. What mass of sodium benzoate should you add to 150.0 mL of a 0.15 M benzoic acid solution to obtain a buffer with a pH of 4.25? (Assume no volume change.)

44. What mass of ammonium chloride should you add to 2.55 L of a 0.155 M NH_3 solution to obtain a buffer with a pH of 9.55? (Assume no volume change.)

45. A 250.0 mL buffer solution is 0.250 M in acetic acid and 0.250 M in sodium acetate.

a. What is the initial pH of this solution?
b. What is the pH after addition of 0.0050 mol of HCl?
c. What is the pH after addition of 0.0050 mol of NaOH?

46. A 100.0 mL buffer solution is 0.175 M in HClO and 0.150 M in NaClO.

a. What is the initial pH of this solution?
b. What is the pH after addition of 150.0 mg of HBr?
c. What is the pH after addition of 85.0 mg of NaOH?

47. For each solution, calculate the initial and final pH after the addition of 0.010 mol of HCl.

a. 500.0 mL of pure water
b. 500.0 mL of a buffer solution that is 0.125 M in $HC_2H_3O_2$ and 0.115 M in $NaC_2H_3O_2$
c. 500.0 mL of a buffer solution that is 0.155 M in $C_2H_5NH_2$ and 0.145 M in $C_2H_5NH_3Cl$

48. For each solution, calculate the initial and final pH after the addition of 0.010 mol of NaOH.

a. 250.0 mL of pure water
b. 250.0 mL of a buffer solution that is 0.195 M in $HCHO_2$ and 0.275 M in $KCHO_2$
c. 250.0 mL of a buffer solution that is 0.255 M in $CH_3CH_2NH_2$ and 0.235 M in $CH_3CH_2NH_3Cl$

49. A 350.0 mL buffer solution is 0.150 M in HF and 0.150 M in NaF. What mass of NaOH does this buffer neutralize before the pH rises above 4.00? If the same volume of the buffer was 0.350 M in HF and 0.350 M in NaF, what mass of NaOH is neutralized before the pH rises above 4.00?

50. A 100.0 mL buffer solution is 0.100 M in NH_3 and 0.125 M in NH_4Br. What mass of HCl does this buffer neutralize before the pH falls below 9.00? If the same volume of the buffer were 0.250 M in NH_3 and 0.400 M in NH_4Br, what mass of HCl is neutralized before the pH falls below 9.00?

51. Determine whether the mixing of each pair of solutions results in a buffer.
   a. 100.0 mL of 0.10 M NH_3; 100.0 mL of 0.15 M NH_4Cl
   b. 50.0 mL of 0.10 M HCl; 35.0 mL of 0.150 M NaOH
   c. 50.0 mL of 0.15 M HF; 20.0 mL of 0.15 M NaOH
   d. 175.0 mL of 0.10 M NH_3; 150.0 mL of 0.12 M NaOH
   e. 125.0 mL of 0.15 M NH_3; 150.0 mL of 0.20 M NaOH

52. Determine whether the mixing of each pair of solutions results in a buffer.
   a. 75.0 mL of 0.10 M HF; 55.0 mL of 0.15 M NaF
   b. 150.0 mL of 0.10 M HF; 135.0 mL of 0.175 M HCl
   c. 165.0 mL of 0.10 M HF; 135.0 mL of 0.050 M KOH
   d. 125.0 mL of 0.15 M CH_3NH_2; 120.0 mL of 0.25 M CH_3NH_3Cl
   e. 105.0 mL of 0.15 M CH_3NH_2; 95.0 mL of 0.10 M HCl

53. Blood is buffered by carbonic acid and the bicarbonate ion. Normal blood plasma is 0.024 M in HCO_3^- and 0.0012 M H_2CO_3 (pK_{a_1} for H_2CO_3 at body temperature is 6.1).
   a. What is the pH of blood plasma?
   b. If the volume of blood in a normal adult is 5.0 L, what mass of HCl can be neutralized by the buffering system in blood before the pH falls below 7.0 (which would result in death)?
   c. Given the volume from part (b), what mass of NaOH can be neutralized before the pH rises above 7.8?

54. The fluids within cells are buffered by $H_2PO_4^-$ and HPO_4^{2-}.
   a. Calculate the ratio of HPO_4^{2-} to $H_2PO_4^-$ required to maintain a pH of 7.1 within a cell.
   b. Could a buffer system employing H_3PO_4 as the weak acid and $H_2PO_4^-$ as the weak base be used as a buffer system within cells? Explain.

55. Which buffer system is the best choice to create a buffer with pH = 7.20? For the best system, calculate the ratio of the masses of the buffer components required to make the buffer.

$HC_2H_3O_2/KC_2H_3O_2$	$HClO_2/KClO_2$
NH_3/NH_4Cl	$HClO/KClO$

56. Which buffer system is the best choice to create a buffer with pH = 9.00? For the best system, calculate the ratio of the masses of the buffer components required to make the buffer.

HF/KF	HNO_2/KNO_2
NH_3/NH_4Cl	$HClO/KClO$

57. A 500.0 mL buffer solution is 0.100 M in HNO_2 and 0.150 M in KNO_2. Determine whether each addition would exceed the capacity of the buffer to neutralize it.
   a. 250.0 mg NaOH
   b. 350.0 mg KOH
   c. 1.25 g HBr
   d. 1.35 g HI

58. A 1.0 L buffer solution is 0.125 M in HNO_2 and 0.145 M in $NaNO_2$. Determine the concentrations of HNO_2 and $NaNO_2$ after the addition of each substance.
   a. 1.5 g HCl
   b. 1.5 g NaOH
   c. 1.5 g HI

Titrations, pH Curves, and Indicators

59. The graphs labeled (a) and (b) are the titration curves for two equal-volume samples of monoprotic acids, one weak and one strong. Both titrations were carried out with the same concentration of strong base.

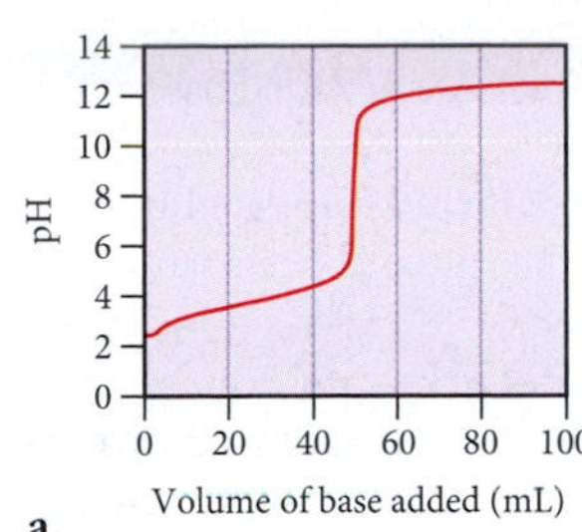

a.

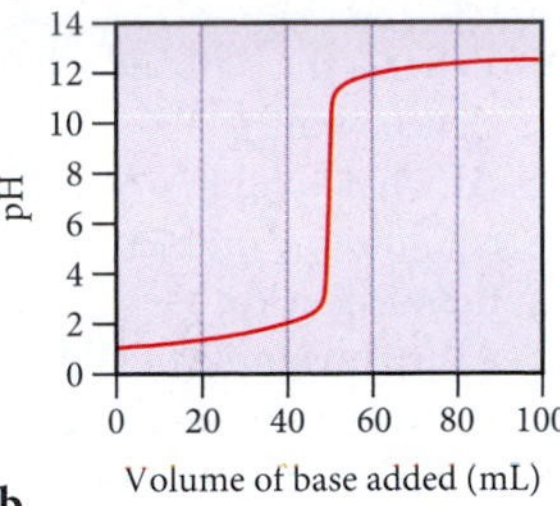

b.

   i. What is the approximate pH at the equivalence point of each curve?
   ii. Which graph corresponds to the titration of the strong acid and which one to the titration of the weak acid?

60. Two 25.0 mL samples, one 0.100 M HCl and the other 0.100 M HF, are titrated with 0.200 M KOH.
   a. What is the volume of added base at the equivalence point for each titration?
   b. Is the pH at the equivalence point for each titration acidic, basic, or neutral?
   c. Which titration curve has the lower initial pH?
   d. Sketch each titration curve.

61. Two 20.0 mL samples, one 0.200 M KOH and the other 0.200 M CH_3NH_2, are titrated with 0.100 M HI.
   a. What is the volume of added acid at the equivalence point for each titration?
   b. Is the pH at the equivalence point for each titration acidic, basic, or neutral?
   c. Which titration curve has the lower initial pH?
   d. Sketch each titration curve.

62. The graphs labeled (a) and (b) are the titration curves for two equal-volume samples of bases, one weak and one strong. Both titrations were carried out with the same concentration of strong acid.

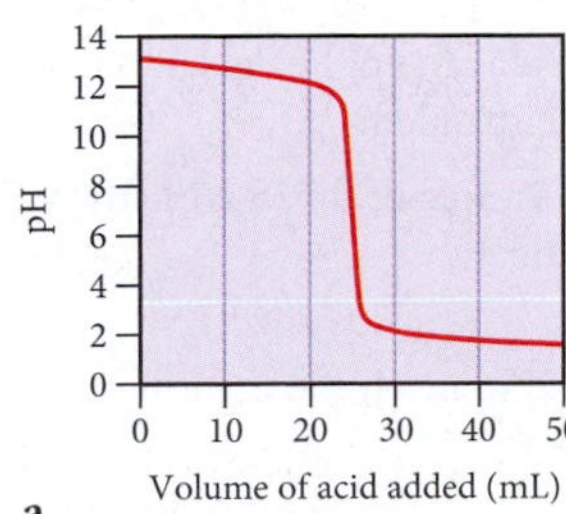

a.

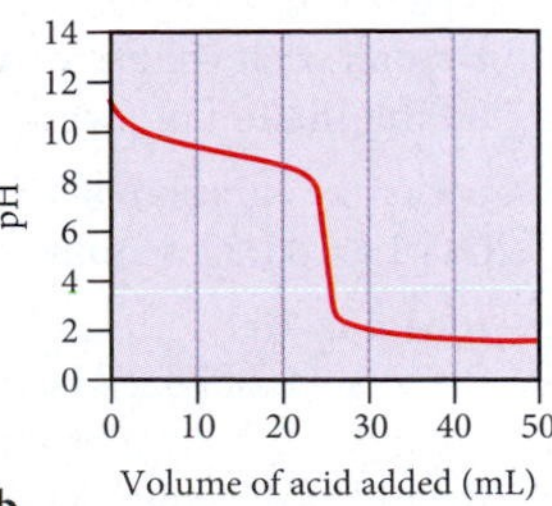

b.

   i. What is the approximate pH at the equivalence point of each curve?
   ii. Which graph corresponds to the titration of the strong base and which one to the weak base?

63. Consider the curve shown here for the titration of a weak monoprotic acid with a strong base and answer each question.

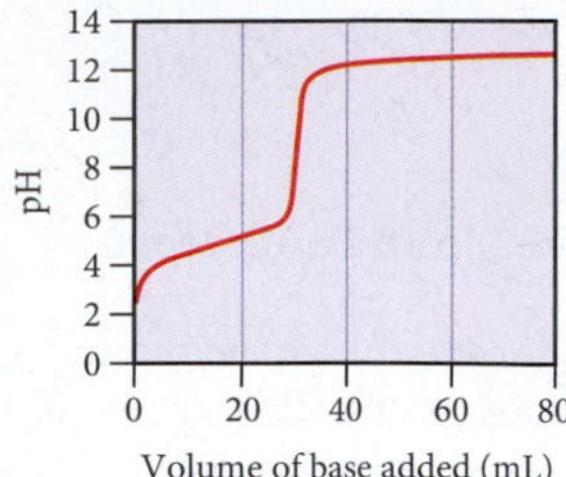

a. What is the pH, and what is the volume of added base at the equivalence point?
b. At what volume of added base is the pH calculated by working an equilibrium problem based on the initial concentration and K_a of the weak acid?
c. At what volume of added base does pH = pK_a?
d. At what volume of added base is the pH calculated by working an equilibrium problem based on the concentration and K_b of the conjugate base?
e. Beyond what volume of added base is the pH calculated by focusing on the amount of excess strong base added?

64. Consider the curve shown here for the titration of a weak base with a strong acid and answer each question.

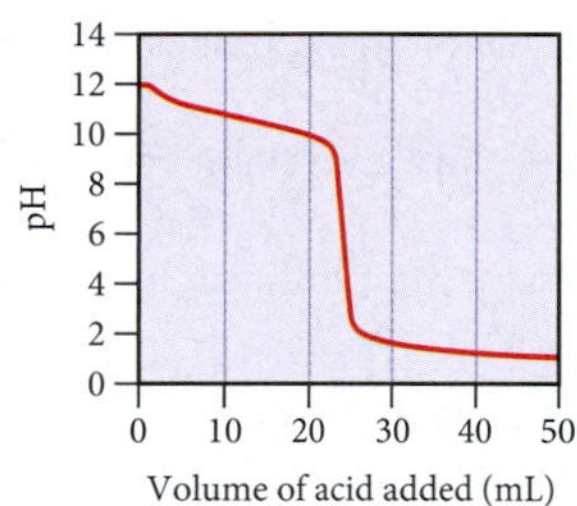

a. What is the pH, and what is the volume of added acid at the equivalence point?
b. At what volume of added acid is the pH calculated by working an equilibrium problem based on the initial concentration and K_b of the weak base?
c. At what volume of added acid does pH = 14 − pK_b?
d. At what volume of added acid is the pH calculated by working an equilibrium problem based on the concentration and K_a of the conjugate acid?
e. Beyond what volume of added acid is the pH calculated by focusing on the amount of excess strong acid added?

65. Consider the titration of a 35.0 mL sample of 0.175 M HBr with 0.200 M KOH. Determine each quantity.

a. the initial pH
b. the volume of added base required to reach the equivalence point
c. the pH at 10.0 mL of added base
d. the pH at the equivalence point
e. the pH after adding 5.0 mL of base beyond the equivalence point

66. A 20.0 mL sample of 0.125 M HNO_3 is titrated with 0.150 M NaOH. Calculate the pH for at least five different points on the titration curve and sketch the curve. Indicate the volume at the equivalence point on your graph.

67. Consider the titration of a 25.0 mL sample of 0.115 M RbOH with 0.100 M HCl. Determine each quantity.

a. the initial pH
b. the volume of added acid required to reach the equivalence point
c. the pH at 5.0 mL of added acid
d. the pH at the equivalence point
e. the pH after adding 5.0 mL of acid beyond the equivalence point

68. A 15.0 mL sample of 0.100 M $Ba(OH)_2$ is titrated with 0.125 M HCl. Calculate the pH for at least five different points on the titration curve and sketch the curve. Indicate the volume at the equivalence point on your graph.

69. Consider the titration of a 20.0 mL sample of 0.105 M $HC_2H_3O_2$ with 0.125 M NaOH. Determine each quantity.

a. the initial pH
b. the volume of added base required to reach the equivalence point
c. the pH at 5.0 mL of added base
d. the pH at one-half of the equivalence point
e. the pH at the equivalence point
f. the pH after adding 5.0 mL of base beyond the equivalence point

70. A 30.0 mL sample of 0.165 M propanoic acid is titrated with 0.300 M KOH. Calculate the pH at each volume of added base: 0 mL, 5 mL, 10 mL, equivalence point, one-half equivalence point, 20 mL, 25 mL. Sketch the titration curve.

71. Consider the titration of a 25.0 mL sample of 0.175 M CH_3NH_2 with 0.150 M HBr. Determine each quantity.

a. the initial pH
b. the volume of added acid required to reach the equivalence point
c. the pH at 5.0 mL of added acid
d. the pH at one-half of the equivalence point
e. the pH at the equivalence point
f. the pH after adding 5.0 mL of acid beyond the equivalence point

72. A 25.0 mL sample of 0.125 M pyridine is titrated with 0.100 M HCl. Calculate the pH at each volume of added acid: 0 mL, 10 mL, 20 mL, equivalence point, one-half equivalence point, 40 mL, 50 mL. Sketch the titration curve.

73. Consider the titration curves (labeled a and b) for equal volumes of two weak acids, both titrated with 0.100 M NaOH.

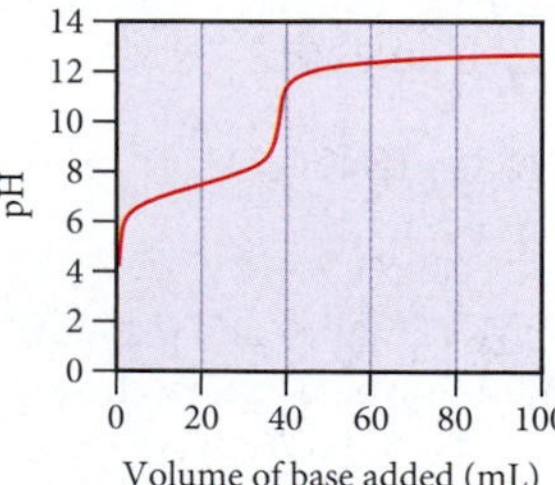

a.

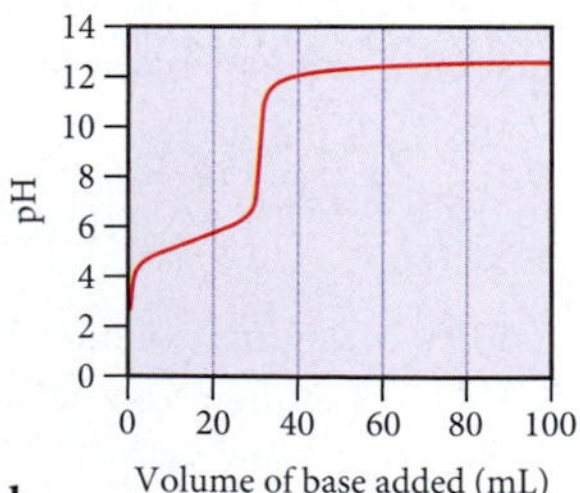

b.

i. Which acid solution is more concentrated?
ii. Which acid has the larger K_a?

74. Consider the titration curves (labeled a and b) for equal volumes of two weak bases, both titrated with 0.100 M HCl.

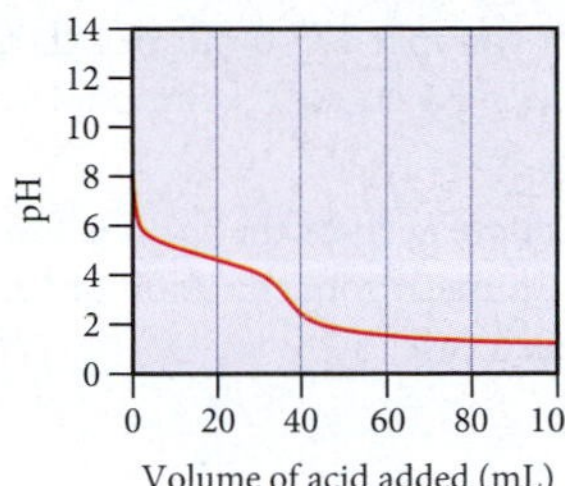

a.

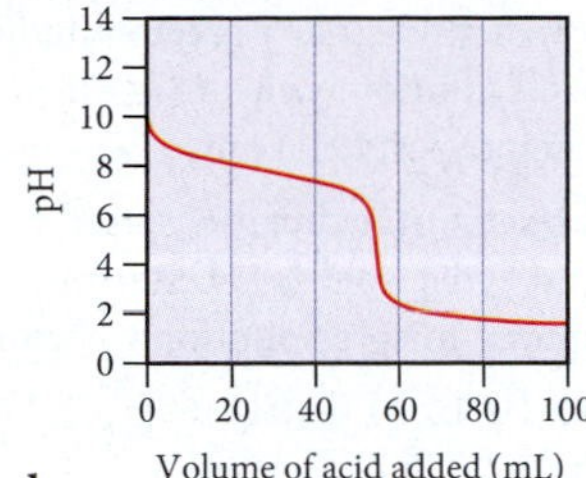

b.

i. Which base solution is more concentrated?
ii. Which base has the larger K_b?

75. A 0.229 g sample of an unknown monoprotic acid is titrated with 0.112 M NaOH. The resulting titration curve is shown here. Determine the molar mass and pK_a of the acid.

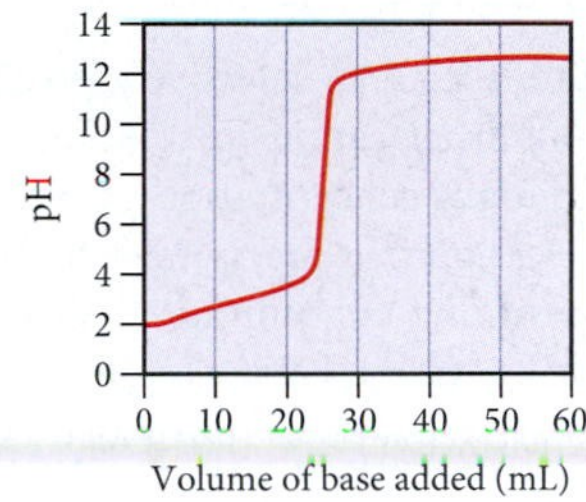

76. A 0.446 g sample of an unknown monoprotic acid is titrated with 0.105 M KOH. The resulting titration curve is shown here. Determine the molar mass and pK_a of the acid.

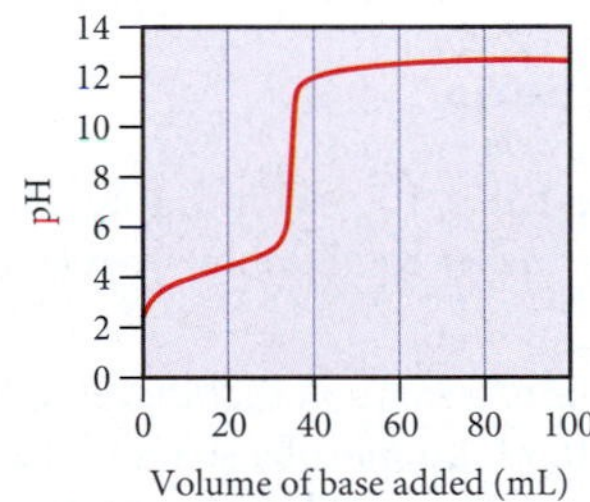

77. A 20.0 mL sample of 0.115 M sulfurous acid (H_2SO_3) solution is titrated with 0.1014 M KOH. At what added volume of base solution does each equivalence point occur?

78. A 20.0 mL sample of a 0.125 M diprotic acid (H_2A) solution is titrated with 0.1019 M KOH. The acid ionization constants for the acid are $K_{a_1} = 5.2 \times 10^{-5}$ and $K_{a_2} = 3.4 \times 10^{-10}$. At what added volume of base does each equivalence point occur?

79. Methyl red has a pK_a of 5.0 and is red in its acid form and yellow in its basic form. If several drops of this indicator are placed in a 25.0 mL sample of 0.100 M HCl, what color does the solution appear? If 0.100 M NaOH is slowly added to the HCl sample, in what pH range will the indicator change color?

80. Phenolphthalein has a pK_a of 9.7. It is colorless in its acid form and pink in its basic form. For each of the pH's, calculate $[In^-]/[HIn]$ and predict the color of a phenolphthalein solution.

a. pH = 2.0
b. pH = 5.0
c. pH = 8.0
d. pH = 11.0

81. Referring to Table 18.1, pick an indicator for use in the titration of each acid with a strong base.

a. HF
b. HCl
c. HCN

82. Referring to Table 18.1, pick an indicator for use in the titration of each base with a strong acid.

a. CH_3NH_2
b. NaOH
c. $C_6H_5NH_2$

Solubility Equilibria

83. Write balanced equations and expressions for K_{sp} for the dissolution of each ionic compound.

a. $BaSO_4$
b. $PbBr_2$
c. Ag_2CrO_4

84. Write balanced equations and expressions for K_{sp} for the dissolution of each ionic compound.

a. $CaCO_3$
b. $PbCl_2$
c. AgI

85. Refer to the K_{sp} values in Table 18.2 to calculate the molar solubility of each compound in pure water.

a. AgBr
b. $Mg(OH)_2$
c. CaF_2

86. Refer to the K_{sp} values in Table 18.2 to calculate the molar solubility of each compound in pure water.

a. MX ($K_{sp} = 1.27 \times 10^{-36}$)
b. Ag_2CrO_4
c. $Ca(OH)_2$

87. Use the given molar solubilities in pure water to calculate K_{sp} for each compound.

a. MX; molar solubility $= 3.27 \times 10^{-11}$ M
b. PbF_2; molar solubility $= 5.63 \times 10^{-3}$ M
c. MgF_2; molar solubility $= 2.65 \times 10^{-4}$ M

88. Use the given molar solubilities in pure water to calculate K_{sp} for each compound.

a. $BaCrO_4$; molar solubility $= 1.08 \times 10^{-5}$ M
b. Ag_2SO_3; molar solubility $= 1.55 \times 10^{-5}$ M
c. $Pd(SCN)_2$; molar solubility $= 2.22 \times 10^{-8}$ M

89. Two compounds with general formulas AX and AX_2 have $K_{sp} = 1.5 \times 10^{-5}$. Which of the two compounds has the higher molar solubility?

90. Consider the compounds with the generic formulas listed and their corresponding molar solubilities in pure water. Which compound has the smallest value of K_{sp}?

a. AX; molar solubility $= 1.35 \times 10^{-4}$ M
b. AX_2; molar solubility $= 2.25 \times 10^{-4}$ M
c. A_2X; molar solubility $= 1.75 \times 10^{-4}$ M

91. Refer to the K_{sp} value from Table 18.2 to calculate the solubility of iron(II) hydroxide in pure water in grams per 100.0 mL of solution.

92. The solubility of copper(I) chloride is 3.91 mg per 100.0 mL of solution. Calculate K_{sp} for CuCl.

93. Calculate the molar solubility of barium fluoride in each liquid or solution.

a. pure water
b. 0.10 M $Ba(NO_3)_2$
c. 0.15 M NaF

94. Calculate the molar solubility of MX ($K_{sp} = 1.27 \times 10^{-36}$) in each liquid or solution.

a. pure water
b. 0.25 M MCl_2
c. 0.20 M Na_2X

95. Calculate the molar solubility of calcium hydroxide in a solution buffered at each pH.

a. pH = 4
b. pH = 7
c. pH = 9

96. Calculate the solubility (in grams per 1.00×10^2 mL of solution) of magnesium hydroxide in a solution buffered at pH = 10. How does this compare to the solubility of $Mg(OH)_2$ in pure water?

97. Is each compound more soluble in acidic solution or in pure water? Explain.

a. $BaCO_3$
b. CuS
c. AgCl
d. PbI_2

98. Is each compound more soluble in acidic solution or in pure water? Explain.

a. Hg_2Br_2
b. $Mg(OH)_2$
c. $CaCO_3$
d. AgI

Precipitation

99. A solution containing sodium fluoride is mixed with one containing calcium nitrate to form a solution that is 0.015 M in NaF and 0.010 M in $Ca(NO_3)_2$. Does a precipitate form in the mixed solution? If so, identify the precipitate.

100. A solution containing potassium bromide is mixed with one containing lead acetate to form a solution that is 0.013 M in KBr and 0.0035 M in $Pb(C_2H_3O_2)_2$. Does a precipitate form in the mixed solution? If so, identify the precipitate.

101. Predict whether a precipitate forms if you mix 75.0 mL of a NaOH solution with pOH = 2.58 with 125.0 mL of a 0.018 M $MgCl_2$ solution. Identify the precipitate, if any.

102. Predict whether a precipitate forms if you mix 175.0 mL of a 0.0055 M KCl solution with 145.0 mL of a 0.0015 M $AgNO_3$ solution. Identify the precipitate, if any.

103. Potassium hydroxide is used to precipitate each of the cations from their respective solution. Determine the minimum concentration of KOH required for precipitation to begin in each case.

a. 0.015 M $CaCl_2$
b. 0.0025 M $Fe(NO_3)_2$
c. 0.0018 M $MgBr_2$

104. Determine the minimum concentration of the precipitating agent on the right you need to cause precipitation of the cation from the solution on the left.

a. 0.035 M $Ba(NO_3)_2$; NaF
b. 0.085 M CaI_2; K_2SO_4
c. 0.0018 M $AgNO_3$; RbCl

105. A solution is 0.010 M in Ba^{2+} and 0.020 M in Ca^{2+}.

a. If sodium sulfate is used to selectively precipitate one of the cations while leaving the other cation in solution, which cation precipitates first? What minimum concentration of Na_2SO_4 will trigger the precipitation of the cation that precipitates first?
b. What is the remaining concentration of the cation that precipitates first, when the other cation begins to precipitate?

106. A solution is 0.022 M in Fe^{2+} and 0.014 M in Mg^{2+}.

a. If potassium carbonate is used to selectively precipitate one of the cations while leaving the other cation in solution, which cation precipitates first? What minimum concentration of K_2CO_3 will trigger the precipitation of the cation that precipitates first?
b. What is the remaining concentration of the cation that precipitates first, when the other cation begins to precipitate?

Complex Ion Equilibria

107. A solution is made 1.1×10^{-3} M in $Zn(NO_3)_2$ and 0.150 M in NH_3. After the solution reaches equilibrium, what concentration of $Zn^{2+}(aq)$ remains?

108. A 120.0 mL sample of a solution that is 2.8×10^{-3} M in $AgNO_3$ is mixed with a 225.0 mL sample of a solution that is 0.10 M in NaCN. After the solution reaches equilibrium, what concentration of $Ag^{+}(aq)$ remains?

109. Use the appropriate values of K_{sp} and K_f to find the equilibrium constant for the reaction.

$$FeS(s) + 6\,CN^-(aq) \rightleftharpoons Fe(CN)_6^{4-}(aq) + S^{2-}(aq)$$

110. Use the appropriate values of K_{sp} and K_f to find the equilibrium constant for the reaction.

$$PbCl_2(s) + 3\,OH^-(aq) \rightleftharpoons Pb(OH)_3^-(aq) + 2\,Cl^-(aq)$$

CUMULATIVE PROBLEMS

111. A 150.0 mL solution contains 2.05 g of sodium benzoate and 2.47 g of benzoic acid. Calculate the pH of the solution.

112. A solution is made by combining 10.0 mL of 17.5 M acetic acid with 5.54 g of sodium acetate and diluting to a total volume of 1.50 L. Calculate the pH of the solution.

113. A buffer is created by combining 150.0 mL of 0.25 M $HCHO_2$ with 75.0 mL of 0.20 M NaOH. Determine the pH of the buffer.

114. A buffer is created by combining 3.55 g of NH_3 with 4.78 g of HCl and diluting to a total volume of 750.0 mL. Determine the pH of the buffer.

115. A 1.0 L buffer solution initially contains 0.25 mol of NH_3 and 0.25 mol of NH_4Cl. In order to adjust the buffer pH to 8.75, should you add NaOH or HCl to the buffer mixture? What mass of the correct reagent should you add?

116. A 250.0 mL buffer solution initially contains 0.025 mol of $HCHO_2$ and 0.025 mol of $NaCHO_2$. In order to adjust the buffer pH to 4.10, should you add NaOH or HCl to the buffer mixture? What mass of the correct reagent should you add?

117. In analytical chemistry, bases used for titrations must often be standardized; that is, their concentration must be precisely determined. Standardization of sodium hydroxide solutions can be accomplished by titrating potassium hydrogen phthalate ($KHC_8H_4O_4$), also known as KHP, with the NaOH solution to be standardized.

a. Write an equation for the reaction between NaOH and KHP.

b. The titration of 0.5527 g of KHP required 25.87 mL of an NaOH solution to reach the equivalence point. What is the concentration of the NaOH solution?

118. A 0.5224 g sample of an unknown monoprotic acid was titrated with 0.0998 M NaOH. The equivalence point of the titration occurs at 23.82 mL. Determine the molar mass of the unknown acid.

119. A 0.25 mol sample of a weak acid with an unknown pK_a is combined with 10.0 mL of 3.00 M KOH, and the resulting solution is diluted to 1.500 L. The measured pH of the solution is 3.85. What is the pK_a of the weak acid?

120. A 5.55 g sample of a weak acid with $K_a = 1.3 \times 10^{-4}$ is combined with 5.00 mL of 6.00 M NaOH, and the resulting solution is diluted to 750 mL. The measured pH of the solution is 4.25. What is the molar mass of the weak acid?

121. A 0.552 g sample of ascorbic acid (vitamin C) is dissolved in water to a total volume of 20.0 mL and titrated with 0.1103 M KOH. The equivalence point occurs at 28.42 mL. The pH of the solution at 10.0 mL of added base was 3.72. From this data, determine the molar mass and K_a for vitamin C.

122. Sketch the titration curve from Problem 121 by calculating the pH at the beginning of the titration, at one-half of the equivalence point, at the equivalence point, and at 5.0 mL beyond the equivalence point. Pick a suitable indicator for this titration from Table 18.1.

123. One of the main components of hard water is $CaCO_3$. When hard water evaporates, some of the $CaCO_3$ is left behind as a white mineral deposit. If a hard water solution is saturated with calcium carbonate, what volume of the solution has to evaporate to deposit 1.00×10^2 mg of $CaCO_3$?

124. Gout—a condition that results in joint swelling and pain—is caused by the formation of sodium urate ($NaC_5H_3N_4O_3$) crystals within tendons, cartilage, and ligaments. Sodium urate precipitates out of blood plasma when uric acid levels become abnormally high. This can happen as a result of eating too many rich foods and consuming too much alcohol, which is why gout is sometimes referred to as the "disease of kings." If the sodium concentration in blood plasma is 0.140 M, and K_{sp} for sodium urate is 5.76×10^{-8}, what minimum concentration of urate results in precipitation?

125. Pseudogout, a condition with symptoms similar to those of gout (see Problem 124), is caused by the formation of calcium diphosphate ($Ca_2P_2O_7$) crystals within tendons, cartilage, and ligaments. Calcium diphosphate precipitates out of blood plasma when diphosphate levels become abnormally high. If the calcium concentration in blood plasma is 9.2 mg/dL and K_{sp} for calcium diphosphate is 8.64×10^{-13}, what minimum concentration of diphosphate results in precipitation?

126. Calculate the solubility of silver chloride in a solution that is 0.100 M in NH_3.

127. Calculate the solubility of CuX in a solution that is 0.150 M in NaCN. K_{sp} for CuX is 1.27×10^{-36}.

128. Aniline, $C_6H_5NH_2$, is an important organic base used in the manufacture of dyes. It has $K_b = 4.3 \times 10^{-10}$. In a certain manufacturing process, it is necessary to keep the concentration of $C_6H_5NH_3^+$ (aniline's conjugate acid, the anilinium ion) below 1.0×10^{-9}M in a solution that is 0.10 M in aniline. Find the concentration of NaOH required for this process.

129. The K_b of hydroxylamine, NH_2OH, is 1.10×10^{-8}. A buffer solution is prepared by mixing 100.0 mL of a 0.36 M hydroxylamine solution with 50.0 mL of a 0.26 M HCl solution. Determine the pH of the resulting solution.

130. A 0.867 g sample of an unknown acid requires 32.2 mL of a 0.182 M barium hydroxide solution for neutralization. Assuming the acid is diprotic, calculate the molar mass of the acid.

131. A 25.0 mL volume of a sodium hydroxide solution requires 19.6 mL of a 0.189 M hydrochloric acid for neutralization. A 10.0 mL volume of a phosphoric acid solution requires 34.9 mL of the sodium hydroxide solution for complete neutralization. Calculate the concentration of the phosphoric acid solution.

132. Determine the mass of sodium formate that must be dissolved in 250.0 cm^3 of a 1.4 M solution of formic acid to prepare a buffer solution with pH = 3.36.

133. What relative masses of dimethyl amine and dimethyl ammonium chloride do you need to prepare a buffer solution of pH = 10.43?

134. You are asked to prepare 2.0 L of a HCN/NaCN buffer that has a pH of 9.8 and an osmotic pressure of 1.35 atm at 298 K. What masses of HCN and NaCN should you use to prepare the buffer? (Assume complete dissociation of NaCN.)

135. What should the molar concentrations of benzoic acid and sodium benzoate be in a solution that is buffered at a pH of 4.55 and has a freezing point of −2.0 °C? (Assume complete dissociation of sodium benzoate and a density of 1.01 g/mL for the solution.)

CHALLENGE PROBLEMS

136. Derive an equation similar to the Henderson–Hasselbalch equation for a buffer composed of a weak base and its conjugate acid. Instead of relating pH to pK_a and the relative concentrations of an acid and its conjugate base (as the Henderson–Hasselbalch equation does), the equation should relate pOH to pK_b and the relative concentrations of a base and its conjugate acid.

137. Soap and detergent action is hindered by hard water so that laundry formulations usually include water softeners—called builders—designed to remove hard water ions (especially Ca^{2+} and Mg^{2+}) from the water. A common builder used in North America is sodium carbonate. Suppose that the hard water used to do laundry contains 75 ppm $CaCO_3$ and 55 ppm $MgCO_3$ (by mass). What mass of Na_2CO_3 is required to remove 90.0% of these ions from 10.0 L of laundry water?

138. A 0.558 g sample of a diprotic acid with a molar mass of 255.8 g/mol is dissolved in water to a total volume of 25.0 mL. The solution is then titrated with a saturated calcium hydroxide solution.

a. Assuming that the pK_a values for each ionization step are sufficiently different to see two equivalence points, determine the volume of added base for the first and second equivalence points.

b. The pH after adding 25.0 mL of the base is 3.82. Find the value of K_{a_1}.

c. The pH after adding 20.0 mL past the first equivalence point is 8.25. Find the value of K_{a_2}.

139. When excess solid $Mg(OH)_2$ is shaken with 1.00 L of 1.0 M NH_4Cl solution, the resulting saturated solution has pH = 9.00. Calculate the K_{sp} of $Mg(OH)_2$.

140. What amount of solid NaOH must be added to 1.0 L of a 0.10 M H_2CO_3 solution to produce a solution with $[H^+] = 3.2 \times 10^{-11}$ M? There is no significant volume change as the result of the addition of the solid.

141. Calculate the solubility of $Au(OH)_3$ in (a) water and (b) 1.0 M nitric acid solution ($K_{sp} = 5.5 \times 10^{-46}$).

142. Calculate the concentration of I^- in a solution obtained by shaking 0.10 M KI with an excess of AgCl(*s*).

143. What volume of 0.100 M sodium carbonate solution is required to precipitate 99% of the Mg from 1.00 L of 0.100 M magnesium nitrate solution?

144. Find the solubility of CuI in 0.40 M HCN solution. The K_{sp} of CuI is 1.1×10^{-12} and the K_f for the $Cu(CN)_2^-$ complex ion is 1×10^{24}.

145. Find the pH of a solution prepared from 1.0 L of a 0.10 M solution of $Ba(OH)_2$ and excess $Zn(OH)_2(s)$. The K_{sp} of $Zn(OH)_2$ is 3×10^{-15} and the K_f of $Zn(OH)_4^{2-}$ is 2×10^{15}.

146. What amount of HCl gas must be added to 1.00 L of a buffer solution that contains [acetic acid] = 2.0 M and [acetate] = 1.0 M in order to produce a solution with pH = 4.00?

CONCEPTUAL PROBLEMS

147. Without doing any calculations, determine if pH = pK_a, pH > pK_a, or pH < pK_a. Assume that HA is a weak monoprotic acid.

a. 0.10 mol HA and 0.050 mol of A^- in 1.0 L of solution
b. 0.10 mol HA and 0.150 mol of A^- in 1.0 L of solution
c. 0.10 mol HA and 0.050 mol of OH^- in 1.0 L of solution
d. 0.10 mol HA and 0.075 mol of OH^- in 1.0 L of solution

148. A buffer contains 0.10 mol of a weak acid and 0.20 mol of its conjugate base in 1.0 L of solution. Determine whether or not each addition exceeds the capacity of the buffer.

a. adding 0.020 mol of NaOH
b. adding 0.020 mol of HCl
c. adding 0.10 mol of NaOH
d. adding 0.010 mol of HCl

149. Consider three solutions:

i. 0.10 M solution of a weak monoprotic acid
ii. 0.10 M solution of strong monoprotic acid
iii. 0.10 M solution of a weak diprotic acid

Each solution is titrated with 0.15 M NaOH. Which quantity is the same for all three solutions?

a. the volume required to reach the final equivalence point
b. the volume required to reach the first equivalence point
c. the pH at the first equivalence point
d. the pH at one-half the first equivalence point

150. Equal volumes of two monoprotic acid solutions (A and B) are titrated with identical NaOH solutions. The volume need to reach the equivalence point for solution A is twice the volume required to reach the equivalence point for solution B, and the pH at the equivalence point of solution A is higher than the pH at the equivalence point for solution B. Which statement is true?

a. The acid in solution A is more concentrated than in solution B and is also a stronger acid than that in solution B.
b. The acid in solution A is less concentrated than in solution B and is also a weaker acid than that in solution B.
c. The acid in solution A is more concentrated than in solution B and is also a weaker acid than that in solution B.
d. The acid in solution A is less concentrated than in solution B and is also a stronger acid than that in solution B.

151. Describe the solubility of CaF_2 in each solution compared to its solubility in water.

a. in a 0.10 M NaCl solution
b. in a 0.10 M NaF solution
c. in a 0.10 M HCl solution

ANSWERS TO CONCEPTUAL CONNECTIONS

Cc 18.1 **(d)** Only this solution contains significant amounts of a weak acid and its conjugate base. (Remember that HNO_3 is a strong acid, but HNO_2 is a weak acid.)

Cc 18.2 **(a)** Since the pH of the buffer is less than the pK_a of the acid, the buffer must contain more acid than base ($[HA] > [A^-]$). In order to raise the pH of the buffer from 4.25 to 4.72, you must add more of the weak base (adding a base will make the buffer solution more basic).

Cc 18.3 **(b)** Since acid is added to the buffer, the pH will become slightly lower (slightly more acidic). Answer **(a)** reflects too large a change in pH for a buffer, and answers **(c)** and **(d)** have the pH changing in the wrong direction.

Cc 18.4 **(a)** Adding 0.050 mol of HCl destroys the buffer because it will react with all of the NaF, leaving no conjugate base in the buffer mixture.

Cc 18.5 **(d)** Because the flask contains 7 H^+ ions, the equivalence point is reached when 7 OH^- ions have been added.

Cc 18.6 **(c)** The pH at the half-equivalence point is the pK_a of the conjugate acid, which is equal to $14.00 - 8.75 = 5.25$.

Cc 18.7 **(c)** Since the volumes and concentrations of all three acids are the same, the volume of NaOH required to reach the first equivalence point (and the only equivalence point for titrations i and iii) is the same for all three titrations.

Cc 18.8 **(c)** The sodium nitrate solution is the only one that has no common ion with barium sulfate. The other two solutions have common ions with barium sulfate; therefore, the solubility of barium sulfate is lower in these solutions.

Cc 18.9 **(c)** Only NaCN contains an anion (CN^-) that forms a complex ion with Cu^{2+} [from Table 18.3 we can see that $K_f = 1.0 \times 10^{25}$ for $Cu(CN)_4{}^{2-}$]. Therefore, the presence of CN^- will drive the dissolution reaction of CuS.

CHAPTER 19

"The law that entropy always increases—the second law of thermodynamics—holds, I think, the supreme position among the laws of Nature."

—Sir Arthur Eddington (1882–1944)

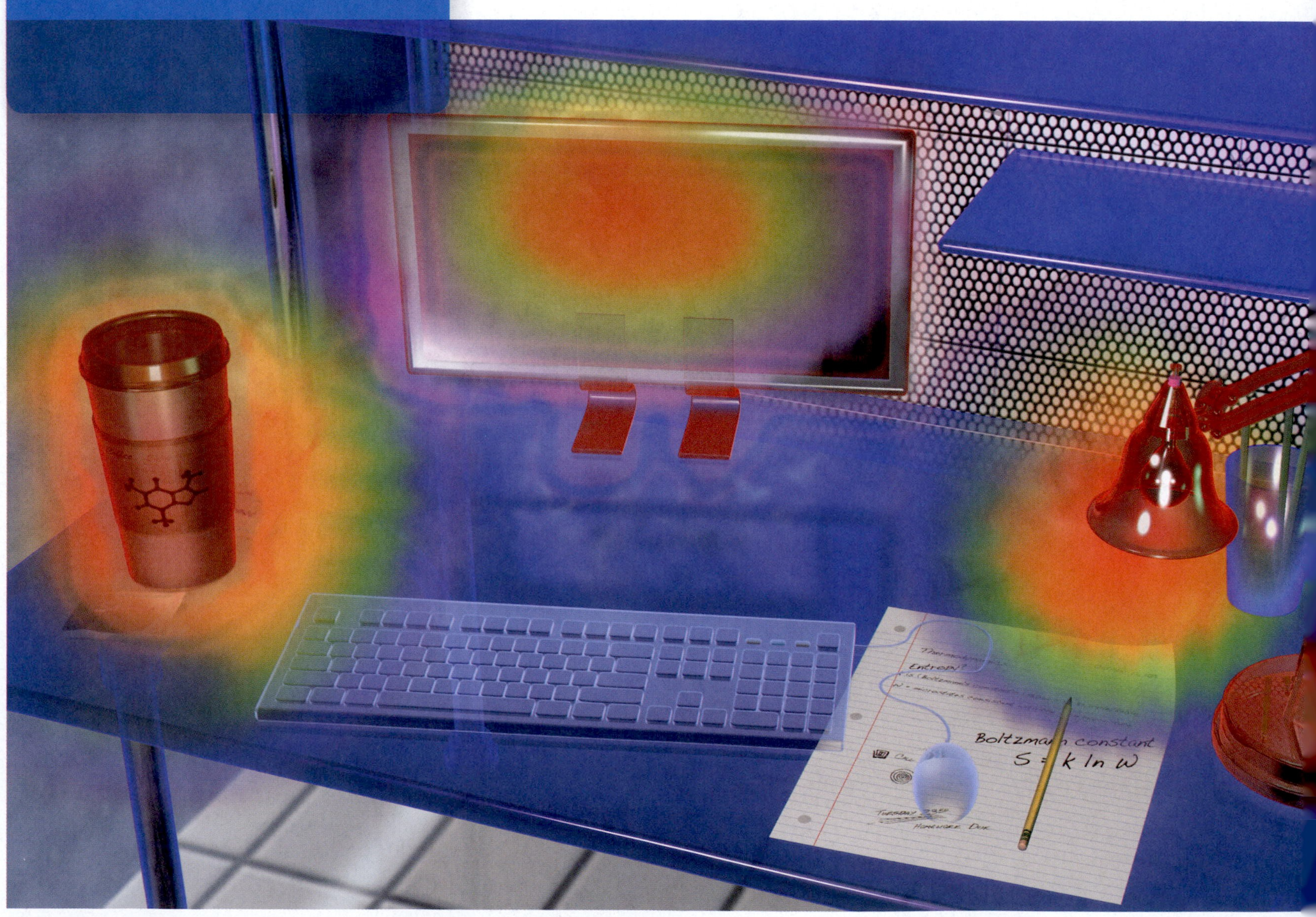

Energy spreads out. Object that are warm disperse energy to their surroundings.

Free Energy and Thermodynamics

19.1 Energy Spreads Out 767

19.2 Spontaneous and Nonspontaneous Processes 768

19.3 Entropy and the Second Law of Thermodynamics 769

19.4 Predicting Entropy and Entropy Changes for Chemical Reactions 774

19.5 Heat Transfer and Entropy Changes of the Surroundings 780

19.6 Gibbs Free Energy 784

19.7 Free Energy Changes in Chemical Reactions: Calculating ΔG°_{rxn} 788

19.8 Free Energy Changes for Nonstandard States: The Relationship between ΔG°_{rxn} and ΔG_{rxn} 794

19.9 Free Energy and Equilibrium: Relating ΔG°_{rxn} to the Equilibrium Constant (K) 797

Key Learning Outcomes 802

THROUGHOUT THIS BOOK, we have examined chemical and physical changes. We have studied how fast chemical changes occur (kinetics) and how to predict how far they will go (through the use of equilibrium constants). We have learned that acids neutralize bases and that gases expand to fill their containers. We now turn to the following question: Why do these changes occur in the first place? What ultimately drives physical and chemical changes in matter? The answer may surprise you. The driving force behind chemical and physical change in the universe is a quantity called *entropy,* which is related to the *dispersion* (or spreading out) of energy. In this chapter, we examine entropy and how we can use it to predict the direction of spontaneous change in chemical processes.

19.1 Energy Spreads Out

We don't have to look very far to observe one of the most fundamental laws of our universe. Sir Arthur Eddington, an early-20th-century astrophysicist and philosopher, called this law "*the most supreme*" among physical laws (see chapter opening quote). He also declared it "*the arrow of time*" and said that any theory found to be inconsistent with it will "*collapse in deepest humiliation.*" Most modern scientists would probably agree with Eddington. The law is called *the second law of thermodynamics*, and we examine a more complete definition of it later in this chapter; however, we can describe it simply in three words: *Energy spreads out.* In all that happens all around us all the time, energy spreads out.

We can see the second law at work by simply observing a hot cup of coffee or a warm computer monitor—they both cool down. The thermal energy concentrated in the hot coffee and the warm monitor *spreads out* into the surroundings. The coffee and the monitor cool down, and the surroundings warm up (just a bit). Energy spreads out (or disperses) from a more concentrated to a less concentrated form. We can also observe the second law by letting go of a book and watching it fall to the floor. As the book falls, the potential energy (due to gravity) that was concentrated in the book gets converted into kinetic energy (energy due to motion) and then dissipates as thermal energy when the book hits the floor. Notice that the potential energy that was concentrated in the book spreads out onto the floor—again, energy spreads out. As another example, consider heating one end of a metal rod. As one end gets hot, the thermal energy spreads out over the entire length of the rod. The second law is so fundamental that all changes—including chemical changes—are governed by it, and no exception has ever been found.

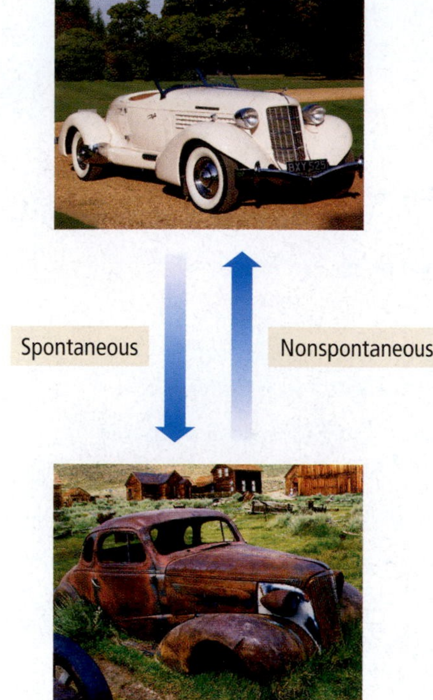

▲ Iron spontaneously rusts when it comes in contact with oxygen.

19.2 Spontaneous and Nonspontaneous Processes

The study of thermodynamics helps us to understand and predict whether a particular process will spontaneously occur. For example, will rust spontaneously form when iron comes into contact with oxygen? Will water spontaneously decompose into hydrogen and oxygen? A **spontaneous process** is one that occurs *without ongoing outside intervention* (such as the performance of work by some external force). For example, as we just discussed, when we drop a book in a gravitational field, it spontaneously drops to the floor. When we place a ball on a slope, it spontaneously rolls down the slope. Both of these processes are spontaneous.

For chemical processes, we must be careful not to confuse the *spontaneity* of the process with its *speed.* In chemical thermodynamics, we study the *spontaneity* of a reaction—the direction in which and the extent to which a chemical reaction proceeds. In kinetics, we study the *speed* of the reaction—how fast it takes place (Figure 19.1 ▼). A reaction may be thermodynamically spontaneous but kinetically slow at a given temperature. For example, the conversion of diamond to graphite is thermodynamically spontaneous. But your diamonds will not become worthless anytime soon because the process is extremely slow kinetically. Although the rate of a spontaneous process can be increased by using a catalyst, a catalyst cannot make a nonspontaneous process spontaneous. Catalysts affect only the rate of a reaction, not the spontaneity.

One last word about nonspontaneity—a nonspontaneous process is not *impossible.* For example, the extraction of iron metal from iron ore is a nonspontaneous process; it does not happen if the iron ore is left to itself, but that does not mean it is impossible. As we will see later in this chapter, a nonspontaneous

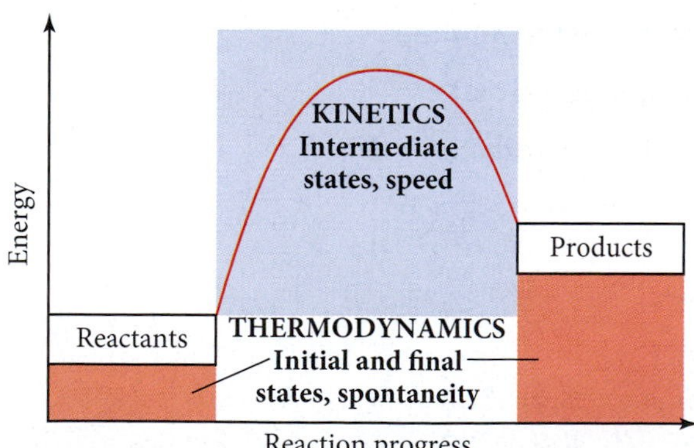

▲ **FIGURE 19.1 Thermodynamics and Kinetics** Thermodynamics deals with the relative chemical potentials of the reactants and products. It enables us to predict whether a reaction will be spontaneous and to calculate how much work it can do. Kinetics involves the chemical potential of intermediate states and enables us to determine why a reaction is slow or fast.

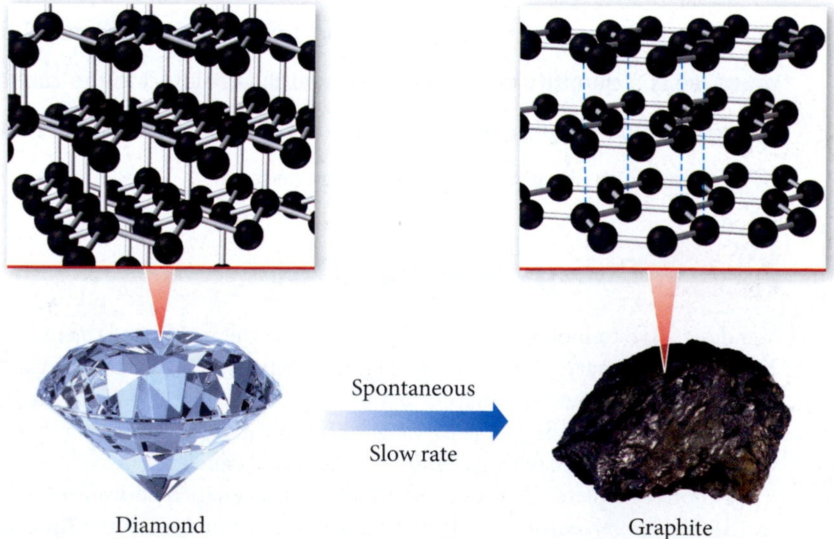

▲ Even though graphite is thermodynamically more stable than diamond, the conversion of diamond to graphite is kinetically so slow that it does not occur at any measurable rate.

process can be made spontaneous by coupling it to another process that is spontaneous or by supplying energy from an external source. Iron can be separated from its ore if external energy is supplied, usually by means of another reaction (that is highly spontaneous).

19.3 Entropy and the Second Law of Thermodynamics

KEY CONCEPT VIDEO Entropy and the Second Law of Thermodynamics

It might initially seem that the way to judge the spontaneity of a chemical process is to examine the change in enthalpy for the process (see Section 10.6). Perhaps, just as a mechanical system proceeds in the direction of lowest potential energy, so a chemical system might proceed in the direction of lowest enthalpy. If this were the case, all exothermic reactions would be spontaneous and all endothermic reactions would not. However, although *most* spontaneous processes are exothermic, some spontaneous processes are *endothermic*. For example, liquid water spilled on the floor at room temperture spontaneously evaporates even though the process is endothermic. Similarly, sodium chloride spontaneously dissolves in water even though the process is slightly endothermic. These examples indicate that enthalpy is not the sole criterion for spontaneity.

The evaporation of water and the dissolution of sodium chloride in water are both endothermic *and* spontaneous. What drives these processes? In each process, energy spreads out (or disperses) as the process occurs. During the evaporation of water, the thermal energy of the water molecules spreads out from a smaller volume (the volume of the liquid) into a much larger volume (the volume of the gas) as the water evaporates. Recall that a gas occupies a much larger volume than its corresponding liquid because the molecules have much more empty space between them.

$H_2O(g)$

Increasing entropy

$H_2O(l)$

▲ When water evaporates, the thermal energy contained in the water molecules spreads out over a larger volume.

In the dissolution of a salt into water, the thermal energy contained in the salt crystals spreads out over the entire volume of the solution as the salt dissolves.

◀ When salt dissolves in water, entropy increases.

In both of these processes, a quantity called *entropy*—related to the spreading out of energy—increases.

Entropy

Entropy is the criterion for spontaneity in all systems, including chemical systems. Formally, **entropy**, abbreviated by the symbol S, has the following definition:

> **Entropy (S) is a thermodynamic function that increases with the number of *energetically equivalent* ways to arrange the components of a system to achieve a particular state.**

Ludwig Boltzmann expressed this definition mathematically in the 1870s as:

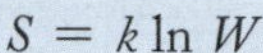

$$S = k \ln W$$

▲ Boltzmann's equation is engraved on his tombstone.

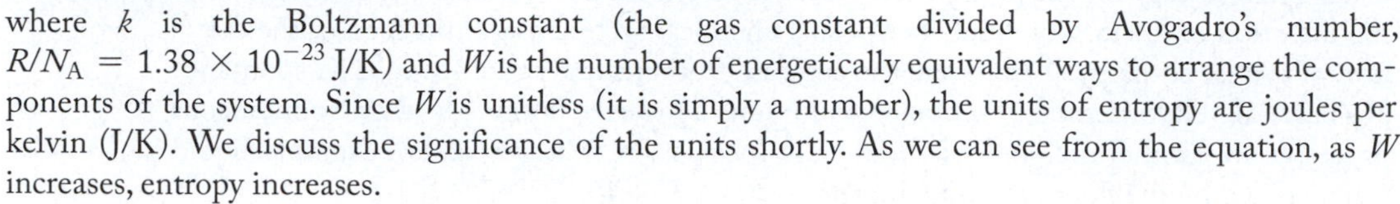

where k is the Boltzmann constant (the gas constant divided by Avogadro's number, $R/N_A = 1.38 \times 10^{-23}$ J/K) and W is the number of energetically equivalent ways to arrange the components of the system. Since W is unitless (it is simply a number), the units of entropy are joules per kelvin (J/K). We discuss the significance of the units shortly. As we can see from the equation, as W increases, entropy increases.

The key to understanding entropy is the quantity W. What does W—the number of energetically equivalent ways to arrange the components of the system—signify? In order to answer this question, we must first understand how the energy of a system can distribute itself among the system's particles.

Imagine a system of particles such as a fixed amount of an ideal gas in a container. The particles have kinetic energy and are colliding with each other and the walls of their container. *A given set of conditions (P, V, and T) defines the* **macrostate** (or state) *of the system.* The overall energy of a macrostate is constant as long as the conditions remain constant. However, the exact distribution of that energy is anything but constant.

At any one instant, a particular gas particle may have lots of kinetic energy, but at the next instant, it may have very little (because it lost its energy through collisions with other particles). *We call the exact internal energy distribution among the particles at any one instant a* **microstate**. We can think of a microstate as a snapshot of the system at a given instant in time. The next instant, the snapshot (the microstate) changes. However, the *macrostate*—defined by P, V, and T—remains constant. Many different microstates can give rise to a particular macrostate. In fact, the microstate (or snapshot) of a given macrostate is generally different from one moment to the next as the energy of the system constantly redistributes itself among the particles of the system.

The quantity, W, is the number of *possible* microstates that can result in a given macrostate. For example, suppose that we have two systems (call them System A and System B) and that each is composed of two particles (one blue and one red). Both systems have a total energy of 4 J, but System A has only one energy level and System B has two:

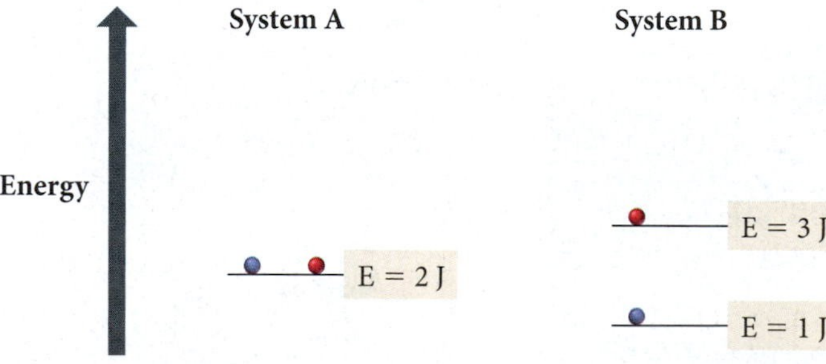

Each system has the same total energy (4 J), but System A has only one possible microstate (the red and the blue particles both occupying the 2 J energy level), while System B has a second possible microstate:

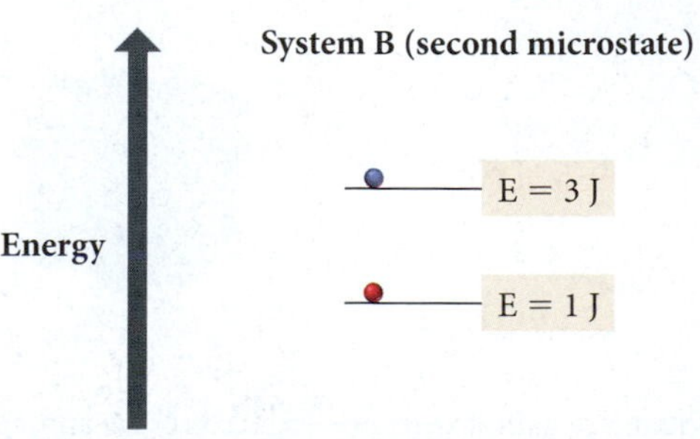

In this second microstate for System B, the blue particle has 3 J and the red one has 1 J (as opposed to System B's first microstate, where the energy of the particles is switched). This second microstate is not possible for System A because it has only one energy level. For System A, $W = 1$, but for System B, $W = 2$. In other words, System B has more microstates in the same 4 J macrostate than System A. Since W is larger for System B than for System A, System B has greater *entropy*; it has more *energetically equivalent ways to arrange the components of the system.*

We can best understand entropy by turning our attention to energy for a moment. The entropy of a macrostate of a system increases with the number of *energetically equivalent* ways to arrange the components of the system to achieve that particular macrostate. This implies that *the state with the highest entropy also has the greatest dispersal of energy.* Returning to our previous example, we find that the energy of System B is dispersed over two energy levels instead of being confined to just one. At the heart of entropy is the concept of energy dispersal or energy randomization. *A state in which a given amount of energy is more highly dispersed (or more highly randomized) has more entropy than a state in which the same energy is more highly concentrated.*

The Second Law of Thermodynamics

We have already alluded to the **second law of thermodynamics**, and now we can formally define it:

For any spontaneous process, the entropy of the *universe* increases ($\Delta S_{univ} > 0$).

The criterion for spontaneity is the entropy of the universe. Processes that increase the entropy of the universe—those that result in greater dispersal or randomization of energy—occur spontaneously. Processes that decrease the entropy of the universe do not occur spontaneously.

See the discussion of state functions in Section 10.3.

Entropy, like enthalpy, is a *state function*—its value depends only on the state of the system, not on how the system arrived at that state. Therefore, for any process, *the change in entropy is the entropy of the final state minus the entropy of the initial state.*

$$\Delta S = S_{final} - S_{initial}$$

Entropy determines the direction of chemical and physical change. *A chemical system proceeds in a direction that increases the entropy of the universe*—it proceeds in a direction that has the largest number of *energetically equivalent* ways to arrange its components.

Macrostates and Microstates

To better understand the second law, let us examine the expansion of an ideal gas into a vacuum (a spontaneous process with no associated change in enthalpy). Consider a flask containing an ideal gas that is connected to an evacuated flask by a tube equipped with a stopcock. When the stopcock is opened, the gas spontaneously expands into the evacuated flask. Since the gas is expanding into a vacuum, the pressure against which it expands is zero, and therefore the work ($w = -P_{ext}\Delta V$) is also zero (See Section 10.4).

However, even though the total energy of the gas does not change during the expansion, the entropy does change. As an illustration, consider a simplified system containing only four gas atoms.

Other states, such as those in which three atoms are in one flask and one atom is in the other, are omitted here to simplify the discussion.

When the stopcock is opened, several possible energetically equivalent final states may result, each with the four atoms distributed in a different way. For example, there could be three atoms in the flask on the left and one in the flask on the right, or vice versa. For simplicity, we consider only the possibilities shown in the right margin: state A, state B, and state C. Since the energy of any one atom is the same in either flask and the atoms do not interact, states A, B, and C are energetically equivalent.

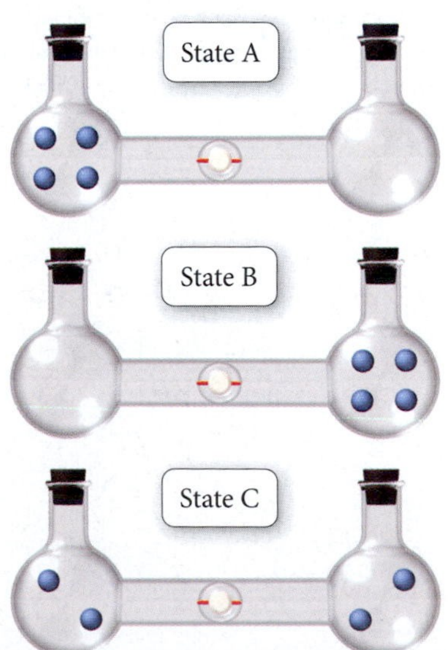

Now we ask the following question for each state: How many microstates give rise to the same macrostate? To keep track of the microstates we label the atoms 1–4. However, even though the atoms have different numbered labels, they are all identical. For states A and B, only one microstate results in the specified macrostate—atoms 1–4 on the left side or the right side, respectively.

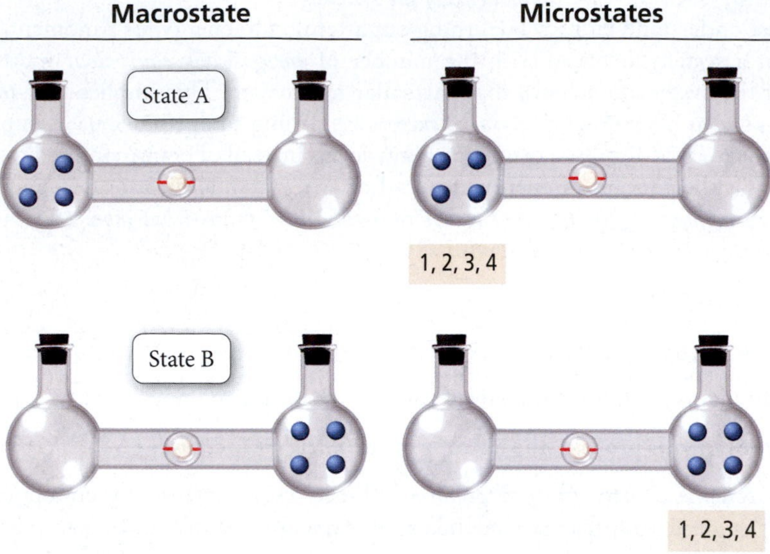

For state C, however, six different possible microstates all result in the same macrostate (two atoms on each side).

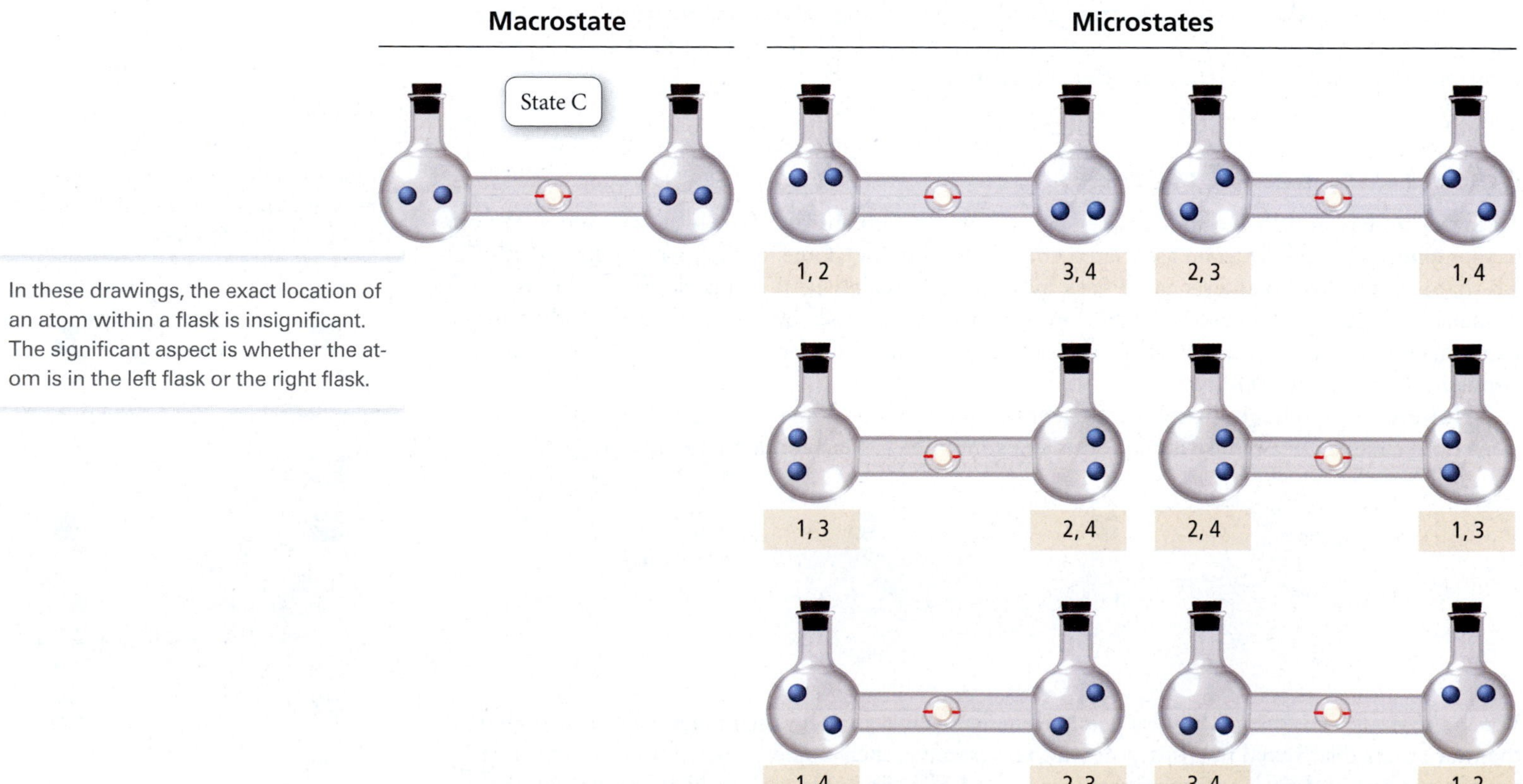

In these drawings, the exact location of an atom within a flask is insignificant. The significant aspect is whether the atom is in the left flask or the right flask.

This means that the statistical probability of finding the atoms in state C is six times greater than the probability of finding the atoms in state A or state B. Consequently, even for a simple system consisting of only four atoms, the atoms are most likely to be found in state C. State C has the greatest entropy—it has the greatest number of energetically equivalent ways to distribute its components.

As the number of atoms increases, the number of microstates that leads to the atoms being equally distributed between the two flasks increases dramatically. For example, with 10 atoms, the number of microstates leading to an equal distribution of atoms between two flasks is 252 and with 20 atoms the number of microstates is 184,756. Yet, the number of microstates that leads to all of the atoms being only in the left flask (or all only in the right flask) does not increase—it is always only 1. In other words, the arrangement in which the atoms are equally distributed between the two flasks has a much larger number of possible microstates and therefore much greater entropy. The system thus tends toward that state.

For n particles, the number of ways to put r particles in one flask and $n - r$ particles in the other flask is $n!/[(n - r)!r!]$. For 10 atoms, $n = 10$ and $r = 5$.

The *change in entropy* in transitioning from a state in which all of the atoms are in the left flask to the state in which the atoms are evenly distributed between both flasks is *positive* because the final state has greater entropy than the initial state:

$$\Delta S = S_{\text{final}} - S_{\text{initial}}$$

Entropy of state in which atoms are distributed between both flasks

Entropy of state in which atoms are all in one flask

Since S_{final} is greater than S_{initial}, ΔS is positive and the process is spontaneous according to the second law. Notice that when the atoms are confined to one flask, their energy is also confined to that one flask; however, when the atoms are evenly distributed between both flasks, their energy is spread out over a greater volume. As the gas expands into the empty flask, energy is dispersed.

The second law explains many phenomena not explained by the first law of thermodynamics. In Section 19.1, we saw that heat travels from a substance at higher temperature (such as a hot cup of coffee) to one at lower temperature (such as the surrounding cooler air). Why? The first law of thermodynamics would not prohibit some heat from flowing the other way—from the surrounding cooler air into the hot cup of coffee. The surroundings could lose 100 J of heat (cooling even more), and the coffee could gain 100 J of heat (warming even more). The first law would not be violated by such a heat transfer. Imagine having your cup of coffee get warmer as it absorbed thermal energy from the surroundings! It will never happen because heat transfer from cold to hot violates the second law of thermodynamics. According to the second law, energy is dispersed, not concentrated. The transfer of heat from a substance of higher temperature to one of lower temperature results in greater energy randomization—the energy that was concentrated in the hot substance becomes dispersed into the surroundings. The second law accounts for this pervasive tendency.

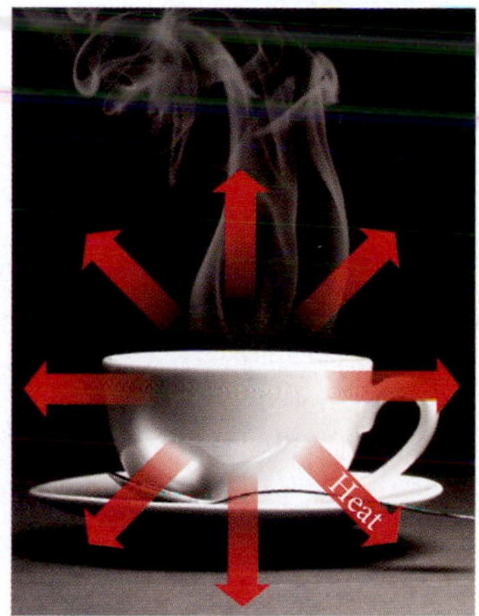

The Units of Entropy

We can understand entropy better by understanding its units—joules per kelvin (J/K). The numerator, joules, is a unit of energy. Since entropy is a measure of energy dispersion, the joules in the numerator represent the amount of energy that is dispersed. As we can see, entropy also depends on temperature in K. However, the temperature unit is in the denominator, so entropy is not just energy dispersion, but energy dispersion per unit temperature. In other words, *entropy is a measure of energy dispersal (joules) per unit temperature (kelvins)*. The higher the temperature, the lower the amount of entropy for a given amount of energy dispersed. For example, if 100 J of energy disperses at a constant temperature of 100 K (and constant pressure), the entropy associated with that energy dispersal is 1 J/K. However, if that same energy disperses at a constant temperature of 10 K (and constant pressure), the energy dispersal is 10 J/K. The same amount of energy has a greater impact at a lower temperature. We examine this concept more closely in Section 19.5. For now, simply remember that entropy is a measure of energy dispersal in a system per unit temperature.

Recall from Section 10.3 that the first law of thermodynamics states that energy is neither created nor destroyed.

19.1 Cc Conceptual Connection

Entropy

Consider these three changes in the possible distributions of six gaseous particles within three interconnected boxes. Which change has a positive ΔS?

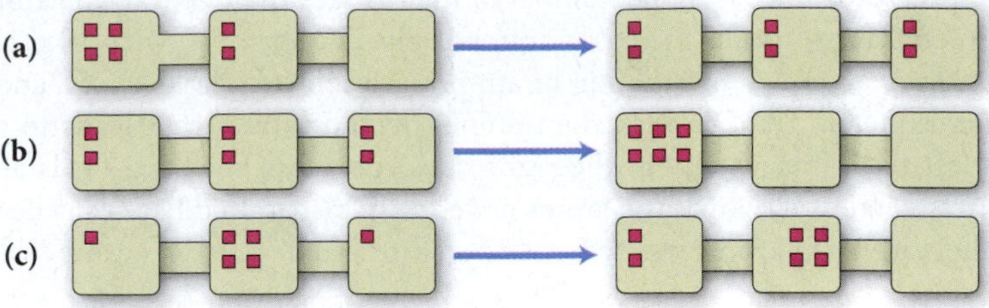

19.4 Predicting Entropy and Entropy Changes for Chemical Reactions

We now turn our attention to predicting and quantifying entropy and entropy changes in a sample of matter. As we examine this topic, we again encounter the theme of this book: *Structure determines properties.* In this case, the property we are interested in is entropy. In this section we see how the structure of the particles that compose a particular sample of matter determines the entropy that the sample possesses at a given temperature and pressure.

The Entropy Change Associated with a Change in State

The entropy of a sample of matter *increases* as it changes state from a solid to a liquid or from a liquid to a gas (Figure 19.2 ▼). The different structures of the three states of matter result in different numbers of energetically equivalent ways of arranging the particles in each state—more ways in the gas than in the liquid, and more in the liquid than in the solid.

A gas has more energetically equivalent ways of arranging its particles because it has more ways to distribute its energy than a solid. The energy in a molecular solid consists largely of the vibrations between its molecules—the molecules vibrate back and forth about relatively fixed positions. If the same substance is vaporized, however, the energy takes the form of straight-line motions of the molecules

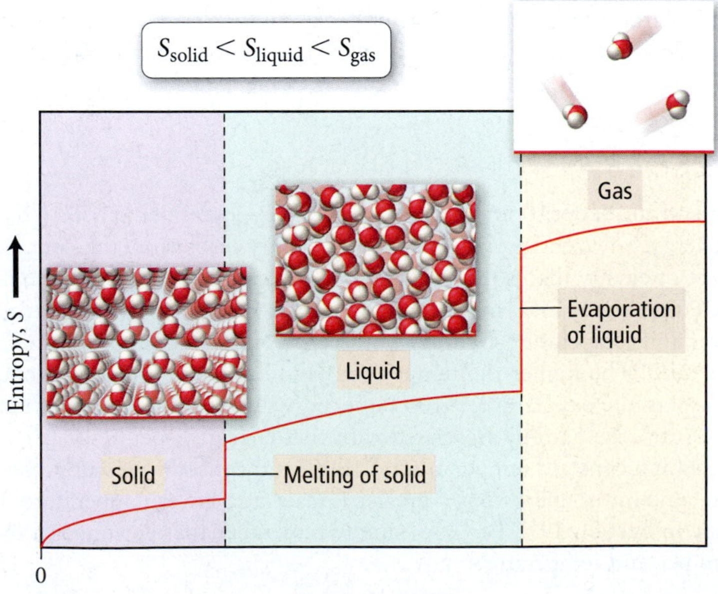

▶ **FIGURE 19.2 Entropy and State Change** Entropy increases when matter changes from a solid to a liquid and from a liquid to a gas.

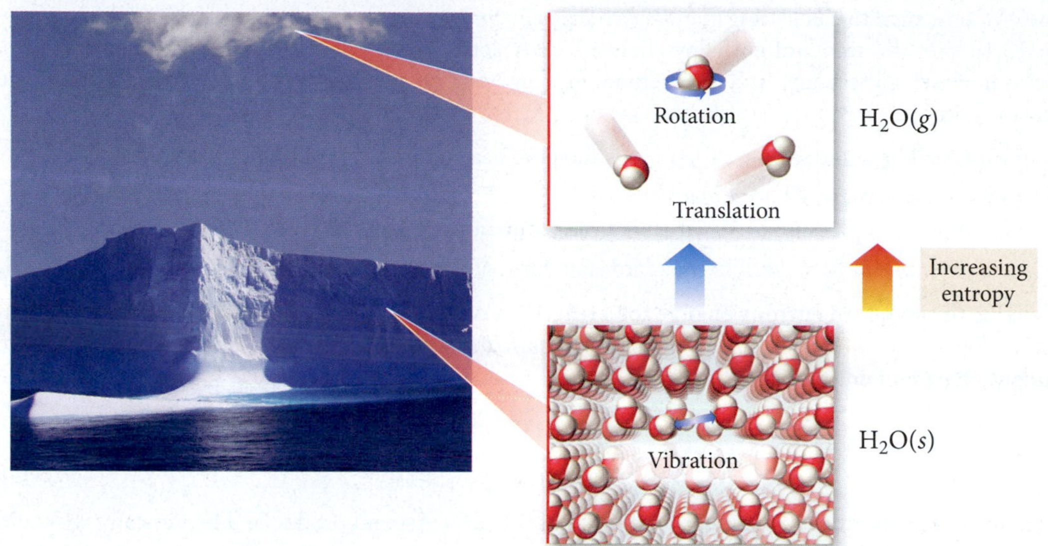

◀ **FIGURE 19.3 "Places" for Energy** In the solid state, energy is contained largely in the vibrations between molecules. In the gas state, energy is contained in both the straight-line motion of molecules (translational energy) and the rotation of molecules (rotational energy).

(called translational energy) and rotations of the molecules (called rotational energy). In other words, when a solid vaporizes into a gas, there are new "places" to put energy (Figure 19.3 ▲). The gas thus has more possible microstates (more energetically equivalent configurations) than the solid and therefore greater entropy.

We can now predict the sign of ΔS for processes involving changes of state (or phase). In general, entropy increases ($\Delta S > 0$) for each of the following:

- the transition from a solid to a liquid
- the transition from a solid to a gas
- the transition from a liquid to a gas
- an increase in the number of moles of a gas during a chemical reaction

EXAMPLE 19.1
Predicting the Sign of Entropy Change

Predict the sign of ΔS for each process.

(a) $H_2O(g) \longrightarrow H_2O(l)$
(b) Solid carbon dioxide sublimes
(c) $2\ N_2O(g) \longrightarrow 2\ N_2(g) + O_2(g)$

SOLUTION

(a) A gas has a greater entropy than a liquid, so the entropy decreases and ΔS is negative.
(b) A solid has a lower entropy than a gas, so the entropy increases and ΔS is positive.
(c) The number of moles of gas increases, so the entropy increases and ΔS is positive.

FOR PRACTICE 19.1

Predict the sign of ΔS for each process.

(a) water boils
(b) $I_2(g) \longrightarrow I_2(s)$
(c) $CaCO_3(s) \longrightarrow CaO(s) + CO_2(g)$

The Entropy Change Associated with a Chemical Reaction ($\Delta S°_{rxn}$)

In Chapter 10, we learned how to calculate standard changes in enthalpy ($\Delta H°_{rxn}$) for chemical reactions. We now turn to calculating standard changes in *entropy* for chemical reactions. Recall from Section 10.10 that the standard enthalpy change for a reaction ($\Delta H°_{rxn}$) is the change in enthalpy for a process in which all reactants and products are in their standard states. Recall also the definitions of the standard state.

The standard state has recently been changed to a pressure of 1 bar, which is very close to 1 atm (1 atm = 1.013 bar). Both standards are now in common use.

- *For a Gas:* The standard state for a gas is the pure gas at a pressure of exactly 1 atm.
- *For a Liquid or Solid:* The standard state for a liquid or solid is the pure substance in its most stable form at a pressure of 1 atm and at the temperature of interest (often taken to be 25 °C).
- *For a Substance in Solution:* The standard state for a substance in solution is a concentration of 1 M.

We define the **standard entropy change for a reaction ($\Delta S°_{rxn}$)** as the change in *entropy* for a process in which all reactants and products are in their standard states. Since entropy is a function of state, the standard change in entropy is the standard entropy of the products minus the standard entropy of the reactants:

$$\Delta S°_{rxn} = S°_{products} - S°_{reactants}$$

But how do we determine the standard entropies of the reactants and products? Theoretically, we could use Boltzmann's definition ($S = k \ln W$); however, practically we rely on tables that come from experimental measurements. Recall from Chapter 10 that we defined *standard molar enthalpies of formation* ($\Delta H°_f$) to use in calculating $\Delta H°_{rxn}$. In this chapter, we define **standard molar entropies ($S°$)** to use in calculating $\Delta S°_{rxn}$.

Standard Molar Entropies ($S°$) and the Third Law of Thermodynamics

In Chapter 10, we defined a *relative* zero for enthalpy. To do this, we assigned a value of zero to the standard enthalpy of formation for an element in its standard state. This was necessary because absolute values of enthalpy cannot be determined. In other words, for enthalpy, there is no absolute zero against which to measure all other values; therefore, we always have to rely on enthalpy changes from an arbitrarily assigned standard (the elements in their standard states and most stable forms). For entropy, however, *there is an absolute zero*. The absolute zero of entropy is established by the **third law of thermodynamics**, which states:

The entropy of a perfect crystal at absolute zero (0 K) is zero.

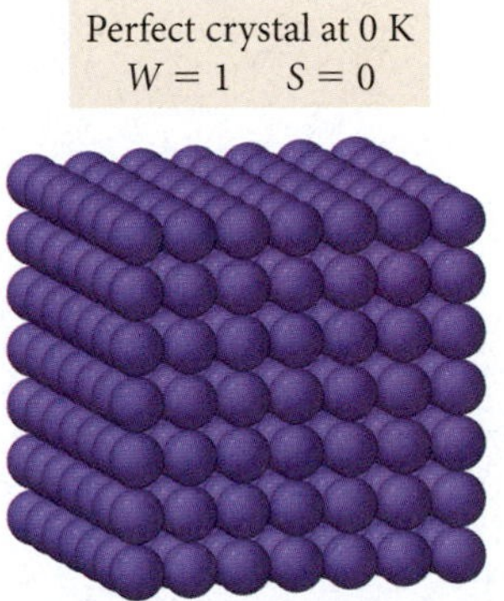

▲ **FIGURE 19.4 Zero Entropy** A perfect crystal at 0 K has only one possible way to arrange its components.

A perfect crystal at a temperature of absolute zero has only one possible way ($W = 1$) to arrange its components (Figure 19.4 ◀). Based on Boltzmann's definition of entropy ($S = k \ln W$), its entropy is zero ($S = k \ln 1 = 0$).

We can measure all entropy values against the absolute zero of entropy as defined by the third law. Table 19.1 (on the next page) lists values of standard entropies at 25 °C for selected substances. A more complete list can be found in Appendix IVB. Standard entropy values are listed per mole of substance, in units of joules per mole per kelvin (J/mol · K). The values are listed per mole because *entropy is an extensive property*—it depends on the amount of the substance.

Some elements exist in two or more forms, called *allotropes*, within the same state.

At 25 °C the standard entropy of any substance is the energy dispersed into one mole of that substance at 25 °C, which depends on the number of "places" to put energy within the substance. The number of "places" to put energy depends on the structure of the substance. In particular, the state of the substance, the molar mass of the substance, the particular allotrope, the molecular complexity of the substance, and its extent of dissolution each affect the entropy contained in the substance (at a given temperature). Let's examine each of these separately.

TABLE 19.1 Standard Molar Entropy Values ($S°$) for Selected Substances at 298 K

Substance	$S°$(J/mol · K)	Substance	$S°$(J/mol · K)	Substance	$S°$(J/mol · K)
Gases		**Liquids**		**Solids**	
$H_2(g)$	130.7	$H_2O(l)$	70.0	$MgO(s)$	27.0
$Ar(g)$	154.8	$CH_3OH(l)$	126.8	$Fe(s)$	27.3
$CH_4(g)$	186.3	$Br_2(l)$	152.2	$Li(s)$	29.1
$H_2O(g)$	188.8	$C_6H_6(l)$	173.4	$Cu(s)$	33.2
$N_2(g)$	191.6			$Na(s)$	51.3
$NH_3(g)$	192.8			$K(s)$	64.7
$F_2(g)$	202.8			$NaCl(s)$	72.1
$O_2(g)$	205.2			$CaCO_3(s)$	91.7
$Cl_2(g)$	223.1			$FeCl_3(s)$	142.3
$C_2H_4(g)$	219.3				

Relative Standard Entropies: Gases, Liquids, and Solids As we saw earlier in this section, the entropy of a gas is generally greater than the entropy of a liquid, which is in turn greater than the entropy of a solid. We can see these trends in the tabulated values of standard entropies. For example, consider the relative standard entropies of liquid water and gaseous water at 25 °C:

	$S°$(J/mol · K)
$H_2O(l)$	70.0
$H_2O(g)$	188.8

Gaseous water has a much greater standard entropy because a gas has more energetically equivalent ways to arrange its components, which in turn results in greater energy dispersal at 25 °C.

Relative Standard Entropies: Molar Mass Consider the standard entropies of the noble gases at 25 °C:

	$S°$ (J/mol · K)
$He(g)$	126.2
$Ne(g)$	146.1
$Ar(g)$	154.8
$Kr(g)$	163.8
$Xe(g)$	169.4

The more massive the noble gas, the greater its entropy at 25 °C. A complete explanation of why entropy increases with increasing molar mass is beyond the scope of this book. Briefly, the energy states associated with the motion of heavy atoms are more closely spaced than those of lighter atoms. The more closely spaced energy states allow for greater dispersal of energy at a given temperature and therefore greater entropy. This trend holds only for elements in the same state. (The effect of a state change—from a liquid to a gas, for example—is far greater than the effect of molar mass.)

Relative Standard Entropies: Allotropes Some elements can exist in two or more forms with different structures—called **allotropes**—in the same state of matter. For example, the allotropes of carbon include diamond and graphite—both solid forms of carbon. Since the arrangement of atoms within these forms is different, their standard molar entropies are different:

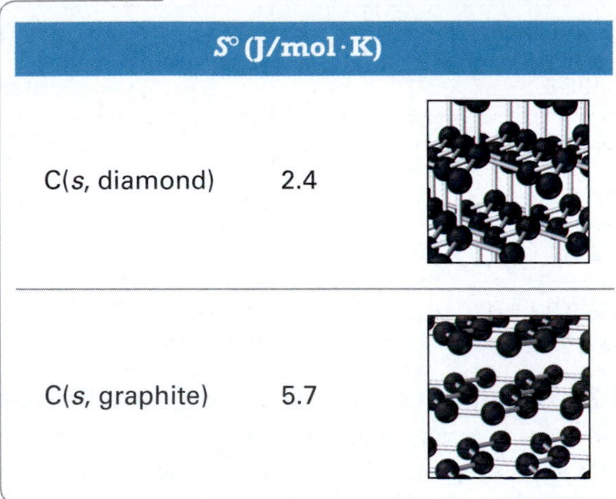

	$S°$ (J/mol·K)
C(*s*, diamond)	2.4
C(*s*, graphite)	5.7

In diamond the atoms are constrained by chemical bonds in a highly restricted three-dimensional crystal structure. In graphite the atoms bond together in sheets, but the sheets have freedom to slide past each other. The less constrained structure of graphite results in more "places" to put energy and therefore greater entropy compared to diamond.

Relative Standard Entropies: Molecular Complexity For a given state of matter, entropy generally increases with increasing molecular complexity. For example, consider the standard entropies of gaseous argon and nitrogen monoxide:

	Molar Mass (g/mol)	$S°$(J/mol·K)
Ar(*g*)	39.948	154.8
NO(*g*)	30.006	210.8

Ar has a greater molar mass than NO, yet it has less entropy at 25 °C. Why? Molecules generally have more "places" to put energy than do atoms. In a gaseous sample of argon, the only form that energy can take is the translational motion of the atoms. In a gaseous sample of NO, on the other hand, energy can take the form of translational motion, rotational motion, and (at high enough temperatures) vibrational motions of the molecules (Figure 19.5 ▼). Therefore, for a given state, molecules generally have

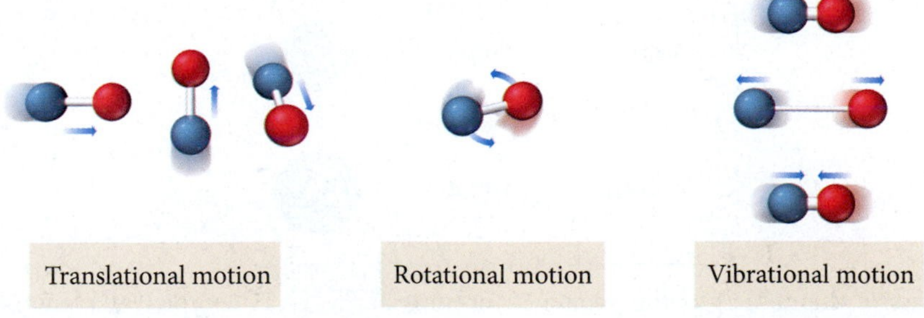

▶ **FIGURE 19.5 "Places" for Energy in Gaseous NO** Energy can be contained in translational motion, rotational motion, and (at high enough temperatures) vibrational motion.

a greater entropy than free atoms. Similarly, more complex molecules generally have more entropy than simpler ones. For example, consider the standard entropies of carbon monoxide and ethene gas:

	Molar Mass (g/mol)	$S°$(J/mol·K)
$CO(g)$	28.01	197.7
$C_2H_4(g)$	28.05	219.3

These two substances have nearly the same molar mass, but the greater complexity of C_2H_4 results in a greater molar entropy. When molecular complexity and molar mass both increase (as is often the case), molar entropy also increases, as demonstrated by the oxides of nitrogen:

	$S°$(J/mol·K)
$NO(g)$	210.8
$NO_2(g)$	240.1
$N_2O_4(g)$	304.4

The increasing molecular complexity as we move down this list, as well as the increasing molar mass, results in more "places" to put energy and therefore greater entropy.

Relative Standard Entropies: Dissolution The dissolution of a crystalline solid into solution usually results in an increase in entropy. For example, consider the standard entropies of solid and aqueous potassium chlorate:

	$S°$(J/mol·K)
$KClO_3(s)$	143.1
$KClO_3(aq)$	265.7

The standard entropies for aqueous solutions are for the solution in its standard state, which is defined as having a concentration of 1 M.

When solid potassium chlorate dissolves in water, the thermal energy that was concentrated within the crystal becomes dispersed throughout the entire solution. The greater energy dispersal results in greater entropy.

Standard Entropies

19.2 Cc Conceptual Connection

Arrange these gases in order of increasing standard molar entropy: SO_3, Kr, Cl_2.

Calculating the Standard Entropy Change ($\Delta S°_{rxn}$) for a Reaction

Since entropy is a state function, and since standard entropies for many common substances are tabulated, we can calculate the standard entropy change for a chemical reaction by calculating the difference in entropy between the products and the reactants. More specifically,

To calculate $\Delta S°_{rxn}$, subtract the standard entropies of the reactants multiplied by their stoichiometric coefficients from the standard entropies of the products multiplied by their stoichiometric coefficients. In the form of an equation:

$$\Delta S°_{rxn} = \sum n_p S°(\text{products}) - \sum n_r S°(\text{reactants}) \qquad [19.1]$$

In Equation 19.1, n_p represents the stoichiometric coefficients of the products, n_r represents the stoichiometric coefficients of the reactants, and $S°$ represents the standard entropies. Keep in mind when using this equation that, *unlike enthalpies of formation, which are zero for elements in their standard states, standard entropies are always nonzero at* 25 °C. Example 19.2 demonstrates the application of Equation 19.1.

EXAMPLE 19.2

Calculating Standard Entropy Changes (ΔS°_{rxn})

Calculate ΔS°_{rxn} for this balanced chemical equation.

$$4\ NH_3(g) + 5\ O_2(g) \longrightarrow 4\ NO(g) + 6\ H_2O(g)$$

SOLUTION

Begin by looking up the standard entropy for each reactant and product in Appendix IVB. Always note the correct state—(*g*), (*l*), (*aq*), or (*s*)—for each reactant and product.

Reactant or product	S°(J/mol · K)
$NH_3(g)$	192.8
$O_2(g)$	205.2
$NO(g)$	210.8
$H_2O(g)$	188.8

Calculate ΔS°_{rxn} by substituting the appropriate values into Equation 19.1. Remember to include the stoichiometric coefficients in your calculation. (The units of S° become J/K when multiplied by the stoichiometric coeffecients in moles.)

$$\begin{aligned}
\Delta S^\circ_{rxn} &= \sum n_p S^\circ(\text{products}) - \sum n_r S^\circ(\text{reactants}) \\
&= [4(S^\circ_{NO(g)}) + 6(S^\circ_{H_2O(g)})] - [4(S^\circ_{NH_3(g)}) + 5(S^\circ_{O_2(g)})] \\
&= [4(210.8\ \text{J/K}) + 6(188.8\ \text{J/K})] - [4(192.8\ \text{J/K}) + 5(205.2\ \text{J/K})] \\
&= 1976.0\ \text{J/K} - 1797.2\ \text{J/K} \\
&= 178.8\ \text{J/K}
\end{aligned}$$

CHECK Notice that ΔS°_{rxn} is positive, as you would expect for a reaction in which the number of moles of gas increases.

FOR PRACTICE 19.2

Calculate ΔS°_{rxn} for this balanced chemical equation.

$$2\ H_2S(g) + 3\ O_2(g) \longrightarrow 2\ H_2O(g) + 2\ SO_2(g)$$

19.5 Heat Transfer and Entropy Changes of the Surroundings

We have now learned how to calculate entropy changes for chemical processes. We also know that the criterion for spontaneity is an *increase* in the entropy of the universe. So, it might seem that, to determine the spontaneity of a chemical process, we simply have to calculate the change in entropy for the process. This is not the case. The criterion for spontaneity is an increase in the entropy of the *universe.* In Chapter 10, we distinguished between a thermodynamic *system* and its *surroundings*. The same distinction is useful in our discussion of entropy. When we calculate the change in entropy for a chemical process, we calculate the change in entropy for the *system* not the *universe*. Several familiar processes result in a decrease in entropy for the system, but are still spontaneous because they increase the entropy of the universe.

For example, when water freezes at temperatures below 0 °C, the entropy of the water decreases, yet the process is spontaneous. Similarly, when water vapor in air condenses into fog on a cold night, the entropy of the water also decreases. Why are these processes spontaneous? Even though the entropy *of the water* decreases during freezing and condensation, the entropy *of the universe* must somehow increase in order for these processes to be spontaneous. For the freezing of water, let us consider the system to be the water. The surroundings are then the rest of the universe. Accordingly, ΔS_{sys} is the entropy change for the water itself, ΔS_{surr} is the entropy change for the surroundings, and ΔS_{univ} is the entropy change for the universe. The entropy change for the universe is the sum of the entropy changes for the system and the surroundings:

$$\Delta S_{univ} = \Delta S_{sys} + \Delta S_{surr}$$

The second law states that the entropy of the universe must increase ($\Delta S_{univ} > 0$) for a process to be spontaneous. The entropy of the *system* can decrease ($\Delta S_{sys} < 0$) as long as the entropy of the *surroundings* increases by a greater amount ($\Delta S_{surr} > -\Delta S_{sys}$), so that the overall entropy of the *universe* undergoes a net increase.

For liquid water freezing or water vapor condensing, we know that the change in entropy for the system ΔS_{sys} is negative (because state changes from liquid to solid and from gas to liquid both result in less entropy). For ΔS_{univ} to be positive, therefore, ΔS_{surr} must be positive and greater in absolute value (or magnitude) than ΔS_{sys} as shown graphically here:

$$\Delta S_{univ} = \Delta S_{sys} + \Delta S_{surr}$$

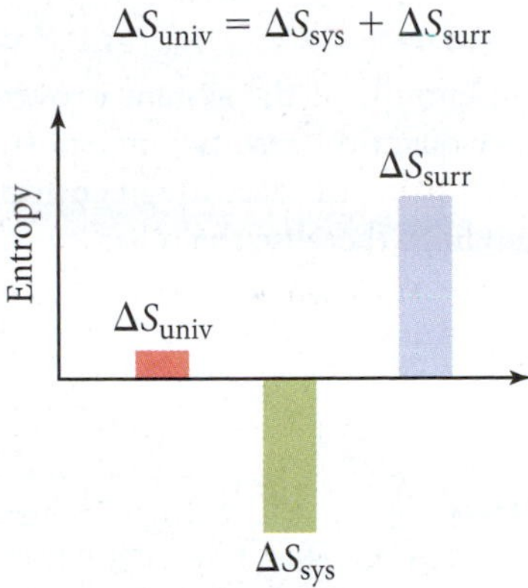

But why does the freezing of ice or the condensation of water increase the entropy of the surroundings? Because both processes are *exothermic:* They give off heat to the surroundings. Because entropy is the dispersal or randomization of energy, *the release of heat energy by the system disperses that energy into the surroundings, increasing the entropy of the surroundings.* The freezing of water below 0 °C and the condensation of water vapor on a cold night both increase the entropy of the universe because the heat given off to the surroundings increases the entropy of the surroundings to a sufficient degree to overcome the entropy decrease in the system.

Even though (as we saw earlier) enthalpy by itself cannot determine spontaneity, the increase in the entropy of the surroundings caused by the release of heat explains why exothermic processes are so *often* spontaneous.

Summarizing Entropy Changes in the Surroundings:

- An exothermic process increases the entropy of the surroundings.
- An endothermic process decreases the entropy of the surroundings.

The Temperature Dependence of ΔS_{surr}

We just discussed how the freezing of water increases the entropy of the surroundings by dispersing heat energy into the surroundings. Yet we know that the freezing of water is not spontaneous at all temperatures. The freezing of water becomes *nonspontaneous* above 0 °C. Why? Because, as we saw in Section 19.3, entropy represents the energy dispersed into a sample of matter per unit temperature—it has the units of joules per kelvin (J/K). The magnitude of the increase in entropy of the surroundings due to the dispersal of energy into the surroundings is *temperature dependent.*

The greater the temperature, the smaller the increase in entropy for a given amount of energy dispersed into the surroundings. The higher the temperature, the lower the amount of entropy for a given amount of energy dispersed. You can understand the temperature dependence of entropy changes due to heat flow with a simple analogy. Imagine that you have \$1000 to give away. If you gave the \$1000 to a rich man, the impact on his net worth would be negligible (because he already has so much money). If you gave the same \$1000 to a poor man, however, his net worth would change substantially (because he has so little money). Similarly, if you disperse 1000 J of energy into surroundings that are hot, the entropy increase is small (because the impact of the 1000 J is small on surroundings that already contain a lot of energy). If you disperse the same 1000 J of energy into surroundings that are cold, however, the entropy increase is large (because the impact of the 1000 J is great on surroundings that contain little energy). For this same reason, the impact of the heat released to the surroundings by the freezing of water depends on the temperature of the surroundings—the higher the temperature, the smaller the impact.

We can now see why water spontaneously freezes at low temperature but not at high temperature. For the freezing of liquid water into ice, the change in entropy of the system is negative at all temperatures.

$$\Delta S_{univ} = \Delta S_{sys} + \Delta S_{surr}$$

Negative (ΔS_{sys})

Positive and large at low temperature
Positive and small at high temperature (ΔS_{surr})

At low temperatures, the decrease in entropy of the system is overcome by the large increase in the entropy of the surroundings (a positive quantity), resulting in a positive ΔS_{univ} and a spontaneous process. At high temperatures, on the other hand, the decrease in entropy of the system is not overcome by the increase in entropy of the surroundings (because the magnitude of the positive ΔS_{surr} is smaller at higher temperatures), resulting in a negative ΔS_{univ}; therefore, the freezing of water is not spontaneous at high temperature as shown graphically here:

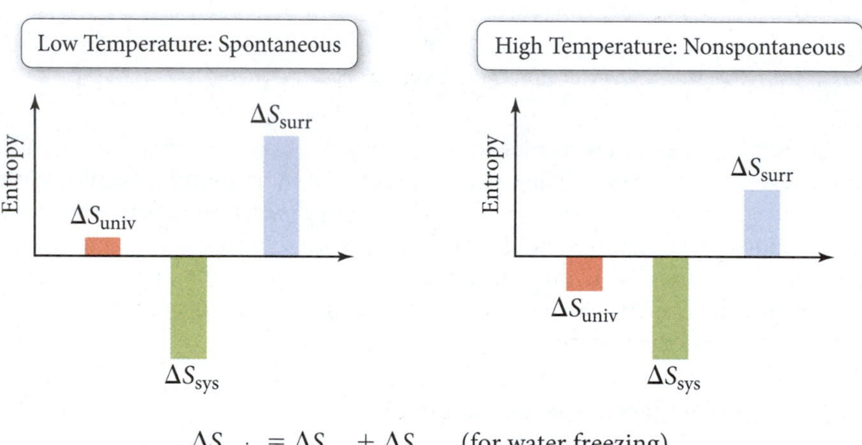

$$\Delta S_{univ} = \Delta S_{sys} + \Delta S_{surr} \quad \text{(for water freezing)}$$

Quantifying Entropy Changes in the Surroundings

We have seen that when a system exchanges heat with the surroundings, it changes the entropy of the surroundings. At constant pressure, we can use q_{sys} to quantify the change in *entropy* for the surroundings (ΔS_{surr}). In general,

- a process that emits heat into the surroundings (q_{sys} negative) *increases* the entropy of the surroundings (positive ΔS_{surr}).
- a process that absorbs heat from the surroundings (q_{sys} positive) *decreases* the entropy of the surroundings (negative ΔS_{surr}).
- the magnitude of the change in entropy of the surroundings is proportional to the magnitude of q_{sys} and inversely proportional to the temperature in kelvins.

We can quantify these relationships with the following proportionalities:

$$\Delta S_{surr} \propto -q_{sys} \qquad [19.2]$$

$$\Delta S_{surr} \propto \frac{1}{T} \qquad [19.3]$$

Combining the proportionalities in Equations 19.2 and 19.3, we get the following general expression at constant temperature:

$$\Delta S_{surr} = \frac{-q_{sys}}{T}$$

For any chemical or physical process occurring at constant temperature and pressure, the entropy change of the surroundings is equal to the energy dispersed into the surroundings ($-q_{sys}$) divided by the temperature of the surroundings in kelvins.

This equation provides insight into why exothermic processes have a tendency to be spontaneous at low temperatures—they increase the entropy of the surroundings. As temperature increases, however, a given negative q produces a smaller positive ΔS_{surr}; thus, exothermicity becomes less of a determining factor for spontaneity as temperature increases.

Under conditions of constant pressure $q_{sys} = \Delta H_{sys}$; therefore,

$$\Delta S_{surr} = \frac{-\Delta H_{sys}}{T} \text{ (constant } P, T) \quad [19.4]$$

EXAMPLE 19.3

Calculating Entropy Changes in the Surroundings

Consider the combustion of propane gas:

$$C_3H_8(g) + 5\,O_2(g) \longrightarrow 3\,CO_2(g) + 4\,H_2O(g) \qquad \Delta H_{rxn} = -2044 \text{ kJ}$$

(a) Calculate the entropy change in the surroundings associated with this reaction occurring at 25 °C.
(b) Determine the sign of the entropy change for the system.
(c) Determine the sign of the entropy change for the universe. Is the reaction spontaneous?

SOLUTION

(a) The entropy change of the surroundings is given by Equation 19.4. Substitute the value of ΔH_{rxn} and the temperature in kelvins and calculate ΔS_{surr}.	$T = 273 + 25 = 298$ K $\Delta S_{surr} = \frac{-\Delta H_{rxn}}{T} = \frac{-(-2044 \text{ kJ})}{298 \text{ K}}$ $= +6.86$ kJ/K $= +6.86 \times 10^3$ J/K
(b) Determine the number of moles of gas on each side of the reaction. An increase in the number of moles of gas implies a positive ΔS_{sys}.	$C_3H_8(g) + 5\,O_2(g) \longrightarrow 3\,CO_2(g) + 4\,H_2O(g)$ 6 mol gas → 7 mol gas ΔS_{sys} is positive.
(c) The change in entropy of the universe is the sum of the entropy changes of the system and the surroundings. If the entropy changes of the system and surroundings are both the same sign, the entropy change for the universe also has the same sign.	$\Delta S_{univ} = \Delta S_{sys} + \Delta S_{surr}$ Positive, Positive Therefore, ΔS_{univ} is positive and the reaction is spontaneous.

FOR PRACTICE 19.3

Consider the reaction between nitrogen and oxygen gas to form dinitrogen monoxide:

$$2\,N_2(g) + O_2(g) \longrightarrow 2\,N_2O(g) \qquad \Delta H_{rxn} = +163.2 \text{ kJ}$$

(a) Calculate the entropy change in the surroundings associated with this reaction occurring at 25 °C.
(b) Determine the sign of the entropy change for the system.
(c) Determine the sign of the entropy change for the universe. Is the reaction spontaneous?

FOR MORE PRACTICE 19.3

A reaction has $\Delta H_{rxn} = -107$ kJ and $\Delta S_{rxn} = 285$ J/K. At what temperature is the change in entropy for the reaction equal to the change in entropy for the surroundings?

19.3 Cc Conceptual Connection

Entropy and Biological Systems

Do biological systems contradict the second law of thermodynamics? By taking energy from their surroundings and synthesizing large, complex biological molecules, plants and animals tend to concentrate energy, not disperse it. How can this be so?

19.6 Gibbs Free Energy

Equation 19.4 in Section 19.5 establishes a relationship between the enthalpy change in a system and the entropy change in the surroundings. Recall that for any process the entropy change of the universe is the sum of the entropy change of the system and the entropy change of the surroundings:

$$\Delta S_{univ} = \Delta S_{sys} + \Delta S_{surr} \quad [19.5]$$

Combining Equation 19.5 with Equation 19.4 gives us the following relationship at constant temperature and pressure:

$$\Delta S_{univ} = \Delta S_{sys} - \frac{\Delta H_{sys}}{T} \quad [19.6]$$

Using Equation 19.6, we can calculate ΔS_{univ} while focusing only on the *system*. If we multiply Equation 19.6 by $-T$, we arrive at the expression:

$$-T\Delta S_{univ} = -T\Delta S_{sys} + T\frac{\Delta H_{sys}}{T}$$

$$= \Delta H_{sys} - T\Delta S_{sys} \quad [19.7]$$

If we drop the subscript *sys*—from now on ΔH and ΔS without subscripts mean ΔH_{sys} and ΔS_{sys}—we get the expression:

$$-T\Delta S_{univ} = \Delta H - T\Delta S \quad [19.8]$$

The right-hand side of Equation 19.8 represents the change in a thermodynamic function called *Gibbs free energy*. The formal definition of **Gibbs free energy (*G*)** is:

$$G = H - TS \quad [19.9]$$

where H is enthalpy, T is the temperature in kelvins, and S is entropy. The *change* in Gibbs free energy, symbolized by ΔG, is expressed as follows (at constant temperature):

$$\Delta G = \Delta H - T\Delta S \quad [19.10]$$

If we combine Equations 19.8 and 19.10, we have an equation that makes clear the significance of ΔG:

$$\Delta G = -T\Delta S_{univ} \quad \text{(constant } T, P) \quad [19.11]$$

The change in Gibbs free energy for a process occurring at constant temperature and pressure is proportional to the negative of ΔS_{univ}. Since ΔS_{univ} is a criterion for spontaneity, ΔG is also a criterion for spontaneity (although opposite in sign). In fact, Gibbs free energy is also called *chemical potential* because it is analogous to mechanical potential energy discussed earlier. Just as mechanical systems tend toward lower potential energy, so chemical systems tend toward lower Gibbs free energy (toward lower chemical potential) (Figure 19.6 ▶).

Summarizing Gibbs Free Energy (at Constant Temperature and Pressure):

- ΔG is proportional to the negative of ΔS_{univ}.
- A decrease in Gibbs free energy ($\Delta G < 0$) corresponds to a spontaneous process.
- An increase in Gibbs free energy ($\Delta G > 0$) corresponds to a nonspontaneous process.

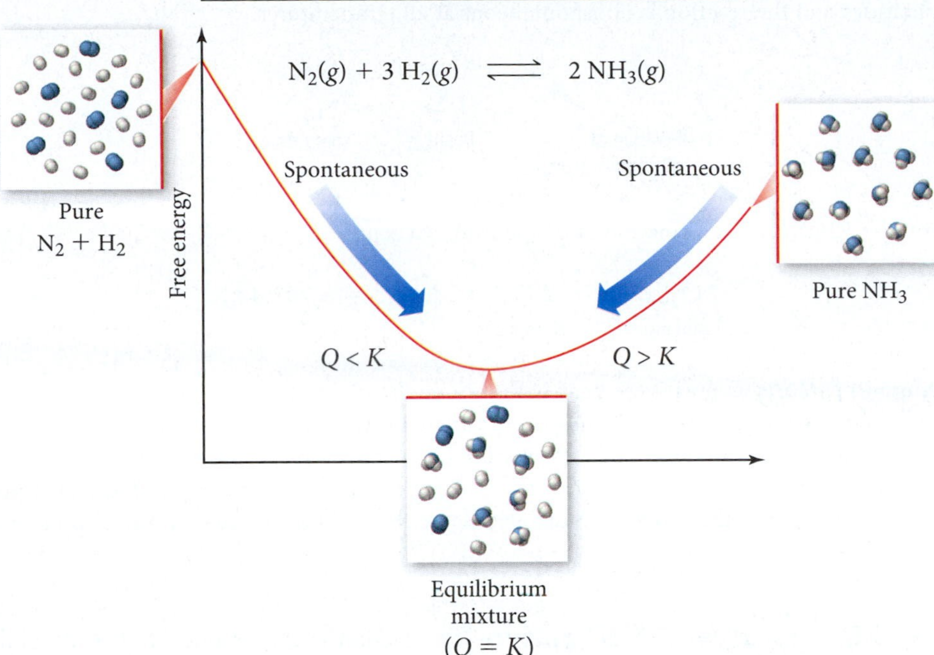

◀ **FIGURE 19.6 Gibbs Free Energy** Gibbs free energy is also called chemical potential because it determines the direction of spontaneous change for chemical systems.

Notice that we can calculate changes in Gibbs free energy solely with reference to the system. So, to determine whether a process is spontaneous, we only have to find the change in *entropy* for the system (ΔS) and the change in *enthalpy* for the system (ΔH). We can then predict the spontaneity of the process at any temperature. In Chapter 10, we learned how to calculate changes in enthalpy (ΔH) for chemical reactions. In Section 19.4, we learned how to calculate changes in entropy (ΔS) for chemical reactions. In Section 19.7, we use those two quantities to calculate changes in free energy (ΔG) for chemical reactions and predict their spontaneity. Before we move on to doing this, however, we examine some examples that demonstrate how ΔH, ΔS, and T affect the spontaneity of chemical processes.

The Effect of ΔH, ΔS, and T on Spontaneity

Case 1: ΔH Negative, ΔS Positive If a reaction is exothermic ($\Delta H < 0$), and if the change in entropy for the reaction is positive ($\Delta S > 0$), then the change in free energy is negative at all temperatures and the reaction is spontaneous at all temperatures.

$$\Delta G = \Delta H - T\,\Delta S$$

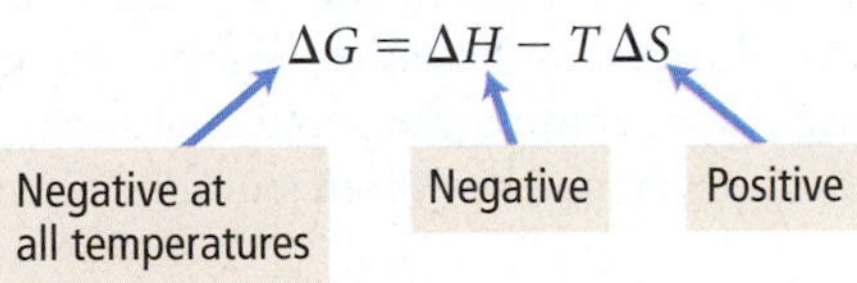

As an example, consider the dissociation of N_2O:

$$\underset{\text{2 mol gas}}{2\,N_2O(g)} \longrightarrow \underset{\text{3 mol gas}}{2\,N_2(g) + O_2(g)} \qquad \Delta H^\circ_{\text{rxn}} = -163.2\text{ kJ}$$

The change in *enthalpy* is negative—heat is emitted, increasing the entropy of the surroundings. The change in *entropy* for the reaction is positive, which means that the entropy of the system increases. (We can see that the change in entropy is positive from the balanced equation—the number of moles of gas increases.) Since the entropy of both the system and the surroundings increases, the entropy of the universe must also increase, making the reaction spontaneous at all temperatures.

Recall from Chapter 10 that ΔH° represents the standard enthalpy change. The definition of the standard state was first given in Section 10.10 and is summarized in Section 19.4.

Case 2: ΔH Positive, ΔS Negative If a reaction is endothermic ($\Delta H > 0$), and if the change in entropy for the reaction is negative ($\Delta S < 0$), then the change in free energy is positive at all temperatures and the reaction is nonspontaneous at all temperatures.

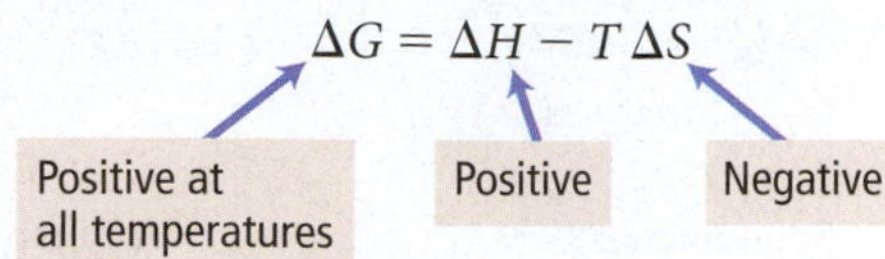

As an example, consider the formation of ozone from oxygen:

$$\underset{3\text{ mol gas}}{3\,O_2(g)} \longrightarrow \underset{2\text{ mol gas}}{2\,O_3(g)} \quad \Delta H^\circ_{rxn} = +285.4\text{ kJ}$$

The change in *enthalpy* is positive—heat is therefore absorbed, *decreasing* the entropy of the surroundings. The change in *entropy* is negative, which means that the entropy of the system decreases. (We can see that the change in entropy is negative from the balanced equation—the number of moles of gas decreases.) Since the entropy of both the system and the surroundings decreases, the entropy of the universe must also decrease, making the reaction nonspontaneous at all temperatures.

Case 3: ΔH Negative, ΔS Negative If a reaction is exothermic ($\Delta H < 0$), and if the change in entropy for the reaction is negative ($\Delta S < 0$), then the sign of the change in free energy depends on temperature. The reaction is spontaneous at low temperature, but nonspontaneous at high temperature.

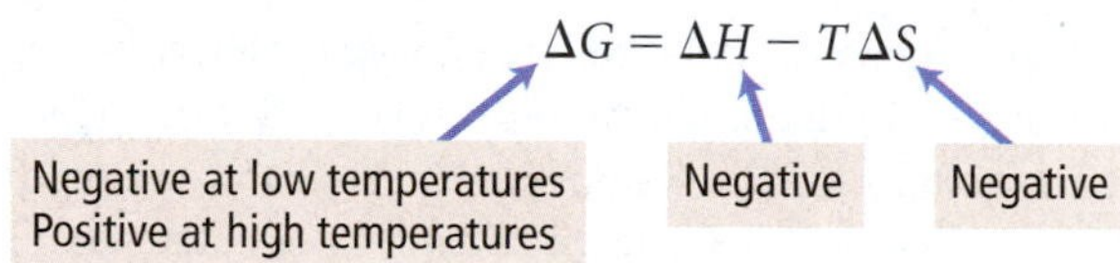

As an example, consider the freezing of liquid water to form ice:

$$H_2O(l) \longrightarrow H_2O(s) \qquad \Delta H^\circ = -6.01\text{ kJ}$$

The change in *enthalpy* is negative—heat is emitted, increasing the entropy of the surroundings. The change in *entropy* is negative, which means that the entropy of the system decreases. (We can see that the change in entropy is negative from the balanced equation—a liquid turns into a solid.)

Unlike the two previous cases, where the changes in *entropy* of the system and of the surroundings had the same sign, the changes here are opposite in sign. Therefore, the sign of the change in free energy depends on the relative magnitudes of the two changes. At a low enough temperature, the heat emitted into the surroundings causes a large entropy change in the surroundings, making the process spontaneous. At high temperature, the same amount of heat is dispersed into warmer surroundings, so the positive entropy change in the surroundings is smaller, resulting in a nonspontaneous process.

Case 4: ΔH Positive, ΔS Positive If a reaction is endothermic ($\Delta H > 0$), and if the change in entropy for the reaction is positive ($\Delta S > 0$), then the sign of the change in free energy again depends on temperature. The reaction is nonspontaneous at low temperature but spontaneous at high temperature.

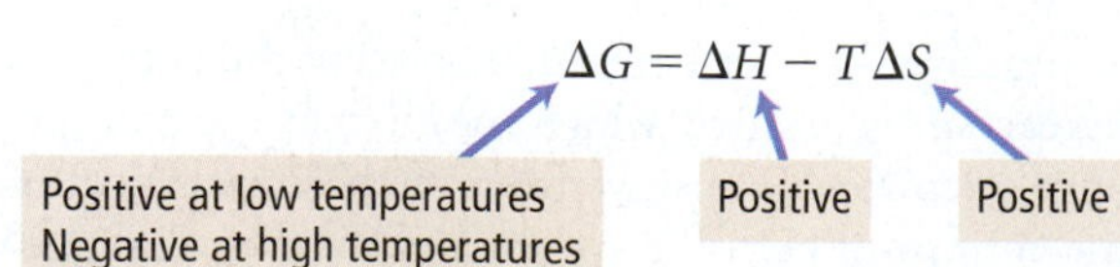

As an example, consider the vaporizing of liquid water to gaseous water:

$$H_2O(l) \longrightarrow H_2O(g) \qquad \Delta H^\circ = +40.7 \text{ kJ (at 100 °C)}$$

The change in *enthalpy* is positive—heat is absorbed from the surroundings, so the entropy of the surroundings decreases. The change in *entropy* is positive, which means that the entropy of the system increases. (We can see that the change in entropy is positive from the balanced equation—a liquid turns into a gas.) The changes in entropy of the system and the surroundings again have opposite signs, only this time the entropy of the surroundings decreases while the entropy of the system increases. In cases such as this, high temperature favors spontaneity because the absorption of heat from the surroundings has less effect on the entropy of the surroundings as temperature increases.

The results of this section are summarized in Table 19.2. Notice that when ΔH and ΔS have opposite signs, the spontaneity of the reaction does not depend on temperature. *When ΔH and ΔS have the same sign, however, the spontaneity does depend on temperature.* The temperature at which the reaction changes from being spontaneous to being nonspontaneous (or vice versa) is the temperature at which ΔG changes sign, which we can determine by setting $\Delta G = 0$ and solving for T, as shown in part b of Example 19.4.

EXAMPLE 19.4

Calculating Gibbs Free Energy Changes and Predicting Spontaneity from ΔH and ΔS

Consider the reaction for the decomposition of carbon tetrachloride gas:

$$CCl_4(g) \longrightarrow C(s, \text{graphite}) + 2\, Cl_2(g) \qquad \Delta H = +95.7 \text{ kJ}; \Delta S = +142.2 \text{ J/K}$$

(a) Calculate ΔG at 25 °C and determine whether the reaction is spontaneous.

(b) If the reaction is not spontaneous at 25 °C, determine at what temperature (if any) the reaction becomes spontaneous.

SOLUTION

(a) Use Equation 19.10 to calculate ΔG from the given values of ΔH and ΔS. The temperature must be in kelvins. *Be sure to express both ΔH and ΔS in the same units (usually joules).*

$$T = 273 + 25 = 298 \text{ K}$$

$$\begin{aligned} \Delta G &= \Delta H - T\Delta S \\ &= 95.7 \times 10^3 \text{ J} - (298 \text{ K})142.2 \text{ J/K} \\ &= 95.7 \times 10^3 \text{ J} - 42.4 \times 10^3 \text{ J} \\ &= +53.3 \times 10^3 \text{ J} \end{aligned}$$

The reaction is not spontaneous.

(b) Since ΔS is positive, ΔG becomes more negative with increasing temperature. To determine the temperature at which the reaction becomes spontaneous, use Equation 19.10 to find the temperature at which ΔG changes from positive to negative (set $\Delta G = 0$ and solve for T). The reaction is spontaneous above this temperature.

$$\Delta G = \Delta H - T\Delta S$$

$$0 = 95.7 \times 10^3 \text{ J} - (T)142.2 \text{ J/K}$$

$$T = \frac{95.7 \times 10^3 \text{ J}}{142.2 \text{ J/K}}$$

$$= 673 \text{ K}$$

FOR PRACTICE 19.4

Consider the reaction:

$$C_2H_4(g) + H_2(g) \longrightarrow C_2H_6(g) \qquad \Delta H = -137.5 \text{ kJ}; \Delta S = -120.5 \text{ J/K}$$

Calculate ΔG at 25 °C and determine whether the reaction is spontaneous. Does ΔG become more negative or more positive as the temperature increases?

TABLE 19.2 The Effect of ΔH, ΔS, and T on Spontaneity

°H	°S	Low Temperature	High Temperature	Example
−	+	Spontaneous ($\Delta G < 0$)	Spontaneous ($\Delta G < 0$)	$2\ N_2O(g) \longrightarrow 2\ N_2(g) + O_2(g)$
+	−	Nonspontaneous ($\Delta G > 0$)	Nonspontaneous ($\Delta G > 0$)	$3\ O_2(g) \longrightarrow 2\ O_3(g)$
−	−	Spontaneous ($\Delta G < 0$)	Nonspontaneous ($\Delta G > 0$)	$H_2O(l) \longrightarrow H_2O(s)$
+	+	Nonspontaneous ($\Delta G > 0$)	Spontaneous ($\Delta G < 0$)	$H_2O(l) \longrightarrow H_2O(g)$

19.4 Cc Conceptual Connection

ΔH, ΔS, and ΔG

Which statement is true regarding the sublimation of dry ice (solid CO_2)?

(a) ΔH is positive, ΔS is positive, and ΔG is positive at low temperature and negative at high temperature.

(b) ΔH is negative, ΔS is negative, and ΔG is negative at low temperature and positive at high temperature.

(c) ΔH is negative, ΔS is positive, and ΔG is negative at all temperatures.

(d) ΔH is positive, ΔS is negative, and ΔG is positive at all temperatures.

19.7 Free Energy Changes in Chemical Reactions: Calculating ΔG°_{rxn}

Standard conditions represent a very specific reaction mixture. For a reaction mixture with reactants and products in their standard states, $Q = 1$.

We have just seen that the criterion for spontaneity is the change in Gibbs free energy (ΔG). When the reactants and products of a reaction are in their standard states, this becomes the **standard change in free energy (ΔG°_{rxn}).** Recall from Section 10.10 that the symbol ° specifies standard states, which means that the *concentrations of all aqueous reactants and products is 1 M* and that *the partial pressure of all gaseous reactants and products is 1 atm*. We will refer to these conditions as standard state conditions, or more simply as standard conditions. In this section, we examine three methods to calculate the standard change in free energy for a reaction (ΔG°_{rxn}). In the first method, we calculate ΔH°_{rxn} and ΔS°_{rxn} from tabulated values of ΔH°_f and S°, and then use the relationship $\Delta G^\circ_{rxn} = \Delta H^\circ_{rxn} - T\Delta S^\circ_{rxn}$ to calculate ΔG°_{rxn}. In the second method, we use tabulated values of free energies of formation to calculate ΔG°_{rxn} directly. In the third method, we determine the free energy change for a stepwise reaction from the free energy changes of each step. At the end of this section, we discuss how to make a nonspontaneous process spontaneous and what is "free" about free energy. Remember that ΔG°_{rxn} is extremely useful because it tells us about the spontaneity of a process at standard conditions. The more negative ΔG°_{rxn} is, the more spontaneous the process—the further it will go toward products to reach equilibrium.

Calculating Standard Free Energy Changes with $\Delta G^\circ_{rxn} = \Delta H^\circ_{rxn} - T\Delta S^\circ_{rxn}$

In Chapter 10 (Section 10.10), we learned how to use tabulated values of standard enthalpies of formation to calculate ΔH°_{rxn}. In Section 19.4, we learned how to use tabulated values of standard entropies to calculate ΔS°_{rxn}. We can use these calculated values of ΔH°_{rxn} and ΔS°_{rxn} to determine the standard free energy change for a reaction by using the equation:

$$\Delta G^\circ_{rxn} = \Delta H^\circ_{rxn} - T\Delta S^\circ_{rxn} \qquad [19.12]$$

Since tabulated values of standard enthalpies of formation (ΔH°_f) and standard entropies (S°) are usually applicable at 25 °C, the equation should (strictly speaking) be valid only when $T = 298$ K (25 °C). However, the changes in ΔH°_{rxn} and ΔS°_{rxn} over a limited temperature range are small when compared to the changes in the value of the temperature itself. Therefore, we can use Equation 19.12 to estimate changes in free energy at temperatures other than 25 °C.

EXAMPLE 19.5

Calculating the Standard Change in Free Energy for a Reaction Using $\Delta G^\circ_{rxn} = \Delta H^\circ_{rxn} - T\Delta S^\circ_{rxn}$

One possible initial step in the formation of acid rain is the oxidation of the pollutant SO_2 to SO_3 by the reaction:

$$SO_2(g) + \frac{1}{2}O_2(g) \longrightarrow SO_3(g)$$

Calculate ΔG°_{rxn} at 25 °C and determine whether the reaction is spontaneous.

SOLUTION

Begin by looking up (in Appendix IVB) the standard enthalpy of formation and the standard entropy for each reactant and product.

Reactant or product	ΔH°_f(kJ/mol)	S°(J/mol·K)
$SO_2(g)$	−296.8	248.2
$O_2(g)$	0	205.2
$SO_3(g)$	−395.7	256.8

Calculate ΔH°_{rxn} using Equation 10.14.

$$\begin{aligned}\Delta H^\circ_{rxn} &= \sum n_p \Delta H^\circ_f(\text{products}) - \sum n_r \Delta H^\circ_f(\text{reactants})\\ &= [\Delta H^\circ_{f,\,SO_3(g)}] - [\Delta H^\circ_{f,\,SO_2(g)} + \tfrac{1}{2}(\Delta H^\circ_{f,\,O_2(g)})]\\ &= -395.7\text{ kJ} - (-296.8\text{ kJ} + 0.0\text{ kJ})\\ &= -98.9\text{ kJ}\end{aligned}$$

Calculate ΔS°_{rxn} using Equation 19.1.

$$\begin{aligned}\Delta S^\circ_{rxn} &= \sum n_p S^\circ(\text{products}) - \sum n_r S^\circ(\text{reactants})\\ &= [S^\circ_{SO_3(g)}] - [S^\circ_{SO_2(g)} + \tfrac{1}{2}(S^\circ_{O_2(g)})]\\ &= 256.8\text{ J/K} - [248.2\text{ J/K} + \tfrac{1}{2}(205.2\text{ J/K})]\\ &= -94.0\text{ J/K}\end{aligned}$$

Calculate ΔG°_{rxn} using the calculated values of ΔH°_{rxn} and ΔS°_{rxn} and Equation 19.12. Convert the temperature to kelvins.

$$T = 25 + 273 = 298\text{ K}$$

$$\begin{aligned}\Delta G^\circ_{rxn} &= \Delta H^\circ_{rxn} - T\Delta S^\circ_{rxn}\\ &= -98.9 \times 10^3\text{ J} - 298\text{ K}(-94.0\text{ J/K})\\ &= -70.9 \times 10^3\text{ J}\\ &= -70.9\text{ kJ}\end{aligned}$$

The reaction is spontaneous at this temperature because ΔG°_{rxn} is negative.

FOR PRACTICE 19.5

Consider the oxidation of NO to NO_2:

$$NO(g) + \frac{1}{2}O_2(g) \longrightarrow NO_2(g)$$

Calculate ΔG°_{rxn} at 25 °C and determine whether the reaction is spontaneous at standard conditions.

EXAMPLE 19.6

Estimating the Standard Change in Free Energy for a Reaction at a Temperature Other than 25 °C Using $\Delta G^\circ_{rxn} = \Delta H^\circ_{rxn} - T\Delta S^\circ_{rxn}$

For the reaction in Example 19.5, estimate the value of ΔG°_{rxn} at 125 °C. Is the reaction more or less spontaneous at this elevated temperature; that is, is the value of ΔG°_{rxn} more negative (more spontaneous) or more positive (less spontaneous)?

SOLUTION

Estimate ΔG°_{rxn} at the new temperature using the calculated values of ΔH°_{rxn} and ΔS°_{rxn} from Example 19.5. For T, convert the given temperature to kelvins. Make sure to use the same units for ΔH°_{rxn} and ΔS°_{rxn} (usually joules).

$$T = 125 + 273 = 398\ \text{K}$$

$$\begin{aligned}\Delta G^\circ_{rxn} &= \Delta H^\circ_{rxn} - T\Delta S^\circ_{rxn} \\ &= -98.9 \times 10^3\ \text{J} - 398\ \text{K}(-94.0\ \text{J/K}) \\ &= -61.5 \times 10^3\ \text{J} \\ &= -61.5\ \text{kJ}\end{aligned}$$

Since the value of ΔG°_{rxn} at this elevated temperature is less negative (or more positive) than the value of ΔG°_{rxn} at 25 °C (which is −70.9 kJ), the reaction is less spontaneous.

FOR PRACTICE 19.6

For the reaction in For Practice 19.5, calculate the value of ΔG°_{rxn} at −55 °C. Is the reaction more spontaneous (more negative ΔG°_{rxn}) or less spontaneous (more positive ΔG°_{rxn}) at the lower temperature?

Calculating ΔG°_{rxn} with Tabulated Values of Free Energies of Formation

Because ΔG°_{rxn} is the *change* in free energy for a chemical reaction—the difference in free energy between the products and the reactants—and because free energy is a state function, we can calculate ΔG°_{rxn} by subtracting the free energies of the reactants of the reaction from the free energies of the products. Also, since we are interested only in *changes* in free energy (and not in absolute values of free energy), we are free to define the *zero* of free energy as conveniently as possible. By analogy with our definition of enthalpies of formation, we define the **free energy of formation (ΔG°_f)** as follows:

> **The free energy of formation (ΔG°_f) is the change in free energy when 1 mol of a compound in its standard state forms from its constituent elements in their standard states. The free energy of formation of pure elements in their standard states is zero.**

We can measure all changes in free energy relative to pure elements in their standard states. To calculate ΔG°_{rxn}, we subtract the free energies of formation of the reactants multiplied by their stoichiometric coefficients from the free energies of formation of the products multiplied by their stoichiometric coefficients. In the form of an equation:

$$\Delta G^\circ_{rxn} = \sum n_p \Delta G^\circ_f(\text{products}) - \sum n_r \Delta G^\circ_f(\text{reactants}) \quad [19.13]$$

In Equation 19.13, n_p represents the stoichiometric coefficients of the products, n_r represents the stoichiometric coefficients of the reactants, and ΔG°_f represents the standard free energies of formation. Table 19.3 lists ΔG°_f values for selected substances. You can find a more complete list in Appendix IVB. Notice that, by definition, *elements* have standard free energies of formation of zero. Notice also that most *compounds* have negative standard free energies of formation. This means that those compounds spontaneously form from their elements in their standard states. Compounds with positive free energies of formation do not spontaneously form from their elements and are therefore less common.

TABLE 19.3 Standard Molar Free Energies of Formation (ΔG°_f) for Selected Substances at 298 K

Substance	ΔG°_f (kJ/mol)	Substance	ΔG°_f (kJ/mol)
$H_2(g)$	0	$CH_4(g)$	−50.5
$O_2(g)$	0	$H_2O(g)$	−228.6
$N_2(g)$	0	$H_2O(l)$	−237.1
C(*s*, graphite)	0	$NH_3(g)$	−16.4
C(*s*, diamond)	2.900	$NO(g)$	+87.6
$CO(g)$	−137.2	$NO_2(g)$	+51.3
$CO_2(g)$	−394.4	$NaCl(s)$	−384.1

Example 19.7 demonstrates the calculation of ΔG°_{rxn} from ΔG°_f values. This method of calculating ΔG°_{rxn} works only at the temperature for which the free energies of formation are tabulated, namely, 25 °C. To estimate ΔG°_{rxn} at other temperatures, we must use $\Delta G^\circ_{rxn} = \Delta H^\circ_{rxn} - T\Delta S^\circ_{rxn}$, as demonstrated previously.

EXAMPLE 19.7

Calculating ΔG°_{rxn} from Standard Free Energies of Formation

Ozone in the lower atmosphere is a pollutant that can form by the following reaction involving the oxidation of unburned hydrocarbons:

$$CH_4(g) + 8\,O_2(g) \longrightarrow CO_2(g) + 2\,H_2O(g) + 4\,O_3(g)$$

Use the standard free energies of formation to determine ΔG°_{rxn} for this reaction at 25 °C.

SOLUTION

Begin by looking up (in Appendix IVB) the standard free energies of formation for each reactant and product. Remember that the standard free energy of formation of a pure element in its standard state is zero.

Reactant/product	ΔG°_f (in kJ/mol)
$CH_4(g)$	−50.5
$O_2(g)$	0.0
$CO_2(g)$	−394.4
$H_2O(g)$	−228.6
$O_3(g)$	163.2

Calculate ΔG°_{rxn} by substituting into Equation 19.13.

$$\begin{aligned}\Delta G^\circ_{rxn} &= \sum n_p \Delta G^\circ_f(\text{products}) - \sum n_r \Delta G^\circ_f(\text{reactants}) \\ &= [\Delta G^\circ_{f,\,CO_2(g)} + 2(\Delta G^\circ_{f,\,H_2O(g)}) + 4(\Delta G^\circ_{f,\,O_3(g)})] - [\Delta G^\circ_{f,\,CH_4(g)} + 8(\Delta G^\circ_{f,\,O_2(g)})] \\ &= [-394.4\text{ kJ} + 2(-228.6\text{ kJ}) + 4(163.2\text{ kJ})] - [-50.5\text{ kJ} + 8(0.0\text{ kJ})] \\ &= -198.8\text{ kJ} + 50.5\text{ kJ} \\ &= -148.3\text{ kJ}\end{aligned}$$

FOR PRACTICE 19.7

One reaction that occurs within a catalytic converter in the exhaust pipe of a car is the simultaneous oxidation of carbon monoxide and reduction of nitrogen monoxide (both of which are harmful pollutants).

$$2\,CO(g) + 2\,NO(g) \longrightarrow 2\,CO_2(g) + N_2(g)$$

Use standard free energies of formation to determine ΔG°_{rxn} for this reaction at 25 °C. Is the reaction spontaneous at standard conditions?

FOR MORE PRACTICE 19.7

In For Practice 19.7, you calculated ΔG°_{rxn} for the simultaneous oxidation of carbon monoxide and reduction of nitrogen monoxide using standard free energies of formation. Calculate ΔG°_{rxn} for that reaction again at 25 °C, only this time use $\Delta G^\circ_{rxn} = \Delta H^\circ_{rxn} - T\Delta S^\circ_{rxn}$. How do the two values compare? Use your results to calculate ΔG°_{rxn} at 500.0 K and explain why you could not calculate ΔG°_{rxn} at 500.0 K using tabulated standard free energies of formation.

Calculating ΔG°_{rxn} for a Stepwise Reaction from the Changes in Free Energy for Each of the Steps

Recall from Section 10.6 that since enthalpy is a state function, we can calculate ΔH°_{rxn} for a stepwise reaction from the sum of the changes in enthalpy for each step (according to Hess's law). Since free energy is also a state function, the same relationships that we covered in Chapter 10 for enthalpy also apply to free energy:

1. If a chemical equation is multiplied by some factor, ΔG_{rxn} is also multiplied by the same factor.
2. If a chemical equation is reversed, ΔG_{rxn} changes sign.
3. If a chemical equation can be expressed as the sum of a series of steps, ΔG_{rxn} for the overall equation is the sum of the free energies of reactions for each step.

Example 19.8 illustrates the use of these relationships to calculate ΔG°_{rxn} for a stepwise reaction.

EXAMPLE 19.8
Calculating ΔG°_{rxn} for a Stepwise Reaction

Find ΔG°_{rxn} for the reaction:

$$3\,C(s) + 4\,H_2(g) \longrightarrow C_3H_8(g)$$

Use the following reactions with known ΔG's:

$$C_3H_8(g) + 5\,O_2(g) \longrightarrow 3\,CO_2(g) + 4\,H_2O(g) \qquad \Delta G^\circ_{rxn} = -2074\text{ kJ}$$
$$C(s) + O_2(g) \longrightarrow CO_2(g) \qquad \Delta G^\circ_{rxn} = -394.4\text{ kJ}$$
$$2\,H_2(g) + O_2(g) \longrightarrow 2\,H_2O(g) \qquad \Delta G^\circ_{rxn} = -457.1\text{ kJ}$$

SOLUTION

To work this problem, manipulate the reactions with known values of ΔG°_{rxn} in such a way as to get the reactants of interest on the left, the products of interest on the right, and other species to cancel.

Since the first reaction has C_3H_8 as a reactant, and the reaction of interest has C_3H_8 as a product, reverse the first reaction and change the sign of ΔG°_{rxn}.	$3\,CO_2(g) + 4\,H_2O(g) \longrightarrow C_3H_8(g) + 5\,O_2(g)$ $\quad \Delta G^\circ_{rxn} = +2074\text{ kJ}$
The second reaction has C as a reactant and CO_2 as a product, as required in the reaction of interest. However, the coefficient for C is 1, and in the reaction of interest, the coefficient for C is 3. Therefore, multiply this equation and its ΔG°_{rxn} by 3.	$3 \times [C(s) + O_2(g) \longrightarrow CO_2(g)]$ $\quad \Delta G^\circ_{rxn} = 3 \times (-394.4\text{ kJ}) = -1183\text{ kJ}$
The third reaction has $H_2(g)$ as a reactant, as required. However, the coefficient for H_2 is 2, and in the reaction of interest, the coefficient for H_2 is 4. Multiply this reaction and its ΔG°_{rxn} by 2.	$2 \times [2\,H_2(g) + O_2(g) \longrightarrow 2\,H_2O(g)]$ $\quad \Delta G^\circ_{rxn} = 2 \times (-457.1\text{ kJ}) = -914.2$
Lastly, rewrite the three reactions after multiplying by the indicated factors and show how they sum to the reaction of interest. ΔG°_{rxn} for the reaction of interest is then the sum of the ΔG's for the steps.	$3\,\cancel{CO_2}(g) + 4\,\cancel{H_2O}(g) \longrightarrow C_3H_8(g) + 5\,\cancel{O_2}(g)$ $\quad \Delta G^\circ_{rxn} = +2074\text{ kJ}$ $3\,C(s) + 3\,\cancel{O_2}(g) \longrightarrow 3\,\cancel{CO_2}(g)$ $\quad \Delta G^\circ_{rxn} = -1183\text{ kJ}$ $4\,H_2(g) + 2\,\cancel{O_2}(g) \longrightarrow 4\,\cancel{H_2O}(g)$ $\quad \Delta G^\circ_{rxn} = -914.2\text{ kJ}$ $3\,C(s) + 4\,H_2(g) \longrightarrow C_3H_8(g)$ $\quad \Delta G^\circ_{rxn} = -23\text{ kJ}$

FOR PRACTICE 19.8

Find ΔG°_{rxn} for the reaction:

$$N_2O(g) + NO_2(g) \longrightarrow 3\,NO(g)$$

Use the following reactions with known ΔG values:

$$2\,NO(g) + O_2(g) \longrightarrow 2\,NO_2(g) \qquad \Delta G^\circ_{rxn} = -71.2\text{ kJ}$$
$$N_2(g) + O_2(g) \longrightarrow 2\,NO(g) \qquad \Delta G^\circ_{rxn} = +175.2\text{ kJ}$$
$$2\,N_2O(g) \longrightarrow 2\,N_2(g) + O_2(g) \qquad \Delta G^\circ_{rxn} = -207.4\text{ kJ}$$

Making a Nonspontaneous Process Spontaneous

As we mentioned in Section 19.2, a process that is nonspontaneous can be made spontaneous by coupling it with another process that is highly spontaneous. For example, hydrogen gas is a potential future fuel because it can be used in a fuel cell (a type of battery in which the reactants are constantly supplied—see Chapter 20) to generate electricity. The main problem with switching to hydrogen is securing a source. Where can we get the huge amounts of hydrogen gas that we need to meet our world's energy demands? Earth's oceans and lakes contain vast amounts of hydrogen. But that hydrogen is locked up in water molecules, and the decomposition of water into hydrogen and oxygen has a positive ΔG°_{rxn} and is therefore nonspontaneous:

$$H_2O(g) \longrightarrow H_2(g) + \tfrac{1}{2}O_2(g) \qquad \Delta G^\circ_{rxn} = +228.6 \text{ kJ}$$

To obtain hydrogen from water, we need to find another reaction with a highly negative ΔG°_{rxn} that can couple with the decomposition reaction to give an overall reaction with a negative ΔG°_{rxn}. For example, the oxidation of carbon monoxide to carbon dioxide has a large negative ΔG°_{rxn} and is highly spontaneous:

$$CO(g) + \tfrac{1}{2}O_2(g) \longrightarrow CO_2(g) \qquad \Delta G^\circ_{rxn} = -257.2 \text{ kJ}$$

If we add the two reactions together, we get a negative ΔG°_{rxn}:

Nonspontaneous

$$\begin{array}{ll} H_2O(g) \longrightarrow H_2(g) + \cancel{\tfrac{1}{2}O_2}(g) & \Delta G^\circ_{rxn} = +228.6 \text{ kJ} \\ CO(g) + \cancel{\tfrac{1}{2}O_2}(g) \longrightarrow CO_2(g) & \Delta G^\circ_{rxn} = -257.2 \text{ kJ} \\ \hline H_2O(g) + CO(g) \longrightarrow H_2(g) + CO_2(g) & \Delta G^\circ_{rxn} = -28.6 \text{ kJ} \end{array}$$

Spontaneous

The reaction between water and carbon monoxide is thus a spontaneous way to generate hydrogen gas.

The coupling of nonspontaneous reactions with highly spontaneous ones is also important in biological systems. The synthesis reactions that create the complex biological molecules (such as proteins and DNA) needed by living organisms, for example, are themselves nonspontaneous. Living systems grow and reproduce by coupling these nonspontaneous reactions to highly spontaneous ones. The main spontaneous reaction that ultimately drives the nonspontaneous ones is the metabolism of food. The oxidation of glucose, for example, is highly spontaneous:

$$C_6H_{12}O_6(s) + 6\,O_2(g) \longrightarrow 6\,CO_2(g) + 6\,H_2O(l) \qquad \Delta G^\circ_{rxn} = -2880 \text{ kJ}$$

Spontaneous reactions such as these ultimately drive the nonspontaneous reactions necessary to sustain life.

Why Free Energy Is "Free"

We often want to use the energy released by a chemical reaction to do work. For example, in an automobile, we use the energy released by the combustion of gasoline to move the car forward. The change in free energy of a chemical reaction represents the maximum amount of energy available, or *free*, to do work (if ΔG°_{rxn} is negative). For many reactions, the amount of free energy change is less than the change in enthalpy for the reaction. Consider the reaction between carbon and hydrogen occurring at 25 °C:

$$C(s, \text{graphite}) + 2\,H_2(g) \longrightarrow CH_4(g)$$

$$\Delta H^\circ_{rxn} = -74.6 \text{ kJ}$$
$$\Delta S^\circ_{rxn} = -80.8 \text{ J/K}$$
$$\Delta G^\circ_{rxn} = -50.5 \text{ kJ}$$

The reaction is exothermic and gives off 74.6 kJ of heat energy. However, the maximum amount of energy available for useful work is only 50.5 kJ (Figure 19.7 ▶). Why? We can see that the change in

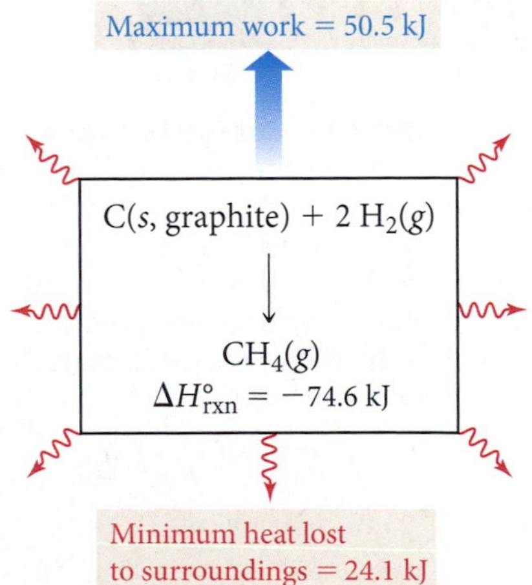

▲ **FIGURE 19.7 Free Energy** Although the reaction produces 74.6 kJ of heat, only a maximum of 50.5 kJ is available to do work. The rest of the energy is lost to the surroundings.

▶ **FIGURE 19.8 A Reversible Process** In a reversible process, the free energy is drawn out in infinitesimally small increments that exactly match the amount of energy that the process is producing in that increment. In this case, grains of sand are removed one at a time, resulting in a series of small expansions in which the weight of sand exactly matches the pressure of the expanding gas. This process is close to reversible—each sand grain would need to have an infinitesimally small mass for the process to be fully reversible.

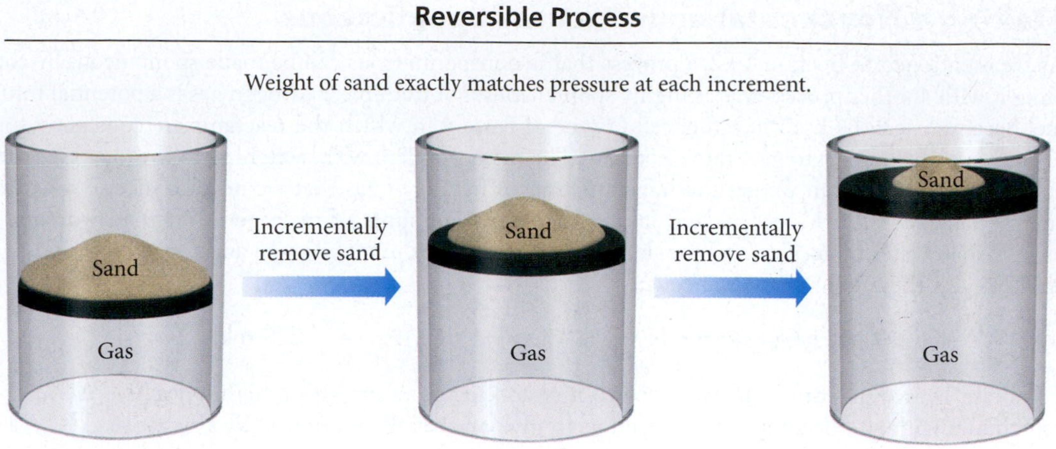

Reversible Process

Weight of sand exactly matches pressure at each increment.

Sand
Gas
Incrementally remove sand
Sand
Gas
Incrementally remove sand
Sand
Gas

entropy of the *system* is negative. Nevertheless, the reaction is spontaneous. This is possible only if some of the emitted heat goes to increase the entropy of the surroundings by an amount sufficient to make the change in entropy of the *universe* positive. The amount of energy available to do work (the free energy) is what remains after accounting for the heat that must be lost to the surroundings.

The change in free energy for a chemical reaction represents a *theoretical limit* as to how much work can be done by the reaction. In a *real* reaction, the amount of energy available to do work is even *less* than ΔG°_{rxn} because additional energy is lost to the surroundings as heat. A reaction that achieves the theoretical limit with respect to free energy is called a **reversible reaction**. A reversible reaction occurs infinitesimally slowly, and the free energy can only be drawn out in infinitesimally small increments that exactly match the amount of energy that the reaction produces during that increment (Figure 19.8 ▲).

More formally, a reversible reaction is one that will change direction upon an infinitesimally small change in a variable (such as temperature or pressure) related to the reaction.

All real reactions are **irreversible reactions** and therefore do not achieve the theoretical limit of available free energy. Consider the process of charging and discharging a battery. A battery contains chemical reactants configured in such a way that, upon spontaneous reaction, they produce an electrical current. The free energy released by the reaction is then harnessed to do work. For example, an electric motor can be wired to the battery. Flowing electrical current makes the motor turn (Figure 19.9 ◀). Owing to resistance in the wire, the flowing electrical current produces some heat, which is lost to the surroundings and is not available to do work. The amount of free energy lost as heat can be lowered by slowing down the rate of current flow. The slower the rate of current flow, the less free energy is lost as heat and the more is available to do work. However, only in the theoretical case of infinitesimally slow current flow is the maximum amount of work (equal to ΔG°_{rxn}) done. Any real rate of current flow results in some loss of energy as heat. This lost energy is sometimes called the "heat tax." Recharging the battery necessarily requires more energy than was obtained as work because some of the energy was lost as heat. Such a heat tax must always be paid in any real process involving energy.

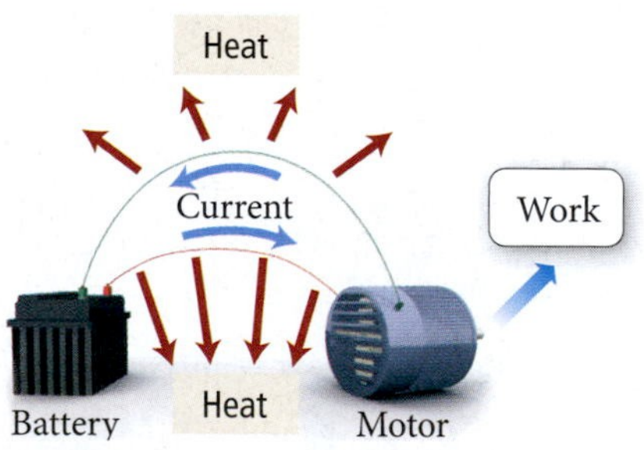

▲ **FIGURE 19.9 Energy Loss in a Battery** When current is drawn from a battery to do work, some energy is lost as heat due to resistance in the wire. Consequently, the quantity of energy required to recharge the battery is more than the quantity of work done.

If the change in free energy of a chemical reaction is positive, then ΔG°_{rxn} represents *the minimum amount of energy required to make the reaction occur*. Again, ΔG°_{rxn} represents a theoretical limit. Making a real nonspontaneous reaction occur always requires more energy than the theoretical limit.

19.8 Free Energy Changes for Nonstandard States: The Relationship between ΔG°_{rxn} and ΔG_{rxn}

We have learned how to calculate the *standard* free energy change for a reaction (ΔG°_{rxn}). However, the standard free energy change applies only to a very narrow set of conditions, namely, those conditions in which the reactants and products are in their standard states. Consider the standard free energy change for the evaporation of liquid water to gaseous water:

$$H_2O(l) \rightleftharpoons H_2O(g) \qquad \Delta G^\circ_{rxn} = +8.59 \text{ kJ/mol}$$

The standard free energy change for this process is positive, so the process is nonspontaneous. But you know that if you spill water onto the floor under ordinary conditions, it spontaneously evaporates. Why? *Because ordinary conditions are not standard conditions* and ΔG°_{rxn} applies only to standard conditions. For a gas (such as the water vapor in the reaction just given), standard conditions are those in which the pure gas is present at a partial pressure of 1 atmosphere. In a flask containing liquid water and water vapor under standard conditions ($P_{H_2O} = 1$ atm) at 25 °C, the water does not vaporize. In fact, since ΔG°_{rxn} is negative for the reverse reaction, the reaction spontaneously occurs in reverse—water vapor condenses.

In open air under ordinary circumstances, however, the partial pressure of water vapor is much less than 1 atm. The conditions are not standard, and therefore the value of ΔG°_{rxn} does not apply. For nonstandard conditions, we must calculate ΔG_{rxn} (as opposed to ΔG°_{rxn}) to predict spontaneity.

We can calculate the **free energy change of a reaction under nonstandard conditions (ΔG_{rxn})** from ΔG°_{rxn} using the relationship:

$$\Delta G_{rxn} = \Delta G^\circ_{rxn} + RT \ln Q \qquad [19.14]$$

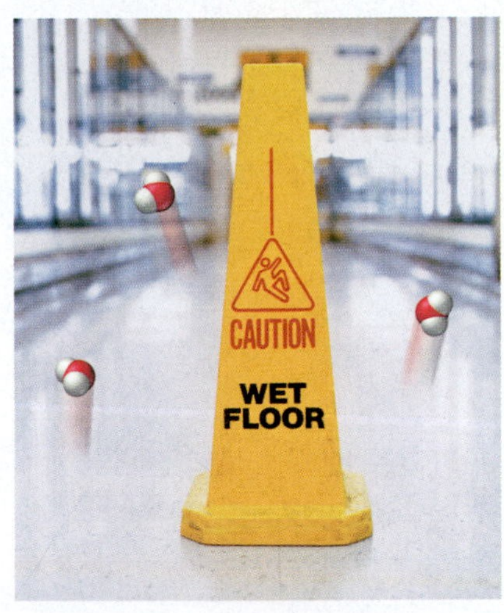

▲ Spilled water spontaneously evaporates even though ΔG° for the vaporization of water is positive. Why?

where Q is the reaction quotient (defined in Section 16.7), T is the temperature in kelvins, and R is the gas constant in the appropriate units (8.314 J/mol • K). In Equation 19.14 and all subsequent thermodynamic equations, we use R = 8.314 J/mol • K. In Equation 19.14 and all subsequent thermodynamic equations, we use Q_p for reactions involving gases, and use Q_c for reactions involving substances dissolved in solution. We can demonstrate the use of this equation by applying it to the liquid–vapor water equilibrium under several different conditions, as shown in Figure 19.10 ▼. Note that by the law of mass action, for this equilibrium, $Q = P_{H_2O}$ (where the pressure is expressed in atmospheres):

$$H_2O(l) \longrightarrow H_2O(g) \qquad Q = P_{H_2O}$$

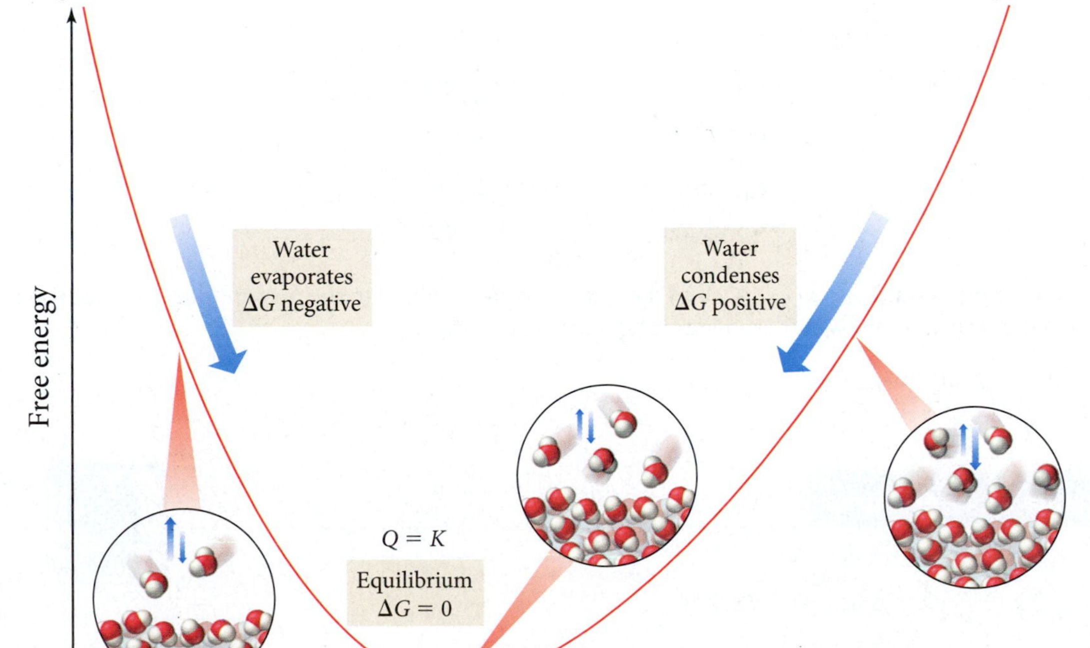

◀ **FIGURE 19.10 Free Energy versus Pressure for Water** The free energy change for the vaporization of water is a function of pressure.

Standard Conditions Under standard conditions, $P_{H_2O} = 1$ atm and therefore $Q = 1$. Substituting, we get the expression:

$$\begin{aligned}\Delta G_{rxn} &= \Delta G^\circ_{rxn} + RT \ln Q \\ &= +8.59 \text{ kJ/mol} + RT\ln(1) \\ &= +8.59 \text{ kJ/mol}\end{aligned}$$

Under standard conditions, Q is always equal to 1, and since $\ln(1) = 0$, the value of ΔG_{rxn} is equal to ΔG°_{rxn}, as expected. For the liquid–vapor water equilibrium, because $\Delta G^\circ_{rxn} > 0$, the reaction is not spontaneous in the forward direction but is spontaneous in the reverse direction. As stated previously, under standard conditions water vapor condenses into liquid water.

Equilibrium Conditions At 25.00 °C, liquid water is in equilibrium with water vapor at a pressure of 0.0313 atm; therefore, $Q = K_p = 0.0313$. Substituting:

$$\begin{aligned}\Delta G_{rxn} &= \Delta G^\circ_{rxn} + RT \ln(0.0313) \\ &= +8.59 \text{ kJ/mol} + 8.314 \frac{\text{J}}{\text{mol} \cdot \cancel{\text{K}}} (298.15 \cancel{\text{K}}) \ln(0.0313) \\ &= +8.59 \text{ kJ/mol} + (-8.59 \times 10^3 \text{ J/mol}) \\ &= +8.59 \text{ kJ/mol} - 8.59 \text{ kJ/mol} \\ &= 0\end{aligned}$$

Under equilibrium conditions, the value of $RT \ln Q$ is always equal in magnitude but opposite in sign to the value of ΔG°_{rxn}. Therefore, the value of ΔG_{rxn} is zero. Because $\Delta G_{rxn} = 0$, the reaction is not spontaneous in either direction, as expected for a reaction at equilibrium.

Other Nonstandard Conditions To calculate the value of ΔG_{rxn} under any other set of nonstandard conditions, calculate Q and substitute the calculated value into Equation 19.14. For example, imagine that the partial pressure of water vapor in the air on a dry (nonhumid) day is 5.00×10^{-3} atm, so $Q = 5.00 \times 10^{-3}$. Substituting:

A water partial pressure of 5.00×10^{-3} atm corresponds to a relative humidity of 16% at 25 °C.

$$\begin{aligned}\Delta G_{rxn} &= \Delta G^\circ_{rxn} + RT \ln(5.00 \times 10^{-3}) \\ &= +8.59 \text{ kJ/mol} + 8.314 \frac{\text{J}}{\text{mol} \cdot \text{K}} (298 \text{ K}) \ln(5.00 \times 10^{-3}) \\ &= +8.59 \text{ kJ/mol} + (-13.1 \times 10^3 \text{ J/mol}) \\ &= +8.59 \text{ kJ/mol} - 13.1 \text{ kJ/mol} \\ &= -4.5 \text{ kJ/mol}\end{aligned}$$

Under these conditions, $\Delta G_{rxn} < 0$, so the reaction is spontaneous in the forward direction, consistent with our experience of water evaporating when spilled on the floor.

EXAMPLE 19.9

Calculating ΔG_{rxn} under Nonstandard Conditions

Consider the reaction at 298 K:

$$2\,NO(g) + O_2(g) \longrightarrow 2\,NO_2(g) \qquad \Delta G^\circ_{rxn} = -71.2 \text{ kJ}$$

Calculate ΔG_{rxn} under these conditions:

$$P_{NO} = 0.100 \text{ atm}; \qquad P_{O_2} = 0.100 \text{ atm}; \qquad P_{NO_2} = 2.00 \text{ atm}$$

Is the reaction more or less spontaneous under these conditions than under standard conditions?

—*Continued on the next page*

Continued from the previous page—

SOLUTION

Use the law of mass action to calculate Q.	$Q = \frac{P^2_{NO_2}}{P^2_{NO}P_{O_2}} = \frac{(2.00)^2}{(0.100)^2(0.100)} = 4.00 \times 10^3$
Substitute Q, T, and ΔG°_{rxn} into Equation 19.14 to calculate ΔG_{rxn}. (Since the units of R include joules, write ΔG°_{rxn} in joules.)	$\Delta G_{rxn} = \Delta G^\circ_{rxn} + RT \ln Q$ $= -71.2 \times 10^3 \text{ J} + 8.314 \frac{\text{J}}{\text{mol} \cdot \text{K}}(298 \text{ K}) \ln (4.00 \times 10^3)$ $= -71.2 \times 10^3 \text{ J} + 20.5 \times 10^3 \text{ J}$ $= -50.7 \times 10^3 \text{ J}$ $= -50.7 \text{ kJ}$ The reaction is spontaneous under these conditions, but less spontaneous than it would be under standard conditions (because ΔG_{rxn} is less negative than ΔG°_{rxn}).

CHECK The calculated result is consistent with what you would expect based on Le Châtelier's principle (see Section 16.9); increasing the concentration of the products and decreasing the concentration of the reactants relative to standard conditions should make the reaction less spontaneous than it was under standard conditions.

FOR PRACTICE 19.9

Consider the reaction at 298 K:

$$2\,H_2S(g) + SO_2(g) \longrightarrow 3\,S(s, \text{rhombic}) + 2\,H_2O(g) \qquad \Delta G^\circ_{rxn} = -102 \text{ kJ}$$

Calculate ΔG_{rxn} under these conditions:

$$P_{H_2S} = 2.00 \text{ atm};\ P_{SO_2} = 1.50 \text{ atm};\quad P_{H_2O} = 0.0100 \text{ atm}$$

Is the reaction more or less spontaneous under these conditions than under standard conditions?

Free Energy Changes and Le Châtelier's Principle

19.5 Cc Conceptual Connection

According to Le Châtelier's principle and the dependence of free energy on reactant and product concentrations, which statement is true? (Assume that both the reactants and products are gaseous.)

(a) A high concentration of reactants relative to products results in a more spontaneous reaction than one in which the reactants and products are in their standard states.

(b) A high concentration of products relative to reactants results in a more spontaneous reaction than one in which the reactants and products are in their standard states.

(c) A reaction in which the reactants are in standard states, but in which no products have formed, has a ΔG_{rxn} that is more positive than ΔG°_{rxn}.

19.9 Free Energy and Equilibrium: Relating ΔG°_{rxn} to the Equilibrium Constant (K)

We have seen that ΔG°_{rxn} determines the spontaneity of a reaction when the reactants and products are in their standard states. In Chapter 16, we learned that the equilibrium constant (K) determines how far a reaction goes toward products, a measure of spontaneity. Therefore, as you might expect, the standard free energy change of a reaction and the equilibrium constant are related—the equilibrium constant becomes larger as the standard free energy change becomes more negative. In other words, if, on one hand, the reactants in a particular reaction undergo a large *negative* free energy change as they become products, then the reaction has a large equilibrium constant, with products strongly favored at equilibrium. If, on the other hand, the reactants in a particular reaction undergo a large *positive* free energy change as they become products, then the reaction has a small equilibrium constant, with reactants strongly favored at equilibrium. We can derive a relationship between ΔG°_{rxn} and K from Equation 19.14. We know that at equilibrium $Q = K$ and $\Delta G_{rxn} = 0$. Making these substitutions:

$$\Delta G_{rxn} = \Delta G^\circ_{rxn} + RT \ln Q$$

$$0 = \Delta G^\circ_{rxn} + RT \ln K$$

$$\Delta G^\circ_{rxn} = -RT \ln K \qquad [19.15]$$

In Equation 19.15 and all subsequent thermodynamic equations, we use K_p for reactions involving gases, and we use K_c for reactions involving substances dissolved in solution.

We can better understand the relationship between ΔG°_{rxn} and K by considering the following ranges of values for K, as summarized in Figure 19.11 ▼:

The relationship between ΔG°_{rxn} and K is logarithmic—small changes in ΔG°_{rxn} have a large effect on K.

- When $K < 1$, $\ln K$ is negative and ΔG°_{rxn} is positive. Under standard conditions (when $Q = 1$), the reaction is spontaneous in the reverse direction.
- When $K > 1$, $\ln K$ is positive and ΔG°_{rxn} is negative. Under standard conditions (when $Q = 1$), the reaction is spontaneous in the forward direction.
- When $K = 1$, $\ln K$ is zero and ΔG°_{rxn} is zero. The reaction happens to be at equilibrium under standard conditions.

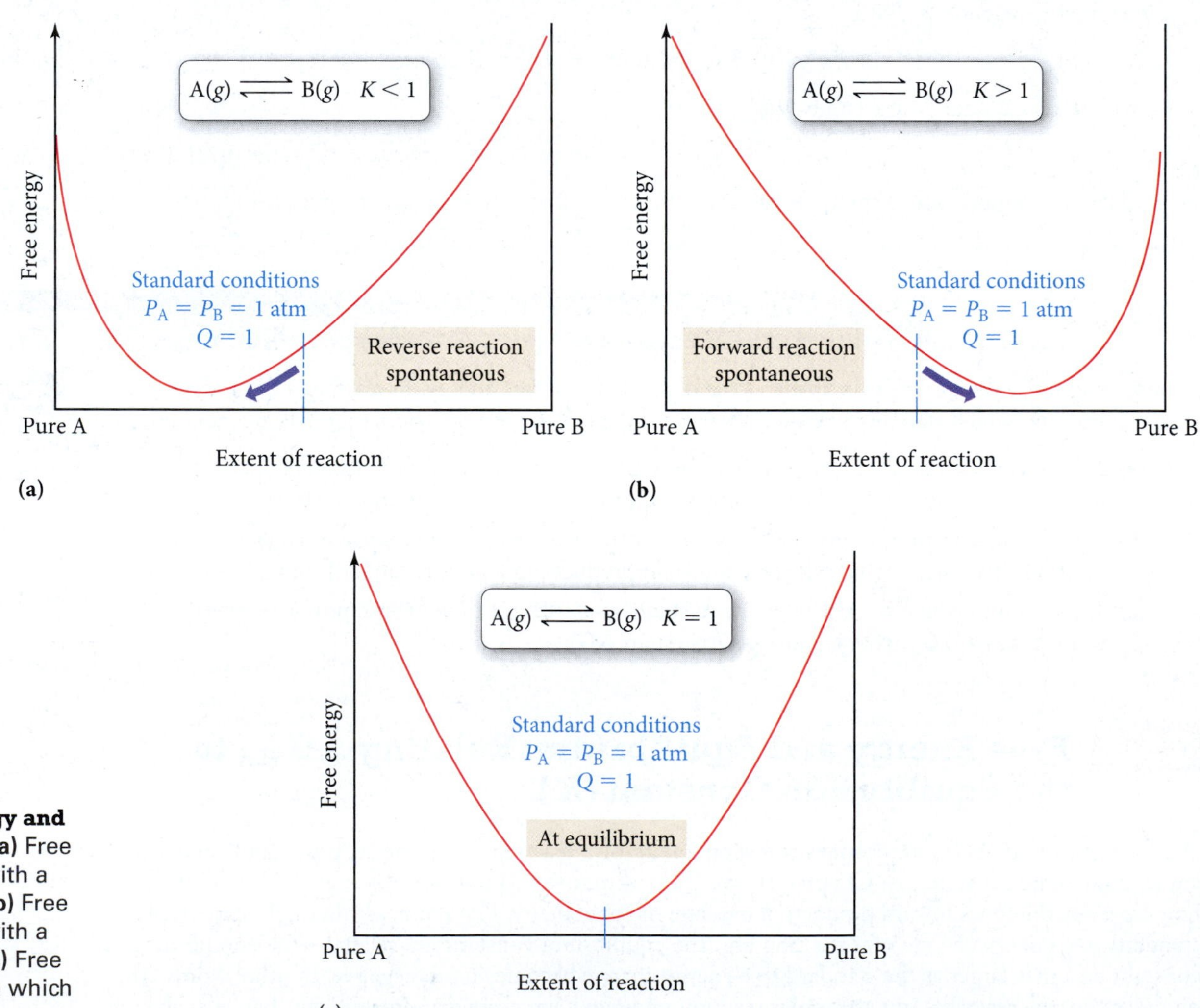

▶ **FIGURE 19.11 Free Energy and the Equilibrium Constant** **(a)** Free energy curve for a reaction with a small equilibrium constant. **(b)** Free energy curve for a reaction with a large equilibrium constant. **(c)** Free energy curve for a reaction in which $K = 1$.

EXAMPLE 19.10
The Equilibrium Constant and ΔG°_{rxn}

Use tabulated free energies of formation to calculate the equilibrium constant for the reaction at 298 K.

$$N_2O_4(g) \rightleftharpoons 2\,NO_2(g)$$

SOLUTION

Look up (in Appendix IVB) the standard free energies of formation for each reactant and product.

Reactant or product	ΔG°_f (in kJ/mol)
$N_2O_4(g)$	99.8
$NO_2(g)$	51.3

Calculate ΔG°_{rxn} by substituting into Equation 19.13.

$$\begin{aligned}\Delta G^\circ_{rxn} &= \sum n_p \Delta G^\circ_f(\text{products}) - \sum n_r \Delta G^\circ_f(\text{reactants}) \\ &= 2[\Delta G^\circ_{f,\,NO_2(g)}] - \Delta G^\circ_{f,\,N_2O_4(g)} \\ &= 2(51.3\text{ kJ}) - 99.8\text{ kJ} \\ &= 2.8\text{ kJ}\end{aligned}$$

Calculate K from ΔG°_{rxn} by solving Equation 19.15 for K and substituting the values of ΔG°_{rxn} and temperature.

$$\Delta G^\circ_{rxn} = -RT \ln K$$

$$\begin{aligned}\ln K &= \frac{-\Delta G^\circ_{rxn}}{RT} \\ &= \frac{-2.8 \times 10^3\text{ J/mol}}{8.314\frac{\text{J}}{\text{mol}\cdot\text{K}}(298\text{ K})} \\ &= -1.13\end{aligned}$$

$$\begin{aligned}K &= e^{-1.13} \\ &= 0.32\end{aligned}$$

FOR PRACTICE 19.10

Calculate ΔG°_{rxn} at 298 K for the reaction:

$$I_2(g) + Cl_2(g) \rightleftharpoons 2\,ICl(g) \qquad K_p = 81.9$$

K and ΔG°_{rxn}

19.6 Cc Conceptual Connection

The equilibrium constant for the reaction $A(g) \rightleftharpoons B(g)$ is less than one. What can you conclude about ΔG°_{rxn} for the reaction?

(a) $\Delta G^\circ_{rxn} = 0$
(b) $\Delta G^\circ_{rxn} < 0$
(c) $\Delta G^\circ_{rxn} > 0$

The Temperature Dependence of the Equilibrium Constant

We now have an equation that relates the standard free energy change for a reaction (ΔG°_{rxn}) to the equilibrium constant for a reaction (K):

$$\Delta G^\circ_{rxn} = -RT \ln K \qquad [19.16]$$

We also have an equation for how the free energy change for a reaction (ΔG°_{rxn}) depends on temperature (T):

$$\Delta G^\circ_{rxn} = \Delta H^\circ_{rxn} - T\Delta S^\circ_{rxn} \quad [19.17]$$

We can combine these two equations to obtain an equation that indicates how the equilibrium constant depends on temperatures. Combining Equations 19.16 and 19.17, we get:

$$-RT \ln K = \Delta H^\circ_{rxn} - T\Delta S^\circ_{rxn} \quad [19.18]$$

We can then divide both sides of Equation 19.18 by the quantity RT:

$$-\ln K = \frac{\Delta H^\circ_{rxn}}{RT} - \frac{T\Delta S^\circ_{rxn}}{RT}$$

Canceling and rearranging, we get this important result:

$$\ln K = -\frac{\Delta H^\circ_{rxn}}{R}\left(\frac{1}{T}\right) + \frac{\Delta S^\circ_{rxn}}{R} \quad [19.19]$$

$$y = mx + b$$

Equation 19.19 is in the form of a straight line. A plot of the natural log of the equilibrium constant (ln K) versus the inverse of the temperature in kelvins ($1/T$) yields a straight line with a slope of $-\Delta H^\circ_{rxn}/R$ and a y-intercept of $\Delta S^\circ_{rxn}/R$. Such a plot is useful for obtaining thermodynamic data (namely, ΔH°_{rxn} and ΔS°_{rxn}) from measurements of K as a function of temperature. However, since ΔH°_{rxn} and ΔS°_{rxn} can themselves be slightly temperature dependent, this analysis works only over a relatively limited temperature range. We can also express the equation in a two-point form:

$$\ln \frac{K_2}{K_1} = -\frac{\Delta H^\circ_{rxn}}{R}\left(\frac{1}{T_2} - \frac{1}{T_1}\right) \quad [19.20]$$

We can use this equation to find ΔH°_{rxn} from a measurement of the equilibrium constant at two different temperatures or to find the equilibrium constant at some other temperature if we know the equilibrium constant at a given temperature and ΔH°_{rxn}.

SELF-ASSESSMENT Quiz

1. Which reaction is most likely to have a positive ΔS_{sys}?
   a) $SiO_2(s) + 3\,C(s) \longrightarrow SiC(s) + 2\,CO(g)$
   b) $6\,CO_2(g) + 6\,H_2O(g) \longrightarrow C_6H_{12}O_6(s) + 6\,O_2(g)$
   c) $CO(g) + Cl_2(g) \longrightarrow COCl_2(g)$
   d) $3\,NO_2(g) + H_2O(l) \longrightarrow 2\,HNO_3(l) + NO(g)$

2. Consider the signs for ΔH_{rxn} and ΔS_{rxn} for several different reactions. In which case is the reaction spontaneous at all temperatures?
   a) $\Delta H_{rxn} < 0$; $\Delta S_{rxn} < 0$
   b) $\Delta H_{rxn} > 0$; $\Delta S_{rxn} > 0$
   c) $\Delta H_{rxn} < 0$; $\Delta S_{rxn} > 0$
   d) $\Delta H_{rxn} > 0$; $\Delta S_{rxn} < 0$

3. Arrange the gases—F_2, Ar, and CH_3F—in order of increasing standard entropy ($S°$) at 298 K.
   a) $F_2 < Ar < CH_3F$
   b) $CH_3F < F_2 < Ar$
   c) $CH_3F < Ar < F_2$
   d) $Ar < F_2 < CH_3F$

4. For a certain reaction $\Delta H_{rxn} = 54.2$ kJ. Calculate the change in entropy for the surroundings (ΔS_{surr}) for the reaction at 25.0 °C. (Assume constant pressure and temperature.)
   a) 2.17×10^3 J/K
   b) -2.17×10^3 J/K
   c) -182 J/K
   d) 182 J/K

5. For a certain reaction $\Delta H°_{rxn} = -255$ kJ and $\Delta S°_{rxn} = 211$ J/K. Calculate $\Delta G°_{rxn}$ at 55 °C.
   a) 11.9×10^3 kJ
   b) 69.5×10^3 kJ
   c) -267 kJ
   d) -324 kJ

6. Use standard entropies to calculate $\Delta S°_{rxn}$ for the balanced chemical equation:

$$2\,PCl_3(l) + O_2(g) \longrightarrow 2\,POCl_3(l)$$

Substance	$S°$ (J/mol·K)
$POCl_3(l)$	222.5
$POCl_3(g)$	325.5
$PCl_3(l)$	217.1
$PCl_3(g)$	311.8
$O_2(g)$	205.2

   a) -194.4 J/K
   b) -199.8 J/K
   c) 10.8 J/K
   d) 1084.4 J/K

7. Use standard free energies of formation to calculate $\Delta G°_{rxn}$ for the balanced chemical equation:

$$Mg(s) + N_2O(g) \longrightarrow MgO(s) + N_2(g)$$

Substance	$\Delta G°_f$ (kJ/mol)
$N_2O(g)$	103.7
$MgO(s)$	−569.3

   a) 673.0 kJ
   b) -673.0 kJ
   c) -465.6 kJ
   d) 465.6 kJ

8. Find $\Delta G°_{rxn}$ for the reaction $2\,A + B \longrightarrow 2\,C$ from the given data.

$$A \longrightarrow B \quad \Delta G°_{rxn} = 128\text{ kJ}$$
$$C \longrightarrow 2\,B \quad \Delta G°_{rxn} = 455\text{ kJ}$$
$$A \longrightarrow C \quad \Delta G°_{rxn} = -182\text{ kJ}$$

   a) -401 kJ
   b) 509 kJ
   c) 401 kJ
   d) -509 kJ

9. For the following reaction, $\Delta G°_{rxn} = 9.4$ kJ at 25 °C. Find ΔG_{rxn} when $P_{NO_2} = 0.115$ atm and $P_{NO} = 9.7$ atm at 25 °C.

$$3\,NO_2(g) + H_2O(l) \longrightarrow 2\,HNO_3(l) + NO(g)$$

   a) -12.3 kJ
   b) 21.7×10^3 kJ
   c) 31.1 kJ
   d) 18.8 kJ

10. The reaction $A(g) \rightleftharpoons B(g)$ has an equilibrium constant of $K_p = 2.3 \times 10^{-5}$. What can you conclude about the sign of $\Delta G°_{rxn}$ for the reaction?
   a) $\Delta G°_{rxn} = 0$
   b) $\Delta G°_{rxn}$ is negative
   c) $\Delta G°_{rxn}$ is positive
   d) Nothing can be concluded about the sign of $\Delta G°_{rxn}$ for the reaction.

11. A reaction has an equilibrium constant of $K_p = 0.018$ at 25 °C. Find $\Delta G°_{rxn}$ for the reaction at this temperature.
   a) -835 J
   b) -4.32 kJ
   c) -9.95 kJ
   d) 9.96 kJ

—Continued

Continued—

12. Which distribution of six particles into three interconnected boxes has the highest entropy?

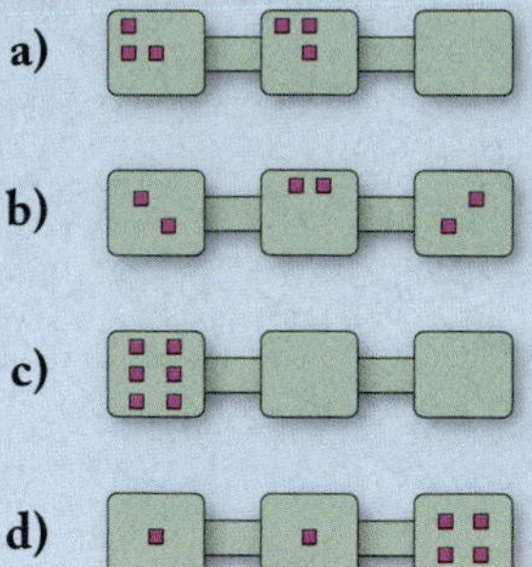

13. Which process results in the increase in entropy of the universe?
    a) the cooling of a hot cup of coffee in room temperature air
    b) the evaporation of water from a desk at room temperature
    c) the melting of snow above 0 °C
    d) all of the above

14. Under which set of conditions is ΔG_{rxn} for the reaction $A(g) \longrightarrow B(g)$ most likely to be negative?
    a) $P_A = 10.0$ atm; $P_B = 10.0$ atm
    b) $P_A = 10.0$ atm; $P_B = 0.010$ atm
    c) $P_A = 0.010$ atm; $P_B = 10.0$ atm
    d) $P_A = 0.010$ atm; $P_B = 0.010$ atm

15. Which statement is true for the freezing of liquid water below 0 °C?
    a) ΔH is positive; ΔS is negative; ΔG is negative
    b) ΔH is negative; ΔS is negative; ΔG is negative
    c) ΔH is positive; ΔS is positive; ΔG is positive
    d) ΔH is positive; ΔS is negative; ΔG is positive

Answers: 1:a; 2:c; 3:d; 4:c; 5:d; 6:a; 7:b; 8:d; 9:c; 10:c; 11:d; 12:b; 13:d; 14:b; 15:b

CHAPTER SUMMARY 19

REVIEW

KEY LEARNING OUTCOMES

CHAPTER OBJECTIVES	ASSESSMENT
Predicting the Sign of Entropy Change (19.4)	• Example 19.1 For Practice 19.1 Exercises 25–26, 29–30, 41–42, 45–46
Calculating Standard Entropy Changes (ΔS°_{rxn}) (19.4)	• Example 19.2 For Practice 19.2 Exercises 37–40
Calculating Entropy Changes in the Surroundings (19.5)	• Example 19.3 For Practice 19.3 For More Practice 19.3 Exercises 41–44
Calculating Gibbs Free Energy Changes and Predicting Spontaneity from ΔH and ΔS (19.6)	• Example 19.4 For Practice 19.4 Exercises 47–52
Calculating the Standard Change in Free Energy for a Reaction Using $\Delta G^\circ_{rxn} = \Delta H^\circ_{rxn} - T\Delta S^\circ_{rxn}$ (19.7)	• Examples 19.5, 19.6 For Practice 19.5, 19.6 Exercises 53–56, 59–60
Calculating ΔG°_{rxn} from Standard Free Energies of Formation (19.7)	• Example 19.7 For Practice 19.7 For More Practice 19.7 Exercises 57–58
Determining ΔG°_{rxn} for a Stepwise Reaction (19.7)	• Example 19.8 For Practice 19.8 Exercises 61–62
Calculating ΔG_{rxn} under Nonstandard Conditions (19.8)	• Example 19.9 For Practice 19.9 Exercises 63–66, 69–70
Relating the Equilibrium Constant and ΔG°_{rxn} (19.9)	• Example 19.10 For Practice 19.10 Exercises 67–68, 71–76

KEY TERMS

Section 19.2
spontaneous process (768)

Section 19.3
entropy (S) (770)
macrostate (770)
microstate (770)
second law of thermodynamics (771)

Section 19.4
standard entropy change for a reaction (ΔS°_{rxn}) (776)
standard molar entropies (S°) (776)
third law of thermodynamics (776)
allotropes (778)

Section 19.6
Gibbs free energy (G) (784)

Section 19.7
standard change in free energy (ΔG°_{rxn}) (788)
free energy of formation (ΔG°_f) (790)

reversible reaction (794)
irreversible reaction (794)

Section 19.8
free energy change of a reaction under nonstandard conditions (ΔG_{rxn}) (795)

KEY CONCEPTS

Energy Spreads Out (19.1)

- Energy has a pervasive tendency to spread out unless prevented from doing so.

Spontaneous and Nonspontaneous Processes (19.2)

- Both spontaneous and nonspontaneous processes can occur, but only spontaneous processes can take place without outside intervention.
- Thermodynamics is the study of the *spontaneity* of reactions, *not* to be confused with kinetics, which is the study of the *rate* of reactions.

Entropy and the Second Law of Thermodynamics (19.3)

- The second law of thermodynamics states that for *any* spontaneous process, the entropy of the universe increases.
- Entropy (S) is proportional to the number of energetically equivalent ways in which the components of a system can be arranged and is a measure of energy dispersal per unit temperature.

Predicting Entropy and Entropy Changes for Chemical Reactions (19.4)

- We calculate the standard change in entropy for a reaction similarly to the way we calculate the standard change in enthalpy for a reaction: by subtracting the sum of the standard entropies of the reactants multiplied by their stoichiometric coefficients from the sum of the standard entropies of the products multiplied by their stoichiometric coefficients.
- Standard entropies are *absolute*: An entropy of zero is established by the third law of thermodynamics as the entropy of a perfect crystal at absolute zero.
- The entropy of a substance at a given temperature depends on its structure, which determines the number of energetically equivalent arrangements of the substance; these include the state, size, and molecular complexity of the substance.

Heat Transfer and Changes in the Entropy of the Surroundings (19.5)

- For a process to be spontaneous, the total entropy of the universe (system plus surroundings) must increase.
- The entropy of the surroundings increases when the change in *enthalpy* of the system (ΔH_{sys}) is negative (i.e., for exothermic reactions).
- The change in entropy of the surroundings for a given ΔH_{sys} depends inversely on temperature—the greater the temperature, the lower the magnitude of ΔS_{surr}.

Gibbs Free Energy (19.6)

- Gibbs free energy, G, is a thermodynamic function that is proportional to the negative of the change in the entropy of the universe. A negative ΔG represents a spontaneous reaction and a positive ΔG represents a nonspontaneous reaction.
- We can calculate the value of ΔG for a reaction from the values of ΔH and ΔS for the *system* using the equation $\Delta G = \Delta H - T\Delta S$.

Free Energy Changes in Chemical Reactions: Calculating ΔG°_{rxn} (19.7)

- There are three ways to calculate ΔG°_{rxn}: (1) from ΔH° and ΔS°, (2) from free energies of formations (only at 25 °C), and (3) from the ΔG°'s of reactions that sum to the reaction of interest.
- The magnitude of a negative ΔG°_{rxn} represents the theoretical amount of energy available to do work, while a positive ΔG°_{rxn} represents the theoretical minimum amount of energy required to make a nonspontaneous process occur.

Free Energy Changes for Nonstandard States: The Relationship between ΔG°_{rxn} and ΔG_{rxn} (19.8)

- The value of ΔG°_{rxn} applies only to standard conditions, and most real conditions are not standard.
- Under nonstandard conditions, we can calculate ΔG_{rxn} from the equation $\Delta G_{rxn} = \Delta G^\circ_{rxn} + RT \ln Q$.

Free Energy and Equilibrium: Relating ΔG°_{rxn} to the Equilibrium Constant (K) (19.9)

- Under standard conditions, the free energy change for a reaction is directly proportional to the negative of the natural log of the equilibrium constant, K; the more negative the free energy change (i.e., the more spontaneous the reaction), the larger the equilibrium constant.
- We can use the temperature dependence of ΔG°_{rxn}, as given by $\Delta G^\circ = \Delta H^\circ - T\Delta S^\circ$, to derive an expression for the temperature dependence of the equilibrium constant.

KEY EQUATIONS AND RELATIONSHIPS

The Definition of Entropy (19.3)

$$S = k \ln W \qquad k = 1.38 \times 10^{-23}\ \text{J/K}$$

Change in Entropy (19.3)

$$\Delta S = S_{final} - S_{initial}$$

Standard Change in Entropy (19.4)

$$\Delta S^\circ_{rxn} = \sum n_p S^\circ(\text{products}) - \sum n_r S^\circ(\text{reactants})$$

Change in the Entropy of the Universe (19.5)

$$\Delta S_{univ} = \Delta S_{sys} + \Delta S_{surr}$$

Change in the Entropy of the Surroundings (19.5)

$$\Delta S_{surr} = \frac{-\Delta H_{sys}}{T} \text{ (constant } T, P)$$

Change in Gibbs Free Energy (19.6)

$$\Delta G = \Delta H - T\Delta S$$

The Relationship between Spontaneity and ΔH, ΔS, and T (19.6)

ΔH	ΔS	Low Temperature	High Temperature
−	+	Spontaneous	Spontaneous
+	−	Nonspontaneous	Nonspontaneous
−	−	Spontaneous	Nonspontaneous
+	+	Nonspontaneous	Spontaneous

Methods for Calculating the Free Energy of Formation (ΔG°_{rxn}) (19.7)

1. $\Delta G^\circ_{rxn} = \Delta H^\circ_{rxn} - T\Delta S^\circ_{rxn}$
2. $\Delta G^\circ_{rxn} = \sum n_p \Delta G^\circ_f(\text{products}) - \sum n_r \Delta G^\circ_f(\text{rectants})$
3. $\Delta G^\circ_{rxn(overall)} = \Delta G^\circ_{rxn(step\ 1)} + \Delta G^\circ_{rxn(step\ 2)} + \Delta G^\circ_{rxn(step\ 3)} + \ldots$

The Relationship between ΔG°_{rxn} and ΔG_{rxn} (19.8)

$$\Delta G_{rxn} = \Delta G^\circ_{rxn} + RT \ln Q \qquad R = 8.314 \text{ J/mol} \cdot \text{K}$$

The Relationship between ΔG°_{rxn} and K (19.9)

$$\Delta G^\circ_{rxn} = -RT \ln K$$

The Temperature Dependence of the Equilibrium Constant (19.9)

$$\ln K = -\frac{\Delta H^\circ_{rxn}}{R}\left(\frac{1}{T}\right) + \frac{\Delta S^\circ_{rxn}}{R}$$

$$\ln \frac{K_2}{K_1} = -\frac{\Delta H^\circ_{rxn}}{R}\left(\frac{1}{T_2} - \frac{1}{T_1}\right)$$

EXERCISES

REVIEW QUESTIONS

1. Explain the idea that energy spreads out and give two examples.
2. What is a spontaneous process? Provide an example.
3. What is a nonspontaneous process? Give an example.
4. Explain the difference between the spontaneity of a reaction (which depends on thermodynamics) and the speed at which the reaction occurs (which depends on kinetics). Can a catalyst make a nonspontaneous reaction spontaneous?
5. What is the precise definition of entropy? What is the significance of entropy being a state function?
6. Why does the entropy of a gas increase when it expands into a vacuum?
7. Explain the difference between macrostates and microstates.
8. Based on its fundamental definition, explain why entropy is a measure of energy dispersion.
9. Provide the definition of the second law of thermodynamics. How does the second law explain why heat travels from a substance at higher temperature to one at lower temperature?
10. What happens to the entropy of a sample of matter when it changes state from a solid to a liquid? From a liquid to a gas?
11. State the third law of thermodynamics and explain its significance.
12. Why is the standard entropy of a substance in the gas state greater than its standard entropy in the liquid state?
13. How does the standard entropy of a substance depend on its molar mass? On its molecular complexity?
14. How can you calculate the standard entropy change for a reaction from tables of standard entropies?
15. Explain why water spontaneously freezes to form ice below 0 °C even though the entropy of the water decreases during the state transition. Why is the freezing of water not spontaneous above 0 °C?
16. Why do exothermic processes tend to be spontaneous at low temperatures? Why does their tendency toward spontaneity decrease with increasing temperature?
17. What is the significance of the change in Gibbs free energy (ΔG) for a reaction?
18. Predict the spontaneity of a reaction (and the temperature dependence of the spontaneity) for each possible combination of signs for ΔH and ΔS (for the system).
    a. ΔH negative, ΔS positive
    b. ΔH positive, ΔS negative
    c. ΔH negative, ΔS negative
    d. ΔH positive, ΔS positive
19. Describe the three different methods to calculate ΔG° for a reaction. Which method would you choose to calculate ΔG° for a reaction at a temperature other than 25 °C?

20. Why is free energy "free"?

21. Explain the difference between $\Delta G°$ and ΔG.

22. Why does water spilled on the floor evaporate even though $\Delta G°$ for the evaporation process is positive at room temperature?

23. How do you calculate the change in free energy for a reaction under nonstandard conditions?

24. How does the value of $\Delta G°$ for a reaction relate to the equilibrium constant for the reaction? What does a negative $\Delta G°$ for a reaction imply about K for the reaction? A positive $\Delta G°$?

PROBLEMS BY TOPIC

Entropy, the Second Law of Thermodynamics, and the Direction of Spontaneous Change

25. Which of these processes is spontaneous?
    a. the combustion of natural gas
    b. the extraction of iron metal from iron ore
    c. a hot drink cooling to room temperature
    d. drawing heat energy from the ocean's surface to power a ship

26. Which of these processes are nonspontaneous? Are the nonspontaneous processes impossible?
    a. a bike going up a hill
    b. a meteor falling to Earth
    c. obtaining hydrogen gas from liquid water
    d. a ball rolling down a hill

27. Suppose that two systems, each composed of two particles represented by circles, have 20 J of total energy. Which system, A or B, has the greater entropy? Why?

System A

10 J

System B

12 J

8 J

28. Suppose two systems, each composed of three particles represented by circles, have 30 J of total energy. In how many energetically equivalent ways can you distribute the particles in each system? Which system has greater entropy?

System A

10 J

System B

12 J

10 J

8 J

Standard Molar Entropy and Changes in Entropy

29. Without doing any calculations, determine the sign of ΔS_{sys} for each chemical reaction.
    a. $2\,KClO_3(s) \longrightarrow 2\,KCl(s) + 3\,O_2(g)$
    b. $CH_2{=}CH_2(g) + H_2(g) \longrightarrow CH_3CH_3(g)$
    c. $Na(s) + \frac{1}{2}Cl_2(g) \longrightarrow NaCl(s)$
    d. $N_2(g) + 3\,H_2(g) \longrightarrow 2\,NH_3(g)$

30. Without doing any calculations, determine the sign of ΔS_{sys} for each chemical reaction.
    a. $Mg(s) + Cl_2(g) \longrightarrow MgCl_2(s)$
    b. $2\,H_2S(g) + 3\,O_2(g) \longrightarrow 2\,H_2O(g) + 2\,SO_2(g)$
    c. $2\,O_3(g) \longrightarrow 3\,O_2(g)$
    d. $HCl(g) + NH_3(g) \longrightarrow NH_4Cl(s)$

31. How does the molar entropy of a substance change with increasing temperature?

32. What is the molar entropy of a pure crystal at 0 K? What is the significance of the answer to this question?

33. For each pair of substances, choose the one that you expect to have the higher standard molar entropy ($S°$) at 25 °C. Explain the reasons for your choices.
    a. $CO(g)$; $CO_2(g)$
    b. $CH_3OH(l)$; $CH_3OH(g)$
    c. $Ar(g)$; $CO_2(g)$
    d. $CH_4(g)$; $SiH_4(g)$
    e. $NO_2(g)$; $CH_3CH_2CH_3(g)$
    f. $NaBr(s)$; $NaBr(aq)$

34. For each pair of substances, choose the one that you expect to have the higher standard molar entropy ($S°$) at 25 °C. Explain the reasons for your choices.
    a. $NaNO_3(s)$; $NaNO_3(aq)$
    b. $CH_4(g)$; $CH_3CH_3(g)$
    c. $Br_2(l)$; $Br_2(g)$
    d. $Br_2(g)$; $F_2(g)$
    e. $PCl_3(g)$; $PCl_5(g)$
    f. $CH_3CH_2CH_2CH_3(g)$; $SO_2(g)$

35. Rank each set of substances in order of increasing standard molar entropy ($S°$). Explain your reasoning.
    a. $NH_3(g)$; $Ne(g)$; $SO_2(g)$; $CH_3CH_2OH(g)$; $He(g)$
    b. $H_2O(s)$; $H_2O(l)$; $H_2O(g)$
    c. $CH_4(g)$; $CF_4(g)$; $CCl_4(g)$

36. Rank each set of substances in order of increasing standard molar entropy ($S°$). Explain your reasoning.
    a. $I_2(s)$; $F_2(g)$; $Br_2(g)$; $Cl_2(g)$
    b. $H_2O(g)$; $H_2O_2(g)$; $H_2S(g)$
    c. $C(s,$ graphite$)$; $C(s,$ diamond$)$; $C(s,$ amorphous$)$

37. Use data from Appendix IVB to calculate ΔS°_{rxn} for each of the reactions. In each case, try to rationalize the sign of ΔS°_{rxn}.
   a. $C_2H_4(g) + H_2(g) \longrightarrow C_2H_6(g)$
   b. $C(s) + H_2O(g) \longrightarrow CO(g) + H_2(g)$
   c. $CO(g) + H_2O(g) \longrightarrow H_2(g) + CO_2(g)$
   d. $2\,H_2S(g) + 3\,O_2(g) \longrightarrow 2\,H_2O(l) + 2\,SO_2(g)$

38. Use data from Appendix IVB to calculate ΔS°_{rxn} for each of the reactions. In each case, try to rationalize the sign of ΔH°_{rxn}.
   a. $3\,NO_2(g) + H_2O(l) \longrightarrow 2\,HNO_3(aq) + NO(g)$
   b. $Cr_2O_3(s) + 3\,CO(g) \longrightarrow 2\,Cr(s) + 3\,CO_2(g)$
   c. $SO_2(g) + \frac{1}{2}O_2(g) \longrightarrow SO_3(g)$
   d. $N_2O_4(g) + 4\,H_2(g) \longrightarrow N_2(g) + 4\,H_2O(g)$

39. Find ΔS° for the formation of $CH_2Cl_2(g)$ from its gaseous elements in their standard states. Rationalize the sign of ΔS°.

40. Find ΔS° for the reaction between nitrogen gas and fluorine gas to form nitrogen trifluoride gas. Rationalize the sign of ΔS°.

Entropy Changes in the Surroundings

41. Without doing any calculations, determine the sign of ΔS_{sys} and ΔS_{surr} for each chemical reaction. In addition, predict under what temperatures (all temperatures, low temperatures, or high temperatures), if any, the reaction is spontaneous.
   a. $C_3H_8(g) + 5\,O_2(g) \longrightarrow 3\,CO_2(g) + 4\,H_2O(g)$ $\quad \Delta H^\circ_{rxn} = -2044\text{ kJ}$
   b. $N_2(g) + O_2(g) \longrightarrow 2\,NO(g) \quad \Delta H^\circ_{rxn} = +182.6\text{ kJ}$
   c. $2\,N_2(g) + O_2(g) \longrightarrow 2\,N_2O(g) \quad \Delta H^\circ_{rxn} = +163.2\text{ kJ}$
   d. $4\,NH_3(g) + 5\,O_2(g) \longrightarrow 4\,NO(g) + 6\,H_2O(g)$ $\quad \Delta H^\circ_{rxn} = -906\text{ kJ}$

42. Without doing any calculations, determine the sign of ΔS_{sys} and ΔS_{surr} for each chemical reaction. In addition, predict under what temperatures (all temperatures, low temperatures, or high temperatures), if any, the reaction is spontaneous.
   a. $2\,CO(g) + O_2(g) \longrightarrow 2\,CO_2(g) \quad \Delta H^\circ_{rxn} = -566.0\text{ kJ}$
   b. $2\,NO_2(g) \longrightarrow 2\,NO(g) + O_2(g) \quad \Delta H^\circ_{rxn} = +113.1\text{ kJ}$
   c. $2\,H_2(g) + O_2(g) \longrightarrow 2\,H_2O(g) \quad \Delta H^\circ_{rxn} = -483.6\text{ kJ}$
   d. $CO_2(g) \longrightarrow C(s) + O_2(g) \quad \Delta H^\circ_{rxn} = +393.5\text{ kJ}$

43. Calculate ΔS_{surr} at the indicated temperature for each reaction.
   a. $\Delta H^\circ_{rxn} = -385\text{ kJ}$; 298 K
   b. $\Delta H^\circ_{rxn} = -385\text{ kJ}$; 77 K
   c. $\Delta H^\circ_{rxn} = +114\text{ kJ}$; 298 K
   d. $\Delta H^\circ_{rxn} = +114\text{ kJ}$; 77 K

44. A reaction has $\Delta H^\circ_{rxn} = -112\text{ kJ}$ and $\Delta S^\circ_{rxn} = 354\text{ J/K}$. At what temperature is the change in entropy for the reaction equal to the change in entropy for the surroundings?

45. Given the values of ΔH°_{rxn}, ΔS°_{rxn}, and T, determine ΔS_{univ} and predict whether each reaction is spontaneous.
   a. $\Delta H^\circ_{rxn} = +115\text{ kJ}$; $\Delta S^\circ_{rxn} = -263\text{ J/K}$; $T = 298\text{ K}$
   b. $\Delta H^\circ_{rxn} = -115\text{ kJ}$; $\Delta S^\circ_{rxn} = +263\text{ J/K}$; $T = 298\text{ K}$
   c. $\Delta H^\circ_{rxn} = -115\text{ kJ}$; $\Delta S^\circ_{rxn} = -263\text{ J/K}$; $T = 298\text{ K}$
   d. $\Delta H^\circ_{rxn} = -115\text{ kJ}$; $\Delta S^\circ_{rxn} = -263\text{ J/K}$; $T = 615\text{ K}$

46. Given the values of ΔH_{rxn}, ΔS_{rxn}, and T, determine ΔS_{univ} and predict whether each reaction is spontaneous.
   a. $\Delta H^\circ_{rxn} = -95\text{ kJ}$; $\Delta S^\circ_{rxn} = -157\text{ J/K}$; $T = 298\text{ K}$
   b. $\Delta H^\circ_{rxn} = -95\text{ kJ}$; $\Delta S^\circ_{rxn} = -157\text{ J/K}$; $T = 855\text{ K}$
   c. $\Delta H^\circ_{rxn} = +95\text{ kJ}$; $\Delta S^\circ_{rxn} = -157\text{ J/K}$; $T = 298\text{ K}$
   d. $\Delta H^\circ_{rxn} = -95\text{ kJ}$; $\Delta S^\circ_{rxn} = +157\text{ J/K}$; $T = 398\text{ K}$

Gibbs Free Energy

47. Calculate the change in Gibbs free energy for each set of ΔH_{rxn}, ΔS_{rxn}, and T given in Problem 45. Predict whether each reaction is spontaneous at the temperature indicated.

48. Calculate the change in Gibbs free energy for each set of ΔH_{rxn}, ΔS_{rxn}, and T given in Problem 46. Predict whether each reaction is spontaneous at the temperature indicated.

49. Calculate the free energy change for this reaction at 25 °C. Is the reaction spontaneous?

$$C_3H_8(g) + 5\,O_2(g) \longrightarrow 3\,CO_2(g) + 4\,H_2O(g)$$
$$\Delta H^\circ_{rxn} = -2217\text{ kJ};\quad \Delta S^\circ_{rxn} = 101.1\text{ J/K}$$

50. Calculate the free energy change for this reaction at 25 °C. Is the reaction spontaneous?

$$2\,Ca(s) + O_2(g) \longrightarrow 2\,CaO(s)$$
$$\Delta H^\circ_{rxn} = -1269.8\text{ kJ};\quad \Delta S^\circ_{rxn} = -364.6\text{ J/K}$$

51. Fill in the blanks in the table. Both ΔH and ΔS refer to the system.

ΔH	ΔS	ΔG	Low Temperature	High Temperature
−	+	−	Spontaneous	______
−	−	Temperature dependent	______	______
+	+	______	______	Spontaneous
______	−	______	Nonspontaneous	Nonspontaneous

52. Predict the conditions (high temperature, low temperature, all temperatures, or no temperatures) under which each reaction is spontaneous.
   a. $H_2O(g) \longrightarrow H_2O(l)$
   b. $CO_2(s) \longrightarrow CO_2(g)$
   c. $H_2(g) \longrightarrow 2\,H(g)$
   d. $2\,NO_2(g) \longrightarrow 2\,NO(g) + O_2(g)$ (endothermic)

53. Methanol burns in oxygen to form carbon dioxide and water. Write a balanced equation for the combustion of liquid methanol and calculate ΔH°_{rxn}, ΔS°_{rxn}, and ΔG°_{rxn} at 25 °C. Is the combustion of methanol spontaneous?

54. In photosynthesis, plants form glucose ($C_6H_{12}O_6$) and oxygen from carbon dioxide and water. Write a balanced equation for photosynthesis and calculate ΔH°_{rxn}, ΔS°_{rxn}, and ΔG°_{rxn} at 25 °C. Is photosynthesis spontaneous?

55. For each reaction, calculate ΔH°_{rxn}, ΔS°_{rxn}, and ΔG°_{rxn} at 25 °C and state whether the reaction is spontaneous. If the reaction is not spontaneous, would a change in temperature make it spontaneous? If so, should the temperature be raised or lowered from 25 °C?
   a. $N_2O_4(g) \longrightarrow 2\,NO_2(g)$
   b. $NH_4Cl(s) \longrightarrow HCl(g) + NH_3(g)$
   c. $3\,H_2(g) + Fe_2O_3(s) \longrightarrow 2\,Fe(s) + 3\,H_2O(g)$
   d. $N_2(g) + 3\,H_2(g) \longrightarrow 2\,NH_3(g)$

56. For each reaction, calculate ΔH°_{rxn}, ΔS°_{rxn}, and ΔG°_{rxn} at 25 °C and state whether the reaction is spontaneous. If the reaction is not spontaneous, would a change in temperature make it spontaneous? If so, should the temperature be raised or lowered from 25 °C?

a. $2\ CH_4(g) \longrightarrow C_2H_6(g) + H_2(g)$
b. $2\ NH_3(g) \longrightarrow N_2H_4(g) + H_2(g)$
c. $N_2(g) + O_2(g) \longrightarrow 2\ NO(g)$
d. $2\ KClO_3(s) \longrightarrow 2\ KCl(s) + 3\ O_2(g)$

57. Use standard free energies of formation to calculate $\Delta G°$ at 25 °C for each reaction in Problem 55. How do the values of $\Delta G°$ calculated this way compare to those calculated from $\Delta H°$ and $\Delta S°$? Which of the two methods could be used to determine how $\Delta G°$ changes with temperature?

58. Use standard free energies of formation to calculate $\Delta G°$ at 25 °C for each reaction in Problem 56. How well do the values of $\Delta G°$ calculated this way compare to those calculated from $\Delta H°$ and $\Delta S°$? Which of the two methods could be used to determine how $\Delta G°$ changes with temperature?

59. Consider the reaction:

$$2\ NO(g) + O_2(g) \longrightarrow 2\ NO_2(g)$$

Estimate $\Delta G°$ for this reaction at each temperature and predict whether the reaction is spontaneous. (Assume that $\Delta H°$ and $\Delta S°$ do not change too much within the given temperature range.)

a. 298 K
b. 715 K
c. 855 K

60. Consider the reaction:

$$CaCO_3(s) \longrightarrow CaO(s) + CO_2(g)$$

Estimate $\Delta G°$ for this reaction at each temperature and predict whether the reaction is spontaneous. (Assume that $\Delta H°$ and $\Delta S°$ do not change too much within the given temperature range.)

a. 298 K
b. 1055 K
c. 1455 K

61. Determine $\Delta G°$ for the reaction:

$$Fe_2O_3(s) + 3\ CO(g) \longrightarrow 2\ Fe(s) + 3\ CO_2(g)$$

Use the following reactions with known $\Delta G°_{rxn}$ values:

$$2\ Fe(s) + \tfrac{3}{2}O_2(g) \longrightarrow Fe_2O_3(s) \quad \Delta G°_{rxn} = -742.2\ kJ$$
$$CO(g) + \tfrac{1}{2}O_2(g) \longrightarrow CO_2(g) \quad \Delta G°_{rxn} = -257.2\ kJ$$

62. Calculate $\Delta G°_{rxn}$ for the reaction:

$$CaCO_3(s) \longrightarrow CaO(s) + CO_2(g)$$

Use the following reactions and given $\Delta G°_{rxn}$ values:

$$Ca(s) + CO_2(g) + \tfrac{1}{2}O_2(g) \longrightarrow CaCO_3(s) \quad \Delta G°_{rxn} = -734.4\ kJ$$
$$2\ Ca(s) + O_2(g) \longrightarrow 2\ CaO(s) \quad \Delta G°_{rxn} = -1206.6\ kJ$$

Free Energy Changes, Nonstandard Conditions, and the Equilibrium Constant

63. Consider the sublimation of iodine at 25.0 °C:

$$I_2(s) \longrightarrow I_2(g)$$

a. Find $\Delta G°_{rxn}$ at 25.0 °C.
b. Find ΔG_{rxn} at 25.0 °C under the following nonstandard conditions:
   i. $P_{I_2} = 1.00$ mmHg
   ii. $P_{I_2} = 0.100$ mmHg
c. Explain why iodine spontaneously sublimes in open air at 25 °C.

64. Consider the evaporation of methanol at 25.0 °C.

$$CH_3OH(l) \longrightarrow CH_3OH(g)$$

a. Find $\Delta G°_{rxn}$ at 25.0 °C.
b. Find ΔG_{rxn} at 25.0 °C under the following nonstandard conditions:
   i. $P_{CH_3OH} = 150.0$ mmHg
   ii. $P_{CH_3OH} = 100.0$ mmHg
   iii. $P_{CH_3OH} = 10.0$ mmHg
c. Explain why methanol spontaneously evaporates in open air at 25.0 °C.

65. Consider the reaction:

$$CH_3OH(g) \rightleftharpoons CO(g) + 2\ H_2(g)$$

Calculate ΔG for this reaction at 25 °C under the following conditions:

$P_{CH_3OH} = 0.855$ atm
$P_{CO} = 0.125$ atm
$P_{H_2} = 0.183$ atm

66. Consider the reaction:

$$CO_2(g) + CCl_4(g) \rightleftharpoons 2\ COCl_2(g)$$

Calculate ΔG for this reaction at 25 °C under the following conditions:

$P_{CO_2} = 0.112$ atm
$P_{CCl_4} = 0.174$ atm
$P_{COCl_2} = 0.744$ atm

67. Use data from Appendix IVB to calculate the equilibrium constant at 25 °C for each reaction.

a. $2\ CO(g) + O_2(g) \rightleftharpoons 2\ CO_2(g)$
b. $2\ H_2S(g) \rightleftharpoons 2\ H_2(g) + S_2(g)$

68. Use data from Appendix IVB to calculate the equilibrium constant at 25 °C for each reaction. $\Delta G°_f$ for $BrCl(g)$ is −1.0 kJ/mol.

a. $2\ NO_2(g) \rightleftharpoons N_2O_4(g)$
b. $Br_2(g) + Cl_2(g) \rightleftharpoons 2\ BrCl(g)$

69. Consider the reaction:

$$CO(g) + 2\ H_2(g) \rightleftharpoons CH_3OH(g) \quad K_p = 2.26 \times 10^4 \text{ at } 25\ °C$$

Calculate ΔG_{rxn} for the reaction at 25 °C under the following conditions:

a. standard conditions
b. at equilibrium
c. $P_{CH_3OH} = 1.0$ atm; $P_{CO} = P_{H_2} = 0.010$ atm

70. Consider the reaction:

$$I_2(g) + Cl_2(g) \rightleftharpoons 2\ ICl(g) \quad K_p = 81.9 \text{ at } 25\ °C$$

Calculate ΔG_{rxn} for the reaction at 25 °C under the following conditions:

a. standard conditions
b. at equilibrium
c. $P_{ICl} = 2.55$ atm; $P_{I_2} = 0.325$ atm; $P_{Cl_2} = 0.221$ atm

71. Estimate the value of the equilibrium constant at 525 K for each reaction in Problem 67.

72. Estimate the value of the equilibrium constant at 655 K for each reaction in Problem 68. ($\Delta H°_f$ for BrCl is 14.6 kJ/mol.)

73. Consider the reaction:

$$H_2(g) + I_2(g) \rightleftharpoons 2\ HI(g)$$

The following data show the equilibrium constant for this reaction measured at several different temperatures. Use the data to find ΔH°_{rxn} and ΔS°_{rxn} for the reaction.

Temperature	K_p
150 K	1.4×10^{-6}
175 K	4.6×10^{-4}
200 K	3.6×10^{-2}
225 K	1.1
250 K	15.5

74. Consider the reaction:

$$2\,NO(g) + O_2(g) \rightleftharpoons 2\,NO_2(g)$$

The following data show the equilibrium constant for this reaction measured at several different temperatures. Use the data to find ΔH°_{rxn} and ΔS°_{rxn} for the reaction.

Temperature	K_p
170 K	3.8×10^{-3}
180 K	0.34
190 K	18.4
200 K	681

75. The change in enthalpy (ΔH°_{rxn}) for a reaction is -25.8 kJ/mol. The equilibrium constant for the reaction is 1.4×10^3 at 298 K. What is the equilibrium constant for the reaction at 655 K?

76. A reaction has an equilibrium constant of 8.5×10^3 at 298 K. At 755 K, the equilibrium constant is 0.65. Find ΔH°_{rxn} for the reaction.

CUMULATIVE PROBLEMS

77. Determine the sign of ΔS_{sys} for each process.

a. water boiling

b. water freezing

c.

78. Determine the sign of ΔS_{sys} for each process.

a. dry ice subliming

b. dew forming

c.

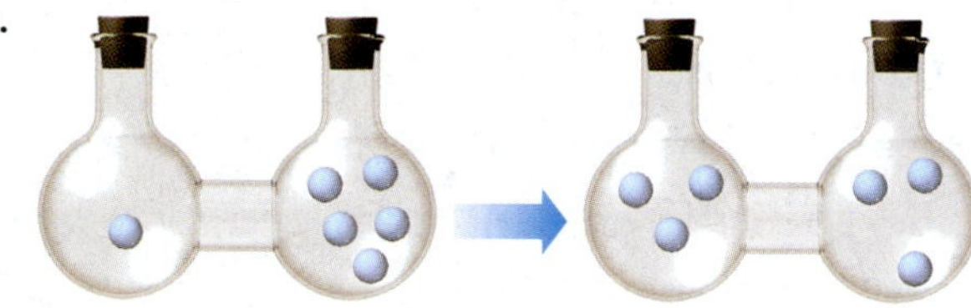

79. Our atmosphere is composed primarily of nitrogen and oxygen, which coexist at 25 °C without reacting to any significant extent. However, the two gases can react to form nitrogen monoxide according to the reaction:

$$N_2(g) + O_2(g) \rightleftharpoons 2\,NO(g)$$

a. Calculate ΔG° and K_p for this reaction at 298 K. Is the reaction spontaneous?

b. Estimate ΔG° at 2000 K. Does the reaction become more spontaneous as temperature increases?

80. Nitrogen dioxide, a pollutant in the atmosphere, can combine with water to form nitric acid. One of the possible reactions is shown here. Calculate ΔG° and K_p for this reaction at 25 °C and comment on the spontaneity of the reaction.

$$3\,NO_2(g) + H_2O(l) \longrightarrow 2\,HNO_3(aq) + NO(g)$$

81. Ethene (C_2H_4) can be halogenated by the reaction:

$$C_2H_4(g) + X_2(g) \rightleftharpoons C_2H_4X_2(g)$$

where X_2 can be Cl_2, Br_2, or I_2. Use the thermodynamic data given to calculate ΔH°, ΔS°, ΔG°, and K_p for the halogenation reaction by each of the three halogens at 25 °C. Which reaction is most spontaneous? Least spontaneous? What is the main factor responsible for the difference in the spontaneity of the three reactions? Does higher temperature make the reactions more spontaneous or less spontaneous?

Compound	ΔH°_f (kJ/mol)	S° (J/mol · K)
$C_2H_4Cl_2(g)$	-129.7	308.0
$C_2H_4Br_2(g)$	$+38.3$	330.6
$C_2H_4I_2(g)$	$+66.5$	347.8

82. H_2 reacts with the halogens (X_2) according to the reaction:

$$H_2(g) + X_2(g) \rightleftharpoons 2\,HX(g)$$

where X_2 can be Cl_2, Br_2, or I_2. Use the thermodynamic data in Appendix IVB to calculate ΔH°, ΔS°, ΔG°, and K_p for the reaction between hydrogen and each of the three halogens. Which reaction is most spontaneous? Least spontaneous? What is the main factor responsible for the difference in the spontaneity of the three reactions? Does higher temperature make the reactions more spontaneous or less spontaneous?

83. Consider this reaction occurring at 298 K:

$$N_2O(g) + NO_2(g) \rightleftharpoons 3\,NO(g)$$

a. Show that the reaction is not spontaneous under standard conditions by calculating ΔG°_{rxn}.

b. If a reaction mixture contains only N_2O and NO_2 at partial pressures of 1.0 atm each, the reaction will be spontaneous until some NO forms in the mixture. What maximum partial pressure of NO builds up before the reaction ceases to be spontaneous?

c. Can the reaction be made more spontaneous by an increase or decrease in temperature? If so, what temperature is required to make the reaction spontaneous under standard conditions?

84. Consider this reaction occurring at 298 K:

$$BaCO_3(s) \rightleftharpoons BaO(s) + CO_2(g)$$

a. Show that the reaction is not spontaneous under standard conditions by calculating ΔG°_{rxn}.
b. If $BaCO_3$ is placed in an evacuated flask, what is the partial pressure of CO_2 when the reaction reaches equilibrium?
c. Can the reaction be made more spontaneous by an increase or decrease in temperature? If so, at what temperature is the partial pressure of carbon dioxide 1.0 atm?

85. Living organisms use energy from the metabolism of food to create an energy-rich molecule called adenosine triphosphate (ATP). The ATP then acts as an energy source for a variety of reactions that the living organism must carry out to survive. ATP provides energy through its hydrolysis, which can be symbolized as follows:

$$ATP(aq) + H_2O(l) \longrightarrow ADP(aq) + P_i(aq) \quad \Delta G^\circ_{rxn} = -30.5 \text{ kJ}$$

where ADP represents adenosine diphosphate and P_i represents an inorganic phosphate group (such as HPO_4^{2-}).

a. Calculate the equilibrium constant, K, for the given reaction at 298 K.
b. The free energy obtained from the oxidation (reaction with oxygen) of glucose ($C_6H_{12}O_6$) to form carbon dioxide and water can be used to re-form ATP by driving the given reaction in reverse. Calculate the standard free energy change for the oxidation of glucose and estimate the maximum number of moles of ATP that can be formed by the oxidation of one mole of glucose.

86. The standard free energy change for the hydrolysis of ATP was given in Problem 85. In a particular cell, the concentrations of ATP, ADP, and P_i are 0.0031 M, 0.0014 M, and 0.0048 M, respectively. Calculate the free energy change for the hydrolysis of ATP under these conditions. (Assume a temperature of 298 K.)

87. These reactions are important in catalytic converters in automobiles. Calculate ΔG° for each at 298 K. Predict the effect of increasing temperature on the magnitude of ΔG°.

a. $2\,CO(g) + 2\,NO(g) \longrightarrow N_2(g) + 2\,CO_2(g)$
b. $5\,H_2(g) + 2\,NO(g) \longrightarrow 2\,NH_3(g) + 2\,H_2O(g)$
c. $2\,H_2(g) + 2\,NO(g) \longrightarrow N_2(g) + 2\,H_2O(g)$
d. $2\,NH_3(g) + 2\,O_2(g) \longrightarrow N_2O(g) + 3\,H_2O(g)$

88. Calculate ΔG° at 298 K for these reactions and predict the effect on ΔG° of lowering the temperature.

a. $NH_3(g) + HBr(g) \longrightarrow NH_4Br(s)$
b. $CaCO_3(s) \longrightarrow CaO(s) + CO_2(g)$
c. $CH_4(g) + 3\,Cl_2(g) \longrightarrow CHCl_3(g) + 3\,HCl(g)$
(ΔG°_f for $CHCl_3(g)$ is -70.4 kJ/mol.)

89. All the oxides of nitrogen have positive values of ΔG°_f at 298 K, but only one common oxide of nitrogen has a positive ΔS°_f. Identify that oxide of nitrogen without reference to thermodynamic data and explain.

90. The values of ΔG°_f for the hydrogen halides become less negative with increasing atomic number. The ΔG°_f of HI is slightly positive. On the other hand, the trend in ΔS°_f is to become more positive with increasing atomic number. Explain.

91. Consider the reaction $X_2(g) \longrightarrow 2\,X(g)$. When a vessel initially containing 755 torr of X_2 comes to equilibrium at 298 K, the equilibrium partial pressure of X is 103 torr. The same reaction is repeated with an initial partial pressure of 748 torr of X_2 at 755 K; the equilibrium partial pressure of X is 532 torr. Find ΔH° for the reaction.

92. Dinitrogen tetroxide decomposes to nitrogen dioxide:

$$N_2O_4(g) \longrightarrow 2\,NO_2(g) \quad \Delta H^\circ_{rxn} = 55.3 \text{ kJ}$$

At 298 K, a reaction vessel initially contains 0.100 atm of N_2O_4. When equilibrium is reached, 58% of the N_2O_4 has decomposed to NO_2. What percentage of N_2O_4 decomposes at 388 K? Assume that the initial pressure of N_2O_4 is the same (0.100 atm).

93. Indicate and explain the sign of ΔS_{univ} for each process.

a. $2\,H_2(g) + O_2(g) \longrightarrow 2\,H_2O(l)$ at 298 K
b. the electrolysis of $H_2O(l)$ to $H_2(g)$ and $O_2(g)$ at 298 K
c. the growth of an oak tree from an acorn

94. The Haber process is very important for agriculture because it converts $N_2(g)$ from the atmosphere into bound nitrogen, which can be taken up and used by plants. The Haber process reaction is $N_2(g) + 3\,H_2(g) \rightleftharpoons 2\,NH_3(g)$. The reaction is exothermic but is carried out at relatively high temperatures. Why?

95. A metal salt with the formula MCl_2 crystallizes from water to form a solid with the composition $MCl_2 \cdot 6\,H_2O$. The equilibrium vapor pressure of water above this solid at 298 K is 18.3 mmHg. What is the value of ΔG for the reaction $MCl_2 \cdot 6\,H_2O(s) \rightleftharpoons MCl_2(s) + 6\,H_2O(g)$ when the pressure of water vapor is 18.3 mmHg? When the pressure of water vapor is 760.0 mmHg?

96. The solubility of AgCl(s) in water at 25 °C is 1.33×10^{-5} mol/L, and its ΔH° of solution is 65.7 k/mol. What is its solubility at 50.0 °C?

CHALLENGE PROBLEMS

97. Review the subsection in this chapter entitled *Making a Nonspontaneous Process Spontaneous* in Section 19.7. The hydrolysis of ATP, shown in Problem 85, is often used to drive nonspontaneous processes—such as muscle contraction and protein synthesis—in living organisms. The nonspontaneous process to be driven must be coupled to the ATP hydrolysis reaction. For example, suppose the nonspontaneous process is $A + B \longrightarrow AB$ (ΔG° positive). The coupling of a nonspontaneous reaction such as this one to the hydrolysis of ATP is often accomplished by the mechanism:

$$A + ATP + H_2O \longrightarrow A{-}P_i + ADP$$
$$A{-}P_i + B \longrightarrow AB + P_i$$
$$\overline{A + B + ATP + H_2O \longrightarrow AB + ADP + P_i}$$

As long as ΔG_{rxn} for the nonspontaneous reaction is less than 30.5 kJ, the reaction can be made spontaneous by coupling in this way to the hydrolysis of ATP. Suppose that ATP is to drive the reaction between glutamate and ammonia to form glutamine:

O^- O
$O{=}C$ CH_2 CH_2 CH $C{=}O$ O^- $+ NH_3 \longrightarrow$
NH_3^+

NH_2 O
$O{=}C$ CH_2 CH_2 CH $C{=}O$ O^- $+ H_2O$
NH_3^+

a. Calculate K for the reaction between glutamate and ammonia. (The standard free energy change for the reaction is $+14.2$ kJ/mol. Assume a temperature of 298 K.)

b. Write a set of reactions such as those given showing how the glutamate and ammonia reaction can couple with the hydrolysis of ATP. What is ΔG°_{rxn} and K for the coupled reaction?

98. Calculate the entropy of each state and rank the states in order of increasing entropy.

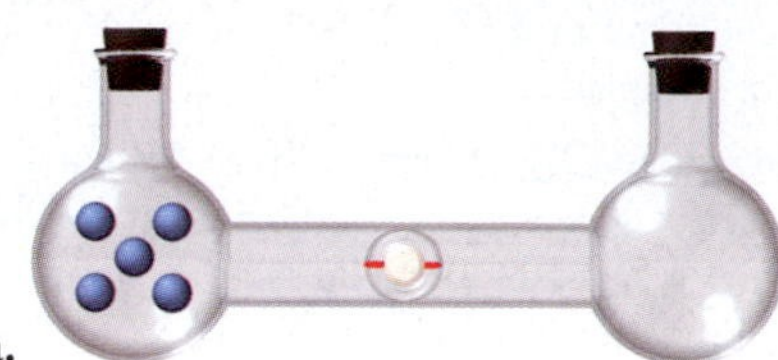
a.

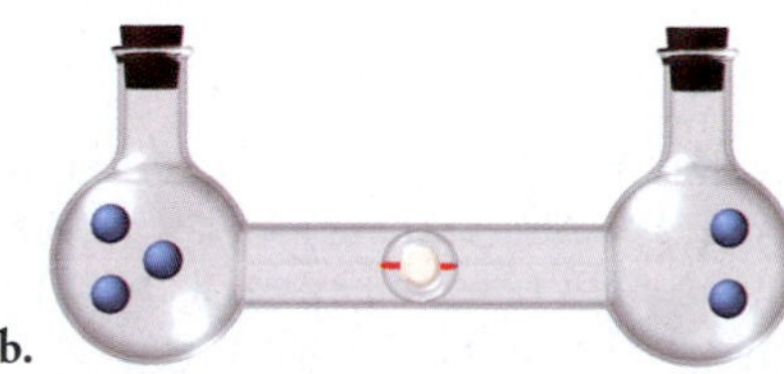
b.

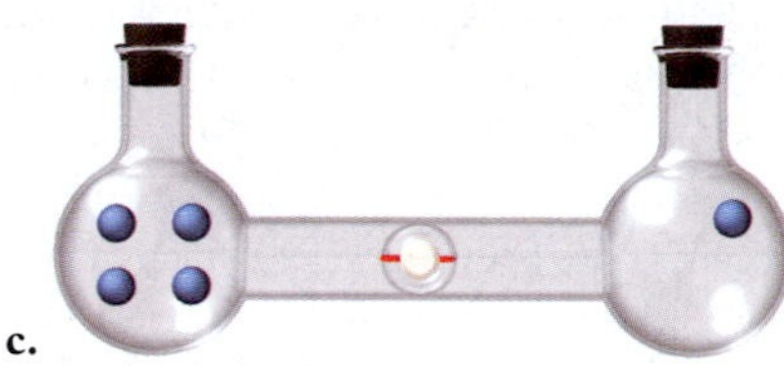
c.

99. Suppose we redefine the standard state as $P = 2$ atm. Find the new standard ΔG°_f values of each substance.

a. $HCl(g)$
b. $N_2O(g)$
c. $H(g)$

Explain the results in terms of the relative entropies of reactants and products of each reaction.

100. The ΔG for the freezing of $H_2O(l)$ at -10 °C is -210 J/mol, and the heat of fusion of ice at this temperature is 5610 J/mol. Find the entropy change of the universe when 1 mol of water freezes at -10 °C.

101. Consider the reaction that occurs during the Haber process:

$$N_2(g) + 3\,H_2(g) \longrightarrow 2\,NH_3(g)$$

The equilibrium constant is 3.9×10^5 at 300 K and 1.2×10^{-1} at 500 K. Calculate ΔH°_{rxn} and ΔS°_{rxn} for this reaction.

102. The salt ammonium nitrate can follow three modes of decomposition: (a) to $HNO_3(g)$ and $NH_3(g)$, (b) to $N_2O(g)$ and $H_2O(g)$, and (c) to $N_2(g)$, $O_2(g)$, and $H_2O(g)$. Calculate ΔG°_{rxn} for each mode of decomposition at 298 K. Explain in light of these results how it is still possible to use ammonium nitrate as a fertilizer and the precautions that should be taken when it is used.

103. Given the tabulated data, calculate ΔS_{vap} for each of the first four liquids. ($\Delta S_{vap} = \Delta H_{vap}/T$, where T is in K)

Compound	Name	bp(°C)	ΔH_{vap} (kJ/mol) at bp
$C_4H_{10}O$	Diethyl ether	34.6	26.5
C_3H_6	Acetone	56.1	29.1
C_6H_6	Benzene	79.8	30.8
$CHCl_3$	Chloroform	60.8	29.4
C_2H_5OH	Ethanol	77.8	38.6
H_2O	Water	100.0	40.7

All four values should be close to each other. Predict whether the last two liquids in the table have ΔS_{vap} in this same range. If not, predict whether it is larger or smaller and explain. Verify your prediction.

CONCEPTUAL PROBLEMS

104. Which statement is true?

a. A spontaneous reaction is always a fast reaction.
b. A spontaneous reaction is always a slow reaction.
c. The spontaneity of a reaction is not necessarily related to the speed of a reaction.

105. Which process is necessarily driven by an increase in the entropy of the surroundings?

a. the condensation of water
b. the sublimation of dry ice
c. the freezing of water

106. Consider the changes in the distribution of nine particles into three interconnected boxes shown here. Which has the most negative ΔS?

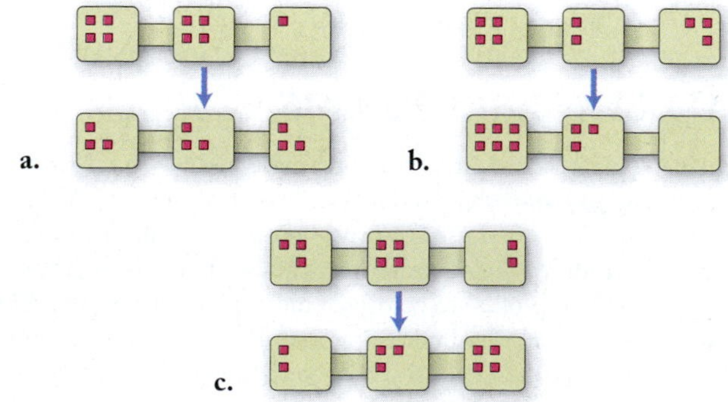

107. Which statement is true?

a. A reaction in which the entropy of the system increases can be spontaneous only if it is exothermic.
b. A reaction in which the entropy of the system increases can be spontaneous only if it is endothermic.
c. A reaction in which the entropy of the system decreases can be spontaneous only if it is exothermic.

108. Which process is spontaneous at 298 K?

a. $H_2O(l) \longrightarrow H_2O(g, 1\text{ atm})$
b. $H_2O(l) \longrightarrow H_2O(g, 0.10\text{ atm})$
c. $H_2O(l) \longrightarrow H_2O(g, 0.010\text{ atm})$

109. The free energy change of the reaction $A(g) \longrightarrow B(g)$ is zero under certain conditions. The *standard* free energy change of the reaction is -42.5 kJ. Which statement must be true about the reaction?

a. The concentration of the product is greater than the concentration of the reactant.
b. The reaction is at equilibrium.
c. The concentration of the reactant is greater than the concentration of the product.

110. The reaction $A(g) \rightleftharpoons B(g)$ has an equilibrium constant of 5.8 and under certain conditions has $Q = 336$. What can you conclude about the sign of ΔG°_{rxn} and ΔG_{rxn} for this reaction under these conditions?

ANSWERS TO CONCEPTUAL CONNECTIONS

Cc 19.1 **(a)** The more spread out the particles are between the three boxes, the greater the entropy. Therefore, the entropy change is positive only in scheme **(a)**.

Cc 19.2 $Kr < Cl_2 < SO_3$. Because krypton is a monoatomic gas, it has the least entropy. Because SO_3 is the most complex molecule, it has the most entropy. The molar masses of the three gases vary slightly, but not enough to overcome the differences in molecular complexity.

Cc 19.3 Biological systems do not violate the second law of thermodynamics. The key to understanding this concept is realizing that entropy changes in the system can be negative as long as the entropy change of the universe is positive. Biological systems can decrease their own entropy, but only at the expense of creating more entropy in the surroundings (which they do primarily by emitting the heat they generate by their metabolic processes). Thus, for any biological process, ΔS_{univ} is positive.

Cc 19.4 **(a)** Sublimation is endothermic (it requires energy to overcome the intermolecular forces that hold solid carbon dioxide together), so ΔH is positive. The number of moles of gas increases when the solid turns into a gas, so the entropy of the carbon dioxide increases and ΔS is positive. Since $\Delta G = \Delta H - T\Delta S$, ΔG is positive at low temperature and negative at high temperature.

Cc 19.5 **(a)** A high concentration of reactants relative to products will lead to $Q < 1$, making the term $RT \ln Q$ in Equation 19.14 negative. ΔG_{rxn} is more negative than ΔG°_{rxn} and the reaction is more spontaneous.

Cc 19.6 **(c)** Since the equilibrium constant is less than 1, the reaction proceeds toward reactants under standard conditions (when $Q = 1$). Therefore, ΔG°_{rxn} is positive.

CHAPTER 20

> "...each metal has a certain power, which is different from metal to metal, of setting the electric fluid in motion..."
>
> —Alessandro Volta (1745–1827)

Lightning is a massive flow of electrical charge from the base of a thundercloud to the ground. In a battery, charge flows in a more controlled fashion but is driven by the same principle.

Electrochemistry

20.1 Lightning and Batteries 813

20.2 Balancing Oxidation–Reduction Equations 814

20.3 Voltaic (or Galvanic) Cells: Generating Electricity from Spontaneous Chemical Reactions 817

20.4 Standard Electrode Potentials 822

20.5 Cell Potential, Free Energy, and the Equilibrium Constant 829

20.6 Cell Potential and Concentration 833

20.7 Batteries: Using Chemistry to Generate Electricity 838

20.8 Electrolysis: Driving Nonspontaneous Chemical Reactions with Electricity 841

20.9 Corrosion: Undesirable Redox Reactions 848

Key Learning Outcomes 852

SOME CHEMICAL REACTIONS RESULT IN THE TRANSFER OF ELECTRONS from one substance to another. We first encountered these kinds of reactions—called oxidation–reduction or redox reactions—in Chapter 9. In an oxidation–reduction reaction, one substance loses electrons and another substance gains them. If we physically separate the reactants in an oxidation–reduction reaction from one another, we can force the electrons to travel through a metal wire in order to get from one reactant to the other. The moving electrons constitute an electrical current. In this way, we can employ the electron-gaining tendency of one substance and the electron-losing tendency of another to force electrons to move through a wire to create electricity. The end result is a battery—a portable source of electrical current. The generation of electricity from spontaneous redox reactions (such as those that occur in a battery) and the use of electricity to drive nonspontaneous redox reactions (such as those that occur in gold or silver plating) are examples of electrochemistry, the subject of this chapter.

20.1 Lightning and Batteries

Lightning dramatically demonstrates the power of the flow of electrical charge. Many of the same principles, although in a much more controlled environment, are at work in a standard flashlight battery. The driving force for both lightning and the battery is the same, and we have encountered it before: Electrons flow *away* from negative charge and toward *positive* charge.

In a thundercloud, violent air currents cause water droplets and ice particles to collide. The collisions knock electrons off of molecules, creating positive and negative charges. The positive charges accumulate on small ice crystals that travel to the top of the thundercloud on rising air currents. The wet slushy bottom of the thundercloud becomes negatively charged. The resulting charge separation exists until a conductive path between the bottom of the cloud (negatively charged) and the top of the cloud (positively charged) can form. The conductive path forms when the charge separation is so great that a channel of ionized air develops. This channel acts like a conductive wire, allowing a massive amount of charge to flow through it in order to equalize the charge separation. This massive flow of electrical charge is lightning. Most lightning occurs within the thundercloud itself or from one thundercloud to another. However, if the thundercloud gets close enough to the ground, the earth underneath the cloud develops a positive charge in response to the negative charge at the base of the cloud. The channel of ionized air can then form between the cloud and the ground, resulting in the flow of charge from the base of the cloud to the earth in what is called cloud-to-ground lightning. Cloud-to-ground lightning is the most visible and dramatic to observers on the ground.

Batteries operate on many of the same principles at work in lightning. A battery is composed of substances that have different affinities for electrons. The substances are separated so that one end of the battery develops a positive charge and the other end a negative charge. The charge separation exists until a conductive path connects the two ends, providing a path through which charge can flow. A metal wire, with a light bulb in line, can provide such a path. When the wire is connected, electrons flow from the negative end of the battery—through the wire and through the light bulb—to the positive end. As the electrons flow through the filament of the light bulb, they create heat and light, much like the flow of electrons from a thundercloud to the earth produces heat and light in a much more dramatic form.

20.2 Balancing Oxidation–Reduction Equations

The reactions that create the flow of electric charge within a battery are oxidation–reduction (redox) reactions. Recall from Section 9.9 that *oxidation* is the loss of electrons, and *reduction* is the gain of electrons. Recall also that we identify oxidation–reduction reactions through changes in oxidation states: *Oxidation corresponds to an increase in oxidation state, and reduction corresponds to a decrease in oxidation state.* For example, consider the reaction between calcium and water:

Review assigning of oxidation states in Section 9.9.

$$Ca(s) + 2\,H_2O(l) \longrightarrow Ca(OH)_2(aq) + H_2(g)$$

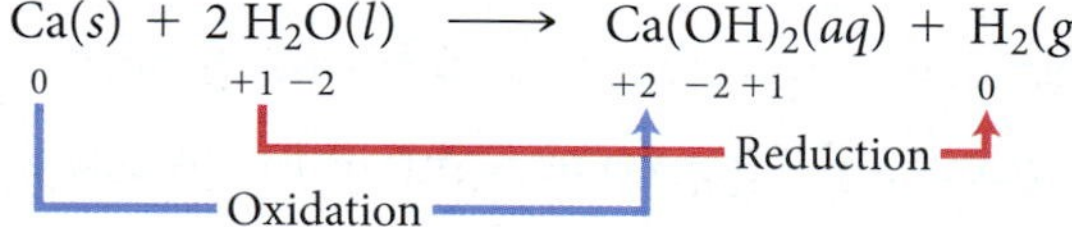

Because calcium increases in oxidation state from 0 to +2, it is oxidized. Because hydrogen decreases in oxidation state from +1 to 0, it is reduced.

Balancing redox reactions can be more complicated than balancing other types of reactions because we must balance both the mass (or number of each type of atom) and the *charge*. We can balance redox reactions occurring in aqueous solutions with a special procedure called the *half-reaction method of balancing*. In this procedure, we break down the overall equation into two half-reactions: one for oxidation and one for reduction. We then balance the half-reactions individually and add them together. The steps differ slightly for reactions occurring in acidic and in basic solution. Examples 20.1 and 20.2 demonstrate the method for an acidic solution, and Example 20.3 demonstrates the method for a basic solution.

PROCEDURE FOR ▼ **Half-Reaction Method of Balancing Aqueous Redox Equations in Acidic Solution**	**EXAMPLE 20.1** **Half-Reaction Method of Balancing Aqueous Redox Equations in Acidic Solution**	**EXAMPLE 20.2** **Half-Reaction Method of Balancing Aqueous Redox Equations in Acidic Solution**
GENERAL PROCEDURE	Balance the redox equation. $Al(s) + Cu^{2+}(aq) \longrightarrow Al^{3+}(aq) + Cu(s)$	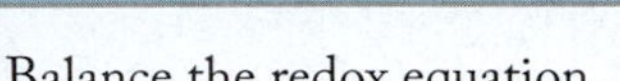 Balance the redox equation. $Fe^{2+}(aq) + MnO_4^-(aq) \longrightarrow Fe^{3+}(aq) + Mn^{2+}(aq)$
Step 1 *Assign oxidation states* to all atoms and identify the substances being oxidized and reduced.	$\underset{0}{Al(s)} + \underset{+2}{Cu^{2+}(aq)} \longrightarrow \underset{+3}{Al^{3+}(aq)} + \underset{0}{Cu(s)}$ Reduction Oxidation	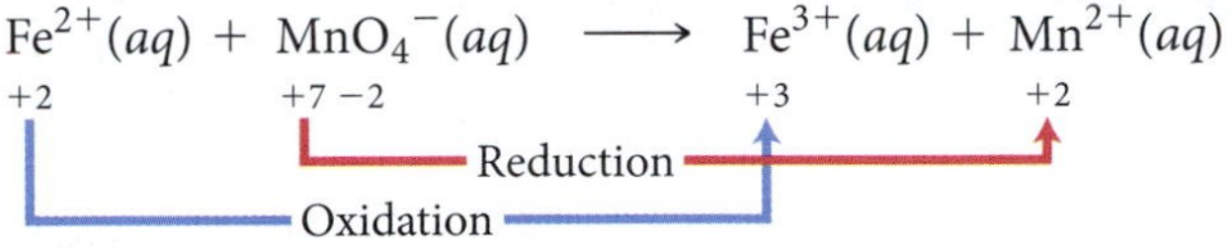
Step 2 *Separate the overall reaction into two half-reactions*: one for oxidation and one for reduction.	***Oxidation:*** $Al(s) \longrightarrow Al^{3+}(aq)$ ***Reduction:*** $Cu^{2+}(aq) \longrightarrow Cu(s)$	***Oxidation:*** $Fe^{2+}(aq) \longrightarrow Fe^{3+}(aq)$ ***Reduction:*** $MnO_4^-(aq) \longrightarrow Mn^{2+}(aq)$
Step 3 *Balance each half-reaction with respect to mass* in the following order: • Balance all elements other than H and O. • Balance O by adding H_2O. • Balance H by adding H^+.	All elements are balanced, so proceed to the next step.	All elements other than H and O are balanced, so proceed to balance H and O. $Fe^{2+}(aq) \longrightarrow Fe^{3+}(aq)$ $MnO_4^-(aq) \longrightarrow Mn^{2+}(aq) + 4\,H_2O(l)$ $8\,H^+(aq) + MnO_4^-(aq) \longrightarrow Mn^{2+}(aq) + 4\,H_2O(l)$
Step 4 *Balance each half-reaction with respect to charge* by adding electrons. (Make the sum of the charges on both sides of the equation equal by adding as many electrons as necessary.)	$Al(s) \longrightarrow Al^{3+}(aq) + 3\,e^-$ $2\,e^- + Cu^{2+}(aq) \longrightarrow Cu(s)$	$Fe^{2+}(aq) \longrightarrow Fe^{3+}(aq) + 1\,e^-$ $5\,e^- + 8\,H^+(aq) + MnO_4^-(aq) \longrightarrow Mn^{2+}(aq) + 4\,H_2O(l)$
Step 5 *Make the number of electrons in both half-reactions equal* by multiplying one or both half-reactions by a small whole number.	$2[Al(s) \longrightarrow Al^{3+}(aq) + 3\,e^-]$ $2\,Al(s) \longrightarrow 2\,Al^{3+}(aq) + 6\,e^-$ $3[2\,e^- + Cu^{2+}(aq) \longrightarrow Cu(s)]$ $6\,e^- + 3\,Cu^{2+}(aq) \longrightarrow 3\,Cu(s)$	$5[Fe^{2+}(aq) \longrightarrow Fe^{3+}(aq) + 1\,e^-]$ $5\,Fe^{2+}(aq) \longrightarrow 5\,Fe^{3+}(aq) + 5\,e^-$ $5\,e^- + 8\,H^+(aq) + MnO_4^-(aq) \longrightarrow Mn^{2+}(aq) + 4\,H_2O(l)$
Step 6 *Add the two half-reactions together*, canceling electrons and other species as necessary.	$2\,Al(s) \longrightarrow 2\,Al^{3+}(aq) + \cancel{6\,e^-}$ $\cancel{6\,e^-} + 3\,Cu^{2+}(aq) \longrightarrow 3\,Cu(s)$ $2\,Al(s) + 3\,Cu^{2+}(aq) \longrightarrow 2\,Al^{3+}(aq) + 3\,Cu(s)$	$5\,Fe^{2+}(aq) \longrightarrow 5\,Fe^{3+}(aq) + \cancel{5\,e^-}$ $\cancel{5\,e^-} + 8\,H^+(aq) + MnO_4^-(aq) \longrightarrow Mn^{2+}(aq) + 4\,H_2O(l)$ $5\,Fe^{2+}(aq) + 8\,H^+(aq) + MnO_4^-(aq) \longrightarrow 5\,Fe^{3+}(aq) + Mn^{2+}(aq) + 4\,H_2O(l)$

—Continued on the next page

Continued from the previous page—

Step 7 *Verify that the reaction is balanced* both with respect to mass and with respect to charge.

Reactants	Products
2 Al	2 Al
3 Cu	3 Cu
6+ charge	6+ charge

Reactants	Products
5 Fe	5 Fe
8 H	8 H
1 Mn	1 Mn
4 O	4 O
17+ charge	17+ charge

FOR PRACTICE 20.1

Balance the redox reaction in acidic solution.

$$H^+(aq) + Cr(s) \longrightarrow H_2(g) + Cr^{2+}(aq)$$

FOR PRACTICE 20.2

Balance the redox reaction in acidic solution.

$$Cu(s) + NO_3^-(aq) \longrightarrow Cu^{2+}(aq) + NO_2(g)$$

When a redox reaction occurs in basic solution, we balance the reaction in a similar manner, except that we add an additional step to neutralize any H^+ with OH^-. The H^+ and the OH^- combine to form H_2O as shown in Example 20.3.

EXAMPLE 20.3

Balancing Redox Reactions Occurring in Basic Solution

Balance the equation occurring in basic solution.

$$I^-(aq) + MnO_4^-(aq) \longrightarrow I_2(aq) + MnO_2(s)$$

SOLUTION

To balance redox reactions occurring in basic solution, you follow the half-reaction method outlined in Examples 20.1 and 20.2, but add an extra step to neutralize the acid with OH^- as shown in Step 3 of this example.

1. Assign oxidation states.

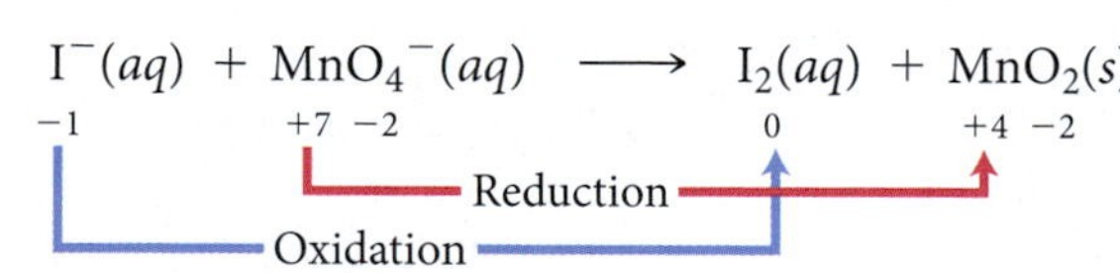

2. Separate the overall reaction into two half-reactions.

Oxidation: $I^-(aq) \longrightarrow I_2(aq)$

Reduction: $MnO_4^-(aq) \longrightarrow MnO_2(s)$

3. Balance each half-reaction with respect to mass.

- Balance all elements other than H and O.

$$\begin{cases} 2\,I^-(aq) \longrightarrow I_2(aq) \\ MnO_4^-(aq) \longrightarrow MnO_2(s) \end{cases}$$

- Balance O by adding H_2O.

$$\begin{cases} 2\,I^-(aq) \longrightarrow I_2(aq) \\ MnO_4^-(aq) \longrightarrow MnO_2(s) + 2\,H_2O(l) \end{cases}$$

- Balance H by adding H^+.

$$\begin{cases} 2\,I^-(aq) \longrightarrow I_2(aq) \\ 4\,H^+(aq) + MnO_4^-(aq) \longrightarrow MnO_2(s) + 2\,H_2O(l) \end{cases}$$

- Neutralize H^+ by adding enough OH^- to neutralize each H^+. Add the same number of OH^- ions to each side of the equation.

$$\begin{cases} 2\,I^-(aq) \longrightarrow I_2(aq) \\ \underbrace{4\,H^+(aq) + 4\,OH^-(aq)}_{4\,H_2O(l)} + MnO_4^-(aq) \longrightarrow MnO_2(s) + 2\,H_2O(l) + 4\,OH^-(aq) \end{cases}$$

—Continued on the next page

Continued from the previous page—

4. Balance each half-reaction with respect to charge.	$2\,I^-(aq) \longrightarrow I_2(aq) + 2\,e^-$ $4\,H_2O(l) + MnO_4^-(aq) + 3\,e^- \longrightarrow MnO_2(s) + 2\,H_2O(l) + 4\,OH^-(aq)$
5. Make the number of electrons in both half-reactions equal.	$3\,[2\,I^-(aq) \longrightarrow I_2(aq) + 2\,e^-]$ $6\,I^-(aq) \longrightarrow 3\,I_2(aq) + 6\,e^-$ $2\,[4\,H_2O(l) + MnO_4^-(aq) + 3\,e^- \longrightarrow MnO_2(s) + 2\,H_2O(l) + 4\,OH^-(aq)]$ $8\,H_2O(l) + 2\,MnO_4^-(aq) + 6\,e^- \longrightarrow 2\,MnO_2(s) + 4\,H_2O(l) + 8\,OH^-(aq)$
6. Add the half-reactions together.	$6\,I^-(aq) \longrightarrow 3\,I_2(aq) + 6\,e^-$ $4\,8\,H_2O(l) + 2\,MnO_4^-(aq) + 6\,e^- \longrightarrow 2\,MnO_2(s) + 4\,H_2O(l) + 8\,OH^-(aq)$ $6\,I^-(aq) + 4\,H_2O(l) + 2\,MnO_4^-(aq) \longrightarrow 3\,I_2(aq) + 2\,MnO_2(s) + 8\,OH^-(aq)$
7. Verify that the reaction is balanced.	(see table below)

Reactants	Products
6 I	6 I
8 H	8 H
2 Mn	2 Mn
12 O	12 O
8– charge	8– charge

FOR PRACTICE 20.3

Balance the following redox reaction occurring in basic solution.

$$ClO^-(aq) + Cr(OH)_4^-(aq) \longrightarrow CrO_4^{2-}(aq) + Cl^-(aq)$$

20.3 Voltaic (or Galvanic) Cells: Generating Electricity from Spontaneous Chemical Reactions

Electrical current is the flow of electric charge. Electrons flowing through a wire or ions flowing through a solution both constitute electrical current. Since redox reactions involve the transfer of electrons from one substance to another, they have the potential to generate electrical current as we discussed in Section 20.1. For example, consider the spontaneous redox reaction:

$$Zn(s) + Cu^{2+}(aq) \longrightarrow Zn^{2+}(aq) + Cu(s)$$

When Zn metal is placed in a Cu^{2+} solution, the greater tendency of zinc to lose electrons results in Zn being oxidized and Cu^{2+} being reduced. Electrons are transferred directly from the Zn to the Cu^{2+} (Figure 20.1 ▶). Although the actual process is more complicated, we can imagine that—on the atomic scale—a zinc atom within the zinc metal transfers two electrons to a copper ion in solution. The zinc atom then becomes a zinc ion dissolved in the solution. The copper(II) ion accepts the two electrons and is deposited on the zinc as solid copper.

Suppose we could separate the zinc atoms and copper(II) ions and force the electron transfer to occur another way—not directly from the zinc atom to the copper(II) ion, but through a wire connecting the two half-reactions. The flowing electrons would constitute an electrical current and could be used to do electrical work.

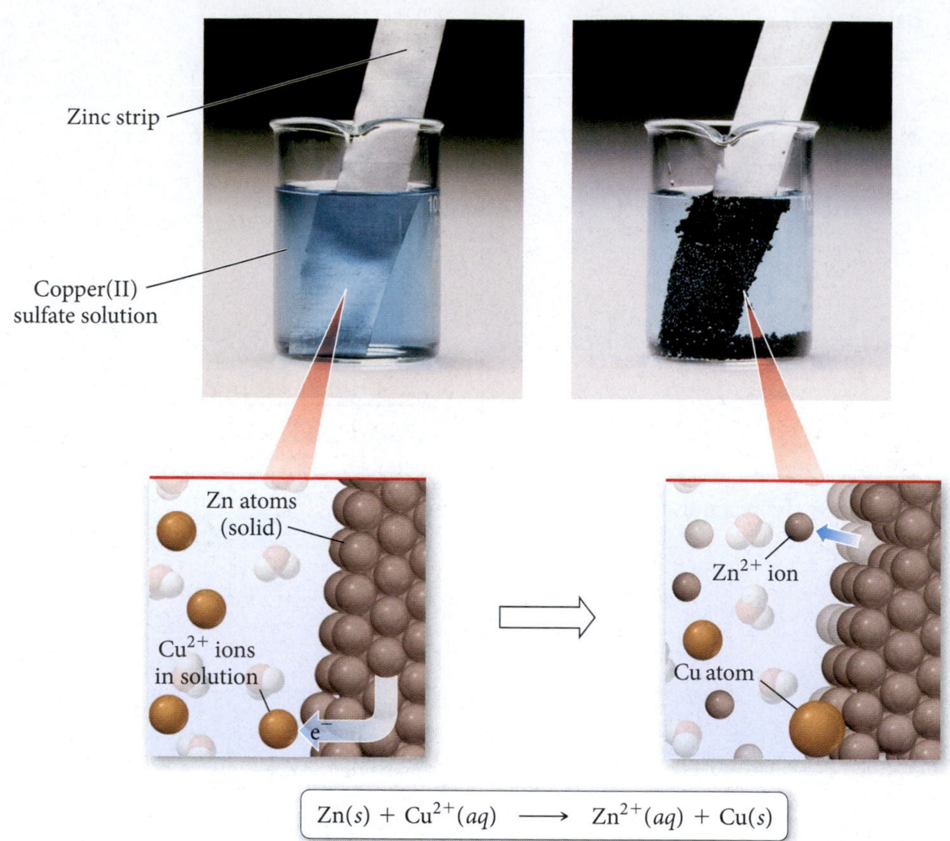

▶ **FIGURE 20.1 A Spontaneous Oxidation–Reduction Reaction** When zinc is immersed in a solution containing copper(II) ions, the zinc atoms transfer electrons to the copper(II) ions. The zinc atoms are oxidized and dissolve in the solution. The copper(II) ions are reduced and are deposited on the eletrode.

The generation of electricity through redox reactions is normally carried out in a device called an **electrochemical cell**. A **voltaic (or galvanic) cell**, is an electrochemical cell that *produces* electrical current from a *spontaneous* chemical reaction. A second type of electrochemical cell, called an **electrolytic cell**, *consumes* electrical current to drive a *nonspontaneous* chemical reaction. We discuss voltaic cells in this section and electrolytic cells in Section 20.8.

In the voltaic cell in Figure 20.2 ▶, a solid strip of zinc is placed in a $Zn(NO_3)_2$ solution to form a **half-cell**. A solid strip of copper placed in a $Cu(NO_3)_2$ solution forms a second half-cell. The strips act as **electrodes**, conductive surfaces through which electrons can enter or leave the half-cells. Each metal strip reaches equilibrium with its ions in solution according to these half-reactions:

$$Zn(s) \rightleftharpoons Zn^{2+}(aq) + 2\,e^-$$

$$Cu(s) \rightleftharpoons Cu^{2+}(aq) + 2\,e^-$$

However, the position of these equilibria—which depends on the potential energy of the electrons in each metal—is not the same for both metals. The electrons in zinc have a higher potential energy and therefore zinc has a greater tendency to ionize than copper, so the zinc half-reaction lies further to the right. As a result, the zinc electrode becomes negatively charged relative to the copper electrode.

If the two half-cells are connected by a wire running from the zinc—through a light bulb or other electrical device—to the copper, electrons spontaneously flow from the zinc electrode (where they have higher potential energy) to the copper electrode (where they have lower potential energy). As the electrons flow away from the zinc electrode, the Zn/Zn^{2+} equilibrium shifts to the right (according to Le Châtelier's principle) and oxidation occurs. As electrons flow to the copper electrode, the Cu/Cu^{2+} equilibrium shifts to the left, and reduction occurs. The flowing electrons constitute an electrical current that lights the bulb.

The *continual* flow of electrical current in a voltaic cell requires a pathway by which counterions can flow to neutralize charge buildup; we discuss this later in the chapter.

The idea that one electrode in a voltaic cell becomes more negatively charged relative to the other electrode due to differences in ionization tendencies is central to understanding how a voltaic cell works.

A Voltaic Cell

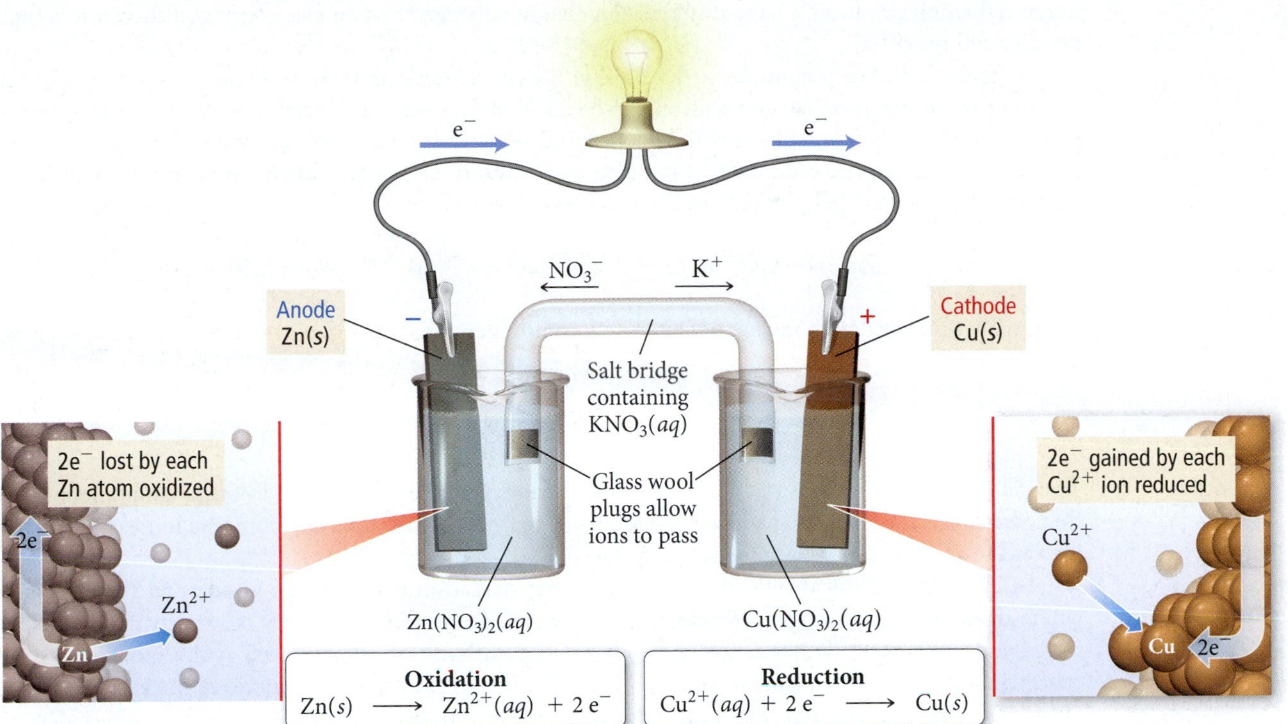

▲ **FIGURE 20.2 A Voltaic Cell** The tendency of zinc to transfer electrons to copper results in a flow of electrons through the wire that lights the bulb. The movement of electrons from the zinc anode to the copper cathode creates a positive charge buildup at the zinc half-cell and a negative charge buildup at the copper half-cell. The flow of ions within the salt bridge (which we describe later in this section) neutralizes this charge buildup, allowing the reaction to continue.

We can understand electrical current and why it flows by analogy with water current in a stream (Figure 20.3 ▶). The *rate of electrons flowing* through a wire is analogous to the *rate of water moving* through a stream. We measure electrical current in units of **amperes (A)** also called *amps*. One ampere represents the flow of one coulomb (a measure of electrical charge) per second.

$$1 \text{ A} = 1 \text{ C/s}$$

Because an electron has a charge of 1.602×10^{-19} C, 1 A corresponds to the flow of 6.242×10^{18} electrons per second.

The *driving force* for electrical current is analogous to the driving force for water current. Water current is driven by a difference in gravitational potential energy. Streams flow downhill, from higher to lower potential energy. Electrical current is also driven by a difference in potential energy. Electrons flow from the electrode in which they have higher potential energy to the electrode in which they have lower potential energy. The difference in potential energy between the the two electrodes is called **potential difference**. *Potential difference is a measure of the difference in potential energy (usually in joules) per unit of charge (coulombs).* The SI unit of potential difference is the **volt (V)**, which is equal to one joule per coulomb.

$$1 \text{ V} = 1 \text{ J/C}$$

In other words, a potential difference of one volt indicates that a charge of one coulomb experiences an energy difference of one joule between the two electrodes.

A large potential difference corresponds to a large difference in potential energy between the two electrodes and therefore a strong tendency for electron flow (analogous to a steeply descending streambed). Potential difference, because it gives rise to the force that results in the motion of electrons, is also referred to as **electromotive force (emf)**. In a voltaic cell, the potential difference between the two electrodes is the **cell potential (E_{cell})** or **cell emf**. The cell potential depends on the relative tendencies

▲ **FIGURE 20.3 An Analogy for Electrical Current** Just as water flows downhill in response to a difference in gravitational potential energy, electrons flow through a conductor in response to an electrical potential difference, creating an electrical current.

of the reactants to undergo oxidation and reduction. Combining the oxidation of a substance with a strong tendency to undergo oxidation and the reduction of a substance with a strong tendency to undergo reduction produces a large difference in charge between the two electrodes and therefore a high positive cell potential.

Standard states, indicated by the symbol °, represent a very specific reaction mixture. For a reaction mixture with reactants and products in their standard states, $Q = 1$.

In general, the cell potential also depends on the concentrations of the reactants and products in the cell and the temperature (which we assume to be 25 °C unless otherwise noted). When the reactants and products are in their standard states (1 M concentration for substances in solution and 1 atm pressure for gaseous substances), the cell potential is called the **standard cell potential (E°_{cell})** or **standard emf**. For example, the standard cell potential in the zinc and copper cell described previously is 1.10 volts.

$$Zn(s) + Cu^{2+}(aq) \longrightarrow Zn^{2+}(aq) + Cu(s) \quad E^\circ_{cell} = +1.10\ V$$

If the zinc is replaced with nickel (which has a lower tendency to be oxidized) the standard cell potential is lower.

$$Ni(s) + Cu^{2+}(aq) \longrightarrow Ni^{2+}(aq) + Cu(s) \quad E^\circ_{cell} = +0.62\ V$$

The cell potential is a measure of the overall tendency of the redox reaction to occur spontaneously—the lower the cell potential, the lower the tendency to occur. A negative cell potential indicates that the forward reaction is not spontaneous.

Note that the anode and cathode need not actually be negatively and positively charged, respectively. The anode is the electrode with the relatively *more* negative (or less positive) charge.

In all electrochemical cells, the electrode where oxidation occurs is the **anode** and the electrode where reduction occurs is the **cathode**. In a voltaic cell, the anode is the more negatively charged electrode, and we label it with a negative (−) sign. The cathode of a voltaic cell is the more positively charged electrode, and we label it with a (+) sign. Electrons flow from the anode to the cathode (from negative to positive) through the wires connecting the electrodes.

As electrons flow out of the anode, positive ions (Zn^{2+} in the preceding example) form in the oxidation half-cell, resulting in a buildup of *positive charge* in the *solution*. As electrons flow into the cathode, positive ions (Cu^{2+} in the preceding example) are reduced at the reduction half-cell, resulting in a buildup of *negative charge* in the *solution*.

If the movement of electrons from anode to cathode were the only flow of charge, the buildup of the opposite charge in the solution would stop electron flow almost immediately. The cell needs a pathway by which counterions can flow between the half-cells without the solutions in the half-cells totally mixing. One such pathway is a **salt bridge**, an inverted, U-shaped tube that contains a strong electrolyte such as KNO_3 and connects the two half-cells (see Figure 20.2). The electrolyte is usually suspended in a gel and held within the tube by permeable stoppers. The salt bridge allows a flow of ions that neutralizes the charge buildup in the solution. *The negative ions within the salt bridge flow to neutralize the accumulation of positive charge at the anode, and the positive ions flow to neutralize the accumulation of negative charge at the cathode.* In other words, the salt bridge completes the circuit, allowing electrical current to flow.

20.1 Cc Conceptual Connection

Voltaic Cells

In a voltaic cell, in which direction do electrons flow?

(a) from higher potential energy to lower potential energy

(b) from the cathode to the anode

(c) from lower potential energy to higher potential energy

Electrochemical Cell Notation

We represent electrochemical cells with a compact notation called a *cell diagram* or *line notation*. For example, we can represent the electrochemical cell in which Zn is oxidized to Zn^{2+} and Cu^{2+} is reduced to Cu as follows:

$$Zn(s)\,|\,Zn^{2+}(aq)\,||\,Cu^{2+}(aq)\,|\,Cu(s)$$

In this representation,

- we write the oxidation half-reaction on the left and the reduction on the right. A double vertical line, indicating the salt bridge, separates the two half-reactions.
- substances in different phases are separated by a single vertical line, which represents the boundary between the phases.
- for some redox reactions, the reactants and products of one or both of the half-reactions may be in the same phase. In these cases (which we explain further next), we separate the reactants and products from each other with a comma in the line diagram. Such cells use an inert electrode, such as platinum (Pt) or graphite, as the anode or cathode (or both).

Consider the redox reaction in which $Fe(s)$ is oxidized and $MnO_4^-(aq)$ is reduced:

$$5\,Fe(s) + 2\,MnO_4^-(aq) + 16\,H^+(aq) \longrightarrow 5\,Fe^{2+}(aq) + 2\,Mn^{2+}(aq) + 8\,H_2O(l)$$

The half-reactions for this overall reaction are:

Oxidation: $Fe(s) \longrightarrow Fe^{2+}(aq) + 2\,e^-$

Reduction: $MnO_4^-(aq) + 5\,e^- + 8\,H^+(aq) \longrightarrow Mn^{2+}(aq) + 4\,H_2O(l)$

Notice that in the reduction half-reaction the principal species are all in the aqueous phase. In this case, the electron transfer needs an electrode on which to occur. An inert platinum electrode is employed, and the electron transfer takes place at its surface. Using line notation, we represent the electrochemical cell corresponding to the above reaction as:

$$Fe(s)\,|\,Fe^{2+}(aq)\,||\,MnO_4^-(aq), H^+(aq), Mn^{2+}(aq)\,|\,Pt(s)$$

The $Pt(s)$ on the far right indicates that an inert platinum electrode acts as the cathode in this reaction, as depicted in Figure 20.4 ▼.

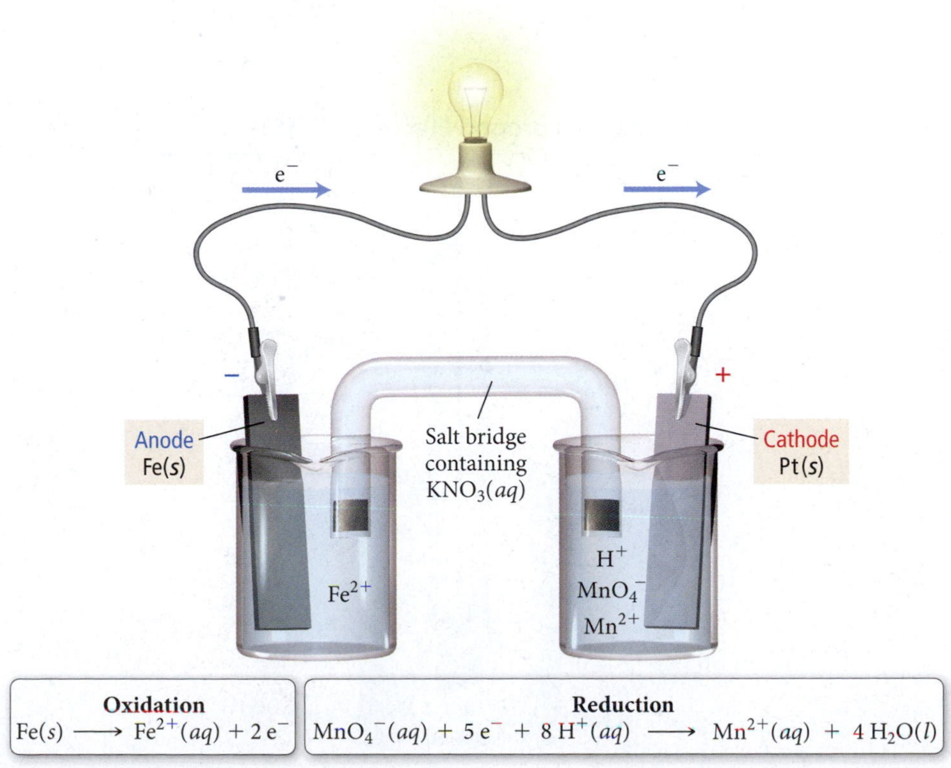

◀ **FIGURE 20.4 Inert Platinum Electrode** When the participants in a half-reaction are all in the aqueous phase, a conductive surface is needed for electron transfer to take place. In such cases an inert electrode of graphite or platinum is often used. In this electrochemical cell, an iron strip acts as the anode and a platinum strip acts as the cathode. Iron is oxidized at the anode, and MnO_4^- is reduced at the cathode.

KEY CONCEPT VIDEO
Standard Electrode Potentials

20.4 Standard Electrode Potentials

As we have just seen, the standard cell potential (E°_{cell}) for an electrochemical cell depends on the specific half-reactions occurring in the half-cells and is a measure of the potential energy difference (per unit charge) between the two electrodes. We can think of the electrode in each half-cell as having its own individual potential, called the **standard electrode potential**. The overall standard cell potential (E°_{cell}) is the difference between the two standard electrode potentials.

We can better understand this idea with an analogy. Consider two water tanks with different water levels connected by a common pipe, as shown in Figure 20.5 ◀. The water in each tank has its own level and corresponding potential energy. When the tanks are connected, water flows from the tank with the higher water level (higher potential energy) to the tank with a lower water level (lower potential energy). Similarly, each half-cell in an electrochemical cell has its own charge and corresponding electrode potential. *When the cells are connected, electrons flow from the electrode with greater potential energy (more negatively charged) to the electode with less potential energy (more positively charged).*

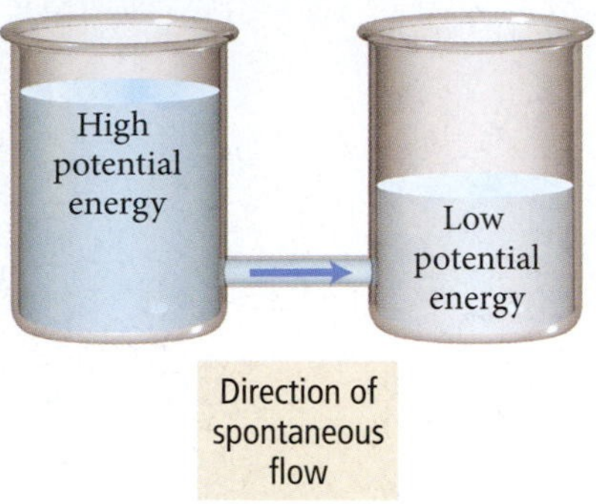

▲ **FIGURE 20.5 An Analogy for Electrode Potential**

One limitation to this analogy is that, unlike the water level in a tank, we cannot measure the electrode potential in a half-cell directly—we can only measure the overall potential that occurs when two half-cells are combined in a whole cell. However, we can arbitrarily assign a potential of zero to the electrode in a *particular* type of half-cell and then measure all other electrode potentials relative to that zero.

The half-cell electrode that is normally chosen to have a potential of zero is the **standard hydrogen electrode (SHE)**. This half-cell consists of an inert platinum electrode immersed in 1 M HCl with hydrogen gas at 1 atm bubbling through the solution, as shown in Figure 20.6 ▼. When the standard hydrogen electrode acts as the cathode, the following half-reaction occurs:

$$2\,H^+(aq) + 2\,e^- \longrightarrow H_2(g) \qquad E^\circ_{cathode} = 0.00\ V$$

If we connect the standard hydrogen electrode to an electrode in another half-cell of interest, we can measure the potential difference (or voltage) between the two electrodes. Since we assigned the standard hydrogen electrode zero voltage, we can determine the electrode potential of the other half-cell.

Consider the electrochemical cell shown in Figure 20.7 ▶. In this electrochemical cell, Zn is oxidized to Zn^{2+} and H^+ is reduced to H_2 under standard conditions (all solutions are 1 M in concentration and all gases are 1 atm in pressure) and at 25 °C. Electrons travel from the anode (where oxidation

Standard Hydrogen Electrode (SHE)

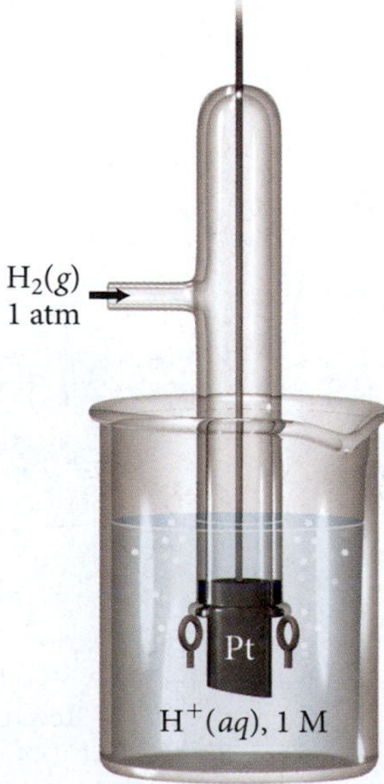

▶ **FIGURE 20.6 The Standard Hydrogen Electrode** The standard hydrogen electrode (SHE) is arbitrarily assigned an electrode potential of zero. All other electrode potentials are then measured relative to the SHE.

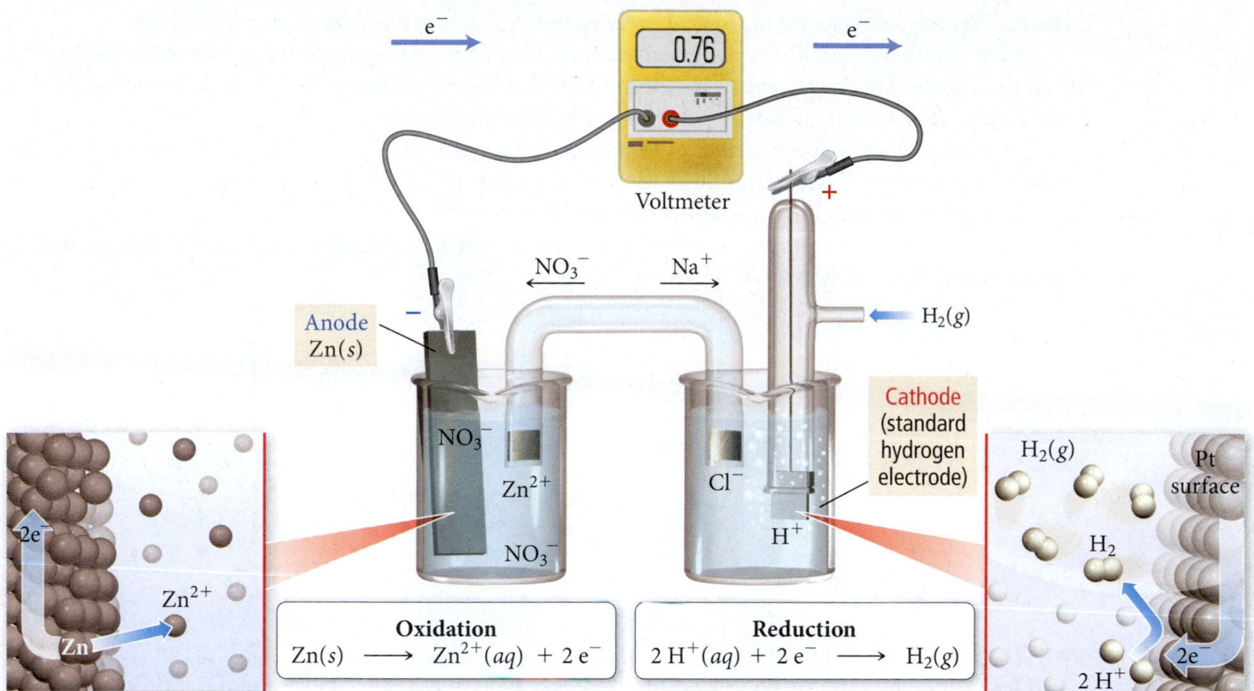

▲ **FIGURE 20.7 Measuring Electrode Potential** Because the electrode potential of the SHE is zero, the electrode potential for the oxidation of Zn is equal to the cell potential.

occurs) to the cathode (where reduction occurs), so we define E°_{cell} as *the difference in voltage between the cathode (final state) and the anode (initial state).*

$$E^{\circ}_{cell} = E^{\circ}_{final} - E^{\circ}_{initial}$$
$$= E^{\circ}_{cathode} - E^{\circ}_{anode}$$

The measured cell potential for this cell is +0.76 V. The anode (in this case, Zn/Zn^{2+}) is at a more negative voltage (higher potential energy) than the cathode (in this case, the SHE). Therefore, electrons spontaneously flow from the anode to the cathode. We can diagram the potential energy and the voltage as follows:

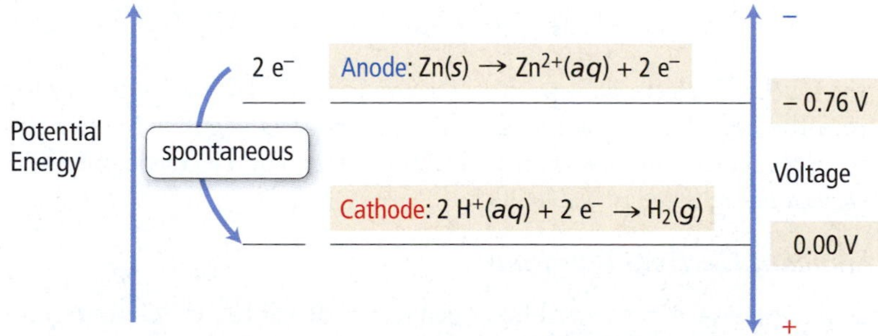

Referring back to our water tank analogy, the zinc half-cell is like the water tank with the higher water level, and electrons therefore flow from the zinc electrode to the standard hydrogen electrode.

Because we assigned the SHE a potential of zero (0.00 V), we can determine the electrode potential for the Zn/Zn^{2+} half-cell (the anode) from the measured cell potential (E°_{cell}).

$$E^{\circ}_{cell} = E^{\circ}_{cathode} - E^{\circ}_{anode}$$
$$0.76\text{ V} = 0.00\text{ V} - E^{\circ}_{anode}$$
$$E^{\circ}_{anode} = -0.76\text{ V}$$

The potential for the Zn/Zn^{2+} electrode is *negative*. The negative potential indicates that an electron at the Zn/Zn^{2+} electrode has greater potential energy than it does at the SHE. *Remember that the more negative the electrode potential is, the greater the potential energy of an electron at that electrode.*

What would happen if we connected an electrode in which the electron has *more positive* potential (than the standard hydrogen electrode) to the standard hydrogen electrode? That electrode would then have a more *positive* voltage (lower potential energy for an electron).

For example, suppose we connect the standard hydrogen electrode to a Cu electrode immersed in a 1 M Cu^{2+} solution. The measured cell potential for this cell is −0.34 V. The anode (Cu/Cu^{2+}) is at a more positive voltage (lower potential energy) than the cathode (the SHE). Therefore, electrons do *not* spontaneously flow from the anode to the cathode. We can diagram the potential energy and the voltage of this cell as follows:

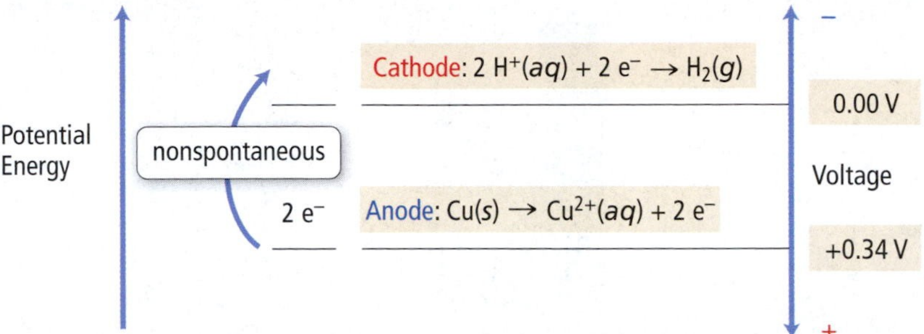

The copper half-cell is like the water tank with the lower water level, and electrons *do not* spontaneously flow from the copper electrode to the standard hydrogen electrode.

We can again determine the electrode potential for the Cu/Cu^{2+} half-cell (the anode) from the measured cell potential.

$$\begin{aligned} E^\circ_{\text{cell}} &= E^\circ_{\text{cathode}} - E^\circ_{\text{anode}} \\ -0.34\text{ V} &= 0.00\text{ V} - E^\circ_{\text{anode}} \\ E^\circ_{\text{anode}} &= +0.34\text{ V} \end{aligned}$$

The potential for the Cu/Cu^{2+} electrode is *positive*. The positive potential indicates that an electron at the Cu/Cu^{2+} electrode has *lower* potential energy than it does at the SHE. The more positive the electrode potential, the lower the potential energy of an electron at that electrode.

By convention, standard electrode potentials are written for *reduction* half-reactions. We write the standard electrode potentials for the two half-reactions just discussed as:

$$Cu^{2+}(aq) + 2\,e^- \longrightarrow Cu(s) \quad E^\circ = +0.34\text{ V}$$

$$Zn^{2+}(aq) + 2\,e^- \longrightarrow Zn(s) \quad E^\circ = -0.76\text{ V}$$

We can see that the Cu/Cu^{2+} electrode is positive relative to the SHE (and will therefore tend to draw electrons *away* from the SHE) and that the Zn/Zn^{2+} electrode is negative relative to the SHE (and will therefore tend to repel electrons toward the SHE). The standard electrode potentials for a number of common half-reactions are listed in Table 20.1.

Summarizing Standard Electrode Potentials:

- The electrode potential of the standard hydrogen electrode (SHE) is exactly zero.
- The electrode in any half-cell with a greater tendency to undergo reduction is positively charged relative to the SHE and therefore has a positive E°.
- The electrode in any half-cell with a lesser tendency to undergo reduction (or greater tendency to undergo oxidation) is negatively charged relative to the SHE and therefore has a negative E°.
- The cell potential of any electrochemical cell (E°_{cell}) is the difference between the electrode potentials of the cathode and the anode ($E^\circ_{\text{cell}} = E^\circ_{\text{cat}} - E^\circ_{\text{an}}$).
- E°_{cell} is positive for spontaneous reactions and negative for nonspontaneous reactions.

TABLE 20.1 Standard Electrode Potentials at 25 °C

	Reduction Half-Reaction			$E°$(V)	
Stronger oxidizing agent	$F_2(g) + 2\,e^-$	⟶	$2\,F^-(aq)$	2.87	Weaker reducing agent
	$H_2O_2(aq) + 2\,H^+(aq) + 2\,e^-$	⟶	$2\,H_2O(l)$	1.78	
	$PbO_2(s) + 4\,H^+(aq) + SO_4^{2-}(aq) + 2\,e^-$	⟶	$PbSO_4(s) + 2\,H_2O(l)$	1.69	
	$MnO_4^-(aq) + 4\,H^+(aq) + 3\,e^-$	⟶	$MnO_2(s) + 2\,H_2O(l)$	1.68	
	$MnO_4^-(aq) + 8\,H^+(aq) + 5\,e^-$	⟶	$Mn^{2+}(aq) + 4\,H_2O(l)$	1.51	
	$Au^{3+}(aq) + 3\,e^-$	⟶	$Au(s)$	1.50	
	$PbO_2(s) + 4\,H^+(aq) + 2\,e^-$	⟶	$Pb^{2+}(aq) + 2\,H_2O(l)$	1.46	
	$Cl_2(g) + 2\,e^-$	⟶	$2\,Cl^-(aq)$	1.36	
	$Cr_2O_7^{2-}(aq) + 14\,H^+(aq) + 6\,e^-$	⟶	$2\,Cr^{3+}(aq) + 7\,H_2O(l)$	1.33	
	$O_2(g) + 4\,H^+(aq) + 4\,e^-$	⟶	$2\,H_2O(l)$	1.23	
	$MnO_2(s) + 4\,H^+(aq) + 2\,e^-$	⟶	$Mn^{2+}(aq) + 2\,H_2O(l)$	1.21	
	$IO_3^-(aq) + 6\,H^+(aq) + 5\,e^-$	⟶	$\frac{1}{2}I_2(aq) + 3\,H_2O(l)$	1.20	
	$Br_2(l) + 2\,e^-$	⟶	$2\,Br^-(aq)$	1.09	
	$VO_2^+(aq) + 2\,H^+(aq) + e^-$	⟶	$VO^{2+}(aq) + H_2O(l)$	1.00	
	$NO_3^-(aq) + 4\,H^+(aq) + 3\,e^-$	⟶	$NO(g) + 2\,H_2O(l)$	0.96	
	$ClO_2(g) + e^-$	⟶	$ClO_2^-(aq)$	0.95	
	$Ag^+(aq) + e^-$	⟶	$Ag(s)$	0.80	
	$Fe^{3+}(aq) + e^-$	⟶	$Fe^{2+}(aq)$	0.77	
	$O_2(g) + 2\,H^+(aq) + 2\,e^-$	⟶	$H_2O_2(aq)$	0.70	
	$MnO_4^-(aq) + e^-$	⟶	$MnO_4^{2-}(aq)$	0.56	
	$I_2(s) + 2\,e^-$	⟶	$2\,I^-(aq)$	0.54	
	$Cu^+(aq) + e^-$	⟶	$Cu(s)$	0.52	
	$O_2(g) + 2\,H_2O(l) + 4\,e^-$	⟶	$4\,OH^-(aq)$	0.40	
	$Cu^{2+}(aq) + 2\,e^-$	⟶	$Cu(s)$	0.34	
	$SO_4^{2-}(aq) + 4\,H^+(aq) + 2\,e^-$	⟶	$H_2SO_3(aq) + H_2O(l)$	0.20	
	$Cu^{2+}(aq) + e^-$	⟶	$Cu^+(aq)$	0.16	
	$Sn^{4+}(aq) + 2\,e^-$	⟶	$Sn^{2+}(aq)$	0.15	
	$2\,H^+(aq) + 2\,e^-$	⟶	$H_2(g)$	0	
	$Fe^{3+}(aq) + 3\,e^-$	⟶	$Fe(s)$	–0.036	
	$Pb^{2+}(aq) + 2\,e^-$	⟶	$Pb(s)$	–0.13	
	$Sn^{2+}(aq) + 2\,e^-$	⟶	$Sn(s)$	–0.14	
	$Ni^{2+}(aq) + 2\,e^-$	⟶	$Ni(s)$	–0.23	
	$Cd^{2+}(aq) + 2\,e^-$	⟶	$Cd(s)$	–0.40	
	$Fe^{2+}(aq) + 2\,e^-$	⟶	$Fe(s)$	–0.45	
	$Cr^{3+}(aq) + e^-$	⟶	$Cr^{2+}(aq)$	–0.50	
	$Cr^{3+}(aq) + 3\,e^-$	⟶	$Cr(s)$	–0.73	
	$Zn^{2+}(aq) + 2\,e^-$	⟶	$Zn(s)$	–0.76	
	$2\,H_2O(l) + 2\,e^-$	⟶	$H_2(g) + 2\,OH^-(aq)$	–0.83	
	$Mn^{2+}(aq) + 2\,e^-$	⟶	$Mn(s)$	–1.18	
	$Al^{3+}(aq) + 3\,e^-$	⟶	$Al(s)$	–1.66	
	$Mg^{2+}(aq) + 2\,e^-$	⟶	$Mg(s)$	–2.37	
	$Na^+(aq) + e^-$	⟶	$Na(s)$	–2.71	
	$Ca^{2+}(aq) + 2\,e^-$	⟶	$Ca(s)$	–2.76	
	$Ba^{2+}(aq) + 2\,e^-$	⟶	$Ba(s)$	–2.90	
Weaker oxidizing agent	$K^+(aq) + e^-$	⟶	$K(s)$	–2.92	Stronger reducing agent
	$Li^+(aq) + e^-$	⟶	$Li(s)$	–3.04	

Example 20.4 illustrates how to calculate the standard potential of an electrochemical cell from the standard electrode potentials of the half-reactions.

EXAMPLE 20.4

Calculating Standard Potentials for Electrochemical Cells from Standard Electrode Potentials of the Half-Reactions

Use tabulated standard electrode potentials to calculate the standard cell potential for this reaction occurring in an electrochemical cell at 25 °C. (The equation is balanced.)

$$Al(s) + NO_3^-(aq) + 4\,H^+(aq) \longrightarrow Al^{3+}(aq) + NO(g) + 2\,H_2O(l)$$

SOLUTION

Begin by separating the reaction into oxidation and reduction half-reactions. (In this case, you can readily see that Al(*s*) is oxidized. In cases where it is not so apparent, you may want to assign oxidation states to determine the correct half-reactions.)	***Oxidation***: $Al(s) \longrightarrow Al^{3+}(aq) + 3\,e^-$ ***Reduction***: $NO_3^-(aq) + 4\,H^+(aq) + 3\,e^- \longrightarrow NO(g) + 2\,H_2O(l)$
Look up the standard electrode potentials for each half-reaction in Table 20.1. Add the half-cell reactions together to obtain the overall redox equation. Calculate the standard cell potential by subtracting the electrode potential of the anode from the electrode potential of the cathode.	***Oxidation (Anode)***: $Al(s) \longrightarrow Al^{3+}(aq) + \cancel{3\,e^-}$ $\quad E° = -1.66\text{ V}$ ***Reduction (Cathode)***: $NO_3^-(aq) + 4\,H^+(aq) + \cancel{3\,e^-} \longrightarrow NO(g) + 2\,H_2O(l)$ $\quad E° = 0.96\text{ V}$ $Al(s) + NO_3^-(aq) + 4\,H^+(aq) \longrightarrow Al^{3+}(aq) + NO(g) + 2\,H_2O(l)$ $E°_{cell} = E°_{cat} - E°_{an}$ $= 0.96\text{ V} - (-1.66\text{ V})$ $= 2.62\text{ V}$

FOR PRACTICE 20.4

Use tabulated standard electrode potentials to calculate the standard cell potential for this reaction occurring in an electrochemical cell at 25 °C. (The equation is balanced.)

$$3\,Pb^{2+}(aq) + 2\,Cr(s) \longrightarrow 3\,Pb(s) + 2\,Cr^{3+}(aq)$$

20.2 Cc Conceptual Connection

Standard Electrode Potentials

An electrode has a negative electrode potential. Which statement is correct regarding the potential energy of an electron at this electrode?

(a) An electron at this electrode has a lower potential energy than it has at a standard hydrogen electrode.

(b) An electron at this electrode has a higher potential energy than it has at a standard hydrogen electrode.

(c) An electron at this electrode has the same potential energy as it has at a standard hydrogen electrode.

Predicting the Spontaneous Direction of an Oxidation–Reduction Reaction

To determine the spontaneous direction of an oxidation–reduction reaction, we examine the electrode potentials of the two relevant half-reactions in Table 20.1. The half-reaction with the more *negative* electrode potential tends to lose electrons and therefore undergo oxidation. (Remember that negative charge repels electrons.) The half-reaction having the more *positive* electrode potential tends to gain electrons and therefore undergo reduction. (Remember that positive charge attracts electrons.)

Consider the two reduction half-reactions:

$$Ni^{2+}(aq) + 2\,e^- \longrightarrow Ni(s) \qquad E° = -0.23\text{ V}$$
$$Mn^{2+}(aq) + 2\,e^- \longrightarrow Mn(s) \qquad E° = -1.18\text{ V}$$

Because the manganese half-reaction has a more negative electrode potential, it repels electrons and proceeds in the reverse direction (oxidation). Because the nickel half-reaction has the more positive (or less negative) electrode potential, it attracts electrons and proceeds in the forward direction.

We can confirm this by calculating the standard electrode potential for manganese acting as the anode (oxidation) and nickel acting as the cathode (reduction).

Oxidation (Anode): $$Mn(s) \longrightarrow Mn^{2+}(aq) + \cancel{2\,e^-} \qquad E° = -1.18\text{ V}$$

Reduction (Cathode): $$Ni^{2+}(aq) + \cancel{2\,e^-} \longrightarrow Ni(s) \qquad E° = -0.23\text{ V}$$

$$Ni^{2+}(aq) + Mn(s) \longrightarrow Ni(s) + Mn^{2+}(aq) \qquad E°_{cell} = E°_{cat} - E°_{an}$$
$$= -0.23\text{ V} - (-1.18\text{ V})$$
$$= 0.95\text{ V}$$

The overall cell potential is positive, indicating a spontaneous reaction. Consider the electrochemical cell corresponding to this spontaneous redox reaction in Figure 20.8 ▶. We draw the manganese half-cell on the left as the anode and the nickel half-cell on the right as the cathode. Electrons flow from the anode to the cathode.

Another way to predict the spontaneity of a redox reaction is to note the relative positions of the two half-reactions in Table 20.1. The table lists half-reactions in order of *decreasing* electrode potential, so the half-reactions near the top of the table—those having large *positive* electrode potentials—attract electrons and therefore tend to occur in the forward direction. Half-reactions near the bottom of the table—those having large *negative* electrode potentials—repel electrons and therefore tend to occur in the reverse direction. In other words, as we move down Table 20.1, the half-reactions become less likely to occur in the forward direction and more likely to occur in the reverse direction. As a result, *any reduction half-reaction listed is spontaneous when paired with the reverse of a half-reaction that appears below it in Table 20.1.*

For example, if we return to our two previous half-reactions involving manganese and nickel, we see that the manganese half-reaction is listed below the nickel half-reaction in Table 20.1.

$$Ni^{2+}(aq) + 2\,e^- \longrightarrow Ni(s) \qquad E° = -0.23\text{ V}$$
$$Mn^{2+}(aq) + 2\,e^- \longrightarrow Mn(s) \qquad E° = -1.18\text{ V}$$

Therefore, the nickel reaction occurs in the forward direction (reduction) and the manganese reaction occurs in the reverse direction (oxidation).

The following mnemonics (NIO and PIR) can help you predict the spontaneous direction of redox reactions:
N.I.O.–More Negative Is Oxidation
P.I.R.–More Positive Is Reduction

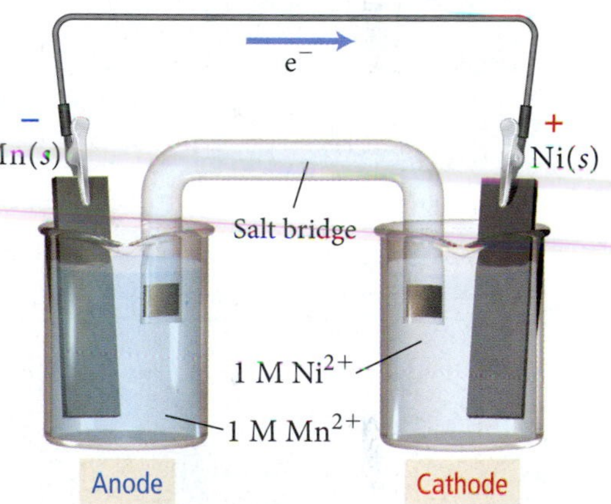

▲ **FIGURE 20.8 Mn/Ni²⁺ Electrochemical Cell** Since the reduction of Mn^{2+} is listed below the reduction of Ni^{2+} in Table 20.1, the reduction of Ni^{2+} is spontaneous when paired with the oxidation of Mn.

Summarizing the Prediction of Spontaneous Direction for Redox Reactions:

- The half-reaction with the more *positive* electrode potential attracts electrons more strongly and undergoes reduction. (Substances listed at the top of Table 20.1 tend to undergo reduction; they are good oxidizing agents.)
- The half-reaction with the more *negative* electrode potential repels electrons more strongly and undergoes oxidation. (Substances listed near the bottom of Table 20.1 tend to undergo oxidation; they are good reducing agents.)
- Any reduction reaction in Table 20.1 is spontaneous when paired with the *reverse* of any of the reactions listed below it on the table.

Recall from Section 9.9 that an *oxidizing* agent causes the oxidation of another substance (and is itself reduced) and that a *reducing* agent causes the reduction of another substance (and is itself oxidized).

EXAMPLE 20.5

Predicting Spontaneous Redox Reactions and Sketching Electrochemical Cells

Without calculating E°_{cell}, predict whether each of these redox reactions is spontaneous (when the reactants and products are in their standard states). If the reaction is spontaneous as written, make a sketch of the electrochemical cell in which the reaction could occur. If the reaction is not spontaneous as written, write an equation for the direction in which the spontaneous reaction occurs and sketch the corresponding electrochemical cell. In your sketches, make sure to label the anode (which should be drawn on the left), the cathode, and the direction of electron flow.

(a) $Fe(s) + Mg^{2+}(aq) \longrightarrow Fe^{2+}(aq) + Mg(s)$

(b) $Fe(s) + Pb^{2+}(aq) \longrightarrow Fe^{2+}(aq) + Pb(s)$

SOLUTION

(a) $Fe(s) + Mg^{2+}(aq) \longrightarrow Fe^{2+}(aq) + Mg(s)$

This reaction involves the reduction of Mg^{2+}:

$$Mg^{2+}(aq) + 2\,e^- \longrightarrow Mg(s) \qquad E^\circ = -2.37\text{ V}$$

and the oxidation of Fe:

$$Fe(s) \longrightarrow Fe^{2+}(aq) + 2\,e^- \qquad E^\circ = -0.45\text{ V}$$

The magnesium half-reaction has the more negative electrode potential and therefore repels electrons more strongly and undergoes oxidation. The iron half-reaction has the more positive electrode potential and therefore attracts electrons more strongly and undergoes reduction. So the reaction as written is *not* spontaneous. (The reaction pairs the reduction of Mg^{2+} of with the reverse of a half-reaction *above it* in Table 20.1—such pairings are not spontaneous.)

However, the reverse reaction is spontaneous.

$$Fe^{2+}(aq) + Mg(s) \longrightarrow Fe(s) + Mg^{2+}(aq)$$

The corresponding electrochemical cell is shown in Figure 20.9 ◀.

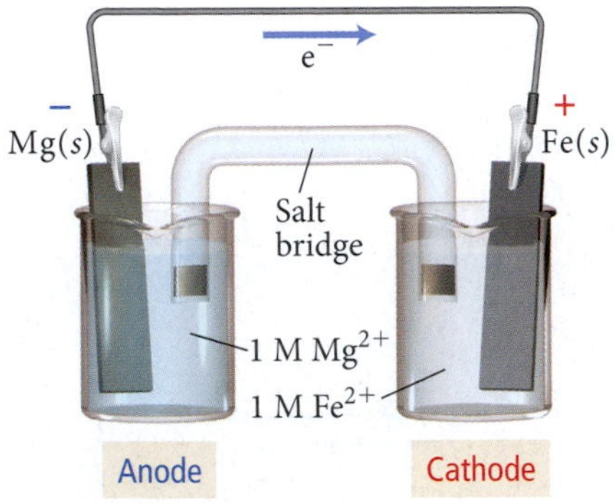

▲ **FIGURE 20.9 Mg/Fe^{2+} Electrochemical Cell**

(b) $Fe(s) + Pb^{2+}(aq) \longrightarrow Fe^{2+}(aq) + Pb(s)$

This reaction involves the reduction of Pb^{2+}:

$$Pb^{2+}(aq) + 2\,e^- \longrightarrow Pb(s) \qquad E^\circ = -0.13\text{ V}$$

and the oxidation of iron:

$$Fe(s) \longrightarrow Fe^{2+}(aq) + 2\,e^- \qquad E^\circ = -0.45\text{ V}$$

The iron half-reaction has the more negative electrode potential and therefore repels electrons and undergoes oxidation. The lead half-reaction has the more positive electrode potential and therefore attracts electrons and undergoes reduction. The reaction *is* spontaneous as written. (The reaction pairs the reduction of Pb^{2+} with the reverse of a half-reaction *below it* in Table 20.1—such pairings are always spontaneous.) The corresponding electrochemical cell is shown in Figure 20.10 ◀.

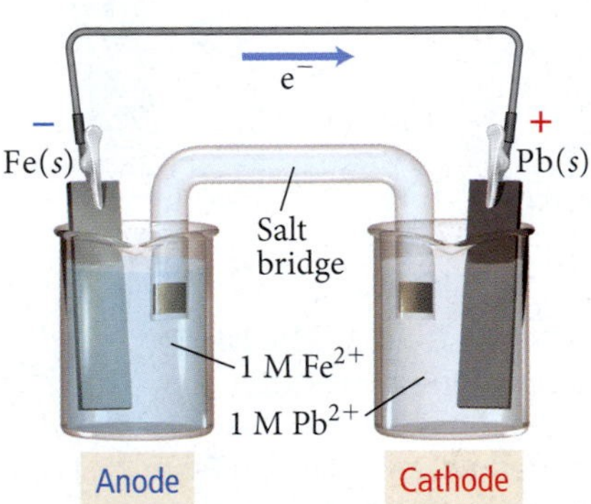

▲ **FIGURE 20.10 Fe/Pb^{2+} Electrochemical Cell**

FOR PRACTICE 20.5

Are these redox reactions spontaneous under standard conditions?

(a) $Zn(s) + Ni^{2+}(aq) \longrightarrow Zn^{2+}(aq) + Ni(s)$

(b) $Zn(s) + Ca^{2+}(aq) \longrightarrow Zn^{2+}(aq) + Ca(s)$

Selective Oxidation — 20.3 Cc Conceptual Connection

A solution contains both NaI and NaBr. Which oxidizing agent could you add to the solution to selectively oxidize $I^-(aq)$ but not $Br^-(aq)$?

(a) Cl_2 **(b)** H_2O_2 **(c)** $CuCl_2$ **(d)** HNO_3

Predicting Whether a Metal Will Dissolve in Acid

Recall from Chapter 17 that acids dissolve some metals. Most acids dissolve metals by the reduction of H^+ ions to hydrogen gas and the corresponding oxidation of the metal to its ion. For example, if solid Zn is submersed into hydrochloric acid, the following reaction occurs:

$$\begin{aligned} 2\,H^+(aq) + \cancel{2\,e^-} &\longrightarrow H_2(g) \\ Zn(s) &\longrightarrow Zn^{2+}(aq) + \cancel{2\,e^-} \\ \hline Zn(s) + 2\,H^+(aq) &\longrightarrow Zn^{2+}(aq) + H_2(g) \end{aligned}$$

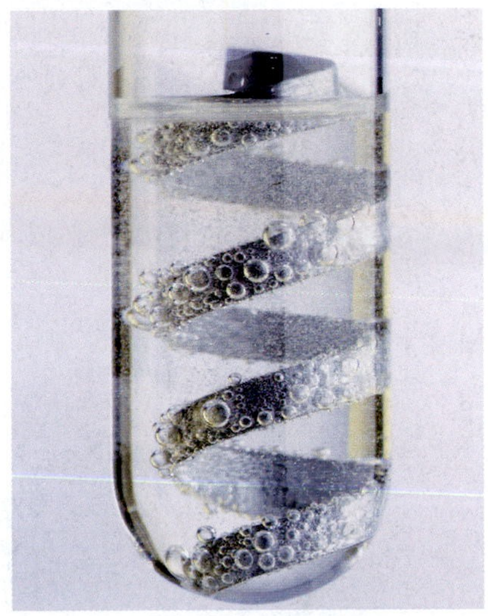

$Zn(s) + 2\,H^+(aq) \longrightarrow Zn^{2+}(aq) + H_2(g)$

▲ When zinc is immersed in hydrochloric acid, the zinc is oxidized, forming ions that become solvated in the solution. Hydrogen ions are reduced, forming bubbles of hydrogen gas.

We observe the reaction as the dissolving of the zinc and the bubbling of hydrogen gas. The zinc is oxidized and the H^+ions are reduced. Notice that this reaction involves the pairing of a reduction half-reaction (the reduction of H^+) with the reverse of a half-reaction that is listed below it in Table 20.1. Therefore, this reaction is spontaneous. What happens, however, if we pair the reduction of H^+ with the oxidation of Cu? The reaction is not spontaneous, because it involves pairing the reduction of H^+ with the reverse of a half-reaction that is listed *above it* in the table. Consequently, copper does not react with H^+ and does not dissolve in acids such as HCl. In general, *metals whose reduction half-reactions are listed below the reduction of* H^+ *to* H_2 *in Table 20.1 dissolve in acids, while metals listed above it do not.*

An important exception to this rule is nitric acid (HNO_3), which can oxidize metals through the reduction half-reaction:

$$NO_3^-(aq) + 4\,H^+(aq) + 3\,e^- \longrightarrow NO(g) + 2\,H_2O(l) \quad E^\circ = 0.96\text{ V}$$

Since this half-reaction is above the reduction of H^+ in Table 20.1, HNO_3 can oxidize metals (such as copper) that can't be oxidized by HCl.

Metals Dissolving in Acids — 20.4 Cc Conceptual Connection

Which metal dissolves in HNO_3 but not in HCl?

(a) Fe **(b)** Au **(c)** Ag

20.5 Cell Potential, Free Energy, and the Equilibrium Constant

We have seen that a positive standard cell potential (E°_{cell}) corresponds to a spontaneous oxidation–reduction reaction when the reactants and products are in their standard states (standard conditions). And we know (from Chapter 19) that the spontaneity of a reaction under standard conditions is determined by the sign of ΔG°. Therefore, E°_{cell} and ΔG° must be related. We also know from Section 19.9 that ΔG° for a reaction is related to the equilibrium constant (K) for the reaction. Since E°_{cell} and ΔG° are related, then E°_{cell} and K must also be related.

Remember that standard conditions (indicated by the symbol °) represent a very specific reaction mixture. For a reaction mixture under standard conditions, Q = 1.

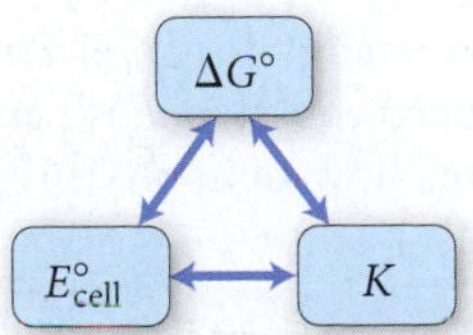

Before we look at the nature of each of these relationships in detail, let's consider the following generalizations.

For a spontaneous redox reaction (one that will proceed in the forward direction when all reactants and products are in their standard states):

- $\Delta G°$ is negative (< 0)
- $E°_{cell}$ is positive (> 0)
- $K > 1$

For a nonspontaneous reaction (one that will proceed in the reverse direction when all reactants and products are in their standard states):

- $\Delta G°$ is positive (> 0)
- $E°_{cell}$ is negative (< 0)
- $K < 1$

The Relationship between $\Delta G°$ and $E°_{cell}$

We can derive a relationship between $\Delta G°$ and $E°_{cell}$ by briefly returning to the definition of potential difference from Section 20.3—a potential difference is a measure of the difference of potential energy per unit charge (q):

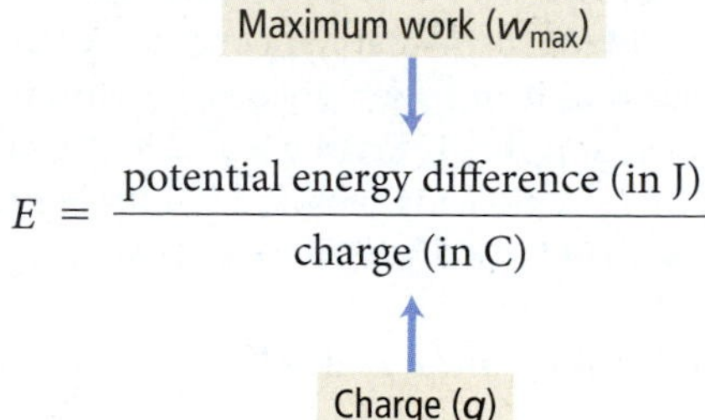

Since the potential energy difference represents the maximum amount of work that can be done by the system on the surroundings, we can write:

$$w_{max} = -qE°_{cell} \qquad [20.1]$$

The negative sign follows the convention used throughout this book that work done by the system on the surroundings is negative.

We can quantify the charge (q) that flows in an electrochemical reaction by using **Faraday's constant (*F*),** which represents the charge in coulombs of 1 mol of electrons.

$$F = \frac{96{,}485 \text{ C}}{\text{mol e}^-}$$

The total charge is $q = nF$, where n is the number of moles of electrons from the balanced chemical equation and F is Faraday's constant. Substituting $q = nF$ into Equation 20.1:

$$\begin{aligned} w_{max} &= -qE°_{cell} \\ &= -nFE°_{cell} \end{aligned} \qquad [20.2]$$

Finally, recall from Chapter 19 that the standard change in free energy for a chemical reaction ($\Delta G°$) represents the maximum amount of work that can be done by the reaction. Therefore, $w_{max} = \Delta G°$. Making this substitution into Equation 20.2, we arrive at the following important result:

$$\Delta G° = -nFE°_{cell} \qquad [20.3]$$

where $\Delta G°$ is the standard change in free energy for an electrochemical reaction, n is the number of moles of electrons transferred in the balanced equation, F is Faraday's constant, and $E°_{cell}$ is the standard cell potential. Example 20.6 demonstrates how to apply this equation to calculate the standard free energy change for an electrochemical cell.

EXAMPLE 20.6

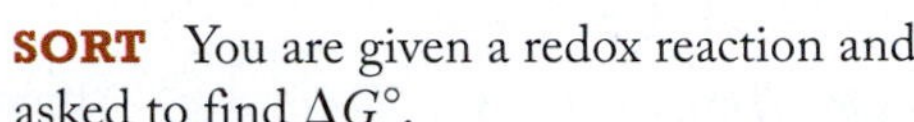

Relating ΔG° and E°_{cell}

Use the tabulated electrode potentials to calculate ΔG° for the reaction.

$$I_2(s) + 2\,Br^-(aq) \longrightarrow 2\,I^-(aq) + Br_2(l)$$

Is the reaction spontaneous under standard conditions?

SORT You are given a redox reaction and asked to find ΔG°.

GIVEN: $I_2(s) + 2\,Br^-(aq) \longrightarrow 2\,I^-(aq) + Br_2(l)$

FIND: ΔG°

STRATEGIZE Refer to the values of electrode potentials in Table 20.1 to calculate E°_{cell}. Then use Equation 20.3 to calculate ΔG° from E°_{cell}.

CONCEPTUAL PLAN

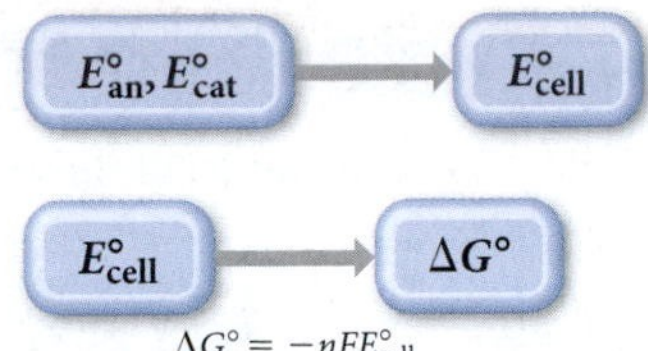

SOLVE Separate the reaction into oxidation and reduction half-reactions and find the standard electrode potentials for each. Determine E°_{cell} by subtracting E°_{an} from E°_{cat}.

SOLUTION

Oxidation (Anode): $2\,Br^-(aq) \longrightarrow Br_2(l) + 2\,e^-$ $\quad E^\circ = 1.09\ V$

Reduction (Cathode): $I_2(s) + 2\,e^- \longrightarrow 2\,I^-(aq)$ $\quad E^\circ = 0.54\ V$

$$I_2(s) + 2\,Br^-(aq) \longrightarrow 2\,I^-(aq) + Br_2(l) \qquad E^\circ_{cell} = E^\circ_{cat} - E^\circ_{an} = -0.55\ V$$

Calculate ΔG° from E°_{cell}. The value of n (the number of moles of electrons) corresponds to the number of electrons that are canceled in the half-reactions. Remember that $1\ V = 1\ J/C$.

$$\begin{aligned}\Delta G^\circ &= -nFE^\circ_{cell}\\ &= -2\ \text{mol e}^-\left(\frac{96{,}485\ C}{\text{mol e}^-}\right)\left(-0.55\frac{J}{C}\right)\\ &= +1.1 \times 10^5\ J\end{aligned}$$

Since ΔG° is positive, the reaction is not spontaneous under standard conditions.

CHECK The answer is in the correct units (joules) and seems reasonable in magnitude ($\approx$110 kJ). You have seen (in Chapter 19) that values of ΔG° typically range from plus or minus tens to hundreds of kilojoules. The sign is positive, as expected for a reaction in which E°_{cell} is negative.

FOR PRACTICE 20.6

Use tabulated electrode potentials to calculate ΔG° for the reaction.

$$2\,Na(s) + 2\,H_2O(l) \longrightarrow H_2(g) + 2\,OH^-(aq) + 2\,Na^+(aq)$$

Is the reaction spontaneous under standard conditions?

Periodic Trends and the Direction of Spontaneity for Redox Reactions

20.5 Cc Conceptual Connection

Consider the result of Example 20.6. The calculation revealed that the reaction is not spontaneous. Based on conceptual reasoning, which statement best explains why I_2 does not oxidize Br^-?

(a) Br is more electronegative than I; therefore, we do not expect Br^- to give up an electron to I_2.

(b) I is more electronegative than Br; therefore, we do not expect I_2 to give up an electron to Br^-.

(c) Br^- is in solution and I_2 is a solid. Solids do not gain electrons from substances in solution.

The Relationship between E°_{cell} and K

We can derive a relationship between the standard cell potential (E°_{cell}) and the equilibrium constant for the redox reaction occurring in the cell (K) by returning to the relationship between ΔG° and K that we learned in Chapter 19. Recall from Section 19.9 that:

$$\Delta G^\circ = -RT \ln K \qquad [20.4]$$

By setting Equations 20.3 and 20.4 equal to each other, we get:

$$-nFE^\circ_{cell} = -RT \ln K$$

$$E^\circ_{cell} = \frac{RT}{nF} \ln K \qquad [20.5]$$

Equation 20.5 is usually simplified for use at 25 °C with the following substitutions:

$$R = 8.314 \frac{\text{J}}{\text{mol} \cdot \text{K}};\ T = 298.15\ \text{K};\ F = \left(\frac{96{,}485\ \text{C}}{\text{mol e}^-}\right);\ \text{and}\ \ln K = 2.303 \log K$$

Substituting into Equation 20.5, we get the following important result:

$$E^\circ_{cell} = \frac{0.0592\ \text{V}}{n} \log K \qquad [20.6]$$

where E°_{cell} is the standard cell potential, n is the number of moles of electrons transferred in the redox reaction, and K is the equilibrium constant for the balanced redox reaction at 25 °C. Example 20.7 demonstrates the use of Equation 20.6.

EXAMPLE 20.7
Relating E°_{cell} and K

Refer to tabulated electrode potentials to calculate K for the oxidation of copper by H^+ (at 25 °C).

$$Cu(s) + 2\,H^+(aq) \longrightarrow Cu^{2+}(aq) + H_2(g)$$

SORT You are given a redox reaction and asked to find K.

GIVEN: $Cu(s) + 2\,H^+(aq) \longrightarrow Cu^{2+}(aq) + H_2(g)$

FIND: K

STRATEGIZE Refer to the values of electrode potentials in Table 20.1 to calculate E°_{cell}. Then use Equation 20.6 to calculate K from E°_{cell}.

CONCEPTUAL PLAN

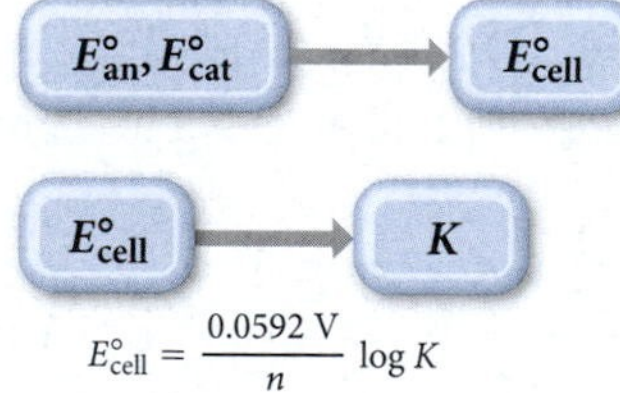

$$E^\circ_{cell} = \frac{0.0592\ \text{V}}{n} \log K$$

SOLVE Separate the reaction into oxidation and reduction half-reactions and find the standard electrode potentials for each. Find E°_{cell} by subtracting E°_{an} from E°_{cat}.

SOLUTION

Oxidation (Anode): $Cu(s) \longrightarrow Cu^{2+}(aq) + 2e^-$ $\quad E^\circ = 0.34\ \text{V}$

Reduction (Cathode): $2\,H^+(aq) + 2e^- \longrightarrow H_2(g)$ $\quad E^\circ = 0.00\ \text{V}$

$Cu(s) + 2\,H^+(aq) \longrightarrow Cu^{2+}(aq) + H_2(g)$ $\quad E^\circ_{cell} = E^\circ_{cat} - E^\circ_{an} = -0.34\ \text{V}$

Calculate K from E°_{cell}. The value of n (the number of moles of electrons) corresponds to the number of electrons that are canceled in the half-reactions.

$$E^\circ_{cell} = \frac{0.0592\ \text{V}}{n} \log K$$

$$\log K = E^\circ_{cell} \frac{n}{0.0592\ \text{V}}$$

$$\log K = -0.34\ \text{V} \frac{2}{0.0592\ \text{V}} = -11.48$$

$$K = 10^{-11.48} = 3.3 \times 10^{-12}$$

—*Continued on the next page*

Continued from the previous page—

CHECK The answer has no units, as expected for an equilibrium constant. The magnitude of the answer is small, indicating that the reaction lies far to the left at equilibrium, as expected for a reaction in which E°_{cell} is negative.

FOR PRACTICE 20.7
Use the tabulated electrode potentials to calculate K for the oxidation of iron by H^+ (at 25 °C).

$$2\,Fe(s) + 6\,H^+(aq) \longrightarrow 2\,Fe^{3+}(aq) + 3\,H_2(g)$$

Notice that the fundamental quantity in the given relationships is the standard change in free energy for a chemical reaction (ΔG°_{rxn}). From that quantity, we can calculate both E°_{cell} and K. The following diagram summarizes the relationships between these three quantities:

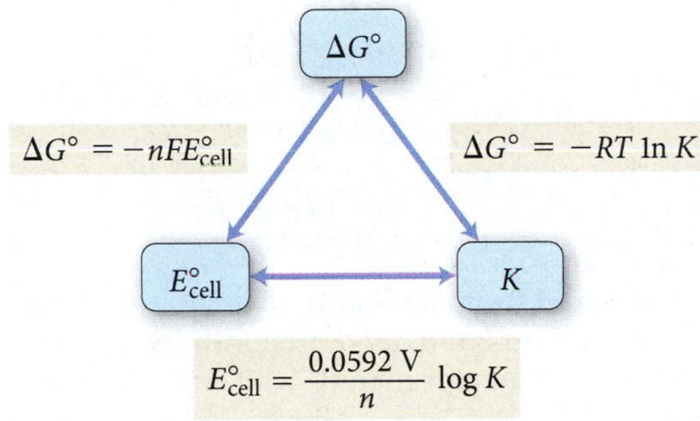

Relating K, ΔG°_{rxn}, and E°_{cell}

20.6 Cc Conceptual Connection

A redox reaction has an equilibrium constant of $K = 1.2 \times 10^3$. Which statement is true regarding ΔG°_{rxn} and E°_{cell} for this reaction?

(a) E°_{cell} is positive and ΔG°_{rxn} is positive.
(b) E°_{cell} is negative and ΔG°_{rxn} is negative.
(c) E°_{cell} is positive and ΔG°_{rxn} is negative.
(d) E°_{cell} is negative and ΔG°_{rxn} is positive.

20.6 Cell Potential and Concentration

We have learned how to find E°_{cell} under standard conditions. For example, we know that when $[Cu^{2+}] = 1\ M$ and $[Zn^{2+}] = 1\ M$, the following reaction produces a potential of 1.10 V:

$$Zn(s) + Cu^{2+}(aq, 1\ M) \longrightarrow Zn^{2+}(aq, 1\ M) + Cu(s) \quad E^\circ_{cell} = 1.10\ V$$

But what if $[Cu^{2+}] > 1\ M$ and $[Zn^{2+}] < 1\ M$? For example, how is the cell potential for the following conditions different from the potential under standard conditions?

$$Zn(s) + Cu^{2+}(aq, 2\ M) \longrightarrow Zn^{2+}(aq, 0.010\ M) + Cu(s) \quad E_{cell} = ?$$

Since the concentration of a reactant is greater than standard conditions, and since the concentration of product is less than standard conditions, we can use Le Châtelier's principle to predict that the reaction has an even stronger tendency to occur in the forward direction and that E_{cell} is therefore greater than +1.10 V (Figure 20.11 ▶).

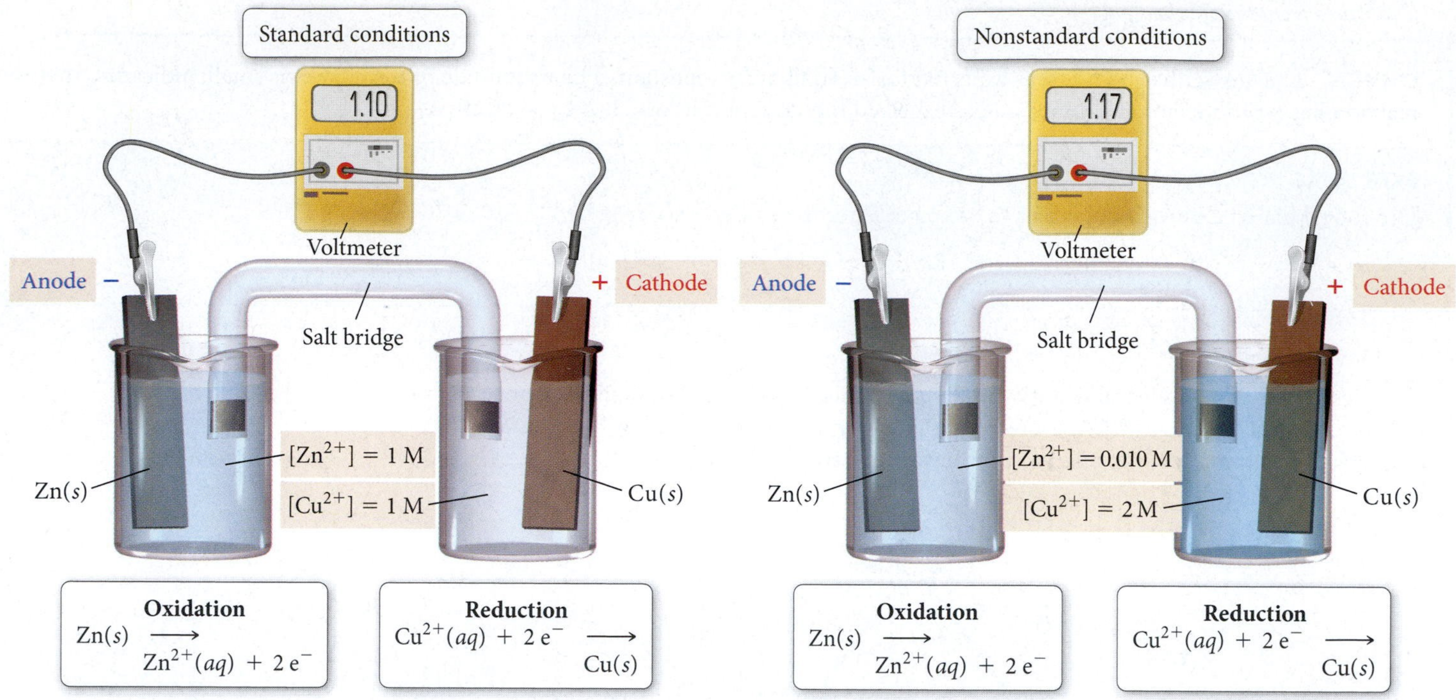

▲ **FIGURE 20.11 Cell Potential and Concentration** This figure compares the Zn/Cu^{2+} electrochemical cell under standard and nonstandard conditions. In this case, the nonstandard conditions consist of a higher Cu^{2+} concentration ($[Cu^{2+}] > 1$ M) at the cathode and a lower Zn^{2+} concentration at the anode ($[Zn^{2+}] < 1$ M). According to Le Châtelier's principle, the forward reaction has a greater tendency to occur, resulting in a greater overall cell potential than the potential under standard conditions.

We can derive an exact relationship between E_{cell} (under nonstandard conditions) and E°_{cell} by considering the relationship between the change in free energy (ΔG) and the *standard* change in free energy (ΔG°) from Section 19.8:

$$\Delta G = \Delta G^\circ + RT \ln Q \qquad [20.7]$$

where R is the gas constant (8.314 J/mol·K), T is the temperature in kelvins, and Q is the reaction quotient corresponding to the nonstandard conditions. Since we know the relationship between ΔG and E_{cell} (Equation 20.3), we can substitute into Equation 20.7:

$$\Delta G = \Delta G^\circ + RT \ln Q$$

$$-nFE_{cell} = -nFE^\circ_{cell} + RT \ln Q$$

We can then divide each side by $-nF$ to arrive at:

$$E_{cell} = E^\circ_{cell} - \frac{RT}{nF} \ln Q \qquad [20.8]$$

As we have seen, R and F are constants; at $T = 25$ °C, $\frac{RT}{nF} \ln Q = \frac{0.0592\text{ V}}{n} \log Q$.

Substituting into Equation 20.8, we arrive at the **Nernst equation:**

$$E_{cell} = E^\circ_{cell} - \frac{0.0592\text{ V}}{n} \log Q \qquad [20.9]$$

where E_{cell} is the cell potential in volts, E°_{cell} is the *standard* cell potential in volts, n is the number of moles of electrons transferred in the redox reaction, and Q is the reaction quotient. Notice that, under standard conditions, $Q = 1$, and (since $\log 1 = 0$) $E_{cell} = E^\circ_{cell}$, as expected. Example 20.8 demonstrates how to calculate the cell potential under nonstandard conditions.

EXAMPLE 20.8

Calculating E_{cell} under Nonstandard Conditions

Determine the cell potential for an electrochemical cell based on the following two half-reactions:

Oxidation: $Cu(s) \longrightarrow Cu^{2+}(aq, 0.010\ M) + 2\ e^-$

Reduction: $MnO_4^-(aq, 2.0\ M) + 4\ H^+(aq, 1.0\ M) + 3\ e^- \longrightarrow MnO_2(s) + 2\ H_2O(l)$

SORT You are given the half-reactions for a redox reaction and the concentrations of the aqueous reactants and products. You are asked to find the cell potential.

GIVEN: $[MnO_4^-] = 2.0\ M; [H^+] = 1.0\ M; [Cu^{2+}] = 0.010\ M$

FIND: E_{cell}

STRATEGIZE Use the tabulated values of electrode potentials to calculate E°_{cell}. Then use Equation 20.9 to calculate E_{cell}.

CONCEPTUAL PLAN

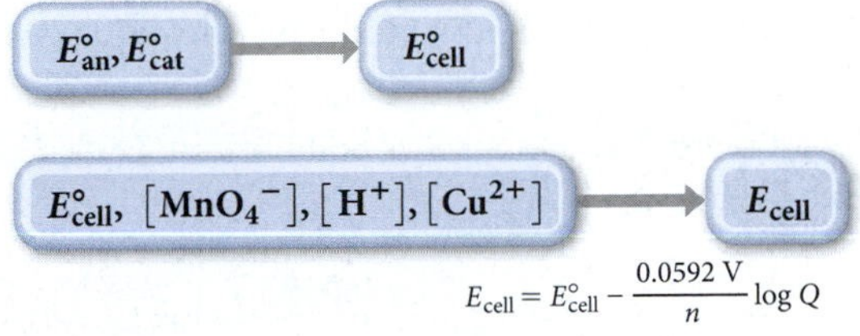

SOLVE Write the oxidation and reduction half-reactions, multiplying by the appropriate coefficients to cancel the electrons. Find the standard electrode potentials for each. Find E°_{cell}.

SOLUTION

Oxidation (Anode): $3[Cu(s) \longrightarrow Cu^{2+}(aq) + 2\ e^-] \quad E^\circ = 0.34\ V$

Reduction (Cathode): $2\ [MnO_4^-(aq) + 4\ H^+(aq) + 3\ e^- \longrightarrow MnO_2(s) + 2\ H_2O(l)] \quad E^\circ = 1.68\ V$

$$3\ Cu(s) + 2\ MnO_4^-(aq) + 8\ H^+(aq) \longrightarrow 3\ Cu^{2+}(aq) + 2\ MnO_2(s) + 4\ H_2O(l)$$

$$E^\circ_{cell} = E^\circ_{cat} - E^\circ_{an} = 1.34\ V$$

Calculate E_{cell} from E°_{cell}. The value of n (the number of moles of electrons) corresponds to the number of electrons (6 in this case) canceled in the half-reactions. Determine Q based on the overall balanced equation and the given concentrations of the reactants and products. (Note that pure liquid water, solid MnO_2, and solid copper are omitted from the expression for Q.)

$$E_{cell} = E^\circ_{cell} - \frac{0.0592\ V}{n} \log Q$$

$$= E^\circ_{cell} - \frac{0.0592\ V}{n} \log \frac{[Cu^{2+}]^3}{[MnO_4^-]^2[H^+]^8}$$

$$= 1.34\ V - \frac{0.0592\ V}{6} \log \frac{(0.010)^3}{(2.0)^2(1.0)^8}$$

$$= 1.34\ V - (-0.065\ V)$$

$$= 1.41\ V$$

CHECK The answer has the correct units (V). The value of E_{cell} is larger than E°_{cell}, as expected based on Le Châtelier's principle because one of the aqueous reactants has a concentration greater than standard conditions and the one aqueous product has a concentration less than standard conditions. Therefore, the reaction has a greater tendency to proceed toward products and a greater cell potential.

FOR PRACTICE 20.8

Determine the cell potential of an electrochemical cell based on the following two half-reactions:

Oxidation: $Ni(s) \longrightarrow Ni^{2+}(aq, 2.0\ M) + 2\ e^-$

Reduction: $VO_2^+(aq, 0.010\ M) + 2\ H^+(aq, 1.0\ M) + e^- \longrightarrow VO^{2+}(aq, 2.0\ M) + H_2O(l)$

From Equation 20.9, we can conclude the following:

- When a redox reaction within a voltaic cell occurs under standard conditions $Q = 1$; therefore, $E_{cell} = E^\circ_{cell}$.

$$E_{cell} = E^\circ_{cell} - \frac{0.0592\ V}{n}\log Q$$

$$= E^\circ_{cell} - \frac{0.0592\ V}{n}\log 1 \quad (\log 1 = 0)$$

$$= E^\circ_{cell}$$

- When a redox reaction within a voltaic cell occurs under conditions in which $Q < 1$, the greater concentration of reactants relative to products drives the reaction to the right, resulting in $E_{cell} > E^\circ_{cell}$.
- When a redox reaction within an electrochemical cell occurs under conditions in which $Q > 1$, the greater concentration of products relative to reactants drives the reaction to the left, resulting in $E_{cell} < E^\circ_{cell}$.
- When a redox reaction reaches equilibrium, $Q = K$. The redox reaction has no tendency to occur in either direction and $E_{cell} = 0$.

$$E_{cell} = E^\circ_{cell} - \frac{0.0592\ V}{n}\log Q$$

$$= E^\circ_{cell} - \frac{0.0592\ V}{n}\log K \quad \left(E^\circ_{cell}\ \text{(see Equation 20.6)}\right)$$

$$= E^\circ_{cell} - E^\circ_{cell}$$

$$= 0$$

This last point explains why batteries do not last forever—as the reactants are depleted, the reaction proceeds toward equilibrium and the potential tends toward zero.

20.7 Cc Conceptual Connection

Relating Q, K, E_{cell}, and E°_{cell}

In an electrochemical cell, $Q = 0.0010$ and $K = 0.10$. What can you conclude about E_{cell} and E°_{cell}?

(a) E_{cell} is positive and E°_{cell} is negative.

(b) E_{cell} is negative and E°_{cell} is positive.

(c) Both E_{cell} and E°_{cell} are positive.

(d) Both E_{cell} and E°_{cell} are negative.

Concentration Cells

Since cell potential depends not only on the half-reactions occurring in the cell, but also on the *concentrations* of the reactants and products in those half-reactions, we can construct a voltaic cell in which both half-reactions are the same, but in which *a difference in concentration drives the current flow.* For example, consider the electrochemical cell shown in Figure 20.12 ▶, in which copperis oxidized at the anode and copper(II) ions are reduced at the cathode. The second part of Figure 20.12 depicts this cell under nonstandard conditions, with $[Cu^{2+}] = 2.0$ M in one half-cell and $[Cu^{2+}] = 0.010$ M in the other:

$$Cu(s) + Cu^{2+}(aq, 2.0\ M) \longrightarrow Cu^{2+}(aq, 0.010\ M) + Cu(s)$$

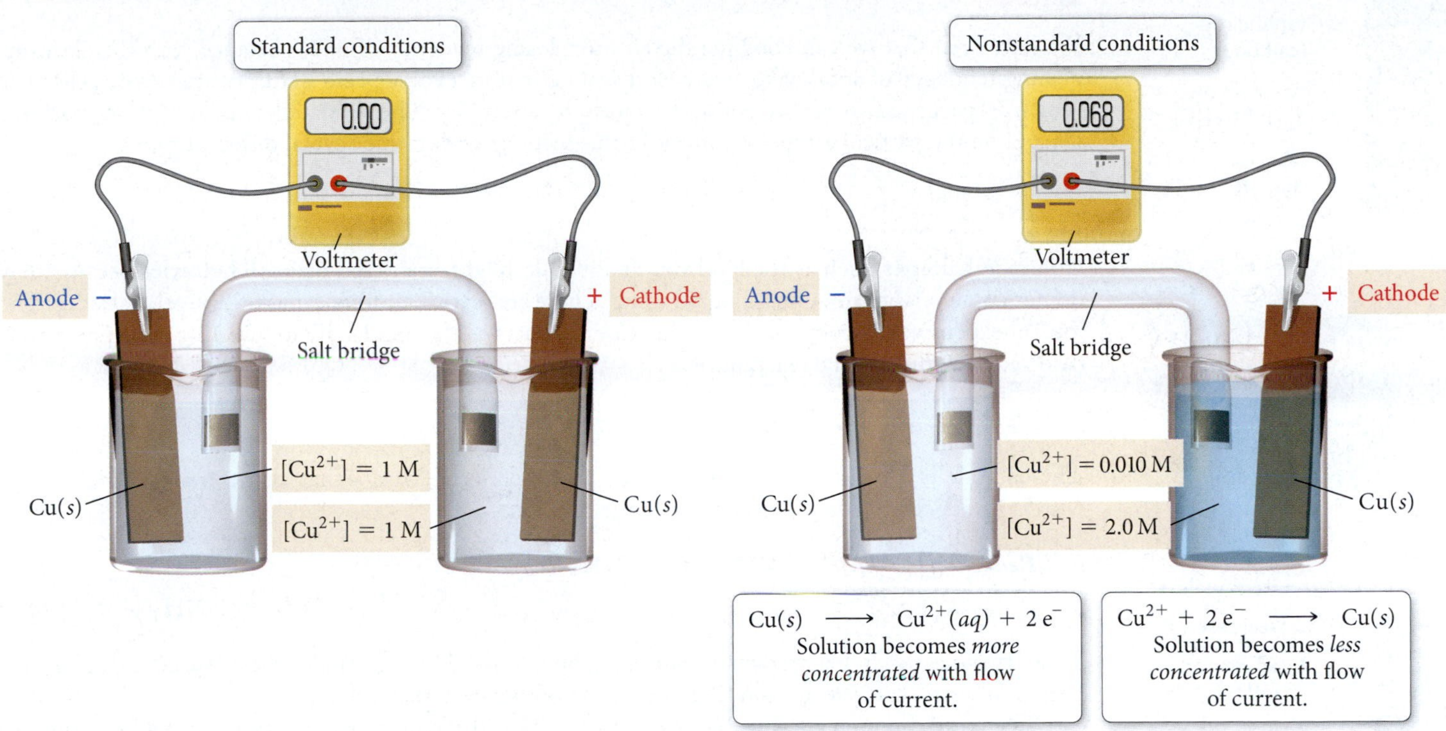

▲ **FIGURE 20.12** **Cu/Cu^{2+} Concentration Cell** If two half-cells have the same Cu^{2+} concentration, the cell potential is zero. If one half-cell has a greater Cu^{2+} concentration than the other, a spontaneous reaction occurs. In the reaction, Cu^{2+} ions in the more concentrated cell are reduced (to solid copper), while Cu^{2+} ions in the more dilute cells are formed (from solid copper). The concentration of copper(II) ions in the two half-cells tends toward equality.

The half-reactions are identical, so the *standard* cell potential is therefore zero.

$$\begin{array}{llll} \textit{Reduction (Cathode):} & Cu^{2+}(aq) + \cancel{2e^-} \longrightarrow Cu(s) & & E^\circ = 0.34\text{ V} \\ \textit{Oxidation (Anode):} & Cu(s) \longrightarrow Cu^{2+}(aq) + \cancel{2e^-} & & E^\circ = 0.34\text{ V} \\ \hline & Cu^{2+}(aq) + Cu(s) \longrightarrow Cu(s) + Cu^{2+}(aq) & & E^\circ_{cell} = E^\circ_{cat} - E^\circ_{an} \\ & & & \phantom{E^\circ_{cell}} = +0.00\text{ V} \end{array}$$

Because of the different concentrations in the two half-cells, we must calculate the cell potential using the Nernst equation.

$$\begin{aligned} E_{cell} &= E^\circ_{cell} - \frac{0.0592\text{ V}}{2}\log\frac{0.010}{2.0} \\ &= 0.000\text{ V} + 0.068\text{ V} \\ &= 0.068\text{ V} \end{aligned}$$

The cell produces a potential of 0.068 V. *Electrons spontaneously flow from the half-cell with the lower copper ion concentration to the half-cell with the higher copper ion concentration.* We can imagine a concentration cell in the same way we think about any concentration gradient. If we mix a concentrated solution of Cu^{2+} with a dilute solution, the Cu^{2+} ions flow from the concentrated solution to the dilute one. Similarly, in a concentration cell, the transfer of electrons *from* the dilute half-cell results in the formation of Cu^{2+} ions in the dilute half-cell. The electrons flow to the concentrated cell, where they react with Cu^{2+} ions and reduce them to Cu(*s*). Therefore, *the flow of electrons has the effect of increasing the concentration of Cu^{2+} in the dilute cell and decreasing the concentration of Cu^{2+} in the concentrated half-cell.*

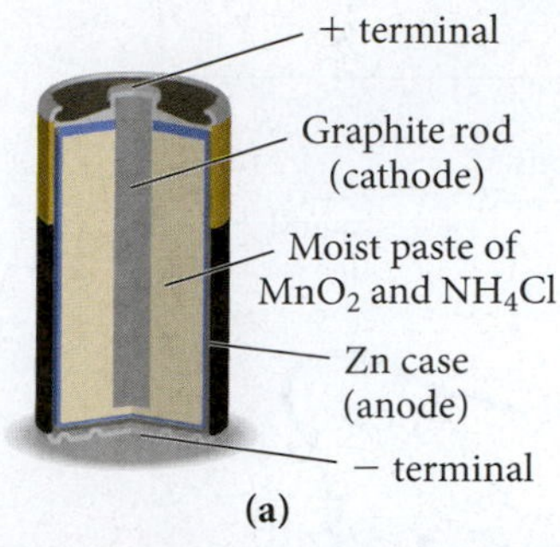

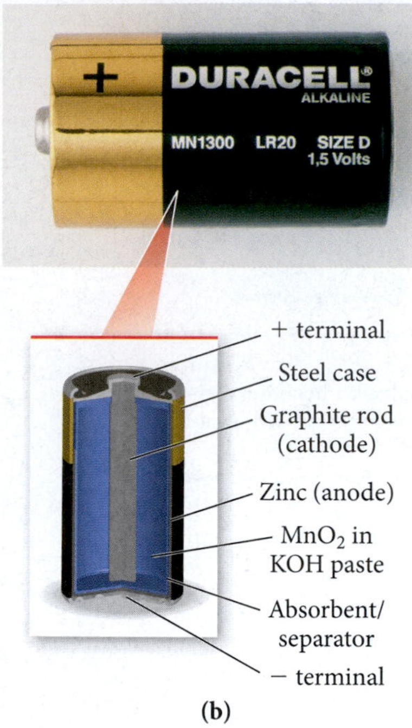

▲ **FIGURE 20.13 Dry-Cell Battery** **(a)** In a common dry-cell battery, the zinc case acts as the anode and a graphite rod immersed in a moist, slightly acidic paste of MnO_2 and NH_4Cl acts as the cathode. **(b)** The longer-lived alkaline batteries employ a graphite cathode immersed in a paste of MnO_2 and a base.

20.7 Batteries: Using Chemistry to Generate Electricity

We have seen that we can combine the electron-losing tendency of one substance with the electron-gaining tendency of another to create electrical current in a voltaic cell. Batteries are voltaic cells conveniently packaged to act as portable sources of electricity. The oxidation and reduction reactions depend on the particular type of battery. In this section, we examine several different types.

Dry-Cell Batteries

Common batteries, such as the kind you find in a flashlight, are called **dry-cell batteries** because they do not contain large amounts of liquid water. There are several familiar types of dry-cell batteries. The most inexpensive are composed of a zinc case that acts as the anode (Figure 20.13(a) ◀). The zinc is oxidized according to the reaction:

Oxidation (Anode): $Zn(s) \longrightarrow Zn^{2+}(aq) + 2\,e^-$

The cathode is a carbon rod immersed in a moist paste of MnO_2 that also contains NH_4Cl. The MnO_2 is reduced to Mn_2O_3 according to the reaction:

Reduction (Cathode): $2\,MnO_2(s) + 2\,NH_4^+(aq) + 2\,e^- \longrightarrow Mn_2O_3(s) + 2\,NH_3(g) + H_2O(l)$

These two half-reactions produce a voltage of about 1.5 V. Two or more of these batteries can be connected in series (cathode-to-anode connection) to produce higher voltages.

The more common **alkaline batteries** (Figure 20.13(b)) employ slightly different half-reactions in a basic medium (therefore the name alkaline). In an alkaline battery, the zinc is oxidized in a basic environment:

Oxidation (Anode): $Zn(s) + \cancel{2\,OH^-(aq)} \longrightarrow Zn(OH)_2(s) + \cancel{2\,e^-}$

Reduction (Cathode): $2\,MnO_2(s) + 2\,H_2O(l) + \cancel{2\,e^-} \longrightarrow 2\,MnO(OH)(s) + \cancel{2\,OH^-(aq)}$

Overall reacion: $Zn(s) + 2\,MnO_2(s) + 2\,H_2O(l) \longrightarrow Zn(OH)_2(s) + 2\,MnO(OH)(s)$

Alkaline batteries have a longer working life and a longer shelf life than their nonalkaline counterparts.

Lead–Acid Storage Batteries

The batteries in most automobiles are **lead–acid storage batteries.** These batteries consist of six electrochemical cells wired in series (Figure 20.14 ▼). Each cell produces 2 V for a total of 12 V. Each cell

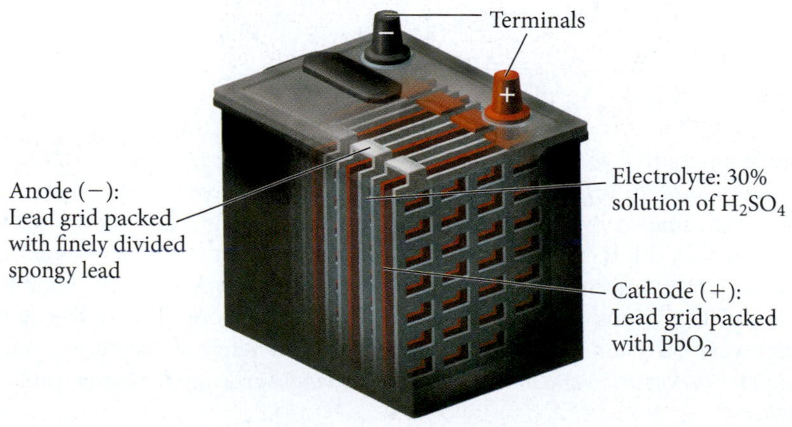

▶ **FIGURE 20.14 Lead–Acid Storage Battery** A lead–acid storage battery consists of six cells wired in series. Each cell contains a porous lead anode and a lead oxide cathode, both immersed in sulfuric acid.

contains a porous lead anode where oxidation occurs and a lead(IV) oxide cathode where reduction occurs according to the reactions:

***Oxidation** (Anode):* $Pb(s) + HSO_4^-(aq) \longrightarrow PbSO_4(s) + H^+(aq) + 2\,e^-$

***Reduction** (Cathode):* $PbO_2(s) + HSO_4^-(aq) + 2\,3H^+(aq) + 2\,e^- \longrightarrow PbSO_4(s) + 2\,H_2O(l)$

Overall reacion: $Pb(s) + PbO_2(s) + 2\,HSO_4^-(aq) + 2\,H^+(aq) \longrightarrow 2\,PbSO_4(s) + 2\,H_2O(l)$

Both the anode and the cathode are immersed in sulfuric acid (H_2SO_4). As electrical current is drawn from the battery, both electrodes become coated with $PbSO_4(s)$. If the battery is run for a long time without recharging, too much $PbSO_4(s)$ builds up on the surface of the electrodes and the battery goes dead. The lead–acid storage battery can be recharged by an electrical current (which must come from an external source such as an alternator in a car). The current causes the preceding reaction to occur in reverse, converting the $PbSO_4(s)$ back to $Pb(s)$ and $PbO_2(s)$.

Other Rechargeable Batteries

The ubiquity of electronic products such as laptops, tablets, cell phones, and digital cameras, as well as the growth in popularity of hybrid electric vehicles, drive the need for efficient, long-lasting, rechargeable batteries. Common types include the **nickel–cadmium (NiCad) battery**, the **nickel–metal hydride (NiMH) battery**, and the **lithium ion battery**.

The Nickel–Cadmium (NiCad) Battery Nickel–cadmium batteries consist of an anode composed of solid cadmium and a cathode composed of $NiO(OH)(s)$. The electrolyte is usually $KOH(aq)$. During operation, the cadmium is oxidized and the NiO(OH) is reduced according to these equations:

***Oxidation** (Anode):* $Cd(s) + 2\,OH^-(aq) \longrightarrow Cd(OH)_2(s) + 2\,e^-$

***Reduction** (Cathode):* $2\,NiO(OH)(s) + 2\,H_2O(l) + 2\,e^- \longrightarrow 2\,Ni(OH)_2(s) + 2\,OH^-(aq)$

The overall reaction produces about 1.30 V. As current is drawn from the NiCad battery, solid cadmium hydroxide accumulates on the anode and solid nickel(II) hydroxide accumulates on the cathode. However, by running current in the opposite direction, the reactants can be regenerated from the products. A common problem in recharging NiCad and other rechargeable batteries is knowing when to stop. Once all of the products of the reaction are converted back to reactants, the charging process should ideally terminate—otherwise the electrical current will drive other, usually unwanted, reactions such as the electrolysis of water to form hydrogen and oxygen gas. These reactions typically damage the battery and may sometimes even cause an explosion. Consequently, most commercial battery chargers have sensors that measure when the charging is complete. These sensors rely on the small changes in voltage or increases in temperature that occur once the products have all been converted back to reactants.

▲ Several types of batteries, including NiCad, NiMH, and lithium ion batteries, are recharged by chargers that use household current.

The Nickel–Metal Hydride (NiMH) Battery Although NiCad batteries were the standard rechargeable battery for many years, they are being replaced by other types of rechargeable batteries, in part because of the toxicity of cadmium and the resulting disposal problems. One of these replacements is the nickel–metal hydride or NiMH battery. The NiMH battery employs the same cathode reaction as the NiCad battery but a different anode reaction. In the anode of a NiMH battery, hydrogen atoms held in a metal alloy are oxidized. If we let M represent the metal alloy, we can write the half-reactions as follows:

***Oxidation** (Anode):* $M \cdot H(s) + OH^-(aq) \longrightarrow M(s) + H_2O(l) + e^-$

***Reduction** (Cathode):* $NiO(OH)(s) + H_2O(l) + e^- \longrightarrow Ni(OH)_2(s) + OH^-(aq)$

In addition to being more environmentally friendly than NiCad batteries, NiMH batteries also have a greater energy density (energy content per unit battery mass), as we can see in Table 20.2. In some cases, a NiMH battery can carry twice the energy of a NiCad battery of the same mass, making NiMH batteries the most common choice for hybrid electric vehicles.

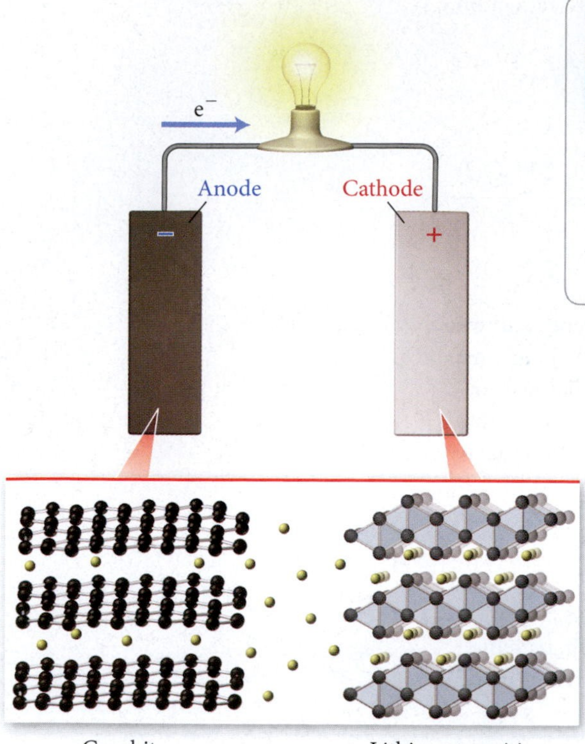

▲ **FIGURE 20.15 Lithium Ion Battery** In the lithium ion battery, the spontaneous flow of lithium ions from the graphite anode to the lithium transition metal oxide cathode causes a corresponding flow of electrons in the external circuit.

TABLE 20.2 Energy Density and Overcharge Tolerance of Several Rechargeable Batteries

Battery Type	Energy Density (W•h/kg)	Overcharge Tolerance
NiCad	45–80	Moderate
NiMH	60–120	Low
Li ion	110–160	Low
Pb storage	30–50	High

The Lithium Ion Battery The newest and most expensive common type of rechargeable battery is the lithium ion battery. Since lithium is the least dense metal (0.53 g/cm^3), lithium batteries have high-energy densities (see Table 20.2). The lithium battery works differently than the other batteries we have examined so far, and the details of its operation are beyond the scope of our current discussion. Briefly, we can describe the operation of the lithium battery as being due primarily to the motion of lithium ions from the anode to the cathode. The anode is composed of graphite into which lithium ions are incorporated between layers of carbon atoms. Upon discharge, the lithium ions spontaneously migrate to the cathode, which consists of a lithium transition metal oxide such as $LiCoO_2$ or $LiMn_2O_4$. The transition metal is reduced during this process. Upon recharging, the transition metal is oxidized, forcing the lithium to migrate back into the graphite (Figure 20.15 ◀). The flow of lithium ions from the anode to the cathode causes a corresponding flow of electrons in the external circuit. Lithium ion batteries are commonly used in applications where light weight and high-energy density are important. These include cell phones, laptop computers, and digital cameras.

Fuel Cells

Fuel cells are like batteries; the key difference is that a battery is self-contained, while in a fuel cell the reactants need to be constantly replenished from an external source. With use, normal batteries lose their ability to generate voltage because the reactants become depleted as electrical current is drawn from the battery. In a **fuel cell**, the reactants—the fuel provided from an external source—constantly flow through the battery, generating electrical current as they undergo a redox reaction. Fuel cells may one day replace—or at least work in combination with—centralized power grid electricity. In addition, vehicles powered by fuel cells may one day usurp vehicles powered by internal combustion engines.

The most common fuel cell is the hydrogen–oxygen fuel cell (Figure 20.16 ▶). In this cell, hydrogen gas flows past the anode (a screen coated with platinum catalyst) and undergoes oxidation:

$$\textit{Oxidation (Anode):}\ 2\,H_2(g) + 4\,OH^-(aq) \longrightarrow 4\,H_2O(l) + 4\,e^-$$

Oxygen gas flows past the cathode (a similar screen) and undergoes reduction:

$$\textit{Reduction (Cathode):}\ O_2(g) + 2\,H_2O(l) + 4\,e^- \longrightarrow 4\,OH^-(aq)$$

The half-reactions sum to the following overall reaction:

$$\textit{Overall reaction:}\ 2\,H_2(g) + O_2(g) \longrightarrow 2\,H_2O(l)$$

Notice that the only product is water. In the space shuttle program, hydrogen–oxygen fuel cells consume hydrogen to provide electricity and astronauts drink the water that is produced by the reaction. In order for hydrogen-powered fuel cells to become more widely used, a more readily available source of hydrogen must be developed.

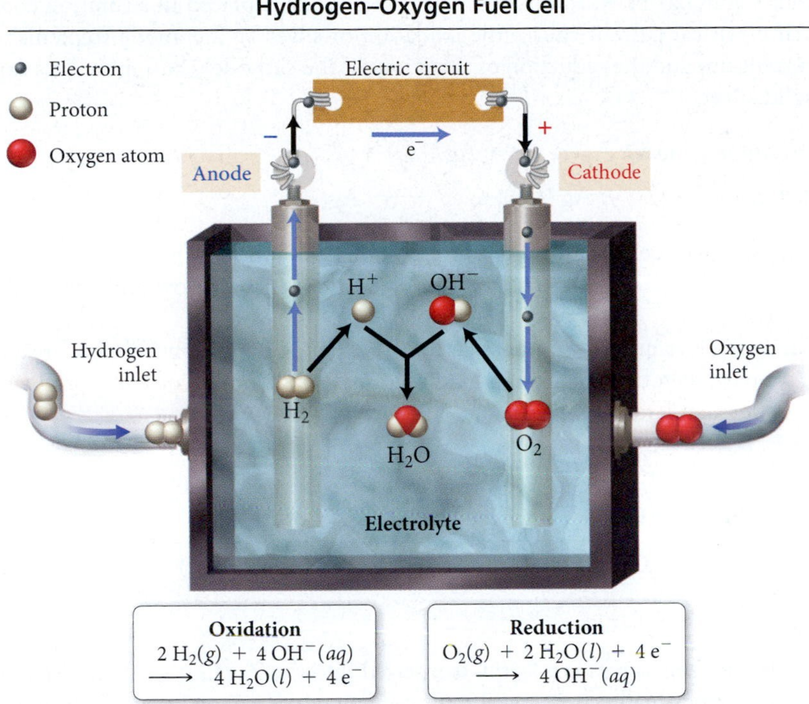

◀ **FIGURE 20.16 Hydrogen–Oxygen Fuel Cell** In this fuel cell, hydrogen and oxygen combine to form water.

20.8 Electrolysis: Driving Nonspontaneous Chemical Reactions with Electricity

In a voltaic cell, a spontaneous redox reaction produces electrical current. In an *electrolytic cell*, electrical current drives an otherwise nonspontaneous redox reaction through a process called **electrolysis**. We have seen that the reaction of hydrogen with oxygen to form water is spontaneous and can be used to produce an electrical current in a fuel cell. Conversely, by *supplying* electrical current, we can cause the reverse reaction to occur, separating water into hydrogen and oxygen (Figure 20.17 ▶).

$$2\,H_2(g) + O_2(g) \longrightarrow 2\,H_2O(l)$$ (spontaneous—produces electrical current; occurs in a voltaic cell)

$$2\,H_2O(l) \longrightarrow 2\,H_2(g) + O_2(g)$$ (nonspontaneous—consumes electrical current; occurs in an electrolytic cell)

Recall from the previous section that one of the problems prohibiting the widespread adoption of hydrogen fuel cells is the scarcity of hydrogen. Where will the hydrogen to power these fuel cells come from? One possible answer is to obtain hydrogen from water through solar-powered electrolysis. A solar-powered electrolytic cell can produce hydrogen from water when the sun is shining. The hydrogen made in this way could be converted back to water to generate electricity and could also be used to power fuel-cell vehicles.

Electrolysis also has numerous other applications. For example, most metals are found in Earth's crust as metal oxides. Converting an oxide to a pure metal requires that the metal be reduced, a nonspontaneous process. Electrolysis can be used to produce these metals. Thus, sodium is produced by the electrolysis of molten sodium chloride (discussed in the following subsection). Electrolysis is also used to plate metals onto other metals. For example, silver can be plated onto a less expensive metal using the electrolytic cell

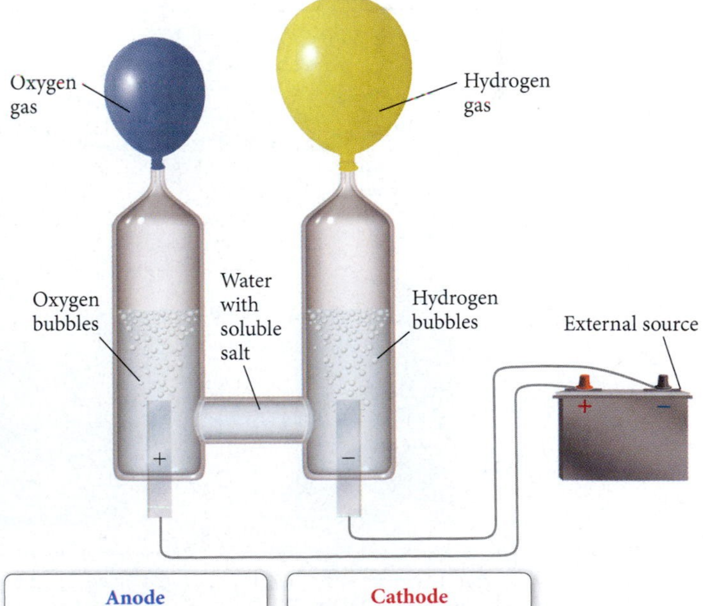

▲ **FIGURE 20.17 Electrolysis of Water** Electrical current can decompose water into hydrogen and oxygen gas.

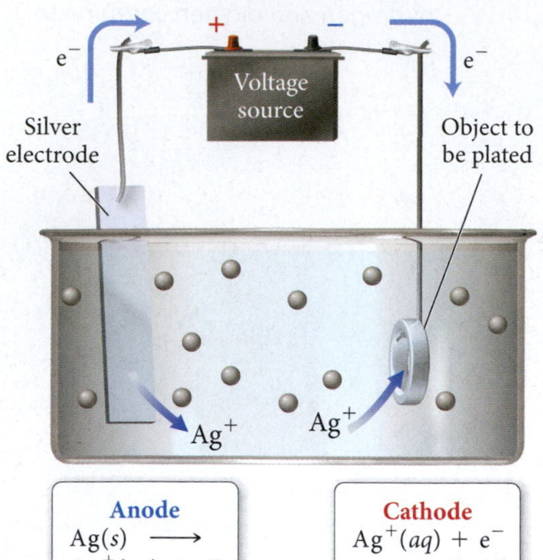

▲ **FIGURE 20.18 Silver Plating** Silver can be plated from a solution of silver ions onto metallic objects in an electrolytic cell.

shown in Figure 20.18 ◀. In this cell, a silver electrode is placed in a solution containing silver ions. An electrical current causes the oxidation of silver at the anode (replenishing the silver ions in solution) and the reduction of silver ions at the cathode (coating the less expensive metal with solid silver).

***Oxidation** (Anode)*: $Ag(s) \longrightarrow Ag^+(aq) + e^-$

***Reduction** (Cathode)*: $Ag^+(aq) + e^- \longrightarrow Ag(s)$

Since the standard cell potential of the reaction is zero, the reaction is not spontaneous under standard conditions. An external power source drives current flow and causes the reaction to occur.

The voltage required to cause electrolysis depends on the specific half-reactions. For example, we have seen that the oxidation of zinc and the reduction of Cu^{2+} produces a voltage of 1.10 V under standard conditions.

$$\text{Reduction (Cathode):}\quad Cu^{2+}(aq) + \cancel{2e^-} \longrightarrow Cu(s) \qquad E^\circ = 0.34\ V$$

$$\text{Oxidation (Anode):}\quad Zn(s) \longrightarrow Zn^{2+}(aq) + \cancel{2e^-} \qquad E^\circ = -0.76\ V$$

$$Cu^{2+}(aq) + Zn(s) \longrightarrow Cu(s) + Zn^{2+}(aq) \qquad E^\circ_{cell} = E^\circ_{cat} - E^\circ_{an} = +1.10\ V$$

If a power source producing *more than 1.10 V* is inserted into the Zn/Cu^{2+} voltaic cell, electrons can be forced to flow in the opposite direction, causing the reduction of Zn^{2+} and the oxidation of Cu, as shown in Figure 20.19 ▼. Notice that in the electrolytic cell, the anode has become the cathode (oxidation always occurs at the anode) and the cathode has become the anode.

In a *voltaic cell*, the anode is the source of electrons and is therefore labeled with a negative charge. The cathode draws electrons and is therefore labeled with a positive charge. In an *electrolytic cell*, however, the source of the electrons is the external power source. The external power source must *draw electrons away* from the anode; thus, the anode must be connected to the positive terminal of the battery (as shown in Figure 20.19). Similarly, the power source drives electrons toward the cathode (where they will be used in reduction), so the cathode must be connected to the

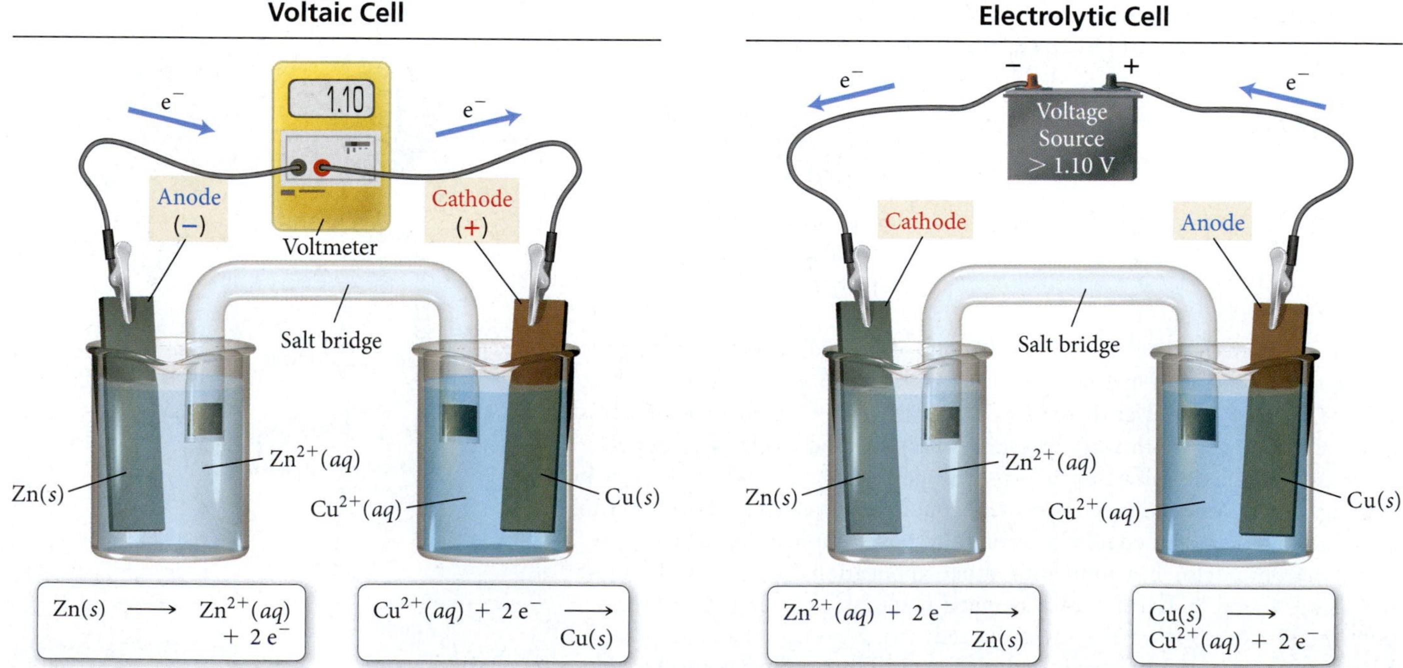

▲ **FIGURE 20.19 Voltaic versus Electrolytic Cells** In a Zn/Cu^{2+} voltaic cell, the reaction proceeds in the spontaneous direction. In a Zn^{2+}/Cu electrolytic cell, electrical current drives the reaction in the nonspontaneous direction.

negative terminal of the battery. The charge labels (+ and −) on an electrolytic cell are the opposite of what they are in a voltaic cell.

Summarizing Characteristics of Electrochemical Cell Types:

In all electrochemical cells:

- Oxidation occurs at the anode.
- Reduction occurs at the cathode.

In voltaic cells:

- The anode is the source of electrons and has a negative charge (anode −).
- The cathode draws electrons and has a positive charge (cathode +).

In electrolytic cells:

- Electrons are drawn away from the anode, which must be connected to the positive terminal of the external power source (anode +).
- Electrons are driven to the cathode, which must be connected to the negative terminal of the power source (cathode −).

Predicting the Products of Electrolysis

Predicting the products of an electrolysis reaction is in some cases relatively straightforward and in other cases more complex. We cover the simpler cases first and then discuss the more complex ones.

Pure Molten Salts Consider the electrolysis of a molten salt such as sodium chloride, shown in Figure 20.20 ▶. Na^+ and Cl^- are the only species present in the cell. The chloride ion cannot be further reduced (−1 is its lowest oxidation state), so it must be oxidized. The sodium ion cannot be further oxidized (+1 is its highest oxidation state), so it must be reduced. Thus, we can write the half-reactions:

Oxidation (Anode):	$2\,Cl^-(l) \longrightarrow Cl_2(g) + \cancel{2\,e^-}$
Reduction (Cathode):	$2\,Na^+(l) + \cancel{2\,e^-} \longrightarrow 2\,Na(s)$
Overall:	$2\,Na^+(l) + 2\,Cl^-(l) \longrightarrow 2\,Na(s) + Cl_2(g)$

Although the reaction as written is not spontaneous, it can be driven to occur in an electrolytic cell by an external power source. We can generalize as follows:

- In the electrolysis of a pure molten salt, the anion is oxidized and the cation is reduced.

Electrolysis of a Molten Salt

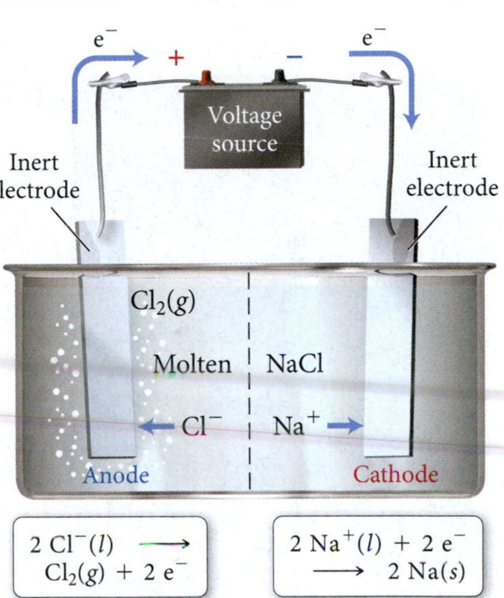

▲ **FIGURE 20.20 Electrolysis of Molten NaCl** In the electrolysis of a pure molten salt, the anion (in this case Cl^-) is oxidized and the cation (in this case Na^+) is reduced.

Mixtures of Cations or Anions What if a molten salt contains more than one anion or cation? For example, suppose the electrolysis cell we just introduced contains both NaCl and KCl. Which of the two cations is reduced at the cathode? In order to answer this question, we must determine which of the two cations is more easily reduced. Although the values of electrode potentials for aqueous solutions given in Table 20.1 do not apply to molten salts, the relative ordering of the electrode potentials does reflect the relative ease with which the metal cations are reduced. We can see from the table that the reduction of Na^+ is *above* the reduction of K^+; that is, Na^+ has a more positive electrode potential.

Throughout this discussion "more positive" means the same thing as "less negative."

$$Na^+(aq) + e^- \longrightarrow Na(s) \qquad E^\circ = -2.71\text{ V} \qquad \text{(for aqueous solution)}$$

$$K^+(aq) + e^- \longrightarrow K(s) \qquad E^\circ = -2.92\text{ V} \qquad \text{(for aqueous solution)}$$

Therefore, Na^+ is easier to reduce than K^+. Consequently, in a mixture of NaCl and KCl, Na^+ has a greater tendency to be reduced at the cathode.

Similarly, what if a mixture of molten salts contained more than one anion? For example, in a mixture of NaBr and NaCl, which of the two anions is oxidized at the cathode? The answer is similar; the anion that is more easily oxidized (the one with the more negative electrode potential).

Throughout this discussion "more negative" means the same thing as "less positive."

$$2\,Cl^-(aq) \longrightarrow Cl_2(g) + 2\,e^- \qquad E^\circ = 1.36\text{ V} \qquad \text{(for aqueous solution)}$$

$$2\,Br^-(aq) \longrightarrow Br_2(l) + 2\,e^- \qquad E^\circ = 1.09\text{ V} \qquad \text{(for aqueous solution)}$$

Since the electrode potential for the bromine half-reaction is more negative, electrons are more easily extracted from it. The bromide ion is therefore oxidized at the anode.

We can generalize as follows:

- The cation that is most easily reduced (the one with the more positive electrode potential) is reduced first.
- The anion that is most easily oxidized (the one with the more negative electrode potential) is oxidized first.

Remember NIO and PIR:
N.I.O.–More Negative Is Oxidation
P.I.R.–More Positive Is Reduction

Aqueous Solutions Electrolysis in an aqueous solution is complicated by the possibility of the electrolysis of water itself. Recall that water can be either oxidized or reduced according to the following half-reactions:

Oxidation (Anode): $2\,H_2O(l) \longrightarrow O_2(g) + 4\,H^+(aq) + 4\,e^-$ $\quad E° = 1.23\text{ V (standard conditions)}$

$E = 0.82\text{ V}([H^+] = 10^{-7}\text{ M})$

Reduction (Cathode): $2\,H_2O(l) + 2\,e^- \longrightarrow H_2(g) + 2\,OH^-(aq)$ $\quad E° = -0.83\text{ V (standard conditions)}$

$E = -0.41\text{ V}([OH^-] = 10^{-7}\text{ M})$

The electrode potentials under standard conditions are shown to the right of each half-reaction. However, in pure water at room temperature, the concentrations of H^+ and OH^- are not standard. The electrode potentials for $[H^+] = 10^{-7}$ M and $[OH^-] = 10^{-7}$ M are shown in blue. Using those electrode potentials, we can calculate E_{cell} for the electrolysis of water as follows:

$$E_{cell} = E_{cat} - E_{an} = -0.41\text{ V} - 0.82\text{ V} = -1.23\text{ V}$$

▲ Pure water is a poor conductor of electrical current, but the addition of an electrolyte allows electrolysis to take place, producing hydrogen and oxygen gas in a stoichiometric ratio.

When a battery with a potential of several volts is connected to an electrolysis cell containing pure water, no reaction occurs, because the concentration of ions in pure water is too low to conduct any significant electrical current. When an electrolyte such as Na_2SO_4 is added to the water, however, electrolysis occurs readily.

In any aqueous solution in which electrolysis is to take place, the electrolysis of water is also possible. For example, consider the electrolysis of a sodium iodide solution, shown in Figure 20.21 ▼. For the electrolysis of *molten* NaI, we can readily predict that I^- is oxidized at the anode and that Na^+ is reduced at the cathode. In an aqueous solution, however, two different oxidation half-reactions are possible at the anode, the oxidation of I^- and the oxidation of water.

Oxidation: $2\,I^-(aq) \longrightarrow I_2(s) + 2\,e^-$ $\quad E° = 0.54\text{ V}$

Oxidation: $2\,H_2O(l) \longrightarrow O_2(g) + 4\,H^+(aq) + 4\,e^-$ $\quad E = 0.82\text{ V } ([H^+] = 10^{-7}\text{ M})$

Electrolysis of an Aqueous Salt Solution

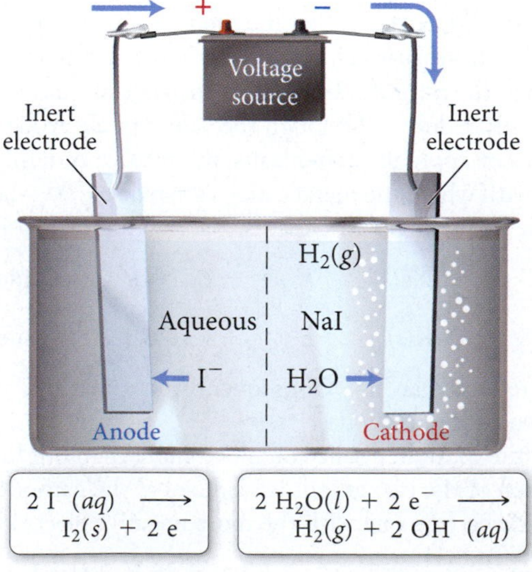

▶ **FIGURE 20.21 Electrolysis of Aqueous NaI** In this cell, I^- is oxidized to I_2 at the anode and H_2O is reduced to H_2 at the cathode. Sodium ions are not reduced, because their electrode potential is more negative than the electrode potential of water.

Similarly, two different reduction half-reactions are possible at the cathode, the reduction of Na^+ and the reduction of water.

$$\textit{Reduction: } 2\,Na^+(aq) + 2\,e^- \longrightarrow 2\,Na(s) \qquad E^\circ = -2.71\ V$$

$$\textit{Reduction: } 2\,H_2O(l) + 2\,e^- \longrightarrow H_2(g) + 2\,OH^-(aq) \qquad E = -0.41\ V\ ([OH^-] = 10^{-7}\ M)$$

How do we know which reactions actually occur? In both cases, the answer is the same: *the half-reaction that occurs more easily*. For oxidation, the half-reaction with the more negative electrode potential is the easier one from which to extract electrons. In this case, therefore, the iodide ion is oxidized at the anode. For reduction, the half-reaction with the more positive electrode potential is the easier one to get to accept electrons. In this case, therefore, water is reduced at the cathode. Notice that Na^+ cannot be reduced in an aqueous solution—water is reduced before Na^+. We can make the following generalization:

- The cations of active metals—those that are not easily reduced, such as Li^+, K^+, Na^+, Mg^{2+}, Ca^{2+}, and Al^{3+}—cannot be reduced from aqueous solutions by electrolysis because water is reduced at a lower voltage.

The Electrolysis of Aqueous Sodium Chloride and Overvoltage An additional complication that we must consider when predicting the products of electrolysis is *overvoltage*—an additional voltage that must be applied in order to get some nonspontaneous reactions to occur. We can demonstrate this concept by considering the electrolysis of a sodium chloride solution, shown in Figure 20.22 ▼. In order to predict the product of the electrolysis, we consider the two possible oxidation half-reactions:

$$\textit{Oxidation: } 2\,Cl^-(aq) \longrightarrow Cl_2(g) + 2\,e^- \qquad E^\circ = 1.36\ V$$

$$\textit{Oxidation: } 2\,H_2O(l) \longrightarrow O_2(g) + 4\,H^+(aq) + 4\,e^- \qquad E = 0.82\ V\ ([H^+] = 10^{-7}\ M)$$

and the two possible reduction half-reactions:

$$\textit{Reduction: } 2\,Na^+(aq) + 2\,e^- \longrightarrow 2\,Na(s) \qquad E^\circ = -2.71\ V$$

$$\textit{Reduction: } 2\,H_2O(l) + 2\,e^- \longrightarrow H_2(g) + 2\,OH^-(aq) \qquad E = -0.41\ V\ ([OH^-] = 10^{-7}\ M)$$

Since the oxidation of water has a more negative electrode potential than the oxidation of Cl^-, we would initially predict that it would be easier to remove electrons from water, and thus water should be oxidized at the anode. Similarly, since the reduction of water has a more positive electrode potential than the reduction of Na^+, we would expect that it would be easier to get water to accept electrons, so water should be reduced at the cathode. In other words, we initially predict that a sodium chloride solution would result in the electrolysis of water, producing oxygen gas at the anode and hydrogen gas at the cathode. If we construct such a cell, however, we find that, although hydrogen gas is indeed formed at the cathode (as predicted), oxygen gas is *not* formed at the anode—chlorine gas is formed instead.

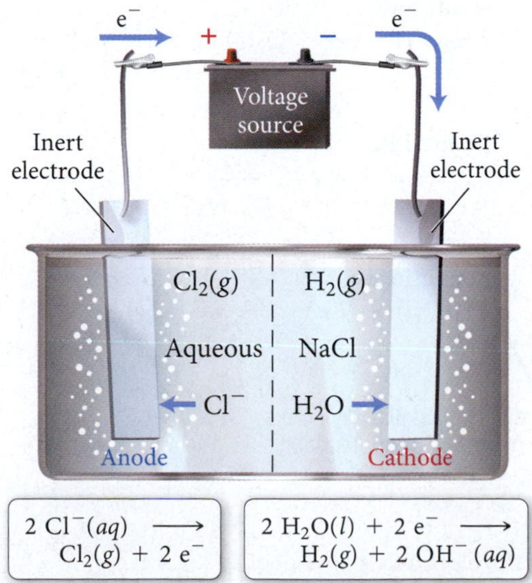

◀ **FIGURE 20.22 Electrolysis of Aqueous NaCl: The Effect of Overvoltage** Because of overvoltage, the anode reaction of this cell is the oxidation of Cl^- to Cl_2 gas rather than the oxidation of water to H^+ and O_2 gas.

Why? The answer is that even though the electrode potential for the oxidation of water is 0.82 V, the reaction actually requires a voltage greater than 0.82 V in order to occur. (The reasons for this behavior are related to kinetic factors that are beyond the scope of our current discussion.) This additional voltage, the *overvoltage*, increases the voltage required for the oxidation of water to about 1.4 V. The result is that the chloride ion oxidizes more easily than water and $Cl_2(g)$ forms at the anode.

EXAMPLE 20.9

Predicting the Products of Electrolysis Reactions

Predict the half-reaction occurring at the anode and the cathode for electrolysis for each reaction.

(a) a mixture of molten $AlBr_3$ and $MgBr_2$

(b) an aqueous solution of LiI

SOLUTION

(a) In the electrolysis of a molten salt, the anion is oxidized and the cation is reduced. However, this mixture contains two cations. Start by writing the possible oxidation and reduction half-reactions that might occur.

Since Br^- is the only anion, write the equation for its oxidation, which occurs at the anode.

At the cathode, both the reduction of Al^{3+} and the reduction of Mg^{2+} are possible. The one that actually occurs is the one that occurs more easily. Since the reduction of Al^{3+} has a more positive electrode potential in aqueous solution, this ion is more easily reduced. Therefore, the reduction of Al^{3+} occurs at the cathode.

Oxidation: $2\,Br^-(l) \longrightarrow Br_2(g) + 2\,e^-$

Reduction: $Al^{3+}(l) + 3\,e^- \longrightarrow Al(s)$ $E° = -1.66$ V (for aqueous solution)

$Mg^{2+}(l) + 2\,e^- \longrightarrow Mg(s)$ $E° = -2.37$ V (for aqueous solution)

Reduction that actually occurs (more positive potential)

(b) Since LiI is in an aqueous solution, two different oxidation half-reactions are possible at the anode, the oxidation of I^- and the oxidation of water. Write half-reactions for each including the electrode potential. Remember to use the electrode potential of water under conditions in which $[H^+] = 10^{-7}$ M. Since the oxidation of I^- has the more negative electrode potential, it will be the half-reaction to occur at the anode.

Similarly, write half-reactions for the two possible reduction half-reactions that might occur at the cathode, the reduction of Li^+ and the reduction of water. Since the reduction of water has the more positive electrode potential (even when considering overvoltage, which would raise the necessary voltage by about 0.4 – 0.6 V), it is the half-reaction that occurs at the cathode.

Oxidation that actually occurs (more negative potential)

Oxidation:
$2\,I^-(aq) \longrightarrow I_2(s) + 2\,e^-$
$E° = 0.54$ V

Oxidation:
$2\,H_2O(l) \longrightarrow O_2(g) + 4\,H^+(aq) + 4\,e^-$
$E° = 0.82$ V ($[H^+] = 10^{-7}$ M)

Reduction:
$2\,Li^+(aq) + 2\,e^- \longrightarrow 2\,Li(s)$
$E° = -3.04$ V

Reduction:
$2\,H_2O(l) + 2\,e^- \longrightarrow H_2(g) + 2\,OH^-(aq)$
$E = -0.41$ V ($[OH^-] = 10^{-7}$ M)

Reduction that actually occurs (more positive potential)

FOR PRACTICE 20.9

Predict the half-reactions occurring at the anode and the cathode for the electrolysis of aqueous Na_2SO_4.

Stoichiometry of Electrolysis

In an electrolytic cell, electrical current drives a particular chemical reaction. In a sense, the electrons act as a reactant and therefore have a stoichiometric relationship with the other reactants and products. Unlike ordinary reactants, for which we usually measure quantity as mass, for electrons we measure quantity as charge. For example, consider an electrolytic cell used to coat copper onto metals, as shown in Figure 20.23 ▼. The half-reaction by which copper is deposited onto the metal is:

$$Cu^{2+}(aq) + 2\,e^- \longrightarrow Cu(s)$$

For every 2 mol of electrons that flow through the cell, 1 mol of solid copper is plated. We can write the stoichiometric relationship:

$$2 \text{ mol } e^- : 1 \text{ mol Cu}(s)$$

We can determine the number of moles of electrons that flow in a given electrolytic cell by measuring the total charge that flows through the cell, which in turn depends on the *magnitude* of the current and on the *time* that the current runs. Recall from Section 20.3 that the unit of current is the ampere.

$$1\text{ A} = 1\frac{\text{C}}{\text{s}}$$

If we multiply the amount of current (in A) flowing through the cell by the time (in s) that the current flowed, we find the total charge that passed through the cell in that time:

$$\text{Current}\left(\frac{\text{C}}{\cancel{\text{s}}}\right) \times \text{time}(\cancel{\text{s}}) = \text{charge (C)}$$

The relationship between charge and the number of moles of electrons is given by Faraday's constant, which, as we saw previously, corresponds to the charge in coulombs of 1 mol of electrons.

$$F = \frac{96{,}485 \text{ C}}{\text{mol } e^-}$$

We can use these relationships to solve problems involving the stoichiometry of electrolytic cells, as demonstrated in Example 20.10.

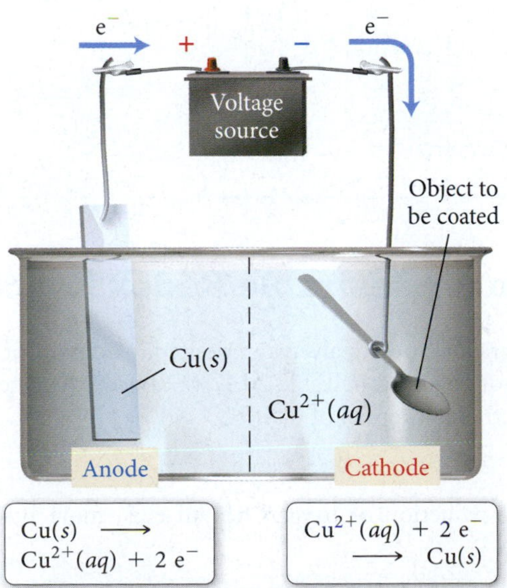

◀ **FIGURE 20.23 Electrolytic Cell for Copper Plating** In this cell, copper ions are plated onto other metals. It takes two moles of electrons to plate one mole of copper atoms.

EXAMPLE 20.10
Stoichiometry of Electrolysis

Gold can be plated out of a solution containing Au^{3+} according to the half-reaction:

$$Au^{3+}(aq) + 3\,e^- \longrightarrow Au(s)$$

What mass of gold (in grams) is plated by a 25-minute flow of 5.5 A current?

SORT You are given the half-reaction for the plating of gold, which shows the stoichiometric relationship between moles of electrons and moles of gold. You are also given the current and duration. You must find the mass of gold that is deposited in that time.

GIVEN: 3 mol e^- : 1 mol Au
5.5 amps
25 min

FIND: g Au

STRATEGIZE You need to find the amount of gold, which is related stoichiometrically to the number of electrons that have flowed through the cell. Begin with time in minutes and convert to seconds. Then, since current is a measure of charge per unit time, use the given current and the time to find the number of coulombs. You can use Faraday's constant to calculate the number of moles of electrons and the stoichiometry of the reaction to find the number of moles of gold. Finally, use the molar mass of gold to convert to mass of gold.

CONCEPTUAL PLAN

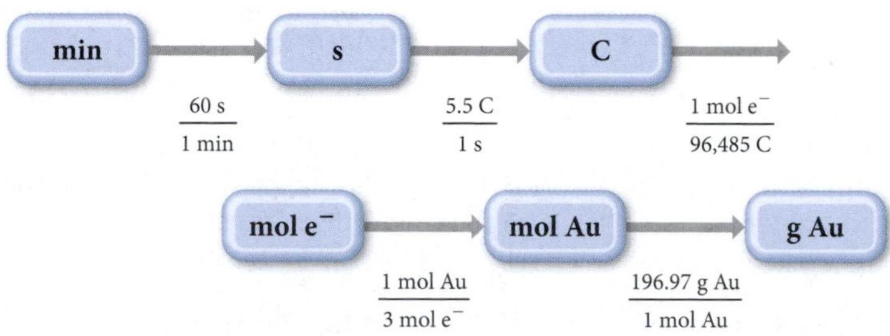

SOLVE Follow the conceptual plan to solve the problem, canceling units to arrive at the mass of gold.

SOLUTION

$$25\ \text{min} \times \frac{60\ \text{s}}{1\ \text{min}} \times \frac{5.5\ \text{C}}{1\ \text{s}} \times \frac{1\ \text{mol e}^-}{96{,}485\ \text{C}} \times \frac{1\ \text{mol Au}}{3\ \text{mol e}^-} \times \frac{196.97\ \text{g Au}}{1\ \text{mol Au}} = 5.6\ \text{g Au}$$

CHECK The answer has the correct units (g Au). The magnitude of the answer is reasonable if you consider that 10 amps of current for 1 hour is the equivalent of about 1/3 mol of electrons (check for yourself), which would produce 1/9 mol (or about 20 g) of gold.

FOR PRACTICE 20.10
Silver can be plated out of a solution containing Ag^+ according to the half-reaction:

$$Ag^+(aq) + e^- \longrightarrow Ag(s)$$

How much time (in minutes) does it take to plate 12 g of silver using a current of 3.0 A?

20.9 Corrosion: Undesirable Redox Reactions

▲ A metal must usually be reduced to extract it from its ore. In corrosion, the metal is oxidized back to its more natural state.

Corrosion is the (usually) gradual, nearly always undesired, oxidation of metals that are exposed to oxidizing agents in the environment. From Table 20.1, we can see that the reduction of oxygen in the presence of water has an electrode potential of 0.40 V.

$$O_2(g) + 2\,H_2O(l) + 4\,e^- \longrightarrow 4\,OH^-(aq) \quad E^\circ = 0.40\ \text{V}$$

In the presence of acid, the reduction of oxygen has an even more positive electrode potential of 1.23 V.

$$O_2(g) + 4\,H^+(aq) + 4\,e^- \longrightarrow 2\,H_2O(l) \quad E^\circ = 1.23\ \text{V}$$

The reduction of oxygen, therefore, has a strong tendency to occur and can bring about the oxidation of other substances, especially metals. Notice that the half-reactions for the reduction of most metal ions are listed *below* the half-reactions for the reduction of oxygen in Table 20.1. Consequently, the oxidation (or corrosion) of those metals is spontaneous when paired with the reduction of oxygen. Corrosion is the opposite of the process by which metals are extracted from their ores. In extraction, the free metal is reduced out from its ore. In corrosion, the metal is oxidized.

Given the ease with which metals oxidize in the presence of oxygen, acid, and water, why are metals used so frequently as building materials in the first place? Many metals form oxides that coat the surface of the metal and prevent further corrosion. For example, bare aluminum metal, with an electrode potential of -1.66 V, is quickly oxidized in the presence of oxygen. However, the oxide that forms at the surface of aluminum is Al_2O_3. In its crystalline form, Al_2O_3 is *sapphire*, a highly inert and structurally solid substance. The Al_2O_3 coating acts to protect the underlying aluminum metal, preventing further corrosion.

▲ Aluminum is stable because its oxide forms a protective film over the underlying metal, preventing further oxidation.

The oxides of iron, however, are not structurally stable, and they tend to flake away from the underlying metal, exposing it to further corrosion. A significant part of the iron produced each year is used to replace rusted iron. Rusting is a redox reaction in which iron is oxidized according to the following half-reaction:

$$Fe(s) \longrightarrow Fe^{2+}(aq) + 2\,e^- \qquad E^\circ = -0.45\text{ V}$$

This oxidation reaction tends to occur at defects on the surface of the iron—known as *anodic regions* because oxidation is occurring at these locations—as shown in Figure 20.24 ▼. The electrons produced at the anodic region travel through the metal to areas called *cathodic regions* where they react with oxygen and H^+ ions dissolved in moisture. (The H^+ ions come from carbonic acid, which naturally forms in water from carbon dioxide in air.)

$$O_2(g) + 4\,H^+(aq) + 4\,e^- \longrightarrow 2\,H_2O(l) \qquad E^\circ = 1.23\text{ V}$$

The overall reaction has a cell potential of 1.68 Vand is highly spontaneous.

$$2\,Fe(s) + O_2(g) + 4\,H^+(aq) \longrightarrow 2\,H_2O(l) + 2\,Fe^{2+}(aq) \qquad E^\circ_{\text{cell}} = 1.68\text{ V}$$

The Fe^{2+} ions formed in the anodic regions can migrate through moisture on the surface of the iron to cathodic regions, where they are further oxidized by reaction with more oxygen.

$$4\,Fe^{2+}(aq) + O_2(g) + (4 + 2n)\,H_2O(l) \longrightarrow \underset{\text{rust}}{2\,Fe_2O_3 \cdot nH_2O(s)} + 8\,H^+(aq)$$

Rust is a hydrated form of iron(III) oxide whose exact composition depends on the conditions under which it forms. Consider each of the following important components in the formation of rust:

- *Moisture must be present for rusting to occur.* The presence of water is necessary because water is a reactant in the last reaction and because charge (either electrons or ions) must be free to flow between the anodic and cathodic regions.
- *Additional electrolytes promote rusting.* The presence of an electrolyte (such as sodium chloride) on the surface of iron promotes rusting because it enhances current flow. This is why cars rust so quickly in cold climates where roads are salted and in coastal areas where salt water mist is present.
- *The presence of acids promotes rusting.* Since H^+ ions are involved in the reduction of oxygen, lower pH enhances the cathodic reaction and leads to faster rusting.

▼ **FIGURE 20.24 Corrosion of Iron: Rusting** The oxidation of iron occurs at anodic regions on the metal surface. The iron ions migrate to cathodic regions, where they react with oxygen and water to form rust.

The Rusting of Iron

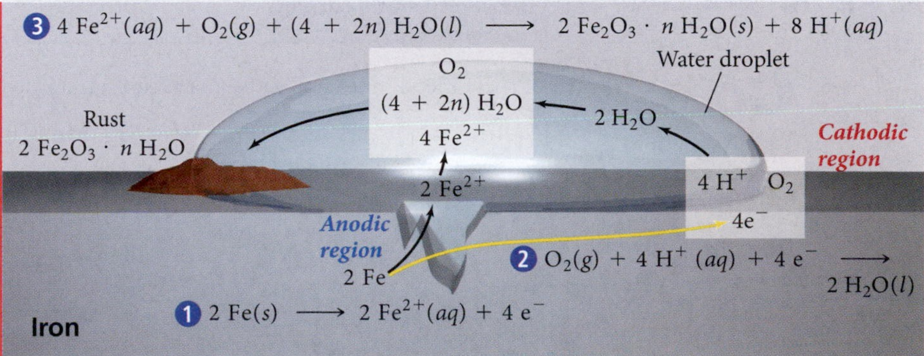

Preventing the rusting of iron is a major industry. The most obvious way to prevent rust is to keep iron dry. Without water, the redox reaction cannot occur. Another way to prevent rust is to coat iron with a substance that is impervious to water. Cars, for example, are painted and sealed to prevent rust. A scratch in the paint can lead to rusting of the underlying iron.

Rust can also be prevented by placing a *sacrificial electrode* in electrical contact with the iron. The sacrificial electrode must be composed of a metal that oxidizes more easily than iron (that is, it must be below iron in Table 20.1). The sacrificial electrode oxidizes in place of the iron (just as the more easily oxidizable species in a mixture is the one to oxidize), protecting the iron from oxidation. A related way to protect iron from rusting is to coat it with a metal that oxidizes more easily than iron. Galvanized nails, for example, are coated with a thin layer of zinc. Since zinc has a more negative electrode potential than iron, it will oxidize in place of the underlying iron (just as a sacrificial electrode does). The oxide of zinc remains on the nail as a protective coating.

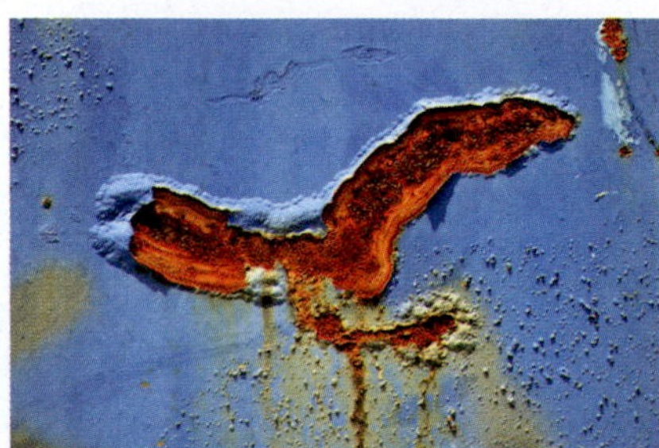

▲ A scratch in paint often allows the underlying iron to rust.

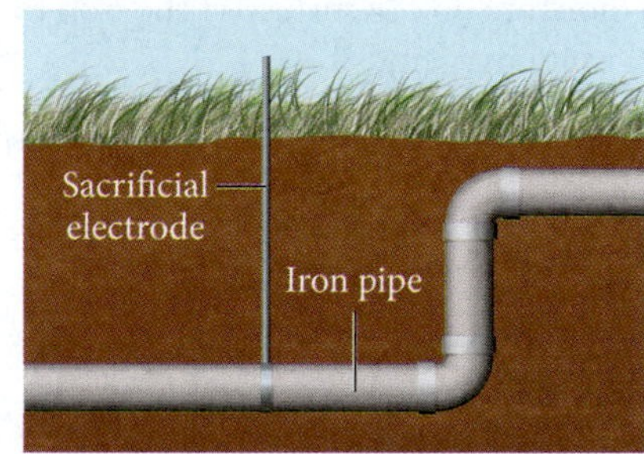

▲ If a metal more active than iron, such as magnesium or aluminum, is in electrical contact with iron, the metal rather than the iron will be oxidized. This principle underlies the use of sacrificial electrodes to prevent the corrosion of iron.

▲ In galvanized nails, a layer of zinc prevents the underlying iron from rusting. The zinc oxidizes in place of the iron, forming a protective layer of zinc oxide.

20.8 Cc Conceptual Connection — Sacrificial Electrodes

Which of these metals does not act as a sacrificial electrode for iron?

Zn, Mg, Mn, Cu

SELF-ASSESSMENT Quiz

1. Balance the redox reaction equation (occurring in acidic solution) and choose the correct coefficients for each reactant and product.

$$\_VO_2^+(aq) + \_Sn(s) + \_H^+(aq) \longrightarrow \_VO^{2+}(aq) + \_Sn^{2+}(aq) + \_H_2O(l)$$

   a) 2,1,4 ⟶ 2,1,2 **b)** 1,1,2 ⟶ 1,1,1
   c) 2,1,2 ⟶ 2,1,1 **d)** 2,1,2 ⟶ 2,1,2

2. Which statement is true for voltaic cells?
   a) Electrons flow from the anode to the cathode.
   b) Electrons flow from the more negatively charged electrode to the more positively charged electrode.
   c) Electrons flow from higher potential energy to lower potential energy.
   d) All of the above

3. Refer to Table 20.1 to calculate E°_{cell} for the reaction.

$$2\ ClO_2(g) + Pb(s) \longrightarrow 2\ ClO_2^-(aq) + Pb^{2+}(aq)$$

   a) 1.77 V **b)** 2.03 V **c)** 0.82 V **d)** 1.08 V

4. Refer to Table 20.1 to determine which statement is true of the voltaic cell pictured here.

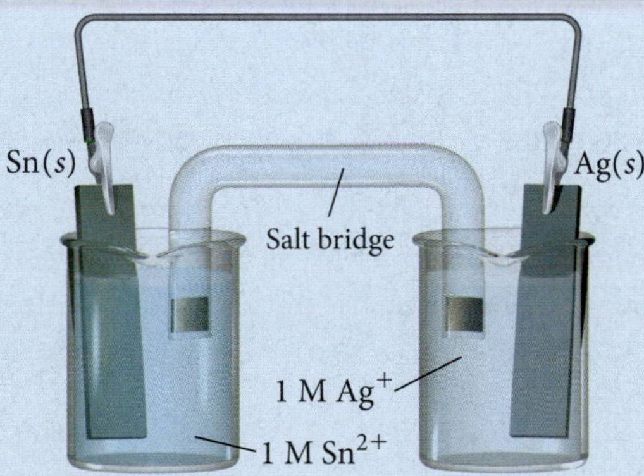

   a) Sn is the anode; Ag is the cathode; electrons flow from left to right
   b) Sn is the cathode; Ag is the anode; electrons flow from left to right
   c) Sn is the anode; Ag is the cathode; electrons flow from right to left
   d) Sn is the cathode; Ag is the anode; electrons flow from right to left

5. Refer to Table 20.1 to determine which metal *does not* dissolve in hydrochloric acid (HCl).
   a) Zn **b)** Cd **c)** Cu **d)** Fe

6. The Zn/Zn^{2+} electrode has a standard electrode potential of $E^\circ = -0.76$ V. How does the relative potential energy of an electron at the Zn/Zn^{2+} electrode compare to the potential energy of an electron at the standard hydrogen electrode?
   a) An electron at the Zn/Zn^{2+} electrode has a higher potential energy than an electron at the standard hydrogen electrode.
   b) An electron at the Zn/Zn^{2+} electrode has a lower potential energy than an electron at the standard hydrogen electrode.
   c) An electron at the Zn/Zn^{2+} electrode has the same potential energy as an electron at the standard hydrogen electrode.
   d) Nothing can be concluded about the relative potential energy of an electron at the standard electrode potential.

7. Refer to Table 20.1 to calculate ΔG° for the reaction.
   $2\ MnO_4^-(aq) + Cd(s) \longrightarrow 2\ MnO_4^{2-}(aq) + Cd^{2+}(aq)$
   a) +30.9 kJ **b)** −30.9 kJ
   c) −185 kJ **d)** +185 kJ

8. A redox reaction has an $E^\circ_{cell} = -0.56$ V. What can you conclude about the equilibrium constant (K) for the reaction?
   a) $K < 1$
   b) $K > 1$
   c) $K = 0$
   d) Nothing can be concluded about K from E°_{cell}.

9. Find E_{cell} for an electrochemical cell based on the following reaction. $[MnO_4^-] = 2.0$ M, $[H^+] = 1.0$ M, and $[Ag^+] = 0.010$ M. E°_{cell} for the reaction is +0.88 V.

$$MnO_4^-(aq) + 4\ H^+(aq) + 3\ Ag(s) \longrightarrow MnO_2(s) + 2\ H_2O(l) + 3\ Ag^+(aq)$$

   a) 0.83 V **b)** 1.00 V **c)** 0.76 V **d)** 0.93 V

10. In an electrochemical cell, $Q = 0.010$ and $K = 855$. What can you conclude about E_{cell} and E°_{cell}?
   a) E_{cell} is positive and E°_{cell} is negative.
   b) E_{cell} is negative and E°_{cell} is positive.
   c) E_{cell} and E°_{cell} are both negative.
   d) E_{cell} and E°_{cell} are both positive.

11. Which reaction occurs at the *anode* of a lead storage battery?
   a) $Zn(s) + 2\ OH^-(aq) \longrightarrow Zn(OH)_2(s) + 2\ e^-$
   b) $PbO_2(s) + HSO_4^-(aq) + 3\ H^+(aq) + 2\ e^- \longrightarrow PbSO_4(s) + 2\ H_2O(l)$
   c) $Pb(s) + HSO_4^-(aq) \longrightarrow PbSO_4(s) + H^+(aq) + 2\ e^-$
   d) $O_2(g) + 2\ H_2O(l) + 4\ e^- \longrightarrow 4\ OH^-(aq)$

—*Continued*

Continued—

12. Which reaction could be used to generate electricity in a voltaic electrochemical cell?
    a) $Pb^{2+}(aq) + Mg(s) \longrightarrow Mg^{2+}(aq) + Pb(s)$
    b) $Zn^{2+}(aq) + Sn(s) \longrightarrow Sn^{2+}(aq) + Zn(s)$
    c) $NaCl(aq) + AgNO_3(aq) \longrightarrow AgCl(s) + NaNO_3(aq)$
    d) None of the above

13. Which reaction occurs at the cathode of an electrolytic cell containing a mixture of molten KCl and $ZnCl_2$?
    a) $K(s) \longrightarrow K^+(l) + e^-$
    b) $K^+(l) + e^- \longrightarrow K(s)$
    c) $Zn^{2+}(l) + 2\,e^- \longrightarrow Zn(s)$
    d) $2\,Cl^-(l) \longrightarrow Cl_2(g) + 2\,e^-$

14. Copper is plated onto the cathode of an electrolytic cell containing $CuCl_2(aq)$. How long does it take to plate 111 mg of copper with a current of 3.8 A?
    a) 1.3×10^3 s
    b) 44 s
    c) 89 s
    d) 22 s

15. Which metal can be used as a sacrificial electrode to prevent the rusting of an iron pipe?
    a) Au **b)** Ag
    c) Cu **d)** Mn

Answers: 1:a; 2:d; 3:d; 4:a; 5:c; 6:a; 7:c; 8:a; 9:b; 10:d; 11:c; 12:a; 13:c; 14:c; 15:d

CHAPTER SUMMARY 20

REVIEW

KEY LEARNING OUTCOMES

CHAPTER OBJECTIVES	ASSESSMENT
Half-Reaction Method of Balancing Aqueous Redox Equations in Acidic Solution (20.2)	• Examples 20.1, 20.2 For Practice 20.1, 20.2 Exercises 33–36
Balancing Redox Reactions Occurring in Basic Solution (20.2)	• Example 20.3 For Practice 20.3 Exercises 37, 38
Calculating Standard Potentials for Electrochemical Cells from Standard Electrode Potentials of the Half-Reactions (20.4)	• Example 20.4 For Practice 20.4 Exercises 41, 42, 57, 58
Predicting Spontaneous Redox Reactions and Sketching Electrochemical Cells (20.4)	• Example 20.5 For Practice 20.5 Exercises 39, 40, 43, 44, 47–50
Relating ΔG° and E°_{cell} (20.5)	• Example 20.6 For Practice 20.6 Exercises 61, 62
Relating E°_{cell} and K (20.5)	• Example 20.7 For Practice 20.7 Exercises 63–68
Calculating E_{cell} under Nonstandard Conditions (20.6)	• Example 20.8 For Practice 20.8 Exercises 69–74
Predicting the Products of Electrolysis Reactions (20.8)	• Example 20.9 For Practice 20.9 Exercises 87–92
Stoichiometry of Electrolysis (20.8)	• Example 20.10 For Practice 20.10 Exercises 95–98

KEY TERMS

Section 20.3
electrical current (817)
electrochemical cell (818)
voltaic (galvanic) cell (818)
electrolytic cell (818)
half-cell (818)
electrode (818)
ampere (A) (819)
potential difference (819)
volt (V) (819)
electromotive force (emf) (819)
cell potential (cell emf) (E_{cell}) (819)
standard cell potential (standard emf) (E°_{cell}) (820)
anode (820)
cathode (820)
salt bridge (820)

Section 20.4
standard electrode potential (822)
standard hydrogen electrode (SHE) (822)

Section 20.5
Faraday's constant (F) (830)

Section 20.6
Nernst equation (834)

Section 20.7
dry-cell battery (838)
alkaline battery (838)
lead–acid storage battery (838)
nickel–cadmium (NiCad) battery (839)
nickel–metal hydride (NiMH) battery (839)
lithium ion battery (839)
fuel cell (840)

Section 20.8
electrolysis (841)

Section 20.9
corrosion (848)

KEY CONCEPTS

Lightning and Batteries (20.1)

- Lightning is the massive flow of electrons from the base of a thundercloud (which is negatively charged) to the ground (which is positively charged).
- In a battery, electrons flow from the negatively charged end to the positively charged end.

Balancing Oxidation–Reduction Equations (20.2)

- Oxidation is the loss of electrons and corresponds to an increase in oxidation state; reduction is the gain of electrons and corresponds to a decrease in oxidation state.
- We can balance redox reactions using the half-reaction method, in which the oxidation and reduction reactions are balanced separately and then added. This method differs slightly for redox reactions in acidic and in basic solutions.

Voltaic (or Galvanic) Cells: Generating Electricity from Spontaneous Chemical Reactions (20.3)

- A voltaic electrochemical cell separates the reactants of a spontaneous redox reaction into two half-cells that are connected by a wire and a means to exchange ions so that electricity is generated.
- In an electrochemical cell, the electrode where oxidation occurs is the anode and the electrode where reduction occurs is the cathode; electrons flow from the anode to the cathode.
- The rate of electrons flowing through a wire is measured in amperes (A), and the cell potential is measured in volts (V).
- A salt bridge allows ions to flow between the half-cell solutions and prevents the buildup of charge.
- Cell diagram or line notation is a technique for symbolizing electrochemical cells concisely by separating the components of the reaction using lines or commas.

Standard Electrode Potentials (20.4)

- The electrode potentials of half-cells are measured in relation to that of a standard hydrogen electrode, which is assigned an electrode potential of zero under standard conditions (solution concentrations of 1 M, gas pressures of 1 atm, and 25 °C).
- A species with a highly positive E° has a strong tendency to attract electrons and undergo reduction (and is therefore an excellent oxidizing agent).
- A species with a highly negative E° has a strong tendency to repel electrons and undergo oxidation (and is therefore an excellent reducing agent).

Cell Potential, Free Energy, and the Equilibrium Constant (20.5)

- In a spontaneous reaction, E°_{cell} is positive, the change in free energy (ΔG°) is negative, and the equilibrium constant (K) is greater than 1.
- In a nonspontaneous reaction, E°_{cell} is negative, ΔG° is positive, and K is less than 1.
- Because E°_{cell}, ΔG°, and K all relate to spontaneity, we can derive equations relating all three quantities.

Cell Potential and Concentration (20.6)

- The standard cell potential (E°_{cell}) is related to the cell potential (E_{cell}) by the Nernst equation, $E_{cell} = E^{\circ}_{cell} - (0.0592\ \text{V}/n) \log Q$.
- As shown by the Nernst equation, E_{cell} is related to the reaction quotient (Q); E_{cell} equals zero when Q equals K.
- In a concentration cell, the reactions at both electrodes are identical and electrons flow because of a difference in concentration.

Batteries: Using Chemistry to Generate Electricity (20.7)

- Batteries are packaged voltaic cells.
- Dry-cell batteries, including alkaline batteries, do not contain large amounts of water.
- The reactions in rechargeable batteries, such as lead–acid storage, nickel–cadmium, nickel–metal hydride, and lithium ion batteries, can be reversed.
- Fuel cells are similar to batteries except that fuel-cell reactants must be continually replenished from an external source.

Electrolysis: Driving Nonspontaneous Chemical Reactions with Electricity (20.8)

- An electrolytic electrochemical cell differs from a voltaic cell in that (1) an electrical charge is used to drive the reaction, and (2) although the anode is still the site of oxidation and the cathode the site of reduction, they are represented with signs opposite those of a voltaic cell (anode +, cathode −).
- In electrolysis reactions, the anion is oxidized; if there is more than one anion, the anion with the more negative E° is oxidized.
- We can use stoichiometry to calculate the quantity of reactants consumed or products produced in an electrolytic cell.

Corrosion: Undesirable Redox Reactions (20.9)

- Corrosion is the undesired oxidation of metal by environmental oxidizing agents.
- When some metals, such as aluminum, oxidize they form a stable compound that prevents further oxidation. Iron, however, does not form a structurally stable compound when oxidized and therefore rust flakes off and exposes more iron to corrosion.
- Iron corrosion can be prevented by preventing water contact, minimizing the presence of electrolytes and acids, or using a sacrificial electrode.

KEY EQUATIONS AND RELATIONSHIPS

Definition of an Ampere (20.3)

$$1\ \text{A} = 1\ \text{C/s}$$

Definition of a Volt (20.3)

$$1\ \text{V} = 1\ \text{J/C}$$

Standard Hydrogen Electrode (20.4)

$$2\ \text{H}^+(aq) + 2\ \text{e}^- \longrightarrow \text{H}_2(g) \qquad E^\circ = 0.00\ \text{V}$$

Equation for Cell Potential (20.4)

$$E^\circ_{\text{cell}} = E^\circ_{\text{cathode}} - E^\circ_{\text{anode}}$$

Relating ΔG° and E°_{cell} (20.5)

$$\Delta G^\circ = -nFE^\circ_{\text{cell}} \qquad F = \frac{96{,}485\ \text{C}}{\text{mol e}^-}$$

Relating E°_{cell} and K (20.5)

$$E^\circ_{\text{cell}} = \frac{0.0592\ \text{V}}{n} \log K \quad (\text{at } 25\ ^\circ\text{C})$$

The Nernst Equation (20.6)

$$E_{\text{cell}} = E^\circ_{\text{cell}} - \frac{0.0592\ \text{V}}{n} \log Q \quad (\text{at } 25\ ^\circ\text{C})$$

EXERCISES

REVIEW QUESTIONS

1. Define oxidation and reduction and explain the basic procedure for balancing redox reactions.
2. Explain the difference between a voltaic (or galvanic) electrochemical cell and an electrolytic one.
3. Which reaction (oxidation or reduction) occurs at the anode of a voltaic cell? What is the sign of the anode? Do electrons flow toward or away from the anode?
4. Which reaction (oxidation or reduction) occurs at the cathode of a voltaic cell? What is the sign of the cathode? Do electrons flow toward or away from the cathode?
5. Explain the purpose of a salt bridge in an electrochemical cell.
6. Which unit is used to measure the magnitude of electrical current? Which unit is used to measure the magnitude of a potential difference? Explain how electrical current and potential difference differ.
7. What is the definition of the standard cell potential (E°_{cell})? What does a large positive standard cell potential imply about the spontaneity of the redox reaction occurring in the cell? What does a negative standard cell potential imply about the reaction?
8. Describe the basic features of a cell diagram (or line notation) for an electrochemical cell.
9. Why do some electrochemical cells employ inert electrodes such as platinum?
10. Describe the standard hydrogen electrode (SHE) and explain its use in determining standard electrode potentials.
11. How is the cell potential of an electrochemical cell (E°_{cell}) related to the potentials of the half-cells?
12. Does a large positive electrode potential indicate a strong oxidizing agent or a strong reducing agent? What about a large negative electrode potential?
13. Is a spontaneous redox reaction obtained by pairing any reduction half-reaction with one listed above it or with one listed below it in Table 20.1?
14. How can Table 20.1 be used to predict whether or not a metal will dissolve in HCl? In HNO_3?
15. Explain why E°_{cell}, $\Delta G^\circ_{\text{rxn}}$, and K are all interrelated.
16. Does a redox reaction with a small equilibrium constant ($K < 1$) have a positive or a negative E°_{cell}? Does it have a positive or a negative $\Delta G^\circ_{\text{rxn}}$?
17. How does E_{cell} depend on the concentrations of the reactants and products in the redox reaction occurring in the cell? What effect does increasing the concentration of a reactant have on E_{cell}? Increasing the concentration of a product?
18. Use the Nernst equation to show that $E_{\text{cell}} = E^\circ_{\text{cell}}$ under standard conditions.
19. What is a concentration electrochemical cell?
20. What are the anode and cathode reactions in a common dry-cell battery? In an alkaline battery?
21. What are the anode and cathode reactions in a lead–acid storage battery? What happens when the battery is recharged?
22. What are the three common types of portable rechargeable batteries, and how does each one work?
23. What is a fuel cell? What is the most common type of fuel cell, and what reactions occur at its anode and cathode?
24. The anode of an electrolytic cell must be connected to which terminal—positive or negative—of the power source?
25. What species is oxidized, and what species is reduced in the electrolysis of a pure molten salt?
26. If an electrolytic cell contains a mixture of species that can be oxidized, how do you determine which species will actually be oxidized? If it contains a mixture of species that can be reduced, how do you determine which one will actually be reduced?
27. Why does the electrolysis of an aqueous sodium chloride solution produce hydrogen gas at the cathode?
28. What is overvoltage in an electrochemical cell? Why is it important?
29. How is the amount of current flowing through an electrolytic cell related to the amount of product produced in the redox reaction?
30. What is corrosion? Why is corrosion only a problem for some metals (such as iron)?
31. Explain the role of each of the following in promoting corrosion: moisture, electrolytes, and acids.
32. How can the corrosion of iron be prevented?

PROBLEMS BY TOPIC

Balancing Redox Reactions

33. Balance each redox reaction occurring in acidic aqueous solution.
    a. $K(s) + Cr^{3+}(aq) \longrightarrow Cr(s) + K^+(aq)$
    b. $Al(s) + Fe^{2+}(aq) \longrightarrow Al^{3+}(aq) + Fe(s)$
    c. $BrO_3^-(aq) + N_2H_4(g) \longrightarrow Br^-(aq) + N_2(g)$
34. Balance each redox reaction occurring in acidic aqueous solution.
    a. $Zn(s) + Sn^{2+}(aq) \longrightarrow Zn^{2+}(aq) + Sn(s)$
    b. $Mg(s) + Cr^{3+}(aq) \longrightarrow Mg^{2+}(aq) + Cr(s)$
    c. $MnO_4^-(aq) + Al(s) \longrightarrow Mn^{2+}(aq) + Al^{3+}(aq)$
35. Balance each redox reaction occurring in acidic aqueous solution.
    a. $PbO_2(s) + I^-(aq) \longrightarrow Pb^{2+}(aq) + I_2(s)$
    b. $SO_3^{2-}(aq) + MnO_4^-(aq) \longrightarrow SO_4^{2-}(aq) + Mn^{2+}(aq)$
    c. $S_2O_3^{2-}(aq) + Cl_2(g) \longrightarrow SO_4^{2-}(aq) + Cl^-(aq)$
36. Balance each redox reaction occurring in acidic aqueous solution.
    a. $I^-(aq) + NO_2^-(aq) \longrightarrow I_2(s) + NO(g)$
    b. $ClO_4^-(aq) + Cl^-(aq) \longrightarrow ClO_3^-(aq) + Cl_2(g)$
    c. $NO_3^-(aq) + Sn^{2+}(aq) \longrightarrow Sn^{4+}(aq) + NO(g)$
37. Balance each redox reaction occurring in basic aqueous solution.
    a. $H_2O_2(aq) + ClO_2(aq) \longrightarrow ClO_2^-(aq) + O_2(g)$
    b. $Al(s) + MnO_4^-(aq) \longrightarrow MnO_2(s) + Al(OH)_4^-(aq)$
    c. $Cl_2(g) \longrightarrow Cl^-(aq) + ClO^-(aq)$
38. Balance each redox reaction occurring in basic aqueous solution.
    a. $MnO_4^-(aq) + Br^-(aq) \longrightarrow MnO_2(s) + BrO_3^-(aq)$
    b. $Ag(s) + CN^-(aq) + O_2(g) \longrightarrow Ag(CN)_2^-(aq)$
    c. $NO_2^-(aq) + Al(s) \longrightarrow NH_3(g) + AlO_2^-(aq)$

Voltaic Cells, Standard Cell Potentials, and Direction of Spontaneity

39. Sketch a voltaic cell for each redox reaction. Label the anode and cathode and indicate the half-reaction that occurs at each electrode and the species present in each solution. Also indicate the direction of electron flow.
    a. $2\,Ag^+(aq) + Pb(s) \longrightarrow 2\,Ag(s) + Pb^{2+}(aq)$
    b. $2\,ClO_2(g) + 2\,I^-(aq) \longrightarrow 2\,ClO_2^-(aq) + I_2(s)$
    c. $O_2(g) + 4\,H^+(aq) + 2\,Zn(s) \longrightarrow 2\,H_2O(l) + 2\,Zn^{2+}(aq)$
40. Sketch a voltaic cell for each redox reaction. Label the anode and cathode and indicate the half-reaction that occurs at each electrode and the species present in each solution. Also indicate the direction of electron flow.
    a. $Ni^{2+}(aq) + Mg(s) \longrightarrow Ni(s) + Mg^{2+}(aq)$
    b. $2\,H^+(aq) + Fe(s) \longrightarrow H_2(g) + Fe^{2+}(aq)$
    c. $2\,NO_3^-(aq) + 8\,H^+(aq) + 3\,Cu(s) \longrightarrow 2\,NO(g) + 4\,H_2O(l) + 3\,Cu^{2+}(aq)$
41. Calculate the standard cell potential for each of the electrochemical cells in Problem 39.
42. Calculate the standard cell potential for each of the electrochemical cells in Problem 40.
43. Consider the voltaic cell:

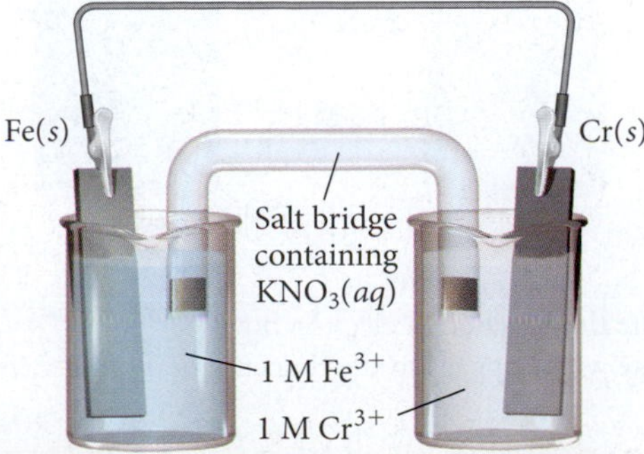

    a. Determine the direction of electron flow and label the anode and the cathode.
    b. Write a balanced equation for the overall reaction and calculate E°_{cell}.
    c. Label each electrode as negative or positive.
    d. Indicate the direction of anion and cation flow in the salt bridge.
44. Consider the voltaic cell:

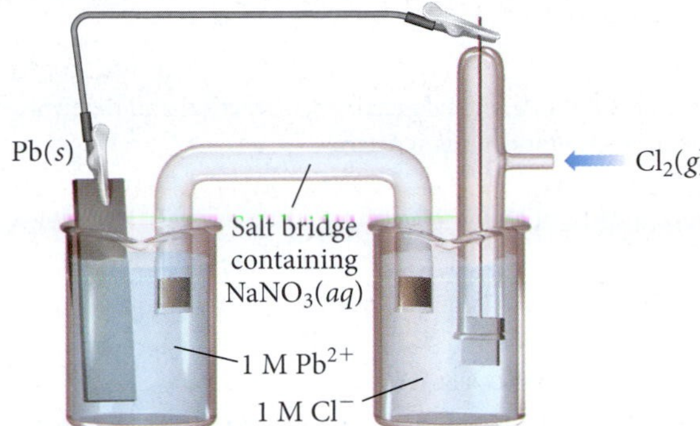

    a. Determine the direction of electron flow and label the anode and the cathode.
    b. Write a balanced equation for the overall reaction and calculate E°_{cell}.
    c. Label each electrode as negative or positive.
    d. Indicate the direction of anion and cation flow in the salt bridge.
45. Use line notation to represent each electrochemical cell in Problem 39.
46. Use line notation to represent each electrochemical cell in Problem 40.
47. Make a sketch of the voltaic cell represented by the line notation. Write the overall balanced equation for the reaction and calculate E°_{cell}.

$$Sn(s)\,|\,Sn^{2+}(aq)\,||\,NO(g)\,|\,NO_3^-(aq), H^+(aq)\,|\,Pt(s)$$

48. Make a sketch of the voltaic cell represented by the line notation. Write the overall balanced equation for the reaction and calculate E°_{cell}.

$$Mn(s)\,|\,Mn^{2+}(aq)\,||\,ClO_2^-(aq)\,|\,ClO_2(g)\,|\,Pt(s)$$

49. Determine whether or not each redox reaction occurs spontaneously in the forward direction.
    a. $Ni(s) + Zn^{2+}(aq) \longrightarrow Ni^{2+}(aq) + Zn(s)$
    b. $Ni(s) + Pb^{2+}(aq) \longrightarrow Ni^{2+}(aq) + Pb(s)$
    c. $Al(s) + 3\,Ag^+(aq) \longrightarrow Al^{3+}(aq) + 3\,Ag(s)$
    d. $Pb(s) + Mn^{2+}(aq) \longrightarrow Pb^{2+}(aq) + Mn(s)$
50. Determine whether each redox reaction occurs spontaneously in the forward direction.
    a. $Ca^{2+}(aq) + Zn(s) \longrightarrow Ca(s) + Zn^{2+}(aq)$
    b. $2\,Ag^+(aq) + Ni(s) \longrightarrow 2\,Ag(s) + Ni^{2+}(aq)$
    c. $Fe(s) + Mn^{2+}(aq) \longrightarrow Fe^{2+}(aq) + Mn(s)$
    d. $2\,Al(s) + 3\,Pb^{2+}(aq) \longrightarrow 2\,Al^{3+}(aq) + 3\,Pb(s)$
51. Which metal could you use to reduce Mn^{2+} ions but not Mg^{2+} ions?
52. Which metal can be oxidized with an Sn^{2+} solution but not with an Fe^{2+} solution?
53. Determine whether or not each metal dissolves in 1 M HCl. For those metals that dissolve, write a balanced redox reaction showing what happens when the metal dissolves.
    a. Al   b. Ag   c. Pb
54. Determine whether or not each metal dissolves in 1 M HCl. For those metals that dissolve, write a balanced redox reaction showing what happens when the metal dissolves.
    a. Cu   b. Fe   c. Au

55. Determine whether or not each metal dissolves in 1 M HNO_3. For those metals that dissolve, write a balanced redox reaction showing what happens when the metal dissolves.

a. Cu b. Au

56. Determine whether or not each metal dissolves in 1 M HIO_3. For those metals that dissolve, write a balanced redox equation for the reaction that occurs.

a. Au b. Cr

57. Calculate E°_{cell} for each balanced redox reaction and determine if the reaction is spontaneous as written.

a. $2\,Cu(s) + Mn^{2+}(aq) \longrightarrow 2\,Cu^{+}(aq) + Mn(s)$
b. $MnO_2(s) + 4\,H^{+}(aq) + Zn(s) \longrightarrow Mn^{2+}(aq) + 2\,H_2O(l) + Zn^{2+}(aq)$
c. $Cl_2(g) + 2\,F^{-}(aq) \longrightarrow F_2(g) + 2\,Cl^{-}(aq)$

58. Calculate E°_{cell} for each balanced redox reaction and determine if the reaction is spontaneous as written.

a. $O_2(g) + 2\,H_2O(l) + 4\,Ag(s) \longrightarrow 4\,OH^{-}(aq) + 4\,Ag^{+}(aq)$
b. $Br_2(l) + 2\,I^{-}(aq) \longrightarrow 2\,Br^{-}(aq) + I_2(s)$
c. $PbO_2(s) + 4\,H^{+}(aq) + Sn(s) \longrightarrow Pb^{2+}(aq) + 2\,H_2O(l) + Sn^{2+}(aq)$

59. Which metal cation is the best oxidizing agent?

a. Pb^{2+} b. Cr^{3+} c. Fe^{2+} d. Sn^{2+}

60. Which metal is the best reducing agent?

a. Mn b. Al c. Ni d. Cr

Cell Potential, Free Energy, and the Equilibrium Constant

61. Use tabulated electrode potentials to calculate ΔG°_{rxn} for each reaction at 25 °C.

a. $Pb^{2+}(aq) + Mg(s) \longrightarrow Pb(s) + Mg^{2+}(aq)$
b. $Br_2(l) + 2\,Cl^{-}(aq) \longrightarrow 2\,Br^{-}(aq) + Cl_2(g)$
c. $MnO_2(s) + 4\,H^{+}(aq) + Cu(s) \longrightarrow Mn^{2+}(aq) + 2\,H_2O(l) + Cu^{2+}(aq)$

62. Use tabulated electrode potentials to calculate ΔG°_{rxn} for each reaction at 25 °C.

a. $2\,Fe^{3+}(aq) + 3\,Sn(s) \longrightarrow 2\,Fe(s) + 3\,Sn^{2+}(aq)$
b. $O_2(g) + 2\,H_2O(l) + 2\,Cu(s) \longrightarrow 4\,OH^{-}(aq) + 2\,Cu^{2+}(aq)$
c. $Br_2(l) + 2\,I^{-}(aq) \longrightarrow 2\,Br^{-}(aq) + I_2(s)$

63. Calculate the equilibrium constant for each of the reactions in Problem 61.

64. Calculate the equilibrium constant for each of the reactions in Problem 62.

65. Calculate the equilibrium constant for the reaction between $Ni^{2+}(aq)$ and $Cd(s)$ (at 25 °C).

66. Calculate the equilibrium constant for the reaction between $Fe^{2+}(aq)$ and $Zn(s)$ (at 25 °C).

67. Calculate ΔG°_{rxn} and E°_{cell} for a redox reaction with $n = 2$ that has an equilibrium constant of $K = 25$ (at 25 °C).

68. Calculate ΔG°_{rxn} and E°_{cell} for a redox reaction with $n = 3$ that has an equilibrium constant of $K = 0.050$ (at 25 °C).

Nonstandard Conditions and the Nernst Equation

69. A voltaic cell employs the following redox reaction:

$$Sn^{2+}(aq) + Mn(s) \longrightarrow Sn(s) + Mn^{2+}(aq)$$

Calculate the cell potential at 25 °C under each set of conditions.

a. standard conditions
b. $[Sn^{2+}] = 0.0100$ M; $[Mn^{2+}] = 2.00$ M
c. $[Sn^{2+}] = 2.00$ M; $[Mn^{2+}] = 0.0100$ M

70. A voltaic cell employs the redox reaction:

$$2\,Fe^{3+}(aq) + 3\,Mg(s) \longrightarrow 2\,Fe(s) + 3\,Mg^{2+}(aq)$$

Calculate the cell potential at 25 °C under each set of conditions.

a. standard conditions
b. $[Fe^{3+}] = 1.0 \times 10^{-3}$ M; $[Mg^{2+}] = 2.50$ M
c. $[Fe^{3+}] = 2.00$ M; $[Mg^{2+}] = 1.5 \times 10^{-3}$ M

71. An electrochemical cell is based on these two half-reactions:

Ox: $Pb(s) \longrightarrow Pb^{2+}(aq, 1.10\ M) + 2\,e^{-}$

Red: $MnO_4^{-}(aq, 1.50\ M) + 4\,H^{+}(aq, 2.0\ M) + 3\,e^{-} \longrightarrow MnO_2(s) + 2\,H_2O(l)$

Calculate the cell potential at 25 °C.

72. An electrochemical cell is based on these two half-reactions:

Ox: $Sn(s) \longrightarrow Sn^{2+}(aq, 2.00\ M) + 2\,e^{-}$

Red: $ClO_2(g, 1.100\ atm) + e^{-} \longrightarrow ClO_2^{-}(aq, 2.00\ M)$

Calculate the cell potential at 25 °C.

73. A voltaic cell consists of a Zn/Zn^{2+} half-cell and a Ni/Ni^{2+} half-cell at 25 °C. The initial concentrations of Ni^{2+} and Zn^{2+} are 1.50 M and 0.100 M, respectively.

a. What is the initial cell potential?
b. What is the cell potential when the concentration of Ni^{2+} has fallen to 0.500 M?
c. What are the concentrations of Ni^{2+} and Zn^{2+} when the cell potential falls to 0.45 V?

74. A voltaic cell consists of a Pb/Pb^{2+} half-cell and a Cu/Cu^{2+} half-cell at 25 °C. The initial concentrations of Pb^{2+} and Cu^{2+} are 0.0500 M and 1.50 M, respectively.

a. What is the initial cell potential?
b. What is the cell potential when the concentration of Cu^{2+} has fallen to 0.200 M?
c. What are the concentrations of Pb^{2+} and Cu^{2+} when the cell potential falls to 0.35 V?

75. Make a sketch of a concentration cell employing two Zn/Zn^{2+} half-cells. The concentration of Zn^{2+} in one of the half-cells is 2.0 M, and the concentration in the other half-cell is 1.0×10^{-3} M. Label the anode and the cathode and indicate the half-reaction occurring at each electrode. Also indicate the direction of electron flow.

76. Consider the concentration cell:

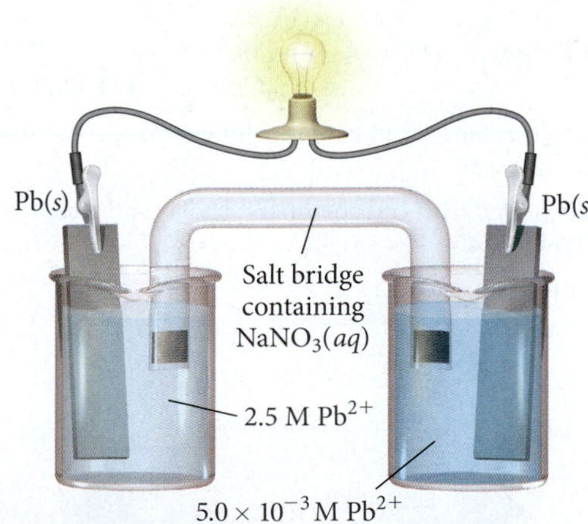

a. Label the anode and cathode.
b. Indicate the direction of electron flow.
c. Indicate what happens to the concentration of Pb^{2+} in each half-cell.

77. A concentration cell consists of two Sn/Sn^{2+} half-cells. The cell has a potential of 0.10 V at 25 °C. What is the ratio of the Sn^{2+} concentrations in the two half-cells?

78. A Cu/Cu^{2+} concentration cell has a voltage of 0.22 V at 25 °C. The concentration of Cu^{2+} in one of the half-cells is 1.5×10^{-3} M. What is the concentration of Cu^{2+} in the other half-cell? (Assume the concentration in the unknown cell to be the *lower* of the two concentrations.)

Batteries, Fuel Cells, and Corrosion

79. Determine the optimum mass ratio of Zn to MnO_2 in an alkaline battery.

80. What mass of lead sulfate is formed in a lead–acid storage battery when 1.00 g of Pb undergoes oxidation?

81. Refer to the tabulated values of ΔG°_f in Appendix IVB to calculate E°_{cell} for a fuel cell that employs the reaction between methane gas (CH_4) and oxygen to form carbon dioxide and gaseous water.

82. Refer to the tabulated values of ΔG°_f in Appendix IVB to calculate E°_{cell} for a fuel cell that employs the following reaction: (ΔG°_f for $HC_2H_3O_2(g) = -374.2$ kJ/mol.)

$$CH_3CH_2OH(g) + O_2(g) \longrightarrow HC_2H_3O_2(g) + H_2O(g)$$

83. Determine whether or not each metal, if coated onto iron, would prevent the corrosion of iron.

a. Zn **b.** Sn **c.** Mn

84. Determine whether or not each metal, if coated onto iron, would prevent the corrosion of iron.

a. Mg **b.** Cr **c.** Cu

Electrolytic Cells and Electrolysis

85. Consider the electrolytic cell:

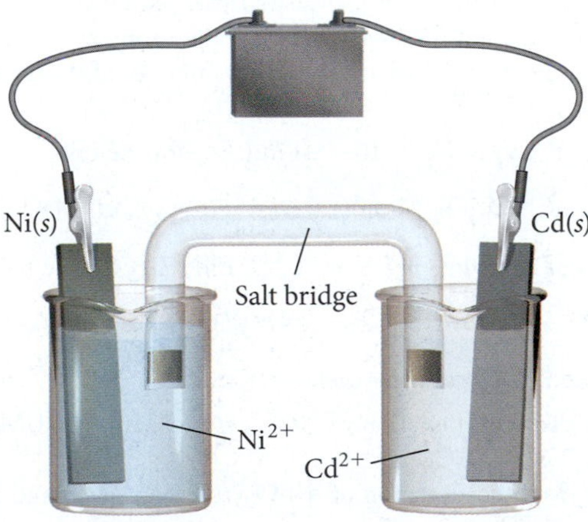

a. Label the anode and the cathode and indicate the half-reactions occurring at each.
b. Indicate the direction of electron flow.
c. Label the terminals on the battery as positive or negative and calculate the minimum voltage necessary to drive the reaction.

86. Draw an electrolytic cell in which Mn^{2+} is reduced to Mn and Sn is oxidized to Sn^{2+}. Label the anode and cathode, indicate the direction of electron flow, and write an equation for the half-reaction occurring at each electrode. What minimum voltage is necessary to drive the reaction?

87. Write equations for the half-reactions that occur in the electrolysis of molten potassium bromide.

88. Which products are obtained in the electrolysis of molten NaI?

89. Write equations for the half-reactions that occur in the electrolysis of a mixture of molten potassium bromide and molten lithium bromide.

90. Which products are obtained in the electrolysis of a molten mixture of KI and KBr?

91. Write equations for the half-reactions that occur at the anode and cathode for the electrolysis of each aqueous solution:

a. $NaBr(aq)$ **b.** $PbI_2(aq)$ **c.** $Na_2SO_4(aq)$

92. Write equations for the half-reactions that occur at the anode and cathode for the electrolysis of each aqueous solution:

a. $Ni(NO_3)_2(aq)$ **b.** $KCl(aq)$ **c.** $CuBr_2(aq)$

93. Make a sketch of an electrolytic cell that electroplates copper onto other metal surfaces. Label the anode and the cathode and indicate the reactions that occur at each.

94. Make a sketch of an electrolytic cell that electroplates nickel onto other metal surfaces. Label the anode and the cathode and indicate the reactions that occur at each.

95. Copper can be electroplated at the cathode of an electrolytic cell by the half-reaction:

$$Cu^{2+}(aq) + 2\,e^- \longrightarrow Cu(s)$$

How much time does it take for 325 mg of copper to be plated at a current of 5.6 A?

96. Silver can be electroplated at the cathode of an electrolytic cell by the half-reaction:

$$Ag^+(aq) + e^- \longrightarrow Ag(s)$$

What mass of silver plates onto the cathode if a current of 6.8 A flows through the cell for 72 min?

97. A major source of sodium metal is the electrolysis of molten sodium chloride. What magnitude of current produces 1.0 kg of sodium metal in 1 hour?

98. What mass of aluminum metal can be produced per hour in the electrolysis of a molten aluminum salt by a current of 25 A?

CUMULATIVE PROBLEMS

99. Consider the unbalanced redox reaction:

$$MnO_4^-(aq) + Zn(s) \longrightarrow Mn^{2+}(aq) + Zn^{2+}(aq)$$

Balance the equation and determine the volume of a 0.500 M $KMnO_4$ solution required to completely react with 2.85 g of Zn.

100. Consider the unbalanced redox reaction:

$$Cr_2O_7^{2-}(aq) + Cu(s) \longrightarrow Cr^{3+}(aq) + Cu^{2+}(aq)$$

Balance the equation and determine the volume of a 0.850 M $K_2Cr_2O_7$ solution required to completely react with 5.25 g of Cu.

101. Consider the molecular view of an Al strip and Cu^{2+} solution. Draw a similar sketch showing what happens to the atoms and ions after the Al strip is submerged in the solution for a few minutes.

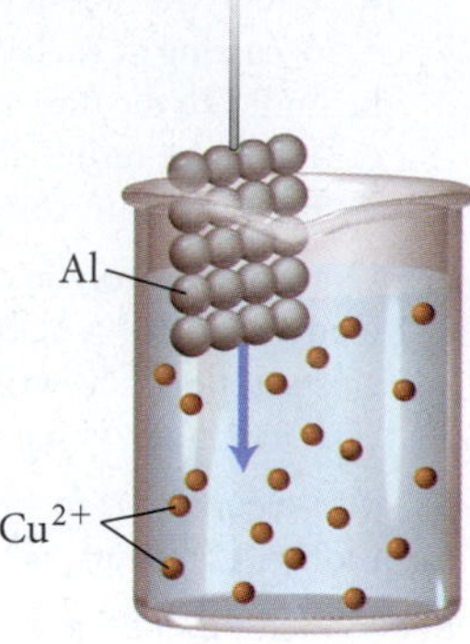

102. Consider the molecular view of an electrochemical cell involving the overall reaction:

$$Zn(s) + Ni^{2+}(aq) \longrightarrow Zn^{2+}(aq) + Ni(s)$$

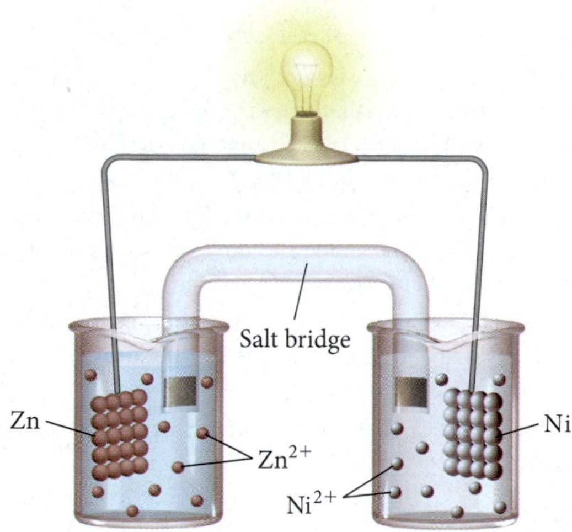

Draw a similar sketch of the cell after it has generated a substantial amount of electrical current.

103. Determine whether HI can dissolve each metal sample. If it can, write a balanced chemical reaction showing how the metal dissolves in HI and determine the minimum volume of 3.5 M HI required to completely dissolve the sample.

a. 2.15 g Al **b.** 4.85 g Cu **c.** 2.42 g Ag

104. Determine if HNO_3 can dissolve each metal sample. If it can, write a balanced chemical reaction showing how the metal dissolves in HNO_3 and determine the minimum volume of 6.0 M HNO_3 required to completely dissolve the sample.

a. 5.90 g Au **b.** 2.55 g Cu **c.** 4.83 g Sn

105. The cell potential of this electrochemical cell depends on the pH of the solution in the anode half-cell.

$$Pt(s)\,|\,H_2(g,\ 1\ atm)\,|\,H^+(aq,\ ?\ M)\,||\,Cu^{2+}(aq,\ 1.0\ M)\,|\,Cu(s)$$

What is the pH of the solution if E_{cell} is 355 mV?

106. The cell potential of this electrochemical cell depends on the gold concentration in the cathode half-cell.

$$Pt(s)\,|\,H_2(g,\ 1.0\ atm)\,|\,H^+(aq,\ 1.0\ M)\,||\,Au^{3+}(aq,\ ?\ M)\,|\,Au(s)$$

What is the concentration of Au^{3+} in the solution if E_{cell} is 1.22 V?

107. A friend wants you to invest in a new battery she has designed that produces 24 V in a single voltaic cell. Why should you be wary of investing in such a battery?

108. What voltage can theoretically be achieved in a battery in which lithium metal is oxidized and fluorine gas is reduced? Why might such a battery be difficult to produce?

109. A battery relies on the oxidation of magnesium and the reduction of Cu^{2+}. The initial concentrations of Mg^{2+} and Cu^{2+} are 1.0×10^{-4} M and 1.5 M, respectively, in 1.0-liter half-cells.

a. What is the initial voltage of the battery?

b. What is the voltage of the battery after delivering 5.0 A for 8.0 h?

c. How long can the battery deliver 5.0 A before going dead?

110. A rechargeable battery is constructed based on a concentration cell constructed of two Ag/Ag^+ half-cells. The volume of each half-cell is 2.0 L, and the concentrations of Ag^+ in the half-cells are 1.25 M and 1.0×10^{-3} M.

a. How long can this battery deliver 2.5 A of current before it dies?

b. What mass of silver is plated onto the cathode by running at 3.5 A for 5.5 h?

c. Upon recharging, how long would it take to redissolve 1.00×10^2 g of silver at a charging current of 10.0 amps?

111. If a water electrolysis cell operates at a current of 7.8 A, how long will it take to generate 25.0 L of hydrogen gas at a pressure of 25.0 atm and a temperature of 25 °C?

112. One type of breathalyzer employs a fuel cell to measure the quantity of alcohol in the breath. When a suspect blows into the breathalyzer, ethyl alcohol is oxidized to acetic acid at the anode:

$$\underset{\text{ethyl alcohol}}{CH_3CH_2OH(g)} + 4OH^-(aq) \longrightarrow \underset{\text{acetic acid}}{HC_2H_3O_2(g)} + 3H_2O(l) + 4e^-$$

At the cathode, oxygen is reduced:

$$O_2(g) + 2\,H_2O(l) + 4\,e^- \longrightarrow 4\,OH^-(aq)$$

The overall reaction is the oxidation of ethyl alcohol to acetic acid and water. When a suspected drunk driver blows 188 mL of his breath through this breathalyzer, the breathalyzer produces an average of 324 mA of current for 10 s. Assuming a pressure of 1.0 atm and a temperature of 25 °C what percent (by volume) of the driver's breath is ethanol?

113. The K_{sp} of CuI is 1.1×10^{-12}. Find E_{cell} for the cell:

$$Cu(s)\,|\,CuI(s)\,|\,I^-(aq)(1.0\ M)\,||\,Cu^+(aq)(1.0\ M)\,|\,Cu(s)$$

114. The K_{sp} of $Zn(OH)_2$ is 1.8×10^{-14}. Find E_{cell} for the half-reaction:

$$Zn(OH)_2(s) + 2\,e^- \rightleftharpoons Zn(s) + 2\,OH^-(aq)$$

115. Calculate ΔG°_{rxn} and K for each reaction.

a. The disproportionation of $Mn^{2+}(aq)$ to $Mn(s)$ and $MnO_2(s)$ in acid solution at 25 °C.

b. The disproportionation of $MnO_2(s)$ to $Mn^{2+}(aq)$ and $MnO_4^-(aq)$ in acid solution at 25 °C.

116. Calculate ΔG°_{rxn} and K for each reaction.

a. The reaction of $Cr^{2+}(aq)$ with $Cr_2O_7^{2-}(aq)$ in acid solution to form $Cr^{3+}(aq)$.

b. The reaction of $Cr^{3+}(aq)$ and $Cr(s)$ to form $Cr^{2+}(aq)$. [The electrode potential of $Cr^{2+}(aq)$ to $Cr(s)$ is −0.91 V.]

117. The molar mass of a metal (M) is 50.9 g/mol; it forms a chloride of unknown composition. Electrolysis of a sample of the molten chloride with a current of 6.42 A for 23.6 minutes produces 1.20 g of M at the cathode. Determine the empirical formula of the chloride.

118. A metal forms the fluoride MF_3. Electrolysis of the molten fluoride by a current of 3.86 A for 16.2 minutes deposits 1.25 g of the metal. Calculate the molar mass of the metal.

119. A sample of impure tin of mass 0.535 g is dissolved in strong acid to give a solution of Sn^{2+}. The solution is then titrated with a 0.0448 M solution of NO_3^-, which is reduced to $NO(g)$. The equivalence point is reached upon the addition of 0.0344 L of the NO_3^- solution. Find the percent by mass of tin in the original sample, assuming that it contains no other reducing agents.

120. A 0.0251 L sample of a solution of Cu^+ requires 0.0322 L of 0.129 M $KMnO_4$ solution to reach the equivalence point. The products of the reaction are Cu^{2+} and Mn^{2+}. What is the concentration of the Cu^{2+} solution?

121. A current of 11.3 A is applied to 1.25 L of a solution of 0.552 M HBr converting some of the H^+ to $H_2(g)$, which bubbles out of the solution. What is the pH of the solution after 73 minutes?

122. A 215 mL sample of a 0.500 M NaCl solution with an initial pH of 7.00 is subjected to electrolysis. After 15.0 minutes, a 10.0 mL portion (or aliquot) of the solution is removed from the cell and titrated with 0.100 M HCl solution. The endpoint in the titration is reached upon addition of 22.8 mL of HCl. Assuming constant current, how much current (in A) was running through the cell?

123. An $MnO_2(s)/Mn^{2+}(aq)$ electrode in which the pH is 10.24 is prepared. Find the $[Mn^{2+}]$ necessary to lower the potential of the half-cell to 0.00 V (at 25 °C).

124. To what pH should you adjust a standard hydrogen electrode to achieve an electrode potential of −0.122 V? (Assume that the partial pressure of hydrogen gas remains at 1 atm.)

CHALLENGE PROBLEMS

125. Suppose a hydrogen–oxygen fuel-cell generator produces electricity for a house. Use the balanced redox reactions and the standard cell potential to predict the volume of hydrogen gas (at STP) required each month to generate the electricity. Assume the home uses 1.2×10^3 kWh of electricity per month.

126. A voltaic cell designed to measure $[Cu^{2+}]$ is constructed of a standard hydrogen electrode and a copper metal electrode in the Cu^{2+} solution of interest. If you want to construct a calibration curve for how the cell potential varies with the concentration of copper(II), what do you plot in order to obtain a straight line? What is the slope of the line?

127. The surface area of an object to be gold plated is 49.8 cm^2 and the density of gold is 19.3 g/cm^3. A current of 3.25 A is applied to a solution that contains gold in the +3 oxidation state. Calculate the time required to deposit an even layer of gold 1.00×10^{-3} cm thick on the object.

128. To electrodeposit all the Cu and Cd from a solution of $CuSO_4$ and $CdSO_4$ required 1.20 F of electricity (1 F = 1 mol e^-). The mixture of Cu and Cd that was deposited had a mass of 50.36 g. What mass of $CuSO_4$ was present in the original mixture?

129. Sodium oxalate, $Na_2C_2O_4$, in solution is oxidized to $CO_2(g)$ by MnO_4^-, which is reduced to Mn^{2+}. A 50.1 mL volume of a solution of MnO_4^- is required to titrate a 0.339 g sample of sodium oxalate. This solution of MnO_4^- is used to analyze uranium-containing samples. A 4.62 g sample of a uranium-containing material requires 32.5 mL of the solution for titration. The oxidation of the uranium can be represented by the change $UO^{2+} \longrightarrow UO_2^{2+}$. Calculate the percentage of uranium in the sample.

130. Three electrolytic cells are connected in a series. The electrolytes in the cells are aqueous copper(II) sulfate, gold(III) sulfate, and silver nitrate. A current of 2.33 A is applied, and after some time 1.74 g Cu is deposited. How long was the current applied? What mass of gold and silver were deposited?

131. The cell $Pt(s)|Cu^+(1\ M), Cu^{2+}(1\ M)||Cu^+(1\ M)|Cu(s)$ has $E^\circ_{cell} = 0.364$ V. The cell $Cu(s)|Cu^{2+}(1\ M)||Cu^{+1}(1\ M)|Cu(s)$ has $E^\circ_{cell} = 0.182$ V. Write the cell reaction for each cell and explain the differences in E°_{cell}. Calculate ΔG° for each cell reaction to help explain these differences.

CONCEPTUAL PROBLEMS

132. An electrochemical cell has a positive standard cell potential but a negative cell potential. Which statement is true?

a. $K > 1$; $Q > K$ **b.** $K < 1$; $Q > K$
c. $K > 1$; $Q < K$ **d.** $K < 1$; $Q < K$

133. Which oxidizing agent oxidizes Br^- but not Cl^-?

a. $K_2Cr_2O_7$(in acid) **b.** $KMnO_4$(in acid) **c.** HNO_3

134. A redox reaction employed in an electrochemical cell has a negative ΔG°_{rxn}. Which statement is true?

a. E°_{cell} is positive; $K < 1$ **b.** E°_{cell} is positive; $K > 1$
c. E°_{cell} is negative; $K > 1$ **d.** E°_{cell} is negative; $K < 1$

135. A redox reaction has an equilibrium constant of $K = 0.055$. What is true of ΔG°_{rxn} and E°_{cell} for this reaction?

ANSWERS TO CONCEPTUAL CONNECTIONS

Cc 20.1 **(a)** In a voltaic cell, electrons flow from higher potential energy to lower potential energy (from the anode to the cathode).

Cc 20.2 **(b)** A negative electrode potential indicates that an electron at that electrode has greater potential energy than it has at a standard hydrogen electrode.

Cc 20.3 **(d)** The reduction of HNO_3 is listed below the reduction of Br_2 and above the reduction of I_2 in Table 20.1. Since any reduction half-reaction is spontaneous when paired with the reverse of a half-reaction below it in the table, the reduction of HNO_3 is spontaneous when paired with the oxidation of I^- but is not spontaneous when paired with the oxidation of Br^-.

Cc 20.4 **(c)** Ag falls *above* the half-reaction for the reduction of H^+ but *below* the half-reaction for the reduction of NO_3^- in Table 20.1.

Cc 20.5 **(a)** Br is more electronegative than I. If the two atoms were in competition for the electron, the electron would go to the more electronegative atom (Br). Therefore, I_2 does not spontaneously gain electrons from Br^-.

Cc 20.6 **(c)** Since $K > 1$, the reaction is spontaneous under standard conditions (when $Q = 1$, the reaction proceeds toward the products). Therefore, E°_{cell} is positive and ΔG°_{rxn} is negative.

Cc 20.7 **(a)** Since $K < 1$, E°_{cell} is negative (under standard conditions, the reaction is not spontaneous). Since $Q < K$, E_{cell} is positive (the reaction is spontaneous under the nonstandard conditions of the cell).

Cc 20.8 Cu. The electrode potential for Fe is more negative than that of Cu. Therefore, Fe oxidizes more easily than Cu.

CHAPTER 21

"I am among those who think that science has great beauty. A scientist in his laboratory is not only a technician; he is also a child placed before natural phenomena which impress him like a fairy tale."

—Marie Curie (1867–1934)

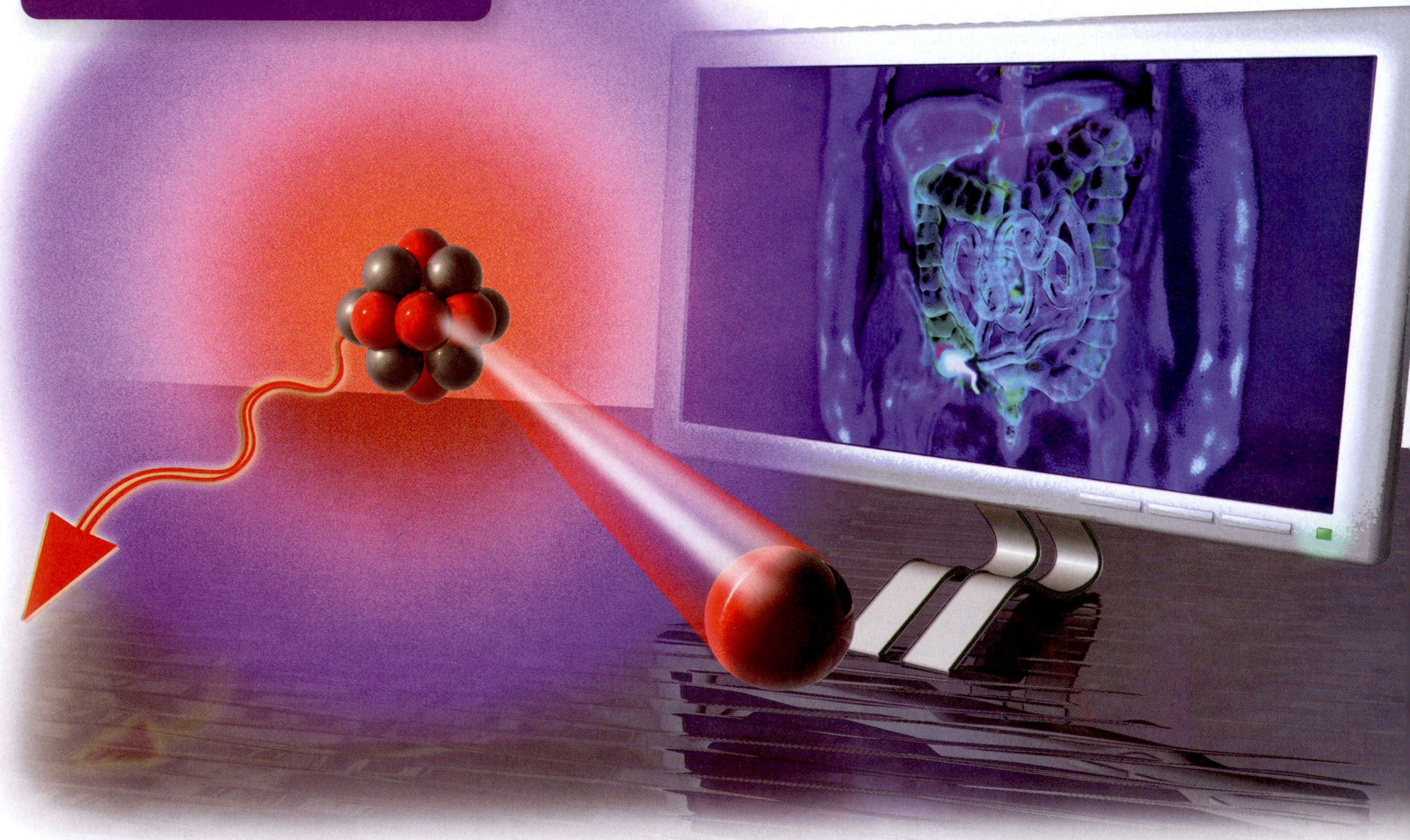

Antibodies labeled with radioactive atoms help to diagnose an infected appendix.

Radioactivity and Nuclear Chemistry

21.1 Diagnosing Appendicitis 861

21.2 The Discovery of Radioactivity 862

21.3 Types of Radioactivity 863

21.4 The Valley of Stability: Predicting the Type of Radioactivity 869

21.5 Detecting Radioactivity 871

21.6 The Kinetics of Radioactive Decay and Radiometric Dating 872

21.7 The Discovery of Fission: The Atomic Bomb and Nuclear Power 879

21.8 Converting Mass to Energy: Mass Defect and Nuclear Binding Energy 883

21.9 Nuclear Fusion: The Power of the Sun 886

21.10 Nuclear Transmutation and Transuranium Elements 887

21.11 The Effects of Radiation on Life 888

21.12 Radioactivity in Medicine and Other Applications 891

Key Learning Outcomes 895

IN THIS CHAPTER, WE EXAMINE RADIO-ACTIVITY and nuclear chemistry, both of which involve changes within the *nuclei* of atoms. Unlike typical chemical processes, in which elements retain their identity, nuclear processes often result in one element changing into another, frequently emitting tremendous amounts of energy. Radioactivity has numerous applications, including the diagnosis and treatment of medical conditions such as cancer, thyroid disease, abnormal kidney and bladder function, and heart disease. Naturally occurring radioactivity allows us to estimate the age of fossils, rocks, and ancient artifacts. And the study of radioactivity, perhaps most famously, led to the discovery of nuclear fission, used for electricity generation and nuclear weapons. In this chapter, we discuss radioactivity—how it was discovered, what it is, and how we use it. Radioactivity—like other chemical phenomena we have explored—also depends on structure, but here it is the structure of the nucleus that determines the radioactive properties. A complete exploration of how nuclear structure affects radioactive properties is beyond our scope, but we will at least get a glimpse of some of the important factors.

21.1 Diagnosing Appendicitis

One morning a few years ago I awoke with a dull pain on the lower right side of my abdomen that grew worse by early afternoon. Since pain in this area can indicate appendicitis (inflammation of the appendix), and since I know that appendicitis can be dangerous if left untreated, I went to the hospital emergency room. The doctor who examined me recommended a simple blood test to determine my white blood cell count. Patients with appendicitis usually have a high white blood cell count because the body is trying to fight the infection. My blood test was negative—I had a normal white blood cell count.

Although my symptoms were consistent with appendicitis, the negative blood test clouded the diagnosis. The doctor said that I could elect to have my appendix removed anyway (even though it might be healthy) or I could submit to another test that might confirm the appendicitis. I chose the additional test, which involved *nuclear medicine*, an area of medical practice that employs *radioactivity* to diagnose and treat disease. **Radioactivity** is the emission of subatomic particles or high-energy electromagnetic radiation by the nuclei of certain atoms. Such atoms are said to be **radioactive**. Most radioactive emissions can pass through many types of matter (such as skin and muscle, in this case).

During the test, antibodies (naturally occurring molecules that fight infection) labeled with radioactive atoms were injected into my bloodstream. Since antibodies attack infection, they migrate to areas of the body where infection is present. If my appendix was infected, the antibodies would accumulate there. I waited about an hour, and then the technician took me to a room and laid me on a table. She inserted a photographic film in a panel above me and removed the covering that prevents exposure of the film. Radioactivity is invisible to the eye, but it exposes photographic film. If my appendix had been infected, it would have (by then) contained a high concentration of the radioactively labeled antibodies. The antibodies would emit radiation and expose the film. The test, however, was negative. No radioactivity was emanating from my appendix. After several hours, the pain in my abdomen subsided and I went home. I never did find out what caused the pain.

21.2 The Discovery of Radioactivity

▲ The greenish light emitted from glow-in-the-dark toys is phosphorescence.

Radioactivity was discovered in 1896 by a French scientist named Antoine-Henri Becquerel (1852–1908). Becquerel was interested in the newly discovered X-rays (see Section 3.2), which became a hot topic of physics research in his time. He hypothesized that X-rays were emitted in conjunction with **phosphorescence**, the long-lived *emission* of light that sometimes follows the absorption of light by certain atoms and molecules. Phosphorescence is probably most familiar to you as the *glow* in glow-in-the-dark products (such as toys or stickers). After such a product is exposed to light, it reemits some of that light, usually at slightly longer wavelengths. If you turn off the room lights or put the glow-in-the-dark product in the dark, you see the greenish glow of the emitted light. Becquerel hypothesized that this visible greenish glow was associated with the emission of X-rays (which are invisible).

To test his hypothesis, Becquerel placed crystals—composed of potassium uranyl sulfate, a compound known to phosphoresce—on top of a photographic plate wrapped in black cloth. He then exposed the crystals to sunlight. He knew the crystals had phosphoresced because he could see the emitted light when he brought them back into the dark. If the crystals had also emitted X-rays, the X-rays would have passed through the black cloth and exposed the underlying photographic plate. Becquerel performed the experiment several times and always got the same result—the photographic plate showed a dark exposure spot where the crystals had been (Figure 21.1 ▼). Becquerel believed his hypothesis was correct and presented the results—that phosphorescence and X-rays were linked—to the French Academy of Sciences.

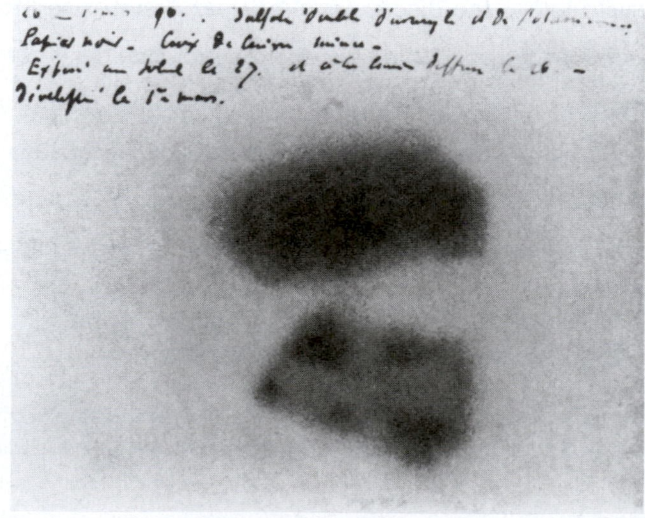

▶ **FIGURE 21.1 The Discovery of Radioactivity** This photographic plate (with Becquerel's original comments at top) played a key role in the discovery of radioactivity. Becquerel placed a uranium-containing compound on the plate (which was wrapped in black cloth to shield it from visible light). He found that the plate was darkened by some unknown form of penetrating radiation that was produced continuously, independently of phosphorescence.

Becquerel later retracted his results, however, when he discovered that a photographic plate with the same crystals showed a dark exposure spot even when the plate and the crystals were stored in a drawer and not exposed to sunlight. Becquerel realized that the crystals themselves were constantly emitting something that exposed the photographic plate, regardless of whether or not they phosphoresced. Becquerel concluded that it was the uranium within the crystals that was the source of the emissions, and he named the emissions *uranic rays*.

Soon after Becquerel's discovery, a young graduate student named Marie Sklodowska Curie (1867–1934) (one of the first women in France to pursue doctoral work) decided to study uranic rays for her doctoral thesis. Her first task was to determine whether any other substances besides uranium (the heaviest known element at the time) emitted these rays. In her search, Curie discovered two new elements, both of which also emitted uranic rays. Curie named one of her newly discovered elements polonium, after her home country of Poland. The other element she named radium, because of its high level of radioactivity. Radium is so radioactive that it gently glows in the dark and emits significant amounts of heat. Since it was clear that these rays were not unique to uranium, Curie changed the name of uranic rays to radioactivity. In 1903, Curie, and her husband, Pierre Curie, as well as Becquerel were all awarded the Nobel Prize in physics for the discovery of radioactivity. In 1911, Curie received a second Nobel Prize, this time in chemistry, for her discovery of the two new elements.

▲ Marie Curie, one of the first women in France to pursue a doctoral degree, was twice awarded the Nobel Prize, in 1903 and 1911. She is seen here with her daughters, in about 1905. Irène (left) became a distinguished nuclear physicist in her own right, winning a Nobel Prize in 1935. Eve (right) wrote a highly acclaimed biography of her mother.

Element 96 is named curium in honor of Marie Curie and her contributions to our understanding of radioactivity.

21.3 Types of Radioactivity

While Curie focused her work on discovering the different kinds of radioactive elements, Ernest Rutherford (1871–1937) and others focused on characterizing the radioactivity itself. These scientists found that the emissions are produced by the nuclei of radioactive atoms. Such nuclei are unstable and spontaneously decompose, emitting small pieces of themselves to gain stability. These fragments are the radioactivity that Becquerel and Curie detected. Natural radioactivity can be categorized into several different types, including *alpha* (α) *decay*, *beta* (β) *decay*, *gamma* (γ) *ray emission*, and *positron emission*. In addition, some unstable atomic nuclei can attain greater stability by absorbing an electron from one of the atom's own orbitals, a process called *electron capture*.

In order to understand these different types of radioactivity, we must briefly review the notation for symbolizing isotopes from Section 1.8. Recall that we can represent any isotope with the following notation:

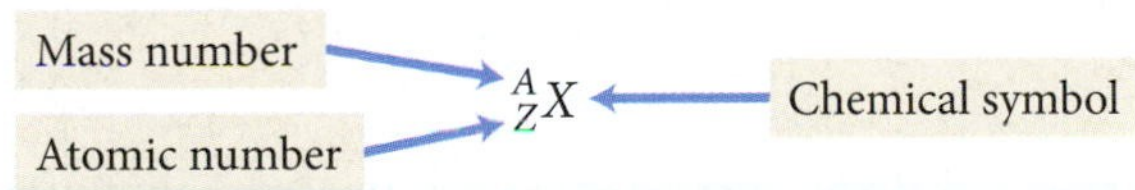

Mass number (A) = the sum of the number of protons and number of neutrons in the nucleus

Atomic number (Z) = the number of protons in the nucleus

Since A represents the sum of the number of protons and neutrons, and since Z represents the number of protons, the number of neutrons in the nucleus (N) is $A - Z$.

$$N = A - Z$$

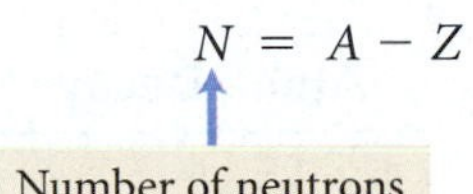

▲ Radium, discovered by Marie Curie, is so radioactive that it glows visibly and emits heat.

For example, the symbol $^{21}_{10}Ne$ represents the neon isotope containing 10 protons and 11 neutrons. The symbol $^{20}_{10}Ne$ represents the neon isotope containing 10 protons and 10 neutrons. Remember that most elements have several different isotopes. When we are discussing nuclear properties, we often refer to a particular isotope (or species) of an element as a **nuclide**.

We represent the main subatomic particles—protons, neutrons, and electrons—with similar notation.

Proton symbol $^{1}_{1}\text{p}$ Neutron symbol $^{1}_{0}\text{n}$ Electron symbol $^{0}_{-1}\text{e}$

The 1 in the lower left of the proton symbol represents 1 proton, and the 0 in the lower left corner of the neutron symbol represents 0 protons. The −1 in the lower left corner of the electron symbol is a bit different from the other atomic numbers; it will make sense when we see it in the context of nuclear decay a bit later in this section.

Alpha (α) Decay

As we discuss in Section 21.4, nuclei are unstable when they are too large or when they contain an unbalanced ratio of neutrons to protons.

Alpha (α) decay occurs when an unstable nucleus emits a particle composed of two protons and two neutrons (Figure 21.2 ▼). Since two protons and two neutrons combined are identical to a helium-4 nucleus, the symbol for alpha radiation is the symbol for helium-4:

Alpha (α) particle $^{4}_{2}\text{He}$

When an element emits an alpha particle, the number of protons in its nucleus changes, transforming the element into a different element. We symbolize this phenomenon with a **nuclear equation**, an equation that represents nuclear processes such as radioactivity. For example, the nuclear equation for the alpha decay of uranium-238 is:

Parent nuclide Daughter nuclide

$$^{238}_{92}\text{U} \longrightarrow {}^{234}_{90}\text{Th} + {}^{4}_{2}\text{He}$$

In nuclear chemistry, we are primarily interested in changes within the nucleus; therefore, the 2+ charge that we would normally write for a helium nucleus is omitted for an alpha particle.

We call the original atom the *parent nuclide* and the product of the decay the *daughter nuclide*. In this case, uranium-238 (the parent nuclide) becomes thorium-234 (the daughter nuclide). Unlike a chemical reaction, in which elements retain their identities, in a nuclear reaction elements often change their identities. Like a chemical equation, however, a nuclear equation must be balanced. *The sum of the atomic numbers on both sides of a nuclear equation must be equal, and the sum of the mass numbers on both sides must also be equal.*

$$^{238}_{92}\text{U} \longrightarrow {}^{234}_{90}\text{Th} + {}^{4}_{2}\text{He}$$

Reactants	Products
Sum of mass numbers = 238	Sum of mass numbers = 234 + 4 = 238
Sum of atomic numbers = 92	Sum of atomic numbers = 90 + 2 = 92

We can deduce the identity and symbol of the daughter nuclide in any alpha decay from the mass and atomic number of the parent nuclide. During alpha decay, the mass number decreases by 4 and the atomic number decreases by 2, as shown in Example 21.1.

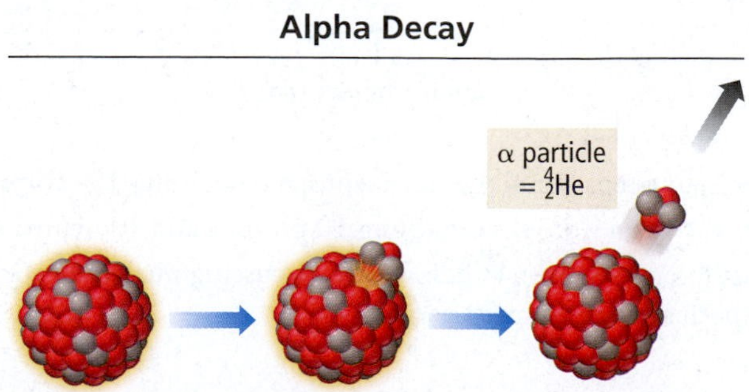

▶ **FIGURE 21.2 Alpha Decay** In alpha decay, a nucleus emits a particle composed of two protons and two neutrons (a helium-4 nucleus).

EXAMPLE 21.1
Writing Nuclear Equations for Alpha Decay

Write the nuclear equation for the alpha decay of Ra-224.

SOLUTION

Begin with the symbol for Ra-224 on the left side of the equation and the symbol for an alpha particle on the right side.	$^{224}_{88}\text{Ra} \longrightarrow {}^{?}_{?}? + {}^{4}_{2}\text{He}$
Equalize the sum of the mass numbers and the sum of the atomic numbers on both sides of the equation by writing the appropriate mass number and atomic number for the unknown daughter nuclide.	$^{224}_{88}\text{Ra} \longrightarrow {}^{220}_{86}? + {}^{4}_{2}\text{He}$
Refer to the periodic table to deduce the identity of the unknown daughter nuclide from its atomic number and write its symbol. Since the atomic number is 86, the daughter nuclide is radon (Rn).	$^{224}_{88}\text{Ra} \longrightarrow {}^{220}_{86}\text{Rn} + {}^{4}_{2}\text{He}$

FOR PRACTICE 21.1
Write the nuclear equation for the alpha decay of Po-216.

Alpha radiation is the 18-wheeler truck of radioactivity. The alpha particle is by far the most massive of all particles emitted by radioactive nuclei. Consequently, alpha radiation has the most potential to interact with and damage other molecules, including biological ones. Highly energetic radiation interacts with other molecules and atoms by ionizing them. When radiation ionizes molecules within the cells of living organisms, those molecules may undergo damaging chemical reactions, and the cells can die or begin to reproduce abnormally. The ability of radiation to ionize other molecules and atoms is called its **ionizing power**. Of all types of radioactivity, alpha radiation has the highest ionizing power.

However, alpha particles, because of their large size, have the lowest **penetrating power**—the ability to penetrate matter. (Imagine a semitruck trying to get through a traffic jam.) In order for radiation to damage important molecules within living cells, it must penetrate into the cell. Alpha radiation does not easily penetrate into cells because it can be stopped by a sheet of paper, by clothing, or even by air. Consequently, a low-level alpha emitter that remains outside the body is relatively safe. If an alpha emitter is ingested, however, it becomes very dangerous because the alpha particles then have direct access to the molecules that compose organs and tissues.

Beta (β) Decay

Beta (β) decay occurs when an unstable nucleus emits an electron (Figure 21.3 ▶). How does a nucleus, which contains only protons and neutrons, emit an electron? In some unstable nuclei, a neutron changes into a proton and emits an electron.

Beta decay Neutron ⟶ proton + emitted electron

The symbol for a beta (β) particle in a nuclear equation is:

Beta (β) particle $^{0}_{-1}e$ •

We can represent beta decay with a nuclear equation:

$$^{1}_{0}\text{n} \longrightarrow {}^{1}_{1}\text{p} + {}^{0}_{-1}\text{e}$$

The −1 reflects the charge of the electron, which is equivalent to an atomic number of −1 in a nuclear equation. When an atom emits a beta particle, its atomic number increases by 1 because it now has an additional proton. For example, the nuclear equation for the beta decay of radium-228 is:

$$^{228}_{88}\text{Ra} \longrightarrow {}^{228}_{89}\text{Ac} + {}^{0}_{-1}\text{e}$$

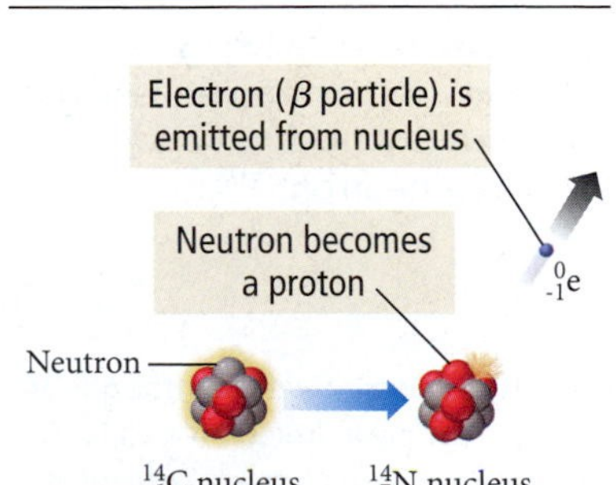

▲ **FIGURE 21.3 Beta Decay** In beta decay, a neutron emits an electron and becomes a proton.

This kind of beta radiation is also called beta minus (β^-) radiation due to its negative charge.

Notice that the nuclear equation is balanced—the sum of the mass numbers on both sides is equal, and the sum of the atomic numbers on both sides is equal.

Beta radiation is the four-door sedan of radioactivity. Beta particles are much less massive than alpha particles and consequently have a lower ionizing power. However, because of their smaller size, beta particles have a higher penetrating power and only something as substantive as a sheet of metal or a thick piece of wood stops them. Consequently, a beta emitter outside of the body poses a higher risk than an alpha emitter. If ingested, however, the beta emitter does less damage than an alpha emitter.

Gamma (γ) Ray Emission

Gamma (γ) ray emission is significantly different from alpha or beta radiation. Gamma radiation is a form of *electromagnetic* radiation (see Section 3.2). Gamma rays are high-energy (short-wavelength) photons. The symbol for a gamma ray is:

Gamma (γ) ray $\quad {}^{0}_{0}\gamma$

A gamma ray has no charge and no mass. When a gamma-ray photon is emitted from a radioactive atom, it does not change the mass number or the atomic number of the element. Gamma rays, however, are usually emitted in conjunction with other types of radiation. For example, the alpha emission of U-238 (discussed previously) is accompanied by the emission of a gamma ray.

$${}^{238}_{92}\text{U} \longrightarrow {}^{234}_{90}\text{Th} + {}^{4}_{2}\text{He} + {}^{0}_{0}\gamma$$

Gamma rays are the motorbikes of radioactivity. They have the lowest ionizing power, but the highest penetrating power. (Imagine a motorbike zipping through a traffic jam.) Stopping gamma rays requires several inches of lead shielding or thick slabs of concrete.

Positron Emission

Positron Emission

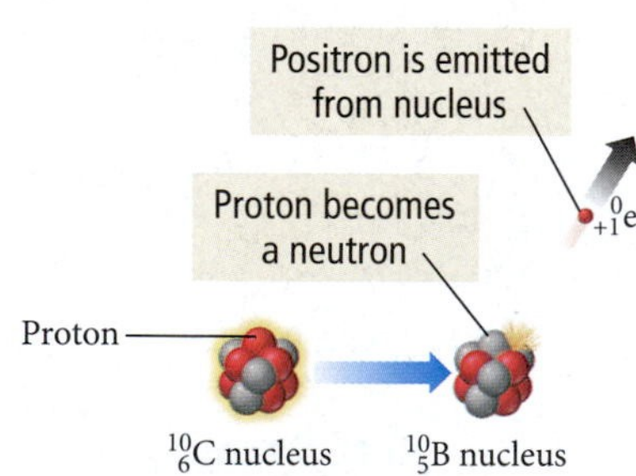

▲ **FIGURE 21.4 Positron Emission** In positron emission, a proton emits a positron and becomes a neutron.

Positron emission occurs when an unstable nucleus emits a positron (Figure 21.4 ◀). A **positron** is the *antiparticle* of the electron; it has the same mass as an electron, but the opposite charge. If a positron collides with an electron, the two particles annihilate each other, releasing energy in the form of gamma rays. In positron emission, a proton becomes a neutron and emits a positron.

Positron emission $\quad$ Proton $\longrightarrow$ neutron + emitted positron

The symbol for a positron in a nuclear equation is:

Positron $\quad {}^{0}_{+1}\text{e}$

We can represent positron emission with this nuclear equation:

$${}^{1}_{1}\text{p} \longrightarrow {}^{1}_{0}\text{n} + {}^{0}_{+1}\text{e}$$

When an atom emits a positron, its atomic number *decreases* by 1 because it has one less proton after emission. Consider the nuclear equation for the positron emission of phosphorus-30 as an example:

$${}^{30}_{15}\text{P} \longrightarrow {}^{30}_{14}\text{Si} + {}^{0}_{+1}\text{e}$$

Positron emission can be thought of as a type of beta emission and is sometimes referred to as beta plus emission (β^+).

We can determine the identity and symbol of the daughter nuclide in any positron emission in a manner similar to that used for alpha and beta decay, as shown in Example 21.2. Positrons are similar to beta particles in their ionizing and penetrating power.

Electron Capture

Unlike the forms of radioactive decay discussed so far, electron capture involves a particle being *absorbed by* instead of *emitted from* an unstable nucleus. **Electron capture** occurs when a nucleus assimilates an electron from an inner orbital of its electron cloud. Like positron emission, the net effect of electron capture is the conversion of a proton into a neutron.

Electron capture Proton + electron $\longrightarrow$ neutron

We can represent electron capture with this nuclear equation:

$$^{1}_{1}\text{p} + {}^{0}_{-1}\text{e} \longrightarrow {}^{1}_{0}\text{n}$$

When an atom undergoes electron capture, its atomic number decreases by 1 because it has one less proton. For example, when Ru-92 undergoes electron capture, its atomic number changes from 44 to 43:

$$^{92}_{44}\text{Ru} + {}^{0}_{-1}\text{e} \longrightarrow {}^{92}_{43}\text{Tc}$$

Table 21.1 summarizes the different kinds of radiation.

EXAMPLE 21.2

Writing Nuclear Equations for Beta Decay, Positron Emission, and Electron Capture

Write the nuclear equation for each type of decay.

(a) beta decay in Bk-249
(b) positron emission in O-15
(c) electron capture in I-111

SOLUTION

(a) In beta decay, the atomic number *increases* by 1 and the mass number remains unchanged. The daughter nuclide is element number 98, californium.	$^{249}_{97}\text{Bk} \longrightarrow {}^{249}_{98}? + {}^{0}_{-1}\text{e}$ $^{249}_{97}\text{Bk} \longrightarrow {}^{249}_{98}\text{Cf} + {}^{0}_{-1}\text{e}$
(b) In positron emission, the atomic number *decreases* by 1 and the mass number remains unchanged. The daughter nuclide is element number 7, nitrogen.	$^{15}_{8}\text{O} \longrightarrow {}^{15}_{7}? + {}^{0}_{+1}\text{e}$ $^{15}_{8}\text{O} \longrightarrow {}^{15}_{7}\text{N} + {}^{0}_{+1}\text{e}$
(c) In electron capture, the atomic number also *decreases* by 1 and the mass number remains unchanged. The daughter nuclide is element number 52, tellurium.	$^{111}_{53}\text{I} + {}^{0}_{-1}\text{e} \longrightarrow {}^{111}_{52}?$ $^{111}_{53}\text{I} + {}^{0}_{-1}\text{e} \longrightarrow {}^{111}_{52}\text{Te}$

FOR PRACTICE 21.2

(a) Write three nuclear equations to represent the nuclear decay sequence that begins with the alpha decay of U-235 followed by a beta decay of the daughter nuclide and then another alpha decay.
(b) Write the nuclear equation for the positron emission of Na-22.
(c) Write the nuclear equation for electron capture in Kr-76.

FOR MORE PRACTICE 21.2

Potassium-40 decays to produce Ar-40. What is the method of decay? Write the nuclear equation for this decay.

TABLE 21.1 Modes of Radioactive Decay

Decay Mode	Process	Change in: A	Z	N/Z*	Example
α	Parent nuclide → Daughter nuclide + ^4_2He (α particle)	−4	−2	Increase	$^{238}_{92}U \longrightarrow {}^{234}_{90}Th + {}^{4}_{2}He$
β	Parent nuclide (Neutron) → Daughter nuclide (Neutron becomes a proton) + $^{0}_{-1}e$ (β particle)	0	+1	Decrease	$^{228}_{88}Ra \longrightarrow {}^{228}_{89}Ac + {}^{0}_{-1}e$
γ	Excited nuclide → Stable nuclide + $^0_0\gamma$ (Photon)	0	0	None	$^{234}_{90}Th \longrightarrow {}^{234}_{90}Th + {}^{0}_{0}\gamma$
Positron emission	Parent nuclide (Proton) → Daughter nuclide (Proton becomes a neutron) + $^{0}_{+1}e$ (Positron)	0	−1	Increase	$^{30}_{15}P \longrightarrow {}^{30}_{14}Si + {}^{0}_{+1}e$
Electron capture	Parent nuclide (Proton) + $^{0}_{-1}e$ → Daughter nuclide (Proton becomes a neutron)	0	−1	Increase	$^{92}_{44}Ru + {}^{0}_{-1}e \longrightarrow {}^{92}_{43}Tc$

* Neutron-to-proton ratio

21.1 Cc Conceptual Connection — Alpha and Beta Decay

Consider the graphical representation of a series of decays shown below. Which kinds of decay do the arrows labeled *x* and *y* represent?

(a) *x* corresponds to alpha decay and *y* corresponds to positron emission.
(b) *x* corresponds to positron emission and *y* corresponds to alpha decay.
(c) *x* corresponds to alpha decay and *y* corresponds to beta decay.
(d) *x* corresponds to beta decay and *y* corresponds to alpha decay.

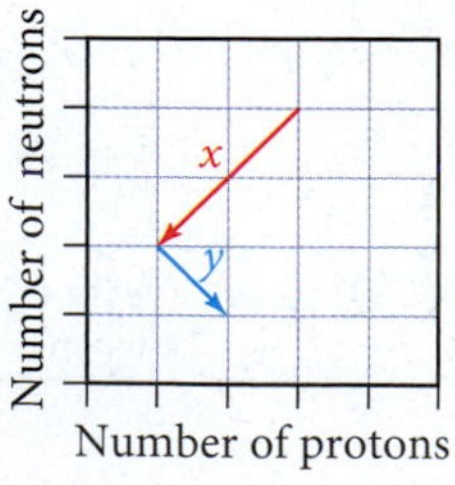

21.4 The Valley of Stability: Predicting the Type of Radioactivity

So far, we have described various different types of radioactivity. But what causes a particular nuclide to be radioactive in the first place? And why do some nuclides decay via alpha decay, while others decay via beta decay or positron emission? Nuclear properties—like all properties—depend on structure, but in this case the relevant structure is not that of an atom or molecule but of the nucleus. The particles that compose the nucleus—protons and neutrons—occupy energy levels that are similar to the energy levels occupied by electrons. A full examination of nuclear structure is beyond the scope of this text, but we can examine a couple of simple factors that influence the stability of the nucleus and the nature of its decay.

A nucleus is a collection of **nucleons**—protons (positively charged) and neutrons (uncharged). We know that positively charged particles such as protons repel one another. So what binds the nucleus together? The binding is provided by a fundamental force of physics known as the **strong force**. All nucleons are attracted to one another by the strong force. However, the strong force acts only at very short distances. We can think of the stability of a nucleus as a balance between the *repulsive* coloumbic force among protons and the *attractive* strong force among all nucleons. The neutrons in a nucleus, therefore, play an important role in stabilizing the nucleus because they attract other nucleons (through the strong force) but lack the repulsive force associated with positive charge. It might seem that adding more neutrons would *always* lead to greater stability, so that the more neutrons the better. This is not the case, however, because as we just discussed protons and neutrons occupy energy levels in a nucleus. As you add more neutrons, they must occupy increasingly higher energy levels within the nucleus. At some point, the energy payback from the strong force is not enough to compensate for the high energy state that the neutrons must occupy.

An important number in determining nuclear stability is the *ratio* of neutrons to protons (N/Z). Figure 21.5 ▶ shows a plot of the number of neutrons versus the number of protons for all known stable nuclei. The green dots along the diagonal of the graph represent stable nuclei; this region is known as the *valley* (or *island*) *of stability*. Notice that for the lighter elements, the N/Z ratio of stable isotopes is about one (equal numbers of neutrons and protons). For example, the most abundant isotope of carbon ($Z = 6$) is carbon-12, which contains six protons and six neutrons. However, beyond about $Z = 20$, the N/Z ratio of stable nuclei begins to get larger. For example, at $Z = 40$, stable nuclei have an N/Z ratio of about 1.25 and at $Z = 80$, the N/Z ratio reaches about 1.5. Above $Z = 83$, stable nuclei do not exist—bismuth ($Z = 83$) is the heaviest element with stable (nonradioactive) isotopes.

The type of radioactivity emitted by a nuclide depends in part on the N/Z ratio.

N/Z too high: Nuclides that lie above the valley of stability have too many neutrons and tend to convert neutrons to protons via beta decay. The process of undergoing beta decay moves the nuclide down in the plot in Figure 21.5 and closer to (or into) the valley of stability.

N/Z too low: Nuclides that lie below the valley of stability have too many protons and tend to convert protons to neutrons via positron emission or electron capture. This moves the nuclide up in the plot in Figure 21.5 and closer to (or into) the valley of stability. (Alpha decay also raises the N/Z ratio for nuclides in which $N/Z > 1$, but the effect is smaller than for positron emission or electron capture.)

One way to decide whether a particular nuclide has an N/Z that is too high, too low, or about right is to consult Figure 21.5. Those nuclides that lie within the valley of stability are stable. Alternatively, we can also compare the mass number of the nuclide to the atomic mass listed in the periodic table for the corresponding element. The atomic mass is an average of the masses of the most stable nuclides for

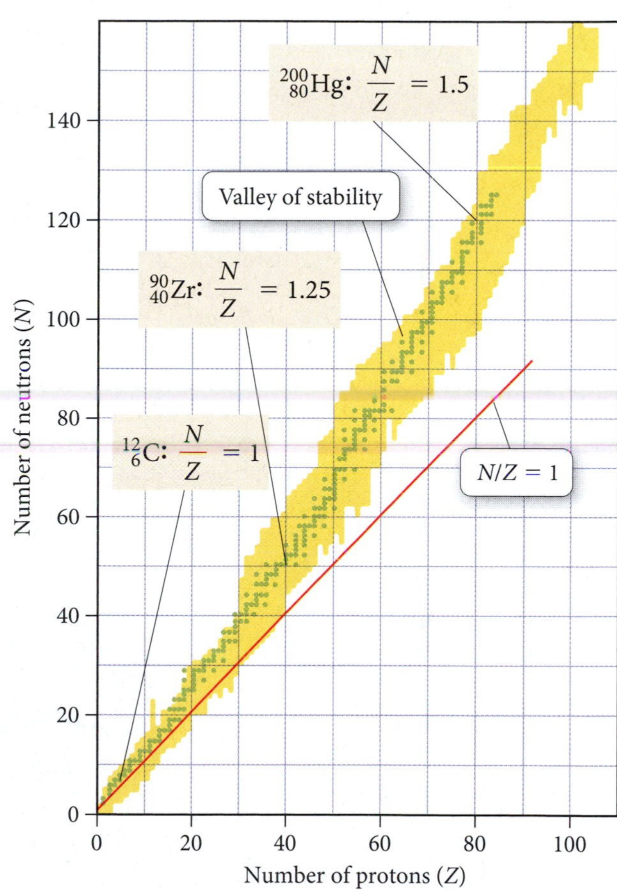

▲ **FIGURE 21.5 Stable and Unstable Nuclei** A plot of N (the number of neutrons) versus Z (the number of protons) for all known stable nuclei—represented by green dots on this graph—shows that these nuclei cluster together in a region known as the valley (or island) of stability. Nuclei with an N/Z ratio that is too high tend to undergo beta decay. Nuclei with an N/Z ratio that is too low tend to undergo positron emission or electron capture.

an element (which is why they occur naturally) and thus represents an N/Z that is about right. For example, suppose we want to evaluate N/Z for Ru-112. Ruthenium has an atomic mass of 101.07 so we know that the nuclide with a mass number of 112 must contain too many neutrons and therefore its N/Z is too high. Example 21.3 demonstrates how to apply these considerations in predicting the mode of decay for a nucleus.

EXAMPLE 21.3 Predicting the Type of Radioactive Decay

Predict whether each nuclide is more likely to decay via beta decay or positron emission.

(a) Mg-28 **(b)** Mg-22 **(c)** Mo-102

SOLUTION

(a) Magnesium-28 has 16 neutrons and 12 protons, so $N/Z = 1.33$. However, for $Z = 12$, you can see from Figure 21.5 that stable nuclei should have an N/Z of about 1. Alternatively, you can see from the periodic table that the atomic mass of magnesium is 24.31. Therefore, a nuclide with a mass number of 28 is too heavy to be stable because the N/Z ratio is too high, so Mg-28 undergoes *beta decay*, resulting in the conversion of a neutron to a proton.

(b) Magnesium-22 has 10 neutrons and 12 protons, so $N/Z = 0.83$ (too low). Alternatively, you can see from the periodic table that the atomic mass of magnesium is 24.31. A nuclide with a mass number of 22 is too light; the N/Z ratio is too low. Therefore, Mg-22 undergoes *positron emission*, resulting in the conversion of a proton to a neutron. (Electron capture would accomplish the same thing as positron emission, but in Mg-22, positron emission is the only decay mode observed.)

(c) Molybdenum-102 has 60 neutrons and 42 protons, so $N/Z = 1.43$. However, for $Z = 42$, you can see from Figure 21.5 that stable nuclei should have an N/Z ratio of about 1.3. Alternatively, you can see from the periodic table that the atomic mass of molybdenum is 95.94. A nuclide with a mass number of 102 is too heavy to be stable; the N/Z ratio is too high. Therefore, Mo-102 undergoes *beta decay*, resulting in the conversion of a neutron to a proton.

FOR PRACTICE 21.3

Predict whether each nuclide is more likely to decay via beta decay or positron emission.

(a) Pb-192 **(b)** Pb-212 **(c)** Xe-114

TABLE 21.2 Number of Stable Nuclides with Even and Odd Numbers of Nucleons

Z	N	Number of Nuclides
Even	Even	157
Even	Odd	53
Odd	Even	50
Odd	Odd	5

Magic Numbers

In addition to the N/Z ratio, the *actual number* of protons and neutrons also affects the stability of the nucleus. Table 21.2 shows the number of nuclei with different possible combinations of even or odd nucleons. Notice that a large number of stable nuclides have both an even number of protons and an even number of neutrons. Only five stable nuclides have an odd number of protons and odd number of neutrons.

The reason for this is related to how nucleons occupy energy levels within the nucleus. Just as atoms with certain numbers of electrons are uniquely stable (in particular, the number of electrons associated with the noble gases: 2, 10, 18, 36, 54, etc.), so nuclei with certain numbers of nucleons (N or $Z = 2, 8, 20, 28, 50, 82$, and $N = 126$), are uniquely stable. These numbers are often referred to as **magic numbers**. Nuclei containing a magic number of protons or neutrons are particularly stable. Note that the magic numbers are even; this accounts in part for the abundance of stable nuclides with even numbers of nucleons. Moreover, nucleons also have a tendency to pair together (much as electrons pair). This tendency and the resulting stability of paired nucleons also contribute to the abundance of stable nuclides with even numbers of nucleons.

Radioactive Decay Series

Atoms with $Z > 83$ are radioactive and decay in one or more steps involving primarily alpha and beta decay (with some gamma decay to carry away excess energy). For example, uranium (atomic number 92) is the heaviest naturally occurring element. Its most common isotope is U-238, an alpha emitter that decays to Th-234.

$$^{238}_{92}\text{U} \longrightarrow ^{234}_{90}\text{Th} + ^{4}_{2}\text{He}$$

The daughter nuclide, Th-234, is itself radioactive—it is a beta emitter that decays to Pa-234.

$$^{234}_{90}\text{Th} \longrightarrow ^{234}_{91}\text{Pa} + ^{0}_{-1}\text{e}$$

Protactinium-234 is also radioactive, decaying to U-234 via beta emission. Radioactive decay continues until a stable nuclide, Pb-206, is reached. Figure 21.6 ▶ illustrates the entire uranium-238 decay series.

A Decay Series

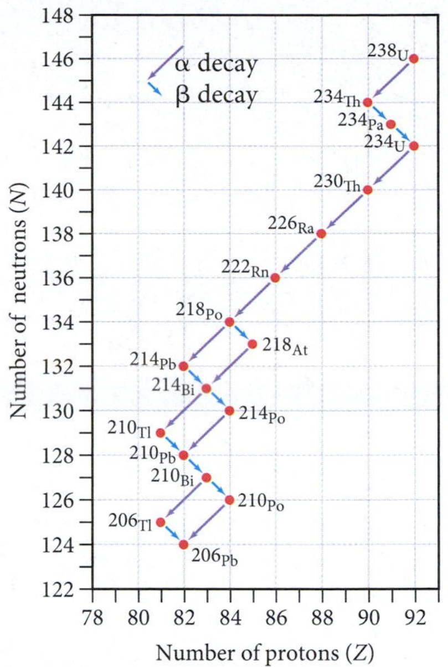

▲ **FIGURE 21.6 The Uranium-238 Radioactive Decay Series** Uranium-238 decays via a series of steps ending in Pb-206, a stable element. Each diagonal line to the left represents an alpha decay, and each diagonal line to the right represents a beta decay.

21.5 Detecting Radioactivity

The particles emitted by radioactive nuclei are highly energetic and can therefore be readily detected. A *radiation detector* detects particles through their interactions with atoms or molecules. The simplest radiation detectors are pieces of photographic film that become exposed when radiation passes through them. **Film-badge dosimeters**—which consist of photographic film held in a small case that is pinned to clothing—are issued to most people working with or near radioactive substances (Figure 21.7 ▼). These badges are collected and processed (or developed) regularly as a way to monitor a person's exposure. The more exposed the film has become in a given period of time, the more radioactivity the person has been exposed to during that period.

Radioactivity can be instantly detected with devices such as a **Geiger-Müller counter** (Figure 21.8 ▼). In this instrument (commonly referred to as a Geiger counter), particles emitted by

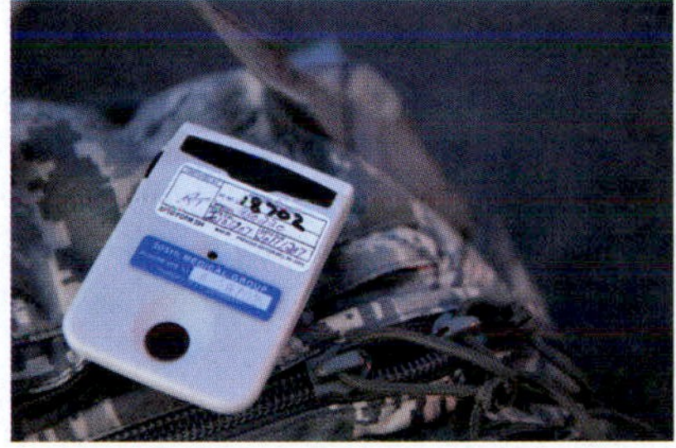

◀ **FIGURE 21.7 Film-Badge Dosimeter** A film-badge dosimeter consists of a piece of photographic film in a light-resistant container. The film's exposure in a given time is proportional to the amount of radiation it receives.

▼ **FIGURE 21.8 Geiger-Müller Counter** When ionizing radiation passes through the argon-filled chamber, it ionizes the argon atoms, giving rise to a brief, tiny pulse of electrical current that is transduced onto a meter or into an audible click.

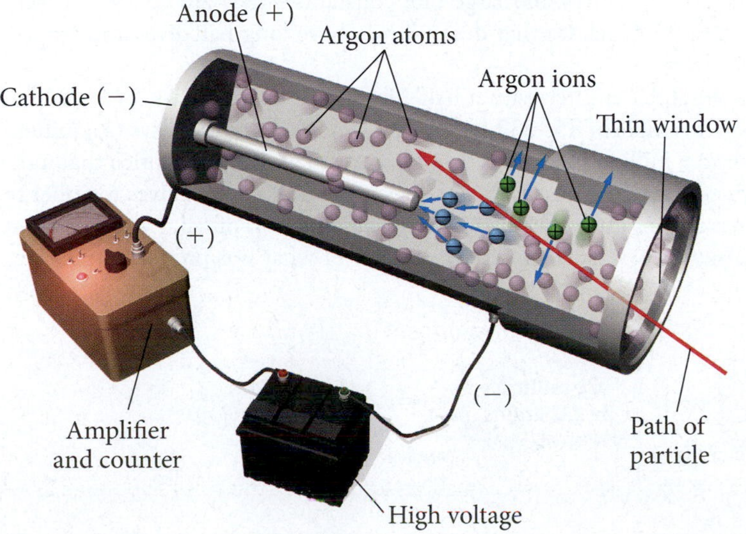

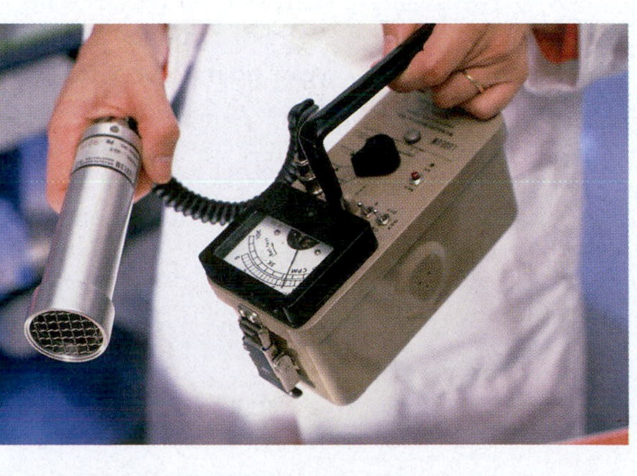

radioactive nuclei pass through an argon-filled chamber. The energetic particles create a trail of ionized argon atoms. High voltage applied between a wire within the chamber and the chamber itself causes these newly formed ions to produce an electrical signal that is displayed on a meter or turned into an audible click. Each click corresponds to a radioactive particle passing through the argon gas chamber. This clicking is the stereotypical sound most people associate with a radiation detector.

A second type of device commonly used to detect radiation instantly is a **scintillation counter**. In a scintillation counter, radioactive emissions pass through a material (such as NaI or CsI) that emits ultraviolet or visible light in response to excitation by energetic particles. The radioactivity excites the atoms to a higher energy state. The atoms release this energy as light, which is detected and turned into an electrical signal that is read on a meter.

21.6 The Kinetics of Radioactive Decay and Radiometric Dating

Radioactivity is a natural component of our environment. The ground beneath you most likely contains radioactive atoms that emit radiation. The food you eat contains a residual quantity of radioactive atoms that are absorbed into your body fluids and incorporated into tissues. Small amounts of radiation from space make it through the atmosphere to constantly bombard Earth. Humans and other living organisms have evolved in this environment and have adapted to survive in it.

One reason for the radioactivity in our environment is the instability of all atomic nuclei beyond atomic number 83 (bismuth). Every element with more than 83 protons in its nucleus is unstable and therefore radioactive. In addition, some isotopes of elements with fewer than 83 protons are also unstable and radioactive. Radioactive nuclides *persist* in our environment because new ones are constantly being formed and because many of the existing ones decay only very slowly.

All radioactive nuclei decay via first-order kinetics, so the rate of decay in a particular sample is directly proportional to the number of nuclei present as indicated in this equation:

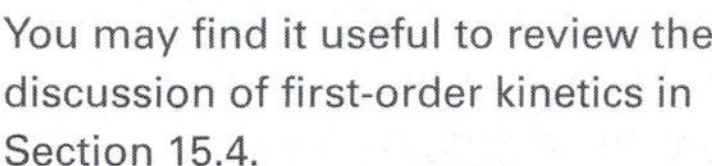

You may find it useful to review the discussion of first-order kinetics in Section 15.4.

$$\text{Rate} = kN$$

where N is the number of radioactive nuclei and k is the rate constant. Different radioactive nuclides decay into their daughter nuclides with different rate constants. Some nuclides decay quickly (large rate constant), while others decay slowly (small rate constant).

The time it takes for one-half of the parent nuclides in a radioactive sample to decay to the daughter nuclides is the *half-life* and is identical to the concept of half-life for chemical reactions that we discussed in Chapter 15. Thus, the relationship between the half-life of a nuclide and its rate constant is given by the same expression that we derived for a first-order reaction in Section 15.5:

$$t_{1/2} = \frac{0.693}{k} \qquad [21.1]$$

Nuclides that decay quickly have short half-lives and large rate constants—they are considered very active (many decay events per unit time). Nuclides that decay slowly have long half-lives and are less active (fewer decay events per unit time).

For example, thorium-232 is an alpha emitter with a half-life of 1.4×10^{10} years, or 14 billion years, so it is not particularly active. A sample of Th-232 containing 1 million atoms decays to $\frac{1}{2}$ million atoms in 14 billion years and then to $\frac{1}{4}$ million in another 14 billion years and so on. Notice that a radioactive sample does *not* decay to *zero* atoms in two half-lives—we can't add two half-lives together to get a "whole" life. The amount that remains after one half-life is always one-half of what was present at the start. The amount that remains after two half-lives is one-quarter of what was present at the start, and so on.

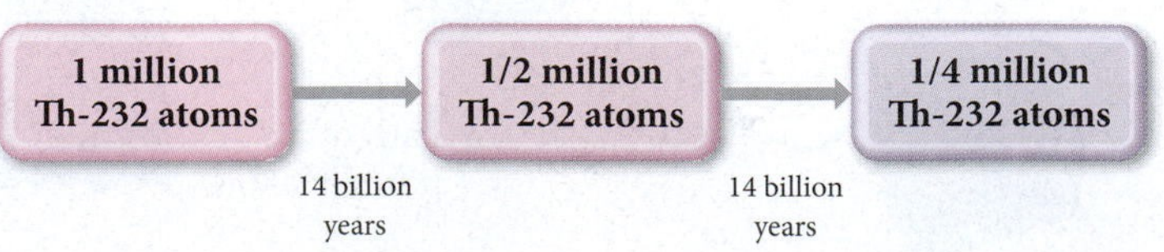

Decay of Radon-220

◀ **FIGURE 21.9 The Decay of Radon-220** Radon-220 decays with a half-life of approximately 1 minute.

Some nuclides have very short half-lives. For example, radon-220 has a half-life of approximately 1 minute (Figure 21.9 ▲). A 1-million-atom sample of radon-220 decays to $\frac{1}{4}$ million radon-220 atoms in just 2 minutes and to approximately 1000 atoms in 10 minutes. Table 21.3 lists several nuclides and their half-lives.

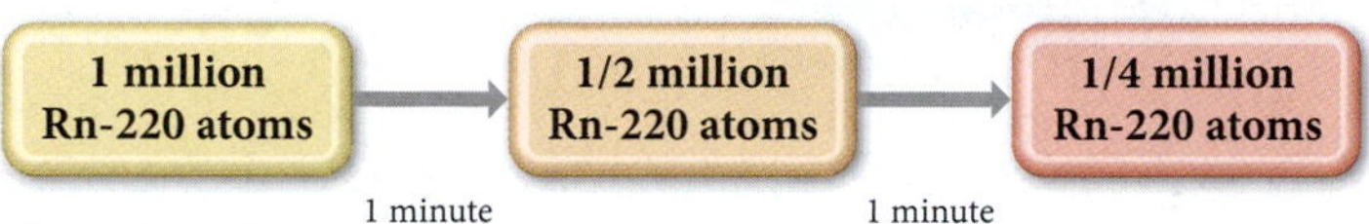

TABLE 21.3 Selected Nuclides and Their Half-Lives

Nuclide	Half-Life	Type of Decay
$^{232}_{90}Th$	1.4×10^{10} yr	Alpha
$^{238}_{92}U$	4.5×10^{9} yr	Alpha
$^{14}_{6}C$	5730 yr	Beta
$^{220}_{86}Rn$	55.6 s	Alpha
$^{219}_{90}Th$	1.05×10^{-6} s	Alpha

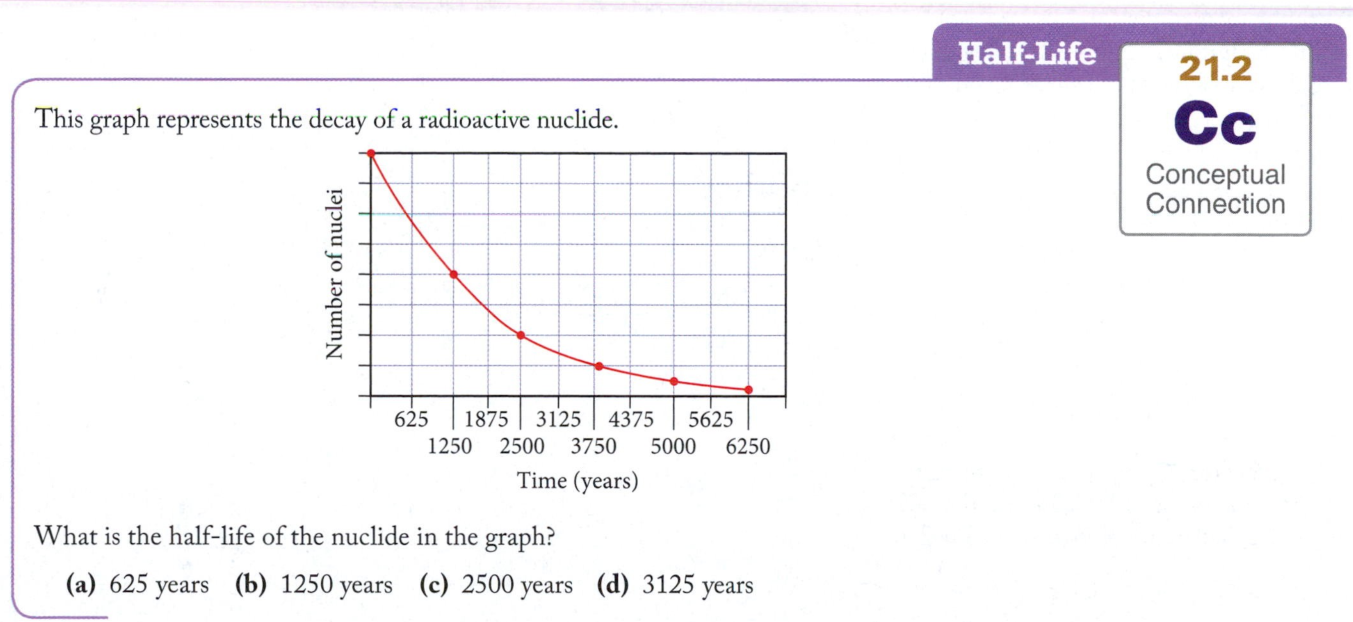

Half-Life

21.2 Cc Conceptual Connection

This graph represents the decay of a radioactive nuclide.

What is the half-life of the nuclide in the graph?

(a) 625 years **(b)** 1250 years **(c)** 2500 years **(d)** 3125 years

The Integrated Rate Law

Recall from Chapter 15 that for first-order chemical reactions, the concentration of a reactant as a function of time is given by the integrated rate law.

$$\ln \frac{[A]_t}{[A]_0} = -kt \qquad [21.2]$$

Since nuclear decay follows first-order kinetics, we can substitute the number of nuclei for concentration to arrive at the equation:

$$\ln\frac{N_t}{N_0} = -kt \qquad [21.3]$$

where N_t is the number of radioactive nuclei at time t and N_0 is the initial number of radioactive nuclei. Example 21.4 demonstrates the use of this equation.

EXAMPLE 21.4

Radioactive Decay Kinetics

Plutonium-236 is an alpha emitter with a half-life of 2.86 years. If a sample initially contains 1.35 mg of Pu-236, what mass of Pu-236 is present after 5.00 years?

SORT You are given the initial mass of Pu-236 in a sample and asked to find the mass after 5.00 years.

GIVEN: $m_{\text{Pu-236}}(\text{initial}) = 1.35\text{ mg}$; $t = 5.00\text{ yr}$; $t_{1/2} = 2.86\text{ yr}$

FIND: $m_{\text{Pu-236}}(\text{final})$

STRATEGIZE Use the integrated rate law (Equation 21.3) to solve this problem. You can determine the value of the rate constant (k) from the half-life expression (Equation 21.1).

CONCEPTUAL PLAN

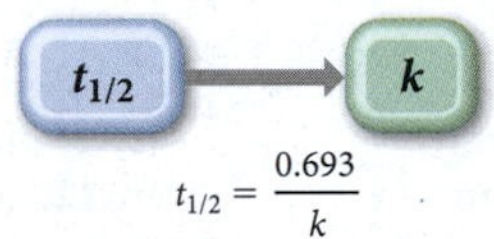

Use the value of the rate constant, the initial mass of Pu-236, and the time along with integrated rate law to find the final mass of Pu-236. Since the mass of the Pu-236 ($m_{\text{Pu-236}}$) is directly proportional to the number of atoms (N), and since the integrated rate law contains the ratio (N_t/N_0), you can substitute the initial and final masses.

$k, m_{\text{Pu-236}}(\text{initial}), t \rightarrow m_{\text{Pu-236}}(\text{final})$

$$\ln\frac{N_t}{N_0} = -kt$$

SOLVE Follow the conceptual plan. Begin by determining the rate constant from the half-life.

SOLUTION

$$t_{1/2} = \frac{0.693}{k}$$

$$k = \frac{0.693}{t_{1/2}} = \frac{0.693}{2.86\text{ yr}} = 0.2423/\text{yr}$$

Solve the integrated rate law for N_t and substitute the values of the rate constant, the initial mass of Pu-236, and the time into the solved equation. Calculate the final mass of Pu-236.

$$\ln\frac{N_t}{N_0} = -kt$$

$$\frac{N_t}{N_0} = e^{-kt}$$

$$N_t = N_0 e^{-kt}$$

$$N_t = 1.35\text{ mg}\,[e^{-(0.2423/\text{yr})(5.00\text{ yr})}]$$

$$N_t = 0.402\text{ mg}$$

CHECK The units of the answer (mg) are correct. The magnitude of the answer (0.402 mg) is about one-third of the original mass (1.35 mg), which seems reasonable given that the amount of time is between one and two half-lives. (One half-life would result in one-half of the original mass, and two half-lives would result in one-fourth of the original mass.)

FOR PRACTICE 21.4

How long will it take for the 1.35 mg sample of Pu-236 in Example 21.4 to decay to 0.100 mg?

Because radioactivity is a first-order process, the rate of decay is linearly proportional to the number of nuclei in the sample. Therefore, we can use the initial rate of decay (rate_0) and the rate of decay at time t (rate_t) in the integrated rate law.

$$\text{Rate}_t = kN_t \qquad \text{Rate}_0 = kN_0$$

$$\frac{N_t}{N_0} = \frac{\text{rate}_t/k}{\text{rate}_0/k} = \frac{\text{rate}_t}{\text{rate}_0}$$

Substituting into Equation 21.3, we get the following result:

$$\ln \frac{\text{rate}_t}{\text{rate}_0} = -kt \qquad [21.4]$$

We can use Equation 21.4 to predict how the rate of decay of a radioactive sample will change with time or how much time has passed based on how the rate has changed (see Examples 21.5 and 21.6 later in this section).

The radioactive isotopes in our environment and their predictable decay with time can therefore be used to estimate the age of rocks or artifacts containing those isotopes. The technique is known as **radiometric dating**, and here we examine two different types individually.

Half-Life and the Amount of Radioactive Sample

21.3 Cc Conceptual Connection

A sample initially contains 1.6 moles of a radioactive isotope. How much of the sample remains after four half-lives?

(a) 0.0 mol

(b) 0.10 mol

(c) 0.20 mol

(d) 0.40 mol

Radiocarbon Dating: Using Radioactivity to Measure the Age of Fossils and Artifacts

Archeologists, geologists, anthropologists, and other scientists use **radiocarbon dating**, a technique devised in 1949 by Willard Libby (1908–1980) at the University of Chicago, to estimate the ages of fossils and artifacts. For example, in 1947, young shepherds searching for a stray goat near the Dead Sea (east of Jerusalem) entered a cave and discovered ancient scrolls that had been stuffed into jars. These scrolls—now named the Dead Sea Scrolls—are 2000-year-old texts of the Hebrew Bible, predating other previously discovered manuscripts by almost a thousand years.

The Dead Sea Scrolls, like other ancient artifacts, contain a radioactive signature that reveals their age. This signature results from the presence of carbon-14 (which is radioactive) in the environment. Carbon-14 is constantly formed in the upper atmosphere by the neutron bombardment of nitrogen.

$$^{14}_{7}\text{N} + ^{1}_{0}\text{n} \longrightarrow ^{14}_{6}\text{C} + ^{1}_{1}\text{H}$$

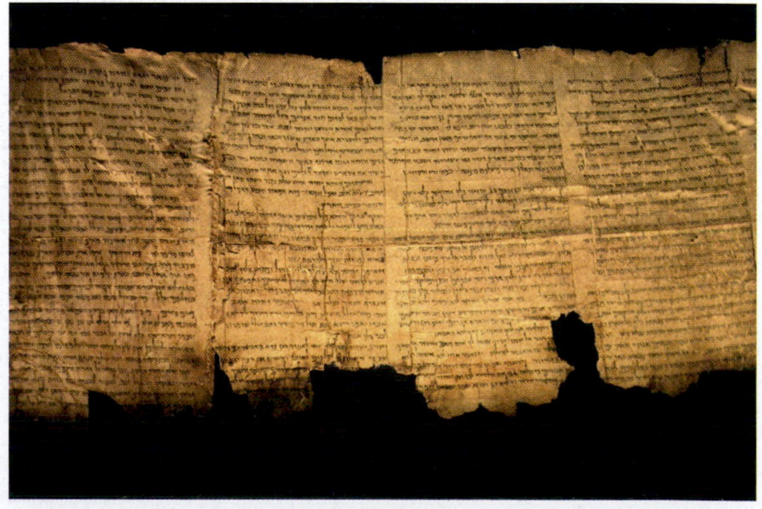

▲ The Dead Sea Scrolls are 2000-year-old biblical manuscripts. Their age was determined by radiocarbon dating.

Libby received the Nobel Prize in 1960 for the development of radiocarbon dating.

After it forms, carbon-14 decays back to nitrogen by beta emission with a half-life of 5730 years.

$$^{14}_{6}\text{C} \longrightarrow {}^{14}_{7}\text{N} + {}^{0}_{-1}\text{e} \qquad t_{1/2} = 5730 \text{ yr}$$

The continuous formation of carbon-14 in the atmosphere and its continuous decay to nitrogen-14 produce a nearly constant equilibrium amount of atmospheric carbon-14. The atmospheric carbon-14 is oxidized to carbon dioxide and incorporated into plants by photosynthesis. The C-14 then makes its way up the food chain and ultimately into all living organisms. As a result, the tissues in all living plants, animals, and humans contain the same ratio of carbon-14 to carbon-12 ($^{14}\text{C}:^{12}\text{C}$) as that found in the atmosphere. When a living organism dies, however, it stops incorporating new carbon-14 into its tissues. The $^{14}\text{C}:^{12}\text{C}$ ratio then begins to decrease with a half-life of 5730 years.

Since many artifacts, including the Dead Sea Scrolls, are made from materials that were once living—such as papyrus, wood, or other plant and animal derivatives—the $^{14}\text{C}:^{12}\text{C}$ ratio in these artifacts indicates their age. For example, suppose an ancient artifact has a $^{14}\text{C}:^{12}\text{C}$ ratio that is 25% of that found in living organisms. How old is the artifact? Since it contains one-quarter as much carbon-14 as a living organism, it must be two half-lives or 11,460 years old. The maximum age that we can estimate from carbon-14 dating is about 50,000 years—beyond that, the amount of carbon-14 becomes too low to measure accurately.

The accuracy of carbon-14 dating can be checked against objects whose ages are known from historical sources. These kinds of comparisons reveal that ages obtained from C-14 dating may deviate from the actual ages by up to about 5%. For a 6000-year-old object, that is a margin of error of about 300 years. The reason for the deviations is the variance of atmospheric C-14 levels over time.

In order to make C-14 dating more accurate, scientists have studied the carbon-14 content of western bristlecone pine trees, which can live up to 5000 years. Each tree trunk contains growth rings corresponding to each year of the tree's life, and the wood in each ring incorporates carbon derived from the carbon dioxide in the atmosphere at that time. The rings thus provide a record of the historical atmospheric carbon-14 content. In addition, the rings of living trees can be correlated with the rings of dead trees (if part of the lifetimes of the trees overlapped), allowing the record to be extended back about 11,000 years. Using the data from the bristlecone pine, the 5% deviations from historical dates can be corrected. In this way, the known ages of bristlecone pine trees are used to calibrate C-14 dating, resulting in more accurate results.

◀ Some western bristlecone pine trees live up to 5000 years; scientists can precisely determine the age of a tree by counting the annual rings in its trunk. The trees can therefore be used to calibrate the time scale for radiocarbon dating.

EXAMPLE 21.5
Radiocarbon Dating

A skull believed to belong to an ancient human being has a carbon-14 decay rate of 4.50 disintegrations per minute per gram of carbon (4.50 dis/min · gC). If living organisms have a decay rate of 15.3 dis/min · gC, how old is the skull? (The decay rate is directly proportional to the amount of carbon-14 present.)

SORT You are given the current rate of decay for the skull and the assumed initial rate. You are asked to determine the age of the skull, which is the time that must have passed in order for the rate to have reached its current value.	**GIVEN:** $\text{rate}_t = 4.50$ dis/min · gC; $\text{rate}_0 = 15.3$ dis/min · gC; **FIND:** t

—*Continued on the next page*

Continued from the previous page—

STRATEGIZE Use the expression for half-life (Equation 21.1) to find the rate constant (k) from the half-life for C-14, which is 5730 yr (Table 21.3).

CONCEPTUAL PLAN

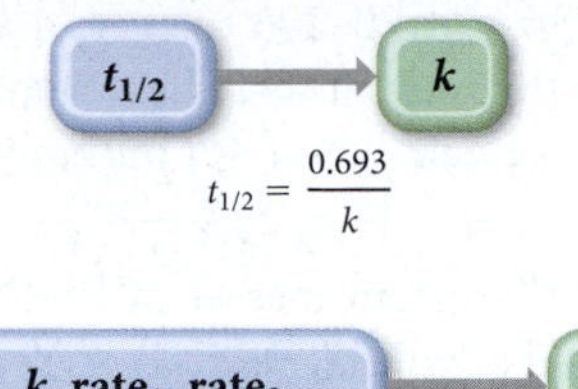

$$t_{1/2} = \frac{0.693}{k}$$

Use the value of the rate constant and the initial and current rates to find t from the integrated rate law (Equation 21.4).

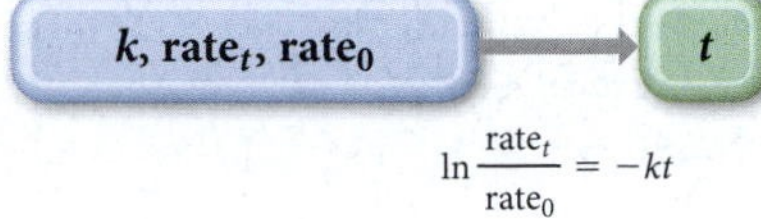

$$\ln \frac{\text{rate}_t}{\text{rate}_0} = -kt$$

SOLVE Follow the conceptual plan. Begin by finding the rate constant from the half-life.

SOLUTION

$$t_{1/2} = \frac{0.693}{k}$$

$$k = \frac{0.693}{t_{1/2}} = \frac{0.693}{5730 \text{ yr}}$$

$$= 1.2\underline{0}9 \times 10^{-4}/\text{yr}$$

Substitute the rate constant and the initial and current rates into the integrated rate law and solve for t.

$$\ln \frac{\text{rate}_t}{\text{rate}_0} = -kt$$

$$t = -\frac{\ln \frac{\text{rate}_t}{\text{rate}_0}}{k} = -\frac{\ln \frac{4.50 \text{ dis/min} \cdot \text{gC}}{15.3 \text{ dis/min} \cdot \text{gC}}}{1.2\underline{0}9 \times 10^{-4}/\text{yr}}$$

$$= 1.0 \times 10^{4} \text{ yr}$$

CHECK The units of the answer (yr) are correct. The magnitude of the answer is about 10,000 years, which is a little less than two half-lives. This value is reasonable given that two half-lives would result in a decay rate of about 3.8 dis/min • gC.

FOR PRACTICE 21.5

A researcher claims that an ancient scroll originated from Greek scholars in about 500 B.C. A measure of its carbon-14 decay rate gives a value that is 89% of that found in living organisms. How old is the scroll and could it be authentic?

Uranium/Lead Dating

Radiocarbon dating can only measure the ages of objects that were once living and that are relatively young (< 50,000 years). Other radiometric dating techniques can measure the ages of prehistoric objects that were never alive. The most dependable technique relies on the ratio of uranium-238 to lead-206 within igneous rocks (rocks of volcanic origin). This technique measures the time that has passed since the rock solidified (at which point the "radiometric clock" was reset).

Because U-238 decays into Pb-206 with a half-life of 4.5×10^{9} years, the relative amounts of U-238 and Pb-206 in a uranium-containing rock reveal its age. For example, if a rock originally contained U-238 and currently contains equal amounts of U-238 and Pb-206, the rock is 4.5 billion years old, assuming that it did not contain any Pb-206 when it was formed. The latter assumption can be tested because the lead that results from the decay of uranium has a different isotopic composition than the lead that is deposited in rocks at the time of their formation. Example 21.6 shows how we can use the relative amounts of Pb-206 and U-238 in a rock to estimate its age.

EXAMPLE 21.6

Using Uranium/Lead Dating to Estimate the Age of a Rock

A meteor contains 0.556 g of Pb-206 to every 1.00 g of U-238. Assuming that the meteor did not contain any Pb-206 at the time of its formation, determine the age of the meteor. Uranium-238 decays to lead-206 with a half-life of 4.5 billion years.

SORT You are given the current masses of Pb-206 and U-238 in a rock and asked to find its age. You are also given the half-life of U-238.

GIVEN: $m_{\text{U-238}} = 1.00\text{ g}$; $m_{\text{Pb-206}} = 0.556\text{ g}$; $t_{1/2} = 4.5 \times 10^9\text{ yr}$

FIND: t

STRATEGIZE Use the integrated rate law (Equation 21.3) to solve this problem. To do so, you must first determine the value of the rate constant (k) from the half-life expression (Equation 21.1).

CONCEPTUAL PLAN

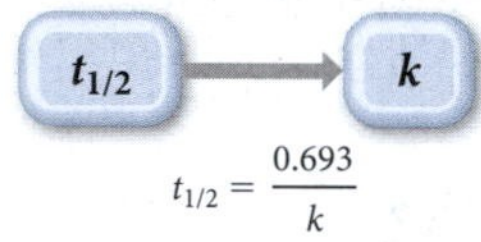

$$t_{1/2} = \frac{0.693}{k}$$

Before substituting into the integrated rate law, you need the ratio of the current amount of U-238 to the original amount (N_t/N_0). The current mass of uranium is 1.00 g. The initial mass includes the current mass (1.00 g) plus the mass that has decayed into lead-206, which you can determine from the current mass of Pb-206.

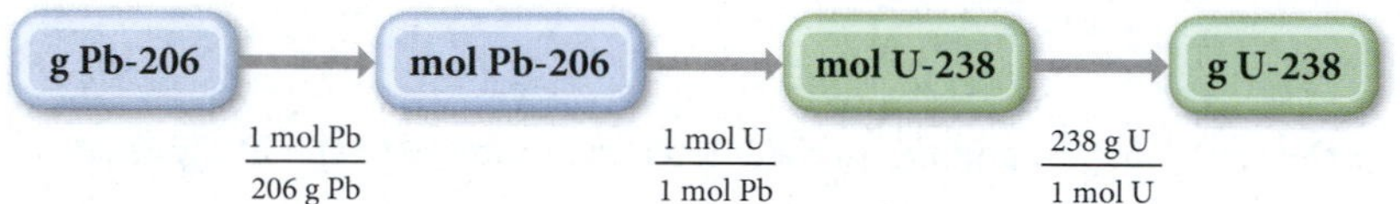

$$\frac{1\text{ mol Pb}}{206\text{ g Pb}} \quad \frac{1\text{ mol U}}{1\text{ mol Pb}} \quad \frac{238\text{ g U}}{1\text{ mol U}}$$

Use the value of the rate constant and the initial and current amounts of U-238 along with the integrated rate law to find t.

$k, N_t, N_0 \rightarrow t$

$$\ln\frac{N_t}{N_0} = -kt$$

SOLVE Follow your plan. Begin by finding the rate constant from the half-life.

SOLUTION

$$t_{1/2} = \frac{0.693}{k}$$

$$k = \frac{0.693}{t_{1/2}} = \frac{0.693}{4.5 \times 10^9\text{ yr}} = 1.5\underline{4} \times 10^{-10}/\text{yr}$$

Determine the mass in grams of U-238 required to form the given mass of Pb-206.

$$0.556\text{ g Pb-206} \times \frac{1\text{ mol Pb-206}}{206\text{ g Pb-206}} \times \frac{1\text{ mol U-238}}{1\text{ mol Pb-206}} \times \frac{238\text{ g U-238}}{1\text{ mol U-238}} = 0.64\underline{2}4\text{ g U-238}$$

Substitute the rate constant and the initial and current masses of U-238 into the integrated rate law and solve for t. (The initial mass of U-238 is the sum of the current mass and the mass that is required to form the given mass of Pb-206.)

$$\ln\frac{N_t}{N_0} = -kt$$

$$t = \frac{\ln\dfrac{N_t}{N_0}}{k} = -\frac{\ln\dfrac{1.00\text{ g}}{1.00\text{ g} + 0.64\underline{2}4\text{ g}}}{1.5\underline{4} \times 10^{-10}/\text{yr}} = 3.2 \times 10^9\text{ yr}$$

CHECK The units of the answer (yr) are correct. The magnitude of the answer is about 3.2 billion years, which is less than one half-life. This value is reasonable given that less than half of the uranium in the meteor has decayed into lead.

FOR PRACTICE 21.6

A rock contains a Pb-206 to U-238 mass ratio of 0.145:1.00. Assuming that the rock did not contain any Pb-206 at the time of its formation, determine its age.

The uranium/lead radiometric dating technique and other radiometric dating techniques (such as the decay of potassium-40 to argon-40) have been widely used to measure the ages of rocks on Earth and have produced highly consistent results. Rocks with ages greater than 3.5 billion years have been found on every continent. The oldest rocks have an age of approximately 4.0 billion years, establishing a lower limit for Earth's age (Earth must be at least as old as its oldest rocks). The ages of about 70 meteorites that have struck Earth have also been extensively studied and have been found to be about 4.5 billion years old. Since the meteorites were formed at the same time as our solar system (which includes Earth), the best estimate for Earth's age is about 4.5 billion years. That age is consistent with the estimated age of our universe—about 13.7 billion years.

The age of the universe is estimated from its expansion rate, which is measured by examining changes in the wavelength of light from distant galaxies.

21.7 The Discovery of Fission: The Atomic Bomb and Nuclear Power

In the mid-1930s Enrico Fermi (1901–1954), an Italian physicist, attempted to synthesize a new element by bombarding uranium—the heaviest known element at that time—with neutrons. Fermi speculated that if a neutron could be incorporated into the nucleus of a uranium atom, the nucleus might undergo beta decay, converting a neutron into a proton. If that happened, a new element, with atomic number 93, would be synthesized for the first time. The nuclear equation for the process is shown here:

The element with atomic number 100 is named fermium in honor of Enrico Fermi.

$$ {}^{238}_{92}\text{U} + {}^{1}_{0}\text{n} \longrightarrow {}^{239}_{92}\text{U} \longrightarrow {}^{239}_{93}\text{X} + {}^{0}_{-1}\text{e} $$

Neutron (${}^{1}_{0}\text{n}$); Newly synthesized element (${}^{239}_{93}\text{X}$)

Fermi performed the experiment and detected the emission of beta particles. However, his results were inconclusive. Had he synthesized a new element? Fermi never chemically examined the products to determine their composition and therefore could not say with certainty that he had.

The element with atomic number 109 is named meitnerium in honor of Lise Meitner.

Three researchers in Germany—Lise Meitner (1878–1968), Fritz Strassmann (1902–1980), and Otto Hahn (1879–1968)—repeated Fermi's experiments and performed careful chemical analysis of the products. What they found in the products—several elements *lighter* than uranium—changed the world forever. On January 6, 1939, Meitner, Strassmann, and Hahn reported that the neutron bombardment of uranium resulted in **nuclear fission**—the splitting of the uranium atom. The nucleus of the neutron-bombarded uranium atom had been split into barium, krypton, and other smaller products. They also determined that the process emits enormous amounts of energy. A nuclear equation for a fission reaction, showing how uranium breaks apart into the daughter nuclides, is shown here:

$$ {}^{235}_{92}\text{U} + {}^{1}_{0}\text{n} \longrightarrow {}^{140}_{56}\text{Ba} + {}^{93}_{36}\text{Kr} + 3\,{}^{1}_{0}\text{n} + \text{energy} $$

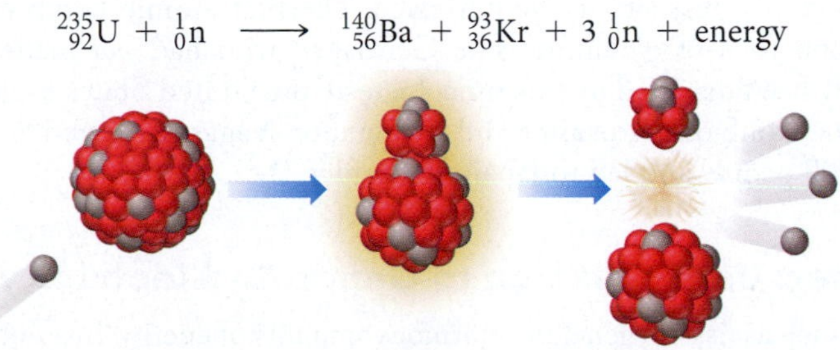

▲ This photograph is of Lise Meitner in Otto Hahn's Berlin laboratory. Together with Hahn and Fritz Strassmann, Meitner determined that U-235 could undergo nuclear fission.

Notice that the initial uranium atom is the U-235 isotope, which constitutes less than 1% of all naturally occurring uranium. U-238, the most abundant uranium isotope, does not undergo fission. Notice also that the process produces three neutrons, which have the potential to initiate fission in three other U-235 atoms.

▶ **FIGURE 21.10 A Self-Amplifying Chain Reaction** The fission of one U-235 nucleus emits neutrons that can then initiate fission in other U-235 nuclei, resulting in a chain reaction that releases enormous amounts of energy.

Fission Chain Reaction

The Atomic Bomb

Scientists quickly realized that a sample rich in U-235 can undergo a **chain reaction** in which neutrons produced by the fission of one uranium nucleus induce fission in other uranium nuclei (Figure 21.10 ▲). This self-amplifying reaction is capable of producing an enormous amount of energy. This is the energy that is harnessed in an atomic bomb. However, to make a bomb, a **critical mass** of U-235—enough U-235 to produce a self-sustaining reaction—is necessary. Fearing that Nazi Germany would develop such a bomb, several U.S. scientists persuaded Albert Einstein (1879–1955), the most famous scientist of the time, to write a letter to President Franklin Roosevelt in 1939 warning of this possibility. Einstein wrote, ". . . and it is conceivable—though much less certain—that extremely powerful bombs of a new type may thus be constructed. A single bomb of this type, carried by boat and exploded in a port, might very well destroy the whole port together with some of the surrounding territory."

▲ On July 16, 1945, in the New Mexico desert, the world's first atomic bomb was detonated. It had the power of 18,000 tons of dynamite.

Einstein's letter convinced Roosevelt, and in 1941 the president assembled the resources to begin the costliest scientific project ever attempted. The top-secret endeavor was called the *Manhattan Project,* and its main goal was to build an atomic bomb before the Germans did. The project was led by physicist J. R. Oppenheimer (1904–1967) at a high-security research facility in Los Alamos, New Mexico. Four years later, Oppenheimer's group successfully detonated the world's first nuclear weapon at a test site in New Mexico. The first atomic bomb exploded with a force equivalent to 18,000 tons of dynamite. The Germans—who had *not* made a successful nuclear bomb—had already been defeated by this time. Instead, the United States used the atomic bomb on Japan, dropping one bomb on Hiroshima and a second on Nagasaki. Together, the bombs killed approximately 200,000 people and led to Japan's surrender.

Nuclear Power: Using Fission to Generate Electricity

Nuclear reactions, such as fission, generate enormous amounts of energy. In a nuclear bomb, the energy is released all at once. The energy can also be released more slowly and used for peaceful purposes such as electricity generation. In the United States, nuclear fission generates about 20% of electricity. In some other countries, nuclear fission generates as much as 70% of electricity. To get an idea of the amount of energy released during fission, imagine a hypothetical nuclear-powered car. Suppose the fuel for such a car was a uranium cylinder about the size of a pencil. How often would you have to refuel the

Albert Einstein
Old Grove Rd.
Nassau Point
Peconic, Long Island

August 2nd, 1939

F.D. Roosevelt,
President of the United States,
White House
Washington, D.C.

Sir:

Some recent work by E.Fermi and L. Szilard, which has been communicated to me in manuscript, leads me to expect that the element uranium may be turned into a new and important source of energy in the immediate future. Certain aspects of the situation which has arisen seem to call for watchfulness and, if necessary, quick action on the part of the Administration. I believe therefore that it is my duty to bring to your attention the following facts and recommendations:

In the course of the last four months it has been made probable - through the work of Joliot in France as well as Fermi and Szilard in America - that it may become possible to set up a nuclear chain reaction in a large mass of uranium,by which vast amounts of power and large quantities of new radium-like elements would be generated. Now it appears almost certain that this could be achieved in the immediate future.

This new phenomenon would also lead to the construction of bombs, and it is conceivable - though much less certain - that extremely powerful bombs of a new type may thus be constructed. A single bomb of this type, carried by boat and exploded in a port, might very well destroy the whole port together with some of the surrounding territory. However, such bombs might very well prove to be too heavy for transportation by air.

◀ Einstein's letter (part of which is shown here) helped persuade Franklin Roosevelt to begin research into the building of a fission bomb.

car? The energy content of the uranium cylinder is equivalent to about 1000 twenty-gallon tanks of gasoline. If you refuel your gasoline-powered car once a week, your nuclear-powered car could go 1000 weeks—almost 20 years—before refueling.

Similarly, a nuclear-powered electricity generation plant can produce a substantial amount of electricity from a small amount of fuel. Such plants exploit the heat created by fission, using it to boil water and make steam, which then turns the turbine on a generator to produce electricity (Figure 21.11 ▶). The fission reaction occurs in the nuclear core of the power plant. The core consists of uranium fuel rods—enriched to about 3.5% U-235—interspersed between retractable neutron-absorbing control rods. When the control rods are fully retracted from the fuel rod assembly, the chain reaction can occur. When the control rods are fully inserted into the fuel assembly, however, they absorb the neutrons that would otherwise induce fission, shutting down the chain reaction. By retracting or inserting the control rods, the operator can increase or decrease the rate of fission. In this way, the fission reaction is controlled to produce the right amount of heat needed for electricity generation. In case of a power failure, the control rods automatically drop into the fuel rod assembly, shutting down the fission reaction.

A typical nuclear power plant generates enough electricity for a city of about 1 million people and uses about 50 kg of fuel per day. In contrast, a coal-burning power plant uses about 2,000,000 kg of fuel to generate the same amount of electricity. Furthermore, a nuclear power plant generates no air pollution and no greenhouse gases. A coal-burning power plant emits pollutants such as carbon monoxide, nitrogen oxides, and sulfur oxides. Coal-burning power plants also emit carbon dioxide, a greenhouse gas.

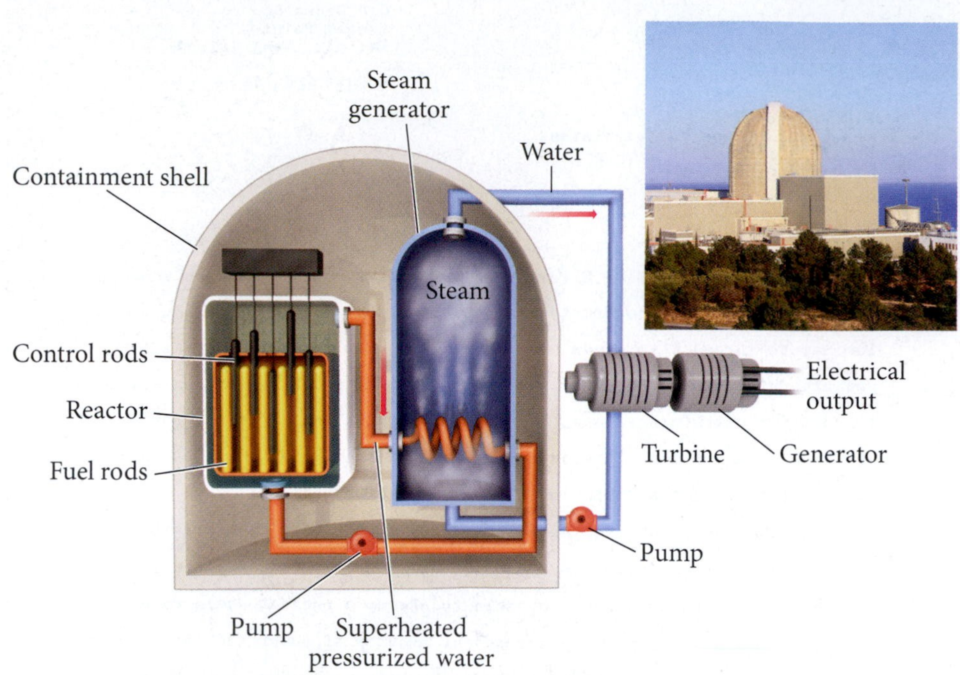

▶ **FIGURE 21.11 A Nuclear Reactor** The fission of U-235 in the core of a nuclear power plant generates heat that creates steam and turns a turbine on an electrical generator. Control rods are raised or lowered to control the fission reaction. (Note that the water carrying heat away from the reactor core is contained within its own pipes and does not come into direct contact with the steam that drives the turbines.)

Nuclear power generation, however, is not without problems. Foremost among them is the danger of nuclear accidents. In spite of safety precautions, the fission reaction occurring in a nuclear power plant can overheat. The most famous examples of this overheating occurred in Chernobyl, in the former Soviet Union, on April 26, 1986, and at the Fukushima Daiichi Nuclear Power Plant in Japan in March of 2011.

Reactor cores in the United States are not made of graphite and cannot burn in the way that the Chernobyl core did.

In the Chernobyl incident, operators of the plant were performing an experiment designed to reduce maintenance costs. In order to perform the experiment they had to disable many of the safety features of the reactor core. The experiment failed with disastrous results. The nuclear core, composed partly of graphite, overheated and began to burn. The accident caused 31 immediate deaths and produced a fire that scattered radioactive debris into the atmosphere, making much of the surrounding land (within about a 32-kilometer radius) uninhabitable. As bad as the accident was, however, it was a not a nuclear detonation. A nuclear power plant *cannot* become a nuclear bomb. The uranium fuel used in electricity generation is not sufficiently enriched in U-235 to produce a nuclear detonation. U.S. nuclear power plants have additional safety features designed to prevent similar accidents. For example, U.S. nuclear power plants have large containment structures to contain radioactive debris in the event of an accident.

▲ In 2011, the Fukushima Daiichi Nuclear Power Plant in Japan overheated as a result of a 9.0 magnitude earthquake that triggered a tsunami that flooded the coastal plant and caused the cooling system pumps to fail.

In the 2011 Japanese accident, a 9.0 magnitude earthquake triggered a tsunami that flooded the coastal plant and caused the cooling system pumps to fail. Several of the nuclear cores within the plant dramatically overheated, and at least one of the cores experienced a partial meltdown (in which the fuel gets so hot that it melts). The accident was intensified by the loss of water in the fuel storage ponds (pools of water used to keep spent fuel as well as future fuel cool), which caused the fuel stored in the ponds to also overheat. The release of radiation into the environment, however, while significant, was lower in Japan than at Chernobyl. No immediate radioactivity-related deaths were reported at the Fukushima plant or the surrounding area. The cleanup of the site, however, will continue for many years.

A second problem associated with nuclear power is waste disposal. Although the amount of nuclear fuel used in electricity generation is small compared to other fuels, the products of the reaction are radioactive and have long half-lives. What do we do with this waste? Currently, in the United States, nuclear waste is stored on site at the nuclear power plants. A single permanent disposal site was being developed in Yucca Mountain, Nevada, to store U.S. waste. However, in the spring of 2010, the Obama administration halted the development of this project and formed the Blue Ribbon Commission on America's Nuclear Future to explore alternatives. In 2012, the committee submitted its final report. Among the recommendations was the immediate development of new disposal and consolidated storage facilities. However, actual locations for such facilities were not part of the recommendations.

21.8 Converting Mass to Energy: Mass Defect and Nuclear Binding Energy

Nuclear fission produces large amounts of energy. But where does the energy come from? The energy comes from the conversion of mass to energy, as described by Einstein's famous equation $E = mc^2$. Here we first look at the conversion of mass to energy in general; then we turn to the topics of mass defect and nuclear binding energy.

The Conversion of Mass to Energy

When a fission reaction occurs, the products have a slightly different mass than the reactants. For example, examine the masses of the reactants and products in the fission equation from Section 21.7.

$$^{235}_{92}U + ^{1}_{0}n \longrightarrow ^{140}_{56}Ba + ^{93}_{36}Kr + 3^{1}_{0}n$$

Mass Reactants		Mass Products	
$^{235}_{92}U$	235.04392 amu	$^{140}_{56}Ba$	139.910581 amu
$^{1}_{0}n$	1.00866 amu	$^{93}_{36}Kr$	92.931130 amu
		$3^{1}_{0}n$	3(1.00866) amu
Total	**236.05258 amu**		**235.86769 amu**

The products of the nuclear reaction have *less mass* than the reactants. The missing mass is converted to energy. In Chapter 1, we learned that matter is conserved in chemical reactions. In nuclear reactions matter can be converted to energy. The relationship between the amount of matter that is lost and the amount of energy formed is given by Einstein's famous equation relating the two quantities:

In a chemical reaction, there are also mass changes associated with the emission or absorption of energy. Because the energy involved in chemical reactions is so much smaller than that of nuclear reactions, however, these mass changes are negligible.

$$E = mc^2$$

where E is the energy produced, m is the mass lost, and c is the speed of light. For example, in the fission reaction just shown, we calculate the quantity of energy produced as follows:

$$\begin{aligned}
\text{Mass lost } (m) &= 236.05258 \text{ amu} - 235.86769 \text{ amu} \\
&= 0.18489 \text{ amu} \times \frac{1.66054 \times 10^{-27} \text{ kg}}{1 \text{ amu}} \\
&= 3.0702 \times 10^{-28} \text{ kg} \\
\text{Energy produced } (E) &= mc^2 \\
&= 3.0702 \times 10^{-28} \text{ kg } (2.9979 \times 10^{8} \text{ m/s})^2 \\
&= 2.7593 \times 10^{-11} \text{ J}
\end{aligned}$$

The result (2.7593×10^{-11} J) is the energy produced when one nucleus of U-235 undergoes fission. This may not seem like much energy, but it is only the energy produced by the fission of a *single* nucleus. If we calculate the energy produced *per mole* of U-235, we can compare it to a chemical reaction.

$$2.7593 \times 10^{-11} \frac{\text{J}}{\text{U-235 atom}} \times \frac{6.0221 \times 10^{23} \text{ U-235 atoms}}{1 \text{ mol U-235}}$$

$$= 1.6617 \times 10^{13} \text{ J/mol U-235}$$

The energy produced by the fission of 1 mol of U-235 is about 17 billion kJ. In contrast, a highly exothermic chemical reaction produces 1000 kJ per mole of reactant. Fission produces over a million times more energy per mole than chemical processes.

Mass Defect and Nuclear Binding Energy

We can examine the formation of a stable nucleus from its component particles as a nuclear reaction in which mass is converted to energy. For example, consider the formation of helium-4 from its components:

The electrons are contained on the left side in the 2 $^{1}_{1}H$, and on the right side in $^{4}_{2}He$. If we write the equation using only two protons on the left ($^{1}_{1}p$), we must also add two electrons to the left.

$$2\,^{1}_{1}H + 2\,^{1}_{0}n \longrightarrow \,^{4}_{2}He$$

	Mass Reactants		Mass Products
$2\,^{1}_{1}H$	2(1.00783) amu	$^{4}_{2}He$	4.00260 amu
$2\,^{1}_{0}n$	2(1.00866) amu		
Total	**4.03298 amu**		**4.00260 amu**

A helium-4 atom has less mass than the sum of the masses of its separate components. This difference in mass, known as the **mass defect**, exists in all stable nuclei. The energy corresponding to the mass defect—obtained by substituting the mass defect into the equation $E = mc^2$ — is the **nuclear binding energy**, the amount of energy required to break apart the nucleus into its component nucleons.

An electron volt is defined as the kinetic energy of an electron that has been accelerated through a potential difference of 1 V.

Although chemists typically report energies in joules, nuclear physicists often use the electron volt (eV) or megaelectron volt (MeV): 1 MeV $= 1.602 \times 10^{-13}$ J. Unlike energy in joules, which is usually reported per mole, energy in electron volts is usually reported per nucleus. A particularly useful conversion for calculating and reporting nuclear binding energies is the relationship between amu (mass units) and MeV (energy units).

$$1 \text{ amu} = 931.5 \text{ MeV}$$

A mass defect of 1 amu, when substituted into the equation $E = mc^2$, gives an energy of 931.5 MeV. Using this conversion factor, we can readily calculate the binding energy of the helium nucleus.

$$\begin{aligned}\text{Mass defect} &= 4.03298 \text{ amu} - 4.00260 \text{ amu} \\ &= 0.03038 \text{ amu}\end{aligned}$$

$$\begin{aligned}\text{Nuclear binding energy} &= 0.03038 \cancel{\text{amu}} \times \frac{931.5 \text{ MeV}}{1 \cancel{\text{amu}}} \\ &= 28.30 \text{ MeV}\end{aligned}$$

The binding energy of the helium nucleus is 28.30 MeV. In order to compare the binding energy of one nucleus to that of another, we calculate the *binding energy per nucleon*, which is the nuclear binding energy of a nuclide divided by the number of nucleons in the nuclide. For helium-4, we calculate the binding energy per nucleon as follows:

$$\begin{aligned}\text{Binding energy per nucleon} &= \frac{28.30 \text{ MeV}}{4 \text{ nucleons}} \\ &= 7.075 \text{ MeV per nucleon}\end{aligned}$$

We can calculate the binding energy per nucleon for other nuclides in the same way. For example, the nuclear binding energy of carbon-12 is 7.680 MeV per nucleon. Since the binding energy per nucleon of carbon-12 is greater than that of helium-4, we conclude that the carbon-12 nuclide is more *stable* (it has lower potential energy).

EXAMPLE 21.7

Mass Defect and Nuclear Binding Energy

Calculate the mass defect and nuclear binding energy per nucleon (in MeV) for C-16, a radioactive isotope of carbon with a mass of 16.014701 amu.

SOLUTION

Calculate the mass defect as the difference between the mass of one C-16 atom and the sum of the masses of 6 hydrogen atoms and 10 neutrons.	$\text{Mass defect} = 6(\text{mass } ^1_1\text{H}) + 10(\text{mass } ^1_0\text{n}) - \text{mass } ^{16}_{6}\text{C}$ $= 6(1.00783 \text{ amu}) + 10(1.00866 \text{ amu}) - 16.014701 \text{ amu}$ $= 0.118879 \text{ amu}$
Calculate the nuclear binding energy by converting the mass defect (in amu) into MeV. (Use 1 amu = 931.5 MeV.)	$0.118879 \text{ amu} \times \dfrac{931.5 \text{ MeV}}{\text{amu}} = 110.74 \text{ MeV}$
Determine the nuclear binding energy per nucleon by dividing by the number of nucleons in the nucleus.	$\text{Nuclear binding energy per nucleon} = \dfrac{110.74 \text{ MeV}}{16 \text{ nucleons}} = 6.921 \text{ MeV/nucleon}$

FOR PRACTICE 21.7

Calculate the mass defect and nuclear binding energy per nucleon (in MeV) for U-238, which has a mass of 238.050784 amu.

Figure 21.12 ▼ shows the binding energy per nucleon plotted as a function of mass number (A). The binding energy per nucleon is relatively low for small mass numbers and increases until about $A = 60$, where it reaches a maximum. Nuclides with mass numbers of about 60, therefore, are among the most stable. Beyond $A = 60$, the binding energy per nucleon decreases again. The figure illustrates why nuclear fission is a highly exothermic process. When a heavy nucleus, such as U-235, breaks up into smaller nuclei, such as Ba-140 and Kr-93, the binding energy per nucleon increases. This is analogous to a chemical reaction in which weak bonds break and strong bonds form. In both cases, the process is exothermic. Figure 21.12 also reveals that the *combining* of two lighter nuclei (below $A = 60$) to form a heavier nucleus should be exothermic as well. This process is called *nuclear fusion*, which we discuss in the next section of this chapter.

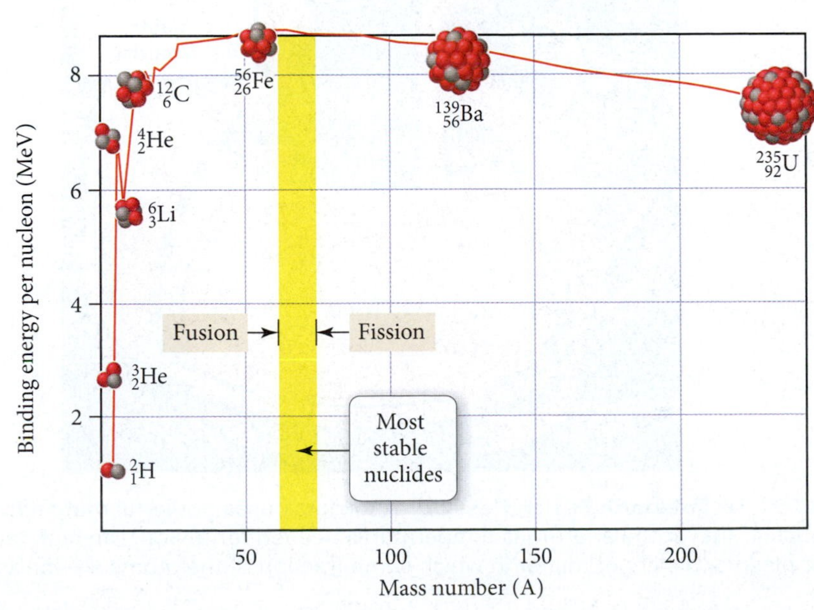

◀ **FIGURE 21.12 Nuclear Binding Energy per Nucleon** The nuclear binding energy per nucleon (a measure of the stability of a nucleus) reaches a maximum at about $A = 60$. Energy can be obtained either by breaking a heavy nucleus up into lighter ones (fission) or by combining lighter nuclei into heavier ones (fusion).

21.9 Nuclear Fusion: The Power of the Sun

Deuterium-Tritium Fusion Reaction

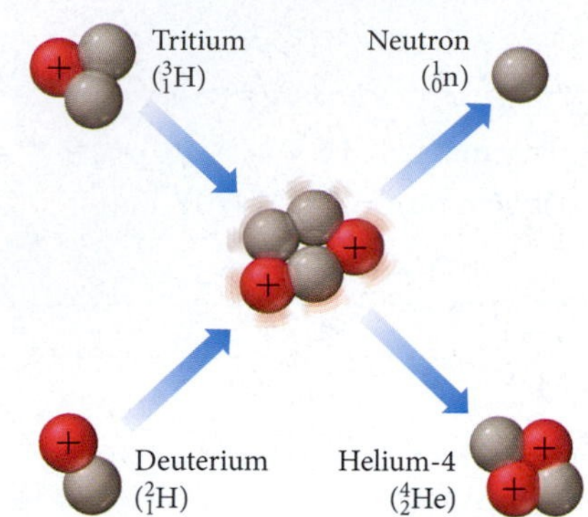

▲ **FIGURE 21.13 A Nuclear Fusion Reaction** In this reaction, two heavy isotopes of hydrogen, deuterium (hydrogen-2) and tritium (hydrogen-3), fuse to form helium-4 and a neutron.

Nuclear fission is the *splitting* of a heavy nucleus to form two or more lighter ones. **Nuclear fusion**, by contrast, is the *combination* of two light nuclei to form a heavier one. Both fusion and fission emit large amounts of energy because, as we have just seen, they both form daughter nuclides with greater binding energies per nucleon than the parent nuclides. Nuclear fusion is the energy source of stars, including our sun. In stars, hydrogen atoms fuse together to form helium atoms, emitting energy in the process.

Nuclear fusion is also the basis of modern nuclear weapons called hydrogen bombs. A modern hydrogen bomb has up to 1000 times the explosive force of the first atomic bombs. These bombs employ the fusion reaction shown here:

$$^2_1H + ^3_1H \longrightarrow ^4_2He + ^1_0n$$

In this reaction, deuterium (the isotope of hydrogen with one neutron) and tritium (the isotope of hydrogen with two neutrons) combine to form helium-4 and a neutron (Figure 21.13 ◀). Because fusion reactions require two positively charged nuclei (which repel each other) to fuse together, extremely high temperatures are required. In a hydrogen bomb, a small fission bomb is detonated first, creating temperatures high enough for fusion to proceed.

Nuclear fusion has been intensely investigated as a way to produce electricity. Because of the higher energy density—fusion provides about 10 times more energy per gram of fuel than does fission—and because the products of the reaction are less problematic than those of fission, fusion holds promise as a future energy source. However, despite concerted efforts, the generation of electricity by fusion remains elusive. One of the main problems is the high temperature required for fusion to occur—no material can withstand those temperatures. Using powerful magnetic fields or laser beams, scientists have succeeded in compressing and heating nuclei to the point where fusion has been initiated and even sustained for brief periods of time (Figure 21.14 ▼). To date, however, the amount of energy generated by fusion reactions has been less than the amount required to get it to occur. After years of allocating billions of dollars on fusion research, the U.S. Congress has reduced funding for these projects. Whether fusion will ever be a viable energy source remains uncertain.

Tokamak Fusion Reactor

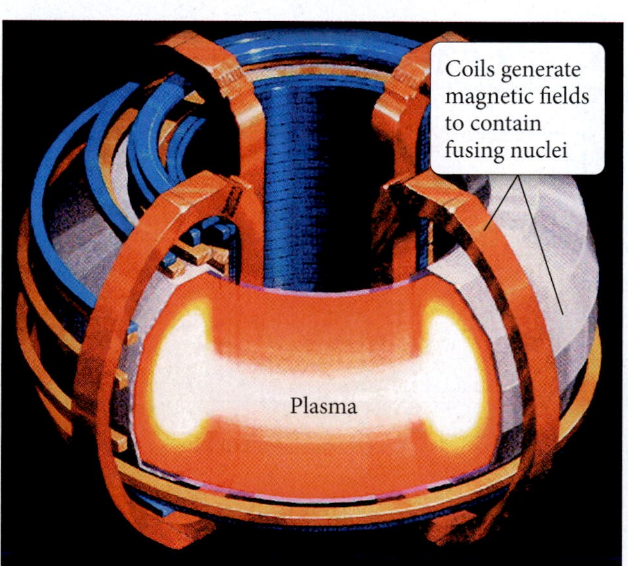

▲ **FIGURE 21.14 Tokamak Fusion Reactor** A tokamak uses powerful magnetic fields to confine nuclear fuel at the enormous temperatures needed for fusion. The high temperatures produce a plasma, a state of matter in which some fraction of the atoms are ionized.

21.10 Nuclear Transmutation and Transuranium Elements

One of the goals of the early chemists of the Middle Ages, who were known as *alchemists*, was to transform ordinary metals into gold. Many alchemists hoped to turn low-cost metals, such as lead or tin, into precious metals and in this way become wealthy. These alchemists were never successful because their attempts were merely chemical—they mixed different metals together or tried to get them to react with other substances in order to turn them into gold. In a chemical reaction, an element retains its identity, so a less valuable metal—such as lead—always remains lead, even when it forms a compound with another element.

Nuclear reactions, by contrast, result in the transformation of one element into another, a process known as **transmutation**. We have already seen how this occurs in radioactive decay, in fission, and in fusion. In addition, other nuclear reactions that transmute elements are possible. For example, in 1919 Ernest Rutherford bombarded nitrogen-17 with alpha particles to form oxygen:

$$^{14}_{7}\text{N} + {}^{4}_{2}\text{He} \longrightarrow {}^{17}_{8}\text{O} + {}^{1}_{1}\text{H}$$

Iréne Joliot-Curie (1897–1956), who was the daughter of Marie Curie, and her husband Frédéric Joliot (1900–1958), bombarded aluminum-27 with alpha particles to form phosphorus:

$$^{27}_{13}\text{Al} + {}^{4}_{2}\text{He} \longrightarrow {}^{30}_{15}\text{P} + {}^{1}_{0}\text{n}$$

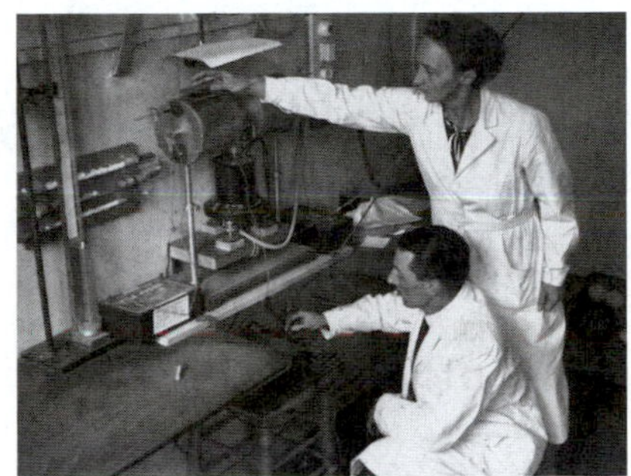

▲ The Joliot-Curies won the 1935 Nobel Prize in Chemistry for their work on nuclear transmutation.

In the 1930s, scientists began building devices that accelerate particles to high velocities, opening the door to even more possibilities. These devices are generally of two types, the **linear accelerator** and the **cyclotron**.

In a *single-stage linear accelerator,* a charged particle such as a proton is accelerated in an evacuated tube. The accelerating force is provided by a potential difference (or voltage) between the ends of the tube. In *multistage linear accelerators,* such as the Stanford Linear Accelerator (SLAC) at Stanford University, a series of tubes of increasing length is connected to a source of alternating voltage, as shown in Figure 21.15 ▼. The voltage alternates in such a way that, as a positively charged particle leaves a particular tube, that tube becomes positively charged, repelling the particle to the next tube. At the same time, the tube that the particle is now approaching becomes negatively charged, pulling the particle toward it. This continues throughout the linear accelerator, allowing the particle to be accelerated to velocities up to 90% of the speed of light. When particles of this speed collide with a target, they produce a shower of subatomic particles that can be studied. For example, researchers using the Stanford Linear Accelerator were awarded the 1990 Nobel Prize in physics for discovering evidence that protons and neutrons were composed of still smaller subatomic particles called *quarks*.

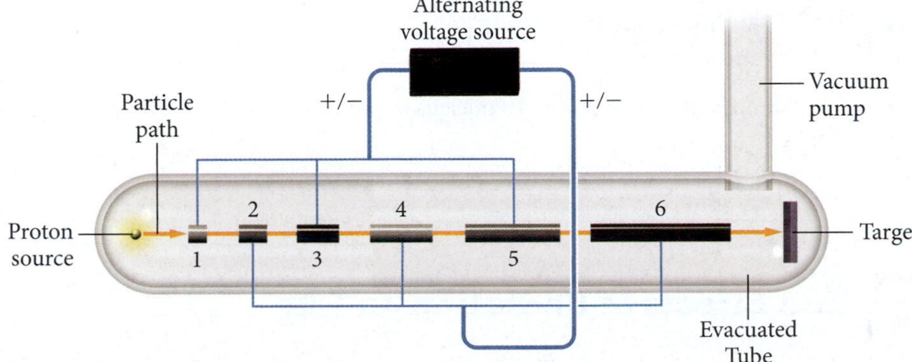

▲ **FIGURE 21.15 The Linear Accelerator** In a multistage linear accelerator, the charge on successive tubes (numbered 1–6 in this diagram) is rapidly alternated in such a way that as a positively charged particle leaves a particular tube, that tube becomes positively charged, repelling the particle toward the next tube. At the same time, the tube that the particle is now approaching becomes negatively charged, pulling the particle toward it. This process repeats itself through a number of tubes until the particle has been accelerated to a high velocity.

▲ The Stanford Linear Accelerator is located at Stanford University in California.

▲ The Fermi National Accelerator Laboratory complex in Batavia, Illinois, includes two cyclotrons in a figure-8 configuration.

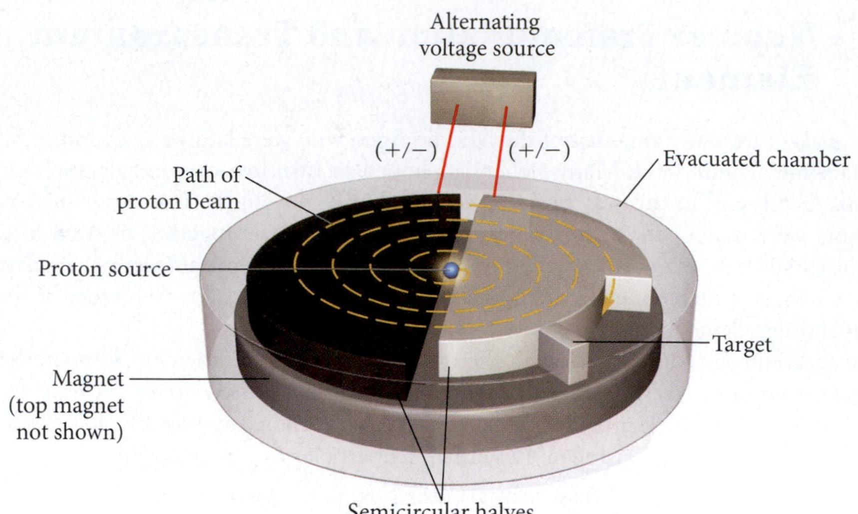

▲ **FIGURE 21.16 The Cyclotron** In a cyclotron, two semicircular D-shaped structures are subjected to an alternating voltage. A charged particle (such as a proton), starting from a point between the two, is accelerated back and forth between them, while additional magnets cause the particle to move in a spiral path.

In a cyclotron, a similarly alternating voltage is used to accelerate a charged particle, only this time the alternating voltage is applied between the two semicircular halves of the cyclotron (Figure 21.16 ▲). A charged particle originally in the middle of the two semicircles is accelerated back and forth between them. Additional magnets cause the particle to move in a spiral path. As the charged particle spirals out from the center, it gains speed and eventually exits the cyclotron aimed at the target.

With linear accelerators or cyclotrons, all sorts of nuclear transmutations can be achieved. In this way, scientists have made nuclides that don't normally exist in nature. For example, uranium-238 can be made to collide with carbon-12 to form an element with atomic number 98:

$$^{238}_{92}\text{U} + ^{12}_{6}\text{C} \longrightarrow ^{244}_{98}\text{Cf} + 6\,^{1}_{0}\text{n}$$

This element was named californium (Cf) because it was first produced (by a slightly different nuclear reaction) at the University of California at Berkeley. Many other nuclides with atomic numbers larger than that of uranium have been synthesized since the 1940s. These synthetic elements—called transuranium elements—have been added to the periodic table.

Most synthetic elements are unstable and have very short half-lives. Some exist for only fractions of a second after they are made.

21.4 Cc Conceptual Connection

Nuclear Transformations

Californium-252 is bombarded with a boron-10 nucleus to produce another nuclide and six neutrons. Which nuclide forms?

21.11 The Effects of Radiation on Life

As we discussed in Section 21.3, the energy associated with radioactivity can ionize molecules. When radiation ionizes important molecules in living cells, problems can develop. The ingestion of radioactive materials—especially alpha and beta emitters—is particularly dangerous because the radioactivity, once inside the body, can do even more damage. We divide the effects of radiation into three different types: acute radiation damage, increased cancer risk, and genetic effects.

Acute Radiation Damage

Acute radiation damage results from exposure to large amounts of radiation in a short period of time. The main sources of this kind of exposure are nuclear bombs and exposed nuclear reactor cores. These high levels of radiation kill large numbers of cells. Rapidly dividing cells, such as those in the immune system and the intestinal lining, are most susceptible. Consequently, people exposed to high levels of radiation have weakened immune systems and a lowered ability to absorb nutrients from food. In milder cases, recovery is possible with time. In more extreme cases, death results, often from infection.

Increased Cancer Risk

Lower doses of radiation over extended periods of time can increase cancer risk. Radiation increases cancer risk because it can damage DNA, the molecules in cells that carry instructions for cell growth and replication. When the DNA within a cell is damaged, the cell normally dies. Occasionally, however, changes in DNA cause cells to grow abnormally and to become cancerous. These cancerous cells grow into tumors that can spread and, in some cases, cause death. Cancer risk increases with increasing radiation exposure. However, cancer is so prevalent and has so many convoluted causes that determining an exact exposure level for increased cancer risk from radiation exposure is difficult.

Genetic Defects

Another possible effect of radiation exposure is genetic defects in future generations. If radiation damages the DNA of reproductive cells—such as eggs or sperm—then the offspring that develop from those cells may have genetic abnormalities. Genetic defects of this type have been observed in laboratory animals exposed to high levels of radiation. However, such genetic defects—with a clear causal connection to radiation exposure—have yet to be verified in humans, even in studies of Hiroshima survivors.

Measuring Radiation Exposure

We can measure radiation exposure in a number of different ways. One method is to measure the number of decay events to which a person is exposed. The unit used for this type of exposure measurement is the *curie* (Ci), defined as 3.7×10^{10} decay events per second. A person exposed to a curie of radiation from an alpha emitter is bombarded by 3.7×10^{10} alpha particles per second. However, recall that different kinds of radiation produce different effects. For example, we know that alpha radiation has a much greater ionizing power than beta radiation. Consequently, a certain number of alpha decays occurring within a person's body (due to the ingestion of an alpha emitter) would do more damage than the same number of beta decays. If the alpha emitter and beta emitter were external to the body, however, the radiation from the alpha emitter would largely be stopped by clothing or the skin (due to the low penetrating power of alpha radiation), while the radiation from the beta emitter could penetrate the skin and cause more damage. Consequently, the curie is not an effective measure of how much biological tissue damage the radiation actually does.

A better way to assess radiation exposure is to measure the amount of energy actually absorbed by body tissue. The units used for this type of exposure measurement are the *gray* (Gy), which corresponds to 1 J of energy absorbed per kilogram of body tissue, and the *rad* (for *r*adiation *a*bsorbed *d*ose), which corresponds to 0.01 Gy.

$$1 \text{ gray (Gy)} = 1 \text{ J/kg body tissue}$$

$$1 \text{ rad} = 0.01 \text{ J/kg body tissue}$$

TABLE 21.4 Exposure by Source for Persons Living in the United States

Source	Dose
Natural Radiation	
A 5-hour jet airplane ride	2.5 mrem/trip (0.5 mrem/hr at 39,000 feet) (whole body dose)
Cosmic radiation from outer space	27 mrem/yr (whole body dose)
Terrestrial radiation	28 mrem/yr (whole body dose)
Natural radionuclides in the body	35 mrem/yr (whole body dose)
Radon gas	200 mrem/yr (lung dose)
Diagnostic Medical Procedures	
Chest X-ray	8 mrem (whole body dose)
Dental X-rays (panoramic)	30 mrem (skin dose)
Dental X-rays (two bitewings)	80 mrem (skin dose)
Mammogram	138 mrem per image
Barium enema (X-ray portion only)	406 mrem (bone marrow dose)
Upper gastrointestinal tract test (X-ray portion only)	244 mrem (bone marrow dose)
Thallium heart scan	500 mrem (whole body dose)
Consumer Products	
Building materials	3.5 mrem/yr (whole body dose)
Luminous watches (H-3 and Pm-147)	0.04–0.1 mrem/yr (whole body dose)
Tobacco products (to smokers of 30 cigarettes per day)	16,000 mrem/yr (bronchial epithelial dose)

Source: Department of Health and Human Services, National Institutes of Health.

Although these units measure the actual energy absorbed by bodily tissues, they still do not account for the amount of damage to biological molecules caused by that energy absorption, which differs from one type of radiation to another and from one type of biological tissue to another. For example, when a gamma ray passes through biological tissue, the energy absorbed is spread out over the long distance that the radiation travels through the body, resulting in a low ionization density within the tissue. When an alpha particle passes through biological tissue, in contrast, the energy is absorbed over a much shorter distance, resulting in a much higher ionization density. The higher ionization density results in greater damage, even though the amount of energy absorbed by the tissue might be the same.

Consequently, a correction factor, called the **biological effectiveness factor**, or **RBE** (for *R*elative *B*iological *E*ffectiveness), is usually multiplied by the dose in rads to obtain the dose in a unit called the **rem** for *r*oentgen *e*quivalent *m*an.

A *roentgen* is the amount of radiation that produces 2.58×10^{-1} C of charge per kg of air.

$$\text{Dose in rads} \times \text{biological effectiveness factor} = \text{dose in rems}$$

The biological effectiveness factor for alpha radiation, for example, is much higher than that for gamma radiation.

The SI unit that corresponds to the rem is the sievert (Sv). However, the rem is still commonly used in the United States. The conversion factor is 1 rem = 0.01 Sv.

On average, each of us is exposed to approximately 310 mrem of radiation per year from the natural sources listed in Table 21.4. The majority of this exposure comes from radon, one of the products in the uranium decay series. As you can see from Table 21.4, however, some medical procedures also involve exposure levels similar to those received from natural sources. The increased use of computed tomography (CT) scans over the last decade—which have associated exposures of 200–1600 mrem—has raised some concerns about the overuse of that technology in medicine.

It takes much more than the average natural radiation dose or the dose from a medical diagnostic procedure to produce significant health effects in humans. The first measurable effect, a decreased white blood cell count, occurs at instantaneous exposures of approximately 20 rem (Table 21.5). Exposures of 100 rem produce a definite increase in cancer risk, and exposures of over 500 rem often result in death.

TABLE 21.5 Effects of Radiation Exposure

Approximate Dose (rem)	Probable Outcome
20–100	Decreased white blood cell count; possible increase in cancer risk
100–400	Radiation sickness including vomiting and diarrhea; skin lesions; increase in cancer risk
500	Death (often within 2 months)
1000	Death (often within 2 weeks)
2000	Death (within hours)

Radiation Exposure

21.5 Cc Conceptual Connection

Suppose a person ingests equal amounts of two nuclides, both of which are beta emitters (of roughly equal energy). Nuclide A has a half-life of 8.5 hours and Nuclide B has a half-life of 15.0 hours. Both nuclides are eliminated from the body within 24 hours of ingestion. Which of the two nuclides produces the greater radiation exposure?

21.12 Radioactivity in Medicine and Other Applications

Radioactivity is often perceived as dangerous; however, it is also immensely useful to physicians in the diagnosis and treatment of disease and has numerous other valuable applications. We can broadly divide the use of radioactivity in medicine into *diagnostic techniques* (which diagnose disease) and *therapeutic techniques* (which treat disease).

Diagnosis in Medicine

The use of radioactivity in diagnosis usually involves a **radiotracer**, a radioactive nuclide attached to a compound or introduced into a mixture in order to track the movement of the compound or mixture within the body. Tracers are useful in the diagnosis of disease because of two main factors: (1) the sensitivity with which radioactivity can be detected, and (2) the identical chemical behavior of a radioactive nucleus and its nonradioactive counterpart. For example, the thyroid gland naturally concentrates iodine. When a patient is given small amounts of iodine-131 (a radioactive isotope of iodine), the radioactive iodine accumulates in the thyroid, just as nonradioactive iodine does. However, the radioactive iodine emits radiation, which can then be detected with great sensitivity and used to measure the rate of iodine uptake by the thyroid, and thus to image the gland.

Different elements are taken up preferentially by different organs or tissues, so various radiotracers are used to monitor metabolic activity and image a variety of organs and structures, including the kidneys, heart, brain, gallbladder, bones, and arteries, as shown in Table 21.6. Radiotracers can also be employed to locate infections or cancers within the body. To locate an infection, antibodies are labeled (or tagged) with a radioactive nuclide, such as technetium-99m (where "m" means metastable), and

TABLE 21.6 Common Radiotracers

Nuclide	Type of Emission	Half-Life	Part of Body Studied
Technetium-99m	Gamma (primarily)	6.01 hours	Various organs, bones
Iodine-131	Beta	8.0 days	Thyroid
Iron-59	Beta	44.5 days	Blood, spleen
Thallium-201	Electron capture	3.05 days	Heart
Fluorine-18	Positron emission	1.83 hours	PET studies of heart, brain
Phosphorus-32	Beta	14.3 days	Tumors in various organs

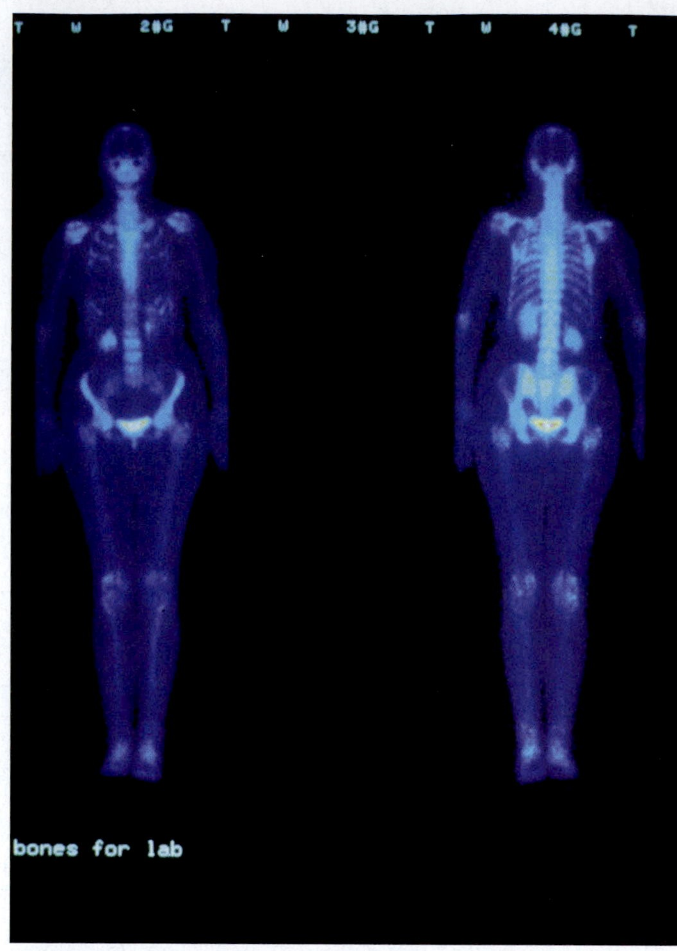

▲ **FIGURE 21.17 A Bone Scan** These images, front and rear views of the human body, were created by the gamma ray emissions of Tc-99m. Such scans are often used to locate cancer that has metastasized (spread) to the bones from a primary tumor elsewhere.

administered to the patient. The tagged antibodies aggregate at the infected site, as described in Section 21.1. Cancerous tumors can be detected because they naturally concentrate phosphorus. When a patient is given phosphorus-32 (a radioactive isotope of phosphorus) or a phosphate compound incorporating another radioactive isotope such as Tc-99m, the tumors concentrate the radioactive substance and become sources of radioactivity that can be detected (Figure 21.17 ▲).

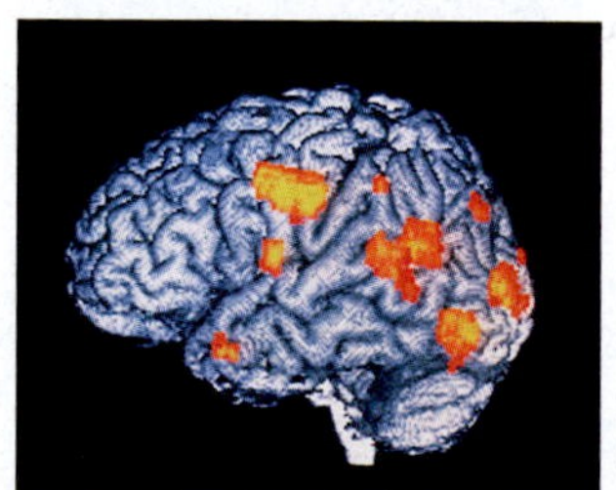

▲ **FIGURE 21.18 A PET Scan** The colored areas indicate regions of high metabolic activity in the brain of a schizophrenic patient experiencing hallucinations.

A specialized imaging technique known as **positron emission tomography (PET)** employs positron-emitting nuclides, such as fluorine-18, synthesized in cyclotrons. The fluorine-18 is attached to a metabolically active substance such as glucose and administered to the patient. As the glucose travels through the bloodstream and to the heart and brain, it carries the radioactive fluorine, which decays with a half-life of just under 2 hours. When a fluorine-18 nuclide decays, it emits a positron that immediately combines with any nearby electrons. Since a positron and an electron are antiparticles, they annihilate one other, producing two gamma rays that travel in exactly opposing directions. The gamma rays can be detected by an array of detectors that can locate the point of origin of the rays with great accuracy. The result is a set of highly detailed images that show both the rate of glucose metabolism and structural features of the imaged organ (Figure 21.18 ◀).

Radiotherapy in Medicine

Because radiation kills cells and is particularly effective at killing rapidly dividing cells, it is often used as a therapy for cancer (cancer cells reproduce much faster than normal cells). Gamma rays are focused on internal tumors to kill them. The gamma ray beam is usually moved in a circular path around the

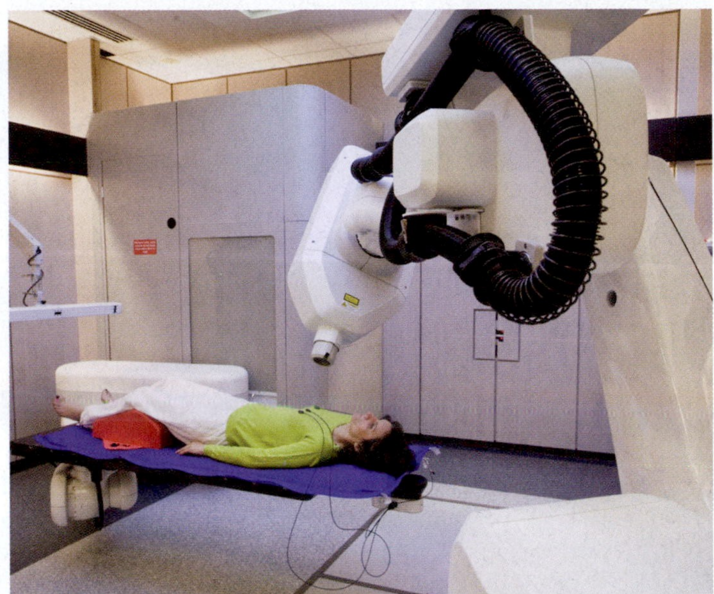

◀ **FIGURE 21.19 Radiotherapy for Cancer** This treatment involves exposing a malignant tumor to gamma rays generated by nuclides such as cobalt-60. The beam is moved in a circular pattern around the tumor to maximize exposure of the tumor to radiation while minimizing the exposure of healthy tissues.

tumor (Figure 21.19 ▲), maximizing the exposure of the tumor while minimizing the exposure of the surrounding healthy tissue. Nonetheless, cancer patients receiving such treatment usually develop the symptoms of radiation sickness, which include vomiting, skin burns, and hair loss.

Why is radiation—which is known to cause cancer—also used to treat cancer? The answer lies in risk analysis. A cancer patient is normally exposed to radiation doses of about 100 rem. Such a dose increases cancer risk by about 1%. However, if the patient has a 100% chance of dying from the cancer that he already has, such a risk becomes acceptable, especially since there is a significant chance of curing the cancer.

Other Applications

Radioactivity is often used to kill microorganisms. For example, physicians use radiation to sterilize medical devices that are to be surgically implanted. The radiation kills bacteria that might otherwise lead to infection. Similarly, radiation is used to kill bacteria and parasites in foods. The irradiation of foods makes them safer to consume and gives them a longer shelf life (Figure 21.20 ▼). The irradiation of raw meat and poultry kills *E. coli* and *Salmonella*, bacteria that can lead to serious illness and even death when consumed. The irradiation of food does not, however, make the food itself radioactive, nor does it decrease the nutritional value of the food. In the United States, the irradiation of many different types of foods—including beef, poultry, potatoes, flour, and fruit—has been approved by the U.S. Food and Drug Administration (FDA) and the U.S. Department of Agriculture (USDA).

Radioactivity is also used to control the populations of harmful insects. For example, fruit flies can be raised in large numbers in captivity and sterilized with radiation. When these fruit flies are released, they mate with wild fruit flies but do not produce offspring. The efforts of the wild fruit flies, which might otherwise lead to reproduction, are wasted and the next generation of flies is smaller than it would otherwise have been. Similar strategies have been employed to control the populations of disease-carrying mosquitoes.

◀ **FIGURE 21.20 Irradiation of Food** Irradiation kills microbes that cause food to decay, allowing for longer and safer storage. The food is not made radioactive and its properties are unchanged in the process. These strawberries were picked at the same time, but those on the bottom were irradiated before storage.

SELF-ASSESSMENT Quiz

1. What daughter nuclide forms when polonium-214 undergoes alpha decay?
   a) $^{218}_{86}Rn$ **b)** $^{214}_{85}At$ **c)** $^{214}_{83}Bi$ **d)** $^{210}_{82}Pb$

2. Which nuclear equation accurately represents the beta decay of Xe-133?
   a) $^{133}_{54}Xe \longrightarrow {}^{133}_{55}Cs + {}^{0}_{-1}e$
   b) $^{133}_{54}Xe \longrightarrow {}^{133}_{53}I + {}^{0}_{+1}e$
   c) $^{133}_{54}Xe \longrightarrow {}^{0}_{-1}e + {}^{133}_{53}I$
   d) $^{133}_{54}Xe \longrightarrow {}^{129}_{52}Cs + {}^{4}_{2}He$

3. Which nuclide is most likely to undergo beta decay?
   a) Si-22 **b)** Rb-91 **c)** Ar-35 **d)** Co-52

4. Which form of radioactive decay would you be most likely to detect if it were happening in the room next to the one you are currently in?
   a) alpha
   b) beta
   c) gamma
   d) positron emission

5. The chart below shows the mass of a decaying nuclide versus time. What is the half-life of the decay?

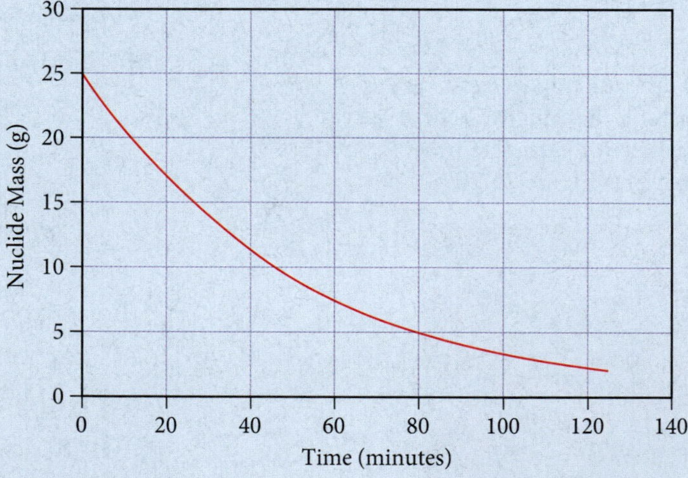

   a) 15 min
   b) 25 min
   c) 35 min
   d) 70 min

6. Iron-59 is a beta emitter with a half-life of 44.5 days. If a sample initially contains 132 mg of iron-59, how much iron-59 is left in the sample after 265 days?
   a) 0.00 mg
   b) 2.13 mg
   c) 33.2 mg
   d) 66.0 mg

7. An artifact has a carbon-14 decay rate of 8.55 disintegrations per minute per gram of carbon (8.55 dis/min • g C). Living organisms have a carbon-14 decay rate of 15.3 dis/min • g C. How old is the artifact? (The half-life of carbon-14 is 5730 yr.)
   a) 4.81×10^{3} yr
   b) 2.10×10^{3} yr
   c) 3.20×10^{3} yr
   d) 1.21×10^{-4} yr

8. An igneous rock contains a Pb-206/U-238 mass ratio of 0.372. How old is the rock? (U-238 decays into Pb-206 with a half-life of 4.5×10^{9} yr.)
   a) 4.50×10^{9} yr
   b) 6.42×10^{9} yr
   c) 2.05×10^{9} yr
   d) 2.32×10^{9} yr

9. Calculate the nuclear binding energy per nucleon for cobalt-59, the only stable isotope of cobalt. The mass of cobalt-59 is 58.933198 amu. (The mass of H^{1}_{1} is 1.00783 amu, and the mass of a neutron is 1.00866 amu.)
   a) 517.3 MeV
   b) 8.768 MeV
   c) 19.16 MeV
   d) 1.011×10^{-5} MeV

10. Which problem is not associated with nuclear power generation?
   a) danger of overheated nuclear core
   b) waste disposal
   c) global warming
   d) none of the above (All of the above are problems associated with nuclear power generation.)

Answers: 1:d; 2:a; 3:b; 4:c; 5:c; 6:b; 7:a; 8:d; 9:b; 10:c

CHAPTER SUMMARY 21

REVIEW

KEY LEARNING OUTCOMES

CHAPTER OBJECTIVES	ASSESSMENT		
Writing Nuclear Equations for Alpha Decay (21.3)	• Example 21.1	For Practice 21.1	Exercises 31–36
Writing Nuclear Equations for Beta Decay, Positron Emission, and Electron Capture (21.3)	• Example 21.2 Exercises 31–36	For Practice 21.2	For More Practice 21.2
Predicting the Type of Radioactive Decay (21.4)	• Example 21.3	For Practice 21.3	Exercises 41, 42
Radioactive Decay Kinetics (21.6)	• Example 21.4	For Practice 21.4	Exercises 45–52
Radiocarbon Dating (21.6)	• Example 21.5	For Practice 21.5	Exercises 53, 54
Using Uranium/Lead Dating to Estimate the Age of a Rock (21.6)	• Example 21.6	For Practice 21.6	Exercises 55, 56
Determining the Mass Defect and Nuclear Binding Energy (21.8)	• Example 21.7	For Practice 21.7	Exercises 65–72

KEY TERMS

Section 21.1
radioactivity (862)
radioactive (862)

Section 21.2
phosphorescence (862)

Section 21.3
nuclide (863)
alpha (α) decay (864)
alpha (α) particle (864)
nuclear equation (864)
ionizing power (865)
penetrating power (865)
beta (β) decay (865)
beta (β) particle (865)
gamma (γ) ray emission (866)
gamma (γ) ray (866)
positron emission (866)
positron (866)
electron capture (867)

Section 21.4
nucleons (869)
strong force (869)
magic numbers (870)

Section 21.5
film-badge dosimeter (871)
Geiger-Müller counter (871)
scintillation counter (872)

Section 21.6
radiometric dating (875)
radiocarbon dating (875)

Section 21.7
nuclear fission (879)
chain reaction (880)
critical mass (880)

Section 21.8
mass defect (884)
nuclear binding energy (884)

Section 21.9
nuclear fusion (886)

Section 21.10
transmutation (887)
linear accelerator (887)
cyclotron (887)

Section 21.11
biological effectiveness factor (RBE) (890)
rem (890)

Section 21.12
radiotracer (891)
positron emission tomography (PET) (892)

KEY CONCEPTS

Diagnosing Appendicitis (21.1)

- Radioactivity is the emission of subatomic particles or energetic electromagnetic radiation by the nuclei of certain atoms.
- Because some of these emissions can pass through matter, radioactivity is useful in medicine and many other areas of study.

The Discovery of Radioactivity (21.2)

- Antoine-Henri Becquerel discovered radioactivity when he found that uranium causes a photographic exposure in the absence of light.
- Marie Sklodowska Curie later determined that this phenomenon was not unique to uranium, and she began calling the rays that produced the exposure radioactivity. Curie also discovered two new elements, polonium and radium.

Types of Radioactivity (21.3)

- The major types of natural radioactivity are alpha (α) decay, beta (β) decay, gamma (γ) ray emission, and positron emission.
- Alpha radiation is helium nuclei. Beta particles are electrons. Gamma rays are electromagnetic radiation of very high energy. Positrons are the antiparticles of electrons.
- A nucleus may absorb one of its orbital electrons in a process called electron capture.

- We can represent each radioactive process with a nuclear equation that illustrates how the parent nuclide changes into the daughter nuclide. In a nuclear equation, although the specific types of atoms may not balance, the atomic numbers and mass numbers must.
- Each type of radioactivity has a different ionizing and penetrating power. These values are inversely related; a particle with a higher ionizing power has a lower penetrating power. Alpha particles are the most massive and have the highest ionizing power, followed by beta particles and positrons, which are equivalent in their ionizing power. Gamma rays have the lowest ionizing power.

The Valley of Stability: Predicting the Type of Radioactivity (21.4)

- The stability of a nucleus, and therefore the probability that it will undergo radioactive decay, depends largely on two factors. The first is the ratio of neutrons to protons (N/Z), because neutrons provide a strong force that overcomes the electromagnetic repulsions between the positive protons. This ratio is one for smaller elements, but becomes greater than one for larger elements. The second factor related to nuclei stability is a concept known as magic numbers; certain numbers of nucleons are more stable than others.

Detecting Radioactivity (21.5)

- Radiation detectors determine the quantity of radioactivity in an area or sample.
- Film-badge dosimeters utilize photographic film for radiation detection; such detectors do not provide an instantaneous response.
- Two detectors that instantly register the amount of radiation are the Geiger-Müller counter, which uses the ionization of argon by radiation to produce an electrical signal, and the scintillation counter, which uses the emission of light induced by radiation.

The Kinetics of Radioactive Decay and Radiometric Dating (21.6)

- All radioactive elements decay according to first-order kinetics (Chapter 15); the half-life equation and the integrated rate law for radioactive decay are derived from the first-order rate laws.
- The kinetics of radioactive decay is used to date objects and artifacts. The age of materials that were once part of living organisms is measured by carbon-14 dating. The age of ancient rocks and even Earth itself is determined by uranium/lead dating.

The Discovery of Fission: The Atomic Bomb and Nuclear Power (21.7)

- Fission is the splitting of an atom, such as uranium-235, into two atoms of lesser atomic weight.
- Because the fission of one uranium-235 atom releases enormous amounts of energy and produces neutrons that can split other uranium-235 atoms, the energy from these collective reactions can be harnessed in an atomic bomb or nuclear reactor.
- Nuclear power produces no air pollution and requires little mass to release lots of energy; however, there is always a danger of accidents, and it is difficult to dispose of nuclear waste.

Converting Mass to Energy: Mass Defect and Nuclear Binding Energy (21.8)

- In a nuclear fission reaction, mass is converted into energy.
- The difference in mass between a nuclide and the individual protons and neutrons that compose it is the mass defect, and the corresponding energy, calculated from Einstein's equation $E = mc^2$, is the nuclear binding energy.
- The stability of a nucleus is determined by the binding energy per nucleon, which increases up to mass number 60 and then decreases.

Nuclear Fusion: The Power of the Sun (21.9)

- Stars produce their energy by a process that is the opposite of fission: nuclear fusion, the combination of two light nuclei to form a heavier one.
- Modern nuclear weapons employ fusion. Although fusion has been examined as a possible method to produce electricity, experiments with hydrogen fusion have been more costly than productive.

Nuclear Transmutation and Transuranium Elements (21.10)

- Nuclear transmutation, the changing of one element to another element, has been used to create the transuranium elements, elements with atomic numbers greater than that of uranium.
- Two devices that accelerate particles to the high speeds necessary for transmutation reactions are the linear accelerator and the cyclotron. Both use alternating voltage to propel particles by electromagnetic forces.

The Effects of Radiation on Life (21.11)

- Acute radiation damage is caused by a large exposure to radiation for a short period of time. Lower radiation exposures may result in increased cancer risk because of damage to DNA. Genetic defects are caused by damage to the DNA of reproductive cells.
- The most effective unit of measurement for the amount of radiation absorbed is the rem, which takes into account the different penetrating and ionizing powers of the various types of radiation.

Radioactivity in Medicine and Other Applications (21.12)

- Radioactivity is central to the diagnosis of medical problems by means of radiotracers and positron emission tomography (PET). Both of these techniques can provide data about the appearance and metabolic activity of an organ, or help locate a tumor.
- Radiation is employed to treat cancer because it kills cells. Radiation is also used to kill bacteria in foods and to control harmful insect populations.

KEY EQUATIONS AND RELATIONSHIPS

The First-Order Rate Law (21.6)

$$\text{Rate} = kN$$

The Half-Life Equation (21.6)

$$t_{1/2} = \frac{0.693}{k} \qquad k = \text{rate constant}$$

The Integrated Rate Law (21.6)

$$\ln \frac{N_t}{N_0} = -kt \qquad N_t = \text{number of radioactive nuclei at time } t$$

$$N_0 = \text{initial number of radioactive nuclei}$$

Einstein's Energy–Mass Equation (21.8)

$$E = mc^2$$

EXERCISES

REVIEW QUESTIONS

1. What is radioactivity? Who discovered it? How was it discovered?
2. Explain Marie Curie's role in the discovery of radioactivity.
3. Define *A*, *Z*, and *X* in the notation used to specify a nuclide: ${}^{A}_{Z}X$.
4. Use the notation from Question 3 to write symbols for a proton, a neutron, and an electron.
5. What is an alpha particle? What happens to the mass number and atomic number of a nuclide that emits an alpha particle?
6. What is a beta particle? What happens to the mass number and atomic number of a nuclide that emits a beta particle?
7. What is a gamma ray? What happens to the mass number and atomic number of a nuclide that emits a gamma ray?
8. What is a positron? What happens to the mass number and atomic number of a nuclide that emits a positron?
9. Describe the process of electron capture. What happens to the mass number and atomic number of a nuclide that undergoes electron capture?
10. Rank alpha particles, beta particles, positrons, and gamma rays in terms of: (a) increasing ionizing power; (b) increasing penetrating power.
11. Explain why the ratio of neutrons to protons (*N/Z*) is important in determining nuclear stability. How can you use the *N/Z* ratio of a nuclide to predict the kind of radioactive decay that it might undergo?
12. What are magic numbers? How are they important in determining the stability of a nuclide?
13. Describe the basic way that each device detects radioactivity: (a) film-badge dosimeter; (b) Geiger-Müller counter; and (c) scintillation counter.
14. Explain the concept of half-life with respect to radioactive nuclides. What rate law is characteristic of radioactivity?
15. Explain the main concepts behind the technique of radiocarbon dating. How is radiocarbon dating corrected for changes in atmospheric concentrations of C-14? What range of ages is reliably determined by C-14 dating?
16. How is the uranium to lead ratio in a rock used to estimate its age? How does this dating technique provide an estimate for Earth's age? How old is Earth according to this dating method?
17. Describe fission. Include the concepts of chain reaction and critical mass in your description. How and by whom was fission discovered? Explain how fission is used to generate electricity.
18. What was the Manhattan Project? Briefly describe its development and culmination.
19. Describe the advantages and disadvantages of using fission to generate electricity.
20. The products of a nuclear reaction usually have a different mass than the reactants. Why?
21. Explain the concepts of mass defect and nuclear binding energy. At what mass number does the nuclear binding energy per nucleon peak? What is the significance of this?
22. What is fusion? Why can fusion and fission both produce energy? Explain.
23. What are some of the challenges associated with using fusion to generate electricity?
24. Explain transmutation and provide one or two examples.
25. How does a linear accelerator work? For what purpose is it used?
26. Explain the basic principles of cyclotron function.
27. How does radiation affect living organisms?
28. Explain why different kinds of radiation affect biological tissues differently, even though the amount of radiation exposure may be the same.
29. Explain the significance of the biological effectiveness factor in measuring radiation exposure. What types of radiation would you expect to have the highest biological effectiveness factor?
30. Describe some of the medical uses, both in diagnosis and in treatment of disease, of radioactivity.

PROBLEMS BY TOPIC

Radioactive Decay and Nuclide Stability

31. Write a nuclear equation for the indicated decay of each nuclide.
    - **a.** U-234 (alpha)
    - **b.** Th-230 (alpha)
    - **c.** Pb-214 (beta)
    - **d.** N-13 (positron emission)
    - **e.** Cr-51 (electron capture)
32. Write a nuclear equation for the indicated decay of each nuclide.
    - **a.** Po-210 (alpha)
    - **b.** Ac-227 (beta)
    - **c.** Tl-207 (beta)
    - **d.** O-15 (positron emission)
    - **e.** Pd-103 (electron capture)
33. Write a partial decay series for Th-232 undergoing the sequential decays: $\alpha, \beta, \beta, \alpha$.
34. Write a partial decay series for Rn-220 undergoing the sequential decays: $\alpha, \beta, \beta, \alpha$.
35. Fill in the missing particle in each nuclear equation.
    - **a.** $\_\_\_\_ \longrightarrow {}^{217}_{85}\text{At} + {}^{4}_{2}\text{He}$
    - **b.** ${}^{241}_{94}\text{Pu} \longrightarrow {}^{241}_{95}\text{Am} + \_\_\_\_$
    - **c.** ${}^{19}_{11}\text{Na} \longrightarrow {}^{19}_{10}\text{Ne} + \_\_\_\_$
    - **d.** ${}^{75}_{34}\text{Se} + \_\_\_\_ \longrightarrow {}^{75}_{33}\text{As}$
36. Fill in the missing particle in each nuclear equation.
    - **a.** ${}^{241}_{95}\text{Am} \longrightarrow {}^{237}_{93}\text{Np} + \_\_\_\_$
    - **b.** $\_\_\_\_ \longrightarrow {}^{233}_{92}\text{U} + {}^{0}_{-1}\text{e}$
    - **c.** ${}^{237}_{93}\text{Np} \longrightarrow \_\_\_\_ + {}^{4}_{2}\text{He}$
    - **d.** ${}^{75}_{35}\text{Br} \longrightarrow \_\_\_\_ + {}^{0}_{+1}\text{e}$

37. Determine whether or not each nuclide is likely to be stable. State your reasons.
 a. Mg-26
 b. Ne-25
 c. Co-51
 d. Te-124

38. Determine whether or not each nuclide is likely to be stable. State your reasons.
 a. Ti-48
 b. Cr-63
 c. Sn-102
 d. Y-88

39. The first six elements of the first transition series have the following number of stable isotopes:

Element	Number of Stable Isotopes
Sc	1
Ti	5
V	1
Cr	3
Mn	1
Fe	4

Explain why Sc, V, and Mn each has only one stable isotope while the other elements have several.

40. Neon and magnesium each have three stable isotopes, while sodium and aluminum each have only one. Explain why this might be so.

41. Predict a likely mode of decay for each unstable nuclide.
 a. Mo-109
 b. Ru-90
 c. P-27
 d. Rn-196

42. Predict a likely mode of decay for each unstable nuclide.
 a. Sb-132
 b. Te-139
 c. Fr-202
 d. Ba-123

43. Which one of each pair of nuclides would you expect to have the longer half-life?
 a. Cs-113 or Cs-125
 b. Fe-62 or Fe-70

44. Which one of each pair of nuclides would you expect to have the longer half-life?
 a. Cs-149 or Cs-139
 b. Fe-45 or Fe-52

The Kinetics of Radioactive Decay and Radiometric Dating

45. One of the nuclides in spent nuclear fuel is U-235, an alpha emitter with a half-life of 703 million years. How long will it take for the amount of U-235 to reach 10.0% of its initial amount?

46. A patient is given 0.050 mg of technetium-99m, a radioactive isotope with a half-life of about 6.0 hours. How long does it take for the radioactive isotope to decay to 1.0×10^{-3} mg? (Assume the nuclide is not excreted from the body.)

47. A radioactive sample contains 1.55 g of an isotope with a half-life of 3.8 days. What mass of the isotope remains after 5.5 days?

48. At 8:00 A.M., a patient receives a 58-mg dose of I-131 to obtain an image of her thyroid. If the nuclide has a half-life of 8 days, what mass of the nuclide remains in the patient at 5:00 P.M. the next day? (Assume the nuclide is not excreted from the body.)

49. A sample of F-18 has an initial decay rate of 1.5×10^5 dis/s. How long will it take for the decay rate to fall to 2.5×10^3 dis/s? (F-18 has a half-life of 1.83 hours.)

50. A sample of Tl-201 has an initial decay rate of 5.88×10^4 dis/s. How long will it take for the decay rate to fall to 287 dis/s? (Tl-201 has a half-life of 3.042 days.)

51. A wooden boat discovered just south of the Great Pyramid in Egypt has a carbon-14/carbon-12 ratio that is 72.5% of that found in living organisms. How old is the boat?

52. A layer of peat beneath the glacial sediments of the last ice age has a carbon-14/carbon-12 ratio that is 22.8% of that found in living organisms. How long ago was this ice age?

53. An ancient skull has a carbon-14 decay rate of 0.85 disintegrations per minute per gram of carbon (0.85 dis/ min • g C). How old is the skull? (Assume that living organisms have a carbon-14 decay rate of 15.3 dis/ min • g C and that carbon-14 has a half-life of 5730 yr.)

54. A mammoth skeleton has a carbon-14 decay rate of 0.48 disintegrations per minute per gram of carbon (0.48 dis/ min • g C). When did the mammoth live? (Assume that living organisms have a carbon-14 decay rate of 15.3 dis/min • g C and that carbon-14 has a half-life of 5730 yr.)

55. A rock from Australia contains 0.438 g of Pb-206 to every 1.00 g of U-238. Assuming that the rock did not contain any Pb-206 at the time of its formation, how old is the rock?

56. A meteor has a Pb-206:U-238 mass ratio of 0.855:1.00. What is the age of the meteor? (Assume that the meteor did not contain any Pb-206 at the time of its formation.)

Fission, Fusion, and Transmutation

57. Write the nuclear reaction for the neutron-induced fission of U-235 to form Xe-144 and Sr-90. How many neutrons are produced in the reaction?

58. Write the nuclear reaction for the neutron-induced fission of U-235 to produce Te-137 and Zr-97. How many neutrons are produced in the reaction?

59. Write the nuclear equation for the fusion of two H-2 atoms to form He-3 and one neutron.

60. Write the nuclear equation for the fusion of H-3 with H-1 to form He-4.

61. A breeder nuclear reactor is a reactor in which nonfissionable (nonfissile) U-238 is converted into fissionable (fissile) Pu-239. The process involves bombardment of U-238 by neutrons to form U-239, which then undergoes two sequential beta decays. Write nuclear equations for this process.

62. Write the series of nuclear equations to represent the bombardment of Al-27 with a neutron to form a product that subsequently undergoes an alpha decay followed by a beta decay.

63. Rutherfordium-257 was synthesized by bombarding Cf-249 with C-12. Write the nuclear equation for this reaction.

64. Element 107, now named bohrium, was synthesized by German researchers by colliding bismuth-209 with chromium-54 to form a bohrium isotope and one neutron. Write the nuclear equation to represent this reaction.

Energetics of Nuclear Reactions, Mass Defect, and Nuclear Binding Energy

65. If 1.0 g of matter is converted to energy, how much energy is formed?

66. A typical home uses approximately 1.0×10^3 kWh of energy per month. If the energy came from a nuclear reaction, what mass would have to be converted to energy per year to meet the energy needs of the home?

67. Calculate the mass defect and nuclear binding energy per nucleon of each nuclide.
   a. O-16 (atomic mass = 15.994915 amu)
   b. Ni-58 (atomic mass = 57.935346 amu)
   c. Xe-129 (atomic mass = 128.904780 amu)

68. Calculate the mass defect and nuclear binding energy per nucleon of each nuclide.
   a. Li-7 (atomic mass = 7.016003 amu)
   b. Ti-48 (atomic mass = 47.947947 amu)
   c. Ag-107 (atomic mass = 106.905092 amu)

69. Calculate the quantity of energy produced per gram of U-235 (atomic mass = 235.043922 amu) for the neutron-induced fission of U-235 to form Xe-144 (atomic mass = 143.9385 amu) and Sr-90 (atomic mass = 89.907738 amu) (discussed in Problem 57).

70. Calculate the quantity of energy produced per mole of U-235 (atomic mass = 235.043922 amu) for the neutron-induced fission of U-235 to produce Te-137 (atomic mass = 136.9253 amu) and Zr-97 (atomic mass = 96.910950 amu) (discussed in Problem 58).

71. Calculate the quantity of energy produced per gram of reactant for the fusion of two H-2 (atomic mass = 2.014102 amu) atoms to form He-3 (atomic mass = 3.016029 amu) and one neutron (discussed in Problem 59).

72. Calculate the quantity of energy produced per gram of reactant for the fusion of H-3 (atomic mass = 3.016049 amu) with H-1 (atomic mass = 1.007825 amu) to form He-4 (atomic mass = 4.002603 amu) (discussed in Problem 60).

Effects and Applications of Radioactivity

73. A 75-kg man is exposed to 32.8 rad of radiation. How much energy is absorbed by his body? Compare this energy to the amount of energy absorbed by his body if he jumps from a chair to the floor (assume that the chair is 0.50 m from the ground and that all of the energy from the fall is absorbed by the man).

74. If a 55-g laboratory mouse is exposed to 20.5 rad of radiation, how much energy is absorbed by the mouse's body?

75. PET studies require fluorine-18, which is produced in a cyclotron and decays with a half-life of 1.83 hours. Assuming that the F-18 can be transported at 60.0 miles/hour, how close must the hospital be to the cyclotron if 65% of the F-18 produced makes it to the hospital?

76. Suppose a patient is given 155 mg of I-131, a beta emitter with a half-life of 8.0 days. Assuming that none of the I-131 is eliminated from the person's body in the first 4.0 hours of treatment, what is the exposure (in Ci) during those first four hours?

CUMULATIVE PROBLEMS

77. Complete each nuclear equation and calculate the energy change (in J/mol of reactant) associated with each. (Be-9 = 9.012182 amu, Bi-209 = 208.980384 amu, He-4 = 4.002603 amu, Li-6 = 6.015122 amu, Ni-64 = 63.927969 amu, Rg-272 = 272.1535 amu, Ta-179 = 178.94593 amu, and W-179 = 178.94707 amu).
   a. $\_\_\_\_ + {}^{9}_{4}\text{Be} \longrightarrow {}^{6}_{3}\text{Li} + {}^{4}_{2}\text{He}$
   b. ${}^{209}_{83}\text{Bi} + {}^{64}_{28}\text{Ni} \longrightarrow {}^{272}_{111}\text{Rg} + \_\_\_\_$
   c. ${}^{179}_{74}\text{W} + \_\_\_\_ \longrightarrow {}^{179}_{73}\text{Ta}$

78. Complete each nuclear equation and calculate the energy change (in J/mol of reactant) associated with each. (Al-27 = 26.981538 amu, Am-241 = 241.056822 amu, He-4 = 4.002603 amu, Np-237 = 237.048166 amu, P-30 = 29.981801 amu, S-32 = 31.972071 amu, and Si-29 = 28.976495 amu).
   a. ${}^{27}_{13}\text{Al} + {}^{4}_{2}\text{He} \longrightarrow {}^{30}_{15}\text{P} + \_\_\_\_$
   b. ${}^{32}_{16}\text{S} + \_\_\_\_ \longrightarrow {}^{29}_{14}\text{Si} + {}^{4}_{2}\text{He}$
   c. ${}^{241}_{95}\text{Am} \longrightarrow {}^{237}_{93}\text{Np} + \_\_\_\_$

79. Write the nuclear equation for the most likely mode of decay for each unstable nuclide.
   a. Ru-114 b. Ra-216 c. Zn-58 d. Ne-31

80. Write the nuclear equation for the most likely mode of decay for each unstable nuclide.
   a. Kr-74
   b. Th-221
   c. Ar-44
   d. Nb-85

81. Bismuth-210 is a beta emitter with a half-life of 5.0 days. If a sample contains 1.2 g of Bi-210 (atomic mass = 209.984105 amu), how many beta emissions occur in 13.5 days? If a person's body intercepts 5.5% of those emissions, to what dose of radiation (in Ci) is the person exposed?

82. Polonium-218 is an alpha emitter with a half-life of 3.0 minutes. If a sample contains 55 mg of Po-218 (atomic mass = 218.008965 amu), how many alpha emissions occur in 25.0 minutes? If the polonium is ingested by a person, to what dose of radiation (in Ci) is the person exposed?

83. Radium-226 (atomic mass = 226.025402 amu) decays to radon-224 (a radioactive gas) with a half-life of 1.6×10^3 years. What volume of radon gas (at 25.0 °C and 1.0 atm) does 25.0 g of radium produce in 5.0 days? (Report your answer to two significant digits.)

84. In one of the neutron-induced fission reactions of U-235 (atomic mass = 235.043922 amu), the products are Ba-140 and Kr-93 (a radioactive gas). What volume of Kr-93 (at 25.0 °C and 1.0 atm) is produced when 1.00 g of U-235 undergoes this fission reaction?

85. When a positron and an electron annihilate one another, the resulting mass is completely converted to energy. Calculate the energy associated with this process in kJ/mol.

86. A typical nuclear reactor produces about 1.0 MW of power per day. What is the minimum rate of mass loss required to produce this much energy?

87. Find the binding energy in an atom of ${}^{3}\text{He}$, which has a mass of 3.016030 amu.

88. The overall hydrogen burning reaction in stars can be represented as the conversion of four protons to one α particle. Use the data for the mass of H-1 and He-4 to calculate the energy released by this process.

89. The nuclide ^{247}Es is made by bombardment of ^{238}U in a reaction that emits five neutrons. Identify the bombarding particle.

90. The nuclide ^{6}Li reacts with ^{2}H to form two identical particles. Identify the particles.

91. The half-life of ^{238}U is 4.5×10^{9} yr. A sample of rock of mass 1.6 g produces 29 dis/s. Assuming all the radioactivity is due to ^{238}U, find the percent by mass of ^{238}U in the rock.

92. The half-life of ^{232}Th is 1.4×10^{10} yr. Find the number of disintegrations per hour emitted by 1.0 mol of ^{232}Th.

93. A 1.50-L gas sample at 745 mm Hg and 25.0 °C contains 3.55% radon-220 by volume. Radon-220 is an alpha-emitter with a half-life of 55.6 s. How many alpha particles are emitted by the gas sample in 5.00 minutes?

94. A 228-mL sample of an aqueous solution contains 2.35% $MgCl_2$ by mass. Exactly half of the magnesium ions are Mg-28, a beta emitter with a half-life of 21 hours. What is the decay rate of Mg-28 in the solution after 4.00 days? (Assume a density of 1.02 g/mL for the solution.)

95. When a positron and an electron collide and annihilate each other, two photons of equal energy are produced. Find the wavelength of these photons.

96. The half-life of ^{235}U, an alpha emitter, is 7.1×10^{8} yr. Calculate the number of alpha particles emitted by 1.0 mg of this nuclide in 1.0 minute.

97. Given that the energy released in the fusion of two deuterium atoms to a ^{3}He and a neutron is 3.3 MeV, and in the fusion to tritium and a proton it is 4.0 MeV. Calculate the energy change for the process $^{3}He + ^{1}n \longrightarrow ^{3}H + ^{1}p$. Suggest an explanation for why this process occurs at much lower temperatures than either of the first two.

98. The nuclide ^{18}F decays by both electron capture and β^{+} decay. Find the difference in the energy released by these two processes. The atomic masses are $^{18}F = 18.000950$ and $^{18}O = 17.9991598$.

CHALLENGE PROBLEMS

99. The space shuttle carries about 72,500 kg of solid aluminum fuel, which is oxidized with ammonium perchlorate according to the reaction shown here:

$$10\ Al(s) + 6\ NH_4ClO_4(s) \longrightarrow 4\ Al_2O_3(s) + 2\ AlCl_3(s) + 12\ H_2O(g) + 3\ N_2(g)$$

The space shuttle also carries about 608,000 kg of oxygen (which reacts with hydrogen to form gaseous water).

a. Assuming that aluminum and oxygen are the limiting reactants, determine the total energy produced by these fuels. (ΔH_f° for solid ammonium perchlorate is −295 kJ/mol.)

b. Suppose that a future space shuttle is powered by matter–antimatter annihilation. The matter is normal hydrogen (containing a proton and an electron), and the antimatter is antihydrogen (containing an antiproton and a positron). What mass of antimatter is required to produce the energy equivalent of the aluminum and oxygen fuel currently carried on the space shuttle?

100. Suppose that an 85.0-gram laboratory animal ingests 10.0 mg of a substance that contained 2.55% by mass Pu-239, an alpha emitter with a half-life of 24,110 years.

a. What is the animal's initial radiation exposure in curies?

b. If all of the energy from the emitted alpha particles is absorbed by the animal's tissues, and if the energy of each emission is 7.77×10^{-12} J, what is the dose in rads to the animal in the first 4.0 hours following the ingestion of the radioactive material? Assuming a biological effectiveness factor of 20, what is the 4.0-hour dose in rems?

101. In addition to the natural radioactive decay series that begins with U-238 and ends with Pb-206, there are natural radioactive decay series that begin with U-235 and Th-232. Both of these series end with nuclides of Pb. Predict the likely end product of each series and the number of α decay steps that occur.

102. The hydride of an unstable nuclide of a Group IIA metal, $MH_2(s)$, decays by α-emission. A 0.025-mol sample of the hydride is placed in an evacuated 2.0 L container at 298 K. After 82 minutes, the pressure in the container is 0.55 atm. Find the half-life of the nuclide.

103. The nuclide ^{38}Cl decays by beta emission with a half-life of 40.0 min. A sample of 0.40 mol of $H^{38}Cl$ is placed in a 6.24-L container. After 80.0 min the pressure is 1650 mmHg. What is the temperature of the container?

104. When BF_3 is bombarded with neutrons, the boron undergoes an α decay, but the F is unaffected. A 0.20-mol sample of BF_3 contained in a 3.0-L container at 298 K is bombarded with neutrons until half of the BF_3 has reacted. What is the pressure in the container at 298 K?

CONCEPTUAL PROBLEMS

105. Closely examine the diagram representing the beta decay of fluorine-21 and draw in the missing nucleus.

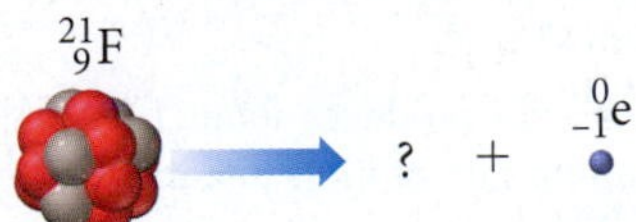

106. Approximately how many half-lives must pass for the amount of radioactivity in a substance to decrease to below 1% of its initial level?

107. A person is exposed for 3 days to identical amounts of two different nuclides that emit positrons of roughly equal energy. The half-life of nuclide A is 18.5 days and the half-life of nuclide B is 255 days. Which of the two nuclides poses the greater health risk?

108. Identical amounts of two different nuclides, an alpha emitter and a gamma emitter, with roughly equal half-lives are spilled in a building adjacent to your bedroom. Which of the two nuclides poses the greater health threat to you while you sleep in your bed? If you accidentally wander into the building and ingest equal amounts of the two nuclides, which one poses the greater health threat?

109. Drugstores in many areas now carry tablets, under such trade names as Iosat and NoRad, to be taken in the event of an accident at a nuclear power plant or a terrorist attack that releases radioactive material. These tablets contain potassium iodide (KI). Can you explain the nature of the protection that they provide?

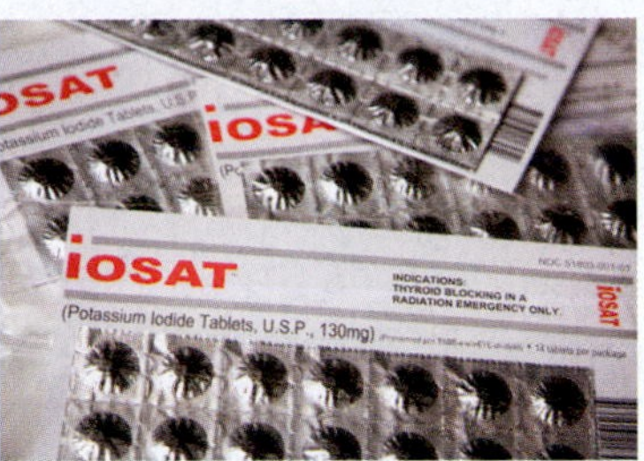

ANSWERS TO CONCEPTUAL CONNECTIONS

Cc 21.1 **(c)** The arrow labeled *x* represents a decrease of two neutrons and two protons, indicative of alpha decay. The arrow labeled *y* represents a decrease of one neutron and an increase of one proton, indicative of beta decay.

Cc 21.2 **(b)** The half-life is the time it takes for the number of nuclei to decay to one-half of their original number.

Cc 21.3 **(b)** 0.10 mol. The sample loses one-half of the number of moles per half-life; so over the course of four half-lives, the amount falls to 0.10 mol.

Cc 21.4 Lawrencium-256

Cc 21.5 Nuclide A. Because nuclide A has a shorter half-life, more of the nuclides will decay, and therefore produce radiation, before they exit the body.

CHAPTER 22

"Organic chemistry just now is enough to drive one mad. It gives one the impression of a primeval, tropical forest full of the most remarkable things. . . ."

—Friedrich Wöhler (1800–1882)

About half of all men's colognes contain at least some patchouli alcohol ($C_{15}H_{26}O$)*, an organic compound (pictured here) derived from the patchouli plant. Patchouli alcohol has a pungent, musty, earthy fragrance.*

Organic Chemistry

22.1 Fragrances and Odors 903
22.2 Carbon: Why It Is Unique 904
22.3 Hydrocarbons: Compounds Containing Only Carbon and Hydrogen 905
22.4 Alkanes: Saturated Hydrocarbons 912
22.5 Alkenes and Alkynes 916
22.6 Hydrocarbon Reactions 921
22.7 Aromatic Hydrocarbons 924
22.8 Functional Groups 928
22.9 Alcohols 929
22.10 Aldehydes and Ketones 931
22.11 Carboxylic Acids and Esters 934
22.12 Ethers 936
22.13 Amines 937
22.14 Polymers 937
Key Learning Outcomes 941

ORGANIC CHEMISTRY IS THE STUDY of carbon-containing compounds. Carbon is unique in the sheer number of compounds that it forms. Millions of organic compounds are known, and researchers discover new ones every day. Carbon is also unique in the diversity of compounds that it forms. In most cases, a fixed number of carbon atoms can combine with a fixed number of atoms of another element to form many different compounds. For example, 10 carbon atoms and 22 hydrogen atoms can form 75 distinctly different compounds. With carbon as the backbone, nature can take the same combination of atoms and bond them together in slightly different ways to produce a huge variety of substances. In other words, in organic chemistry, we see the theme that structure determines properties played out over and over again. It is not surprising that life is based on the chemistry of carbon because life needs diversity to exist, and organic chemistry is nothing if not diverse. In this chapter, we peer into Friedrich Wöhler's "primeval tropical forest" (see chapter-opening quotation) and discover the most remarkable things.

22.1 Fragrances and Odors

Have you ever ridden an elevator with someone wearing too much cologne? Or found yourself too close to a skunk? Or caught a whiff of rotting fish? What causes these fragrances and odors? When we inhale certain molecules called odorants, they bind with olfactory receptors in our noses. This interaction is

$CH_3CH{=}CHCH_2SH$

2-Butene-1-thiol

$$\begin{array}{l}\quad CH_3 \\ \quad\ | \\ CH_3CHCH_2CH_2SH\end{array}$$

3-Methyl-1-butanethiol

▲ The smell of skunk is due primarily to the molecules shown here.

largely determined by the kind of structure-dependent lock and key mechanism that we discussed in the beginning of Chapter 6. When the odorant binds to its receptor, a nerve signal is sent to the brain that we experience as a smell. Some smells, such as that of cologne, are pleasant (when not overdone). Other smells, such as that of the skunk or rotting fish, are unpleasant. Our sense of smell helps us identify food, people, and other organisms, and it alerts us to dangers such as polluted air or spoiled food. Smell (olfaction) is one way we probe the environment around us.

Odorants, if they are to reach our noses, must be volatile. However, many volatile substances have no scent at all. Nitrogen, oxygen, water, and carbon dioxide molecules, for example, constantly pass through our noses, yet they produce no smell because they do not bind to olfactory receptors. Most common smells are caused by **organic molecules,** molecules containing carbon combined with several other elements such as hydrogen, nitrogen, oxygen, and sulfur. Organic molecules are responsible for the smells of roses, vanilla, cinnamon, almond, jasmine, body odor, and rotting fish. When you wander into a rose garden, you experience the sweet smell caused in part by geraniol, an organic compound emitted by roses. Men's colognes often contain patchouli alcohol, an earthy-smelling organic compound extracted from the patchouli plant. If you have been in the vicinity of skunk spray (or have been unfortunate enough to be sprayed yourself), you are familiar with the unpleasant smell of 2-butene-1-thiol and 3-methyl-1-butanethiol, two particularly odoriferous compounds present in the secretion that skunks use to defend themselves.

The study of compounds containing carbon combined with one or more of the elements mentioned previously (hydrogen, nitrogen, oxygen, and sulfur), including their properties and their reactions, is known as **organic chemistry**. Besides composing much of what we smell, organic compounds are prevalent in foods, drugs, petroleum products, and pesticides. Organic chemistry is also the basis for living organisms. Life has evolved based on carbon-containing compounds, making organic chemistry of utmost importance to any person interested in understanding living organisms.

22.2 Carbon: Why It Is Unique

Why did life evolve based on the chemistry of carbon? Why is life not based on some other element? The answer may not be simple, but we know that life—in order to exist—must entail complexity, and carbon chemistry is clearly complex. The number of compounds containing carbon is greater than the number of compounds containing all of the other elements combined. The reasons for carbon's unique and versatile behavior include its ability to form four covalent bonds, its ability to form double and triple bonds, and its tendency to *catenate* (that is, to form chains).

Carbon's Tendency to Form Four Covalent Bonds

Carbon—with its four valence electrons—forms four covalent bonds. Consider the Lewis structure and space-filling models of two simple carbon compounds, methane and ethane.

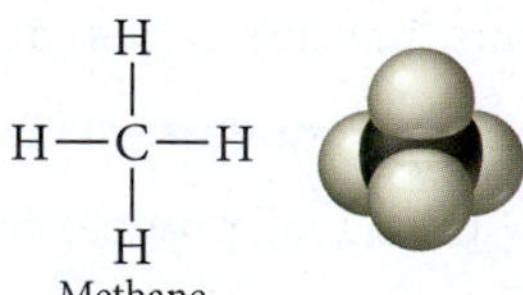

Methane

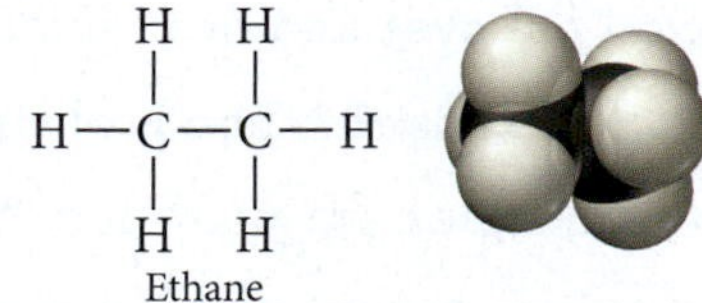

Ethane

The geometry about a carbon atom forming four single bonds is tetrahedral, as we can see in the figure for methane. Carbon's ability to form four bonds, and to form those bonds with a number of different elements, results in its potential to form many different compounds. As you learn to draw structures for organic compounds, always remember to draw carbon with four bonds.

Carbon's Ability to Form Double and Triple Bonds

Carbon atoms also form double bonds (trigonal planar geometry) and triple bonds (linear geometry), adding even more diversity to the number of compounds that carbon forms.

$H_2C{=}CH_2$

Ethene

H—C≡C—H

Ethyne

In contrast, silicon (the element in the periodic table with properties closest to that of carbon) does not readily form double or triple bonds because the greater size of silicon atoms results in a Si—Si bond that is too long for much overlap between nonhybridized *p* orbitals.

Carbon's Tendency to Catenate

Carbon, more than any other element, can bond to itself to form chain, branched, and ring structures.

$CH_3CH_2CH_3$

Propane

$(CH_3)_3CH$

Isobutane

C_6H_{12}

Cyclohexane

Although other elements can form chains, none surpasses carbon at this ability. Silicon, for example, can form chains with itself. However, silicon's affinity for oxygen (the Si—O bond is stronger than the Si—Si bond) coupled with the prevalence of oxygen in our atmosphere means that silicon–silicon chains are readily oxidized to form silicates (the silicon–oxygen compounds that compose a significant proportion of minerals). By contrast, the C—C bond (347 kJ/mol) and the C—O bond (359 kJ/mol) are nearly the same strength, allowing carbon chains to exist relatively peacefully in an oxygen-rich environment. Silicon's affinity for oxygen robs it of the rich diversity that catenation provides to carbon.

22.3 Hydrocarbons: Compounds Containing Only Carbon and Hydrogen

Hydrocarbons—compounds that contain only carbon and hydrogen—are the simplest organic compounds. However, because of carbon's versatility, many different kinds of hydrocarbons exist. We use hydrocarbons as fuels. Candle wax, oil, gasoline, LP (liquefied petroleum) gas, and natural gas are all composed of hydrocarbons. Hydrocarbons are also the starting materials in the synthesis of many different consumer products including fabrics, soaps, dyes, cosmetics, drugs, plastic, and rubber.

As shown in Figure 22.1 ▼, we classify hydrocarbons into four different types: **alkanes**, **alkenes**, **alkynes**, and **aromatic hydrocarbons**. Alkanes, alkenes, and alkynes—also called **aliphatic**

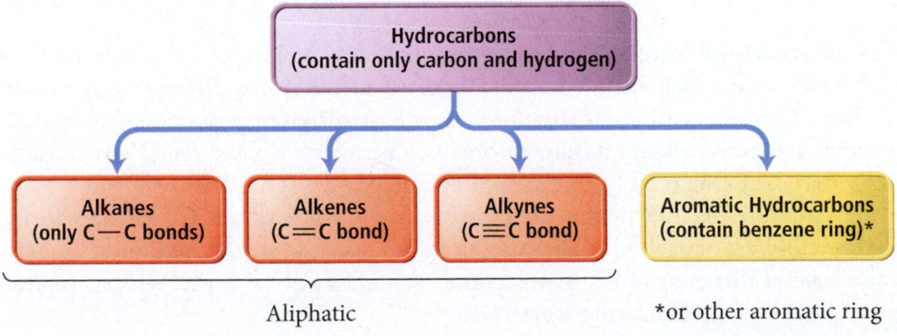

◀ **FIGURE 22.1 Four Types of Hydrocarbons**

hydrocarbons—are differentiated based on the kinds of bonds between carbon atoms. (We discuss aromatic hydrocarbons in detail in Section 22.7.) As we can see in Table 22.1, alkanes have only single bonds between carbon atoms, alkenes have a double bond, and alkynes have a triple bond.

TABLE 22.1 Alkanes, Alkenes, Alkynes

Type of Hydrocarbon	Type of Bonds	Generic Formula*	Example
Alkane	All single	C_nH_{2n+2}	H H H—C—C—H H H Ethane
Alkenes	One (or more) double	C_nH_{2n}	H H C=C H H Ethene
Alkynes	One (or more) triple	C_nH_{2n-2}	H—C≡C—H Ethyne

* n is the number of carbon atoms. These formulas apply only to noncyclic structures containing no more than one multiple bond.

Drawing Hydrocarbon Structures

Throughout this book, we have relied primarily on molecular formulas as the simplest way to represent compounds. In organic chemistry, however, molecular formulas are insufficient because, as we have already discussed, the same atoms can bond together in different ways to form different compounds. For example, consider an alkane with 4 carbon atoms and 10 hydrogen atoms. Two different structures, named butane and isobutane, are possible:

```
                                     H
                                     |
   H  H  H  H                  H H—C—H H
   |  |  |  |                  |     |     |
H—C—C—C—C—H             H—C———C———C—H
   |  |  |  |                  |     |     |
   H  H  H  H                  H     H     H
     Butane                      Isobutane
```

Butane and isobutane are **structural isomers,** molecules with the same molecular formula but different structures. More specifically, notice that the atoms are connected together in a different way in isobutene than in butane. Because of their different structures, they have different properties—indeed, they are different compounds. Isomerism is ubiquitous in organic chemistry. Butane has 2 structural isomers. Pentane (C_5H_{12}) has 3, hexane (C_6H_{14}) has 5, and decane ($C_{10}H_{22}$) has 75!

We represent the structure of a particular hydrocarbon with a **structural formula,** a formula that shows not only the numbers of each kind of atoms, but also how the atoms are bonded together. Organic chemists use several different kinds of structural formulas. For example, we can represent butane and isobutane in each of the following ways:

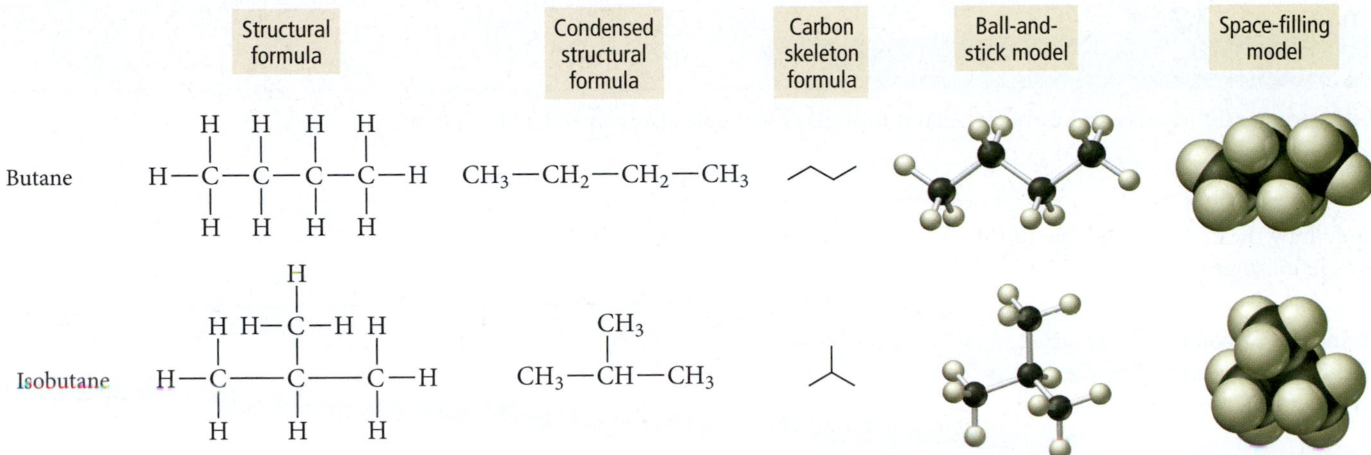

The structural formula shows all of the carbon and hydrogen atoms in the molecule and how they are bonded together. The *condensed structural formula* groups the hydrogen atoms with the carbon atom to which they are bonded. Condensed structural formulas may show some of the bonds (as the previous examples do) or none at all. For example, the condensed structural formula for butane can also be written as $CH_3CH_2CH_2CH_3$. The *carbon skeleton formula* (also called a line formula) shows the carbon–carbon bonds only as lines. Each end or bend of a line represents a carbon atom bonded to as many hydrogen atoms as necessary to form a total of four bonds. Carbon skeleton formulas allow us to draw complex structures quickly.

Note that structural formulas are generally not three-dimensional representations of the molecule, as space-filling or ball-and-stick models are. Instead, they are two-dimensional representations that show how atoms are bonded together. As such, the most important feature of a structural formula is the *connectivity* of the atoms, not the exact way the formula is drawn. For example, consider the two condensed structural formulas for butane and the corresponding space-filling models below them:

$CH_3—CH_2—CH_2—CH_3$ $\quad$ $CH_3—CH_2—CH_2$ (with CH_3 bonded below the last CH_2)

Same molecule

Since rotation about single bonds is relatively unhindered at room temperature, the two structural formulas are identical, even though they are drawn differently.

We represent double and triple bonds in structural formulas with double or triple lines. For example, we draw the structural formulas for C_3H_6 (propene) and C_3H_4 (propyne) as follows:

	Structural formula	Condensed structural formula	Carbon skeleton formula	Ball-and-stick model	Space-filling model
Propene	H—C=C—C—H (with H above each carbon and H below the third carbon)	$CH_2{=}CH—CH_3$			
Propyne	H—C≡C—C—H (with H above and below the third carbon)	$CH{\equiv}C—CH_3$			

The kind of structural formula we use depends on how much information we want to portray. Example 22.1 illustrates how to write structural formulas for a compound.

EXAMPLE 22.1

Writing Structural Formulas for Hydrocarbons

Write the structural formulas and carbon skeleton formulas for the five isomers of C_6H_{14} (hexane).

SOLUTION

To begin, draw the carbon backbone of the straight-chain isomer.

```
C—C—C—C—C—C
```

Determine the carbon backbone structure of the other isomers by arranging the carbon atoms in four other unique ways.

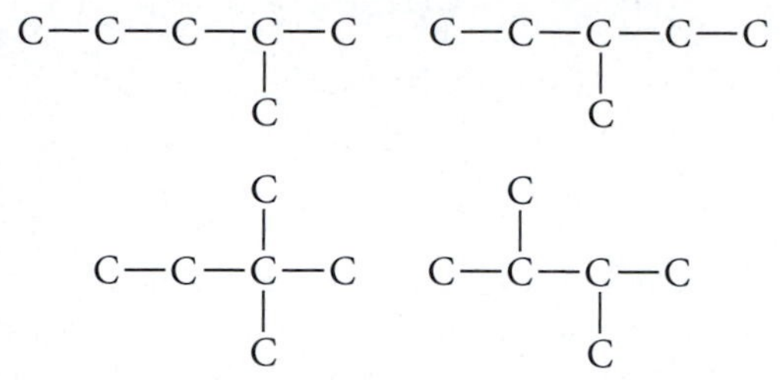

```
C—C—C—C—C      C—C—C—C—C
      |            |
      C            C

      C          C
      |          |
  C—C—C—C      C—C—C—C
      |            |
      C            C
```

Fill in all the hydrogen atoms so that each carbon forms four bonds.

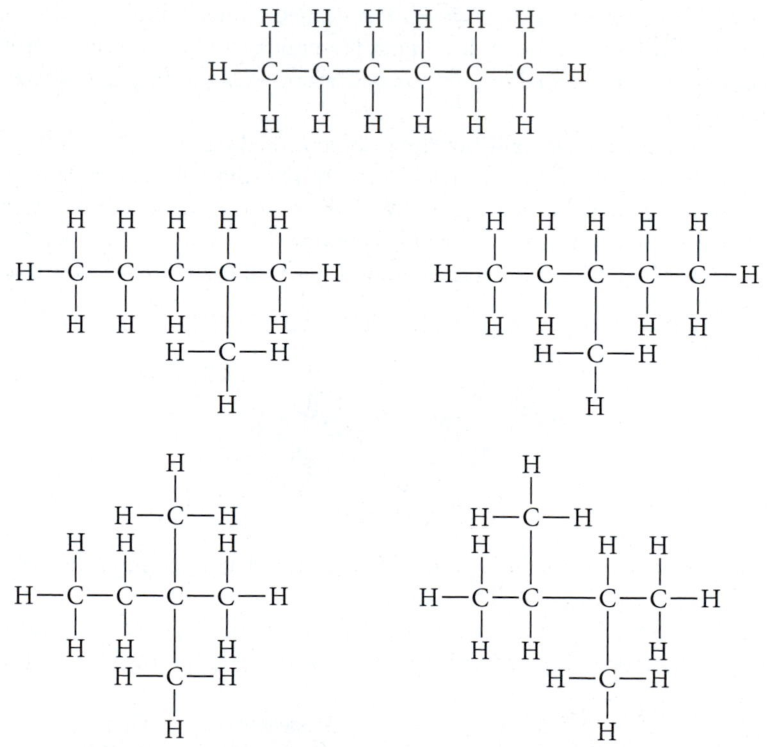

```
    H   H   H   H   H   H
    |   |   |   |   |   |
H—C—C—C—C—C—C—H
    |   |   |   |   |   |
    H   H   H   H   H   H

    H   H   H   H   H              H   H   H   H   H
    |   |   |   |   |              |   |   |   |   |
H—C—C—C—C—C—H          H—C—C—C—C—C—H
    |   |   |   |   |              |   |   |   |   |
    H   H   H   |   H              H   H   |   H   H
            H—C—H                      H—C—H
                |                          |
                H                          H

            H                          H
            |                          |
          H—C—H                      H—C—H
    H   H   |   H              H     |       H   H
    |   |   |   |              |     |       |   |
H—C—C—C—C—H            H—C—C———C—C—H
    |   |   |   |              |     |       |   |
    H   H   |   H              H     H       |   H
          H—C—H                          H—C—H
            |                              |
            H                              H
```

Write the carbon skeleton formulas by using lines to represent each carbon–carbon bond. Remember that each end or bend represents a carbon atom.

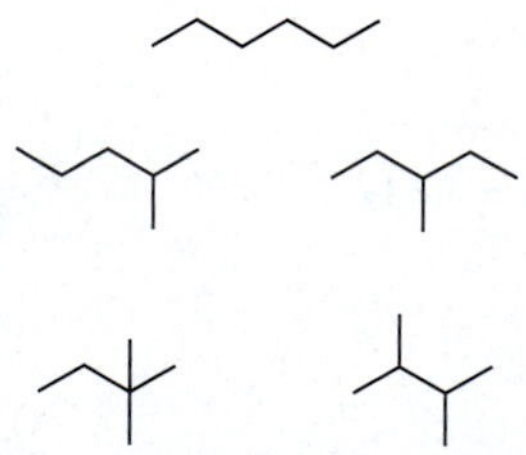

FOR PRACTICE 22.1

Write the structural formulas and carbon skeleton formulas for the three isomers of C_5H_{12} (pentane).

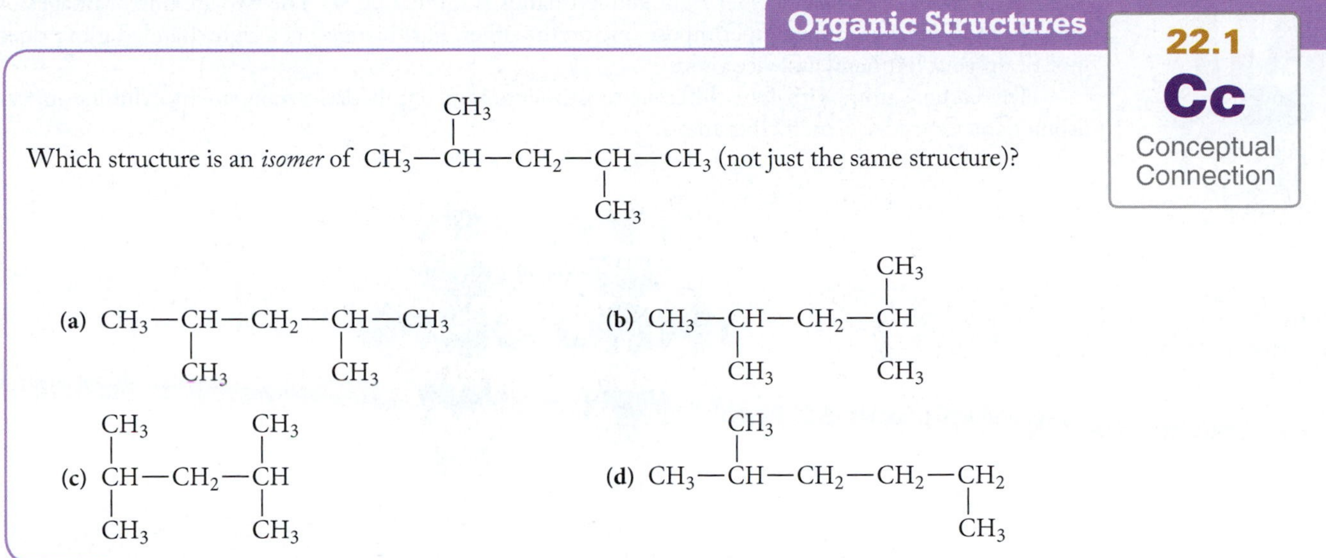

Stereoisomerism and Optical Isomerism

Stereoisomers are molecules in which the atoms have the same connectivity but a different spatial arrangement. We categorize stereoisomers into two types: geometric (or cis–trans) isomers and optical isomers. We discuss geometric isomers in Section 22.5. **Optical isomers** are two molecules that are nonsuperimposable mirror images of one another. Consider the molecule shown here with its mirror image.

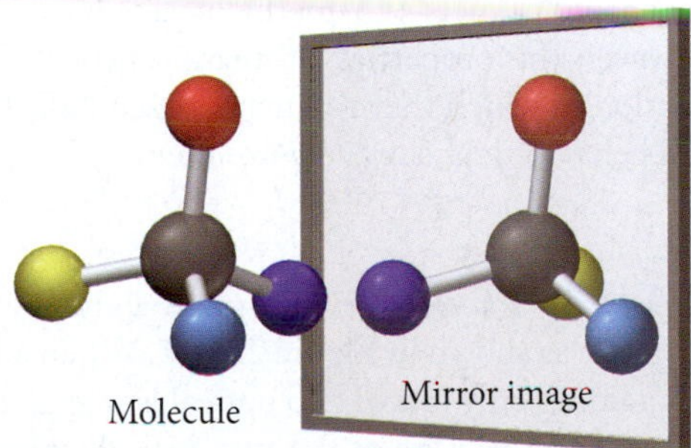

The molecule cannot be superimposed onto its mirror image. If we swing the mirror image around to try to superimpose the two, we find that there is no way to get all four substituent atoms to align together. (A substituent is an atom or group of atoms that is substituted for a hydrogen atom in an organic compound.)

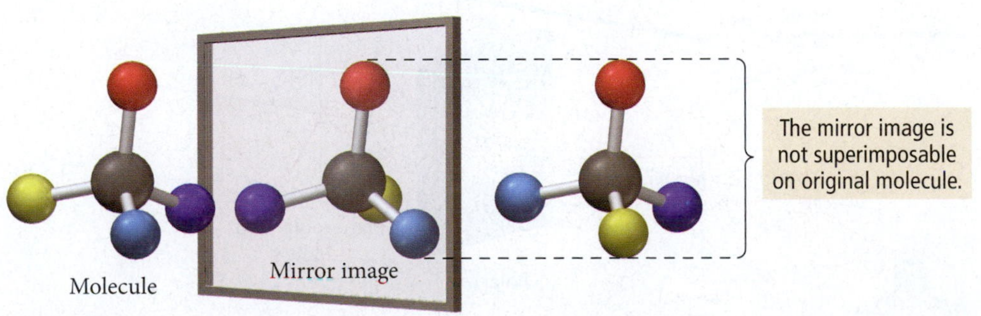

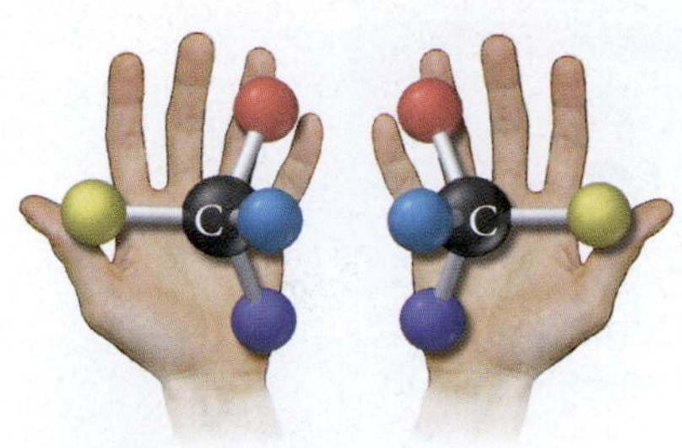

▲ **FIGURE 22.2 Mirror Images** The left and right hands are nonsuperimposable mirror images, just as are optical isomers.

Optical isomers are similar to your right and left hands (Figure 22.2 ◀). The two are mirror images of one another, but you cannot superimpose one on the other. For this reason, a right-handed glove does not fit on your left hand and vice versa.

Any carbon atom with four different substituents in a tetrahedral arrangement exhibits optical isomerism. Consider 3-methylhexane:

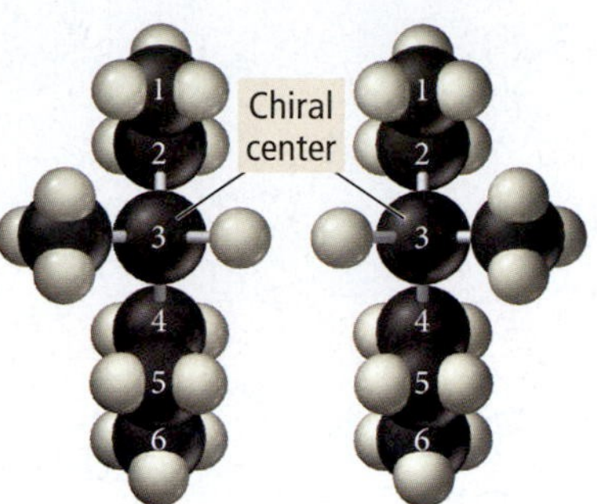

Optical isomers of 3-methylhexane

These molecules are nonsuperimposable mirror images and are optical isomers of one another. Optical isomers are also called **enantiomers.** Any molecule, such as 3-methylhexane, that exhibits optical isomerism is said to be **chiral**, from the Greek word *cheir*, which means "hand." Optical isomerism is important, not only to organic chemistry, but also to biology and biochemistry. Most biological molecules are chiral and usually only one or the other enantiomer is active in biological systems. For example, glucose, the primary fuel of cells, is chiral. Only one of the enantiomers of glucose has that familiar sweet taste and only that enantiomer can fuel our cellular functioning; the other enantiomer is not even metabolized by the body.

Some of the physical and chemical properties of enantiomers are indistinguishable from one another. For example, both of the optical isomers of 3-methylhexane have identical freezing points, melting points, and densities. However, the properties of enantiomers differ from one another in two important ways: (1) in the direction in which they rotate polarized light (which is one way to tell them apart) and (2) in their chemical behavior in a chiral environment.

Rotation of Polarized Light *Plane-polarized light* is light that is made up of electric field waves that oscillate in only one plane as shown in Figure 22.3 ▼. When a beam of plane-polarized light is directed through a sample containing only one of two optical isomers, the plane of polarization of the light is rotated as shown in Figure 22.4 ▶. One of the two optical isomers rotates the polarization of

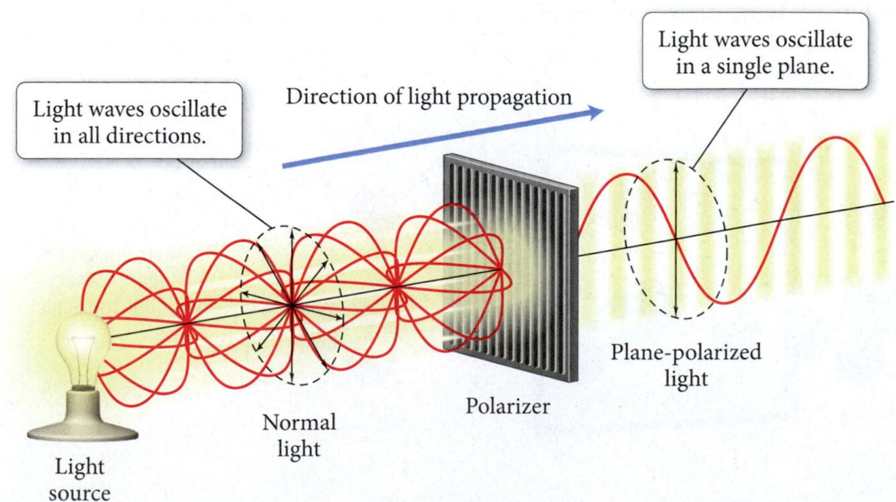

▶ **FIGURE 22.3 Plane-Polarized Light** The electric field of plane-polarized light oscillates in one plane.

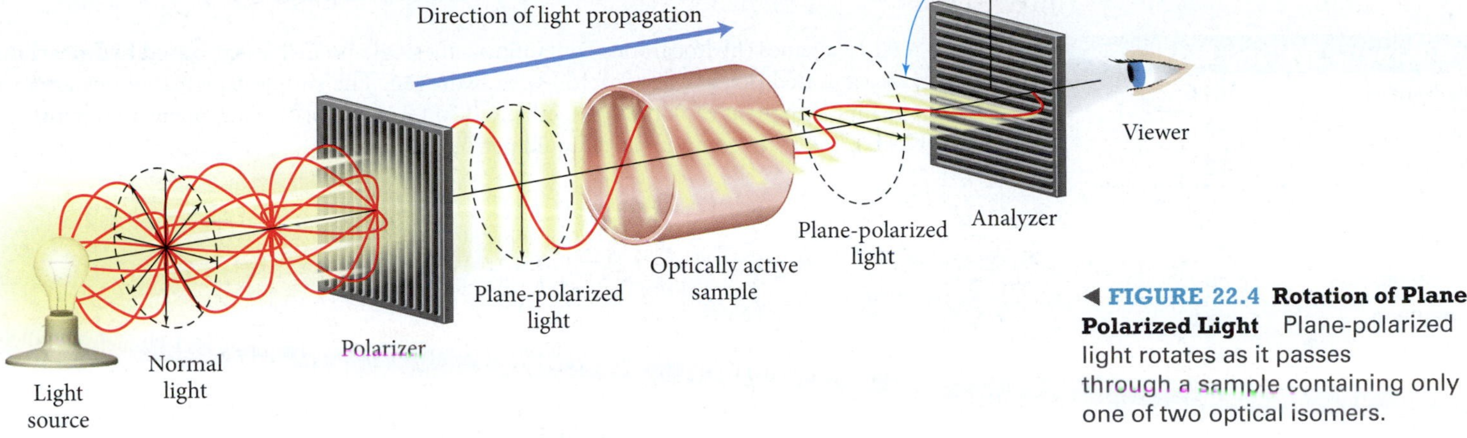

◀ **FIGURE 22.4 Rotation of Plane-Polarized Light** Plane-polarized light rotates as it passes through a sample containing only one of two optical isomers.

the light clockwise and is called the **dextrorotatory** isomer (or the *d* isomer). The other isomer rotates the polarization of the light counterclockwise and is called the **levorotatory** isomer (or the *l* isomer). An equimolar mixture of both optical isomers does not rotate the polarization of light at all and is called a **racemic mixture**.

Dextrorotatory means turning clockwise or to the right. *Levorotatory* means turning counterclockwise or to the left.

Chemical Behavior in a Chiral Environment Optical isomers also exhibit different chemical behavior when they are in a chiral environment (a chiral environment is one that is not superimposable on its mirror image). Enzymes are large biological molecules that catalyze reactions in living organisms and provide chiral environments. Consider the following simplified picture of two enantiomers in a chiral environment:

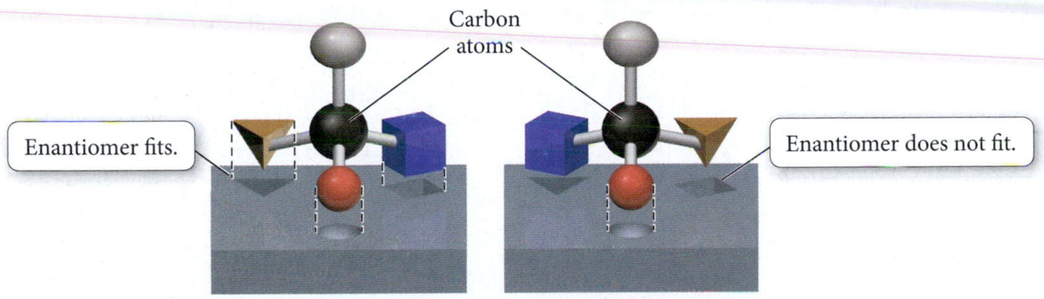

One of the enantiomers fits the template, but the other does not, no matter how it is rotated. In this way, an enzyme is able to catalyze the reaction of one enantiomer because that particular enantiomer fits the "template." As we have already seen, most biological molecules are chiral, and usually only one or the other enantiomer is active in biological systems. Even subtle differences in structure, such as the difference between one enantiomer and the other, affect properties.

Optical Isomers — 22.2 Cc Conceptual Connection

Which molecule exhibits optical isomerism?

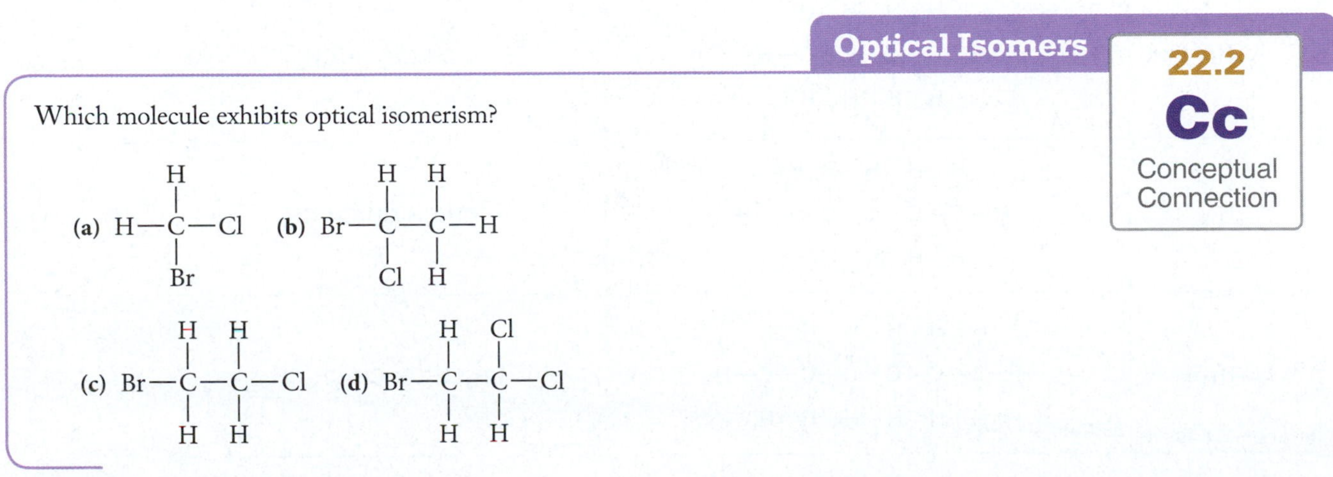

TABLE 22.2 *n*-Alkane Boiling Points

n-Alkane	Boiling Point (°C)
Methane	–161.5
Ethane	–88.6
Propane	–42.1
n-Butane	–0.5
n-Pentane	36.0
n-Hexane	68.7
n-Heptane	98.5
n-Octane	125.6

22.4 Alkanes: Saturated Hydrocarbons

We often refer to alkanes (hydrocarbons containing only single bonds) as **saturated hydrocarbons** because they are saturated (loaded to capacity) with hydrogen. The simplest hydrocarbons are methane (CH_4), the main component of natural gas; ethane (C_2H_6), a minority component in natural gas; and propane (C_3H_8), the main component of liquid petroleum (LP) gas.

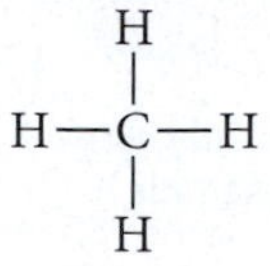

```
     H             H  H              H  H  H
     |             |  |              |  |  |
  H—C—H        H—C—C—H          H—C—C—C—H
     |             |  |              |  |  |
     H             H  H              H  H  H
  Methane         Ethane            Propane
```

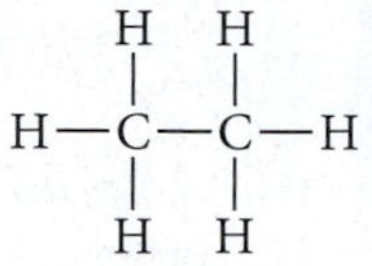

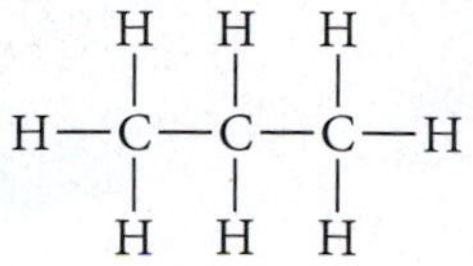

Alkanes containing four or more carbon atoms may be straight or branched (as we have already seen). The straight-chain isomers are often called normal alkanes, or *n*-alkanes. As the number of carbon atoms increases in the *n*-alkanes, so does their boiling point (as shown in Table 22.2). The increase is due to the increasing dispersion force with increasing molar mass (see Section 12.3). Methane, ethane, propane, and *n*-butane are all gases at room temperature, but the next *n*-alkane in the series, pentane, is a liquid at room temperature. Pentane is a component of gasoline. Table 22.3 summarizes the *n*-alkanes through decane, which contains ten carbon atoms. Like pentane, hexane through decane are all components of gasoline. Table 22.4 summarizes the uses of hydrocarbons.

TABLE 22.3 *n*-Alkanes

<table>
<tr><th>n</th><th>Name</th><th>Molecular Formula C_nH_{2n+2}</th><th>Structural Formula</th><th>Condensed Structural Formula</th></tr>
<tr><td>1</td><td>Methane</td><td>CH_4</td><td>H
|
H—C—H
|
H</td><td>CH_4</td></tr>
<tr><td>2</td><td>Ethane</td><td>C_2H_6</td><td>H H
| |
H—C—C—H
| |
H H</td><td>CH_3CH_3</td></tr>
<tr><td>3</td><td>Propane</td><td>C_3H_8</td><td>H H H
| | |
H—C—C—C—H
| | |
H H H</td><td>$CH_3CH_2CH_3$</td></tr>
<tr><td>4</td><td>n-Butane</td><td>C_4H_{10}</td><td>H H H H
| | | |
H—C—C—C—C—H
| | | |
H H H H</td><td>$CH_3CH_2CH_2CH_3$</td></tr>
<tr><td>5</td><td>n-Pentane</td><td>C_5H_{12}</td><td>H H H H H
| | | | |
H—C—C—C—C—C—H
| | | | |
H H H H H</td><td>$CH_3CH_2CH_2CH_2CH_3$</td></tr>
<tr><td>6</td><td>n-Hexane</td><td>C_6H_{14}</td><td>H H H H H H
| | | | | |
H—C—C—C—C—C—C—H
| | | | | |
H H H H H H</td><td>$CH_3CH_2CH_2CH_2CH_2CH_3$</td></tr>
</table>

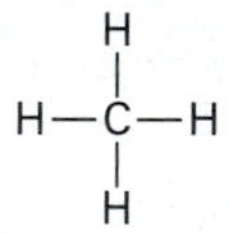

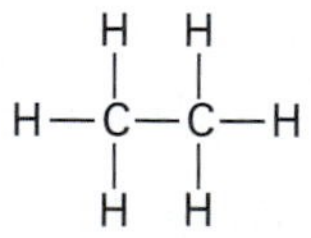

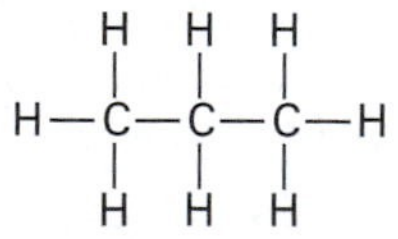

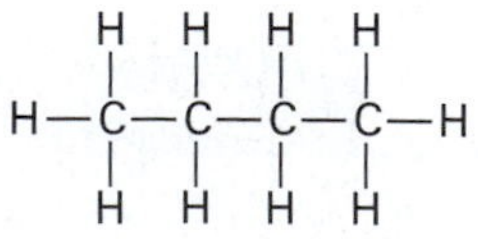

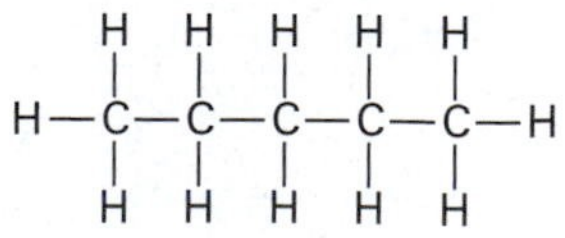

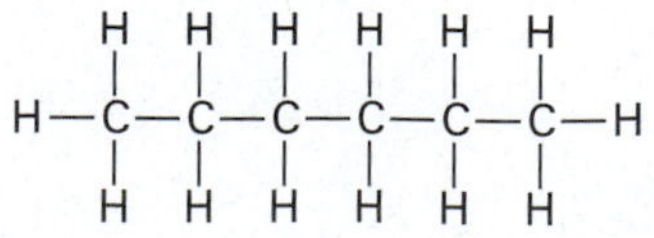

—*Continued*

Continued—

7	*n*-Heptane	C_7H_{16}	H H H H H H H H—C—C—C—C—C—C—C—H H H H H H H H	$CH_3CH_2CH_2CH_2CH_2CH_2CH_3$
8	*n*-Octane	C_8H_{18}	H H H H H H H H H—C—C—C—C—C—C—C—C—H H H H H H H H H	$CH_3CH_2CH_2CH_2CH_2CH_2CH_2CH_3$
9	*n*-Nonane	C_9H_{20}	H H H H H H H H H H—C—C—C—C—C—C—C—C—C—H H H H H H H H H H	$CH_3CH_2CH_2CH_2CH_2CH_2CH_2CH_2CH_3$
10	*n*-Decane	$C_{10}H_{22}$	H H H H H H H H H H H—C—C—C—C—C—C—C—C—C—C—H H H H H H H H H H H	$CH_3CH_2CH_2CH_2CH_2CH_2CH_2CH_2CH_2CH_3$

TABLE 22.4 Uses of Hydrocarbons

Number of Carbon Atoms	State (at 25 °C)	Major Uses
1–4	Gas	Heating fuel, cooking fuel
5–7	Low-boiling liquid	Solvents, gasoline
6–18	Liquid	Gasoline
12–24	Liquid	Jet fuel, portable-stove fuel
18–50	High-boiling liquid	Diesel fuel, lubircants, heating oil
50+	Solid	Petroleum jelly, paraffin wax

Naming Alkanes

Many organic compounds have common names that we can learn only through familiarity. Because of the sheer number of organic compounds, however, chemists need a systematic method of nomenclature. In this book, we adopt the nomenclature system recommended by the IUPAC (International Union of Pure and Applied Chemistry), which is used throughout the world. In this system, the longest continuous chain of carbon atoms—called the base chain—determines the base name of the compound. The root of the base name depends on the number of carbon atoms in the base chain, as shown in Table 22.5. Base names for alkanes always have the ending *-ane*. Groups of carbon atoms branching off the base chain are alkyl groups and are named as substituents. Recall from Section 22.3 that a *substituent* is an atom or group of atoms that has been substituted for a hydrogen atom in an organic compound. Common alkyl groups are shown in Table 22.6.

TABLE 22.5 Prefixes for Base Names of Alkane Chains

Number of Carbon Atoms	Prefix
1	meth-
2	eth-
3	prop-
4	but-
5	pent-
6	hex-
7	hept-
8	oct-
9	non-
10	dec-

TABLE 22.6 Common Alkyl Groups

Condensed Structural Formula	Name	Condensed Structural Formula	Name
$—CH_3$	Methyl	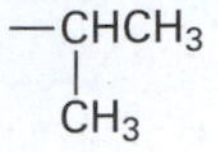 $—CHCH_3$ with CH_3 on the CH	Isopropyl
$—CH_2CH_3$	Ethyl	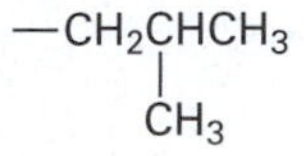 $—CH_2CHCH_3$ with CH_3 on the CH	Isobutyl
$—CH_2CH_2CH_3$	Propyl	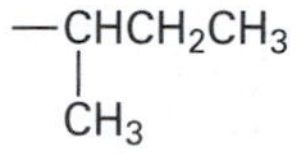 $—CHCH_2CH_3$ with CH_3 on the first CH	*sec*-Butyl
$—CH_2CH_2CH_2CH_3$	Butyl	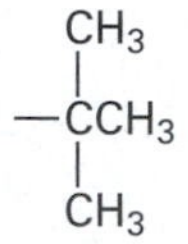 $—CCH_3$ with two CH_3 on the C	*tert*-Butyl

The procedure demonstrated in Examples 22.2 and 22.3 allows us to systematically name many alkanes. The procedure is presented in the left column and two examples of applying the procedure are shown in the center and right columns.

PROCEDURE FOR Naming Alkanes

EXAMPLE 22.2 Naming Alkanes

Name this alkane.

$CH_3—CH_2—CH(—CH_2—CH_3)—CH_2—CH_3$ (the ethyl group, $CH_2—CH_3$, is drawn hanging below the CH)

EXAMPLE 22.3 Naming Alkanes

Name this alkane.

$CH_3—CH(CH_3)—CH_2—CH(CH_2CH_3)—CH_2—CH_2—CH(CH_3)—CH_3$ (the branches are drawn hanging below the CH groups)

1. Count the number of carbon atoms in the longest continuous carbon chain to determine the base name of the compound (which may not be the horizontal chain). Locate the prefix corresponding to this number of atoms in Table 22.5 and add the ending *-ane* to form the base name.

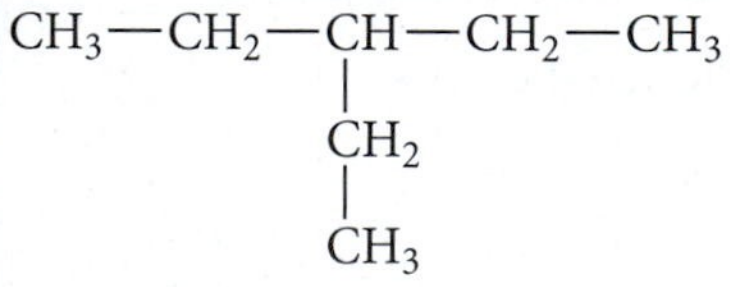

SOLUTION (Example 22.2)

This compound has five carbon atoms in its longest continuous chain.

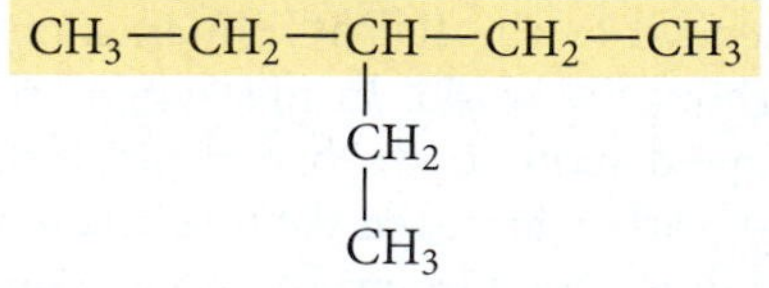

$CH_3—CH_2—CH—CH_2—CH_3$ (highlighted), with $CH_2—CH_3$ hanging below the CH

The correct prefix from Table 22.5 is *pent-*. The base name is pentane.

SOLUTION (Example 22.3)

This compound has eight carbon atoms in its longest continuous chain.

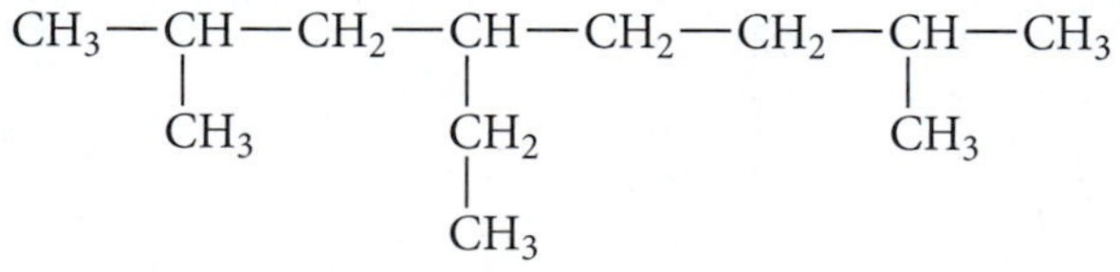

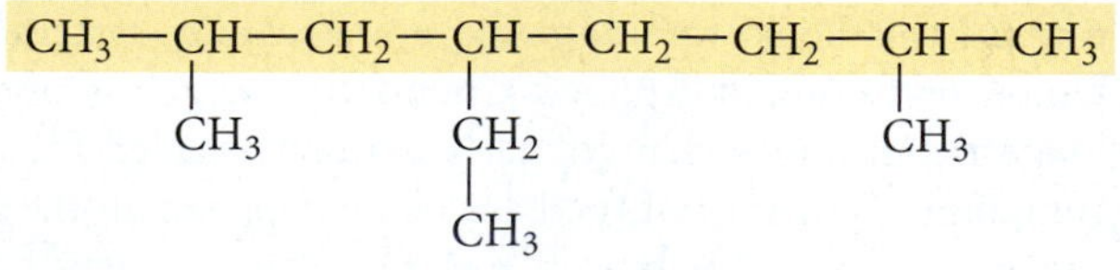

$CH_3—CH—CH_2—CH—CH_2—CH_2—CH—CH_3$ (highlighted), with CH_3, CH_2CH_3 and CH_3 hanging below the three CH groups

The correct prefix from Table 22.5 is *oct-*. The base name is octane.

—Continued on the next page

Continued from the previous page—

2. Consider every branch from the base chain to be a substituent. Name each substituent according to Table 22.6.	This compound has one substituent named *ethyl*. $CH_3-CH_2-\underset{\substack{\vert\\CH_2\\\vert\\CH_3}}{CH}-CH_2-CH_3$ ethyl (CH_2-CH_3)	This compound has one substituent named *ethyl* and two named *methyl.* $CH_3-\underset{\substack{\vert\\CH_3}}{CH}-CH_2-\underset{\substack{\vert\\CH_2\\\vert\\CH_3}}{CH}-CH_2-CH_2-\underset{\substack{\vert\\CH_3}}{CH}-CH_3$ ethyl (CH_2-CH_3); methyl (CH_3, CH_3)
3. Beginning with the end closest to the branching, number the base chain and assign a number to each substituent. (If two substituents occur at equal distances from each end, go to the next substituent to determine from which end to start numbering.)	Number the base chain as follows: $\overset{1}{CH_3}-\overset{2}{CH_2}-\underset{\substack{\vert\\CH_2\\\vert\\CH_3}}{\overset{3}{CH}}-\overset{4}{CH_2}-\overset{5}{CH_3}$ Assign the number 3 to the ethyl substituent.	Number the base chain as follows: $\overset{1}{CH_3}-\underset{\substack{\vert\\CH_3}}{\overset{2}{CH}}-\overset{3}{CH_2}-\underset{\substack{\vert\\CH_2\\\vert\\CH_3}}{\overset{4}{CH}}-\overset{5}{CH_2}-\overset{6}{CH_2}-\underset{\substack{\vert\\CH_3}}{\overset{7}{CH}}-\overset{8}{CH_3}$ Assign the number 4 to the ethyl substituent and the numbers 2 and 7 to the two methyl substituents.
4. Write the name of the compound in the following format: (substituent number)-(substituent name) (base name) If there are two or more substituents, give each one a number and list them alphabetically with hyphens between words and numbers.	The name of the compound is: 3-ethylpentane	The basic form of the name of the compound is: 4-ethyl-2,7-methyloctane List ethyl before methyl because substituents are listed in alphabetical order.
5. If a compound has two or more identical substituents, indicate the number of identical substituents with the prefix *di-* (2), *tri-* (3), or *tetra-* (4) before the substituent's name. Separate the numbers indicating the positions of the substituents relative to each other with a comma. Do not take the prefixes into account when alphabetizing.	This step does not apply to this compound.	This compound has two methyl substituents; therefore, the name of the compound is: 4-ethyl-2,7-dimethyloctane
	FOR PRACTICE 22.2 Name this alkane. $CH_3-CH_2-\underset{\substack{\vert\\CH_3}}{CH}-CH_2-CH_2-CH_3$	**FOR PRACTICE 22.3** Name this alkane. $CH_3-CH_2-\underset{\substack{\vert\\CH_3}}{CH}-CH_2-\underset{\substack{\vert\\CH_3}}{CH}-CH_2-CH_3$

EXAMPLE 22.4
Naming Alkanes

Name this alkane. $CH_3-CH(CH_3)-CH_2-CH(CH_3)-CH_3$

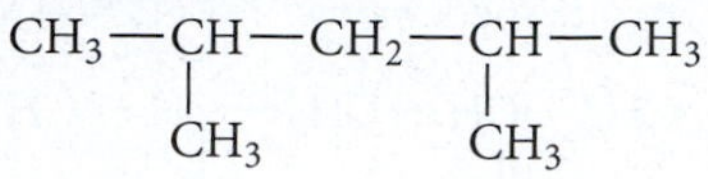

SOLUTION

1. The longest continuous carbon chain has five atoms. Therefore, the base name is pentane.	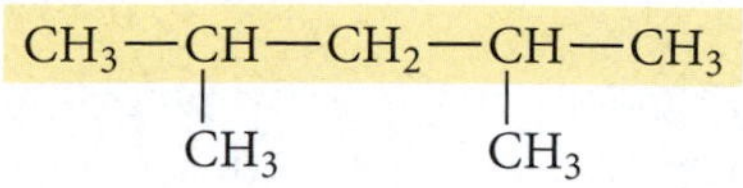 $CH_3-CH(CH_3)-CH_2-CH(CH_3)-CH_3$
2. This compound has two substituents, both named methyl.	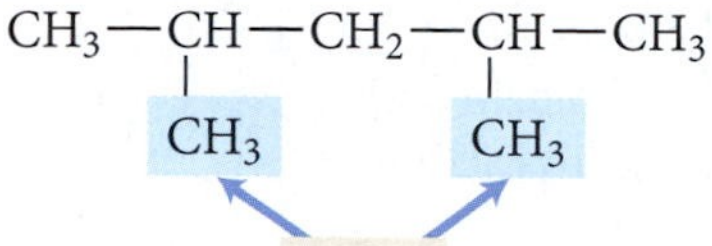 $CH_3-CH(CH_3)-CH_2-CH(CH_3)-CH_3$ (both CH_3 substituents: methyl)
3. Since both substituents are equidistant from the ends, it does not matter from which end you start numbering.	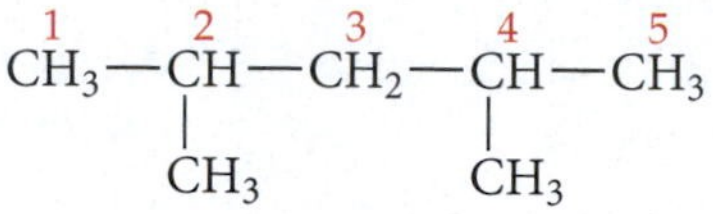 $\overset{1}{CH_3}-\overset{2}{CH}(CH_3)-\overset{3}{CH_2}-\overset{4}{CH}(CH_3)-\overset{5}{CH_3}$
4, 5. Use the general form for the name: (substituent number)-(substituent name)(base name) Because this compound contains two identical substituents, step 5 from the naming procedure applies and you use the prefix *di-*. Indicate the position of each substituent with a number separated by a comma.	2,4-dimethylpentane

FOR PRACTICE 22.4

Name this alkane. $CH_3-CH(CH_3)-CH_2-CH(CH_3)-CH(CH_3)-CH_3$

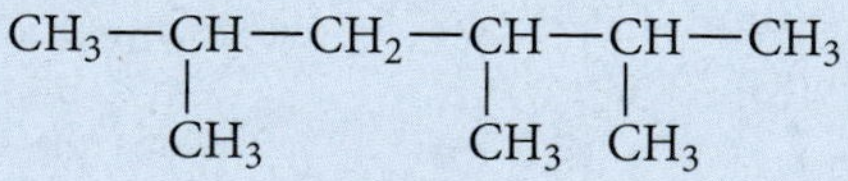

22.5 Alkenes and Alkynes

Alkenes are hydrocarbons containing at least one double bond between carbon atoms. Alkynes contain at least one triple bond. Because of the double or triple bond, alkenes and alkynes have fewer hydrogen atoms than the corresponding alkanes and are therefore **unsaturated hydrocarbons** because they are not loaded to capacity with hydrogen. Recall that noncyclic alkenes have the formula C_nH_{2n} and noncyclic alkynes have the formula C_nH_{2n-2}. The simplest alkene is ethene (C_2H_4), also called ethylene.

The general formulas shown here for alkenes and alkynes assume only one multiple bond.

	Formula	Structural formula	Space-filling model
Ethene or ethylene	C_2H_4	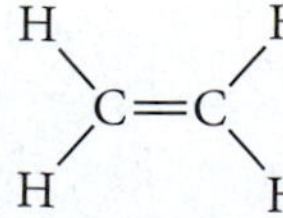	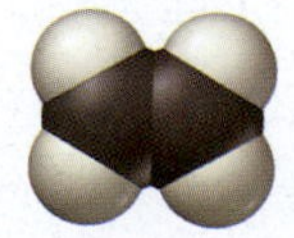

The geometry about each carbon atom in ethene is trigonal planar (see Example 7.3 for the valence bond model of ethene); this makes ethene a flat, rigid molecule. Ethene is a ripening agent in fruit such as bananas. When a banana within a cluster of bananas begins to ripen, it emits ethene. The ethene causes other bananas in the cluster to ripen. Banana farmers usually pick bananas green for ease of shipping. When the bananas arrive at their destination, they are often "gassed" with ethene to initiate ripening. Table 22.7 lists the names and structures of several other alkenes. Most of them do not have familiar uses except as minority components in fuels.

TABLE 22.7 Alkenes

n	Name	Molecular Formula C_nH_{2n}	Structural Formula	Condensed Structural Formula
2	Ethene	C_2H_4	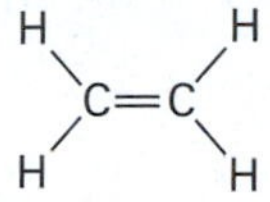 $H_2C{=}CH_2$	$CH_2{=}CH_2$
3	Propene	C_3H_6	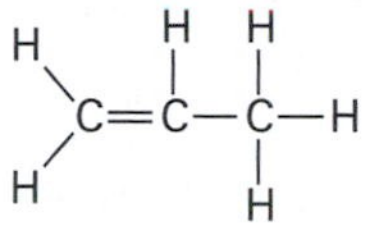 $H_2C{=}CH{-}CH_3$	$CH_2{=}CHCH_3$
4	1-Butene*	C_4H_8	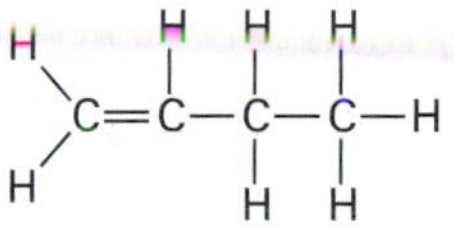 $H_2C{=}CH{-}CH_2{-}CH_3$	$CH_2{=}CHCH_2CH_3$
5	1-Pentene*	C_5H_{10}	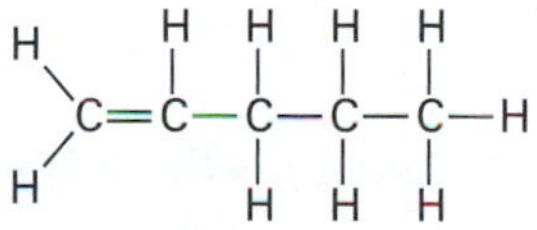 $H_2C{=}CH{-}CH_2{-}CH_2{-}CH_3$	$CH_2{=}CHCH_2CH_2CH_3$
6	1-Hexene*	C_6H_{12}	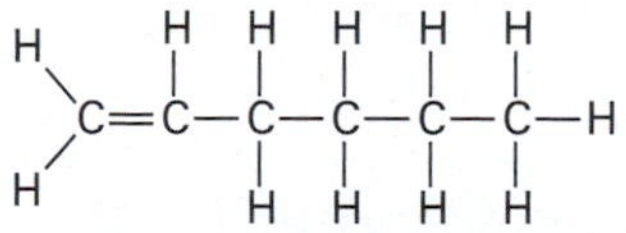 $H_2C{=}CH{-}CH_2{-}CH_2{-}CH_2{-}CH_3$	$CH_2{=}CHCH_2CH_2CH_2CH_3$

* These alkenes have one or more isomers depending on the position of the double bond. The isomers shown here have the double bond in the 1 position, meaning the first carbon–carbon bond of the chain.

The simplest alkyne is ethyne, C_2H_2, also known as acetylene.

	Formula	Structural formula	Space-filling model
Ethyne or acetylene	C_2H_2	$H{-}C{\equiv}C{-}H$	

TABLE 22.8 Alkynes

n	Name	Molecular Formula C_nH_{2n-2}	Structural Formula	Condensed Structural Formula
2	Ethyne	C_2H_2	H—C≡C—H	$CH{\equiv}CH$
3	Propyne	C_3H_4	H | H—C≡C—C—H | H	$CH{\equiv}CCH_3$
4	1-Butyne*	C_4H_6	H H | | H—C≡C—C—C—H | | H H	$CH{\equiv}CCH_2CH_3$
5	1-Pentyne*	C_5H_8	H H H | | | H—C≡C—C—C—C—H | | | H H H	$CH{\equiv}CCH_2CH_2CH_3$
6	1-Hexyne*	C_6H_{10}	H H H H | | | | H—C≡C—C—C—C—C—H | | | | H H H H	$CH{\equiv}CCH_2CH_2CH_2CH_3$

* These alkynes have one or more isomers depending on the position of the triple bond. The isomers shown here have the triple bond in the 1 position, meaning the first carbon–carbon bond of the chain.

▲ Welding torches burn ethyne in pure oxygen to produce the very hot flame needed for melting metals.

The geometry about each carbon atom in ethyne is linear, making ethyne a linear molecule. Ethyne (or acetylene) is commonly used as fuel for welding torches. Table 22.8 shows the names and structures of several other alkynes. Like alkenes, the alkynes do not have familiar uses other than as minority components of gasoline.

Naming Alkenes and Alkynes

We name alkenes and alkynes in the same way we name alkanes with the following exceptions:

- The base chain is the longest continuous carbon chain *that contains the double or triple bond.*
- The base name has the ending *-ene* for alkenes and *-yne* for alkynes.
- We number the base chain to give the double or triple bond the lowest possible number.
- We insert a number indicating the position of the double or triple bond (lowest possible number) just before the base name.

For example, the alkene and alkyne shown here are 2-methyl-2-pentene and 1-butyne:

$$CH_3CH_2CH{=}\underset{\displaystyle |\atop \displaystyle CH_3}{C}CH_3$$

2-methyl-2-pentene

$$CH{\equiv}CCH_2CH_3$$

1-butyne

EXAMPLE 22.5
Naming Alkenes and Alkynes

Name each compound.

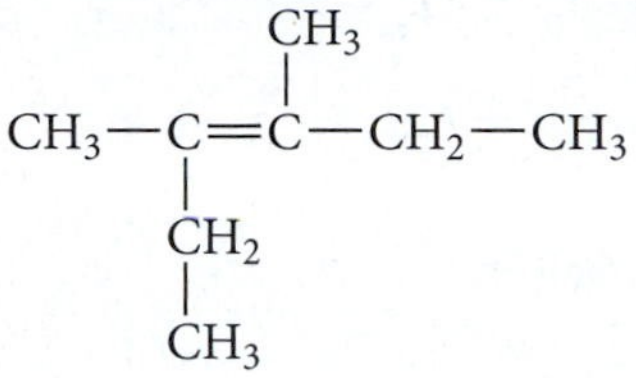

(a)

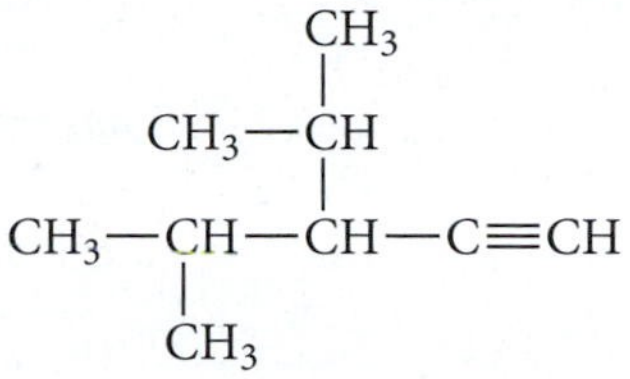

```
            CH3
            |
CH3—C=C—CH2—CH3
       |
       CH2
       |
       CH3
```

(b)

```
                  CH3
                  |
            CH3—CH
                  |
CH3—CH—CH—C≡CH
       |
       CH3
```

SOLUTION

(a) 1. The longest continuous carbon chain containing the double bond has six carbon atoms. The base name is therefore *hexene*.

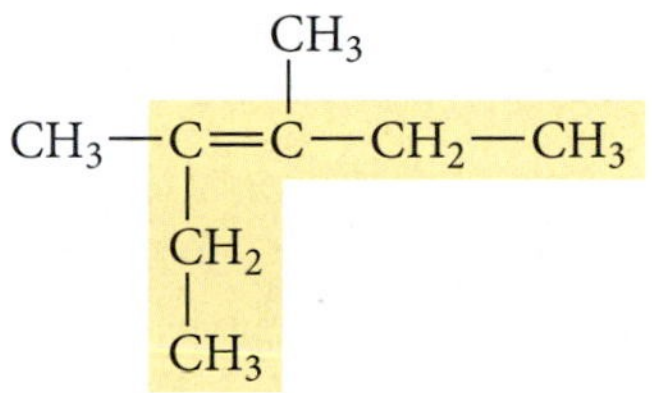

```
            CH3
            |
CH3—C=C—CH2—CH3
       |
       CH2
       |
       CH3
```

2. The two substituents are both methyl.

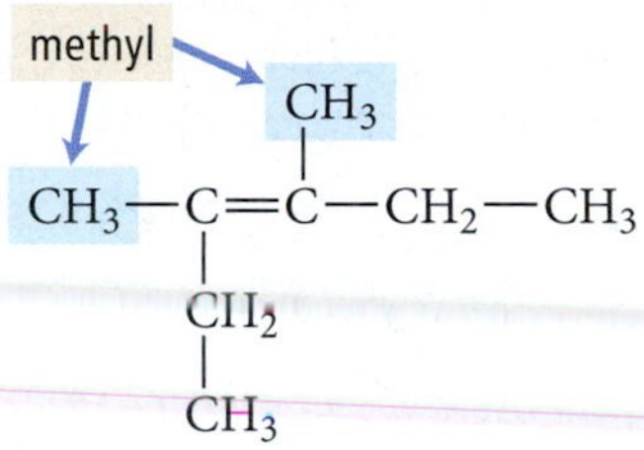

```
methyl
            CH3
            |
CH3—C=C—CH2—CH3
       |
       CH2
       |
       CH3
```

3. Recall that one of the exceptions listed previously states that, in naming alkenes, you should number the chain so that the double bond has the lowest number. In this case, the *double bond* is equidistant from the ends. Assign the double bond the number 3. The two methyl groups are then at positions 3 and 4.

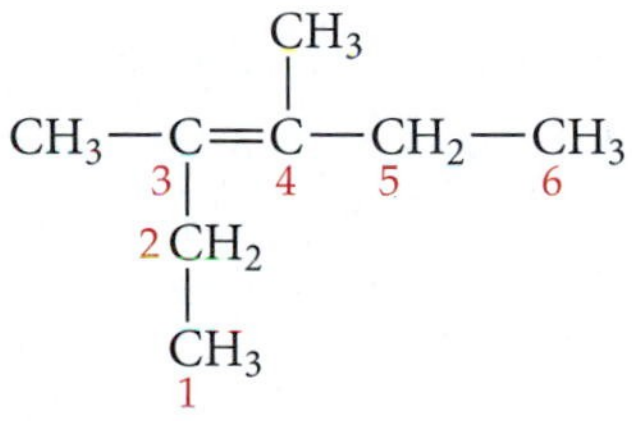

```
            CH3
            |
CH3—C=C—CH2—CH3
      3|  4    5      6
     2CH2
       |
       CH3
       1
```

4, 5. Use the general form for the name:

(substituent number)-(substituent name)(base name)

Because this compound contains two identical substituents, step 5 of the naming procedure applies. Use the prefix *di*-. In addition, indicate the position of each substituent with a number separated by a comma. Because this compound is an alkene, specify the position of the double bond, isolated by hyphens, just before the base name.

3,4-dimethyl-3-hexene

(b) 1. The longest continuous carbon chain containing the triple bond is five carbon atoms long; therefore, the base name is *pentyne*.

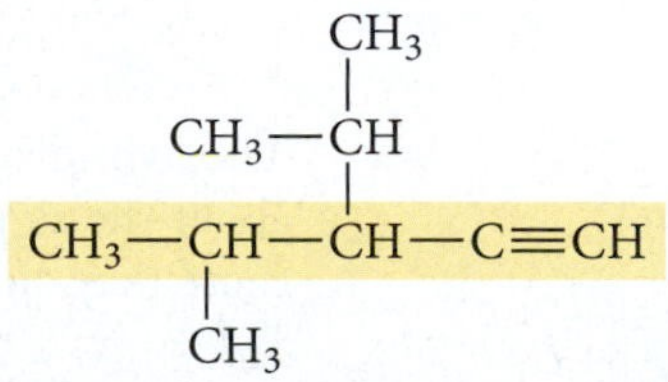

```
                  CH3
                  |
            CH3—CH
                  |
CH3—CH—CH—C≡CH
       |
       CH3
```

—*Continued on the next page*

Continued from the previous page—

2. There are two substituents; one is a methyl group and the other an isopropyl group.	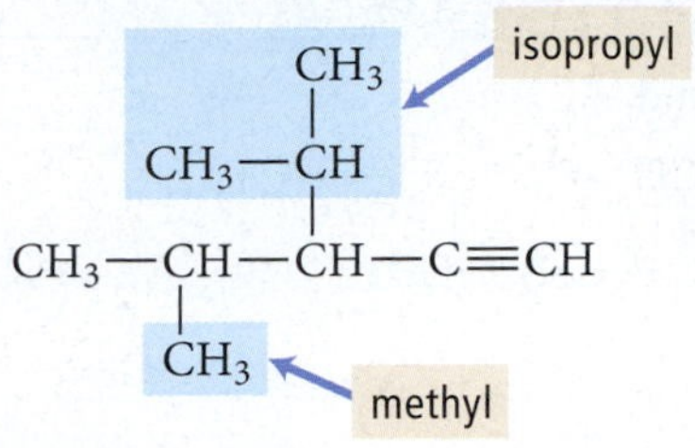 $CH_3-CH(CH_3)-CH(CH(CH_3)_2)-C\equiv CH$ (isopropyl; methyl)
3. Number the base chain, giving the triple bond the lowest number (1). Assign the isopropyl and methyl groups the numbers 3 and 4, respectively.	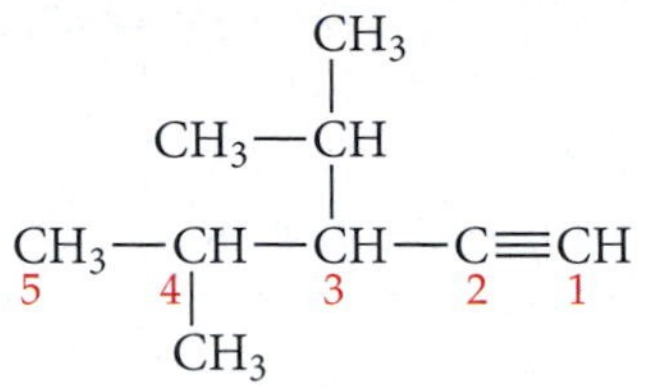 $\underset{5}{CH_3}-\underset{4}{CH}(CH_3)-\underset{3}{CH}(CH(CH_3)_2)-\underset{2}{C}\equiv\underset{1}{CH}$
4. Use the general form for the name: (substituent number)-(substituent name)(base name) Since there are two substituents, list both of them in alphabetical order. Since this compound is an alkyne, specify the position of the triple bond with a number isolated by hyphens just before the base name.	3-isopropyl-4-methyl-1-pentyne

FOR PRACTICE 22.5

Name each compound.

(a) 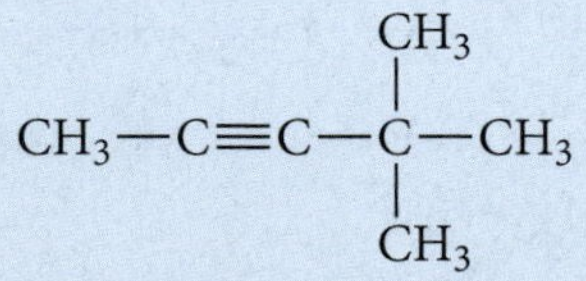 $CH_3-C\equiv C-C(CH_3)_2-CH_3$

(b) 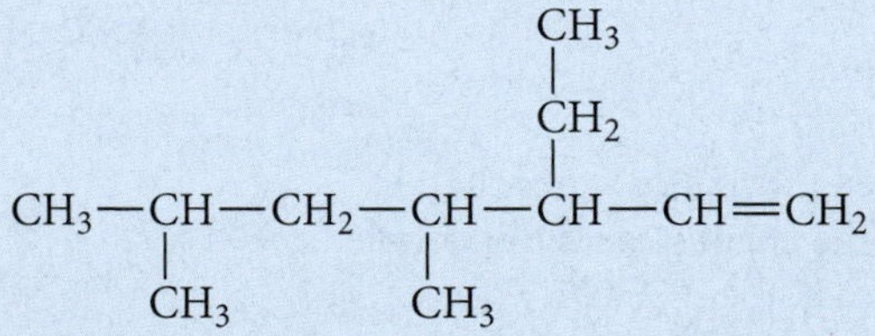 $CH_3-CH(CH_3)-CH_2-CH(CH_3)-CH(CH_2CH_3)-CH=CH_2$

Geometric (Cis–Trans) Isomerism in Alkenes

A major difference between a single bond and a double bond is the degree to which rotation occurs about the bond. As discussed in Section 7.3, rotation about a double bond is highly restricted due to the overlap between unhybridized *p* orbitals on the adjacent carbon atoms. Consider the difference between 1,2-dichloroethane and 1,2-dichloroethene.

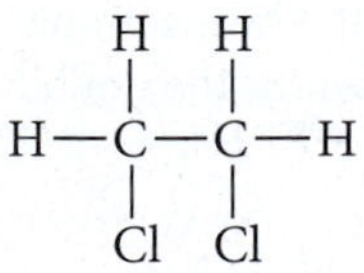

$H-CH(Cl)-CH(Cl)-H$

1,2-Dichloroethane

$ClHC=CHCl$

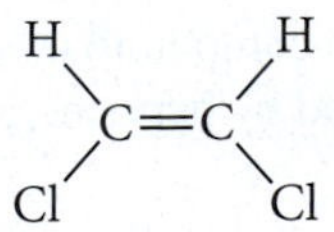

1,2-Dichloroethene

The hybridization of the carbon atoms in 1,2-dichloroethane is sp^3, resulting in relatively free rotation about the sigma single bond. Consequently, the two structures are identical at room temperature because they quickly interconvert.

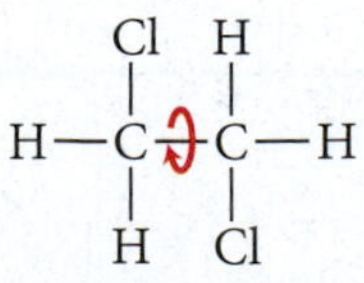

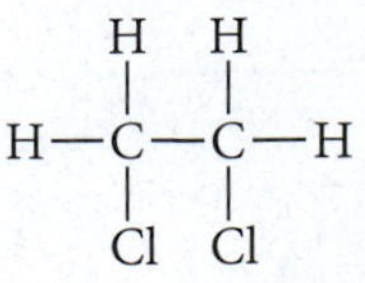

$H-C(Cl)(H)-C(H)(Cl)-H$ ⇌ $H-C(H)(Cl)-C(H)(Cl)-H$

TABLE 22.9 Physical Properties of *cis*- and *trans*-1,2-Dichloroethene

Name	Structure	Space-filling Model	Density (g/mL)	Melting Point (°C)	Boiling Point (°C)
cis-1,2-Dichloroethene	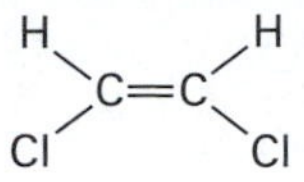 H H C=C Cl Cl		1.284	−80.5	60.1
trans-1,2-Dichloroethene	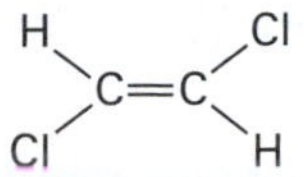 H Cl C=C Cl H	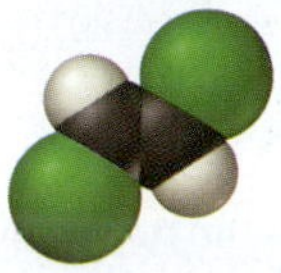	1.257	−49.4	47.5

In contrast, rotation about the double bond (sigma + pi) in 1,2-dichloroethene is restricted, so at room temperature, 1,2-dichloroethene exists in two isomeric forms.

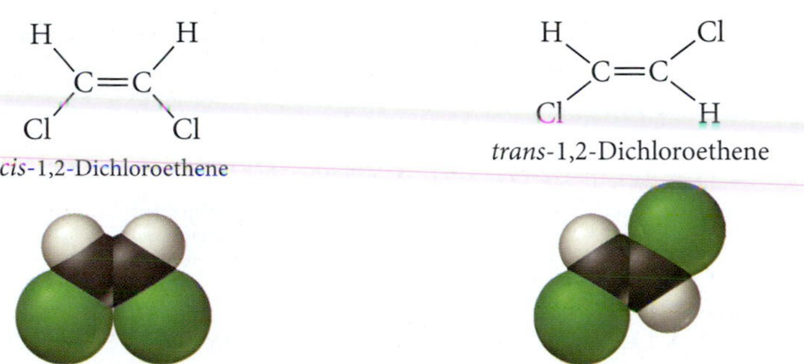

cis-1,2-Dichloroethene *trans*-1,2-Dichloroethene

These two forms of 1,2-dichloroethene are different compounds with different properties as shown in Table 22.9. This kind of isomerism is a type of stereoisomerism (see Section 22.3) called **geometric** (or **cis–trans**) **isomerism**. We distinguish between the two isomers with the designations *cis* (meaning "same side") and *trans* (meaning "opposite sides"). Cis–trans isomerism is common in alkenes. As another example, consider *cis*- and *trans*-2-butene. Like the two isomers of 1,2-dichloroethene, these two isomers have different physical properties. For example, *cis*-2-butene boils at 3.7 °C, and *trans*-2-butene boils at 0.9 °C.

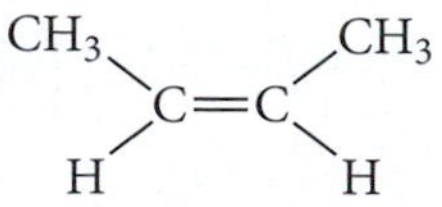

CH_3 CH_3
C=C
H H

cis-2-Butene

CH_3 H
C=C
H CH_3

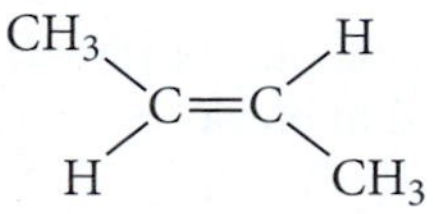

trans-2-Butene

22.6 Hydrocarbon Reactions

One of the most common hydrocarbon reactions is combustion, the burning of hydrocarbons in the presence of oxygen. Hydrocarbon combustion reactions are highly exothermic and are commonly used to warm homes and buildings, to generate electricity, and to power the engines of cars, ships, and airplanes. It is not an exaggeration to say that hydrocarbon combustion makes our current way of life possible. Approximately 90% of the energy produced in the United States is generated by hydrocarbon combustion. Alkanes, alkenes, and alkynes all undergo combustion. In a combustion reaction, the hydrocarbon reacts with oxygen to form carbon dioxide and water.

$$CH_3CH_2CH_3(g) + 5\,O_2(g) \longrightarrow 3\,CO_2(g) + 4\,H_2O(g) \quad \text{Alkane combustion}$$

$$CH_2{=}CHCH_2CH_3(g) + 6\,O_2(g) \longrightarrow 4\,CO_2(g) + 4\,H_2O(g) \quad \text{Alkene combustion}$$

$$CH{\equiv}CCH_3(g) + 4\,O_2(g) \longrightarrow 3\,CO_2(g) + 2\,H_2O(g) \quad \text{Alkyne combustion}$$

Reactions of Alkanes

In addition to combustion reactions, alkanes also undergo **substitution reactions,** in which one or more hydrogen atoms on an alkane are replaced by one or more other atoms. The most common substitution reaction is *halogen substitution* (also referred to as *halogenation*). For example, methane can react with chlorine gas in the presence of heat or light to form chloromethane.

$$\underset{\text{Methane}}{CH_4(g)} + \underset{\text{Chlorine}}{Cl_2(g)} \xrightarrow{\text{heat or light}} \underset{\text{Chloromethane}}{CH_3Cl(g)} + HCl(g)$$

Ethane reacts with chlorine gas to form chloroethane.

$$\underset{\text{Ethane}}{CH_3CH_3(g)} + \underset{\text{Chlorine}}{Cl_2(g)} \xrightarrow{\text{heat or light}} \underset{\text{Chloromethane}}{CH_3CH_2Cl(g)} + HCl(g)$$

Multiple halogenation reactions can occur because halogens can replace more than one of the hydrogen atoms on an alkane. For example, chloromethane can react with chlorine, and the product of that reaction can react again (and so on).

$$\underset{\text{Chloromethane}}{CH_3Cl(g)} + \underset{\text{Chlorine}}{Cl_2(g)} \xrightarrow{\text{heat or light}} CH_2Cl_2(g) + HCl(g)$$

Dichloromethane (also known as methylene chloride)

$$\underset{\text{Dichloromethane}}{CH_2Cl_2(g)} + \underset{\text{Chlorine}}{Cl_2(g)} \xrightarrow{\text{heat or light}} CHCl_3(g) + HCl(g)$$

Trichloromethane (also known as chloroform)

$$\underset{\text{Trichloromethane}}{CHCl_3(g)} + \underset{\text{Chlorine}}{Cl_2(g)} \xrightarrow{\text{heat or light}} CCl_4(g) + HCl(g)$$

Tetrachloromethane (also known as carbon tetrachloride)

The general form for halogen substitution reactions is:

$$\underset{\text{Alkane}}{R{-}H} + \underset{\text{Halogen}}{X_2} \xrightarrow{\text{heat or light}} \underset{\text{Haloalkane}}{R{-}X} + \underset{\text{Hydrogen halide}}{HX}$$

Notice that the halogenation of hydrocarbons requires initiation with heat or light, which causes the chlorine–chlorine bond to break.

$$Cl{-}Cl \xrightarrow{\text{heat or light}} Cl\cdot + Cl\cdot$$

The resulting chlorine atoms are *free radicals* (see Section 6.5), as the dot that represents each chlorine atom's unpaired electron indicates. Chlorine radicals are highly reactive and attack the C—H bond in hydrocarbons. The subsequent reaction proceeds by this mechanism:

$$Cl\cdot + R{-}H \longrightarrow R\cdot + HCl$$

$$R\cdot + Cl_2 \longrightarrow R{-}Cl + Cl\cdot$$

Notice that a chlorine free radical is produced as a product of the last step. This free radical can go on to react again, unless it encounters another chlorine free radical, in which case it reacts with it to re-form Cl_2.

Reactions of Alkenes and Alkynes

Alkenes and alkynes undergo **addition reactions** in which molecules add across (on either side of) the multiple bond. For example, ethene reacts with chlorine gas to form dichloroethane.

$$H_2C{=}CH_2 + Cl{-}Cl \longrightarrow H{-}CH(Cl){-}CH(Cl){-}H$$

The addition of chlorine converts the carbon–carbon double bond into a single bond because each carbon atom bonds to a chlorine atom. Alkenes and alkynes can also add hydrogen in hydrogenation reactions. For example, in the presence of an appropriate catalyst, propene reacts with hydrogen gas to form propane.

$$H{-}CH_2{-}CH{=}CH_2 + H{-}H \xrightarrow{\text{catalyst}} H{-}CH_2{-}CH_2{-}CH_2{-}H$$

We often indicate the presence of a catalyst by adding a label over the reaction arrow.

INGREDIENTS: SOYBEAN OIL, FULLY HYDROGENATED PALM OIL, PARTIALLY HYDROGENATED PALM AND SOYBEAN OILS, MONO AND DIGLYCERIDES, TBHQ AND CITRIC ACID (ANTIOXIDANTS).
MANUFACTURED BY ©/® THE J.M. SMUCKER COMPANY ORRVILLE, OH 44667 U.S.A.

▲ Partially hydrogenated vegetable oil is a saturated fat that is made by hydrogenating unsaturated fats.

Hydrogenation reactions convert unsaturated hydrocarbons into saturated hydrocarbons. For example, hydrogenation reactions convert unsaturated vegetable oils into saturated fats. Most vegetable oils are unsaturated because their carbon chains contain double bonds. The double bonds put bends into the carbon chains that result in less efficient packing of molecules; thus vegetable oils are liquids at room temperature, while saturated fats are solids at room temperature. When food manufacturers add hydrogen to the double bonds of vegetable oil, the unsaturated fat is converted into a saturated fat, turning the liquid oil into a solid at room temperature. As we have seen so many times, structure determines properties. The words "partially hydrogenated vegetable oil" on a label indicate a food product that contains saturated fats made via hydrogenation reactions.

Alkenes can also add unsymmetrical reagents across the double bond. For example, ethene reacts with hydrogen chloride to form chloroethane.

$$H_2C{=}CH_2 + HCl \longrightarrow H{-}CH_2{-}CH(Cl){-}H$$

Chloroethane

If the alkene itself is also unsymmetrical, then the addition of an unsymmetrical reagent leads to the potential for two different products. For example, when HCl reacts with propene, two products are possible:

$$CH_3{-}CH{=}CH_2 + HCl \longrightarrow CH_3{-}CH_2{-}CH(Cl){-}H \quad \text{1-chloropropane (not observed)}$$

$$CH_3{-}CH{=}CH_2 + HCl \longrightarrow CH_3{-}CH(Cl){-}CH_2{-}H \quad \text{2-chloropropane}$$

When this reaction is carried out in the lab, however, only the 2-chloropropane forms. We can predict the product of the addition of an unsymmetrical reagent to an unsymmetrical alkene with *Markovnikov's rule,* which states the following:

> **When a polar reagent is added to an unsymmetrical alkene, the positive end (the least electronegative part) of the reagent adds to the carbon atom that has the most hydrogen atoms.**

In most reactions of this type, the positive end of the reagent is hydrogen; therefore, the hydrogen atom bonds to the carbon atom that already contains the most hydrogen atoms.

EXAMPLE 22.6
Alkene Addition Reactions

Determine the products of the reactions.

(a) $CH_3CH_2CH{=}CH_2 + Br_2 \longrightarrow$ **(b)** $CH_3CH_2CH{=}CH_2 + HBr \longrightarrow$

SOLUTION

(a) The reaction of 1-butene with bromine is an example of a symmetric addition. The bromine adds across the double bond, and each carbon forms a single bond to a bromine atom.

$$\begin{array}{ccccccccccc} & & H & & H & & H & & H & & \\ & & | & & | & & | & & | & & \\ H & - & C & - & C & - & C & = & C & - & H \\ & & | & & | & & & & & & \\ & & H & & H & & & & & & \end{array} + Br-Br \longrightarrow$$

$$\begin{array}{ccccccccccc} & & H & & H & & H & & Br & & \\ & & | & & | & & | & & | & & \\ H & - & C & - & C & - & C & - & C & - & H \\ & & | & & | & & | & & | & & \\ & & H & & H & & Br & & H & & \end{array}$$

(b) The reaction of 1-butene with hydrogen bromide is an example of an unsymmetrical addition. Apply Markovnikov's rule to determine which carbon the hydrogen bonds with and which carbon the bromine atom bonds with. Markovnikov's rule predicts that the hydrogen bonds to the end carbon in this case.

$$\begin{array}{ccccccccccc} & & H & & H & & H & & H & & \\ & & | & & | & & | & & | & & \\ H & - & C & - & C & - & C & = & C & - & H \\ & & | & & | & & & & & & \\ & & H & & H & & & & & & \end{array} + H-Br \longrightarrow$$

$$\begin{array}{ccccccccccc} & & H & & H & & H & & H & & \\ & & | & & | & & | & & | & & \\ H & - & C & - & C & - & C & - & C & - & H \\ & & | & & | & & | & & | & & \\ & & H & & H & & Br & & H & & \end{array}$$

FOR PRACTICE 22.6

Determine the products of the reactions.

(a) $$\begin{array}{ccccccc} & & CH_3 & & & & \\ & & | & & & & \\ CH_3 & - & C & - & C & = & CH_2 \\ & & | & & | & & \\ & & H & & H & & \end{array} + H_2 \xrightarrow{\text{catalyst}}$$

(b) $$\begin{array}{ccccccc} & & CH_3 & & & & \\ & & | & & & & \\ CH_3 & - & C & - & C & = & CH_2 \\ & & | & & | & & \\ & & H & & H & & \end{array} + HCl \longrightarrow$$

22.7 Aromatic Hydrocarbons

As you might imagine, determining the structure of organic compounds has not always been easy. During the mid-1800s chemists were working to determine the structure of a particularly stable organic compound named benzene (C_6H_6). In 1865, Friedrich August Kekulé (1829–1896) had a dream in which he envisioned chains of carbon atoms as snakes. The snakes danced before him, and one of them twisted around and bit its tail. Based on that vision, Kekulé proposed the following structure for benzene:

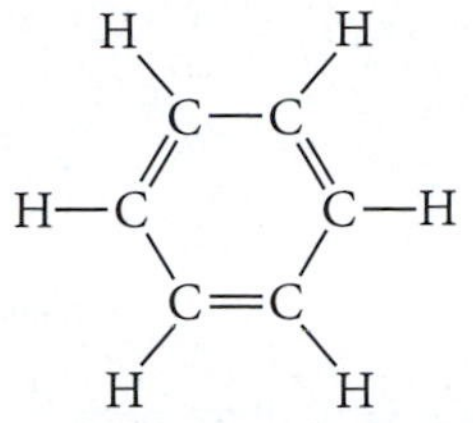

This structure has alternating single and double bonds. When we examine the carbon–carbon bond lengths in benzene, however, we find that all the bonds are the same length, which indicates that the following resonance structures are a more accurate representation of benzene:

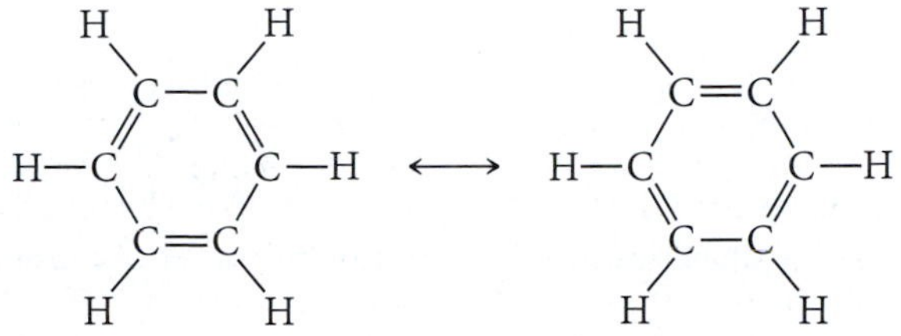

Recall from Section 6.4 that the actual structure of a molecule represented by resonance structures is intermediate between the two resonance structures and is called a *resonance hybrid.*

The true structure of benzene is a hybrid of the two resonance structures. We often represent benzene with the following carbon skeletal formula (or line formula):

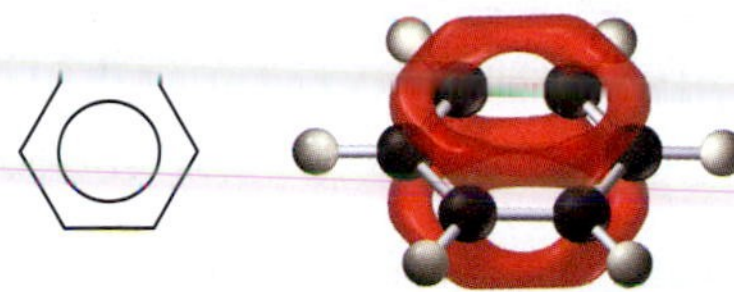

The ring represents the delocalized π electrons that occupy the molecular orbital shown superimposed on the ball-and-stick model. When drawing benzene rings, either by themselves or as parts of other compounds, organic chemists use either this diagram or just one of the resonance structures with alternating double bonds. Both representations indicate the same thing—a benzene ring.

The benzene ring structure occurs in many organic compounds. An atom or group of atoms can substitute for one or more of the six hydrogen atoms on the ring to form compounds referred to as *substituted benzenes,* such as chlorobenzene and phenol.

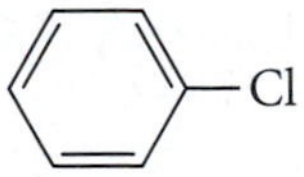

Chlorobenzene 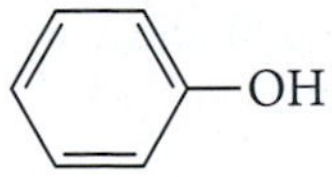 Phenol

Because many compounds containing benzene rings have pleasant aromas, benzene rings are also called *aromatic rings*, and compounds containing them are called *aromatic compounds*. Aromatic compounds are responsible for the pleasant smells of cinnamon, vanilla, and jasmine.

Naming Aromatic Hydrocarbons

Monosubstituted benzenes—benzenes in which only one of the hydrogen atoms has been substituted—are often named as derivatives of benzene.

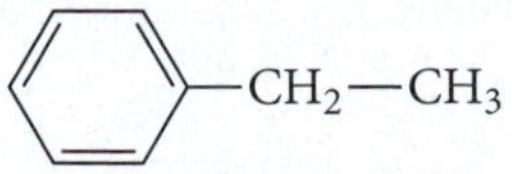

Ethylbenzene 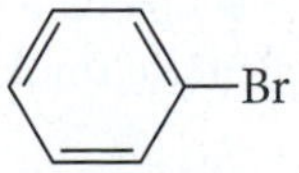 Bromobenzene

These names take the general form:

(name of substituent)benzene

However, many monosubstituted benzenes have names that can only be learned through familiarity. Some common ones are shown here.

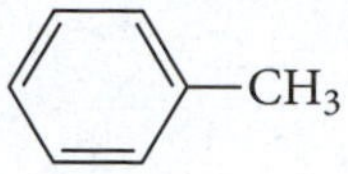

Toluene ($C_6H_5-CH_3$) Aniline ($C_6H_5-NH_2$) Phenol (C_6H_5-OH) Styrene ($C_6H_5-CH=CH_2$)

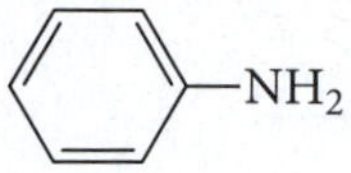

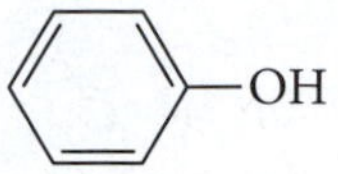

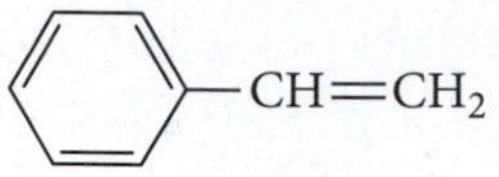

We name some substituted benzenes, especially those with large substituents, by treating the benzene ring as the substituent. In these cases, we refer to the benzene substituent as a **phenyl group**.

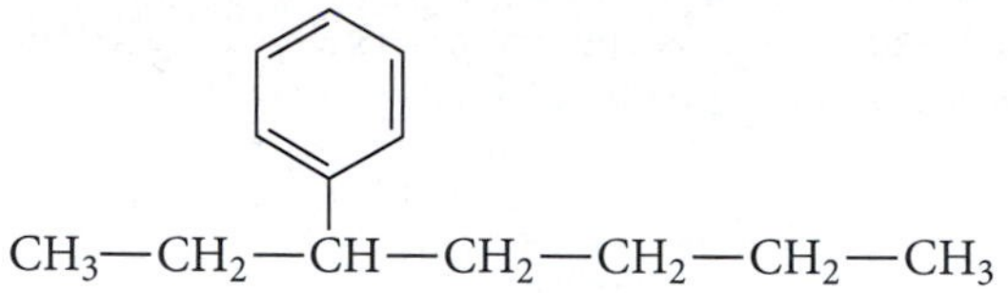

$CH_3-CH_2-CH(C_6H_5)-CH_2-CH_2-CH_2-CH_3$
3-Phenylheptane

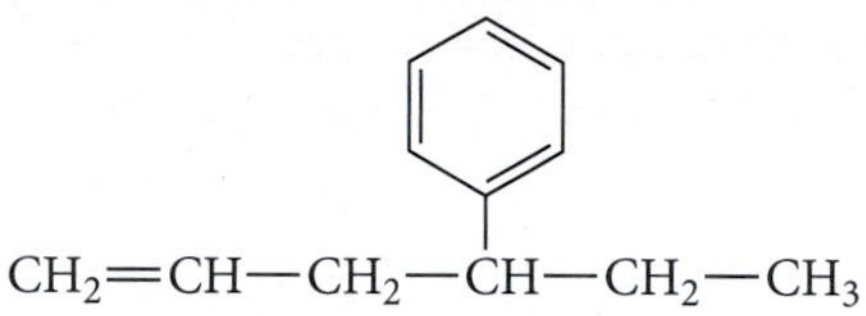

$CH_2=CH-CH_2-CH(C_6H_5)-CH_2-CH_3$
4-Phenyl-1-hexene

Disubstituted benzenes—benzenes in which two hydrogen atoms have been substituted—are numbered, and the substituents are listed alphabetically. We determine the order of numbering on the ring by the alphabetical order of the substituents.

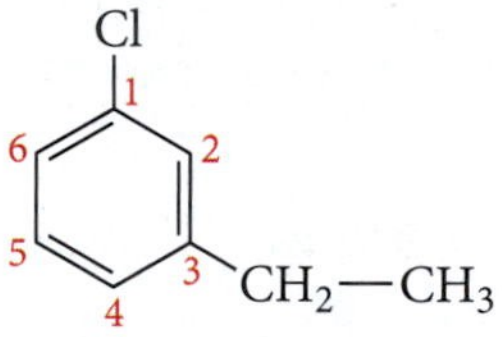

1-Chloro-3-ethylbenzene

1-Bromo-2-chlorobenzene

When the two substituents are identical, we use the prefix *di-*.

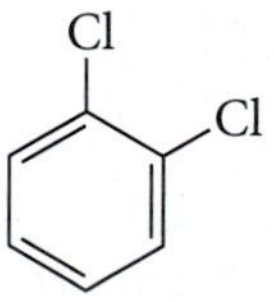

1,2-Dichlorobenzene 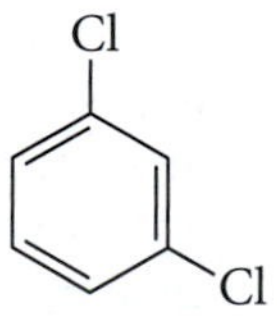 1,3-Dichlorobenzene  1,4-Dichlorobenzene

Also in common use, in place of numbering, are the prefixes *ortho* (1,2 disubstituted), *meta* (1,3 disubstituted), and *para* (1,4 disubstituted).

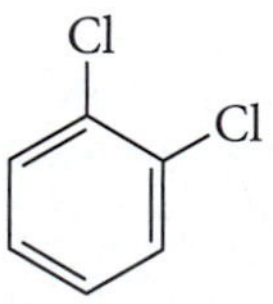

ortho-Dichlorobenzene or *o*-Dichlorobenzene

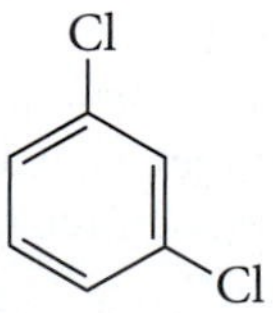

meta-Dichlorobenzene or *m*-Dichlorobenzene

para-Dichlorobenzene or *p*-Dichlorobenzene

Compounds containing fused aromatic rings are called *polycyclic aromatic hydrocarbons*. Some common examples (shown in Figure 22.5 ▶) include naphthalene, the substance that composes mothballs, and pyrene, a carcinogen found in cigarette smoke.

Reactions of Aromatic Compounds

We might expect benzene to react similarly to the alkenes, readily undergoing addition reactions across its double bonds. However, because of electron delocalization around the ring and the

Naphthalene

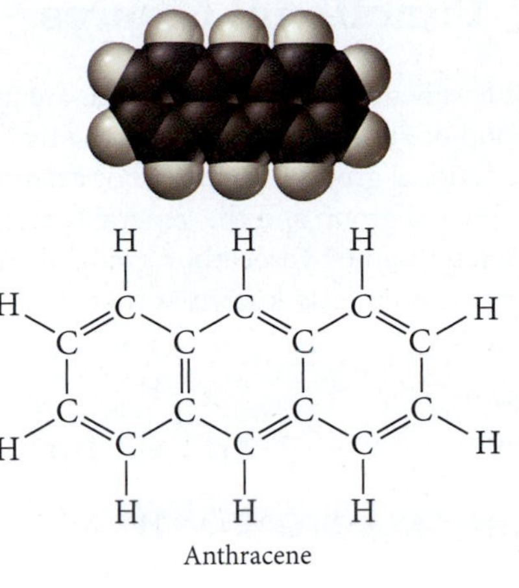
Anthracene

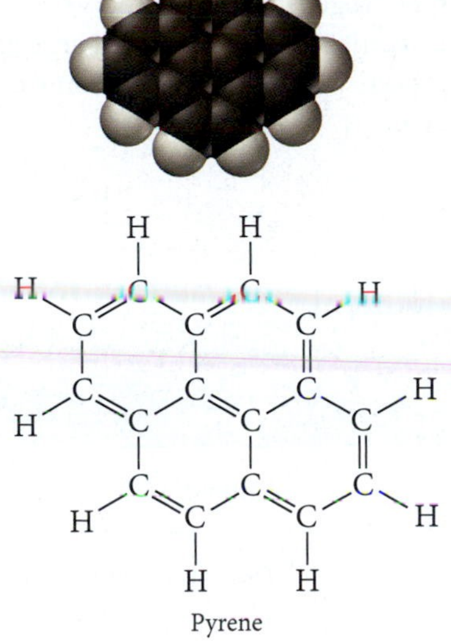
Pyrene

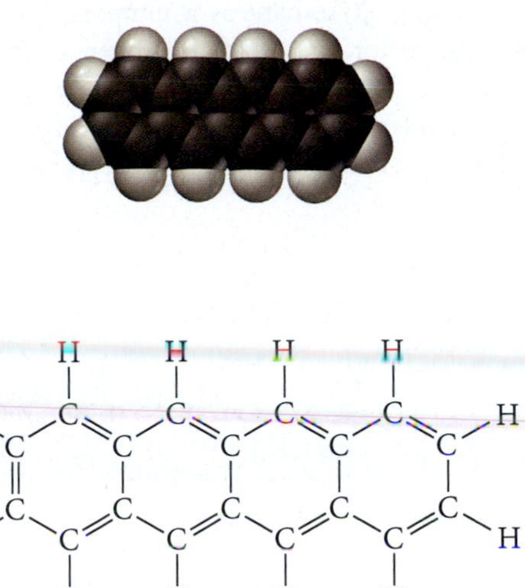
Tetracene

◀ **FIGURE 22.5 Polycyclic Aromatic Compounds** The structures of some common polycyclic aromatic compounds contain fused rings.

resulting greater stability, benzene does not typically undergo addition reactions. Instead, benzene undergoes substitution reactions in which the hydrogen atoms are replaced by other atoms or groups of atoms as shown in these examples:

$$C_6H_6 + Cl{-}Cl \xrightarrow{FeCl_3} C_6H_5Cl + HCl$$

Chlorobenzene

$$C_6H_6 + CH_3{-}CH_2Cl \xrightarrow{AlCl_3} C_6H_5{-}CH_2{-}CH_3 + HCl$$

Ethylbenzene

The substances shown over the arrows are catalysts needed to increase the rate of the reaction.

22.8 Functional Groups

Most other families of organic compounds are hydrocarbons with a **functional group**—a characteristic atom or group of atoms—inserted into the hydrocarbon. A group of organic compounds that all have the same functional group is a *family*. For example, the members of the family of alcohols each have an —OH functional group and the general formula R—OH, where R represents a hydrocarbon group. (That is, we refer to the hydrocarbon group as an "R group.") Some specific examples include methanol and isopropyl alcohol (also known as rubbing alcohol).

$$\begin{array}{c} \mathrm{H} \\ | \\ \mathrm{H{-}C{-}OH} \\ | \\ \mathrm{H} \end{array} \qquad \begin{array}{c} \mathrm{H}\quad \mathrm{CH_3} \\ |\qquad | \\ \mathrm{H{-}C{-}C{-}OH} \\ |\qquad | \\ \mathrm{H}\quad \mathrm{H} \end{array}$$

Methanol — Isopropyl alcohol

The presence of a functional group in a hydrocarbon alters the properties of the compound significantly. For example, methane is a nonpolar gas. By contrast, methanol—methane with an —OH group substituted for one of the hydrogen atoms—is a polar, hydrogen-bonded liquid at room temperature. Although each member of a family is unique and different, their common functional group causes some similarities in both their physical and chemical properties. Table 22.10 lists some common functional groups, their general formulas, and an example of each.

TABLE 22.10 Some Common Functional Groups

Family	General Formula*	Condensed General Formula	Example	Name
Alcohols	R—OH	ROH	CH_3CH_2OH	Ethanol (ethyl alcohol)
Ethers	R—O—R	ROR	CH_3OCH_3	Dimethyl ether
Aldehydes	$\mathrm{R{-}\overset{\Large O}{\overset{\|\|}{C}}{-}H}$	RCHO	$\mathrm{CH_3{-}\overset{\Large O}{\overset{\|\|}{C}}{-}H}$	Ethanal (acetaldehyde)
Ketones	$\mathrm{R{-}\overset{\Large O}{\overset{\|\|}{C}}{-}R}$	RCOR	$\mathrm{CH_3{-}\overset{\Large O}{\overset{\|\|}{C}}{-}CH_3}$	Propanone (acetone)
Carboxylic acids	$\mathrm{R{-}\overset{\Large O}{\overset{\|\|}{C}}{-}OH}$	RCOOH	$\mathrm{CH_3{-}\overset{\Large O}{\overset{\|\|}{C}}{-}OH}$	Ethanoic acid (acetic acid)
Esters	$\mathrm{R{-}\overset{\Large O}{\overset{\|\|}{C}}{-}OR}$	RCOOR	$\mathrm{CH_3{-}\overset{\Large O}{\overset{\|\|}{C}}{-}OCH_3}$	Methyl acetate
Amines	$\mathrm{R{-}\overset{R}{\overset{\|}{N}}{-}R}$	R_3N	$\mathrm{CH_3CH_2{-}\overset{H}{\overset{\|}{N}}{-}H}$	Ethylamine

*In ethers, ketones, esters, and amines, the R groups may be the same or different.

22.9 Alcohols

As we discussed in Section 22.8, **alcohols** are organic compounds containing the —OH functional group, or **hydroxyl group**, and they have the general formula R—OH. In addition to methanol and isopropyl alcohol, ethanol and 1-butanol (shown here) are also common alcohols.

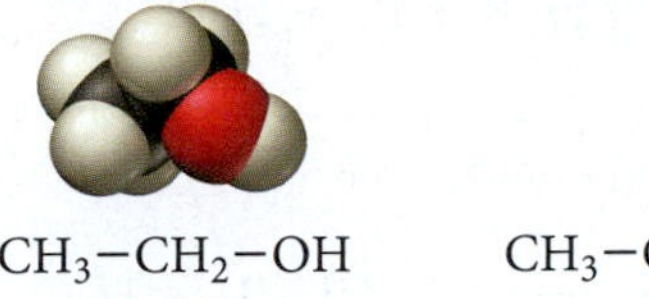

CH_3-CH_2-OH
Ethanol

$CH_3-CH_2-CH_2-CH_2-OH$
1-Butanol

Naming Alcohols

The names of alcohols are like the names of alkanes with the following differences:

- The base chain is the longest continuous carbon chain that contains the —OH functional group.
- The base name has the ending *-ol.*
- We number the base chain to assign the —OH group the lowest possible number.
- We insert a number indicating the position of the —OH group just before the base name. For example:

$CH_3CH_2CH_2CH(OH)CH_3$
2-Pentanol

$CH_2(OH)-CH_2-CH(CH_3)-CH_3$
3-Methyl-1-butanol

About Alcohols

The familiar alcohol in alcoholic beverages, ethanol, is most commonly formed by the yeast fermentation of sugars, such as glucose, from fruits and grains.

$$\underset{\text{Glucose}}{C_6H_{12}O_6} \xrightarrow{\text{yeast}} \underset{\text{Ethanol}}{2\,CH_3CH_2OH} + 2\,CO_2$$

Alcoholic beverages contain ethanol, water, and a few other components that impart flavor and color. Beer usually contains 3–6% ethanol. Wine contains about 12–14% ethanol, and spirits—beverages like whiskey, rum, or tequila—range from 40% to 80% ethanol, depending on their *proof.* The proof of an alcoholic beverage is twice the percentage of its ethanol content, so an 80-proof whiskey contains 40% ethanol. Ethanol is used as a gasoline additive because it increases the octane rating of gasoline and fosters its complete combustion, reducing the levels of certain pollutants such as carbon monoxide and the precursors of ozone.

Isopropyl alcohol (or 2-propanol) is available at any drug store under the name of rubbing alcohol. It is commonly used as a disinfectant for wounds and to sterilize medical instruments. Isopropyl alcohol should never be consumed internally, as it is highly toxic. Four ounces of isopropyl alcohol can cause death. A third common alcohol is methanol, also called wood alcohol. Methanol is commonly used as a laboratory solvent and as a fuel additive. Like isopropyl alcohol, methanol is toxic and should never be consumed.

Alcohol Reactions

Alcohols undergo a number of reactions including substitution, elimination (or dehydration), and oxidation. Alcohols also react with active metals to form strong bases.

Substitution Alcohols react with acids such as HBr to form halogenated hydrocarbons.

$$ROH + HBr \longrightarrow R{-}Br + H_2O$$

In these reactions, the halogen replaces the hydroxyl group on the alcohol. For example, ethanol reacts with hydrobromic acid to form bromoethane and water:

$$CH_3CH_2OH + HBr \longrightarrow CH_3CH_2Br + H_2O$$

Elimination (or Dehydration) In the presence of concentrated acids such as H_2SO_4, alcohols react and eliminate water, forming alkenes. These kinds of reactions are known as **elimination reactions**. For example, ethanol eliminates water to form ethene according to the reaction:

$$\underset{\text{H}}{\underset{|}{CH_2}}{-}\underset{\text{OH}}{\underset{|}{CH_2}} \xrightarrow{H_2SO_4} CH_2{=}CH_2 + H_2O$$

Oxidation In organic chemistry, we think of oxidation and reduction in terms of the changes to the carbon atoms in the molecule. Thus, oxidation is the gaining of oxygen or the losing of hydrogen by a carbon atom. Reduction is the loss of oxygen or the gaining of hydrogen by a carbon atom. We can draw a series showing relative states of oxidation:

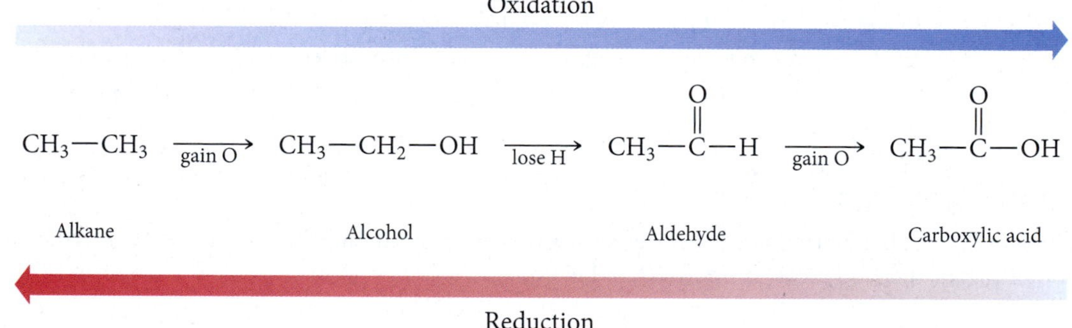

In this view, an alcohol is a partially oxidized hydrocarbon; it can be further oxidized to form an aldehyde or carboxylic acid, or it can be reduced to form a hydrocarbon (but this is rare). For example, ethanol can be oxidized to acetic acid according to the reaction:

$$CH_3CH_2OH \xrightarrow[H_2SO_4]{Na_2Cr_2O_7} CH_3COOH$$

Reaction with Active Metals Alcohols react with active metals, such as sodium, much as water does. For example, methanol reacts with sodium to form *sodium methoxide* and hydrogen gas:

$$CH_3OH + Na \longrightarrow \underset{\text{Sodium methoxide}}{CH_3ONa} + \tfrac{1}{2}H_2$$

The reaction of *water* with sodium produces *sodium hydroxide* and hydrogen gas:

$$H_2O + Na \longrightarrow NaOH + \tfrac{1}{2}H_2$$

In both cases, a strong base forms (OH^- in the case of water and CH_3O^- in the case of methanol).

EXAMPLE 22.7 Alcohol Reactions

Determine the type of reaction (substitution, dehydration, oxidation, or reaction with an active metal) that occurs in each case, and write formulas for the products.

(a) $CH_3-\underset{}{\overset{CH_3}{\overset{|}{C}H}}-CH_2-CH_2-OH + HBr \longrightarrow$

(b) $CH_3-\overset{CH_3}{\overset{|}{C}H}-CH_2-CH_2-OH \xrightarrow[H_2SO_4]{Na_2Cr_2O_7}$

SOLUTION

(a) An alcohol reacting with an acid is an example of a *substitution reaction*. The product of the substitution reaction is a halogenated hydrocarbon and water.

$$CH_3-\overset{CH_3}{\overset{|}{C}H}-CH_2-CH_2-OH + HBr \longrightarrow$$

$$CH_3-\overset{CH_3}{\overset{|}{C}H}-CH_2-CH_2-Br + H_2O$$

(b) An alcohol in solution with sodium dichromate and acid undergoes an *oxidation reaction*. The product of the oxidation reaction is a carboxylic acid functional group. (We discuss carboxylic acid functional groups in detail in Section 22.11.)

$$CH_3-\overset{CH_3}{\overset{|}{C}H}-CH_2-CH_2-OH \xrightarrow[H_2SO_4]{Na_2Cr_2O_7}$$

$$CH_3-\overset{CH_3}{\overset{|}{C}H}-CH_2-\overset{O}{\overset{\|}{C}}-OH$$

FOR PRACTICE 22.7

Determine the type of reaction (substitution, dehydration, oxidation, or reaction with an active metal) that occurs in each case, and write formulas for the products.

(a) $CH_3CH_2OH + Na \longrightarrow$

(b) $CH_3-\overset{CH_3}{\overset{|}{C}H}-CH_2-OH \xrightarrow{H_2SO_4}$

22.10 Aldehydes and Ketones

Aldehydes and **ketones** have the following general formulas:

$$\underset{\text{Aldehyde}}{R-\overset{O}{\overset{\|}{C}}-H} \qquad \underset{\text{Ketone}}{R-\overset{O}{\overset{\|}{C}}-R}$$

The condensed structural formula for aldehydes is RCHO; for ketones it is RCOR.

The functional group for both aldehydes and ketones is the **carbonyl group**:

$$\overset{O}{\overset{\|}{\diagup C \diagdown}}$$

Ketones have an R group attached to both sides of the carbonyl, while aldehydes have one R group and a hydrogen atom. (An exception is formaldehyde, which is an aldehyde with two H atoms bonded to the carbonyl group.)

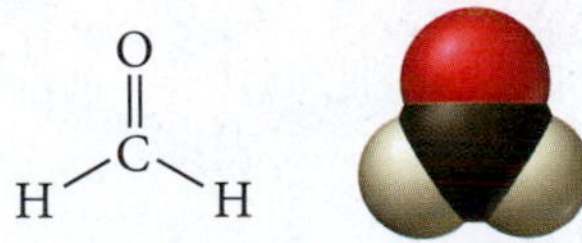

Formaldehyde or methanal

Figure 22.6 ▼ shows other common aldehydes and ketones.

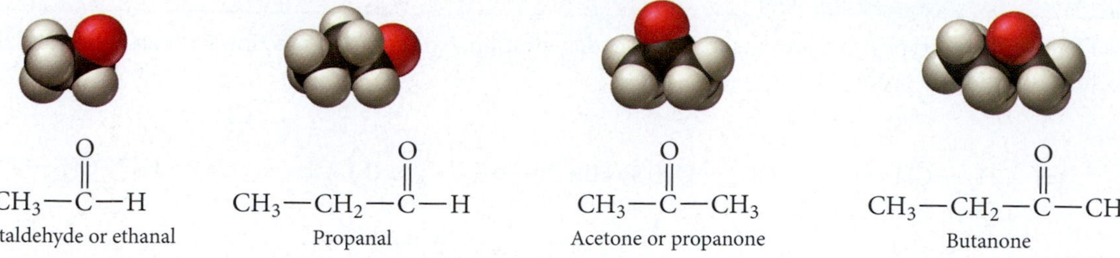

▶ **FIGURE 22.6 Common Aldehydes and Ketones**

Naming Aldehydes and Ketones

Many aldehydes and ketones have common names that we can learn only by becoming familiar with them, but we can systematically name simple aldehydes according to the number of carbon atoms in the longest continuous carbon chain that contains the carbonyl group. We form the base name from the name of the corresponding alkane by dropping the *-e* and adding the ending *-al*.

$$CH_3-CH_2-CH_2-\overset{O}{\overset{\|}{C}}-H$$

Butanal

$$CH_3-CH_2-CH_2-CH_2-\overset{O}{\overset{\|}{C}}-H$$

Pentanal

We name simple ketones according to the longest continuous carbon chain containing the carbonyl group, forming the base name from the name of the corresponding alkane by dropping the letter *-e* and adding the ending *-one*. For ketones, we number the chain to give the carbonyl group the lowest possible number.

$$CH_3-CH_2-CH_2-\overset{O}{\overset{\|}{C}}-CH_3$$

2-Pentanone

$$CH_3-CH_2-CH_2-\overset{O}{\overset{\|}{C}}-CH_2-CH_3$$

3-Hexanone

About Aldehydes and Ketones

The most familiar aldehyde is probably formaldehyde, shown earlier in this section. Formaldehyde is a gas with a pungent odor. It is often mixed with water to make formalin, a preservative and disinfectant. Formaldehyde is also found in wood smoke, which is one reason smoking is an effective method of food preservation—the formaldehyde kills bacteria. Aromatic aldehydes, those that also contain an aromatic ring, have pleasant aromas. For example, vanillin causes the smell of vanilla, cinnamaldehyde is the sweet-smelling component of cinnamon, and benzaldehyde accounts for the smell of almonds (Figure 22.7 ▼).

O, C, H, H_3C, O, HO

Vanillin

H, O, C, C=C, H, H

Cinnamaldehyde

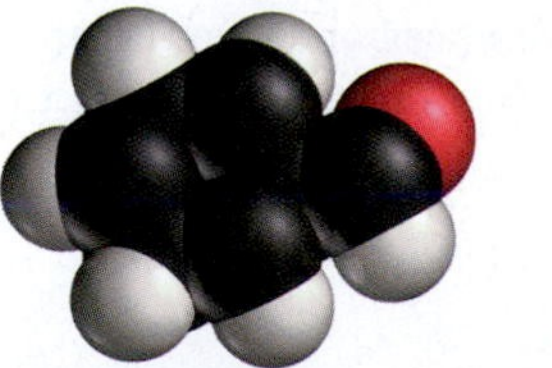

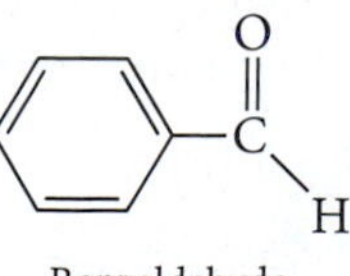

Benzaldehyde

▲ **FIGURE 22.7 The Nutty Aroma of Almonds** Benzaldehyde is partly responsible for the smell of almonds.

▲ **FIGURE 22.8 The Fragrance of Raspberries** Ionone is partly responsible for the smell of raspberries.

The most familiar ketone is acetone, the main component of nail polish remover. Other ketones have more pleasant aromas. For example, carvone is largely responsible for the smell of spearmint, 2-heptanone (among other compounds) for the smell of cloves, and ionone for the smell of raspberries (Figure 22.8▲).

Carvone

$$CH_3-CH_2-CH_2-CH_2-CH_2-\overset{\displaystyle O}{\overset{\|}{C}}-CH_3$$

2-Heptanone

Aldehyde and Ketone Reactions

Aldehydes and ketones can form from the *oxidation* of alcohols. For example, ethanol can be oxidized to ethanal, and 2-propanol can be oxidized to 2-propanone (or acetone).

$$\underset{\text{Ethanol}}{H-CH_2-CH_2-OH} \xrightarrow{\text{oxidation}} \underset{\text{Ethanal}}{H-CH_2-CHO} \qquad \underset{\text{2-Propanol}}{H-CH_2-CH(OH)-CH_2-H} \xrightarrow{\text{oxidation}} \underset{\text{2-Propanone}}{H-CH_2-C(=O)-CH_2-H}$$

In the reverse reaction, an aldehyde or ketone is reduced to an alcohol. For example, 2-butanone can be reduced to 2-butanol in the presence of a reducing agent.

$$\underset{\text{2-Butanone}}{CH_3-CH_2-C(=O)-CH_3} \xrightarrow{\text{reduction}} \underset{\text{2-Butanol}}{CH_3-CH_2-CH(OH)-CH_3}$$

The carbonyl group in aldehydes and ketones is unsaturated, much like the double bond in an alkene. Because of this feature, the most common reactions of aldehydes and ketones are addition reactions. However, in contrast to the carbon–carbon double bond in alkenes, which is nonpolar, the double bond in the carbonyl group is highly polar (Figure 22.9 ▶). Consequently, additions across the double bond result in the more electronegative part of the reagent bonding to the carbon atom and the less electronegative part (often hydrogen) bonding to the oxygen atom. For example, HCN adds across the carbonyl double bond in formaldehyde.

$$H_2C=O + H-C\equiv N \xrightarrow{\text{NaCN}} N\equiv C-CH_2-O-H$$

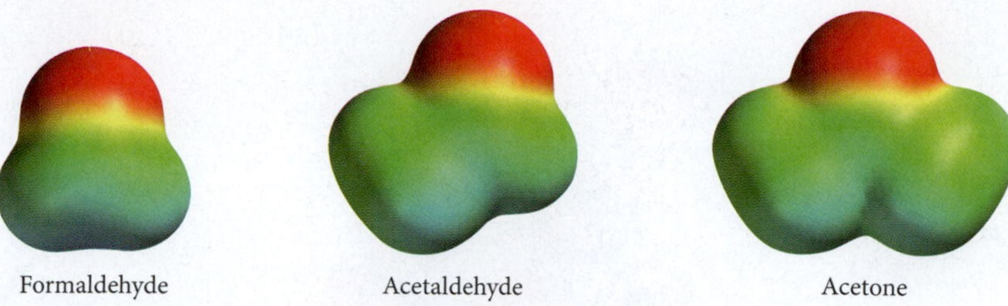

▲ **FIGURE 22.9 Electrostatic Potential Maps of the Carbonyl Group** Members of the carbonyl group are highly polar, as shown in these plots of electrostatic potential.

22.11 Carboxylic Acids and Esters

Carboxylic acids and **esters** have the general formulas:

The condensed structural formula for carboxylic acids is RCOOH; for esters it is RCOOR.

$$\underset{\text{Carboxylic acid}}{R-\overset{\overset{\displaystyle O}{\|}}{C}-OH} \qquad \underset{\text{Ester}}{R-\overset{\overset{\displaystyle O}{\|}}{C}-OR}$$

Figure 22.10 ◀ shows the structures of some common carboxylic acids and esters.

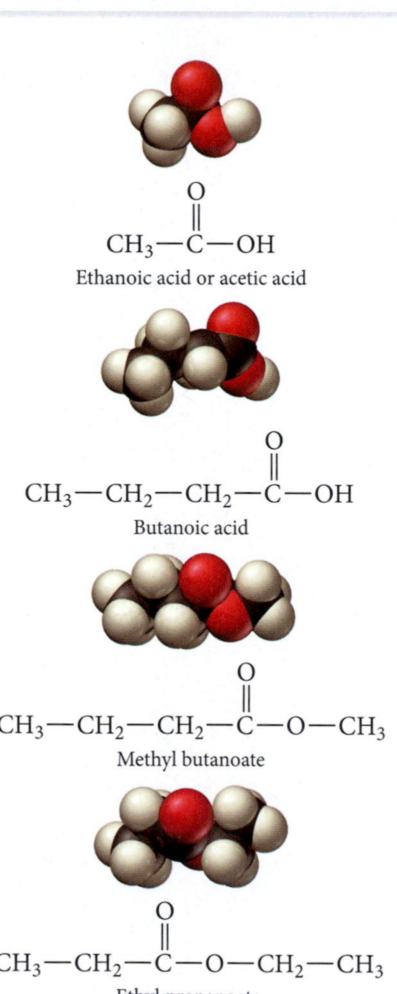

▲ **FIGURE 22.10 Common Carboxylic Acids and Esters**

Naming Carboxylic Acids and Esters

We name carboxylic acids according to the number of carbon atoms in the longest chain containing the —COOH functional group. We form the base name by dropping the *-e* from the name of the corresponding alkane and adding the ending *–oic acid.*

$$\underset{\text{Propanoic acid}}{CH_3-CH_2-\overset{\overset{\displaystyle O}{\|}}{C}-OH} \qquad \underset{\text{Pentanoic acid}}{CH_3-CH_2-CH_2-CH_2-\overset{\overset{\displaystyle O}{\|}}{C}-OH}$$

We name esters as if they were derived from a carboxylic acid by replacing the H on the OH with an alkyl group. The R group from the parent acid forms the base name of the compound. We change the *-ic* on the name of the corresponding carboxylic acid to *-ate*, and drop *acid*, naming the R group that replaced the H on the carboxylic acid as an alkyl group with the ending *-yl*, as shown in the following examples:

$$\underset{\text{Propanoic acid}}{CH_3-CH_2-\overset{\overset{\displaystyle O}{\|}}{C}-OH} \qquad \underset{\text{Pentanoic acid}}{CH_3-CH_2-CH_2-CH_2-\overset{\overset{\displaystyle O}{\|}}{C}-OH}$$

$$\underset{\text{Methyl propanoate}}{CH_3-CH_2-\overset{\overset{\displaystyle O}{\|}}{C}-OCH_3} \qquad \underset{\text{Ethyl pentanoate}}{CH_3-CH_2-CH_2-CH_2-\overset{\overset{\displaystyle O}{\|}}{C}-OCH_2CH_3}$$

About Carboxylic Acids and Esters

Like all acids, carboxylic acids taste sour. The most familiar carboxylic acid is ethanoic acid, better known by its common name, acetic acid. Acetic acid is the component in vinegar that imparts its characteristic flavor and aroma. It can form by the oxidation of ethanol, which is why wines left open to air become sour. Some yeasts and bacteria also form acetic acid when they metabolize sugars in bread dough. These are added to bread dough to make sourdough bread. Other common carboxylic acids

$$H-\overset{\overset{\displaystyle O}{\|}}{C}-OH$$

Formic or methanoic acid

$$CH_3-\overset{\overset{\displaystyle OH}{|}}{CH}-\overset{\overset{\displaystyle O}{\|}}{C}-OH$$

Lactic acid

include methanoic acid (formic acid), present in ant bites; lactic acid, which collects in muscles after intense exercise causing soreness; and citric acid, found in limes, lemons, and oranges (Figure 22.11▼).

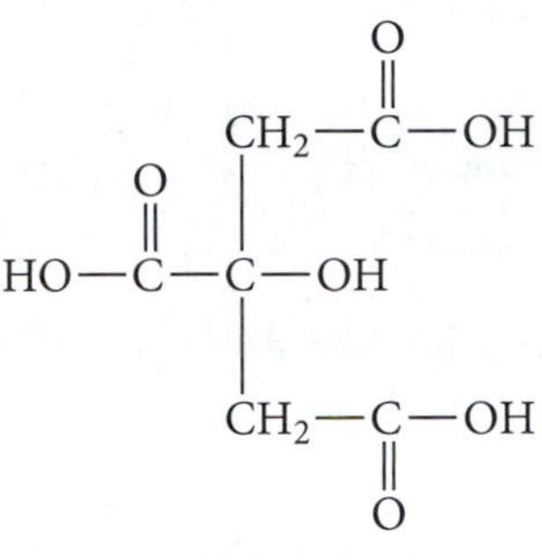

Citric acid

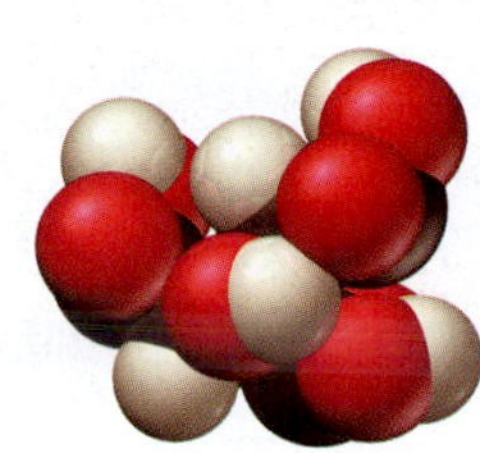

◀ **FIGURE 22.11 The Tart Taste of Limes** Citric acid is partly responsible for the sour taste of limes.

Esters are best known for their sweet smells. Methyl butanoate is largely responsible for the smell and taste of apples, and ethyl butanoate is largely responsible for the smell and taste of pineapples (see Figure 22.12 ▼).

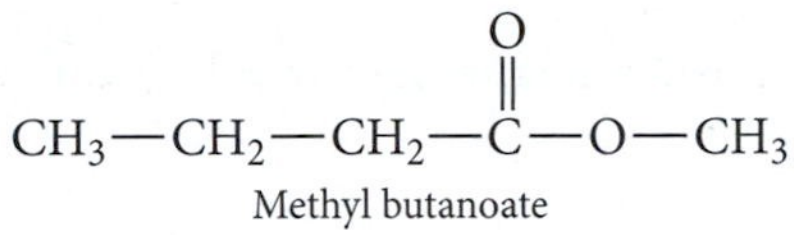

Methyl butanoate

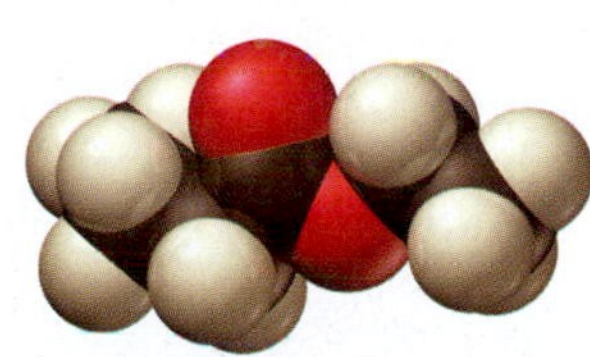

$$CH_3-CH_2-CH_2-\overset{\overset{\displaystyle O}{\|}}{C}-O-CH_2-CH_3$$

Ethyl butanoate

▲ **FIGURE 22.12 The Aroma of Pineapple** Ethyl butanoate is partly responsible for the aroma of pineapples.

Carboxylic Acid and Ester Reactions

Carboxylic acids act as weak acids in solution.

$$RCOOH(aq) + H_2O(l) \rightleftharpoons H_3O^+(aq) + RCOO^-(aq)$$

Like all acids, carboxylic acids react with strong bases via neutralization reactions. For example, propanoic acid reacts with sodium hydroxide to form sodium propanoate and water.

$$CH_3CH_2COOH(aq) + NaOH(aq) \longrightarrow CH_3CH_2COO^-Na^+(aq) + HOH(l)$$

A carboxylic acid reacts with an alcohol to form an ester via a **condensation reaction**—a reaction in which two (or more) organic compounds join, often with the loss of water (or some other small molecule).

$$\underset{\text{Acid}}{R-\overset{\overset{\displaystyle O}{\|}}{C}-OH} + \underset{\text{Alcohol}}{HO-R'} \xrightarrow{H_2SO_4} \underset{\text{Ester}}{R-\overset{\overset{\displaystyle O}{\|}}{C}-O-R'} + \underset{\text{Water}}{H_2O}$$

An important example of this reaction is the formation of acetylsalicylic acid (aspirin) from ethanoic acid (acetic acid) and salicylic acid (originally obtained from the bark of the willow tree).

$$CH_3-\overset{\overset{\displaystyle O}{\|}}{C}-OH + \text{HO}-\overset{\overset{\displaystyle O}{\|}}{C}-C_6H_4-OH \longrightarrow \text{HO}-\overset{\overset{\displaystyle O}{\|}}{C}-C_6H_4-O-\underset{\underset{\displaystyle O}{\|}}{C}-CH_3 + H_2O$$

Acetic acid — Salicylic acid — Acetylsalicylic acid

If we subject a carboxylic acid to high temperatures, it undergoes a condensation reaction with itself to form an acid anhydride (*anhydride* means "without water").

$$\text{RCOOH}(aq) + \text{HOOCR}(aq) \longrightarrow \underset{\text{Acid anhydride}}{\text{RCOOOCR}(aq)} + \text{HOH}(aq)$$

We can add water to an acid anhydride to reverse the reaction just shown and regenerate the carboxylic acid molecules.

22.3 Cc Conceptual Connection

Oxidation

Arrange the compounds from least oxidized to most oxidized.

(a) $CH_3-\overset{\overset{\displaystyle O}{\|}}{C}-CH_3$

(b) $CH_3-\overset{\overset{\displaystyle O}{\|}}{C}-O-CH_3$

(c) $CH_3-CH_2-CH_3$

(d) $CH_3-\underset{\underset{\displaystyle OH}{|}}{CH}-CH_3$

22.12 Ethers

Ethers are organic compounds with the general formula ROR. The two R groups may be identical or they may be different. Some common ethers are shown in Figure 22.13 ◀.

CH_3-O-CH_3
Dimethyl ether

$CH_3-O-CH_2-CH_3$
Ethyl methyl ether

$CH_3-CH_2-O-CH_2-CH_3$
Diethyl ether

▲ **FIGURE 22.13 Ethers**

Naming Ethers

Common names for ethers have the format:

(R group 1) (R group 2) ether

If the two R groups differ, we use each of their names in alphabetical order. If the two R groups are the same, we use the prefix *di-*. Some examples include:

$$\underset{\text{Dipropyl ether}}{H_3C-CH_2-CH_2-O-CH_2-CH_2-CH_3}$$

$$\underset{\text{Ethyl propyl ether}}{H_3C-CH_2-O-CH_2-CH_2-CH_3}$$

About Ethers

The most common ether is diethyl ether. Diethyl ether is a useful laboratory solvent because it can dissolve many organic compounds and it has a low boiling point (34.6 °C). The low boiling point allows for easy removal of the solvent. Diethyl ether was used as a general anesthetic for many years. When inhaled, diethyl ether depresses the central nervous system, causing unconsciousness and insensitivity to pain. Its use as an anesthetic has decreased in recent years because other compounds have the same anesthetic effect with fewer side effects (such as nausea).

22.13 Amines

The simplest nitrogen-containing compound is ammonia (NH_3). **Amines** are organic compounds containing nitrogen that are derived from ammonia, with one or more of the hydrogen atoms replaced by alkyl groups. Like ammonia, amines are weak bases. We systematically name amines according to the hydrocarbon groups attached to the nitrogen and assign the ending *-amine*.

$$CH_3-CH_2-\underset{\displaystyle H}{\underset{|}{N}}-H$$

Ethylamine

$$CH_3-CH_2-\underset{\displaystyle CH_3}{\underset{|}{N}}-H$$

Ethylmethylamine

Amines are most commonly known for their awful odors. When a living organism dies, the bacteria that feast on its proteins emit amines. For example, trimethylamine causes the smell of rotten fish, and cadaverine causes the smell of decaying animal flesh.

$$CH_3-\overset{\displaystyle CH_3}{\overset{|}{N}}-CH_3$$

Trimethylamine

$$NH_2-CH_2-CH_2-CH_2-CH_2-CH_2-NH_2$$

Cadaverine

Amine Reactions

Just as carboxylic acids act as weak acids, so amines act as weak bases.

$$RNH_2(aq) + H_2O(l) \rightleftharpoons RNH_3^+(aq) + OH^-(aq)$$

Like all bases, amines react with strong acids to form salts called ammonium salts. For example, methylamine reacts with hydrochloric acid to form methylammonium chloride.

$$CH_3NH_2(aq) + HCl(aq) \longrightarrow \underset{\text{Methylammonium chloride}}{CH_3NH_3{}^+Cl^-(aq)}$$

A biochemically important amine reaction is the condensation reaction between a carboxylic acid and an amine.

$$CH_3COOH(aq) + HNHR(aq) \longrightarrow CH_3CONHR(aq) + HOH(l)$$

This reaction is responsible for the formation of proteins from amino acids.

22.14 Polymers

Polymers are long, chainlike molecules composed of repeating units called **monomers**. In this section, we learn about *synthetic* polymers, which we encounter daily in plastic products such as PVC tubing, styrofoam coffee cups, nylon rope, and plexiglass windows. Polymer materials are common in our everyday lives, found in everything from computers to toys to packaging materials. Most polymers are durable, partly because of the length of their molecules. In general, the longer the length of a molecule, the greater the intermolecular

▲ **FIGURE 22.14 Polyethylene** Soda and juice bottles are made from polyethylene.

▲ **FIGURE 22.15 Polyvinyl Chloride** Polyvinyl chloride is used for many plastic plumbing supplies, such as pipes and connectors.

forces between molecules, and the higher the melting point and boiling point of the substance. Because breaking or tearing a polymeric material involves either overcoming the intermolecular forces between chains, or actually breaking the covalent bonds between monomers, polymers tend to be durable materials.

One of the simplest synthetic polymers is polyethylene. The polyethylene monomer is ethene (also called ethylene).

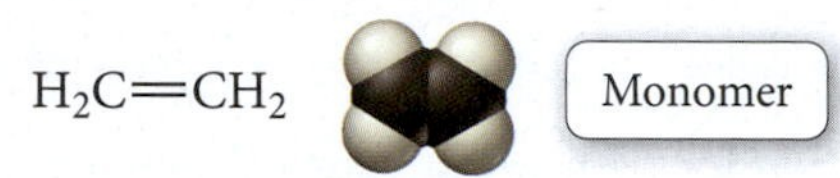

Ethene or ethylene

Ethene monomers can react with each other, breaking the double bond between carbons and bonding together to form a long polymer chain.

$$\cdots CH_2{-}CH_2{-}CH_2{-}CH_2{-}CH_2{-}CH_2{-}CH_2{-}CH_2{-}CH_2\cdots$$

Polymer

Polyethylene

Polyethylene is the plastic that is used for soda bottles, juice containers, and garbage bags (Figure 22.14 ▲). It is an example of an **addition polymer**, a polymer in which the monomers link together without the elimination of any atoms.

Substituted polyethylenes make up an entire class of polymers. For example, polyvinyl chloride (PVC)—the plastic used to make certain kinds of pipes and plumbing fixtures—is composed of monomers in which a chlorine atom has been substituted for one of the hydrogen atoms in ethene (Figure 22.15 ▲). These monomers (shown here) react to form PVC:

$$\underset{\displaystyle Cl}{\underset{|}{HC}}{=}CH_2$$

Monomer

Chloroethene

$$\cdots\underset{\displaystyle Cl}{\underset{|}{CH}}{-}CH_2{-}\underset{\displaystyle Cl}{\underset{|}{CH}}{-}CH_2{-}\underset{\displaystyle Cl}{\underset{|}{CH}}{-}CH_2{-}\underset{\displaystyle Cl}{\underset{|}{CH}}{-}CH_2{-}\underset{\displaystyle Cl}{\underset{|}{CH}}\cdots$$

Polymer

Polyvinyl chloride (PVC)

Table 22.11 lists several other substituted polyethylene polymers.

Some polymers—called copolymers—consist of two different kinds of monomers. For example, the monomers that compose nylon 6,6 are hexamethylenediamine and adipic acid. These two monomers add together via the condensation reaction shown here:

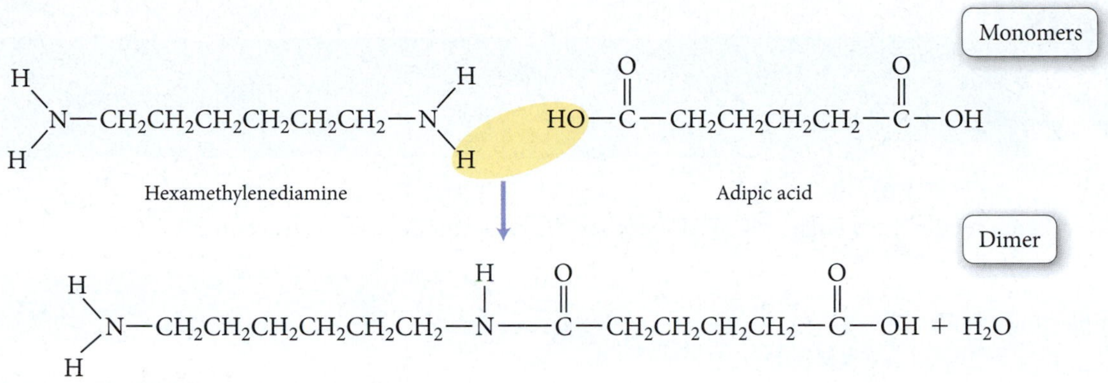

Monomers

$$\mathrm{H_2N{-}CH_2CH_2CH_2CH_2CH_2CH_2{-}NH_2 \quad + \quad HO{-}\overset{O}{\overset{\|}{C}}{-}CH_2CH_2CH_2CH_2{-}\overset{O}{\overset{\|}{C}}{-}OH}$$

Hexamethylenediamine   Adipic acid

Dimer

$$\mathrm{H_2N{-}CH_2CH_2CH_2CH_2CH_2CH_2{-}\underset{}{\overset{H}{\overset{|}{N}}}{-}\overset{O}{\overset{\|}{C}}{-}CH_2CH_2CH_2CH_2{-}\overset{O}{\overset{\|}{C}}{-}OH + H_2O}$$

The product that forms in the reaction of two monomers is called a **dimer**. The polymer (nylon 6,6) forms as the dimer continues to add more monomers. Polymers that eliminate an atom or a small group of atoms during polymerization are **condensation polymers**. Nylon 6,6 and other similar nylons can be drawn into fibers and used to make consumer products such as panty hose, carpet fibers, and fishing line. Table 22.11 shows other condensation polymers.

TABLE 22.11 Polymers of Commercial Importance

Polymer	Structure	Uses
Addition Polymers		
Polyethylene	$-\!(CH_2-CH_2)_n\!-$	Films, packaging, bottles
Polypropylene	$\left[CH_2-\underset{CH_3}{\underset{\vert}{CH}}\right]_n$	Kitchenware, fibers, appliances
Polystyrene	$\left[CH_2-\underset{C_6H_5}{\underset{\vert}{CH}}\right]_n$	Packaging, disposable food containers, insulation
Polyvinyl chloride	$\left[CH_2-\underset{Cl}{\underset{\vert}{CH}}\right]_n$	Pipe fittings, clear film for meat packaging
Condensation Polymers		
Polyurethane	$\left[\underset{O}{\underset{\|}{C}}-NH-R-NH-\underset{O}{\underset{\|}{C}}-O-R'-O\right]_n$ R, R′ = $-CH_2-CH_2-$ (*for example*)	"Foam" furniture stuffing, spray-on insulation, automotive parts, footwear, water-protective coatings
Polyethylene terephthalate (a polyester)	$\left[O-CH_2-CH_2-O-\underset{O}{\underset{\|}{C}}-C_6H_4-\underset{O}{\underset{\|}{C}}\right]_n$	Tire cord, magnetic tape, apparel, soda bottles
Nylon 6,6	$\left[NH-(CH_2)_6-NH-\underset{O}{\underset{\|}{C}}-(CH_2)_4-\underset{O}{\underset{\|}{C}}\right]_n$	Home furnishings, apparel, carpet fibers, fish line, polymer blends

SELF-ASSESSMENT Quiz

1. Which property of carbon is related to its ability to form a large number of compounds?
   a) its tendency to form four covalent bonds
   b) its ability to form double and triple bonds
   c) its tendency to catenate
   d) all of the above

2. What is the correct formula for the alkane (noncyclic) containing eight carbon atoms?
   a) C_8H_{16} **b)** C_8H_{18}
   c) C_8H_{14} **d)** C_8H_8

3. Which structure is not an isomer of $CH_3CH_2CH_2CH_2CH_3$?
   a) 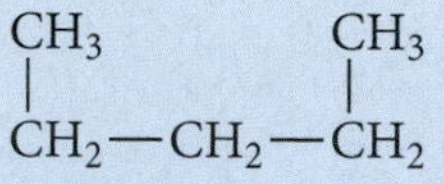 $CH_3-CH_2-CH_2-CH_2-CH_3$ (drawn as $CH_2-CH_2-CH_2$ with CH_3 on each end carbon)
   b) 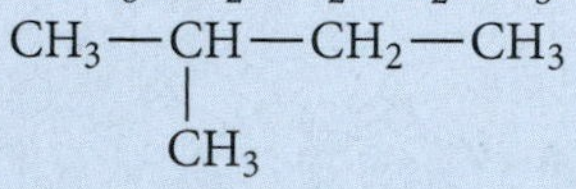 $CH_3-CH(CH_3)-CH_2-CH_3$
   c) 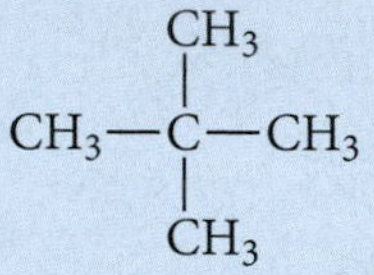 $CH_3-C(CH_3)_2-CH_3$
   d) None of the above (all are isomers).

4. Which structure can exhibit optical isomerism?
   a) 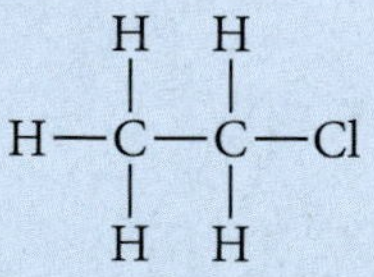 $H-CH_2-CH_2-Cl$ (H—C(H)(H)—C(H)(H)—Cl)
   b) H—C(H)(Cl)—C(H)(H)—C(Cl)(H)—H
   c) H—C(Br)(H)—C(Br)(H)—C(H)(H)—H
   d) 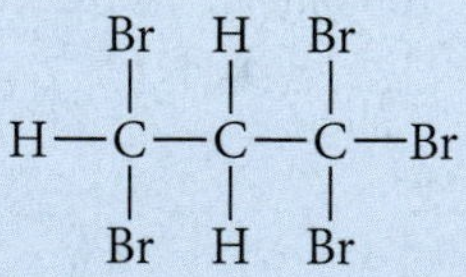 H—C(Br)(Br)—C(H)(H)—C(Br)(Br)—Br

5. Name the compound.

$$CH_3-CH_2-CH_2-CH(CH_2CH_3)-CH(CH_3)-CH_3$$

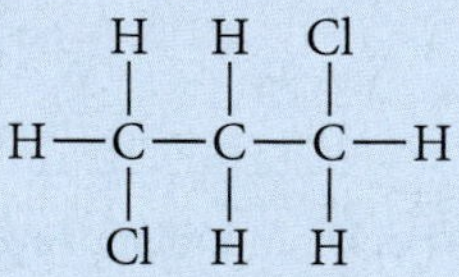

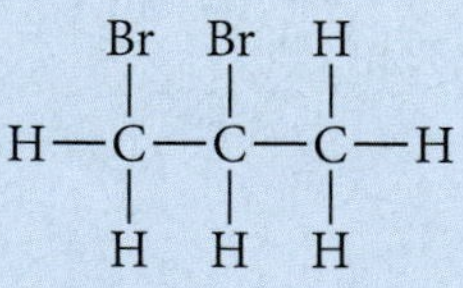

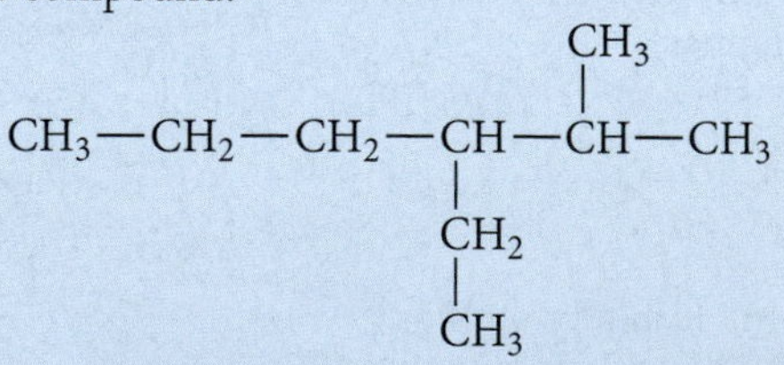

   a) 4-ethyl-5-methylhexane **b)** 3-ethyl-2-methylhexane
   c) 3-ethyl-2-methylnonane **d)** 4-methyl-5ethylhexane

6. Name the compound.

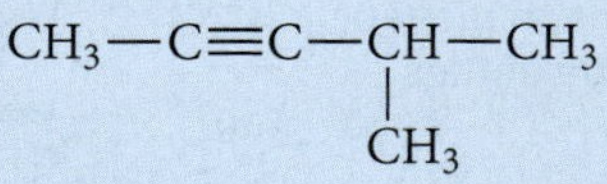

$$CH_3-C{\equiv}C-CH(CH_3)-CH_3$$

   a) 4-methyl-2-pentyne **b)** 2-methyl-3-pentyne
   c) 2-methyl-3-hexyne **d)** 4-methyl-2-hexyne

7. Determine the major product of the reaction.

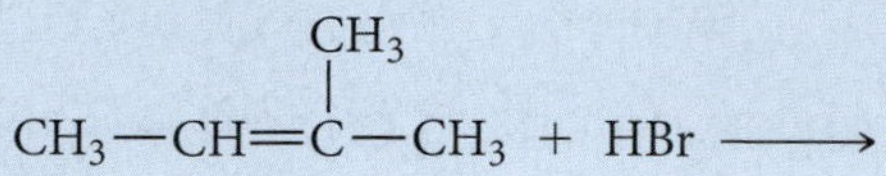

$$CH_3-CH{=}C(CH_3)-CH_3 + HBr \longrightarrow$$

   a) $CH_2Br-CH{=}C(CH_3)-CH_3$ **b)** $CH_3-CH_2-CH(CH_3)-CH_3$
   c) $CH_3-CH(Br)-CH(CH_3)-CH_3$ **d)** 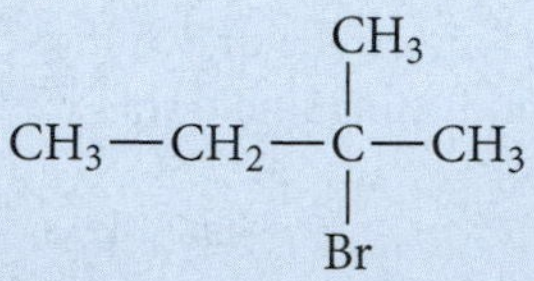 $CH_3-CH_2-C(CH_3)(Br)-CH_3$

8. Determine the product of the reaction.

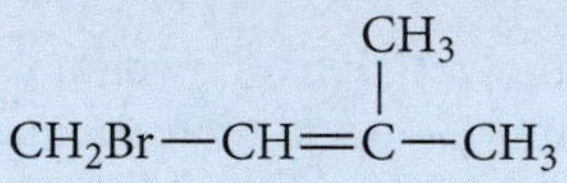

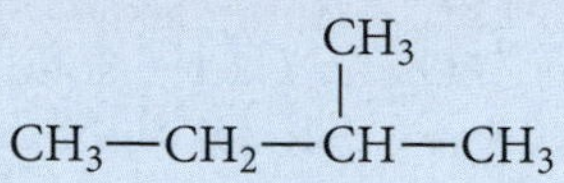

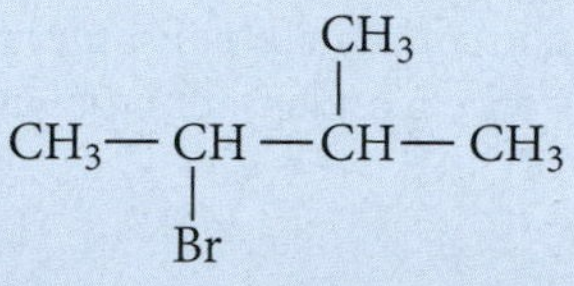

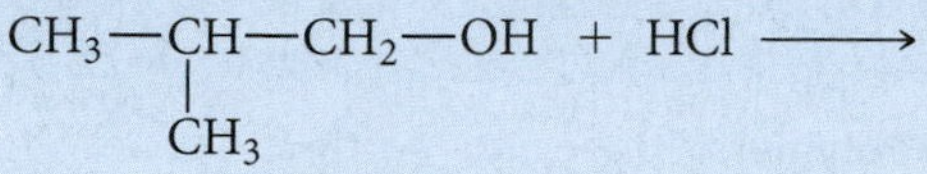

$$CH_3-CH(CH_3)-CH_2-OH + HCl \longrightarrow$$

   a) 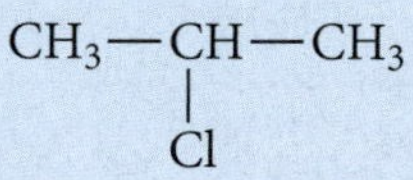 $CH_3-CH(Cl)-CH_3$ **b)** 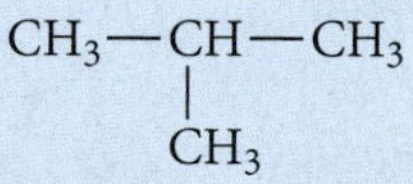 $CH_3-CH(CH_3)-CH_3$
   c) 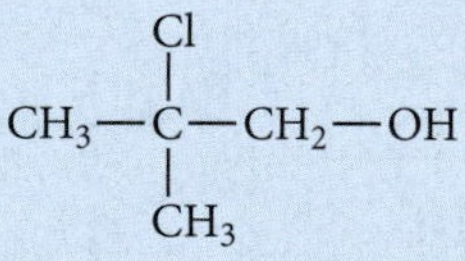 $CH_3-C(Cl)(CH_3)-CH_2-OH$ **d)** 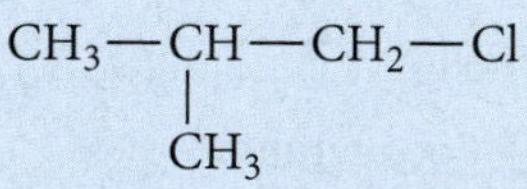 $CH_3-CH(CH_3)-CH_2-Cl$

9. Which compound is an ester?
   a) $CH_3-CH_2-O-CH_3$
   b) 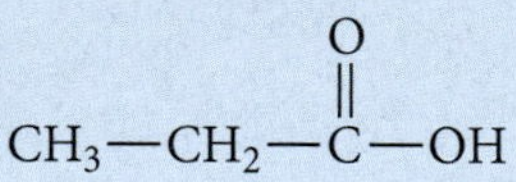 $CH_3-CH_2-C(=O)-OH$
   c) $CH_3-C(=O)-CH_3$
   d) 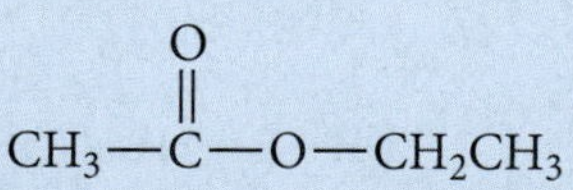 $CH_3-C(=O)-O-CH_2CH_3$

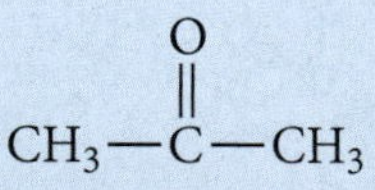

10. Which compound is most likely to have a foul odor?
   a) 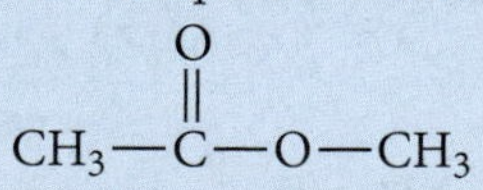 $CH_3-C(=O)-O-CH_3$
   b) $CH_3-CH_2-NH_2$
   c) $CH_3-CH_2-O-CH_3$
   d) $CH_3-CH_2-CH_2-OH$

Answers: 1:d; 2:b; 3:a; 4:c; 5:b; 6:a; 7:d; 8:d; 9:d; 10:b

CHAPTER SUMMARY 22

REVIEW

KEY LEARNING OUTCOMES

CHAPTER OBJECTIVES	ASSESSMENT
Writing Structural Formulas for Hydrocarbons (22.3)	• Example 22.1 For Practice 22.1 Exercises 37, 38
Naming Alkanes (22.4)	• Examples 22.2, 22.3, 22.4 For Practice 22.2, 22.3, 22.4 Exercises 43, 44
Naming Alkenes and Alkynes (22.5)	• Example 22.5 For Practice 22.5 Exercises 53–56
Writing Reactions: Addition Reactions (22.6)	• Example 22.6 For Practice 22.6 Exercises 59–62
Writing Reactions: Alcohols (22.9)	• Example 22.7 For Practice 22.7 Exercises 75, 76

KEY TERMS

Section 22.1
organic molecule (904)
organic chemistry (904)

Section 22.3
alkane (905)
alkene (905)
alkyne (905)
aromatic hydrocarbon (905)
aliphatic hydrocarbon (905)
structural isomers (906)
structural formula (906)
stereoisomers (909)
optical isomers (909)
enantiomers (910)
chiral (910)
dextrorotatory (911)
levorotatory (911)
racemic mixture (911)

Section 22.4
saturated hydrocarbon (912)

Section 22.5
unsaturated hydrocarbon (916)
geometric (cis–trans) isomerism (921)

Section 22.6
substitution reaction (922)
addition reaction (923)

Section 22.7
phenyl group (926)

Section 22.8
functional group (928)

Section 22.9
alcohol (929)
hydroxyl group (929)
elimination reaction (930)

Section 22.10
aldehyde (931)
ketone (931)
carbonyl group (931)

Section 22.11
carboxylic acid (934)
esters (934)
condensation reaction (935)

Section 22.12
ether (936)

Section 22.13
amine (937)

Section 22.14
polymer (937)
monomer (937)
addition polymer (938)
dimer (939)
condensation polymer (939)

KEY CONCEPTS

Fragrances and Odors (22.1)

- Organic chemistry is the study of organic compounds, which contain carbon (and other elements including hydrogen, oxygen, and nitrogen).

Carbon (22.2)

- Carbon forms more compounds than all the other elements combined.
- Carbon's four valence electrons (in conjunction with its size) allow carbon to form four bonds (in the form of single, double, or triple bonds).
- Carbon also has the capacity to catenate, to form long chains, because of the strength of the carbon–carbon bond.

Hydrocarbons (22.3)

- Organic compounds containing only carbon and hydrogen are called hydrocarbons, the key components of fuels.
- Hydrocarbons are divided into four different types: alkanes, alkenes, alkynes, and aromatic hydrocarbons.
- Stereoisomers are molecules that feature the same atoms bonded in the same order but arranged differently in space. Optical isomerism, a type of stereoisomerism, occurs when two molecules are nonsuperimposable mirror images of one another.

Alkanes (22.4)

- Alkanes are saturated hydrocarbons—they contain only single bonds and are therefore represented by the generic formula C_nH_{2n+2}. Alkane names always end in *-ane*.

Alkenes and Alkynes (22.5)

- Alkenes and alkynes are unsaturated hydrocarbons—they contain double bonds (alkenes) or triple bonds (alkynes) and are represented by the generic formulas C_nH_{2n} and C_nH_{2n-2}, respectively.
- Alkene names end in *-ene* and alkynes end in *-yne*.
- Because rotation about a double bond is severely restricted, geometric (or cis–trans) isomerism occurs in alkenes.

Hydrocarbon Reactions (22.6)

- The most common hydrocarbon reaction is probably combustion, in which hydrocarbons react with oxygen to form carbon dioxide and water; this reaction is exothermic and is used to provide most of our society's energy.
- Alkanes can also undergo substitution reactions, where heat or light causes another atom, commonly a halogen such as bromine, to be substituted for a hydrogen atom.
- Unsaturated hydrocarbons undergo addition reactions. If the addition reaction is between two unsymmetrical molecules, Markovnikov's rule predicts that the positive end of the polar reagent adds to the carbon with the most hydrogen atoms.

Aromatic Hydrocarbons (22.7)

- Aromatic hydrocarbons contain six-membered benzene rings represented with alternating single and double bonds that become equivalent through resonance. These compounds are called aromatic because they often produce pleasant fragrances.
- Because of the stability of the aromatic ring, benzene is more stable than a straight-chain alkene, and it undergoes substitution rather than addition reactions.

Functional Groups (22.8)

- Characteristic groups of atoms, such as hydroxyl (—OH), are called functional groups. Molecules that contain the same functional group have similar chemical and physical properties, and they are referred to as families.

Alcohols (22.9)

- The family of alcohols contains the —OH group and is named with the suffix *-ol*.
- Alcohols are commonly used in gasoline, in alcoholic beverages, and in sterilization procedures.
- Alcohols undergo substitution reactions, in which a substituent such as a halogen replaces the hydroxyl group.
- Alcohols undergo elimination reactions, in which water is eliminated across a bond to form an alkene, and oxidation or reduction reactions.
- Alcohols also react with active metals to form alkoxide ions and hydrogen gas.

Aldehydes and Ketones (22.10)

- Aldehydes and ketones both contain a carbonyl group (a carbon atom double-bonded to oxygen).
- In aldehydes, the carbonyl group is at the end of a carbon chain, while in ketones it is between two other carbon atoms.
- Aldehydes are named with the suffix *-al* and ketones with the suffix *-one*.
- A carbonyl can be formed by the oxidation of an alcohol or reverted to an alcohol by reduction.
- Like alkenes, carbonyls undergo addition reactions; however, because the carbon–oxygen bond is highly polar, the electronegative component of the reagent always adds to the carbon atom, and the less electronegative part adds to the oxygen.

Carboxylic Acids and Esters (22.11)

- Carboxylic acids contain a carbonyl group and a hydroxide on the same carbon and are named with the suffix *-oic acid*.
- Esters contain a carbonyl group bonded to an oxygen atom that is in turn bonded to an R group; they are named with the suffix *-oate*.
- Carboxylic acids taste sour, such as acetic acid in vinegar, while esters smell sweet.
- Carboxylic acids react as weak acids but can also form esters through condensation reactions with alcohols.

Ethers (22.12)

- The family of ethers contains an oxygen atom between two R groups.
- Ethers are named with the ending *-yl ether*.

Amines (22.13)

- Amines are organic compounds that contain nitrogen and are named with the suffix *-amine*.
- They are known for their terrible odors; the smell of decaying animal flesh is produced by cadaverine.
- Amines act as weak bases and produce a salt when mixed with a strong acid.
- The combination of an amine with a carboxylic acid leads to a condensation reaction; this reaction is used by our bodies to produce proteins from amino acids.

Polymers (22.14)

- Polymers are long, chainlike molecules that consist of repeating units called monomers. They can be natural or synthetic.
- Polyethylene is an addition polymer, a polymer formed without the elimination of any atoms.
- Condensation polymers, such as nylon, are formed by the elimination of small groups of atoms.

KEY EQUATIONS AND RELATIONSHIPS

Halogen Substitution Reactions in Alkanes (22.6)

$$\underset{\text{alkane}}{R{-}H} + \underset{\text{halogen}}{X_2} \xrightarrow{\text{heat or light}} \underset{\text{haloalkane}}{R{-}X} + \underset{\text{hydrogen halide}}{HX}$$

Common Functional Groups (22.8)

Family	General Formula	Condensed General Formula	Example	Name
Alcohols	R—OH	ROH	CH_3CH_2OH	Ethanol (ethyl alcohol)
Ethers	R—O—R	ROR	CH_3OCH_3	Dimethyl ether
Aldehydes	R—C(=O)—H	RCHO	H_3C—C(=O)—H	Ethanal (acetaldehyde)
Ketones	R—C(=O)—R	RCOR	H_3C—C(=O)—CH_3	Propanone (acetone)
Carboxylic acids	R—C(=O)—OH	RCOOH	H_3C—C(=O)—OH	Acetic acid
Esters	R—C(=O)—OR	RCOOR	H_3C—C(=O)—OCH_3	Methyl acetate
Amines	R—N(—R)—R	R_3N	H_3CH_2C—N(—H)—H	Ethylamine

Alcohol Reactions (22.9)

Substitution $ROH + HBr \longrightarrow R\text{—}Br + H_2O$

Oxidation $R\text{—}CH_2\text{—}CH_2(\text{—}OH) \xrightarrow[H_2SO_4]{Na_2Cr_2O_7} R\text{—}CH_2\text{—}C(=O)\text{—}OH$

Alcohol → Carboxylic acid

Elimination $R\text{—}CH_2\text{—}CH_2(\text{—}OH) \xrightarrow{H_2SO_4} R\text{—}CH\text{=}CH_2 + H_2O$

Carboxylic Acid Condensation Reactions (22.11)

$$\underset{\text{Acid}}{R\text{—}C(=O)\text{—}OH} + \underset{\text{Alcohol}}{HO\text{—}R'} \xrightarrow{H_2SO_4} \underset{\text{Ester}}{R\text{—}C(=O)\text{—}O\text{—}R'} + \underset{\text{Water}}{H_2O}$$

Amine Acid–Base Reactions (22.13)

$$RNH_2(aq) + H_2O(l) \longrightarrow RNH_3^+(aq) + OH^-(aq)$$

Amine–Carboxylic Acid Condensation Reactions (22.13)

$$CH_3COOH(aq) + HNHR(aq) \longrightarrow CH_3CONHR(aq) + H_2O(l)$$

EXERCISES

REVIEW QUESTIONS

1. What kinds of molecules often trigger our sense of smell?
2. What is organic chemistry?
3. What is unique about carbon and carbon-based compounds? Why did life evolve around carbon?
4. Why does carbon form such a large diversity of compounds?
5. Why does silicon exhibit less diversity of compounds than carbon does?
6. Describe the geometry and hybridization about a carbon atom that forms the following:
   a) four single bonds.
   b) two single bonds and one double bond.
   c) one single bond and one triple bond.
7. What are hydrocarbons? What are their main uses?

8. What are the main classifications of hydrocarbons? What are their generic molecular formulas?

9. Explain the differences between a structural formula, a condensed structural formula, a carbon skeleton formula, a ball-and-stick model, and a space-filling model.

10. What are structural isomers? How do the properties of structural isomers differ from one another?

11. What are optical isomers? How do the properties of optical isomers differ from one another?

12. Define each term related to optical isomerism: enantiomers, chiral, dextrorotatory, levorotatory, racemic mixture.

13. What is the difference between saturated and unsaturated hydrocarbons?

14. What are the key differences in the way that alkanes, alkenes, and alkynes are named?

15. Explain geometric isomerism in alkenes. How do the properties of geometric isomers differ from one another?

16. Describe and provide an example of a hydrocarbon combustion reaction.

17. What kinds of reactions are common to alkanes? List an example of each.

18. Describe each kind of reaction.
a) substitution reaction **b)** addition reaction
c) elimination reaction

19. What kinds of reactions are common to alkenes? List an example of each.

20. Explain Markovnikov's rule and give an example of a reaction to which it applies.

21. What is the structure of benzene? What are the different ways in which this structure is represented?

22. What kinds of reactions are common to aromatic compounds? Provide an example of each.

23. What is a functional group? List some examples.

24. What is the generic structure of alcohols? Write the structures of two specific alcohols.

25. Explain oxidation and reduction with respect to organic compounds.

26. Which kinds of reactions are common to alcohols? Provide an example of each.

27. What are the generic structures for aldehydes and ketones? Write a structure for a specific aldehyde and ketone.

28. Which kinds of reactions are common to aldehydes and ketones? List an example of each.

29. What are the generic structures for carboxylic acids and esters? Write a structure for a specific carboxylic acid and ester.

30. Which kinds of reactions are common to carboxylic acids and esters? Provide an example of each.

31. What is the generic structure of ethers? Write the structures of two specific ethers.

32. What is the generic structure of amines? Write the structures of two specific amines.

33. What is a polymer? What is the difference between a polymer and a copolymer?

34. How do an addition polymer and a condensation polymer differ from each other?

PROBLEMS BY TOPIC

Hydrocarbons

35. Based on the molecular formula, determine whether each compound is an alkane, alkene, or alkyne. (Assume that the hydrocarbons are noncyclic and there is no more than one multiple bond.)
a. C_5H_{12} **b.** C_3H_6 **c.** C_7H_{12} **d.** $C_{11}H_{22}$

36. Based on the molecular formula, determine whether each compound is an alkane, alkene, or alkyne. (Assume that the hydrocarbons are noncyclic and there is no more than one multiple bond.)
a. C_8H_{16} **b.** C_4H_6 **c.** C_7H_{16} **d.** C_2H_2

37. Write structural formulas for each of the nine structural isomers of heptane.

38. Write structural formulas for any 6 of the 18 structural isomers of octane.

39. Determine whether each compound exhibits optical isomerism.
a. CCl_4
b. 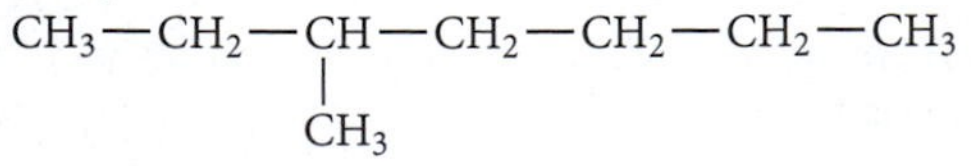

c. 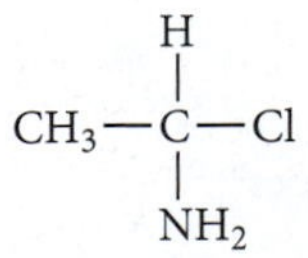

d. $CH_3CHClCH_3$

40. Determine whether each compound exhibits optical isomerism.
a. $CH_3CH_2CHClCH_3$ **b.** $CH_3CCl_2CH_3$
c. 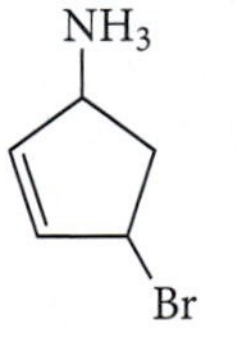

d. 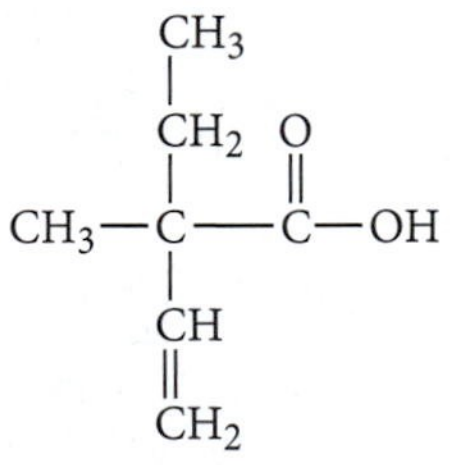

41. Determine whether the molecules in each pair are the same or enantiomers.
a. 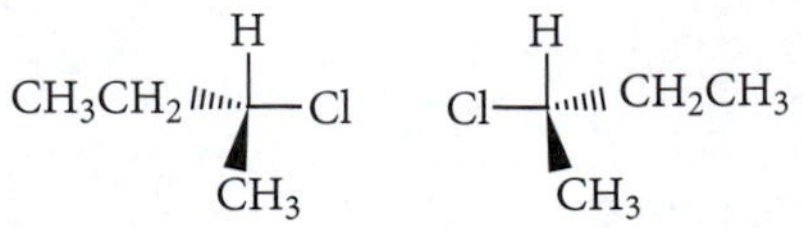

b. 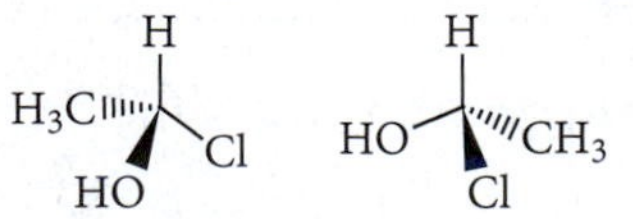

c. 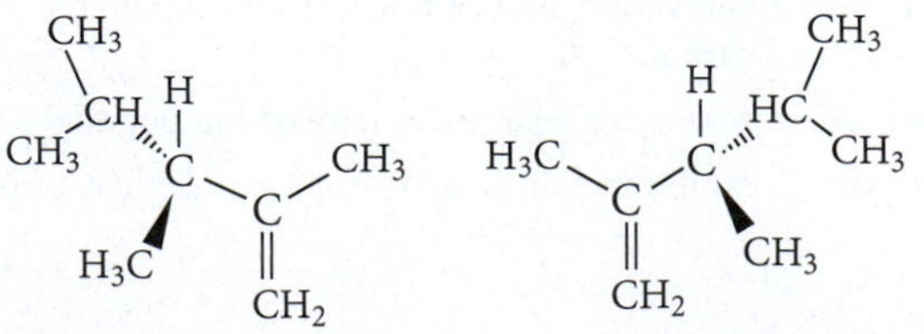

42. Determine whether the molecules in each pair are the same or enantiomers.

a. CH_3, H_3C, Cl, HO CH_3, CH_3, Cl, OH

b. H, H_3C, Br, H_2N H, Br, CH_3, NH_2

c. H, H_3C, C, CCl_3, CH_3—CH, CH_3 CH_3, H, HC, CH_3, C, Cl_3C, CH_3

Alkanes

43. Name each alkane.

a. $CH_3—CH_2—CH_2—CH_2—CH_3$

b. $CH_3—CH_2—CH—CH_3$
   (with CH_3 on the CH)

c.
$$\begin{array}{l} \qquad\qquad\qquad\qquad CH_3 \\ \quad CH_3 \qquad\quad\;\; CH—CH_3 \\ CH_3—CH—CH_2—CH—CH_2—CH_2—CH_3 \end{array}$$

d.
$$\begin{array}{l} CH_3—CH—CH_2—CH—CH_2—CH_3 \\ \quad\;\; CH_3 \qquad\quad\; CH_2—CH_3 \end{array}$$

44. Name each alkane.

a. $CH_3—CH—CH_3$
   (with CH_3 on the CH)

b.
$$\begin{array}{l} \quad\;\; CH_3 \qquad\quad\;\; CH_3 \\ CH_3—CH—CH_2—CH—CH_2 \\ \qquad\qquad\qquad\qquad\quad\;\; CH_3 \end{array}$$

c.
$$\begin{array}{l} \quad\;\; CH_3\, CH_3 \\ CH_3—C—C—CH_3 \\ \quad\;\; CH_3\, CH_3 \end{array}$$

d.
$$\begin{array}{l} \qquad\qquad\qquad\qquad\quad\; CH_3 \\ \quad\;\; CH_3 \qquad\qquad\quad\; CH_2 \\ CH_3—CH—CH_2—CH—CH—CH_2—CH_2—CH_3 \\ \qquad\qquad\qquad\;\; CH_3 \end{array}$$

45. Draw a structure for each alkane.

a. 3-ethylhexane
b. 3-ethyl-3-methylpentane
c. 2,3-dimethylbutane
d. 4,7-diethyl-2,2-dimethylnonane

46. Draw a structure for each alkane.

a. 2,2-dimethylpentane
b. 3-isopropylheptane
c. 4-ethyl-2,2-dimethylhexane
d. 4,4-diethyloctane

47. Complete and balance each hydrocarbon combustion reaction.

a. $CH_3CH_2CH_3 + O_2 \longrightarrow$
b. $CH_3CH_2CH{=}CH_2 + O_2 \longrightarrow$
c. $CH{\equiv}CH + O_2 \longrightarrow$

48. Complete and balance each hydrocarbon combustion reaction.

a. $CH_3CH_2CH_2CH_3 + O_2 \longrightarrow$
b. $CH_2{=}CHCH_3 + O_2 \longrightarrow$
c. $CH{\equiv}CCH_2CH_3 + O_2 \longrightarrow$

49. List all the possible products for each alkane substitution reaction. (Assume monosubstitution.)

a. $CH_3CH_3 + Br_2 \longrightarrow$
b. $CH_3CH_2CH_3 + Cl_2 \longrightarrow$
c. $CH_2Cl_2 + Br_2 \longrightarrow$
d. $CH_3—CH—CH_3 + Cl_2 \longrightarrow$
   (with CH_3 on the CH)

50. List all the possible products for each alkane substitution reaction. (Assume monosubstitution.)

a. $CH_4 + Cl_2 \longrightarrow$
b. $CH_3CH_2Br + Br_2 \longrightarrow$
c. $CH_3CH_2CH_2CH_3 + Cl_2 \longrightarrow$
d. $CH_3CHBr_2 + Br_2 \longrightarrow$

Alkenes and Alkynes

51. Write structural formulas for each of the possible isomers of *n*-hexene that are formed by moving the position of the double bond.

52. Write structural formulas for each of the possible isomers of *n*-pentyne that are formed by moving the position of the triple bond.

53. Name each alkene.

a. $CH_2{=}CH—CH_2—CH_3$

b.
$$\begin{array}{l} \qquad\;\; CH_3\;\; CH_3 \\ CH_3—CH—C{=}CH—CH_3 \end{array}$$

c.
$$\begin{array}{l} CH_2{=}CH—CH—CH_2—CH_2—CH_3 \\ \qquad CH_3—CH \\ \qquad\qquad\;\; CH_3 \end{array}$$

d.
$$\begin{array}{l} \qquad\;\; CH_3 \\ CH_3—CH—CH{=}C—CH_3 \\ \qquad\qquad\qquad\;\; CH_2—CH_3 \end{array}$$

54. Name each alkene.

a. $CH_3-CH_2-CH=CH-CH_2-CH_3$

b.
$$\begin{array}{lll} CH_3- & CH & -CH=CH-CH_3 \\ & | & \\ & CH_3 & \end{array}$$

c.
$$\begin{array}{lllllll} & & & CH_3 & & & \\ & & & | & & & \\ & CH_3 & & CH_2 & & & \\ & | & & | & & & \\ CH_3- & CH & -CH= & C & - & CH & -CH_3 \\ & & & & & | & \\ & & & & & CH_3 & \end{array}$$

d.
$$\begin{array}{lllll} & CH_3 & & CH_3 & \\ & | & & | & \\ CH_3- & C & -CH= & C & -CH_2-CH_3 \\ & | & & & \\ & CH_3 & & & \end{array}$$

55. Name each alkyne.

a. $CH_3-C\equiv C-CH_3$

b.
$$\begin{array}{lll} & CH_3 & \\ & | & \\ CH_3-C\equiv C- & C & -CH_2-CH_3 \\ & | & \\ & CH_3 & \end{array}$$

c.
$$\begin{array}{lll} CH\equiv C- & CH & -CH_2-CH_2-CH_3 \\ & | & \\ & CH-CH_3 & \\ & | & \\ & CH_3 & \end{array}$$

d.
$$\begin{array}{lllll} & & & CH_3 & \\ & & & | & \\ CH_3- & CH & -C\equiv C- & CH & -CH_2 \\ & | & & & \ \ | \\ & CH_2 & & & \ \ CH_2 \\ & | & & & \ \ | \\ & CH_3 & & & \ \ CH_3 \end{array}$$

56. Name each alkyne.

a.
$$\begin{array}{lll} CH\equiv C- & CH & -CH_3 \\ & | & \\ & CH_3 & \end{array}$$

b.
$$\begin{array}{lllll} & & & CH_3 & \\ & & & | & \\ CH_3-C\equiv C- & CH & - & CH & -CH_2-CH_3 \\ & | & & & \\ & CH_3 & & & \end{array}$$

c.
$$\begin{array}{lll} & CH_3 & \\ & | & \\ CH\equiv C- & C & -CH_2-CH_3 \\ & | & \\ & CH_2 & \\ & | & \\ & CH_3 & \end{array}$$

d.
$$\begin{array}{lllll} & & & CH_3 & \\ & & & | & \\ CH_3-C\equiv C- & CH & - & C & -CH_3 \\ & | & & | & \\ & CH_2 & & CH_3 & \\ & | & & & \\ & CH_3 & & & \end{array}$$

57. Draw the correct structure for each compound.

a. 4-octyne
b. 3-nonene
c. 3,3-dimethyl-1-pentyne
d. 5-ethyl-3,6-dimethyl-2-heptene

58. Draw the correct structure for each compound.

a. 2-hexene
b. 1-heptyne
c. 4,4-dimethyl-2-hexene
d. 3-ethyl-4-methyl-2-pentene

59. List the products of each alkene addition reaction.

a. $CH_3-CH=CH-CH_3 + Cl_2 \longrightarrow$

b.
$$\begin{array}{lll} CH_3- & CH & -CH=CH-CH_3 + HBr \longrightarrow \\ & | & \\ & CH_3 & \end{array}$$

c. $CH_3-CH_2-CH=CH-CH_3 + Br_2 \longrightarrow$

d.
$$\begin{array}{lllll} & & & CH_3 & \\ & & & | & \\ CH_3- & CH & -CH= & C & -CH_3 + HCl \longrightarrow \\ & | & & & \\ & CH_3 & & & \end{array}$$

60. What are the products of each alkene addition reaction?

a.
$$\begin{array}{lll} CH_3- & CH & -CH=CH_2 + Br_2 \longrightarrow \\ & | & \\ & CH_3 & \end{array}$$

b. $CH_2=CH-CH_3 + Cl_2 \longrightarrow$

c.
$$\begin{array}{lll} & CH_3 & \\ & | & \\ CH_3- & C & -CH=CH_2 + HCl \longrightarrow \\ & | & \\ & CH_3 & \end{array}$$

d.
$$\begin{array}{lllll} & & & CH_3 & \\ & & & | & \\ CH_3- & CH & -CH= & C & -CH_3 + HBr \longrightarrow \\ & | & & & \\ & CH_2-CH_3 & & & \end{array}$$

61. Complete each hydrogenation reaction.

a. $CH_2=CH-CH_3 + H_2 \xrightarrow{\text{catalyst}}$

b.
$$\begin{array}{lll} CH_3- & CH & -CH=CH_2 + H_2 \xrightarrow{\text{catalyst}} \\ & | & \\ & CH_3 & \end{array}$$

c.
$$\begin{array}{lllll} CH_3- & CH & - & C & =CH_2 + H_2 \xrightarrow{\text{catalyst}} \\ & | & & | & \\ & CH_3 & & CH_3 & \end{array}$$

62. Complete each hydrogenation reaction.

a. $CH_3-CH_2-CH=CH_2 + H_2 \xrightarrow{\text{catalyst}}$

b. 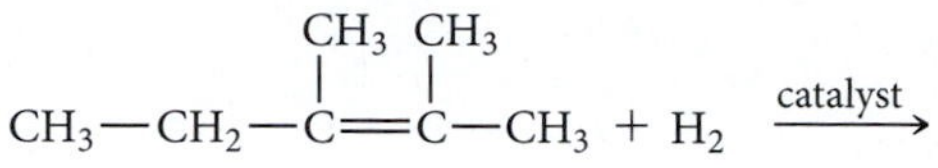 $CH_3-CH_2-\underset{}{\overset{CH_3}{\overset{|}{C}}}=\overset{CH_3}{\overset{|}{C}}-CH_3 + H_2 \xrightarrow{\text{catalyst}}$

c. $CH_3-CH_2-\underset{CH_3}{\underset{|}{C}}=CH_2 + H_2 \xrightarrow{\text{catalyst}}$

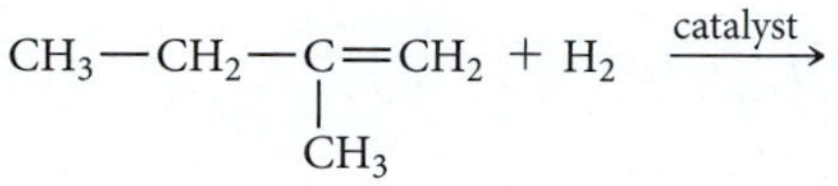

Aromatic Hydrocarbons

63. Name each monosubstituted benzene.

a. 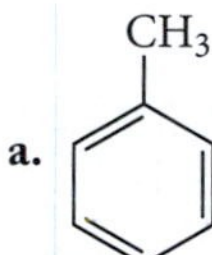

b. 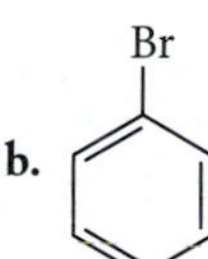

c. 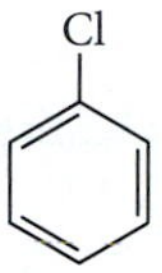 Cl (chlorine on a benzene ring)

64. Name each monosubstituted benzene.

a. 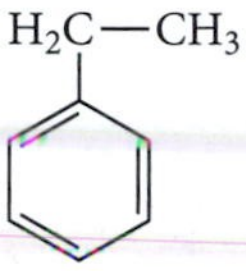 H_2C-CH_3 (on a benzene ring)

b. 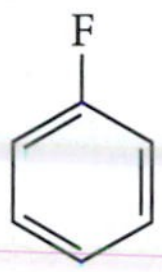 F (on a benzene ring)

c. 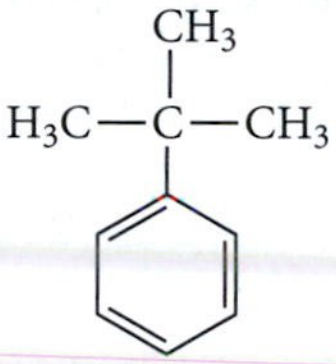 $H_3C-\overset{CH_3}{\overset{|}{C}}-CH_3$ (on a benzene ring)

65. Name each compound in which the benzene ring is best treated as a substituent.

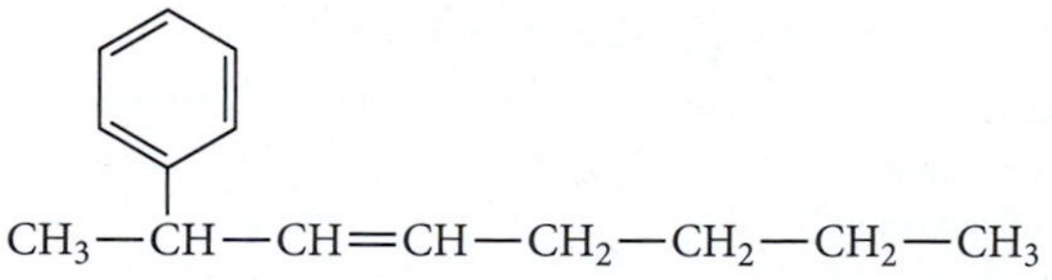

a. 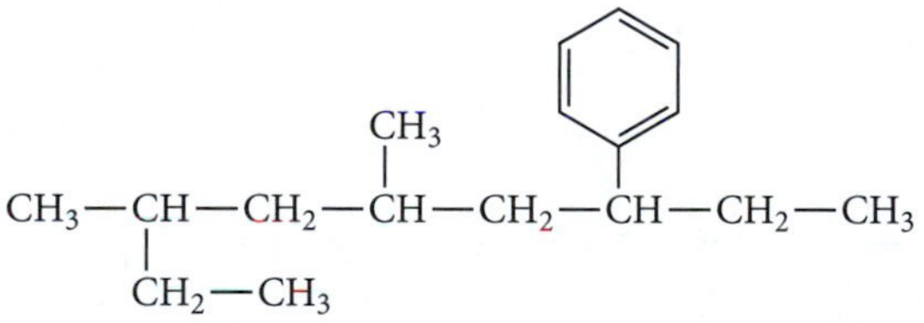 $CH_3-\underset{CH_2-CH_3}{\underset{|}{CH}}-CH_2-\overset{CH_3}{\overset{|}{CH}}-CH_2-\overset{C_6H_5}{\overset{|}{CH}}-CH_2-CH_3$

b. $CH_3-\overset{C_6H_5}{\overset{|}{CH}}-CH=CH-CH_2-CH_2-CH_2-CH_3$

c. $CH_3-C\equiv C-\underset{CH_3}{\underset{|}{CH}}-\underset{CH_3}{\underset{|}{CH}}-\overset{C_6H_5}{\overset{|}{CH}}-CH_2-CH_3$

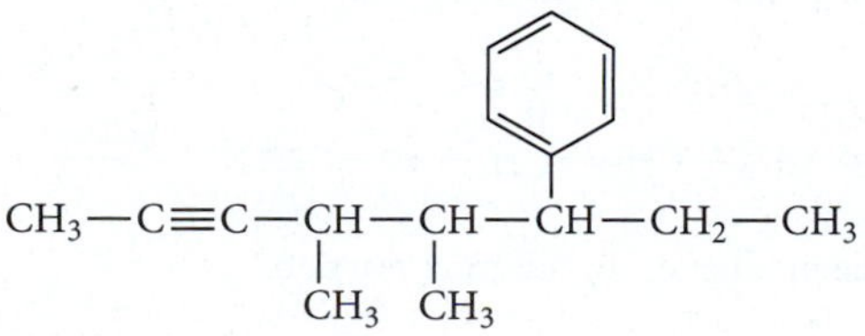

66. Name each compound in which the benzene ring is best treated as a substituent.

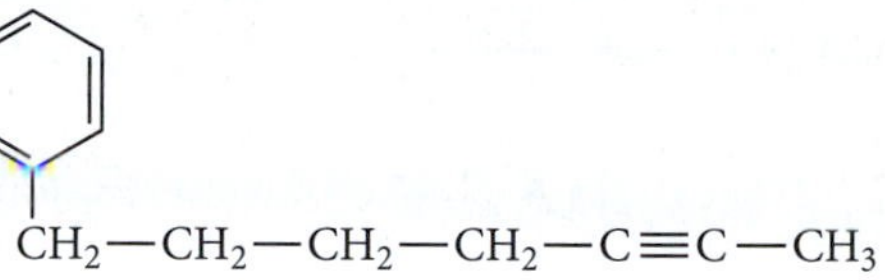

a. 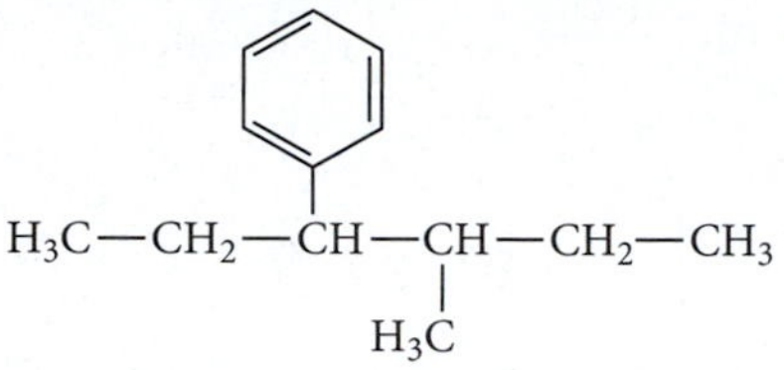 $H_3C-CH_2-\overset{C_6H_5}{\overset{|}{CH}}-\underset{H_3C}{\underset{|}{CH}}-CH_2-CH_3$

b. $\overset{C_6H_5}{\overset{|}{CH_2}}-CH_2-CH_2-CH_2-C\equiv C-CH_3$

c. $CH_3-\underset{CH_3}{\underset{|}{CH}}-\underset{CH_3}{\underset{|}{CH}}-\overset{CH_3}{\overset{|}{C}}=CH-\overset{C_6H_5}{\overset{|}{CH}}-CH_2-CH_3$

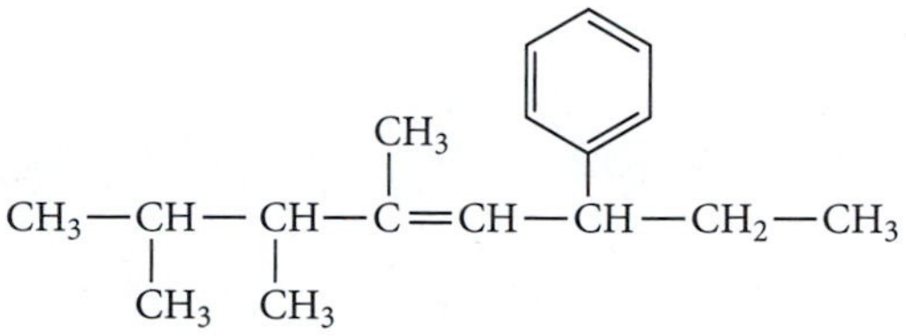

67. Name each disubstituted benzene.

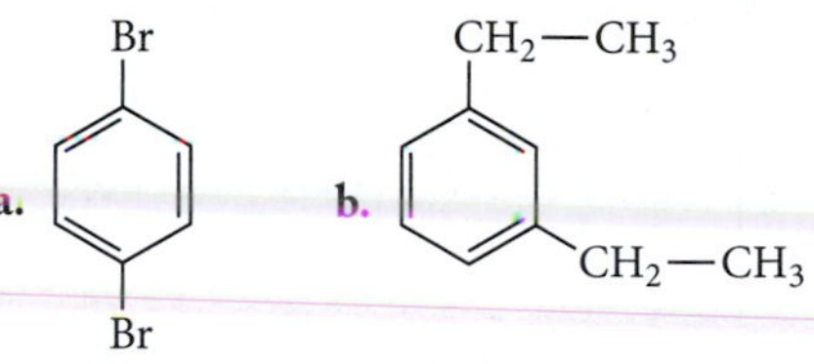

c. 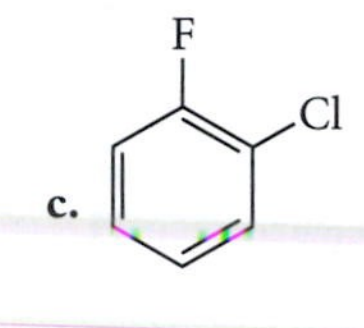

68. Name each disubstituted benzene.

a. 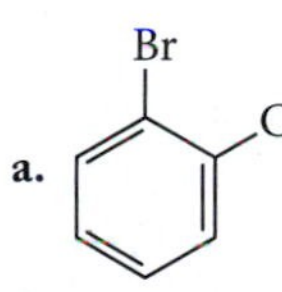

b. 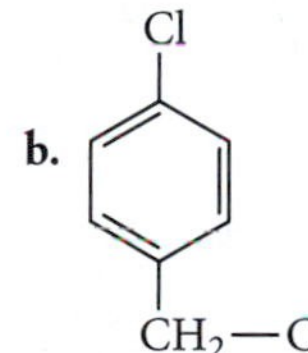

c. 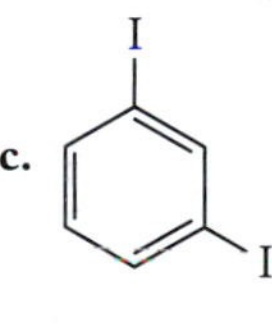

69. Draw the structure for each compound.

a. isopropylbenzene
b. *meta*-dibromobenzene
c. 1-chloro-4-methylbenzene

70. Draw the structure for each compound.

a. ethylbenzene
b. 1-iodo-2-methylbenzene
c. *para*-diethylbenzene

71. What are the products of each aromatic substitution reaction?

a. 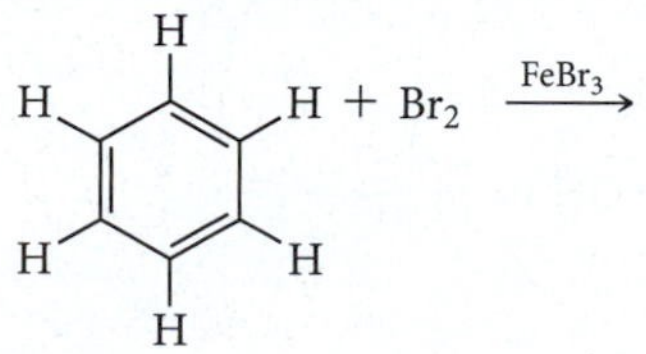 C_6H_6 (benzene ring with six H) $+ Br_2 \xrightarrow{FeBr_3}$

b. 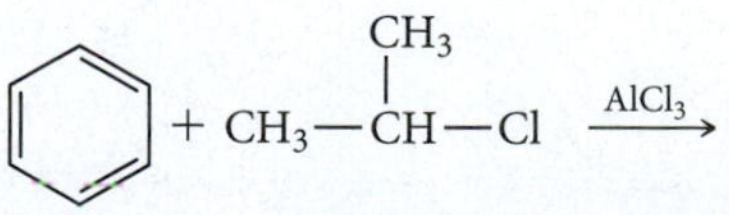 (benzene) $+ CH_3-\overset{CH_3}{\overset{|}{CH}}-Cl \xrightarrow{AlCl_3}$

72. What are the products of each aromatic substitution reaction?

a. 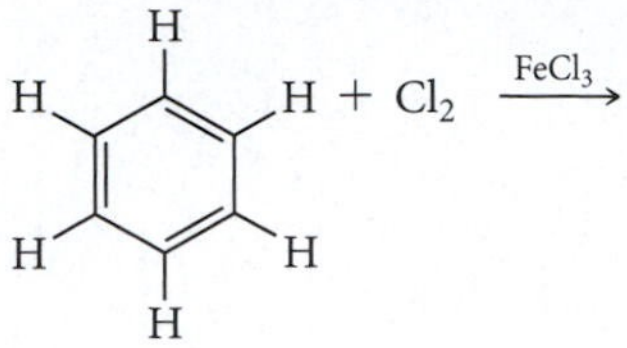 Benzene (C_6H_6, each ring carbon bearing H) $+ Cl_2 \xrightarrow{FeCl_3}$

b. 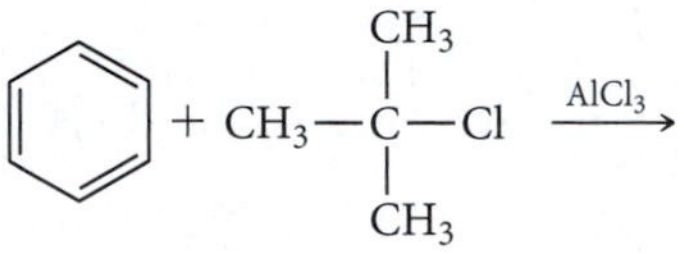 Benzene $+ CH_3-C(CH_3)_2-Cl \xrightarrow{AlCl_3}$

Alcohols

73. Name each alcohol.

a. $CH_3-CH_2-CH_2-OH$

b. 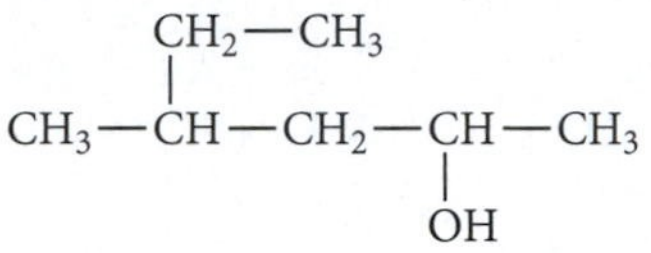 $CH_3-CH(CH_2-CH_3)-CH_2-CH(OH)-CH_3$

c. 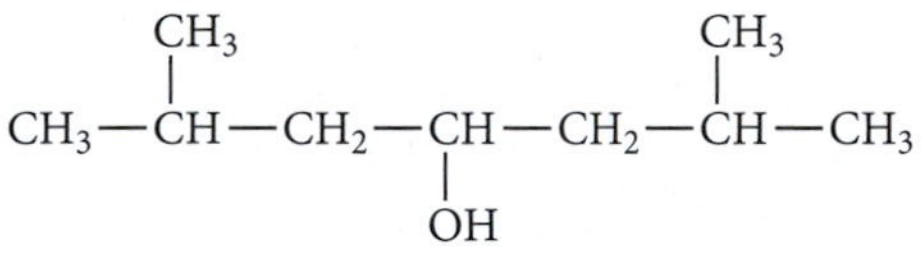 $CH_3-CH(CH_3)-CH_2-CH(OH)-CH_2-CH(CH_3)-CH_3$

d. 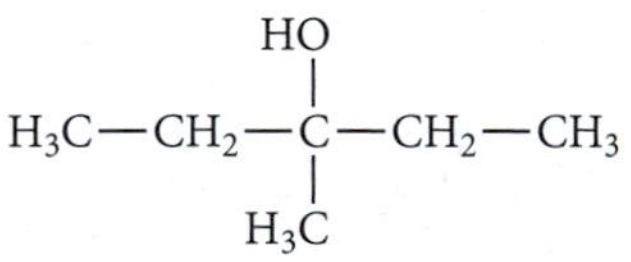 $H_3C-CH_2-C(HO)(H_3C)-CH_2-CH_3$

74. Draw the structure for each alcohol.

a. 2-butanol
b. 2-methyl-1-propanol
c. 3-ethyl-1-hexanol
d. 2-methyl-3-pentanol

75. List the products of each alcohol reaction.

a. $CH_3-CH_2-CH_2-OH + HBr \longrightarrow$

b. 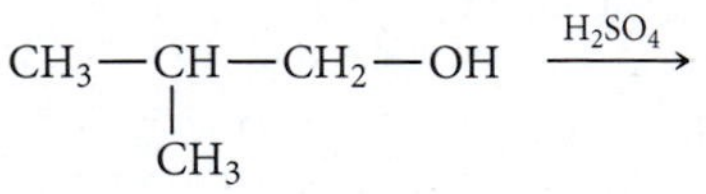 $CH_3-CH(CH_3)-CH_2-OH \xrightarrow{H_2SO_4}$

c. $CH_3-CH_2-OH + Na \longrightarrow$

d. 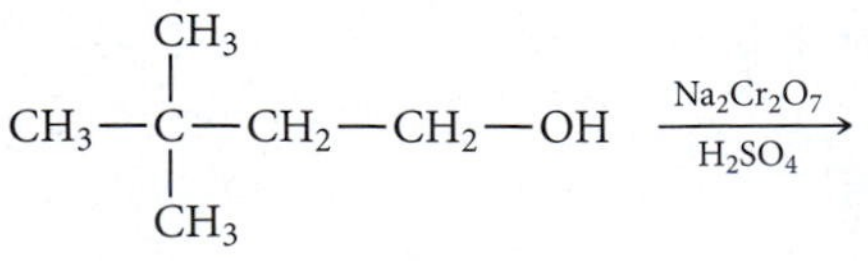 $CH_3-C(CH_3)_2-CH_2-CH_2-OH \xrightarrow[H_2SO_4]{Na_2Cr_2O_7}$

76. List the products of each alcohol reaction.

a. 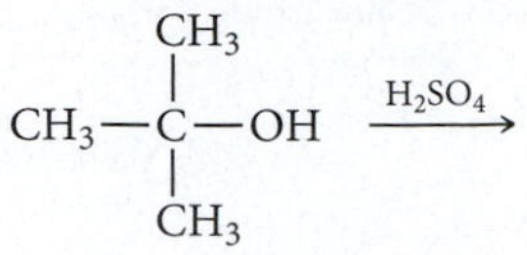 $CH_3-C(CH_3)_2-OH \xrightarrow{H_2SO_4}$

b. 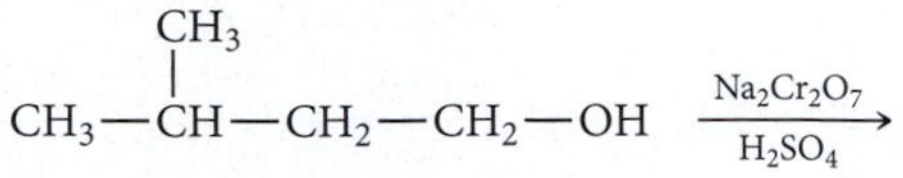 $CH_3-CH(CH_3)-CH_2-CH_2-OH \xrightarrow[H_2SO_4]{Na_2Cr_2O_7}$

c. $CH_3-CH_2-OH + HCl \longrightarrow$

d. 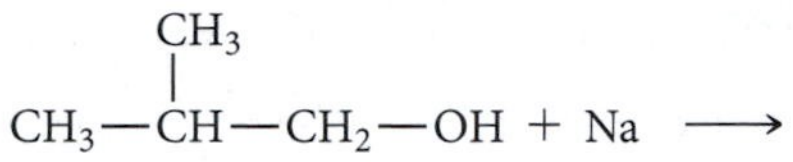 $CH_3-CH(CH_3)-CH_2-OH + Na \longrightarrow$

Aldehydes and Ketones

77. Name each aldehyde or ketone.

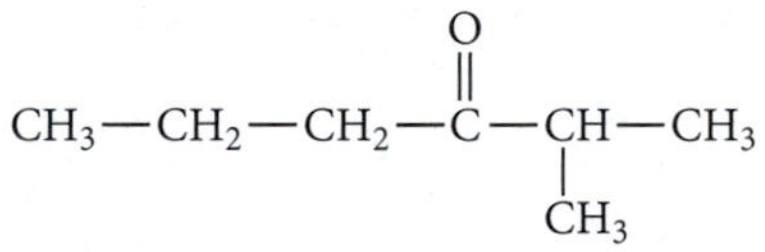

a. 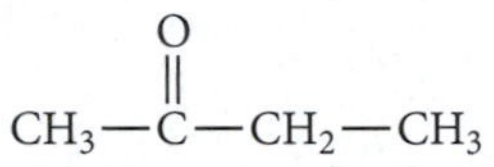 $CH_3-C(=O)-CH_2-CH_3$

b. $CH_3-CH_2-CH_2-C(=O)-CH(CH_3)-CH_3$

c. $CH_3-C(CH_3)_2-CH_2-CH(CH_3)-CH_2-C(=O)-H$

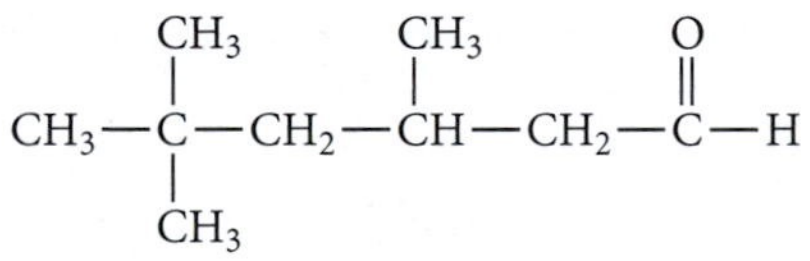

d. 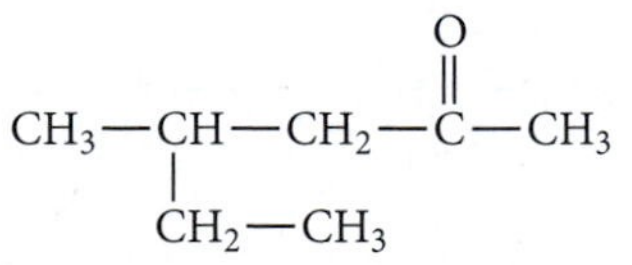 $CH_3-CH(CH_2-CH_3)-CH_2-C(=O)-CH_3$

78. Draw the structure of each aldehyde or ketone.

a. hexanal
b. 2-pentanone
c. 2-methylbutanal
d. 4-heptanone

79. Determine the product of the addition reaction.

$CH_3-CH_2-CH_2-CH(=O) + H-C\equiv N \xrightarrow{NaCN}$

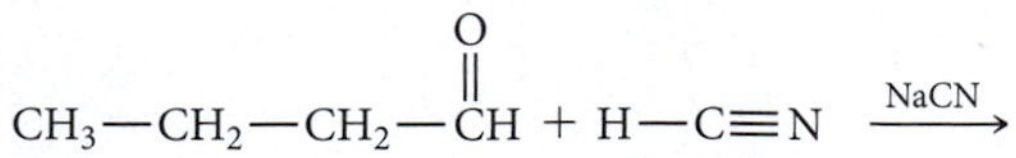

80. Determine the product of the addition reaction.

$CH_3-C(=O)-CH_2-CH_3 + HCN \xrightarrow{NaCN}$

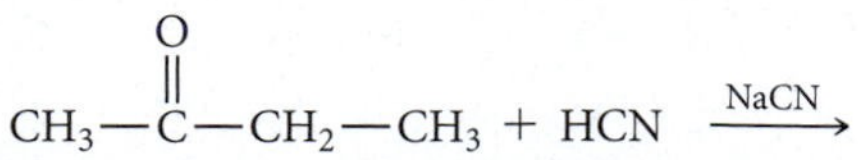

Carboxylic Acids and Esters

81. Name each carboxylic acid or ester.

a. 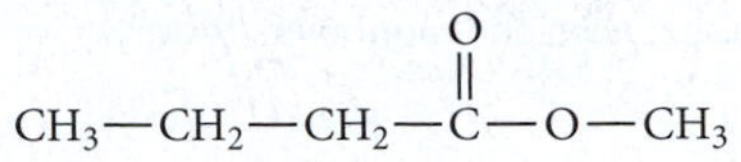 $CH_3-CH_2-CH_2-\overset{\overset{\displaystyle O}{\|}}{C}-O-CH_3$

b. $CH_3-CH_2-\overset{\overset{\displaystyle O}{\|}}{C}-OH$

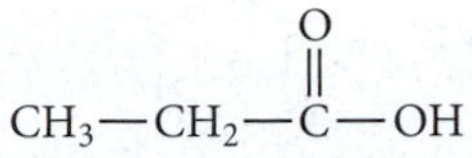

c. $CH_3-\underset{\underset{\displaystyle CH_3}{|}}{CH}-CH_2-CH_2-CH_2-\overset{\overset{\displaystyle O}{\|}}{C}-OH$

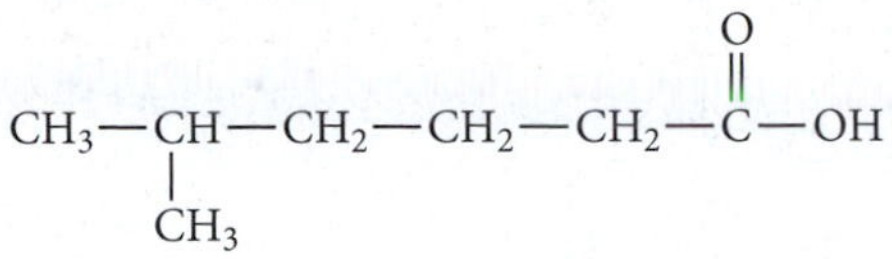

d. 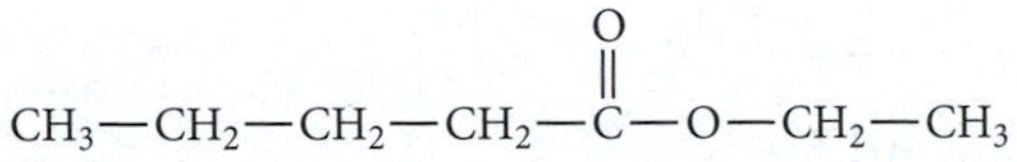 $CH_3-CH_2-CH_2-CH_2-\overset{\overset{\displaystyle O}{\|}}{C}-O-CH_2-CH_3$

82. Draw the structure of each carboxylic acid or ester.

a. pentanoic acid
b. methyl hexanoate
c. 3-ethylheptanoic acid
d. butyl ethanoate

83. Determine the products of each carboxylic acid reaction.

a. 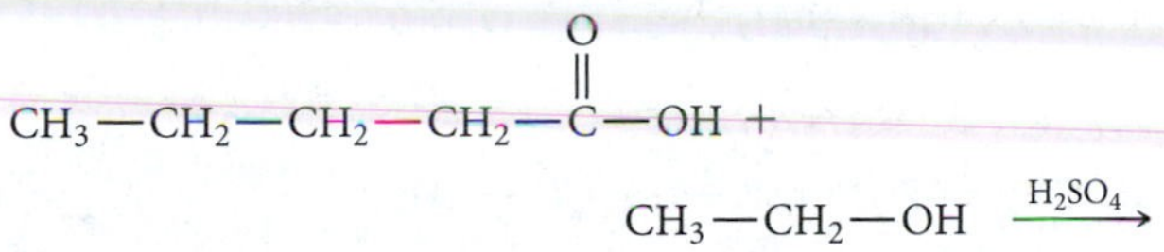 $CH_3-CH_2-CH_2-CH_2-\overset{\overset{\displaystyle O}{\|}}{C}-OH +$
$CH_3-CH_2-OH \xrightarrow{H_2SO_4}$

b. 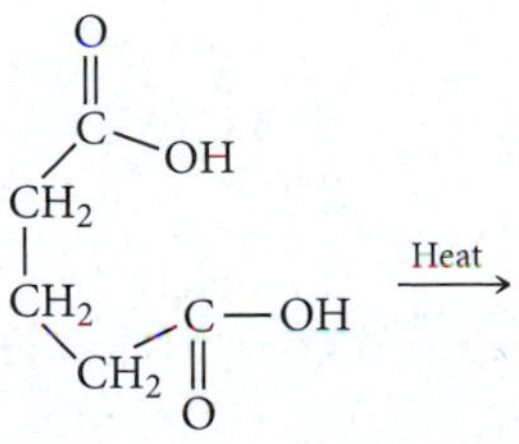 $HO-\overset{\overset{\displaystyle O}{\|}}{C}-CH_2-CH_2-CH_2-\overset{\overset{\displaystyle O}{\|}}{C}-OH \xrightarrow{Heat}$

84. Determine the products of each carboxylic acid reaction.

a. 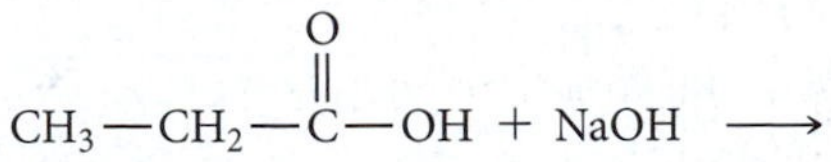 $CH_3-CH_2-\overset{\overset{\displaystyle O}{\|}}{C}-OH + NaOH \longrightarrow$

b. 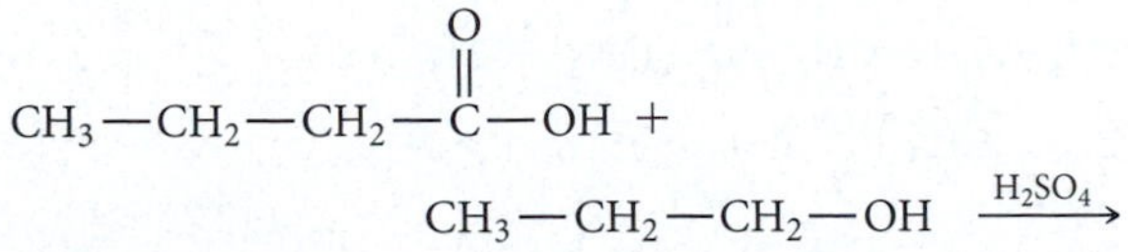 $CH_3-CH_2-CH_2-\overset{\overset{\displaystyle O}{\|}}{C}-OH +$
$CH_3-CH_2-CH_2-OH \xrightarrow{H_2SO_4}$

Ethers

85. Name each ether.

a. $CH_3-CH_2-CH_2-O-CH_2-CH_3$
b. $CH_3-CH_2-CH_2-CH_2-CH_2-O-CH_2-CH_3$
c. $CH_3-CH_2-CH_2-O-CH_2-CH_2-CH_3$
d. $CH_3-CH_2-O-CH_2-CH_2-CH_2-CH_3$

86. Draw the structure for each ether.

a. ethyl propyl ether
b. dibutyl ether
c. methyl hexyl ether
d. dipentyl ether

Amines

87. Name each amine.

a. 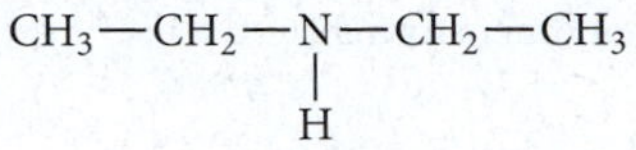 $CH_3-CH_2-\underset{\underset{\displaystyle H}{|}}{N}-CH_2-CH_3$

b. 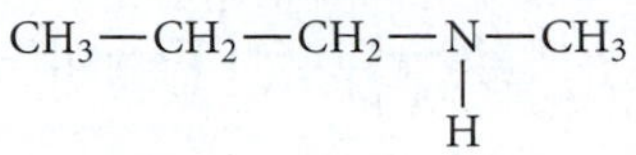 $CH_3-CH_2-CH_2-\underset{\underset{\displaystyle H}{|}}{N}-CH_3$

c. 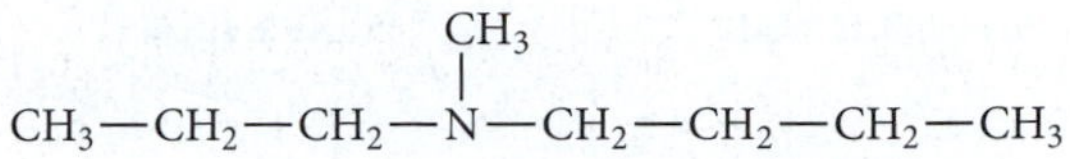 $CH_3-CH_2-CH_2-\overset{\overset{\displaystyle CH_3}{|}}{N}-CH_2-CH_2-CH_2-CH_3$

88. Draw the structure for each amine.

a. isopropylamine
b. triethylamine
c. butylethylamine

89. Classify each amine reaction as acid–base or condensation and list its products.

a. $CH_3NHCH_3 + HCl \longrightarrow$
b. $CH_3CH_2NH_2 + CH_3CH_2COOH \longrightarrow$
c. $CH_3NH_2 + H_2SO_4 \longrightarrow$

90. List the products of each amine reaction.

a. $N(CH_2CH_3)_3 + HNO_3 \longrightarrow$

b. 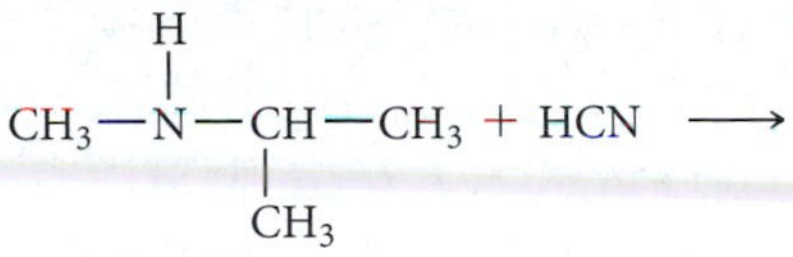 $CH_3-\overset{\overset{\displaystyle H}{|}}{N}-\underset{\underset{\displaystyle CH_3}{|}}{CH}-CH_3 + HCN \longrightarrow$

c. 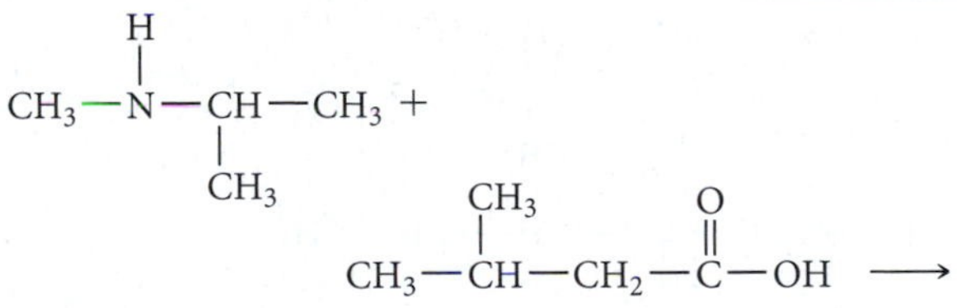 $CH_3-\overset{\overset{\displaystyle H}{|}}{N}-\underset{\underset{\displaystyle CH_3}{|}}{CH}-CH_3 +$
$CH_3-\overset{\overset{\displaystyle CH_3}{|}}{CH}-CH_2-\overset{\overset{\displaystyle O}{\|}}{C}-OH \longrightarrow$

Polymers

91. Teflon is an addition polymer formed from the monomer shown here. Draw the structure of the polymer.

$F_2C=CF_2$

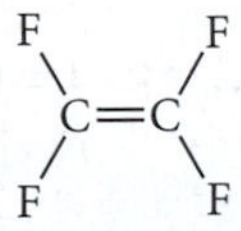

92. Saran, the polymer used to make clingy plastic food wrap, is an addition polymer formed from two monomers—vinylidene chloride and vinyl chloride. Draw the structure of the polymer. (*Hint:* The monomers alternate.)

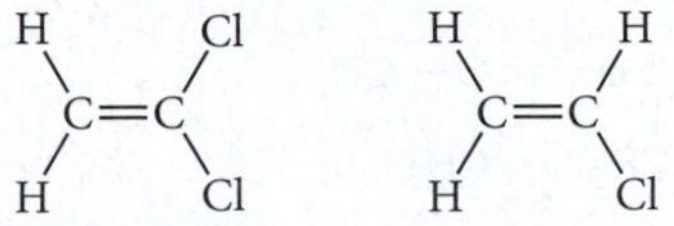

$H_2C=CCl_2$ — Vinylidene chloride

$H_2C=CHCl$ — Vinyl chloride

93. One kind of polyester is a condensation copolymer formed from terephthalic acid and ethylene glycol. Draw the structure of the dimer and circle the ester functional group. [*Hint:* Water (circled) is eliminated when the bond between the monomers forms.]

Terephthalic acid: $HO-C(=O)-C_6H_4-C(=O)-OH$ Ethylene glycol: $HO-CH_2-CH_2-OH$

Terephthalic acid   Ethylene glycol

94. Nomex, a condensation copolymer useful to firefighters because of its flame-resistant properties, forms from isophthalic acid and *m*-aminoaniline. Draw the structure of the dimer. (*Hint:* Water is eliminated when the bond between the monomers forms.)

Isophthalic acid: $HO-C(=O)-C_6H_4-C(=O)-OH$ *m*-Aminoaniline: $H_2N-C_6H_4-NH_2$

Isophthalic acid   *m*-Aminoaniline

CUMULATIVE PROBLEMS

95. Identify each organic compound as an alkane, alkene, alkyne, aromatic hydrocarbon, alcohol, ether, aldehyde, ketone, carboxylic acid, ester, or amine, and provide a name for the compound.

a. $H_3C-HC(CH_3)-CH_2-C(=O)-O-CH_3$

b. $CH_3-CH_2-CH(CH_3)-CH_2-O-CH_2-CH_3$

c. Benzene ring bearing H_2C-CH_3 and CH_3 (meta)

d. $CH_3-C{\equiv}C-CH(CH_3)-CH(CH_2-CH_3)-CH_2-CH_3$

e. $H_3C-CH_2-CH_2-CH(=O)$

f. $H_3C-CH(H_3C)-CH_2(OH)$

c. $CH_3-CH_2-CH(CH_3)-CH_2-C(=O)-OH$

d. $CH_3-CH(CH_3)-N(H)-CH_2-CH_2-CH_2-CH_3$

e. $CH_3-CH(CH_2-CH_3)-CH_2-CH(CH_2-OH)-CH_3$

f. $CH_3-CH_2-CH_2-C(=O)-CH(CH_3)-CH_3$

96. Identify each organic compound as an alkane, alkene, alkyne, aromatic hydrocarbon, alcohol, ether, aldehyde, ketone, carboxylic acid, ester, or amine, and provide a name for the compound.

a. $H_3C-HC(CH_3)-C(H_3C)=C(CH_3)-CH_3$

b. $CH_3-C(CH_3)_2-CH_2-CH(CH_3)-CH_2-CH_3$

97. Name each compound.

a. $CH_3-CH_2-CH(CH_3)-CH_2-CH(HC(CH_3)-CH_2-CH_3)-CH_2-CH_2-CH_2-CH_3$

b. $CH_3-CH(CH_3)-CH_2-C(=O)-CH_2-CH_3$

c. $CH_3-CH(CH_3)-CH(OH)-CH_3$

d. $CH_3-CH(CH_3)-CH(CH_2-CH_3)-CH(CH_3)-C{\equiv}C-H$

98. Name each compound.

a.
```
                   CH3 CH3
                    |   |
CH3—CH=CH—C—CH—CH2—CH3
                    |
                   CH2
                    |
                   CH3
```

b. (benzene ring with Br and CH_2—CH_3 substituents)
```
Br
[benzene ring]
      CH2—CH3
```

c.
```
            CH3         O        CH3
             |          ||        |
CH3—CH2—CH—CH2—C—O—CH—CH3
```

d.
```
                      O
                      ||
CH3—CH—CH2—CH
      |
     CH3
```

99. Determine whether each pair of structures are isomers or the same molecule drawn in two different ways.

a.
```
              O
              ||
CH3—CH2—C—O—CH3

       O
       ||
CH3—C—O—CH2—CH3
```

b. (two benzene rings, each with two I substituents)
```
I                I
[ring]—I     [ring]
                    —I
```

c.
```
CH3—HC—CH—CH3
      |    |
     CH3 CH3

         CH3
          |
CH3—HC—CH—CH3
      |
     CH3
```

100. Determine whether each pair of structures are isomers or the same molecule drawn two different ways.

a.
```
CH3—CH2—CH—CH2—CH3
             |
            CH3

            CH3
             |
            CH2
             |
CH3—CH2—CH—CH3
```

b.
```
       CH3
        |
CH3—CH—CH2—O—CH2—CH3

                                    CH3
                                     |
CH3—CH2—CH2—O—CH2—CH2
```

c.
```
       CH3         O
        |          ||
CH3—CH—CH2—C—CH2—CH3

                   O
                   ||
CH3—CH—CH2—C—CH2
       |            |
      CH3          CH3
```

101. What minimum amount of hydrogen gas, in grams, is required to completely hydrogenate 15.5 kg of 2-butene?

102. How many kilograms of CO_2 does the complete combustion of 3.8 kg of *n*-octane produce?

103. Classify each organic reaction as combustion, alkane substitution, alkene addition or hydrogenation, aromatic substitution, or alcohol substitution, elimination, or oxidation.

a. $2\ CH_3CH{=}CH_2 + 9\ O_2 \longrightarrow 6\ CO_2 + 6\ H_2O$

b. $CH_3CH_2CH_3 + Cl_2 \longrightarrow CH_3CH_2CH_2Cl + HCl$

c.
```
                                   H2SO4
CH3—CH2—CH—CH2—OH   ——→
              |
             CH3           CH3—CH2—C=CH2
                                      |
                                     CH3
```

d. (benzene) $+ I_2 \xrightarrow{FeI_3}$ (iodobenzene, ring with I) $+ HI$

104. Determine the products of each reaction.

a.
```
CH3—CH2—C=CH2 + H2 ——→
              |
             CH3
```

b. $CH_3—CH_2—CH_2—CH_2—OH + HCl \longrightarrow$

c.
```
             CH3         O
              |          ||
CH3—CH2—CH—CH2—C—OH +
                               CH3CH2OH ——→
```

d.
```
              H
              |
CH3—CH2—N—CH2—CH3 + HCl ——→
```

105. Draw the structure that corresponds to each name and indicate which structures can exist as stereoisomers.

a. 3-methyl-1-pentene
b. 3,5-dimethyl-2-hexene
c. 3-propyl-2-hexene

106. Identify the two compounds that display stereoisomerism and draw their structures.

a. 3-methyl-3-pentanol
b. 2-methyl-2-pentanol
c. 3-methyl-2-pentanol
d. 2-methyl-3-pentanol
e. 2,4-dimethyl-3-pentanol

107. There are 11 structures (ignoring stereoisomerism) with the formula C_4H_8O that have no carbon branches. Draw the structures and identify the functional groups in each.

108. There are eight structures with the formula C_3H_7NO in which the O is part of a carbonyl group. Draw the structures and identify the functional groups in each.

109. Explain why carboxylic acids are much stronger acids than alcohols.

110. The hydrogen at C-1 of 1-butyne is much more acidic than the hydrogens at C-1 in 1-butene. Explain.

CHALLENGE PROBLEMS

111. Determine the one or two steps it takes to get from the starting material to the product using reactions found in this chapter.

a.
$$\begin{array}{l} H_2C-C(=O)-OH \\ \;| \\ H_2C-C(=O)-OH \end{array} \longrightarrow \begin{array}{l} H_2C-C(=O) \\ \;|\qquad\quad \diagdown \\ \;|\qquad\qquad O \\ \;|\qquad\quad \diagup \\ H_2C-C(=O) \end{array}$$

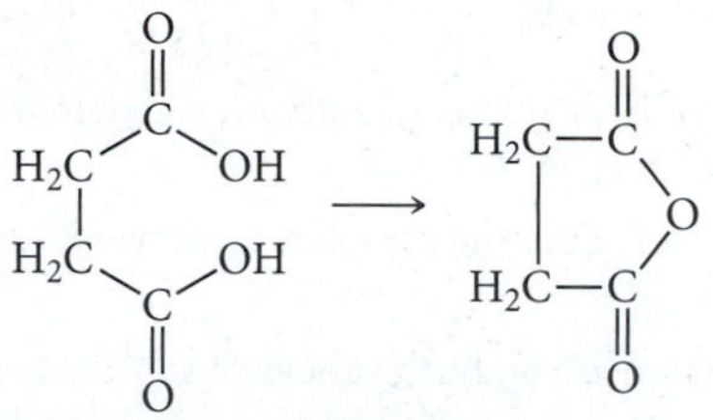

b.
$$\begin{array}{c} \qquad\qquad\qquad\quad CH_2-OH \\ \qquad\qquad\qquad\quad | \\ CH_3-CH-CH_2-CH-CH_3 \\ | \qquad\qquad\qquad \\ CH_2-CH_3 \qquad\qquad \end{array} \longrightarrow$$

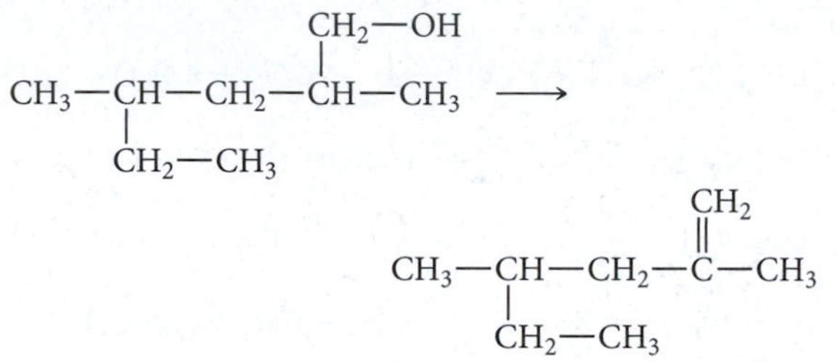

$$\begin{array}{c} \qquad\qquad\qquad\quad CH_2 \\ \qquad\qquad\qquad\quad \| \\ CH_3-CH-CH_2-C-CH_3 \\ | \qquad\qquad\qquad \\ CH_2-CH_3 \qquad\qquad \end{array}$$

c.
$$\begin{array}{c} CH_3-CH_2-C=CH_2 \\ \qquad\quad | \\ \qquad\quad CH_3 \end{array} \longrightarrow \begin{array}{c} \qquad\qquad Br \\ \qquad\qquad | \\ CH_3-CH_2-C-CH_3 \\ \qquad\qquad | \\ \qquad\qquad CH_3 \end{array}$$

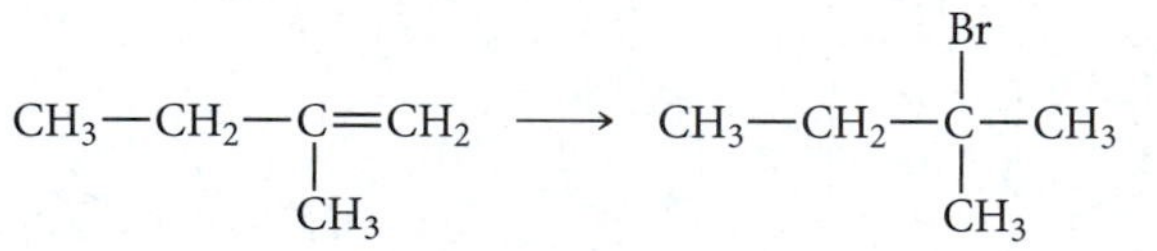

112. Given the following synthesis of ethyl 3-chloro-3-methylbutanoate, fill in the missing intermediates or reactants.

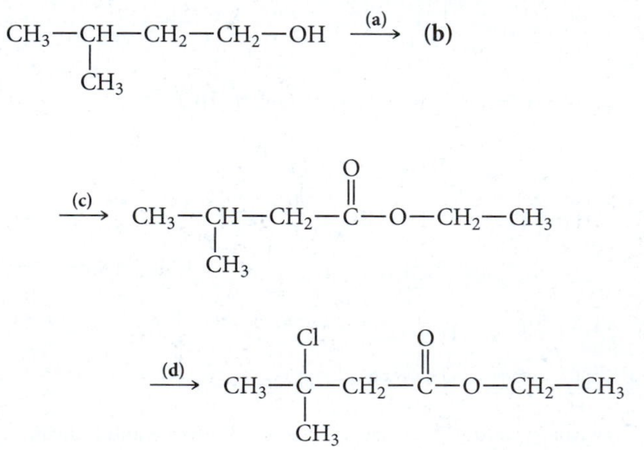

$$\begin{array}{l} CH_3-CH-CH_2-CH_2-OH \\ \qquad\;\; | \\ \qquad\;\; CH_3 \end{array} \xrightarrow{(a)} (b)$$

$$\xrightarrow{(c)} \begin{array}{l} \qquad\qquad\qquad\qquad\quad O \\ \qquad\qquad\qquad\qquad\quad \| \\ CH_3-CH-CH_2-C-O-CH_2-CH_3 \\ \qquad\;\; | \\ \qquad\;\; CH_3 \end{array}$$

$$\xrightarrow{(d)} \begin{array}{l} \qquad\quad\; Cl \qquad\qquad O \\ \qquad\quad\; | \qquad\qquad\;\; \| \\ CH_3-C-CH_2-C-O-CH_2-CH_3 \\ \qquad\quad\; | \\ \qquad\quad\; CH_3 \end{array}$$

113. For the chlorination of propane, the two isomers shown here are possible.

$$CH_3CH_2CH_3 + Cl_2 \longrightarrow$$

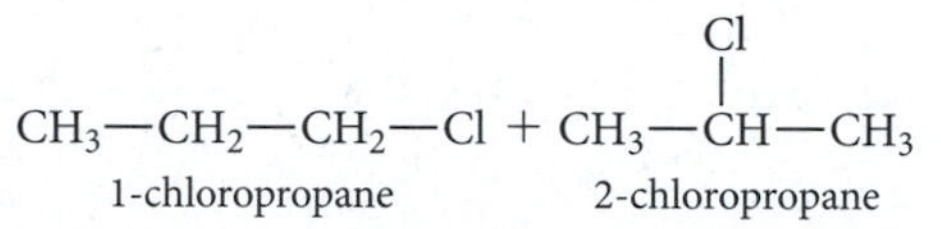

$$\underset{\text{1-chloropropane}}{CH_3-CH_2-CH_2-Cl} + \underset{\text{2-chloropropane}}{\begin{array}{c} Cl \\ | \\ CH_3-CH-CH_3 \end{array}}$$

Propane has six hydrogen atoms on terminal carbon atoms—called primary (1°) hydrogen atoms—and two hydrogen atoms on the interior carbon atom—called secondary (2°) hydrogen atoms.

a. If the two different types of hydrogen atoms were equally reactive, what ratio of 1-chloropropane to 2-chloropropane would we expect as monochlorination products?

b. The result of a reaction yields 55% 2-chloropropane and 45% 1-chloropropane. What can we conclude about the relative reactivity of the two different kinds of hydrogen atoms? Determine a ratio of the reactivity of one type of hydrogen atom to the other.

114. There are two isomers of C_4H_{10}. Suppose that each isomer is treated with Cl_2 and the products that have the composition $C_4H_8Cl_2$ are isolated. Find the number of different products that form from each of the original C_4H_{10} compounds. Do not consider optical isomerism.

115. Identify the compounds formed in the previous problem that are chiral.

116. Nitromethane has the formula CH_3NO_2, with the N bonded to the C and without O—O bonds. Draw its two most important contributing structures.

a. What is the hybridization of the C, and how many hybrid orbitals are in the molecule?

b. What is the shortest bond?

c. Between which two atoms is the strongest bond found?

d. Predict whether the HCH bond angles are greater or less than 109.5° and justify your prediction.

117. Free radical fluorination of methane is uncontrollably violent, and free radical iodination of methane is a very poor reaction. Explain these observations in light of bond energies.

118. There are two compounds with the formula C_3H_6, one of which does not have a multiple bond. Draw its structure and explain why it is much less stable than the isomer with the double bond.

119. Consider molecules that have two carbons and two chlorines. Draw the structures of three of these with no dipole moment and two with a dipole moment.

CONCEPTUAL PROBLEMS

120. Pick the more oxidized structure from each pair.

a. $CH_3-\overset{\overset{\displaystyle O}{\|}}{C}H$ or CH_3-CH_2-OH

b. CH_3-CH_2-OH or CH_3-CH_3

c. $CH_3-CH_2-\overset{\overset{\displaystyle O}{\|}}{C}H$ or $CH_3-CH_2-\overset{\overset{\displaystyle O}{\|}}{C}-OH$

121. Draw the structure and name a compound with the formula C_8H_{18} that forms only one product with the formula $C_8H_{17}Br$ when it is treated with Br_2.

122. Determine whether each structure is chiral.

a. $H\overset{\overset{\displaystyle Cl}{|}}{\underset{\underset{\displaystyle CH_3}{|}}{C}}-CH_3$

b. $CH_3-\underset{\underset{\displaystyle Cl}{|}}{CH}-\underset{\underset{\displaystyle CH_3}{|}}{CH}-CH_3$

c. CH_3-CH_2-OH

d. $H\overset{\overset{\displaystyle Cl}{|}}{\underset{\underset{\displaystyle CH_3}{|}}{C}}-\underset{\underset{\displaystyle Br}{|}}{CH_2}$

ANSWERS TO CONCEPTUAL CONNECTIONS

Cc 22.1 **(d)** The others are simply the same structure drawn in slightly different ways.

Cc 22.2 **(b)** This structure is the only one that contains a carbon atom (the one on the left) with four different substituent groups attached (a Br atom, a Cl atom, an H atom, and a CH_3 group).

Cc 22.3 **(c, d, a, b)** Oxidation includes the gain of oxygen or the loss of hydrogen.

CHAPTER 23

"Chemistry must become the astronomy of the molecular world."

—Alfred Werner (1866–1919)

The color of ruby is caused by a splitting of the d-orbital energy levels in Cr^{3+} by the host crystal.

Transition Metals and Coordination Compounds

23.1 The Colors of Rubies and Emeralds 955

23.2 Properties of Transition Metals 956

23.3 Coordination Compounds 960

23.4 Structure and Isomerization 965

23.5 Bonding in Coordination Compounds 970

23.6 Applications of Coordination Compounds 975

Key Learning Outcomes 980

IN THIS CHAPTER, WE EXAMINE the chemistry of the transition metals and an important class of their compounds called coordination compounds. We discuss how coordination compounds form all of the types of isomers that we have studied so far, as well as some new types. In our examination of the transition metals, we draw on much of what we learned in Chapters 3 and 4 about electronic structure and periodicity. We also briefly revisit valence bond theory to explain bonding in coordination compounds, but we quickly shift to a different theory—crystal field theory—that better explains many of the properties of these compounds. Transition metals and coordination compounds are important, not only because of their interesting chemistry, but because of their numerous applications. Coordination compounds are the basis for a number of therapeutic drugs, chemical sensors, and coloring agents. In addition, many biological molecules contain transition metals that bond in ways that are similar to coordination compounds.

23.1 The Colors of Rubies and Emeralds

Rubies are deep red and emeralds are brilliant green, yet the color of both gemstones is caused by the same ion, Cr^{3+}. The difference lies in the crystal that hosts the ion. Rubies are crystals of aluminum oxide (Al_2O_3) in which about 1% of the Al^{3+} ions are replaced by Cr^{3+} ions. Emeralds, by contrast, are crystals of beryllium aluminum silicate [$Be_3Al_2(SiO_3)_6$] in which a similar percentage of the Al^{3+} ions is replaced by Cr^{3+}. The embedded Cr^{3+} ion is red in the aluminum oxide crystal but green in the beryllium aluminum silicate crystal. Why?

The answer to this question lies in the effect that the host crystal has on the energies of the atomic orbitals in Cr^{3+}. Atoms in the crystal create a field around the ion—sometimes called the *crystal field*—that splits the five normally degenerate *d* orbitals into two or more levels. The color of the gemstone is caused by electron transitions between these levels. In rubies, the crystal field is stronger (and the

▲ **Ruby and Emerald** The Cr^{3+} ion is responsible for both the red color of ruby and the green color of emerald.

corresponding splitting of the *d* orbitals greater) than it is in emeralds. Recall from Chapter 3 that the color of a substance depends on the colors *absorbed* by that substance, which in turn depends on the energy differences between the orbitals involved in the absorption. The greater splitting in ruby results in a greater energy difference between the *d* orbitals of Cr^{3+} and consequently, the absorption of a different color of light than in emerald. Here we see the theme of this book repeated yet again—the structure that surrounds the Cr^{3+} ion determines its color.

The splitting of the *d* orbitals in transition metal ions embedded within host crystals also causes the colors of several other gemstones. For example, the red in garnet, which has $Mg_3Al_2(SiO_4)_3$ as a host crystal, and the yellow-green of peridot, which has Mg_2SiO_4 as a host crystal, are both caused by electron transitions between *d* orbitals in Fe^{2+}. Similarly, the blue in turquoise, which has $[Al_6(PO_4)_4(OH)_8 \cdot 4\,H_2O]^{2-}$ as a host crystal, is caused by transitions between the *d* orbitals in Cu^{2+}.

In this chapter, we examine the properties of the transition metals and their ions more closely. We also examine the properties of coordination compounds in some detail. We first discussed this common type of transition metal compound in Chapter 18 (see Section 18.7). In a coordination compound, bonds to a central metal ion split the *d* orbitals much as they are split in the crystals of gemstones. The theory that explains these splittings and the corresponding colors is **crystal field theory**, which we also explore in this chapter.

▲ **Garnet, Peridot, and Turquoise** The red in garnet and the yellow-green of peridot are both caused by Fe^{2+}. The blue of turquoise is caused by Cu^{2+}.

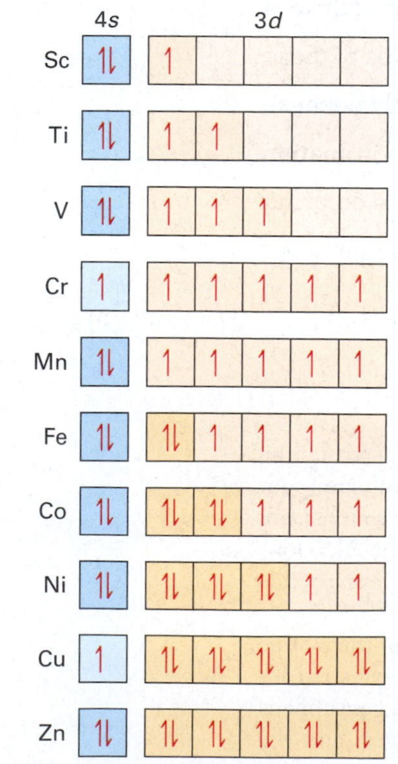

TABLE 23.1 First-Row Transition Metal Orbital Occupancy

	4s	3d				
Sc	↑↓	↑				
Ti	↑↓	↑	↑			
V	↑↓	↑	↑	↑		
Cr	↑	↑	↑	↑	↑	↑
Mn	↑↓	↑	↑	↑	↑	↑
Fe	↑↓	↑↓	↑	↑	↑	↑
Co	↑↓	↑↓	↑↓	↑	↑	↑
Ni	↑↓	↑↓	↑↓	↑↓	↑	↑
Cu	↑	↑↓	↑↓	↑↓	↑↓	↑↓
Zn	↑↓	↑↓	↑↓	↑↓	↑↓	↑↓

23.2 Properties of Transition Metals

Transition metals have more uniform properties than the main-group elements. For example, almost all transition metals have moderate to high densities, good electrical conductivity, high melting points, and moderate to extreme hardness. Their similar properties are related to their similar electron configurations: They all have electrons in *d* orbitals that can be involved in metallic bonding. In spite of their similarities, however, each element is also unique, and they exhibit a wide variety of chemical behavior. Before we examine some of the periodic properties of the transition metals, let's review the electron configurations of these elements, first discussed in Chapter 4.

Electron Configurations

Recall from Section 4.4 that, as we move to the right across a row of transition elements, electrons are added to $(n-1)d$ orbitals (where n is the row number in the periodic table and also the quantum number of the highest occupied principal level). For example, as we move across the fourth-period transition metals, electrons are added to the $3d$ orbitals, as shown in Table 23.1.

In general, the ground state electron configuration for the first two rows of transition elements is [noble gas] $ns^2(n-1)d^x$ and for the third and fourth rows is [noble gas] $ns^2(n-2)f^{14}(n-1)d^x$, where x ranges from 1 to 10. Recall from Section 4.4, however, that because the ns and $(n-1)d$ sublevels are close in energy, many exceptions occur. For example, in the first transition series of the *d* block, the outer configuration is $4s^23d^x$ with two exceptions: Cr is $4s^13d^5$ and Cu is $4s^13d^{10}$. This behavior is related to the closely spaced $3d$ and $4s$ energy levels and the stability associated with a half-filled or completely filled d sublevel.

Recall from Section 4.7 that the transition metals form ions by losing electrons from the ns orbital *before* losing electrons from the $(n-1)d$ orbitals. For example, Fe^{2+} has an electron configuration of [Ar] $3d^6$, because it has lost both of the $4s$ electrons to form the 2+ charge. Examples 23.1 and 23.2 review the steps in writing electron configurations for transition metals and their ions.

PROCEDURE FOR Writing Electron Configurations	EXAMPLE 23.1 Writing Electron Configurations for Transition Metals	EXAMPLE 23.2 Writing Electron Configurations for Transition Metals
	Write the ground state electron configuration for Zr.	Write the ground state electron configuration for Co^{3+}.
Identify the noble gas that precedes the element and write it in square brackets.	**SOLUTION** [Kr]	**SOLUTION** [Ar]
Count down the periods to determine the outer principal quantum level—this is the quantum level for the s orbital. Subtract one to obtain the quantum level for the d orbital. If the element is in the third or fourth transition series, include $(n-2)f^{14}$ electrons in the configuration.	Zr is in the fifth period, so the orbitals used are [Kr] $5s4d$	Co is in the fourth period, so the orbitals used are [Ar] $4s3d$
Count across the row to see how many electrons are in the neutral atom and fill the orbitals accordingly.	Zr has four more electrons than Kr. [Kr] $5s^24d^2$	Co has nine more electrons than Ar. [Ar] $4s^23d^7$
For an cation, remove the required number of electrons, first from the s and then from the d orbitals.		Co^{3+} has lost three electrons relative to the Co atom. [Ar] $4s^03d^6$ or [Ar] $3d^6$
	FOR PRACTICE 23.1 Write the ground state electron configuration for Os.	**FOR PRACTICE 23.2** Write the ground state electron configuration for Nb^{2+}.

Atomic Size

As we discussed in Section 4.6, for main-group elements, the size of atoms decreases across a period and increases down a column. For transition metals, however, there is little variation in size across a row (other than for the first two elements in each transition metal row, such as Sc and Ti in the first row), as shown in Figure 23.1 ▶. The reason for the difference is that, across a row of transition elements, the number of electrons in the outermost principal energy level (highest n value) is nearly constant. As another proton is added to the nucleus with each successive element, another electron is added as well, but the electron goes into an $n-1$ orbital. The number of outermost electrons thus stays the same, and the electrons experience a roughly constant effective nuclear charge as we move across the row, keeping the radii approximately constant.

Looking down a group, we see a small but expected increase in size from the first transition metal row to the second, but the size of elements in the third row is about the same as it is for those in the second row. This pattern is also different from that of the main-group elements, especially when we consider that in any given column, the third transition row has 32 more electrons than the second row. The third transition row elements are not larger because 14 of the 32 electrons are in a $(n-2)f$ sublevel,

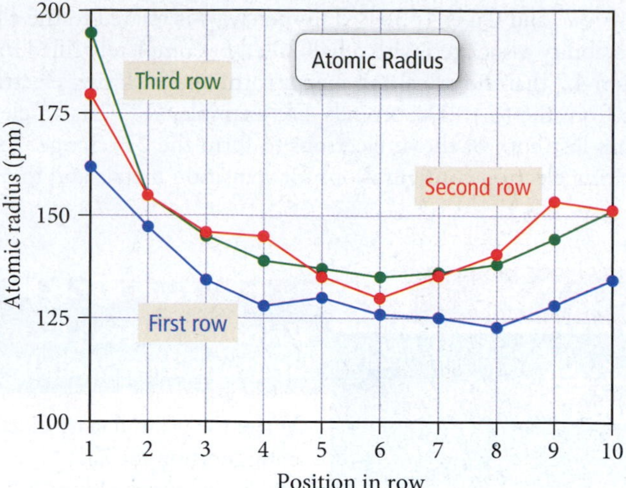

▶ **FIGURE 23.1 Trends in Atomic Radius** With the exception of a decrease in radius from the first to the second element, there is only a small variation in atomic radius across each transition metal row. There is a small and expected increase in radius from the first to the second transition metal row but virtually no difference in radius from the second to the third.

and while electrons in *f* orbitals are in lower principal quantum levels, they are not very effective at shielding the outer electrons from nuclear charge. Consequently, the outer electrons are held more tightly by the nucleus, offsetting the typical increase in size between the periods. This effect is called the **lanthanide contraction**.

23.1 Cc Conceptual Connection

Atomic Size

Which element has the larger atomic radius, Fe or W?

Ionization Energy

The first ionization energies of transition elements follow the expected main-group periodic trend and slowly increase across each transition metal row (Figure 23.2 ▼), but the increase is smaller than for main-group elements. As we move down a group, we see that the third transition row generally has a higher ionization energy than do the first two rows, a trend counter to that observed in the main-group elements. In the transition elements the charge of the nucleus increases substantially from one row to the next, but there is only a small increase in atomic size between the first and second rows, and no increase in size is observed between the second and third rows. The outer electrons are therefore held more tightly in the third transition row than in the first two rows.

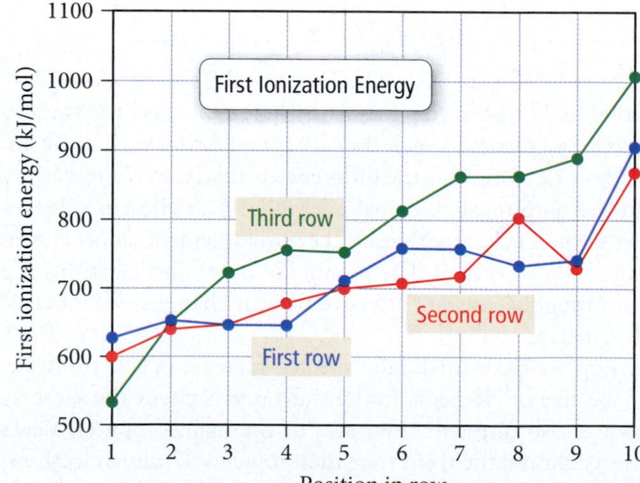

▶ **FIGURE 23.2 Trends in First Ionization Energy** First ionization energy generally increases across each transition metal row, following the main-group trend. However, in contrast to the main-group trend, the third transition metal row has a greater ionization energy than the first and second rows.

Electronegativity

The electronegativity values of the transition metals, like their ionization energies, follow the main-group trend and slowly increase across a row, as shown in Figure 23.3 ▼. The increase is smaller than the increase that occurs in the main-group elements, but we would expect that given the similarity in the sizes of the atoms across the transition metal rows. The trend in electronegativity values down a group (or column) is another example of the transition metals behaving differently from the main-group elements. The electronegativity values generally increase from the first transition metal row to the second, but there is no further increase for the third row. In the main-group elements, by contrast, we see a *decrease* in electronegativity down a group. The difference is again caused by the relatively small change in atomic size as we move down a column for the transition elements, accompanied by a large increase in nuclear charge. One of the heaviest metals, gold (Au), is also the most electronegative metal. Its electronegativity value (EN = 2.4) is even higher than that of some nonmetals (EN of P is 2.1), and compounds of an Au^- ion have been observed.

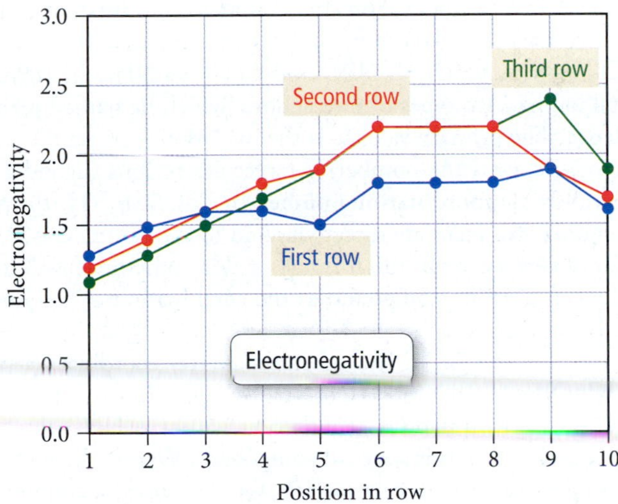

◀ **FIGURE 23.3 Trends in Electronegativity** The electronegativity of the transition metal elements generally increases across a row, following the main-group trend. However, in contrast to the main-group trend, electronegativity increases from the first transition metal row to the second. There is little electronegativity difference between the second and third transition rows.

Oxidation States

Unlike main-group metals, which tend to exhibit only one oxidation state, the transition metals often exhibit a variety of oxidation states (Figure 23.4 ▼). The highest oxidation state for a transition metal

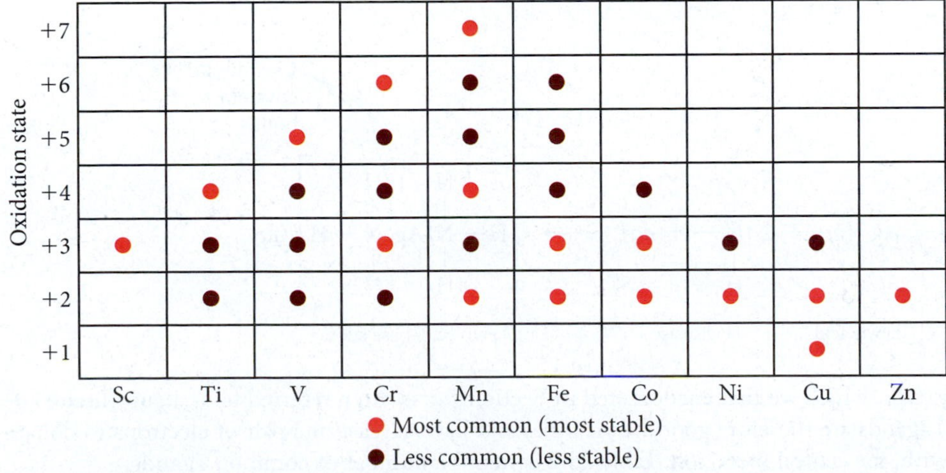

▲ **FIGURE 23.4 First-Row Transition Metal Oxidation States** The transition metals exhibit many more oxidation states than the main-group elements. These oxidation states range from +7 to +1.

In some rare cases, oxidation states higher than +7 have been observed.

Metals in high-oxidation states, such as +7, exist only when the metal is bound to a highly electronegative element such as oxygen; they do not exist as bare ions.

is +7 for manganese (Mn). The electron configuration of manganese in this oxidation state corresponds to the loss of all the electrons in the 4*s* and 3*d* orbitals, leaving a noble gas electron configuration ([Ar]). This is the same configuration we see for all of the highest oxidation states of the elements to the left of Mn. To the right of manganese, the oxidation states are all lower, mostly +2 or +3. A +2 oxidation state for a transition metal is not surprising, since 4*s* electrons are readily lost.

23.3 Coordination Compounds

We discussed at the end of Chapter 18 that transition metals tend to form *complex ions*. A **complex ion** contains a central metal ion bound to one or more *ligands*. A **ligand** is a Lewis base (or electron donor) that forms a bond with the metal. When a complex ion combines with one or more *counterions* (ions of opposite charge that are not acting as ligands), the resulting neutral compound is a **coordination compound**. The first coordination compounds were discovered in the early eighteenth century, but their nature was not understood until nearly 200 years later. Swiss chemist Alfred Werner (1866–1919) studied coordination compounds extensively—especially a series of cobalt(III) compounds with ammonia, whose formulas were then written as $CoCl_3 \cdot 6\,NH_3$, $CoCl_3 \cdot 5\,NH_3$, and $CoCl_3 \cdot 4\,NH_3$. In 1893, he proposed that the central metal ion has two types of interactions, which he named **primary valence** and **secondary valence**. The primary valence is the oxidation state on the central metal atom, and the secondary valence is the number of molecules or ions (ligands) directly bound to the metal atom, called the **coordination number**. In $CoCl_3 \cdot 6\,NH_3$ the primary valence is +3, and the ammonia molecules are directly bound to the central cobalt, giving a coordination number of 6. Today we write the formula of this compound as $[Co(NH_3)_6]Cl_3$ to better represent the coordination compound as the combination of a complex ion, $Co(NH_3)_6^{3+}$, and three Cl^- counterions.

We write the formulas of the other cobalt(III) coordination compounds studied by Werner as $[Co(NH_3)_5Cl]Cl_2$ and $[Co(NH_3)_4Cl_2]Cl$. In these two cases, the complex ions are $Co(NH_3)_5Cl^{2+}$ (with two Cl^- counterions) and $Co(NH_3)_4Cl_2^+$ (with one Cl^- counterion), respectively. With this series of compounds, Werner demonstrated that the Cl^- can replace NH_3 in the secondary valence. In other words, Cl^- can act as a counterion, or it can bond directly to the metal as a ligand and, therefore, be part of the complex ion.

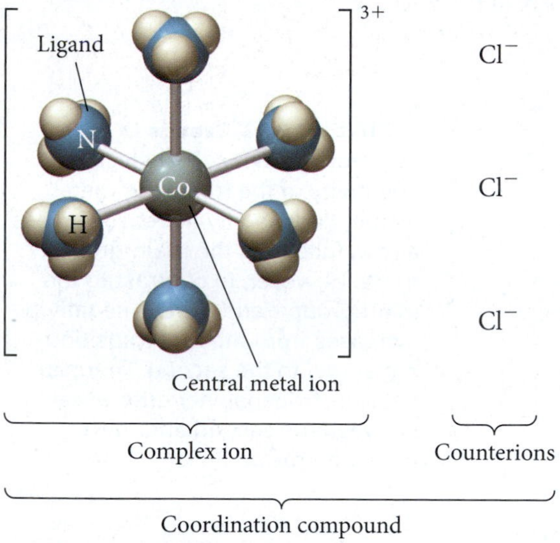

▲ **Coordination Compound** A coordination compound contains a complex ion and corresponding counterions. The complex ion contains a central metal atom coordinated to several ligands. The compound shown here is $[Co(NH_3)_6]Cl_3$.

Ligands

The complex ion itself contains the metal ion in the center and the ligands—which can be neutral molecules or ions—arranged around it. We can think of the metal–ligand complex as a Lewis acid–base adduct (see Section 17.11) because the bond is formed when the ligand donates a pair of electrons to an empty orbital on the metal. For example, consider the reaction between the silver ion and ammonia:

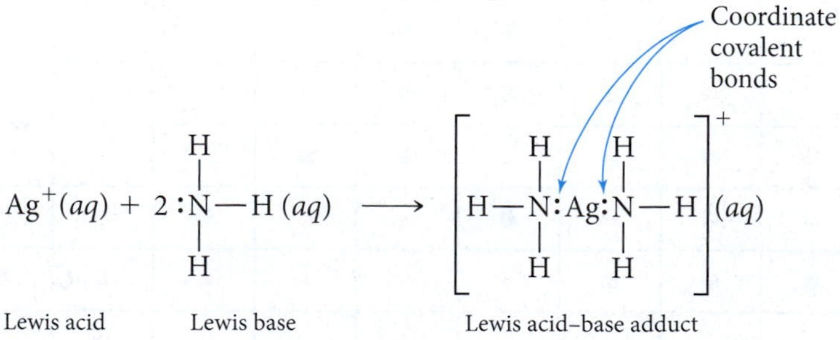

$$\underset{\text{Lewis acid}}{Ag^+(aq)} + \underset{\text{Lewis base}}{2\,:NH_3\,(aq)} \longrightarrow \underset{\text{Lewis acid–base adduct}}{[H_3N\!:\!Ag\!:\!NH_3]^+(aq)}$$

A bond of this type, which we first encountered in Section 7.2, is often referred to as a **coordinate covalent bond**. Ligands are therefore good Lewis bases and have at least one pair of electrons to donate to, and bond with, the central metal ion. Table 23.2 contains a number of common ligands.

Ligands that donate only one electron pair to the central metal are **monodentate**. Some ligands, however, have the ability to donate two pairs of electrons (from two different atoms) to the metal; these

TABLE 23.2 Common Ligands

Name	Lewis structure
Water	H—Ö—H
Ammonia	H—N̈—H, with H on N
Chloride ion	$[:\ddot{Cl}:]^-$
Carbon monoxide	:C≡O:
Cyanide ion	$[:C\equiv N:]^-$
Thiocyanate ion	$[:\ddot{S}=C=\ddot{N}:]^-$
Oxalate ion (ox)	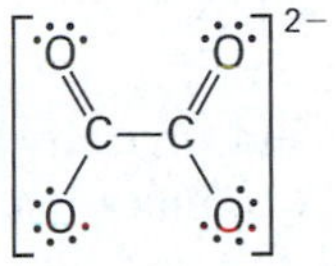 $[(^-O)(O=)C-C(=O)(O^-)]^{2-}$
Ethylenediamine (en)	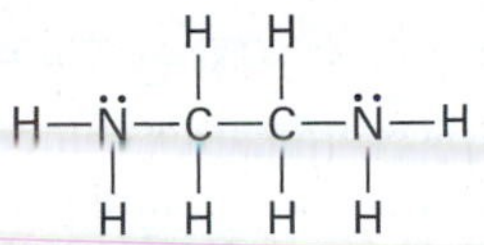 $H_2N-CH_2-CH_2-NH_2$ (H—N̈—C—C—N̈—H with H's)
Ethylenediaminetetraacetate (EDTA)	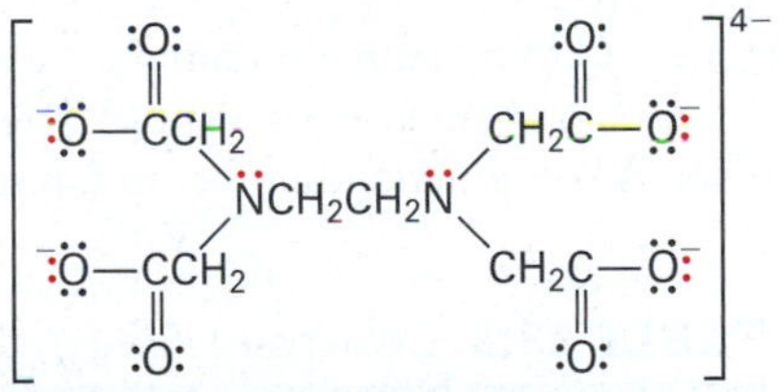 $[(^-OOCCH_2)_2NCH_2CH_2N(CH_2COO^-)_2]^{4-}$

are **bidentate**. Examples of bidentate ligands include the oxalate ion (abbreviated ox) and the ethylenediamine molecule (abbreviated en), which we show in the right margin. The ethylenediamine ligand is shown bonded to Co^{3+} in Figure 23.5(a) ▶ on p. 962.

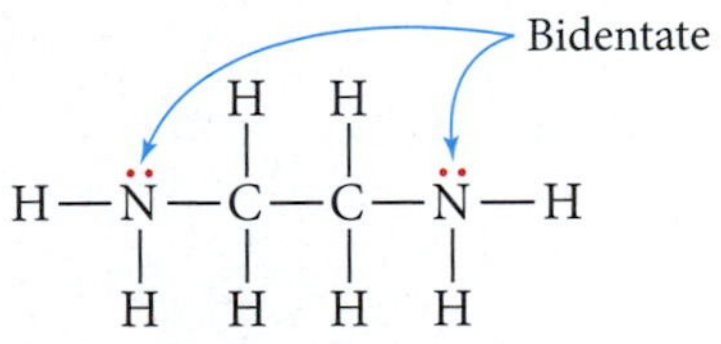

Bidentate

Ethylenediamine

Some ligands, called **polydentate** ligands, can donate even more than two electron pairs (from more than two atoms) to the metal. The most common polydentate ligand is the ethylenediaminetetraacetate ion ($EDTA^{4-}$) shown below.

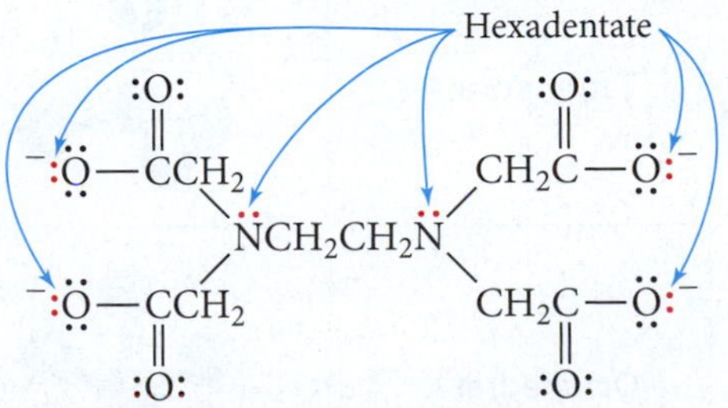

Hexadentate

$EDTA^{4-}$

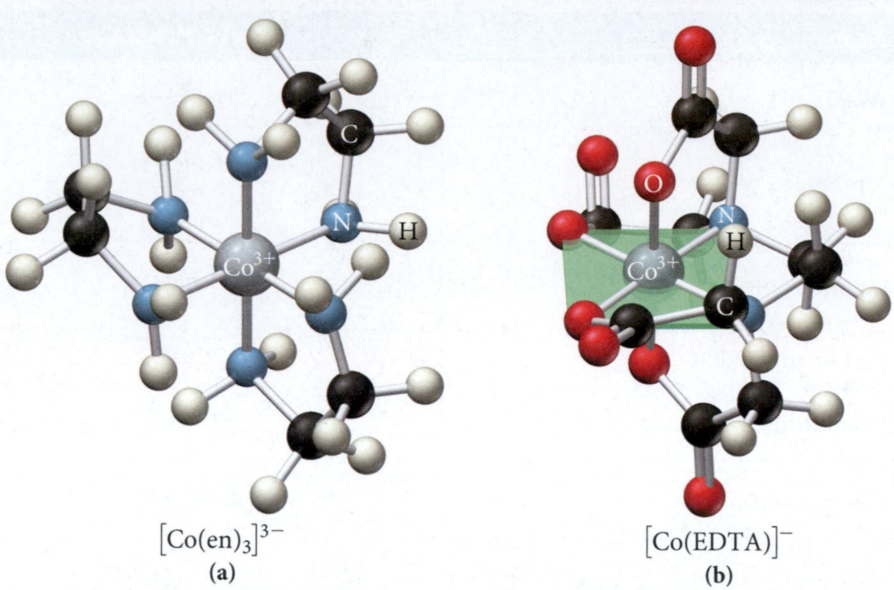

▶ **FIGURE 23.5 Bidentate and Polydentate Ligands Coordinated to Co(III)** **(a)** Ethylenediamine is a bidentate ligand; **(b)** EDTA is a hexadentate ligand.

The EDTA ligand wraps itself completely around the metal, donating up to six pairs of electrons (Figure 23.5b ▲). A complex ion that contains either a bidentate or polydentate ligand is a **chelate** (pronounced "key-late"), and the coordinating ligand is known as a **chelating agent**.

Coordination Numbers and Geometries

A survey of many coordination compounds shows that coordination numbers can vary from as low as 2 to as high as 12. The most common coordination numbers are 6, as occurs in $[Co(NH_3)_6]^{3+}$, and 4, as occurs in $[PdCl_4]^{2-}$. Coordination numbers greater than 6 are rarely observed for the first-row transition metals. Typically, only 1+ metal ions have a coordination number as low as 2, as occurs in $[Ag(NH_3)_2]^+$. Odd coordination numbers exist, but they are rare.

The common geometries of complex ions, shown in Table 23.3, depend in part on their coordination number. A coordination number of 2 results in a linear geometry, and a coordination number of 6

TABLE 23.3 Common Geometries of Complex Ions

Coordination Number	Shape	Model	Example
2	Linear		$[Ag(NH_3)_2]^+$
4	Square planar		$[PdCl_4]^{2-}$
4	Tetrahedral		$[Zn(NH_3)_4]^{2+}$
6	Octahedral		$[Fe(H_2O)_6]^{3+}$

results in an octahedral geometry. A coordination number of 4 can have either a tetrahedral geometry or a square planar geometry, depending on the number of *d* electrons in the metal ion. Metal ions with a d^8 electron configuration (such as $[PdCl_4]^{2-}$) exhibit square planar geometry, and metal ions with a d^{10} electron configuration (such as $[Zn(NH_3)_4]^{2+}$) exhibit tetrahedral geometry.

Naming Coordination Compounds

To name coordination compounds, we follow a series of general rules based on the system originally proposed by Werner. As with ionic compounds, the name of the cation goes before the name of the anion.

Guidelines for Naming Complex Ions	Examples
1. Name the ligands. • Name neutral ligands as molecules with the following notable exceptions:	$NH_2CH_2CH_2NH_2$ is ethylenediamine.
H_2O (aqua)	H_2O is aqua.
NH_3 (ammine)	
CO (carbonyl)	
• Name anionic ligands with the name of the ion plus an ending modified as follows:	
-ide becomes -o	Cl^- is chloro.
-ate becomes -ato	SO_4^{2-} is sulfato.
-ite becomes -ito	SO_3^{2-} is sulfito.
Table 23.4 lists the names of some common ligands.	
2. List the names of the ligands in alphabetical order before the name of the metal cation.	Ammine (NH_3) is listed before chloro (Cl^-), which is listed before nitrito (NO_2^-).
3. Use a prefix to indicate the number of ligands (when there is more than one of a particular type): *di-* (2), *tris-* (3), *tetra-* (4), *penta-* (5), or *hexa-* (6).	Trichloro indicates three Cl^- ligands. Tetraammine indicates four NH_3 ligands.
If the name of the ligand already contains a prefix, such as ethylenediamine, place parentheses around the ligand name and use *bis-* (2), *tris-* (3), or *tetrakis-* (4) to indicate the number. Prefixes do not affect the order in which you list the ligands.	Tris(ethylenediamine) indicates three ethylenediamine ligands.
4. Name the metal. **a.** When the complex ion is a cation, use the name of the metal followed by the oxidation state written with a Roman numeral.	In cations: Co^{3+} is cobalt(III). Pt^{2+} is platinum(II). Cu^+ is copper(I).
b. If the complex ion is an anion, drop the ending of the metal and add *–ate* followed by the oxidation state written with a Roman numeral. Some metals use the Latin root with the *ate* ending. Table 23.5 lists the names for some common metals in anionic complexes.	In anions: Co^{3+} is cobaltate(III). Pt^{2+} is platinate(II). Cu^+ is cuprate(I).
5. Write the entire name of the complex ion by listing the ligands first followed by the metal.	$[Pt(NH_3)_2Cl_4]^{2-}$ diamminetetrachloroplatinate(II). $[Co(NH_3)_6]^{3+}$ is hexaamminecobalt(III).

TABLE 23.4 Names and Formulas of Common Ligands

Ligand	Name in Complex Ion
Anions	
Bromide, Br^-	Bromo
Chloride, Cl^-	Chloro
Hydroxide, OH^-	Hydroxo
Cyanide, CN^-	Cyano
Nitrite, NO_2^-	Nitro
Oxalate, $C_2O_4^{2-}$ (ox)	Oxalato
Ethylenediaminetetraacetate ($EDTA^{4-}$)	Ethylenediaminetetraacetato
Neutral molecules	
Water, H_2O	Aqua
Ammonia, NH_3	Ammine
Carbon monoxide, CO	Carbonyl
Ethylenediamine (en)	Ethylenediamine

TABLE 23.5 Names of Common Metals When Found in Anionic Complex Ions

Metal	Name in Anionic Complex
Chromium	Chromate
Cobalt	Cobaltate
Copper	Cuprate
Gold	Aurate
Iron	Ferrate
Lead	Plumbate
Manganese	Manganate
Molybdenum	Molybdate
Nickel	Nickelate
Platinum	Platinate
Silver	Argentate
Tin	Stannate
Zinc	Zincate

When you write the *formula* of a complex ion, write the symbol for the metal first, followed by neutral molecules and then anions. If there is more than one anion or neutral molecule acting as a ligand, list them in alphabetical order based on the chemical symbol.

PROCEDURE FOR ▼ Naming Coordination Compounds | **EXAMPLE 23.3 Naming Coordination Compounds** | **EXAMPLE 23.4 Naming Coordination Compounds**

PROCEDURE FOR Naming Coordination Compounds	EXAMPLE 23.3	EXAMPLE 23.4
	Name the compound. $[Cr(H_2O)_5Cl]Cl_2$.	Name the compound. $K_3[Fe(CN)_6]$.
Identify the cation and anion and first name the simple ion (i.e., not the complex one).	**SOLUTION** $[Cr(H_2O)_5Cl]^{2+}$ is a complex cation. Cl^- is chloride.	**SOLUTION** K^+ is potassium. $[Fe(CN)_6]^{3-}$ is a complex anion.
Give each ligand a name and list them in alphabetical order.	H_2O is aqua. Cl^- is chloro.	CN^- is cyano.
Name the metal ion.	Cr^{3+} is chromium(III).	Fe^{3+} is ferrate(III) because the complex is anionic.
Name the complex ion by adding prefixes to indicate the number of each ligand followed by the name of each ligand followed by the name of the metal ion.	$[Cr(H_2O)_5Cl]^{2+}$ is pentaaquachloro-chromium(III).	$[Fe(CN)_6]^{3-}$ is hexacyanoferrate(III).
Name the compound by writing the name of the cation before the anion. The only space is between ion names.	$[Cr(H_2O)_5Cl]Cl_2$ is pentaaquachloro-chromium(III) chloride.	$K_3[Fe(CN)_6]$ is potassium hexacyano-ferrate(III).
	FOR PRACTICE 23.3 Name the following compound: $[Mn(CO)(NH_3)_5]SO_4$.	**FOR PRACTICE 23.4** Name the following compound: $Na_2[PtCl_4]$.

23.4 Structure and Isomerization

Isomerism is common in coordination compounds. We broadly divide the isomerism observed in coordination compounds into two categories, each with subcategories, as shown in Figure 23.6 ▼. In **structural isomers**, atoms are connected to one another in different ways, whereas in **stereoisomers**, atoms are connected in the same way but the ligands have a different spatial arrangement about the metal atom.

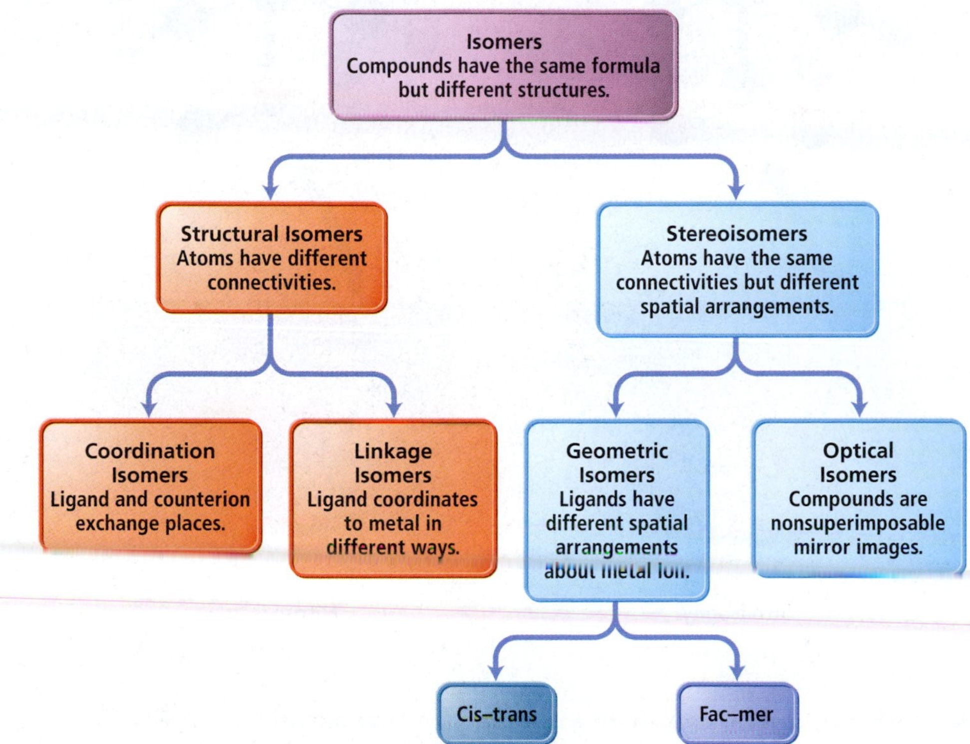

▲ **FIGURE 23.6 Types of Isomers**

Structural Isomerism

We subdivide the broad category of structural isomers into two types: coordination isomers and linkage isomers. **Coordination isomers** occur when a coordinated ligand exchanges places with the uncoordinated counterion. For example, two different compounds have the general formula $Co(NH_3)_5BrCl$. In one of them, the bromine coordinates to the metal and chloride is a counterion, pentaamminebromocobalt(II) chloride, $[Co(NH_3)_5Br]Cl$; in the other one, the chlorine coordinates to the metal and bromide is the counterion, pentaamminechlorocobalt(II) bromide, $[Co(NH_3)_5Cl]Br$.

Linkage isomers have ligands that coordinate to the metal in different orientations. For example, the nitrite ion (NO_2^-) has a lone pair on the N atom as well as lone pairs on the O atoms—either of the two atoms can form coordinate covalent bonds with the metal. When the nitrite ion coordinates through the N atom it is a *nitro* ligand and is represented as NO_2^-, but when it coordinates through the O atom, it is a *nitrito* ligand and is usually represented as ONO^-. We can see an example of linkage isomerization in the yellow-orange complex ion pentaamminenitrocobalt(III), $[Co(NH_3)_5NO_2]^{2+}$, which contrasts with the red-orange complex ion pentaamminenitritocobalt(III), $[Co(NH_3)_5ONO]^{2+}$, as shown in Figure 23.7 ▶. Table 23.6 lists other ligands capable of linkage isomerization.

▶ **FIGURE 23.7 Linkage Isomers** In $[Co(NH_3)_5NO_2]^{2+}$, the NO_2 ligand bonds to the central metal atom through the nitrogen atom. In $[Co(NH_3)_5ONO]^{2+}$, the NO_2 ligand bonds through the oxygen atoms. The different isomers have different colors.

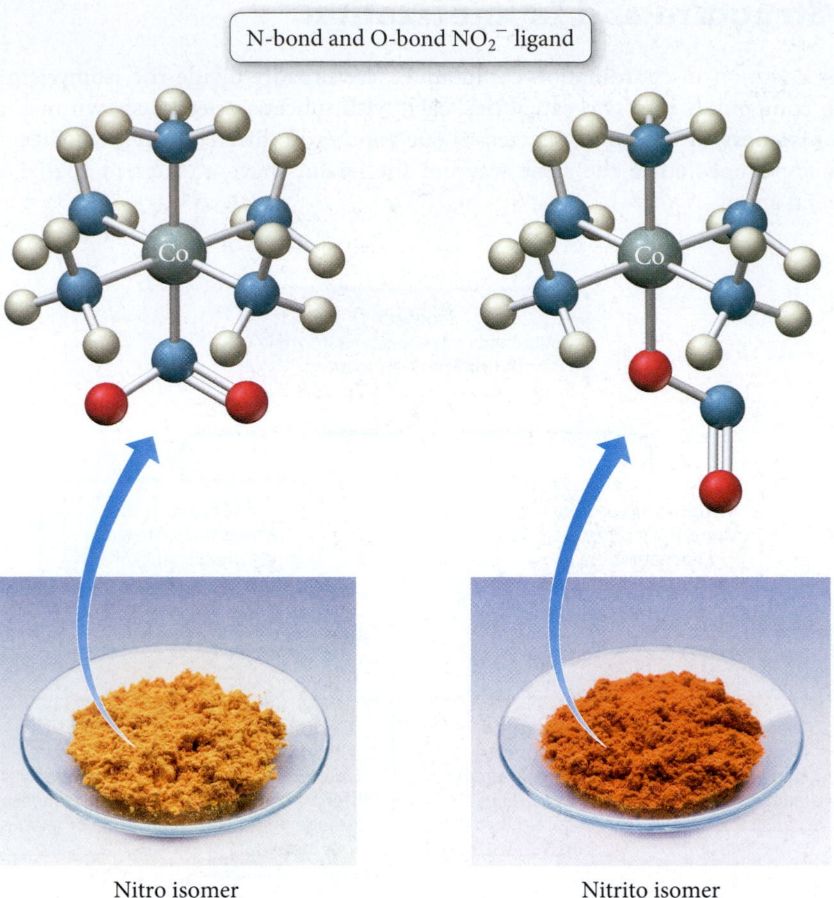

TABLE 23.6 Ligands Capable of Linkage Isomerization

Ligand	Bonding through left atom	Bonding through right atom
$[:\ddot{O}-\ddot{N}=\ddot{O}]^-$	nitrito (O)	nitro (N)
$[:C\equiv N:]^-$	cyano (C)	isocyano (N)
$[:\ddot{O}=C=\ddot{N}:]^-$	cyanato (O)	isocyanato (N)
$[:\ddot{S}=C=\ddot{N}:]^-$	thiocyanato (S)	isothiocyanato (N)

Stereoisomerism

We can also subdivide the broad category of stereoisomers into two types: geometric isomers and optical isomers. **Geometric isomers** result when the ligands bonding to the metal have a different spatial arrangement. One type of geometric isomerism, as we saw in Section 22.5, is cis–trans isomerism, which in complex ions occurs in square planar complexes of the general formula MA_2B_2 or octahedral complexes of the general formula MA_4B_2. For example, cis–trans isomerism occurs in the square planar complex $Pt(NH_3)_2Cl_2$. Figure 23.8(a) ▶ shows the two distinct ways that ligands can be oriented around the metal. In one complex, the Cl^- ligands are next to each other on one side of the molecule—this is the cis isomer. In the other complex, the Cl^- ligands are on opposite sides of the molecule—this is the trans isomer. Geometric isomerism also exists in the octahedral complex ion $[Co(NH_3)_4Cl_2]^+$. As shown in Figure 23.8(b) ▶, the ligands in this complex ion arrange themselves around the metal in two ways, one with the Cl^- ligands on the same side (the cis isomer) and another with the Cl^- ligands

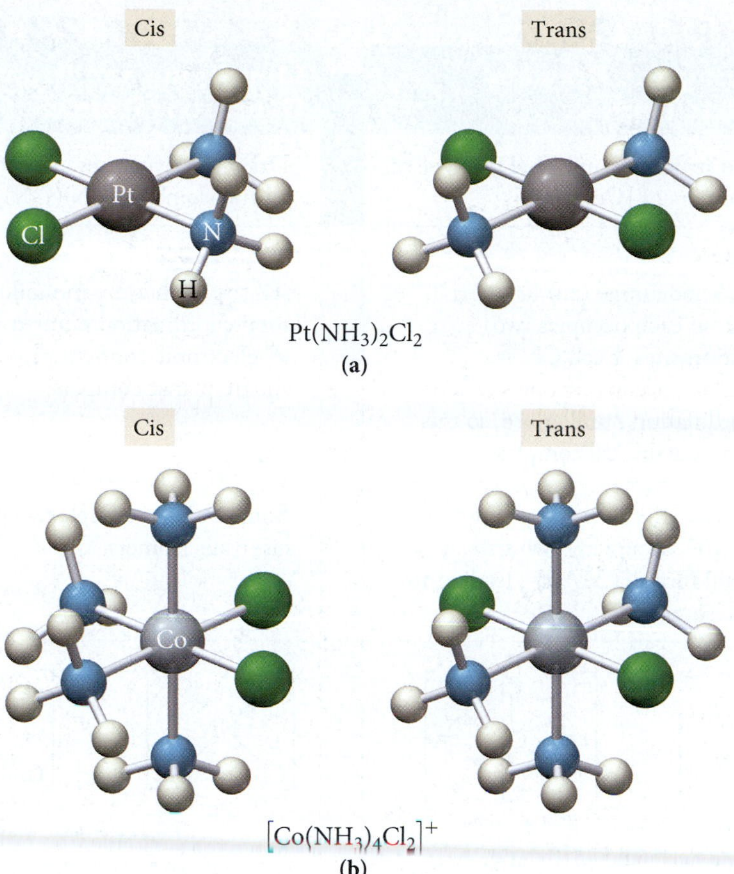

▲ **FIGURE 23.8 Cis–trans Isomerism** **(a)** Cis–trans isomerism in square planar $Pt(NH_3)_2Cl_2$. In the cis isomer, the Cl^- ligands are next to each other on one side of the molecule. In the trans isomer, the Cl^- ligands are on opposite sides of the molecule. **(b)** Cis–trans isomerism in octahedral $[Co(NH_3)_4Cl_2]^+$. In the cis isomer, the Cl^- ligands are on the same side. In the trans isomer, the Cl^- ligands are on opposite sides.

on opposite sides of the metal (the trans isomer). Note that cis–trans isomerism does not occur in tetrahedral complexes because all bond angles around the metal are 109.5°, and each corner of a tetrahedron is considered to be adjacent to all three other corners.

Another type of geometric isomerism is fac–mer isomerism, which occurs in octahedral complexes of the general formula MA_3B_3. For example, in $Co(NH_3)_3Cl_3$, the ligands arrange themselves around the metal in two ways (Figure 23.9 ▼). In the fac isomer the three Cl^- ligands are all on one side of the molecule and make up one face of the octahedron (fac is short for facial). In the mer isomer, the three ligands form an arc around the middle of the octahedron (mer is short for meridional).

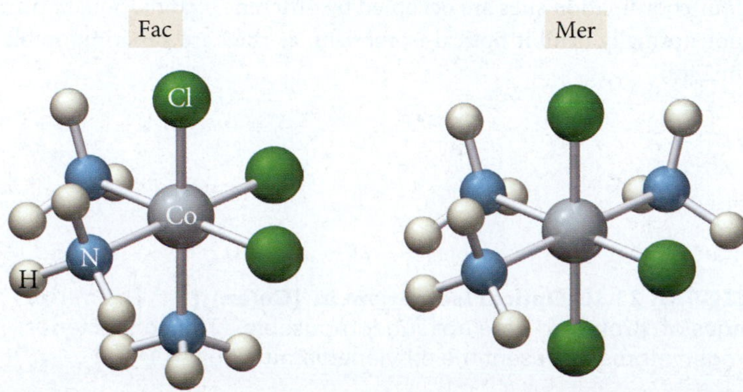

◀ **FIGURE 23.9 Fac–Mer Isomerism in $Co(NH_3)_3Cl_3$** In the fac isomer, the three Cl^- ligands are all on one side of the molecule and make up one face of the octahedron. In the mer isomer, the three ligands inscribe an arc around the middle (or meridian) of the octahedron.

PROCEDURE FOR **Identifying and Drawing Geometric Isomers**	**EXAMPLE 23.5** **Identifying and Drawing Geometric Isomers**	**EXAMPLE 23.6** **Identifying and Drawing Geometric Isomers**
	Draw the structures and label the type of all the isomers of $[Co(en)_2Cl_2]^+$.	Draw the structures and label the type for all the isomers of $[Ni(CN)_2Cl_2]^{2-}$.
Identify the coordination number and the geometry around the metal.	**SOLUTION** The ethylenediamine (en) ligand is bidentate, so each occupies two coordination sites. Each Cl^- is monodentate, occupying one site. The total coordination number is 6, so this must be an octahedral complex.	**SOLUTION** All the ligands are monodentate, so the total coordination number is 4. Ni^{2+} is a d^8 electronic configuration, so we expect a square planar complex.
Identify whether this is a cis–trans or fac–mer isomerism.	With ethylenediamine occupying four sites and Cl^- occupying two sites, it fits the general formula MA_4B_2, leading to cis–trans isomers.	Square planar complexes can only have cis–trans isomers.
Draw and label the two isomers.	Cis: $[Co(en)_2Cl_2]^+$ with both Cl adjacent; Trans: $[Co(en)_2Cl_2]^+$ with Cl above and below Co Cis Trans	Cis: $[NiCl_2(CN)_2]^{2-}$ with Cl, Cl on one side and CN, CN on the other; Trans: Cl opposite Cl, CN opposite CN Cis Trans
	FOR PRACTICE 23.5 Draw the structures and label the type of all the isomers of $[Cr(H_2O)_3Cl_3]^+$.	**FOR PRACTICE 23.6** Draw the structures and label the type of all the isomers of $[Co(NH_3)_2Cl_2(ox)]^-$.

The second category of stereoisomerism is optical isomerism. Recall from Section 22.3 that **optical isomers** are nonsuperimposable mirror images of one another. If you hold your right hand up to a mirror, the image in the mirror looks like your left hand. No matter how you rotate or flip your left hand, you cannot superimpose it on your right hand. Molecules or ions that exhibit this quality are *chiral*. The isomers are *enantiomers*, and they exhibit the property of optical activity (the rotation of polarized light). The complex ion $[Co(en)_3]^{3+}$ is nonsuperimposable on its mirror image, so it is a chiral complex (Figure 23.10 ◀).

The compounds in Example 23.7 demonstrate optical isomerism in octahedral complexes. Tetrahedral complexes can also exhibit optical isomerism, but only if all four coordination sites are occupied by different ligands. Square planar complexes do not normally exhibit optical isomerism, as they are superimposable on their mirror images.

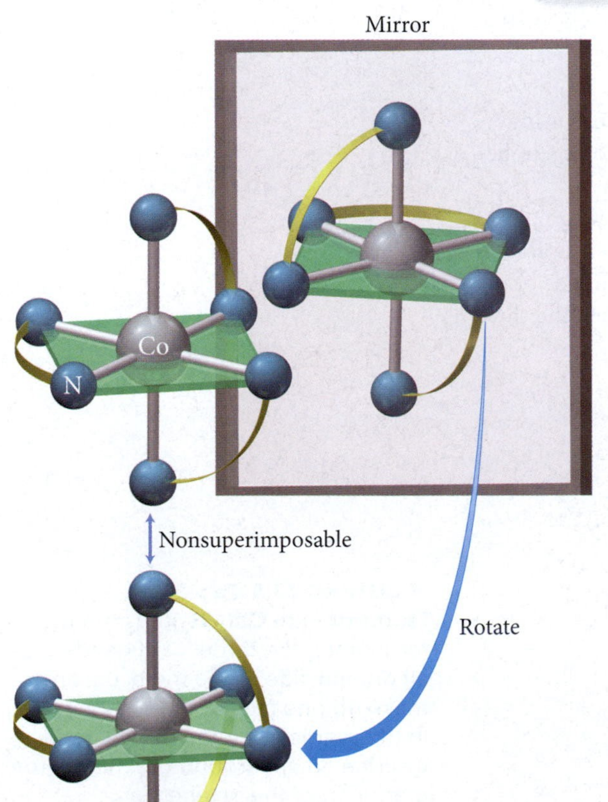

◀ **FIGURE 23.10 Optical Isomerism in $[Co(en)_3]^{3+}$** The mirror images of $[Co(en)_3]^{3+}$ are not superimposable. (The connected nitrogen atoms represent the ethylenediamine ligand.)

EXAMPLE 23.7
Recognizing and Drawing Optical Isomers

Determine whether the cis or trans isomers in Example 23.5 are optically active (demonstrate optical isomerism).

SOLUTION

Draw the trans isomer of $[Co(en)_2Cl_2]^+$ and its mirror image. Check to see if they are superimposable by rotating one isomer 180°.

Mirror

Trans

Superimposable

In this case the two are identical, so there is no optical activity.

Draw the cis isomer and its mirror image. Check to see if they are superimposable by rotating one isomer 180°.

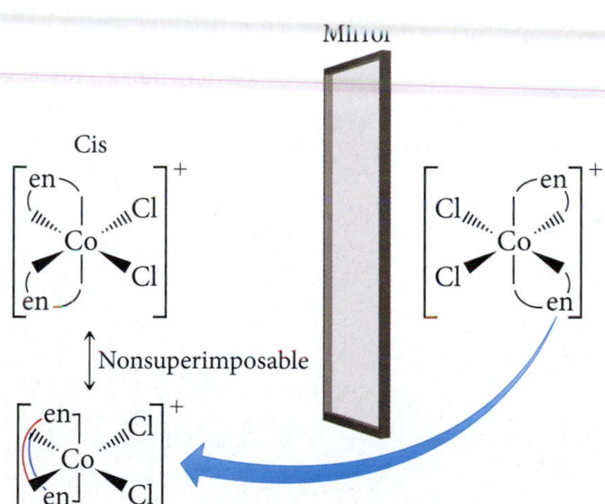

In this case the two structures are not superimposable, so the cis isomer does exhibit optical activity.

FOR PRACTICE 23.7

Determine whether the fac or mer isomers of $[Cr(H_2O)_3Cl_3]^+$ are optically active.

23.5 Bonding in Coordination Compounds

The bonding in complex ions can be described by valence bond theory (first introduced in Chapter 7). However, many of the properties of these ions are better described by a different model known as crystal field theory. We examine each model separately in this section.

Valence Bond Theory

The bonding in complex ions, particularly the geometries of the ions, can be described by one of our previous bonding models, valence bond theory. Recall from Section 7.2 that in valence bond theory, a coordinate covalent bond is the overlap between a completely filled atomic orbital and an empty atomic orbital. In complex ions, the filled orbital is on the ligand, and the empty orbital is on the metal ion. The metal ion orbitals are hybridized according to the geometry of the complex ion. The common hybridization schemes are shown in Figure 23.11 ▼. An octahedral complex ion requires six empty orbitals in an octahedral arrangement on the metal ion. A full set of d^2sp^3 hybrid orbitals results in the exact orbitals needed for this geometry. A set of sp^3 hybrid orbitals results in a tetrahedral arrangement of orbitals; a set of dsp^2 hybrid orbitals results in a square planar arrangement; and a set of sp hybrid orbitals results in a linear arrangement of orbitals. In each case, the coordinate covalent bond is formed by the overlap between the orbitals on the ligands and the hybridized orbitals on the metal ion.

▶ **FIGURE 23.11 Common Hybridization Schemes in Complex Ions** The valence bond model hybridization schemes are deduced from the geometry of the complex ion.

Geometry	Hybridization	Orbitals
Linear	sp	
Tetrahedral	sp^3	
Square planar	dsp^2	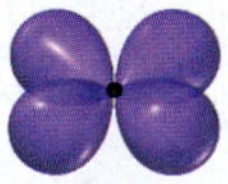
Octahedral	d^2sp^3	

Crystal Field Theory

Valence bond theory, while useful for describing the geometries of the complex ions, cannot explain other properties of complex ions such as color and magnetism. Crystal field theory (CFT), a bonding model for transition metal complexes, does account for these properties. To illustrate the basic principles of CFT, we first examine the central metal atom's *d* orbitals in an octahedral complex.

Octahedral Complexes and *d* Orbital Splitting The basic premise of CFT is that complex ions form because of attractions between the electrons on the ligands and the positive charge on the metal ion. However, the electrons on the ligands also repel the electrons in the *unhybridized* metal *d* orbitals. CFT focuses on these repulsions. Figure 23.12 ▶ shows how the ligand positions superimpose on the *d* orbitals in an octahedral complex. Notice that the ligands in an octahedral complex are located in the same space as the lobes of the $d_{x^2-y^2}$ and d_{z^2} orbitals. The repulsions *between* electron pairs in the ligands and any potential electrons in the *d* orbitals result in an increase in the energies of these orbitals. In contrast, the d_{xy}, d_{xz}, and d_{yz} orbitals lie *between* the axes and have nodes directly on

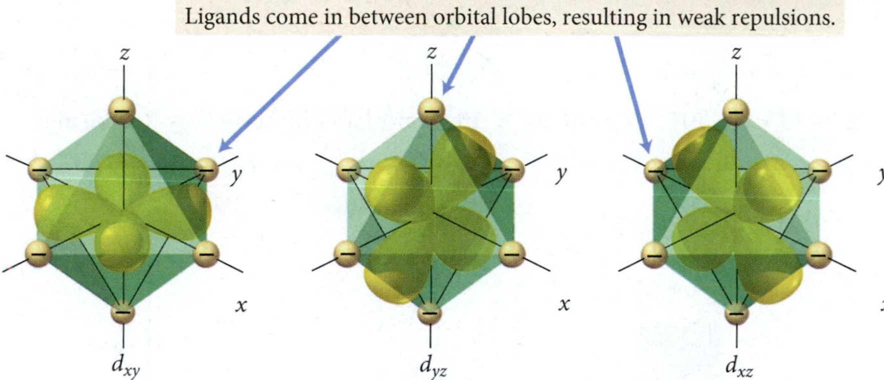

◀ **FIGURE 23.12 Relative Positions of *d* Orbitals and Ligands in an Octahedral Complex** The ligands in an octahedral complex (represented here as spheres of negative charge) interact most strongly with the d_{z^2} and $d_{x^2-y^2}$ orbitals.

the axes, which results in less repulsion and lower energies for these three orbitals. In other words, the *d* orbitals—which are degenerate in the bare metal ion—are split into higher and lower energy levels because of the spatial arrangement of the ligands (Figure 23.13 ▼). The difference in energy between these split *d* orbitals is known as the crystal field splitting energy (Δ). The magnitude of the splitting depends on the particular complex. In **strong-field complexes**, the splitting is large; and in **weak-field complexes**, the splitting is small.

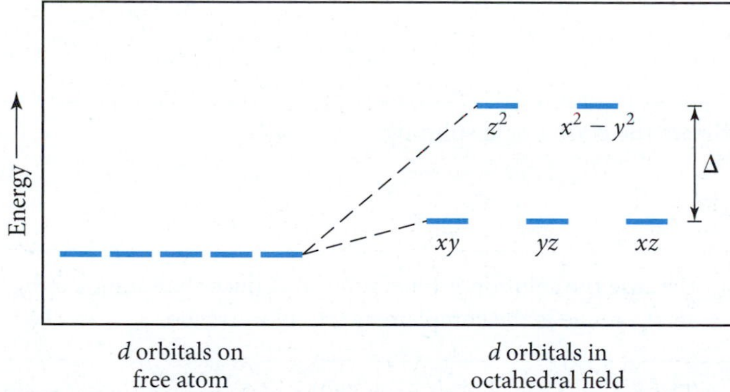

▲ **FIGURE 23.13 *d* Orbital Splitting in an Octahedral Field** The otherwise degenerate *d* orbitals are split into two energy levels by the ligands in an octahedral complex ion.

The Color of Complex Ions and Crystal Field Strength In Section 23.1 we saw that transition metals in host crystals often show brilliant colors because of the crystal field splitting of their *d* orbitals. Solutions of complex ions display brilliant colors because of similar splittings. For example, an $[Fe(CN)_6]^{3-}$ solution is deep red, and an $[Ni(NH_3)_6]^{2+}$ solution is blue (Figure 23.14 ▶). Recall from Chapter 3 that the color of an object is related to the absorption of light energy by its electrons. If a substance absorbs all of the visible wavelengths, it appears black. If it transmits (or reflects) all the wavelengths (absorbs no light), it appears colorless. A substance appears to be a particular color if it absorbs some visible light but also transmits (or reflects) the wavelengths associated with that color. A substance also appears to be a given color if it transmits (or reflects) most wavelengths but absorbs the

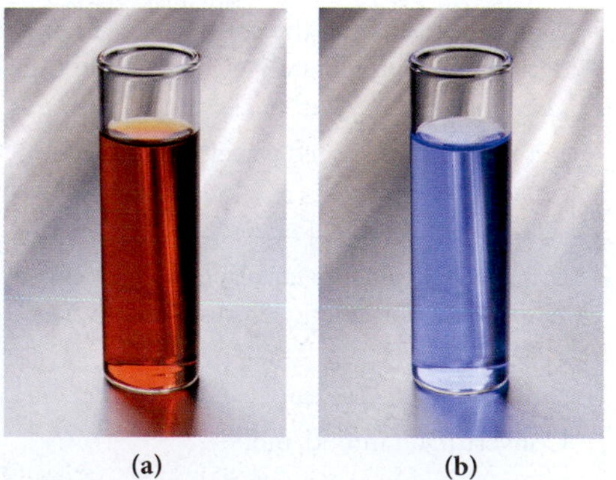

(a) (b)

▲ **FIGURE 23.14 Colors of Complex Ions** **(a)** The complex ion $[Fe(CN)_6]^{3-}$ forms a deep red solution, and **(b)** $[Ni(NH_3)_6]^{2+}$ is blue.

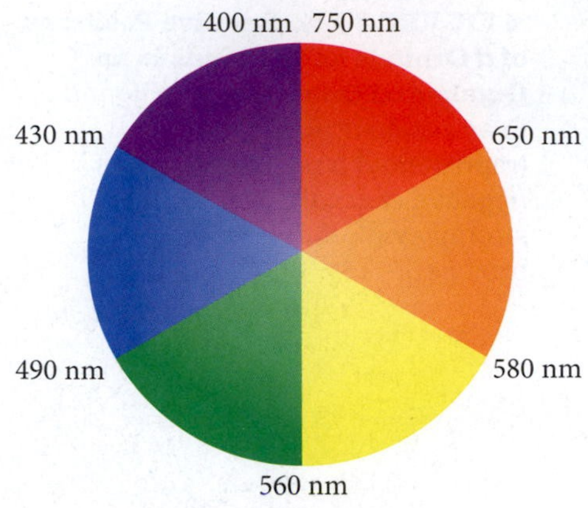

▲ **FIGURE 23.15 The Color Wheel** Colors across from one another on the color wheel are complementary. A substance that absorbs a color on the wheel appears to be its complementary color.

complementary color on a color wheel (Figure 23.15 ◀). For example, a substance that absorbs green light (the complement of red) appears red. A solution of $[Ti(H_2O)_6]^{3+}$ is purple because it absorbs strongly between 490 and 580 nm, the yellow-green region of the visible spectrum (Figure 23.16a ▼).

The easiest way to measure the energy difference between the *d* orbitals in a complex ion is to use spectroscopy to determine the wavelength of light absorbed when an electron makes a transition from the lower energy *d* orbitals to the higher energy ones. With that information we can calculate the crystal field splitting energy, Δ:

$$E_{\text{photon}} = h\nu = \frac{hc}{\lambda} = \Delta$$

Consider the $[Ti(H_2O)_6]^{3+}$ absorption spectrum shown in Figure 23.16b. The maximum absorbance is at 498 nm. Using this wavelength, we calculate Δ:

$$\Delta = hc/\lambda = (6.626 \times 10^{-34}\ \text{J}\cdot\cancel{\text{s}})(3.00 \times 10^{8}\ \cancel{\text{m}}/\cancel{\text{s}})/(498\ \cancel{\text{nm}} \times 1 \times 10^{-9}\ \cancel{\text{m}}/\cancel{\text{nm}})$$

$$\Delta = 3.99 \times 10^{-19}\ \text{J}$$

This energy corresponds to a single $[Ti(H_2O)_6]^{3+}$ ion. We can convert to kilojoules per mole:

$$\Delta = (3.99 \times 10^{-19}\ \cancel{\text{J}}/\cancel{\text{ion}})(6.02 \times 10^{23}\ \cancel{\text{ion}}/\text{mol})(1\ \text{kJ}/1000\ \cancel{\text{J}}) = 240\ \text{kJ/mol}$$

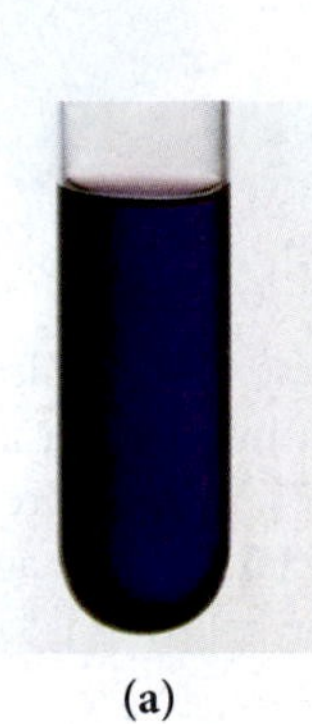

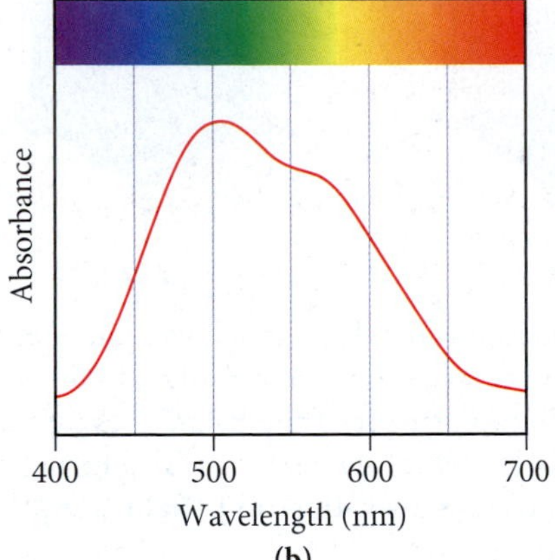

(a) (b)

▶ **FIGURE 23.16 The Color and Absorption Spectrum of $[Ti(H_2O)_6]^{3+}$** **(a)** A solution containing $[Ti(H_2O)_6]^{3+}$ is purple. **(b)** The absorption spectrum of $[Ti(H_2O)_6]^{3+}$ extends across the green-yellow region of the spectrum.

EXAMPLE 23.8
Crystal Field Splitting Energy

The complex ion $[Cu(NH_3)_6]^{2+}$ is blue in aqueous solution. Estimate the crystal field splitting energy (in kJ/mol) for this ion.

SOLUTION

Begin by consulting the color wheel (Figure 23.15) to determine approximately what wavelength is being absorbed.	Because the solution is blue, you can deduce that orange light is absorbed since orange is the complementary color to blue.
Estimate the absorbed wavelength.	The color orange ranges from 580 to 650 nm, so you can estimate the average wavelength as 615 nm.
Calculate the energy corresponding to this wavelength, using $E = hc/\lambda$. This energy corresponds to Δ.	$E = \frac{(6.626 \times 10^{-34}\ \text{J}\cdot\cancel{\text{s}})(3.00 \times 10^{8}\ \cancel{\text{m}}/\cancel{\text{s}})}{(615\ \cancel{\text{nm}})(1 \times 10^{-9}\ \cancel{\text{m}}/\cancel{\text{nm}})}$ $E = 3.23 \times 10^{-19}\ \text{J} = \Delta$
Convert J/ion into kJ/mol.	$\Delta = \frac{(3.23 \times 10^{-19}\ \cancel{\text{J}}/\cancel{\text{ion}})(6.02 \times 10^{23}\ \cancel{\text{ion}}/\text{mol})}{(1000\ \cancel{\text{J}}/\text{kJ})}$ $\Delta = 195\ \text{kJ/mol}$

FOR PRACTICE 23.8
The complex ion $[Co(NH_3)_5NO_2]^{2+}$ is yellow. Estimate the crystal field splitting energy (in kJ/mol) for this ion.

The magnitude of the crystal field splitting in a complex ion—and, therefore whether it is a strong-field or a weak-field complex—depends in large part on the ligands attached to the central metal ion. Spectroscopic studies of various ligands attached to the same metal allow us to arrange different ligands in order of their ability to split the *d* orbitals. This list is known as the *spectrochemical series* and is arranged from ligands that result in the largest Δ to those that result in the smallest:

$$CN^- > NO_2^- > en > NH_3 > H_2O > OH^- > F^- > Cl^- > Br^- > I^-$$

large Δ — small Δ

typically strong-field ligands — typically weak-field ligands

Ligands that produce large values of Δ are *strong-field ligands* and those that give small values of Δ are *weak-field ligands*.

The metal ion also has an effect on the magnitude of Δ. If we examine different metal ions with the same ligand, we find that Δ increases as the charge on the metal ion increases. A greater charge on the metal draws the ligands closer, causing greater repulsion with the *d* orbitals and therefore a larger Δ. An example of this behavior occurs in the complex ions between NH_3 (a ligand in the middle of the spectrochemical series) and the +2 or +3 oxidation states of cobalt. Hexaamminecobalt(II) ion, $[Co(NH_3)_6]^{2+}$, has a weak crystal field (small Δ) and hexaamminecobalt(III) ion, $[Co(NH_3)_6]^{3+}$, has a strong field (large Δ).

Weak- and Strong-Field Ligands

23.2 Cc Conceptual Connection

Two ligands, A and B, both form complexes with a particular metal ion. When the metal ion complexes with ligand A, the resulting solution is red. When the metal ion complexes with ligand B, the resulting solution is yellow. Which of the two ligands produces the larger Δ?

Magnetic Properties and Crystal Field Strength The strength of the crystal field can affect the magnetic properties of a transition metal complex. Recall that, according to Hund's rule, electrons occupy degenerate orbitals singly as long as an empty orbital is available. When the energies of the *d* orbitals are split by ligands, the lower energy orbitals fill first. Once they are half-filled, the next electron can either: (1) pair with an electron in one of the lower energy half-filled orbitals by overcoming the electron–electron repulsion associated with having two electrons in the same orbital; or (2) go into an empty orbital of higher energy by overcoming the energy difference between the orbitals—in this case, the crystal field splitting energy, Δ. The magnitude of Δ compared to the electron–electron repulsions determines which of these two actually occurs.

Recall from Section 4.7 that a paramagnetic species contains unpaired electrons and a diamagnetic one does not.

We can compare two iron(II) complexes to see the difference in behavior under strong- and weak-field conditions. $[Fe(CN)_6]^{4-}$ is known to be diamagnetic and $[Fe(H_2O)_6]^{2+}$ is known to be paramagnetic. Both of these complexes contain Fe^{2+}, which has an electron configuration of [Ar] $3d^6$. In the case of $[Fe(CN)_6]^{4-}$, CN^- is a strong-field ligand that generates a large Δ, so it takes more energy to occupy the higher energy level than it does to pair the electrons in the lower energy level. The result is that all six electrons are paired and the compound is diamagnetic, as shown in the upper figure in the right margin.

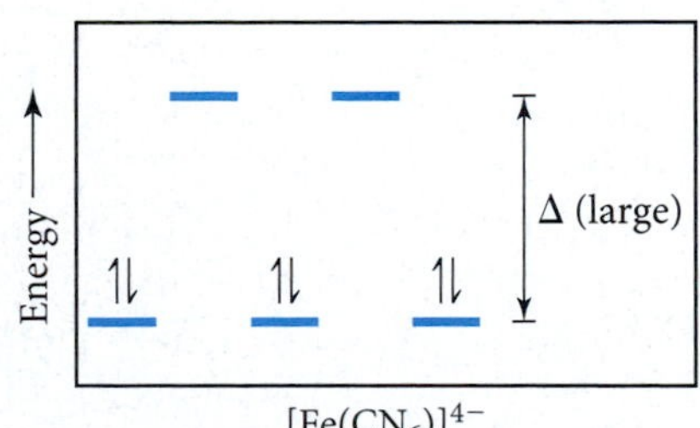

$[Fe(CN_6)]^{4-}$

In $[Fe(H_2O)_6]^{2+}$, H_2O is a weak-field ligand that generates a small Δ, so the electron pairing energy is greater than Δ. Consequently, the first five electrons occupy the five *d* orbitals singly and only the sixth pairs up, resulting in a paramagnetic compound with four unpaired electrons, as shown in the lower figure in the right margin.

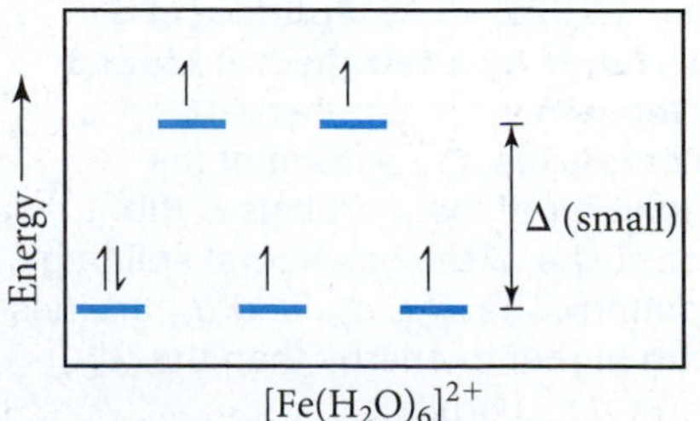

$[Fe(H_2O)_6]^{2+}$

In general, complexes with strong-field ligands have fewer unpaired electrons relative to the free metal ion and are therefore called **low-spin complexes**. Complexes with weak-field ligands, by contrast, have the same number of unpaired electrons as the free metal ion and are known as **high-spin complexes**.

When we examine the orbital diagrams of the d^1 through d^{10} metal ions in octahedral complexes, we find that only d^4, d^5, d^6, and d^7 metal ions have low- and high-spin possibilities. Since there are three lower energy *d* orbitals, the d^1, d^2, and d^3 metal ions always have unpaired electrons, independent of Δ. In the d^8, d^9, and d^{10} metal ions, the three lower energy orbitals are completely filled, so the remaining electrons fill the two higher orbitals (as expected according to Hund's rule), also independent of Δ.

PROCEDURE FOR ▼	**EXAMPLE 23.9**	**EXAMPLE 23.10**
Determining the Number of Unpaired Electrons in Octahedral Complexes	**High- and Low-Spin Octahedral Complexes**	**High- and Low-Spin Octahedral Complexes**
	How many unpaired electrons are there in the complex ion $[CoF_6]^{3-}$?	How many unpaired electrons are there in the complex ion $[Co(NH_3)_5NO_2]^{2+}$?
Begin by determining the charge and number of *d* electrons on the metal.	**SOLUTION** The metal is Co^{3+} and has a d^6 electronic configuration.	**SOLUTION** The metal is Co^{3+} and has a d^6 electronic configuration.
Look at the spectrochemical series to determine whether the ligand is a strong-field or a weak-field ligand.	F^- is a weak-field ligand, so Δ is relatively small.	NH_3 and NO_2^- are both strong-field ligands, so Δ is relatively large.
Decide if the complex is high- or low-spin and draw the electron configuration.	Weak-field ligands yield high-spin configurations.	Strong-field ligands yield low-spin configurations. 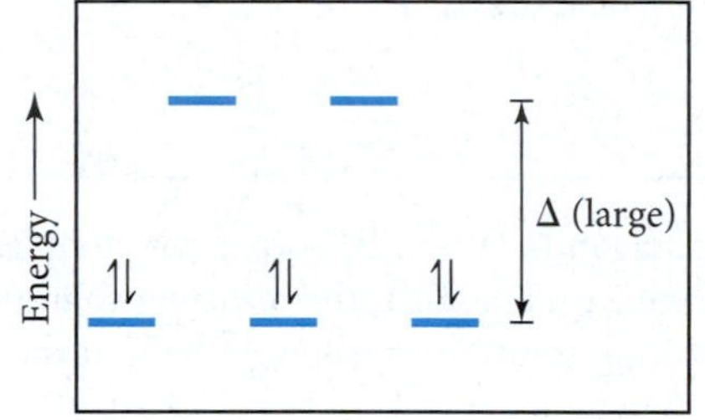
Count the unpaired electrons.	This configuration has four unpaired electrons.	This configuration has no unpaired electrons.
	FOR PRACTICE 23.9 How many unpaired electrons are there in the complex ion $[FeCl_6]^{3-}$?	**FOR PRACTICE 23.10** How many unpaired electrons are there in the complex ion $[Co(CN)_6]^{4-}$?

Tetrahedral and Square Planar Complexes and *d* Orbital Splitting So far, we have examined the *d* orbital energy changes only for octahedral complexes, but transition metal complexes can have other geometries, such as tetrahedral and square planar. We use crystal field theory to determine the *d* orbital splitting pattern for these geometries as well. For a tetrahedral complex, the *d* orbital splitting pattern is the opposite of the octahedral splitting pattern: three *d* orbitals (d_{xy}, d_{xz}, and d_{yz}) are higher in energy, and two *d* orbitals ($d_{x^2-y^2}$ and d_{z^2}) are lower in energy (Figure 23.17 ▼). Almost all tetrahedral complexes are high-spin because of reduced ligand–metal interactions. The *d* orbitals in a tetrahedral complex interact with only four ligands, as opposed to six in the octahedral complex, so the value of Δ is generally smaller.

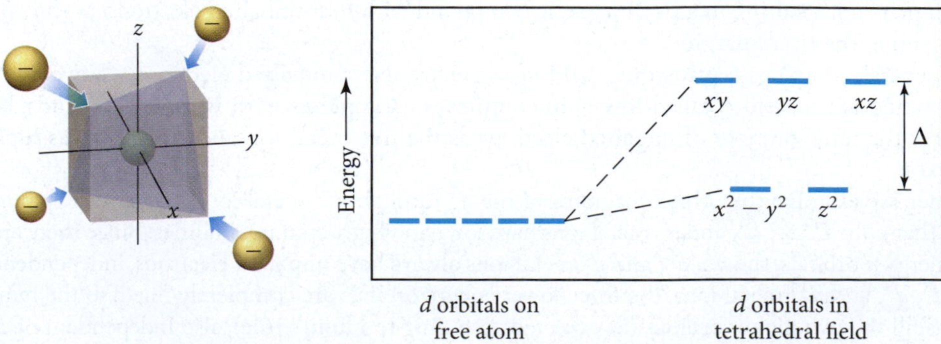

▶ **FIGURE 23.17 Splitting of *d* Orbitals by a Tetrahedral Ligand Geometry** In tetrahedral complexes, the pattern of the splitting of the *d* orbitals is the opposite of the octahedral splitting pattern. The d_{xy}, d_{yz}, and d_{xz} orbitals are higher in energy than the d_{z^2} and $d_{x^2-y^2}$ orbitals.

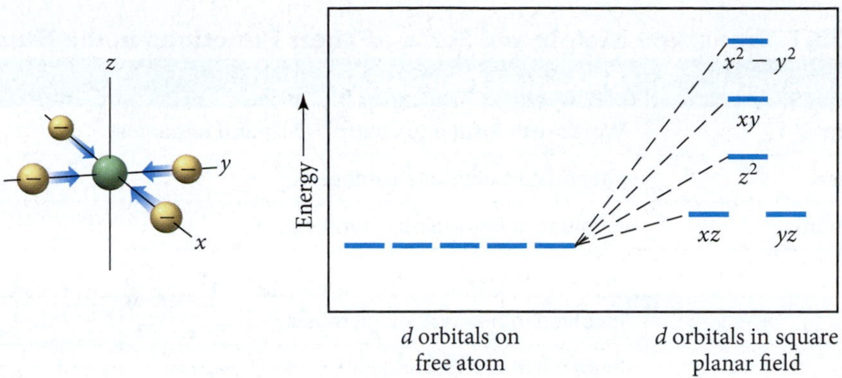

◀ **FIGURE 23.18 Splitting of *d* Orbitals by a Square Planar Ligand Geometry** Square planar complexes produce the *d* orbital energy pattern shown here.

A square planar complex results in the most complex splitting pattern of the three geometries (Figure 23.18 ▲). As we discussed previously, square planar complexes occur in d^8 metal ions, such as Pt^{2+}, Pd^{2+}, Ir^+, or Au^{3+}, and in nearly all cases they are low-spin.

23.6 Applications of Coordination Compounds

Coordination compounds are found in living systems, in industry, and even in household products. In this section we describe a few applications of coordination compounds.

Chelating Agents

In Section 23.3, we introduced the chelating agent ethylenediaminetetraacetate ion ($EDTA^{4-}$). This ligand has lone pairs on six different donor atoms that can interact with a metal ion to form very stable metal complexes. EDTA is used to treat the victims of heavy metal poisoning such as lead poisoning. The patient is given $[Ca(EDTA)]^{2-}$ and since the lead complex ($K_f = 2 \times 10^{18}$) is more stable than the calcium complex ($K_f = 4 \times 10^{10}$), the lead displaces the calcium. The body excretes the lead complex and leaves behind the calcium, which is nontoxic (and is in fact a nutrient).

Chemical Analysis

Some ligands are selective in their binding, preferring specific metal ions; these ligands can be used in chemical analysis. For example, dimethylglyoxime (dmg) is used to chemically analyze a sample for Ni^{2+} or Pd^{2+}. In the presence of Ni^{2+}, an insoluble red precipitate forms, and in the presence of Pd^{2+}, an insoluble yellow precipitate forms. Similarly, the SCN^- ligand is used to test for Co^{2+} or Fe^{3+}. In the presence of Co^{2+} a blue solution forms, and in the presence of Fe^{3+} a deep red solution forms (Figure 23.19 ▶).

(a) (b)

▲ **FIGURE 23.19 Chemical Analysis with SCN^-** **(a)** Blue indicates Co^{2+}. **(b)** Red indicates Fe^{3+}.

Coloring Agents

Because of the wide variety of colors found in coordination complexes, they are often used as coloring agents. For example, a commercially available agent, iron blue, is a mixture of the hexacyano complexes of iron(II) and iron(III). Iron blue is used in ink, paint, cosmetics (eye shadow), and blueprints.

Biomolecules

Living systems contain many molecules based on metal complexes. Hemoglobin (involved in oxygen transport), cytochrome c (involved in electron transport), carbonic anhydrase (involved in respiration), and chlorophyll (involved in photosynthesis) all have coordinated metal ions that are critical to their structure and function. Table 23.7 summarizes the biological significance of many of the other first-row transition metals.

TABLE 23.7 Transition Metals and Some of Their Functions in the Human Body

Transition Metal	Biological Function
Chromium	Works with insulin to control utilization of glucose
Manganese	Fat and carbohydrate synthesis
Molybdenum	Involved in hemoglobin synthesis
Iron	Oxygen transport
Copper	Involved in hemoglobin synthesis
Zinc	Involved in cell reproduction and tissue growth; part of more than 70 enzymes; assists in the utilization of carbohydrate, protein, and fat

Hemoglobin and Cytochrome C In hemoglobin and in cytochrome c, an iron complex called a heme is connected to a protein, as shown in the figure below. A heme is an iron ion coordinated to a flat, polydentate ligand called a porphyrin (Figure 23.20 ▼). The porphyrin ligand has a planar ring structure with four nitrogen atoms that can coordinate to the metal ion. Different porphyrins have different substituent groups connected around the outside of the ring.

In hemoglobin, the iron complex is octahedral, with the four nitrogen atoms of the porphyrin in a square planar arrangement around the metal. A nitrogen atom from the protein occupies the fifth

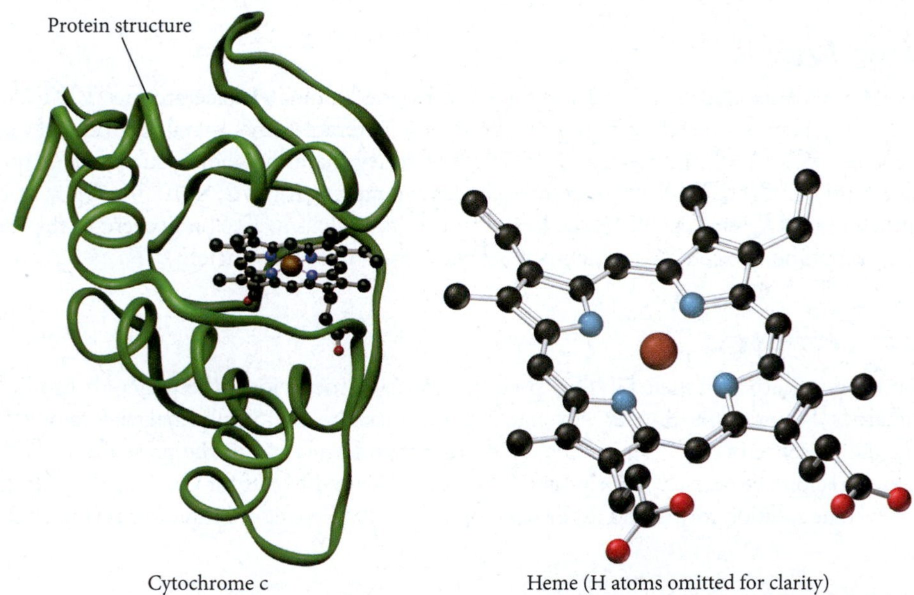

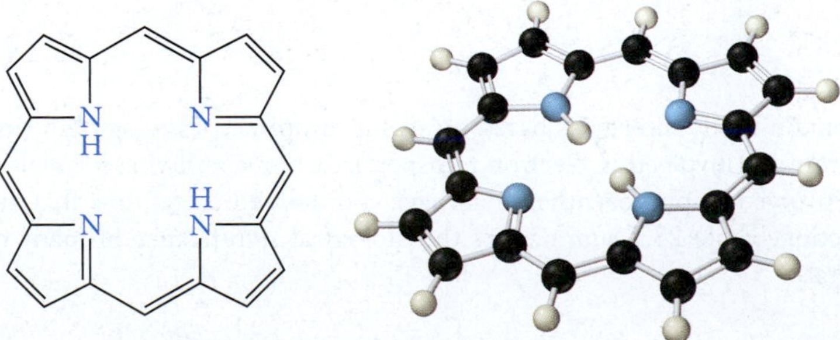

▶ **FIGURE 23.20 Porphyrin** A porphyrin has four nitrogen atoms that can coordinate to a central metal atom.

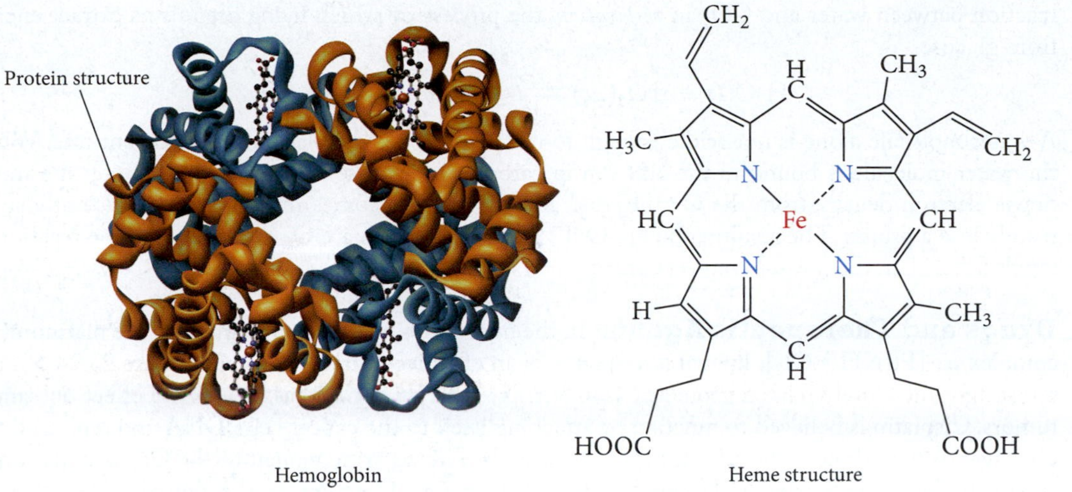

Protein structure

Hemoglobin

Heme structure

◀ **FIGURE 23.21 Hemoglobin** In hemoglobin, the iron complex is octahedral, with the four nitrogen atoms of the porphyrin in a square planar arrangement around the metal. A nitrogen atom from a nearby amino acid of the protein occupies the fifth coordination site, and either O_2 or H_2O occupies the last coordination site.

coordination site, and either O_2 or H_2O occupies the last coordination site (Figure 23.21 ▲). In the lungs, where the oxygen content is high, the hemoglobin coordinates to an O_2 molecule. The oxygen-rich hemoglobin is carried by the bloodstream to areas throughout the body that are depleted in oxygen, where oxygen is released and replaced by a water molecule. The hemoglobin then travels back to the lungs to repeat the cycle.

Chlorophyll Chlorophyll, shown in Figure 23.22 ▼, is another porphyrin-based biomolecule, but in chlorophyll the porphyrin is not surrounded by a protein and the coordinated metal is magnesium (which is not a transition metal). Chlorophyll is essential for the *photosynthesis* process performed by plants, in which light energy from the sun is converted to chemical energy to fuel the plant's growth.

Carbonic Anhydrase In carbonic anhydrase, the zinc ion is bound in a tetrahedral complex, with three of the coordination sites occupied by nitrogen atoms from the surrounding protein and the fourth site available to bind a water molecule (Figure 23.23 ▼). Carbonic anhydrase catalyzes the

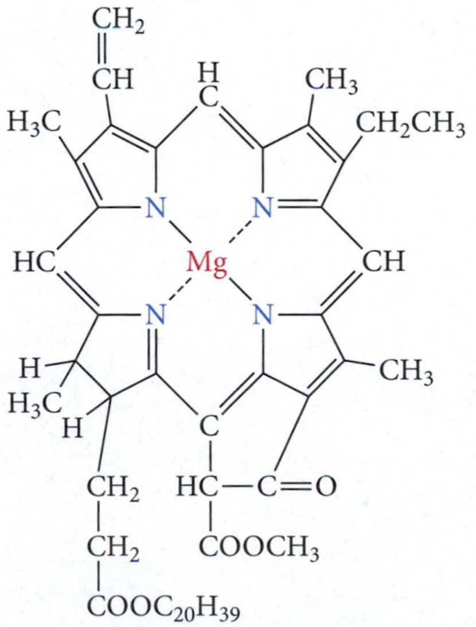

▲ **FIGURE 23.22 Chlorophyll** Chlorophyll, involved in photosynthesis in plants, contains magnesium coordinated to a porphyrin.

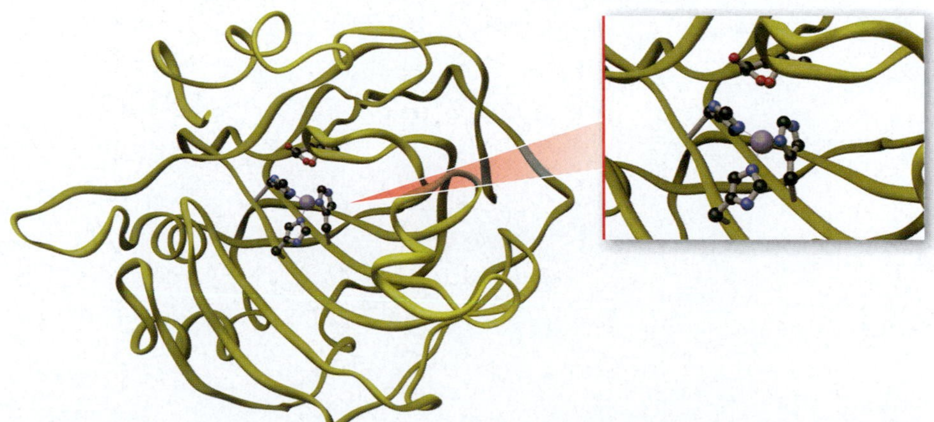

▲ **FIGURE 23.23 Carbonic Anhydrase** Carbonic anhydrase contains a zinc ion that is bound in a tetrahedral complex, with three of the coordination sites occupied by nitrogen atoms from surrounding amino acids. The fourth site is available to bind a water molecule.

reaction between water and CO_2 in *respiration*, the process by which living organisms extract energy from glucose.

$$H_2O(l) + CO_2(aq) \rightleftharpoons H^+(aq) + HCO_3^-(aq)$$

A water molecule alone is not acidic enough to react with a CO_2 molecule at a sufficient rate. When the water molecule is bound to the zinc ion in carbonic anhydrase, the positive charge on the metal draws electron density from the O—H bond and the H_2O becomes more acidic—sufficiently so to readily lose a proton. The resulting bound OH^- easily reacts with a CO_2 molecule, and the reaction is much faster than the uncatalyzed version.

Drugs and Therapeutic Agents In the mid-1960s, researchers found that the platinum(II) complex *cis*-$[Pt(NH_3)_2Cl_2]$, known as cisplatin, is an effective anticancer agent (Figure 23.24 ▼). Interestingly, the closely related geometric isomer *trans*-$[Pt(NH_3)_2Cl_2]$ has little or no effect on cancer tumors. Cisplatin is believed to function by attaching itself to the cancer cell's DNA and replacing the Cl^- ligands with donor atoms from the DNA strands. The cis arrangement of the Cl^- ligands corresponds to the geometry required to bind to the DNA strands. The trans isomer, though closely related, cannot bind properly due to the arrangement of the Cl^- ligands and is therefore not an effective agent. Cisplatin and other closely related platinum(II) complexes are used in chemotherapy for certain types of cancer and are among the most effective anticancer agents available for these cases.

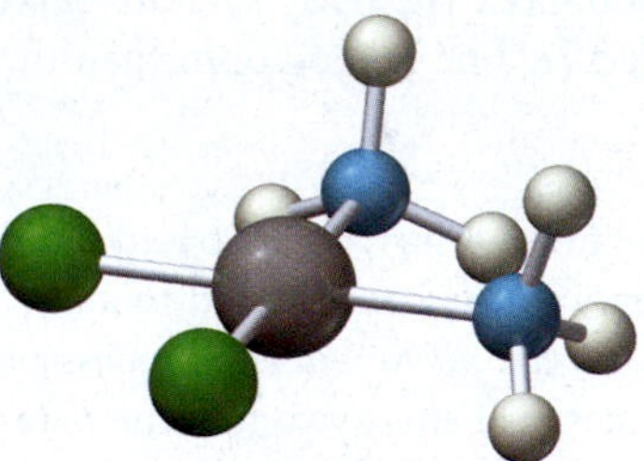

◀ **FIGURE 24.24 Cisplatin** Cisplatin is an effective anticancer agent.

SELF-ASSESSMENT Quiz

1. What is the electron configuration of the Cu^+ ion?
   a) $[Ar]4s^23d^9$
   b) $[Ar]4s^23d^8$
   c) $[Ar]4s^13d^9$
   d) $[Ar]4s^03d^{10}$

2. Which metal has the highest first ionization energy?
   a) Ti
   b) Mn
   c) Ru
   d) Au

3. What is the name of the compound $[CoCl(NH_3)_5]Cl_2$?
   a) pentaamminetrichlorocobalt(III) chloride
   b) pentaamminechlorocobalt(III) chloride
   c) pentaamminechlorocobalt(II) chloride
   d) pentaamminetrichlorocobalt(II) chloride

4. What is the formula of hexaaquamanganese(II) sulfate?
   a) $[Mn(OH)_6]SO_4$
   b) $[Mn(H_2O)_6]SO_4$
   c) $[Mn(H_2O)_6]_2SO_4$
   d) $[Mn_2(H_2O)_6]_2SO_4$

5. Which complex ion can exhibit geometric isomerism? Assume that M is the metal ion, A and B are ligands, and the geometry is octahedral.
   a) $[MA_6]^{2+}$
   b) $[MA_5B]^{2+}$
   c) $[MA_4B_2]^{2+}$
   d) $[MAB_5]^{2+}$

6. Pick the optical isomer of the complex ion represented here.

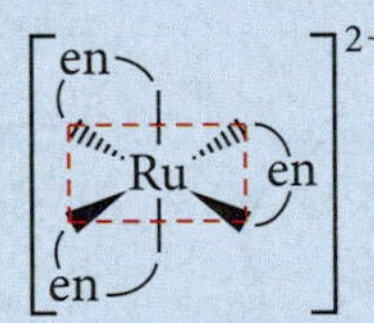

a)

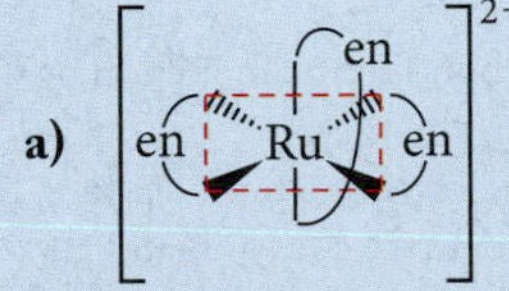

b)

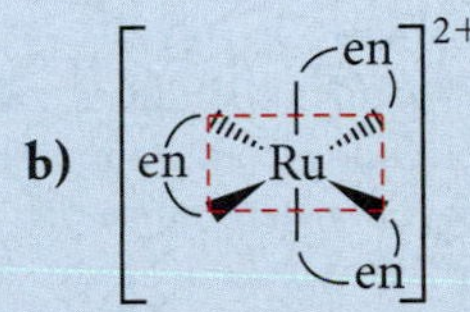

c)

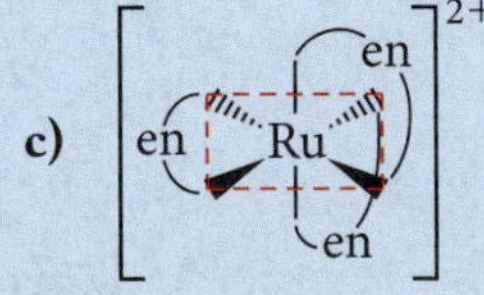

d)

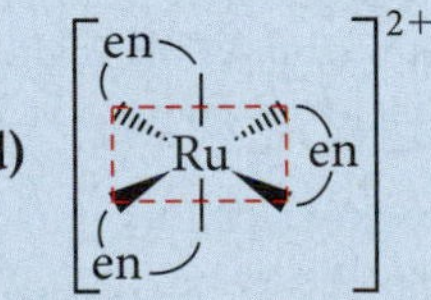

7. According to valence bond theory, what is the hybridization of the central metal ion in an octahedral complex ion?
   a) sp
   b) sp^3
   c) dsp^2
   d) d^2sp^3

8. Estimate the crystal field splitting energy (in kJ/mol) for a complex ion that is red in solution.
   a) 228 kJ/mol
   b) 171 kJ/mol
   c) 2.84×10^{-19} kJ/mol
   d) 3.79×10^{-19} kJ/mol

9. Use crystal field theory to determine the number of unpaired electrons in the complex ion $[Fe(CN)_6]^{4-}$.
   a) 0
   b) 2
   c) 3
   d) 4

10. Which complex ion is diamagnetic?
   a) $[Cr(H_2O)_4Cl_2]^+$
   b) $[Fe(H_2O)_6]^{2+}$
   c) $[Co(NH_3)_6]^{3+}$
   d) $[CoCl_6]^{3-}$

Answers: 1:d; 2:d; 3:b; 4:b; 5:c; 6:b; 7:d; 8:a; 9:a; 10:c

CHAPTER SUMMARY 23

REVIEW

KEY LEARNING OUTCOMES

CHAPTER OBJECTIVES	ASSESSMENT
Writing Electronic Configurations for Transition Metals and Their Ions (23.2)	• Examples 23.1, 23.2 For Practice 23.1, 23.2 Exercises 19–20, 57, 58
Naming Coordination Compounds (23.3)	• Examples 23.3, 23.4 For Practice 23.3, 23.4 Exercises 23–28
Identifying and Drawing Geometric Isomers (23.4)	• Examples 23.5, 23.6 For Practice 23.5, 23.6 Exercises 5–40, 61–62
Recognizing and Drawing Optical Isomers (23.4)	• Example 23.7 For Practice 23.7 Exercises 39–40, 61–62
Crystal Field Splitting Energy (23.5)	• Example 23.8 For Practice 23.8 Exercises 43–46
Determining the Number of Unpaired Electrons in Octahedral Complexes (23.5)	• Examples 23.9, 23.10 For Practice 23.9, 23.10 Exercises 49–52, 65

KEY TERMS

Section 23.1
crystal field theory (956)

Section 23.2
lanthanide contraction (958)

Section 23.3
complex ion (960)
ligand (960)
coordination compound (960)
primary valence (960)
secondary valence (960)
coordination number (960)
coordinate covalent bond (960)
monodentate (960)
bidentate (961)
polydentate (961)
chelate (962)
chelating agent (962)

Section 23.4
structural isomers (965)
stereoisomers (965)
coordination isomers (965)
linkage isomers (965)
geometric isomers (966)
optical isomers (968)

Section 23.5
strong-field complex (971)
weak-field complex (971)
low-spin complex (973)
high-spin complex (973)

KEY CONCEPTS

Electron Configurations (23.2)

- As we move across a row of transition elements, we add electrons to the $(n-1)d$ orbitals, resulting in a general electron configuration for first- and second-row transition elements of [noble gas] $ns^2(n-1)d^x$ and for the third and fourth rows of [noble gas] $ns^2(n-2)f^{14}(n-1)d^x$, where x ranges from 1 to 10.
- A transition element forms a cation by losing electrons from the ns orbitals before losing electrons from the $(n-1)d$ orbitals.

Periodic Trends (23.2)

- The variations in atomic size, ionization energy, and electronegativity across a row in the periodic table for transition metals are similar to those of main-group elements (although the trends are less pronounced and less regular). As we move down a group, however, atomic size increases from the first row of transition metals to the second but stays roughly constant from the second row to the third because of the lanthanide contraction. This contraction results in ionization energy and electronegativity trends as we move down a column for transition metals that are opposite of the main-group elements.

Composition and Naming of Coordination Compounds (23.3)

- A coordination compound is composed of a complex ion and a counterion.
- A complex ion contains a central metal ion bound to one or more ligands. The number of ligands directly bound to the metal ion is called the coordination number.
- The ligand forms a coordinate covalent bond to the metal ion by donating a pair of electrons to an empty orbital on the metal.
- Ligands that donate a single pair of electrons are monodentate. A ligand that donates two pairs of electrons is bidentate, and a ligand that donates more than two pairs is polydentate.
- In naming coordination compounds, we use the name of the cation followed by the name of the anion. To name a complex ion, we use the guidelines outlined in Section 23.3.

Types of Isomers (23.4)

- We broadly divide the isomerism observed in coordination compounds into two categories: structural isomers, in which atoms are connected differently to one another, and stereoisomers, in which atoms are connected in the same way but the ligands have a different spatial arrangement about the metal atom.

- Structural isomers are either coordination isomers (a coordinated ligand exchanges places with an uncoordinated counterion) or linkage isomers (a particular ligand has the ability to coordinate to the metal in different ways).
- Stereoisomers are either geometric isomers (the ligands bonded to the metal have a different spatial arrangement relative to each other, leading to cis–trans or fac–mer isomers) or optical isomers (nonsuperimposable mirror images of one another).

Bonding in Coordination Compounds (23.5)

- Crystal field theory is a bonding model for transition metal complex ions. The model describes how the degeneracy of the *d* orbitals is broken by the repulsive forces between the electrons on the ligands around the metal ion and the *d* orbitals in the metal ion.
- The energy difference between the split *d* orbitals is the crystal field splitting energy (Δ). The magnitude of Δ depends in large part on the ligands bound to the metal.
- Octahedral complexes with a d^4, d^5, d^6, or d^7 metal ion can have two possible electronic configurations with different numbers of unpaired electrons. The first, called high-spin, has the same number of unpaired electrons as the free metal ion and is usually the result of a weak crystal field. The second, called low-spin, has fewer unpaired electrons than the free metal ion and is usually the result of a strong crystal field.

KEY EQUATIONS AND RELATIONSHIPS

Crystal Field Splitting Energy (23.5)

$\Delta = hc/\lambda$ (where λ is the wavelength of maximum absorption)

EXERCISES

REVIEW QUESTIONS

1. When a transition metal atom forms an ion, which electrons are lost first?
2. Explain why transition metals exhibit multiple oxidation states instead of a single oxidation state (which most of the main-group metals do).
3. Why is the +2 oxidation state so common for transition metals?
4. Explain why atomic radii of elements in the third row of the transition metals are no larger than those of elements in the second row.
5. Gold is the most electronegative transition metal. Explain.
6. Briefly define each term.
   a) coordination number
   b) ligand
   c) bidentate and polydentate
   d) complex ion
   e) chelating agent
7. Using the Lewis acid–base definition, how would you categorize a ligand? How would you categorize a transition metal ion?
8. Explain the differences between each pair of isomer types.
   a) structural isomer and stereoisomer
   b) linkage isomer and coordination isomer
   c) geometric isomer and optical isomer
9. Which complex ion geometry has the potential to exhibit cis–trans isomerism: linear, tetrahedral, square planar, octahedral?
10. How can you tell whether a complex ion is optically active?
11. Explain the differences between weak-field and strong-field metal complexes.
12. Explain why compounds of Sc^{3+} are colorless, but compounds of Ti^{3+} are colored.
13. Explain why compounds of Zn^{2+} are white, but compounds of Cu^{2+} are often blue or green.
14. Explain the differences between high-spin and low-spin metal complexes.
15. Why are almost all tetrahedral complexes high-spin?
16. Many transition metal compounds are colored. How does crystal field theory account for this?

PROBLEMS BY TOPIC

Properties of Transition Metals

17. Write the ground state electron configuration for each atom and ion pair.
    a. Ni, Ni^{2+} b. Mn, Mn^{4+} c. Y, Y^{+} d. Ta, Ta^{2+}
18. Write the ground state electron configuration for each atom and ion pair.
    a. Zr, Zr^{2+} b. Co, Co^{2+} c. Tc, Tc^{3+} d. Os, Os^{4+}
19. Determine the highest possible oxidation state for each element.
    a. V b. Re c. Pd
20. Which first-row transition metal(s) has the following highest possible oxidation state?
    a. +3 b. +7 c. +4

Coordination Compounds

21. Determine the oxidation state and coordination number of the metal ion in each complex ion.
    a. $[Cr(H_2O)_6]^{3+}$ b. $[Co(NH_3)_3Cl_3]^-$
    c. $[Cu(CN)_4]^{2-}$ d. $[Ag(NH_3)_2]^+$
22. Determine the oxidation state and coordination number of the metal ion in each complex ion.
    a. $[Co(NH_3)_5Br]^{2+}$ b. $[Fe(CN)_6]^{4-}$
    c. $[Co(ox)_3]^{4-}$ d. $[PdCl_4]^{2-}$
23. Name each complex ion or coordination compound.
    a. $[Cr(H_2O)_6]^{3+}$ b. $[Cu(CN)_4]^{2-}$
    c. $[Fe(NH_3)_5Br]SO_4$ d. $[Co(H_2O)_4(NH_3)(OH)]Cl_2$

24. Name each complex ion or coordination compound.

a. $[Cu(en)_2]^{2+}$
b. $[Mn(CO)_3(NO_2)_3]^{2+}$
c. $Na[Cr(H_2O)_2(ox)_2]$
d. $[Co(en)_3][Fe(CN)_6]$

25. Write the formula for each complex ion or coordination compound.

a. hexaamminechromium(III)
b. potassium hexacyanoferrate(III)
c. ethylenediaminedithiocyanatocopper(II)
d. tetraaquaplatinum(II) hexachloroplatinate(IV)

26. Write the formula for each complex ion or coordination compound.

a. hexaaquanickel(II) chloride
b. pentacarbonylchloromanganese(I)
c. ammonium diaquatetrabromovanadate(III)
d. tris(ethylenediamine)cobalt(III) trioxalatoferrate(III)

27. Write the formula and the name of each complex ion.

a. a complex ion with Co^{3+} as the central ion and three NH_3 molecules and three CN^- ions as ligands
b. a complex ion with Cr^{3+} as the central ion and a coordination number of 6 with ethylenediamine ligands

28. Write the formula and the name of each complex ion or coordination compound.

a. a complex ion with four water molecules and two ONO^- ions connected to an iron(III) ion
b. a coordination compound made of two complex ions: one a complex of vanadium(III) with two ethylenediamine molecules and two Cl^- ions as ligands and the other a complex of nickel(II) having a coordination number of 4 with Cl^- ions as ligands

Structure and Isomerism

29. Draw two linkage isomers of $[Mn(NH_3)_5(NO_2)]^{2+}$.

30. Draw two linkage isomers of $[PtCl_3(SCN)]^{2-}$.

31. Write the formulas and names for the coordination isomers of $[Fe(H_2O)_6]Cl_2$.

32. Write the formulas and names for the coordination isomers of $[Co(en)_3][Cr(ox)_3]$.

33. Which complexes exhibit geometric isomerism?

a. $[Cr(NH_3)_5(OH)]^{2+}$
b. $[Cr(en)_2Cl_2]^+$
c. $[Cr(H_2O)(NH_3)_3Cl_2]^+$
d. $[Pt(NH_3)Cl_3]^-$
e. $[Pt(H_2O)_2(CN)_2]$

34. Which complexes exhibit geometric isomerism?

a. $[Co(H_2O)_2(ox)_2]^-$
b. $[Co(en)_3]^{3+}$
c. $[Co(H_2O)_2(NH_3)_2(ox)]^+$
d. $[Ni(NH_3)_2(en)]^{2+}$
e. $[Ni(CO)_2Cl_2]$

35. If W, X, Y, and Z are different monodentate ligands, how many geometric isomers are there for each ion?

a. square planar $[NiWXYZ]^{2+}$
b. tetrahedral $[ZnWXYZ]^{2+}$

36. How many geometric isomers are there for each species?

a. $[Fe(CO)_3Cl_3]$
b. $[Mn(CO)_2Cl_2Br_2]^+$

37. Draw the structures and label the type for all the isomers of each ion.

a. $[Cr(CO)_3(NH_3)_3]^{3+}$
b. $[Pd(CO)_2(H_2O)Cl]^+$

38. Draw the structures and label the type for all the isomers of each species.

a. $[Fe(CO)_4Cl_2]^+$
b. $[Pt(en)Cl_2]$

39. Determine if either isomer of $[Cr(NH_3)_2(ox)_2]^-$ is optically active.

40. Determine if either isomer of $[Fe(CO)_3Cl_3]$ is optically active.

Bonding in Coordination Compounds

41. Draw the octahedral crystal field splitting diagram for each metal ion.

a. Zn^{2+}
b. Fe^{3+} (high- and low-spin)
c. V^{3+}
d. Co^{2+} (high-spin)

42. Draw the octahedral crystal field splitting diagram for each metal ion.

a. Cr^{3+}
b. Cu^{2+}
c. Mn^{3+} (high- and low-spin)
d. Fe^{2+} (low-spin)

43. The $[CrCl_6]^{3-}$ ion has a maximum absorbance in its absorption spectrum at 735 nm. Calculate the crystal field splitting energy (in kJ/mol) for this ion.

44. The absorption spectrum of the complex ion $[Rh(NH_3)_6]^{3+}$ has maximum absorbance at 295 nm. Calculate the crystal field splitting energy (in kJ/mol) for this ion.

45. Three complex ions of cobalt(III), $[Co(CN)_6]^{3-}$, $[Co(NH_3)_6]^{3+}$, and $[CoF_6]^{3-}$, absorb light at wavelengths of (in no particular order) 290 nm, 440 nm, and 770 nm. Match each complex ion to the appropriate wavelength absorbed. What color would you expect each solution to be?

46. Three bottles of aqueous solutions are discovered in an abandoned lab. The solutions are green, yellow, and purple. It is known that three complex ions of chromium(III) were commonly used in that lab: $[Cr(H_2O)_6]^{3+}$, $[Cr(NH_3)_6]^{3+}$, and $[Cr(H_2O)_4Cl_2]^+$. Determine the likely identity of each of the colored solutions.

47. The $[Mn(NH_3)_6]^{2+}$ ion is paramagnetic with five unpaired electrons. The NH_3 ligand is usually a strong-field ligand. Is NH_3 acting as a strong-field in this case?

48. The complex $[Fe(H_2O)_6]^{2+}$ is paramagnetic. Is the H_2O ligand inducing a strong or weak field?

49. How many unpaired electrons do you expect each complex ion to have?

a. $[RhCl_6]^{3-}$
b. $[Co(OH)_6]^{4-}$
c. *cis*- $[Fe(en)_2(NO_2)_2]^+$

50. How many unpaired electrons do you expect each complex ion to have?

a. $[Cr(CN)_6]^{4-}$
b. $[MnF_6]^{4-}$
c. $[Ru(en)_3]^{2+}$

51. How many unpaired electrons do you expect the complex ion $[CoCl_4]^{2-}$ to have if it is a tetrahedral shape?

52. The complex ion $[PdCl_4]^{2-}$ is known to be diamagnetic. Use this information to determine if it is a tetrahedral or square planar structure.

Applications of Coordination Compounds

53. What structural feature do hemoglobin, cytochrome c, and chlorophyll have in common?

54. Identify the central metal atom in each complex.

a. hemoglobin
b. carbonic anhydrase
c. chlorophyll
d. iron blue

55. Hemoglobin exists in two predominant forms in our bodies. One form, known as oxyhemoglobin, has O_2 bound to the iron and the other, known as deoxyhemoglobin, has a water molecule bound instead. Oxyhemoglobin is a low-spin complex that gives arterial blood its red color, and deoxyhemoglobin is a high-spin complex that gives venous blood its darker color. Explain these observations in terms of crystal field splitting. Would you categorize O_2 as a strong- or weak-field ligand?

56. Carbon monoxide and the cyanide ion are both toxic because they bind more strongly than oxygen to the iron in hemoglobin (Hb).

$$\text{Hb} + O_2 \rightleftharpoons \text{HbO}_2 \qquad K = 2 \times 10^{12}$$
$$\text{Hb} + \text{CO} \rightleftharpoons \text{HbCO} \qquad K = 1 \times 10^{14}$$

Calculate the equilibrium constant value for this reaction.

$$\text{HbO}_2 + \text{CO} \rightleftharpoons \text{HbCO} + O_2$$

Does the equilibrium favor reactants or products?

CUMULATIVE PROBLEMS

57. Recall from Chapter 4 that Cr and Cu are exceptions to the normal orbital filling, resulting in a $[\text{Ar}]4s^1 3d^x$ configuration. Write the ground state electron configuration for each species.

a. Cr, Cr^+, Cr^{2+}, Cr^{3+} **b.** Cu, Cu^+, Cu^{2+}

58. Most of the second-row transition metals do not follow the normal orbital filling pattern. Five of them—Nb, Mo, Ru, Rh, and Ag—have a $[\text{Kr}]\ 5s^1 4d^x$ configuration and Pd has a $[\text{Kr}]\ 4d^{10}$ configuration. Write the ground state electron configuration for each species.

a. Mo, Mo^+, Ag, Ag^+ **b.** Ru, Ru^{3+}
c. Rh, Rh^{2+} **d.** Pd, Pd^+, Pd^{2+}

59. Draw the Lewis diagrams for each ligand. Indicate the lone pair(s) that may be donated to the metal. Indicate any you expect to be bidentate or polydentate.

a. NH_3 **b.** SCN^- **c.** H_2O

60. Draw the Lewis diagrams for each ligand. Indicate the lone pair(s) that may be donated to the metal. Indicate any you expect to be bidentate or polydentate.

a. CN^-
b. bipyridyl (bipy), which has the following structure:

N N

c. NO_2^-

61. List all the different formulas for an octahedral complex made from a metal (M) and three different ligands (A, B, and C). Describe any isomers for each complex.

62. Amino acids, such as glycine (gly), form complexes with the trace metal ions found in the bloodstream. Glycine, whose structure is shown here, acts as a bidentate ligand coordinating with the nitrogen atom and one of the oxygen atoms.

```
      H   O
      |   ||
H2N—C—C—OH
      |
      H
```

Draw all the possible isomers of:

a. square planar $[\text{Ni(gly)}_2]$
b. tetrahedral $[\text{Zn(gly)}_2]$
c. octahedral $[\text{Fe(gly)}_3]$

63. Oxalic acid solutions remove rust stains. Draw a complex ion that is likely responsible for this effect. Does it have any isomers?

64. W, X, Y, and Z are different monodentate ligands.

a. Is the square planar $[\text{NiWXYZ}]^{2+}$ optically active?
b. Is the tetrahedral $[\text{ZnWXYZ}]^{2+}$ optically active?

65. Hexacyanomanganate(III) ion is a low-spin complex. Draw the crystal field splitting diagram with electrons filled in appropriately. Is this complex paramagnetic or diamagnetic?

66. Determine the color and approximate wavelength absorbed most strongly by each solution.

a. blue solution **b.** red solution **c.** yellow solution

67. Draw the structures of all the geometric isomers of $[\text{Ru}(H_2O)_2(NH_3)_2Cl_2]^+$. Draw the mirror images of any that are chiral.

68. A 0.32 mol amount of NH_3 is dissolved in 0.47 L of a 0.38 M silver nitrate solution. Calculate the equilibrium concentrations of all species in the solution.

CHALLENGE PROBLEMS

69. When a solution of $PtCl_2$ reacts with the ligand trimethylphosphine, $P(CH_3)_3$, two compounds are produced. The compounds share the same elemental analysis: 46.7% Pt; 17.0% Cl; 14.8% P; 17.2% C; 4.34% H. Determine the formula, draw the structure, and give the systematic name for each compound.

70. Draw a crystal field splitting diagram for a trigonal planar complex ion. Assume the plane of the molecule is perpendicular to the *z* axis.

71. Draw a crystal field splitting diagram for a trigonal bipyramidal complex ion. Assume the axial positions are on the *z* axis.

72. Explain why $[\text{Ni}(NH_3)_4]^{2+}$ is paramagnetic, while $[\text{Ni(CN)}_4]^{2-}$ is diamagnetic.

73. Sulfide (S^{2-}) salts are notoriously insoluble in aqueous solution.

a. Calculate the molar solubility of nickel(II) sulfide in water. $K_{sp}(\text{NiS}) = 3 \times 10^{-16}$
b. Nickel(II) ions form a complex ion in the presence of ammonia with a formation constant (K_f) of 2.0×10^8: $Ni^{2+} + 6\,NH_3 \rightleftharpoons [\text{Ni}(NH_3)_6]^{2+}$. Calculate the molar solubility of NiS in 3.0 M NH_3.
c. Explain any differences between the answers to parts a and b.

74. Calculate the solubility of $\text{Zn(OH)}_2(s)$ in 2.0 M NaOH solution. (*Hint:* You must take into account the formation of Zn(OH)_4^{2-}, which has a $K_f = 2 \times 10^{15}$.)

75. Halide complexes of metal M of the form $[MX_6]^{3-}$ are found to be stable in aqueous solution. But it is possible that they undergo rapid ligand exchange with water (or other ligands) that is not detectable because the complexes are less stable. This property is referred to as their *lability*. Suggest an experiment to measure the lability of these complexes that does not employ radioactive labels.

76. The K_f for $[\text{Cu(en)}_2]^{2+}$ is much larger than the one for $[\text{Cu}(NH_3)_4]^{2+}$. This difference is primarily an entropy effect. Explain why and calculate the difference between the $\Delta S°$ values at 298 K for the complete dissociation of the two complex ions. (*Hint:* The value of ΔH is about the same for both systems.)

77. When solid Cd(OH)_2 is added to a solution of 0.10 M NaI, some of it dissolves. Calculate the pH of the solution at equilibrium.

CONCEPTUAL PROBLEMS

78. Two ligands, A and B, both form complexes with a particular metal ion. When the metal ion complexes with ligand A, the solution is green. When the metal ion complexes with ligand B, the solution is violet. Which of the two ligands results in the larger Δ?

79. Which element has the higher first ionization energy, Cu or Au?

80. The complexes of Fe^{3+} have magnetic properties that depend on whether the ligands are strong or weak field. Explain why this observation supports the idea that electrons are lost from the 4*s* orbital before the 3*d* orbitals in the transition metals.

ANSWERS TO CONCEPTUAL CONNECTIONS

Cc 23.1 The element W has the larger radius because it is in the third transition row and Fe is in the first. Atomic radii increase from the first to the second transition row and stay roughly constant from the second to the third.

Cc 23.2 Ligand B forms a yellow solution, which means that the complex absorbs in the violet region. Ligand A forms a red solution, which means that the complex absorbs in the green. Since the violet region of the electromagnetic spectrum is of shorter wavelength (higher energy) than the green region, ligand B produces a higher Δ.

APPENDIX I

The Units of Measurement

The two most common unit systems are the *metric system*, used in most of the world, and the *English System*, used in the United States. Scientists use the International System of Units (SI), which is based on the metric system.

The abbreviation *SI* comes from the French, *Système International d'Unités.*

The Standard Units

Table A1.1 shows the standard SI base units. In this appendix, we focus on the first four of these units: the *meter*, the standard unit of length; the *kilogram*, the standard unit of mass; the *second*, the standard unit of time; and the *kelvin*, the standard unit of temperature.

TABLE A1.1 SI Base Units

Quantity	Unit	Symbol
Length	Meter	m
Mass	Kilogram	kg
Time	Second	s
Temperature	Kelvin	K
Amount of substance	Mole	mol
Electric current	Ampere	A
Luminous intensity	Candela	cd

The velocity of light in a vacuum is 3.00×10^8 m/s.

Scientific notation is reviewed in Appendix IIIA.

The Meter: A Measure of Length

A **meter (m)** is slightly longer than a yard (1 yard is 36 inches while 1 meter is 39.37 inches). Thus, a 100-yard football field measures only 91.4 meters. The meter was originally defined as 1/10,000,000 of the distance from the equator to the North Pole (through Paris). The International Bureau of Weights and Measures now defines it more precisely as the distance light travels through a vacuum in a designated period of time, 1/299,792,458 second. Scientists commonly deal with a wide range of lengths and distances. The separation between the sun and the closest star (Proxima Centauri) is about 3.8×10^{16} m, while many chemical bonds measure about 1.5×10^{-10} m.

The Kilogram: A Measure of Mass

The **kilogram (kg)**, defined as the mass of a metal cylinder kept at the International Bureau of Weights and Measures at Sèvres, France, is a measure of *mass*, a quantity different from *weight*. The **mass** of an object is a measure of the quantity of matter within it, while the weight of an object is a measure of the *gravitational pull* on its matter. If you could weigh yourself on the moon, for example, its weaker gravity would pull on you with less force than does Earth's gravity, resulting in a lower weight. A 130-pound (lb) person on Earth would weigh only 21.5 lb on the moon. However, the person's mass—the quantity of matter in his or her body—remains the same on every planet. One kilogram of mass is the equivalent of 2.205 lb of weight on Earth, so if we express mass in kilograms, a 130-lb person has a mass of approximately 59 kg and this book has a mass of about 2.5 kg. Another common unit of mass is the gram (g). One gram is 1/1000 kg. A nickel (5¢) has a mass of about 5 g.

▲ A nickel (5 cents) weighs about 5 grams.

The Second: A Measure of Time

If you live in the United States, the **second (s)** is perhaps the most familiar SI unit. The International Bureau of Weights and Measures originally defined the second in terms of the day and the year, but a second is now defined more precisely as the duration of 9,192,631,770 periods of the radiation emitted from a certain transition in a cesium-133 atom. (We discuss transitions and the emission of radiation by atoms in Chapter 3.) Scientists measure time on a large range of scales. The human heart beats about once every second; the age of the universe is estimated to be about 4.32×10^{17} s (13.7 billion years); and some molecular bonds break or form in time periods as short as 1×10^{-15} s.

The Kelvin: A Measure of Temperature

The **kelvin (K)** is the SI unit of **temperature**. The temperature of a sample of matter is a measure of the amount of average kinetic energy—the energy due to motion—of the atoms or molecules that compose the matter. The molecules in a *hot* glass of water are, on average, moving faster than the molecules in a *cold* glass of water. Temperature is a measure of this molecular motion.

Temperature also determines the direction of thermal energy transfer, or what we commonly call *heat.* Thermal energy transfers from hot objects to cold ones. For example, when you touch another person's warm hand (and yours is cold), thermal energy flows *from their hand to yours*, making your hand feel warmer. However, if you touch an ice cube, thermal energy flows *out of your hand* to the ice, cooling your hand (and possibly melting some of the ice cube).

Figure A1.1 shows the three temperature scales. The most common in the United States is the *Fahrenheit scale* (°F), shown on the left. On the Fahrenheit scale, water freezes at 32 °F and boils at 212 °F at sea level. Room temperature is approximately 72 °F. The Fahrenheit scale was originally determined by assigning 0 °F to the freezing point of a concentrated saltwater solution and 96 °F to normal body temperature.

Normal body temperature was later measured more accurately to be 98.6 °F.

Scientists and citizens of most countries other than the United States typically use the *Celsius* (°C) *scale*, shown in the middle of Figure A1.1 ▼. On this scale, pure water freezes at 0 °C and boils at 100 °C (at sea level). Room temperature is approximately 22 °C. The Fahrenheit scale and the Celsius scale differ both in the size of their respective degrees and the temperature each designates as "zero." Both the Fahrenheit and Celsius scales allow for negative temperatures.

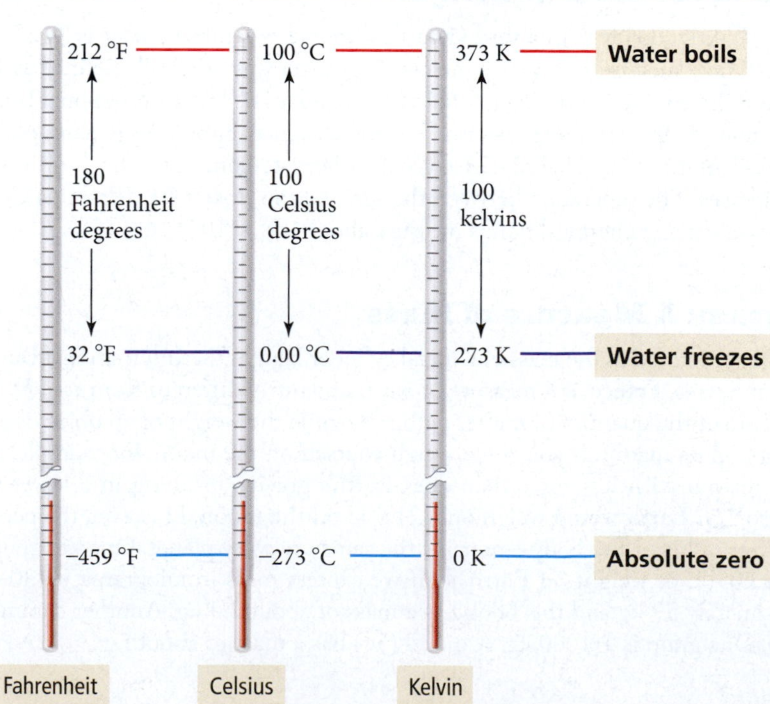

▶ **FIGURE A1.1 Comparison of the Fahrenheit, Celsius, and Kelvin Temperature Scales** The Fahrenheit degree is five-ninths the size of the Celsius degree and the kelvin. The zero point of the Kelvin scale is absolute zero (the lowest possible temperature), whereas the zero point of the Celsius scale is the freezing point of water.

The SI unit for temperature, as we have seen, is the kelvin, shown on the right in Figure A1.1. The *Kelvin scale* (sometimes also called the *absolute scale)* avoids negative temperatures by assigning 0 K to the coldest temperature possible, absolute zero. Absolute zero (−273 °C or −459 °F) is the temperature at which molecular motion virtually stops. Lower temperatures do not exist. The size of the kelvin is identical to that of the Celsius degree—the only difference is the temperature that each designates as zero. You can convert between the temperature scales with these formulas:

Molecular motion does not *completely* stop at absolute zero because of the uncertainty principle in quantum mechanics, which we discuss in Chapter 3.

$$°\text{C} = \frac{(°\text{F} - 32)}{1.8}$$

$$\text{K} = °\text{C} + 273.15$$

Note that we refer to Kelvin temperatures in kelvins (*not* "degrees Kelvin") or K (*not* °K).

The Celsius Temperature Scale

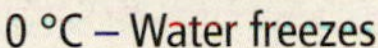

0 °C – Water freezes

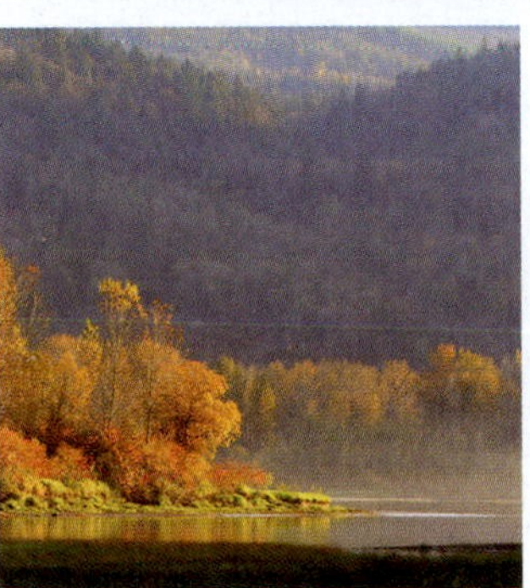

10 °C – Brisk fall day

22 °C – Room temperature

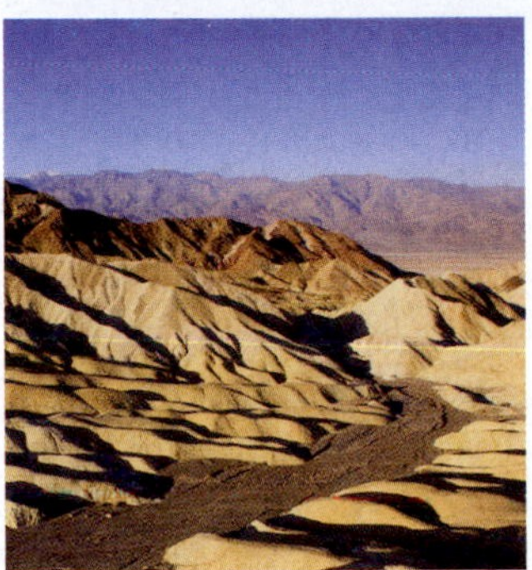

45 °C – Summer day in Death Valley

EXAMPLE A1.1
Converting between Temperature Scales

A sick child has a temperature of 40.00 °C. What is the child's temperature in **(a)** K and **(b)** °F?

SOLUTION

(a) Begin by finding the equation that relates the quantity that is given (°C) and the quantity you are trying to find (K).	$\text{K} = °\text{C} + 273.15$
Since this equation gives the temperature in K directly, substitute in the correct value for the temperature in °C and calculate the answer.	$\text{K} = °\text{C} + 273.15$ $\text{K} = 40.00 + 273.15 = 313.15\ \text{K}$
(b) To convert from °C to °F, find the equation that relates these two quantities.	$°\text{C} = \frac{(°\text{F} - 32)}{1.8}$
Since this equation expresses °C in terms of °F, solve the equation for °F.	$°\text{C} = \frac{(°\text{F} - 32)}{1.8}$ $1.8(°\text{C}) = (°\text{F} - 32)$ $°\text{F} = 1.8(°\text{C}) + 32$
Now substitute °C into the equation and calculate the answer. *Note: The number of digits reported in this answer follows significant figure conventions, covered in Appendix II.*	$°\text{F} = 1.8(°\text{C}) + 32$ $°\text{F} = 1.8(40.00\ °\text{C}) + 32 = 104.00\ °\text{F}$

FOR PRACTICE A1.1

Gallium is a solid metal at room temperature, but will melt to a liquid in your hand. The melting point of gallium is 85.6 °F. What is this temperature on (a) the Celsius scale and (b) the Kelvin scale?

Prefix Multipliers

Scientific notation (see Appendix IIIA) allows us to express very large or very small quantities in a compact manner by using exponents. For example, we write the diameter of a hydrogen atom as 1.06×10^{-10} m. The International System of Units uses the **prefix multipliers** shown in Table A1.2 with the standard units. These multipliers change the value of the unit by powers of 10 (just like an exponent does in scientific notation). For example, the kilometer has the prefix "kilo" meaning 1000 or 10^3. Therefore,

$$1 \text{ kilometer} = 1000 \text{ meters} = 10^3 \text{ meters}$$

TABLE A1.2 SI Prefix Multipliers

Prefix	Symbol	Multiplier	
exa	E	1,000,000,000,000,000,000	(10^{18})
peta	P	1,000,000,000,000,000	(10^{15})
tera	T	1,000,000,000,000	(10^{12})
giga	G	1,000,000,000	(10^{9})
mega	M	1,000,000	(10^{6})
kilo	k	1000	(10^{3})
deci	d	0.1	(10^{-1})
centi	c	0.01	(10^{-2})
milli	m	0.001	(10^{-3})
micro	μ	0.000001	(10^{-6})
nano	n	0.000000001	(10^{-9})
pico	p	0.000000000001	(10^{-12})
femto	f	0.000000000000001	(10^{-15})
atto	a	0.000000000000000001	(10^{-18})

Similarly, the millimeter has the prefix "milli" meaning 0.001 or 10^{-3}.

$$1 \text{ millimeter} = 0.001 \text{ meters} = 10^{-3} \text{ meters}$$

When we report a measurement, we choose a prefix multiplier close to the size of the quantity we are measuring. For example, to state the diameter of a hydrogen atom, which is 1.06×10^{-10} m, we use picometers (106 pm) or nanometers (0.106 nm) rather than micrometers or millimeters. We choose the prefix multiplier that is most convenient for a particular number.

Units of Volume

Many scientific measurements can be expressed in a single unit but require combinations of units. For example, velocities are often reported in units such as km/s, and densities are often reported in units of g/cm^3. Both of these units are *derived units*, combinations of other units. An important SI-derived unit for chemistry is the m^3, used to report measurements of volume.

Volume is a measure of space. Any unit of length, when cubed (raised to the third power), becomes a unit of volume. The cubic meter (m^3), cubic centimeter (cm^3), and cubic millimeter (mm^3) are all units of volume. The cubic nature of volume is not always intuitive, and studies have shown that our brains are not naturally wired to think abstractly, which we need to do in order to think about volume. For example, consider this question: How many small cubes measuring 1 cm on each side are required to construct a large cube measuring 10 cm (or 1 dm) on a side?

The answer to this question, as we can see by carefully examining the unit cube in Figure A1.2 ▼, is 1000 small cubes. When we go from a linear, one-dimensional distance to a three-dimensional volume, we must raise both the linear dimension *and* its unit to the third power (not just multiply by 3). The volume of a cube is equal to the length of its edge cubed:

$$\text{volume of cube} = (\text{edge length})^3$$

A cube with a 10-cm edge length has a volume of $(10 \text{ cm})^3$ or 1000 cm^3, and a cube with a 100-cm edge length has a volume of $(100 \text{ cm})^3 = 1{,}000{,}000 \text{ cm}^3$. Other common units of volume in chemistry are the **liter (L)** and the **milliliter (mL).** One milliliter (10^{-3} L) is equal to 1 cm^3. A gallon of gasoline contains 3.785 L. Table A1.3 lists some common units for volume and their equivalents.

Relationship between Length and Volume

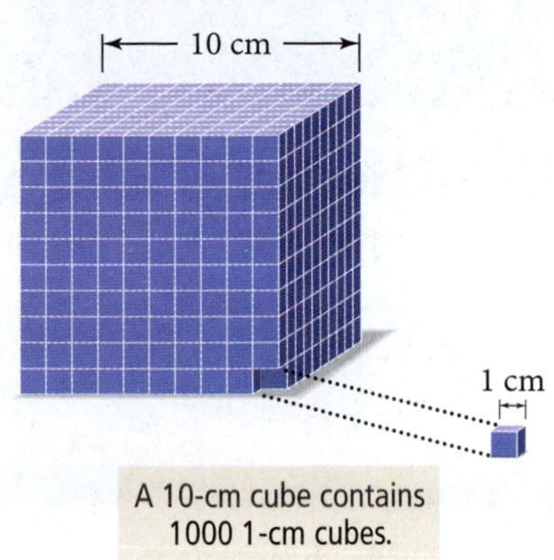

▲ **FIGURE A1.2 The Relationship between Length and Volume**

TABLE A1.3 Common units for volume and their equivalents

1 liter (L) = 1000 mL = 1000 cm^3
1 liter (L) = 1.057 quarts (qt)
1 U.S. gallon (gal) = 3.785 liters (L)

Exercises

1. Convert each temperature.
   a. 32 °F to °C (temperature at which water freezes)
   b. 77 K to °F (temperature of liquid nitrogen)
   c. −109 °F to °C (temperature of dry ice)
   d. 98.6 °F to K (body temperature)
2. Convert each temperature.
   a. 212 °F to °C (temperature of boiling water at sea level)
   b. 22 °C to K (approximate room temperature)
   c. 0.00 K to °F (coldest temperature possible, also known as absolute zero)
   d. 2.735 K to °C (average temperature of the universe as measured from background black body radiation)
3. The coldest temperature ever measured in the United States is −80 °F on January 23, 1971, in Prospect Creek, Alaska. Convert that temperature to °C and K. (Assume that −80 °F is accurate to two significant figures.)
4. The warmest temperature ever measured in the United States is 134 °F on July 10, 1913, in Death Valley, California. Convert that temperature to °C and K.
5. Use the prefix multipliers to express each measurement without any exponents.
   a. 1.2×10^{-9} m **b.** 22×10^{-15} s
   c. 1.5×10^{9} g **d.** 3.5×10^{6} L
6. Use prefix multipliers to express each measurement without any exponents.
   a. 38.8×10^{5} g **b.** 55.2×10^{-10} s
   c. 23.4×10^{11} m **d.** 87.9×10^{-7} L
7. Use scientific notation to express each quantity with only the base units (no prefix multipliers).
   a. 4.5 ns **b.** 18 fs **c.** 128 pm **d.** 35 μm
8. Use scientific notation to express each quantity with only the base units (no prefix multipliers).
   a. 35 μL **b.** 225 Mm **c.** 133 Tg **d.** 1.5 cg
9. Complete the table.

a. 1245 kg	1.245×10^{6} g	1.245×10^{9} mg
b. 515 km	________ dm	________ cm
c. 122.355 s	________ ms	________ ks
d. 3.345 kJ	________ J	________ mJ

10. Complete the table.

a. 355 km/s	________ cm/s	________ m/ms
b. 1228 g/L	________ g/mL	________ kg/ML
c. 554 mK/s	________ K/s	________ μK/ms
d. 2.554 mg/mL	________ g/L	________ μg/mL

11. Express the quantity 254,998 m in each unit.
    a. km **b.** Mm **c.** mm **d.** cm
12. Express the quantity 556.2×10^{-12} s in each unit.
    a. ms **b.** ns **c.** ps **d.** fs
13. How many 1-cm squares does it take to construct a square that is 1 m on each side?
14. How many 1-cm cubes does it take to construct a cube that is 4 cm on edge?
15. Convert 15.0 L to each unit.
    a. mL **b.** cm^3 **c.** gal **d.** qt
16. Convert $4.58 \times 10^3 \text{ cm}^3$ to each unit.
    a. L **b.** mL **c.** gal **d.** qt

APPENDIX II

Significant Figure Guidelines

Counting Significant Figures

In Section 2.2, we discussed how the precision of a measurement depends on the instrument used to make the measurement. This precision must be preserved not only when recording the measurement, but also when performing calculations that use the measurement. We preserve this precision by using *significant figures*. In any reported measurement, the non–place-holding digits—those that are not simply marking the decimal place—are called *significant figures* (or *significant digits*). For example, the number 23.5 has three significant figures while the number 23.56 has four. *The greater the number of significant figures, the greater the certainty of the measurement.*

To determine the number of significant figures in a number containing zeroes, we distinguish between zeroes that are significant and those that simply mark the decimal place. For example, in the number 0.0008, the leading zeroes (zeroes to the left of the first nonzero digit) mark the decimal place but *do not* add to the certainty of the measurement and are therefore not significant; this number has only one significant figure. In contrast, the trailing zeroes (zeroes at the end of a number) in the number 0.000800 *do add* to the certainty of the measurement and are therefore counted as significant; this number has three significant figures.

To determine the number of significant figures in a number, follow these rules. (See examples on the right.)

Significant Figure Rules	Examples	
1. All nonzero digits are significant.	28.03	0.0540
2. Interior zeroes (zeroes between two nonzero digits) are significant.	408	7.0301
3. Leading zeroes (zeroes to the left of the first nonzero digit) are not significant. They only serve to locate the decimal point.	0.0032 (0.00 not significant)	0.00006 (0.0000 not significant)
4. Trailing zeroes (zeroes at the end of a number) are categorized as follows:		
• Trailing zeroes after a decimal point are always significant.	45.000	3.5600
• Trailing zeroes before a decimal point (and after a nonzero number) are always significant.	140.00	2500.55
• Trailing zeroes before an *implied* decimal point are ambiguous and should be avoided by using scientific notation.	1200	ambiguous
	1.2×10^3	2 significant figures
	1.20×10^3	3 significant figures
	1.200×10^3	4 significant figures
• Some textbooks put a decimal point after one or more trailing zeroes if the zeroes are to be considered significant. We avoid that practice in this book, but you should be aware of it.	1200.	4 significant figures (common in some textbooks)

Exact Numbers

Exact numbers have no uncertainty and thus do not limit the number of significant figures in any calculation. We regard an exact number as having an unlimited number of significant figures. Exact numbers originate from three sources:

- From the accurate counting of discrete objects. For example, 3 atoms means 3.00000… atoms.
- From defined quantities, such as the number of centimeters in 1 m. Because 100 cm is defined as 1 m,

$$100 \text{ cm} = 1 \text{ m} \quad \textit{means} \quad 100.00000\ldots \text{ cm} = 1.0000000\ldots \text{ m}$$

- From integral numbers that are part of an equation. For example, in the equation, $radius = \dfrac{diameter}{2}$, the number 2 is exact and therefore has an unlimited number of significant figures.

EXAMPLE A2.1

Determining the Number of Significant Figures in a Number

How many significant figures are in each number?

(a) 0.04450 m
(b) 5.0003 km
(c) 10 dm = 1 m
(d) 1.000×10^5 s
(e) 0.00002 mm
(f) 10,000 m

SOLUTION

(a) 0.04450 m	*Four significant figures.* The two 4s and the 5 are significant (Rule 1). The trailing zero is after a decimal point and is therefore significant (Rule 4). The leading zeroes only mark the decimal place and are therefore not significant (Rule 3).
(b) 5.0003 km	*Five significant figures.* The 5 and 3 are significant (Rule 1), as are the three interior zeroes (Rule 2).
(c) 10 dm = 1 m	*Unlimited significant figures.* Defined quantities have an unlimited number of significant figures.
(d) 1.000×10^5 s	*Four significant figures.* The 1 is significant (Rule 1). The trailing zeroes are after a decimal point and therefore significant (Rule 4).
(e) 0.00002 mm	*One significant figure.* The 2 is significant (Rule 1). The leading zeroes only mark the decimal place and are therefore not significant (Rule 3).
(f) 10,000 m	*Ambiguous.* The 1 is significant (Rule 1), but the trailing zeroes occur before an implied decimal point and are therefore ambiguous (Rule 4). Without more information, you would assume one significant figure. It is better to write this as 1×10^5 to indicate one significant figure or as 1.0000×10^5 to indicate five (Rule 4).

FOR PRACTICE A2.1

How many significant figures are in each number?

(a) 554 km
(b) 7 pennies
(c) 1.01×10^5 m
(d) 0.00099 s
(e) 1.4500 km
(f) 21,000 m

Significant Figures in Calculations

When we use measured quantities in calculations, the results of the calculation must reflect the precision of the measured quantities. We should not lose or gain precision during mathematical operations. Follow these rules when carrying significant figures through calculations.

Rules for Calculations	Examples
1. In multiplication or division, the result carries the same number of significant figures as the factor with the fewest significant figures.	1.052 (4 sig. figures) × 12.054 (5 sig. figures) × 0.53 (2 sig. figures) = 6.7208 = 6.7 (2 sig. figures) 2.0035 (5 sig. figures) ÷ 3.20 (3 sig. figures) = 0.626094 = 0.626 (3 sig. figures)
2. In addition or subtraction, the result carries the same number of decimal places as the quantity with the fewest decimal places.	2.345 + 0.07 + 2.9975 = 5.4125 = 5.41 5.9 − 0.221 = 5.679 = 5.7 In addition and subtraction, it is helpful to draw a line next to the number with the fewest decimal places. This line determines the number of decimal places in the answer.
3. When rounding to the correct number of significant figures, round down if the last (or leftmost) digit dropped is 4 or less; round up if the last (or leftmost) digit dropped is 5 or more.	To two significant figures: 5.37 rounds to 5.4 5.34 rounds to 5.3 5.35 rounds to 5.4 5.349 rounds to 5.3 Notice in the last example that only the *last (or leftmost) digit being dropped* determines in which direction to round—ignore all digits to the right of it.
4. To avoid rounding errors in multistep calculations round only the final answer—do not round intermediate steps. If you write down intermediate answers, keep track of significant figures by underlining the least significant digit.	$6.78 \times 5.903 \times (5.489 - 5.01)$ $= 6.78 \times 5.903 \times 0.4\underline{7}9$ $= 19.1707$ $= 19$ (underline least significant digit)

A few books recommend a slightly different rounding procedure for in cases where the last digit is 5. However, the procedure presented here is consistent with electronic calculators, and we use it throughout this book.

Notice that for multiplication or division, the quantity with the fewest *significant figures* determines the number of *significant figures* in the answer, but for addition and subtraction, the quantity with the fewest *decimal places* determines the number of *decimal places* in the answer. In multiplication and division, we focus on significant figures, but in addition and subtraction we focus on decimal places. When a problem involves addition or subtraction, the answer may have a different number of significant figures than the initial quantities. Keep this in mind in problems that involve both addition or subtraction and multiplication or division. For example,

$$\frac{1.002 - 0.999}{3.754} = \frac{0.003}{3.754}$$
$$= 7.99 \times 10^{-4}$$
$$= 8 \times 10^{-4}$$

The answer has only one significant figure, even though the initial numbers had three or four significant figures.

EXAMPLE A2.2
Significant Figures in Calculations

Perform each calculation to the correct number of significant figures.

(a) $1.10 \times 0.5120 \times 4.0015 \div 3.4555$

(b)
$$\begin{array}{r} 0.355 \\ +105.1 \\ \underline{-100.5820} \end{array}$$

(c) $4.562 \times 3.99870 \div (452.6755 - 452.33)$

(d) $(14.84 \times 0.55) - 8.02$

SOLUTION

(a) Round the intermediate result (in blue) to three significant figures to reflect the three significant figures in the least precisely known quantity (1.10).	$1.10 \times 0.5120 \times 4.0015 \div 3.4555$ $= 0.65219$ $= 0.652$
(b) Round the intermediate answer (in blue) to one decimal place to reflect the quantity with the fewest decimal places (105.1). Notice that 105.1 is *not* the quantity with the fewest significant figures, but it has the fewest decimal places and therefore determines the number of decimal places in the answer.	0.355 $+105.1$ $\underline{-100.5820}$ $4.8730 = 4.9$
(c) Mark the intermediate result to two decimal places to reflect the number of decimal places in the quantity within the parentheses having the smallest number of decimal places (452.33). Round the final answer to two significant figures to reflect the two significant figures in the least precisely known quantity (0.3455).	$4.562 \times 3.99870 \div (452.6755 - 452.33)$ $= 4.562 \times 3.99870 \div 0.3455$ $= 52.79904$ $= 53$ 2 places of the decimal
(d) Mark the intermediate result to two significant figures to reflect the number of significant figures in the quantity within the parentheses having the smallest number of significant figures (0.55). Round the final answer to one decimal place to reflect the one decimal place in the least precisely known quantity (8.162).	$(14.84 \times 0.55) - 8.02 = 8.162 - 8.02$ $= 0.142$ $= 0.1$

FOR PRACTICE A2.2

Perform each calculation to the correct number of significant figures.

(a) $3.10007 \times 9.441 \times 0.0301 \div 2.31$

(b)
$$\begin{array}{r} 0.881 \\ +132.1 \\ \underline{-12.02} \end{array}$$

(c) $2.5110 \times 21.20 \div (44.11 + 1.223)$

(d) $(12.01 \times 0.3) + 4.811$

EXERCISES

1. For each number, underline the zeroes that are significant and draw an x through the zeroes that are not.
   a. 1,050,501 km
   b. 0.0020 m
   c. 0.000000000000002 s
   d. 0.001090 cm
2. For each number, underline the zeroes that are significant and draw an x through the zeroes that are not.
   a. 180,701 mi
   b. 0.001040 m
   c. 0.005710 km
   d. 90,201 m
3. How many significant figures are in each number?
   a. 0.000312 m
   b. 312,000 s
   c. 3.12×10^5 km
   d. 13,127 s
   e. 2000
4. How many significant figures are in each number?
   a. 0.1111 s
   b. 0.007 m
   c. 108,700 km
   d. 1.563300×10^{11} m
   e. 30,800
5. Which numbers are exact (and therefore have an unlimited number of significant figures)?
   a. $\pi = 3.14$
   b. 12 inches = 1 foot
   c. EPA gas mileage rating of 26 miles per gallon
   d. 1 gross = 144
6. Indicate the number of significant figures in each number. If the number is an exact number, indicate that it has an unlimited number of significant figures.
   a. 305,435,087 (2008 U.S. population)
   b. 2.54 cm = 1 in
   c. 11.4 g/cm^3 (density of lead)
   d. 12 = 1 dozen
7. Round each number to four significant figures.
   a. 156.852
   b. 156.842
   c. 156.849
   d. 156.899
8. Round each number to three significant figures.
   a. 79,845.82
   b. 1.548937×10^7
   c. 2.3499999995
   d. 0.000045389
9. Calculate to the correct number of significant figures.
   a. $9.15 \div 4.970$
   b. $1.54 \times 0.03060 \times 0.69$
   c. $27.5 \times 1.82 \div 100.04$
   d. $(2.290 \times 10^6) \div (6.7 \times 10^4)$
10. Calculate to the correct number of significant figures.
   a. $89.3 \times 77.0 \times 0.08$
   b. $(5.01 \times 10^5) \div (7.8 \times 10^2)$
   c. $4.005 \times 74 \times 0.007$
   d. $453 \div 2.031$
11. Calculate to the correct number of significant figures.
   a. $43.7 - 2.341$
   b. $17.6 + 2.838 + 2.3 + 110.77$
   c. $19.6 + 58.33 - 4.974$
   d. $5.99 - 5.572$
12. Calculate to the correct number of significant figures.
   a. $0.004 + 0.09879$
   b. $1239.3 + 9.73 + 3.42$
   c. $2.4 - 1.777$
   d. $532 + 7.3 - 48.523$
13. Calculate to the correct number of significant figures.
   a. $(24.6681 \times 2.38) + 332.58$
   b. $(85.3 - 21.489) \div 0.0059$
   c. $(512 \div 986.7) + 5.44$
   d. $[(28.7 \times 10^5) \div 48.533] + 144.99$
14. Calculate to the correct number of significant figures.
   a. $[(1.7 \times 10^6) \div (2.63 \times 10^5)] + 7.33$
   b. $(568.99 - 232.1) \div 5.3$
   c. $(9443 + 45 - 9.9) \times 8.1 \times 10^6$
   d. $(3.14 \times 2.4367) - 2.34$

APPENDIX III

Common Mathematical Operations in Chemistry

A. Scientific Notation

A number written in scientific notation consists of a **decimal part**, a number that is usually between 1 and 10, and an **exponential part**, 10 raised to an **exponent**, *n*.

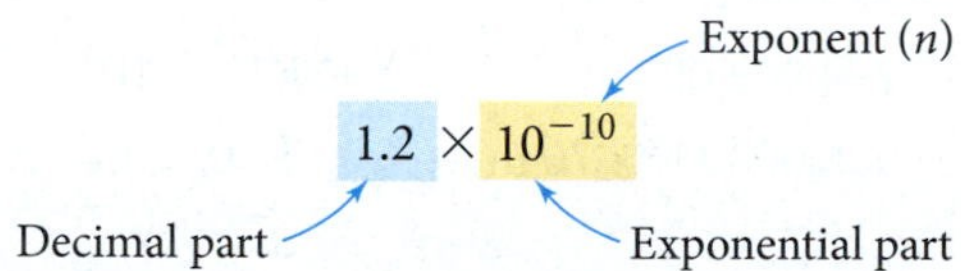

Each of the following numbers is written in both scientific and decimal notation:

$$1.0 \times 10^{5} = 100{,}000 \qquad 1.0 \times 10^{-6} = 0.000001$$

$$6.7 \times 10^{3} = 6700 \qquad 6.7 \times 10^{-3} = 0.0067$$

A positive exponent means 1 multiplied by 10 *n* times.

$$10^{0} = 1$$

$$10^{1} = 1 \times 10$$

$$10^{2} = 1 \times 10 \times 10 = 100$$

$$10^{3} = 1 \times 10 \times 10 \times 10 = 1000$$

A negative exponent ($-n$) means 1 divided by 10 *n* times.

$$10^{-1} = \frac{1}{10} = 0.1$$

$$10^{-2} = \frac{1}{10 \times 10} = 0.01$$

$$10^{-3} = \frac{1}{10 \times 10 \times 10} = 0.001$$

To convert a number to scientific notation, we move the decimal point to obtain a number between 1 and 10 and then multiply by 10 raised to the appropriate power. For example, to write 5983 in scientific notation, we move the decimal point to the left three places to get 5.983 (a number between 1 and 10) and then multiply by 1000 to make up for moving the decimal point.

$$5983 = 5.983 \times 1000$$

Since 1000 is 10^3, we write:

$$5983 = 5.983 \times 10^{3}$$

We can do this in one step by counting how many places we move the decimal point to obtain a number between 1 and 10 and then writing the decimal part multiplied by 10 raised to the number of places we moved the decimal point.

$$5983 = 5.983 \times 10^{3}$$

If the decimal point is moved to the left, as in the previous example, the exponent is positive. If the decimal is moved to the right, the exponent is negative.

$$0.00034 = 3.4 \times 10^{-4}$$

To express a number in scientific notation:

1. **Move the decimal point to obtain a number between 1 and 10.**
2. **Write the result from step 1 multiplied by 10 raised to the number of places you moved the decimal point.**
   - *The exponent is positive if you moved the decimal point to the left.*
   - *The exponent is negative if you moved the decimal point to the right.*

Consider the following additional examples:

$$290{,}809{,}000 = 2.90809 \times 10^{8}$$

$$0.000000000070 \text{ m} = 7.0 \times 10^{-11} \text{ m}$$

Multiplication and Division

To multiply numbers expressed in scientific notation, multiply the decimal parts and add the exponents.

$$(A \times 10^{m})(B \times 10^{n}) = (A \times B) \times 10^{m+n}$$

To divide numbers expressed in scientific notation, divide the decimal parts and subtract the exponent in the denominator from the exponent in the numerator.

$$\frac{(A \times 10^{m})}{(B \times 10^{n})} = \left(\frac{A}{B}\right) \times 10^{m-n}$$

Consider the following example involving multiplication:

$$(3.5 \times 10^{4})(1.8 \times 10^{6}) = (3.5 \times 1.8) \times 10^{4+6}$$

$$= 6.3 \times 10^{10}$$

Consider the following example involving division:

$$\frac{(5.6 \times 10^{7})}{(1.4 \times 10^{3})} = \left(\frac{5.6}{1.4}\right) \times 10^{7-3}$$

$$= 4.0 \times 10^{4}$$

Addition and Subtraction

To add or subtract numbers expressed in scientific notation, rewrite all the numbers so that they have the same exponent, then add or subtract the decimal parts of the numbers. The exponents remained unchanged.

$$\begin{array}{r} A \times 10^{n} \\ \pm B \times 10^{n} \\ \hline (A \pm B) \times 10^{n} \end{array}$$

Notice that the numbers *must have* the same exponent. Consider the following example involving addition:

$$\begin{array}{r} 4.82 \times 10^{7} \\ +3.4 \times 10^{6} \\ \hline \end{array}$$

First, express both numbers with the same exponent. In this case, we rewrite the lower number and perform the addition as follows:

$$\begin{array}{r} 4.82 \times 10^{7} \\ +0.34 \times 10^{7} \\ \hline 5.16 \times 10^{7} \end{array}$$

Consider the following example involving subtraction:

$$\begin{array}{r} 7.33 \times 10^{5} \\ -1.9 \times 10^{4} \\ \hline \end{array}$$

First, express both numbers with the same exponent. In this case, we rewrite the lower number and perform the subtraction as follows:

$$\begin{array}{r} 7.33 \times 10^{5} \\ -0.19 \times 10^{5} \\ \hline 7.14 \times 10^{5} \end{array}$$

Powers and Roots

To raise a number written in scientific notation to a power, raise the decimal part to the power and multiply the exponent by the power:

$$\begin{aligned} (4.0 \times 10^{6})^{2} &= 4.0^{2} \times 10^{6\times 2} \\ &= 16 \times 10^{12} \\ &= 1.6 \times 10^{13} \end{aligned}$$

To take the nth root of a number written in scientific notation, take the nth root of the decimal part and divide the exponent by the root:

$$\begin{aligned} (4.0 \times 10^{6})^{1/3} &= 4.0^{1/3} \times 10^{6/3} \\ &= 1.6 \times 10^{2} \end{aligned}$$

B. Logarithms

Common (or Base 10) Logarithms

The common or base 10 logarithm (abbreviated log) of a number is the exponent to which 10 must be raised to obtain that number. For example, the log of 100 is 2 because 10 must be raised to the second power to get 100. Similarly, the log of 1000 is 3 because 10 must be raised to the third power to get 1000. The logs of several multiples of 10 are shown below:

$$\begin{aligned} \log 10 &= 1 \\ \log 100 &= 2 \\ \log 1000 &= 3 \\ \log 10{,}000 &= 4 \end{aligned}$$

Because $10^{0} = 1$ by definition, $\log 1 = 0$.

The log of a number smaller than one is negative because 10 must be raised to a negative exponent to get a number smaller than one. For example, the log of 0.01 is −2 because 10 must be raised to −2 to get 0.01. Similarly, the log of 0.001 is −3 because 10 must be raised to −3 to get 0.001. The logs of several fractional numbers are shown below:

$$\begin{aligned} \log 0.1 &= -1 \\ \log 0.01 &= -2 \end{aligned}$$

$$\log 0.001 = -3$$

$$\log 0.0001 = -4$$

The logs of numbers that are not multiples of 10 can be computed on your calculator. See your calculator manual for specific instructions.

Inverse Logarithms

The inverse logarithm or invlog function is exactly the opposite of the log function. For example, the log of 100 is 2 and the inverse log of 2 is 100. The log function and the invlog function undo one another.

$$\log 100 = 2$$

$$\text{invlog}\, 2 = 100$$

$$\text{invlog}(\log 100) = 100$$

The inverse log of a number is 10 rasied to that number.

$$\text{invlog}\, x = 10^x$$

$$\text{invlog}\, 3 = 10^3 = 1000$$

The inverse logs of numbers can be computed on your calculator. See your calculator manual for specific instructions.

Natural (or Base *e*) Logarithms

The natural (or base e) logarithm (abbreviated ln) of a number is the exponent to which e (which has the value of 2.71828...) must be raised to obtain that number. For example, the ln of 100 is 4.605 because e must be raised to 4.605 to get 100. Similarly, the ln of 10.0 is 2.303 because e must be raised to 2.303 to get 10.0.

The inverse natural logarithm or invln function is exactly the opposite of the ln function. For example, the ln of 100 is 4.605 and the inverse ln of 4.605 is 100. The inverse ln of a number is simply e raised to that number.

$$\text{invln}\, x = e^x$$

$$\text{invln}\, 3 = e^3 = 20.1$$

The invln of a number can be computed on your calculator. See your calculator manual for specific instructions.

Mathematical Operations Using Logarithms

Because logarithms are exponents, mathematical operations involving logarithms are similar to those involving exponents as follows:

$$\log(a \times b) = \log a + \log b \qquad \ln(a \times b) = \ln a + \ln b$$

$$\log \frac{a}{b} = \log a - \log b \qquad \ln \frac{a}{b} = \ln a - \ln b$$

$$\log a^n = n \log a \qquad \ln a^n = n \ln a$$

C. Quadratic Equations

A quadratic equation contains at least one term in which the variable x is raised to the second power (and no terms in which x is raised to a higher power). A quadratic equation has the following general form:

$$ax^2 + bx + c = 0$$

A quadratic equation can be solved for x using the quadratic formula:

$$x = \frac{-b \pm \sqrt{b^2 - 4ac}}{2a}$$

Quadratic equations are often encountered when solving equilibrium problems. Below we show how to use the quadratic formula to solve a quadratic equation for x.

$$3x^2 - 5x + 1 = 0 \quad (\textit{quadratic equation})$$

$$x = \frac{-b \pm \sqrt{b^2 - 4ac}}{2a}$$

$$= \frac{-(-5) \pm \sqrt{(-5)^2 - 4(3)(1)}}{2(3)}$$

$$= \frac{5 \pm 3.6}{6}$$

$$x = 1.43 \quad \text{or} \quad x = 0.233$$

As you can see, the solution to a quadratic equation usually has two values. In any real chemical system, one of the values can be eliminated because it has no physical significance. (For example, it may correspond to a negative concentration, which does not exist.)

D. Graphs

Graphs are often used to visually show the relationship between two variables. For example, in Chapter 11 we show the following relationship between the volume of a gas and its pressure:

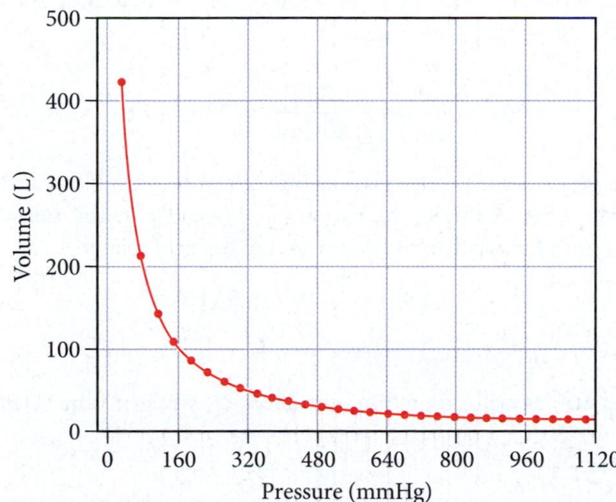

◀ **Volume versus Pressure** A plot of the volume of a gas sample—as measured in a J-tube—versus pressure. The plot shows that volume and pressure are inversely related.

The horizontal axis is the x-axis and is normally used to show the independent variable. The vertical axis is the y-axis and is normally used to show how the other variable (called the dependent variable) varies with a change in the independent variable. In this case, the graph shows that as the pressure of a gas sample increases, its volume decreases.

Many relationships in chemistry are *linear*, which means that if you change one variable by a factor of n the other variable will also change by a factor of n. For example, the volume of a gas is linearly related to the number of moles of gas. When two quantities are linearly related, a graph of one versus the other produces a straight line. For example, the graph below shows how the volume of an ideal gas sample depends on the number of moles of gas in the sample:

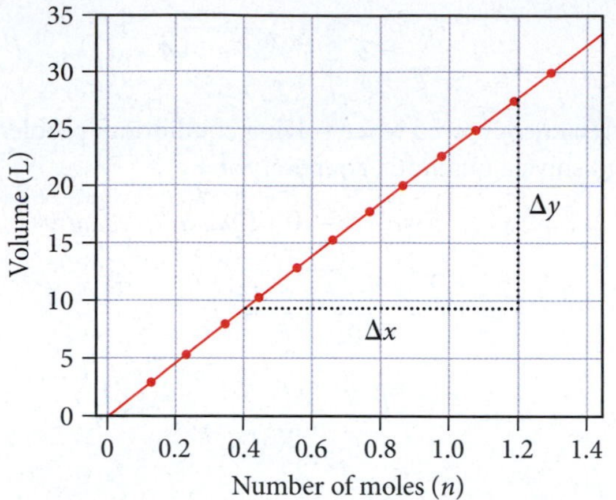

▶ **Volume versus Number of Moles** The volume of a gas sample increases linearly with the number of moles of gas in the sample.

A linear relationship between any two variables x and y can be expressed by the following equation:

$$y = mx + b$$

where m is the slope of the line and b is the y-intercept. The slope is the change in y divided by the change in x.

$$m = \frac{\Delta y}{\Delta x}$$

For the graph above, we can estimate the slope by simply estimating the changes in y and x for a given interval. For example, between $x = 0.4$ mol and 1.2 mol, $\Delta x = 0.80$ mol and we can estimate that $\Delta y = 18$ L. Therefore the slope is

$$m = \frac{\Delta y}{\Delta x} = \frac{18\text{ L}}{0.80\text{ mol}} = 23\text{ L/mol}$$

In several places in this book, logarithmic relationships between variables can be plotted in order to obtain a linear relationship. For example, the variables $[\text{A}]_t$ and t in the following equation are not linearly related, but the natural logarithm of $[\text{A}]_t$ and t are linearly related.

$$\ln[\text{A}]_t = -kt + \ln[\text{A}]_0$$

$$y = mx + b$$

A plot of $\ln[\text{A}]_t$ versus t will therefore produce a straight line with slope $= -k$ and y-intercept $= \ln[\text{A}]_0$.

APPENDIX IV

Useful Data

A. Atomic Colors

Atomic number:	1	4	5	6	7	8	9
Atomic symbol:	**H**	**Be**	**B**	**C**	**N**	**O**	**F**

Atomic number:	11	12	14	15	16	17	19
Atomic symbol:	**Na**	**Mg**	**Si**	**P**	**S**	**Cl**	**K**

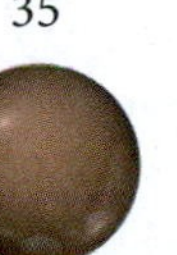

Atomic number:	20	29	30	35	53	54
Atomic symbol:	**Ca**	**Cu**	**Zn**	**Br**	**I**	**Xe**

B. Standard Thermodynamic Quantities for Selected Substances at 25 °C

Substance	ΔH_f° (kJ/mol)	ΔG_f° (kJ/mol)	S° (J/mol · K)
Aluminum			
$Al(s)$	0	0	28.32
$Al(g)$	330.0	289.4	164.6
$Al^{3+}(aq)$	−538.4	−483	−325
$AlCl_3(s)$	−704.2	−628.8	109.3
$Al_2O_3(s)$	−1675.7	−1582.3	50.9
Barium			
$Ba(s)$	0	0	62.5
$Ba(g)$	180.0	146.0	170.2
$Ba^{2+}(aq)$	−537.6	−560.8	9.6
$BaCO_3(s)$	−1213.0	−1134.4	112.1
$BaCl_2(s)$	−855.0	−806.7	123.7
$BaO(s)$	−548.0	−520.3	72.1
$Ba(OH)_2(s)$	−944.7		
$BaSO_4(s)$	−1473.2	−1362.2	132.2
Beryllium			
$Be(s)$	0	0	9.5
$BeO(s)$	−609.4	−580.1	13.8

—*Continued on the next page*

Substance	ΔH_f° (kJ/mol)	ΔG_f° (kJ/mol)	S° (J/mol · K)
$Be(OH)_2(s)$	−902.5	−815.0	45.5
Bismuth			
$Bi(s)$	0	0	56.7
$BiCl_3(s)$	−379.1	−315.0	177.0
$Bi_2O_3(s)$	−573.9	−493.7	151.5
$Bi_2S_3(s)$	−143.1	−140.6	200.4
Boron			
$B(s)$	0	0	5.9
$B(g)$	565.0	521.0	153.4
$BCl_3(g)$	−403.8	−388.7	290.1
$BF_3(g)$	−1136.0	−1119.4	254.4
$B_2H_6(g)$	36.4	87.6	232.1
$B_2O_3(s)$	−1273.5	−1194.3	54.0
$H_3BO_3(s)$	−1094.3	−968.9	90.0
Bromine			
$Br(g)$	111.9	82.4	175.0
$Br_2(l)$	0	0	152.2
$Br_2(g)$	30.9	3.1	245.5
$Br^-(aq)$	−121.4	−102.8	80.71
$HBr(g)$	−36.3	−53.4	198.7
Cadmium			
$Cd(s)$	0	0	51.8
$Cd(g)$	111.8	77.3	167.7
$Cd^{2+}(aq)$	−75.9	−77.6	−73.2
$CdCl_2(s)$	−391.5	−343.9	115.3
$CdO(s)$	−258.4	−228.7	54.8
$CdS(s)$	−161.9	−156.5	64.9
$CdSO_4(s)$	−933.3	−822.7	123.0
Calcium			
$Ca(s)$	0	0	41.6
$Ca(g)$	177.8	144.0	154.9
$Ca^{2+}(aq)$	−542.8	−553.6	−53.1
$CaC_2(s)$	−59.8	−64.9	70.0
$CaCO_3(s)$	−1207.6	−1129.1	91.7
$CaCl_2(s)$	−795.4	−748.8	108.4
$CaF_2(s)$	−1228.0	−1175.6	68.5
$CaH_2(s)$	−181.5	−142.5	41.4

Substance	ΔH_f° (kJ/mol)	ΔG_f° (kJ/mol)	S° (J/mol · K)
$Ca(NO_3)_2(s)$	−938.2	−742.8	193.2
$CaO(s)$	−634.9	−603.3	38.1
$Ca(OH)_2(s)$	−985.2	−897.5	83.4
$CaSO_4(s)$	−1434.5	−1322.0	106.5
$Ca_3(PO_4)_2(s)$	−4120.8	−3884.7	236.0
Carbon			
$C(s$, graphite$)$	0	0	5.7
$C(s$, diamond$)$	1.88	2.9	2.4
$C(g)$	716.7	671.3	158.1
$CH_4(g)$	−74.6	−50.5	186.3
$CH_3Cl(g)$	−81.9	−60.2	234.6
$CH_2Cl_2(g)$	−95.4		270.2
$CH_2Cl_2(l)$	−124.2	−63.2	177.8
$CHCl_3(l)$	−134.1	−73.7	201.7
$CCl_4(g)$	−95.7	−62.3	309.7
$CCl_4(l)$	−128.2	−66.4	216.4
$CH_2O(g)$	−108.6	−102.5	218.8
CH_2O_2 (l, formic acid)	−425.0	−361.4	129.0
CH_3NH_2 (g, methylamine)	−22.5	32.7	242.9
$CH_3OH(l)$	−238.6	−166.6	126.8
$CH_3OH(g)$	−201.0	−162.3	239.9
$C_2H_2(g)$	227.4	209.9	200.9
$C_2H_4(g)$	52.4	68.4	219.3
$C_2H_6(g)$	−84.68	−32.0	229.2
$C_2H_5OH(l)$	−277.6	−174.8	160.7
$C_2H_5OH(g)$	−234.8	−167.9	281.6
C_2H_3Cl (g, vinyl chloride)	37.2	53.6	264.0
$C_2H_4Cl_2$ (l, dichloroethane)	−166.8	−79.6	208.5
C_2H_4O (g, acetaldehyde)	−166.2	−133.0	263.8
$C_2H_4O_2$ (l, acetic acid)	−484.3	−389.9	159.8
$C_3H_8(g)$	−103.85	−23.4	270.3
C_3H_6O (l, acetone)	−248.4	−155.6	199.8
C_3H_7OH (l, isopropanol)	−318.1		181.1

Substance	ΔH_f° (kJ/mol)	ΔG_f° (kJ/mol)	S° (J/mol · K)
$C_4H_{10}(l)$	−147.3	−15.0	231.0
$C_4H_{10}(g)$	−125.7	−15.71	310.0
$C_6H_6(l)$	49.1	124.5	173.4
$C_6H_5NH_2$ (l, aniline)	31.6	149.2	191.9
C_6H_5OH (s, phenol)	−165.1	−50.4	144.0
$C_6H_{12}O_6$ (s, glucose)	−1273.3	−910.4	212.1
$C_{10}H_8$ (s, naphthalene)	78.5	201.6	167.4
$C_{12}H_{22}O_{11}$ (s, sucrose)	−2226.1	−1544.3	360.24
$CO(g)$	−110.5	−137.2	197.7
$CO_2(g)$	−393.5	−394.4	213.8
$CO_2(aq)$	−413.8	−386.0	117.6
$CO_3^{2-}(aq)$	−677.1	−527.8	−56.9
$HCO_3^-(aq)$	−692.0	−586.8	91.2
$H_2CO_3(aq)$	−699.7	−623.2	187.4
$CN^-(aq)$	151	166	118
$HCN(l)$	108.9	125.0	112.8
$HCN(g)$	135.1	124.7	201.8
$CS_2(l)$	89.0	64.6	151.3
$CS_2(g)$	116.7	67.1	237.8
$COCl_2(g)$	−219.1	−204.9	283.5
$C_{60}(s)$	2327.0	2302.0	426.0
Cesium			
$Cs(s)$	0	0	85.2
$Cs(g)$	76.5	49.6	175.6
$Cs^+(aq)$	−258.0	−292.0	132.1
$CsBr(s)$	−400	−387	117
$CsCl(s)$	−438	−414	101.2
$CsF(s)$	−553.5	−525.5	92.8
$CsI(s)$	−342	−337	127
Chlorine			
$Cl(g)$	121.3	105.3	165.2
$Cl_2(g)$	0	0	223.1
$Cl^-(aq)$	−167.1	−131.2	56.6

Substance	ΔH_f° (kJ/mol)	ΔG_f° (kJ/mol)	S° (J/mol · K)
$HCl(g)$	−92.3	−95.3	186.9
$HCl(aq)$	−167.2	−131.2	56.5
$ClO_2(g)$	102.5	120.5	256.8
$Cl_2O(g)$	80.3	97.9	266.2
Chromium			
$Cr(s)$	0	0	23.8
$Cr(g)$	396.6	351.8	174.5
$Cr^{3+}(aq)$	−1971		
$CrO_4^{2-}(aq)$	−872.2	−717.1	44
$Cr_2O_3(s)$	−1139.7	−1058.1	81.2
$Cr_2O_7^{2-}(aq)$	−1476	−1279	238
Cobalt			
$Co(s)$	0	0	30.0
$Co(g)$	424.7	380.3	179.5
$CoO(s)$	−237.9	−214.2	53.0
$Co(OH)_2(s)$	−539.7	−454.3	79.0
Copper			
$Cu(s)$	0	0	33.2
$Cu(g)$	337.4	297.7	166.4
$Cu^+(aq)$	51.9	50.2	−26
$Cu^{2+}(aq)$	64.9	65.5	−98
$CuCl(s)$	−137.2	−119.9	86.2
$CuCl_2(s)$	−220.1	−175.7	108.1
$CuO(s)$	−157.3	−129.7	42.6
$CuS(s)$	−53.1	−53.6	66.5
$CuSO_4(s)$	−771.4	−662.2	109.2
$Cu_2O(s)$	−168.6	−146.0	93.1
$Cu_2S(s)$	−79.5	−86.2	120.9
Fluorine			
$F(g)$	79.38	62.3	158.75
$F_2(g)$	0	0	202.79
$F^-(aq)$	−335.35	−278.8	−13.8
$HF(g)$	−273.3	−275.4	173.8
Gold			
$Au(s)$	0	0	47.4
$Au(g)$	366.1	326.3	180.5

—*Continued on the next page*

Substance	ΔH_f°(kJ/mol)	ΔG_f°(kJ/mol)	S°(J/mol · K)
Helium			
He(*g*)	0	0	126.2
Hydrogen			
H(*g*)	218.0	203.3	114.7
H^+(*aq*)	0	0	0
H^+(*g*)	1536.3	1517.1	108.9
H_2(*g*)	0	0	130.7
Iodine			
I(*g*)	106.76	70.2	180.79
I_2(*s*)	0	0	116.14
I_2(*g*)	62.42	19.3	260.69
I^-(*aq*)	−56.78	−51.57	106.45
HI(*g*)	26.5	1.7	206.6
Iron			
Fe(*s*)	0	0	27.3
Fe(*g*)	416.3	370.7	180.5
Fe^{2+}(*aq*)	−87.9	−84.94	−113.4
Fe^{3+}(*aq*)	−47.69	−10.54	−293.3
$FeCO_3$(*s*)	−740.6	−666.7	92.9
$FeCl_2$(*s*)	−341.8	−302.3	118.0
$FeCl_3$(*s*)	−399.5	−334.0	142.3
FeO(*s*)	−272.0	−255.2	60.75
$Fe(OH)_3$(*s*)	−823.0	−696.5	106.7
FeS_2(*s*)	−178.2	−166.9	52.9
Fe_2O_3(*s*)	−824.2	−742.2	87.4
Fe_3O_4(*s*)	−1118.4	−1015.4	146.4
Lead			
Pb(*s*)	0	0	64.8
Pb(*g*)	195.2	162.2	175.4
Pb^{2+}(*aq*)	0.92	−24.4	18.5
$PbBr_2$(*s*)	−278.7	−261.9	161.5
$PbCO_3$(*s*)	−699.1	−625.5	131.0
$PbCl_2$(*s*)	−359.4	−314.1	136.0
PbI_2(*s*)	−175.5	−173.6	174.9
$Pb(NO_3)_2$(*s*)	−451.9		
PbO(*s*)	−217.3	−187.9	68.7
PbO_2(*s*)	−277.4	−217.3	68.6

Substance	ΔH_f°(kJ/mol)	ΔG_f°(kJ/mol)	S°(J/mol · K)
PbS(*s*)	−100.4	−98.7	91.2
$PbSO_4$(*s*)	−920.0	−813.0	148.5
Lithium			
Li(*s*)	0	0	29.1
Li(*g*)	159.3	126.6	138.8
Li^+(*aq*)	−278.47	−293.3	12.24
LiBr(*s*)	−351.2	−342.0	74.3
LiCl(*s*)	−408.6	−384.4	59.3
LiF(*s*)	−616.0	−587.7	35.7
LiI(*s*)	−270.4	−270.3	86.8
$LiNO_3$(*s*)	−483.1	−381.1	90.0
LiOH(*s*)	−487.5	−441.5	42.8
Li_2O(*s*)	−597.9	−561.2	37.6
Magnesium			
Mg(*s*)	0	0	32.7
Mg(*g*)	147.1	112.5	148.6
Mg^{2+}(*aq*)	−467.0	−455.4	−137
$MgCl_2$(*s*)	−641.3	−591.8	89.6
$MgCO_3$(*s*)	−1095.8	−1012.1	65.7
MgF_2(*s*)	−1124.2	−1071.1	57.2
MgO(*s*)	−601.6	−569.3	27.0
$Mg(OH)_2$(*s*)	−924.5	−833.5	63.2
$MgSO_4$(*s*)	−1284.9	−1170.6	91.6
Mg_3N_2(*s*)	−461	−401	88
Manganese			
Mn(*s*)	0	0	32.0
Mn(*g*)	280.7	238.5	173.7
Mn^{2+}(*aq*)	−219.4	−225.6	−78.8
MnO(*s*)	−385.2	−362.9	59.7
MnO_2(*s*)	−520.0	−465.1	53.1
MnO_4^-(*aq*)	−529.9	−436.2	190.6
Mercury			
Hg(*l*)	0	0	75.9
Hg(*g*)	61.4	31.8	175.0
Hg^{2+}(*aq*)	170.21	164.4	−36.19
Hg_2^{2+}(*aq*)	166.87	153.5	65.74
$HgCl_2$(*s*)	−224.3	−178.6	146.0

Substance	ΔH_f°(kJ/mol)	ΔG_f°(kJ/mol)	S°(J/mol · K)
$HgO(s)$	−90.8	−58.5	70.3
$HgS(s)$	−58.2	−50.6	82.4
$Hg_2Cl_2(s)$	−265.4	−210.7	191.6
Nickel			
$Ni(s)$	0	0	29.9
$Ni(g)$	429.7	384.5	182.2
$NiCl_2(s)$	−305.3	−259.0	97.7
$NiO(s)$	−239.7	−211.7	37.99
$NiS(s)$	−82.0	−79.5	53.0
Nitrogen			
$N(g)$	472.7	455.5	153.3
$N_2(g)$	0	0	191.6
$NF_3(g)$	−132.1	−90.6	260.8
$NH_3(g)$	−45.9	−16.4	192.8
$NH_3(aq)$	−80.29	−26.50	111.3
$NH_4^+(aq)$	−133.26	−79.31	111.17
$NH_4Br(s)$	−270.8	−175.2	113.0
$NH_4Cl(s)$	−314.4	−202.9	94.6
$NH_4CN(s)$	0.4		
$NH_4F(s)$	−464.0	−348.7	72.0
$NH_4HCO_3(s)$	−849.4	−665.9	120.9
$NH_4I(s)$	−201.4	−112.5	117.0
$NH_4NO_3(s)$	−365.6	−183.9	151.1
$NH_4NO_3(aq)$	−339.9	−190.6	259.8
$HNO_3(g)$	−133.9	−73.5	266.9
$HNO_3(aq)$	−207	−110.9	146
$NO(g)$	91.3	87.6	210.8
$NO_2(g)$	33.2	51.3	240.1
$NO_3^-(aq)$	−206.85	−110.2	146.70
$NOBr(g)$	82.2	82.4	273.7
$NOCl(g)$	51.7	66.1	261.7
$N_2H_4(l)$	50.6	149.3	121.2
$N_2H_4(g)$	95.4	159.4	238.5
$N_2O(g)$	81.6	103.7	220.0
$N_2O_4(l)$	−19.5	97.5	209.2
$N_2O_4(g)$	9.16	99.8	304.4
$N_2O_5(s)$	−43.1	113.9	178.2
$N_2O_5(g)$	13.3	117.1	355.7
Oxygen			
$O(g)$	249.2	231.7	161.1
$O_2(g)$	0	0	205.2
$O_3(g)$	142.7	163.2	238.9
$OH^-(aq)$	−230.02	−157.3	−10.90
$H_2O(l)$	−285.8	−237.1	70.0
$H_2O(g)$	−241.8	−228.6	188.8
$H_2O_2(l)$	−187.8	−120.4	109.6
$H_2O_2(g)$	−136.3	−105.6	232.7
Phosphorus			
P(*s*, white)	0	0	41.1
P(*s*, red)	−17.6	−12.1	22.8
$P(g)$	316.5	280.1	163.2
$P_2(g)$	144.0	103.5	218.1
$P_4(g)$	58.9	24.4	280.0
$PCl_3(l)$	−319.7	−272.3	217.1
$PCl_3(g)$	−287.0	−267.8	311.8
$PCl_5(s)$	−443.5		
$PCl_5(g)$	−374.9	−305.0	364.6
$PF_5(g)$	−1594.4	−1520.7	300.8
$PH_3(g)$	5.4	13.5	210.2
$POCl_3(l)$	−597.1	−520.8	222.5
$POCl_3(g)$	−558.5	−512.9	325.5
$PO_4^{3-}(aq)$	−1277.4	−1018.7	−220.5
$HPO_4^{2-}(aq)$	−1292.1	−1089.2	−33.5
$H_2PO_4^-(aq)$	−1296.3	−1130.2	90.4
$H_3PO_4(s)$	−1284.4	−1124.3	110.5
$H_3PO_4(aq)$	−1288.3	−1142.6	158.2
$P_4O_6(s)$	−1640.1		
$P_4O_{10}(s)$	−2984	−2698	228.9
Platinum			
$Pt(s)$	0	0	41.6
$Pt(g)$	565.3	520.5	192.4
Potassium			
$K(s)$	0	0	64.7

—Continued on the next page

Substance	ΔH_f° (kJ/mol)	ΔG_f° (kJ/mol)	S° (J/mol · K)
$K(g)$	89.0	60.5	160.3
$K^+(aq)$	−252.14	−283.3	101.2
$KBr(s)$	−393.8	−380.7	95.9
$KCN(s)$	−113.0	−101.9	128.5
$KCl(s)$	−436.5	−408.5	82.6
$KClO_3(s)$	−397.7	−296.3	143.1
$KClO_4(s)$	−432.8	−303.1	151.0
$KF(s)$	−567.3	−537.8	66.6
$KI(s)$	−327.9	−324.9	106.3
$KNO_3(s)$	−494.6	−394.9	133.1
$KOH(s)$	−424.6	−379.4	81.2
$KOH(aq)$	−482.4	−440.5	91.6
$KO_2(s)$	−284.9	−239.4	116.7
$K_2CO_3(s)$	−1151.0	−1063.5	155.5
$K_2O(s)$	−361.5	−322.1	94.14
$K_2O_2(s)$	−494.1	−425.1	102.1
$K_2SO_4(s)$	−1437.8	−1321.4	175.6
Rubidium			
$Rb(s)$	0	0	76.8
$Rb(g)$	80.9	53.1	170.1
$Rb^+(aq)$	−251.12	−283.1	121.75
$RbBr(s)$	−394.6	−381.8	110.0
$RbCl(s)$	−435.4	−407.8	95.9
$RbClO_3(s)$	−392.4	−292.0	152
$RbF(s)$	−557.7		
$RbI(s)$	−333.8	−328.9	118.4
Scandium			
$Sc(s)$	0	0	34.6
$Sc(g)$	377.8	336.0	174.8
Selenium			
$Se(s, gray)$	0	0	42.4
$Se(g)$	227.1	187.0	176.7
$H_2Se(g)$	29.7	15.9	219.0
Silicon			
$Si(s)$	0	0	18.8
$Si(g)$	450.0	405.5	168.0
$SiCl_4(l)$	−687.0	−619.8	239.7

Substance	ΔH_f° (kJ/mol)	ΔG_f° (kJ/mol)	S° (J/mol · K)
$SiF_4(g)$	−1615.0	−1572.8	282.8
$SiH_4(g)$	34.3	56.9	204.6
$SiO_2(s, quartz)$	−910.7	−856.3	41.5
$Si_2H_6(g)$	80.3	127.3	272.7
Silver			
$Ag(s)$	0	0	42.6
$Ag(g)$	284.9	246.0	173.0
$Ag^+(aq)$	105.79	77.11	73.45
$AgBr(s)$	−100.4	−96.9	107.1
$AgCl(s)$	−127.0	−109.8	96.3
$AgF(s)$	−204.6	−185	84
$AgI(s)$	−61.8	−66.2	115.5
$AgNO_3(s)$	−124.4	−33.4	140.9
$Ag_2O(s)$	−31.1	−11.2	121.3
$Ag_2S(s)$	−32.6	−40.7	144.0
$Ag_2SO_4(s)$	−715.9	−618.4	200.4
Sodium			
$Na(s)$	0	0	51.3
$Na(g)$	107.5	77.0	153.7
$Na^+(aq)$	−240.34	−261.9	58.45
$NaBr(s)$	−361.1	−349.0	86.8
$NaCl(s)$	−411.2	−384.1	72.1
$NaCl(aq)$	−407.2	−393.1	115.5
$NaClO_3(s)$	−365.8	−262.3	123.4
$NaF(s)$	−576.6	−546.3	51.1
$NaHCO_3(s)$	−950.8	−851.0	101.7
$NaHSO_4(s)$	−1125.5	−992.8	113.0
$NaI(s)$	−287.8	−286.1	98.5
$NaNO_3(s)$	−467.9	−367.0	116.5
$NaNO_3(aq)$	−447.5	−373.2	205.4
$NaOH(s)$	−425.8	−379.7	64.4
$NaOH(aq)$	−470.1	−419.2	48.2
$NaO_2(s)$	−260.2	−218.4	115.9
$Na_2CO_3(s)$	−1130.7	−1044.4	135.0
$Na_2O(s)$	−414.2	−375.5	75.1
$Na_2O_2(s)$	−510.9	−447.7	95.0
$Na_2SO_4(s)$	−1387.1	−1270.2	149.6

Substance	ΔH_f° (kJ/mol)	ΔG_f° (kJ/mol)	S° (J/mol · K)
$Na_3PO_4(s)$	−1917	−1789	173.8
Strontium			
$Sr(s)$	0	0	55.0
$Sr(g)$	164.4	130.9	164.6
$Sr^{2+}(aq)$	−545.51	−557.3	−39
$SrCl_2(s)$	−828.9	−781.1	114.9
$SrCO_3(s)$	−1220.1	−1140.1	97.1
$SrO(s)$	−592.0	−561.9	54.4
$SrSO_4(s)$	−1453.1	−1340.9	117.0
Sulfur			
$S(s$, rhombic)	0	0	32.1
$S(s$, monoclinic)	0.3	0.096	32.6
$S(g)$	277.2	236.7	167.8
$S_2(g)$	128.6	79.7	228.2
$S_8(g)$	102.3	49.7	430.9
$S^{2-}(aq)$	41.8	83.7	22
$SF_6(g)$	−1220.5	−1116.5	291.5
$HS^-(aq)$	−17.7	12.4	62.0
$H_2S(g)$	−20.6	−33.4	205.8
$H_2S(aq)$	−39.4	−27.7	122
$SOCl_2(l)$	−245.6		
$SO_2(g)$	−296.8	−300.1	248.2
$SO_3(g)$	−395.7	−371.1	256.8
$SO_4^{2-}(aq)$	−909.3	−744.6	18.5
$HSO_4^-(aq)$	−886.5	−754.4	129.5
$H_2SO_4(l)$	−814.0	−690.0	156.9
$H_2SO_4(aq)$	−909.3	−744.6	18.5
$S_2O_3^{2-}(aq)$	−648.5	−522.5	67
Tin			
$Sn(s$, white)	0	0	51.2
$Sn(s$, gray)	−2.1	0.1	44.1
$Sn(g)$	301.2	266.2	168.5
$SnCl_4(l)$	−511.3	−440.1	258.6
$SnCl_4(g)$	−471.5	−432.2	365.8
$SnO(s)$	−280.7	−251.9	57.2
$SnO_2(s)$	−577.6	−515.8	49.0
Titanium			
$Ti(s)$	0	0	30.7
$Ti(g)$	473.0	428.4	180.3
$TiCl_4(l)$	−804.2	−737.2	252.3
$TiCl_4(g)$	−763.2	−726.3	353.2
$TiO_2(s)$	−944.0	−888.8	50.6
Tungsten			
$W(s)$	0	0	32.6
$W(g)$	849.4	807.1	174.0
$WO_3(s)$	−842.9	−764.0	75.9
Uranium			
$U(s)$	0	0	50.2
$U(g)$	533.0	488.4	199.8
$UF_6(s)$	−2197.0	−2068.5	227.6
$UF_6(g)$	−2147.4	−2063.7	377.9
$UO_2(s)$	−1085.0	−1031.8	77.0
Vanadium			
$V(s)$	0	0	28.9
$V(g)$	514.2	754.4	182.3
Zinc			
$Zn(s)$	0	0	41.6
$Zn(g)$	130.4	94.8	161.0
$Zn^{2+}(aq)$	−153.39	−147.1	−109.8
$ZnCl_2(s)$	−415.1	−369.4	111.5
$ZnO(s)$	−350.5	−320.5	43.7
ZnS (s, zinc blende)	−206.0	−201.3	57.7
$ZnSO_4(s)$	−982.8	−871.5	110.5

C. Aqueous Equilibrium Constants

1. Ionization Constants for Acids at 25 °C

Name	Formula	K_{a_1}	K_{a_2}	K_{a_3}
Acetic	$HC_2H_3O_2$	1.8×10^{-5}		
Acetylsalicylic	$HC_9H_7O_4$	3.3×10^{-4}		

—Continued on the next page

Name	Formula	K_{a_1}	K_{a_2}	K_{a_3}
Adipic	$H_2C_6H_8O_4$	3.9×10^{-5}	3.9×10^{-6}	
Arsenic	H_3AsO_4	5.5×10^{-3}	1.7×10^{-7}	5.1×10^{-12}
Arsenous	H_3AsO_3	5.1×10^{-10}		
Ascorbic	$H_2C_6H_6O_6$	8.0×10^{-5}	1.6×10^{-12}	
Benzoic	$HC_7H_5O_2$	6.5×10^{-5}		
Boric	H_3BO_3	5.4×10^{-10}		
Butanoic	$HC_4H_7O_2$	1.5×10^{-5}		
Carbonic	H_2CO_3	4.3×10^{-7}	5.6×10^{-11}	
Chloroacetic	$HC_2H_2O_2Cl$	1.4×10^{-3}		
Chlorous	$HClO_2$	1.1×10^{-2}		
Citric	$H_3C_6H_5O_7$	7.4×10^{-4}	1.7×10^{-5}	4.0×10^{-7}
Cyanic	HCNO	2×10^{-4}		
Formic	$HCHO_2$	1.8×10^{-4}		
Hydrazoic	HN_3	2.5×10^{-5}		
Hydrocyanic	HCN	4.9×10^{-10}		
Hydrofluoric	HF	3.5×10^{-4}		
Hydrogen chromate ion	$HCrO_4^-$	3.0×10^{-7}		
Hydrogen peroxide	H_2O_2	2.4×10^{-12}		
Hydrogen selenate ion	$HSeO_4^-$	2.2×10^{-2}		
Hydrosulfuric	H_2S	8.9×10^{-8}	1×10^{-19}	
Hydrotelluric	H_2Te	2.3×10^{23}	1.6×10^{-11}	

Name	Formula	K_{a_1}	K_{a_2}	K_{a_3}
Hypobromous	HBrO	2.8×10^{-9}		
Hypochlorous	HClO	3.5×10^{-8}		
Hypoiodous	HIO	2.3×10^{-11}		
Iodic	HIO_3	1.7×10^{-1}		
Lactic	$HC_3H_5O_3$	1.4×10^{-4}		
Maleic	$H_2C_4H_2O_4$	1.2×10^{-2}	5.9×10^{-7}	
Malonic	$H_2C_3H_2O_4$	1.5×10^{-3}	2.0×10^{-6}	
Nitrous	HNO_2	4.6×10^{-4}		
Oxalic	$H_2C_2O_4$	6.0×10^{-2}	6.1×10^{-5}	
Paraperiodic	H_5IO_6	2.8×10^{-2}	5.3×10^{-9}	
Phenol	HC_6H_5O	1.3×10^{-10}		
Phosphoric	H_3PO_4	7.5×10^{-3}	6.2×10^{-8}	4.2×10^{-13}
Phosphorous	H_3PO_3	5×10^{-2}	2.0×10^{-7}	
Propanoic	$HC_3H_5O_2$	1.3×10^{-5}		
Pyruvic	$HC_3H_3O_3$	4.1×10^{-3}		
Pyrophosphoric	$H_4P_2O_7$	1.2×10^{-1}	7.9×10^{-3}	2.0×10^{-7}
Selenous	H_2SeO_3	2.4×10^{-3}	4.8×10^{-9}	
Succinic	$H_2C_4H_4O_4$	6.2×10^{-5}	2.3×10^{-6}	
Sulfuric	H_2SO_4	Strong acid	1.2×10^{-2}	
Sulfurous	H_2SO_3	1.6×10^{-2}	6.4×10^{-8}	
Tartaric	$H_2C_4H_4O_6$	1.0×10^{-3}	4.6×10^{-5}	
Trichloroacetic	$HC_2Cl_3O_2$	2.2×10^{-1}		
Trifluoroacetic acid	$HC_2F_3O_2$	3.0×10^{-1}		

2. Dissociation Constants for Hydrated Metal Ions at 25 °C

Cation	Hydrated Ion	K_a
Al^{3+}	$Al(H_2O)_6^{3+}$	1.4×10^{-5}
Be^{2+}	$Be(H_2O)_6^{2+}$	3×10^{-7}
Co^{2+}	$Co(H_2O)_6^{2+}$	1.3×10^{-9}
Cr^{3+}	$Cr(H_2O)_6^{3+}$	1.6×10^{-4}
Cu^{2+}	$Cu(H_2O)_6^{2+}$	3×10^{-8}
Fe^{2+}	$Fe(H_2O)_6^{2+}$	3.2×10^{-10}

Cation	Hydrated Ion	K_a
Fe^{3+}	$Fe(H_2O)_6^{3+}$	6.3×10^{-3}
Ni^{2+}	$Ni(H_2O)_6^{2+}$	2.5×10^{-11}
Pb^{2+}	$Pb(H_2O)_6^{2+}$	3×10^{-8}
Sn^{2+}	$Sn(H_2O)_6^{2+}$	4×10^{-4}
Zn^{2+}	$Zn(H_2O)_6^{2+}$	2.5×10^{-10}

3. Ionization Constants for Bases at 25 °C

Name	Formula	K_b
Ammonia	NH_3	1.76×10^{-5}
Aniline	$C_6H_5NH_2$	3.9×10^{-10}

Name	Formula	K_b
Bicarbonate ion	HCO_3^-	2.3×10^{-8}
Carbonate ion	CO_3^{2-}	1.8×10^{-4}

Name	Formula	K_b
Codeine	$C_{18}H_{21}NO_3$	1.6×10^{-6}
Diethylamine	$(C_2H_5)_2NH$	6.9×10^{-4}
Dimethylamine	$(CH_3)_2NH$	5.4×10^{-4}
Ethylamine	$C_2H_5NH_2$	5.6×10^{-4}
Ethylenediamine	$C_2H_8N_2$	8.3×10^{-5}
Hydrazine	H_2NNH_2	1.3×10^{-6}
Hydroxylamine	$HONH_2$	1.1×10^{-8}
Ketamine	$C_{13}H_{16}ClNO$	3×10^{-7}
Methylamine	CH_3NH_2	4.4×10^{-4}
Morphine	$C_{17}H_{19}NO_3$	1.6×10^{-6}
Nicotine	$C_{10}H_{14}N_2$	1.0×10^{-6}
Piperidine	$C_5H_{10}NH$	1.33×10^{-3}
Propylamine	$C_3H_7NH_2$	3.5×10^{-4}
Pyridine	C_5H_5N	1.7×10^{-9}
Strychnine	$C_{21}H_{22}N_2O_2$	1.8×10^{-6}
Triethylamine	$(C_2H_5)_3N$	5.6×10^{-4}
Trimethylamine	$(CH_3)_3N$	6.4×10^{-5}

4. Solubility Product Constants for Compounds at 25 °C

Compound	Formula	K_{sp}
Aluminum hydroxide	$Al(OH)_3$	1.3×10^{-33}
Aluminum phosphate	$AlPO_4$	9.84×10^{-21}
Barium carbonate	$BaCO_3$	2.58×10^{-9}
Barium chromate	$BaCrO_4$	1.17×10^{-10}
Barium fluoride	BaF_2	2.45×10^{-5}
Barium hydroxide	$Ba(OH)_2$	5.0×10^{-3}
Barium oxalate	BaC_2O_4	1.6×10^{-6}
Barium phosphate	$Ba_3(PO_4)_2$	6×10^{-39}
Barium sulfate	$BaSO_4$	1.07×10^{-10}
Cadmium carbonate	$CdCO_3$	1.0×10^{-12}
Cadmium hydroxide	$Cd(OH)_2$	7.2×10^{-15}
Cadmium sulfide	CdS	8×10^{-28}
Calcium carbonate	$CaCO_3$	4.96×10^{-9}
Calcium chromate	$CaCrO_4$	7.1×10^{-4}
Calcium fluoride	CaF_2	1.46×10^{-10}
Calcium hydroxide	$Ca(OH)_2$	4.68×10^{-6}
Calcium hydrogen phosphate	$CaHPO_4$	1×10^{-7}
Calcium oxalate	CaC_2O_4	2.32×10^{-9}
Calcium phosphate	$Ca_3(PO_4)_2$	2.07×10^{-33}
Calcium sulfate	$CaSO_4$	7.10×10^{-5}
Chromium(III) hydroxide	$Cr(OH)_3$	6.3×10^{-31}
Cobalt(II) carbonate	$CoCO_3$	1.0×10^{-10}
Cobalt(II) hydroxide	$Co(OH)_2$	5.92×10^{-15}
Cobalt(II) sulfide	CoS	5×10^{-22}
Copper(I) bromide	$CuBr$	6.27×10^{-9}
Copper(I) chloride	$CuCl$	1.72×10^{-7}
Copper(I) cyanide	$CuCN$	3.47×10^{-20}
Copper(II) carbonate	$CuCO_3$	2.4×10^{-10}
Copper(II) hydroxide	$Cu(OH)_2$	2.2×10^{-20}
Copper(II) phosphate	$Cu_3(PO_4)_2$	1.40×10^{-37}
Copper(II) sulfide	CuS	1.27×10^{-36}
Iron(II) carbonate	$FeCO_3$	3.07×10^{-11}
Iron(II) hydroxide	$Fe(OH)_2$	4.87×10^{-17}
Iron(II) sulfide	FeS	3.72×10^{-19}
Iron(III) hydroxide	$Fe(OH)_3$	2.79×10^{-39}
Lanthanum fluoride	LaF_3	2×10^{-19}
Lanthanum iodate	$La(IO_3)_3$	7.50×10^{-12}
Lead(II) bromide	$PbBr_2$	4.67×10^{-6}
Lead(II) carbonate	$PbCO_3$	7.40×10^{-14}
Lead(II) chloride	$PbCl_2$	1.17×10^{-5}
Lead(II) chromate	$PbCrO_4$	2.8×10^{-13}
Lead(II) fluoride	PbF_2	3.3×10^{-8}
Lead(II) hydroxide	$Pb(OH)_2$	1.43×10^{-20}
Lead(II) iodide	PbI_2	9.8×10^{-9}
Lead(II) phosphate	$Pb_3(PO_4)_2$	1×10^{-54}
Lead(II) sulfate	$PbSO_4$	1.82×10^{-8}
Lead(II) sulfide	PbS	9.04×10^{-29}
Magnesium carbonate	$MgCO_3$	6.82×10^{-6}

—Continued on the next page

Compound	Formula	K_{sp}
Magnesium fluoride	MgF_2	5.16×10^{-11}
Magnesium hydroxide	$Mg(OH)_2$	2.06×10^{-13}
Magnesium oxalate	MgC_2O_4	4.83×10^{-6}
Manganese(II) carbonate	$MnCO_3$	2.24×10^{-11}
Manganese(II) hydroxide	$Mn(OH)_2$	1.6×10^{-13}
Manganese(II) sulfide	MnS	2.3×10^{-13}
Mercury(I) bromide	Hg_2Br_2	6.40×10^{-23}
Mercury(I) carbonate	Hg_2CO_3	3.6×10^{-17}
Mercury(I) chloride	Hg_2Cl_2	1.43×10^{-18}
Mercury(I) chromate	Hg_2CrO_4	2×10^{-9}
Mercury(I) cyanide	$Hg_2(CN)_2$	5×10^{-40}
Mercury(I) iodide	Hg_2I_2	5.2×10^{-29}
Mercury(II) hydroxide	$Hg(OH)_2$	3.1×10^{-26}
Mercury(II) sulfide	HgS	1.6×10^{-54}
Nickel(II) carbonate	$NiCO_3$	1.42×10^{-7}
Nickel(II) hydroxide	$Ni(OH)_2$	5.48×10^{-16}
Nickel(II) sulfide	NiS	3×10^{-20}
Silver bromate	$AgBrO_3$	5.38×10^{-5}
Silver bromide	AgBr	5.35×10^{-13}

Compound	Formula	K_{sp}
Silver carbonate	Ag_2CO_3	8.46×10^{-12}
Silver chloride	AgCl	1.77×10^{-10}
Silver chromate	Ag_2CrO_4	1.12×10^{-12}
Silver cyanide	AgCN	5.97×10^{-17}
Silver iodide	AgI	8.51×10^{-17}
Silver phosphate	Ag_3PO_4	8.89×10^{-17}
Silver sulfate	Ag_2SO_4	1.20×10^{-5}
Silver sulfide	Ag_2S	6×10^{-51}
Strontium carbonate	$SrCO_3$	5.60×10^{-10}
Strontium chromate	$SrCrO_4$	3.6×10^{-5}
Strontium phosphate	$Sr_3(PO_4)_2$	1×10^{-31}
Strontium sulfate	$SrSO_4$	3.44×10^{-7}
Tin(II) hydroxide	$Sn(OH)_2$	5.45×10^{-27}
Tin(II) sulfide	SnS	1×10^{-26}
Zinc carbonate	$ZnCO_3$	1.46×10^{-10}
Zinc hydroxide	$Zn(OH)_2$	3×10^{-17}
Zinc oxalate	ZnC_2O_4	2.7×10^{-8}
Zinc sulfide	ZnS	2×10^{-25}

5. Complex Ion Formation Constants in Water at 25 °C

Complex Ion	K_f
$[Ag(CN)_2]^-$	1×10^{21}
$[Ag(EDTA)]^{3-}$	2.1×10^{7}
$[Ag(en)_2]^+$	5.0×10^{7}
$[Ag(NH_3)_2]^+$	1.7×10^{7}
$[Ag(SCN)_4]^{3-}$	1.2×10^{10}
$[Ag(S_2O_3)_2]^-$	2.8×10^{13}
$[Al(EDTA)]^-$	1.3×10^{16}
$[AlF_6]^{3-}$	7×10^{19}
$[Al(OH)_4]^-$	3×10^{33}
$[Al(ox)_3]^{3-}$	2×10^{16}
$[CdBr_4]^{2-}$	5.5×10^{3}
$[Cd(CN)_4]^{2-}$	3×10^{18}
$[CdCl_4]^{2-}$	6.3×10^{2}
$[Cd(en)_3]^{2+}$	1.2×10^{12}

Complex Ion	K_f
$[CdI_4]^{2-}$	2×10^{6}
$[Co(EDTA)]^{2-}$	2.0×10^{16}
$[Co(EDTA)]^-$	1×10^{36}
$[Co(en)_3]^{2+}$	8.7×10^{13}
$[Co(en)_3]^{3+}$	4.9×10^{48}
$[Co(NH_3)_6]^{2+}$	1.3×10^{5}
$[Co(NH_3)_6]^{3+}$	2.3×10^{33}
$[Co(OH)_4]^{2-}$	5×10^{9}
$[Co(ox)_3]^{4-}$	5×10^{9}
$[Co(ox)_3]^{3-}$	1×10^{20}
$[Co(SCN)_4]^{2-}$	1×10^{3}
$[Cr(EDTA)]^-$	1×10^{23}
$[Cr(OH)_4]^-$	8.0×10^{29}
$[CuCl_3]^{2-}$	5×10^{5}

Complex Ion	K_f
$[Cu(CN)_4]^{2-}$	1.0×10^{25}
$[Cu(EDTA)]^{2-}$	5×10^{18}
$[Cu(en)_2]^{2+}$	1×10^{20}
$[Cu(NH_3)_4]^{2+}$	1.7×10^{13}
$[Cu(ox)_2]^{2-}$	3×10^{8}
$[Fe(CN)_6]^{4-}$	1.5×10^{35}
$[Fe(CN)_6]^{3-}$	2×10^{43}
$[Fe(EDTA)]^{2-}$	2.1×10^{14}
$[Fe(EDTA)]^{-}$	1.7×10^{24}
$[Fe(en)_3]^{2+}$	5.0×10^{9}
$[Fe(ox)_3]^{4-}$	1.7×10^{5}
$[Fe(ox)_3]^{3-}$	2×10^{20}
$[Fe(SCN)]^{2+}$	8.9×10^{2}
$[Hg(CN)_4]^{2-}$	1.8×10^{41}
$[HgCl_4]^{2-}$	1.1×10^{16}
$[Hg(EDTA)]^{2-}$	6.3×10^{21}
$[Hg(en)_2]^{2+}$	2×10^{23}
$[HgI_4]^{2-}$	2×10^{30}
$[Hg(ox)_2]^{2-}$	9.5×10^{6}
$[Ni(CN)_4]^{2-}$	2×10^{31}
$[Ni(EDTA)]^{2-}$	3.6×10^{18}
$[Ni(en)_3]^{2+}$	2.1×10^{18}
$[Ni(NH_3)_6]^{2+}$	2.0×10^{8}
$[Ni(ox)_3]^{4-}$	3×10^{8}
$[PbCl_3]^{-}$	2.4×10^{1}
$[Pb(EDTA)]^{2-}$	2×10^{18}
$[PbI_4]^{2-}$	3.0×10^{4}
$[Pb(OH)_3]^{-}$	8×10^{13}
$[Pb(ox)_2]^{2-}$	3.5×10^{6}
$[Pb(S_2O_3)_3]^{4-}$	2.2×10^{6}
$[PtCl_4]^{2-}$	1×10^{16}
$[Pt(NH_3)_6]^{2+}$	2×10^{35}
$[Sn(OH)_3]^{-}$	3×10^{25}
$[Zn(CN)_4]^{2-}$	2.1×10^{19}
$[Zn(EDTA)]^{2-}$	3×10^{16}
$[Zn(en)_3]^{2+}$	1.3×10^{14}
$[Zn(NH_3)_4]^{2+}$	2.8×10^{9}
$[Zn(OH)_4]^{2-}$	2×10^{15}
$[Zn(ox)_3]^{4-}$	1.4×10^{8}

D. Standard Electrode Potentials at 25 °C

Half-Reaction	$E°(V)$
$F_2(g) + 2\,e^- \longrightarrow 2\,F^-(aq)$	2.87
$O_3(g) + 2\,H^+(aq) + 2\,e^- \longrightarrow O_2(g) + H_2(l)$	2.08
$Ag^{2+}(aq) + e^- \longrightarrow Ag^+(aq)$	1.98
$Co^{3+}(aq) + e^- \longrightarrow Co^{2+}(aq)$	1.82
$H_2O_2(aq) + 2\,H^+(aq) + 2\,e^- \longrightarrow 2\,H_2O(l)$	1.78
$PbO_2(s) + 4\,H^+(aq) + SO_4^{2-}(aq) + 2\,e^- \longrightarrow PbSO_4(s) + 2\,H_2O(l)$	1.69
$MnO_4^-(aq) + 4\,H^+(aq) + 3\,e^- \longrightarrow MnO_2(s) + 2\,H_2O(l)$	1.68
$2\,HClO(aq) + 2\,H^+(aq) + 2\,e^- \longrightarrow Cl_2(g) + 2\,H_2O(l)$	1.61
$MnO_4^-(aq) + 8\,H^+(aq) + 5\,e^- \longrightarrow Mn^{2+}(aq) + 4\,H_2O(l)$	1.51
$Au^{3+}(aq) + 3\,e^- \longrightarrow Au(s)$	1.50
$2\,BrO_3^-(aq) + 12\,H^+(aq) + 10\,e^- \longrightarrow Br_2(l) + 6\,H_2O(l)$	1.48
$PbO_2(s) + 4\,H^+(aq) + 2\,e^- \longrightarrow Pb^{2+}(aq) + 2\,H_2O(l)$	1.46
$Cl_2(g) + 2\,e^- \longrightarrow 2\,Cl^-(aq)$	1.36
$Cr_2O_7^{2-}(aq) + 14\,H^+(aq) + 6\,e^- \longrightarrow 2\,Cr^{3+}(aq) + 7\,H_2O(l)$	1.33
$O_2(g) + 4\,H^+(aq) + 4\,e^- \longrightarrow 2\,H_2O(l)$	1.23
$MnO_2(s) + 4\,H^+(aq) + 2\,e^- \longrightarrow Mn^{2+}(aq) + 2\,H_2O(l)$	1.21
$IO_3^-(aq) + 6\,H^+(aq) + 5\,e^- \longrightarrow \frac{1}{2}\,I_2(aq) + 3\,H_2O(l)$	1.20
$Br_2(l) + 2\,e^- \longrightarrow 2\,Br^-(aq)$	1.09
$AuCl_4^-(aq) + 3\,e^- \longrightarrow Au(s) + 4\,Cl^-(aq)$	1.00
$VO_2^+(aq) + 2\,H^+(aq) + e^- \longrightarrow VO^{2+}(aq) + H_2O(l)$	1.00
$HNO_2(aq) + H^+(aq) + e^- \longrightarrow NO(g) + 2\,H_2O(l)$	0.98
$NO_3^-(aq) + 4\,H^+(aq) + 3\,e^- \longrightarrow NO(g) + 2\,H_2O(l)$	0.96
$ClO_2(g) + e^- \longrightarrow ClO_2^-(aq)$	0.95
$2\,Hg^{2+}(aq) + 2\,e^- \longrightarrow 2\,Hg_2^{2+}(aq)$	0.92

—Continued on the next page

Half-Reaction	$E°$(V)
$Ag^+(aq) + e^- \longrightarrow Ag(s)$	0.80
$Hg_2^{2+}(aq) + 2\,e^- \longrightarrow 2\,Hg(l)$	0.80
$Fe^{3+}(aq) + e^- \longrightarrow Fe^{2+}(aq)$	0.77
$PtCl_4^{2-}(aq) + 2\,e^- \longrightarrow Pt(s) + 4\,Cl^-(aq)$	0.76
$O_2(g) + 2\,H^+(aq) + 2\,e^- \longrightarrow H_2O_2(aq)$	0.70
$MnO_4^-(aq) + e^- \longrightarrow Mno_4^{2-}(aq)$	0.56
$I_2(s) + 2\,e^- \longrightarrow 2\,I^-(aq)$	0.54
$Cu^+(aq) + e^- \longrightarrow Cu(s)$	0.52
$O_2(g) + 2\,H_2O(l) + 4\,e^- \longrightarrow 4\,OH^-(aq)$	0.40
$Cu^{2+}(aq) + 2\,e^- \longrightarrow Cu(s)$	0.34
$BiO^+(aq) + 2\,H^+(aq) + 3\,e^- \longrightarrow Bi(s) + H_2O(l)$	0.32
$Hg_2Cl_2(s) + 2\,e^- \longrightarrow 2\,Hg(l) + 2\,Cl^-(aq)$	0.27
$AgCl(s) + e^- \longrightarrow Ag(s) + Cl^-(aq)$	0.22
$SO_4^{2-}(aq) + 4\,H^+(aq) + 2\,e^- \longrightarrow H_2SO_3(aq) + H_2O(l)$	0.20
$Cu^{2+}(aq) + e^- \longrightarrow Cu^+(aq)$	0.16
$Sn^{4+}(aq) + 2\,e^- \longrightarrow Sn^{2+}(aq)$	0.15
$S(s) + 2\,H^+(aq) + 2\,e^- \longrightarrow H_2S(g)$	0.14
$AgBr(s) + e^- \longrightarrow Ag(s) + Br^-(aq)$	0.071
$2\,H^+(aq) + 2\,e^- \longrightarrow H_2(g)$	0.00
$Fe^{3+}(aq) + 3\,e^- \longrightarrow Fe(s)$	−0.036
$Pb^{2+}(aq) + 2\,e^- \longrightarrow Pb(s)$	−0.13
$Sn^{2+}(aq) + 2\,e^- \longrightarrow Sn(s)$	−0.14

Half-Reaction	$E°$(V)
$AgI(s) + e^- \longrightarrow Ag(s) + I^-(aq)$	−0.15
$N_2(g) + 5\,H^+(aq) + 4\,e^- \longrightarrow N_2H_5^+(aq)$	−0.23
$Ni^{2+}(aq) + 2\,e^- \longrightarrow Ni(s)$	−0.23
$Co^{2+}(aq) + 2\,e^- \longrightarrow Co(s)$	−0.28
$PbSO_4(s) + 2\,e^- \longrightarrow Pb(s) + SO_4^{2-}(aq)$	−0.36
$Cd^{2+}(aq) + 2\,e^- \longrightarrow Cd(s)$	−0.40
$Fe^{2+}(aq) + 2\,e^- \longrightarrow Fe(s)$	−0.45
$2\,CO_2(g) + 2\,H^+(aq) + 2\,e^- \longrightarrow H_2C_2O_4(aq)$	−0.49
$Cr^{3+}(aq) + e^- \longrightarrow Cr^{2+}(aq)$	−0.50
$Cr^{3+}(aq) + 3\,e^- \longrightarrow Cr(s)$	−0.73
$Zn^{2+}(aq) + 2\,e^- \longrightarrow Zn(s)$	−0.76
$2\,H_2O(l) + 2\,e^- \longrightarrow H_2(g) + 2\,OH^-(aq)$	−0.83
$Mn^{2+}(aq) + 2\,e^- \longrightarrow Mn(s)$	−1.18
$Al^{3+}(aq) + 3\,e^- \longrightarrow Al(s)$	−1.66
$H_2(g) + 2\,e^- \longrightarrow 2\,H^-(aq)$	−2.23
$Mg^{2+}(aq) + 2\,e^- \longrightarrow Mg(s)$	−2.37
$La^{3+}(aq) + 3\,e^- \longrightarrow La(s)$	−2.38
$Na^+(aq) + e^- \longrightarrow Na(s)$	−2.71
$Ca^{2+}(aq) + 2\,e^- \longrightarrow Ca(s)$	−2.76
$Ba^{2+}(aq) + 2\,e^- \longrightarrow Ba(s)$	−2.90
$K^+(aq) + e^- \longrightarrow K(s)$	−2.92
$Li^+(aq) + e^- \longrightarrow Li(s)$	−3.04

E. Vapor Pressure of Water at Various Temperatures

T (°C)	P (torr)	T (°C)	P (torr)	T (°C)	P (torr)	T (°C)	P (torr)
0	4.58	21	18.65	35	42.2	92	567.0
5	6.54	22	19.83	40	55.3	94	610.9
10	9.21	23	21.07	45	71.9	96	657.6
12	10.52	24	22.38	50	92.5	98	707.3
14	11.99	25	23.76	55	118.0	100	760.0
16	13.63	26	25.21	60	149.4	102	815.9
17	14.53	27	26.74	65	187.5	104	875.1
18	15.48	28	28.35	70	233.7	106	937.9
19	16.48	29	30.04	80	355.1	108	1004.4
20	17.54	30	31.82	90	525.8	110	1074.6

APPENDIX V

Answers to Selected End-of-Chapter Problems

Chapter 1

33. a. pure substance **b.** pure substance
c. homogeneous mixture **d.** heterogeneous mixture

35. a. homogeneous mixture
b. pure substance, compound
c. pure substance, element
d. heterogeneous mixture

37.

Substance	Pure or Mixture	Type
Aluminum	Pure	Element
Apple juice	Mixture	Homogeneous
Hydrogen peroxide	Pure	Compound
Chicken soup	Mixture	Heterogeneous

39. a. pure substance, compound
b. mixture, heterogeneous
c. mixture, homogeneous
d. pure substance, element

41. a. theory **b.** observation
c. law **d.** observation

43. Several answers possible.

45. 13.5 g

47. These results are not consistent with the law of definite proportions because sample 1 is composed of 11.5 parts Cl to 1 part C and sample 2 is composed of 9.05 parts Cl to 1 part C. The law of definite proportions states that a given compound always contains exactly the same proportion of elements by mass.

49. 23.8 g

51. For the law of multiple proportions to hold, the ratio of the masses of O combining with 1 g of O's in the compound should be a small whole number. $0.3369/0.168 = 2.00$

53. Sample 1: $1.00 \text{ g } O_2/1.00 \text{ g S}$;
sample 2: $1.50 \text{ g } O_2/1.00 \text{ g S}$
Sample 2 / sample 1 $= 1.50/1.00 = 1.50$
3 O atoms/2 O atoms $= 1.5$

55. a. not consistent
b. consistent: Dalton's atomic theory states that the atoms of a given element are identical.
c. consistent: Dalton's atomic theory states that atoms combine in simple whole-number ratios to form compounds.
d. not consistent

57. a. consistent: Rutherford's nuclear model states that the atom is largely empty space.
b. consistent: Rutherford's nuclear model states that most of the atom's mass is concentrated in a tiny region called the nucleus.
c. not consistent
d. not consistent

59. -2.3×10^{-19} C

61. a, b, c

63. a. Ag-107 **b.** Ag-109
c. U-238 **d.** H-2

65. a. 7 ${}^{1}_{1}p$ and 7 ${}^{0}_{1}n$ **b.** 11 ${}^{1}_{1}p$ and 12 ${}^{0}_{1}n$
c. 86 ${}^{1}_{1}p$ and 136 ${}^{0}_{1}n$ **d.** 82 ${}^{1}_{1}p$ and 126 ${}^{0}_{1}n$

67. 6 ${}^{1}_{1}p$ and 8 ${}^{1}_{0}n$, ${}^{14}_{6}C$

69. a. 28 ${}^{1}_{1}p$ and 26 e^- **b.** 16 ${}^{1}_{1}p$ and 18 e^-
c. 35 ${}^{1}_{1}p$ and 36 e^- **d.** 24 ${}^{1}_{1}p$ and 21 e^-

71.

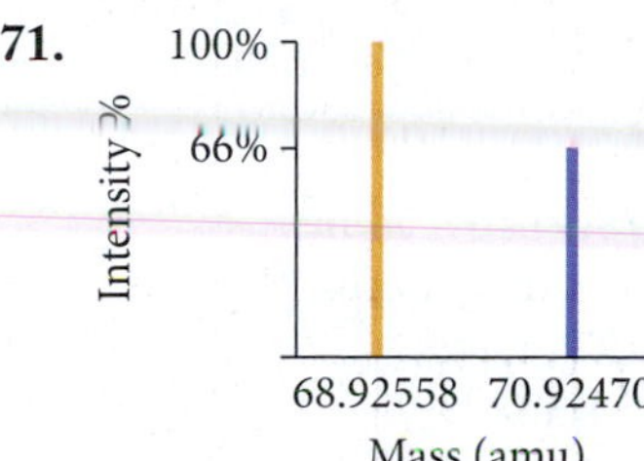

73. The fluorine-19 isotope must have a large percent abundance, which would make fluorine produce a large peak at this mass. Chlorine has two isotopes (Cl-35 and Cl-37). The atomic mass is simply the weighted average of these two, which means that there is no chlorine isotope with a mass of 35.45 amu.

75. 121.8 amu, Sb

77. Br-79 78.92 amu 50.96%

79. 152 amu

81. 1.50 g

83. 207 amu

85. ${}^{237}Pa$, ${}^{238}U$, ${}^{239}Np$, ${}^{240}Pu$, ${}^{235}Ac$, ${}^{234}Ra$, etc.

87. 106.91 amu

89. 0.423

91. 63.67 g/mol

93. 25.06 g/mol

95. c

97. a. Law
b. Theory
c. Observation
d. Law

Chapter 2

15. c
17. a. 73.0 mL **b.** 88.2 °C **c.** 645 mL
19. no
21. 1.26 g/cm^3
23. a. 463 g **b.** 3.7 L
25. 201. $\times 10^3$ g
27. a. 2.78×10^4 cm^3 **b.** 1.898×10^{-3} kg
c. 1.98×10^7 cm
29. a. 60.6 in **b.** 3.14×10^3 g
c. 3.7 qt **d.** 4.29 in
31. 5.0×10^1 min
33. 4.0×10^1 mi/gal
35. a. 1.95×10^{-4} km^2 **b.** 1.95×10^4 dm^2
c. 1.95×10^6 cm^2
37. 0.680 mi^2
39. 0.95 mL
41. a. 1.92×10^9 J **b.** 5.14×10^4 cal
c. 2.37×10^6 J **d.** 0.681 Cal
43. a. 9.987×10^6 J
b. 9.987×10^3 kJ
c. 2.78 kWh
45. 4.35×10^9 J
47. 3.32×10^{24} atoms
49. a. 0.295 mol Ar **b.** 0.0543 mol Zn
c. 0.144 mol Ta **d.** 0.0304 mol Li
51. 2.11×10^{22} atoms
53. a. 1.01×10^{23} atoms **b.** 6.78×10^{21} atoms
c. 5.39×10^{21} atoms **d.** 5.6×10^{20} atoms
55. a. 36 grams **b.** 0.187 grams
c. 62 grams **d.** 3.1 grams
57. 2.6×10^{21} atoms
59. 3.239×10^{-22} g
61. a. mass of can of gold = 1.9×10^4 g
mass of can of sand = 3.0×10^3 g
b. Yes, the thief sets off the trap because the can of sand is lighter than the gold cylinder.
63. 22 in^3
65. 7.6 g/cm^3
67. 3.11×10^5 lb
69. 3.3×10^2 km
71. 6.8×10^{-15}
73. 2.4×10^{19} km
75. 488 grams
77. 0.661 Ω
79. 0.492
81. $V_n = 8.2 \times 10^{-8}$ pm^3, $V_a = 1.4 \times 10^6$ pm^3, 5.9×10^{-12}%
83. 6.022×10^{21} dollars total, 9.3×10^{11} dollars per person, billionaires
85. 4.76×10^{24} atoms
87. 75.0% gold
89. 2.4×10^{13} atoms
91. 7.3×10^{11} g/cm^3
93. a. 1.6×10^{-20} L
b. 1.3×10^{-18} pressurized O_2/nanocontainers
c. 1.7×10^2 g oxygen
d. 1.3×10^{20} nanocontainers
e. 2.0 L Not feasible
95. 50% of the spheres are copper
97. 1×10^{78} atoms/universe
99. Substance A
101. If the amu and mole were not based on the same isotope, the numerical values obtained for an atom of material and a mole of material would not be the same. If, for example, the mole was based on the number of particles in C-12 but the amu was changed to a fraction of the mass of an atom of Ne-20, the number of particles and the number of amu that make up one mole of material would no longer be the same. We would no longer have the relationship where the mass of an atom in amu is numerically equal to the mass of a mole of those atoms in grams.

Chapter 3

35. 499 s
37. (i) d, c, b, a
(ii) a, b, c, d
39. a. 4.74×10^{14} Hz **b.** 5.96×10^{14} Hz
c. 5.8×10^{18} Hz
41. a. 3.14×10^{-19} J **b.** 3.95×10^{-19} J
c. 3.8×10^{-15} J
43. 1.03×10^{16} photons
45. a. 79.8 kJ/mol **b.** 239 kJ/mol
c. 798 kJ/mol
47.
49. 3.6×10^6 m/s
51. 5.39 nm
53. 1.1×10^{-34} m. The wavelength of a baseball is negligible with respect to its size.
55. $\Delta v = 1.04 \times 10^5$ m/s
57. 2*s*
59. a. $l = 0$ **b.** $l = 0, 1$
c. $l = 0, 1, 2$ **d.** $l = 0, 1, 2, 3$
61. c
63. See Figures 7.25 and 7.26. The 2*s* and 3*p* orbitals would, on average, be farther from the nucleus and have more nodes than the 1*s* and 2*p* orbitals.
65. $n = 1$
67. $2p \longrightarrow 1s$
69. a. 122 nm, UV **b.** 103 nm, UV
c. 486 nm, visible **d.** 434 nm, visible
71. n = 2
73. 344 nm
75. 6.4×10^{17} photons/s
77. 0.0547 nm
79. 91.2 nm
81. a. 4 **b.** 9 **c.** 16
83. $n = 4 \longrightarrow n = 3$, $n = 5 \longrightarrow n = 3$, $n = 6 \longrightarrow n = 3$, respectively
85. 4.84×10^{14} s^{-1}

87. 11 m
89. 6.78×10^{-3} J
91. 632 nm
93. **a.** $E_1 = 2.51 \times 10^{-18}$ J, $E_2 = 1.00 \times 10^{-17}$ J, $E_3 = 2.26 \times 10^{-17}$ J
b. 26.5 nm, UV; 15.8 nm, UV
95. 1*s*:

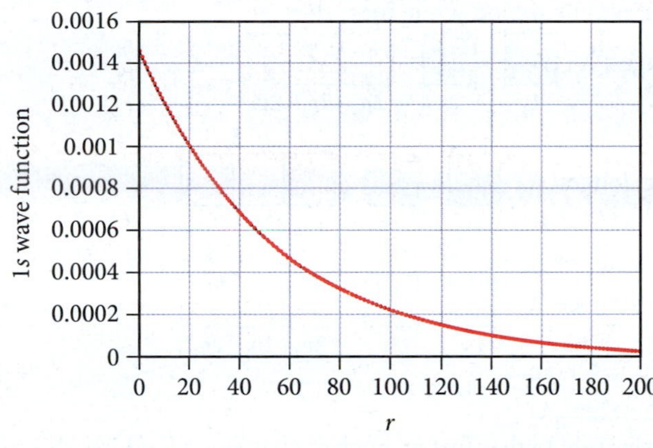

2*s*:

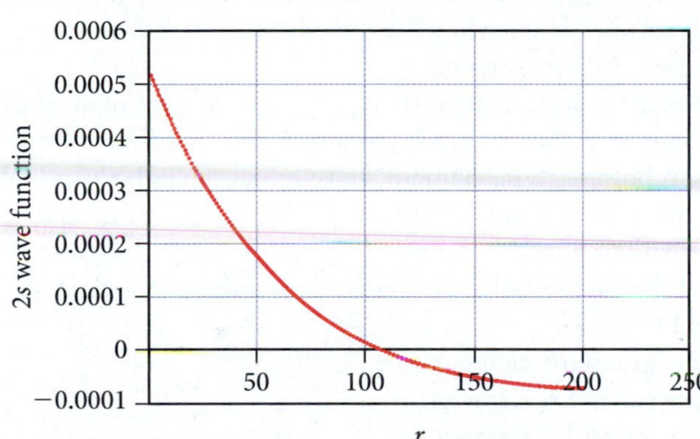

The plot for the 2*s* wave function extends below the *x*-axis. The *x*-intercept represents the radial node of the orbital.
97. 7.39×10^5 m/s
99. $\Delta E = 1.1 \times 10^{-20}$ J, 7.0×10^2 nm
101. 11 m
103. In the Bohr model, electrons exist in specific orbits encircling the atom. In the quantum-mechanical model, electrons exist in orbitals that are really probability density maps of where the electron is likely to be found. The Bohr model is inconsistent with Heisenberg's uncertainty principle.
105. **a.** yes **b.** no **c.** yes **d.** no

Chapter 4

41. **a.** potassium, metal **b.** barium, metal
c. iodine, nonmetal **d.** oxygen, nonmetal
e. antimony, metalloid
43. **a** and **b** are main-group elements.
45. **a.** $1s^22s^22p^63s^23p^2$
b. $1s^22s^22p^4$
c. $1s^22s^22p^63s^23p^64s^1$
d. $1s^22s^22p^6$
47. **a.** 1*s* [↑↓] 2*s* [↑↓] 2*p* [↑][↑][↑]
b. 1*s* [↑↓] 2*s* [↑↓] 2*p* [↑↓][↑↓][↑]
c. 1*s* [↑↓] 2*s* [↑↓] 2*p* [↑↓][↑↓][↑↓] 3*s* [↑↓]
d. 1*s* [↑↓] 2*s* [↑↓] 2*p* [↑↓][↑↓][↑↓] 3*s* [↑↓] 3*p* [↑][ ][ ]
49. **a.** [Ne] $3s^23p^3$ **b.** [Ar] $4s^23d^{10}4p^2$
c. [Kr] $5s^24d^2$ **d.** [Kr] $5s^24d^{10}5p^5$
51. **a.** 1 **b.** 10 **c.** 5 **d.** 2
53. **a.** V, As **b.** Se **c.** V **d.** Kr
55. **a.** 2 **b.** 1 **c.** 10 **d.** 6
57. reactive metal: **a**, reactive nonmetal: **c**
59. **a.** 1 valence electron, alkali metal
b. 7 valance electrons, halogen
c. 2 valence electrons, alkaline earth metal
d. 2 valence electrons, alkaline earth metal
e. 8 valence electrons, noble gas
61. Cl and F because they are in the same group or family. Elements in the same group or family have similar chemical properties.
63. **a.** 2− [Ne] **b.** 1+ [Ar]
c. 3+ [Ne] **d.** 1+ [Kr]
65. **c**
67. The valence electrons of nitrogen will experience a greater effective nuclear charge. The valence electrons of both atoms are screened by two core electrons, but N has a greater number of protons and therefore a greater net nuclear charge.
69. **a.** 1+ **b.** 2+ **c.** 6+ **d.** 4+
71. **a.** In **b.** Si **c.** Pb **d.** C
73. F, S, Si, Ge, Ca, Rb
75. **a.** [Ne] **b.** [Kr] **c.** [Kr]
d. [Ar] $3d^6$ **e.** [Ar] $3d^9$
77. **a.** [Ar] Diamagnetic
b. [Ar] 3*d* [↑][↑][↑][ ][ ] Paramagnetic
c. [Ar] 3*d* [↑↓][↑↓][↑↓][↑][↑] Paramagnetic
d. [Ar] 3*d* [↑][↑][↑][↑][↑] Paramagnetic
79. **a.** Li **b.** I^- **c.** Cr **d.** O^{2-}
81. O^{2-}, F^-, Ne, Na^+, Mg^{2+}
83. **a.** Br **b.** Na
c. cannot tell based on periodic trends **d.** P
85. In, Si, N, F
87. **a.** second and third **b.** fifth and sixth
c. sixth and seventh **d.** first and second

89. a. Na **b.** S **c.** C **d.** F

91. a. Sr **b.** Bi
c. cannot tell based on periodic trends **d.** As

93. S, Se, Sb, In, Ba, Fr

95. Br: $1s^2 2s^2 2p^6 3s^2 3p^6 4s^2 3d^{10} 4p^5$
Kr: $1s^2 2s^2 2p^6 3s^2 3p^6 4s^2 3d^{10} 4p^6$
Krypton's outer electron shell is filled, giving it chemical stability. Bromine is missing an electron from its outer shell and subsequently has a high electron affinity. Bromine tends to be easily reduced by gaining an electron, giving the bromide ion stability due to the filled *p* subshell which corresponds to krypton's chemically stable electron configuration.

97. V: [Ar] $4s^2 3d^3$
V^{3+}:[Ar] $3d^2$
Both V and V^{3+} contain unpaired electrons in their $3d$ orbitals.

99. A substitute for K^+ would need to exhibit a 1+ electric charge and have similar mass and atomic radius. Na^+ or Rb^+ might be good substitutes, but their radii are significantly smaller and larger, respectively. Based on mass, Ca^+ and Ar^+ are the closest to K^+. Because the first ionization energy of Ca^+ is closest to that of K^+, Ca^+ might be a good choice for a substitute. The difficulty lies in Ca's low second ionization energy, making it easily oxidized to form Ca^{2+}.

101. Si, Ge

103. a. N: [He] $2s^2 2p^3$, Mg: [Ne] $3s^2$, O: [He] $2s^2 2p^4$, F: [He] $2s^2 2p^5$, Al: [Ne] $3s^2 3p^1$
b. Mg, Al, O, F, N
c. Al, Mg, O, N, F
d. Aluminum's first ionization energy is lower than Mg because its $3p$ electron is shielded by the $3s$ orbital. Oxygen's first ionization energy is lower than that of N because its fourth $2p$ electron experiences electron–electron repulsion by the other electron in its orbital.

105. For main-group elements, atomic radii decrease across a period because the addition of a proton in the nucleus and an electron in the outermost energy level increases Z_{eff}. This does not happen in the transition metals because the electrons are added to the $n_{highest-1}$ orbital and the Z_{eff} stays roughly the same.

107. Noble gases are exceptionally unreactive due to the stability of their completely filled outer quantum levels and their high ionization energies. The ionization energies of Kr, Xe, and Rn are low enough to form some compounds.

109. 6A: ns^2np^4, 7A: ns^2np^5, group 7A elements require only one electron to achieve a noble gas configuration. Since group 6A elements require two electrons, their affinity for one electron is less negative because one electron will merely give them an np^5 configuration.

111. 85

113. a. One If By Land (O, Ne, I, F, B, Y, La, Nd)
b. Atoms Are Fun (N, U, Fe, Ra, S, Mo, Ta backwards)

115. 1.390×10^3 kJ/mol, 86.14 nm

117. a. $d_{Ar} \approx 2$ g/L, $d_{Xe} \approx 6.5$ g/L
b. $d_{118} \approx 13$ g/L
c. mass $= 3.35 \times 10^{-23}$ g/Ne atom, density of Ne atom $= 2.3 \times 10^4$ g/L. The separation of Ne atoms relative to their size is immense.
d. Kr: 2.69×10^{22} atoms/L, Ne: 2.69×10^{22} atoms/L. It seems Ar will also have 2.69×10^{22} atoms/L.
$d_{Ar} = 1.78$ g/L. This corresponds to accepted values.

119. Density increases to the right because, though electrons are added successively across the period, they are added to the $3d$ subshell, which is not a part of the outermost principal energy level. As a result, the atomic radius does not increase significantly across the period while mass does.

121.

	$1s$	$2s$	$2p_x$	$2p_y$	$2p_z$
Longest λ:	↑↓	↑↓	↑	↓	
Next longest λ:	↑↓	↑↓	↑↓		
Next longest λ:	↑↓	↑	↑	↑	↑

123. 168, noble gas

125. A relatively high effective nuclear charge is found in gallium with its completed $3d$ subshell and in thallium with its completed $4f$ subshell, accounting for the relatively high first ionization energies of these elements.

127. The second electron affinity requires the addition of an electron to something that is already negatively charged. The monoanions of both of these elements have relatively high electron density in a relatively small volume. As we shall see in Chapter 9, the dianions of these elements do exist in many compounds because they are stabilized by chemical bonding.

129. 120, 170

131. a. any group 6A element
b. any group 5A element
c. any group 1A element

133. a. true **b.** true
c. false **d.** true

135. Since Ca has valence electrons of $4s^2$, it has a relatively low ionization energy to lose 2 electrons. In contrast, F has a highly exothermic electron affinity when gaining 1 electron, but not a second electron because of its $2s^2 2p^5$ valence electrons. Therefore, calcium and fluoride combine in a 2:1 ratio.

Chapter 5

29. a. molecular **b.** ionic
c. ionic **d.** molecular

31. a. NO_2 **b.** C_5H_{12} **c.** C_2H_5

33. a. 3 Mg, 2 P, 8 O **b.** 1 Ba, 2 Cl
c. 1 Fe, 2 N, 4 O **d.** 1 Ca, 2 O, 2 H

35. a. NH_3 **b.** C_2H_6 **c.** SO_3

37. $1s^2\underline{2s^2 2p^3}$ ·N̈:

39. a. ·Ȧl· **b.** Na^+
c. :C̤̈l: **d.** $[:\ddot{\underset{..}{Cl}}:]^-$

41. a. $Na^+[:\ddot{\underset{..}{F}}:]^-$ **b.** $Ca^{2+}[:\ddot{\underset{..}{O}}:]^{2-}$
c. $Sr^{2+}\ 2[:\ddot{\underset{..}{Br}}:]^-$ **d.** $2\ K^+[:\ddot{\underset{..}{O}}:]^{2-}$

43. a. SrSe **b.** $BaCl_2$
c. Na_2S **d.** Al_2O_3

45. One factor of lattice energy is the product of the charges of the two ions. The product of the ion charges for CsF is +1, while that for BaO is +4. Because this product is four times greater, the lattice energy is also four times greater.

47. a. CaO **b.** ZnS
c. RbBr **d.** Al_2O_3

49. a. $Ca(OH)_2$ **b.** $CaCrO_4$
c. $Ca_3(PO_4)_2$ **d.** $Ca(CN)_2$

51. a. magnesium nitride **b.** potassium fluoride
c. sodium oxide **d.** lithium sulfide
e. cesium fluoride **f.** potassium iodide

53. a. tin(II) oxide **b.** chromium(III) sulfide
c. rubidium iodide **d.** barium bromide

55. a. copper(I) nitrite **b.** magnesium acetate
c. barium nitrate **d.** lead(II) acetate

57. a. $NaHSO_3$ **b.** $LiMnO_4$
c. $AgNO_3$ **d.** K_2SO_4
e. $RbHSO_4$ **f.** $KHCO_3$

59. a. cobalt(II) sulfate heptahydrate
b. $IrBr_3 \cdot 4\ H_2O$
c. Magnesium bromate hexahydrate
d. $K_2CO_3 \cdot 2\ H_2O$

61. a. H:H, filled duets **b.** $:\ddot{\underset{..}{Cl}}:\ddot{\underset{..}{Cl}}:$, filled octets
c. $\ddot{\underset{..}{O}}=\ddot{\underset{..}{O}}$, filled octets **d.** $:N\equiv N:$, filled octets

63. a. carbon monoxide **b.** nitrogen triiodide
c. silicon tetrachloride **d.** tetranitrogen tetraselenide

65. a. PCl_3 **b.** ClO
c. S_2F_4 **d.** PF_5

67. a. Strontium chloride
b. Tin(IV) oxide
c. Diphosphorus pentasulfide

69. a. Potassium chlorate
b. Diiodine pentoxide
c. Lead(II) sulfate

71. a. 46.01 amu **b.** 58.12 amu
c. 180.16 amu **d.** 238.03 amu

73. a. 0.471 mol **b.** 0.0362 mol
c. 968 mol **d.** 0.279 mol

75. a. 0.554 mol **b.** 28.4 mol
c. 0.378 mol **d.** 1093 mol

77. a. 2.2×10^{23} molecules
b. 7.06×10^{23} molecules
c. 4.16×10^{23} molecules
d. 1.09×10^{23} molecules

79. a. 0.0790 g **b.** 0.84 g **c.** 2.992×10^{-22} g

81. 0.10 mg

83. a. 74.87% C **b.** 79.88% C
c. 92.24% C **d.** 37.23% C

85. NH_3: 82.27% N
$CO(NH_2)_2$: 46.65% N
NH_4NO_3: 35.00% N
$(NH_4)_2SO_4$: 21.20% N
NH_3 has the highest N content

87. 20.8 g F

89. 196 μg KI

91. a. 2 : 1 **b.** 4 : 1 **c.** 6 : 2 : 1

93. a. 0.885 mol H **b.** 5.2 mol H
c. 29 mol H **d.** 33.7 mol H

95. a. 3.3 g Na **b.** 3.6 g Na
c. 1.4 g Na **d.** 1.7 g Na

97. a. Ag_2O **b.** $Co_3As_2O_8$ **c.** $SeBr_4$

99. a. C_5H_7N **b.** $C_4H_5N_2O$

101. $C_{13}H_{18}O_2$

103. NCl_3

105. a. $C_{12}H_{14}N_2$ **b.** $C_6H_3Cl_3$
c. $C_{10}H_{20}N_2S_4$

107. CH_2

109. C_2H_4O

111. a. inorganic **b.** organic
c. organic **d.** inorganic

113. a. functionalized hydrocarbon, alcohol
b. hydrocarbon
c. functionalized hydrocarbon, ketone
d. functionalized hydrocarbon, amine

115. 1.50×10^{24} molecules EtOH

117. a. K_2CrO_4, 40.27% K, 26.78% Cr, 32.95% O
b. $Pb_3(PO_4)_2$, 76.60% Pb, 7.63% P, 15.77% O
c. $CoBr_2$, 26.94% Co, 73.06% Br

119. 1.80×10^2 g Cl/yr

121. M = Fe

123. estradiol = $C_{18}H_{24}O_2$

125. $C_{18}H_{20}O_2$

127. $7\ H_2O$

129. C_6H_9BrO

131. 1.87×10^{21} atoms

133. 92.93 amu

135. $x = 1, y = 2$

137. 41.7 mg

139. 0.224 g

141. 22.0% by mass

143. 1.6×10^7 kg Cl

145. 7.8×10^3 kg rock

147. $C_5H_{10}SI$

149. X_2Y_3

151. The sphere in the molecular models represents the electron cloud of the atom. On this scale, the nucleus would be too small to see.

153. The statement is incorrect because a chemical formula is based on the ratio of atoms combined, not the ratio of grams combined. The statement should read: "The chemical formula for ammonia (NH_3) indicates that ammonia contains three hydrogen atoms to each nitrogen atom."

Chapter 6

23. a. pure covalent **b.** polar covalent
c. pure covalent **d.** ionic bond

25. 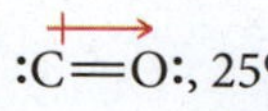 :C=O:, 25%

27. 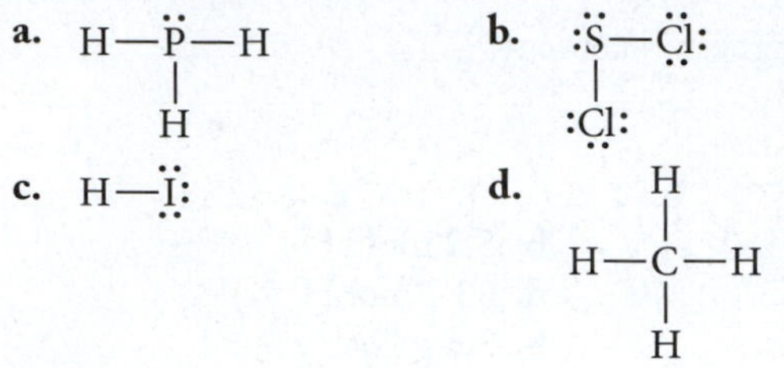 a. H—P—H
 |
 H

b. :S—Cl:
 |
 :Cl:

c. H—I:

d. H
 |
H—C—H
 |
 H

29.

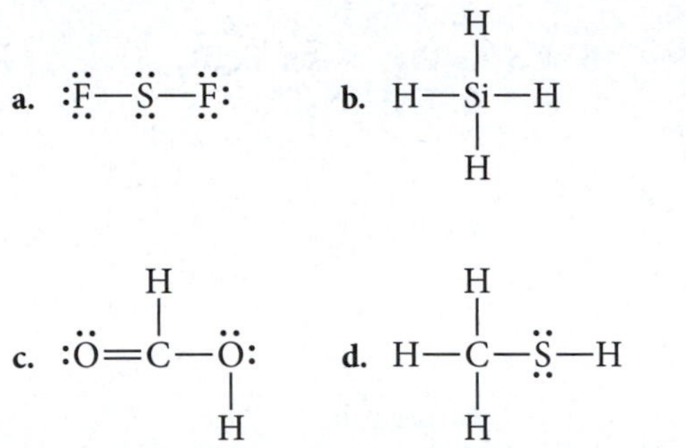

a. :F—S—F:

b. H
 |
H—Si—H
 |
 H

c. H
 |
:O=C—O:
 |
 H

d. H
 |
H—C—S—H
 |
 H

31. 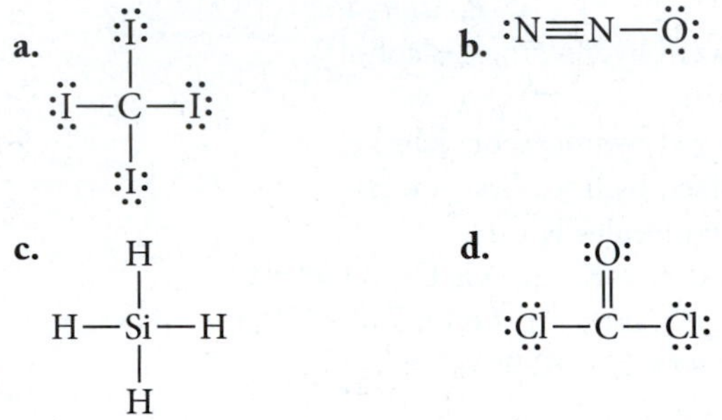 a. :I:
 |
:I—C—I:
 |
 :I:

b. :N≡N—O:

c. H
 |
H—Si—H
 |
 H

d. :O:
 ‖
:Cl—C—Cl:

33. 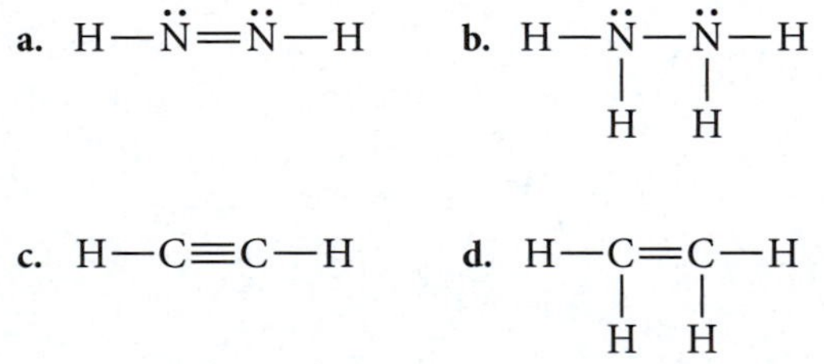 a. H—N=N—H

b. H—N—N—H
 | |
 H H

c. H—C≡C—H

d. H—C=C—H
 | |
 H H

35. 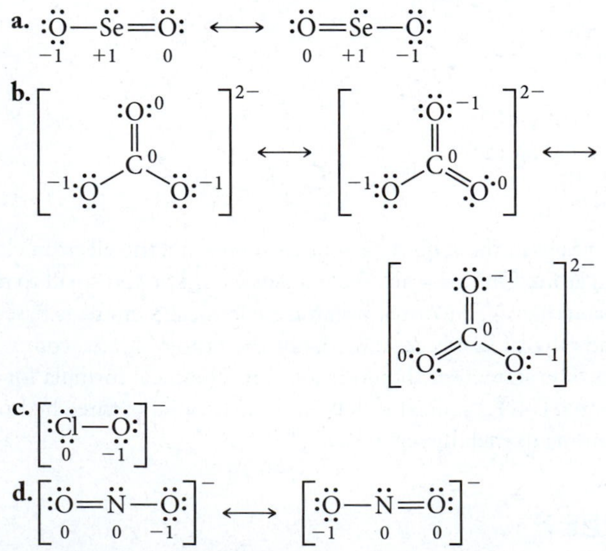 a. :O—Se=O: ⟷ :O=Se—O:
-1 $+1$ 0 0 $+1$ -1

b. [:O:^{0} ‖ C^{0} $^{-1}$:O: :O:$^{-1}$]$^{2-}$ ⟷ [:O:$^{-1}$ ‖ C^{0} $^{-1}$:O: O:^{0}]$^{2-}$ ⟷ [:O:$^{-1}$ ‖ C^{0} ^{0}O: O:$^{-1}$]$^{2-}$

c. [:Cl—O:]$^{-}$
0 -1

d. [:O=N—O:]$^{-}$ ⟷ [:O—N=O:]$^{-}$
0 0 -1 -1 0 0

37. 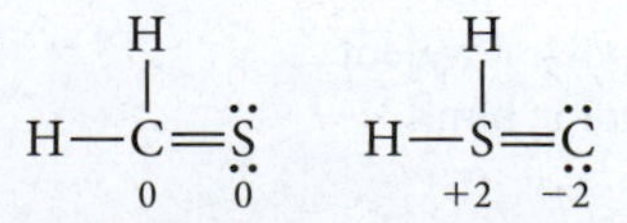 H H
 | |
H—C=S H—S=C
0 0 $+2$ -2

H_2CS is the better structure

39. 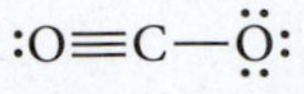 :O≡C—O:

does not provide a significant contribution to the resonance hybrid as it has a +1 formal charge on a very electronegative atom (oxygen).

41. 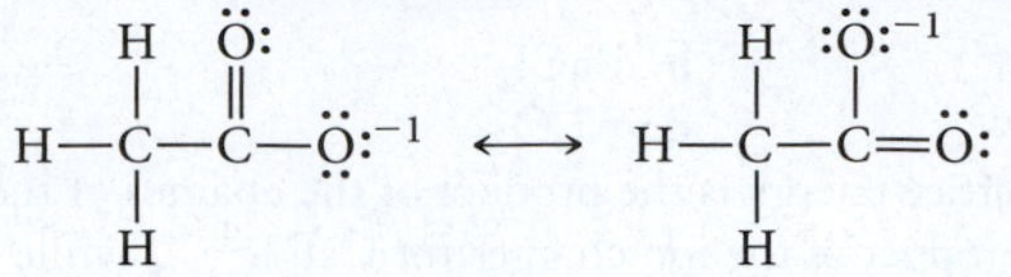 H O: H :O:$^{-1}$
 | ‖ | |
H—C—C—O:$^{-1}$ ⟷ H—C—C=O:
 | |
 H H

43. N has a formal charge of +1; O has a formal charge of −1.

45. 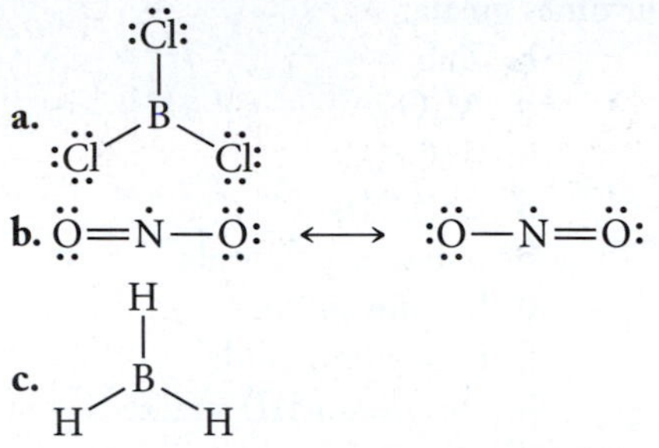 **a.** :Cl:
 |
 B
:Cl Cl:

b. O=N—O: ⟷ :O—N=O:

c. H
 |
 B
H H

47. 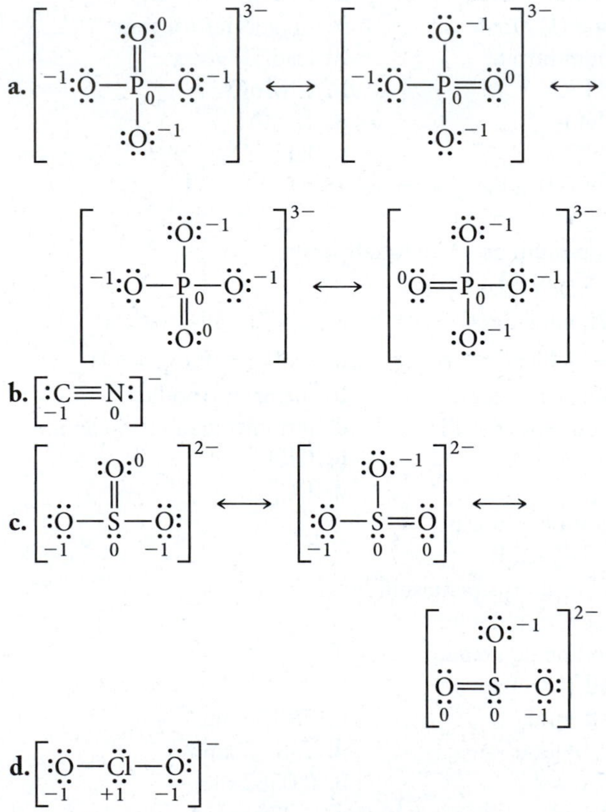 a. [:O:^{0} ‖ $^{-1}$:O—P$_{0}$—O:$^{-1}$ | :O:$^{-1}$]$^{3-}$ ⟷ [:O:$^{-1}$ | $^{-1}$:O—P$_{0}$=O^{0} | :O:$^{-1}$]$^{3-}$ ⟷

[:O:$^{-1}$ | $^{-1}$:O—P$_{0}$—O:$^{-1}$ ‖ :O:^{0}]$^{3-}$ ⟷ [:O:$^{-1}$ | ^{0}O=P$_{0}$—O:$^{-1}$ | :O:$^{-1}$]$^{3-}$

b. [:C≡N:]$^{-}$
-1 0

c. [:O:^{0} ‖ :O—S—O:]$^{2-}$ ⟷ [:O:$^{-1}$ | :O—S=O]$^{2-}$ ⟷ [:O:$^{-1}$ | O=S—O:]$^{2-}$
-1 0 -1 / -1 0 0 / 0 0 -1

d. [:O—Cl—O:]$^{-}$
-1 $+1$ -1

49. 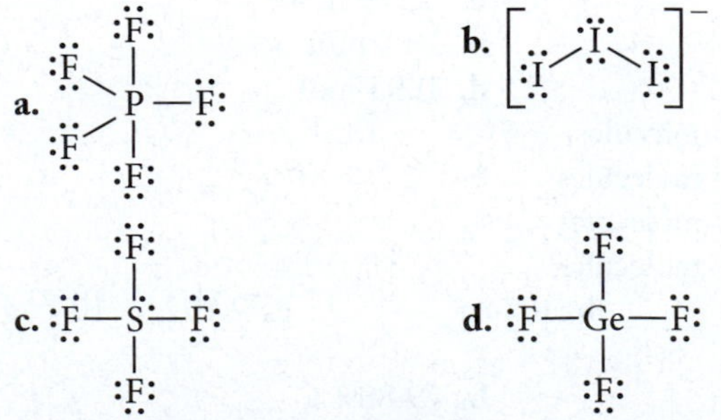 a. 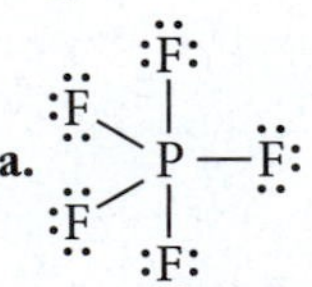

b. [:I—I—I:]$^{-}$

c. :F:
 |
:F—S—F:
 |
 :F:

d. :F:
 |
:F—Ge—F:
 |
 :F:

51. H_3CCH_3, H_2CCH_2, HCCH

53. 4

55. a. 4 e^{-} groups, 4 bonding groups, 0 lone pair

b. 5 e^{-} groups, 3 bonding groups, 2 lone pairs

c. 6 e^{-} groups, 5 bonding groups, 1 lone pair

57. a. e^- geometry: tetrahedral
molecular geometry: trigonal pyramidal
idealized bond angle: 109.5°, deviation

b. e^- geometry: tetrahedral
molecular geometry: bent
idealized bond angle: 109.5°, deviation

c. e^- geometry: tetrahedral
molecular geometry: tetrahedral
idealized bond angle: 109.5°, deviation (due to large size of Cl compared to H)

d. e^- geometry: linear
molecular geometry: linear
idealized bond angle: 180°

59. H_2O has a smaller bond angle due to lone pair–lone pair repulsions, the strongest electron group repulsion.

61. a. seesaw, (S with four F)
b. T-shape, (F—Cl—F with F)
c. linear, F—I—F
d. square planar, (Br—I—Br with Br above and below)

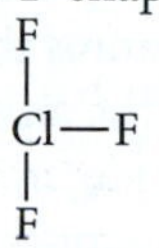

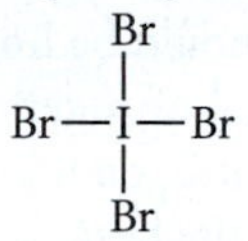

63. a. linear, H—C≡C—H
b. Trigonal planar, (H)(H)C=C(H)(H)
c. tetrahedral, $H_3C—CH_3$

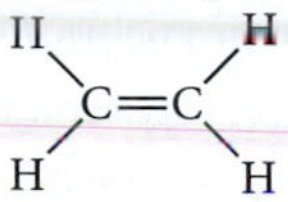

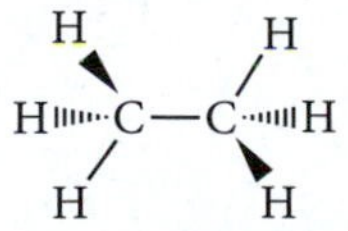

65. a. The lone pair will cause lone pair–bonding pair repulsions, pushing the three bonding pairs out of the same plane. The correct molecular geometry is trigonal pyramidal.
b. The lone pair should take an equatorial position to minimize 90° bonding pair interactions. The correct molecular geometry is seesaw.
c. The lone pairs should take positions on opposite sides of the central atom to reduce lone pair–lone pair interactions. The correct molecular geometry is square planar.

67. a. C: tetrahedral
O: bent 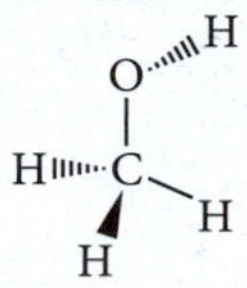
(H—O—CH_3)
b. C's: tetrahedral
O: bent 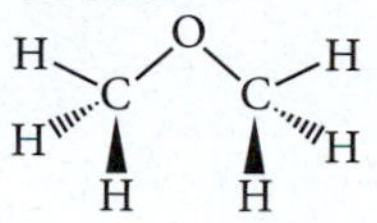
(H_3C—O—CH_3)
c. O's: bent
H—O—O—H

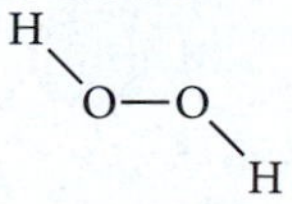

69. The vectors of the polar bonds in both CO_2 and CCl_4 oppose each other with equal magnitude and sum to 0.

71. PF_3, polar
SBr_2, nonpolar
$CHCl_3$, polar
CS_2, nonpolar

73. a. polar **b.** polar **c.** polar **d.** nonpolar

75. a. Ba^{2+} $[:\ddot{O}—C(=O)—\ddot{O}:]^{2-}$ ⟷ Ba^{2+} $[:\ddot{O}—C(—\ddot{O}:)=\ddot{O}]^{2-}$ ⟷ Ba^{2+} $[\ddot{O}=C(—\ddot{O}:)—\ddot{O}:]^{2-}$

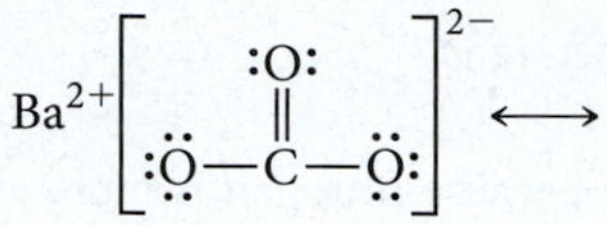

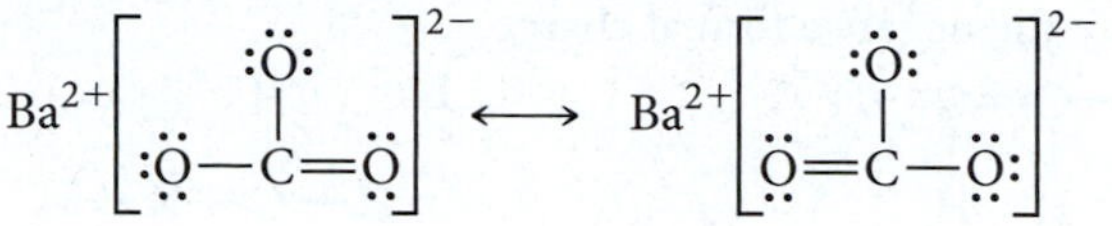

b. Ca^{2+} $2[:\ddot{O}—H]^-$

c. K^+ $[:\ddot{O}—N(=O)—\ddot{O}:]^-$ ⟷ K^+ $[:\ddot{O}—N(—\ddot{O}:)=\ddot{O}]^-$ ⟷ K^+ $[\ddot{O}=N(—\ddot{O}:)—\ddot{O}:]^-$

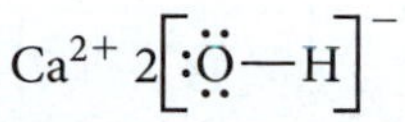

d. Li^+ $[:\ddot{I}—\ddot{O}:]^-$

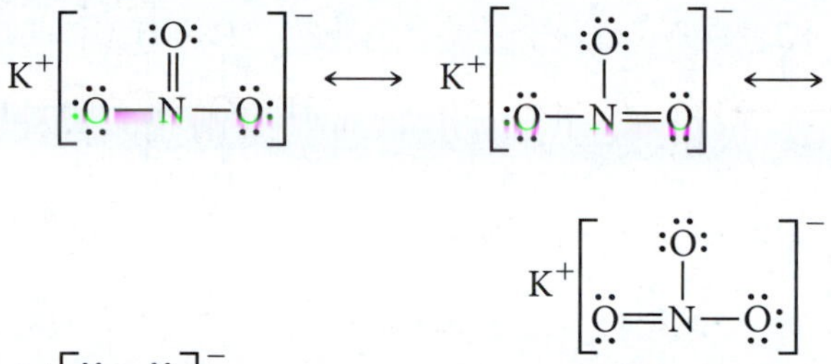

77. a. (cyclobutane, C_4H_8)
b. (cyclobutadiene, C_4H_4, two resonance structures) ⟷
c. (cyclohexane, C_6H_{12})
d. (benzene, C_6H_6, two resonance structures) ⟷

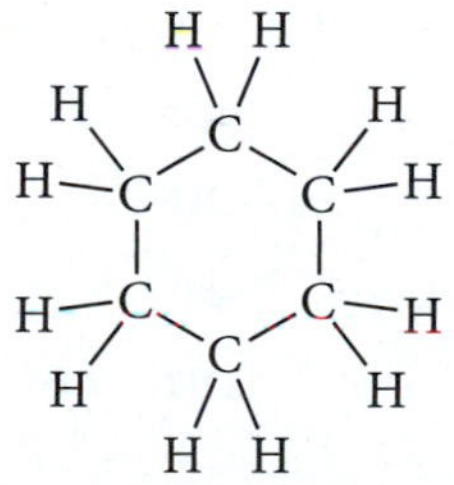

79. CH_2O_2,
H—C(=O)—$\ddot{O}$—H

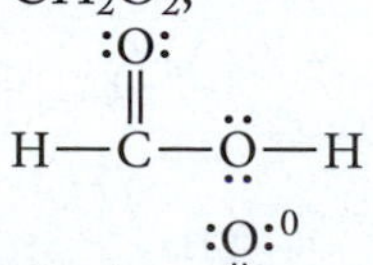

81.
H—$\ddot{O}$—N(=O)—$\ddot{O}$: ⟷ H—$\ddot{O}$—N(—$\ddot{O}$:)=$\ddot{O}$ ⟷ H—$\ddot{O}$=N(—$\ddot{O}$:)—$\ddot{O}$:
formal charges: O 0 (=O); H 0, O 0, N +1, O −1 ⟷ O −1; H 0, O 0, N +1, O 0 ⟷ O −1; H 0, O +1, N +1, O −1

Most important (first two structures)

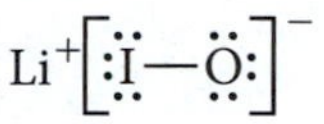

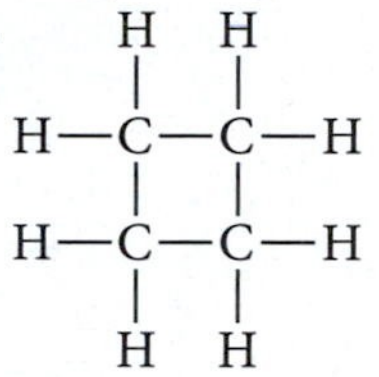

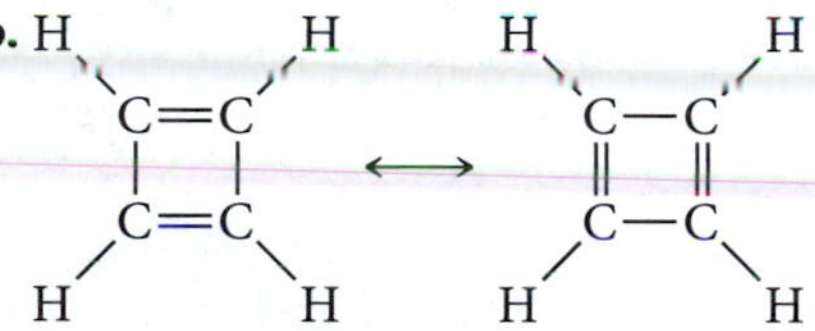

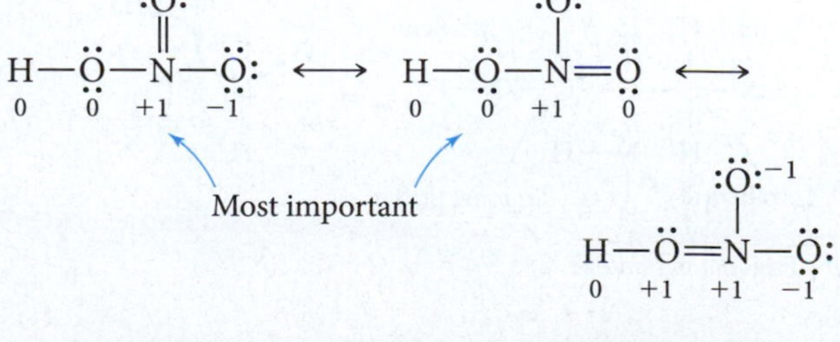

83. $[\ddot{C}{=}N{=}\ddot{O}]^-$ (formal charges −2, +1, 0) ⟷ $[:C{\equiv}\ddot{N}{-}\ddot{O}:]^-$ (formal charges −1, +1, −1)

The fulminate ion is less stable because nitrogen is more electronegative than carbon and should therefore be terminal to accommodate the negative formal charge.

85. **a.** H—C=C—H (each C bonded to H below)
b. H—C(H)(H)—N̈(H)—H
c. :O: double-bonded to C; H—C—H
d. H—C(H)(H)—C(H)(H)—Ö—H
e. :O: double-bonded to C; H—C—Ö—H

87. :S: double-bonded to C; H—C—N̈(H)—H. Labels: Nonpolar (C=S), Polar (N—H), Nonpolar (H—C), Polar (C—N)

89. **a.** $[\ddot{O}{=}\ddot{O}]^-$ **b.** $[:\ddot{O}:]^-$
c. :Ö—H **d.** H—C(H)(H)—Ö—Ö·

c. Polar
d. Nonpolar

91. H—C≡C—H

93. **a.** :Ö—C(H)(H)—C(H)(:N(H)—H)—C(=:O:)—Ö—H
Labels: sp^3, Bent; sp^3, Tetrahedral; sp^3, Tetrahedral; sp^3, bent; sp^2, Trigonal planar; sp^3, Trigonal pyramidal
(3D structure: H, OH, H_2N, O, OH)

b. 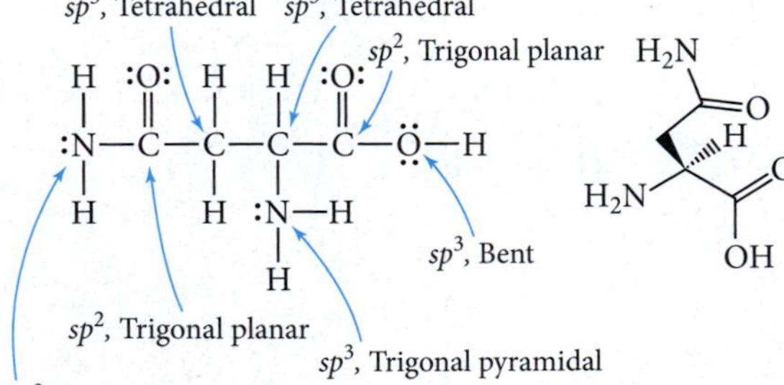

(3D structure: H_2N, O, H, H_2N, O, OH)

c. :S̈(H)—C(H)(H)—C(H)(:N(H)—H)—C(=:O:)—Ö—H
Labels: sp^3, Bent; sp^3, Tetrahedral; sp^3, Bent; sp^3, Tetrahedral; sp^2, Trigonal planar; sp^3, Trigonal pyramidal
(3D structure: SH, H, N, O, H, H, OH)

95. **a.** water soluble **b.** fat soluble
c. water soluble **d.** fat soluble

97. $[:N{\equiv}N{-}\ddot{\underset{..}{N}}:]^-$ ⟷ $[:\ddot{\underset{..}{N}}{-}N{\equiv}N:]^-$ ⟷ $[\ddot{N}{=}N{=}\ddot{N}]^-$

99. $r_{HCl} = 113$ pm
$r_{HF} = 84$ pm

101. (cubane structure: eight C atoms at the corners of a cube, each bonded to one H)

103. As you move down the column from F to Cl to Br to I, the atomic radius of the atoms increases. Because of this, the larger atoms cannot be accommodated with the smaller bond angle. The attached atoms themselves would begin to overlap their orbitals. So, as the size of the attached atoms increases, the bond angle becomes larger, approaching the hybridized 109.5° angle.

105. The bond angle for the nitrogen closest to the C atom should be bent. The bond angle for the nitrogen closest to the terminal nitrogen should be linear. The nitrogen nitrogen bond closest to the terminal nitrogen atom should be shorter than the other nitrogen nitrogen bond (due to resonance).

107. **a.** This is the best.
b. This statement is similar to a. but leaves out nonbonding lone-pair electron groups.
c. Molecular geometries are not determined by overlapping orbitals, but rather by the number and type of electron groups around each central atom.

Chapter 7

31. **a.** 0 **b.** 3 **c.** 1

33.

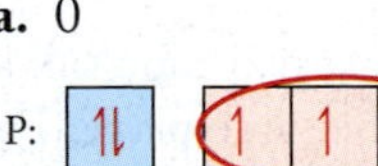

H1: 1s

H2: 1s

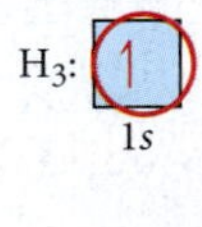

Expected bond angle = 90°

Valence bond theory is compatible with experimentally determined bond angle of 93.3° without hybrid orbitals.

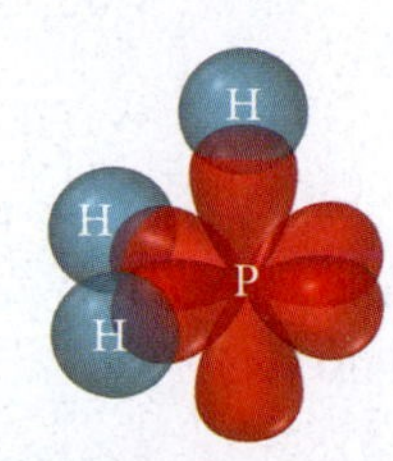

35. 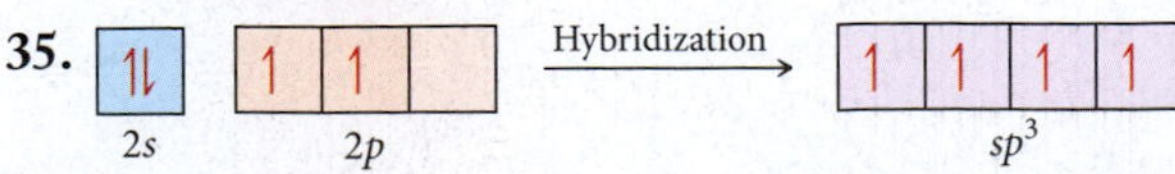

37. sp^2

39. a. sp^3

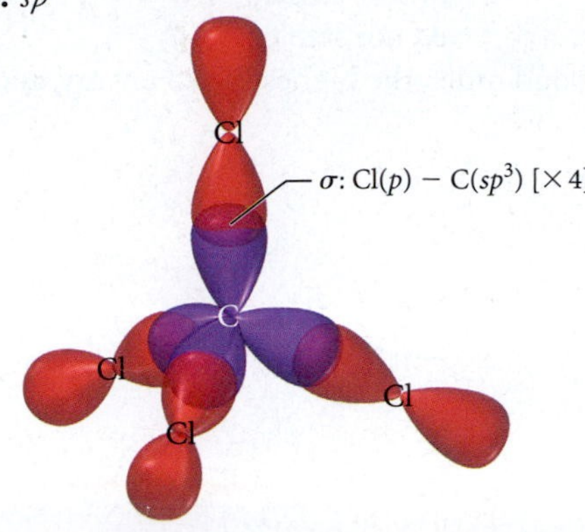

b. sp^3

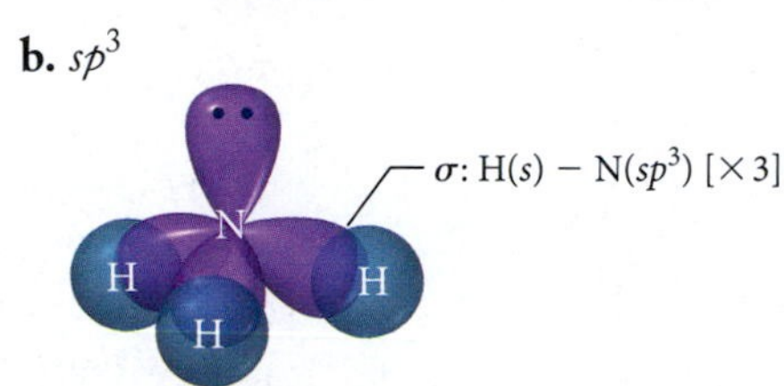

c. sp^3

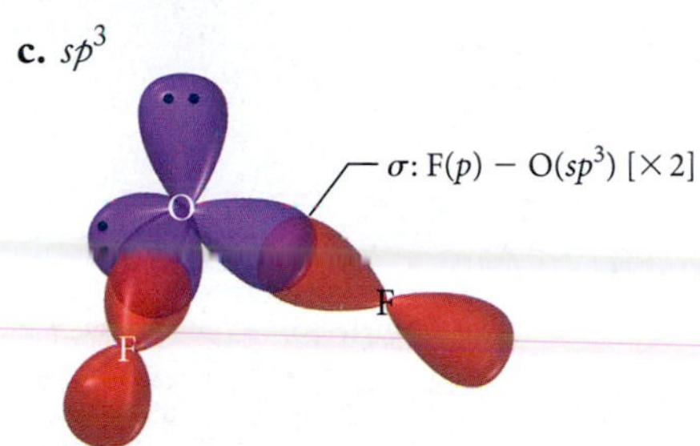

d. sp

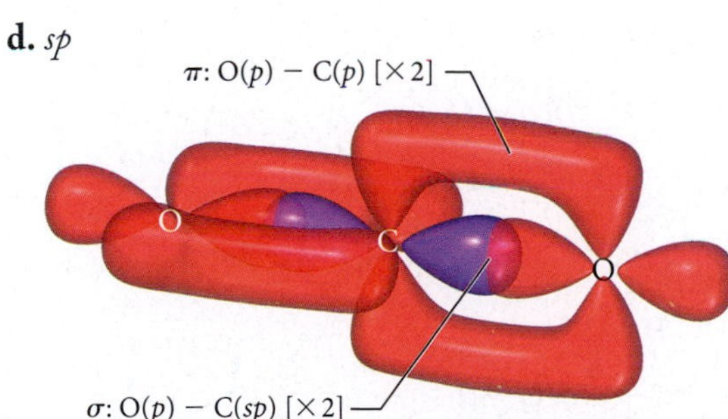

41. a. sp^2

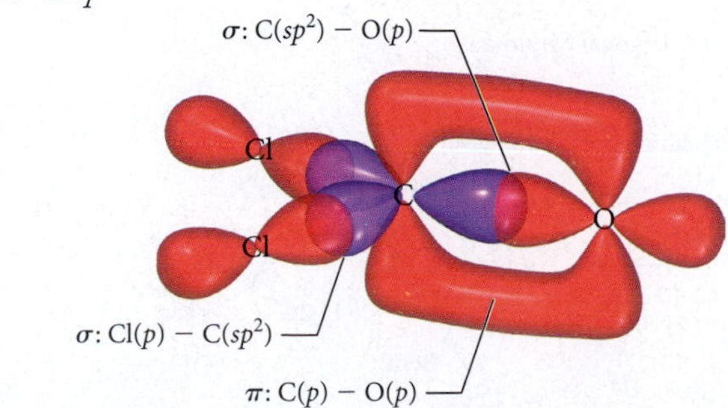

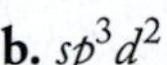

b. sp^3d^2

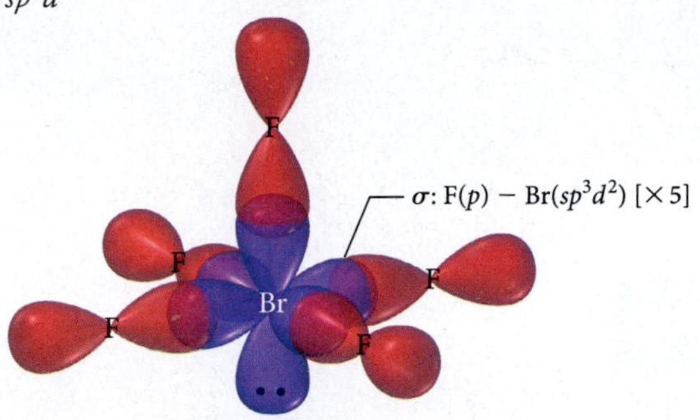

c. sp^3d

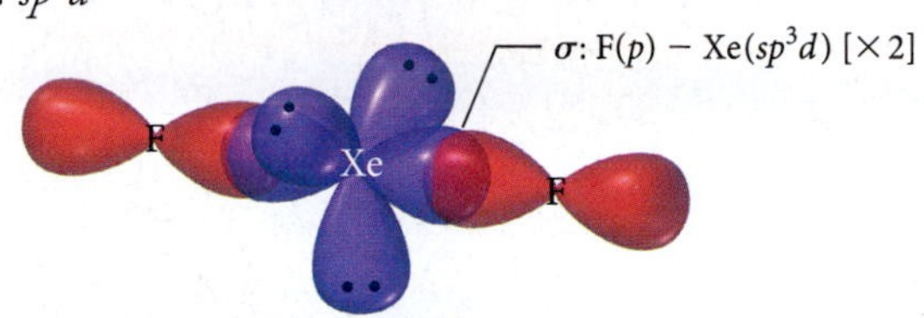

d. sp^3d

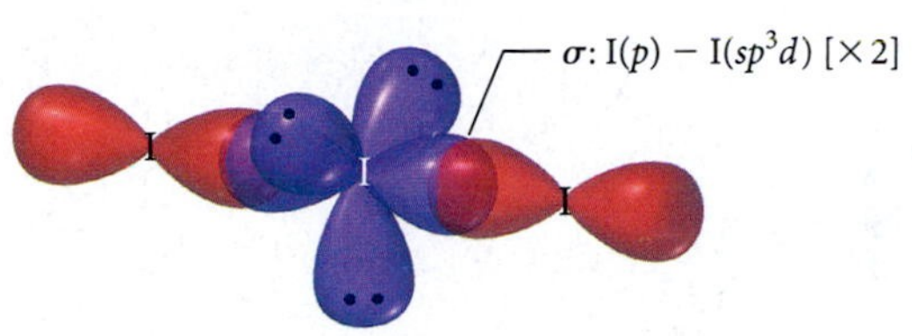

43. a. N's: sp^2

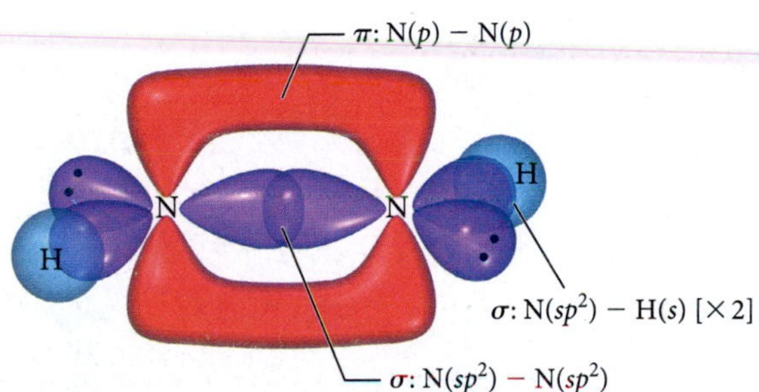

b. N's: sp^3

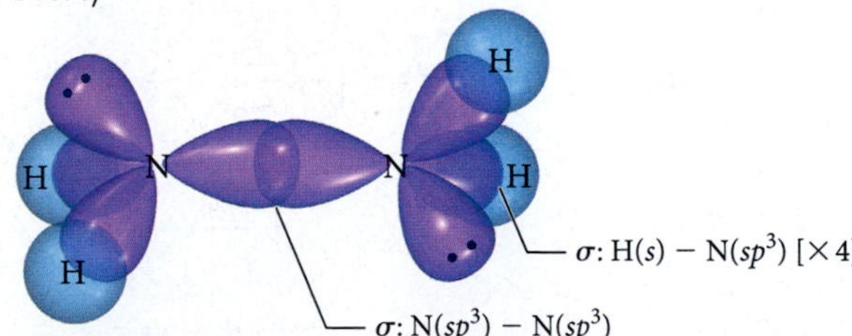

c. C: sp^3
N: sp^3

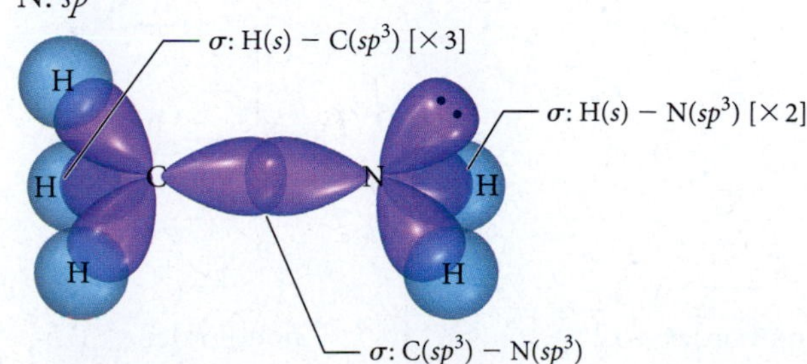

45. Structure of $CH_3CH(NH_2)COOH$ (H–C(H)(H)–C(H)(NH₂)–C(=O)–O–H), with labels: the CH_3 carbon, the central C and the N are sp^3; the carbonyl C is sp^2; the –OH oxygen is sp^3.

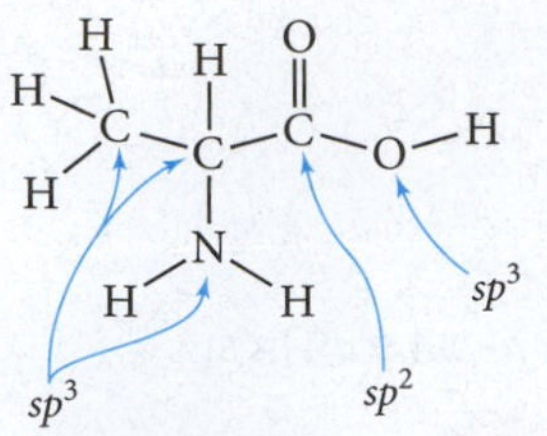

47.

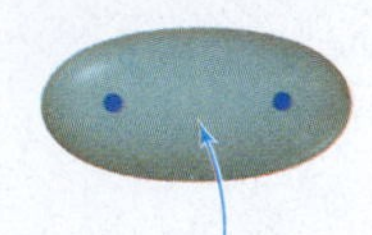

49. Be_2^+

σ^*_{2s}: 1 electron (↑)

σ_{2s}: 2 electrons (↑↓)

Be_2^-

σ_{2p}: 1 electron (↑)

σ^*_{2s}: 2 electrons (↑↓)

σ_{2s}: 2 electrons (↑↓)

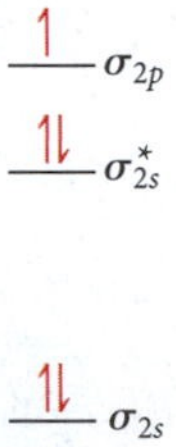

bond order $Be_2^+ = 1/2$

bond order $Be_2^- = 1/2$

Both will exist in gas phase.

51. Bonding

Antibonding

53. a. σ^*_{2p}: empty; π^*_{2p}: empty, empty; σ_{2p}: empty; π_{2p}: empty, empty; σ^*_{2s}: ↑↓; σ_{2s}: ↑↓

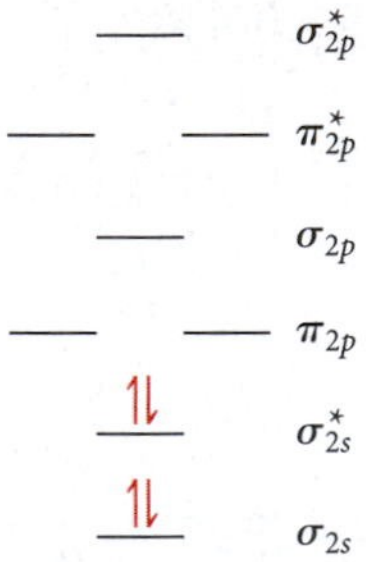

bond order = 0

diamagnetic

b. σ^*_{2p}: empty; π^*_{2p}: empty, empty; σ_{2p}: empty; π_{2p}: ↑, ↑; σ^*_{2s}: ↑↓; σ_{2s}: ↑↓

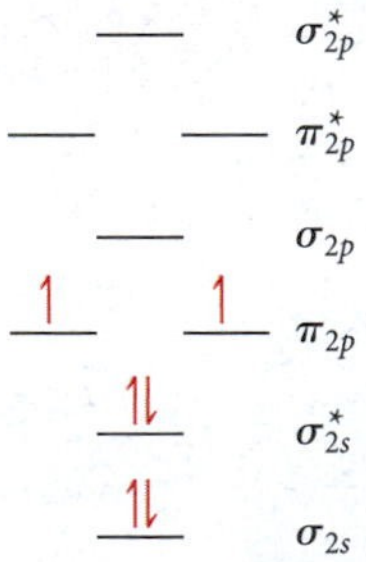

bond order = 1

paramagnetic

c. σ^*_{2p}: empty; π^*_{2p}: empty, empty; σ_{2p}: empty; π_{2p}: ↑↓, ↑↓; σ^*_{2s}: ↑↓; σ_{2s}: ↑↓

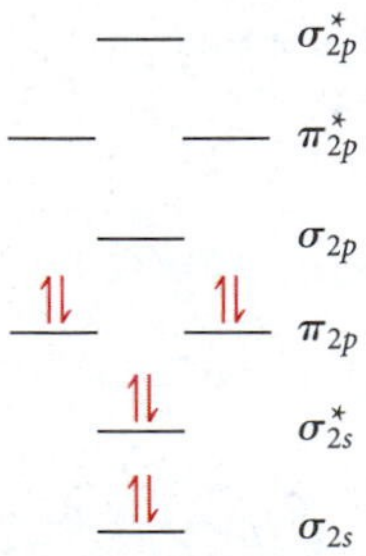

bond order = 2

diamagnetic

d. σ^*_{2p}: empty; π^*_{2p}: empty, empty; σ_{2p}: ↑; π_{2p}: ↑↓, ↑↓; σ^*_{2s}: ↑↓; σ_{2s}: ↑↓

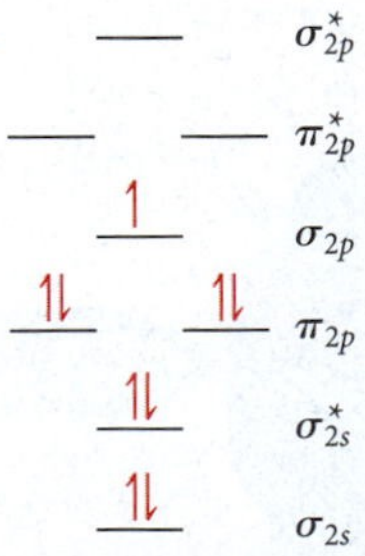

bond order = 2.5

paramagnetic

55. a. not stable **b.** not stable
c. stable **d.** not stable

57. C_2^- has the highest bond order, the highest bond energy, and the shortest bond length.

59. σ^*_{2p}: empty; π^*_{2p}: empty, empty; π_{2p}: ↑↓, ↑↓; σ_{2p}: ↑↓; σ^*_{2s}: ↑↓; σ_{2s}: ↑↓

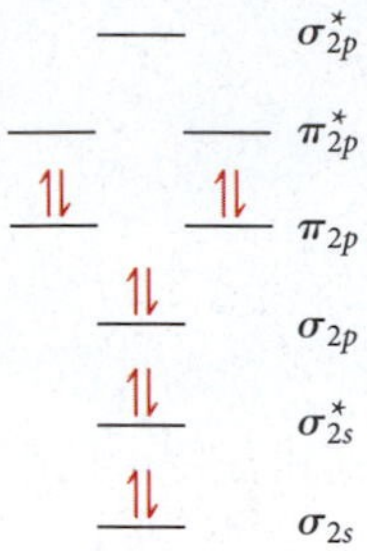

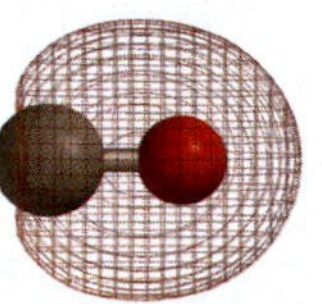

bond order = 3

61. 2.81×10^{23} e^-

63. a

65. a. p-type
b. n-type

67. a. :Ö: double-bonded to C, with :F̤: on each side (:F–C–F:)

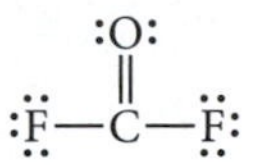

trigonal planar polar

C: sp^2

b. :C̤l–S̤–S̤–C̤l:

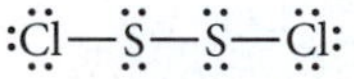

bent polar

S's: sp^3

c. :F̤: above and below S, :F̤: on each side (:F–S–F: with lone pair on S)

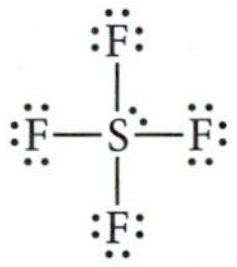

seesaw polar

S: sp^3d

69. a. H–Ö–C(H)(H)–C(H)(NH₂)–C(=O)–Ö–H

Labels: sp^3, Bent (hydroxyl O); sp^3, Tetrahedral; sp^3, Tetrahedral; sp^2, Trigonal planar; sp^3, bent; sp^3, Trigonal pyramidal

Skeletal structure: H, OH, O, H_2N, OH

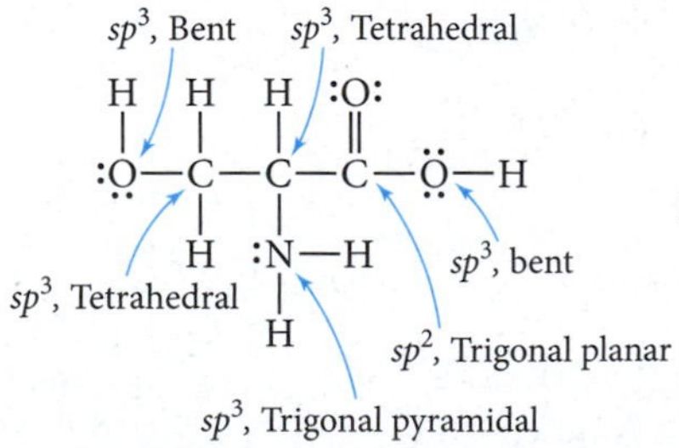

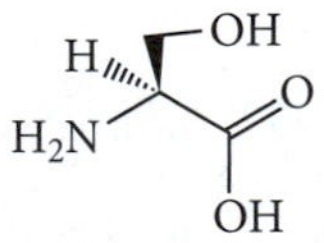

b. H₂N̈–C(=O)–C(H)(H)–C(H)(NH₂)–C(=O)–Ö–H

Labels: sp^3, Tetrahedral; sp^3, Tetrahedral; sp^2, Trigonal planar; sp^3, Bent; sp^2, Trigonal planar; sp^3, Trigonal pyramidal; sp^3, Trigonal pyramidal

Skeletal structure: H_2N, O, H, O, H_2N, OH

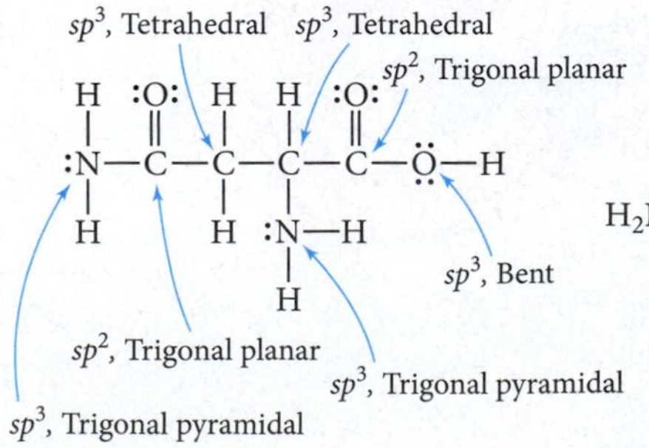

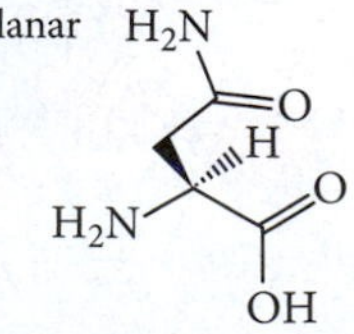

c.

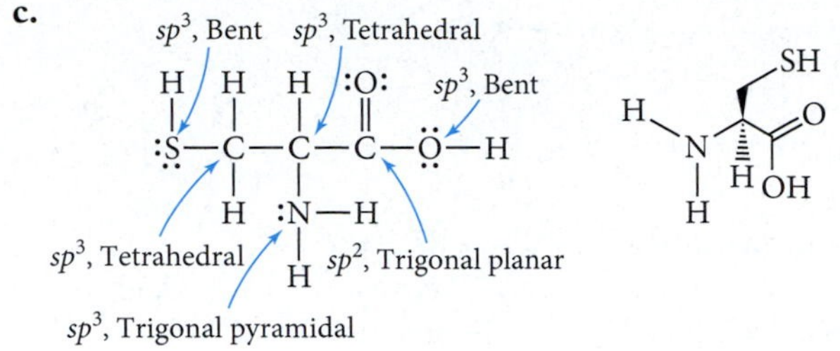

sp^3, Bent sp^3, Tetrahedral sp^3, Bent

H H H :O:
:S—C—C—C—O—H
H :N—H
H

sp^3, Tetrahedral sp^2, Trigonal planar

sp^3, Trigonal pyramidal

SH, H, N, H, O, OH, H

71. σ bonds: 25
π bonds: 4
lone pairs: on O's and N (without methyl group): sp^2 orbitals
on N's (with methyl group): sp^3 orbitals

73.

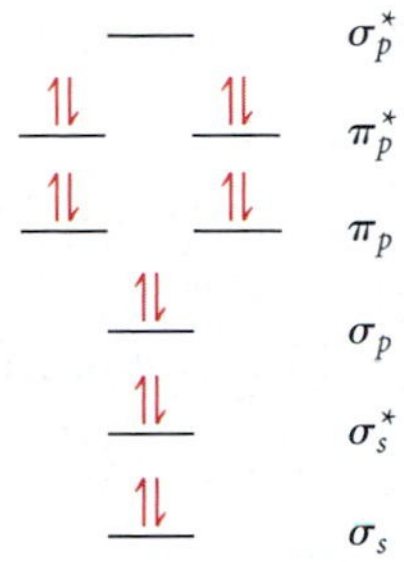

σ_p^*: —
π_p^*: ⥮ ⥮
π_p: ⥮ ⥮
σ_p: ⥮
σ_s^*: ⥮
σ_s: ⥮

bond order = 1

75. BrF, unhybridized, linear
:Br—F:

BrF_2^- has two bonds and three lone pairs on the central atom. The hybridization is sp^3d. The electron geometry is trigonal bipyramidal with the three lone pairs equatorial. The molecular geometry is linear.

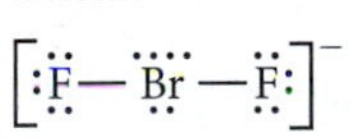

[:F—Br—F:]$^-$

BrF_3 has three bonds and two lone pairs on the central atom. The hybridization is sp^3d. The electron geometry is trigonal bipyramidal with the two lone pairs equatorial. The molecular geometry is T-shaped.

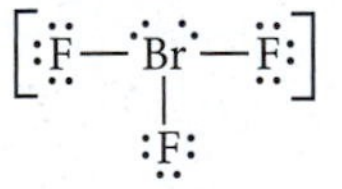

[:F—Br—F:] with :F: below Br

BrF_4^- has four bonds and two lone pairs on the central atom. The hybridization is sp^3d^2. The electron geometry is octahedral with the two lone pairs on the same axis. The molecular geometry is square planar.

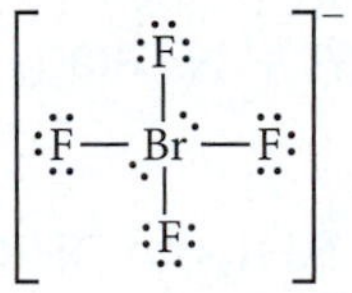

[:F—Br—F: with :F: above and :F: below Br]$^-$

BrF_5 has five bonds and one lone pair on the central atom. The hybridization is sp^3d^2. The electron geometry is octahedral. The molecular geometry is square pyramidal.

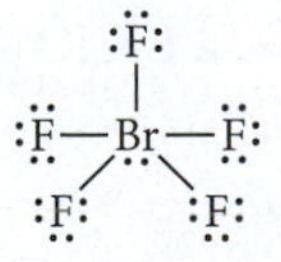

:F: above; :F—Br—F:; :F: :F: below

77. **a.** 10 **b.** 14 **c.** 2

79. According to valence bond theory, CH_4, NH_3, and H_2O are all sp^3 hybridized. This hybridization results in a tetrahedral electron group configuration with a 109.5° bond angle. NH_3 and H_2O deviate from this idealized bond angle because their lone electron pairs exist in their own sp^3 orbitals. The presence of lone pairs lowers the tendency for the central atom's orbitals to hybridize. As a result, as lone pairs are added, the bond angle moves further from the 109.5° hybrid angle and closer to the 90° unhybridized angle.

81. NH_3 is stable due to its bond order of 3.

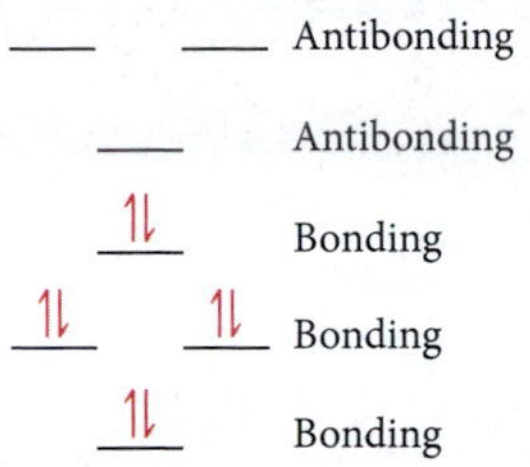

— — Antibonding
— Antibonding
⥮ Bonding
⥮ ⥮ Bonding
⥮ Bonding

83. In addition to the 2*s* and the three 2*p* orbitals, one more orbital is required to make 5 hybrid orbitals. The closest in energy is the 3*s* orbital. So the hybridization is s^2p^3. VSEPR predicts trigonal bipyramidal geometry for five identical substituents.

85. Lewis theory defines a single bond, double bond, and triple bond as a sharing of two electrons, four electrons, and six electrons, respectively, between two atoms. Valence bond theory defines a single bond as a sigma overlap of two orbitals, a double bond as a single sigma bond combined with a pi bond, and a triple bond as a double bond with an additional pi bond. Molecular orbital theory defines a single bond, double bond, and triple bond as a bond order of 1, 2, or 3, respectively, between two atoms.

87. Metals conduct heat well because the free electrons in the electron sea can efficiently transfer heat energy throughout the metal.

Chapter 8

15. **a.** chemical **b.** physical **c.** chemical **d.** chemical

17. **a.** physical **b.** chemical **c.** physical

19. physical, chemical, physical, physical, physical

21. **a.** chemical **b.** physical **c.** physical **d.** chemical

23. $2\,SO_2(g) + O_2(g) + 2\,H_2O(l) \longrightarrow 2\,H_2SO_4(aq)$

25. $2\,Na(s) + 2\,H_2O(l) \longrightarrow H_2(g) + 2\,NaOH(aq)$

27. $C_{12}H_{22}O_{11}(s) + H_2O(l) \longrightarrow 4\,C_2H_5OH(aq) + 4\,CO_2(g)$

29. **a.** $PbS(s) + 2\,HBr(aq) \longrightarrow PbBr_2(s) + H_2S(g)$
b. $CO(g) + 3\,H_2(g) \longrightarrow CH_4(g) + H_2O(l)$
c. $4\,HCl(aq) + MnO_2(s) \longrightarrow MnCl_2(aq) + 2\,H_2O(l) + Cl_2(g)$
d. $C_5H_{12}(l) + 8\,O_2(g) \longrightarrow 5\,CO_2(g) + 6\,H_2O(g)$

31. $Na_2CO_3(aq) + CuCl_2(aq) \longrightarrow CuCO_3(s) + 2\,NaCl(aq)$

33. **a.** $2\,CO_2(g) + CaSiO_3(s) + H_2O(l) \longrightarrow SiO_2(s) + Ca(HCO_3)_2(aq)$
b. $2\,Co(NO_3)_3(aq) + 3\,(NH_4)_2S(aq) \longrightarrow Co_2S_3(s) + 6\,NH_4NO_3(aq)$

c. $Cu_2O(s) + C(s) \longrightarrow 2Cu(s) + CO(g)$
d. $H_2(g) + Cl_2(g) \longrightarrow 2HCl(g)$

35. $2\,C_6H_{14}(g) + 19\,O_2(g) \longrightarrow 12\,CO_2(g) + 14\,H_2O(g)$, 68 mol O_2

37. **a.** 5.0 mol NO_2 **b.** 14. mol NO_2 **c.** 0.281 mol NO_2 **d.** 53.1 mol NO_2

39.

mol SiO_2	mol C	mol SiC	mol CO
3	9	3	6
2	6	2	4
5	15	5	10
2.8	8.4	2.8	5.6
0.517	1.55	0.517	1.03

41. **a.** 9.3 g HBr, 0.12 g H_2

43. **a.** 5.56 g $BaCl_2$ **b.** 6.55 g $CaCO_3$ **c.** 6.09 g MgO **d.** 6.93 g Al_2O_3

45. **a.** Na **b.** Na **c.** Br_2 **d.** Na

47. 3 molecules Cl_2

49. **a.** 2 mol **b.** 7 mol **c.** 9.40 mol

51. 0.5 mol O_2

53. **a.** 2.5 g **b.** 31.1 g **c.** 1.16 g

55. 2.91 grams CO_2 remaining

57. limiting reactant: Pb^{2+}, theoretical yield: 34.5 g $PbCl_2$, percent yield: 85.3%

59. limiting reactant: NH_3, theoretical yield: 240.5 kg CH_4N_2O, percent yield: 70.01%

61. **a.** $S(s) + O_2(g) \longrightarrow SO_2(g)$
b. $2\,C_3H_6(g) + 9\,O_2(g) \longrightarrow 6\,CO_2(g) + 6\,H_2O(g)$
c. $2\,Ca(s) + O_2(g) \longrightarrow 2\,CaO(g)$
d. $C_5H_{12}S(l) + 9\,O_2(g) \longrightarrow 5\,CO_2(g) + SO_2(g) + 6\,H_2O(g)$

63. $Sr(s) + I_2(g) \longrightarrow SrI_2(s)$

65. $2\,Li(s) + 2\,H_2O(l) \longrightarrow 2\,Li^+(aq) + 2\,OH^-(aq) + H_2(g)$

67. $H_2(g) + Br_2(g) \longrightarrow 2\,HBr(g)$

69. 3.1 kg

71. limiting reactant: $C_7H_6O_3$, theoretical yield: 1.63 g $C_9H_8O_4$, percent yield: 74.8%

73. b

75. 0.333 g PH_3

77. 30.8 kg CO_2

79. 1.6 g C_2H_2

81. 2.0 mg

83. 96.6 g Mn

85. d

87. a

Chapter 9

21. **a.** 1.17 M LiCl
b. 0.123 M $C_6H_{12}O_6$
c. 0.00453 M NaCl

23. **a.** 0.150 M NO_3^-
b. 0.300 M NO_3^-
c. 0.450 M NO_3^-

25. **a.** 1.3 mol
b. 1.5 mol
c. 0.211 mol

27. 37 g

29. 0.27 M

31. 6.0 L

33. 37.1 mL

35. 2.1 L

37. barium nitrate, 2.81g $Ba(NO_3)_2$, 87.1%

39. **a.** yes **b.** no **c.** yes **d.** no

41. **a.** soluble Ag^+, NO_3^- **b.** soluble Pb^{2+}, $C_2H_3O_2^-$ **c.** soluble K^+, NO_3^- **d.** soluble NH_4^+, S^{2-}

43. **a.** NO REACTION **b.** NO REACTION
c. $CrBr_2(aq) + Na_2CO_3(aq) \longrightarrow CrCO_3(s) + 2\,NaBr(aq)$
d. $3\,NaOH(aq) + FeCl_3(aq) \longrightarrow Fe(OH)_3(s) + 3\,NaCl(aq)$

45. **a.** $K_2CO_3(aq) + Pb(NO_3)_2(aq) \longrightarrow PbCO_3(s) + 2\,KNO_3(aq)$
b. $Li_2SO_4(aq) + Pb(C_2H_3O_2)_2(aq) \longrightarrow PbSO_4(s) + 2\,LiC_2H_3O_2(aq)$
c. $Cu(NO_3)_2(aq) + MgS(aq) \longrightarrow CuS(s) + Mg(NO_3)_2(aq)$
d. NO REACTION

47. **a.** Complete:
$H^+(aq) + Cl^-(aq) + Li^+(aq) + OH^-(aq) \longrightarrow H_2O(l) + Li^+(aq) + Cl^-(aq)$
Net: $H^+(aq) + OH^-(aq) \longrightarrow H_2O(l)$
b. Complete:
$Mg^{2+}(aq) + S^{2-}(aq) + Cu^{2+}(aq) + 2\,Cl^-(aq) \longrightarrow CuS(s) + Mg^{2+}(aq) + 2Cl^-(aq)$
Net: $Cu^{2+}(aq) + S^{2-}(aq) \longrightarrow CuS(s)$
c. Complete:
$Na^+(aq) + OH^-(aq) + HC_2H_3O_2(aq) \longrightarrow H_2O(l) + Na^+(aq) + C_2H_3O_2^-(aq)$
Net: $OH^-(aq) + HC_2H_3O_2(aq) \longrightarrow H_2O(l) + C_2H_3O_2^-(aq)$
d. Complete:
$6Na^+(aq) + 2PO_4^{3-}(aq) + 3Ni^{2+}(aq) + 6Cl^-(aq) \longrightarrow Ni_3(PO_4)_2(s) + 6Na^+(aq) + 6Cl^-(aq)$
Net: $3Ni^{2+}(aq) + 2PO_4^{3-}(aq) \longrightarrow Ni_3(PO_4)_2(s)$

49. Complete:
$Hg_2^{2+}(aq) + 2NO_3^-(aq) + 2Na^+(aq) + 2Cl^-(aq) \longrightarrow Hg_2Cl_2(s) + 2Na^+(aq) + 2NO_3^-(aq)$
Net: $Hg_2^{2+}(aq) + 2Cl^-(aq) \longrightarrow Hg_2Cl_2(s)$

51. **a.** hydroiodic acid **b.** nitric acid **c.** carbonic acid

53. **a.** HF **b.** HBr **c.** H_2SO_3

55. Molecular: $HBr(aq) + KOH(aq) \longrightarrow H_2O(l) + KBr(aq)$
Net ionic: $H^+(aq) + OH^-(aq) \longrightarrow H_2O(l)$

57. **a.** $H_2SO_4(aq) + Ca(OH)_2(aq) \longrightarrow 2\,H_2O(l) + CaSO_4(s)$
b. $HClO_4(aq) + KOH(aq) \longrightarrow H_2O(l) + KClO_4(aq)$
c. $H_2SO_4(aq) + 2\,NaOH(aq) \longrightarrow 2\,H_2O(l) + Na_2SO_4(aq)$

59. 0.1810 M $HClO_4$

61. **a.** $2\,HBr(aq) + NiS(s) \longrightarrow H_2S(g) + NiBr_2(aq)$
b. $NH_4I(aq) + NaOH(aq) \longrightarrow H_2O(l) + NH_3(g) + NaI(aq)$
c. $2\,HBr(aq) + Na_2S(aq) \longrightarrow H_2S(g) + 2\,NaBr(aq)$
d. $2\,HClO_4(aq) + Li_2CO_3(aq) \longrightarrow H_2O(l) + CO_2(g) + 2\,LiClO_4(aq)$

63. $2\,HClO_4(aq) + Li_2CO_3(aq) \longrightarrow H_2O(l) + CO_2(g) + 2\,LiClO_4(aq)$

65. **a.** Ag: 0 **b.** Ag: +1 **c.** Ca: +2, F: −1 **d.** H: +1, S: −2 **e.** C: +4, O: −2 **f.** Cr: +6, O: −2

67. **a.** +2 **b.** +6 **c.** +3

69. **a.** redox reaction, oxidizing agent: O_2, reducing agent: Li
b. redox reaction, oxidizing agent: Fe^{2+}, reducing agent: Mg
c. not a redox reaction
d. not a redox reaction
71. 3.32 M
73. 1.1 g
75. **b**
77. **a.** $2\ HCl(aq) + Hg_2(NO_3)_2(aq) \longrightarrow Hg_2Cl_2(s) + 2\ HNO_3(aq)$
b. $KHSO_3(aq) + HNO_3(aq) \longrightarrow H_2O(l) + SO_2(g) + KNO_3(aq)$
c. $2\ NH_4Cl(aq) + Pb(NO_3)_2(aq) \longrightarrow PbCl_2(s) + 2\ NH_4NO_3(aq)$
d. $2\ NH_4Cl(aq) + Ca(OH)_2(aq) \longrightarrow 2\ NH_3(g) + 2\ H_2O(g) + CaCl_2(aq)$
79. 22 g
81. 6.9 g
83. 0.531 L HCl
85. Br is the oxidizing agent, Au is the reducing agent, 38.8 g $KAuF_4$
87. 11.8 g AgI
89. 5.5% by mass
91. Ca^{2+} and Cu^{2+} present in the original solution
Net ionic for first precipitate:
$Ca^{2+}(aq) + SO_4^{\ 2-}(aq) \longrightarrow CaSO_4(s)$
Net ionic for second precipitate:
$Cu^{2+}(aq) + CO_3^{\ 2-}(aq) \longrightarrow CuCO_3(s)$
93.

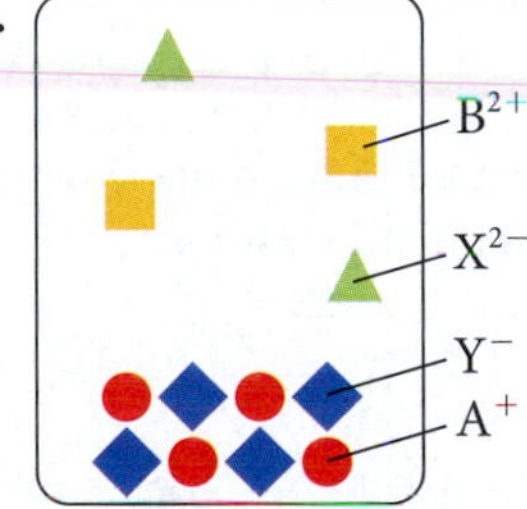

95. The charge of an ion is that actual charge it has due to gaining or losing electrons. The oxidation state of an atom is the charge *it would have* in a compound if all of the bonding electrons were assigned to the more electronegative atom.

Chapter 10

31. **d**
33. **a.** heat, + **b.** work, − **c.** heat, +
35. -7.27×10^2 kJ
37. 311 kJ
39. The drinks that went into cooler B had more thermal energy than the refrigerated drinks that went into cooler A. The temperature difference between the drinks in cooler B and the ice was greater than the difference between the drinks and the ice in cooler A. More thermal energy was exchanged between the drinks and the ice in cooler B, which resulted in more melting.
41. 4.7×10^5 J
43. **a.** 7.6×10^2 °C **b.** 4.3×10^2 °C
c. 1.3×10^2 °C **d.** 49 °C
45. -2.8×10^2 J
47. 489 J
49. $\Delta E = -3463$ J, $\Delta H = -3452$ kJ
51. **a.** exothermic, − **b.** endothermic, +
c. exothermic, −
53. -4.30×10^3 kJ
55. 6.46×10^4 kJ
57. 9.5×10^2 g CO_2
59. mass of silver 77.1 grams
61. Final temperature 28.4 °C
63. Specific heat capacity of substance A 1.10 J/g · °C
65. Measurement B corresponds to conditions of constant pressure. Measurement A corresponds to conditions of constant volume. When a fuel is burned under constant pressure, some of the energy released does work on the atmosphere by expanding against it. Less energy is manifest as heat due to this work. When a fuel is burned under constant volume, all of the energy released by the combustion reaction is evolved as heat.
67. -6.3×10^3 kJ/mol
69. -1.6×10^5 J
71. **a.** $-\Delta H_1$
b. $2\ \Delta H_1$
c. $-\frac{1}{2}\Delta H_1$
73. −23.9 kJ
75. −173.2 kJ
77. −128 kJ
79. −614 kJ
81. **a.** $N_2(g) + 3\ H_2(g) \longrightarrow 2\ NH_3(g)$, $\Delta H_f^\circ = -45.9$ kJ/mol
b. $C(s, \text{graphite}) + O_2(g) \longrightarrow CO_2(g)$, $\Delta H_f^\circ = -393.5$ kJ/mol
c. $2\ Fe(s) + 3/2\ O_2(g) \longrightarrow Fe_2O_3(s)$, $\Delta H_f^\circ = -824.2$ kJ/mol
d. $C(s, \text{graphite}) + 2\ H_2(g) \longrightarrow CH_4(g)$, $\Delta H_f^\circ = -74.6$ kJ/mol
83. −380.2 kJ/mol
85. **a.** −137.1 kJ **b.** −41.2 kJ
c. −137 kJ **d.** 290.7 kJ
87. $6\ CO_2(g) + 6\ H_2O(l) \longrightarrow C_6H_{12}O_6(s) + 6\ O_2(g)$, $\Delta H_{rxn}^\circ = 2803$ kJ
89. −113.0 kJ/mol
91. As the size of the alkaline earth metal ions increases, so does the distance between the metal cations and oxygen anions. Therefore, the magnitude of the lattice energy decreases accordingly because the potential energy decreases as the distance increases.
93. One factor of lattice energy is the product of the charges of the two ions. The product of the ion charges for CsF is −1 while that for BaO is −4. Because this product is four times greater, the lattice energy is also four times greater.
95. −708 kJ/mol
97. $\Delta E = -1.7$ J, $q = -0.5$ J, $w = -1.2$ J
99. 78 g
101. $\Delta H = 6.0$ kJ/mol, 1.1×10^2 g
103. 26.1 °C
105. palmitic acid: 9.9378 Cal/g, sucrose: 3.938 Cal/g, fat contains more Cal/g than sugar
107. 5.7 Cal/g

109. $\Delta E = 0$, $\Delta H = 0$, $q = -w = 3.0 \times 10^3$ J
111. −294 kJ/mol
113. 23.9 °C
115. The reaction is exothermic due to the energy released when the Al_2O_3 lattice forms.
117. $\Delta H_{rxn(H_2)} = -243$ kJ/mol $= -121$ kJ/g
$\Delta H_{rxn(CH_4)} = -802$ kJ/mol $= -50.0$ kJ/g
CH_4 yields more energy per mole, while H_2 yields more energy per gram.
119. 333 kJ/mol
121. 7.3×10^3 g H_2SO_4
123. 7.2×10^2 g
125. 78.2 °C
127. $q = 1030$ kJ, $\Delta H = 1030$ kJ, $\Delta E = 952$ kJ, $w = -78$ kJ
129.

$$\text{O=S=O} + \text{O—H} \longrightarrow \text{H—O—S(=O)—O}$$

$$\text{H—O—S(=O)—O} + \text{O=O} \longrightarrow$$

$$\text{O—S(=O)—O} + \text{H—O—O}$$

$$\text{O—S(=O)—O} + \text{H—O—H} \longrightarrow \text{(H—O)(H—O)S(=O)=O}$$

$\Delta H_{rxn} = -172$ kJ
131. −2162 kJ/mol
133. **d**
135. Refrigerator A contains only air, which will cool quickly, but will not stabilize the temperature. Refrigerator B contains containers of water, which require a great deal of energy to cool on day 1 but will remain stable at a cold temerature on day 2.
137. Substance A
139. The internal energy of a chemical system is the sum of its kinetic energy and its potential energy. It is this potential energy that is the energy source in an exothermic chemical reaction. Under normal circumstances, chemical potential energy (or simply chemical energy) arises primarily from the electrostatic forces between the protons and electrons that compose the atoms and molecules within the system. In an exothermic reaction, some bonds break and new ones form, and the protons and electrons go from an arrangement of high potential energy to one of lower potential energy. As they rearrange, their potential energy is converted into kinetic energy. Heat is emitted in the reaction, and so it feels hot to the touch.
141. **a**

Chapter 11

25. **a.** 0.832 atm **b.** 632 mmHg
c. 12.2 psi **d.** 8.43×10^4 Pa
27. **a.** 809.0 mmHg **b.** 1.064 atm
c. 809.0 torr **d.** 107.9 kPa
29. **a.** 832 mmHg **b.** 718 mmHg
31. 4.4×10^2 mmHg
33. 58.9 mL
35. 4.22 L
37. 3.0 L The volume would not be different if the gas was argon.
39. 1.16 atm
41. 2.1 mol
43. Yes, the final gauge pressure is 43.5 psi, which exceeds the maximum rating.
45. 16.2 L
47. 286 atm, 17.5 bottles purged
49. **b**
51. 4.76 atm
53. 37.3 L
55. 9.43 g/L
57. 44.0 g/mol
59. 4.00 g/mol
61. $P_{tot} = 434$ torr, $\text{mass}_{N_2} = 0.437$ g, $\text{mass}_{O_2} = 0.237$ g, $\text{mass}_{He} = 0.0340$ g
63. 1.84 atm
65. $\chi_{N_2} = 0.627$, $\chi_{O_2} = 0.373$, $P_{N_2} = 0.687$ atm, $P_{O_2} = 0.409$ atm
67. $P_{H_2} = 0.921$ atm, $\text{mass}_{H_2} = 0.0539$ g
69. 7.47×10^{-2} g
71. **a.** yes
b. no
c. No. Even though the argon atoms are more massive than the helium atoms, both have the same kinetic energy at a given temperature. The argon atoms therefore move more slowly, and so exert the same pressure as the helium atoms.
d. He
73. F_2: $u_{rms} = 442$ m/s, $KE_{avg} = 3.72 \times 10^3$ J;
Cl_2: $u_{rms} = 324$ m/s, $KE_{avg} = 3.72 \times 10^3$ J;
Br_2: $u_{rms} = 216$ m/s, $KE_{avg} = 3.72 \times 10^3$ J;
rankings: u_{rms}: $Br_2 < Cl_2 < F_2$, KE_{avg}: $Br_2 = Cl_2 = F_2$,
rate of effusion: $Br_2 < Cl_2 < F_2$
75. rate $^{238}UF_6$/rate $^{235}UF_6 = 0.99574$
77. krypton
79. A has the higher molar mass, B has the higher rate of effusion.
81. 38 L
83. $V_{H_2} = 48.2$ L, $V_{CO} = 24.1$ L
85. 22.8 g NaN_3
87. 60.4%
89. F_2, 2.84 g ClF_3
91. That the volume of gas particles is small compared to the space between them breaks down under conditions of high pressure. At high pressure the particles themselves occupy a significant portion of the total gas volume.
93. 0.05826 L (ideal); 0.0708 L (V.D.W.); Difference because of high pressure, at which Ne no longer acts ideally.
95. 97.8%
97. 27.8 g/mol
99. C_4H_{10}
101. 4.70 L
103. $2\,HCl(aq) + K_2S(s) \longrightarrow H_2S(g) + 2\,KCl(aq)$, 0.191 g $K_2S(s)$

105. 11.7 L
107. $mass_{air} = 8.56$ g, $mass_{He} = 1.20$ g, mass difference $= 7.36$ g
109. 4.76 L/s
111. total force $= 6.15 \times 10^3$ pounds; no, the can cannot withstand this force.
113. 5.8×10^3 balloons
115. 4.0 cm
117. 77.7%
119. 0.32 grams
121. 311 K
123. 5.0 g
125. C_3H_8
127. 0.39 g Ar
129. 74.0 mmHg
131. 25% N_2H_4
133. 25%
135. $P_{CH_4} = 7.30 \times 10^{-2}$ atm, $P_{O_2} = 4.20 \times 10^{-1}$ atm, $P_{NO} = 2.79 \times 10^{-3}$ atm, $P_{CO_2} = 5.03 \times 10^{-3}$ atm, $P_{H_2O} = 5.03 \times 10^{-3}$ atm, $P_{NO_2} = 2.51 \times 10^{-2}$ atm, $P_{OH} = 1.01 \times 10^{-2}$ atm, $P_{tot} = 0.542$ atm
137. 0.42 atm
139. Because helium is less dense than air, the balloon moves in a direction opposite the direction in which the air inside the car is moving due to the acceleration and deceleration of the car.
141. −29%
143. **a.** false **b.** false **c.** false **d.** true
145. Four times the intial pressure
147. Although the velocity "tails" have different lengths, the average length of the tails on the helium atoms is longer than the average length of the tails on the neon atoms, which is in turn longer than the average length of the tails on the krypton atoms. The lighter the atom, the faster they must move on average to have the same kinetic energy.

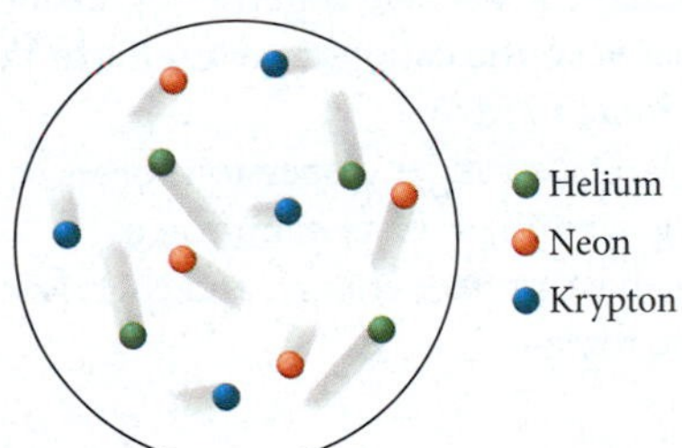

Chapter 12

33. **a.** dispersion
b. dispersion, dipole–dipole, hydrogen bonding
c. dispersion, dipole–dipole
d. dispersion
35. **a.** dispersion, dipole–dipole
b. dispersion, dipole–dipole, hydrogen bonding
c. dispersion
d. dispersion
37. **a, b, c, d,** Boiling point increases with increasing intermolecular forces. The molecules increase in their intermolecular forces as follows:
a. dispersion forces;
b. stronger dispersion forces (broader electron cloud);
c. dispersion forces and dipole–dipole interactions;
d. dispersion forces, dipole–dipole interactions, and hydrogen bonding.
39. **a.** CH_3OH, hydrogen bonding
b. CH_3CH_2OH, hydrogen bonding
c. CH_3CH_3, greater mass, broader electron cloud causes greater dispersion forces
41. **a.** Br_2, smaller mass results in weaker dispersion forces
b. H_2S, lacks hydrogen bonding
c. PH_3, lacks hydrogen bonding
43. **a.** not homogeneous
b. homogeneous, dispersion, dipole–dipole, hydrogen bonding, ion–dipole
c. homogeneous, dispersion
d. homogeneous, dispersion, dipole–dipole, hydrogen bonding
45. Water. Surface tension increases with increasing intermolecular forces, and water can hydrogen bond while acetone cannot.
47. compound A
49. When the tube is clean, water experiences adhesive forces with glass that are stronger than its cohesive forces, causing it to climb the surface of a glass tube. Water does not experience strong intermolecular forces with oil, so if the tube is coated in oil, the water's cohesive forces will be greater and it will not be attracted to the surface of the tube.
51. The water in the 12-cm dish will evaporate more quickly. The vapor pressure does not change but the surface area does. The water in the dish evaporates more quickly because the greater surface area allows for more molecules to obtain enough energy at the surface and break free.
53. Water is more volatile than vegetable oil. When the water evaporates, the endothermic process results in cooling.
55. 0.405 L
57. 91 °C
59. $\Delta H_{vap} = 24.7$ kJ/mol, bp $= 239$ K
61. 41 torr
63. 22.0 kJ
65. 2.7 °C
67. 30.5 kJ
69. Water has strong intermolecular forces. It is polar and experiences hydrogen bonding.
71. Water's exceptionally high specific heat capacity has a moderating effect on Earth's climate. Also, its high ΔH_{vap} causes water evaporation and condensation to have a strong effect on temperature.
73. The general trend is that melting point increases with increasing mass. This is due to the fact that the electrons of the larger molecules are held more loosely and a stronger dipole moment can be induced more easily. HF is the exception to the rule. It has a relatively high melting point due to hydrogen bonding.
75. yes, 1.22 g
77. 26 °C

79.

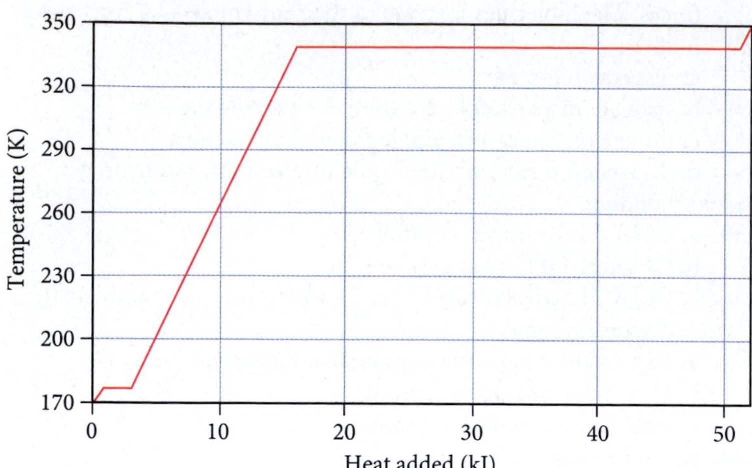

81. 3.4×10^3 g H_2O
83. 26 mmHg
85. Decreasing the pressure will decrease the temperature of liquid nitrogen. Because the nitrogen is boiling, its temperature must be constant at a given pressure. As the pressure decreases, the boiling point decreases, and therefore so does the temperature. If the pressure drops below the pressure of the triple point, the phase change will shift from vaporization to sublimation and the liquid nitrogen will become solid.
87. 70.7 L
89. The melting of an ice cube in a glass of water will not raise or lower the level of the liquid in the glass as long as the ice is always floating in the liquid. This is because the ice will displace a volume of water based on its mass. By the same logic, melting floating icebergs will not raise ocean levels (assuming that the dissolved solids content, and thus the density, will not change when the icebergs melt). Dissolving ice formations that are supported by land will raise the ocean levels, just as pouring more water into the glass will raise the liquid level in the glass.
91. Substance A
93. The liquid segment will have the least steep slope because it takes the most kJ/mol to raise the temperature of the phase.
95. There are substantial intermolecular attractions in the liquid, but virtually none in the gas.
97.

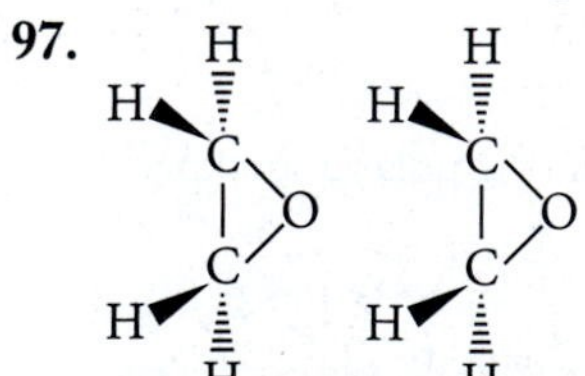

Chapter 13

19. **a.** solid
b. liquid
c. gas
d. supercritical fluid
e. solid/liquid
f. liquid/gas
g. solid/liquid/gas

21. N_2 has a stable liquid phase at 1 atm.

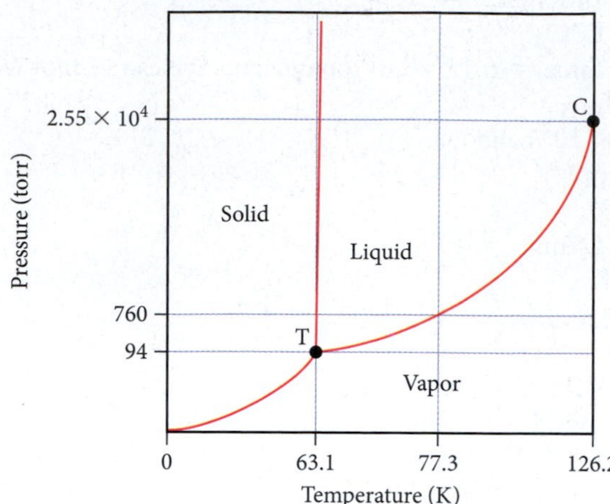

23. **a.** 0.027 mmHg
b. rhombic
25. 162 pm
27. **a.** 1
b. 2
c. 4
29. $l = 393$ pm, $d = 21.3$ g/cm^3
31. 129 pm
33. 134.5 pm
35. 6.0×10^{23} atoms/mol
37. **a.** atomic
b. molecular
c. ionic
d. atomic
39. LiCl(*s*). The other three solids are held together by intermolecular forces, while LiCl is held together by stronger coulombic interactions between the cations and anions of the crystal lattice.
41. **a.** $TiO_2(s)$, ionic solid
b. $SiCl_4(s)$, larger, stronger dispersion forces
c. Xe(*s*), larger, stronger dispersion forces
d. CaO, ions have greater charge, and therefore stronger coulombic forces.
43. CuI
45. TiO_2
47. Cs: $1(1) = 1$
Cl: $8(1/8) = 1$
1:1
CsCl
Ba: $8(1/8) + 6(1/2) = 4$
Cl: $8(1) = 8$
$4:8 = 1:2$
$BaCl_2$
49. Graphite consists of covalently bonded sheets that are held to each other by weak interactions, allowing them to slip past each other. Diamond is not a good lubricant because it is an extremely strong network covalent solid, where all of the carbon atoms are covalently bonded.

51. gas ⟶ liquid ⟶ solid

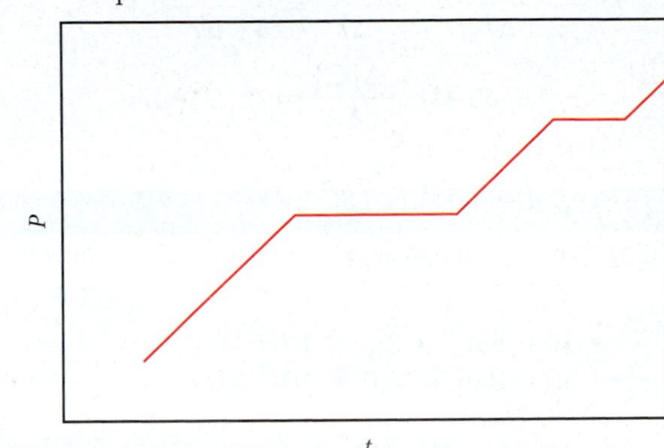

53. CsCl has a higher melting point than AgI because of its higher coordination number. In CsCl, one anion bonds to eight cations (and vice versa) while in AgI, one anion bonds only to four cations.

55. **a.** $4r$

b. $c^2 = a^2 + b^2$ $c = 4r$, $a = l$, $b = l$

$(4r)^2 = l^2 + l^2$

$16r^2 = 2l^2$

$8r^2 = l^2$

$l = \sqrt[3]{8r^2}$

$l = 2\sqrt{2}r$

57. 8 atoms/unit

59. **a.** $CO_2(s) \longrightarrow CO_2(g)$ at 195 K

b. $CO_2(s) \longrightarrow$ triple point at 216 K $\longrightarrow CO_2(g)$ just above 216 K

c. $CO_2(s) \longrightarrow CO_2(l)$ at somewhat above 216 K $\longrightarrow CO_2(g)$ at around 250 K

d. $CO_2(s) \longrightarrow CO_2(g) \longrightarrow$ supercritical fluid

61. 55.843 g/mol

63. 2.00 g/cm^3

65. body diagonal $= \sqrt{6}r$, radius $= (\sqrt{3} - \sqrt{2})\, r/\sqrt{2} = 0.2247r$

67. solid and gas

Chapter 14

25. **a.** hexane, toluene, or CCl_4; dispersion forces

b. water, methanol; dispersion, dipole–dipole, hydrogen bonding

c. hexane, toluene, or CCl_4; dispersion forces

d. water, acetone, methanol, ethanol; dispersion, ion–dipole

27. $HOCH_2CH_2CH_2OH$

29. **a.** water; dispersion, dipole–dipole, hydrogen bonding

b. hexane; dispersion

c. water; dispersion, dipole–dipole

d. water; dispersion, dipole–dipole, hydrogen bonding

31. **a.** endothermic

b. The lattice energy is greater in magnitude than the heat of hydration.

c.

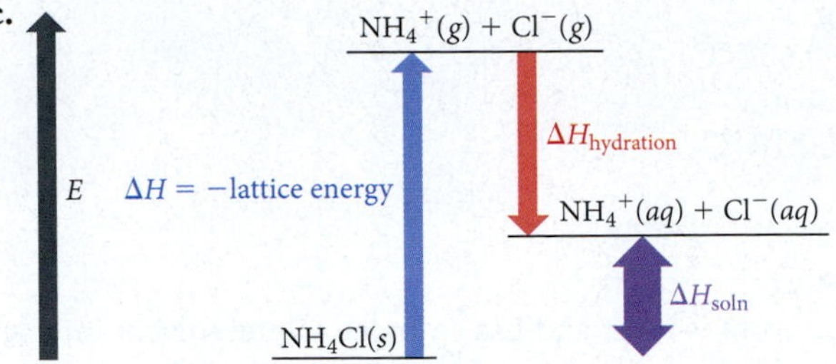

d. The solution forms because chemical systems tend toward greater entropy.

33. -797 kJ/mol

35. $\Delta H_{soln} = -6 \times 10^1$ kJ/mol, $-$ 7 kJ of energy evolved

37. unsaturated

39. About 31 g will precipitate.

41. Boiling water releases any O_2 dissolved in it. The solubility of gases decreases with increasing temperature.

43. As pressure increases, nitrogen will more easily dissolve in blood. To reverse this process, divers should ascend to lower pressures.

45. 1.1 g

47. 1.92 M, 2.0 *m*, 10.4%

49. 0.340 L

51. 1.6×10^2 g Ag

53. 1.4×10^4 g

55. Add water to 7.31 mL of concentrated solution until a total volume of 1.15 L is acquired.

57. **a.** Add water to 3.73 g KCl to a volume of 100 mL.

b. Add 3.59 g KCl to 96.41 g H_2O.

c. Add 5.0 g KCl to 95 g H_2O.

59. **a.** 0.417 M

b. 0.444 *m*

c. 7.41% by mass

d. 0.00794

e. 0.794% by mole

61. 0.89 M

63. 15 *m*, 0.22

65. The level has decreased more in the beaker filled with pure water. The dissolved salt in the seawater decreases the vapor pressure and subsequently lowers the rate of vaporization.

67. 30.7 torr

69. **a.** $P_{hep} = 24.4$ torr, $P_{oct} = 5.09$ torr

b. 29.5 torr

c. 80.8% heptane by mass, 19.2% octane by mass

d. The vapor is richer in the more volatile component.

71. $P_{chl} = 51.9$ torr, $P_{ace} = 274$ torr, $P_{tot} = 326$ torr. The solution is not ideal. The chloroform–acetone interactions are stronger than the chloroform–chloroform and acetone–acetone interactions.

73. freezing point (fp) $= -1.27$ °C, bp $= 100.349$ °C

75. freezing point (fp) $= 1.0$ °C, boiling point (bp) $= 82.4$ °C

77. 1.8×10^2 g/mol

79. 26.1 atm

81. 6.36×10^3 g/mol

83. **a.** fp $= -0.558$ °C, bp $= 100.154$ °C

b. fp $= -1.98$ °C, bp $= 100.546$ °C

c. fp $= -2.5$ °C, bp $= 100.70$ °C

85. 160 g

87. a. −0.632 °C
b. 5.4 atm
c. 100.18 °C

89. 2.3

91. 3.4

93. 23.0 torr

95. Chloroform is polar and has stronger solute–solvent interactions than nonpolar carbon tetrachloride.

97. $\Delta H_{soln} = 51$ kJ/mol, −8.7 °C

99. 2.2×10^{-3} M/atm

101. 1.3×10^4 L

103. 0.24 g

105. −24 °C

107. a. 1.1% by mass/V
b. 1.6% by mass/V
c. 5.3% by mass/V

109. 2.484

111. 0.229 atm

113. χ_{CHCl_3}(original) = 0.657, P_{CHCl_3}(condensed) = 0.346 atm

115. 1.74 M

117. $C_6H_{14}O_2$

119. 12 grams

121. 6.4×10^{-3} L

123. 22.4% glucose by mass, 77.6% sucrose by mass

125. $P_{iso} = 0.131$ atm, $P_{pro} = 0.068$ atm. The major intermolecular attractions are between the OH groups. The OH group at the end of the chain in propyl alcohol is more accessible than the one in the middle of the chain in isopropyl alcohol. In addition, the molecular shape of propyl alcohol is a straight chain of carbon atoms, while that of isopropyl alcohol is a branched chain and is more like a ball. The contact area between two ball-like objects is smaller than that of two chain-like objects. The smaller contact area in isopropyl alcohol means the molecules don't attract each other as strongly as do those of propyl alcohol. As a result of both of these factors, the vapor pressure of isopropyl alcohol is higher.

127. 0.0097 m

129. Na_2CO_3 0.050 M, $NaHCO_3$ 0.075 M

131. The water should not be immediately cycled back into the river. As the water was warmed, dissolved oxygen would have been released, since the amount of a gas able to be dissolved into a liquid decreases as the temperature of the liquid increases. As such, the water returned to the river would lack dissolved oxygen if it was still hot. To preserve the dissolved oxygen necessary for the survival of fish and other aquatic life, the water must first be cooled.

133. b. NaCl

Chapter 15

27. a. $\text{Rate} = -\frac{1}{2}\frac{\Delta[\text{HBr}]}{\Delta t} = \frac{\Delta[\text{H}_2]}{\Delta t} = \frac{\Delta[\text{Br}_2]}{\Delta t}$
b. 1.8×10^{-3} M/s
c. 0.040 mol Br_2

29. a. $\text{Rate} = -\frac{1}{2}\frac{\Delta[\text{A}]}{\Delta t} = -\frac{\Delta[\text{B}]}{\Delta t} = \frac{1}{3}\frac{\Delta[\text{C}]}{\Delta t}$
b. $\frac{\Delta[\text{B}]}{\Delta t} = -0.0500$ M/s, $\frac{\Delta[\text{C}]}{\Delta t} = 0.150$ M/s

31.

$\Delta[Cl_2]/\Delta t$	$\Delta[F_2]/\Delta t$	$\Delta[ClF_3]/\Delta t$	Rate
−0.012 M/s	−0.036 M/s	0.024 M/s	0.012 M/s

33. a. 0 ⟶ 10 s: Rate = 8.7×10^{-3} M/s
40 ⟶ 50 s: Rate = 6.0×10^{-3} M/s
b. 1.4×10^{-2} M/s

35. a. (i) 1.0×10^{-2} M/s
(ii) 8.5×10^{-3} M/s
(iii) 0.013 M/s
b.

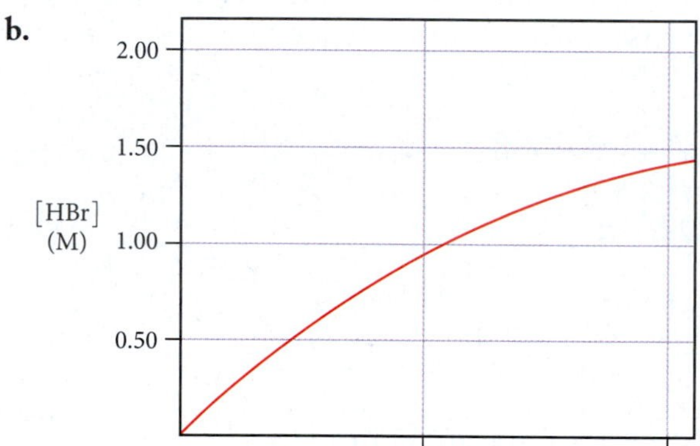

37. a. first order
b.

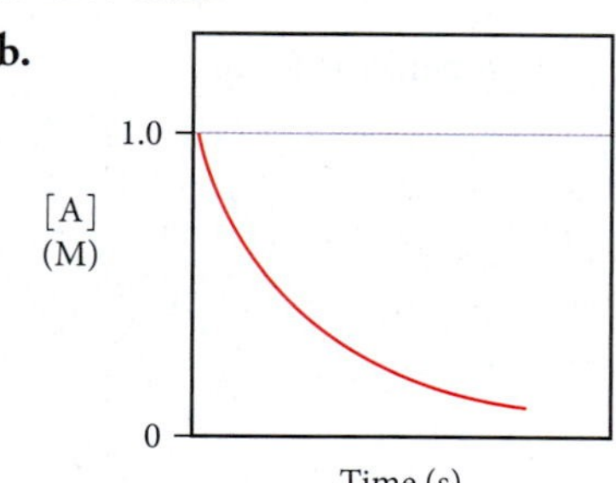

c. Rate = $k[\text{A}]^1$, $k = 0.010\ \text{s}^{-1}$

39. a. s^{-1} **b.** $\text{M}^{-1}\text{s}^{-1}$ **c.** $\text{M} \cdot \text{s}^{-1}$

41. a. Rate = $k[\text{A}][\text{B}]^2$ **b.** third order
c. 2 **d.** 4
e. 1 **f.** 8

43. second order, Rate = $5.25\ \text{M}^{-1}\text{s}^{-1}[\text{A}]^2$

45. Rate = $k[\text{NO}_2][\text{F}_2]$, $k = 2.57\ \text{M}^{-1}\text{s}^{-1}$, second order

47. a. zero order **b.** first order **c.** second order

49. second order, $k = 2.25 \times 10^{-2}\ \text{M}^{-1}\text{s}^{-1}$, [AB] at 25 s = 0.619 M

51. first order, $k = 1.12 \times 10^{-2}\ \text{s}^{-1}$, Rate = 2.8×10^{-3} M/s

53. a. $4.5 \times 10^{-3}\ \text{s}^{-1}$ **b.** Rate = $4.5 \times 10^{-3}\ \text{s}^{-1}[\text{A}]$
c. 1.5×10^2 s **d.** [A] = 0.0908 M

55. a. 4.88×10^3 s **b.** 9.8×10^3 s
c. 1.7×10^3 s **d.** 0.146 M at 200 s, 0.140 M at 500 s

57. 6.8×10^8 yrs; 1.8×10^{17} atoms

59.

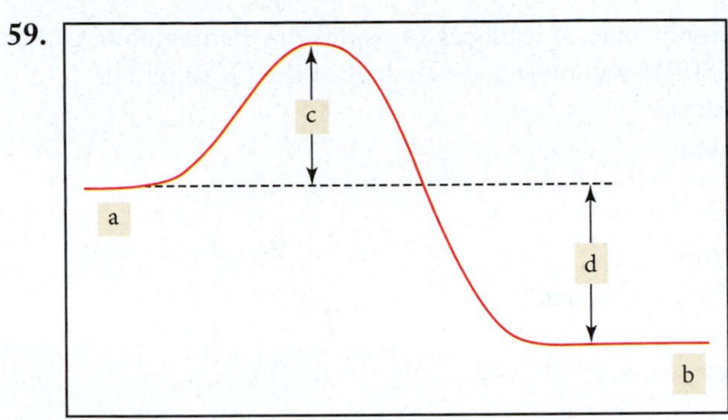

61. 17 s^{-1}

63. 61.90 kJ/mol

65. $E_a = 251$ kJ/mol, $A = 7.93 \times 10^{11}\ s^{-1}$

67. $E_a = 23.0$ kJ/mol, $A = 8.05 \times 10^{10}\ s^{-1}$

69. **a.** 122 kJ/mol
b. 0.101 s^{-1}

71. 47.85 kJ/mol

73. **a**

75. The mechanism is valid.

77. **a.** $Cl_2(g) + CHCl_3(g) \longrightarrow HCl(g) + CCl_4(g)$
b. $Cl(g)$, $CCl_3(g)$
c. Rate $= k[Cl_2]^{1/2}[CHCl_3]$

79. Heterogeneous catalysts require a large surface area because catalysis can only happen at the surface. A greater surface area means greater opportunity for the substrate to react, which results in a faster reaction.

81. 10^{12}

83. **a.** first order, $k = 0.0462\ hr^{-1}$
b. 15 hr
c. 5.0×10^1 hr

85. 0.0531 M/s

87. rate $= 4.5 \times 10^{-4}\,[CH_3CHO]^2$, $k = 4.5 \times 10^{-4}$, 0.37 atm

89. 219 torr

91. 1×10^{-7} s

93. 1.6×10^2 seconds

95. **a.** 2
b.

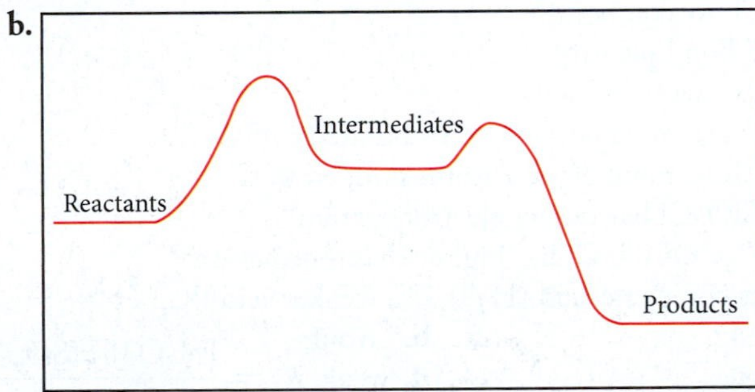

c. first step
d. exothermic

97. **a.** 5.41 s
b. 2.2 s for 25%, 5.4 s for 50%
c. 0.28 at 10 s, 0.077 at 20 s

99. **a.** $E_a = 89.5$ kJ/mol, $A = 4.22 \times 10^{11}\ s^{-1}$
b. $2.5 \times 10^{-5}\ M^{-1}\ s^{-1}$
c. 6.0×10^{-4} M/s

101. **a.** No
b. No bond is broken, and the two radicals attract each other.
c. Formation of diatomic gases from atomic gases

103. 1.35×10^4 years

105. **a.** Both are valid. For both, all steps sum to overall reaction, and the predicted rate law is consistent with experimental data.
b. Buildup of $I(g)$

107. rate $= k_2[(k_1/k_{-1})[Br_2]]^{1/2}[H_2]$
The rate law is 3/2 order overall.

109. **a.** 0% **b.** 25% **c.** 33%

111. 174 kJ

113. **a.** second order
b. $CH_3NC + CH_3NC \underset{k_2}{\overset{k_1}{\rightleftharpoons}} CH_3NC^* + CH_3NC$ (fast)

$CH_3NC^* \xrightarrow{k_3} CH_3CN$ (slow)

Rate $= k_3[CH_3NC^*]$

$k_1[CH_3NC]^2 = k_2[CH_3NC^*][CH_3NC]$

$[CH_3NC^*] = \frac{k_1}{k_2}[CH_3NC]$

Rate $= k_3 \times \frac{k_1}{k_2}[CH_3NC]$

Rate $= k[CH_3NC]$

115. Rate $= k[A]^2$

$$\text{Rate} = -\frac{d[A]}{dt}$$

$$\frac{d[A]}{dt} = -k[A]^2$$

$$-\frac{d[A]}{[A]^2} = k\,dt$$

$$\int_{[A]_0}^{[A]} -\frac{1}{[A]^2}d[A] = \int_0^t k\,dt$$

$$\left[\frac{1}{[A]}\right]_{[A]_0}^{[A]} = k\,[t]_0^t$$

$$\frac{1}{[A]} - \frac{1}{[A]_0} = kt$$

$$\frac{1}{[A]} = kt + \frac{1}{[A]_0}$$

117. Rate $= k[CO][Cl_2]^{\frac{3}{2}}$

119. $[Cl_2] = 0.0084$ mol/L, $[NO] = 0.017$ mol/L

121. B is first order and A is second order.
B will be linear if you plot ln[B] vs. time; A will be linear if you plot 1/[A] vs. time.

Chapter 16

21. **a.** $K = \frac{[SbCl_3][Cl_2]}{[SbCl_5]}$ **b.** $K = \frac{[NO]^2[Br_2]}{[BrNO]^2}$

c. $K = \frac{[CS_2][H_2]^4}{[CH_4][H_2S]^2}$ **d.** $K = \frac{[CO_2]^2}{[CO]^2[O_2]}$

23. The concentration of the reactants will be greater. No, this is not dependent on initial concentrations; it is dependent on the value of K_c.

25. **a.** figure v
b. The change in the decrease of reactants and increase of products would be faster.
c. No, catalysts affect kinetics, not equilibrium.

27. **a.** 4.42×10^{-5}, reactants favored
b. 1.50×10^{2}, products favored
c. 1.96×10^{-9}, reactants favored

29. 1.3×10^{-29}

31. **a.** 2.56×10^{-23}
b. 1.3×10^{22}
c. 81.9

33. **a.** $K_c = \frac{[HCO_3^-][OH^-]}{[CO_3^{2-}]}$ **b.** $K_c = [O_2]^3$
c. $K_c = \frac{[H_3O^+][F^-]}{[HF]}$ **d.** $K_c = \frac{[NH_4^+][OH^-]}{[NH_3]}$

35. 136

37.

T(K)	$[N_2]$	$[H_2]$	$[NH_3]$	K_c
500	0.115	0.105	0.439	1.45×10^{-3}
575	0.110	0.249	0.128	9.6
775	0.120	0.140	4.39×10^{-3}	0.0584

39. 234 torr

41. 18

43. 3.3×10^{2}

45. 764

47. More solid will form.

49. Additional solid will not dissolve.

51. **a.** [A] = 0.20 M, [B] = 0.80 M
b. [A] = 0.33 M, [B] = 0.67 M
c. [A] = 0.38 M, [B] = 1.2 M

53. $[N_2O_4]$ = 0.0115 M, $[NO_2]$ = 0.0770 M

55. 0.199 M

57. 1.9×10^{-3} M

59. 7.84 torr

61. **a.** [A] = 0.38 M, [B] = 0.62 M, [C] = 0.62 M
b. [A] = 0.90 M, [B] = 0.095 M, [C] = 0.095 M
c. [A] = 1.0 M, [B] = 3.2×10^{-3} M, [C] = 3.2×10^{-3} M

63. **a.** shift left **b.** shift right **c.** shift right

65. **a.** shift right **b.** no effect
c. no effect **d.** shift left

67. **a.** shift right **b.** shift left **c.** no effect

69. Increase temperature $\longrightarrow$ shift right, decrease temperature $\longrightarrow$ shift left. Increasing the temperature will increase the equilibrium constant.

71. **b, d**

73. **a.** 1.7×10^{2}
b. $\frac{[Hb-CO]}{[Hb-O_2]} = 0.85$ or 17/20

CO is highly toxic, as it blocks O_2 uptake by hemoglobin. CO at a level of 0.1% will replace nearly half of the O_2 in blood.

75. **a.** 1.68 atm
b. 1.41 atm

77. 0.406 g

79. **b, c, d**

81. 0.0144 atm

83. 3.1×10^{2} g, 20% yield

85. 0.12 atm

87. 0.72 atm

89. 0.017 g

91. 0.226

93. **a.** 29.3
b. 86.3 torr

95. $P_{NO} = P_{Cl_2} = 429$ torr

97. 1.27×10^{-2}

99. $K_P = 5.1 \times 10^{-2}$

101. Yes, because the volume affects Q.

103. a = 1, b = 2

Chapter 17

31. **a.** acid, $HNO_3(aq) \longrightarrow H^+(aq) + NO_3^-(aq)$
b. acid, $NH_4^+(aq) \rightleftharpoons H^+(aq) + NH_3(aq)$
c. base, $KOH(aq) \longrightarrow K^+(aq) + OH^-(aq)$
d. acid, $HC_2H_3O_2(aq) \rightleftharpoons H^+(aq) + C_2H_3O_2^-(aq)$

33. **a.** $H_2CO_3(aq) + H_2O(l) \rightleftharpoons H_3O^+(aq) + HCO_3^-(aq)$
acid base conj.acid conj.base
b. $NH_3(aq) + H_2O(l) \rightleftharpoons NH_4^+(aq) + OH^-(aq)$
base acid conj.acid conj.base
c. $HNO_3(aq) + H_2O(l) \rightleftharpoons H_3O^+(aq) + NO_3^-(aq)$
acid base conj.acid conj.base
d. $C_5H_5N(aq) + H_2O(l) \rightleftharpoons C_5H_5NH^+(aq) + OH^-(aq)$
base acid conj.acid conj.base

35. **a.** Cl^- **b.** HSO_3^- **c.** CHO_2^- **d.** F^-

37. $H_2PO_4^-(aq) + H_2O(l) \rightleftharpoons HPO_4^{2-}(aq) + H_3O^+(aq)$
$H_2PO_4^-(aq) + H_2O(l) \rightleftharpoons H_3PO_4(aq) + OH^-(aq)$

39. **a.** HCl, weaker bond
b. HF, bond polarity
c. H_2Se, weaker bond

41. **a.** H_2SO_4, more oxygen atoms bonded to S
b. $HClO_2$, more oxygen atoms bonded to Cl
c. HClO, Cl has higher electronegativity
d. CCl_3COOH, Cl has higher electronegativity

43. S^{2-}, its conjugate acid (H_2S), is a weaker acid than H_2S

45. **a.** strong **b.** strong
c. strong **d.** weak, $K_a = \frac{[H_3O^+][HSO_3^-]}{[H_2SO_3]}$

47. **a, b, c**

49. **a.** 8.3×10^{-7}, basic
b. 1.2×10^{-10}, acidic
c. 2.9×10^{-13}, acidic

51. **a.** pH = 7.77, pOH = 6.23
b. pH = 7.00, pOH = 7.00
c. pH = 5.66, pOH = 8.34

53.

$[H_3O^+]$	$[OH^-]$	pH	Acidic or Basic
7.1×10^{-4}	1.4×10^{-11}	3.15	Acidic
3.7×10^{-9}	2.7×10^{-6}	8.43	Basic
7.9×10^{-12}	1.3×10^{-3}	11.1	Basic
6.3×10^{-4}	1.6×10^{-11}	3.20	Acidic

55. $[H_3O^+] = 1.5 \times 10^{-7}$ M, pH = 6.81

57. pH = 1.36, 1.35, 1.34 A difference of 1 in the second significant digit in a concentration value produces a difference of 0.01 in pH. Therefore the second significant digit in value of the concentration corresponds to the hundredths place in a pH value.

59. **a.** $[H_3O^+] = 0.25$ M, $[OH^-] = 4.0 \times 10^{-14}$ M, pH = 0.60
b. $[H_3O^+] = 0.015$ M, $[OH^-] = 6.7 \times 10^{-13}$ M, pH = 1.82
c. $[H_3O^+] = 0.072$ M, $[OH^-] = 1.4 \times 10^{-13}$ M, pH = 1.14
d. $[H_3O^+] = 0.105$ M, $[OH^-] = 9.5 \times 10^{-14}$ M, pH = 0.979

61. **a.** 1.8 g **b.** 0.57 g **c.** 0.045 g

63. 2.21

65. $[H_3O^+] = 2.5 \times 10^{-3}$ M, pH = 2.59

67. **a.** 1.82 (approximation valid)
b. 2.18 (approximation breaks down)
c. 2.72 (approximation breaks down)

69. 2.75

71. 6.8×10^{-6}

73. 0.0063%

75. **a.** 0.42% **b.** 0.60% **c.** 1.3% **d.** 1.9%

77. 3.61×10^{-5}

79. **a.** pH = 2.03, percent ionization = 3.7%
b. pH = 2.24, percent ionization = 5.7%
c. pH = 2.40, percent ionization = 8.0%

81. **a.** 0.939 **b.** 1.07 **c.** 2.19 **d.** 3.02

83. **a.** $[OH^-] = 0.15$ M, $[H_3O^+] = 6.7 \times 10^{-14}$ M, pH = 13.17, pOH = 0.83
b. $[OH^+] = 0.003$ M, $[H_3O^+] = 3.3 \times 10^{-12}$ M, pH = 11.48, pOH = 2.52
c. $[OH^-] = 9.6 \times 10^{-4}$ M, $[H_3O^+] = 1.0 \times 10^{-11}$ M, pH = 10.98, pOH = 3.02
d. $[OH^-] = 8.7 \times 10^{-5}$ M, $[H_3O^+] = 1.1 \times 10^{-10}$ M, pH = 9.93, pOH = 4.07

85. 13.842

87. 0.104 L

89. **a.** $NH_3(aq) + H_2O(l) \rightleftharpoons NH_4^+(aq) + OH^-(aq)$,

$$K_b = \frac{[NH_4^+][OH^-]}{[NH_3]}$$

b. $HCO_3^-(aq) + H_2O(l) \rightleftharpoons$

$$H_2CO_3(aq) + OH^-(aq),\ K_b = \frac{[H_2CO_3][OH^-]}{[HCO_3^-]}$$

c. $CH_3NH_2(aq) + H_2O(l) \rightleftharpoons$

$$CH_3NH_3^+(aq) + OH^-(aq),\ K_b = \frac{[CH_3NH_3^+][OH^-]}{[CH_3NH_2]}$$

91. $[OH^-] = 1.6 \times 10^{-3}$ M, pOH = 2.79, pH = 11.21

93. 7.48

95. 6.7×10^{-7}

97. **a.** neutral
b. basic,
$ClO^-(aq) + H_2O(l) \rightleftharpoons HClO(aq) + OH^-(aq)$
c. basic,
$CN^-(aq) + H_2O(l) \rightleftharpoons HCN(aq) + OH^-(aq)$
d. neutral

99. $[OH^-] = 2.0 \times 10^{-6}$ M, pH = 8.30

101. **a.** acidic,
$NH_4^+(aq) + H_2O(l) \rightleftharpoons NH_3(aq) + H_3O^+(aq)$
b. neutral
c. acidic, $Co(H_2O)_6^{3+}(aq) + H_2O(l) \rightleftharpoons Co(H_2O)_5(OH)^{2+}(aq) + H_3O^+(aq)$
d. acidic, $CH_2NH_3^+(aq) + H_2O(l) \rightleftharpoons CH_2NH_2(aq) + H_3O^+(aq)$

103. **a.** acidic **b.** basic
c. neutral **d.** acidic
e. acidic

105. NaOH, $NaHCO_3$, NaCl, NH_4ClO_2, NH_4Cl

107. **a.** 5.13 **b.** 8.87 **c.** 7.0

109. $[K^+] = 0.15$ M, $[F^-] = 0.15$ M, $[HF] = 2.1 \times 10^{-6}$ M, $[OH^-] = 2.1 \times 10^{-6}$ M; $[H_3O^+] = 4.8 \times 10^{-9}$ M

111. **a.** F^- **b.** NO_2^- **c.** ClO^-

113. $H_3PO_4(aq) + H_2O(l) \rightleftharpoons H_2PO_4^-(aq) + H_3O^+(aq)$,

$$K_{a_1} = \frac{[H_3O^+][H_2PO_4^-]}{[H_3PO_4]}$$

$H_2PO_4^-(aq) + H_2O(l) \rightleftharpoons HPO_4^{2-}(aq) + H_3O^+(aq)$,

$$K_{a_2} = \frac{[H_3O^+][HPO_4^{2-}]}{[H_2PO_4^-]}$$

$HPO_4^{2-}(aq) + H_2O(l) \rightleftharpoons PO_4^{3-}(aq) + H_3O^+(aq)$,

$$K_{a_3} = \frac{[H_3O^+][PO_4^{3-}]}{[HPO_4^{2-}]}$$

115. **a.** $[H_3O^+] = 0.048$ M, pH = 1.32
b. $[H_3O^+] = 0.12$ M, pH = 0.92

117. $[H_2SO_3] = 0.418$ M
$[HSO_3^-] = 0.082$ M
$[SO_3^{2-}] = 6.4 \times 10^{-8}$ M
$[H_3O^+] = 0.082$ M

119. **a.** $[H_3O^+] = 0.50$ M, pH = 0.30
b. $[H_3O^+] = 0.11$ M, pH = 0.96 (*x is small* approximation breaks down)
c. $[H_3O^+] = 0.059$ M, pH = 1.23

121. **a.** Lewis acid **b.** Lewis acid
c. Lewis base **d.** Lewis base

123. **a.** acid: Fe^{3+}, base: H_2O **b.** acid: Zn^{2+}, base: NH_3
c. acid: BF_3, base: $(CH_3)_3N$

125. **a.** weak **b.** strong **c.** weak **d.** strong

127. If blood became acidic, the H^+ concentration would increase. According to Le Châtelier's principle, equilibrium would be shifted to the left and the concentration of oxygenated Hb would decrease.

129. All acid will be neutralized.

131. $[H_3O^+]$(Great Lakes) $= 3 \times 10^{-5}$ M, $[H_3O^+]$(West Coast) $= 4 \times 10^{-6}$ M. The rain over the Great Lakes is about 8 times more concentrated.

133. 2.7

135. **a.** 2.000 **b.** 1.52 **c.** 12.95 **d.** 11.12 **e.** 5.03

137. **a.** 1.260 **b.** 8.22 **c.** 0.824 **d.** 8.57 **e.** 1.171

139. **a.** $CN^-(aq) + H^+(aq) \rightleftharpoons HCN(aq)$
b. $NH_4^+(aq) + OH^-(aq) \rightleftharpoons NH_3(aq) + H_2O(l)$
c. $CN^-(aq) + NH_4^+(aq) \rightleftharpoons HCN(aq) + NH_3(aq)$
d. $HSO_4^-(aq) + C_2H_3O_2^-(aq) \rightleftharpoons SO_4^{2-}(aq) + HC_2H_3O_2(aq)$
e. no reaction between the major species

141. 0.794

143. $K_a = 8.3 \times 10^{-4}$

145. The student forgot to account for the dissociation of water. Correct pH is 6.79.

147. 2.14

149. $[A^-] = 4.5 \times 10^{-5}$ M
$[H^+] = 2.2 \times 10^{-4}$ M
$[HA_2^-] = 1.8 \times 10^{-4}$ M

151. 9.28

153. 50.1 g $NaHCO_3$

155. **b**

157. $CH_3COOH < CH_2ClCOOH < CHCl_2COOH < CCl_3COOH$

Chapter 18

25. **d**

27. **a.** 3.62 **b.** 9.11

29. pure water: 2.1%, in $NaC_7H_5O_2$: 0.065%. The percent ionization in the sodium benzoate solution is much smaller because the presence of the benzoate ion shifts the equilibrium to the left.

31. **a.** 2.14 **b.** 8.32 **c.** 3.46

33. $HCl + NaC_2H_3O_2 \longrightarrow HC_2H_3O_2 + NaCl$
$NaOH + HC_2H_3O_2 \longrightarrow NaC_2H_3O_2 + H_2O$

35. **a.** 3.62 **b.** 9.11

37. **a.** 7.60 **b.** 11.18 **c.** 4.61

39. **a.** 3.86 **b.** 8.95

41. 3.5

43. 3.7 g

45. **a.** 4.74 **b.** 4.68 **c.** 4.81

47. **a.** initial 7.00, after 1.70
b. initial 4.71, after 4.56
c. initial 10.78, after 10.66

49. 1.2 g; 2.7 g

51. **a.** yes **b.** no **c.** yes **d.** no **e.** no

53. **a.** 7.4 **b.** 0.3 g **c.** 0.14 g

55. $KClO/HClO = 0.79$

57. **a.** does not exceed capacity **b.** does not exceed capacity **c.** does not exceed capacity **d.** does not exceed capacity

59. **(i)** **a.** pH = 8, **b.** pH = 7
(ii) **a.** weak acid, **b.** strong acid

61. **a.** 40.0 mL HI for both
b. KOH: neutral, CH_3NH_2: acidic
c. CH_3NH_2
d. Titration of KOH with HI:

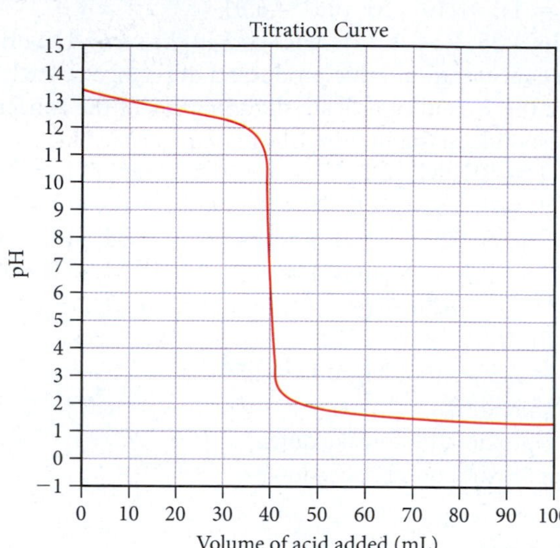

Titration of CH_3NH_2 with HI:

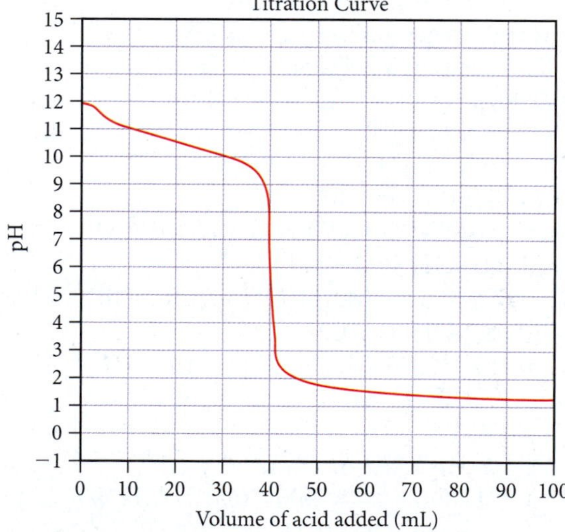

63. **a.** pH = 9, added base = 30 mL
b. 0 mL **c.** 15 mL **d.** 30 mL **e.** 30 mL

65. **a.** 0.757 **b.** 30.6 mL **c.** 1.038 **d.** 7 **e.** 12.15

67. **a.** 13.06 **b.** 28.8 mL **c.** 12.90 **d.** 7 **e.** 2.07

69. **a.** 2.86 **b.** 16.8 mL **c.** 4.37 **d.** 4.74 **e.** 8.75 **f.** 12.17

71. **a.** 11.94 **b.** 29.2 mL **c.** 11.33 **d.** 10.64 **e.** 5.87 **f.** 1.90

73. **(i)** (a)
    (ii) (b)
75. $pK_a = 3$, 82 g/mol
77. First equivalence: 22.7 mL
    Second equivalence: 45.4 mL
79. The indicator will appear red. The pH range is 4 to 6.
81. **a.** phenol red, *m*-nitrophenol
    b. alizarin, bromothymol blue, phenol red
    c. alizarin yellow R
83. **a.** $BaSO_4(s) \rightleftharpoons Ba^{2+}(aq) + SO_4^{2-}(aq)$, $K_{sp} = [Ba^{2+}][SO_4^{2-}]$
    b. $PbBr_2(s) \rightleftharpoons Pb^{2+}(aq) + 2\,Br^-(aq)$, $K_{sp} = [Pb^{2+}][Br^-]^2$
    c. $Ag_2CrO_4(s) \rightleftharpoons 2\,Ag^+(aq) + CrO_4^{2-}(aq)$, $K_{sp} = [Ag^+]^2[CrO_4^{2-}]$
85. **a.** 7.31×10^{-7} M **b.** 3.72×10^{-5} M **c.** 3.32×10^{-4} M
87. **a.** 1.07×10^{-21} **b.** 7.14×10^{-7} **c.** 7.44×10^{-11}
89. AX_2
91. 2.07×10^{-5} g/100 mL
93. **a.** 0.0183 M **b.** 0.00755 M **c.** 0.00109 M
95. **a.** 5×10^{14} M **b.** 5×10^{8} M **c.** 5×10^{4} M
97. **a.** more soluble, CO_3^{2-} is basic
    b. more soluble, S^{2-} is basic
    c. not, neutral
    d. not, neutral
99. precipitate will form, CaF_2
101. precipitate will form, $Mg(OH)_2$
103. **a.** 0.018 M **b.** 1.4×10^{-7} M **c.** 1.1×10^{-5} M
105. **a.** $BaSO_4$, 1.1×10^{-8} M **b.** 3.0×10^{-8} M
107. 8.7×10^{-10} M
109. 5.6×10^{16}
111. 4.03
113. 3.57
115. HCl, 4.7 g
117. **a.** $NaOH(aq) + KHC_8H_4O_4(aq) \longrightarrow Na^+(aq) + K^+(aq) + C_8H_4O_4^{2-}(aq) + H_2O(l)$
     b. 0.1046 M
119. 4.73
121. 176 g/mol; 1.0×10^{-4}
123. 14.2 L
125. 1.6×10^{-7} M
127. 8.0×10^{-8} M
129. 6.29
131. 0.172 M
133. The ratio by mass of dimethyl ammonium chloride to dimethyl amine needed is 3.6.
135. 0.18 M benzoic acid, 0.41 M sodium benzoate
137. 51.6 g
139. 1.8×10^{-11} (based on this data)
141. **a.** 5.5×10^{-25} M **b.** 5.5×10^{-4} M
143. 1.38 L
145. 12.97
147. **a.** $pH < pK_a$ **b.** $pH > pK_a$
     c. $pH = pK_a$ **d.** $pH > pK_a$
149. **b**
151. **a.** no difference **b.** less soluble
     c. more soluble

Chapter 19

25. **a, c**
27. System B has the greatest entropy. There is only one energetically equivalent arrangement for System A. However, the particles of System B may exchange positions for a second energetically equivalent arrangement.
29. **a.** $\Delta S > 0$ **b.** $\Delta S < 0$
    c. $\Delta S < 0$ **d.** $\Delta S < 0$
31. It increases.
33. **a.** $CO_2(g)$, greater molar mass and complexity
    b. $CH_3OH(g)$, gas phase
    c. $CO_2(g)$, greater molar mass and complexity
    d. $SiH_4(g)$, greater molar mass
    e. $CH_3CH_2CH_3(g)$, greater molar mass and complexity
    f. NaBr(*aq*), aqueous
35. **a.** He, Ne, SO_2, NH_3, CH_3CH_2OH. From He to Ne there is an increase in molar mass; beyond that, the molecules increase in complexity.
    b. $H_2O(s)$, $H_2O(l)$, $H_2O(g)$; increase in entropy in going from solid to liquid to gas phase
    c. CH_4, CF_4, CCl_4; increasing entropy with increasing molar mass
37. **a.** −120.8 J/K, decrease in moles of gas
    b. 133.9 J/K, increase in moles of gas
    c. −42.0 J/K, small change because moles of gas stay constant
    d. −390.8 J/K, decrease in moles of gas
39. −89.3 J/K, decrease in moles of gas
41. **a.** $\Delta S_{sys} > 0$, $\Delta S_{surr} > 0$, spontaneous at all temperatures
    b. $\Delta S_{sys} < 0$, $\Delta S_{surr} < 0$, nonspontaneous at all temperatures
    c. $\Delta S_{sys} < 0$, $\Delta S_{surr} < 0$, nonspontaneous at all temperatures
    d. $\Delta S_{sys} > 0$, $\Delta S_{surr} > 0$, spontaneous at all temperatures
43. **a.** 1.29×10^3 J/K
    b. 5.00×10^3 J/K
    c. -3.83×10^2 J/K
    d. -1.48×10^3 J/K
45. **a.** −649 J/K >, nonspontaneous
    b. 649 J/K, spontaneous
    c. 123 J/K, spontaneous
    d. −76 J/K, nonspontaneous
47. **a.** 1.93×10^5 J, nonspontaneous
    b. -1.93×10^5 J, spontaneous
    c. -3.7×10^4 J, spontaneous
    d. 4.7×10^4 J, nonspontaneous
49. -2.247×10^6 J, spontaneous
51.

ΔH	ΔS	ΔG	Low Temperature	High Temperature
−	+	−	Spontaneous	Spontaneous
−	−	Temperature dependent	Spontaneous	Nonspontaneous
+	+	Temperature dependent	Nonspontaneous	Spontaneous
+	−	+	Nonspontaneous	Nonspontaneous

53. $\Delta H^\circ_{rxn} = -1277$ kJ, $\Delta S^\circ_{rxn} = 313.6$ J/K, $\Delta G^\circ_{rxn} = -1.370 \times 10^3$ kJ; yes

55. a. $\Delta H^\circ_{rxn} = 57.2$ kJ, $\Delta S^\circ_{rxn} = 175.8$ J/K, $\Delta G^\circ_{rxn} = 4.8 \times 10^3$ J/mol; nonspontaneous, becomes spontaneous at high temperatures
b. $\Delta H^\circ_{rxn} = 176.2$ kJ, $\Delta S^\circ_{rxn} = 285.1$ J/K, $\Delta G^\circ_{rxn} = 91.2$ kJ; nonspontaneous, becomes spontaneous at high temperatures
c. $\Delta H^\circ_{rxn} = 98.8$ kJ, $\Delta S^\circ_{rxn} = 141.5$ J/K, $\Delta G^\circ_{rxn} = 56.6$ kJ; nonspontaneous, becomes spontaneous at high temperatures
d. $\Delta H^\circ_{rxn} = -91.8$ kJ, $\Delta S^\circ_{rxn} = -198.1$ J/K, $\Delta G^\circ_{rxn} = -32.8$ kJ; spontaneous

57. a. 2.8 kJ **b.** 91.2 kJ **c.** 56.4 kJ **d.** −32.8 kJ
Values are comparable. The method using ΔH° and ΔS° can be used to determine how ΔG° changes with temperature.

59. a. −72.5 kJ, spontaneous
b. −11.4 kJ, spontaneous
c. 9.1 kJ, nonspontaneous

61. −29.4 kJ

63. a. 19.3 kJ
b. (i) 2.9 kJ
(ii) −2.9 kJ
c. The partial pressure of iodine is very low.

65. 11.9 kJ

67. a. 1.48×10^{90} **b.** 2.09×10^{-26}

69. a. −24.8 kJ **b.** 0 **c.** 9.4 kJ

71. a. 1.90×10^{47} **b.** 1.51×10^{-13}

73. $\Delta H^\circ = 50.6$ kJ
$\Delta S^\circ = 226$ J/K

75. 4.8

77. a. + **b.** − **c.** −

79. a. $\Delta G^\circ = 175.2$ kJ, $K = 1.95 \times 10^{-31}$, nonspontaneous
b. 133 kJ, yes

81. Cl_2: $\Delta H^\circ_{rxn} = -182.1$ kJ, $\Delta S^\circ_{rxn} = -134.4$ J/K, $\Delta G^\circ_{rxn} = -142.0$ kJ $K = 7.94 \times 10^{24}$
Br_2: $\Delta H^\circ_{rxn} = -121.6$ kJ, $\Delta S^\circ_{rxn} = -134.2$ J/K, $\Delta G^\circ_{rxn} = -81.6$ kJ $K = 2.02 \times 10^{14}$
I_2: $\Delta H^\circ_{rxn} = -48.3$ kJ, $\Delta S^\circ_{rxn} = -132.2$ J/K, $\Delta G^\circ_{rxn} = -8.9$ kJ $K = 37$
Cl_2 is the most spontaneous, I_2 is the least. Spontaneity is determined by the standard enthalpy of formation of the dihalogenated ethane. Higher temperatures make the reactions less spontaneous.

83. a. 107.8 kJ **b.** 5.0×10^{-7} atm
c. spontaneous at higher temperatures, $T = 923.4$ K

85. a. 2.22×10^5 **b.** 94.4 mol

87. a. $\Delta G^\circ = -689.6$ kJ, ΔG° becomes less negative
b. $\Delta G^\circ = -665.2$ kJ, ΔG° becomes less negative
c. $\Delta G^\circ = -632.4$ kJ, ΔG° becomes less negative
d. $\Delta G^\circ = -549.3$ kJ, ΔG° becomes less negative

89. With one exception, the formation of any oxide of nitrogen at 298 K requires more moles of gas as reactants than are formed as products. For example, 1 mol of N_2O requires 0.5 mol of O_2 and 1 mol of N_2, 1 mol of N_2O_3 requires 1 mol of N_2 and 1.5 mol of O_2, and so on. The exception is NO, where 1 mol of NO requires 0.5 mol of O_2 and 0.5 mol of N_2:

$$\frac{1}{2}N_2(g) + \frac{1}{2}O_2(g) \longrightarrow NO(g)$$

This reaction has a positive ΔS because what is essentially mixing of the N and O has taken place in the product.

91. 15.0 kJ

93. a. Positive, the process is spontaneous. It is slow unless a spark is applied.
b. Positive, although the change in the system is not spontaneous; the overall change, which includes such processes as combustion or water flow to generate electricity, is spontaneous.
c. Positive, the acorn oak/tree system is becoming more ordered, so the processes associated with growth are not spontaneous. But they are driven by spontaneous processes such as the generation of heat by the sun and the reactions that produce energy in the cell.

95. At 18.3 mmHg $\Delta G = 0$, At 760 mmHg $\Delta G^\circ = 55.4$ kJ

97. a. 3.24×10^{-3}
b.
$$NH_3 + ATP + H_2O \longrightarrow NH_3{-}P_i + ADP$$
$$NH_3{-}P_i + C_5H_8O_4N^- \longrightarrow C_5H_9O_3N_2 + P_i + H_2O$$
$$NH_3 + C_5H_8O_4N^- + ATP \longrightarrow C_5H_9O_3N_2 + ADP + P_i$$
$\Delta G^\circ = -16.3$ kJ, $K = 7.20 \times 10^2$

99. a. −95.3 kJ/mol. Since the number of moles of reactants and products are the same, the decrease in volume affects the entropy of both equally, so there is no change in ΔG.
b. 102.8 kJ/mol. The entropy of the reactants (1.5 mol) is decreased more than the entropy of the product (1 mol). Since the product is relatively more favored at lower volume, ΔG is less positive.
c. 204.2 kJ/mol. The entropy of the product (1 mol) is decreased more than the entropy of the reactant (0.5 mol). Since the product is relatively less favored, ΔG is more positive.

101. $\Delta H^\circ = -93$ kJ, $\Delta S^\circ = -2.0 \times 10^2$ J/K

103. ΔS_{vap} diethyl ether $= 86.1$ J/mol K, ΔS_{vap} acetone $= 88.4$ J/mol K, ΔS_{vap} benzene $= 87.3$ J/mol K, ΔS_{vap} chloroform $= 88.0$ J/mol K. Because water and ethanol hydrogen bond, they are more ordered in the liquid and we expect ΔS_{vap} to be more positive. Ethanol 38600/351.0 = 110 J/mol K, H_2O = 40700/373.2 = 109 J/mol K

105. a and **c** will both increase the entropy of the surroundings because they are both exothermic reactions (adding thermal energy to the sourroundings).

107. c. If entropy of a system is increasing, the enthalpy of a reaction can be overcome (if necessary) by the entropy change as long as the temperature is high enough. If the entropy change of the system is decreasing, the reaction must be exothermic in order to be spontaneous since the entropy is working against spontaneity.

109. a and **b** are both true. Since $\Delta G_{rxn} = \Delta G^\circ_{rxn} + RT \ln Q$ and $\Delta G^\circ_{rxn} = -42.5$ kJ, in order for $\Delta G_{rxn} = 0$ the second term must be positive. This necessitates that $Q > 1$ or that we have more product than reactant. Any reaction at equilibrium has $\Delta G_{rxn} = 0$.

Chapter 20

33. a. $3\,K(s) + Cr^{3+}(aq) \longrightarrow Cr(s) + 3\,K^+(aq)$
b. $2\,Al(s) + 3\,Fe^{2+}(aq) \longrightarrow 2\,Al^{3+}(aq) + 3\,Fe(s)$
c. $2\,BrO_3^-(aq) + 3\,N_2H_4(g) \longrightarrow 2\,Br^-(aq) + 3\,N_2(g) + 6\,H_2O(l)$

35. a. $PbO_2(s) + 2\,I^-(aq) + 4\,H^+(aq) \longrightarrow Pb^{2+}(aq) + I_2(s) + 2\,H_2O(l)$

b. $5\,SO_3^{\,2-}(aq) + 2\,MnO_4^{\,-}(aq) + 6\,H^+(aq) \longrightarrow 5\,SO_4^{\,2-}(aq) + 2\,Mn^{2+}(aq) + 3\,H_2O(l)$

c. $S_2O_3^{\,2-}(aq) + 4\,Cl_2(g) + 5\,H_2O(l) \longrightarrow 2\,SO_4^{\,2-}(aq) + 8\,Cl^-(aq) + 10\,H^+(aq)$

37. a. $H_2O_2(aq) + 2\,ClO_2(aq) + 2\,OH^-(aq) \longrightarrow O_2(g) + 2\,ClO_2^{\,-}(aq) + 2\,H_2O(l)$

b. $Al(s) + MnO_4^{\,-}(aq) + 2\,H_2O(l) \longrightarrow Al(OH)_4^{\,-}(aq) + MnO_2(s)$

c. $Cl_2(g) + 2\,OH^-(aq) \longrightarrow Cl^-(aq) + ClO^-(aq) + H_2O(l)$

39. a.

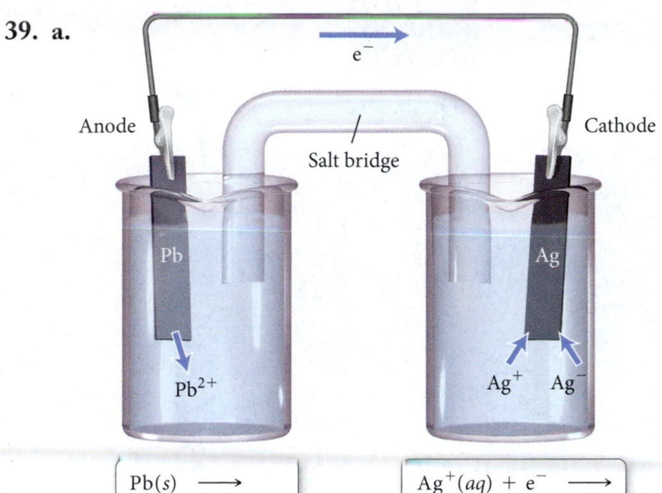

b.

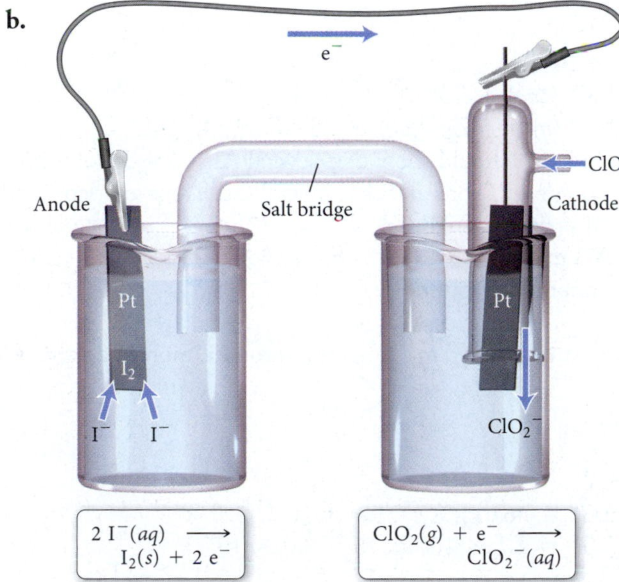

c.

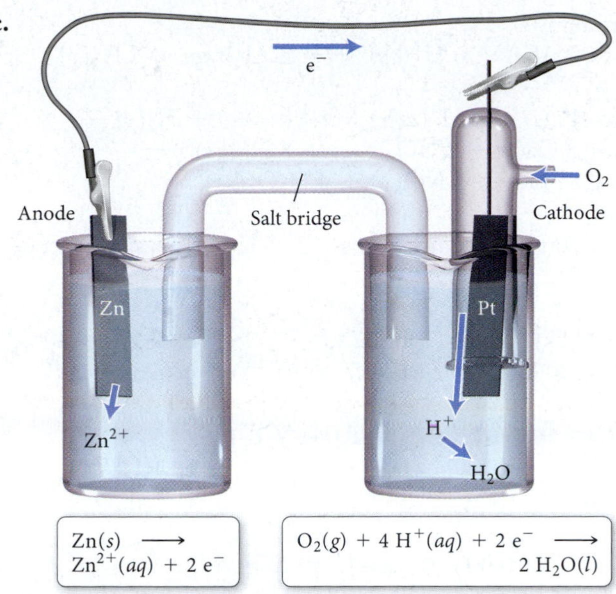

41. a. 0.93 V **b.** 0.41 V **c.** 1.99 V

43. a, c, d

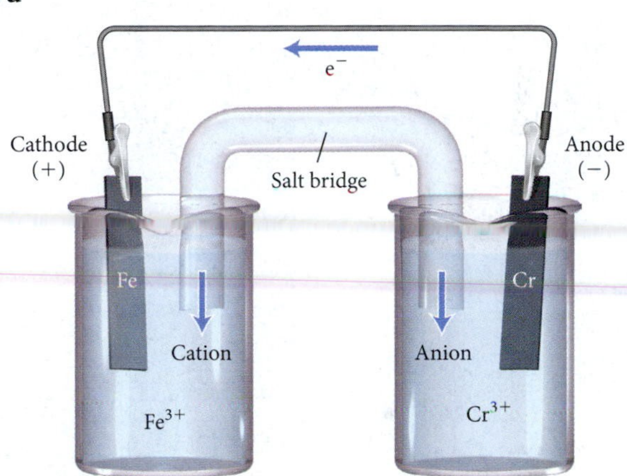

b. $Cr(s) + Fe^{3+}(aq) \longrightarrow Cr^{3+}(aq) + Fe(s)$, $E^\circ_{cell} = 0.69$ V

45. a. $Pb(s)\,|\,Pb^{2+}(aq)\,||\,Ag^+(aq)\,|\,Ag(s)$

b. $Pt(s), I_2(s)\,|\,I^-(aq)\,||\,ClO_2^{\,-}(aq)\,|\,ClO_2(g)\,|\,Pt(s)$

c. $Zn(s)\,|\,Zn^{2+}(aq)\,||\,H_2O(l)\,|\,H^+(aq)\,|\,O_2(g)\,|\,Pt(s)$

47.

$3\,Sn(s) + 2\,NO_3^{\,-}(aq) + 8\,H^+(aq) \longrightarrow 3\,Sn^{2+}(aq) + 2\,NO(g) + 4\,H_2O(l)$, $E^\circ_{cell} = 1.10$ V

49. b, c occur spontaneously in the forward direction

51. aluminum

53. a. yes, $2\,Al(s) + 6\,H^+(aq) \longrightarrow 2\,Al^{3+}(aq) + 3\,H_2(g)$
b. no
c. yes, $Pb(s) + 2\,H^+(aq) \longrightarrow Pb^{2+}(aq) + H_2(g)$

55. a. yes, $3\,Cu(s) + 2\,NO_3^-(aq) + 8\,H^+(aq) \longrightarrow 3\,Cu^{2+}(aq) + 2\,NO(g) + 4\,H_2O(l)$
b. no

57. a. −1.70 V, nonspontaneous **b.** 1.97 V, spontaneous
c. −1.51, nonspontaneous

59. a

61. a. −432 kJ **b.** 52 kJ **c.** -1.7×10^2 kJ

63. a. 5.31×10^{75} **b.** 7.7×10^{-10} **c.** 6.3×10^{29}

65. 5.6×10^5

67. $\Delta G^\circ = -7.\underline{9}7$ kJ, $E^\circ_{cell} = 0.041$ V

69. a. 1.04 V **b.** 0.97 V **c.** 1.11 V

71. 1.87 V

73. a. 0.56 V **b.** 0.52 V
c. $[Ni^{2+}] = 0.003$ M, $[Zn^{2+}] = 1.60$ M

75.

e⁻
Anode
Cathode
Salt bridge
Zn
Zn
Anions
Cations
Zn^{2+}
Zn^{2+}
1.0×10^{-3} M
2.0 M
$Zn(s) \longrightarrow Zn^{2+}(aq) + 2\,e^-$
$Zn^{2+}(aq) + 2\,e^- \longrightarrow Zn(s)$

77. $\dfrac{[Sn^{2+}](ox)}{[Sn^{2+}](red)} = 4.2 \times 10^{-4}$

79. 0.3762

81. 1.038 V

83. a, c would prevent the corrosion of iron

85.

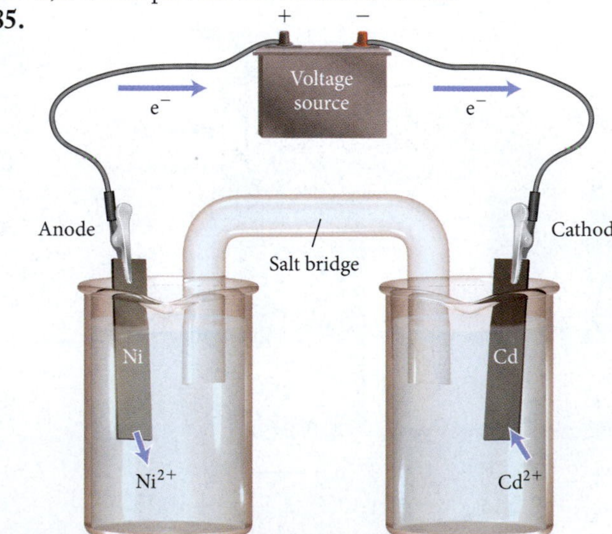

minimum voltage = 0.17 V

87. oxidation: $2\,Br^-(l) \longrightarrow Br_2(g) + 2\,e^-$
reduction: $K^+(l) + e^- \longrightarrow K(l)$

89. oxidation: $2\,Br^-(l) \longrightarrow Br_2(g) + 2\,e^-$
reduction: $K^+(l) + e^- \longrightarrow K(l)$

91. a. anode: $2\,Br^- \longrightarrow Br_2(l) + 2\,e^-$
cathode: $2\,H_2O(l) + 2\,e^- \longrightarrow H_2(g) + 2\,OH^-(aq)$
b. anode: $2\,I^-(aq) \longrightarrow I_2(s) + 2\,e^-$
cathode: $Pb^{2+}(aq) + 2\,e^- \longrightarrow Pb(s)$
c. anode: $2\,H_2O(l) \longrightarrow O_2(g) + 4\,H^+(aq) + 4\,e^-$
cathode: $2\,H_2O(l) + 2\,e^- \longrightarrow H_2(g) + 2\,OH^-(aq)$

93.

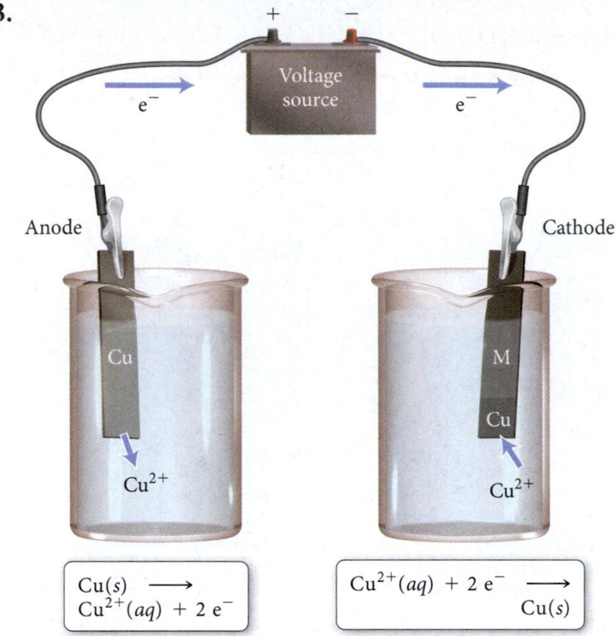

95. 1.8×10^2 s

97. 1.2×10^3 A

99. $2\,MnO_4^-(aq) + 5\,Zn(s) + 16\,H^+(aq) \longrightarrow 2\,Mn^{2+}(aq) + 5\,Zn^{2+}(aq) + 8\,H_2O(l)$ 34.9 mL

101. The drawing should show that several Al atoms dissolve into solution as Al^{3+} ions and that several Cu^{2+} ions are deposited on the Al surface as solid Cu.

103. a. 68.3 mL **b.** cannot be dissolved
c. cannot be dissolved

105. 0.25

107. There are no paired reactions that produce more than about 5 or 6 V.

109. a. 2.83 V **b.** 2.71 V **c.** 16 hr

111. 176 h

113. 0.71 V

115. a. $\Delta G^\circ = 461$ kJ, $K = 1.4 \times 10^{-81}$
b. $\Delta G^\circ = 2.7 \times 10^2$ kJ, $K = 2.0 \times 10^{-48}$

117. MCl_4

119. 51.3%

121. pH = 0.85

123. 0.83 M

125. 4.1×10^5 L

127. 435 s

129. 8.39% U

131. The overall cell reaction for both cells is $2\,Cu^+(aq) \longrightarrow Cu^{2+}(aq) + Cu(s)$. The difference in E° is because n = 1 for the first cell and n = 2 for the second cell. For both cells, $\Delta G^\circ = -35.1$ kJ.

133. a

135. ΔG°_{rxn} is positive and E°_{cell} is negative.

Chapter 21

31. a. ${}^{234}_{92}U \longrightarrow {}^{4}_{2}He + {}^{230}_{90}Th$
b. ${}^{230}_{90}Th \longrightarrow {}^{4}_{2}He + {}^{226}_{88}Ra$
c. ${}^{214}_{82}Pb \longrightarrow {}^{0}_{-1}e + {}^{214}_{83}Bi$
d. ${}^{13}_{7}N \longrightarrow {}^{0}_{+1}e + {}^{13}_{6}C$
e. ${}^{51}_{24}Cr + {}^{0}_{-1}e \longrightarrow {}^{51}_{23}V$

33. ${}^{232}_{90}Th \longrightarrow {}^{4}_{2}He + {}^{228}_{88}Ra$
${}^{228}_{88}Ra \longrightarrow {}^{0}_{-1}e + {}^{228}_{89}Ac$
${}^{228}_{89}Ac \longrightarrow {}^{0}_{-1}e + {}^{228}_{90}Th$
${}^{228}_{90}Th \longrightarrow {}^{4}_{2}He + {}^{224}_{88}Ra$

35. a. ${}^{221}_{87}Fr$ **b.** ${}^{0}_{-1}e$ **c.** ${}^{0}_{+1}e$ **d.** ${}^{0}_{-1}e$

37. a. stable, N/Z ratio is close to 1, acceptable for low Z atoms
b. not stable, N/Z ratio much too high for low Z atom
c. not stable, N/Z ratio is less than 1, much too low
d. stable, N/Z ratio is acceptable for this Z

39. Sc, V, and Mn, each have odd numbers of protons. Atoms with an odd number of protons typically have less stable isotopes than those with an even number of protons.

41. a. beta decay **b.** positron emission
c. positron emission **d.** positron emission

43. a. Cs-125 **b.** Fe-62

45. 2.34×10^9 years

47. 0.57 g

49. 10.8 hrs

51. 2.66×10^3 yr

53. 2.4×10^4 yr

55. 2.7×10^9 yr

57. ${}^{235}_{92}U + {}^{1}_{0}n \longrightarrow {}^{144}_{54}Xe + {}^{90}_{38}Sr + 2\ {}^{1}_{0}n$

59. ${}^{2}_{1}H + {}^{2}_{1}H \longrightarrow {}^{3}_{2}He + {}^{1}_{0}n$

61. ${}^{238}_{92}U + {}^{1}_{0}n \longrightarrow {}^{239}_{92}U$
${}^{239}_{92}U \longrightarrow {}^{239}_{93}Np + {}^{0}_{-1}e$
${}^{239}_{93}Np \longrightarrow {}^{239}_{94}Pu + {}^{0}_{-1}e$

63. ${}^{249}_{98}Cf + {}^{12}_{6}C \longrightarrow {}^{257}_{104}Rf + 4\ {}^{1}_{0}n$

65. 9.0×10^{13} J

67. a. mass defect = 0.13701 amu
binding energy = 7.976 MeV/ nucleon
b. mass defect = 0.54369 amu
binding energy = 8.732 MeV/ nucleon
c. mass defect = 1.16754 amu
binding energy = 8.431 MeV/ nucleon

69. 7.228×10^{10} J/g U-235

71. 7.84×10^{10} J/g H-2

73. radiation: 25 J, fall: 370 J

75. 68 mi

77. a. ${}^{1}_{1}p + {}^{9}_{4}Be \longrightarrow {}^{6}_{3}Li + {}^{4}_{2}He$
1.03×10^{11} J/mol
b. ${}^{209}_{83}Bi + {}^{64}_{28}Ni \longrightarrow {}^{272}_{111}Rg + {}^{1}_{0}n$
1.141×10^{13} J/mol
c. ${}^{179}_{74}W + {}^{0}_{-1}e \longrightarrow {}^{179}_{73}Ta$
7.59×10^{10} J/mol

79. a. ${}^{114}_{44}Ru \longrightarrow {}^{0}_{-1}e + {}^{114}_{45}Rh$
b. ${}^{216}_{88}Ra \longrightarrow {}^{0}_{+1}e + {}^{216}_{87}Fr$
c. ${}^{58}_{30}Zn \longrightarrow {}^{0}_{+1}e + {}^{58}_{29}Cu$
d. ${}^{31}_{10}Ne \longrightarrow {}^{0}_{-1}e + {}^{31}_{11}Na$

81. 2.9×10^{21} beta emissions, 3700 Ci

83. 1.6×10^{-5} L

85. 4.94×10^7 kJ/mol

87. 7.72 MeV

89. ${}^{14}N$

91. 0.15%

93. 1.24×10^{21} atoms

95. 2.42×10^{-12} m

97. −0.7 MeV, there is no coulombic barrier for collision with a neutron.

99. a. 1.164×10^{10} kJ **b.** 0.1299 g

101. U-235 forms Pb-207 in seven α-decays and four β-decays, and Th-232 forms Pb-208 in six α-decays and four β-decays.

103. 3.0×10^2 K

105. ${}^{21}_{9}F \longrightarrow {}^{21}_{10}Ne + {}^{0}_{-1}e$

107. Nuclide A is more dangerous because the half-life is shorter (18.5 days) and so it decays faster.

109. Iodine is used by the thyroid gland to make hormones. Normally we ingest iodine in foods, especially iodized salt. The thyroid gland cannot tell the difference between stable and radioactive iodine and will absorb both. KI tablets work by blocking radioactive iodine from entering the thyroid. When a person takes KI, the stable iodine in the tablet gets absorbed by the thyroid. Because KI contains so much stable iodine, the thyroid gland becomes "full" and cannot absorb any more iodine—either stable or radioactive—for the next 24 hours.

Chapter 22

35. a. alkane **b.** alkene
c. alkyne **d.** alkene

37.
```
CH3—CH2—CH2—CH2—CH2—CH2—CH3

CH3—CH—CH2—CH2—CH2—CH3
    |
    CH3

CH3—CH2—CH—CH2—CH2—CH3
        |
        CH3

    CH3
    |
CH3—CH—CH—CH2—CH3
        |
        CH3

        CH3
        |
CH3—CH2—C—CH2—CH3
        |
        CH3
```

$$\begin{array}{ccccccccc} & & \mathrm{CH_3} & & & & & & \\ & & | & & & & & & \\ \mathrm{H_3C} & - & \mathrm{C} & - & \mathrm{CH_2} & - & \mathrm{CH_2} & - & \mathrm{CH_3} \\ & & | & & & & & & \\ & & \mathrm{CH_3} & & & & & & \end{array}$$

$$\begin{array}{ccccccccc} \mathrm{H_3C} & - & \mathrm{CH} & - & \mathrm{CH_2} & - & \mathrm{CH} & - & \mathrm{CH_3} \\ & & | & & & & | & & \\ & & \mathrm{CH_3} & & & & \mathrm{CH_3} & & \end{array}$$

$$\begin{array}{ccccccc} & & \mathrm{CH_3} & & & & \\ & & | & & & & \\ \mathrm{H_3C} & - & \mathrm{C} & - & \mathrm{CH} & - & \mathrm{CH_3} \\ & & | & & | & & \\ & & \mathrm{CH_3} & & \mathrm{CH_3} & & \end{array}$$

$$\begin{array}{ccccccccc} \mathrm{H_3C} & - & \mathrm{CH_2} & - & \mathrm{CH} & - & \mathrm{CH_2} & - & \mathrm{CH_3} \\ & & & & | & & & & \\ & & & & \mathrm{CH_2} & & & & \\ & & & & | & & & & \\ & & & & \mathrm{CH_3} & & & & \end{array}$$

39. a. no **b.** yes **c.** yes **d.** no

41. a. enantiomers **b.** same **c.** enantiomers

43. a. pentane
b. 2-methylbutane
c. 4-isopropyl-2-methylheptane
d. 4-ethyl-2-methylhexane

45. a.
$$\begin{array}{ccccccccccc} \mathrm{CH_3} & - & \mathrm{CH_2} & - & \mathrm{CH} & - & \mathrm{CH_2} & - & \mathrm{CH_2} & - & \mathrm{CH_3} \\ & & & & | & & & & & & \\ & & & & \mathrm{CH_2{-}CH_3} & & & & & & \end{array}$$

b.
$$\begin{array}{ccccccccc} & & & & \mathrm{CH_3} & & & & \\ & & & & | & & & & \\ \mathrm{CH_3} & - & \mathrm{CH_2} & - & \mathrm{C} & - & \mathrm{CH_2} & - & \mathrm{CH_3} \\ & & & & | & & & & \\ & & & & \mathrm{CH_2{-}CH_3} & & & & \end{array}$$

c.
$$\begin{array}{ccccccc} \mathrm{CH_3} & - & \mathrm{CH} & - & \mathrm{CH} & - & \mathrm{CH_3} \\ & & | & & | & & \\ & & \mathrm{CH_3} & & \mathrm{CH_3} & & \end{array}$$

d.
$$\begin{array}{ccccccccccccccccc} & & \mathrm{CH_3} & & & & & & & & & & \mathrm{CH_2{-}CH_3} & & & & \\ & & | & & & & & & & & & & | & & & & \\ \mathrm{CH_3} & - & \mathrm{C} & - & \mathrm{CH_2} & - & \mathrm{CH} & - & \mathrm{CH_2} & - & \mathrm{CH_2} & - & \mathrm{CH} & - & \mathrm{CH_2} & - & \mathrm{CH_3} \\ & & | & & & & | & & & & & & & & & & \\ & & \mathrm{CH_3} & & & & \mathrm{CH_2{-}CH_3} & & & & & & & & & & \end{array}$$

47. a. $\mathrm{CH_3CH_2CH_3} + 5\,\mathrm{O_2} \longrightarrow 3\,\mathrm{CO_2} + 4\,\mathrm{H_2O}$
b. $\mathrm{CH_3CH_2CH{=}CH_2} + 6\,\mathrm{O_2} \longrightarrow 4\,\mathrm{CO_2} + 4\,\mathrm{H_2O}$
c. $2\,\mathrm{CH{\equiv}CH} + 5\,\mathrm{O_2} \longrightarrow 4\,\mathrm{CO_2} + 2\,\mathrm{H_2O}$

49. a. $\mathrm{CH_3CH_2Br}$
b. $\mathrm{CH_3CH_2CH_2Cl}$, $\mathrm{CH_3CHClCH_3}$
c. $\mathrm{CHCl_2Br}$
d.
$$\begin{array}{ccccccc} & & \mathrm{H} & & & & \\ & & | & & & & \\ \mathrm{CH_3} & - & \mathrm{C} & - & \mathrm{CH_2} & - & \mathrm{Cl} \\ & & | & & & & \\ & & \mathrm{CH_3} & & & & \end{array}$$

$$\begin{array}{ccccc} & & \mathrm{Cl} & & \\ & & | & & \\ \mathrm{CH_3} & - & \mathrm{C} & - & \mathrm{CH_3} \\ & & | & & \\ & & \mathrm{CH_3} & & \end{array}$$

51. $\mathrm{CH_2{=}CH{-}CH_2{-}CH_2{-}CH_2{-}CH_3}$
$\mathrm{CH_3{-}CH{=}CH{-}CH_2{-}CH_2{-}CH_3}$
$\mathrm{CH_3{-}CH_2{-}CH{=}CH{-}CH_2{-}CH_3}$

53. a. 1-butene
b. 3,4-dimethyl-2-pentene
c. 3-isopropyl-1-hexene
d. 2,4-dimethyl-3-hexene

55. a. 2-butyne
b. 4,4-dimethyl-2-hexyne
c. 3-isopropyl-1-hexyne
d. 3,6-dimethyl-4-nonyne

57. a. $\mathrm{CH_3{-}CH_2{-}CH{-}C{\equiv}C{-}CH_2{-}CH_2{-}CH_3}$
b.
$$\begin{array}{lcccccccccccc} \mathrm{CH_3{-}CH_2{-}} & \mathrm{CH} & & & & & & & & & & \\ & \| & & & & & & & & & & \\ & \mathrm{CH} & - & \mathrm{CH_2} & - & \mathrm{CH_2} & - & \mathrm{CH_2} & - & \mathrm{CH_2} & - & \mathrm{CH_3} \end{array}$$

c.
$$\begin{array}{ccccccc} & & \mathrm{CH_3} & & & & \\ & & | & & & & \\ \mathrm{CH{\equiv}C} & - & \mathrm{CH_2} & - & \mathrm{CH_2} & - & \mathrm{CH_3} \\ & & | & & & & \\ & & \mathrm{CH_3} & & & & \end{array}$$

d.
$$\begin{array}{ccccccccccccc} & & & & \mathrm{CH_3} & & & & & & \mathrm{CH_3} & & \\ & & & & | & & & & & & | & & \\ \mathrm{CH_3} & - & \mathrm{CH} & = & \mathrm{C} & - & \mathrm{CH_2} & - & \mathrm{CH} & - & \mathrm{CH} & - & \mathrm{CH_3} \\ & & & & & & & & | & & & & \\ & & & & & & & & \mathrm{CH_2{-}CH_3} & & & & \end{array}$$

59. a.
$$\begin{array}{ccccccc} \mathrm{CH_3} & - & \mathrm{CH} & - & \mathrm{CH} & - & \mathrm{CH_3} \\ & & | & & | & & \\ & & \mathrm{Cl} & & \mathrm{Cl} & & \end{array}$$

b.
$$\begin{array}{cccccccccc} \mathrm{CH_3} & - & \mathrm{CH} & - & \mathrm{CH_2} & - & \mathrm{CH} & - & \mathrm{CH_3} & + \\ & & | & & & & | & & & \\ & & \mathrm{CH_3} & & & & \mathrm{Br} & & & \end{array}$$

$$\begin{array}{ccccccccc} \mathrm{CH_3} & - & \mathrm{CH} & - & \mathrm{CH} & - & \mathrm{CH_2} & - & \mathrm{CH_3} \\ & & | & & | & & & & \\ & & \mathrm{CH_3} & & \mathrm{Br} & & & & \end{array}$$

c.
$$\begin{array}{ccccccccc} \mathrm{CH_3} & - & \mathrm{CH_2} & - & \mathrm{CH} & - & \mathrm{CH} & - & \mathrm{CH_3} \\ & & & & | & & | & & \\ & & & & \mathrm{Br} & & \mathrm{Br} & & \end{array}$$

d.
$$\begin{array}{ccccccccc} & & & & & & \mathrm{CH_3} & & \\ & & & & & & | & & \\ \mathrm{CH_3} & - & \mathrm{CH} & - & \mathrm{CH_2} & - & \mathrm{C} & - & \mathrm{CH_3} \\ & & | & & & & | & & \\ & & \mathrm{CH_3} & & & & \mathrm{Cl} & & \end{array}$$

61. a. $\mathrm{CH_2{=}CH{-}CH_3} + \mathrm{H_2} \longrightarrow \mathrm{CH_3{-}CH_2{-}CH_3}$
b.
$$\begin{array}{ccccccccc} \mathrm{CH_3} & - & \mathrm{CH} & - & \mathrm{CH} & = & \mathrm{CH_2} & + \; \mathrm{H_2} & \longrightarrow \\ & & | & & & & & & \\ & & \mathrm{CH_3} & & & & & & \end{array}$$

$$\begin{array}{ccccccc} \mathrm{CH_3} & - & \mathrm{CH} & - & \mathrm{CH_2} & - & \mathrm{CH_3} \\ & & | & & & & \\ & & \mathrm{CH_3} & & & & \end{array}$$

c.
$$\begin{array}{ccccccccc} \mathrm{CH_3} & - & \mathrm{CH} & - & \mathrm{C} & = & \mathrm{CH_2} & + \; \mathrm{H_2} & \longrightarrow \\ & & | & & | & & & & \\ & & \mathrm{CH_3} & & \mathrm{CH_3} & & & & \end{array}$$

$$\begin{array}{ccccccc} \mathrm{CH_3} & - & \mathrm{CH} & - & \mathrm{CH} & - & \mathrm{CH_3} \\ & & | & & | & & \\ & & \mathrm{CH_3} & & \mathrm{CH_3} & & \end{array}$$

63. a. methylbenzene or toluene
b. bromobenzene
c. chlorobenzene

65. a. 3,5-dimethyl-7-phenylnonane
b. 2-phenyl-3-octene
c. 4,5-dimethyl-6-phenyl-2-octyne

67. a. 1,4-dibromobenzene or *p*-dibromobenzene
b. 1,3-diethylbenzene or *m*-diethylbenzene
c. 1-chloro-2-fluorobenzene or *o*-chlorofluorobenzene

69. a. 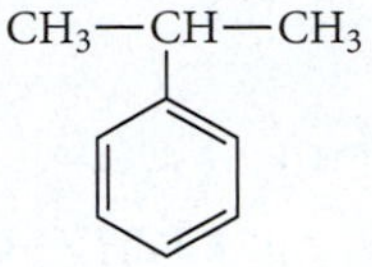 $CH_3—CH—CH_3$ (with benzene ring on central CH)

b. 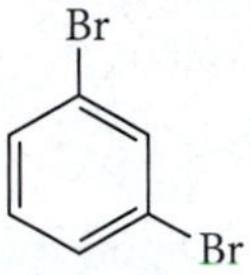 Br, Br (1,3-dibromobenzene)

c.  Cl, CH_3 (4-chlorotoluene)

71. a. 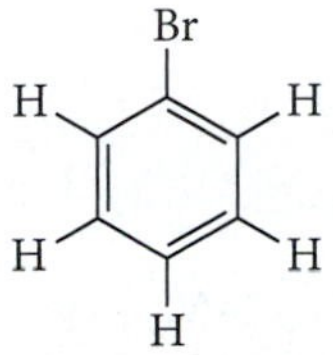 Br, H, H, H, H, H (bromobenzene) + HBr

b. 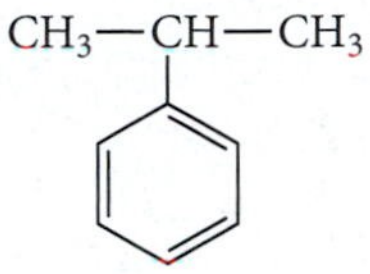 $CH_3—CH—CH_3$ (with benzene ring on central CH) + HCl

73. a. 1-propanol
b. 4-methyl-2-hexanol
c. 2,6-dimethyl-4-heptanol
d. 3-methyl-3-pentanol

75. a. $CH_3CH_2CH_2Br + H_2O$

b. 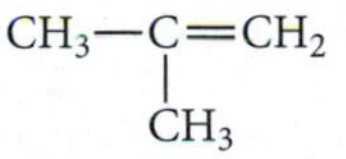 $CH_3—C(CH_3){=}CH_2 + H_2O$

c. $CH_3CH_2ONa + \frac{1}{2}H_2$

d. 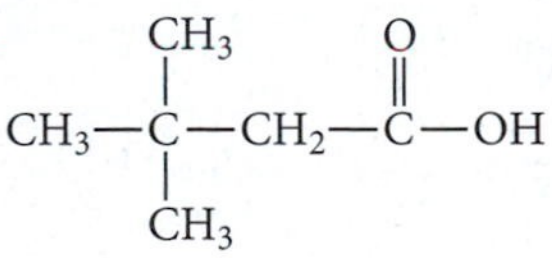 $CH_3—C(CH_3)_2—CH_2—C(=O)—OH$

77. a. butanone **b.** pentanal
c. 3,5,5-trimethylhexanal **d.** 4-methyl-2-hexanone

79. 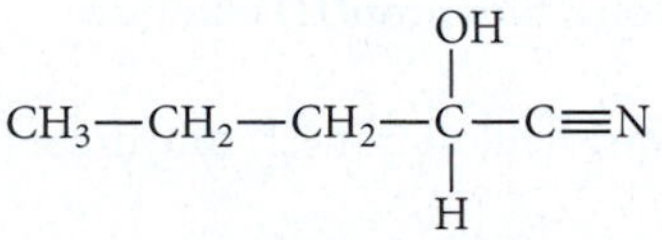 $CH_3—CH_2—CH_2—C(OH)(H)—C{\equiv}N$

81. a. methylbutanoate **b.** propanoic acid
c. 5-methylhexanoic acid **d.** ethylpentanoate

83. a. 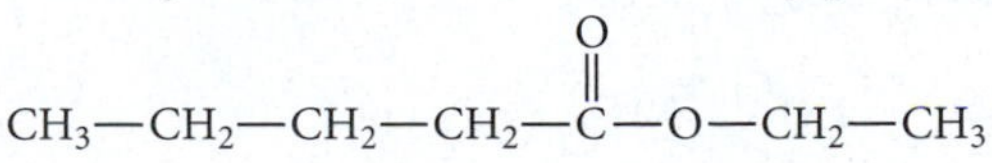 $CH_3—CH_2—CH_2—CH_2—C(=O)—O—CH_2—CH_3 + H_2O$

b. 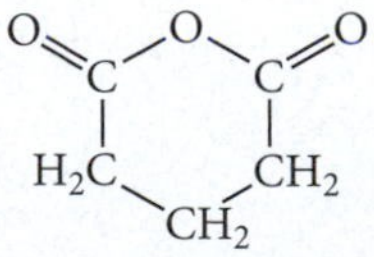 O=C–O–C=O ring with H_2C, CH_2, CH_2 (cyclic anhydride) + H_2O

85. a. ethyl propyl ether **b.** ethyl pentyl ether
c. dipropyl ether **d.** butyl ethyl ether

87. a. diethylamine **b.** methylpropylamine
c. butylmethylpropylamine

89. a. acid–base, $(CH_3)_2NH_2^+(aq) + Cl^-(aq)$
b. condensation, $CH_3CH_2CONHCH_2CH_3(aq) + H_2O$
c. acid–base, $CH_3NH_3^+(aq) + HSO_4^-(aq)$

91. 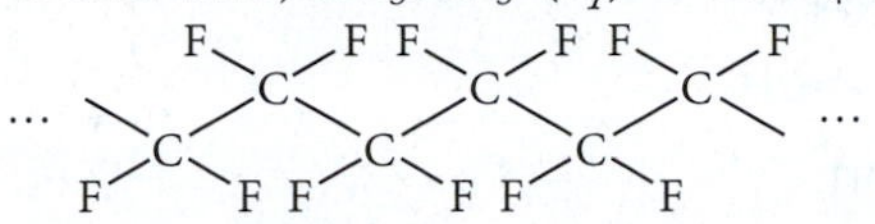 $\cdots —CF_2—CF_2—CF_2—CF_2—CF_2—CF_2— \cdots$

93. 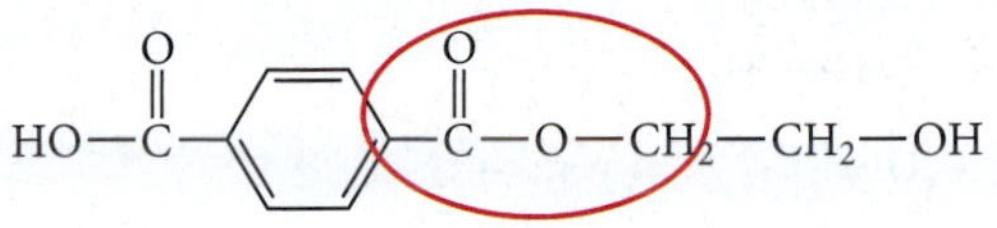 $HO—C(=O)—C_6H_4—C(=O)—O—CH_2—CH_2—OH$

95. a. ester, methyl 3-methylbutanoate
b. ether, ethyl 2-methylbutyl ether
c. aromatic, 1-ethyl-3-methylbenzene or *m*-ethylmethylbenzene
d. alkyne, 5-ethyl-4-methyl-2-heptyne
e. aldehyde, butanal
f. alcohol, 2-methyl-1-propanol

97. a. 5-isobutyl-3-methylnonane
b. 5-methyl-3-hexanone
c. 3-methyl-2-butanol
d. 4-ethyl-3,5-dimethyl-1-hexyne

99. a. isomers **b.** isomers **c.** same

101. 558 g

103. a. combustion **b.** alkane substitution
c. alcohol elimination **d.** aromatic substitution

105. a. 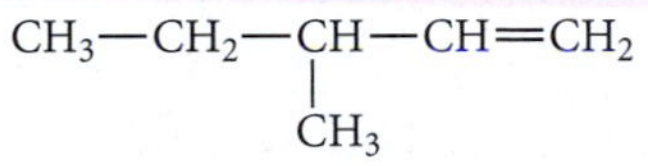 $CH_3—CH_2—CH(CH_3)—CH{=}CH_2$
Can exist as a stereoisomer

b. 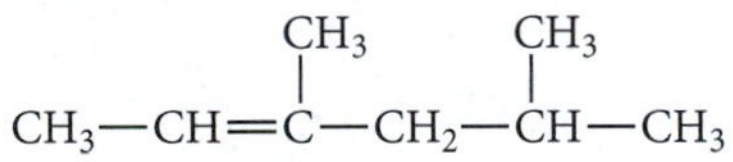 $CH_3—CH{=}C(CH_3)—CH_2—CH(CH_3)—CH_3$
Can exist as a stereoisomer

c. 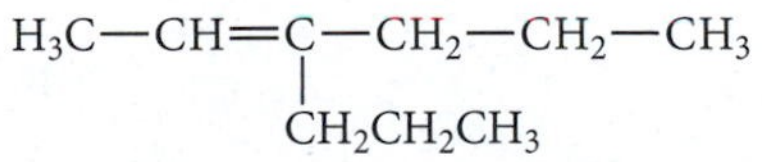 $H_3C—CH{=}C(CH_2CH_2CH_3)—CH_2—CH_2—CH_3$
Can exist as a stereoisomer

107.
a. 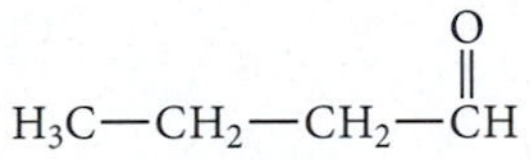 $H_3C—CH_2—CH_2—CH(=O)$
Aldehyde

b. 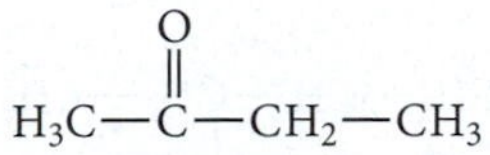 $H_3C—C(=O)—CH_2—CH_3$
Ketone

c. $H_3C—CH{=}CH—O—CH_3$
Alkene, ether

d. $H_2C{=}CH—O—CH_2—CH_3$
Alkene, ether

e. $H_2C{=}CH—CH_2—O—CH_3$
Alkene, ether

f. $H_3C—CH{=}CH—CH_2—OH$
Alkene, alcohol

g. 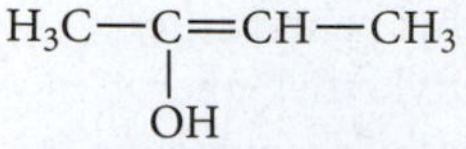 $H_3C-C(OH)=CH-CH_3$

Alkene, alcohol

h. 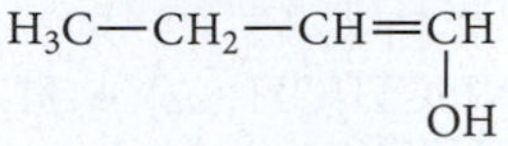 $H_3C-CH_2-CH=CH(OH)$

Alkene, alcohol

i. 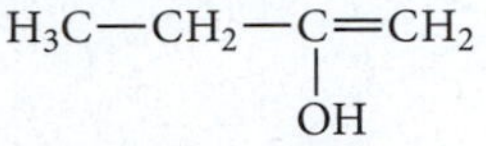 $H_3C-CH_2-C(OH)=CH_2$

Alkene, alcohol

j. 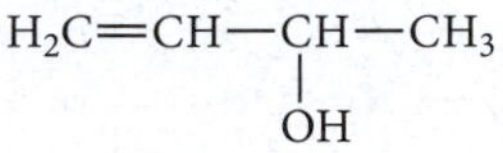 $H_2C=CH-CH(OH)-CH_3$

Alkene, alcohol

k. $H_2C=CH-CH_2-CH_2-OH$

Alkene, alcohol

109. In the acid form of the carboxylic acid, electron withdrawal by the C=O enhances acidity. The conjugate base, the carboxylate anion, is stabilized by resonance so the two O atoms are equivalent and bear the negative charge equally.

111. a. 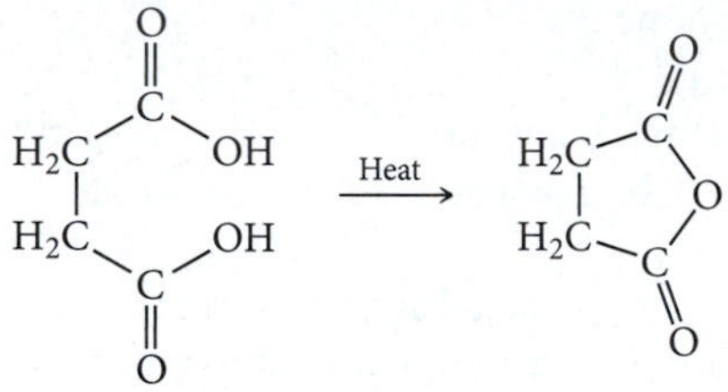 $H_2C(COOH)-H_2C(COOH)$ $\xrightarrow{\text{Heat}}$ cyclic anhydride ($H_2C-C(=O)-O-C(=O)-CH_2$ ring)

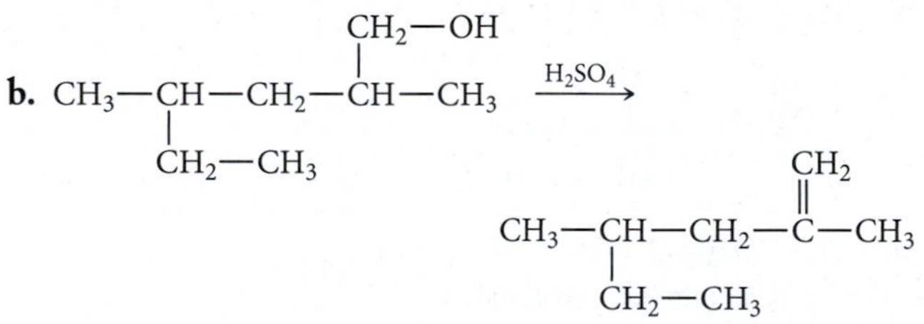

b. $CH_3-CH(CH_2-CH_3)-CH_2-CH(CH_2-OH)-CH_3$ $\xrightarrow{H_2SO_4}$ $CH_3-CH(CH_2-CH_3)-CH_2-C(=CH_2)-CH_3$

c. $CH_3-CH_2-C(CH_3)=CH_2 + HBr \longrightarrow CH_3-CH_2-CBr(CH_3)-CH_3$

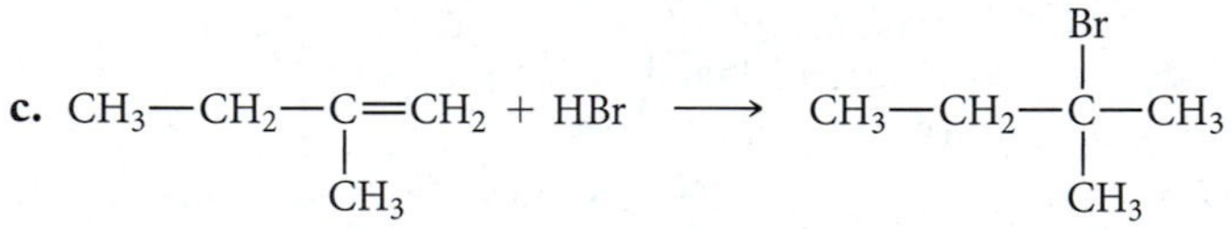

113. a. 3:1

b. 2° hydrogen atoms are more reactive. The reactivity of 2° hydrogens to 1° hydrogens is 11:3.

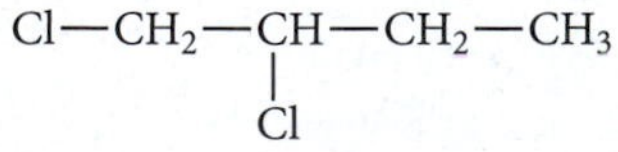

115. 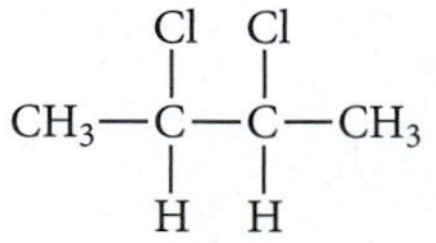 $CH_3-CHCl-CHCl-CH_3$ Chiral

$Cl-CH_2-CHCl-CH_2-CH_3$ Chiral

$Cl-CH_2-CH_2-CHCl-CH_3$ Chiral

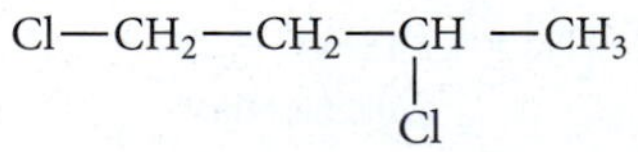

117. The first propagation step for F is very rapid and exothermic because of the strength of the H—F bond that forms. For I the first propagation step is endothermic and slow because the H—I bond that forms is relatively weak.

119. $Cl-C\equiv C-Cl$ trans $ClHC=CHCl$ $ClCH_2-CH_2Cl$

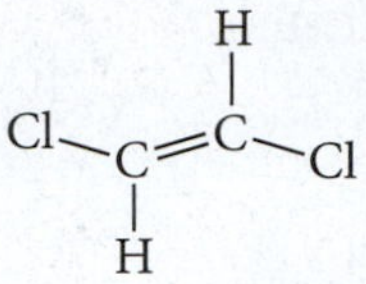

No dipole moment

cis $ClHC=CHCl$ Cl_2CHCH_3 Dipole moment

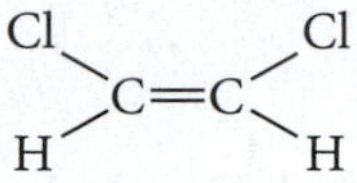

121. 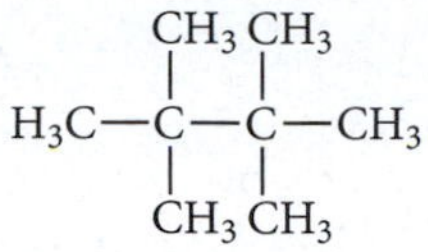 $H_3C-C(CH_3)_2-C(CH_3)_2-CH_3$

2,2,3,3-tetramethylbutane

Chapter 23

17. a. [Ar] $4s^23d^8$, [Ar] $3d^8$
b. [Ar] $4s^23d^5$, [Ar] $3d^3$
c. [Kr] $5s^24d^1$, [Kr] $5s^14d^1$
d. [Xe] $6s^24f^{14}5d^3$, [Xe] $4f^{14}5d^3$

19. a. +5 **b.** +7
c. +4

21. a. +3, 6 **b.** +2, 6
c. +2, 4 **d.** +1, 2

23. a. hexaaquachromium(III)
b. tetracyanocuprate(II)
c. pentaaminebromoiron(III) sulfate
d. amminetetraaquahydroxycobalt(III) chloride

25. a. $[Cr(NH_3)_6]^{3+}$
b. $K_3[Fe(CN)_6]$
c. $[Cu(en)(SCN)_2]$
d. $[Pt(H_2O)_4][PtCl_6]$

27. a. $[Co(NH_3)_3(CN)_3]$, triamminetricyanocobalt(III)
b. $[Cr(en)_3]^{3+}$, tris(ethylenediamine)chromium(III)

29. 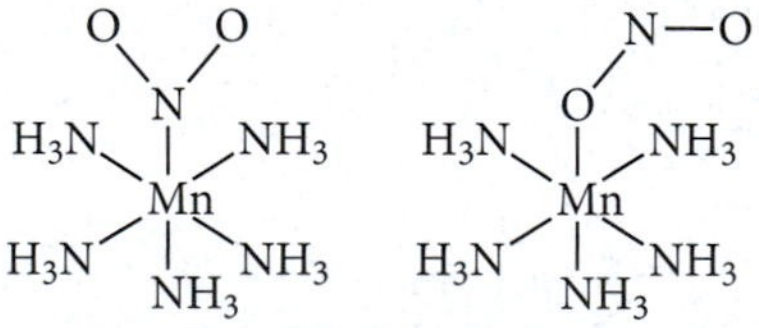 $[Mn(NH_3)_5(NO_2)]$ (N-bonded nitro) $[Mn(NH_3)_5(ONO)]$ (O-bonded nitrito)

31. $[Fe(H_2O)_5Cl]Cl \cdot H_2O$, pentaaquachloroiron(II) chloride monohydrate
$[Fe(H_2O)_4Cl_2] \cdot 2\ H_2O$, tetraaquadichloroiron(II) dihydrate

33. **b, c, e**

35. a. 3 **b.** No geometric isomers.

37. a. 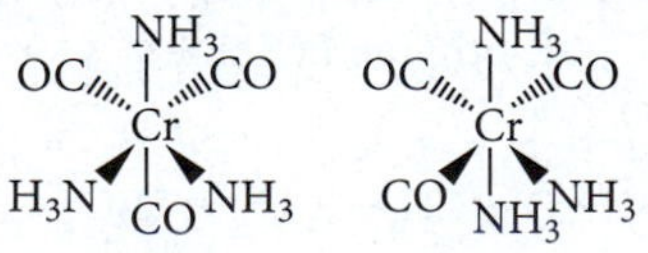 $[Cr(NH_3)_3(CO)_3]$ Fac   Mer

b. 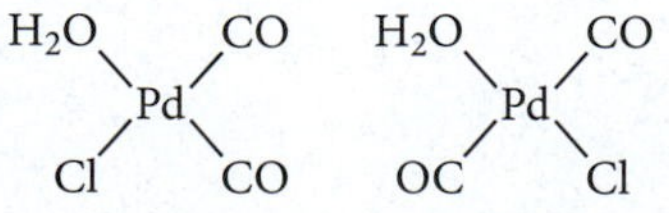 $[Pd(H_2O)(CO)_2Cl]$ Cis   Trans

39. *cis* isomer is optically active

41. a.

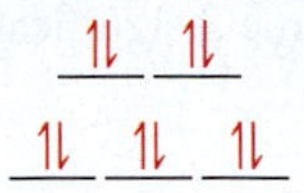

b.

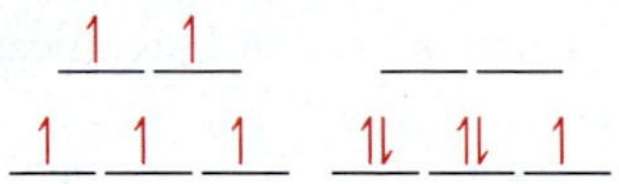

c.

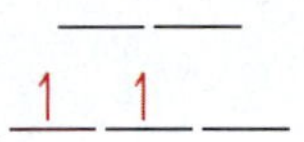

d.

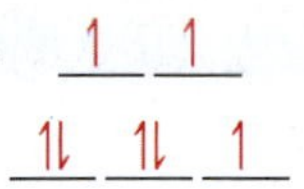

43. 163 kJ/mol

45. $[Co(CN)_6]^{3-}$ ⟶ 290 nm, colorless
$[Co(NH_3)_6]^{3+}$ ⟶ 440 nm, yellow
$[CoF_6]^{3-}$ ⟶ 770 nm, green

47. weak

49. a. 4 **b.** 3 **c.** 1

51. 3

53. porphyrin

55. Water is a weak field ligand that forms a high-spin complex with hemoglobin. Because deoxyhemoglobin is a weak field, it absorbs large wavelength light and appears blue. Oxyhemoglobin is a low-spin complex and absorbs small wavelength light, so O_2 must be a strong field ligand.

57. a. [Ar] $4s^1 3d^5$, [Ar] $3d^5$, [Ar] $3d^4$, [Ar] $3d^3$
b. [Ar] $4s^1 3d^{10}$, [Ar] $3d^{10}$, [Ar] $3d^9$

59. a.

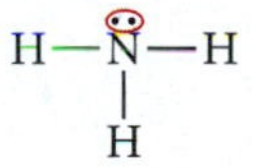

b.

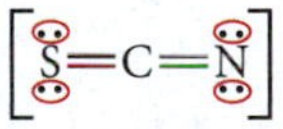

c.

61. $[MA_2B_2C_2]$ all cis; A trans and B and C cis; B trans and A and C cis; C trans and A and B cis; all trans.
$[MA_2B_3C]$ will have fac–mer isomers.
$[MAB_2C_3]$ will have fac–mer isomers.
$[MAB_3C_2]$ will have fac–mer isomers.
$[MA_3B_2C]$ will have fac–mer isomers.
$[MA_2BC_3]$ will have fac–mer isomers.
$[MA_3BC_2]$ will have fac–mer isomers.
$[MABC_2]$ will have AB cis–trans isomers.
$[MAB_4C]$ will have AC cis–trans isomers.
$[MA_4BC]$ will have BC cis–trans isomers.
$[MABC_4]$ will have AB cis–trans isomers.

63.

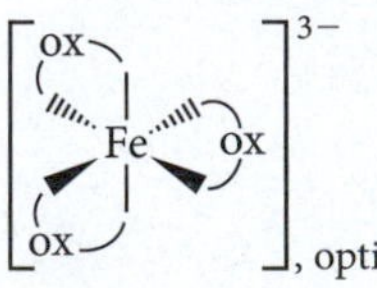

, optical isomers

65.

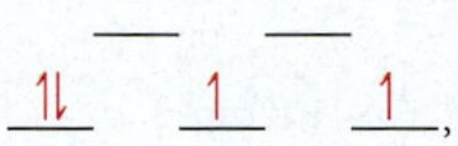

, paramagnetic

67. 1.

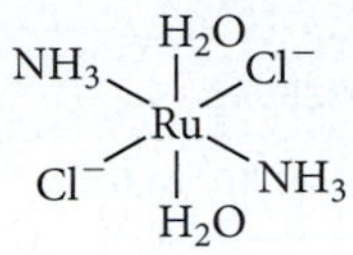

2.

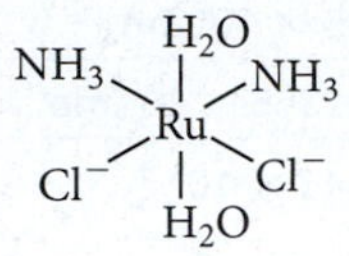

3.

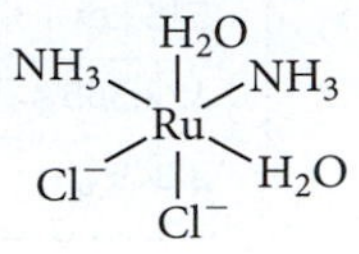

4.

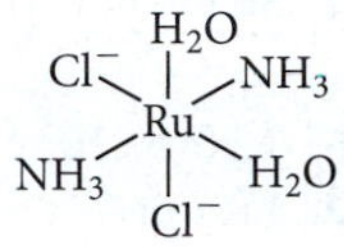

5.

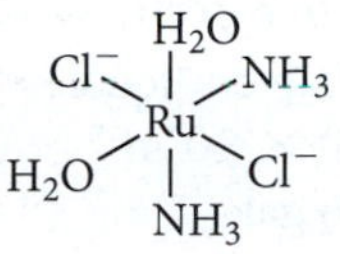

Only structure 3 is chiral. This is its mirror image.

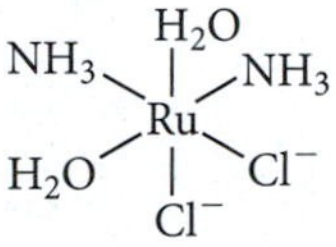

69.

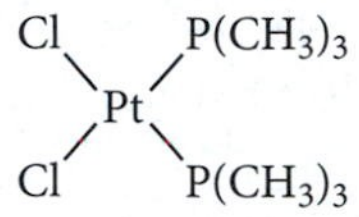

cis-dichlorobis (trimethyl phosphine) platinum(II)

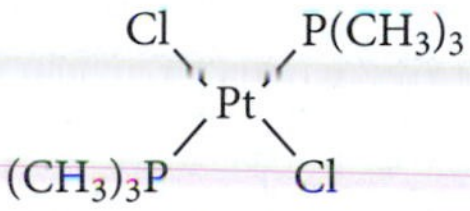

trans-dichlorobis (trimethyl phosphine) platinum(II)

71.

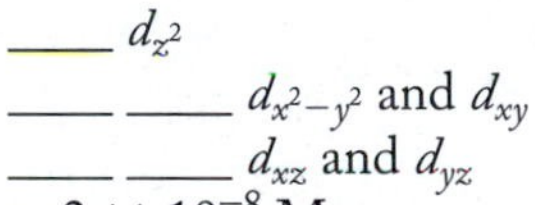

73. a. 2×10^{-8} M
b. 6.6×10^{-3} M
c. NiS will dissolve more easily in the ammonia solution because the formation of the complex ion is favorable, removing Ni^{2+} ions from the solution allowing more NiS to dissolve.

75. Prepare a solution that contains both $[MCl_6]^{3-}$ and $[MBr_6]^{3-}$ and see if any complex ions that contain both Cl and Br form. If they do, it would demonstrate that these complexes are labile.

77. pH = 10.1

79. Au

Appendix I

1. a. 0 °C **b.** −321 °
c. −78.3 ° **d.** 310.2 K

3. −62.2 °C, 210.9 K

5. a. 1.2 nm **b.** 22 fs
c. 1.5 Gg **d.** 3.5 ML

7. a. 4.5×10^{-9} s **b.** 1.8×10^{-14} s
c. 1.28×10^{-10} m **d.** 3.5×10^{-5} m

9.

1245 kg	1.245×10^6 g	1.245×10^9 mg
515 km	5.15×10^6 dm	5.15×10^7 cm
122.355 s	1.22355×10^5 ms	0.122355 ks
3.345 kJ	3.345×10^3 J	3.345×10^6 mJ

11. **a.** 254.998 km **b.** 2.54998×10^{-1} Mm
c. 254998×10^3 mm **d.** 254998×10^2 cm

13. 10,000 1-cm squares

15. **a.** 1.50×10^3 mL **b.** 1.50×10^3 cm^3
c. 3.96 gal **d.** 15.9 qt

Appendix II

1. **a.** 1,050,501 **b.** 0.0020
c. 0.000000000000002 **d.** 0.001090

3. **a.** 3
b. ambiguous, without more information assume 3 significant figures
c. 3
d. 5
e. ambiguous, without more information assume 1 significant figure

5. **a.** not exact **b.** exact
c. not exact **d.** exact

7. **a.** 156.9 **b.** 156.8
c. 156.8 **d.** 156.9

9. **a.** 1.84 **b.** 0.033
c. 0.500 **d.** 34

11. **a.** 41.4 **b.** 133.5
c. 73.0 **d.** 0.42

13. **a.** 391.3 **b.** 1.1×10^4
c. 5.96 **d.** 5.93×10^4

APPENDIX VI

Answers to In Chapter Practice Problems

Chapter 1

1.1. For the first sample:

$$\frac{\text{mass of oxygen}}{\text{mass of carbon}} = \frac{17.2 \text{ g O}}{12.9 \text{ g C}} = 1.33 \text{ or } 1.33{:}1$$

For the second sample:

$$\frac{\text{mass of oxygen}}{\text{mass of carbon}} = \frac{10.5 \text{ g O}}{7.88 \text{ g C}} = 1.33 \text{ or } 1.33{:}1$$

The ratios of oxygen to carbon are the same in the two samples of carbon monoxide, so these results are consistent with the law of definite proportions.

1.2. $$\frac{\text{mass of hydrogen to 1 g of oxygen in water}}{\text{mass of hydrogen to 1 g of oxygen in hydrogen peroxide}} = \frac{0.125}{0.0625} = 2.00$$

The ratio of the mass of hydrogen from one compound to the mass of hydrogen in the other is equal to 2. This is a simple whole number and therefore consistent with the law of multiple proportions.

1.3. **a.** $Z = 6, A = 13, {}^{13}_{6}\text{C}$

b. 19 protons, 20 neutrons

1.4. 24.31 amu

1.4. **For More Practice**

70.92 amu

Chapter 2

2.1. The thermometer shown has markings every 1 °F; thus, the first digit of uncertainty is 0.1. The answer is 103.4 °F.

2.2. 21.4 g/cm^3 This matches the density of platinum.

2.2. **For More Practice**

4.50 g/cm^3 The metal is titanium.

2.3. 3.15 yd

2.4. 2.445 gal

2.5. 1.61×10^6 cm^3

2.5. **For More Practice**

3.23×10^3 kg

2.6. 1.03 kg

2.6. **For More Practice**

2.9×10^{-2} cm^3

2.7. 0.855 cm

2.8. 2.70 g/cm^3

2.9. 4.65×10^{-2} mol Ag

2.10. 0.563 mol Cu

2.10. For More Practice

22.6 g Ti

2.11. 1.3×10^{22} C atoms

2.11. For More Practice

6.87 g W

2.12. $l = 1.72$ cm

2.12. For More Practice

2.90×10^{24} Cu atoms

Chapter 3

3.1. $5.83 \times 10^{14}\ s^{-1}$

3.2. 2.64×10^{20} photons

3.2. For More Practice

435 nm

3.3. **a.** blue < green < red

b. red < green < blue

c. red < green < blue

3.4. 6.1×10^6 m/s

3.5. For the 5*d* orbitals:

$n = 5$

$l = 2$

$m_l = -2, -1, 0, 1, 2$

The five integer values for m_l signify that there are five 5*d* orbitals.

3.6. **a.** l cannot equal 3 if $n = 3$. $l = 2$

b. m_l cannot equal -2 if $l = -1$. Possible values for $m_l = -1, 0,$ or 1

c. l cannot be 1 if $n = 1$. $l = 0$

3.7. 397 nm

3.7. For More Practice

$n = 1$

Chapter 4

4.1. **a.** Cl $1s^22s^22p^63s^23p^5$ or [Ne] $3s^23p^5$

b. Si $1s^22s^22p^63s^23p^2$ or [Ne] $3s^23p^2$

c. Sr $1s^22s^22p^63s^23p^64s^23d^{10}4p^65s^2$ or [Kr] $5s^2$

d. O $1s^22s^22p^4$ or [He] $2s^22p^4$

4.2. There are no unpaired electrons.

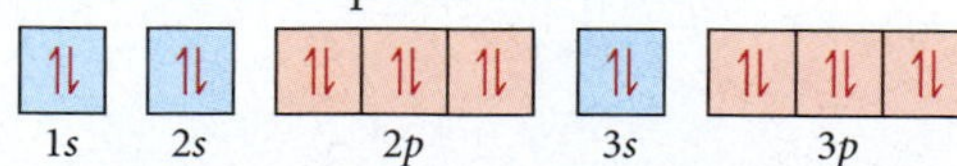

4.3. $1s^22s^22p^63s^23p^3$ or [Ne] $3s^23p^3$. The five electrons in the $3s^23p^3$ orbitals are the valence electrons, while the 10 electrons in the $1s^22s^22p^6$ orbitals belong to the core.

4.4. Bi [Xe] $6s^2 4f^{14} 5d^{10} 6p^3$

4.4. For More Practice

I [Kr] $5s^2 4d^{10} 5p^5$

4.5. **a.** N^{3-}

b. Rb^+

4.6. **a.** Sn

b. cannot predict

c. W

d. Se

4.6. For More Practice

Rb > Ca > Si > S > F

4.7. **a.** $[Ar]4s^0 3d^7$. Co^{2+} is paramagnetic.

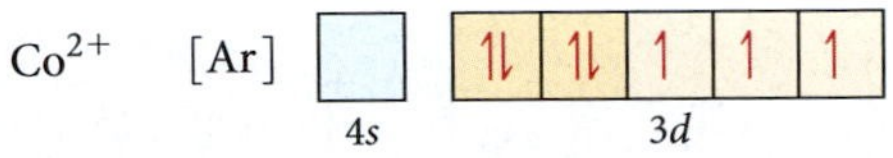

Co^{2+} [Ar] 4s: [ ] 3d: [⇅][⇅][↑][↑][↑]

b. $[He]2s^2 2p^6$. N^{3-} is diamagnetic.

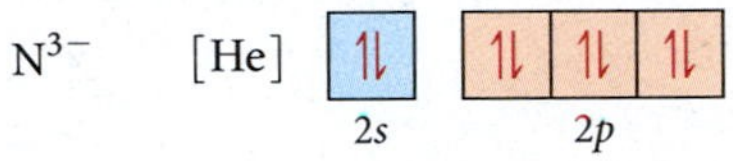

N^{3-} [He] 2s: [⇅] 2p: [⇅][⇅][⇅]

c. $[Ne]3s^2 3p^6$. Ca^{2+} is diamagnetic.

Ca^{2+} [Ne] 3s: [⇅] 3p: [⇅][⇅][⇅]

4.8. **a.** K

b. F^-

c. Cl^-

4.8. For More Practice

$Cl^- > Ar > Ca^{2+}$

4.9. **a.** I

b. Ca

c. cannot predict

d. F

4.9. For More Practice

F > S > Si > Ca > Rb

4.10. **a.** Sn

b. cannot predict based on simple trends (Po is larger)

c. Bi

d. B

4.10. For More Practice

Cl < Si < Na < Rb

Chapter 5

5.1. **a.** C_5H_{12}

b. HgCl

c. CH_2O

5.2. Mg_3N_2

5.3. K_2S

5.4. AlN

5.5. silver nitride

5.5. For More Practice

Rb_2S

5.6. iron(II) sulfide

5.6. For More Practice

RuO_2

5.7. tin(II) chlorate

5.7. For More Practice

$Co_3(PO_4)_2$

5.8. dinitrogen pentoxide

5.8. For More Practice

PBr_3

5.9. 164.10 amu

5.10. 5.839×10^{20} $C_{13}H_{18}O_2$ molecules

5.10. For More Practice

1.06 g H_2O

5.11. 53.29%

5.11. For More Practice

74.19% Na

5.12. 83.9 g Fe_2O_3

5.12. For More Practice

8.6 g Na

5.13. 4.0 g O

5.13. For More Practice

3.60 g C

5.14. CH_2O

5.15. $C_{13}H_{18}O_2$

5.16. C_6H_6

5.16. For More Practice

$C_2H_8N_2$

5.17. C_2H_5

5.18. C_2H_4O

Chapter 6

6.1. **a.** pure covalent

b. ionic

c. polar covalent

6.2. $:C{\equiv}O:$

6.3. H—C(=O)—H

6.4. $[:\ddot{Cl}:\ddot{O}:]^-$

6.5. $[:O=N-O:]^- \longleftrightarrow [:O-N=O:]^-$

6.6.

Structure	A			B			C		
	:N=N=O:			:N≡N—O:			:N—N≡O:		
Number of valence e⁻	5	5	6	5	5	6	5	5	6
Number of nonbonding e⁻	4	0	4	2	0	6	6	0	2
1/2(number of bonding e⁻)	2	4	2	3	4	1	1	4	3
Formal charge	**−1**	**+1**	**0**	**0**	**+1**	**−1**	**−2**	**+1**	**+1**

Structure B contributes the most to the correct overall structure of N_2O.

6.6. For More Practice

The nitrogen is +1, the singly bonded oxygen atoms are −1, and the double-bonded oxygen atom has no formal charge.

6.7. $H_2C^{-1}-N^{+1}\equiv N: \longleftrightarrow H_2C=N^{+1}=N^{-1}:$

6.8. XeF_4: :F—Xe—F: with :F: above and below Xe (Xe with two lone pairs)

6.8. For More Practice

H_3PO_4: H—O—P⁰(=O⁰)(—O—H)—O—H

6.9. tetrahedral

CCl_4: :Cl—C—Cl: with :Cl: above and below C

6.10. bent

6.11. linear

6.12.

Atom	Number of Electron Groups	Number of Lone Pairs	Molecular Geometry
Carbon (left)	4	0	Tetrahedral
Carbon (right)	3	0	Trigonal planar
Oxygen	4	2	Bent

6.13. The molecule is nonpolar.

Chapter 7

7.1. The xenon atom has six electron groups and therefore has an octahedral electron geometry. An octahedral electron geometry corresponds to sp^3d^2 hybridization (refer to Table 7.1).

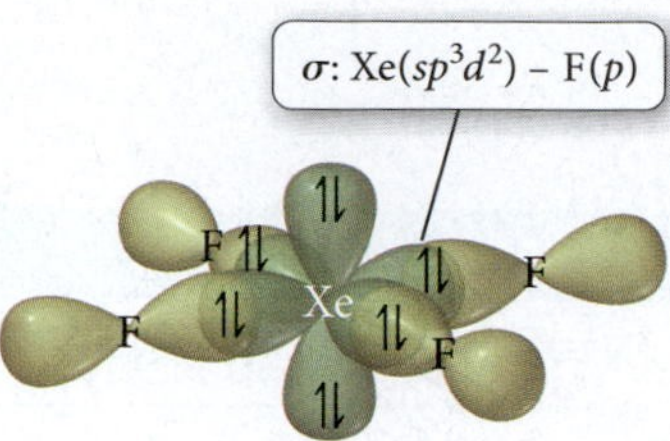

7.2. Since there are only two electron groups around the central atom (C), the electron geometry is linear. According to Table 7.1, the corresponding hybridization on the carbon atom is *sp*.

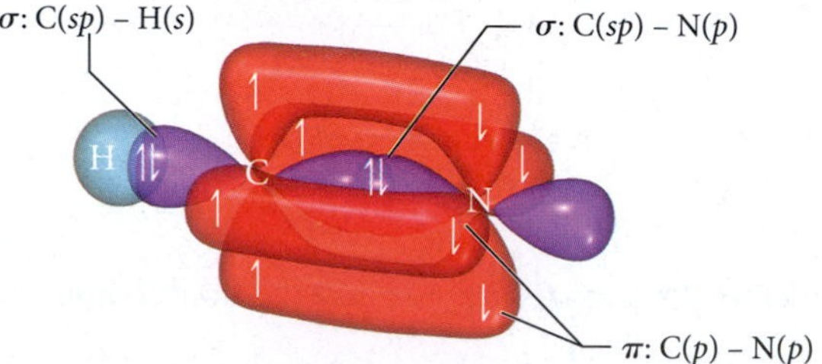

7.3. Since there are only two electron groups about the central atom (C), the electron geometry is linear. The hybridization on C is *sp* (refer to Table 7.1).

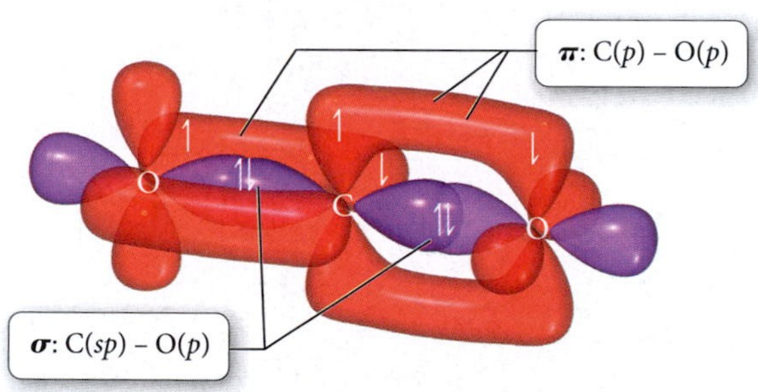

7.3. For More Practice

There are five electron groups about the central atom (I); therefore the electron geometry is trigonal bipyramidal, and the corresponding hybridization of I is sp^3d (refer to Table 7.1).

7.4. H_2^+ bond order $= +\frac{1}{2}$

Since the bond order is positive, the H_2^+ ion should be stable; however, the bond order of H_2^+ is lower than the bond order of H_2 (bond order = 1). Therefore, the bond in H_2^+ is weaker than in H_2.

7.5. The bond order of N_2^+ is 2.5, which is lower than that of the N_2 molecule (bond order = 3); therefore the bond is weaker. The MO diagram shows that the N_2^+ ion has one unpaired electron and is therefore paramagnetic.

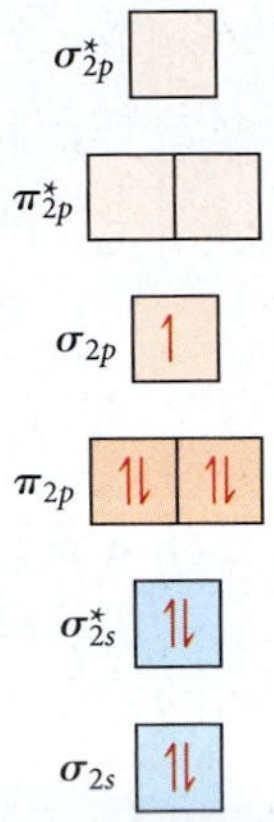

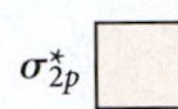

7.5. For More Practice

The bond order of Ne_2 is 0, which indicates that dineon does not exist.

7.6. The bond order of NO is +2.5. The MO diagram shows that the ion has one unpaired electron and is therefore paramagnetic.

Chapter 8

8.1. $SiO_2(s) + 3\ C(s) \longrightarrow SiC(s) + 2\ CO(g)$

8.2. $2\ C_2H_6(g) + 7\ O_2(g) \longrightarrow 4\ CO_2(g) + 6\ H_2O(g)$

8.3. $Pb(NO_3)_2(aq) + 2\ KCl(aq) \longrightarrow PbCl_2(s) + 2\ KNO_3(aq)$

8.4. 4.08 g HCl

8.5. 22 kg HNO_3

8.6. H_2 is the limiting reagent, since it produces the least amount of NH_3. Therefore, 29.4 kg NH_3 is the theoretical yield.

8.7. CO is the limiting reagent, since it only produces 114 g Fe. Therefore, 114 g Fe is the theoretical yield: percentage yield = 63.4% yield.

8.8. $2\ C_2H_5SH(l) + 9\ O_2(g) \longrightarrow 4\ CO_2(g) + 2\ SO_2(g) + 6\ H_2O(g)$

8.9. **a.** $2\ Al(s) + 3\ Cl_2(g) \longrightarrow 2\ AlCl_3(s)$

b. $2\ Li(s) + 2\ H_2O(l) \longrightarrow 2\ Li^+(aq) + 2\ OH^-(aq) + H_2(g)$

c. $H_2(g) + Br_2(l) \longrightarrow 2\ HBr(g)$

Chapter 9

9.1. 0.214 M $NaNO_3$

9.1. For More Practice

44.6 g KBr

9.2. 402 g $C_{12}H_{22}O_{11}$

9.2. For More Practice

221 mL of KCl solution

9.3. 667 mL

9.3. For More Practice

0.105 L

9.4. 51.4 mL HNO_3 solution

9.4. For More Practice

0.170 g CO_2

9.5. **a.** Insoluble.

b. Insoluble.

c. Soluble.

d. Soluble.

9.6. $NH_4Cl(aq) + Fe(NO_3)_3(aq) \longrightarrow$ NO REACTION

9.7. $2\,NaOH(aq) + CuBr_2(aq) \longrightarrow Cu(OH)_2(s) + 2\,NaBr(aq)$

9.8. $2\,H^+(aq) + 2\,I^-(aq) + Ba^{2+}(aq) + 2\,OH^-(aq) \longrightarrow 2\,H_2O(l) + Ba^{2+}(aq) + 2\,I^-(aq)$

$H^+(aq) + OH^-(aq) \longrightarrow H_2O(l)$

9.9. hydrofluoric acid

9.10. nitrous acid

9.10. **For More Practice**

$HClO_4$

9.11. $H_2SO_4(aq) + 2\,LiOH(aq) \longrightarrow 2\,H_2O(l) + Li_2SO_4(aq)$

$H^+(aq) + OH^-(aq) \longrightarrow H_2O(aq)$

9.12. $9.03 \times 10^{-2}\,M\,H_2SO_4$

9.12. **For More Practice**

24.5 mL NaOH solution

9.13. $2\,HBr(aq) + K_2SO_3(aq) \longrightarrow H_2O(l) + SO_2(g) + 2\,KBr(aq)$

9.13. **For More Practice**

$2\,H^+(aq) + S^{2-}(aq) \longrightarrow H_2S(g)$

9.14. **a.** Cr = 0.

b. $Cr^{3+} = +3$.

c. $Cl^- = -1$, C = +4.

d. Br = −1, Sr = +2.

e. O = −2, S = +6.

f. O = −2, N = +5.

9.15. Sn is oxidized and N is reduced.

9.15. **For More Practice**

b. Reaction b is the only redox reaction. Al is oxidized and O is reduced.

9.16. **a.** This is a redox reaction in which Li is the reducing agent (it is oxidized) and Cl_2 is the oxidizing reagent (it is reduced).

b. This is a redox reaction in which Al is the reducing agent and Sn^{2+} is the oxidizing agent.

c. This is not a redox reaction because no oxidation states change.

d. This is a redox reaction in which C is the reducing agent and O_2 is the oxidizing agent.

Chapter 10

10.1. $\Delta E = 71\ J$

10.2. $C_s = 0.38 \dfrac{J}{g \cdot {}^\circ C}$

The specific heat capacity of gold is 0.128 J/g · °C; therefore the rock cannot be pure gold.

10.2. **For More Practice**

$T_f = 42.1\ ^\circ C$

10.3. 37.8 g Cu

10.4. −122 J

10.4. For More Practice

$\Delta E = -998$ J

10.5. $\Delta E_{reaction} = -3.91 \times 10^3$ kJ/mol C_6H_{14}

10.5. For More Practice

$$C_{cal} = 4.55 \frac{kJ}{°C}$$

10.6. **a.** endothermic, positive ΔH.

b. endothermic, positive ΔH.

c. exothermic, negative ΔH.

10.7. -2.06×10^3 kJ

10.7. For More Practice

33 g C_4H_{10}

99 g CO_2

10.8. $\Delta H_{rxn} = -68$ kJ

10.9. $N_2O(g) + NO_2(g) \longrightarrow 3\ NO(g),\ \Delta H_{rxn} = +157.6$ kJ

10.9. For More Practice

$3\ H_2(g) + O_3(g) \longrightarrow 3\ H_2O(g),\ \Delta H_{rxn} = -868.1$ kJ

10.10. $CH_3OH(g) + \frac{3}{2} O_2(g) \longrightarrow CO_2(g) + 2\ H_2O(g)$

$\Delta H_{rxn} = -641$ kJ

10.10. For More Practice

$\Delta H_{rxn} = -8.0 \times 10^1$ kJ

10.11. **a.** $Na(s) + \frac{1}{2} Cl_2(g) \longrightarrow NaCl(s),\ \Delta H_f^\circ = -411.2$ kJ/mol

b. $Pb(s) + N_2(g) + 3\ O_2(g) \longrightarrow Pb(NO_3)_2(s),\ \Delta H_f^\circ = -451.9$ kJ/mol

10.12. $\Delta H^\circ_{rxn} = -851.5$ kJ

10.13. $\Delta H^\circ_{rxn} = -1648.4$ kJ

111 kJ emitted (−111 kJ)

10.14. KI < LiBr < CaO

10.14. For More Practice

$MgCl_2$

Chapter 11

11.1. 15.0 psi

11.1. For More Practice

80.6 kPa

11.2. 2.1 atm at a depth of approximately 11 m.

11.3. 123 mL

11.4. 11.3 L

11.5. 1.63 atm, 23.9 psi

11.6. 16.1 L

11.6. For More Practice

976 mmHg

11.7. $d = 4.91$ g/L

11.7. For More Practice

44.0 g/mol

11.8. 70.7 g/mol

11.9. 0.0610 mol H_2

11.10. 4.2 atm

11.11. 12.0 mg H_2

11.12. $u_{rms} = 238$ m/s

11.13. $\frac{\text{rate}_{H_2}}{\text{rate}_{Kr}} = 6.44$

11.14. 82.3 g Ag_2O

11.14. For More Practice

7.10 g Ag_2O

11.15. 6.53 L O_2

Chapter 12

12.1. b, c

12.2. HF has a higher boiling point than HCl because, unlike HCl, HF is able to form hydrogen bonds. The hydrogen bond is the strongest of the intermolecular forces and requires more energy to break.

12.3. 5.83×10^3 kJ

12.3. For More Practice

49 °C

12.4. 33.8 kJ/mol

12.5. 7.04×10^3 torr

Chapter 13

13.1. No phase transition occurs.

13.2. 29.4°

13.3. 3.24×10^{-23} cm^3

13.4. $7.18 \frac{\text{g}}{\text{cm}^3}$

13.5. Metallic

Chapter 14

14.1. **a.** not soluble

b. soluble

c. not soluble

d. not soluble

14.2. 2.7×10^{-4} M

14.3. 42.5 g $C_{12}H_{22}O_{11}$

14.3. For More Practice

3.3×10^4 L

14.4. **a.** $M = 0.415$ M

b. $m = 0.443$ m

c. % by mass = 13.2%

d. $\chi C_{12}H_{22}O_{11} = 0.00793$

e. mole percent = 0.793%

14.5. 0.600 M

14.5. For More Practice

0.651 m

14.6. 22.5 torr

14.6. For More Practice

0.144

14.7. **a.** $P_{\text{benzene}} = 26.6$ torr

$P_{\text{toluene}} = 20.4$ torr

b. 47.0 torr

c. 52.5% benzene; 47.5% toluene

The vapor will be richer in the more volatile component, which in this case is benzene.

14.8. $T_f = -4.8$ °C

14.9. 101.84 °C

14.10. 11.8 atm

14.11. −0.60 °C

14.12. 0.014 mol NaCl

Chapter 15

15.1. $\frac{\Delta[H_2O_2]}{\Delta t} = -4.40 \times 10^{-3}$ M/s

$\frac{\Delta[I_3^-]}{\Delta t} = 4.40 \times 10^{-3}$ M/s

15.2. **a.** Rate $= k[CHCl_3][Cl_2]^{1/2}$. (Fractional-order reactions are not common but are occasionally observed.)

b. $3.5\ M^{-1/2} \cdot s^{-1}$

15.3. 5.78×10^{-2} M

15.4. 0.0277 M

15.5. 1.64×10^{-3} M

15.6. 79.2 s

15.7. $2.07 \times 10^{-5} \frac{L}{mol \cdot s}$

15.8. $6.13 \times 10^{-4} \frac{L}{mol \cdot s}$

15.9. $2\ A + B \longrightarrow A_2B$

Rate $= k[A]^2$

Chapter 16

16.1. $K = \frac{[CO_2]^3[H_2O]^4}{[C_3H_8][O_2]^5}$

16.2. 2.1×10^{-13}

16.2. For More Practice

1.4×10^2

16.3. 6.2×10^2

16.4. $K_c = \frac{[Cl_2]^2}{[HCl]^4[O_2]}$

16.5. 9.4

16.6. 1.1×10^{-6}

16.7. $Q_c = 0.0196$

Reaction proceeds to the left.

16.8. 0.033 M

16.9. $[N_2] = 4.45 \times 10^{-3}$ M

$[O_2] = 4.45 \times 10^{-3}$ M

$[NO] = 1.1 \times 10^{-3}$ M

16.10. $[N_2O_4] = 0.005$ M

$[NO_2] = 0.041$ M

16.11. $P_{I_2} = 0.0027$ atm

$P_{CI_2} = 0.0027$ atm

$P_{ICI_2} = 0.246$ atm

16.12. 1.67×10^{-7} M

16.13. 6.78×10^{-6} M

16.14. Adding Br_2 increases the concentration of Br_2, causing a shift to the left (away from the Br_2). Adding BrNO increases the concentration of BrNO, causing a shift to the right.

16.15. Decreasing the volume causes the reaction to shift right. Increasing the volume causes the reaction to shift left.

16.16. If we increase the temperature, the reaction shifts to the left. If we decrease the temperature, the reaction shifts to the right.

Chapter 17

17.1. a. H_2O donates a proton to C_5H_5N, making it the acid. The conjugate base is therefore OH^-. Since C_5H_5N accepts the proton, it is the base and becomes the conjugate acid $C_5H_5NH^+$.

b. Since HNO_3 donates a proton to H_2O it is the acid, making NO_3^- the conjugate base. Since H_2O is the proton acceptor, it is the base and becomes the conjugate acid, H_3O^+.

17.2. a. $[H_3O^+] = 6.7 \times 10^{-13}$ M

Since $[H_3O^+] < [OH^-]$ the solution is basic.

b. $[H_3O^+] = 1.0 \times 10^{-7}$ M

Neutral solution.

c. $[H_3O^+] = 1.2 \times 10^{-5}$ M

Since $[H_3O^+] > [OH^-]$ the solution is acidic.

17.3. **a.** 8.02 (basic)
b. 11.85 (basic)

17.4. 4.3×10^{-9} M

17.5. 9.4×10^{-3} M

17.6. 3.28

17.7. 2.72

17.8. 1.8×10^{-6}

17.9. 0.85%

17.10. 4.0×10^{-7} M

17.11. $[OH^-] = 0.020$ M
pH = 12.30

17.12. $[OH^-] = 1.2 \times 10^{-2}$ M
pH = 12.08

17.13. **a.** weak base
b. pH-neutral

17.14. 9.07

17.15. **a.** pH-neutral
b. weak acid
c. weak acid

17.16. **a.** basic
b. acidic
c. pH-neutral
d. acidic

17.17. 3.83

17.18. $[SO_4^{2-}] = 0.00386$ M
pH = 1.945

17.19. 5.6×10^{-11} M

Chapter 18

18.1. 4.44

18.1. **For More Practice**
3.44

18.2. 9.14

18.3. 4.87

18.3. **For More Practice**
4.65

18.4. 9.68

18.4. **For More Practice**
9.56

18.5. hypochlorous acid (HClO); 2.4 g NaClO

18.6. 1.74

18.7. 8.08

18.8. 2.30×10^{-6} M

18.9. 5.3×10^{-13}

18.10. 1.21×10^{-5} M

18.11. $FeCO_3$ will be more soluble in an acidic solution than $PbBr_2$ because the CO_3^{2-} ion is a basic anion, whereas Br^- is the conjugate base of a strong acid (HBr) and is therefore pH-neutral.

18.12. $Q > K_{sp}$; therefore, a precipitate forms.

18.13. 2.9×10^{-6} M

18.14. **a.** AgCl precipitates first; [NaCl] $= 7.1 \times 10^{-9}$ M

b. $[Ag^+]$ is 1.5×10^{-8} M when $PbCl_2$ begins to precipitate, and $[Pb^{2+}]$ is 0.085 M.

18.15. 9.6×10^{-6} M

Chapter 19

19.1. **a.** positive

b. negative

c. positive

19.2. -153.2 J/K

19.3. **a.** -548 J/K

b. ΔS_{sys} is negative.

c. ΔS_{univ} is negative, and the reaction is not spontaneous.

19.3. For More Practice

375 K

19.4. $\Delta G = -101.6 \times 10^3$ J

Therefore the reaction is spontaneous. Since both ΔH and ΔS are negative, as the temperature increases ΔG will become more positive.

19.5. $\Delta G^\circ_{rxn} = -36.3$ kJ

Since ΔG°_{rxn} is negative, the reaction is spontaneous at this temperature.

19.6. $\Delta G^\circ_{rxn} = -42.1$ kJ

Since the value of ΔG°_{rxn} at the lowered temperature is more negative (or less positive) (which is -36.3 kJ), the reaction is more spontaneous.

19.7. $\Delta G^\circ_{rxn} = -689.6$ kJ

Since ΔG°_{rxn} is negative, the reaction is spontaneous at this temperature.

19.7. For More Practice

$\Delta G^\circ_{rxn} = -689.7$ kJ (at 25°)

The value calculated for ΔG°_{rxn} from the tabulated values (-689.6 kJ) is the same, to within 1 in the least significant digit, as the value calculated using the equation for ΔG°_{rxn}.

$\Delta G^\circ_{rxn} = -649.7$ kJ (at 500.0 K)

You could not calculate ΔG°_{rxn} at 500.0 K using tabulated ΔG°_f values because the tabulated values of free energy are calculated at a standard temperature of 298 K, much lower than 500 K.

19.8. $+107.1$ kJ

19.9. $\Delta G_{rxn} = -129$ kJ

The reaction is more spontaneous under these conditions than under standard conditions because ΔG_{rxn} is more negative than ΔG°_{rxn}.

19.10. -10.9 kJ

Chapter 20

20.1. $2\ Cr(s) + 4\ H^+(aq) \longrightarrow 2\ Cr^{2+}(aq) + 2\ H_2(g)$

20.2. $Cu(s) + 4\ H^+(aq) + 2\ NO_3^-(aq) \longrightarrow Cu^{2+}(aq) + 2\ NO_2(g) + 2\ H_2O(l)$

20.3. $3\ ClO^-(aq) + 2\ Cr(OH)_4^-(aq) + 2\ OH^-(aq) \longrightarrow 3\ Cl^-(aq) + 2\ CrO_4^{2-}(aq) + 5\ H_2O(l)$

20.4. $+0.60$ V

20.5. **a.** The reaction *will* be spontaneous as written.

b. The reaction *will not* be spontaneous as written.

20.6. $\Delta G° = -3.63 \times 10^5$ J

Since $\Delta G°$ is negative, the reaction is spontaneous.

20.7. 4.5×10^3

20.8. 1.08 V

20.9. *Anode:* $2\ H_2O(l) \longrightarrow O_2(g) + 4\ H^+(aq) + 4\ e^-$

Cathode: $2\ H_2O(l) + 2\ e^- \longrightarrow H_2(g) + 2\ OH^-(aq)$

20.10. 6.0×10^1 min

Chapter 21

21.1. $^{216}_{84}Po \longrightarrow {}^{212}_{82}Pb + {}^{4}_{2}He$

21.2. **a.** $^{235}_{92}U \longrightarrow {}^{231}_{90}Th + {}^{4}_{2}He$

$^{231}_{90}Th \longrightarrow {}^{231}_{91}Pa + {}^{0}_{-1}e$

$^{231}_{91}Pa \longrightarrow {}^{227}_{89}Ac + {}^{4}_{2}He$

b. $^{22}_{11}Na \longrightarrow {}^{22}_{10}Ne + {}^{0}_{+1}e$

c. $^{76}_{36}Kr + {}^{0}_{-1}e \longrightarrow {}^{76}_{35}Br$

21.2. For More Practice

Positron emission $\left({}^{40}_{19}K \longrightarrow {}^{40}_{18}Ar + {}^{0}_{+1}e\right)$ or electron capture $\left({}^{40}_{19}K + {}^{0}_{-1}e \longrightarrow {}^{40}_{18}Ar\right)$

21.3. **a.** positron emission

b. beta decay

c. positron emission

21.4. 10.7 yr

21.5. $t = 964$ yr

No, the C-14 content suggests that the scroll is from about A.D. 1000, not 500 B.C.

21.6. 1.0×10^9 yr

21.7. Mass defect $= 1.934$ amu

Nuclear binding energy $= 7.569$ MeV/nucleon

Chapter 22

22.1. or

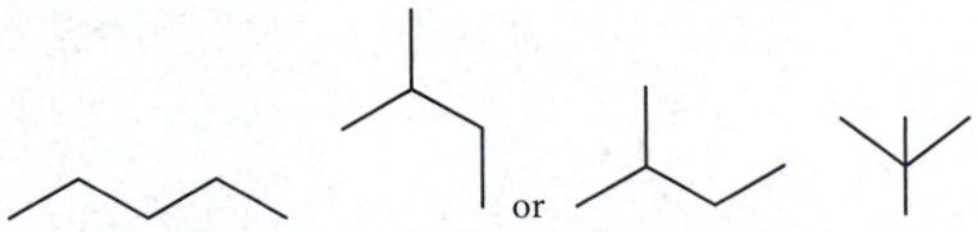

22.2. 3-methylhexane

22.3. 3,5-dimethylheptane

22.4. 2,3,5-trimethylhexane

22.5. **a.** 4,4-dimethyl-2-pentyne

b. 3-ethyl-4,6-dimethyl-1-heptene

22.6. **a.** 2-methylbutane

```
   H   CH3                                  H   CH3 H   H
   |    |                                   |    |  |   |
H—C—C—C=C—H + H2   ——→   H—C—C—C—C—H
   |    |   |   |                            |    |  |   |
   H   H   H   H                            H   H  H   H
```

b. 2-chloro-3-methylbutane

```
   H   CH3                                   H   CH3 H   H
   |    |                                    |    |  |   |
H—C—C—C=C—H + HCl   ——→   H—C—C—C—C—H
   |    |   |   |                             |    |  |   |
   H   H   H   H                             H   H  Cl  H
```

22.7. **a.** Alcohol reacting with an active metal.

$CH_3CH_2OH + Na \longrightarrow CH_3CH_2ONa + \frac{1}{2}H_2$

b. dehydration reaction

```
   H   CH3 H                      H   CH3
   |    |   |                     |    |
H—C—C—C—OH  —H2SO4→  H—C—C=C—H + H2O
   |    |   |                     |        |
   H   H   H                      H        H
```

Chapter 23

23.1. $[Xe]\ 6s^2 4f^{14} 5d^6$

23.2. $[Kr]\ 5s^0 4d^3$ or $[Kr]\ 4d^3$

23.3. pentaamminecarbonylmanganese(II) sulfate

23.4. sodium tetrachloroplatinate(II)

23.5. The complex ion $[Cr(H_2O)_3Cl_3]^+$ fits the general formula MA_3B_3, which results in fac and mer isomers.

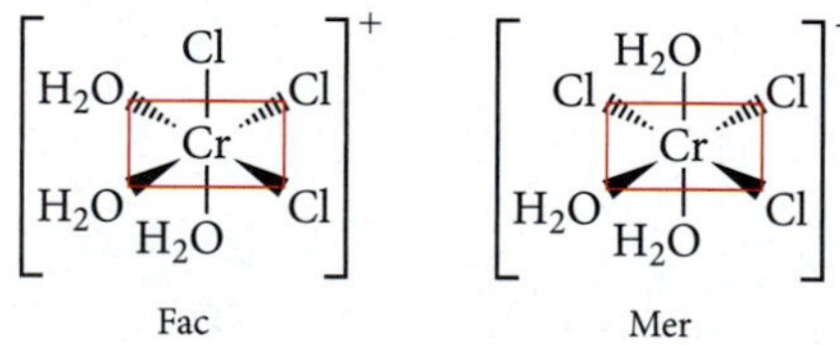

Fac   Mer

23.6. The oxalate ligand is a small bidentate ligand so it will have to occupy two adjacent (cis) positions of the octahedron. There are three ways to arrange the two NH_3 and two Cl^- ligands in the four remaining positions. One has both NH_3 and both Cl^- in cis positions (cis isomer). Another has the NH_3 ligands in a trans arrangement with both Cl^- in cis positions (*trans*-ammine isomer). The third has both NH_3 ligands cis and the Cl^- ligands trans (*trans*-chloro isomer).

Cl, Cl, Co, ox, H_3N, NH_3 —

Cis isomer

NH_3, Cl, Co, ox, Cl, NH_3 —   Cl, H_3N, Co, ox, H_3N, Cl —

Trans (in NH_3)   Trans (in Cl^-)

23.7. Both the fac and mer isomers are superimposable (by rotating 180°) on their mirror images, so neither one is optically active.

23.8. 288 kJ/mol

23.9. 5 unpaired electrons

23.10. 1 unpaired electron

Appendix I

A1.1. **a.** 29.8 °C

b. 302.9 K

Appendix II

A2.1. **a.** Each figure in this number is significant by rule 1: three significant figures.

b. This is a defined quantity that has an unlimited number of significant figures.

c. Both 1's are significant (rule 1), and the interior zero is significant as well (rule 2): three significant figures.

d. Only the two 9's are significant; the leading zeroes are not (rule 3): two significant figures.

e. There are five significant figures because the 1, 4, and 5 are nonzero (rule 1) and the trailing zeroes are after a decimal point, so they are significant as well (rule 4).

f. The number of significant figures is ambiguous because the trailing zeroes occur before an implied decimal point (rule 4). Assume two significant figures.

A2.2. **a.** 0.381

b. 121.0

c. 1.174

d. 8

GLOSSARY

absorption spectrum A plot of the absorption of light of a sample of matter as a function of wavelength. (3.3)

accuracy A term that refers to how close a measured value is to the actual value. (2.2)

acid See *Arrhenius definitions (of acids and bases), Brønsted-Lowry definitions (of acids and bases), and Lewis acid.* (9.4)

acid ionization constant (K_a) The equilibrium constant for the ionization reaction of a weak acid; used to compare the relative strengths of weak acids. (17.5)

acid–base reaction (neutralization reaction) A reaction in which an acid reacts with a base and the two neutralize each other, producing water. (9.7)

acid–base titration A laboratory procedure in which a basic (or acidic) solution of unknown concentration is reacted with an acidic (or basic) solution of known concentration, in order to determine the concentration of the unknown. (18.4)

acidic solution A solution containing an acid that creates additional H_3O^+ ions, causing $[H_3O^+]$ to increase. (17.6)

activated complex (transition state) A high-energy intermediate state between reactant and product. (15.6)

activation energy An energy barrier in a chemical reaction that must be overcome for the reactants to be converted into products. (15.6)

active site The specific area of an enzyme at which catalysis occurs. (15.8)

actual yield The amount of product actually produced by a chemical reaction. (8.5)

addition polymer A polymer in which the monomers simply link together without the elimination of any atoms. (22.14)

addition reaction A type of organic reaction in which two substituents are added across a double bond. (22.6)

alcohol A member of the family of organic compounds that contain a hydroxyl functional group (—OH). (22.9)

aldehyde A member of the family of organic compounds that contain a carbonyl functional group (C═O) bonded to two R groups, one of which is a hydrogen atom. (22.10)

aliphatic hydrocarbon An organic compound consisting of hydrogen and carbon atoms and containing no benzene rings; alkanes, alkenes, and alkynes are aliphatic hydrocarbons. (22.3)

alkali metals Highly reactive metals in group 1A of the periodic table. (4.5)

alkaline battery A dry-cell battery that employs slightly different half-reactions in a basic medium. (20.7)

alkaline earth metals Fairly reactive metals in group 2A of the periodic table. (4.5)

alkaloid Organic bases found in plants; they are often poisonous. (17.2)

alkane A hydrocarbon containing only single bonds. (22.3)

alkene A hydrocarbon containing one or more carbon–carbon double bonds. (22.3)

alkyne A hydrocarbon containing one or more carbon–carbon triple bonds. (22.3)

allotrope One of two or more forms of the same element; each form has a different structure. (19.4)

alpha (α) decay The form of radioactive decay that occurs when an unstable nucleus emits a particle composed of two protons and two neutrons. (21.3)

alpha (α) particle A low-energy particle released during alpha decay; equivalent to a He-4 nucleus. (21.3)

amine A member of a family of organic compounds containing nitrogen that are derived from ammonia with one or more of the hydrogen atoms replaced by alkyl groups. (22.3)

amorphous solid A solid in which atoms or molecules do not have any long-range order. (12.2)

ampere (A) The SI unit for electrical current; $1\ A = 1\ C/s$. (20.3)

amphoteric Able to act as either an acid or a base. (17.3)

amplitude The vertical height of a crest (or depth of a trough) of a wave; a measure of wave intensity. (3.2)

angular momentum quantum number (l) An integer that determines the shape of an orbital. (3.5)

anion A negatively charged ion. (1.8)

anode The electrode in an electrochemical cell where oxidation occurs; electrons flow away from the anode. (20.3)

antibonding orbital A molecular orbital that is higher in energy than any of the atomic orbitals from which it was formed. (7.4)

aqueous solution A solution in which water acts as the solvent. (9.2, 14.2)

aromatic hydrocarbon A hydrocarbon containing an aromatic (or benzene) ring. (22.3)

Arrhenius definitions (of acids and bases) The definitions of an acid as a substance that produces H^+ ions in aqueous solution and a base as a substance that produces OH^- ions in aqueous solution. (9.7, 17.3)

Arrhenius equation An equation that relates the rate constant of a reaction to the temperature, the activation energy, and the frequency factor; $k = Ae^{\frac{-E_a}{RT}}$. (15.6)

Arrhenius plot A plot of the natural log of the rate constant ($\ln k$) versus the inverse of the temperature in kelvins ($1/T$) that yields a straight line with a slope of $-E_a/R$ and a y-intercept of $\ln A$. (15.6)

atmosphere (atm) A unit of pressure based on the average pressure of air at sea level; $1\ atm = 101{,}325\ Pa$. (11.2)

atom A submicroscopic particle that constitutes the fundamental building block of ordinary matter; the smallest identifiable unit of an element. (1.1)

atomic mass (atomic weight) The average mass in amu of the atoms of a particular element based on the relative abundance of the various isotopes; it is numerically equivalent to the mass in grams of one mole of the element. (1.9)

atomic mass unit (amu) A unit used to express the masses of atoms and subatomic particles; defined as 1/12 the mass of a carbon atom containing 6 protons and 6 neutrons. (1.8)

atomic number (Z) The number of protons in an atom; the atomic number defines the element. (1.8)

atomic orbital (AO) A mathematical function that represents a state of an electron in an atom. (7.2)

atomic radius A set of average bonding radii determined from measurements on a large number of elements and compounds. (4.6)

atomic solids Solids whose composite units are atoms; they include nonbonding atomic solids, metallic atomic solids, and network covalent solids. (13.5)

atomic theory The theory that each element is composed of tiny indestructible particles called atoms, that all atoms of a given element have the same mass and other properties, and that atoms combine in simple, whole-number ratios to form compounds. (1.3, 1.5)

aufbau principle The principle that indicates the pattern of orbital filling in an atom. (4.3)

autoionization The process by which water acts as an acid and a base with itself. (17.6)

Avogadro's law The law that states that the volume of a gas is directly proportional to its amount in moles ($V \propto n$). (11.3)

Avogadro's number The number of ^{12}C atoms in exactly 12 g of ^{12}C; equal to 6.0221421×10^{23}. (2.8)

ball-and-stick molecular model A representation of the arrangement of atoms in a molecule that shows how the atoms are bonded to each other and the overall shape of the molecule. (5.3)

band gap An energy gap that exists between the valence band and conduction band of semiconductors and insulators. (7.6)

band theory A model for bonding in atomic solids that comes from molecular orbital theory in which atomic orbitals combine and become delocalized over the entire crystal. (7.6)

barometer An instrument used to measure atmospheric pressure. (11.2)

base See *Arrhenius definitions (of acids and bases), Brønsted-Lowry definitions (of acids and bases), and Lewis base.* (9.7)

base ionization constant (K_b) The equilibrium constant for the ionization reaction of a weak base; used to compare the relative strengths of weak bases. (17.8)

basic solution A solution containing a base that creates additional OH^- ions, causing the $[OH^-]$ to increase. (17.6)

bent geometry A molecular geometry in which three atoms bond in a nonlinear arrangement. (6.8)

beta (β) decay The form of radioactive decay that occurs when an unstable nucleus emits an electron. (21.3)

beta (β) particle A medium-energy particle released during beta decay; equivalent to an electron. (21.3)

bidentate Describes ligands that donate two electron pairs to the central metal. (23.3)

bimolecular An elementary step in a reaction that involves two particles, either the same species or different, that collide and go on to form products. (15.7)

binary acid An acid composed of hydrogen and a nonmetal. (9.7)

binary compound A compound that contains only two different elements. (5.6)

biological effectiveness factor (RBE) A correction factor multiplied by the dose of radiation exposure in rad to obtain the dose in rem. (21.11)

body-centered cubic A unit cell that consists of a cube with one atom at each corner and one atom at the center of the cube. (13.4)

boiling point The temperature at which the vapor pressure of a liquid equals the external pressure. (12.5)

boiling point elevation The effect of a solute that causes a solution to have a higher boiling point than the pure solvent. (14.6)

bomb calorimeter A piece of equipment designed to measure ΔE_{rxn} for combustion reactions at constant volume. (10.5)

bond energy The energy required to break 1 mol of the bond in the gas phase. (6.6)

bond length The average length of a bond between two particular atoms in a variety of compounds. (6.6)

bond order For a molecule, the number of electrons in bonding orbitals minus the number of electrons in nonbonding orbitals divided by two; a positive bond order implies that the molecule is stable. (7.4)

bonding orbital A molecular orbital that is lower in energy than any of the atomic orbitals from which it was formed. (7.4)

bonding pair A pair of electrons shared between two atoms. (5.7)

Born–Haber cycle A hypothetical series of steps based on Hess's law that represents the formation of an ionic compound from its constituent elements. (10.11)

Boyle's law The law that states that volume of a gas is inversely proportional to its pressure $\left(V \propto \frac{1}{P}\right)$. (11.3)

Brønsted-Lowry definitions (of acids and bases) The definitions of an acid as a proton (H^+ ion) donor and a base as a proton acceptor. (17.3)

buffer A solution containing significant amounts of both a weak acid and its conjugate base (or a weak base and its conjugate acid) that resists pH change by neutralizing added acid or added base. (18.2)

buffer capacity The amount of acid or base that can be added to a buffer without destroying its effectiveness. (18.3)

calorie (cal) A unit of energy defined as the amount of energy required to raise one gram of water 1 °C; equal to 4.184 J. (2.4, 10.2)

calorimetry The experimental procedure used to measure the heat evolved in a chemical reaction. (10.5)

capillary action The ability of a liquid to flow against gravity up a narrow tube due to adhesive and cohesive forces. (12.4)

carbonyl group A functional group consisting of a carbon atom double-bonded to an oxygen atom ($C{=}O$). (22.10)

carboxylic acid An organic acid containing the functional group $-COOH$. (17.2, 22.11)

catalyst A substance that is not consumed in a chemical reaction but increases the rate of the reaction by providing an alternate mechanism in which the rate-determining step has a smaller activation energy. (15.8)

cathode The electrode in an electrochemical cell where reduction occurs; electrons flow toward the cathode. (20.3)

cathode rays A stream of electrons produced when a high electrical voltage is applied between two electrodes within a partially evacuated tube. (1.6)

cathode ray tube A partially evacuated tube equipped with electrodes to produce cathode rays. (1.6)

cation A positively charged ion. (1.8)

cell potential (cell emf) (E_{cell}) The potential difference between the cathode and the anode in an electrochemical cell. (20.3)

Celsius (°C) scale The temperature scale most often used by scientists (and by most countries other than the United States), on which pure water freezes at 0 °C and boils at 100 °C (at atmospheric pressure). (Appendix I)

chain reaction A series of reactions in which previous reactions cause future ones; in a fission bomb, neutrons produced by the fission of one uranium nucleus induce fission in other uranium nuclei. (21.7)

Charles's law The law that states that the volume of a gas is directly proportional to its temperature ($V \propto T$). (11.3)

chelate A complex ion that contains either a bi- or polydentate ligand. (23.3)

chelating agent The coordinating ligand of a chelate. (23.3)

chemical bond The sharing or transfer of electrons to attain stable electron configurations for the bonding atoms. (5.2)

chemical change A change that alters the molecular composition of a substance; see also *chemical reaction.* (8.2)

chemical energy The energy associated with the relative positions of electrons and nuclei in atoms and molecules. (10.2)

chemical equation A symbolic representation of a chemical reaction; a balanced equation contains equal numbers of the atoms of each element on both sides of the equation. (8.3)

chemical formula A symbolic representation of a compound that indicates the elements present in the compound and the relative number of atoms of each. (5.3)

chemical property A property that a substance displays only by changing its composition via a chemical change. (8.2)

chemical reaction A process by which one or more substances are converted to one or more different substances; see also *chemical change.* (1.5, 8.3)

chemical symbol A one- or two-letter abbreviation for an element that is listed directly below its atomic number on the periodic table. (1.8)

chemistry The science that seeks to understand the properties of matter by studying the structure of the particles that compose matter. (1.1)

chiral Describes a molecule that is not superimposable on its mirror image. (22.3)

Clausius–Clapeyron equation The equation that displays the exponential relationship between vapor pressure and temperature; $\ln (P_{vap}) = \frac{-\Delta H_{vap}}{R}\left(\frac{1}{T}\right) + \ln \beta$. (12.5)

coffee-cup calorimeter A piece of equipment designed to measure ΔH_{rxn} for reactions at constant pressure. (10.7)

colligative property A property that depends on the amount of a solute but not on the type. (14.6)

collision frequency In the gas phase, the number of collisions that occur per unit time between gaseous particles. (15.6)

collision model A model of chemical reactions in which a reaction occurs after a sufficiently energetic collision between two reactant molecules. (15.6)

combustion analysis A method of obtaining empirical formulas for unknown compounds, especially those containing carbon and hydrogen, by burning a sample of the compound in pure oxygen and analyzing the products of the combustion reaction. (5.11)

combustion reaction A type of chemical reaction in which a substance combines with oxygen to form one or more oxygen-containing compounds; the reaction often causes the evolution of heat and light in the form of a flame. (8.6)

common ion effect The tendency for a common ion to decrease the solubility of an ionic compound or to decrease the ionization of a weak acid or weak base. (18.2)

common name A traditional name of a compound that gives little or no information about its chemical structure; for example, the common name of $NaHCO_3$ is "baking soda." (5.6)

complementary properties Properties that exclude one another; that is, the more you know about one, the less you know about the other. For example, the wave nature and particle nature of the electron are complementary. (3.4)

complete ionic equation An equation that lists individually all of the ions present as either reactants or products in a chemical reaction. (9.6)

complex ion An ion that contains a central metal ion bound to one or more ligands. (18.7, 23.3)

composition Refers to the type of particles that compose matter; by composition is one way to classify matter. (1.2)

compound A substance composed of two or more elements in fixed, definite proportions. (1.2)

concentrated solution A solution that contains a large amount of solute relative to the amount of solvent. (9.2, 14.5)

condensation The phase transition from gas to liquid. (12.5)

condensation polymer A polymer formed by elimination of an atom or small group of atoms (usually water) between pairs of monomers during polymerization. (22.14)

condensation reaction A reaction in which two or more organic compounds are joined, often with the loss of water or some other small molecule. (22.11)

conjugate acid A base to which a proton has been added. (17.3)

conjugate acid–base pair Two substances related to each other by the transfer of a proton. (17.3)

conjugate base An acid from which a proton has been removed. (17.3)

constructive interference The interaction of waves from two sources that align with overlapping crests, resulting in a wave of greater amplitude. (3.2)

conversion factor A factor used to convert between two different units; a conversion factor can be constructed from any two quantities known to be equivalent. (2.5)

coordinate covalent bond The bond formed when a ligand donates electrons to an empty orbital of a metal in a complex ion. (23.3)

coordination compound A neutral compound made when a complex ion combines with one or more counterions. (23.3)

coordination isomers Isomers of complex ions that occur when a coordinated ligand exchanges places with the uncoordinated counterion. (23.4)

coordination number The number of atoms with which each atom in a crystal lattice is in direct contact. (13.4)

coordination number (secondary valence) The number of molecules or ions directly bound to the metal atom in a complex ion. (23.3)

core electrons Those electrons in a complete principal energy level and those in complete *d* and *f* sublevels. (4.4)

corrosion The gradual, nearly always undesired, oxidation of metals that occurs when metals are exposed to oxidizing agents in the environment. (20.9)

Coulomb's law The law that states that the potential energy (E) of two charged particles depends on their charges (q_1 and q_2) and on their separation (r): $E = \frac{1}{4\pi\varepsilon_0}\frac{q_1 q_2}{r}$. (4.3)

covalent bond A chemical bond in which two atoms share electrons that interact with the nuclei of both atoms, lowering the potential energy of each through electrostatic interactions. (5.2)

covalent radius (bonding atomic radius) In nonmetals, one-half the distance between two atoms bonded together, and in metals, one-half the distance between two adjacent atoms in a crystal of the metal. (4.6)

critical mass The necessary amount of a radioactive isotope required to produce a self-sustaining fission reaction. (21.7)

critical point The temperature and pressure above which a supercritical fluid exists. (13.2)

critical pressure The pressure required to bring about a transition to a liquid at the critical temperature. (12.5)

critical temperature The temperature above which a liquid cannot exist, regardless of pressure. (12.5)

crystalline lattice The regular arrangement of atoms in a crystalline solid. (12.2, 13.4)

crystalline solid (crystal) A solid in which atoms, molecules, or ions are arranged in patterns with long-range, repeating order. (12.2)

cubic closest packing A closest-packed arrangement in which the third layer of atoms is offset from the first; the same structure as the face-centered cubic. (13.4)

cyclotron A particle accelerator in which a charged particle is accelerated in an evacuated ring-shaped tube by an alternating voltage applied to each semicircular half of the ring. (21.10)

Dalton's law of partial pressures The law that states that the sum of the partial pressures of the components in a gas mixture must equal the total pressure. (11.6)

de Broglie relation The observation that the wavelength of a particle is inversely proportional to its momentum $\lambda = \frac{h}{mv}$. (3.4)

degenerate Describes two or more electron orbitals with the same value of *n* that have the same energy. (4.3)

density (*d*) The ratio of an object's mass to its volume. (2.3)

deposition The phase transition from gas to solid. (12.6)

destructive interference The interaction of waves from two sources that are aligned so that the crest of one overlaps the trough of the other, resulting in cancellation. (3.2)

deterministic A characteristic of the classical laws of motion, which imply that present circumstances determine future events. (3.4)

dextrorotatory Capable of rotating the plane of polarization of light clockwise. (22.3)

diamagnetic The state of an atom or ion that contains only paired electrons and is, therefore, slightly repelled by an external magnetic field. (4.7)

diffraction The phenomena by which a wave emerging from an aperture spreads out to form a new wave front. (3.2)

diffusion The process by which a gas spreads through a space occupied by another gas. (11.9)

dilute solution A solution that contains a very small amount of solute relative to the amount of solvent. (9.2, 14.5)

dimensional analysis The use of units as a guide to solving problems. (2.5)

dimer The product that forms from the reaction of two monomers. (22.14)

diode A device that allows the flow of electrical current in only one direction. (7.6)

dipole moment A measure of the separation of positive and negative charge in a molecule. (6.2)

dipole–dipole force An intermolecular force exhibited by polar molecules that results from the uneven charge distribution. (12.3)

diprotic acid An acid that contains two ionizable protons. (9.7, 17.5)

dispersion force (London force) An intermolecular force exhibited by all atoms and molecules that results from fluctuations in the electron distribution. (12.3)

double bond The bond that forms when two electrons are shared between two atoms. (5.7)

dry-cell battery A battery that does not contain a large amount of liquid water, often using the oxidation of zinc and the reduction of MnO_2 to provide the electrical current. (20.7)

duet A Lewis structure with two dots, signifying a filled outer electron shell for the elements H and He. (5.4)

dynamic equilibrium The point at which the rate of the reverse reaction or process equals the rate of the forward reaction or process. (12.5, 14.4, 16.2)

effective nuclear charge (Z_{eff}) The actual nuclear charge experienced by an electron, defined as the charge of the nucleus plus the charge of the shielding electrons. (4.3)

effusion The process by which a gas escapes from a container into a vacuum through a small hole. (11.9)

electrical charge A fundamental property of certain particles that causes them to experience a force in the presence of electric fields. (1.6)

electrical current The flow of electric charge. (20.3)

electrochemical cell A device in which a chemical reaction either produces or is carried out by an electrical current. (20.3)

electrolysis The process by which electrical current is used to drive an otherwise nonspontaneous redox reaction. (20.8)

electrolyte A substance that dissolves in water to form solutions that conduct electricity. (9.4)

electrolytic cell An electrochemical cell that uses electrical current to drive a nonspontaneous chemical reaction. (20.3)

electromagnetic radiation A form of energy embodied in oscillating electric and magnetic fields. (3.2)

electromagnetic spectrum The range of the wavelengths of all possible electromagnetic radiation. (3.2)

electromotive force (emf) The force that results in the motion of electrons due to a difference in potential. (20.3)

electron A negatively charged, low-mass particle found outside the nucleus of all atoms that occupies most of the atom's volume but contributes almost none of its mass. (1.6)

electron affinity (EA) The energy change associated with the gaining of an electron by an atom in its gaseous state. (4.8)

electron capture The form of radioactive decay that occurs when a nucleus assimilates an electron from an inner orbital. (21.3)

electron configuration A notation that shows the particular orbitals that are occupied by electrons in an atom. (4.3)

electron geometry The geometrical arrangement of electron groups in a molecule. (6.8)

electron groups A general term for lone pairs, single bonds, multiple bonds, or lone electrons in a molecule. (6.7)

electron sea model A model for bonding in metals in which the metal is viewed as an array of positive ions immersed in a sea of electrons. (7.6)

electron spin A fundamental property of electrons; spin can have a value of $\pm 1/2$. (3.5)

electronegativity The ability of an atom to attract electrons to itself in a covalent bond. (6.2)

element A substance that cannot be chemically broken down into simpler substances. (1.2)

elementary step An individual step in a reaction mechanism. (15.7)

elimination reaction A reaction in which two molecules join together while eliminating a small molecule such as water. (22.9)

emission spectrum The range of wavelengths emitted by a particular element; used to identify the element. (3.3)

empirical formula A chemical formula that shows the simplest whole-number ratio of atoms in the compound. (5.3)

empirical formula molar mass The sum of the masses of all the atoms in an empirical formula. (5.11)

enantiomers (optical isomers) Two molecules that are nonsuperimposable mirror images of one another. (22.3)

endothermic Describes a process that absorbs heat from its surroundings. (2.4)

endothermic reaction A chemical reaction that absorbs heat from its surroundings; for an endothermic reaction, $\Delta H > 0$. (10.6)

endpoint The point of pH change where an indicator changes color. (18.4)

energy The capacity to do work. (2.4, 10.2)

English system The system of units used in the United States and various other countries in which the inch is the unit of length, the pound is the unit of force, and the ounce is the unit of mass. (1.3)

enthalpy (*H*) The sum of the internal energy of a system and the product of its pressure and volume. (10.6)

enthalpy (heat) of reaction (ΔH_{rxn}) The heat that is emitted or absorbed during a chemical reaction under conditions of constant pressure. (10.6)

enthalpy of solution (ΔH_{soln}) The heat that is emitted or absorbed during a solution formation under conditions of constant pressure. (14.3)

entropy (*S*) A thermodynamic function that is proportional to the number of energetically equivalent ways to arrange the components of a system to achieve a particular state; a measure of the energy randomization or energy dispersal in a system. (14.2)

enzyme A biochemical catalyst made of protein that increases the rates of biochemical reactions. (15.8)

equilibrium constant (*K*) The ratio, at equilibrium, of the concentrations of the products of a reaction raised to their stoichiometric coefficients to the concentrations of the reactants raised to their stoichiometric coefficients. (16.3)

equivalence point The point in a titration at which the added solute completely reacts with the solute present in the solution; for acid–base titrations, the point at which the amount of acid is stoichiometrically equal to the amount of base in solution. (9.7, 18.4)

ester A member of the family of organic compounds with the general structure R—COO—R. (22.11)

ether A member of the family of organic compounds of the form R—O—R. (22.12)

exact numbers Numbers that have no uncertainty and thus do not limit the number of significant figures in any calculation. (1.7)

exothermic Describes a process that releases heat to its surroundings. (2.4)

exothermic reaction A chemical reaction that releases heat to its surroundings; for an exothermic reaction, $\Delta H < 0$. (10.6)

experiment A highly controlled procedure designed to generate observations that may support a hypothesis or prove it wrong. (1.3)

exponential factor A number between zero and one that represents the fraction of molecules that have enough energy to make it over the activation barrier on a given approach. (15.6)

extensive property A property that depends on the amount of a given substance, such as mass. (2.3)

face-centered cubic A crystal structure in which the unit cell consists of a cube with one atom at each corner and one atom in the center of every face. (13.4)

family A group of organic compounds with the same functional group. (3.11)

family (group) One of the columns within the main group elements in the periodic table that contain elements that exhibit similar chemical properties. (4.2)

Faraday's constant (*F*) The charge in coulombs of 1 mol of electrons: $F = \frac{96{,}485\text{ C}}{\text{mol e}^-}$. (20.5)

film-badge dosimeter A device for monitoring exposure to radiation consisting of photographic film held in a small case that is pinned to clothing. (21.5)

first law of thermodynamics The law that states that the total energy of the universe is constant. (10.3)

formal charge The charge that an atom in a Lewis structure would have if all the bonding electrons were shared equally between the bonded atoms. (6.4)

formation constant (K_f) The equilibrium constant associated with reactions for the formation of complex ions. (18.7)

formula mass The average mass of a molecule of a compound in amu. (5.9)

formula unit The smallest, electrically neutral collection of ions in an ionic compound. (5.5)

free energy of formation (ΔG_f°) The change in free energy when 1 mol of a compound forms from its constituent elements in their standard states. (19.7)

free radical A molecule or ion with an odd number of electrons in its Lewis structure. (6.5)

freezing The phase transition from liquid to solid. (12.6)

freezing point depression The effect of a solute that causes a solution to have a lower melting point than the pure solvent. (14.6)

frequency (ν) For waves, the number of cycles (or complete wavelengths) that pass through a stationary point in one second. (3.2)

frequency factor (*A*) The number of times that reactants approach the activation energy per unit time. (15.6)

fuel cell A voltaic cell that uses the oxidation of hydrogen and the reduction of oxygen, forming water, to provide electrical current. (20.7)

fullerenes Carbon clusters, such as C_{60}, bonded in roughly spherical shapes containing from 36 to over 100 carbon atoms. (13.7)

functional group A characteristic atom or group of atoms that imparts certain chemical properties to an organic compound. (22.8)

gamma (γ) ray emission The form of radioactive decay that occurs when an unstable nucleus emits extremely high-frequency electromagnetic radiation. (3.2, 21.3)

gamma (γ) rays The form of electromagnetic radiation with the shortest wavelength and highest energy. (7.2, 19.3)

gas The state of matter in which atoms or molecules have a great deal of space between them and are free to move relative to one another; lacking a definite shape or volume, a gas conforms to the shape and volume of its container. (1.2)

gas-evolution reaction A reaction in which two aqueous solutions are mixed and a gas forms, resulting in bubbling. (9.8)

Geiger-Müller counter A device used to detect radioactivity that uses argon atoms that become ionized in the presence of energetic particles to produce an electrical signal. (21.5)

geometric (cis-trans) isomerism A form of stereoisomerism involving the orientation of functional groups in a molecule that contains bonds incapable of rotating. (22.5)

geometric isomers For complex ions, isomers that result when the ligands bonded to the metal have a different spatial arrangement. (23.4)

Gibbs free energy (*G*) A thermodynamic state function related to enthalpy and entropy by the equation $G = H - TS$; chemical systems tend toward lower Gibbs free energy, also called the *chemical potential*. (19.5)

Graham's law of effusion The law that states the ratio of effusion of two different gases is inversely proportional to the square root of the ratio of their molar masses. (11.9)

graphite An elemental form of carbon consisting of flat sheets of carbon atoms, bonded together as interconnected hexagonal rings held together by intermolecular forces, that can easily slide past each other. (13.7)

ground state The lowest energy state of an atom or molecule. (4.3)

half-cell One half of an electrochemical cell in which either oxidation or reduction occurs. (20.3)

half-life ($t_{1/2}$) The time required for the concentration of a reactant or the amount of a radioactive isotope to fall to one-half of its initial value. (13.4)

halogen One of the highly reactive nonmetals in group 7A of the periodic table. (4.5)

heat (q) The flow of energy caused by a temperature difference. (10.2)

heat capacity (C) The quantity of heat required to change a system's temperature by 1 °C. (10.4)

heat of fusion (ΔH_{fus}) The amount of heat required to melt 1 mole of a solid. (12.6)

heat of hydration ($\Delta H_{hydration}$) The enthalpy change that occurs when 1 mole of gaseous solute ions is dissolved in water. (14.3)

heat (or enthalpy) of reaction (ΔH_{rxn}) The enthalpy change for a chemical reaction. (10.6)

heat (or enthalpy) of vaporization (ΔH_{vap}) The amount of heat required to vaporize one mole of a liquid to a gas. (12.5)

Heisenberg's uncertainty principle The principle stating that due to the wave–particle duality, it is fundamentally impossible to precisely determine both the position and velocity of a particle at a given moment in time. (3.4)

Henderson–Hasselbalch equation The equation used to easily calculate the pH of a buffer solution from the initial concentrations of the buffer components, assuming that the "*x is small*" approximation is valid; $\text{pH} = \text{p}K_a + \log \frac{[\text{base}]}{[\text{acid}]}$. (18.2)

Henry's law The equation that expresses the relationship between solubility of a gas and pressure; $S_{gas} = k_H P_{gas}$. (14.4)

Hess's law The law stating that if a chemical equation can be expressed as the sum of a series of steps, then ΔH_{rxn} for the overall equation is the sum of the heats of reactions for each step. (10.8)

heterogeneous catalysis Catalysis in which the catalyst and the reactants exist in different phases. (15.8)

heterogeneous mixture A mixture in which the composition varies from one region to another. (1.3)

hexagonal closest packing A closest-packed arrangement in which the atoms of the third layer align exactly over those in the first layer. (13.4)

high-spin complex A complex ion with weak field ligands that have the same number of unpaired electrons as the free metal ion. (23.5)

homogeneous catalysis Catalysis in which the catalyst exists in the same phase as the reactants. (15.8)

homogeneous mixture A mixture with the same composition throughout. (1.3)

Hund's rule The principle stating that when electrons fill degenerate orbitals, they first fill them singly with parallel spins. (4.3)

hybrid orbitals Orbitals that form from the combination of standard atomic orbitals and that correspond more closely to the actual distribution of electrons in a chemically bonded atom. (7.3)

hybridization A mathematical procedure in which standard atomic orbitals are combined to form new, hybrid orbitals. (7.3)

hydrate An ionic compound that contains a specific number of water molecules associated with each formula unit. (5.6)

hydrocarbon An organic compound that contains only carbon and hydrogen. (5.12)

hydrogen bond A strong dipole–dipole attractive force between a hydrogen bonded to O, N, or F and one of these electronegative atoms on a neighboring molecule. (11.3)

hydrogenation The catalyzed addition of hydrogen to alkene double bonds to make single bonds. (15.8)

hydronium ion H_3O^+, the ion formed from the association of a water molecule with an H^+ ion donated by an acid. (9.7.17.3)

hydroxyl group In organic chemistry, an —OH group. (22.9)

hypothesis A tentative interpretation or explanation of an observation. A good hypothesis is *falsifiable*. (1.3)

hypoxia A physiological condition caused by low levels of oxygen, marked by dizziness, headache, shortness of breath, and eventually unconsciousness or even death in severe cases. (11.6)

ideal gas A gas in which interactions between particles and particles size are both negligible; a gas that behaves as described by the ideal gas law. (11.4)

ideal gas constant The proportionality constant of the ideal gas law, R, equal to 8.314 J/mol · K or 0.08206 L · atm/mol · K. (5.4)

ideal gas law The law that combines the relationships of Boyle's, Charles's, and Avogadro's laws into one comprehensive equation of state with the proportionality constant R in the form $PV = nRT$. (5.4)

ideal solution A solution that follows Raoult's law at all concentrations for both solute and solvent. (14.6)

indeterminacy The principle that present circumstances do not necessarily determine future events in the quantum-mechanical realm. (3.4)

indicator A dye whose color depends on the pH of the solution in which it is dissolved; often used to detect the endpoint of a titration. (9.7, 18.4)

infrared (IR) radiation Electromagnetic radiation emitted from warm objects, with wavelengths slightly larger than those of visible light. (3.2)

insoluble Describes a compound that is incapable of dissolving in water or is extremely difficult to dissolve in water. (9.4)

integrated rate law A relationship between the concentrations of the reactants in a chemical reaction and time. (15.5)

intensive property A property such as density that is independent of the amount of a given substance. (2.3)

interference The superposition of two or more waves overlapping in space, resulting in either an increase in amplitude (constructive interference) or a decrease in amplitude (destructive interference). (3.2)

internal energy (E) The sum of the kinetic and potential energies of all the particles that compose a system. (10.3)

International System of Units (SI) The standard unit system used by scientists; based on the metric system. (1.3)

ion An atom or molecule with a net charge caused by the loss or gain of electrons. (1.8)

ion product constant for water (K_w) The equilibrium constant for the autoionization of water. (17.6)

ion–dipole force An intermolecular force between an ion and the oppositely charged end of a polar molecule. (12.3)

ionic bond A chemical bond formed between two oppositely charged ions, generally a metallic cation and a nonmetallic anion, that are attracted to one another by electrostatic forces. (5.2)

ionic compound A compound composed of cations and anions bound together by electrostatic attraction. (5.2)

ionic solid A solid whose composite units are ions; ionic solids generally have high melting points. (13.5)

ionization energy (IE) The energy required to remove an electron from an atom or ion in its gaseous state. (4.7)

ionizing power The ability of radiation to ionize molecules and atoms. (21.3)

irreversible reaction A reaction that does not achieve the theoretical limit of available free energy. (19.7)

isomer One of two or more molecules with the same chemical formula, but with a different structure. (7.3)

isotope One of two or more atoms of the same element with the same number of protons but different numbers of neutrons and consequently different masses. (1.8)

joule (J) The SI unit for energy: equal to 1 kg · m^2/s^2. (2.4, 10.2)

ketone A member of the family of organic compounds that contain a carbonyl functional group (C=O) bonded to two R groups, neither of which is a hydrogen atom. (22.10)

ketose A sugar that is a ketone. (21.3)

kilogram (kg) The SI standard unit of mass defined as the mass of a block of metal kept at the International Bureau of Weights and Measures at Sèvres, France. (1.6)

kilowatt-hour (kWh) An energy unit used primarily to express large amounts of energy produced by the flow of electricity; equal to 3.60×10^6 J. (2.4)

kinetic energy The energy associated with motion of an object. (2.4, 10.2)

kinetic molecular theory A model of an ideal gas as a collection of point particles in constant motion undergoing completely elastic collisions. (11.7)

lanthanide contraction The trend toward leveling off in size of the atoms in the third and fourth transition rows due to the ineffective shielding of the *f* sublevel electrons. (23.2)

lattice energy The energy associated with forming a crystalline lattice from gaseous ions. (5.5)

law See *scientific law*

law of conservation of energy The law stating that energy can neither be created nor destroyed, only converted from one form to another. (2.4, 10.2)

law of conservation of mass The law stating that matter is neither created nor destroyed in a chemical reaction. (1.3)

law of definite proportions The law stating that all samples of a given compound have the same proportions of their constituent elements. (1.5)

law of mass action The relationship between the balanced chemical equation and the expression of the equilibrium constant. (16.3)

law of multiple proportions The law stating that when two elements (A and B) form two different compounds, the masses of element B that combine with one gram of element A can be expressed as a ratio of small whole numbers. (1.5)

Le Châtelier's principle The principle that when a chemical system at equilibrium is disturbed, the system shifts in a direction that minimizes the disturbance. (16.9)

lead–acid storage battery A battery that uses the oxidation of lead and the reduction of lead(IV) oxide in sulfuric acid to provide electrical current. (20.7)

levorotatory Capable of rotating the polarization of light counterclockwise. (22.3)

Lewis acid An atom, ion, or molecule that is an electron pair acceptor. (15.11)

Lewis base An atom, ion, or molecule that is an electron pair donor. (15.11)

Lewis electron-dot structure (Lewis structure) A drawing of a molecule that represents chemical bonds between atoms as shared or transferred electrons; the valence electrons of atoms are represented as dots. (5.4)

Lewis model A simple model of chemical bonding, which uses diagrams to represent bonds between atoms as lines or pairs of dots. In this model, atoms bond together to obtain stable octets (eight valence electrons). (5.4)

Lewis symbol A symbol of an element in which dots represent valence electrons. (5.4)

ligand A neutral molecule or an ion that acts as a Lewis base with the central metal ion in a complex ion. (18.7, 23.3)

limiting reactant The reactant that has the smallest stoichiometric amount in a reactant mixture and consequently limits the amount of product in a chemical reaction. (8.5)

linear accelerator A particle accelerator in which a charged particle is accelerated in an evacuated tube by a potential difference between the ends of the tube or by alternating charges in sections of the tube. (21.10)

linear geometry The molecular geometry of three atoms that form a 180° bond angle which results from the repulsion of two electron groups. (6.7)

linkage isomers Isomers of complex ions that occur when some ligands coordinate to the metal in different ways. (23.4)

liquid The state of matter in which atoms or molecules pack about as closely as they do in solid matter but are free to move relative to each other, resulting in a fixed volume but not a fixed shape. (1.2)

lithium ion battery A battery that produces electrical current in the form of motion of lithium ions from the anode to the cathode. (20.7)

lone pair A pair of electrons associated with only one atom. (5.7)

low-spin complex A complex ion that has strong field ligands and has fewer unpaired electrons relative to the free metal ion. (23.5)

macrostate The overall state of a system as defined by a given set of conditions (such as P, V, and T). (19.3)

magic numbers Certain numbers of nucleons (N or $Z = 2, 8, 20, 28, 50, 82$, and $N = 126$) that confer unique stability. (21.4)

magnetic quantum number (m_l) An integer that specifies the orientation of an orbital. (3.5)

main-group element One of the elements found in the *s* or *p* blocks of the periodic table, whose properties tend to be predictable based on their position in the table. (4.2)

manometer An instrument used to determine the pressure of a gaseous sample, consisting of a liquid-filled U-shaped tube with one end exposed to the ambient pressure and the other end connected to the sample. (5.2)

mass A measure of the quantity of matter making up an object. (1.6)

mass defect The difference in mass between the nucleus of an atom and the sum of the separated particles that make up that nucleus. (21.8)

mass number (A) The sum of the number of protons and neutrons in an atom. (1.8)

mass percent composition (mass percent) An element's percentage of the total mass of a compound containing the element. (5.10)

mass spectrometry An experimental method of determining the precise mass and relative abundance of isotopes in a given sample using an instrument called a *mass spectrometer*. (1.9)

matter Anything that occupies space and has mass. (1.1)

mean free path The average distance that a molecule in a gas travels between collisions. (11.9)

melting (fusion) The phase transition from solid to liquid. (12.6)

melting point The temperature at which the molecules of a solid have enough thermal energy to overcome intermolecular forces and become a liquid. (12.6)

metal A member of a large class of elements that are generally good conductors of heat and electricity, are malleable, ductile, and lustrous, and tend to lose electrons during chemical changes. (4.5)

metallic atomic solid An atomic solid held together by metallic bonds; metallic atomic solids have variable melting points. (13.5)

metalloid A member of a category of elements found on the boundary between the metals and nonmetals of the periodic table, with properties intermediate between those of both groups; metalloids are also called *semimetals*. (4.5)

metric system The system of measurements used in most countries in which the meter is the unit of length, the kilogram is the unit of mass, and the second is the unit of time. (1.3)

microstate The exact distribution of internal energy at any one instant among the particles that compose a system. (19.3)

microwaves Electromagnetic radiation with wavelengths slightly longer than those of infrared radiation; used for radar and in microwave ovens. (3.2)

millimeter of mercury (mmHg) A common unit of pressure referring to the air pressure required to push a column of mercury to a height of 1 mm in a barometer; 760 mmHg = 1 atm. (11.2)

miscibility The ability of substances to mix without separating into two phases. (12.3)

miscible The ability of two or more substances to be soluble in each other in all proportions. (14.2)

mixture A substance composed of two or more different types of atoms or molecules that can be combined in variable proportions. (1.3)

molality (*m*) A means of expressing solution concentration as the number of moles of solute per kilogram of solvent. (14.5)

molar heat capacity The amount of heat required to raise the temperature of one mole of a substance by 1 °C. (10.4)

molar mass The mass in grams of one mole of atoms of an element; numerically equivalent to the atomic mass of the element in amu. (2.8)

molar solubility The solubility of a compound in units of moles per liter. (18.5)

molar volume The volume occupied by one mole of a gas; the molar volume of an ideal gas at STP is 22.4 L. (11.5)

molarity (M) A means of expressing solution concentration as the number of moles of solute per liter of solution. (9.2, 14.5)

mole (mol) A unit defined as the amount of material containing 6.0221421×10^{23} (Avogadro's number) particles. (2.8)

mole fraction (χ_A) The number of moles of a component in a mixture divided by the total number of moles in the mixture. (5.6); a means of expressing solution concentration as the amount of solute in moles per total amount of solute and solvent in moles. (11.6, 14.5)

mole percent (mol %) A means of expressing solution concentration as the mole fraction multiplied by 100%. (14.5)

molecular compound A compound composed of two or more covalently bonded nonmetals. (5.2)

molecular element One of a group of elements that exist in nature with diatomic or polyatomic molecules as their basic unit. (3.4)

molecular equation An equation showing the complete neutral formula for each compound in a reaction. (9.6)

molecular formula A chemical formula that shows the actual number of atoms of each element in a molecule of a compound. (5.3)

molecular geometry The geometrical arrangement of atoms in a molecule. (6.8)

molecular orbital A mathematical function that represents a state of an electron in a molecule.(7.4)

molecular orbital diagram An energy diagram showing the atomic orbitals of the atoms that compose a molecule, the molecular orbitals of the molecule, their relative energies, and the placement of the valence electrons in the molecular orbitals.(7.4)

molecular orbital theory An advanced model of chemical bonding in which electrons reside in molecular orbitals delocalized over the entire molecule. In the simplest version, the molecular orbitals are simply linear combinations of atomic orbitals. (7.4)

molecular solid A solid whose composite units are molecules; molecular solids generally have low melting points. (11.5)

molecularity The number of reactant particles involved in an elementary step. (15.7)

molecule Two or more atoms joined chemically in a specific geometrical arrangement. (1.1)

monodentate Describes ligands that donate only one electron pair to the central metal. (23.3)

monoprotic acid An acid that contains only one ionizable proton. (17.5)

nanotube A long, tubular structure consisting of interconnected C_6 rings. (13.7)

natural abundance The relative percentage of a particular isotope in a naturally occurring sample with respect to other isotopes of the same element. (1.8)

Nernst equation The equation relating the cell potential of an electrochemical cell to the standard cell potential and the reaction quotient; $E_{cell} = E^{\circ}_{cell} - \frac{0.0592\ \text{V}}{n} \log Q$. (20.6)

net ionic equation An equation that shows only the species that actually change during a reaction. (9.6)

network covalent atomic solid An atomic solid held together by covalent bonds; network covalent atomic solids have high melting points. (13.5)

neutral Describes the state of a solution in which the concentrations of H_3O^+ and OH^- are equal. (17.6)

neutron An electrically neutral subatomic particle found in the nucleus of an atom, with a mass almost equal to that of a proton. (1.7)

nickel–cadmium (NiCad) battery A battery that consists of an anode composed of solid cadmium and a cathode composed of NiO(OH)(*s*) in a KOH solution. (20.7)

nickel–metal hydride (NiMH) battery A battery that uses the same cathode reaction as the NiCad battery but a different anode reaction, the oxidation of hydrogens in a metal alloy. (20.7)

nitrogen narcosis A physiological condition caused by an increased partial pressure of nitrogen, resulting in symptoms similar to those of intoxication. (11.6)

noble gas One of the group 8A elements, which are largely unreactive (inert) due to their stable filled *p* orbitals. (2.7)

node A point where the wave function (ψ), and therefore the probability density (ψ^2) and radial distribution function, all go through zero (3.6)

nonbonding atomic solid An atomic solid held together by dispersion forces; nonbonding atomic solids have low melting points. (13.5)

nonbonding orbital An orbital whose electrons remain localized on an atom. (7.4)

nonelectrolyte A compound that does not dissociate into ions when dissolved in water. (9.4)

nonmetal A member of a class of elements that tend to be poor conductors of heat and electricity and usually gain electrons during chemical reactions. (4.5)

nonvolatile Not easily vaporized. (12.5)

normal boiling point The temperature at which the vapor pressure of a liquid equals 1 atm. (12.5)

n-type semiconductor A semiconductor that employs negatively charged electrons in the conduction band as the charge carriers. (7.6)

nuclear binding energy The amount of energy required to break apart the nucleus into its component nucleons. (21.8)

nuclear equation An equation that represents nuclear processes such as radioactivity. (21.3)

nuclear fission The splitting of the nucleus of an atom, which results in a tremendous release of energy. (21.7)

nuclear fusion The combination of two light nuclei to form a heavier one. (21.9)

nuclear theory The theory that most of the atom's mass and all of its positive charge are contained in a small, dense nucleus. (1.7)

nucleon One of the particles that compose the nucleus, which are protons and neutrons. (21.4)

nucleus The very small, dense core of the atom that contains most of the atom's mass and all of its positive charge; the nucleus is composed of protons and neutrons. (1.7)

nuclide A particular isotope of an atom. (21.3)

octahedral geometry The molecular geometry of seven atoms with 90° bond angles. (6.7)

octahedral hole A space that exists in the middle of six atoms on two adjacent close-packed sheets of atoms in a crystal lattice. (23.4)

octet A Lewis symbol with eight dots, signifying a filled outer electron shell for *s* and *p* block elements. (5.4)

octet rule The tendency for most bonded atoms to possess or share eight electrons in their outer shell in order to obtain stable electron configurations and lower their potential energy. (5.4)

optical isomers Two molecules that are nonsuperimposable mirror images of one another. (22.3, 23.4)

orbital A probability distribution map, based on the quantum-mechanical model of the atom, used to describe the likely position of an electron in an atom; also an allowed energy state for an electron. (3.5)

orbital diagram A diagram that gives information similar to an electron configuration but symbolizes an electron as an arrow in a box representing an orbital, with the arrow's direction denoting the electron's spin. (4.3)

organic chemistry The study of carbon-based compounds. (22.1)

organic compound A compound containing carbon combined with several other elements including hydrogen, nitrogen, oxygen, or sulfur. (5.12)

organic molecule A molecule containing carbon combined with several other elements including hydrogen, nitrogen, oxygen, or sulfur. (22.1)

orientation factor In collision theory, a variable that is a measure of how specific the orientation of the colliding molecules must be. A large orientation factor (near 1) indicates that the colliding molecules can have virtually any orientation and the reaction will still occur. A small orientation factor indicates that the colliding molecules must have a highly specific orientation for the reaction to occur. (15.6)

osmosis The flow of solvent from a solution of lower solute concentration to one of higher solute concentration. (14.6)

osmotic pressure The pressure required to stop osmotic flow. (14.6)

overall order The sum of the orders of all reactants in a chemical reaction. (13.3)

oxidation The loss of one or more electrons; also the gaining of oxygen or the loss of hydrogen. (9.9)

oxidation state (oxidation number) A positive or negative whole number that represents the "charge" an atom in a compound would have if all shared electrons were assigned to the atom with a greater attraction for those electrons. (9.9)

oxidation–reduction (redox) reaction A reaction in which electrons are transferred from one reactant to another and the oxidation states of certain atoms are changed. (9.9)

oxidizing agent A substance that causes the oxidation of another substance; an oxidizing agent gains electrons and is reduced. (9.9)

oxyacid An acid composed of hydrogen and an oxyanion. (9.7)

oxyanion A polyatomic anion containing a nonmetal covalently bonded to one or more oxygen atoms. (5.6)

oxygen toxicity A physiological condition caused by an increased level of oxygen in the blood, resulting in muscle twitching, tunnel vision, and convulsions. (11.6)

packing efficiency The percentage of volume of a unit cell occupied by the atoms, assumed to be spherical. (13.4)

paramagnetic The state of an atom or ion that contains unpaired electrons and is, therefore, attracted by an external magnetic field. (4.7)

partial pressure (P_n) The pressure due to any individual component in a gas mixture. (11.6)

parts by mass A unit for expressing solution concentration as the mass of the solute divided by the mass of the solution multiplied by a multiplication factor. (14.5)

parts by volume A unit for expressing solution concentration as the volume of the solute divided by the volume of the solution multiplied by a multiplication factor. (14.5)

parts per billion (ppb) A unit for expressing solution concentration in parts by mass where the multiplication factor is 10^9. (14.5)

parts per million (ppm) A unit for expressing solution concentration in parts by mass where the multiplication factor is 10^6. (14.5)

pascal (Pa) The SI unit of pressure, defined as 1 N/M^2. (11.2)

Pauli exclusion principle The principle that no two electrons in an atom can have the same four quantum numbers. (4.3)

penetrating power The ability of radiation to penetrate matter. (21.3)

penetration The phenomenon of some higher-level atomic orbitals having significant amounts of probability within the space occupied by orbitals of lower energy level. For example, the 2*s* orbital penetrates into the 1*s* orbital. (4.3)

percent by mass A unit for expressing solution concentration in parts by mass with a multiplication factor of 100%. (14.5)

percent ionic character The ratio of a bond's actual dipole moment to the dipole moment it would have if the electron were transferred completely from one atom to the other, multiplied by 100%. (6.2)

percent ionization The concentration of ionized acid in a solution divided by the initial concentration of acid multiplied by 100%. (17.7)

percent yield The percentage of the theoretical yield of a chemical reaction that is actually produced; the ratio of the actual yield to the theoretical yield multiplied by 100%. (8.5)

periodic law A law based on the observation that when the elements are arranged in order of increasing mass, certain sets of properties recur periodically. (4.2)

periodic property A property of an element that is predictable based on an element's position in the periodic table. (4.1)

periodic table A table that arranges all known elements in order of increasing atomic number; elements with similar properties generally fall into columns on the periodic table. (1.8)

permanent dipole A permanent separation of charge; a molecule with a permanent dipole always has a slightly negative charge at one end and a slightly positive charge at the other. (12.3)

pH The negative log of the concentration of H_3O^+ in a solution; the pH scale is a compact way to specify the acidity of a solution. (17.6)

phase With regard to waves and orbitals, the sign of the amplitude of the wave, which can be positive or negative. (3.6)

phase diagram A map of the state (or phase) of a substance as a function of pressure and temperature. (13.2)

phenyl group A benzene ring treated as a substituent. (22.7)

phosphorescence The long-lived emission of light that sometimes follows the absorption of light by certain atoms and molecules. (19.2)

photoelectric effect The observation that many metals emit electrons when light falls upon the metal. (3.2)

photon (quantum) The smallest possible packet of electromagnetic radiation with an energy equal to $h\nu$. (3.2)

physical change A change that alters the state or appearance of a substance but not its chemical composition. (8.2)

physical property A property that a substance displays without changing its chemical composition. (8.2)

pi (π) bond The bond that forms between two *p* orbitals that overlap side to side. (7.3)

p–n junctions Tiny areas in electronic circuits that have p-type semiconductors on one side and n-type on the other. (7.6)

polar covalent bond A covalent bond between two atoms with significantly different electronegativities, resulting in an uneven distribution of electron density. (6.2)

polyatomic ion An ion composed of two or more atoms. (5.6)

polydentate Describes ligands that donate more than one electron pair to the central metal. (23.3)

polymer A long chain-like molecule composed of many repeating units. (22.14)

polyprotic acid An acid that contains more than one ionizable proton and releases them sequentially. (9.7, 17.10)

positron The particle released in positron emission; equal in mass to an electron but opposite in charge. (21.3)

positron emission The form of radioactive decay that occurs when an unstable nucleus emits a positron. (21.3)

positron emission tomography (PET) A specialized imaging technique that employs positron-emitting nuclides, such as fluorine-18, as a radiotracer. (21.12)

potential difference A measure of the difference in potential energy (usually in joules) per unit of charge (coulombs). (20.3)

potential energy The energy associated with the position or composition of an object. (2.4, 10.2)

precipitate A solid, insoluble ionic compound that forms in, and separates from, a solution. (9.5)

precipitation reaction A reaction in which a solid, insoluble product forms upon mixing two solutions. (9.5)

precision A term that refers to how close a series of measurements are to one another or how reproducible they are. (2.2)

pressure A measure of force exerted per unit area; in chemistry, most commonly the force exerted by gas molecules as they strike the surfaces around them. (10.4, 11.1)

pressure–volume work The work that occurs when a volume change takes place against an external pressure. (10.4)

primary valence The oxidation state on the central metal atom in a complex ion. (23.3)

principal level (shell) The group of orbitals with the same value of *n*. (3.5)

principal quantum number (*n*) An integer that specifies the overall size and energy of an orbital. The higher the quantum number *n*, the greater the average distance between the electron and the nucleus and the higher its energy. (3.5)

probability density The probability (per unit volume) of finding the electron at a point in space as expressed by a three-dimensional plot of the wave function squared (ψ^2). (3.6)

product A substance produced in a chemical reaction; products appear on the right-hand side of a chemical equation. (8.3)

proton A positively charged subatomic particle found in the nucleus of an atom. (1.7)

p-type semiconductor A semiconductor that employs positively charged "holes" in the valence band as the charge carriers. (7.6)

pure substance A substance composed of only one type of atom or molecule. (1.3)

quantum number One of four interrelated numbers that determine the shape and energy of orbitals, as specified by a solution of the Schrödinger equation. (3.5)

quantum-mechanical model A model that explains the behavior of absolutely small particles such as electrons and photons. (3.1)

quartz A silicate crystal that has a formula unit of SiO_2. (13.7)

racemic mixture An equimolar mixture of two optical isomers that does not rotate the plane of polarization of light at all. (22.3)

radial distribution function A mathematical function that represents the total probability of finding the electron within a thin spherical shell at a distance r from the nucleus in an atom. (3.6)

radio waves The form of electromagnetic radiation with the longest wavelengths and smallest energy. (3.2)

radioactive Describes the state of unstable atoms that emit subatomic particles or high-energy electromagnetic radiation. (21.1)

radioactivity The emission of subatomic particles or high-energy electromagnetic radiation by the unstable nuclei of certain atoms. (1.7, 21.1)

radiocarbon dating A form of radiometric dating based on the C-14 isotope. (21.6)

radiometric dating A technique used to estimate the age of rocks, fossils, or artifacts that depends on the presence of radioactive isotopes and the predictable decay of those isotopes over time. (21.6)

radiotracer A radioactive nuclide attached to a compound or introduced into a mixture in order to track the movement of the compound or mixture within the body. (21.12)

random error Error that has equal probability of being too high or too low. (2.2)

Raoult's law The equation used to determine the vapor pressure of a solution; $P_{soln} = X_{solv}P^{\circ}_{solv}$. (14.6)

rate constant (*k*) A constant of proportionality in the rate law. (15.4)

rate law A relationship between the rate of a reaction and the concentration of the reactants. (15.4)

rate-determining step The step in a reaction mechanism that occurs more slowly than any of the other steps. (15.7)

reactant A starting substance in a chemical reaction; reactants appear on the left-hand side of a chemical equation. (8.3)

reaction intermediates Species that are formed in one step of a reaction mechanism and consumed in another. (15.7)

reaction mechanism A series of individual chemical steps by which an overall chemical reaction occurs. (15.7)

reaction order (*n*) A value in the rate law that determines how the rate depends on the concentration of the reactants. (15.5)

reaction quotient (Q_c) The ratio, at any point in the reaction, of the concentrations of the products of a reaction raised to their stoichiometric coefficients to the concentrations of the reactants raised to their stoichiometric coefficients. (16.7)

recrystallization A technique used to purify solids in which the solid is put into hot solvent until the solution is saturated; when the solution cools, the purified solute comes out of solution. (14.4)

reducing agent A substance that causes the reduction of another substance; a reducing agent loses electrons and is oxidized. (9.9)

reduction The gaining of one or more electrons; also the gaining of hydrogen or the loss of oxygen. (9.9)

rem A unit of radiation exposure that stands for roentgen equivalent man, where a roentgen is defined as the amount of radiation that produces 2.58×10^{-4} C of charge per kg of air. (21.1)

resonance hybrid The actual structure of a molecule that is intermediate between two or more resonance structures. (6.4)

resonance structures Two or more valid Lewis structures that are shown with double-headed arrows between them to indicate that the actual structure of the molecule is intermediate between them. (6.4)

reversible As applied to a reaction, describes the ability to proceed in either the forward or the reverse direction. (16.2)

reversible reaction A reaction that achieves the theoretical limit with respect to free energy and changes direction upon an infinitesimally small change in a variable (such as temperature or pressure) related to the reaction. (19.7)

salt An ionic compound formed in a neutralization reaction by the replacement of an H^+ ion from the acid with a cation from the base. (9.7)

salt bridge An inverted, U-shaped tube containing a strong electrolyte such as KNO_3 that connects two half-cells, allowing a flow of ions that neutralizes charge build-up. (20.3)

saturated hydrocarbon A hydrocarbon containing no double bonds in the carbon chain. (22.4)

saturated solution A solution in which the dissolved solute is in dynamic equilibrium with any undissolved solute; any added solute will not dissolve. (14.4)

scientific law A brief statement or equation that summarizes past observations and predicts future ones. (1.3)

scintillation counter A device for the detection of radioactivity that includes a material that emits ultraviolet or visible light in response to excitation by energetic particles. (21.5)

second law of thermodynamics The law stating that for any spontaneous process, the entropy of the universe increases ($\Delta S_{univ} > 0$). (23.3)

secondary valence The number of molecules or ions directly bound to the metal atom in a complex ion; also called the *coordination number*. (23.3)

seesaw geometry The molecular geometry of a molecule with trigonal bipyramidal electron geometry and one lone pair in an axial position. (6.8)

selective precipitation A process involving the addition of a reagent to a solution that forms a precipitate with one of the dissolved ions but not the others. (18.6)

semiconductor A material with intermediate electrical conductivity that can be changed and controlled. (4.5)

semipermeable membrane A membrane that selectively allows some substances to pass through but not others. (14.6)

shielding The effect on an electron of repulsion by electrons in lower-energy orbitals that screen it from the full effects of nuclear charge. (4.3)

sigma (σ) bond The bond that forms between a combination of any two *s*, *p*, or hybridized orbitals that overlap end to end. (7.3)

silica A silicate crystal that has a formula unit of SiO_2, also called *quartz*. (13.7)

silicate A covalent atomic solid that contains silicon, oxygen, and various metal atoms. (13.7)

simple cubic A unit cell that consists of a cube with one atom at each corner. (13.4)

solid A state of matter in which atoms or molecules are packed close to one another in fixed locations with definite volume. (1.2)

solubility The amount of a substance that will dissolve in a given amount of solvent. (14.2)

solubility product constant (K_{sp}) The equilibrium expression for a chemical equation representing the dissolution of a slightly to moderately soluble ionic compound. (16.5)

soluble Describes a compound that is able to dissolve to a significant extent, usually in water. (9.4)

solute The minority component of a solution. (9.2, 14.1)

solution A homogeneous mixture of two substances. (9.2, 14.1)

solvent The majority component of a solution. (9.2, 14.1)

space-filling molecular model A representation of a molecule that shows how the atoms fill the space between them. (5.3)

specific heat capacity (C_s) The amount of heat required to raise the temperature of 1 g of a substance by 1 °C (10.4)

spectator ion An ion in a complete ionic equation that does not participate in the reaction and therefore remains in solution. (9.6)

spin quantum number (m_s) The fourth quantum number, which denotes the electron's spin as either 1/2 (up arrow) or −1/2 (down arrow). (3.5)

spontaneous process A process that occurs without ongoing outside intervention. (19.2)

square planar geometry The molecular geometry of a molecule with octahedral electron geometry and two lone pairs. (6.8)

square pyramidal geometry The molecular geometry of a molecule with octahedral electron geometry and one lone pair. (10.3)

standard cell potential (standard emf) (E°_{cell}) The cell potential for a system in standard states (solute concentration of 1 M and gaseous reactant partial pressure of 1 atm). (20.3)

standard change in free energy (ΔG°_{rxn}) The change in free energy for a process when all reactants and products are in their standard states. (19.7)

standard electrode potential A measure of the potential energy experienced by charged particles at an electrode in an electrochemical cell; the standard cell potential is the difference between the standard electrode potentials of the anode and cathode. (20.4)

standard enthalpy change (ΔH°) The change in enthalpy for a process when all reactants and products are in their standard states. (10.10)

standard enthalpy (or heat) of formation (ΔH°_f) The change in enthalpy when 1 mol of a compound forms from its constituent elements in their standard states. (10.10)

standard entropy change (ΔS_{rxn}) The change in entropy for a process when all reactants and products are in their standard states. (19.4)

standard entropy change for a reaction (ΔS°_{rxn}) The change in entropy for a process in which all reactants and products are in their standard states. (17.6)

Standard Hydrogen Electrode (SHE) A half-cell consisting of an inert platinum electrode immersed in 1 M HCl with hydrogen gas at 1 atm bubbling through the solution; used as the standard of a cell potential of zero. (20.4)

standard molar entropy (S°) A measure of the energy dispersed into one mole of a substance at a particular temperature. (19.4)

standard state For a gas the standard state is the pure gas at a pressure of exactly 1 atm; for a liquid or solid the standard state is the pure substance in its most stable form at a pressure of 1 atm and the temperature of interest (often taken to be 25°C); for a substance in solution the standard state is a concentration of exactly 1 M. (10.10)

standard temperature and pressure (STP) Conditions of $T = 0$ °C (273 K) and $P = 1$ used primarily in reference to a gas. (11.5)

state A classification of a form of matter as a solid, liquid, or gas. (1.3)

state function A function whose value depends only on the state of the system, not on how the system got to that state. (10.3)

stereoisomers Molecules in which the atoms are bonded in the same order but have a different spatial arrangement. (22.3, 23.4)

stock solution A highly concentrated form of a solution used in laboratories to make less concentrated solutions via dilution. (9.2)

stoichiometry The numerical relationships between amounts of reactants and products in a balanced chemical equation. (8.4)

strong acid An acid that completely ionizes in solution. (9.4, 17.3)

strong base A base that completely dissociates in solution. (17.3)

strong electrolyte A substance that completely dissociates into ions when dissolved in water. (9.4)

strong force Of the four fundamental forces of physics, the one that is the strongest but acts over the shortest distance; the strong force is responsible for holding the protons and neutrons together in the nucleus of an atom. (21.4)

strong-field complex A complex ion in which the crystal field splitting is large. (23.5)

structural formula A molecular formula that shows how the atoms in a molecule are connected or bonded to each other. (5.3, 22.3)

structural isomers Molecules with the same molecular formula but different structures. (22.3, 23.4)

sublevel (subshell) Those orbitals in the same principal level with the same value of n and l. (3.5)

sublimation The phase transition from solid to gas. (12.6)

substance A specific instance of matter. (1.2)

substitution reaction A chemical reaction in which one atom or group of atoms takes the place of another atom or group of atoms. (22.6)

substrate The reactant molecule of a biochemical reaction that binds to an enzyme at the active site. (15.8)

supersaturated solution An unstable solution in which more than the equilibrium amount of solute is dissolved. (14.4)

surface tension The energy required to increase the surface area of a liquid by a unit amount; responsible for the tendency of liquids to minimize their surface area, giving rise to a membrane-like surface. (12.4)

surroundings In thermodynamics, everything in the universe that exists outside the system under investigation. (10.2)

system In thermodynamics, the portion of the universe that is singled out for investigation. (10.2)

systematic error Error that tends toward being consistently either too high or too low. (2.2)

systematic name An official name based on well-established rules for a compound, which can be determined by examining its chemical structure. (5.6)

termolecular Describes an elementary step of a reaction in which three particles collide and go on to form products. (15.7)

tetrahedral geometry The molecular geometry of five atoms with 109.5° bond angles. (6.7)

tetrahedral hole A space that exists directly above the center point of three closest-packed metal atoms in one plane and a fourth metal located directly above the center point in the adjacent plane in a crystal lattice. (23.4)

theoretical yield The greatest possible amount of product that can be produced in a chemical reaction based on the amount of limiting reactant. (8.5)

theory A proposed explanation for observations and laws, based on well-established and tested hypotheses; a theory presents a model of the way nature works and predicts behavior beyond the observations and laws on which it was based. (1.3)

thermal energy A type of kinetic energy associated with the temperature of an object, arising from the motion of individual atoms or molecules in the object; see also *heat*. (2.4, 10.2)

thermal equilibrium The point at which there is no additional net transfer of heat between a system and its surroundings. (10.4)

thermochemistry The study of the relationship between chemistry and energy. (10.1)

thermodynamics The general study of energy and its interconversions. (10.3)

third law of thermodynamics The law stating that the entropy of a perfect crystal at absolute zero (0 K) is zero. (19.4)

titration A laboratory procedure in which a substance in a solution of known concentration is reacted with another substance in a solution of unknown concentration in order to determine the unknown concentration; see also *acid–base titration*. (9.7)

transition element (transition metal) One of the elements found in the d block of the periodic table whose properties tend to be less predictable based simply on their position in the table. (4.2)

transmutation The transformation of one element into another as a result of nuclear reactions. (21.10)

trigonal bipyramidal geometry The molecular geometry of six atoms with 120° bond angles between the three equatorial electron groups and 90° bond angles between the two axial electron groups and the trigonal plane. (6.7)

trigonal planar geometry The molecular geometry of four atoms with 120° bond angles in a plane. (6.7)

trigonal pyramidal geometry The molecular geometry of a molecule with tetrahedral electron geometry and one lone pair. (6.8)

triple bond The bond that forms when three electron pairs are shared between two atoms. (5.7)

triple point The unique set of conditions at which all three phases of a substance are equally stable and in equilibrium. (13.2)

triprotic acid An acid that contains three ionizable protons. (15.4)

T-shaped geometry The molecular geometry of a molecule with trigonal bipyramidal electron geometry and two lone pairs in axial positions. (6.8)

ultraviolet (UV) radiation Electromagnetic radiation with slightly smaller wavelengths than visible light. (3.2)

unimolecular Describes a reaction that involves only one particle that goes on to form products. (15.7)

unit A standard quantity used to specify measurements. (1.3)

unit cell The smallest divisible unit of a crystal that, when repeated in three dimensions, reproduces the entire crystal lattice. (13.4)

unsaturated fat A triglyceride with one or more double bonds in the hydrocarbon chain; unsaturated fats tend to be liquid at room temperature. (21.2)

unsaturated hydrocarbon A hydrocarbon that includes one or more double or triple bonds. (22.5)

unsaturated solution A solution containing less than the equilibrium amount of solute; any added solute will dissolve until equilibrium is reached. (14.4)

valence bond theory An advanced model of chemical bonding in which electrons reside in quantum-mechanical orbitals localized on individual atoms that are a hybridized blend of standard atomic orbitals; chemical bonds result from an overlap of these orbitals. (7.2)

valence electrons The electrons that are important in chemical bonding. For main-group elements, the valence electrons are those in the outermost principal energy level. (4.4)

valence shell electron pair repulsion (VSEPR) theory A theory that allows prediction of the shapes of molecules based on the idea that electrons—either as lone pairs or as bonding pairs—repel one another. (6.7)

van der Waals equation The extrapolation of the ideal gas law that considers the effects of intermolecular forces and particle volume in a nonideal gas; $P + a\left(\frac{n}{V}\right)^2 \times (V - nb) = nRT$. (5.9)

van der Waals radius (nonbonding atomic radius) One-half the distance between the centers of adjacent, nonbonding atoms in a crystal. (4.6)

van't Hoff factor (*i*) The ratio of moles of particles in a solution to moles of formula units dissolved. (14.7)

vapor pressure The partial pressure of a vapor that is in dynamic equilibrium with its liquid. (11.6, 12.5)

vaporization The phase transition from liquid to gas. (12.5)

viscosity A measure of the resistance of a liquid to flow. (12.4)

visible light Electromagnetic radiation with frequencies that can be detected by the human eye. (3.2)

volatile Tending to vaporize easily. (12.5)

voltaic (galvanic) cell An electrochemical cell that produces electrical current from a spontaneous chemical reaction. (20.3)

volt (V) The SI unit used to measure potential difference; equivalent to 1 J/C. (20.3)

wave function (ψ) The mathematical function that describes the wavelike nature of the electron. (3.5)

wavelength (λ) The distance between adjacent crests of a wave. (3.2)

weak acid An acid that does not completely ionize in water. (9.4, 17.3)

weak base A base that only partially ionizes in water. (17.3)

weak electrolyte A substance that does not completely ionize in water and only weakly conducts electricity in solution. (9.4)

weak-field complex A complex ion in which the crystal field splitting is small. (23.5)

work (w) The result of a force acting through a distance. (2.4, 10.2)

X-rays Electromagnetic radiation with wavelengths slightly longer than those of gamma rays; used to image bones and internal organs. (3.2)

X-ray diffraction A powerful laboratory technique that allows for the determination of the arrangement of atoms in a crystal and the measuring of the distance between them. (13.3)

CREDITS

Photo Credits

Chapter 1 Page 6 left: Jeff Greenberg "0 people images"/Alamy. Page 6 bottom center: cdrcom/Fotolia. Page 6 center right: YinYang/iStockphoto/Getty Images. Page 6 right: Денис Ларкин/iStockphoto/Getty Images. Page 7: Tomas Abad/Alamy. Page 10 center left: Charles D. Winters/Science Source. Page 10 center: Charles D. Winters/Science Source. Page 10 center right: Joseph Calev/Shutterstock. Page 13: Richard Megna/Fundamental Photographs. Page 19 top left: Steve Cole/Photodisc/Getty Images. Page 19 top center: Pearson Education. Page 20: Library of Congress, Prints & Photographs Division. Page 25: NASA Images. **Chapter 2** Page 37 center: Richard Megna/Fundamental Photographs. Page 37 top right: Richard Megna/Fundamental Photographs. Page 37 center right: Richard Megna/Fundamental Photographs. Page 37 bottom center: Pearson. Page 51 center right: Maxwell Art and Photo/Pearson. Page 51 bottom right: Marek/Fotolia. Page 58 center left: Pearson. Page 58 center: Pearson. Page 58 center right: Pearson. Page 58 bottom left: Pearson. Page 58 bottom center: Warren Rosenberg/Fundamental Photographs. Page 58 bottom right: Warren Rosenberg/Fundamental Photographs. Page 59: IBM Research, Almaden Research Center. Page 60: NASA Images. **Chapter 3** Page 65: Yahor Piaskouski/iStockphoto/Getty Images. Page 66: Mopic/Shutterstock. Page 67 top right: Poznyakov/Shutterstock. Page 67 center right: Image Courtesy of SPI Corp/www.x20.org. Page 68: Bonita R. Cheshier/Shutterstock. Page 73 bottom center: Tom Bochsler/Pearson. Page 73 bottom right: Karin Hildebrand Lau/Shutterstock. Page 76 bottom left: Michael Smith/iStockphoto/Getty Images. Page 76 top left: Jerry Mason/Science Source. Page 76 top center: Andrew Lambert Photography/Science Source. Page 76 top right: Andrew Lambert Photography/Science Source. Page 76 right: Andrew Lambert Photography/Science Source. Page 80: Segre Collection/AIP/Science Source. Page 81 top right: Stephen Dunn/Getty Images. Page 81 bottom left: REUTERS/XXSTRINGERXX xxxxx/Reuters/ Corbis. **Chapter 4** Page 102: Popova Olga/Fotolia. Page 103 bottom left: Charles D. Winters/Science Source. Page 103 bottom center: Richard Megna/Fundamental Photographs. Page 117 left: SPL/Science Source. Page 117 top left: Charles D. Winters/Science Source. Page 117 center left: Joe Belanger/Alamy. Page 117 center: David J. Green/Alamy. Page 117 bottom center: studiomode/Alamy. Page 117 top left: Jeff J Daly/Alamy. Page 117 top center: Harry Taylor/DK Images. Page 117 top right: Richard Megna/Fundamental Photographs. Page 117 right: Steve Gorton/DK Images. Page 117 bottom right: Charles D. Winters/Science Source. Page 117 bottom: Perennou Nuridsany/Science. Page 134 sodium: sciencephotos/Alamy. Page 134 magnesium: Richard Megna/Fundamental Photographs. Page 134 aluminium: amana images inc./Alamy. Page 134 silicon: Jeff J Daly/Alamy. Page 134 phosphorus: Charles D. Winters/Science Source/Photo Researchers, Inc. Page 134 sulphur: Steve Gorton/Dorling Kindersley. Page 134 cholorine: Charles D. Winters/Science Source/Photo Researchers, Inc. Page 134 nitrogen: DK Images. Page 134 phosphorus: Charles D. Winters/Science Source/Photo Researchers, Inc. Page 134 arsenic: Harry Taylor/DK Images. Page 134 antimony: Manamana/Shutterstock. Page 134 bismuth: Ted Kinsman/Science Source. **Chapter 5** Page 147 top right: artjazz/Shutterstock. Page 147 top left: Madlen/Shutterstock. Page 147 bottom left: Richard Megna/Fundamental Photographs. Page 147 bottom center: Charles Falco/Science Source. Page 147 bottom right: Charles D. Winters/Science Source. Page 154 top left: Richard Megna/Fundamental Photographs. Page 154 bottom left: Richard Megna/Fundamental Photographs. Page 158: Pearson. Page 160 center left: Richard Megna/Fundamental Photographs. Page 160 center right: Richard Megna/Fundamental Photographs. Page 167: NASA Images. Page 169: Richard Megna/Fundamental Photographs. Page 177: Sinelyov/Shutterstock. **Chapter 6** Page 188: Vaclav Volrab/Shutterstock. Page 222 top: Kip Peticolas/Fundamental Photographs. Page 222 bottom: Richard Megna/Fundamental Photographs. **Chapter 7** Page 256: Richard Megna/Fundamental Photographs. Page 261: mariusz szczygieł/Fotolia. **Chapter 8** Page 273: Michael Dalton/Fundamental Photographs. Page 290 top left: Andrew Lambert Photography/Science Source. Page 290 center left: Richard Megna/Fundamental Photographs. Page 290 center: Richard Megna/Fundamental Photographs. Page 290 center right: Richard Megna/Fundamental Photographs. Page 291: Richard Megna/Fundamental Photographs. Page 294 center: Richard Megna/Fundamental Photographs. Page 294 center right: Richard Megna/Fundamental Photographs. Page 294 bottom right: Roman Sigaev/iStockphoto/Getty Images. **Chapter 9** Page 301: Javier Impelluso/Shutterstock. Page 311 top left: Richard Megna/Fundamental Photographs. Page 311 top center: Richard Megna/Fundamental Photographs. Page 311 top right: Richard Megna/Fundamental Photographs. Page 312: Richard Megna/Fundamental Photographs. Page 313: Stephanie Weiler/Corbis. Page 314: Mike Dunning/DK Images. Page 315: Richard Megna/Fundamental Photographs. Page 320 center left: Shutterstock. Page 320 bottom right: Richard Megna/Fundamental Photographs. Page 321: Pearson. Page 323: Richard Megna/Fundamental Photographs. Page 325 bottom left: Richard Megna/Fundamental Photographs. Page 325 bottom center: Richard Megna/Fundamental Photographs. Page 325 bottom right: Richard Megna/Fundamental Photographs. Page 327 top right: Chip Clark/Fundamental Photographs. Page 327 bottom left: Richard Megna/Fundamental Photographs. Page 329 top left: Tom Bochsler/Pearson Education/PH College. Page 329 top right: Charles D. Winters/Science Source. Page 329 bottom left: Richard Megna/Fundamental Photographs. **Chapter 10** Page 350: Palette7/Shutterstock. Page 359: Tom Bochsler/Pearson. Page 373: Richard Megna/Fundamental Photographs. Page 387: md8speed/Shutterstock. **Chapter 11** Page 398: Carlos Caetano/Shutterstock. Page 402: Michael Dalton/Fundamental Photographs. Page 437 center left: PH College/Pearson Education. Page 437 center right: PH College/Pearson Education. **Chapter 12** Page 449: Richard Megna/Fundamental Photographs. Page 454 top left: Douglas Allen/iStockphoto/Getty Images. Page 454 bottom left: Andrei Kuzmik/Shutterstock. Page 455 top right: Johnson Space Center/NASA Images. Page 455 bottom right: Image courtesy of IRIS International, Inc. Page 456 top left: Richard Megna/Fundamental Photographs. Page 456 top center: Richard Megna/Fundamental Photographs. Page 457: Maridav/Fotolia. Page 461: Fundamental Photographs, NYC. Page 466: Reika/Shutterstock. Page 468 bottom left: Pearson Science/Pearson. Page 468 center: Dorling Kindersley Media Library/DK Images. Page 468 center right: Can Balcioglu/Shutterstock. Page 470: NG Images/Alamy. Page 471: Pearson Science/Pearson. Page 476 center left: Daniel Taeger/Shutterstock. Page 476 top right: Professor Nivaldo Jose Tro. Page 476 center: Harry Taylor/DK Images. **Chapter 13** Page 485 center right: Karl J Blessing/Shutterstock. Page 485 bottom right: Ted Kinsman/Science Source. Page 487: Science Source. Page 495 left: Volodymyr Goinyk/Shutterstock. Page 495 center left: Andrew Syred/Science Source. Page 495 right: Mirka Moksha/Shutterstock. Page 495 center right: Sashkin/Shutterstock. Page 496 left: Charles D. Winters/Science Source. Page 496 right: Barış Muratoglu/iStockphoto/Getty Images. Page 499 left: Siim Sepp/Shutterstock. Page 499 right: Tim Parmenter/The Natural History Museum, London/DK Images. Page 500: Lijuan Guo/Fotolia. **Chapter 14** Page 510 center left: Biology Pics/Science Source. Page 510 bottom left: Charles D. Winters/Getty Images. Page 510 top left: Masonjar/Shutterstock. Page 519 top left: Richard Megna/Fundamental Photographs. Page 519 top center: Richard Megna/Fundamental Photographs. Page 519 top right: Richard Megna/Fundamental Photographs. Page 520 top left: Richard Megna/Fundamental Photographs. Page 520 center left: Pearson. Page 528: ollo/iStockphoto/Getty Images. Page 534: Clark Brennan/Alamy. Page 541 bottom right: Pearson. Page 541 bottom center: Pearson. Page 541 bottom left: Pearson. Page 542 left: RubberBall/Alamy. Page 542 right: dtimiraos/iStockphoto/Getty Images. **Chapter 15** Page 590: NASA Images. Page 591 top left: NASA Images. Page 591 top right: NASA Images. **Chapter 16** Page 610 top left: Martin McCarthy/Getty Images. Page 610 top left: Kenneth Eward Illustration. Page 643 top left: Richard Megna/Fundamental Photographs. Page 643 top right: Richard Megna/Fundamental Photographs. **Chapter 17** Page 657 top: Clive Streeter/Dorling Kindersley. Page 657 bottom left: IgorDutina/iStockphoto. Page 657 bottom right: Mitch Hrdlicka/Getty. Page 658: Stacey Stambaugh/Pearson Education. Page 706: Eco Images/Universal Image Group/Getty Images. **Chapter 18** Page 712 top left: Richard Megna/Fundamental Photographs. Page 712 top center: Richard Megna/Fundamental Photographs. Page 712 top right: Richard Megna/Fundamental Photographs. Page 725 bottom left: Richard Megna/Fundamental Photographs. Page 725 bottom center: Richard Megna/Fundamental Photographs. Page 725 bottom right: Richard Megna/Fundamental Photographs. Page 738: Pearson. Page 739: Pearson. Page 744: National Park Service. Page 745: Richard Megna/Fundamental Photographs. Page 746 top left: Charles D. Winters/Science Source. Page 746 top center: Charles D. Winters/Science Source. Page 751 top left: Richard Megna/Fundamental Photographs. Page 751 bottom left: Richard Megna/Fundamental Photographs. Page 753 top left: Richard Megna/Fundamental Photographs. Page 753 center right: Richard Megna/Fundamental Photographs. Page 753 bottom left: Richard Megna/Fundamental Photographs. **Chapter 19** Page 768 center left: Motoring Picture Library/Alamy. Page 768 center right: Shubroto Chattopadhyay/Spirit/Corbis. Page 768 bottom left: BlackJack3D/iStockphoto/Getty Images. Page 768 bottom right: Siim Sepp/Shutterstock. Page 769 center: Paradoxdes/Fotolia. Page 769 bottom left: Richard Megna/Fundamental Photographs. Page 770 left: Johan Kocur. Page 773: Svetlana Lukienko/Shutterstock. Page 775 top left: Jan Will/Fotolia. Page 795 top right: Ryan McVay/Photodisc/Getty Images. **Chapter 20** Page 818 top center: Richard Megna/Fundamental Photographs. Page 818 top right: Richard Megna/Fundamental Photographs. Page 819: Alejandro Díaz Díez/AGE Fotostock. Page 829: Richard Megna/Fundamental Photographs. Page 838: Dave King/Courtesy of Duracell Ltd/DK Images. Page 839: Paul Mogford/Alamy. Page 844: Charles D. Winters/Science Source. Page 848: Lusoimages/Shutterstock. Page 849 top right: Donovan Reese/Riser/Getty Images. Page 849 bottom left: Alan Pappe/Stockbyte/Getty Images. Page 850 center left: busypix/iStockphoto/Getty Images. Page 850 right: Pearson. **Chapter 21** Page 862 center left: photovideostock/istockphoto/Getty Images. Page 862 bottom center: SPL/Science Source. Page 863 top right: AFP/Getty Images. Page 863 bottom right: Klaus Guldbrandsen/Science Source. Page 871 center left: United States Air Force. Page 871 bottom right: Hank Morgan/Science Source. Page 875: Baz Ratner/Reuters/Corbis. Page 876: sierrarat/istockphoto/Getty Images. Page 879: American Institute of Physics/Emilio Segre Visual ArchivesMeitner C2. Page 880: Stocktrek/Photodisc/Getty Images. Page 881: Franklin D. Roosevelt Library. Page 882 top right: Marlee90/iStockphoto/Getty Images. Page 882 center left: AP Photo/Kyodo. Page 886: General Atomics. Page 887: Fox Photos/Hulton Archive/Getty Images. Page 888 top left: David Parker/Science Source. Page 888 top left: Fermilab/

Science Source. Page 892 top right: Custom Medical Stock Photo/Newscom. Page 892 bottom left: D.Silbersweig/Science Source. Page 893 top left: Centre Oscar Lambret/Science Source. Page 893 bottom center: Cordelia Molloy/Science Source. Page 901: Craig Ellenwood/Alamy. **Chapter 22** Page 904 top left: Pearson. Page 917: Chad Koski/Shutterstock. Page 918: John Casey/Fotolia. Page 923: Pearson. Page 932: Lyudmila Suvorova/Shutterstock. Page 933: Andreas Von Einsiedel/ DK Images. Page 935 top left: felinda/istockphoto/Getty Images. Page 935 bottom left: Yasonya/ Shutterstock. Page 938 top left: PackShot/Fotolia. Page 938 top right: Comstock Images/ Comstock/JupiterImages/Getty Images. **Chapter 23** Page 956 top left: Tim Parmenter/DK Images. Page 956 left: Tim Parmenter/DK Images. Page 956 right: Asya Babushkina/Shutterstock. Page 956 center right: Harry Taylor/DK Images. Page 956 center: Harry Taylor/DK Images. Page 966 top left: Richard Megna/Fundamental Photographs. Page 966 top right: Richard Megna/Fundamental Photographs. Page 971 bottom center: Richard Megna/Fundamental Photographs. Page 971 bottom left: Richard Megna/Fundamental Photographs. Page 972: Richard Megna/Fundamental Photographs. Page 975: Richard Megna/Fundamental Photographs. **Appendix** Page A1: Richard Megna/Fundamental Photographs. Page A3 left: yui/Shutterstock. Page A3 center left: iStockphoto/Getty Images. Page A3 center right: iStockphoto/Getty Images. Page A3 right: Byron W. Moore/ Shutterstock.

Text Credits

Chapter 1 Page 7: Antoine Laurent Laviosier, Law of Conservation of Mass, 1789. Page 9: Thomas Kuhn, *The Structure of Scientific Revolutions*, 1962. Page 11: Joseph Proust, Law of Definite Proportions, 1798-1804. Page 12: John Dalton, 1804. Page 13: John Dalton, 1808. Page 14: J. J. Thomson, 1897. Page 17: E. Rutherford as quoted in S. Glasstone, *Principles of Nuclear Reactor Engineering*, 1958, p. 93. **Chapter 3** Page 63: E. Schrödinger, *The Interpretation of Quantum Mechanics: Dublin Seminars* (1949-1955) *and Other Unpublished Essays*, Ox Bow Press (1995). Page 63: From A. Einstein and I. Born, "Letter to Max Born (4 December 1926)"; *The Born-Einstein Letters*, Walker and Company, 1971. Page 64: E. Schrödinger as translated by John D. Trimmer in "The Present Situation in Quantum Mechanics," American Philosophical Society, 1935. Page 75: From N. Bohr, *The Correspondence Principle*, 1923. Page 79: P.A.M. Dirac, *The Principles of Quantum Mechanics*, Oxford University Press, 4th edition, 1958 (first published 1930), pp. 3-4. **Chapter 6** Page 191 Figure 6.3: Data from Linus Pauling, *The Nature of the Chemical Bond*. **Chapter 8** Page 272 Figure 8.1: Tro, Nivaldo J., *Chemistry: A Molecular Approach*, 2nd Ed., ©2011, p. 128. Reprinted and Electronically reproduced by permission of Pearson Education, Inc., Upper Saddle River, New Jersey. Page 272 Figure 8.2: From the National Climate Data Center, National Environmental Satellite, Data and Information Service, and the National Oceanic and Atmospheric Administration. **Chapter 11** Page 392 Figure 11.2: Based on "Surface Fronts," The Weather Channel, June 7, 2010, available at http://www.nc-climate.ncsu.edu/edu/k12/.fronts. **Chapter 15** Page 590 Table 15.4: U.S. Environmental Protection Agency, Our Nation's Air: Status and Trends through 2010. Page 591 Figure 15.19: NASA Ozone Watch, OMI instrument (KNMI/NASA) onboard the Aura satellite. **Chapter 21** Page 881: Letter, Albert Einstein to F.D. Roosevelt, August 2, 1939, Franklin D. Roosevelt Library, Hyde Park, New York. Page 890 Table 21.4: Department of Health and Human Services, National Institutes of Health.

INDEX

A

A. *See* **Frequency factor; Mass number**
A (ampere), 9, 819
Absolute concentration, 722–723
Absolute zero, 776
Absorption, 867, 956
Absorption spectra, 76, 972
Accuracy, 37–38
Acetaldehyde (ethanal), 247, 932, 933
Acetic acid
- in buffers, 711
- formula, 656, 664
- naming, 322
- from oxidation of ethanol, 930
- pH of, 671
- in solution, 311
- sources of, 657
- space-filling model, 934
- structure, 178
- uses, 936

Acetone
- and carbon disulfide, 531–533
- dipole–dipole forces, 448
- model and uses, 932–934
- vaporization of, 457

Acetonitrile, 450, 576–577
Acetylaldehyde, 932, 934
Acetylene (ethyne)
- formula and models of, 151, 917
- hybridization of, 243–244
- triple bond in, 178, 905
- uses, 918

Acetylsalicylic acid, 166, 936
Acid(s). *See also specific acids*
- acid ionization constant, 663–665
- amino, 937
- Arrhenius, 320, 658–660
- binary, 321, 661–662
- Brønsted-Lowry, 659–661, 688
- buffer effectiveness and amounts of, 722
- concentrations of, in buffers, 722–723
- conjugate, 660–661, 688, 720–721
- defined, 310, 658–661
- dilution of, 304
- diprotic, 320, 664, 695–696, 736
- hydronium ion concentration of, 670–680
- ions and salts as, 684, 688–691
- Lewis, 696–697, 748
- molecular structure of, 661–663
- monoprotic, 664, 665
- nature of, 656–657
- oxyacids, 322, 662–663
- percent ionization of, 676–678
- pH of, 670–680
- polyprotic. *See* **Polyprotic acids**
- properties of, 320–321
- and rusting, 849
- strength of, 661–665
- strong. *See* **Strong acid(s)**
- triprotic, 664
- weak. *See* **Weak acid(s)**

Acid anhydride, 936
Acid–base chemistry, 654–700
- acid strength, 661–665
- and autoionization of water, 666–667
- in Batman comic book series, 654–656
- definitions of acids and bases, 658–661
- hydronium ion concentration/pH of acids, 670–680
- hydroxide ion concentration/pH of bases, 680–684
- ions and salts, 684–691
- Lewis acids and bases, 696–697
- nature of acids and bases, 656–657
- pH scale, 668–669
- pOH scale, 669–670
- polyprotic acids, 691–696

Acid–base reactions, 319–326
- binary acids in, 321
- hydrochloric acid and sodium hydroxide, 322–324
- oxyacids in, 322
- and properties of acids/bases, 320–321
- titrations, 324–326

Acid–base titrations. *See* **Titrations**
Acidic cations, 688–689
Acidic solutions
- balancing redox reactions in, 815–816
- defined, 666
- hydronium ion concentration and pH of, 671–680
- pH of, 668
- on pOH scale, 670
- salt solutions as, 689–691
- solubility of ionic compounds in, 744
- solubility of metal hydroxides in, 752

Acid ionization constant (K_a), 663–665, 692, A-23–A-24
Acidosis, 710
Actinides, 115–116
Activated complex, 577
Activation energy (E_a)
- in Arrhenius equation, 577
- from Arrhenius plots, 578–580
- defined, 557, 576

Active nuclear charge, 121
Active site, enzyme, 592
Actual yield, 284
Acute radiation damage, 889
Addition
- of chemical equations, 616
- and scientific notation, A-12–A-13
- vector, 221–222

Addition polymers, 938
Addition reactions, 923–924
Adhesive forces, 455–456
Adipic acid, 938–939
Aerosols, 401–402
Age, measuring, 875–879
Air, 404, 407, 420
Alchemists, 887
Alcohols, 929–931
- and aldehydes/ketones, 933
- and carboxylic acids, 935–936
- solubility of, 514

Aldehydes, 931–934
Aliphatic hydrocarbons, 905–906
Aliquots, 562
Alkali metal chlorides, 377
Alkali metals, 118, 290, 291
Alkaline batteries, 838
Alkaline earth metals, 118
Alkaloids, 658
Alkanes, 905, 912–916
- alkynes and alkenes vs., 906
- naming, 913–916
- reactions of, 922–923

***n*-Alkanes, boiling points of, 447, 912–913**
Alkenes, 905, 916–921
- alkynes and alkanes vs., 906
- geometric isomerism in, 920–921
- naming, 918–920
- reactions of, 923–924
- unsymmetrical, 923–924

Alkyl groups, 914
Alkynes, 905, 916–920
alkanes and alkenes vs., 906
naming, 918–920
reactions of, 923–924
Allotropes, 776, 778
Alpha (α) decay, 863–865, 868
Alpha (α) particles, 16, 864–865
Altitude, 346, 393, 461
Aluminum, 5
density, 39, 100–102, 492
electron configuration, 111
and iron(III) oxide, 373
isotopes of, 20
molar mass, 53
and oxygen, 155
sacrificial electrodes of, 850
and water, 352–353
Aluminum-27 isotope, 887
Aluminum acetate, 690
Aluminum ions, 119, 697, 752–753
Aluminum nitrate, 689
Aluminum oxide, 849, 955
Amines, 681, 937
Amino acids, 937
Ammonia, 163, 696
and boron trifluoride, 696
and cobalt(III) compounds, 960
formation of, 285, 287, 424, 617, 641
formula and models, 151
hybrid orbitals, 238–239
nitrogen-to-hydrogen mass ratio of, 11
and silver, 748–749, 960
and silver chloride, 750–751
as weak base, 681–682
Ammonium bromide, 689
Ammonium nitrate, 159, 514–515, 517
Ammonium nitrite, 690
Ammonium salts, 937
Amorphous solids, 444
Ampere (A), 9, 819
Amphotericity, 660, 752
Amphoteric metal hydroxides, 752–753
Amplitude, 65
amu (atomic mass unit), 18, 884
Angular momentum quantum number (*l*), 82–83
Aniline, 926
Anion(s)
defined, 22
from electrolysis, 843–844
and ionic bonds, 147
ionic radii of, 127
monoatomic, 156
as weak bases, 684–688
in weak diprotic acid solution, 695–696
Anodes, 820, 823
Anodic regions, 849
Antacids, 322
Antarctica, ozone hole over, 590, 591
Antibodies, radioactively labeled, 860, 862
Antibonding orbitals, 250–252
Antifluorite structure, 498
Antifreeze, 534
glucose as, 508–510
toxicity of, 709–710
Antimony, 117
Antiparticles, 866
AOs. *See* **Atomic orbitals**
Appendicitis, diagnosis of, 860–862
Approximations, 632–635. *See also* ***x* is small approximation**
Aqueous ionic equilibrium, 708–757
and amphoteric metal hydroxides, 752–753
buffers, 710–724
complex ion equilibria, 748–753
in precipitation reactions, 745–748
solubility-product constant, 739–745
titrations, 725–739
and toxicity of antifreeze, 709–710
Aqueous reactions, 313–334
acid–base, 319–326
gas-evolution, 327–328
oxidation–reduction, 328–334
precipitation, 313–318
representing, 318–319
Aqueous solutions
defined, 302, 510
electrolysis in, 844–846
and heats of hydration, 516–518
standard entropies of, 779
types of, 308–311
Arbitrary elements, in science, 9
Area, pressure and, 392
Argon, 20, 118, 496
atomic radius, 414
electron configuration, 115
in Geiger-Müller counter, 872
ideal vs. real behavior, 425
and neon, 511
standard entropy, 778
Argon-40 isotope, 879
Aristotle, 7, 9
Aromatic compounds, 926–927
Aromatic hydrocarbons, 905, 924–927
Arrhenius, Svante, 576
Arrhenius acids, 320, 658–660
Arrhenius bases, 658–660
Arrhenius equation, 576–581
activation energy in, 577
and Arrhenius plots, 578–580
exponential factor in, 577–578
frequency factor in, 577
two-point form, 580
Arrhenius plots, 578–580
Arsenic, 117, 244
Arsenic pentafluoride, 203
Ascorbic acid, 693, 695
Aspirin, 166, 173, 936
Atmosphere (atm), 394
Atom(s), 2–28. *See also* **Quantum-mechanical model of atom**
atomic radii, 119–123
average mass, 22–25
as building blocks of matter, 9
and classification of matter, 4–7
converting between moles and number of, 52
defined, 4
discovery of electrons, 13–16
modern atomic theory, 10–13
orientation factors for reactions of, 582
origins of, 25
radioactive, 862
and scientific approach to knowledge, 7–9
shape of, 92
structure of, 16–18
and structure/properties of matter, 2–4
subatomic particles, 18–22
Atomic bomb, 880, 881
Atomic colors, A-17
Atomic mass, 22–25, 165
Atomic mass unit (amu), 18, 884
Atomic number (*Z*), 18, 19, 103, 120
Atomic orbitals (AOs)
defined, 82
degenerate, 106
d orbitals, 90, 91
filling of, 105–112
f orbitals, 90, 91
hybridization of, 236–248, 970
of Lewis acids, 696
linear combinations of, 249–252
for multi-electron atoms, 108–112
overlap of, 234–236
and periodic table, 113–114
phase of, 92
p orbitals, 90
shapes of, 87–92
s orbitals, 87–90
sp^2 hybridization, 239–243
sp^3d^2 hybridization, 245
sp^3d hybridization, 244
sp^3 hybridization, 237–239
sp hybridization, 243–244
Atomic radii, 119–123, 490, 958
Atomic solids, 495–496
Atomic spectroscopy, 73–76, 84–86
Atomic theory, 8, 10–13
Atomic weight. *See* **Atomic mass**
Atomos, 9
Aufbau principle, 109
Autoionization of water, 666–667
Average bond energy, 205
Average bond length, 206

Average kinetic energy, 414, 417–418
Average rate of reaction, 559–560
Avogadro, Amedeo, 51, 400
Avogadro's law, 400, 401, 415
Avogadro's number, 51. *See also* **Mole(s)**
Axial positions, 209

B

Bacon, Francis, 9
Baking soda, 156, 301. *See also* **Sodium bicarbonate**
Balancing equations
 general procedure, 275–278
 half-reaction method, 814–817
Ball-and-stick molecular models, 150
Band gap, 262
Band theory, 261–262
Barium, 74, 76, 118, 879
Barium chloride hexahydrate, 160
Barometers, 393
Basal sliding, 480
Base(s). *See also specific bases*
 Arrhenius, 658–660
 Brønsted-Lowry, 659–661
 and buffer effectiveness, 722
 conjugate, 660–661
 and conjugate acids, 720–721
 defined, 658–661
 hydroxide ion concentration of, 680–684
 ions and salts as, 684–691
 Lewis, 696, 748
 nature of, 657
 pH of, 680–684
 properties, 320–321
Base 10 logarithms, A-13–A-14
Base *e* logarithms, A-14
Base ionization constant (K_b), 681, A-24–A-25
Basic anions, 684–687
Basic solutions
 balancing redox reactions in, 816–817
 defined, 667
 metal hydroxides in, 752
 pH of, 668, 682–684
 pOH of, 670
 salt solutions as, 689–691
Batman comic book, 654–656
Battery(-ies), 838–841
 alkaline, 838
 dry-cell, 838
 energy loss in, 794
 flow of electrical charge in, 812–814
 fuel cells vs., 840–841
 irreversible reactions in, 794
 lead–acid storage, 838–839
 lithium ion, 840
 nickel–cadmium, 839
 nickel–metal hydride, 839–840
 rechargeable, 839–840
 redox reactions in, 328
Baumgartner, Felix, 390–393
Becquerel, Antoine-Henri, 16, 862–863
Bent geometry, 4, 212
Benzaldehyde, 932
Benzene
 formula and models, 151
 molecular orbital model, 260
 structure, 924–925
 surface tension, 454
Benzoic acid, 715
Berkelium, 20
Beryllium
 effective nuclear charge, 122
 incomplete octet formation by, 202, 207
 isotopes, 20
 Lewis structure, 151
Beryllium aluminum silicate, 955
Beta (β) decay, 863, 865–866, 868
Beta minus (β^-) decay, 866
Beta (β) particles, 865–866
Bicarbonate ions, 684, 710
Bidentate ligands, 961, 962
Bimolecular steps, 583–584
Binary acids, 321, 661–662
Binary compounds, 156–158
Binary ionic compounds, 156–158
Binding energy, 70
Biological effectiveness factor (RBE), 890
Biomolecules, 975–978
Bismuth, 872
Blood, buffers in, 708, 710
Blood drawing, 455
Blue Ribbon Commission on America's Nuclear Future, 882
Body-centered cubic unit cell, 487–489
Bohr, Niels, 62, 63, 74–75, 88
Bohr model, 74–75, 77, 88
Boiling, 273. *See also* **Vaporization (evaporation)**
Boiling point
 defined, 461
 and dipole moment, 449
 and dispersion force, 447
 and intermolecular forces, 453
 of main-group hydrides, 470
 and temperature, 461–465
Boiling point elevation, 533–537
Boltzmann, Ludwig, 770, 776
Boltzmann constant, 770
Bomb calorimeter, 356–357
Bombs
 atomic, 880, 881
 of bomb calorimeter, 356
 hydrogen, 886
Bond(s). *See also* **Covalent bonds**
 in compounds, 146–148
 coordinate covalent, 960
 defined, 146
 directional, 163
 energy from breaking, 368–369
 hydrogen, 450–452
 ionic, 147, 148, 152–154
 metallic, 496
 nondirectional, 163
 nonpolar covalent, 192
 pi, 240–241
 polar covalent, 191
 polarity of, 190–194
 sigma, 240–241
 single, 161–162, 241–243
 triple, 162, 243–244, 905
Bond dissociation energy. *See* **Bond energies**
Bond energies, 204–205, 367–370
Bond enthalpy. *See* **Bond energies**
Bonding, 188–226, 233–265. *See also* **Ionic bonding**
 and bond energies, 204–205
 and bond length, 206
 in coordination compounds, 970–975
 covalent, 147, 161–163
 and electron delocalization, 248–259
 electronegativity and bond polarity, 190–194
 energy required for, 368–369
 exceptions to octet rule, 201–204
 and formal charge, 199–201
 hybridization and bonding schemes, 215–240
 and hybridization of atomic orbitals, 236–248
 hydrogen, 450–452, 470
 and ionization energies, 132
 Lewis theory, 194–204
 magnetic property of liquid oxygen, 232–234
 in metals, 261
 and molecular geometry, 215–219
 molecular orbital theory, 248–260
 and molecular shape/polarity, 219–223
 in morphine, 188–190
 orbital overlap in, 234–236
 in polyatomic molecules, 259–260
 resonance in, 196–198
 in semiconductors, 261–262
 valence bond theory, 234–248
 VSEPR theory, 207–219
Bonding atomic radius (covalent radius), 119
Bonding electron pairs, 161, 211
Bonding orbitals, 249–252, 258
Bond length, 206
Bond order, 251, 252
Bond polarity, 662
Bond strength, 662
Bone scans, 892
Born–Haber cycle, 154, 375–377
Boron, 151, 202–203
Boron trifluoride, 696
Boscovich, Roger Joseph, 2
Boyle, Robert, 9, 395

Boyle's law, 395–397
and Charles's law, 399
and ideal gas law, 401
and kinetic molecular theory, 415
Bragg's law, 486, 487
Bridging hydrogens, 194
Bristlecone pine trees, 876
Bromate, 160
Bromine
atomic radii, 119
and 1-butene, 924
electron configuration, 111
family of, 118
and fluorine, 291
and potassium, 291, 582
Bromine trifluoride, 247
Bromobenzene, 925
1-Bromo-2-chlorobenzene, 926
Bromoethane, 930
Brønsted-Lowry acids, 659–661, 688
Brønsted-Lowry bases, 659–661
Buckminsterfullerene, 500
Buckyballs, 500
Buffer capacity, 724
Buffer range, 723–724
Buffers, 710–724
action of, 718
bases and conjugate acids as, 720–721
in blood, 708, 710
calculating pH changes in, 716–720
characteristics, 710–711
effectiveness, 722–725
Henderson–Hasselbalch equation for, 713–716
pH, 712–713
Burning bill demonstration, 342–344
Burns, steam, 457
Butanal, 932
Butane
combustion of, 276–277, 358
isobutane vs., 906, 907
***n*-Butane, 912**
Butanedione, 174–175
Butanoic acid, 934
1-Butanol, 929
2-Butanol, 933
Butanone, 932
2-Butanone, 933
1-Butene, 924
***cis*-2-Butene, 921**
***trans*-2-Butene, 921**
2-Butene-1-thiol, 904
1-Butyne, 918

C

***C* (heat capacity), 350–352**
Cadaverine, 937
Cadmium, in batteries, 839
CaF_2 (fluorite) structure, 498
Cal (Calorie), 42, 344
cal (calorie), 41–42, 344
Calcium
absorption spectrometry, 76
and chelating agents, 975
and chlorine, 153
emission spectra, 76
family of, 118
and oxygen, 155
Calcium acetate, 689
Calcium bromide, 157
Calcium carbonate, 159, 744
Calcium chloride, 302
Calcium fluoride, 495
dissolution of, 739–740
relative solubility, 742
solubility-product constant, 745
unit cell of, 498
Calcium hydroxide, 747–748
Calcium ions, 76, 747
Calcium nitrate, 689
Calcium oxide, 156
Calcium sulfate hemihydrate, 160
Calcium sulfide, 497
Californium, 888
Calorie (Cal), 42, 344
calorie (cal), 41–42, 344
Calorimetry
constant-pressure, 362–364
constant-volume, 355–357, 364
defined, 356
Cancer, 889, 892–893, 978
Candela (cd), 9
Capillary action, 455–456
Carbon
atomic mass, 24
in carbon dioxide, 11, 12
chemical symbol, 20
and cobalt(III) oxide, 276–277
from combustion analysis, 175–177
combustion of, 289
electron configuration, 109
electronegativity, 200
hybridization in, 237, 238
and hydrogen, 793–794
isotopes of, 21
Lewis structure of, 151
molar mass, 53
as network covalent atomic solid, 499–501
nuclear charge, 18
in organic chemistry, 177–178, 903
origin of, 25
and oxygen, 163, 347
properties, 904–905
standard entropy, 778
and sulfur, 332
Carbon-12 isotope, 51
binding energy per nucleon, 884
collision of uranium-238 and, 888
molar mass, 52
and radiocarbon dating, 876
Carbon-14 dating, 875–877
Carbonate ions, 313
Carbon dioxide, 6
chemical formula, 149
from combustion, 279–281, 289, 922
decomposition of, 11, 347
from electricity generation, 881
formula mass of, 165
as greenhouse gas, 272
molar mass, 165–166
from oxidation of carbon monoxide, 793
oxygen-to-carbon mass ratio, 12
phase diagram, 484, 485
polarity, 220
solubility, 520–522
supercritical, 465
and water, 620, 621, 697, 978
Carbon disulfide, 495, 531–533
Carbonic acid, 710, 849
Carbonic anhydrase, 975, 977–978
Carbon monoxide
from electricity generation, 881
and ethanol in gasoline, 929
and nitrogen dioxide, 566–567, 580, 584–585
oxidation of, 793
oxygen-to-carbon mass ratio, 12
standard entropy, 779
Carbon skeleton formula, 907
Carbon tetrachloride, 787
Carbonyl group, 931
Carboxylic acids, 657, 934–937
Carvone, 933
Catalysis, 588–592
Catalysts, 588–592, 923, 927
Catalytic converters, 589–590
Catenation of carbon, 904, 905
Cathode rays, 13–14
Cathode ray tubes, 13–14
Cathodes, 820, 823
Cathodic regions, 849
Cation(s)
defined, 22
from electrolysis, 843–844
and ionic bonds, 147
ionic radii of, 126
as Lewis acids, 697
of metals, 156–158, 688–689
as weak acids, 688–689
cd (Candela), 9
Cell diagrams, 820–821
Cell emf. *See* **Cell potential (E_{cell})**
Cell potential (E_{cell}), 829–837
and concentration, 833–837
defined, 819–820
and free energy/equilibrium constant, 829–833

under nonstandard conditions, 833–837
Celsius scale, A-2
Certainty of measurements, 36–37
Cesium chloride, 497
CFCs. *See* **Chlorofluorocarbons**
CFT. *See* **Crystal field theory**
Chadwick, James, 17
Chain reactions, 880
Charges
active nuclear, 121
effective nuclear, 107, 121–122
electrical, 14–15, 18, 812–814
formal, 199–201
ionic, 330
and lattice energy, 377–378
metals with invariant, 156
in voltaic cells, 820
Charles, J. A. C., 398
Charles's law, 397–399
and ideal gas law, 401
and kinetic molecular theory, 414, 415
Chelate, 962
Chelating agents, 962, 975
Chemical analysis, coordination compounds in, 975
Chemical bonds. *See* **Bond(s); Bonding**
Chemical changes, 273–274
Chemical energy, 41, 344
Chemical equations
for aqueous reactions, 318–319
balancing, 275–278, 814–817
and equilibrium constants, 615–617
for precipitation reactions, 317
writing, 274
Chemical equilibrium. *See* **Equilibrium**
Chemical formula(s)
and composition, 172
of compounds, 148–149, 172–177
conversion factors from, 170–171
experimental determination of, 172–177
of ionic compounds, 153, 155
and Lewis structures, 194
and mass percent composition, 168
Chemical kinetics. *See* **Reaction rate(s)**
Chemical potential, 785
Chemical properties, 273
Chemical reactions. *See* **Reaction(s)**
Chemical symbols, 20
Chemistry, 4
Chernobyl nuclear accident (1986), 882
χ. *See* **Mole fraction**
Chirality, 910, 911, 968–969
Chloride ions, 127, 843
Chlorine, 20
atomic mass, 22–23
atomic radii, 127
and calcium, 153
chemical symbol, 20
from electrolysis, 845–846
electron affinity, 132, 133
electron configuration, 114, 117
electronegativity, 663
and ethene, 923
family of, 118
and hydrogen, 192, 330
and iodine, 291
ionization energies, 129
and iron, 290
isotopes of, 20, 21
Lewis structure, 161
mass spectrum, 24
and methane, 367, 368, 922
and potassium, 152–153
and sodium, 10, 192, 329
Chlorobenzene, 925, 927
Chloroethane, 923
1-Chloro-3-ethylbenzene, 926
Chlorofluorocarbons (CFCs), 167, 568, 588, 590
Chloromethane, 922
Chlorophyll, 975, 977
1-Chloropropane, 923
2-Chloropropane, 923–924
Chlorous acid, 724
Chromium, 115–116, 157
Chromium(III) bromide, 157
Chromium(III) ions, 157
Ci (Curie), 889
Cinnamaldehyde, 177, 932
***cis* isomers**
of alkenes, 920–921
of coordination compounds, 966–967
properties of, 242–243
Cisplatin, 978
Cis–trans isomerism
in alkenes, 920–921
in coordination compounds, 966–967
and properties of compounds, 242–243
Citric acid, 935
Classical (term), 68
Clausius–Clayperon equation, 462–465
Climate change, 270–273
Closest-packed structures, 493–494, 496
Club soda, 510, 520, 522
Coal, 881
Cobalt(II) chloride, 160
Cobalt(III) coordination compounds, 960, 962
Cobalt(III) hexahydrate, 160
Cobalt(III) oxide, 276–277
Coefficients, in chemical equations, 614, 616
Coffee-cup calorimeters, 362
Cohesive forces, 455–456
Colligative properties of solutions, 528–542
boiling point elevation, 533–537
freezing point depression, 533–535
and medical uses, 541–542
osmotic pressure, 537–538
with strong electrolytes, 539–542
vapor pressure lowering, 528–532, 540–541
with volatile solutes, 530–532
Collision frequency, 581
Collision model, 556–557, 581–582
Color(s)
and absorption of light, 66
atomic, A-17
complementary, 972
of complex ions, 971–973
of rubies and emeralds, 954–956
of solutions with indicators, 737–739
Coloring agents, 975
Combustion analysis, 175–177
Combustion reactions (combustion), 274, 289
in bomb calorimeters, 356–357
and burning bill demonstration, 343–344
of butane, 276–277, 358
of ethanol, 343–344
of fossil fuels, 270–273, 279–281
of gasoline, 42, 353–354
of hydrocarbons, 921–922
of hydrogen, 329
of methane, 284–285, 372
of natural gas, 274–275
of octane, 270, 279, 329, 374–375
of petroleum, 280–281
of propane, 360, 361, 783
of sucrose, 356–357
water in, 360
Common ion effect, 712–713, 743–744
Common logarithms, A-13–A-14
Common names, 156
Complementary color, 972
Complementary properties, 80
Complete ionic equations, 318, 319
Complex ion equilibria, 748–753
and amphoteric metal hydroxides, 752–753
and formation constant, 749
and solubility, 750–752
Complex ion formation constants, A-26–A-27
Complex ions, 960
crystal field strength and color of, 971–973
defined, 748
formation of, 749, 751
geometries, 962–963
hybridization schemes, 970
naming, 963–964
Composition of matter, 4, 6–7, 344
Compounds, 144–181. *See also* **Coordination compounds; Ionic compounds**
aromatic, 926–927
binary, 156–158
chemical bonds in, 146–148
chemical formulas of, 148–149, 172–177
composition of, 167–172
in composition of matter, 6
covalent, 195

covalent bonding in, 161–163
formula mass for, 165
insoluble, 311, 316, 510, 745
interhalogen, 291
ionic bonding in, 152–154
Lewis models, 150–152
Lewis structures, 194–196
mole concept for, 165–167
molecular, 148, 163–165, 194–195
molecular models, 150
nonpolar, 449
organic, 177–178, 200–201
polar, 449
polyatomic, 196
properties of molecules vs., 144–146
pure, 370
soluble, 311, 510, 745
standard enthalpy of formation, 370
standard free energy of formation, 790
Compressibility of gases, 5–6, 443
Concentrated buffers, 723
Concentrated solutions, 302, 522
Concentration. *See also* **Equilibrium concentration**
absolute, 722–723
of acids and conjugate bases in buffers, 722–723
of anions in weak diprotic acid solutions, 695–696
and cell potential, 833–837
and electrical current, 836
and half-life of reaction, 573–575
of hydronium ions in acids, 670–680
of hydroxide ions in bases, 680–684
initial, 627–630, 633–635, 713–716
and integrated rate law, 568–575
and Le Châtelier's principle, 636–638
and percent ionization, 676–677
in rate laws, 563–575
and rate of reaction, 556
and reaction order, 563–567
and reaction rate, 558–559, 563–575
and selective precipitation, 747–748
of solutions, 302–306, 522–527
time dependence of, 568–576
Concentration cells, 836–837
Conceptual plan, 44, 45
Condensation, 457
Condensation polymers, 938–939
Condensation reactions, 935–936
Condensed states, 441, 445–456. *See also* **Liquid(s); Solid(s)**
Condensed structural formulas, 907
Conduction bands, 262
Coniine, 658
Conjugate acid-base pairs, 660–661, 685
Conjugate acids
in Brønsted–Lowry definition, 660–661
in buffers, 720–721
cations as, 688
Conjugate bases
in Brønsted–Lowry definition, 660–661
in buffer solutions, 710–711
concentrations of, 722–723
Connectivity, in structural formula, 907
Conservation of energy, law of, 40, 344–349
Conservation of mass, law of, 7, 10
Constant composition, law of, 11
Constant-pressure calorimetry, 362–364
Constant-volume calorimetry, 355–357, 364
Constructive interference, 68, 485
Continuous spectrum, 74
Conversion factor(s), 43–44
from chemical formulas, 170–171
concentrations as, 524–525
mass percent composition as, 168–169
molarity as, 303–304
Coordinate covalent bonds, 960
Coordination compounds
applications of, 975–978
bonding in, 970–975
coordination numbers of, 962–963
geometries of, 962–963
isomerism of, 965–969
ligands in, 960–962
naming of, 963–964
Coordination isomers, 965
Coordination numbers, 488, 962–963
Copernicus, Nicolaus, 9
Copper, 51, 54
atomic mass, 23
bonding in, 261
in concentration cells, 836–837
density, 40
electron configuration, 115–116
emission spectra, 76
heat capacity, 351
and H^+ ions, 829, 832–833
molar mass, 53
and zinc, 817–818, 820, 842
Copper(II) ions, 157
Copper(II) oxide, 157
Copper plating, 847
Copper(II) sulfate pentahydrate, 160
Core electrons, 112–113, 252
Corrosion, 848–850. *See also* **Rusting**
Coulomb's law
and intermolecular forces, 445
and lattice energy, 154, 377
and sublevel energy splitting, 106–107
and types of chemical bonds, 146–148
Counterions, 688, 960
Covalent bonding, 147, 161–163
Covalent bonds
carbon in, 904
coordinate, 960
defined, 148
ionic vs., 147
and Lewis model, 152
Covalent compounds, 195
Covalent radius (bonding atomic radius), 119
Creativity, in science, 8–9
Crick, Francis H.C., 487
Critical mass, 880
Critical point, 465, 483
Critical pressure (P_c), 465
Critical temperature (T_c), 465
Crystal field, 955–956
Crystal field strength
and color of complex ions, 971–973
and magnetic properties, 973–974
Crystal field theory (CFT), 956, 970–975
color of complex ions and crystal field strength, 971–973
magnetic properties and crystal field strength, 973–974
octahedral complexes and *d* orbital splitting, 970–971
tetrahedral/square planar complexes and *d* orbital splitting, 974–975
Crystalline lattice, 487
Crystalline solids, 480–504
closest-packed structures of, 493–494
defined, 444
fundamental types, 495–497
and glacial movement, 480–482
ionic, 497–498
network covalent atomic, 498–501
and phase diagrams, 482–485
unit cells of, 488–492
X-ray crystallography of, 485–487
Crystallography, X-ray, 485–487
Crystals, diffraction from, 486
C_s (specific heat capacity), 350–351, 471
Cubic closest packing, 494
Cubic unit cells, 488
curie (Ci), 889
Curie, Eve, 863
Curie, Irène, 863
Curie, Marie Sklodowska, 16, 20, 860, 863, 887
Curie, Pierre, 863
Curium, 20, 863
Cyanate ion, 200
Cyclohexane, 177, 905
Cyclotron, 887, 888
Cytochrome c, 975, 976

D

d. *See* **Density**
Dalton, John, 8–10, 12, 13, 20, 22
Dalton's law of partial pressures, 407–408, 415
Daughter nuclide, 864
Davisson-Germer experiment, 77
***d* block, 114, 115**
Dead Sea Scrolls, 875, 876

De Broglie, Louis, 63, 77
De Broglie relation, 78
De Broglie wavelength, 78–79
Decane, 906, 912
Decay
  alpha, 863–865, 868
  beta, 863, 865–866, 868
  beta minus, 866
  radioactive, 872–875
  radioactive decay series, 871
Decimal part, scientific notation, A-11
Decomposition
  of carbon tetrachloride, 787
  of ozone, 579, 587–589
  standard heat of formation for, 372
  of water, 11, 172, 841
Deep-sea diving, 409–411
Definite proportions, law of, 11
Degenerate orbitals, 106
Dehydration reactions, 930
Delocalized electrons, 197, 248–259
  in linear combinations of atomic orbitals, 249–252
  of second-period heteronuclear diatomic molecules, 258–259
  of second-period homonuclear diatomic molecules, 252–257
Democritus, 9
Density (*d*)
  of aluminum, 100–102, 492
  calculating, 38–40
  as conversion factor, 48
  and crystal structure, 492
  of gases, 393, 401, 404–406
  as periodic property, 102
  probability, 87–89
Deoxyribonucleic acid (DNA), 487, 889
Deposition, 466
Derived units, A-4–A-5
Destructive interference, 68, 485–486
Deterministic laws, 80
Deuterium, 886
Dextrorotatory isomers, 911
Diagnostic medical procedures, radioactivity in, 891–892
Diamagnetism, 124–125, 973
Diamminetetrachloroplatinate(III), 963
Diamond, 5, 442, 496
  graphite from, 768
  properties, 499
  standard entropy, 778
Diatomic elements, 161, 206
Diatomic molecules
  second-period heteronuclear, 258–259
  second-period homonuclear, 252–257
Diborane, 194
1,2-Dichlorobenzene, 926
1,3-Dichlorobenzene, 926
1,4-Dichlorobenzene, 926
***meta*-Dichlorobenzene, 926**
***ortho*-Dichlorobenzene, 926**
***para*-Dichlorobenzene, 926**
Dichloroethane, 923
1,2-Dichloroethane, 241–242, 920
1,2-Dichloroethene, 241–242, 920–921
***cis*-1,2,-Dichloroethene, 921**
***trans*-1,2,-Dichloroethene, 921**
Dichloromethane, 463, 922
Diethyl ether, 462, 936, 937
Differential rate law, 569
Diffraction
  electron, 77
  light, 68, 69
  X-ray, 485–487
Diffraction patterns, 486
Diffusion of gases, 420
Dihydrogen sulfide, 235–236
Dilute buffers, 723
Dilute solutions, 302, 522
Dilutions, 304–306
Dimensional analysis, 43–44
Dimers, 939
Dimethyl ether, 440–442, 451, 936
Dimethylglyoxime (dmg), 975
3,4-Dimethyl-3-hexene, 919
2,4-Dimethylpentane, 916
Dinitrogen monoxide, 12, 164, 562
Diodes, 262
Dipole(s)
  ion–dipole forces, 453
  ion–dipole interactions, 517
  permanent, 448
  temporary, 446
Dipole–dipole forces, 448–450
Dipole moment (μ)
  and boiling point, 449
  and bond polarity, 192–193
  of polar molecules, 220–223
Dipropyl ether, 936
Diprotic acid(s)
  anions in weak, 695–696
  defined, 664
  properties of, 320
  titrations of, 736
Dirac, P.A.M., 63, 79
Directional bonds, 163
Dispersal of energy, 511, 766–768, 773
Dispersion forces, 446–448
Dissociation, 311
  photodissociation, 588
  of polyprotic acids, 696
  and strength of acids/bases, 659
  of weak bases, 681
Dissociation constants, A-24
Dissolution, 779, 829
Disubstituted benzenes, 926
Division, A-12
dmg (dimethylglyoxime), 975
DNA (deoxyribonucleic acid), 487, 889
Döbereiner, Johann, 102
DOE (U.S. Department of Energy), 280
Dopants, 262
Doping, 262
***d* orbitals**
  and colors of rubies and emeralds, 954–956
  hybridization of, 244–245
  maximum electrons in, 110
  shape of, 90, 91
***d* orbital splitting**
  and octahedral complexes, 970–971
  and tetrahedral/square planar complexes, 974–975
Double bond(s)
  carbon in, 905
  covalent, 162
  and electron groups, 207
  and sp^2 hybridization, 239–243
Drugs, coordination compounds as, 978
Dry-cell batteries, 838
Dry ice, 166, 442, 466, 484, 495
Ductility, 261
Duet, 152
Dynamic equilibrium
  defined, 611–612
  in solutions, 519
  and vapor pressure, 459–460

E

ΔE. *See* Internal energy change
***E* (internal energy), 346–349**
E_a. *See* Activation energy
EA (electron affinity), 132–133
Eagle Nebula, 25
Ears, pressure imbalance in, 393
E_{cell}. *See* Cell potential
$E°_{cell}$. *See* Standard cell potential
Ectotherms, reaction rates in, 554–556
Eddington, Sir Arthur, 766, 767
Edge length, 490
EDTA^{4-} ion. *See* Ethylenediaminetetraacetate ion
Effective nuclear charge (Z_{eff}), 107, 121–122
Effusion of gases, 420–421
Eigenstates, 82
Einstein, Albert, 20, 63, 70–72, 346, 880, 881, 883
Einsteinium, 20
Eka-aluminum, 103
Eka-silicon, 103
Elastic collisions, 414
elBulli, 301
Electrical charge
  in batteries and lightning, 812–814
  of electrons, 14–15, 18
  of neutrons, 18
  of protons, 18

Electrical current, 817
and concentration, 836
driving force for, 819
in voltaic cells, 818
Electric field, 14, 64
Electricity
from batteries, 838–841
driving nonspontaneous reactions with, 841–848
from nuclear power, 880–881
from spontaneous reactions, 817–821
Electrochemical cells
concentration cells, 836–837
defined, 818
drawing, 827–828
notation for, 820–821
Electrochemistry, 812–854
balancing redox reactions, 814–817
batteries, 812–814, 838–841
cell potential, 829–837
corrosion, 848–850
electrolysis, 841–848
lightning, 812–814
standard electrode potentials, 822–829
voltaic cells, 817–821
Electrodes, 818. *See also* **Standard electrode potentials**
platinum, 821, 822
sacrificial, 850
standard hydrogen, 822–824
Electrolysis, 841–848
predicting products of, 843–846
stoichiometry of, 847–848
voltaic cells vs. electrolytic cells, 842–843
Electrolyte(s)
dissociation of, 659
and rusting, 849
strong, 310, 539–542, 659
weak, 311, 659
Electrolyte solutions, 309–311, 539–542
Electrolytic cells
for copper plating, 847
current in, 841
defined, 818
voltaic vs., 842–843
Electromagnetic radiation, 64–65, 866
Electromagnetic spectrum, 66–67
Electromotive force (emf), 819
Electron(s)
absorption of, 867
core, 112–113, 252
delocalized, 197, 248–259
discovery of, 13–16
energy of, 81–82
of ions, 22
localized, 197
nonbonding, 161, 211
positron as antiparticle of, 866
properties of, 18
in redox reactions, 330
in resonance structures, 196–198
transitions of, 75, 85–86
unpaired, 233–234, 256
valence, 112–116, 150–152
wave nature of, 77–78
Electron affinity (EA), 132–133
Electron capture, 863, 867
Electron configurations, 105–119
for families of elements, 117–118
of ions, 118–119, 124–125
of metals and nonmetals, 116–117
and orbital filling, 105–112
of transition metals, 956–957
of valence electrons, 112–116
Electron diffraction, 77
Electronegativity
and bond polarity, 191–192
and percent ionic character, 193
and strength of oxyacids, 662–663
of transition metals, 959
Electron geometry
basic shapes, 207–210
hybridization and bonding schemes from, 246
and lone pairs, 211–215
and molecular geometry, 216
Electron groups
defined, 207
and geometry, 207–210, 215
and lone pairs, 211–215
Electron orbitals, 81
Electron pairs
bonding, 161, 211
and Lewis acids/bases, 696
lone pairs, 161, 211–215, 682
Electron sea model, 261, 496
Electron spin, 83, 105–106, 109
Electron transfer, 152–153
Electron volt (eV), 884
Electrostatic forces, 14
Electrostatic potential maps, 191, 308, 330, 450
Element(s). *See also* **Main-group elements; Transition metals (transition elements)**
atomic masses, 22–25
atomic spectroscopy and identification of, 75–76
compounds vs., 144–146
defined, 6
diatomic, 161, 206
electronegativities, 191
families of, 117–118
groups of, 104, 117–118
inner transition, 115–116
mass percent, 167
molar masses, 52
octaves of, 102
origins of, 25
proton numbers, 18–20
standard enthalpies of formation, 370
standard free energies of formation, 790
transuranium, 888
triads of, 102
Elementary steps, 583–584
Elimination reactions, 930
Emeralds, color of, 955–956
emf (electromotive force), 819
Emission
gamma ray, 863, 866
of light, 862
positron, 863, 866
Emission spectra, 74–76, 85
Empirical formula molar mass, 174
Empirical formulas
from combustion analysis, 176–177
defined, 148
from experimental data, 173
from molecular formulas, 149
Enantiomers, 910–911, 968–969
Endogenous (term), 190
Endorphins, 190
Endothermic processes
defined, 42
melting, 467
particle view of, 360
in solution, 515
spontaneous, 769
vaporization, 457
Endothermic reactions
and bond energies, 368
defined, 358
Le Châtelier's principle for, 642
spontaneity of, 786–787
Endpoint of titration, 737
Energy, 40–43, 343
activation, 557, 576–580
in batteries, 794
binding, 70
for bonding/breaking bonds, 368–369
chemical, 41, 344
conservation of mass–energy, 346
converting mass to, 883–885
defined, 40
dispersal of, 511, 766–768, 773
of electrons, 81–82
exchange of, 344–345, 414
free. *See* **Free energy**
interaction, 234–235
internal, 346–349
ionization, 958
kinetic, 344, 345, 414, 417–418
lattice, 153–154, 375–378
law of conservation of, 40, 344–349
of melting and freezing, 467
nature of, 40–41, 344–345
nuclear binding, 884–885
and octet rule, 163

of orbitals, 249
potential. *See* **Potential energy**
quantifying changes in, 42–43
quantization of, 70–73
sign conventions for, 349
in solution formation, 514–518
sublimation, 375
and temperature, 456–457
thermal. *See* **Thermal energy**
total, 40
units of, 41–42
and vaporization, 457–459

Energy diagrams, 347, 585, 589
English system of measurement, 8
Enthalpy (H), 358–361
Enthalpy change (ΔH), 362–366
in chemical reactions, 362–366
in endothermic vs. exothermic processes, 358–360
internal energy change vs., 358
and spontaneity, 785–788
standard, 370, 372–373
stoichiometry with, 360–361
Enthalpy of reaction (ΔH_{rxn}), 367–375
from bond energies, 367–370
in coffee-cup calorimeter, 362–363
and Hess's law, 364–366
measuring, 362–364
relationships involving, 364–366
standard change in free energy from, 788–789
from standard enthalpies of formation, 370–375
stoichiometry involving, 360–361
Enthalpy of solution (ΔH_{soln}), 515–518
Enthalpy of vaporization. *See* **Heat of vaporization (ΔH_{vap})**
Entropy (S), 769–784
defined, 770–771
in microstates and macrostates, 771–773
predicting, 774–780
and second law of thermodynamics, 771
and solutions, 511
standard molar, 776–779
units of, 773
Entropy change (ΔS), 774–780
and changes in state, 774–775
in chemical reactions, 776
defined, 773
and spontaneity, 785–788
standard, 776, 779–780
Entropy change for surroundings (ΔS_{surr}), 780–784
and entropy change in universe, 780–781
quantifying, 782–783
temperature dependence of, 781–782
Enzymes, 591–592, 911
Equations
chemical. *See* **Chemical equations**
problems involving, 45, 49–50
Equatorial positions, 209
Equilibrium, 608–647. *See also* **Aqueous ionic equilibrium**
complex ion, 748–753
concentration change and, 636–638
dynamic, 459–460, 519, 611–612
and fetal hemoglobin, 608–611
free energy change of reactions in, 796
and Gibbs free energy, 797–800
heterogeneous, 620–621
Le Châtelier's principle for, 636–643
and pH changes in buffers, 717–718
pressure changes and, 638–641
and reaction quotient, 623–626
solubility, 739–745
in solutions, 518–522
temperature changes and, 641–643
thermal, 350
volume changes and, 638–641
Equilibrium approach to calculating pH, 715–716
Equilibrium concentration, 626–635
from K and concentrations of products and reactants, 626–627
from K and initial concentrations/pressures, 627–631
K from, 621–623
and pH of weak acids, 671
simplifying approximations of, 632–635
with small K, 633–635
Equilibrium constant (K), 609, 612–623, 632–635, A-23–A-27
approximations in problems with, 632–635
and chemical equations, 615–617
for chemical reactions, 614
defined, 613
and direction of change, 624–626
and equilibrium concentrations, 621–623, 626–631, 633
for fetal hemoglobin, 610
for heterogeneous equilibria, 620–621
for ionization of weak acid, 664–665
and law of mass action, 613
from pH, 675
and pressure, 617–620
significance of, 614–615
small, 631–635
and standard cell potential, 832–833
and standard change in free energy, 797–800
temperature dependence of, 799–800
units of, 619–620
Equivalence point
defined, 325, 725
in titration of strong acids, 726, 727
in titration of weak acids, 730, 732
Error, 38
Ester(s), 934–936
Estimations, 37, 49
Ethanal, 247, 932, 933
Ethanal (acetaldehyde), 247, 932, 933
Ethane, 448–449, 904, 912
Ethanoic acid, 934, 936. *See also* **Acetic acid**
Ethanol (ethyl alcohol), 5, 700
combustion of, 289, 343–344
dilution of, 306
dimethyl ether vs., 440–442, 451
from glucose, 929
hydrogen bonding in, 451
oxidization of, 933
reactions of, 930
in solution with water, 510
uses of, 929
Ethene
and chlorine, 923
double bond in, 905
formula and models of, 916
hybridization and bonding scheme for, 248
and hydrogen, 582
hydrogenation of, 590, 591
and hydrogen chloride, 923
polymers of, 938
standard entropy, 779
triple bond in, 178
uses, 917
Ethers, 936–937
Ethyl alcohol. *See* **Ethanol**
Ethylamine, 937
Ethylbenzene, 925, 927
Ethyl butanoate, 935
Ethylene, 916, 917, 938. *See also* **Ethene**
Ethylenediamine ligand, 961
Ethylenediaminetetraacetate ion ($EDTA^{4-}$), 961, 962, 975
Ethylene glycol
in antifreeze, 534, 709–710
boiling point elevation with, 536
freezing point of, 535
solution of, 526
Ethylmethylamine, 937
Ethyl methyl ether, 936
4-Ethyl-2,7-methyloctane, 914–915
3-Ethylpentane, 914–915
Ethyl pentanoate, 934
Ethyl propanoate, 934
Ethyl propyl ether, 936
Ethyne (acetylene)
formula and models of, 151, 917
hybridization of, 243–244
triple bond in, 178, 905
uses, 918
Europium, 20
eV (electron volt), 884
Evaporation. *See* **Vaporization**

Exact numbers, A-7
Excitation, radiation and, 84–85
Exothermic processes
condensation, 457
defined, 42
freezing, 467
particle view of, 360
in solution, 514, 515
Exothermic reactions
and bond energies, 368
defined, 359
and lattice energies, 153–154
Le Châtelier's principle for, 641
spontaneity of, 785, 786
Expanded octets, 194, 203–204
Experiments, 7
Exponential factors, 577–578
Exponential part, scientific notation, A-11
Exponents, 614, A-11
Extensive properties, 39, 776
Extrapolation, 397

F

Face-centered cubic unit cell, 487, 491
Fac–mer isomerism, 967
Fahrenheit scale, A-2
Falsifiability, 7
Family(-ies)
of elements, 104, 117–118
of organic compounds, 928
Faraday's constant, 830, 847
***f* block, 114, 116**
FDA (U.S. Food and Drug Administration), 169, 893
Fermi, Enrico, 879
Fermi National Accelerator Laboratory, 888
Fermium, 879
Fetal hemoglobin, 608–611
Feynman, Richard P., 342
Film-badge dosimeters, 871
Find information (in problem solving), 45, 49
Fireworks, 76
First ionization energy (IE_1), 128–129, 131, 958
First law of thermodynamics, 346–349, 773
First-order reactions
defined, 564
half-life of, 573–574
identifying, from experimental data, 564
integrated rate law for, 569–571
Fishing flies, 454
Fission, nuclear, 879–882
Flame tests, 76
Flash freezing, 471
Fluids
intravenous, 542
supercritical, 465
Fluoride ions, 124
Fluorine
and boron, 202–203
and bromine, 291
electron configuration, 124
electronegativity, 191
family of, 118
formal charge, 199
hydrogen bonding with, 450
Lewis structure, 151
and magnesium, 155
and potassium, 155
Fluorine-18 isotope, 892
Fluorine ions, 22
Fluorite (CaF_2) structure, 498
Fluoromethane, 452
Food, irradiation of, 893
***f* orbitals, 90, 91, 110**
Force(s). *See also* **Intermolecular forces**
adhesive, 455–456
cohesive, 455–456
from collisions of gas particles, 415, 416
electromotive, 819
electrostatic, 14
intramolecular, 163
and pressure, 392
in solutions, 512
strong, 869
total, 416
Formal charge, 199–201
Formaldehyde
addition reaction with, 933–934
dipole–dipole forces, 448–449
formula and model of, 931
geometry, 208
intermolecular forces in, 452
uses, 932
Formalin, 932
Formation constant (K_f), 749
Formic acid, 664, 724, 935
Formula mass, 165
Formula unit, 152
Fossil fuels, combustion of, 270–273, 279–281
4*f* orbitals, 91
Francium, 191
Franklin, Rosalind, 487
Free energy
Gibbs, 784–785
of reversible reactions, 793–794
Free energy change (ΔG), 784–800
defined, 784
and equilibrium, 797–800
in nonstandard states, 794–797
and spontaneity, 785–788
standard, 788–800
theoretical limit on, 794
Free energy change of reaction under non-standard conditions (ΔG_{rxn}), 794–797
Free energy of formation (ΔG°_f), 790–791
Free radicals, 202, 922–923
Freezing, 466, 467, 471
Freezing point depression, 533–535, 539–540
Freon-112 refrigerant, 168
Frequency (ν), 65–66, 70, 72
Frequency factor (A)
in Arrhenius equation, 577
from Arrhenius plots, 578–580
and collision model, 581–582
defined, 576
Fructose, 174
Fuel cells, 840–841
Fukushima Daiichi Nuclear Power Plant accident (2011), 882
Fuller, R. Buckminster, 500
Fullerenes, 500
Functional groups, 928. *See also specific types*
Fusion (melting), 466–467
Fusion (nuclear), 885–886

G

G (Gibbs free energy), 784–785
ΔG (free energy change), 784–800
Gadolinium, 116
Galileo Galilei, 9
Gallium, 103, 262
Galvanic (voltaic) cells, 817–821, 842–843
Gamma (γ) ray emission, 863, 866
Gamma (γ) rays, 66, 866, 892–893
Garnet, 956
Gas(es), 370, 390–432
Avogadro's law, 400
Boyle's law, 395–397
Charles's law, 397–399
collecting, over water, 412–413
defined, 5–6
density, 404–406
diffusion of, 420
dipole moments, 193
effusion of, 420–421
entropy and state changes in, 774–775
greenhouse, 270–273
on heating curves, 468–470
ideal gas law, 401–407, 618, 638
kinetic molecular theory of, 414–417
mean free path, 420
molar mass, 406–407
molar volume, 404
noble, 116–118, 447, 496, 777
partial pressure, 407–413
pressure, 392–395
properties of, 443
real, 425–428
solids and liquids vs., 442–445
solubility of, 520–521
standard molar entropies, 777
stoichiometry with, 422–424
and sublimation, 466
and supersonic skydiving, 390–392

temperature and molecular velocities, 417–419
and vaporization, 456–465
Gas chromatographs, 562
Gas-evolution reactions, 327–328
Gasoline, 5, 442
alkanes in, 912
combustion of, 42, 273, 353–354
ethanol in, 929
Gauge pressure, 403
Gay-Lussac's law, 402
Geiger-Müller counter, 871–872
Genetic defects, 889
Geodesic dome, 500
Geometric isomerism, 920–921, 966–967
Geometry. *See* **Electron geometry; Molecular geometry**
Germanium, 103, 113
ΔG°_f (free energy of formation), 790–791
Gibbs free energy (G), 784–785. *See also* **Free energy change (ΔG)**
Given information (in problem solving), 45, 49
Glaciers, movement of, 480–482
Global warming, 272
Glowing, 862
Glucose
alcohols from, 929
formation of, 273, 281
formula and models, 131
formula mass of, 165
oxidation of, 793
in respiration, 978
in solution, 527
in wood frogs, 508–510
Glycine, 218
Glycolic acid, 710
Gold, 959
Gold foil experiment, 16–17
Gold plating, 848
Gore, George, 188
Graham's law of effusion, 421
Graphite, 496
conversion of diamond to, 768
in lithium–ion batteries, 840
methane from, 370
properties, 499
standard entropy of, 778
Gravitational pull, A-1
Gray (Gy), 889
Grease, solubility in water, 510
Greenhouse effect, 272
Greenhouse gases, 270–273
Ground state, 105
Groups of elements, 104, 117–118
ΔG°_{rxn}. *See* **Standard change in free energy**
ΔG_{rxn} (free energy change of reaction under nonstandard conditions), 794–797
Gy (gray), 889

H

ΔH. *See* **Enthalpy change**
H (enthalpy), 358–361
ΔH° (standard enthalpy change), 370, 372–373
H_3O^+ ions. *See* **Hydronium ions**
Hahn, Otto, 879
Hair, bleaching of, 328
Half-cell, 818
Half-life ($t_{1/2}$) of reaction, 573–575, 872–875
Half-reaction method of balancing, 814–817
Halides, 160
Halogenation, 922–923
Halogens, 118, 290–291
Halogen substitution reactions, 922–923
Hand warmers, chemical, 375
Hard water, 313
Harpoon mechanism, 582
Heat
and energy, 344
and entropy changes for surroundings, 781
and internal energy, 348–349
quantification of, 349–353
transfer of, 780–784
Heat at constant volume (q_V), 356
Heat capacity (C), 350–352
Heating curve for water, 468–470
Heat of fusion (ΔH_{fus}), 467
Heat of hydration ($\Delta H_{hydration}$), 516–518
Heat of reaction. *See* **Enthalpy of reaction (ΔH_{rxn})**
Heat of vaporization (ΔH_{vap}), 463, 467, 472–473
Heat tax, 794
Heisenberg, Werner, 63, 80
Heisenberg's uncertainty principle, 79–80
Helium, 6, 20, 117, 442
density, 404
dispersion forces in, 446
effective nuclear charge, 121
electron configuration, 105–106, 152
emission spectra, 73, 74
intermolecular forces, 428
Lewis structure, 151, 152
mass, 17
molar mass, 53
molecular orbital diagram, 251
origin of, 25
structure and behavior, 3
valence electrons, 114, 116
Helium-4 isotope, 864, 884, 886
Heme, 976
Hemlock, 658
Hemoglobin
as coordination compound, 975–977
fetal, 608–611
and oxygen, 710
Henderson–Hasselbalch equation, 713–716, 720–721
Henry's law, 521–522
Henry's law constant, 521
Heptane, 512
2-Heptanone, 933
Hess's law, 364–366, 372, 375, 377, 515, 791
Heterogeneous catalysis, 590–591
Heterogeneous equilibrium, 620–621
Heterogeneous mixtures, 7
Heteronuclear diatomic molecules, 258–259
Hexaamminecobalt(III), 963, 973
Hexaamminecobalt(II) ion, 973
Hexagonal closest packing, 493–494
Hexamethylenediamine, 938–939
Hexane, 912
as solvent, 512, 514
structural formula, 908
structural isomers of, 906
3-Hexanone, 932
ΔH_f° (standard enthalpy of formation), 370–375, 776
ΔH_{fus} (heat of fusion), 467
$\Delta H_{hydration}$ (heat of hydration), 516–518
High-spin complexes, 973–974
Hinshelwood, Cyril N., 554, 555
H^+ ions, copper and, 829, 832–833
Hiroshima, Japan, 880
HIV protease, 487
Hoffmann, Roald, 440, 456
Homogeneous catalysis, 590
Homogeneous mixtures, 7
Homonuclear diatomic molecules, 252–257
Hooke, Robert, 395
Hot-air balloon, 398
ΔH_{rxn}. *See* **Enthalpy of reaction**
ΔH_{soln} (enthalpy of solution), 515–518
Hund's rule, 109–110, 238
ΔH_{vap}. *See* **Heat of vaporization**
Hybridization
of atomic orbitals, 234, 236–248
resonance, 197, 198
sp, 243–244
sp^2, 239–243
sp^3, 237–239
sp^3d, 244
sp^3d^2, 245
Hybridization and bonding schemes, 245–248
Hybrid orbitals, 236–237, 970
Hybrids, 197, 198, 237
Hydrated aluminum ions, 752–753
Hydrated metal ions, A-24
Hydrates, 160
Hydrobromic acid, 321, 930
Hydrocarbons, 178, 905–927
aliphatic, 905–906
aromatic, 905, 924–927
classification, 905–906
isomerism in, 909–911
oxidation of, 791
reactions of, 921–924

saturated, 912–916
structures, 906–909
unsaturated, 916–921
uses, 913
viscosity, 455
Hydrochloric acid, 321
acid ionization constant, 663–664
dilution of, 304
and limestone, 327
properties, 320
and sodium bicarbonate, 327
and sodium hydroxide, 322–324
in solution, 310
titration of, 726–728
uses, 656
and zinc, 320, 413, 829
Hydrofluoric acid, 685
Hydrogen
from alcohol and metal reaction, 930
in ammonia, 11
and carbon, 793–794
and chlorine, 192, 330
collecting, over water, 412
from combustion analysis, 175–177
combustion of, 289, 329
density, 404
effective nuclear charge, 121
from electrolysis of aqueous sodium chloride, 845
electron configuration, 105
emission spectrum, 73, 74
and ethene, 582
formal charge, 199
formation of, 365, 369, 413, 793
interaction energy, 234–235
and iodine, 558–560, 565–566, 621
and iodine monochloride, 583
Lewis structure, 162
mass, 17
molecular orbital diagram, 250–251
and nitrogen, 424, 641
and nitrogen monoxide, 566, 586
origin of, 25
and oxygen, 163, 369
and propene, 923
properties of water vs., 144–146
as reducing agent, 333
Rydberg constant, 82
Schrödinger equation, 82–84
standard hydrogen electrode, 822–824
transitions in, 75, 85–86
from water, 4, 11, 171, 172, 841
Hydrogenation, 590, 591
Hydrogen bombs, 886
Hydrogen bonding, 450–452, 470
Hydrogen bromide, 924
Hydrogen chloride, 658, 923
Hydrogen fluoride, 190–191
Hydrogen halides, 291, 662
Hydrogen–oxygen fuel cell, 840–841
Hydrogen peroxide, 148–149, 162, 452
Hydrogen sulfide (dihydrogen sulfide), 235–236
Hydroiodic acid, 321
Hydronium ions (H_3O^+)
in acidic solutions, 666
and Arrhenius definition of acids, 320
concentration of, 670–680
defined, 659
and percent ionization, 676–677
and pH, 668–669
water vs., 162
Hydroxide ions (OH^-)
in basic solutions, 666
concentration of, 680–684
and pH, 669
and pOH, 669–670
precipitation with, 747
Hydroxyl group, 929
Hyperosmotic solutions, 541
Hypochlorite ion, 158–159
Hypochlorous acid, 724
***Hypo-* prefix, 160**
Hyposmotic solutions, 542
Hypothesis, 7
Hypoxia, 409–410

I

***i* (van't Hoff factor), 539–540**
Ice, 5, 442, 495
cooling water with, 470
formation of, 786
melting of, 466
sublimation of, 466
ICE table, 622–623, 671
Ideal gas(es), 401
real gases vs., 427–428
solution of, 511
Ideal gas constant, 401
Ideal gas law, 401–407, 618, 638
defined, 401–403
and density of gases, 404–406
and kinetic molecular theory, 416–417
molar mass of gas, 406–407
molar volume at STP, 404
Ideal solutions, 530–531
IE (ionization energies), 128–132, 958
IE_1. *See* First ionization energy
IE_2 (second ionization energy), 128, 131–132
IE_3 (third ionization energy), 128
Incomplete octets, 202–203
Indeterminacy, 80–81
Indicators, 325, 725, 737–739
Inelastic collisions, 414
Infrared (IR) radiation, 67
Initial concentration
equilibrium concentration and, 627–630, 633–635
and pH of buffer solution, 713–716
Initial rates, method of, 564–565
Inner electron configuration, 110, 114
Inner transition elements (actinides), 115–116
In phase (term), 68
Insect control, with radioactivity, 893
Insoluble compounds, 311, 316, 510, 745
Instantaneous dipoles, 446
Instantaneous rate of reaction, 560
Integrated rate law, 568–575
first-order, 569–571
and half-life of reaction, 573–575
for radioactive decay, 873–875
second-order, 571–572
zero-order, 573
Intensity, of light, 65, 71, 72
Intensive properties, 39
Interaction energy, 234–235
Interference
constructive vs. destructive, 485–486
electron, 77
light, 68, 69
Interference patterns, 68, 486
Interhalogen compounds (interhalides), 291
Intermediates, reaction, 583
Intermolecular forces, 445–456
and capillary action, 455–456
defined, 442
dipole–dipole, 448–450
dispersion, 446–448
and hydrogen bonding, 450–452
of ideal gases, 401
ion–dipole, 453
in molecular compounds, 163
of real gases, 426–427
in solutions, 511–514
and surface tension, 454–455
and viscosity, 455
Internal energy (*E*), 346–349
Internal energy change (ΔE)
for chemical reactions, 355–357
enthalpy change vs., 358
and first law of thermodynamics, 347–348
International System of Units (SI), 8, 9, A-1–A-3
International Union of Pure and Applied Chemistry (IUPAC), 23, 913
Intramolecular forces, 163
Intravenous fluids, 542
Inverse logarithms, A-14
Inverse relationships, 395
Iodate, 160
Iodine
and chlorine, 291
family of, 118
and hydrogen, 558–560, 565–566, 621
on periodic table, 103
phase diagrams, 484

Iodine-131 isotope, 891
Iodine monochloride, 583
Ion(s), 124–132. *See also specific ions; specific types*
acid–base properties of, 684–691
after selective precipitation, 747–748
charge on, 377–378
common ion effect, 712–713, 743–744
electron configurations of, 118–119, 124–125
electrons of, 22
formation of, 118–119
ionic radii, 126–127
ionization energy, 128–132
magnetic properties of, 124–125
size of, 377
in strong electrolyte solutions, 539
Ion–dipole forces, 453
Ion–dipole interactions, 517
Ionic bond(s), 147, 148, 152
Ionic bonding, 152–154
covalent vs., 147
and electron transfer, 152–153
and lattice energy, 153–154
models vs. reality of, 154
Ionic charge, 330
Ionic compounds, 155–160
balancing equations with, 278
binary, 156–158
chemical formulas of, 153, 155
defined, 148
formation of, 147
hydrated, 160
lattice energies for, 375–378
naming, 156–160
solubility of, 311–313, 739–745
Ionic equations, 318–319
Ionic radii, 126–127
Ionic solids, 154, 495, 497–498
Ionizable protons, 664
Ionization, 311
percent, 676–678
of polyprotic acids, 691–692
of strong acids, 663
of strong bases, 680
of weak acids, 664
of weak bases, 681
Ionization energies (IE), 128–132, 958
Ionizing power, 865
Ionone, 933
Ion product constant for water (K_w), 666–667
Iron, 495
and chlorine, 290
from iron ore, 768
and lead, 828
and magnesium, 828
as metallic atomic solid, 496
oxides of, 849
rusting of, 273, 328, 768, 849–850
Iron blue, 975
Iron(III) chloride, 689
Iron(III) fluoride, 690
Iron(II) ions, 157
Iron(III) ions, 157
Iron ore, 768
Iron(III) oxide, 373
Iron(II) sulfate, 159
IR (infrared) radiation, 67
Irradiation of foods, 893
Irregular tetrahedron geometry, 213
Irreversible reactions, 794
Island (valley) of stability, 869–871
Isobutane, 177, 905–907
Isomerism
cis–trans, 242–243, 920–921, 966–967
of coordination compounds, 965–969
fac–mer, 967
geometric, 920–921, 966–967
optical, 909–911, 968–969
stereo-, 909–911, 966–969
Isomers
cis, 242–243, 920–921, 966–967
coordination, 965
dextrorotatory, 911
levorotatory, 911
linkage, 965–966
structural, 906, 965–966
trans, 242–243, 920–921, 966–967
Isopropyl alcohol (rubbing alcohol), 442, 928, 929
3-Isopropyl-4-methyl-1-pentyne, 919–920
Isosmotic solutions, 541, 542
Isotonic solutions, 541
Isotope(s), 20–21
atomic masses of, 22
ionic radii of, 128
natural abundance of, 20, 22
notation for, 863–864
IUPAC (International Union of Pure and Applied Chemistry), 23, 913

J

J. *See* **Joule**
Joliot-Curie, Frédéric, 887
Joliot-Curie, Irène, 887
Joule (J), 41, 42, 344
Joule, James, 41

K

K. *See* **Kelvin**
K. *See* **Equilibrium constant**
K_a. *See* **Acid ionization constant**
K_b (base ionization constant), 681, A-24–A-25
Kekulé, Friedrich August, 924–925
Kelvin (K), 8, 9, 399, A-2–A-3
Kepler, Johannes, 9
Ketones, 931–934
K_f (formation constant), 749
Kilogram (kg), 8, 9, A-1
Kilowatt-hour (kWh), 42
Kinetic energy
average, 414, 417–418
and motion, 344
transformation of potential and, 345
Kinetic molecular theory (KMT) of gases, 414–417
Kinetics
chemical. *See* **Reaction rate(s)**
of radioactive decay, 872–875
and thermodynamics, 768–769
KMT (kinetic molecular theory) of gases, 414–417
Krypton, 118, 119, 879
K_{sp}. *See* **Solubility-product constant**
Kuhn, Thomas, 9
K_w (ion product constant for water), 666–667
kWh (kilowatt-hour), 42

L

***l* (angular momentum quantum number), 82–83**
L (liter), A-5
Lactic acid, 935
Lag time, 70
λ. *See* **Wavelength**
Lanthanide contraction, 958
Lattice energy(-ies), 153–154, 375–378
Lattice point, 488
Lattices, 148
Lavoisier, Antoine, 7, 10
Lavoisier, Marie, 7
Law of conservation of energy, 40, 344–349
Law of conservation of mass, 7, 10
Law of constant composition, 11
Law of definite proportions, 11
Law of mass action, 613, 615, 618, 749
Law of multiple proportions, 12
Laws, scientific, 7–8
LCAO (linear combination of atomic orbitals), 249–252
Lead, 156, 828, 975
Lead-206 isotope, 877–879
Lead–acid storage batteries, 838–839
Lead(IV) chloride, 158
Lead(II) nitrate, 313–314, 746
Lead(IV) oxide, 839
Le Châtelier, Henri, 608
Le Châtelier's principle, 636–643
and acid–base chemistry, 670, 677, 678, 693
and buffers, 712
and concentration change, 636–638
and free energy changes, 797
and solubility, 712
and temperature change, 641–643

and volume/pressure change, 638–641
Length
bond, 260
unit of measurement, A-1
and volume, A-5
Leucippus, 9
Levorotatory isomers, 911
Lewis, Gilbert N., 144, 150, 654, 696
Lewis acids, 696–697, 748
Lewis bases, 696, 748
Lewis electron-dot structures (Lewis structures), 150–152
and bond polarity, 190–191
and covalent bonding, 161–163
for molecular compounds, 194–195
for polyatomic ions, 196
Lewis models, 150–152
Lewis symbols, 151
Lewis theory
exceptions to octet rule, 201–204
formal charge in, 199–201
for molecular compounds, 194–195
for polyatomic compounds, 196
resonance in, 196–198
Libby, Willard, 875, 876
Life, effects of radiation on, 888–891
Ligands, 748, 960–962
bidentate, 961, 962
monodentate, 960–961
names and formulas of, 964
nitrito and nitro, 965, 966
strong-field, 973
weak-field, 973
Light, 64–73
diffraction of, 68, 69
in electromagnetic spectrum, 66–67
emission of, 862
intensity of, 65, 71, 72
interference, 68, 69
particle nature of, 68, 70–73
plane-polarized, 910–911
visible, 66, 67
wave nature of, 64–66
white, 66, 74
Lightning, 65, 812–814
Limestone, 327, 744
Limiting reactant, 283–289
Linear accelerators, 887
Linear combination of atomic orbitals (LCAO), 249–252
Linear geometry, 207, 214
Linear relationships, 397
Line notation, 820–821
Linkage isomers, 965–966
Liquefied petroleum (LP), 905, 912
Liquid(s)
defined, 5
entropy and state changes for, 774–775
and fusion, 466–467
on heating curve, 468–470
in heterogeneous equilibria, 620–621
intermolecular forces in, 445–456
properties of water, 470–471
solids and gases vs., 442–445
standard molar entropies of, 777
standard state for, 370
structure and properties of, 440–442
vaporization of, 456–465
vapor pressure, 459–465
Liter (L), A-5
Lithium
effective nuclear charge, 121
electron configuration, 109, 124
energy levels of molecular orbitals, 261–262
family of, 118
flame tests for, 76
Lewis structure, 151
molecular orbital theory, 253
and water, 290
Lithium bromide, 517
Lithium ion batteries, 840
Lithium ions, 22, 124
Lithium phosphate, 278
Lithium sulfate, 317
Litmus paper, 658
Lizards, reaction rates in, 554–556
Lobes, atomic orbital, 90
Localized electrons, 197
Logarithmic scale, 668
London, Fritz W., 446
London forces (dispersion forces), 446–448
Lone pairs, 161, 211–215, 682
Low-spin complexes, 973–974
LP (liquefied petroleum), 905, 912
Lung volume, 397

M

M. *See* **Molarity**
m. *See* **Meter**
m **(molality), 524**
M16 (Eagle Nebula), 25
Macrostates, 770–773
Magic numbers, 870
Magnesium
electron configuration, 111
family of, 118
and fluorine, 155
ionization energy, 131
and iron, 828
sacrificial electrodes of, 850
Magnesium-22 isotope, 870
Magnesium-28 isotope, 870
Magnesium hydroxide, 742, 747, 748
Magnesium ions, 747
Magnetic properties
and crystal field strength, 973–974
of ions, 124–125
of liquid oxygen, 232–234
Magnetic quantum number (m_l), 82, 83
Main-group elements, 104
atomic radii, 122
electron affinities, 132–133
electron configurations, 112–115
ionization energy, 129
valence electrons, 112
Main-group hydrides, boiling points of, 470
Malleability, 261
Manganese, 827, 960
Manhattan Project, 880
Manometer, 394–395
Markovnikov's rule, 924
***Mars Climate Orbiter*, 34–36, 44**
***Mars Curiosity Rover*, 470**
Mars Polar Lander, 35
Mass. *See also* **Molar mass**
atomic, 22–25, 165
concentration as parts by, 524, 525
critical, 880
empirical formula molar, 174
and energy, 883–885
formula, 165
law of conservation of mass, 7, 10
and moles, 52–55, 170–171
and number of molecules, 166
parts by, 524, 525
percent by, 524
unit of measurement, A-1
Mass action, law of, 613, 615, 618, 749
Mass defect, 884–885
Mass–energy, conservation of, 346
Mass number (*A*), 20–21
Mass percent composition, 167–169
Mass spectrometry, 24–25
Mass-to-mass conversions, 280–281
Mathematical operations, A-11–A-16
and graphs, A-15–A-16
with logarithms, A-13–A-14
and quadratic equations, A-15
and scientific notation, A-11–A-13
Mathematical operators, 82
Matter. *See also* **Particulate nature of matter; States of matter**
building blocks of, 9
classification of, 4–7
composition of, 4, 6–7, 344
defined, 3
properties and structure of, 2–4
wave nature of, 77–81
Maxwell, James Clerk, 390
Mean free path, 420
Measurement. *See also* **Units of measurement**
of density, 38–40
of energy, 40–43
of pressure, 394–395
of reaction rates, 561–562
reliability of, 36–38

in science, 8
and unit conversion, 43–44
Medical solutions, 541–542
Medicine, nuclear, 860, 891–893
Megaelectron volt (MeV), 884
Meitner, Lise, 879
Meitnerium, 879
Melting, 466–467
Melting points, 154, 466
Membranes, semipermeable, 537
Mendeleev, Dmitri, 100–104, 112, 116
Meniscus, 456
Mercury, 20, 496
absorption spectra, 76
emission spectra, 73, 76
measuring pressure with, 393–395
meniscus of, 456
Metal(s). *See also* **Transition metals (transition elements)**
and acids, 320
and alcohols, 930
alkali, 118, 290, 291
alkaline earth, 118
in anionic complex ions, 964
bonds in, 261
cations of, 688–689
corrosion of, 848–850
with invariant charges, 156
with multiple types of cations, 157–158
with one type of cation, 156–157
from ores, 848, 849
periodic properties of, 116–117
predicting dissolution of, 829
as reducing agents, 333
semimetals, 117
Metal halides, 290–291
Metal hydride, 839–840
Metallic atomic solids, 495, 496
Metallic bonds, 496
Metallic character, 133–135
Metalloids, 117
Meter (m), 8, 9, A-1
Methane, 177, 912
and chlorine, 367, 368, 922
combustion of, 284–285, 372
formula and models, 150
geometry of, 209
hybrid orbitals, 236
Lewis structure, 904
standard enthalpy of formation, 370
Methanoic acid (formic acid), 664, 724, 935
Methanol, 928, 929
geometry, 219
and sodium methoxide, 930
synthesis of, 422–423
uses, 929
vapor pressure, 464–465
Method of initial rates, 564–565
Method of successive approximations, 633, 671
Methyl alcohol, 289
Methylamine, 682
Methylammonium chloride, 937
3-Methyl-1-butanethiol, 904
Methyl butanoate, 934, 935
3-Methyl-1-Butanol, 929
3-Methylhexane, 910
Methyl isonitrile, 576–577
2-Methyl-2-pentene, 918
Methyl propanoate, 934
Metric system, 8
MeV (megaelectron volt), 884
Meyer, Julius Lothar, 102–103
Microstates, 770–773
Microwaves, 67
Milk of magnesia, 322, 744
Millikan, Robert, 14–15, 18
Milliliter (mL), A-5
Millimeter of mercury (mmHg), 393
Miscibility, 449, 512
Mixtures, 6–7
compounds vs., 146
electrolysis of, 843–844
heterogeneous, 7
homogeneous, 7
racemic, 911
m_l (magnetic quantum number), 82, 83
mL (milliliter), A-5
mmHg (millimeter of mercury), 393
MO diagrams. *See* **Molecular orbital diagrams**
Moisture, rusting and, 849
mol % (mole percent), 526
Molality (*m*), 524
Molar heat capacity, 350
Molarity (M)
in calculations, 303–304
and concentration of solutions, 523–524
defined, 302–303
Molar mass, 165–167
defined, 52–53
and dispersion force, 446–447
and empirical formula, 174
of gases, 406–407
and standard molar entropies, 777–778
and velocity distribution, 418
Molar solubility, 720–743
Molar volume
of real gases, 425–426
at standard temperature and pressure, 404
and stoichiometry, 423–424
Mole(s), 51–55
of compounds, 165–167
in ideal gas law, 401–407
and mass, 52–55, 170–171
and number of atoms, 52
and pressure, 424
as SI base unit, 8, 9
and volume, 400
Mole concept, 51, 54, 55
Molecular complexity, standard molar entropy and, 778–779
Molecular compounds, 148
formulas and names for, 163–165
Lewis structures of, 194–195
Molecular equations, 318
Molecular formulas, 148, 149, 174
Molecular gastronomy, 300–302
Molecular geometry
basic shapes, 207–211
of coordination compounds, 962–963
and electron geometry, 216
and electron groups, 208, 209, 215
and hybridization, 238
and lone pairs, 211–215
and polarity, 219–223
representing, on paper, 218
and VSEPR, 215–219
Molecular imposters, 190
Molecularity, 583–584
Molecular mass, 165
Molecular models, 150, 167
Molecular orbital (MO) diagrams, 250–251, 255, 256
Molecular orbitals (MOs), 249
antibonding, 250–252
bonding, 249–252, 258
nonbonding, 259
Molecular orbital (MO) theory, 248–260
electron delocalization in, 248–259
linear combinations of atomic orbitals, 249–252
for polyatomic molecules, 259–260
for second-period heteronuclear diatomic molecules, 252–259
Molecular shape, 219–223, 447
Molecular solids, 495
Molecular structure, acid strength and, 661–663
Molecular velocities, of gases, 417–419
Molecular weight, 165
Molecules, 4
of compounds, 148
mass and number of, 166
of molecular solids, 495
polarity and shape of, 219–223
predicting geometries of, 215–219
properties of compounds vs., 144–146
Mole fraction (χ), 408, 410–411, 525
Mole percent (mol %), 526
Mole-to-mole conversions, 279–280
Molten salts, electrolysis of, 843
Molybdenum-102 isotope, 870
Monatomic anions, 156
Monodentate ligands, 960–961
Monomers, 937–938
Monoprotic acids, 664, 665

Monosubstituted benzenes, 925–926
Monoxide, 164
Morphinan, 188
Morphine, 188–190
MOs. *See* **Molecular orbitals**
Moseley, Henry, 103
MO theory. *See* **Molecular orbital theory**
Motion, kinetic energy and, 344
m_s (spin quantum number), 82, 83
μ. *See* **Dipole moment**
Multi-electron atoms
  electron configurations, 109–112
  sublevel energy splitting, 106–109
Multiple proportions, law of, 12
Multiplication, 616, A-12
Multistage linear accelerator, 887

N

***n*.** *See* **Principal quantum number; Reaction order**
Nagasaki, Japan, 880
Nanoribbons, 500
Nanotubes, 500–501
Natural abundance of isotopes, 20, 22
Natural gas
  alkanes in, 912
  combustion of, 274–275, 289
  hydrocarbons in, 905
  molecular models, 150
Natural logarithms, A-14
***Nature of Chemical Bond, The* (Pauling), 191**
Negative charge, in voltaic cells, 820
Neon
  and argon, 511
  electron configuration, 110, 114, 116
  emission spectra, 73, 75
  ion formation, 118
  isotopes of, 20–21
  Lewis structure, 151
  as noble gas, 118
Neopentane, 447
Nernst equation, 834, 837
Net ionic equations, 318–319
Network covalent atomic solids, 495, 496, 498–501
Neutral anions, 684, 685
Neutral cations, 688, 689
Neutralization reactions. *See* **Acid–base reactions**
Neutral solutions, 666
  pH of, 668
  pOH of, 670
  salt solutions as, 689–691
Neutron(s), 17
  in isotopes, 20–21
  properties of, 18
  and valley of stability, 869–870
Newlands, John, 102
Newton, Isaac, 9, 480
Newton's laws of motion, 80, 416
Nickel, 496, 591, 827
Nickel–cadmium (NiCad) batteries, 839
Nickel(II) chloride, 317
Nickel–metal hydride (NiMH) batteries, 839–840
Nitrate, 160
Nitrate ions, 322
Nitric acid
  formula, 322
  oxidation of metals by, 829
  and PAN, 202
  and sodium carbonate, 328
  titrations with, 729
  uses, 657
Nitrite, 160
Nitrite ions, 965
Nitrito ligands, 965, 966
Nitrogen, 6, 71, 442
  in ammonia, 11
  and boron, 203
  and carbon-14 isotope, 875
  density, 405
  electronegativity, 200
  and hydrogen, 424, 641
  hydrogen bonding with, 450
  ionization energy, 131
  Lewis structure, 151, 162
  and oxygen, 12, 163, 173
  partial pressure in air, 408
  and potassium bromide, 520
  velocity distribution, 418
Nitrogen-14 isotope, 876
Nitrogen-17 isotope, 887
Nitrogen dioxide
  and carbon monoxide, 566–567, 580, 584–585
  oxygen to nitrogen ratio in, 12
  structure of, 164
  and water, 289
Nitrogen ions, 118
Nitrogen monoxide
  and hydrogen, 566, 586
  Lewis structure, 202
  oxidation of, 619
  standard entropy, 778
Nitrogen narcosis, 410
Nitrogen oxides, 881
Nitrogen triiodide, 164
Nitro ligands, 965, 966
Nitromethane, 200–201
Nitrous acid, 724
Noble gases, 116–118
  boiling points, 447
  nonbonding atomic solids of, 496
  standard entropies, 777
Nodes, orbital, 88, 250
Nonbonding atomic radius (van der Waals radius), 119
Nonbonding atomic solids, 495, 496
Nonbonding electrons, 161, 211
Nonbonding orbitals, 259
Nondirectional bonds, 163
Nonelectrolyte(s), 310–311, 529–533
Nonelectrolyte solutions, 310–311
Nonideal solutions, 531
Nonmetal(s), 117
Nonpolar compounds, 449
Nonpolar covalent bonds, 192
Nonpolar molecules, 220
Nonspontaneous processes
  and electrolytic cells, 818, 841–848
  made spontaneous, 793
  spontaneous vs., 768–769
Nonstandard conditions, cell potentials under, 833–837
Nonstandard states, energy changes in, 794–797
Nonvolatile liquids, 457
Nonvolatile solutes
  and boiling point elevation, 535–536
  and freezing point depression, 534–535
  and vapor pressure lowering, 528–530
Normal boiling point, 461
Normalization, 25
Normal science, 9
n-type semiconductors, 262
Nuclear binding energy, 884–885
Nuclear chemistry
  converting mass to energy, 883–885
  fission, 879–882
  fusion, 886
  transmutation, 887–888
Nuclear equation(s), 864, 865
Nuclear fission, 879–882
Nuclear fusion, 885
Nuclear medicine, 860, 891–893
Nuclear power, 880–881
Nuclear reactors, 882
Nuclear theory of atom, 17
Nucleon(s), 869
  binding energy per, 884, 885
  magic numbers of, 870
Nucleus, 17, 861, 864
Nuclides, 863, 864, 873
Nylon 6,6 polymer, 938–939

O

Obama, Barack, 882
Observation, 7
Octahedral complexes, 970–971
Octahedral geometry, 210
Octane, combustion of, 270, 279, 329, 374–375

Octaves of elements, 102
Octet(s), 152, 202–204
Octet rule, 152, 163, 201–204
Odd-electron species, 202–203
Odors, of organic molecules, 902–904
OH^- ions. *See* **Hydroxide ions**
Oil
hydrocarbons in, 905
and water, 222, 449
Oil drop experiment, 14–15
Olive, spherical, 300–302
orbital 1*s*, 87–88, 92, 108
***On the Revolution of the Heavenly Orbs* (Copernicus), 9**
Operators (mathematical), 82
Opium poppy, 188, 190
Oppenheimer, J. R., 880
Optical isomerism, 909–911, 968–969
Orbital(s). *See* **Atomic orbitals (AOs); Molecular orbitals (MOs)**
Orbital blocks, periodic table, 113–116
Orbital diagrams, 105, 106, 109–111
Order of magnitude estimations, 49
Order of reaction. *See* **Reaction order**
Ores, metals from, 848, 849
Organic chemistry, 902–943
alcohols, 929–931
aldehydes and ketones, 931–934
alkanes, 912–916
alkenes, 916–921
alkynes, 916–920
amines, 937
aromatic hydrocarbons, 924–927
carboxylic acids and esters, 934–936
defined, 904
ethers, 936–937
functional groups, 928
hydrocarbons, 905–927
odors of organic molecules, 902–904
polymers, 937–939
properties of carbon, 904–905
reactions of hydrocarbons, 921–924
Organic compounds
properties of, 177–178
resonance structures and formal charge for, 200–201
Organic molecules, odors of, 902–904
Orientation factor, 581–582
Orientation of particles, reaction rate and, 557, 581
Osmosis, 537, 541
Osmosis cell, 537
Osmotic pressure, 537–538
Outer electron configuration, 114
Overall order, 565–567
Overlapping orbitals, 238
Overvoltage, 845–846
Oxalate ion, 961
Oxidation
of alcohols, 930, 933
of carbon monoxide, 793
defined, 814
definition of, 330, 332
of glucose, 793
of hydrocarbons, 791
of nitrogen monoxide, 619
and oxidation number, 332–334
Oxidation number. *See* **Oxidation states**
Oxidation–reduction (redox) reactions, 328–334
balancing, 814–817
corrosion, 848–850
identifying, 332–334
and oxidation states, 330–332
predicting direction of, 827–829, 831
spontaneous, 817–818
Oxidation states, 330–332, 959–960
Oxidizing agents, 333, 827
Oxyacids, 322, 662–663
Oxyanions, 160
Oxygen, 20, 442
and aluminum, 155
and calcium, 155
and carbon, 163, 347
in carbon dioxide, 11, 12
collecting, over water, 413
in combustion analysis, 175–177
combustion of, 289
electron configuration, 151
emission spectrum, 75
in fetal hemoglobin, 608–611
in formation of ozone, 786
and hemoglobin, 710, 977
from hydrocarbon combustion, 922
and hydrogen, 163, 369
hydrogen bonding with, 450
hydrogen–oxygen fuel cell, 840–841
ionization energy, 131
and iron, 768
Lewis structure, 151, 162
magnetic property of liquid, 232–234
and nitrogen, 12, 163, 173
and nitrogen monoxide, 202
oxidation state, 331, 332
as oxidizing agent, 333
oxyacid strength and bonding of, 663
paramagnetism of, 256–257
partial pressure limits of, 410
properties of water vs., 144–146
redox reactions without, 329
reduction of, 848–849
root mean square velocity, 419
and sodium, 330
from water, 4, 11, 172, 841
Oxygen toxicity, 410
Ozone, 167, 568
decomposition of, 579, 587–589
and ethanol in gasoline, 929
formation of, 786, 791
heterogeneous catalysis and depletion of, 590, 591
molecular orbital model, 260

P

Pa (pascal), 394
Packing efficiency, 488
Palladium, 589, 591
PAN (peroxyacetylnitrate), 202
Paramagnetic (term), 124, 233, 256–257, 973
Parent nuclide, 864
Partially hydrogenated vegetable oil, 923
Partial pressures (P_n), 407–413
collecting gases over water, 412–413
Dalton's law, 407–408, 415
and deep-sea diving, 409–411
defined, 407
and equilibrium concentration, 630–631
equilibrium constant in terms of, 617–620
and total pressure, 408–409
of water, 796
Particle nature
of electron, 80
of light, 68, 70–73
Particulate nature of matter
and behavior of gases, 391, 414–417
and Boyle's law, 396
and Charles's law, 396
and exothermic/endothermic processes, 360
and matter classification, 4–7
and the mole, 51
and pressure, 392–395
and reaction rates, 556–557
and structure/properties, 3–4
Parts by mass, 524, 525
Parts by volume, 524–525
Parts per billion (ppb), 524
Parts per million (ppm), 524
Pascal (Pa), 394
Patchouli alcohol, 902, 904
Pauli, Wolfgang, 105
Pauli exclusion principle, 105–106
Pauling, Linus, 144, 191, 232, 233, 270
***p* block, 114**
P_c (critical pressure), 465
Penetrating power, 865
Penetration, electron, 107, 109
Pentaaminebromocobalt(II) chloride, 965
Pentaaminechlorocobalt(II) bromide, 965
Pentaamminenitrocobalt(III), 965
Pentaaquachlorochromium(III) chloride, 964
Pentanal, 932
Pentane, 512, 906, 912
***n*-Pentane, 447, 460, 465**
Pentanoic acid, 934
2-Pentanol, 929
2-Pentanone, 932

Percent, 524
Percent by mass, 524
Percent ionic character, 193
Percent ionization, acids, 676–678
Percent yield, 283–289
Perfect crystals, 776
Peridot, 956
Periodic law, 102–103
Periodic property(-ies), 100–138
atomic radii, 119–123
defined, 102
density of aluminum, 100–102
effective nuclear charge, 121–122
electron affinity, 132–133
electron configurations and, 105–119, 124–125
for families of elements, 117–118
of ions, 118–119, 124–132
metallic character, 133–135
of metals and nonmetals, 116–117
orbital filling and, 105–112
valence electrons, 112–116
Periodic table
creation of, 101–104
electron configurations from, 114–115
orbital blocks in, 113–114
organization of, 18, 19
Permanent dipoles, 448
Peroxyacetylnitrate (PAN), 202
Perpetual motion machines, 346
***Per-* prefix, 160**
Perturbation theory, 234
PET (positron emission tomography), 892
Petroleum. *See also* **Gasoline**
combustion of, 280–281
liquefied, 905, 912
pH
of acids, 670–680
of bases, 680–684
of basic solutions, 682–684
buffers and changes in, 716–720
of buffer solutions, 712–716
equilibrium constant from, 675
of mixtures of acids, 678–680
and percent ionization, 676–678
and pK_a, 731
of polyprotic acids, 693–695
of salt solutions, 689–691
and solubility, 744–745
of strong acids, 670
of strong bases, 680–682
of weak acids, 671–676
of weak bases, 681–684
Phase (of orbitals), 92
Phase diagrams, 482–485, 533–534
pH curves (titration curves), 725, 728, 733–734, 736
Phenol, 925, 926
Phenolphthalein, 325, 738
Phenyl group, 926
3-Phenylheptane, 926
4-Phenyl-1-hexene, 926
pH meter, 737
Phosphorescence, 862
Phosphoric acid, 691
Phosphorus, 111, 262, 887
Phosphorus-30 isotope, 866
Phosphorus-32 isotope, 892
Phosphorus pentachloride, 164
Photodissociation, 588
Photoelectric effect, 68, 70–73
Photons, 71
Photosynthesis, 977
pH scale, 668–669
Physical changes, 273–274
Physical properties, 273
Pi (π) bonds, 240–241
pK_a scale, 670, 731
Planck, Max, 63, 70
Plane-polarized light, 910–911
Platinum
in catalytic converters, 589
density, 39
electrodes of, 821, 822
and hydrogenation of ethene, 591
in hydrogen–oxygen fuel cell, 840
Plato, 7, 9
Plutonium-236 isotope, 874
P_n. *See* **Partial pressures**
p–n junctions, 262
pOH scale, 669–670
Polar compounds, 449
Polar covalent bonds, 191
Polarimetry, 562
Polarity
and acid strength, 662
of bonds, 190–194
and dispersion forces, 446
and molecular shape, 219–223
and solubility, 308
of water, 470
Polarized light, rotation of, 910–911
Polar molecules, 219–223
Polar stratospheric clouds (PSCs), 590
Pollutant(s), 590
Polonium, 20, 863
Polyatomic ion(s)
balancing equations with, 278
in ionic compounds, 158–160
Lewis structures for, 196
oxidation numbers in, 332
Polyatomic molecules, 259–260
Polycyclic aromatic hydrocarbons, 926, 927
Polydentate ligands, 961, 962
Polyethylene, 938
Polymers, 937–939
Polyprotic acids, 691–696
anions in weak solutions of, 695–696
pH of, 693–695
properties, 320
titrations of, 736–737
Poly vinyl chloride, 938
Popper, Karl, 300
***p* orbitals, 90, 110**
hybridization of, 237–245
mixing of *s* and, 255
Porphyrin, 976
Position
of electrons, 80
and potential energy, 344
Positive charge, in voltaic cells, 820
Positron emission, 863, 866
Positron emission tomography (PET), 892
Positrons, 866
Potassium
and bromine, 291, 582
and chlorine, 152–153
family of, 118
flame tests for, 76
and fluorine, 155
oxidation state, 331
and water, 290
Potassium-40 isotope, 879
Potassium bromide, 520, 689
Potassium carbonate, 317
Potassium chlorate, 779
Potassium chloride, 156
Potassium hexacyanoferrate(III), 964
Potassium hydroxide, 324, 517, 658, 747–748
Potassium iodide, 313–315, 746
Potassium nitrite, 689
Potassium uranyl sulfate, 862–863
Potential difference, 819
Potential energy
changes in systems to lower, 40–41
defined, 40, 344
and entropy, 511
and intermolecular forces, 445
and standard electrode potential, 823
transformation of kinetic and, 345
Power (for work)
ionizing, 865
nuclear, 880–881
penetrating, 865
Power (mathematical)
raising units to a, 47–48
and scientific notation, A-13
ppb (parts per billion), 524
ppm (parts per million), 524
Precipitate, 313
Precipitation reactions, 313–318
aqueous ionic equilibrium in, 745–748
in molecular gastronomy, 301–302
selective precipitation, 747–748
from supersaturated solution, 519
Precision, 37–38
Pre-exponential factor, 578

Prefixes, 160, 913
Prefix multipliers, A-4
Premise, 3
Pressure(s). *See also* **Vapor pressure**
and Avogadro's law, 400
and Charles's law, 398
constant-pressure calorimetry, 362–364
critical, 465
defined, 353, 392
equilibrium and change in, 638–641
equilibrium concentration from, 627–631
equilibrium constant in terms of, 617–620
free energy vs., 795
in gases, 392–395
gauge, 403
in ideal gas law, 401–407
and kinetic molecular theory, 415
and Le Châtelier's principle, 638–641
and number of moles of gases, 424
osmotic, 537–538
partial, 407–413
on phase diagrams, 483–484
and reaction rate, 562
and solubility, 520–522
and states of matter, 444
total, 403, 408–409
and volume in gases, 395–397, 638
and weather, 392
Pressure–volume work, 353–355
Primary valence, 960
Principal level (principal shell), 83, 112, 114
Principal quantum number (*n*), 82, 83, 115
Principles, 8
Proactinium-234 isotope, 871
Probability density, 87–89
Probability distribution maps, 81
Problem solving
with equations, 49–50
four-step procedure, 45–46
order of magnitude estimations, 49
unit conversion problems, 43–44
units in, 35–36
with units raised to a power, 47–48
Products
in chemical equations, 275
of electrolysis, 843–846
equilibrium concentrations of, 626–627
gaseous, 422–424
reaction rate and concentration of, 559
states of, 274
Proof, of alcoholic beverages, 929
Propanal, 932
Propane
carbon bonds in, 177
catenation of carbon in, 905
combustion of, 360, 361, 783
formation of, 923
formula, 912
state of, 444
Propanoic acid, 934, 935
2-Propanol, 933
2-Propanone, 933
Propene, 907, 923
Properties. *See also* **Colligative properties of solutions; Periodic property(-ies)**
chemical vs. physical, 273
complementary, 80
extensive, 39
intensive, 39
magnetic, 124–125, 232–234, 973–974
Proportionality, 396
Propylene glycol, 534
Propyne, 907
Protein(s), 190, 937
Proton(s), 17
in Brønsted-Lowry acids and bases, 659–660
of elements, 18–20
ionizable, 664
and valley of stability, 869–870
Proton acceptors, 659
Proton donors, 659
Proust, Joseph, 11, 12
PSCs (polar stratospheric clouds), 590
p-type semiconductors, 262
Pure compounds, 370
Pure substances, 6, 7

Q

Q **(reaction quotient), 623–626, 745–746**
Quadratic equations, 628
Quantification, 35
Quantum (of light), 71
Quantum-mechanical model of atom, 62–95
and atomic spectroscopy, 73–76, 84–86
defined, 64
and diffraction, 68, 69
and electromagnetic spectrum, 66–67
and interference, 68, 69
nature of light, 64–73
and periodic properties, 102
Schrödinger equation for hydrogen atom, 82–84
and Schrödinger's cat, 62–64
shapes of atomic orbitals, 87–92
wave nature of matter, 77–81
Quantum numbers, 82–84, 112
Quantum particles, behavior of, 64
Quarks, 887
Quartz, 501
q_V (heat at constant volume), 356

R

Racemic mixtures, 911
rad, 889
Radial distribution function, 88, 89, 108
Radiation
effects on life, 888–891
electromagnetic, 64–65, 866
and excitation, 84–86
infrared, 67
ultraviolet, 67
Radiation detectors, 871–872
Radiation exposure, measuring, 889–890
Radicals. *See* **Free radicals**
Radioactive (term), 862
Radioactive decay, 868, 872–875
alpha, 863–865, 868
beta, 865–866, 868
electron capture, 866
positron emission, 866
Radioactive decay series, 871
Radioactivity, 16
applications of, 891–893
defined, 862
detection of, 871–872
in diagnosis of appendicitis, 860–862
discovery of, 862–863
effects of radiation on life, 888–891
kinetics of radioactive decay, 872–875
measuring age with, 875–879
predicting type of, 869–871
types of, 863–868
and valley of stability, 869–871
Radiocarbon dating, 875–877
Radiometric dating, 875–879
Radiotherapy, 892–893
Radiotracers, 891
Radio waves, 67
Radium, 863
Radium-228 isotope, 865–866
Radon, 890
Radon-220 isotope, 873
Rana sylvatica **(wood frogs), 508–510**
Random error, 38
Raoult, François-Marie, 508
Raoult's law, 529–531, 533
Rapture of the deep, 410
Rate constant(s) (*k*), 563, 566–567
Rate-determining step, 584–585
Rate laws, 563–575
defined, 563
for elementary steps, 583–584
first-order reaction, 564, 569–571, 573–574
and half-life of reaction, 573–575
integrated, 568–575
for overall reaction, 584–585
and reaction order, 563–567
second-order reaction, 564, 571–572, 574
zero-order reaction, 563–564, 573–575
Rate-limiting step, 585–586
RBE (biological effectiveness factor), 890
Reactant(s)
in chemical equations, 275
equilibrium concentrations of, 626–627
gaseous, 422–424
limiting, 283–289

reaction order for multiple, 565–567
reaction rate and concentration of, 556, 558–559. *See also* **Rate laws**
states of, 274

Reaction(s), 175, 270–293
acid–base, 319–326
addition, 923–924
of alcohols, 930
of aldehydes and ketones, 933–934
of alkali metals, 290
of amines, 937
aqueous. *See* **Aqueous reactions**
of aromatic compounds, 926–927
of carboxylic acids and esters, 935–936
chain, 880
and chemical change, 273–274
chemical equations for, 274–278
climate change and combustion of fossil fuels, 270–273
combustion. *See* **Combustion reactions (combustion)**
condensation, 935–936
defined, 10
dehydration, 930
direction of change in, 623–626
elimination, 930
endothermic, 358, 368, 642, 786–787
enthalpy of. *See* **Enthalpy of reaction (ΔH_{rxn})**
entropy change for, 776
equilibrium constant for, 614
exothermic, 153–154, 359, 368, 641, 785, 786
first-order, 564, 569–571, 573–574
gases in, 422–424
gas-evolution, 327–328
half-life of, 573–575, 872–875
of halogens, 290–291, 922–923
of hydrocarbons, 921–924
internal energy change for, 355–357
irreversible, 794
limiting reactant of, 283–289
mass and energy in, 883
oxidation–reduction. *See* **Oxidation–reduction (redox) reactions**
precipitation, 313–318, 747–748
reversible, 611, 793–794
second-order, 564, 565, 571–572, 574
standard enthalpy change for, 372–373
standard entropy change for, 776, 779–780, 789–790
stepwise, 791–792
stoichiometry of, 279–283, 422–424
substitution, 922–923, 930
thermite, 373
yield of, 283–289
zero-order, 563–565, 573–575

Reaction intermediates, 583
Reaction mechanisms, 583–588
with fast initial step, 585–587
rate-determining steps and overall reaction rate laws, 584–585
rate laws for elementary steps, 583–584
two-step, 584–585

Reaction order (*n*), 563–567
defined, 563–564
determining, 564–565
for multiple reactants, 565–567

Reaction quotient (*Q*), 623–626, 745–746
Reaction rate(s), 554–597
Arrhenius equation, 576–581
and catalysis, 588–592
and collision model of temperature, 581–582
and concentration, 563–575
defined, 557–559
in dynamic equilibrium, 611
in ectotherms, 554–556
and half-life of reaction, 573–575
integrated rate law, 568–575
measuring, 559–562
and order of reaction, 563–567
and particulate nature of matter, 556–557
for radioactive decay, 872–875
rate laws, 563–575
and reaction mechanism, 583–588
spontaneity of reaction vs., 768–769
and temperature, 576–582

Reagents, unsymmetrical, 923–924
Real gases, 425–428
Rechargeable batteries, 839–840
Recrystallization, 520
Red blood cells, 541
Redox reactions. *See* **Oxidation–reduction reactions**
Reducing agents, 333, 827
Reduction, 330, 332, 814
of aldehydes and ketones, 933
and oxidation number, 332–334
of oxygen, 848–849
standard electrode potentials for, 824

Relative solubility, 742
Reliability of measurements, 36–38
rem, 890
Repulsion
between electron groups, 207, 208, 215
by lone pairs, 211–212
strong force vs., 869

Resonance, 196–198
Resonance hybrids, 197, 198
Resonance stabilization, 197
Resonance structures, 197–198, 200–201
Respiration, 978
Reversible reactions, 611, 793–794
Revolutions, scientific, 9
Rhodium, 589
Rock candy, 520
Rock salt structure, 497
Roentgen, 890
Roosevelt, Franklin, 880, 881
Root mean square velocity, 417–419
Roots (mathematical), A-13
Rotation, of polarized light, 910–911
***R* proportionality constant, 417**
Rubbing alcohol, 442, 928, 929
Rubidium, 118, 291
Rubies, color of, 954–956
Rusting, 273, 328, 768, 849–850
Ruthenium-112 isotope, 870
Rutherford, Ernest, 16–18, 20, 863, 887
Rutherfordium, 20
Rydberg, Johannes, 74
Rydberg equation, 74, 85

S

***S*. *See* Entropy**
ΔS. *See* Entropy change
$S°$ (standard molar entropies), 776–779
Sacrificial electrodes, 850
Salicylic acid, 936
Salt bridges, 820
Salts. *See also* **Table salt**
acid–base properties, 684–691
defined, 322, 324, 684
electrolysis of, 843
freezing point depression with, 528
in solutions, 308, 309, 510, 689–691, 844–846

Sapphire, 849
Saturated hydrocarbons, 912–916. *See also* **Alkanes**
Saturated solutions, 519, 745
***s* block, 114**
Schrödinger, Erwin, 63, 64
Schrödinger equation, 82–84, 249
Schrödinger's cat, 62–64
Science, 8–9
Scientific approach to knowledge, 7–9
Scientific laws, 7–8
Scientific notation, A-11–A-13
Scientific revolution, 9
Scintillation counters, 872
Screening (shielding), 107, 109, 121–122
Scuba diving, 396
Seawater, 537, 747
Second (s), 8, 9, A-1, A-2
Secondary valence, 960
Second ionization energy (IE_2), 128, 131–132
Second law of thermodynamics, 350, 767, 771–773
Second-order reactions
defined, 564
half-life of, 574
identifying, from experimental data, 565
integrated rate law for, 571–572

Second-period heteronuclear diatomic molecules, 258–259

Second-period homonuclear diatomic molecules, 252–257
Seesaw geometry, 213
Selective precipitation, 747–748
Selenium, 20, 115
Semiconductors, 117, 261–262
Semimetals, 117
Semipermeable membranes, 537
SHE (standard hydrogen electrode), 822–824
Shielding (screening), 107, 109, 121–122
SI. *See* **International System of Units**
Sievert (Sv), 890
Sigma (σ) bonds, 240–241
Significant figures, A-6–A-9
in calculations, A-8–A-9
counting, A-6, A-7
and exact numbers, A-7
Silica, 501
Silicates, 501
Silicon, 112, 117, 262, 905
Silicon dioxide, 495, 496
Silver, 24–25, 156, 960
Silver chloride
and ammonia, 750–751
molar solubility of, 740–741
solubility of, 311–312
Silver ions, 124, 748–749
Silver nitrate, 311–312, 517, 746
Silver plating, 841–842
Simple cubic unit cell, 488–489
Single bonds
covalent, 161–162
double bond vs., 241–243
Single-stage linear accelerators, 887
Skydiving, supersonic, 390–392
SLAC (Stanford Linear Accelerator), 887
Snowflakes, 485
Socrates, 658
Soda pop, 520, 521, 525
Sodium
atomic radii, 126
chemical symbol, 20
and chlorine, 10, 192, 329
from electrolysis of sodium chloride, 841
electron affinity, 133
electron configuration, 116
electron sea model for, 261
emission spectrum, 76
family of, 118
flame tests for, 76
ionization energies, 129, 131
and oxygen, 330
and sulfur, 153
and water, 290, 930
Sodium acetate, 519, 711, 746
Sodium alginate, 302
Sodium benzoate, 715
Sodium bicarbonate, 156, 684
and hydrochloric acid, 327
polyatomic ions in, 159
uses, 658
and vinegar, 301
Sodium bromide, 746
Sodium carbonate, 313, 328
Sodium chloride, 6
Born-Haber cycle for production of, 375–377
consumption of, 169
electrolysis of, 841, 843, 845–846
enthalpy of solution for, 517
formation of, 10, 290
freezing point depression with, 539
ionic bonding in, 152
as ionic solid, 495
lattice energy, 153–154
neutral solutions of, 689
and potassium iodide, 315
solubility, 308–310, 510
in solution, 515, 518–519, 524–525, 528, 769
structure, 497
vapor pressure lowering with, 540
Sodium fluoride, 517, 689
Sodium hydroxide, 320, 658
and hydrochloric acid, 322–324
from metal and water reaction, 930
and propanoic acid, 935
titrations with, 726–734, 736
in water, 514
Sodium hypochlorite, 158
Sodium ions, 22, 126, 312, 843
Sodium methoxide, 930
Sodium nitrate, 317
Sodium nitrite, 159
Sodium oxide, 165, 282–283
Sodium propanoate, 935
Solid(s). *See also* **Crystalline solids**
amorphous, 444
atomic, 495–496
defined, 5
entropy and state changes for, 774–775
fusion of, 466–467
on heating curve, 468–470
in heterogeneous equilibria, 620–621
intermolecular forces in, 445–456
ionic, 154, 495, 497–498
liquids and gases vs., 442–445
metallic atomic, 495, 496
molecular, 495
network covalent atomic, 495, 496, 498–501
nonbonding atomic, 495, 496
solubility of, 519–520
standard molar entropies of, 777
standard state for, 370
structure and properties of, 440–442
sublimation of, 466
Solubility, 510–514
of amphoteric metal hydroxides, 752–753
and common ion effect, 743–744
and complex ion equilibria, 750–752
of gases, 520–521
of ionic compounds, 311–313, 739–745
molar, 720–743
and pH, 744–745
and pressure, 520–522
relative, 742
of solids, 519–520
and temperature, 519–520
of vitamins, 513–514
Solubility equilibria, 739–745
Solubility-product constant (K_{sp}), 739–745
and common ion effect, 743–744
and molar solubility, 720–742
and precipitation, 745–746
and relative solubility, 742
for selected substances, A-25–A-26
and selective precipitation, 747
Solubility rules, 312, 739
Soluble compounds, 311, 510, 745
Solutes, 302, 510
interactions between, 512
solvent interactions with, 308, 512
volatile, 530–532
Solution(s), 300–313, 508–546. *See also* **Acidic solutions; Aqueous solutions; Basic solutions**
colligative properties of, 528–542
concentrated, 302, 522
concentration of, 302–306, 522–527
defined, 510
dilute, 302, 522
electrolyte, 309–311, 539–542
energetics for formation of, 514–518
enthalpy of, 515–518
entropy in, 511
equilibrium in, 518–522
hyperosmotic, 541
hyposmotic, 542
ideal, 530–531
intermolecular forces in, 511–514
isosmotic, 541, 542
isotonic, 541
medical, 541–542
in molecular gastronomy, 300–302
neutral, 666, 668, 670, 689–691
nonelectrolyte, 310–311
nonideal, 531
saturated, 519, 745
and solubility, 510–514
standard state for, 370
stock, 304–306
stoichiometry of, 307–308
supersaturated, 519, 746
thirsty, 537
types of, 510
unsaturated, 519, 745
in voltaic cells, 820
in wood frog cells, 508–510

Solvent(s), 510. *See also specific solvents*
boiling point elevation in, 535
common laboratory, 512
defined, 302
freezing point depression in, 535
interactions between, 512
mole fraction for, 525
solute interactions with, 308, 512
***s* orbitals, 87–89, 110**
hybridization of, 237–245
mixing of *p* and, 255
sp^2 hybridization, 239–243
sp^3d^2 hybridization, 245
sp^3d hybridization, 244
sp^3 hybridization, 237–239
Space-filling molecular models, 150
Specific heat capacity (C_s), 350–351, 471
Spectator ions, 318
Spectrochemical series, 973
Spectrometers, 562
Spectrometry, 24–25, 562
Spectroscopy, 73–76, 84–86
Spherification, 301–302
***sp* hybridization, 243–244**
Spin-pairing, 235–236
Spontaneity
and entropy changes for surroundings, 781
and Gibbs free energy, 785–788
and second law of thermodynamics, 771
and standard change in free energy, 793–794
Spontaneous processes
electricity from, 817–821
making, 793
nonspontaneous vs., 768–769
and predicting direction of redox reactions, 827–829, 831
Square planar complexes, 975
Square planar geometry, 214
Square pyramidal geometry, 214
ΔS°_{rxn}. *See* **Standard entropy change for a reaction**
ΔS_{surr}. *See* **Entropy change for surroundings**
Stability
and nuclear binding energy, 884
of nuclei, 864
and potential energy, 41
valley of, 869–871
Stalactites, 744
Stalagmites, 744
Standard atomic weight. *See* **Atomic mass**
Standard cell potential (E°_{cell}), 820
in concentration cells, 837
and equilibrium constant, 832–833
and free energy, 830–831
and standard change in free energy, 830–831
Standard change in free energy (ΔG°_{rxn}), 788–800
and equilibrium constant, 797–800
from free energies of formation, 790–791
free energy changes in nonstandard states vs., 794–797
from ΔH°_{rxn} and ΔS°_{rxn}, 788–789
and spontaneity, 793–794
and standard cell potential, 830–831
for stepwise reaction, 791–792
Standard electrode potentials, 822–829
from potentials of half-reactions, 826
predicting direction of redox reaction with, 827–829
predicting dissolution of metals with, 829
of selected substances, 825, A-27–A-28
and standard hydrogen electrode, 822–824
Standard emf. *See* **Standard cell potential (E°_{cell})**
Standard enthalpy change (ΔH°), 370, 372–373
Standard enthalpy of formation (ΔH°_f), 370–375, 776
Standard entropy change, 779–780
Standard entropy change for a reaction (ΔS°_{rxn}), 776
calculating, 779–780
standard change in free energy from, 788–789
Standard heat of formation. *See* **Standard enthalpy of formation (ΔH°_f)**
Standard hydrogen electrode (SHE), 822–824
Standard molar entropies (S°), 776–779
Standard state, 370
Standard temperature and pressure (STP), molar volume at, 404
Stanford Linear Accelerator (SLAC), 887
State functions, 346, 771
States of matter, 4–6. *See also specific states*
changes between, 444–445
in chemical equations, 274
entropy and changes in, 774–775
properties of, 443
and standard molar entropies, 777
Stationary states, 75
Steam burn, 457
Stepwise reactions, standard change in free energy for, 791–792
Stereoisomerism, 909–911, 966–969
Stern–Gerlach experiment, 124
Stock solutions, 304–306
Stoichiometry, 279–283
defined, 271, 279
for electrolysis, 847–848
with ΔH, 360–361
for pH changes in buffers, 716–717
for reactions with gases, 422–424
for solutions, 307–308
STP (standard temperature and pressure), molar volume at, 404
Strassmann, Fritz, 879
Strong acid(s)
and buffer solutions, 716–719
defined, 659
as electrolytes, 310–311
hydronium ion concentration and pH of, 670
ionization constants, 663–664
mixtures of weak and, 678–679
in titrations, 725–729, 735–736
Strong base(s)
and buffer solutions, 717, 719
counterions of, 688
defined, 659
hydroxide ion concentration and pH of, 680–682
in titrations, 725–734, 730–734, 736
Strong electrolyte(s), 310
colligative properties of solutions with, 539–542
dissociation of, 659
and vapor pressure, 540–541
Strong-field complexes, 971
Strong-field ligands, 973
Strong force, 869
Strontium, 76, 118
Strontium chloride, 278
Structural formulas, 148–149, 906–909
Structural isomers, 906, 965–966
***Structure of Scientific Revolutions, The* (Kuhn), 9**
Styrene, 926
Subatomic particles, 18–22
Subjectivity, in science, 8–9
Sublevels (subshells), 83, 106–109
Sublimation, 466, 564
Sublimation energy, 375
Substance(s)
defined, 4
pure, 6, 7
Substituents, 913
Substituted benzenes, 925
Substitution reactions, 922–923, 930
Substrates, enzyme, 592
Subtraction, A-13
Successive approximations, method of, 633, 671
Sucrase, 592
Sucrose, 525
combustion of, 356–357
enzyme-catalyzed reaction of, 592
freezing point depression by, 539
hydrolysis of, 556, 561–562
vapor pressure lowering by, 529–530
Sugar(s)
alcohol from, 929
burning of, 273
as organic compound, 177
solubility of, 519
in solutions, 308, 310, 510

Sulfate ions, 322
Sulfite ions, 322
Sulfur, 20
and carbon, 332
electron configuration, 111
hybrid orbitals, 237
ion formation, 119
and sodium, 153
Sulfur hexafluoride, 203, 245
Sulfuric acid, 657
formation of, 282
ionization constant, 692
ionization of, 320
in lead–acid storage batteries, 839
and lithium sulfide, 327
pH of, 693–695
and potassium hydroxide, 324
Sulfurous acid, 322, 691–692, 736
Sulfur oxides, 881
Sun, nuclear fusion on, 886
Supercritical fluids, 465
Supersaturated solutions, 519, 746
Supersonic skydiving, 390–392
Surface tension, 454–455
Surroundings
entropy changes for, 780–784
and systems, 42, 345, 348
transfer of energy to, 347–348
Sv (sievert), 890
Sweating, 457
Synthetic polymers, 937
Systematic error, 38
Systematic names, 156
Systems
energy flow in, 345
entropy changes in surroundings and, 780–781, 784
internal energy change of, 346–348
and surroundings, 42, 345, 348

T

$t_{1/2}$ (half-life) of reaction, 573–575, 872–875
Table salt, 156, 177, 495. *See also* **Sodium chloride**
bonding theories and structure of, 150
ionic bonding in, 152
T_c (critical temperature), 465
Technetium-99m isotope, 891, 892
Tellurium, 103
Temperature
and Arrhenius equation, 576–581
and average kinetic energy, 414
and boiling point, 461–465
and Boyle's law, 396
collision model of, 581–582
and energy, 456–457
and entropy change for surroundings, 781–782
equilibrium and changes in, 641–643
and equilibrium constants, 621, 799–800
global, 272
heat capacity and changes in, 350–352
in ideal gas law, 401–407
and Le Châtelier's principle, 641–643
and melting, 466
and molecular velocity of gases, 417–419
on phase diagram, 483
and reaction rate, 555–557, 576–582
and solubility, 519–520
and spontaneity, 785–788
and state of matter, 5, 444
and thermal energy, 344
unit of measurement, A-2–A-3
and units of entropy, 773
and vapor pressure, 412, 461–465
and viscosity, 455
and volume in gases, 397–399
Temporary dipoles, 446
Terminal atoms, 194
Termolecular steps, 583
Tetrachloromethane, 922
Tetrahedral complexes, 974
Tetrahedral geometry, 208–209
Tetrahedron, 150, 208
Tetraphosphorus decasulfide, 164
Theoretical yield, 283–289
Theories, 8, 9
Therapeutic agents, 978
Therapeutic techniques, radioactivity in, 891–893
Thermal energy
defined, 40, 344
distribution of, 456, 578
in endothermic and exothermic processes, 360
and intermolecular forces, 442
transfer of, 352–353
Thermal equilibrium, 350
Thermite reaction, 373
Thermochemistry, 342–382
burning bill demonstration, 342–344
changes in enthalpy, 362–366
changes in internal energy, 355–357
defined, 344
enthalpy, 358–361, 367–375
first law of thermodynamics, 346–349
heat, 349–353
lattice energies for ionic compounds, 375–378
and nature of energy, 344–345
work, 353–355
Thermodynamics, 766–804
defined, 346
and dispersion of energy, 766–768
entropy, 769–784
and equilibrium, 797–800
first law of, 346–349, 773
free energy, 784–800
heat transfer, 780–784
microstates and macrostates in, 771–773
in nonstandard states, 794–797
second law of, 350, 767, 771–773
spontaneous vs. nonspontaneous processes, 768–769
standard quantities, A-17–A-23
third law of, 776
Third ionization energy (IE_3), 128
Third law of thermodynamics, 776
Thirsty solutions, 537
Thomson, J.J., 13–14, 16
Thorium-232 isotope, 872
3*d* orbitals, 90, 91, 108
3*p* orbitals, 108
3*s* orbitals, 108
Threshold frequency, 70
Time
and concentration, 568–576
unit of measurement, A-2
Tin, 20, 156
Titanium, 288
Titration curves. *See* **pH curves**
Titrations, 725–739
as acid–base reactions, 324–326
indicators for, 737–739
of polyprotic acids, 736–737
of strong acids, 726–729
of weak acids, 730–734
of weak bases, 735–736
Tokamak fusion reactor, 886
Toluene, 926
Torr, 393
Torricelli, Evangelista, 393
Total energy, 40
Total force, 416
Total pressure, 403, 408–409
Trajectory, 80–81
***trans* isomers**
of alkenes, 920–921
of coordination compounds, 966–967
properties of, 242–243
Transition metals (transition elements), 156. *See also* **Coordination compounds**
atomic radii, 122–123, 957–958
and color of rubies and emeralds, 954–956
electron configurations, 115–116, 956–957
electronegativity, 959
ionization energy, 958
oxidation states, 959–960
on periodic table, 104
Transitions, electron, 75, 85–86
Transition state, 577
Transmutation, 887–888
Transuranium elements, 888
Triads of elements, 102
Trichloromethane, 922
Trigonal bipyramidal geometry, 209

Trigonal planar geometry, 208
Trigonal pyramidal geometry, 211
Trimethylamine, 937
Triple bond(s), 162, 243–244, 905
Triple point, 483
Triprotic acids, 664
Tritium, 886
Truth, of theories, 9
T-shaped geometry, 213
Turquoise, 956
orbitals 2*p*, 90, 92, 108
orbital 2*s*, 89, 108
orbital 3*s*, 89
Two-point forms of equations
Arrhenius equation, 580
Clausius–Clapeyron equation, 464–465
Two-step reaction mechanisms, 584–585

U

Ultraviolet (UV) radiation, 67
Uncertainty principle, 79–80
Unimolecular steps, 583
Unit cells, 488–492, 497–498
Unit conversion problems, 45–48
United States, radiation exposure in, 890
Units of measurement, 8, 9, A-1–A-5
converting, 43–44
defined, 35
derived, A-4–A-5
energy, 41–42
entropy, 773
equilibrium constant, 619–620
prefix multipliers for, A-4
pressure, 393–394
in problem solving, 35–36
raising, to a power, 47–48
reaction rate, 558–559
SI base units, A-1–A-3
volume, A-4–A-5
Universe
age of, 879
entropy changes in, 780–781, 784
Unpaired electrons, 233–234, 256
Unsaturated hydrocarbons, 916–921. *See also* **Alkenes; Alkynes**
cis–trans isomerism in, 920–921
naming, 918–920
Unsaturated solutions, 519, 745
Unsymmetrical alkenes, 923–924
Uranic rays, 863
Uranium, 20, 879, 881, 882
Uranium-235 isotope, 880
Uranium-238 isotope
alpha decay of, 864
radioactive decay series, 871
radiometric dating with, 877–879
transmutation of, 888
U.S. Department of Agriculture (USDA), 893
U.S. Department of Energy (DOE), 280
U.S. Food and Drug Administration (FDA), 169, 893
UV (ultraviolet) radiation, 67

V

v. *See* **Frequency**
V (volt), 819
Valence, coordination compound, 960
Valence bands, 262
Valence bond theory, 234–248
for coordination compounds, 970
hybridization of atomic orbitals, 236–248
orbital overlap in, 234–236
sp^2 hybridization, 239–243
sp^3d^2 hybridization, 245
sp^3d hybridization, 244
sp^3 hybridization, 237–239
sp hybridization, 243–244
writing hybridization and bonding schemes, 245–248
Valence electrons, 112–116, 150–152
Valence shell electron pair repulsion (VSEPR) theory, 189, 206–219
five basic shapes of atoms, 207–210
lone pairs in, 211–215
predicting molecular geometries with, 215–219
Valley of stability, 869–871
Van der Waals, Johannes, 426, 427
Van der Waals constants, 426
Van der Waals equation, 427
Van der Waals radius (nonbonding atomic radius), 119
Vandium, 124
Vandium ion, 124
Vanillin, 932
Van't Hoff factor (*i*), 539–540
Vaporization (evaporation), 456–465
energetics of, 457–459
process of, 456–457
of water, 344, 457, 769, 787, 794–796
Vapor pressure, 459–465
and critical point, 465
and dynamic equilibrium, 459–460
lowering of, 528–532, 540–541
and temperature, 461–465
of water, 412, 459, A-28
Variational principle, 249
Vector quantities, 220–222
Vegetable oil, 923
Velocities
of electrons, 80
of gases, 417–419
root mean square, 417–419
Vinegar, 301, 657. *See also* **Acetic acid**
Viscosity, 455
Visible light, 66, 67
Vitamin A, 513
Vitamin B_5, 513, 514
Vitamin C, 513
Vitamin K_3, 513
Vitamins, solubility of, 513–514
Volatile liquids, 457, 530–533
Volt (V), 819
Volta, Allesandro, 812
Voltaic (galvanic) cells, 817–821, 842–843
Volume
and amount of gas in moles, 400
constant-volume calorimetry, 356, 364
equilibrium and change in, 638–641
heat at constant volume, 356
in ideal gas law, 401–407
and Le Châtelier's principle, 638–641
lung, 397
molar, 404, 423–426
parts by, 524–525
and pressure, 395–397, 638
of real gases, 425–426
and temperature, 397–399
unit cell, 490
units of measurement, A-4–A-5
and work, 353–355
VSEPR theory. *See* **Valence shell electron pair repulsion theory**

W

W (watt), 42
***War Crimes 4: Judgment at Gotham* (comic book series), 655**
Water, 5, 6–7, 51, 163, 442. *See also* **Aqueous reactions; Aqueous solutions**
from acid–base reactions, 324, 935
and alkali metals, 290
amphotericity, 660
atoms in, 2, 4
autoionization of, 666–667
boiling point, 461
boiling point elevation, 536
and carbon dioxide, 620, 621, 697, 978
chemical formula, 148
collecting gases over, 412–413
from combustion, 289, 360, 922
decomposition of, 11, 172, 841
electrolysis of, 844
electrostatic potential map, 308
elimination of, 930
evaporation/vaporization of, 344, 457, 769, 787, 794–796
formation of, 424, 659
freezing of, 780–781, 786
freezing point depression, 534
from fuel cells, 840, 841
geometry, 212
heat capacity, 352
heating curve, 468–470

heat of fusion, 467
hydrogen bonding in, 452
hydronium ion vs., 162
intermolecular forces in, 445–446
Lewis structure, 161
meniscus of, 456
and oil, 222, 449
partial pressure, 796
phase diagram, 482–484
physical changes in, 273, 274
polarity, 220
properties, 4, 144–146, 470–471
and rubidium, 291
and sodium, 930
solubility in, 510, 520–521
as solvent, 302, 512, 514–515, 518–519, 528
specific heat capacity, 350
standard molar entropy, 777
states of, 442–443
structure, 4, 150
from substitution reactions, 930
surface tension, 454–455
vapor pressure, 412, 459, A-28
Waters of hydration, 160
Watson, James, 487
Watt (W), 42
Wave function, 82, 249
Wavelength (λ), 72
de Broglie, 78–79
of light, 65, 66
Wave nature
of electrons, 77–78, 80
of light, 64–66
of matter, 77–81
Wave–particle duality, 64, 73
Wax, 905
Weak acid(s)
acid ionization constant, 664
and buffer solutions, 710–711
cations as, 688–689
defined, 659
diprotic, 695–696
as electrolytes, 311
equilibrium constant for ionization, 664–665
in mixtures, 678–680
percent ionization, 676–678
pH of, 671–675
titrations of, 730–734
Weak base(s)
anions as, 684–688
in buffer solutions, 720–721
cations as conjugate acids of, 688
defined, 659
pH of, 681–684
titrations of, 735–736
Weak electrolytes, 311, 659
Weak-field complexes, 971
Weak-field ligands, 973
Weather, pressure and, 392
Weighing, estimation in, 37
Weight, 165, A-1. *See also* **Atomic mass**
Weighted linear sum, 249
Werner, Alfred, 954, 960, 963
White light, 66, 74
Wilhelmy, Ludwig, 556, 561
Wilkins, Maurice, 487
Witt, Otto N., 708, 709
Wöhler, Friedrich, 902, 903
Wood alcohol, 929. *See also* **Methanol**
Wood frogs, 508–510
Work
defined, 40
and energy, 344
and internal energy, 348–349
quantification of, 353–355
World War II, 880

X

Xenon, 118, 426, 427, 495
***x is small* approximation**
and equilibrium concentration, 633–635
and Henderson–Hasselbalch equation, 714, 716
and pH in buffer solutions, 717
and pH of polyprotic acids, 696
and pH of weak acids, 671–673
X-ray crystallography, 485–487
X-ray diffraction, 485–487
X-rays, 67

Y

Yield
actual, 284
percent, 283–289
theoretical, 283–289
Yucca Mountain, Nevada, 882

Z

Z. *See* **Atomic number**
Z_{eff} (effective nuclear charge), 107, 121–122
Zero-order reaction
defined, 563–564
half-life of, 574–575
identifying, from experimental data, 565
integrated rate law for, 573
Zinc, 156, 496
and carbonic anhydrase, 977
and copper, 817–818, 820, 842
in dry-cell batteries, 838
galvanized nails coated with, 850
and hydrochloric acid, 320, 413, 829
Zinc blende structure, 498
Zinc ion, 124–125
Zinc sulfide, 498